Handbook of Ecological Parameters and Ecotoxicology

HANDBOOK OF ECOLOGICAL PARAMETERS AND ECOTOXICOLOGY

Sven Erik Jørgensen

Royal Danish School of Pharmacy, Department of General Chemistry, Section of Environmental Chemistry, Universitetsparken 2, DK 2100, Copenhagen Ø, Denmark

Søren Nors Nielsen

National Environmental Research Institute, Division of Policy Analysis, Frederiksborgvej 399, DK 4000, Roskilde, Denmark

Leif Albert Jørgensen

National Environmental Research Institute, Division of Policy Analysis, Frederiksborgvej 399, DK 4000, Roskilde, Denmark

ELSEVIER

Amsterdam — London — New York — Tokyo 1991

ELSEVIER SCIENCE PUBLISHERS B.V.
Molenwerf 1
P.O. Box 181, 1000 AD Amsterdam, The Netherlands

Distributors for the United States and Canada:

ELSEVIER SCIENCE PUBLISHING COMPANY INC.
655, Avenue of the Americas
New York, NY 10010, U.S.A.

ISBN 0-444-88604-4

Printed in The Netherlands

INTRODUCTION

Many handbooks are available in chemistry and physics that makes it feasible for a scientist to find the chemical and physical data he needs to carry out estimations and calculations. Chemistry and physics have been quantitative sciences for more than a century and they have therefore a tradition for handbooks, datacollections and databases.

Environmental science and ecology have during the last decade rapidly become at least semiquantitative sciences due to a number of reasons. First of all there is a need for quantitative estimations of environmental impacts and pollution effects. Secondly, our environmental concerns have intensified the development of environmental science and ecology. Thirdly, the development of computer science has rendered it feasible to overview such complex systems as ecosystems with many interacting variables and processes. Environmental modelling has therefore become an important discipline within the family of environmental and ecological sciences. Fourthly, we have realized that it is not possible to solve the environmental problems by simple means. Zero discharge is not possible for economical reasons and environmental technology can not solve the problems alone, but must work hand in hand with engineering methods applied in the ecosystem - it may be called ecological engineering or ecotechnology. Consequently, we must attempt to optimize very complex issues in very complex systems which is not feasible without quantification.

Several handbooks in environmental science have already been published, but most of them have not been handbooks of data and parameters, as we know them from chemistry and physics. This handbook attempts to fill this gap. The idea behind the handbook is to present a collection of data and parameters needed for everyone who deals with environmental and ecological issues, either she or he is a modeller, a theoretical scientist or an environmental manager.

It is difficult to determine constants and parameters in ecology as accurate as in chemistry and physics, or expressed in another way: it is much more difficult to transfer laboratory data to the environment. The accuracy of the data in this handbook is therefore less than in handbooks of chemistry and physics, and it is more uncertain to transfer ecological data from one situation to another. Most of the data in this handbook are therefore accompanied with information on the conditions that were valid for the analysis and determinations to enable the user to evaluate the possibilities to transfer the data to his conditions. As the literature sources for all data are indicated, further details on the conditions can easily be found if needed.

This handbook contains *ecological and ecotoxicological* data available for a quantitative evaluation of environmental impacts and effects. The number of purely *toxicological* data is, however, rather limited. Only toxicological data of direct interest for environmental problems have been included.

There are two reasons for this: 1) the book would be too large and unwieldy, 2) data bases already exist in Europe and U.S.A. with pure toxicological data.

Environmental modelling has developed very rapidly during the last 10-15 years, and models are used more and more in environmental management. The handbook has therefore put emphasis on the parameters used in environmental modelling. This does not imply that calibration of model parameters is redundant, eventhough it may be the case for a limited number of parameters, but it facilitates the calibration significantly because the calibration is so much easier when it is feasible to limit the parameters to a certain (realistic) interval.

The data and parameters are organized in 7 chapters:

The first chapter covers **Composition and Ecological Parameters of Living Organisms**. In this chapter it is possible to find the concentrations of various compounds and elements in living organisms as they are found in nature. These informations are of use in setting up mass balances and in a comparison between a given set of analytical data, for instance your own and usually obtained analytical data. Furthermore, the chapter contains data on size volume and/or weight of organisms and on rate parameters such as growth rates, uptake rates, half saturation constants etc. of use in estimation of the ecological parameters that are of great importance in ecological modelling as underlined above.

Chapter 2 is named **The Ecosphere and the Chemical Compounds**. It contains data about the physico-chemical processes such as biological decomposition, run off and sorption in the ecosphere. The chapter considers the big scale processes and presents therefore the global cycling of elements and the processes characteristic for the entire ecosphere. Climatic data and information about the spheres including productivity are given in this chapter as well.

Data about light and light processes are given in the last part of the chapter, followed by information on environmental processes of pesticides. The last part of chapter 2 gives data about sediment and sediment processes.

Chapter 3 is devoted to **Equations for Environmental Processes**. It does not include submodels or theoretically well established equations for biological processes - a good overview of these equations may be found in S.E. Jørgensen, Fundamentals of Ecological Modelling (1988) - but only empirical equations, mainly found on basis of regression analysis. The equations listed in this chapter may be useful in modelling to replace tables which usually are cumbersome to use in computer programs.

Chapter 4 focuses on **Biological Effects**. All tables in this chapter are headed as "Effect of .....", the dots being a chemical compound. The effects includes mortality, i.e. lethal concentrations are indicated, but also sublethal effects may be found. The tables are easily applied as they are ordered alphabetically on basis of the chemical compounds. Notice that elements are given as the chemical symbols, for instance Cd for Cadmium.

Chapter 5 is entitled **Chemical Compounds and the Living Organisms**. It attempts to cover the interaction between chemical compounds and the biosphere. The chapter gives information on

1) background concentrations to be able to compare a given concentration with what is otherwise found in nature, particularly in uncontaminated nature,

2) concentration factors, i.e. the ratio of concentrations in organisms and their environment,

3) concentrations of various compounds found in typical parts of the environment and

4) excretion and uptake rates for chemical compounds by various organisms.

The sixth chapter contains information on **Processes in the Environment**. It gives equilibrium and rate constants for environmentally important processes.

The last chapter is named **Pesticides** and is devoted to *effects* of the pesticides. Chemical companies producing pesticides may have more information, not given in the accessible literature. It has only been possible for us to summarize the data that can be found in the literature, but this should anyhow facilitate the search for data for those who do not have access to the data of the chemical producers.

Some of the tables throughout the book needs some additional information. They are given as footnotes which can be found last in the volume.

Original references are given for all data as the users may wish to critically evaluate the basis for the data. A comprehensive index has been included to facilitate the use of the handbook.

CONTENTS

LIST OF TABLES

LIST OF TABLES

Chapter 1

COMPOSITION AND ECOLOGICAL PARAMETERS OF LIVING ORGANISMS

Algae
Bacteria
Detritus
Fish
Fungi
Invertebrates
Kormophytes
Vertebrates
Zooplankton

1-001 Algae
ATP - C / biomass ratio

Species	Value	Condition
Trichodesmium spp.	3.3 microgram ATP / Colony	ATP-Carbon/colony, N Pacific, N-fixing algae, n=6[1]

Source: 1.Mague, T.H., F.C. Mague and O. Holm-Hansen. 1977. Mar. Biol. 41(3), pp 213-227.

1-002 Algae
ATP / (C + N) ratio

Species	Value	Condition
Chaetoceros lauderi	0.00457 by weight	Cultivated, marine[1]
Chlamydomonas magnusii	0.0018 by weight	Cultivated, marine[1]
Coscinodiscus sp.	0.00319 by weight	Cultivated, marine[1]
Cryptomonas pseudobaltica	0.00357 by weight	Cultivated, marine[1]
Heterothrix sp.	0.00149 by weight	Cultivated, marine[1]
Phaeodactylum tricornutum	0.0024 by weight	Cultivated, marine[1]
Tetraselmis stricia	0.00153 by weight	Cultivated, marine[1]

Source: 1.Berland, B.R., P.J. Bonin, P.L. Laborde and S.Y. Maestrini. 1972. Mar. biol., 13, pp 338-345.

1-003 Algae
ATP / C ratio

Species	Value	Condition
Chaetoceros lauderi	0.00557 by weight	Cultivated, marine[1]
Chlamydomonas magnusii	0.00213 by weight	Cultivated, marine[1]
Coscinodiscus sp.	0.00389 by weight	Cultivated, marine[1]
Cryptomonas pseudobaltica	0.00447 by weight	Cultivated, marine[1]
Heterothrix sp.	0.00169 by weight	Cultivated, marine[1]
Marine phytoplankton	0.00345 by weight	Urea as source of N, stationary phase, 293 K, Southern Calif[2]
Marine phytoplankton	0.00357 by weight	NO3 as source of N, log phase, 293 K, Southern Calif[2]
Marine phytoplankton	0.0037 by weight	NO3 as source of N, stationary phase, 293 K, Southern Calif[2]
Marine phytoplankton	0.00385 by weight	NH4 as source of N, log phase, 293 K, Southern Calif[2]
Marine phytoplankton	0.00385 by weight	Urea as source of N, log phase, 293 K, Southern Calif[2]
Marine phytoplankton	0.00435 by weight	Nh4 as source of N, stationary phase, 293 K, Southern Calif[2]
Peridinium cinctum fa. westii	2374 by weight	Grown without phosphorus[3]
Peridinium cinctum fa. westii	297 by weight	Grown with phosphorus[3]
Peridinium cinctum fa. westii	310 by weight	None[3]
Phaeodactylum tricornutum	0.003 by weight	Cultivated, marine[1]
Tetraselmis stricia	0.00197 by weight	Cultivated, marine[1]
Thalassiosira pseudonana	0.000478 by weight	P-limited chemostat, growth=0.017 1/h, 295 K, 0.07 cal/cm2*min[4]
Thalassiosira pseudonana	0.00123 by weight	P-limited chemostat, growth=0.041 1/h, 295 K, 0.07 cal/cm2*min[4]
Thalassiosira pseudonana	0.00324 by weight	N-limited chemostat, growth=0.0085 1/h, 295 K, 0.07 cal/cm2*min[4]
Thalassiosira pseudonana	0.00555 by weight	N-limited chemostat, growth=0.017 1/h, 295 K, 0.07 cal/cm2*min[4]
Thalassiosira pseudonana	0.00719 by weight	N-limited chemostat, growth=0.041 1/h, 295 K, 0.07 cal/cm2*min[4]

Source: 1.Berland, B.R., P.J. Bonin, P.L. Laborde and S.Y. Maestrini. 1972. Mar. biol., 13, pp 338-345.
2.Eppley, R.W., A.F. Carlucci, O.H.-Hansen, D. Kiefer, J.J. McCarthy, E. Venrick and P.M. Williams. 1971. Limnol. Oceanogr., 16(5), pp 741-757.
3.Cavari, B. 1976. Limnol. Oceanogr. 21(2), pp 231-236.
4.Perry, M.J. 1976. Limnol. Oceanogr. 21(1), pp 88-107.

1-004 Algae
ATP / Chlorophyll a ratio

Species	Value	Condition
Chlorella sp.	0.29 mg/mg	200 - 300 FTC, r=0.725[1]
Freshwater algae	0.09 mg/mg	Plankton from Bivins Arm, hypereutrophic, r=0.869[1]
Freshwater algae	0.11 mg/mg	Plankton from Newnans Lake, eutrophic, r=0.718[1]
Freshwater algae	0.69 mg/mg	Plankton from Anderson-Cue Lake, oligotrophic, r=0.790[1]
Microcystis aeruginosa	0.09 mg/mg	200 - 300 FTC, r=0.683[1]
Selenastrum capricornatum	0.35 mg/mg	200 - 300 FTC, r=0.539[1]
Thalassiosira pseudonana	0.037 by weight	P-limited chemostat, growth=0.041 1/h, 295 K, 0.07 cal/cm2*min[2]
Thalassiosira pseudonana	0.038 by weight	P-limited chemostat, growth=0.017 1/h, 295 K, 0.07 cal/cm2*min[2]
Thalassiosira pseudonana	0.2 by weight	N-limited chemostat, growth=0.041 1/h, 295 K, 0.07 cal/cm2*min[2]
Thalassiosira pseudonana	0.25 by weight	N-limited chemostat, growth=0.017 1/h, 295 K, 0.07 cal/cm2*min[2]
Thalassiosira pseudonana	0.33 by weight	N-limited chemostat, growth=0.0085 1/h, 295 K, 0.07 cal/cm2*min[2]

Source: 1.Brungs, W.A., J.R. Geckler and M. Gast. 1976. Water Res. 10(1), pp 37-44.
2.Perry, M.J. 1976. Limnol. Oceanogr. 21(1), pp 88-107.

1-005 Algae
ATP / N ratio

Species	Value	Condition
Chaetoceros lauderi	0.0262 by weight	Cultivated, marine[1]
Chlamydomonas magnusii	0.013 by weight	Cultivated, marine[1]
Coscinodiscus sp.	0.0178 by weight	Cultivated, marine[1]
Cryptomonas pseudobaltica	0.0153 by weight	Cultivated, marine[1]
Heterothrix sp.	0.0127 by weight	Cultivated, marine[1]
Phaeodactylum tricornutum	0.0125 by weight	Cultivated, marine[1]
Tetraselmis stricia	0.00661 by weight	Cultivated, marine[1]
Thalassiosira pseudonana	0.00441 by weight	P-limited chemostat, growth=0.017 1/h, 295 K, 0.07 cal/cm2*min[2]
Thalassiosira pseudonana	0.0053 by weight	P-limited chemostat, growth=0.041 1/h, 295 K, 0.07 cal/cm2*min[2]
Thalassiosira pseudonana	0.0357 by weight	N-limited chemostat, growth=0.041 1/h, 295 K, 0.07 cal/cm2*min[2]
Thalassiosira pseudonana	0.04 by weight	N-limited chemostat, growth=0.0085 1/h, 295 K, 0.07 cal/cm2*min[2]
Thalassiosira pseudonana	0.0476 by weight	N-limited chemostat, growth=0.017 1/h, 295 K, 0.07 cal/cm2*min[2]

Source: 1.Berland, B.R., P.J. Bonin, P.L. Laborde and S.Y. Maestrini. 1972. Mar. biol., 13, pp 338-345.
2.Perry, M.J. 1976. Limnol. Oceanogr. 21(1), pp 88-107.

1-006 Algae
ATP / P ratio

Species	Value	Condition
Thalassiosira pseudonana	0.114 by weight	N-limited chemostat, growth=0.0085 1/h, 295 K, 0.07 cal/cm2*min[1]
Thalassiosira pseudonana	0.115 by weight	P-limited chemostat, growth=0.041 1/h, 295 K, 0.07 cal/cm2*min[1]
Thalassiosira pseudonana	0.123 by weight	P-limited chemostat, growth=0.017 1/h, 295 K, 0.07 cal/cm2*min[1]
Thalassiosira pseudonana	0.175 by weight	N-limited chemostat, growth=0.041 1/h, 295 K, 0.07 cal/cm2*min[1]
Thalassiosira pseudonana	0.182 by weight	N-limited chemostat, growth=0.017 1/h, 295 K, 0.07 cal/cm2*min[1]

Source: 1.Perry, M.J. 1976. Limnol. Oceanogr. 21(1), pp 88-107.

1-007 Algae
ATP / biomass ratio

Species	Value	Condition
Anabaena flos-aquae	0.26 mg/g dry weight	200 - 300 FTC, r=0.607[1]
Chaetoceros lauderi	0.269 microgram ATP / mm**3	Cultivated, marine[2]
Chlamydomonas magnusii	0.402 microgram ATP / mm**3	Cultivated, marine[2]
Chlorella sp.	0.38 mg/g dry weight	200 - 300 FTC, r=0.860[1]
Coscinodiscus sp.	0.206 microgram ATP / mm**3	Cultivated, marine[2]
Cryptomonas pseudobaltica	0.666 microgram ATP / mm**3	Cultivated, marine[2]
Freshwater algae	0.12 mg/g dry weight	Plankton from Newnans lake, eutrophic, r=0.673[1]
Freshwater algae	0.13 mg/g dry weight	Plankton from Bivins arm, hypereutrophic, r=0.649[1]
Freshwater algae	0.96 mg/g dry weight	Plankton from Anderson-Cue Lake, oligotrophic, r=0.790[1]
Heterothrix sp.	0.597 microgram ATP / mm**3	Cultivated, marine[2]
Marine phytoplankton	1E-12 g/cell	NH4 as source of N, stationary phase, 293 K, Southern Calif[3]
Marine phytoplankton	5.3E-13 g/cell	Urea as source of[3]
Marine phytoplankton	8.2E-13 g/cell	NO3 as source of N, stationary phase, 293 K, Southern Calif[3]
Microcystis aeruginosa	0.36 mg/g dry weight	200 - 300 FTC, r=0.718[1]
Phaeodactylum tricornutum	0.469 microgram ATP / mm**3	Cultivated, marine[2]
Selenastrum capricornutum	0.24 mg/g dry weight	200 - 300 FTC, r=0.244[1]
Tetraselmis stricia	0.159 microgram ATP / mm**3	Cultivated, marine[2]
Thalassiosira sp.	2.6E-15 gram/cell	Growth rate=0.017 1/h, cell density=6.7E8 cell/l, P-limited chemostat[4]
Thalassiosira sp.	3.9E-15 gram/cell	Growth rate=0.041 1/h, cell density=4.0E8 cell/l, P-limited chemostat[4]
Trichodesmium spp.	13 ng/colony	ATP / Colony, N Pacific, N-fixing algae, n=6[5]

Source: 1.Brungs, W.A., J.R. Geckler and M. Gast. 1976. Water Res. 10(1), pp 37-44.
2.Berland, B.R., P.J. Bonin, P.L. Laborde and S.Y. Maestrini. 1972. Mar. biol., 13, pp 338-345.
3.Eppley, R.W., A.F. Carlucci, O.H.-Hansen, D. Kiefer, J.J. McCarthy, E. Venrick and P.M. Williams. 191. Limnol. Oceanogr., 16(5), pp 741-757.
4.Perry, M.J. 1976. Limnol. Oceanogr. 21(1), pp 88-107.
5.Mague, T.H., F.C. Mague and O. Holm-Hansen. 1977. Mar. Biol. 41(3), pp 213-227.

1-008 Algae
ATP / chlorophyll a ratio

Species	Value	Condition
Anabaena flos-aquae	0.09 by weight	200 - 300 ftc, r = 0.707[1]

Source: 1.Brungs, W.A., J.R. Geckler and M. Gast. 1976. Water Res. 10(1), pp 37-44.

1-009 Algae
Ash / biomass ratio

Species	Value	Condition
Anabaena sp.	5.19 % of dry weight	Algae from ponds and lakes in southeastern United States[1]
Aphanizomenon sp.	7.21 % of dry weight	Algae from ponds and lakes in southeastern United States[1]
Bacillariophyceae	58.5 % of dry weight	Plankton, 8 species[2]
Chara sp.	43.4 % of dry weight	Algae from ponds and lakes in southeastern United States[1]
Cladophora sp.	23.38 % of dry weight	Algae from ponds and lakes in southeastern United States[1]
Euglena sp.	4.12 % of dry weight	Algae from ponds and lakes in southeastern United States[1]
Hydrodictyon sp.	17.94 % of dry weight	Algae from ponds and lakes in southeastern United States[1]
Lyngbya sp.	17.2 % of dry weight	Algae from ponds and lakes in southeastern United States[1]
Microcystis sp.	6.2 % of dry weight	Algae from ponds and lakes in southeastern United States[1]
Mougeotia sp.	14.54 % of dry weight	Algae from ponds and lakes in southeastern United States[1]
Nitella sp.	19.11 % of dry weight	Algae from ponds and lakes in southeastern United States[1]
Oedogonium sp.	12.69 % of dry weight	Algae from ponds and lakes in southeastern United States[1]
Phaeophyceae	28.8 % of dry weight	16 species[2]
Pithophora sp.	27.77 % of dry weight	Algae from ponds and lakes in southeastern United States[1]
Rhizoclonium sp.	17.36 % of dry weight	Algae from ponds and lakes in southeastern United States[1]
Spirogyra sp.	13.06 % of dry weight	Algae from ponds and lakes in southeastern United States[1]
Spirogyra sp., giant	13.86 % of dry weight	Algae from ponds and lakes in southeastern United States[1]

Source: 1.Water Quality Criteria. 1968. Federal Water Pollution Control Administration Washington D.C., pp 234
2.Vinogradov, A.P. 1953. The Elementary Chemical Composition of Marine Organisms, Sears Foundation NeHaven.

1-010 Algae
Assimilation

Species	Value	Condition
Algae from Pacific Ocean	1.15 micro g C/(micro g chlorophyll * ho	Chlorophyll=0.092, inorg N-0.46, inorg P=0.33, Si=3.2 mg/m3[1]
Algae from Pacific Ocean	1.36 micro g C/(micro g chlorophyll * ho	Chlorophyll=0.155, inorg N = .00, inorg P=0.26, Si=1.8 mg/m3[1]
Algae from Pacific Ocean	10.6 micro g C/(microgram chl*hour)	Chlorophyll=0.155, inorg N=0.00, inorg P=0.26, Si=1.8 mg/m3[1]
Algae from Pacific Ocean	2.91 micro g C/(micro g chlorophyll * ho	Chlorophyll=0.174, inorg N-0.49, inorg P=0.22, Si=2.1 mg/m3[1]
Algae from Pacific Ocean	3.15 micro g C/(micro g chlorophyll * ho	Mean of 8 nitrogen poor stations[1]
Algae from Pacific Ocean	3.24 micro g C/(micro g chlorophyll * ho	Chlorophyll=0.099, inorg N-0.46, inorg P=0.24, Si=1.5 mg/m3[1]

Algae from Pacific Ocean	3.4 micro g C/(micro g chlorophyll * ho	Chlorophyll=0.135, inorg N-0.34, inorg P=0.23, Si=3.8 mg/m3[1]
Algae from Pacific Ocean	3.53 micro g C/(micro g chlorophyll * ho	Chlorophyll=0.099, inorg N=7.81, inorg P=0.71, Si=--- mg/m3[1]
Algae from Pacific Ocean	3.81 micro g C/(micro g chlorophyll * ho	Chlorophyll=0.142, inorg N-0.45, inorg P=0.30, Si=3.1 mg/m3[1]
Algae from Pacific Ocean	4.04 micro g C/(micro g chlorophyll * ho	Chlorophyll=0.102, inorg N=7.16, inorg P=0.62, Si=3.5 mg/m3[1]
Algae from Pacific Ocean	4.12 micro g C/(micro g chlorophyll * ho	Chlorophyll=0.224, inorg N-0.48, inorg P=0.27, Si=2.0 mg/m3[1]
Algae from Pacific Ocean	4.55 micro g C/(micro g chlorophyll * ho	Chlorophyll=0.263, inorg N=8.35, inorg P=0.83, Si=5.2 mg/m3[1]
Algae from Pacific Ocean	4.93 micro g C/(micro g chlorophyll * ho	Chlorophyll=0.211, inorg N=5.71, inorg P=0.27, Si=5.0 mg/m3[1]
Algae from Pacific Ocean	4.95 micro g C/(micro g chlorophyll * ho	Mean of 9 nitrogen rich stations[1]
Algae from Pacific Ocean	4.98 micro g C/(micro g chlorophyll * ho	Chlorophyll=0.105, inorg N=7.42, inorg P=0.67, Si=3.6 mg/m3[1]
Algae from Pacific Ocean	5.18 micro g C/(micro g chlorophyll * ho	Chlorophyll=0.158, inorg N-0.40, inorg P=0.24, Si=2.5 mg/m3[1]
Algae from Pacific Ocean	5.22 micro g C/(micro g chlorophyll * ho	Chlorophyll=0.151, inorg N=8.61, inorg P=0.75, Si=3.8 mg/m3[1]
Algae from Pacific Ocean	5.23 micro g C/(micro g chlorophyll * ho	Chlorophyll=0.201, inorg N=2.50, inorg P=0.35, Si=3.1 mg/m3[1]
Algae from Pacific Ocean	5.84 micro g C/(micro g chlorophyll * ho	Chlorophyll=0.041, inorg N=8.11, inorg P=0.69, Si=1.6 mg/m3[1]
Algae from Pacific Ocean	6.19 micro g C/(micro g chlorophyll * ho	Chlorophyll=0.181, inorg N=6.09, inorg P=0.58, Si=3.2 mg/m3[1]

Source: 1.Thomas, W.H. 1970. Limnol. Oceanogr. 15(3), pp 380-385.

1-011 Algae
Assimilation number

Species	Value	Condition
Upwelling area	2.59 mg C (mg Chl a)-1	average of number 1.85-3.15, NW-Africa[1]
Upwelling area	5.15 mg C (mg Chl a)-1	average of number 2.86-6.75, off Peru[1]

Source: 1.Minas, H.J., Minas, M. and Packard, T.T. Limnol. Oceanogr., 31(6), 1182, 1986.

1-012 Algae
Assimilation of nitrogen

Species	Value	Condition
Natural population	0.053-1.1 umol l-1 d-1	Hawaiian coastal waters, from incorporation of 14C into protein[1]

Source: 1.Laws, E.A., et al., Limnol.Oceanogr., 29(6), 1161, 1984.

1-013 Algae
B / biomass ratio

Species	Value	Condition
Chara sp.	6.7 microgram/l	Algae from ponds and lakes in southeastern United States[1]
Cladophora sp.	84.6 microgram/l	Algae from ponds and lakes in southeastern United States[1]
Euglena sp.	3.8 microgram/l	Algae from ponds and lakes in southeastern United States[1]
Lyngbya sp.	112 microgram/l	Algae from ponds and lakes in southeastern United States[1]

Microcystis sp.	3.6 microgram/l	Algae from ponds and lakes in southeastern United States[1]
Mougeotia sp.	8 microgram/l	Algae from ponds and lakes in southeastern United States[1]
Nitella sp.	9.8 microgram/l	Algae from ponds and lakes in southeastern United States[1]
Oedogonium sp.	8.1 microgram/l	Algae from ponds and lakes in southeastern United States[1]
Pithophora sp.	65 microgram/l	Algae from ponds and lakes in southeastern United States[1]
Rhizoclonium sp.	1.8 microgram/l	Algae from ponds and lakes in southeastern United States[1]
Spirogyra sp.	4.2 microgram/l	Algae from ponds and lakes in southeastern United States[1]
Spirogyra sp., giant	4.3 microgram/l	Algae from ponds and lakes in southeastern United States[1]

Source: 1.Water Quality Criteria. 1968. Federal Water Pollution Control Administration Washington D.C., pp 234

1-014 Algae
Biomass

Species	Value	Condition
Algae	1600 kcal/m**2	Max standing crop, initial value, thermal ecosystem[1]
Chlorella vulgaris	3.4E-11 g	Green algae[2]
Cladophora sp.	14.5 g dry weight/m**2	Annual mean:t.inor.P = 2.3 mg/l, NO3 = 7.4, NH4 = 0.7 mg N/l, River[3]
Cladophora sp.	15.5 g dry weight/m**2	Annual mean:t.inor.P = 0.8 mg/l, NO3 = 3.5, NH4 = 0.4 mg N/l, River[3]
Cladophora sp.	20.3 g dry weight/m**2	Annual mean:t.inor.P = 0.8 mg/l, NO3 = 4.4, NH4 = 0.4 mg N/l, River[3]
Cladophora sp.	23 g dry weight/m**2	Annual mean:t.inor.P = 1.1 mg/l, NO3 = 4.7, NH4 = 0.5 mg N/l, River[3]
Cladophora sp.	31.6 g dry weight/m**2	Annual mean:t.inor.P = 2.3 mg/l, NO3 = 7.4, NH4 = 0.7 mg N/l, River[3]
Cladophora sp.	34.6 g dry weight/m**2	Annual mean:t.inor.P = 1.1 mg/l, NO3 = 2.7, NH4 = 0.6 mg N/l, River[3]
Cladophora sp.	37.4 g dry weight/m**2	Annual mean:t.inor.P = 1.7 mg/l, NO3 = 6.7, NH4 = 0.05 mg N/l, R. Gr.[3]
Cladophora sp.	5 g dry weight/m**2	Annual mean:t.inor.P = 0.7 mg/l, NO3 = 4.9, NH4 = 0.3 mg N/l, River[3]
Cladophora sp.	5 g dry weight/m**2	Annual mean:t.inor.P = 1.3 mg/l, NO3 = 7.0, NH4 = 0.2 mg N/l, R. Gr.[3]
Cladophora sp.	50.1 g dry weight/m**2	Annual mean:t.inor.P = 1.8 mg/l, NO3 = 4.3, NH4 = 0.3 mg N/l, River[3]
Cladophora sp.	52.6 g dry weight/m**2	Annual mean:t.inor.P = 2.9 mg/l, NO3 = 6.7, NH4 = 0.9 mg N/l, River[3]
Cladophora sp.	54.8 g dry weight/m**2	Annual mean:t.inor.P = 2.8 mg/l, NO3 = 6.6, NH4 = 0.9 mg N/l, River[3]
Cladophora sp.	60.3 g dry weight/m**2	Annual mean:t.inor.P = 1.5 mg/l, NO3 = 4.3, NH4 = 0.9 mg N/l, River[3]
Cladophora sp.	62.9 g dry weight/m**2	Annual mean:t.inor.P = 1.0 mg/l, NO3 = 4.8, NH4 = 0.05 mg N/l, River
Cladophora sp.	66.4 g dry weight/m**2	Annual mean:t.inor.P = 4.6 mg/l, NO3 = 12.5, NH4 = 2.8 mg N/l, River[3]
Cladophora sp.	73.7 g dry weight/m**2	Annual mean:t.inor.P = 2.1 mg/l, NO3 = 2.7, NH4 = 0.4 mg N/l, River[3]
Cladophora sp.	74.7 g dry weight/m**2	Annual mean:t.inor.P = 1.6 mg/l, NO3[3]
Valonia ventriculosa	68 g	The largest unicellular algae[2]

Source: 1.Wiegert, R.G. 1975. In: B.C. Patten (ed.), Systems Analysis and Simulation in Ecology, Vol. 3, pp 15-181. 2.Brock, T.D. 1966. Principles of Microbial Ecology. Prentice-Hall. New Jersey. 3.Pitcairn, C.E.R. and H.A. Hawkes. 1973. Water Res. 7(1/2), pp 159-171.

1-015 Algae
Biosynthesis of polycyclic aromatic hydrocarbon, PAH

Species	Value	Condition
Synthesis of benz(a)anthracene by Chlorella vulgaris	5.7 ug/kg dry algae	19 d, substrate = acetate as sole C source, C-14 labeled[1]
Synthesis of benzo(a)fluoranthene by Chlorella vulgaris	3.7 ug/kg dry algae	19 d, substrate = acetate as sole C source, C-14 labeled[1]
Synthesis of benzo(a)pyrene by Chlorella vulgaris	0.7 ug/kg dry algae	19 d, substrate = acetate as sole C source, C-14 labeled[1]
Synthesis of benzo(ghi)perylene by Chlorella vulgaris	2 ug/kg dry algae	19 d, substrate = acetate as sole C source, C-14 labeled[1]
Synthesis of benzo(ghi)perylene by Chlorella vulgaris	2.0 ug/kg dry algae	19 d, substrate = acetate as sole C source, C-14 labeled[1]
Synthesis of benzo(k)fluoranthene by Chlorella vulgaris	1.3 ug/kg dry algae	19 d, substrate = acetate as sole C source, C-14 labeled[1]
Synthesis of fluoranthene by Chlorella vulgaris	54.6 ug/kg dry algae	19 d, substrate = acetate as sole C source, C-14 labeled[1]
Synthesis of indeno(1,2,3-cd)pyrene by Chlorella vulgaris	1.4 ug/kg dry algae	19 d, substrate = acetate as sole C source, C-14 labeled[1]

Source: 1.Schmelts, N. and D. Hoffman. 1976. In: R. Freudenthal and P.W. Jones (eds.): Carcinogenes - Comprehesive Survey. Vol 1. Polynuclear Aromatic Hydrocarbones. NY. Raven Press.

1-016 Algae
Biotin / biomass ratio

Species	Value	Condition
Marine phytoplankton	1.4E-16 g/cell	Nh4 as source of N, stationary phase, 293 K, Southern Calif[1]
Marine phytoplankton	1.9E-16 g/cell	Urea as source of N, stationary phase, 293 K, Southern Calif[1]
Marine phytoplankton	3.2E-16 g/cell	NO3 as source of N, stationary phase, 293 K, Southern Calif[1]

Source: 1.Eppley, R.W., A.F. Carlucci, O.H.-Hansen, D. Kiefer, J.J. McCarthy, E. Venrick and P.M. Williams. 1971. Limnol. Oceanogr., 16(5), pp 741-757.

1-017 Algae
C / N / P ratio

Species	Value	Condition
Trichodesmium spp.	4.05 C / N ratio by weight	N Pacific, N-fixing algae, n=6, C:N:P = 81:20:1[1]

Source: 1.Mague, T.H., F.C. Mague and O. Holm-Hansen. 1977. Mar. Biol. 41(3), pp 213-227.

1-018 Algae
C / N ratio

Species	Value	Condition
Algae	12.9 by weight	Used in models, rg = 20 - 5.88[1]
Algae	5 by weight	Used in models, Lake Michigan[2]
Algae	5 by weight	Used in models, Lake Ontario[3]

Algae	5 by weight	Used in models, Potomac Estuary[3]
Algae	5.14 by weight	Used in models, Lake Erie[3]
Algae	5.88 by weight	Used in models, Grays Harbor, Washington[4]
Algae	6.25 by weight	Used in models[5]
Algae	7.14 by weight	Used in models, San Joaquin River Estuary, Calif[3]
Algae	7.14 by weight	Used in models, San Joaquin River, Calif[3]
Asterionella formosa	5.78	algal culture[6]
Chaetoceros debilis	11.4 by atoms	Nitrogen limited, dilution rate=0.097-0.008/hr, rg = 9.8 - 12.9[7]
Chaetoceros debilis	3.2 by atoms	Not limited, cont light : 0.08 ly/min, dilution=1/3-1/2 of[8]
Chaetoceros debilis	4 by atoms	NH4 starved, 72 h cont light : 0.08 ly/min, dilution=1/3-1/2 of[8]
Chaetoceros debilis	4.1 by atoms	Si starved, 72 h cont light : 0.08 ly/min, dilution=1/3-1/2 of[8]
Chaetoceros debilis	5.4 by atoms	Si limited, cont light : 0.08 ly/min, dilution=1/3-1/2 of max.gr.rate[8]
Chaetoceros debilis	6.4 by atoms	NH4 limited, cont light : 0.08 ly/min, dilution=1/3-1/2 of[8]
Chaetoceros lauderi	4.79 by weight	Cultivated, marine[9]
Chaetoceros lauderi	5.6 by atoms	Grown under optimal condition[9]
Chaetoceros sp.	6.6 by atoms	Grown under optimal condition[10]
Chlamydomonas magnusii	6.43 by weight	Cultivated, marine[9]
Chlorella sp.	4.82	algal culture[6]
Coscinodiscus sp.	4.56 by weight	Cultivated, marine[9]
Coscinodiscus sp.	5.3 by atoms	Grown under optimal condition[9]
Cryptomonas erosa	5.12	algal culture[6]
Cryptomonas pseudobaltica	3.52 by weight	Cultivated, marine[9]
Cyclotella nana	13.9 by atoms	Nitrate limited, dilution rate=0.077-0.009/hr, rg = 7.7 - 20.0[11]
Ditylum brightwellii	5.2 by atoms	Grown under optimal condition[12]
Fragilaria crotonensis	6.82	algal culture[6]
Heterothrix sp.	7.51 by weight	Cultivated, marine[9]
Isochrysis galbana	6.0 mol mol-1	light conditions: 320 umol quanta m-2 s-1[13]
Isochrysis galbana	6.4 mol mol-1	light conditions: 70 umol quanta m-2 s-1[13]
Isochrysis galbana	6.6 mol mol-1	light conditions: 150 umol quanta m-2 s-1[13]
Isochrysis galbana	7.3 mol mol-1	light conditions: 30 umol quanta m-2 s-1[13]
Isochrysis galbana	8.7 mol mol-1	light conditions: 600 umol quanta m-2 s-1[13]
Marine phytoplankton	3 by weight	Vigorously growing, excess nitrate in water[14]
Marine phytoplankton	3.9 by weight	NH4 as source of N, log phase, 293 K, Southern Calif[15]
Marine phytoplankton	4.3 by weight	NH4 as source of N, stationary phase, 293 K, Southern Calif[15]
Marine phytoplankton	4.5 by weight	Unhealthy phytoplankton, nitrate depleted water[14]
Marine phytoplankton	5.2 by weight	NO3 as source of N, stationary phase, 293 K, Southern Calif[15]
Marine phytoplankton	5.6 by weight	Urea as source of N, log phase, 293 K, Southern Calif[15]
Marine phytoplankton	5.9 by weight	NO3 as source of N, log phase, 293 K, Southern Calif[15]
Marine phytoplankton	6 by weight	Range = 4 - 8[16]
Marine phytoplankton	7 by weight	Urea as source of N, stationary phase, 293 K, Southern Calif[15]

Melosira italica	3.38 pg cell-1	algal culture[6]
Melosira italica tenuissima	4.87	algal culture[6]
Phaeodactylum tricornutum	4.23 by weight	Cultivated, marine[9]
Phaeodactylum tricornutum	4.9 by atoms	Grown under optimal condition[9]
Prorocentrum micans	4.3 mol mol-1	light conditions: 70 umol quanta m-2 s-1[13]
Prorocentrum micans	4.6 mol mol-1	light conditions: 600 umol quanta m-2 s-1[13]
Prorocentrum micans	4.8 mol mol-1	light conditions: 150 umol quanta m-2 s-1[13]
Prorocentrum micans	4.8 mol mol-1	light conditions: 320 umol quanta m-2 s-1[13]
Skeletonema costatum	13.2 by atoms	Si starved, 72 h cont light : 0.08 ly/min, dilution=1/3-1/2 of max.gr.rate[8]
Skeletonema costatum	14.8 by atoms	NH4 starved, 72 h cont light : 0.08 ly/min, dilution=1/3-1/2 of max.gr.rate8
Skeletonema costatum	5.1 by atoms	Not limited, cont light : 0.08 ly/min, dilution=1/3-1/2 of max.gr.rate[8]
Skeletonema costatum	5.3 by atoms	Grown under optimal condition[10]
Skeletonema costatum	6.8 by atoms	Si limited, cont light : 0.08 ly/min, dilution=1/3-1/2 of max.gr.rate[8]
Skeletonema costatum	7.1 by atoms	Nitrogen limited, batch culture, 72-115 hrs starvation[17]
Skeletonema costatum	9.5 by atoms	NH4 limited, cont light : 0.08 ly/min, dilution=1/3-1/2 of max.gr.rate[8]
Stephanodiscus astraea minitula	3.99 pg cell-1	algal culture[6]
Stephanodiscus hantzschii	4.89 pg cell-1	algal culture[6]
Stephanodiscus niagarae	6.24	algal culture[6]
Tabellaria fenestrata	4.29 pg cell-1	algal culture[6]
Tetraselmis stricia	3.78 by weight	Cultivated, marine[9]
Thalassiosira fluviatilis	9.15 by atoms	Nitrate limited, batch culture, 72-115 hrs starvation[18]
Thalassiosira gravida	12.3 by atoms	NH4 limited, cont light : 0.08 ly/min, dilution=1/3-1/2 of max.gr.rate[8]
Thalassiosira gravida	2.7 by atoms	Not limited, cont light : 0.08 ly/min, dilution=1/3-1/2 of max.gr.rate[8]
Thalassiosira gravida	3.5 by atoms	NH4 starved, 72 h cont light : 0.08 ly/min, dilution=1/3-1/2 of max.gr.rate[8]
Thalassiosira gravida	4.1 by atoms	Si limited, cont light : 0.08 ly/min, dilution=1/3-1/2 of max.gr.rate[8]
Thalassiosira gravida	4.6 by atoms	Si starved, 72 h cont light : 0.08 ly/min, dilution=1/3-1/2 of max.gr.rate[8]
Thalassiosira pseudonana	10 by atoms	N limited chemostat, growth=0.017 1/h, 295 K, 0.07 cal/cm2*min[19]
Thalassiosira pseudonana	10.25 by atoms	Nitrogen limited, dilution rate=0.041-0.008/hr, rg = 5.7 - 14.8[20]
Thalassiosira pseudonana	10.7 by atoms	P limited chemostat, growth=0.017 1/h, 295 K, 0.07 cal/cm2*min[19]
Thalassiosira pseudonana	14.8 by atoms	N limited chemostat, growth=0.0085 1/h, 295 K, 0.07 cal/cm2*min[19]
Thalassiosira pseudonana	5.2 by atoms	Grown under optimal condition[11]
Thalassiosira pseudonana	5.7 by atoms	N limited chemostat, growth=0.041 1/h, 295 K, 0.07 cal/cm2*min[19]
Thalassiosira pseudonana	5.8 by atoms	P limited chemostat, growth=0.041 1/h, 295 K, 0.07 cal/cm2*min[19]
Thalassiosira pseudonana	7.4 by atoms	Nitrate limited, batch culture, 72-115 hrs starvation, rg = 4.8 -[21]

Thalassiosira weisflogii	7.4 mol mol-1	light conditions: 30 umol quanta m-2 s-1[13]
Thalassiosira weisflogii	7.6 mol mol-1	light conditions: 150 umol quanta m-2 s-1[13]
Thalassiosira weisflogii	7.7 mol mol-1	light conditions: 70 umol quanta m-2 s-1[13]
Thalassiosira weisflogii	9.4 mol mol-1	light conditions: 600 umol quanta m-2 s-1[13]
Thalassiosira weisflogii	9.7 mol mol-1	light conditions: 320 umol quanta m-2 s-1[13]
Trichodesmium spp.	4.1 by weight	N Pacific, N-fixing algae, n=6[22]

Source: 1.Baca and Arnett. 1976. A Limnological Model For Eutrophic lakes and Impoundments. Battle Pacific NWLab. EPA.
2.Canale, R.P., L.M. De Palma and A.H. Vogel. 1976. in: Canale, R.P. (ed.); Mathematical modelling of biochemical processes in Aquatic Ecosystems. Ann Arbor Science Press.
3.O'Connor, F.B. 1963. in: Doeksen, J. and J. van dẽr Drift (eds.); Soil Organisms. pp 32-48. North-Holland publ. Co. Amsterdam.
4.Battelle. 1974. Development of a Mathematical Water Quality Model for Gray's Harbor and the ChehalisRiver, Wash. Documentation Report. Pacific NW Lab. Richland, Wash.
5.O'Connor, D.J., R.V. Thomann and D.M. DiToro. 1973. Dynamic Water Quality Forecasting and Management. Office of Research and Development U.S. Environmental Protection Agency, Wash. 20460.
6.Infante, A. and Litt, A.H., Limnol.Oceanogr., 30(5), 1053, 1985.
7.Thomas, W.H. and A.N. Dodson. 1972. Limnol. Oceanogr. 17, pp 515-523.
8.Harrison, P.J., H.L. Conway, R.W. Holmes and C.O. Davis. 1977. Mar. Biol., 43(1), pp 19-31.
9.Berland, B.R., P.J. Bonin, P.L. Laborde and S.Y. Maestrini. 1972. Mar. biol., 13, pp 338-345.
10.Parsons, T.R., K. Stephens and J.D.H. Strickland. 1961. J. Fish. Res. Bd. Can., 18, pp 1001-1016.
11.Caperon, J. and J. Meyer. 1972. Deep Sea Res., 19, pp 601-632.
12.Strickland, J.D.H., O. Holm-Hansen, R.W. Eppley and R. J. Linn. 1969. Limnol. Oceanogr., 14, pp 23-2.
13.Falkowski, P.G., Dubinsky, Z. and Wyman, K.., Limnol.Oceanogr., 30(2), 311, 1985.
14.Antia, N.J., C.D. McAllister, T.R. Parsons, K. Stephens and J.H.D. Strickland. 1963. Limnol. Oceanog., 8, pp 166-183.
15.Eppley, R.W., A.F. Carlucci, O.H.-Hansen, D. Kiefer, J.J. McCarthy, E. Venrick and P.M. Williams. 1971. Limnol. Oceanogr., 16(5), pp 741-757.
16.Strickland, J.D.H. 1960. Fish. Res. Bd. Can. Bull., 122, pp 172.
17.Harrison, P.J., H.L. Conway and R.C. Dugdale. 1976. Mar. biol. 35, pp 177-186.
18.Holm-Hansen, O., W.H. Sutcliffe, Jr. and J. Sharp. 1968. Limnol. Oceanogr., 13, pp 507-514.
19.Perry, M.J. 1976. Limnol. Oceanogr. 21(1), pp 88-107.
20.Eppley, R.W. and E.H. Renger. 1974. J. Phycol., 10, pp 15-23.
21.Hobson, L.A. and R.J. Pariser. 1971. J. exp. mar. biol. Ecol. 6, pp 71-78.
22.Mague, T.H., F.C. Mague and O. Holm-Hansen. 1977. Mar. Biol. 41(3), pp 213-227.

1-019 Algae
C / P ratio

Species	Value	Condition
Asterionella japonica	200 by atoms	Batch culture, cell vol=12 % Carbon[1]
Chaetoceros debilis	24 by atoms	NH4 starved, 72 h cont light : 0.08 ly/min, dilution=1/3-1/2 of[2]
Chaetoceros debilis	29 by atoms	NH4 limited, cont light : 0.08 ly/min, dilution=1/3-1/2 of[2]
Chaetoceros debilis	33 by atoms	Si starved, 72 h cont light : 0.08 ly/min, dilution=1/3-1/2 of[2]
Chaetoceros debilis	39 by atoms	Not limited, cont light : 0.08 ly/min, dilution=1/3-1/2 of[2]
Chaetoceros debilis	43 by atoms	Si limited, cont light : 0.08 ly/min, dilution=1/3-1/2 of max.gr.rate[2]
Chaetoceros gracilis	1000 by atoms	Batch culture, cell vol=12 % Carbon[1]
Marine phytoplankton	20 by weight	Vigorously growing, excess nitrate in water[3]
Marine phytoplankton	33 by weight	Unhealthy phytoplankton, nitrate depleted water[3]
Marine phytoplankton	40 by weight	Range = 25 - 55[4]
Monochrysis lutheri	143 by atoms	Batch culture, cell vol=12 % Carbon[1]
Phaeodactylum tricornutum	416 by atoms	Batch culture, dry cell=50 % Carbon[1]
Skeletonema costatum	104 by atom	Si starved, 72 h cont light : 0.08 ly/min, dilution=1/3-1/2 of max.gr.rate[2]
Skeletonema costatum	46 by atom	NH4 limited, cont light : 0.08 ly/min, dilution=1/3-1/2 of max.gr.rate[2]

Skeletonema costatum	51 by atom	Not limited, cont light : 0.08 ly/min, dilution=1/3-1/2 of max.gr.rate[2]
Skeletonema costatum	57 by atom	Si limited, cont light : 0.08 ly/min, dilution=1/3-1/2 of max.gr.rate[2]
Skeletonema costatum	62 by atoms	NH4 starved, 72 h cont light : 0.08 ly/min, dilution=1/3-1/2 of max.gr.rate[2]
Thalassiosira gravida	18 by atoms	NH4 starved, 72 h cont light : 0.08 ly/min, dilution=1/3-1/2 of max.gr rate[2]
Thalassiosira gravida	24 by atoms	NH4 limited, cont light : 0.08 ly/min, dilution=1/3-1/2 of max.gr.rate[2]
Thalassiosira gravida	24 by atoms	Not limited, cont light : 0.08 ly/min, dilution=1/3-1/2 of max.gr.rate[2]
Thalassiosira gravida	40 by atoms	Si limited, cont light : 0.08 ly/min, dilution=1/3-1/2 of max.gr.rate[2]
Thalassiosira gravida	51 by atoms	Si starved, 72 h cont light : 0.08 ly/min, dilution=1/3-1/2 of max.gr.rate[2]
Thalassiosira pseudonana	220 by atoms	P limited chemostat, growth=0.041 1/h, 295 K, 0.07 cal/cm2*min[5]
Thalassiosira pseudonana	63.4 by atoms	N limited chemostat, growth=0.041 1/h, 295 K, 0.07 cal/cm2*min[5]
Thalassiosira pseudonana	665 by atoms	P limited chemostat, growth=0.017 1/h, 295 K, 0.07 cal/cm2*min[5]
Thalassiosira pseudonana	77 by atoms	Nitrogen limited, dilution rate=0.041-0.008/hr, rg = 63.4 - 90.5[6]
Thalassiosira pseudonana	84.3 by atoms	N limited chemostat, growth=0.017 1/h, 295 K, 0.07 cal/cm2*min[5]
Thalassiosira pseudonana	90.5 by atoms	N limited chemostat, growth=0.0085 1/h, 295 K, 0.07 cal/cm2*min[5]

Source: 1.Tett, P., J.C. Cottrelll, D.O. Trew and B.J.B. Wood. 1975. Limnol. Oceanogr. 20(4), pp 587-603.
2.Harrison, P.J., H.L. Conway, R.W. Holmes and C.O. Davis. 1977. Mar. Biol., 43(1), pp 19-31.
3.Antia, N.J., C.D. McAllister, T.R. Parsons, K. Stephens and J.H.D. Strickland. 1963. Limnol. Oceanogr., 8, pp 166-183.
4.Strickland, J.D.H. 1960. Fish. Res. Bd. Can. Bull., 122, pp 172.
5.Perry, M.J. 1976. Limnol. Oceanogr. 21(1), pp 88-107.
6.Eppley, R.W. and E.H. Renger. 1974. J. Phycol., 10, pp 15-23.

1-020 Algae
C / Si ratio

Species	Value	Condition
Bacillariophyceae	5 by weight	None[1]
Chaetoceros debilis	19.6 by atoms	Si starved, 72 h cont light : 0.08 ly/min, dilution=1/3-1/2 of max.gr.rate[2]
Chaetoceros debilis	22.5 by atoms	Si limited, cont light : 0.08 ly/min, dilution=1/3-1/2 of max.gr.rate[2]
Chaetoceros debilis	3 by atoms	NH4 limited, cont light : 0.08 ly/min, dilution=1/3-1/2 of max.gr.rate[2]
Chaetoceros debilis	3.6 by atoms	NH4 starved, 72 h cont light : 0.08 ly/min, dilution=1/3-1/2 of max.gr.rate[2]
Chaetoceros debilis	5.8 by atoms	Not limited, cont light : 0.08 ly/min, dilution=1/3-1/2 of max.gr.rate[2]
Marine phytoplankton	1 by weight	Unhealthy phytoplankton, nitrate depleted water[3]
Marine phytoplankton	1.25 by weight	See ref.[4]
Marine phytoplankton	1.67 by weight	Vigorously growing, excess nitrate in water[3]
Skeletonema costatum	11.9 by atoms	NH4 starved, 72 h cont light : 0.08 ly/min, dilution=1/3-1/2 of max.gr.rate[2]
Skeletonema costatum	27.7 by atoms	Si limited, cont light : 0.08 ly/min, dilution=1/3-1/2 of max.gr.rate[2]
Skeletonema costatum	36.5 by atoms	Si starved, 72 h cont light : 0.08 ly/min, dilution=1/3-1/2 of max.gr.rate[2]

Skeletonema costatum	7.8 by atoms	NH4 limited, cont light : 0.08 ly/min, dilution=1/3-1/2 of max.gr.rate[2]
Skeletonema costatum	9.4 by atoms	Not limited, cont light : 0.08 ly/min, dilution=1/3-1/2 of max.gr.rate[2]
Thalassiosira gravida	24.4 by atoms	Si limited, cont light : 0.08 ly/min, dilution=1/3-1/2 of max.gr.rate[2]
Thalassiosira gravida	3.5 by atoms	NH4 starved, 72 h cont light : 0.08 ly/min, dilution=1/3-1/2 of max.gr.rate[2]
Thalassiosira gravida	39.2 by atoms	Si starved, 72 h cont light : 0.08 ly/min, dilution=1/3-1/2 of max.gr.rate[2]
Thalassiosira gravida	6.3 by atoms	Not limited, cont light : 0.08 ly/min, dilution=1/3-1/2 of max.gr.rate[2]
Thalassiosira gravida	7.6 by atoms	NH4 limited, cont light : 0.08 ly/min, dilution=1/3-1/2 of[2]

Source: 1.Canale, R.P., L.M. De Palma and A.H. Vogel. 1976. in: Canale, R.P. (ed.); Mathematical modelling of biochemical processes in Aquatic Ecosystems. Ann Arbor Science Press.
2.Harrison, P.J., H.L. Conway, R.W. Holmes and C.O. Davis. 1977. Mar. Biol., 43(1), pp 19-31.
3.Antia, N.J., C.D. McAllister, T.R. Parsons, K. Stephens and J.H.D. Strickland. 1963. Limnol. Oceanog., 8, pp 166-183.
4.Strickland, J.D.H. 1960. Fish. Res. Bd. Can. Bull., 122, pp 172.

1-021 Algae
C / biomass ratio

Species	Value	Condition
Agmenellum quadruplicatum	42 % of dry weight	Marine algae, two diff. methods of sampling, also = 44.9[1]
Algae	22.5 % of dry weight	Mean elementary composition[2]
Algae	45 % of dry weight	Range = 40 - 50 %[3]
Amphidinium carteri	40.7 % of dry weight	Marine algae, two diff. methods of sampling, also = 47.1 %[1]
Amphipleura rutilans	46.42 % of dry weight	Marine algae[4]
Anabaena sp.	49.7 % of dry weight	Algae from ponds and lakes in southeastern United States[5]
Anabaenopsis sp.	49.57 % of dry weight	Freshwater algae[4]
Aphanizomenon sp.	47.65 % of dry weight	Algae from ponds and lakes in southeastern United States[5]
Asterionella formosa	59.14 pg cell-1	algal culture[6]
Chaetoceros lauderi	49.3 microgram C/mm**3	Cultivated, marine[7]
Chaetoceros sp.	31 % of dry weight	Marine algae[1]
Chara sp.	29.3 % of dry weight	Algae from ponds and lakes in southeastern United States[5]
Chlamydomonas magnusii	200 microgram C/mm**3	Cultivated, marine[7]
Chlamydomonas sp.	47.46 % of dry weight	Freshwater algae[4]
Chlorella pyrenoidosa	49.51 % of dry weight	Freshwater algae[4]
Chlorella pyrenoidosa	54.89 % of dry weight	Freshwater algae[8]
Chlorella pyrenoidosa	70.17 % of dry weight	Freshwater algae[4]
Chlorella sp.	0.376 pg cell-1	algal culture[6]
Chlorella vulgaris	53.37 % of dry weight	Freshwater algae[8]
Cladophora sp.	35.27 % of dry weight	Algae from ponds and lakes in southeastern United States[5]
Coscinodiscus angstii	0.00084 microgram C/cell	Diameter = 35 micro m[9]
Coscinodiscus angstii var. granulomarginatus	0.003334 microgram C/cell	Diameter = 87 micro m[9]

Coscinodiscus excentricus	0.001644 microgram C/cell	Diameter = 75 micro m[9]
Coscinodiscus sp.	15.9 % of dry weight	Marine algae, two diff. methods of sampling, also = 18.8 %[1]
Coscinodiscus sp.	53.5 microgram C/mm**3	Cultivated, marine[7]
Cryptomonas erosa	125.3 pg cell-1	algal culture[6]
Cryptomonas pseudobaltica	167 microgram C/mm**3	Cultivated, marine[7]
Cyclotella meneghiniana	26 %	of dry weight, biomass est. by conv. from morphometric data[10]
Cyclotella sp.	11 %	of dry weight, biomass est. by conv. from morphometric data[10]
Dunaliella salina	39.6 % of dry weight	Marine algae, two diff. methods of sampling, also = 45.4 %[1]
Egregia menziesii	49.97 % of dry weight	Marine algae[4]
Euglena sp.	48.14 % of dry weight	Algae from ponds and lakes in southeastern United States[5]
Exuviaella sp.	44.5 % of dry weight	Marine algae, two diff. methods of sampling, also = 47.4 %[1]
Fragilaria capucina	15 %	of dry weight, biomass est. by conv. from morphometric data[10]
Fragilaria capucina	18 %	of dry weight, biomass est. by conv. from morphometric data[10]
Fragilaria crotonensis	79.70 pg cell-1	algal culture[6]
Fragilaria intermedia	12 %	of dry weight, biomass est. by conv. from morphometric data[10]
Gigartina agardhii	42.9 % of dry weight	Marine algae[4]
Gonyaulax polyedra	154.9 microgram C/mm**3	N = 10, SD = 12.7[11]
Gymnodinium splendens	129.2 microgram C/mm**3	N = 5, SD = 5.3[11]
Heterothrix sp.	351 microgram C/mm**3	Cultivated, marine[7]
Hydrodictyon sp.	39.96 % of dry weight	Algae from ponds and lakes in southeastern United States[5]
Isochrysis galbana	10.3 pg C cell-1	light conditions: 30 umol quanta m-2 s-1[12]
Isochrysis galbana	12.4 pg C cell-1	light conditions: 70 umol quanta m-2 s-1[12]
Isochrysis galbana	15.0 pg C cell-1	light conditions: 150 umol quanta m-2 s-1[12]
Isochrysis galbana	16.5 pg C cell-1	light conditions: 320 umol quanta m-2 s-1[12]
Isochrysis galbana	20.0 pg C cell-1	light conditions: 600 umol quanta m-2 s-1[12]
Lauderia borealis	33 microgram C/mm**3	Cell diameter = 36 micro m, n = 6, SD = 3.8[11]
Lauderia borealis	85 microgram C/mm**3	Cell diameter = 19 micro m, n = 66, SD = 13.8[11]
Lyngbya sp.	40.23 % of dry weight	Algae from ponds and lakes in southeastern United States[5]
Macrocystis pyrifera	46.02 % of dry weight	Marine algae[4]
Marine phytoplankton	1.54E-10 g/cell	Urea as source of N, stationary phase, 293 K, Southern Calif[13]
Marine phytoplankton	2.25E-10 g/cell	Nh4 as source of N, stationary phase, 293 K, Southern Calif[13]
Marine phytoplankton	2.2E-10 g/cell	NO3 as source of N, stationary phase, 293 K, Southern Calif[13]
Meliosira granulata	14 %	of dry weight, biomass est. by conv. from morphometric data[10]
Meliosira islandica	10 %	of dry weight, biomass est. by conv. from morphometric data[10]
Melosira italica	117.1 pg cell-1	algal culture[6]
Melosira italica tenuissima	94.58 pg cell-1	algal culture[6]
Microcystis sp.	46.46 % of dry weight	Algae from ponds and lakes in southeastern United States[5]

Monochrysis lutheri	53.2 % of dry weight	Marine algae, two diff. methods of sampling, also = 44.6 %[1]
Mougeotia sp.	40.74 % of dry weight	Algae from ponds and lakes in southeastern United States[5]
Navicula torquatum	48.45 % of dry weight	Marine algae[4]
Nitella sp.	38.43 % of dry weight	Algae from ponds and lakes in southeastern United States[5]
Nitzschia closterium	54.98 % of dry weight	Freshwater algae[8]
Oedogonium sp.	40.84 % of dry weight	Algae from ponds and lakes in southeastern United States[5]
Oicomonas termo	51.76 % of dry weight	Freshwater algae[4]
Phaeodactylum tricornutum	160 microgram C/mm**3	Cultivated, marine[7]
Phaeodactylum tricornutum	37.5 % of dry weight	Marine algae, two diff. methods of sampling, also = 49.5 %[1]
Phaeophyceae	34.5 % of dry weight	Mean elementary composition[2]
Pithophora sp.	35.38 % of dry weight	Algae from ponds and lakes in southeastern United States[5]
Prorocentrum micans	1,068 pg C cell-1	light conditions: 70 umol quanta m-2 s-1[12]
Prorocentrum micans	1,096 pg C cell-1	light conditions: 150 umol quanta m-2 s-1[12]
Prorocentrum micans	1,117 pg C cell-1	light conditions: 320 umol quanta m-2 s-1[12]
Prorocentrum micans	1,178 pg C cell-1	light conditions: 600 umol quanta m-2 s-1[12]
Prorocentrum micans	213.8 migrogram C/mm**3	N = 11, SD = 17.7[11]
Rhizoclonium sp.	39.1 % of dry weight	Algae from ponds and lakes in southeastern United States[5]
Scenedesmus basilensis	54.82 % of dry weight	Freshwater algae[8]
Scenedesmus obliquus	54.55 % of dry weight	Freshwater algae[8]
Scenedesmus obliquus	54.87 % of dry weight	Freshwater algae[8]
Skeletonema costatum	26 % of dry weight	Marine algae, two diff. methods of sampling, also 31.3 %[1]
Spirogyra sp.	42.4 % of dry weight	Algae from ponds and lakes in southeastern United States[5]
Stephanodiscus alpinus	17 %	of dry weight, biomass est. by conv. from morphometric data[10]
Stephanodiscus astraea minitula	34.42 pg cell-1	algal culture[6]
Stephanodiscus binderanus	30 %	of dry weight, biomass est. by conv. from morphometric data[10]
Stephanodiscus hantzschii	29.18 pg cell-1	algal culture[6]
Stephanodiscus minutus	21 %	of dry weight, biomass est. by conv. from morphometric data[10]
Stephanodiscus niagarae	1,748 pg cell-1	algal culture[6]
Stephanodiscus niagarae	17 %	of dry weight, biomass est. by conv. from morphometric data[10]
Stephanodiscus subtilis	29 %	of dry weight, biomass est. by conv. from morphometric data[10]
Stephanodiscus tenuis	27 %	of dry weight, biomass est. by conv. from morphometric data[10]
Stichococcus bacillaris	52.65 % of dry weight	Freshwater algae[4]
Stichococcus bacillaris	53.82 % of dry weight	Freshwater algae[8]
Stichococcus bacillaris	57.14 % of dry weight	Freshwater algae[4]
Syracosphaera carteri	34.6 % of dry weight	Marine algae, two diff. methods of sampling, also 35.2 %[1]
Tabellaria fenestrata	221.1 pg cell-1	algal culture[6]
Tetraselmis maculata	35.1 % of dry weight	Marine algae, two diff. methods of sampling, also 36.8 %[1]
Tetraselmis stricia	116 microgram C/mm**3	Cultivated, marine[7]

Thalassiosira fluviatilis	0.000094 microgram C/cell	Diameter = 11 u[9]
Thalassiosira pseudonana	8E-6 microgram C/cell	Diameter = 3.8 u[9]
Thalassiosira sp.	2.5E-13 moles/cell	Growth rate=0.041 1/h, cell density=4.0E8 cell/l, P limited chemostat[14]
Thalassiosira sp.	4.5E-13 moles/cell	Growth rate=0.017 1/h, cell density=6.7E8 cell/l, P limited chemostat[14]
Thalassiosira weisflogii	275 pg C cell-1	light conditions: 30 umol quanta m-2 s-1[12]
Thalassiosira weisflogii	277 pg C cell-1	light conditions: 70 umol quanta m-2 s-1[12]
Thalassiosira weisflogii	291 pg C cell-1	light conditions: 150 umol quanta m-2 s-1[12]
Thalassiosira weisflogii	326 pg C cell-1	light conditions: 600 umol quanta m-2 s-1[12]
Thalassiosira weisflogii	328 pg C cell-1	light conditions: 320 umol quanta m-2 s-1[12]
Trichodesmium spp.	9.7 microgram C/colony	C/colony, N Pacific, N-fixing algae, n=6[15]
Ulva sp.	44.36 % of dry weight	Marine algae.[4]

Source: 1.Parsons, T. and M. Takahashi. 1973. Biological Oceanographic Processesp. p 186 Pergamon Press.
2.Bowen, H.J.M. 1966. Trace Elements in Biochemistry, Ac. Press.
3.The Hydrological Engineering Center Corps of Engineers U.S. Army. 1975. Water Quality for river-resevoir systems, Davis, Cal. 9 5 6 1 6 (916) 499-2329.
4.Milner, H.W. 1953. in: Burlew, J.S. (ed.): Algal Culture from Laboratory to Pilot Plant. Publ. 600.arnegie Inst of Wash. Wash. D.C.
5.Water Quality Criteria. 1968. Federal Water Pollution Control Administration Washington D.C., pp 234
6.Infante, A. and Litt, A.H., Limnol.Oceanogr., 30(5), 1053, 1985.
7.Berland, B.R., P.J. Bonin, P.L. Laborde and S.Y. Maestrini. 1972. Mar. biol., 13, pp 338-345.
8.Ketchum, B.H. 1939. Am. J. Bot. 26. pp 399-407.
9.Frost, B.W. 1972. Limnol. Oceanogr., 17(6), pp 805-815.
10.Sicko-Goad, L.M., Schelske, C.L. and Stoermer, E.F., Limnol.Oceanogr., 29(6), 1170, 1984.
11.Paffenhoefer, G.-A. 1976. Limnol. Oceanogr. 21(1), pp 39-50.
12.Falkowski, P.G., Dubinsky, Z. and Wyman, K.., Limnol.Oceanogr., 30(2), 311, 1985.
13.Eppley, R.W., A.F. Carlucci, O.H.-Hansen, D. Kiefer, J.J. McCarthy, E. Venrick and P.M. Williams. 1971. Limnol. Oceanogr., 16(5), pp 741-757.
14.Perry, M.J. 1976. Limnol. Oceanogr. 21(1), pp 88-107.
15.Mague, T.H., F.C. Mague and O. Holm-Hansen. 1977. Mar. Biol. 41(3), pp 213-227.

1-022 Algae
C / biotin ratio

Species	Value	Condition
Marine phytoplankton	1600000 by weight	NH4 as source of N, stationary phase, 293 K, Southern Calif[1]
Marine phytoplankton	540000 by weight	NH4 as source of N, log phase, 293 K, Southern Calif[1]
Marine phytoplankton	590000 by weight	NO3 as source of N, log phase, 293 K, Southern Calif[1]
Marine phytoplankton	590000 by weight	Urea as source of N, log phase, 293 K, Southern Calif[1]
Marine phytoplankton	680000 by weight	NO3 as source of N, stationary phase, 293 K, Southern Calif[1]
Marine phytoplankton	820000 by weight	Urea as source of N, stationary phase, 293 K, Southern Calif[1]

Source: 1.Eppley, R.W., A.F. Carlucci, O.H.-Hansen, D. Kiefer, J.J. McCarthy, E. Venrick and P.M. Williams. 1971. Limnol. Oceanogr., 16(5), pp 741-757.

1-023 Algae
C / chlorophyll a ratio

Species	Value	Condition
Algae	0.05 mg C/microgram chl	Lake Ontario[1]
Algae	0.05 mg C/microgram chl	Potomac River[2]
Chaetoceros debilis	32 by weight	Not limited, cont light : 0.08 ly/min, dilution=1/3-1/2 of max.gr.rate[3]

Chaetoceros debilis	43 by weight	NH4 starved, 72 h cont light : 0.08 ly/min, dilution=1/3-1/2 of max.gr.rate[3]
Chaetoceros debilis	47 by weight	NH4 limited, cont light : 0.08 ly/min, dilution=1/3-1/2 of max.gr.rate[3]
Chaetoceros debilis	55 by weight	Nitrogen limited, dilution rate=0.097-0.008/hr, rg = 57.4-53.6[4]
Chaetoceros debilis	55 by weight	Si limited, cont light : 0.08 ly/min, dilution=1/3-1/2 of max.gr.rate[3]
Chaetoceros debilis	76 by weight	Si starved, 72 h cont light : 0.08 ly/min, dilution=1/3-1/2 of max.gr.rate[3]
Cyclotella nana	288 by weight	Nitrate limited, dilution rate=0.077-0.009/hr, rg = 76-500[5]
Marine phytoplankton	112 by weight	NH4 as source of N, log phase, 293 K, Southern Calif[6]
Marine phytoplankton	25 by weight	Vigorously growing, excess nitrate in water[7]
Marine phytoplankton	30 by weight	See ref.[8]
Marine phytoplankton	46 by weight	NO3 as source of N, stationary phase, 293 K, Southern Calif[6]
Marine phytoplankton	58 by weight	NH4 as source of N, stationary phase, 293 K, Southern Calif[6]
Marine phytoplankton	60 by weight	Unhealthy phytoplankton, nitrate depleted water[7]
Marine phytoplankton	77 by weight	Urea as source of N, stationary phase, 293 K, Southern Calif[6]
Marine phytoplankton	90 by weight	NO3 as source of N, log phase, 293 K, Southern Calif[6]
Marine phytoplankton	96 by weight	Urea as source of N, log phase, 293 K, Southern Calif[6]
Skeletonema costatum	145 by weight	Nitrogen limited, batch culture, 72-115 hrs starvation, rg = 70-220[9]
Skeletonema costatum	178 by weight	Si starved, 72 h cont light : 0.08 ly/min, dilution=1/3-1/2 ofmax.gr.rate[3]
Skeletonema costatum	61 by weight	Not limited, cont light : 0.08 ly/min, dilution=1/3-1/2 of max.gr.rate[3]
Skeletonema costatum	65 by weight	Si limited, cont light : 0.08 ly/min, dilution=1/3-1/2 of max.gr.rate[3]
Skeletonema costatum	72 by weight	NH4 limited, cont light : 0.08 ly/min, dilution=1/3-1/2 of max.gr.rate[3]
Skeletonema costatum	98 by weight	NH4 starved, 72 h cont light : 0.08 ly/min, dilution=1/3-1/2 of max.gr.rate[3]
Thalassiosira fluviatilis	22 by weight	Nitrate limited, batch culture, 72-115 hrs starvation, rg = 18-26[10]
Thalassiosira gravida	106 by weight	NH4 starved, 72 h cont light : 0.08 ly/min, dilution=1/3-1/2 of max.gr.rate[3]
Thalassiosira gravida	26 by weight	Not limited, cont light : 0.08 ly/min, dilution=1/3-1/2 of max.gr.rate[3]
Thalassiosira gravida	34 by weight	Si limited, cont light : 0.08 ly/min, dilution=1/3-1/2 of max.gr.rate[3]
Thalassiosira gravida	46 by weight	Si starved, 72 h cont light : 0.08 ly/min, dilution=1/3-1/2 of max.gr.rate[3]
Thalassiosira gravida	55 by weight	NH4 limited, cont light : 0.08 ly/min, dilution=1/3-1/2 of max.gr.rate[3]
Thalassiosira pseudonana	155 by weight	Nitrate limited, batch culture, 72-115 hrs starvation, rg = 111-200[10]
Thalassiosira pseudonana	28.6 by weight	P limited chemostat, growth=0.041 1/h, 295 K, 0.07 cal/cm2*min[11]
Thalassiosira pseudonana	28.7 by weight	N limited chemostat, growth=0.041 1/h, 295 K, 0.07 cal/cm2*min[11]
Thalassiosira pseudonana	44.3 by weight	N limited chemostat, growth=0.017 1/h, 295 K, 0.07 cal/cm2*min[11]
Thalassiosira pseudonana	60 by weight	Nitrogen limited, dilution rate=0.041-0.008/hr, rg = 28.7-91.4[12]
Thalassiosira pseudonana	66.7 by weight	P limited chemostat, growth=0.017 1/h, 295 K, 0.07 cal/cm2*min[11]

Thalassiosira pseudonana	91.4 by weight	N limited chemostat, growth=0.0085 1/h, 295 K, 0.07 cal/cm2*min[11]

Source: 1.Thomann, R.V., D.M. Di Toro, R.P. Winfield and D.J. O'Connor. 1975. Mathematical Modeling Of Phytoplankton in Lake Otario: 1. National Environmental Reseach Center. EPA. Oregon.
2.O'Connor, D.J., R.V. Thomann and D.M. DiToro. 1973. Dynamic Water Quality Forecasting and Management.Office of Research and Development U.S. Environmental Protection Agency, Wash. 20460.
3.Harrison, P.J., H.L. Conway, R.W. Holmes and C.O. Davis. 1977. Mar. Biol., 43(1), pp 19-31.
4.Thomas, W.H. and A.N. Dodson. 1972. Limnol. Oceanogr. 17, pp 515-523.
5.Caperon, J. and J. Meyer. 1972. Deep Sea Res., 19, pp 601-632.
6.Eppley, R.W., A.F. Carlucci, O.H.-Hansen, D. Kiefer, J.J. McCarthy, E. Venrick and P.M. Williams. 1971. Limnol. Oceanogr., 16(5), pp 741-757.
7.Antia, N.J., C.D. McAllister, T.R. Parsons, K. Stephens and J.H.D. Strickland. 1963. Limnol. Oceanogr., 8, pp 166-183.
8.Strickland, J.D.H. 1960. Fish. Res. Bd. Can. Bull., 122, pp 172.
9.Holm-Hansen, O., W.H. Sutcliffe, Jr. and J. Sharp. 1968. Limnol. Oceanogr., 13, pp 507-514.
10.Hobson, L.A. and R.J. Pariser. 1971. J. exp. mar. biol. Ecol. 6, pp 71-78.
11.Perry, M.J. 1976. Limnol. Oceanogr. 21(1), pp 88-107.
12.Eppley, R.W. and E.H. Renger. 1974. J. Phycol., 10, pp 15-23.

1-024 Algae
C / vitamin b1 ratio

Species	Value	Condition
Marine phytoplankton	27000 by weight	NO3 as source of N, stationary phase, 293 K, Southern Calif[1]
Marine phytoplankton	35000 by weight	NH4 as source of N, stationary phase, 293 K, Southern Calif[1]
Marine phytoplankton	41000 by weight	Urea as source of N, stationary phase, 293 K, Southern Calif[1]
Marine phytoplankton	58000 by weight	NO3 as source of N, log phase, 293 K, Southern Calif[1]
Marine phytoplankton	60000 by weight	NH4 as source of N, log phase, 293 K, Southern Calif[1]
Marine phytoplankton	67000 by weight	Urea as source of N, log phase, 293 K, Southern Calif[1]

Source: 1.Eppley, R.W., A.F. Carlucci, O.H.-Hansen, D. Kiefer, J.J. McCarthy, E. Venrick and P.M. Williams. 1971. Limnol. Oceanogr., 16(5), pp 741-757.

1-025 Algae
C / vitamin b12 ratio

Species	Value	Condition
Marine phytoplankton	1400000 by weight	NH4 as source of N, stationary phase, 293 K, Southern Calif[1]
Marine phytoplankton	680000 by weight	NO3 as source of N, stationary phase, 293 K, Southern Calif[1]
Marine phytoplankton	750000 by weight	NH4 as source of N, log phase, 293 K, Southern Calif[1]
Marine phytoplankton	770000 by weight	Urea as source of N, stationary phase, 293 K, Southern Calif[1]
Marine phytoplankton	780000 by weight	Urea as source of N, log phase, 293 K, Southern Calif[1]
Marine phytoplankton	870000 by weight	NO3 as source of N, log phase, 293 K, Southern Calif[1]

Source: 1.Eppley, R.W., A.F. Carlucci, O.H.-Hansen, D. Kiefer, J.J. McCarthy, E. Venrick and P.M. Williams. 1971. Limnol. Oceanogr., 16(5), pp 741-757.

1-026 Algae
C2H2 reduction / N2 fixation ratio

Species	Value	Condition
Trichodesmium spp.	1.9 by mole	N Pacific, N-fixing algae, incubation in laboratory[1]

Source: 1.Mague, T.H., F.C. Mague and O. Holm-Hansen. 1977. Mar. Biol. 41(3), pp 213-227.

1-027 Algae
C2H2 reduction rate

Species	Value	Condition
Trichodesmium spp.	251 pmoles C2H4/(microgram algal N)*hr	N Pacific, N-fixing algae, incubation in laboratory[1]
Trichodesmium spp.	620 pmoles C2H4/hr	N Pacific, N-fixing algae, max rate[1]

Source: 1.Mague, T.H., F.C. Mague and O. Holm-Hansen. 1977. Mar. Biol. 41(3), pp 213-227.

1-028 Algae
Ca / biomass ratio

Species	Value	Condition
Anabaena sp.	0.36 % of dry weight	Algae from ponds and lakes in southeastern United States[1]
Aphanizomenon sp.	0.73 % of dry weight	Algae from ponds and lakes in southeastern United States[1]
Chara sp.	8.03 % of dry weight	Algae from ponds and lakes in southeastern United States[1]
Chlorophyta	0.157 mg/(gram of acid-soluble fraction)	Average of 7 marine species, Puerto Rico[2]
Cladophora sp.	1.69 % of dry weight	Algae from ponds and lakes in southeastern United States[1]
Euglena sp.	0.05 % of dry weight	Algae from ponds and lakes in southeastern United States[1]
Hydrodictyon sp.	0.69 % of dry weight	Algae from ponds and lakes in southeastern United States[1]
Lyngbya sp.	0.45 % of dry weight	Algae from ponds and lakes in southeastern United States[1]
Microcystis sp.	0.53 % of dry weight	Algae from ponds and lakes in southeastern United States[1]
Mougeotia sp.	1.68 % of dry weight	Algae from ponds and lakes in southeastern United States[1]
Nitella sp.	1.89 % of dry weight	Algae from ponds and lakes in southeastern United States[1]
Oedogonium sp.	0.44 % of dry weight	Algae from ponds and lakes in southeastern United States[1]
Phaeophyceae	0.099 gram/(gram of acid-soluble fractio	Average of 4 marine species, Puerto Rico[2]
Pithophora sp.	3.82 % of dry weight	Algae from ponds and lakes in southeastern United States[1]
Rhizoclonium sp.	0.52 % of dry weight	Algae from ponds and lakes in southeastern United States[1]
Rhodophyta	0.106 gram/(gram of acid-soluble fractio	Average of 9 marine species, Puerto Rico[2]
Spirogyra sp.	0.57 % of dry weight	Algae from ponds and lakes in southeastern United States[1]
Spirogyra sp., giant	0.84 % of dry weight	Algae from ponds and lakes in southeastern United States[1]

Source: 1.Water Quality Criteria. 1968. Federal Water Pollution Control Administration Washington D.C., pp 234
2.Edgington, D.N., S.A. Gordon, M.M. Thommes and L.R. Almodovar. 1970 Limnol. Oceanogr.,, 15(6), pp 94-955.

1-029 Algae
Caloric content

Species	Value	Condition
Achnantes taeniata	2.604 kcal/g dry weight	Estuarine, bombcal., pure culture, log-phase[1]
Anabaena solitaria	5.14 kcal/g ash free dry weight	Estuarine, bombcal.[1]
Anabaena solitaria	5.140 kcal/g ash free dry weight	Estuarine, bombcal.[1]
Anacystis spp.	4.781 kcal/g dry weight	Freshwater, bombcal[1]
Chara baltica	1.383 kcal/g dry weight	Estuarine, bombcal., as detritus[1]

Chara baltica	1.776 kcal/g dry weight	Estuarine, bombcal., also C. aspera[1]
Chara baltica	2.95 kcal/g ash free dry weight	Estuarine, bombcal., as detritus[1]
Chara baltica	3.56 kcal/g ash free dry weight	Estuarine, bombcal., also C. aspera[1]
Chara spp.	0.415 kcal/g wet weight	Freshwater, bombcal[1]
Chara spp.	2.255 kcal/g dry weight	Freshwater, bombcal[1]
Chara spp.	4.798 kcal/g ash free dry weight	Freshwater, bombcal[1]
Cladophora spp.	2.12 kcal/g dry weight	Marine, bombcal.[1]
Cladophora spp.	3.22 kcal/g dry weight	Marine, oxydation[1]
Cladophora spp.	5.17 kcal/g ash free dry weight	Marine, bombcal.[1]
Delesseria decipiens	3.7 kcal/g dry weight	Marine, bombcal.[1]
Delesseria decipiens	3.70 kcal/g dry weight	Marine, bombcal.[1]
Delesseria decipiens	5.44 kcal/g ash free dry weight	Marine, bombcal.[1]
Ectocarpus dimorphus	3.82 kcal/g dry weight	Marine, bombcal.[1]
Ectocarpus dimorphus	5.09 kcal/g ash free dry weight	Marine, bombcal.[1]
Elachistea fucicola	3.92 kcal/g dry weight	Marine, bombcal.[1]
Elachistea fucicola	5.16 kcal/g ash free dry weight	Marine, bombcal.[1]
Enteromorpha intestinalis	2.65 kcal/g dry weight	Estuarine, bombcal., 12 h as fine particles before mesurement[1]
Enteromorpha intestinalis	2.650 kcal/g dry weight	Estuarine, bombcal., 12 h as fine particles before mesurement[1]
Enteromorpha intestinalis	2.714 kcal/g dry weight	Estuarine, bombcal.[1]
Enteromorpha intestinalis	3.41 kcal/g ash free dry weight	Estuarine, bombcal., 12 h as fine particles before mesurement[1]
Enteromorpha intestinalis	3.52 kcal/g ash free dry weight	Estuarine, bombcal.[1]
Enteromorpha spp,	4.7 kcal/g ash free dry weight	Marine, bombcal.[1]
Enteromorpha spp,	4.70 kcal/g ash free dry weight	Marine, bombcal.[1]
Enteromorpha spp.	3.01 kcal/g dry weight	Marine, bombcal.[1]
Fucus distichus	0.65 kcal/g wet weight	Marine, bombcal.[1]
Fucus distichus	3.43 kcal/g dry weight	Marine, bombcal.[1]
Fucus distichus	4.64 kcal/g ash free dry weight	Marine, bombcal.[1]
Gymnodinium splendens	5708 kcal/kg ash-free dry weight	N=5, SD = 166.1[2]
Laminaria saccharina	0.65 kcal/g wet weight	Marine, bombcal.[1]
Laminaria saccharina	2.84 kcal/g dry weight	Marine, bombcal.[1]
Laminaria saccharina	4.37 kcal/g ash free dry weight	Marine, bombcal.[1]

Lauderia borealis	5418 kcal/kg ash-free dry weight	N=7, SD = 57.3[2]
Lithothamnion sp.	0.67 kcal/g dry weight	Marine, bombcal.[1]
Lithothamnion sp.	3.94 kcal/g ash free dry weight	Marine, bombcal.[1]
Marine phytoplankton	314 cal/100 mg ash-free dry weight	Adriatic sea[3]
Melosira sp.	5.15 kcal/g ash free dry weight	Freshwater, bombcal[1]
Melosira sp.	5.150 kcal/g ash free dry weight	Freshwater, bombcal[1]
Microcystis firma	5.06 kcal/g ash free dry weight	Freshwater, bombcal, pure culture, washed[1]
Microcystis spp.	4.781 kcal/g ash free dry weight	Marine, bombcal., rg = 0.812[1]
Monostroma fuscum	4.56 kcal/g dry weight	Marine, bombcal.[1]
Monostroma fuscum	5.18 kcal/g ash free dry weight	Marine, bombcal.[1]
Mycrocystis firma	4.596 kcal/g dry weight	Estuarine, bombcal., pure cultur, washed[1]
Navicula minima	3.218 kcal/g dry weight	Freshwater, bombcal[1]
Nitella spp.	4.413 kcal/g ash free dry weight	Freshwater, oxydation[1]
Nitzschia paradoxa	3.28 kcal/g dry weight	Marine, bombcal.[1]
Nitzschia paradoxa	5.47 kcal/g ash free dry weight	Marine, bombcal.[1]
Oscillatoria terebriformis	1.249 kcal/g dry weight	Freshwater, bombcal[1]
Oscillatoria terebriformis	4.405 kcal/g ash free dry weight	Freshwater, bombcal[1]
Polisiphonia brodiae	3.5 kcal/g dry weight	Marine, bombcal.[1]
Polisiphonia brodiae	3.50 kcal/g dry weight	Marine, bombcal.[1]
Polysiphonia brodiae	5 kcal/g ash free dry weight	Marine, bombcal.[1]
Polysiphonia brodiae	5.00 kcal/g ash free dry weight	Marine, bombcal.[1]
Prorocentrum micans	5218 kcal/kg ash-free dry weight	N=4, SD = 52.0[2]
Scenedesmus quadricauda	4.172 kcal/g dry weight	Freshwater, oxydationr, pure culture, log-phase[1]
Scytosiphon lomentaria	2.8 kcal/g dry weight	Marine, bombcal.[1]
Scytosiphon lomentaria	2.80 kcal/g dry weight	Marine, bombcal.[1]
Scytosiphon lomentaria	4.67 kcal/g ash free dry weight	Marine, bombcal.[1]
Thalassiosira rotula	5479 kcal/kg ash-free dry weight	N=2, SD = 57.3[2]
Ulva lactuca	3.75 kcal/g dry weight	Marine, bombcal.[1]
Ulva lactuca	4.75 kcal/g ash free dry weight	Marine, bombcal.[1]
Ulva spp.	4 kcal/g dry weight	Marine, bombcal.[1]

Ulva spp.	4.00 kcal/g dry weight	Marine, bombcal.[1]
Ulva spp.	4.94 kcal/g ash free dry weight	Marine, bombcal.[1]

Source: 1.Oertzen, J.-A. 1973. Wiss. Z. Univ. Rostock. 22, pp 1153-1158.
2.Paffenhoefer, G.-A. 1976. Limnol. Oceanogr. 21(1), pp 39-50.
3.Baric, J. 1974. Acta Adriat., Inst. Oceanogr. Ribar. 15(10). pp 1-12.

1-030 Algae
Carbohydrate / biomass ratio

Species	Value	Condition
Agmenellum quadruplicatum	38 % of ash free dry weight	Blue - green algae, marine, anthrone reaction[1]
Amphidinium carteri	39 % of ash free dry weight	Dinophyceae, marine, anthrone reaction[1]
Chlorella sp.	23.5 % of dry weight	Unfavorable environment[2]
Chlorella sp.	33.4 % of dry weight	Favorable environment[2]
Dunaliella salina	32 % of ash free dry weight	Chlorophyceae, marine, anthrone reaction[1]
Exuviaella sp.	42 % of ash free dry weight	Dinophyceae, marine, anthrone reaction[1]
Marine phytoplankton	0.11 by weight	Unhealthy phytoplankton, nitrate depleted water[3]
Marine phytoplankton	0.2 by weight	Vigorously growing, excess nitrate in water[3]
Monochrysis lutheri	34 % of ash free dry weight	Chrysophyceae, marine, anthrone reaction[1]
Skeletonema costatum	33 % of ash free dry weight	Bacillariophyceae, marine, anthrone reaction[1]
Tetraselmis maculata	20 % of ash free dry weight	Prasinophyceae, marine, anthrone reaction[1]

Source: 1.Parsons, T.R., K. Stephens and J.D.H. Strickland. 1961. J. Fish. Res. Bd. Can., 18, pp 1001-1016.
2.Milner, H.W. 1953. in: Burlew, J.S. (ed.): Algal Culture from Laboratory to Pilot Plant. Publ. 600.arnegie Inst of Wash. Wash. D.C.
3.Antia, N.J., C.D. McAllister, T.R. Parsons, K. Stephens and J.H.D. Strickland. 1963. Limnol. Oceanog., 8, pp 166-183.

1-031 Algae
Carbon fixation/nitrogen fixation ratio

Species	Value	Condition
Trichodesmium spp.	1.65 by weight	N Pacific, N-fixing algae, range = 1.2 - 2.1[1]

Source: 1.Mague, T.H., F.C. Mague and O. Holm-Hansen. 1977. Mar. Biol. 41(3), pp 213-227.

1-032 Algae
Carbon, intracellular

Species	Value	Condition
Bacillariophyceae	33 % of dry weight	Range=19-50 %[1]
Blue - green algae	36 % of dry weight	Range=28-45 %[1]
Chlorophyceae	43 % of dry weight	Range=35-48 %[1]
Chrysophyceae	40 % of dry weight	Range=35-45 %[1]
Dinophyceae	43 % of dry weight	Range=37-47 %[1]

Source: 1.Strickland, J.D.H. 1965. Chemical Oceanog., 1, pp 503.

1-033 Algae
Carotenoids / chlorophyll a ratio

Species	Value	Condition
Marine phytoplankton	0.9 by weight	See ref.[1]
Marine phytoplankton	0.95 by weight	Vigorously growing, excess nitrate in water[2]
Marine phytoplankton	1.15 by weight	Unhealthy phytoplankton, nitrate depleted water[2]

Source: 1.Strickland, J.D.H. 1960. Fish. Res. Bd. Can. Bull., 122, pp 172.
2.Antia, N.J., C.D. McAllister, T.R. Parsons, K. Stephens and J.H.D. Strickland. 1963. Limnol. Oceanogr., 8, pp 166-183.

1-034 Algae
Cell diameter

Species	Value	Condition
Coccolithus huxleyi	5 micro meters	None[1]
Ditylum brightwellii	30 micro meters	None[1]
Thalassiosira pseudonana	6.25 micro meters	Clone 3h, range = 2.5 - 10[2]
Thalassiosira pseudonana	7.5 micro meters	Clone 13-1, range = 4.0 - 11.0[2]

Source: 1.Parsons, T.R. and M. Takahashi. 1973. Limnol. Oceanogr., 18(4), pp 511-515.
2.Hecky, R.E. and P. Kilham. 1974. Limnol. Oceanogr., 19(2), pp 361-365.

1-035 Algae
Cell volume

Species	Value	Condition
Agmenellum quadruplicatum	1.5 micro m**3	Blue - green algae, marine[1]
Amphidinium carteri	740 micro m**3	Dinophyceae, marine[1]
Amphiphora ornata	17650 micro m**3	None[2]
Anabaena flos-aquae	80000 micro m**3	Col.[2]
Anabaena flos-aquae	80000 micro m**3	Freshwater plankton, calculated volume[3]
Anabaena spp.	125 micro m**3	Lake Kinneret, rg = 100 - 150[4]
Ankistrodesmus falcatus	250 micro m**3	Freshwater plankton, calculated volume[3]
Ankistrodesmus falcatus	250 micro m**3	None[2]
Aphanocapsa delicatissima	4 micro m**3	None[2]
Aphanothece clathrata	10 micro m**3	None[2]
Aphanothece nidulans	5 micro m**3	None[2]
Asterionella formosa	350 micro m**3	Michigan[2]
Asterionella formosa	480 micro m**3	Lake Tahoe, surface area/volume=1.7[5]
Asterionella formosa	700 micro m**3	Europe[2]
Asterionella formosa	700 micro m**3	Freshwater plankton, calculated volume[3]
Botryococcus braunii	10000 micro m**3	Col.[2]
Botryococcus braunii	10000 micro m**3	Freshwater plankton, calculated volume, col[3]
Ceratium hirundinella	4000 micro m**3	None[2]
Ceratium hirundinella	70000 micro m**3	Freshwater plankton, calculated volume[3]
Ceratium hirundinella	75000 micro m**3	Lake Kinneret[4]
Chaetoceros lauderi	2275 micro m**3	Cultivated, marine[6]

Chlamydomonas magnusii	65.3 micro m**3	Cultivated, marine[6]
Chlamydomonas subcompleta	250 micro m**3	None[2]
Chlorella sp.	24 micro**3	Count by microscope[7]
Chlorella sp.	30 micro m**3	Lake Kinneret[4]
Chlorella vulgaris	200 micro m**3	Freshwater plankton, calculated volume[3]
Chlorella vulgaris	200 micro m**3	None[2]
Chodatella spp.	400 micro m**3	Lake Kinneret, rg = 200 - 600[4]
Chromulina pyriformis	50 micro m**3	None[2]
Chroococcus limneticus	400 micro m**3	Col.[2]
Chroococcus spp.	280 micro m**3	Lake Kinneret, range = 60 - 500[4]
Chroococcus turgidus	1000 micro m**3	Col.[2]
Chroomonas nordstedtii	35 micro m**3	None[2]
Closterium acicularis	4000 micro m**3	Freshwater plankton, calculated volume[3]
Closterium acicularis	4000 micro m**3	None[2]
Closterium acicularis var. subpronum	3000 micro m**3	Lake Kinneret[4]
Coelosphaerium naegelianum	15000 micro m**3	Col.[2]
Coscinodiscus angstii	26000 micro m**3	Diameter = 35 u[8]
Coscinodiscus angstii var. granulomarginatus	160000 micro m**3	Diameter = 87 u, centric[8]
Coscinodiscus excentricus	63000 micro m**3	Diameter = 75 u[8]
Coscinodiscus sp.	60400 micro m**3	Cultivated, marine[6]
Cosmarium phaseoluss	3000 micro m**3	Freshwater plankton, calculated volume[3]
Cosmarium phaseoluss	3000 micro m**3	None[2]
Cosmarium reniforme	30000 micro m**3	Freshwater plankton, calculated volume[3]
Cosmarium reniforme	30000 micro m**3	None[2]
Cosmarium sp.	500 micro m**3	Lake Kinneret[4]
Cryptomonas erosa	1000 micro m**3	Michigan[2]
Cryptomonas erosa	250 micro m**3	Europe[2]
Cryptomonas erosa	250 micro m**3	Freshwater plankton, calculated volume[3]
Cryptomonas ovata	250 micro m**3	Freshwater plankton, calculated volume[3]
Cryptomonas ovata	250 micro m**3	None[2]
Cryptomonas pseudobaltica	410 micro m**3	Cultivated, marine[6]
Cryptomonas spp.	3600 micro m**3	Lake Kinneret, three typical values: 250 3600 6200[4]
Cyclotella bodanica	10000 micro m**3	Freshwater plankton, calculated volume[3]
Cyclotella bodanica	10000 micro m**3	None[2]
Cyclotella comensis	400 micro m**3	Freshwater plankton, calculated volume[3]
Cyclotella sp.	600 micro m**3	Lake Kinneret, range = 200 - 1000[4]
Cyclotella stelligera	77 micro m**3	Lake Tahoe, surface area/volume=1.6[5]
Cymatopleura solea	80000 micro m**3	Freshwater plankton, calculated volume[3]
Cymatopleura solea	80000 micro m**3	None[2]

Dactylococpsis smithii	1500 micro m**3	Col.[2]
Diatoma vulgare	4350 micro m**3	None[2]
Dinobryon borgei	1500 micro m**3	None[2]
Dinobryon divergens	800 micro m**3	None[2]
Dinobryon divergens and Dinobryan sociale	800 micro m**3	Freshwater plankton, calculated volume[3]
Dinobryon sociale	800 micro m**3	None[2]
Dinobryon sp.	900 micro m**3	Lake Tahoe, surface area/volume=0.5[5]
Diplopsalis acuta	11000 micro m**3	Lake Kinneret[4]
Dunaliella salina	400 micro m**3	Chlorophyceae, marine[1]
Elakatothrix gelatinosa	195 micro m**3	Lake Kinneret, range = 90 - 300[4]
Erkenia subaequiciliata	30 micro m**3	Lake Kinneret[4]
Exuviaella sp.	780 micro m**3	Dinophyceae, marine[1]
Fragilaria crotonensis	200000 micro m**3	1 mm of a thread[2]
Fragilaria crotonensis	200000 micro m**3	Freshwater plankton, calculated volume, 1 mm of a thread[3]
Fragilaria crotonensis	510 micro m**3	Lake Tahoe, surface area/volume=1.5[5]
Gloeocapsa rupestris	18 micro m**3	None[2]
Gloeocococcus schroeteri	5000 micro m**3	Col.[2]
Gloeocococcus schroeteri	5000 micro m**3	Freshwater plankton, calculated volume, col[3]
Gomphosphaeria lacustris	2000 micro m**3	Col.[2]
Gomphosphaeria lacustris	2000 micro m**3	Freshwater plankton, calculated volume, col[3]
Gymnodinium fuscum	10000 micro m**3	None[2]
Gymnodinium helveticum	20000 micro m**3	Freshwater plankton, calculated volume[3]
Gymnodinium helveticum	20000 micro m**3	None[2]
Gymnodinium ordinatum	400 micro m**3	None[2]
Heterothrix sp.	101 micro m**3	Cultivated, marine[6]
Isochrysis galbana	37 um3	light conditions: 30 umol quanta m-2 s-1[9]
Isochrysis galbana	50 um3	light conditions: 70 umol quanta m-2 s-1[9]
Isochrysis galbana	52 um3	light conditions: 320 umol quanta m-2 s-1[9]
Isochrysis galbana	55 um3	light conditions: 150 umol quanta m-2 s-1[9]
Isochrysis galbana	57 um3	light conditions: 600 umol quanta m-2 s-1[9]
Kephyrion sp.	50 micro m**3	Freshwater plankton, calculated volume[3]
Mallomonas caudata	10000 micro m**3	Freshwater plankton, calculated volume[3]
Mallomonas caudata	12000 micro m**3	None[2]
Mallomonas urnaformis	1200 micro m**3	None[2]
Melosira granulata	1100 micro m**3	Lake Kinneret, three typical values: 320 1100 1500[4]
Melosira granulata	60000 micro m**3	1 mm of a thread[2]
Melosira granulata	60000 micro m**3	Freshwater plankton, calculated volume, 1 mm of a thread[3]
Melosira islandica	80000 micro m**3	1 mm of a thread[2]
Melosira islandica	80000 micro m**3	Freshwater plankton, calculated volume, 1 mm of a thread[3]

Merismopedia tenuissima	8 micro m**3	None[2]
Microcystis aeruginosa	100000 micro m**3	Col.[2]
Microcystis aeruginosa	100000 micro m**3	Freshwater plankton, calculated volume, col[3]
Microcystis flos-aquae	50 micro m**3	None[2]
Microcystis spp.	60 micro m**3	Lake Kinneret, range = 20 - 100[4]
Monochrysis lutheri	28 micro m**3	Chrysophyceae, marine[1]
Navicula spp.	850 micro m**3	Lake Kinneret, range = 300 - 1400[4]
Nitzschia gracilis	240 micro m**3	None[2]
Oocystis solitaria	400 micro m**3	Freshwater plankton, calculated volume[3]
Oocystis solitaria	400 micro m**3	None[2]
Oocystis sp.	250 micro m**3	Lake Kinneret, range = 200 - 300[4]
Oscillatoria limnetica	17500 micro m**3	1 mm of a thread[2]
Oscillatoria rubescens	30000 micro m**3	1 mm of a thread[2]
Oscillatoria rubescens	30000 micro m**3	Freshwater plankton, calculated volume, 1 mm of a thread[3]
Pandorina morum	4000 micro m**3	Col.[2]
Pandorina morum	4000 micro m**3	Freshwater plankton, calculated volume[3]
Pediastrum spp.	100 micro m**3	Lake Kinneret, three typical values: 30 100 200[4]
Peridinium cinctum	40000 micro m**3	None[2]
Peridinium cinctum fa. westii	1000 micro m**3	Lake Kinneret, range = 750 - 1250[4]
Peridinium spp.	8000 micro m**3	Lake Kinneret, range = 5400 - 10600[4]
Peridinium willei	40000 micro m**3	Freshwater plankton, calculated volume[3]
Peridinium willei	40000 micro m**3	None[2]
Phaeodactylum tricornutum	555 micro m**3	Cultivated, marine[6]
Prorocentrum micans	4,340 um3	light conditions: 70 umol quanta m-2 s-1[9]
Prorocentrum micans	5,096 um3	light conditions: 150 umol quanta m-2 s-1[9]
Prorocentrum micans	5,122 um3	light conditions: 320 umol quanta m-2 s-1[9]
Prorocentrum micans	5,350 um3	light conditions: 600 umol quanta m-2 s-1[9]
Radiocystis geminata	7 micro m**3	Lake Kinneret[4]
Rhizochrysis limnetica	1200 micro m**3	None[2]
Rhodomonas lacustris	175 micro m**3	Michigan[2]
Rhodomonas lacustris	200 micro m**3	Europe[2]
Rhodomonas lacustris	200 micro m**3	Freshwater plankton, calculated volume[3]
Rhodomonas minuta	175 micro m**3	Lake Kinneret, range = 100 - 250[4]
Rhodomonas minuta	200 micro m**3	None[2]
Scenedesmus quadricauda	1000 micro m**3	Freshwater plankton, calculated volume[3]
Scenedesmus quadricauda	1000 micro m**3	None[2]
Scenedesmus spp.	100 micro m**3	Lake Kinneret, three typical values: 30 100 200[4]
Selenastrum capricornutum	29 micro**3	Particle counter[7]
Selenastrum capricornutum	30 micro**3	Particle counter[7]
Selenastrum capricornutum	60 micro**3	Particle counter[7]

Selenastrum capricornutum	80 micro**3	Particle counter[7]
Skeletonema costatum	1390 micro m**3	Bacillariophyceae, marine[1]
Staurastrum paradoxum	20000 micro m**3	Freshwater plankton, calculated volume[3]
Staurastrum paradoxum	20000 micro m**3	None[2]
Stephanodiscus astraea	2000 micro m**3	Freshwater plankton, calculated volume[3]
Stephanodiscus astraea	2000 micro m**3	None[2]
Stephanodiscus hantzschii var. pusillus	155 micro m**3	Lake Kinneret, range = 110 - 200[4]
Stephanodiscus hantzschii var. pusillus	200 micro m**3	Freshwater plankton, calculated volume[3]
Stephanodiscus hantzschii var. pusillus	200 micro m**3	None[2]
Stephanodiscus niagarea	5000 micro m**3	None[2]
Synechococcus aeruginosus	350 micro m**3	None[2]
Synedra acus	250 micro m**3	None[2]
Synedra acus var. angustissima	1000 micro m**3	Freshwater plankton, calculated volume[3]
Synedra acus var. angustissima	1000 micro m**3	None[2]
Synedra capitata	950 micro m**3	None[2]
Synedra delicatissima	300 micro m**3	None[2]
Synedra ulna	50 micro m**3	None[2]
Tabellaria fenestrata	3000 micro m**3	Michigan[2]
Tabellaria fenestrata	4000 micro m**3	Europe[2]
Tabellaria fenestrata	4000 micro m**3	Freshwater plankton, calculated volume[3]
Tetraedron minimum	40 micro m**3	Freshwater plankton, calculated volume[3]
Tetraedron minimum	40 micro m**3	None[2]
Tetraedron spp.	85 micro m**3	Lake Kinneret, range = 30 - 140[4]
Tetraselmis maculata	310 micro m**3	Prasinophyceae, marine[1]
Tetraselmis stricia	376 micro m**3	Cultivated, marine[6]
Thalassiosira fluviatilis	1450 micro m**3	Diameter = 11 u[8]
Thalassiosira pseudonana	55 micro m**3	Diameter = 3.8 u[8]
Thalassiosira weisflogii	1,172 um3	light conditions: 30 umol quanta m-2 s-1[9]
Thalassiosira weisflogii	1,364 um3	light conditions: 320 umol quanta m-2 s-1[9]
Thalassiosira weisflogii	1,460 um3	light conditions: 70 umol quanta m-2 s-1[9]
Thalassiosira weisflogii	1,480 um3	light conditions: 600 umol quanta m-2 s-1[9]
Thalassiosira weisflogii	1,675 um3	light conditions: 150 umol quanta m-2 s-1[9]
Trachelomonas hispida	4200 micro m**3	None[2]
Trachelomonas volvocina	1800 micro m**3	None[2]
Ulothrix zonata	6000 micro m**3	1 mm of a thread[2]
Uroglena americana	90000 micro m**3	Col.[2]
Uroglena americana	90000 micro**3	Freshwater plankton, calculated volume, col[3]

Source: 1. Parsons, T.R., K. Stephens and J.D.H. Strickland. 1961. J. Fish. Res. Bd. Can. 18, pp 1001-1016.
2. Wetzel, R.G. 1975. Limnology. Saunders. Philadelphia.
3. IBP Handbook No 12. 1974. in: R.A. Vollenweider (ed.=: A Manual for Measuring Primary Production in Aquatic Environments, 2 edition.
4. Berman, T. and U. Pollingher. 1974. Limnol. Oceanogr. 19)1=, pp 31-54.
5. Tilzer, M.M., H.W. Pearl and C.R. Golterman. 1977. Limnol. Oceanogr. 21(1), pp 84-91.
6. Berland, B.R., P.J. Bonin, P.L. Laborde and S.Y. Maestrini. 1972. Mar. Biol 13, pp 338-345.
7. Claesson, A. 1975. Vatten 31(4), pp 333-338.
8. Frost, B.W. 1972. Limnol. Oceanogr., 17(6), pp 805-815.

1-036 Algae
Chlorophyll a / biomass ratio

Species	Value	Condition
Algae	1.75 microgram chl/l	Conc of algae = 100 cell/ml[1]
Marine phytoplankton	2E-12 g/cell	Urea as source of N, stationary phase, 293 K, Southern Calif[2]
Marine phytoplankton	3.9E-12 g/cell	NH4 as source of N, stationary phase, 293 K, Southern Calif[2]
Marine phytoplankton	4.8E-12 g/cell	NO3as source of N, stationary phase, 293 K, Southern Calif[2]
Oscillatoria redekei	0.25 % of dry weight	Grown at 293 K[3]
Thalassiosira sp.	1.05E-13 gram/cell	Growth rate=0.041 1/h, cell density=4.0E8 cell/l, P limited chemostat[4]
Thalassiosira sp.	8.1E-14 gram/cell	Growth rate=0.017 1/h, cell density=6.7E8 cell/l, P limited chemostat[4]
Trichodesmium spp.	34 ng/colony	Chl a/colony, N Pacific, N-fixing algae, n=6[5]

Source: 1.O'Connor, D.J., R.V. Thomann and D.M. DiToro. 1973. Dynamic Water Quality Forecasting and Management.Office of Re search and Development U.S. Environmental Protection Agency, Wash. 20460.
2.Eppley, R.W., A.F. Carlucci, O.H.-Hansen, D. Kiefer, J.J. McCarthy, E. Venrick and P.M. Williams. 1971. Limnol. Oceanogr., 16(5), pp 741-757.
3.Gibson, C.E. 1975. A Field and Laboratory study of Oxygen uptake by Planktonic Blue-green Algae.
4.Perry, M.J. 1976. Limnol. Oceanogr. 21(1), pp 88-107.
5.Mague, T.H., F.C. Mague and O. Holm-Hansen. 1977. Mar. Biol. 41(3), pp 213-227.

1-037 Algae
Chlorophyll / C ratio

Species	Value	Condition
Isochrysis galbana	13.70 mg mg-1 * 10e3	light conditions: 150 umol quanta m-2 s-1[1]
Isochrysis galbana	17.24 mg mg-1 * 10e3	light conditions: 70 umol quanta m-2 s-1[1]
Isochrysis galbana	23.81 mg mg-1 * 10e3	light conditions: 30 umol quanta m-2 s-1[1]
Isochrysis galbana	5.78 mg mg-1 * 10e3	light conditions: 600 umol quanta m-2 s-1[1]
Isochrysis galbana	8.77 mg mg-1 * 10e3	light conditions: 320 umol quanta m-2 s-1[1]
Prorocentrum micans	2.42 mg mg-1 * 10e3	light conditions: 600 umol quanta m-2 s-1[1]
Prorocentrum micans	2.89 mg mg-1 * 10e3	light conditions: 320 umol quanta m-2 s-1[1]
Prorocentrum micans	4.74 mg mg-1 * 10e3	light conditions: 150 umol quanta m-2 s-1[1]
Prorocentrum micans	6.21 mg mg-1 * 10e3	light conditions: 70 umol quanta m-2 s-1[1]
Thalassiosira weisflogii	15.53 mg mg-1 * 10e3	light conditions: 600 umol quanta m-2 s-1[1]
Thalassiosira weisflogii	21.10 mg mg-1 * 10e3	light conditions: 320 umol quanta m-2 s-1[1]
Thalassiosira weisflogii	34.60 mg mg-1 * 10e3	light conditions: 150 umol quanta m-2 s-1[1]
Thalassiosira weisflogii	42.74 mg mg-1 * 10e3	light conditions: 70 umol quanta m-2 s-1[1]
Thalassiosira weisflogii	50.00 mg mg-1 * 10e3	light conditions: 30 umol quanta m-2 s-1[1]

Source: 1.Falkowski, P.G., Dubinsky, Z. and Wyman, K.., Limnol.Oceanogr., 30(2), 311, 1985.

1-038 Algae
Chlorophyll a

Species	Value	Condition
Natural population	0.47-1.8 ug l-1	Hawaiian coastal waters[1]

Source: 1.Laws, E.A., et al., Limnol.Oceanogr., 29(6), 1161, 1984.

1-039 Algae
Cu / biomass ratio

Species	Value	Condition
Chara sp.	19 microgram/l	Algae from ponds and lakes in southeastern United States[1]
Cladophora sp.	190 microgram/l	Algae from ponds and lakes in southeastern United States[1]
Euglena sp.	290 microgram/l	Algae from ponds and lakes in southeastern United States[1]
Hydrodictyon sp.	114 microgram/l	Algae from ponds and lakes in southeastern United States[1]
Lyngbya sp.	101 microgram/l	Algae from ponds and lakes in southeastern United States[1]
Microcystis sp.	37 microgram/l	Algae from ponds and lakes in southeastern United States[1]
Mougeotia sp.	143 microgram/l	Algae from ponds and lakes in southeastern United States[1]
Nitella sp.	39 microgram/l	Algae from ponds and lakes in southeastern United States[1]
Oedogonium sp.	75 microgram/l	Algae from ponds and lakes in southeastern United States[1]
Pithophora sp.	23 microgram/l	Algae from ponds and lakes in southeastern United States[1]
Rhizoclonium sp.	75 microgram/l	Algae from ponds and lakes in southeastern United States[1]
Spirogyra sp.	47 microgram/l	Algae from ponds and lakes in southeastern United States[1]
Spirogyra sp., giant	34 microgram/l	Algae from ponds and lakes in southeastern United States[1]

Source: 1.Water Quality Criteria. 1968. Federal Water Pollution Control Administration Washington D.C., pp 234

1-040 Algae
Dark uptake

Species	Value	Condition
Algae from pacific Ocean	1.3 % of photosynthesis	Chlorophyll=0.211, inorg N=5.71, inorg P =0.27, Si =5.0 mg/m3[1]
Algae from pacific Ocean	1.5 % of photosynthesis	Chlorophyll=0.201, inorg N=2.50, inorg P =0.35, Si =3.1 mg/m3[1]
Algae from pacific Ocean	13.6 % of photosynthesis	Chlorophyll=0.105, inorg N=7.42, inorg P =0.67, Si =3.6 mg/m3[1]
Algae from pacific Ocean	15.4 % of photosynthesis	Chlorophyll=0.092, inorg N=0.46, inorg P =0.33, Si =3.2 mg/m3[1]
Algae from pacific Ocean	2.2 % of photosynthesis	Chlorophyll=0.135, inorg N=0.34, inorg P =0.23, Si =3.8 mg/m3[1]
Algae from pacific Ocean	2.7 % of photosynthesis	Chlorophyll=0.224, inorg N=0.48, inorg P =0.27, Si =2.0 mg/m3[1]
Algae from pacific Ocean	2.8 % of photosynthesis	Chlorophyll=0.151, inorg N=8.61, inorg P =0.75, Si =3.8 mg/m3[1]
Algae from pacific Ocean	3 % of photosynthesis	Chlorophyll=0.158, inorg N=0.40, inorg P =0.24, Si =2.5 mg/m3[1]
Algae from pacific Ocean	3.1 % of photosynthesis	Chlorophyll=0.263, inorg N=8.35, inorg P =0.83, Si =5.2 mg/m3[1]
Algae from pacific Ocean	3.3 % of photosynthesis	Chlorophyll=0.102, inorg N=7.16, inorg P =0.62, Si =3.5 mg/m3[1]
Algae from pacific Ocean	3.3 % of photosynthesis	Chlorophyll=0.142, inorg N=0.45, inorg P =0.30, Si =3.1 mg/m3[1]
Algae from pacific Ocean	3.6 % of photosynthesis	Chlorophyll=0.174, inorg N=0.49, inorg P =0.22, Si =2.1 mg/m3[1]
Algae from pacific Ocean	4.8 % of photosynthesis	Mean of 8 nitrogen rich stations[1]
Algae from pacific Ocean	5 % of photosynthesis	Chlorophyll=0.041, inorg N=8.11, inorg P =0.69, Si =1.6 mg/m3[1]

Algae from pacific Ocean	5.2 % of photosynthesis	Chlorophyll=0.099, inorg N=0.46, inorg P =0.24, Si =1.5 mg/m3[1]
Algae from pacific Ocean	5.7 % of photosynthesis	Mean of 8 nitrogen poor stations[1]
Algae from pacific Ocean	8 % of photosynthesis	Chlorophyll=0.099, inorg N=7.81, inorg P =0.71, Si =--- mg/m3[1]

Source: 1.Thomas, W.H. 1970. Limnol. Oceanogr. 15(3), pp 380-385.

1-041 Algae
Dry / wet weight ratio

Species	Value	Condition
Agmenellum quadruplicatum	0.563 mg dry weight/mm**3	None[1]
Amphidinium carteri	0.346 mg dry weight/mm**3	None[1]
Aphanizomenon flos-aquae f. gracile	0.35 mg dry weight/mm**3	None[1]
Chaetoceros sp.	0.352 mg dry weight/mm**3	None[1]
Coenococcus planktonicus	0.508 mg dry weight/mm**3	None[1]
Coscinodiscus sp.	0.214 mg dry weight/mm**3	None[1]
Cyclotella meneghiniana	0.31 mg dry weight/mm**3	None[1]
Dictyosphaerium pulchellum	0.611 mg dry weight/mm**3	None[1]
Dunaliella salina	0.333 mg dry weight/mm**3	None[1]
Exuviaella sp.	0.341 mg dry weight/mm**3	None[1]
Melosira granulata	1.02 mg dry weight/mm**3	None[1]
Melosira italica subsp. subartica	0.857 mg dry weight/mm**3	None[1]
Monochrysis lutheri	1.25 mg dry weight/mm**3	None[1]
Oscillatoria agardhii var isothrix	0.425 mg dry weight/mm**3	None[1]
Oscillatoria bourrellyi	0.62 mg dry weight/mm**3	None[1]
Paulschulzia tenera	0.667 mg dry weight/mm**3	None[1]
Phaeodactylum tricornutum	0.181 mg dry weight/mm**3	None[1]
Skeletonema costatum	0.484 mg dry weight/mm**3	None[1]
Stephanodiscus astraea	0.889 mg dry weight/mm**3	None[1]
Stephanodiscus hantzschii	0.23 mg dry weight/mm**3	None[1]
Syracosphaera carteri	0.286 mg dry weight/mm**3	None[1]
Tetraselmis maculata	0.738 mg dry weight/mm**3	None[1]

Source: 1.Lund, J.W.G. 1964. Verh. Int. Ver. Limnol. 15, pp 37-56.

1-042 Algae
Excretion of P

Species	Value	Condition
Freshwater algae	0.84 1/hour	Steady state solution to Castle Lake, autolysis of diss organic P[1]
Seston	0.0002 microgram P/hour	Stonehouse pond, may 17, 1973[2]
Seston	0.01 microgram P/hour	Stonehouse pond, feb 22, 1973[2]
Seston	0.0512 microgram P/hour	Stonehouse pond, june 17, 1973[2]
Seston	0.4608 microgram P/hour	Stonehouse pond, april 6, 1973[2]

Source: 1.Richey, J.E. 1977. in: Hall, C.A.S. and J.W. Day, Jr. (eds.); Ecosystem Modeling in theory and practice. John Wiley and Sons. New York.
2.Ferrante, J.G. 1976. Hydrobiologia, 49(3), pp 203-204.

1-043 Algae
Excretion of dissolved organic C

Species	Value	Condition
Anabaena cylindrica	16.9 % of gross photosynthesis	12 h, old culture, stationary phase[1]
Anabaena cylindrica	3 % of gross photosynthesis	n=2, 12 h, stationary phase, rg = 2.0 - 4.0[1]
Chlorella sp.	1.65 % of photosynthesis	3 h, full sunlight, 305 K, 117500 cell/ml, chl a = 47 ng/ml, rg = 1.5[2]
Coccolithus huxleyi	3.65 % of gross photosynthesis	n=3, 12 h, stationary phase, rg = 2.1 - 5.2[1]
Coccolithus huxleyi	33.5 % of gross photosynthesis	12 h, old culture, stationary phase, rg = 32 - 35[1]
Dicrateria sp.	4.35 % of photosynthesis	3 h, full sunlight, 305 K, 270000 cell/ml, chl a = 88 ng/ml, rg = 4.1[2]
Dunaliella tertiolecta	3.05 % of gross photosynthesis	n=7, 12 h, stationary phase, rg = 1.0 - 5.1[1]
Dunaliella tertiolecta	8 % of gross photosynthesis	12 h, old culture, stationary phase[1]
Freshwater phytoplankton	26 % of photosynthesis	4 h, Lake Windermere, rg = 3 - 49 %[3]
Freshwater phytoplankton	31 % of photosynthesis	4 h, british resevoirs, rg = 2 - 60 %[3]
Freshwater phytoplankton	35.5 % of photosynthesis	6 h, swedish lakes, rg = 1 - 70 %[3]
Freshwater phytoplankton	44.5 % of photosynthesis	5 h, Lake Ontario, rg = 22 - 67 %[3]
Gonyaulax polyedra	32 % of gross photosynthesis	12 h, old culture, stationary phase[1]
Gonyaulax polyedra	4.3 % of gross photosynthesis	n=8, 12 h, stationary phase, rg = 2.3 - 6.3 %[1]
Gymnodinium splendens	3.05 % of gross photosynthesis	n=10, 12 h, stationary phase, rg = 0.6 - 5.5 %[1]
Lauderia sp.	16 % of gross photosynthesis	n=5, 12 h, stationary phase, rg = 13.0 - 19.0 %[1]
Marine phytoplankton	0.2 % of photosynthesis	2 h, inshore N Atlantic[3]
Marine phytoplankton	10 % of photosynthesis	Neritic N Atlantic, 3 - 24 h, rg = 4 - 16 %[3]
Marine phytoplankton	12 % of photosynthesis	6 h, coastal N Pacific, rg = 7 - 17 %[3]
Marine phytoplankton	13.5 % of photosynthesis	2 h, coastral N Atlantic, rg = 0 - 27 %[3]
Marine phytoplankton	14 % of photosynthesis	6 h, Gulf of California, rg = 10 - 18 %[3]
Marine phytoplankton	17.5 % of photosynthesis	4 h, Strait of Georgia, rg = 9 - 26 %[3]
Marine phytoplankton	19.5 % of photosynthesis	2 h, inshore Indian Ocean, rg = 10 - 29 %[3]
Marine phytoplankton	19.5 % of photosynthesis	6 h, offshore N Pacific, rg = 13 - 26 %[3]
Marine phytoplankton	19.5 % of photosynthesis	Inshore AntArtic, 4 h, rg = 1 - 38 %[3]
Marine phytoplankton	22.5 % of photosynthesis	6 h, offshore N Atlantic, rg = 0 - 45 %[3]
Marine phytoplankton	23 % of photosynthesis	6 h, Scottish Sea Lochs, rg = 0 - 46 %[3]
Marine phytoplankton	23.5 % of photosynthesis	Offshore N Pacific, 12 h, rg = 20 - 27 %[3]
Marine phytoplankton	41.5 % of photosynthesis	Western N Atlantic, 4 h, rg = 13 - 70 %[3]
Marine phytoplankton	42.5 % of photosynthesis	6 h, Menai Strait, rg = 10 - 75 %[3]
Marine phytoplankton	8.3 % of gross photosynthesis	n=3, 12 h, stationary phase, rg = 6.9 - 9.7 %[1]

Monochrysis lutheri	29.8 % of gross photosynthesis	12 h, old culture, stationary phase[1]
Nostoc muscorum	3.5 % of gross photosynthesis	n=2, 12 h, stationary phase, rg = 2.0 - 5.0 %[1]
Nostoc muscorum	39.2 % of gross photosynthesis	12 h, old culture, stationary phase[1]
Skeletonema costatum	38 % of photosynthesis	5 h, light = 120000 lux[3]
Skeletonema costatum	4.6 % of photosynthesis	48 h, light = 10000 lux[3]
Skeletonema costatum	5.6 % of photosynthesis	96 h, light = 25000 lux[3]
Skeletonema costatum	6.2 % of photosynthesis	96 h, light = 3000 lux[3]
Skeletonema costatum	9 % of photosynthesis	48 h, light = 3000 lux[3]
Tetraselmis carteriformis	1.43 % of photosynthesis	3 h, full sunlight, 305 K, 21300 cell/ml, chl a = 93 ng/ml, rg = 1.05[2]
Tetraselmis gracilis	3.1 % of photosynthesis	3 h, full sunlight, 305 K, 43500 cell/ml, chl a = 107 ng/ml, rg = 2.8[2]
Thalassiosira fluviatilis	13 % of photosynthesis	5 h, light = 120000 lux[3]
Thalassiosira fluviatilis	3.6 % of photosynthesis	96 h, light = 25000 lux[3]
Thalassiosira fluviatilis	4.5 % of photosynthesis	48 h, light = 10000 lux[3]
Thalassiosira fluviatilis	4.9 % of photosynthesis	96 h, light = 3000 lux[3]
Thalassiosira fluviatilis	6.2 % of photosynthesis	48 h, light = 3000 lux[3]
Thalassiosira pseudonana	27 % of photosynthesis	5 h, light = 120000 lux[3]
Thalassiosira pseudonana	4 % of photosynthesis	48 h, light = 10000 lux[3]
Thalassiosira pseudonana	4.6 % of photosynthesis	48 h, light = 3000 lux[3]

Source: 1.Berman, T. and O. Holm-Hansen. 1974. Mar. biol., 28(4), pp 305-310.
2.Samuels, S., N.M. Shah and G.E. Fogg. 1971. J. mar. biol. Ass. U.K. 51(4), pp 793-798.
3.Sharp, J.H. 1977. Limnol. Oceanogr. 22(3), pp 381-399.

1-044 Algae
Fat / biomass ratio

Species	Value	Condition
Agmenellum quadruplicatum	16 % of ash free dry weight	Blue - green algae, marine, saponifiable fraction only[1]
Amphidinium carteri	23 % of ash free dry weight	Dinophyceae, marine, saponifiable fraction only[1]
Dunaliella salina	7 % of ash free dry weight	Chlorophyceae, marine, saponifiable fraction only[1]
Exuviaella sp.	17 % of ash free dry weight	Dinophyceae, marine, saponifiable fraction only[1]
Monochrysis lutheri	13 % of ash free dry weight	Chrysophyceae, marine, saponifiable fraction only[1]
Skeletonema costatum	7 % of ash free dry weight	Bacillariophyceae, marine, saponifiable fraction only[1]
Tetraselmis maculata	4 % of ash free dry weight	Prasinophyceae, marine, saponifiable fraction only[1]

Source: 1.Parsons, T.R., K. Stephens and J.D.H. Strickland. 1961. J. Fish. Res. Bd. Can., 18, pp 1001-1016.

1-045 Algae
Fe / biomass ratio

Species	Value	Condition
Anabaena sp.	80 microgram/l	Algae from ponds and lakes in southeastern USA[1]
Aphanizomenon sp.	833 microgram/l	Algae from ponds and lakes in southeastern USA[1]
Chara sp.	2520 microgram/l	Algae from ponds and lakes in southeastern USA[1]
Cladophora sp.	2300 microgram/l	Algae from ponds and lakes in southeastern USA[1]
Euglena sp.	1545 microgram/l	Algae from ponds and lakes in southeastern USA[1]
Hydrodictyon sp.	1963 microgram/l	Algae from ponds and lakes in southeastern USA[1]
Lyngbya sp.	3866 microgram/l	Algae from ponds and lakes in southeastern USA[1]
Microcystis sp.	322 microgram/l	Algae from ponds and lakes in southeastern USA[1]
Mougeotia sp.	60 microgram/l	Algae from ponds and lakes in southeastern USA[1]
Nitella sp.	2180 microgram/l	Algae from ponds and lakes in southeastern USA[1]
Oedogonium sp.	1645 microgram/l	Algae from ponds and lakes in southeastern USA[1]
Pithophora sp.	2836 microgram/l	Algae from ponds and lakes in southeastern USA[1]
Rhizoclonium sp.	1820 microgram/l	Algae from ponds and lakes in southeastern USA[1]
Spirogyra sp.	1368 microgram/l	Algae from ponds and lakes in southeastern USA[1]
Spirogyra sp., giant	1793 microgram/l	Algae from ponds and lakes in southeastern USA[1]

Source: 1.Water Quality Criteria. 1968. Federal Water Pollution Control Administration Washington D.C., pp 234

1-046 Algae
Fe quota

Species	Value	Condition
Thalassiosira weissflogii	2-25 10e5 mol Fe l-1 cellvolume	values increases with increasing concentration in environment[1]

Source: 1.Harrison, G.I. and Morel, F.M.M. Limnol. Oceanogr., 31(5), 989, 1986.

1-047 Algae
Gross photosynthesis

Species	Value	Condition
Isochrysis galbana	11.8 umol O2 cell-1 min-1 * 10e10	light conditions: 150 umol quanta m-2 s-1[1]
Isochrysis galbana	15.0 umol O2 cell-1 min-1 * 10e10	light conditions: 320 umol quanta m-2 s-1[1]
Isochrysis galbana	16.2 umol O2 cell-1 min-1 * 10e10	light conditions: 600 umol quanta m-2 s-1[1]
Isochrysis galbana	2.9 umol O2 cell-1 min-1 * 10e10	light conditions: 30 umol quanta m-2 s-1[1]
Isochrysis galbana	5.9 umol O2 cell-1 min-1 * 10e10	light conditions: 70 umol quanta m-2 s-1[1]
Prorocentrum micans	133 umol O2 cell-1 min-1 * 10e10	light conditions: 70 umol quanta m-2 s-1[1]
Prorocentrum micans	165 umol O2 cell-1 min-1 * 10e10	light conditions: 150 umol quanta m-2 s-1[1]
Prorocentrum micans	230 umol O2 cell-1 min-1 * 10e10	light conditions: 320 umol quanta m-2 s-1[1]
Prorocentrum micans	280 umol O2 cell-1 min-1 * 10e10	light conditions: 600 umol quanta m-2 s-1[1]
Thalassiosira weisflogii	130 umol O2 cell-1 min-1 * 10e10	light conditions: 70 umol quanta m-2 s-1[1]
Thalassiosira weisflogii	220 umol O2 cell-1 min-1 * 10e10	light conditions: 150 umol quanta m-2 s-1[1]
Thalassiosira weisflogii	360 umol O2 cell-1 min-1 * 10e10	light conditions: 320 umol quanta m-2 s-1[1]

Thalassiosira weisflogii	38 umol O2 cell-1 min-1 * 10e10	light conditions: 30 umol quanta m-2 s-1[1]
Thalassiosira weisflogii	490 umol O2 cell-1 min-1 * 10e10	light conditions: 600 umol quanta m-2 s-1[1]

Source: 1.Falkowski, P.G., Dubinsky, Z. and Wyman, K.., Limnol.Oceanogr., 30(2), 311, 1985.

1-048 Algae
Growth rate

Species	Value	Condition
Amphidinium carteri	8.8 hours	Doubling time, continuous saturating light, 291 K[1]
Amphidinium sp.	0 growth	2E-3 g atom N/l added as ala, marine, batch, 293 K, F001[2]
Amphidinium sp.	0 growth	2E-3 g atom N/l added as asp, marine, batch, 293 K, F001[2]
Amphidinium sp.	0 growth	2E-3 g atom N/l added as glu, marine, batch, 293 K, F001[2]
Amphidinium sp.	0 growth	2E-3 g atom N/l added as gly, marine, batch, 293 K, F001[2]
Amphidinium sp.	0 growth	2E-3 g atom N/l added as mix of amino acids, marine, batch, 293 K,[2]
Amphidinium sp.	0 growth	2E-3 g atom N/l added as ser, marine, batch, 293 K, F001[2]
Amphidinium sp.	0 growth	2E-3 g atom N/l added as thr, marine, batch, 293 K, F001[2]
Amphidinium sp.	3 days	2E-3 g atom N/l added as NO3, marine, batch, 293 K, F001[2]
Amphidinium sp.	3.3 days	2E-3 g atom N/l added as orn, marine, batch, 293 K, F001[2]
Amphidinium sp.	4.5 days	2E-3 g atom N/l added as lys, marine, batch, 293 K, F001[2]
Amphidinium sp.	9.2 days	2E-3 g atom N/l added as val, marine, batch, 293 K, F001[2]
Anabaena cylindrica	0.96 doublings/day	298 K, saturating light, synthetic medium, blue green alga[3]
Anabaena cylindrica	10.6 hours	Doubling time, continuous saturating light, 298 K[1]
Anacystis nidulans	11.5 doublings/day	314 K, saturating light, synthetic medium, blue green alga[3]
Anacystis nidulans	2 hours	Doubling time, continuous saturating light, 314 K[1]
Asterionella formosa	1.9 doublings/day	291.5 K, saturating light, nonsynthetic medium[3]
Asterionella formosa	9.6 hours	Doubling time, continuous saturating light, 293 K[1]
Asterionella japonica	0 doublings/day	Culture in log phase, 20 % sunlight, no added N, 293-298 K[4]
Asterionella japonica	0.9 doublings/day	Culture in log phase, 20 % sunlight, 75 ug atom N as urea, 293-298 K[4]
Asterionella japonica	1.1 doublings/day	Culture in log phase, 20 % sunlight, 75 ug atom N as NO3, 293-298 K[4]
Asterionella japonica	13.9 hours	Doubling time, continuous saturating light, 293-298 K[1]
Bacillariophyceae	0.29 doublings/day	Culture in log phase, 20 % sunlight, no added N, 293-298 K[4]
Bacillariophyceae	1.16 doublings/day	Culture in log phase, 20 % sunlight, 75 ug atom N/l as NH4, 293-298 K[4]
Bacillariophyceae	1.33 doublings/day	Culture in log phase, 20 % sunlight, no added N, 293-298 K[4]
Bacillariophyceae	1.35 doublings/day	Culture in log phase, 20 % sunlight, 75 ug atom N/l as NO3, 293-298 K[4]
Cachonina sp.	0 growth	2E-3 g atom N/l added as asp, marine, batch, 293 K, F001[2]
Cachonina sp.	0 growth	2E-3 g atom N/l added as glu, marine, batch, 293 K, F001[2]
Cachonina sp.	0 growth	2E-3 g atom N/l added as gly, marine, batch, 293 K, F001[2]
Cachonina sp.	0 growth	2E-3 g atom N/l added as lys, marine, batch, 293 K, F001[2]
Cachonina sp.	0 growth	2E-3 g atom N/l added as orn, marine, batch, 293 K, F001[2]
Cachonina sp.	0 growth	2E-3 g atom N/l added as ser, marine, batch, 293 K, F001[2]
Cachonina sp.	0 growth	2E-3 g atom N/l added as thr, marine, batch, 293 K, F001[2]
Cachonina sp.	0 growth	2E-3 g atom N/l added as val, marine, batch, 293 K, F001[2]

Cachonina sp.	2.6 days	2E-3 g atom N/l added as NO3, marine, batch, 293 K, F001[2]
Cachonina sp.	8 days	2E-3 g atom N/l added as mix of amino acids, marine, batch, 293 K,[2]
Ceratium tripos	82.8 hours	Doubling time, continuous saturating light, 293 K[1]
Chaetoceros lauderi	33 hours	Generation time, grown under optimal condition[5]
Chaetoceros sp.	0 growth	2E-3 g atom N/l added as asp, marine, batch, 293 K, F001[2]
Chaetoceros sp.	0 growth	2E-3 g atom N/l added as glu, marine, batch, 293 K, F001[2]
Chaetoceros sp.	0 growth	2E-3 g atom N/l added as lys, marine, batch, 293 K, F001[2]
Chaetoceros sp.	0 growth	2E-3 g atom N/l added as ser, marine, batch, 293 K, F001[2]
Chaetoceros sp.	0 growth	2E-3 g atom N/l added as thr, marine, batch, 293 K, F001[2]
Chaetoceros sp.	0.31 doublings/day	Culture in log phase, 20 % sunlight, no added N, 293-298 K[4]
Chaetoceros sp.	1.13 doublings/day	Culture in log phase, 20 % sunlight, 75 ug atom N as NH4, 293-298 K[4]
Chaetoceros sp.	1.21 doublings/day	Culture in log phase, 20 % sunlight, 75 ug atom N as urea, 293-298 K[4]
Chaetoceros sp.	1.45 doublings/day	Culture in log phase, 20 % sunlight, 75 ug atom N as NO3, 293-298 K[4]
Chaetoceros sp.	17 hours	Generation time, grown under optimal condition[6]
Chaetoceros sp.	2.5 days	2E-3 g atom N/l added as NO3, marine, batch, 293 K, F001[2]
Chaetoceros sp.	3.6 days	2E-3 g atom N/l added as ala, marine, batch, 293 K, F001[2]
Chaetoceros sp.	3.8 days	2E-3 g atom N/l added as orn, marine, batch, 293 K, F001[2]
Chaetoceros sp.	4.2 days	2E-3 g atom N/l added as mix of amino acids, marine, batch, 293 K,[2]
Chaetoceros sp.	8.2 days	2E-3 g atom N/l added as gly, marine, batch, 293 K, F001[2]
Chlamydomonas reinhardtii	3.8 doublings/day	298 K, saturating light, synthetic medium, green alga[3]
Chlamydomonas sp.	0 growth	2E-3 g atom N/l added as glu, marine, batch, 293 K, F001[2]
Chlamydomonas sp.	0 growth	2E-3 g atom N/l added as gly, marine, batch, 293 K, F001[2]
Chlamydomonas sp.	0 growth	2E-3 g atom N/l added as lys, marine, batch, 293 K, F001[2]
Chlamydomonas sp.	0 growth	2E-3 g atom N/l added as ser, marine, batch, 293 K, F001[2]
Chlamydomonas sp.	2.1 days	2E-3 g atom N/l added as mix of amino acids, marine, batch, 293 K,[2]
Chlamydomonas sp.	3.4 days	2E-3 g atom N/l added as NO3, marine, batch, 293 K, F001[2]
Chlamydomonas sp.	3.4 days	2E-3 g atom N/l added as ala, marine, batch, 293 K, F001[2]
Chlorella ellipsoidea	3.6 doublings/day	298 K, saturating light, synthetic medium, green alga[3]
Chlorella pyrenoidosa	19.6 hours	Doubling time, continuous saturating light, 293 K, planktonic strain[1]
Chlorella pyrenoidosa	2.6 hours	Doubling time, continuous saturating light, 312 K, 7-11-05 strain[1]
Chlorella pyrenoidosa	60.2 hours	Doubling time, continuous saturating light, 283 K, planktonic strain[1]
Chlorella pyrenoidosa	7.75 hours	Doubling time, continuous saturating light, 298 K[1]
Chlorella pyrenoidosa	8 hours	Doubling time, continuous saturating light, 298 K, Emerson strain[1]
Chlorella sp.	0 growth	2E-3 g atom N/l added as ala, marine, batch, 293 K, F001[2]
Chlorella sp.	0 growth	2E-3 g atom N/l added as asp, marine, batch, 293 K, F001[2]
Chlorella sp.	0 growth	2E-3 g atom N/l added as glu, marine, batch, 293 K, F001[2]
Chlorella sp.	0 growth	2E-3 g atom N/l added as gly, marine, batch, 293 K[2]
Chlorella sp.	0 growth	2E-3 g atom N/l added as lys, marine, batch, 293 K, F001[2]
Chlorella sp.	0 growth	2E-3 g atom N/l added as ser, marine, batch, 293 K, F001[2]
Chlorella sp.	0 growth	2E-3 g atom N/l added as val, marine, batch, 293 K, F001[2]

Chlorella sp.	1.5 days	2E-3 g atom N/l added as NO3, marine, batch, 293 K, F001[2]
Chlorella sp.	11 days	2E-3 g atom N/l added as thr, marine, batch, 293 K, F001[2]
Chlorella sp.	23.5 days	2E-3 g atom N/l added as mix of amino acids, marine, batch, 293 K,[2]
Chlorella sp.	5.5 days	2E-3 g atom N/l added as orn, marine, batch, 293 K, F001[2]
Coccolithus huxleyi	0.8 doublings/day	With coccoliths, coastal estuarine, spring[7]
Coccolithus huxleyi	1.37 doublings/day	With coccoliths, subtropical, summer[7]
Coccolithus huxleyi	1.46 doublings/day	No coccoliths, subarctic Pacific, spring[7]
Coccolithus huxleyi	1.59 doublings/day	With coccoliths, tropical upwelling[7]
Coccolithus huxleyi	1.61 doublings/day	No coccoliths, Antarctic upwelling, summer[7]
Coccolithus sp.	0 growth	2E-3 g atom N/l added as asp, marine, batch, 293 K, F001[2]
Coccolithus sp.	0 growth	2E-3 g atom N/l added as glu, marine, batch, 293 K, F001[2]
Coccolithus sp.	0 growth	2E-3 g atom N/l added as lys, marine, batch, 293 K, F001[2]
Coccolithus sp.	0 growth	2E-3 g atom N/l added as mix of amino acids, marine, batch, 293 K,[2]
Coccolithus sp.	0 growth	2E-3 g atom N/l added as orn, marine, batch, 293 K, F001[2]
Coccolithus sp.	0 growth	2E-3 g atom N/l added as ser, marine, batch, 293 K, F001[2]
Coccolithus sp.	0 growth	2E-3 g atom N/l added as thr, marine, batch, 293 K, F001[2]
Coccolithus sp.	0 growth	2E-3 g atom N/l added as val, marine, batch, 293 K, F001[2]
Coccolithus sp.	2 days	2E-3 g atom N/l added as NO3, marine, batch, 293 K, F001[2]
Coccolithus sp.	3.3 days	2E-3 g atom N/l added as ala, marine, batch, 293 K, F001[2]
Coccolithus sp.	3.5 days	2E-3 g atom N/l added as gly, marine, batch, 293 K, F001[2]
Coscinodiscus sp.	29 hours	Generation time, grown under optimal condition[5]
Coscinodiscus sp.	30 hours	Doubling time, continuous saturating light, 291 K[1]
Cricosphaera galbana	20.1 hours	Doubling time, continuous saturating light, 291 K[1]
Cricosphaera sp.	0 growth	2E-3 g atom N/l added as asp, marine, batch, 293 K, F001[2]
Cricosphaera sp.	0 growth	2E-3 g atom N/l added as glu, marine, batch, 293 K, F001[2]
Cricosphaera sp.	0 growth	2E-3 g atom N/l added as gly, marine, batch, 293 K, F001[2]
Cricosphaera sp.	0 growth	2E-3 g atom N/l added as ser, marine, batch, 293 K, F001[2]
Cricosphaera sp.	0 growth	2E-3 g atom N/l added as thr, marine, batch, 293 K, F001[2]
Cricosphaera sp.	2.2 days	2E-3 g atom N/l added as NO3, marine, batch, 293 K, F001[2]
Cricosphaera sp.	2.4 days	2E-3 g atom N/l added as mix of amino acids, marine, batch, 293 K,[2]
Cricosphaera sp.	2.6 days	2E-3 g atom N/l added as orn, marine, batch, 293 K, F001[2]
Cricosphaera sp.	3 days	2E-3 g atom N/l added as lys, marine, batch, 293 K, F001[2]
Cricosphaera sp.	3.5 days	2E-3 g atom N/l added as val, marine, batch, 293 K, F001[2]
Cricosphaera sp.	3.7 days	2E-3 g atom N/l added as ala, marine, batch, 293 K, F001[2]
Cyclotella sp.	0 growth	2E-3 g atom N/l added as ala, marine, batch, 293 K, F001[2]
Cyclotella sp.	0 growth	2E-3 g atom N/l added as asp, marine, batch, 293 K, F001[2]
Cyclotella sp.	0 growth	2E-3 g atom N/l added as glu, marine, batch, 293 K, F001[2]
Cyclotella sp.	0 growth	2E-3 g atom N/l added as gly, marine, batch, 293 K, F001[2]
Cyclotella sp.	0 growth	2E-3 g atom N/l added as lys, marine, batch, 293 K, F001[2]
Cyclotella sp.	0 growth	2E-3 g atom N/l added as mix of amino acids, marine, batch, 293 K,[2]

Cyclotella sp.	0 growth	2E-3 g atom N/l added as orn, marine, batch, 293 K, F001[2]
Cyclotella sp.	0 growth	2E-3 g atom N/l added as ser, marine, batch, 293 K, F001[2]
Cyclotella sp.	0 growth	2E-3 g atom N/l added as thr, marine, batch, 293 K, F001[2]
Cyclotella sp.	0 growth	2E-3 g atom N/l added as val, marine, batch, 293 K, F001[2]
Cyclotella sp.	2.1 days	2E-3 g atom N/l added as NO3, marine, batch, 293 K, F001[2]
Cylindrotheca closterium	0 doublings/day	Culture in log phase, 20 % sunlight, no added N, 293-298 K[4]
Cylindrotheca closterium	1.1 doublings/day	Culture in log phase, 20 % sunlight, 75 ug atom N/l as NH4, 293-298 K[4]
Cylindrotheca closterium	1.1 doublings/day	Culture in log phase, 20 % sunlight, 75 ug atom N/l as urea, 293-298[4]
Cylindrotheca closterium	1.13 doublings/day	Culture in log phase, 20 % sunlight, 75 ug atom N/l as NO3, 293-298 K[4]
Ditylum brightwellii	0.84 doublings/day	Senescent cells, subtropical, summer[7]
Ditylum brightwellii	1.22 doublings/day	Healthy cells, subarctic Pacific, spring[7]
Ditylum brightwellii	1.48 doublings/day	Healthy cells, coastal estuarine, spring[7]
Ditylum brightwellii	1.66 doublings/day	Healthy cells, Antarctic upwelling, summer[7]
Ditylum brightwellii	1.9 doublings/day	Healthy cells, tropical upwelling[7]
Ditylum brightwellii	15 hours	Generation time, grown under optimal condition[8]
Dunaliella sp.	0 growth	2E-3 g atom N/l added as ala, marine, batch, 293 K, F001[2]
Dunaliella sp.	0 growth	2E-3 g atom N/l added as asp, marine, batch, 293 K, F001[2]
Dunaliella sp.	0 growth	2E-3 g atom N/l added as glu, marine, batch, 293 K, F001[2]
Dunaliella sp.	0 growth	2E-3 g atom N/l added as gly, marine, batch, 293 K, F001[2]
Dunaliella sp.	0 growth	2E-3 g atom N/l added as lys, marine, batch, 293 K, F001[2]
Dunaliella sp.	0 growth	2E-3 g atom N/l added as mix of amino acids, marine, batch, 293 K,[2]
Dunaliella sp.	0 growth	2E-3 g atom N/l added as orn, marine, batch, 293 K, F001[2]
Dunaliella sp.	0 growth	2E-3 g atom N/l added as ser, marine, batch, 293 K, F001[2]
Dunaliella sp.	0 growth	2E-3 g atom N/l added as thr, marine, batch, 293 K, F001[2]
Dunaliella sp.	0 growth	2E-3 g atom N/l added as val, marine, batch, 293 K, F001[2]
Dunaliella sp.	2.2 days	2E-3 g atom N/l added as NO3, marine, batch, 293 K, F001[2]
Dunaliella tertiolecta	0.018 1/hour	NO3=4.46, NO2=0.20, NH4=0.19 ug atom N/l, specific growth rate, n=3[9]
Dunaliella tertiolecta	0.045 1/hour	NO3=4.03, NO2=0.29, NH4=0.31 ug atom N/l, specific growth rate, n=3[9]
Dunaliella tertiolecta	0.057 1/hour	NO3=6.60, NO2=0.16, NH4=0.14 ug atom N/l, specific growth rate, n=3[9]
Dunaliella tertiolecta	0.065 1/hour	NO3=19.53, NO2=0.05, NH4=0.45 ug atom N/l, specific growth rate, n=3[9]
Dunaliella tertiolecta	0.073 1/hour	NO3=17.16, NO2=0.24, NH4=0.47 ug atom N/l, specific growth rate, n=3[9]
Dunaliella tertiolecta	0.073 1/hour	NO3=2.92, NO2=0.36, NH4=23.70 ug atom N/l, specific growth rate, n=3[9]
Dunaliella tertiolecta	0.073 1/hour	NO3=4.05, NO2=0.78, NH4=14.98 ug atom N/l, specific growth rate, n=3[9]
Dunaliella tertiolecta	0.073 1/hour	NO3=6.42, NO2=0.28, NH4=23.80 ug atom N/l, specific growth rate, n=3[9]
Dunaliella tertiolecta	0.073 1/hour	NO3=7.91, NO2=0.14, NH4=13.50 ug atom N/l, specific growth rate, n=3[9]
Dunaliella tertiolecta	0.074 1/hour	NO3=12.75, NO2=0.04, NH4=13.10 ug atom N/l, specific growth rate, n=3[9]
Dunaliella tertiolecta	0.074 1/hour	NO3=16.20, NO2=0.22, NH4=5.13 ug atom N/l, specific growth rate, n=3[9]
Dunaliella tertiolecta	0.074 1/hour	NO3=19.00, NO2=0.00, NH4=0.10 ug atom N/l, specific growth rate, n=3[9]
Dunaliella tertiolecta	0.074 1/hour	NO3=2.88, NO2=0.06, NH4=27.50 ug atom N/l, specific growth rate, n=3[9]
Dunaliella tertiolecta	0.076 1/hour	NO3=16.79, NO2=0.02, NH4=0.00 ug atom N/l, specific growth rate, n=3[9]

Euglena gracilis	2.2 doublings/day	298 K, saturating light, synthetic medium, euglenoid alga[3]
Hemiaulus sinensis	0 doublings/day	Culture in log phase, 20 % sunlight, no added N, 293-298 K[4]
Hemiaulus sinensis	0.7 doublings/day	Culture in log phase, 20 % sunlight, 75 ug atom N/l as NH4, 293-298 K[4]
Hemiaulus sinensis	1.16 doublings/day	Culture in log phase, 20 % sunlight, 75 ug atom N/l as urea, 293-298[4]
Hemiaulus sinensis	1.3 doublings/day	Culture in log phase, 20 % sunlight, 75 ug atom N/l as NO3, 293-298 K[4]
Isochrysis galbana	0.30 d-1	light conditions: 30 umol quanta m-2 s-1[10]
Isochrysis galbana	0.70 d-1	light conditions: 70 umol quanta m-2 s-1[10]
Isochrysis galbana	1.10 d-1	light conditions: 150 umol quanta m-2 s-1[10]
Isochrysis galbana	1.20 d-1	light conditions: 320 umol quanta m-2 s-1[10]
Isochrysis galbana	1.20 d-1	light conditions: 600 umol quanta m-2 s-1[10]
Isochrysis galbana	30.2 hours	Doubling time, continuous saturating light, 293 K[1]
Isochrysis sp.	0 growth	2E-3 g atom N/l added as asp, marine, batch, 293 K, F001[2]
Isochrysis sp.	0 growth	2E-3 g atom N/l added as glu, marine, batch, 293 K, F001[2]
Isochrysis sp.	0 growth	2E-3 g atom N/l added as thr, marine, batch, 293 K, F001[2]
Isochrysis sp.	1.5 days	2E-3 g atom N/l added as NO3, marine, batch, 293 K, F001[2]
Isochrysis sp.	3.6 days	2E-3 g atom N/l added as gly, marine, batch, 293 K, F001[2]
Isochrysis sp.	5 days	2E-3 g atom N/l added as orn, marine, batch, 293 K, F001[2]
Isochrysis sp.	5.1 days	2E-3 g atom N/l added as lys, marine, batch, 293 K, F001[2]
Isochrysis sp.	5.5 days	2E-3 g atom N/l added as val, marine, batch, 293 K, F001[2]
Isochrysis sp.	6.5 days	2E-3 g atom N/l added as mix of amino acids, marine, batch, 293 K,[2]
Isochrysis sp.	6.6 days	2E-3 g atom N/l added as ser, marine, batch, 293 K, F001[2]
Isochrysis sp.	8.2 days	2E-3 g atom N/l added as ala, marine, batch, 293 K, F001[2]
Laminaria longicruris	0.25 1/month	Stripe length = 0-24 cm, biomass = 122 g, Nova Scotia, rg = 0.21 -[11]
Leptocylindrus danicus	0.4 doublings/day	Culture in log phase, 20 % sunlight, no added N, 293-298 K[4]
Leptocylindrus danicus	0.95 doublings/day	Culture in log phase, 20 % sunlight, 75 ug atom N as NO3, 293-298 K[4]
Leptocylindrus danicus	0.96 doublings/day	Culture in log phase, 20 % sunlight, 75 ug atom N as NH4, 293-298 K[4]
Leptocylindrus danicus	1.34 doublings/day	Culture in log phase, 20 % sunlight, 75 ug atom N as urea, 293-298 K[4]
Marine phytoplankton	0.26 doublings/day	Average, nutrient-poor water, Sargasso Sea[12]
Marine phytoplankton	0.33 doublings/day	Average, nutrient-poor water, off Southern California, 293 K, rg =[12]
Marine phytoplankton	0.35 doublings/day	Average, nutrient-poor water, off Montauk Pt.[12]
Marine phytoplankton	0.37 doublings/day	Average, nutrient-poor water, off the Carolinas[12]
Marine phytoplankton	0.45 doublings/day	Average, nutrient-poor water, Florida Strait[12]
Marine phytoplankton	0.67 doublings/day	Average, nutrient-rich water, Peru current, 290-293 K, April 1966[12]
Marine phytoplankton	0.7 doublings/day	Average, nutrient-poor water, off Southern California, 285-294 K[12]
Marine phytoplankton	0.73 doublings/day	Average, nutrient-rich water, Peru current, 291-292 K, June 1969[12]
Marine phytoplankton	1 doublings/day	Average, nutrient-rich water, W Arabian Sea, 300-301 K, less than[12]
Marine phytoplankton	1 doublings/day	Average, nutrient-rich water, off SW Africa[12]
Melosira sp.	0 growth	2E-3 g atom N/l added as NO3, marine, batch, 293 K, F001[2]
Melosira sp.	0 growth	2E-3 g atom N/l added as ala, marine, batch, 293 K, F001[2]
Melosira sp.	0 growth	2E-3 g atom N/l added as asp, marine, batch, 293 K, F001[2]

Melosira sp.	0 growth	2E-3 g atom N/l added as glu, marine, batch, 293 K, F001[2]
Melosira sp.	0 growth	2E-3 g atom N/l added as gly, marine, batch, 293 K, F001[2]
Melosira sp.	0 growth	2E-3 g atom N/l added as lys, marine, batch, 293 K, F001[2]
Melosira sp.	0 growth	2E-3 g atom N/l added as mix of amino acids, marine, batch, 293 K,[2]
Melosira sp.	0 growth	2E-3 g atom N/l added as orn, marine, batch, 293 K, F001[2]
Melosira sp.	0 growth	2E-3 g atom N/l added as ser, marine, batch, 293 K, F001[2]
Melosira sp.	0 growth	2E-3 g atom N/l added as thr, marine, batch, 293 K, F001[2]
Melosira sp.	0 growth	2E-3 g atom N/l added as val, marine, batch, 293 K, F001[2]
Monodus subterraneus	24.3 hours	Doubling time, continuous saturating light, 298 K[1]
Monodus subterraneus	37.8 hours	Doubling time, continuous saturating light, 293 K[1]
Monodus subterraneus	42.7 hours	Doubling time, continuous saturating light, 303 K[1]
Monodus subterraneus	97.7 hours	Doubling time, continuous saturating light, 288 K[1]
Navicula incerta	13.8 days	2E-3 g atom N/l added as orn, marine, batch, 293 K, F001[2]
Navicula incerta	3 days	2E-3 g atom N/l added as asp, marine, batch, 293 K, F001[2]
Navicula incerta	3 days	2E-3 g atom N/l added as glu, marine, batch, 293 K, F001[2]
Navicula incerta	3.8 days	2E-3 g atom N/l added as NO3, marine, batch, 293 K, F001[2]
Navicula incerta	3.8 days	2E-3 g atom N/l added as thr, marine, batch, 293 K, F001[2]
Navicula incerta	4 days	2E-3 g atom N/l added as ala, marine, batch, 293 K, F001[2]
Navicula incerta	4 days	2E-3 g atom N/l added as mix of amino acids, marine, batch, 293 K,[2]
Navicula incerta	4 days	2E-3 g atom N/l added as val, marine, batch, 293 K, F001[2]
Navicula incerta	4.7 days	2E-3 g atom N/l added as ser, marine, batch, 293 K, F001[2]
Navicula incerta	7.6 days	2E-3 g atom N/l added as lys, marine, batch, 293 K, F001[2]
Navicula incerta	9.4 days	2E-3 g atom N/l added as gly, marine, batch, 293 K, F001[2]
Navicula minima	1.4 doublings/day	298 K, saturating light, nonsynthetic medium, diatom[3]
Navicula salinarum	0 growth	2E-3 g atom N/l added as asp, marine, batch, 293 K, F001[2]
Navicula salinarum	0 growth	2E-3 g atom N/l added as glu, marine, batch, 293 K, F001[2]
Navicula salinarum	0 growth	2E-3 g atom N/l added as gly, marine, batch, 293 K, F001[2]
Navicula salinarum	0 growth	2E-3 g atom N/l added as lys, marine, batch, 293 K, F001[2]
Navicula salinarum	0 growth	2E-3 g atom N/l added as ser, marine, batch, 293 K, F001[2]
Navicula salinarum	0 growth	2E-3 g atom N/l added as thr, marine, batch, 293 K, F001[2]
Navicula salinarum	0 growth	2E-3 g atom N/l added as val, marine, batch, 293 K, F001[2]
Navicula salinarum	11.3 days	2E-3 g atom N/l added as NO3, marine, batch, 293 K, F001[2]
Navicula salinarum	7.5 days	2E-3 g atom N/l added as ala, marine, batch, 293 K, F001[2]
Navicula salinarum	8 days	2E-3 g atom N/l added as mix of amino acids, marine, batch, 293 K,[2]
Nitzschia frustulum	3.4 days	2E-3 g atom N/l added as NO3, marine, batch, 293 K, F001[2]
Nitzschia frustulum	4.3 days	2E-3 g atom N/l added as ala, marine, batch, 293 K, F001[2]
Nitzschia frustulum	4.7 days	2E-3 g atom N/l added as lys, marine, batch, 293 K, F001[2]
Nitzschia frustulum	5.2 days	2E-3 g atom N/l added as orn, marine, batch, 293 K, F001[2]
Nitzschia frustulum	5.7 days	2E-3 g atom N/l added as val, marine, batch, 293 K, F001[2]
Nitzschia frustulum	5.8 days	2E-3 g atom N/l added as asp, marine, batch, 293 K, F001[2]

Nitzschia frustulum	5.8 days	2E-3 g atom N/l added as glu, marine, batch, 293 K, F001[2]
Nitzschia frustulum	6 days	2E-3 g atom N/l added as thr, marine, batch, 293 K, F001[2]
Nitzschia frustulum	6.5 days	2E-3 g atom N/l added as mix of amino acids, marine, batch, 293 K,[2]
Nitzschia frustulum	6.5 days	2E-3 g atom N/l added as ser, marine, batch, 293 K, F001[2]
Nitzschia frustulum	7.3 days	2E-3 g atom N/l added as gly, marine, batch, 293 K, F001[2]
Nitzschia ovalis	1.7 days	2E-3 g atom N/l added as asp, marine, batch, 293 K, F001[2]
Nitzschia ovalis	1.7 days	2E-3 g atom N/l added as glu, marine, batch, 293 K, F001[2]
Nitzschia ovalis	2 days	2E-3 g atom N/l added as mix of amino acids, marine, batch, 293 K, F001[2]
Nitzschia ovalis	3.3 days	2E-3 g atom N/l added as ala, marine, batch, 293 K, F001[2]
Nitzschia ovalis	3.5 days	2E-3 g atom N/l added as NO3, marine, batch, 293 K, F001[2]
Nitzschia ovalis	4 days	2E-3 g atom N/l added as orn, marine, batch, 293 K, F001[2]
Nitzschia ovalis	4.4 days	2E-3 g atom N/l added as lys, marine, batch, 293 K, F001[2]
Nitzschia ovalis	4.4 days	2E-3 g atom N/l added as val, marine, batch, 293 K, F001[2]
Nitzschia ovalis	4.8 days	2E-3 g atom N/l added as ser, marine, batch, 293 K, F001[2]
Nitzschia ovalis	5.4 days	2E-3 g atom N/l added as gly, marine, batch, 293 K, F001[2]
Nitzschia ovalis	9.5 days	2E-3 g atom N/l added as thr, marine, batch, 293 K, F001[2]
Nitzschia sp.	0.3 doublings/day	Culture in log phase, 20 % sunlight, no added N, 293-298 K[4]
Nitzschia sp.	1.36 doublings/day	Culture in log phase, 20 % sunlight, 75 ug atom N as NH4, 293-298 K[4]
Nitzschia sp.	1.48 doublings/day	Culture in log phase, 20 % sunlight, 75 ug atom N as urea, 293-298 K[4]
Nitzschia sp.	1.52 doublings/day	Culture in log phase, 20 % sunlight, 75 ug atom N as NO3, 293-298 K[4]
Nitzschia sp.	1.8 days	2E-3 g atom N/l added as NO3, marine, batch, 293 K, F001[2]
Nitzschia sp.	2 days	2E-3 g atom N/l added as gly, marine, batch, 293 K, F001[2]
Nitzschia sp.	2 days	2E-3 g atom N/l added as ser, marine, batch, 293 K, F001[2]
Nitzschia sp.	2.7 days	2E-3 g atom N/l added as orn, marine, batch, 293 K, F001[2]
Nitzschia sp.	2.9 days	2E-3 g atom N/l added as lys, marine, batch, 293 K, F001[2]
Nitzschia sp.	3 days	2E-3 g atom N/l added as asp, marine, batch, 293 K, F001[2]
Nitzschia sp.	3 days	2E-3 g atom N/l added as glu, marine, batch, 293 K, F001[2]
Nitzschia sp.	3 days	2E-3 g atom N/l added as mix of amino acids, marine, batch, 293 K,[2]
Nitzschia sp.	3 days	2E-3 g atom N/l added as thr, marine, batch, 293 K, F001[2]
Nitzschia sp.	3.7 days	2E-3 g atom N/l added as val, marine, batch, 293 K, F001[2]
Nitzschia sp.	4.2 days	2E-3 g atom N/l added as ala, marine, batch, 293 K, F001[2]
Phaeodactylum tricornutum	0 growth	2E-3 g atom N/l added as thr, marine, batch, 293 K, F001[2]
Phaeodactylum tricornutum	1 days	2E-3 g atom N/l added as ala, marine, batch, 293 K, F001[2]
Phaeodactylum tricornutum	1.5 days	2E-3 g atom N/l added as NO3, marine, batch, 293 K, F001[2]
Phaeodactylum tricornutum	1.7 days	2E-3 g atom N/l added as orn, marine, batch, 293 K, F001[2]
Phaeodactylum tricornutum	1.7 days	2E-3 g atom N/l added as val, marine, batch, 293 K, F001[2]
Phaeodactylum tricornutum	1.8 days	2E-3 g atom N/l added as ser, marine, batch, 293 K, F001[2]
Phaeodactylum tricornutum	10 hours	Doubling time, continuous saturating light, 298 K[1]
Phaeodactylum tricornutum	13 hours	Generation time, grown under optimal condition[5]
Phaeodactylum tricornutum	2 days	2E-3 g atom N/l added as mix of amino acids, marine, batch, 293 K,[2]

Phaeodactylum tricornutum	2.1 days	2E-3 g atom N/l added as gly, marine, batch, 293 K, F001[2]
Phaeodactylum tricornutum	2.2 days	2E-3 g atom N/l added as asp, marine, batch, 293 K, F001[2]
Phaeodactylum tricornutum	2.2 days	2E-3 g atom N/l added as glu, marine, batch, 293 K, F001[2]
Phaeodactylum tricornutum	2.3 days	2E-3 g atom N/l added as lys, marine, batch, 293 K, F001[2]
Phaeodactylum tricornutum	2.4 doublings/day	281 K, nonsynthetic medium[3]
Phaeodactylum tricornutum	2.5 days	2E-3 g atom N/l added as asp, marine, batch, 293 K, F001[2]
Phaeodactylum tricornutum	2.5 days	2E-3 g atom N/l added as glu, marine, batch, 293 K, F001[2]
Phaeodactylum tricornutum	3 days	2E-3 g atom N/l added as val, marine, batch, 293 K, F001[2]
Phaeodactylum tricornutum	3.5 days	2E-3 g atom N/l added as NO3, marine, batch, 293 K, F001[2]
Phaeodactylum tricornutum	3.6 days	2E-3 g atom N/l added as lys, marine, batch, 293 K, F001[2]
Phaeodactylum tricornutum	4.1 days	2E-3 g atom N/l added as ala, marine, batch, 293 K, F001[2]
Phaeodactylum tricornutum	4.4 days	2E-3 g atom N/l added as ser, marine, batch, 293 K, F001[2]
Phaeodactylum tricornutum	5.6 days	2E-3 g atom N/l added as gly, marine, batch, 293 K, F001[2]
Phaeodactylum tricornutum	5.7 days	2E-3 g atom N/l added as orn, marine, batch, 293 K, F001[2]
Phaeodactylum tricornutum	5.9 days	2E-3 g atom N/l added as mix of amino acids, marine, batch, 293 K,[2]
Platymonas sp.	0 growth	2E-3 g atom N/l added as lys, marine, batch, 293 K, F001[2]
Platymonas sp.	1.2 days	2E-3 g atom N/l added as ala, marine, batch, 293 K, F001[2]
Platymonas sp.	1.5 days	2E-3 g atom N/l added as mix of amino acids, marine, batch, 293 K,[2]
Platymonas sp.	1.5 days	2E-3 g atom N/l added as ser, marine, batch, 293 K, F001[2]
Platymonas sp.	1.6 days	2E-3 g atom N/l added as NO3, marine, batch, 293 K, F001[2]
Platymonas sp.	1.7 days	2E-3 g atom N/l added as gly, marine, batch, 293 K, F001[2]
Platymonas sp.	23.5 days	2E-3 g atom N/l added as val, marine, batch, 293 K, F001[2]
Platymonas sp.	27.5 days	2E-3 g atom N/l added as thr, marine, batch, 293 K, F001[2]
Platymonas sp.	3.6 days	2E-3 g atom N/l added as asp, marine, batch, 293 K, F001[2]
Platymonas sp.	3.6 days	2E-3 g atom N/l added as glu, marine, batch, 293 K, F001[2]
Platymonas sp.	30 hour	Doubling time, Newport Bay[13]
Platymonas sp.	6.9 days	2E-3 g atom N/l added as orn, marine, batch, 293 K, F001[2]
Platymonas subcordiformis	0 growth	2E-3 g atom N/l added as lys, marine, batch, 293 K, F001[2]
Platymonas subcordiformis	1.4 days	2E-3 g atom N/l added as gly, marine, batch, 293 K, F001[2]
Platymonas subcordiformis	1.7 days	2E-3 g atom N/l added as ser, marine, batch, 293 K, F001[2]
Platymonas subcordiformis	2.2 days	2E-3 g atom N/l added as NO3, marine, batch, 293 K, F001[2]
Platymonas subcordiformis	2.2 days	2E-3 g atom N/l added as ala, marine, batch, 293 K, F001[2]
Platymonas subcordiformis	2.5 days	2E-3 g atom N/l added as asp, marine, batch, 293 K, F001[2]
Platymonas subcordiformis	2.5 days	2E-3 g atom N/l added as glu, marine, batch, 293 K, F001[2]
Platymonas subcordiformis	3.1 days	2E-3 g atom N/l added as mix of amino acids, marine, batch, 293 K,[2]
Platymonas subcordiformis	3.2 days	2E-3 g atom N/l added as thr, marine, batch, 293 K, F001[2]
Platymonas subcordiformis	6.5 days	2E-3 g atom N/l added as val, marine, batch, 293 K, F001[2]
Platymonas subcordiformis	9.1 days	2E-3 g atom N/l added as orn, marine, batch, 293 K, F001[2]
Porphyridium aerugineum	1.4 doublings/day	294 K, saturating light, synthetic medium, red alga[3]
Porphyridium sp.	2.5 days	2E-3 g atom N/l added as NO3, marine, batch, 293 K, F001[2]

Prorocentrum micans	0 doublings/day	Culture in log phase, 20 % sunlight, no added N, 293-298 K[4]
Prorocentrum micans	0.075 d-1	light conditions: 70 umol quanta m-2 s-1[10]
Prorocentrum micans	0.108 d-1	light conditions: 150 umol quanta m-2 s-1[10]
Prorocentrum micans	0.164 d-1	light conditions: 320 umol quanta m-2 s-1[10]
Prorocentrum micans	0.178 d-1	light conditions: 600 umol quanta m-2 s-1[10]
Prorocentrum micans	0.63 doublings/day	Culture in log phase, 20 % sunlight, 75 ug atom N/l as NH4, 293-298 K[4]
Prorocentrum micans	0.63 doublings/day	Culture in log phase, 20 % sunlight, no added N, 293-298 K[4]
Prorocentrum micans	0.7 doublings/day	Culture in log phase, 20 % sunlight, 75 ug atom N/l as NO3, 293-298 K[4]
Prorocentrum micans	55.5 hours	Doubling time, continuous saturating light, 293 K[1]
Scenedesmus obliquus	2.2 doublings/day	298 K, saturating light, synthetic medium, green alga[3]
Schizothrix calcicola	3.4 doublings/day	303 K, saturating light, synthetic medium, blue green alga[3]
Selenastrum capricornutum	0.6 +/- 0.3 doublings d-1	culture with 8.0 uM TP and Zn 1.0 uM[14]
Selenastrum capricornutum	0.7 +/- 0.2 doublings d-1	culture with 10.0 uM TP and Zn 1.0 uM[14]
Selenastrum capricornutum	0.8 +/- 0.3 doublings d-1	culture with 12.0 uM TP and Zn 1.0 uM[14]
Selenastrum capricornutum	1.5 +/- 0.3 doublings d-1	culture with 8.0 uM TP and Zn 0.5 uM[14]
Selenastrum capricornutum	1.6 +/- 0.2 doublings d-1	culture with 10.0 uM TP and Zn 0.5 uM[14]
Selenastrum capricornutum	1.6 +/- 0.3 doublings d-1	culture with 12.0 uM TP and Zn 0.5 uM[14]
Selenastrum capricornutum	1.7 +/- 0.4 doublings d-1	culture with 12.0 uM TP and Zn < 0.1 uM[14]
Selenastrum capricornutum	1.8 +/- 0.2 doublings d-1	culture with 10.0 uM TP and Zn < 0.1 uM[14]
Selenastrum capricornutum	1.9 +/- 0.3 doublings d-1	culture with 8.0 uM TP and Zn < 0.1 uM[14]
Skeletonema costatum	0.25 doublings/day	Culture in log phase, 20 % sunlight, no added N, 293-298 K[4]
Skeletonema costatum	0.88 doublings/day	Culture in log phase, 20 % sunlight, 75 ug atom N/l as NH4, 293-298 K[4]
Skeletonema costatum	1.3 doublings/day	Culture in log phase, 20 % sunlight, 75 ug atom N/l as NO3, 293-298 K[4]
Skeletonema costatum	1.44 doublings/day	Culture in log phase, 20 % sunlight, 75 ug atom N/l as urea, 293-298[4]
Skeletonema costatum	13 hours	Generation time, grown under optimal condition[6]
Skeletonema costatum	13.1 hours	Doubling time, continuous saturating light, 291 K[1]
Skeletonema sp.	0 growth	2E-3 g atom N/l added as ala, marine, batch, 293 K, F001[2]
Skeletonema sp.	0 growth	2E-3 g atom N/l added as asp, marine, batch, 293 K, F001[2]
Skeletonema sp.	0 growth	2E-3 g atom N/l added as glu, marine, batch, 293 K, F001[2]
Skeletonema sp.	0 growth	2E-3 g atom N/l added as lys, marine, batch, 293 K, F001[2]
Skeletonema sp.	0 growth	2E-3 g atom N/l added as mix of amino acids, marine, batch, 293 K,[2]
Skeletonema sp.	0 growth	2E-3 g atom N/l added as orn, marine, batch, 293 K, F001[2]
Skeletonema sp.	0 growth	2E-3 g atom N/l added as ser, marine, batch, 293 K, F001[2]
Skeletonema sp.	0 growth	2E-3 g atom N/l added as thr, marine, batch, 293 K, F001[2]
Skeletonema sp.	0 growth	2E-3 g atom N/l added as val, marine, batch, 293 K, F001[2]

Skeletonema sp.	5.5 days	2E-3 g atom N/l added as NO3, marine, batch, 293 K, F001[2]
Skeletonema sp.	6 days	2E-3 g atom N/l added as gly, marine, batch, 293 K, F001[2]
Stephanopyxis sp.	0 growth	2E-3 g atom N/l added as ala, marine, batch, 293 K, F001[2]
Stephanopyxis sp.	0 growth	2E-3 g atom N/l added as asp, marine, batch, 293 K, F001[2]
Stephanopyxis sp.	0 growth	2E-3 g atom N/l added as glu, marine, batch, 293 K, F001[2]
Stephanopyxis sp.	0 growth	2E-3 g atom N/l added as gly, marine, batch, 293 K, F001[2]
Stephanopyxis sp.	0 growth	2E-3 g atom N/l added as lys, marine, batch, 293 K, F001[2]
Stephanopyxis sp.	0 growth	2E-3 g atom N/l added as mix of amino acids, marine, batch, 293 K,[2]
Stephanopyxis sp.	0 growth	2E-3 g atom N/l added as orn, marine, batch, 293 K, F001[2]
Stephanopyxis sp.	0 growth	2E-3 g atom N/l added as ser, marine, batch, 293 K, F001[2]
Stephanopyxis sp.	0 growth	2E-3 g atom N/l added as thr, marine, batch, 293 K, F001[2]
Stephanopyxis sp.	0 growth	2E-3 g atom N/l added as val, marine, batch, 293 K, F001[2]
Stephanopyxis sp.	1.7 days	2E-3 g atom N/l added as NO3, marine, batch, 293 K, F001[2]
Tabellaria flocculosa var. flocculosa	1.4 doublings/day	293 K, saturating light, nonsynthetic medium[3]
Thalassiosira pseudonana	15 hours	Generation time, grown under optimal condition[15]
Thalassiosira rotula	10.8 hours	Generation time, 2400 lux, 99 % CL = 9.1 - 13.3 hrs, 291 K[16]
Thalassiosira rotula	11.6 hours	Generation time, 4000 lux, 99 % CL = 9.6 - 14.4 hrs, 285 K[16]
Thalassiosira rotula	15.5 hours	Generation time, 2400 lux, 99 % CL = 14.9 - 16.3 hrs, 285 K[16]
Thalassiosira rotula	20.6 hours	Generation time, 1200 lux, 99 % CL = 17.7 - 24.6 hrs, 291 K[16]
Thalassiosira rotula	27.6 hours	Generation time, 1200 lux, 99 % CL = 26.1 - 29.2 hrs, 285 K[16]
Thalassiosira rotula	28.5 hours	Generation time, 2400 lux, 99 % CL = 27.0 - 30.0 hrs, 278 K[16]
Thalassiosira rotula	37.7 hours	Generation time, 1200 lux, 99 % CL = 34.3 - 41.9 hrs, 278 K[16]
Thalassiosira rotula	38.9 hours	Generation time, 600 lux, 99 % CL = 35.7 - 42.7 hrs, 285 K[16]
Thalassiosira rotula	39.7 hours	Generation time, 2400 lux, 99 % CL = 34.6 - 46.7 hrs, 277 K[16]
Thalassiosira rotula	9.2 hours	Generation time, 2400 lux, 99 % CL = 8.4 - 10.1 hrs, 295 K[16]
Thalassiosira rotula	9.7 hours	Generation time, 5500 lux, 99 % CL = 9.2 - 10.3 hrs, 285 K[16]
Thalassiosira sp.	0 growth	2E-3 g atom N/l added as ala, marine, batch, 293 K, F001[2]
Thalassiosira sp.	0 growth	2E-3 g atom N/l added as asp, marine, batch, 293 K, F001[2]
Thalassiosira sp.	0 growth	2E-3 g atom N/l added as glu, marine, batch, 293 K, F001[2]
Thalassiosira sp.	0 growth	2E-3 g atom N/l added as gly, marine, batch, 293 K, F001[2]
Thalassiosira sp.	0 growth	2E-3 g atom N/l added as lys, marine, batch, 293 K, F001[2]
Thalassiosira sp.	0 growth	2E-3 g atom N/l added as thr, marine, batch, 293 K, F001[2]
Thalassiosira sp.	0 growth	2E-3 g atom N/l added as val, marine, batch, 293 K, F001[2]
Thalassiosira sp.	1.6 days	2E-3 g atom N/l added as NO3, marine, batch, 293 K, F001[2]
Thalassiosira sp.	2.5 days	2E-3 g atom N/l added as orn, marine, batch, 293 K, F001[2]
Thalassiosira sp.	4.1 days	2E-3 g atom N/l added as mix of amino acids, marine, batch, 293 K,[2]
Thalassiosira sp.	5.3 days	2E-3 g atom N/l added as ser, marine, batch, 293 K, F001[2]
Thalassiosira weisflogii	0.25 d-1	light conditions: 30 umol quanta m-2 s-1[10]
Thalassiosira weisflogii	0.62 d-1	light conditions: 70 umol quanta m-2 s-1[10]
Thalassiosira weisflogii	1.15 d-1	light conditions: 150 umol quanta m-2 s-1[10]

Thalassiosira weisflogii	1.73 d-1	light conditions: 320 umol quanta m-2 s-1[10]
Thalassiosira weisflogii	1.80 d-1	light conditions: 600 umol quanta m-2 s-1[10]
Trichodesmium spp.	177 hours	Doubling time, based on ATP carbon, N Pacific, N-fixing algae[17]
Trichodesmium spp.	185 hours	Doubling time, from N assimilation, N Pacific, N-fixing algae[17]
Trichodesmium spp.	520 hours	Doubling time, from C assim.(tot.part.carb.), N Pacific, N-fixing[17]

Source: 1.Fogg, G.E. 1966. Algal Cultures and Phytoplankton Ecology. Univ. of Wisconsin Press. Wis.
2.Wheeler, P.A., B.B. North and G.C. Stephens. 1974. Limnol. Oceanogr. 19(2), pp 249-259.
3.Hoogenhant, H. and J. Amesz. 1965. Arch. Microbiol. 50, pp 10-24.
4.Eppley, R.W., A.F. Carlucci, O.H.-Hansen, D. Kiefer, J.J. McCarthy, E. Venrick and P.M. Williams. 1971. Limnol. Oceanogr., 16(5), pp 741-757.
5.Berland, B.R., P.J. Bonin, P.L. Laborde and S.Y. Maestrini. 1972. Mar. biol., 13, pp 338-345.
6.Parsons, T.R., K. Stephens and J.D.H. Strickland. 1961. J. Fish. Res. Bd. Can., 18, pp 1001-1016.
7.Parsons, T.R. and M. Takahashi. 1973. Limnol. Oceanogr., 18(4), pp 511-515.
8.Strickland, J.D.H., O. Holm-Hansen, R.W. Eppley and R. J. Linn. 1969. Limnol. Oceanogr., 14, pp 23-2.
9.Bienfang, P.K. 1975. Limnol. Oceanogr., 20(3), pp 402-411.
10.Falkowski, P.G., Dubinsky, Z. and Wyman, K.., Limnol.Oceanogr., 30(2), 311, 1985.
11.Mann, K.H. 1972. Mar. Biol. 14, pp 199-209.
12.Eppley, R.W. 1972. Fish Bull., 70, pp 1063-1085.
13.North, B.B. 1975. Limnol. Oceanogr. 20(1), pp 20-27.
14.Kuwabara, J.S., Davis, J.A. and Chang, C.C.Y. Limnol. Oceanogr., 31, 503, 1986.
15.Caperon, J. and J. Meyer. 1972. Deep Sea Res., 19, pp 601-632.
16.Schoene, H.K. 1972. Mar. biol. 13, pp 284-291.
17.Mague, T.H., F.C. Mague and O. Holm-Hansen. 1977. Mar. Biol. 41(3), pp 213-227.

1-049 Algae
H / biomass ratio

Species	Value	Condition
Algae	4.6 % of dry weight	Mean elementary composition[1]
Phaeophyceae	4.1 % of dry weight	Mean elementary composition[1]

Source: 1.Bowen, H.J.M. 1966. Trace Elements in Biochemistry, Ac. Press.

1-050 Algae
Half saturation constant for Fe uptake

Species	Value	Condition
Thalassiosira weissflogii	3 10e-19 M	free ferric Fe[1]

Source: 1.Harrison, G.I. and Morel, F.M.M. Limnol. Oceanogr., 31(5), 989, 1986.

1-051 Algae
Half saturation constant for N uptake

Species	Value	Condition
Oscillatoria agardhii gomomt	35 micromoles/l	Light int.=3.6w/m**2, pH=7.3-7.8, N limited, NO3, batch cult.[1]
Oscillatoria agardhii gomomt	36 micromoles/l	Light int.=0.1w/m**2, pH=7.1-7.6, N limited, NO3, batch cult.[1]
Oscillatoria agardhii gomomt	36 micromoles/l	Light int.=5.5w/m**2, pH=7.1-7.6, N limited, NO3, batch cult.[1]
Thalassiosira pseudonana	5E-7 M	Growth rate = 0.017 1/h, chemostat, oligotrof isolate[2]
Thalassiosira pseudonana	5E-7 M	Growth rate = 0.041 1/h, chemostat, oligotrof isolate[2]
Thalassiosira pseudonana	5E-7 M	Growth rate =0.0085 1/h, chemostat, oligotrof isolate[2]

Source: 1.Liere, L. van, W. Zevenboom and L.R. Mur. 1975. in: Conf. on nitrogen as a water Pollutant, Vol 2, Copenhagen.
2.Perry, M.J. 1976. Limnol. Oceanogr. 21(1), pp 88-107.

1-052 Algae
Half saturation constant for NH4 - N uptake

Species	Value	Condition
Asterionella japonica	1 micromoles/l	None[1]
Chaetoceros debilis	0.5 ug-atom/l	Chemostat, uptake=0 for conc = 0.2 ug-atom/l, NH4-limitation[2]
Chaetoceros debilis	2.2 ug-atom/l	Chemostat, uptake=0 for conc = 0.3 ug-atom/l, Si-limitation[2]
Chaetoceros gracilis	0.4 micromoles/l	None[1]
Coccolithus huxleyi	0.1 micromoles/l	None[1]
Coscinodiscus lineatus	2 micromoles/l	None[1]
Coscinodiscus wailesii	4.9 micromoles/l	None[1]
Cyclotella nana	0.4 micromoles/l	None[1]
Ditylum brightwellii	1.1 micromoles/l	None[1]
Dunaliella tertiolecta	0.17 micromoles/l	None[3]
Dunaliella tertiolecta	0.6 micromoles/l	None[1]
Gonyaulax polyedra	5.5 micromoles/l	None[1]
Gymnodinium splendens	1.1 micromoles/l	None[1]
Leptocylindrus danicus	0.7 micromoles/l	None[1]
Marine phytoplankton	0.15 microgram atom/l	North Pacific[4]
Monochrysis lutheri	0.29 micromoles/l	None[3]
Monochrysis lutheri	0.4 micromoles/l	None[1]
Neritic bacillariophyceae	63.5 microgram/l	Range = 7.0 - 120[5]
Neritic or littoral flagellate	42 microgram/l	Range = 7.0 - 77.0[5]
Oceanic species	3.5 microgram/l	Range = 1.4 - 5.6[5]
Oscillatoria thiebautii	6.74 microgram/l	Sargasso Sea[6]
Rhizosolenia robusta	7.5 micromoles/l	None[1]
Rhizosolenia stolterfothii	0.5 micromoles/l	None[1]
Skeletonema costatum	0.5 ug-atom/l	Chemostat, uptake=0 for conc = 0.2 ug-atom/l, NH4-limitation[2]
Skeletonema costatum	0.8 micromoles/l	None[1]
Skeletonema costatum	1.3 ug-atom/l	Chemostat, uptake=0 for conc = 0.2 ug-atom/l, Si-limitation[2]
Thalassiosira gravida	0.3 ug-atom/l	Chemostat, uptake=0 for conc = 0.5 ug-atom/l, Si-limitation[2]
Thalassiosira gravida	0.5 ug-atom/l	Chemostat, uptake=0 for conc = 0.2 ug-atom/l, NH4-limitation[2]

Source: 1.Eppley, R.W. and J.N. Rogers and J.J. McCarthy. 1969. Limnol. Oceanogr. 14. pp 912-920.
2.Conway, H.L. and P.J. Harrison. 1977. Mar. Biol., 43(1), pp 33-43.
3.Caperon, J. and J. Meyer. 1972. Deep Sea Res., 19, pp 601-632.
4.Eppley, R.W., E.H. Renger, E.L. Venrick and M.M. Mullin. 1973. Limnol. Oceanogr., 18(4), pp 534-551.
5.Eppley, R.W. and J.L. Coatsworth. 1968. J. Phycol., 4. pp 151-156.
6.Carpenter, E.J. and J.J. McCarthy. 1975. Limnol. Oceanogr., 20(3), pp 389-401.

1-053 Algae
Half saturation constant for NO3 - N uptake

Species	Value	Condition
Anabaena cylindrica	70 micromoles/l	None[1]
Asterionella japonica	1 micromoles/l	None[2]

Bellochia sp.	0.5 micromoles/l	Range = 0.1 - 0.9[3]
Chaetoceros gracilis	0.2 micromoles/l	None[2]
Coccochloris stagnina	0.31 micromoles/l	None[4]
Coccolithus huxleyi	0.1 microgram atom/l	None[5]
Coccolithus huxleyi	0.1 micromoles/l	None[2]
Coscinodiscus lineatus	2.6 micromoles/l	None[2]
Coscinodiscus wailesii	3.6 micromoles/l	None[2]
Cyclotella nana	0.35 micromoles/l	None[4]
Cyclotella nana	0.5 micromoles/l	None[2]
Cyclotella nana	1.15 micromoles/l	Range = 0.4 - 1.9[3]
Cyclotella nana	1.8 micromoles/l	None[6]
Ditylum brightwellii	0.6 microgram atom/l	None[5]
Ditylum brightwellii	0.6 micromoles/l	None[2]
Dunaliella sp.	2.8E-7 M	288 K, 95% CL = +- 0.68, light=9500 lux, batch culture[7]
Dunaliella sp.	9.5E-7 M	298 K, 95% CL = +- 0.72, light=9500 lux, batch culture[7]
Dunaliella tertiolecta	0.21 micromoles/l	None[4]
Dunaliella tertiolecta	1.4 micromoles/l	None[2]
Fragilaria pinnata	1.1 micromoles/l	Range = 0.6 - 1.6[3]
Gonyaulax polyedra	9.5 micromoles/l	None[2]
Gymnodinium splendens	1.02E-6 M	291 K, 95% CL = +- 0.99, light=9500 lux, batch culture[7]
Gymnodinium splendens	3.8 micromoles/l	None[2]
Gymnodinium splendens	6.55E-6 M	298 K, 95% CL = +- 0.85, light=9500 lux, batch culture[7]
Isochrysis galbana	0.1 micromoles/l	None[2]
Leptocylindrus danicus	1.25 micromoles/l	None[2]
Monochrysis lutheri	0.42 micromoles/l	None[4]
Monochrysis lutheri	0.6 micromoles/l	None[2]
Neritic bacillariophyceae	17.15 microgram/l	Range = 6.3 - 28[8]
Oceanic species	4.2 microgram/l	Range = 1.4 - 7.0[8]
Phaeodactylum tricornutum	2.6 micromoles/l	None[9]
Rhizosolenia robusta	3 micromoles/l	None[2]
Rhizosolenia stolterfothii	1.7 micromoles/l	None[2]
Skeletonema costatum	0.45 micromoles/l	None[2]
Thalassiosira pseudonana	0.38 micro M	Clone 13-1[10]
Thalassiosira pseudonana	1.87 micro M	Clone 3h[10]

Source: 1.Hattori, A. 1962. Plant Cell Physiol. 3, pp 355-369.
2.Eppley, R.W. and J.N. Rogers and J.J. McCarthy. 1969. Limnol. Oceanogr. 14. pp 912-920.
3.Carpenter, E.J. and R.R.L. Guillard. 1971. Ecology, 52, pp 183-185.
4.Caperon, J. and J. Meyer. 1972. Deep Sea Res., 19, pp 601-632.
5.Parsons, T.R. and M. Takahashi. 1973. Limnol. Oceanogr., 18(4), pp 511-515.
6.MacIsaac, J.J. and R.C. Dagdale. 1972. Deep Sea Res. 16. pp 415-422.
7.Thomas, W.H. and A.N. Dodson. 1974. Mar biol. 24, pp 213-217.
8.Eppley, R.W. and J.L. Coatsworth. 1968. J. Phycol., 4. pp 151-156.
9.Ketchum, B.H. 1939. Am. J. Bot. 26. pp 399-407.
10.Hecky, R.E. and P. Kilham. 1974. Limnol. Oceanogr., 19(2), pp 361-365.

1-054 Algae
Half saturation constant for P uptake

Species	Value	Condition
Asterionella japonica	0.12 microgram atom P/l	Tropical ocean, batch culture[1]
Biddulphia sinensis	0.17 microgram atom P/l	Tropical estuary, batch culture[2]
Cerataulina bergonii	3.33 microgram P/l	Volume=2712 u**3, P content=1700E-15 g-atom/cell[3]
Ceratium furca	0.15 microgram atom P/l	Tropical estuary, batch culture[3]
Chaetoceros curvisetus	0.944 microgram P/l	Volume=2170 u**3, P content=100-310E-15 g-atom/cell[3]
Chaetoceros gracilis	3.7 mg/m3	None[4]
Chaetoceros socialis	0.75 microgram P/l	Volume=100 u**3, P content=3.2-7.0E-15 g-atom/cell, rg = 0.389 - 1.11[3]
Chlorella pyrenoidosa	4.5 micromoles/l	Range = 4.0 - 5.0[5]
Coscinodiscus gigas	2.84 microgram P/l	Volume=5150805 u**3, rg = 1.78 - 3.89[3]
Coscinodiscus granii	0.627 microgram P/l	Volume=172292 u**3, P content=1000-4000E-15 g-atom/cell[3]
Cyclotella nana	0.58 micromoles/l	None[6]
Cyclotella nana	1 mg/m3	None[4]
Dinobryon cylindricum	0.72 micromoles P/l	New Jersey[7]
Dinobryon cylindricum	0.8 micromoles/l	None[8]
Dinobryon sociale var. americanum	0.39 micromoles P/l	New Jersey, in p-free medium for 2 d[7]
Dinobryon sociale var. americanum	0.42 micromoles/l	New Jersey, directly from culture[7]
Dinobryon sociale var. americanum	0.5 micromoles/l	None[8]
Ditylum brightwellii	3.33 microgram P/l	Volume=175472 u**3, P content=10000E-15 g-atom/cell[3]
Euglena gracilis	0.3 microgram atom P/l	Continuous culture[9]
Euglena gracilis	16 micromoles/l	None[10]
Freshwater algae	0.011 mg ortho P/l	Determined from winter PO4-P conc and max primary production, rg=[11]
Freshwater phytoplankton	0.18 micro M	Jordan River, upstream L Kinneret, 289 K, P =0.44 micro M[12]
Freshwater phytoplankton	0.21 micro M	Southern Lake Kinneret, 301 K, P=0.11 micro M[12]
Freshwater phytoplankton	0.24 micro M	Central Lake Kinneret, 287.5 K, P =0.14 micro M[12]
Freshwater phytoplankton	0.25 micro M	Southern Lake Kinneret, 304 K, P =0.037 micro M[12]
Freshwater phytoplankton	0.79 micro M	Southern Lake Kinneret, 290 K, P =0.10 micro M[12]
Marine phytoplankton	0.905 microgram atom P/l	Chesapeake Bay, natural phytoplankton, 5 stations, rg = 0.09 - 1.72[13]
Microcystis aeruginosa	10 microgram/l	Estimated value[14]
Navicula arenaria	0.1 microgram atom P/l	Estuary, improved batch culture[15]
Nitzschia actinastroides	0.014 microgram atom P/l	Oligotrophic lake, continuous culture[16]
Nitzschia actinastroides	1 micromoles/l	None[17]
Pediastrum duplex	1.1 micromoles/l	None[8]

Phaeodactylum tricornutum	10 microgram/l	None[18]
Scenedesmus obliquus	1.5 mg/m3	None[4]
Scenedesmus obliquus	48.4 microgram atom P/l	Fresh water, continuous culture, too high value[19]
Scenedesmus sp.	0.03 microgram atom P/l	Fresh water, continuous culture[20]
Scenedesmus sp.	0.6 micromoles/l	None[21]
Selenastrum capricornutum	5 mg/m3	Batch culture[22]
Skeletonema costatum	0.722 microgram P/l	Volume=144 u**3, P content=8.0-28.0E-15 g-atom/cell, rg = 0.244 - 1.2[3]
Thalassiosira fluviatilis	1 mg/m3	None[4]
Thalassiosira fluviatilis	1.72 micromoles/l	None[6]
Thalassiosira pseudonana	6E-7 M	Growth rate = 0.041 1/h, chemostat, oligotrof isolate[23]
Thalassiosira pseudonana	7E-7 M	Growth rate = 0.017 1/h, chemostat, oligotrof isolate[23]
Thalassiosira pseudonana	7E-7 M	Growth rate = 0.035 1/h, chemostat, oligotrof isolate[23]
Thalassiosira pseudonana	7E-7 M	Growth rate = 0.041 1/h, chemostat, isolate from moriches bay[23]

Source: 1.Thomas, W.H. and A.N. Dodson. 1968. Biol. Bull. Mar. Biol. Lab. , Woods Hole 134, pp 199-208.
2.Quasim, S.Z., P.M.A. Bhattathiri and V.P. Devassy. 1973. Mar. Biol. 21, pp 299-304.
3.Finenko, Z.Z. and D.K. Krupatkina-Aki Nina. 1974. Mar. Biol., 26, pp 193-201.
4.Fogg, G.E. 1973. Water Res., 7(1/2), pp 77-91.
5.Jeanjean, R. 1969. Bull. Soc. Fr. Phy siol. reg.15, pp 150-171.
6.Fuhs, G.W., S.P. Demmerle, E. Canelli and M. Chen. 1972. in; G.E. Likens (ed); Nutrients and eutrophcation. Amer. Soc. Limnol. Oceangr. Spec. Symp. 1, pp 113-132.
7.Lehman, J.T., J. Sarvala, I. Hakala and M.-L. Viljanen. 1976. Limnol. Oceanogr. 21(5), pp 724-730.
8.Livingston, H.D. and G. Thompson. 1971. Limnol. Oceanogr. 16(5), pp 786-796.
9.Chisholm, S.W., R.G. Struss and P.A. Nobbs. 1975. J. Phycol., 11, pp 307-373.
10.Blum, J.J. 1966. J. Gen. Physiol., 49, pp 1125-1137.
11.Clendenning, K. A., T.E. Brown and H.C. Eyster. 1956. Can. J. Bot., 34, pp 943-966.
12.Halmann, M. and M. Stiller. 1974. Limnol. Oceanogr., 19(5), pp 774-783.
13.Joergensen, S.E., J.H. Henriksen, M. Friis, L.A. Joergensen and H.F. Mejer. Handbook of Environmental. Data and Ecological Parameters. 1979. ISEM. Copenhagen.
14.Riley, G.A. 1965. Limnol. Oceanogr., 10(suppl.), pp 202-215.
15.Admiraal, W. 1977. Mar. Biol., 42(1), pp 1-8.
16.Muller, H. von. 1972. Arch. Hydrobiol. Suppl. 38(4). pp 399-484.
17.Muller, H. von. 1972. Arch. Hydrobiol. Suppl. 38(4). pp 399-484.
18.Gerloff, S. 1957 Ecology 38, pp 556-561.
19.Eichhorn, M. 1969. Flora, Jena (Abt.A.) 159, pp 494-506.
20.Yull Rhee, G. 1973. J. Phycol. 9, pp 495-506.
21.Rhee, G.-Y. 1973. J. Phycol. 9. pp 495-506.
22.Toerien, D.F. and C.H. Huang. Water Res. 7(11), pp 1673-1681.
23.Perry, M.J. 1976. Limnol. Oceanogr. 21(1), pp 88-107.

1-055 Algae
Half saturation constant for ammonium growth

Species	Value	Condition
Selenastrum capricornutum	5.237 microgram N/l	293 K, chemost., ammonium growth limited, r=0.975, monod eq.[1]
Selenastrum capricornutum	5.237 microgram N/l	293 K, chemostat[2]
Selenastrum capricornutum	5.246 microgram N/l	300 K, chemost., ammonium growth limited, r=0.989, monod eq.[1]
Selenastrum capricornutum	5.248 microgram N/l	301 K, chemost., ammonium growth limited, r=0.799, monod eq.[1]
Selenastrum capricornutum	5.251 microgram N/l	297 K, chemost., ammonium growth limited, r=0.996, monod eq.[1]
Selenastrum capricornutum	5.505 microgram N/l	306 K, chemost., ammonium growth limited, r=0.949, monod eq.[1]

Source: 1.Reynolds, J.H., E.J. Middlebrooks, D.B. Porcella and W.J. Grenney, 1975. J. Wat. Poll. Control Fed.7(10), pp 2420-2436.
2.Pitter, P. 1976. Water Res., 10(3), pp 231-236.

1-056 Algae
Half saturation constant for growth

Species	Value	Condition
Thalassiosira weissflogii	1.1 10e-21 M	free ferric Fe[1]

Source: 1.Harrison, G.I. and Morel, F.M.M., Limnol. Oceanogr., 31(5), 989, 1986.

1-057 Algae
Half saturation constant for light

Species	Value	Condition
Algae	0.005 kcal/(m**2*sec)	Range = 0.004 - 0.006[1]
Coccolithus huxleyi	0.002 ly/min	None[2]
Ditylum brightwellii	0.009 ly/min	None[2]
Sargassum sp.	0.014 ly/min	Incubator, comm.resp.=20-30% of max photosynt.rate =0.45 mg c/g[3]
Skeletonema costatum	0.0003 ly/min	N as nitrate, sun adapted, estaurine phytoplankton[4]
Skeletonema costatum	0.0004 ly/min	N as nitrate, shade adapted, estaurine phytoplankton[4]
Skeletonema costatum	0.003 ly/min	N as nitrate and NH4, sun adapted, estaurine phytoplankton[4]
Skeletonema costatum	0.007 ly/min	N as nitrate and NH4, shade adapted, estaurine phytoplankton[4]
chlorophyta	0.0017 ly/min	N as nitrate, shade adapted, estaurine phytoplankton[4]
chlorophyta	0.0024 ly/min	N as nitrate, sun adapted, estaurine phytoplankton[4]
chlorophyta	0.0026 ly/min	N as nitrate and NH4, shade adapted, estaurine phytoplankton[4]
chlorophyta	0.0066 ly/min	N as nitrate and NH4, sun adapted, estaurine phytoplankton[4]

Source: 1.The Hydrological Engineering Center Corps of Engineers U.S. Army. 1975. Water Quality for river-resevoir systems, Davis, Cal. 9 5 6 1 6 (916) 499-2329.
2.Parsons, T.R. and M. Takahashi. 1973. Limnol. Oceanogr., 18(4), pp 511-515.
3.Carpenter, E.J. and J.L. Cox. 1974. Lim nol. Oceanogr., 19(3), pp 429-436.
4.Bates, S.S. 1976. Limnol. Oceanogr., 21(2), pp 212-218.

1-058 Algae
Half saturation constant for nitrate growth

Species	Value	Condition
Neritic or littoral flagellate	69.2 microgram/l	Flagellate, rg = 8.4 - 130[1]

Source: 1.Eppley, R.W. and J.L. Coatsworth. 1968. J. Phycol., 4. pp 151-156.

1-059 Algae
Half saturation constant for nitrite uptake

Species	Value	Condition
Anabaena cylindrica	40 micromoles/l	None[1]
Chlorella pyrenoidosa	25 micromoles/l	None[2]

Source: 1.Hattori, A. 1962. Plant Cell Physiol. 3, pp 355-369.
2.Eppley, R.W. and J.N. Rogers and J.J. McCarthy. 1969. Limnol. Oceanogr. 14. pp 912-920.

1-060 Algae
Half saturation constant for nitrogen growth

Species	Value	Condition
Algae	0.025 mg/l	Lake Ontario[1]
Algae	0.025 mg/l	Potomac River[2]
Algae	0.075 mg/l	Range = 0.05 - 0.1[3]
Algae	0.15 mg/l	Range = 0.1 - 0.2[3]
Phytoplankton	25 microgram/l	Delta Estuary[4]
Phytoplankton	25 microgram/l	Lake Erie[4]
Phytoplankton	25 microgram/l	Lake Ontario[4]
Phytoplankton	25 microgram/l	Potomac Estuary[4]
Phytoplankton	25 microgram/l	San Joaquin River[4]

Source: 1.Thomann, R.V., D.M. Di Toro, R.P. Winfield and D.J. O'Connor. 1975. Mathematical Modeling of Phytoplankton in Lake Ontario: 1. National Environmental Reseach Center. EPA. Oregon.
2.O'Connor, D.J., R.V. Thomann and D.M. DiToro. 1973. Dynamic Water Quality Forecasting and Management. Office of Research and Development U.S. Environmental Protection Agency, Wash. 20460.
3.The Hydrological Engineering Center Corps of Engineers U.S. Army. 1975. Water Quality for river-resevoir systems, Davis, Cal. 9 5 6 1 6 (916) 499-2329.
4.O'Connor, D.J., D.M. DiToro and R.V. Thomann. 1975. in: Russel, C.S. (ed.); Ecological Modeling in a Resource Management Framework. pp 149-209.

1-061 Algae
Half saturation constant for phosphorus growth

Species	Value	Condition
Algae	0.002 mg/l	Lake Ontario[1]
Algae	0.005 mg/l	Potomac River[2]
Algae	0.035 mg/l	Range = 0.02 - 0.05[3]
Algae	0.06 mg/l	Range = 0.04 - 0.08 mg/l[3]
Chaetoceros gracilis	25 microgram/l	None[1]
Phytoplankton	10 microgram/l	Lake Erie[4]
Phytoplankton	2 microgram/l	Lake Ontario[4]
Phytoplankton	5 microgram/l	Potomac Estuary[4]

Source: 1.Thomann, R.V., D.M. Di Toro, R.P. Winfield and D.J. O'Connor. 1975. Mathematical Modeling of Phytoplankton in Lake Ontario: 1. National Environmental Reseach Center. EPA. Oregon.
2.O'Connor, D.J., R.V. Thomann and D.M. DiToro. 1973. Dynamic Water Quality Forecasting and Management. Office of Research and Development U.S. Environmental Protection Agency, Wash. 20460.
3.The Hydrological Engineering Center Corps of Engineers U.S. Army. 1975. Water Quality for river-resevoir systems, Davis, Cal. 9 5 6 1 6 (916) 499-2329.
4.O'Connor, D.J., D.M. DiToro and R.V. Thomann. 1975. in: Russel, C.S. (ed.); Ecological Modeling in a Resource Management Framework. pp 149-209.

1-062 Algae
Half saturation constant for silicium uptake

Species	Value	Condition
Ditylum brightwellii	2.96 microgram atom Si/l	Uptake, no uptake for =2.96 ug-at/l, 25000 erg/(cm2*sec), 293 K, F077[1]
Licmophora sp.	2.58 microgram atom Si/l	Uptake, no uptake for =1.30 ug-at/l, 25000 erg/(cm2*sec), 293 K, F077[1]
Nitzschia actinastroides	3.5 micromoles/l	None[2]

Skeletonema costatum	0.8 micromoles/l	None[3]
Thalassiosira decipiens	3.37 microgram atom Si/l	Uptake, no uptake for =1.33 ug-at/l, 25000 erg/(cm2*sec), 293 K, F077[1]
Thalassiosira pseudonana	0.19 micro M	Clone 13-1[4]
Thalassiosira pseudonana	0.98 micro M	Clone 3h[4]
Thalassiosira pseudonana	1.39 microgram atom Si/l	Uptake, no uptake for =0.67 ug-at/l, 25000 erg/(cm2*sec), 293 K, F077[1]
Thalassiosira pseudonana	2.15 micromoles/l	Range = 1.4 - 2.9[3]

Source: 1.Paasche, F. 1973. Mar. Biol. 19. pp 262-269.
2.Muller, H. von. 1972. Arch. Hydrobiol. Suppl. 38(4). pp 399-484.
3.Paasche, F. 1973. Mar. Biol. 19. pp 262-269.
4.Hecky, R.E. and P. Kilham. 1974. Limnol. Oceanogr., 19(2), pp 361-365.

1-063 Algae
Half saturation constant for total nitrogen growth

Species	Value	Condition
Scenedesmus gracile	150 microgram/l	Total nitrogen considered as limiting[1]

Source: 1.O'Connor, D.J., R.V. Thomann and D.M. DiToro. 1973. Dynamic Water Quality Forecasting and Management. Office of Research and Development U.S. Environmental Protection Agency, Wash. 20460.

1-064 Algae
Half saturation constant for total phosphorus growth

Species	Value	Condition
Scenedesmus gracile	10 microgram/l	Total phosphorus considered as limiting[1]

Source: 1.O'Connor, D.J., R.V. Thomann and D.M. DiToro. 1973. Dynamic Water Quality Forecasting and Management. Office of Research and Development U.S. Environmental Protection Agency, Wash. 20460.

1-065 Algae
K / biomass ratio

Species	Value	Condition
Anabaena sp.	1.2 % of dry weight	Algae from ponds and lakes in southeastern United States[1]
Aphanizomenon sp.	0.68 % of dry weight	Algae from ponds and lakes in southeastern United States[1]
Chara sp.	2.3 % of dry weight	Algae from ponds and lakes in southeastern United States[1]
Cladophora sp.	6.08 % of dry weight	Algae from ponds and lakes in southeastern United States[1]
Euglena sp.	0.34 % of dry weight	Algae from ponds and lakes in southeastern United States[1]
Hydrodictyon sp.	4.21 % of dry weight	Algae from ponds and lakes in southeastern United States[1]
Lyngbya sp.	0.42 % of dry weight	Algae from ponds and lakes in southeastern United States[1]
Microcystis sp.	0.79 % of dry weight	Algae from ponds and lakes in southeastern United States[1]
Mougeotia sp.	1.2 % of dry weight	Algae from ponds and lakes in southeastern United States[1]
Nitella sp.	3.73 % of dry weight	Algae from ponds and lakes in southeastern United States[1]
Oedogonium sp.	3.03 % of dry weight	Algae from ponds and lakes in southeastern United States[1]

Pithophora sp.	3.06 % of dry weight	Algae from ponds and lakes in southeastern United States[1]
Rhizoclonium sp.	1.9 % of dry weight	Algae from ponds and lakes in southeastern United States[1]
Spirogyra sp.	0.92 % of dry weight	Algae from ponds and lakes in southeastern United States[1]
Spirogyra sp., giant	0.99 % of dry weight	Algae from ponds and lakes in southeastern United States[1]

Source: 1.Water Quality Criteria. 1968. Federal Water Pollution Control Administration Washington D.C., pp 234.

1-066 Algae
Lipid / biomass ratio

Species	Value	Condition
Chlorella sp.	20.2 % of dry weight	Favorable environment[1]
Chlorella sp.	63.4 % of dry weight	Unfavorable environment[1]
Marine phytoplankton	0.15 by weight	Vigorously growing, excess nitrate in water[2]
Marine phytoplankton	0.35 by weight	Unhealthy phytoplankton, nitrate depleted water[2]

Source: 1.Milner, H.W. 1953. in: Burlew, J.S. (ed.): Algal Culture from Laboratory to Pilot Plant. Publ. 600. Carnegie Inst. of Wash. Wash. D.C.
2.Antia, N.J., C.D. McAllister, T.R. Parsons, K. Stephens and J.H.D. Strickland. 1963. Limnol. Oceanogr., 8, pp 166-183.

1-067 Algae
Max growth rate

Species	Value	Condition
Acrosiphonia centralis	15 mg O2/(gram dry weight*hour)	Spring, 283 K, light=14.4 mW/cm2, benthic, Baltic Sea[1]
Algae	0.1 1/day	Potomac River[2]
Algae	0.1 1/day	Used in models[2]
Algae	0.51 1/day	275.6 K, natural association[3]
Algae	0.58 1/day	Temp factor is 1 at 273 K, Lake Ontario[4]
Algae	0.605 1/day	Range = 0.01 - 1.2[5]
Algae	0.63 1/day	277 K, natural association[3]
Algae	1.03 1/day	Range = 0.06 - 2.0[6]
Algae	1.03 1/day	Range = 0.06 - 2.0[7]
Algae	1.5 1/day	Temp factor is 1 at 293 K, range = 1.0 - 2.0[8]
Algae	2 1/day	Temp factor is 1 at 293 K, range = 1.0 - 3.0[8]
Ceramium strictum	17 mg O2/(gram dry weight*hour)	Autumn, 288 K, light=16.3 mW/cm2, benthic algae, Baltic Sea[1]
Cerataulina bergonii	1.47 divisions/day	Volume=2712 u**3, P content=1700E-15 g-atom/cell[9]
Chaetoceros curvisetus	1.83 divisions/day	Volume=2170 u**3, P content=100-310E-15 g-atom/cell, rg = 1.45 - 2.20[9]
Chaetoceros gracilis	0.107 1/hour	N limited, NH4, 298 K, continuous light : 0.05ly/min[10]
Chaetoceros socialis	1.84 divisions/day	Volume=100 u**3, P content=3.2-7.0E-15 g-atom/cell, rg = 1.53 - 2.13[9]
Chaetomorpha sp.	14 micromoles C/(mg chlorophyll a*hour)	297 K, 1.3E17 quanta/(s*cm2), bubbled with air[11]
Chlamydomonas reinhardtii	2.64 1/day	298 K[12]
Chlamydomonas reinhardtii	2.64 1/day	298 K[3]

Chlorella ellipsoidea	1.2 1/day	288 K[3]
Chlorella ellipsoidea	3.14 1/day	298 K[3]
Chlorella ellipsoidea	3.14 1/day	298 K[13]
Chlorella pyrenoidosa	0.18 1/hour	N limited, NO3, 308 K, continuous light : 0.05cal/(cm**2)*min[14]
Chlorella pyrenoidosa	0.18 1/hour	Specific growth rate, 308 K, substrate NO3, ligth=0.05 cal/(cm2*min)[15]
Chlorella pyrenoidosa	0.2 1/day	283 K[16]
Chlorella pyrenoidosa	0.2 1/day	283 K, high-temperature strain[3]
Chlorella pyrenoidosa	0.9 mg dw/ml	Determinated at end of exponential fase after 65 hrs[17]
Chlorella pyrenoidosa	1.1 1/day	288 K[16]
Chlorella pyrenoidosa	1.1 1/day	288 K, high-temperature strain[3]
Chlorella pyrenoidosa	1.36 1/day	292 K, NH4 limiting, continuous cult., freshwater[18]
Chlorella pyrenoidosa	1.45 1/day	292 K, NO3 limiting, continuous cult., freshwater[19]
Chlorella pyrenoidosa	1.84 1/day	301.5 K, NH4 limiting, continuous cult., freshwater[18]
Chlorella pyrenoidosa	1.95 1/day	298 K, P limiting, continuous cult., freshwater[20]
Chlorella pyrenoidosa	1.96 1/day	298 K[21]
Chlorella pyrenoidosa	1.96 1/day	298 K[3]
Chlorella pyrenoidosa	2.14 1/day	298 K, P limiting, continuous cult., freshwater[20]
Chlorella pyrenoidosa	2.15 1/day	298 K[12]
Chlorella pyrenoidosa	2.15 1/day	298 K[3]
Chlorella pyrenoidosa	2.22 1/day	301.5 K, NO3 limiting, continuous cult., freshwater[19]
Chlorella pyrenoidosa	2.4 1/day	293 K[16]
Chlorella pyrenoidosa	2.4 1/day	293 K, high-temperature strain[3]
Chlorella pyrenoidosa	3.9 1/day	298 K[16]
Chlorella pyrenoidosa	3.94 1/day	308 K, NH4 limiting, continuous cult., freshwater[18]
Chlorella pyrenoidosa	4.26 1/day	312.2 K, NH4 limiting, continuous cult., freshwater[18]
Chlorella pyrenoidosa	4.32 1/day	308.5 K, NO3 limiting, continuous cult., freshwater[18]
Chlorella pyrenoidosa	5.65 1/day	312.2 K, NO3 limiting, continuous cult., freshwater[19]
Chlorella pyrenoidosa	9.9 micromoles C/(mg chlorophyll a*hour)	297 K, 1.3E17 quanta/(s*cm2), bubbled with air[11]
Chlorella sp.	1.88 1/day	298 K, NO3 limiting, continuous cult., freshwater[22]
Chlorella vulgaris	1.8 1/day	298 K[12]
Chlorella vulgaris	1.8 1/day	298 K[3]
Cladophora glomerata	27 mg O2/(gram dry weight*hour)	Summer, 293 K, light=18.4 mW/cm2, benthic algae, Baltic Sea[1]
Coccochloris sp.	2.16 1/day	298 K, NO3 limiting, continuous cult., marine[23]
Coccochloris stagnina	0.086 1/hour	N limited, NO3, 298 K, continuous light : 5.8cal/(cm**2)*hr[24]
Coccochloris stagnina	0.086 1/hour	Specific growth rate, 298 K, substrate NO3, ligth= 5.8 cal/(cm2*h)[15]
Coccolithus huxleyi	1.75 doublings/day	Optimal conditions[25]
Coscinodiscus gigas	0.62 divisions/day	Volume = 5150805 u**3, rg = 0.51 - 0.73[9]

Coscinodiscus granii	0.71 divisions/day	Volume=172292 u**3, P content=1000-4000E-15 g-atom/cell[9]
Cyclotella nana	0.086 1/hour	N limited, NO3, 298 K, continuous light : 5.8cal/(cm**2)*hr[24]
Delesseria sanguinea	2.2 mg O2/(gram dry weight*hour)	Winter, 278 K, light=10 mW/cm2, benthic algae, Baltic Sea[1]
Delesseria sanguinea	3 mg O2/(gram dry weight*hour)	Autumn, 288 K, light=8.0 mW/cm2, benthic algae, Baltic Sea[1]
Delesseria sanguinea	4 mg O2/(gram dry weight*hour)	Summer, 293 K, light=4.9 mW/cm2, benthic algae, Baltic Sea[1]
Delesseria sanguinea	8.8 mg O2/(gram dry weight*hour)	Spring, 283 K, light=10 mW/cm2, benthic algae, Baltic Sea[1]
Ditylum brightwellii	1.2 divisions/day	Volume=175472 u**3, P content=10000E-15 g-atom/cell[9]
Ditylum brightwellii	2.32 doublings/day	Optimal conditions[25]
Dumontia incrassata	17 mg O2/(gram dry weight*hour)	Spring, 283 K, light=18.4 mW/cm2, benthic algae, Baltic Sea[1]
Dumontia incrassata	9 mg O2/(gram dry weight*hour)	Winter, 278 K, light=5.5 mW/cm2, benthic algae, Baltic Sea[1]
Dunaliella sp.	1.85 doublings/day	298 K, 95% CL = +- 0.24, light=9500 lux, batch culture[26]
Dunaliella sp.	1.93 doublings/day	288 K, 95% CL = +- 0.26, light=9500 lux, batch culture[26]
Dunaliella tertiolecta	0.07 1/hour	N limited, NH4, 298 K, continuous light : 5.8cal/(cm**2)*hr[24]
Dunaliella tertiolecta	0.8 1/day	288 K, Fe limiting, continuous cult., marine[27]
Dunaliella tertiolecta	1.83 1/day	298 K, NH4 limiting, continuous cult., marine[23]
Ectocarpus confervoides	14 mg O2/(gram dry weight*hour)	Spring, 283 K, light=26.7 mW/cm2, benthic algae, Baltic Sea[1]
Enteromorpha intestinalis	28 mg O2/(gram dry weight*hour)	Winter, 278 K, light=13.5 mW/cm2, benthic algae, Baltic Sea[1]
Enteromorpha intestinalis	30 mg O2/(gram dry weight*hour)	Spring, 283 K, light=17.5 mW/cm2, benthic algae, Baltic Sea[1]
Enteromorpha linza	22 mg O2/(gram dry weight*hour)	Autumn, 288 K, light=18.4 mW/cm2, benthic algae, Baltic Sea[1]
Enteromorpha prolifera	13 mg O2/(gram dry weight*hour)	Summer, 293 K, light=5.6 mW/cm2, benthic algae, Baltic Sea[1]
Enteromorpha prolifera	42 mg O2/(gram dry weight*hour)	Autumn, 288 K, light=18.4 mW/cm2, benthic algae, Baltic Sea[1]
Enteromorpha sp.	79.6 micromoles C/(mg chlorophyll a*hour	297 K, 1.3E17 quanta/(s*cm2), bubbled with air[11]
Fucus serratus	2.4 mg O2/(gram dry weight*hour)	Spring, 283 K, light=23.6 mW/cm2, benthic algae, Baltic Sea[1]
Fucus serratus	3.1 mg O2/(gram dry weight*hour)	Autumn, 288 K, light=3.1 mW/cm2, benthic algae, Baltic Sea[1]
Fucus serratus	5.4 mg O2/(gram dry weight*hour)	Winter, 278 K, light=43.6 mW/cm2, benthic algae, Baltic Sea[1]
Fucus vesiculosus	5.4 mg O2/(gram dry weight*hour)	Winter, 278 K, light=30.2 mW/cm2, benthic algae, Baltic Sea[1]
Fucus vesiculosus	6 mg O2/(gram dry weight*hour)	Spring, 283 K, light=28.4 mW/cm2, benthic algae, Baltic Sea[1]
Fucus vesiculosus	6.6 mg O2/(gram dry weight*hour)	Autumn, 288 K, light=23.6 mW/cm2, benthic algae, Baltic Sea[1]

Furcellaria fastigiata	1.8 mg O2/(gram dry weight*hour)	Autumn, 288 K, light=4.3 mW/cm2, benthic algae, Baltic Sea[1]
Furcellaria fastigiata	2.3 mg O2/(gram dry weight*hour)	Winter, 278 K, light=10 mW/cm2, benthic algae, Baltic Sea[1]
Furcellaria fastigiata	3.2 mg O2/(gram dry weight*hour)	Spring, 283 K, light=10 mW/cm2, benthic algae, Baltic Sea[1]
Glenodinium sp.	61.7 micromoles C/(mg chlorophyll a*hour	297 K, 1.3E17 quanta/(s*cm2), bubbled with air[11]
Gymnodinium splendens	0.48 doublings/day	291 K, 95% CL = +-0.06, light=9500 lux, batch culture[26]
Gymnodinium splendens	0.83 doublings/day	298 K, 95% CL = +-0.42, light=9500 lux, batch culture[26]
Laminaria digitata	1.3 mg O2/(gram dry weight*hour)	Winter, 278 K, light=1.3 mW/cm2, benthic algae, Baltic Sea[1]
Laminaria digitata	3.8 mg O2/(gram dry weight*hour)	Spring, 283 K, light=26.7 mW/cm2, benthic algae, Baltic Sea[1]
Laminaria digitata	3.9 mg O2/(gram dry weight*hour)	Autumn, 288 K, light=6.3 mW/cm2, benthic algae, Baltic Sea[1]
Laminaria saccharina	2 mg O2/(gram dry weight*hour)	Autumn, 288 K, light=23.6 mW/cm2, benthic algae, Baltic Sea[1]
Laminaria saccharina	2 mg O2/(gram dry weight*hour)	Winter, 278 K, light=2.8 mW/cm2, benthic algae, Baltic Sea[1]
Marine phytoplankton	1.2 mg c/(m**3*h)	Low nutrient level, central north Pacific[28]
Marine phytoplankton	3.7 mg c/(m**3*h)	Low nutrient level, 15 miles from coast[28]
Monochrysis lutheri	0.84 1/day	292 K, b12 limiting, continuous cult., marine[29]
Nannochloris atomus	1.54 1/day	283 K[3]
Nannochloris atomus	1.54 1/day	283 K, flagellate[30]
Nannochloris atomus	2.16 1/day	293 K[3]
Nannochloris atomus	2.16 1/day	293 K, flagellate, flagellate[30]
Nitzschia actinastroides	2.06 1/day	296 K, P limiting, continuous cult., marine[31]
Nitzschia closterium	0.67 1/day	283 K[32]
Nitzschia closterium	0.67 1/day	283 K[3]
Nitzschia closterium	1.19 1/day	288.5 K[32]
Nitzschia closterium	1.19 1/day	288.5 K[3]
Nitzschia closterium	1.55 1/day	292 K[32]
Nitzschia closterium	1.55 1/day	292 K[3]
Nitzschia closterium	1.75 1/day	300 K[32]
Nitzschia closterium	1.75 1/day	300 K[3]
Oscillatoria agardhii	0.036 1/hour	Specific growth rate, 293 K, substrate NO3, ligth saturated[15]
Oscillatoria agardhii gomomt	0.017 1/hour	Light =0.1w/m**2, pH = 7.1-7.6, N limited, NO3, batch cult.[33]
Oscillatoria agardhii gomomt	0.02 1/hour	Light =3.6w/m**2, pH = 7.3-7.8, N limited, NO3, batch cult.[33]
Oscillatoria agardhii gomomt	0.02 1/hour	Light =5.5w/m**2, pH = 7.1-7.6, N limited, NO3, batch cult.[33]
Oscillatoria agardhii gomomt	0.036 1/hour	N limited, NO3, 293 K, cont.sat.light[34]

Petalonia fascia	34 mg O2/(gram dry weight*hour)	Spring, 283 K, light=43.6 mW/cm2, benthic algae, Baltic Sea[1]
Phycodrys rubens	2 mg O2/(gram dry weight*hour)	Winter, 278 K, light=1.4 mW/cm2, benthic algae, Baltic Sea[1]
Phycodrys rubens	3.7 mg O2/(gram dry weight*hour)	Summer, 293 K, light=4.3 mW/cm2, benthic algae, Baltic Sea[1]
Phycodrys rubens	4.1 mg O2/(gram dry weight*hour)	Autumn, 288 K, light=4.3 mW/cm2, benthic algae, Baltic Sea[1]
Phycodrys rubens	4.8 mg O2/(gram dry weight*hour)	Spring, 283 K, light=10 mW/cm2, benthic algae, Baltic Sea[1]
Phyllophora brodiaei	1.2 mg O2/(gram dry weight*hour)	Autumn, 288 K, light=6.3 mW/cm2, benthic algae, Baltic Sea[1]
Phyllophora brodiaei	1.9 mg O2/(gram dry weight*hour)	Spring, 283 K, light=9.12 mW/cm2, benthic algae, Baltic Sea[1]
Phyllophora brodiaei	2 mg O2/(gram dry weight*hour)	Winter, 278 K, light=4.9 mW/cm2, benthic algae, Baltic Sea[1]
Phytoplankton	1.3 1/day	Lake Erie, 293 K[35]
Phytoplankton	2 1/day	Potomac Estuary, 293 K[35]
Phytoplankton	2 1/day	San Joaquin River, 293 K[35]
Phytoplankton	2.1 1/day	Lake Ontario, 293 K[35]
Phytoplankton	2.5 1/day	Delta Estuary, 293 K[35]
Polysiphonia nigrescens	2.3 mg O2/(gram dry weight*hour)	Winter, 278 K, light=26.7 mW/cm2, benthic algae, Baltic Sea[1]
Polysiphonia nigrescens	4.3 mg O2/(gram dry weight*hour)	Spring, 283 K, light=18.5 mW/cm2, benthic algae, Baltic Sea[1]
Polysiphonia nigrescens	5.3 mg O2/(gram dry weight*hour)	Autumn, 288 K, light=14.5 mW/cm2, benthic algae, Baltic Sea[1]
Porphyra leucosticta	29 mg O2/(gram dry weight*hour)	Winter, 278 K, light=4.3 mW/cm2, benthic algae, Baltic Sea[1]
Sargassum sp.	44 micromol C/(mg chlorophyll a*hour)	297 K, 1.3E17 quanta/(s*cm2), bubbled with air[11]
Scenedesmus obliquus	1.52 1/day	298 K[12]
Scenedesmus obliquus	1.52 1/day	298 K[3]
Scenedesmus quadricauda	2.02 1/day	298 K[21]
Scenedesmus quadricauda	2.02 1/day	298 K[3]
Scytosiphon lomentaria	12.5 mg O2/(gram dry weight*hour)	Spring, 283 K, light=26.7 mW/cm2, benthic algae, Baltic Sea[1]
Selenastrum capricornutum	1.274 1/day	306 K, chemost., ammonium-N limited, monod eq., most prob.spec.rate[36]
Selenastrum capricornutum	1.274 1/day	306 K, chemost., ammonium-N limited, r=0.949, monod eq.[36]
Selenastrum capricornutum	1.365 1/day	293 K, chemost., ammonium-N limited, monod eq., most prob.spec.rate[36]
Selenastrum capricornutum	1.39 1/day	301 K, chemost., ammonium-N limited, monod eq., most prob.spec.rate[36]
Selenastrum capricornutum	1.398 1/day	293 K, chemost., ammonium-N limited, r=0.975, monod eq.[36]
Selenastrum capricornutum	1.412 1/day	300 K, chemost., ammonium-N limited, monod eq., most prob.spec.rate[36]
Selenastrum capricornutum	1.54 1/day	300 K, chemost., ammonium-N limited, r=0.989, monod eq.[36]

Selenastrum capricornutum	1.544 1/day	297 K, chemost., ammonium-N limited, r=0.996, monod eq.[36]
Selenastrum capricornutum	1.596 1/day	301 K, chemost., ammonium-N limited, r=0.799, monod eq.[36]
Selenastrum capricornutum	1.85 1/day	297 K, P limiting, continuous cult., freshwater[37]
Selenastrum capricornutum	1.85 1/day	Batch culture, specific growth rate[38]
Selenastrum capricornutum	1.992 1/day	297 K, chemost., ammonium-N limited, monod eq., most prob.spec.rate[36]
Skeletonema costatum	1.27 1/day	292 K, b12 limiting, continuous cult., marine[29]
Skeletonema costatum	3.26 divisions/day	Volume=144 u**3, P content=8.0-28.0E-15 g-atom/cell, rg = 1.51 - 5.00[9]
Thalassiosira pseudonana	1.14 1/day	291 K, P limiting, continuous cult., marine[39]
Thalassiosira pseudonana	1.27 1/day	286.5 K, P limiting, continuous cult., marine[39]
Thalassiosira pseudonana	1.34 1/day	289 K, b12 limiting, continuous cult., marine[40]
Thalassiosira pseudonana	1.46 1/day	297 K, P limiting, continuous cult., marine[39]
Thalassiosira pseudonana	2.09 1/day	298 K, NO3 limiting, continuous cult., marine[23]
Thalassiosira pseudonana	2.1 doublings/day	Clone 13-1[41]
Thalassiosira pseudonana	2.77 1/day	293 K, Si limiting, continuous cult., marine[42]
Thalassiosira pseudonana	3.6 doublings/day	Clone 3h[41]
Thalassiosira pseudonana	74.6 micromol C/(mg chlorophyll a*hour)	297 K, 1.3E17 quanta/(s*cm2), bubbled with air[11]
Ulotrix speciosa	35 mg O2/(gram dry weight*hour)	Spring, 283 K, light=23.6 mW/cm2, benthic algae, Baltic Sea[1]
Ulvopsis grevillei	13 mg O2/(gram dry weight*hour)	Winter, 278 K, light=14.5 mW/cm2, benthic algae, Baltic Sea[1]
Ulvopsis grevillei	47 mg O2/(gram dry weight*hour)	Spring, 283 K, light=26.7 mW/cm2, benthic algae, Baltic Sea[1]
Zooxanthellae	16.5 micromol C/(mg chlorophyll a*hour)	297 K, 1.3E17 quanta/(s*cm2), bubbled with air[11]

Source: 1.King, R.J. and W. Schramm. 1976. Mar. Biol. 37, pp 209-213.
2.O'Connor, D.J., R.V. Thomann and D.M. DiToro. 1973. Dynamic Water Quality Forecasting and Management. Office of Research and Development U.S. Environmental Protection Agency, Wash. 20460.
3.Di Toro, D.M., D.J. O'Connor and R.V. Thomann. 1971. A Dynamic Model of the Phytoplankton Population in the Sacramento-San Joaquin Delta. Advances in Chemistry Series. Vol 106.
4.Thomann, R.V., D.M. Di Toro, R.P. Winfield and D.J. O'Connor. 1975. Mathematical Modeling of Phytoplankton in Lake Ontario: 1. National Environmental Research Center. EPA. Oregon.
5.Lung, W.S., R.P. Canele and P.L. Freedman. 1976. Water Res., 10(12), pp 1101-1114.
6.Imboden, D.M. 1974. Limnol. Oceanogr., 19, pp 297-304.
7.Snodgrass, W.J. and C.R. O'Melia. 1975. Environ. Sci. Technol., 9, pp 937-944.
8.The Hydrological Engineering Center Corps of Engineers U.S. Army. 1975. Water Quality for river-resevoir systems, Davis, Cal. 9 5 6 1 6 (916) 499-2329.
9.Finenko, Z.Z. and D.K. Krupatkina-Aki Nina. 1974. Mar. Biol., 26, pp 193-201.
10.Thomas, W.H. and A.N. Dodson. 1972. Limnol. Oceanogr. 17, pp 515-523.
11.Burris, J.E. 1977. Mar. Biol., 39(4), pp 371-379.
12.Sorokin, C. and R.W. Krauss. 1958. The effects of light intensity on the growth rates of Green Algae. Plant Physiol, 33, pp 109-113.
13.Tamiya, H., E. Hase, K. Shibata et. al. 1964. Algal Culture from Laboratory to Pilot Plant. J.S. Burew(ed), Publ 600, Carnegie Inst. Wash DC. pp 204-234.
14.Shelef, G., W.J. Oswald and C.C. Golueke. 1972. Adv. Wat. Poll. Res., Proc. of the 5th Int. Conf. onWater Poll., Paper III, 25, pp 1-9.
15.Nyholm,N. 1977. Prog. Wat. Tech. 8(4/5), pp 347-358.
16.Sorokin, C. and R.W. Krauss. 1962. Effects of temperature and Illuminance on Chorella growth uncoupled from cell division, Plant Physiol, 37, pp 109-113.
17.Wright, S.J.L. 1972. Chemosphere 1(1), pp 11-14.
18.Shelef, G., W.J. Oswald and C.C. Golveke. 1970. In S.H. Jenkins(ed) Proc. Int. Conf. Water Pollut. Res. 5th, pp 13-25. Pergamon.
19.Sheler, G. 1968. Kinetics in Algal Systems in Waste Treatment. Light Intensity and Nitrogen Concentration as Growth limiting Factors. Ph. D. Thesis. Univ. Calif. Berkeley.
20.Zabat, M. 1970. Kinetics of Phosphorus removal by Algae Ph. D. Thesis. Univ. Calif. Berkeley.

21.Myers, J. 1964. in: Burlew, J.S. (ed.): Dynamic water quality, Forecasting and Manegement, Carnegienst. of Wash. D.C., Publ. 600, pp 37-54.
22.Williams, F.M. 1965. Population Growth and regulation in continuously Culture Algae. Ph. D. Thesis. Univ. Yale.
23.Caperon, J. and J. Meyer. 1972. Deep Sea Res., 19, pp 601-632.
24.Caperon, J. and J. Meyer. 1972. Deep Sea Res., 19, pp 601-632.
25.Parsons, T.R. and M. Takahashi. 1973. Limnol. Oceanogr., 18(4), pp 511-515.
26.Thomas, W.H. and A.N. Dodson. 1974. Mar biol. 24, pp 213-217.
27.Davies, A.G. 1970. J. mar. biol. Ass. U.K., 50, pp 65-86.
28.Bienfang, P. and K. Gundersen. 1977. Mar Biol., 43(3), pp 187-199.
29.Droop, M.R. 1970. Helgol Wiss. Meeresunters., 20, pp 629-636.
30.Yentsch, C.S. and R.W. Lee. 1966. Mar. Res., 24, pp 319-337.
31.Soeder, C.J., H. Muller, H.D. Payer and H. Schulle. 1971. Mitt. Int. Ver. Theor. Angew. Limnol., 19, pp 39-58.
32.Spencer, C.P. 1954. Studies of the culture of a Marine Diatom, J. Marine Biol. Assoc. U.K., Quoted in Harvey H.W., 1966. The Chemistry and Fertility of Sea Water, Cambridge Univ. pp 94.
33.Liere, L. van, W. Zevenboom and L.R. Mur. 1975. in: Conf. on nitrogen as a water Pollutant, Vol 2, session B2. Copenhagen.
34.Lisemu. 1975. in: Conf. on Nitrogen as a water pollutant, Vol 2, session B2. Copenhagen.
35.O'Connor, D.J., D.M. DiToro and R.V. Thomann. 1975. in: Russel, C.S. (ed.); Ecological Modeling in a Resource Management Framework. pp 149-209.
36.Reynolds, J.H., E.J. Middlebrooks, D.B. Porcella and W.J. Grenney, 1975. J. Wat. Poll. Control Fed.7(10), pp 2420-2436.
37.Toerien, D.F., C.H. Huang, J. Radimsky, E.A. Pearson and J. Scherfig. 1971. Final Report: Provisional Algal assay Procedures Sanit. Eng. Res. Lab., SERL rep., Univ. Calif. Berkeley, pp 71-76.
38.Toerien, D.F. and C.H. Huang. Water Res. 7(11), pp 1673-1681.
39.Fuhs, G.W. 1969. J. Phycol., 5, pp 312-321.
40.Swift, D.G. 1967. Growth of Vitamin B(12) limited Cultures: Cyclotella Nana, Monochrysis Lutheri and Isochrysis Galbana. MS. Thesis, John Hopkins Univ., Baltimore.
41.Hecky, R.E. and P. Kilham. 1974. Limnol. Oceanogr., 19(2), pp 361-365.
42.Paasche, E. 1973. Mar. Biol., 19, pp 117-126.

1-068 Algae
Max photosynthetic efficiency

Species	Value	Condition
Anabaena cylindrica	1.7 mg C/((mg chl)*hour*klux)	Culture:293 K;exp.:303 K, light = 10klux[1]
Chaetoceros sp.	1.5 mg C/((mg chl)*hour*klux)	Culture:293 K, light = 8klux;exp.:293 K, light = 20klux[2]
Chlorella ellipsoidea	5.5 mg C/((mg chl)*hour*klux)	Culture:293 K;exp.:303 K, light = 20klux[1]
Chlorella vulgaris	1.1 mg C/(mg chl(a+b))*hour*klux	Culture: light int.=3 klux;exp.:light int.=6 klux[3]
Chlorella vulgaris	3.8 mg C/((mg chl a)*hour*klux)	Culture:light int.=30 klux;exp.:light int.=30 klux[3]
Coccolithus huxleyi	2.2 mg C/(mg chl(a+b))*hour*klux	Culture:287 K, light = 38 klux;exp.:light int.:>38 klux, less than[4]
Cyclotella meneghiniana	2.75 mg C/((mg chl a)*hour*klux)	Culture:light int.=3 klux, rg = 2.1 - 3.4[5]
Cyclotella meneghiniana	4 mg C/((mg chl a)*hour*klux)	Culture:light int.=30 klux, rg = 3.6 - 4.4[5]
Hymenomonas sp.	1.9 mg C/(mg chl(a+b))*hour*klux	Culture:287 K, light = 38 klux;exp.:light int.:>38 klux, less than[4]
Ice flora	0.4 mg C/((mg chl a)*hour*klux)	Antarctic, 271.5 K[6]
Ice flora	2.6 mg C/((mg chl a)*hour*klux)	Antarctic, 271.4 K[7]
Phytoplankton	0.5 mg C/((mg chl)*hour*klux)	Kuroshio, generel val, summer, rg = 0.3 - 0.7[8]
Phytoplankton	0.55 mg C/((mg chl)*hour*klux)	Oligotrophic lake, generel val, rg = 0.1 - 1.0[8]
Phytoplankton	1.25 mg C/((mg chl a)*hour*klux)	Depth = 0 m, arctic, summer, rg = 1.0 - 1.5[9]
Phytoplankton	1.5 mg C/((mg chl)*hour*klux)	Depth = 0 m, temperate, winter[9]
Phytoplankton	1.5 mg C/((mg chl)*hour*klux)	Mesotrophic lake, generel val, rg = 1 - 2
Phytoplankton	1.7 mg C/((mg chl a)*hour*klux)	Depth = 10 m, 281-293 K, western north Pacific, summer, rg = 1.4 -[10]
Phytoplankton	12.95 mg C/((mg chl)*hour*klux)	293-298 K, Tokyo Bay, Skeletonema bloom, summer, rg = 9.0 - 16.9[11]
Phytoplankton	2.3 mg C/((mg chl a)*hour*klux)	Antarctic[7]
Phytoplankton	3.15 mg C/((mg chl a)*hour*klux)	296-300 K, tropical Pacific, autumn, rg = 1.1 - 5.2[10]

Phytoplankton	3.15 mg C/((mg chl a)*hour*klux)	Tropical Pacific, generel val, spring, nitrogen-poor water[12]
Phytoplankton	3.15 mg C/((mg chl)*hour*klux)	Northern, generel val, no vertical stability, rg = 2.9 - 3.4[9]
Phytoplankton	3.5 mg C/((mg chl)*hour*klux)	Depth=10m, 293 K, tropical Atlantic, rg = 3.0 - 4.0[13]
Phytoplankton	4 mg C/((mg chl)*hour*klux)	Bays and coastal waters near Japan, generel val, rg = 2 - 6[8]
Phytoplankton	4 mg C/((mg chl)*hour*klux)	Eutrophic lake, generel val, rg = 2 - 6[8]
Phytoplankton	4.1 mg C/((mg chl)*hour*klux)	Depth = 0 m, temperate, summer, rg = 4.0 - 4.2[9]
Phytoplankton	4.5 mg C/((mg chl)*hour*klux)	Oyashio, generel val, summer, rg = 3 - 6[8]
Phytoplankton	4.95 mg C/((mg chl a)*hour*klux)	Tropical Pacific, generel val, spring, nitrogen-rich water[12]
Phytoplankton	8 mg C/((mg chl)*hour*klux)	Tropical, generel val[9]
Scenedesmus sp.	5 mg C/((mg chl)*hour*klux)	Culture:293 K;exp.:303 K[1]
Skeletonema costatum	3.6 mg C/((mg chl)*hour*klux)	Culture:293 K;exp.:293 K, light = 13 klux[14]
Skeletonema costatum	6.2 mg C/((mg chl)*hour*klux)	Culture:293 K, light = 8klux;exp.:293 K, light = 20klux[2]
Synedra sp.	2 mg C/((mg chl)*hour*klux)	Culture:293 K;exp.:293 K, light = 15 klux[1]

Source: 1.Aruga, Y. 1965. Bot. Mag., Tokyo, 78, pp 360-365.
2.Nakaniski, M. and M. Monji. 1965. J. Fac. Sci., Univ. Tokyo, Sec III Botany, 9, pp 19-42.
3.Steemann Nielsen, E. 1961. Plant Physiol., 14, pp 868-876.
4.Jeffrey, S.W. and M.B. Allen. 1964. J. Gen. Microbiol., 36, pp 277-288.
5.Joergensen, E.G. 1964. Physiol. Plant.,17, pp 136-145.
6.Bunt, J.S. 1964. Antarctic Res. Ser., 1, pp 27-31.
7.Burkholder, P.R. and E.F. Mandelli. 1965. Science, 149, pp 872-874.
8.Ichimura, S. and Y. Aruga. 1964. in; Y. Miyake and T. Koyoma (eds.); Recent Researches in the Fields of Hydrosphere, Atmosphere and Nuclear Geochemistry, Tokyo, pp 13-37.
9.Steemann Nielsen, E. and V.K. Hansen. 1975. Plant Physiol., 12, pp 353-370.
10.Takahaski, M., K. Satake and N. Nakamoto. 1972. J. Oceanogr. Soc. Japan, 28, pp 27-36.
11.Hogetsu, K., M. Sakamoto and H. Sami kawa. 1959. Bot. Mag., Tokyo, 72, pp 421-422.
12.Thompson, G.E. and R.W. Davies. 1976. Trans. Am. Fish. Soc. 105(2), pp 208-219.
13.Yentsch, C.S. and R.W. Lee. 1966. Mar. Res., 24, pp 319-337.
14.Aruga, Y. 1965. Bot. Mag., Tokyo, 78, pp 280-288.

1-069 Algae
Max rate of NH4 - N uptake

Species	Value	Condition
Biddulphia aurita	2.1E-8 micromoles /(cell*hour)	None[1]
Chaetoceros debilis	0.16 1/hour	Chemostat, uptake=0 for conc = 0.2 ug-atom/l, NH4 limitation[2]
Chaetoceros debilis	0.18 1/hour	Chemostat, uptake=0 for conc = 0.3 ug-atom/l, Si limitation[2]
Dunaliella tertiolecta	6.6E-8 micromoles /(cell*hour)	Range = 2.6 - 10.6e-8[3]
Marine phytoplankton	57 ng-atoms/(l*day)	North Pacific[4]
Monochrysis lutheri	2.95E-8 micromoles /(cell*hour)	Range = 2.1 - 3.8E-8[3]
Oscillatoria thiebautii	0.59 picogram n/(cell*hour)	Sargasso Sea[5]
Skeletonema costatum	0.1 1/hour	Chemostat, uptake=0 for conc = 0.2 ug-atom/l, Si limitation[2]
Skeletonema costatum	0.16 1/hour	Chemostat, uptake=0 for conc = 0.2 ug-atom/l, NH4 limitation[2]
Thalassiosira gravida	0.03 1/hour	Chemostat, uptake=0 for conc = 0.5 ug-atom/l, Si limitation[2]

Thalassiosira gravida	0.16 1/hour	Chemostat, uptake=0 for conc = 0.2 ug-atom/l, NH4 limitation[2]

Source: 1.Lui, N.S.T. and O.A. Roels. 1972. J. Phycol. 8, pp 259-263.
2.Conway, H.L. and P.J. Harrison. 1977. Mar. Biol., 43(1), pp 33-43.
3.Caperon, J. and J. Meyer. 1972. Deep Sea Res., 19, pp 601-632.
4.Eppley, R.W., E.H. Renger, E.L. Venrick and M.M. Mullin. 1973. Limnol. Oceanogr., 18(4), pp 534-551.
5.Carpenter, E.J. and J.J. McCarthy. 1975. Limnol. Oceanogr., 20(3), pp 389-401.

1-070 Algae
Max rate of NO3 - N uptake

Species	Value	Condition
Biddulphia aurita	9E-9 micromoles /(cell*hour)	None[1]
Coccochloris stagnina	6.5E-10 micromoles /(cell*hour)	Range = 4.0 - 9.0E-10[2]
Cyclotella nana	6.5E-9 micromoles /(cell*hour)	Range = 4.0 - 9.0E-9[3]
Cyclotella nana	9.5E-9 micromoles /(cell*hour)	Range = 0.3 - 1.6E-8[2]
Ditylum brightwellii	1.2E-6 micromoles /(cell*hour)	None[4]
Dunaliella tertiolecta	2.95E-8 micromoles /(cell*hour)	Range = 2.2 - 3.7E-8[2]
Monochrysis lutheri	1.4E-8 micromoles /(cell*hour)	None[2]
Skeletonema costatum	1.52 microgram-atoms NO3-N/(E6 cells*hou	N as nitrate and NH4, shade adapted, estaurine phytoplankton[5]
Skeletonema costatum	2.78 microgram-atoms NO3-N/(E6 cells*hou	N as nitrate and NH4, sun adapted, estaurine phytoplankton[5]
Skeletonema costatum	7.14 microgram-atoms NO3-N/(E6 cells*hou	N as nitrate, sun adapted, estaurine phytoplankton[5]
Skeletonema costatum	8.33 microgram-atoms NO3-N/(E6 cells*hou	N as nitrate, shade adapted, estaurine phytoplankton[5]
chlorophyta	2.14 microgram-atoms NO3-N/(E6 cells*hou	N as nitrate and NH4, shade adapted, estaurine phytoplankton[5]
chlorophyta	4.35 microgram-atoms NO3-N/(E6 cells*hou	N as nitrate and NH4, sun adapted, estaurine phytoplankton[5]
chlorophyta	4.78 microgram-atoms NO3-N/(E6 cells*hou	N as nitrate, sun adapted, estaurine phytoplankton[5]
chlorophyta	5.56 microgram-atoms NO3-N/(E6 cells*hou	N as nitrate, shade adapted, estaurine phytoplankton[5]

Source: 1.Lui, N.S.T. and O.A. Roels. 1972. J. Phycol. 8, pp 259-263.
2.Caperon, J. and J. Meyer. 1972. Deep Sea Res., 19, pp 601-632.
3.Carpenter, E.J. and R.R.L. Guillard. 1971. Ecology, 52, pp 183-185.
4.Eppley, R.W. and J.L. Coatsworth. 1968. J. Phycol., 4. pp 151-156.
5.Bates, S.S. 1976. Limnol. Oceanogr., 21(2), pp 212-218.

1-071 Algae
Max rate of P uptake

Species	Value	Condition
Asterionella formosa	8E-9 micromoles /(cell*hour)	None[1]
Dinobryon cylindricum	2.01E-9 micro moles/(cell*hour)	New Jersey[2]
Dinobryon sociale var. americanum	1.39E-8 micro moles/(cell*hour)	New Jersey, in P-free medium for 2 d[2]
Dinobryon sociale var. americanum	3.66E-9 micro moles/(cell*hour)	New Jersey, directly from culture[2]
Euglena gracilis	2.4E-7 micromoles /(cell*hour)	None[3]
Freshwater phytoplankton	0.04 micromoles/hour	Central Lake Kinneret, 287.5 K, P =0.14 micro m[4]

Freshwater phytoplankton	0.06 micromoles/hour	Jordan river, upstream L. Kinneret, 289 K, P =0.44 micro m[4]
Freshwater phytoplankton	0.066 micromoles/hour	Southern Lake Kinneret, 301 K, P =0.11 micro m[4]
Freshwater phytoplankton	0.18 micromoles/hour	Southern Lake Kinneret, 290.0 K, P =0.10 micro m[4]
Freshwater phytoplankton	1.11 micromoles/hour	Southern Lake Kinneret, 304 K, P =0.037 micro m[4]
Marine phytoplankton	1.98 microgram-atom p/(l*h)	Chesapeake Bay, natural phytoplankton, 5 stations, rg = 0.02 - 2.95[5]
Pediastrum duplex	4E-8 micromoles /(cell*hour)	None[6]
Phaeodactylum tricornutum	1.1E-9 micromoles /(cell*hour)	Range = 0.7 - 1.4E-9[7]
Scenedesmus sp.	1.2E-8 micromoles /(cell*hour)	None[8]

Source: 1.Mackkereth, F.J. 1953. J. Expt. Bot. 4. pp 296-313.
2.Lehman, J.T., J. Sarvala, I. Hakala and M.-L. Viljanen. 1976. Limnol. Oceanogr. 21(5), pp 724-730.
3.Blum, J.J. 1966. J. Gen. Physiol., 49, pp 1125-1137.
4.Halmann, M. and M. Stiller. 1974. Limnol. Oceanogr., 19(5), pp 774-783.
5.Taguchi, S. and H. Ishii. 1972. in: Takenouti, A.Y.(ed). Biological Oceanogra phy of the northern Noth Pacific Ocean. pp 419-431. Idemitsu Shoten. Tokyo.
6.Livingston, H.D. and G. Thompson. 1971. Limnol. Oceanogr. 16(5), pp 786-796.
7.Kuenzler, E.J. and B.H. Ketchum. 1962. Biol. Bull. 125, pp 134-145.
8.Rhee, G.-Y. 1973. J. Phycol. 9. pp 495-506.

1-072 Algae
Max rate of photosynthesis

Species	Value	Condition
Anabaena sp. and Oscillatoria sp.	0.66 mg C/(mg chlorophyll a*hour)	283-304 K, photosynthetic capacity = 13.6 mg C/(m3*h)[1]
Asterionella sp.	1.54 mg C/(mg chlorophyll a*hour)	291 K, photosynthetic capacity = 6.0 mg C/(m3*h)[1]
Melosira sp.	2.01 mg C/(mg chlorophyll a*hour)	301.5 K, photosynthetic capacity = 5.0 mg C/(m3*h)[1]
Melosira sp.	2.28 mg C/(mg chlorophyll a*hour)	291 K, photosynthetic capacity = 19.5 mg C/(m3*h)[1]
Microcystis sp.	6.4 mg C/(mg chlorophyll a*hour)	301.9 K, photosynthetic capacity = 10.5 mg C/(m3*h)[1]
Nitzschia sp.	3.96 mg C/(mg chlorophyll a*hour)	295-299 K, photosynthetic capacity = 8.9 mg C/(m3*h)[1]
Trichodesmium spp.	0.38 microgram C/(microgram chl.a)*hr	N Pacific, N-fixing algae[2]

Source: 1.Maljkovic, D. and M. Branica, 1971. Limnol. Oceanogr. 16(5), pp 779-785.
2.Mague, T.H., F.C. Mague and O. Holm-Hansen. 1977. Mar. Biol. 41(3), pp 213-227.

1-073 Algae
Max rate of silicium uptake

Species	Value	Condition
Ditylum brightwellii	26.6 picogram Si/(cell*hour)	Uptake, no uptake for =2.96 ug-at/l, 25000 erg/(cm2*sec), 293 K, F077[1]
Ditylum brightwellii	9.5E-7 micromoles /(cell*hour)	None[2]
Licmophora sp.	2.15 picogram Si/(cell*hour)	Uptake, no uptake for =1.30 ug-at/l, 25000 erg/(cm2*sec), 293 K, F077[1]
Licmophora sp.	7.7E-8 micromoles /(cell*hour)	None[2]
Navicula pelliculosa	1E-8 micromoles /(cell*hour)	None[3]
Nitzschia alba	4.1E-9 micromoles /(cell*hour)	Range = 2.5 - 5.7E-9[4]
Skeletonema costatum	0.095 picogram Si/(cell*hour)	Uptake, no uptake for =0.32 ug-at/l, 25000 erg/(cm2*sec), 293 K, F077[1]
Skeletonema costatum	3.4E-9 micromoles /(cell*hour)	None[2]

Thalassiosira decipiens	1.46E-7 micromoles /(cell*hour)	None[2]
Thalassiosira decipiens	4.09 picogram Si/(cell*hour)	Uptake, no uptake for =1.33 ug-at/l, 25000 erg/(cm2*sec), 293 K, F077[1]
Thalassiosira pseudonana	0.073 picogram Si/(cell*hour)	Uptake, no uptake for =0.67 ug-at/l, 25000 erg/(cm2*sec), 293 K, F077[1]
Thalassiosira pseudonana	2.6E-9 micromoles /(cell*hour)	None[2]

Source: 1.Paasche, F. 1973. Mar. Biol. 19. pp 262-269.
2.Paasche, F. 1973. Mar. Biol. 19. pp 262-269.
3.Busby, W.F. and J. Lewin. 1967. J. Phydrocol. 3, pp 127-131.
4.Lewin, J. and C. Chen. 1968. J. Phycol. 4, pp 161-166.

1-074 Algae
Max yield

Species	Value	Condition
Selenastrum capricornutum	805 mg biomass/mg P utilized	Batch culture[1]

Source: 1.Toerien, D.F. and C.H. Huang. Water Res. 7(11), pp 1673-1681.

1-075 Algae
Max yield coefficient

Species	Value	Condition
Selenastrum capricornutum	22.01 mg cell/mg ammonium-N	293 K, chemost.[1]
Selenastrum capricornutum	25.57 mg cell/mg ammonium-N	306 K, chemost.[1]
Selenastrum capricornutum	29.94 mg cell/mg ammonium-N	301 K, chemost.[1]
Selenastrum capricornutum	33.44 mg cell/mg ammonium-N	300 K, chemost.[1]
Selenastrum capricornutum	61.2 mg cell/mg ammonium-N	297 K, chemost.[1]

Source: 1.Reynolds, J.H., E.J. Middlebrooks, D.B. Porcella and W.J. Grenney, 1975. J. Wat. Poll. Control Fed.7(10), pp 2420-2436.

1-076 Algae
Maximum uptake rate

Species	Value	Condition
Ammonia	1.44 - 76.0 ng-atom l-1 h-1	different stations in the Pacific ocean, nov. 1982 - feb. 1983[1]
Nitrate	0.23 - 1.44 ng-atom l-1 h-1	different stations in the Pacific ocean, nov. 1982 - feb. 1983[1]
Urea	0.42 - 4.00 ng-atom l-1 h-1	different stations in the Pacific ocean, nov. 1982 - feb. 1983[1]

Source: 1.Kanda, J., Saino, T. and Hattori, A., Limnol. Oceanogr., 30(5), 987, 1985.

1-077 Algae
Mg / biomass ratio

Species	Value	Condition
Anabaena sp.	0.42 % of dry weight	Algae from ponds and lakes in southeastern United States[1]
Aphanizomenon sp.	0.21 % of dry weight	Algae from ponds and lakes in southeastern United States[1]
Chara sp.	0.92 % of dry weight	Algae from ponds and lakes in southeastern United States[1]
Cladophora sp.	0.23 % of dry weight	Algae from ponds and lakes in southeastern United States[1]

Euglena sp.	0.07 % of dry weight	Algae from ponds and lakes in southeastern United States[1]
Hydrodictyon sp.	0.17 % of dry weight	Algae from ponds and lakes in southeastern United States[1]
Lyngbya sp.	0.14 % of dry weight	Algae from ponds and lakes in southeastern United States[1]
Microcystis sp.	0.17 % of dry weight	Algae from ponds and lakes in southeastern United States[1]
Mougeotia sp.	0.57 % of dry weight	Algae from ponds and lakes in southeastern United States[1]
Nitella sp.	0.95 % of dry weight	Algae from ponds and lakes in southeastern United States[1]
Oedogonium sp.	0.16 % of dry weight	Algae from ponds and lakes in southeastern United States[1]
Pithophora sp.	0.2 % of dry weight	Algae from ponds and lakes in southeastern United States[1]
Rhizoclonium sp.	0.21 % of dry weight	Algae from ponds and lakes in southeastern United States[1]
Spirogyra sp.	0.45 % of dry weight	Algae from ponds and lakes in southeastern United States[1]
Spirogyra sp., giant	0.3 % of dry weight	Algae from ponds and lakes in southeastern United States[1]

Source: 1.Water Quality Criteria. 1968. Federal Water Pollution Control Administration Washington D.C., pp 234

1-078 Algae
Min cell N

Species	Value	Condition
Anabaena sp.	1E-7 micromoles/cell	None[1]
Asterionella formosa	6E-7 micromoles/cell	None[1]
Dinobryon sp.	1.8E-8 micromoles/cell	None[1]
Gymnodinium sp.	3.9E-7 micromoles/cell	None[1]
Isochrysis galbana	3E-8 micromoles/cell	None[2]

Source: 1.Grim, J. 1939. Int. Rev. Hydrobiol. 39. pp 139-315.
2.Droop, M.R. 1973. J. Phycol., 9, pp 264-272.

1-079 Algae
Min cell P

Species	Value	Condition
Anabaena sp.	2.5E-9 micromoles/cell	None[1]
Asterionella formosa	2.25E-9 micromoles/cell	Range = 1.5 - 3.0E-9[2]
Asterionella formosa	2E-9 micromoles/cell	None[3]
Asterionella formosa	3E-8 micromoles/cell	None[1]
Asterionella japonica	1.5E-9 micromoles/cell	None[2]
Chlorella pyrenoidosa	3E-9 micromoles/cell	None[2]
Cyclotella nana	9E-10 micromoles/cell	None[4]
Dinobryon sp.	5E-10 micromoles/cell	None[1]
Gymnodinium sp.	1.1E-8 micromoles/cell	None[1]
Nitzschia actinastroides	3E-9 micromoles/cell	Diameter = 35 u[2]

Nitzschia actinastroides	4E-9 micromoles/cell	Diameter = 50 - 55 u[2]
Phaeodactylum tricornutum	2E-9 micromoles/cell	None[2]
Scenedesmus quadricauda	4.5E-9 micromoles/cell	None[2]
Scenedesmus sp.	1.7E-9 micromoles/cell	None[5]
Thalassiosira fluviatilis	1.22E-8 micromoles/cell	None[4]

Source: 1.Grim, J. 1939. Int. Rev. Hydrobiol. 39. pp 139-315.
2.Muller, H. von. 1972. Arch. Hydrobiol. Suppl. 38(4). pp 399-484.
3.Mackkereth, F.J. 1953. J. Expt. Bot. 4. pp 296-313.
4.Fuhs, G.W., S.P. Demmerle, E. Canelli and M. Chen. 1972. in; G.E. Likens (ed); Nutrients and eutrophication. Amer. Soc. Limnol. Oceangr. Spec. Symp. 1, pp 113-132.
5.Rhee, G.-Y. 1973. J. Phycol. 9. pp 495-506.

1-080 Algae
Min cell Si

Species	Value	Condition
Asterionella formosa	1.8E-6 micromoles/cell	None[1]
Asterionella formosa	2E-6 micromoles/cell	None[2]
Fragilaria crotonensis	4E-6 micromoles/cell	None[1]
Navicula pelliculosa	5E-8 micromoles/cell	None[3]
Nitzschia alba	3E-7 micromoles/cell	None[4]
Thalassiosira pseudonana	2E-8 micromoles/cell	None[5]

Source: 1.Grim, J. 1939. Int. Rev. Hydrobiol. 39. pp 139-315.
2.Hughes, J.C. and J.W.G. Lund. 1962. Arch. Microbiol. 42. pp 117-129.
3.Busby, W.F. and J. Lewin. 1967. J. Phydrocol. 3, pp 127-131.
4.Lewin, J. and C. Chen. 1968. J. Phycol. 4, pp 161-166.
5.Paasche, F. 1973. Mar. Biol. 19. pp 262-269.

1-081 Algae
Minimal phosphorus requirements

Species	Value	Condition
Asterionella sp.	0.2 microgram P/(mm**3 cell volume)	Less than[1]
Fragilaria sp.	0.275 microgram P/(mm**3 cell volume)	Range = 0.2 - 0.35[1]
Microcystis sp.	0.5 microgram P/(mm**3 cell volume)	Less than[1]
Oscillatoria sp.	0.5 microgram P/(mm**3 cell volume)	Less than[1]
Scenedesmus sp.	0.5 microgram P/(mm**3 cell volume)	Less than[1]
Tabellaria sp.	0.525 microgram P/(mm**3 cell volume)	Range = 0.45 - 0.6[1]

Source: 1.Vollenweider, R.A. 1968. Scientific Fundamentals of the Eutrophication of Lakes and Flowing Waters, with Particular Reference to N and P as Factors in Eutroph. OECD. DAS/CSI/68.27. Paris.

1-082 Algae
Mn / biomass ratio

Species	Value	Condition
Anabaena sp.	800 microgram/l	Algae from ponds and lakes in southeastern United States[1]
Aphanizomenon sp.	167 migrogram/l	Algae from ponds and lakes in southeastern United States[1]

Chara sp.	2926 microgram/l	Algae from ponds and lakes in southeastern United States[1]
Cladophora sp.	1040 microgram/l	Algae from ponds and lakes in southeastern United States[1]
Euglena sp.	240 microgram/l	Algae from ponds and lakes in southeastern United States[1]
Hydrodictyon sp.	1373 microgram/l	Algae from ponds and lakes in southeastern United States[1]
Lyngbya sp.	5230 microgram/l	Algae from ponds and lakes in southeastern United States[1]
Microcystis sp.	2751 microgram/l	Algae from ponds and lakes in southeastern United States[1]
Mougeotia sp.	1080 microgram/l	Algae from ponds and lakes in southeastern United States[1]
Nitella sp.	2388 microgram/l	Algae from ponds and lakes in southeastern United States[1]
Oedogonium sp.	1729 microgram/l	Algae from ponds and lakes in southeastern United States[1]
Pithophora sp.	829 microgram/l	Algae from ponds and lakes in southeastern United States[1]
Rhizoclonium sp.	1687 microgram/l	Algae from ponds and lakes in southeastern United States[1]
Spirogyra sp.	1641 microgram/l	Algae from ponds and lakes in southeastern United States[1]
Spirogyra sp., giant	1658 microgram/l	Algae from ponds and lakes in southeastern United States[1]

Source: 1.Water Quality Criteria. 1968. Federal Water Pollution Control Administration Washington D.C., pp 234

1-083 Algae
Mortality

Species	Value	Condition
Freshwater phytoplankton	0.004 1/hour	Steady state solution to Castle Lake.[1]

Source: 1.Richey, J.E. 1977. in: Hall, C.A.S. and J.W. Day, Jr. (eds.); Ecosystem Modeling in theory and practice. John Wiley and Sons. New York.

1-084 Algae
N / P ratio

Species	Value	Condition
Chaetoceros debilis	12 by atoms	Not limited, cont.light:0.08ly/min, dilution=1/3-1/2 of max.gr.rate[1]
Chaetoceros debilis	4.6 by atoms	NH4 limited, cont.light:0.08ly/min, dilution=1/3-1/2 of max.gr.rate[1]
Chaetoceros debilis	6 by atoms	NH4 starved, 72 h cont.light:0.08ly/min, dilution=1/3-1/2 of ma.gr.rate[1]
Chaetoceros debilis	7.9 by atoms	Si limited, cont.light:0.08ly/min, dilution=1/3-1/2 of max.gr.rate[1]
Chaetoceros debilis	8 by atoms	Si starved, 72 h cont.light:0.08ly/min, dilution=1/3-1/2 of max.gr.rate[1]
Freshwater algae	104 by weight	Lake Mergozzo, 0.004 mg P/l, 0.524 mg N/l[2]
Freshwater algae	104 by weight	Lake Piano, 0.01 mg P/l, 1.045 mg N/l[2]
Freshwater algae	116 by weight	Lake Ghirla, 0.005 mg P/l, 0.582 mg N/l[2]

Freshwater algae	126 by weight	Lake Monate, 0.001 mg P/l, 0.177 mg N/l[2]
Freshwater algae	15	Mean atomic ratio, rg = 10 - 20[3]
Freshwater algae	15 by weight	Lake Como (Brienno), 0.053 mg P/l, 0.78 ng n/l[2]
Freshwater algae	156 by weight	Lake Mezzola, 0.003 mg P/l, 0.468 mg N/l[2]
Freshwater algae	16 by weight	Lake Candia, 0.005 mg P/l, 0.083 mg N/l[2]
Freshwater algae	16 by weight	Lake Pusiano, 0.068 mg P/l, 1.116 mg N/l[2]
Freshwater algae	189 by weight	Lake Annona, 0.002 mg P/l, 0.472 mg N/l[2]
Freshwater algae	19 by weight	Lake Alserio, 0.061 mg P/l, 1.165 mg N/l
Freshwater algae	19 by weight	Lake Como (Lecco), 0.035 mg P/l, 0.653 mg N/l[2]
Freshwater algae	19 by weight	Lake Como, 0.035 mg P/l, 0.67 mg N/l[2]
Freshwater algae	20 by weight	Lake Endine, 0.005 mg P/l, 0.1 mg N/l[2]
Freshwater algae	24 by weight	Lake Olginate, 0.029 mg P/l, 0.707 mg N/l[2]
Freshwater algae	247 by weight	Lake Ganna, 0.002 mg P/l, 0.494 mg N/l[2]
Freshwater algae	26 by weight	Lake Garlate, 0.029 mg P/l, 0.746 mg N/l[2]
Freshwater algae	291 by weight	Lake Segrino, 0.005 mg P/l, 1.454 mg N/l[2]
Freshwater algae	35 by weight	Lake Iseo (South), 0.019 mg P/l, 0.669 mg N/l[2]
Freshwater algae	36 by weight	Lake Iseo (north), 0.017 mg P/l, 0.695 mg N/l[2]
Freshwater algae	4.1 by weight	Lake Varese, 0.363 mg P/l, 1.494 mg N/l[2]
Freshwater algae	43 by weight	Lake Maggiore (Pallanza), 0.018 mg P/l, 0.754 mg N/l[2]
Freshwater algae	5.1 by weight	Lake Lugano, 0.122 mg P/l, 0.616 mg N/l[2]
Freshwater algae	5.6 by weight	Lake Viverone, 0.117 mg P/l, 0.652 mg N/l[2]
Freshwater algae	5.7 by weight	Lake Oggiono , 0.161 mg P/l, 0.91 mg N/l[2]
Freshwater algae	53 by weight	Lake Garda (Rive), 0.006 mg P/l. 0.322 mg N/l[2]
Freshwater algae	58 by weight	Lake Montorfano, 0.004 mg P/l, 0.524 mg N/l[2]
Freshwater algae	6.7 by weight	Lake Lugano, 0.091 mg P/l, 0.616 mg N/l[2]
Freshwater algae	6.75	Mean weight ratio, rg = 4.5 - 9[3]
Freshwater algae	63 by weight	Lake Garda (Sirmione), 0.005 mg P/l, 0.317 mg N/l[2]
Freshwater algae	65 by weight	Lake Idro, 0.006 mg P/l, 0.389 mg N/l[2]
Freshwater algae	68 by weight	Lake Comabbio, 0.004 mg P/l, 0.274 mg N/l[2]
Freshwater algae	7.9 by weight	from 33 stations in lakes[3]
Freshwater algae	79 by weight	Lake Garda (Gargnano), 0.004 mg P/l, o.319 mg N/l[2]
Freshwater algae	9.3 by weight	Lake Avigliana, 0.033 mg P/l, 0.3 mg N/l[2]
Freshwater algae	90 by weight	Lake Maggiore (Meina), 0.019 mg P/l, 0.717 mg N/l[2]
Marine phytoplankton	13.5 by atoms	Vigorously growing, excess nitrate in water, rg = 12 - 15[4]
Marine phytoplankton	14.5 by atoms	See ref.[5]
Marine phytoplankton	16.5 by atoms	Unhealthy phytoplankton, nitrate depleted water, rg = 15 - 18[4]
Skeletonema costatum	10 by atoms	Not limited, cont.light:0.08ly/min, dilution=1/3-1/2 of max.gr.rate[1]
Skeletonema costatum	10.15 by atoms	Ammonium limited, dilution rate=0.110-0.042/hr, rg = 5.1 - 15.2[6]

Skeletonema costatum	4.2 by atoms	NH4 starved, 72 h cont.light:0.08ly/min, dilution=1/3-1/2 of ma.gr.rate[1]
Skeletonema costatum	4.8 by atoms	NH4 limited, cont.light:0.08ly/min, dilution=1/3-1/2 of max.gr.rate[1]
Skeletonema costatum	7.7 by atoms	Si starved, 72 h cont.light:0.08ly/min, dilution=1/3-1/2 of[1]
Skeletonema costatum	8.4 by atoms	Si limited, cont.light:0.08ly/min, dilution=1/3-1/2 of max.gr.rate[1]
Skeletonema costatum	9.9 by atoms	Si limited, dilution rate=0.061-0.035/hr, rg = 6.1 - 10.0[7]
Skeletonema costatum	9.9 by atoms	Si limited, dilution rate=0.086-0.022/hr, rg = 9.8 - 10.0[6]
Thalassiosira gravida	11.1 by atoms	Si starved, 72 h cont.light:0.08ly/min, dilution=1/3-1/2 of max.gr.rate[1]
Thalassiosira gravida	2 by atoms	NH4 limited, cont.light:0.08ly/min, dilution=1/3-1/2 of max.gr.rate[1]
Thalassiosira gravida	5.1 by atoms	NH4 starved, 72 h cont.light:0.08ly/min, dilution=1/3-1/2 of max.gr.rate[1]
Thalassiosira gravida	8.8 by atoms	Not limited, cont.light:0.08ly/min, dilution=1/3-1/2 of max.gr.rate[1]
Thalassiosira gravida	9.7 by atoms	Si limited, cont.light:0.08ly/min, dilution=1/3-1/2 of max.gr.rate[1]
Thalassiosira pseudonana	37.2 by weight	P limited chemostat, growth=0.041 1/h, 295 K, 0.07 cal/cm2*min[8]
Thalassiosira pseudonana	6.4 by mole	N limited chemostat, growth=0.0085 1/h, 295 K, 0.07 cal/cm2*min[8]
Thalassiosira pseudonana	61.1 by weight	P limited chemostat, growth=0.017 1/h, 295 K, 0.07 cal/cm2*min[8]
Thalassiosira pseudonana	8.05 by atoms	Nitrogen limited, dilution rate=0.041-0.008/hr, rg = 6.4 - 9.7[9]
Thalassiosira pseudonana	8.3 by weight	N limited chemostat, growth=0.017 1/h, 295 K, 0.07 cal/cm2*min[8]
Thalassiosira pseudonana	9.7 by weight	N limited chemostat, growth=0.041 1/h, 295 K, 0.07 cal/cm2*min[8]

Source: 1.Harrison, P.J., H.L. Conway, R.W. Holmes and C.O. Davis. 1977. Mar. Biol., 43(1), pp 19-31.
2.Joergensen, S.E., J.H. Henriksen, M. Friis, L.A. Joergensen and H.F. Mejer. Handbook of Environmental Data and Ecological Parameters. 1979. ISEM. Copenhagen.
3.Clendenning, K. A., T.E. Brown and H.C. Eyster. 1956. Can. J. Bot., 34, pp 943-966.
4.Antia, N.J., C.D. McAllister, T.R. Parsons, K. Stephens and J.H.D. Strickland. 1963. Limnol. Oceanogr., 8, pp 166-183.
5.Strickland, J.D.H. 1960. Fish. Res. Bd. Can. Bull., 122, pp 172.
6.Harrison, P.J., H.L. Conway and R.C. Dugdale. 1976. Mar. biol. 35, pp 177-186.
7.Davis, C.O. 1973. Effects of changes in light intensity and photoperiod on the Si-limited cont. culture of the marine diatom Skeletomena costatum. 122 p. Ph. D. Univ Wash. Seattle.
8.Perry, M.J. 1976. Limnol. Oceanogr. 21(1), pp 88-107.
9.Eppley, R.W. and E.H. Renger. 1974. J. Phycol., 10, pp 15-23.

1-085 Algae
N / Si ratio

Species	Value	Condition
Chaetoceros debilis	0.5 by atoms	NH4 limited, cont.light:0.08ly/min, dilution=1/3-1/2 of max.gr.rate[1]
Chaetoceros debilis	0.9 by atoms	NH4 starved, 72 h cont.light:0.08ly/min, dilution=1/3-1/2 of max.gr.rate[1]
Chaetoceros debilis	1.8 by atoms	Not limited, cont.light:0.08ly/min, dilution=1/3-1/2 of max.gr.rate[1]

Chaetoceros debilis	4.1 by atoms	Si limited, cont.light:0.08ly/min, dilution=1/3-1/2 of max.gr.rate[1]
Chaetoceros debilis	4.8 by atoms	Si starved, 72 h cont.light:0.08ly/min, dilution=1/3-1/2 of max.gr.rate[1]
Skeletonema costatum	0.8 by atoms	NH4 limited, cont.light:0.08ly/min, dilution=1/3-1/2 of max.gr.rate[1]
Skeletonema costatum	0.8 by atoms	NH4 starved, 72 h cont.light:0.08ly/min, dilution=1/3-1/2 of max.gr.rate[1]
Skeletonema costatum	1.9 by atoms	Not limited, cont.light:0.08ly/min, dilution=1/3-1/2 of max.gr.rate[1]
Skeletonema costatum	2.19 by atoms	Ammonium limited, dilution rate=0.110-0.042/hr, rg = 0.58 - 3.8[2]
Skeletonema costatum	2.8 by atoms	Si starved, 72 h cont.light:0.08ly/min, dilution=1/3-1/2 of max.gr.rate[1]
Skeletonema costatum	4.1 by atoms	Si limited, cont.light:0.08ly/min, dilution=1/3-1/2 of max.gr.rate[1]
Skeletonema costatum	4.5 by atoms	Si limited, dilution rate=0.086-0.022/hr, rg = 3.3 - 5.7[2]
Thalassiosira gravida	0.6 by atoms	NH4 limited, cont.light:0.08ly/min, dilution=1/3-1/2 of max.gr.rate[1]
Thalassiosira gravida	1.4 by atoms	NH4 starved, 72 h cont.light:0.08ly/min, dilution=1/3-1/2 of max.gr.rate[1]
Thalassiosira gravida	2.3 by atoms	Not limited, cont.light:0.08ly/min, dilution=1/3-1/2 of max.gr.rate[1]
Thalassiosira gravida	6 by atoms	Si limited, cont.light:0.08ly/min, dilution=1/3-1/2 of max.gr.rate[1]
Thalassiosira gravida	8.5 by atoms	Si starved, 72 h cont.light:0.08ly/min, dilution=1/3-1/2 of max.gr.rate[1]

Source: 1.Harrison, P.J., H.L. Conway, R.W. Holmes and C.O. Davis. 1977. Mar. Biol., 43(1), pp 19-31. 2.Harrison, P.J., H.L. Conway and R.C. Dugdale. 1976. Mar. biol. 35, pp 177-186.

1-086 Algae
N / biomass ratio

Species	Value	Condition
Algae	0.08 fraction of dry weight	Range = 0.07 - 0.09[1]
Algae	0.11 by weight	Used in models, rg = 0.05 - 0.17[2]
Algae	0.14 by weight	Used in models, Lake Erie[3]
Algae	0.14 by weight	Used in models, San Joaquin River estuary, Calif[3]
Algae	0.14 by weight	Used in models, San Joaquin River, Calif[3]
Algae	0.2 by weight	Used in models, Lake Michigan[4]
Algae	0.2 by weight	Used in models, Lake Ontario[3]
Algae	0.2 by weight	Used in models, Potomac estuary[3]
Algae	3.8 % of dry weight	Mean elementary composition[5]
Anabaena sp.	9.43 % of dry weight	Algae from ponds and lakes in southeastern United States[6]
Aphanizomenon sp.	8.57 % of dry weight	Algae from ponds and lakes in southeastern United States[6]
Asterionella formosa	10.22 pg cell-1	algal culture[7]
Chaetoceros lauderi	10.3 microgram N/mm**3	Cultivated, marine[8]

Chara sp.	2.46 % of dry weight	Algae from ponds and lakes in southeastern United States[6]
Chlamydomonas magnusii	31.4 microgram N/mm**3	Cultivated, marine[8]
Chlorella sp.	0.078 pg cell-1	algal culture[7]
Chlorophyta	9.54 mg/(gram of acid-soluble fraction)	Average of 7 marine species, Puerto Rico[9]
Cladophora sp.	2.3 % of dry weight	Algae from ponds and lakes in southeastern United States[6]
Coscinodiscus sp.	11.7 microgram N/mm**3	Cultivated, marine[8]
Cryptomonas erosa	24.47 pg cell-1	algal culture[7]
Cryptomonas pseudobaltica	47.6 microgram N/mm**3	Cultivated, marine[8]
Euglena sp.	5.14 % of dry weight	Algae from ponds and lakes in southeastern United States[6]
Fragilaria crotonensis	11.68 pg cell-1	algal culture[7]
Heterothrix sp.	46.6 microgram N/mm**3	Cultivated, marine[8]
Hydrodictyon sp.	3.87 % of dry weight	Algae from ponds and lakes in southeastern United States[6]
Lyngbya sp.	5.01 % of dry weight	Algae from ponds and lakes in southeastern United States[6]
Marine phytoplankton	2.2E-11 g/cell	Urea as source of N, stationary phase, 293 K, Southern Calif[10]
Marine phytoplankton	4.2E-11 g/cell	NO3 as source of N, stationary phase, 293 K, Southern Calif[10]
Marine phytoplankton	5.3E-11 g/cell	NH4 as source of N, stationary phase, 293 K, Southern Calif[10]
Melosira italica	34.58 pg cell-1	Algal culture[7]
Melosira italica tenuissima	19.42 pg cell-1	Algal culture[7]
Microcystis sp.	8.08 % of dry weight	Algae from ponds and lakes in southeastern United States[6]
Mougeotia sp.	1.77 % of dry weight	Algae from ponds and lakes in southeastern United States[6]
Nitella sp.	2.7 % of dry weight	Algae from ponds and lakes in southeastern United States[6]
Oedogonium sp.	2.64 % of dry weight	Algae from ponds and lakes in southeastern United States[6]
Phaeodactylum tricornutum	37.6 microgram N/mm**3	Cultivated, marine[8]
Phaeophyceae	1.5 % of dry weight	Mean elementary composition[5]
Phaeophyceae	8.87 mg/(gram of acid-soluble fraction)	Average of 4 marine species, Puerto Rico[9]
Pithophora sp.	2.57 % of dry weight	Algae from ponds and lakes in southeastern United States[6]
Rhizoclonium sp.	3.46 % of dry weight	Algae from ponds and lakes in southeastern United States[6]
Rhodophyta	10.5 mg/(gram of acid-soluble fraction)	Average of 9 marine species, Puerto Rico[9]
Spirogyra sp.	3.01 % of dry weight	Algae from ponds and lakes in southeastern United States[6]
Spirogyra sp., giant	2.35 % of dry weight	Algae from ponds and lakes in southeastern United States[6]

Stephanodiscus astraea minitula	8.62 pg cell-1	Algal culture[7]
Stephanodiscus hantzschii	5.96 pg cell-1	Algal culture[7]
Stephanodiscus niagarae	280.0 pg cell-1	Algal culture[7]
Tabellaria fenestrata	51.45 pg cell-1	Algal culture[7]
Tetraselmis stricia	34.6 microgram N/mm**3	Cultivated, marine[8]
Thalassiosira sp.	4.2E-14 moles/cell	Growth rate=0.017 1/h, cell density=6.7E8 cell/l, P limited chemostat[11]
Thalassiosira sp.	4.3E-14 moles/cell	Growth rate=0.041 1/h, cell density=4.0E8 cell/l, P limited chemostat[11]
Trichodesmium spp.	2.4 microgram N/colony	N/colony, N Pacific, N-fixing algae, n=6[12]

Source: 1.The Hydrological Engineering Center Corps of Engineers U.S. Army. 1975. Water Quality for river-resevoir systems, Davis, Cal. 9 5 6 1 6 (916) 499-2329.
2.Baca and Arnett. 1976. A Limnological Model For Eutrophic lakes and Impound ments. Battle Pacific NW. Lab. EPA.
3.O'Connor, F.B. 1963. in: Doeksen, J. and J. van der Drift (eds.); Soil Organisms. pp 32-48. North-Holand publ. Co. Amsterdam.
4.Canale, R.P., L.M. De Palma and A.H. Vogel. 1976. in: Canale, R.P. (ed.); Mathematical modelling of Biochemical processes in Aquatic Ecosystems. Ann Arbor Science Press.
5.Bowen, H.J.M. 1966. Trace Elements in Biochemistry, Ac. Press.
6.Water Quality Criteria. 1968. Federal Water Pollution Control Administration Washington D.C., pp 234
7.Infante, A. and Litt, A.H., Limnol. Oceanogr., 30(5), 1053, 1985.
8.Berland, B.R., P.J. Bonin, P.L. Laborde and S.Y. Maestrini. 1972. Mar. biol., 13, pp 338-345.
9.Edgington, D.N., S.A. Gordon, M.M. Thommes and L.R. Almodovar. 1970 Limnol. Oceanogr.,, 15(6), pp 94-955.
10.Eppley, R.W., A.F. Carlucci, O.H.-Hansen, D. Kiefer, J.J. McCarthy, E. Venrick and P.M. Williams. 191. Limnol. Oceanogr., 16(5), pp 741-757.
11.Perry, M.J. 1976. Limnol. Oceanogr. 21(1), pp 88-107.
12.Mague, T.H., F.C. Mague and O. Holm-Hansen. 1977. Mar. Biol. 41(3), pp 213-227.

1-087 Algae
N / chlorophyll a ratio

Species	**Value**	**Condition**
Algae	0.01 mg N/microgram chl	Potomac River[1]
Algae	10 by weight	Used in models, Lake Ontario[2]
Algae	10 by weight	Used in models, Potomac Estuary[2]
Algae	7 by weight	Used in models, Lake Erie[2]
Algae	7 by weight	Used in models, San Joaquin River Estuary, Calif[2]
Algae	7 by weight	Used in models, San Joaquin River, Calif[2]
Chaetoceros debilis	11.6 by weight	Not limited, cont.light:0.08ly/min, dilution=1/3-1/2 of max.gr.rate[3]
Chaetoceros debilis	11.7 by weight	Si limited, cont.light:0.08ly/min, dilution=1/3-1/2 of max.gr.rate[3]
Chaetoceros debilis	12.6 by weight	NH4 starved, 72 h cont.light:0.08ly/min, dilution=1/3-1/2 of max.gr.rate[3]
Chaetoceros debilis	21.9 by weight	Si starved, 72 h cont.light:0.08ly/min, dilution=1/3-1/2 of max.gr.rate[3]
Chaetoceros debilis	3.85 by weight	Nitrogen limited, dilution rate=0.097-0.008/hr, rg = 2.9 - 4.8[4]
Chaetoceros debilis	8.5 by weight	NH4 limited, cont.light:0.08ly/min, dilution=1/3-1/2 of max.gr.rate[3]
Cyclotella nana	22.7 by weight	Nitrate limited, dilution rate=0.077-0.009/hr, rg = 11.7 - 35.6[5]
Marine phytoplankton	11 by weight	Urea as source of N, stationary phase, 293 K, Southern Calif[6]
Marine phytoplankton	14 by weight	NH4 as source of N, stationary phase, 293 K, Southern Calif[6]

Marine phytoplankton	15 by weight	NO3 as source of N, log phase, 293 K, Southern Calif[6]
Marine phytoplankton	17 by weight	Urea as source of N, log phase, 293 K, Southern Calif[6]
Marine phytoplankton	29 by weight	NH4 as source of N, log phase, 293 K, Southern Calif[6]
Marine phytoplankton	8.8 by weight	NO3 as source of N, stationary phase, 293 K, Southern Calif[6]
Skeletonema costatum	11.2 by weight	Si limited, cont.light:0.08ly/min, dilution=1/3-1/2 of max.gr.rate[3]
Skeletonema costatum	14 by weight	Not limited, cont.light:0.08ly/min, dilution=1/3-1/2 of max.gr.rate[3]
Skeletonema costatum	15.7 by weight	Si starved, 72 h cont.light:0.08ly/min, dilution=1/3-1/2 of max.gr.rate[3]
Skeletonema costatum	15.75 by weight	Nitrogen limited, batch culture, 72-115 hrs starvation, rg = 11.5 -[7]
Skeletonema costatum	17.25 by weight	Ammonium limited, dilution rate=0.110-0.042/hr[8]
Skeletonema costatum	42.65 by weight	Si limited, dilution rate=0.086-0.022/hr, rg = 10.7 - 74.6[8]
Skeletonema costatum	7.7 by weight	NH4 starved, 72 h cont.light:0.08ly/min, dilution=1/3-1/2 of max.gr.rate[3]
Skeletonema costatum	8.8 by weight	NH4 limited, cont.light:0.08ly/min, dilution=1/3-1/2 of max.gr.rate[3]
Thalassiosira fluviatilis	2.55 by weight	Nitrate limited, batch culture, 72-115 hrs starvation, rg = 2.2 - 2.9[9]
Thalassiosira gravida	11.4 by weight	Not limited, cont.light:0.08ly/min, dilution=1/3-1/2 of max.gr.rate[3]
Thalassiosira gravida	15.1 by weight	NH4 starved, 72 h cont.light:0.08ly/min, dilution=1/3-1/2 of max.gr.rate[3]
Thalassiosira gravida	17.2 by weight	Si starved, 72 h cont.light:0.08ly/min, dilution=1/3-1/2 of max.gr.rate[3]
Thalassiosira gravida	5.3 by weight	NH4 limited, cont.light:0.08ly/min, dilution=1/3-1/2 of max.gr.rate[3]
Thalassiosira gravida	9.8 by weight	Si limited, cont.light:0.08ly/min, dilution=1/3-1/2 of max.gr.rate[3]
Thalassiosira pseudonana	21.7 by weight	Nitrate limited, batch culture, 72-115 hrs starvation, rg = 20 - 23.3
Thalassiosira pseudonana	5.2 by weight	N limited chemostat, growth=0.017 1/h, 295 K, 0.07 cal/cm2*min[10]
Thalassiosira pseudonana	5.7 by weight	P limited chemostat, growth=0.041 1/h, 295 K, 0.07 cal/cm2*min[10]
Thalassiosira pseudonana	5.8 by weight	N limited chemostat, growth=0.041 1/h, 295 K, 0.07 cal/cm2*min[10]
Thalassiosira pseudonana	6.65 by weight	Nitrogen limited, dilution rate=0.041-0.008/hr, rg = 5.8 - 7.5[11]
Thalassiosira pseudonana	7.3 by weight	P limited chemostat, growth=0.017 1/h, 295 K, 0.07 cal/cm2*min[10]
Thalassiosira pseudonana	7.5 by weight	N limited chemostat, growth=0.0085 1/h, 295 K, 0.07 cal/cm2*min[10]

Source: 1.O'Connor, D.J., R.V. Thomann and D.M. DiToro. 1973. Dynamic Water Quality Forecasting and Management. Office of Research and Development U.S. Environmental Protection Agency, Wash. 20460.
2.O'Connor, F.B. 1963. in: Doeksen, J. and J. van der Drift (eds.); Soil Organisms. pp 32-48. North-Holland publ. Co. Amsterdam.
3.Harrison, P.J., H.L. Conway, R.W. Holmes and C.O. Davis. 1977. Mar. Biol., 43(1), pp 19-31.
4.Thomas, W.H. and A.N. Dodson. 1972. Limnol. Oceanogr. 17, pp 515-523.
5.Caperon, J. and J. Meyer. 1972. Deep Sea Res., 19, pp 601-632.

6.Eppley, R.W., A.F. Carlucci, O.H.-Hansen, D. Kiefer, J.J. McCarthy, E. Venrick and P.M. Williams. 1917. Limnol. Oceanogr., 16(5), pp 741-757.
7.Holm-Hansen, O., W.H. Sutcliffe, Jr. and J. Sharp. 1968. Limnol. Oceanogr., 13, pp 507-514.
8.Harrison, P.J., H.L. Conway and R.C. Dugdale. 1976. Mar. biol. 35, pp 177-186.
9.Hobson, L.A. and R.J. Pariser. 1971. J. exp. mar. biol. Ecol. 6, pp 71-78.
10.Perry, M.J. 1976. Limnol. Oceanogr. 21(1), pp 88-107.
11.Eppley, R.W. and E.H. Renger. 1974. J. Phycol., 10, pp 15-23.

1-088 Algae
Na / biomass ratio

Species	Value	Condition
Anabaena sp.	0.18 % of dry weight	Algae from ponds and lakes in southeastern United States[1]
Aphanizomenon sp.	0.19 % of dry weight	Algae from ponds and lakes in southeastern United States[1]
Chara sp.	0.13 % of dry weight	Algae from ponds and lakes in southeastern United States[1]
Cladophora sp.	0.18 % of dry weight	Algae from ponds and lakes in southeastern United States[1]
Euglena sp.	0.02 % of dry weight	Algae from ponds and lakes in southeastern United States[1]
Hydrodictyon sp.	0.38 % of dry weight	Algae from ponds and lakes in southeastern United States[1]
Lyngbya sp.	0.06 % of dry weight	Algae from ponds and lakes in southeastern United States[1]
Microcystis sp.	0.04 % of dry weight	Algae from ponds and lakes in southeastern United States[1]
Mougeotia sp.	0.49 % of dry weight	Algae from ponds and lakes in southeastern United States[1]
Nitella sp.	0.28 % of dry weight	Algae from ponds and lakes in southeastern United States[1]
Oedogonium sp.	0.06 % of dry weight	Algae from ponds and lakes in southeastern United States[1]
Pithophora sp.	0.07 % of dry weight	Algae from ponds and lakes in southeastern United States[1]
Rhizoclonium sp.	0.09 % of dry weight	Algae from ponds and lakes in southeastern United States[1]
Spirogyra sp.	1.42 % of dry weight	Algae from ponds and lakes in southeastern United States[1]
Spirogyra sp., giant	1.43 % of dry weight	Algae from ponds and lakes in southeastern United States[1]

Source: 1.Water Quality Criteria. 1968. Federal Water Pollution Control Administration Washington D.C., pp 234

1-089 Algae
Nitrogen fixation

Species	Value	Condition
Anabaena dominance	0.5 microgram N/(l*hour)	Smith, Alaska, 1963, rg = 0 - 1[1]
Anabaena scheremetievi	1.12E-9 mg N/(cell*day)	Chernoye, USSR, 1937[1]
Blue - green algae dominance	0.5025 microgram N/(l*hour)	Wingra, Wisc., 1961, rg = 0.005 - 1[1]
Blue - green algae dominance	21.53 microgram N/(l*hour)	Lake Mendota, Wisc., 1960, 1967, rg = 0.07 - 43[1]
Blue - green algae dominance	3 microgram N/(l*hour)	Sanctuary, Penn., 1959, rg = 0 - 6[1]
Freshwater algae	3 mg N/(g carbohydrate)	Assuming respiration rate of 24 mg O2/(mg chlorophyll*day)[2]

Freshwater algae	33 % of total input	Lake George[3]
Freshwater algae	36.4 microgram N/(mg chlorophyll*day)	Assuming 10 % N and 1.5 % chlorophyll, rg = 2.8 - 70[2]
Freshwater algae	5 microgram N/(mg n*day)	Lake Tjeukemeer[4]
Trichodesmium spp.	0.27 microgram N/((microgram chl a)*hr)	N Pacific, N fixing algae, using C2H4:N2 conversion ratio of 1.9:1,[5]
Trichodesmium spp.	134 pmoles N2/((microgram algal N)*hr)	N Pacific, N fixing algae, incubation in laboratory[5]

Source: 1.Wetzel, R.G. 1975. Limnology. Saunders. Philadelphia.
2.Golterman, H.L. 1975. Physiological Limnol., Elsevier Amsterdam.
3.Viner, A.B. and Horne, A.J. 1971. Nature 232, pp 417-418.
4.Horne, A.J. and G.E. Fogg. 1970. Proc. R. ent Soc., London, Ser. B. 175, pp 351-366.
5.Mague, T.H., F.C. Mague and O. Holm-Hansen. 1977. Mar. Biol. 41(3), pp 213-227.

1-090 Algae
Nitrogen, intracellular

Species	Value	Condition
Bacillariophyceae	4.9 % of average dry weight	Range=2.7-5.9 %[1]
Blue - green algae	4.9 % of average dry weight	Range=4.5-5.8 %[1]
Chlorophyceae	7.8 % of average dry weight	Range=6.6-9.1 %[1]
Chrysophyceae	4.4 % of average dry weight	Range=3.3-5.0%[1]

Source: 1.Strickland, J.D.H. 1965. Chemical Oceanogr., 1, pp 503.

1-091 Algae
Nutrient utilization constant

Species	Value	Condition
Selenastrum capricornutum	0.427 days	303 K, chemost., r=-0.472, eq. F004, net yield coeff.=max.yield[1]
Selenastrum capricornutum	0.558 days	293 K, chemost., r=0.931, eq. F004, net yield coeff.=max.yield[1]
Selenastrum capricornutum	0.75 days	301 K, chemost., r=0.804, eq. F004, net yield coeff.=max.yield[1]
Selenastrum capricornutum	0.825 days	300 K, chemost., r=0.442, eq. F004, net yield coeff.=max.yield[1]
Selenastrum capricornutum	0.84 days	306 K, chemost., r=-0.459, eq. F003, ka=1/k1, k1=nutrient util.rate[1]
Selenastrum capricornutum	1.05 days	293 K, chemost., r=0.934, eq. F003, ka=1/k1, k1=nutrient util.rate[1]
Selenastrum capricornutum	1.3 days	300 K, chemost., r=0.470, eq. F003, ka=1/k1, k1=nutrient util.rate[1]
Selenastrum capricornutum	1.36 days	301 K, chemost., r=0.800, eq. F003, ka=1/k1, k1=nutrient util.rate[1]
Selenastrum capricornutum	2.64 days	297 K, chemost., r=0.899, eq. F004, net yield coeff.=max.yield[1]
Selenastrum capricornutum	3.16 days	297 K, chemost., r=0.908, eq. F003, ka=1/k1, k1=nutrient util.rate[1]

Source: 1.Reynolds, J.H., E.J. Middlebrooks, D.B. Porcella and W.J. Grenney, 1975. J. Wat. Poll. Control Fed.7(10), pp 2420-2436.

1-092 Algae
O / biomass ratio

Species	Value	Condition
Algae	44 % of dry weight	Mean elementary composition[1]
Phaeophyceae	37.7 % of dry weight	Mean elementary composition[1]

Source: 1.Bowen, H.J.M. 1966. Trace Elements in Biochemistry, Ac. Press.

1-093 Algae
Organic content

Species	Value	Condition
Chlorophyta	0.304 gram/(gram of acid-soluble fractio	Average of 7 marine species, Puerto Rico[1]
Phaeophyceae	0.313 gram/(gram of acid-soluble fractio	Average of 4 marine species, Puerto Rico, total organic content[1]
Rhodophyta	0.406 gram/(gram of acid-soluble fractio	Average of 9 marine species, Puerto Rico, total organic content[1]

Source: 1.Edgington, D.N., S.A. Gordon, M.M. Thommes and L.R. Almodovar. 1970 Limnol. Oceanogr.,, 15(6), pp 94-955.

1-094 Algae
P / biomass ratio

Species	Value	Condition
Algae	0.011 by weight	Range = 0.01 - 0.012[1]
Anabaena sp.	0.77 % of dry weight	Algae from ponds and lakes in southeastern United States[2]
Aphanizomenon sp.	1.17 % of dry weight	Algae from ponds and lakes in southeastern United States[2]
Chara sp.	0.25 % of dry weight	Algae from ponds and lakes in southeastern United States[2]
Cladophora sp.	0.56 % of dry weight	Algae from ponds and lakes in southeastern United States[2]
Euglena sp.	0.67 % of dry weight	Algae from ponds and lakes in southeastern United States[2]
Hydrodictyon sp.	0.24 % of dry weight	Algae from ponds and lakes in southeastern United States[2]
Lyngbya sp.	0.31 % of dry weight	Algae from ponds and lakes in southeastern United States[2]
Microcystis sp.	0.68 % of dry weight	Algae from ponds and lakes in southeastern United States[2]
Mougeotia sp.	0.25 % of dry weight	Algae from ponds and lakes in southeastern United States[2]
Nitella sp.	0.23 % of dry weight	Algae from ponds and lakes in southeastern United States[2]
Oedogonium sp.	0.08 % of dry weight	Algae from ponds and lakes in southeastern United States[2]
Pithophora sp.	0.3 % of dry weight	Algae from ponds and lakes in southeastern United States[2]
Rhizoclonium sp.	0.43 % of dry weight	Algae from ponds and lakes in southeastern United States[2]
Spirogyra sp.	0.2 % of dry weight	Algae from ponds and lakes in southeastern United States[2]

Spirogyra sp., giant	0.23 % of dry weight	Algae from ponds and lakes in southeastern United States[2]
Thalassiosira sp.	1.1E-15 moles/cell	Growth rate=0.041 1/h, cell density=4.0E8 cell/l, P limited chemostat[3]
Thalassiosira sp.	6.8E-16 moles/cell	Growth rate=0.017 1/h, cell density=6.7E8 cell/l, P limited chemostat[3]
Trichodesmium spp.	0.12 microgram	P/colony, N Pacific, N fixing algae, n=6[4]

Source: 1.The Hydrological Engineering Center Corps of Engineers U.S. Army. 1975. Water Quality for river-resevoir systems, Davis, Cal. 9 5 6 1 6 (916) 499-2329.
2.Water Quality Criteria. 1968. Federal Water Pollution Control Administration Washington D.C., pp 234
3.Perry, M.J. 1976. Limnol. Oceanogr. 21(1), pp 88-107.
4.Mague, T.H., F.C. Mague and O. Holm-Hansen. 1977. Mar. Biol. 41(3), pp 213-227.

1-095 Algae
P / chlorophyll a ratio

Species	Value	Condition
Algae	0.001 mg P/microgram chl	Potomac River[1]
Chaetoceros debilis	2.2 by weight	Not limited, cont.light:0.08ly/min, dilution=1/3-1/2 of max.gr.rate[2]
Chaetoceros debilis	2.6 by weight	Si limited, cont.light:0.08ly/min, dilution=1/3-1/2 of max.gr.rate[2]
Chaetoceros debilis	4.3 by weight	NH4 limited, cont.light:0.08ly/min, dilution=1/3-1/2 of max.gr.rate[2]
Chaetoceros debilis	4.8 by weight	NH4 starved, 72 h cont.light:0.08ly/min, dilution=1/3-1/2 of max.gr.rate[2]
Chaetoceros debilis	6.2 by weight	Si starved, 72 h cont.light:0.08ly/min, dilution=1/3-1/2 of[2]
Skeletonema costatum	10.85 by weight	Si limited, dilution rate=0.086-0.022/hr, rg = 2.5 - 19.2[3]
Skeletonema costatum	2.8 by weight	Not limited, cont.light:0.08ly/min, dilution=1/3-1/2 of max.gr.rate[2]
Skeletonema costatum	3 by weight	Si limited, cont.light:0.08ly/min, dilution=1/3-1/2 of max.gr.rate[2]
Skeletonema costatum	4.2 by weight	NH4 limited, cont.light:0.08ly/min, dilution=1/3-1/2 of max.gr.rate[2]
Skeletonema costatum	4.35 by weight	NH4 limited, dilution rate=0.110-0.042/hr, rg = 2.6 - 6.1[3]
Skeletonema costatum	4.6 by weight	NH4 starved, 72 h cont.light:0.08ly/min, dilution=1/3-1/2 of max.gr.rate[2]
Skeletonema costatum	4.6 by weight	Si starved, 72 h cont.light:0.08ly/min, dilution=1/3-1/2 of max.gr.rate[2]
Thalassiosira gravida	15.7 by weight	NH4 starved, 72 h cont.light:0.08ly/min, dilution=1/3-1/2 of max.gr.rate[2]
Thalassiosira gravida	2.1 by weight	Si starved, 72 h cont.light:0.08ly/min, dilution=1/3-1/2 of max.gr.rate[2]
Thalassiosira gravida	2.3 by weight	Si limited, cont.light:0.08ly/min, dilution=1/3-1/2 of max.gr.rate[2]
Thalassiosira gravida	3 by weight	Not limited, cont.light:0.08ly/min, dilution=1/3-1/2 of max.gr.rate[2]
Thalassiosira gravida	6.1 by weight	NH4 limited, cont.light:0.08ly/min, dilution=1/3-1/2 of max.gr.rate[2]
Thalassiosira pseudonana	0.27 by weight	P limited chemostat, growth=0.017 1/h, 295 K, 0.07 cal/cm2*min[4]

Thalassiosira pseudonana	0.32 by weight	P limited chemostat, growth=0.041 1/h, 295 K, 0.07 cal/cm2*min[4]
Thalassiosira pseudonana	1.17 by weight	N limited chemostat, growth=0.041 1/h, 295 K, 0.07 cal/cm2*min[4]
Thalassiosira pseudonana	1.36 by weight	N limited chemostat, growth=0.017 1/h, 295 K, 0.07 cal/cm2*min[4]
Thalassiosira pseudonana	1.9 by weight	Nitrogen limited, dilution rate=0.041-0.008/hr, rg = 1.2 - 2.6[5]
Thalassiosira pseudonana	2.61 by weight	N limited chemostat, growth=0.0085 1/h, 295 K, 0.07 cal/cm2*min[4]

Source: 1.O'Connor, D.J., R.V. Thomann and D.M. DiToro. 1973. Dynamic Water Quality Forecasting and Management. Office of Research and Development U.S. Environmental Protection Agency, Wash. 20460.
2.Harrison, P.J., H.L. Conway, R.W. Holmes and C.O. Davis. 1977. Mar. Biol., 43(1), pp 19-31.
3.Harrison, P.J., H.L. Conway and R.C. Dugdale. 1976. Mar. biol. 35, pp 177-186.
4.Perry, M.J. 1976. Limnol. Oceanogr. 21(1), pp 88-107.
5.Eppley, R.W. and E.H. Renger. 1974. J. Phycol., 10, pp 15-23.

1-096 Algae
Particulate organic C

Species	Value	Condition
Natural population	2.75-19.8 uM	conc. Hawaiian coastal waters[1]

Source: 1.Laws, E.A., et al., Limnol.Oceanogr., 29(6), 1161, 1984.

1-097 Algae
Particulate organic N

Species	Value	Condition
Natural population	0.39-2.59 uM	conc. Hawaiian coastal waters[1]

Source: 1.Laws, E.A., et al., Limnol.Oceanogr., 29(6), 1161, 1984.

1-098 Algae
Particulate organic P

Species	Value	Condition
Natural population	14-193 uM	conc. Hawaiian coastal waters, from not detectable[1]

Source: 1.Laws, E.A., et al., Limnol.Oceanogr., 29(6), 1161, 1984.

1-099 Algae
Phosphorus, intracellular

Species	Value	Condition
Bacillariophyceae	1.1 % of dry weight	Range=0.4-2.0 %[1]
Blue - green algae	1.1 % of dry weight	Range=0.8-1.4 %[1]
Chlorophyceae	2.9 % of dry weight	Range=2.4-3.3 %[1]
Chrysophyceae	2.1 % of dry weight	Range=1.2-3.0 %[1]
Dinophyceae	1 % of dry weight	Range=0.6-1.1 %[1]
Dinophyceae	4.4 % of dry weight	Range=3.3-5.0%[1]

Source: 1.Strickland, J.D.H. 1965. Chemical Oceanogr., 1, pp 503.

1-100 Algae
Photosynthesis

Species	Value	Condition
Fucus vesiculosus	1.2 mg O2/(g dw*h)	288 K, 21.5 mW/cm2=saturation, thallus[1]
Fucus vesiculosus	9.27 mg O2/(g dw*h)	288 K, 21.5 mW/cm2=saturation, apices[1]
Laminaria digitata	1.19 mg O2/(g dw*h)	288 K, 21.5 mW/cm2=saturation, thallus[1]
Laminaria digitata	3.97 mg O2/(g dw*h)	288 K, 21.5 mW/cm2=saturation, several cm from thallus margin[1]

Source: 1.King, R.J. and W. Schramm. 1976. Mar. Biol. 37, pp 209-213.

1-101 Algae
Photosynthetic efficiency

Species	Value	Condition
Anabaena cylindrica	0.25 mg C/((mg chl)*hr*klux)	Initial slope, natural population, dominating specie, 283-303 K, pond[1]
Asterionella japonica	14.5 kilolux	Saturation point, tropical marine algae, 2 h[2]
Asterionella sp.	2.8 klux	Ik, linear saturation level of photosynthesis[3]
Biddulphia sinensis	17.75 kilolux	Saturation point, tropical marine algae, 2 h[2]
Ceratium furca	13 kilolux	Saturation point, tropical marine algae, 2 h[2]
Chaetoceros lorenzianus	11.75 kilolux	Saturation point, tropical marine algae, 2 h[2]
Chaetoceros sp.	0.13 mg C/((mg chl)*hr*klux)	Initial slope, culture, 293k[4]
Chlorella ellipsoidea	0.5 mg C/((mg chl)*hr*klux)	Initial slope, culture, 278-303k[1]
Chlorella pyrenoidosa	0.48 mg C/((mg chl a)*hr*klux)	Initial slope, culture[5]
Chlorella vulgaris	0.5 mg C/(mg chl(a+b))*hr*klux	Initial slope, culture, 293k[6]
Coccolithus huxleyi	0.13 mg C/(mg chl(a+b))*hr*klux	Initial slope, culture[7]
Coscinodiscus radiatus	11.5 kilolux	Saturation point, tropical marine algae, 2 h[2]
Cryptomonas sp.	0.8 mg C/((mg chl a)*hr*klux)	Initial slope, natural population, dominating specie, lake, app value[8]
Dictyocha sp.	15.5 kilolux	Saturation point, tropical marine algae, 2 h[2]
Dinophysis miles	34.5 kilolux	Saturation point, tropical marine algae, 2 h[2]
Freshwater algae	15.5 kerg/sec	Ik, linear saturation level of photosynthesis, rg = 14 - 17
Hymenomonas sp.	0.1 mg C/(mg chl(a+b))*hr*klux	Initial slope, culture[7]
Marine nannoplankton	1.1 mg C/(mg chl a*hour)	0 NO3, 298-302 K, Calif current, evening, rg = +-0.4[10]
Marine nannoplankton	12.1 mg C/(mg chl a*hour)	5-12 ug-atom NO3/l, 298-301 K, Calif current, evening, rg = +-2.8[10]
Marine nannoplankton	18.3 mg C/(mg chl a*hour)	5-12 ug-atom NO3/l, 298-301 K, Calif current, morning, rg = +-2.6[10]
Marine nannoplankton	2.1 mg C/(mg chl a*hour)	0 NO3, 298-302 K, Calif current, morning, rg = +-0.5[10]
Marine nannoplankton	6.4 mg C/(mg chl a*hour)	0.1-1 ug-atom NO3/l, 288-291 K, Calif current, evening, rg = +-0.4[10]
Marine nannoplankton	6.7 mg C/(mg chl a*hour)	0.1-1 ug-atom NO3/l, 288-291 K, Calif current, morning, rg = +-0.6[10]
Marine netplankton	0.8 mg C/(mg chl a*hour)	0 NO3, 298-302 K, Calif current, morning, rg = +-0.4[10]

Marine netplankton	11.1 mg C/(mg chl a*hour)	5-12 ug-atom NO3/l, 298-301 K, Calif current, evening, rg = +- 1.6[10]
Marine netplankton	2.6 mg C/(mg chl a*hour)	0 NO3, 298-302 K, Calif current, evening, rg = +-0.2[10]
Marine netplankton	3.4 mg C/(mg chl a*hour)	0.1-1 ug-atom NO3/l, 288-291 K, Calif current, evening, rg = +-0.6[10]
Marine netplankton	3.8 mg C/(mg chl a*hour)	0.1-1 ug-atom NO3/l, 288-291 K, Calif current, morning, rg = +-0.9[10]
Marine netplankton	6.3 mg C/(mg chl a*hour)	5-12 ug-atom NO3/l, 298-301 K, Calif current, morning, rg = +-1.1[10]
Phytoplankton	0.05 mg C/((mg chl a)*hr*klux)	Initial slope, natural population, Kuroshio(general)[11]
Phytoplankton	0.17 mg C/((mg chl a)*hr*klux)	Initial slope, natural population, Kuroshio & Oyashio(general)[11]
Phytoplankton	0.35 mg C/((mg chl a)*hr*klux)	Initial slope, natural population, Oyashio(general)[11]
Phytoplankton	0.35 mg C/((mg chl a)*hr*klux)	Initial slope, natural population, northern and arctic(general)[12]
Phytoplankton	0.35 mg C/((mg chl a)*hr*klux)	Initial slope, natural population, tropical, temperate(summer,[12]
Phytoplankton	0.36 mg C/((mg chl a)*hr*klux)	Initial slope, natural population, 277.5-279 K, arctic[12]
Phytoplankton	0.42 mg C/((mg chl a)*hr*klux)	Initial slope, natural population, Antarctic[13]
Planktoniella sol	14.5 kilolux	Saturation point, tropical marine algae, 2 h[2]
Quantum yield	0.1-0.005 mol C Einstein-1	varies with light intensity, form curves[14]
Rhizosolenia styliformis	23.5 kilolux	Saturation point, tropical marine algae, 2 h[2]
Scenedesmus sp.	0.65 mg C/((mg chl)*hr*klux)	Initial slope, culture, 283-313k[1]
Skeletonema costatum	0.44 mg C/((mg chl)*hr*klux)	Initial slope, culture, 293k[4]
Skeletonema costatum	0.52 mg C/((mg chl a)*hr*klux)	Initial slope, culture[5]
Skeletonema costatum	0.56 mg C/((mg chl)*hr*klux)	Initial slope, natural population, dominating specie, 293 K, Tokyo[1]
Synedra sp.	0.65 mg C/((mg chl)*hr*klux)	Initial slope, natural population, dominating specie, 283-303 K, pond[1]
Triceratium favus	17.25 kilolux	Saturation point, tropical marine algae, 2 h[2]

Source: 1.Aruga, Y. 1965. Bot. Mag., Tokyo, 78, pp 360-365.
2.Qasim, S.Z., P.M.A. Bhattathiri and V.P. Devassy. 1972. Mar. Biol. 16, pp 22-27.
3.Talling, J.F. 1966. J. Ecol. 54, pp 99-127.
4.Nakaniski, M. and M. Monji. 1965. J. Fac. Sci., Univ. Tokyo, Sec III Botany, 9, pp 19-42.
5.Steemann Nielsen, E. 1968. Plant Physiol., 21, pp 401-413.
6.Steemann Nielsen, E. 1961. Plant Physiol., 14, pp 868-876.
7.Jeffrey, S.W. and M.B. Allen. 1964. J. Gen. Microbiol., 36, pp 277-288.
8.Ichimura, S., S. Nagasawa and T. Tanaka. 1968. Bot. Mag., Tokyo, 81, pp 1-10.
9.Vollenweider, R.A. 1960. Mem. Ist. Ital. Idrobiol., 12 pp 201-244.
10.Malone, T.C. 1971. Mar. Biol. 10(4), pp 285-289.
11.Ichimura, S, Y. Saijo and Y. Aruga. 1962. Bot. Mag., Tokyo, 75, pp 212-220.
12.Steemann Nielsen, E. and V.K. Hansen. 1975. Plant Physiol., 12, pp 353-370.
13.Burkholder, P.R. and E.F. Mandelli. 1965. Science, 149, pp 872-874.
14.Kishino, M., Okami, N., Takahashi, M. and Ichimura, S. Limnol. Oceanogr., 31, 557, 1986.

1-102 Algae
Primary productivity

Species	Value	Condition
Upwelling area	2.312 g C m-2 d-1	14C method, avg. of range 0.663-4.735[1]
Upwelling area	2.759 g C m-2 d-1	14C method, avg. of range 0.715-4.413[1]

Source: 1.Minas, H.J., Minas, M. and Packard, T.T. Limnol. Oceanogr., 31(6), 1182, 1986.

1-103 Algae
Production

Species	Value	Condition
Natural population	0.5-11 C umol l-1 d-1	Hawaiian coastal waters[1]

Source: *1.Laws, E.A., et al., Limnol. Oceanogr., 29(6), 1161, 1984.*

1-104 Algae
Production / biomass ratio

Species	Value	Condition
Agarum cribrosum	4.2	Annual production, initial biomass[1]
Bossiella sp.	0.5 1/year	Yearly turnover, Southern Calif[2]
Calliarthron sp.	0.3 1/year	Yearly turnover, Southern Calif[2]
Corallina sp.	0.5 1/year	Yearly turnover, Southern Calif[2]
Freshwater lakes	0.09 mg C assimilated/(mg biomass C*day)	Lake Zurich, rg = 0.05 - 0.13[3]
Freshwater lakes	0.19 mg C assimilated/(mg biomass C*day)	Lake Erken, rg = 0.08 - 0.3[3]
Freshwater lakes	0.32 mg C assimilated/(mg biomass C*day)	Czech resevoirs, rg = 0.2 - 0.44[3]
Freshwater lakes	0.35 mg C assimilated/(mg biomass C*day)	Swedish lakes, rg = 0.1 - 0.6[3]
Freshwater lakes	0.385 mg C assimilated/(mg biomass C*day	Obersee, rg = 0.31 - 0.46[3]
Freshwater lakes	0.39 mg C assimilated/(mg biomass C*day)	Rotsee[3]
Freshwater lakes	0.45 mg C assimilated/(mg biomass C*day)	Devils Lake, rg = 0.2 - 0.44[3]
Freshwater lakes	0.515 mg C assimilated/(mg biomass C*day	Lake Kinneret, rg = 0.01 - 1.02[3]
Freshwater lakes	0.7 mg C assimilated/(mg biomass C*day)	Untersee, rg = 0.2 - 1.2[3]
Freshwater lakes	1.085 mg C assimilated/(mg biomass C*day	Polish lakes, rg = 0.67 - 1.5[3]
Freshwater lakes	1.8 mg C assimilated/(mg biomass C*day)	Tystrup-Bavelse lakes, rg = 0.2 - 3.4[3]
Freshwater lakes	2.15 mg C assimilated/(mg biomass C*day)	Lake Maggiore, rg = 0.28 - 4.11[3]
Freshwater lakes	5.78 mg C assimilated/(mg biomass C*day)	Lake Norrviken, rg = 0.26-11.3
Laminaria digitata	9.8	Annual production, initial biomass[1]
Laminaria longicruris	7.2	Annual production, initial biomass[1]
Lithothamnioid algae	1.5 1/year	Yearly turnover, Southern Calif[2]
Melobesia sp.	2 1/year	Yearly turnover, Southern Calif[2]
Miscellaneous Articulated algae	0.5 1/year	Yearly turnover, Southern Calif[2]

Source: *1.Mann, K.H. 1972. Mar. Biol. 14, pp 199-209.*
2.Smith, K.L. Jr. 1974. Limnol. Oceanogr. 19(6), pp 939-944.
3.Berman, T. and U. Pollingher. 1974. Limnol. Oceanogr., 19(1), pp 31-54.

1-105 Algae
Productivity index

Species	Value	Condition
Natural population	6.3-15.5 g C g-1 Chla h-1	Hawaiian coastal waters[1]

Source: 1.Laws, E.A., et al., Limnol. Oceanogr., 29(6), 1161, 1984.

1-106 Algae
Protein / biomass ratio

Species	Value	Condition
Agmenellum quadruplicatum	44 % of ash free dry weight	Blue - green algae, marine, nitrogen*6.25 = protein[1]
Amphidinium carteri	36 % of ash free dry weight	Dinophyceae, marine, nitrogen*6.25 = protein[1]
Chlorella sp.	13.1 % of dry weight	Unfavorable environment[2]
Chlorella sp.	46.4 % of dry weight	Favorable environment[2]
Dunaliella salina	58 % of ash free dry weight	Chlorophyceae, marine, nitrogen*6.25 = protein[1]
Exuviaella sp.	35 % of ash free dry weight	Dinophyceae, marine, nitrogen*6.25 = protein[1]
Monochrysis lutheri	53 % of ash free dry weight	Chrysophyceae, marine, nitrogen*6.25 = protein[1]
Skeletonema costatum	58 % of ash free dry weight	Bacillariophyceae, marine, nitrogen*6.25 = protein[1]
Tetraselmis maculata	68 % of ash free dry weight	Prasinophyceae, marine, nitrogen*6.25 = protein[1]

Source: 1.Parsons, T.R., K. Stephens and J.D.H. Strickland. 1961. J. Fish. Res. Bd. Can., 18, pp 1001-1016.
2.Milner, H.W. 1953. in: Burlew, J.S. (ed.): Algal Culture from Laboratory to Pilot Plant. Publ. 600. Carnegie Inst of Wash. Wash. D.C.

1-107 Algae
Quantum requirement

Species	Value	Condition
Isochrysis galbana	11 quanta O2-1	light conditions: 30 umol quanta m-2 s-1[1]
Isochrysis galbana	11 quanta O2-1	light conditions: 70 umol quanta m-2 s-1[1]
Isochrysis galbana	14 quanta O2-1	light conditions: 150 umol quanta m-2 s-1[1]
Isochrysis galbana	19 quanta C-1	light conditions: 70 umol quanta m-2 s-1[1]
Isochrysis galbana	20 quanta O2-1	light conditions: 320 umol quanta m-2 s-1[1]
Isochrysis galbana	21 quanta C-1	light conditions: 150 umol quanta m-2 s-1[1]
Isochrysis galbana	22 quanta C-1	light conditions: 30 umol quanta m-2 s-1[1]
Isochrysis galbana	33 quanta C-1	light conditions: 320 umol quanta m-2 s-1[1]
Isochrysis galbana	43 quanta O2-1	light conditions: 600 umol quanta m-2 s-1[1]
Isochrysis galbana	62 quanta C-1	light conditions: 600 umol quanta m-2 s-1[1]
Prorocentrum micans	179 quanta C-1	light conditions: 600 umol quanta m-2 s-1[1]
Prorocentrum micans	21 quanta O2-1	light conditions: 70 umol quanta m-2 s-1[1]
Prorocentrum micans	25 quanta O2-1	light conditions: 320 umol quanta m-2 s-1[1]
Prorocentrum micans	27 quanta O2-1	light conditions: 150 umol quanta m-2 s-1[1]
Prorocentrum micans	57 quanta O2-1	light conditions: 600 umol quanta m-2 s-1[1]
Prorocentrum micans	66 quanta C-1	light conditions: 70 umol quanta m-2 s-1[1]
Prorocentrum micans	81 quanta C-1	light conditions: 320 umol quanta m-2 s-1[1]
Prorocentrum micans	82 quanta C-1	light conditions: 150 umol quanta m-2 s-1[1]

Thalassiosira weisflogii	11 quanta O2-1	light conditions: 70 umol quanta m-2 s-1[1]
Thalassiosira weisflogii	12 quanta O2-1	light conditions: 150 umol quanta m-2 s-1[1]
Thalassiosira weisflogii	14 quanta O2-1	light conditions: 320 umol quanta m-2 s-1[1]
Thalassiosira weisflogii	22 quanta C-1	light conditions: 150 umol quanta m-2 s-1[1]
Thalassiosira weisflogii	22 quanta C-1	light conditions: 70 umol quanta m-2 s-1[1]
Thalassiosira weisflogii	23 quanta C-1	light conditions: 30 umol quanta m-2 s-1[1]
Thalassiosira weisflogii	23 quanta C-1	light conditions: 320 umol quanta m-2 s-1[1]
Thalassiosira weisflogii	27 quanta O2-1	light conditions: 600 umol quanta m-2 s-1[1]
Thalassiosira weisflogii	42 quanta C-1	light conditions: 600 umol quanta m-2 s-1[1]
Thalassiosira weisflogii	9 quanta O2-1	light conditions: 30 umol quanta m-2 s-1[1]

Source: 1.Falkowski, P.G., Dubinsky, Z. and Wyman, K.., Limnol. Oceanogr., 30(2), 311, 1985.

1-108 Algae
Ra / biomass ratio

Species	Value	Condition
Chlorophyta	0.0487 pico curie/(gram of acid-soluble	Average of 7 marine species, Puerto Rico[1]
Phaeophyceae	0.095 pico curie/(gram of acid-soluble f	Average of 4 marine species, Puerto Rico[1]
Rhodophyta	0.047 pico curie/(gram of acid-soluble f	Average of 9 marine species, Puerto Rico[1]

Source: 1.Edgington, D.N., S.A. Gordon, M.M. Thommes and L.R. Almodovar. 1970 Limnol. Oceanogr.,, 15(6), pp 94-955.

1-109 Algae
Rate of N uptake

Species	Value	Condition
Dissodinium sp.	0.002 pg-atoms N/(cell*min)	Inital 6 uM N as NH4, 2972 cell/ml, 4 h, dark, 296 K, log phase[1]
Dissodinium sp.	0.003 pg-atoms N/(cell*min)	Inital 10 uM N as NO3, 3147 cell/ml, 4 h, dark, 296 K, log phase[1]
Dissodinium sp.	0.003 pg-atoms N/(cell*min)	Inital 6 uM N as NH4, 3215 cell/ml, 4 h, light, 296 K, log phase[1]
Dissodinium sp.	0.005 pg-atoms N/(cell*min)	Inital 9 uM N as NO3, 3087 cell/ml, 4 h, light, 296 K, log phase[1]
Pyrocystis fusiformis	0.11 pg-atoms N/(cell*min)	Inital 10 uM N as NO3, 110 cell/ml, 4 h, dark, 296 K, log phase[1]
Pyrocystis fusiformis	0.11 pg-atoms N/(cell*min)	Inital 6 uM N as NH4, 129 cell/ml, 4 h, dark, 296 K, log phase[1]
Pyrocystis fusiformis	0.12 pg-atoms N/(cell*min)	Inital 9 uM N as NH4, 118 cell/ml, 4 h, light, 296 K, log phase[1]
Pyrocystis fusiformis	0.14 pg-atoms N/(cell*min)	Inital 9 uM N as NO3, 195 cell/ml, 4 h, light, 296 K, log phase[1]
Pyrocystis noctiluca	0.03 pg-atoms N/(cell*min)	Inital 4 uM N as NH4, 173 cell/ml, 4 h, dark, 296 K, log phase[1]
Pyrocystis noctiluca	0.04 pg-atoms N/(cell*min)	Inital 6 uM N as NH4, 91 cell/ml, 4 h, dark, 296 K, log phase[1]
Pyrocystis noctiluca	0.05 pg-atoms N/(cell*min)	Inital 4 uM N as NO3, 134 cell/ml, 4 h, dark, 296 K, log phase[1]

Pyrocystis noctiluca	0.05 pg-atoms N/(cell*min)	Inital 9 uM N as NH4, 70 cell/ml, 4 h, light, 296 K, log phase[1]
Pyrocystis noctiluca	0.06 pg-atoms N/(cell*min)	Inital 10 uM N as NH4, 85 cell/ml, 4 h, light, 296 K, log phase[1]
Pyrocystis noctiluca	0.06 pg-atoms N/(cell*min)	Inital 4 uM N as NO3, 238 cell/ml, 4 h, dark, 296 K, log phase[1]
Pyrocystis noctiluca	0.07 pg-atoms N/(cell*min)	Inital 9 uM N as NO3, 303 cell/ml, 4 h, light, 296 K, log phase[1]
Pyrocystis noctiluca	0.08 pg-atoms N/(cell*min)	Inital 9 uM N as NO3, 23 cell/ml, 4 h, light, 296 K, log phase[1]

Source: 1.Bhovichitra, M. and E. Swift, 1977. Limnol. Oceanogr., 22(1), pp 73-83.

1-110 Algae
Rate of NH4 - N uptake

Species	Value	Condition
Chaetoceros debilis	0.019 1/hour	Internally controlled uptake, chemostat, Si starvation[1]
Chaetoceros debilis	0.063 1/hour	Surge uptake, chemostat, Si starvation[1]
Chaetoceros debilis	0.076 1/hour	Internally controlled uptake, chemostat, NH4 starvation[1]
Chaetoceros debilis	0.085 1/hour	Internally controlled uptake, chemostat, NH4 limitation[1]
Chaetoceros debilis	0.085 1/hour	Surge uptake, chemostat, NH4 limitation[1]
Chaetoceros debilis	0.125 1/hour	Internally controlled uptake, chemostat, Si limitation[1]
Chaetoceros debilis	0.167 1/hour	Surge uptake, chemostat, Si limitation[1]
Chaetoceros debilis	0.253 1/hour	Surge uptake, chemostat, NH4 starvation[1]
Marine bacillariophyceae	0.0309 1/hour	Stationary phase, N added as 75 ug-atoms NH4 N, 293 K[2]
Marine bacillariophyceae	0.0318 1/hour	Log phase, N added as 75 ug-atoms NH4 N, 293 K[2]
Phytoplankton	0.1 1/day	Stady state solution to model of Corpus Christi Bay[3]
Skeletonema costatum	0.01 1/hour	Internally controlled uptake, chemostat, Si starvation[1]
Skeletonema costatum	0.042 1/hour	Surge uptake, chemostat, Si starvation[1]
Skeletonema costatum	0.06 1/hour	Internally controlled uptake, chemostat, Si limitation[1]
Skeletonema costatum	0.088 1/hour	Internally controlled uptake, chemostat, NH4 starvation[1]
Skeletonema costatum	0.124 1/hour	Internally controlled uptake, chemostat, NH4 limitation[1]
Skeletonema costatum	0.17 1/hour	Surge uptake, chemostat, Si limitation[1]
Skeletonema costatum	0.172 1/hour	Surge uptake, chemostat, NH4 starvation[1]
Skeletonema costatum	0.268 1/hour	Surge uptake, chemostat, NH4 limitation[1]
Thalassiosira gravida	0.04 1/hour	Internally controlled uptake, chemostat, Si limitation[1]
Thalassiosira gravida	0.095 1/hour	Surge uptake, chemostat, Si limitation[1]
Thalassiosira gravida	0.11 1/hour	Internally controlled uptake, chemostat, NH4 limitation[1]
Thalassiosira gravida	0.206 1/hour	Surge uptake, chemostat, NH4 limitation[1]

Source: 1.Conway, H.L. and P.J. Harrison. 1977. Mar. Biol., 43(1), pp 33-43.
2.McCarthy, J.J. and R.W. Eppley. 1972. Limnol. Oceanogr. 17(3), pp 371-382.
3.Penumalli, B.R. R.H. Flake and E.G. Fruh. 1976 Ecol. Modell. 2(2), pp 101-115.

1-111 Algae
Rate of NO3 - N uptake

Species	Value	Condition
Marine bacillariophyceae	0.0236 1/hour	Stationary phase, N added as 75 ug-atoms NO3-N, 293 K[1]
Marine bacillariophyceae	0.0278 1/hour	Log phase, N added as 75 ug-atoms NO3-N, 293 K[1]
Phytoplankton	0.4 1/day	Stady state solution to model of Corpus Christi bay[2]

Source: 1.McCarthy, J.J. and R.W. Eppley. 1972. Limnol. Oceanogr. 17(3), pp 371-382.
2.Penumalli, B.R. R.H. Flake and E.G. Fruh. 1976 Ecol. Modell. 2(2), pp 101-115.

1-112 Algae
Rate of P uptake

Species	Value	Condition
Freshwater phytoplankton	0.319 1/hour	Steady state solution to Castle Lake.[1]
Seston	0.037 microgram P/hour	Stonehouse Pond, Feb 22, 1973[2]
Seston	0.064 microgram P/hour	Stonehouse Pond, April 6, 1973[2]
Seston	0.1438 microgram P/hour	Stonehouse Pond, May 17, 1973[2]
Seston	3.182 microgram P/hour	Stonehouse Pond, June 17, 1973[2]

Source: 1.Richey, J.E. 1977. in: Hall, C.A.S. and J.W. Day, Jr. (eds.); Ecosystem Modeling in theory and practice. John Wiley and Sons. New York.
2.Ferrante, J.G. 1976. Hydrobiologia, 49(3), pp 203-204.

1-113 Algae
Rate of urea - N uptake

Species	Value	Condition
Marine bacillariophyceae	0.0327 1/hour	Stationary phase, N added as 75 ug-atoms urea-N, 293 K[1]
Marine bacillariophyceae	0.0418 1/hour	Log phase, N added as 75 ug-atoms urea-N, 293 K[1]

Source: 1.McCarthy, J.J. and R.W. Eppley. 1972. Limnol. Oceanogr. 17(3), pp 371-382.

1-114 Algae
Reflectance ratio

Species	Value	Condition
Anabaena sp.	0.326 625 to 650 nm, conc of algae=8 mg/	625 to 650 nm, conc of algae=8 mg/l[1]
Anabaena sp.	0.62 625 to 650 nm, conc of algae=4 mg/l	625 to 650 nm, conc of algae=4 mg/l[1]
Chlorella sp.	1.52 625 to 650 nm, conc of algae=4 mg/l	625 to 650 nm, conc of algae=4 mg/l[1]
Chlorella sp.	2.22 625 to 650 nm, conc of algae=8 mg/l	625 to 650 nm, conc of algae=8 mg/l[1]
Microcystis sp.	0.636 625 to 650 nm, conc of algae=8 mg/	625 to 650 nm, conc of algae=8 mg/l[1]
Microcystis sp.	0.824 625 to 650 nm, conc of algae=4 mg/	625 to 650 nm, conc of algae=4 mg/l[1]
Selenastrum sp.	1.153 625 to 650 nm, conc of algae=4 mg/	625 to 650 nm, conc of algae=4 mg/l[1]
Selenastrum sp.	1.51 625 to 650 nm, conc of algae=8 mg/l	625 to 650 nm, conc of algae=8 mg/l[1]

Source: 1.Gramms, L.C. and W.C. Boyle. 1972. Water Res., 6(12), pp 1433-1444.

1-115 Algae
Respiration

Species	Value	Condition
Isochrysis galbana	1.1 umol O2 cell-1 min-1 * 10e10	light conditions: 30 umol quanta m-2 s-1[1]
Isochrysis galbana	1.3 umol O2 cell-1 min-1 * 10e10	light conditions: 70 umol quanta m-2 s-1[1]
Isochrysis galbana	2.0 umol O2 cell-1 min-1 * 10e10	light conditions: 150 umol quanta m-2 s-1[1]
Isochrysis galbana	2.2 umol O2 cell-1 min-1 * 10e10	light conditions: 320 umol quanta m-2 s-1[1]
Isochrysis galbana	2.3 umol O2 cell-1 min-1 * 10e10	light conditions: 600 umol quanta m-2 s-1[1]
Prorocentrum micans	125 umol O2 cell-1 min-1 * 10e10	light conditions: 320 umol quanta m-2 s-1[1]
Prorocentrum micans	150 umol O2 cell-1 min-1 * 10e10	light conditions: 600 umol quanta m-2 s-1[1]
Prorocentrum micans	70 umol O2 cell-1 min-1 * 10e10	light conditions: 70 umol quanta m-2 s-1[1]
Prorocentrum micans	85 umol O2 cell-1 min-1 * 10e10	light conditions: 150 umol quanta m-2 s-1[1]
Thalassiosira weisflogii	20 umol O2 cell-1 min-1 * 10e10	light conditions: 30 umol quanta m-2 s-1[1]
Thalassiosira weisflogii	47 umol O2 cell-1 min-1 * 10e10	light conditions: 70 umol quanta m-2 s-1[1]
Thalassiosira weisflogii	72 umol O2 cell-1 min-1 * 10e10	light conditions: 150 umol quanta m-2 s-1[1]
Thalassiosira weisflogii	75 umol O2 cell-1 min-1 * 10e10	light conditions: 320 umol quanta m-2 s-1[1]
Thalassiosira weisflogii	82 umol O2 cell-1 min-1 * 10e10	light conditions: 600 umol quanta m-2 s-1[1]

Source: 1.Falkowski, P.G., Dubinsky, Z. and Wyman, K.., Limnol. Oceanogr., 30(2), 311, 1985.

1-116 Algae
Respiration rate

Species	Value	Condition
Algae	0.005 1/day-celcius	Potomac River, end. resp[1]
Algae	0.005 1/day-celcius	Used in models, end. resp[1]
Algae	0.0055 1/day	Temp factor is 1 at 293 K, temp.<278 K, end. resp, rg = 0.001 - 0.01[2]
Algae	0.031 kcal/kcal*day	Initial values, thermal ecosystem, transfer coeff.[3]
Algae	0.1 1/day	Temp factor is 1 at 273 K, Lake Ontario, end. resp[4]
Algae	0.125 1/day	Temp factor is 1 at 293 K, temp.>278 K, end. resp, rg = 0.05 - 2.0[2]
Anabaena sp.	4.5 microliter/((mg dry weight)*h)	298 K, no other indication, end. resp, rg = 4.2 - 4.8[5]
Anabaena variabilis	12.3 microliter/((mg dry weight)*h)	Growing cell (starving cell = 3.08) 307 K, min. salt medium, end.[5]
Anabaena variabilis	3.08 microliter/((mg dry weight)*h)	Growing cell, 0.1m potassium phosphate buffer pH = 7., end. resp[5]
Anabaena variabilis	8.4 microliter/((mg dry weight)*h)	Growing cell (starving cell = 1.7), 298 K, end. resp[5]
Anacystis nidulans	1.6 microliter/((mg dry weight)*h)	Growing cell (starving cell = 0.3) 298 K, end. resp[5]
Anacystis nidulans	4.7 microliter/((mg dry weight)*h)	Growing cell (starving cell = 1.9) 312 K, end. resp[5]
Anacystis nidulans	5 microliter/((mg dry weight)*h)	298 K, no other indication, end. resp, rg = 4 - 6[5]
Anacystis nidulans	6.8 microliter/((mg dry weight)*h)	Growing cell, 307 K, 0.1m potassium phosphate buffer pH = 7., end.[5]
Anacystis nidulans	7.5 microliter/((mg dry weight)*h)	Growing cell (starving cell = 2.9) none temp indication, end. resp[5]

Coscinodiscus excentricus	0.075 1/day	289 K, end. resp[6]
Coscinodiscus excentricus	0.11 1/day	289 K, end. resp[6]
Mixed naturel	1.3 microgram/((microgram chl)*h)	290-307 K, laboratory expiriment, end. resp, rg = 0.4 - 2.2[5]
Mixed naturel	1.52 microgram/((microgram chl)*h)	277-288 K, field observation, end. resp, rg = 0.1 - 2.95[5]
Mixed naturel	2.1 microgram/((microgram chl)*h)	296-308 K, field observation, end. resp, rg = 0.2 - 4.0[5]
Natural association	0.024 1/day	275 K, end. resp, rg = +- 0.012[6]
Natural association	0.03 1/day	275 K, end. resp[6]
Natural association	0.11 1/day	291 K, end. resp, rg = +- 0.007[6]
Natural association	0.12 1/day	291 K, end. resp[6]
Nitzschia closterium	0.035 1/day	279 K, end. resp[6]
Nitzschia closterium	0.08 1/day	293 K, end. resp[6]
Nitzschia closterium	0.17 1/day	308 K, end. resp[6]
Nostoc muscorum	4.4 microliter/((mg dry weight)*h)	Growing cell (starving cell = 1.1), 298 K, end. resp[5]
Oscillatoria redekei	3.23 microgram/((microgram chl)*h)	293 K, laboratory expiriment, end. resp, rg = 0.16 - 3.23[5]
Phormidium luridum	5 microliter/((mg dry weight)*h)	298 K, no other indication, end. resp, rg = 4 - 6[5]
Phytoplankton	0.08 1/day	Lake Erie, 293 K, end. resp[7]
Phytoplankton	0.1 1/day	Delta Estuary, 293 K, end. resp[7]
Phytoplankton	0.1 1/day	Lake Ontario, 293 K, end. resp[7]
Phytoplankton	0.1 1/day	Potomac Estuary, 293 K, end. resp[7]
Phytoplankton	0.1 1/day	San Joaquin River, 293 K, end. resp[7]

Source: 1.O'Connor, D.J., R.V. Thomann and D.M. DiToro. 1973. Dynamic Water Quality Forecasting and Management. Office of Research and Development U.S. Environmental Protection Agency, Wash. 20460.
2.The Hydrological Engineering Center Corps of Engineers U.S. Army. 1975. Water Quality for river-resevoir systems, Davis, Cal. 9 5 6 1 6 (916) 499-2329.
3.Wiegert, R.G. 1975. In: B.C. Patten (ed.), Systems Analysis and Simulation in Ecology, Vol. 3, pp 15-181.
4.Thomann, R.V., D.M. Di Toro, R.P. Winfield and D.J. O'Connor. 1975. Mathematical Modeling of Phytoplankton in Lake Ontario: 1. National Environmental Reseach Center. EPA. Oregon.
5.Gibson, C.E. 1975. A Field and Laboratory study of Oxygen uptake by Planktonic Blue-green Algae.
6.Riley, G.A. and R. von Arx. 1949. J. marine res., 8(11), pp 60-72.
7.O'Connor, D.J., D.M. DiToro and R.V. Thomann. 1975. in: Russel, C.S. (ed.); Ecological Modeling in a Resource Management Framework. pp 149-209.

1-117 Algae
S / biomass ratio

Species	Value	Condition
Algae	0.6 % of dry weight	Mean elementary composition, interpolated[1]
Anabaena sp.	0.53 % of dry weight	Algae from ponds and lakes in southeastern United States, end resp[2]
Aphanizomenon sp.	1.18 % of dry weight	Algae from ponds and lakes in southeastern United States[2]
Chara sp.	0.55 % of dry weight	Algae from ponds and lakes in southeastern United States[2]
Cladophora sp.	1.58 % of dry weight	Algae from ponds and lakes in southeastern United States[2]
Euglena sp.	0.19 % of dry weight	Algae from ponds and lakes in southeastern United States[2]
Hydrodictyon sp.	1.41 % of dry weight	Algae from ponds and lakes in southeastern United States[2]

Lyngbya sp.	0.28 % of dry weight	Algae from ponds and lakes in southeastern United States[2]
Microcystis sp.	0.27 % of dry weight	Algae from ponds and lakes in southeastern United States[2]
Mougeotia sp.	0.36 % of dry weight	Algae from ponds and lakes in southeastern United States[2]
Nitella sp.	0.34 % of dry weight	Algae from ponds and lakes in southeastern United States[2]
Oedogonium sp.	0.15 % of dry weight	Algae from ponds and lakes in southeastern United States[2]
Phaeophyceae	1.2 % of dry weight	Mean elementary composition[1]
Pithophora sp.	1.42 % of dry weight	Algae from ponds and lakes in southeastern United States[2]
Rhizoclonium sp.	0.27 % of dry weight	Algae from ponds and lakes in southeastern United States[2]
Spirogyra sp.	0.27 % of dry weight	Algae from ponds and lakes in southeastern United States[2]
Spirogyra sp., giant	0.24 % of dry weight	Algae from ponds and lakes in southeastern United States[2]

Source: 1.Bowen, H.J.M. 1966. Trace Elements in Biochemistry, Ac. Press.
2.Water Quality Criteria. 1968. Federal Water Pollution Control Administration Washington D.C., pp 234

1-118 Algae
Saturating light intensity

Species	Value	Condition
Phytoplankton	300 langleys/day	Delta Estuary[1]
Phytoplankton	300 langleys/day	Potomac Estuary[1]
Phytoplankton	300 langleys/day	San Joaquin River[1]
Phytoplankton	350 langleys/day	Lake Erie[1]
Phytoplankton	350 langleys/day	Lake Ontario[1]

Source: 1.O'Connor, D.J., D.M. DiToro and R.V. Thomann. 1975. in: Russel, C.S. (ed.); Ecological Modeling in a Resource Management Framework. pp 149-209.

1-119 Algae
Settling rate

Species	Value	Condition
Algae	0.1 m/day	Lake Ontario[1]
Algae	0.185 m/day	Range = 0.1 - 0.27[2]
Algae	0.25 m/day	Range = 0.1 - 0.4[3]
Algae	0.4 m/day	Range = 0.3 - 0.5[4]
Algae	1 m/day	Range = 0.0 - 2.0[5]
Asterionella formosa	0.2 m/day	Exponential phase, 95% CL = +- 0.06, radius=12.5 U, 293 K, n=6[6]
Asterionella formosa	0.22 m/day	1-4 cells/chain, 51E3 cell/ml, 11 d culture, 293 K, live[7]
Asterionella formosa	0.23 m/day	Living cells, 1-4 cells per colony[7]
Asterionella formosa	0.25 m/day	5-8 cells/chain, 51E3 cell/ml, 11 d culture, 293 K, live[7]
Asterionella formosa	0.26 m/day	1-4 cells/chain, 24E3 cell/ml, 9 d culture, 293 K, live[7]

Asterionella formosa	0.26 m/day	5-8 cells/chain, 24E3 cell/ml, 9 d culture, 293 K, live[7]
Asterionella formosa	0.27 m/day	9-12 cells/chain, 116E3 cell/ml, 15 d culture, 293 K, live[7]
Asterionella formosa	0.29 m/day	Living cells, 5-8 cells per colony[7]
Asterionella formosa	0.31 m/day	1500 Microgram CuSO4/l, >8 cells[7]
Asterionella formosa	0.31 m/day	9-12 cells/chain, 215E3 cell/ml, 17 d culture, 293 K, live[7]
Asterionella formosa	0.31 m/day	Living cells, total colony[7]
Asterionella formosa	0.32 m/day	9-12 cells/chain, 51E3 cell/ml, 11 d culture, 293 K, live[7]
Asterionella formosa	0.32 m/day	Alive, 293 K[7]
Asterionella formosa	0.33 m/day	9-12 cells/chain, 24E3 cell/ml, 9 d culture, 293 K, live[7]
Asterionella formosa	0.34 m/day	1-4 cells/chain, 116E3 cell/ml, 15 d culture, 293 K, live[7]
Asterionella formosa	0.34 m/day	Living cells, >8 cells per colony[7]
Asterionella formosa	0.35 m/day	500 ug CuSO4/l, 293 K[7]
Asterionella formosa	0.36 m/day	1-4 cells/chain, 215E3 cell/ml, 17 d culture, 293 K, live[7]
Asterionella formosa	0.36 m/day	500 Microgram CuSO4/l, >8 cells[7]
Asterionella formosa	0.37 m/day	5-8 cells/chain, 215E3 cell/ml, 17 d culture, 293 K, live[7]
Asterionella formosa	0.38 m/day	500 Microgram CuSO4/l, 1-4 cells[7]
Asterionella formosa	0.38 m/day	500 Microgram CuSO4/l, 5-8 cells[7]
Asterionella formosa	0.38 m/day	500 Microgram CuSO4/l, total colony[7]
Asterionella formosa	0.4 m/day	1-4 cells/chain, 240E3 cell/ml, 19 d culture, 293 K, live[7]
Asterionella formosa	0.4 m/day	5-8 cells/chain, 116E3 cell/ml, 15 d culture, 293 K, live[7]
Asterionella formosa	0.41 m/day	1500 Microgram CuSO4/l, total colony[7]
Asterionella formosa	0.42 m/day	In lugols solution, 293 K[7]
Asterionella formosa	0.43 m/day	5-8 cells/chain, 240E3 cell/ml, 19 d culture, 293 K, live[7]
Asterionella formosa	0.44 m/day	1500 Microgram CuSO4/l, 5-8 cells[7]
Asterionella formosa	0.44 m/day	In formalin, 293 K[7]
Asterionella formosa	0.45 m/day	1500 Microgram CuSO4/l, 1-4 cells[7]
Asterionella formosa	0.5 m/day	9-12 cells/chain, 240E3 cell/ml, 19 d culture, 293 K, live[7]
Asterionella formosa	0.52 m/day	Heat killed, 293 K[7]
Asterionella formosa	0.59 m/day	5-8 cells/chain, 219E3 cell/ml, 22 d culture, 293 K, live[7]
Asterionella formosa	0.61 m/day	1-4 cells/chain, 219E3 cell/ml, 22 d culture, 293 K, live[7]
Asterionella formosa	0.76 m/day	9-12 cells/chain, 219E3 cell/ml, 22 d culture, 293 K, live[7]
Asterionella formosa	1.48 m/day	Stationary phase, 95% CL = +- 1.05, radius=12.5 U, 293 K, n=7[6]

Bacteriastrum hyalinum	0.815 m/day	Area/vol ratio=0.29-0.33, dia=10*30 U, centric , chain-forming[8]
Chaetoceros lauderi	1 m/day	Area/vol ratio=0.19-0.41, dia=20*34 U, centric , chain-forming[8]
Coccolithus huxleyi	0.3 m/day	No coccoliths[9]
Coccolithus huxleyi	1.5 m/day	With coccoliths[9]
Coscinodiscus lineatus	1.9 m/day	Exponential phase, radius=25 U[10]
Coscinodiscus lineatus	6.8 m/day	Stationary phase, radius=25 U[10]
Cyclotella meneghiniana	0.08 m/day	Exponential phase, 95% CL = +- 0.10, radius=1.0 U, 293 K, n=4[6]
Cyclotella meneghiniana	0.24 m/day	Stationary phase, 95% CL = +-0.31, radius=1.0 U, 293 K, n=4[6]
Ditylum brightwellii	0.5 m/day	Healthy cells[9]
Ditylum brightwellii	0.6 m/day	Exponential phase, radius=75 U[10]
Ditylum brightwellii	1.3 m/day	Mean of 7 values dur.a 8 h light-16 hdark cyclus[11]
Ditylum brightwellii	2 m/day	Senescent cells[9]
Ditylum brightwellii	3.1 m/day	Stationary phase, radius=75 U[10]
Melosira agassizii	0.67 m/day	Exponential phase, 95% CL = +-0.48, radius=27.4 U, 293 K, n=3[6]
Melosira agassizii	1.87 m/day	Stationary phase, 95% CL = +-0.38, radius=27.4 U, 293 K, n=15[6]
Nitzschia seriata	0.38 m/day	Area/vol ratio=1.18-1.65, dia=4*40 U, pennate, chain-forming[12]
Phytoplankton	0.1 m/day	Lake Ontario[13]
Phytoplankton	15 m/day	Living cells, rg = 0 - 30[14]
Rhizosolenia robusta	1.1 m/day	Exponential phase, radius=42 U[7]
Rhizosolenia robusta	4.7 m/day	Stationary phase, radius=42 U[7]
Rhizosolenia setigera	0.315 m/day	Cell size= 6*245 U, centric, elongate, normal pre-auxospore[15]
Rhizosolenia setigera	0.425 m/day	Area/volume ratio=0.62-0.75, spineless pre-auxospore, centric[15]
Rhizosolenia setigera	0.78 m/day	Total population, rg = 0.63 - 0.95[15]
Rhizosolenia setigera	1.37 m/day	Area/vol ratio=0.10-0.16, cell size=33*363 U, post-auxospore[15]
Scenedesmus quadricauda	0.27 m/day	Exponential phase, 95% CL = +-0.04, radius=4.2 U, 293 K, n=8[6]
Scenedesmus quadricauda	0.89 m/day	Stationary phase, 95% CL = +-0.06, radius=4.2 U, 293 K, n=14[6]
Skeletonema costatum	0.83 m/day	Area/vol ratio=0.81-1.01, dia=5*20 U, centric , chain-forming[15]
Tabellaria flocculosa	0.22 m/day	Living cells[7]
Tabellaria flocculosa	0.39 m/day	Alive, 293 K[7]
Tabellaria flocculosa	0.46 m/day	In formalin, 293 K[7]
Tabellaria flocculosa	1.11 m/day	500 ug CuSO4/l, 293 K[7]
Tabellaria flocculosa	1.29 m/day	In lugols solution[7]
Tabellaria flocculosa	1.5 m/day	Heat killed, 293 K[7]

Thalassiosira nana	0.19 m/day	Area/vol ratio=0.88-1.20, dia=4.3-5.2 U, centric, unicellular[12]
Thalassiosira pseudonana	0.15 m/day	Clone 13-1, healthy[16]
Thalassiosira pseudonana	0.85 m/day	Clone 13-1, senescent[16]
Thalassiosira rotula	1.245 m/day	Area/vol ratio=0.23-0.29, dia=19-34 U, centric , chain-forming[12]
Thalassiosira rotula	17.1 m/day	5500 lux, 285 K[17]
Thalassiosira rotula	5.5 m/day	1200 lux, 285 K[17]
Thalassiosira rotula	9.2 m/day	2400 lux, 285 K[17]

Source: 1.Thomann, R.V., D.M. Di Toro, R.P. Winfield and D.J. O'Connor. 1975. Mathematical Modeling of Phytoplankton in Lake Ontario: 1. National Environmental Reseach Center. EPA. Oregon.
2.Snodgrass, W.J. and C.R. O'Melia. 1975. Environ. Sci. Technol., 9, pp 937-944.
3.Imboden, D.M. 1974. Limnol. Oceanogr., 19, pp 297-304.
4.Lung, W.S., R.P. Canele and P.L. Freedman. 1976. Water Res., 10(12), pp 1101-1114.
5.The Hydrological Engineering Center Corps of Engineers U.S. Army. 1975. Water Quality for river-resevoir systems, Davis, Cal. 9 5 6 1 6 (916) 499-2329.
6.Titman, D. and P. Kilham. 1976. Limnol. Oceanogr. 21(3), pp 409-417.
7.Smayda, T.J. 1974. Limnol. Oceanogr. 19(4), pp 628-635.
8.Smayda, T.J. and B.J. Boleyn. 1966. Limnol. Oceanogr. 11, pp 35-43.
9.Parsons, T.R. and M. Takahashi. 1973. Limnol. Oceanogr., 18(4), pp 511-515.
10.Eppley, R.W., R.W. Holmes and J.D.H. Strickland. 1967. J. Explor. Mar. Biol. Ecol., 1, pp 191-208.
11.Anderson, L.W.J. and B.M. Sweeney. 1977. Limnol. Oceanogr., 22(3), pp 539-551.
12.Smayda, T.J. and B.J. Boleyn. 1965. Limnol. Oceanogr. 10, pp 499-509.
13.O'Connor, D.J., D.M. DiToro and R.V. Thomann. 1975. in: Russel, C.S. (ed.); Ecological Modeling in a Resource Management Framework. pp 149-209.
14.Smayda, T.J. 1970. Oceanogr. Mar. Biol. Ann. Rev., 8, pp 353-414.
15.Smayda, T.J. and B.J. Boleyn. 1966. Limnol. Oceanogr. 11, pp 18-34.
16.Hecky, R.E. and P. Kilham. 1974. Limnol. Oceanogr., 19(2), pp 361-365.
17.Schoene, H.K. 1972. Mar. biol. 13, pp 284-291.

1-120 Algae
Si / P ratio

Species	Value	Condition
Chaetoceros debilis	1.7 by atoms	Si starved, 72 h cont.light:0.08ly/min, dilution=1/3-1/2 of max.gr.rate[1]
Chaetoceros debilis	1.9 by atoms	Si limited, cont.light:0.08ly/min, dilution=1/3-1/2 of max.gr.rate[1]
Chaetoceros debilis	6.6 by atoms	NH4 starved, 72 h cont.light:0.08ly/min, dilution=1/3-1/2 of max.gr.rate[1]
Chaetoceros debilis	6.7 by atoms	Not limited, cont.light:0.08ly/min, dilution=1/3-1/2 of max.gr.rate[1]
Chaetoceros debilis	9.7 by atoms	NH4 limited, cont.light:0.08ly/min, dilution=1/3-1/2 of max.gr.rate[1]
Skeletonema costatum	2 by atoms	Si limited, cont.light:0.08ly/min, dilution=1/3-1/2 of max.gr.rate[1]
Skeletonema costatum	2.25 by atoms	Si limited, dilution rate=0.061-0.035/hr, rg = 1.5 - 3.0[2]
Skeletonema costatum	2.35 by atoms	Silicate limited, dil rate=0.086-0.022/hr, rg = 1.7 - 3.0[3]
Skeletonema costatum	2.8 by atoms	Si starved, 72 h cont.light:0.08ly/min, dilution=1/3-1/2 of max.gr.rate[1]
Skeletonema costatum	5.2 by atoms	NH4 starved, 72 h cont.light:0.08ly/min, dilution=1/3-1/2 of max.gr.rate[1]
Skeletonema costatum	5.4 by atoms	Not limited, cont.light:0.08ly/min, dilution=1/3-1/2 of max.gr.rate[1]
Skeletonema costatum	5.9 by atoms	NH4 limited, cont.light:0.08ly/min, dilution=1/3-1/2 of max.gr.rate[1]
Skeletonema costatum	6.4 by atoms	NH4 limited, dilution rate=0.110-0.042/hr, rg = 4.0 - 8.8[3]

Thalassiosira gravida	1.3 by atoms	Si starved, 72 h, cont.light = 0.08 ly/min, dil=1/3-1/2 of max.gr.rate[1]
Thalassiosira gravida	1.6 by atoms	Si limited, cont.light = 0.08 ly/min, dil=1/3-1/2 of max gr rate[1]
Thalassiosira gravida	3.2 by atoms	NH4 limited, cont.light = 0.08 ly/min, dil=1/3-1/2 of max gr rate[1]
Thalassiosira gravida	3.7 by atoms	NH4 starved, 72 h, cont.light = 0.08 ly/min, dil=1/3-1/2 of max gr rate[1]
Thalassiosira gravida	3.8 by atoms	Non limited, cont.light = 0.08 ly/min, dil=1/3-1/2 of max gr rate[1]

Source: 1.Harrison, P.J., H.L. Conway, R.W. Holmes and C.O. Davis. 1977. Mar. Biol., 43(1), pp 19-31.
2.Davis, C.O. 1973. Effects of changes in light intensity and photoperiod on the Si-limited cont. culture of the marine diatom Skeletomena costatum. 122 p. Ph. D. Univ Wash. Seattle.
3.Harrison, P.J., H.L. Conway and R.C. Dugdale. 1976. Mar. biol. 35, pp 177-186.

1-121 Algae
Si / biomass ratio

Species	Value	Condition
Asterionella formosa	137 microgram Si/(E6 cells)	England, n=28, rg = 100 - 175[1]
Asterionella formosa	139 microgram Si/(E6 cells)	Germany, n=5, rg = 97 - 171[1]
Asterionella formosa	153 microgram Si/(E6 cells)	England, n=2, rg = 140 - 165[1]
Asterionella formosa	178 microgram Si/(E6 cells)	USA[1]
Asterionella formosa	50 microgram Si/(E6 cells)	Culture, rg = 43 - 59[1]
Cyclotella bodanica	32000 microgram Si/(E6 cells)	Germany[1]
Cyclotella comta	750 microgram Si/(E6 cells)	Germany, n=1[1]
Cyclotella glomerata	220 microgram Si/(E6 cells)	Germany, n=1[1]
Cyclotella meneghiniana	49 %	of dry weight, biomass est. by conv. from morphometric data[2]
Cyclotella socialis	1000 microgram Si/(E6 cells)	Germany, n=2, rg = 850 - 1150[1]
Cyclotella sp.	78 %	of dry weight, biomass est. by conv. from morphometric data[2]
Diatoma elongatum	250 microgram Si/(E6 cells)	Germany[1]
Fragilaria capucina	64 %	of dry weight, biomass est. by conv. from morphometric data[2]
Fragilaria capucina	69 %	of dry weight, biomass est. by conv. from morphometric data[2]
Fragilaria crotonensis	189 microgram Si/(E6 cells)	England, n=2, rg = 188 - 190[1]
Fragilaria crotonensis	189 microgram Si/(E6 cells)	Germany, n=4, rg = 170 - 215[1]
Fragilaria intermedia	76 %	of dry weight, biomass est. by conv. from morphometric data[2]
Meliosira granulata	72 %	of dry weight, biomass est. by conv. from morphometric data[2]
Meliosira islandica	79 %	of dry weight, biomass est. by conv. from morphometric data[2]
Melosira granulata	130 microgram Si/(E6 cells)	Germany, n=1[1]
Melosira granulata	620 microgram Si/(E6 cells)	Sudan, n=1[1]
Melosira islandica subsp. helvetica	280 microgram Si/(E6 cells)	Germany, n=1[1]
Melosira italica	320 microgram Si/(E6 cells)	Germany, n=1[1]

Melosira italica subsp. subArtica	237 microgram Si/(E6 cells)	England, n=3, rg = 165 - 274[1]
Stephanodiscus alpinus	66 %	of dry weight, biomass est. by conv. from morphometric data[2]
Stephanodiscus astraea	8500 microgram Si/(E6 cells)	Germany, n=1[1]
Stephanodiscus binderanus	39 %	of dry weight, biomass est. by conv. from morphometric data[2]
Stephanodiscus hantzschii	41 microgram Si/(E6 cells)	England, n=3, rg = 32 - 47[1]
Stephanodiscus hantzschii var. pusillus	100 microgram Si/(E6 cells)	Germany, n=1[1]
Stephanodiscus minutus	59 %	of dry weight, biomass est. by conv. from morphometric data[2]
Stephanodiscus niagarae	66 %	of dry weight, biomass est. by conv. from morphometric data[2]
Stephanodiscus subtilis	41 %	of dry weight, biomass est. by conv. from morphometric data[2]
Stephanodiscus tenuis	45 %	of dry weight, biomass est. by conv. from morphometric data[2]
Synedra acus var. angustissima	1175 microgram Si/(E6 cells)	Germany, n=2, rg = 1100 - 1250[1]
Synedra acus var. radians	286 microgram Si/(E6 cells)	England, n=1[1]
Tabellaria flocculosa var. asterionelloides	310 microgram Si/(E6 cells)	England, n=5, rg = 250 - 420[1]
Tabellaria flocculosa var. asterionelloides	395 microgram Si/(E6 cells)	Germany / Switzerland, n=2, rg = 250 - 420[1]

Source: 1.Lund, J.W.G. 1965. Biol Rev. 40, pp 231-293.
2.Sicko-Goad, L.M., Schelske, C.L. and Stoermer, E.F., Limnol. Oceanogr., 29(6), 1170, 1984.

1-122 Algae
Si / chlorophyll a ratio

Species	Value	Condition
Chaetoceros debilis	13 by weight	Not limited, cont.light:0.08ly/min, dilution=1/3-1/2 of max.gr.rate[1]
Chaetoceros debilis	28.1 by weight	NH4 starved, 72 h cont.light:0.08ly/min, dilution=1/3-1/2 of max.gr.rate[1]
Chaetoceros debilis	36.3 by weight	NH4 limited, cont.light:0.08ly/min, dilution=1/3-1/2 of max.gr.rate[1]
Chaetoceros debilis	5.6 by weight	Si limited, cont.light:0.08ly/min, dilution=1/3-1/2 of max.gr.rate[1]
Chaetoceros debilis	9 by weight	Si starved, 72 h cont.light:0.08ly/min, dilution=1/3-1/2 of max.gr.rate[1]
Skeletonema costatum	11.4 by weight	Si starved, 72 h cont.light:0.08ly/min, dilution=1/3-1/2 of max.gr.rate[1]
Skeletonema costatum	15 by weight	Not limited, cont.light:0.08ly/min, dilution=1/3-1/2 of max.gr.rate[1]
Skeletonema costatum	20.8 by weight	NH4 starved, 72 h cont.light:0.08ly/min, dilution=1/3-1/2 of[1]
Skeletonema costatum	21.6 by weight	NH4 limited, cont.light:0.08ly/min, dilution=1/3-1/2 of max.gr.rate[1]
Skeletonema costatum	21.9 by weight	NH4 limited, dilution rate=0.110-0.042/hr, rg = 21.0 - 22.7[2]
Skeletonema costatum	23.8 by weight	Si limited, dilution rate=0.086-0.022/hr, rg = 3.7 - 43.9[2]

Skeletonema costatum	5.5 by weight	Si limited, cont.light:0.08ly/min, dilution=1/3-1/2 of max.gr.rate[1]
Thalassiosira gravida	17.1 by weight	NH4 limited, cont.light:0.08ly/min, dilution=1/3-1/2 of max.gr.rate[1]
Thalassiosira gravida	2.4 by weight	Si starved, 72 h cont.light:0.08ly/min, dilution=1/3-1/2 of max.gr.rate[1]
Thalassiosira gravida	3.3 by weight	Si limited, cont.light:0.08ly/min, dilution=1/3-1/2 of max.gr.rate[1]
Thalassiosira gravida	50.7 by weight	NH4 starved, 72 h cont.light:0.08ly/min, dilution=1/3-1/2 of max.gr.rate[1]
Thalassiosira gravida	9.8 by weight	Not limited, cont.light:0.08ly/min, dilution=1/3-1/2 of max.gr.rate[1]

Source: 1.Harrison, P.J., H.L. Conway, R.W. Holmes and C.O. Davis. 1977. Mar. Biol., 43(1), pp 19-31. 2.Harrison, P.J., H.L. Conway and R.C. Dugdale. 1976. Mar. biol. 35, pp 177-186.

1-123 Algae
Specific gravity

Species	Value	Condition
Anabaena flos-aquae	1.0065	Normal cells[1]
Anabaena flos-aquae	1.0085	Cells exposed to a sudden pressure to destroy vacuoles[1]

Source: 1.Klebahn, H. 1922. Jh. wiss. Bot. 61, p 535.

1-124 Algae
Specific growth rate

Species	Value	Condition
Natural population	1-2 d-1	Hawaiian coastal waters[1]

Source: 1.Laws, E.A., et al., Limnol. Oceanogr., 29(6), 1161, 1984.

1-125 Algae
Temperature coefficient

Species	Value	Condition
Algae	1.066	Growth, Lake Ontario[1]
Algae	1.08	Respiration, Lake Ontario[1]

Source: 1.Thomann, R.V., D.M. Di Toro, R.P. Winfield and D.J. O'Connor. 1975. Mathematical Modeling of Phytoplankton in Lake Ontario: 1. National Environmental Reseach Center. EPA. Oregon.

1-126 Algae
Temperature for max growth rate

Species	Value	Condition
Synechococcus lividus	65 degrees celcius	One strain, Chroococcales[1]

Source: 1.Meeks, J.C. and R.W. Castenholz. 1971. Arch. Mikrobiol. 78, 25-41.

1-127 Algae
Th / biomass ratio

Species	Value	Condition
Chlorophyta	0.122 microgram/(gram acid soluble fract	Average of 7 marine species, Puerto Rico[1]
Phaeophyceae	0.17 microgram/(gram acid soluble fracti	Average of 4 marine species, Puerto Rico[1]
Rhodophyta	0.205 microgram/(gram acid soluble fract	Average of 9 marine species, Puerto Rico[1]

Source: 1.Edgington, D.N., S.A. Gordon, M.M. Thommes and L.R. Almodovar. 1970 Limnol. Oceanogr.,, 15(6), pp 94-955.

1-128 Algae
Threshold density

Species	Value	Condition
Algae	100 kcal/m**2	Initial values, thermal ecosystem, limit-egestion proportion[1]

Source: 1.Wiegert, R.G. 1975. In: B.C. Patten (ed.), Systems Analysis and Simulation in Ecology, Vol. 3, pp 15-181.

1-129 Algae
U / biomass ratio

Species	Value	Condition
Chlorophyta	0.88 microgram/(gram acid soluble fraction)	Average of 7 marine species, Puerto Rico[1]
Phaeophyceae	0.79 microgram/(gram acid soluble fraction)	Average of 4 marine species, Puerto Rico[1]
Rhodophyta	0.39 microgram/(gram acid soluble fraction)	Average of 9 marine species, Puerto Rico[1]

Source: 1.Edgington, D.N., S.A. Gordon, M.M. Thommes and L.R. Almodovar. 1970 Limnol. Oceanogr.,, 15(6), pp 94-955.

1-130 Algae
Upper temperature limit for growth rate

Species	Value	Condition
Algae	57.5 degrees celcius	Rg = 55 - 60 cel deg, app. value[1]
Aphanocapsa thermalis	55 degrees celcius	More than value, Chroococcales[2]
Blue-green algae	71.5 degrees celcius	Rg = 70 - 73 cel deg, app. value[1]
Calothrix sp.	53 degrees celcius	Rg = 52-54 deg cel., Nostocales[2]
Mastigocladus laminosus	63.5 degrees celcius	Rg = 63-64 deg cel., Nostocales[2]
Oscillatoria amphibia	57 degrees celcius	Oscillatoriales[2]
Oscillatoria germinata	55 degrees celcius	Oscillatoriales[2]
Oscillatoria okenii	60 degrees celcius	More then value, Oscillatoriales[2]
Oscillatoria terebriformis	53 degrees celcius	Oscillatorialeso[2]
Phormidium laminosum	58.5 degrees celcius	Rg = 57-60 deg cel., Oscillatoriales[2]
Phormidium purpurasiens	46.5 degrees celcius	Rg = 46-47 deg cel., Oscillatoriales[2]
Pleurocapsa sp.	53 degrees celcius	Rg = 52-52 deg cel., Chamaesiphonales[2]
Spirulina sp.	57.5 degrees celcius	Rg = 57-60 deg cel., Oscillatoriales[2]
Symploca thermalis	46 degrees celcius	Rg = 45-47 deg cel., Oscillatoriales[2]

Synechococcus elongatus	68 degrees celcius	Chroococcales, rg = 66 - 70 deg cel.[2]
Synechococcus lividus	74 degrees celcius	One strain, Chroococcales[1]
Synechococcus minervae	60 degrees celcius	Chroococcales[2]
Synechocystis aquatilus	47.5 degrees celcius	Chroococcales, rg = 45-50 deg cel.[2]

Source: 1.Meeks, J.C. and R.W. Castenholz. 1971. Arch. Mikrobiol. 78, 25-41. 2.Castenholz, R.W. 1969. Bact. Rev. 33, 476-504.

1-131 Algae
Uptake

Species	Value	Condition
Nitrogen	109 ug N g-1 ash-free dry weight day-1	Epilithic periphyton, Oct. 1981[1]
Nitrogen	20 ug N g-1 ash-free dry weight day-1	Epilithic periphyton, Dec. 1980[1]
Nitrogen	22 ug N g-1 ash-free dry weight day-1	Epilithic periphyton, Mar. 1981[1]
Nitrogen	23 ug N g-1 ash-free dry weight day-1	Epilithic periphyton, Apr. 1981[1]
Nitrogen	24 ug N g-1 ash-free dry weight day-1	Epilithic periphyton, Feb. 1981[1]
Nitrogen	24 ug N g-1 ash-free dry weight day-1	Epilithic periphyton, Nov. 1980[1]
Nitrogen	38 ug N g-1 ash-free dry weight day-1	Epilithic periphyton, Oct. 1980[1]
Nitrogen	39 ug N g-1 ash-free dry weight day-1	Epilithic periphyton, July 1980[1]
Nitrogen	40 ug N g-1 ash-free dry weight day-1	Epilithic periphyton, May 1981[1]
Nitrogen	55 ug N g-1 ash-free dry weight day-1	Epilithic periphyton, Sep. 1980[1]
Nitrogen	60 ug N g-1 ash-free dry weight day-1	Epilithic periphyton, Jul. 1981[1]
Nitrogen	62 ug N g-1 ash-free dry weight day-1	Epilithic periphyton, Jun. 1981[1]
Nitrogen	90 ug N g-1 ash-free dry weight day-1	Epilithic periphyton, Aug. 1981[1]
Nitrogen	92 ug N g-1 ash-free dry weight day-1	Epilithic periphyton, Sep. 1981[1]

Source: 1.Reuter, J.E., Loeb, S.L. and Goldman C.R. Limnol. Oceanogr., 31, 149, 1986.

1-132 Algae
Uptake of Ammonia

Species	Value	Condition
Natural population	0.12-1.2 umol l-1 d-1	Hawaiian coastal waters[1]

Source: 1.Laws, E.A., et al., Limnol.Oceanogr., 29(6), 1161, 1984.

1-133 Algae
Uptake of phosphate

Species	Value	Condition
Natural population	0.0071-0.22 umol l-1 d-1	Hawaiian coastal waters[1]

Source: 1.Laws, E.A., et al., Limnol.Oceanogr., 29(6), 1161, 1984.

1-134 Algae
Uptake rate

Species	Value	Condition
Ammonia	0.21 - 0.85 mg-atoms N m-3 h-1 * 10	arctic surface waters, fraction < 1 um[1]
Ammonia	0.46 - 1.33 mg-atoms N m-3 h-1 * 10	arctic surface waters, fraction < 15 um[1]
Ammonia	0.64 - 1.92 mg-atoms N m-3 h-1 * 10	arctic surface waters, fraction < 200 um[1]
Nitrate	0.55 - 1.24 mg-atoms N m-3 h-1 * 10	arctic surface waters, fraction < 1 um[1]
Nitrate	0.56 - 1.94 mg-atoms N m-3 h-1 * 10	arctic surface waters, fraction < 15 um[1]
Nitrate	0.88 - 3.65 mg-atoms N m-3 h-1 * 10	arctic surface waters, fraction < 200 um[1]
carbon	19.6 - 76.0 ng-atom l-1 h-1	different stations in the Pacific ocean, nov. 1982 - feb. 1983[2]
urea	0.12 - 1.29 mg-atoms N m-3 h-1 * 10	arctic surface waters, fraction < 1 um[1]
urea	0.16 - 1.53 mg-atoms N m-3 h-1 * 10	arctic surface waters, fraction < 15 um[1]
urea	0.22 - 2.27 mg-atoms N m-3 h-1 * 10	arctic surface waters, fraction < 200 um[1]

Source: 1.Probyn, T.A. and Painting, S.J., Limnol. Oceanogr., 30, 1327, 1985.
2.Kanda, J., Saino, T. and Hattori, A., Limnol.Oceanogr., 30(5), 987, 1985.

1-135 Algae
Vitamin b1 / biomass ratio

Species	Value	Condition
Phytoplankton	3.8E-15 g/cell	Urea as source of N, stationary phase, 293 K, Southern Calif, marine[1]
Phytoplankton	6.4E-15 g/cell	NH4 as source of N, stationary phase, 293 K, Southern Calif, marine[1]
Phytoplankton	8.2E-15 g/cell	NO3 as source of N, stationary phase, 293 K, Southern Calif, marine[1]

Source: 1.Eppley, R.W., A.F. Carlucci, O.H.-Hansen, D. Kiefer, J.J. McCarthy, E. Venrick and P.M. Williams. 1971. Limnol. Oceanogr., 16(5), pp 741-757.

1-136 Algae
Vitamin b12 / biomass ratio

Species	Value	Condition
Phytoplankton	1.6E-16 g/cell	NH4 as source of N, stationary phase, 293 K, Southern Calif, marine[1]
Phytoplankton	2E-16 g/cell	Urea as source of N, stationary phase, 293 K, Southern Calif, marine[1]
Phytoplankton	3.2E-16 g/cell	NO3 as source of N, stationary phase, 293 K, Southern Calif, marine[1]

Source: 1.Eppley, R.W., A.F. Carlucci, O.H.-Hansen, D. Kiefer, J.J. McCarthy, E. Venrick and P.M. Williams. 1971. Limnol. Oceanogr., 16(5), pp 741-757.

1-137 Algae
Volume

Species	Value	Condition
Asterionella formosa	1,122 um3	algal culture[1]
Chlorella sp.	2.6 um3	algal culture[1]
Cryptomonas erosa	1,478 um3	algal culture[1]
Fragilaria crotonensis	1,082 um3	algal culture[1]
Melosira italica	923 um3	algal culture[1]
Melosira italica tenuissima	1,180 um3	algal culture[1]
Society	14.39-17.31 10e5 um3 ml-1	range from several years, beginning of spring bloom, Lake Zurich[2]
Society	16.05-25.02 10e5 um3 ml-1	range from several years, beginning of late summer turb.max., Lake Zurich[2]
Society	16.05-25.02 10e5 um3 ml-1	range from several years, beginning of spring bloom, Lake Zurich[2]
Society	17.55-40.78 10e5 um3 ml-1	range from several years, end of late summer turb.max., Lake Zurich[2]
Society	6.76-34.10 10e5 um3 ml-1	range from several years, autumn bloom, Lake Zurich[2]
Stephanodiscus Hantzschii	213 um3	algal culture[1]
Stephanodiscus astraea minutula	509 um3	algal culture[1]
Stephanodiscus niagarae	11,213 um3	algal culture[1]
Tabellaria fenestrata	4,942 um3	algal culture[1]

Source: 1.Infante, A. and Litt, A.H., Limnol. Oceanogr., 30(5), 1053, 1985.
2.Schanz, F. Limnol.Oceanogr., 30(2), 299, 1985.

1-138 Algae
Zn / biomass ratio

Species	Value	Condition
Aphanizomenon sp.	120 microgram/l	Algae from ponds and lakes in southeastern United States[1]
Chara sp.	89 microgram/l	Algae from ponds and lakes in southeastern United States[1]
Cladophora sp.	10 microgram/l	Algae from ponds and lakes in southeastern United States[1]
Euglena sp.	73 microgram/l	Algae from ponds and lakes in southeastern United States[1]
Hydrodictyon sp.	129 microgram/l	Algae from ponds and lakes in southeastern United States[1]
Lyngbya sp.	171 microgram/l	Algae from ponds and lakes in southeastern United States[1]
Microcystis sp.	48 microgram/l	Algae from ponds and lakes in southeastern United States[1]
Mougeotia sp.	520 microgram/l	Algae from ponds and lakes in southeastern United States[1]
Nitella sp.	240 microgram/l	Algae from ponds and lakes in southeastern United States[1]
Oedogonium sp.	119 microgram/l	Algae from ponds and lakes in southeastern United States[1]

Pithophora sp.	29 microgram/l	Algae from ponds and lakes in southeastern United States[1]
Rhizoclonium sp.	89 microgram/l	Algae from ponds and lakes in southeastern United States[1]
Spirogyra sp.	72 microgram/l	Algae from ponds and lakes in southeastern United States[1]
Spirogyra sp., giant	46 microgram/l	Algae from ponds and lakes in southeastern United States[1]

Source: 1.Water Quality Criteria. 1968. Federal Water Pollution Control Administration Washington D.C., pp 234

1-139 Bacteria
ATP / biomass ratio

Species	Value	Condition
Aerobacter aerogenes	2.8E-11 microgram ATP/cell	Grown at optimal conditions[1]
Bacillus cereus	1.1E-10 microgram ATP/cell	Grown at optimal conditions[1]
Bacillus coagulans	1.7E-10 microgram ATP/cell	Grown at optimal conditions[1]
Bacillus globigii	5.4E-10 microgram ATP/cell	Grown at optimal conditions[1]
Bacillus subtilis	2.82E-10 microgram ATP/cell	Mean value[2]
Brevibacterium helvolum	3.7E-11 microgram ATP/cell	Grown at optimal conditions[1]
Erwinia carotovora	4.4E-11 microgram ATP/cell	Grown at optimal conditions[1]
Escherichia coli	1.45E-10 microgram ATP/cell	Mean value[2]
Escherichia coli	1E-10 microgram ATP/cell	Grown at optimal conditions[1]
Flavobacterium arborescens	1.5E-10 microgram ATP/cell	Grown at optimal conditions[1]
Gaffkya tetragena	6.1E-11 microgram ATP/cell	Grown at optimal conditions[1]
Klebsiella pneumoniae	5E-10 microgram ATP/cell	Grown at optimal conditions[1]
Micrococcus lysodeikticus	1.3E-10 microgram ATP/cell	Grown at optimal conditions[1]
Mycobacterium phlei	1.9E-10 microgram ATP/cell	Grown at optimal conditions[1]
Mycobacterium smegmatis	8.9E-10 microgram ATP/cell	Grown at optimal conditions[1]
Proteus vulgaris	1.8E-10 microgram ATP/cell	Grown at optimal conditions[1]
Pseudomonas aeruginosa	1E-10 microgram ATP/cell	Grown at optimal conditions[1]
Pseudomonas fluorescens	1.6E-10 microgram ATP/cell	Mean value[2]
Pseudomonas fluorescens	3.1E-10 microgram ATP/cell	Grown at optimal conditions[1]
Sarcina lutea	3.7E-11 microgram ATP/cell	Grown at optimal conditions[1]
Serratia marcescens	1E-10 microgram ATP/cell	Grown at optimal conditions[1]
Staphylococcus auerus	6.4E-11 microgram ATP/cell	Grown at optimal conditions[1]

Source: 1.Chappelle, E.W. and G.V. Levin. 1968. Biochem. Med. 2, pp 41-52.
2.Deustachio, A.J. and G.V. Levin. 1967. Bact. Proc., 20, pp 121.

1-140 Bacteria
Ash / biomass ratio

Species	Value	Condition
Bacteria	6.97 % of dry weight	21 species[1]

Source: 1.Spector, W.S. 1956. Handbook of Biological Data, Saunders.

1-141 Bacteria
Biosynthesis of Benzo(a)pyrene

Species	Value	Condition
Bacillus badius on peptone broth + betha-carotene	0.098 ug/l	Substrate=10 mg/l, rg=0.023, control=0.084, n=4, 186 h culture[1]
Bacillus badius on peptone broth + lycopene	0.124 ug/l	Substrate=10 mg/l, rg=0.028, control=0.084, n=4, 186 h culture[1]
Bacillus badius on peptone broth + naphthalene acid	0.135 ug/l	Substrate=10 mg/l, rg=0.005, control=0.084, n=4, 186 h culture[1]
Bacillus badius on peptone broth + sodium acetate	0.059 ug/l	Substrate=10 mg/l, rg=0.001, control=0.084, n=4, 186 h culture[1]
Bacillus badius on peptone broth + vitamin K11	0.119 ug/l	Substrate=10 mg/l, rg=0.027, control=0.084, n=4, 186 h culture[1]

Source: 1.Naiussat, P., C. Auger and L. Mallet. 1970. C.R. Acad. Sci (Paris) Ser. D. 270, 1042-1045.

1-142 Bacteria
Biosynthesis of pyrylene

Species	Value	Condition
Bacillus badius on peptone broth + betha-carotene	0.31 ug/l	Substrate=10 mg/l, rg=0.075, control=0.30, n=4, 186 h culture[1]
Bacillus badius on peptone broth + lycopene	0.474 ug/l	Substrate=10 mg/l, rg=0.064, control=0.30, n=4, 186 h culture[1]
Bacillus badius on peptone broth + naphthalene acid	0.415 ug/l	Substrate=10 mg/l, rg=0.088, control=0.30, n=4, 186 h culture[1]
Bacillus badius on peptone broth + sodium acetate	0.259 ug/l	Substrate=10 mg/l, rg=0.015, control=0.30, n=4, 186 h culture[1]
Bacillus badius on peptone broth + vitamin K11	0.365 ug/l	Substrate=10 mg/l, rg=0.063, control=0.30, n=4, 186 h culture[1]

Source: 1.Naiussat, P., C. Auger and L. Mallet. 1970. C.R. Acad. Sci (Paris) Ser. D. 270, 1042-1045.

1-143 Bacteria
C / biomass ratio

Species	Value	Condition
Bacteria	48.5 % of dry weight	Mean elementary composition[1]
Decomposer	49 % of dry weight	Mean elementary composition, mean value[1]
Escherichia coli	12.14 %	Atomic composition[2]

Source: 1.Bowen, H.J.M. 1966. Trace Elements in Biochemistry, Ac. Press.
2.Morowitz, H.J., M.E. Tourtellotte, W.R. Guild, E. Castro and C. Woese. 1962. J. Mol. Boil., 4, p 93.

1-144 Bacteria
Ca / biomass ratio

Species	Value	Condition
Escherichia coli	0.03 %	Atomic composition[1]

Source: 1.Morowitz, H.J., M.E. Tourtellotte, W.R. Guild, E. Castro and C. Woese. 1962. J. Mol. Boil., 4, p 93.

1-145 Bacteria
Carbon production

Species	Value	Condition
Marine Bacteria	0.5 ng C l-1 h-1	on marine snow, Atlantic, calculated from thymidine incorporation rates[1]
Marine Bacteria	1.0 ng C l-1 h-1	on marine snow, Atlantic, calculated from thymidine incorporation rates[1]
Marine Bacteria	1.4 ng C l-1 h-1	on marine snow, Atlantic, calculated from thymidine incorporation rates[1]
Marine Bacteria	18.9 ng C l-1 h-1	in seawater, Atlantic, calculated from thymidine incorporation rates[1]
Marine Bacteria	38.6 ng C l-1 h-1	in seawater, Atlantic, calculated from thymidine incorporation rates[1]
Marine bacteria	0.3 ng C l-1 h-1	on marine snow, Atlantic, calculated from thymidine incorporation rates[1]
Marine bacteria	10.2 ng C l-1 h-1	in seawater, Atlantic, calculated from thymidine incorporation rates[1]
Marine bacteria	12.9 ng C l-1 h-1	in seawater, Atlantic, calculated from thymidine incorporation rates[1]
Marine bacteria	19 ng C l-1 h-1	in seawater, California, calculated from thymidine incorporation rates[1]
Marine bacteria	2 ng C l-1 h-1	on marine snow, California, calculated from thymidine incorporation rates[1]
Marine bacteria	201 ng C l-1 h-1	in seawater, California, calculated from thymidine incorporation rates[1]
Marine bacteria	223 ng C l-1 h-1	in seawater, California, calculated from thymidine incorporation rates[1]
Marine bacteria	3 ng C l-1 h-1	on marine snow, California, calculated from thymidine incorporation rates[1]
Marine bacteria	3.6 ng C l-1 h-1	on marine snow, Atlantic, calculated from thymidine incorporation rates[1]
Marine bacteria	31 ng C l-1 h-1	on marine snow, California, calculated from thymidine incorporation rates[1]
Marine bacteria	7.8 ng C l-1 h-1	in seawater, Atlantic, calculated from thymidine incorporation rates[1]
Marine bacteria	89 ng C l-1 h-1	in seawater, California, calculated from thymidine incorporation rates[1]
marine bacteria	5 ng C l-1 h-1	on marine snow, California, calculated from thymidine incorporation rates[1]

Source: 1.Alldredge, A.L., Cole, J.J. and Caron, D.A. Limnol. Oceanogr., 31, 68, 1986.

1-146 Bacteria
Cell dry weight

Species	Value	Condition
Nitrosomonas sp.	1E-12 g	None[1]

Source: 1.Curtis, E.J.C., K. Durrant and M.M.I. Harman. 1975. Water Res., 9(3), pp 255-268.

1-147 Bacteria
Cell production rate

Species	Value	Condition
Cells	0.44-1.98 10E8 cells l-1 h-1	estimated from thymidine incorporation rate, Lake Michigan[1]

Source: 1.Scavia, D., Laird, G.A. and Fahnenstiel, G.L. Limnol. Oceanogr., 31, 612, 1986.

1-148 Bacteria
Concentration for optimal growth

Species	Value	Condition
Nitrifying bacteria	0.0002 mg Mg/l	See ref.[1]
Nitrifying bacteria	0.0115 mg P/l	Range = 0.003 - 0.02[2]
Nitrifying bacteria	1 mg Ca/l	See ref.[3]
Nitrifying bacteria	6 mg Fe/l	See ref.[4]
Nitrobacter sp.	0.3 mg Fe/l	See ref.[4]
Nitrobacter sp.	1705 mg P/l	Range = 310 - 3100[5]
Nitrosomonas sp.	0.1 mg Fe/l	See ref.[4]
Nitrosomonas sp.	620 mg P/l	Range = 310 - 930[5]

Source: 1.Meiklejohn, J. 1952. Nature 170, p 1132.
2.Meiklejohn, J. and G. Wimmer. 1952. Z. Hyg., 48, p 135.
3.Kingma Boltjes, T.Y. 1936. Arch. Mikrobiol. 6, p 79.
4.Meiklejohn, J. 1954. Some Aspects of the Physiology of the Nitrifying Bacteria. In B.A. Fry and J.L.Peel(ed), The fourth Symposium of the Society for General Microbiology
5.Van Droogenbroeck, R. and H. Laudelout. 1967. Antonie van Leeuwenhoek, 33, p 287.

1-149 Bacteria
Covalent bond distribution

Species	Value	Condition
Escherichia coli	11280000000 EV	Contribution to total bond energy from C-C (single bond)[1]
Escherichia coli	1160000000 EV	Contribution to total bond energy from C-N (double bond)[1]
Escherichia coli	1470000000 EV	Contribution to total bond energy from C-C (double bond)[1]
Escherichia coli	2300000000 EV	Contribution to total bond energy from H-O (single bond)[1]
Escherichia coli	28620000000 EV	Contribution to total bond energy from H-C (single bond)[1]
Escherichia coli	3770000000 EV	Contribution to total bond energy from C-O (single bond)[1]
Escherichia coli	5840000000 EV	Contribution to total bond energy from H-N (single bond)[1]
Escherichia coli	6620000000 EV	Contribution to total bond energy from C-N (single bond)[1]
Escherichia coli	8300000000 EV	Contribution to total bond energy from C-O (double bond)[1]

Source: 1.Robinson, E. and R.C. Robbins, 1975. In: S.F. Singer (ed.), The Changing Global Environment, pp 111-23. D. Reidel Publ. Comp.

1-150 Bacteria
Density

Species	Value	Condition
Carex walteriana	0.27 10e7 AODC	initial value for exp., fraction < 1um[1]
Carex walteriana	15.23 10e7 AODC	initial value for exp., unfractionated[1]
Carex walteriana	16.82 10e7 AODC	final value for exp., fraction < 1 um, candicidin added[1]
Carex walteriana	7.52 10e7 AODC	final value for exp., fraction < 1 um[1]
Carex walteriana	90.95 10e7 AODC	final value for exp., unfractionated[1]
Carex walteriana	99.56 10e7 AODC	final value for exp., unfractionated, candicidin added[1]
Marine Bacteria	0.007 10e8 bact. l-1	on marine snow, Atlantic, 0.1 % of total[2]
Marine Bacteria	0.018 10e8 bact. l-1	on marine snow, Atlantic, 0.2 % of total[2]
Marine Bacteria	0.018 10e8 bact. l-1	on marine snow, California, 0.9 % of total[2]
Marine Bacteria	0.032 10e8 bact. l-1	on marine snow, California, 1.3 % of total[2]
Marine Bacteria	0.055 10e8 bact. l-1	on marine snow, California, 3.0 % of total[2]
Marine Bacteria	0.068 10e8 bact. l-1	on marine snow, Atlantic, 2.3 % of total[2]
Marine Bacteria	0.068 10e8 bact. l-1	on marine snow, California, 2.8 % of total[2]
Marine Bacteria	0.089 10e8 bact. l-1	on marine snow, Atlantic, 0.9 % of total[2]
Marine Bacteria	0.150 10e8 bact. l-1	on marine snow, Atlantic, 2.8 % of total[2]
Marine Bacteria	0.280 10e8 bact. l-1	on marine snow, Atlantic, 4.4 % of total[2]
Marine Bacteria	1.84 10e8 bact. l-1	in seawater, California[2]
Marine Bacteria	10.4 10e8 bact. l-1	in seawater, Atlantic[2]
Marine Bacteria	2.08 10e8 bact. l-1	in seawater, California[2]
Marine Bacteria	2.39 10e8 bact. l-1	in seawater, California[2]
Marine Bacteria	2.92 10e8 bact. l-1	in seawater, Atlantic[2]
Marine Bacteria	5.32 10e8 bact. l-1	in seawater, Atlantic[2]
Marine Bacteria	5.51 10e8 bact. l-1	in seawater, Atlantic[2]
Marine Bacteria	6.4 10e8 bact. l-1	in seawater, Atlantic[2]
Marine Bacteria	9.13 10e8 bact. l-1	in seawater, Atlantic[2]
Rhizophora mangle	0.07 10e7 AODC	initial value for exp., fraction < 1 um[1]
Rhizophora mangle	0.20 10e7 AODC	initial value for exp., unfractionated[1]
Rhizophora mangle	12.43 10e7 AODC	final value for exp., unfractionated[1]
Rhizophora mangle	32.95 10e7 AODC	final value for exp., fraction < 1 um[1]
Spartina alterniflora	0.16 10e7 AODC	initial value for exp., fraction < 1 um[1]
Spartina alterniflora	11.60 10e7 AODC	final value for exp., fraction < 1 um, candicidin added[1]
Spartina alterniflora	16.89 10e7 AODC	final value for exp., unfractionated, candicidin added[1]
Spartina alterniflora	6.30 10e7 AODC	initial value for exp., unfractionated[1]
Spartina alterniflora	7.13 10e7 AODC	final value for exp., fraction < 1 um[1]
Spartina alterniflora	9.74 10e7 AODC	final value for exp., unfractionated[1]
Taxodium ascendens	0.15 10e7 AODC	initial value for exp., fraction < 1 um[1]

Taxodium ascendens	10.17 10e7 AODC	final value for exp., fraction < 1 um, candicidin added[1]
Taxodium ascendens	26.80 10e7 AODC	final value for exp., unfractionated, candicidin added[1]
Taxodium ascendens	4.88 10e7 AODC	final value for exp., fraction < 1 um[1]
Taxodium ascendens	6.9 10e7 AODC	initial value for exp., unfractionated[1]
Taxodium ascendens	8.30 10e7 AODC	final value for exp., unfractionated[1]

Source: 1.Benner, R., Moran, M.A. and Hodson, R.E. Limnol. Oceanogr., 31, 89, 1986.
2.Alldredge, A.L., Cole, J.J. and Caron, D.A. Limnol. Oceanogr., 31, 68, 1986.

1-151 Bacteria
Growth rate

Species	Value	Condition
Filaments bacteria	2.8 h for generation time	White Clay Creek, in situ, pool, summer, 289.5-294 K[1]
Filaments bacteria	4 h for generation time	White Clay Creek, in situ, riffle, summer, 289.5-294 K[1]
Filaments bacteria	4.0 h for generation time	White Clay Creek, in situ, riffle, summer, 289.5-294 K[1]
Filaments bacteria	42 h for generation time	White Clay Creek, in situ, riffle, winter, 273-278 K[1]
Filaments bacteria	45 h for generation time	White Clay Creek, in situ, pool, winter, 273-278 K[1]
Filaments bacteria	8.4 h for generation time	White Clay Creek, in situ, pool, spring, 281-293 K[1]
Filaments bacteria	9.65 h for generation time	White Clay Creek, in situ, riffle, spring, 281-293 K[1]
Nitrosomonas sp.	34.5 hours	Mean generation time[2]
Unicellular bacteria	10.8 h for generation time	White Clay Creek, in situ, pool, spring, 281-293 K[1]
Unicellular bacteria	11.5 h for generation time	White Clay Creek, in situ, riffle, spring, 281-293 K[1]
Unicellular bacteria	4.8 h for generation time	White Clay Creek, in situ, riffle, summer, 289.5-294 K[1]
Unicellular bacteria	48 h for generation time	White Clay Creek, in situ, riffle, winter, 273-278 K[1]
Unicellular bacteria	58 h for generation time	White Clay Creek, in situ, pool, winter, 273-278 K[1]
Unicellular bacteria	7 h for generation time	White Clay Creek, in situ, pool, summer, 289.5-294 K[1]
Unicellular bacteria	7.0 h for generation time	White Clay Creek, in situ, pool, summer, 289.5-294 K[1]

Source: 1.Boot, T.L. 1975. Limnol. Oceanogr. 20, 191-197.
2.Curtis, E.J.C., K. Durrant and M.M.I. Harman. 1975. Water Res., 9(3), pp 255-268.

1-152 Bacteria
Growth using aliphatic hydrocarbon as substrate

Species	Value	Condition
Acinetobacter	50 % growth occured	Sub = alkanes = 9,14, 16,18, 32 and sub alkanes, tot fatty acid exam.[1]
Bacterium	50 % growth occured	Sub = alkanes 10,11,14,15,17,18, tot. fatty acids found, B. stratin H[1]
Corynebacterium simplex	50 % growth occured	Sub = alkanes and alk-1-enes, 12-18. tot fatty acids found[1]
Micrococcus freundenreichii	50 % growth occured	Sub = various alkanes, phospholipids found[1]
Mycobacterium vaccae	50 % growth occured	Sub = alkanes 2-8, tot fatty acids and phospholipids found[1]
Mycobacterium vaccae	50 % growth occured	Sub = alkanes, alk-1-ene, branched-chain alkanes, tot fatty acids found[1]

Bacteria sp.	50 % growth occured	Sub = alkanes 14-17, fatty acids found, A. HO1N[1]
Bacteria sp.	50 % growth occured	Sub = alkanes 16,17, phospholipids found, A. HO1N[1]
Bacteria sp.	50 % growth occured	Sub = hexadecane, phospholipids found, A. HO1N[1]

Source: 1.Watkinson, R.J. 1978. Development in Biodegradation of Hydrocarbons - 1. Applied Science publishers.London.

1-153 Bacteria
H / biomass ratio

Species	Value	Condition
Bacteria	7.4 % of dry weight	Mean elementary composition[1]
Decomposer	7 % of dry weight	Mean elementary composition, mean value[1]
Escherichia coli	9.94 %	Atomic composition[2]

Source: 1.Bowen, H.J.M. 1966. Trace Elements in Biochemistry, Ac. Press.
2.Morowitz, H.J., M.E. Tourtellotte, W.R. Guild, E. Castro and C. Woese. 1962. J. Mol. Boil., 4, p 93.

1-154 Bacteria
Half saturation constant for N uptake

Species	Value	Condition
Nitrosomonas europaea	0.32 mg NH3-N/l	298 K, pH=7[1]
Nitrosomonas europaea	0.36 mg NH3-N/l	298 K, pH=8[1]
Nitrosomonas europaea	0.81 mg NH3-N/l	298 K, pH=9[1]
Nitrosomonas europaea	2 mg (NH4(+) + NH3)-N/l	298 K, pH=9[1]
Nitrosomonas europaea	5.6 mg (NH4(+) + NH3)-N/l	298 K, pH=8[1]
Nitrosomonas europaea	56 mg (NH4(+) + NH3)-N/l	298 K, pH=7[1]

Source: 1.Suzucki, I., U. Dular and S.C. Kwok. 1974. J. Bact. 120, pp 556-558.

1-155 Bacteria
Half saturation constant for P uptake

Species	Value	Condition
Escherichia coli	0.7 Umol	Minimal medium, 293 K, uptake study[1]
Escherichia coli	17.35 Umol	Minimal medium, 303 K[1]

Source: 1.Mitchell, R. and C. Chamberlin. 1978. in: Berg. G. (ed.); Indicators of Enteric Contamination in Natural Waters.

1-156 Bacteria
Half saturation constant for acetate uptake

Species	Value	Condition
Ancalomicrobium sp.	0.762 microgram/ml	Oxidation lagoon, 290 K[1]
Ancalomicrobium sp.	1.13 microgram/ml	Oxidation lagoon, 290 K[1]
Bacteria community	0.387 microgram/ml	Oxidation lagoon, 290 K[1]
Bacteria community	1.45 microgram/ml	Oxidation lagoon, 290 K[1]
Elongated rod	0.166 microgram/ml	Oxidation lagoon, 290 K[1]

Source: 1.Stanley, P.M. and J.T. Stanley. 1977. Limnol. Oceanogr. 22(1), pp 26-37.

1-157 Bacteria
Half saturation constant for acetic acid

Species	Value	Condition
Sulfate reducing bacteria	5.7 mg/l	304 K, experimental data according to volumetric reaction[1]

Source: 1.Middleton, A.C. and A.W. Lawrence. 1977. J. Water Pollut. Control, 49(7), pp 1659-1670.

1-158 Bacteria
Half saturation constant for ammonium uptake

Species	Value	Condition
Pseudomonad	0.26 mmol	Chemostat, late log phase, 288 K, marine[1]

Source: 1.Brown, C.M., P.S. MacDonald-Brown and S.O Stanley. 1975. Mar. Biol., 31(1), pp 7-13.

1-159 Bacteria
Half saturation constant for glucose

Species	Value	Condition
Escherichia coli	0.378 Umol	303 K[1]
Escherichia coli	19.4 Umol	Minimal medium, 1942[1]
Escherichia coli	22 Umol	Minimal medium, 1942[1]
Escherichia coli	405 Umol	303 K, 1964, minimal medium[1]
Escherichia coli	41.7 Umol	Minimal medium, 1958[1]
Escherichia coli	44 Umol	Seawater, 293 K[1]
Escherichia coli	550 Umol	303 K, minimal medium[1]

Source: 1.Mitchell, R. and C. Chamberlin. 1978. in: Berg. G. (ed.); Indicators of Enteric Contamination in Natural Waters.

1-160 Bacteria
Half saturation constant for growth

Species	Value	Condition
Acinetobacter sp. on 2-hydroxycyclohexanone	11.2 Um	80 uM NADPH and 200 uM O2, 1,2-mono-oxygenase from A. NCIB 9871[1]
Acinetobacter sp. on 2-methylcyclohexanone	12 Um	80 uM NADPH and 200 uM O2, 1,2-mono-oxygenase from A. NCIB 9871[1]
Acinetobacter sp. on 2-methylcyclohexanone	12.0 Um	80 uM NADPH and 200 uM O2, 1,2-mono-oxygenase from A. NCIB 9871[1]
Acinetobacter sp. on 4-hydroxycyclohexanone	390 Um	80 uM NADPH and 200 uM O2, 1,2-mono-oxygenase from A. NCIB 9871[1]
Acinetobacter sp. on 4-methylcyclohexanone	9 Um	80 uM NADPH and 200 uM O2, 1,2-mono-oxygenase from A. NCIB 9871[1]
Acinetobacter sp. on 4-methylcyclohexanone	9.0 Um	80 uM NADPH and 200 uM O2, 1,2-mono-oxygenase from A. NCIB 9871[1]
Acinetobacter sp. on camphor	1000 Um	80 uM NADPH and 200 uM O2, 1,2-mono-oxygenase from A. NCIB 9871[1]
Acinetobacter sp. on cycloheptanone	200 Um	80 uM NADPH and 200 uM O2, 1,2-mono-oxygenase from A. NCIB 9871[1]
Acinetobacter sp. on cyclohexan-1,2-dione	170 Um	80 uM NADPH and 200 uM O2, 1,2-mono-oxygenase from A. NCIB 9871[1]
Acinetobacter sp. on cyclohexanone	6.3 Um	80 uM NADPH and 200 uM O2, 1,2-mono-oxygenase from A. NCIB 9871[1]

Acinetobacter sp. on cyclooctanone	1800 Um	80 uM NADPH and 200 uM O2, 1,2-mono-oxygenase from A. NCIB 9871[1]
Acinetobacter sp. on cyclopentanone	3600 Um	80 uM NADPH and 200 uM O2, 1,2-mono-oxygenase from A. NCIB 9871[1]
Acinetobacter sp. on dihydrocarvone	640 Um	80 uM NADPH and 200 uM O2, 1,2-mono oxygenase from NCIB[1]
Acinetobacter sp. on norcamphor	140 Um	80 uM NADPH and 200 uM O2, 1,2-mono-oxygenase from A. NCIB 9871[1]

Source: 1.Yamamoto, S., M. Katagiri, H. Maeno and O. Hayaishi. 1965. J. Biol. Chem. 240, 3408.

1-161 Bacteria
Half saturation constant for lactose

Species	Value	Condition
Escherichia coli	111 Umol	Minimal medium, 293 K[1]
Escherichia coli	50 Umol	Seawater, 293 K[1]

Source: 1.Mitchell, R. and C. Chamberlin. 1978. in: Berg. G. (ed.); Indicators of Enteric Contamination in Natural Waters.

1-162 Bacteria
Half saturation constant for nitrate uptake

Species	Value	Condition
Pseudomonad	0.26 mm	Chemostat, late log phase, 288 K, marine[1]

Source: 1.Brown, C.M., P.S. MacDonald-Brown and S.O Stanley. 1975. Mar. Biol., 31(1), pp 7-13.

1-163 Bacteria
Half saturation constant for nitrogen oxidation

Species	Value	Condition
Nitrobacter sp.	5 mg N/l	298 K[1]
Nitrobacter sp.	8 mg N/l	305 K[1]
Nitrosomonas sp.	1.2 mg N/l	293 K[1]
Nitrosomonas sp.	10 mg N/l	303 K[1]
Nitrosomonas sp.	3.5 mg N/l	298 K[1]

Source: 1.Painter, H.A. 1975. Microbial Transformations of inorganic Nitrogen. in: Conf. on Nitrogen as a Water Poll. Vol 1, Session A1. Copenhagen.

1-164 Bacteria
Half saturation constant for oxygen uptake

Species	Value	Condition
Larger marine bacteria	6.2 microgram-atom/l	Length=4.0 U, 293 K, O2 limiting[1]
Nitrobacter sp.	0.25 mg/l	291 K[2]
Nitrobacter sp.	0.5 mg/l	305 K[2]
Nitrobacter sp.	1 mg/l	303 K[2]
Nitrosomonas sp.	0.3 mg/l	293 K[2]
Nitrosomonas sp.	0.5 mg/l	303 K[2]
Small marine bacteria	0.022 microgram-atom/l	Length=0.5 U, 293 K, O2 limiting[1]

Source: 1.Cline, J.D. and F.A. Richards. 1972. Limnol. Oceanogr., 17(6), pp 885-900.
2.Painter, H.A. 1975. Microbial Transformations of inorganic Nitrogen. in: Conf. on Nitrogen as a Water Poll. Vol 1, Session A1. Copenhagen.

1-165 Bacteria
K / biomass ratio

Species	Value	Condition
Escherichia coli	0.05 %	Atomic composition[1]

Source: 1.Morowitz, H.J., M.E. Tourtellotte, W.R. Guild, E. Castro and C. Woese. 1962. J. Mol. Boil., 4, p 93.

1-166 Bacteria
Lipid / biomass ratio

Species	Value	Condition
Mycoplasma gallisepticum	11 % of dry weight	None[1]

Source: 1.Morowitz, H.J., M.E. Tourtellotte, W.R. Guild, E. Castro and C. Woese. 1962. J. Mol. Boil., 4, p 93.

1-167 Bacteria
Max growth rate

Species	Value	Condition
Ammonia oxidation bacteria	0.33 1/day	Max specific growth rate, ammonia oxid. in activated sludge, 298 K[1]
Azotobacter sp.	0.35 1/hr	Specific growth rate, rg = 0.3 - 0.4[2]
Bacteria	3.7 1/day	Max specific growth rate, aerobic treatment of waste water, 298 K[3]
Denitrifying bacteria	1.8 1/day	Max specific growth rate, denitrification using methanol, 298 K[4]
Methane fermenting bacteria	0.13 1/day	Max specific growth rate, methane ferm. of acetic acid, 298 K[5]
Nitrite oxidation bacteria	0.65 1/day	Max specific growth rate, nitrite oxidation in estuary water, 298 K[6]
Nitrobacter sp.	0.14 1/day	In sludge, 293 K[7]
Nitrobacter sp.	1 1/day	In Thames water, 293 K[7]
Nitrobacter sp.	1.39 1/day	305 K[8]
Nitrobacter sp.	1.39 1/day	Pure culture, 293 K[7]
Nitrosomonas sp.	0.33 1/day	293 K, pH=7.5, activated sludge treated sewage, val.usual.much less[9]
Nitrosomonas sp.	0.59 1/day	293 K, activated sludge treated sewage[10]
Nitrosomonas sp.	0.7 1/day	293 K, pH=7.5, pure culture and in Thames River water[9]
Nitrosomonas sp.	1 1/day	Isolated from activated sludge, 300 K, doubling time=16 hrs[11]
Nitrosomonas sp.	1.5 1/day	Clear medium, continuous culture, 303 K, doubling time=11 hrs[12]
Nitrosomonas sp.	2.2 1/day	Clear medium, batch culture, 303 K, doubling time=8 hrs[12]
Sulfate reducing bacteria	0.33 1/day	Max specific growth rate, using acetic acid, 298 K[13]
Sulfolobus sp.	4.9 hrs	Generation time, in nature, 348 K[14]
Sulfolobus sp.	5.8 hrs	Generation time, in nature, 348 K[14]
Sulfolobus sp.	6.4 hrs	Generation time, in nature, 348 K[14]
Sulfolobus sp.	9.9 hrs	Generation time, in nature, 353 K[14]

Source: 1.Downing, A.L., H.A. Painter and G. Knowles. 1964. J. Proc. Inst. Sew. Purif., pp 130-153.

2.Alexander, M. and P.W. Wilson. 1954. Appl. Microbiol., 2, pp 135-140.
3.Benedek, P. and F. Horvath. 1967. Water Res., 1, p 663.
4.Stensel, H.D. et. al. 1973. Water Pollut. Control, 45, pp 243.
5.Lawrence, A.W. and P.L. McCarty. 1969. Water Pollut. Control, 41, R1.(12)
6.Knowles, C., A.L. Downing and M.J. Barrett. 1965. J. Gen. Microbiol. 38, pp 263-278.
7.Department of Scientific and Industrial Research. 1964. Water Pollution research. 1963. H.M.S.O. London.
8.Boon, B. and H. Laudelout. 1962. Biochem. J. 85, pp 440-447.
9.Downing, A.L., H.A. Painter and G. Knowles. 1964. J. Proc. Inst. Sew. Purif., pp 130-153.
10.Hall, I. 1974. Water Pollut Control, 3, pp 538-547.
11.Loveless, J.E. and H.A. Painter. 1968. J. Gen. Microbiol. 52, pp 1-14.
12.Skinner, M. and N. Walker. 1961. Arch. Microbiol. 38, pp 339-349.
13.Middleton, A.C. and A.W. Lawrence. 1977. J. Water Pollut. Control, 49(7), pp 1659-1670.
14.Weiss, R.L. 1973. Limnol. Oceanogr. 18(6), pp 877-883.

1-168 Bacteria
Max rate of acetate uptake

Species	Value	Condition
Ancalomicrobium sp.	25.7 microgram/(ml*min)	Oxidation lagoon, 290 K[1]
Ancalomicrobium sp.	374 microgram/(ml*min)	Oxidation lagoon, 290 K[1]
Bacteria Lake Erken	0.0365 ug/(l*h)	Under ice, rg=1.9-5.4E-2, no bac=50E6/l, phyt prod=500 mgC/(m2*d), 277 K[2]
Bacteria Lake Erken	3.65E-2 ug/(l*h)	Under ice, rg=1.9-5.4E-2, no bac=50E6/l, phyt prod=500 mgC/(m2*d), 277 K[2]
Bacteria community	10900 microgram/(ml*min)	Oxidation lagoon, 290 K[1]
Bacteria community	2210 microgram/(ml*min)	Oxidation lagoon, 290 K[1]
Bacteria from Lotsjoen	0.6 ug/(l*h)	Winter, rg=3.3-8.7E-1, 273.4-273.8 K, Sweden[3]
Bacteria from Lotsjoen	6.0E-1 ug/(l*h)	Winter, rg=3.3-8.7E-1, 273.4-273.8 K, Sweden[3]
Bacteria from Lotsjoen	80 ug/(l*h)	Summer, Sweden[3]
Bacteria from Marion Lake	4 ug C/(g sediment*h)	Rg=3-5, winter, Canada, 276 K, no bac=5E4/(cm3 sediment)[4]
Bacteria from Marion Lake	45 ug C/(g sediment*h)	Rg=40-50, summer, Canada, 283 K, no bac=2E6/(cm3 sediment)[4]
Elongated rod	388 microgram/(ml*min)	Oxidation lagoon, 290 K[1]
Freshwater bacteria	17 microgram/(l*day)	Upper Klamath Lake I, rg = 2 - 55[5]

Source: 1.Stanley, P.M. and J.T. Stanley. 1977. Limnol. Oceanogr. 22(1), pp 26-37.
2.Chimiklis, P.E. and E.P. Karlander. 1973. Plant. Physiol. 51, 48-56.
3.Allen, H.L. 1968. Ecology 49, 346-349.
4.Hall, K.J. and K.O. Hyatt. 1974. J. Fish. Res. bd Canada 31, 893-911.
5.Wright, R.T. 1975. Limnol. Oceanogr. 20(4), pp 626-633.

1-169 Bacteria
Max rate of acetic acid uptake

Species	Value	Condition
Sulfate reducing bacteria	8.3 1/day	304 K, experimental data according to volumetric reaction[1]

Source: 1.Middleton, A.C. and A.W. Lawrence. 1977. J. Water Pollut. Control, 49(7), pp 1659-1670.

1-170 Bacteria
Max rate of acetic uptake

Species	Value	Condition
Bacteria from West Blue Lake, Manitoba	0.031 ug/(l*h)	No bac=5.8 /ml, ice surface, winter, phyt prod=18-67 mg C/(m2*d)(summer)[1]
Bacteria from West Blue Lake, Manitoba	3.1E-2 ug/(l*h)	No bac=5.8 /ml, ice surface, winter, phyt prod=18-67 mg C/(m2*d)(summer)[1]

Source: 1.Ward, F.J. and G.G.C. Robinson. 1974. J. Fish. Res. bd Canada 31, 977-1005.

1-171 Bacteria
Max rate of alanine uptake

Species	Value	Condition
Bacteria from Upper Klamath Lake	2.5 ug/(l*h)	Feb, 28 K, natural eutrophic, ice surface during winter, rg=2-3[1]

Source: 1.Burnison, B.K. and R.Y. Morita. 1974. Appl. Microbiol. 27, 488-495.

1-172 Bacteria
Max rate of asparagine uptake

Species	Value	Condition
Bacteria from Upper Klamath Lake	5 ug/(l*h)	Feb, 28 K, natural eutrophic, ice surface during winter[1]

Source: 1.Burnison, B.K. and R.Y. Morita. 1974. Appl. Microbiol. 27, 488-495.

1-173 Bacteria
Max rate of aspartate uptake

Species	Value	Condition
Bacteria from Upper Klamath Lake	4.5 ug/(l*h)	Feb, 28 K, natural eutrophic, ice surface during winter, rg=4-5[1]

Source: 1.Burnison, B.K. and R.Y. Morita. 1974. Appl. Microbiol. 27, 488-495.

1-174 Bacteria
Max rate of citric uptake

Species	Value	Condition
Bacteria from West Blue Lake, Manitoba	0.008 ug/(l*h)	No bac=2.4 /ml, ice surface, winter, phyt prod=18-67 mg C/(m2*d)(summer)[1]
Bacteria from West Blue Lake, Manitoba	8.0E-3 ug/(l*h)	No bac=2.4 /ml, ice surface, winter, phyt prod=18-67 mg C/(m2*d)(summer)[1]

Source: 1.Ward, F.J. and G.G.C. Robinson. 1974. J. Fish. Res. bd Canada 31, 977-1005.

1-175 Bacteria
Max rate of fumaric uptake

Species	Value	Condition
Bacteria from West Blue Lake, Manitoba	0.048 ug/(l*h)	No bac=13,1/ml, ice surface, winter, phyt prod=18-67 mg C/(m2*d)(summer)[1]
Bacteria from West Blue Lake, Manitoba	4.8E-2 ug/(l*h)	No bac=13,1/ml, ice surface, winter, phyt prod=18-67 mg C/(m2*d)(summer)[1]

Source: 1.Ward, F.J. and G.G.C. Robinson. 1974. J. Fish. Res. bd Canada 31, 977-1005.

1-176 Bacteria
Max rate of glucose uptake

Species	Value	Condition
Bacteria Lake Erken	0.034 ug/(l*h)	Under ice, rg=2.3-4.5E-2, no bac=50E6/l, phyt prod=500 mgC/(m2*d), 277 K[1]
Bacteria Lake Erken	3.4E-2 ug/(l*h)	Under ice, rg=2.3-4.5E-2, no bac=50E6/l, phyt prod=500 mgC/(m2*d), 277 K[1]
Bacteria from Char lake, artic	0.0045 ug/(l*h)	Rg = 1-8E-3 ug/(l*h), no. bac=0.1-2E8/l, phyt prod=25 mgC/(m2*d)[1]
Bacteria from Char lake, artic	4.5E-3 ug/(l*h)	Rg = 1-8E-3 ug/(l*h), no. bac=0.1-2E8/l, phyt prod=25 mgC/(m2*d)[1]

Bacteria from Lapland lakes	0.0015 ug/(l*h)	Phyt prod=50 mg C/(m2*d), rg=1-2E-3, ice+snow covered most of year, n=6[2]
Bacteria from Lapland lakes	1.5E-3 ug/(l*h)	Phyt prod=50 mg C/(m2*d), rg=1-2E-3, ice+snow covered most of year, n=6[2]
Bacteria from Marion Lake	15 ug C/(g sediment*h)	Rg=10-20, summer, Canada, 283 K, no bac=2E6/(cm3 sediment)[3]
Bacteria from Marion Lake	2 ug C/(g sediment*h)	Rg=1-3, winter, Canada, 276 K, no bac=5E4/(cm3 sediment)[3]
Bacteria from Meretta Lake, artic	0.38 ug/(l*h)	Rg = 0.1-7.5E-1 ug/(l*h), no. bac=2-80E8/l, phyt prod=125 mg C/(m2*d)[1]
Bacteria from Meretta Lake, artic	3.8E-1 ug/(l*h)	Rg = 0.1-7.5E-1 ug/(l*h), no. bac=2-80E8/l, phyt prod=125 mg C/(m2*d)[1]
Freshwater bacteria	9 microgram/(l*day)	Upper Klamath Lake I, rg = 2 - 31[4]

Source: 1.Chimiklis, P.E. and E.P. Karlander. 1973. Plant. Physiol. 51, 48-56.
2.Batterton, J.C. Jr and C. van Baalen. 1971. Arch. Mikrobiol. 76, 151-165.
3.Hall, K.J. and K.O. Hyatt. 1974. J. Fish. Res. bd Canada 31, 893-911.
4.Wright, R.T. 1975. Limnol. Oceanogr. 20(4), pp 626-633.

1-177 Bacteria
Max rate of glutamate uptake

Species	Value	Condition
Bacteria from Upper Klamath Lake	5 ug/(l*h)	Feb, 28 K, natural eutrophic, ice surface during winter, rg=4-6[1]

Source: 1.Burnison, B.K. and R.Y. Morita. 1974. Appl. Microbiol. 27, 488-495.

1-178 Bacteria
Max rate of glutamic uptake

Species	Value	Condition
Bacteria from antarctic seawater	0.0007 ug/(l*h)	Respiration = 4 %, 273 K[1]
Bacteria from antarctic seawater	0.0011 ug/(l*h)	273 K, in situ, 1975[2]
Bacteria from antarctic seawater	0.004 ug/(l*h)	Respiration, 273 K[1]
Bacteria from antarctic seawater	0.0092 ug/(l*h)	Rg = 0.05-1.8, respiration=21-36 % of prod, 273 K[1]
Bacteria from antarctic seawater	0.0445 ug/(l*h)	Rg = 0.2-8.7, respiration=51-77 % of prod, 273 K[1]
Bacteria from antarctic seawater	0.07E-2 ug/(l*h)	Respiration = 4 %, 273 K[1]
Bacteria from antarctic seawater	0.11E-2 ug/(l*h)	273 K, in situ, 1975[2]
Bacteria from antarctic seawater	0.4E-2 ug/(l*h)	Respiration, 273 K[1]
Bacteria from antarctic seawater	0.92E-2 ug/(l*h)	Rg = 0.05-1.8, respiration=21-36 % of prod, 273 K[1]
Bacteria from antarctic seawater	4.45E-2 ug/(l*h)	Rg = 0.2-8.7, respiration=51-77 % of prod, 273 K[1]
Bacteria from artic ice	0.0115 ug/(l*h)	Rg = 0.1-2.2, respiration=60-86 % of prod, 273 K[3]
Bacteria from artic ice	1.15E-2 ug/(l*h)	Rg = 0.1-2.2, respiration=60-86 % of prod, 273 K[3]
Bacteria from artic seawater	0.085 ug/(l*h)	Rg = 0.1-17, respiration=44-76 % of prod, 273 K[3]
Bacteria from artic seawater	8.5E-2 ug/(l*h)	Rg = 0.1-17, respiration=44-76 % of prod, 273 K[3]

Bacteria from artic sediment	0.86 ug/(l*h)	Rg = 2 - 170, respiration=32-65 % of prod, 273 K[3]
Bacteria from artic sediment	86E-2 ug/(l*h)	Rg = 2 - 170, respiration=32-65 % of prod, 273 K[3]

Source: 1.Morita, R.Y. 1975. Bacteriol. Rev. 39, 144-167.
2.Gillespie, P.A., R.Y. Morita and L.P. Jones. 1976. J. Oceanogr. Soc. Japan. 32, 811-818.
3.Morita, R.Y., R.P. Griffiths and S.S. Hayasaka. 1977. 3rd SCAR/IUBS Symp on Antarctic Biology. N.A.S.Heterotrophic potential of microorganisms in Antarctic waters.

1-179 Bacteria
Max rate of glycine uptake

Species	Value	Condition
Bacteria from Marion Lake	0.3 ug C/(g sediment*h)	Rg=0.1-0.5, winter, Canada, 276 K, no bac=5E4/(cm3 sediment)[1]
Bacteria from Marion Lake	2.5 ug C/(g sediment*h)	Rg=2-3, summer, Canada, 283 K, no bac=2E6/(cm3 sediment)[1]
Freshwater bacteria	7 microgram/(l*day)	Upper Klamath Lake I, range = 0.04 - 23[2]

Source: 1.Hall, K.J. and K.O. Hyatt. 1974. J. Fish. Res. bd Canada 31, 893-911.
2.Wright, R.T. 1975. Limnol. Oceanogr. 20(4), pp 626-633.

1-180 Bacteria
Max rate of glycolate uptake

Species	Value	Condition
Freshwater bacteria	10 microgram/(l*day)	Upper Klamath Lake I, rg = 1 - 37[1]

Source: 1.Wright, R.T. 1975. Limnol. Oceanogr. 20(4), pp 626-633.

1-181 Bacteria
Max rate of glycolic uptake

Species	Value	Condition
Bacteria from West Blue Lake, Manitoba	0.009 ug/(l*h)	No bac=9.0 /ml, ice surface, winter, phyt prod=18-67 mg C/(m2*d)(summer)[1]
Bacteria from West Blue Lake, Manitoba	9.0E-3 ug/(l*h)	No bac=9.0 /ml, ice surface, winter, phyt prod=18-67 mg C/(m2*d)(summer)[1]

Source: 1.Ward, F.J. and G.G.C. Robinson. 1974. J. Fish. Res. bd Canada 31, 977-1005.

1-182 Bacteria
Max rate of lactic uptake

Species	Value	Condition
Bacteria from West Blue Lake, Manitoba	0.12 ug/(l*h)	No bac=6.0 /ml, ice surface, winter, phyt prod=18-67 mg C/(m2*d)(summer)[1]
Bacteria from West Blue Lake, Manitoba	1.2E-1 ug/(l*h)	No bac=6.0 /ml, ice surface, winter, phyt prod=18-67 mg C/(m2*d)(summer)[1]

Source: 1.Ward, F.J. and G.G.C. Robinson. 1974. J. Fish. Res. bd Canada 31, 977-1005.

1-183 Bacteria
Max rate of lysine uptake

Species	Value	Condition
Bacteria from Upper Klamath Lake	2 ug/(l*h)	Feb, 28 K, natural eutrophic, ice surface during winter[1]

Source: 1.Burnison, B.K. and R.Y. Morita. 1974. Appl. Microbiol. 27, 488-495.

1-184 Bacteria
Max rate of malic uptake

Species	Value	Condition
Bacteria from West Blue Lake, Manitoba	0.036 ug/(l*h)	No bac=12.0/ml, ice surface, winter, phyt prod=18-67 mg C/(m2*d)(summer)[1]
Bacteria from West Blue Lake, Manitoba	3.6E-2 ug/(l*h)	No bac=12.0/ml, ice surface, winter, phyt prod=18-67 mg C/(m2*d)(summer)[1]

Source: 1.Ward, F.J. and G.G.C. Robinson. 1974. J. Fish. Res. bd Canada 31, 977-1005.

1-185 Bacteria
Max rate of photosynthesis

Species	Value	Condition
Chlorobium sp.	0.38 mg C/(mg chlorophyll a*hour)	283-304.5 K, photosynthetic capacity = 13.5 mg C/(m3*h)[1]
Chromatium sp.	0.26 mg C/(mg chlorophyll a*hour)	291 K, photosynthetic capacity = 5.6 mg C/(m3*h)[1]
Chromatium sp.	0.3 mg C/(mg chlorophyll a*hour)	291 K, photosynthetic capacity = 6.0 mg C/(m3*h)[1]
Chromatium sp.	2.14 mg C/(mg chlorophyll a*hour)	295-299 K, photosynthetic capacity = 61.7 mg C/(m3*h)[1]
Chromatium sp. and green sulfur bacteria	0.17 mg C/(mg chlorophyll a*hour)	301.9 K, photosynthetic capacity = 172 mg C/(m3*h)[1]
Purple and green sulfur bacteria	0.25 mg C/(mg chlorophyll a*hour)	301.5 K, photosynthetic capacity = 45.3 mg C/(m3*h)[1]

Source: 1.Maljkovic, D. and M. Branica, 1971. Limnol. Oceanogr. 16(5), pp 779-785.

1-186 Bacteria
Max rate of proline uptake

Species	Value	Condition
Bacteria from Upper Klamath Lake	5 ug/(l*h)	Feb, 28 K, natural eutrophic, ice surface during winter, more than value[1]

Source: 1.Burnison, B.K. and R.Y. Morita. 1974. Appl. Microbiol. 27, 488-495.

1-187 Bacteria
Max rate of pyruvic uptake

Species	Value	Condition
Bacteria from West Blue Lake, Manitoba	0.077 ug/(l*h)	No bac=21.7/ml, ice surface, winter, phyt prod=18-67 mg C/(m2*d)(summer)[1]
Bacteria from West Blue Lake, Manitoba	7.7E-2 ug/(l*h)	No bac=21.7/ml, ice surface, winter, phyt prod=18-67 mg C/(m2*d)(summer)[1]

Source: 1.Ward, F.J. and G.G.C. Robinson. 1974. J. Fish. Res. bd Canada 31, 977-1005.

1-188 Bacteria
Max rate of succinate uptake

Species	Value	Condition
Bacteria from West Blue Lake, Manitoba	0.028 ug/(l*h)	No bac=24.6/ml, ice surface, winter, phyt prod=18-67 mg C/(m2*d)(summer)[1]
Bacteria from West Blue Lake, Manitoba	2.8E-2 ug/(l*h)	No bac=24.6/ml, ice surface, winter, phyt prod=18-67 mg C/(m2*d)(summer)[1]

Source: 1.Ward, F.J. and G.G.C. Robinson. 1974. J. Fish. Res. bd Canada 31, 977-1005.

1-189 Bacteria
Max rate of uptake

Species	Value	Condition
Acinetobacter sp. on 2-hydroxycyclohexanone	26 nmol/min	80 uM NADPH and 200 uM O2, 1,2-mono-oxygenase from A. NCIB 9871[1]
Acinetobacter sp. on 2-hydroxycyclohexanone	26.0 nmol/min	80 uM NADPH and 200 uM O2, 1,2-mono-oxygenase from A. NCIB 9871[1]
Acinetobacter sp. on 2-methylcyclohexanone	39.8 nmol/min	80 uM NADPH and 200 uM O2, 1,2-mono-oxygenase from A. NCIB 9871[1]
Acinetobacter sp. on 4-hydroxycyclohexanone	29.7 nmol/min	80 uM NADPH and 200 uM O2, 1,2-mono-oxygenase from A. NCIB 9871[1]
Acinetobacter sp. on 4-methylcyclohexanone	38.9 nmol/min	80 uM NADPH and 200 uM O2, 1,2-mono-oxygenase from A. NCIB 9871[1]
Acinetobacter sp. on camphor	5.2 nmol/min	80 uM NADPH and 200 uM O2, 1,2-mono-oxygenase from A. NCIB 9871[1]
Acinetobacter sp. on cycloheptanone	24.9 nmol/min	80 uM NADPH and 200 uM O2, 1,2-mono-oxygenase from A. NCIB 9871[1]
Acinetobacter sp. on cyclohexan-1,2-dione	12.7 nmol/min	80 uM NADPH and 200 uM O2, 1,2-mono-oxygenase from A. NCIB 9871[1]
Acinetobacter sp. on cyclohexanone	42.9 nmol/min	80 uM NADPH and 200 uM O2, 1,2-mono-oxygenase from A. NCIB 9871[1]
Acinetobacter sp. on cyclooctanone	5.5 nmol/min	80 uM NADPH and 200 uM O2, 1,2-mono-oxygenase from A. NCIB 9871[1]
Acinetobacter sp. on cyclopentanone	27.4 nmol/min	80 uM NADPH and 200 uM O2, 1,2-mono-oxygenase from A. NCIB 9871[1]
Acinetobacter sp. on dihydrocarbone	10.3 nmol/min	80 uM NADPH and 200 uM O2, 1,2-mono-oxygenase from A. NCIB 9871[1]
Acinetobacter sp. on norcamphor	35.7 nmol/min	80 uM NADPH and 200 uM O2, 1,2-mono-oxygenase from A. NCIB 9871[1]

Source: 1.Yamamoto, S., M. Katagiri, H. Maeno and O. Hayaishi. 1965. J. Biol. Chem. 240, 3408.

1-190 Bacteria
Mg / biomass ratio

Species	Value	Condition
Escherichia coli	0.03 %	Atomic composition[1]

Source: 1.Morowitz, H.J., M.E. Tourtellotte, W.R. Guild, E. Castro and C. Woese. 1962. J. Mol. Boil., 4, p 93.

1-191 Bacteria
Mortality

Species	Value	Condition
Bacteria	0.038 1/hour	Steady state solution to Castle Lake.[1]
Coliform bacteria	0.0167 1/hour	Wastewater lagoon, 280.9 - 298.5 K, rg = 0.0083 - 0.029[2]
Coliform bacteria	0.017 1/hour	Leaf River, Miss[2]
Coliform bacteria	0.02 1/hour	Boise River, Idaho[3]
Coliform bacteria	0.02 1/hour	Lake Washington[3]
Coliform bacteria	0.02 1/hour	Missouri River, winter[2]
Coliform bacteria	0.02 1/hour	San Franscico bay estuary[3]
Coliform bacteria	0.021 1/hour	Groundwater stream, 283 K[2]

Coliform bacteria	0.024 1/hour	Upper Illinois River, Dec - Mar[2]
Coliform bacteria	0.026 1/hour	Lower Illinois River, Dec - Mar[2]
Coliform bacteria	0.029 1/hour	Lower Illinois River, Apr and Nov.[2]
Coliform bacteria	0.037 1/hour	Lower Illinois River, Oct and May[2]
Coliform bacteria	0.042 1/hour	North Fork Kings River, Calif[3]
Coliform bacteria	0.043 1/hour	Tennessee River, Knoxville[2]
Coliform bacteria	0.043 1/hour	Tennessee River, Knoxville, summer[2]
Coliform bacteria	0.043 1/hour	Upper Illinois River, Apr and Nov.[2]
Coliform bacteria	0.045 1/hour	Ohio River, winter, 278 K[2]
Coliform bacteria	0.049 1/hour	Ohio River, summer, 293 K[2]
Coliform bacteria	0.053 1/hour	Lake Ontario, rg = 0.02 - 0.083[3]
Coliform bacteria	0.055 1/hour	Tennessee River, Chattanooga[2]
Coliform bacteria	0.055 1/hour	Tennessee River, Chattanooga, summer[2]
Coliform bacteria	0.07 1/hour	Maturation ponds[2]
Coliform bacteria	0.072 1/hour	Sacramento River, summer[2]
Coliform bacteria	0.0732 1/hour	All streams, rg = 0.0004 - 0.146[3]
Coliform bacteria	0.083 1/hour	Maturation ponds, 280.9 - 298.5 K[2]
Coliform bacteria	0.0835 1/hour	All streams, rg = 0.042 - 0.125[3]
Coliform bacteria	0.085 1/hour	Lower Illinois River, June - Sept[2]
Coliform bacteria	0.085 1/hour	Upper Illinois River, June - Sept[2]
Coliform bacteria	0.105 1/hour	Upper Illinois River, Oct and May[2]
Coliform bacteria	0.136 cm**2/cal	Ligth induced death rate, n=61, first order reaction, lab data[3]
Coliform bacteria	0.168 cm**2/cal	Ligth induced death rate, n=24, first order reaction, field data[3]
Coliform bacteria	0.177 1/hour	Long Island estuary, New York, rg = 0.02 - 0.333[3]
Coliform bacteria	0.23 1/hour	Cumberland River, summer[2]
Coliform bacteria	0.481 cm**2/cal	Ligth induced death rate, n=14, first order reaction, field data[3]
Coliform bacteria	0.63 1/hour	Shallow turbulent stream[2]
Coliform bacteria	1.1 1/hour	Glatt River[2]
Escherichia coli	0.354 cm**2/cal	Ligth induced death rate, n= 4, first order reaction, lab data[3]
Escherichia coli	0.362 cm**2/cal	Ligth induced death rate, n= 4, first order reaction, field data[3]
Fecal coliform	0.0076 1/hour	Average for stormwater runoff, 293 K, 8th-14th day[4]
Fecal coliform	0.0105 1/hour	Average for stormwater runoff, 283 K, 14 d[4]
Fecal coliform	0.0134 1/hour	Average for stormwater runoff, 293 K, 4th-7th day[4]
Fecal coliform	0.0259 1/hour	Average for stormwater runoff, 293 K, 3rd day[4]
Fecal coliform	0.0413 1/hour	Average for stormwater runoff, 293 K, 2nd day[4]
Fecal coliform	0.0419 1/hour	Many small secondary treatment effluents, sugar beet, June[4]
Fecal coliform	0.0481 1/hour	Stormwater runoff, 24 h after initial die-away[4]

Fecal coliform	0.0542 1/hour	Many small secondary treatment effluents, sugar beet, Aug[4]
Fecal coliform	0.0577 1/hour	Stormwater runoff, initial die-away[4]
Fecal coliform	0.0654 1/hour	Average for stormwater runoff, 293 K, first day[4]
Fecal coliform	0.0668 1/hour	Industrial and primary treatment discharge, June[4]
Fecal coliform	0.0962 1/hour	Stormwater runoff, 48 h after initial die-away[4]
Fecal coliform	0.13 1/hour	Unchlorinated primary domestic effluent, June[4]
Fecal streptococci	0 cm**2/cal	Ligth induced death rate, n= 3, first order reaction, field data[3]
Fecal streptococci	0.0019 1/hour	Average stormwater runoff, 283 K, 14 d[4]
Fecal streptococci	0.0044 1/hour	Average stormwater runoff, 293 K, 4th-14th d[4]
Fecal streptococci	0.0065 1/hour	Average stormwater runoff, 293 K, 3rd day[4]
Fecal streptococci	0.007 cm**2/cal	Ligth induced death rate, n= 1, first order reaction, field data[3]
Fecal streptococci	0.0125 1/hour	Average stormwater runoff, 293 K, 2 d[4]
Fecal streptococci	0.0625 1/hour	Stormwater runoff, initial die-away[4]
Fecal streptococci	0.086 cm**2/cal	Ligth induced death rate, n=3 , first order reaction, lab data, rg=[3]
Fecal streptococci	0.091 cm**2/cal	Ligth induced death rate, n=12, first order reaction, field, initial[3]
Salmonella typhimurium	1.48 cm**2/cal	Ligth induced death rate, n= 2, first order reaction, lab data[3]
Serratia marcescens	0.002 cm**2/cal	Ligth induced death rate, n=1 , first order reaction, lab data[3]
Total coliform	0.0162 1/hour	Clean rivers and lakes, rg = 0.0288 - 0.0096[4]
Total coliform	0.0182 1/hour	Industrial and primary treatment discharge, Oct[4]
Total coliform	0.0211 1/hour	Stormwater runoff, initial die-away[4]
Total coliform	0.0223 1/hour	Industrial and primary treatment discharge, June[4]
Total coliform	0.0264 1/hour	Polluted rivers and lakes, rg = 0.0384 - 0.0144[4]
Total coliform	0.0293 1/hour	Many small secondary treatment effluents, sugar beet, initial Oct[4]
Total coliform	0.0374 1/hour	Illinois River, 31-50 h after peak, summer[4]
Total coliform	0.0384 1/hour	Stormwater runoff, initial regrowth[4]
Total coliform	0.04 1/hour	Unchlorinated primary domestic effluent, Oct[4]
Total coliform	0.0425 1/hour	Many small secondary treatment effluents, sugar beet, June[4]
Total coliform	0.051 1/hour	Many small secondary treatment effluents, sugar beet, Aug[4]
Total coliform	0.0761 1/hour	Illinois River, 21-30 h after peak, summer[4]
Total coliform	0.0889 1/hour	Illinois River, first 20 h after peak, summer[4]
Total coliform	0.1053 1/hour	Unchlorinated primary domestic effluent, June[4]

Source: 1.Richey, J.E. 1977. in: Hall, C.A.S. and J.W. Day, Jr. (eds.); Ecosystem Modeling in theory and practice. John Wiley and Sons. New York.
2.Mitchell, R. and C. Chamberlin. 1978. in: Berg. G. (ed.); Indicators of Enteric Contamination in Natural Waters.
3.Tetra Tech, Inc. 1977. Ecological constants. San F. Calif.
4.Lombardo, P.S. 1972. Hydrocomp, Inc. Mathematical Model of Water Quality in Rivers and Impoundsments

1-192 Bacteria
N / biomass ratio

Species	Value	Condition
Bacteria	10.7 % of dry weight	Mean elementary composition[1]
Decomposer	8 % of dry weight	Mean elementary composition[1]
Escherichia coli	3.04 %	Atomic composition[2]

Source: 1.Bowen, H.J.M. 1966. Trace Elements in Biochemistry, Ac. Press.
2.Morowitz, H.J., M.E. Tourtellotte, W.R. Guild, E. Castro and C. Woese. 1962. J. Mol. Boil., 4, p 93.

1-193 Bacteria
Na / biomass ratio

Species	Value	Condition
Escherichia coli	0.06 %	Atomic composition[1]

Source: 1.Morowitz, H.J., M.E. Tourtellotte, W.R. Guild, E. Castro and C. Woese. 1962. J. Mol. Boil., 4, p 93.

1-194 Bacteria
Nitrogen fixation

Species	Value	Condition
Azotobacter sp.	1.6E-11 mg N/(cell*day)	Chernoye, USSR, 1937[1]
Heterotrophic bacteria	0.026 microgram N/(l*hour)	Mary, Wisc, 1968, rg = 0.003 - 0.47[1]
Heterotrophic bacteria	0.042 microgram N/(l*hour)	Mize, FL, August 1968, rg = 0.000 - 0.083[1]
Heterotrophic bacteria	0.196 microgram N/(l*hour)	Mize, FL, July 1968, rg = 0.083 - 0.308[1]

Source: 1.Wetzel, R.G. 1975. Limnology. Saunders. Philadelphia.

1-195 Bacteria
Numbers

Species	Value	Condition
Bacteria	1000000 per ml	Total bac., saproph.=650/ml, prim prod=35 mg C/(m2*d), Gross. Ploener[1]
Bacteria	1100000 per ml	Total bac., saproph.=1000/ml, prim prod=57 mg C/(m2*d), Pluss See[1]
Bacteria	500000 per ml	Total bac., saproph.=300/ml, prim prod=16 mg C/(m2*d), Schoehsee[1]
Bacteria	500000 per ml	Total bac., saproph.=350/ml, prim prod=17 mg C/(m2*d), Schluensee[1]

Source: 1.Overbeck, J. 1965. Naturwissenschaften 51, pp 145.

1-196 Bacteria
O / biomass ratio

Species	Value	Condition
Bacteria	27 % of dry weight	Mean elementary composition[1]
Decomposer	30 % of dry weight	Mean elementary composition[1]
Escherichia coli	73.68 %	Atomic composition[2]

Source: 1.Bowen, H.J.M. 1966. Trace Elements in Biochemistry, Ac. Press.
2.Morowitz, H.J., M.E. Tourtellotte, W.R. Guild, E. Castro and C. Woese. 1962. J. Mol. Boil., 4, p 93.

1-197 Bacteria
Oxidation rate

Species	Value	Condition
Nitrosomonas sp.	0.286 mg N/((mg dry cells)*hour)	279.5 K, March, River Tame[1]
Nitrosomonas sp.	0.292 mg N/((mg dry cells)*hour)	280 K, Feb, River Trent[1]
Nitrosomonas sp.	0.292 mg N/((mg dry cells)*hour)	Calc.from val.at 293 K, 280 K[1]
Nitrosomonas sp.	0.299 mg N/((mg dry cells)*hour)	280.3 K, March, River Trent[1]
Nitrosomonas sp.	0.325 mg N/((mg dry cells)*hour)	281 K, Feb, River Tame[1]
Nitrosomonas sp.	0.325 mg N/((mg dry cells)*hour)	281 K, November, River Tame[1]
Nitrosomonas sp.	0.338 mg N/((mg dry cells)*hour)	281.5 K, December, River Tame[1]
Nitrosomonas sp.	0.364 mg N/((mg dry cells)*hour)	282.3 K, November, River Trent[1]
Nitrosomonas sp.	0.37 mg N/((mg dry cells)*hour)	282.5 K, December, River Trent[1]
Nitrosomonas sp.	0.39 mg N/((mg dry cells)*hour)	283 K, April, River Tame[1]
Nitrosomonas sp.	0.39 mg N/((mg dry cells)*hour)	283 K, April, River Trent[1]
Nitrosomonas sp.	0.61 mg N/((mg dry cells)*hour)	287.8 K, June, River Tame[1]
Nitrosomonas sp.	0.662 mg N/((mg dry cells)*hour)	288.5 K, June, River Trent[1]
Nitrosomonas sp.	0.67 mg N/((mg dry cells)*hour)	Cell dry wt.=E-12 g NH4-N conc.=10mg/l, 293 K, River Tame[1]
Nitrosomonas sp.	0.753 mg N/((mg dry cells)*hour)	290 K, September, River Trent[1]
Nitrosomonas sp.	0.831 mg N/((mg dry cells)*hour)	291 K, May, River Trent[1]
Nitrosomonas sp.	0.89 mg N/((mg dry cells)*hour)	Cell dry wt.=E-12 g NH4-N conc.=100mg/l, 293 K, River Tame[1]
Nitrosomonas sp.	0.909 mg N/((mg dry cells)*hour)	292 K, August, River Tame[1]
Nitrosomonas sp.	0.961 mg N/((mg dry cells)*hour)	292.5 K, August, River Trent[1]
Nitrosomonas sp.	0.98 mg N/((mg dry cells)*hour)	293 K[1]
Nitrosomonas sp.	1.08 mg N/((mg dry cells)*hour)	Growth rate=1, yield=0.04, 293 K, according to volumetric reaction[1]
Nitrosomonas sp.	1.104 mg N/((m d*c*h)	294 K, July, River Tame[1]
Nitrosomonas sp.	1.16 mg N/((mg d*c*h)	Cell dry wt.=E-12 g NH4-N conc.=10mg/l, 293 K, River Trent[1]
Nitrosomonas sp.	1.2 mg N/((mg d*c*h)	Cell dry wt.=E-12 g NH4-N conc.=10mg/l, 293 K, River Trent[1]
Nitrosomonas sp.	1.273 mg N/((mg dry cells)*hour)	295.5 K, July, River Trent[1]

Source: 1.Curtis, E.J.C., K. Durrant and M.M.I. Harman. 1975. Water Res., 9(3), pp 255-268.

1-198 Bacteria
P / biomass ratio

Species	Value	Condition
Escherichia coli	0.6 %	Atomic composition[1]

Source: 1.Morowitz, H.J., M.E. Tourtellotte, W.R. Guild, E. Castro and C. Woese. 1962. J. Mol. Boil., 4, p 93.

1-199 Bacteria
Production

Species	Value	Condition
Bacteria	2.6 mg C l-1 3d-1	growth in experimental flasks on macrophyte leachate[1]
Natural population	0.52-3.60 C umol l-1 d-1	Hawaiian coastal waters, from DNA-synthesis[2]

Source: 1.Findlay, S., Carlough, L., Crocker, M.T., Gill, H.K., Meyer, J.L. and Smith, P.J. Limnol. Oceanogr.1(6), 1335, 1986.
2.Laws, E.A., et al., Limnol.Oceanogr., 29(6), 1161, 1984.

1-200 Bacteria
Production / biomass ratio

Species	Value	Condition
Bacteria	0.45 per day	Per day, southern lakes of USSR, rg = 0.3 - 0.6[1]

Source: 1.Greze, V.N. 1973. Okeanologiya, 13(1), pp 107-113.

1-201 Bacteria
Protein / biomass ratio

Species	Value	Condition
Mycoplasma gallisepticum	80 % of dry weight	None[1]

Source: 1.Morowitz, H.J., M.E. Tourtellotte, W.R. Guild, E. Castro and C. Woese. 1962. J. Mol. Boil., 4, p 93.

1-202 Bacteria
Rate of P uptake

Species	Value	Condition
Bacteria	0.05 1/hour	Steady state solution to Castle Lake.[1]

Source: 1.Richey, J.E. 1977. in: Hall, C.A.S. and J.W. Day, Jr. (eds.); Ecosystem Modeling in theory and practice. John Wiley and Sons. New York.

1-203 Bacteria
Rate of alkane oxidation

Species	Value	Condition
Pseudomonas oleovorans with 1,6-hexanediol	2.3 units of CO2 production	All rates are relative to control and shows possible degradation[1]
Pseudomonas oleovorans with 1,7-octadiene	18.1 units of CO2 production	All rates are relative to control and shows possible degradation[1]
Pseudomonas oleovorans with 1-butanol	0.7 units of CO2 production	All rates are relative to control and shows possible degradation[1]
Pseudomonas oleovorans with 1-dodecanol	0.4 units of CO2 production	All rates are relative to control and shows possible degradation[1]
Pseudomonas oleovorans with 1-hexanol	21 units of CO2 production	All rates are relative to control and shows possible degradation[1]

Pseudomonas oleovorans with 1-hexanol	21.0 units of CO2 production	All rates are relative to control and shows possible degradation[1]
Pseudomonas oleovorans with 1-pentanol	12.2 units of CO2 production	All rates are relative to control and shows possible degradation[1]
Pseudomonas oleovorans with 1-propanol	1.1 units of CO2 production	All rates are relative to control and shows possible degradation[1]
Pseudomonas oleovorans with 2-hexanol	2.3 units of CO2 production	All rates are relative to control and shows possible degradation[1]
Pseudomonas oleovorans with 2-methylhexane	7 units of CO2 production	All rates are relative to control and shows possible degradation[1]
Pseudomonas oleovorans with 2-methylhexane	7.0 units of CO2 production	All rates are relative to control and shows possible degradation[1]
Pseudomonas oleovorans with 3-hexanol	21 units of CO2 production	All rates are relative to control and shows possible degradation[1]
Pseudomonas oleovorans with 3-hexanol	21.0 units of CO2 production	All rates are relative to control and shows possible degradation[1]
Pseudomonas oleovorans with dicyclopropyl ketone	23.3 units of CO2 production	All rates are relative to control and shows possible degradation[1]
Pseudomonas oleovorans with dicyclopropyl methanol	18.2 units of CO2 production	All rates are relative to control and shows possible degradation[1]
Pseudomonas oleovorans with n-heptane	24.5 units of CO2 production	All rates are relative to control and shows possible degradation[1]
Pseudomonas oleovorans with n-undecane	0.6 units of CO2 production	All rates are relative to control and shows possible degradation[1]
Pseudomonas oleovorans with no added compound	1 units of CO2 production	All rates are relative to control and shows possible degradation[1]
Pseudomonas oleovorans with no added compound	1.0 units of CO2 production	All rates are relative to control and shows possible degradation[1]

Source: 1.Grund, J. Shapiro, Fennewald, Bacha, Leahy, Markbreiter, Neider. 1975. also M. Toepfer. J. Bact. 123546-556.

1-204 Bacteria
Rate of production of tetramethyl - Pb

Species	Value	Condition
Bacteria from lake sediments	125 ug tetramethyl Pb/week	First week, Pb added as Me3PbOAc, 10 mg/50 g sediment[1]
Bacteria from lake sediments	256 ug tetramethyl Pb/week	Fourth week, Pb added as Me3PbOAc, 10 mg/50 g sediment[1]
Bacteria from lake sediments	555 ug tetramethyl Pb/week	Third week, Pb added as Me3PbOAc, 10 mg/50 g sediment[1]
Bacteria from lake sediments	642 ug tetramethyl Pb/week	Second week, Pb added as Me3PbOAc, 10 mg/50 g sediment[1]

Source: 1.Wong, P.T.S., Y.K. Chau and P.L. Luxon. 1975. Nature 253(5489), pp 263-264.

1-205 Bacteria
Respiration rate

Species	Value	Condition
Azotobacter chroococcum	750 ml O2/((g wet weight)*hour)	301 K, rg = 500 - 1000[1]
Bacillus fluorescens non liquefaciens	4100 ml O2/((g wet weight)*hour)	Temp not stated[1]
Bacillus mesentericus vulgatus	12.1 ml O2/((g wet weight)*hour)	289 K[1]
Bacteria	7E-9 microlitre O2/(hour*individual)	Grassland soil, 293 K, equivalent to 575 gcal/g*day at 289 K[2]

Source: 1.Giese, A.C. 1962. Cell Physiology. 2nd edition. Saunders. Philadelphia.
2.MacFadyen, A. 1963. Animal Ecology, Aims and Methods. 2nd edition. Pitman. London.

1-206 Bacteria
S / biomass ratio

Species	Value	Condition
Bacteria	0.61 % of dry weight	Mean elementary composition[1]
Decomposer	0.5 % of dry weight	Mean elementary composition[1]
Escherichia coli	0.32 %	Atomic composition[2]

Source: 1.Bowen, H.J.M. 1966. Trace Elements in Biochemistry, Ac. Press.
2.Morowitz, H.J., M.E. Tourtellotte, W.R. Guild, E. Castro and C. Woese. 1962. J. Mol. Boil., 4, p 93.

1-207 Bacteria
Temperature coefficient

Species	Value	Condition
Bacteria	1.02	Ammonia decay, temperature factor is 1 at 293 K[1]
Bacteria	1.02	Nitrite decay, temperature factor is 1 at 293 K[1]
Bacteria	1.04	Coliform dieoff, temperature factor is 1 at 293 K[1]
Bacteria	1.04	Detritus decay, temperature factor is 1 at 293 K[1]
Bacteria	1.047	BOD decay, temperature factor is 1 at 293 K[1]

Source: 1.The Hydrological Engineering Center Corps of Engineers U.S. Army. 1975. Water Quality for river-resevoir systems, Davis, Cal. 9 5 6 1 6 (916) 499-2329.

1-208 Bacteria
Temperature for max growth

Species	Value	Condition
Clostridium sp.	281.2 K	Origin = coast and Puget sound, sp 54, marine psychrophillic bac[1]
Clostridium sp.	281.3 K	Origin = coast and Puget sound, sp 19, marine psychrophillic bac[1]
Clostridium sp.	282.4	Origin = coast and Puget sound, sp 40, marine psychrophillic bac[1]
Clostridium sp.	282.4 Origin = coast and Puget sound, sp	Origin = coast and Puget sound, sp 40, marine psychrophillic bac[1]
Clostridium sp.	282.8 K	Origin = coast and Puget sound, sp 5, marine psychrophillic bac[1]
Clostridium sp.	283 K	Origin = coast and Puget sound, sp 41, marine psychrophillic bac[1]
Clostridium sp.	283.0 K	Origin = coast and Puget sound, sp 41, marine psychrophillic bac[1]
Clostridium sp.	283.7 K	Origin = coast and puget sound, sp 2, marine psychrophillic bac[2]
Cytophaga psychrophila	290.5 K	Origin = silver salmon, rg=288-293 K, marine psychrophillic bac[3]

Marine bacteria	279.5 K	Origin = water, Antarctic, E5-4-4, rg=279-280 K, marine psychrophillic bac[4]
Vibrio marinus	288 K	Origin = water off Oregon coast, marine psychrophillic bac[5]
Vibrio sp.	277 K	Origin = water, Antarctic, Vibrio AP-2-24, marine psychrophillic bac[6]
Vibrio sp.	280 K	Origin = water, Antarctic, Vibrio 18-300, marine psychrophillic bac[7]
Vibrio sp.	283 K	Origin = water, Antarctic, Vibrio 18-500, marine psychrophillic bac[7]

Source: 1.Finne, W.H. and J.R. Matches. 1974. Can. J. Microbiol. 20, 1639-1645.
2.Matches, J.R. and J. Liston. 1973. In: L.H. Stevenson and R.R. Colwell: Estuarine Microbial Ecology.Univ. S Carolina Press. Columbia, 345-361.
3.Borg, A.F. 1960. Wildlife Dis. 8, 1-85.
4.Christian, R.R. and W.J. Wiebe. 1974. Can. J. Microbiol. 20, 1341-1345.
5.Morita, R.Y. and R.D. Haight. 1964. Limnol. Oceanogr. 9, 102-106.
6.Morita, R.Y. 1975. Bacteriol. Rev. 39, 144-167.
7.Baross, J.A., F.J. Hanus and R.Y. Morita. 1974. In: R.R. Colwell and R.Y. Morita (eds.): Effect of the Ocean Environment on Microbial Activity. Univ. Park Press. Baltimore. 180-202.

1-209 Bacteria
Temperature for max growth rate

Species	Value	Condition
Actinobifida chromogena	56.5 degrees celcius	Rg = 55-58 deg cel., Actinomycetes, non-photosyntic bac.[1]
Actinobifida dichotomica	54 degrees celcius	Rg = 50-58 deg cel., Ac, non-photosyntic bac.[1]
Bacillus acidocaldarius	62.5 degrees celcius	Acidophile, spore former, non-photosyntic bac., rg = 60-65 deg cel.[2]
Bacillus coagulans	41.5 degrees celcius	Acidophile, spore former, non-photosyntic bac., rg=37-45 deg cel.[3]
Bacillus sp.	57.5 degrees celcius	Hydrocarbon oxidizer, spore former, non-photosyntic bac., rg=55-60[4]
Bacillus sp.	72 degrees celcius	Proteolytic, B YT-P, spore former, non-photosyntic bac.[5]
Bacillus sp.	80 degrees celcius	Uncertain, B. YT-G, spore former, non-photosyntic bac.[6]
Bacillus stearothermophilus	57.5 degrees celcius	Versatile, common, spore former, non-photosyntic bac., rg=50-65 deg cel.[7]
Bifidobacterium thermophilum	46.5 degrees celcius	Lactic acid bac., non-photosyntic bac.[8]
Chloroflexus aurantiacus	55 degrees celcius	Chlorobiaceae, photosynthetic bac[9]
Chloroflexus aurantiacus	55 degrees celcius	Chlorobiaceae, photosynthetic bacteria[9]
Clostridium sp.	283.4 K	Origin = sediment off Washington, Pacific, marine psychrophillic bac[10]
Clostridium sp.	60 degrees celcius	Cellulose digester, spore former, non-photosyntic bac.[11]
Clostridium sp.	62.5 degrees celcius	Rg = 50-75 deg cel., butyric formation, spore former, non-photosyntic bac.[11]
Clostridium spp.	40 degrees celcius	Rg = 35-45 deg cel., many species, spore former, non-photosyntic bac.[1]
Clostridium thermoaceticum	57.5 degrees celcius	SF, non-photosyntic bac., rg=55-60 deg cel.[13]
Clostridium thermocellulaseum	57.5 degrees celcius	Cellulose digester, spore former, non-photosyntic bac., rg=55-60[13]
Clostridium thermocellum	60 degrees celcius	Cellulose digester, spore former, non-photosyntic bac.[13]
Clostridium thermosaccharolyticum	55 degrees celcius	SF, non-photosyntic bac.[14]
Desulfotomaculum nigrificans	55 degrees celcius	Sulphate reducer, spore former, non-photosyntic bac.[15]
Desulfovibrio thermophilus	65 degrees celcius	Sulfate reducing bac., non-photosyntic bac.[16]

Hydrogenomonas thermophilus	50 degrees celcius	Pseudomonad, non-photosyntic bac.[17]
Lactobacillus bulgaricus	40 degrees celcius	Lactic acid bac., non-photosyntic bac.[18]
Lactobacillus sp.	35 degrees celcius	Lactic acid bac., non-photosyntic bac., rg=30-40 deg cel.[8]
Lactobacillus thermophilus	56.5 degrees celcius	Lactic acid bac., non-photosyntic bac., rg=50-63 deg cel[19]
Methanobacterium thermoautotrophicum	67.5 degrees celcius	Methane producing bac., non-photosyntic bac., rg=65-70 deg cel.[20]
Methylococcus capsulatus	40 degrees celcius	Methane oidozing, non-photosyntic bac., rg=30-50 deg cel.[21]
Micropolyspora caesia	36.5 degrees celcius	Actinomycetes, non-photosyntic bac., rg=28-45 deg cel.[1]
Micropolyspora faeni	50 degrees celcius	Actinomycetes, non-photosyntic bac.[1]
Micropolyspora rectivirgula	50 degrees celcius	Actinomycetes, non-photosyntic bac., rg=45-55 deg cel.[1]
Micropolyspora rubrobrunea	50 degrees celcius	Actinomycetes, non-photosyntic bac., rg=45-55 deg cel.[1]
Micropolyspora thermovirida	45 degrees celcius	Actinomycetes, non-photosyntic bac., rg=40-50 deg cel.[1]
Micropolyspora virdinigra	50 degrees celcius	Actinomycetes, non-photosyntic bac., rg=45-55 deg cel.[1]
Pseudonocardia thermophila	45 degrees celcius	Actinomycetes, non-photosyntic bac., rg=40-50 deg cel.[22]
Streptococcus thermophilus	42.5 degrees celcius	Lactic acid bac., non-photosyntic bac., rg=40-45 deg cel.[23]
Streptomyces fragmentosporus	55 degrees celcius	Rg = 50-60 deg cel., Actinomycetes, non-photosyntic bac.[24]
Streptomyces thermonitrificans	47.5 degrees celcius	Rg = 45-50 deg cel., Actinomycetes, non-photosyntic bac.[25]
Streptomyces thermoviolaceus	50 degrees celcius	Actinomycetes, non-photosyntic bac.[25]
Streptosporangium album	52.5 degrees celcius	S. a. var thermophilum, Actinoplanes, non-photosyntic bac., rg=50-55[26]
Sulfolobus acidocaldarius	72.5 degrees celcius	Acidophile, Sulfur oxidizing, non-photosyntic bac., rg=70-75 deg cel.[27]
Thermoactinomyces sacchari	57.5 degrees celcius	Actinomycetes, non-photosyntic bac., rg=55-60 deg cel.[28]
Thermoactinomyces vulgaris	60 degrees celcius	Actinomycetes, non-photosyntic bac.[28]
Thermomicrobium roseum	72.5 degrees celcius	Gram negative, non-photosyntic bac., rg=70-75 deg cel.[29]
Thermomonospora citrina	57.5 degrees celcius	Actinomycetes, non-photosyntic bac., rg=55-60 deg cel.[26]
Thermomonospora curvata	50 degrees celcius	Actinomycetes, non-photosyntic bac.[26]
Thermomonospora viridis	50 degrees celcius	Actinomycetes, non-photosyntic bac.[28]
Thermoplasma acidophilum	59 degrees celcius	Acidophile, Mycoplasma, non-photosyntic bac.[30]
Thermus aquaticus	70 degrees celcius	Gram negative, non-photosyntic bac.[31]
Thermus flavus	72.5 degrees celcius	Gram negative, non-photosyntic bac.[32]
Thermus ruber	60 degrees celcius	Gram negative, non-photosyntic bac.[33]
Thermus sp.	70 degrees celcius	Rg = 69-71 deg cel., Thermus X-1, gram negative, non-photosyntic bac.[33]
Thermus thermophilus	70 degrees celcius	Gram negative, non-photosyntic bac.[34]
Thiobacillus sp.	50 degrees celcius	Rg = 55-60 del cel., Sulfur oxidizing, non-photosyntic bac.[35]
Thiobacillus thermophilica	57.5 degrees celcius	Spore former, Sulfur oxidizing, non-photosyntic bac., rg=55-60 deg cel.[36]

Source: 1.Cross, T. 1974. In: R.E. Buchanan and N.E. Gibbons (eds.): Bergeys Manual of Determinative Bacteriology. Williams and Williams. Baltimore, 861-863.
2.Darland, W.H. and T.D. Brock. 1971. J. Gen. Microbiol. 67, 9-15.
3.Belly, R.T. and T.D. Brock. 1974. J. Appl. Bact. 37, 175-177.
4.Mateless, R.I., J.N. Baruah and S.R. Tannenbaum. 1967. Science 157, 1322-1323.
5.Heinen, U.J. and W. Heinen. 1972. Arch. Mikrobiol. 82, 1-23.
6.Heinen, W. 1971. Arch. Mikrobiol. 76, 2-17.
7.Gibson, T. and R.E. Gordon. 1974. In: R.E. Buchanan and N.E. Gibbons (eds.): Bergeys Manual of Determinative Bacteriology. Williams and Williams. Baltimore, 529-550.

8.Rogosa, M. 1974. In: R.E. Buchanan and N.E. Gibbons (eds.): Bergeys Manual of Determinative Bacteriology. Williams and Williams. Baltimore, 669-676.
9.Bauld, J. and T.D. Brock. 1973. Arch. Microbiol. 92, 267-284.
10.Matches, J.R. and J. Liston. 1973. In: L.H. Stevenson and R.R. Colwell: Estuarine Microbial Ecology.Univ. S Carolina Press. Columbia, 345-361.
11.Loginova, L.G., A.E. Kosmachev, R.S. Golovacheva and Seregina. 1962. Mikrobiologiya 31, 1082-1086.
12.Smith, L.D.S. and G. Hobbs. 1974. In: R.E. Buchanan and N.E. Gibbons (eds.): Bergeys Manual of Determinative Bacteriology. Williams and Williams. Baltimore, 551-572.
13.Breed, R.S., E.G.D. Murray and N.R. Smith. 1974. In: R.E. Buchanan and N.E. Gibbons (eds.): Bergeys Manual of Determinative Bacteriology. Williams and Williams. Baltimore.
14.Hollaus, F. and U. Steytr. 1972. Arch. Mikrobiol. 86, 129-146.
15.Campbell, L.L. 1974. In: R.E. Buchanan and N.E. Gibbons (eds.): Bergeys Manual of Determinative Bacteriology. Williams and Williams. Baltimore, 572-573.
16.Rozanova, E.P. and A.I. Khudyakova. 1974. Mikrobiologiya 43, 1069-1075.
17.McGee, J.M., L.R. Bowen and R.G. Tischer. 1967. Nature 214, 715-716.
18.Orla-Jensen, S. 1942. The Lactic Acid Bacteria. 2nd ed. Copenhagen. 197 pp.
19.Gaughran, E.R.L. 1947. Bact. Rev. 11, 189-225.
20.Zeikus, J.G. and R.S. Wolfe. 1972. J. Bact. 109, 707-713.
21.Leadbetter, E.R. 1974. In: R.E. Buchanan and N.E. Gibbons (eds.): Bergeys Manual of Determinative Bacteriology. Williams and Williams. Baltimore, 267-269.
22.Henssen, A. 1974. In: R.E. Buchanan and N.E. Gibbons (eds.): Bergeys Manual of Determinative Bacteriology. Williams and Williams. Baltimore, 746-747.
23.Deibel, R.H. and H.W. Seeley, Jr. 1974. In: R.E. Buchanan and N.E. Gibbons (eds.): Bergeys Manual of Determinative Bacteriology. Williams and Williams. Baltimore, 490-509.
24.Henssen, A. 1969. Arch. Mikrobiol. 67, 21-27.
25.Pridham, T.G. and H.D. Tresner. 1974. In: R.E. Buchanan and N.E. Gibbons (eds.): Bergeys Manual of Determinative Bacteriology. Williams and Williams. Baltimore, 748-829.
26.Manachini, P., A. Craveri and R. Craveri. 1966. Annal. Microbiol. 16, 83-90.
27.Brock, T.D., K.M. Brock, R.T. Belly and R.L. Weiss. 1972. Arch. Mikrobiol. 84, 54-68.
28.Kuster, E. 1974. In: R.E. Buchanan and N.E. Gibbons (eds.): Bergeys Manual of Determinative Bacteriology. Williams and Williams. Baltimore, 858-859.
29.Jackson, T.J., R.F. Ramaley and W.G. Meinschein. 1973. Int. J. Syst. Bacteriol. 23, 28-36.
30.Belly, R.T., M.R. Tansey and T.D. Brock. 1973. J. Phycol. 9, 123-127.
31.Brock, T.D. and H. Freeze. 1969. J. Bact. 98, 289-297.
32.Oshima, T. and K. Imahori. 1974. Int. J. Syst. Bactiol. 24, 102-112.
33.Ramaley, R.F. and J. Hixson. 1970. J. Bact. 103, 527-528.
34.Loginova, L.G. and L.A. Egorova. 1975. Mikrobiologiya 44, 661-665.
35.Williams, R.A.D. and D.S. Hoare. 1972. J. Gen. Microbiol. 70, 555-566.
36.Egorova, A.A. and Z. Derygina. 1963. Mikrobiologiya 32, 437-446.

1-210 Bacteria
Thymidine incorporation

Species	Value	Condition
Freshwater bacteria	5-25 10-8 nmole ml-1 min-1	Lake Michigan[1]
Marine Bacteria	0.002 pmol l-1 h-1	on marine snow, Atlantic[2]
Marine Bacteria	0.0036 pmol l-1 h-1	on marine snow, Atlantic[2]
Marine Bacteria	0.004 pmol l-1 h-1	on marine snow, Atlantic[2]
Marine Bacteria	0.0073 pmol l-1 h-1	on marine snow, Atlantic[2]
Marine Bacteria	0.0099 pmol l-1 h-1	on marine snow, Atlantic[2]
Marine Bacteria	0.017 pmol l-1 h-1	on marine snow, California[2]
Marine Bacteria	0.026 pmol l-1 h-1	on marine snow, Atlantic[2]
Marine Bacteria	0.029 pmol l-1 h-1	on marine snow, California[2]
Marine Bacteria	0.188 pmol l-1 h-1	on marine snow, California[2]
Marine Bacteria	0.27 10e-21 mol cell-1 h-1	on marine snow, Atlantic[2]
Marine Bacteria	0.31 10e-21 mol cell-1 h-1	on marine snow, Atlantic[2]
Marine Bacteria	0.625 pmol l-1 h-1	in seawater, California[2]
Marine Bacteria	1.12 10e-21 mol cell-1 h-1	on marine snow, Atlantic[2]
Marine Bacteria	1.41 pmol l-1 h-1	in seawater, Atlantic[2]

Marine Bacteria	1.67 10e-21 mol cell-1 h-1	on marine snow, Atlantic[2]
Marine Bacteria	1.84 pmol l-1 h-1	in seawater, Atlantic[2]
Marine Bacteria	12.38 10e-21 mol cell-1 h-1	in seawater, California[2]
Marine Bacteria	2.27 10e-21 mol cell-1 h-1	on marine snow, Atlantic[2]
Marine Bacteria	2.32 pmol l-1 h-1	in seawater, Atlantic[2]
Marine Bacteria	2.5 10e-21 mol cell-1 h-1	on marine snow, California[2]
Marine Bacteria	2.56 10e-21 mol cell-1 h-1	in seawater, Atlantic[2]
Marine Bacteria	2.80 pmol l-1 h-1	in seawater, Atlantic[2]
Marine Bacteria	2.88 10e-21 mol cell-1 h-1	in seawater, Atlantic[2]
Marine Bacteria	2.96 pmol l-1 h-1	in seawater, California[2]
Marine Bacteria	3.01 10e-21 mol cell-1 h-1	in seawater, Atlantic[2]
Marine Bacteria	3.03 10e-21 mol cell-1 h-1	in seawater, California[2]
Marine Bacteria	3.40 pmol l-1 h-1	in seawater, Atlantic[2]
Marine Bacteria	30.75 10e-21 mol cell-1 h-1	in seawater, California[2]
Marine Bacteria	34.31 10e-21 mol cell-1 h-1	on marine snow, California[2]
Marine Bacteria	36.14 10e-21 mol cell-1 h-1	in seawater, California[2]
Marine Bacteria	5.31 10e-21 mol cell-1 h-1	in seawater, Atlantic[2]
Marine Bacteria	5.58 10e-21 mol cell-1 h-1	on marine snow, Atlantic[2]
Marine Bacteria	6.65 pmol l-1 h-1	in seawater, California[2]
Marine Bacteria	6.71 10e-21 mol cell-1 h-1	in seawater, Atlantic[2]
Marine Bacteria	6.98 pmol l-1 h-1	in seawater, Atlantic[2]
Marine Bacteria	7.35 pmol l-1 h-1	in seawater, California[2]
Marine Bacteria	7.95 10e-21 mol cell-1 h-1	in seawater, Atlantic[2]
Marine Bacteria	9.06 10e-21 mol cell-1 h-1	on marine snow, California[2]
Marine Bacteria	9.44 10e-21 mol cell-1 h-1	on marine snow, California[2]

Source: 1.Scavia, D., Laird, G.A. and Fahnenstiel, G.L. Limnol. Oceanogr., 31, 612, 1986. 2.Alldredge, A.L., Cole, J.J. and Caron, D.A. Limnol. Oceanogr., 31, 68, 1986.

1-211 Bacteria
Upper temperature limit for growth rate

Species	Value	Condition
Bacillus acidocaldarius	70 degrees Celcius	The genera as a whole, non-photosynthetic, acid pH[1]
Bacillus acidocaldarius	70 degrees celcius	Acidophile, spore former, non-photosyntic bac.[2]
Bacillus brevis	50 degrees celcius	Rg = 40-60 deg cel., non-photosyntic bac., spore former[3]
Bacillus circulans	42.5 degrees celcius	Rg = 35-50 deg cel., non-photosyntic bac., spore former[3]
Bacillus coagulans	60 degrees celcius	Acidophile, spore former, non-photosyntic bac.[4]
Bacillus laterosporus	42.5 degrees celcius	Rg = 35-50 deg cel., non-photosyntic bac., spore former[3]
Bacillus licheniformis	52.5 degrees celcius	Rg = 50-55 deg cel, non-photosyntic bac., spore former[5]
Bacillus macerans	45 degrees celcius	Rg = 40-50 deg cel., non-photosyntic bac., spore former[3]
Bacillus pumilus	47.5 degrees celcius	Rg = 45-50 deg cel, non-photosyntic bac., spore former[3]
Bacillus sp.	70 degrees celcius	Hydrocarbon oxidizer, spore former, non-photosyntic bac.[6]
Bacillus sp.	82 degrees Celcius	The genera as a whole, non-photosynthetic, neutral pH[1]
Bacillus sp.	82 degrees celcius	Proteolytic, B YT-P, spore former, non-photosyntic bac.[7]
Bacillus sp.	85 degrees celcius	Uncertain, B. YT-G, spore former, non-photosyntic bac.[8]
Bacillus sphaericus	67.5 degrees celcius	Rg = 65-70, carboxylic acid util., many strains, non-photosyntic , spore[9]
Bacillus stearothermophilus	72.5 degrees celcius	Versatile, common, rg=70-75 deg cel., spore former, non-photosyntic bac.[3]
Bacillus stearothermophilus	75 degrees Celcius	The genera as a whole, non-photosynthetic, neutral pH[1]
Bacillus subtilis	62.5 degrees celcius	Rg = 55-70 deg cel., alcohol utilizer, many strains, non-photosyntic bac.[9]
Chemolithotrophic bacteria	90 degrees celcius	More than value, app. value[10]
Chloroflexus aurantiacus	71.5 degrees celcius	Chlorobiaceae, photosynthetic bac, rg = 70-73 deg cel.[11]
Chromatium sp.	58.5 degrees celcius	Natural obs, Chromatiaceae, photosynthetic bacteria[12]
Clostridium sp.	289.3 K	Origin = coast and Puget sound, sp 19, marine psychrophillic bac[13]
Clostridium sp.	289.3 K	Origin = coast and Puget sound, sp 54, marine psychrophillic bac[13]
Clostridium sp.	289.4 K	Origin = coast and Puget sound, sp 41, marine psychrophillic bac[13]
Clostridium sp.	290 K	Origin = coast and Puget sound, sp 5, marine psychrophillic bac[13]
Clostridium sp.	290.0 K	Origin = coast and Puget sound, sp 5, marine psychrophillic bac[13]
Clostridium sp.	290.5	Origin = coast and Puget sound, sp 40, marine psychrophillic bac[13]
Clostridium sp.	290.5 K	Origin = sediment off Washington, Pacific, marine psychrophillic bac[14]
Clostridium sp.	290.5 Origin = coast and Puget sound, sp	Origin = coast and Puget sound, sp 40, marine psychrophillic bac[13]
Clostridium sp.	291.3 K	Origin = coast and puget sound, sp 2, marine psychrophillic bac[14]
Clostridium sp.	75 degrees Celcius	The genera as a whole, non-photosynthetic, rg=74-76, neutral pH[1]
Clostridium sp.	75 degrees celcius	Cellulose digester, spore former, non-photosyntic bac.[15]
Clostridium tartarivorum	67 degrees celcius	SF, non-photosyntic bac.[16]

Clostridium thermoaceticum	65 degrees celcius	SF, non-photosyntic bac.[17]
Clostridium thermocellulaseum	65 degrees celcius	Cellulose digester, spore former, non-photosyntic bac.[17]
Clostridium thermocellum	68 degrees celcius	Cellulose digester, spore former, non-photosyntic bac.[17]
Clostridium thermohydrosulfuricum	75 degrees celcius	Rg = 74-76, reduces sulphite[16]
Clostridium thermosaccharolyticum	67 degrees celcius	SF, non-photosyntic bac.[16]
Cytophaga psychrophila	293 K	Origin = silver salmon, marine psychrophillic bac[18]
Desulfotomaculum nigrificans	70 degrees celcius	Sulphate reducer, spore former, non-photosyntic bac.[19]
Desulfotomaculum sp.	70 degrees Celcius	The genera as a whole, non-photosynthetic, neutral pH[1]
Desulfovibrio thermophilus	85 degrees celcius	Sulfate reducing bac., non-photosyntic bac.[20]
Heterotrophic bacteria	90 degrees celcius	More than value, app. value[10]
Hydrogenomonas sp.	60 degrees Celcius	The genera as a whole, non-photosynthetic, neutral pH[1]
Hydrogenomonas thermophilus	60 degrees celcius	Pseudomonad, non-photosyntic bac.[21]
Lactobacillus bulgaricus	52.5 degrees celcius	Lactic acid bac., non-photosyntic bac.[22]
Lactobacillus sp.	53 degrees Celcius	The genera as a whole, non-photosynthetic, neutral pH[1]
Lactobacillus sp.	53 degrees celcius	Lactic acid bac., non-photosyntic bac.[23]
Lactobacillus thermophilus	65 degrees celcius	Lactic acid bac., non-photosyntic bac.[24]
Leptospira sp.	54 degrees Celcius	The genera as a whole, non-photosynthetic, neutral pH[1]
Marine bacteria	283 K	Origin = water, Antarctic, E5-4-4, marine psychrophillic bac[25]
Methanobacterium sp.	75 degrees Celcius	The genera as a whole, non-photosynthetic, neutral pH[1]
Methanobacterium thermoautotrophicum	75 degrees celcius	Methane producing bac., non-photosyntic bac.[26]
Methylococcus capsulatus	55 degrees celcius	Methane oidozing, non-photosyntic bac.[27]
Microbispora aerata	55 degrees celcius	Actinomycetes, non-photosyntic bac.[28]
Microbispora bispora	60 degrees celcius	Actinomycetes, non-photosyntic bac.[28]
Microbispora thermodiastatica	55 degrees celcius	Actinomycetes, non-photosyntic bac.[28]
Micropolyspora caesia	55 degrees celcius	Actinomycetes, non-photosyntic bac.[28]
Micropolyspora faeni	60 degrees celcius	Actinomycetes, non-photosyntic bac.[28]
Micropolyspora rectivirgula	65 degrees celcius	Actinomycetes, non-photosyntic bac.[28]
Micropolyspora rubrobrunea	65 degrees celcius	Actinomycetes, non-photosyntic bac.[28]
Micropolyspora sp.	65 degrees Celcius	The genera as a whole, non-photosynthetic, neutral pH[1]
Micropolyspora thermovirida	57 degrees celcius	Actinomycetes, non-photosyntic bac.[28]
Micropolyspora virdinigra	65 degrees celcius	Actinomycetes, non-photosyntic bac.[28]
Microspora sp.	60 degrees Celcius	The genera as a whole, non-photosynthetic, neutral pH[1]
Photosynthetic bacteria	71.5 degrees celcius	Rg = 70 - 73 cel deg, app. value[10]
Pseudonocardia	60 degrees Celcius	The genera as a whole, non-photosynthetic, neutral pH[1]
Pseudonocardia thermophila	60 degrees celcius	Actinomycetes, non-photosyntic bac.[29]
Streptococcus thermophilus	50 degrees celcius	Lactic acid bac., non-photosyntic bac.[30]
Streptomyces sp.	60 degrees Celcius	The genera as a whole, non-photosynthetic, neutral pH[1]
Streptomyces thermoviolaceus	60 degrees celcius	Lactic acid bac., non-photosyntic bac.[31]
Streptomyces thermovulgaris	60 degrees celcius	Actinomycetes, non-photosyntic bac.[31]

Streptosporangium album	70 degrees celcius	S. a. var thermophilum, Actinoplanes, non-photosyntic bac.[32]
Streptosporangium sp.	70 degrees Celcius	The genera as a whole, non-photosynthetic, neutral pH[1]
Sulfolobus acidocaldarius	87.5 degrees celcius	Acidophile, Sulfur oxidizing, non-photosyntic bac., rg=85-90 del cel.[33]
Sulfolobus sp.	87.5 degrees Celcius	The genera as a whole, non-photosynthetic, rg=85-90, acid pH[1]
Thermoactinomyces candidus	60 degrees celcius	Actinomycetes, non-photosyntic bac.[34]
Thermoactinomyces citrina	72.5 degrees celcius	Actinomycetes, non-photosyntic bac., rg = 70-75 deg cel.[32]
Thermoactinomyces sacchari	65 degrees celcius	Actinomycetes, non-photosyntic bac.[35]
Thermoactinomyces sp.	70 degrees Celcius	The genera as a whole, non-photosynthetic, neutral pH[1]
Thermoactinomyces vulgaris	70 degrees celcius	Actinomycetes, non-photosyntic bac.[35]
Thermomicrobium roseum	85 degrees celcius	Gram negative, non-photosyntic bac.[36]
Thermomicrobium sp.	85 degrees Celcius	The genera as a whole, non-photosynthetic, neutral pH[1]
Thermomonospora curvata	65 degrees celcius	Actinomycetes, non-photosyntic bac.[32]
Thermomonospora sp.	72.5 degrees Celcius	The genera as a whole, non-photosynthetic, rg=70-75, neutral pH[1]
Thermomonospora viridis	60 degrees celcius	Actinomycetes, non-photosyntic bac.[35]
Thermoplasma acidophilum	65 degrees celcius	Acidophile, Mycoplasma, non-photosyntic bac.[37]
Thermoplasma sp.	59 degrees Celcius	The genera as a whole, non-photosynthetic, acid pH[1]
Thermus aquaticus	79 degrees celcius	Gram negative, non-photosyntic bac.[38]
Thermus flavus	80 degrees celcius	Gram negative, non-photosyntic bac.[39]
Thermus ruber	80 degrees celcius	Gram negative, non-photosyntic bac.[40]
Thermus sp.	85 degrees Celcius	The genera as a whole, non-photosynthetic, neutral pH[1]
Thermus thermophilus	85 degrees celcius	Gram negative, non-photosyntic bac.[41]
Thiobacillus sp.	55 degrees Celcius	The genera as a whole, non-photosynthetic, acid pH[1]
Thiobacillus sp.	57.5 degrees celcius	Rg = 55-60 del cel., Sulfur oxidizing, non-photosyntic bac.[42]
Thiobacillus thermophilica	80 degrees celcius	Spore former, Sulfur oxidizing, non-photosyntic bac.[43]
Thiobacillus thiooxidans	55 degrees celcius	Acidophile, Sulfur oxidizing bac, non-photosyntic bac.[44]
Vibrio marinus	293 K	Origin = water off Oregon coast, marine psychrophillic bac[45]
Vibrio sp.	282 K	Origin = water, Antarctic, Vibrio AP-2-24, marine psychrophillic bac[46]
Vibrio sp.	286 K	Origin = water, Antarctic, Vibrio 18-300, marine psychrophillic bac[47]
Vibrio sp.	289 K	Origin = water, Antarctic, Vibrio 18-500, marine psychrophillic bac[47]

Source: 1.Tansey, M.R. and T.D. Brock. 1978. In: D.J. Kushner: Microbial Life in Extreme Environments. Academic Press. London. 159-216.
2.Darland, W.H. and T.D. Brock. 1971. J. Gen. Microbiol. 67, 9-15.
3.Gibson, T. and R.E. Gordon. 1974. In: R.E. Buchanan and N.E. Gibbons (eds.): Bergeys Manual of Determinative Bacteriology. Williams and Williams. Baltimore, 529-550.
4.Belly, R.T. and T.D. Brock. 1974. J. Appl. Bact. 37, 175-177.
5.Golovacheva, R.S., Loginova, Salikhov, Kolesnikov and Zaltseva. 1975. Mikrobiologiya 44, 265-268.
6.Mateless, R.I., J.N. Baruah and S.R. Tannenbaum. 1967. Science 157, 1322-1323.
7.Heinen, U.J. and W. Heinen. 1972. Arch. Mikrobiol. 82, 1-23.
8.Heinen, W. 1971. Arch. Mikrobiol. 76, 2-17.
9.Allen, M.B. 1953. Bact. Rev. 17, 125-173.
10.Meeks, J.C. and R.W. Castenholz. 1971. Arch. Mikrobiol. 78, 25-41.
11.Bauld, J. and T.D. Brock. 1973. Arch. Microbiol. 92, 267-284.
12.Castenholz, R.W. 1969. Bact. Rev. 33, 476-504.
13.Finne, W.H. and J.R. Matches. 1974. Can. J. Microbiol. 20, 1639-1645.
14.Matches, J.R. and J. Liston. 1973. In: L.H. Stevenson and R.R. Colwell: Estuarine Microbial Ecology.Univ. S Carolina Press. Columbia, 345-361.
15.Loginova, L.G., A.E. Kosmachev, R.S. Golovacheva and Seregina. 1962. Mikrobiologiya 31, 1082-1086.

16.Hollaus, F. and U. Steytr. 1972. Arch. Mikrobiol. 86, 129-146.
17.Breed, R.S., E.G.D. Murray and N.R. Smith. 1974. In: R.E. Buchanan and N.E. Gibbons (eds.): Bergeys Manual of Determinative Bacteriology. Williams and Williams. Baltimore.
18.Borg, A.F. 1960. Wildlife Dis. 8, 1-85.
19.Campbell, L.L. 1974. In: R.E. Buchanan and N.E. Gibbons (eds.): Bergeys Manual of Determinative Bacteriology. Williams and Williams. Baltimore, 572-573.
20.Rozanova, E.P. and A.I. Khudyakova. 1974. Mikrobiologiya 43, 1069-1075.
21.McGee, J.M., L.R. Bowen and R.G. Tischer. 1967. Nature 214, 715-716.
22.Orla-Jensen, S. 1942. The Lactic Acid Bacteria. 2nd ed. Copenhagen. 197 pp.
23.Rogosa, M. 1974. In: R.E. Buchanan and N.E. Gibbons (eds.): Bergeys Manual of Determinative Bacteriology. Williams and Williams. Baltimore, 669-676.
24.Gaughran, E.R.L. 1947. Bact. Rev. 11, 189-225.
25.Christian, R.R. and W.J. Wiebe. 1974. Can. J. Microbiol. 20, 1341-1345.
26.Zeikus, J.G. and R.S. Wolfe. 1972. J. Bact. 109, 707-713.
27.Leadbetter, E.R. 1974. In: R.E. Buchanan and N.E. Gibbons (eds.): Bergeys Manual of Determinative Bacteriology. Williams and Williams. Baltimore, 267-269.
28.Cross, T. 1974. In: R.E. Buchanan and N.E. Gibbons (eds.): Bergeys Manual of Determinative Bacteriology. Williams and Williams. Baltimore, 861-863.
29.Henssen, A. 1974. In: R.E. Buchanan and N.E. Gibbons (eds.): Bergeys Manual of Determinative Bacteriology. Williams and Williams. Baltimore, 746-747.
30.Deibel, R.H. and H.W. Seeley, Jr. 1974. In: R.E. Buchanan and N.E. Gibbons (eds.): Bergeys Manual of Determinative Bacteriology. Williams and Williams. Baltimore, 490-509.
31.Pridham, T.G. and H.D. Tresner. 1974. In: R.E. Buchanan and N.E. Gibbons (eds.): Bergeys Manual of Determinative Bacteriology. Williams and Williams. Baltimore, 748-829.
32.Manachini, P., A. Craveri and R. Craveri. 1966. Annal. Microbiol. 16, 83-90.
33.Brock, T.D., K.M. Brock, R.T. Belly and R.L. Weiss. 1972. Arch. Mikrobiol. 84, 54-68.
34.Kurup, V.P., J.J. Barboriak, J.N. Fink and M.P. Lechevalier. 1975. Int. J. Syst. Bacteriol. 25, 150-54.
35.Kuster, E. 1974. In: R.E. Buchanan and N.E. Gibbons (eds.): Bergeys Manual of Determinative Bacteriology. Williams and Williams. Baltimore, 858-859.
36.Jackson, T.J., R.F. Ramaley and W.G. Meinschein. 1973. Int. J. Syst. Bacteriol. 23, 28-36.
37.Belly, R.T., M.R. Tansey and T.D. Brock. 1973. J. Phycol. 9, 123-127.
38.Brock, T.D. and H. Freeze. 1969. J. Bact. 98, 289-297.
39.Oshima, T. and K. Imahori. 1974. Int. J. Syst. Bactiol. 24, 102-112.
40.Ramaley, R.F. and J. Hixson. 1970. J. Bact. 103, 527-528.
41.Loginova, L.G. and L.A. Egorova. 1975. Mikrobiologiya 44, 661-665.
42.Williams, R.A.D. and D.S. Hoare. 1972. J. Gen. Microbiol. 70, 555-566.
43.Egorova, A.A. and Z. Derygina. 1963. Mikrobiologiya 32, 437-446.
44.Fliermans, C.B. and T.D. Brock. 1972. J. Bact. 111, 343-350.
45.Morita, R.Y. and R.D. Haight. 1964. Limnol. Oceanogr. 9, 102-106.
46.Morita, R.Y. 1975. Bacteriol. Rev. 39, 144-167.
47.Baross, J.A., F.J. Hanus and R.Y. Morita. 1974. In: R.R. Colwell and R.Y. Morita (eds.): Effect of te Ocean Environment on Microbial Activity. Univ. Park Press. Baltimore. 180-202.

1-212 Bacteria
Uptake

Species	Value	Condition
Nitrate	0.001 umol l-1 h-1	ambient conc. 1.30 uM, 1 hour exp.[1]
Nitrate	0.001 umol l-1 h-1	ambient conc. 1.30 uM, 4 hour exp.[1]
Nitrate	0.002 umol l-1 h-1	ambient conc. 1.30 uM, 8 hour exp.[1]
alanine	0.005 umol l-1 h-1	ambient conc. 1.01 uM, 8 hour exp.[1]
alanine	0.008 umol l-1 h-1	ambient conc. 1.01 uM, 4 hour exp.[1]
alanine	0.014 umol l-1 h-1	ambient conc. 1.01 uM, 1 hour exp.[1]
amino acids	0.062 umol l-1 h-1	mixture ambient conc. 1.30 uM, 4 hour exp.[1]
amino acids	0.068 umol l-1 h-1	mixture ambient conc. 1.30 uM, 8 hour exp.[1]
amino acids	0.167 umol l-1 h-1	mixture ambient conc. 1.30 uM, 1 hour exp.[1]
ammonium	0.039 umol l-1 h-1	ambient conc. 1.5 uM, 8 hour exp.[1]
ammonium	0.050 umol l-1 h-1	ambient conc. 1.5 uM, 4 hour exp.[1]
ammonium	0.064 umol l-1 h-1	ambient conc. 1.5 uM, 1 hour exp.[1]
glycine	0.008 umol l-1 h-1	ambient conc. 1.01 uM, 1 hour exp.[1]
glycine	0.015 umol l-1 h-1	ambient conc. 1.01 uM, 4 hour exp.[1]
glycine	0.016 umol l-1 h-1	ambient conc. 1.01 uM, 8 hour exp.[1]

urea	-0.002 umol l-1 h-1	ambient conc. 2.0 uM, 4 hour exp.[1]
urea	0.000 umol l-1 h-1	ambient conc. 2.0 uM, 8 hour exp.[1]
urea	0.002 umol l-1 h-1	ambient conc. 2.0 uM, 1 hour exp.[1]

Source: 1.Wheeler, P.A. and Kirchman, D.L. Limnol. Oceanogr., 31(5), 998, 1986.

1-213 Bacteria
Volume

Species	Value	Condition
Freshwater Bacteria	0.042	rods, electron and flurescence microscopy, Lake Michigan, 9. September[1]
Freshwater Bacteria	0.063	rods, electron and flurescence microscopy, Lake Michigan, 9. April[1]
Freshwater Bacteria	0.070	rods, electron and flurescence microscopy, Lake Michigan, 16. August[1]
Freshwater Bacteria	0.070	rods, electron and flurescence microscopy, Lake Michigan, 22. May[1]
Freshwater Bacteria	0.087	rods, electron and flurescence microscopy, Lake Michigan, 1. July[1]

Source: 1.Scavia, D., Laird, G.A. and Fahnenstiel, G.L. Limnol. Oceanogr., 31, 612, 1986.

1-214 Bacteria
Water activity for limiting growth rate

Species	Value	Condition
Bacillus subtilis	0.949	Zero growth at this water activity[1]
Bacillus subtilis	0.949 Zero growth at this water activity	Zero growth at this water activity[1]
Escherichia coli	0.932	Zero growth at this water activity[1]
Escherichia coli	0.932 Zero growth at this water activity	Zero growth at this water activity[1]
Halobacteria and halococci	0.805	Zero growth at this water activity, rg=0.75-0.88, in NaCl[1]
Halobacteria and halococci	0.805 Zero growth at this water activity	Zero growth at this water activity, rg=0.75-0.88, in NaCl[1]
Lactobacillus sp.	0.9 Zero growth at this water activity	Zero growth at this water activity[1]
Lactobacillus sp.	0.90	Zero growth at this water activity[1]
Paracoccus halodenitrificans	0.92	Zero growth at this water activity, rg=0.86-0.98, in NaCl[1]
Paracoccus halodenitrificans	0.92 Zero growth at this water activity,	Zero growth at this water activity, rg=0.86-0.98, in NaCl[1]
Pseudomonas aeruginosa	0.945	Zero growth at this water activity[1]
Pseudomonas aeruginosa	0.945 Zero growth at this water activity	Zero growth at this water activity[1]
Staphyllococcus aureus	0.87	Zero growth at this water activity, rg=0.86-0.88[1]
Staphyllococcus aureus	0.87 Zero growth at this water activity,	Zero growth at this water activity, rg=0.86-0.88[1]
Vibrio costicola	0.92	Zero growth at this water activity, rg=0.86-0.98, in NaCl[1]
Vibrio costicola	0.92 Zero growth at this water activity,	Zero growth at this water activity, rg=0.86-0.98, in NaCl[1]
Vibrio metchnikovii	0.97	Zero growth at this water activity, rg=0.96-0.995[1]

Vibrio metchnikovii	0.97 Zero growth at this water activity,	Zero growth at this water activity, rg=0.96-0.995[1]

Source: 1.Kushner, D.J. 1978. In: D.J. Kushner: Microbial Life in Extreme Environments. AP. London 317-368.

1-215 Bacteria
Wet weight

Species	Value	Condition
Escherichia coli	9.6E-13 g	None[1]
Influenza bacillus	2.4E-14 g	None[1]
Pneumococcus bacterium	1.9E-15 g	None[1]
T-even bacteriophage	3.3E-16 g	None[1]
Typhoid bacillus	1.7E-12 g	None[1]

Source: 1.Brock, T.D. 1966. Principles of Microbial Ecology. Prentice-Hall. New Jersey.

1-216 Bacteria
Yield

Species	Value	Condition
Nitrobacter sp.	0.02 Yield=wt.of cells/wt.of nitrogen ox	Yield=wt.of cells/wt.of nitrogen oxidized[1]
Nitrosomonas sp.	0.06 Yield=wt.of cells/wt.of nitrogen ox	Yield=wt.of cells/wt.of nitrogen oxidized[1]

Source: 1.Painter, H.A. 1975. Microbial Transformations of inorganic Nitrogen. in: Conf. on Nitrogen as a Water Poll. Vol 1, Session A1. Copenhagen.

1-217 Detritus
C / biomass ratio

Species	Value	Condition
Detritus	45 % of dry weight	Range = 40 - 50 %[1]

Source: 1.The Hydrological Engineering Center Corps of Engineers U.S. Army. 1975. Water Quality for river-resevoir systems, Davis, Cal. 9 5 6 1 6 (916) 499-2329.

1-218 Detritus
N / biomass ratio

Species	Value	Condition
Detritus	0.08	Range = 0.07 - 0.09[1]

Source: 1.The Hydrological Engineering Center Corps of Engineers U.S. Army. 1975. Water Quality for river-resevoir systems, Davis, Cal. 9 5 6 1 6 (916) 499-2329.

1-219 Detritus
N / chlorophyll a ratio

Species	Value	Condition
Detritus	0.01 mg N/mg chl	Lake Ontario[1]

Source: 1.Thomann, R.V., D.M. Di Toro, R.P. Winfield and D.J. O'Connor. 1975. Mathematical Modeling of Phytoplankton in Lake Ontario: 1. National Environmental Reseach Center. EPA. Oregon.

1-220 Detritus
Oxygen uptake

Species	Value	Condition
Detritus	0.28 mg O2/(g*hour)	297 K, detritus associated with Thalassia, part. diameter=1.4 mm[1]
Detritus	0.3 mg O2/(g*hour)	297 K, detritus associated with Thalassia, part. diameter=0.28 mm[1]
Detritus	0.32 mg O2/(g*hour)	297 K, detritus associated with Thalassia, part. diameter=0.43 mm[1]
Detritus	0.48 mg O2/(g*hour)	297 K, detritus associated with Thalassia, part. diameter=0.2 mm[1]
Detritus	1.03 mg O2/(g*hour)	297 K, detritus associated with Thalassia, 5 d, particle diameter =0.1 mm[1]

Source: 1.Fenchel, T. 1970. Limnol. Oceanogr. 15(1), pp 14-20.

1-221 Detritus
P / biomass ratio

Species	Value	Condition
Detritus	0.85 % of dry weight	Range = 0.5 - 1.2 % of dry weight[1]

Source: 1.The Hydrological Engineering Center Corps of Engineers U.S. Army. 1975. Water Quality for river-resevoir systems, Davis, Cal. 9 5 6 1 6 (916) 499-2329.

1-222 Detritus
P / chlorophyll a ratio

Species	Value	Condition
Detritus	0.001 mg P/microgram chl	Lake Ontario[1]

Source: 1.Thomann, R.V., D.M. Di Toro, R.P. Winfield and D.J. O'Connor. 1975. Mathematical Modeling of Phytoplankton in Lake Ontario: 1. National Environmental Reseach Center. EPA. Oregon.

1-223 Detritus
Settling rate

Species	Value	Condition
Detritus	0.001 1/day	Lake Ontario[1]
Detritus	1 m/day	293 K, rg = 0.0 - 2.0[2]
Fecal pellets from 3 species	232 m/day	Food = F006, fecal pel.vol.=(1.6-15.5)E6 U**3, n=75, F005, rg = 53 -[3]
Fecal pellets from Acartia clausii	116 m/day	Food=micro-flagellate, fecal pel.vol.=0.21E6 U**3, n=7, rg = 72 - 210[4]
Fecal pellets from Euphausia krohnii	240 m/day	Food=natural diet, fecal pel.vol.=(3-110)E6 U**3, n=113[3]
Fecal pellets from Euphausia pacifica	43 m/day	Food = Skeletonema costatum[5]
Fecal pellets from Pontella meadii	51 m/day	Food = dunaliella salina, fecal pel.vol.=3.7E6 U**3, n=28, rg = 23 -[6]
Fecal pellets from Pontella meadii	54 m/day	Food = Skeletonema costatum, fecal pel.vol.=8.2E6 U**3, n=22, rg = 22[6]
Fecal pellets from Pontella meadii	77 m/day	Food = gonyaulax monilata, fecal pel.vol.=8.3E6 U**3, n=8, rg = 15 -[6]
Fecal pellets from Pontella meadii	88 m/day	Food = F006, fecal pel.vol.=11.7E6 U**3, n=27, rg = 18 - 153[6]
Fecal pellets from unknown species	158.8 m/day	Fecal pel.vol.=(0.2-2.5)E6 U**3, n=41, rg = 36 - 376[7]

Fecal pellets from unknown species	666 m/day	Fecal pel.vol.=38.6E6 U**3, n=6, rg = 550 - 860[6]
Nitrogen detritus	0.001 1/day	Used in models, Lake Ontario[8]
Nitrogen detritus	0.028 1/day	Used in models, Lake Erie[8]
Nitrogen detritus	0.05 1/day	Used in models, Lake Michigan[9]
Nitrogen detritus	0.1 1/day	Used in models, Potomac estuary[8]

Source: 1.Thomann, R.V., D.M. Di Toro, R.P. Winfield and D.J. O'Connor. 1975. Mathematical Modeling of Phytoplankton in Lake Ontario: 1. National Environmental Reseach Center. EPA. Oregon.
2.The Hydrological Engineering Center Corps of Engineers U.S. Army. 1975. Water Quality for river-resevoir systems, Davis, Cal. 9 5 6 1 6 (916) 499-2329.
3.Fowler, S.W. and L.F. Small. 1972. Limnol. Oceanogr., 17, pp 293-296.
4.Smayda, T.J. 1974. Limnol. Oceanogr. 19(4), pp 628-635.
5.Osterberg, C., A.G. Carey and H. Curl, Jr. 1963. Nature. 200, pp 1276-1277.
6.Turner, J.T. 1977. Mar. Biol. 40(3), pp 249-259.
7.Smayda, T.J. 1969. Limnol. Oceanogr. 14, pp 621-625.
8.O'Connor, F.B. 1963. in: Doeksen, J. and J. van der Drift (eds.); Soil Organisms. pp 32-48. North-Holland publ. Co. Amsterdam.
9.Canale, R.P., L.M. De Palma and A.H. Vogel. 1976. in: Canale, R.P. (ed.); Mathematical modelling of Biochemical processes in Aquatic Ecosystems. Ann Arbor Science Press.

1-224 Fish
Assimilation

Species	Value	Condition
Ctenopharyngodon sp.	14 % of ingestion	None[1]
Fundulus heteroclitus	0.05 % of ingested Zn	Diet = plant detritus, measured after 24 h[2]
Fundulus heteroclitus	0.3 % of ingested Zn	Diet = shrimp, measured after 24 h[2]
Perca fluviatilis	35.8 % of ingestion	None[1]
Pleuronectes platessa	0.1 % of ingested Zn	Diet = mussel or Nereis, measured after 4 d[2]
Pleuronectes platessa	0.28 % of ingested Zn	Diet = shrimp, measured after 4 d[2]
Pleuronectes platessa	0.36 % of ingested Zn	Diet = Neries, measured after 4 d[2]

Source: 1.Wetzel, R.G. 1975. Limnology. Saunders. Philadelphia.
2.Wolfe, 1975. In: Cronin, L.E. (ed.); Estuarine Research, Vol. 1. Ac. Press N.Y.

1-225 Fish
C / biomass ratio

Species	Value	Condition
Fish	45 % of dry weight	Range = 40 - 50 %[1]
Fish	47.5 % of dry weight	Mean elementary composition[2]
Mola mola	48.2 % of dry weight	None[3]

Source: 1.The Hydrological Engineering Center Corps of Engineers U.S. Army. 1975. Water Quality for river-resevoir systems, Davis, Cal. 9 5 6 1 6 (916) 499-2329.
2.Bowen, H.J.M. 1966. Trace Elements in Biochemistry, Ac. Press.
3.Leidy, G.R. and R.M. Jenkins. 1977. The development of Fishery Compartments and Population Rate Coef. for Use in Resevoir Ecosystem Model. USDI Fish and Wildlife Service. US Army. COE.

1-226 Fish
Caloric content

Species	Value	Condition
Fish	1 kcal/g wet weight	None[1]
Solea solea	0.849 kcal/(g wet weight)	1.5 - 5.0 g, rg = 0.626 - 1.073[2]

Source: 1.Waters, T.F. 1977. Adv. Ecol. Res. 10, pp 91-164.
2.Bromley, P.J. 1971. J. Cons. Int. Explor. Mer. 34(1), pp 131-134.

1-227 Fish
Conversion efficiency

Species	Value	Condition
Clupea harengus	12.25 % of intake	Food = 1.3 % of body wet weight, 279.2 K[1]
Clupea harengus	6.99 % of intake	Food at saturation, 277.5 K[1]
Clupea harengus	9.78 % of intake	Food at saturation, 279.5 K[1]
Fish	45 % of dry weight	Range = 40 - 50 %[2]

Source: 1.De Silva, S.S. and F. Balbontin. 1974. J. Fish. Biol., 6, pp 645-658.
2.The Hydrological Engineering Center Corps of Engineers U.S. Army. 1975. Water Quality for river-resevoir systems, Davis, Cal. 9 5 6 1 6 (916) 499-2329.

1-228 Fish
Density

Species	Value	Condition
Rutilus rutilus	< 3 individulals m-2	average for several years, from diagram[1]

Source: 1.Cryer, M., Peirson, G. and Townsend, C.R. Limnol. Oceanogr., 31(5), 1022, 1986.

1-229 Fish
Development of larvae

Species	Value	Condition
Achirus lineatus	10.8 %	Larvae = 32.0 no/l, dw = 169.9 ug, 16 first days after hatch[1]
Achirus lineatus	14.5 %	Larvae = 2.0 no/l, dw = 141.9 ug, 16 first days after hatch[1]
Achirus lineatus	19.6 %	Larvae = 8.0 no/l, dw = 161.3 ug, 16 first days after hatch[1]
Achirus lineatus	31.5 %	Larvae = 0.5 no/l, dw = 235.8 ug, 16 first days after hatch[1]
Anchoa mitchilli	22 %	Larvae = 8.0 no/l, dw = 402.7 ug, 16 first days after hatch[1]
Anchoa mitchilli	31 %	Larvae = 32.0 no/l, dw = 334.2 ug, 16 first days after hatch[1]
Anchoa mitchilli	31.3 %	Larvae = 2.0 no/l, dw = 518.7 ug, 16 first days after hatch[1]
Anchoa mitchilli	32.9 %	Larvae = 0.5 no/l, dw = 578.0 ug, 16 first days after hatch[1]

Source: 1.Houde, E.D. 1977. Mar. Biol., 43(4), pp 333-341.

1-230 Fish
Development time of larvae

Species	Value	Condition
Achirus lineatus	11.5 hours	Hours after yolk absorption before which larvae must be fed, 297 K[1]
Achirus lineatus	11.5 hours	Hours after yolk absorption before which larvae must be fed, 301 K[1]
Achirus lineatus	12 hours	Hours after eye pigmentation when yolk is absorbed, 305 K[1]
Achirus lineatus	12.5 hours	Hours after eye pigmentation when yolk is absorbed, 301 K[1]
Achirus lineatus	20.5 hours	Hours after eye pigmentation when yolk is absorbed, 297 K[1]
Achirus lineatus	24 hours	Good survival if fed before this hour after eye pigmentation, 301 K[1]
Achirus lineatus	32 hours	Good survival if fed before this hour after eye pigmentation, 297 K[1]
Achirus lineatus	58.2 hours	Hours after hatch when yolk is absorbed, 305 K[1]
Achirus lineatus	59.7 hours	Hours after hatch when yolk is absorbed, 301 K[1]

Achirus lineatus	79.8 hours	Hours after hatch when yolk is absorbed, 297 K[1]
Anchoa mitchilli	1 hours	Hours after eye pigmentation when yolk is absorbed, 305 K[1]
Anchoa mitchilli	15 hours	Hours after yolk absorption before which larvae must be fed, 305 K[1]
Anchoa mitchilli	16 hours	Good survival if fed before this hour after eye pigmentation, 305 K[1]
Anchoa mitchilli	22.5 hours	Hours after yolk absorption before which larvae must be fed, 301 K[1]
Anchoa mitchilli	27.4 hours	Hours after hatch when yolk is absorbed, 305 K[1]
Anchoa mitchilli	32 hours	Good survival if fed before this hour after eye pigmentation, 301 K[1]
Anchoa mitchilli	32.5 hours	Hours after yolk absorption before which larvae must be fed, 297 K[1]
Anchoa mitchilli	36.6 hours	Hours after hatch when yolk is absorbed, 301 K[1]
Anchoa mitchilli	40 hours	Good survival if fed before this hour after eye pigmentation, 297 K[1]
Anchoa mitchilli	41 hours	Hours after hatch when yolk is absorbed, 297 K[1]
Anchoa mitchilli	7.5 hours	Hours after eye pigmentation when yolk is absorbed, 297 K[1]
Anchoa mitchilli	9.5 hours	Hours after eye pigmentation when yolk is absorbed, 301 K[1]
Archosargus rhomboidalis	10.5 hours	Hours after eye pigmentation when yolk is absorbed, 295 K[1]
Archosargus rhomboidalis	16 hours	Hours after eye pigmentation when yolk is absorbed, 303 K[1]
Archosargus rhomboidalis	16 hours	Hours after yolk absorption before which larvae must be fed, 303 K[1]
Archosargus rhomboidalis	19.5 hours	Hours after yolk absorption before which larvae must be fed, 299 K[1]
Archosargus rhomboidalis	20.5 hours	Hours after eye pigmentation when yolk is absorbed, 299 K[1]
Archosargus rhomboidalis	29.5 hours	Hours after yolk absorption before which larvae must be fed, 295 K[1]
Archosargus rhomboidalis	32 hours	Good survival if fed before this hour after eye pigmantation, 303 K[1]
Archosargus rhomboidalis	40 hours	Good survival if fed before this hour after eye pigmentation, 295 K[1]
Archosargus rhomboidalis	40 hours	Good survival if fed before this hour after eye pigmentation, 299 K[1]
Archosargus rhomboidalis	40.9 hours	Hours after hatch when yolk is absorbed, 303 K[1]
Archosargus rhomboidalis	50 hours	Hours after hatch when yolk is absorbed, 299 K[1]
Archosargus rhomboidalis	52.9 hours	Hours after hatch when yolk is absorbed, 295 K[1]

Source: 1.Houde, E.D. 1977. Mar. Biol., 43(4), pp 333-341.

1-231 Fish
Efficiency

Species	Value	Condition
Fish	10 %	Fish production/primary production, oceanic area[1]
Fish	15 %	Fish production/primary production, coastal area[1]
Fish	20 %	Fish production/primary production, upwelling area, marine[1]

Source: 1.Ryther, J.H. 1969. Science 166, pp 72-76.

1-232 Fish
Egestion

Species	Value	Condition
Ctenopharyngodon sp.	86 % of ingestion	None[1]
Perca fluviatilis	64.2 % of ingestion	None[1]

Source: 1.Wetzel, R.G. 1975. Limnology. Saunders. Philadelphia.

1-233 Fish
Food items

Species	Value	Condition
Black bass	6 %	Calculated by volume, terrestrial origin of food in % of total[1]
Black bass	8 %	Calculated by volume, invertebrates in % of total food[1]
Black bass	86 %	Calculated by volume, fish in % of total food[1]
Bowfin	100 %	Calculated by volume, fish in % of total food[1]
Buffalofish	40 %	Calculated by volume, detritus in % of total food[1]
Buffalofish	5 %	Calculated by volume, invertebrates in % of total food[1]
Buffalofish	5 %	Calculated by volume, plant in % of total food[1]
Buffalofish	50 %	Calculated by volume, zooplankton in % of total food[1]
Bullhead	10 %	Calculated by volume, plant in % of total food[1]
Bullhead	15 %	Calculated by volume, fish in % of total food[1]
Bullhead	25 %	Calculated by volume, detritus in % of total food[1]
Bullhead	50 %	Calculated by volume, invertebrates in % of total food[1]
Carp	10 %	Calculated by volume, zooplankton in % of total food[1]
Carp	20 %	Calculated by volume, invertebrates in % of total food[1]
Carp	30 %	Calculated by volume, plant in % of total food[1]
Carp	40 %	Calculated by volume, detritus in % of total food[1]
Carpsuckers	15 %	Calculated by volume, plant in % of total food[1]
Carpsuckers	15 %	Calculated by volume, zooplankton in % of total food[1]
Carpsuckers	5 %	Calculated by volume, invertebrates in % of total food[1]
Carpsuckers	65 %	Calculated by volume, detritus in % of total food[1]
Catfish	10 %	Calculated by volume, invertebrates in % of total food[1]
Catfish	10 %	Calculated by volume, plant in % of total food[1]
Catfish	80 %	Calculated by volume, fish in % of total food[1]
Freshwater drum	34 %	Calculated by volume, fish in % of total food[1]
Freshwater drum	58 %	Calculated by volume, invertebrates in % of total food[1]
Freshwater drum	8 %	Calculated by volume, detritus in % of total food[1]
Gars	100 %	Calculated by volume, fish in % of total food[1]
Gizzard shad	10 %	Calculated by volume, plant in % of total food[1]
Gizzard shad	5 %	Calculated by volume, invertebrates in % of total food[1]
Gizzard shad	5 %	Calculated by volume, zooplankton in % of total food[1]
Gizzard shad	80 %	Calculated by volume, detritus in % of total food[1]

Haddock	10 %	% weight of stomach content, 31 - 35 cm, molluscs in % of total food[2]
Haddock	11 %	% weight of stomach content, 31 - 35 cm, annelids in % of total food[2]
Haddock	11 %	% weight of stomach content, 31 - 35 cm, fish in % of total food[2]
Haddock	27 %	% weight of stomach content, 31 - 35 cm, echinoderms in % of total[2]
Haddock	41 %	% weight of stomach content, 31 - 35 cm, crustacea in % of total food[2]
Hog suckers	15 %	Calculated by volume, zooplankton in % of total food[1]
Hog suckers	5 %	Calculated by volume, invertebrates in % of total food[1]
Hog suckers	80 %	Calculated by volume, detritus in % of total food[1]
Madtoms	18 %	Calculated by voiume, fish in % of total food[1]
Madtoms	27 %	Calculated by volume, detritus in % of total food[1]
Madtoms	55 %	Calculated by volume, invertebrates in % of total food[1]
Minnows	20 %	Calculated by volume, invertebrates in % of total food[1]
Minnows	20 %	Calculated by volume, plant in % of total food[1]
Minnows	60 %	Calculated by volume, zooplankton in % of total food[1]
Perch	20 %	Calculated by volume, invertebrates in % of total food[1]
Perch	20 %	Calculated by volume, zooplankton in % of total food[1]
Perch	60 %	Calculated by volume, fish in % of total food[1]
Pickerels	100 %	Calculated by volume, fish in % of total food[1]
Redhorses	100 %	Calculated by volume, invertebrates in % of total food[1]
Salmo gairdneri	10 %	Calculated by volume, fish in % of total food[1]
Salmo gairdneri	10 %	Calculated by volume, terrestrial origin of food in % of total[1]
Salmo gairdneri	15 %	Calculated by volume, zooplankton in % of total food[1]
Salmo gairdneri	5 %	Calculated by volume, plant in % of total food[1]
Salmo gairdneri	60 %	Calculated by volume, invertebrates in % of total food[1]
Salvelinus fontinalis	5 %	Calculated by volume, terrestrial origin of food in % of total[1]
Salvelinus fontinalis	5 %	Calculated by volume, zooplankton in % of total food[1]
Salvelinus fontinalis	90 %	Calculated by volume, invertebrates in % of total food[1]
Silversides	20 %	Calculated by volume, invertebrates in % of total food[1]
Silversides	80 %	Calculated by volume, zooplankton in % of total food[1]
Suckers	15 %	Calculated by volume, plant in % of total food[1]
Suckers	15 %	Calculated by volume, zooplankton in % of total food[1]
Suckers	5 %	Calculated by volume, invertebrates in % of total food[1]
Suckers	65 %	Calculated by volume, detritus in % of total food[1]
Sunfishes	10 %	Calculated by volume, plant in % of total food[1]
Sunfishes	15 %	Calculated by volume, terrestrial origin of food in % of total[1]
Sunfishes	5 %	Calculated by volume, detritus in % of total food[1]
Sunfishes	5 %	Calculated by volume, fish in % of total food[1]
Sunfishes	65 %	Calculated by volume, invertebrates in % of total food[1]

Temperate basses	10 %	Calculated by volume, zooplankton in % of total food[1]
Temperate basses	20 %	Calculated by volume, invertebrates in % of total food[1]
Temperate basses	70 %	Calculated by volume, fish in % of total food[1]
Threadfin shad	10 %	Calculated by volume, invertebrates in % of total food, young[1]
Threadfin shad	10 %	Calculated by volume, zooplankton in % of total food, young[1]
Threadfin shad	15 %	Calculated by volume, invertebrates in % of total food, adult[1]
Threadfin shad	15 %	Calculated by volume, zooplankton in % of total food, adult[1]
Threadfin shad	30 %	Calculated by volume, plant in % of total food, adult[1]
Threadfin shad	30 %	Calculated by volume, plant in % of total food, young[1]
Threadfin shad	5 %	Calculated by volume, detritus in % of total food, adult[1]
Threadfin shad	50 %	Calculated by volume, detritus in % of total food, young[1]
Threadfin shad	55 %	Calculated by volume, zooplankton in % of total food, adult[1]
Whitling	16 %	% weight of stomach content, 21 - 25 cm, annelids in % of total food[2]
Whitling	32 %	% weight of stomach content, 21 - 25 cm, fish in % of total food[2]
Whitling	43 %	% weight of stomach content, 21 - 25 cm, crustacea in % of total food[2]
Whitling	9 %	% weight of stomach content, 21 - 25 cm, molluscs in % of total food[2]
Crappie	15 %	Calculated by volume, zooplankton in % of total food[1]
Crappie	20 %	Calculated by volume, invertebrates in % of total food[1]
Crappie	5 %	Calculated by volume, detritus in % of total food[1]
Crappie	5 %	Calculated by volume, plant in % of total food[1]
Crappie	55 %	Calculated by volume, fish in % of total food[1]

Source: 1.Leidy, G.R. and R.M. Jenkins. 1977. The development of Fishery Compartments and Population Rate Coef. for Use in Reservoir Ecosystem Model. USDI Fish and Wildlife Service. US Army. COE.
2.Jones,R. 1954. Mar. Res. 2.

1-234 Fish
Growth and reproduction

Species	Value	Condition
Ctenopharyngodon sp.	1.9 % of ingestion	None[1]
Ctenopharyngodon sp.	14 % of ingestion	None[1]
Perca fluviatilis	19.5 % of ingestion	None[1]
Perca fluviatilis	54.5 % of assimilation	None[1]

Source: 1.Wetzel, R.G. 1975. Limnology. Saunders. Philadelphia.

1-235 Fish
H / biomass ratio

Species	Value	Condition
Fish	6.8 % of dry weight	Mean elementary composition[1]

Source: 1.Bowen, H.J.M. 1966. Trace Elements in Biochemistry, Ac. Press.

1-236 Fish
Half saturation constant

Species	Value	Condition
Channel catfish	3.1 % of body weight consumed/day	Weight = 4 g, 303 K, mixed diet[1]
Fish	0.075 mg/l	Value for grazing on zooplankton, rg = 0.05 - 0.1[2]
Fish	1025 mg/m**2	Grazing on benthos, rg = 50 - 2000[2]
Micropterus salmoides	4.6 % of body weight consumed/day	Lenght = 24.5 cm, 294 K, food = Minnow[1]
Muskellunge	5.6 % of body weight consumed/day	Length = 17.0 cm, 17.0 g, 292.5 K, food=Minnow[1]
Reticulate sculpin	4.4 % of body weight consumed/day	Weight = 1.2 g, 284.1 K, food = midge larvae[1]
Smallmouth bass	7.2 % of body weight consumed/day	Length = 8.3 - 20.2 cm, 293.3 K, food = Minnow[1]
Sockeye salmon	3.9 % of body weight consumed/day	Weight = 6.9 g, 283 K, mixed diet[1]
Sockeye salmon	7.9 % of body weight consumed/day	Weight = 7.1 g, 288 K, mixed diet[1]

Source: 1.Leidy, G.R. and R.M. Jenkins. 1977. The development of Fishery Compartments and Population Rate Coef. for Use in Reservoir Ecosystem Model. USDI Fish and Wildlife Service. US Army. COE.
2.The Hydrological Engineering Center Corps of Engineers U.S. Army. 1975. Water Quality for river-resevoir systems, Davis, Cal. 9 5 6 1 6 (916) 499-2329.

1-237 Fish
Lipid / biomass ratio

Species	Value	Condition
Eel	39.4 % of dry weight	Yellow[1]
Eel	63.8 % of dry weight	Silver[1]
Herring	49.5 % of dry weight	None[1]
Plaice	9.4 % of dry weight	None[1]

Source: 1.Special Rept. Ser. Med. Res. Com. London No. 297. 1960. The Composition of Foods.

1-238 Fish
Max growth rate

Species	Value	Condition
Channel catfish	3.4 % of body weight/day	Weight = 4 g, 303 K, mixed diet[1]
Fish	0.025 1/day	Temp factor is 1 at 293 K, rg = 0.02 - 0.03[2]
Micropterus salmoides	3.9 % of body weight/day	Lenght = 24.5 cm, 294 K, food = Minnow[1]
Muskellunge	3.9 % of body weight/day	Length = 17.0 cm, 17.0 g, 292.5 K, food=Minnow[1]
Reticulate sculpin	1.7 % of body weight/day	Weight = 1.2 g, 284.1 K, food = midge larvae[1]
Smallmouth bass	4.7 % of body weight/day	Length = 8.3 - 20.2 cm, 293.3 K, food = Minnow[1]
Sockeye salmon	1.8 % of body weight/day	Weight = 6.9 g, 283 K, mixed diet[1]
Sockeye salmon	4.2 % of body weight/day	Weight = 7.1 g, 288 K, mixed diet[1]

Source: 1.Leidy, G.R. and R.M. Jenkins. 1977. The development of Fishery Compartments and Population Rate Coef. for Use in Reservoir Ecosystem Model. USDI Fish and Wildlife Service. US Army. COE.
2.The Hydrological Engineering Center Corps of Engineers U.S. Army. 1975. Water Quality for river-resevoir systems, Davis, Cal. 9 5 6 1 6 (916) 499-2329.

1-239 Fish
Mortality

Species	Value	Condition
Engraulis japonicus	1.6 per year	Average mortality value, yearly[1]
Engraulis ringens	1.15 per year	Average mortality value, yearly[1]
Eopsetta jordani	0.2 per year	Average mortality value, yearly[1]
Eopsetta jordani	0.24 per year	Average mortality value, yearly[1]
Fish	0.003 per day	293 K, rg = 0.001 - 0.005[2]
Gadus macrocephalus	0.75 per year	Average mortality value, rg = 0.6 - 0.9[1]
Lepidopsetta bilineata	0.25 per year	Average mortality value, yearly[1]
Lepidopsetta bilineata	0.5 per year	Average mortality value, yearly[1]
Merluccius productus	0.3 per year	Average mortality value, yearly[1]
Merluccius productus	0.4 per year	Average mortality value, yearly[1]
Sebastodes alutus	0.1 per year	Average mortality value, yearly[1]
Thunnus albacores	0.8 per year	Average mortality value, yearly[1]

Source: 1.Alverson, D.L. and M.J. Carney. 1975. J. Cons. int. Explor. Mer., 36(2), pp 133-143.
2.The Hydrological Engineering Center Corps of Engineers U.S. Army. 1975. Water Quality for river-resevoir systems, Davis, Cal. 9 5 6 1 6 (916) 499-2329.

1-240 Fish
N / biomass ratio

Species	Value	Condition
Bluegill	16.7 % of dry weight	None[1]
Bluegill	2.72 % of wet weight	None[1]
Carp	2.6 % of wet weight	None[1]
Channel catfish	2.35 % of dry weight	None[1]
Fish	11.4 % of dry weight	Mean elementary composition[2]
Fish	16.3 % of dry weight	General average[1]
Fish	8 % of dry weight	Range = 7 - 9 %[3]
Fish and fish larvae	9.5 % of dry weight	Dried at 333 K, Sargasso Sea, rg = 8.3 - 10.7[4]
Largescale sucker	2.4 % of wet weight	None[1]
Mola mola	17.4 % of dry weight	Range = 16.6 - 18.2 %[1]
Northern squawfish	2.5 % of wet weight	Range = +- 0.1[1]
Salmo gairdneri	2.9 % of wet weight	None[1]
Solea solea	21.34 mg N/g	1.5 - 5.0 g, rg = 16.53 - 26.14 mg N/g[5]

Source: 1.Leidy, G.R. and R.M. Jenkins. 1977. The development of Fishery Compartments and Population Rate Coef. for Use in Reservoir Ecosystem Model. USDI Fish and Wildlife Service. US Army. COE.
2.Bowen, H.J.M. 1966. Trace Elements in Biochemistry, Ac. Press.
3.The Hydrological Engineering Center Corps of Engineers U.S. Army. 1975. Water Quality for river-resevoir systems, Davis, Cal. 9 5 6 1 6 (916) 499-2329.
4.Beers, J.R. 1966. Limnol. Oceanogr., 11, pp 520-528.
5.Bromley, P.J. 1971. J. Cons. Int. Ex plor. Mer. 34(1), pp 131-134.

1-241 Fish
O / biomass ratio

Species	Value	Condition
Fish	29 % of dry weight	Mean elementary composition[1]

Source: 1.Bowen, H.J.M. 1966. Trace Elements in Biochemistry, Ac. Press.

1-242 Fish
P / biomass ratio

Species	Value	Condition
Bluegill	4.75 % of dry weight	Range = +-0.70[1]
Carp	0.5 % of wet weight	Range = +-0.01[1]
Channel catfish	0.86 % of wet weight	None[1]
Cod	0.6 % of wet weight	None[1]
Eel	0.68 % of wet weight	None[1]
Fish	0.22 % of wet weight	Range = 0.1 - 0.4, general average for fish flesh[1]
Fish	1.1 % of wet weight	Range = 1 - 1.2 %, used in models[2]
Haddock	0.97 % of wet weight	None[1]
Halibut	0.44 % of wet weight	None[1]
Herring	0.56 % of wet weight	None[1]

Largescale sucker	0.3 % of wet weight	None[1]
Mackerel	0.56 % of wet weight	None[1]
Northern squawfish	0.4 % of wet weight	None[1]
Salmo gairdneri	0.4 % of wet weight	None[1]
Salmon	0.59 % of wet weight	None[1]
Trout	0.81 % of wet weight	None[1]
Turbot	0.48 % of wet weight	None[1]

Source: 1.Leidy, G.R. and R.M. Jenkins. 1977. The development of Fishery Compartments and Population Rate Coef. for Use in Reservoir Ecosystem Model. USDI Fish and Wildlife Service. US Army. COE.
2.The Hydrological Engineering Center Corps of Engineers U.S. Army. 1975. Water Quality for river-resevoir systems, Davis, Cal. 9 5 6 1 6 (916) 499-2329.

1-243 Fish
Production

Species	Value	Condition
Elassoma evergladei	34.7 kg wet wt. ha-1	estimated by instantaneous growth rate method, Okefenokee[1]
Elassoma evergladei	75.1 kg wet wt. ha-1	estimated by size-frequency method, Okefenokee[1]
Enneacanthus obesus	66.2 kg wet wt. ha-1	estimated by instantaneous growth rate method, Okefenokee[1]
Enneacanthus obesus	78.0 kg wet wt. ha-1	estimated by size-frequency method, Okefenokee[1]
Leptolucania ommata	17.6 kg wet wt. ha-1	estimated by size-frequency method, Okefenokee[1]
Leptolucania ommata	8.9 kg wet wt. ha-1	estimated by instantaneous growth rate method, Okefenokee[1]

Source: 1.Freeman, B.J. and Freeman, M.C., Limnol.Oceanogr., 30(3), 686, 1985.

1-244 Fish
Production / biomass ratio

Species	Value	Condition
Alburnus alburnus	1.92 on yearly basis	Prod=915 kg dw/(ha*y), River Thames, Engl, max 2 y[1]
Aspius aspius	0.55 on yearly basis	Prod=1.54 kg dw/(ha*y), Vistual River, Poland[1]
Cottus carolinae	1.8 on yearly basis	Prod=8.0 kg dw/(ha*y), Steeles Run, 3nd order stream, KY[1]
Cottus carolinae	5.5 on yearly basis	Prod=77.0 kg dw/(ha*y), Steeles Run, 2nd order stream, KY[1]
Cottus cognatus	1.2 on yearly basis	Prod=59.4 kg dw/(ha*y), Vally Lake, Minn[1]
Elassoma evergladei	1.9	estimated by instantaneous growth rate method, Okefenokee[2]
Elassoma evergladei	3.0	estimated by size-frequency method, Okefenokee[2]
Enneacanthus obesus	3.3	estimated by instantaneous growth rate method, Okefenokee[2]
Enneacanthus obesus	3.4	estimated by size-frequency method, Okefenokee[2]
Esox lucius	0.32 on yearly basis	Prod=0.75 kg dw/(ha*y), reservoir, Czech.[1]
Esox lucius	0.7 on yearly basis	Prod=1.5 kg dw/(ha*y), Vistula River, Poland[1]
Etheostoma spectabile	1.7 on yearly basis	Prod=23 kg dw/(ha*y), Steeles Run, 3nd order stream, KY[1]
Etheostoma spectabile	2.4 on yearly basis	Prod=52 kg dw/(ha*y), Steeles Run, 2nd order stream, KY[1]
Fish	0.01 per day	Southern lakes of USSR, less than[3]
Fish fauna	0.3 on yearly basis	Prod = 14 kg dw/(ha*y), Myastro Lake, USSR, piscivores[1]
Fish fauna	0.4 on yearly basis	Prod = 14 kg dw/(ha*y), Batorin Lake, USSR, piscivores[1]
Fish fauna	0.4 on yearly basis	Prod = 26 kg dw/(ha*y), Myastro Lake, USSR, benthivores[1]

Fish fauna	0.4 on yearly basis	Prod = 9 kg dw/(ha*y), Naroch Lake, USSR, piscivores[1]
Fish fauna	0.5 on yearly basis	Prod = 13 kg dw/(ha*y), Naroch Lake, USSR, benthivores[1]
Fish fauna	0.5 on yearly basis	Prod = 60 kg dw/(ha*y), Batorin Lake, USSR, benthivores[1]
Fish fauna	0.7 on yearly basis	Prod = 22 kg dw/(ha*y), Naroch Lake, USSR, planktivores[1]
Fish fauna	0.7 on yearly basis	Prod = 24 kg dw/(ha*y), Myastro Lake, USSR, planktivores[1]
Fish fauna	0.8 on yearly basis	Prod = 9 kg dw/(ha*y), Batorin Lake, USSR, planktivores[1]
Fish fauna	0.85 on yearly basis	Prod = 244 kg dw/(ha*y), Rybinsk River, USSR, all trophic levels[1]
Gobio gobio	1.94 on yearly basis	Prod=289 kg dw/(ha*y), River Thames, Engl, max 2 y[1]
Leptolucania ommata	2.0	estimated by instantaneous growth rate method, Okefenokee[2]
Leptolucania ommata	3.6	estimated by size-frequency method, Okefenokee[2]
Leuciscus cephalus	0.27 on yearly basis	Prod=2.1 kg dw/(ha*y), reservoir, Czech.[1]
Leuciscus cephalus	0.43 on yearly basis	Prod=0.10 kg dw/(ha*y), Vistula River, Poland[1]
Leuciscus leuciscus	1.75 on yearly basis	Prod=52 kg dw/(ha*y), River Thames, Engl, 1 y[1]
Oncorhynchus kisutch	2.4 on yearly basis	Prod=86 kg dw/(ha*y), n=3 streams, juveniles, Oregon[1]
Perca fluviatilis	0.35 on yearly basis	Prod=7.3 kg dw/(ha*y), reservoir, Czech., n=3 y[1]
Perca fluviatilis	0.41 on yearly basis	Prod=0.04 kg dw/(ha*y), Vistula River, Poland[1]
Rutilus rutilus	0.24 on yearly basis	Prod=19.2 kg dw/(ha*y), resevoir, Czech.[1]
Rutilus rutilus	1.12 on yearly basis	Prod=281 kg dw/(ha*y), River Thames, Engl, max 2 y[1]
Salmo clarki	1 on yearly basis	Prod=41 kg dw/(ha*y), n=3 streams, Oregon[1]
Salmo gairdneri	2 on yearly basis	Prod=47.6 kg dw/(ha*y), Whistler Bend, Oregon[1]
Salmo gairdneri	2.4 on yearly basis	Prod=132 kg dw/(ha*y), juveniles, Bothwells Creek, Ontario[1]
Salmo trutta	0.5 on yearly basis	Prod=23 kg dw/(ha*y), Great Dodgen Pot, slow growth, Engl[1]
Salmo trutta	0.5 on yearly basis	Prod=38 kg dw/(ha*y), River Yarty, fry prod. in tribs., Engl[1]
Salmo trutta	1 on yearly basis	Prod=106 kg dw/(ha*y), Black Earth Creek, polluted, Wisc.[1]
Salmo trutta	1.1 on yearly basis	Prod=132 kg dw/(ha*y), Walla Brook, n=2 y, Engl[1]
Salmo trutta	1.1 on yearly basis	Prod=22 kg dw/(ha*y), n=3 streams, Engl, normal growth[1]
Salmo trutta	1.2 on yearly basis	Prod=328 kg dw/(ha*y), Black Earth River, poll. abated, Wisc.[1]
Salmo trutta	2 on yearly basis	Prod=544 kg dw/(ha*y), Horokiwi Stream, n=6 places, NZ[1]
Salmo trutta	4.9 on yearly basis	Prod=85 kg dw/(ha*y), nursery stream, Hinau main water, NZ[1]
Salmo trutta	5 on yearly basis	Prod=89 kg dw/(ha*y), nursery stream, Hinaki main water, NZ[1]
Salvelinus fontinalis	0.62 on yearly basis	Prod = 0.21 kg dw/(ha*y), Bill Lake, Quebec[1]
Salvelinus fontinalis	0.7 on yearly basis	Prod = 38 kg dw/(ha*y), Gallienne Creek, n=2 y, Quebec[1]
Salvelinus fontinalis	0.71 on yearly basis	Prod=2.2 kg dw/(ha*y), Matamek Lake, Quebec[1]
Salvelinus fontinalis	1.05 on yearly basis	Prod = 25 kg dw/(ha*y), Sherry Creek, n=3 y, Quebec[1]
Salvelinus fontinalis	1.2 on yearly basis	Prod = 16 kg dw/(ha*y), Tchinicaman River, n=2 y, Quebec[1]
Salvelinus fontinalis	1.3 on yearly basis	Prod = 145 kg dw/(ha*y), Valley Creek, post-flood, Minn[1]
Salvelinus fontinalis	1.6 on yearly basis	Prod = 116 kg dw/(ha*y), Lawrence Creek, n=6 y, Minn[1]
Salvelinus fontinalis	1.6 on yearly basis	Prod = 63 kg dw/(ha*y), Kailhosru Creek, n=2 y, Quebec[1]
Stizostedion lucioperca	0.57 on yearly basis	Prod = 5.5 kg dw/(ha*y), Lake Balaton, Hungary[1]
Stizostedion lucioperca	0.72 on yearly basis	Prod = 0.91 kg dw/(ha*y), Vistula River, Poland[1]

Stizostedion vitreum vitreum	0.43 on yearly basis	Prod = 2.1 kg dw/(ha*y), West Blue Lake, Manitoba[1]
Tinca tinca	0.18 on yearly basis	Prod = 0.17 kg dw/(ha*y), resevoir, Czech.[1]

Source: 1.Waters, T.F. 1977. Adv. Ecol. Res. 10, pp 91-164.
2.Freeman, B.J. and Freeman, M.C., Limnol.Oceanogr., 30(3), 686, 1985.
3.Greze, V.N. 1973. Okeanologiya, 13(1), pp 107-113.

1-245 Fish
Protein / biomass ratio

Species	Value	Condition
Eel	33.6 % of dry weight	Silver eel[1]
Eel	58 % of dry weight	Yellow eel[1]
Herring	45.7 % of dry weight	None[1]
Plaice	79.5 % of dry weight	None[1]

Source: 1.Special Rept. Ser. Med. Res. Com. London No. 297. 1960. The Composition of Foods.

1-246 Fish
Respiration rate

Species	Value	Condition
Ctenopharyngodon sp.	12.2 % of ingestion	None[1]
Ctenopharyngodon sp.	86 % of assimilation	None[1]
Fish	0.00055 1/day	End resp., temp factor is 1 at 293 K, temp.<278 K, rg = 0.0001-0.001[2]
Fish	0.0055 1/day	Temp factor is 1 at 293 K, temp>278 K, end resp, rg = 0.001 - 0.01[2]
Perca fluviatilis	16.2 % of ingestion	None[1]
Perca fluviatilis	45.5 % of assimilation	None[1]
Salvelinus fontinalis	0.17 ml O2/hour	Weight = 5 g, 8 cm, basal metabolism, oxygen not limiting, 278 K[3]
Salvelinus fontinalis	0.36 ml O2/hour	Weight = 5 g, 8 cm, basal metabolism, oxygen not limiting, 283 K[3]
Salvelinus fontinalis	0.52 ml O2/hour	Oxygen conc = 3.2 mg O2/l, weight = 5 g, 8 cm[3]
Salvelinus fontinalis	0.6 ml O2/hour	Weight = 5 g, 8 cm, basal metabolism, oxygen not limiting, 288 K[3]
Salvelinus fontinalis	0.68 ml O2/hour	Weight = 5 g, 8 cm, basal metabolism, oxygen not limiting, 293 K[3]
Salvelinus fontinalis	0.85 ml O2/hour	Oxygen conc = 5.0 mg O2/l, weight = 5 g, 8 cm[3]
Salvelinus fontinalis	0.97 ml O2/hour	Weight = 5 g, 8 cm, active metabolism, oxygen not limiting, 278 K[3]
Salvelinus fontinalis	1 ml O2/hour	Oxygen conc = 6.4 mg O2/l, weight = 5 g, 8 cm[3]
Salvelinus fontinalis	1.38 ml O2/hour	Weight = 5 g, 8 cm, active metabolism, oxygen not limiting, 283 K[3]
Salvelinus fontinalis	1.9 ml O2/hour	Weight = 5 g, 8 cm, active metabolism, oxygen not limiting, 288 K[3]
Salvelinus fontinalis	10 ml O2/hour	Oxygen conc = 3.4 mg O2/l, weight = 135 g, 23 cm[3]
Salvelinus fontinalis	102 ml O2/hour	Oxygen conc = 9.3 mg O2/l, weight = 1000 g, 45 cm[3]
Salvelinus fontinalis	11.3 ml O2/hour	Weight = 135 g, 23 cm, basal metabolism, oxygen not limiting, 293 K[3]
Salvelinus fontinalis	12.6 ml O2/hour	Oxygen conc = 5.0 mg O2/l, weight = 135 g, 23 cm[3]
Salvelinus fontinalis	124 ml O2/hour	Weight = 1 kg, 45 cm, active metabolism, oxygen not limiting, 283 K[3]

Salvelinus fontinalis	127 ml O2/hour	Weight = 1 kg, 45 cm, active metabolism, oxygen not limiting, 278 K[3]
Salvelinus fontinalis	149 ml O2/hour	Weight = 1 kg, 45 cm, active metabolism, oxygen not limiting, 293 K[3]
Salvelinus fontinalis	158 ml O2/hour	Weight = 1 kg, 45 cm, active metabolism, oxygen not limiting, 288 K[3]
Salvelinus fontinalis	17.2 ml O2/hour	Oxygen conc = 6.4 mg O2/l, weight = 135 g, 23 cm[3]
Salvelinus fontinalis	18.3 ml O2/hour	Weight = 1 kg, 45 cm, basal metabolism, oxygen not limiting, 278 K[3]
Salvelinus fontinalis	2.8 ml O2/hour	Weight = 5 g, 8 cm, active metabolism, oxygen not limiting, 293 K[3]
Salvelinus fontinalis	2.9 ml O2/hour	Oxygen conc = 9.3 mg O2/l, weight = 5 g, 8 cm[3]
Salvelinus fontinalis	20.7 ml O2/hour	Weight = 135 g, 23 cm, active metabolism, oxygen not limiting, 278 K[3]
Salvelinus fontinalis	23.5 ml O2/hour	Oxygen conc = 9.3 mg O2/l, weight = 135 g, 23 cm[3]
Salvelinus fontinalis	23.9 ml O2/hour	Weight = 135 g, 23 cm, active metabolism, oxygen not limiting, 283 K[3]
Salvelinus fontinalis	29.1 ml O2/hour	Weight = 135 g, 23 cm, active metabolism, oxygen not limiting, 288 K
Salvelinus fontinalis	3.4 ml O2/hour	Weight = 135 g, 23 cm, basal metabolism, oxygen not limiting, 278 K[3]
Salvelinus fontinalis	31.6 ml O2/hour	Weight = 135 g, 23 cm, active metabolism, oxygen not limiting, 293 K[3]
Salvelinus fontinalis	37 ml O2/hour	Weight = 1 kg, 45 cm, basal metabolism, oxygen not limiting, 283 K[3]
Salvelinus fontinalis	47.7 ml O2/hour	Oxygen conc = 3.6 mg O2/l, weight = 1000 g, 45 cm[3]
Salvelinus fontinalis	48.9 ml O2/hour	Weight = 1 kg, 45 cm, basal metabolism, oxygen not limiting, 288 K[3]
Salvelinus fontinalis	55.2 ml O2/hour	Weight = 1 kg, 45 cm, basal metabolism, oxygen not limiting, 293 K[3]
Salvelinus fontinalis	67.4 ml O2/hour	Oxygen conc = 5.0 mg O2/l, weight = 1000 g, 45 cm[3]
Salvelinus fontinalis	7 ml O2/hour	Weight = 135 g, 23 cm, basal metabolism, oxygen not limiting, 283 K[3]
Salvelinus fontinalis	9.7 ml O2/hour	Weight = 135 g, 23 cm, basal metabolism, oxygen not limiting, 288 K[3]
Salvelinus fontinalis	92.1 ml O2/hour	Oxygen conc = 6.4 mg O2/l, weight = 1000 g, 45 cm[3]

Source: 1.Wetzel, R.G. 1975. Limnology. Saunders. Philadelphia.
2.The Hydrological Engineering Center Corps of Engineers U.S. Army. 1975. Water Quality for river-resevoir systems, Davis, Cal. 9 5 6 1 6 (916) 499-2329.
3.Madsen, B.L., P.N. Markmann and G.Rasmussen. 1977. Vand, 8(1), pp 4-8.

1-247 Fish
S / biomass ratio

Species	Value	Condition
Pisces	1 % of dry weight	Mean elementary composition[1]

Source: 1.Bowen, H.J.M. 1966. Trace Elements in Biochemistry, Ac. Press.

1-248 Fish
Wet / dry

Species	Value	Condition
Fish	5 by weight	None[1]

Source: 1.Waters, T.F. 1977. Adv. Ecol. Res. 10, pp 91-164.

1-249 Fungi
Ash / biomass ratio

Species	Value	Condition
Fungi	6.32 % of dry weight	11 species[1]

Source: 1.Baumeister, W. 1958. In; Ruhland, W. (ed); Encyclopedia of Plant Physiology, 4, p 5.

1-250 Fungi
C / biomass ratio

Species	Value	Condition
Fungi	49.4 % of dry weight	Mean elementary composition[1]

Source: 1.Bowen, H.J.M. 1966. Trace Elements in Biochemistry, Ac. Press.

1-251 Fungi
Carbohydrate / biomass ratio

Species	Value	Condition
Saccharomyces cerevisiae	37 % of dry weight	None[1]
Torulopsis utilis	30 % of dry weight	None[1]

Source: 1.Grylls, F.S.M. 1961. in: Long, C. (ed.): Biochemists Handbook. Van Nostrand Princeton. New Jersey.

1-252 Fungi
Density

Species	Value	Condition
Carex walteriana	0 colonies ml-1	fraction < 1 um[1]
Carex walteriana	0 colonies ml-1	fraction < 1 um, candicidin added[1]
Carex walteriana	60 +/- 40 colonies ml-1	unfractionated, candicidin added[1]
Carex walteriana	880 +/- 260 colonies ml-1	unfractionated[1]
Spartina alterniflora	0 colonies ml-1	fraction < 1 um, candicidin added[1]
Spartina alterniflora	0 colonies ml-1	unfractionated, candicidin added[1]
Spartina alterniflora	2 +/- 8 colonies ml-1	fraction < 1 um[1]
Spartina alterniflora	82 +/- 86 colonies ml-1	unfractionated[1]
Taxodium ascendens	0 colonies ml-1	fraction < 1 um[1]
Taxodium ascendens	0 colonies ml-1	fraction < 1 um, candicidin added[1]
Taxodium ascendens	0 colonies ml-1	unfractionated, candicidin added[1]
Taxodium ascendens	112 +/- 22 colonies ml-1	unfractionated[1]

Source: 1.Benner, R., Moran, M.A. and Hodson, R.E. Limnol. Oceanogr., 31, 89, 1986.

1-253 Fungi
Growth rate

Species	Value	Condition
Basisporium gallarum	0.45 doublings/day	298 K, liquid shaken culture[1]
Monilinia fructicola	0.8 doublings/day	298 K, liquid shaken culture[1]
Saccharomyces cerevisiae	12 doublings/day	303 K, liquid shaken culture[1]

Source: 1.Brock, T.D. 1966. Principles of Microbial Ecology. Prentice-Hall. New Jersey.

1-254 Fungi
Growth using aliphatic hydrocarbon as substrate

Species	Value	Condition
Candida guilliermondii	50 % growth occured	Sub = alkane fractions, tot fatty acids found, yeast[1]
Candida lipolytica	50 % growth occured	Sub = alkane 12-17, tot fatty acids found, yeast[1]
Candida lipolytica	50 % growth occured	Sub = alkanes 11-18, tot fatty acids and phospholipids found, yeast[1]
Candida lipolytica	50 % growth occured	Sub = alkanes 12+16 in deparaffinated gas oil, tot fatty acids found, yeast[1]
Candida lipolytica	50 % growth occured	Sub = hexadecane, tot fatty acids found, yeast[1]
Candida sp.	50 % growth occured	Sub = alkanes 12-16, fatty acids and triglycerides found, C. 107, yeast[1]
Candida sp.	50 % growth occured	Sub = alkanes 14-16, 18 and mix, tot fatty acids found, C. strain H17, yeast[1]
Candida tropicalis	50 % growth occured	Sub = alkanes 11-18, tot fatty acids and phospholipids found, yeast[1]
Candida tropicalis	50 % growth occured	Sub = alkanes 12-16, fatty acids and triglycerids found,, yeast[1]
Candida tropicalis	50 % growth occured	Sub = alkanes 12-17, tot fatty acids, yeast[1]
Candida tropicalis	50 % growth occured	Sub = mixed alkanes, fatty acids and phospholipids fiund, yeast[1]
Cladosporium resinae	50 % growth occured	Sub = Aviation fuel, alkanes 10-16, total fatty acids found[1]
Cladosporium resinae	50 % growth occured	Sub = alkanes 10-16, hydrocarbons in lipid found[1]
Cunninghamella elegans	50 % growth occured	Sub = alkanes 13-15, alkenes 14-18, tot fatty acids found[1]
Endomycopsis lipolytica	50 % growth occured	Sub = alkanes 10,12-16, 17, C-16-1-ene and -ol, tot fatty acids found, yeast[1]
Paecilomyces persicinus	50 % growth occured	Sub = hexadecane, polar lipids found[1]
Penicillium zonatum	50 % growth occured	Sub = alkanes 13-15, tot fatty acids found[1]
Rhodotorula gracilis	50 % growth occured	Sub = alkane mixture, total fatty acids found, yeast[1]

Source: 1.Watkinson, R.J. 1978. Development in Biodegradation of Hydrocarbons - 1. Applied Science publishers.London.

1-255 Fungi
H / biomass ratio

Species	Value	Condition
Fungi	5.5 % of dry weight	Mean elementary composition, interpolated[1]

Source: 1.Bowen, H.J.M. 1966. Trace Elements in Biochemistry, Ac. Press.

1-256 Fungi
Lipid / biomass ratio

Species	Value	Condition
Saccharomyces cerevisiae	2 % of dry weight	None[1]
Torulopsis utilis	3 % of dry weight	None[1]

Source: 1.Grylls, F.S.M. 1961. in: Long, C. (ed.): Biochemists Handbook. Van Nostrand Princeton. New Jersey.

1-257 Fungi
Lower temperature limit for growth rate

Species	Value	Condition
Absidia corymbifera	14 degrees Celcius	Zygomycetes[1]
Acremonium alabamensis	25 degrees Celcius	Deuteromycetes, less than value[2]
Acrophialophora fusispora	14 degrees Celcius	Deuteromycetes[1]
Allescheria terrestris	22 degrees Celcius	Also Cephalosporium sp., Ascomycetes[1]
Aspergillus candidus	12.5 degrees Celcius	Rg = 50-55, Deuteromycetes, rg=10-15[3]
Aspergillus fumigatus	16 degrees Celcius	Rg = 52-55, Deuteromycetes, rg=12-20[1]
Burgoa-Papulaspora sp.	20 degrees Celcius	Mycelia sterilia, less than value[4]
Byssochlamys verrucosa	15 degrees Celcius	Also Paecilomyces, Ascomycetes[5]
Calcarisporium thermophile	16 degrees Celcius	Deuteromycetes[6]
Cephalosporium sp.	25 degrees Celcius	Deuteromycetes[1]
Chaetomium sp.	14 degrees Celcius	Chaetomium sp. A, Ascomycetes[1]
Chaetomium sp.	20 degrees Celcius	Ascomycetes, more than value[7]
Chaetomium thermophile	26 degrees Celcius	Rg = 58-61, Ascomycetes, rg=25-27[1]
Chaetomium thermophile	26.5 degrees Celcius	Chaetomium thermophile var. coprophile, Ascomycetes, rg=25-28[1]
Chaetomium thermophile	26.5 degrees Celcius	Chaetomium thermophile var. dissitum, Ascomycetes, rg=25-28[1]
Coprinus sp.	20 degrees Celcius	Basidiomycetes[1]
Emiricella nidulans	10 degrees Celcius	E. n. stat. conid Aspergillus, Ascomycetes[8]
Humicola grisea	22 degrees Celcius	Rg = 55-56, H.g. var thermoidea, Deuteromycetes, rg=20-24[1]
Humicola insolens	21.5 degrees Celcius	Deuteromycetes, rg=20-23[1]
Humicola langinosa	29 degrees Celcius	Deuteromycetes, rg=28-30[1]
Humicola stellata	24 degrees Celcius	Deuteromycetes, less than value[1]
Malbranchea pulchella	27.5 degrees Celcius	Rg = 53057, M.p. var. sulfurea, Deuteromycetes, rg=25-30[1]
Mocor pusillus	23.5 degrees Celcius	Rg = 55-60, Zygomycetes,, rg=24-25[1]
Mortierella turficola	25 degrees Celcius	Zygomycetes[9]
Mucor miehei	24.5 degrees Celcius	Rg = 55-57, Zygomycetes[1]
Mucor sp.	25 degrees Celcius	Mucor sp. I, Zygomycetes[1]
Mucor sp.	25 degrees Celcius	Mucor sp. II, Zygomycetes[1]
Myriococcum albomyces	25.5 degrees Celcius	Rg=55-57, equal to Melanocarpus a., Ascomycetes, rg=25-26[10]
Paecilomyces puntonii	30 degrees Celcius	Deuteromycetes, more than value[7]
Paecilomyces spp.	30 degrees Celcius	Rg = 55-60, P. spp. group b, 7 strains, Deuteromycetes, less than value[1]

Paecilomyces spp.	30 degrees Celcius	Rg = 55-60, P. spp. group c, 6 strains, Deuteromycetes, less than value[1]
Paecilomyces spp.	30 degrees Celcius	Rg = 55-60, P. spp., group d, 4 strains, Deuteromycetes, less than value[1]
Paecilomyces variotii	5 degrees Celcius	Deuteromycetes[11]
Papulaspora thermophila	29.5 degrees Celcius	Mycelia sterilia[1]
Penicillium agrillaceum	15 degrees Celcius	Deuteromycetes[12]
Penicillium sp.	20 degrees Celcius	Deuteromycetes, more than value[7]
Penicillium sp.	20 degrees Celcius	P sp. A, Deuteromycetes[1]
Phanerochaete chrysosporium	12 degrees Celcius	Basidiomycetes, less than value[13]
Rhizomucor sp.	27.5 degrees Celcius	Rg = 60-61, Zygomycetes, rg=25-30[1]
Rhizopus arrhizus	15 degrees Celcius	Zygomycetes, more than value[14]
Rhizopus cohnii	10 degrees Celcius	Zygomycetes[1]
Rhizopus microsporus	12 degrees Celcius	Zygomycetes[1]
Rhizopus sp.	16 degrees Celcius	Rhizopus sp. I, Zygomycetes[1]
Rhizopus sp.	16 degrees Celcius	Rhizopus sp. II, Zygomycetes[1]
Rhizopus sp.	25 degrees Celcius	Rhizopus sp. A, Zygomycetes[1]
Rhizopus sp.	28 degrees Celcius	Rhizopus sp. III, Zygomycetes[1]
Scolecobasidium sp.	15 degrees Celcius	Equal to Diplorhinotrichum gallopavum, Deuteromycetes, rg=14-16[1]
Sphaerospora saccata	12 degrees Celcius	Ascomycetes[1]
Sporotricum thermophile	21 degrees Celcius	Deuteromycetes, rg=18-24[15]
Stilbella thermophila	25 degrees Celcius	Deuteromycetes[1]
Talaromyces byssochlamydoides	25 degrees Celcius	T.b. stat. conid. Paecilomyces, Ascomycetes[11]
Talaromyces emersonii	27.5 degrees Celcius	Rg=55-60, T.e. stat. conid. Penicillium, Ascomycetes, rg=25-30[1]
Talaromyces leycettanus	18 degrees Celcius	T.l. sta. conid. Penicillium, Ascomycetes[1]
Talaromyces thermophilus	27.5 degrees Celcius	Rg = 57-60, T.t. stat. conid. Penicillium, Ascomycetes, rg=25-30[1]
Theilavia sp.	16 degrees Celcius	Theilavis sp. A, Ascomycetes[1]
Theilavia thermophila	20 degrees Celcius	T.t. stat. conid. Sporotrichum, Ascomycetes[1]
Thermoascus aurantiacus	27.5 degrees Celcius	Rg = 55-62, Ascomycetes, rg=25-30[1]
Thermoascus crustaceus	20 degrees Celcius	T.c. sta. conid. Paecilomyces, Ascomycetes[1]
Thermoascus thermophilus	20 degrees Celcius	T.t. stat. conid. Polypaecilum, Ascomycetes, more than value[10]
Thermomyces ibadanesis	33 degrees Celcius	Deuteromycetes, rg=31-35[1]
Thielavia australiensis	20 degrees Celcius	T.a. sta. conid. Chrysosporium, Ascomycetes[16]
Torula thermophila	23 degrees Celcius	Deuteromycetes[1]
Tritirachium sp.	16 degrees Celcius	Equal to Nodulisporium cylindroconium, Deuteromycetes[17]

Source: 1.Crisan, E.V. 1973. Mycologia 65, 1171-1198.
2.Morgan-Jones, G. 1974. Can. J. Bot. 52, 429-431.
3.Christensen, C.M. and H,H, Kaufmann. 1974. In: C.M. Christensen (ed.): Storage of Cereal Grains and Cereal Chemists. Amer. Ass. of Cereal Chemists. St. Paul. Minnesota. 158-192.
4.Tansey, M.R. 1973. Mycologia 65, 594-601.
5.Samson, R.A. and M.R. Tansey. 1975. Trans. Br. Mycol. Soc. 65, 512-514.
6.Hoog, G.S. de 1975. Studies in Mycology No. 7. Centraalbureau voor Schimmelcultures, Baarn, Netherlads. Eicker, A. 1972. S. Afr. J. Sci. 68, 150-155.
8.Evans, H.C. 1971. Trans. Br. Mycol. Soc. 57, 267-272.
9.Hayes, W.A. 1969. Mushr. Sci. 7, 173-186
10.Arx, J.A. 1975. Studies in Mycology No. 8. Centraalbureau voor Schimmelcultures, Baarn, Netherlands.
11.Stolk, A.C. and R.A. Samson. 1972. Studies in Mycology No. 2. Centraalbureau voor Schimmelcultures, Baarn, Netherlands.

12.Minoura, K., M. Yokoe, T. Kizima and T. Nehira. 1973. Trans. Mycol. Soc. Japan 14, 352-361.
13.Burdsall, H.H. Jr. and W. Eslyn. 1974. Mycotaxon 1, 123-133.
14.Nilsson, T. 1973. Stud. For. Suec. 104, 1-40.
15.Klopotek, A. 1974. Arch. Microbiol. 98, 365-369.
16.Tansey, M.R. and M.A. Jack. 1975. Can. J. Bot. 53, 81-83.
17.Schmelts, N. and D. Hoffman. 1976. In: R. Freudenthal and P.W. Jones (eds.): Carcinogenes - Comprehesive Survey. Vol 1. Polynuclear Aromatic Hydrocarbones. NY. Raven Press.

1-258 Fungi
N / biomass ratio

Species	Value	Condition
Fungi	5.1 % of dry weight	Mean elementary composition[1]

Source: 1.Bowen, H.J.M. 1966. Trace Elements in Biochemistry, Ac. Press.

1-259 Fungi
O / biomass ratio

Species	Value	Condition
Fungi	34 % of dry weight	Mean elementary composition[1]

Source: 1.Bowen, H.J.M. 1966. Trace Elements in Biochemistry, Ac. Press.

1-260 Fungi
Protein / biomass ratio

Species	Value	Condition
Saccharomyces cerevisiae	40 % of dry weight	None[1]
Torulopsis utilis	54.4 % of dry weight	None[1]

Source: 1.Grylls, F.S.M. 1961. in: Long, C. (ed.): Biochemists Handbook. Van Nostrand Princeton. New Jersey.

1-261 Fungi
Respiration rate

Species	Value	Condition
Fungi	161 gcal/(g*day)	Grassland soil, 293 K[1]
Neurospora crassa	6.4 ml O2/((g wet weight)*hour)	299 K[2]
Saccharomyces cerevisiae	11.25 ml O2/((g wet weight)*hour)	299 K, rg = 8 - 14.5[2]

Source: 1.MacFadyen, A. 1963. Animal Ecology, Aims and Methods. 2nd edition. Pitman. London.
2.Giese, A.C. 1962. Cell Physiology. 2nd edition. Saunders. Philadelphia.

1-262 Fungi
S / biomass ratio

Species	Value	Condition
Fungi	0.4 % of dry weight	Mean elementary composition, interpolated[1]

Source: 1.Bowen, H.J.M. 1966. Trace Elements in Biochemistry, Ac. Press.

1-263 Fungi
Temperature for max growth rate

Species	Value	Condition
Absidia corymbifera	40 degrees Celcius	Zygomycetes[1]
Acrophialophora fusispora	40 degrees Celcius	Deuteromycetes[1]
Allescheria terrestris	43.5 degrees Celcius	Also Cephalosporium sp., Ascomycetes, rg=42-45[1]

Aspergillus candidus	47.5 degrees Celcius	Deuteromycetes, rg=45-50[2]
Aspergillus fumigatus	40 degrees Celcius	Deuteromycetes, rg=37-43[1]
Byssochlamys verrucosa	40 degrees Celcius	Also Paecilomyces, Ascomycetes[3]
Calcarisporium thermophile	40 degrees Celcius	Deuteromycetes[4]
Cephalosporium sp.	42.5 degrees Celcius	Deuteromycetes[1]
Chaetomium sp.	40 degrees Celcius	Chaetomium sp. A, Ascomycetes[1]
Chaetomium sp.	41.5 degrees Celcius	Ascomycetes, rg=40-43[5]
Chaetomium thermophile	47.5 degrees Celcius	Chaetomium thermophile var. dissitum, Ascomycetes, rg=45-50[1]
Chaetomium thermophile	50 degrees Celcius	Ascomycetes[1]
Chaetomium thermophile	50 degrees Celcius	Chaetomium thermophile var. coprophile, Ascomycetes, rg=45-55[1]
Coprinus sp.	45 degrees Celcius	Basidiomycetes[1]
Emiricella nidulans	36 degrees Celcius	E. n. stat. conid Aspergillus, Ascomycetes, rg=35-37[6]
Hansenula polymorpha	39.5 degrees Celcius	Ascomycetes, rg=37-42[7]
Humicola grisea	42 degrees Celcius	H.g. var thermoidea, Deuteromycetes, rg=40-45[1]
Humicola insolens	40 degrees Celcius	Deuteromycetes, rg=35-45[1]
Humicola langinosa	50 degrees Celcius	Deuteromycetes, rg=45-55[1]
Humicola stellata	40 degrees Celcius	Deuteromycetes[1]
Malbranchea pulchella	45.5 degrees Celcius	M.p. var. sulfurea, Deuteromycetes, rg=45-46[1]
Mocor pusillus	45 degrees Celcius	Zygomycetes, rg=35-55[1]
Mortierella turficola	47 degrees Celcius	Zygomycetes[8]
Mucor miehei	40 degrees Celcius	Zygomycetes, rg=35-45[1]
Mucor sp.	43.5 degrees Celcius	Mucor sp. II, Zygomycetes, rg=42-45[1]
Mucor sp.	45 degrees Celcius	Mucor sp. I, Zygomycetes[1]
Myriococcum albomyces	39 degrees Celcius	equal to Melanocarpus a., Ascomycetes, rg=3542[9]
Paecilomyces spp.	47.5 degrees Celcius	P. spp. group b, 7 strains, Deuteromycetes, rg=45-50[1]
Paecilomyces spp.	47.5 degrees Celcius	P. spp. group c, 6 strains, Deuteromycetes, rg=45-50[1]
Paecilomyces spp.	47.5 degrees Celcius	P. spp., group d, 4 strains, Deuteromycetes, rg=45-50[1]
Paecilomyces variotii	37.5 degrees Celcius	Deuteromycetes[10]
Papulaspora thermophila	45 degrees Celcius	Mycelia sterilia[1]
Penicillium agrillaceum	35 degrees Celcius	Deuteromycetes[11]
Penicillium sp.	42 degrees Celcius	P sp. A, Deuteromycetes[1]
Penicillium sp.	45 degrees Celcius	Deuteromycetes[5]
Phanerochaete chrysosporium	38 degrees Celcius	Basidiomycetes, rg=36-40[12]
Rhizomucor sp.	49 degrees Celcius	Zygomycetes, rg=45-53[1]
Rhizopus arrhizus	42.5 degrees Celcius	Zygomycetes, rg=40-45[13]
Rhizopus cohnii	42 degrees Celcius	Zygomycetes[1]
Rhizopus microsporus	40 degrees Celcius	Zygomycetes[1]
Rhizopus sp.	40 degrees Celcius	Rhizopus sp. II, Zygomycetes[1]
Rhizopus sp.	42.5 degrees Celcius	Rhizopus sp. I, Zygomycetes, rg=40-45[1]
Rhizopus sp.	48.5 degrees Celcius	Rhizopus sp. A, Zygomycetes, rg=45-52[1]

Rhizopus sp.	49 degrees Celcius	Rhizopus sp. III, Zygomycetes, rg=48-50[1]
Scolecobasidium sp.	40 degrees Celcius	equal to Diplorhinotrichum gallopavum, Deuteromycetes[1]
Sphaerospora saccata	38.5 degrees Celcius	Ascomycetes, rg=35-42[1]
Sporotricum thermophile	45 degrees Celcius	Deuteromycetes, rg=40-50[14]
Stilbella thermophila	42.5 degrees Celcius	Deuteromycetes, rg=35-50[1]
Talaromyces byssochlamydoides	42.5 degrees Celcius	T.b. stat. conid. Paecilomyces, Ascomycetes, rg=40-45[10]
Talaromyces emersonii	42.5 degrees Celcius	T.e. stat. conid. Penicillium, Ascomycetes, rg=40-45[1]
Talaromyces leycettanus	42 degrees Celcius	T.l. sta. conid. Penicillium, Ascomycetes[1]
Talaromyces thermophilus	47.5 degrees Celcius	T.t. stat. conid. Penicillium, Ascomycetes, rg=45-50[1]
Theilavia sp.	40 degrees Celcius	Theilavis sp. A, Ascomycetes[1]
Theilavia thermophila	45 degrees Celcius	T.t. stat. conid. Sporotrichum, Ascomycetes[1]
Thermoascus aurantiacus	43 degrees Celcius	Ascomycetes, rg=40-46[1]
Thermoascus crustaceus	37 degrees Celcius	T.c. sta. conid. Paecilomyces, Ascomycetes[1]
Thermomyces ibadanesis	44.5 degrees Celcius	Deuteromycetes, rg=42-47[1]
Torula thermophila	40 degrees Celcius	Deuteromycetes, rg=35-45[1]
Torulopsis candida	36.5 degrees Celcius	Deuteromycetes, rg=36-37[15]
Tritirachium sp.	40 degrees Celcius	Equal to Nodulisporium cylindroconium, Deuteromycetes[15]

Source: 1.Crisan, E.V. 1973. Mycologia 65, 1171-1198.
2.Christensen, C.M. and H,H, Kaufmann. 1974. In: C.M. Christensen (ed.): Storage of Cereal Grains and Cereal Chemists. Amer. Ass. of Cereal Chemists. St. Paul. Minnesota. 158-192.
3.Samson, R.A. and M.R. Tansey. 1975. Trans. Br. Mycol. Soc. 65, 512-514.
4.Hoog, G.S. de 1975. Studies in Mycology No. 7. Centraalbureau voor Schimmelcultures, Baarn, Netherlands.
5.Eicker, A. 1972. S. Afr. J. Sci. 68, 150-155.
6.Evans, H.C. 1971. Trans. Br. Mycol. Soc. 57, 267-272.
7.Levine, D.W. and C.L. Cooney. 1973. Appl. Microbiol. 26, 981-990.
8.Hayes, W.A. 1969. Mushr. Sci. 7, 173-186
9.Arx, J.A. 1975. Studies in Mycology No. 8. Centraalbureau voor Schimmelcultures, Baarn, Netherlands.
10.Stolk, A.C. and R.A. Samson. 1972. Studies in Mycology No. 2. Centraalbureau voor Schimmelcultures, Baarn, Netherlands.
11.Minoura, K., M. Yokoe, T. Kizima and T. Nehira. 1973. Trans. Mycol. Soc. Japan 14, 352-361.
12.Burdsall, H.H. Jr. and W. Eslyn. 1974. Mycotaxon 1, 123-133.
13.Nilsson, T. 1973. Stud. For. Suec. 104, 1-40.
14.Klopotek, A. 1974. Arch. Microbiol. 98, 365-369.
15.Schmelts, N. and D. Hoffman. 1976. In: R. Freudenthal and P.W. Jones (eds.): Carcinogenes - Comprehesive Survey. Vol 1. Polynuclear Aromatic Hydrocarbones. NY. Raven Press.

1-264 Fungi
Upper temperature limit for growth rate

Species	Value	Condition
Absidia corymbifera	50 degrees Celcius	Zygomycetes[1]
Acremonium alabamensis	50 degrees Celcius	More than value, Deuteromycetes[2]
Acrophialophora fusispora	50 degrees Celcius	Deuteromycetes[1]
Allescheria terrestris	55 degrees Celcius	Also Cephalosporium sp., Ascomycetes[1]
Aspergillus candidus	52.5 degrees Celcius	Rg = 50-55, Deuteromycetes[3]
Aspergillus fumigatus	53.5 degrees Celcius	Rg = 52-55, Deuteromycetes[1]
Burgoa-Papulaspora sp.	53 degrees Celcius	Mycelia sterilia[4]
Byssochlamys verrucosa	55 degrees Celcius	Also Paecilomyces, Ascomycetes[5]
Calcarisporium thermophile	50 degrees Celcius	Deuteromycetes[6]
Cephalosporium sp.	50 degrees Celcius	More than value, Deuteromycetes[1]
Chaetomium sp.	50 degrees Celcius	Chaetomium sp. A, Ascomycetes[1]

Chaetomium sp.	50 degrees Celcius	More than value, Ascomycetes[7]
Chaetomium thermophile	59 degrees Celcius	Chaetomium thermophile var. coprophile, rg=58-60, Ascomycetes[1]
Chaetomium thermophile	59 degrees Celcius	Chaetomium thermophile var. dissitum, rg=58-60, Ascomycetes[1]
Chaetomium thermophile	59.5 degrees Celcius	Rg = 58-61, Ascomycetes[1]
Coprinus sp.	55 degrees Celcius	Basidiomycetes[1]
Emiricella nidulans	51 degrees Celcius	E. n. stat. conid Aspergillus, Ascomycetes[8]
Fungi	61 degrees celcius	Rg = 60 - 62 cel deg, app. value[9]
Hansenula polymorpha	50 degrees Celcius	Ascomycetes[10]
Humicola grisea	55.5 degrees Celcius	Rg = 55-56, H.g. var thermoidea, Deuteromycetes[1]
Humicola insolens	55 degrees Celcius	Deuteromycetes[1]
Humicola langinosa	60 degrees Celcius	Deuteromycetes[1]
Humicola stellata	50 degrees Celcius	Deuteromycetes[1]
Malbranchea pulchella	55 degrees Celcius	Rg = 53057, M.p. var. sulfurea, Deuteromycetes[1]
Mocor pusillus	57.5 degrees Celcius	Rg = 55-60, Zygomycetes[1]
Mortierella turficola	55 degrees Celcius	Zygomycetes[11]
Mucor miehei	56 degrees Celcius	Rg = 55-57, Zygomycetes[1]
Mucor sp.	55 degrees Celcius	Mucor sp. II, Zygomycetes[1]
Mucor sp.	56 degrees Celcius	Mucor sp. I, Zygomycetes[1]
Myriococcum albomyces	56 degrees Celcius	Rg=55-57, equal to Melanocarpus a., Ascomycetes[12]
Paecilomyces puntonii	50 degrees Celcius	More than value, Deuteromycetes[7]
Paecilomyces spp.	57.5 degrees Celcius	Rg = 55-60, P. spp. group b, 7 strains, Deuteromycetes[1]
Paecilomyces spp.	57.5 degrees Celcius	Rg = 55-60, P. spp. group c, 6 strains, Deuteromycetes[1]
Paecilomyces spp.	57.5 degrees Celcius	Rg = 55-60, P. spp., group d, 4 strains, Deuteromycetes[1]
Paecilomyces variotii	50 degrees Celcius	Deuteromycetes[13]
Papulaspora thermophila	52 degrees Celcius	Mycelia sterilia[1]
Penicillium agrillaceum	50 degrees Celcius	Deuteromycetes[14]
Penicillium sp.	50 degrees Celcius	More than value, Deuteromycetes[7]
Penicillium sp.	55 degrees Celcius	P sp. A, Deuteromycetes[1]
Phanerochaete chrysosporium	50 degrees Celcius	Basidiomycetes[15]
Rhizomucor sp.	60.5 degrees Celcius	Rg = 60-61, Zygomycetes[1]
Rhizopus arrhizus	55 degrees Celcius	Zygomycetes[16]
Rhizopus cohnii	55 degrees Celcius	Zygomycetes[1]
Rhizopus microsporus	50 degrees Celcius	Zygomycetes[1]
Rhizopus sp.	50 degrees Celcius	Rhizopus sp. II, Zygomycetes[1]
Rhizopus sp.	55 degrees Celcius	Rhizopus sp. I, Zygomycetes[1]
Rhizopus sp.	60 degrees Celcius	Rhizopus sp. A, Zygomycetes[1]
Rhizopus sp.	60 degrees Celcius	Rhizopus sp. III, Zygomycetes[1]
Scolecobasidium sp.	51 degrees Celcius	Rg=50-52, equal to Diplorhinotrichum gallopavum, Deuteromycetes[1]
Sphaerospora saccata	50 degrees Celcius	Ascomycetes[1]
Sporotricum thermophile	55 degrees Celcius	Deuteromycetes[17]

Stilbella thermophila	55 degrees Celcius	Deuteromycetes[1]
Talaromyces byssochlamydoides	50 degrees Celcius	More than value, T.b. stat. conid. Paecilomyces, Ascomycetes[13]
Talaromyces emersonii	57.5 degrees Celcius	Rg=55-60, T.e. stat. conid. Penicillium, Ascomycetes[1]
Talaromyces leycettanus	55 degrees Celcius	T.l. sta. conid. Penicillium, Ascomycetes[1]
Talaromyces thermophilus	58.5 degrees Celcius	Rg = 57-60, T.t. stat. conid. Penicillium, Ascomycetes[1]
Theilavia sp.	52 degrees Celcius	Theilavis sp. A, Ascomycetes[1]
Theilavia thermophila	56 degrees Celcius	T.t. stat. conid. Sporotrichum, Ascomycetes[1]
Thermoascus aurantiacus	58.5 degrees Celcius	Rg = 55-62, Ascomycetes[1]
Thermoascus crustaceus	55 degrees Celcius	T.c. sta. conid. Paecilomyces, Ascomycetes[1]
Thermoascus thermophilus	50 degrees Celcius	More than value, T.t. stat. conid. Polypaecilum, Ascomycetes[12]
Thermomyces ibadanesis	60.5 degrees Celcius	Rg = 60-61, Deuteromycetes[1]
Thielavia australiensis	50 degrees Celcius	More than value, T.a. sta. conid. Chrysosporium, Ascomycetes[18]
Torula thermophila	58 degrees Celcius	Deuteromycetes[1]
Torulopsis candida	50 degrees Celcius	More than value, Deuteromycetes[19]
Tritirachium sp.	55 degrees Celcius	Equal to Nodulisporium cylindroconium, Deuteromycetes[20]

Source: 1.Crisan, E.V. 1973. Mycologia 65, 1171-1198.
2.Morgan-Jones, G. 1974. Can. J. Bot. 52, 429-431.
3.Christensen, C.M. and H,H, Kaufmann. 1974. In: C.M. Christensen (ed.): Storage of Cereal Grains and Cereal Chemists. Amer. Ass. of Cereal Chemists. St. Paul. Minnesota. 158-192.
4.Tansey, M.R. 1973. Mycologia 65, 594-601.
5.Samson, R.A. and M.R. Tansey. 1975. Trans. Br. Mycol. Soc. 65, 512-514.
6.Hoog, G.S. de 1975. Studies in Mycology No. 7. Centraalbureau voor Schimmelcultures, Baarn, Netherlands.
7.Eicker, A. 1972. S. Afr. J. Sci. 68, 150-155.
8.Evans, H.C. 1971. Trans. Br. Mycol. Soc. 57, 267-272.
9.Meeks, J.C. and R.W. Castenholz. 1971. Arch. Mikrobiol. 78, 25-41.
10.Levine, D.W. and C.L. Cooney. 1973. Appl. Microbiol. 26, 981-990.
11.Hayes, W.A. 1969. Mushr. Sci. 7, 173-186
12.Arx, J.A. 1975. Studies in Mycology No. 8. Centraalbureau voor Schimmelcultures, Baarn, Netherlands.
13.Stolk, A.C. and R.A. Samson. 1972. Studies in Mycology No. 2. Centraalbureau voor Schimmelcultures, Baarn, Netherlands.
14.Minoura, K., M. Yokoe, T. Kizima and T. Nehira. 1973. Trans. Mycol. Soc. Japan 14, 352-361.
15.Burdsall, H.H. Jr. and W. Eslyn. 1974. Mycotaxon 1, 123-133.
16.Nilsson, T. 1973. Stud. For. Suec. 104, 1-40.
17.Klopotek, A. 1974. Arch. Microbiol. 98, 365-369.
18.Tansey, M.R. and M.A. Jack. 1975. Can. J. Bot. 53, 81-83.
19.Andrusenko, M. Y. and G.M. Lakhonina. 1969. Uzbek. Biol. Zh. 21-23.
20.Schmelts, N. and D. Hoffman. 1976. In: R. Freudenthal and P.W. Jones (eds.): Carcinogenes - Comprehesive Survey. Vol 1. Polynuclear Aromatic Hydrocarbones. NY. Raven Press.

1-265 Fungi
Water activity for limiting growth rate

Species	Value	Condition
Aspergillus flavus	0.90	Zero growth at this water activity[1]
Aspergillus niger	0.84	Zero growth at this water activity[1]
Candida pseudotropicalis	0.93	Zero growth at this water activity, in NaCl[1]
Penicillium spp.	0.85	Zero growth at this water activity, rg=0.8-0.9, 8 species[1]
Rhizopus nigricans	0.93	Zero growth at this water activity[1]
Saccharomyces cerevisiae	0.90	Zero growth at this water activity, in agar[1]
Saccharomyces cerevisiae	0.92	Zero growth at this water activity, in glucose[1]
Saccharomyces cerevisiae	0.94	Zero growth at this water activity, in NaCl[1]
Saccharomyces rouxii	0.845	Zero growth at this water activity, in glucose[1]

Saccharomyces rouxii	0.857	Zero growth at this water activity, in sucrose[1]
Saccharomyces rouxii	0.860	Zero growth at this water activity, in NaCl[1]
Sporendonema sebi	0.77	Zero growth at this water activity[1]
Xeromyces bisporus	0.65	Zero growth at this water activity[1]

Source: 1.Kushner, D.J. 1978. In: D.J. Kushner: Microbial Life in Extreme Environments. Academic Press. London317-368.

1-266 Fungi
Wet weight

Species	Value	Condition
Saccharomyces cerevisiae	7.1E-11 g	None[1]

Source: 1.Brock, T.D. 1966. Principles of Microbial Ecology. Prentice-Hall. New Jersey.

1-267 Invertebrates
Ash / biomass ratio

Species	Value	Condition
Acheta domesticus	5.7 % of dry weight	None[1]
Agrypnia obsoleta	4.72 % of dry weight	Trichoptera, 95 % CL = +-0.02, 12 h at 873 K[2]
Annelida	22.4 % of dry weight	46 species[3]
Asellus aquaticus	28.62 % of dry weight	Isopoda, 95 % CL = +-0.22, 12 h at 873 K[2]
Brachoria initialis	54.9 % of dry weight	None[1]
Camponotus americanus	2.8 % of dry weight	None[1]
Centroptilum luteolum	8.14 % of dry weight	Ephemeroptera, 95 % CL = +- 0.17, 12 h at 873 K[2]
Chaoborus flavicans	6.93 % of dry weight	Diptera, 95 % CL = +- 0.04, 12 h at 873 K[2]
Chironomus plumosus	5.16 % of dry weight	Diptera, 95 % CL = +- 0.11, 12 h at 873 K[2]
Coelenterata	48.8 % of dry weight	16 species[3]
Cordulia aenea	4.52 % of dry weight	Odonata, 95 % CL = +- 0.07, 12 h at 873 K[2]
Crustacea	21.7 % of dry weight	25 species[3]
Cylisticus convexus	45.2 % of dry weight	None[4]
Drassyllus virginiana	3.8 % of dry weight	None[5]
Echinodermata	60.8 % of dry weight	9 species[3]
Erpobdella octoculata	4.47 % of dry weight	Hirudinea, 95 % CL = 0.08, 12 h at 873 K[2]
Evarthrus sodalis	3.2 % of dry weight	None[1]
Gammaracanthus lacustris	20.69 % of dry weight	Amphipoda, 95 % CL = +- 0.21, 12 h at 873 K[2]
Heptagenia fuscogrisea	5.39 % of dry weight	Ephemeroptera, 95 % CL = +- 0.13, 12 h at 873 K[2]
Invertebrates and zooplankton	0.9 by weight	Invertebrates found as zooplankton[6]
Leptophlebia sp.	6.24 % of dry weight	Ephemeroptera, 95 % CL = +- 0.07, 12 h at 873 K[2]
Limnephilidae sp.	9.02 % of dry weight	Trichoptera, 95 % CL = +- 0.12, 12 h at 873 K[2]
Lumbricus terrestris	23 % of dry weight	None[7]
Mesodon thyroides	22.8 % of dry weight	None[1]
Mollusca	11 % of dry weight	70 species[3]
Mysis relicta	8.61 % of dry weight	Mysidacea, 95 % CL = +- 0.02, 12 h at 873 K[2]

Notonecta glauca	3.86 % of dry weight	Heterpotera, 95 % CL = +- 0.08, 12 h at 873 K^2
Otocryptops sexspinosus	7.2 % of dry weight	None[1]
Pallacea quadrispinosa	30.37 % of dry weight	Amphipoda, 95 % CL = +_0.76, 12 h at 873 K^2
Parcoblatta sp.	4 % of dry weight	None[5]
Sialis sp.	4.28 % of dry weight	Megaloptera, 95 % CL = +- 0.04, 12 h at 873 K^2
Tomocerus sp.	8.7 % of dry weight	None[5]

Source: 1.Edwards, C.A., D.E. Reichle and D.A. Crossley, Jr. 1970. in: Reichle, D.E. (ed); Analysis of Temperae Forest Ecosystems. pp 147-172. Springer-Verlag. Berlin.
2.Salonen, K., J. Sarvala, I. Hakala and M.-L. Viljanen. 1976. Limnol. Oceanogr., 21(5), pp 724-730.
3.Vinogradov, A.P. 1953. The Elementary Chemical Composition of Marine Organisms, Sears Foundation New Haven.
4.Reichle, D.E. 1967. Ecology 48, pp 351-366.
5.Moulder, B.C., D.E. Reichle and S.I. Auerbach. 1970. Oak Ridge National Lab. USA, ORNL 4452.
6.Waters, T.F. 1977. Adv. Ecol. Res. 10, pp 91-164.
7.French, C.E., S.A. Liscinsky and D.R. Miller. 1957. J. Wildl. Mgmt. 21, pp 348.

1-268 Invertebrates
Assimilation

Species	Value	Condition
Asellus aquaticus	30.3 % of ingestion	None[1]
Blood worm	60 % of Hg intake	Food = algae detritus[2]
Calanus helgolandicus and naphthalene	58.1 % of naphthalene intake	Female, soluble release = 26.9 %[3]
Calanus helgolandicus and naphthalene	60.7 % of naphthalene intake	Male, soluble release = 33.9 %[3]
Diplacodes trivialis	91 % of ingestion	46 mg animal, ingestion = 61.5 cal/(kcal*anim.*d)[4]
Lanice conchilega	73.9 %	Based on calories, rg = 70.6 - 77.2[5]
Lestes sponsa	36.6 % of ingestion	None[1]
Meganyctiphanes norvegica	0.15 % of ingested Zn	Diet = artemia[6]
Philaenus spumarius	43.5 % of ingestion	4rd instar, 1.16 mg, 298 K, rg = 30 - 57[7]
Philaenus spumarius	44 % of ingestion	3rd instar, 0.39 mg, 296 K, rg = 30 - 58[7]

Source: 1.Wetzel, R.G. 1975. Limnology. Saunders. Philadelphia.
2.Joergensen, S.E., J.H. Henriksen, M. Friis, L.A. Joergensen and H.F. Mejer. Handbook of Environmental Data and Ecological Parameters. 1979. ISEM. Copenhagen.
3.Harris, R.P., V. Berdugo, E.D.S. Corner, Kilvington and Ohara. 1977. In: D.A. Wolfe (ed.): Fate and Effects of Petroleum Hydrocarbons in Marine Ecosystems and Organisms. NY. Pergamon Press.
4.Pandian, T.J. and S. Mathavan. 1974. Odonatologica, 3(4), pp 241-248.
5.Buhr, K.-J. 1976. Mar. Biol., 38(4), pp 373-383.
6.Wolfe, 1975. In: Cronin, L.E. (ed.); Estuarine Research, Vol. 1. Ac. Press N.Y.
7.Wiegert, R.G. 1964. Ecol. Monogr. 34, pp 225-241.

1-269 Invertebrates
Biomass, mean annual

Species	Value	Condition
Araneae	115.5 mg dry weight/m**2	From a mesic deciduous forest, yellow poplar, Arthropoda[1]
Chilopoda	32.32 mg dry weight/m**2	From a mesic deciduous forest, yellow poplar, Arthropoda[1]
Coleoptera	134.4 mg dry weight/m**2	From a mesic deciduous forest, yellow poplar, Insecta, larvae[1]
Diplopoda	249.6 mg dry weight/m**2	From a mesic deciduous forest, yellow poplar, Arthropoda[1]
Diplura	58.3 mg dry weight/m**2	From a mesic deciduous forest, yellow poplar, Insecta, Arthropoda[1]

Diptera	849.6 mg dry weight/m**2	From a mesic deciduous forest, yellow poplar, Insecta, larvae[1]
Enchytraeidae	500 mg dry weight/m**2	From a mesic deciduous forest, yellow poplar, Annelida[1]
Entomobryidae	295.6 mg dry weight/m**2	From a mesic deciduous forest, yellow poplar, Insecta[1]
Gamasina	2836.3 mg dry weight/m**2	From a mesic deciduous forest, yellow poplar, Arthropoda[1]
Hymenoptera	16.92 mg dry weight/m**2	From a mesic deciduous forest, yellow poplar, Insecta[1]
Isotomidae	150.4 mg dry weight/m**2	From a mesic deciduous forest, yellow poplar, Insecta[1]
Lepidoptera	5.73 mg dry weight/m**2	From a mesic deciduous forest, yellow poplar, Insecta, larvae[1]
Lumbricidae	10640 mg dry weight/m**2	From a mesic deciduous forest, yellow poplar, Annelida[1]
Microflora	124000 mg dry weight/m**2	From a mesic deciduous forest, yellow poplar[1]
Nematoda	950 mg dry weight/m**2	From a mesic deciduous forest, yellow poplar[1]
Onychiuridae	37.6 mg dry weight/m**2	From a mesic deciduous forest, yellow poplar, Insecta[1]
Oribatei	3678.2 mg dry weight/m**2	From a mesic deciduous forest, yellow poplar, Arthropoda[1]
Orthoptera	26.91 mg dry weight/m**2	From a mesic deciduous forest, yellow poplar, Insecta[1]
Pauropoda	10.3 mg dry weight/m**2	From a mesic deciduous forest, yellow poplar, Arthropoda[1]
Phalangida	5.79 mg dry weight/m**2	From a mesic deciduous forest, yellow poplar, Arthropoda[1]
Poduridae	12.4 mg dry weight/m**2	From a mesic deciduous forest, yellow poplar, Insecta[1]
Prostigmata	59.3 mg dry weight/m**2	From a mesic deciduous forest, yellow poplar, Arthropoda[1]
Protura	8.3 mg dry weight/m**2	From a mesic deciduous forest, yellow poplar, Arthropoda[1]
Pseudoscorpionida	0.3 mg dry weight/m**2	From a mesic deciduous forest, yellow poplar, Arthropoda[1]
Psocoptera	2.3 mg dry weight/m**2	From a mesic deciduous forest, yellow poplar, Insecta[1]
Pulmonata	222.46 mg dry weight/m**2	From a mesic deciduous forest, yellow poplar[1]
Sminthuridae	6.3 mg dry weight/m**2	From a mesic deciduous forest, yellow poplar, Insecta[1]
Symphyla	104.1 mg dry weight/m**2	From a mesic deciduous forest, yellow poplar, Arthropoda[1]
Uropodina	237.7 mg dry weight/m**2	From a mesic deciduous forest, yellow poplar, Arthropoda[1]

Source: 1.Reichle, D.E. 1977. In: U. Lohm and T. Persson (eds.): Soil Organisms as Components of Ecosystems. Ecol. Bull (Stockholm) 25, 145-156.

1-270 Invertebrates
Body weight

Species	Value	Condition
Araneae	0.199 mg dry weight/animal	From a mesic deciduous forest, yellow poplar, Arthropoda[1]
Chilopoda	0.69 mg dry weight/animal	From a mesic deciduous forest, yellow poplar, Arthropoda[1]
Coleoptera	0.209 mg dry weight/animal	From a mesic deciduous forest, yellow poplar, Insecta, larvae[1]
Diplopoda	8.757 mg dry weight/animal	From a mesic deciduous forest, yellow poplar, Arthropoda[1]
Diplura	0.081 mg dry weight/animal	From a mesic deciduous forest, yellow poplar, Insecta, Arthropoda[1]
Diptera	0.564 mg dry weight/animal	From a mesic deciduous forest, yellow poplar, Insecta, larvae[1]
Enchytraeidae	0.08 mg dry weight/animal	From a mesic deciduous forest, yellow poplar, Annelida[1]
Entomobryidae	0.045 mg dry weight/animal	From a mesic deciduous forest, yellow poplar, Insecta[1]
Gamasina	0.28 mg dry weight/animal	From a mesic deciduous forest, yellow poplar, Arthropoda[1]
Hymenoptera	0.11 mg dry weight/animal	From a mesic deciduous forest, yellow poplar, Insecta[1]
Isotomidae	0.017 mg dry weight/animal	From a mesic deciduous forest, yellow poplar, Insecta[1]
Lepidoptera	0.103 mg dry weight/animal	From a mesic deciduous forest, yellow poplar, Insecta, larvae[1]
Lumbricidae	118 mg dry weight/animal	From a mesic deciduous forest, yellow poplar, Annelida[1]
Nematoda	0.0003 mg dry weight/animal	From a mesic deciduous forest, yellow poplar[1]
Onychiuridae	0.005 mg dry weight/animal	From a mesic deciduous forest, yellow poplar, Insecta[1]
Oribatei	0.099 mg dry weight/animal	From a mesic deciduous forest, yellow poplar, Arthropoda[1]
Orthoptera	7.69 mg dry weight/animal	From a mesic deciduous forest, yellow poplar, Insecta[1]
Pauropoda	0.006 mg dry weight/animal	From a mesic deciduous forest, yellow poplar, Arthropoda[1]
Phalangida	0.3 mg dry weight/animal	From a mesic deciduous forest, yellow poplar, Arthropoda[1]
Poduridae	0.005 mg dry weight/animal	From a mesic deciduous forest, yellow poplar, Insecta[1]
Prostigmata	0.004 mg dry weight/animal	From a mesic deciduous forest, yellow poplar, Arthropoda[1]
Protura	0.003 mg dry weight/animal	From a mesic deciduous forest, yellow poplar, Arthropoda[1]
Pseudoscorpionida	0.413 mg dry weight/animal	From a mesic deciduous forest, yellow poplar, Arthropoda[1]
Psocoptera	0.068 mg dry weight/animal	From a mesic deciduous forest, yellow poplar, Insecta[1]

Pulmonata	37.98 mg dry weight/animal	From a mesic deciduous forest, yellow poplar[1]
Sminthuridae	0.008 mg dry weight/animal	From a mesic deciduous forest, yellow poplar, Insecta[1]
Symphyla	0.089 mg dry weight/animal	From a mesic deciduous forest, yellow poplar, Arthropoda[1]
Uropodina	0.182 mg dry weight/animal	From a mesic deciduous forest, yellow poplar, Arthropoda[1]

Source: 1.Reichle, D.E. 1977. In: U. Lohm and T. Persson (eds.): Soil Organisms as Components of Ecosystems. Ecol. Bull (Stockholm) 25, 145-156.

1-271 Invertebrates
C / biomass ratio

Species	Value	Condition
Agrypnia obsoleta	47.3 % of dry weight	Trichoptera, 95 % CL = +- 1.4, total C[1]
Agrypnia obsoleta	49.7 % of ash free dry weight	Trichoptera, organic carbon, 95 % CL = +- 1.3[1]
Annelida	40.2 % of dry weight	Mean elementary composition[2]
Arthropoda	44.6 % of dry weight	Mean elementary composition, land[2]
Asellus aquaticus	34.3 % of dry weight	Isopoda, 95 % CL = +- 0.9, total C[1]
Asellus aquaticus	44.7 % of ash free dry weight	Isopoda, organic carbon, 95 % CL = +- 1.1[1]
Benthos	0.4 by weight	Used in models, N Fork Kings River, Calif[3]
Benthos	0.45 based on dry weight	Range = 0.4 - 0.5[4]
Benthos	0.5 by weight	Used in models, Boise River, Idaho[5]
Centroptilum luteolum	49.7 % of dry weight	Ephemeroptera, 95 % CL = +- 1.8, total C[1]
Centroptilum luteolum	54.1 % of ash free dry weight	Ephemeroptera, organic carbon, 95 % CL = +- 1.9[1]
Chaoborus flavicans	47.3 % of dry weight	Diptera, 95 % CL = +- 0.7, total C[1]
Chaoborus flavicans	50.8 % of ash free dry weight	Diptera, organic carbon, 95 % CL = +- 0.7[1]
Chironomus plumosus	45.1 % of dry weight	Diptera, 95 % CL = +- 0.8, total C[1]
Chironomus plumosus	47.5 % of ash free dry weight	Diptera, organic carbon, 95 % CL = +- 0.9[1]
Coelenterata	43.6 % of dry weight	Mean elementary composition[2]
Cordulia aenea	47.4 % of dry weight	Odonata, 95 % CL = +- 0.8, total C[1]
Cordulia aenea	49.6 % of ash free dry weight	Odonata, organic carbon, 95 % CL = +- 0.8[1]
Echinodermata	40 % of dry weight	Mean elementary composition, interpolated[2]
Erpobdella octoculata	48.3 % of dry weight	Hirudinea, 95 % CL = +- 2.3, total C[1]
Erpobdella octoculata	50.6 % of ash free dry weight	Hirudinea , organic carbon, 95 % CL = +- 2.4[1]
Gammaracanthus lacustris	41.7 % of dry weight	Amphipoda, 95 % CL = +- 0.8, total C[1]
Gammaracanthus lacustris	51.1 % of ash free dry weight	Amphipoda, organic carbon, 95 % CL = +- 0.9[1]
Heptagenia fuscogrisea	52.2 % of dry weight	Ephemeroptera, 95 % CL = +- 1.2, total C[1]

Heptagenia fuscogrisea	55.1 % of ash free dry weight	Ephemeroptera, organic carbon, 95 % CL = +- 1.2[1]
Invertebrate	51.8 % of ash free dry weight	SD = +- 4.0, mean of 19 species[1]
Invertebrates and zooplankton	0.5 g C/gram dry weight	Invertebrates found as zplankton[6]
Leptophlebia sp.	49.3 % of dry weight	Ephemeroptera, 95 % CL = +- 1.2, total C[1]
Leptophlebia sp.	52.6 % of ash free dry weight	Ephemeroptera, organic carbon, 95 % CL = +- 1.3[1]
Limnephilidae sp.	46.4 % of dry weight	Trichoptera, 95 % CL = +- 0.7, total C[1]
Limnephilidae sp.	51 % of ash free dry weight	Trichoptera, organic carbon, 95 % CL = +- 0.8[1]
Mollusca	39.9 % of dry weight	Mean elementary composition, soft parts[2]
Mysis relicta	50 % of dry weight	Mysidacea, 95 % CL = +- 1.0, total C[1]
Mysis relicta	54.6 % of ash free dry weight	Mysidacea, organic carbon, 95 % CL = +- 1.1[1]
Nematoda	40 % of dry weight	Mean elementary composition, interpolated[2]
Notonecta glauca	50 % of dry weight	Heteroptera, 95 % CL = +- 0.9, total C[1]
Notonecta glauca	52 % of ash free dry weight	Heteroptera, organic carbon, 95 % CL = +- 1.0[1]
Pallacea quadrispinosa	35.4 % of dry weight	Amphipoda, 95 % CL = +- 1.2, total C[1]
Pallacea quadrispinosa	47 % of ash free dry weight	Amphipoda, organic carbon, 95 % CL = +- 1.7[1]
Sialis sp.	49.2 % of dry weight	Megaloptera, 95 % CL = +- 0.6, total C[1]
Sialis sp.	51.1 % of ash free dry weight	Megaloptera, organic carbon, 95 % CL = +- 0.6[1]
Tintinnopsis acuminata	822 pg	Ciliate, grown on Isochrysis galbana at 25 oC[7]
Tintinnopsis acuminata	847 pg	Ciliate, grown on Isochrysis galbana at 20 oC[7]
Tintinnopsis acuminata	866 pg	Ciliate, grown on Isochrysis galbana at 15 oC[7]
Tintinnopsis vasculum	8,337 pg	Ciliate, grown on Dicrateria inornata at 15 oC[7]
Tintinnopsis vasculum	8,679 pg	Ciliate, grown on Dicrateria inornata at 10 oC[7]
Tintinnopsis vasculum	8,940 pg	Ciliate, grown on Dicrateria inornata at 5 oC[7]

Source: 1.Salonen, K., J. Sarvala, I. Hakala and M.-L. Viljanen. 1976. Limnol. Oceanogr., 21(5), pp 724-730.
2.Bowen, H.J.M. 1966. Trace Elements in Biochemistry, Ac. Press.
3.Tetra Tech, Inc. 1976. Estuary Water Quality Models, Lond Island, New York User's Guide. For Nassau-Suffolk Regional Planning Board. Hauppauge, New York. 11787.
4.The Hydrological Engineering Center Corps of Engineers U.S. Army. 1975. Water Quality for river-resevoir systems, Davis, Cal. 9 5 6 1 6 (916) 499-2329.
5.Chen, C.W. and Wells. 1975. Boise River Water Quality-Ecological Model for Urban Planning Study. US Army Eng. Walla Walla, Idaho Water Res. Board.
6.Waters, T.F. 1977. Adv. Ecol. Res. 10, pp 91-164.
7.Verity, P.G., Limnol. Oceanogr., 30, 1268, 1985.

1-272 Invertebrates
CO2, annual production

Species	Value	Condition
Annelida, total	55.6 g CO2/(m**2*year)	From a mesic deciduous forest, yellow poplar[1]
Araneae	1.73 g CO2/(m**2*year)	From a mesic deciduous forest, yellow poplar, Arthropoda[1]
Arthropoda, total	139.76 g CO2/(m**2*year)	From a mesic deciduous forest, yellow poplar[1]
Chilopoda	0.38 g CO2/(m**2*year)	From a mesic deciduous forest, yellow poplar, Arthropoda[1]

Coleoptera	1.99 g CO2/(m**2*year)	From a mesic deciduous forest, yellow poplar, Insecta, larvae[1]
Diplopoda	1.8 g CO2/(m**2*year)	From a mesic deciduous forest, yellow poplar, Arthropoda[1]
Diplura	1.03 g CO2/(m**2*year)	From a mesic deciduous forest, yellow poplar, Insecta, Arthropoda[1]
Diptera	10.4 g CO2/(m**2*year)	From a mesic deciduous forest, yellow poplar, Insecta, larvae[1]
Enchytraeidae	8.9 g CO2/(m**2*year)	From a mesic deciduous forest, yellow poplar, Annelida[1]
Entomobryidae	5.88 g CO2/(m**2*year)	From a mesic deciduous forest, yellow poplar, Insecta[1]
Gamasina	39.7 g CO2/(m**2*year)	From a mesic deciduous forest, yellow poplar, Arthropoda[1]
Hymenoptera	0.28 g CO2/(m**2*year)	From a mesic deciduous forest, yellow poplar, Insecta[1]
Isotomidae	3.6 g CO2/(m**2*year)	From a mesic deciduous forest, yellow poplar, Insecta[1]
Lepidoptera	0.1 g CO2/(m**2*year)	From a mesic deciduous forest, yellow poplar, Insecta, larvae[1]
Lumbricidae	46.7 g CO2/(m**2*year)	From a mesic deciduous forest, yellow poplar, Annelida[1]
Microflora, total	2291 g CO2/(m**2*year)	From a mesic deciduous forest, yellow poplar[1]
Micrpflora	2291 g CO2/(m**2*year)	From a mesic deciduous forest, yellow poplar[1]
Nematoda	49.43 g CO2/(m**2*year)	From a mesic deciduous forest, yellow poplar[1]
Onychiuridae	1.14 g CO2/(m**2*year)	From a mesic deciduous forest, yellow poplar, Insecta[1]
Oribatei	62.9 g CO2/(m**2*year)	From a mesic deciduous forest, yellow poplar, Arthropoda[1]
Orthoptera	0.2 g CO2/(m**2*year)	From a mesic deciduous forest, yellow poplar, Insecta[1]
Pauropoda	0.3 g CO2/(m**2*year)	From a mesic deciduous forest, yellow poplar, Arthropoda[1]
Phalangida	0.08 g CO2/(m**2*year)	From a mesic deciduous forest, yellow poplar, Arthropoda[1]
Poduridae	0.34 g CO2/(m**2*year)	From a mesic deciduous forest, yellow poplar, Insecta[1]
Prostigmata	1.87 g CO2/(m**2*year)	From a mesic deciduous forest, yellow poplar, Arthropoda[1]
Protura	0.28 g CO2/(m**2*year)	From a mesic deciduous forest, yellow poplar, Arthropoda[1]
Pseudoscorpionida	0.12 g CO2/(m**2*year)	From a mesic deciduous forest, yellow poplar, Arthropoda[1]
Psocoptera	0.04 g CO2/(m**2*year)	From a mesic deciduous forest, yellow poplar, Insecta[1]
Pulmonata	1.22 g CO2/(m**2*year)	From a mesic deciduous forest, yellow poplar[1]
Sminthuridae	0.17 g CO2/(m**2*year)	From a mesic deciduous forest, yellow poplar, Insecta[1]
Symphyla	1.82 g CO2/(m**2*year)	From a mesic deciduous forest, yellow poplar, Arthropoda[1]
Uropodina	3.61 g CO2/(m**2*year)	From a mesic deciduous forest, yellow poplar, Arthropoda[1]

Source: 1.Reichle, D.E. 1977. In: U. Lohm and T. Persson (eds.): Soil Organisms as Components of Ecosystems. Ecol. Bull (Stockholm) 25, 145-156.

1-273 Invertebrates Ca / biomass ratio

Species	Value	Condition
Arctiidae	4.7 mg/(g ash free dry weight)	Soil or litter invert.[1]
Camponotus pennsylvanicus	2.9 mg/(g ash free dry weight)	Soil or litter invert.[1]
Ceuthophilus gracilipes	2.1 mg/(g ash free dry weight)	Soil or litter invert.[1]
Dixidesmus erasus	294.9 mg/(g ash free dry weight)	Soil or litter invert.[1]

Geotrupes sp.	0.8 mg/(g ash free dry weight)	Soil or litter invert.[1]
Leiobunum flavum	7.2 mg/(g ash free dry weight)	Soil or litter invert.[1]
Ligidium blueridgensis	108.9 mg/(g ash free dry weight)	Soil or litter invert.[1]
Mesomphix vulgatus	16.7 mg/(g ash free dry weight)	Soil or litter invert.[1]
Otocryptops sexspinosus	0.6 mg/(g ash free dry weight)	Soil or litter invert.[1]
Pachylomerides audouini	2.1 mg/(g ash free dry weight)	Soil or litter invert.[1]
Sciomyzidae	2 mg/(g ash free dry weight)	Soil or litter invert.[1]
Scytonotus granulatus	103.8 mg/(g ash free dry weight)	Soil or litter invert.[1]
Sphaeroderus stenostomus	0.5 mg/(g ash free dry weight)	Soil or litter invert.[1]

Source: 1.Reichle, D.E., M.H. Shanks and D.A. Crossey, Jr. 1969. Ann. Entomol. Soc. Amer. 62, pp 57-62.

1-274 Invertebrates
Caloric content

Species	Value	Condition
Abra alba	4.491 kcal/g dry weight	Marine, bombcal., adult, Kiel Bay[1]
Abra alba	5.228 kcal/g ash free dry weight	Marine, bombcal., Kiel bay[1]
Acheta domesticus	5896 g cal/(g ash free dry weight)	Ash = 5.7 % of dry weight[2]
Aeschna cyanea	5432 cal/g dry weight	Germany[3]
Aeschna cyanea	5837 cal/g ash free dry weight	Germany[3]
Agrypnia obsoleta	19.6 kJ/g ash free dry weight	Trichoptera, 95 % CL = +- 1.1, correction for nitrogen[4]
Agrypnia obsoleta	21.5 kJ/g ash free dry weight	Trichoptera, 95 % CL = +- 0.4, no correction for nitrogen[4]
Agrypnia obsoleta	39.2 kJ/g organic carbon	Trichoptera, 95 % CL = +- 2.9, correction for nitrogen[4]
Agrypnia obsoleta	43.3 kJ/g organic carbon	Trichoptera, 95 % CL = +- 1.3, no corection for nitrogen[4]
Aphrodita hastata	0.486 kcal/g wet weight	Marine, bombcal.[1]
Aphrodita hastata	2638 gcal/(g dry weight)	Incl ash, water=80+-4%, 20 anim, 6 samples, all seasons[5]
Aphrodita hastata	3.438 kcal/g dry weight	Marine, bombcal.[1]
Arachnida	5.62 kcal/g ash free dry weight	None[6]
Arachnida	5.82 kcal/g ash free dry weight	None[7]
Arctica islandica	4031 gcal/(g dry weight)	Shell-free, incl ash, wat=86+-5%, 72 anim, 11 samp, all seasons[5]
Asellus aquaticus	17.6 kJ/g ash free dry weight	Isopoda, 95 % CL = +- 0.6, correction for nitrogen[4]

Asellus aquaticus	19.9 kJ/g ash free dry weight	Isopoda, 95 % CL = +-0.6, no correction for nitrogen[4]
Asellus aquaticus	2.957 kcal/g dry weight	Freshwater, bombcal, lab culture[1]
Asellus aquaticus	3493 cal/g dry weight	Germany[3]
Asellus aquaticus	39.3 kJ/g organic carbon	Isopoda, 95 % CL = +- 1.9, correction for nitrogen[4]
Asellus aquaticus	4.586 kcal/g ash free dry weight	Freshwater, bombcal, lab culture[1]
Asellus aquaticus	44.6 kJ/g organic carbon	Isopoda, 95 % CL = +- 2.5, no corection for nitrogen[4]
Asellus aquaticus	5521 cal/g ash free dry weight	Germany[3]
Asplanchna sp.	1.11 kcal/g wet weight	Freshwater, oxydation[1]
Asplanchna sp.	4.27 kcal/g dry weight	Freshwater, oxydation[1]
Asplanchna sp.	4.93 kcal/g ash free dry weight	Freshwater, oxydation[1]
Astarte undata	4234 gcal/(g dry weight)	Shell-free, incl ash, wat=87+-3%, 154 anim, 10 samp, all seasons[5]
Asterias vulgaris	0.633 kcal/g wet weight	Marine, bombcal., with CaCO3 correction[1]
Asterias vulgaris	2.551 kcal/g dry weight	Marine, bombcal., with CaCO3 correction[1]
Baetis rhodani	6044 cal/g dry weight	Germany[3]
Brachoria initialis	5453 g cal/(g ash free dry weight)	Ash = 7.2 % of dry weight[2]
Camponotus americanus	6247 g cal/(g ash free dry weight)	Ash = 2.8 % of dry weight[2]
Cardium lamarcki	3.701 kcal/g dry weight	Estuarine, bombcal., juv, Grabower Bodden[1]
Centroptilum luteolum	23.7 kJ/g ash free dry weight	Ephemeroptera, 95 % CL = +- 0.5, correction for nitrogen[4]
Centroptilum luteolum	26.1 kJ/g ash free dry weight	Ephemeroptera, 95 % CL = +- 0.4, no correction for nitrogen[4]
Centroptilum luteolum	43.8 kJ/g organic carbon	Ephemeroptera, 95 % CL = +- 1.7, correction for nitrogen[4]
Centroptilum luteolum	48.4 kJ/g organic carbon	Ephemeroptera, 95 % CL = +- 1.7, no corection for nitrogen[4]
Chaetopsis sp.	5.52 kcal/g ash free dry weight	None[8]
Chaoborus flavicans	20.4 kJ/g ash free dry weight	Diptera, 95 % CL = +- 0.4, correction for nitrogen[4]
Chaoborus flavicans	22.7 kJ/g ash free dry weight	Diptera, 95 % CL = +- 0.3, no correction for nitrogen[4]
Chaoborus flavicans	40.1 kJ/g organic carbon	Diptera, 95 % CL = +- 0.6, correction for nitrogen[4]
Chaoborus flavicans	44.7 kJ/g organic carbon	Diptera, 95 % CL = +- 0.8, no corection for nitrogen[4]
Chironomus gregarius	5.43 kcal/g dry weight	Freshwater, oxydation, larvae[1]
Chironomus gregarius	5.430 kcal/g dry weight	Freshwater, oxydation, larvae[1]

Chironomus gregarius	5.779 kcal/g ash free dry weight	Freshwater, oxydation, larvae[1]
Chironomus plumosus	18.6 kJ/g ash free dry weight	Diptera, 95 % CL = 2.0, correction for nitrogen[4]
Chironomus plumosus	20.8 kJ/g ash free dry weight	Diptera, 95 % CL = 2.2, no correction for nitrogen[4]
Chironomus plumosus	39.1 kJ/g organic carbon	Diptera, 95 % CL = 4.1, correction for nitrogen[4]
Chironomus plumosus	4.686 kcal/g dry weight	Freshwater, oxydation, larvae[1]
Chironomus plumosus	43.7 kJ/g organic carbon	Diptera, 95 % CL = 4.6, no corection for nitrogen[4]
Chironomus plumosus	5.372 kcal/g ash free dry weight	Freshwater, oxydation, larvae[1]
Chironomus sp.	4.157 kcal/g dry weight	Estuarine, bombcal., larvae[1]
Chironomus sp.	5.054 kcal/g dry weight	Estuarine, bombcal., larvae, Bodstedter Bodden[1]
Chironomus teniforceps	5.89 kcal/g dry weight	Freshwater, oxydation, larvae[1]
Chironomus teniforceps	5.890 kcal/g dry weight	Freshwater, oxydation, larvae[1]
Clava multicornis	5.5 kcal/g ash free dry weight	Marine, bombcal.[1]
Clava multicornis	5.500 kcal/g ash free dry weight	Marine, bombcal.[1]
Colus stimpsoni	4587 cal/g dry weight	Shell-free, incl ash, water = 68 %, 4 anim, n = 1, July to Sept[5]
Cordulia aenea	20.2 kJ/g ash free dry weight	Odonata, 95 % CL = 0.9, correction for nitrogen[4]
Cordulia aenea	22.4 kJ/g ash free dry weight	Odonata, 95 % CL = 0.8, no correction for nitrogen[4]
Cordulia aenea	40.7 kJ/g organic carbon	Odonata, 95 % CL = 2.4, correction for nitrogen[4]
Cordulia aenea	45.2 kJ/g organic carbon	Odonata, 95 % CL = 1.2, no corection for nitrogen[4]
Crangon septemspinosa	4088 cal/g dry weight	Incl ash, water = 76 %, 15 anim, n = 1, November[5]
Crangon septemspinosa	4272 cal/g wet weight	Incl ash, water=74+-2%, 43 animals, 4samples, Jan to Aug[5]
Crematogaster clara	6.13 kcal/g ash free dry weight	None[8]
Crenobia alpina	6055 cal/g dry weight	Germany[3]
Crenobia alpina	6426 cal/g ash free dry weight	Germany[3]
Cyclops strenuus	4.764 kcal/g dry weight	Freshwater, bombcal[1]
Cyclops vicinus	0.65 kcal/g wet weight	Freshwater, oxydation[1]
Cyclops vicinus	4.36 kcal/g dry weight	Freshwater, oxydation[1]
Cyclops vicinus	4.82 kcal/g ash free dry weight	Freshwater, oxydation[1]
Cylisticus convexus	6123 cal/(g ash free dry wt)	Ash = 45.2 % of dry weight, SE = +-9.7[9]
Cyprina islandica	4.547 kcal/g dry weight	Marine, bombcal., adult, Kiel Bay[1]

Cyprina islandica	5.086 kcal/g ash free dry weight	Marine, bombcal., adult, Kiel Bay[1]
Dentalium entale	4560 cal/g dry weight	Shell-free, incl ash, water = 73 %, 100 animals, 1 sample, April[5]
Diastylis rathkei	3.953 kcal/g dry weight	Marine, bombcal., adult, Kiel Bay[1]
Diastylis rathkei	6.017 kcal/g ash free dry weight	Marine, bombcal., adult, Kiel Bay[1]
Drassyllus virginiana	5794 cal/(g ash free dry wt)	Ash = 3.8 % of dry weight, SE = +-89.4[10]
Dreissena polymorpha	0.69 kcal/g wet weight	Freshwater, oxydation, without shell[1]
Dreissena polymorpha	0.690 kcal/g wet weight	Freshwater, oxydation, without shell[1]
Dreissena polymorpha	3.99 kcal/g dry weight	Freshwater, oxydation, without shell[1]
Dreissena polymorpha	3.990 kcal/g dry weight	Freshwater, oxydation, without shell[1]
Dreissena polymorpha	4.41 kcal/g ash free dry weight	Freshwater, oxydation, without shell[1]
Dreissena polymorpha	4.410 kcal/g ash free dry weight	Freshwater, oxydation, without shell[1]
Dugesia gonocephala	6043 cal/g dry weight	Germany[3]
Dugesia gonocephala	6453 cal/g ash free dry weight	Germany[3]
Erpobdella octoculata	19.8 kJ/g ash free dry weight	Hirudinea, 95 % CL = 0.6, correction for nitrogen[4]
Erpobdella octoculata	22.1 kJ/g ash free dry weight	Hirudinea, 95 % CL = 0.6, no correction for nitrogen[4]
Erpobdella octoculata	40.8 kJ/g organic carbon	Hirudinea, 95 % CL = 2.8, correction for nitrogen[4]
Erpobdella octoculata	43.9 kJ/g organic carbon	Hirudinea, 95 % CL = 2.5, no corection for nitrogen[4]
Erpobdella octoculata	5607 cal/g dry weight	Germany[3]
Erpobdella octoculata	5995 cal/g ash free dry weight	Germany[3]
Eusimulium sp.	4510 cal/g dry weight	Germany[3]
Eusimulium sp.	5562 cal/g ash free dry weight	Germany[3]
Evarthrus sodalis	5672 cal/(g ash free dry wt)	Ash = 3.2 % of dry weight[2]
Gammaracanthus lacustris	21.7 kJ/g ash free dry weight	Amphipoda, 95 % CL = 0.2, correction for nitrogen[4]
Gammaracanthus lacustris	24.1 kJ/g ash free dry weight	Amphipoda, 95 % CL = 0.2, no correction for nitrogen[4]
Gammaracanthus lacustris	45.5 kJ/g organic carbon	Amphipoda, 95 % CL = 1.1, correction for nitrogen[4]
Gammaracanthus lacustris	47.1 kJ/g organic carbon	Amphipoda, 95 % CL = 0.8, no corection for nitrogen[4]
Gammarus fossarum	3578 cal/g dry weight	Germany, 10 - 13 mm[3]
Gammarus fossarum	3735 cal/g dry weight	Germany, 2 - 5 mm[3]
Gammarus fossarum	3896 cal/g dry weight	Germany, 6 - 9 mm[3]
Gammarus fossarum	5201 cal/g ash free dry weight	Germany, 10 - 13 mm[3]

Gammarus fossarum	5354 cal/g ash free dry weight	Germany, 6 - 9 mm[3]
Gammarus fossarum	5452 cal/g ash free dry weight	Germany, 2 - 5 mm[3]
Gammarus pulex	4.035 kcal/g dry weight	Freshwater, oxydation[1]
Gammerus duebeni	4.412 kcal/g dry weight	Estuarine, bombcal., rg = +- 0.175[1]
Gammerus pulex	5.695 kcal/g ash free dry weight	Freshwater, oxydation[1]
Glossiphonia complanata	5544 cal/g dry weight	Germany[3]
Glossiphonia complanata	5713 cal/g ash free dry weight	Germany[3]
Heptagenia fuscogrisea	24.1 kJ/g ash free dry weight	Ephemeroptera, 95 % CL = 0.5, correction for nitrogen[4]
Heptagenia fuscogrisea	26.2 kJ/g ash free dry weight	Ephemeroptera, 95 % CL = 0.6, no correction for nitrogen[4]
Heptagenia fuscogrisea	43.6 kJ/g organic carbon	Ephemeroptera, 95 % CL = 1.3, correction for nitrogen[4]
Heptagenia fuscogrisea	47.5 kJ/g organic carbon	Ephemeroptera, 95 % CL = 1.7, no corection for nitrogen[4]
Hyas araneus	4958 cal/g dry weight	Shell-free, incl ash, water = 81 %, 1 animal, 1 sample, March[5]
Hydropsyche sp.	5348 cal/g dry weight	Germany[3]
Hydropsyche sp.	5814 cal/g ash free dry weight	Germany[3]
Idothea baltica	2.06 kcal/g dry weight	Marine, oxydation, youngs[1]
Idothea baltica	2.060 kcal/g dry weight	Marine, oxydation, youngs[1]
Idothea baltica	2.44 kcal/g dry weight	Marine, oxydation, 20 d old[1]
Invertebrate	21.5 kJ/g ash free dry weight	SD = +-2.7, mean of 19 species, correction for nitrogen[4]
Invertebrate	23.7 kJ/g ash free dry weight	SD = +-2.6, mean of 19 species, no correction for nitrogen[4]
Invertebrate	41.4 kJ/g organic carbon	SD = +-1.9, mean of 19 species, correction for nitrogen[4]
Invertebrate	45.8 kJ/g organic carbon	SD = +-1.7, mean of 19 species, no correction for nitrogen[4]
Invertebrates and zooplankton	5 kcal/g dry weight	Invertebrates found as zooplankton[11]
Isoperla gortzi	5038 cal/g dry weight	Germany[3]
Leptocheirus pinguis	2740 cal/g dry weight	Incl ash, water = 83 %, 1500 anim, 8 samples, all seasons, rg =[5]
Leptodora kindtii	5.3 kcal/g dry weight	Freshwater, bombcal[1]
Leptodora kindtii	5.300 kcal/g dry weight	Freshwater, bombcal[1]
Leptodora kindtii	5.749 kcal/g ash free dry weight	Freshwater, bombcal[1]
Leptophlebia marginata	5647 cal/g dry weight	Germany[3]
Leptophlebia sp.	21.9 kJ/g ash free dry weight	Ephemeroptera, 95 % CL = 0.8, correction for nitrogen[4]
Leptophlebia sp.	24.4 kJ/g ash free dry weight	Ephemeroptera, 95 % CL = 0.8, no correction for nitrogen[4]

Leptophlebia sp.	41.6 kJ/g organic carbon	Ephemeroptera, 95 % CL = 1.4, correction for nitrogen[4]
Leptophlebia sp.	46.4 kJ/g organic carbon	Ephemeroptera, 95 % CL = 1.3, no corection for nitrogen[4]
Limacina retroversa	3.543 kcal/g dry weight	Marine, bombcal.[1]
Limacina retroversa	5.026 kcal/g ash free dry weight	Marine, bombcal., rg = 0.278[1]
Limnephilidae sp.	20.5 kJ/g ash free dry weight	Trichoptera, 95 % CL = 1.3, correction for nitrogen[4]
Limnephilidae sp.	22 kJ/g ash free dry weight	Trichoptera, 95 % CL = 0.3, no correction for nitrogen[4]
Limnephilidae sp.	40.2 kJ/g organic carbon	Trichoptera, 95 % CL = 0.5, correction for nitrogen[4]
Limnephilidae sp.	43.2 kJ/g organic carbon	Trichoptera, 95 % CL = 0.4, no corection for nitrogen[4]
Limnodrilus hoffmeisteri	0.76 kcal/g wet weight	Freshwater, oxydation, rg = +- 0.026[1]
Limnodrilus hoffmeisteri	0.760 kcal/g wet weight	Freshwater, oxydation, rg = +- 0.026[1]
Lumbricus terrestris	5745.7 cal/(g ash free dry wt)	Ash = 23.0 % of dry weight[12]
Lumbrinereis fragilis	4565 cal/g dry weight	Incl ash, water = 80 %, 200 anim, 12 samples, all seasons, rg =[5]
Macrocyclops albidus	0.6 kcal/g wet weight	Freshwater, oxydation[1]
Macrocyclops albidus	0.60 kcal/g wet weight	Freshwater, oxydation[1]
Macrocyclops albidus	3.07 kcal/g dry weight	Freshwater, oxydation[1]
Macrocyclops albidus	3.94 kcal/g ash free dry weight	Freshwater, oxydation[1]
Meganyctiphanes norvegica	4945 cal/g dry weight	Incl ash, water = 81. 855 anim, 11 samples, all seasons[5]
Meganyctiphanes norvegica	5040 cal/g dry weight	Incl ash, water = 81 %, 767 anim, 12 samples, al seasons[5]
Mesodon thyroides	5970.9 cal/(g ash free dry wt)	Ash = 22.8 % of dry weight[2]
Mya arenaria	4.46 kcal/g dry weight	Estuarine, bombcal., juv, Grabower Boden[1]
Mya arenaria	4.460 kcal/g dry weight	Estuarine, bombcal., juv, Grabower Boden[1]
Myrmica sp.	6.309 kcal/g dry weight	Equal to 1.92 kcal/g fresh weight[13]
Mysis relicta	23.3 kJ/g ash free dry weight	Mysidacea, 95 % CL = 0.2, correction for nitrogen[4]
Mysis relicta	25.7 kJ/g ash free dry weight	Mysidacea, 95 % CL = 0.1, no correction for nitrogen[4]
Mysis relicta	42.5 kJ/g organic carbon	Mysidacea, 95 % CL = 1.0, correction for nitrogen[4]
Mysis relicta	47.3 kJ/g organic carbon	Mysidacea, 95 % CL = 1.2, no corection for nitrogen[4]
Mysis stenolepis	4714 cal/g dry weight	Incl ash, water = 79 %, 300 anim, 3 samples, fall-winter[5]
Mytilus edulis	5.05 kcal/g dry weight	Estuarine, bombcal., juv and adult, rg = 0.05[1]
Nemoura sp.	5488 cal/g dry weight	Germany[3]
Neomysis mirabilis	1.11 kcal/g wet weight	Marine, oxydation, female = 11.5 mm[1]

Neomysis mirabilis	1.110 kcal/g wet weight	Marine, oxydation, female = 11.5 mm[1]
Neomysis mirabilis	5.47 kcal/g dry weight	Marine, oxydation, female = 11.5 mm[1]
Neomysis mirabilis	5.470 kcal/g dry weight	Marine, oxydation, female = 11.5 mm[1]
Nephtys ciliata	0.747 kcal/g wet weight	Marine, bombcal.[1]
Nephtys ciliata	4.0661 kcal/g dry weight	Marine, bombcal.[1]
Nephtys incisa	3984 cal/g dry weight	Incl ash, water = 81 %, 156 anim, 12 samples, all seasons, rg = 3773[5]
Nephtys sp.	4.184 kcal/g dry weight	Marine, bombcal., adult, Kiel Bay[1]
Nephtys sp.	5.294 kcal/g ash free dry weight	Marine, bombcal., adult, Kiel Bay[1]
Nereis diversicolor	3.993 kcal/g dry weight	Estuarine, bombcal., S = 1.2 %[1]
Nereis diversicolor	4.179 kcal/g dry weight	Estuarine, bombcal., S = 0.7 %[1]
Nereis diversicolor	5.46 kcal/g ash free dry weight	Estuarine, bombcal., S = 0.7 %[1]
Notonecta glauca	21.3 kJ/g ash free dry weight	Heteroptera, 95 % CL =, correction for nitrogen[4]
Notonecta glauca	23.8 kJ/g ash free dry weight	Heteroptera, 95 % CL = 0.4, no correction for nitrogen[4]
Notonecta glauca	41 kJ/g organic carbon	Heteroptera, 95 % CL = 1.6, correction for nitrogen[4]
Notonecta glauca	45.9 kJ/g organic carbon	Heteroptera, 95 % CL = 1.7, no corection for nitrogen[4]
Nuculana sp.	4184 cal/g dry weight	Shell-free, incl ash, water = 68 %, 100 anim, 1 sample, June - Aug[5]
Orchelimum fidicinium	5.432 kcal/g dry weight	None[14]
Otocryptops sexspinosus	5344 cal/(g ash free dry wt)	Ash = 6.9 % of dry weight[2]
Palaemonetes varians var. microgenitor	4.47 kcal/g dry weight	Estuarine, bombcal., male, adult, rg = 0.142[1]
Palaemonetes varians var. microgenitor	4.470 kcal/g dry weight	Estuarine, bombcal., male, adult, rg = 0.142[1]
Palaemonetes varians var. microgenitor	5.06 kcal/g dry weight	Estuarine, bombcal., female, adult, rg = +- 0.98[1]
Palaemonetes varians var. microgenitor	5.060 kcal/g dry weight	Estuarine, bombcal., female, adult, rg = +- 0.98[1]
Palaemonetes varians var. microgenitor	5.073 kcal/g ash free dry weight	Estuarine, bombcal., male, adult, rg = 0.164[1]
Palaemonetes varians var. microgenitor	5.142 kcal/g dry weight	Estuarine, bombcal., 1. larvae stage, rg = 0.63[1]
Palaemonetes varians var. microgenitor	5.41 kcal/g ash free dry weight	Estuarine, bombcal., female adult, rg = 0.120[1]
Palaemonetes varians var. microgenitor	5.410 kcal/g ash free dry weight	Estuarine, bombcal., female adult, rg = 0.120[1]
Palaemonetes varians var. microgenitor	5.817 kcal/g ash free dry weight	Estuarine, bombcal., 1. larvae stage, rg = 0.85[1]

Pallacea quadrispinosa	19 kJ/g ash free dry weight	Amphipoda, 95 % CL =0.4, correction for nitrogen[4]
Pallacea quadrispinosa	21.3 kJ/g ash free dry weight	Amphipoda, 95 % CL = 0.4, no correction for nitrogen[4]
Pallacea quadrispinosa	40.4 kJ/g organic carbon	Amphipoda, 95 % CL = 1.8, correction for nitrogen[4]
Pallacea quadrispinosa	45.3 kJ/g organic carbon	Amphipoda, 95 % CL = 2.1, no corection for nitrogen[4]
Pandalus montagui	4610 cal/g dry weight	Incl ash, water = 72 %, 67 anim, 9 samples, all seasons, rg = 4390[5]
Paramecium caudatum	3.827 kcal/g dry weight	Freshwater, oxydation[1]
Parcoblatta sp.	6063 cal/(g ash free dry wt)	Ash = 8.7 % of dry weight[10]
Pectinaria hypoborea	0.484 kcal/g wet weight	Marine, bombcal., Mar - Apr 1967[1]
Pectinaria hypoborea	0.624 kcal/g wet weight	Marine, bombcal., Jun - Oct 1967[1]
Pectinaria hypoborea	3.242 kcal/g dry weight	Marine, bombcal., Jun - Oct 1967[1]
Pectinaria hypoborea	3.242 kcal/g dry weight	Marine, bombcal., Mar - Apr 1967[1]
Pectinaria koreni	3.735 kcal/g dry weight	Marine, bombcal.[1]
Pectinaria koreni	5.102 kcal/g ash free dry weight	Marine, bombcal.[1]
Philaenus spumarius	5.807 kcal/g dry weight	None[15]
Plectrocnemia conspersa	5703 cal/g dry weight	Germany[3]
Plectrocnemia conspersa	5958 cal/g ash free dry weight	Germany[3]
Polyphemus pediculus	0.74 kcal/g wet weight	Freshwater, oxydation[1]
Polyphemus pediculus	4.55 kcal/g dry weight	Freshwater, oxydation[1]
Polyphemus pediculus	5.9 kcal/g ash free dry weight	Freshwater, oxydation[1]
Polyphemus pediculus	5.90 kcal/g ash free dry weight	Freshwater, oxydation[1]
Praxillella praetermissa	2365 cal/g dry weight	Incl ash, water = 79 %, 300 anim, 1 sample, April - Nov[5]
Prokelisia marginata	5.38 kcal/g ash free dry weight	None[8]
Radix peregra ovata	4608 cal/g dry weight	Germany[3]
Radix peregra ovata	5957 cal/g ash free dry weight	Germany[3]
Rhithrogena semicolorata	4955 cal/g dry weight	Germany[3]
Rhithrogena semicolorata	5579 cal/g ash free dry weight	Germany[3]
Rhyacophila fasciata	5869 cal/g dry weight	Germany[3]
Rhyacophila fasciata	6181 cal/g ash free dry weight	Germany[3]
Sarcophaga bullata	5464 cal/(g ash free dry wt)	SE = +- 243[16]

Schistocerca sp.	5.363 kcal/g dry weight	None[14]
Scrobicularia plana	5.097 kcal/g ash free dry weight	Marine, bombcal., rg = 0.52[1]
Sialis sp.	21.2 kJ/g ash free dry weight	Megaloptera, 95 % CL = 0.4, correction for nitrogen[4]
Sialis sp.	23.5 kJ/g ash free dry weight	Megaloptera, 95 % CL = 0.3, no correction for nitrogen[4]
Sialis sp.	41.5 kJ/g organic carbon	Megaloptera, 95 % CL = 1.0, correction for nitrogen[4]
Sialis sp.	46 kJ/g organic carbon	Megaloptera, 95 % CL = 0.8, no corection for nitrogen[4]
Sphaeroma pulchellum	1.66 kcal/g dry weight	Marine, oxydation, youngs[1]
Sphaeroma pulchellum	1.9 kcal/g dry weight	Marine, oxydation, 14 d old[1]
Sphaeroma pulchellum	1.90 kcal/g dry weight	Marine, oxydation, 14 d old[1]
Sphaeroma rugicauda	3.004 kcal/g dry weight	Estuarine, bombcal., adult male, rg = +- 0.121[1]
Sphaeroma rugicauda	4.553 kcal/g ash free dry weight	Estuarine, bombcal., adult male, rg = +- 0.231[1]
Spirontocaris sp.	4065 cal/g dry weight	Incl ash, water = 78 %, 18 anim, 1 sample, April[5]
Spirontocaris sp.	4423 cal/g dry weight	Incl ash, water = 73 %, 20 anim, 1 sample, October[5]
Tenebrio molitor	6.314 kcal/g ash free dry weight	None[6]
Tetrahymena pyriformis	5.938 kcal/g ash free dry weight	Freshwater, bombcal, rg = +- 0.026[1]
Tomocerus sp.	6063 cal/(g ash free dry wt)	Ash = 3.8 % of dry weight[10]
Tribolium castaneum	6 kcal/g dry weight	None[17]
Trigonotylus sp.	5.3 kcal/g ash free dry weight	None[8]
Tubifex tubifex	5.28 kcal/g dry weight	Freshwater, bombcal[1]
Tubifex tubifex	5.280 kcal/g dry weight	Freshwater, bombcal[1]
Tubifex tubifex	5.547 kcal/g ash free dry weight	Freshwater, bombcal[1]
Tyroglyphus interna	5808 cal/(g ash free dry wt)	95 % CL = 446[18]
Unciola leucopis	2147 cal/g dry weight	Incl ash, water = 80 %, 1800 anim, 7 samples, all seasons, rg = 1845[5]
Venericardia borealis	3678 cal/g dry weight	Shell-free, incl ash, wat=82 %, 188 anim, n=10, all seasons, rg=[5]
Yoldia sapotilla	4216 cal/g dry weight	Shell-free, incl ash, wat = 78 %, 100 anim, n = 4, all seasons, rg =[5]

Source: 1.Oertzen, J.-A. 1973. Wiss. Z. Univ. Rostock. 22, pp 1153-1158.
2.Edwards, C.A., D.E. Reichle and D.A. Crossley, Jr. 1970. in: Reichle, D.E. (ed); Analysis of Temperate Forest Ecosystems. pp 147-172. Springer-Verlag. Berlin.
3.Caspers, N. 1975. Arch. Hydrobiol., 75(4), pp 484-489.
4.Salonen, K., J. Sarvala, I. Hakala and M.-L. Viljanen. 1976. Limnol. Oceanogr., 21(5), pp 724-730.
5.Tyler, A.V. 1973. Mar biol. 19, pp 258-261.
6.Slobodkin, L.B. and S. Richman. 1961. Nature 191, pp 299.
7.Gibb, J. 1957. Bird study, 4, pp 207-215.
8.Kale, H.W. 1965. Publ. Nutt. Ornithol. Club. Cambridge. Massachusettes.
9.Reichle, D.E. 1967. Ecology 48, pp 351-366.
10.Moulder, B.C., D.E. Reichle and S.I. Auerbach. 1970. Oak Ridge National Lab. USA, ORNL 4452.
11.Waters, T.F. 1977. Adv. Ecol. Res. 10, pp 91-164.
12.French, C.E., S.A. Liscinsky and D.R. Miller. 1957. J. Wildl. Mgmt. 21, pp 348.
13.Petal, J. 1964. in: Petrusewicz, K. (ed.); Secondary Productivity of Terrestrial Ecosystems. pp 841-55. Krakow. Warszawa.

14.Golley, F.B. 1959. Mimeogr. Univ. Georgia, pp 7.
15.Wiegert, R.G. 1965. BioScience 15, pp 543-545.
16.Slobodkin, L.B. 1962. Advan. Ecol. Res. 1, pp 69-101.
17.Klekowsky, R.Z., T. Prus and H. Zyromska-Rudzka. 1967. in: Petrusewitz, K. (ed). Secondary Productivity of Terrestrial Ecosystems, pp 859-879, PWN, Warszawa.
18.Slobodkin, L.B. and S. Richman. 1961. Nature 191, pp 299.

1-275 Invertebrates
Carbohydrate / biomass ratio

Species	Value	Condition
Oyster	28.2 % of dry weight	None[1]
Sea urchin	3.4 % of dry weight	40 h larva[1]
Sea urchin	5.4 % of dry weight	Unfertilized egg[1]
Silkworm	1.8 % of dry weight	Larva[1]
Silkworm	6.5 % of dry weight	Imago[1]

Source: 1.Florkin, M. 1949. Biochemical Evolution. Academic Press. New York.

1-276 Invertebrates
Conversion efficiency

Species	Value	Condition
Benthos	60 % of intake	Range = 40 - 80 %[1]

Source: 1.The Hydrological Engineering Center Corps of Engineers U.S. Army. 1975. Water Quality for river-resevoir systems, Davis, Cal. 9 5 6 1 6 (916) 499-2329.

1-277 Invertebrates
Density

Species	Value	Condition
Ampharete acutifrons	0.3 pr. 0.01 m-1	Benthos under natural conditions[1]
Ampharete americana	0.3 pr. 0.01 m-1	Benthos under natural conditions[1]
Amphitrite sp.	0.3 pr. 0.01 m-1	Benthos under natural conditions[1]
Arabella iricolor	0.6 pr. 0.01 m-1	Benthos under natural conditions[1]
Axiothella mucosa	0.6 pr. 0.01 m-1	Benthos under natural conditions[1]
Capitella capitata	8 pr. 0.01 m-1	Benthos under natural conditions[1]
Cirratulidae spp.	8.3 pr. 0.01 m-1	Benthos under natural conditions[1]
Clymenella torquata	1 pr. 0.01 m-1	Benthos under natural conditions[1]
Eteone heteropoda	0.3 pr. 0.01 m-1	Benthos under natural conditions[1]
Glycera dibranchiata	4.7 pr. 0.01 m-1	Benthos under natural conditions[1]
Glycinde solitaria	0.3 pr. 0.01 m-1	Benthos under natural conditions[1]
Haploscoloplos fragilis	7.7 pr. 0.01 m-1	Benthos under natural conditions[1]
Haploscoloplos robustus	0.6 pr. 0.01 m-1	Benthos under natural conditions[1]
Heteromastus filiformis	117 pr. 0.01 m-1	Benthos under natural conditions[1]
Idotea triloba	0.3 pr. 0.01 m-1	Benthos under natural conditions[1]
Lumbrinereis tenuis	21 pr. 0.01 m-1	Benthos under natural conditions[1]
Macoma nasuta	1 pr. 0.01 m-1	Benthos under natural conditions[1]
Magelona rosea	3.7 pr. 0.01 m-1	Benthos under natural conditions[1]
Mediomastus californiensis	3 pr. 0.01 m-1	Benthos under natural conditions[1]

Mercinaria mercinaria	0.3 pr. 0.01 m-1	Benthos under natural conditions[1]
Nereis succinea	22 pr. 0.01 m-1	Benthos under natural conditions[1]
Notomastus lobatus	0.6 pr. 0.01 m-1	Benthos under natural conditions[1]
Oligochaeta	13.7 pr. 0.01 m-1	Benthos under natural conditions[1]
Paraonis fulgens	0.3 pr. 0.01 m-1	Benthos under natural conditions[1]
Pectinaria gouldii	0.3 pr. 0.01 m-1	Benthos under natural conditions[1]
Petaloproctus socialis	0.6 pr. 0.01 m-1	Benthos under natural conditions[1]
Phloe minuta	0.6 pr. 0.01 m-1	Benthos under natural conditions[1]
Prionospio cirrobranchiata	0.3 pr. 0.01 m-1	Benthos under natural conditions[1]
Sigambra bussi	0.3 pr. 0.01 m-1	Benthos under natural conditions[1]
Spiophanes bombyx	4 pr. 0.01 m-1	Benthos under natural conditions[1]
Streblospio benedictii	16 pr. 0.01 m-1	Benthos under natural conditions[1]
Tagelus plebeius	0.3 pr. 0.01 m-1	Benthos under natural conditions[1]
Tellina texana	0.6 pr. 0.01 m-1	Benthos under natural conditions[1]
Tharyx marioni	30.3 pr. 0.01 m-1	Benthos under natural conditions[1]

Source:1.Luckenbach, M.W., Limnol. Oceanogr., 31, 867, 1986.

1-278 Invertebrates
Development time

Species	Value	Condition
Fly	2 days	Initial values, thermal ecosystem, 1 and 2 instar larvae[1]
Fly	2 days	Initial values, thermal ecosystem, to hatching, eggs[1]
Fly	4 days	Initial values, thermal ecosystem, 3 instar larvae[1]
Fly	6 days	Initial values, thermal ecosystem, pupae[1]

Source: 1.Wiegert, R.G. 1975. In: B.C. Patten (ed.), Systems Analysis and Simulation in Ecology, Vol. 3, pp 15-181.

1-279 Invertebrates
Dry weight

Species	Value	Condition
Ablabesmyia sp.	0.005 mg	Benthos, dried at 343 K, 0 - 1 mm[1]
Ablabesmyia sp.	0.007 mg	Benthos, dried at 343 K, 1 - 2 mm[1]
Ablabesmyia sp.	0.01 mg	Benthos, dried at 343 K, 2 - 3 mm[1]
Ablabesmyia sp.	0.06 mg	Benthos, dried at 343 K , more than 3 mm[1]
Ablabesmyia sp.	0.154 mg	Emergent insects, dried at 343 K[1]
Caenis simulans	0.01 mg	Benthos, dried at 343 K, 0 - 1 mm[1]
Caenis simulans	0.044 mg	Benthos, dried at 343 K, 1 - 2 mm[1]
Caenis simulans	0.13 mg	Benthos, dried at 343 K, 2 - 3 mm[1]
Caenis simulans	0.154 mg	Emergent insects, dried at 343 K[1]
Caenis simulans	0.32 mm	Benthos, dried at 343 K, 3 - 4 mm[1]
Caenis simulans	0.573 mg	Benthos, dried at 343 K, 4 - 5 mm[1]
Callibaetis ferrugineus	0.02 mm	Benthos, dried at 343 K, 0 - 1 mm[1]
Callibaetis ferrugineus	0.05 mg	Benthos, dried at 343 K, 1 - 2 mm[1]

Callibaetis ferrugineus	0.08 mg	Benthos, dried at 343 K, 2 - 3 mm[1]
Callibaetis ferrugineus	0.14 mg	Benthos, dried at 343 K, 3 - 4 mm[1]
Callibaetis ferrugineus	0.31 mg	Benthos, dried at 343 K, 4 - 5 mm[1]
Callibaetis ferrugineus	0.63 mg	Benthos, dried at 343 K, 5 - 6 mm[1]
Callibaetis ferrugineus	0.85 mg	Benthos, dried at 343 K, 6 - 7 mm[1]
Callibaetis ferrugineus	1.25 mg	Emergent insects, dried at 343 K[1]
Callibaetis ferrugineus	1.68 mm	Benthos, dried at 343 K, more than 7 mm[1]
Ceratopogonidae	0.025 mg	Benthos, dried at 343 K, 0 - 5 mm[1]
Ceratopogonidae	0.14 mg	Benthos, dried at 343 K, 5 - 10 mm[1]
Ceratopogonidae	0.26 mg	Benthos, dried at 343 K, 10 - 15 mm[1]
Chaoborus sp.	0.75 mg	Emergent insects, dried at 343 K[1]
Chironomus sp.	0.007 mg	Benthos, dried at 343 K, instar 2[1]
Chironomus sp.	0.2 mg	Benthos, dried at 343 K, instar 3[1]
Chironomus sp.	1.5 mg	Benthos, dried at 343 K, instar 4[1]
Chironomus tentans	0.04 mg	Benthos, dried at 343 K, instar 2[1]
Chironomus tentans	0.45 mg	Benthos, dried at 343 K, instar 3[1]
Chironomus tentans	1.22 mg	Emergent insects, benthos, dried at 343 K[1]
Chironomus tentans	1.3 mg	Benthos, dried at 343 K, instar 4[1]
Chironomus tentans	2.1 mg	Benthos, dried at 343 K, pupae[1]
Corixidae	0.6 mg	Benthos, dried at 343 K, 0 - 10 mm[1]
Corynoneura sp.	0.007 mg	Benthos, dried at 343 K, 1 - 2 mm[1]
Diptera	0.005 mg	Benthos, dried at 343 K, 0 - 1 mm, miscellananeous[1]
Diptera	0.007 mg	Benthos, dried at 343 K, 1 - 2 mm, miscellananeous[1]
Diptera	0.01 mg	Benthos, dried at 343 K, 2 - 3 mm, miscellananeous[1]
Diptera	0.02 mg	Benthos, dried at 343 K, 3 - 4 mm, miscellananeous[1]
Diptera	0.04 mg	Benthos, dried at 343 K, 4 - 5 mm, miscellananeous[1]
Diptera	0.065 mg	Benthos, dried at 343 K, 5 - 6 mm, miscellananeous[1]
Diptera	0.1 mg	Benthos, dried at 343 K, pupae, miscellananeous[1]
Diptera	0.146 mg	Emergent insects, dried at 343 K, miscellananeous[1]
Diptera	1.5 mg	Benthos, dried at 343 K, more than 6 mm, miscellananeous[1]
Dytiscidae	0.04 mg	Benthos, dried at 343 K, instar 1[1]
Dytiscidae	0.17 mg	Benthos, dried at 343 K, instar 2[1]
Dytiscidae	0.97 mg	Benthos, dried at 343 K, instar 3[1]
Glyptotendipes barbipes	0.002 mg	Instar 1, lab, 297 K[2]
Glyptotendipes barbipes	0.025 mg	Instar 2, lab, 297 K[2]
Glyptotendipes barbipes	0.16 mg	Instar 3, lab, 297 K[2]
Glyptotendipes barbipes	0.65 mg	Instar 4, lab, 297 K[2]
Glyptotendipes sp.	0.005 mg	Benthos, dried at 343 K, instar 1[1]
Glyptotendipes sp.	0.02 mg	Benthos, dried at 343 K, instar 2[1]
Glyptotendipes sp.	0.15 mg	Benthos, dried at 343 K, instar 3[1]

Glyptotendipes sp.	0.45 mg	Benthos, dried at 343 K, instar 4[1]
Glyptotendipes sp.	0.85 mg	Emergent insects, dried at 343 K[1]
Haliplidae	0.47 mg	Benthos, dried at 343 K, 0 - 5 mm[1]
Haliplidae	0.94 mg	Benthos, dried at 343 K, more than 5 mm[1]
Hirudinea	1.43 mg	Benthos, dried at 343 K, 0 - 10 mm[1]
Hirudinea	5.7 mg	Benthos, dried at 343 K, more than 10 mm[1]
Hyalella azteca	0.015 mg	Benthos, dried at 343 K, 0 - 1 mm[1]
Hyalella azteca	0.036 mg	Benthos, dried at 343 K, 1 - 2 mm[1]
Hyalella azteca	0.08 mg	Benthos, dried at 343 K, 2 - 3 mm[1]
Hyalella azteca	0.133 mg	Benthos, dried at 343 K, 3 - 4 mm[1]
Hyalella azteca	0.2 mg	Benthos, dried at 343 K, 4 - 5 mm[1]
Hyalella azteca	0.41 mg	Benthos, dried at 343 K, 5 - 6 mm[1]
Hyalella azteca	0.6 mg	Benthos, dried at 343 K, more than 6 mm[1]
Hydrophilidae	1.04 mg	Benthos, dried at 343 K, 0 - 5 mm[1]
Hydrophilidae	2.14 mg	Benthos, dried at 343 K, 5 - 10 mm[1]
Hydrophilidae	4.42 mg	Benthos, dried at 343 K, 10 - 15 mm[1]
Microtendipes sp.	0.005 mg	Benthos, dried at 343 K, instar 1[1]
Microtendipes sp.	0.01 mg	Benthos, dried at 343 K, instar 2[1]
Microtendipes sp.	0.02 mg	Benthos, dried at 343 K, instar 3[1]
Microtendipes sp.	0.1 mg	Benthos, dried at 343 K, instar 4[1]
Miscellaneous trichoptera	0.06 mg	Emergent insects, dried at 343 K[1]
Phryganea sp.	18.05 mg	Emergent insects, dried at 343 K[1]
Polycentropus sp.	0.91 mg	Emergent insects, dried at 343 K[1]
Procladius sp.	0.007 mg	Benthos, dried at 343 K, 1 - 2 mm[1]
Procladius sp.	0.01 mg	Benthos, dried at 343 K, 2 - 3 mm[1]
Procladius sp.	0.02 mg	Benthos, dried at 343 K, 3 - 4 mm[1]
Procladius sp.	0.035 mg	Benthos, dried at 343 K, more than 4 mm[1]
Tanytarsini	0.007 mg	Benthos, dried at 343 K, 1 - 2 mm[1]
Tanytarsini	0.01 mg	Benthos, dried at 343 K, 2 - 3 mm[1]
Tanytarsini	0.02 mg	Benthos, dried at 343 K, 3 - 4 mm[1]
Tanytarsini	0.035 mg	Benthos, dried at 343 K, more than 4 mm[1]
Trichoptera	0.026 mg	Benthos, dried at 343 K, 1 - 3 mm[1]
Trichoptera	0.074 mg	Benthos, dried at 343 K, 3 - 5 mm[1]
Trichoptera	0.768 mg	Benthos, dried at 343 K, more than 5 mm[1]
Zygoptera	0.35 mg	Benthos, dried at 343 K, 0 - 5 mm[1]
Zygoptera	2.19 mg	Benthos, dried at 343 K, 5 - 10 mm[1]
Zygoptera	5.61 mg	Emergent insects, dried at 343 K[1]
Zygoptera	5.83 mg	Benthos, dried at 343 K, 11 - 15 mm[1]

Source: 1.Hall, D.J., W.E. Cooper and E.E. Werner. 1970. Limnol. Oceanogr., 15(3), pp 838-928.
2.Kimerle. R.A. and N.H. Anderson. 1971. Limnol. Oceanogr. 16(4), pp 646-659.

1-280 Invertebrates
Egestion

Species	Value	Condition
Asellus aquaticus	69.7 % of ingestion	None[1]
Fly	0.3 fraction of ingestion	Initial values, thermal ecosystem, limit-egestion proportion, 1 + 2 instar[2]
Fly	0.3 fraction of ingestion	Initial values, thermal ecosystem, limit-egestion proportion, 3 instar[2]
Fly	0.3 fraction of ingestion	Initial values, thermal ecosystem, limit-egestion proportion, adult[2]
Lestes sponsa	63.4 % of ingestion	None[1]
Philaenus spumarius	0.25 cal/(animal*day)	Adult, 3.87 mg, 294 K, excreting urea[3]
Philaenus spumarius	0.41 cal/(animal*day)	Adult, 3.87 mg, 294 K, excreting uric acid[3]
Philaenus spumarius	0.89 cal/(animal*day)	3rd instar, 0.39 mg, 296 K, rg = 0.43 - 1.35[3]
Philaenus spumarius	2.3 cal/(animal*day)	Adult, 3.87 mg, 294 K, excreting uric acid, rg = 1.12 - 3.48[3]

Source: 1.Wetzel, R.G. 1975. Limnology. Saunders. Philadelphia.
2.Wiegert, R.G. 1975. In: B.C. Patten (ed.), Systems Analysis and Simulation in Ecology, Vol. 3, pp 15-181.
3.Wiegert, R.G. 1964. Ecol. Monogr. 34, pp 225-241.

1-281 Invertebrates
Emigration rate

Species	Value	Condition
Fly	0.007 kcal/(kcal*day)	Initial values, thermal ecosystem, transfer coeff., incl DOM[1]

Source: 1.Wiegert, R.G. 1975. In: B.C. Patten (ed.), Systems Analysis and Simulation in Ecology, Vol. 3, pp 15-181.

1-282 Invertebrates
Excretion

Species	Value	Condition
Calanus helgolandicus and naphthalene	39.3 % of naphthalene intake	Male, soluble release = 33.9 %[1]
Calanus helgolandicus and naphthalene	41.9 % of naphthalene intake	Female, soluble release = 26.9 %[1]
Tintinnopsis acuminata	13 pg N tintinnid-1 h-1	at 20 oC, ciliate[2]
Tintinnopsis acuminata	21 pg N tintinnid-1 h-1	at 25 oC, ciliate[2]
Tintinnopsis acuminata	9 pg N tintinnid-1 h-1	at 15 oC, ciliate[2]
Tintinnopsis vasculum	32 pg N tintinnid-1 h-1	at 5 oC, ciliate[2]
Tintinnopsis vasculum	58 pg N tintinnid-1 h-1	at 10 oC, ciliate[2]
Tintinnopsis vasculum	82 pg N tintinnid-1 h-1	at 15 oC, ciliate[2]

Source: 1.Harris, R.P., V. Berdugo, E.D.S. Corner, Kilvington and Ohara. 1977. In: D.A. Wolfe (ed.): Fate and Effects of Petroleum Hydrocarbons in Marine Ecosystems and Organisms. NY. Pergamon Press.
2.Verity, P.G., Limnol. Oceanogr., 30, 1268, 1985.

1-283 Invertebrates
Filtration rate

Species	Value	Condition
Arctica islandica	2 l/hr	277 K[1]
Arctica islandica	3.5 l/hr	283 K[1]

Arctica islandica	4.1 l/hr	287 K[1]
Arctica islandica	4.3 l/hr	293 K[1]
Hiatella arctica	0.0055 l/hr	278 K[2]
Hiatella arctica	0.0121 l/hr	293 K[2]
Hiatella arctica	0.0165 l/hr	283 K[2]
Hiatella arctica	0.0341 l/hr	288 K[2]
Hydroides norvegica	900 ml/(hour*g wet weight)	289-290 K, anim wt = 12.3 mg, 2-3 hrs, n=44, graphite[3]
Lanice conchilega	105 ml/(hour*g wet weight)	285 K, anim wt = 248 mg, 5 exp, 40 days detritus, F078[4]
Lanice conchilega	1240 ml/(hour*g wet weight)	285 K, anim wt = 7.3 mg, 1 exp, 21 days detritus, F078[4]
Lanice conchilega	175 ml/(hour*g wet weight)	285 K, anim wt = 77.1 mg, 4 exp, 46 days detritus, F078[4]
Lanice conchilega	180 ml/(hour*g wet weight)	285 K, anim wt = 72.4 mg, 3 exp, 20 days detritus, F078[4]
Lanice conchilega	610 ml/(hour*g wet weight)	285 K, anim wt = 18.5 mg, 2 exp, 24 days detritus, F078[4]
Mytilus edulis	0.106 l/hr	303 K[5]
Mytilus edulis	0.353 l/hr	278 K[5]
Mytilus edulis	0.62 l/hr	277 K[6]
Mytilus edulis	1.1 l/hr	283 K[6]
Mytilus edulis	1.24 l/hr	293 K[6]
Mytilus edulis	1.33 l/hr	287 K[6]
Mytilus edulis	1.468 l/hr	283 K[5]
Mytilus edulis	1.751 l/hr	288 K[5]
Mytilus edulis	1.796 l/hr	293 K[5]
Mytilus edulis	1.798 l/hr	298 K[5]
Myxicola infundibulum	100 ml/(hour*g wet weight)	290-293 K, anim wt = 2695 mg, 2-3 hrs, n=45, graphite[3]
Oysters	250 l/day	Wet weight = 5 g, rg = 200 - 300[7]
Pomatoceros triqueter	1400 ml/(hour*g wet weight)	289-290 K, anim wt = 18.5 mg, 2-3 hrs, n=45, graphite[3]
Pomatoceros triqueter	900 ml/(hour*g wet weight)	285 K, anim wt = 11.8 mg, 0.5-1hr, n=12, Dunaliella sp. (22E6cells/l)[8]
Pomatoceros triqueter	970 ml/(hour*g wet weight)	291 K, anim wt = 11.8 mg, 0.5-1hr, n=11, Dunaliella sp. (24E6cells/l)[8]
Sabella pavonina	390 ml/(hour*g wet weight)	291-293 K, anim wt = 187 mg, 2-3 hrs, n=50, graphite[3]
Salmacina dysteri	2090 ml/(hour*g wet weight)	290-291 K, anim wt = 0.14 mg, 2-3 hrs, n=50, graphite[3]
Schizobranchia insignis	70 ml/(hour*g wet weight)	285-286 K, anim wt = 500-2000 mg, 3 hrs, n=10, graphite, mean val[9]
Spirorbis borealis	950 ml/(hour*g wet weight)	290-292 K, anim wt = 0.24 mg, 2-3 hrs, n=99, graphite[3]

Source: 1.Winter, J.E. 1969. Mar. Biol., 4, pp 87-135.
2.Schultze, E.H. 1975. Mar. biol., 30(4), pp 331-341.
3.Dales, R.P. 1957. J. mar. biol. Ass. U.K., 36, pp 309-316.
4.Buhr, K.-J. 1976. Mar. Biol., 38(4), pp 373-383.
5.Schultze, E.H. 1975. Mar. biol., 30(4), pp 331-341.

6.Theede, H. 1963. Kieler Meeresforsch., 19, pp 20-41.
7.Moore, S.F. and R.L. Dwyer. 1974. Water Res. 8(10), pp 819-827
8.Kloeckner, K. 1976. Oekologie von Pomatoceros triqueter (L.) (Serpulidae, Polycheata), Dissertation,Universitaet Tubingen.
9.Dales, R.P. 1961. Biol. Bull. mar. Biol. Lab., Woods Hole, 121, pp 82-91.

1-284 Invertebrates
Grazing rate

Species	Value	Condition
Fly	0.751 kcal/(kcal*day)	Initial values, thermal ecosystem, transfer coeff., adult[1]
Fly	1.456 kcal/(kcal*day)	Initial values, thermal ecosystem, transfer coeff., 3 instar[1]
Fly	2.94 kcal/(kcal*day)	Initial values, thermal ecosystem, transfer coeff., 1 + 2 instar[1]

Source: 1.Wiegert, R.G. 1975. In: B.C. Patten (ed.), Systems Analysis and Simulation in Ecology, Vol. 3, pp 15-181.

1-285 Invertebrates
Growth and reproduction

Species	Value	Condition
Asellus aquaticus	18.2 % of assimilation	None[1]
Asellus aquaticus	5.6 % of ingestion	None[1]
Lestes sponsa	23.5 % of ingestion	None[1]
Lestes sponsa	64 % of assimilation	None[1]

Source: 1.Wetzel, R.G. 1975. Limnology. Saunders. Philadelphia.

1-286 Invertebrates
Growth efficiency

Species	Value	Condition
Alburnus alburnus	8.8 % of ingestion	Prod = 30.5 kcal/(m2*y)[1]
Allolobophora rosea	17.58 % of ingestion	Prod = 2.717 kcal/(m2*y)[1]
Anatopynia dyari	15.2 % of ingestion	Prod = 37.4 kcal/(m2*y)[1]
Asellus aquaticus	24.1 % of ingestion	Prod = 16.296 kcal/(m2*y)[1]
Asellus miliaris	18.2 % of ingestion	Prod = 108.3 kcal/(m2*y)[1]
Calopsetra dives	25.1 % of ingestion	Prod = 130.7 kcal/(m2*y)[1]
Chorthippus parallellus	55.3 % of ingestion	Prod = 0.420 kcal/(m2*y)[1]
Chorthippus parallellus	58.9 % of ingestion	Prod = 0.273 kcal/(m2*y)[1]
Diplacodes trivialis	48.1 % of assimilation	46 mg animal, ingestion = 61.5 cal/(kcal*anim.*d)[2]
Gammarus pseudolimnaeus	15 % of ingestion	Prod = 183.0 kcal/(m2*y)[1]
Gobio gobio	6.6 % of ingestion	Prod = 2.6 kcal/(m2*y)[1]
Japonaria laminata armigera	46.5 % of ingestion	Prod = 26.9 kcal/(m2*y)[1]
Japonaria laminata armigera	50 % of ingestion	Prod = 1.5 kcal/(m2*y)[1]
Leptoterna dolobrata	52.3 % of ingestion	Prod = 0.096 kcal/(m2*y)[1]
Leptoterna dolobrata	58.1 % of ingestion	Prod = 0.140 kcal/(m2*y)[1]
Leuciscus leuciscus	5.96 % of ingestion	Prod = 1.5 kcal/(m2*y)[1]
Ligidium japonica	18 % of ingestion	Prod = 3.5 kcal/(m2*y)[1]
Limnodrilus hoffmeisteri claraparede	24.9 % of ingestion	Prod = 160.5 kcal/(m2*y)[1]

Lithobius crassipes	5.51 % of ingestion	Prod = 0.074 kcal/(m2*y)[1]
Lithobius fortificatus	4.08 % of ingestion	Prod = 0.085 kcal/(m2*y)[1]
Littorina irrorata	14 % of ingestion	Prod = 40.6 kcal/(m2*y)[1]
Metapenaeus monoceros	21.6 % of ingestion	Gross eff., food = estuarine detritus[3]
Metapenaeus monoceros	24.1 % of assimilation	Net effic., food = estuarine detritus[3]
Modiolus demissus	30 % of ingestion	Prod = 16.7 kcal/(m2*y)[1]
Onychiurus procampatus	46.2 % of ingestion	Prod = 2.61 kcal/(m2*y)[1]
Orchelimum fidicinium	36.7 % of ingestion	Prod = 10.8 kcal/(m2*y)[1]
Oribatid mites	21.4 % of ingestion	Prod = 0.43 kcal/(m2*y)[1]
Orthoptera	15.6 % of ingestion	Prod = 4.0 kcal/(m2*y), 3 species[1]
Orthoptera	34.4 % of ingestion	Prod = 4.573 kcal/(m2*y)[1]
Orthoptera	36.3 % of ingestion	Prod = 0.638 kcal/(m2*y), most melanopus[1]
Orthoptera	38.5 % of ingestion	Prod = 0.373 kcal/(m2*y)[1]
Perca fluviatilis	7.2 % of ingestion	Prod = 1.8 kcal/(m2*y)[1]
Philaenus spumarius	41.2 % of ingestion	Prod = 15963 kcal/(m2*y)[1]
Physa pseudolimnaeus	27.9 % of ingestion	Prod = 170.6 kcal/(m2*y)[1]
Pimpla instigator	65.4 % of ingestion	Prod = 0.017 kcal/(m2*y), larvae[1]
Pisidium virginicum	47.3 % of ingestion	Prod = 81.8 kcal/(m2*y)[1]
Pteromalus puparum	56.9 % of ingestion	Prod = 0.074 kcal/(m2*y), larvae[1]
Pyrrhosoma nymphula	51.3 % of ingestion	Prod = 3.889 kcal/(m2*y)[1]
Pyrrhosoma nymphula	58.2 % of ingestion	Prod = 4.436 kcal/(m2*y)[1]
Rutilus rutilus	8.3 % of ingestion	Prod = 10.7 kcal/(m2*y)[1]
Scrobicularia plana	20.65 % of ingestion	Prod = 124.0 kcal/(m2*y)[1]
Scrobicularia plana	24.3 % of ingestion	Prod = 17.5 kcal/(m2*y)[1]
Trichoptera	23.7 % of ingestion	Prod = 20.9 kcal/(m2*y)[1]

Source: 1.McNeill, S. and J.H. Lawton. 1970. Nature 225, pp 472-474
2.Pandian, T.J. and S. Mathavan. 1974. Odonatologica, 3(4), pp 241-248.
3.Qasim, S.Z. and D.C.V. Easterson. 1974. Indian J. Mar. Sci. 3(2), pp 131-134.

1-287 Invertebrates
Growth rate

Species	Value	Condition
Calandra oryzae	113.6 rate	Net production rate, lenght of 1 generation = 58.1 d[1]
Daphnia pulex	20.7 rate	Net production rate, lenght of 1 generation = 15.4 d[1]
Daphnia pulex	21.4 rate	Net production rate, lenght of 1 generation = 10.6 d[1]
Dugesia tigrina	2 rate	Net production rate, lenght of 1 generation = 10.3 d[1]
Eurostus hilleri	3.1 rate	Net production rate, lenght of 1 generation = 115.6 d[1]
Gibbium psylloides	75.2 rate	Net production rate, lenght of 1 generation = 182.6 d[1]
Glyptotendipes barbipes	100 mg/(gram*day)	Instar 3, lab, 297 K[2]
Glyptotendipes barbipes	116 mg/(gram*day)	Age = 20 d, 297 K, lab, relative growth[2]
Glyptotendipes barbipes	172 mg/(gram*day)	Age = 14.7 d, 297 K, lab, relative growth[2]
Glyptotendipes barbipes	200 mg/(gram*day)	Instar 2, lab, 297 K[2]

Glyptotendipes barbipes	25 mg/(gram*day)	Age = 52 d, 297 K, lab, relative growth[2]
Glyptotendipes barbipes	294 mg/(gram*day)	Age = 8.7 d, 297 K, lab, relative growth[2]
Glyptotendipes barbipes	363 mg/(gram*day)	Age = 5.5 d, 297 K, lab, relative growth[2]
Glyptotendipes barbipes	40 mg/(gram*day)	Age = 24.5 d, 297 K, lab, relative growth[2]
Glyptotendipes barbipes	400 mg/(gram*day)	Instar 1, lab, 297 K[2]
Glyptotendipes barbipes	44 mg/(gram*day)	Age = 32.1 d, 297 K, lab, relative growth[2]
Glyptotendipes barbipes	45 mg/(gram*day)	Age = 42 d, 297 K, lab, relative growth[2]
Glyptotendipes barbipes	469 mg/(gram*day)	Age = 1.6 d, 297 K, lab, relative growth[2]
Glyptotendipes barbipes	50 mg/(gram*day)	Instar 4, lab, 297 K[2]
Lasioderma serricorne	13.2 rate	Net production rate, lenght of 1 generation = 35.6 d[1]
Mezium affine	64.2 rate	Net production rate, lenght of 1 generation = 287.3 d[1]
Microtus agrestis	5.9 rate	Net production rate, lenght of 1 generation = 170.4 d[1]
Nezara viridula	1.6 rate	Net production rate, lenght of 1 generation = 10.3 d[1]
Niptus hololeucus	2.3 rate	Net production rate, lenght of 1 generation = 138.1 d[1]
Octopus vulgaris	0.19 % per day	288 K, weight=664g, mature female, optimal condition[3]
Octopus vulgaris	0.52 % per day	288 K, weight=587g, mature female, optimal condition[3]
Octopus vulgaris	0.63 % per day	288 K, weight=580g, immature female, optimal condition[3]
Octopus vulgaris	0.79 % per day	288 K, weight=675g, immature female, optimal condition[3]
Octopus vulgaris	1.11 % per day	288 K, weight=530g, mature female, optimal condition[3]
Octopus vulgaris	1.35 % per day	288 K, weight=500g, immature female, optimal condition[3]
Pavona clivosa	2.5 cm/year	Radial growth rate, coral[4]
Pavona gigantea	2 cm/year	Radial growth rate, coral[4]
Pediculus humanus	30.9 rate	Net production rate, lenght of 1 generation = 30.9 d[1]
Philaenus spumarius	0.25 cal/(animal*day)	3rd instar, 0.39 mg, 296 K[5]
Philaenus spumarius	0.74 cal/(animal*day)	4rd instar, 1.16 mg, 298 K[5]
Phyllopertha horticola	2.9 rate	Net production rate, lenght of 1 generation = 365.0 d[1]
Physa gyrina	515.3 rate	Net production rate, lenght of 1 generation = 142.0 d[1]
Physa gyrina	642.7 rate	Net production rate, lenght of 1 generation = 242.7 d[1]
Pocillopora damicornis	10 cm/year	Radial growth rate, coral[4]
Pocillopora robusta	5 cm/year	Radial growth rate, coral[4]
Porites panamensis	1 cm/year	Radial growth rate, coral[4]
Psammocora stellata	1 cm/year	Radial growth rate, coral[4]
Ptinus fur	10.8 rate	Net production rate, lenght of 1 generation = 182.3 d[1]
Ptinus sexpunctatus	4.1 rate	Net production rate, lenght of 1 generation = 227.5 d[1]
Ptinus tectus	116.1 rate	Net production rate, lenght of 1 generation = 149.6 d[1]
Rhabditis marina	11.3 micro m/hour	298 K, 2.5 % salinitet, male, max rate[6]
Rhabditis marina	11.7 micro m/hour	298 K, 2.5 % salinitet, female, max rate[6]
Stethomezium squamosum	41 rate	Net production rate, lenght of 1 generation = 190.6 d[1]
Tribolium castaneum	275 rate	Net production rate, lenght of 1 generation = 80.0 d[1]
Trigonogenius globulus	49.8 rate	Net production rate, lenght of 1 generation = 171.8 d[1]

Source: 1.Laughlin, R. 1965. J. Anim. Ecol. 34, pp 77-91.
2.Kimerle. R.A. and N.H. Anderson. 1971. Limnol. Oceanogr. 16(4), pp 646-659.
3.Manguld, K. and S. von Boletzky. 1973. Mar. biol. 19, pp 7-12.
4.Maguire, L.A. and J.J. Porter. 1977. Ecol. Modell. 3(4), pp 249-271.
5.Wiegert, R.G. 1964. Ecol. Monogr. 34, pp 225-241.
6.Tietjen, J.H., J.J. Lee, J. Rullman, A. Greengart and J. Trompeter. 1970. Limnol. Oceanogr. 15(4), p535-543.

1-288 Invertebrates
H / biomass ratio

Species	Value	Condition
Annelida	5.9 % of dry weight	Mean elementary composition[1]
Arthropoda	7.3 % of dry weight	Mean elementary composition, land[1]
Coelenterata	4.5 % of dry weight	Mean elementary composition, interpolated[1]
Echinodermata	4.5 % of dry weight	Mean elementary composition, interpolated[1]
Mollusca	6 % of dry weight	Mean elementary composition, soft parts[1]
Nematoda	6 % of dry weight	Mean elementary composition, interpolated[1]

Source: 1.Bowen, H.J.M. 1966. Trace Elements in Biochemistry, Ac. Press.

1-289 Invertebrates
Half saturation constant for sediment

Species	Value	Condition
Benthos	100 mg/m**2	Used in models, San Franscico Bay estuary[1]
Benthos	1025 mg/m**2	Grazing on sediment, rg = 50 - 2000[2]
Benthos	50 mg/m**2	Used in models, Boise River, Idaho[3]

Source: 1.Chen, C.W. 1975. in: Russell, C.S.; Ecological Modeling, in a Resource Management Framework. NOAA.
2.The Hydrological Engineering Center Corps of Engineers U.S. Army. 1975. Water Quality for river-resevoir systems, Davis, Cal. 9 5 6 1 6 (916) 499-2329.
3.Chen, C.W. and Wells. 1975. Boise River Water Quality-Ecological Model for Urban Planning Study. USrmy Eng. Walla Walla, Idaho Water Res. Board.

1-290 Invertebrates
Lipid / biomass ratio

Species	Value	Condition
Mussel	11.9 % of dry weight	None[1]
Oyster	11.1 % of dry weight	None[2]
Sea urchin	17.4 % of dry weight	40 h larva[2]
Sea urchin	31.2 % of dry weight	Unfertilized egg[2]
Silkworm	13.3 % of dry weight	Larva[2]
Silkworm	24.3 % of dry weight	Imago[2]

Source: 1.Special Rept. Ser. Med. Res. Com. London No. 297. 1960. The Composition of Foods.
2.Florkin, M. 1949. Biochemical Evolution. Academic Press. New York.

1-291 Invertebrates
Max standing crop

Species	Value	Condition
Fly	150 kcal/m**2	Initial values, thermal ecosystem, 1 + 2 instar[1]
Fly	150 kcal/m**2	Initial values, thermal ecosystem, 3 instar[1]
Fly	315 kcal/m**2	Initial values, thermal ecosystem, adult[1]

Source: 1.Wiegert, R.G. 1975. In: B.C. Patten (ed.), Systems Analysis and Simulation in Ecology, Vol. 3, pp 15-181.

1-292 Invertebrates
Mortality

Species	Value	Condition
Benthos	0.001 1/day	None[1]
Benthos	0.001 1/day	Used in models[2]
Benthos	0.003 1/day	Used in models, N Fork Kings River, Calif[3]
Fly	0.056 kcal/(kcal*day)	Initial values, thermal ecosystem, transfer coeff., pupae[4]
Fly	0.122 kcal/(kcal*day)	Initial values, thermal ecosystem, transfer coeff., base rate, 3 instar[4]
Fly	0.148 kcal/(kcal*day)	Initial values, thermal ecosystem, transfer coeff., adult[4]
Fly	0.215 kcal/(kcal*day)	Initial values, thermal ecosystem, transfer coeff., egg[4]
Fly	0.221 kcal/(kcal*day)	Initial values, thermal ecosystem, transfer coeff., 1 + 2 instar[4]

Source: 1.The Hydrological Engineering Center Corps of Engineers U.S. Army. 1975. Water Quality for river-resevoir systems, Davis, Cal. 9 5 6 1 6 (916) 499-2329.
2.U.S. Army Corps of Engineers. 1974. Water Quality for River-Resevoirs Systems.
3.Tetra Tech, Inc. 1976. Estuary Water Quality Models, Long Island, New York User's Guide. For Nassau-Suffolk Regional Planning Board. Hauppauge, New York. 11787.
4.Wiegert, R.G. 1975. In: B.C. Patten (ed.), Systems Analysis and Simulation in Ecology, Vol. 3, pp 15-181.

1-293 Invertebrates
N / biomass ratio

Species	Value	Condition
Agrypnia obsoleta	8 % of ash free dry weight	Trichoptera, 95 % CL = +-1.9[1]
Annelida	9.9 % of dry weight	Mean elementary composition[2]
Arthropoda	12.3 % of dry weight	Mean elementary composition, land[2]
Asellus aquaticus	9.7 % of ash free dry weight	Isopoda, 95 % CL = +-1.4[1]
Benthos	0.08 by weight	Range = 0.07 - 0.09[3]
Benthos	0.08 by weight	Used in models, N Fork Kings River, Calif[4]
Benthos	0.09 by weight	Used in models, Boise River, Idaho[5]
Centroptilum luteolum	9.9 % of ash free dry weight	Ephemeroptera, 95 % CL = +-1.2[1]
Chaoborus flavicans	9.5 % of ash free dry weight	Diptera, 95 % CL = +-0.4[1]
Chironomus plumosus	8.7 % of ash free dry weight	Diptera, 95 % CL = +-1.4[1]
Coelenterata	6.3 % of dry weight	Mean elementary composition[2]
Cordulia aenea	9 % of ash free dry weight	Odonata, 95 % CL = +-0.7[1]

Echinodermata	4.4 % of dry weight	Mean elementary composition[2]
Erpobdella octoculata	9.4 % of ash free dry weight	Hirudinea, 95 % CL = +-0.6[1]
Gammaracanthus lacustris	9.6 % of ash free dry weight	Amphipoda, 95 % CL = +-0.7[1]
Heptagenia fuscogrisea	8.8 % of ash free dry weight	Ephemeroptera, 95 % CL = +-0.8[1]
Invertebrate	9 % of ash free dry weight	Standard derivation = +-1.3, mean of 19 species[1]
Leptophlebia sp.	10.1 % of ash free dry weight	Ephemeroptera, 95 % CL = +-0.8[1]
Limnephilidae sp.	6.2 % of ash free dry weight	Trichoptera, 95 % CL = +-1.0[1]
Mollusca	8.5 % of dry weight	Mean elementary composition, soft parts[2]
Mysis relicta	10 % of ash free dry weight	Mysidacea, 95 % CL = +-0.4[1]
Nematoda	11 % of dry weight	Mean elementary composition[2]
Notonecta glauca	10.3 % of ash free dry weight	Heterpotera, 95 % CL = +-0.6[1]
Pallacea quadrispinosa	9.4 % of ash free dry weight	Amphipoda, 95 % CL = +-0.7[1]
Sialis sp.	9.3 % of ash free dry weight	Megaloptera, 95 % CL = +-1.1[1]
Tintinnopsis acuminata	175 pg	Ciliate, grown on Dicrateria inornata at 25 oC[6]
Tintinnopsis acuminata	182 pg	Ciliate, grown on Dicrateria inornata at 20 oC[6]
Tintinnopsis acuminata	188 pg	Ciliate, grown on Dicrateria inornata at 15 oC[6]
Tintinnopsis vasculum	1,730 pg	Ciliate, grown on Dicrateria inornata at 15 oC[6]
Tintinnopsis vasculum	1,875 pg	Ciliate, grown on Dicrateria inornata at 10 oC[6]
Tintinnopsis vasculum	1,901 pg	Ciliate, grown on Dicrateria inornata at 5 oC[6]

Source: 1.Salonen, K., J. Sarvala, I. Hakala and M.-L. Viljanen. 1976. Limnol. Oceanogr., 21(5), pp 724-730.
2.Bowen, H.J.M. 1966. Trace Elements in Biochemistry, Ac. Press.
3.The Hydrological Engineering Center Corps of Engineers U.S. Army. 1975. Water Quality for river-resevoir systems, Davis, Cal. 9 5 6 1 6 (916) 499-2329.
4.Tetra Tech, Inc. 1976. Estuary Water Quality Models, Long Island, New York User's Guide. For Nassau-Suffolk Regional Planning Board. Hauppauge, New York. 11787.
5.Chen, C.W. and Wells. 1975. Boise River Water Quality-Ecological Model for Urban Planning Study. US Army Eng. Walla Walla, Idaho Water Res. Board.
6.Verity, P.G., Limnol. Oceanogr., 30, 1268, 1985.

1-294 Invertebrates
Na / biomass ratio

Species	Value	Condition
Arctiidae	9.2 mg/(g ash free dry wt)	Soil or litter invert.[1]
Camponotus pennsylvanicus	9.4 mg/(g ash free dry wt)	Soil or litter invert.[1]
Ceuthophilus gracilipes	5.7 mg/(g ash free dry wt)	Soil or litter invert.[1]
Dixidesmus erasus	3.7 mg/(g ash free dry wt)Soil or litter invert.[1]	
Geotrupes sp.	3.4 mg/(g ash free dry wt)	Soil or litter invert.[1]
Leiobunum flavum	5.7 mg/(g ash free dry wt)	Soil or litter invert.[1]
Ligidium blueridgensis	5.1 mg/(g ash free dry wt)	Soil or litter invert.[1]
Mesomphix vulgatus	2.2 mg/(g ash free dry wt)	Soil or litter invert.[1]
Mesomphix vulgatus	8.8 mg/(g ash free dry wt)	Soil or litter invert.[1]
Otocryptops sexspinosus	9.2 mg/(g ash free dry wt)	Soil or litter invert.[1]
Pachylomerides audouini	7.3 mg/(g ash free dry wt)	Soil or litter invert.[1]
Sciomyzidae	1.2 mg/(g ash free dry wt)	Soil or litter invert.[1]
Scytonotus granulatus	1.4 mg/(g ash free dry wt)	Soil or litter invert.[1]
Sphaeroderus stenostomus	3.8 mg/(g ash free dry wt)	Soil or litter invert.[1]

Source: 1.Reichle, D.E., M.H. Shanks and D.A. Crossey, Jr. 1969. Ann. Entomol. Soc. Amer. 62, pp 57-62.

1-295 Invertebrates
Numbers of eggs

Species	Value	Condition
Amphipholis squamata	6 eggs/individual	Ophiuroid, direct development, rg = 2 - 10[1]
Amphiura angularis	5 eggs/individual	Ophiuroid, direct development[1]
Amphiura annulifera	2 eggs/individual	Ophiuroid, direct development[1]
Amphiura belgicae	105 eggs/individual	Ophiuroid, direct development, rg = 10 - 200[1]
Amphiura filiformis	5000 eggs/individual	Ophiuroid, indirect development, app value[1]
Amphiura microplax	100 eggs/individual	Ophiuroid, direct development[1]
Amphiura monorima	7.5 eggs/individual	Ophiuroid, direct development, rg = 5 - 10[1]
Astrochlamys bruneus	2000 eggs/individual	Ophiuroid, direct development[1]
Kirks ophiuroid	55 eggs/individual	Ophiuroid, direct development, rg = 10 - 100[1]
Ophiacantha densispina	20 eggs/individual	Ophiuroid, direct development[1]
Ophiacantha vivipara	4 eggs/individual	Ophiuroid, direct development, rg = 2 - 6[1]

Ophioceres incipiens	405 eggs/individual	Ophiuroid, direct development, rg = 10 - 800[1]
Ophiochondrus stelliger	5 eggs/individual	Ophiuroid, direct development[1]
Ophiocoma echinata	717000 eggs/individual	Ophiuroid, indirect development[1]
Ophiocoma erinaceus	253000 eggs/individual	Ophiuroid, indirect development[1]
Ophiolebella biscutifera	3 eggs/individual	Ophiuroid, direct development[1]
Ophiomitrella clavigera	15 eggs/individual	Ophiuroid, direct development, rg = 10 - 20[1]
Ophiomitrella falklandica	35 eggs/individual	Ophiuroid, direct development, rg = 20 - 50[1]
Ophiomitrella ingrata	10 eggs/individual	Ophiuroid, direct development[1]
Ophiomyxa brevirima	600 eggs/individual	Ophiuroid, direct development, rg = 2 - 1200[1]
Ophionotus hexactis	55 eggs/individual	Ophiuroid, direct development, rg = 30 - 80[1]
Ophiopholis aculeata	352000 eggs/individual	Ophiuroid, indirect development[1]
Ophioplocus esmarki	200 eggs/individual	Ophiuroid, direct development[1]
Ophioscolex nutrix	15 eggs/individual	Ophiuroid, direct development, rg = 10 - 20[1]
Ophiosteira antarctica	35 eggs/individual	Ophiuroid, direct development, rg = 20 - 50[1]
Ophiozonella falklandi	25 eggs/individual	Ophiuroid, direct development, rg = 20 - 30[1]
Ophiura meridionalis	70 eggs/individual	Ophiuroid, direct development, rg = 60 - 80[1]
Ophiura rouchi	100 eggs/individual	Ophiuroid, direct development[1]
Ophiurolepis martensi	55 eggs/individual	Ophiuroid, direct development, rg = 60 - 80[1]
Pectinura cylindrica	10 eggs/individual	Ophiuroid, direct development, less than[1]
Pectinura gracilis	10 eggs/individual	Ophiuroid, direct development, less than[1]
Stegophiura nodosa	120 eggs/individual	Ophiuroid, direct development[1]
Stegophiura sculpta	700 eggs/individual	Ophiuroid, direct development[1]

Source: 1.Schoener, A. 1972. Limnol. Oceanogr. 17(2), pp 193-199.

1-296 Invertebrates
O / biomass ratio

Species	Value	Condition
Annelida	34 % of dry weight	Mean elementary composition[1]
Arthropoda	32.3 % of dry weight	Mean elementary composition, land[1]
Coelenterata	27.1 % of dry weight	Mean elementary composition[1]
Echinodermata	45.6 % of dry weight	Mean elementary composition[1]
Mollusca	39 % of dry weight	Mean elementary composition, soft parts[1]
Nematoda	35.4 % of dry weight	Mean elementary composition[1]

Source: 1.Bowen, H.J.M. 1966. Trace Elements in Biochemistry, Ac. Press.

1-297 Invertebrates
Oviposition rate

Species	Value	Condition
Fly	0.056 kcal/kcal*day	Initial values, thermal ecosystem, transfer coeff.[1]

Source: 1.Wiegert, R.G. 1975. In: B.C. Patten (ed.), Systems Analysis and Simulation in Ecology, Vol. 3, pp 15-181.

1-298 Invertebrates
Oxygen consumption

Species	Value	Condition
Araneae	26.1 cm**3 O2/(g dry weight*day)	From a mesic deciduous forest, yellow poplar, 288 K, Arthropoda[1]
Chilopoda	20.5 cm**3 O2/(g dry weight*day)	From a mesic deciduous forest, yellow poplar, 288 K, Arthropoda[1]
Coleoptera	25.8 cm**3 O2/(g dry weight*day)	From a mesic deciduous forest, yellow poplar, 288 K, Insecta, larvae[1]
Diplopoda	12.6 cm**3 O2/(g dry weight*day)	From a mesic deciduous forest, yellow poplar, 288 K, Arthropoda[1]
Diplura	31 cm**3 O2/(g dry weight*day)	From a mesic deciduous forest, yellow poplar, 288 K, Insecta, Arthropoda[1]
Diptera	21.3 cm**3 O2/(g dry weight*day)	From a mesic deciduous forest, yellow poplar, 288 K, Insecta, larvae[1]
Enchytraeidae	31 cm**3 O2/(g dry weight*day)	From a mesic deciduous forest, yellow poplar, 288 K, Annelida[1]
Entomobryidae	34.7 cm**3 O2/(g dry weight*day)	From a mesic deciduous forest, yellow poplar, 288 K, Insecta[1]
Gamasina	24.4 cm**3 O2/(g dry weight*day)	From a mesic deciduous forest, yellow poplar, 288 K, Arthropoda[1]
Hymenoptera	29.2 cm**3 O2/(g dry weight*day)	From a mesic deciduous forest, yellow poplar, 288 K, Insecta[1]
Isotomidae	41.8 cm**3 O2/(g dry weight*day)	From a mesic deciduous forest, yellow poplar, 288 K, Insecta[1]
Lepidoptera	29.6 cm**3 O2/(g dry weight*day)	From a mesic deciduous forest, yellow poplar, 288 K, Insecta, larvae[1]
Lumbricidae	7.65 cm**3 O2/(g dry weight*day)	From a mesic deciduous forest, yellow poplar, 288 K, Annelida[1]
Nematoda	90.7 cm**3 O2/(g dry weight*day)	From a mesic deciduous forest, yellow poplar, 288 K[1]
Onychiuridae	52.9 cm**3 O2/(g dry weight*day)	From a mesic deciduous forest, yellow poplar, 288 K, Insecta[1]
Oribatei	29.8 cm**3 O2/(g dry weight*day)	From a mesic deciduous forest, yellow poplar, 288 K, Arthropoda[1]
Orthoptera	12.9 cm**3 O2/(g dry weight*day)	From a mesic deciduous forest, yellow poplar, 288 K, Insecta[1]
Pauropoda	51 cm**3 O2/(g dry weight*day)	From a mesic deciduous forest, yellow poplar, 288 K, Arthropoda[1]
Phalangida	24.1 cm**3 O2/(g dry weight*day)	From a mesic deciduous forest, yellow poplar, 288 K, Arthropoda[1]
Poduridae	48.3 cm**3 O2/(g dry weight*day)	From a mesic deciduous forest, yellow poplar, 288 K, Insecta[1]
Prostigmata	55.2 cm**3 O2/(g dry weight*day)	From a mesic deciduous forest, yellow poplar, 288 K, Arthropoda[1]
Protura	58.3 cm**3 O2/(g dry weight*day)	From a mesic deciduous forest, yellow poplar, 288 K, Arthropoda[1]
Pseudoscorpionida	22.6 cm**3 O2/(g dry weight*day)	From a mesic deciduous forest, yellow poplar, 288 K, Arthropoda[1]
Psocoptera	23 cm**3 O2/(g dry weight*day)	From a mesic deciduous forest, yellow poplar, 288 K, Insecta[1]

Pulmonata	9.55 cm**3 O2/(g dry weight*day)	From a mesic deciduous forest, yellow poplar, 288 K[1]
Sminthuridae	48.3 cm**3 O2/(g dry weight*day)	From a mesic deciduous forest, yellow poplar, 288 K, Insecta[1]
Symphyla	30.4 cm**3 O2/(g dry weight*day)	From a mesic deciduous forest, yellow poplar, 288 K, Arthropoda[1]
Uropodina	26.5 cm**3 O2/(g dry weight*day)	From a mesic deciduous forest, yellow poplar, 288 K, Arthropoda[1]

Source: 1.Reichle, D.E. 1977. In: U. Lohm and T. Persson (eds.): Soil Organisms as Components of Ecosystems. Ecol. Bull (Stockholm) 25, 145-156.

1-299 Invertebrates P / biomass ratio

Species	Value	Condition
Benthos	0.011 by weight	Range = 0.01 - 0.012[1]
Benthos	0.015 by weight	Used in models, Boise River, Idaho[2]
Benthos	0.015 by weight	Used in models, N Fork Kings River, Calif[3]

Source: 1.The Hydrological Engineering Center Corps of Engineers U.S. Army. 1975. Water Quality for river-resevoir systems, Davis, Cal. 9 5 6 1 6 (916) 499-2329.
2.Chen, C.W. and Wells. 1975. Boise River Water Quality-Ecological Model for Urban Planning Study. US Army Eng. Walla Walla, Idaho Water Res. Board.
3.Tetra Tech, Inc. 1976. Estuary Water Quality Models, Long Island, New York User's Guide. For Nassau-Suffolk Regional Planning Board. Hauppauge, New York. 11787.

1-300 Invertebrates Production

Species	Value	Condition
Gammarus mucronatus	12.4 - 15.8 g dry wt. m-2 y-1	mean cohort turnover rate, different methods[1]

Source: 1.LaFrance, K. and Ruber, E., Limnol. Oceanogr., 30(5), 1109, 1985.

1-301 Invertebrates Production / biomass ratio

Species	Value	Condition
Acanthina spinata	0.3 on yearly basis	Southern Calif macroinvertebrates[1]
Acmaea sp.	1 on yearly basis	Southern Calif macroinvertebrates[1]
Aletes squamigerus	0.3 on yearly basis	Southern Calif macroinvertebrates[1]
Alloperla mediana	2.8 on yearly basis	Prod = 0.17 kg dw/(ha*y), Walker Branch, voltinism=2 y, Tenn.[2]
Amphipoda	0.024 per day	Per day, southern lakes of USSR, rg = 0.008 - 0.04[3]
Ancylus lacustris	3.3 on yearly and shell-free basis	Prod = 0.09 kg dw/(ha*y), R Thames, voltinism=uni, Engl[2]
Ancylus lacustris	3.32 on yearly basis	Prod = 0.83 g wet weight/yr, Allan curve method[4]
Anodonta anatina	0.2 on yearly and shell-free basis	Prod = 22.0 kg dw/(ha*y), R Thames, voltinism=2+ y, Engl[2]
Anodonta anatina	0.2 on yearly basis	Prod = 211.0 g wet weight/yr, less shells[4]
Anodonta piscinalis	0.33 on yearly and shell-free basis	Prod = 4.9 kg dw/(ha*y), Danube R pools, voltinism=2+ y, Romania[2]
Antropora sp.	2 on yearly basis	Southern Calif macroinvertebrates[1]
Archynopteryx curvata	3.8 on yearly basis	Sagehen creek, voltinism=uni, Calif[2]
Asellus aquaticus	2 on yearly basis	Prod = 27.8 kg dw/(ha*y), L Pajep Maske., voltinism=2 y, Sweden[2]

Asellus aquaticus	2 on yearly basis	Prod = 52.7 kg dw/(ha*y), L Erken, voltinism=uni, Sweden[2]
Asellus aquaticus	3.17 on yearly basis	Prod = 13.30 g wet weight/yr, Allan curve method[4]
Asellus aquaticus	3.2 on yearly basis	Prod = 1.38 kg dw/(ha*y), R Thames, voltinism=multi, Engl[2]
Asellus aquaticus	4.7 on yearly basis	Prod = 9.4 kg dw/(ha*y), Egwys Nunydd R., voltinism=bi, Wales, UK[2]
Astraea undosa	0.1 on yearly basis	Southern Calif macroinvertebrates[1]
Astrometis sertulifera	0.3 on yearly basis	Southern Calif macroinvertebrates[1]
Astropecten armatus	0.3 on yearly basis	Southern Calif macroinvertebrates[1]
Athripsoides ancylus	5.8 on yearly basis	Prod = 0.099 kg dw/(ha*y), Brasheeas Creek, voltinism=uni, Kentucky[2]
Baetis rhodani	8.4 on yearly basis	Prod = 8.9 kg dw/(ha*y), trout streams, voltinism=bi, Czech.[2]
Baetis vagans	9.7 on yearly basis	Prod = 21 kg dw/(ha*y), Valley Creek, voltinism=tri, Minn[2]
Balanophyllia sp.	0.1 on yearly basis	Southern Calif macroinvertebrates[1]
Balanus sp.	1 on yearly basis	Southern Calif macroinvertebrates[1]
Barnacles	1 on yearly basis	Southern Calif macroinvertebrates, miscellaneous[1]
Bellamya unicolor	5.5 on yearly and shell-free basis	Prod = 156.3 kg dw/(ha*y), Lake Chad, voltinism=uni, max value[2]
Bithynia tentaculata	1.6 on yearly and shell-free basis	Prod = 1.1 kg dw/(ha*y), R Thames, voltinism=2-3 y, Engl[2]
Bithynia tentaculata	1.63 on yearly basis	Prod = 10.60 g wet weight/yr, Allan curve method[4]
Brachiopods	1 on yearly basis	Southern Calif macroinvertebrates, miscellaneous[1]
Bryozoans	2 on yearly basis	Southern Calif macroinvertebrates, miscellaneous[1]
Bursa californicus	0.1 on yearly basis	Southern Calif macroinvertebrates[1]
Chama pellucida	0.3 on yearly basis	Southern Calif macroinvertebrates[1]
Chaoborus flavicans	1.7 on yearly basis	Prod = 29.2 kg dw/(ha*y), Lake Esrom, voltinism=uni, Denm[2]
Chione undatella	0.2 on yearly basis	Southern Calif macroinvertebrates[1]
Chironomidae	4.97 on yearly basis	Prod = 8.65 g wet weight/yr, 7 spp., Allan curve method[4]
Chironomus anthracinus	3.8 on yearly basis	Prod = 128.9 kg dw/(ha*y), Lake Esrom, voltinism=2 y, Denm[2]
Chironomus commutatus	1.49 on yearly basis	Max biomass used, yearly value[5]
Chironomus plumosus	6.4 on yearly basis	Prod = 35.1 kg dw/(ha*y), Egwys Nunydd R., UK, voltinism=bi[2]
Chitons	0.3 on yearly basis	Southern Calif macroinvertebrates[1]
Choroterpes mexicanus	15.4 on yearly basis	Prod = 13.7 kg dw/(ha*y), Brazos River, voltinism=tri, Texas[2]
Chthamalus sp.	2 on yearly basis	Southern Calif macroinvertebrates[1]
Cladotanytarsus mancus	5.2 on yearly basis	Prod =[2]
Cleopatra bulimoides	3.5 on yearly and shell-free basis	Prod = 119.0 kg dw/(ha*y), Lake Chad, voltinism=1+ y, max value[2]
Coelenterata	0.1 on yearly basis	Southern Calif macroinvertebrates, miscellaneous[1]
Conus californicus	0.2 on yearly basis	Southern Calif macroinvertebrates[1]
Corbicula africana	2.9 on yearly and shell-free basis	Prod = 106.3 kg dw/(ha*y), Lake Chad, voltinism=1-2 y, max value[2]
Crangonyx pseudogracilis	5.8 on yearly basis	Prod = 0.22 kg dw/(ha*y), R Thames, voltinism=multi, Engl[2]
Crangonyx pseudogracilis	5.83 on yearly basis	Prod = 2.10 g wet weight/yr, Allan curve method[4]
Crangonyx richmondensis	2 on yearly basis	Prod = 3.5 kg dw/(ha*y), Marion Lake, voltinism=1.5 y, B.C.[2]

Crepidula sp.	0.2 on yearly basis	Southern Calif macroinvertebrates[1]
Cypraea spadicea	1 on yearly basis	Southern Calif macroinvertebrates[1]
Cyrnus trimaculatus	4.6 on yearly basis	Prod = 0.38 kg dw/(ha*y), R Thames, voltinism=uni, Engl[2]
Cyrnus trimaculatus	4.63 on yearly basis	Prod = 3.70 g wet weight/yr, Allan curve method[4]
Deleatidium sp.	3.5 on yearly basis	Prod = 42.3 kg dw/(ha*y), Selwyn River, voltinism=bi, NZ[2]
Dermasterias imbricata	0.3 on yearly basis	Southern Calif macroinvertebrates[1]
Diaperoccia californica	0.3 on yearly basis	Southern Calif macroinvertebrates[1]
Ecdyonurus sp.	8.4 on yearly basis	Prod = 14.1 kg dw/(ha*y), trout streams, voltinism=uni, Czech.[2]
Echinodermata	0.3 on yearly basis	Southern Calif macroinvertebrates, miscellaneous[1]
Ephemera strigata	2.3 on yearly basis	Prod = 7.1 kg dw/(ha*y), Yoshino River, voltinism=uni, Japan[2]
Ephemerella subvaria	5.8 on yearly basis	Prod = 44.5 kg dw/(ha*y), Luxemburg Creek, voltinism=uni, Minn[2]
Epitonium sp.	0.2 on yearly basis	Southern Calif macroinvertebrates[1]
Erpobdella octoculata	1.5 on yearly basis	Prod = 36.0 kg dw/(ha*y), Wilfin Beck, voltinism=2 y, n=3 y, Engl[2]
Erpobdella octoculata	3.5 on yearly basis	Prod = 3.4 kg dw/(ha*y), R Thames, voltinism=2-3 y, Engl[2]
Erpobdella octoculata	3.52 on yearly basis	Prod = 32.80 g wet weight/yr, Allan curve method[4]
Eupomatus sp.	1 on yearly basis	Southern Calif macroinvertebrates[1]
Fissurella volcano	0.3 on yearly basis	Southern Calif macroinvertebrates[1]
Gammarus mucronatus	4.3	mean cohort turnover rate, different methods[6]
Gammarus pulex	1.22 on yearly basis	Old cohort, production=1.2 g dw/(m2*y)[7]
Gammarus pulex	3.35 on yearly basis	Young cohort, production=2.6 g dw/(m2*y)[7]
Gammarus pulex	6.66 on yearly basis	Prod = 1.40 g wet weight/yr, Allan curve method[4]
Gammarus pulex	6.7 on yearly basis	Prod = 0.14 kg dw/(ha*y), R Thames, voltinism=multi, Engl[2]
Gammarus tigrinus	6.2 on yearly basis	Prod = 130 kg dw/(ha*y), Tjeukemeer River, voltinism=uni, Netherlands[2]
Glossiphonia complanata	2.07 on yearly basis	Prod = 5.57 g wet weight/yr, Allan curve method[4]
Glossiphonia complanata	2.1 on yearly basis	Prod = 0.58 kg dw/(ha*y), R Thames, voltinism=2-3 y, Engl[2]
Glossiphonia heteroclita	5.25 on yearly basis	Prod = 1.68 g wet weight/yr, Allan curve method[4]
Glossiphonia heteroclita	5.3 on yearly basis	Prod = 0.17 kg dw/(ha*y), R Thames, voltinism=uni, Engl[2]
Glyptotendipes barbipes	8.5 on yearly basis	Prod = 1616 kg dw/(ha*y), waste lagoon, near shore, voltinism=multi,[2]
Glyptotendipes barbipes	8.5 on yearly basis	Prod = 74 kg dw/(ha*y), waste lagoon, entire lag., voltinism=multi,[2]
Glyptotendipes paripes	5.9 on yearly basis	Prod = 30.8 kg dw/(ha*y), Egwys Nunydd R., UK, voltinism=bi[2]
Glyptotendipes sp.	5 on yearly basis	Prod = 405 kg dw/(ha*y), Loch Leven, max 2 y, voltinism=multi,[2]
Haliotis rufescens	0.05 on yearly basis	Southern Calif macroinvertebrates[1]
Helobdella stagnalis	3 on yearly basis	Prod = 11.9 kg dw/(ha*y), Egwys Nunydd R., voltinism=bi, max 2 y, UK[2]
Helobdella stagnalis	3.5 on yearly basis	Prod = 8.7 kg dw/(ha*y), Egwys Nunydd R., voltinism=bi, UK[2]
Helobdella stagnalis	4.8 on yearly basis	Prod = 3.5 kg dw/(ha*y), R Thames, voltinism=2-3 y, Engl[2]
Helobdella stagnalis	4.84 on yearly basis	Prod = 21.7 g wet weight/yr, Allan curve method[4]
Henricia leviuscula	0.3 on yearly basis	Southern Calif macroinvertebrates[1]

Hexagenia limbata	2.8 on yearly basis	Prod = 15.0 kg dw(ha*y), Clark Lake, volt=1.7 y, Neb, incl Hexagenia[2]
Hexagenia limbata	4.4 on yearly basis	Prod = 1.2 kg dw/(ha*y), Tuttle Creek River, n=4 y, voltinism=uni,[2]
Hyalella azteca	4.2 on yearly basis	Prod = 11.5 kg dw/(ha*y), Marion Lake, voltinism=uni, B.C.[2]
Hyalella azteca	4.8 on yearly basis	Prod = 19.3 kg dw/(ha*y), Sugarloaf Lake, voltinism=bi, Mich.[2]
Ilyodrilus hammoniensis	0.7 on yearly basis	Prod = 11.7 kg dw/(ha*y), Lake Esrom, Denmark[2]
Kellettia kellettii	0.1 on yearly basis	Southern Calif macroinvertebrates[1]
Limnaea pereger	11.52 on yearly basis	Prod = 48.40 g wet weight/yr, Allan curve method[4]
Limnochironomus pulsus	6.3 on yearly basis	Prod = 11.0 kg dw(ha*y), Egwys Nunydd R., UK, voltinism=bi[2]
Limnochironomus sp.	3.6 on yearly basis	Prod = 5.9 kg dw/(h*y), Loch Leven, max 2 y, voltinism=multi,[2]
Limnodrilus spp.	12.5 on yearly basis	Prod = 53.6 kg dw/(ha*y), Egwys Nunydd R., voltinism=multi, Wales, UK[2]
Linckia columbiae	0.3 on yearly basis	Southern Calif macroinvertebrates[1]
Lithophaga plumula	0.2 on yearly basis	Southern Calif macroinvertebrates[1]
Littorina saxatilis	4.1 on yearly and shell-free basis	Prod = 32.5 kg dw/(ha*y), Nova Scotia, voltinism=2+ y, intertidal[2]
Lottia gigantea	0.2 on yearly basis	Southern Calif macroinvertebrates[1]
Lovenia cordiformis	0.3 on yearly basis	Southern Calif macroinvertebrates[1]
Lymnaea peregra	11.5 on yearly and shell-free basis	Prod = 5.02 kg dw/(ha*y), R Thames, voltinism=multi, Engl[2]
Lytechinus anamesus	1 on yearly basis	Southern Calif macroinvertebrates[1]
Macoma baltica	1.5 on yearly and shell-free basis	Prod = 19.3 kg dw/(h*y), Nova Scotia, voltinism=2+ y, intertidal[2]
Megathura crenulata	0.3 on yearly basis	Southern Calif macroinvertebrates[1]
Melania tuberculata	4.4 on yearly and shell-free basis	Prod = 154.5 kg dw/(ha*y), Lake Chad, voltinism=1-2 y, max value[2]
Membranipora sp.	2 on yearly basis	Southern Calif macroinvertebrates[1]
Mesoplanktonic group	0.055 per day	Per day, southern lakes of USSR, rg = 0.01 - 0.1[3]
Microtendipes chloris	12.7 on yearly basis	Prod = 0.24 kg dw/(ha*y), Mikolajskie Lake, voltinism=multi, Poland[2]
Microtendipes sp.	5.7 on yearly basis	Prod = 16.2 kg dw/(ha*y), Egwys Nunydd R., UK, voltinism=bi[2]
Mitella polymerus	1 on yearly basis	Southern Calif macroinvertebrates[1]
Mollusca	0.0125 per day	Per day, southern lakes of USSR, rg = 0.005 - 0.02[3]
Mollusca	0.3 on yearly basis	Southern Calif macroinvertebrates, miscellaneous[1]
Muricea sp.	0.1 on yearly basis	Southern Calif macroinvertebrates[1]
Mya arenaria	2.5 on yearly and shell-free basis	Prod = 116 kg dw/(ha*y), Nova Scotia, voltinism=2+ y, intertidal[2]
Mytilus sp.	0.3 on yearly basis	Southern Calif macroinvertebrates[1]
Neanthes virens	1.6 on yearly basis	Prod = 90.4 kg dw/(ha*y), R Thames, voltinism=2 y, estuary, Engl[2]
Nuttallina californica	0.3 on yearly basis	Southern Calif macroinvertebrates[1]
Oligochaeta	7.4 on yearly basis	Prod = 12-97 kg dw/(ha*y), peat land, Calif, rg = 5.5 - 9.3[2]
Ophiuroidea	0.2 on yearly basis	Southern Calif macroinvertebrates[1]
Orconectes virilis	0.94 on yearly basis	Prod = 22.3 kg dw/(ha*y), West Lost Lake, voltinism=3.5 y, Mich,[2]

Orconectes virilis	1.1 on yearly basis	Prod = 8.2 kg dw/(ha*y), North Twin Lake, voltinism=3.5 y, n=9 y,[2]
Orconectes virilis	2.3 on yearly basis	Prod = 28.3 kg dw/(ha*y), West Lost Lake, voltinism=3.5 y, Mich.,[2]
Ostrea lurida	0.1 on yearly basis	Southern Calif macroinvertebrates[1]
Parachironomus tener	5.8 on yearly basis	Prod =[2]
Patiria miniata	0.3 on yearly basis	Southern Calif macroinvertebrates[1]
Penitella conradi	0.3 on yearly basis	Southern Calif macroinvertebrates[1]
Phidolopora sp.	0.5 on yearly basis	Southern Calif macroinvertebrates[1]
Pisater sp.	0.3 on yearly basis	Southern Calif macroinvertebrates[1]
Pisidium casertanum	0.2 on yearly and shell-free basis	Prod = 4.38 kg dw/(ha*y), Lake Esrom, voltinism=2 y, Denmark[2]
Planorbis contortus	0.3 on yearly and shell-free basis	Prod = 0.017 kg dw/(ha*y), R Thames, voltinism=uni, Engl[2]
Planorbis contortus	0.32 on yearly basis	Prod = 0.16 g wet weight/yr, Allan curve method[4]
Planorbis vortex	6.47 on yearly basis	Prod = 1.10 g wet weight/yr, Allan curve method[4]
Planorbis vortex	6.5 on yearly and shell-free basis	Prod = 0.11 kg dw/(ha*y), R Thames, voltinism=uni, Engl[2]
Pododesmus pernoides	0.7 on yearly basis	Southern Calif macroinvertebrates[1]
Pontoporeia affinis	4.1 on yearly basis	Prod = 5.2 kg dw/(ha*y), Lake Krasnoye, voltinism=uni, USSR[2]
Potamophylax cingulatus	4.4 on yearly basis	Prod = 7.7 kg dw/(ha*y), Stampes Lake, voltinism=uni, tissue only,[2]
Potamothrix hammoniensis	0.86 on yearly basis	Prod = 53.6 kg dw/(ha*y), Lake Esrom, voltinism=4 y, Denm[2]
Procladius choreus	6.5 on yearly basis	Prod = 34.1 kg dw/(ha*y), Eglwys Nunydd River, UK voltinism=multi[2]
Procladius pectinatus	1.9 on yearly basis	Prod = 5.2 kg dw/(ha*y), Lake Esrom, voltinism=1 y, Denm[2]
Procladius rufovittatus	3.2 on yearly basis	Prod = 20.1 kg dw/(ha*y), Eglwys Nunydd River, UK voltinism=uni[2]
Protothara staminea	0.3 on yearly basis	Southern Calif macroinvertebrates[1]
Pyrrhosoma nymphula	3.2 on yearly basis	Prod = 7.1 kg dw/(ha*y), ponds in Engl, voltinism=2 y[2]
Rhithrogena semicolorata	8.4 on yearly basis	Prod = 20.8 kg dw/(ha*y), truot streams, voltinism=uni, Czech[2]
Scrupocellaria sp.	2 on yearly basis	Southern Calif macroinvertebrates[1]
Semele sp.	0.3 on yearly basis	Southern Calif macroinvertebrates[1]
Serpula sp.	1 on yearly basis	Southern Calif macroinvertebrates[1]
Sialis mitsuhashii	3.9 on yearly basis	Prod = 2.5 kg dw/(ha*y), Lake Tatsu-Nama, voltinism=3 y, Japan[2]
Simulium brevicaule	6.2 on yearly basis	Prod = 0.60 kg dw/(h*y), Lissuraga River, voltinism=multi, n=2 places[2]
Simulium monticola	11.1 on yearly basis	Prod = 1.12 kg dw/(ha*y), Lissuraga River, voltinism=multi, n=4[2]
Simulium ornatum	12.9 on yearly basis	Prod = 1.29 kg dw/(h*y), Lissuraga River, n=2 places, voltinism=multi[2]
Sphaerium corneum	3.5 on yearly and shell-free basis	Prod = 2.85 kg dw/(ha*y), River Thames, voltinism=2-3 y, Engl[2]
Sphaerium corneum	3.53 on yearly basis	Prod = 27.50 g wet weight/yr, Allan curve method[4]
Spirorbis sp.	4 on yearly basis	Southern Calif macroinvertebrates[1]
Stenoperla prasina	2.5 on yearly basis	Prod = 20.8 kg dw/(ha*y), Selwyn River, voltinism=uni, NZ[2]
Stictochironomus sp.	9.8 on yearly basis	Prod = 102 kg dw/(ha*y), Loch Leven, max 2 y, voltinism=multi,[2]

Strongylocentrotus franciscanus	1 on yearly basis	Southern Calif macroinvertebrates[1]
Strongylocentrotus purpuratus	1 on yearly basis	Southern Calif macroinvertebrates[1]
Surculites sp.	0.2 on yearly basis	Southern Calif macroinvertebrates[1]
Tanytarsus holochlorus	6.6 on yearly basis	Prod = 15.5 kg dw/(ha*y), Eglwys Nunydd River, UK voltinism=bi[2]
Tanytarsus inopertus	5.2 on yearly basis	Prod =[2]
Tanytarsus lugens	7.6 on yearly basis	Prod = 24.0 kg dw/(ha*y), Eglwys Nunydd River, UK voltinism=tri[2]
Tegula funebralis	0.1 on yearly basis	Southern Calif macroinvertebrates[1]
Terebratalia sp.	1 on yearly basis	Southern Calif macroinvertebrates[1]
Terebratulina sp.	1 on yearly basis	Southern Calif macroinvertebrates[1]
Tetraclita sp.	1 on yearly basis	Southern Calif macroinvertebrates[1]
Tubificidae	1 on yearly basis	Prod = 71.0 g wet weight/yr[4]
Unio pictorum	0.14 on yearly and shell-free basis	Prod = 8.8 kg dw/(ha*y), River Thames, voltinism=10-12 y, Engl[2]
Unio pictorum	0.14 on yearly basis	Prod = 84.6 g wet weight/yr[4]
Unio pictorum	0.15 on yearly and shell-free basis	Prod = 3.7 kg dw/(ha*y), Danube River pools, voltinism=2+ y, Romania[2]
Unio tumidus	0.13 on yearly and shell-free basis	Prod = 3.3 kg dw/(ha*y), River Thames, voltinism=10-12 y, Engl[2]
Unio tumidus	0.25 on yearly and shell-free basis	Prod = 3.4 kg dw/(ha*y), Danube River pools, voltinism=2+ y, Romania[2]
Viviparus viviparus	1.59 on yearly basis	Prod = 42.94 g wet weight/yr, Allan curve method[4]
Viviparus viviparus	1.6 on yearly and shell-free basis	Prod = 4.46 kg dw/(ha*y), River Thames, voltinism=2-3 y, Engl[2]
annelida	1 on yearly basis	Southern Calif macroinvertebrates, miscellaneous[1]

Source: 1.Smith, K.L. Jr. 1974. Limnol. Oceanogr. 19(6), pp 939-944.
2.Waters, T.F. 1977. Adv. Ecol. Res. 10, pp 91-164.
3.Greze, V.N. 1973. Okeanologiya, 13(1), pp 107-113.
4.Mann, K.H. 1971. in: Edmondson, W.T. (ed.); A Manual on Methods for the Assessment of Secondary Productivity of Fresh Waters. pp 204. IBP Handbook No 17.
5.Laville, H. 1975. Ann. Limnol., 11(1), pp 67-77.
6.LaFrance, K. and Ruber, E., Limnol. Oceanogr., 30(5), 1109, 1985.
7.Iversen, T.M. and J.Jessen. 1977. Fresh. Biol. 7, pp 287-298.

1-302 Invertebrates
Protein / biomass ratio

Species	Value	Condition
Mussel	73.7 % of dry weight	None[1]
Oyster	51.2 % of dry weight	None[2]
Sea urchin	60.6 % of dry weight	40 h larva[2]
Sea urchin	66.9 % of dry weight	Unfertilized egg[2]
Silkworm	55.5 % of dry weight	None[2]
Silkworm	63.4 % of dry weight	Imago[2]

Source: 1.Special Rept. Ser. Med. Res. Com. London No. 297. 1960. The Composition of Foods.
2.Florkin, M. 1949. Biochemical Evolution. Academic Press. New York.

1-303 Invertebrates
Respiration

Species	Value	Condition
Tintinnopsis acuminata	114 pg C tintinnid-1 h-1	at 25 oC, ciliate[1]
Tintinnopsis acuminata	43 pg C tintinnid-1 h-1	at 15 oC, ciliate[1]
Tintinnopsis acuminata	78 pg C tintinnid-1 h-1	at 20 oC, ciliate[1]
Tintinnopsis vasculum	114 pg C tintinnid-1 h-1	at 5 oC, ciliate[1]
Tintinnopsis vasculum	314 pg C tintinnid-1 h-1	at 10 oC, ciliate[1]
Tintinnopsis vasculum	462 pg C tintinnid-1 h-1	at 15 oC, ciliate[1]

Source: 1. Verity, P.G., Limnol. Oceanogr., 30, 1268, 1985.

1-304 Invertebrates
Respiration rate

Species	Value	Condition
Araneae	3 microlitre O2/(hour*animal)	Grassland soil, 292 K, equivalent to 27 gcal/(g*day) at 289 K[1]
Asellus aquaticus	24.7 % of ingestion	None[2]
Asellus aquaticus	81.8 % of assimilation	None[2]
Benthic foraminifera	2.5 microlitre O2/(mg wet weight*hour)	Optimum lab. conditions, 278 - 308 K, rg = 0.5 - 4.5[3]
Benthos	0.00055 1/day	Temp factor is 1 at 293 K, temp.<278 K, rg = 0.0001 - 0.001[4]
Benthos	0.00055 1/day	Used in models, inactive, rg = 0.0001 - 0.001[5]
Benthos	0.001 1/day	Used in models, Lake Washington[6]
Benthos	0.001 1/day	Used in models, N Fork Kings river, Calif, inactive[7]
Benthos	0.001 1/day	Used in models, San Francisco Bay estuary[8]
Benthos	0.002 1/day	Used in models, N Fork Kings river, Calif, active[7]
Benthos	0.0055 1/day	Temp factor is 1 at 293 K, temp>278 K, rg = 0.001 - 0.01[4]
Benthos	0.0055 1/day	Used in models, active, rg = 0.001 - 0.01[5]
Benthos	0.01 1/day	Used in models, Boise river, Idaho[9]
Ceuthophilus gracilipes	263 microlitre O2/(g wet weight*hour)	Mean wet weight = 259 mg, 293 K, soil invert.[10]
Cognettia cognettii	625 microlitre O2/(g wet weight*hour)	Mean wet weight = 0.19 mg, 293 K, soil invert.[11]
Coleoptera	60 microlitre O2/(hour*animal)	Grassland soil, 286 K, equivalent to 39 gcal/(g*day) at 289 K[1]
Collembola	0.55 microlitre O2/(hour*animal)	Grassland soil, 289 K, equivalent to 144 gcal/(g*day) at 289 K[1]
Cylisticus convexus	205 microlitre O2/(g wet weight*hour)	Mean wet weight = 35 mg, 293 K, soil invert.[12]
Diplacodes trivialis	51.9 % of assimilation	46 mg animal, ingestion = 61.5 cal/(kcal*anim.*d)[13]
Diptera	105 microlitre O2/(hour*animal)	Grassland soil, 286 K, equivalent to 29 gcal/(g*day) at 289 K[1]

Dixidesmus erasus	140 microlitre O2/(g wet weight*hour)	Mean wet weight = 78 mg, 293 K, soil invert.[10]
Enchytraeidae	0.117 microlitre O2/(hour*individual)	Grassland soil, 289 K, equivalent to 100 gcal/g*day at 289 K[1]
Fly	0.03 kcal/(kcal*day)	Initial values, thermal ecosystem, transfer coeff., pupae[14]
Fly	0.06 kcal/(kcal*day)	Initial values, thermal ecosystem, transfer coeff., egg[14]
Fly	0.33 kcal/(kcal*day)	Initial values, thermal ecosystem, transfer coeff., 3 instar larvae[14]
Fly	0.36 kcal/(kcal*day)	Initial values, thermal ecosystem, transfer coeff., 1 and 2 instar[14]
Fly	0.47 kcal/(kcal*day)	Initial values, thermal ecosystem, transfer coeff., adult[14]
Isopoda	9 microlitre O2/(hour*animal)	Grassland soil, 292 K, equivalent to 36 gcal/(g*day) at 289 K[1]
Isoptera	199 microlitre O2/(g wet weight*hour)	Mean wet weight = 2.1 mg, 293 K, soil invert.[10]
Lestes sponsa	13.2 % of ingestion	None[2]
Lestes sponsa	36 % of assimilation	None[2]
Limnodrilus hoffmeisteri	0.18 microlitre O2/(mg*hr)	278 K, respiration rate, mean of 3 exp.[15]
Limnodrilus hoffmeisteri	0.21 microlitre O2/(mg*hr)	283 K, respiration rate, mean of 3 exp.[15]
Limnodrilus hoffmeisteri	0.31 microlitre O2/(mg*hr)	288 K, respiration rate, mean of 3 exp.[15]
Limnodrilus hoffmeisteri	0.48 microlitre O2/(mg*hr)	293 K, respiration rate, mean of 3 exp.[15]
Lumbricidae	300 microlitre O2/(hour*animal)	Grassland soil, 289 K, equivalent to 7 gcal/(g*day) at 289 K[1]
Lumbricus terrestris	36 microlitre O2/(g wet weight*hour)	Mean wet weight = 2700 mg, 288 K, soil invert.[16]
Mesodon thyroides	91 microlitre O2/(g wet weight*hour)	Mean wet weight = 2108 mg(inc shell), 293 K, soil invert.[17]
Mitopus morio	785 microlitre O2/(g wet weight*hour)	Mean wet weight = 3.7 - 56.1 mg, 289 K, soil invert., rg = 610 - 960[18]
Myriapoda	10 microlitre O2/(hour*animal)	Grassland soil, 292 K, equivalent to 36 gcal/(g*day) at 289 K[1]
Nematoda	0.0012 microlitre O2/(hour*individual)	Grassland soil, 289 K, equivalent to 144 gcal/g*day at 289 K[1]
Nematoda	468 microlitre O2/(g wet weight*hour)	Mean wet weight = 0.000135-0.000196 mg, 289 K, soil invert.[19]
Onychiurus procampatus	660 microlitre O2/(g wet weight*hour)	Mean wet weight = 0.01 - 0.10 mg, 288 K, soil invert., rg = 320 -[20]
Opiliones	10.5 microlitre O2/(hour*animal)	Grassland soil, 292 K, equivalent to 53 gcal/(g*day) at 289 K[1]
Orbited mites	0.12 microlitre O2/(hour*animal)	Grassland soil, 287 K, equivalent to 72 gcal/(g*day) at 289 K[1]
Otocryptops sexspinosus	73 microlitre O2/(g wet weight*hour)	Mean wet weight = 432 mg, 293 K, soil invert.[17]
Paramecium sp.	0.5 ml o2/(g wet wt.)*hr	293 K, ciliate[21]
Parasitic mites	0.5 microlitre O2/(hour*animal)	Grassland soil, 289 K, equivalent to 280 gcal/(g*day) at 289 K[1]

Parcoblatta sp.	158 microlitre O2/(g wet weight*hour)	Mean wet weight = 173 mg, 293 K, soil invert.[10]
Parhyalla whelpleyi	2.95 mg O2/(g dry weight*hour)	297 K, 92 h[22]
Peloscolex multisetosus	0.35 microlitre O2/(mg*hr)	283 K, respiration rate, mean of 3 exp.[15]
Peloscolex multisetosus	0.39 microlitre O2/(mg*hr)	278 K, respiration rate, mean of 3 exp.[15]
Peloscolex multisetosus	0.42 microlitre O2/(mg*hr)	288 K, respiration rate, mean of 3 exp.[15]
Peloscolex multisetosus	0.56 microlitre O2/(mg*hr)	293 K, respiration rate, mean of 3 exp.[15]
Philaenus spumarius	0.34 cal/(animal*day)	3rd instar, 0.39 mg, 296 K[23]
Philaenus spumarius	0.76 cal/(animal*day)	4rd instar, 1.16 mg, 298 K[23]
Philaenus spumarius	1 cal/(animal*day)	Adult, 3.87 mg, 294 K, excreting urea[23]
Philaenus spumarius	1 cal/(animal*day)	Adult, 3.87 mg, 294 K, excreting uric acid[23]
Platynorthrus peltifer	925 microlitre O2/(g wet weight*hour)	Mean wet weight = 0.063 mg, 288 K, soil invert.[24]
Popilius disjunctus	124 microlitre O2/(g wet weight*hour)	Mean wet weight = 1630 mg, 293 K, soil invert.[10]
Protozoa	0.01 microlitre O2/(hour*individual)	Grassland soil, 295.5 K, equivalent to 14 gcal/g*day at 289 K[1]
Ptyoiulus impressus	56 microlitre O2/(g wet weight*hour)	Mean wet weight = 432 mg, 293 K, soil invert.[10]
Rhincalanus gigas	0.47 microlitre O2/(mg dry weight*hour)	271.7 K, respiration artic pond, SD =+- 0.05[25]
Rhincalanus gigas	3.07 microlitre O2/(aminal*hour)	271.7 K, respiration artic pond, SD =+0.40[25]
Schizocosa sp.	251 microlitre O2/(g wet weight*hour)	Mean wet weight = 23 mg, 293 K, soil invert.[26]
Snails	1000 microlitre O2/(hour*individual)	Grassland soil, 301 K, equivalent to 29 gcal/g*day at 289 K[1]
Sphaeroderus stenostomus	194 microlitre O2/(g wet weight*hour)	Mean wet weight = 169 mg, 293 K, soil invert.[10]
Tipula sp.	146 microlitre O2/(g wet weight*hour)	Mean wet weight = 277 - 607 mg, 286 K, soil invert.[27]
Tubifex tubifex	0.21 microlitre O2/(mg*hr)	283 K, respiration rate, mean of 3 exp.[15]
Tubifex tubifex	0.24 microlitre O2/(mg*hr)	278 K, respiration rate, mean of 3 exp.[15]
Tubifex tubifex	0.4 microlitre O2/(mg*hr)	288 K, respiration rate, mean of 3 exp.[15]
Tubifex tubifex	0.51 microlitre O2/(mg*hr)	293 K, respiration rate, mean of 3 exp.[15]

Source: 1.MacFadyen, A. 1963. Animal Ecology, Aims and Methods. 2nd edition. Pitman. London.
2.Wetzel, R.G. 1975. Limnology. Saunders. Philadelphia.
3.Lee, J.J. and W.A. Muller. 1973. Am. Zool. 13(1), pp 215-223.
4.The Hydrological Engineering Center Corps of Engineers U.S. Army. 1975. Water Quality for river-reservoir systems, Davis, Cal. 9 5 6 1 6 (916) 499-2329.
5.U.S. Army Corps of Engineers. 1974. Water Quality for River-Reservoirs Systems.
6.Chen, C.W. and G.T. Orlob. 1975. in: Patten, B.C. (ed.); Systems Analysis and Simulation in Ecology,vol 3. pp 476-537. Acad. Press. N.Y.
7.Tetra Tech, Inc. 1976. Estuary Water Quality Models, Lond Island, New York User's Guide. For Nassau-Suffolk Regional Planning Board. Hauppauge, New York. 11787.
8.Chen, C.W. 1975. in: Russell, C.S.; Ecological Modeling, in a Resource Management Framework. NOAA.

9.Chen, C.W. and Wells. 1975. Boise River Water Quality-Ecological Model for Urban Planning Study. US Army Eng. Walla Walla, Idaho Water Res. Board.
10.Reichle, D.E. 1968. Ecology 49, pp 538-542.
11.O'Connor, D.J. and D.M. Di Toro. 1970. J. San. Eng. Div, ASCE. 96, SA2.
12.Reichle, D.E. 1967. Ecology 48, pp 351-366.
13.Pandian, T.J. and S. Mathavan. 1974. Odonatologica, 3(4), pp 241-248.
14.Wiegert, R.G. 1975. In: B.C. Patten (ed.), Systems Analysis and Simulation in Ecology, Vol. 3, pp 15-181.
15.Brinkhurst, R.O., K.E. Chua and N.K. Kaushik. 1972. Limnol. Oceanogr., 17(1), pp 122-133.
16.Satchell, J.E. 1967. in: Burges, A. and F. Raw. (eds.); Soil Biology. pp 259-322. Academic Press. Lodon.
17.Edwards, C.A., D.E. Reichle and D.A. Crossley, Jr. 1970. in: Reichle, D.E. (ed); Analysis of Temperate Forest Ecosystems. pp 147-172. Springer-Verlag. Berlin.
18.Philipson, J. 1964. Oikos 15, pp 130-139.
19.Nielsen, C.O. 1961. Oikos 12, pp 17-35.
20.Healey, I.N. 1967. in: Petrusewicz, K. (ed.); Secondary Productivity of Temperate Terrestrial Ecosysems. pp 695-708. Panstwowe Wydawnictwo Naukowe. Warsaw.
21.Giese, A.C. 1962. Cell Physiology. 2nd edition. Saunders. Philadelphia.
22.Fenchel, T. 1970. Limnol. Oceanogr. 15(1), pp 14-20.
23.Wiegert, R.G. 1964. Ecol. Monogr. 34, pp 225-241.
24.Berthet, P. 1963. In: Doeksen, J. and J. van der Drift. (eds); Soil Organism. North-Holland Publ. Co. Amsterdam.
25.Rakusa-Suszcewski, S., M.A. McWhite and M.O. Cahoon. 1976. Limnol. Oceanogr. 21(5), pp 763-765.
26.Moulder, B.C., D.E. Reichle and S.I. Auerbach. 1970. Oak Ridge National Lab. USA, ORNL 4452.
27.Bornebusch, C.H. 1930. Forstl. Forsoegsvaesen Copenhagen. 11, pp 1-224.

1-305 Invertebrates
Respiratory coefficient

Species	Value	Condition
Asellus aquaticus	4.49	P/R, 0 - 116 d[1]
Ctenopharyngodon idella	1.47	P/R, on aminal food, 6 m, 40-120 g[1]
Ctenopharyngodon idella	5.9	P/R, on plant food, 6 m, 20-70 g[1]
Lestes sponsa	0.56	P/R, 0 - 37 d[1]
Rhizoglyphus echinopus	0.43	P/R, average adult + eggs, 0-72 d[1]
Rhizoglyphus echinopus	0.6	P/R, trichonympha active, 0-20 d[1]
Rhizoglyphus echinopus	0.67	P/R, average new hatched adult, 0-22 d[1]
Rhizoglyphus echinopus	0.69	P/R, protonympha active, 0-16 d[1]
Rhizoglyphus echinopus	0.8	P/R, triconympha resting, 0-21 d[1]
Rhizoglyphus echinopus	1.02	P/R, protonympha resting, 0-17 d[1]
Rhizoglyphus echinopus	1.84	P/R, larvae active, 0-11 d[1]
Rhizoglyphus echinopus	4.35	P/R, larvae resting, 0-13 d[1]
Simocephalus vetulus	0.37	P/R, female, 0-7 d, immature[1]
Simocephalus vetulus	0.54	P/R, female, 0-20 d, reproducing[1]
Tribolium castaneum	1.26	P/R, average adult + eggs, 0-116 d[1]
Tribolium castaneum	2.4	P/R, prepupa, 0-24 d[1]
Tribolium castaneum	3.8	P/R, new hatched adult, average, 0-36 d[1]

Source: 1.McNeill, S. and J.H. Lawton. 1970. Nature 225, pp 472-474

1-306 Invertebrates
S / biomass ratio

Species	Value	Condition
Annelida	1.4 % of dry weight	Mean elementary composition[1]
Arthropoda	0.44 % of dry weight	Mean elementary composition, land[1]
Coelenterata	1.9 % of dry weight	Mean elementary composition[1]
Echinodermata	0.5 % of dry weight	Mean elementary composition[1]

Mollusca	1.5 % of dry weight	Mean elementary composition, soft parts[1]
Nematoda	1.3 % of dry weight	Mean elementary composition, interpolated[1]

Source: 1.Bowen, H.J.M. 1966. Trace Elements in Biochemistry, Ac. Press.

1-307 Invertebrates Settling rate

Species	Value	Condition
Foraminiferal shells	0.3 cm/sec	62-125 U, rg = 0.24 - 0.35[1]
Foraminiferal shells	0.6 cm/sec	125-177 U, rg = 0.56 - 0.73[1]
Foraminiferal shells	1 cm/sec	177-250 U, rg = 0.8 - 1.3[1]
Foraminiferal shells	2.3 cm/sec	>250 U, rg = 2.1 - 2.4[1]

Source: 1.Berger, W.H. and D.J.W. Piper. 1972. Limnol. Oceanogr., 17(2), pp 275-287.

1-308 Invertebrates Threshold density

Species	Value	Condition
Fly	284 kcal/m**2	Initial values, thermal ecosystem, limit-egestion proportion, adult[1]
Fly	50 kcal/m**2	Initial values, thermal ecosystem, limit-egestion proportion, 1 and 2 instar[1]

Source: 1.Wiegert, R.G. 1975. In: B.C. Patten (ed.), Systems Analysis and Simulation in Ecology, Vol. 3, pp 15-181.

1-309 Invertebrates Turnover time of iodine

Species	Value	Condition
Amphipoda	34.4 hours	Living in freshwater sediment[1]

Source: 1.McNaughton, S.J. and L.L. Wolf. 1975. General Ecology. Holt, Rinehart and Winston, Int., New York.

1-310 Invertebrates Upper temperature limit for growth rate

Species	Value	Condition
Protozoa	56 degrees celcius	App. value[1]

Source: 1.Meeks, J.C. and R.W. Castenholz. 1971. Arch. Mikrobiol. 78, 25-41.

1-311 Invertebrates Wet / dry

Species	Value	Condition
Invertebrates and zooplankton	6 by weight	Invertebrates found as zooplankton[1]

Source: 1.Waters, T.F. 1977. Adv. Ecol. Res. 10, pp 91-164.

1-312 Invertebrates Wet weight

Species	Value	Condition
Paramecium	4.6E-7 g	Ciliate, typical size[1]
Tetrahymena pyriformis	1.9E-8 g	None[1]

Source: 1.Brock, T.D. 1966. Principles of Microbial Ecology. Prentice-Hall. New Jersey.

1-313 Invertebrates growth efficiency

Species	Value	Condition
Gammarus mucronatus	0.34 - 0.56	mean cohort turnover rate, different methods[1]

Source: 1.LaFrance, K. and Ruber, E., Limnol. Oceanogr., 30(5), 1109, 1985.

1-314 Kormophytes, macrophytes Ash / biomass ratio

Species	Value	Condition
Amblystegium riparium	29.24 % of dry weight	Musci, n = 4, San Marcos River, TX, SD =+- 0.15[1]
Angiosperms	17.2 % of dry weight	Mean of 15 species, San Marcos River, TX[1]
Cabomba caroliniana	22.8 % of dry weight	N = 4, Angiosperm, San Marcos River, TX, SD =+- 0.85[1]
Ceratophyllum demersum	8.78 % of dry weight	N = 4, Angiosperm, San Marcos River, TX, SD =+- 0.20[1]
Elodea densa	16.77 % of dry weight	N = 4, Angiosperm, San Marcos River, TX, SD =+- 0.97[1]
Emergent	9 % of dry weight	Range = 5 - 12[2]
Free floating	18.28 % of dry matter	Amazon, SE = +- 1.805[3]
Heteranthera dubia	19.71 % of dry weight	N = 4, Angiosperm, San Marcos River, TX, SD =+- 2.77[1]
Ludwigia palustra	25.32 % of dry weight	N = 4, Angiosperm, San Marcos River, TX, SD =+- 1.60[1]
Myriophyllum brasiliense	10.08 % of dry weight	N = 4, Angiosperm, San Marcos River, TX, SD =+- 1.17[1]
Myriophyllum heterophyllum	8.89 % of dry weight	N = 4, Angiosperm, San Marcos River, TX, SD =+- 0.40[1]
Najas guadalupensis	18.28 % of dry weight	N = 4, Angiosperm, San Marcos River, TX, SD =+- 1.37[1]
Plant with floating leaves	16 % of dry weight	Range = 10 - 25[2]
Potamogeton crispus	19.27 % of dry weight	N = 4, Angiosperm, San Marcos River, TX, SD =+- 0.76[1]
Potamogeton illinoeasis	15.85 % of dry weight	N = 4, Angiosperm, San Marcos River, TX, SD =+- 1.04[1]
Potamogeton pectinatus	15.35 % of dry weight	N = 4, Angiosperm, San Marcos River, TX, SD =+- 1.08[1]
Rhizoclonium hieroglyphicum	29.17 % of dry weight	N = 4, Chlorophyceae, San Marcos River, TX, SD =+- 0.47[1]
Riccia fluitans	27.06 % of dry weight	N = 4, Hepaticae, San Marcos River, TX, SD =+- 0.13[1]
Rooted	10.57 % of dry matter	Amazon, SE = +- 2.352[3]
Sagittaria platyphylla	22.6 % of dry weight	N = 4, Angiosperm, San Marcos River, TX, SD =+- 0.07[1]
Submersed	21 % of dry weight	Range = 9 - 25[2]
Utricularia subulata	19.92 % of dry weight	N = 4, Angiosperm, San Marcos River, TX, SD =+- 0.17[1]
Vallisneria americana	24.33 % of dry weight	N = 4, Angiosperm, San Marcos River, TX, SD =+- 2.38[1]
Zannichellia palustrus	10.6 % of dry weight	N = 4, Angiosperm, San Marcos River, TX, SD =+- 0.37[1]

Source: 1.Hannan, H.H. and T.C. Dorris. 1970. Limnol. Oceanogr., 15(3) pp 442-453.
2.Strabskraba, M. 1968. Mitt. Int. Ver. Limnol. 14, pp 212-230.
3.Howard-Williams, C. and W.J. Junk. 1977. Arch. Hydrobiol., 79(4), pp 446-464.

1-315 Kormophytes, macrophytes Biomass

Species	Value	Condition
Berula sp.	500 g dry weight/m**2	Submergents domination, max biomass, River lvel, Engl[1]
Carex sp.	850 g dry weight/m**2	Emergents dominating, max biomass, Minn wetlands[2]

Chara sp.	110 g dry weight/m**2	Submergents dominating, max biomass, Lawrence Lake, hardwater, Mich[3]
Eichhorina sp.	1051 g dry weight/m**2	Floating plant, max biomass, New Orleans, LA, rg = 630 - 1472[4]
Emergents dominating	16.5 g dry weight/m**2	Max biomass, Onega Lake, USSR, rg = 0.1 - 33[5]
Emergents dominating	5.55 g dry weight/m**2	Max biomass, Ladoga Lake, USSR, rg = 0.4 - 10.7[5]
Emergents dominating	635 g dry weight/m**2	Max biomass, polish lakes, rg = 440 - 830[6]
Emergents dominating	950 g dry weight/m**2	Max biomass, Surlingham Broad, Engl, rg = 800 - 1100[7]
Phragmites sp.	1650 g dry weight/m**2	Emergents dominating, aerial, max biomass, Opatovicky p., Czec[8]
Phragmites sp.	7280 g dry weight/m**2	Emergents, below ground, max biomass, Opatovicky p., Czec[8]
Potamogeton sp.	380 g dry weight/m**2	Submergents dominating, max biomass, River Yare, Engl[9]
Ranunculus sp.	250 g dry weight/m**2	Submergents dominating, max biomass, River Test, Engl, rg = 100 - 400[10]
Ruppia sp.	60 g dry weight/m**2	Submergents dominating, max biomass, Borax Lake, Calif[11]
Scirpus subterminalis	338 g dry weight/m**2	Submergents dominating, max biomass, Lawrence Lake, hardwater, Mich[3]
Submergents dominating	0.07 g dry weight/m**2	Max biomass, Trout Lake, softwater, Wisc[12]
Submergents dominating	1.73 g dry weight/m**2	Max biomass, Sweeney Lake, softwater, Wisc[13]
Submergents dominating	100 g dry weight/m**2	Max biomass, Spiggie Lake, dystrophic, Scotland[14]
Submergents dominating	130 g dry weight/m**2	Annuals, max biomass, Lawrence Lake, hardwater, Mich[3]
Submergents dominating	16.8 g dry weight/m**2	Max biomass, Weber Lake, softwater, Wisc[15]
Submergents dominating	202 g dry weight/m**2	Max biomass, Lake Mendota, hardwater, Wisc.[16]
Submergents dominating	32 g dry weight/m**2	Max biomass, Lowes Lake, dystrophic, Scotland[14]
Submergents dominating	400 g dry weight/m**2	Max biomass, Croispol, Scotland[14]
Thalassia sp.	4000 g dry weight/m**2	Submergents dom., max biomass, saline channals, Puerto Rico[17]
Typha sp.	4640 g dry weight/m**2	Emergents dominating, max biomass, Cedar Creek, Minn[18]

Source: 1.Wetzel, R.G. and M. Owens. 1960. J. Ecol. 48, pp 151-160.
2.Bernard, J.M. 1973. Pol. Arch. Hydro biol., 20, pp 207-214.
3.Rich, P.H., R.G. Wetzel and N.V. Thury. 1971. Freshwater Biol. 1, pp 3-21.
4.Penfound, W.T. and T.T. Earle. 1948. Ecol. Monogr. 18, pp 447-472.
5.Raspopov, I.M. 1971. Hidrobiologia 12, pp 241-247.
6.Szczepanska, W. 1973. Pol. Arch. Hydrobiol. 15, pp 51-77.
7.Buttery, B.R. and J.M. Lambert. 1973. J. Ecol., 53, pp 163-181.
8.Dykyjova, D. and D. Hradecka. 1973. Pol. Arch. Hydrobiol., 20, pp 111-119.
9.Owens, M. and R.W. Edwards. 1962. J. Ecol. 50, pp 157-162.
10.Owens, M. and R.W. Edwards. 1961. J. Ecol. 49, pp 119-126.
11.Wetzel, R.G. Int. Rev. ges. Hydrobiol 49, pp 1-64.
12.Wilson, L.R. 1941. Trans. Wis. Acad. Sci. Arts lett. 33, pp 135-146.
13.Wilson, L.R. 1937. Bull. Torr. Bot. Club 64, pp 199-208.
14.Spence, D.H.N., R.M. Campbell and J. Chrystal. 1971. Hydrobiologia 12, pp 169-176.
15.Potzger, J.E. and W.A. Engel. 1942. Trans. Wis. Acad. Sci. Arts lett. 34, pp 149-166.
16.Rickett, H.W. 1921. Trans. Wis. Acad. Sci. Arts Lett. 20, pp 501-527.
17.Burkholder, P.R. and G.H. Bornside. 1957. Bull. Torrey. bot. Club, 84, pp 366-383.
18.Bray. J.R., D.B. Lawrence and L.C. Pearson. 1959. Oikos, 10, pp 38-49.

1-316 Kormophytes, macrophytes
Biomass in roots

Species	Value	Condition
Alisma plantago aquatica	40 % of total biomass	Emergent[1]
Ceratophyllum sp.	5 % of total biomass	Submersed, less than[1]
Chara sp.	10 % of total biomass	Submersed, less than[1]
Cyperus fuscus	7.5 % of total biomass	Emergent, rg = 7 - 8[1]
Eichhornia crassipes	29 % of total biomass	Submersed roots, floating, rg = 10 - 48[1]
Elodea canadensis	2.6 % of total biomass	Submersed[1]
Equisetum fluviatile	61 % of total biomass	Emergent, rg = 40 - 83[1]
Glyceria maxima	53.5 % of total biomass	Emergent, rg = less than 30 - 67[1]
Littorella uniflora	50.5 % of total biomass	Submersed, rg = 46 - 55[1]
Nuphar luteum	65 % of total biomass	Emergent, rg = 50 - 80[1]
Nuphar pumilum	64 % of total biomass	Emergent, rg = 48 - 80[1]
Phragmites communis	66 % of total biomass	Emergent, rg = more than 36 - 96[1]
Schoenoplectus lacustris	68 % of total biomass	Emergent, rg = more than 46 - 90[1]
Sparganium spp.	45.5 % of total biomass	Emergent, rg = more than 25 - 66[1]
Thalassia testudinum	80 % of total biomass	Marine, submersed, rg = 75 - 85[1]
Typha angustifolia	45.5 % of total biomass	Emergent, rg = less than 32 - 59[1]
Typha latifolia	46.5 % of total biomass	Emergent, rg = 43 - 50[1]
Typha sp.	64 % of total biomass	Hybrid, emergent[1]
Zizania aquatica	7.5 % of total biomass	Floating-leaved, rg = 7 - 8[1]

Source: 1.Westlake, D.F. 1968. Methods of Productivity Studies in Roots Systems and Rhizosphere Organisms. Pub. House Nauka. Leningrad.

1-317 Kormophytes, macrophytes
C / biomass ratio

Species	Value	Condition
Amblystegium riparium	27.12 % of dry weight	Musci, n = 5, San Marcos River, TX, SD = +-0.24[1]
Angiosperms	31.7 % of dry weight	Mean of 15 species, San Marcos River, TX[1]
Cabomba caroliniana	28.36 % of dry weight	N = 5, Angiosperm, San Marcos River, TX, SD = +-0.74[1]
Ceratophyllum demersum	34.34 % of dry weight	N = 5, Angiosperms, San Marcos River, TX, SD = +-0.81[1]
Elodea densa	30.84 % of dry weight	N = 5, Angiosperms, San Marcos River, TX, SD = +-0.74[1]
Heteranthera dubia	27.18 % of dry weight	N = 5, Angiosperms, San Marcos River, TX, SD = +-0.59[1]
Ludwigia palustra	29.54 % of dry weight	N = 5, Angiosperms, San Marcos River, TX, SD = +-0.67[1]
Myriophyllum brasiliense	35.28 % of dry weight	N = 5, Angiosperms, San Marcos River, TX, SD = +-1.19[1]
Myriophyllum heterophyllum	37.61 % of dry weight	N = 5, Angiosperms, San Marcos River, TX, SD = +-0.61[1]
Najas guadalupensis	31.92 % of dry weight	N = 5, Angiosperms, San Marcos River, TX, SD = +-0.79[1]
Potamogeton crispus	30.74 % of dry weight	N = 5, Angiosperms, San Marcos River, TX, SD = +-0.74[1]
Potamogeton illinoeasis	32.73 % of dry weight	N = 5, Angiosperms, San Marcos River, TX, SD = +-0.76[1]
Potamogeton pectinatus	30.49 % of dry weight	N = 5, Angiosperms, San Marcos River, TX, SD = +-0.86[1]

Rhizoclonium hieroglyphicum	25.84 % of dry weight	N = 5, Chlorophyceae, San Marcos River, TX, SD = +-0.41[1]
Riccia fluitans	28.66 % of dry weight	N = 5, Hepaticae, San Marcos River, TX, SD = +-0.24[1]
Sagittaria platyphylla	33.3 % of dry weight	N = 5, Angiosperms, San Marcos River, TX, SD = +-0.56[1]
Utricularia subulata	29.69 % of dry weight	N = 5, Angiosperms, San Marcos River, TX, SD = +-0.79[1]
Vallisneria americana	28.6 % of dry weight	N = 5, Angiosperms, San Marcos River, TX, SD = +-0.77[1]
Zannichellia palustrus	34.83 % of dry weight	N = 5, Angiosperms, San Marcos River, TX, SD = +-0.58[1]

Source: 1.Hannan, H.H. and T.C. Dorris. 1970. Limnol. Oceanogr., 15(3) pp 442-453.

1-318 Kormophytes, macrophytes Ca / biomass ratio

Species	Value	Condition
Brasenia schreberi	1.79 % of dry weight	Par Pond, SC, May 69, soft wat, water = 2.64 Ca ppm, n=4, SD = +-0.05[1]
Ceratophyllum demersum	0.77 % of dry weight	Par Pond, SC, May 69, soft wat, water = 2.64 Ca ppm, n=4, SD = +-0.06[1]
Eleocharis acicularis	0.53 % of dry weight	Par Pond, SC, May 69, soft wat, water = 2.64 Ca ppm, n=4, SD = +-0.13[1]
Eleocharis quadrangulata	0.2 % of dry weight	Par Pond, SC, May 69, soft wat, water = 2.64 Ca ppm, n=1[1]
Free floating	14.16 g/kg dry weight	Amazon, SE = +-3.149[2]
Halophila ovalis	1.06 % of dry weight	Par Pond, SC, May 69, soft wat, water = 2.64 Ca ppm, n=5, SD = +-0.01[1]
Hydrocotyle sp.	1.85 % of dry weight	Par Pond, SC, May 69, soft wat, water = 2.64 Ca ppm, n=4, SD = +-0.26[1]
Juncus effusus	0.38 % of dry weight	Par Pond, SC, May 69, soft wat, water = 2.64 Ca ppm, n=5, SD = +-0.11[1]
Myriophyllum heterophyllum	1.47 % of dry weight	Par Pond, SC, May 69, soft wat, water = 2.64 Ca ppm, n=4, SD = +-0.11[1]
Najas guadalupensis	0.98 % of dry weight	Par Pond, SC, May 69, soft wat, water = 2.64 Ca ppm, n=4, SD = +-0.14[1]
Nelumbo lutea	1.56 % of dry weight	Par Pond, SC, May 69, soft wat, water = 2.64 Ca ppm, n=2, SD = +-0.18[1]
Nuphar advena	1.08 % of dry weight	Par Pond, SC, May 69, soft wat, water = 2.64 Ca ppm, n=2[1]
Panicum hemitonium	0.38 % of dry weight	Par Pond, SC, May 69, soft wat, water = 2.64 Ca ppm, n=3, SD = +-0.04[1]
Pontederia cordata	0.96 % of dry weight	Par Pond, SC, May 69, soft wat, water = 2.64 Ca ppm, n=1[1]
Potamogeton diversifolius	1.14 % of dry weight	Par Pond, SC, May 69, soft wat, water = 2.64 Ca ppm, n=1[1]
Rooted	4.18 g/kg dry weight	Amazon, SE = +-1.597[2]
Sagittaria latifolia	0.55 % of dry weight	Par Pond, SC, May 69, soft wat, water = 2.64 Ca ppm, n=2[1]
Scirpus americanus	0.5 % of dry weight	Par Pond, SC, May 69, soft wat, water = 2.64 Ca ppm, n=5, SD = +-0.01[1]
Typha latifolia	0.76 % of dry weight	Par Pond, SC, May 69, soft wat, water = 2.64 Ca ppm, n=4, SD = +-0.18[1]
Utricularia inflata	0.67 % of dry weight	Par Pond, SC, May 69, soft wat, water = 2.64 Ca ppm, n=4, SD = +-0.07[1]

Source: 1.Boyd, C.E. 1970. Arch. Hydrobiol., 67(1), pp 78-85.
2.Howard-Williams, C. and W.J. Junk. 1977. Arch. Hydrobiol., 79(4), pp 446-464.

1-319 Kormophytes, macrophytes
Caloric content

Species	Value	Condition
Emergent	4480 cal/g	Range = 4207 - 4987[1]
Free floating	4615 cal/(g ash free dry weight)	Amazon, SE = +-0.087[2]
Juncus effusus	4.455 kcal/g ash free dry weight	Freshwater, oxydation[3]
Phragmites communis	4.581 kcal/g ash free dry weight	Freshwater, bombcal[3]
Plant with floating leaves	4770 cal/g	Range = 4560 - 5140[1]
Rooted	4479 cal/(g ash free dry weight)	Amazon, SE = +-0.039[2]
Scirpus lacustris	4.556 kcal/g ash free dry weight	Freshwater, oxydation[3]
Submersed	4580 cal/g	Range = 4165 - 5201[1]
Typha spp.	4.34 kcal/g dry weight	Freshwater, bombcal[3]
Typha spp.	4.340 kcal/g dry weight	Freshwater, bombcal[3]

Source: 1.Straskraba, M. 1968. Mitt. Int. Ver. Limnol. 14, pp 212-230.
2.Howard-Williams, C. and W.J. Junk. 1977. Arch. Hydrobiol., 79(4), pp 446-464.
3.Oertzen, J.-A. 1973. Wiss. Z. Univ. Rostock. 22, pp 1153-1158.

1-320 Kormophytes, macrophytes
Carbohydrate / biomass ratio

Species	Value	Condition
Emergent	50 % of organic matter	Range = 23 - 69[1]
Plant with floating leaves	42 % of organic matter	Range = 31 - 64[1]
Submersed	51 % of organic matter	Range = 20 - 70[1]

Source: 1.Straskraba, M. 1968. Mitt. Int. Ver. Limnol. 14, pp 212-230.

1-321 Kormophytes, macrophytes
Chlorophyll a / biomass ratio

Species	Value	Condition
Brasenia schreberi	3.26 mg/g dry weight	Par Pond, SC, May 69, soft water, SD = +-0.32[1]
Hydrocotyle sp.	5.13 mg/g dry weight	Par Pond, SC, May 69, soft water, SD = +-0.15[1]
Juncus effusus	1.26 mg/g dry weight	Par Pond, SC, May 69, soft water, SD = +-0.08[1]
Myriophyllum heterophyllum	7.74 mg/g dry weight	Par Pond, SC, May 69, soft water, SD = +-0.75[1]
Najas guadalupensis	8.52 mg/g dry weight	Par Pond, SC, May 69, soft water, SD = +-0.52[1]
Nelumbo lutea	3.62 mg/g dry weight	Par Pond, SC, May 69, soft water, SD = +-0.54[1]
Nuphar advena	3.86 mg/g dry weight	Par Pond, SC, May 69, soft water, SD = +-0.58[1]
Nymphaea odorata	3.38 mg/g dry weight	Par Pond, SC, May 69, soft water, SD = +-0.14[1]
Pontederia cordata	3.26 mg/g dry weight	Par Pond, SC, May 69, soft water, SD = +-0.07[1]
Potamogeton diversifolius	5.74 mg/g dry weight	Par Pond, SC, May 69, soft water, SD = +-0.13[1]
Scirpus americanus	2.32 mg/g dry weight	Par Pond, SC, May 69, soft water, SD = +-0.21[1]

Typha latifolia	2.61 mg/g dry weight	Par Pond, SC, May 69, soft water, SD = +-0.21[1]
Utricularia inflata	6.19 mg/g dry weight	Par Pond, SC, May 69, soft water, SD = +-0.33[1]

Source: 1.Boyd, C.E. 1970. Arch. Hydrobiol., 67(1), pp 78-85.

1-322 Kormophytes, macrophytes
Chlorophyll b / biomass ratio

Species	Value	Condition
Brasenia schreberi	0.1 mg/g dry weight	Par Pond, SC, May 69, soft water, SD = +-0.06[1]
Hydrocotyle sp.	1.23 mg/g dry weight	Par Pond, SC, May 69, soft water, SD = +-0.05[1]
Juncus effusus	0.23 mg/g dry weight	Par Pond, SC, May 69, soft water, SD = +-0.02[1]
Myriophyllum heterophyllum	2.21 mg/g dry weight	Par Pond, SC, May 69, soft water, SD = +-0.09[1]
Najas guadalupensis	2.51 mg/g dry weight	Par Pond, SC, May 69, soft water, SD = +-0.19[1]
Nelumbo lutea	0.82 mg/g dry weight	Par Pond, SC, May 69, soft water, SD = +-0.16[1]
Nuphar advena	0.79 mg/g dry weight	Par Pond, SC, May 69, soft water, SD = +-0.15[1]
Nymphaea odorata	0.43 mg/g dry weight	Par Pond, SC, May 69, soft water, SD = +-0.19[1]
Pontederia cordata	0.77 mg/g dry weight	Par Pond, SC, May 69, soft water, SD = +-0.15[1]
Potamogeton diversifolius	1.46 mg/g dry weight	Par Pond, SC, May 69, soft water, SD = +-0.01[1]
Scirpus americanus	0.41 mg/g dry weight	Par Pond, SC, May 69, soft water, SD = +-0.07[1]
Typha latifolia	0.65 mg/g dry weight	Par Pond, SC, May 69, soft water, SD = +-0.06[1]
Utricularia inflata	0.65 mg/g dry weight	Par Pond, SC, May 69, soft water, SD = +-0.11[1]

Source: 1.Boyd, C.E. 1970. Arch. Hydrobiol., 67(1), pp 78-85.

1-323 Kormophytes, macrophytes
Crude fiber / biomass ratio

Species	Value	Condition
Emergent	32 % of organic matter	Range = 17.6 - 45.4[1]
Plant with floating leaves	27 % of organic matter	Range = 14 - 38[1]
Submersed	27 % of organic matter	Range = 14.0 - 61.6[1]

Source: 1.Straskraba, M. 1968. Mitt. Int. Ver. Limnol. 14, pp 212-230.

1-324 Kormophytes, macrophytes
Cu / biomass ratio

Species	Value	Condition
Brasenia schreberi	32 mg/kg dry weight	Par Pond, SC, May 69, soft wat, water = trace Cu, n=4, SD = +-5[1]
Ceratophyllum demersum	30 mg/kg dry weight	Par Pond, SC, May 69, soft wat, water = trace Cu, n=4, SD = +-7[1]
Eleocharis acicularis	42 mg/kg dry weight	Par Pond, SC, May 69, soft wat, water = trace Cu, n=4, SD = +-6[1]
Eleocharis quadrangulata	20 mg/kg dry weight	Par Pond, SC, May 69, soft wat, water = trace Cu, n=2[1]
Hydrocotyle sp.	53 mg/kg dry weight	Par Pond, SC, May 69, soft wat, water = trace Cu, n=4, SD = +-8[1]
Myriophyllum heterophyllum	44 mg/kg dry weight	Par Pond, SC, May 69, soft wat, water = trace Cu, n=4, SD = +-9[1]
Najas guadalupensis	48 mg/kg dry weight	Par Pond, SC, May 69, soft wat, water = trace Cu, n=4, SD = +-8[1]
Nelumbo lutea	40 mg/kg dry weight	Par Pond, SC, May 69, soft wat, water = trace Cu, n=4, SD = +-7[1]
Nuphar advena	35 mg/kg dry weight	Par Pond, SC, May 69, soft wat, water = trace Cu, n=2[1]
Nymphaea odorata	36 mg/kg dry weight	Par Pond, SC, May 69, soft wat, water = trace Cu, n=5, SD = +-7[1]

Panicum hemitonium	26 mg/kg dry weight	Par Pond, SC, May 69, soft wat, water = trace Cu, n=3, SD = +-2[1]
Pontederia cordata	60 mg/kg dry weight	Par Pond, SC, May 69, soft wat, water = trace Cu, n=1[1]
Potamogeton diversifolius	36 mg/kg dry weight	Par Pond, SC, May 69, soft wat, water = trace Cu, n=1[1]
Sagittaria latifolia	57 mg/kg dry weight	Par Pond, SC, May 69, soft wat, water = trace Cu, n=2[1]
Typha latifolia	37 mg/kg dry weight	Par Pond, SC, May 69, soft wat, water = trace Cu, n=4, SD = +-7[1]
Utricularia inflata	47 mg/kg dry weight	Par Pond, SC, May 69, soft wat, water = trace Cu, n=4, SD = +-6[1]

Source: 1.Boyd, C.E. 1970. Arch. Hydrobiol., 67(1), pp 78-85.

1-325 Kormophytes, macrophytes
Dark respiration

Species	Value	Condition
Lemna minor	1.25 %	% of initial internal 14C h-1, 5 Oct 1976[1]
Lemna minor	1.44 %	% of initial internal 14C h-1, 18 Aug 1976[1]
Lemna minor	1.63 %	% of initial internal 14C h-1, 15 Sep 1976[1]
Lemna minor	2.60 %	% of initial internal 14C h-1, 4 Sep 1976[1]

Source: 1.Filbin, G.J. and Hough, R.A., Limnol. Oceanogr., 30(2), 322, 1985.

1-326 Kormophytes, macrophytes
Dry / wet weight ratio

Species	Value	Condition
Amblystegium riparium	30.57 % of wet weight	N=3, Musci, San Marcos River, SD = +-0.82[1]
Angiosperms	11.5 Mean for 15 species, San Marcos Riv	Mean for 15 species, San Marcos River, TX[1]
Cabomba caroliniana	7.47 % of wet weight	N = 3, Angiosperm, San Marcos River, TX, SD = +-0.56[1]
Ceratophyllum demersum	8.9 % of wet weight	N = 3, Angiosperm, San Marcos River, TX, SD = +-0.36[1]
Elodea densa	10.47 % of wet weight	N = 3, Angiosperm, San Marcos River, TX, SD = +-1.75[1]
Emergent	21 % of dry weight	Range = 0.15 - 0.30[2]
Free floating	7.15 % of wet weight	Amazon, SE = +-0.520[3]
Heteranthera dubia	6.53 % of wet weight	N = 3, Angiosperm, San Marcos River, TX, SD = +-0.41[1]
Ludwigia palustra	13.5 % of wet weight	N = 3, Angiosperm, San Marcos River, TX, SD = +-0.50[1]
Myriophyllum brasiliense	20 % of wet weight	N = 3, Angiosperm, San Marcos River, TX, SD = +-0.20[1]
Myriophyllum heterophyllum	12.77 % of wet weight	N = 3, Angiosperm, San Marcos River, TX, SD = +-0.22[1]
Najas guadalupensis	10.83 % of wet weight	N = 3, Angiosperm, San Marcos River, TX, SD = +-1.22[1]
Plant with floating leaves	9 % of dry weight	Range = 0.16 - 0.20[2]
Potamogeton crispus	13.66 % of wet weight	N = 3, Angiosperm, San Marcos River, TX, SD = +-0.31[1]
Potamogeton illinoeasis	14.2 % of wet weight	N = 3, Angiosperm, San Marcos River, TX, SD = +-0.26[1]
Potamogeton pectinatus	15.56 % of wet weight	N = 3, Angiosperm, San Marcos River, TX, SD = +-0.20[1]
Rhizoclonium hieroglyphicum	27.43 % of wet weight	N = 3, Chlorophyceae, San Marcos River, TX, SD = +-1.80[1]
Riccia fluitans	8.87 % of wet weight	N = 3, Angiosperm, San Marcos River, TX, SD = +-0.42[1]
Rooted	19.81 % of wet weight	Amazon, SE = +-2.937[3]
Sagittaria platyphylla	6.4 % of wet weight	N = 3, Angiosperm, San Marcos River, TX, SD = +-0.26[1]
Submersed	12 % of dry weight	Range = 0.08 - 0.15[2]

Utricularia subulata	13.47 % of wet weight	N = 3, Angiosperm, San Marcos River, TX, SD = +-0.16[1]
Vallisneria americana	6.5 % of wet weight	N = 3, Angiosperm, San Marcos River, TX, SD = +-0.45[1]
Zannichellia palustrus	12.86 % of wet weight	N = 3, Angiosperm, San Marcos River, TX, SD = +-0.16[1]

Source: 1.Hannan, H.H. and T.C. Dorris. 1970. Limnol. Oceanogr., 15(3) pp 442-453.
2.Straskraba, M. 1968. Mitt. Int. Ver. Limnol. 14, pp 212-230.
3.Howard-Williams, C. and W.J. Junk. 1977. Arch. Hydrobiol., 79(4), pp 446-464.

1-327 Kormophytes, macrophytes
Ether extract / biomass ratio

Species	Value	Condition
Emergent	2.1 % of organic matter	Range = 0.4 - 6.1[1]
Plant with floating leaves	4 % of organic matter	Range = 2.8 - 5.7[1]

Source: 1.Straskraba, M. 1968. Mitt. Int. Ver. Limnol. 14, pp 212-230.

1-328 Kormophytes, macrophytes
Fe / biomass ratio

Species	Value	Condition
Brasenia schreberi	500 mg/kg dry weight	Par Pond, SC, May 69, soft wat, water = 0.029 Fe ppm, n=4, SD = +-40[1]
Ceratophyllum demersum	1053 mg/kg dry weight	Par Pond, SC, May 69, soft wat, water = 0.029 Fe ppm, n=4, SD = +-344[1]
Eleocharis acicularis	2920 mg/kg dry weight	Par Pond, SC, May 69, soft wat, water = 0.029 Fe ppm, n=4, SD = +-613[1]
Eleocharis quadrangulata	460 mg/kg dry weight	Par Pond, SC, May 69, soft wat, water = 0.029 Fe ppm, n=2[1]
Hydrocotyle sp.	1245 mg/kg dry weight	Par Pond, SC, May 69, soft wat, water = 0.029 Fe ppm, n=4, SD = +-326[1]
Myriophyllum heterophyllum	2000 mg/kg dry weight	Par Pond, SC, May 69, soft wat, water = 0.029 Fe ppm, n=4, SD = +-286[1]
Najas guadalupensis	712 mg/kg dry weight	Par Pond, SC, May 69, soft wat, water = 0.029 Fe ppm, n=4, SD = +-215[1]
Nelumbo lutea	126 mg/kg dry weight	Par Pond, SC, May 69, soft wat, water = 0.029 Fe ppm, n=4, SD = +-46[1]
Nuphar advena	740 mg/kg dry weight	Par Pond, SC, May 69, soft wat, water = 0.029 Fe ppm, n=2[1]
Nymphaea odorata	600 mg/kg dry weight	Par Pond, SC, May 69, soft wat, water = 0.029 Fe ppm, n=5, SD = +-128[1]
Panicum hemitonium	560 mg/kg dry weight	Par Pond, SC, May 69, soft wat, water = 0.029 Fe ppm, n=3[1]
Pontederia cordata	200 mg/kg dry weight	Par Pond, SC, May 69, soft wat, water = 0.029 Fe ppm, n=1[1]
Potamogeton diversifolius	2140 mg/kg dry weight	Par Pond, SC, May 69, soft wat, water = 0.029 Fe ppm, n=1[1]
Sagittaria latifolia	460 mg/kg dry weight	Par Pond, SC, May 69, soft wat, water = 0.029 Fe ppm, n=2[1]
Scirpus americanus	133 mg/kg dry weight	Par Pond, SC, May 69, soft wat, water = 0.029 Fe ppm, n=3, SD = +-19[1]
Typha latifolia	120 mg/kg dry weight	Par Pond, SC, May 69, soft wat, water = 0.029 Fe ppm, n=4, SD = +-40[1]
Utricularia inflata	2112 mg/kg dry weight	Par Pond, SC, May 69, soft wat, water = 0.029 Fe ppm, n=4, SD = +-222[1]

Source: 1.Boyd, C.E. 1970. Arch. Hydrobiol., 67(1), pp 78-85.

1-329 Kormophytes, macrophytes
Gross photosynthesis

Species	Value	Condition
Lotic community	10.3 gram O2/(m2*day)	Comm resp = 6.7 - 15.4 g O2/(m2*d), P/R=1.13, Ivel River, rg = 3.2[1]
Lotic community	14.95 gram O2/(m2*day)	Comm resp=4.1-19.9 g O2/(m2*d), P/R=0.59-1.38, San Marcos River[2]
Lotic community	16.6 gram O2/(m2*day)	Comm resp = 14.6 g O2/(m2*d), P/R=1.1, headwater area, Silver[3]
Lotic community	21.3 gram O2/(m2*day)	Comm resp = 12.6 g O2/(m2*d), P/R=1.70, granite bed, Blue River[4]
Lotic community	3 gram O2/(m2*day)	Comm resp = 7.7 g O2/(m2*d), P/R=0.39, sand bed, Blue River, OKL[4]
Lotic community	35 gram O2/(m2*day)	Comm resp = 5.0 g O2/(m2*d), P/R=7.0, spring, Silver Springs, FL[3]
Lotic community	5.05 gram O2/(m2*day)	Comm resp = 0.7 - 21.5 g O2/(m2*d), Neuse River, NC, rg = 0.3 - 9.8[5]
Lotic community	6.8 gram O2/(m2*day)	Comm resp = 11.0 g O2/(m2*d), P/R=0.64, limestone bed, Blue River[4]
Lotic community	7.2 gram O2/(m2*day)	Comm resp = 4.2 - 20.2 g O2/(m2*d), P/R=0.1 - 1.1, Itchen River[3]
Lotic community	8 gram O2/(m2*day)	Comm resp = 2.8 g O2/(m2*d), P/R=2.9, winter, Silver Springs, FL[6]

Source: 1.Edwards, R.W. and M. Owens. 1960. J. Ecol., 48, pp 151-180.
2.Hannan, H.H. and T.C. Dorris. 1970. Limnol. Oceanogr., 15(3) pp 442-453.
3.Odum, H.T. 1956. Limnol. Oceanogr. 1, pp 102-117.
4.Duffer, W.R. and T.C. Dorris. 1966. Limnol. Oceanogr., 11, pp 143-151.
5.Hoskin, C.M. 1959. Publ. Inst. Mar. Sci. Univ Tex. 6, pp 186-192.
6.Odum, H.T. 1957. Ecol. Monogr. 27, pp 55-112.

1-330 Kormophytes, macrophytes
K / biomass ratio

Species	Value	Condition
Brasenia schreberi	0.99 % of dry weight	Par Pond, SC, May 69, soft wat, water = 1.61 K ppm, n=4, SD = +-0.01[1]
Ceratophyllum demersum	4.01 % of dry weight	Par Pond, SC, May 69, soft wat, water = 1.61 K ppm, n=4, SD = +-0.59[1]
Eleocharis acicularis	2.86 % of dry weight	Par Pond, SC, May 69, soft wat, water = 1.61 K ppm, n=4, SD = +-0.71[1]
Eleocharis quadrangulata	1.81 % of dry weight	Par Pond, SC, May 69, soft wat, water = 1.61 K ppm, n=1[1]
Free floating	36.75 g/kg dry weight	Amazon, SE = +-4.239[2]
Hydrocotyle sp.	1.73 % of dry weight	Par Pond, SC, May 69, soft wat, water = 1.61 K ppm, n=4, SD = +-0.42[1]
Juncus effusus	0.89 % of dry weight	Par Pond, SC, May 69, soft wat, water = 1.61 K ppm, n=5, SD = +-0.06[1]
Myriophyllum heterophyllum	1.25 % of dry weight	Par Pond, SC, May 69, soft wat, water = 1.61 K ppm, n=4, SD = +-0.08[1]
Najas guadalupensis	3.49 % of dry weight	Par Pond, SC, May 69, soft wat, water = 1.61 K ppm, n=4, SD = +-0.53[1]
Nelumbo lutea	2.27 % of dry weight	Par Pond, SC, May 69, soft wat, water = 1.61 K ppm, n=2, SD = +-0.25[1]
Nuphar advena	1.88 % of dry weight	Par Pond, SC, May 69, soft wat, water = 1.61 K ppm, n=2[1]
Nymphaea odorata	1.28 % of dry weight	Par Pond, SC, May 69, soft wat, water = 1.61 K ppm, n=5, SD = +-0.13[1]

Panicum hemitonium	1.06 % of dry weight	Par Pond, SC, May 69, soft wat, water = 1.61 K ppm, n=3, SD = +-0.13[1]
Pontederia cordata	2.58 % of dry weight	Par Pond, SC, May 69, soft wat, water = 1.61 K ppm, n=1[1]
Potamogeton diversifolius	3.08 % of dry weight	Par Pond, SC, May 69, soft wat, water = 1.61 K ppm, n=1[1]
Rooted	25.43 g/kg dry weight	Amazon, SE = +-4.583[2]
Sagittaria latifolia	4.04 % of dry weight	Par Pond, SC, May 69, soft wat, water = 1.61 K ppm, n=2[1]
Scirpus americanus	2.83 % of dry weight	Par Pond, SC, May 69, soft wat, water = 1.61 K ppm, n=5, SD = +-0.17[1]
Typha latifolia	2.65 % of dry weight	Par Pond, SC, May 69, soft wat, water = 1.61 K ppm, n=4, SD = +-0.50[1]
Utricularia inflata	1.98 % of dry weight	Par Pond, SC, May 69, soft wat, water = 1.61 K ppm, n=4, SD = +-0.13[1]

Source: 1.Boyd, C.E. 1970. Arch. Hydrobiol., 67(1), pp 78-85.
2.Howard-Williams, C. and W.J. Junk. 1977. Arch. Hydrobiol., 79(4), pp 446-464.

1-331 Kormophytes, macrophytes
Light respiration

Species	Value	Condition
Lemna minor	0.82 %	% of initial internal 14C h-1, 18 Aug 1976[1]
Lemna minor	1.22 %	% of initial internal 14C h-1, 5 Oct 1976[1]
Lemna minor	1.83 %	% of initial internal 14C h-1, 15 Sep 1976[1]
Lemna minor	3.82 %	% of initial internal 14C h-1, 4 Sep 1976[1]

Source: 1.Filbin, G.J. and Hough, R.A., Limnol.Oceanogr., 30(2), 322, 1985.

1-332 Kormophytes, macrophytes
Max standing crop

Species	Value	Condition
Submersed macrophytes	123 g dry weight/m2	Maple Lodge Channel, Engl, polluted, freshwater[1]
Submersed macrophytes	320 g dry weight/m2	River Ivel, Engl, freshwater[2]
Submersed macrophytes	322 g dry weight/m2	River Chess, Engl, freshwater[2]
Submersed macrophytes	381 g dry weight/m2	River Yare, Engl, freshwater[2]
Submersed macrophytes	385 g dry weight/m2	River test, Engl, freshwater[2]
Submersed macrophytes	412 g dry weight/m2	Three Florida Rivers, freshwater[3]
Submersed macrophytes	520 g dry weight/m2	River Ivel, Engl, freshwater[4]
Submersed macrophytes	526 g dry weight/m2	Three Florida springs, freshwater[3]
Submersed macrophytes	621 g dry weight/m2	Silver Springs, FL, freshwater[5]
Submersed macrophytes	638 g dry weight/m2	San Marcos River, TX, freshwater[6]
Submersed macrophytes	809 g dry weight/m2	Silver Springs, FL, macrophytes and associated aufwuchs, freshwater[5]

Source: 1.Westlake, D.F. 1961. Int. Ver. Theor. Angew. Limnol. Verh. 15, pp 499-504.
2.Owens, M. and R.W. Edwards. 1962. J. Ecol. 50, pp 157-162.
3.Penfound, W.T. 1956. Limnol. Oceanogr. 1(1), pp 92-101.
4.Edwards, R.W. and M. Owens. 1960. J. Ecol., 48, pp 151-180.
5.Odum, H.T. 1957. Ecol. Monogr. 27, pp 55-112.
6.Hannan, H.H. and T.C. Dorris. 1970. Limnol. Oceanogr., 15(3) pp 442-453.

1-333 Kormophytes, macrophytes
Mg / biomass ratio

Species	Value	Condition
Brasenia schreberi	0.26 % of dry weight	Par Pond, SC, May 69, soft wat, water = 1.04 Mg ppm, n=4, SD = +-0.01[1]
Ceratophyllum demersum	0.42 % of dry weight	Par Pond, SC, May 69, soft wat, water = 1.04 Mg ppm, n=4, SD = +-0.10[1]
Eleocharis acicularis	0.33 % of dry weight	Par Pond, SC, May 69, soft wat, water = 1.04 Mg ppm, n=4, SD = +-0.06[1]
Eleocharis quadrangulata	0.1 % of dry weight	Par Pond, SC, May 69, soft wat, water = 1.04 Mg ppm, n=1[1]
Free floating	3.9 g/kg dry weight	Amazon, SE = +- 0.602[2]
Hydrocotyle sp.	0.47 % of dry weight	Par Pond, SC, May 69, soft wat, water = 1.04 Mg ppm, n=4[1]
Juncus effusus	0.11 % of dry weight	Par Pond, SC, May 69, soft wat, water = 1.04 Mg ppm, n=5, SD = +-0.01[1]
Myriophyllum heterophyllum	0.26 % of dry weight	Par Pond, SC, May 69, soft wat, water = 1.04 Mg ppm, n=4[1]
Najas guadalupensis	0.47 % of dry weight	Par Pond, SC, May 69, soft wat, water = 1.04 Mg ppm, n=4[1]
Nelumbo lutea	0.23 % of dry weight	Par Pond, SC, May 69, soft wat, water = 1.04 Mg ppm, n=2[1]
Nuphar advena	0.27 % of dry weight	Par Pond, SC, May 69, soft wat, water = 1.04 Mg ppm, n=2[1]
Nymphaea odorata	0.14 % of dry weight	Par Pond, SC, May 69, soft wat, water = 1.04 Mg ppm, n=5[1]
Panicum hemitonium	0.25 % of dry weight	Par Pond, SC, May 69, soft wat, water = 1.04 Mg ppm, n=3[1]
Pontederia cordata	0.15 % of dry weight	Par Pond, SC, May 69, soft wat, water = 1.04 Mg ppm, n=1[1]
Potamogeton diversifolius	0.19 % of dry weight	Par Pond, SC, May 69, soft wat, water = 1.04 Mg ppm, n=1[1]
Rooted	1.93 g/kg dry weight	Amazon, SE = +- 0.426[2]
Sagittaria latifolia	0.18 % of dry weight	Par Pond, SC, May 69, soft wat, water = 1.04 Mg ppm, n=2[1]
Scirpus americanus	0.22 % of dry weight	Par Pond, SC, May 69, soft wat, water = 1.04 Mg ppm, n=5[1]
Typha latifolia	0.15 % of dry weight	Par Pond, SC, May 69, soft wat, water = 1.04 Mg ppm, n=4[1]
Utricularia inflata	0.21 % of dry weight	Par Pond, SC, May 69, soft wat, water = 1.04 Mg ppm, n=4[1]

Source: 1.Boyd, C.E. 1970. Arch. Hydrobiol., 67(1), pp 78-85.
2.Howard-Williams, C. and W.J. Junk. 1977. Arch. Hydrobiol., 79(4), pp 446-464.

1-334 Kormophytes, macrophytes
Mn / biomass ratio

Species	Value	Condition
Brasenia schreberi	265 mg/kg dry weight	Par Pond, SC, May 69, soft wat, water = 0.002 Mn ppm, n=4, SD = +- 72[1]
Ceratophyllum demersum	486 mg/kg dry weight	Par Pond, SC, May 69, soft wat, water = 0.002 Mn ppm, n=4, SD = +- 62[1]
Eleocharis acicularis	192 mg/kg dry weight	Par Pond, SC, May 69, soft wat, water = 0.002 Mn ppm, n=4, SD = +- 30[1]
Eleocharis quadrangulata	460 mg/kg dry weight	Par Pond, SC, May 69, soft wat, water = 0.002 Mn ppm, n=2[1]
Hydrocotyle sp.	196 mg/kg dry weight	Par Pond, SC, May 69, soft wat, water = 0.002 Mn ppm, n=4, SD = +- 71[1]
Myriophyllum heterophyllum	473 mg/kg dry weight	Par Pond, SC, May 69, soft wat, water = 0.002 Mn ppm, n=4, SD = +- 64[1]
Najas guadalupensis	201 mg/kg dry weight	Par Pond, SC, May 69, soft wat, water = 0.002 Mn ppm, n=4, SD = +- 62[1]

Nelumbo lutea	607 mg/kg dry weight	Par Pond, SC, May 69, soft wat, water = 0.002 Mn ppm, n=4[1]
Nuphar advena	300 mg/kg dry weight	Par Pond, SC, May 69, soft wat, water = 0.002 Mn ppm, n=2[1]
Nymphaea odorata	128 mg/kg dry weight	Par Pond, SC, May 69, soft wat, water = 0.002 Mn ppm, n=5, SD = +- 13[1]
Panicum hemitonium	560 mg/kg dry weight	Par Pond, SC, May 69, soft wat, water = 0.002 Mn ppm, n=4[1]
Pontederia cordata	200 mg/kg dry weight	Par Pond, SC, May 69, soft wat, water = 0.002 Mn ppm, n=1[1]
Potamogeton diversifolius	160 mg/kg dry weight	Par Pond, SC, May 69, soft wat, water = 0.002 Mn ppm, n=1[1]
Sagittaria latifolia	460 mg/kg dry weight	Par Pond, SC, May 69, soft wat, water = 0.002 Mn ppm, n=2[1]
Scirpus americanus	292 mg/kg dry weight	Par Pond, SC, May 69, soft wat, water = 0.002 Mn ppm, n=3, SD = +- 60[1]
Typha latifolia	412 mg/kg dry weight	Par Pond, SC, May 69, soft wat, water = 0.002 Mn ppm, n=4, SD = +- 67[1]
Utricularia inflata	480 mg/kg dry weight	Par Pond, SC, May 69, soft wat, water = 0.002 Mn ppm, n=4[1]

Source: 1.Boyd, C.E. 1970. Arch. Hydrobiol., 67(1), pp 78-85.

1-335 Kormophytes, macrophytes N / biomass ratio

Species	Value	Condition
Free floating	22.4 g/kg dry weight	Amazon, SE = +- 1.950[1]
Rooted	15.13 g/kg dry weight	Amazon, SE = +- 2.986[1]

Source: 1.Howard-Williams, C. and W.J. Junk. 1977. Arch. Hydrobiol., 79(4), pp 446-464.

1-336 Kormophytes, macrophytes Na / biomass ratio

Species	Value	Condition
Brasenia schreberi	0.66 % of dry weight	Par Pond, SC, May 69, soft wat, water = 6.70 Na ppm, n=4[1]
Ceratophyllum demersum	1.16 % of dry weight	Par Pond, SC, May 69, soft wat, water = 6.70 Na ppm, n=4[1]
Eleocharis acicularis	0.54 % of dry weight	Par Pond, SC, May 69, soft wat, water = 6.70 Na ppm, n=4[1]
Eleocharis quadrangulata	0.12 % of dry weight	Par Pond, SC, May 69, soft wat, water = 6.70 Na ppm, n=1[1]
Free floating	6.36 g/kg dry weight	Amazon, SE = +- 1.729[2]
Hydrocotyle sp.	0.98 % of dry weight	Par Pond, SC, May 69, soft wat, water = 6.70 Na ppm, n=4[1]
Juncus effusus	0.4 % of dry weight	Par Pond, SC, May 69, soft wat, water = 6.70 Na ppm, n=5[1]
Myriophyllum heterophyllum	1.87 % of dry weight	Par Pond, SC, May 69, soft wat, water = 6.70 Na ppm, n=4[1]
Najas guadalupensis	0.61 % of dry weight	Par Pond, SC, May 69, soft wat, water = 6.70 Na ppm, n=4[1]
Nelumbo lutea	0.28 % of dry weight	Par Pond, SC, May 69, soft wat, water = 6.70 Na ppm, n=2[1]
Nuphar advena	1.47 % of dry weight	Par Pond, SC, May 69, soft wat, water = 6.70 Na ppm, n=2[1]
Nymphaea odorata	1.35 % of dry weight	Par Pond, SC, May 69, soft wat, water = 6.70 Na ppm, n=5[1]
Panicum hemitonium	0.19 % of dry weight	Par Pond, SC, May 69, soft wat, water = 6.70 Na ppm, n=3[1]
Pontederia cordata	0.83 % of dry weight	Par Pond, SC, May 69, soft wat, water = 6.70 Na ppm, n=1[1]
Potamogeton diversifolius	0.44 % of dry weight	Par Pond, SC, May 69, soft wat, water = 6.70 Na ppm, n=1[1]
Rooted	2.32 g/kg dry weight	Amazon, SE = +- 2.032[2]
Sagittaria latifolia	0.14 % of dry weight	Par Pond, SC, May 69, soft wat, water = 6.70 Na ppm, n=2[1]
Scirpus americanus	0.09 % of dry weight	Par Pond, SC, May 69, soft wat, water = 6.70 Na ppm, n=5[1]

Typha latifolia	0.28 % of dry weight	Par Pond, SC, May 69, soft wat, water = 6.70 Na ppm, n=4[1]
Utricularia inflata	1.52 % of dry weight	Par Pond, SC, May 69, soft wat, water = 6.70 Na ppm, n=4[1]

Source: 1.Boyd, C.E. 1970. Arch. Hydrobiol., 67(1), pp 78-85.
2.Howard-Williams, C. and W.J. Junk. 1977. Arch. Hydrobiol., 79(4), pp 446-464.

1-337 Kormophytes, macrophytes
Net photosynthesis

Species	Value	Condition
Lemna minor	0.08 mg g-1 h-1	p.m., 16 Nov, 17.2 klux[1]
Lemna minor	0.22 mg g-1 h-1	p.m., 21 May, 8.4 klux[1]
Lemna minor	0.28 mg g-1 h-1	noon, 16 Nov, 20.0 klux[1]
Lemna minor	0.32 mg g-1 h-1	a.m., 16 Nov, 21.2 klux[1]
Lemna minor	0.38 mg g-1 h-1	noon, 21 May, 10.9 klux[1]
Lemna minor	0.50 mg g-1 h-1	a.m., 21 May, 16 klux[1]
Lemna minor	0.70 mg g-1 h-1	p.m., 1 July, 128 klux[1]
Lemna minor	0.92 mg g-1 h-1	a.m., 1 July, 64 klux[1]
Lemna minor	1.02 mg g-1 h-1	noon, 1 July, 104 klux[1]
Lemna minor	2.18 mg g-1 h-1	p.m., 30 Sep, 78.0 klux[1]
Lemna minor	2.52 mg g-1 h-1	noon, 30 Sep, 72.0 klux[1]
Lemna minor	2.62 mg g-1 h-1	a.m., 30 Sep, 61.0 klux[1]
Lemna minor	2.67 mg g-1 h-1	p.m., 1 Aug, 101 klux[1]
Lemna minor	3.50 mg g-1 h-1	noon, 1 Aug, 136 klux[1]
Lemna minor	3.80 mg g-1 h-1	a.m., 1 Aug, 74 klux[1]
Lemna minor	5.49 mg g-1 h-1	p.m., 27 Aug, 76.0 klux[1]
Lemna minor	6.46 mg g-1 h-1	noon, 27 Aug, 101 klux[1]
Lemna minor	6.74 mg g-1 h-1	a.m., 27 Aug, 91.0 klux[1]
Lemna minor	6.90 mg g-1 h-1	p.m., 14 Aug, 118 klux[1]
Lemna minor	8.05 mg g-1 h-1	noon, 14 Aug, 108 klux[1]
Lemna minor	8.72 mg g-1 h-1	a.m., 14 Aug, 48.8 klux[1]

Source: 1.Filbin, G.J. and Hough, R.A., Limnol.Oceanogr., 30(2), 322, 1985.

1-338 Kormophytes, macrophytes
P / biomass ratio

Species	Value	Condition
Brasenia schreberi	0.14 % of dry weight	Par Pond, SC, May 69, soft wat, water = 0.008 P ppm, n=4[1]
Ceratophyllum demersum	0.26 % of dry weight	Par Pond, SC, May 69, soft wat, water = 0.008 P ppm, n=4[1]
Eleocharis acicularis	0.24 % of dry weight	Par Pond, SC, May 69, soft wat, water = 0.008 P ppm, n=4[1]
Eleocharis quadrangulata	0.1 % of dry weight	Par Pond, SC, May 69, soft wat, water = 0.008 P ppm, n=1[1]
Free floating	2.11 g/kg dry weight	Amazon, SE = +- 0.295[2]
Hydrocotyle sp.	0.18 % of dry weight	Par Pond, SC, May 69, soft wat, water = 0.008 P ppm, n=4[1]
Juncus effusus	0.27 % of dry weight	Par Pond, SC, May 69, soft wat, water = 0.008 P ppm, n=5[1]
Myriophyllum heterophyllum	0.16 % of dry weight	Par Pond, SC, May 69, soft wat, water = 0.008 P ppm, n=4[1]

Najas guadalupensis	0.15 % of dry weight	Par Pond, SC, May 69, soft wat, water = 0.008 P ppm, n=4[1]
Nelumbo lutea	0.19 % of dry weight	Par Pond, SC, May 69, soft wat, water = 0.008 P ppm, n=2[1]
Nuphar advena	0.4 % of dry weight	Par Pond, SC, May 69, soft wat, water = 0.008 P ppm, n=2[1]
Nymphaea odorata	0.18 % of dry weight	Par Pond, SC, May 69, soft wat, water = 0.008 P ppm, n=5[1]
Panicum hemitonium	0.14 % of dry weight	Par Pond, SC, May 69, soft wat, water = 0.008 P ppm, n=3[1]
Pontederia cordata	0.24 % of dry weight	Par Pond, SC, May 69, soft wat, water = 0.008 P ppm, n=1[1]
Potamogeton diversifolius	0.27 % of dry weight	Par Pond, SC, May 69, soft wat, water = 0.008 P ppm, n=1[1]
Rooted	1.5 g/kg dry weight	Amazon, SE = +- 0.371[2]
Sagittaria latifolia	0.3 % of dry weight	Par Pond, SC, May 69, soft wat, water = 0.008 P ppm, n=2[1]
Scirpus americanus	0.18 % of dry weight	Par Pond, SC, May 69, soft wat, water = 0.008 P ppm, n=5[1]
Typha latifolia	0.14 % of dry weight	Par Pond, SC, May 69, soft wat, water = 0.008 P ppm, n=4[1]
Utricularia inflata	0.12 % of dry weight	Par Pond, SC, May 69, soft wat, water = 0.008 P ppm, n=4[1]

Source: 1.Boyd, C.E. 1970. Arch. Hydrobiol., 67(1), pp 78-85.
2.Howard-Williams, C. and W.J. Junk. 1977. Arch. Hydrobiol., 79(4), pp 446-464.

1-339 Kormophytes, macrophytes
Phosphorus uptake

Species	Value	Condition
Myriophyllum spicatum	0.8 g Productionm-2 y-1	Uptake by shoot, Lake Wingra, Wisconsin[1]
Myriophyllum spicatum	2.2 g Productionm-2 y-1	Uptake by root, Lake Wingra, Wisconsin[1]
Myriophyllum spicatum	3.0 g Productionm-2 y-1	total uptake by whole plant, Lake Wingra, Wisconsin[1]

Source: 1.Smith, C.S. and Adams, M.S. Limnol. Oceanogr., 31(6), 1312, 1986.

1-340 Kormophytes, macrophytes
Production

Species	Value	Condition
Emergents dominating	180 g dw/(year*m**2)	Sphagnum, on hummocks, blancet bog, Engl[1]
Emergents dominating	2500 g dw/(year*m**2)	Typha, Cedar Creek, Minn[2]
Emergents dominating	290 g dw/(year*m**2)	Sphagnum, in pools, Blanket Bog, Engl[1]
Emergents dominating	340 g dw/(year*m**2)	Sphagnum, on lawns, Blanket Bog, Engl[1]
Emergents dominating	738 g dw/(year*m**2)	Carex, Minn wetlands[3]
Floating	2950 g dw/(year*m**2)	Eichhorina, New Orleans, LA., rg = 1500 - 4400[4]
Submergents dominating	155 g dw/(year*m**2)	Chara, Lake Lawrence, hardwater[5]
Submergents dominating	199 g dw/(year*m**2)	Annuals, Lake Lawrence, hardwater[5]
Submergents dominating	595 g dw/(year*m**2)	Scirpus subterminalis, Lake Lawrence, hardwater[5]
Submergents dominating	64 g dw/(year*m**2)	Ruppia, Borax Lake, Calif, saline[6]

Source: 1.Clymo, R.S. and E.J.F. Reddaway. 1974. J. Ecol., 62, pp 191-196.
2.Bray. J.R., D.B. Lawrence and L.C. Pearson. 1959. Oikos, 10, pp 38-49.
3.Bernard, J.M. 1973. Pol. Arch. Hydro biol., 20, pp 207-214.
4.Penfound, W.T. and T.T. Earle. 1948. Ecol. Monogr. 18, pp 447-472.
5.Rich, P.H., R.G. Wetzel and N.V. Thury. 1971. Freshwater Biol. 1, pp 3-21.
6.Wetzel, R.G. Int. Rev. ges. Hydrobiol 49, pp 1-64.

1-341 Kormophytes, macrophytes
Protein / biomass ratio

Species	Value	Condition
Emergent	13 % of organic matter	Range = 3.5 - 26.7 %[1]
Plant with floating leaves	26.5 % of organic matter	Range = 18 - 44 %[1]
Submersed	22 % of organic matter	Range = 7.5 - 34.7 %[1]

Source: 1.Straskraba, M. 1968. Mitt. Int. Ver. Limnol. 14, pp 212-230.

1-342 Kormophytes, macrophytes
Rate of N uptake

Species	Value	Condition
Ceratophyllum demersum	0.01 microgram NO3-N/(mg plant N*hour)	20 cm plant, 4 h start at 0240 cdt, light = 0.1 ly/4 h, 294.9 K[1]
Ceratophyllum demersum	0.015 microgram NO3-N/(mg plant N*hour)	20 cm plant, 4 h start at 2240 cdt, light = 0 ly/4 h, 294.6 K[1]
Ceratophyllum demersum	0.237 microgram NH4-N/(mg plant N*hour)	20 cm plant, 4 h start at 0240 cdt, light = 0.1 ly/4 h, 294.9 K[1]
Ceratophyllum demersum	0.238 microgram NH4-N/(mg plant N*hour)	20 cm plant, 4 h start at 2240 cdt, light = 0 ly/4 h, 294.6 K[1]
Ceratophyllum demersum	0.257 microgram NH4-N/(mg plant N*hour)	20 cm plant, 4 h start at 1040 cdt, light = 192 ly/4 h, 297.4 K[1]
Ceratophyllum demersum	0.291 microgram NH4-N/(mg plant N*hour)	20 cm plant, 4 h start at 1440 cdt, light = 57 ly/4 h, 297.4 K[1]
Ceratophyllum demersum	0.298 microgram NH4-N/(mg plant N*hour)	20 cm plant, 4 h start at 1840 cdt, light = 1.6 ly/4 h, 296 K[1]
Ceratophyllum demersum	0.32 microgram NO3-N/(mg plant N*hour)	20 cm plant, 4 h start at 1840 cdt, light = 1.6 ly/4 h, 296 K[1]
Ceratophyllum demersum	0.394 microgram NH4-N/(mg plant N*hour)	20 cm plant, 4 h start at 0640 cdt, light = 23 ly/4 h, 294.7 K[1]
Ceratophyllum demersum	0.399 microgram NO3-N/(mg plant N*hour)	20 cm plant, 4 h start at 1440 cdt, light = 57 ly/4 h, 297.4 K[1]
Ceratophyllum demersum	0.504 microgram NO3-N/(mg plant N*hour)	20 cm plant, 4 h start at 0640 cdt, light = 23 ly/4 h, 294.7 K[1]
Ceratophyllum demersum	0.591 microgram NO3-N/(mg plant N*hour)	20 cm plant, 4 h start at 1040 cdt, light = 192 ly/4 h, 297.4 K[1]
Eichhornia crassipes	11.4 g N/(plant*day)	Static water, 25 % hoaglands solution, NC[2]
Eichhornia crassipes	18.4 g N/(plant*day)	Flowing water, 25 % hoaglands solution, NC[2]
Eichhornia crassipes	19.8 g N/(plant*day)	Static water, 50 % hoaglands solution, NC[2]
Eichhornia crassipes	20.8 g N/(plant*day)	Flowing water, 50 % hoaglands solution, NC[2]
Eichhornia crassipes	5.3 g N/(plant*day)	Static water, 10 % hoaglands solution, NC[2]
Eichhornia crassipes	6.6 g N/(plant*day)	Static water, sewage effluent, NC[2]
Eichhornia crassipes	9.9 g N/(plant*day)	Flowing water, 10 % hoaglands solution, NC[2]
Eichhornia crassipes	96 kg N/ton of macrophyte	Macrophyte in aeration pond[3]

Source: 1.Toetz, D.W. 1971. Limnol. Oceanogr. 16(5), pp 819-821.
2.Rigers, H.E. and D.E. Davis. 1972. Weed Sci. 20(5), pp 423-428.
3.Cornwell, D.A., J. Zoltek Jr., C.D. Patrinely, S. Furman and J.I. Kim. 1977. Water Pollut. Control, 49(1), pp 57-65.

1-343 Kormophytes, macrophytes
Rate of P uptake

Species	Value	Condition
Eichhornia crassipes	1.1 g P/(plant*day)	Static water, 10 % hoaglands solution, NC[1]
Eichhornia crassipes	1.6 g P/(plant*day)	Static water, sewage effluent, NC[1]
Eichhornia crassipes	1.7 g P/(plant*day)	Flowing water, 10 % hoaglands solution, NC[1]
Eichhornia crassipes	18 kg PO4-P/ton of macrophyte	Macrophyte in aeration pond[2]
Eichhornia crassipes	2.1 g P/(plant*day)	Static water, 25 % hoaglands solution, NC[1]
Eichhornia crassipes	2.5 g P/(plant*day)	Flowing water, 25 % hoaglands solution, NC[1]
Eichhornia crassipes	3.1 g P/(plant*day)	Static water, 50 % hoaglands solution, NC[1]
Eichhornia crassipes	3.3 g P/(plant*day)	Flowing water, 50 % hoaglands solution, NC[1]
Zostera marina	0.0016 microgram PO4-P/plant	Dark, 3000 min, from upper water to lower water[3]
Zostera marina	0.057 microgram PO4-P/plant	Dark, 3000 min, from upper water to plant, net transport[3]
Zostera marina	0.088 microgram PO4-P/plant	Dark, 3000 min, from lower water to roots and rhizomes[3]
Zostera marina	0.088 microgram PO4-P/plant	Light, 3000 min, from upper water to lower water[3]
Zostera marina	0.098 microgram PO4-P/plant	Dark, 3000 min, from lower water to plant[3]
Zostera marina	0.116 microgram PO4-P/plant	Dark, 3000 min, from upper water to leaves and stems[3]
Zostera marina	0.16 microgram PO4-P/plant	Dark, 3000 min, from lower water to upper water[3]
Zostera marina	0.16 microgram PO4-P/plant	Light, 3000 min, from lower water to roots and rhizomes[3]
Zostera marina	0.216 microgram PO4-P/plant	Dark, 3000 min, from upper water to plant[3]
Zostera marina	0.256 microgram PO4-P/plant	Dark, 3000 min, from lower water to plant, net transport[3]
Zostera marina	0.34 microgram PO4-P/plant	Light, 3000 min, from lower water to plant[3]
Zostera marina	0.392 microgram PO4-P/plant	Light, 3000 min, from upper water to plant, net transport[3]
Zostera marina	0.66 microgram PO4-P/plant	Light, 3000 min, from lower water to upper water[3]
Zostera marina	0.88 microgram PO4-P/plant	Light, 3000 min, from upper water to leaves and stems[3]
Zostera marina	0.912 microgram PO4-P/plant	Light, 3000 min, from lower water to plant, net transport[3]
Zostera marina	0.964 microgram PO4-P/plant	Light, 3000 min, from upper water to plant[3]

Source: 1.Rigers, H.E. and D.E. Davis. 1972. Weed Sci. 20(5), pp 423-428.
2.Cornwell, D.A., J. Zoltek Jr., C.D. Patrinely, S. Furman and J.I. Kim. 1977. Water Pollut. Control, 49(1), pp 57-65.
3.McRoy, C.P. and R.J. Barsdate. 1970. Limnol. Oceanogr. 15(1) pp 6-13.

1-344 Kormophytes, macrophytes
Release of dissolved organic C

Species	Value	Condition
Lemna minor	0.80 %	% of initial internal 14C h-1, 5 Oct 1976, dark[1]
Lemna minor	0.90 %	% of initial internal 14C h-1, 15 Sep 1976, dark[1]
Lemna minor	1.04 %	% of initial internal 14C h-1, 4 Sep 1976, dark[1]
Lemna minor	1.82 %	% of initial internal 14C h-1, 5 Oct 1976, light[1]
Lemna minor	3.30 %	% of initial internal 14C h-1, 18 Aug 1976, dark[1]
Lemna minor	3.92 %	% of initial internal 14C h-1, 15 Sep 1976, light[1]
Lemna minor	4.05 %	% of initial internal 14C h-1, 4 Sep 1976, light[1]
Lemna minor	5.52 %	% of initial internal 14C h-1, 18 Aug 1976, light[1]

Source: 1.Filbin, G.J. and Hough, R.A., Limnol.Oceanogr., 30(2), 322, 1985.

1-345 Kormophytes, macrophytes
S / biomass ratio

Species	Value	Condition
Brasenia schreberi	0.11 % of dry weight	Par Pond, SC, May 69, soft wat, water = 1.03 S ppm, n=4, SD = +- 0.01[1]
Ceratophyllum demersum	0.3 % of dry weight	Par Pond, SC, May 69, soft wat, water = 1.03 S ppm, n=4, SD = +- 0.02[1]
Eleocharis acicularis	0.28 % of dry weight	Par Pond, SC, May 69, soft wat, water = 1.03 S ppm, n=4, SD = +- 0.04[1]
Eleocharis quadrangulata	0.15 % of dry weight	Par Pond, SC, May 69, soft wat, water = 1.03 S ppm, n=1[1]
Hydrocotyle sp.	0.16 % of dry weight	Par Pond, SC, May 69, soft wat, water = 1.03 S ppm, n=4, SD = +- 0.01[1]
Juncus effusus	0.26 % of dry weight	Par Pond, SC, May 69, soft wat, water = 1.03 S ppm, n=5, SD = +- 0.01[1]
Myriophyllum heterophyllum	0.24 % of dry weight	Par Pond, SC, May 69, soft wat, water = 1.03 S ppm, n=4, SD = +- 0.02[1]
Najas guadalupensis	0.28 % of dry weight	Par Pond, SC, May 69, soft wat, water = 1.03 S ppm, n=4, SD = +- 0.01[1]
Nelumbo lutea	0.16 % of dry weight	Par Pond, SC, May 69, soft wat, water = 1.03 S ppm, n=2, SD = +- 0.01[1]
Nuphar advena	0.32 % of dry weight	Par Pond, SC, May 69, soft wat, water = 1.03 S ppm, n=2[1]
Nymphaea odorata	0.14 % of dry weight	Par Pond, SC, May 69, soft wat, water = 1.03 S ppm, n=5, SD = +- 0.02[1]
Panicum hemitonium	0.23 % of dry weight	Par Pond, SC, May 69, soft wat, water = 1.03 S ppm, n=3, SD = +- 0.02[1]
Pontederia cordata	0.22 % of dry weight	Par Pond, SC, May 69, soft wat, water = 1.03 S ppm, n=1[1]
Potamogeton diversifolius	0.5 % of dry weight	Par Pond, SC, May 69, soft wat, water = 1.03 S ppm, n=1[1]
Sagittaria latifolia	0.15 % of dry weight	Par Pond, SC, May 69, soft wat, water = 1.03 S ppm, n=2[1]
Scirpus americanus	0.59 % of dry weight	Par Pond, SC, May 69, soft wat, water = 1.03 S ppm, n=5, SD = +- 0.02[1]

Typha latifolia	0.15 % of dry weight	Par Pond, SC, May 69, soft wat, water = 1.03 S ppm, n=4, SD = +- 0.02[1]
Utricularia inflata	0.26 % of dry weight	Par Pond, SC, May 69, soft wat, water = 1.03 S ppm, n=4, SD = +- 0.02[1]

Source: 1.Boyd, C.E. 1970. Arch. Hydrobiol., 67(1), pp 78-85.

1-346 Kormophytes, macrophytes Si / biomass ratio

Species	Value	Condition
Free floating	12.35 g/kg dry weight	Amazon, SE = +- 2.752[1]
Rooted	17.2 g/kg dry weight	Amazon, SE = +- 4.785[1]

Source: 1.Howard-Williams, C. and W.J. Junk. 1977. Arch. Hydrobiol., 79(4), pp 446-464.

1-347 Kormophytes, macrophytes Zn / biomass ratio

Species	Value	Condition
Brasenia schreberi	267 mg/kg dry weight	Par Pond, SC, May 69, soft wat, water = 0.008 Zn ppm, n=4, SD = +- 94[1]
Ceratophyllum demersum	100 mg/kg dry weight	Par Pond, SC, May 69, soft wat, water = 0.008 Zn ppm, n=4, SD = +- 24[1]
Eleocharis acicularis	68 mg/kg dry weight	Par Pond, SC, May 69, soft wat, water = 0.008 Zn ppm, n=4, SD = +- 7[1]
Eleocharis quadrangulata	45 mg/kg dry weight	Par Pond, SC, May 69, soft wat, water = 0.008 Zn ppm, n=2[1]
Hydrocotyle sp.	31 mg/kg dry weight	Par Pond, SC, May 69, soft wat, water = 0.008 Zn ppm, n=3, SD = +- 5[1]
Hydrocotyle sp.	53 mg/kg dry weight	Par Pond, SC, May 69, soft wat, water = 0.008 Zn ppm, n=4, SD = +- 8[1]
Myriophyllum heterophyllum	54 mg/kg dry weight	Par Pond, SC, May 69, soft wat, water = 0.008 Zn ppm, n=4, SD = +- 14[1]
Najas guadalupensis	48 mg/kg dry weight	Par Pond, SC, May 69, soft wat, water = 0.008 Zn ppm, n=4, SD = +- 17[1]
Nelumbo lutea	50 mg/kg dry weight	Par Pond, SC, May 69, soft wat, water = 0.008 Zn ppm, n=4, SD = +- 8[1]
Nuphar advena	50 mg/kg dry weight	Par Pond, SC, May 69, soft wat, water = 0.008 Zn ppm, n=2[1]
Nymphaea odorata	32 mg/kg dry weight	Par Pond, SC, May 69, soft wat, water = 0.008 Zn ppm, n=5, SD = +- 5[1]
Pontederia cordata	67 mg/kg dry weight	Par Pond, SC, May 69, soft wat, water = 0.008 Zn ppm, n=1[1]
Potamogeton diversifolius	60 mg/kg dry weight	Par Pond, SC, May 69, soft wat, water = 0.008 Zn ppm, n=1[1]
Sagittaria latifolia	46 mg/kg dry weight	Par Pond, SC, May 69, soft wat, water = 0.008 Zn ppm, n=2[1]
Typha latifolia	30 mg/kg dry weight	Par Pond, SC, May 69, soft wat, water = 0.008 Zn ppm, n=4, SD = +- 5[1]
Utricularia inflata	108 mg/kg dry weight	Par Pond, SC, May 69, soft wat, water = 0.008 Zn ppm, n=4, SD = +- 27[1]

Source: 1.Boyd, C.E. 1970. Arch. Hydrobiol., 67(1), pp 78-85.

1-348 Kormophytes, plants
Al / biomass ratio

Species	Value	Condition
Alfalfa	0.0025 %	Atomic composition[1]

Source: 1.Berthet, P. 1963. In: Doeksen, J. and J. van der Drift. (eds); Soil Organism. North-Holland Publ. Co,Amsterdam.

1-349 Kormophytes, plants
Ash / biomass ratio

Species	Value	Condition
Angiosperms	0.57 % of dry weight	Wood, 9 species[1]
Angiosperms	6.39 % of dry weight	Leaves, 7 species[1]
Brasenia schreberi	7.6 % of dry weight	None[2]
Bryophytes	3.28 % of dry weight	8 species[1]
Ceratophyllum demersum	14.9 % of dry weight	None[2]
Eleocharis acicularis	11.2 % of dry weight	None[2]
Ferns	7.71 % of dry weight	6 species[1]
Gymnosperms	0.166 % of dry weight	Wood, 5 species[1]
Gymnosperms	2.41 % of dry weight	Leaves and twigs, 3 species[1]
Myriophyllum heterophyllum	12.2 % of dry weight	None[2]
Najas guadalupensis	12.8 % of dry weight	None[2]
Nelumbo lutea	8.8 % of dry weight	None[2]
Nuphar advena	10.6 % of dry weight	None[2]
Nymphaea odorata	8.1 % of dry weight	None[2]
Typha latifolia	7.5 % of dry weight	None[2]
Utricularia inflata	14 % of dry weight	None[2]

Source: 1.Baumeister, W. 1958. In; Ruhland, W. (ed); Encyclopedia of Plant Physiology, 4, p 5.
2.Boyd, C.E. and C.P. Goodyear. 1971. Arch. Hydrobiol., 69(2), pp 256-270.

1-350 Kormophytes, plants
Average width of leaves

Species	Value	Condition
Coniferous forest	0.2 m	Input value[1]
Grass	0.005 m	Input value[1]
Maize	0.05 m	Input value[1]

Source: 1.Goudrian, J. 1977. Crop Micrometeorology: A Simulation Study. Chap 4. Pudoc. Wageningen.

1-351 Kormophytes, plants
Br / biomass ratio

Species	Value	Condition
Alfalfa	0.0001 %	Atomic composition[1]

Source: 1.Berthet, P. 1963. In: Doeksen, J. and J. van der Drift. (eds); Soil Organism. North-Holland Publ. Co, Amsterdam.

1-352 Kormophytes, plants
C / biomass ratio

Species	Value	Condition
Alfalfa	11.34 %	Atomic composition[1]
Angiosperms	45.4 % of dry weight	Mean elementary composition[2]
Mean plant	45 % of dry weight	Mean elementary composition[2]

Source: 1.Berthet, P. 1963. In: Doeksen, J. and J. van der Drift. (eds); Soil Organism. North-Holland Publ. Co. Amsterdam.
2.Bowen, H.J.M. 1966. Trace Elements in Biochemistry, Ac. Press.

1-353 Kormophytes, plants
Ca / biomass ratio

Species	Value	Condition
Alfalfa	0.58 %	Atomic composition[1]

Source: 1.Berthet, P. 1963. In: Doeksen, J. and J. van der Drift. (eds); Soil Organism. North-Holland Publ. Co. Amsterdam.

1-354 Kormophytes, plants
Caloric content

Species	Value	Condition
Alpine	4619 cal/g dry matter	Roots or rhizomes[1]
Alpine	4713 cal/g ash free dry weight	Roots or rhizomes[1]
Alpine	5000 cal/g dry matter	Shoots or leaves[1]
Alpine	5102 cal/g ash free dry weight	Shoots or leaves[1]
Brasenia schreberi	4.03 kcal/g dry weight	None[2]
Ceratophyllum demersum	3.91 kcal/g dry weight	None[2]
Conifers	4927 cal/g dry weight	Shoots or leaves[3]
Corn	3.5 kcal/g	Per g harvest weight[4]
Crude sugar	3.7 kcal/g	Per g harvest weight[4]
Cymodocea sp.	2841 cal/g dry weight	Tropical seagras, sheath[5]
Cymodocea sp.	3239 cal/g dry weight	Shoots or leaves[5]
Cymodocea sp.	3240 cal/g dry weight	Tropical seagras, leaf, green, rg = 3228 - 3253[5]
Cymodocea sp.	3253 cal/g dry weight	Tropical seagras, leaf, brown[5]
Cymodocea sp.	3389 cal/g dry weight	Tropical seagras, rhizome, rg = 3266 - 3421[5]
Cymodocea sp.	3999 cal/g ash free dry weight	Shoots or leaves[5]
Deciduous forest	4759 cal/g dry weight	Shoots or leaves[3]
Eleocharis acicularis	4.26 kcal/g dry weight	None[2]
Enhalus sp.	2515 cal/g dry weight	Tropical seagras, leaf, distal half[5]
Enhalus sp.	2890 cal/g dry weight	Shoots or leaves[5]
Enhalus sp.	3078 cal/g dry weight	Tropical seagras, leaf, proximal half, rg = 3032 - 3124[5]
Enhalus sp.	3109 cal/g dry weight	Tropical seagras, sheath[5]
Enhalus sp.	3552 cal/g dry weight	Tropical seagras, rhizome, rg = 3522 - 3581[5]

Enhalus sp.	3802 cal/g ash free dry weight	Shoots or leaves[5]
Halodule sp.	2658 cal/g dry weight	Tropical seagras, sheath, rg = 2507 - 2808[5]
Halodule sp.	3134 cal/g dry weight	Tropical seagras, rhizome, rg = 3018 - 3249[5]
Halodule sp.	3219 cal/g dry weight	Shoots or leaves[5]
Halodule sp.	3220 cal/g dry weight	Tropical seagras, leaf, rg = 3193 - 3246[5]
Halodule sp.	4350 cal/g ash free dry weight	Shoots or leaves[5]
Halophila minor	1452 cal/g dry weight	Tropical seagras, whole plant, rg = 1407 - 1495[5]
Halophila ovalis	1954 cal/g dry weight	Tropical seagras, whole plant[5]
Halophila ovalis	2603 cal/g dry weight	Tropical seagras, leaf[5]
Halophila ovalis	2814 cal/g dry weight	Shoots or leaves[5]
Halophila ovalis	2814 cal/g dry weight	Tropical seagras, leaf[5]
Halophila ovalis	2959 cal/g dry weight	Tropical seagras, rhizome[5]
Halophila ovalis	3124 cal/g dry weight	Tropical seagras, whole plant[5]
Halophila ovalis	6507 cal/g ash free dry weight	Shoots or leaves[5]
Halophila spinulosa	2814 cal/g dry weight	Shoots or leaves[5]
Halophila spinulosa	2814 cal/g dry weight	Tropical seagras, leaf[5]
Halophila spinulosa	3629 cal/g ash free dry weight	Shoots or leaves[5]
Herbs	4056 cal/g dry weight	Shoots or leaves, old field herbs[6]
Hydrotrida carolinensis	4.06 kcal/g dry weight	None[2]
Mangrove	4034 cal/g dry weight	Roots or rhizomes, red mangrove[6]
Mangrove	4190 cal/g dry weight	Shoots or leaves, red mangrove[6]
Myriophyllum heterophyllum	3.96 kcal/g dry weight	None[2]
Najas guadalupensis	3.92 kcal/g dry weight	None[2]
Nelumbo lutea	4.23 kcal/g dry weight	None[2]
Nuphar advena	4.32 kcal/g dry weight	None[2]
Nymphaea odorata	4.18 kcal/g dry weight	None[2]
Potato	0.9 kcal/g	Per g harvest weight[4]
Rice	3.6 kcal/g	Per g harvest weight[4]
Soybean	4 kcal/g	Per g harvest weight[4]
Syringodium sp.	2178 cal/g dry weight	Shoots or leaves[5]
Syringodium sp.	2178 cal/g dry weight	Tropical seagras, leaf[5]
Syringodium sp.	2792 cal/g dry weight	Tropical seagras, sheath[5]
Syringodium sp.	2966 cal/g dry weight	Tropical seagras, rhizome[5]
Syringodium sp.	3630 cal/g ash free dry weight	Shoots or leaves[5]
Thalassia sp.	1990 cal/g dry weight	Shoots or leaves[7]
Thalassia sp.	2618 cal/g ash free dry weight	Shoots or leaves[7]
Tropical forest	4041 cal/g dry weight	Roots or rhizomes, wet tropical forest[8]

Tropical forest	4065 cal/g dry weight	Shoots or leaves, wet tropical forest[8]
Typha latifolia	4.26 kcal/g dry weight	None[2]
Utricularia inflata	4.02 kcal/g dry weight	None[2]
Wheat	3.3 kcal/g	Per g harvest weight[4]
Zostera sp.	2857 cal/g dry weight	Roots or rhizomes[9]
Zostera sp.	3369 cal/g dry weight	Shoots or leaves[9]
Zostera sp.	3571 cal/g ash free dry weight	Roots or rhizomes[9]
Zostera sp.	4211 cal/g ash free dry weight	Shoots or leaves[9]

Source: 1.Hadley, E.B. and L.C. Bliss. 1964. Ecol. Monogr., 34, pp 331-357.
2.Boyd, C.E. and C.P. Goodyear. 1971. Arch. Hydrobiol., 69(2), pp 256-270.
3.Ovington, J.D. and D. Heitkamp, 1960. J. Ecol. 48, pp 639-646.
4.United States Department of Agriculture. 1963. Agriculture Handbook, 8.
5.Birch, W.R. 1975. J. appl. Ecol., 12(1), pp 201-212.
6.Golley, F.B. 1961. Ecology 42, pp 581-584.
7.Burkholder, P.R., L.M. Burkholder and J.A. Rivero. 1969. Bull. Torreya bot. Club, 86, pp 88-93.
8.Golley, F.B. 1969. Ecology 50, pp 517-519.
9.McRoy, C.P. 1970. J. Fish. Res. bd. Can. 27, pp 1811-1829.

1-355 Kormophytes, plants
Cl / biomass ratio

Species	Value	Condition
Alfalfa	0.07 %	Atomic composition[1]
Cymodocea sp.	17 % of ash weight	Tropical seagrasses, leaves, also Halodule sp.[2]
Cymodocea sp.	25 % of ash weight	Tropical seagrasses, rhizomes, also Halodule sp.[2]
Enhalus sp.	26 % of ash weight	Tropical seagrasses, leaves, also Syringodium sp.[2]
Enhalus sp.	27 % of ash weight	Tropical seagrasses, rhizomes, also Syringodium sp.[2]
Halophila decipiens	15 % of ash weight	Tropical seagrasses[2]
Halophila spinulosa	32 % of ash weight	Tropical seagrasses[2]

Source: 1.Berthet, P. 1963. In: Doeksen, J. and J. van der Drift. (eds); Soil Organism. North-Holland Publ. Co, Amsterdam.
2.Birch, W.R. 1975. J. appl. Ecol., 12(1), pp 201-212.

1-356 Kormophytes, plants
Compensation point

Species	Value	Condition
Artemisia herba-alba	40.5 degrees celcius	Upper temp at comp.point of CO2, May 2, n = 4, non-irrigated[1]
Hammada scoparia	53.2 degrees celcius	Upper temp at comp.point of CO2, Aug 16, n = 13, non-irrigated[1]
Hammada scoparia	53.4 degrees celcius	Upper temp at comp.point of CO2, Aug 16, n = 11, irrigated[1]
Prunus armeniaca	48 degrees celcius	Upper temp at comp.point of CO2, July 5, n = 13, irrigated[1]
Prunus armeniaca	48.6 degrees celcius	Upper temp at comp.point of CO2, June 27, n=11, non-irrigated[1]
Reaumuria negevensis	46.4 degrees celcius	Upper temp at comp.point of CO2, May 16, n = 4, non-irrigated[1]
Reaumuria negevensis	47.6 degrees celcius	Upper temp at comp.point of CO2, Aug 16, n = 4, irrigated[1]
Salsola inermis	47.6 degrees celcius	Upper temp at comp.point of CO2, Aug 30, n = 4, non-irrigated[1]

Zygophyllum dumosum	43.9 degrees celcius	Upper temp at comp.point of CO2, May 16, n = 11, non-irrigated[1]
Zygophyllum dumosum	47.8 degrees celcius	Upper temp at comp.point of CO2, Aug 30, n = 5, irrigated[1]

Source: 1.Lange, O.L., E.-D. Schulze, M. Evenari, L. Kappen and U. Buschbom. 1974. Oecologia 17, pp 97-110.

1-357 Kormophytes, plants
Cu / biomass ratio

Species	Value	Condition
Alfalfa	0.0003 %	Atomic composition[1]

Source: 1.Berthet, P. 1963. In: Doeksen, J. and J. van der Drift. (eds); Soil Organism. North-Holland Publ. Co, Amsterdam.

1-358 Kormophytes, plants
Depth containing 63 % of the roots

Species	Value	Condition
Barley	32.7 cm	Density = 50 plants/m2, wet weight = 0.556 kg/m2[1]
Barley	36.4 cm	Density = 800 plants/m2, wet weight = 0.727 kg/m2[1]
Barley	38.4 cm	Low soil moisture, wet weight = 0.963 kg/m2[1]
Barley	57.1 cm	High soil moisture, wet weight = 0.755 kg/m2[1]
Barley	70.4 cm	Density = 200 plants/m2, wet weight = 0.742 kg/m2[1]
Brussel sprouts	70.4 cm	By root number profile, age = 180 d, wet weight = 8.64 kg/m2[2]
Cabbage	14.1 cm	Sandy loam, age = 73 d, wet weight = 5.6 kg/m2[3]
Cabbage	15.1 cm	Sandy loam, age = 73 d, wet weight = 6.8 kg/m2[3]
Cabbage	17.3 cm	Sandy loam, age = 77 d, wet weight = 7.6 kg/m2[3]
Cabbage	17.8 cm	Sandy loam, age = 86 d, wet weight = 8.9 kg/m2[3]
Cabbage	2.8 cm	Sandy loam, age = 41 d, wet weight = 0.69 kg/m2[3]
Cabbage	2.8 cm	Sandy loam, age = 41 d, wet weight = 0.72 kg/m2[3]
Cabbage	41.5 cm	Sandy loam, age = 77 d, wet weight = 9.6 kg/m2[3]
Cabbage	5.5 cm	Sandy loam, age = 56 d, wet weight = 2.6 kg/m2[3]
Cabbage	5.5 cm	Sandy loam, age = 63 d, wet weight = 2.6 kg/m2[3]
Cabbage	6.2 cm	Sandy loam, age = 64 d, wet weight = 3.0 kg/m2[3]
Cabbage	6.4 cm	Sandy loam, age = 86 d, wet weight = 8.2 kg/m2[3]
Cabbage	7.1 cm	Sandy loam, age = 57 d, wet weight = 2.8 kg/m2[3]
Cabbage	9.6 cm	Sandy loam, age = 92 d, wet weight = 11.3 kg/m2[3]
Carrot	11.8 cm	High fertility, age = 59 d, wet weight = 5.5 kg/m2[4]
Carrot	12.5 cm	Low fertility, age = 59 d, wet weight = 3.6 kg/m2[4]
Carrot	14.6 cm	Peat, by soil injection of P-32[5]
Carrot	19.7 cm	Low fertility, age = 79 d, wet weight = 6.6 kg/m2[4]
Carrot	20.9 cm	Low fertility, age = 38 d, wet weight = 1.5 kg/m2[4]
Carrot	22.9 cm	High fertility, age = 79 d, wet weight = 8.9 kg/m2[4]
Carrot	23.1 cm	Peat, by soil injection of P-32[5]
Carrot	23.8 cm	Peat, by soil injection of P-32[5]

Carrot	24.8 cm	High fertility, age = 135 d, wet weight = 12.1 kg/m2[4]
Carrot	24.8 cm	High fertility, age = 38 d, wet weight = 0.9 kg/m2[4]
Carrot	33.9 cm	Low fertility, age = 135 d, wet weight = 9.9 kg/m2[4]
Carrot	38.8 cm	Peat, by soil injection of P-32, wet weight = 8.25 kg/m2[5]
Celeriac	19.6 cm	Age = 108 d[6]
Cotton	31.3 cm	Irrigated, by weight of roots[7]
Cotton	93.5 cm	Irrigated, by numbers of roots[7]
Endive	24.1 cm	Age = 65 d[6]
Hyparrhenia rufa	49 cm	Age = 120 d, by dry weight[8]
Kohl-rabi	21.3 cm	Age = 87 d[6]
Leek	19.8 cm	Age = 136 d[6]
Lettuce	10.4 cm	High fertility, age = 57 d, wet weight = 4.3 kg/m2[4]
Lettuce	16.1 cm	Low fertility, age = 75 d, wet weight = 3.1 kg/m2[4]
Lettuce	16.5 cm	Low fertility, age = 78 d, wet weight = 1.6 kg/m2[4]
Lettuce	16.8 cm	Irrigated, age = 42 d, wet weight = 6.1 kg/m2[9]
Lettuce	24.1 cm	Non-irrigated, age = 42 d, wet weight = 2.8 kg/m2[9]
Lettuce	32.9 cm	Non-irrigated, age = 42 d, wet weight = 2.8 kg/m2[9]
Maize	13.9 cm	High n1 defoliations, 2 y, by injection of Rb-86[10]
Maize	14 cm	High n3 defoliations, 2 y, by injection of Rb-86[10]
Maize	14.1 cm	Low n3 defoliations, 2 y, by root dry weight[10]
Maize	16.2 cm	Low n3 defoliations, 2 y, by injection of Rb-86[10]
Maize	21.6 cm	Low n1 defoliations, 2 y, by injection of Rb-86[10]
Maize	44 cm	Age = 100 d, wet weight = 1.26 kg/m2[11]
Maize	44.8 cm	Control, age = 100 d, wet weight = 0.672 kg/m2[12]
Maize	63.1 cm	Phosphate added, age = 100 d, wet weight = 0.915 kg/m2[12]
Oats	40.6 cm	Age = 105 d, by root weight[7]
Onion	10.9 cm	Silt, by soil injection of P-32[5]
Onion	11.4 cm	Age = 106 d[6]
Onion	11.7 cm	Sandy loam, age = 102 d, wet weight = 5.1 kg/m2[9]
Onion	11.7 cm	Silt, by soil injection of P-32[5]
Onion	16.1 cm	Silt, by soil injection of P-32[5]
Onion	19.1 cm	Peat, by soil injection of P-32, wet weight = 6.31 kg/m2[5]
Onion	5.8 cm	Peat, by soil injection of P-32[5]
Onion	5.9 cm	Peat, by soil injection of P-32[5]
Onion	5.9 cm	Silt, by soil injection of P-32, wet weight = 3.05 kg/m2[5]
Onion	9.5 cm	Peat, by soil injection of P-32[5]
Parsley	27.1 cm	Age = 188 d[6]
Pasture	16.3 cm	Age = 120 d, by dry weight[13]
Pea	32.9 cm	Age = 116 d[6]
Pea	35.7 cm	Irrigated, age = 105 d, by weight[7]

Pea	64.5 cm	Irrigated, age = 105 d, by numbers[7]
Pepper	10.8 cm	By soil injection of P-32[5]
Pepper	14.1 cm	By soil injection of P-32, wet weight = 2.70 kg/m2[5]
Pepper	14.2 cm	By soil injection of P-32[5]
Pepper	14.3 cm	By soil injection of P-32[5]
Rye grass	13.2 cm	Age = 120 d, by dry weight[14]
Rye grass	13.5 cm	Low n1 defoliations, 2 y, by root dry weight[10]
Rye grass	13.9 cm	High n1 defoliations, 2 y, by root dry weight[10]
Rye grass	16.4 cm	Age = 120 d, by total length, single plants[14]
Rye grass	16.7 cm	High n3 defoliations, 2 y, by root dry weight[10]
Rye grass	17.8 cm	Age = 120 d, by total number, single plants[14]
Rye grass	19.3 cm	Low n1 defoliations, 2 y, by root number[10]
Rye grass	20.1 cm	High n3 defoliations, 2 y, by root number[10]
Rye grass	20.5 cm	Low n3 defoliations, 2 y, by root number[10]
Rye grass	21.6 cm	High n1 defoliations, 2 y, by root number[10]
Sorghum sp.	135.1 cm	Increasing soil moisture (2), age = 83 d[15]
Sorghum sp.	144.6 cm	Increasing moisture (1), age = 83 d[15]
Sorghum sp.	59.9 cm	Increasing soil moisture (4), age = 83 d[15]
Sorghum sp.	92.1 cm	Increasing soil moisture (3), age = 83 d[15]
Spring barley	11.1 cm	Cultivation depth = 7.5 cm, age = 34 d, wet weight = 0.154 kg/m2[16]
Spring barley	11.4 cm	Cultivation depth = 15 cm, age = 34 d, wet weight = 0.148 kg/m2[16]
Spring barley	11.4 cm	Cultivation depth = 23 cm, age = 34 d, wet weight = 0.160 kg/m2[16]
Spring barley	12.2 cm	Cultivation depth = 15 cm, age = 69 d, wet weight = 0.929 kg/m2[16]
Spring barley	14.3 cm	Cultivation depth = 0 cm, age = 69 d, wet weight = 0.890 kg/m2[16]
Spring barley	14.6 cm	Cultivation depth = 23 cm, age = 69 d, wet weight = 0.929 kg/m2[16]
Spring barley	20.2 cm	Cultivation depth = 7.5 cm, age = 69 d, wet weight = 0.907 kg/m2[16]
Spring barley	7.2 cm	Cultivation depth = 15 cm, age = 97 d[16]
Spring barley	8.7 cm	Cultivation depth = 0 cm, age = 94 d[16]
Spring barley	9.1 cm	Cultivation depth = 7.5 cm, age = 95 d[16]
Spring barley	9.6 cm	Cultivation depth = 0 cm, age = 34 d, wet weight = 0.114 kg/m2[16]
Tomato	20.8 cm	Irrigated, by weight[7]
Tomato	21.7 cm	Age = 103 d[6]
Tomato	46.1 cm	Irrigated, by numbers[7]
Vetch	45.7 cm	Irrigated, age = 105 d, by weight of roots[7]
Vetch	54.9 cm	Irrigated, age = 105 d, by numbers of roots[7]
Winter wheat	38 cm	Age = 180 d, ranger[17]

Winter wheat	43 cm	Age = 180 d, culture 3[17]
Winter wheat	44.8 cm	Age = 180 d, culture 25[17]

Source: 1.Kirby, E.J.M. and O. Rackham. 1971. J. appl. Ecol. 8, pp 919-924.
2.Verheij, E.W.H. 1970. Neth. J. Agric. Sci. 18, pp 89-104.
3.Gerwits, A. 1972. Rep. Natn. Veg. Res. St. 1971. pp 36.
4.Page, E.R. and A. Gerwitz. 1969. J. Sci. Fd. Agric. 20, pp 85-90.
5.Hamnes, J.K. and J.F. Bartz. 1963. Agron. J. 55, pp 329-333.
6.Hoesslin, R.V. 1955. Festscr. 25 Jaehr. Bestehen Gaern. Hochschulstud., Deutschland. pp 47-62.
7.Bloodworth, M.E., C.A. Burleson and W.R. Cowey. 1958. Agron. J., 50, pp 317-320.
8.Daubenmire, R. 1972. J. appl. Ecol., 9, pp 11-23.
9.Gerwitz, A. and E.R. Page. 1974. J. Appl. Ecol., 11, pp 773-781.
10.Newbould, P., F.B. Ellis, B.T. Barnes, K.R. Howse, T.E. Williams and E.A. Garwood. 1970. Rep. agric.Res. Coun. Let combe lab., 1969, pp 45-48.
11.Foth, H.D. 1962. Agron. J., 54, pp 49-52.
12.Fehrenbacher, J.B. and J.D. Alexander. 1955. Agron. J., 47, pp 468-472.
13.Schuurman, J.J. and M.A.J. Goedewaagen. 1971. Methods for Examination of Roots Systems and Roots. 2e. Wageningen.
14.Rappaport, J. 1938. Meded. Landb-Hoogesch. Opzoek. Stns. Gent. 6, pp 121-204.
15.Nakayama, F.S. and C.H.M.v. Bavel. 1963. Agron. J. 55, pp 271-274.
16.Newbould, P., F.B. Ellis, B.T. Barnes, K.R. Howse, T.G. Elliott and N. Squires. 1970. Rep. agric. Re. Coun. Letcombe Lab., 1969, pp 40-43.
17.Newbould, P., F.B. Ellis, B.T. Barnes, K.R. Howse and F.G.H. Lupton. 1970. Rep. agric. Res. Coun. Lecombe Lab., 1969, pp 38-40.

1-359 Kormophytes, plants
Drag coefficient of the leaves

Species	Value	Condition
Maize	0.2 Input value, also coniferous forest	Input value, also coniferous forest and grass[1]

Source: 1.Goudrian, J. 1977. Crop Micrometeorology: A Simulation Study. Chap 4. Pudoc. Wageningen.

1-360 Kormophytes, plants
Exchange coefficient for heat

Species	Value	Condition
Coniferous forest	0.055 m**2/sec	Equivalent heat flux = 1 J/(m**2)*sec, min value[1]
Coniferous forest	0.122 m**2/sec	Equivalent heat flux = 10 J/(m**2)*sec, min value[1]
Coniferous forest	0.26 m**2/sec	Equivalent heat flux = 100 J/(m**2)*sec, min value[1]
Grass	0.00019 m**2/sec	Equivalent heat flux = 1 J/(m**2)*sec, min value[1]
Grass	0.00041 m**2/sec	Equivalent heat flux = 10 J/(m**2)*sec, min value[1]
Grass	0.00089 m**2/sec	Equivalent heat flux = 100 J/(m**2)*sec, min value[1]
Maize	0.0067 m**2/sec	Equivalent heat flux = 1 J/(m**2)*sec, min value[1]
Maize	0.0145 m**2/sec	Equivalent heat flux = 10 J/(m**2)*sec, min value[1]
Maize	0.0312 m**2/sec	Equivalent heat flux = 100 J/(m**2)*sec, min value[1]

Source: 1.Goudrian, J. 1977. Crop Micrometeorology: A Simulation Study. Chap 4. Pudoc. Wageningen.

1-361 Kormophytes, plants
Extinction factor for wind

Species	Value	Condition
Coniferous forest	1.88 m	Calculated value[1]
Grass	3.54 m	Calculated value[1]
Maize	2.55 m	Calculated value[1]

Source: 1.Goudrian, J. 1977. Crop Micrometeorology: A Simulation Study. Chap 4. Pudoc. Wageningen.

1-362 Kormophytes, plants
Fe / biomass ratio

Species	Value	Condition
Alfalfa	0.027 %	Atomic composition[1]

Source: 1.Berthet, P. 1963. In: Doeksen, J. and J. van der Drift. (eds); Soil Organism. North-Holland Publ. Co, Amsterdam.

1-363 Kormophytes, plants
H / biomass ratio

Species	Value	Condition
Alfalfa	8.72 % of dry weight	Atomic composition[1]
Angiosperms	5.5 % of dry weight	Mean elementary composition[2]
Plant	5 % of dry weight	Mean elementary composition, mean value[2]

Source: 1.Berthet, P. 1963. In: Doeksen, J. and J. van der Drift. (eds); Soil Organism. North-Holland Publ. Co, Amsterdam.
2.Bowen, H.J.M. 1966. Trace Elements in Biochemistry, Ac. Press.

1-364 Kormophytes, plants
Height of canopy

Species	Value	Condition
Coniferous forest	10 m	Input value[1]
Coniferous forest	4.33 m	Calc value, (height of can)-(zero plane displacement)[1]
Grass	0.053 m	Calc value, (height of can)-(zero plane displacement)[1]
Grass	0.2 m	Input value[1]
Maize	0.84 m	Calc value, (height of can)-(zero plane displacement)[1]
Maize	2.5 m	Input value[1]

Source: 1.Goudrian, J. 1977. Crop Micrometeorology: A Simulation Study. Chap 4. Pudoc. Wageningen.

1-365 Kormophytes, plants
I / biomass ratio

Species	Value	Condition
Alfalfa	0 %	Atomic composition[1]

Source: 1.Berthet, P. 1963. In: Doeksen, J. and J. van der Drift. (eds); Soil Organism. North-Holland Publ. Co, Amsterdam.

1-366 Kormophytes, plants
K / biomass ratio

Species	Value	Condition
Alfalfa	0.17 %	Atomic composition[1]
Brasenia schreberi	0.99 % of dry weight	None[2]
Ceratophyllum demersum	4.01 % of dry weight	None[2]
Cymodocea sp.	10 % of ash weight	Tropical seagrasses, rhizomes, also Halodule sp.[3]
Cymodocea sp.	5 % of ash weight	Tropical seagrasses, leaves, also Halodule sp.[3]
Eleocharis acicularis	2.86 % of dry weight	None[2]
Enhalus sp.	10.8 % of ash weight	Tropical seagrasses, rhizomes, also Syringodium sp.[3]
Enhalus sp.	2.5 % of ash weight	Tropical seagrasses, leaves, also Syringodium sp.[3]
Halophila decipiens	1.7 % of ash weight	Tropical seagrasses[3]

Halophila spinulosa	4.1 % of ash weight	Tropical seagrasses[3]
Myriophyllum heterophyllum	1.25 % of dry weight	None[2]
Najas guadalupensis	3.49 % of dry weight	None[2]
Nelumbo lutea	2.27 % of dry weight	None[2]
Nuphar advena	1.88 % of dry weight	None[2]
Nymphaea odorata	1.28 % of dry weight	None[2]
Typha latifolia	2.65 % of dry weight	None[2]
Utricularia inflata	1.98 % of dry weight	None[2]

Source: 1.Berthet, P. 1963. In: Doeksen, J. and J. van der Drift. (eds); Soil Organism. North-Holland Publ. Co, Amsterdam.
2.Boyd, C.E. and C.P. Goodyear. 1971. Arch. Hydrobiol., 69(2), pp 256-270.
3.Birch, W.R. 1975. J. appl. Ecol., 12(1), pp 201-212.

1-367 Kormophytes, plants
Leaf area density

Species	Value	Condition
Coniferous forest	0.2 m**2 leaf/m**3 air	Input value[1]
Grass	25 m**2 leaf/m**3 air	Input value[1]
Maize	1.2 m**2 leaf/m**3 air	Input value[1]

Source: 1.Goudrian, J. 1977. Crop Micrometeorology: A Simulation Study. Chap 4. Pudoc. Wageningen.

1-368 Kormophytes, plants
Leaf area index

Species	Value	Condition
Coniferous forest	2 m**2 leaf/m**2 ground	Input value[1]
Grass	5 m**2 leaf/m**2 ground	Input value[1]
Maize	3 m**2 leaf/m**2 ground	Input value[1]

Source: 1.Goudrian, J. 1977. Crop Micrometeorology: A Simulation Study. Chap 4. Pudoc. Wageningen.

1-369 Kormophytes, plants
Leaf biomass

Species	Value	Condition
Abies spp.	14.9 t/ha	Evergreen, coniferous, 9 samples, Japan[1]
Betula ermanii	2.7 t/ha	Deciduous, broadleaved, 10 samples, Japan[1]
Betula spp.	2.7 t/ha	Deciduous, broadleaved, 21 samples, Japan[1]
Camellia japonica	8.6 t/ha	Evergreen, broadleaved, 4 samples, Japan[1]
Castanopsis cuspidata	8.5 t/ha	Evergreen, broadleaved, 11 samples, Japan[1]
Chamaecyparis obtusa	13.2 t/ha	Evergreen, coniferous, 7 samples, Japan[1]
Fagus crenata	3.1 t/ha	Deciduous, broadleaved, 86 samples, Japan[1]
Larix leptolepis	3.9 t/ha	Deciduous, coniferous, 15 samples, Japan[1]
Pinus densiflora	6.6 t/ha	Evergreen, coniferous, 108 samples, Japan[1]
Pinus thunbergii	9.9 t/ha	Evergreen, coniferous, 12 samples, Japan[1]

Source: 1.Satoo, T. 1970. in: Reichle, D.E. (ed.); Analysis of Temperate Forest Ecosystem. pp 55-72. Springer-Verlag. Berlin.

1-370 Kormophytes, plants
Leaf resistance

Species	Value	Condition
Acer rubrum	0.2 min/cm	Free water surface = 0.0367 min/cm, min res, boundary layer[1]
Adenostoma fasciculatum	0.71 min/cm	Free water surface = 0.0367 min/cm, min res, boundary layer[1]
Adenostoma fasciculatum	1 min/cm	Free water surface = 0.0367 min/cm, min res, boundary layer[1]
Ammophila breviligulata	0.05 min/cm	Free water surface = 0.0367 min/cm, min res, boundary layer[1]
Arctostaphylos uva-ursi	0.0834 min/cm	Free water surface = 0.0367 min/cm, min res, boundary layer[1]
Avicennia tomentosa	0.04 min/cm	Free water surface = 0.0367 min/cm, min res, boundary layer[1]
Betula papyrifera	0.1 min/cm	Free water surface = 0.0367 min/cm, min res, boundary layer[1]
Ceanothus crassifolius	0.35 min/cm	Free water surface = 0.0367 min/cm, min res, boundary layer[1]
Chamaedaphne calyculata	0.2 min/cm	Free water surface = 0.0367 min/cm, min res, boundary layer[1]
Coccoloba uvifera	0.04 min/cm	Free water surface = 0.0367 min/cm, min res, boundary layer[1]
Conocarpus erectus	0.05 min/cm	Free water surface = 0.0367 min/cm, min res, boundary layer[1]
Gossypium sp.	0.02 min/cm	Free water surface = 0.0367 min/cm, res of boundary layer, lower[1]
Gossypium sp.	0.084 min/cm	Free water surface = 0.0367 min/cm, res of boundary layer, mean[1]
Gossypium sp.	0.148 min/cm	Free water surface = 0.0367 min/cm, res of boundary layer, upper[1]
Laguncularia racemosa	0.03 min/cm	Free water surface = 0.0367 min/cm, min res, boundary layer[1]
Pinus resinosa	0.5 min/cm	Free water surface = 0.0367 min/cm, min res, boundary layer[1]
Pinus strobus	0.67 min/cm	Free water surface = 0.0367 min/cm, min res, boundary layer[1]
Populus grandidentata	0.0667 min/cm	Free water surface = 0.0367 min/cm, min res, boundary layer[1]
Populus tremuloides	0.05 min/cm	Free water surface = 0.0367 min/cm, res of boundary layer, wet site[1]
Populus tremuloides	0.0667 min/cm	Free water surface = 0.0367 min/cm, res of boundary layer, dry site[1]
Pteridium aquilinum	3.3 min/cm	Free water surface = 0.0367 min/cm, min res, boundary layer[1]
Quercus dumosa	0.196 min/cm	Free water surface = 0.0367 min/cm, min res, boundary layer[1]
Rhizophora mangle	0.04 min/cm	Free water surface = 0.0367 min/cm, min res, boundary layer[1]
Thuja occidentalis	1.25 min/cm	Free water surface = 0.0367 min/cm, res of boundary layer, shade[1]

Thuja occidentalis	1.25 min/cm	Free water surface = 0.0367 min/cm, res of boundary layer, sune[1]
Vitis sp.	0.25 min/cm	Free water surface = 0.0367 min/cm, min res, boundary layer,[1]

Source: 1.Collier, B.D., G.W. Cox, A.W. Johnson and P.C. Miller. 1974. Dynamic Ecology. Prentice-Hall. New Jersey.

1-371 Kormophytes, plants
Mg / biomass ratio

Species	Value	Condition
Alfalfa	0.08 %	Atomic composition[1]

Source: 1.Berthet, P. 1963. In: Doeksen, J. and J. van der Drift. (eds); Soil Organism. North-Holland Publ. Co, Amsterdam.

1-372 Kormophytes, plants
Mixing length

Species	Value	Condition
Coniferous forest	1.13 m	Calculated value[1]
Grass	0.016 m	Calculated value[1]
Maize	0.23 m	Calculated value[1]

Source: 1.Goudrian, J. 1977. Crop Micrometeorology: A Simulation Study. Chap 4. Pudoc. Wageningen.

1-373 Kormophytes, plants
Mn / biomass ratio

Species	Value	Condition
Alfalfa	0.0004 %	Atomic composition[1]

Source: 1.Berthet, P. 1963. In: Doeksen, J. and J. van der Drift. (eds); Soil Organism. North-Holland Publ. Co, Amsterdam.

1-374 Kormophytes, plants
N / biomass ratio

Species	Value	Condition
Alfalfa	0.825 %	Atomic composition[1]
Angiosperms	3.3 % of dry weight	Mean elementary composition[2]
Plant	3 % of dry weight	Mean elementary composition, mean value[2]

Source: 1.Berthet, P. 1963. In: Doeksen, J. and J. van der Drift. (eds); Soil Organism. North-Holland Publ. Co, Amsterdam.
2.Bowen, H.J.M. 1966. Trace Elements in Biochemistry, Ac. Press.

1-375 Kormophytes, plants
Na / biomass ratio

Species	Value	Condition
Alfalfa	0.03 %	Atomic composition[1]
Cymodocea sp.	11.8 % of ash weight	Tropical seagrasses, leaves, also Halodule sp.[2]
Cymodocea sp.	14.1 % of ash weight	Tropical seagrasses, rhizomes, also Halodule sp.[2]
Enhalus sp.	15.5 % of ash weight	Tropical seagrasses, leaves, also Syringodium sp.[2]
Enhalus sp.	20 % of ash weight	Tropical seagrasses, rhizomes, also Syringodium sp.[2]
Halophila decipiens	9.6 % of ash weight	Tropical seagrasses[2]
Halophila spinulosa	18.6 % of ash weight	Tropical seagrasses[2]

Source: 1.Berthet, P. 1963. In: Doeksen, J. and J. van der Drift. (eds); Soil Organism. North-Holland Publ. Co, Amsterdam.
2.Birch, W.R. 1975. J. appl. Ecol., 12(1), pp 201-212.

1-376 Kormophytes, plants
Nitrogen fixation

Species	Value	Condition
Arachis hypogaea	124 kg N/(ha*year)	Leguminous plant[1]
Aristida argentea	1500 microgram N/(m**2*hour)	During daylight, Jan 75, non-legume, rg = 0 - 3000[2]
Burkea africana	1500 microgram N/(m**2*hour)	During daylight, Jan 75, legume, rg = 1200 - 2200[2]
Cajanus cajan	224 kg N/(ha*year)	Leguminous plant[1]
Calopogonium mucunoides	202 kg N/(ha*year)	Leguminous plant[1]
Canavalia ensiformis	49 kg N/(ha*year)	Leguminous plant[1]
Cassia mimosoides	4820 microgram N/(m**2*hour)	During daylight, Jan 75, legume, rg = 240 - 9600[2]
Centrosema pubescens	259 kg N/(ha*year)	Leguminous plant[1]
Cicer arietinum	103 kg N/(ha*year)	Leguminous plant[1]
Cyamopsis tetragonolobus	130 kg N/(ha*year)	Leguminous plant[1]
Desmodium intortum	897 kg N/(ha*year)	Leguminous plant, also Desmodium canum[1]
Dichapetalum cymosum	820 microgram N/(m**2*hour)	During daylight, Jan 75, non-legume, rg = 160 - 1480[2]
Digitaria eriantha	220 microgram N/(m**2*hour)	During daylight, Jan 75, non-legume, rg = 0 - 440[2]
Elephantorrhiza sp.	1540 microgram N/(m**2*hour)	During daylight, Jan 75, legume, rg = 360 - 2720[2]
Enterolobium saman	150 kg N/(ha*year)	Leguminous plant[1]
Eragrostis pallens	1640 microgram N/(m**2*hour)	During daylight, Jan 75, non-legume, rg = 160 - 3120[2]
Glycine max	103 kg N/(ha*year)	Leguminous plant[1]
Hyperthelia dissoluta	960 microgram N/(m**2*hour)	During daylight, Jan 75, non-legume, rg = 40 - 1880[2]
Indegofera sordida	2140 microgram N/(m**2*hour)	During daylight, Jan 75, legume, rg = 520 - 3760[2]
Indigoferia oxytropis	39500 microgram N/(m**2*hour)	During daylight, Jan 75, legume, rg = 1920 - 77200[2]
Lens culinaris	101 kg N/(ha*year)	Leguminous plant[1]
Lespedeza sp.	95 kg N/(ha*year)	Leguminous plant[1]
Leucaena glauca	277 kg N/(ha*year)	Leguminous plant[1]
Lotonosis bainesii	62 kg N/(ha*year)	Leguminous plant[1]
Lotus corniculatus	116 kg N/(ha*year)	Leguminous plant[1]
Lupinus sp.	176 kg N/(ha*year)	Leguminous plant[1]
Medicago sp.	199 kg N/(ha*year)	Leguminous plant[1]
Melilotus sp.	199 kg N/(ha*year)	Leguminous plant[1]
Mikanea cordata	120 kg N/(ha*year)	Leguminous plant[1]
Ochna pulchra	2020 microgram N/(m**2*hour)	During daylight, Jan 75, non-legume, rg = 800 - 3240[2]
Parinari capense	3640 microgram N/(m**2*hour)	During daylight, Jan 75, non-legume, rg = 40 - 7240[2]

Phaseolus atrapurpurea	291 kg N/(ha*year)	Leguminous plant[1]
Phaseolus aureus	202 kg N/(ha*year)	Leguminous plant, green grain[1]
Phaseolus aureus	61 kg N/(ha*year)	Leguminous plant, mung[1]
Pisum sativum	65 kg N/(ha*year)	Leguminous plant[1]
Pueraria phaseoloides	99 kg N/(ha*year)	Leguminous plant[1]
Pygmaeothamnus zeyheri	960 microgram N/(m**2*hour)	During daylight, Jan 75, non-legume, rg = 80 - 1840[2]
Rhynchosia monophylla	5000 microgram N/(m**2*hour)	During daylight, Jan 75, legume, rg = 920 - 9080[2]
Sesbania cannabina	542 kg N/(ha*year)	Leguminous plant[1]
Stylosanthes sp.	124 kg N/(ha*year)	Leguminous plant[1]
Tephrosia lupiniifolia	14071 microgram N/(m**2*hour)	During daylight, Jan 75, legume, rg = 4240 - 23902[2]
Terminalia sericea	420 microgram N/(m**2*hour)	During daylight, Jan 75, non-legume, 40 - 800[2]
Thalassia sp.	9.45 microgram N/(mg plant-N*hour)	In light, with epiphytes, Redfish Bay, rg = 2.4 - 16.5[3]
Trigonella foenumgraecum	342 kg N/(ha*year)	Leguminous plant[1]
Vicia faba	210 kg N/(ha*year)	Leguminous plant[1]
Vigna sinensis	198 kg N/(ha*year)	Leguminous plant[1]

Source: 1.Nutman, P S.(ed). Symbiotic Nitrogen fixation in Plants, International Biological Program. Cambridge. University Press, Cambridge 1975, vol. 5, pp 211.
2.Hirst, S.M. 1975. Savanna Ecosystem project. Progress report 1974/75.
3.Goering, J.J. and P.L. Parker. 1972. Limnol. Oceanogr., 17(2), pp 320-323.

1-377 Kormophytes, plants
O / biomass ratio

Species	Value	Condition
Alfalfa	77.9 %	Atomic composition[1]
Angiosperms	41 % of dry weight	Mean elementary composition[2]
Plant	40 % of dry weight	Mean elementary composition, mean value[2]

Source: 1.Berthet, P. 1963. In: Doeksen, J. and J. van der Drift. (eds); Soil Organism. North-Holland Publ. Co, Amsterdam.
2.Bowen, H.J.M. 1966. Trace Elements in Biochemistry, Ac. Press.

1-378 Kormophytes, plants
P / biomass ratio

Species	Value	Condition
Alfalfa	0.71 %	Atomic composition[1]
Brasenia schreberi	0.14 % of dry weight	None[2]
Ceratophyllum demersum	0.26 % of dry weight	None[2]
Cymodocea sp.	0.6 % of ash weight	Tropical seagrasses, leaves, also Halodule sp.[3]
Cymodocea sp.	0.6 % of ash weight	Tropical seagrasses, rhizomes, also Halodule sp.[3]
Eleocharis acicularis	0.24 % of dry weight	None[2]
Enhalus sp.	0.7 % of ash weight	Tropical seagrasses, leaves, also Syringodium sp.[3]
Enhalus sp.	0.8 % of ash weight	Tropical seagrasses, rhizomes, also Syringodium sp.[3]
Halophila decipiens	0.5 % of ash weight	Tropical seagrasses[3]

Halophila spinulosa	0.4 % of ash weight	Tropical seagrasses[3]
Myriophyllum heterophyllum	0.16 % of dry weight	None[2]
Najas guadalupensis	0.15 % of dry weight	None[2]
Nelumbo lutea	0.19 % of dry weight	None[2]
Nuphar advena	0.4 % of dry weight	None[2]
Nymphaea odorata	0.18 % of dry weight	None[2]
Typha latifolia	0.14 % of dry weight	None[2]
Utricularia inflata	0.12 % of dry weight	None[2]

Source: 1.Berthet, P. 1963. In: Doeksen, J. and J. van der Drift. (eds); Soil Organism. North-Holland Publ. Co, Amsterdam.
2.Boyd, C.E. and C.P. Goodyear. 1971. Arch. Hydrobiol., 69(2), pp 256-270.
3.Birch, W.R. 1975. J. appl. Ecol., 12(1), pp 201-212.

1-379 Kormophytes, plants Photosynthesis

Species	Value	Condition
Amaranthus edulis	58 mg CO2/(dm**2*hour)	SD = +- 5[1]
Beta vulgaris	24 mg CO2/(dm**2*hour)	SD = +- 2[1]
Gossypium barbadense	29 mg CO2/(dm**2*hour)	SD = +- 2[1]
Gossypium hirsutum	30 mg CO2/(dm**2*hour)	SD = +- 3[1]
Helianthus annus	36 mg CO2/(dm**2*hour)	SD = +- 4[1]
Solidago virgaurea	17.9 mg CO2/(dm**2*hour)	Mean of 6, shaded habitat, saturation value[2]
Solidago virgaurea	2.31 mg CO2/(dm**2*hourE4*-erg/(cm**2*sec	Mean of 6, exposed habitat, initial slope[2]
Solidago virgaurea	25.5 mg CO2/(dm**2*hour)	Mean of 6, exposed habitat, saturation value, rg = +- 2.0[2]
Solidago virgaurea	3.1 mg CO2/(dm**2*hourE4-*erg/(cm**2*sec)	Mean of 6, shaded habitat, initial slope[2]
Sorghum vulgare	55 mg CO2/(dm**2*hour)	SD = +- 6[1]
Zea mays	53 mg CO2/(dm**2*hour)	SD = +- 6[1]

Source: 1.El Sharkawy, M.A., R.S. Loomis and W.A. Williams. 1967. Physiol. Plant., 20, pp 171-186.

1-380 Kormophytes, plants Production

Species	Value	Condition
Betula maximowicziana	0.2 t/(ha*year)	Net prod, temp. deciduous broadleaved forest, Japan, estimat,[1]
Betula maximowicziana	0.3 t/(ha*year)	Net prod, temp. deciduous broadleaved forest, Japan, estimat,[1]
Betula maximowicziana	0.5 t/(ha*year)	Net prod, temp. deciduous broadleaved forest, Japan, estimat,[1]
Betula maximowicziana	1 t/(ha*year)	Net prod, temp. deciduous broadleaved forest, Japan, estimat, branch[1]

Betula maximowicziana	1.1 t/(ha*year)	Net prod, temp. deciduous broadleaved forest, Japan, estimat, branch[1]
Betula maximowicziana	1.8 t/(ha*year)	Net prod, temp. deciduous broadleaved forest, Japan, estimat, leaf[1]
Betula maximowicziana	2.2 t/(ha*year)	Net prod, temp. deciduous broadleaved forest, Japan, estimat, leaf[1]
Betula maximowicziana	2.6 t/(ha*year)	Net prod, temp. deciduous broadleaved forest, Japan, estimat, leaf[1]
Betula maximowicziana	2.8 t/(ha*year)	Net prod, temp. deciduous broadleaved forest, Japan, estimat, stem[1]
Betula maximowicziana	2.9 t/(ha*year)	Net prod, temp. deciduous broadleaved forest, Japan, estimat, stem[1]
Betula maximowicziana	3.7 t/(ha*year)	Net prod, temp. deciduous broadleaved forest, Japan, estimat, stem[1]
Betula maximowicziana	5.5 t/(ha*year)	Net prod, temp. deciduous broadleaved forest, Japan, estimat, total[1]
Betula maximowicziana	6.1 t/(ha*year)	Net prod, temp. deciduous broadleaved forest, Japan, estimat, total[1]
Betula maximowicziana	6.2 t/(ha*year)	Net prod, temp. deciduous broadleaved forest, Japan, estimat, total[1]
Betula maximowicziana	6.3 t/(ha*year)	Net prod, temp. deciduous broadleaved forest, Japan, estimat, total[1]
Betula maximowicziana	7.2 t/(ha*year)	Net prod, temp. deciduous broadleaved forest, Japan, estimat, total[1]
Betula maximowicziana	7.5 t/(ha*year)	Net prod, temp. deciduous broadleaved forest, Japan, estimat, total[1]
Fagus crenata	1.8 t/(ha*year)	Branch, net prod, temp. deciduous broadleaved forest, Japan, estimat[1]
Fagus crenata	2.5 t/(ha*year)	Stem, net prod, temp. deciduous broadleaved forest, Japan, estimat[1]
Fagus crenata	3 t/(ha*year)	Leaf, net prod, temp. deciduous broadleaved forest, Japan, estimat[1]
Fagus crenata	7.3 t/(ha*year)	Total tree, net prod, temp. deciduous broadleaved forest, Japan,[1]
Populus davidiana	0.9 t/(ha*year)	Branch, net prod, temp. deciduous broadleaved forest, Japan, estimat[1]
Populus davidiana	12.4 t/(ha*year)	Total stand, net prod, temp. deciduous broadleaved forest, Japan,[1]
Populus davidiana	2.2 t/(ha*year)	Leaf, net prod, temp. deciduous broadleaved forest, Japan, estimat[1]
Populus davidiana	3.6 t/(ha*year)	Undegrowth, net prod, temp. deciduous broadleaved forest, Japan,[1]
Populus davidiana	5.6 t/(ha*year)	Stem, net prod, temp. deciduous broadleaved forest, Japan, estimat[1]
Populus davidiana	8.7 t/(ha*year)	Total tree, net prod, temp. deciduous broadleaved forest, Japan,[1]

Source: 1.Satoo, T. 1970. in: Reichle, D.E. (ed.); Analysis of Temperate Forest Ecosystem. pp 55-72. Springer-Verlag. Berlin.

1-381 Kormophytes, plants
Production, net primary

These obeservations are from a place in Japan with the same altitudinal zone.

Species	Value	Condition
Abies sachalinensis	13 t/ha	Net aboveground prod, Japan, age=26 y[1]
Larix leptolepis	16.5 t/ha	Net aboveground prod, Japan, age=21 y[1]
Picea abies	11.4 t/ha	Net aboveground prod, Japan, age=46 y[1]
Picea abies	11.7 t/ha	Net aboveground prod, Japan, age=46 y[1]
Picea abies	12.4 t/ha	Net aboveground prod, Japan, age=47 y[1]
Picea abies	7.3 t/ha	Net aboveground prod, Japan, age=45 y[1]
Populus davidiana	8.7 t/ha	Net aboveground prod., Japan, age=40 y[1]

Source: 1.Satoo, T. 1970. in: Reichle, D.E. (ed.); Analysis of Temperate Forest Ecosystem. pp 55-72. Springer-Verlag. Berlin.

1-382 Kormophytes, plants
Protein / biomass ratio

Species	Value	Condition
Brasenia schreberi	10.9 % of dry weight	None[1]
Ceratophyllum demersum	17.1 % of dry weight	None[1]
Eleocharis acicularis	14.1 % of dry weight	None[1]
Eleocharis quadrangulata	11.8 % of dry weight	Crude protein=nitrogen*6.25[1]
Hydrotrida carolinensis	10.5 % of dry weight	None[1]
Juncus effusus	7.8 % of dry weight	Crude protein=nitrogen*6.25[1]
Myriophyllum heterophyllum	13.5 % of dry weight	None[1]
Najas guadalupensis	14.4 % of dry weight	None[1]
Nelumbo lutea	12.1 % of dry weight	None[1]
Nuphar advena	21.6 % of dry weight	None[1]
Nymphaea odorata	14.6 % of dry weight	None[1]
Orontium aquaticum	21.2 % of dry weight	Crude protein=nitrogen*6.25[1]
Pontederia cordata	8.8 % of dry weight	Crude protein=nitrogen*6.25[1]
Scirpus americanus	7.6 % of dry weight	Crude protein=nitrogen*6.25[1]
Scirpus validus	5.8 % of dry weight	Crude protein=nitrogen*6.25[1]
Typha latifolia	4 % of dry weight	None[1]
Utricularia inflata	11.4 % of dry weight	None[1]

Source: 1.Boyd, C.E. and C.P. Goodyear. 1971. Arch. Hydrobiol., 69(2), pp 256-270.

1-383 Kormophytes, plants
Rb / biomass ratio

Species	Value	Condition
Alfalfa	0.0005 %	Atomic composition[1]

Source: 1.Berthet, P. 1963. In: Doeksen, J. and J. van der Drift. (eds); Soil Organism. North-Holland Publ. Co, Amsterdam.

1-384 Kormophytes, plants
Resistance to heat

Species	Value	Condition
Coniferous forest	182 sec/m	Equiv heat flux=1 J/((m**2)*sec), trans betw canopy and ref height[1]
Coniferous forest	38 sec/m	Equiv heat flux=100 J/((m**2)*sec), trans betw canopy and ref height[1]
Coniferous forest	82 sec/m	Equiv heat flux=10 J/((m**2)*sec), trans betw canopy and ref height[1]
Grass	1053 sec/m	Equiv heat flux=1 J/((m**2)*sec), trans betw canopy and ref height[1]
Grass	225 sec/m	Equiv heat flux=100 J/((m**2)*sec), trans betw canopy and ref height[1]
Grass	488 sec/m	Equiv heat flux=10 J/((m**2)*sec), trans betw canopy and ref height[1]
Maize	172 sec/m	Equiv heat flux=10 J/((m**2)*sec), trans betw canopy and ref height[1]
Maize	373 sec/m	Equiv heat flux=1 J/((m**2)*sec), trans betw canopy and ref height[1]
Maize	80 sec/m	Equiv heat flux=100 J/((m**2)*sec), trans betw canopy and ref height[1]

Source: 1.Goudrian, J. 1977. Crop Micrometeorology: A Simulation Study. Chap 4. Pudoc. Wageningen.

1-385 Kormophytes, plants
Respiration rate

Species	Value	Condition
Amaranthus edulis	0 mg CO2/(dm**2*hour)	In light, zero CO2[1]
Amaranthus edulis	5.2 mg CO2/(dm**2*hour)	In dark, normal air, SD = +- 0.8[1]
Amaranthus edulis	5.8 mg CO2/(dm**2*hour)	In dark, zero CO2, SD = +- 1.0[1]
Barley	1 mg CO2/(g*hour)	Specific dark respiration, less than[2]
Barley	12.5 % of daily photosynthesis	Except may with higher values, Japan, dark, rg = 5-20[2]
Beta vulgaris	2.4 mg CO2/(dm**2*hour)	In light, zero CO2, SD = +- 0.3[1]
Beta vulgaris	2.6 mg CO2/(dm**2*hour)	In dark, normal air, SD = +- 0.4[1]
Beta vulgaris	4.9 mg CO2/(dm**2*hour)	In dark, zero CO2, SD = +- 0.8[1]
Gossypium barbadense	1.7 mg CO2/(dm**2*hour)	In light, zero CO2, SD = +- 1.1[1]
Gossypium barbadense	3.5 mg CO2/(dm**2*hour)	In dark, normal air, SD = +- 0.9[1]
Gossypium barbadense	6.1 mg CO2/(dm**2*hour)	In dark, zero CO2, SD = +- 0.9[1]
Gossypium hirsutum	2.2 mg CO2/(dm**2*hour)	In light, zero CO2, SD = +- 0.3[1]
Gossypium hirsutum	3.2 mg CO2/(dm**2*hour)	In dark, normal air, SD = +- 0.9[1]
Gossypium hirsutum	5.6 mg CO2/(dm**2*hour)	In dark, zero CO2, SD = +- 0.8[1]
Helianthus annus	3.1 mg CO2/(dm**2*hour)	In light, zero CO2, SD = +- 0.9[1]
Helianthus annus	5 mg CO2/(dm**2*hour)	In dark, normal air, SD = +- 1.3[1]
Helianthus annus	6.1 mg CO2/(dm**2*hour)	In dark, zero CO2, SD = +- 1.2[1]
Sorghum vulgare	0 mg CO2/(dm**2*hour)	In light, zero CO2[1]
Sorghum vulgare	1.6 mg CO2/(dm**2*hour)	In dark, normal air, SD = +- 0.2[1]
Sorghum vulgare	2.8 mg CO2/(dm**2*hour)	In dark, zero CO2, SD = +- 0.1[1]
Zea mays	0 mg CO2/(dm**2*hour)	In light, zero CO2[1]
Zea mays	2.9 mg CO2/(dm**2*hour)	In dark, normal air, SD = +- 0.3[1]
Zea mays	4.1 mg CO2/(dm**2*hour)	In dark, zero CO2, SD = +- 0.7[1]

Source: 1.El Sharkawy, M.A., R.S. Loomis and W.A. Williams. 1967. Physiol. Plant., 20, pp 171-186. 2.Fortescue, J.A.C. and G.G. Marten. 1970. in: Richle, D.E. (ed.); Analysis of Temperate Forest Ecosystems. pp 173-198. Springer-Verlag. Berlin.

1-386 Kormophytes, plants
Roughness length

Species	Value	Condition
Coniferous forest	1.27 m	Calculated value[1]
Grass	0.018 m	Calculated value[1]
Maize	0.26 m	Calculated value[1]

Source: 1.Goudrian, J. 1977. Crop Micrometeorology: A Simulation Study. Chap 4. Pudoc. Wageningen.

1-387 Kormophytes, plants
Roughness length / height of canopy

Species	Value	Condition
Coniferous forest	0.127 Calculated value	Calculated value[1]
Grass	0.0916 Calculated value	Calculated value[1]
Maize	0.105 Calculated value	Calculated value[1]

Source: 1.Goudrian, J. 1977. Crop Micrometeorology: A Simulation Study. Chap 4. Pudoc. Wageningen.

1-388 Kormophytes, plants
S / biomass ratio

Species	Value	Condition
Alfalfa	0.1 %	Atomic composition[1]
Angiosperms	0.44 % of dry weight	Mean elementary composition[2]
Mean plant	0.5 % of dry weight	Mean elementary composition[2]

Source: 1.Berthet, P. 1963. In: Doeksen, J. and J. van der Drift. (eds); Soil Organism. North-Holland Publ. Co, Amsterdam.
2.Bowen, H.J.M. 1966. Trace Elements in Biochemistry, Ac. Press.

1-389 Kormophytes, plants
Si / biomass ratio

Species	Value	Condition
Alfalfa	0.093 %	Atomic composition[1]

Source: 1.Berthet, P. 1963. In: Doeksen, J. and J. van der Drift. (eds); Soil Organism. North-Holland Publ. Co, Amsterdam.

1-390 Kormophytes, plants
Sn / biomass ratio

Species	Value	Condition
Alfalfa	0 %	Atomic composition[1]

Source: 1.Berthet, P. 1963. In: Doeksen, J. and J. van der Drift. (eds); Soil Organism. North-Holland Publ. Co, Amsterdam.

1-391 Kormophytes, plants
Temperature gradient

Species	Value	Condition
Coniferous forest	0.015 celcius/m	Equivalent heat flux = 1 J/(m**2)*sec, F080[1]
Coniferous forest	0.066 celcius/m	Equivalent heat flux = 10 J/(m**2)*sec, F080[1]
Coniferous forest	0.3 celcius/m	Equivalent heat flux = 100 J/(m**2)*sec, F080[1]
Grass	19.5 celcius/m	Equivalent heat flux = 10 J/(m**2)*sec, F080[1]
Grass	4.2 celcius/m	Equivalent heat flux = 1 J/(m**2)*sec, F080[1]
Grass	90 celcius/m	Equivalent heat flux = 100 J/(m**2)*sec, F080[1]
Maize	0.12 celcius/m	Equivalent heat flux = 1 J/(m**2)*sec, F080[1]
Maize	0.56 celcius/m	Equivalent heat flux = 10 J/(m**2)*sec, F080[1]
Maize	2.56 celcius/m	Equivalent heat flux = 100 J/(m**2)*sec, F080[1]

Source: 1.Goudrian, J. 1977. Crop Micrometeorology: A Simulation Study. Chap 4. Pudoc. Wageningen.

1-392 Kormophytes, plants
Transpiration

Species	Value	Condition
Alisma plantago	1.7	Ratio max transpiration/min transpiration, aquatic[1]
Avena sativa	112 g water/(g dry matter)	Total dw = 349, root dw = 93, shoot dw = 256 g/pot, Israel, n=6, Aug[2]
Avena sativa	170 g water/(g dry matter)	Total dw = 27.3, root dw = 17.0, shoot dw = 10.3 g/pot, Israel, n=4[2]
Avena sativa	194 g water/(g dry matter)	Total dw = 367, root dw = 87, shoot dw = 280 g/pot, Israel, n=2[2]
Avena sativa	211 g water/(g dry matter)	Total dw = 105, root dw = 20, shoot dw = 85 g/pot, Israel, n=6, Mar[2]
Avena sativa	221 g water/(g dry matter)	Total dw = 414, root dw = 73, shoot dw = 341 g/pot, Israel, n=2[2]
Avena sativa	233 g water/(g dry matter)	Total dw = 213, root dw = 57, shoot dw = 156 g/pot, Israel, n=2, Oct[2]
Avena sativa	235 g water/(g dry matter)	Total dw = 175, root dw = 52, shoot dw = 123 g/pot, Israel, n=2, Oct[2]
Avena sativa	79 g water/(g dry matter)	Total dw = 254, root dw = 79, shoot dw = 175 g/pot, Israel, n=8[2]
Avena sterilis	111 g water/(g dry matter)	Total dw = 72, root dw = 29, shoot dw = 43 g/pot, Israel, n=3[2]
Avena sterilis	124 g water/(g dry matter)	Total dw = 140, root dw = 65, shoot dw = 75 g/pot, Israel, n=10, Mar[2]
Avena sterilis	129 g water/(g dry matter)	Total dw = 141, root dw = 45, shoot dw = 96 g/pot, Israel, n=3[2]
Avena sterilis	154 g water/(g dry matter)	Total dw = 263, root dw = 63, shoot dw = 201 g/pot, Israel, n=3[2]
Avena sterilis	164 g water/(g dry matter)	Total dw = 309, root dw = 93, shoot dw = 216 g/pot, Israel, n=2[2]
Beta vulgaris	0.636 g water/(dm**2*hour)	In dark, SD = +- 0.07[3]
Beta vulgaris	3.34 g water/(dm**2*hour)	In light, SD = +- 0.58[3]

Eichhornia crassipes	1.5	Ratio max transpiration/min transpiration, aquatic[1]
Erucaria boveana	207 g water/(g dry matter)	Total dw = 71, root dw = 16, shoot dw = 55 g/pot, Israel, n=2[2]
Gossypium barbadense	0.623 g water/(dm**2*hour)	In dark, SD = +- 0.12[3]
Gossypium barbadense	2.58 g water/(dm**2*hour)	In light, SD = +- 0.35[3]
Gossypium hirsutum	0.661 g water/(dm**2*hour)	In dark, SD = 0.08[3]
Gossypium hirsutum	3 g water/(dm**2*hour)	In light, SD = +- 0.47[3]
Helianthus annus	0.769 g water/(dm**2*hour)	In dark, SD = +- 0.17[3]
Helianthus annus	3.67 g water/(dm**2*hour)	In light, SD = +- 0.08[3]
Hordeum murinum	129 g water/(g dry matter)	Total dw = 78, root dw = 31, shoot dw = 47 g/pot, Israel, n=10[2]
Hordeum murinum	144 g water/(g dry matter)	Total dw = 95, root dw = 26, shoot dw = 69 g/pot, Israel, n=2[2]
Hordeum murinum	159 g water/(g dry matter)	Total dw = 166, root dw = 38, shoot dw = 128 g/pot, Israel, n=2[2]
Hordeum murinum	178 g water/(g dry matter)	Total dw = 288, root dw = 38, shoot dw = 128 g/pot, Israel, n=2[2]
Hordeum sativum	104 g water/(g dry matter)	Total dw = 169, root dw = 71, shoot dw = 98 g/pot, Israel, n=8[2]
Hordeum sativum	118 g water/(g dry matter)	Total dw = 303, root dw = 85, shoot dw = 218 g/pot, Israel, n=7, Aug[2]
Hordeum sativum	173 g water/(g dry matter)	Total dw = 23.4, root dw = 15.7, shoot dw = 7.7 g/pot, Israel, n=4[2]
Hordeum sativum	187 g water/(g dry matter)	Total dw = 118, root dw = 25, shoot dw = 93 g/pot, Israel, n=6, Mar[2]
Hordeum sativum	187 g water/(g dry matter)	Total dw = 360, root dw = 90, shoot dw = 270 g/pot, Israel, n=3[2]
Hordeum sativum	191 g water/(g dry matter)	Total dw = 256, root dw = 98, shoot dw = 167 g/pot, Israel, n=3[2]
Hordeum sativum	217 g water/(g dry matter)	Total dw = 181, root dw = 63, shoot dw = 118 g/pot, Israel, n=3, Oct[2]
Laurus nobilis	8.4	Ratio max transpiration/min transpiration, meso- and xerophytes[1]
Malva sylvestris	142 g water/(g dry matter)	Total dw = 139, root dw = 34, shoot dw = 105 g/pot, Israel, n=6[2]
Medicago hispida	157 g water/(g dry matter)	Total dw = 64, root dw = 17, shoot dw = 47 g/pot, Israel, n=10[2]
Medicago hispida	169 g water/(g dry matter)	Total dw = 60, root dw = 12, shoot dw = 48 g/pot, Israel, n=3[2]
Medicago hispida	192 g water/(g dry matter)	Total dw = 83, root dw = 11, shoot dw = 72 g/pot, Israel, n=3[2]
Medicago hispida	226 g water/(g dry matter)	Total dw = 34, root dw = 7, shoot dw = 27 g/pot, Israel, n=5, Mar[2]
Nymphaea marliacea	3.2	Ratio max transpiration/min transpiration, aquatic[1]
Phalaris minor	157 g water/(g dry matter)	Total dw = 83, root dw = 30, shoot dw = 52 g/pot, Israel, n=2[2]
Phalaris minor	162 g water/(g dry matter)	Tot dw = 283, root dw = 105, shoot dw = 178 g/pot, Israel, n=2[2]

Phalaris minor	190 g water/(g dry matter)	Total dw = 116, root dw = 27, shoot dw = 89 g/pot, Israel, n=2[2]
Potamogeton natans	1.4	Ratio max transpiration/min transpiration, aquatic[1]
Quercus ilex	4.6	Ratio max transpiration/min transpiration, meso- and xerophytes[1]
Reboudia pinnata	146 g water/(g dry matter)	Total dw = 42, root dw = 7, shoot dw = 35 g/pot, Israel, n=6, Oct-Jan[2]
Reboudia pinnata	155 g water/(g dry matter)	Total dw = 65, root dw = 15, shoot dw = 50 g/pot, Israel, n=2[2]
Reboudia pinnata	168 g water/(g dry matter)	Total dw = 41, root dw = 4, shoot dw = 37 g/pot, Israel, n=3, Dec-Feb[2]
Reboudia pinnata	275 g water/(g dry matter)	Total dw = 60, root dw = 6, shoot dw = 54 g/pot, Israel, n=6, Mar[2]
Sorghum vulgare	0.596 g water/(dm**2*hour)	In dark, SD = +- 0.10[3]
Sorghum vulgare	3.26 g water/(dm**2*hour)	In light, SD = +- 0.20[3]
Stipa capensis	165 g water/(g dry matter)	Total dw = 117, root dw = 11, shoot dw = 106 g/pot, Israel, n=2[2]
Stipa capensis	169 g water/(g dry matter)	Total dw = 134, root dw = 24, shoot dw = 110 g/pot, Israel, n=2[2]
Stipa capensis	181 g water/(g dry matter)	Total dw = 63, root dw = 15, shoot dw = 48 g/pot, Israel, n=2[2]
Syringa vulgaris	4.9	Ratio max transpiration/min transpiration, meso- and xerophytes[1]
Trigonella arabica	131 g water/(g dry matter)	Tot dw = 50, root dw = 5.5, shoot dw = 44.5 g/pot, Israel, n=3,[2]
Trigonella arabica	154 g water/(g dry matter)	Total dw = 26, root dw = 4, shoot dw = 22 g/pot, Israel, Dec-Feb[2]
Trigonella arabica	222 g water/(g dry matter)	Total dw = 60, root dw = 8, shoot dw = 92 g/pot, Israel, n=6, Mar[2]
Triticum durum	194 g water/(g dry matter)	Total dw = 113, root dw = 21, shoot dw = 92 g/pot, Israel, n=6, Mar[2]
Triticum sativum	100 g water/(g dry matter)	Total dw = 189, root dw = 61, shoot dw = 128 g/pot, Israel, n=3,[2]
Triticum sativum	129 g water/(g dry matter)	Total dw = 157, root dw = 47, shoot dw = 110 g/pot, Israel, n=6,[2]
Triticum sativum	131 g water/(g dry matter)	Total dw = 212, root dw = 51, shoot dw = 161 g/pot, Israel, n=3,[2]
Triticum sativum	166 g water/(g dry matter)	Total dw = 121, root dw = 41, shoot dw = 80 g/pot, Israel, n=3,[2]
Triticum sativum	190 g water/(g dry matter)	Total dw = 231, root dw = 70, shoot dw = 161 g/pot, Israel, n=3,[2]
Triticum sativum	217 g water/(g dry matter)	Total dw = 166, root dw = 44, shoot dw = 122 g/pot, Israel, n=3,[2]
Triticum sativum	220 g water/(g dry matter)	Total dw = 100, root dw = 19, shoot dw = 81 g/pot, Israel, n=6, Mar[2]
Viola tricolor	7.6	Ratio max transpiration/min transpiration, meso- and xerophytes[1]
Zea mays	0.556 g water/(dm**2*hour)	In dark, SD = +- 0.10[3]
Zea mays	3.71 g water/(dm**2*hour)	In light, SD = +- 0.51[3]

Source: 1.Gessner, F. 1959. Hydrobotanik. II. Stoffhauhalt. VEB.
2.Keulen, H. van. 1975. Simulation of water use and herbage growth in arid regions. Center for Agricutural Publishing and Documentation. Wageningen.
3.El Sharkawy, M.A., R.S. Loomis and W.A. Williams. 1967. Physiol. Plant., 20, pp 171-186.

1-393 Kormophytes, plants
Turbulence intensity

Species	Value	Condition
Coniferous forest	0.5 Input value	Input value[1]
Grass	0.5 Input value	Input value[1]
Maize	0.5 Input value	Input value[1]

Source: 1.Goudrian, J. 1977. Crop Micrometeorology: A Simulation Study. Chap 4. Pudoc. Wageningen.

1-394 Kormophytes, plants
Vein length densities

Species	Value	Condition
Acer campestris	102 cm/cm**2	Terrestrial plant[1]
Fraxinus excelsior	88 cm/cm**2	Terrestrial plant[1]
Nelumbo sp.	104 cm/cm**2	Aerial leaf, edge, aquatic plant[1]
Nelumbo sp.	68 cm/cm**2	Floating leaf, edge, aquatic plant[1]
Nelumbo sp.	71 cm/cm**2	Floating leaf, middle, aquatic plant[1]
Nelumbo sp.	91 cm/cm**2	Aerial leaf, middle, aquatic plant[1]
Nuphar luteum	7 cm/cm**2	Submersed leaf, aquatic plant[1]
Nymphaea mexicana	44 cm/cm**2	Floating leaf, aquatic plant[1]
Potamogeton praelongus	14 cm/cm**2	Submersed leaf, aquatic plant[1]
Quercus cerris	86 cm/cm**2	Terrestrial plant[1]
Rosa canina	108 cm/cm**2	Terrestrial plant[1]
Salix rubrum	52 cm/cm**2	Terrestrial plant[1]
Salvinia auriculata	14 cm/cm**2	Floating leaf, aquatic plant[1]

Source: 1.Gessner, F. 1959. Hydrobotanik. II. Stoffhauhalt. VEB.

1-395 Kormophytes, plants
Von Karmans constant

Species	Value	Condition
Coniferous forest	0.4	Input value[1]
Grass	0.4	Input value[1]
Maize	0.4	Input value[1]

Source: 1.Goudrian, J. 1977. Crop Micrometeorology: A Simulation Study. Chap 4. Pudoc. Wageningen.

1-396 Kormophytes, plants
Wind parameter

Species	Value	Condition
Coniferous forest	0.326	Calculated value, friction vel/wind vel at top of canopy[1]
Coniferous forest	3.41	Calc value, ((canopy height)-(zero plane displacm))/(roughness)[1]
Grass	0.31	Calc value, ((lai)*(average width of leaves)/(mix length)[1]

Grass	0.376	Calculated value, friction vel/wind vel at top of canopy[1]
Grass	2.9	Calc value, ((canopy height)-(zero plane displacm))/(roughness)[1]
Maize	0.22	Calc value, ((lai)*(average width of leaves)/(mix length)[1]
Maize	0.343	Calculated value, friction vel/wind vel at top of canopy[1]
Maize	3.21	Calc value, ((canopy height)-(zero plane displacm))/(roughness)[1]

Source: 1.Goudrian, J. 1977. Crop Micrometeorology: A Simulation Study. Chap 4. Pudoc. Wageningen.

1-397 Kormophytes, plants
Zero plane displacement

Species	Value	Condition
Coniferous forest	5.67 m	Calculated value[1]
Grass	0.147 m	Calculated value[1]
Maize	1.66 m	Calculated value[1]

Source: 1.Goudrian, J. 1977. Crop Micrometeorology: A Simulation Study. Chap 4. Pudoc. Wageningen.

1-398 Kormophytes, plants
Zero plane displacement / height of canopy

Species	Value	Condition
Coniferous forest	0.567	Calculated value[1]
Grass	0.735	Calculated value[1]
Maize	0.664	Calculated value[1]

Source: 1.Goudrian, J. 1977. Crop Micrometeorology: A Simulation Study. Chap 4. Pudoc. Wageningen.

1-399 Kormophytes, plants
Zn / biomass ratio

Species	Value	Condition
Alfalfa	0.0004 %	Atomic composition[1]

Source: 1.Berthet, P. 1963. In: Doeksen, J. and J. van der Drift. (eds); Soil Organism. North-Holland Publ. Co, Amsterdam.

1-400 Kormophytes, trees
B / biomass ratio

Species	Value	Condition
Aspen	0.91 g/complete tree	Tree fresh wet=232.1 kg, height=15.2 m, univ forest, Stillwater,[1]
Balsam fir	1.06 g/complete tree	Tree fresh wet=316.4 kg, height=13.7 m, univ forest, Stillwater,[1]
Hemlock	1.39 g/complete tree	Tree fresh wet=294.6 kg, height=13.4 m, univ forest, Stillwater,[1]
Red maple	1.08 g/complete tree	Tree fresh wet=283.6 kg, height=14.7 m, univ forest, Stillwater,[1]
Red spruce	0.98 g/complete tree	Tree fresh wet=275.9 kg, height=15.8 m, univ forest, Stillwater,[1]
White birch	1.33 g/complete tree	Tree fresh wet=371.9 kg, height=18.7 m, univ forest, Stillwater,[1]
White pine	0.89 g/complete tree	Tree fresh wet=376.9 kg, height=15.8 m, univ forest, Stillwater,[1]

Source: 1.Young, H.E. and V.P. Guinn. 1966. Tappi 49, pp 190-197.

1-401 Kormophytes, trees
C / biomass ratio

Species	Value	Condition
Distribution of C in different pools in a tropical rain forest	15 % in organic layer	Tot C = 37000 g/m2, Middle Magdalena Valley, Columbia[1]
Distribution of C in different pools in a tropical rain forest	35 % in soil	Tot N = 6763 g/m2, Middle Magdalena Valley, Columbia[1]
Distribution of C in different pools in a tropical rain forest	50 % in vegetation	Tot C = 37000 g/m2, Middle Magdalena Valley, Columbia[1]
Distribution of Ca in different pools in a tropical rain forest	78 % in vegetation	Tot Ca = 557 g/m2, Middle Magdalena Valley, Columbia[1]
Distribution of Mg in different pools in a tropical rain forest	21 % in soil	Tot Mg = 204 g/m2, Middle Magdalena Valley, Columbia[1]

Source: 1.De las Salas and H. Foelster. 1976. Turriaba 26, 179-186.

1-402 Kormophytes, trees
Ca / biomass ratio

Species	Value	Condition
Distribution of Ca in different pools in a tropical rain forest	17 % in organic layer	Tot Ca = 557 g/m2, Middle Magdalena Valley, Columbia[1]
Distribution of Ca in different pools in a tropical rain forest	5 % in soil	Tot Ca = 557 g/m2, Middle Magdalena Valley, Columbia[1]
Distribution of Ca in different pools in a tropical rain forest	5 % in soil	Tot Ca = 557 g/m2, Middle Magdalena Valley, Columbia[1]

Source: 1.De las Salas and H. Foelster. 1976. Turriaba 26, 179-186.

1-403 Kormophytes, trees
Compensation point

Species	Value	Condition
Beech	7.5 %	Percent of full winter sunlight in Maine[1]
Eastern hemlock	8.5 %	Percent of full winter sunlight in Maine[1]
Eastern white pine	10 %	Percent of full winter sunlight in Maine[1]
Engelmann spruce	11 %	Percent of full winter sunlight in Maine[1]
Larch	18 %	Percent of full winter sunlight in Maine[1]
Northern white cedar	19 %	Percent of full winter sunlight in Maine[1]
Ponderosa pine	31 %	Percent of full winter sunlight in Maine[1]
Red oak	14 %	Percent of full winter sunlight in Maine[1]
Sugar maple	3.5 %	Percent of full winter sunlight in Maine[1]

Source: 1.Burns, G.P. 1923. Ecology 4.

1-404 Kormophytes, trees
Cu / biomass ratio

Species	Value	Condition
Aspen	1.03 g/complete tree	Tree fresh wet=232.1 kg, height=15.2 m, univ forest, Stillwater,[1]
Balsam fir	1.71 g/complete tree	Tree fresh wet=316.4 kg, height=13.7 m, univ forest, Stillwater,[1]
Hemlock	0.84 g/complete tree	Tree fresh wet=294.6 kg, height=13.4 m, univ forest, Stillwater,[1]

Red maple	1.32 g/complete tree	Tree fresh wet=283.6 kg, height=14.7 m, univ forest, Stillwater,[1]
Red spruce	0.99 g/complete tree	Tree fresh wet=275.9 kg, height=15.8 m, univ forest, Stillwater,[1]
White birch	1.88 g/complete tree	Tree fresh wet=371.9 kg, height=18.7 m, univ forest, Stillwater,[1]
White pine	1.22 g/complete tree	Tree fresh wet=376.9 kg, height=15.8 m, univ forest, Stillwater,[1]

Source: 1.Young, H.E. and V.P. Guinn. 1966. Tappi 49, pp 190-197.

1-405 Kormophytes, trees
Fe / biomass ratio

Species	Value	Condition
Aspen	5.1 g/complete tree	Tree fresh wet=232.1 kg, height=15.2 m, univ forest, Stillwater[1]
Balsam fir	9.64 g/complete tree	Tree fresh wet=316.4 kg, height=13.7 m, univ forest, Stillwater[1]
Hemlock	7.84 g/complete tree	Tree fresh wet=294.6 kg, height=13.4 m, univ forest, Stillwater[1]
Pinus resinosa	0.0096 kg/ha	3-7 cm diam., total weight in veg, Halliday Field, Ont, Canada, stem[2]
Pinus resinosa	0.0272 kg/ha	3-7 cm diam., total weight in veg, Halliday Field, Ont, Canada, stem[2]
Pinus resinosa	0.2604 kg/ha	7 cm-stump, total weight in veg, Halliday Field, Ont, Canada, stem[2]
Pinus resinosa	0.2908 kg/ha	Total weight in veg, Halliday Field, Ont, Canada, living branches[2]
Pinus resinosa	0.4646 kg/ha	7 cm-stump, total weight in veg, Halliday Field, Ont, Canada, stem[2]
Pinus resinosa	0.5602 kg/ha	Total weight in veg, Halliday Field, Ont, Canada, twigs[2]
Pinus resinosa	1.0621 kg/ha	Total weight in veg, Halliday Field, Ont, Canada, dead branches[2]
Pinus resinosa	1.0717 kg/ha	Total weight in veg, Halliday Field, Ont, Canada, needles[2]
Pinus resinosa	3.7466 kg/ha	Total weight in veg, Halliday Field, Ont, Canada, all aerial organs[2]
Red maple	7.28 g/complete tree	Tree fresh wet=283.6 kg, height=14.7 m, univ forest, Stillwater[1]
Red pine	0.005 g/tree	3-7 cm dia., 34 y., n=10, Halliday Field, Ont, Canada, stem wood[2]
Red pine	0.0142 g/tree	3-7 cm dia., 34 y., n=10, Halliday Field, Ont, Canada, stem bark[2]
Red pine	0.2914 g/tree	34 y., n=10, Halliday Field, Ont, Canada, twigs[2]
Red pine	0.5524 g/tree	34 y., n=10, Halliday Field, Ont, Canada, dead branches[2]
Red pine	0.5574 g/tree	34 y., n=10, Halliday Field, Ont, Canada, needles[2]
Red spruce	7.66 g/complete tree	Tree fresh wet=275.9 kg, height=15.8 m, univ forest, Stillwater[1]
White birch	8.08 g/complete tree	Tree fresh wet=371.9 kg, height=18.7 m, univ forest, Stillwater[1]
White pine	7.54 g/complete tree	Tree fresh wet=376.9 kg, height=15.8 m, univ forest, Stillwater[1]

Source: 1.Young, H.E. and V.P. Guinn. 1966. Tappi 49, pp 190-197.
2.Fortescue, J.A.C. and G.G. Marten. 1970. in: Richle, D.E. (ed.); Analysis of Temperate Forest Ecosystems. pp 173-198. Springer-Verlag. Berlin.

1-406 Kormophytes, trees
K / biomass ratio

Species	Value	Condition
Distribution of K in different pools in a tropical rain forest	28 % in soil	Tot K = 428 g/m2, Middle Magdalena Valley, Columbia[1]
Distribution of K in different pools in a tropical rain forest	65 % in vegetation	Tot K = 428 g/m2, Middle Magdalena Valley, Columbia[1]
Distribution of K in different pools in a tropical rain forest	7 % in organic layer	Tot K = 428 g/m2, Middle Magdalena Valley, Columbia[1]
Distribution of K in different pools in a tropical rain forest	7 % in organic layer	Tot K = 428 g/m2, Middle Magdalena Valley, Columbia[1]

Source: 1.De las Salas and H. Foelster. 1976. Turriaba 26, 179-186.

1-407 Kormophytes, trees
Mg / biomass ratio

Species	Value	Condition
Distribution of Mg in different pools in a tropical rain forest	14 % in organic layer	Tot Mg = 204 g/m2, Middle Magdalena Valley, Columbia[1]
Distribution of Mg in different pools in a tropical rain forest	65 % in vegetation	Tot Mg = 204 g/m2, Middle Magdalena Valley, Columbia[1]

Source: 1.De las Salas and H. Foelster. 1976. Turriaba 26, 179-186.

1-408 Kormophytes, trees
Mn / biomass ratio

Species	Value	Condition
Aspen	9.6 g/complete tree	Tree fresh wet=232.1 kg, height=15.2 m, univ forest, Stillwater[1]
Balsam fir	69.3 g/complete tree	Tree fresh wet=316.4 kg, height=13.7 m, univ forest, Stillwater[1]
Hemlock	66.13 g/complete tree	Tree fresh wet=294.6 kg, height=13.4 m, univ forest, Stillwater[1]
Pinus resinosa	0.2246 kg/ha	3-7 cm diam., total weight in veg, Halliday Field, Ont, Canada, stem[2]
Pinus resinosa	0.256 kg/ha	3-7 cm diam., total weight in veg, Halliday Field, Ont, Canada, stem[2]
Pinus resinosa	0.8221 kg/ha	Total weight in veg, Halliday Field, Ont, Canada, dead branches[2]
Pinus resinosa	1.1018 kg/ha	Total weight in veg, Halliday Field, Ont, Canada, twigs[2]
Pinus resinosa	1.3966 kg/ha	Total weight in veg, Halliday Field, Ont, Canada, living branches[2]
Pinus resinosa	22.6457 kg/ha	Total weight in veg, Halliday Field, Ont, Canada, all aerial organs[2]
Pinus resinosa	3.4499 kg/ha	7 cm-stump, total weight in veg, Halliday Field, Ont, Canada, stem[2]
Pinus resinosa	5.717 kg/ha	Total weight in veg, Halliday Field, Ont, Canada, needles[2]
Pinus resinosa	9.6777 kg/ha	7 cm-stump, total weight in veg, Halliday Field, Ont, Canada, stem[2]
Red maple	41.64 g/complete tree	Tree fresh wet=283.6 kg, height=14.7 m, univ forest, Stillwater[1]

Red pine	0.1169 g/tree	3-7 cm dia., 34 y., n=10, Halliday Field, Ont, Canada, stem bark[2]
Red pine	0.1332 g/tree	3-7 cm dia., 34 y., n=10, Halliday Field, Ont, Canada, stem wood[2]
Red pine	0.4276 g/tree	34 y., n=10, Halliday Field, Ont, Canada, dead branches[2]
Red pine	0.5731 g/tree	34 y., n=10, Halliday Field, Ont, Canada, twigs[2]
Red pine	2.9738 g/tree	34 y., n=10, Halliday Field, Ont, Canada, needles[2]
Red spruce	61.03 g/complete tree	Tree fresh wet=275.9 kg, height=15.8 m, univ forest, Stillwater[1]
White birch	21.96 g/complete tree	Tree fresh wet=371.9 kg, height=18.7 m, univ forest, Stillwater[1]
White pine	16.42 g/complete tree	Tree fresh wet=376.9 kg, height=15.8 m, univ forest, Stillwater[1]

Source: 1.Young, H.E. and V.P. Guinn. 1966. Tappi 49, pp 190-197.
2.Fortescue, J.A.C. and G.G. Marten. 1970. in: Richle, D.E. (ed.); Analysis of Temperate Forest Ecosystems. pp 173-198. Springer-Verlag. Berlin.

1-409 Kormophytes, trees
Mo / biomass ratio

Species	Value	Condition
Aspen	0.3 g/complete tree	Tree fresh wet=232.1 kg, height=15.2 m, univ forest, Stillwater[1]
Balsam fir	0.34 g/complete tree	Tree fresh wet=316.4 kg, height=13.7 m, univ forest, Stillwater[1]
Hemlock	0.23 g/complete tree	Tree fresh wet=294.6 kg, height=13.4 m, univ forest, Stillwater[1]
Red maple	0.35 g/complete tree	Tree fresh wet=283.6 kg, height=14.7 m, univ forest, Stillwater[1]
Red spruce	0.35 g/complete tree	Tree fresh wet=275.9 kg, height=15.8 m, univ forest, Stillwater[1]
White birch	0.47 g/complete tree	Tree fresh wet=371.9 kg, height=18.7 m, univ forest, Stillwater[1]
White pine	0.18 g/complete tree	Tree fresh wet=376.9 kg, height=15.8 m, univ forest, Stillwater[1]

Source: 1.Young, H.E. and V.P. Guinn. 1966. Tappi 49, pp 190-197.

1-410 Kormophytes, trees
N / biomass ratio

Species	Value	Condition
Distribution of N in different pools in a tropical rain forest	10 % in organic layer	Tot N = 6763 g/m2, Middle Magdalena Valley, Columbia[1]
Distribution of N in different pools in a tropical rain forest	11 % in vegetation	Tot N = 6763 g/m2, Middle Magdalena Valley, Columbia[1]
Distribution of N in different pools in a tropical rain forest	79 % in soil	Tot N = 6763 g/m2, Middle Magdalena Valley, Columbia[1]

Source: 1.De las Salas and H. Foelster. 1976. Turriaba 26, 179-186.

1-411 Kormophytes, trees
Nitrogen distribution

Species	Value	Condition
Pine forest	0.34 Gallery forest	Gallery forest, leaf/wood ratio[1]
Pine forest	0.6 Temperate british 55 y forest, root/	Temperate british 55 y forest, root/shoot ratio of N[1]
Pine forest	12.4 gram N/m**2	Temperate british 55 y forest, leaves[1]
Pine forest	18.4 gram N/m**2	Temperate british 55 y forest, in roots[1]
Pine forest	18.5 gram N/m**2	Temperate british 55 y forest, wood above ground[1]
Pine forest	3 % of total N	Temperate british 55 y forest, wood above ground[1]
Pine forest	4 Temperate british 55 y forest, turnove	Temperate british 55 y forest, turnover of N[1]
Pine forest	40.9 gram N/m**2	Temperate british 55 y forest, litter[1]
Pine forest	6 % of total N	Temperate british 55 y forest, in biomass[1]
Pine forest	730.8 gram N/m**2	Temperate british 55 y forest, in soil[1]
Tropical forest	0.3 Gallery forest	Gallery forest, root/shoot ratio of N[1]
Tropical forest	0.76 Gallery forest	Gallery forest, leaf/wood ratio[1]
Tropical forest	28.2 gram N/m**2	Gallery forest, in roots[1]
Tropical forest	3.9 gram N/m**2	Gallery forest, litter[1]
Tropical forest	41.2 gram N/m**2	Gallery forest, wood above ground[1]
Tropical forest	44 % of total N	Gallery forest, wood above ground[1]
Tropical forest	52.6 gram N/m**2	Gallery forest, leaves[1]
Tropical forest	57.8 % of total N	Gallery forest, in biomass[1]
Tropical forest	85.3 gram N/m**2	Gallery forest, in soil[1]

Source: 1.Ovington, J.D. 1962. Adv. Ecol. Res. 1, pp 103-192.

1-412 Kormophytes, trees
Pollem dimensions, height

Species	Value	Condition
Cephalonica fir	86.2 um	Specific value[1]
Douglas fir	54.8 um	Specific value[1]
English oak	21.5 um	Specific value[1]
European alder	13.7 um	Specific value[1]
European birch	16.8 um	Specific value[1]
European cut-leaf linden	20.6 um	Specific value[1]
European larch	50 um	Specific value[1]
Horsechestnut	18.2 um	Specific value[1]
Norway spruce	66.3 um	Specific value[1]
Rock elm	17.7 um	Specific value[1]
Scotch pine	36 um	Specific value[1]
Silver fir	62.7 um	Specific value[1]
Sugar maple	24.6 um	Specific value[1]

Source: 1.Stanley, R.G. and H.. Linskens. 1974. Pollen Biology Biochemistry Management. Springer-Verlag. NY. 37 pp.

1-413 Kormophytes, trees
Pollem dimensions

Species	Value	Condition
European beech	41.1 um	Specific value[1]

Source: 1.Stanley, R.G. and H.. Linskens. 1974. Pollen Biology Biochemistry Management. Springer-Verlag. NY. 37 pp.

1-414 Kormophytes, trees
Pollem dimensions, length

Species	Value	Condition
Cephalonica fir	97.1 um	Specific value[1]
Douglas fir	84.8 um	Specific value[1]
English oak	40.8 um	Specific value[1]
European alder	26.4 um	Specific value[1]
European beech	55.1 um	Specific value[1]
European birch	10.1 um	Specific value[1]
European cut-leaf linden	40.5 um	Specific value[1]
European larch	76 um	Specific value[1]
Horsechestnut	31 um	Specific value[1]
Norway spruce	85.8 um	Specific value[1]
Rock elm	33.4 um	Specific value[1]
Scotch pine	41.5 um	Specific value[1]
Silver fir	97.8 um	Specific value[1]
Sugar maple	32.5 um	Specific value[1]

Source: 1.Stanley, R.G. and H.. Linskens. 1974. Pollen Biology Biochemistry Management. Springer-Verlag. NY. 37 pp.

1-415 Kormophytes, trees
Pollem dimensions, volume

Species	Value	Condition
Cephalonica fir	4.226E-7 cm3	Specific value[1]
Douglas fir	2.192E-7 cm3	Specific value[1]
English oak	1.33E-8 cm3	Specific value[1]
European alder	4.4E-9 cm3	Specific value[1]
European beech	5.03E-8 cm3	Specific value[1]
European birch	2.9E-9 cm3	Specific value[1]
European cut-leaf linden	1.5E-8 cm3	Specific value[1]
European larch	1.802E-7 cm3	Specific value[1]
Horsechestnut	4.8E-9 cm3	Specific value[1]
Norway spruce	2.782E-7 cm3	Specific value[1]
Rock elm	1.28E-8 cm3	Specific value[1]
Scotch pine	3.55E-8 cm3	Specific value[1]
Silver fir	4.994E-7 cm3	Specific value[1]
Sugar maple	1.65E-8 cm3	Specific value[1]

Source: 1.Stanley, R.G. and H.. Linskens. 1974. Pollen Biology Biochemistry Management. Springer-Verlag. NY. 37 pp.

1-416 Kormophytes, trees
Pollem dimensions, weight

Species	Value	Condition
Cephalonica fir	2.122E-7 g	Specific value[1]
Douglas fir	1.888E-7 g	Specific value[1]
English oak	5.7E-9 g	Specific value[1]
European alder	1.4E-9 g	Specific value[1]
European beech	2.6E-8 g	Specific value[1]
European birch	8E-10 g	Specific value[1]
European cut-leaf linden	6.5E-9 g	Specific value[1]
European larch	1.763E-7 g	Specific value[1]
Horsechestnut	9E-10 g	Specific value[1]
Norway spruce	1.108E-7 g	Specific value[1]
Rock elm	6.8E-9 g	Specific value[1]
Scotch pine	3.7E-8 g	Specific value[1]
Silver fir	2.516E-7 g	Specific value[1]
Sugar maple	6.6E-9 g	Specific value[1]

Source: 1.Stanley, R.G. and H.. Linskens. 1974. Pollen Biology Biochemistry Management. Springer-Verlag. NY. 37 pp.

1-417 Kormophytes, trees
Pollem dimensions, width

Species	Value	Condition
Cephalonica fir	98.6 um	Specific value[1]
Douglas fir	81.1 um	Specific value[1]
English oak	26.1 um	Specific value[1]
European alder	22.8 um	Specific value[1]
European beech	40.5 um	Specific value[1]
European birch	10.1 um	Specific value[1]
European cut-leaf linden	40.1 um	Specific value[1]
European larch	72 um	Specific value[1]
Horsechestnut	16.4 um	Specific value[1]
Norway spruce	80.5 um	Specific value[1]
Rock elm	32.7 um	Specific value[1]
Scotch pine	45.9 um	Specific value[1]
Silver fir	102.9 um	Specific value[1]
Sugar maple	23.6 um	Specific value[1]

Source: 1.Stanley, R.G. and H.. Linskens. 1974. Pollen Biology Biochemistry Management. Springer-Verlag. NY. 37 pp.

1-418 Kormophytes, trees
Pollen dispersion distance

Species	Value	Condition
Alder	546.7 km	Average distance with windspeed 10 m/sec[1]
English oak	199 km	Average distance with windspeed 10 m/sec[1]
European filbert	267.8 km	Average distance with windspeed 10 m/sec[1]
Norway spruce	22.2 km	Average distance with windspeed 10 m/sec[1]
Scotch pine	267.8 km	Average distance with windspeed 10 m/sec[1]

Source: 1.Stanley, R.G. and H.. Linskens. 1974. Pollen Biology Biochemistry Management. Springer-Verlag. NY. 37 pp.

1-419 Kormophytes, trees
Pollen production

Species	Value	Condition
Alder	50000000 no of pollen/(g biomass of tree	Estimated per reproductive part[1]
Alder	7239000000 no of pollen/(g biomass of tr	Estimated per reproductive part[2]
Alnus sp.	0.04 ml pollen/((g catkin)*year)	Estimated per reproductive part[1]
Apple tree	3800 No of pollen/((g anther)*year)	Estimated per reproductive part[2]
Betula sp.	0.1 ml pollen/((g catkin)*year)	Estimated per reproductive part[1]
Common juniper	400000 No of pollen/((g flower)*year)	Estimated per reproductive part[2]
European alder	4500000 No of pollen/((g catkin)*year)	Estimated per reproductive part[2]
European asp	12500 No of pollen/((g anther)*year)	Estimated per reproductive part[2]
European beach	175000 No of pollen/((g catkin)*year)	Estimated per reproductive part[2]
European beech	409000000 no of pollen/(g biomass of tre	Estimated per reproductive part[2]
European birch	150000000 no of pollen/(g biomass of tre	Estimated per reproductive part[1]
European birch	30000000 no of pollen/(g biomass of tree	Estimated per reproductive part[1]
European birch	5562000000 no of pollen/(g biomass of tr	Estimated per reproductive part[2]
European filbert	50000000 no of pollen/(g biomass of tree	Estimated per reproductive part[1]
European filbert	5603000000 no of pollen/(g biomass of tr	Estimated per reproductive part[2]
European horsebean	3149000000 no of pollen/(g biomass of tr	Estimated per reproductive part[2]

European oak	1250000 No of pollen/((g catkin)*year)	Estimated per reproductive part[2]
European white birch	6000000 No of pollen/((g catkin)*year)	Estimated per reproductive part[2]
Fagus sp.	0.013 ml pollen/((g inflorescence)*year)	Estimated per reproductive part[1]
Larix sp.	0.012 ml pollen/((g strobili)*year)	Estimated per reproductive part[1]
Liquidambar sp.	0.25 ml pollen/((g flower)*year)	Estimated per reproductive part[1]
Littleleaf linden	5603000000 no of pollen/(g biomass of tr	Estimated per reproductive part[2]
Norway maple	1000 No of pollen/((g anther)*year)	Estimated per reproductive part[2]
Norway maple	8000 No of pollen/((g flower)*year)	Estimated per reproductive part[2]
Norway spruce	400000000 no of pollen/(g biomass of tre	Estimated per reproductive part[1]
Norway spruce	5481000000 no of pollen/(g biomass of tr	Estimated per reproductive part[2]
Norway spruce	600000 No of pollen/((g flower)*year)	Estimated per reproductive part[2]
Pinus sp.	0.21 ml pollen/((g strobili)*year)	Estimated per reproductive part[1]
Platanus sp.	0.08 ml pollen/((g head)*year)	Estimated per reproductive part[1]
Populus sp.	0.75 ml pollen/((g catkin)*year)	Estimated per reproductive part[1]
Pseudotsuga sp.	0.022 ml pollen/((g strobili)*year)	Estimated per reproductive part[1]
Scotch pine	12000000 no of pollen/(g biomass of tree	Estimated per reproductive part[1]
Scotch pine	160000 No of pollen/((g flower)*year)	Estimated per reproductive part[2]
Scotch pine	6462000000 no of pollen/(g biomass of tr	Estimated per reproductive part[2]
Sessile oak	654000000 no of pollen/(g biomass of tre	Estimated per reproductive part[2]
Ulmus sp.	0.0025 g pollen/((g flower)*year)	Estimated per reproductive part[1]

Source: 1.Brooks, J. 1971. In: J. Brooks, P.R. Grant, M. Muir, P. van Gijzel and G. Shaw. (eds.): Sporopolleni. Academic Press. NY. 351-407.
2.Erdtman, G. 1969. Handbook of Palynology. Hofner. NY. 486 pp.

1-420 Kormophytes, trees
Pollen production, annual variation

Species	Value	Condition
Alnus sp.	118 % of normal catch	Average catch year : 1948-1948 = 385, year = 1948[1]
Alnus sp.	120 % of normal catch	Average catch year : 1945-1948 = 385, year = 1945[1]
Alnus sp.	226 % of normal catch	Average catch year : 1944-1948 = 385, year = 1944[1]

Alnus sp.	34 % of normal catch	Average catch year : 1947-1948 = 385, year = 1947[1]
Alnus sp.	45 % of normal catch	Average catch year : 1943-1948 = 385, year = 1943[1]
Alnus sp.	55 % of normal catch	Average catch year : 1946-1948 = 385, year = 1946[1]
Betula sp.	106 % of normal catch	Average catch year : 1946-1948 = 620, year = 1946[1]
Betula sp.	160 % of normal catch	Average catch year : 1944-1948 = 620, year = 1944[1]
Betula sp.	214 % of normal catch	Average catch year : 1948-1948 = 620, year = 1948[1]
Betula sp.	22 % of normal catch	Average catch year : 1947-1948 = 620, year = 1947[1]
Betula sp.	42 % of normal catch	Average catch year : 1945-1948 = 620, year = 1945[1]
Betula sp.	57 % of normal catch	Average catch year : 1943-1948 = 620, year = 1943[1]
Fagus sp.	11 % of normal catch	Average catch year : 1945-1948 = 273, year = 1945[1]
Fagus sp.	140 % of normal catch	Average catch year : 1944-1948 = 273, year = 1944[1]
Fagus sp.	143 % of normal catch	Average catch year : 1946-1948 = 273, year = 1946[1]
Fagus sp.	23 % of normal catch	Average catch year : 1943-1948 = 273, year = 1943[1]
Fagus sp.	275 % of normal catch	Average catch year : 1948-1948 = 273, year = 1948[1]
Fagus sp.	8 % of normal catch	Average catch year : 1947-1948 = 273, year = 1947[1]
Fraxinus sp.	159 % of normal catch	Average catch year : 1946-1948 = 675, year = 1946[1]
Fraxinus sp.	24 % of normal catch	Average catch year : 1945-1948 = 675, year = 1945[1]
Fraxinus sp.	25 % of normal catch	Average catch year : 1944-1948 = 675, year = 1944[1]
Fraxinus sp.	330 % of normal catch	Average catch year : 1943-1948 = 675, year = 1943[1]
Fraxinus sp.	54 % of normal catch	Average catch year : 1947-1948 = 675, year = 1947[1]
Fraxinus sp.	8 % of normal catch	Average catch year : 1948-1948 = 675, year = 1948[1]
Pinus sp.	121 % of normal catch	Average catch year : 1944-1948 = 386, year = 1944[1]
Pinus sp.	127 % of normal catch	Average catch year : 1946-1948 = 386, year = 1946[1]
Pinus sp.	146 % of normal catch	Average catch year : 1948-1948 = 386, year = 1948[1]
Pinus sp.	40 % of normal catch	Average catch year : 1947-1948 = 386, year = 1947[1]
Pinus sp.	72 % of normal catch	Average catch year : 1945-1948 = 386, year = 1945[1]
Pinus sp.	92 % of normal catch	Average catch year : 1943-1948 = 386, year = 1943[1]
Quercus sp.	10 % of normal catch	Average catch year : 1947-1948 = 2776, year = 1947[1]
Quercus sp.	117 % of normal catch	Average catch year : 1948-1948 = 2776, year = 1948[1]
Quercus sp.	280 % of normal catch	Average catch year : 1946-1948 = 2776, year = 1946[1]
Quercus sp.	42 % of normal catch	Average catch year : 1945-1948 = 2776, year = 1945[1]
Quercus sp.	58 % of normal catch	Average catch year : 1943-1948 = 2776, year = 1943[1]
Quercus sp.	90 % of normal catch	Average catch year : 1944-1948 = 2776, year = 1944[1]
Ulmus sp.	150 % of normal catch	Average catch year : 1948-1948 = 4579, year = 1948[1]
Ulmus sp.	180 % of normal catch	Average catch year : 1945-1948 = 4579, year = 1945[1]
Ulmus sp.	48 % of normal catch	Average catch year : 1947-1948 = 4579, year = 1947[1]
Ulmus sp.	50 % of normal catch	Average catch year : 1946-1948 = 4579, year = 1946[1]
Ulmus sp.	84 % of normal catch	Average catch year : 1943-1948 = 4579, year = 1943[1]
Ulmus sp.	87 % of normal catch	Average catch year : 1944-1948 = 4579, year = 1944[1]

Source: 1.Sandberg, D.V., S.G. Pickford and E.F. Darley. 1975. J. Air Pollut. Control Asssoc. 21, 629-635.

1-421 Kormophytes, trees
Production / biomass ratio

Species	Value	Condition
oak - pine forest	0.0806 1/year	Total wood tissue, net production[1]
oak - pine forest	0.1 1/year	Below ground, net production[1]
oak - pine forest	0.13 1/year	Above ground, net production[1]
oak - pine forest	0.833 1/year	Leaves, net production[1]

Source: 1.Whittaker, R.A. and G.M. Woodwell. 1969. J. Ecol. 57, pp 103-174.

1-422 Kormophytes, trees
Zn / biomass ratio

Species	Value	Condition
Abies lasiocarpa, foliage	45 mg/kg dry weight	Forest tree, rg = 43 - 47 ppm[1]
Adenostema sp., foliage	52 mg/kg dry weight	Forest tree, rg = 33 - 71 ppm[1]
Arcosta phylos, foliage	52 mg/kg dry weight	Forest tree, rg = 30 - 73 ppm[1]
Aspen	5.62 g/complete tree	Tree fresh wet=232.1 kg, height=15.2 m, univ forest, Stillwater,[2]
Balsam fir	4.65 g/complete tree	Tree fresh wet=316.4 kg, height=13.7 m, univ forest, Stillwater,[2]
Betula glandulosa, foliage	55 mg/kg dry weight	Forest tree, rg=37-308 ppm, soil conc = 25-1400 ppm[1]
Betule lenta, foliage	135 mg/kg dry weight	Forest tree[1]
Eucalyptus deglupa, foliage	28 mg/kg dry weight	Forest tree, rg = 6 - 49 ppm[1]
Hemlock	1.06 g/complete tree	Tree fresh wet=294.6 kg, height=13.4 m, univ forest, Stillwater,[2]
Nyssa sylvatica, foliage	40 mg/kg dry weight	Forest tree[1]
Picea abies, foliage	76 mg/kg dry weight	Forest tree, rg = 48 - 83 ppm[1]
Picea glauca, foliage	50 mg/kg dry weight	Forest tree, rg = 30 - 70 ppm[1]
Picea glauca, foliage	55 mg/kg dry weight	Forest tree[1]
Pinus contorta, foliage	50 mg/kg dry weight	Forest tree, rg = 30 - 70 ppm[1]
Pinus elliottii, foliage	43 mg/kg dry weight	Forest tree, rg = 25 - 61 ppm[1]
Pinus pinaster, foliage	5.5 mg/kg dry weight	Forest tree, rg = 3 - 8 ppm[1]
Pinus radiata, foliage	19 mg/kg dry weight	Forest tree, rg = 1 - 36 ppm[1]
Pinus taeda, foliage	41 mg/kg dry weight	Forest tree, rg = 36 - 45 ppm[1]
Quercus prinus, foliage	55 mg/kg dry weight	Forest tree[1]
Quercus rubra, foliage	90 mg/kg dry weight	Forest tree, rg=70-155 ppm, soil conc = 25-1400 ppm[1]
Quercus sp., foliage	43 mg/kg dry weight	Forest tree, rg = 15 - 70 ppm[1]
Red maple	7.66 g/complete tree	Tree fresh wet=283.6 kg, height=14.7 m, univ forest, Stillwater,[2]
Salix alaxenis, foliage	100 mg/kg dry weight	Forest tree, rg = 97-332 ppm, soil conc = 25 - 1400 ppm[1]
Salix phylicifolia, foliage	100 mg/kg dry weight	Forest tree, rg = 92 - 266 ppm[1]
Sassafras albidum, foliage	100 mg/kg dry weight	Forest tree[1]
White birch	10.98 g/complete tree	Tree fresh wet=371.9 kg, height=18.7 m, univ forest, Stillwater,[2]
White pine	4.91 g/complete tree	Tree fresh wet=376.9 kg, height=15.8 m, univ forest, Stillwater,[2]

Source: 1.Hook, R.I. van, D.W. Johnson and B.P. Spalding. 1980. In: Nriagu, J.O. (ed.): Zinc in the Environmen. Part I: Ecological cycling. J. Wiley and sons. NY. 419-437.
2.Young, H.E. and V.P. Guinn. 1966. Tappi 49, pp 190-197.

1-423 Kormophytes, trees
Zn/biomass

Species	Value	Condition
Red spruce	3.69 g/complete tree	Tree fresh wet=275.9 kg, height=15.8 m, univ forest, Stillwater,[1]

Source: 1.Young, H.E. and V.P. Guinn. 1966. Tappi 49, pp 190-197.

1-424 Vertebrates
Al / biomass ratio

Species	Value	Condition
Homo sapiens	0.0001 %	Atomic composition[1]

Source: 1.Berthet, P. 1963. In: Doeksen, J. and J. van der Drift. (eds); Soil Organism. North-Holland Publ. Co, Amsterdam.

1-425 Vertebrates
Ash / biomass ratio

Species	Value	Condition
Pisces	5.2 % of dry weight	630 species[1]

Source: 1.Vinogradov, A.P. 1953. The Elementary Chemical Composition of Marine Organisms, Sears Foundation New Haven.

1-426 Vertebrates
Assimilation

Species	Value	Condition
Micropterus salmoides	40 % of hg intake	Food = mosquito fish[1]
Mosquito fish	12 % of hg intake	Food = blood worm[2]
Predator	85 % of ingestion	Typical value[3]

Source: 1.Huckabee, J.W. and B.G. Blaylock. 1972. in: Dhar, S.K. (ed.); Metal Ions in Biological Systems. pp 15-160. Plenum Publ. Corp. New York.
2.Joergensen, S.E., J.H. Henriksen, M. Friis, L.A. Joergensen and H.F. Mejer. Handbook of Environmental Data and Ecological Parameters. 1979. ISEM. Copenhagen.
3.Glass, N.R. 1971. in: Patten, B.C. (ed); System Analysis and Simulation in Ecology. 1, pp 325-363. Aademic Press. New York.

1-427 Vertebrates
Biomass at different life time

Species	Value	Condition
Bonasa umbellus	11 g wet weight	Age class in days = 0, female[1]
Bonasa umbellus	12 g wet weight	Age class in days = 0, male[1]
Bonasa umbellus	163 g wet weight	Age class in days = 42, female[1]
Bonasa umbellus	180 g wet weight	Age class in days = 42, male[1]
Bonasa umbellus	20 g wet weight	Age class in days = 7, female[1]
Bonasa umbellus	21 g wet weight	Age class in days = 7, male[1]
Bonasa umbellus	260 g wet weight	Age class in days = 56, female[1]
Bonasa umbellus	295 g wet weight	Age class in days = 56, male[1]
Bonasa umbellus	32 g wet weight	Age class in days = 14, female[1]
Bonasa umbellus	352 g wet weight	Age class in days = 70, female[1]
Bonasa umbellus	39 g wet weight	Age class in days = 14, male[1]
Bonasa umbellus	400 g wet weight	Age class in days = 70, male[1]
Bonasa umbellus	428 g wet weight	Age class in days = 84, female[1]
Bonasa umbellus	479 g wet weight	Age class in days = 84, male[1]
Bonasa umbellus	498 g wet weight	Age class in days = 112, female[1]
Bonasa umbellus	511 g wet weight	Age class in days = 140, female[1]
Bonasa umbellus	550 g wet weight	Age class in days = 112, male[1]
Bonasa umbellus	579 g wet weight	Age class in days = 140, male[1]
Bonasa umbellus	587 g wet weight	Adult, female[1]
Bonasa umbellus	651 g wet weight	Adult, male[1]
Bonasa umbellus	81 g wet weight	Age class in days = 28, female[1]
Bonasa umbellus	95 g wet weight	Age class in days = 28, male[1]
Kobus leche	100.9 kg wet weight	Age class in years = 7, male[2]
Kobus leche	101.8 kg wet weight	Age class in years = 8, male[2]
Kobus leche	103.2 kg wet weight	Age class in years = 4, male[2]
Kobus leche	104.5 kg wet weight	Age class in years = 5, male[2]
Kobus leche	105.9 kg wet weight	Age class in years = 6, male[2]

Kobus leche	4.5 kg wet weight	Age class in years = 0, female[2]
Kobus leche	4.8 kg wet weight	Age class in years = 0, male[2]
Kobus leche	54.6 kg wet weight	Age class in years = 1, male[2]
Kobus leche	60 kg wet weight	Age class in years = 1, female[2]
Kobus leche	67.1 kg wet weight	Age class in years = 2, male[2]
Kobus leche	72.7 kg wet weight	Age class in years = 2, female[2]
Kobus leche	75.4 kg wet weight	Age class in years = 5, female[2]
Kobus leche	78.1 kg wet weight	Age class in years = 3, female[2]
Kobus leche	78.1 kg wet weight	Age class in years = 7, female[2]
Kobus leche	80 kg wet weight	Age class in years = 4, female[2]
Kobus leche	80.9 kg wet weight	Age class in years = 8, female[3]
Kobus leche	83.2 kg wet weight	Age class in years = 6, female[2]
Kobus leche	85.5 kg wet weight	Age class in years = 3, male[2]
Loxodonta africana	1000 kg wet weight	Age class in years = 7, both sexes[4]
Loxodonta africana	1204.5 kg wet weight	Age class in years = 8, both sexes[4]
Loxodonta africana	1409.1 kg wet weight	Age class in years = 9, both sexes[4]
Loxodonta africana	1613.6 kg wet weight	Age class in years = 10, both sexes[4]
Loxodonta africana	204.5 kg wet weight	Age class in years = 2, both sexes[4]
Loxodonta africana	318.1 kg wet weight	Age class in years = 3, both sexes[4]
Loxodonta africana	4090.9 kg wet weight	Age class in years = 25, both sexes[4]
Loxodonta africana	454.5 kg wet weight	Age class in years = 4, both sexes[4]
Loxodonta africana	613.6 kg wet weight	Age class in years = 5, both sexes[4]
Loxodonta africana	795.4 kg wet weight	Age class in years = 6, both sexes[4]
Loxodonta africana	90 kg wet weight	Age class in years = 1, both sexes[4]
Meleagris gallopavo	1220 g wet weight	Age class in days = 70, male[5]
Meleagris gallopavo	1240 g wet weight	Age class in days = 84, female[5]
Meleagris gallopavo	1580 g wet weight	Age class in days = 98, female[5]
Meleagris gallopavo	1640 g wet weight	Age class in days = 84, male[5]
Meleagris gallopavo	1980 g wet weight	Age class in days = 112, female[5]
Meleagris gallopavo	2100 g wet weight	Age class in days = 98, male[5]
Meleagris gallopavo	250 g wet weight	Age class in days = 28, female[5]
Meleagris gallopavo	2520 g wet weight	Age class in days = 126, female[5]
Meleagris gallopavo	2600 g wet weight	Age class in days = 112, male[5]
Meleagris gallopavo	280 g wet weight	Age class in days = 28, male[5]
Meleagris gallopavo	3000 g wet weight	Age class in days = 140, female[5]
Meleagris gallopavo	3320 g wet weight	Age class in days = 126, male[5]
Meleagris gallopavo	3320 g wet weight	Age class in days = 154, female[5]
Meleagris gallopavo	3480 g wet weight	Age class in days = 168, female[5]
Meleagris gallopavo	3620 g wet weight	Age class in days = 182, female[5]
Meleagris gallopavo	3710 g wet weight	Age class in days = 196, female[5]

Meleagris gallopavo	3770 g wet weight	Age class in days = 210, female[5]
Meleagris gallopavo	3910 g wet weight	Age class in days = 252, female[5]
Meleagris gallopavo	3960 g wet weight	Age class in days = 280, female[5]
Meleagris gallopavo	40 g wet weight	Age class in days = 0, female[5]
Meleagris gallopavo	40 g wet weight	Age class in days = 0, male[5]
Meleagris gallopavo	4050 g wet weight	Age class in days = 140, male[5]
Meleagris gallopavo	4620 g wet weight	Age class in days = 154, male[5]
Meleagris gallopavo	480 g wet weight	Age class in days = 42, female[5]
Meleagris gallopavo	5100 g wet weight	Age class in days = 168, male[5]
Meleagris gallopavo	5500 g wet weight	Age class in days = 182, male[5]
Meleagris gallopavo	560 g wet weight	Age class in days = 42, male[5]
Meleagris gallopavo	5780 g wet weight	Age class in days = 196, male[5]
Meleagris gallopavo	5950 g wet weight	Age class in days = 210, male[5]
Meleagris gallopavo	6260 g wet weight	Age class in days = 252, male[5]
Meleagris gallopavo	6350 g wet weight	Age class in days = 280, male[5]
Meleagris gallopavo	690 g wet weight	Age class in days = 56, female[5]
Meleagris gallopavo	80 g wet weight	Age class in days = 14, female[5]
Meleagris gallopavo	80 g wet weight	Age class in days = 14, male[5]
Meleagris gallopavo	850 g wet weight	Age class in days = 56, male[5]
Meleagris gallopavo	960 g wet weight	Age class in days = 70, female[5]
Mus musculus	1.25 g wet weight	Age class in weeks = 0, male[5]
Mus musculus	1.26 g wet weight	Age class in weeks = 0, female[5]
Mus musculus	11.94 g wet weight	Age class in weeks = 4, female[5]
Mus musculus	12.13 g wet weight	Age class in weeks = 4, male[5]
Mus musculus	14.73 g wet weight	Age class in weeks = 5, female[5]
Mus musculus	14.94 g wet weight	Age class in weeks = 5, male[5]
Mus musculus	16.77 g wet weight	Age class in weeks = 6, female[5]
Mus musculus	17.6 g wet weight	Age class in weeks = 6, male[5]
Mus musculus	18.42 g wet weight	Age class in weeks = 7, female[5]
Mus musculus	19.81 g wet weight	Age class in weeks = 8, female[5]
Mus musculus	20.06 g wet weight	Age class in weeks = 7, male[5]
Mus musculus	21.59 g wet weight	Age class in weeks = 8, male[5]
Mus musculus	22.99 g wet weight	Age class in weeks = 12, female[5]
Mus musculus	24.78 g wet weight	Age class in weeks = 12, male[5]
Mus musculus	26.11 g wet weight	Age class in weeks = 16, female[5]
Mus musculus	27.76 g wet weight	Age class in weeks = 16, male[5]
Mus musculus	29.03 g wet weight	Age class in weeks = 20, female[5]
Mus musculus	29.91 g wet weight	Age class in weeks = 20, male[5]
Mus musculus	31 g wet weight	Age class in weeks = 24, female[5]
Mus musculus	31.07 g wet weight	Age class in weeks = 24, male[5]

Mus musculus	33 g wet weight	Age class in weeks = 28, female[5]
Mus musculus	33.07 g wet weight	Age class in weeks = 28, male[5]
Mus musculus	34.07 g wet weight	Age class in weeks = 32, male[5]
Mus musculus	34.52 g wet weight	Age class in weeks = 32, female[5]
Mus musculus	35.49 g wet weight	Age class in weeks = 36, male[5]
Mus musculus	36.38 g wet weight	Age class in weeks = 36, female[5]
Mus musculus	4.03 g wet weight	Age class in weeks = 1, male[5]
Mus musculus	4.04 g wet weight	Age class in weeks = 1, female[5]
Mus musculus	6.52 g wet weight	Age class in weeks = 2, female[5]
Mus musculus	6.54 g wet weight	Age class in weeks = 2, male[5]
Mus musculus	8.72 g wet weight	Age class in weeks = 3, female[5]
Mus musculus	8.76 g wet weight	Age class in weeks = 3, male[5]
Odocoileus hemionus columbianus	3.1 kg wet weight	Age class in years = 0, male[3]
Odocoileus hemionus columbianus	3.4 kg wet weight	Age class in years = 0, female[3]
Odocoileus hemionus columbianus	35.4 kg wet weight	Age class in years = 1, female[3]
Odocoileus hemionus columbianus	36.4 kg wet weight	Age class in years = 1, male[3]
Odocoileus hemionus columbianus	41.4 kg wet weight	Age class in years = 5, female[3]
Odocoileus hemionus columbianus	42.7 kg wet weight	Age class in years = 6, female[3]
Odocoileus hemionus columbianus	49.6 kg wet weight	Age class in years = 2, female[3]
Odocoileus hemionus columbianus	54.9 kg wet weight	Age class in years = 2, male[3]
Odocoileus hemionus columbianus	56.7 kg wet weight	Age class in years = 3, female[3]
Odocoileus hemionus columbianus	63.5 kg wet weight	Age class in years = 4, female[3]
Odocoileus hemionus columbianus	71.8 kg wet weight	Age class in years = 3, male[3]
Odocoileus hemionus columbianus	74.5 kg wet weight	Age class in years = 4, male[3]
Odocoileus hemionus columbianus	87.3 kg wet weight	Age class in years = 5, male[3]
Odocoileus hemionus columbianus	96.4 kg wet weight	Age class in years = 6, male[3]
Passer domesticus	10 g wet weight	Age class in days = 4, both sexes[6]
Passer domesticus	18 g wet weight	Age class in days = 8, both sexes[6]
Passer domesticus	2 g wet weight	Age class in days = 1, both sexes[6]
Passer domesticus	20 g wet weight	Age class in days = 10, both sexes[6]
Passer domesticus	25 g wet weight	Age class in days = 15, both sexes[6]
Passer domesticus	30 g wet weight	Adult, female[6]
Passer domesticus	30.5 g wet weight	Adult, male[6]

Rattus norveginus	113 g wet weight	Age class in weeks = 5, female[5]
Rattus norveginus	138 g wet weight	Age class in weeks = 5, male[5]
Rattus norveginus	147 g wet weight	Age class in weeks = 6, female[5]
Rattus norveginus	16.2 g wet weight	Age class in weeks = 1, female[5]
Rattus norveginus	17.5 g wet weight	Age class in weeks = 1, male[5]
Rattus norveginus	172 g wet weight	Age class in weeks = 7, female[5]
Rattus norveginus	188 g wet weight	Age class in weeks = 6, male[5]
Rattus norveginus	196 g wet weight	Age class in weeks = 8, female[5]
Rattus norveginus	227 g wet weight	Age class in weeks = 10, female[5]
Rattus norveginus	233 g wet weight	Age class in weeks = 7, male[5]
Rattus norveginus	251 g wet weight	Age class in weeks = 12, female[5]
Rattus norveginus	274 g wet weight	Age class in weeks = 15, female[5]
Rattus norveginus	274 g wet weight	Age class in weeks = 8, male[5]
Rattus norveginus	303 g wet weight	Age class in weeks = 20, female[5]
Rattus norveginus	33.5 g wet weight	Age class in weeks = 2, female[5]
Rattus norveginus	335 g wet weight	Age class in weeks = 30, female[5]
Rattus norveginus	339 g wet weight	Age class in weeks = 10, male[5]
Rattus norveginus	358 g wet weight	Age class in weeks = 40, female[5]
Rattus norveginus	36.9 g wet weight	Age class in weeks = 2, male[5]
Rattus norveginus	393 g wet weight	Age class in weeks = 12, male[5]
Rattus norveginus	440 g wet weight	Age class in weeks = 15, male[5]
Rattus norveginus	490 g wet weight	Age class in weeks = 20, male[5]
Rattus norveginus	5.75 g wet weight	Age class in weeks = 0, female[5]
Rattus norveginus	53 g wet weight	Age class in weeks = 3, female[5]
Rattus norveginus	59.6 g wet weight	Age class in weeks = 3, male[5]
Rattus norveginus	6.08 g wet weight	Age class in weeks = 0, male[5]
Rattus norveginus	79.5 g wet weight	Age class in weeks = 4, female[5]
Rattus norveginus	92.9 g wet weight	Age class in weeks = 4, male[5]
Sylvilagus floridanus	1106 g wet weight	Age class in weeks = 21, both sexes[7]
Sylvilagus floridanus	134.5 g wet weight	Age class in weeks = 4, both sexes[7]
Sylvilagus floridanus	234.5 g wet weight	Age class in weeks = 5, both sexes[7]
Sylvilagus floridanus	25.6 g wet weight	Age class in weeks = 0, both sexes[7]
Sylvilagus floridanus	274.5 g wet weight	Age class in weeks = 6, both sexes[7]
Sylvilagus floridanus	374.5 g wet weight	Age class in weeks = 7, both sexes[7]
Sylvilagus floridanus	414 g wet weight	Age class in weeks = 9, both sexes[7]
Sylvilagus floridanus	435.6 g wet weight	Age class in weeks = 8, both sexes[7]
Sylvilagus floridanus	497.4 g wet weight	Age class in weeks = 10, both sexes[7]
Sylvilagus floridanus	517.5 g wet weight	Age class in weeks = 11, both sexes[7]
Sylvilagus floridanus	58.3 g wet weight	Age class in weeks = 1, both sexes[7]
Sylvilagus floridanus	659.7 g wet weight	Age class in weeks = 12, both sexes[7]

Sylvilagus floridanus	68.5 g wet weight	Age class in weeks = 2, both sexes[7]
Sylvilagus floridanus	765.3 g wet weight	Age class in weeks = 13, both sexes[7]
Sylvilagus floridanus	785.4 g wet weight	Age class in weeks = 14, both sexes[7]
Sylvilagus floridanus	92.4 g wet weight	Age class in weeks = 3, both sexes[7]
Sylvilagus floridanus	945 g wet weight	Age class in weeks = 15, both sexes[7]
Sylvilagus floridanus	975.2 g wet weight	Age class in weeks = 16, both sexes[7]
Sylvilagus floridanus	981 g wet weight	Age class in weeks = 19, both sexes[7]

Source: 1.Bump, G., R.W. Darrow, F.C. Edminster and W.F. Crissley. 1947. The Ruffed Grouse: Life History - propagation management. Albany. New York Conservation Dept.
2.Robinette, W.L. and G.F.T. Child. 1964. Puku 2, pp 84-117.
3.Brown, E.R. 1961. Wash. State Game Dept. Biol. Bull, No 13, pp 124.
4.Petrides, G.A. and W.G. Swank. 1966. Proc. Internat. Grassland Cong., Sao Paulo, Brazil. 9, pp 831-92.
5.Altman, P.L. and D.S. Dittmar. 1964. Biology Data Book. Wash. D.C. Federation of Amer. Soc. Exp. Bio.
6.Summers-smith, D. 1963. The House Sparrow. Collins. London.
7.Lord, R.D., Jr. 1963. The Cottentail rabbit in Illinois. Carbondale. State Ill. Univ. Press.

1-428 Vertebrates
Br / biomass ratio

Species	Value	Condition
Homo sapiens	0.0002 %	Atomic composition[1]

Source: 1.Berthet, P. 1963. In: Doeksen, J. and J. van der Drift. (eds); Soil Organism. North-Holland Publ. Co, Amsterdam.

1-429 Vertebrates
C / biomass ratio

Species	Value	Condition
Animals	45 % of dry weight	Mean elementary composition[1]
Homo sapiens	19.37 %	Atomic composition[2]
Mammalia	48.4 % of dry weight	Mean elementary composition[1]

Source: 1.Bowen, H.J.M. 1966. Trace Elements in Biochemistry, Ac. Press.

2.Berthet, P. 1963. In: Doeksen, J. and J. van der Drift. (eds); Soil Organism. North-Holland Publ. Co, Amsterdam.

1-430 Vertebrates
Ca / biomass ratio

Species	Value	Condition
Homo sapiens	1.38 %	Atomic composition[1]

Source: 1.Berthet, P. 1963. In: Doeksen, J. and J. van der Drift. (eds); Soil Organism. North-Holland Publ. Co, Amsterdam.

1-431 Vertebrates
Caloric content

Species	Value	Condition
Apodemus agrarius	5.179 kcal/g dry weight	Equal to 1.64 kcal/g fresh weight[1]
Apodemus flavicollis	5.002 kcal/g dry weight	Equal to 1.40 kcal/g fresh weight[1]
Bufo bufo	5.003 kcal/g dry weight	Equal to 1.25 kcal/g fresh weight[2]
Capreolus capreolus	6.065 kcal/g dry weight	Equal to 2.16 kcal/g fresh weight[3]
Clethrionomys glareolus	4.934 kcal/g dry weight	Equal to 1.45 kcal/g fresh weight[1]
Cyanocitta cristata	5.09 kcal/g dry weight	None[4]

Hylocichla mustellina	5.99 kcal/g dry weight	After migration[4]
Hylocichla mustellina	7.725 kcal/g dry weight	Before migration[4]
Lepus europaeus	7.5 kcal/g dry weight	Equal to 1.87 kcal/g fresh weight[5]
Microtus arvalis	4.856 kcal/g dry weight	Equal to 1.47 kcal/g fresh weight[1]
Microtus pennsylvanicus	4.65 kcal/g dry weight	Equal to 1.40 kcal/g fresh weight[3]
Mus musculus	5.675 kcal/g dry weight	Equal to 1.70 kcal/g fresh weight[3]
Mus musculus	6.205 kcal/g dry weight	Equal to 2.25 kcal/g fresh weight[6]
Oryzomys palustris	5.84 kcal/g dry weight	Equal to 1.90 kcal/g fresh weight[7]
Parula americana	6.075 kcal/g dry weight	After migration[4]
Parula americana	7.68 kcal/g dry weight	Before migration[4]
Passer domesticus	5.512 kcal/g dry weight	Equal to 1.91 kcal/g fresh weight[8]
Passer domesticus	5.535 kcal/g dry weight	None[4]
Perdix perdix	5.253 kcal/g dry weight	Equal to 1.90 kcal/g fresh weight[9]
Phoca hispida	3.8 kcal/g wet weight	2 - 5.5 % ash, rg = 2.3 - 5.3[10]
Piranga olivacea	6.645 kcal/g dry weight	After migration[4]
Piranga olivacea	7.575 kcal/g dry weight	Before migration[4]
Rana arvalis	4.628 kcal/g dry weight	Equal to 1.16 kcal/g fresh weight[2]
Richmodena cardinalis	5.53 kcal/g dry weight	None[4]
Sorex araneus	4.555 kcal/g dry weight	Equal to 1.50 kcal/g fresh weight[1]
Sorex araneus	4.991 kcal/g dry weight	Equal to 1.53 kcal/g fresh weight[11]
Sorex minutus	5.025 kcal/g dry weight	Equal to 1.63 kcal/g fresh weight[11]
Telmatodytes palustris	6.17 kcal/g ash free dry weight	None[12]
Zenoidura macraura	2.61 kcal/g fresh weight	None[13]

Source: 1.Gorecki, A. 1969. Acta theriol. 18, pp 333-352.
2.Mazur, T. 1967. Ekol. pol. 15, pp 607-613.
3.Golley, F.B. 1959. Mimeogr. Univ. Georgia, pp 7.
4.Odum, E.P., G.S. Marshall and T.G. Marples. 1965. Ecology 46, pp 901-904.
5.Myrcha, A. 1967. Small Mammals Newslett. 3, pp 10.
6.Myrcha, A. and W. Walkowa. 1968. Acta theriol 13, pp 391-400.
7.Davis, D.E. and F.B. Golley. 1963. Principles in Mammalogy. Reinh. Publ. Corp. London.
8.Golley, F.B. 1967. in: Petrusewicz, K. (ed.); Secondary Productivity of Terrestrial Ecosystems. pp 9-124. Warszawa. Krakow.
9.Szwykowska, M.M. 1969. Ekol. pol. 17, pp 795-809.
10.Stirling, I. and E.H. McEwan. 1975. Can. J. Zool. 53(8), pp 1021-1027.
11.Myrcha, A. 1969. Acta theriol. 14, pp 211-227.
12.Kale, H.W. 1965. Publ. Nutt. Ornithol. Club. Cambridge. Massachusettes.
13.Brisbin, I.L. 1968. Ecology 49, pp 792-794.

1-432 Vertebrates
Cl / biomass ratio

Species	Value	Condition
Homo sapiens	0.18 %	Atomic composition[1]

Source: 1.Berthet, P. 1963. In: Doeksen, J. and J. van der Drift. (eds); Soil Organism. North-Holland Publ. Co, Amsterdam.

1-433 Vertebrates
Consumption

Species	Value	Condition
Balaenoptera borealis	4.43 % of body wet weigth	Equal to 900 kg food per day (ww)[1]
Clethrionomys glareolus and Apodemus flavicollis	20000 kcal/(ha*year)	Food avai = 1024E3 kcal/(ha*y), cladonia, Poland[2]
Clethrionomys glareolus and Apodemus flavicollis	31500 kcal/(ha*year)	Food avai = 2415-7080E3 kcal/(ha*y), pinewood, Poland, rg = 21-42E3[2]
Clethrionomys glareolus and Apodemus flavicollis	75000 kcal/(ha*year)	Food avai = 1950E3 kcal/(ha*y), beechwood, Poland[2]
Clethrionomys glareolus and Apodemus flavicollis	88500 kcal/(ha*year)	Food avai = 13040E3 kcal/(ha*y), oak-ppine forest, Poland[2]
Clethrionomys glareolus, Apodemus flavicollis and Apodemus agrarius	95000 kcal/(ha*year)	Food avai = 2050E3 kcal/(ha*y), oak-hornbeam forest, Poland[2]
Clethrionomys rutilus and Tamiasciurus sp.	179000 kcal/(ha*year)	Food avai = 1320E3 kcal/(ha*y), taiga forest, Alaska[2]
Dipodomys merriami and Lepus californicus	131000 kcal/(ha*year)	Food avai = 2395E3 kcal/(ha*y), desert shrub, Arizona[2]
Microtus agrestis	515000 kcal/(ha*year)	Food avai = 1950E3 kcal/(ha*y), spruce forest, Sweden, rg = 3.3-7E5[2]
Microtus oeconomus	375000 kcal/(ha*year)	Food avai = 6650E3 kcal/(ha*y), forest on peat-bog, Poland[2]

Source: 1.Kawamura, A. 1974. Sci. Rep. Whales Res. Inst. Tokyo. 26, pp 25-144.
2.Grodzwsky, W. 1975. in: Grodzinsky, W., R.Z. Klekowsky and A. Duncan (eds). Methods for Ecological Boenergetics IBP Handbook No. 24, pp 65-94. Blackwell, Oxford.

1-434 Vertebrates
Cu / biomass ratio

Species	Value	Condition
Homo sapiens	0.0004 %	Atomic composition[1]

Source: 1.Berthet, P. 1963. In: Doeksen, J. and J. van der Drift. (eds); Soil Organism. North-Holland Publ. Co, Amsterdam.

1-435 Vertebrates
Efficiency

Species	Value	Condition
Phallusia mammillata	90 % converted of filtrated amount	Rg = 86-93, pH = 7.8, 24 h, 288 K, O2 = 5 mg/l, s = 3.8%[1]

Source: 1.Finenko, Z.Z. and D.K. Krupatkina-Aki Nina. 1974. Mar. Biol., 26, pp 193-201.

1-436 Vertebrates
Fat / biomass ratio

Species	Value	Condition
Carp	5.5 % of weight	Dependent and sex, age and condition, rg = +-1-10 %[1]
Cod and haddock	0.5 % of weight	Dependent and sex, age and condition, rg = +-0-1 %[1]
Domestic animals and homo sapiens	19 % of weight	Dependent and sex, age and condition, rg = +-13-25 %[1]
Eel	22.5 % of weight	Dependent and sex, age and condition, rg = +-15-30 %[1]
Herring	12.5 % of weight	Dependent and sex, age and condition, rg = +-5-20 %[1]
Insects	10 % of weight	Dependent and sex, age and condition[1]

Lobster	3 % of weight	Dependent and sex, age and condition[1]
Marine mammals	15 % of weight	Dependent and sex, age and condition, more than value[1]
Oyster	1.2 % of weight	Dependent and sex, age and condition[1]
Perch and pike	1.5 % of weight	Dependent and sex, age and condition, rg = +-1-2 %[1]
Salmon	8.5 % of weight	Dependent and sex, age and condition, rg = +-2-15 %[1]
Trout	1 % of weight	Dependent and sex, age and condition[1]
Wild living mammals and birds	2 % of weight	Dependent and sex, age and condition, rg = +-1-3 %[1]

Source: 1.Bruggeman, W.A. 1982. In: O. Hutzinger: Environmental chemistry. Reactions and processes. Vol 2. Part B. Springer-verlag. Berlin. 83-102.

1-437 Vertebrates
Fe / biomass ratio

Species	Value	Condition
Homo sapiens	0.005 %	Atomic composition[1]

Source: 1.Berthet, P. 1963. In: Doeksen, J. and J. van der Drift. (eds); Soil Organism. North-Holland Publ. Co, Amsterdam.

1-438 Vertebrates
Filtration rate

Species	Value	Condition
Phallusia mammillata	4380 ml/(hour*g org dry weight)	Food = Monochrysis lutheri, 24 h, 288 K, pH = 7.8, O2 = 5 mg/l[1]

Source: 1.Finenko, Z.Z. and D.K. Krupatkina-Aki Nina. 1974. Mar. Biol., 26, pp 193-201.

1-439 Vertebrates
Growth efficiency

Species	Value	Condition
Adenota kobthomasi	1.3 % of ingestion	Production = 0.81 kcal/(m2*y)[1]
Clethrionomys rutilus	1.89 % of ingestion	Production = 0.00024 kcal/(m2*y)[1]
Glaucomys sabrinus yukonensis	1.18 % of ingestion	Production = 0.00002 kcal/(m2*y)[1]
Leptoterna dolobrata	50.8 % of ingestion	Production = 0.064 kcal/(m2*y)[1]
Loxodonta africana	1.46 % of ingestion	Production = 0.34 kcal/(m2*y)[1]
Microtus oeconomus	1.82 % of ingestion	Production = 0.00024 kcal/(m2*y)[1]
Microtus pennsylvanicus	2.95 % of ingestion	Production = 0.517 kcal/(m2*y)[1]
Mustela rixosa alleghaniensis	2.34 % of ingestion	Production = 0.013 kcal/(m2*y)[1]
Passerculus sandwichensis	1.11 % of ingestion	Production = 0.34 kcal/(m2*y)[1]
Peromyscus polionotus	1.79 % of ingestion	Production = 0.12 kcal/(m2*y)[1]
Pitymys subterraneus	1.57 % of ingestion	Production = 0.0955 kcal/(m2*y)[1]
Tamiasciurus hudsoni preblei	1.37 % of ingestion	Production = 0.00020 kcal/(m2*y)[1]

Source: 1.McNeill, S. and J.H. Lawton. 1970. Nature 225, pp 472-474

1-440 Vertebrates
Growth rate

Species	Value	Condition
Microtus orcadensis	12.2 rate	Net productation rate, lenght of 1 generation = 332.4 d[1]
Rattus norvegicus	16.5 rate	Net productation rate, lenght of 1 generation = 344.1 d[1]
Rattus norvegicus	16.8 rate	Net productation rate, lenght of 1 generation = 236.5 d[1]

Source: 1.Laughlin, R. 1965. J. Anim. Ecol. 34, pp 77-91.

1-441 Vertebrates
H / biomass ratio

Species	Value	Condition
Animals	6.5 % of dry weight	Mean elementary composition[1]
Homo sapiens	9.31 %	Atomic composition[2]
Mammalia	6.6 % of dry weight	Mean elementary composition[1]

Source: 1.Bowen, H.J.M. 1966. Trace Elements in Biochemistry, Ac. Press.
2.Berthet, P. 1963. In: Doeksen, J. and J. van der Drift. (eds); Soil Organism. North-Holland Publ. Co, Amsterdam.

1-442 Vertebrates
I / biomass ratio

Species	Value	Condition
Homo sapiens	0.0001 %	Atomic composition[1]

Source: 1.Berthet, P. 1963. In: Doeksen, J. and J. van der Drift. (eds); Soil Organism. North-Holland Publ. Co, Amsterdam.

1-443 Vertebrates
K / biomass ratio

Species	Value	Condition
Homo sapiens	0.22 %	Atomic composition[1]

Source: 1.Berthet, P. 1963. In: Doeksen, J. and J. van der Drift. (eds); Soil Organism. North-Holland Publ. Co, Amsterdam.

1-444 Vertebrates
Mg / biomass ratio

Species	Value	Condition
Homo sapiens	0.04 %	Atomic composition[1]

Source: 1.Berthet, P. 1963. In: Doeksen, J. and J. van der Drift. (eds); Soil Organism. North-Holland Publ. Co, Amsterdam.

1-445 Vertebrates
Mn / biomass ratio

Species	Value	Condition
Homo sapiens	0.0001 %	Atomic composition[1]

Source: 1.Berthet, P. 1963. In: Doeksen, J. and J. van der Drift. (eds); Soil Organism. North-Holland Publ. Co, Amsterdam.

1-446 Vertebrates
N / biomass ratio

Species	Value	Condition
Animals	10 % of dry weight	Mean elementary composition[1]
Homo sapiens	5.14 %	Atomic composition[2]
Mammalia	12.8 % of dry weight	Mean elementary composition[1]

Source: 1.Bowen, H.J.M. 1966. Trace Elements in Biochemistry, Ac. Press.
2.Berthet, P. 1963. In: Doeksen, J. and J. van der Drift. (eds); Soil Organism. North-Holland Publ. Co, Amsterdam.

1-447 Vertebrates
Na / biomass ratio

Species	Value	Condition
Homo sapiens	0.26 %	Atomic composition[1]

Source: 1.Berthet, P. 1963. In: Doeksen, J. and J. van der Drift. (eds); Soil Organism. North-Holland Publ. Co, Amsterdam.

1-448 Vertebrates
O / biomass ratio

Species	Value	Condition
Animals	30 % of dry weight	Mean elementary composition[1]
Homo sapiens	62.81 %	Atomic composition[2]
Mammalia	18.6 % of dry weight	Mean elementary composition[1]

Source: 1.Bowen, H.J.M. 1966. Trace Elements in Biochemistry, Ac. Press.
2.Berthet, P. 1963. In: Doeksen, J. and J. van der Drift. (eds); Soil Organism. North-Holland Publ. Co, Amsterdam.

1-449 Vertebrates
P / biomass ratio

Species	Value	Condition
Homo sapiens	0.63 %	Atomic composition[1]

Source: 1.Berthet, P. 1963. In: Doeksen, J. and J. van der Drift. (eds); Soil Organism. North-Holland Publ. Co, Amsterdam.

1-450 Vertebrates
Pb / biomass ratio

Species	Value	Condition
Homo sapiens	0.0001 %	Atomic composition[1]

Source: 1.Berthet, P. 1963. In: Doeksen, J. and J. van der Drift. (eds); Soil Organism. North-Holland Publ. Co, Amsterdam.

1-451 Vertebrates
Rb / biomass ratio

Species	Value	Condition
Homo sapiens	0.0009 %	Atomic composition[1]

Source: 1.Berthet, P. 1963. In: Doeksen, J. and J. van der Drift. (eds); Soil Organism. North-Holland Publ. Co, Amsterdam.

1-452 Vertebrates
S / biomass ratio

Species	Value	Condition
Animals	1.3 % of dry weight	Mean elementary composition[1]
Homo sapiens	0.64 %	Atomic composition[2]
Mammalia	1.6 % of dry weight	Mean elementary composition[1]

Source: 1.Bowen, H.J.M. 1966. Trace Elements in Biochemistry, Ac. Press.
2.Berthet, P. 1963. In: Doeksen, J. and J. van der Drift. (eds); Soil Organism. North-Holland Publ. Co, Amsterdam.

1-453 Vertebrates
Si / biomass ratio

Species	Value	Condition
Homo sapiens	0.004 %	Atomic composition[1]

Source: 1.Berthet, P. 1963. In: Doeksen, J. and J. van der Drift. (eds); Soil Organism. North-Holland Publ. Co, Amsterdam.

1-454 Vertebrates
Sn / biomass ratio

Species	Value	Condition
Homo sapiens	0.0002 %	Atomic composition[1]

Source: 1.Berthet, P. 1963. In: Doeksen, J. and J. van der Drift. (eds); Soil Organism. North-Holland Publ. Co, Amsterdam.

1-455 Vertebrates
Thermal conductivity

Species	Value	Condition
Caribou	0.0049 cal/(cm*min*deg)	Conductivity of winter fur, thickness = 3.5 cm[1]
Dall sheep	0.0081 cal/(cm*min*deg)	Conductivity of winter fur, thickness = 7 cm[1]
Dog	0.007 cal/(cm*min*deg)	Conductivity of winter fur, thickness = 4.0 cm[1]
Polar bear	0.0113 cal/(cm*min*deg)	Conductivity of winter fur, thickness = 6.5 cm[1]
Rabbit	0.0052 cal/(cm*min*deg)	Conductivity of winter fur, thickness = 3.0 cm[1]
Red fox	0.0078 cal/(cm*min*deg)	Conductivity of winter fur, thickness = 4.5 cm[1]
Shrew	0.0035 cal/(cm*min*deg)	Conductivity of winter fur, thickness = 0.5 cm, summer[1]
Spermophilus undulatus	0.0052 cal/(cm*min*deg)	Conductivity of winter fur, thickness = 1.5 cm[1]
Weasel	0.0056 cal/(cm*min*deg)	Conductivity of winter fur, thickness = 0.8 cm[1]
White fox	0.0058 cal/(cm*min*deg)	Conductivity of winter fur, thickness = 5 cm[1]
Wolf	0.009 cal/(cm*min*deg)	Conductivity of winter fur, thickness = 6.5 cm[1]

Source: 1.Scholander, P.F. 1955. Evolution 9, pp 15-26.

1-456 Vertebrates
Zn / biomass ratio

Species	Value	Condition
Homo sapiens	0.0025 %	Atomic composition[1]

Source: 1.Berthet, P. 1963. In: Doeksen, J. and J. van der Drift. (eds); Soil Organism. North-Holland Publ. Co, Amsterdam.

1-457 Zooplankton
ATP / C ratio

Species	Value	Condition
Calanus finmarchicus	0.86 % of total carbon	Starvation in 14 days, stage 4 and 5, SD =+-0.39[1]
Calanus finmarchicus	0.87 % of total carbon	No starvation, stage 4 and 5, SD =+-0.33[1]
Calanus finmarchicus	0.91 % of total carbon	Starvation in 23 days, stage 4 and 5, SD =+-0.33[1]
Calanus finmarchicus	0.92 % of total carbon	Starvation in 8 days, stage 4 and 5, SD =+-0.37[1]
Mesocyclops leuckarti	0.00793 by weight	Copepodite[2]
Mesocyclops leuckarti	0.0106 by weight	Adult female copepod[2]

Source: 1.Balch, N. 1972. Limnol. Oceanogr., 17(6), pp 906-908.
2.Cavari, B. 1976. Limnol. Oceanogr. 21(2), pp 231-236.

1-458 Zooplankton
ATP / biomass ratio

Species	Value	Condition
Calanus finmarchicus	0.9756 microgram ATP / animal	Starvation in 14 days, n=10, stage 4 and 5, SD =+-451.6[1]
Calanus finmarchicus	1.2686 microgram ATP / animal	Starvation in 23 days, n=10, stage 4 and 5, SD =+-544.0[1]
Calanus finmarchicus	1.2711 microgram ATP / animal	Starvation in 8 days, n=10, stage 4 and 5, SD =+-326.1[1]
Calanus finmarchicus	1.6142 microgram ATP / animal	No starvation, n=10, stage 4 and 5, SD =+-507.6[1]
Calanus finmarchicus	1.9 % of nonlipid carbon	Starvation in 14 days, stage 4 and 5, nonlipid C, SD =+-0.87[1]
Calanus finmarchicus	2.04 % of nonlipid carbon	Starvation in 23 days, stage 4 and 5, nonlipid C, SD =+-0.78[1]
Calanus finmarchicus	2.07 % of nonlipid carbon	Starvation in 8 days, stage 4 and 5, nonlipid C, SD =+-0.70[1]
Calanus finmarchicus	2.23 % of nonlipid carbon	No starvation, stage 4 and 5, nonlipid C, SD =+-0.70[1]

Source: 1.Balch, N. 1972. Limnol. Oceanogr., 17(6), pp 906-908.

1-459 Zooplankton
Al / biomass ratio

Species	Value	Condition
Calanus finmarchicus	0 %	Atomic composition[1]

Source: 1.Vernadsky, W. 1933. Z. Krist. Mineral. Petrog. Abt. B Mineral. Petrog. Mitt., 44, p 191.

1-460 Zooplankton
Ash / biomass ratio

Species	Value	Condition
Amphipoda	23.5 % of dry weight	Dried at 333 K, North Pacific, rg = 10 - 37[1]
Chaetognatha	4.5 % of dry weight	Dried at 333 K, North Pacific, rg = 4 - 5[1]
Copepoda	20.5 % of dry weight	Dried at 378 K, continental shelf off New York, 18 - 23[2]
Copepoda	4 % of dry weight	Dried at 333 K, North Pacific, rg = 2.0 - 6.0[1]
Ctenophora	72.5 % of dry weight	Dried at 378 K, continental shelf off New York, rg = 70 - 75[2]
Daphnia cristata	8.07 % of dry weight	Cladocera , 95 % CL = +- 0.03, 12 h at 873 K[3]
Eudiaptomus gracilis	5.16 % of dry weight	Copepoda , 95 % CL = +- 0.05, 12 h at 873 K[3]
Euphausiacea	20.5 % of dry weight	Dried at 378 K, continental shelf off New York, rg = 19 - 22[2]
Euphausiacea	8.5 % of dry weight	Dried at 333 K, North Pacific, rg = 8 - 9[1]

Limnocalanus macrurus	2.92 % of dry weight	Copepoda , 95 % CL = +- 0.05, 12 h at 873 K[3]
Macrocyclops albidus	4.66 % of dry weight	Copepoda , 95 % CL = +- 0.02, 12 h at 873 K[3]
Pteropods	36 % of dry weight	Dried at 333 K, North Pacific, rg = 29 - 43[1]
Pteropods	44 % of dry weight	Dried at 378 K, continental shelf off New York, rg = 24 - 64[2]
Tunicata	74 % of dry weight	Dried at 378 K, continental shelf off New York, rg = 71 - 77[2]

Source: 1.Omori, M. 1969. Mar. Biol., 3, pp 4-10.
2.Curl, H. 1962. Rapp. Proc.-Verb. Cons. Int. Explor. Mer., 153, pp 183-189.
3.Salonen, K., J. Sarvala, I. Hakala and M.-L. Viljanen. 1976. Limnol. Oceanogr., 21(5), pp 724-730.

1-461 Zooplankton
Assimilation

Species	Value	Condition
Ceriodaphnia reticulata	10.6 % of ingestion	None[1]
Daphnia pulex	100 % of ingested food	Food = Anabaena flos-aquae, 298 K[2]
Daphnia pulex	100 % of ingested food	Food = Chlorella vulgaris, 298 K[2]
Daphnia pulex	13.7 % of consumed	Food = 100E-4 cell/ml, Chlamydomoas sp., 40 d, after 18 instars[3]
Daphnia pulex	14 % of energy consumed	0-40 d old, food = 100000 cells/ml, Chlamydomonas reinhardtii[4]
Daphnia pulex	14.2 % of energy consumed	6-34 d old, food = 100000 cells/ml, Chlamydomonas reinhardtii[4]
Daphnia pulex	15.8 % of energy consumed	0-6 d old, food = 50000 cells/ml, Chlaymdomonas reinhardtii[4]
Daphnia pulex	15.8 % of ingested food	Food = Anacystis nidulans, 298 K[2]
Daphnia pulex	16 % of energy consumed	0-40 d old, food = 75000 cells/ml, Chlamydomonas reinhardtii[4]
Daphnia pulex	16.3 % of consumed	Food = 75E-4 cell/ml, Chlamydomoas sp., 40 d, after 18 instars[3]
Daphnia pulex	16.8 % of energy consumed	6-34 d old, food = 75000 cells/ml, Chlamydomonas reinhardtii[4]
Daphnia pulex	20 % of consumed	Food = 50E-4 cell/ml, Chlamydomoas sp., 40 d, after 18 instars[3]
Daphnia pulex	20 % of energy consumed	0-40 d old, food = 50000 cells/ml, Chlamydomonas reinhardtii[4]
Daphnia pulex	20.2 % of energy consumed	6-34 d old, food = 50000 cells/ml, Chlamydomonas reinhardtii[4]
Daphnia pulex	22.5 % of ingestion	Range = 14 - 31 %[1]
Daphnia pulex	23.9 % of energy consumed	0-6 d old, food = 25000 cells/ml, Chlaymdomonas reinhardtii[4]
Daphnia pulex	31 % of energy consumed	0-40 d old, food = 25000 cells/ml, Chlamydomonas reinhardtii[4]
Daphnia pulex	31.3 % of consumed	Food = 25E-4 cell/ml, Chlamydomoas sp., 40 d, after 18 instars[3]
Daphnia pulex	31.7 % of energy consumed	6-34 d old, food = 25000 cells/ml, Chlamydomonas reinhardtii[4]
Daphnia pulex	6.6 % of energy consumed	0-6 d old, food = 100000 cells/ml, Chlamydomonas reinhardtii[4]
Daphnia pulex	67.2 % of ingested food	Food = Synechococcus elongata, 298 K[2]
Daphnia pulex	70.6 % of ingested food	Food = Gloeocapsa alpicola, 298 K[2]
Daphnia pulex	71 % of ingested food	Food = Ankistrodesmus falcatus, 298 K[2]
Daphnia pulex	77.8 % of ingested food	Food = Synechocystis sp., 298 K[2]
Daphnia pulex	8.4 % of energy consumed	0-6 d old, food = 75000 cells/ml, Chlamydomonas reinhardtii[4]
Daphnia pulex	81.3 % of ingested food	Food = Synechococcus cedrorum, 298 K[2]
Daphnia pulex	85.3 % of ingested food	Food = Merismopedia sp., 298 K[2]
Daphnia schodleri	78 % of ingestion	295 K, no eggs, filtering rate = 0.96 ml/(animal*hr), n=5[5]
Daphnia schodleri	81 % of ingestion	295 K, with eggs, filtering rate = 1.06 ml/(animal*hr), n=5[5]
Daphnia schodleri	82 % of ingestion	283 K, no eggs, filtering rate = 0.55 ml/(animal*hr), n=5[5]

Daphnia schodleri	94 % of ingestion	283 K, with eggs, filtering rate = 0.50 ml/(animal*hr), n=5[5]
Leptodora kindtii	40 % of ingestion	None[1]
Mesocyclops albidus	42.5 % of ingestion	Range = 20 - 75 %[1]
Simocephalus vectulus	31.7 % of ingestion	Reproducing female[1]
Simocephalus vectulus	72.4 % of ingestion	Juvenile, female[1]
Zooplankton	0.2 % of ingested Zn	Calculated on relative concentration factor, marine[6]
Zooplankton	0.6 mg C/mg C	Delta Estuary[7]
Zooplankton	0.6 mg C/mg C	Lake Ontario[7]
Zooplankton	0.6 mg C/mg C	San Joaquin River[7]
Zooplankton	0.65 mg C/mg C	Lake Erie[7]

Source: 1.Wetzel, R.G. 1975. Limnology. Saunders. Philadelphia.
2.Arnold, D.E. 1971. Limnol. Oceanogr., 16(6), pp 906-920.
3.Richman, S. 1958. Ecol. Monogr. 28, pp 273-291.
4.Richman, S. 1958. Ecol. Monogr. 28, pp 273-291.
5.Hayward, R.S. and D.N. Gallup. 1976. Arch. Hydrobiol., 77(2), pp 139-163.
6.Wolfe, 1975. In: Cronin, L.E. (ed.); Estuarine Research, Vol. 1. Ac. Press N.Y.
7.O'Connor, D.J., D.M. DiToro and R.V. Thomann. 1975. in: Russel, C.S. (ed.); Ecological Modeling in aResource Management Framework. pp 149-209.

1-462 Zooplankton Biomass

Species	Value	Condition
Calanoids	13.4 ug dry wt. l-1	Brompton, Canada[1]
Calanoids	16.3 ug dry wt. l-1	Argent, Canada[1]
Calanoids	16.9 ug dry wt. l-1	Magog, Canada[1]
Calanoids	18.4 ug dry wt. l-1	Bowker, Canada[1]
Calanoids	18.5 ug dry wt. l 1	Waterloo, Canada[1]
Calanoids	2.5 ug dry wt. l-1	Massawippi, Canada[1]
Calanoids	25.6 ug dry wt. l-1	Orford, Canada[1]
Calanoids	4.4 ug dry wt. l-1	Central basin, Canada[1]
Calanoids	57.6 ug dry wt. l-1	Brome, Canada[1]
Calanoids	6.1 ug dry wt. l-1	Lovering, Canada[1]
Calanoids	6.2 ug dry wt. l-1	North basin, Canada[1]
Calanoids	8.4 ug dry wt. l-1	South basin, Canada[1]
Cladocerans	101.8 ug dry wt. l-1	Waterloo, Canada[1]
Cladocerans	13.5 ug dry wt. l-1	Brompton, Canada[1]
Cladocerans	14.8 ug dry wt. l-1	Central basin, Canada[1]
Cladocerans	2.9 ug dry wt. l-1	Massawippi, Canada[1]
Cladocerans	20.5 ug dry wt. l-1	Lovering, Canada[1]
Cladocerans	29.0 ug dry wt. l-1	Orford, Canada[1]
Cladocerans	29.9 ug dry wt. l-1	Argent, Canada[1]
Cladocerans	34.0 ug dry wt. l-1	South basin, Canada[1]
Cladocerans	35.2 ug dry wt. l-1	North basin, Canada[1]
Cladocerans	48.6 ug dry wt. l-1	Brome, Canada[1]
Cladocerans	5.8 ug dry wt. l-1	Bowker, Canada[1]

Cladocerans	87.4 ug dry wt. l-1	Magog, Canada[1]
Cyclopoids	10.9 ug dry wt. l-1	Orford, Canada[1]
Cyclopoids	23.6 ug dry wt. l-1	Lovering, Canada[1]
Cyclopoids	25.1 ug dry wt. l-1	Brome, Canada[1]
Cyclopoids	27.6 ug dry wt.l-1	Argent, Canada[1]
Cyclopoids	37.2 ug dry wt.l-1	North basin, Canada[1]
Cyclopoids	43.0 ug dry wt. l-1	Central basin, Canada[1]
Cyclopoids	44.2 ug dry wt. l-1	South basin, Canada[1]
Cyclopoids	65.4 ug dry wt. l-1	Waterloo, Canada[1]
Cyclopoids	67.3 ug dry wt. l-1	Massawippi, Canada[1]
Cyclopoids	7.1 ug dry wt. l-1	Brompton, Canada[1]
Cyclopoids	9.1 ug dry wt. l-1	Bowker, Canada[1]
Cyclopoids	91.1 ug dry wt. l-1	Magog, Canada[1]
Invertebrates	12.3 ug dry wt. l-1	Lovering, Canada[1]
Invertebrates	13.5 ug dry wt. l-1	Brompton, Canada[1]
Invertebrates	13.5 ug dry wt. l-1	Orford, Canada[1]
Invertebrates	14.0 ug dry wt. l-1	Bowker, Canada[1]
Invertebrates	18.3 ug dry wt. l-1	North basin, Canada[1]
Invertebrates	18.7 ug dry wt. l-1	Argent, Canada[1]
Invertebrates	22.5 ug dry wt. l-1	Brome, Canada[1]
Invertebrates	29.1 ug dry wt. l-1	Central basin, Canada[1]
Invertebrates	32.5 ug dry wt. l-1	Massawippi, Canada[1]
Invertebrates	33.2 ug dry wt. l-1	South Basin, Canada[1]
Invertebrates	55.8 ug dry wt. l-1	Magog, Canada[1]
Invertebrates	55.8 ug dry wt. l-1	Waterloo, Canada[1]
Nauplii	18.2 ug dry wt. l-1	Brompton, Canada[1]
Nauplii	25.9 ug dry wt. l-1	Central basin, Canada[1]
Nauplii	29.8 ug dry wt. l-1	Argent, Canada[1]
Nauplii	30.7 ug dry wt. l-1	Brome, Canada[1]
Nauplii	34.8 ug dry wt. l-1	Lovering, Canada[1]
Nauplii	40.8 ug dry wt. l-1	North basin, Canada[1]
Nauplii	41.1 ug dry wt. l-1	South basin, Canada[1]
Nauplii	46.0 ug dry wt. l-1	Magog, Canada[1]
Nauplii	50.0 ug dry wt. l-1	Massawippi, Canada[1]
Nauplii	54.4 ug dry wt. l-1	Waterloo, Canada[1]
Nauplii	6.5 ug dry wt. l-1	Bowker, Canada[1]
Nauplii	8.8 ug dry wt. l-1	Orford, Canada[1]
Rotifers	16.3 ug dry wt. l-1	Magog, Canada[1]
Rotifers	17.3 ug dry wt. l-1	Central basin, Canada[1]
Rotifers	19.3 ug dry wt. l-1	Massawippi, Canada[1]

Rotifers	2.5 ug dry wt. l-1	Bowker, Canada[1]
Rotifers	23.9 ug dry wt. l-1	North basin, Canada[1]
Rotifers	28.9 ug dry wt. l-1	South Basin, Canada[1]
Rotifers	3.7 ug dry wt. l-1	Brompton, Canada[1]
Rotifers	4.2 ug dry wt. l-1	Orford, Canada[1]
Rotifers	5.4 ug dry wt. l-1	Lovering, Canada[1]
Rotifers	6.4 ug dry wt. l-1	Argent, Canada[1]
Rotifers	7.0 ug dry wt. l-1	Waterloo, Canada[1]
Rotifers	7.6 ug dry wt. l-1	Brome, Canada[1]
Total standing biomass of society	0.1-10 mg dry wt l-1	from curve[2]
Individual weight	0.2-7 ug ind.-1	mean individual dry wt[2]

Source: 1.Pace, M.L. Limnol. Oceanogr., 31, 45, 1986.
2.Cryer, M., Peirson, G. and Townsend, C.R. Limnol. Oceanogr., 31(5), 1022, 1986.

1-463 Zooplankton
Birth rate

Species	Value	Condition
Bosmina longirostris	0.064 d-1	enclosure experiment with addition of Daphnia pulex[1]
Bosmina longirostris	0.099 d-1	controls for enclosure experiment[1]
Bosmina longirostris	0.128 d-1	enclosure experiment with addition of nutrients and Daphnia[1]
Bosmina longirostris	0.224 d-1	enclosure experiment with addition of nutrients[1]

Source: 1.Vanni, M.J. Limnol. Oceanogr. 31(5), 1039, 1986.

1-464 Zooplankton
Br / biomass ratio

Species	Value	Condition
Calanus finmarchicus	0.0009 %	Atomic composition[1]

Source: 1.Vernadsky, W. 1933. Z. Krist. Mineral. Petrog. Abt. B Mineral. Petrog. Mitt., 44, p 191.

1-465 Zooplankton
C / biomass ratio

Species	Value	Condition
Amphipoda	37 % of dry weight	Dried at 333 K, North Pacific, rg = 26 - 48[1]
Calanus finmarchicus	118.1 microgram C/animal	Starvation in 14 days, n=10, stage 4 and 5, SD =+-53.8[2]
Calanus finmarchicus	153 microgram C/animal	Starvation in 8 days, n=10, stage 4 and 5, SD =+-44.3[2]
Calanus finmarchicus	153.3 microgram C/animal	Starvation in 23 days, n=12, stage 4 and 5, SD =+-26.5[2]
Calanus finmarchicus	187.4 microgram C/animal	No starvation, n=9, stage 4 and 5, SD =+-71.6[2]
Calanus finmarchicus	6.1 %	Atomic composition[3]
Chaetognatha	28 % of dry weight	Dried at 333 K, Sargasso Sea, rg = 22 - 34[4]
Chaetognatha	46 % of dry weight	Dried at 333 K, North Pacific, rg = 44 - 48[1]
Copepoda	37 % of dry weight	Dried at 378 K, continental shelf off New York, 32 - 42[5]
Copepoda	41.5 % of dry weight	Dried at 333 K, Sargasso Sea, rg = 35 - 48[4]
Copepoda	52.5 % of dry weight	Dried at 333 K, North Pacific, rg = 39 - 66[1]

Ctenophora	6.4 % of dry weight	Dried at 378 K, continental shelf off New York[5]
Daphnia cristata	50.7 % of dry weight	Cladocera, 95 % CL = +-0.8, total C[6]
Daphnia cristata	55.2 % of ash free dry weight	Cladocera, 95 % CL +-0.9, organic C[6]
Eudiaptomus gracilis	49.8 % of dry weight	Copepoda, 95 % CL = +-0.4, total C[6]
Eudiaptomus gracilis	52.6 % of ash free dry weight	Copepoda, 95 % CL = +-0.4, organic C[6]
Euphausiacea	35 % of dry weight	Dried at 378 K, continental shelf off New York, rg = 33 - 37[5]
Euphausiacea	43 % of dry weight	Dried at 333 K, North Pacific, rg = 39 - 47[1]
Euphausiacea and Mysidae	39 % of dry weight	Dried at 333 K, Sargasso Sea, rg = 35 - 43[4]
Hydromedusae	7.5 % of dry weight	Dried at 333 K, Sargasso Sea, rg = 5 - 10[4]
Limnocalanus macrurus	62.1 % of dry weight	Copepoda, 95 % CL = +-0.6, total C[6]
Limnocalanus macrurus	64 % of ash free dry weight	Copepoda, 95 % CL = +-0.6, organic C[6]
Macrocyclops albidus	48.2 % of dry weight	Copepoda, 95 % CL = +-0.5, total C[6]
Macrocyclops albidus	50.7 % of ash free dry weight	Copepoda, 95 % CL = +-0.6, organic C[6]
Polychaeta	30 % of dry weight	Dried at 333 K, Sargasso Sea, rg = 16 - 44[4]
Pteropods	23 % of dry weight	Dried at 333 K, North Pacific, rg = 17 - 29[1]
Pteropods	23 % of dry weight	Dried at 333 K, Sargasso Sea, rg = 21 - 25[4]
Pteropods	27 % of dry weight	Dried at 378 K, continental shelf off New York, rg = 26 - 28[5]
Siphonophora	9.5 % of dry weight	Dried at 333 K, Sargasso Sea, rg = 3 - 16[4]
Tunicata	9 % of dry weight	Dried at 378 K, continental shelf off New York, rg = 7 - 11[5]
Zooplankton	0.00013 mg C/animal	Used in model[7]
Zooplankton	45 % of dry weight	Range = 40 - 50 %[8]

Source: 1.Omori, M. 1969. Mar. Biol., 3, pp 4-10.
2.Balch, N. 1972. Limnol. Oceanogr., 17(6), pp 906-908.
3.Vernadsky, W. 1933. Z. Krist. Mineral. Petrog. Abt. B Mineral. Petrog. Mitt., 44, p 191.
4.Beers, J.R. 1966. Limnol. Oceanogr., 11, pp 520-528.
5.Curl, H. 1962. Rapp. Proc.-Verb. Cons. Int. Explor. Mer., 153, pp 183-189.
6.Salonen, K., J. Sarvala, I. Hakala and M.-L. Viljanen. 1976. Limnol. Oceanogr., 21(5), pp 724-730.
7.O'Connor, D.J., R.V. Thomann and D.M. DiToro. 1973. Dynamic Water Quality Forecasting and Management. Office of Re search and Development U.S. Environmental Protection Agency, Wash. 20460.
8.The Hydrological Engineering Center Corps of Engineers U.S. Army. 1975. Water Quality for river-reservoir systems, Davis, Cal. 9 5 6 1 6 (916) 499-2329.

1-466 Zooplankton
Ca / biomass ratio

Species	Value	Condition
Calanus finmarchicus	0.04 %	Atomic composition[1]

Source: 1.Vernadsky, W. 1933. Z. Krist. Mineral. Petrog. Abt. B Mineral. Petrog. Mitt., 44, p 191.

1-467 Zooplankton
Caloric content

Species	Value	Condition
Acartia clausi	5.59 kcal/g dry weight	Marine, oxydation, copepodit III[1]
Acartia clausi	5.590 kcal/g dry weight	Marine, oxydation, copepodit III[1]
Bosmina coregoni	5.137 kcal/g dry weight	Freshwater, bombcal, adult without egg[1]

Bosmina coregoni	5.327 kcal/g dry weight	Estuarine, bombcal., adult with egg[1]
Bosmina coregoni	5.439 kcal/g ash free dry weight	Estuarine, bombcal., adult without egg[1]
Bosmina coregoni	5.629 kcal/g ash free dry weight	Estuarine, bombcal., adult with egg, rg = +- 0.059[1]
Bosmina longirostris	0.6 kcal/g wet weight	Freshwater, oxydation[1]
Bosmina longirostris	0.60 kcal/g wet weight	Freshwater, oxydation[1]
Bosmina longirostris	5.81 kcal/g dry weight	Freshwater, oxydation[1]
Bosmina longirostris	6.47 kcal/g ash free dry weight	Freshwater, oxydation[1]
Calanus finmarchicus	5.531 kcal/g dry weight	Marine, bombcal.[1]
Calanus finmarchicus	6.194 kcal/g ash free dry weight	Marine, bombcal.[1]
Calanus hyperboreus	5.4 kcal/g ash free dry weight	Marine, bombcal., rg = 0.197[1]
Calanus hyperboreus	5.400 kcal/g ash free dry weight	Marine, bombcal., rg = 0.197[1]
Calanus hyperboreus	7.432 kcal/g ash free dry weight	Marine, bombcal., adult female[1]
Ceriodaphnia pulchella	0.73 kcal/g wet weight	Freshwater, oxydation[1]
Ceriodaphnia pulchella	4.36 kcal/g dry weight	Freshwater, oxydation[1]
Ceriodaphnia pulchella	4.69 kcal/g ash free dry weight	Freshwater, oxydation[1]
Chydorus sphaericus	5.407 kcal/g dry weight	Freshwater, bombcal, rg = +- 0.064[1]
Chydorus sphaericus	5.609 kcal/g ash free dry weight	Freshwater, bombcal, rg = +- 0.057[1]
Daphnia cristata	23.4 kJ/g ash free dry weight	Cladocera, 95 % CL = +- 0.6, correction for nitrogen[2]
Daphnia cristata	25.3 kJ/g ash free dry weight	Cladocera, 95 % CL = +-0.6, no correction for nitrogen[2]
Daphnia cristata	42.4 kJ/g organic Carbon	Cladocera, 95 % CL = +- 0.7, correction for nitrogen[2]
Daphnia cristata	45.8 kJ/g organic Carbon	Cladocera, 95 % CL = +- 0.4, no corection for nitrogen[2]
Daphnia magna	3.694 kcal/g dry weight	Freshwater, oxydation[1]
Daphnia magna	4.988 kcal/g ash free dry weight	Freshwater, oxydation[1]
Daphnia pulex	4.395 kcal/g dry weight	Freshwater, oxydation[1]
Daphnia pulex	4.419 kcal/g dry weight	Freshwater, bombcal[1]
Daphnia pulex	5.392 kcal/g ash free dry weight	Freshwater, bombcal[1]
Daphnia pulex	5.674 kcal/g ash free dry weight	Freshwater, oxydation[1]
Diaptomus leptopus	5.416 kcal/g ash free dry weight	Freshwater, bombcal[1]
Eudiaptomus gracilis	22.2 kJ/g ash free dry weight	Copepoda, 95 % CL = +- 0.2, correction for nitrogen[2]
Eudiaptomus gracilis	24.7 kJ/g ash free dry weight	Copepoda, 95 % CL = +- 0.3, no correction for nitrogen[2]

Eudiaptomus gracilis	42.2 kJ/g organic Carbon	Copepoda, 95 % CL = +- 0.7, correction for nitrogen[2]
Eudiaptomus gracilis	47.3 kJ/g organic Carbon	Copepoda, 95 % CL = +- 2.1, no corection for nitrogen[2]
Eurytemora velox	0.51 kcal/g wet weight	Freshwater, oxydation[1]
Eurytemora velox	3.37 kcal/g dry weight	Freshwater, oxydation[1]
Eurytemora velox	4.7 kcal/g ash free dry weight	Freshwater, oxydation[1]
Eurytemora velox	4.70 kcal/g ash free dry weight	Freshwater, oxydation[1]
Limnocalanus macrurus	29.7 kJ/g ash free dry weight	Copepoda, 95 % CL = +-0.3, correction for nitrogen[2]
Limnocalanus macrurus	31.2 kJ/g ash free dry weight	Copepoda, 95 % CL = +- 0.4, no correction for nitrogen[2]
Limnocalanus macrurus	46.4 kJ/g organic Carbon	Copepoda, 95 % CL = +-0.5, correction for nitrogen[2]
Limnocalanus macrurus	48.8 kJ/g organic Carbon	Copepoda, 95 % CL = +-0.4, no corection for nitrogen[2]
Macrocyclops albidus	20.3 kJ/g ash free dry weight	Copepoda, 95 % CL = +-0.4, correction for nitrogen[2]
Macrocyclops albidus	22.8 kJ/g ash free dry weight	Copepoda, 95 % CL = +-0.3, no correction for nitrogen[2]
Macrocyclops albidus	39.9 kJ/g organic Carbon	Copepoda, 95 % CL = +-0.4, correction for nitrogen[2]
Macrocyclops albidus	45 kJ/g organic Carbon	Copepoda, 95 % CL = +-0.4, no corection for nitrogen[2]
Simocephalus vetulus	0.55 kcal/g wet weight	Freshwater, oxydation[1]
Simocephalus vetulus	3.8 kcal/g dry weight	Freshwater, oxydation[1]
Simocephalus vetulus	3.80 kcal/g dry weight	Freshwater, oxydation[1]
Simocephalus vetulus	4.72 kcal/g ash free dry weight	Freshwater, oxydation[1]

Source: 1.Oertzen, J.-A. 1973. Wiss. Z. Univ. Rostock. 22, pp 1153-1158.
2.Salonen, K., J. Sarvala, I. Hakala and M.-L. Viljanen. 1976. Limnol. Oceanogr., 21(5), pp 724-730.

1-468 Zooplankton
Carbon consumption

Species	Value	Condition
Daphnia pulicaria	0.0265 ug C ind-1 d-1	with algal species Chlorella sp.[1]
Daphnia pulicaria	11.50 ug C ind-1 d-1	with algal species Tabellaria fenestrata[1]
Daphnia pulicaria	13.11 ug C ind-1 d-1	with algal species Cryptomonas erosa[1]
Daphnia pulicaria	3.13 ug C ind-1 d-1	with algal species Asterionella formosa[1]
Daphnia pulicaria	3.55 ug C ind-1 d-1	with algal species Fragilaria crotonensis[1]
Daphnia pulicaria	3.58 ug C ind-1 d-1	with algal species Stephanodiscus astraea minutula[1]
Daphnia pulicaria	3.77 ug C ind-1 d-1	with algal species Stephanodiscus hantzschii[1]
Daphnia thorata	0.04 ug C ind-1 d-1	with algal species Chlorella sp.[1]
Daphnia thorata	0.52 ug C ind-1 d-1	with algal species Fragilaria crotonensis[1]
Daphnia thorata	10.95 ug C ind-1 d-1	with algal species Tabellaria fenestrata[1]
Daphnia thorata	13.35 ug C ind-1 d-1	with algal species Cryptomonas erosa[1]
Daphnia thorata	3.12 ug C ind-1 d-1	with algal species Asterionella formosa[1]
Daphnia thorata	3.43 ug C ind-1 d-1	with algal species Stephanodiscus astraea minutula[1]
Daphnia thorata	3.50 ug C ind-1 d-1	with algal species Stephanodiscus hantzschii[1]

Source: 1.Infante, A. and Litt, A.H., Limnol.Oceanogr., 30(5), 1053, 1985.

1-469 Zooplankton
Cl / biomass ratio

Species	Value	Condition
Calanus finmarchicus	1.05 %	Atomic composition[1]

Source: 1.Vernadsky, W. 1933. Z. Krist. Mineral. Petrog. Abt. B Mineral. Petrog. Mitt., 44, p 191.

1-470 Zooplankton
Clearance rate

Species	Value	Condition
Daphnia longispina	34.1 +/- 1.8 ml (mg dry wt)-1 h-1	incubation exp., fed on Rhodomonas lacustris[1]
Daphnia pulex	12.4 +/- 1.6 ml (mg dry wt)-1 h-1	incubation exp., fed on Scenedesmus sp.[1]
Daphnia pulicaria	1.98 ml ind.-1 d-1	with algal species Chlorella sp.[2]
Daphnia pulicaria	12.88 ml ind.-1 d-1	with algal species Cryptomonas erosa[2]
Daphnia pulicaria	19.24 ml ind.-1 d-1	with algal species Stephanodiscus astraea minutula[2]
Daphnia pulicaria	25.00 ml ind.-1 d-1	with algal species Stephanodiscus hantzschii[2]
Daphnia pulicaria	3.19 ml ind.-1 d-1	with algal species Fragilaria crotonensis[2]
Daphnia pulicaria	32.68 ml ind.-1 d-1	with algal species Tabellaria fenestrata[2]
Daphnia pulicaria	5.36 ml ind.-1 d-1	with algal species Asterionella formosa[2]
Daphnia thorata	0.39 ml ind.-1 d-1	with algal species Fragilaria crotonensis[2]
Daphnia thorata	11.80 ml ind.-1 d-1	with algal species Tabellaria fenestrata[2]
Daphnia thorata	13.06 ml ind.-1 d-1	with algal species Stephanodiscus hantzschii[2]
Daphnia thorata	13.59 ml ind.-1 d-1	with algal species Cryptomonas erosa[2]
Daphnia thorata	15.62 ml ind.-1 d-1	with algal species Stephanodiscus astraea minutula[2]
Daphnia thorata	3.68 ml ind.-1 d-1	with algal species Chlorella sp.[2]
Daphnia thorata	5.40 ml ind.-1 d-1	with algal species Asterionella formosa[2]
Thysanoessa raschii	15.4 +/- 3 ml krill-1 h-1	concentration of pigment with zero ingestion[3]

Source: 1.Olsen, Y. and Østgaard, K., Limnol.Oceanogr., 30(4), 844, 1985.
2.Infante, A. and Litt, A.H., Limnol.Oceanogr., 30(5), 1053, 1985.
3.McClatchie, S. Limnol. Oceanogr., 31, 469, 1986.

1-471 Zooplankton
Consumption

Species	Value	Condition
Daphnia pulicaria	0.183 um3 d-1	with algal species Chlorella sp.[1]
Daphnia pulicaria	104,088 cells ind.-1 d-1	with algal species Stephanodiscus astraea minutula[1]
Daphnia pulicaria	104,663 cells ind.-1 d-1	with algal species Cryptomonas erosa[1]
Daphnia pulicaria	129,275 cells ind.-1 d-1	with algal species Stephanodiscus hantzschii[1]
Daphnia pulicaria	154.69 um3 d-1	with algal species Cryptomonas erosa[1]
Daphnia pulicaria	264.11 um3 d-1	with algal species Tabellaria fenestrata[1]
Daphnia pulicaria	27.54 um3 d-1	with algal species Stephanodiscus hantzschii[1]
Daphnia pulicaria	44,561 cells ind.-1 d-1	with algal species Fragilaria crotonensis[1]

Daphnia pulicaria	48.22 um3 d-1	with algal species Fragilaria crotonensis[1]
Daphnia pulicaria	52,027 cells ind.-1 d-1	with algal species Tabellaria fenestrata[1]
Daphnia pulicaria	52,935 cells ind-1 d-1	with algal species Asterionella formosa[1]
Daphnia pulicaria	52.98 um3 d-1	with algal species Stephanodiscus astraea minutula[1]
Daphnia pulicaria	59.39 um3 d-1	with algal species Asterionella formosa[1]
Daphnia pulicaria	70,452 ml ind.-1 d-1	with algal species Chlorella sp.[1]
Daphnia thorata	0.275 um3 d-1	with algal species Chlorella sp.[1]
Daphnia thorata	105,829 cells ind.-1 d-1	with algal species Chlorella sp.[1]
Daphnia thorata	106,423 cells ind.-1 d-1	with algal species Cryptomonas erosa[1]
Daphnia thorata	157.29 um3 d-1	with algal species Cryptomonas erosa[1]
Daphnia thorata	199,878 cells ind.-1 d-1	with algal species Stephanodiscus hantzschii[1]
Daphnia thorata	244.80 um3 d-1	with algal species Tabellaria fenestrata[1]
Daphnia thorata	25.53 um3 d-1	with algal species Stephanodiscus hantzschii[1]
Daphnia thorata	49,537 cells ind.-1 d-1	with algal species Tabellaria fenestrata[1]
Daphnia thorata	50.66 um3 d-1	with algal species Stephanodiscus astraea minutula[1]
Daphnia thorata	52,726 cells ind.-1 d-1	with algal species Asterionella formosa[1]
Daphnia thorata	59.16 um3 d-1	with algal species Asterionella formosa[1]
Daphnia thorata	6,502 cells ind.-1 d-1	with algal species Fragilaria crotonensis[1]
Daphnia thorata	7.04 um3 d-1	with algal species Fragilaria crotonensis[1]
Daphnia thorata	99,531 cells ind.-1 d-1	with algal species Stephanodiscus astraea minutula[1]

Source: 1.Infante, A. and Litt, A.H., Limnol.Oceanogr., 30(5), 1053, 1985.

1-472 Zooplankton
Conversion efficiency

Species	Value	Condition
Zooplankton	0.6 Lake Ontario, carnivorous	Lake Ontario, carnivorous[1]
Zooplankton	0.6 Lake Ontario, herbivorous	Lake Ontario, herbivorous[1]
Zooplankton	0.6 Lake Ontario, upper trophic level 1	Lake Ontario, upper trophic level 1[1]
Zooplankton	0.6 Lake Ontario, upper trophic level 2	Lake Ontario, upper trophic level 2[1]
Zooplankton	0.6 Used in models	Used in models[2]
Zooplankton	0.65 Range = 0.5 - 0.8	Range = 0.5 - 0.8[3]

Source: 1.Thomann, R.V., D.M. Di Toro, R.P. Winfield and D.J. O'Connor. 1975. Mathematical Modeling of Phytoplankton in Lake Ontario: 1. National Environmental Reseach Center. EPA. Oregon.
2.O'Connor, D.J., R.V. Thomann and D.M. DiToro. 1973. Dynamic Water Quality Forecasting and Management. Office of Research and Development U.S. Environmental Protection Agency, Wash. 20460.
3.The Hydrological Engineering Center Corps of Engineers U.S. Army. 1975. Water Quality for river-reservoir systems, Davis, Cal. 9 5 6 1 6 (916) 499-2329.

1-473 Zooplankton
Cu / biomass ratio

Species	Value	Condition
Calanus finmarchicus	0.0004 %	Atomic composition[1]

Source: 1.Vernadsky, W. 1933. Z. Krist. Mineral. Petrog. Abt. B Mineral. Petrog. Mitt., 44, p 191.

1-474 Zooplankton
Death rate

Species	Value	Condition
Bosmina longirostris	0.190 d-1	controls for enclosure experiment[1]
Bosmina longirostris	0.260 d-1	enclosure experiment with addition of nutrients[1]
Bosmina longirostris	0.264 d-1	enclosure experiment with addition of Daphnia[1]
Bosmina longirostris	0.278 d-1	enclosure experiment with addition of nutrients and Daphnia[1]

Source: 1. Vanni, M.J. Limnol. Oceanogr. 31(5), 1039, 1986.

1-475 Zooplankton
Density

Species	Value	Condition
Tropocyclops prasinus	< 10,000 ind. l-1	exp. in enclosure with Daphnia pulex[1]
Polyarthra vulgaris	< 15,000 ind. l-1	competition exp. in enclosure with Daphnia pulex[1]
Keratella cochlearis	< 200,000 ind. l-1	Competition exp. in enclosure with Daphnia pulex[1]
Diaptomus pallidus	< 5,000 ind. l-1	Competition exp. in enclosure with Daphnia pulex[1]
Mesocyclops edax	< 5,000 ind. l-1	Competition exp. in enclosure with Daphnia pulex[1]
Bosmina longirostris	1-1000 individuals l-1	yearly variation[2]
Calanoids	0.1 l-1	Massawippi, Canada[3]
Calanoids	1.3 l-1	Central basin, Canada[3]
Calanoids	1.7 l-1	Lovering, Canada[3]
Calanoids	1.9 l-1	South basin, Canada[3]
Calanoids	2.2 l-1	Magog, Canada[3]
Calanoids	2.2 l-1	Waterloo, Canada[3]
Calanoids	2.3 l-1	North basin, Canada[3]
Calanoids	20.7 l-1	Brome, Canada[3]
Calanoids	3.0 l-1	Argent, Canada[3]
Calanoids	3.0 l-1	Bowker, Canada[3]
Calanoids	5.5 l-1	Brompton, Canada[3]
Calanoids	8.2 l-1	Orford, Canada[3]
Ceriodaphnia quadrangula	0-100 individuals l-1	yearly variation[2]
Cladocerans	10.6 l-1	Brome, Canada[3]
Cladocerans	12.3 l-1	South basin, Canada[3]
Cladocerans	13.7 l-1	North basin, Canada[3]
Cladocerans	2.8 l-1	Bowker, Canada[3]
Cladocerans	3.8 l-1	Massawippi, Canada[3]
Cladocerans	32.6 l-1	Waterloo, Canada[3]
Cladocerans	41.8 l-1	Magog, Canada[3]
Cladocerans	5.7 l-1	Argent, Canada[3]
Cladocerans	5.8 l-1	Brompton, Canada[3]
Cladocerans	6.6 l-1	Orford, Canada[3]

Cladocerans	6.8 l-1	Central basin, Canada[3]
Cladocerans	8.6 l-1	Lovering, Canada[3]
Cyclopoids	1.7 l-1	Bowker, Canada[3]
Cyclopoids	1.9 l-1	Orford, Canada[3]
Cyclopoids	10.4 l-1	Lovering, Canada[3]
Cyclopoids	11.0 l-1	Waterloo, Canada[3]
Cyclopoids	11.8 l-1	Central basin, Canada[3]
Cyclopoids	16.1 l-1	North basin, Canada[3]
Cyclopoids	16.6 l-1	South basin, Canada[3]
Cyclopoids	20.2 l-1	Magog, Canada[3]
Cyclopoids	20.5 l-1	Massawippi, Canada[3]
Cyclopoids	4.5 l-1	Brompton, Canada[3]
Cyclopoids	5.7 l-1	Brome, Canada[3]
Cyclopoids	8.2 l-1	Argent, Canada[3]
Daphnia hyalina	0-1000 individuals l-1	yearly variation[2]
Daphnia hyalina	120.3 +/- 29.0	enclosure experiment without fish, 27 days[2]
Daphnia hyalina	18.8 +/- 9.5 individuals l-1	effect of 2 roach in enclosure experiment, 27 days[2]
Daphnia hyalina	5.6 individuals l-1	open water reference for enclosure experiments[2]
Daphnia pulex	< 25,000 individuals l-1	enclosure experiment, Larimore Pond[1]
Daphnia pulex	< 9,000 individuals l-1	enclosure experiment, Dynamite Pond[1]
Diaptomus pallidus	2,000- 50,000 ind. l-1	level in lake during enclosure experiment[3]
Diaptomus pallidus	< 30,000 ind. l-1	enclosure competition exp. without Daphnia pulex (control)[1]
Invertebrates	13,771 l-1	Magog, Canada[3]
Invertebrates	21,556 l-1	Waterloo, Canada[3]
Invertebrates	3,318 l-1	Bowker, Canada[3]
Invertebrates	4,380 l-1	Orford, Canada[3]
Invertebrates	4,492 l-1	Brompton, Canada[3]
Invertebrates	5,815 l-1	Brome, Canada[3]
Invertebrates	5,996 l-1	Lovering, Canada[3]
Invertebrates	6,718 l-1	Argent, Canada[3]
Invertebrates	7,217 l-1	North basin, Canada[3]
Invertebrates	8,404 l-1	Central basin, Canada[3]
Invertebrates	8,549 l-1	Massawippi, Canada[3]
Invertebrates	8,769 l-1	South basin, Canada[3]
Keratella cochlearis	1,000-200,000 ind. l-1	level in lake during enclosure experiment[4]
Keratella cochlearis	< 250,000 ind. l-1	enclosure competition exp. without Daphnia pulex (control)[1]
Mesocyclops edax	< 22,000 ind. l-1	enclosure competition exp. without Daphnia pulex (control)[1]
Mesocyclops edax	around 10.000 ind. l-1	level in lake during enclosure experiment[3]
Nauplii	12 l-1	Orford, Canada[3]
Nauplii	25 l-1	Brompton, Canada[3]

Nauplii	39 l-1	Argent, Canada[3]
Nauplii	41 l-1	Brome, Canada[3]
Nauplii	48 l-1	Lovering, Canada[3]
Nauplii	54 l-1	South basin, Canada[3]
Nauplii	57 l-1	North basin, Canada[3]
Nauplii	64 l-1	Magog, Canada[3]
Nauplii	68 l-1	Massawippi, Canada[3]
Nauplii	75 l-1	Waterloo, Canada[3]
Nauplii	9 l-1	Bowker, Canada[3]
Polyarthra vulgaris	1.000-11.000 ind. l-1	level in lake during enclosure experiment[4]
Polyarthra vulgaris	< 15.000 ind. l-1	enclosure competition exp. without Daphnia pulex (control)[1]
Rotifers	100 l-1	Bowker, Canada[3]
Rotifers	161 l-1	Orford, Canada[3]
Rotifers	193 l-1	Argent, Canada[3]
Rotifers	212 l-1	Lovering, Canada[3]
Rotifers	255 l-1	Brome, Canada[3]
Rotifers	476 l-1	North basin, Canada[3]
Rotifers	587 l-1	Magog, Canada[3]
Rotifers	683 l-1	Central basin, Canada[3]
Rotifers	703 l-1	Waterloo, Canada[3]
Rotifers	805 l-1	Massawippi, Canada[3]
Rotifers	915 l-1	South basin, Canada[3]
Rotifers	99 l-1	Brompton, Canada[3]
Tropocyclops prasinus	< 50.000 ind. l-1	enclosure competition exp. without Daphnia pulex (control)[1]
Tropocyclops prasinus	around 1000 ind. l-1	level in lake during enclosure experiment[3]

Source: 1.Vanni, M.J. Limnol. Oceanogr. 31(5), 1039, 1986.
2.Cryer, M., Peirson, G. and Townsend, C.R. Limnol. Oceanogr., 31(5), 1022, 1986.
3.Pace, M.L. Limnol. Oceanogr., 31, 45, 1986.

1-476 Zooplankton
Development time

Species	Value	Condition
Acartia clausii	107.4 hours	Median development time to nauplius stage 4, 293 K, autumn[1]
Acartia clausii	130.5 hours	Median development time to nauplius stage 5, 293 K, autumn[1]
Acartia clausii	145.3 hours	Median development time to nauplius stage 6, 293 K, spring[1]
Acartia clausii	155.3 hours	Median development time to nauplius stage 6, 293 K, autumn[1]
Acartia clausii	171.4 hours	Median development time to copepodit stage 1, 293 K, spring[1]
Acartia clausii	180.4 hours	Median development time to copepodit stage 1, 293 K, autumn[1]
Acartia clausii	184.6 hours	Median development time to nauplius stage 3, 283 K, spring[1]
Acartia clausii	196 hours	Median development time to copepodit stage 2, 293 K, spring[1]
Acartia clausii	204.7 hours	Median development time to copepodit stage 2, 293 K, autumn[1]
Acartia clausii	217.6 hours	Median development time to nauplius stage 6, 288 K, summer[1]

Acartia clausii	218.9 hours	Median development time to copepodit stage 3, 293 K, spring[1]
Acartia clausii	230.7 hours	Median development time to copepodit stage 3, 293 K, autumn[1]
Acartia clausii	239 hours	Median development time to nauplius stage 4, 283 K, spring[1]
Acartia clausii	239.6 hours	Median development time to copepodit stage 4, female, 293 K, spring[1]
Acartia clausii	242.1 hours	Median development time to copepodit stage 4, male, 293 K, spring[1]
Acartia clausii	254 hours	Median development time to copepodit stage 4, male, 293 K, autumn[1]
Acartia clausii	260 hours	Median development time to copepodit stage 4, female, 293 K, autumn[1]
Acartia clausii	260.9 hours	Median development time to copepodit stage 1, 288 K, summer[1]
Acartia clausii	263 hours	Median development time to copepodit stage 5, male, 293 K, spring[1]
Acartia clausii	269.2 hours	Median development time to copepodit stage 5, female, 293 K, spring[1]
Acartia clausii	278.7 hours	Median development time to copepodit stage 5, male, 293 K, autumn[1]
Acartia clausii	287 hours	Median development time to copepodit stage 5, female, 293 K, autumn[1]
Acartia clausii	29.3 hours	Median development time to nauplius stage 1, 293 K, spring[1]
Acartia clausii	290 hours	Median development time to male, 293 K, spring[1]
Acartia clausii	295.8 hours	Median development time to nauplius stage 5, 283 K, spring[1]
Acartia clausii	299 hours	Median development time to copepodit stage 2, 288 K, summer[1]
Acartia clausii	300.9 hours	Median development time to female, 293 K, spring[1]
Acartia clausii	31.2 hours	Median development time to nauplius stage 1, 293 K, autumn[1]
Acartia clausii	312 hours	Median development time to male, 293 K, autumn[1]
Acartia clausii	319 hours	Median development time to female, 293 K, autumn[1]
Acartia clausii	339 hours	Median development time to copepodit stage 3, 288 K, summer[1]
Acartia clausii	348.9 hours	Median development time to nauplius stage 6, 283 K, spring[1]
Acartia clausii	363.6 hours	Median development time to copepodit stage 4, male, 288 K, summer
Acartia clausii	376.8 hours	Median development time to copepodit stage 4, female, 288 K, summer[1]
Acartia clausii	402.6 hours	Median development time to copepodit stage 5, male, 288 K, summer[1]
Acartia clausii	409.4 hours	Median development time to copepodit stage 1, 283 K, spring[1]
Acartia clausii	443.1 hours	Median development time to copepodit stage 5, female, 288 K, summer[1]
Acartia clausii	45.8 hours	Median development time to nauplius stage 1, 288 K, summer[1]
Acartia clausii	47 hours	Median development time to nauplius stage 2, 293 K, autumn[1]
Acartia clausii	475.4 hours	Median development time to copepodit stage 2, 283 K, spring[1]
Acartia clausii	485.1 hours	Median development time to male, 288 K, summer[1]
Acartia clausii	72 hours	Median development time to nauplius stage 1, 283 K, spring[1]
Acartia clausii	82.6 hours	Median development time to nauplius stage 3, 293 K, autumn[1]
Ceriodaphnia sp.	1.8 days	Field life tables, 296 K, median nutrient level, egg development[2]
Ceriodaphnia sp.	1.8 days	Grazing on 0.4E5 cell algae/l, 296 K, lab, egg develop, SE =+-0.09[2]

Ceriodaphnia sp.	1.8 days	Grazing on 4E5 cell algae/l, 296 K, lab, egg development[2]
Ceriodaphnia sp.	1.9 days	Field life tables, 296 K, high nutrient level, egg develop[2]
Ceriodaphnia sp.	1.9 days	Field life tables, 296 K, low nutrient level, egg develop[2]
Ceriodaphnia sp.	3 days	Grazing on 4E5 cell algae/l, 296 K, lab maturation time, SE = +-0.11[2]
Ceriodaphnia sp.	3.1 days	Field life tables, 296 K, low nutrient level maturation time[2]
Ceriodaphnia sp.	3.2 days	Field life tables, 296 K, median nutrient level maturation time[2]
Ceriodaphnia sp.	3.8 days	Field life tables, 296 K, high nutrient level maturation time[2]
Ceriodaphnia sp.	5 days	Grazing on 0.4E5 cell algae/l, 296 K, lab maturation time[2]
Keratella valga	1.5 days	296 k maturation time[2]
Keratella valga	4 days	290 k maturation time[3]
Limnocalanus macrurus	11 days	Nauplius stage 5, Char Lake, field data, artic[4]
Limnocalanus macrurus	12 days	Copepodit stage 1, Char Lake, field data, artic, rg = 11 - 13[4]
Limnocalanus macrurus	14 days	Nauplius stage 3, Char Lake, field data, artic[4]
Limnocalanus macrurus	15 days	Nauplius stage 4, Char Lake, field data, artic[4]
Limnocalanus macrurus	15 days	Nauplius stage 6, Char Lake, field data, artic[4]
Limnocalanus macrurus	182.5 days	Copepodit stage 6, Char Lake, field data, artic, rg = 65 - 300[4]
Limnocalanus macrurus	19.5 days	Copepodit stage 2, Char Lake, field data, artic, rg = 19 - 20[4]
Limnocalanus macrurus	20 days	Nauplius stage 1, Char Lake, field data, artic[4]
Limnocalanus macrurus	22 days	Nauplius stage 2, Char Lake, field data, artic[4]
Limnocalanus macrurus	22.5 days	Copepodit stage 4, Char Lake, field data, artic, rg = 20 - 25[4]
Limnocalanus macrurus	23.5 days	Copepodit stage 3, Char Lake, field data, artic, rg = 21 - 26[4]
Limnocalanus macrurus	47.5 days	Copepodit stage 5, Char Lake, field data, artic, rg = 45 - 50[4]
Neodiaptomus schmackeri	2 days	Egg to nauplius, 301 - 303 K, lab, rg = 1 - 3[5]
Neodiaptomus schmackeri	22 days	Copepodite to adult, male, 301 - 303 K, lab, rg = 13 - 29[5]
Neodiaptomus schmackeri	25 days	Copepodite to adult, 301 - 303 K, field observation[5]
Neodiaptomus schmackeri	26 days	Copepodite to adult, female, 301 - 303 K, lab, rg = 22 - 29[5]
Neodiaptomus schmackeri	3 days	Nauplius to copepodite, 301 - 303 K, field observation[5]
Neodiaptomus schmackeri	32 days	One complete generation(egg to egg), 301 - 303 K, lab, rg = 25 - 37[5]
Neodiaptomus schmackeri	4 days	Nauplius to copepodite, 301 - 303 K, lab, rg = 2 - 7[5]
Simocephalus serrulatus	2 days	Field experiment, 296 K, egg develop[2]
Simocephalus serrulatus	6.5 days	Field experiment, 296 k maturation time[2]

Source: 1.Landry, M.R. 1975. Limnol. Oceanog. 20(5), pp 854-857.
2.Hall, D.J., W.E. Cooper and E.E. Werner. 1970. Limnol. Oceanogr., 15(3), pp 838-928.
3.Kolisko, A. 1938. Arch. Hydrobiol. 33, pp 165-207.
4.Roff, J.C. and J.C.H. Carter. 1972. Limnol. Oceanogr., 17(3), pp 363-370.
5.Dunn, I,G. 1970. Limnol. Oceanogr. 15(3), pp 373-379.

1-477 Zooplankton
Dry / wet weight ratio

Species	Value	Condition
Amphipoda	27.5 % of wet weight	Dried at 333 K, North Pacific, rg = 18.4 - 36.6[1]
Chaetognatha	12.85 % of wet weight	Dried at 333 K, North Pacific, rg = 11.6 - 14.1[1]
Chaetognatha	6.7 % of wet weight	Dried at 333 K, Sargasso Sea, rg = 6.0 - 7.4[2]
Copepoda	13 % of wet weight	Dried at 378 K, continental shelf off New York, rg = 10.2 - 15.8[3]
Copepoda	13.95 % of wet weight	Dried at 333 K, Sargasso Sea, rg = 11.6 - 16.3[2]
Copepoda	21.05 % of wet weight	Dried at 333 K, North Pacific, rg = 9.2 - 33.9[1]
Ctenophora	4.85 % of wet weight	Dried at 378 K, continental shelf off New York, rg = 4.7 - 5.0[3]
Euphausiacea	19.5 % of wet weight	Dried at 378 K, continental shelf off New York, rg = 19.0 - 20.0[3]
Euphausiacea	20.75 % of wet weight	Dried at 333 K, North Pacific, rg = 20.2 - 21.3[1]
Euphausiacea and Mysidae	16.25 % of wet weight	Dried at 333 K, Sargasso Sea, rg = 14.5 - 18.0[2]
Fish larvae	13.95 % of wet weight	Dried at 333 K, Sargasso Sea, rg = 11.9 - 16.0, also mature fish[2]
Hydromedusae	5.2 % of wet weight	Dried at 333 K, Sargasso Sea, rg = 0.3 - 10.1[2]
Polychaeta	16.35 % of wet weight	Dried at 333 K, Sargasso Sea, rg = 5.7 - 16.35[2]
Pteropods	11.25 % of wet weight	Dried at 378 K, continental shelf off New York, rg = 3.5 - 19[3]
Pteropods	27 % of wet weight	Dried at 333 K, Sargasso Sea, rg = 22 - 32[2]
Pteropods	30.7 % of wet weight	Dried at 333 K, North Pacific, rg = 25.0 - 36.4[1]
Siphonophora	3.2 % of wet weight	Dried at 333 K, Sargasso Sea, rg = 0.3 - 6.1[2]
Tunicata	4.05 % of wet weight	Dried at 378 K, continental shelf off New York, rg 4.0 - 4.1[3]

Source: 1.Omori, M. 1969. Mar. Biol., 3, pp 4-10.
2.Beers, J.R. 1966. Limnol. Oceanogr., 11, pp 520-528.
3.Curl, H. 1962. Rapp. Proc.-Verb. Cons. Int. Explor. Mer., 153, pp 183-189.

1-478 Zooplankton
Dry weight

Species	Value	Condition
Acartia tonsa	12.9 microgram	Female[1]
Acroperus harpae	3 microgram	Specific gravity = 1.025, juvenile[2]
Acroperus harpae	6 microgram	Specific gravity = 1.025, adult[2]
Alona costata	1 microgram	Specific gravity = 1.025, juvenile[2]
Alona costata	3 microgram	Specific gravity = 1.025, adult[2]
Alona guttata	1 microgram	Specific gravity = 1.025, juvenile[2]
Alona guttata	1.8 microgram	Specific gravity = 1.025, adult[2]
Ascomorpha sp.	0.04 microgram	Specific gravity = 1.025[2]
Asplanchna girodi	1 microgram	Specific gravity = 1.025[2]
Asplanchna priodonta	2 microgram	Specific gravity = 1.025[2]
Bosmina longirostris	1 microgram	Specific gravity = 1.025, juvenile[2]
Bosmina longirostris	1.8 microgram	Specific gravity = 1.025, adult[2]
Brachionus plicatilis	0.06 microgram	Specific gravity = 1.025[2]
Brachyuran larvae	1.17 microgram	Megalops[3]

Calanoid copepodids	3 microgram	Specific gravity = 1.025[2]
Calanoides carinatus	0.13 mg	None[3]
Calanus discaudatus	26.15 microgram	Length=2.04 mm, copepodite stage 4[4]
Calanus finmarchicus	113.8 microgram	Copepodit 5[1]
Calanus pacificus	0.001 mg	Nauplii 1[3]
Calanus pacificus	0.01 mg	Copepodite 2[3]
Calanus pacificus	0.026 mg	Copepodite 4[3]
Calanus pacificus	0.044 mg	Copepodite 5[3]
Calanus pacificus	0.179 mg	Female[3]
Calanus pacificus	0.6 microgram	Length=0.27 mm, nauplii stage 3[4]
Centropages typicus	37.3 microgram	Female[1]
Ceriodaphnia reticulata	2.2 microgram	Specific gravity = 1.025, juvenile[2]
Ceriodaphnia reticulata	4.2 microgram	Specific gravity = 1.025, adult[2]
Chironomus tentans	0.005 mg	Dried at 343 K, instar 1[2]
Chydorus sphaericus	1 microgram	Specific gravity = 1.025, juvenile[2]
Chydorus sphaericus	2 microgram	Specific gravity = 1.025, adult[2]
Comptocerus rectirostris	2 microgram	Specific gravity = 1.025, juvenile[2]
Comptocerus rectirostris	5 microgram	Specific gravity = 1.025, adult[2]
Cyclopoid copepodids	3 microgram	Specific gravity = 1.025[2]
Cyclopoid nauplii	0.01 microgram	Specific gravity = 1.025[2]
Cyclops vericans rubellus	2 microgram	Specific gravity = 1.025[2]
Cyclops vernalis	8.6 microgram	Specific gravity = 1.025[2]
Daphnia pulex	35 microgram	Specific gravity = 1.025, adult[2]
Daphnia pulex	8 microgram	Specific gravity = 1.025, juvenile[2]
Diaphanosoma brachyurum	3 microgram	Specific gravity = 1.025, juvenile[2]
Diaphanosoma brachyurum	7 microgram	Specific gravity = 1.025, adult[2]
Diaptomus leptopus	20 microgram	Specific gravity = 1.025[2]
Diaptomus pallidus	8.1 microgram	Specific gravity = 1.025[2]
Epilabidocera amphitrites	0.5 mg	None[3]
Eucyclops agilis	8 microgram	Specific gravity = 1.025[2]
Euphausia pacifica	6.36 mg	None[3]
Eurycercus lamellatus	10 microgram	Specific gravity = 1.025, juvenile[2]
Eurycercus lamellatus	80 microgram	Specific gravity = 1.025, adult[2]
Gastropus sp.	0.05 microgram	Specific gravity = 1.025[2]
Keratella cochlearis	0.05 microgram	Specific gravity = 1.025[2]
Keratella hiemalis	0.05 microgram	Specific gravity = 1.025[2]
Leydigia acanthoceroides	3 microgram	Specific gravity = 1.025, juvenile[2]
Leydigia acanthoceroides	6 microgram	Specific gravity = 1.025, adult[2]

Leydigia quadrangularis	3 microgram	Specific gravity = 1.025, juvenile[2]
Leydigia quadrangularis	6 microgram	Specific gravity = 1.025, adult[2]
Macrocyclops albidus	40 microgram	Specific gravity = 1.025[2]
Macrothrix laticornis	1 microgram	Specific gravity = 1.025, juvenile[2]
Macrothrix laticornis	2 microgram	Specific gravity = 1.025, adult[2]
Macrothrix rosea	1 microgram	Specific gravity = 1.025, juvenile[2]
Macrothrix rosea	2 microgram	Specific gravity = 1.025, adult[2]
Mesocyclops edax	18 microgram	Specific gravity = 1.025[2]
Mesocyclops leuckarti	15 microgram	Specific gravity = 1.025[2]
Nematoscelis atlantica	0.91 mg	None[3]
Paracyclops fimbriatus poppei	3 microgram	Specific gravity = 1.025[2]
Parathemisto pacifica	1.65 mg	None[3]
Pleurobrachia bacheii	11.4 mg	None[3]
Pleuroncodes planipes	432 mg	None[3]
Pleuroxus denticulatus	1 microgram	Specific gravity = 1.025, juvenile[2]
Pleuroxus denticulatus	2 microgram	Specific gravity = 1.025, adult[2]
Pleuroxus procurvis	2 microgram	Specific gravity = 1.025, juvenile[2]
Pleuroxus procurvis	4 microgram	Specific gravity = 1.025, adult[2]
Polyarthra sp.	0.05 microgram	Specific gravity = 1.025[2]
Pseudocalanus minutus	21.5 microgram	Female[1]
Sagitta elegans	1.43 mg	None[3]
Simocephalus serrulatus	33.5 microgram	Specific gravity = 1.025, small adult[2]
Simocephalus serrulatus	66.2 microgram	Specific gravity = 1.025, large adult[2]
Simocephalus serrulatus	8.5 microgram	Specific gravity = 1.025, juvenile[2]
Synchaeta pectinata	0.2 microgram	Specific gravity = 1.025[2]
Tomopteris septentrionalis	5.61 mg	None[3]
Tortanus discaudatus	36.9 microgram	Total length=1.97 mm, female[4]
Tropocyclops prasinus	3 microgram	Specific gravity of volumes = 1.025[2]

Source: 1.Dagg, M. 1977. Limnol. Oceanogr., 22(1), pp 99-107.
2.Hall, D.J., W.E. Cooper and E.E. Werner. 1970. Limnol. Oceanogr., 15(3), pp 838-928.
3.King, F.D. and T.T. Packard. 1975. Limnol. Oceanogr. 20(5), pp 849-853.
4.Ambler, J.W. and B.W. Frost. 1974. Limnol. Oceanogr. 19(3), pp 446-451.

1-479 Zooplankton Egestion

Species	Value	Condition
Daphnia pulex	77.5 % of ingestion	Range = 69 - 86[1]
Leptodora kindtii	60 % of ingestion	None[1]
Simocephalus vectulus	27.6 % of ingestion	Juvenile, female[1]
Simocephalus vectulus	68.3 % of ingestion	Reproducing female[1]

Source: 1.Wetzel, R.G. 1975. Limnology. Saunders. Philadelphia.

1-480 Zooplankton
Excretion

Species	Value	Condition
Phronima sedentaria	10.8 microgram tot-N/(mg dry weight*day)	12 h starv before exp, 286 K, resp=17.3 ul O2/mg dw, rg = +-0.3[1]
Phronima sedentaria	5.3 microgram org-N/(mg dry weight*day)	12 h starv before exp, 286 K, resp=17.3 ul O2/mg dw[1]
Phronima sedentaria	5.5 microgram NH4-N/(mg dry weight*day)	12 h starv before exp, 286 K, resp=17.3 ul O2/mg dw, 95 % CL = +-1.3[1]
Sagitta setosa	20.1 microgram org-N/(mg dry weight*day)	16 h starv before exp, 283-298 K, 6 h[1]
Sagitta setosa	30.7 microgram NH4-N/(mg dry weight*day)	16 h starv before exp, 283-298 K, 6 h, 95 % CL = +- 4.0[1]
Sagitta setosa	70.8 microgram NH4-N/(mg dry weight*day)	16 h starv before exp, 283-298 K, 6 h, 95 % CL = +- 5.6[1]
Salpa fusiformis	1.4 microgram org-N/(mg dry weight*day)	3 h starv before exp, 288 K, resp=7.7 ul O2/mg dw, 6 h[1]
Salpa fusiformis	2 microgram NH4-N/(mg dry weight*day)	3 h starv before exp, 288 K, resp=7.7 ul O2/mg dw, 6 h, 95 % CL=0.7[1]
Salpa fusiformis	3.4 microgram tot-N/(mg dry weight*day)	3 h starv before exp, 288 K, resp=7.7 ul O2/mg dw, 6 h[1]
Thalia democratica	13.6 microgram tot-N/(mg dry weight*day)	6 h starv before exp, 288 K, resp=6.7 ul O2/mg dw, 6 h[1]
Thalia democratica	4.1 microgram org-N/(mg dry weight*day)	6 h starv before exp, 288 K, resp=6.7 ul O2/mg dw, 6 h[1]
Thalia democratica	9.5 microgram NH4-N/(mg dry weight*day)	6 h starv before exp, 288 K, resp=6.7 ul O2/mg dw, 6 h[1]
Zooplankton	0.0001 microgram P/(mg*hour)	Stonehouse Pond, April 6, 1973, freshwater[2]
Zooplankton	0.0006 microgram P/(mg*hour)	Stonehouse Pond, Feb 22, 1973, freshwater[2]
Zooplankton	0.0172 microgram P/(mg*hour)	Stonehouse Pond, May 17, 1973, freshwater[2]
Zooplankton	0.0203 microgram P/(mg*hour)	Stonehouse Pond, June 17, 1973, freshwater[2]
Zooplankton	0.104 1/hour	Steady state solution to Castle Lake, freshwater, phosphorus[3]
Zooplankton	10.3 % of body P/day	Summer[4]
Zooplankton	2.6 % of body N/day	Winter[4]
Zooplankton	4.6 % of body N/day	Summer[4]
Zooplankton	6.5 % of body P/day	Winter[4]

Source: 1.Mayzaud, P. and S. Pallot. 1973. Mar biol. 19, pp 307-314.
2.Ferrante, J.G. 1976. Hydrobiologia, 49(3), pp 203-204.
3.Richey, J.E. 1977. in: Hall, C.A.S. and J.W. Day, Jr. (eds.); Ecosystem Modeling in theory and practice. John Wiley and Sons. New York.
4.Golterman, H.L. 1975. Physiological Limnol., Elsevier Amsterdam.

1-481 Zooplankton
Excretion of P

Species	Value	Condition
Daphnia longispina	1.16 - 1.50 ug Production(mg dry wt.)-1	incubation experiment, fed on Rhodomonas lacustris[1]
Daphnia longispina	1.16 - 1.52 ug Production(mg dry wt.)-1	incubation experiment, fed on Rhodomonas lacustris[1]
Daphnia pulex	0.41 - 0.56 ug Production(mg dry wt.)-1	incubation experiment, fed on Scenedesmus sp.[1]

Source: 1.Olsen, Y. and Østgaard, K., Limnol. Oceanogr., 30(4), 844, 1985.

1-482 Zooplankton
Excretion of ammonium

Species	Value	Condition
Beroe cucumis	11.58 nmol mg protein-1 h-1	spring[1]
Beroe cucumis	4.76 nmol mg protein-1 h-1	summer[1]
Boreomysis arctica	7.05 nmol mg protein-1 h-1	summer[1]
Calanus finmarchius	10.65 nmol mg protein-1 h-1	spring[1]
Calanus finmarchius	8.10 nmol mg protein-1 h-1	summer[1]
Calanus hyperboreus	12.07 nmol mg protein-1 h-1	spring[1]
Calanus hyperboreus	2.82 nmol mg protein-1 h-1	summer[1]
Chiridius armatus	13.15 nmol mg protein-1 h-1	summer, males[1]
Chiridius armatus	13.75 nmol mg protein-1 h-1	spring, juvenile[1]
Chiridius armatus	3.80 nmol mg protein-1 h-1	spring, males[1]
Chiridius armatus	9.34 nmol mg protein-1 h-1	spring, females[1]
Concoecia sp.	10.80 nmol mg protein-1 h-1	summer[1]
Concoecia sp.	3.36 nmol mg protein-1 h-1	spring[1]
Euchaeata norvegica	11.61 nmol mg protein-1 h-1	spring, CIII[1]
Euchaeata norvegica	13.04 nmol mg protein-1 h-1	spring, CV[1]
Euchaeata norvegica	14.23 nmol mg protein-1 h-1	spring, females[1]
Euchaeata norvegica	5.60 nmol mg protein-1 h-1	spring, males[1]
Euchaeata norvegica	7.82 nmol mg protein-1 h-1	spring, CIV[1]

Euchaeta norvegica	3.75 nmol mg protein-1 h-1	summer, females[1]
Eukrohnia batypelagica	19.38 nmol mg protein-1 h-1	spring[1]
Eukrohnia hamata	5.26 nmol mg protein-1 h-1	spring[1]
Meganyctiphanes norvegica	3.66 nmol mg protein-1 h-1	spring[1]
Meganyctiphanes norvegica	3.85 nmol mg protein-1 h-1	summer[1]
Metridia longa	14.80 nmol mg protein-1 h-1	summer[1]
Metridia longa	20.01 nmol mg protein-1 h-1	spring[1]
Parathemisto abyssorum	4.05 nmol mg protein-1 h-1	summer[1]
Parathemisto abyssorum	9.44 nmol mg protein-1 h-1	spring[1]
Pleurobrachia pileus	22.87 nmol mg protein-1 h-1	spring[1]
Sagitta elegans	79.86 nmol mg protein-1 h-1	spring[1]
Tomopteris helgolandica	9.16 nmol mg protein-1 h-1	spring[1]

Source: 1.Baamstedt, U., Limnol. Oceanogr., 30(3), 607, 1985.

1-483 Zooplankton
Excretion of dissolved organic P

Species	Value	Condition
Beroe cucumis	1.26 nmol mg protein-1 h-1	summer[1]
Beroe cucumis	4.13 nmol mg protein-1 h-1	spring[1]
Boreomysis arctica	0.81 nmol mg protein-1 h-1	summer[1]
Calanus finmarchius	0.17 nmol mg protein-1 h-1	summer[1]
Calanus finmarchius	1.11 nmol mg protein-1 h-1	spring[1]
Calanus hyperboreus	0.14 nmol mg protein-1 h-1	summer[1]
Calanus hyperboreus	1.01 nmol mg protein-1 h-1	spring[1]
Chiridius armatus	0.18 nmol mg protein-1 h-1	spring, males[1]
Chiridius armatus	0.20 nmol mg protein-1 h-1	spring, juvenile[1]
Chiridius armatus	0.29 nmol mg protein-1 h-1	spring, females[1]
Chiridius armatus	0.98 nmol mg protein-1 h-1	summer, males[1]

Concoecia sp.	0.73 nmol mg protein-1 h-1	spring[1]
Concoecia sp.	1.10 nmol mg protein-1 h-1	summer[1]
Euchaeata norvegica	0.14 nmol mg protein-1 h-1	spring, females[1]
Euchaeata norvegica	0.29 nmol mg protein-1 h-1	spring, CIII[1]
Euchaeata norvegica	0.43 nmol mg protein-1 h-1	spring, CIV[1]
Euchaeata norvegica	0.59 nmol mg protein-1 h-1	spring, CV[1]
Euchaeata norvegica	0.66 nmol mg protein-1 h-1	spring, males[1]
Euchaeta norvegica	0.63 nmol mg protein-1 h-1	summer, females[1]
Eukrohnia batypelagica	1.20 nmol mg protein-1 h-1	spring[1]
Eukrohnia hamata	0.32 nmol mg protein-1 h-1	spring[1]
Eutonina indicans	2.57 nmol mg protein-1 h-1	summer[1]
Meganyctiphanes norvegica	0.26 nmol mg protein-1 h-1	summer[1]
Meganyctiphanes norvegica	0.30 nmol mg protein-1 h-1	spring[1]
Metridia longa	0.48 nmol mg protein-1 h-1	summer[1]
Metridia longa	1.17 nmol mg protein-1 h-1	spring[1]
Parathemisto abyssorum	1.02 nmol mg protein-1 h-1	spring[1]
Parathemisto abyssorum	1.61 nmol mg protein-1 h-1	summer[1]
Pasiphaea multidentata	0.23 nmol mg protein-1 h-1	spring[1]
Pleurobrachia pileus	1.27 nmol mg protein-1 h-1	spring[1]
Sagitta elegans	0.58 nmol mg protein-1 h-1	spring[1]
Tomopteris helgolandica	0.86 nmol mg protein-1 h-1	spring[1]

Source: 1.Baamstedt, U., Limnol.Oceanogr., 30(3), 607, 1985.

1-484 Zooplankton
Excretion of phosphate

Species	Value	Condition
Beroe cucumis	1.67 nmol mg protein-1 h-1	summer[1]
Beroe cucumis	1.84 nmol mg protein-1 h-1	spring[1]
Boreomysis arctica	2.56 nmol mg protein-1 h-1	summer[1]

Calanus finmarchius	0.89 nmol mg protein-1 h-1	summer[1]
Calanus finmarchius	2.06 nmol mg protein-1 h-1	spring[1]
Calanus hyperboreus	1.05 nmol mg protein-1 h-1	summer[1]
Calanus hyperboreus	2.67 nmol mg protein-1 h-1	spring[1]
Chiridius armatus	1.21 nmol mg protein-1 h-1	spring, males[1]
Chiridius armatus	1.67 nmol mg protein-1 h-1	spring, juvenile[1]
Chiridius armatus	2.66 nmol mg protein-1 h-1	summer, males[1]
Chiridius armatus	2.97 nmol mg protein-1 h-1	spring, females[1]
Concoecia sp.	0.73 nmol mg protein-1 h-1	spring[1]
Concoecia sp.	1.88 nmol mg protein-1 h-1	summer[1]
Euchaeata norvegica	0.27 nmol mg protein-1 h-1	spring, males[1]
Euchaeata norvegica	1.09 nmol mg protein-1 h-1	spring, CIII[1]
Euchaeata norvegica	1.53 nmol mg protein-1 h-1	spring, CIV[1]
Euchaeata norvegica	1.70 nmol mg protein-1 h-1	spring, CV[1]
Euchaeata norvegica	2.48 nmol mg protein-1 h-1	spring, females[1]
Euchaeta norvegica	1.60 nmol mg protein-1 h-1	summer, females[1]
Eukrohnia batypelagica	3.82 nmol mg protein-1 h-1	spring[1]
Eukrohnia hamata	1.20 nmol mg protein-1 h-1	spring[1]
Eutonina indicans	0.35 nmol mg protein-1 h-1	summer[1]
Meganyctiphanes norvegica	2.49 nmol mg protein-1 h-1	summer[1]
Meganyctiphanes norvegica	3.06 nmol mg protein-1 h-1	spring[1]
Metridia longa	0.86 nmol mg protein-1 h-1	summer[1]
Metridia longa	3.32 nmol mg protein-1 h-1	spring[1]
Parathemisto abyssorum	1.93 nmol mg protein-1 h-1	summer[1]
Parathemisto abyssorum	2.38 nmol mg protein-1 h-1	spring[1]
Pasiphaea multidentata	1.69 nmol mg protein-1 h-1	spring[1]

Pleurobrachia pileus	8.71 nmol mg protein-1 h-1	spring[1]
Sagitta elegans	1.16 nmol mg protein-1 h-1	spring[1]
Tomopteris helgolandica	0.93 nmol mg protein-1 h-1	spring[1]

Source: 1.Baamstedt, U., Limnol.Oceanogr., 30(3), 607, 1985.

1-485 Zooplankton
Excretion of urea

Species	Value	Condition
Beroe cucumis	2.21 nmol mg protein-1 h-1	summer[1]
Boreomysis arctica	0.96 nmol mg protein-1 h-1	summer[1]
Calanus finmarchius	3.52 nmol mg protein-1 h-1	summer[1]
Calanus finmarchius	6.31 nmol mg protein-1 h-1	spring[1]
Calanus hyperboreus	0.66 nmol mg protein-1 h-1	summer[1]
Calanus hyperboreus	1.51 nmol mg protein-1 h-1	spring[1]
Chiridius armatus	1.25 nmol mg protein-1 h-1	summer, males[1]
Chiridius armatus	1.92 nmol mg protein-1 h-1	spring, females[1]
Chiridius armatus	2.72 nmol mg protein-1 h-1	spring, males[1]
Chiridius armatus	3.09 nmol mg protein-1 h-1	spring, juvenile[1]
Concoecia sp.	1.61 nmol mg protein-1 h-1	summer[1]
Concoecia sp.	3.33 nmol mg protein-1 h-1	spring[1]
Euchaeata norvegica	3.03 nmol mg protein-1 h-1	spring, females[1]
Euchaeata norvegica	3.94 nmol mg protein-1 h-1	spring, CV[1]
Euchaeata norvegica	4.99 nmol mg protein-1 h-1	spring, CIV[1]
Euchaeata norvegica	6.30 nmol mg protein-1 h-1	spring, CIII[1]
Euchaeta norvegica	1.26 nmol mg protein-1 h-1	summer, females[1]
Eukrohnia batypelagica	1.29 nmol mg protein-1 h-1	spring[1]
Eukrohnia hamata	0.93 nmol mg protein-1 h-1	spring[1]
Meganyctiphanes norvegica	1.33 nmol mg protein-1 h-1	spring[1]
Meganyctiphanes norvegica	2.92 nmol mg protein-1 h-1	summer[1]

Metridia longa	2.17 nmol mg protein-1 h-1	summer[1]
Metridia longa	7.18 nmol mg protein-1 h-1	spring[1]
Parathemisto abyssorum	0.97 nmol mg protein-1 h-1	summer[1]
Parathemisto abyssorum	1.31 nmol mg protein-1 h-1	spring[1]
Pleurobrachia pileus	7.76 nmol mg protein-1 h-1	spring[1]
Sagitta elegans	3.45 nmol mg protein-1 h-1	spring[1]
Tomopteris helgolandica	1.80 nmol mg protein-1 h-1	spring[1]

Source: 1.Baamstedt, U., Limnol. Oceanogr., 30(3), 607, 1985.

1-486 Zooplankton
Excretion rate of P

Species	Value	Condition
Daphnia pulex	0.40 - 0.52 ug Production(mg dry wt.)-1	incubation experiment, fed on Scenedesmus sp.[1]

Source: 1.Olsen, Y. and Østgaard, K., Limnol. Oceanogr., 30(4), 844, 1985.

1-487 Zooplankton
Fe / biomass ratio

Species	Value	Condition
Calanus finmarchicus	0.007 %	Atomic composition[1]

Source: 1.Vernadsky, W. 1933. Z. Krist. Mineral. Petrog. Abt. B Mineral. Petrog. Mitt., 44, p 191.

1-488 Zooplankton
Filtration rate

Species	Value	Condition
Acartia clausii	0.05 ml/(anim*hour)	2 h, n=3, coulter counter technique[1]
Acartia clausii	0.06 ml/(anim*hour)	4 h, n=3, carbon-14 technique[1]
Acartia clausii	0.07 ml/(anim*hour)	2 h, n=3, carbon-14 technique[1]
Acartia clausii	0.09 ml/(anim*hour)	4 h, n=3, coulter counter technique[1]
Acartia clausii	0.1 ml/(anim*hour)	4 h, n=3, carbon-14 technique[1]
Acartia clausii	0.11 ml/(anim*hour)	6 h, n=3, coulter counter technique[1]
Acartia clausii	0.12 ml/(anim*hour)	6 h, n=3, chlorophyll a technique[1]
Acartia clausii	0.12 ml/(anim*hour)	6 h, n=3, coulter counter technique[1]
Acartia clausii	0.13 ml/(anim*hour)	6 h, n=3, carbon-14 technique[1]
Acartia clausii	0.14 ml/(anim*hour)	4 h, n=3, carbon-14 technique[1]
Acartia clausii	0.14 ml/(anim*hour)	4 h, n=3, chlorophyll a technique[1]
Acartia clausii	0.14 ml/(anim*hour)	6 h, n=3, carbon-14 technique[1]
Acartia clausii	0.15 ml/(anim*hour)	2 h, n=3, chlorophyll a technique[1]
Acartia clausii	0.16 ml/(anim*hour)	4 h, n=3, chlorophyll a technique[1]
Acartia clausii	0.16 ml/(anim*hour)	6 h, n=3, chlorophyll a technique[1]

Acartia clausii	0.17 ml/(anim*hour)	2 h, n=3, chlorophyll a technique[1]
Acartia clausii	0.18 ml/(anim*hour)	4 h, n=3, coulter counter technique[1]
Acartia clausii	0.24 ml/(anim*hour)	2 h, n=3, coulter counter technique[1]
Acartia tonsa	1.63 ml/(anim*day)	Food = 1203 ug C/l as detritus, 293 K, seawater, n=3, SD = +- 1.75[2]
Acartia tonsa	2.64 ml/(anim*day)	Food = 1193 ug C/l as detritus, 293 K, seawater, n=3, SD = +- 0.89[2]
Acartia tonsa	3.65 ml/(anim*day)	Food = 474 ug C/l as detritus, 293 K, seawater, n=3, SD = +- 4.23[2]
Acartia tonsa	5.28 ml/(anim*day)	Food = 737 ug C/l as detritus, 293 K, seawater, n=3, SD = +-3.38[2]
Bosmina longirostris	0.44 ml/(animal*day)	Animal length = 0.4 - 0.6 mm[3]
Bosmina longirostris	0.45 ml/(anim*day)	Sept 1969, Drowned Bog Lake, Ont, % of total filt.=44.8[3]
Bosmina longirostris	0.46 ml/(anim*day)	Sept 1968, Drowned Bog Lake, Ont, % of total filt.=85[3]
Brachionus calyciflorus	1.1 litre/(mg dry weight*day)	Range = 0.6 - 1.5[4]
Calanus finmarchicus	0.05 litre/(mg dry weight*day)	None[5]
Calanus helgolandicus	350 ml/day	100 ug C animal, culture, rg = 200 - 500[6]
Calanus sp.	1.335 litre/(mg dry weight*day)	Range = 0.67 - 2.0[7]
Centropages hamatus	1.13 litre/(mg dry weight*day)	Range = 0.67 - 1.6[8]
Ceriodaphnia quadrangula	4.6 ml/(animal*day)	Animal length = 0.7 - 0.9 mm[3]
Chydorus sphaericus	0.18 ml/(animal*day)	Animal length = 0.1 - 0.2 mm[3]
Daphnia galeata	3.7 ml/(animal*day)	Animal length = 1.5 - 1.7 mm[9]
Daphnia galeata	6.4 ml/(animal*day)	Animal length = 1, 5 - 1.7 mm[3]
Daphnia longispina	2.3 ml/(animal*day)	Animal length = 0.7 - 1.2 mm[10]
Daphnia magna	0.25 litre/(mg dry weight*day)	Range = 0.2 - 0.3[11]
Daphnia magna	0.74 litre/(mg dry weight*day)	None[12]
Daphnia parvula	1.6 ml/(anim*day)	Sept 1969, Drowned Bog Lake, Ont, % of total filt.=7.5[3]
Daphnia parvula	3.8 ml/(animal*day)	Animal length = 0.7 - 1.2 mm[3]
Daphnia rosea	3.6 ml/(animal*day)	Animal length = 1.3 - 1.6 mm[9]
Daphnia rosea	5.5 ml/(animal*day)	Animal length = 1.3 - 1.6 mm[3]
Daphnia schodleri	1 ml/animal*hr	Food = Chlamydomonas sp., max.value, 293 K, n=4[13]
Daphnia schodleri	1.1 ml/animal*hr	Food = Frustulia sp., max.value, 293 K, n=2[13]
Daphnia schodleri	1.2 ml/animal*hr	Food = Ankistrodesmus sp., max.value, 293 K, n=10[13]
Daphnia schodleri	1.2 ml/animal*hr	Food = Frustulia and Ankistrodesmus, max.value, 293 K, n=4[13]
Daphnia sp.	0.9 litre/(mg dry weight*day)	Range = 0.2 - 1.6[14]
Daphnia sp.	1.06 litre/(mg dry weight*day)	None[15]
Diaphanosoma brachyurum	1.2 ml/(anim*day)	Sept 1969, Drowned Bog Lake, Ont, % of total filt.=0.4[3]
Diaphanosoma brachyurum	1.6 ml/(animal*day)	Animal length = 0.9 - 1.4 mm[3]

Diaptomus gracilis	0.02 ml/(anim*day)	Food = < 30E-3 cells/ml, Pediastrum, 293 K[16]
Diaptomus gracilis	0.19 ml/(anim*day)	Food = < 30E-3 cells/ml, bacteria, 293 K[16]
Diaptomus gracilis	0.61 ml/(anim*day)	Food = < 30.0E-3 cells/ml, Chlorella, 278 K[16]
Diaptomus gracilis	0.68 ml/(anim*day)	Food = 198.0E-3 cells/ml, Melosira and Asterionella sp.[3]
Diaptomus gracilis	0.87 ml/(anim*day)	Food = < 30E-3 cells/ml, Carteria, 293 K[16]
Diaptomus gracilis	0.94 ml/(anim*day)	Food = < 30E-3 cells/ml, Scenedesmus, 285 K[16]
Diaptomus gracilis	1.32 ml/(anim*day)	Food = < 30E-3 cells/ml, Scenedesmus, 293 K[16]
Diaptomus gracilis	1.51 ml/(anim*day)	Food = < 30E-3 cells/ml, Chlorella, 285 K[16]
Diaptomus gracilis	1.61 ml/(anim*day)	Food = < 30E-3 cells/ml, Ankistrodesmus, 285 K[16]
Diaptomus gracilis	1.76 ml/(anim*day)	Food = < 30E-3 cells/ml, Diplophaeria, 285 K[16]
Diaptomus gracilis	1.94 ml/(anim*day)	Food = 24.2-52.0E-3 cells/ml, Melosira and Asterionella sp., rg=[3]
Diaptomus gracilis	1.96 ml/(anim*day)	Food = < 30E-3 cells/ml, Nitzschia, 293 K[16]
Diaptomus gracilis	2.16 ml/(anim*day)	Food = < 30E-3 cells/ml, Haematococcus, 293 K[16]
Diaptomus gracilis	2.4 ml/(anim*day)	Food = < 30E-3 cells/ml, Chlorella, 293 K[16]
Diaptomus gracilis	2.45 ml/(anim*day)	Food = < 30E-3 cells/ml, Ankistrodesmus, 293 K[16]
Diaptomus gracilis	2.54 ml/(anim*day)	Food = < 30E-3 cells/ml, Diplosphaeria, 293 K[16]
Diaptomus graciloides	1.55 ml/(anim*day)	Food = natural phytoplankton, rg = 0.3 - 2.8[17]
Diaptomus graciloides	4.1 ml/(anim*day)	Food = 13.6E-3 cells/ml, Scenedesmus sp.[18]
Diaptomus oregonensis	0.7 ml/(anim*day)	Food = in situ phytoplankton, rg = 0.00 - 1.4[3]
Diaptomus oregonensis	0.85 ml/(anim*day)	Food = 52.0-198.0E-3 cells/ml, Chlorella, rg = 0.3 - 1.4[19]
Diaptomus oregonensis	1.95 ml/(anim*day)	Food = 25.0-52.0E-3 cells/ml, Chlamydomonas sp., rg = 1.4 - 2.5[19]
Diaptomus oregonensis	2.1 ml/(anim*day)	Sept 1969, Drowned Bog Lake, Ont, % of total filt.=0.9[3]
Diaptomus siciloides	2 ml/(anim*day)	Food = Pandorina and Chlamydomonas sp.[20]
Diaptomus siciloides	2.5 ml/(anim*day)	Food = Pandorina and Chlamydomonas sp.[19]
Holopedium gibberum	9.4 ml/(anim*day)	Sept 1969, Drowned Bog Lake, Ont, % of total filt.=46.2[3]
Limnocalanus macrurus	0.1 ml/(anim*day)	Food = Rhodoturula (yeast), > 278 K[21]
Limnocalanus macrurus	1.24 ml/(anim*day)	Food = Chlamydomonas, > 278 K[21]
Limnocalanus macrurus	2.45 ml/(anim*day)	Food = Scenedesmus, > 278 K[21]
Rhincalanus nasutus	1.25 litre/(mg dry weight*day)	Range = 0.3 - 2.2[22]
Zooplankton	0.01 1(mg*day-celcius)	Lake Ontario, upper trophic level 1[23]
Zooplankton	0.06 1(mg*day-celcius)	Lake Ontario, carnivorous[23]
Zooplankton	0.06 1(mg*day-celcius)	Lake Ontario, upper trophic level 2[23]
Zooplankton	0.06 1/(mg*day)	Lake Ontario, herbivorous[23]
Zooplankton	0.13 l/((mg carbon)*day)	San Joaquin River, 293 K[24]
Zooplankton	0.13 l/(mg carbon*day)	Used in models[25]
Zooplankton	0.18 l/((mg carbon)*day)	Delta Estuary, 293 K[24]
Zooplankton	0.25 l/((mg carbon)*day)	Lake Erie, 293 K[24]
Zooplankton	0.95 litre/(mg dry weight*day)	Range = 0.8 - 1.1, natural association[7]
Zooplankton	1.2 l/((mg carbon)*day)	Lake Ontario, 293 K[24]

Source: 1.Hargis, J.R. 1977. Limnol. Oceanogr., 22(5), pp 942-944.

2.Roman, M.R. 1977. Mar. Biol. 42(2), pp 149-155.
3.Haney, J.F. 1973. Arch. Hydrobiol., 72, pp 87-132.
4.Hutchinson, G.E. 1957. A Treatise on Limnology, vol II. Introduction of Lake Biology and the Limnoplankton, Wiley, N.Y., pp 306-354.
5.Adams, J.A. and J.H. Steele. 1966. Some contemporary studies in Marine Science. Shipboard Experimentation the feeding of Calanus. H. Barnes(ed), G. Allen and Unwin LTD., London, pp 19-85.
6.Shelef, G., W.J. Oswald and C.C. Golueke. 1972. Adv. Wat. Poll. Res., Proc. of the 5th Int. Conf. on Water Poll., Paper III, 25, pp 1-9.
7.Riley, G.A. and R. von Arx. 1949. J. marine res., 8(11), pp 60-72.
8.Anraku, M. and M. OMori. 1963. Limnol. Oceanogr., 8(1), pp 116-126.
9.Burns, C.W. and F.H. Rigler. 1967. Limnol. Oceanogr. 12. pp 492-502.
10.Nauwerck, A. 1959. Arch. Hydrobiol. Suppl. 25, pp 83-101.
11.McMahon, J.W. and F.H. Rigler. 1965. Limnol. Oceanogr. 10(1), pp 105-113.
12.Ryther, J.H. 1954. Ecology, 35, pp 522-33.
13.Hayward, R.S. and D.N. Gallup. 1976. Arch. Hydrobiol., 77(2), pp 139-163.
14.Burns, C.W. 1967. Limnol. Oceanogr., 14(5), pp 693-700.
15.Wright, J.C. 1958. Limnol. Oceanogr., 3(2), pp 150-159.
16.Kibby. H.V. 1971. Ecol. Monogr. 41, pp 311-327.
17.Nauwerck, A. 1959. Arch. Hydrobiol. Suppl. 25, pp 83-101.
18.Malovitskaia, L.M. and J. Sorekin. 1961. Trudy Inst. Biol. Vodokhranilishch 4, pp 262-272.
19.Richman, S. 1966. Verh. Int. Ver. Limnol. 16, pp 392-398.
20.Comita, G.W. 1964. Verh. Int. Ver. Limnol., 15, pp 646-653.
21.Kibby, H.V. and F.H. Rigler. 1973. Verh Int. Ver. Limnol. 18, pp 1457-1461.
22.Mullin, M.M. 1963. Limnol. Oceanogr., 8(2), pp 239-250.
23.Thomann, R.V., D.M. Di Toro, R.P. Winfield and D.J. O'Connor. 1975. Mathematical Modeling of Phytoplankton in Lake Ontario: 1. National Environmental Reseach Center. EPA. Oregon.
24.O'Connor, D.J., D.M. DiToro and R.V. Thomann. 1975. in: Russel, C.S. (ed.); Ecological Modeling in a Resource Management Framework. pp 149-209.
25.O'Connor, D.J., R.V. Thomann and D.M. DiToro. 1973. Dynamic Water Quality Forecasting and Management. Office of Research and Development U.S. Environmental Protection Agency, Wash. 20460.

1-489 Zooplankton
Grazing rate

Species	Value	Condition
Acartia tonsa	0.06 microgram C/((microgram anim C)*day	Filtration rate = 3.65 ml/(anim*d)(SD=4.23), 293 K, seawater, n=3[1]
Acartia tonsa	0.33 microgram C/((microgram anim C)*day	Filtration rate = 5.28 ml/(anim*d)(SD=3.38), 293 K, seawater, n=3[1]
Acartia tonsa	0.66 microgram C/((microgram anim C)*day	Filtration rate = 2.64 ml/(anim*d)(SD=0.89), 293 K, seawater, n=3[1]
Acartia tonsa	0.81 microgram C/((microgram anim C)*day	Filtration rate = 1.63 ml/(anim*d)(SD=1.75), 293 K, seawater, n=3[1]
Calanus helgolandicus	11.78 microgram C/animal	Food = Laudria borealis = 49 ug C/l, n=2, stage CI to CIII[2]
Calanus helgolandicus	133.9 microgram C/animal	Food=Laudria borealis = 49 ug C/l, n=2, from NII to adult[2]
Calanus helgolandicus	21.55 microgram C/animal	Food = Laudria borealis =100 ug C/l, stage CI to CIII, n=3[2]
Calanus helgolandicus	4.61 microgram C/animal	Food = Laudria borealis = 49 ug C/l, n=2, stage nii to ci[2]
Calanus helgolandicus	40.84 microgram C/animal	Food = Laudria borealis =100 ug C/l, stage cv to adult[2]
Calanus helgolandicus	47.28 microgram C/animal	Food = Laudria borealis = 49 ug C/l, n=2, stage CIII to CV[2]
Calanus helgolandicus	69.23 microgram C/animal	Food = Laudria borealis = 49 ug C/l, n=2, stage CV to adult[2]
Calanus helgolandicus	7.84 microgram C/animal	Food = Laudria borealis =100 ug C/l, stage NII to CI, n=3[2]
Calanus helgolandicus	98 microgram C/animal	Food = Laudria borealis =100 ug C/l, stage CIII to CV, n=3[2]
Calanus pacificus	0.1 microgram C/(animal*hour)	23 cell/ml= 19 ug C/l, cell dia = 35 my, Coscinodiscus angstii[3]
Calanus pacificus	0.11 microgram C/(animal*hour)	11 cell/ml=18 ug C/l, cell dia = 75 U as Coscinodiscus excentricus

Calanus pacificus	0.11 microgram C/(animal*hour)	4 cell/ml=13 ug C/l, cell dia = 87 U as Coscinodiscus angstii[3]
Calanus pacificus	0.14 microgram C/(animal*hour)	200 cell/ml=27 ug C/l, cell dia =13 U as Thalassiosira fluviatilis[3]
Calanus pacificus	0.34 microgram C/(animal*hour)	1010 cell/ml=95 ug C/l, cell dia = 11 U as Thalassiosira fluviatilis[3]
Daphnia magna	250000 cells/hr	Saccharomyces cer.(66 U**3)= cons=0.016 mm**3/hr*anim[4]
Daphnia magna	2800 cells/hr	Tetrahymena pyr.(18000 U**3)= cons=0.051 mm**3/hr*anim[4]
Daphnia magna	500000 cells/hr	Chlorella vulg. (34 U**3)= consump=0.017 mm**3/hr*anim[4]
Daphnia magna	5600000 cells/hr	E.coli(0.9 U**3)= cons=0.005 mm**3/hr*anim[4]
Daphnia pulex	0.47 cal/period	0-6 d old, food = 25000 cells/ml, Chlamydomonas reinhardtii[5]
Daphnia pulex	0.58 cal/period	0-6 d old, food = 50000 cells/ml, Chlamydomonas reinhardtii[5]
Daphnia pulex	1.39 cal/period	0-6 d old, food = 75000 cells/ml, Chlamydomonas reinhardtii[5]
Daphnia pulex	1.91 cal/period	0-6 d old, food = 100000 cells/ml, Chlamydomonas reinhardtii[5]
Daphnia pulex	1010000 cells/(anim*hour)	Food = Synechocystis sp., 298 K[6]
Daphnia pulex	1110000 cells/(anim*hour)	Food = Chlorella vulgaris, 298 K[6]
Daphnia pulex	1120000 cells/(anim*hour)	Food = Synechococcus cedrorum, 298 K[6]
Daphnia pulex	13 cal/period	6-34 d old, food = 50000 cells/ml, Chlamydomonas reinhardtii[5]
Daphnia pulex	13.6 cal/period	0-40 d old, food = 50000 cells/ml, Chlamydomonas reinhardtii[5]
Daphnia pulex	1330000 cells/(anim*hour)	Food = Gloeocapsa alpicola, 298 K[6]
Daphnia pulex	1650000 cells/(anim*hour)	Food = Anacystis nidulans, 298 K[6]
Daphnia pulex	19.3 cal/period	6-34 d old, food = 75000 cells/ml, Chlamydomonas reinhardtii[5]
Daphnia pulex	20.7 cal/period	0-40 d old, food = 75000 cells/ml, Chlamydomonas reinhardtii[5]
Daphnia pulex	2200000 cells/(anim*hour)	Food = Ankistrodesmus falcatus, 298 K[6]
Daphnia pulex	27.3 cal/period	6-34 d old, food = 100000 cells/ml, Chlamydomonas reinhardtii[5]
Daphnia pulex	29.2 cal/period	0-40 d old, food = 100000 cells/ml, Chlamydomonas reinhardtii[5]
Daphnia pulex	330000 cells/(anim*hour)	Food = Synechococcus elongata, 298 K[6]
Daphnia pulex	340000 cells/(anim*hour)	Food = Anabaena flos-aquae, 298 K[6]
Daphnia pulex	5.67 cal/period	6-34 d old, food = 25000 cells/ml, Chlamydomonas reinhardtii[5]
Daphnia pulex	510000 cells/(anim*hour)	Food = Merismopedia sp., 298 K[6]
Daphnia pulex	6.1 cal/period	0-40 d old, food = 25000 cells/ml, Chlamydomonas reinhardtii[5]
Daphnia schodleri	10000 cells/(anim*hour)	Food = Ankistrodesmus sp., max.value, 293 K, n=10[7]
Daphnia schodleri	10000 cells/(anim*hour)	Food = Frustulia and Ankistrodesmus, max value, 293 K, n=4[7]
Daphnia schodleri	30000 cells/(anim*hour)	Food = Chlamydomonas sp., max.value, 293 K, n=4[7]
Daphnia schodleri	9800 cells/(anim*hour)	Food = Frustulia sp., max.value, 293 K, n=2[7]
Zooplankton	0.015 1/hour	Steady state solution to Castle Lake, on phytoplankton[8]
Zooplankton	0.2502 microgram P/(mg*hour)	Food = seston, Stonehouse Pond, May 17, 1973[9]
Zooplankton	2.8917 microgram P/(mg*hour)	Food = seston, Stonehouse Pond, Feb 22, 1973[9]
Zooplankton	88.96 microgram P/(mg*hour)	Food = seston, Stonehouse Pond, April 6, 1973[9]
Zooplankton	88.96 microgram P/(mg*hour)	Food = seston, Stonehouse Pond, June 17, 1973[9]

Source: 1.Roman, M.R. 1977. Mar. Biol. 42(2), pp 149-155.
2.Paffenhoefer, G.-A. 1976. Limnol. Oceanogr. 21(1), pp 39-50.
3.Frost, B.W. 1975. Limnol. Oceanogr., 29(2), pp 263-266.
4.McMahon, J.W. and F.H. Rigler. 1965. Limnol. Oceanogr. 10(1), pp 105-113.
5.Richman, S. 1958. Ecol. Monogr. 28, pp 273-291.
6.Arnold, D.E. 1971. Limnol. Oceanogr., 16(6), pp 906-920.
7.Hayward, R.S. and D.N. Gallup. 1976. Arch. Hydrobiol., 77(2), pp 139-163.
8.Richey, J.E. 1977. in: Hall, C.A.S. and J.W. Day, Jr. (eds.); Ecosystem Modeling in theory and practice. John Wiley and Sons. New York.
9.Ferrante, J.G. 1976. Hydrobiologia, 49(3), pp 203-204.

1-490 Zooplankton
Gross growth efficiency

Species	Value	Condition
Calanus finmarchicus	17.2 %	Food = Thalassiosira= 490 ug C/l, 279-291 K, CV and adults[1]
Calanus finmarchicus	26.75 %	Food = seawater, based on phosphorus, CV to adult, 285 k[1]
Calanus finmarchicus	34 %	Food = seawater, hatching to adulthood, 283 K, based on nitrogen[1]
Calanus finmarchicus	35.7 %	Food = Skeletonema = 200 ug C/l, 283 K, based on nitrogen, cii to cv[1]
Calanus helgolandicus	17.2 %	Food = Thalassiosira= 490 ug C/l, 279-291 K, CV and adult, based on P[1]
Calanus helgolandicus	29.55 %	Food = seawater, based on nitrogen, CV and adult, 285 k[1]
Calanus helgolandicus	35.7 %	Food = Skeletonema= 200 ug C/l, 283 K, based on nitrogen, CII to CV[1]
Calanus hyperboreus	15.15 %	Food=T. fluviatilis=6.5 mg ash-free dw/l, 275-278 K, CV, on ash-fre[1]
Calanus hyperboreus	27.5 %	Food=T. fluviatilis=1.7 mg asg-free dw/l, 275-278 K, CV, on ash-fre[1]
Rhincalanus nasutus	30 %	Food = T. fluviatilis=352 ug C/l, whole life, 283 K, based on carbon[1]
Rhincalanus nasutus	34 %	Food = Ditylum brigh.=200 ug C/l, whole life, 283 K, based on carbon[1]
Rhincalanus nasutus	37 %	Food = T. fluviatilis=196 ug C/l, whole life, 288 K, based on carbon[1]
Rhincalanus nasutus	45 %	Food = Ditylum brigh.=148 ug C/l, whole life, 288 K, based on carbon[1]

Source: 1.Paffenhoefer, G.-A. 1976. Limnol. Oceanogr. 21(1), pp 39-50.

1-491 Zooplankton
Growth and reproduction

Species	Value	Condition
Daphnia pulex	10 % of consumption	Food = 100E-4 cell/ml, Chlamydomonas, 40 d, after 18 instars[1]
Daphnia pulex	11.4 % of consumption	Food = 75E-4 cell/ml, Chlamydomonas, 40 d, after 18 instars[1]
Daphnia pulex	13.1 % of consumption	Food = 50E-4 cell/ml, Chlamydomonas, 40 d, after 18 instars[1]
Daphnia pulex	13.5 % of ingestion	Range = 10 - 17 %[2]
Daphnia pulex	17.4 % of consumption	Food = 25E-4 cell/ml, Chlamydomonas, 40 d, after 18 instars[1]
Daphnia pulex	56 % of assimilation	Food = 25E-4 cell/ml, Chlamydomonas, 40 d, after 18 instars[1]
Daphnia pulex	64.5 % of assimilation	Range = 56 - 73 %[2]
Daphnia pulex	65.3 % of assimilation	Food = 50E-4 cell/ml, Chlamydomonas, 40 d, after 18 instars[1]
Daphnia pulex	69.8 % of assimilation	Food = 75E-4 cell/ml, Chlamydomonas, 40 d, after 18 instars[1]
Daphnia pulex	73 % of assimilation	Food = 100E-4 cell/ml, Chlamydomonas, 40 d, after 18 instars[1]

Leptodora kindtii	7.3 % of assimilation	None[2]
Mesocyclops albidus	25 % of ingestion	App value[2]
Mesocyclops albidus	50 % of assimilation	App value[2]
Simocephalus vectulus	20.5 % of ingestion	Reproducing female[2]
Simocephalus vectulus	52.9 % of ingestion	Juvenile, female[2]
Simocephalus vectulus	64.7 % of assimilation	None[2]
Simocephalus vectulus	73.1 % of assimilation	Reproducing female[2]

Source: 1.Richman, S. 1958. Ecol. Monogr. 28, pp 273-291.
2.Wetzel, R.G. 1975. Limnology. Saunders. Philadelphia.

1-492 Zooplankton
Growth efficiency

Species	Value	Condition
Daphnia pulex	0.43 % of energy consumed	6-34 d old, food = 100000 cells/ml, Chlamydomonas reinhardtii[1]
Daphnia pulex	0.57 % of energy consumed	6-34 d old, food = 75000 cells/ml, Chlamydomonas reinhardtii[1]
Daphnia pulex	0.78 % of energy consumed	6-34 d old, food = 50000 cells/ml, Chlamydomonas reinhardtii[1]
Daphnia pulex	1.25 % of energy consumed	6-34 d old, food = 25000 cells/ml, Chlamydomonas reinhardtii[1]
Daphnia pulex	10 % of energy consumed	0-40 d old, food = 100000 cells/ml, Chlamydomonas reinhardtii[1]
Daphnia pulex	11.4 % of energy consumed	0-40 d old, food = 75000 cells/ml, Chlamydomonas reinhardtii[1]
Daphnia pulex	13.1 % of energy consumed	0-40 d old, food = 50000 cells/ml, Chlamydomonas reinhardtii[1]
Daphnia pulex	13.2 % of energy assimilated	0-6 d old, food = 25000 cells/ml, Chlamydomonas reinhardtii[1]
Daphnia pulex	17.4 % of energy consumed	0-40 d old, food = 25000 cells/ml, Chlamydomonas reinhardtii[1]
Daphnia pulex	2.99 % of energy assimilated	6-34 d old, food = 100000 cells/ml, Chlamydomonas reinhardtii[1]
Daphnia pulex	3.41 % of energy assimilated	6-34 d old, food = 75000 cells/ml, Chlamydomonas reinhardtii[1]
Daphnia pulex	3.85 % of energy assimilated	6-34 d old, food = 50000 cells/ml, Chlamydomonas reinhardtii[1]
Daphnia pulex	3.9 % of energy assimilated	0-6 d old, food = 100000 cells/ml, Chlamydomonas reinhardtii[1]
Daphnia pulex	3.95 % of energy assimilated	6-34 d old, food = 25000 cells/ml, Chlamydomonas reinhardtii[1]
Daphnia pulex	4.8 % of energy assimilated	0-6 d old, food = 75000 cells/ml, Chlamydomonas reinhardtii[1]
Daphnia pulex	56 % of energy assimilated	0-40 d old, food = 25000 cells/ml, Chlamydomonas reinhardtii[1]
Daphnia pulex	65 % of energy assimilated	0-40 d old, food = 50000 cells/ml, Chlamydomonas reinhardtii[1]
Daphnia pulex	70 % of energy assimilated	0-40 d old, food = 75000 cells/ml, Chlamydomonas reinhardtii[1]
Daphnia pulex	73 % of energy assimilated	0-40 d old, food = 100000 cells/ml, Chlamydomonas reinhardtii[1]
Daphnia pulex	9.1 % of energy assimilated	0-6 d old, food = 50000 cells/ml, Chlamydomonas reinhardtii[1]

Source: 1.Richman, S. 1958. Ecol. Monogr. 28, pp 273-291.

1-493 Zooplankton
Growth rate

Species	Value	Condition
Canuella perplexa	0.042 1/day	Spring 1970, field data, 280.6 K[1]
Ceriodaphnia sp.	0.44 1/day	Mean of 18 different field experiment[2]
Ceriodaphnia sp.	0.58 1/day	Near optimal lab conditions, 296 K, grazing on 4E5 algae/l[2]
Ceriodaphnia sp.	7.5 days	4E5 cell algae/l, 296 K, near optimal lab, generation time[2]
Daphnia pulex	0 1/day	Food = Gloeocapsa alpicola, 274 K, net reprod.rate=0[3]
Daphnia pulex	0 1/day	Food = Gloeocapsa alpicola, 298 K, net reprod.rate=0[3]
Daphnia pulex	0 1/day	Food = Merismopedia sp., 274 K, net reprod.rate=0[3]
Daphnia pulex	0 1/day	Food = Merismopedia sp., 298 K, net reprod.rate=0[3]
Daphnia pulex	0 1/day	Food = Synechocystis sp., 274 K, net reprod.rate=0[3]
Daphnia pulex	0 1/day	Food = Synechocystis sp., 298 K, net reprod.rate=0[3]
Daphnia pulex	0.01 1/day	Food = Synechococcus elongata, 274 K, net reprod.rate=0.1, less than[3]
Daphnia pulex	0.01 1/day	Food = Synechococcus elongata, 298 K, net reprod.rate=0.9, less than[3]
Daphnia pulex	0.023 microgram C/(anim*hour)	Detritus = 187.5 mg C/m3, dissolved oxygen=5.6 mg/l[4]
Daphnia pulex	0.032 microgram C/(anim*hour)	Detritus = 194.7 mg C/m3, dissolved oxygen=4.3 mg/l[4]
Daphnia pulex	0.04 1/day	No food, net reprod.rate=1.6[3]
Daphnia pulex	0.04 microgram C/(anim*hour)	Detritus = 238.0 mg C/m3, dissolved oxygen=2.8 mg/l[4]
Daphnia pulex	0.043 microgram C/(anim*hour)	Detritus = 194.7 mg C/m3, dissolved oxygen=5.3 mg/l[4]
Daphnia pulex	0.046 microgram C/(anim*hour)	Detritus = 213.4 mg C/m3, dissolved oxygen=4.5 mg/l[4]
Daphnia pulex	0.05 microgram C/(anim*hour)	Detritus = 228.5 mg C/m3, dissolved oxygen=3.1 mg/l[4]
Daphnia pulex	0.09 1/day	Food = Anacystis nidulans, 298 K, net reprod.rate=2.1[3]
Daphnia pulex	0.16 1/day	Food = Anabaena flos-aquae, 274 K, net reprod.rate=34.1[3]
Daphnia pulex	0.21 1/day	Food = Anabaena flos-aquae, 298 K, net reprod.rate=82.4[3]
Daphnia pulex	0.26 1/day	Food = Anacystis nidulans, 274 K, net reprod.rate=39.9[3]
Daphnia pulex	0.29 1/day	Food = Synechococcus cedrorum, 274 K, net reprod.rate=181.9[3]
Daphnia pulex	0.37 1/day	Food = Ankistrodesmus falcatus, 274 K, net reprod.rate=306.9[3]
Daphnia pulex	0.43 1/day	Food = Synechococcus cedrorum, 298 K, net reprod.rate=225.4[3]
Daphnia pulex	0.57 1/day	Food = yeast, 283 K, net reprod.rate=1289.2[3]
Daphnia pulex	0.74 1/day	Food = Ankistrodesmus falcatus, 298 K, net reprod.rate=2906.5[3]
Daphnia rosea	0.184 ng P/(anim*hour)	Food = 0.5E5 cells Rhodotorula/ml, adult growth rate, n=6[5]
Daphnia rosea	0.229 ng P/(anim*hour)	Food = 1E5 cells Rhodotorula/ml, adult growth rate, n=10[5]

Daphnia rosea	0.44 ng P/(anim*hour)	Food = 0.125E5 cells Rhodotorula/ml, adult growth rate, n=5[5]
Daphnia rosea	0.64 ng P/(anim*hour)	Food = 0.125E5 cells Rhodotorula/ml, reprod.rate, n=5, SD = +-0.16[5]
Daphnia rosea	1.74 ng P/(anim*hour)	Food = 0.5E5 cells Rhodotorula/ml, reprod.rate, n=6, SD = +-0.56[5]
Daphnia rosea	1.89 ng P/(anim*hour)	Food = 1E5 cells Rhodotorula/ml, reprod.rate, n=10, SD = +-0.47[5]
Daphnia rosea	118 ng P/(mg*hour)	Food = 0.5E5 cells Rhodotorula/ml, juvenile growth const., n=7[5]
Daphnia rosea	119 ng P/(mg*hour)	Food = 1E5 cells Rhodotorula/ml, juvenile growth const., n=14[5]
Daphnia rosea	76 ng P/(mg*hour)	Food = 0.125E5 cells Rhodotorula/ml, juvenile growth const., n=7[5]
Euchlanis dilatata	0.76 1/day	Mean value, 295 K, lab, littoral rotifer[6]
Euchlanis dilatata	1.4 1/days	Max value, 295 K, lab, littoral rotifer[6]
Halicyclops magniceps	0.056 1/day	Spring 1969, field data, 282.8 K[1]
Paronychocamptus nanus	0.042 1/day	Spring 1969, field data, 288 K[1]
Paronychocamptus nanus	0.043 1/day	Spring 1971, field data, 288.4 K[1]
Paronychocamptus nanus	0.048 1/day	Autum 1971, field data[1]
Paronychocamptus nanus	0.056 1/day	Spring 1970, field data, 280.6 K[1]
Philodina roseola	0.196 1/day	288 K, median life span=28 days[7]
Philodina roseola	0.238 1/day	293 K, median life span=10.2 days[7]
Philodina roseola	0.322 1/day	298 K, median life span=5.9 days[7]
Philodina roseola	0.5 1/day	303 K, median life span=3.0 days[7]
Simocephalus serrulatus	0.27 1/day	Field experiment, 296 K[2]
Simocephalus serrulatus	0.55 1/day	Feeding on 8E5 cell algae/l, lab, 296 K[2]
Simocephalus serrulatus	10 days	Field experiment, 296 K, generation time[2]
Tachidius discipea	0.045 1/day	Spring 1969, field data, 283.5 K[1]
Tachidius discipea	0.049 1/day	Spring 1969, field data, 282.8 K[1]
Tachidius discipea	0.056 1/day	Spring 1971, field data, 280.8 K[1]
Zooplankton	0.225 1/day	Temp factor is 1 at 293 K, rg = 0.15 - 0.30, max value[8]

Source: 1.Heip, C. 1972. Mar. Biol., 12, pp 219-221.
2.Hall, D.J., W.E. Cooper and E.E. Werner. 1970. Limnol. Oceanogr., 15(3), pp 838-928.
3.Arnold, D.E. 1971. Limnol. Oceanogr., 16(6), pp 906-920.
4.Bell, R.K. and F.J. Ward. 1970. Limnol. Oceanogr., 15(5), pp 715-726.
5.McRoy, C.P. and R.J. Barsdate. 1970. Limnol. Oceanogr. 15(1) pp 6-13.
6.King, C.E. 1967. Ecology 48, pp 111-128.
7.Schaefer, E.D. and W.O. Pipes. 1973. Water Res., 7(12), pp 1781-1790.
8.The Hydrological Engineering Center Corps of Engineers U.S. Army. 1975. Water Quality for river-reservoir systems, Davis, Cal. 9 5 6 1 6 (916) 499-2329.

1-494 Zooplankton
H / biomass ratio

Species	Value	Condition
Amphipoda	6 % of dry weight	Dried at 333 K, North Pacific, rg = 4.4 - 7.6[1]
Calanus finmarchicus	10.21 %	Atomic composition[2]
Chaetognatha	7.4 % of dry weight	Dried at 333 K, North Pacific, rg = 7.2 - 7.6[1]
Copepoda	8.5 % of dry weight	Dried at 333 K, North Pacific, rg = 6.7 - 10.3[1]
Euphausiacea	7.15 % of dry weight	Dried at 333 K, North Pacific, rg = 6.7 - 7.6[1]
Pteropods	2.45 % of dry weight	Dried at 333 K, North Pacific, rg = 1.1 - 3.8[1]

Source: 1.Omori, M. 1969. Mar. Biol., 3, pp 4-10.
2.Vernadsky, W. 1933. Z. Krist. Mineral. Petrog. Abt. B Mineral. Petrog. Mitt., 44, p 191.

1-495 Zooplankton
Half saturation constant for grazing

Species	Value	Condition
Bosmina coregoni	2.05 mg C/l	Range = 0.1 - 4.0[1]
Daphnia magna	15 mg C/l	Food = Chlorella vulgaris = 0.124E-6 mg C/cell[1]
Daphnia magna	9.6 mg C/l	Food = Chlorella vulgaris = 0.124E-6 mg C/cell[1]
Daphnia rosea	0.16 mg C/l	Food = Rhodotorula glutinis = 0.776E-8 mg C/cell, 4.2 u long[1]
Diaptomus oregonensis	1.6 mg C/l	Food = Chlamydomonas and Chlorella = 0.124E-6 mg C/cell[1]
Zooplankton	0.5 mg/l	On algae, rg = 0.4 - 0.6[2]
Zooplankton	10 microgram chl/l	Lake Ontario[3]
Zooplankton	10 microgram chl/l	Lake Ontario, on algae, herbivorous[4]
Zooplankton	50 microgram chl/l	Delta Estuary, used in model[3]
Zooplankton	50 microgram chl/l	Lake Erie[3]
Zooplankton	50 microgram chl/l	San Joaquin River[3]
Zooplankton	60 microgram chl/l	Potomac river on algae[5]
Zooplankton	60 microgram chl/l	Used in models, on algae[5]

Source: 1.Scavia, D. and B.J. Eadie. 1976. Ecol. Modell., 2, pp 315-319.
2.The Hydrological Engineering Center Corps of Engineers U.S. Army. 1975. Water Quality for river-reservoir systems, Davis, Cal. 9 5 6 1 6 (916) 499-2329.
3.O'Connor, D.J., D.M. DiToro and R.V. Thomann. 1975. in: Russel, C.S. (ed.); Ecological Modeling in a Resource Management Framework. pp 149-209.
4.Thomann, R.V., D.M. Di Toro, R.P. Winfield and D.J. O'Connor. 1975. Mathematical Modeling of Phytoplankton in Lake Ontario: 1. National Environmental Reseach Center. EPA. Oregon.
5.O'Connor, D.J., R.V. Thomann and D.M. DiToro. 1973. Dynamic Water Quality Forecasting and Management. Office of Research and Development U.S. Environmental Protection Agency, Wash. 20460.

1-496 Zooplankton
I / biomass ratio

Species	Value	Condition
Calanus finmarchicus	0.0002 %	Atomic composition[1]

Source: 1.Vernadsky, W. 1933. Z. Krist. Mineral. Petrog. Abt. B Mineral. Petrog. Mitt., 44, p 191.

1-497 Zooplankton
Ingestion rate

Species	Value	Condition
Daphnia longispina	32.5 +/- 2.6 ug C (mg dry wt)-1 h-1	incubation exp., fed on Scenedesmus sp.[1]
Daphnia pulex	8.76 +/- 0.78 ug C (mg dry wt)-1 h-1	incubation exp., fed on Scenedesmus sp.[1]
Thysanoessa raschii	0.25 ug pigment l-1	concentration of pigment with zero ingestion[2]
Thysanoessa raschii	0.5-0.9 ug pigment l-1	None[2]
Copepods	0.88 - 2.54 ng Chl copepods-1 h-1	evacuation exp., 22. Dec.,[3]
Copepods	0.89 - 3.51 ng Chl copepods-1 h-1	evacuation exp., 21. Dec.[3]
Copepods	10.42 - 14.68 ng Chl copepods-1 h-1	grazing exp., 7. Oct.[3]
Copepods	11.17 - 12.77 ng Chl copepods-1 h-1	grazing exp., 14. Oct.[3]

Copepods	3.06 - 11.29 ng Chl copepods-1 h-1	grazing exp., 22. Dec.,[3]
Copepods	3.16 - 7.33 ng Chl copepods-1 h-1	grazing exp., 21. Dec.[3]
Copepods	8.42 - 11.98 ng Chl copepods-1 h-1	evacuation exp., 14. Oct.[3]
Copepods	9.77 - 11.50 ng Chl copepods-1 h-1	evacuation exp., 7. Oct.[3]

Source: 1.Olsen, Y. and Østgaard, K., Limnol. Oceanogr., 30(4), 844, 1985.
2.McClatchie, S. Limnol. Oceanogr., 31, 469, 1986.
3.Wang, R. and Conover, R.J. Limnol. Oceanogr., 31, 867, 1986.

1-498 Zooplankton
Ingestion time

Species	Value	Condition
Chaoborus trivittatus	103.6 sec	Time used to ingest prey = Diaptomus = 1.8 mm, n=35, 4. instar[1]
Chaoborus trivittatus	11.4 sec	Time used to ingest prey = Diaptomus = 1.4 mm, n=22, 4. instar[1]
Chaoborus trivittatus	210.5 sec	Time used to ingest prey = Daphnia = 1.8 mm, n=21, 4. instar[1]
Chaoborus trivittatus	24.3 sec	Time used to ingest prey = Daphnia = 1.0 mm, n=28, 4. instar[1]
Chaoborus trivittatus	5 sec	Time used to ingest prey = Diaptomus = 1.0 mm, n=16, 4. instar[1]
Chaoborus trivittatus	74.2 sec	Time used to ingest prey = Daphnia = 1.4 mm, n=24, 4. instar[1]

Source: 1.Swift, M.C. and A.Y. Fedorenko, 1975. Limnol. Oceanogr. 20(3), pp 418-425.

1-499 Zooplankton
K / biomass ratio

Species	Value	Condition
Calanus finmarchicus	0.29 %	Atomic composition[1]

Source: 1.Vernadsky, W. 1933. Z. Krist. Mineral. Petrog. Abt. B Mineral. Petrog. Mitt., 44, p 191.

1-500 Zooplankton
Length

Species	Value	Condition
Acanthocyclops vernalis	1.125 mm	Late copepodits, rg = 1.08 - 1.17 mm[1]
Acanthocyclops vernalis	1.5 mm	Adults[1]
Bosmina longirostris	0.4 mm	Max adult length, Gatun Lake[2]
Bosminopsis deitersi	0.35 mm	Max adult length, Gatun Lake[2]
Ceriodaphnia cornuta	0.4 mm	Max adult length, Gatun Lake, unhorned or horned morph[2]
Chydorus eurynotus	0.5 mm	Max adult length, Gatun Lake[2]
Cyclops vicinus	1.125 mm	Late copepodits, rg = 1.08 - 1.17[1]
Cyclops vicinus	1.5 mm	Adults[1]
Daphnia middendorffiana	0.7 mm	Neonates, rg = 0.32 - 1.08 mm[1]
Daphnia middendorffiana	2.62 mm	Adults, rg = 2.32 - 2.92 mm[1]
Diaphanosoma brachyurum	0.9 mm	Max adult length, Gatun Lake[2]
Diaptomus alaskaensis	1.6 mm	Adults[1]
Diaptomus gatunensis	0.8 mm	Max adult length, Gatun Lake, copepodids[2]

Diaptomus gatunensis	1.5 mm	Max adult length, Gatun Lake, adults[2]
Eubosmina tubicens	0.45 mm	Max adult length, Gatun Lake[2]
Eucyclops agilis	0.8 mm	Max adult length, Gatun Lake, copepodids[2]
Eucyclops agilis	1.5 mm	Adults[1]
Eucyclops agilis	1.5 mm	Max adult length, Gatun Lake, adults[2]
Heterocope septentrionalis	0.73 mm	Nauplii and copepodit I, rg = 0.66 - 0.80 mm[1]
Heterocope septentrionalis	2.56 mm	Copepodits IV and V, rg = 2.20 - 2.92[1]
Moina minutus	0.6 mm	Max adult length, Gatun Lake[2]
Nauplii	0.25 mm	Max adult length, Gatun Lake, total nauplii[2]
Polyartemiella hazeni	1.23 mm	Nauplii[1]
Polyartemiella hazeni	2.55 mm	Early metanauplii, rg = 1.67 - 3.42[1]
Polyartemiella hazeni	5.41 mm	Late metanauplii, rg = 4.16 - 6.66[1]

Source: 1.Dodson, S.I. 1975. Limnol. Oceanogr., 20(3), pp 426-433.
2.Zaret, T.M. and J.S. Suffern. 1976. Limnol. Oceanogr. 21(6), pp 804-813.

1-501 Zooplankton
Max feeding rate

Species	Value	Condition
Centropages typicus	22 % of body weight/6 hours	Prey = Artemia nauplii, 281 K[1]
Centropages typicus	48 % of body weight/6 hours	Prey = Artemia nauplii, 288 K[1]
Labidocera jollae	27 % of body weight/6 hours	Prey = anchovy fish larvae[2]
Metridia lucens	24 % of body weight/6 hours	Prey = Artemia nauplii[3]
Sagitta hispida	16 % of body weight/6 hours	Prey = Artemia nauplii[4]
Tortanus discaudatus	17 % of body weight/6 hours	Prey = calanus nauplii 3[5]
Tortanus discaudatus	24 % of body weight/6 hours	Prey = calanus nauplii 5[5]

Source: 1.Anraku, M. 1964. J. Oceanogr. Soc. Japan, 20(5), pp 19-29.
2.Lillelund, K. and R. Lasker. 1971. Fish. Bull., 69, pp 655-667.
3.Haq, S.M. 1967. Limnol. Oceanogr., 12, pp 40-51.
4.Reeve, M.R. 1964. Nature, 201, pp 211-213.
5.Ambler, J.W. and B.W. Frost. 1974. Lim nol. Oceanogr. 19(3), pp 446-451.

1-502 Zooplankton
Mean length

Species	Value	Condition
Daphnia pulicaria	1,161 um	fed on Stephanociscus astraea minutula[1]
Daphnia pulicaria	1,262 um	fed on Melosira italica tenuissima[1]
Daphnia pulicaria	1,316 um	fed on Asterionella formosa[1]
Daphnia pulicaria	1,328 um	fed on Chlorella sp., density 2 * 10e5 cells l-1[1]
Daphnia pulicaria	1,541 um	fed on Melosira italica[1]
Daphnia pulicaria	1,613 um	fed on Chlorella sp., density 1 * 10e4 cells l-1[1]

Daphnia pulicaria	1,705 um	fed on Fragilaria crotonensis[1]
Daphnia pulicaria	1,716 um	fed on Tabellaria fenestrata[1]
Daphnia pulicaria	1,816 um	fed on Stephanociscus hantzschii[1]
Daphnia pulicaria	1,819 um	fed on Cryptomonas erosa[1]
Daphnia thorata	1,115 um	fed on Chlorella sp., density 2 * 10e5 cells l-1[1]
Daphnia thorata	1,147 um	fed on Fragilaria crotonensis[1]
Daphnia thorata	1,197 um	fed on Melosira italica[1]
Daphnia thorata	1,277 um	fed on Tabellaria fenestrata[1]
Daphnia thorata	1,285 um	fed on Stephanodiscus astraea minutula[1]
Daphnia thorata	1,366 um	fed on Chlorella sp., density 1 * 10e4 cells l-1[1]
Daphnia thorata	1,394 um	fed on Stephanodiscus hantzschii[1]
Daphnia thorata	1,416 um	fed on Asterionella formosa[1]
Daphnia thorata	1,427 um	fed on Cryptomonas erosa[1]
Daphnia thorata	945 um	fed on Melosira italica tenuissima[1]

Source: 1.Infante, A. and Litt, A.H., Limnol.Oceanogr., 30(5), 1053, 1985.

1-503 Zooplankton
Mg / biomass ratio

Species	Value	Condition
Calanus finmarchicus	0.03 %	Atomic composition[1]

Source: 1.Vernadsky, W. 1933. Z. Krist. Mineral. Petrog. Abt. B Mineral. Petrog. Mitt., 44, p 191.

1-504 Zooplankton
Min conc for feeding

Species	Value	Condition
Calanus pacificus	0.016 mg C/l	No feeding below this value[1]
Zooplankton	0.115 mg C/l	No feeding below this value, rg = 0.04 - 0.19, 14 marine species[1]

Source: 1.Scavia, D. and B.J. Eadie. 1976. Ecol. Modell., 2, pp 315-319.

1-505 Zooplankton
Mn / biomass ratio

Species	Value	Condition
Calanus finmarchicus	0 %	Atomic composition[1]

Source: 1.Vernadsky, W. 1933. Z. Krist. Mineral. Petrog. Abt. B Mineral. Petrog. Mitt., 44, p 191.

1-506 Zooplankton
Mortality

Species	Value	Condition
Copepoda	0.004 1/day	Used in model, Lake Michigan, rg = 0.003 - 0.005[1]
Daphnia pulicaria	0 %	during 7 days exp., fed on Cryptomonas erosa[2]
Daphnia pulicaria	0 %	during 7 days exp., fed on Fragilaria crotonensis[2]
Daphnia pulicaria	10 %	during 7 days exp., fed on Stephanodiscus astraea minutula[2]
Daphnia pulicaria	100 %	during 7 days exp., fed on Stephanodiscus niagarae[2]
Daphnia pulicaria	20 %	during 7 days exp., fed on Chlorella sp., density 2 * 10e5 cells l-1[2]

Daphnia pulicaria	25 %	during 7 days exp., fed on Asterionella formosa[2]
Daphnia pulicaria	30 %	during 7 days exp., fed on Melosira italica tenuissima[2]
Daphnia pulicaria	40 %	during 7 days exp., fed on Chlorella sp., density 1 * 10e4 cells l-1[2]
Daphnia pulicaria	5 %	during 7 days exp., fed on Melosira italica[2]
Daphnia pulicaria	5 %	during 7 days exp., fed on Stephanodiscus hantzschii[2]
Daphnia pulicaria	5 %	during 7 days exp., fed on Tabellaria fenestrata[2]
Daphnia thorata	100 %	during 7 days exp., fed on Stephanodiscus niagarae[2]
Daphnia thorata	15 %	during 7 days exp., fed on Cryptomonas erosa[2]
Daphnia thorata	15 %	during 7 days exp., fed on Stephanodiscus hantzschii[2]
Daphnia thorata	20 %	during 7 days exp., fed on Melosira italica[2]
Daphnia thorata	25 %	during 7 days exp., fed on Tabellaria fenestrata[2]
Daphnia thorata	30 %	during 7 days exp., fed on Chlorella sp., density 2 * 10e5 cells l-1[2]
Daphnia thorata	30 %	during 7 days exp., fed on Stephanodiscus astraea minutula[2]
Daphnia thorata	35 %	during 7 days exp., fed on Chlorella sp., density 1 * 10e4 cells l-1[2]
Daphnia thorata	40 %	during 7 days exp., fed on Asterionella formosa[2]
Daphnia thorata	45 %	during 7 days exp., fed on Fragilaria crotonensis[2]
Daphnia thorata	70 %	during 7 days exp., fed on Melosira italica tenuissima[2]
Freshwater zooplankton	0.004 1/hour	Steady state solution to Castle Lake.[3]
Zooplankton	0.005 1/day	293 K[4]
Zooplankton	0.005 1/day	Used in model[1]
Zooplankton	0.02 1/day	Used in model, N Fork Kings River, Calif[1]
Zooplankton	0.075 1/day	Used in models[5]

Source: 1.Tetra Tech, Inc. 1977. Ecological constants. San F. Calif.
2.Infante, A. and Litt, A.H., Limnol.Oceanogr., 30(5), 1053, 1985.
3.Richey, J.E. 1977. in: Hall, C.A.S. and J.W. Day, Jr. (eds.); Ecosystem Modeling in theory and practice. John Wiley and Sons. New York.
4.The Hydrological Engineering Center Corps of Engineers U.S. Army. 1975. Water Quality for river-reservoir systems, Davis, Cal. 9 5 6 1 6 (916) 499-2329.
5.O'Connor, D.J., R.V. Thomann and D.M. DiToro. 1973. Dynamic Water Quality Forecasting and Management. Office of Research and Development U.S. Environmental Protection Agency, Wash. 20460.

1-507 Zooplankton N / P ratio

Species	Value	Condition
Calanus cristatus	19 by atoms	Unfilt water, based on respiration and excretion rates[1]
Calanus finmarchicus	10.8 by atoms	Filt water, based on respiration and excretion rates[2]
Calanus finmarchicus	11 by atoms	Filt water, based on respiration and excretion rates, total N/total P[3]
Calanus helgolandicus	16.5 by atoms	Filt water, based on respiration and excretion rates, total N/total P[4]
Calanus plumchrus	13 by atoms	Unfilt water, based on respiration and excretion rates[1]
Sagitta hispida	11.3 by atoms	Filt water, based on respiration and excretion rates[5]
Zooplankton	10.33 by atoms	Unfilt water, based on respiration and excretion rates, mixed[6]
Zooplankton	6.8 by atoms	Unfilt water, based on respiration and excretion rates, mixed[7]
Zooplankton	7 by atoms	Unfilt water, based on respiration and excretion rates, mixed[8]
Zooplankton	9.98 by atoms	Unfilt water, based on respiration and excretion rates, mixed[9]

Source: 1.Taguchi, S. and H. Ishii. 1972. in: Takenouti, A.Y.(ed). Biological Oceanography of the northern Noth Pacific Ocean. pp 419-431. Idemitsu Shoten. Tokyo.
2.Butler, E.I, E.D.S. Corner and S.M. Marshall. 1969. J. mar. biol. Ass. U.K. 49, pp 977-1001.
3.Butler, E.I., E.D.S. Corner and S.M. Marshall. 1970. J. mar. biol. Ass. U.K. 50, pp 525-560.
4.Corner, E.D.S., R.N. Head and C.C. Kilvington. 1972. J. mar. biol. Ass. U.K., 52, pp 847-861.
5.Beers, J.R. 1964. J. Cons. perm. inst. Explor. Mer., 29, pp 123-129.
6.Le Borgne, R.P. 1973. Mar. Biol. 19, pp 249-257.
7.Mullin, M.M., H.J. Perry, E.H. Renger and P.M. Evans. 1975. Mar. Sci. Commun., 1(1), pp 1-13.
8.Harris, E. 1959. Bull. Bingham oceanogr. Coll., 17, pp 31-65.
9.Martin, J.H. 1968. Limnol. Oceanogr. 13, pp 63-71.

1-508 Zooplankton
N / biomass ratio

Species	Value	Condition
Amphipoda	6.3 % of dry weight	Dried at 333 K, North Pacific, rg = 4.4 - 8.2[1]
Calanus finmarchicus	1.52 %	Atomic composition[2]
Calanus finmarchicus	14.8 microgram N/animal	Starvation in 14 days, n=10, stage 4 and 5, SD = +-4.4[3]
Calanus finmarchicus	18.3 microgram N/animal	Starvation in 23 days, n=12, stage 4 and 5, SD = +-2.7[3]
Calanus finmarchicus	18.7 microgram N/animal	Starvation in 8 days, n=10, stage 4 and 5, SD = +-4.2[3]
Calanus finmarchicus	20.1 microgram N/animal	No starvation, n=9, stage 4 and 5, SD = +-5.6[3]
Chaetognatha	10.9 % of dry weight	Dried at 333 K, North Pacific, rg = 10.7 - 11.1[1]
Chaetognatha	7.85 % of dry weight	Dried at 333 K, Sargasso Sea, rg = 6.3 - 9.4[4]
Copepoda	5.9 % of dry weight	Dried at 378 K, continental shelf off New York, rg = 4.7 - 7.1[5]
Copepoda	9.1 % of dry weight	Dried at 333 K, North Pacific, rg = 5.1 - 13.1[1]
Copepoda	9.7 % of dry weight	Dried at 333 K, Sargasso Sea, rg = 8.2 - 11.2[4]
Ctenophora	0.65 % of dry weight	Dried at 378 K, continental shelf off New York, rg = 0.2 - 1.1[5]
Daphnia cristata	7.4 % of ash free dry weight	Cladocera , 95 % CL = +-1.2[6]
Eudiaptomus gracilis	10.1 % of ash free dry weight	Copepoda , 95 % CL = +-0.4[6]
Euphausiacea	10.35 % of dry weight	Dried at 333 K, North Pacific, rg = 10.0 - 10.7[1]
Euphausiacea	6.15 % of dry weight	Dried at 378 K, continental shelf off New York, rg = 5.2 - 7.1[5]
Euphausiacea and Mysidae	9.95 % of dry weight	Dried at 333 K, Sargasso Sea, rg = 9.4 - 10.5[4]
Hydromedusae	4.15 % of dry weight	Dried at 333 K, Sargasso Sea, rg = 1.4 - 6.9[4]
Limnocalanus macrurus	6.2 % of ash free dry weight	Copepoda , 95 % CL = +-0.3[6]
Macrocyclops albidus	10.2 % of ash free dry weight	Copepoda , 95 % CL = +-1.0[6]
Polychaeta	7.8 % of dry weight	Dried at 333 K, Sargasso Sea, rg = 4.4 - 11.2[4]
Pteropods	3.35 % of dry weight	Dried at 378 K, continental shelf off New York, rg = 2.2 - 5.0[5]
Pteropods	3.45 % of dry weight	Dried at 333 K, Sargasso Sea, rg = 2.7 - 4.2[4]
Pteropods	3.75 % of dry weight	Dried at 333 K, North Pacific, rg = 1.5 - 6.0[1]
Siphonophora	2.7 % of dry weight	Dried at 333 K, Sargasso Sea, rg = 1.0 - 4.4[4]
Tunicata	0.9 % of dry weight	Dried at 378 K, continental shelf off New York, rg = 0.3 - 1.5[5]
Zooplankton	8 % of dry weight	Range = 7 - 9[7]

Source: 1.Omori, M. 1969. Mar. Biol., 3, pp 4-10.
2.Vernadsky, W. 1933. Z. Krist. Mineral. Petrog. Abt. B Mineral. Petrog. Mitt., 44, p 191.
3.Balch, N. 1972. Limnol. Oceanogr., 17(6), pp 906-908.

4.Beers, J.R. 1966. Limnol. Oceanogr., 11, pp 520-528.
5.Curl, H. 1962. Rapp. Proc.-Verb. Cons. Int. Explor. Mer., 153, pp 183-189.
6.Salonen, K., J. Sarvala, I. Hakala and M.-L. Viljanen. 1976. Limnol. Oceanogr., 21(5), pp 724-730.
7.The Hydrological Engineering Center Corps of Engineers U.S. Army. 1975. Water Quality for river-reservoir systems, Davis, Cal. 9 5 6 1 6 (916) 499-2329.

1-509 Zooplankton
Na / biomass ratio

Species	Value	Condition
Calanus finmarchicus	0.54 %	Atomic composition[1]

Source: 1.Vernadsky, W. 1933. Z. Krist. Mineral. Petrog. Abt. B Mineral. Petrog. Mitt., 44, p 191.

1-510 Zooplankton
O / N ratio

Species	Value	Condition
Acartia clausii	1.61 by atoms	Filt water, based on respiration and excretion rates[1]
Boreal zooplankton	103 by atoms	Unfilt water, resp and excretion rates, ninhydrin-N, 10 sp[2]
Calanus cristatus	5.7 by atoms	Unfilt water, based on respiration and excretion rates[3]
Calanus helgolandicus	12.7 by atoms	Filt water, based on resp and excretion rates, ninhydrin-N[4]
Calanus plumchrus	6.8 by atoms	Unfilt water, based on respiration and excretion rates[3]
Meganyctiphanes norvegica	8.45 by atoms	Filt water, based on respiration and excretion rates[1]
Phronima sedentaria	4.27 by atoms	Based on respiration and excretion rates[1]
Sagitta hispida	6.8 by atoms	Filt water, based on respiration and excretion rates[5]
Sagitta setosa	1.75 by atoms	Filt water, based on respiration and excretion rates[1]
Temora stylifera	11 by atoms	Filt water, based on respiration and excretion rates[6]
Zooplankton	13.48 by atoms	For respiration/excretion of NH4-N, 22 hrs, 288 K, n=44, marine[7]
Zooplankton	13.48 by atoms	Unfilt water, based on respiration and excretion rates, mixed[8]
Zooplankton	41 by atoms	Unfilt water, based on respiration and excretion rates, mixed[1]
Zooplankton	59.5 by atoms	Filt wat, resp and excretion rates, 81 sp, rg=4-115, boreal[9]
Zooplankton	6.64 by atoms	For respiration/excretion of total N, 22 hrs, 288 K, n=36, marine[7]
Zooplankton	7.7 by atoms	Unfilt water, based on respiration and excretion rates, mixed[10]

Source: 1.Martin, J.H. 1968. Limnol. Oceanogr. 13, pp 63-71.
2.Conover, R.J. and E.D.S. Corner. 1968. J. mar. biol. Ass. U.K., 48, pp 49-75.
3.Taguchi, S. and H. Ishii. 1972. in: Takenouti, A.Y.(ed). Biological Oceanography of the northern North Pacific Ocean. pp 419-431. Idemitsu Shoten. Tokyo.
4.Corner, E.D.S., C.B. Cowey and S.M. Marshall. 1965. J. mar. biol. Ass. U.K., 45, pp 429-442.
5.Reeve, M.R., J.E.G. Raymont and J.K.B. Raymont. 1970. Mar. Biol. 6, pp 357-364.
6.Nival, P., G. Malara, R. Charra, I. Palazzoli and S. Nival. 1974. J. Exp. Mar. Biol. Ecol. 15, pp 21-260
7.Le Borgne, R.P. 1973. Mar. Biol. 19, pp 249-257.
8.Le Borgne, R.P. 1973. Mar. Biol. 19, pp 249-257.
9.Ikeda, T. 1974. Mem. Fac. Fish. Hokkaido Univ. 22, pp 1-97.
10.Harris, E. 1959. Bull. Bingham oceanogr. Coll., 17, pp 31-65.

1-511 Zooplankton
O / P ratio

Species	Value	Condition
Calanus cristatus	110 by atoms	Unfilt water, based on respiration and excretion rates[1]
Calanus plumchrus	89 by atoms	Unfilt water, based on respiration and excretion rates[1]
Zooplankton	142.4 by atoms	For respiration/excretion of PO4-P, 22 hrs, 288 K, n=41, marine[2]
Zooplankton	142.4 by atoms	Unfilt water, based on respiration and excretion rates, mixed[3]
Zooplankton	222 by atoms	Unfilt water, based on respiration and excretion rates, mixed[4]
Zooplankton	54 by atoms	Unfilt water, based on respiration and excretion rates, mixed[5]
Zooplankton	66.8 by atoms	For respiration/excretion of total P, 22 hrs, 288 K, n=40, marine[2]
Zooplankton	72 by atoms	Unfilt water, based on respiration and excretion rates, mixed[6]

Source: 1.Taguchi, S. and H. Ishii. 1972. in: Takenouti, A.Y.(ed). Biological Oceanography of the northern North Pacific Ocean. pp 419-431. Idemitsu Shoten. Tokyo.
2.Le Borgne, R.P. 1973. Mar. Biol. 19, pp 249-257.
3.Martin, J.H. 1968. Limnol. Oceanogr. 13, pp 63-71.
4.Harris, E. 1959. Bull. Bingham oceanogr. Coll., 17, pp 31-65.
5.Satomi, M. and L.R. Pomeroy. 1965. Ecology 46, pp 877-881.

1-512 Zooplankton
P / biomass ratio

Species	Value	Condition
Calanus finmarchicus	0.13 %	Atomic composition[1]
Chaetognatha	0.6 % of dry weight	Dried at 333 K, Sargasso Sea, rg = 0.5 - 0.7[2]
Copepoda	0.6 % of dry weight	Dried at 378 K, continental shelf off New York, rg = 0.4 - 0.8[3]
Copepoda	0.75 % of dry weight	Dried at 333 K, Sargasso Sea, rg = 0.7 - 0.8[2]
Ctenophora	0.15 % of dry weight	Dried at 378 K, continental shelf off New York, rg = 0.1 - 0.2[3]
Euphausiacea	1.05 % of dry weight	Dried at 378 K, continental shelf off New York, rg = 0.9 - 1.2[3]
Euphausiacea and Mysidae	1.5 % of dry weight	Dried at 333 K, Sargasso Sea, rg = 1.4 - 1.6[2]
Fish larvae	1.35 % of dry weight	Dried at 333 K, Sargasso Sea, rg = 0.9 - 1.8, also fish[2]
Hydromedusae	0.25 % of dry weight	Dried at 333 K, Sargasso Sea, rg = 0.1 - 0.4[2]
Polychaeta	1.1 % of dry weight	Dried at 333 K, Sargasso Sea, rg = 0.4 - 1.8[2]
Pteropods	0.3 % of dry weight	Dried at 333 K, Sargasso Sea, rg = 0.2 - 0.4[2]
Pteropods	0.45 % of dry weight	Dried at 378 K, continental shelf off New York, rg = 0.3 - 0.6[3]
Siphonophora	0.15 % of dry weight	Dried at 333 K, Sargasso Sea, rg = < 0.1 - 0.2[2]
Tunicata	0.2 % of dry weight	Dried at 378 K, continental shelf off New York, rg = 0.1 - 0.3[3]
Zooplankton	1.1 % of dry weight	Range = 1.0 - 1.2 %[4]

Source: 1.Vernadsky, W. 1933. Z. Krist. Mineral. Petrog. Abt. B Mineral. Petrog. Mitt., 44, p 191.
2.Beers, J.R. 1966. Limnol. Oceanogr., 11, pp 520-528.
3.Curl, H. 1962. Rapp. Proc.-Verb. Cons. Int. Explor. Mer., 153, pp 183-189.
4.The Hydrological Engineering Center Corps of Engineers U.S. Army. 1975. Water Quality for river-reservoir systems, Davis, Cal. 9 5 6 1 6 (916) 499-2329.

1-513 Zooplankton
Pb / biomass ratio

Species	Value	Condition
Calanus finmarchicus	0 %	Atomic composition[1]

Source: 1.Vernadsky, W. 1933. Z. Krist. Mineral. Petrog. Abt. B Mineral. Petrog. Mitt., 44, p 191.

1-514 Zooplankton
Production

Species	Value	Condition
Archtodiaptomus sp.	0.0014 g dry weight/(day*m**2)	Filter org, annual, biomass turnover = 162 d, Sevan, S. USSR, 2[1]
Asplanchna priodonta	0.0014 g dry weight/(day*m**2)	Predator May-Oct, biomass turnover = 2.9 d, Naroch Lake, USSR[2]
Asplanchna priodonta	0.0061 g dry weight/(day*m**2)	Predator May-Oct, biomass turnover = 2.5 d, Myastro Lake, USSR[2]
Asplanchna priodonta	0.0105 g dry weight/(day*m**2)	Predator May-Oct, biomass turnover = 3.2 d, Batorin Lake, USSR[2]
Asplanchna sp.	0.0031 g dry weight/(day*m**2)	Predator annual, Severson Lake, Minn, eutrophic[3]
Bosmina longirostris	0.007 g dry weight/(day*m**2)	Filter org, annual, Severson Lake, Minn[4]
Bosmina longirostris	0.067 g dry weight/(day*m**2)	Filter org, May-Nov 1967, Sanctuary Lake, PA, plus Bosmina coregoni[4]
Bosmina longirostris	0.183 g dry weight/(day*m**2)	Filter org, May-Nov 1966, Sanctuary Lake, PA , plus Bosmina coregoni[4]
Brachionus sp.	0.0075 g dry weight/(day*m**2)	Filter org, annual, Severson Lake, Minn, eutrophic[3]
Ceriodaphnia reticulata	0.031 g dry weight/(day*m**2)	Filter org, July-Nov, Sanctuary Lake, PA[4]
Chaoborus punctipennis	0.0001 g dry weight/(day*m**2)	Predator annual, Severson Lake, Minn, eutrophic[3]
Chydorus sphaericus	0.004 g dry weight/(day*m**2)	Filter org, July-Aug 1966, Sanctuary Lake, PA eutrophic res.[4]
Chydorus sphaericus	0.047 g dry weight/(day*m**2)	Filter org, July-Aug 1967, Sanctuary Lake, PA[4]
Cladocera	0.0002 g dry weight/(day*m**2)	Predator May-Oct, biomass turnover = 10.8 d, Batorin Lake, USSR[2]
Cladocera	0.0003 g dry weight/(day*m**2)	Predator May-Oct, biomass turnover = 11.3 d, Naroch Lake, USSR[2]
Cladocera	0.0009 g dry weight/(day*m**2)	Predator May-Oct, biomass turnover = 3.9 d, Myastro Lake, USSR[2]
Cladocera	0.0026 g dry weight/(day*m**2)	Filter org, May-Oct, biomass turnover = 13.7 d, Naroch Lake USSR[5]
Cladocera	0.015 g dry weight/(day*m**2)	Filter org, May-Oct, biomass turnover = 10.9 d, Myastro Lake USSR[5]
Cladocera	0.033 g dry weight/(day*m**2)	Filter org, May-Oct, biomass turnover = 10.5 d, Batorin Lake USSR[5]
Cyclops sp.	0.0008 g dry weight/(day*m**2)	Predator May-Oct, biomass turnover = 9.7 d, Naroch Lake, USSR[2]
Cyclops sp.	0.0023 g dry weight/(day*m**2)	Predator May-Oct, biomass turnover = 19.4 d, Myastro Lake, USSR[2]

Cyclops sp.	0.0094 g dry weight/(day*m**2)	Predator May-Oct, biomass turnover = 14.2 d, Batorin Lake, USSR[2]
Cyclops strenuus	0.0004 g dry weight/(day*m**2)	Filter org, annual, Buttermere, Engl, oligotrophic[6]
Cyclops strenuus	0.0005 g dry weight/(day*m**2)	Filter org, annual, Rydal water, Engl, eutrophic[6]
Cyclops strenuus	0.0006 g dry weight/(day*m**2)	Filter org, annual, Grasmere, Engl, eutrophic[6]
Cyclops strenuus	0.0007 g dry weight/(day*m**2)	Filter org, annual, biomass turnover = 79.3 d, Lake Sevan, S USSR[1]
Cyclops strenuus	0.0017 g dry weight/(day*m**2)	Filter org, annual, Esthwaite water, Engl, eutrophic[6]
Daphnia galeata mendotae	0.03 g dry weight/(day*m**2)	Filter org, May-Nov 1967, Sanctuary Lake, PA, eutrophic[4]
Daphnia galeata mendotae	0.114 g dry weight/(day*m**2)	Filter org, April-Sep, Canyon Ferry, Mont, eutroph, turnover = 10.0 d[7]
Daphnia galeata mendotae	0.407 g dry weight/(day*m**2)	Filter org, May-Nov 1966, Sanctuary Lake, PA, eutrophic[4]
Daphnia hyalina	0.32 g dry weight/(day*m**2)	Filter org, annual 1971, biomass turnover = 15.9 d, eutrophic, Wales[8]
Daphnia hyalina	0.57 g dry weight/(day*m**2)	Filter org, annual, biomass turnover = 21.3 d, eutrophic, Eglweys
Daphnia longispina	0.006 g dry weight/(day*m**2)	Filter org, annual, Lake Sevan, southern USSR, biomass turnover =[1]
Daphnia parvula	0.01 g dry weight/(day*m**2)	Filter org, annual, Severson Lake eutrophic, Minn[3]
Daphnia schodleri	0.227 g dry weight/(day*m**2)	Filter org, April-Sep, Canyon Ferry, Mont, eutroph, turnover = 6.7 d[7]
Diaphanosoma leuchtenbergianum	0.0067 g dry weight/(day*m**2)	Filter org, annual, Severson Lake, Minn, eutrophic[3]
Diaptomus siciloides	0.0342 g dry weight/(day*m**2)	Filter org, annual, Severson Lake, Minn, eutrophic[3]
Eudiaptomus graciloides	0.001 g dry weight/(day*m**2)	Filter org, May-Oct, biomass turnover = 24.7 d, Naroch Lake, USSR[5]
Eudiaptomus graciloides	0.0065 g dry weight/(day*m**2)	Filter org, May-Oct, biomass turnover = 20.2 d, Myastro Lake USSR[5]
Eudiaptomus graciloides	0.007 g dry weight/(day*m**2)	Filter org, May-Oct, biomass turnover = 15.4 d, Batorin Lake USSR[5]
Filinia longiseta	0.0011 g dry weight/(day*m**2)	Filter org, annual, Severson Lake, Minn, eutrophic[3]
Keratella cochlearis	0.0007 g dry weight/(day*m**2)	Filter org, annual, Severson Lake, Minn, eutrophic[3]
Keratella quadrata	0.0021 g dry weight/(day*m**2)	Filter org, annual, Severson Lake, Minn eutrophic[3]
Leptodora kindtii	0.003 g dry weight/(day*m**2)	Predator May-Nov 1966, Sanctuary Lake, PA[4]
Leptodora kindtii	0.013 g dry weight/(day*m**2)	Predator May-Oct 1967, Sanctuary Lake, PA[4]
Mesocyclops edax	0.0046 g dry weight/(day*m**2)	Filter org, annual, Severson Lake, Minn, eutrophic[3]
Polyarthra sp.	0.001 g dry weight/(day*m**2)	Filter org, annual, Severson Lake, Minn, eutrophic[3]
Rotifera	0.0024 g dry weight/(day*m**2)	Filter org, May-Oct, biomass turnover = 3.9 d, Myastro Lake, USSR[5]

Rotifera	0.0027 g dry weight/(day*m**2)	Filter org, May-Oct, biomass turnover = 1.7 d, Naroch Lake, USSR[5]
Rotifera	0.0105 g dry weight/(day*m**2)	Filter org, May-Oct, biomass turnover = 2.6 d, Batorin Lake, USSR[5]
Synchaeta sp.	0.00009 g dry weight/(day*m**2)	Predator annual, Severson Lake, Minn, eutrophic[3]
Zooplankton	11823 mg dry weight/m3	Mean biomass = 1734 ug/l, high nutrient, based on 12 w, 1967[9]
Zooplankton	1316 mg dry weight/m3	Mean biomass = 104 ug/l, median nutrient, based on 8 w, 1965[9]
Zooplankton	3043 mg dry weight/m3	Mean biomass = 134 ug/l, median nutrient, based on 12 w, 1967[9]
Zooplankton	3412 mg dry weight/m3	Mean biomass = 124 ug/l, low nutrient, based on 12 w, 1967[9]
Zooplankton	3514 mg dry weight/m3	Mean biomass = 141 ug/l, median nutrient, based on 12 w, 1966[9]
Zooplankton	3925 mg dry weight/m3	Mean biomass = 259 ug/l, low nutrient, based on 12 w, 1966[9]
Zooplankton	4756 mg dry weight/m3	Mean biomass = 492 ug/l, high nutrient, based on 12 w, 1966[9]
Zooplankton	5762 mg dry weight/m3	Mean biomass = 416 ug/l, high nutrient, based on 8 w, 1965[9]
Zooplankton	963 mg dry weight/m3	Mean biomass = 67 ug/l, low nutrient, based on 8 w, 1965[9]

Source: 1.Winberg, G.G. 1971. Methods for the Estimation of Production of Aquatic Animals. Academic Press.
2.Winberg, G.G. et al. 1973. Hydrobiol. J. 9, pp 77-84.
3.Comita, G.W. 1972. Arch. Hydrobiol. 70, pp 14-66.
4.Cummins, K.W., R.R. Costa, R.E. Rowe, G.A. Moshiri, R.M. Scanlon and R.K. Zajdel. 1969. Oikos, 20, p189-223.
5.Winberg, G.G. et al. 1970. in: Kajak, Z. and A. Hillbricht-Ilkowska (eds.); Productivity Problems of Freshwater. pp 383-404. PWN Polish Scientific Publishers. Warsaw.
6.Smyly, W.J.P. 1973. Oecologia 11, pp 163-186.
7.Wright, J.C. 1965. Limnol. Oceanogr. 10, pp 583-590.
8.George, D.G. and R.W. Edwards. 1974. Freshwater Biol., 4, pp 445-465.
9.Hall, D.J., W.E. Cooper and E.E. Werner. 1970. Limnol. Oceanogr., 15(3), pp 838-928.

1-515 Zooplankton
Production / assimilation

Species	Value	Condition
Asellopsis intermedia	11.8 %	Cumulative C budget to 300 d[1]
Asellopsis intermedia	17.9 %	Cumulative C budget to 126 d[1]
Asellopsis intermedia	35.9 %	Cumulative C budget to 55 d[1]
Asellopsis intermedia	56.7 %	Cumulative C budget to 14 d[1]
Calanus finmarchicus	22.7 %	Cumulative N budget, adult phase, 250 eggs/(female*35 d) = growth[2]
Calanus finmarchicus	37.7 %	Cumulative N budget, total life, 250 eggs/(female*35 d) = growth rate[2]
Calanus finmarchicus	55 %	Cumulative N budget, pre-adult phase[2]
Calanus finmarchicus	61.3 %	Cumulative N budget, total life, 1500 eggs/(female*35 d) = growth[2]
Calanus finmarchicus	63.8 %	Cumulative N budget, adult phase, 1500 eggs/(female*35 d) = growth[2]
Calliopius laeviusculus	24.1 %	Cumulative C budget, total life, 3 diff temp[3]
Calliopius laeviusculus	43.4 %	Cumulative C budget, pre-adult phase, 3 diff temp[3]
Euphausia pacifica	11.8 %	Cumulative C budget over 25 - 70 d, less than 1 mg dry weight[4]
Euphausia pacifica	25.8 %	Cumulative C budget over 70 d, less than 1 mg dry weight[4]
Hyalella azteca	12.6 %	Daily calorie budget for large adult[5]
Menippe mercenaria	28.4 %	Daily calorie budget, stage 1[6]
Menippe mercenaria	38.4 %	Daily calorie budget, stage 2[6]
Menippe mercenaria	43.9 %	Daily calorie budget, megalopa[6]

Menippe mercenaria	46.2 %	Daily calorie budget, stage 3[6]
Menippe mercenaria	46.8 %	Daily calorie budget, stage 4[6]
Menippe mercenaria	47.5 %	Cumulative calorie budget to megalopa[6]
Menippe mercenaria	59.3 %	Daily calorie budget, stage 5[6]
Metamysidopsis elongata	30 %	Cumulative calorie budget to 103 d[7]
Metamysidopsis elongata	38 %	Cumulative calorie budget to 45 d[7]
Metamysidopsis elongata	43 %	Cumulative calorie budget to 68 d[7]
Metamysidopsis elongata	44 %	Cumulative calorie budget to 53 d[7]
Mysis relicta	17 %	Cumulative calorie budget for total life[8]
Tigriopus brevicornis	17.2 %	Cumulative N budget, pre-adult phase[9]
Tigriopus brevicornis	26.5 %	Cumulative N budget, total life, female[9]
Tigriopus brevicornis	29.3 %	Cumulative N budget, adult phase[9]

Source: 1.Lasker, R., J.B.J. Wells and A.D. McIntyre. 1970. J. mar. biol. Ass. U.K. 50, pp 147-160.
2.Corner, E.D.S. and B.S. Newell. 1967. J. mar. biol. Ass. U.K., 47, pp 113-120.
3.Dagg, M.J. 1976. Int. Rev. Gesamt. Hydrobiol. 61(3), pp 297-357.
4.Lasker, R. 1966. J. Fish. Res. bd. Can. 23, pp 1291-1317.
5.Hargrave, B.T. 1970. Limnol. Oceanogr., 15(1), pp 21-30.
6.Mootz, C.A. and C.E. Epifanio. 1974. Biol. Bull. mar. Biol. Lab. Woods Hole 146, pp 44-55.
7.Clutter, R.I. and G.H. Thielacker. 1971. Fish. Bull., U.S. 69, pp 93-115.
8.Lasenby, D.C. and R.R. Langford. 1972. J. Fish. Res. bd. Can. 29, pp 1701-1708.
9.Harris, R.P. 1973. J. mar. biol. Ass. U.K., 53, pp 785-800.

1-516 Zooplankton
Production / biomass ratio

Species	Value	Condition
Cladocera	22 on yearly basis	Prod=19.2 g dw/(m2*y), Taltowisko Lake, Poland, mesotrophic, total[1]
Cladocera	31 on yearly basis	Prod=24.7 g dw/(m2*y), Makolajkie Lake, Poland, eutrophic, total[1]
Copepoda	1.2 on yearly basis	Prod=0.27 g dw/(m2*y), Snowflake Lake, Alberta, total Copepoda[1]
Copepoda	12 on yearly basis	Prod=7.8 g dw/(m2*y), Taltowisko Lake, Poland, mesotrophic, total[1]
Copepoda	17.5 on yearly basis	Prod=9.9 g dw/(m2*y), Mikolajskie Lake, Poland, eutrophic, total[1]
Copepoda	3 on yearly basis	Prod=0.75 g dw/(m2*y), Teardrop Lake, Alberta, total Copepoda[1]
Cyclops vicinus	21 on yearly basis	Prod=10.2 g dw/(m2*y), Eglwys Nunydd River Wales, UK[1]
Daphnia hyalina	23 on yearly basis	Prod=35.3 g dw/(m2*y), Eglwys Nunydd River, Wales, n=2 y, UK[1]
Daphnia longispina	2.5 on yearly basis	Prod=0.56 g dw/(m2*y), Lake Port-Bielh, Fr[1]
Eudiaptomus graciloides	6.8 on yearly basis	Prod=7.7 g dw/(ha*y), Lake Esrom, Denm[1]
Eurycercus lamellatus	20 on yearly basis	Prod=0.09 g dw/(ha*y), Lake Naroch, USSR[1]
Mixodiaptomus laciniatus	2.5 on yearly basis	Prod=1.3 g dw/(m2*y), Lake Port-Bielh, Fr[1]
Rotifera	10.1 on yearly basis	Prod=0.08 g dw/(m2*y), Lake Zelenetskoye, USSR, max 2 y, cold N lake[1]
Rotifera	42 on yearly basis	Prod=6.2 g dw/(m2*y), Taltowisko Lake, Poland, mesotrophic, total[1]
Rotifera	43 on yearly basis	Prod=14.7 g dw/(m2*y), Mikolajkie Lake, Poland, eutrophic, total[1]
Rotifera	60 on yearly basis	Prod=5.4 g dw/(m2*y), Rybinsk River, USSR, total rotifera[1]
Rotifera	64 on yearly basis	Prod=4.6 g dw/(m2*y), Three Lakes, USSR, total rotifera, predator[1]
Rotifera	75 on yearly basis	Prod=3.9 g dw/(m2*y), Three Lakes, USSR, total rotifera, non-predator[1]

Rotifera	8 on yearly basis	Prod=0.05 g dw/(m2*y), Lake Akulkino, USSR, max 2 y, cold N lake, tot[1]
Thermocyclops hyalinus	28.5 on yearly basis	Prod=16.1 g dw/(m2*y), Lake George, Uganda[1]
Zooplankton	11 on yearly basis	Prod=2.0 g dw/(m2*y), Lake Krugloe, USSR, total zooplankton, predator[1]
Zooplankton	13 on yearly basis	Prod=11 g dw/(m2*y), Lake Krugloe, USSR, total zooplankton,[1]
Zooplankton	13 on yearly basis	Prod=3.4 g dw/(m2*y), Lake Krivoe, USSR, total zooplankton,[1]
Zooplankton	13.5 on yearly basis	Prod=16.1 g dw/(m2*y), Lake Baibal, USSR, total zooplankton,[1]
Zooplankton	15 on yearly basis	Prod=2.0 g dw/(m2*y), Red Lake, USSR, total zooplankton, predator[1]
Zooplankton	18 on yearly basis	Prod=17.7 g dw/(m2*y), Red Lake, USSR, total zooplankton,[1]
Zooplankton	2.4 on yearly basis	Prod=1.6 g dw/(m2*y), Lake Baikal, USSR, total zooplankton, predator[1]
Zooplankton	22 on yearly basis	Prod=20.8 g dw/(m2*y), n=9, lakes in USSR, total zooplankton,[1]
Zooplankton	5 on yearly basis	Prod=0.52 g dw/(m2*y), Lake Krivoe, USSR, total zooplankton, predator[1]
Zooplankton	7.6 on yearly basis	Prod=0.70 g dw/(m2*y), L. Zelenetzkoye, USSR, max of 2 y[1]
Zooplankton	8.7 on yearly basis	Prod=0.087 g dw/(m2*y), L. Akulkino, USSR, cold N lake[1]
Zooplankton, Rotifera, Cladocera and Copepoda	13 on yearly basis	Prod=1.60 g dw/(m2*y), Lake Krugloe, USSR, total[1]
Zooplankton, Rotifera, Copepoda and Cladocera	14 on yearly basis	Prod=1.98 g dw/(m2*y), Lake Krivoe, USSR, total[1]
Zooplankton, Rotifera, Copepoda and Cladocera	22 on yearly basis	Prod=33.2 g dw/(m2*y), Taltowisko Lake, Poland, mesotrophic, total[1]
Zooplankton, Rotifera, Copepoda and Cladocera	27 on yearly basis	Prod=49.3 g dw/(m2*y), Mikolajskie Lake, Poland, eutrophic, total[1]

Source: 1.Waters, T.F. 1977. Adv. Ecol. Res. 10, pp 91-164.

1-517 Zooplankton
Production of young

Species	Value	Condition
Daphnia pulex	10 % of energy consumed	6-34 d old, food = 100000 cells/ml, Chlamydomonas reinhardtii[1]
Daphnia pulex	10.9 % of energy consumed	6-34 d old, food = 75000 cells/ml, Chlamydomonas reinhardtii[1]
Daphnia pulex	12.5 % of energy consumed	6-34 d old, food = 50000 cells/ml, Chlamydomonas reinhardtii[1]
Daphnia pulex	132 youngs	6-34 d old, food = 50000 cells/ml, Chlamydomonas reinhardtii[1]
Daphnia pulex	16.5 % of energy consumed	6-34 d old, food = 25000 cells/ml, Chlamydomonas reinhardtii[1]
Daphnia pulex	177 youngs	6-34 d old, food = 75000 cells/ml, Chlamydomonas reinhardtii[1]
Daphnia pulex	223 youngs	6-34 d old, food = 100000 cells/ml, Chlamydomonas reinhardtii[1]
Daphnia pulex	52 % of energy assimilated	6-34 d old, food = 25000 cells/ml, Chlamydomonas reinhardtii[1]
Daphnia pulex	61.8 % of energy assimilated	6-34 d old, food = 50000 cells/ml, Chlamydomonas reinhardtii[1]
Daphnia pulex	66.8 % of energy assimilated	6-34 d old, food = 75000 cells/ml, Chlamydomonas reinhardtii[1]
Daphnia pulex	70.5 % of energy assimilated	6-34 d old, food = 100000 cells/ml, Chlamydomonas reinhardtii[1]

Daphnia pulex	76.3 youngs	6-34 d old, food = 25000 cells/ml, Chlamydomonas reinhardtii[1]

Source: 1.Richman, S. 1958. Ecol. Monogr. 28, pp 273-291.

1-518 Zooplankton
Protein content

Species	Value	Condition
Aurelia aurita	1.01 mg ind.-1	incubation exp., dark, 4-6 C[1]
Balanus sp.	0.003 - 0.006 mg ind.-1	incubation exp., dark, 4-6 C, Nauplii[1]
Beroe cucumis	1.02 - 2.63 mg ind.-1	incubation exp., dark, 4-6 C[1]
Boreomysis arctica	2.61 - 5.23 mg ind.-1	incubation exp., dark, 4-6 C[1]
Calanus finmarchius	0.06 - 0.11 mg ind.-1	incubation exp., dark, 4-6 C[1]
Calanus hyperboreus	0.43 - 1.89 mg ind.-1	incubation exp., dark, 4-6 C[1]
Chiridius armatus	0.16 mg ind.-1	incubation exp., dark, 4-6 C, CV[1]
Chiridius armatus	0.17 - 0.43 mg ind.-1	incubation exp., dark, 4-6 C, females[1]
Chiridius armatus	0.22 - 0.30 mg ind.-1	incubation exp., dark, 4-6 C, males[1]
Concoecia sp.	0.13 - 0.18 mg ind.-1	incubation exp., dark, 4-6 C[1]
Euchaeta norvegica	0.05 mg ind.-1	incubation exp., dark, 4-6 C, CIII[1]
Euchaeta norvegica	0.10 - 0.15 mg ind.-1	incubation exp., dark, 4-6 C, CIV[1]
Euchaeta norvegica	0.41 - 0.47 mg ind.-1	incubation exp., dark, 4-6 C, CV[1]
Euchaeta norvegica	0.48 mg ind.-1	incubation exp., dark, 4-6 C, males[1]
Euchaeta norvegica	0.76 - 1.70 mg ind.-1	incubation exp., dark, 4-6 C, females[1]
Eukrohnia bathypelagica	0.81 - 1.45 mg ind.-1	incubation exp., dark, 4-6 C[1]
Eukrohnia hamata	1.93 - 3.00 mg ind.-1	incubation exp., dark, 4-6 C[1]
Eutonina indicans	0.39 - 0.85 mg ind.-1	incubation exp., dark, 4-6 C[1]
Meganyctiphanes norvegica	1.97 - 35.79 mg ind.-1	incubation exp., dark, 4-6 C[1]
Metridia longa	0.04 - 0.14 mg ind.-1	incubation exp., dark, 4-6 C[1]
Parathemisto abyssorum	0.12 - 2.15 mg ind.-1	incubation exp., dark, 4-6 C[1]
Pasiphaea multidentata	2.26 - 10.09 mg ind.-1	incubation exp., dark, 4-6 C[1]
Pleurobrachia pileus	0.39 - 0.92 mg ind.-1	incubation exp., dark, 4-6 C[1]
Sagitta elegans	0.31 - 1.05 mg ind.-1	incubation exp., dark, 4-6 C[1]
Tomopteris helgolandica	1.08 - 8.99 mg ind.-1	incubation exp., dark, 4-6 C[1]

Source: 1.Baamstedt, U., Limnol.Oceanogr., 30(3), 607, 1985.

1-519 Zooplankton
Rb / biomass ratio

Species	Value	Condition
Calanus finmarchicus	0 %	Atomic composition[1]

Source: 1.Vernadsky, W. 1933. Z. Krist. Mineral. Petrog. Abt. B Mineral. Petrog. Mitt., 44, p 191.

1-520 Zooplankton
Respiration rate

Species	Value	Condition
Acartia clausii	12.4 microlitre O2/(mg dry weight*hour)	Endogenous resp, 286 K[1]
Acartia clausii	13 microlitre O2/(mg dry weight*hour)	Endogenous resp, 283 K[1]
Acartia clausii	15.8 microlitre O2/(mg dry weight*hour)	Endogenous resp, 290 K[1]
Acartia clausii	18.9 microlitre O2/(mg dry weight*hour)	Endogenous resp, 293 K[1]
Brachyuran larvae	1.45 microlitre O2/(animal*hour)	Decapoda, megalops, 289 K, mean of 2 experiments, rg = +- 0.19[2]
Calanoides carinatus	0.472 microlitre O2/(animal*hour)	Copepoda, 289 K, mean of 13 experiments, rg = +-0.324[2]
Calanus finmarchicus	1.3 ml O2/(mg dry weight*day)	Endogenous resp, 277 K[3]
Calanus finmarchicus	1.3 ml O2/(mg dry weight*day)	Endogenous resp, 277 K[4]
Calanus finmarchicus	1.4 ml O2/(mg dry weight*day)	Endogenous resp, 283 K[3]
Calanus finmarchicus	1.4 ml O2/(mg dry weight*day)	Endogenous resp, 283 K[4]
Calanus finmarchicus	2.3 ml O2/(mg dry weight*day)	Endogenous resp, 288 K[3]
Calanus finmarchicus	2.3 ml O2/(mg dry weight*day)	Endogenous resp, 288 K[4]
Calanus finmarchicus	4.2 ml O2/(mg dry weight*day)	Endogenous resp, 293 K[3]
Calanus finmarchicus	4.2 ml O2/(mg dry weight*day)	Endogenous resp, 293 K[4]
Calanus pacificus	0.004 microlitre O2/(animal*hour)	Copepoda, nauplii 1, 288 K, mean of 2 experiments, rg = +- 0.001[2]
Calanus pacificus	0.044 microlitre O2/(animal*hour)	Copepoda, copepodite 2, 288 K, mean of 2 experiments, rg = +- 0.044[2]
Calanus pacificus	0.086 microlitre O2/(animal*hour)	Copepoda, copepodite 4, 288 K, mean of 3 experiments, rg = +- 0.007[2]
Calanus pacificus	0.259 microlitre O2/(animal*hour)	Copepoda, copepodite 5, 288 K, mean of 5 experiments, rg = +- 0.050[2]
Calanus pacificus	0.59 microlitre O2/(animal*hour)	Copepoda, female, 288 K, mean of 3 experiments, rg = +- 0.019[2]
Calanus sp.	0.876 microlitre O2/(animal*hour)	Copepoda, 289 K, mean of 3 experiments, rg = +- 0.105[2]
Centropages hamatus	2.27 microlitre O2/(mg dry weight*hour)	Endogenous resp, 279 K[1]
Centropages hamatus	4.54 microlitre O2/(mg dry weight*hour)	Endogenous resp, 283 K[1]
Centropages hamatus	6.34 microlitre O2/(mg dry weight*hour)	Endogenous resp, 286 K[1]
Centropages hamatus	7.56 microlitre O2/(mg dry weight*hour)	Endogenous resp, 290 K[1]
Ceriodaphnia reticulata	1.8 % of ingestion	Juvenile, female[5]

Ceriodaphnia sp.	14.2 microlitre O2/(mg dry weight*hour)	Endogenous resp, 292 K, also Daphnia sp.[1]
Ceriodaphnia sp.	2.7 microlitre O2/(mg dry weight*hour)	Endogenous resp, 277 K, also Daphnia sp.[1]
Cladocera	14.2 ml O2/(mg dry weight*day)	Endogenous resp, 291 K[6]
Cladocera	14.2 ml O2/(mg dry weight*day)	Endogenous resp, 291 K[4]
Cladocera	2.7 ml O2/(mg dry weight*day)	Endogenous resp, 277 K[6]
Cladocera	2.7 ml O2/(mg dry weight*day)	Endogenous resp, 277 K[4]
Copepoda	0.075 1/day	Used in model, San Joaquin River, Calif[7]
Copepoda	12.2 ml O2/(mg dry weight*day)	Endogenous resp, 291 K[6]
Copepoda	12.2 ml O2/(mg dry weight*day)	Endogenous resp, 291 K[4]
Copepoda	20.9 microlitre O2/(animal*hour)	300 K, mean of 6 experiments, rg = +- 21.2, mixed copepods[2]
Copepoda	3.4 ml O2/(mg dry weight*day)	Endogenous resp, 277 K[6]
Copepoda	3.4 ml O2/(mg dry weight*day)	Endogenous resp, 277 K[4]
Copepoda	3.8 ml O2/(mg dry weight*day)	Endogenous resp, 277 K[6]
Copepoda	3.8 ml O2/(mg dry weight*day)	Endogenous resp, 277 K[4]
Copepoda	4.1 ml O2/(mg dry weight*day)	Endogenous resp, 281 K[6]
Copepoda	4.1 ml O2/(mg dry weight*day)	Endogenous resp, 281 K[4]
Copepoda	5.2 ml O2/(mg dry weight*day)	Endogenous resp, 285 K[6]
Copepoda	5.2 ml O2/(mg dry weight*day)	Endogenous resp, 285 K[4]
Copepoda	6.5 ml O2/(mg dry weight*day)	Endogenous resp, 289 K[6]
Copepoda	6.5 ml O2/(mg dry weight*day)	Endogenous resp, 289 K[4]
Copepoda	8.2 ml O2/(mg dry weight*day)	Endogenous resp, 291 K[6]
Copepoda	8.2 ml O2/(mg dry weight*day)	Endogenous resp, 291 K[4]
Cyclops sp.	12.2 microlitre O2/(mg dry weight*hour)	Endogenous resp, 292 K, also Diaptomus sp.[1]
Cyclops sp.	3.8 microlitre O2/(mg dry weight*hour)	Endogenous resp, 277 K, also Diaptomus sp.[1]
Daphnia magna	3.4 microlitre O2/(mg dry weight*hour)	Endogenous resp, 283 K[1]
Daphnia magna	5.4 microlitre O2/(mg dry weight*hour)	Endogenous resp, 298 K[1]
Daphnia pulex	35.5 % of assimilation	Range = 27 - 44 %[5]

Daphnia pulex	37 % of growth and reproduction	Food = 100E-4 cell/ml, Chlamydomonas, 40 d, after 18 instars[8]
Daphnia pulex	37 % of growth and young production	0-40 d old, food = 100000 cells/ml, Chlamydomonas reinhardtii[9]
Daphnia pulex	43 % of growth and young production	0-40 d old, food = 75000 cells/ml, Chlamydomonas reinhardtii[9]
Daphnia pulex	43.3 % of growth and reproduction	Food = 75E-4 cell/ml, Chlamydomonas, 40 d, after 18 instars[8]
Daphnia pulex	53 % of growth and reproduction	Food = 50E-4 cell/ml, Chlamydomonas, 40 d, after 18 instars[8]
Daphnia pulex	53 % of growth and young production	0-40 d old, food = 50000 cells/ml, Chlamydomonas reinhardtii[9]
Daphnia pulex	78.6 % of growth and reproduction	Food = 25E-4 cell/ml, Chlamydomonas, 40 d, after 18 instars[8]
Daphnia pulex	78.6 % of growth and young production	0-40 d old, food = 25000 cells/ml, Chlamydomonas reinhardtii[9]
Daphnia pulex	9 % of ingestion	Range = 4 - 14 %[5]
Diaptomus arcticus	0.982 microlitre O2/(mg dry weight*hour)	Endogenous resp, 278 K[1]
Diaptomus arcticus	1.48 microlitre O2/(mg dry weight*hour)	Endogenous resp, 283 K[1]
Diaptomus arcticus	2.06 microlitre O2/(mg dry weight*hour)	Endogenous resp, 288 K[1]
Diaptomus arcticus	2.93 microlitre O2/(mg dry weight*hour)	Endogenous resp, 293 K[1]
Diaptomus arcticus	4.38 microlitre O2/(mg dry weight*hour)	Endogenous resp, 298 K[1]
Diaptomus clacipes	1.8 ml O2/(mg dry weight*day)	Endogenous resp, 278 K[10]
Diaptomus clacipes	1.8 ml O2/(mg dry weight*day)	Endogenous resp, 278 K[4]
Diaptomus clacipes	10.6 microlitre O2/(mg dry weight*hour)	Endogenous resp, 298 K[1]
Diaptomus clacipes	12.5 ml O2/(mg dry weight*day)	Endogenous resp, 298 K[10]
Diaptomus clacipes	12.5 ml O2/(mg dry weight*day)	Endogenous resp, 298 K[4]
Diaptomus clacipes	2.14 microlitre O2/(mg dry weight*hour)	Endogenous resp, 278 K[1]
Diaptomus clacipes	2.4 ml O2/(mg dry weight*day)	Endogenous resp, 283 K[10]
Diaptomus clacipes	2.4 ml O2/(mg dry weight*day)	Endogenous resp, 283 K[4]
Diaptomus clacipes	2.85 microlitre O2/(mg dry weight*hour)	Endogenous resp, 283 K[1]
Diaptomus clacipes	4.64 microlitre O2/(mg dry weight*hour)	Endogenous resp, 288 K[1]
Diaptomus clacipes	5.1 ml O2/(mg dry weight*day)	Endogenous resp, 288 K[10]
Diaptomus clacipes	5.1 ml O2/(mg dry weight*day)	Endogenous resp, 288 K[4]
Diaptomus clacipes	7.05 microlitre O2/(mg dry weight*hour)	Endogenous resp, 293 K[1]

Diaptomus clacipes	8.5 ml O2/(mg dry weight*day)	Endogenous resp, 293 K[10]
Diaptomus clacipes	8.5 ml O2/(mg dry weight*day)	Endogenous resp, 293 K[4]
Diaptomus leptopus	11.7 microlitre O2/(mg dry weight*hour)	Endogenous resp, 298 K[1]
Diaptomus leptopus	12.1 ml O2/(mg dry weight*day)	Endogenous resp, 298 K[4]
Diaptomus leptopus	12.5 ml O2/(mg dry weight*day)	Endogenous resp, 298 K[10]
Diaptomus leptopus	2.33 microlitre O2/(mg dry weight*hour)	Endogenous resp, 278 K[1]
Diaptomus leptopus	2.5 ml O2/(mg dry weight*day)	Endogenous resp, 278 K[10]
Diaptomus leptopus	2.5 ml O2/(mg dry weight*day)	Endogenous resp, 278 K[4]
Diaptomus leptopus	2.8 ml O2/(mg dry weight*day)	Endogenous resp, 283 K[10]
Diaptomus leptopus	2.8 ml O2/(mg dry weight*day)	Endogenous resp, 283 K[4]
Diaptomus leptopus	3.06 microlitre O2/(mg dry weight*hour)	Endogenous resp, 283 K[1]
Diaptomus leptopus	5.07 microlitre O2/(mg dry weight*hour)	Endogenous resp, 288 K[1]
Diaptomus leptopus	5.3 ml O2/(mg dry weight*day)	Endogenous resp, 288 K[10]
Diaptomus leptopus	5.3 ml O2/(mg dry weight*day)	Endogenous resp, 288 K[4]
Diaptomus leptopus	7.4 ml O2/(mg dry weight*day)	Endogenous resp, 293 K[10]
Diaptomus leptopus	7.4 ml O2/(mg dry weight*day)	Endogenous resp, 293 K[4]
Diaptomus leptopus	7.76 microlitre O2/(mg dry weight*hour)	Endogenous resp, 293 K[1]
Diaptomus oregonensis	13.8 microlitre O2/(mg dry weight*hour)	Endogenous resp, 293 K[1]
Diaptomus oregonensis	21 microlitre O2/(mg dry weight*hour)	Endogenous resp, 298 K[1]
Diaptomus oregonensis	3.88 microlitre O2/(mg dry weight*hour)	Endogenous resp, 278 K[1]
Diaptomus oregonensis	4.71 microlitre O2/(mg dry weight*hour)	Endogenous resp, 283 K[1]
Diaptomus oregonensis	8.65 microlitre O2/(mg dry weight*hour)	Endogenous resp, 288 K[1]
Diaptomus siciloides	13.5 ml O2/(mg dry weight*day)	Endogenous resp, 293 K[10]
Diaptomus siciloides	13.5 ml O2/(mg dry weight*day)	Endogenous resp, 293 K[4]
Diaptomus siciloides	16 microlitre O2/(mg dry weight*hour)	Endogenous resp, 293 K[1]
Diaptomus siciloides	21 ml O2/(mg dry weight*day)	Endogenous resp, 298 K[10]
Diaptomus siciloides	21 ml O2/(mg dry weight*day)	Endogenous resp, 298 K[4]

Diaptomus siciloides	24.4 microlitre O2/(mg dry weight*hour)	Endogenous resp, 298 K[1]
Diaptomus siciloides	4.42 microlitre O2/(mg dry weight*hour)	Endogenous resp, 278 K[1]
Diaptomus siciloides	4.8 ml O2/(mg dry weight*day)	Endogenous resp, 278 K[10]
Diaptomus siciloides	4.8 ml O2/(mg dry weight*day)	Endogenous resp, 278 K[4]
Diaptomus siciloides	5.26 microlitre O2/(mg dry weight*hour)	Endogenous resp, 283 K[1]
Diaptomus siciloides	5.5 ml O2/(mg dry weight*day)	Endogenous resp, 283 K[10]
Diaptomus siciloides	5.5 ml O2/(mg dry weight*day)	Endogenous resp, 283 K[4]
Diaptomus siciloides	7.8 ml O2/(mg dry weight*day)	Endogenous resp, 288 K[10]
Diaptomus siciloides	7.8 ml O2/(mg dry weight*day)	Endogenous resp, 288 K[4]
Diaptomus siciloides	9.92 microlitre O2/(mg dry weight*hour)	Endogenous resp, 288 K[1]
Diaptomus sp.	1.1 ml O2/(mg dry weight*day)	Endogenous resp, 278 K[10]
Diaptomus sp.	1.1 ml O2/(mg dry weight*day)	Endogenous resp, 278 K[4]
Diaptomus sp.	1.7 ml O2/(mg dry weight*day)	Endogenous resp, 283 K[10]
Diaptomus sp.	1.7 ml O2/(mg dry weight*day)	Endogenous resp, 283 K[4]
Diaptomus sp.	2.1 ml O2/(mg dry weight*day)	Endogenous resp, 288 K[10]
Diaptomus sp.	2.1 ml O2/(mg dry weight*day)	Endogenous resp, 288 K[4]
Diaptomus sp.	3 ml O2/(mg dry weight*day)	Endogenous resp, 293 K[10]
Diaptomus sp.	3 ml O2/(mg dry weight*day)	Endogenous resp, 293 K[4]
Diaptomus sp.	3.3 microlitre O2/(mg dry weight*hour)	Endogenous resp, 277 K, also Cyclops sp.[1]
Diaptomus sp.	4.3 microlitre O2/(mg dry weight*hour)	Endogenous resp, 281 K, also Cyclops sp.[1]
Diaptomus sp.	4.3 ml O2/(mg dry weight*day)	Endogenous resp, 298 K[10]
Diaptomus sp.	4.3 ml O2/(mg dry weight*day)	Endogenous resp, 298 K[4]
Diaptomus sp.	5.7 microlitre O2/(mg dry weight*hour)	Endogenous resp, 285 K, also Cyclops sp.[1]

Diaptomus sp.	6.6 microlitre O2/(mg dry weight*hour)	Endogenous resp, 289 K, also Cyclops sp.[1]
Diaptomus sp.	8.07 microlitre O2/(mg dry weight*hour)	Endogenous resp, 292 K, also Cyclops sp.[1]
Epilabidocera amphitrites	52.6 microlitre O2/(animal*hour)	Copepoda, 283 K, mean of 4 experiments, rg = +- 17.2[2]
Euphausia pacifica	7.57 microlitre O2/(animal*hour)	Euphausiacea, 281 K, mean of 5 experiments, rg = +- 2.55[2]
Leptodora kindtii	92.7 % of assimilation	None[5]
Leuckartiara octona	2.89 microlitre O2/(animal*hour)	Coelenterata, 283 K, mean of 3 experiments, rg = +- 1.34[2]
Mesocyclops albidus	20 % of ingestion	App value[5]
Mesocyclops albidus	50 % of assimilation	App value[5]
Micro zooplankton	0.0001 microlitre O2/(hour*anim)	Animal = E-5 mg, Pacific Ocean[11]
Micro zooplankton	0.5 microlitre/(hour*anim)	Animal = 0.5 mg, Pacific Ocean[11]
Nauplii zooplankton	0.04 1/day	Used in model, Lake Michigan[7]
Nematoscelis atlantica	4.78 microlitre O2/(animal*hour)	Euphausiacea, 289 K, mean of 8 experiments, rg = +-5.31[2]
Parathemisto pacifica	2.49 microlitre O2/(animal*hour)	Amphipoda, 283 K, mean of 7 experiments, rg = +- 1.08[2]
Phialidium gregarium	1.94 microlitre O2/(animal*hour)	Coelenterata, 286 K, mean of 9 experiments, rg = +-0.17[2]
Pleurobrachia bacheii	2.19 microlitre O2/(animal*hour)	Ctenophora, 286 K, mean of 15 experiments, rg = +-1.46[2]
Pleuroncodes planipes	183 microlitre O2/(animal*hour)	Decapoda, 296 K, mean of 5 experiments, rg = +- 12.4[2]
Sagitta elegans	0.896 microlitre O2/(animal*hour)	Chaetognatha, 281 K, mean of 3 experiments, rg = +- 0.117[2]
Simocephalus vectulus	11.2 % of ingestion	Reproducing female[5]
Simocephalus vectulus	19.5 % of ingestion	Juvenile, female[5]
Simocephalus vectulus	26.9 % of assimilation	Juvenile, female[5]
Simocephalus vectulus	35.3 % of assimilation	Reproducing female[5]
Stomotoca artra	10 microlitre O2/(animal*hour)	Coelenterata, 286 K, mean of 36 experiments, rg = +- 2.9[2]
Temora longicornis	10.3 microlitre O2/(mg dw*hour)	Endogenous resp, 293 K[1]
Temora longicornis	5.4 microlitre O2/(mg dw*hour)	Endogenous resp, 283 K[1]
Temora longicornis	5.4 microlitre O2/(mg dw*hour)	Endogenous resp, 286 K[1]
Temora longicornis	7.2 microlitre O2/(mg dw*hour)	Endogenous resp, 290 K[1]
Tomopteris septentrionalis	6.15 microlitre O2/(animal*hour)	Annelida, 281 K, mean of 5 experiments, rg = +- 1.82[2]
Zooplankton	0.001 1/day-celcius	Lake Ontario, endogenous resp, carnivorous[12]
Zooplankton	0.001 1/day-celcius	Lake Ontario, endogenous resp, herbivorous[12]
Zooplankton	0.002 1/day-celcius	Lake Ontario, endogenous resp, upper trophic level 2[12]
Zooplankton	0.0055 1/day	Temp factor is 1 at 293 K, temp.<278 K, endogenous resp[1]3

Zooplankton	0.01 1/day	Used in model, Boise River, Idaho[7]
Zooplankton	0.01 1/day	Used in model, Lake Washington[7]
Zooplankton	0.01 1/day	Used in model, San Francisco Bay estuary[7]
Zooplankton	0.0125 1/day	Used in model, N Fork River, Calif, rg = 0.005 - 0.02[7]
Zooplankton	0.0125 1/day	Used in model, rg = 0.005 - 0.3[7]
Zooplankton	0.02 1/day	Lake Ontario, 293 K, endogenous respiration[14]
Zooplankton	0.02 1/day	Used in model, Lake Ontario[7]
Zooplankton	0.0505 1/day	Used in model, rg = 0.001 - 0.10[7]
Zooplankton	0.06 1/day	Temp factor is 1 at 293 K, temp>278 K, endogenous resp[13]
Zooplankton	0.06 1/day	Used in model, Lake Michigan, adult[7]
Zooplankton	0.075 1/day	San Joaquin River, 293 K, endogenous respiration[14]
Zooplankton	0.1 1/day	Delta Estuary, 293 K, endogenous respiration[14]
Zooplankton	0.1 1/day	Used in model, San Joaquin River, Delta Estuary, Calif[7]
Zooplankton	0.16 1/day	Lake Erie, 293 K, endogenous respiration[14]
Zooplankton	0.16 1/day	Used in model, Lake Erie[7]
Zooplankton	0.36 1/day	Used in model[7]
Zooplankton	1.2 mg O2/(kg dry weight*min)	Fish and crustaceans, n=8, dw rg=0.132-2.02 g, depth=900-1300 m[15]
Zooplankton	12.6 mg O2/(kg dry weight*min)	Fish and crustaceans, n=19, dw rg=0.0041-0.104 g, depth=0-400 m[15]
Zooplankton	31.9 microlitre O2/(animal*hour)	297 K, mean of 7 experiments, rg = +-22.2, mixed zoplankton[2]
Zooplankton	4.5 mg O2/(kg dry weight*min)	Fish and crustaceans, n=53, dw rg=0.0049-2.22 g, depth=400-900 m[15]

Source: 1.Lombardo, P.S. 1972. Hydrocomp, Inc. Mathematical Model of Water Quality in Rivers and Impoundments
2.King, F.D. and T.T. Packard. 1975. Limnol. Oceanogr. 20(5), pp 849-853.
3.Riley, G.A. 1963. in: Hill. M.N. (ed.): The Sea Theory of Foodchains relations in the Ocean. pp 438-63. Interscience. N.Y.
4.Di Toro, D.M., D.J. O'Connor and R.V. Thomann. 1971. A Dynamic Model of the Phytoplankton Population in the Sacramento-San Joaquin Delta. Advances in Chemistry Series. Vol 106.
5.Wetzel, R.G. 1975. Limnology. Saunders. Philadelphia.
6.Bishop, J.W. 1968. Limnol. Oceanogr., 13(1), pp 58-62.
7.Tetra Tech, Inc. 1977. Ecological constants. San F. Calif.
8.Richman, S. 1958. Ecol. Monogr. 28, pp 273-291.
9.Richman, S. 1958. Ecol. Monogr. 28, pp 273-291.
10.Comita, G.W. 1968. Limnol. Oceanogr., 13, pp 51-57.
11.Klekowski, R.Z. 1975. in: Vinogradov, M.E. (ed.); Pelagic Ecosystem of the Pacific Ocean, pp 372-378
12.Thomann, R.V., D.M. Di Toro, R.P. Winfield and D.J. O'Connor. 1975. Mathematical Modeling of Phytoplankton in Lake Ontario: 1. National Environmental Reseach Center. EPA. Oregon.
13.The Hydrological Engineering Center Corps of Engineers U.S. Army. 1975. Water Quality for river-reservoir systems, Davis, Cal. 9 5 6 1 6 (916) 499-2329.
14.O'Connor, D.J., D.M. DiToro and R.V. Thomann. 1975. in: Russel, C.S. (ed.); Ecological Modeling in a Resource Management Framework. pp 149-209.
15.Childress, J.J. 1971. Limnol. Ocean ogr., 16(1), pp 104-106.

1-521 Zooplankton
Respiratory coefficient

Species	Value	Condition
Copepoda	0.7 Living off fat reserves	Living off fat reserves[1]
Copepoda	1 Feeding on phytoplankton	Feeding on phytoplankton[1]

Source: 1.Parsons, T. and M. Takahashi. 1973. Biological Oceanographic Processes. pp 186 Pergamon Press.

1-522 Zooplankton
S / biomass ratio

Species	Value	Condition
Calanus finmarchicus	0.14 %	Atomic composition[1]

Source: 1.Vernadsky, W. 1933. Z. Krist. Mineral. Petrog. Abt. B Mineral. Petrog. Mitt., 44, p 191.

1-523 Zooplankton
Si / biomass ratio

Species	Value	Condition
Calanus finmarchicus	0.007 %	Atomic composition[1]

Source: 1.Vernadsky, W. 1933. Z. Krist. Mineral. Petrog. Abt. B Mineral. Petrog. Mitt., 44, p 191.

1-524 Zooplankton
Sn / biomass ratio

Species	Value	Condition
Calanus finmarchicus	0.0009 %	Atomic composition[1]

Source: 1.Vernadsky, W. 1933. Z. Krist. Mineral. Petrog. Abt. B Mineral. Petrog. Mitt., 44, p 191.

1-525 Zooplankton
Vertical migration

Species	Value	Condition
Bosmina longirostris	1.8 m	Maximum obs amplitude, Silver Lake Col[1]
Bosmina longirostris	12 m/hour	Descent, Lake Michigan[2]
Bosmina longirostris	19 m/hour	Ascent, Lake Michigan[2]
Bosmina longirostris	20 m	Maximum obs amplitude, Lake Michigan[2]
Cyclops bicuspidatus	3.2 m	Maximum obs amplitude, Silver Lake, Col[1]
Daphnia galeata mendotae	1.4 m/hour	Descent and ascent, Mendota, Wisc[3]
Daphnia galeata mendotae	1.5 m	Maximum obs amplitude, Lake Mendota, Wisc[3]
Daphnia galeata mendotae	10 m	Maximum obs amplitude, Lake Michigan[4]
Daphnia longispina	31 m	Maximum obs amplitude, Lake Lucerne[5]
Daphnia longispina	5.8 m	Maximum obs amplitude, Grand Lake, Col[1]
Daphnia longispina	50 m	Maximum obs amplitude, Lake Victoria[5]
Daphnia pulex	2.7 m	Maximum obs amplitude, Lake Summit, Col[1]
Daphnia retrocurva	10.6 m/hour	Ascent, Lake Michigan[2]
Daphnia retrocurva	24.3 m	Maximum obs amplitude, Lake Michigan[2]
Daphnia retrocurva	5 m/hour	Descent, Lake Michigan[2]
Daphnia schodleri	1.4 m/hour	Descent and ascent, Lake Mendota[3]
Daphnia schodleri	1.5 m	Maximum obs amplitude, Lake Mendota[3]
Diaptomus shoshone	8.8 m	Maximum obs amplitude, Lake Summit[1]
Holopedium gibberum	13.7 m	Maximum obs amplitude, Lake Michigan[2]
Holopedium gibberum	6.9 m/hour	Ascent, Lake Michigan[2]

Leptodora kindtii	2 m/hour	Descent, Lake Michigan[2]
Leptodora kindtii	4.2 m/hour	Ascent, Lake Michigan[2]
Leptodora kindtii	8.5 m	Maximum obs amplitude, Lake Michigan[2]
Limnocalanus macrurus	18 m/hour	Ascent, Lake Michigan[2]
Limnocalanus macrurus	24 m	Maximum obs amplitude, Lake Michigan[2]
Limnocalanus macrurus	9.8 m/hour	Descent, Lake Michigan[2]
Mysis relicta	13.9 m/hour	Descent, Lake Michigan[2]
Mysis relicta	18 m/hour	Ascent, Lake Michigan[2]
Mysis relicta	39 m/hour	Descent and ascent, Lake Michigan, rg = 30 - 48[6]
Mysis relicta	40 m	Maximum obs amplitude, Lake Michigan[2]
Mysis relicta	76 m	Maximum obs amplitude, Lake Michigan[6]
Polyphemus pediculus	13 m	Maximum obs amplitude, Lake Michigan[2]
Polyphemus pediculus	2 m/hour	Descent, Lake Michigan[2]
Polyphemus pediculus	3.6 m/hour	Ascent, Lake Michigan[2]
Pontoporeia affinis	11.7 m/hour	Ascent, Lake Michigan[2]
Pontoporeia affinis	13.9 m/hour	Descent, Lake Michigan[2]
Pontoporeia affinis	40 m	Maximum obs amplitude, Lake Michigan[2]

Source: 1.Pennak, R.W. 1944. Ecology 25, pp 387-403.
2.McNaught, D.C. and A.D. Hasler. 1966. Verh. Int. Ver. Limnol., 16, pp 194-203.
3.McNaught, D.C. and A.D. Hasler. 1964. J. Fish. Res. Bd Can. 21, pp 291-318.
4.Wells, L. 1960. Fish. Bull. U.S. Fish. Wildl. Serv. 60(172), pp 343-369.
5.Worthington, E.B. 1931. Int. Rev. ges. Hydrobiol. 25, pp 394-436.
6.Beeton, A.M. 1960. J. Fish. Res. Bd Can., 17, pp 517-539.

Chapter 2

THE ECOSPHERE AND CHEMICAL COMPOUNDS

Biological Half-Life Time
Chemical Compounds
Ecosphere
Light
Pesticides

2-001 Biological half-life time of 2,3,2',3'-TCB

Species	Value	Condition
Mouse	0.9 days	Uptake from food = 10 ppm in 20 d, carcass, initial=0.13 ug/g ww[1]

Source: 1.Mizotani, T., K. Hidaka, T. Ohe and M. Matsumoto. 1977. Bull. Environ. Contam. Toxicol. 18(4), pp 45-461.

2-002 Biological half-life time of 2,4,2',4'-TCB

Species	Value	Condition
Mouse	7.8 days	Uptake from food = 10 ppm in 20 d, liver, initial=4.94 ug/g ww[1]
Mouse	9.2 days	Uptake from food = 10 ppm in 20 d, carcass, initial=9.93 ug/g ww[1]

Source: 1.Mizotani, T., K. Hidaka, T. Ohe and M. Matsumoto. 1977. Bull. Environ. Contam. Toxicol. 18(4), pp 45-461.

2-003 Biological half-life time of 2,4,2',4'-TCB + phenobarbital

Species	Value	Condition
Mouse	4.9 days	Food=10 tcb + 200 ppm pb in 20 d, carcass, initial=2.28 ug/g ww[1]

Source: 1.Mizotani, T., K. Hidaka, T. Ohe and M. Matsumoto. 1977. Bull. Environ. Contam. Toxicol. 18(4), pp 45-461.

2-004 Biological half-life time of 2,5,2',5'-TCB

Species	Value	Condition
Mouse	2.6 days	Uptake from food = 10 ppm in 20 d, liver, initial=0.84 ug/g ww[1]
Mouse	3.4 days	Uptake from food = 10 ppm in 20 d, carcass, initial=4.36 ug/g ww[1]

Source: 1.Mizotani, T., K. Hidaka, T. Ohe and M. Matsumoto. 1977. Bull. Environ. Contam. Toxicol. 18(4), pp 45-461.

2-005 Biological half-life time of 2,5,2',5'-TCB + phenobarbital

Species	Value	Condition
Mouse	2.6 days	Food=10 tcb + 200 ppm pb in 20 d, carcass, initial=0.28 ug/g ww[1]

Source: 1.Mizotani, T., K. Hidaka, T. Ohe and M. Matsumoto. 1977. Bull. Environ. Contam. Toxicol. 18(4), pp 45-461.

2-006 Biological half-life time of 3,4,3',4'-TCB

Species	Value	Condition
Mouse	0.9 days	Uptake from food = 10 ppm in 20 d, carcass, initial=0.25 ug/g ww[1]

Source: 1.Mizotani, T., K. Hidaka, T. Ohe and M. Matsumoto. 1977. Bull. Environ. Contam. Toxicol. 18(4), pp 45-461.

2-007 Biological half-life time of 3,5,3',5'-TCB

Species	Value	Condition
Mouse	2.1 days	Uptake from food = 10 ppm in 20 d, carcass, initial=4.11 ug/g ww[1]
Mouse	2.2 days	Uptake from food = 10 ppm in 20 d, liver, initial=0.53 ug/g ww[1]

Source: 1.Mizotani, T., K. Hidaka, T. Ohe and M. Matsumoto. 1977. Bull. Environ. Contam. Toxicol. 18(4), pp 45-461.

2-008 Biological half-life time of 3,5,3',5'-TCB + phenobarbital

Species	Value	Condition
Mouse	1.5 days	Food=10 tcb + 200 ppm pb in 20 d, carcass, initial=2.57 ug/g ww[1]

Source: 1.Mizotani, T., K. Hidaka, T. Ohe and M. Matsumoto. 1977. Bull. Environ. Contam. Toxicol. 18(4), pp 45-461.

2-009 Biological half-life time of Ag

Species	Value	Condition
Dog	1 day	Whole body, oral intake[1]
Dog	2.4 days	Injection of silver[2]
Homo sapiens, liver	52 day	Inhalation[3]
Homo sapiens, lung	1 day	Inhalation[3]

Monkey	1 day	Whole body, oral intake[1]
Monkey	1.8 days	Injection of silver[2]
Mouse	1 day	Whole body, oral intake[1]
Rat	1 day	Whole body, oral intake[1]
Rat	2.2 days	Injection of $AgNO_3$[2]

Source: 1.Furchner, J.E., C.R. Richmond and G.A. Drake. 1968. Health Phys. 15, pp 505-514.
2.Habighorst, L.V. and W. Buchwald. 1971. Nucl. Med. 9, pp 35-38.
3.Newton, D. and A. Holmes. 1966. Radiat. Res. 29, pp 403-412.

2-010 Biological half-life time of Aliphatic hydrocarbons

Species	Value	Condition
Mussel	3.5 days	Initial = 250 ug/kg dw, lab. experiment[1]

Source: 1.Fossato, V.U. 1975. Mar. Poll. Bull., 6, p 7.

2-011 Biological half-life time of Aromatic content

Species	Value	Condition
Mussel	4.5 weeks	Natural contamin. animals, rg = 4 - 5 weeks[1]

Source: 1.Di Salvo. L.H., H.E. Guard and L. Hunter. 1975. Environ. Sci. Technol., 9, pp 247.

2-012 Biological half-life time of As

Species	Value	Condition
Homo sapiens	10 hours	Inorganic As(III)[1]
Homo sapiens	20 hours	Organic As from seafood[1]
Homo sapiens	30 hours	methylated As[1]
Homo sapiens	45 hours	Mainly urinary excretion, rg = 30 - 60 hours[2]
Rat	60 days	Single exposure, As(III) oral intake[3]

Source: 1.Crecelius, E.A. 1977. Environ. Health Perspect. 103, pp 147-150.
2.Wickstroem, G. 1972. Work-Environmental-Health, 9, pp 2-8.
3.Mealy, J. Jr., G.L. Brownell and W.H. Sweet. 1959. Arch. Neurol. Psychiatr. 81, pp 310-320.

2-013 Biological half-life time of Ba

Species	Value	Condition
Homo sapiens	8.5 days	Clearence of body fluids from the lungs, $BaSO_4$ administrated[1]

Source: 1.Morrow, P.E., F.R. Gibb and L. Johnson. 1964. Health Phys. 10, pp 543-549.

2-014 Biological half-life time of Be

Species	Value	Condition
Cow, milk	19 hours	Be-7, less than 0.002 % recovered in milk of injected dose[1]
Rat, blood	3 hours	Injection of $BeCl_2$, overall half-life time[2]
Rat, bone	147 days	$BeCl_2$, 18 % of administrated dose[3]
Rat, lung	1 day	Be citrate, completly cleared after 4 d[4]
Rat, lung	2 weeks	Overall reaction, a smaller fraction remainds, perhaps incapsulated[5]
Rat, lung	20 days	$BeCl_2$, 82 % of administrated dose[3]

Source: 1.Mullen, A.L., R.E. Stanley, S.R. Lloyd and A.A. Moghissi. 1972. Health Phys. 22, pp 17-22.
2.Vacher, J. and H.B. Stoner. 1968. Brit. J. Exp. Pathol. 49, pp 315-323.
3.Kuznetsov, A.V., O.G. Matveev and G.D. Suntsov. 1974. Gig. Sanit. 10, pp 113-114.
4.Van Cleve, C.D. and C.T. Kaylor. 1955. Arch. Ind. Health 11, pp 375-392.
5.Reeves, A.L. 1967. Cancer Res. 27, pp 1895-1899.

2-015 Biological half-life time of Benzo(a)pyrene

Species	Value	Condition
Mytilus edulis	16 days	Initial = 45 ug/kg dw, seawater, 281 K, natural contamin. animals[1]
Rangia cuneata	3.5 days	Initial = 7.2 ug/kg ww, 24 h exposure, rg = 2 - 5 days[2]

Source: 1.Dunn, B.P. and H.F. Stich. 1976. Bull. Environ. Contam. Toxicol., 15(4), pp 398-401.
2.Neff, J.M. and J.W. Anderson. 1976. Proceedings of the 1975 Conference in Prevention and Control of Oil Pollution. 469 pp, American Petrolium Institute. Wash. D.C.

2-016 Biological half-life time of Bi

Species	Value	Condition
Homo sapiens, bone	13.3 days	Model value[1]
Homo sapiens, kidney	6 days	Model value[1]
Homo sapiens, liver	15 days	Model value[1]
Homo sapiens, spleen	10 days	Model value[1]
Homo sapiens, whole body	5 days	Model value[1]

Source: 1.Environmental Studies Board. 1972. Water Quality Data. National Academy of Engineering. Wash D.C..

2-017 Biological half-life time of Cd

Species	Value	Condition
Acheta domesticus	7 days	Uptake from food, Cd-109[1]
Dog	380 days	Intrav.injection of 115-Cd, rg = 260 - 500 days[2]
Dog	5 years	Single intrav.injection of 115-Cd, only urinary excretion, rg = 3 - 7 years[2]
Dog	589 days	0.32-0.40 mg 115-Cd/kg, single intrav. injection , fecal excretion[2]
Gambusia affinis	61 days	Uptake from food, Cd-109[1]
Lycosa spp.	99 days	Uptake from food, Cd-109[1]
Lysmata seticaudata	378 days	Exposed to Cd in 2 months, Cd-109[3]
Micropterus salmoides	21 days	Uptake from food, Cd-109[1]
Mouse	100 days	Single intraperitonal injection of 109-Cd (17U ci), more than[2]
Mouse	170 days	Single intraperitonal or vene injection , excl 1.day, rg = 40 - 300[2]
Mouse	2 years	O.25 mg Cd/kg day, 5 days/w for 25 w, subcut.injection , male[2]
Mouse	200 days	Single oral dose of 109-Cd, female[2]
Mouse	62.5 days	Single intravenous injection , excl first rapid clearence[2]
Mytilus galloprovincialis	1254 days	Lab. animals, Cd-109[3]
Mytilus galloprovincialis	307 days	Field animals, Cd-109[3]
Rat	300 days	Intramuscle injection of 109-Cd (2U ci/rat), excl initial excretion[2]
Sigmodon hispidus	18 days	Uptake from food, Cd-109[1]
Spizella passerina	99 days	Uptake from food, Cd-109[1]
Squirrel monkeys	2 years	4 weeks after oral exposure, more than value[2]

Source: 1.Hook, R.I. van Jr., B.G. Blaylock, E.A. Bondietti, C.W. Francis, J.W. Huckabee, D.E. Reichle, F.H. Seeton and J.P. Wintherspoon. 1976. Environ. Qual. Saf. 5, pp 167-182.
2.Friberg, L., M. Piscater, G.F. Nordberg and T. Kjellstroem. Cadmium in the Environment 2nd edition, CRC-Press.
3.Fowler, S.W. and G. Benayoun. 1974. in: Comparative Studies of Food and Environ. Cont.IAEA, pp 159-178. Vienna.

2-018 Biological half-life time of Ce

Species	Value	Condition
Euphausia pacifica	7.5 hours	Measured over 2 months, Ce-144[1]

Source: 1.Fowler, S.W., L.F. Small and J.M. Dean. 1971. Mar. Biol., 8, pp 224-231.

2-019 Biological half-life time of Co

Species	Value	Condition
Ahnfeltia paradoxa	20.7 days	27 d, 288 K, Co-60, blade[1]
Ahnfeltia paradoxa	57.6 days	27 d, 288 K, Co-60, stem[1]
Chondrus ocellatus	18.3 days	23 d, 288 K, Co-60, blade[1]
Chondrus ocellatus, stem	41.6 days	23 d, 288 K, Co-60[1]
Clam, soft tissue	30 days	Co-60[2]
Crangon crangon	1.2 days	Uptake from food, fast phase = 78 %, Co-60, SD = +- 0.5[3]
Crangon crangon	9.8 days	Uptake from food, long live compartment = 22 %, Co-60, SD = +- 2.7[3]
Dog, lung	35 days	CoO, inhalation, after fast initial phase[4]
Dog, lung	60 days	Co3O4, inhalation, after fast initial phase[5]
Homo sapiens	2 days	Fast phase, 90 % of dose[6]
Homo sapiens	2 years	Second phase, 10 % of intake[6]
Homo sapiens	800 days	Intravene injection of CoCl2, 9-16% of dose[7]
Laminaria japonica	12.8 days	35 d, 288 K, Co-60, base[1]
Laminaria japonica	24.1 days	35 d, 288 K, Co-60, tip[1]
Laminaria japonica	36.7 days	35 d, 288 K, Co-60, stem[1]
Micropogon undulatus	31 days	Ambient temp and salinity, injection, Co-60[8]
Micropogon undulatus	31 days	Co-57[9]
Micropogon undulatus	31 days	Co-58[9]
Micropogon undulatus	31 days	Co-60[2]
Micropogon undulatus	31 days	Co-60[9]
Mya arenaria	127 days	Slow phase, Co-60, rg = 84 - 170, most soft tissue[10]
Mya arenaria	180 days	Co-60, muscle[10]
Mya arenaria	240 days	Co-60, total soft tissue[10]
Mya arenaria	9 days	Fast phase, Co-60, rg = 3 - 15, most soft tissue[10]
Mytilus edulis	11.2 days	Half-time of accumulation, Co-60, rg = 4.7 - 17.7, most soft tissue[11]
Mytilus edulis	17.7 days	Half-time of accumulation, Co-60, muscle[11]
Mytilus edulis	5.75 days	Fast phase, Co-60, rg = 5.5 - 6, intact[12]
Mytilus edulis	64.5 days	Slow phase, Co-60, rg = 57 - 72, intact[12]
Oyster	54 days	Co-60, soft tissue[2]
Sargassum thunbergii	12.1 days	13 d, 288 K, Co-60, centre[1]
Sargassum thunbergii	7.5 days	13 d, 288 K, Co-60, tip[1]
Sargassum thunbergii	79.6 days	13 d, 288 K, Co-60, base[1]

Source: 1.Gilat, Y., Laichter and N.H. Shafrir. 1975. Impacts of Nuclear releases into the Aquatic Environment. IAEA-SM-198/18, pp 63-75. Vienna.
2.Baptist, J.P., D.E. Hoss and C.W. Lewis. 1970. Health Phys., 18(2), pp 141-148.
3.Weers, A.W. van and Petten. 1975. Impacts of Nuclear Rel. into the Aqua. Environ. IAE-SM-198/44. pp 349. Vienna
4.Barnes, J.E., G.M. Kanapilly and G.J. Newton 1976. Health Phys. 30, 391-398.
5.Smith,T., C.J. Edmonds and C.F. Barnaby. 1972. Health Phys. 22, pp 359-367.
6.Morsy, S.M. and F.M. El-Assaly. 1970. Health Phys. 19, pp 769-773.
7.Smith, T., C.J. Edmonds and C.F. Barnaby. 1972 Health Phys., 22, pp 359-367.
8.Baptist, J.P., D.E. Hoss and C.W. Lewis. 1970. Health Phys., 18(2), pp 141-148.
9.Baptist, J.P., D.E. Hoss and C.W. Lewis. 1970. Health Phys., 18(2), pp 141-148.
10.Harrison, F.L. 1973. In; Radioactive contamination of the Marine Environment. Vienna, IAEA.
11.Pentreath, R.L. 1973. J. Mar. Biol. Ass. U.K., 53, pp 127.
12.Weers, A.W. van. 1973. In Radioactive Contamination of the Marine Environment. Vienna, IAEA.

2-020 Biological half-life time of Cr

Species	Value	Condition
Micropogon undulatus	0.5 days	Fast phase, 94 %, Cr-51[1]
Micropogon undulatus	70 days	Ambient temp and salinity, injection, Cr-51[2]
Micropogon undulatus	70 days	Slow phase, 6 %, Cr-51[1]

Source: 1.Baptist, J.P., D.E. Hoss and C.W. Lewis. 1970. Health Phys., 18(2), pp 141-148.
2.Baptist, J.P., D.E. Hoss and C.W. Lewis. 1970. Health Phys., 18(2), pp 141-148.

2-021 Biological half-life time of Cs

Species	Value	Condition
Aix sponsa	17.3 days	Free living ducks, effective half-life time, Cs-137, rg = 10.2 - 26.8[1]
Arctic fox	1.0 days	Fast phase, 18 %, female, 4.9 kg, Cs-134[2]
Arctic fox	17.5 days	Slow phase, 83 %, female, 4.9 kg, Cs-134[2]
Blennius pholis	203 days	Slow phase, Cs-137[3]
Blennius pholis	203 days	Slow phase, Cs-137, rg = +- 18[4]
Blennius pholis	25 days	Fast phase, Cs-137, rg = 20 - 30[3]
Blennius pholis	25 days	Fast phase, Cs-137, rg = 20 - 30[4]
Cancer pagurus	4 days	Half-time of accumulation, Cs-137[5]
Carcinus maenas	10 days	Half-time of accumulation, Cs-137[5]
Clam, soft tissue	12 days	Slow phase, Cs-137[6]
Clam, soft tissue	3 days	Fast phase, Cs-137[6]
Coyote	1.5 days	Fast phase, 20 %, female, 9.5 kg, Cs-134[2]
Coyote	22.0 days	Slow phase, 87 %, male, 11.5 kg, Cs-134[2]
Coyote	26.0 days	Slow phase, 80 %, female, 9.5 kg, Cs-134[2]
Coyote	3.5 days	Fast phase, 13 %, male, 11.5 kg, Cs-134[2]
Euphausia pacifica	6 days	Measured over 2 m, Cs-137[7]
Homo sapiens	0.1 year	1961-1962, author = Rosoff, cited in ref., Cs-137[8]
Homo sapiens	0.2 year	1961-1962, author = Cohn, cited in ref., Cs-137[8]
Homo sapiens	0.2 year	1965-1968, author = Naversten, cited in ref., Cs-137[8]
Homo sapiens	0.3 year	1956-1963, author = Rondo, cited in ref., Cs-137[8]
Homo sapiens	0.3 year	1961-1962, author = Suguri, cited in ref., Cs-137[8]
Homo sapiens	0.4 year	1965-1968, author = Lloyd, cited in ref., Cs-137[8]
Homo sapiens	1.1 year	1965-1968, author = Godfrey, cited in ref., Cs-137[8]
Homo sapiens	1.3 year	1965-1968, author = Hesp, cited in ref., Cs-137[8]
Homo sapiens	1.4 year	1964-1968, author = Shukla, cited in ref., Cs-137[8]
Homo sapiens	1.4 year	1966-1968, author = Joyet, cited in ref., Cs-137[8]
Homo sapiens	1.5 year	1965-1967, author = MacDonald, cited in ref., Cs-137[8]
Homo sapiens	1.5 year	1965-1968, author = Scott, cited in ref., Cs-137[8]
Homo sapiens	1.6 year	1964-1967, author = Gustafson, cited in ref., Cs-137[8]
Homo sapiens	1.9 year	1964-1966, author = Richmond, cited in ref., Cs-137[8]
Homo sapiens	100 years	1963-1973, author = Rahola, cited in ref., Cs-137, value equal to[8]
Homo sapiens	5.4 year	1963-1970, author = MacDonald, cited in ref., Cs-137[8]

Micropogon undulatus	0.7 days	Fast phase, liver, Cs-137[9]
Micropogon undulatus	100 years	Slow phase, liver, Cs-137, value equal to infinite[9]
Micropogon undulatus	13.4 days	Fast phase, gonad, Cs-137[9]
Micropogon undulatus	24.1 days	Medium II phase, liver, Cs-137[9]
Micropogon undulatus	26.2 days	Medium phase, skin, Cs-137[9]
Micropogon undulatus	290 days	Slow phase, skin, Cs-137[9]
Micropogon undulatus	34.8 days	Fast phase, muscle, Cs-137[9]
Micropogon undulatus	4.2 days	Medium I phase, liver, Cs-137[9]
Micropogon undulatus	6.2 days	Fast phase, skin, Cs-137[9]
Micropogon undulatus	911 days	Slow phase, gonad, Cs-137[9]
Micropogon undulatus	94.7 days	Slow phase, muscle, Cs-137[9]
Micropogon undulatus, muscle	34.8 days	Fast phase, Cs-137[10]
Micropogon undulatus, muscle	94.7 days	Slow phase, Cs-137[10]
Mugil chelo	154 days	Slow phase, Cs-137[3]
Mugil chelo	154 days	Slow phase, Cs-137, rg = +- 24[4]
Mugil chelo	25 days	Fast phase, Cs-137, rg = 20 - 30[3]
Mugil chelo	25 days	Fast phase, Cs-137, rg = 20 - 30[4]
Mya arenaria	2.7 days	Fast phase, Cs-137, rg = 1 - 4.6, most soft tissue[11]
Mya arenaria	3.3 days	Fast phase, Cs-137, muscle[11]
Mya arenaria	3.6 days	Fast phase, Cs-137, total soft tissue[11]
Mya arenaria	33 days	Slow phase, Cs-137, muscle[11]
Mya arenaria	40 days	Slow phase, Cs-137, total soft tissue[11]
Mya arenaria	43 days	Slow phase, Cs-137, most soft tissue, rg = 14 - 82[11]
Mytilus galloprovincialis	2.7 days	Half life time of accumulation, Cs-137, rg = 2.5 - 2.9[12]
Oyster	12 days	Slow phase, Cs-137, soft tissue[10]
Oyster	3 days	Fast phase, Cs-137, soft tissue[10]
Paralichthys dentatus	36.9 days	Long lived compartment, Cs-137[10]
Paralichthys dentatus	36.9 days	Slow phase, whole body, Cs-137[9]
Paralichthys dentatus	5.3 days	Fast phase, Cs-137[10]
Paralichthys dentatus	5.3 days	Fast phase, whole body, Cs-137[9]
Perch	2 days	Fast phase, orally uptake, 5 - 10 %, Cs-137, app value[13]
Perch	200 days	Slow phase, orally uptake, 90 %, Cs-137[13]
Pleuronectes platessa	108 days	Slow phase, Cs-137, rg = 78 - 138, most soft tissue[14]
Pleuronectes platessa	11 days	Fast phase, Cs-137, rg = 7 - 15, most soft tissue[14]
Pleuronectes platessa	130 days	30 g, 283 K, Cs-134, rg = 120 - 140[15]
Pleuronectes platessa	136 days	Marine, Cs-137, muscle[14]
Pleuronectes platessa, muscle	130 days	30 g, 283 K, Cs-134, rg = 120 - 140[15]
Portunas puber	8 days	Half-time of accumulation, Cs-137[5]
Raia clavata	165 days	Slow phase, Cs-137, rg = 150 - 180, most soft tissue[14]
Raia clavata	180 days	Cs-137, muscle[14]

Raia clavata	27.5 days	Fast phase, Cs-137, rg = 7 - 50, most soft tissue[14]
Raia clavata, muscle	185 days	80 g, 283 K, Cs-134, rg = 180 - 190[15]
Roach	78.5 days	Slow phase, orally uptake, 90 %, Cs-137, rg = 57 - 100[13]
Salmo gairdneri	50 days	Slow phase, orally uptake, 90 %, Cs-137, rg = 20 - 80[13]
Salmo trutta	22.1 days	Uptake from water, biological half-life, blood cells, Cs-137[16]
Salmo trutta	22.6 days	Uptake from water, biological half-life, whole blood, Cs-137[16]
Salmo trutta	29.8 days	Uptake from water, biological half-life, gill filament, Cs-137[16]
Salmo trutta	32.9 days	Uptake from water, biological half-life, heart, Cs-137[16]
Salmo trutta	36.1 days	Uptake from water, biological half-life, blood plasma, Cs-137[16]
Salmo trutta	36.7 days	Uptake from water, biological half-life, liver, Cs-137[16]
Salmo trutta	37.1 days	Uptake from water, biological half-life, spleen, Cs-137[16]
Salmo trutta	37.6 days	Whole body, uptake from water, biological half-life, fish = 8.0-40.0[16]
Salmo trutta	38.2 days	Uptake from water, biological half-life, kidney, Cs-137[16]
Salmo trutta	38.5 days	Uptake from water, biological half-life, nerve cord, Cs-137[16]
Salmo trutta	41.4 days	Whole body, uptake from water, biological half-life, fish = 1.5-16.5[16]
Salmo trutta	41.9 days	Uptake from water, biological half-life, gut, Cs-137[16]
Salmo trutta	46.9 days	Uptake from water, biological half-life, skin, Cs-137[16]
Salmo trutta	62.4 days	Uptake from water, biological half-life, bone, Cs-137[16]
Salmo trutta	79.9 days	Whole body, uptake from water, biological half-life, fish = 75-280 g,[16]
Salmo trutta	84.0 days	Uptake from water, biological half-life, whole fish, Cs-137[16]
Salmo trutta	88.0 days	Uptake from water, biological half-life, gonad, Cs-137[16]
Salmo trutta	95 days	Whole body, uptake from water, biological half-life, fish = 600-820[16]
Salmo trutta	99.0 days	Uptake from water, biological half-life, muscle, Cs-137[16]
Silver fox	1.1 days	Fast phase, 15 %, male, 5.3 kg, Cs-134[2]
Silver fox	25.3 days	Slow phase, 85 %, male, 5.3 kg, Cs-134[2]
Wolf	1.8 days	Fast phase, 16 %, female, 31.3 kg, Cs-134[2]
Wolf	23.0 days	Slow phase, 84 %, female, 31.3 kg, Cs-134[2]

Source: 1.Fendley, T.T., M.N. Manlove and I.L. Brisbin, Jr. 1977. Health Phys., 32(5), pp 415-422.
2.Holleman, D.F. and J.R. Luick. 1976. Health Phys., 30(2), pp 243-245.
3.Fraizer, A. and J.C. Ancellin. 1975. Impacts of Nuclear releases into the Aquatic Environment. IAEA-m-198/8, pp 49-61. Vienna.
4.Fraizier, A. and A. Vilquin. 1971. Mar. Biol., 10, pp 154-156.
5.Bryan, G.W. 1963. J. Mar. Biol. Ass. U.K., 43, pp 519.
6.Baptist, J.P., D.E. Hoss and C.W. Lewis. 1970. Health Phys., 18(2), pp 141-148.
7.Fowler, S.W., L.F. Small and J.M. Dean. 1971. Mar. Biol., 8, pp 224-231.
8.Lewis, J.T., L.W. Grossman, J.G.Kereiakes and E.L.Saenger 1976. Health Physics. 30(3), pp 318-320.
9.Baptist, J.P. and T.J. Price. 1962. Fishery Bull. Fish Wildl Serv. U.S., 206, pp 177-187.
10.Baptist, J.P. and T.J. Price. 1962. Fishery Bull. Fish Wildl Serv. U.S., 206, pp 177-187.
11.Harrison, F.L. 1973. In; Radioactive contamination of the Marine Environment. Vienna, IAEA.
12.Bryan, G.W. 1961. J. Mar. Biol. Ass. U.K., 41, p 551.
13.Hasanen, E., S. Kolehmainen and J.K. Miettinen. 1966. Proc. 1st Int. Congr. Radiat. Prot., pp 401-40.
14.Jefferies, P.F. and C.J. Hewett. 1971. J Mar. Biol. Ass. U.K., 51, pp 411.
15.Jefferies, D.F. and C.J. Hewett. 1971. J. mar. biol. Ass. U.K. 51, pp 411-422.
16.Hewett, C.J. and D.F. Jefferies. 1976. J. Fish. Biol. 9, pp 479-489.

2-022 Biological half-life time of Cu

Species	Value	Condition
Homo sapiens	4 weeks	Injection, whole body, normal persons[1]

Source: 1.Dekaban, A. R. Aamodt, W.F. Rumble, G.S. Johnston and O Reilly. 1975. Arch. Neurol. 32, pp 672-675.

2-023 Biological half-life time of DDD

Species	Value	Condition
Beef cattle	50 days	Depletion from body fat[1]
Lactating cows	37 days	Depletion from body fat[2]
Sheep	26 days	Depletion from body fat[3]
Sheep	4 weeks	Depletion from body fat[4]

Source: 1.McCully, K.A., D.C. Villeneuve, W.P. McKinley, W.E.J. Phillips and M. Hidiroglou. 1966. J. Assoc. Oficial Agr. Chem. 49, pp 966.
2.Fries, G.F., G.S. Marrow, Jr. and C.H. Cordon. 1973. J. Agric. Food Chem. 21, pp 117.
3.Reynolds, P., I.L. Lindahl, H.C. Cecil and J. Bitman. 1976. Bull. Environ. Contam. Toxicol. 16(2), p240-247.
4.Hunnego, J.N. and D.L. Harrison. 1971. New Zealand. J. Agr. Res. 14, pp 406.

2-024 Biological half-life time of DDE

Species	Value	Condition
Beef cattle	213 days	Depletion from body fat[1]
Lactating cows	70 days	Depletion from body fat[2]
Sheep	14 weeks	Depletion from body fat[3]
Sheep	223 days	Depletion from body fat[4]

Source: 1.McCully, K.A., D.C. Villeneuve, W.P. McKinley, W.E.J. Phillips and M. Hidiroglou. 1966. J. Assoc. Oficial Agr. Chem. 49, pp 966.
2.Fries, G.F., G.S. Marrow, Jr. and C.H. Cordon. 1973. J. Agric. Food Chem. 21, pp 117.
3.Hunnego, J.N. and D.L. Harrison. 1971. New Zealand. J. Agr. Res. 14, pp 406.
4.Reynolds, P., I.L. Lindahl, H.C. Cecil and J. Bitman. 1976. Bull. Environ. Contam. Toxicol. 16(2), p240-247.

2-025 Biological half-life time of DDT

Species	Value	Condition
Beef cattle	103 days	Depletion from body fat[1]
Lactating cows	140 days	Depletion from body fat[2]
Sheep	9 weeks	Depletion from body fat[3]
Sheep	90 days	Depletion from body fat[4]

Source: 1.McCully, K.A., D.C. Villeneuve, W.P. McKinley, W.E.J. Phillips and M. Hidiroglou. 1966. J. Assoc. Oficial Agr. Chem. 49, pp 966.
2.Fries, G.F., G.S. Marrow, Jr. and C.H. Cordon. 1973. J. Agric. Food Chem. 21, pp 117.
3.Hunnego, J.N. and D.L. Harrison. 1971. New Zealand. J. Agr. Res. 14, pp 406.
4.Reynolds, P., I.L. Lindahl, H.C. Cecil and J. Bitman. 1976. Bull. Environ. Contam. Toxicol. 16(2), p240-247.

2-026 Biological half-life time of Di-2-ethylhexyl phthalate

Species	Value	Condition
Pimephales promelas	11.8 days	1.24 g, uptake from water in 56 d, water = 8.1 ug/l[1]
Pimephales promelas	13.3 days	1.24 g, uptake from water in 56 d, water = 14 ug/l[1]
Pimephales promelas	14.2 days	1.24 g, uptake from water in 56 d, water = 1.9 ug/l[1]
Pimephales promelas	15.4 days	1.24 g, uptake from water in 56 d, water = 30 ug/l[1]
Pimephales promelas	18.3 days	1.24 g, uptake from water in 56 d, water = 62 ug/l[1]
Pimephales promelas	6.2 days	1.24 g, uptake from water in 56 d, water = 2.5 ug/l[1]
Pimephales promelas	6.2 days	1.24 g, uptake from water in 56 d, water = 4.6 ug/l[1]

Source: 1.Mayer, 1976. J. Fish. Res. Bd Can. 33, pp 2610-2613.

2-027 Biological half-life time of Dieldrin

Species	Value	Condition
Rat	3.13 days	Intake from food, adipose tissue[1]

Source: 1.Edwards, C.A. 1975. Persistent Pesticides in the Environment, 2nd edition, CRC-Press.

2-028 Biological half-life time of Endosulfan

Species	Value	Condition
Mytilus edulis	10.2 days	Uptake from water = 0.1 mg/l, digestive gland[1]
Mytilus edulis	11.6 days	Uptake from water = 0.1 mg/l, gonad and mantle[1]
Mytilus edulis	13.9 days	Uptake from water = 0.1 mg/l, whole body[1]
Mytilus edulis	6.0 days	Uptake from water = 0.1 mg/l, remaining tissue[1]
Mytilus edulis	9.5 days	Uptake from water = 0.1 mg/l, gill[1]

Source: 1.Roberts, D. 1975. Bull. Environ. Contam. Toxicol. 31, pp 170-176.

2-029 Biological half-life time of Endrin

Species	Value	Condition
Ictalurus punctatus	10 days	After 147 d uptake from food, 303 K, 17 g, food = 3 % of ww/d, kidney[1]
Ictalurus punctatus	10 days	After 147 d uptake from food, 303 K, 17 g, food = 3 % of ww/d, liver[1]
Ictalurus punctatus	11 days	After 147 d uptake from food, 303 K, 17 g, food = 3 % of ww/d,[1]
Ictalurus punctatus	12 days	After 147 d uptake from food, 303 K, 17 g, food = 3 % of ww/d, whole[1]
Ictalurus punctatus	15 days	After 147 d uptake from food, 303 K, 17 g, food = 3 % of ww/d,[1]

Source: 1.Jackson, G.A. 1976. Bull. Environ. Contam. Toxicol. 16(5), pp 505-507.

2-030 Biological half-life time of Fe

Species	Value	Condition
Homo sapiens	2310 days	Person on 70 kg, 4-5 g Fe/70 kg, excretion = 0.93 mg/day[1]
Littorina obtusata	11.0 days	Intake from food (Fucus s.), 283 K, Fe-59[2]
Micropogon undulatus	215 days	Ambient temp and salinity, injection, Fe-59[3]
Micropogon undulatus	215 days	Slow phase, 96 %, Fe-55[4]
Micropogon undulatus	215 days	Slow phase, 96 %, Fe-59[4]
Micropogon undulatus	4.0 days	Fast phase, 4 %, Fe-55[4]
Micropogon undulatus	4.0 days	Fast phase, 4 %, Fe-59[4]

Source: 1.Florkin and Stutz (ed). 1972. Comprehensive Biochemistry, Elsevier Publ. Comp., 21(2).
2.Young, M.L. 1975. J. Mar. Biol. Ass. U.K., 55, pp 583-610.
3.Baptist, J.P., D.E. Hoss and C.W. Lewis. 1970. Health Phys., 18(2), pp 141-148.
4.Baptist, J.P., D.E. Hoss and C.W. Lewis. 1970. Health Phys., 18(2), pp 141-148.

2-031 Biological half-life time of Hexachlorobenzene

Species	Value	Condition
Lepomis cyanellus	10.0 days	Carcass, uptake from food=100 ppm, one feeding, 296 K[1]
Lepomis cyanellus	19.6 days	Liver, uptake from food=100 ppm, one feeding, 296 K[1]
Lepomis cyanellus	8.0 days	Stomach and pyloric caeca, uptake from food=100 ppm, one feeding, 296[1]
Lepomis cyanellus	8.1 days	Intestine, uptake from food=100 ppm, one feeding, 296 K[1]
Lepomis cyanellus	8.5 days	Muscle, uptake from food=100 ppm, one feeding, 296 K[1]

Source: 1.Sanborn, J.R., W.F. Childers and L.G. Hansen. 1977. J. Agric. Food. Chem. 25(3), pp 551-553.

2-032 Biological half-life time of Hg

Species	Value	Condition
Crassostrea virginica	16.8 days	Seawat during uptake = 10 ppb, 5.4 ppm in anim, 25 d, 295 K[1]
Crassostrea virginica	19.9 days	Seawat during uptake = 100 ppb, 91.6 ppm in anim, 80 d, 295 K[1]
Crassostrea virginica	35.4 days	Seawat during uptake = 10 ppb, 12.1 ppm in anim, 80 d, 298 K[1]
Crassostrea virginica	9.3 days	Seawat during uptake = 100 ppb, 26.8 ppm in anim, 25 d, 295 K[1]

Eel	950 days	Methyl-mercury, fish, rg = 900 - 1000 days[2]
Flounder	550 days	Methyl-mercury, fish, rg = 400 - 700 days[2]
Gambusia affinis	17.5 days	Uptake from water and food[3]
Homo sapiens	31.5 days	Oral intake of inorganic Hg, rg = 3 - 60[4]
Homo sapiens	42 days	Oral dose of 0.1 ug Hg/kg, absorp =15% of dose, 90d, urin.excretion,[5]
Homo sapiens	80 days	Monomethyl-Hg, rg = 70 - 90[4]
Lebistes reticulatus	56.5 days	High desorption rate for high concentration, rg = 38 - 75[6]
Lepomis macrochirus	130 days	Slow phase = 60 %, methyl-Hg, 297 K[7]
Lepomis macrochirus	38.5 days	Fast phase = 40 %, methyl-Hg, 297 K[7]
Neanthes succinea	4.80 hours	Polychaete, 7.9 % of total burden, fast phase, Hg-203[8]
Perch	500 days	Methyl-mercury, fish[2]
Pike	2 days	App value, fast phase, 283 K, 300 g fish, 5 - 10 %[9]
Pike	600 days	Methyl-mercury, fish, rg = 500 - 700 days[2]
Pike	640 days	90-95 %, slow phase, 283 K, rg = +- 120[9]
Pusa hispida	20 days	Fast phase = 45 %, day 30-100[10]
Pusa hispida	500 days	Slow phase = 55 %, day 100-153[10]
Salmo gairdneri	2.5 years	283 K, rg = 2 - 3[7]
Salmo gairdneri	200 days	Intragastric dose of methyl-Hg, 0.5 mg Hg/kg ww, larger than value[11]

Source: 1.Cunningham, P.A. and M.R. Tripp. 1975. Mar. Biol., 31(4), pp 311-319.
2.Grant, N. 1971. Environment 13, 16.
3.EPA. 1976. Fate and Biological Effects of Mercury introduced into Artificial streams. EPA-600/3-76-00.
4.Friberg, L., M. Piscater, G.F. Nordberg and T. Kjellstroem. Cadmium in the Environment 2nd edition, CRC-Press.
5.Friberg, L. and J. Vostal. 1974. Mercury in the Environment. CRC-Press.
6.Kudo, A. 1976. J. Environ. Qual. 5(4), pp 427-430.
7.Burrows, W.D. and P.A. Krenkel. 1973. Environ. Sci. Technol., 7(13), pp 1127-1130.
8.Luoma, S.N. 1977. Mar. Biol. 41(3), pp 269-273.
9.Jarvenpaa, T., M. Tillander and J.K. Miettinen. 1970. Suom. Kemistilehti. 43, pp 439-442.
10.Tillander, M., J.K. Miettinen and I. Koivisto. 1972. in: Ruivo, M. (ed.); Marine Pollution and Sea Lfe. pp 303-305. Fishing Trading news (books) Ltd. London.
11.Giblin, F.J. and E.J. Massaro. 1973. Toxicol. Appl. Pharmacol. 24(1), pp 81-91.

2-033 Biological half-life time of Hydrocarbon

Species	Value	Condition
Mytilus edulis	11.4 days	Uptake from water = 302 ug/l in 80 days, extrac=n-hexane, June - Oct[1]
Mytilus edulis	16.1 days	Uptake from water = 228 ug/l in 40 days, extrac=n-hexane, Nov - Jan[1]
Mytilus edulis	5.94 days	Uptake from water = 381 ug/l in 80 days, extrac=n-hexane, march-june[1]

Source: 1.Fossato, V.U. and W.J. Canzonier. 1976. Mar Biol., 36, pp 243-250.

2-034 Biological half-life time of I

Species	Value	Condition
Micropogon undulatus	0.25 days	Fast phase, 59 %, I-131[1]
Micropogon undulatus	2.25 days	Slow phase, 40 %, I-131[1]
Micropogon undulatus	24 days	Ambient temp and salinity, injection, I-131[2]
Salmo gairdneri	4 days	287 K, biological half-life time, I-131, rg = 1 - 7[3]

Source: 1.Baptist, J.P., D.E. Hoss and C.W. Lewis. 1970. Health Phys., 18(2), pp 141-148.
2.Baptist, J.P., D.E. Hoss and C.W. Lewis. 1970. Health Phys., 18(2), pp 141-148.
3.Huun, J.B. 1963. Uptake, Distribution and excreation Of the radionuclide I-131 by rainbow trout. Theis, Michigan Statee Univ. East Lansing, Mich. 126 p.

2-035 Biological half-life time of In

Species	Value	Condition
Micropogon undulatus	224 days	Ambient temp and salinity, injection, In-114m[1]
Micropogon undulatus	224 days	Slow phase, 90 %, In-114m[2]
Micropogon undulatus	3.5 days	Fast phase, 10 %, In-114m[2]
Mouse	2 days	Intravenous injection, hydrated In oxide, fast phase = 25 %[3]
Mouse	21 days	Eff half-life, I-131[4]
Mouse	69 days	Slow phase, 50 % of total, InCl injected[3]
Mouse	73.8 days	Intravenous injection, hydrated In oxide, slow phase = 75 %[3]
Mouse, lung	1.9 days	Fast phase, 50 % of total, InCl injected[3]
Mouse, lung	3.5 days	Hydrated In oxide, intravenous injection[5]
Mouse, whole body	14.5 days	Hydrated In oxide, intravenous injection[5]
Rat	44 days	Subcutanous injection[6]
Rat	80 days	Intramuscular.injection[6]
Rat, lung	9 days	Inhaled In sesquioxide part., rg = 8 - 10 d[7]

Source: 1.Baptist, J.P., D.E. Hoss and C.W. Lewis. 1970. Health Phys., 18(2), pp 141-148.
2.Baptist, J.P., D.E. Hoss and C.W. Lewis. 1970. Health Phys., 18(2), pp 141-148.
3.Castronovo, F.P. and H.N. Wagner. 1973. J. Nucl. Med. 14, pp 677-682.
4.Brown, R.A., C.M. Nazario, R.S. de Tirado, J. Castrillon and E.T. Agard. 1977. Environ. Res., 13(1),pp 56-61.
5.Stern, H.S., D.A. Goldwin, U. Scheffel, H.N. Wagner and Kramer. 1967. Nucleonics 25, pp 62-68.
6.Smith, G.A., R.G. Thomas and J.K. Scott. 1960. Health Phys., 4, pp 101-108.
7.Morrow, P.E., F.R. Gibb, R. Cloutier, L.J. Casarett and Scott. 1958. Univ. Rochester Atomic Energy Poj. No. UR508. Univ. Roch. NY.

2-036 Biological half-life time of Mn

Species	Value	Condition
Anodonta nuttalliana	1300 days	Freshwater, Mn-54[1]
Homo sapiens	12.5 days	Healty Mn miners[2]
Homo sapiens	35.5 days	Normal person[2]
Homo sapiens	39 days	For 70% of dose, parenteral(no uptake thr.intestine)[3]
Pleuronectes platessa	39.5 days	Uptake from food, 283 K, n = 10, 30 g fish[4]

Source: 1.Harris, R.P. 1973. J. mar. biol. Ass. U.K., 53, pp 785-800.
2.Mena, I., O. Martin, Fuenzalida, and G.C. Cotzias. 1967. Neurology 17, pp 128-136.
3.Mahoney, J.P. and W.J. Small. 1968. J. Clinical Investigation, 47, pp 643.
4.Pentreath, R.J. 1976. J. Exp. mar. Biol. Ecol. 21, pp 179-189.

2-037 Biological half-life time of Na

Species	Value	Condition
Carp	17.5 days	Orally uptake, Na-24, rg = 10 - 25[1]
Perch	11 days	Orally uptake, Na-24, rg = 7 - 15[1]

Source: 1.Hasanen, E., S. Kolehmainen and J.K. Miettinen. 1966. Proc. 1st Int. Congr. Radiat. Prot., pp 401-40.

2-038 Biological half-life time of Naphthalene

Species	Value	Condition
Mytilus edulis	1.3 days	4 hours exposure, initial = 7 mg/kg dw[1]

Source: 1.Lee, R.F., R. Sauerheber and A.A. Benson. 1972. Science 177, pp 344.

2-039 Biological half-life time of Nb

Species	Value	Condition
Micropogon undulatus	465 days	Ambient temp and salinity, injection, Nb-95[1]
Micropogon undulatus	465 days	Slow phase, 46 %, Nb-95[2]
Micropogon undulatus	5.0 days	Fast phase, 54 %, Nb-95[2]

Source: 1.Baptist, J.P., D.E. Hoss and C.W. Lewis. 1970. Health Phys., 18(2), pp 141-148.
2.Baptist, J.P., D.E. Hoss and C.W. Lewis. 1970. Health Phys., 18(2), pp 141-148.

2-040 Biological half-life time of Ni

Species	Value	Condition
Rat	6.3 hours	Plasma, injected Ni salt[1]

Source: 1.Onkelinx, C., J. Becker and F.W. Sunderman. 1973. Res. Commun. Chem. Pathol. Pharmacol. &, pp 664-67.

2-041 Biological half-life time of Pb

Species	Value	Condition
Crassostrea virginica	13.2 days	Seawat during uptake = 25 ppb, 24 ppm in anim, 21 d, 293 K[1]
Crassostrea virginica	143.0 days	Seawat during uptake = 200 ppb, 203 ppm in anim, 21 d, 293 K[1]
Crassostrea virginica	33.3 days	Seawat during uptake = 50 ppb, 32 ppm in anim, 21 d, 293 K[1]
Crassostrea virginica	52.0 days	Seawat during uptake = 100 ppb, 79 ppm in anim, 21 d, 293 K[1]
Homo sapiens, bone	10 years	None[2]
Homo sapiens, bone	20 years	None[3]
Homo sapiens, whole body	5 year	Assuming one compartment model, body burden = 80 mg[4]
Mytilus edulis	102.0 days	Seawat during uptake = 500 ppb, 2480 ppm in anim, 40 d, 285 K[5]
Mytilus edulis	43.2 days	Seawat during uptake = 5 ppb, 33.7 ppm in anim, 40 d, 285 K[5]
Mytilus edulis	48.1 days	Seawat during uptake = 100 ppb, 559 ppm in anim, 40 d, 285 K[5]
Mytilus edulis	62.4 days	Seawat during uptake = 50 ppb, 395 ppm in anim, 40 d, 285 K[5]
Mytilus edulis	66.5 days	Seawat during uptake = 200 ppb, 1344 ppm in anim, 40 d, 285 K[5]
Mytilus edulis	66.8 days	Seawat during uptake = 5000 ppb, 17625 ppm in anim, 40 d, 285 K[5]
Mytilus edulis	68.5 days	Seawat during uptake = 10 ppb, 65.7 ppm in anim, 40 d, 285 K[5]
Mytilus edulis	70.3 days	Seawat during uptake = 1000 ppb, 4972 ppm in anim, 40 d, 285 K[5]
Rat	100 days	Eff half-life, n=150, inj. of 25 U ci/kg, last phase, Pb-210[6]
Rat	109 days	Eff half-life, n=150, inj. of 25 U ci/kg, last phase, Pb-210[6]
Rat	18 days	Eff half-life, n=150, inj. of 25 U ci/kg, last phase, Pb-210[6]
Rat	20 days	Eff half-life, n=150, inj. of 25 U ci/kg, last phase, Pb-210[6]
Rat	21 days	Eff half-life, n=150, inj. of 25 U ci/kg, last phase, Pb-210[6]
Rat	27 days	Eff half-life, n=150, inj. of 25 U ci/kg, last phase, Pb-210[6]
Rat	44 days	Eff half-life, n=150, inj. of 25 U ci/kg, last phase, Pb-210[6]
Rat	63 days	Eff half-life, n=150, inj. of 25 U ci/kg, last phase, Pb-210[6]
Rat	64 days	Eff half-life, n=150, inj. of 25 U ci/kg, last phase, Pb-210[6]
Rat	86 days	Eff half-life, n=150, inj. of 25 U ci/kg, last phase, Pb-210[6]

Source: 1.Pringle, B.H., D.E. Hissong, E.L. Katz and S.T. Mulawka. 1968. J. San. Eng. Div. Am. Soc. Civ. Eng.,94(5970), pp 455-475.
2.Task Group on Metal Accumulation. 1973. Environ. Physiol. Biochem 3, pp 65-107.
3.Task Group on Metal Toxicity. 1976. Effects and Dose-Response Relationships of Toxic Metals. p 111. Elsevier.
4.Tsuchiya, K and M. Sugita. 1971. Nord. Hyg. Tidskr. 53, pp 105-110.
5.Schulz-Baldes, M. 1973.. Mar. Biol., 21, pp 98-102.
6.Torvik, E., E. Pfitzer, J.G.Kereiakes and R. Blanchard. 1974. Health Physics, 26, pp 81-87.

2-042 Biological half-life time of Pentachlorophenol

Species	Value	Condition
Lepomis cyanellus	12.4 days	Liver, uptake from food=100 ppm, one feeding, 296 K[1]
Lepomis cyanellus	12.5 days	Carcass, uptake from food=100 ppm, one feeding, 296 K[1]
Lepomis cyanellus	12.9 days	Intestine, uptake from food=100 ppm, one feeding, 296 K[1]
Lepomis cyanellus	16.6 days	Stomach and pyloric caeca, uptake from food=100 ppm, one feeding, 296[1]
Lepomis cyanellus	8.9 days	Muscle, uptake from food=100 ppm, one feeding, 296 K[1]

Source: 1.Sanborn, J.R., W.F. Childers and L.G. Hansen. 1977. J. Agric. Food. Chem. 25(3), pp 551-553.

2-043 Biological half-life time of Phthalic acid

Species	Value	Condition
Rat	10.1 hours	Oral dose = 0.4 mg/kg, urine=75 %, feces=25 %, calculated from ref[1]
Rat	8.2 hours	Oral dose = 4.0 mg/kg, urine=75 %, feces=25 %, calculated from ref[1]

Source: 1.Williams, D.T., H.M. Cunningham and B.J. Blanchfield. 1972. Bull. Environ. Contam. Toxicol. 7(1), pp57-62.

2-044 Biological half-life time of Po

Species	Value	Condition
Homo sapiens	2.4 days	Excretion with feces, fast phase, Po-210[1]
Homo sapiens	33 days	Excretion with urin, 3 % of total, value for the compartment, Po-210[1]
Homo sapiens	52 days	Excretion with feces, slow phase, Po-210[1]
Meganyctiphanes norvegica	9 days	Excretion soluble po-210, marine, Po-210[2]
Rat	112 days	Eff half-life, n=150, inj. of 25 U ci/kg, last phase, Po-210[3]
Rat	34 days	Eff half-life, n=150, inj. of 25 U ci/kg, last phase, Po-210[3]
Rat	41 days	Eff half-life, n=150, inj. of 25 U ci/kg, last phase, Po-210[3]
Rat	50 days	Eff half-life, n=150, inj. of 25 U ci/kg, last phase, Po-210[3]
Rat	56 days	Eff half-life, n=150, inj. of 25 U ci/kg, last phase, Po-210[3]
Rat	58 days	Eff half-life, n=150, inj. of 25 U ci/kg, last phase, Po-210[3]
Rat	60 days	Eff half-life, n=150, inj. of 25 U ci/kg, last phase, Po-210[3]
Rat	67 days	Eff half-life, n=150, inj. of 25 U ci/kg, last phase, Po-210[3]
Rat	72 days	Eff half-life, n=150, inj. of 25 U ci/kg, last phase, Po-210[3]

Source: 1.Scott, L.M. and C.M. West. 1975. Health Physics, 28, pp 563-565.
2.Hethering, J.A., D.F. Jefferies and M.B. Lovett. 1975. Impacts of Nuclear Releases into the Aquatic Environment. IAEA-SM-198/29, pp 193-210. Vienna.
3.Torvik, E., E. Pfitzer, J.G.Kereiakes and R. Blanchard. 1974. Health Physics, 26, pp 81-87.

2-045 Biological half-life time of Pu

Species	Value	Condition
Ictalurus punctatus	13.7 days	Pu-hydroxide, % uptake = 6.17+-1.39, r=0.992, Pu-237, SE = +- 1.7[1]
Ictalurus punctatus	24.3 days	Pu-fulvate, % uptake = 1.02+-0.18, r=0.972, Pu-237, SE = +- 4.5[1]
Ictalurus punctatus	30.2 days	Pu-citrate, % uptake = 10.5+-2.57, r=0.996, Pu-237, SE = +-2.8[1]
Ictalurus punctatus	30.2 days	Pu-citrate, % uptake = 10.5+-2.57, r=0.996, Pu-237, rg = +- 2.8[1]
Lysmata seticaudata	42 days	Uptake from seawater, Pu(+6), Pu-237[2]
Mytilus galloprovincialis	39 days	Uptake from phytoplankton suspension, 7 days of uptake, Pu(+6), sea.,[2]
Mytilus galloprovincialis	776 days	Uptake from seawater, Pu(+6), Pu-237[2]
Neries	79 days	Uptake from seawater, Pu(+6), between day 4 and 35, Pu-237[2]

Source: 1.Eyman, L.D. and T.R. Trabalka. 1977. Health Phys., 32(6), pp 475-478.

2.Fowler, S.W. and G. Benayoun. 1974. in: Comparative Studies of Food and Environmental Contamination. IAEA, pp 159-178. Vienna.

2-046 Biological half-life time of Sb

Species	Value	Condition
Dog	2 days	Fast phase, 80 %, aerosol intake, whole body measurement[1]
Dog	68 days	Fast phase, 20 %, aerosol intake, whole body measurement[1]

Source: 1.Felicetti, S.W., R.G. Thomas, and R.O. McClellan. 1974. Health Phys. 26, pp 525-531.

2-047 Biological half-life time of Se

Species	Value	Condition
Homo sapiens	103 days	Third slow phase, sodium selenite, oral dose[1]
Lumbricus sp.	64 days	Annelida[2]
Lysmata seticaudata	58 days	Uptake of Se(+4), 293 K, 0.8 g animal, field, Se-75[3]
Lysmata seticaudata	60 days	Uptake of Se(+4), 293 K, 0.8 g animal, lab, Se-75[3]
Mouse	13 days	None[2]
Mytilus galloprovincialis	63 days	Uptake of Se(+4), 293 K, 8 g animal, lab, Se-75[3]
Mytilus galloprovincialis	81 days	Uptake of Se(+4), 293 K, 8 g animal, field, Se-75[3]
Pheasant	10 days	None[2]
Rat	27 days	Third slow phase, intake of 0.99 mg/(kg*d)[4]
Rat	70 days	Third slow phase, intake of 0.04 mg/(kg*d)[4]

Source: 1.Griffiths, N.M. R.D.H. Stewart and M.F. Robinson. 1976. Brit. J. Nutr. 35, pp 373-382.
2.Gissel-Nielsen, G. 1973. Ambio, 2, pp 114-117.
3.Fowler, S.W. and G. Benayoun. 1976. Mar. Biol., 37, pp 59-68.
4.EwanR.C. A.L. Pope and C.A. Baumann. 1967. J. Nutr. 91, pp 547-554.

2-048 Biological half-life time of Sn

Species	Value	Condition
Dog	95 days	Four compartment model, longest given, inj of Sn(II)[1]
Monkey	95 days	Four compartment model, longest given, inj of Sn(II)[1]
Mouse	2 days	Bis-tributyl-Sn oxide, injection[2]
Mouse	25 days	Eff half-life, Sn-113[3]
Mouse	95 days	Four compartment model, longest given, inj of Sn(II)[1]
Rat	15 days	Rg = 10 - 20 d, inj of Sn(II), liver and kidney[4]
Rat	3.5 days	Diethyltin chloride, oral dose, liver and kidney[2]
Rat	95 days	Four compartment model, longest given, inj of Sn(II)[1]
Rat, bone	34 days	Injection of Sn(II)[4]
Rat, bone	34 days	Injection of Sn(IV)[4]
Rat, brain	3 days	Triphenyl-Sn[5]
Rat, brain	40 days	Tricyclohexyl-Sn compounds[6]
Rat, skeleton	100 days	Inj of Sn-113[1]
Rat, whole body	84 days	Inj of Sn-113[7]

Source: 1.Furchner J.E. and G.A. Drake. 1976. Health Phys. 31, pp 219-224.
2.Brown, R.A., Nazario, De Tirado, Castrillon, and Agard. 1977. Environ. Res. 13, pp 56-61.
3.Brown, R.A., C.M. Nazario, R.S. de Tirado, J. Castrillon and E.T. Agard. 1977. Environ. Res., 13(1),pp 56-61.
4.Hiles, R.A. 1974. Toxicol. Appl. Pharmacol. 27, pp 366-379.
5.Heath, D.F. 1963. In: Radiation and Radioisotopes applied to insects and Agricultural Importance, pp188-193. IAEA. Vienna.
6.FAO/WHO. 1971. Evaluation of some pesticide in food, pp 327-366. FAO and WHO. Rome.
7.Durbin, P.W., K.G. Scott and J.G. Hamilton. 1957. Univ. California Publ. Pharmacol. 3, pp 1-34.

2-049 Biological half-life time of Sr

Species	Value	Condition
Lepomis macrochirus	100 years	In blood, 97 % of total, slow phase, Sr-90, value equal to infinite[1]
Lepomis macrochirus	2 hour	In blood, 2 % of total, fast phase, Sr-90[1]
Lepomis macrochirus	35 days	In blood, 1 % of total, medium phase, Sr-90[1]
Micropogon undulatus	1.2 days	Fast phase, 84 %, Sr-85, liver[2]
Micropogon undulatus	1.25 days	Fast phase, 21 %, Sr-85[2]
Micropogon undulatus	1.25 days	Fast phase, 21 %, Sr-85, whole body[2]
Micropogon undulatus	1.25 days	Fast phase, 21 %, Sr-89[2]
Micropogon undulatus	1.25 days	Fast phase, 21 %, Sr-90[2]
Micropogon undulatus	1.8 days	Fast phase, 84 %, Sr-85, gonad[2]
Micropogon undulatus	100 days	Slow phase, 11 %, Sr-85, muscle[2]
Micropogon undulatus	103 days	Slow phase, 4.3 %, Sr-85, gonad[2]
Micropogon undulatus	120.0 days	100 %, Sr-85, scales[2]
Micropogon undulatus	136 days	Slow phase, 2.6 %, Sr-85, liver[2]
Micropogon undulatus	138 days	Ambient temp and salinity, injection, Sr-85[3]
Micropogon undulatus	138 days	Slow phase, 79 %, Sr-85[2]
Micropogon undulatus	138 days	Slow phase, 79 %, Sr-85, whole body[2]
Micropogon undulatus	138 days	Slow phase, 79 %, Sr-89[2]
Micropogon undulatus	138 days	Slow phase, 79 %, Sr-90[2]
Micropogon undulatus	18.0 days	Slow phase, 29 %, Sr-85, spleen[2]
Micropogon undulatus	223.0 days	100 %, Sr-85, bone[2]
Micropogon undulatus	3.6 days	Fast phase, 88 %, Sr-85, skin[2]
Micropogon undulatus	4.8 days	Fast phase, 90 %, Sr-85, kidney[2]
Micropogon undulatus	4.8 days	Medium phase, 14 %, Sr-85, liver[2]
Micropogon undulatus	5.4 days	Fast phase, 89 %, Sr-85, muscle[2]
Micropogon undulatus	51.0 days	Slow phase, 12 %, Sr-85, skin[2]
Micropogon undulatus	6.0 days	Fast phase, 73 %, Sr-85, gills[2]
Micropogon undulatus	8.0 days	100 %, Sr-85, heart[2]
Micropogon undulatus	8.4 days	Fast phase, 71 %, Sr-85, spleen[2]
Micropogon undulatus	8.6 days	Medium phase, 11 %, Sr-85, gonad[2]
Micropogon undulatus	91.0 days	Slow phase, 27 %, Sr-85, gills[2]
Micropogon undulatus	94.0 days	Slow phase, 10 %, Sr-85, kidney[2]
Pomoxis annularis	30 minutes	In fish flesh, Clinch River, Tenn, Sr-90, rg = 12 - 48 minutes[4]

Source: 1.Reed, J.R. and D.J. Nelson. 1969. Proc. 2nd Nat. Symp. on Radioecology. USAEC. Conf. 670503, pp 226-33.
2.Baptist, J.P., D.E. Hoss and C.W. Lewis. 1970. Health Phys., 18(2), pp 141-148.
3.Baptist, J.P., D.E. Hoss and C.W. Lewis. 1970. Health Phys., 18(2), pp 141-148.
4.Nelson, D.J. 1967. Proc. Int. Symp. Ra dioecol. Concent. Process., 1966, Stockholm, pp 843-851.

2-050 Biological half-life time of TCDD

Species	Value	Condition
Rat	11 days	Liver, male[1]
Rat	12 days	Total body, male[1]
Rat	13 days	Liver, female[1]
Rat	15 days	Total body, female[1]
Rat	17 days	Body (except liver), female[1]
Rat	21 days	Body (except liver), male[1]
Rat	25 days	Range = 20 - 30[2]

Source: 1.Fries, G.F. and G.S. Marrow. 1975. J. agric. Fd. Chem. 23(2), pp 265-269.
2.Mercier, M.J. 1977. 2,3,7,8-Tetrachlorodibenzo-p-dioxin, An Overview. Liege. Belgium.

2-051 Biological half-life time of Tc

Species	Value	Condition
Haliotis rufescens	60 days	Uptake only from seawater, conc=50 nci/l, mean of 4, 284 K, Tc-95[1]

Source: 1.Spies, R.B. 1975. Health Physics, 29, pp 695-699.

2-052 Biological half-life time of Te

Species	Value	Condition
Homo sapiens	31 days	Whole body retention model[1]
Rat	14 days	51 - 58 % of dose, rg = +- 1 d, inj of tellorous acid[2]
Rat	19 hours	42 - 49 % of dose, injection of tellurous acid[2]
Rat, blood	9.2 days	Injection of sodium tellurite[3]
Rat, bone	600 days	Injection of sodium tellurite, SE = +- 5000 days[3]
Rat, kidney	23 days	Injection of sodium tellurite[3]
Rat, liver	10.2 days	Injection of sodium tellurite[3]
Rat, liver	17.7 days	Injection of sodium tellurite[3]

Source: 1.Lie, R., R.G. Thomas and J.K. Scott. 1960. Health Phys. 2, pp 334-340.
2.Slouka, V. and J. Hradil. 1970. Sb. Ved. Pr. VLVDU Hradce Kralove 47, pp 21-43.
3.Hollins, J.G. 1969. Health Phys. 17, pp 497-505.

2-053 Biological half-life time of Tl

Species	Value	Condition
Homo sapiens	5.5 days	Range = 3 - 8[1]
Rat	3.3 days	Whole body, independent of adm. route[2]
Rat	4 days	Intravenous dose[3]

Source: 1.Oehme, F.W. 1972. Clinical Toxicology, 5(2), pp 151-167.
2.Lie, R., R.G. Thomas and J.K. Scott. 1960. Health Phys. 2, pp 334-340.
3.Rauws, A.G. 1974. Nauyn-Schmiedebergs Arch. Pharmacol. 284, pp 295-306.

2-054 Biological half-life time of Toxaphene

Species	Value	Condition
Rat	8.3 days	Oral dose = 20 mg/kg, urine = 70 %, feces = 30 %[1]

Source: 1.Crowder, L.A. and E.F. Dindal. 1974. Bull. Environ. Contam. Toxicol., 12(3), pp 320-327.

2-055 Biological half-life time of Tritium

Species	Value	Condition
Homo sapiens	2020 days	Slow compartment = 13 %[1]
Homo sapiens	21 days	Fast compartment = 92 %[1]
Homo sapiens	280 days	Slow compartment = 8 %[1]
Homo sapiens	33 days	Fast compartment = 87 %[1]

Source: 1.Moghissi, A.A., M.W. Carter and R. Lieberman. 1971. Health Physics. 21(1), pp 57-60.

2-056 Biological half-life time of U

Species	Value	Condition
Dog, lung	15 months	Inhalation of UO2 dust[1]
Homo sapiens	270 days	Rg = 180 - 360, normal diet[1]
Monkey, lung	15 months	Inhalation of UO2 dust[1]

Source: 1.Hursh, J.B., W.F. Neuman, T. Toribara, H. Wilson and Waterhouse. 1973. Health Phys. 17, pp 239-258.

2-057 Biological half-life time of V

Species	Value	Condition
Homo sapiens	42 days	Whole body[1]

Source: 1.ICRP 1960. Report of Committiee II on Permissible Dose for Internal Radiation. ICRP Publ 2. PergamonPress. Oxford.

2-058 Biological half-life time of W

Species	Value	Condition
Dog	139 days	Compartment = 1.6 %, inhaled W-181 tungstic oxide, slow phase[1]
Dog	9 hours	Compartment = 94 %, fast phase, inhaled W-181 tungstic oxide[1]
Rat	3.65 days	Compartment = 2 %, inj of sodium W-181 tungstate[1]
Rat	8.8 hours	Compartment = 15 %, inj of sodium W-181 tungstate[1]
Rat	86 min	Compartment = 82 %, inj of sodium W-181 tungstate[1]
Rat	99 days	Compartment = 1 %, inj of sodium W-181 tungstate[1]
Rat, bone	1100 days	Slow compartment[2]
Rat, soft tissue	1 day	None[2]
Rat, spleen	44 days	None[2]
Rat, whole body	10 hours	Fast compartment[2]

Source: 1.Aamodt, R.L. 1973. Health Phys. 24, pp 519-524.
2.Kaye, S.V. 1968. Health Phys. 15, pp 398-417.

2-059 Biological half-life time of Zn

Species	Value	Condition
Anodonta nuttalliana	650 days	Freshwater, Zn-65[1]
Euphausia pacifica	140 days	Measured over 2 m, Zn-65[2]
Gambusia affinis	13.6 days	Medium phase = 4.30 %, 0.19 g, uptake mainly from water, Zn-65[3]
Gambusia affinis	2.47 days	Compartment 1 = 8.72 %, 0.19 g, uptake mainly from water, Zn-65[3]
Gambusia affinis	235 days	Slow phase 3 = 91.1 %, 0.19 g, uptake mainly from water, Zn-65[3]
Homo sapiens	162 days	Uptake from food, Zn-65[4]
Littorina obtusata	14.1 days	Intake from water, 283 K, Zn-65[5]
Littorina obtusata	8.9 days	Intake from food (Fucus s.), 283 K, Zn-65[5]

Micropogon undulatus	138 days	Ambient temp and salinity, injection, Zn-65[6]
Micropogon undulatus	138 days	Slow phase, 90 %, Zn-65[7]
Micropogon undulatus	6.5 days	Fast phase, 10 %, Zn-65[7]
Mytilus californianus	76 days	Zn-65, rg - +- 3.5[8]
Pleuronectes platessa	102.8 days	Uptake from food, 283 K, n = 10, 30 g fish[9]
Starry flounders	139 days	Juvenile, depending on initial concentration, Zn-65[10]

Source: 1.Harris, R.P. 1973. J. mar. biol. Ass. U.K., 53, pp 785-800.
2.Fowler, S.W., L.F. Small and J.M. Dean. 1971. Mar. Biol., 8, pp 224-231.
3.Willis, J.N. and N.Y. Jones. 1977. Health Physics. 32(5), pp 381-387.
4.Foster, R.F. and J.F. Honstead. 1967. Health Physics 13, pp 39-43.
5.Young, M.L. 1975. J. Mar. Biol. Ass. U.K., 55, pp 583-610.
6.Baptist, J.P., D.E. Hoss and C.W. Lewis. 1970. Health Phys., 18(2), pp 141-148.
7.Baptist, J.P., D.E. Hoss and C.W. Lewis. 1970. Health Phys., 18(2), pp 141-148.
8.Young, D.R. and T.R. Folson. 1967. Biol. Bull. 133, pp 438-447.
9.Pentreath, R.J. 1976. J. Exp. mar. Biol. Ecol. 21, pp 179-189.
10.Renfro, W.C. 1973. Mar. biol. 21, pp 305-316.

2-060 Chemical Compounds
Half-life time in water

Species	Value	Condition
Ribose	19 days	Half-life time, 278 K, 5 stations at Southampton, rg = 8 to more than value[1]
Ribose	36 days	Half-life time, 300 K, Porto Novo, River Vellar, more than value[1]
Ribose	6 days	Half-life time, 295 K, 5 stations at Southampton, rg = 3 - 9 d[1]

Source: 1.Austin, J. and J.D. Redmond. 1972. Mar. Poll. Bull. 3(3), pp 45-47.

2-061 Chemical compounds
Absorption maxima

Species	Value	Condition
Bacterio - chlorophyll a	365 nm	Occurrence = in purple and green bacteria, in organic solvents[1]
Bacterio - chlorophyll a	605 nm	Occurrence = in purple and green bacteria, in organic solvents[1]
Bacterio - chlorophyll a	770 nm	Occurrence = in purple and green bacteria, in organic solvents[1]
Bacterio - chlorophyll b	368 nm	Occurrence = found in a strain of Rhodopseudomonas, in organic solvents[1]
Bacterio - chlorophyll b	583 nm	Occurrence = found in a strain of Rhodopseudomonas, in organic solvents[1]
Bacterio - chlorophyll b	795 nm	Occurrence = found in a strain of Rhodopseudomonas, in organic solvents[1]
Bacterio - chlorophyll c	432 nm	Occurrence = in green bacteria, in organic solvents[1]
Bacterio - chlorophyll c	660 nm	Occurrence = in green bacteria, in organic solvents[1]
Beta-Carotene	450 nm	Occurrence = in higher plants and most algae, in organic solvents[1]
Chlorophyll a	420 nm	Occurrence = in all higher plants and in algae, in organic solvents[1]
Chlorophyll a	660 nm	Occurrence = in all higher plants and in algae, in organic solvents[1]
Chlorophyll b	435 nm	Occurrence = in all higher plants and in algae, in organic solvents[1]
Chlorophyll b	643 nm	Occurrence = in all higher plants and in algae, in organic solvents[1]
Chlorophyll c	445 nm	Occurrence = in diatoms and brown algae, in organic solvents[1]
Chlorophyll c	625 nm	Occurrence = in diatoms and in brown algae, in organic solvents[1]
Chlorophyll d	450 nm	Occurrence = in red algae, in organic solvents[1]
Chlorophyll d	690 nm	Occurrence = in red algae, in organic solvents[1]
Fucoxanthol	425 nm	Occurrence = in diatoms and brown algae, in organic solvents[1]
Fucoxanthol	450 nm	Occurrence = in diatoms and brown algae, in organic solvents[1]
Fucoxanthol	475 nm	Occurrence = in diatoms and brown algae, in organic solvents[1]

Luteol	425 nm	Occurrence = in green algae, red algae and higher plants, in organic solvents[1]
Luteol	445 nm	Occurrence = in green algae, red algae and higher plants, in organic solvents[1]
Luteol	475 nm	Occurrence = in green algae, red algae and higher plants, in organic solvents[1]
Phycocyanins	618 nm	Occurrence = blue-green algae and in some red algae, in organic solvents[1]
Phycoerthrins	490 nm	Occurrence = in red algae and in some blue-green algae, in organic solvents[1]
Phycoerthrins	546 nm	Occurrence = in red algae and in some blue-green algae, in organic solvents[1]
Phycoerthrins	576 nm	Occurrence = in red algae and in some blue-green algae, in organic solvents[1]
Violaxanthol	425 nm	Occurrence = in higher plants, in organic solvents[1]
Violaxanthol	450 nm	Occurrence = in higher plants, in organic solvents[1]
Violaxanthol	475 nm	Occurrence = in higher plants, in organic solvents[1]
alpha-Carotene	420 nm	Occurrence = in most plants and some algae, in organic solvents[1]
alpha-Carotene	440 nm	Occurrence = in most algae and some algae, in organic solvents[1]
alpha-Carotene	470 nm	Occurrence = in most algae and some algae, in organic solvents[1]
beta-Carotene	425 nm	Occurrence = in higher plants and most algae, in organic solvents[1]
beta-Carotene	480 nm	Occurrence = in higher plants and most algae, in organic solvents[1]

Source: 1.Hall, D.O. and K.K. Rao. 1978. Phytosynthesis. 2nd ed. Camelot Press. Southampton.

2-062 Chemical compounds
Adsorption

Species	Value	Condition
Diquat on Aldrich, commercial humic acid	0.0 meq/100 g	Added: 0 paraquat and 0 meq/100 g Diquat, pH=5.8, competition[1]
Diquat on Aldrich, commercial humic acid	0.0 meq/100 g	Added: 160 paraquat and 0 meq/100 g Diquat, pH=5.3, competition[1]
Diquat on Aldrich, commercial humic acid	43.1 meq/100 g	Added: 80 paraquat and 80 meq/100 g Diquat, pH=5.3, competition[1]
Diquat on Aldrich, commercial humic acid	87.6 meq/100 g	Added: 0 paraquat and 160 meq/100 g Diquat, pH=5.3, competition[1]
Diquat on Humic fraction of histosol	0.0 meq/100 g	Added: 0 paraquat and 0 meq/100 g Diquat, pH=4.9, competition[1]
Diquat on Humic fraction of histosol	0.0 meq/100 g	Added: 160 paraquat and 0 meq/100 g Diquat, pH=3.5, competition[1]
Diquat on Humic fraction of histosol	35.8 meq/100 g	Added: 80 paraquat and 80 meq/100 g Diquat, pH=3.7, competition[1]
Diquat on Humic fraction of histosol	76.4 meq/100 g	Added: 0 paraquat and 160 meq/100 g Diquat, pH=3.5, competition[1]
Diquat on Humin fraction of histosol	0.0 meq/100 g	Added: 0 paraquat and 0 meq/100 g Diquat, pH=7.0, competition[1]
Diquat on Humin fraction of histosol	0.0 meq/100 g	Added: 160 paraquat and 0 meq/100 g Diquat, pH=6.0, competition[1]
Diquat on Humin fraction of histosol	36.1 meq/100 g	Added: 80 paraquat and 80 meq/100 g Diquat, pH=6.0, competition[1]
Diquat on Humin fraction of histosol	77.3 meq/100 g	Added: 0 paraquat and 160 meq/100 g Diquat, pH=6.0, competition[1]
Diquat on untreated histosol	0.0 meq/100 g	Added: 0 paraquat and 0 meq/100 g Diquat, pH=5.7, competition[1]
Diquat on untreated histosol	0.0 meq/100 g	Added: 160 paraquat and 0 meq/100 g Diquat, pH=4.8, competition[1]
Diquat on untreated histosol	27.3 meq/100 g	Added: 80 paraquat and 80 meq/100 g Diquat, pH=4.8, competition[1]
Diquat on untreated histosol	55.2 meq/100 g	Added: 0 paraquat and 160 meq/100 g Diquat, pH=4.8, competition[1]
Diquat on washed histosol	0.0 meq/100 g	Added: 0 paraquat and 0 meq/100 g Diquat, pH=3.8, competition[1]
Diquat on washed histosol	0.0 meq/100 g	Added: 160 paraquat and 0 meq/100 g Diquat, pH=3.4, competition[1]
Diquat on washed histosol	22.4 meq/100 g	Added: 80 paraquat and 80 meq/100 g Diquat, pH=3.4, competition[1]

Diquat on washed histosol	47.7 meq/100 g	Added: 0 paraquat and 160 meq/100 g Diquat, pH=3.4, competition[1]
Paraquat and diquat on Aldrich, commercial humic acid	0.0 meq/100 g	Added: 0 paraquat and 0 meq/100 g Diquat, pH=5.8, competition[1]
Paraquat and diquat on Aldrich, commercial humic acid	87.2 meq/100 g	Added: 80 paraquat and 80 meq/100 g Diquat, pH=5.3, competition[1]
Paraquat and diquat on Aldrich, commercial humic acid	87.6 meq/100 g	Added: 0 paraquat and 160 meq/100 g Diquat, pH=5.3, competition[1]
Paraquat and diquat on Aldrich, commercial humic acid	88.0 meq/100 g	Added: 160 paraquat and 0 meq/100 g Diquat, pH=5.3, competition[1]
Paraquat and diquat on Humic fraction of histosol	0.0 meq/100 g	Added: 0 paraquat and 0 meq/100 g Diquat, pH=4.9, competition[1]
Paraquat and diquat on Humic fraction of histosol	75.0 meq/100 g	Added: 160 paraquat and 0 meq/100 g Diquat, pH=3.5, competition[1]
Paraquat and diquat on Humic fraction of histosol	76.4 meq/100 g	Added: 0 paraquat and 160 meq/100 g Diquat, pH=3.5, competition[1]
Paraquat and diquat on Humic fraction of histosol	76.6 meq/100 g	Added: 80 paraquat and 80 meq/100 g Diquat, pH=3.7, competition[1]
Paraquat and diquat on Humin fraction of histosol	0.0 meq/100 g	Added: 0 paraquat and 0 meq/100 g Diquat, pH=7.0, competition[1]
Paraquat and diquat on Humin fraction of histosol	77.3 meq/100 g	Added: 0 paraquat and 160 meq/100 g Diquat, pH=6.0, competition[1]
Paraquat and diquat on Humin fraction of histosol	78.1 meq/100 g	Added: 160 paraquat and 0 meq/100 g Diquat, pH=6.0, competition[1]
Paraquat and diquat on Humin fraction of histosol	78.2 meq/100 g	Added: 80 paraquat and 80 meq/100 g Diquat, pH=6.0, competition[1]
Paraquat and diquat on untreated histosol	0.0 meq/100 g	Added: 0 paraquat and 0 meq/100 g Diquat, pH=5.7, competition[1]
Paraquat and diquat on untreated histosol	55.2 meq/100 g	Added: 0 paraquat and 160 meq/100 g Diquat, pH=4.8, competition[1]
Paraquat and diquat on untreated histosol	55.7 meq/100 g	Added: 80 paraquat and 80 meq/100 g Diquat, pH=4.8, competition[1]
Paraquat and diquat on untreated histosol	56.2 meq/100 g	Added: 160 paraquat and 0 meq/100 g Diquat, pH=4.8, competition[1]
Paraquat and diquat on washed histosol	0.0 meq/100 g	Added: 0 paraquat and 0 meq/100 g Diquat, pH=3.8, competition[1]
Paraquat and diquat on washed histosol	47.2 meq/100 g	Added: 160 paraquat and 0 meq/100 g Diquat, pH=3.4, competition[1]
Paraquat and diquat on washed histosol	47.7 meq/100 g	Added: 0 paraquat and 160 meq/100 g Diquat, pH=3.4, competition[1]
Paraquat and diquat on washed histosol	49.7 meq/100 g	Added: 80 paraquat and 80 meq/100 g Diquat, pH=3.4, competition[1]
Paraquat on Aldrich, commercial humic acid	0.0 meq/100 g	Added: 0 paraquat and 0 meq/100 g Diquat, pH=5.8, competition[1]
Paraquat on Aldrich, commercial humic acid	0.0 meq/100 g	Added: 0 paraquat and 160 meq/100 g Diquat, pH=5.3, competition[1]
Paraquat on Aldrich, commercial humic acid	44.1 meq/100 g	Added: 80 paraquat and 80 meq/100 g Diquat, pH=5.3, competition[1]
Paraquat on Aldrich, commercial humic acid	88.0 meq/100 g	Added: 160 paraquat and 0 meq/100 g Diquat, pH=5.3, competition[1]
Paraquat on Humic fraction of histosol	0.0 meq/100 g	Added: 0 paraquat and 0 meq/100 g Diquat, pH=4.9, competition[1]
Paraquat on Humic fraction of histosol	0.0 meq/100 g	Added: 0 paraquat and 160 meq/100 g Diquat, pH=3.5, competition[1]

Paraquat on Humic fraction of histosol	40.8 meq/100 g	Added: 80 paraquat and 80 meq/100 g Diquat, pH=3.7, competition[1]
Paraquat on Humic fraction of histosol	75.0 meq/100 g	Added: 160 paraquat and 0 meq/100 g Diquat, pH=3.5, competition[1]
Paraquat on Humin fraction of histosol	0.0 meq/100 g	Added: 0 paraquat and 0 meq/100 g Diquat, pH=7.0, competition[1]
Paraquat on Humin fraction of histosol	0.0 meq/100 g	Added: 0 paraquat and 160 meq/100 g Diquat, pH=6.0, competition[1]
Paraquat on Humin fraction of histosol	42.1 meq/100 g	Added: 80 paraquat and 80 meq/100 g Diquat, pH=6.0, competition[1]
Paraquat on Humin fraction of histosol	78.1 meq/100 g	Added: 160 paraquat and 0 meq/100 g Diquat, pH=6.0, competition[1]
Paraquat on untreated histosol	0.0 meq/100 g	Added: 0 paraquat and 0 meq/100 g Diquat, pH=5.7, competition[1]
Paraquat on untreated histosol	0.0 meq/100 g	Added: 0 paraquat and 160 meq/100 g Diquat, pH=4.8, competition[1]
Paraquat on untreated histosol	28.4 meq/100 g	Added: 80 paraquat and 80 meq/100 g Diquat, pH=4.8, competition[1]
Paraquat on untreated histosol	56.2 meq/100 g	Added: 160 paraquat and 0 meq/100 g Diquat, pH=4.8, competition[1]
Paraquat on washed histosol	0.0 meq/100 g	Added: 0 paraquat and 0 meq/100 g Diquat, pH=3.8, competition[1]
Paraquat on washed histosol	0.0 meq/100 g	Added: 0 paraquat and 160 meq/100 g Diquat, pH=3.4, competition[1]
Paraquat on washed histosol	27.3 meq/100 g	Added: 80 paraquat and 80 meq/100 g Diquat, pH=3.4, competition[1]
Paraquat on washed histosol	47.2 meq/100 g	Added: 160 paraquat and 0 meq/100 g Diquat, pH=3.4, competition[1]

Source: 1.Duursma, E.D. 1965. In: J.P. Riley and G. Skirrow (eds.): Chemical Oceanography. Vol 1. Academic Pres. 433.

2-063 Chemical compounds
Adsorption on carbon

Species	Value	Condition
1,2 Dichlorobenzene	129 mg/g carbon	On carbon surface[1]
1,2,4 Trichlorobenzene	151 mg/g carbon	On carbon surface[1]
1,3 Dichlorobenzene	118 mg/g carbon	On carbon surface[1]
1,4 Dichlorobenzene	121 mg/g carbon	On carbon surface[1]
1-Chlorobenzene	130 mg/g carbon	On carbon surface[1]
Chlorobenzene	91 mg/g carbon	On carbon surface[1]
Hexachlorobenzene	450 mg/g carbon	On carbon surface[1]

Source: 1.Dobbs, R.A. and J.M. Cohen. 1980. Carbon adsorption isotherms for Toxic organics. EPA 600/8-80-023.

2-064 Chemical compounds
Adsorption on sand

Species	Value	Condition
Cu from waste water	0.02 mg Cu/l per gram sand	Cu conc = 9.5 mg/l, pH = 4.0[1]
Cu from waste water	0.26 mg Cu/l per gram sand	Cu conc = 9.5 mg/l, pH = 5.0[1]
Cu from waste water	0.32 mg Cu/l per gram sand	Cu conc = 9.5 mg/l, pH = 5.6[1]
Cu from waste water	0.33 mg Cu/l per gram sand	Cu conc = 8.7 mg/l, pH = 6.0[1]
Cu from waste water	0.38 mg Cu/l per gram sand	Cu conc = 5.1 mg/l, pH = 11.0[1]
Cu from waste water	0.45 mg Cu/l per gram sand	Cu conc = 5.5 mg/l, pH = 7.0[1]

Source: 1.Netzer, A. and P. Wilkinson. 1974. Proceedings of the 29th Industrial Waste Conference, Purde Univerity, 841-845.

2-065 Chemical compounds
Cation exchange capacity, CEC

Species	Value	Condition
Chlorite	35 m**2/g	Sediment material, rg = 20 - 50[1]
Fe hydroxide	18 meq/100 g	Sediment material, rg = 10 - 25[1]
Illite	25 meq/100 g	Sediment material, rg = 10 - 40[1]
Kaolinite	9 meq/100 g	Sediment material, rg = 3 - 15[1]
Montmorillonite	100 meq/100 g	Sediment material, rg = 80 - 120[1]
Silicic acid	23 meq/100 g	Sediment material, rg = 11 - 34, amorphous[1]

Source: 1.Foestner, U. and G.T.W. Wittmann. 1979. Metal Pollution in the Aquatic Environment. Springer-Verlag.Berlin.

2-066 Chemical compounds
Evaporation rate

Species	Value	Condition
CCl2=CCl2	25.7 min	50 % evaporation, dilute water sol., initial = 1 ppm, 298 K[1]
CCl3CCl3	45 min	50 % evaporation, dilute water sol., initial = 1 ppm, 298 K[1]
CCl4	29 min	50 % evaporation, dilute water sol., initial = 1 ppm, 298 K[1]
CH2=CCl2	22 min	50 % evaporation, dilute water sol., initial = 1 ppm, 298 K[1]
CH2=CClCH2Cl	20 min	50 % evaporation, dilute water sol., initial = 1 ppm, 298 K[1]
CH2=CClCH3	29 min	50 % evaporation, dilute water sol., initial = 1 ppm, 298 K[1]
CH2=CHCH2Cl	27 min	50 % evaporation, dilute water sol., initial = 1 ppm, 298 K[1]
CH2=CHCl	26 min	50 % evaporation, dilute water sol., initial = 1 ppm, 298 K[1]
CH2Cl2	20.7 min	50 % evaporation, dilute water sol., initial = 1 ppm, 298 K[1]
CH2ClCCl2CH2Cl	47 min	50 % evaporation, dilute water sol., initial = 1 ppm, 298 K[1]
CH2ClCCl3	43 min	50 % evaporation, dilute water sol., initial = 1 ppm, 298 K[1]
CH2ClCH2Cl	29 min	50 % evaporation, dilute water sol., initial = 1 ppm, 298 K[1]
CH2ClCHCl2	21 min	50 % evaporation, dilute water sol., initial = 1 ppm, 298 K[1]
CH2ClCHClCHCl2	51 min	50 % evaporation, dilute water sol., initial = 1 ppm, 298 K[1]
CH3CCl3	20 min	50 % evaporation, dilute water sol., initial = 1 ppm, 298 K[1]
CH3CH2Cl	21 min	50 % evaporation, dilute water sol., initial = 1 ppm, 298 K[1]
CH3CHCl2	22 min	50 % evaporation, dilute water sol., initial = 1 ppm, 298 K[1]
CH3Cl	27 min	50 % evaporation, dilute water sol., initial = 1 ppm, 298 K[1]
CHCl2CCl3	45 min	50 % evaporation, dilute water sol., initial = 1 ppm, 298 K[1]
CHCl2CHCl2	56 min	50 % evaporation, dilute water sol., initial = 1 ppm, 298 K[1]
CHCl3	21 min	50 % evaporation, dilute water sol., initial = 1 ppm, 298 K[1]
CHCl=CCl2	21.3 min	50 % evaporation, dilute water sol., initial = 1 ppm, 298 K[1]
CHCl=CClCH2Cl	49 min	50 % evaporation, dilute water sol., initial = 1 ppm, 298 K[1]
CHCl=CHCH2Cl	31 min	50 % evaporation, dilute water sol., initial = 1 ppm, 298 K[1]
CHCl=CHCH3	16 min	50 % evaporation, dilute water sol., initial = 1 ppm, 298 K[1]
CHCl=CHCl	18 min	50 % evap., dilute water sol., initial = 1 ppm, 298 K, cis[1]
CHCl=CHCl	24 min	50 % evap., dilute water sol., initial = 1 ppm, 298 K, trans[1]
PCB	0.004 mg/((cm**2)*hour)	Surface = 12.28 cm2, 373 K, 72 h, aroclor 5442[2]

PCB	0.004 mg/((cm**2)*hour)	Surface = 12.28 cm2, 373 K, 72 h, aroclor 5460[2]
PCB	0.007 mg/((cm**2)*hour)	Surface = 12.28 cm2, 373 K, 72 h, aroclor 4465[2]
PCB	0.009 mg/((cm**2)*hour)	Surface = 12.28 cm2, 373 K, 24 h, aroclor 1260[2]
PCB	0.013 mg/((cm**2)*hour)	Surface = 12.28 cm2, 373 K, 24 h, aroclor 1262[2]
PCB	0.053 mg/((cm**2)*hour)	Surface = 12.28 cm2, 373 K, 24 h, aroclor 1254[2]
PCB	0.152 mg/((cm**2)*hour)	Surface = 12.28 cm2, 373 K, 24 h, aroclor 1448[2]
PCB	0.338 mg/((cm**2)*hour)	Surface = 12.28 cm2, 373 K, 24 h, aroclor 1242[2]
PCB	0.874 mg/((cm**2)*hour)	Surface = 12.28 cm2, 373 K, 24 h, aroclor 1232[2]
PCB	1.74 mg/((cm**2)*hour)	Surface = 12.28 cm2, 373 K, 24 h, aroclor 1221[2]

Source: 1.Dilling, W.L., N.B. Tefertiller and G.J. Kallos. 1975. Environ. Sci. Technol. 9(9), pp 833-837. 2.Hutzinger, O., S. Safe and V. Zitko. 1974. The Chemistry of PCBs, CRC-Press.

2-067 Chemical compounds Fulvic acid composition

Species	Value	Condition
Acidic amino acids	25.7 mg/g	Amino acid content[1]
Aromatic amino acids	9.5 mg/g	Amino acid content[1]
Basic amino acids	7.5 mg/g	Amino acid content[1]
C	43.2 % by atoms	Rg = 38.4 - 48.7, elemental composition[1]
C / H ratio	7.9 by atoms	Rg = 6.9 - 8.9, elemental composition[1]
C / N ratio	12.1 by atoms	Rg = 9.4 - 14.8, elemental composition[1]
Carbonyl	4.5 meq/g dry organic matter	Rg = 3.0 - 6.0, functional group content[1]
Carboxyl	2.5 meq/g dry organic matter	Rg = 1.0 - 4.0, functional group content[1]
H	5.5 % by atoms	Rg = 4.3 - 6.6, elemental composition[1]
Molecular weight range	100 % of dry organic matter	Weight = less than 10000[1]
N	3.9 % by atoms	Rg = 2.6 - 5.2, elemental composition[1]
Neutral amino acids	29.2 mg/g	Amino acid content[1]
O	46.2 % by atoms	Rg = 41.8 - 54.5, elemental composition[1]
Phenolic hydroxyl	0.5 meq/g dry organic matter	Rg = 0.0 - 1.0, functional group content[1]
Quinone	3.6 meq/g dry organic matter	Rg = 2.4 - 4.7, functional group content[1]
Sulfur amino acids	4.6 mg/g	Amino acid content[1]
Total acidity	3.8 meq/g dry organic matter	Rg = 2.0 - 5.5, functional group content[1]
Total amino acids	76.5 mg/g	Amino acid content[1]

Source: 1.Rashid, M.A. 1974. In: Offshore geology of eastern Canada. Geol. Surv. Can. paper 74-30, I, 123-132.

2-068 Chemical compounds of Half-life in water

Species	Value	Condition
Atrazine	25.7 min	Init = 1.5E-4 M,without fulvic acid, 253.7 nm, photolysis, pH = 11.0[1]
Atrazine	31.6 min	Init = 1.5E-4 M,without fulvic acid, 253.7 nm, photolysis, pH = 7.0[1]
Atrazine	35.0 min	Init = 1.5E-4 M,without fulvic acid, 253.7 nm, photolysis, pH = 3.5[1]
Atrazine	81.3 min	Init = 1.5E-4 M, with 0.01 % fulvic acid, 253.7 nm, photolysis, pH = 11.[1]
Atrazine	92.8 min	Init = 1.5E-4 M, with 0.01 % fulvic acid, 253.7 nm, photolysis, pH = 7.0[1]
Atrazine	94.4 min	Init = 1.5E-4 M, with 0.01 % fulvic acid, 253.7 nm, photolysis, pH = 3.5[1]

Source: 1.Khan, S.U. and M. Schnitzer. 1978. Environ. Sci. Health B13(3), 299.

2-069 Chemical compounds
Half-life time in atmosphere

Species	Value	Condition
1,1,2-trichloroethane	15.9 hours	Init = 10 ppm, humid = 35 %, uv = 2.6*normal, 300 K, with 5 ppm NO[1]
1-butene oxide	15.2 hours	Init = 10 ppm, humid = 35 %, uv = 2.6*normal, 300 K, with 5 ppm NO2[1]
1-butene oxide	15.9 hours	Init = 10 ppm, humid = 35 %, uv = 2.6*normal, 300 K, with 5 ppm NO[1]
1-methoxy-2-propanol	3.1 hours	Init = 10 ppm, humid = 35 %, uv = 2.6*normal, 300 K, with 5 ppm NO[1]
2,4,4-trimethyl-1--pentene	0.60 hours	Inital = 10 ppm, humid = 35 %, uv = 2.6*normal, 300 K, with 5 ppm NO[1]
Allyl chloride	0.05 hours	Reaction products = chloracetic acid, chloracetaldehyde[2]
Anthracene	38000 hours	Degradation by RO2* radical[3]
Benz(a)anthracene	370 hours	Degradation by O3 = 2E-9 M[3]
Benzo(a)pyrene	2.5E5 hours	Degradation by RO2* radical[3]
Benzo(a)pyrene	870 hours	Degradation by O3 = 2E-9 M[3]
CCl2=CCl2	14.2 hours	Init = 10 ppm, humid = 35 %, uv = 2.6*normal, 300 K, with 5 ppm NO[1]
CCl2=CCl2	8.3 hours	Init = 10 ppm, humid = 35 %, uv = 2.6*normal, 300 K, with 16.8 ppm[1]
CH2Cl2	10 % per d	Init = 10 ppm, humid = 35 %, uv = 2.6*norm, 300 K, with 16.8 ppm NO2,[1]
CH2Cl2	6 % per day	Init = 10 ppm, humid = 35 %, uv = 2.6*normal, 300 K, with 5 ppm NO,[1]
CH3CCl3	10 % per d	Init = 10 ppm, humid = 35 %, uv = 2.6*norm, 300 K, with 16.8 ppm NO2,[1]
CH3CCl3	6 % per day	Init = 10 ppm, humid = 35 %, uv = 2.6*normal, 300 K, with 5 ppm NO,[1]
CHCl=CCl2	2.9 hours	Init = 10 ppm, humid = 35 %, uv = 2.6*normal, 300 K, with 16.8 ppm[1]
CHCl=CCl2	3.5 hours	Init = 10 ppm, humid = 35 %, uv = 2.6*normal, 300 K, with 5 ppm NO[1]
Carbon tetrachloride	1000 hours	More than value, reaction product = phosgene, chlorine[2]
Chlorobenzene	8.7 hours	Init = 10 ppm, humid = 35 %, uv = 2.6*normal, 300 K, with 5 ppm NO[1]
Chloroform	15 hours	Reaction product = phosgene, chlorine[2]
Cis - dichloroethylene	2.8 hours	Init = 10 ppm, humid = 35 %, uv = 2.6*normal, 300 K, with 16.8 ppm[1]
Cis - dichloroethylene	3.0 hours	Init = 10 ppm, humid = 35 %, uv = 2.6*normal, 300 K, with 5 ppm NO[1]
Cyclohexane	6.5 hours	Init = 10 ppm, humid = 35 %, uv = 2.6*normal, 300 K, with 16.8 ppm[1]
Cyclohexane	6.9 hours	Init = 10 ppm, humid = 35 %, uv = 2.6*normal, 300 K, with 5 ppm NO[1]
Cyclohexane	7.5 hours	Init = 10 ppm, humid = 35 %, uv = 2.6*normal, 300 K, with 5 ppm NO2[1]
Cyclohexene	0.19 hours	Init = 10 ppm, humid = 35 %, uv = 2.6*normal, 300 K, with 5 ppm NO2[1]
Cyclohexene	0.87 hours	Init = 10 ppm, humid = 35 %, uv = 2.6*normal, 300 K, with 5 ppm NO[1]
Dibenzanthracene	340 hours	Degradation by O3 = 2E-9 M[3]
Dimethyldibenzanthra-cene	140 hours	Degradation by O3 = 2E-9 M, less than value[3]
Dioxane	3.4 hours	Init = 10 ppm, humid = 35 %, uv = 2.6*normal, 300 K, with 5 ppm NO[1]
Epichlorohydrin	16.0 hours	Init = 10 ppm, humid = 35 %, uv = 2.6*normal, 300 K, with 5 ppm NO[1]
Ethyl acetate	14.6 hours	Init = 10 ppm, humid = 35 %, uv = 2.6*normal, 300 K, with 5 ppm NO[1]
Ethyl benzene	5.0 hours	Init = 10 ppm, humid = 35 %, uv = 2.6*normal, 300 K, with 5 ppm NO[1]
Ethylene	2.5 hours	Init = 10 ppm, humid = 35 %, uv = 2.6*normal, 300 K, with 16.8 ppm[1]
Ethylene	2.9 hours	Init = 10 ppm, humid = 35 %, uv = 2.6*normal, 300 K, with 5 ppm NO[1]

Ethylene dichloride	8 hours	Products = chloroacetaldehyde, chloroacetyl, chloride, formyl chloride[2]
Isobutyl alcohol	3.5 hours	Init = 10 ppm, humid = 35 %, uv = 2.6*normal, 300 K, with 5 ppm NO[1]
M-xylene	2.9 hours	none[1]
Methyl chloride	12 hours	Product = phosgene, chlorine, formyl chloride[2]
Methyl chloroform	140 hours	Reaction products = phosgene, chlorine[2]
Methyl ethyl ketone	9.8 hours	Init = 10 ppm, humid = 35 %, uv = 2.6*normal, 300 K, with 5 ppm NO[1]
Methyl isobutyl ketone	3.5 hours	Init = 10 ppm, humid = 35 %, uv = 2.6*normal, 300 K, with 5 ppm NO[1]
Methylene dichloride	15 hours	None[2]
N-butyl alcohol	6.5 hours	Init = 10 ppm, humid = 35 %, uv = 2.6*normal, 300 K, with 5 ppm NO[1]
N-methylpyrrole	0.16 hours	Init = 10 ppm, humid = 35 %, uv = 2.6*normal, 300 K, with 5 ppm NO[1]
Nitromethane	9.2 hours	Init = 10 ppm, humid = 35 %, uv = 2.6*normal, 300 K, with 5 ppm NO[1]
P-xylene	3.1 hours	none[1]
Perchloroethylene	10 hours	Reaction prod. = phosgene, Cl2C(OH)COCl, chlorine, trichloroacetic acid[2]
Perylene	38000 hours	Degradation by RO2* radical[3]
Phenanthrene	2E8 hours	Degradation by RO2* radical[3]
Pyrene	2.4E5 hours	Degradation by RO2* radical[3]
Pyrene	560 hours	Degradation by O3 = 2E-9 M[3]
S-butyl alcohol	4.0 hours	Init = 10 ppm, humid = 35 %, uv = 2.6*normal, 300 K, with 5 ppm NO[1]
T-butyl alcohol	34.5 hours	Init = 10 ppm, humid = 35 %, uv = 2.6*normal, 300 K, with 5 ppm NO[1]
Tetracene	96 hours	Degradation by RO2* radical[3]
Toluene	6.8 hours	Init = 10 ppm, humid = 35 %, uv = 2.6*normal, 300 K, with 5 ppm NO[1]
Trans-2-butene	0.17 hours	Init = 10 ppm, humid = 35 %, uv = 2.6*normal, 300 K, with 5 ppm NO2[1]
Trans-2-butene	0.30 hours	Init = 10 ppm, humid = 35 %, uv = 2.6*normal, 300 K, with 5 ppm NO[1]
Trans-dichloroethylene	2.8 hours	Init = 10 ppm, humid = 35 %, uv = 2.6*normal, 300 K, with 16.8 ppm[1]
Trans-dichloroethylene	2.9 hours	Init = 10 ppm, humid = 35 %, uv = 2.6*normal, 300 K, with 5 ppm NO[1]
Trichloroethylene	1 hours	Reaction products = phosgene, formyl chloride and dichloroacetic acid[2]
Trioxane	4.7 hours	Init = 10 ppm, humid = 35 %, uv = 2.6*normal, 300 K, with 5 ppm NO[1]
Vinyl chloride	1 hours	None[2]
Vinyl chloride	4.3 hours	Init = 10 ppm, humid = 35 %, uv = 2.6*normal, 300 K, with 5 ppm NO[1]
Vinylidene chloride	1 hours	None[2]
Vinylidene chloride	2.1 hours	Init = 10 ppm, humid = 35 %, uv = 2.6*normal, 300 K, with 5 ppm NO[1]

Source: 1.Dilling, W.L., C.J. Bredeweg and N.B. Tefertiller. 1976. Environ. Sci. Tech nol., 10(4), pp 351-356. 2.Pearson, C.R. 1982. In: O. Hutzinger (ed.): Environmental Chemistry. Springer-Verlag. Volume 3, partB. Berlin. 3.Radding, S.B., T. Mill, Gould, Liu, Johnson, Bomberger and Fojo. 1976. The Environmental Fate of Selcted Polynuclear Aromatic Hdrocarbons. EPA 560/5-75-009. Wash DC.

2-070 Chemical compounds
Half-life time in soil

Species	Value	Condition
Cd	1.3 y used for a 10 % decreased in metal conc	Control Organic spruce, Sweden, artificially rain = 150 l/m2, pH=2.8[1]
Cd	1.7 y used for a 10 % decreased in metal conc	Polluted organic spruce, Sweden, artificially rain = 150 l/m2, pH=2.8[1]
Cd	20 y used for a 10 % decreased in metal conc	Polluted organic spruce, Sweden, artificially rain = 150 l/m2, pH=4.2[1]

Cd	3 y used for a 10 % decreased in metal conc	Control Organic spruce, Sweden, artificially rain = 150 l/m2, pH=3.2[1]
Cd	4.5 y used for a 10 % decreased in metal conc	Polluted organic spruce, Sweden, artificially rain = 150 l/m2, pH=3.2[1]
Cd	6 y used for a 10 % decreased in metal conc	Control Organic spruce, Sweden, artificially rain = 150 l/m2, pH=4.2[1]
Cr	125 y used for a 10 % decreased in metal conc	Polluted organic spruce, Sweden, artificially rain = 150 l/m2, pH=4.2[1]
Cr	15 y used for a 10 % decreased in metal conc	Control Organic spruce, Sweden, artificially rain = 150 l/m2, pH=2.8[1]
Cr	19 y used for a 10 % decreased in metal conc	Control Organic spruce, Sweden, artificially rain = 150 l/m2, pH=3.2[1]
Cr	20 y used for a 10 % decreased in metal conc	Control Organic spruce, Sweden, artificially rain = 150 l/m2, pH=4.2[1]
Cr	60 y used for a 10 % decreased in metal conc	Polluted organic spruce, Sweden, artificially rain = 150 l/m2, pH=2.8[1]
Cr	60 y used for a 10 % decreased in metal conc	Polluted organic spruce, Sweden, artificially rain = 150 l/m2, pH=3.2[1]
Cu	100 y used for a 10 % decreased in metal conc	Polluted organic spruce, Sweden, artificially rain = 150 l/m2, pH=4.2[1]
Cu	11 y used for a 10 % decreased in metal conc	Control Organic spruce, Sweden, artificially rain = 150 l/m2, pH=3.2[1]
Cu	13 y used for a 10 % decreased in metal conc	Control Organic spruce, Sweden, artificially rain = 150 l/m2, pH=4.2[1]
Cu	19 y used for a 10 % decreased in metal conc	Polluted organic spruce, Sweden, artificially rain = 150 l/m2, pH=3.2[1]
Cu	6 y used for a 10 % decreased in metal conc	Polluted organic spruce, Sweden, artificially rain = 150 l/m2, pH=2.8[1]
Cu	9 y used for a 10 % decreased in metal conc	Control Organic spruce, Sweden, artificially rain = 150 l/m2, pH=2.8[1]
Mn	0.5 y used for a 10 % decreased in metal conc	Control Organic spruce, Sweden, artificially rain = 150 l/m2, pH=2.8[1]
Mn	1.5 y used for a 10 % decreased in metal conc	Control Organic spruce, Sweden, artificially rain = 150 l/m2, pH=3.2[1]
Mn	1.5 y used for a 10 % decreased in metal conc	Polluted organic spruce, Sweden, artificially rain = 150 l/m2, pH=2.8[1]
Mn	3 y used for a 10 % decreased in metal conc	Control Organic spruce, Sweden, artificially rain = 150 l/m2, pH=4.2[1]
Mn	35 y used for a 10 % decreased in metal conc	Polluted organic spruce, Sweden, artificially rain = 150 l/m2, pH=4.2[1]
Mn	4 y used for a 10 % decreased in metal conc	Polluted organic spruce, Sweden, artificially rain = 150 l/m2, pH=3.2[1]
Ni	15 y used for a 10 % decreased in metal conc	Polluted organic spruce, Sweden, artificially rain = 150 l/m2, pH=4.2[1]
Ni	2 y used for a 10 % decreased in metal conc	Control Organic spruce, Sweden, artificially rain = 150 l/m2, pH=2.8[1]
Ni	2 y used for a 10 % decreased in metal conc	Control Organic spruce, Sweden, artificially rain = 150 l/m2, pH=3.2[1]
Ni	2 y used for a 10 % decreased in metal conc	Polluted organic spruce, Sweden, artificially rain = 150 l/m2, pH=2.8[1]
Ni	4.5 y used for a 10 % decreased in metal conc	Polluted organic spruce, Sweden, artificially rain = 150 l/m2, pH=3.2[1]

Ni	5 y used for a 10 % decreased in metal conc	Control Organic spruce, Sweden, artificially rain = 150 l/m2, pH=4.2[1]
Pb	100 y used for a 10 % decreased in metal conc	Polluted organic spruce, Sweden, artificially rain = 150 l/m2, pH=3.2[1]
Pb	17 y used for a 10 % decreased in metal conc	Polluted organic spruce, Sweden, artificially rain = 150 l/m2, pH=2.8[1]
Pb	20 y used for a 10 % decreased in metal conc	Control Organic spruce, Sweden, artificially rain = 150 l/m2, pH=2.8[1]
Pb	200 y used for a 10 % decreased in metal conc	Polluted organic spruce, Sweden, artificially rain = 150 l/m2, pH=4.2[1]
Pb	45 y used for a 10 % decreased in metal conc	Control Organic spruce, Sweden, artificially rain = 150 l/m2, pH=3.2[1]
Pb	80 y used for a 10 % decreased in metal conc	Control Organic spruce, Sweden, artificially rain = 150 l/m2, pH=4.2[1]
V	17 y used for a 10 % decreased in metal conc	Control Organic spruce, Sweden, artificially rain = 150 l/m2, pH=4.2[1]
V	2 y used for a 10 % decreased in metal conc	Polluted organic spruce, Sweden, artificially rain = 150 l/m2, pH=4.2[1]
V	27.5 y used for a 10 % decreased in metal conc	Control Organic spruce, Sweden, artificially rain = 150 l/m2, pH=3.2[1]
V	6.5 y used for a 10 % decreased in metal conc	Polluted organic spruce, Sweden, artificially rain = 150 l/m2, pH=3.2[1]
V	9 y used for a 10 % decreased in metal conc	Control Organic spruce, Sweden, artificially rain = 150 l/m2, pH=2.8[1]
V	9 y used for a 10 % decreased in metal conc	Polluted organic spruce, Sweden, artificially rain = 150 l/m2, pH=2.8[1]
Zn	0.8 y used for a 10 % decreased in metal conc	Control Organic spruce, Sweden, artificially rain = 150 l/m2, pH=2.8[1]
Zn	1.2 y used for a 10 % decreased in metal conc	Polluted organic spruce, Sweden, artificially rain = 150 l/m2, pH=2.8[1]
Zn	2 y used for a 10 % decreased in metal conc	Control Organic spruce, Sweden, artificially rain = 150 l/m2, pH=3.2[1]
Zn	3 y used for a 10 % decreased in metal conc	Polluted organic spruce, Sweden, artificially rain = 150 l/m2, pH=3.2[1]
Zn	7 y used for a 10 % decreased in metal conc	Control Organic spruce, Sweden, artificially rain = 150 l/m2, pH=4.2[1]
Zn	9 y used for a 10 % decreased in metal conc	Polluted organic spruce, Sweden, artificially rain = 150 l/m2, pH=4.2[1]

Source: 1.Inman, R.E., R.B. Ingersoll and E.A. Levy. 1971. Science 172, 1229-1231.

2-071 Chemical compounds
Half-life time in stratosphere

Species	Value	Condition
Chlorflouromethan	1.3 months	Altitude 30 km, f-11[1]
Chlorflouromethan	1.3 months	Altitude 40 km, f-12[1]
Chlorflouromethan	11 months	Altitude 30 km, f-12[1]
Chlorflouromethan	15 hours	Top of atmosphere, f-11[1]
Chlorflouromethan	16 days	Altitude 50 km, f-12[1]
Chlorflouromethan	2.1 days	Altitude 50 km, f-11[1]
Chlorflouromethan	3.5 days	Top of atmosfere, f-12[1]
Chlorflouromethan	3E5 years	Altitude 10 km, f-11[1]

Chlorflouromethan	4.7 days	Altitude 40 km, f-11[1]
Chlorflouromethan	4E6 years	Altitude 10 km, f-12[1]
Chlorflouromethan	6.6 years	Altitude 20 km, f-11[1]
Chlorflouromethan	66 years	Altitude 20 km, f-12[1]

Source: 1.Rowland, D.F.S. and M.J. Molina. 1975. Rev. Geophys. Spach. Phys. 13, pp 1-35.

2-072 Chemical compounds
Half-life time in water

Species	Value	Condition
Acetate	0.5 hours	Bac uptake, Lotsjoen, Sweden, summer[1]
Acetate	13 hours	Bac uptake, Lotsjoen, Sweden, winter, rg=4-22, 273.5 K[1]
Acetate	315 hours	Bac uptake, Erken, Sweden, 276 K, rg=200-430 h, prod=500 mg C/(m2*d)[2]
Acetate	416 hours	Bac uptake, W Blue L., Manitoba, ice surface, winter, prod=40 mg C/(m2*d)[3]
Alanine	40 hours	Bac uptake, rg=20-60, Upper Klamath L, ice surf, winter[4]
All polyaromatic hydrocarbons, PAH	0.5 hours	Degradation by Cl2, less than value[5]
All polyaromatic hydrocarbons, PAH	10 hours	Degradation by HO*radical[5]
Anthracene	32.6 mg compound remaining after 5 h	Initial=100 mg, natural light, water+acetone (or tetrachloride)[6]
Anthracene	5 hours	Degradation by singlet oxygen[5]
Anthracene	52.9 mg compound remaining after 5 h	Initial=100 mg, natural light, water+acetone (or tetrachloride)[6]
Asparginc	30 hours	Bac uptake, rg=20-30, Upper Klamath L, ice surf, winter[4]
Aspartate	4.5 hours	Bac uptake, rg=4-5, Upper Klamath L, ice surf, winter[4]
Atrazine	0.70 days	313 K, 0.2 g humic acid as catalyst, hydrolysis, init atra=60 umol/l[7]
Atrazine	1.73 days	313 K, 0.2 g humic acid as catalyst, hydrolysis, init atra=60 umol/l[7]
Atrazine	11.34 days	313 K, 0.2 g humic acid as catalyst, hydrolysis, init atra=60 umol/l[7]
Atrazine	1352 days	285 K, hydrogen ion as catalyst, pH=4, hydrolysis, init atra=60 umol/l[7]
Atrazine	138 days	298 K, 0.04 M acetate buffer as catalyst, hydrolysis, init atra=60 umol/l[7]
Atrazine	18.42 days	303 K, 0.08 M acetate buffer as catalyst, hydrolysis, init atra=60 umol/l[7]
Atrazine	244 days	298 K, hydrogen ion as catalyst, pH=4, hydrolysis, init atra=60 umol/l[7]
Atrazine	302.5 days	285 K, 0.08 M acetate buffer as catalyst, hydrolysis, init atra=60 umol/l[7]
Atrazine	34.17 days	303 K, 0.04 M acetate buffer as catalyst, hydrolysis, init atra=60 umol/l[7]
Atrazine	48.6 days	303 K, hydrogen ion as catalyst, pH=4, hydrolysis, init atra=60 umol/l[7]

Atrazine	661 days	285 K, 0.04 M acetate buffer as catalyst, hydrolysis, init atra=60 umol/l[7]
Atrazine	71.02 days	298 K, 0.08 M acetate buffer as catalyst, hydrolysis, init atra=60 umol/l[7]
Benz(a)anthracene	0.0 mg compound remaining after 5 h	Initial=100 mg, natural light, water+acetone (or tetrachloride)[6]
Benz(a)anthracene	0.45 hours	Degradation by O3 = 10E-4 M[5]
Benz(a)anthracene	10 hours	Degradation by singlet oxygen[5]
Benz(a)anthracene	27 min	Water with O3 = 1E-4 m, 298 K[5]
Benz(a)anthracene	45.5 mg compound remaining after 5 h	Initial=100 mg, natural light, water+acetone (or tetrachloride)[6]
Benzo(a)pyrene	1.05 hours	Degradation by O3 = 10E-4 M[5]
Benzo(a)pyrene	5 hours	Degradation by singlet oxygen[5]
Benzo(a)pyrene	63 min	Water with O3 = 1E-4 m, 298 K[5]
Benzo(a)pyrene + 1-Methylnaphthalene	90.5 mg compound remaining after 5 h	Initial=100 mg, natural light, water+acetone (or tetrachloride)[6]
Benzo(a)pyrene + 1-Methylnaphthalene	93.6 mg compound remaining after 5 h	Initial=100 mg, natural light, water+acetone (or tetrachloride)[6]
Benzo(a)pyrene degradated by all processes	1.5 days	Oligotrophic lake, summer sunlight, except dilution[8]
Benzo(a)pyrene degradated by all processes	2.9 days	River water, summer sunlight, except dilution[8]
Benzo(a)pyrene degradated by all processes	7.3 days	Eutrophic pond, summer sunlight, except dilution[8]
Benzo(a)pyrene degradated by all processes	7.4 days	Eutrophic lake, summer sunlight, except dilution[8]
Benzo(a)pyrene degradated by oxidation	340 days	Eutrophic lake, summer sunlight[8]
Benzo(a)pyrene degradated by oxidation	340 days	Eutrophic pond, summer sunlight[8]
Benzo(a)pyrene degradated by oxidation	340 days	Oligotrophic lake, summer sunlight[8]
Benzo(a)pyrene degradated by oxidation	340 days	River water, summer sunlight[8]
Benzo(a)pyrene degradated by photolysis	1.5 days	Oligotrophic lake, summer sunlight[8]
Benzo(a)pyrene degradated by photolysis	3.0 days	River water, summer sunlight[8]
Benzo(a)pyrene degradated by photolysis	7.5 days	Eutrophic lake, summer sunlight[8]
Benzo(a)pyrene degradated by photolysis	7.5 days	Eutrophic pond, summer sunlight[8]
Benzo(a)pyrene degradated by volatilization	140 days	River water, summer sunlight[8]
Benzo(a)pyrene degradated by volatilization	350 days	Eutrophic pond, summer sunlight[8]
Benzo(a)pyrene degradated by volatilization	700 days	Eutrophic lake, summer sunlight[8]
Benzo(a)pyrene degradated by volatilization	700 days	Oligotrophic lake, summer sunlight[8]

Benzothiophene + 1-Methylnaphthalene	0.0 mg naphthalene rem. after 5 h	Initial=100 + 100 mg, natural light, water+acetone (or tetrachloride)[6]
Benzothiophene + 1-Methylnaphthalene	181.0 mg benzothiphene rem. after 5 h	Initial=100 + 100 mg, natural light, water+acetone (or tetrachloride)[6]
Benzothiophene + 1-Methylnaphthalene	36.30 mg benzothiphene rem. after 5 h	Initial=100 + 100 mg, natural light, water+acetone (or tetrachloride)[6]
Carbaryl	51.7 hours	Direct photolysis, midday, midsummer, latitude 40 N[9]
Carbon tetrachloride	1E8	Hydrolysis, more than value[10]
Carbon tetrachloride	1E8	Solvolysis, more than value[10]
Chloroform	10E4	Hydrolysis[10]
Chloroform	14E6	Solvolysis[10]
Chrysene	94.0 mg compound remaining after 5 h	Initial=100 mg, natural light, water+acetone (or tetrachloride)[6]
Chrysene	96.0 mg compound remaining after 5 h	Initial=100 mg, natural light, water+acetone (or tetrachloride)[6]
Citric acid	100 hours	Bac uptake, W Blue L., Manitoba, ice surface, winter, prod=40 mg C/(m2*d)[3]
Dibenzanthracene	0.42 hours	Degradation by O3 = 10E-4 M[5]
Dibenzanthracene	25 min	Water with O3 = 1E-4 m, 298 K[5]
Dibenzothiophene + 1-methylnaphthalene	0.0 mg naphthalene rem. after 5 h	Initial=100 + 100 mg, natural light, water+acetone (or tetrachloride)[6]
Dibenzothiophene + 1-methylnaphthalene	16.2 mg dibenzothiophene rem. after 5 h	Initial=100 + 100 mg, natural light, water+acetone (or tetrachloride)[6]
Dibenzothiophene + 1-methylnaphthalene	20.6 mg dibenzothiophene rem. after 5 h	Initial=100 + 100 mg, natural light, water+acetone (or tetrachloride)[6]
Dibenzothiophene + 1-methylnaphthalene	40.1 mg naphthalene rem. after 5 h	Initial=100 + 100 mg, natural light, water+acetone (or tetrachloride)[6]
Dimethylanthracene	0.05 hours	Degradation by singlet oxygen[5]
Dimethylbenz(a)anthracene	5 hours	Degradation by singlet oxygen, less than value[5]
Dimethylbenzanthracene	10 min	Water with O3 = 1E-4 m, 298 K, less than value[5]
Dimethyldibenzanthracene	0.17 hours	Degradation by O3 = 10E-4 M[5]
Dimethyldibenzanthracene	5 hours	Degradation by singlet oxygen, less than value[5]
Diphenylanthracene	0.6 hours	Degradation by singlet oxygen[5]
Ethylene dichloride	1E3	Hydrolysis, by dehydrochlorination, more than value[10]
Ethylene dichloride	1E3	Solvolysis, more than value, by dehydrochlorination[10]
Fluorene	91.9 mg compound remaining after 5 h	Initial=100 mg, natural light, water+acetone (or tetrachloride)[6]
Fluorene	94.3 mg compound remaining after 5 h	Initial=100 mg, natural light, water+acetone (or tetrachloride)[6]
Fumaric acid	50 hours	Bac uptake, W Blue L., Manitoba, ice surface, winter, prod=40 mg C/(m2*d)[3]
Glucose	11 days	300 K, Porto Novo, Kille Backwater[11]
Glucose	17 days	278 K, 5 stations at Southampton, rg = 8 to > 26 days[11]
Glucose	17 days	300 K, Porto Novo, Bay of Bengal, more than value[11]
Glucose	25.5 days	300 K, Porto Novo, 2 st.from River Vellar, rg = 21 to > 30 days[11]

Glucose	45 days	300 K, distilled water[11]
Glucose	6.5 days	295 K, 5 stations at Southampton, rg = 3 - 10[11]
Glucose	80 hours	Bac uptake, Erken, phyt prod=500 mg C/(m2*d), 276 K, rg=60-100[2]
Glucose	9.5 days	300 K, 2 stations from Kille Backwater and mangrove swamp, rg = 9 -[11]
Glucose	90 hours	Bac uptake, Meretta Lake, phyt prod=125 mg C/(m2*d), artic, rg=5-175[12]
Glucose	900 hours	Bac uptake, Char lake, artic, phyt prod = 25 mg C/(m2*d), rg=43-1700 h[12]
Glutamate	12.5 hours	Bac uptake, rg=10-15, Upper Klamath L, ice surf, winter[4]
Glycollic acid	950 hours	Bac uptake, W Blue L., Manitoba, ice surface, winter, prod=40 mg C/(m2*d)[13]
Lactic acid	303 hours	Bac uptake, W Blue L., Manitoba, ice surface, winter, prod=40 mg C/(m2*d)[3]
Lysine	12.5 hours	Bac uptake, rg=10-15, Upper Klamath L, ice surf, winter[4]
Malic acid	8 hours	Bac uptake, W Blue L., Manitoba, ice surface, winter, prod=40 mg C/(m2*d)[3]
Methyl chloride	200	Solvolysis[10]
Methyl chloride	230	Hydrolysis[10]
Methyl chloroform	25	Solvolysis, by dehydrochlorination[10]
Methylene dichloride	22E4	Solvolysis[10]
Methylene dichloride	23E7	Hydrolysis[10]
Phenanthrene	35.9 mg compound remaining after 5 h	Initial=100 mg, natural light, water+acetone (or tetrachloride)[8]
Phenanthrene	57.0 mg compound remaining after 5 h	Initial=100 mg, natural light, water+acetone (or tetrachloride)[8]
Proline	80 hours	Bac uptake, rg=25-125, Upper Klamath L, ice surf, winter[4]
Pyrene	0.68 hours	Degradation by O3 = 10E-4 M[5]
Pyrene	41 min	Water with O3 = 1E-4 m, 298 K[5]
Pyrene	89.1 mg compound remaining after 5 h	Initial=100 mg, natural light, water+acetone (or tetrachloride)[8]
Pyrene	94.6 mg compound remaining after 5 h	Initial=100 mg, natural light, water+acetone (or tetrachloride)[8]
Pyruvic acid	300 hours	Bac uptake, W Blue L., Manitoba, ice surface, winter, prod=40 mg C/(m2*d)[3]
Rubrene	0.02 hours	Degradation by singlet oxygen[5]
Succinic acid	33 hours	Bac uptake, W Blue L., Manitoba, ice surface, winter, prod=40 mg C/(m2*d)[3]
Sulfur	10000 min	pH = 7.5-7.7, 281-282 K, no order of reaction[14]
Sulfur	130 min	pH > 11, order of reaction with respect to sulfide = 1[14]
Sulfur	175 min	pH = 7.5-7.8, 283 K, order of reaction with respect to sulfide = 1[14]
Sulfur	180 min	pH = 8, 296-297 K, order of reaction with respect to sulfide = 1[14]

Sulfur	24 min	pH = 8.2, 298 K, order of reaction with respect to sulfide = i[14]
Sulfur	280 min	pH = 8, 296-297 K, order of reaction with respect to sulfide = 1[14]
Sulfur	3000 min	pH = 7.94, 298 K, order of reaction with respect to sulfide = 1.34[14]
Sulfur	336 min	pH = 7.5-7.8, 283 K, order of reaction with respect to sulfide = 1[14]
Sulfur	370 min	pH = 8, 283 K, order of reaction with respect to sulfide = 1[14]
Sulfur	570 min	pH = 8, 283 K, order of reaction with respect to sulfide = 1[14]
Sulfur	600 min	pH = 7.6-7.7, 281-282 K, no order of reaction[14]
Sulfur	861 min	pH = 7.5-.78, 283 K, order of reaction with respect to sulfide = 1[14]
Tetracene	5 hours	Degradation by singlet oxygen, less than value[5]

Source: 1.Allen, H.L. 1968. Ecology 49, 346-349.
2.Batterton, J.C. Jr and C. van Baalen. 1971. Arch. Mikrobiol. 76, 151-165.
3.Ward, F.J. and G.G.C. Robinson. 1974. J. Fish. Res. bd Canada 31, 977-1005.
4.Burnison, B.K. and R.Y. Morita. 1974. Appl. Microbiol. 27, 488-495.
5.Radding, S.B., T. Mill, Gould, Liu, Johnson, Bomberger and Fojo. 1976. The Environmental Fate of Selected Polynuclear Aromatic Hydrocarbons. EPA 560/5-75-009. Wash DC.
6.Nagata, S. and G. Kondo. 1977. Proc of 1977 Oil Spill Conf. (Prevention, Behavior, Control, Cleanup) Wash D.C. American Petroleum Institute.
7.Li, G-C. and G.T Felbeck. 1972. Soil Sci. 113, 140.
8.Keckes, S. and J.K. Miettinen. 1970. Proc FAO Conf. Rome. Dec 9-18.
9.Paris, D.F., D.L. Lewis and N.L. Wolfe. 1975. Environ Sci Technol. 9, 135.
10.Pearson, C.R. 1982. In: O. Hutzinger (ed.): Environmental Chemistry. Springer-Verlag. Volume 3, part B. Berlin.
11.Austin, J. and J.D. Redmond. 1972. Mar. Poll. Bull. 3(3), pp 45-47.
12.Chimiklis, P.E. and E.P. Karlander. 1973. Plant. Physiol. 51, 48-56.
13.Schrpensel, H.W., F. Pietig and E. Kruse. 1976. In: J.O. Nriagu (ed.): Environmental Biochemistry, Vl 2. Ann Arbor Science. Ann Arbor. Mich. 597-607.
14.Almgren, T. and I. Hagstroem. 1974. Water Res. 8(7), pp 395-400.

2-073 Chemical compounds
Humic acid composition

Species	Value	Condition
Acidic amino acids	25.5 mg/g	Rg = 25.5 - 25.6, amino acid content[1]
Alcoholic hydroxyl	1.5 meq/g dry org. matter	Rg = 0.0 - 3.0, functional group content[1]
Aromatic amino acids	9.3 mg/g	Rg = 8.7 - 9.8, amino acid content[1]
Basic amino acids	16.8 mg/g	Rg = 15.2 - 18.3, amino acid content[1]
C	58.8 % by atoms	Rg = 50.2 - 63.8, elemental composition[1]
C / H ratio	9.3 by atoms	Rg = 7.8 - 10.8, elemental composition[1]
C / N ratio	12.7 by atoms	Rg = 11.4 - 13.9, elemental composition[1]
Carbonyl	4.5 meq/g dry org. matter	Rg = 3.0 - 6.0, functional group content[1]
Carboxyl	3.5 meq/g dry org. matter	Rg = 2.0 - 5.0, functional group content[1]
H	6.7 % by atoms	Rg = 5.5 - 7.0, elemental composition[1]
Molecular weight range	13.5 % of dry organic matter	Rg = 5 - 22, weight = 10000 - 100000[1]
Molecular weight range	18.5 % of dry organic matter	rg = 12 - 25, weight = 100000 - 200000[1]
Molecular weight range	34 % of dry organic matter	Rg = 27 - 41, weight = less than 10000[1]
Molecular weight range	37.5 % of dry organic matter	Rg = 25 - 40, weight more than 200000[1]
N	4.6 % by atoms	Rg = 3.6 - 5.6, elemental composition[1]
Neutral amino acids	43.2 mg/g	Rg = 41.6 - 44.8, amino acid content[1]

O	31.5 % by atoms	Rg = 24.1 - 40.6, elemental composition[1]
Phenolic hydroxyl	1.3 meq/g dry organic matter	Rg = 0.0 - 2.5, functional group content[1]
Quinone	2.3 meq/g dry organic matter	Rg = 1.8 - 2.8, functional group content[1]
Sulfur amino acids	3.7 mg/g	Rg = 3.7 - 3.8, amino acid content[1]
Total acidity	4.5 meq/g dry organic matter	Rg = 2.0 - 7.0, functional group content[1]
Total amino acids	99 mg/g	Rg = 98 - 100, amino acid content[1]

Source: 1.Rashid, M.A. 1974. In: Offshore geology of eastern Canada. Geol. Surv. Can. paper 74-30, I, 123-132.

2-074 Chemical compounds
Isoelectric point

Species	Value	Condition
Al(OH)3	5.0	pH value for zero point of charge, alpha-Al(OH)3[1]
Al2O3	9.1	pH value for zero point of charge, alpha-Al2O3[1]
Albite	2.0	pH value for zero point of charge[1]
CuO	9.5	pH value for zero point of charge[1]
Fe(OH)3	8.5	pH value for zero point of charge, amorph[1]
Fe2O3	6.7	pH value for zero point of charge, gamma-Fe2O3[1]
Fe3O4	6.5	pH value for zero point of charge[1]
Kaolinite	4.6	pH value for zero point of charge[1]
MgO	12.4	pH value for zero point of charge[1]
MnO2	3.25	pH value for zero point of charge, rg = 2 - 4.5[1]
Montmorillonite	2.5	pH value for zero point of charge[1]
SiO2	2.0	pH value for zero point of charge[1]
ZrSiO4	5	pH value for zero point of charge[1]

Source: 1.Stumm, W. and J.J. Morgan, 1970. Aquatic Chemistry. Wiley N.Y.

2-075 Chemical compounds
Loss by runoff

Species	Value	Condition
Carbon	0.26 kg/ha	Runoff = 17 cm, Jun 69-Oct 69, undisturbed forest, fine part organic[1]
Carbon	0.82 kg/ha	Runoff = 36 cm, Jun 69-Oct 69, deforested, fine part organic C[1]
Carbon	1.13 kg/ha	Runoff = 86 cm, Oct 67-May 68, undisturbed forest, fine part organic[1]
Carbon	1.45 kg/ha	Runoff = 90 cm, Jun 68-May 69, undisturbed forest, fine part organic[1]
Carbon	1.94 kg/ha	Runoff = 92 cm, Oct 67-May 68, deforested, fine part organic C[1]
Carbon	2.58 kg/ha	Runoff = 116 cm, Jun 68-May 69, deforested, fine part organic C[1]
Dissolved organic matter	4.18 kg/ha	Runoff = 17 cm, Jun 69-Oct 69, undisturbed forest[1]
Dissolved organic matter	4.84 kg/ha	Runoff = 116 cm, Jun 68-May 69, deforested[1]
Dissolved organic matter	5.89 kg/ha	Runoff = 36 cm, Jun 69-Oct 69, deforested[1]
Dissolved organic matter	7.76 kg/ha	Runoff = 86 cm, Oct 67-May 68, undisturbed forest[1]
Dissolved organic matter	8.51 kg/ha	Runoff = 90 cm, Jun 68-May 69, undisturbed forest[1]
Dissolved organic matter	9.84 kg/ha	Runoff = 92 cm, Oct 67-May 68, deforested[1]
Inorganic particles	133 g P/ha	Particles > 1 mm, 2 July 68-30 Sep 68, deforested[1]
Inorganic particles	47 g P/ha	Particles > 1 mm, 9 Oct 67-22 May 68, deforested[1]

Inorganic particles	6 g P/ha	Particles > 1 mm, 9 Oct 67-22 May 68, undisturbed forest[1]
Inorganic particles	8.8 g P/ha	Particles > 1 mm, 2 July 68-30 Sep 68, undisturbed forest[1]
Nitrate	0.90 kg N/ha	Citrus, 1973, 0.53 % of supply, in drainage, FL[2]
Nitrate	12.30 kg N/ha	Citrus, 1975, 7.26 % of supply, in drainage, FL[2]
Nitrate	2.95 kg N/ha	Citrus, 1974, 1.74 % of supply, in drainage, FL[2]
Organic matter from Imperata grassland	9.00 g organic matter/(m2*year)	Mindano in Philippines, 1968[3]
Organic matter from corn swidden	89.07 g organic matter/(m2*year)	Mindano in Philippines, 1968, cropping period, 2 y old corn[3]
Organic matter from logged-over forest	83.61 g organic matter/(m2*year)	Mindano in Philippines, 1968[3]
Organic matter from new Abaca plantation	10.57 g organic matter/(m2*year)	Mindano in Philippines, 1968[3]
Organic matter from new corn swidden	11.67 g organic matter/(m2*year)	Mindano in Philippines, 1968, inter-cropping period[3]
Organic matter from new corn swidden	33.07 g organic matter/(m2*year)	Mindano in Philippines, 1968, cropping period[3]
Organic matter from new corn swidden	44.74 g organic matter/(m2*year)	Mindano in Philippines, 1968, total loss[3]
Organic matter from old Abaca plantation	13.28 g organic matter/(m2*year)	Mindano in Philippines, 1968, 10 y forest[3]
Organic matter from primary forest	4.50 g organic matter/(m2*year)	Mindano in Philippines, 1968[3]
Organic matter from rice swidden	1172.69 g organic matter/(m2*year)	Mindano in Philippines, 1968, cropping period, 12 y old rice swidden[3]
Organic matter from rice swidden	1212.74 g organic matter/(m2*year)	Mindano in Philippines, 1968, total loss, 12 y old rice swidden[3]
Organic matter from rice swidden	20.88 g organic matter/(m2*year)	Mindano in Philippines, 1968, cropping period[3]
Organic matter from rice swidden	27.86 g organic matter/(m2*year)	Mindano in Philippines, 1968, total loss[3]
Organic matter from rice swidden	39.74 g organic matter/(m2*year)	Mindano in Philippines, 1968, inter-cropping period, 12 y old rice swidden[3]
Organic matter from rice swidden	6.98 g organic matter/(m2*year)	Mindano in Philippines, 1968, inter-cropping period[3]
Organic matter from softwood tree fallow	6.52 g organic matter/(m2*year)	Mindano in Philippines, 1968[3]
Organic particles	0.6 g P/ha	Particles > 1 mm, 2 July 68-30 Sep 68, undisturbed forest[1]
Organic particles	1.2 g P/ha	Particles > 1 mm, 1 Oct 68-31 May 69, undisturbed[1]
Organic particles	1.7 g P/ha	Particles > 1 mm, 22 May 68-2 July 68, undisturbed forest[1]
Organic particles	15 g P/ha	Particles > 1 mm, 1 Oct 68-31 May 69, deforested[1]
Organic particles	28 g P/ha	Particles > 1 mm, 22 May 68-2 July 68, deforested[1]
Organic particles	3.5 g P/ha	Particles > 1 mm, 9 Oct 67-22 May 68, undisturbed forest[1]
Organic particles	37 g P/ha	Particles > 1 mm, 9 Oct 67-22 May 68, deforested[1]

Organic particles	7 g P/ha	Particles > 1 mm, 2 July 68-30 Sep 68, deforested[1]
Phosphate	0.51 kg P/ha	Citrus, 1975, 0.81 % of supply, in drainage, FL[2]
Phosphate	0.82 kg P/ha	Citrus, 1973, 4.44 % of supply, in drainage, FL[2]
Phosphate	0.97 kg P/ha	Citrus, 1974, 5.25 % of supply, in drainage, FL[2]
Total dissolved PO4 - P	15.62 g/ha	Runoff = 92 cm, Oct 67-May 68, deforested, incl fine part P[1]
Total dissolved PO4 - P	20.49 g/ha	Runoff = 116 cm, Jun 68-May 69, deforested, incl fine part P[1]
Total dissolved PO4 - P	9.22 g/ha	Runoff = 90 cm, Jun 68-May 69, undisturbed, incl fine part P[1]
Total dissolved PO4 - P	9.86 g/ha	Runoff = 86 cm, Oct 67-May 68, undisturbed, incl fine part P[1]

Source: 1.Hobbie, J.E. and G.E. Likens. 1973. Limnol. Oceanogr., 734-742.
2.Mansell, R.S. et al. 1977. Fertilizer and Pesticide Movements from Citrus groves in Florida flatwood Soils. EPA-600/2-77-177.
3.Kellman, M.C. 1969. J. Trop. Geogr. 28, 40-56.

2-076 Chemical compounds
Maximum light scattering

Species	Value	Condition
1,5-Hexadiene	40E4 1/meter	Aerosol formation from hydrocarbons[1]
1,6-Heptadiene	160E4 1/meter	Aerosol formation from hydrocarbons[1]
1,7-Octadiene	180E4 1/meter	Aerosol formation from hydrocarbons[1]
1-Heptene	1E4 1/meter	Aerosol formation from hydrocarbons[1]
1-Octene	1E4 1/meter	Aerosol formation from hydrocarbons[1]
2,6-Dimethylheptane	1E4 1/meter	Aerosol formation from hydrocarbons[1]
2,6-octadiene	1E4 1/meter	Aerosol formation from hydrocarbons[1]
2-Methyl-1,5-hexadiene	110E4 1/meter	Aerosol formation from hydrocarbons[1]
5-Methyl-1-hexene	1E4 1/meter	Aerosol formation from hydrocarbons[1]
Cyclohexene	90E4 1/meter	Aerosol formation from hydrocarbons[1]
Ethylbenzene	1E4 1/meter	Aerosol formation from hydrocarbons[1]
Glutaraldehyde	0E4 1/meter	Aerosol formation from hydrocarbons[2]
Mesitylene	1E4 1/meter	Aerosol formation from hydrocarbons[1]
alpha-Pinene	180E4 1/meter	Aerosol formation from hydrocarbons[1]
alpha-Xylene	8E4 1/meter	Aerosol formation from hydrocarbons[1]
trans-4-Octene	1E4 1/meter	Aerosol formation from hydrocarbons[1]

Source: 1.Sandberg, D.V., S.G. Pickford and E.F. Darley. 1975. J. Air Pollut. Control Asssoc. 21, 629-635. 2.OBrian, R.J, J.R. Holmes and A.H. Bockian. 1975. Environ. Sci. Technol 9, 568-576.

2-077 Chemical compounds
Overall mass transfer coefficient

Species	Value	Condition
2,2,4-trimethylpentane	0.124 meter/hour	In water, 298 K, liquid[1]
Aldrin	3.72E-3 meter/hour	In water, 298 K, liquid[1]
Benzene	0.144 meter/hour	In water, 298 K, liquid[1]
Biphenyl	0.092 meter/hour	In water, 298 K, liquid[1]
Cumene	0.119 meter/hour	In water, 298 K, liquid[1]
DDT	9.34E-3 meter/hour	In water, 298 K, liquid[1]
Dieldrin	5.33E-5 meter/hour	In water, 298 K, liquid[1]
Hg	0.092 meter/hour	In water, 298 K, liquid[1]
Lindane	1.5E-4 meter/hour	In water, 298 K, liquid[1]
N-octane	0.124 meter/hour	In water, 298 K, liquid[1]
Naphthalene	0.096 meter/hour	In water, 298 K, liquid[1]
O-xylene	0.123 meter/hour	In water, 298 K, liquid[1]
PCB	0.057 meter/hour	In water, 298 K, liquid, aroclor 1242[1]
PCB	0.067 meter/hour	In water, 298 K, liquid, aroclor 1254[1]
PCB	0.067 meter/hour	In water, 298 K, liquid, aroclor 1260[1]
PCB	0.072 meter/hour	In water, 298 K, liquid, aroclor 1248[1]
Toluene	0.133 meter/hour	In water, 298 K, liquid[1]

Source: 1.Mackay, D. and P.J. Leinonen, 1975. Environ. Sci. Tech. 9(13), pp 1178-1180.

2-078 Chemical compounds
Partion coefficient

Species	Value	Condition
1,2 Dichlorobenzene	3.55 = log P	Water / air, log P (O/w)[1]
1,2,3 Trichlorobenzene	4.11 = log P	Water / air, log P (O/w)[1]
1,2,3,4 Trteachlorobenzene	4.46 = log P	Water / air, log P (O/w)[1]
1,2,3,5 Tetrachlorobenzene	4.71 = log P	Water / air, log P (O/w)[1]
1,2,4 Trichlorobenzene	4.08 = log P	Water / air, log P (O/w)[1]
1,2,4,5 Tetrachlorobenzene	4.52 = log P	Water / air, log P (O/w)[1]
1,3 dichlorobenzene	3.44 = log P	Water / air, log P (O/w)[1]
1,3,5 Trichlorobenzene	4.15 = log P	Water / air, log P (O/w)[1]
1,4 dichlorobenzene	3.38 = log P	Water / air, log P (O/w)[1]
1-Chloro 4-Nitrobenzene	2.4 = log P	Water / air, log P (O/w)[1]
2,3-Dichloro aniline	2.8 = log P	Water / air, log P (O/w)[1]
2-Chloro aniline	1.9 = log P	Water / air, log P (O/w)[1]
2-Chloro toluene	3.42 = log P	Water / air, log P (O/w)[1]
3,4-Dichloro aniline	2.7 = log P	Water / air, log P (O/w)[1]
3-Chloro aniline	1.9 = log P	Water / air, log P (O/w)[1]
4-Chloro aniline	1.8 = log P	Water / air, log P (O/w)[1]
Chlorobenzene	2.84 = log P	Water / air, log P (O/w)[1]
Hexachlorobenzene	5.5 = log P	Water / air, log P (O/w)[1]
Pentachlorobenzene	5.28 = log P	Water / air, log P (O/w)[1]

Source: 1.Dobbs, R.A. and J.M. Cohen. 1980. Carbon adsorption isotherms for Toxic organics. EPA 600/8-80-023.

2-079 Chemical compounds
Partition coefficient

Species	Value	Condition
1,4 Dichlorobenzene	3.38 = log(kow)	Octanol/water, log value, 298 K[1]
1-methyl-naphthalene	3.86 = log(kow)	Octanol/water, log value, 298 K[1]
2,6-dimethyl-naphthalene	4.38 = log(kow)	Octanol/water, log value, 298 K[1]
2-methyl-naphthalene	4.11 = log(kow)	Octanol/water, log value, 298 K[1]
Benzene	2.13 = log(kow)	Octanol/water, log value, 298 K[1]
Carbon tetachloride	1.1 water/air	298 K, water/air, w/v per w/v, CCl4[2]
Chlorobenzene	2.84 = log(kow)	Octanol/water, log value, 298 K[1]
Chloroform	8.6 water/air	298 K, water/air, w/v per w/v, CHCl3[2]
Diphenyl	4.03 = log(kow)	Octanol/water, log value, 298 K[1]
Ethyl benzene	3.15 = log(kow)	Octanol/water, log value, 298 K[1]
Ethylene dichloride	26.4 water/air	298 K, water/air, w/v per w/v, CH2ClCH2Cl[2]
Hexachlorobutadiene	0.97 water/air	298 K, water/air, w/v per w/v, C4Cl6[2]
Isopropyl benzene	3.43 = log(kow)	Octanol/water, log value, 298 K[1]
Methyl chloride	3.3 water/air	298 K, water/air, w/v per w/v, CH3Cl[2]

Methyl chloroform	0.71 water/air	298 K, water/air, w/v per w/v, $CH3CCl3$[2]
Methylene chloride	8.1 water/air	298 K, water/air, w/v per w/v, $CH2Cl2$[2]
Naphthalene	3.37 = log(kow)	Octanol/water, log value, 298 K[1]
Nitrobenzene	1.85 = log(kow)	Octanol/water, log value, 298 K[1]
Pentachloroethane	11.8 water/air	298 K, water/air, w/v per w/v, C2HCl5[2]
Perchloroethylene	1.22 water/air	298 K, water/air, w/v per w/v, C2Cl4[2]
Tetrachloroethane	6.3 water/air	298 K, water/air, w/v per w/v, C2H2Cl4 (sym)[2]
Toluene	2.69 = log(kow)	Octanol/water, log value, 298 K[1]
Trichloroethylene	2.74 water/air	298 K, water/air, w/v per w/v, CHClCCl2[2]
Vinyl chloride	0.02 water/air	298 K, water/air, w/v per w/v, CH2CHCl[2]
Vinylidene chloride	0.16 water/air	298 K, water/air, w/v per w/v, CH2CCl2[2]
o-Xylene	3.15 = log(kow)	Octanol/water, log value, 298 K[1]

Source: 1.D'Ans-Lax. 1967. Taschenbuch fur Chemiker und Fysiker. Springer-Verlag. Berlin.
2.McConnell, G., D.M. Ferguson and C.R. Pearson. 1975. Endaevour 34, 13.

2-080 Chemical compounds
Residence time in stratosphere

Species	Value	Condition
Aerosols	1.5 years	Varies greatly, sink; diffusion into troposphere, rg = 1 - 2[1]
CO2	2 years	Varies greatly, sink; diffusion into troposphere[1]
Ozone	1.05 years	Varies greatly, sink; reduction to oxygen, rg = 0.1 - 2[1]
Water vapor	2 years	Varies greatly, sink; diffusion into tropopause, form.of nitric acid[1]

Source: 1.Almquist, E. 1974. Ambio, 3(5), pp 161-167.

2-081 Chemical compounds
Residence time in troposphere

Species	Value	Condition
Aerosols	10 days	Varies greatly, sink; rain, snow, hail, large aerosols[1]
Aerosols	30 days	Varies greatly, Small aerosols, sinks: Rain, snow, hail[1]
CO	1.55 years	Varies greatly, sink; prob. oxidation to CO2, rg = 0.1 - 3[1]
CO2	4 years	Varies greatly, sink; biol.absorption, photosynt., absorption in[1]
Hydrogen sulfide	2 days	Varies greatly, sink; oxid.to sulphur dioxide, removal through[1]
Methane	57.5 years	Varies, sink; photochem.react. (carbon monoxide major sink?)[1]
NH3	3.5 days	Varies greatly, sink; NH3+sulf.diox. ->ammoniumsulf., ox.to nitrate,[1]
Nitric oxide	5 days	Varies greatly, sink; oxid.to nitrate, source:combust., biol.act.in[1]
Nitrogen dioxide	5 days	Varies greatly, sink; oxidation to nitrate[1]
Nitrous oxide	2.5 years	Varies greatly, sink; photodissipat.in strato., biol.action in soils,[1]
Ozone	2 months	Varies greatly, sink; reduction to oxygen, rg = 1 - 3[1]
SO2	4 days	Varies greatly, sink; oxidation to sulfates[1]
Water vapor	10 days	Varies greatly, sink; tropopause acts as a cold-trap[1]

Source: 1.Almquist, E. 1974. Ambio, 3(5), pp 161-167.

2-082 Chemical compounds
Solubility

Species	Value	Condition
(Hg)2SO4	0.06 g/100 g water	In water, 298 K, Hg(I) compound[1]
1,2 Dichlorobenzene	100 mg/l	In water, 298 K[2]
1,3 Dichlorobenzene	123 mg/l	In water, 298 K[2]
1,3,5 Trichlorobenzene	30 mg/l	o1[2]
1,4 Dichlorobenzene	87.2 mg/l	In water, 298 K[1]
1,5-Dimethylnaphthalene	2740 ug/liter	Distilled water at 298-300 K, No. aromatic rings=2[3]
1-Methylanthracene	11.1 ug/liter	Distilled water, 286.9 K, rg = +- 0.3[4]
1-Methylanthracene	14.5 ug/liter	Distilled water, 291.3 K, rg = +- 0.1[4]
1-Methylanthracene	19.1 ug/liter	Distilled water, 296.1 K, rg = +- 0.6[4]
1-Methylanthracene	24.2 ug/liter	Distilled water, 300.0 K, rg = +- 0.1[4]
1-Methylanthracene	32.1 ug/liter	Distilled water, 304.1 K, rg = +- 0.3[4]
1-Methylanthracene	7.06 ug/liter	Distilled water, 279.3 K, rg = +- 0.18[4]
1-Methylanthracene	8.48 ug/liter	Distilled water, 282.1 K, rg = +- 0.09[4]
1-Methylanthracene	9.43 ug/liter	Distilled water, 283.8 K, rg = +- 0.37[4]
1-Methylnaphthalene	25800 ug/liter	Distilled water at 298-300 K, No. aromatic rings=2[3]
1-Methylphenanthrene	0.3 mg/liter	Seawater, S=3.2 %, 295-298 K, No. aromatic rings=3, rg=+-0.1[4]
1-Methylphenanthrene	114 ug/liter	Distilled water, 281.9 K, rg = +- 4[4]
1-Methylphenanthrene	147 ug/liter	Distilled water, 287.0 K, rg = +- 1[4]
1-Methylphenanthrene	193 ug/liter	Distilled water, 292.2 K, rg = +- 1[4]
1-Methylphenanthrene	255 ug/liter	Distilled water, 297.1 K, rg = +- 5[4]
1-Methylphenanthrene	269 ug/liter	Distilled water at 298-300 K, No. aromatic rings=3[3]
1-Methylphenanthrene	304 ug/liter	Distilled water, 299.9 K, rg = +- 1[4]
1-Methylphenanthrene	355 ug/liter	Distilled water, 302.9 K, rg = +- 2[4]
1-Methylphenanthrene	95.2 ug/liter	Distilled water, 279.6 K, rg = +- 0.2[4]
1-methyl-naphthalene	28.5 mg/l	In water, 298 K[1]
10-Amylbenz(a)anthracene	0.8 ug/liter	Distilled water at 298-300 K, No. aromatic rings=4[3]
10-Butylbenz(a)anthracene	8 ug/liter	Distilled water at 298-300 K, No. aromatic rings=4[3]
10-Ethylbenz(a)anthracene	45 ug/liter	Distilled water at 298-300 K, No. aromatic rings=4[3]
10-Methylbenz(a)anthracene	55 ug/liter	Distilled water at 298-300 K, No. aromatic rings=4[3]
2',3,4-trichlorobiphenyl	0.078 ng/l	In water[5]
2,2',3,3',4,4',5,5'-octachloro-biphenyl	0.0070 mg/l	In water[5]
2,2',3,3'-tetrachlorobiphenyl	0.034 mg/l	In water[5]
2,2',3,5'-tetrachlorobiphenyl	0.170 mg/l	In water[5]
2,2',4,4',5,5'--hexachlorobiphenyl	0.0088 mg/l	In water[5]
2,2',4,4'-tetrachlorobiphenyl	0.068 mg/l	In water[5]

2,2',4,5,5'--pentachlorobiphenyl	0.031 mg/l	In water[5]
2,2',5,5'-tetrachlorobiphenyl	0.046 mg/l	In water[5]
2,2'-dichlorobiphenyl	1.50 mg/l	In water[5]
2,2'3,4,5'-pentachlorobiphenyl	0.022 mg/l	In water[5]
2,2,4-trimethylpentane	2.44 mg/l	In water, 298 K[6]
2,3',4',5-tetrachlorobiphenyl	0.041 mg/l	In water[5]
2,3',4,4'-tetrachlorobiphenyl	0.058 mg/l	In water[5]
2,3,6-Trimethylnaphthalene	1.7 mg/liter	Seawater, S=3.2 %, 295-298 K, rg=+-0.6[4]
2,4'-dichlorobiphenyl	1.88 mg/l	In water[5]
2,4,4'-trichlorobiphenyl	0.085 mg/l	In water[5]
2,4-dichlorobiphenyl	1.40 mg/l	In water[5]
2,6-Dimethylnaphthalene	2.4 mg/liter	Seawater, S=3.2 %, 295-298 K, rg =+-0.5[4]
2,6-diisopropyl-naphthalene	0.1 mg/l	In water, 298 K[1]
2,6-dimethyl-naphthalene	2.0 mg/l	In water, 298 K[1]
2-Ethylnaphthalene	8000 ug/liter	Distilled water at 298-300 K, No. aromatic rings=2[3]
2-Methylanthracene	21.3 ug/liter	Distilled water at 298-300 K, No. aromatic rings=3[3]
2-methyl-naphthalene	25.4 mg/l	In water, 298 K[1]
2-monochlorobiphenyl	5.9 mg/l	In water[5]
3,3',4,4'-tetrachlorobiphenyl	0.175 mg/l	In water[5]
3-monochlorobiphenyl	3.5 mg/l	In water[5]
4,4'-dichlorobiphenyl	0.07 mg/l	In water with humic acid extract[5]
4,4'-dichlorobiphenyl	0.08 mg/l	In water[5]
4,4'-dichlorobiphenyl	10.0 mg/l	In water with 1 % tween 80, F010, larger than value[5]
4,4'-dichlorobiphenyl	5.9 mg/l	In water with 0.1 % tween 80, F010[5]
4,5-Methylenephenanthrene	1100 ug/liter	Distilled water at 298-300 K, No. aromatic rings=3[3]
4-Chloro aniline	5300 mg/l	In water, 298 K[2]
4-monochlorobiphenyl	1.19 mg/l	In water[5]
5-Methylbenzo(a)pyrene	0.8 ug/liter	Distilled water at 298-300 K, No. aromatic rings=5[3]
9,10-Dimethylbenz(a)-anthracene	43 ug/liter	Distilled water at 298-300 K, No. aromatic rings=4[3]
Acenaphthalene	3470 ug/liter	Distilled water at 298-300 K, No. aromatic rings=2[3]
Adenine	0.0068 moles/liter	In water[7]
Alanine	1.665 moles/liter	In water[7]
Aldrin	0.105 ppm	293-298 K, rg = 0.01 - 0.2, in water[8]
Aldrin	0.2 mg/l	In water, 298 K[6]
Allyl chloride	100 mg/l	298 K, in water, CH2ClCHCH2[9]
Anthracene	12.7 ug/liter	Distilled water, 278.2 K, rg = +- 0.4[4]
Anthracene	17.5 ug/liter	Distilled water, 283.0 K, rg = +- 0.3[4]
Anthracene	22.2 ug/liter	Distilled water, 287.1 K, rg = +- 0.1[4]
Anthracene	29.1 ug/liter	Distilled water, 291.3 K, rg = +- 0.6[4]

Anthracene	37.2 ug/liter	Distilled water, 295.4 K, rg = +- 1.1[4]
Anthracene	43.4 ug/liter	Distilled water, 297.6 K, rg = +- 0.1[4]
Anthracene	45 ug/liter	Distilled water at 298-300 K, No. aromatic rings=3, rg=30-75[3]
Anthracene	55.7 ug/liter	Distilled water, 301.7 K, rg = +- 0.7[4]
Anthracene	75 mg/m3	In water at 298 K[10]
Asparagine	0.186 moles/liter	In water[7]
Aspartic acid	0.038 moles/liter	In water[7]
Asulam	5000.0 ppm	Purity = 40.3 %, in water, sodium salt[11]
Benz(a)anthracene	10 ug/liter	Distilled water at 298-300 K, No. aromatic rings=4, rg=9.4-11[3]
Benzene	1780 mg/l	In water, 298 K[6]
Benzene	1780 mg/l	In water, 298 K[1]
Benzo(a)anthracene	10 mg/m3	In water at 298 K[10]
Benzo(a)pyrene	2.0 ug/liter	Distilled water at 298-300 K, No. aromatic rings=5, rg=0.172-6.05[3]
Benzo(e)pyrene	4 ug/liter	Distilled water at 298-300 K, No. aromatic rings=5[3]
Biphenyl	7.48 mg/l	In water, 298 K[6]
Butralin	1.0 ppm	Purity = 61.9 %, in water, emulsifiable concentrate, as soduim salt[11]
CCl3F	110 mg/100 ml	293 K, in water[12]
Camphechlor	1.7 ppm	293-298 K, rg = 0.4 - 3.0, in water[8]
Carbon tetachloride	785 mg/l	298 K, in water, CCl4[9]
Chlordane	0 ppm	293-298 K, very low, in water[8]
Chlorobenzene	472 mg/l	In water, 298 K[1]
Chlorobenzene	600 mg/l	In water, 298 K[2]
Chloroform	8200 mg/l	298 K, in water, CHCl3[9]
Chrysene	1.8 ug/liter	Distilled water at 298-300 K, No. aromatic rings=4, rg=1.5-6[3]
Chrysene	6 mg/m3	In water at 298 K[10]
Crude oil with aromatic rings	0.0009 mg crude oil/liter water	Mixing 1 part crude Kuwait oil with 9 parts seawater, 3 ring[13]
Crude oil with aromatic rings	0.001 mg crude oil/liter water	Mixing 1 part crude S Louisiana oil with 9 parts seawater, 3 ring[13]
Crude oil with aromatic rings	0.02 mg crude oil/liter water	Mixing 1 part crude S Louisiana oil with 9 parts seawater, 2 ring[13]
Crude oil with aromatic rings	0.11 mg crude oil/liter water	Mixing 1 part crude Kuwait oil with 9 parts seawater, 2 ring[13]
Crude oil with aromatic rings	3 mg crude oil/liter water	Mixing 1 part crude S Louisiana oil with 9 parts seawater, 1 ring[13]
Crude oil with aromatic rings	5 mg crude oil/liter water	Mixing 1 part crude Kuwait oil with 9 parts seawater, 1 ring[13]
Cumene	50 mg/l	In water, 298 K[6]
Cytosine	0.064 moles/liter	In water[7]
DDT	0.0205 ppm	293-298 K, rg = 0.001 - 0.04, in water[8]
DDT	1.2E-3 mg/l	In water, 298 K[6]
Decachlorobiphenyl	0.015 mg/l	In water[5]
Dibenz(ac)anthracene	0.55 ug/liter	Distilled water at 298-300 K, No. aromatic rings=5, rg=0.5-0.6[3]

Dicamba	4500.0 ppm	Purity = 49.5 %, in water, dimethylamine salt[11]
Dieldrin	0.175 ppm	293-298 K, rg = 0.1 - 0.25, in water[8]
Dieldrin	0.25 mg/l	In water, 298 K[6]
Diphenyl	7.0 mg/l	In water, 298 K[1]
Endrin	0.175 ppm	293-298 K, rg = 0.1 - 0.25, in water[8]
Ethyl benzene	152 mg/l	In water, 298 K[1]
Ethyl bromide	9100 mg/l	298 K, in water, C2H5Br[9]
Ethyl chloride	5740 mg/l	298 K, in water, C2H5Cl[9]
Ethylene dibromide	4000 mg/l	298 K, in water, CH2BrCH2Br[9]
Ethylene dichloride	8800 mg/l	298 K, in water, CH2ClCH2Cl[9]
Fatty acid	0.0000011 moles/liter	In water[7]
Flouranthene	240 ug/liter	Distilled water at 298-300 K, No. aromatic rings=3, rg=206-265[3]
Fluometuron	85.0 ppm	Purity = 83.5 %, in water, wettable powder[11]
Fluoranthene	0.1 mg/liter	Seawater, S=3.2 %, 295-298 K, No. aromatic rings=3, rg=+-0.06[4]
Fluoranthene	265 mg/m3	In water at 298 K[10]
Fluorene	0.8 mg/liter	Seawater, S=3.2 %, 295-298 K, No. aromatic rings=2, rg=+-0.2[4]
Fluorene	1685 ug/liter	Distilled water at 298-300 K, No. aromatic rings=2[3]
Glucose	3.452 moles/liter	In water[7]
Glutamic acid	0.058 moles/liter	In water[7]
Glycine	2.900 moles/liter	In water[7]
Guanine	0.000033 moles/liter	In water[7]
Heptachlor	0 ppm	293-298 K, in water, very low[8]
Hexachlorobenzene	0.007 mg/l	In water, 298 K[2]
Hexachlorobutadiene	2 mg/l	298 K, in water, C4Cl6[9]
Hexachloroethane	50 mg/l	298 K, in water, C2Cl6[9]
Hg	3E-2 mg/l	In water, 298 K[6]
Hg2Br2	4E-6 g/100 g water	In water, 298 K, Hg(I) compound[1]
Hg2Cl2	2E-4 g/100 g water	In water, 298 K, Hg(I) compound[1]
Hg2I2	2E-8 g/100 g water	In water, 298 K, Hg(I) compound[1]
HgBr2	0.62 g/100 g water	In water, 298 K, Hg(II) compound[1]
HgCl2	6.6 g/100 g water	In water, 293 K, Hg(II) compound[1]
HgI2	6E-3 g/100 g water	In water, 298 K, Hg(II) compound[1]
HgO	5.3E-3 g/100 g water	In water, 298 K, Hg(II) compound, yellow and red HgO[1]
HgS	1E-6 g/100 g water	In water, 291 K, Hg(II) compound, alpha-HgS[1]
Histidine	0.262 moles/liter	In water[7]
Isoleucine	0.303 moles/liter	In water[7]
Isopropyl benzene	50 mg/l	In water, 298 K[1]
Leptophos	0.03 ppm	Purity = 0.321 g/ml, in water, wettable powder[11]
Leucine	0.181 moles/liter	In water[7]
Lindane	7.3 mg/l	In water, 298 K[6]

Lindane	8.5 ppm	293-298 K, in water, rg = 7.0 - 10.0[8]
Methazole	1.5 ppm	Purity = 88.5 %, in water, wettable powder[11]
Methyl bromide	900 mg/l	298 K, in water, CHBr3[9]
Methyl chloride	7250 mg/l	298 K, in water, CH3Cl[9]
Methyl chloroform	480 mg/l	298 K, in water, CH3CCl3[9]
Methylene chloride	13200 mg/l	298 K, in water, CH2Cl2[9]
N-octane	0.66 mg/l	In water, 298 K[6]
Naphthacene	1.25 ug/liter	Distilled water at 298-300 K, No. aromatic rings=4, rg=1.0-1.5[3]
Naphthalene	1.5 mg/m3	In water at 298 K[10]
Naphthalene	12500 mg/m3	In water at 298 K[10]
Naphthalene	21.0 mg/liter	Seawater, S=3.2 %, 295-298 K, No. aromatic rings=2, rg=20-22[3]
Naphthalene	22000 ug/liter	Distilled water at 298-300 K, No. aromatic rings=2, rg=12500-34000[3]
Naphthalene	31.7 mg/l	In water, 298 K[1]
Naphthalene	33 mg/l	In water, 298 K[6]
Nitrobenzene	205.7 mg/l	In water, 298 K[1]
Norflurazon	28.0 ppm	Purity = 88.5 %, in water, wettable powder[11]
O-xylene	175 mg/l	In water, 298 K[6]
Oxadiazon	0.7 ppm	Purity = 0.223 g/ml, in water, emulsifiable concentrate[11]
Oxygen	23.1 ml O2/liter	313 K, volume O2 (ml) dissolved in 1 litre of water[10]
Oxygen	24.4 ml O2/liter	308 K, volume O2 (ml) dissolved in 1 litre of water[10]
Oxygen	26.1 ml O2/liter	303 K, volume O2 (ml) dissolved in 1 litre of water[10]
Oxygen	28.3 ml O2/liter	298 K, volume O2 (ml) dissolved in 1 litre of water[10]
Oxygen	31.0 ml O2/liter	293 K, volume O2 (ml) dissolved in 1 litre of water[10]
Oxygen	34.2 ml O2/liter	288 K, volume O2 (ml) dissolved in 1 litre of water[10]
Oxygen	38.0 ml O2/liter	283 K, volume O2 (ml) dissolved in 1 litre of water[10]
Oxygen	42.9 ml O2/liter	278 K, volume O2 (ml) dissolved in 1 litre of water[10]
Oxygen	48.9 ml O2/liter	273 K, volume O2 (ml) dissolved in 1 litre of water[10]
PBB	20 ug/l	In water, firemaster BP-6, rg = 10 - 30 ug/l[14]
PCB	14 ug/l	In water, aroclor 1260, rg = 3 - 25 ug/l[14]
PCB	41 ug/l	In water, aroclor 1254, rg = 12 - 70 ug/l[14]
PCB	4250 ug/l	In water, aroclor 1221, rg = 3500 - 5000 ug/l[14]
PCB	450 ug/l	In water, aroclor 1242, rg = 200 - 700 ug/l[14]
PCB	560 ug/l	In water, aroclor 1016, rg = 220 - 900 ug/l[14]
PCB	77 ug/l	In water, aroclor 1248, rg = 54 - 100 ug/l[14]
Pb(OH)2PbCl2	-68 = log(K)	298 K, water, M units, solubility product, mineral = fiedlerite[15]
Pb(UO2)2(PO4)2*4H2O	-47.7 = log(K)	298 K, water, M units, solubility product, mineral = przhevalskite[15]
Pb2(UO2)3(PO4)2(OH)4*3H20	-35.3 = log(K)	298 K, water, M units, solubility product, mineral = demonite[15]
Pb2Cu(PO4)(OH)3*3H2O	-9.3 = log(K)	298 K, water, M units, solubility product, mineral = tsumebite[15]
Pb2UO2(PO4)*2H2O	-45.5 = log(K)	298 K, water, M units, solubility product, mineral = parsonite[15]
Pb3(CO3)2(OH)2	-17.0 = log(K)	298 K, water, M units, solubility product, mineral = hydrocerussite[15]

Pb3(PO4)2	-44.3 = log(K)	298 K, water, M units, solubility product[15]
Pb4O(PO4)2	-36.8 = log(K)	298 K, water, M units, solubility product[15]
Pb5(PO4)3Cl	-84.4 = log(K)	298 K, water, M units, solubility product, mineral = Chloropyromorphite[15]
PbAl3(PO4)(OH)6*SO4	-15.0 = log(K)	298 K, water, M units, solubility product, mineral = hindalite[15]
PbAl3(PO4)2(OH)5*H20	-29.2 = log(K)	298 K, water, M units, solubility product, mineral = plumbogummite[15]
PbCO3	-12.8 = log(K)	298 K, water, M units, solubility product, mineral = cerussite[15]
PbCO3*PbCl2	-19.8 = log(K)	298 K, water, M units, solubility product, mineral = phosgenite[15]
PbCO3*PbO	-0.5 = log(K)	298 K, water, M units, solubility product[15]
PbCl2	-4.77 = log(K)	298 K, water, M units, solubility product, mineral = cotunnite[15]
PbClOH	-13.7 = log(K)	298 K, water, M units, solubility product, mineral = paralaurionite[15]
PbF2	6.15 = log(K)	298 K, water, M units, solubility product[15]
PbFCl	8.62 = log(K)	298 K, water, M units, solubility product. mineral = matlockite[15]
PbFe3(PO4)(OH)6*SO4	-28.6 = log(K)	298 K, water, M units, solubility product, mineral = corkite[15]
PbHPO4	-11.4 = log(K)	298 K, water, M units, solubility product, mineral = phosphate-schulterite[15]
PbS	-28.1 = log(K)	298 K, water, M units, solubility product, mineral = galena[15]
PbSO4	-7.72 = log(K)	298 K, water, M units, solubility product, mineral = anglesite[15]
PbSO4*2 PbO	10.9 = log(K)	298 K, water, M units, solubility product[15]
PbSO4*3 PbO	22.1 = log(K)	298 K, water, M units, solubility product[15]
PbSO4*PbO	-0.26 = log(K)	298 K, water, M units, solubility product, mineral = lanarkite[15]
PbSeO4	-0.26 = log(K)	298 K, water, M units, solubility product, mineral = kerstenite[15]
PbSiO3	7.13 = log(K)	298 K, water, M units, solubility product, mineral = alamosire[15]
PbSiO4	19.8 = log(K)	298 K, water, M units, solubility product[15]
Pentachloroethane	340 mg/l	298 K, in water, C2HCl5[9]
Perchloroethylene	150 mg/l	298 K, in water, C2Cl4[9]
Perylene	0.5 ug/liter	Distilled water at 298-300 K, No. aromatic rings=5, less than value[3]
Phenanthrene	0.65 mg/liter	Seawater, S=3.2 %, 295-298 K, No. aromatic rings=3, rg=0.6-0.71[4]
Phenanthrene	1200 ug/liter	Distilled water at 298-300 K, No. aromatic rings=3, rg=1002-1600[3]
Phenanthrene	1277 ug/liter	Distilled water, 302.9 K, rg = +- 11[4]
Phenanthrene	1600 mg/m3	In water at 298 K[10]
Phenanthrene	432 ug/liter	Distilled water, 281.5 K, rg = +- 4[4]
Phenanthrene	468 ug/liter	Distilled water, 283.0 K, rg = +- 2[4]
Phenanthrene	512 ug/liter	Distilled water, 285.5 K, rg = +- 1[4]
Phenanthrene	601 ug/liter	Distilled water, 288.0 K, rg = +- 7[4]
Phenanthrene	816 ug/liter	Distilled water, 294.0 K, rg = +- 8[4]
Phenanthrene	995 ug/liter	Distilled water, 297.3 K, rg = +- 1[4]
Phenylalanine	0.175 moles/liter	In water[7]
Picene	2.5 ug/liter	Distilled water at 298-300 K, No. aromatic rings=5[3]
Profluralin	1.0 ppm	Purity = 80.4 %, in water, wettable powder, as sodium salt[11]
Proline	6.572 moles/liter	In water[7]

Prometryn	48.0 ppm	Purity = 80.4 %, in water, wettable powder[11]
Propylene dichloride	2700 mg/l	298 K, in water, CH2ClCHClCH3[9]
Pyrene	165 ug/liter	Distilled water at 298-300 K, No. aromatic rings=4, rg=132-175[3]
Pyrene	175 mg/m3	In water at 298 K[10]
Serine	0.464 moles/liter	In water[7]
Styrene	250 mg/l	In water, 298 K[1]
Tetrachloroethane	2900 mg/l	298 K, in water, C2H2Cl4 (sym)[9]
Th 6040	0.3 ppm	Purity = 25.9 %, in water, wettable powder[11]
Thymine	0.032 moles/liter	In water[7]
Toluene	515 mg/l	In water, 298 K[6]
Toluene	515 mg/l	In water, 298 K[1]
Trichloroethylene	1100 mg/l	298 K, in water, CHClCCl2[9]
Trophenylene	40 ug/liter	Distilled water at 298-300 K, No. aromatic rings=4, rg=38-43[3]
Tryptophan	0.055 moles/liter	In water[7]
Tyrosine	0.0025 moles/liter	In water[7]
Uracil	0.032 moles/liter	In water[7]
Valine	0.740 moles/liter	In water[7]
Vinyl chloride	60 mg/l	298 K, in water, CH2CHCl[9]
Vinylidene chloride	400 mg/l	298 K, in water, CH2CCl2[9]
alpha - PbO	12.7 = log(K)	298 K, water, M units, solubility product, mineral = litharge[15]
beta - PbO	12.9 = log(K)	298 K, water, M units,solubility product, mineral = massicot[15]
m-Xylene	162 mg/l	In water, 298 K[1]
o-Xylene	175 mg/l	In water, 298 K[1]
p-Xylene	185 mg/l	In water, 298 K[1]

Source: 1.D'Ans-Lax. 1967. Taschenbuch fur Chemiker und Fysiker. Springer-Verlag. Berlin.
2.Dobbs, R.A. and J.M. Cohen. 1980. Carbon adsorption isotherms for Toxic organics. EPA 600/8-80-023.
3.May, W.E., S.P. Wasik and D.H. Freeman. 1978. Anal. Chem. 50, 175-179.
4.Neff, J.M. 1979. Polycyclic Aromatic Hydrocarbons in the Aquatic Environment. Applied Sci. Publ. London.
5.Hutzinger, O., S. Safe and V. Zitko. 1974. The Chemistry of PCBs, CRC-Press.
6.Mackay, D. and P.J. Leinonen, 1975. Environ. Sci. Tech. 9(13), pp 1178-1180.
7.Morowitz, H.J., M.E. Tourtellotte, W.R. Guild, E. Castro and C. Woese. 1962. J. Mol. Boil., 4, p 93.
8.Edwards, C.A. 1975. Persistent Pesticides in the Environment, 2nd edition, CRC-Press.
9.McConnell, G., D.M. Ferguson and C.R. Pearson. 1975. Endaevour 34, 13.
10.Handbook of Chemistry and Physics., 1967. The Chemical Rubber Company.
11.Carringer, R.D., J.B. Weber and T.J. Monaco. 1975. J. agric. Fd Chem., 23(3), pp 568-572.
12.McConnell, G., D.M. Ferguson and G.R. Pearson. 1975. Endeavour 13.
13.Andersen, J.W.R., R.C. Clark and J.J. Stegeman. 1974. In: Marine Bioassays Workshop Proceedings. APIEPA and Marine Tech. Soc. Wash. DC. Marine Technology Society. 36-75.
14.Howard, P.H. 1973. Preliminary Hazard Assesment of Chlorinated Naphthalenes. EPA 560/2/74-001.
15.Wagman, D.D., W.H. Evans, Parker, Hallow, Bailey and Schumm. 1969. Selected values of chemical thermodynamic properties. Nat. Bur. Stand. Tech. Note, 270-273.

2-083 Chemical compounds
Sorption

Species	Value	Condition
2,4,5-trichlorophenol	63.6 % of added amount	0.5 mg/l per 5 g bentonite (bentone 18c), pH = 7.4, 295 K[1]
2,4,5-trichlorophenol	97.9 % of added amount	0.5 mg/l per 5 g bentonite (bentone 24), pH = 7.6, 295 K[1]
2,4,6-trichlorophenol	55.0 % of added amount	0.5 mg/l per 5 g bentonite (bentone 18c), pH = 7.4, 295 K[1]
2,4,6-trichlorophenol	91.8 % of added amount	0.5 mg/l per 5 g bentonite (bentone 24), pH = 7.5, 295 K[1]
2,4-dibromophenol	55.4 % of added amount	0.5 mg/l per 5 g bentonite (bentone 18c), pH = 7.4, 295 K[1]
2,4-dibromophenol	97.9 % of added amount	0.5 mg/l per 5 g bentonite (bentone 24), pH = 7.6, 295 K[1]
2,4-dichlorophenol	38 % of added amount	0.5 mg/l per 5 g bentonite (bentone 18c), pH = 7.5, 295 K[1]
2,4-dichlorophenol	96.4 % of added amount	0.5 mg/l per 5 g bentonite (bentone 24), pH = 7.7, 295 K[1]
2,5-dichlorophenol	38 % of added amount	0.5 mg/l per 5 g bentonite (bentone 18c), pH = 7.5, 295 K[1]
2,5-dichlorophenol	93.9 % of added amount	0.5 mg/l per 5 g bentonite (bentone 24), pH = 7.9, 295 K[1]
2,6-dichlorophenol	32 % of added amount	0.5 mg/l per 5 g bentonite (bentone 18c), pH = 7.4, 295 K[1]
2,6-dichlorophenol	78.6 % of added amount	0.5 mg/l per 5 g bentonite (bentone 24), pH = 7.8, 295 K[1]
2,6-dimethylphenol	72.8 % of added amount	0.5 mg/l per 5 g bentonite (bentone 24), pH = 7.9, 295 K[1]
2,6-dimethylphenol	8 % of added amount	0.5 mg/l per 5 g bentonite (bentone 18c), pH = 7.6, 295 K[1]
3,4-dichlorophenol	34 % of added amount	0.5 mg/l per 5 g bentonite (bentone 18c), pH = 7.6, 295 K[1]
3,4-dichlorophenol	96.9 % of added amount	0.5 mg/l per 5 g bentonite (bentone 24), pH = 7.9, 295 K[1]
3,4-dimethylphenol	13 % of added amount	0.5 mg/l per 5 g bentonite (bentone 18c), pH = 7.7, 295 K[1]
3,4-dimethylphenol	65.2 % of added amount	0.5 mg/l per 5 g bentonite (bentone 24), pH = 7.8, 295 K[1]
3,5-dichlorophenol	42 % of added amount	0.5 mg/l per 5 g bentonite (bentone 18c), pH = 7.6, 295 K[1]
3,5-dichlorophenol	96.4 % of added amount	0.5 mg/l per 5 g bentonite (bentone 24), pH = 7.9, 295 K[1]
Alkyl benzene sulfonate	0.0 mg/l	F081, init = 20 mg/l, 21 d, algae = 30 mg/l, on Oscillatoria chlybia[2]
Alkyl benzene sulfonate	0.0 mg/l	F081, init = 20 mg/l, 21 d, algae = 35 mg/l, on Chlorella pyrenoidosa[2]
Alkyl benzene sulfonate	0.0 mg/l	F081, init = 20 mg/l, 21 d, algae = 40 mg/l, on Oscillatoria formosa[2]
Alkyl benzene sulfonate	0.0 mg/l	F081, init = 20 mg/l, 21 d, algae = 72 mg/l, on Oscillatoria tenuis[2]
Alkyl benzene sulfonate	0.0 mg/l	F081, init = 20 mg/l, 21 d, on activated sludge[2]
Alkyl benzene sulfonate	0.0 mg/l	F081, init = 20 mg/l, 21 d, on soil[2]
Alkyl benzene sulfonate	0.2 mg/l	F081, init = 20 mg/l, 21 d, algae = 40 mg/l, on Chlorella vulgaris[2]
Alkyl benzene sulfonate	0.4 mg/l	F081, init = 20 mg/l, 21 d, algae = 50 mg/l, on Scenedesmus obliquus[2]
Alkyl benzene sulfonate	0.5 mg/l	F081, init = 20 mg/l, 21 d, algae = 41 mg/l, on Ankistrodesmus[2]
Alkyl benzene sulfonate	0.5 mg/l	F081, init = 20 mg/l, 21 d, algae = 58 mg/l, on Anabaena cylindrica[2]
Alkyl benzene sulfonate	0.5 mg/l	F081, init = 20 mg/l, 21 d, algae = 87 mg/l, on Gloeocapsa alpicola[2]
Alkyl benzene sulfonate	0.6 mg/l	F081, init = 20 mg/l, 21 d, on raw wastewater[2]
Alkyl benzene sulfonate	0.8 mg/l	F081, init = 20 mg/l, 21 d, algae = 55 mg/l, on Cylindrospermum sp.[2]
Alkyl benzene sulfonate	0.9 mg/l	F081, init = 20 mg/l, 21 d, algae = 71 mg/l, on Anabaena variabilis[2]
Alkyl benzene sulfonate	1.0 mg/l	F081, init = 20 mg/l, 21 d, algae = 50 mg/l, on Oscillatoria borneti[2]

Alkyl benzene sulfonate	1.5 mg/l	F081, init = 20 mg/l, 21 d, algae = 47 mg/l, on Calothrix parietina[2]
Alkyl benzene sulfonate	1.5 mg/l	F081, init = 20 mg/l, 21 d, on pond water[2]
Alkyl benzene sulfonate	2.0 mg/l	F081, init = 20 mg/l, 21 d, algae = 20 mg/l, on Scenedesmus[2]
Alkyl benzene sulfonate	2.0 mg/l	F081, init = 20 mg/l, 21 d, algae = 90 mg/l, on Ulothrix fimbriata[2]
Alkyl benzene sulfonate	2.9 mg/l	F081, init = 20 mg/l, 21 d, algae = 90 mg/l, on Phormidium faveolarum[2]
Alkyl benzene sulfonate	4.0 mg/l	F081, init = 20 mg/l, 21 d, algae = 121 mg/l, on Anacystis nidulans[2]
Alkyl polyethoxylate	0.0 mg/l	F081, init = 20 mg/l, 21 d, on soil[2]
Alkyl polyethoxylate	0.2 mg/l	F081, init = 20 mg/l, 21 d, algae = 20 mg/l, on Scenedesmus[2]
Alkyl polyethoxylate	0.4 mg/l	F081, init = 20 mg/l, 21 d, algae = 90 mg/l, on Phormidium faveolarum[2]
Alkyl polyethoxylate	0.7 mg/l	F081, init = 20 mg/l, 21 d, algae = 35 mg/l, on Chlorella pyrenoidosa[2]
Alkyl polyethoxylate	0.9 mg/l	F081, init = 20 mg/l, 21 d, algae = 58 mg/l, on Anabaena cylindrica[2]
Alkyl polyethoxylate	1.0 mg/l	F081, init = 20 mg/l, 21 d, on raw wastewater[2]
Alkyl polyethoxylate	1.1 mg/l	F081, init = 20 mg/l, 21 d, algae = 47 mg/l, on Calothrix parietina[2]
Alkyl polyethoxylate	1.6 mg/l	F081, init = 20 mg/l, 21 d, algae = 50 mg/l, on Scenedesmus obliquus[2]
Alkyl polyethoxylate	2.1 mg/l	F081, init = 20 mg/l, 21 d, algae = 40 mg/l, on Oscillatoria formosa[2]
Alkyl polyethoxylate	2.7 mg/l	F081, init = 20 mg/l, 21 d, algae = 71 mg/l, on Anabaena variabilis[2]
Alkyl polyethoxylate	3.5 mg/l	F081, init = 20 mg/l, 21 d, algae = 72 mg/l, on Oscillatoria tenuis[2]
Alkyl polyethoxylate	4.4 mg/l	F081, init = 20 mg/l, 21 d, algae = 40 mg/l, on Nostoc muscorum[2]
Alkyl polyethoxylate	4.9 mg/l	F081, init = 20 mg/l, 21 d, algae = 87 mg/l, on Gloeocapsa alpicola[2]
Alkyl polyethoxylate	5.3 mg/l	F081, init = 20 mg/l, 21 d, algae = 30 mg/l, on Oscillatoria chlybia[2]
Alkyl polyethoxylate	5.5 mg/l	F081, init = 20 mg/l, 21 d, algae = 40 mg/l, on Chlorella vulgaris[2]
Alkyl polyethoxylate	5.5 mg/l	F081, init = 20 mg/l, 21 d, algae = 50 mg/l, on Oscillatoria borneti[2]
Alkyl polyethoxylate	5.6 mg/l	F081, init = 20 mg/l, 21 d, algae = 55 mg/l, on Cylindrospermum sp.[2]
Alkyl polyethoxylate	6.3 mg/l	F081, init = 20 mg/l, 21 d, on pond water[2]
Alkyl polyethoxylate	7.0 mg/l	F081, init = 20 mg/l, 21 d, algae = 90 mg/l, on Ulothrix fimbriata[2]
Alkyl polyethoxylate	7.2 mg/l	F081, init = 20 mg/l, 21 d, algae = 121 mg/l, on Anacystis nidulans[2]
Alkyl polyethoxylate	8.0 mg/l	F081, init = 20 mg/l, 21 d, on activated sludge[2]
Alkyl polyethoxylate	8.9 mg/l	F081, init = 20 mg/l, 21 d, algae = 41 mg/l, on Ankistrodesmus[2]
Linear alkyl sulfonate	0.0 mg/l	F081, init = 20 mg/l, 21 d, algae = 30 mg/l, on Oscillatoria chlybia[2]
Linear alkyl sulfonate	0.0 mg/l	F081, init = 20 mg/l, 21 d, algae = 35 mg/l, on Chlorella pyrenoidosa[2]
Linear alkyl sulfonate	0.0 mg/l	F081, init = 20 mg/l, 21 d, algae = 40 mg/l, on Oscillatoria formosa[2]
Linear alkyl sulfonate	0.0 mg/l	F081, init = 20 mg/l, 21 d, algae = 55 mg/l, on Cylindrospermum sp.[2]
Linear alkyl sulfonate	0.1 mg/l	F081, init = 20 mg/l, 21 d, algae = 50 mg/l, on Oscillatoria borneti[2]
Linear alkyl sulfonate	0.1 mg/l	F081, init = 20 mg/l, 21 d, algae = 50 mg/l, on Scenedesmus obliquus[2]

Linear alkyl sulfonate	0.1 mg/l	F081, init = 20 mg/l, 21 d, on pond water[2]
Linear alkyl sulfonate	0.3 mg/l	F081, init = 20 mg/l, 21 d, algae = 87 mg/l, on Gloeocapsa alpicola[2]
Linear alkyl sulfonate	0.6 mg/l	F081, init = 20 mg/l, 21 d, algae = 40 mg/l, on Chlorella vulgaris[2]
Linear alkyl sulfonate	1.1 mg/l	F081, init = 20 mg/l, 21 d, algae = 72 mg/l, on Oscillatoria tenuis[2]
Linear alkyl sulfonate	1.3 mg/l	F081, init = 20 mg/l, 21 d, algae = 71 mg/l, on Anabaena variabilis[2]
Linear alkyl sulfonate	1.4 mg/l	F081, init = 20 mg/l, 21 d, algae = 58 mg/l, on Anabaena cylindrica[2]
Linear alkyl sulfonate	1.4 mg/l	F081, init = 20 mg/l, 21 d, on soil[2]
Linear alkyl sulfonate	1.5 mg/l	F081, init = 20 mg/l, 21 d, algae = 47 mg/l, on Calothrix parietina[2]
Linear alkyl sulfonate	1.7 mg/l	F081, init = 20 mg/l, 21 d, algae = 20 mg/l, on Scenedesmus[2]
Linear alkyl sulfonate	1.9 mg/l	F081, init = 20 mg/l, 21 d, algae = 90 mg/l, on Phormidium faveolarum[2]
Linear alkyl sulfonate	2.0 mg/l	F081, init = 20 mg/l, 21 d, algae = 40 mg/l, on Nostoc muscorum[2]
Linear alkyl sulfonate	2.2 mg/l	F081, init = 20 mg/l, 21 d, algae = 121 mg/l, on Anacystis nidulans[2]
Linear alkyl sulfonate	2.6 mg/l	F081, init = 20 mg/l, 21 d, algae = 90 mg/l, on Ulothrix fimbriata[2]
Linear alkyl sulfonate	3.5 mg/l	F081, init = 20 mg/l, 21 d, algae = 41 mg/l, on Ankistrodesmus[2]
Linear alkyl sulfonate	4.9 mg/l	F081, init = 20 mg/l, 21 d, on activated sludge[2]
Linear alkyl sulfonate	6.0 mg/l	F081, init = 20 mg/l, 21 d, on raw wastewater[2]
M-aminophenol	0.8 % of added amount	0.5 mg/l per 5 g bentonite (bentone 18c), pH = 7.5, 295 K[1]
M-aminophenol	12 % of added amount	0.5 mg/l per 5 g bentonite (bentone 24), pH = 7.7, 295 K[1]
M-chlorophenol	13 % of added amount	0.5 mg/l per 5 g bentonite (bentone 18c), pH = 7.6, 295 K[1]
M-chlorophenol	81.0 % of added amount	0.5 mg/l per 5 g bentonite (bentone 24), pH = 7.7, 295 K[1]
Nonyl phenol polyethoxylate	0.0 mg/l	F081, init = 20 mg/l, 21 d, on soil[2]
Nonyl phenol polyethoxylate	0.8 mg/l	F081, init = 20 mg/l, 21 d, algae = 90 mg/l, on Phormidium faveolarum[2]
Nonyl phenol polyethoxylate	0.9 mg/l	F081, init = 20 mg/l, 21 d, algae = 40 mg/l, on Chlorella vulgaris[2]
Nonyl phenol polyethoxylate	1.9 mg/l	F081, init = 20 mg/l, 21 d, algae = 50 mg/l, on Oscillatoria borneti[2]
Nonyl phenol polyethoxylate	10.6 mg/l	F081, init = 20 mg/l, 21 d, on pond water[2]
Nonyl phenol polyethoxylate	11.1 mg/l	F081, init = 20 mg/l, 21 d, algae = 55 mg/l, on Cylindrospermum sp.[2]
Nonyl phenol polyethoxylate	13.1 mg/l	F081, init = 20 mg/l, 21 d, algae = 71 mg/l, on Anabaena variabilis[2]
Nonyl phenol polyethoxylate	14.8 mg/l	F081, init = 20 mg/l, 21 d, algae = 90 mg/l, on Ulothrix fimbriata[2]
Nonyl phenol polyethoxylate	2.1 mg/l	F081, init = 20 mg/l, 21 d, algae = 47 mg/l, on Calothrix parietina[2]
Nonyl phenol polyethoxylate	2.5 mg/l	F081, init = 20 mg/l, 21 d, on activated sludge[2]
Nonyl phenol polyethoxylate	2.7 mg/l	F081, init = 20 mg/l, 21 d, algae = 40 mg/l, on Nostoc muscorum[2]
Nonyl phenol polyethoxylate	2.8 mg/l	F081, init = 20 mg/l, 21 d, algae = 40 mg/l, on Oscillatoria formosa[2]
Nonyl phenol polyethoxylate	2.9 mg/l	F081, init = 20 mg/l, 21 d, algae = 20 mg/l, on Scenedesmus[2]
Nonyl phenol polyethoxylate	3.3 mg/l	F081, init = 20 mg/l, 21 d, algae = 50 mg/l, on Scenedesmus obliquus[2]
Nonyl phenol polyethoxylate	3.6 mg/l	F081, init = 20 mg/l, 21 d, algae = 58 mg/l, on Anabaena cylindrica[2]

Nonyl phenol polyethoxylate	5.1 mg/l	F081, init = 20 mg/l, 21 d, algae = 121 mg/l, on Anacystis nidulans[2]
Nonyl phenol polyethoxylate	5.7 mg/l	F081, init = 20 mg/l, 21 d, algae = 35 mg/l, on Chlorella pyrenoidosa[2]
Nonyl phenol polyethoxylate	6.2 mg/l	F081, init = 20 mg/l, 21 d, algae = 41 mg/l, on Ankistrodesmus[2]
Nonyl phenol polyethoxylate	6.5 mg/l	F081, init = 20 mg/l, 21 d, algae = 30 mg/l, on Oscillatoria chlybia[2]
Nonyl phenol polyethoxylate	7.0 mg/l	F081, init = 20 mg/l, 21 d, algae = 72 mg/l, on Oscillatoria tenuis[2]
Nonyl phenol polyethoxylate	7.0 mg/l	F081, init = 20 mg/l, 21 d, algae = 87 mg/l, on Gloeocapsa alpicola[2]
Nonyl phenol polyethoxylate	7.5 mg/l	F081, init = 20 mg/l, 21 d, on raw wastewater[2]
O-aminophenol	25.5 % of added amount	0.5 mg/l per 5 g bentonite (bentone 24), pH = 8.4, 295 K[1]
O-aminophenol	29 % of added amount	0.5 mg/l per 5 g bentonite (bentone 18c), pH = 7.8, 295 K[1]
O-bromophenol	18 % of added amount	0.5 mg/l per 5 g bentonite (bentone 18c), pH = 7.4, 295 K[1]
O-bromophenol	85.6 % of added amount	0.5 mg/l per 5 g bentonite (bentone 24), pH = 7.6, 295 K[1]
O-chlorophenol	15 % of added amount	0.5 mg/l per 5 g bentonite (bentone 18c), pH = 7.6, 295 K[1]
O-chlorophenol	76.6 % of added amount	0.5 mg/l per 5 g bentonite (bentone 24), pH = 7.8, 295 K[1]
O-chlorophenoxyacetic acid	17.0 % of added amount	0.5 mg/l per 5 g bentonite (bentone 18c), pH = 7.3, 295 K[1]
O-chlorophenoxyacetic acid	43.6 % of added amount	0.5 mg/l per 5 g bentonite (bentone 24), pH = 7.4, 295 K[1]
O-phenyl phenol	47 % of added amount	0.5 mg/l per 5 g bentonite (bentone 18c), pH = 7.6, 295 K[1]
O-phenyl phenol	96.4 % of added amount	0.5 mg/l per 5 g bentonite (bentone 24), pH = 7.3, 295 K[1]
P-chlorophenol	48.5 % of added amount	0.5 mg/l per 5 g bentonite (bentone 24), pH = 8.0, 295 K[1]
P-chlorophenol	6 % of added amount	0.5 mg/l per 5 g bentonite (bentone 18c), pH = 7.7, 295 K[1]
P-chlorophenoxyacetic acid	11.0 % of added amount	0.5 mg/l per 5 g bentonite (bentone 18c), pH = 7.4, 295 K[1]
P-chlorophenoxyacetic acid	45.6 % of added amount	0.5 mg/l per 5 g bentonite (bentone 24), pH = 7.4, 295 K[1]
Zn sorption on clay, bentonite	10 % fixed of add	30 d wet incubation, pH=6.2, desorb 1 M NH4OAc (pH=7), add Zn=500/kg clay[3]
Zn sorption on clay, bentonite	11 % fixed of add	30 d wet incubation, pH=8.8, desorb 1 M NH4OAc (pH=7), add Zn=1000/kg clay[3]
Zn sorption on clay, bentonite	13 % fixed of add	30 d wet incubation, pH=4.9, desorb 1 M NH4OAc (pH=7), add Zn=40/kg clay[3]
Zn sorption on clay, bentonite	13 % fixed of add	30 d wet incubation, pH=7.6, desorb 1 M NH4OAc (pH=7), add Zn=500/kg clay[3]
Zn sorption on clay, bentonite	13 % fixed of add	Alt. wet and dry cycles, pH=4.9, desorb 1 M NH4OAc (pH=7), add Zn=1000/kg clay[3]
Zn sorption on clay, bentonite	14 % fixed of add	30 d wet incubation, pH=6.2, desorb 1 M NH4OAc (pH=7), add Zn=250/kg clay[3]
Zn sorption on clay, bentonite	14 % fixed of add	Alt. wet and dry cycles, pH=4.9, desorb 1 M NH4OAc (pH=7), add Zn=125/kg clay[3]
Zn sorption on clay, bentonite	14 % fixed of add	Alt. wet and dry cycles, pH=4.9, desorb 1 M NH4OAc (pH=7), add Zn=250/kg clay[3]
Zn sorption on clay, bentonite	14 % fixed of add	Alt. wet and dry cycles, pH=4.9, desorb 1 M NH4OAc (pH=7), add Zn=500/kg clay[3]

Zn sorption on clay, bentonite	14 % fixed of add	Alt. wet and dry cycles, pH=6.2, desorb 1 M NH4OAc (pH=7), add Zn=1000/kg clay^{3}
Zn sorption on clay, bentonite	15 % fixed of add	30 d wet incubation, pH=6.2, desorb 1 M NH4OAc (pH=7), add Zn=40/kg clay^{3}
Zn sorption on clay, bentonite	17 % fixed of add	30 d wet incubation, pH=8.8, desorb 1 M NH4OAc (pH=7), add Zn=500/kg clay^{3}
Zn sorption on clay, bentonite	17 % fixed of add	Alt. wet and dry cycles, pH=6.2, desorb 1 M NH4OAc (pH=7), add Zn=500/kg clay^{3}
Zn sorption on clay, bentonite	18 % fixed of add	30 d wet incubation, pH=7.6, desorb 1 M NH4OAc (pH=7), add Zn=250/kg clay^{3}
Zn sorption on clay, bentonite	20 % fixed of add	Alt. wet and dry cycles, pH=7.6, desorb 1 M NH4OAc (pH=7), add Zn=1000/kg clay^{3}
Zn sorption on clay, bentonite	20 % fixed of add	Alt. wet and dry cycles, pH=7.6, desorb 1 M NH4OAc (pH=7), add Zn=500/kg clay^{3}
Zn sorption on clay, bentonite	22 % fixed of add	Alt. wet and dry cycles, pH=6.2, desorb 1 M NH4OAc (pH=7), add Zn=250/kg clay^{3}
Zn sorption on clay, bentonite	23 % fixed of add	Alt. wet and dry cycles, pH=8.8, desorb 1 M NH4OAc (pH=7), add Zn=1000/kg clay^{3}
Zn sorption on clay, bentonite	25 % fixed of add	30 d wet incubation, pH=7.6, desorb 1 M NH4OAc (pH=7), add Zn=40/kg clay^{3}
Zn sorption on clay, bentonite	26 % fixed of add	30 d wet incubation, pH=8.8, desorb 1 M NH4OAc (pH=7), add Zn=250/kg clay^{3}
Zn sorption on clay, bentonite	27 % fixed of add	Alt. wet and dry cycles, pH=8.8, desorb 1 M NH4OAc (pH=7), add Zn=500/kg clay^{3}
Zn sorption on clay, bentonite	29 % fixed of add	Alt. wet and dry cycles, pH=7.6, desorb 1 M NH4OAc (pH=7), add Zn=250/kg clay^{3}
Zn sorption on clay, bentonite	34 % fixed of add	Alt. wet and dry cycles, pH=6.2, desorb 1 M NH4OAc (pH=7), add Zn=125/kg clay^{3}
Zn sorption on clay, bentonite	38 % fixed of add	30 d wet incubation, pH=8.8, desorb 1 M NH4OAc (pH=7), add Zn=40/kg clay^{3}
Zn sorption on clay, bentonite	39 % fixed of add	Alt. wet and dry cycles, pH=7.6, desorb 1 M NH4OAc (pH=7), add Zn=125/kg clay^{3}
Zn sorption on clay, bentonite	39 % fixed of add	Alt. wet and dry cycles, pH=8.8, desorb 1 M NH4OAc (pH=7), add Zn=250/kg clay^{3}
Zn sorption on clay, bentonite	4 % fixed of add	30 d wet incubation, pH=4.9, desorb 1 M NH4OAc (pH=7), add Zn=1000/kg clay^{3}
Zn sorption on clay, bentonite	46 % fixed of add	Alt. wet and dry cycles, pH=8.8, desorb 1 M NH4OAc (pH=7), add Zn=125/kg clay^{3}
Zn sorption on clay, bentonite	6 % fixed of add	30 d wet incubation, pH=4.9, desorb 1 M NH4OAc (pH=7), add Zn=500/kg clay^{3}
Zn sorption on clay, bentonite	7 % fixed of add	30 d wet incubation, pH=4.9, desorb 1 M NH4OAc (pH=7), add Zn=250/kg clay^{3}
Zn sorption on clay, bentonite	7 % fixed of add	30 d wet incubation, pH=6.2, desorb 1 M NH4OAc (pH=7), add Zn=1000/kg clay^{3}
Zn sorption on clay, bentonite	9 % fixed of add	30 d wet incubation, pH=7.6, desorb 1 M NH4OAc (pH=7), add Zn=1000/kg clay^{3}
Zn sorption on clay, illite	0 % fixed of add	30 d wet incubation, pH=4.9, desorb 1 M NH4OAc (pH=7), add Zn=40/kg clay^{3}
Zn sorption on clay, illite	0 % fixed of add	30 d wet incubation, pH=6.2, desorb 1 M NH4OAc (pH=7), add Zn=40/kg clay^{3}
Zn sorption on clay, illite	10 % fixed of add	30 d wet incubation, pH=6.2, desorb 1 M NH4OAc (pH=7), add Zn=320/kg clay^{3}
Zn sorption on clay, illite	13 % fixed of add	30 d wet incubation, pH=4.9, desorb 1 M NH4OAc (pH=7), add Zn=80/kg clay^{3}

Zn sorption on clay, illite	14 % fixed of add	30 d wet incubation, pH=7.6, desorb 1 M NH4OAc (pH=7), add Zn=320/kg clay[3]
Zn sorption on clay, illite	15 % fixed of add	30 d wet incubation, pH=6.2, desorb 1 M NH4OAc (pH=7), add Zn=160/kg clay[3]
Zn sorption on clay, illite	16 % fixed of add	30 d wet incubation, pH=8.8, desorb 1 M NH4OAc (pH=7), add Zn=320/kg clay[3]
Zn sorption on clay, illite	19 % fixed of add	30 d wet incubation, pH=6.2, desorb 1 M NH4OAc (pH=7), add Zn=80/kg clay[3]
Zn sorption on clay, illite	22 % fixed of add	30 d wet incubation, pH=7.6, desorb 1 M NH4OAc (pH=7), add Zn=160/kg clay[3]
Zn sorption on clay, illite	22 % fixed of add	Alt. wet and dry cycles, pH=4.9, desorb 1 M NH4OAc (pH=7), add Zn=320/kg clay[3]
Zn sorption on clay, illite	23 % fixed of add	Alt. wet and dry cycles, pH=4.9, desorb 1 M NH4OAc (pH=7), add Zn=40/kg clay[3]
Zn sorption on clay, illite	24 % fixed of add	Alt. wet and dry cycles, pH=4.9, desorb 1 M NH4OAc (pH=7), add Zn=160/kg clay[3]
Zn sorption on clay, illite	25 % fixed of add	Alt. wet and dry cycles, pH=4.9, desorb 1 M NH4OAc (pH=7), add Zn=80/kg clay[3]
Zn sorption on clay, illite	26 % fixed of add	Alt. wet and dry cycles, pH=6.2, desorb 1 M NH4OAc (pH=7), add Zn=320/kg clay[3]
Zn sorption on clay, illite	28 % fixed of add	30 d wet incubation, pH=7.6, desorb 1 M NH4OAc (pH=7), add Zn=80/kg clay[3]
Zn sorption on clay, illite	28 % fixed of add	30 d wet incubation, pH=8.8, desorb 1 M NH4OAc (pH=7), add Zn=160/kg clay[3]
Zn sorption on clay, illite	29 % fixed of add	Alt. wet and dry cycles, pH=6.2, desorb 1 M NH4OAc (pH=7), add Zn=80/kg clay[3]
Zn sorption on clay, illite	34 % fixed of add	Alt. wet and dry cycles, pH=6.2, desorb 1 M NH4OAc (pH=7), add Zn=160/kg clay[3]
Zn sorption on clay, illite	35 % fixed of add	Alt. wet and dry cycles, pH=6.2, desorb 1 M NH4OAc (pH=7), add Zn=40/kg clay[3]
Zn sorption on clay, illite	38 % fixed of add	30 d wet incubation, pH=8.8, desorb 1 M NH4OAc (pH=7), add Zn=80/kg clay[3]
Zn sorption on clay, illite	39 % fixed of add	Alt. wet and dry cycles, pH=7.6, desorb 1 M NH4OAc (pH=7), add Zn=320/kg clay[3]
Zn sorption on clay, illite	4 % fixed of add	30 d wet incubation, pH=4.9, desorb 1 M NH4OAc (pH=7), add Zn=320/kg clay[3]
Zn sorption on clay, illite	40 % fixed of add	Alt. wet and dry cycles, pH=7.6, desorb 1 M NH4OAc (pH=7), add Zn=40/kg clay[3]
Zn sorption on clay, illite	43 % fixed of add	Alt. wet and dry cycles, pH=7.6, desorb 1 M NH4OAc (pH=7), add Zn=80/kg clay[3]
Zn sorption on clay, illite	44 % fixed of add	Alt. wet and dry cycles, pH=8.8, desorb 1 M NH4OAc (pH=7), add Zn=320/kg clay[3]
Zn sorption on clay, illite	45 % fixed of add	Alt. wet and dry cycles, pH=8.8, desorb 1 M NH4OAc (pH=7), add Zn=40/kg clay[3]
Zn sorption on clay, illite	48 % fixed of add	Alt. wet and dry cycles, pH=7.6, desorb 1 M NH4OAc (pH=7), add Zn=160/kg clay[3]
Zn sorption on clay, illite	55 % fixed of add	Alt. wet and dry cycles, pH=8.8, desorb 1 M NH4OAc (pH=7), add Zn=80/kg clay[3]
Zn sorption on clay, illite	60 % fixed of add	Alt. wet and dry cycles, pH=8.8, desorb 1 M NH4OAc (pH=7), add Zn=160/kg clay[3]
Zn sorption on clay, illite	8 % fixed of add	30 d wet incubation, pH=4.9, desorb 1 M NH4OAc (pH=7), add Zn=160/kg clay[3]

Zn sorption on clay, illite	8 % fixed of add	30 d wet incubation, pH=7.6, desorb 1 M NH4OAc (pH=7), add Zn=40/kg clay[3]
Zn sorption on clay, illite	8 % fixed of add	30 d wet incubation, pH=8.8, desorb 1 M NH4OAc (pH=7), add Zn=40/kg clay[3]
Zn sorption on clay, kaolinite	0 % fixed of add	Alt. wet and dry cycles, pH=4.9, desorb 1 M NH4OAc (pH=7), add Zn=160/kg clay[3]
Zn sorption on clay, kaolinite	0 % fixed of add	Alt. wet and dry cycles, pH=4.9, desorb 1 M NH4OAc (pH=7), add Zn=40 meq/kg cla[3]
Zn sorption on clay, kaolinite	0 % fixed of add	Alt. wet and dry cycles, pH=4.9, desorb 1 M NH4OAc (pH=7), add Zn=40 meq/kg clay[3]
Zn sorption on clay, kaolinite	0 % fixed of add	Alt. wet and dry cycles, pH=4.9, desorb 1 M NH4OAc (pH=7), add Zn=80/kg clay[3]
Zn sorption on clay, kaolinite	0 % fixed of add	Alt. wet and dry cycles, pH=6.2, desorb 1 M NH4OAc (pH=7), add Zn=160/kg clay[3]
Zn sorption on clay, kaolinite	0 % fixed of add	Alt. wet and dry cycles, pH=6.2, desorb 1 M NH4OAc (pH=7), add Zn=40 meq/kg cla[3]
Zn sorption on clay, kaolinite	0 % fixed of add	Alt. wet and dry cycles, pH=6.2, desorb 1 M NH4OAc (pH=7), add Zn=40 meq/kg clay[3]
Zn sorption on clay, kaolinite	0 % fixed of add	Alt. wet and dry cycles, pH=6.2, desorb 1 M NH4OAc (pH=7), add Zn=80/kg clay[3]
Zn sorption on clay, kaolinite	1 % fixed of add	Alt. wet and dry cycles, pH=7.6, desorb 1 M NH4OAc (pH=7), add Zn=40 meq/kg cla[3]
Zn sorption on clay, kaolinite	1 % fixed of add	Alt. wet and dry cycles, pH=7.6, desorb 1 M NH4OAc (pH=7), add Zn=40 meq/kg clay[3]
Zn sorption on clay, kaolinite	10 % fixed of add	30 d wet incubation, pH=4.9, desorb 1 M NH4OAc (pH=7), add Zn=160/kg clay[3]
Zn sorption on clay, kaolinite	11 % fixed of add	30 d wet incubation, pH=6.2, desorb 1 M NH4OAc (pH=7), add Zn=160/kg clay[3]
Zn sorption on clay, kaolinite	11 % fixed of add	30 d wet incubation, pH=7.6, desorb 1 M NH4OAc (pH=7), add Zn=160/kg clay[3]
Zn sorption on clay, kaolinite	11 % fixed of add	30 d wet incubation, pH=8.8, desorb 1 M NH4OAc (pH=7), add Zn=160/kg clay[3]
Zn sorption on clay, kaolinite	12 % fixed of add	30 d wet incubation, pH=6.2, desorb 1 M NH4OAc (pH=7), add Zn=125/kg clay[3]
Zn sorption on clay, kaolinite	12 % fixed of add	Alt. wet and dry cycles, pH=8.8, desorb 1 M NH4OAc (pH=7), add Zn=160/kg clay[3]
Zn sorption on clay, kaolinite	21 % fixed of add	30 d wet incubation, pH=7.6, desorb 1 M NH4OAc (pH=7), add Zn=125/kg clay[3]
Zn sorption on clay, kaolinite	38 % fixed of add	30 d wet incubation, pH=8.8, desorb 1 M NH4OAc (pH=7), add Zn=125/kg clay[3]
Zn sorption on clay, kaolinite	4 % fixed of add	Alt. wet and dry cycles, pH=7.6, desorb 1 M NH4OAc (pH=7), add Zn=80/kg clay[3]
Zn sorption on clay, kaolinite	7 % fixed of add	30 d wet incubation, pH=4.9, desorb 1 M NH4OAc (pH=7), add Zn=125/kg clay[3]
Zn sorption on clay, kaolinite	7 % fixed of add	30 d wet incubation, pH=4.9, desorb 1 M NH4OAc (pH=7), add Zn=80/kg clay[3]
Zn sorption on clay, kaolinite	7 % fixed of add	Alt. wet and dry cycles, pH=8.8, desorb 1 M NH4OAc (pH=7), add Zn=80/kg clay[3]
Zn sorption on clay, kaolinite	8 % fixed of add	30 d wet incubation, pH=6.2, desorb 1 M NH4OAc (pH=7), add Zn=80/kg clay[3]
Zn sorption on clay, kaolinite	8 % fixed of add	Alt. wet and dry cycles, pH=7.6, desorb 1 M NH4OAc (pH=7), add Zn=160/kg clay[3]
Zn sorption on clay, kaolinite	8 % fixed of add	Alt. wet and dry cycles, pH=8.8, desorb 1 M NH4OAc (pH=7), add Zn=40 meq/kg cla[3]

Zn sorption on clay, kaolinite	8 % fixed of add	Alt. wet and dry cycles, pH=8.8, desorb 1 M NH4OAc (pH=7), add Zn=40 meq/kg clay[3]
Zn sorption on clay, kaolinite	9 % fixed of add	30 d wet incubation, pH=7.6, desorb 1 M NH4OAc (pH=7), add Zn=80/kg clay[3]
Zn sorption on clay, kaolinite	9 % fixed of add	30 d wet inc., pH=8.8, desorb 1 M NH4OAc (pH=7), add Zn=80/kg [3]

Source: 1.Miller, R.W. and S.D. Faust. 1973. Environ. Lett. 4(3), pp 211-223.
2.Davis, E.M. and E.F. Gloyna. 1969. Water Pollut. Control. 41 (8), pp 1494-1504.
3.Reddy, M.R. and H.F. Perkins. 1974. Soil Sci. Soc. Am. Proc. 38, 229-231.

2-084 Chemical compounds
Sorption capacity

Species	Value	Condition
H2S and soil	11.0 mg H2S/g dry soil	Air = 100 ppm H2S, moist air, soil = thurman[1]
H2S and soil	15.4 mg H2S/g dry soil	Air = 100 ppm H2S, dry air, soil = thurman[1]
H2S and soil	40.6 mg H2S/g dry soil	Air = 100 ppm H2S, moist air, soil = harpster[1]
H2S and soil	44.5 mg H2S/g dry soil	Air = 100 ppm H2S, moist air, soil = okoboji[1]
H2S and soil	44.7 mg H2S/g dry soil	Air = 100 ppm H2S, moist air, soil = mean of 6 soils[1]
H2S and soil	46.6 mg H2S/g dry soil	Air = 100 ppm H2S, dry air, soil = harpster[1]
H2S and soil	50.7 mg H2S/g dry soil	Air = 100 ppm H2S, dry air, soil = mean of 6 soils[1]
H2S and soil	51.6 mg H2S/g dry soil	Air = 100 ppm H2S, moist air, soil = atroria[1]
H2S and soil	52.9 mg H2S/g dry soil	Air = 100 ppm H2S, dry air, soil = okoboji[1]
H2S and soil	58.1 mg H2S/g dry soil	Air = 100 ppm H2S, moist air, soil = weller[1]
H2S and soil	61.0 mg H2S/g dry soil	Air = 100 ppm H2S, dry air, soil = weller[1]
H2S and soil	62.5 mg H2S/g dry soil	Air = 100 ppm H2S, moist air, soil = regina[1]
H2S and soil	62.9 mg H2S/g dry soil	Air = 100 ppm H2S, dry air, soil = atroria[1]
H2S and soil	65.2 mg H2S/g dry soil	Air = 100 ppm H2S, dry air, soil = regina[1]
SO2 and soil	1.1 mg SO2/g dry soil	Air = 100 ppm SO2, dry air, soil = thurman[1]
SO2 and soil	10.2 mg SO2/g dry soil	Air = 100 ppm SO2, dry air, soil = harpster[1]
SO2 and soil	10.2 mg SO2/g dry soil	Air = 100 ppm SO2, dry air, soil = okoboji[1]
SO2 and soil	13.2 mg SO2/g dry soil	Air = 100 ppm SO2, dry air, soil = regina[1]
SO2 and soil	15.3 mg SO2/g dry soil	Air = 100 ppm SO2, dry air, soil = weller[1]
SO2 and soil	31.4 mg SO2/g dry soil	Air = 100 ppm SO2, moist air, soil = okoboji[1]
SO2 and soil	31.9 mg SO2/g dry soil	Air = 100 ppm SO2, moist air, soil = weller[1]
SO2 and soil	37.1 mg SO2/g dry soil	Air = 100 ppm SO2, moist air, soil = atroria[1]
SO2 and soil	37.8 mg SO2/g dry soil	Air = 100 ppm SO2, moist air, soil = mean of 6 soils[1]
SO2 and soil	50.4 mg SO2/g dry soil	Air = 100 ppm SO2, moist air, soil = regina[1]
SO2 and soil	66.8 mg SO2/g dry soil	Air = 100 ppm SO2, moist air, soil = harpster[1]
SO2 and soil	8.9 mg SO2/g dry soil	Air = 100 ppm SO2, dry air, soil = atroria[1]
SO2 and soil	9.3 mg SO2/g dry soil	Air = 100 ppm SO2, moist air, soil = thurman[1]
SO2 and soil	9.8 mg SO2/g dry soil	Air = 100 ppm SO2, dry air, soil = mean of 6 soils[1]

Source: 1.Smith, K.A., J.M. Bremner and M.A. Tabatabai. 1973. Soil Sci. 116, 313-319.

2-085 Chemical compounds
Specific surface area

Species	Value	Condition
Calcite	12.5 m**2/g	Less than 2 um[1]
Fe hydroxide	300 m**2/g	Sediment material, freshly precipitated[1]
Humic acids from soils	1900 m**2/g	Sediment material[1]
Humic acids from soils	0.38 eq/100g	Less than 2 um, rg = 170 - 590[1]
Illite	55 m**2/g	Sediment material, rg = 30 - 80[1]
Kaolinite	30 m**2/g	Sediment material, rg = 10 - 50[1]
Montmorillonite	100 m**2/g	Sediment material, rg = 50 - 150[1]

Source: 1.Foestner, U. and G.T.W. Wittmann. 1979. Metal Pollution in the Aquatic Environment. Springer-Verlag.Berlin.

2-086 Chemical compounds
Vapour pressure

Species	Value	Condition
1,2 Dichlorobenzene	1 mm Hg	298 K[1]
1,2,3 Trichlorobenzene	1 mm Hg	Less than value[1]
1,2,4 Trichlorobenzene	1 mm Hg	Less than value[1]
1,2,4,5 Tetrachlorobenzene	1 mm Hg	Less than value[1]
1,3 Dichlorobenzene	0.6 mm Hg	298 K[1]
1,3 Dichlorobenzene	3 mm Hg	298 K[1]
1,3,5 Trichlorobenzene	1 mm Hg	Less than value[1]
1,4 Dichlorobenzene	0.5 mm Hg	In water, 298 K[2]
1,4 Dichlorobenzene	79 mg/l	In water, 298 K[1]
2-Chloro toluene	2.7 mm Hg	298 K[1]
3-Chloro aniline	0.1 mm Hg	Less than value[1]
4-Chloro aniline	0.015 mm Hg	298 K[1]
Allyl chloride	340 mm Hg	298 K, CH2ClCHCH2[3]
Benzene	74.3 mm Hg	In water, 298 K[2]
Carbon tetachloride	90 mm Hg	298 K, CCl4[3]
Chlorobenzene	8.8 mm Hg	298 K[1]
Chlorobenzene	8.8 mm Hg	In water, 298 K[2]
Chloroform	151 mm Hg	298 K, CHCl3[3]
Chloronaphthalenes	1E-1 mm HG	Halowax 1031[4]
Chloronaphthalenes	1E-4 mm HG	Halowax 1014[4]
Chloronaphthalenes	5E-2 mm HG	Halowax 1000[4]
Chloronaphthalenes	5E-4 mm HG	Halowax 1099[4]
Chloroprene	190 mm Hg	298 K, CH2CHCClCH2[3]
Ethyl benzene	7.1 mm Hg	In water, 298 K[2]
Ethyl bromide	38 mm Hg	298 K, C2H5Br[3]
Ethyl chloride	1000 mm Hg	298 K, C2H5Cl[3]
Ethylene dibromide	11 mm Hg	298 K, CH2BrCH2Br[3]

Ethylene dichloride	64 mm Hg	298 K, CH2ClCH2Cl[3]
Hexachlorobenzene	1E-5 mm Hg	298 K[1]
Hexachlorobutadiene	0.15 mm Hg	298 K, C4Cl6[3]
Hexachloroethane	0.4 mm Hg	298 K, C2Cl6[3]
Isopropyl benzene	4.0 mm Hg	In water, 298 K[2]
Methyl bromide	1420 mm Hg	298 K, CHBr3[3]
Methyl chloride	3756 mm Hg	298 K, CH3Cl[3]
Methyl chloroform	96 mm Hg	298 K, CH3CCl3[3]
Methylene chloride	362 mm Hg	298 K, CH2Cl2[3]
Naphthalene	0.05 mm Hg	In water, 298 K[2]
Nitrobenzene	0.27 mm Hg	In water, 298 K[2]
PBB	1E-6 mm Hg	Firemaster BP-6[4]
PCB	12.7E-5 mm Hg	Aroclor 1254, rg = 1.8E-4 - 7.7E-5 mm Hg[4]
PCB	1E-2 mm Hg	Aroclor 1221,[4]
PCB	2E-4 mm Hg	Aroclor 1016[4]
PCB	6.2E-4 mm Hg	Aroclor 1248, rg = 8.3E-4 - 5E-4 mm Hg[4]
PCB	6.5E-4 mm Hg	Aroclor 1242, rg = 9E-4 - 4.1E-4 mm Hg[4]
PCB	6.6E-5 mm Hg	Aroclor 1260, rg = 0.9E-4 - 4.1E-5 mm Hg[4]
PCB	7.5E-3 mm Hg	Aroclor 1232, rg = 5E-3 - 1E-3 mm Hg[4]
Pentachloroethane	3.4 mm Hg	298 K, C2HCl5[3]
Perchloroethylene	14 mm Hg	298 K, C2Cl4[3]
Propylene dichloride	0.42 mm Hg	298 K, CH2ClCHClCH3[3]
Tetrachloroethane	5 mm Hg	298 K, C2H2Cl4 (sym)[3]
Toluene	22.4 mm Hg	In water, 298 K[2]
Trichloroethylene	58 mm Hg	298 K, CHClCCl2[3]
Vinyl chloride	2320 mm Hg	298 K, CH2CHCl[3]
Vinylidene chloride	497 mm Hg	298 K, CH2CCl2[3]
m-Xylene	6.4 mm Hg	In water, 298 K[2]
o-Xylene	5.2 mm Hg	In water, 298 K[2]
p-Xylene	6.8 mm Hg	In water, 298 K[2]

Source: 1.Dobbs, R.A. and J.M. Cohen. 1980. Carbon adsorption isotherms for Toxic organics. EPA 600/8-80-023.
2.D'Ans-Lax. 1967. Taschenbuch fur Chemiker und Fysiker. Springer-Verlag. Berlin.
3.McConnell, G., D.M. Ferguson and C.R. Pearson. 1975. Endaevour 34, 13.
4.Howard, P.H. 1973. Preliminary Hazard Assesment of Chlorinated Naphthalenes. EPA 560/2/74-001.

2-087 Chemical compounds
Size of Zn compounds

Species	Value	Condition
Size of Zn adsorbed on inorganic colloids in water	260 nm	Rg = 10 - 500 nm, example = Zn(2+)-Fe2O3 or Zn(2+)-SiO2, diameter given[1]
Size of Zn adsorbed on organic colloids in water	260 nm	Rg=10-500 nm, example=Zn(2+)-humic acid, Zn(2+)-organic detritus, diameter given[1]
Size of organic complexes in water	1.5 nm	Rg = 1-2 nm, example = Zn-citrate or Zn-glycinate, diameter given[1]
Size of particulate matter in water	450 nm	Retained by 0.45 um filter, diameter given[1]
Size of simple hydrated metal ion in water	0.8 nm	Example = Zn(H2O)6(2+), diameter given[1]
Size of simple inorganic complexe in water	1 nm	Example = Zn(H2O)5Cl(+), Zn(H2O)5)H(+), diameter given[1]
Size of stable inorganic complexes in water	1.5 nm	Rg = 1-2 nm, example = ZnS, ZnCO3, Zn2SiO4, diameter given[1]
Size of stable organic complexes in water	3 nm	Rg = 2-4 nm, Zn-humate or Zn-cysteinate, diameter given[1]

Source: 1.Ngiagu, J.O. 1980. Zinc in the Environment. Part 1: Ecological cycling. Wiley and Sons. NY.

2-088 Ecosphere
Animal biomass

Species	Value	Condition
Algal beds and reef	12E12 g dry matter	Area = 0.6E12 m2[1]
Boreal forest	57E12 g dry matter	Global value[1]
Continental shelf	160E12 g dry matter	Area = 26.6E12 m2[1]
Cultivated land	6E12 g dry matter	Global value[1]
Desert and semi desert shrubs	8E12 g dry matter	Global value[1]
Estuaries	21E12 g dry matter	Excluding marsh, area = 1.4E12 m2[1]
Extreme desert, rock and sand	0.02E12 g dry matter	Global value[1]
Lake and stream	10E12 g dry matter	Global value[1]
Open ocean	800E12 g dry matter	Area = 332E12 m2[1]
Savanna	220E12 g dry matter	Global value[1]
Swamp and marsh	20E12 g dry matter	Global value[1]
Temperate deciduous forest	110E12 g dry matter	Global value[1]
Temperate evergreen forest	50E12 g dry matter	Global value[1]
Temperate grassland	60E12 g dry matter	Global value[1]
Total biosphere	2002E12 g dry matter	Area = 510E12 m2[1]
Total marine	997E12 g dry matter	Area = 361E12 m2[1]
Total, terrestrial	1005E12 g dry matter	Global value[1]
Tropical rain forest	330E12 g dry matter	Global value[1]
Tropical seasonal forest	90E12 g dry matter	Global value[1]
Tundra and alpine	3.5E12 g dry matter	Global value[1]
Upwelling zones	4E12 g dry matter	Area = 0.4E12 m2[1]
Woodland and shrubland	40E12 g dry matter	Global value[1]

Source: 1.Whittaker, R.H. and G.E. Likens. 1973. in: WoodwellG.M. and E.V. Pecan (eds.): Carbon and Biosphere.AEC Symposium Series 30, 281-302. NTIS. Springfield. Virginia.

2-089 Ecosphere
Animal consumption

Species	Value	Condition
Algal beds and reef	240E12 g dry matter/year	Area = 0.6E12 m2[1]
Boreal forest	380E12 g dry matter/year	Global value[1]
Continental shelf	3000E12 g dry matter/year	Area = 26.6E12 m2[1]
Cultivated land	90E12 g dry matter/year	Global value[1]
Desert and semi desert shrubs	48E12 g dry matter/year	Global value[1]
Estuaries	320E12 g dry matter/year	Excluding marsh, area = 1.4E12 m2[1]
Extreme desert, rock and sand	0.2E12 g dry matter/year	Global value[1]
Lake and stream	100E12 g dry matter/year	Global value[1]
Open ocean	16600E12 g dry matter/year	Area = 332E12 m2[1]
Savanna	2000E12 g dry matter/year	Global value[1]
Swamp and marsh	320E12 g dry matter/year	Global value[1]
Temperate deciduous forest	420E12 g dry matter/year	Global value[1]
Temperate evergreen forest	260E12 g dry matter/year	Global value[1]
Temperate grassland	540E12 g dry matter/year	Global value[1]
Total biosphere	28040E12 g dry matter/year	Area = 510E12 m2[1]
Total marine	20230E12 g dry matter/year	Area = 361E12 m2[1]
Total, terrestrial	7810E12 g dry matter/year	Global value[1]
Tropical rain forest	2600E12 g dry matter/year	Global value[1]
Tropical seasonal forest	720E12 g dry matter/year	Global value[1]
Tundra and alpine	33E12 g dry matter/year	Global value[1]
Upwelling zones	70E12 g dry matter/year	Area = 0.4E12 m2[1]
Woodland and shrubland	300E12 g dry matter/year	Global value[1]

Source: 1.Whittaker, R.H. and G.E. Likens. 1973. in: WoodwellG.M. and E.V. Pecan (eds.): Carbon and Biosphere.AEC Symposium Series 30, 281-302. NTIS. Springfield. Virginia.

2-090 Ecosphere
Animal production

Species	Value	Condition
Algal beds and reef	36E12 g dry matter/year	Area = 0.6E12 m2[1]
Boreal forest	38E12 g dry matter/year	Global value[1]
Continental shelf	430E12 g dry matter/year	Area = 26.6E12 m2[1]
Cultivated land	9E12 g dry matter/year	Global value[1]
Desert and semi desert shrubs	0.076E12 g dry matter/year	Global value[1]
Estuaries	48E12 g dry matter/year	Excluding marsh, area = 1.4E12 m2[1]
Extreme desert, rock and sand	0.02E12 g dry matter/year	Global value[1]
Lake and stream	10E12 g dry matter/year	Global value[1]
Open ocean	2500E12 g dry matter/year	Area = 332E12 m2[1]
Savanna	300E12 g dry matter/year	Global value[1]
Swamp and marsh	32E12 g dry matter/year	Global value[1]

Temperate deciduous forest	42E12 g dry matter/year	Global value[1]
Temperate evergreen forest	26E12 g dry matter/year	Global value[1]
Temperate grassland	80E12 g dry matter/year	Global value[1]
Total biosphere	3934E12 g dry matter/year	Area = 510E12 m2[1]
Total marine	3025E12 g dry matter/year	Area = 361E12 m2[1]
Total, terrestrial	909E12 g dry matter/year	Global value[1]
Tropical rain forest	260E12 g dry matter/year	Global value[1]
Tropical seasonal forest	72E12 g dry matter/year	Global value[1]
Tundra and alpine	3E12 g dry matter/year	Global value[1]
Upwelling zones	11E12 g dry matter/year	Area = 0.4E12 m2[1]
Woodland and shrubland	30E12 g dry matter/year	Global value[1]

Source: 1.Whittaker, R.H. and G.E. Likens. 1973. in: WoodwellG.M. and E.V. Pecan (eds.): Carbon and Biosphere.AEC Symposium Series 30, 281-302. NTIS. Springfield. Virginia.

2-091 Ecosphere
Area

Species	Value	Condition
Arable land	0.23E8 km**2	None[1]
Brown algae zone	0.009E8 km**2	Depth = 0-20m[1]
Coniferous forest	0.10E8 km**2	None[1]
Continental shelf	0.27E8 km**2	Depth = 0-200 m[1]
Deciduous forest	0.049E8 km**2	None[1]
Deep seas	3.34E8 km**2	Depth: > = 200 m[1]
Desert	0.33E8 km**2	None[1]
Earth - surface	5.11E14 m**2	None[2]
Forest	0.144E8 km**2	Other forest, excl tropical, deciduous and confiferous forest[1]
Forest	0.44E8 km**2	Total forest area[1]
Freshwater	0.0024E8 km**2	None[1]
Grassland	0.27E8 km**2	None[1]
Ice	0.11E8 km**2	None[1]
Land	1.48E14 m**2	None[2]
Land	1.49E8 km**2	Total area[1]
Littoral zone	0.000144E8 km**2	None[1]
Ocean	3.61E8 km**2	Total area[1]
Tropical forest	0.147E8 km**2	None[1]
Tundra	0.11E8 km**2	None[1]
Whole earth	5.10E8 km**2	None[1]

Source: 1.Bowen, H.J.M. 1966. Trace Elements in Biochemistry, Ac. Press.
2.Singer, S.F. 1975. In: S.F. Singer (ed.), The Changing Global Environment, pp 25-43. D. Reidel Publ.Comp.

2-092 Ecosphere
Atmospheric turbidity

Species	Value	Condition
Davos	0.026 unit of Schuepp coefficient	1914-26[1]
Davos	0.046 unit of Schuepp coefficient	1957-59[1]
Jerusalem	0.043 unit of Schuepp coefficient	1930-34[1]
Jerusalem	0.050 unit of Schuepp coefficient	1961-68[1]
McMurdo	0.016 unit of Schuepp coefficient	1949-52, Nov[1]
McMurdo	0.022 unit of Schuepp coefficient	1966, Nov[1]
McMurdo	0.024 unit of Schuepp coefficient	1966, Dec[1]
McMurdo	0.026 unit of Schuepp coefficient	1949-52, Dec[1]
Mexico city	0.087 unit of Schuepp coefficient	1911-28[1]
Mexico city	0.181 unit of Schuepp coefficient	1957-62[1]
Mt. Montezuma	0.0076 unit of Schuepp coefficient	1940-48, Chile[1]
Mt. Montezuma	0.0171 unit of Schuepp coefficient	1920-30, Chile[1]
Table Mountain	0.0081 unit of Schuepp coefficient	1940-50, Calif[1]
Table Mountain	0.0185 unit of Schuepp coefficient	1925-30, Calif[1]
Washington DC	0.105 unit of Schuepp coefficient	1903-07[1]
Washington DC	0.165 unit of Schuepp coefficient	1962-66[1]

Source: 1.Ellsaesser, H.W. 1975. In: S.F. Singer (ed.): The Changing Global Environment, pp 235-269. D. ReidelPubl. Comp.

2-093 Ecosphere
Biomass

Species	Value	Condition
Algal beds and reef	2E3 g dry matter/m2	Rg = 0.04 - 4E3 g/m2, total biomass = 1.2E15 g dm[1]
Angiosperms on land	1144549E12 g dry matter in ecosystem typ	Mean biomass = 1800 - 7600 g DM/m2[2]
Animals	0.0050 kg dry weight/m**2	Total = 0.0183E14 kg, incl bacteria, marine animals[3]
Animals	0.014 kg dry weight/m**2	Total = 0.021E14 kg, land animals[3]
Annelida in sea bed	200E12 g dry matter in ecosystem type	Mean biomass = 7.1 g DM/m2[2]
Annelida in soil	915E12 g dry matter in ecosystem type	Mean biomass = 2.3 - 13 g DM/m2[2]
Arable land	0.67 kg dry weight/m**2	Total = 0.15E14 kg, n = 15[3]
Arthropoda in sea bed	0.85E12 g dry matter in ecosystem type	Mean biomass = 0.03 g DM/m2[2]
Arthropoda in soil	839E12 g dry matter in ecosystem type	Mean biomass = 4 - 11 g DM/m2[2]
Atmosphere	5.3E21 g	Value per unit area=1030 g/cm**2, total mass
Bacteria in ocean	720E12 g dry matter in ecosystem type	Mean biomass = 2o g DM/m2[2]
Bacteria in sea bed	16E12 g dry matter in ecosystem type	Mean biomass = 0.004 - 0.2 g DM/m2[2]
Bacteria in soil	4421E12 g dry matter in ecosystem type	Mean biomass = 30 - 240 g DM/m2[2]
Bird on land, tame	0.09E12 g dry matter in ecosystem type	Mean biomass = 0.00062 g DM/m2[2]

Bird on land, wild	0.15E12 g dry matter in ecosystem type	Mean biomass = 0.001 g DM/m2[2]
Brown algae	3.4 kg dry weight/m**2	Total = 0.0305E14 kg, n = 8[3]
Brown algae on continental shelf	3050E12 g dry matter in ecosystem type	Mean biomass = 4300 g DM/m2[2]
Coelenterata in sea bed	100E12 g dry matter in ecosystem type	Mean biomass = 3.7 g DM/m2[2]
Commercial fishery catch	0.0063E15 g C in total	None[4]
Coniferous forest	26.2 kg dry weight/m**2	Total = 2.62E14 kg, n = 18[3]
Continental shelf	0.001E3 g dry matter/m2	Rg = 0.001 - 0.04E3 g/m2, total biomass = 0.27E15 g dm[1]
Coral reef	703 gram dry weight/m**2	Typical value, Eniwetok[5]
Deciduous forest	12.2 kg dry weight/m**2	Total = 0.60E14 kg, n = 9[3]
Decomposers, total	7000E12 g dry matter in ecosystem type	Mean biomass = 13.7 g DM/m2[2]
Desert	0.225 kg dry weight/m**2	Total = 0.074E14 kg, estimated values[3]
Dry land	7.60 kg dry weight/m**2	Total = 11.36E14 kg, in total[3]
Echinodermata in sea bed	76E12 g dry matter in ecosystem type	Mean biomass = 2.7 g DM/m2[2]
Estuaries	1E3 g dry matter/m2	Excluding marsh, rg = 0.01-4E3 g/m2, total biomass = 1.4E15 g dm[1]
Fish in ocean	650E12 g dry matter in ecosystem type	Mean biomass = 1.8 g DM/m2[2]
Forest	6.7 kg dry weight/m**2	Total = 0.97E14 kg, est val, other forest, excl tropical, decid, conif forest[3]
Freshwater	96 gram dry weight/m**2	Typical value, Lake Wisconsin[5]
Fungi in soil	1843E12 g dry matter in ecosystem type	Mean biomass = 10 - 100 g DM/m2[2]
Grassland	0.87 kg dry weight/m**2	Total = 0.235E14 kg, n = 6[3]
Homo sapiens	30E12 g dry matter in ecosystem type	Mean biomass = 0.2 g DM/m2[2]
Mammalia in ocean	55E12 g dry matter in ecosystem type	Mean biomass = 0.15 g DM/m2[2]
Mammalia on land, tame	170E12 g dry matter in ecosystem type	Mean biomass = 1.1 g DM/m2[2]
Mammalia on land, wild	9E12 g dry matter in ecosystem type	Mean biomass = 0.12 g DM/m2[2]
Mollusca in sea bed	21E12 g dry matter in ecosystem type	Mean biomass = 0.75 g DM/m2[2]
Mollusca in soil	191.41E12 g dry matter in ecosystem type	Mean biomass = 0 - 3.3 g DM/m2[2]
Nekton total in oceans	0.12E15 g C in total	None[6]
Nematoda in sea bed	2.3E12 g dry matter in ecosystem type	Mean biomass = 0.08 g DM/m2[2]
Nematoda in soil	240E12 g dry matter in ecosystem type	Mean biomass = 0.2 - 4 g DM/m2[2]
Ocean	0.0201 kg dry weight/m**2	Total = 0.0728E14 kg, in total[3]
Ocean plankton	0.0067 kg dry weight/m**2	Total = 0.024E14 kg, n = many[3]
Ocean water	4 gram dry weight/m**2	Typical value, English Channel[5]
Old field	500 gram dry weight/m**2	Typical value, Georgia[5]
Open ocean	0.003E3 g dry matter/m2	Rg = 0-0.005E3 g/m2, total biomass = 1.0E15 g dm[1]
Phytobenthos in oceans	0.01E15 g C in total	None[7]
Phytobenthos in oceans	0.02E15 g C in total	None[6]
Phytobenthos in oceans	0.06E15 g C in total	None[4]
Phytoplankton in oceans	0.07E15 g C in total	None[7]
Phytoplankton in oceans	0.14E15 g C in total	None[6]

Phytoplankton in oceans	0.59E15 g C in total	None[1]
Plankton in ocean	2400E12 g dry matter in ecosystem type	Mean biomass = 6.7 g DM/m2[2]
Plankton on seabed	0.81E12 g dry matter in ecosystem type	Mean biomass = 0.03 g DM/m2[2]
Protozoa in ocean	19E12 g dry matter in ecosystem type	Mean biomass = 0.052 g DM/m2[2]
Protozoa in sea bed	7.2E12 g dry matter in ecosystem type	Mean biomass = 0.02 g DM/m2[2]
Protozoa in soil	404E12 g dry matter in ecosystem type	Mean biomass = 0.4 - 8 g DM/m2[2]
Soil decomposers	0.047 kg dry weight/m**2	Total = 0.070E14 kg[3]
Total animals	3930E12 g dry matter in ecosystem type	Mean biomass = 7.6 g DM/m2[2]
Total biosphere	3.6E3 g dry matter/m2	Total biomass both marine continental = 1841E15 g[1]
Total global	1160930E12 g dry matter in ecosystem type	Equal to 522698.5E12 g C[2]
Total marine	0.01E3 g dry matter/m2	Total production = 3.9E15 g dm[1]
Total plants	1150000E12 g dry matter in ecosystem type	Mean biomass = 2250 g DM/m2[2]
Tropical forest	44.5 kg dry weight/m**2	Total = 6.55E14 kg, estimated values, n = 1[3]
Tropical rainforest	40000 gram dry weight/m**2	Typical value, Panama[5]
Tundra	0.68 kg dry weight/m**2	Total = 0.074E14 kg, estimated values, n = 1, sphagnum[3]
Upwelling zones	0.02E3 g dry matter/m2	Rg = 0.005 - 0.1E3 g/m2, total biomass = 0.008E15 g dm[1]
Whole earth	2.25 kg dry weight/m**2	Total = 11.43E14 kg[3]
Zoobenthos in oceans	1.67E15 g C in total	None[6]
Zooplankton in oceans	1.55E15 g C in total	None[4]
Zooplankton in oceans	1.94E15 g C in total	None[6]

Source: 1.Whittaker, R.H. and G.E. Likens. 1973. in: WoodwellG.M. and E.V. Pecan (eds.): Carbon and Biosphere.AEC Symposium Series 30, 281-302. NTIS. Springfield. Virginia.
2.Ajtay, G.L., P. Ketner and P. Duvigneaud. 1979. in: Bolin, B., E.T. Degens, S. Kempe and P. Ketner.(ds.) The Global Carbon Cycle. 129-181. SCOPE. John Wiley and Sons. NY.
3.Bowen, H.J.M. 1966. Trace Elements in Biochemistry, Ac. Press.
4.Moiseev, P.A. 1969. The living Resources of the World Ocean. Moscow. Engl by Israel Program of Scientfic Translations Ltd., Jerusalem 1971.
5.Odum, E.P. 1971. Fundamentals of Ecology. Third edition. W.B. Saunders, Philadelphia.
6.Bogorov, V.G. 1969. Znaniye. Biol. Ser. Moscow 6, 3-5.
7.Bazilevich, , N.I., L.Y. Rodin and N.N/ Rozov. 1971. Sov. Geogr. Rev. Transl. 12, 293-317.

2-094 Ecosphere
Biomass and surface area

Species	Value	Condition
Cold desert, grey or sandy soils	0.4 kg dry organic matter/m2	Total area = 1.0E6 km2, Total biomass on earth = 0.2E15 g C, on land[1]
Cold desert, mountainous	0.7 kg dry organic matter/m2	Total area = 1.5E6 km2, Total biomass on earth = 0.5E15 g C, on land[1]
Cultivated land, subtropical	3.0 kg dry organic matter/m2	Total area = 0.3E6 km2, Total biomass on earth = 0.4E15 g C, on land[1]
Cultivated land, temperate	1.2 kg dry organic matter/m2	Total area = 6.7E6 km2, Total biomass on earth = 4E15 g C, on land[1]
Cultivated land, tropical	1.4 kg dry organic matter/m2	Total area = 8.5E6 km2, Total biomass on earth = 5E15 g C, on land[1]
Forest, equatorial rain	35 kg dry organic matter/m2	Total area = 11.0E6 km2, Total biomass on earth = 175E15 g C, on land[1]

Forest, tropical seasonal	18 kg dry organic matter/m2	Total area = 5.0E6 km2, Total biomass on earth = 41E15 g C, on land[1]
Glacier	0 kg dry organic matter/m2	Total area = 13.9E6 km2, Total biomass on earth = 0.0E15 g C, on land[1]
Halophytes	0.2 kg dry organic matter/m2	Total area = 0.5E6 km2, Total biomass on earth = 0.05E15 g C, on land[1]
Lake and stream	0.4 kg dry organic matter/m2	Total area = 2.0E6 km2, Total biomass on earth = 0.4E15 g C, on land[1]
Mangroves	7 kg dry organic matter/m2	Total area = 0.3E6 km2, Total biomass on earth = 1.0E15 g C, on land[1]
Mixed forest	28 kg dry organic matter/m2	Total area = 1.5E6 km2, Total biomass on earth = 20E15 g C, on land[1]
Peatland	3.5 kg dry organic matter/m2	Total area = 1.5E6 km2, Total biomass on earth = 2.4E15 g C, on land
Polar desert	0.5 kg dry organic matter/m2	Total area = 1.5E6 km2, Total biomass on earth = 0.3E15 g C, on land[1]
Savanna, derived grass	3.0 kg dry organic matter/m2	Total area = 4.5E6 km2, Total biomass on earth = 6E15 g C, on land[1]
Savanna, derived low trees	7.0 kg dry organic matter/m2	Total area = 4.5E6 km2, Total biomass on earth = 14E15 g C, on land[1]
Savanna, spiny forest	11 kg dry organic matter/m2	Total area = 2.0E6 km2, Total biomass on earth = 10E15 g C, on land[1]
Savanna, spiny shrub and tree savanna	5 kg dry organic matter/m2	Total area = 8.0E6 km2, Total biomass on earth = 18E15 g C, on land[1]
Sclerophyll forest	15 kg dry organic matter/m2	Total area = 1.3E6 km2, Total biomass on earth = 9E15 g C, on land[1]
Sclerophyll forest, chaparrals and macchia	7 kg dry organic matter/m2	Total area = 2.7E6 km2, Total biomass on earth = 9E15 g C, on land[1]
Semi desert, scrub and grass, subtropical and mediterranean	1.3 kg dry organic matter/m2	Total area = 6.9E6 km2, Total biomass on earth = 4E15 g C, on land[1]
Semi desert, scrub and grass, temperate	1.1 kg dry organic matter/m2	Total area = 2.3E6 km2, Total biomass on earth = 1.1E15 g C, on land[1]
Swamp and marsh, temperate	5 kg dry organic matter/m2	Total area = 0.6E6 km2, Total biomass on earth = 1.4E15 g C, on land[1]
Swamp and marsh, tropical	12 kg dry organic matter/m2	Total area = 1.4E6 km2, Total biomass on earth = 8E15 g C, on land[1]
Taiga, middle	23 kg dry organic matter/m2	Total area = 6.0E6 km2, Total biomass on earth = 63E15 g C, on land[1]
Taiga, mountainous	17 kg dry organic matter/m2	Total area = 5.7E6 km2, Total biomass on earth = 43E15 g C, on land[1]
Taiga, nothern	11 kg dry organic matter/m2	Total area = 2.1E6 km2, Total biomass on earth = 11E15 g C, on land[1]
Taiga, southern	26 kg dry organic matter/m2	Total area = 1.6E6 km2, Total biomass on earth = 19E15 g C, on land[1]
Temperate forest, coniferous plantations	18 kg dry organic matter/m2	Total area = 2.0E6 km2, Total biomass on earth = 16E15 g C, on land
Temperate forest, deciduous	28 kg dry organic matter/m2	Total area = 3.0E6 km2, Total biomass on earth = 38E15 g C, on land[1]
Temperate forest, evergreen	30 kg dry organic matter/m2	Total area = 4.0E6 km2, Total biomass on earth = 55E15 g C, on land[1]
Temperate grassland, chernozems steppe	2.0 kg dry organic matter/m2	Total area = 3.5E6 km2, Total biomass on earth = 3.2E15 g C, on land[1]

Temperate grassland, evergreen	2.0 kg dry organic matter/m2	Total area = 2.4E6 km2, Total biomass on earth = 2.2E15 g C, on land[1]
Temperate grassland, kastanozems steppe	1.4 kg dry organic matter/m2	Total area = 2.8E6 km2, Total biomass on earth = 1.8E15 g C, on land[1]
Temperate grassland, solonetz steppe	1.3 kg dry organic matter/m2	Total area = 0.3E6 km2, Total biomass on earth = 0.2E15 g C, on land[1]
Towns	3.5 kg dry organic matter/m2	Total area = 1.0E6 km2, Total biomass on earth = 1.6E15 g C, on land[1]
Tundra, grey soil	2.8 kg dry organic matter/m2	Total area = 3.8E6 km2, Total biomass on earth = 5E15 g C, on land[1]
Tundra, mountainous	0.5 kg dry organic matter/m2	Total area = 2.9E6 km2, Total biomass on earth = 0.7E15 g C, on land[1]
Warm desert, extreme	0.01 kg dry organic matter/m2	Total area = 8.0E6 km2, Total biomass on earth = 0.03E15 g C, on land[1]
Warm desert, subtropical and mediterranean	0.3 kg dry organic matter/m2	Total area = 4.2E6 km2, Total biomass on earth = 0.5E15 g C, on land[1]
Warm desert, tropical	0.1 kg dry organic matter/m2	Total area = 5.6E6 km2, Total biomass on earth = 0.3E15 g C, on land[1]

Source: 1.Duvigneaud, P. 1972. La pensee et les hommes 7, 286-303.

2-095 Ecosphere
Carbon content

Species	Value	Condition
Arable land	0.30 kg C/m**2	Total = 0.069E14 kg, n = 15[1]
Brown algae	1.0 kg C/m**2	Total = 0.0090E14 kg, n = 8[1]
Coniferous forest	11.8 kg C/m**2	Total = 1.18E14 kg, n = 18[1]
Deciduous forest	5.5 kg C/m**2	Total = 0.27E14 kg, n = 9[1]
Desert	0.1 kg C/m**2	Total = 0.033E14 kg, estimated values[1]
Dry land	3.42 kg C/m**2	Total = 5.10E14 kg, in total[1]
Forest	3.0 kg C/m**2	Total = 0.43E14 kg, est values, other forest, excl decid, tropical, coniferous forest[1]
Grassland	0.39 kg C/m**2	Total = 0.106E14 kg, n = 6[1]
Land animals	0.0064 kg C/m**2	Total = 0.0095E14 kg[1]
Marine animals	0.0023 kg C/m**2	Total = 0.0083E14 kg, incl bacteria[1]
Ocean	0.0063 kg C/m**2	Total = 0.0227E14 kg, in total[1]
Ocean plankton	0.0015 kg C/m**2	Total = 0.0054E14 kg, n = many[1]
Soil decomposers	0.021 kg C/m**2	Total = 0.0315E14 kg[1]
Tropical forest	20.0 kg C/m**2	Total = 2.94E14 kg, estimated value, n = 1[1]
Tundra	0.3 kg C/m**2	Total = 0.033E14 kg, estimated values, n = 1, sphagnum[1]
Whole earth	1.004 kg C/m**2	Total = 5.12E14 kg[1]

Source: 1.Bowen, H.J.M. 1966. Trace Elements in Biochemistry, Ac. Press.

2-096 Ecosphere
Change in surface area of livestock units

Species	Value	Condition
Africa	45.6 % increase in 24 years	Period start at 1951-52 with 121.8E6 livestock units[1]
Asia	62.6 % increase in 24 years	Period start at 1951-52 with 344.7E6 livestock units[1]
Europe	19.3 % increase in 24 years	Period start at 1951-52 with 136.0E6 livestock units[1]
North and Central America	46.0 % increase in 24 years	Period start at 1951-52 with 135.0E6 livestock units[1]
Oceania	55.1 % increase in 24 years	Period start at 1951-52 with 36.5E6 livestock units[1]
South America	39.2 % increase in 24 years	Period start at 1951-52 with 159.7E6 livestock units[1]
USSR	58.2 % increase in 24 years	Period start at 1951-52 with 78.8E6 livestock units[1]
World total	48.2 % increase in 24 years	Period start at 1951-52 with 1500.8E6 livestock units[1]

Source: 1.Ajtay, G.L., P. Ketner and P. Duvigneaud. 1979. in: Bolin, B., E.T. Degens, S. Kempe and P. Ketner.(ds.) The Global Carbon Cycle. 129-181. SCOPE. John Wiley and Sons. NY.

2-097 Ecosphere
Change in surface area of permanent pasture

Species	Value	Condition
Africa	27.3 % increase in 23 years	Period start at 1952 with 6230E9 m2[1]
Asia	27.0 % increase in 23 years	Period start at 1952 with 4330E9 m2[1]
Europe	10.9 % decrease in 23 years	Period start at 1952 with 980E9 m2[1]
North and Central America	15.2 % decrease in 23 years	Period start at 1952 with 3840E9 m2[1]
Oceania	20.7 % increase in 23 years	Period start at 1952 with 3760E9 m2[1]
South America	45.0 % increase in 23 years	Period start at 1952 with 3070E9 m2[1]
USSR	202.5 % increase in 23 years	Period start at 1952 with 1240E9 m2[1]
World total	29.2 % increase in 23 years	Period start at 1952 with 23450E9 m2[1]

Source: 1.Ajtay, G.L., P. Ketner and P. Duvigneaud. 1979. in: Bolin, B., E.T. Degens, S. Kempe and P. Ketner.(ds.) The Global Carbon Cycle. 129-181. SCOPE. John Wiley and Sons. NY.

2-098 Ecosphere
Changes in production of cultivated plants

Species	Value	Condition
Cereals	46 % increase in 20 years	Total production = 2550E12 g C/y in 1973, period start in 1953[1]
Fibre crop	62 % increase in 20 years	Total production = 190E12 g C/y in 1973, period start in 1953[1]
Oil crop	11 % decrease in 20 years	Total production = 128E12 g C/y in 1973, period start in 1953[1]
Other then given by same ref	77 % increase in 20 years	Total production = 204E12 g C/y in 1973, period start in 1953[1]
Pulses, vegetables	353 % increase in 20 years	Total production = 4126E12 g C/y in 1973, period start in 1953[1]
Roots, tubers	74 % increase in 20 years	Total production = 219E12 g C/y in 1973, period start in 1953[1]
Shrubs	83 % increase in 20 years	Total production = 119E12 g C/y in 1973, period start in 1953[1]
Total for all cultivated plants	58 % increase in 20 years	Total prod = 3966E12 g C/y in 1973, per. start 53[1]
Trees	30 % increase in 20 years	Total prod = 144E12 g C/y in 1973, period start 53[1]

Source: 1.Ajtay, G.L., P. Ketner and P. Duvigneaud. 1979. in: Bolin, B., E.T. Degens, S. Kempe and P. Ketner.(ds.) The Global Carbon Cycle. 129-181. SCOPE. John Wiley and Sons. NY.

2-099 Ecosphere
Changes in production of cultivated plants, area

Species	Value	Condition
Cereals	18 % increase in 20 years	Total area = 7098E9 m2 in 1973, period start in 1953[1]
Fibre crop	68 % decrease in 20 years	Total area = 137E9 m2 in 1973, period start in 1953[1]
Oil crop	27 % decrease in 20 years	Total area = 527E9 m2 in 1973, period start in 1953[1]
Other then given by same ref	50 % increase in 20 years	Total area = 154E9 m2 in 1973, period start in 1953[1]
Pulses, vegetables	135 % increase in 20 years	Total area = 1346E9 m2 in 1973, period start in 1953[1]
Roots, tubers	51 % increase in 20 years	Total area = 596E9 m2 in 1973, period start in 1953[1]
Shrubs	16 % increase in 20 years	Total area = 346E9 m2 in 1973, period start in 1953[1]
Total for all cultivated plants	21 % increase in 20 years	Total area = 10577E9 m2 in 1973, period start in 1953[1]
Trees	65 % increase in 20 years	Total area = 373E9 m2 in 1973, period start in 1953[1]

Source: 1.Ajtay, G.L., P. Ketner and P. Duvigneaud. 1979. in: Bolin, B., E.T. Degens, S. Kempe and P. Ketner.(ds.) The Global Carbon Cycle. 129-181. SCOPE. John Wiley and Sons. NY.

2-100 Ecosphere
Changes in production of cultivated plants, biomass

Species	Value	Condition
Cereals	60 % increase in 20 years	Total biomass = 2436E12 g C in 1973, period start in 1953[1]
Fibre crop	61 % increase in 20 years	Total biomass = 187E12 g C in 1973, period start in 1953[1]
Oil crop	11 % decrease in 20 years	Total biomass = 116E12 g C in 1973, period start in 1953[1]
Other then given by same ref	78 % increase in 20 years	Total biomass = 189E12 g C in 1973, period start in 1953[1]
Pulses, vegetables	334 % increase in 20 years	Total biomass = 382E12 g C in 1973, period start in 1953[1]
Roots, tubers	17 % increase in 20 years	Total biomass = 190E12 g C in 1973, period start in 1953[1]
Shrubs	77 % increase in 20 years	Total biomass = 539E12 g C in 1973, period start in 1953[1]
Total for all cultivated plants	26 % increase in 20 years	Total biomass = 5399E12 g C in 1973, period start 1953[1]
Trees	27 % decrease in 20 years	Total biomass = 1360E12 g C in 1973, period start 1953[1]

Source: 1.Ajtay, G.L., P. Ketner and P. Duvigneaud. 1979. in: Bolin, B., E.T. Degens, S. Kempe and P. Ketner.(ds.) The Global Carbon Cycle. 129-181. SCOPE. John Wiley and Sons. NY.

2-101 Ecosphere
Classification of lakes

Species	Value	Condition
Cultural eutrophic	550 gram C/(m**2*year)	Net primary productivity, rg = 300 - 800[1]
Mesotrophic	40 gram C/(m**2*year)	Net primary productivity, rg = 10 - 70[1]
Natural eutrophic	235 gram C/(m**2*year)	Net primary productivity, rg = 70 - 400[1]
Oligotrophic	5.05 gram C/(m**2*year)	Net primary productivity, rg = 0.1 - 10[1]

Source: 1.Rodhe, W. 1969. Nat. Acad. Sci. (U.S.) Symp. Eutrophication, pp 50-64.

2-102 Ecosphere
Density

Species	Value	Condition
Air	1.25E-3 gram/cm**3	Air, still[1]
Clay	1.95 gram/cm**3	Range = 1.7 - 2.2, wet clay[1]
Ice	2.0 gram/cm**3	Range = 1.7 - 2.3[1]

Sand	1.55 gram/cm**3	Range = 1.4 - 1.7, dry sand[1]
Soil organic matter	1.3 gram/cm**3	None[1]
Water	1.0 gram/cm**3	Still water[1]

Source: 1.Collier, B.D., G.W. Cox, A.W. Johnson and P.C. Miller. 1974. Dynamic Ecology. Prentice-Hall. New Jerey.

2-103 Ecosphere
Diurnal damping depth

Species	Value	Condition
Air	72.7 cm	Range = 69.7 - 75.7, still air[1]
Clay	16.0 cm	Range = 13.5 - 18.5, wet clay[1]
Clay	8.8 cm	Range = 7.4 - 10.2, dry clay[1]
Concrete	24.6 cm	None[1]
Fe	265.5 cm	None[1]
Ice	17.3 cm	Range = 14.1 - 20.5[1]
Rock	18.9 cm	Range = 16.0 - 21.8[1]
Sand	16.6 cm	Wet sand[1]
Sand	8.3 cm	Range = 5.8 - 10.5, dry sand[1]
Silver	215 cm	None[1]
Soil organic matter	5.5 cm	None[1]
Water	6.2 cm	Range = 6.0 - 6.4, still water[1]
Wood	7.85 cm	Range = 7.4 - 8.3, dry wood[1]

Source: 1.Collier, B.D., G.W. Cox, A.W. Johnson and P.C. Miller. 1974. Dynamic Ecology. Prentice-Hall. New Jerey.

2-104 Ecosphere
Energy flow, biomass and number

Species	Value	Condition
Acartia sp.	2.5 kcal/(m**2*day)	Marine copepod, biomass = 2.0 g/m2, density=E5 anim/m2[1]
Littorina sp.	1.0 kcal/(m**2*day)	Intertidal snail, biomass = 10.0 g/m2, density=200 anim/m2[1]
Microtus sp.	0.7 kcal/(m**2*day)	Meadow mice, biomass = 0.6 g/m2, density=E-2 anim/m2[1]
Odocoileus sp.	0.5 kcal/(m**2*day)	Deer, biomass = 1.1 g/m2, density=E(-5) anim/m2[1]
Orchelimum sp.	0.4 kcal/(m**2*day)	Salt marsh grasshoppers, biomass = 1.0 g/m2, density=10 anim/m2[1]
Soil bacteria	1.0 kcal/(m**2*day)	Biomass = 0.001 g/m2, density=E12 anim/m2[1]

Source: 1.Odum, E.P. 1968. Amer. Zool. 8, pp 11-18.

2-105 Ecosphere
Fluxes within the Cd cycle

Species	Value	Condition
Accumulation in sediments	2.0E8 g/y	Hydrogenous precipitation[1]
Accumulation in sediments	2.2E10 g/y	Conc = 1.5 mg/kg, Material flux (g/year) = 1.5E16, lithogenous depos.[1]
Accumulation in sediments	3E9 g/y	Conc = 1.0 mg/kg, Material flux (g/year) = 3E15, skeletal input[1]
Atmospheric emission, anthropogenic sources	7.3E9 g/y	none[1]
Atmospheric emission, natural sources	0.8E9 g/y	none[1]
Atmospheric fallout	2.4E9 g/y	On oceans[1]

Atmospheric fallout	5.7E9 g/y	On land[1]
Biological uptake, freshwater	2E9 g/y	Conc = 2.0 mg/kg, Material flux (g/year) = 1.0E15[1]
Biological uptake, marine	2.4E11 g/y	Conc = 4.0 mg/kg, Material flux (g/year) = 6E16[1]
Biological uptake, terrestrial	3.6E10 g/y	Conc = 0.3 mg/kg, Material flux (g/year) = 12E16[1]
Continental denudation	1.2E10 g/y	Conc = 0.5 mg/kg, Material flux (g/year) = 2.5E16[1]
Fertilizers	6.6E8 g/y	Conc = 7 mg/kg, Material flux (g/year) = 9.4E13[1]
Human consumption	2.2E8 g/y	Uptake = 0.15 mg/person/day, total population = 4E9[1]
Litter fall	3.1E10 g/y	Conc = 0.6 mg/kg, Material flux (g/year) = 5.2E16[1]
Rivers to Oceans, part. Cd	2.25E10 g/y	Conc = 1.5 mg/kg, Material flux (g/year) = 1.5E16[1]
Rivers to oceans, diss. Cd	1.6E9 g/y	Conc = 0.05 mg/kg, Material flux (g/year) = 0.32E10[1]
Waste disposal	2.2E8 g/y	Conc = 0.8 mg/kg, Material flux (g/year) = 2.8E14, fly ash on land[1]
Waste disposal	4.8E8 g/y	Conc = 23 mg/kg, Material flux (g/year) = 2.1E13, sewage on land[1]
Waste disposal	5.8E8 g/y	Conc = 23 mg/kg, Material flux (g/year) = 2.5E13, in oceans[1]

Source: 1.Nriagu, J.O. 1979. in: Nriagu, J.O. (ed): Copper in the Environment. Part 1A. Elsevier. Amsterdam. p137-183.

2-106 Ecosphere
Fluxes within the Pb cycles

Species	Value	Condition
From aerosols to oceans	37000 t/year	Pb from gasoline, industrial sources, total ocean[1]
From aerosols to Oceans	*1000 t/year*	*From neolithic sources, total ocean*[1]
From aerosols to oceans	3000 t/year	Pb from smelters and forest fires, industrial sources, total ocean[1]
From rivers	60000 t/year	Pb as soluble, mainly from aerosols, industrial sources, total ocean[1]
From rivers and sewers to oceans	200000 t/year	Pb as solids, industrial sources, total ocean[1]
From rivers to Oceans	13000 t/year	Soluble Pb, Neolithic sources, total ocean[1]
From rivers to oceans	100000 t/year	Solid Pb, neolithic sources, total ocean[1]

Source: 1.Patterson, CC, D. Settle, B. Schaule and M. Burnett. 1976. in: H.L. Windom and R.A. Duce (eds.): Transport of pollutant lead to the ocean and within the ocean ecosystem. Lexington. D.C. Heath. pp 23-28

2-107 Ecosphere
Forest cutting and industrial wood

Species	Value	Condition
Africa	309E6 m3	Industrial wood = 14 %, 1973[1]
Asia, except U.S.S.R.	691E6 m3	Industrial wood = 29 %, 1973[1]
Europe, except U.S.S.R.	3328E6 m3	Industrial wood = 83 %, 1973[1]
Latin America	278.1E6 m3	Industrial wood = 19 %, 1973[1]
North America	480E6 m3	Industrial wood = 96 %, 1973[1]
Oceania	27E6 m3	Industrial wood = 78 %, 1973[1]
Total world	2500E6 m3	Industrial wood = 54 %, 1973[1]
USSR	383E6 m3	Industrial wood = 78 %, 1973[1]

Source: 1.Skogsstyrelsen 1976. Skogstatistisj AArsbok 1974. Sveriges officiella statistik. Skogsstyrelsen. Joekoeking.

2-108 Ecosphere
Forest cutting, industrial and fuel wood

Species	Value	Condition
Argentina, both coniferous and broadleaved forest	11.5E9 m3 solid vol.	Value is year 1973, in 1972 = 12.7E6 m3[1]
Argentina, broadleaved forest	10.5E9 m3 solid vol.	Industrial wood = 2.5E6 m3, Fuel wood = 8.0E6 m3, 1973[1]
Argentina, coniferous forest	0.6E9 m3 solid vol.	Industrial wood = 0.6E6 m3, Fuel wood = 0E6 m3, 1973[1]
Australia, both coniferous and broadleaved forest	12.1E9 m3 solid vol.	Value is year 1973, in 1972 = 14.1E6 m3[1]
Australia, broadleaved forest	9.4E9 m3 solid vol.	Industrial wood = 8.7E6 m3, Fuel wood = 0.7E6 m3, 1973[1]
Australia, coniferous forest	2.7E9 m3 solid vol.	Industrial wood = 2.7E6 m3, Fuel wood = 0E6 m3, 1973[1]
Austria, both coniferous and broadleaved forest	11.9E9 m3 solid vol.	Value is year 1973, in 1972 = 12.5E6 m3[1]
Austria, broadleaved forest	2.1E9 m3 solid vol.	Industrial wood = 1.4E6 m3, Fuel wood = 0.7E6 m3, 1973[1]
Austria, coniferous forest	9.8E9 m3 solid vol.	Industrial wood = 9.6E6 m3, Fuel wood = 0.2E6 m3, 1973[1]
Brazil, both coniferous and broadleaved forest	163.8E9 m3 solid vol.	Value is year 1973, in 1972 = 163.8E6 m3[1]
Brazil, broadleaved forest	137.7E9 m3 solid vol.	Industrial wood = 12.7E6 m3, Fuel wood = 125.0E6 m3, 1973[1]
Brazil, coniferous forest	26.1E9 m3 solid vol.	Industrial wood = 11.1E6 m3, Fuel wood = 15.0E6 m3, 1973[1]
Canada, both coniferous and broadleaved forest	124.2E9 m3 solid vol.	Value is year 1973, in 1972 = 119.7E6 m3[1]
Canada, broadleaved forest	11.6E9 m3 solid vol.	Industrial wood = 9.4E6 m3, Fuel wood = 2.2E6 m3, 1973[1]
Canada, coniferous forest	112.6E9 m3 solid vol.	Industrial wood = 111.4E6 m3, Fuel wood = 1.2E6 m3, 1973[1]
China, both coniferous and broadleaved forest	183.2E9 m3 solid vol.	Value is year 1973, in 1972 = 179.0E6 m3[1]
China, broadleaved forest	100.7E9 m3 solid vol.	Industrial wood = 18.5E6 m3, Fuel wood = 82.2E6 m3, 1973[1]
China, coniferous forest	82.5E9 m3 solid vol.	Industrial wood = 28.5E6 m3, Fuel wood = 54.0E6 m3, 1973[1]
Columbia, both coniferous and broadleaved forest	24.9E9 m3 solid vol.	Value is year 1973, in 1972 = 26.8E6 m3[1]
Columbia, broadleaved forest	24.9E9 m3 solid vol.	Industrial wood = 4.9E6 m3, Fuel wood = 20.0E6 m3, 1973[1]
Columbia, coniferous forest	0.0E9 m3 solid vol.	Industrial wood = 0.0E6 m3, Fuel wood = 0.0E6 m3, 1973[1]
Czechoslovakia, both coniferous and broadleaved forest	15.1E9 m3 solid vol.	Value is year 1973, in 1972 = 14.7E6 m3[1]
Czechoslovakia, broadleaved forest	3.8E9 m3 solid vol.	Industrial wood = 3.0E6 m3, Fuel wood = 0.8E6 m3, 1973[1]
Czechoslovakia, coniferous forest	11.3E9 m3 solid vol.	Industrial wood = 10.6E6 m3, Fuel wood = 0.7E6 m3, 1973[1]
DDR, both coniferous and broadleaved forest	9.4E9 m3 solid vol.	Value is year 1973, in 1972 = 7.9E6 m3[1]
DDR, broadleaved forest	1.4E9 m3 solid vol.	Industrial wood = 1.4E6 m3, Fuel wood = 0.0E6 m3, 1973[1]
DDR, coniferous forest	8.0E9 m3 solid vol.	Industrial wood = 7.2E6 m3, Fuel wood = 0.8E6 m3, 1973[1]
Ethiopia, both coniferous and broadleaved forest	24.3E9 m3 solid vol.	Value is year 1973, in 1972 = 24.2E6 m3[1]
Ethiopia, broadleaved forest	21.3E9 m3 solid vol.	Industrial wood = 1.1E6 m3, Fuel wood = 20.2E6 m3, 1973[1]
Ethiopia, coniferous forest	3.0E9 m3 solid vol.	Industrial wood = 0.2E6 m3, Fuel wood = 2.8E6 m3, 1973[1]
Finland, both coniferous and broadleaved forest	42.9E9 m3 solid vol.	Value is year 1973, in 1972 = 42.9E6 m3[1]

Finland, broadleaved forest	11.6E9 m3 solid vol.	Industrial wood = 5.7E6 m3, Fuel wood = 5.9E6 m3, 1973[1]
Finland, coniferous forest	31.3E9 m3 solid vol.	Industrial wood = 29.9E6 m3, Fuel wood = 1.4E6 m3, 1973[1]
France, both coniferous and broadleaved forest	33.9E9 m3 solid vol.	Value is year 1973, in 1972 = 33.9E6 m3[1]
France, broadleaved forest	19.2E9 m3 solid vol.	Industrial wood = 14.4E6 m3, Fuel wood = 4.8E6 m3, 1973[1]
France, coniferous forest	14.7E9 m3 solid vol.	Industrial wood = 13.7E6 m3, Fuel wood = 1.0E6 m3, 1973[1]
India, both coniferous and broadleaved forest	116.6E9 m3 solid vol.	Value is year 1973, in 1972 = 116.6E6 m3[1]
India, broadleaved forest	112.0E9 m3 solid vol.	Industrial wood = 9.2E6 m3, Fuel wood = 102.8E6 m3, 1973[1]
India, coniferous forest	4.6E9 m3 solid vol.	Industrial wood = 1.4E6 m3, Fuel wood = 3.2E6 m3, 1973[1]
Indonesia, both coniferous and broadleaved forest	131.2E9 m3 solid vol.	Value is year 1973, in 1972 = 120.0E6 m3[1]
Indonesia, broadleaved forest	131.1E9 m3 solid vol.	Industrial wood = 27.1E6 m3, Fuel wood = 104.0E6 m3, 1973[1]
Indonesia, coniferous forest	0.1E9 m3 solid vol.	Industrial wood = 0.1E6 m3, Fuel wood = 0.1E6 m3, 1973[1]
Italy, both coniferous and broadleaved forest	11.7E9 m3 solid vol.	Value is year 1973, in 1972 = 13.1E6 m3[1]
Italy, broadleaved forest	10.3E9 m3 solid vol.	Industrial wood = 5.0E6 m3, Fuel wood = 5.3E6 m3, 1973[1]
Italy, coniferous forest	1.4E9 m3 solid vol.	Industrial wood = 1.1E6 m3, Fuel wood = 0.3E6 m3, 1973[1]
Japan, both coniferous and broadleaved forest	44.6E9 m3 solid vol.	Value is year 1973, in 1972 = 46.2E6 m3[1]
Japan, broadleaved forest	19.0E9 m3 solid vol.	Industrial wood = 17.5E6 m3, Fuel wood = 1.5E6 m3, 1973[1]
Japan, coniferous forest	25.6E9 m3 solid vol.	Industrial wood = 25.6E6 m3, Fuel wood = 0.0E6 m3, 1973[1]
Mexico, both coniferous and broadleaved forest	14.1E9 m3 solid vol.	Value is year 1973, in 1972 = 14.2E6 m3[1]
Mexico, broadleaved forest	6.5E9 m3 solid vol.	Industrial wood = 0.5E6 m3, Fuel wood = 6.0E6 m3, 1973[1]
Mexico, coniferous forest	7.6E9 m3 solid vol.	Industrial wood = 5.0E6 m3, Fuel wood = 2.6E6 m3, 1973[1]
New Zealand, both coniferous and broadleaved forest	8.8E9 m3 solid vol.	Value is year 1973, in 1972 = 8.8E6 m3[1]
New Zealand, broadleaved forest	0.3E9 m3 solid vol.	Industrial wood = 0.2E6 m3, Fuel wood = 0.1E6 m3, 1973[1]
New Zealand, coniferous forest	8.5E9 m3 solid vol.	Industrial wood = 8.0E6 m3, Fuel wood = 0.5E6 m3, 1973[1]
Nigeria, both coniferous and broadleaved forest	59.8E9 m3 solid vol.	Value is year 1973, in 1972 = 59.8E6 m3[1]
Nigeria, broadleaved forest	59.8E9 m3 solid vol.	Industrial wood = 3.0E6 m3, Fuel wood = 56.8E6 m3, 1973[1]
Norway, both coniferous and broadleaved forest	8.6E9 m3 solid vol.	Value is year 1973, in 1972 = 8.3E6 m3[1]
Norway, broadleaved forest	0.9E9 m3 solid vol.	Industrial wood = 0.4E6 m3, Fuel wood = 0.5E6 m3, 1973[1]
Norway, coniferous forest	7.7E9 m3 solid vol.	Industrial wood = 7.5E6 m3, Fuel wood = 0.2E6 m3, 1973[1]
Philippines, both coniferous and broadleaved forest	34.9E9 m3 solid vol.	Value is year 1973, in 1972 = 33.1E6 m3[1]
Philippines, broadleaved forest	34.9E9 m3 solid vol.	Industrial wood = 13.8E6 m3, Fuel wood = 21.1E6 m3, 1973[1]
Philippines, coniferous forest	0.0E9 m3 solid vol.	Industrial wood = 0.0E6 m3, Fuel wood = 0.0E6 m3, 1973[1]
Poland, both coniferous and broadleaved forest	21.9E9 m3 solid vol.	Value is year 1973, in 1972 = 18.8E6 m3[1]
Poland, broadleaved forest	5.4E9 m3 solid vol.	Industrial wood = 4.5E6 m3, Fuel wood = 0.9E6 m3, 1973[1]
Poland, coniferous forest	16.5E9 m3 solid vol.	Industrial wood = 15.6E6 m3, Fuel wood = 0.9E6 m3, 1973[1]

Portugal, both coniferous and broadleaved forest	7.4E9 m3 solid vol.	Value is year 1973, in 1972 = 7.0E6 m3[1]
Portugal, broadleaved forest	2.2E9 m3 solid vol.	Industrial wood = 2.0E6 m3, Fuel wood = 0.2E6 m3, 1973[1]
Portugal, coniferous forest	5.2E9 m3 solid vol.	Industrial wood = 4.9E6 m3, Fuel wood = 0.3E6 m3, 1973[1]
Romania, both coniferous and broadleaved forest	21.4E9 m3 solid vol.	Value is year 1973, in 1972 = 21.2E6 m3[1]
Romania, broadleaved forest	14.5E9 m3 solid vol.	Industrial wood = 9.6E6 m3, Fuel wood = 4.9E6 m3, 1973[1]
Romania, coniferous forest	6.9E9 m3 solid vol.	Industrial wood = 6.5E6 m3, Fuel wood = 0.4E6 m3, 1973[1]
South Africa, both coniferous and broadleaved forest	10.5E9 m3 solid vol.	Value is year 1973, in 1972 = 9.6E6 m3[1]
South Africa, broadleaved forest	6.0E9 m3 solid vol.	Industrial wood = 5.1E6 m3, Fuel wood = 0.9E6 m3, 1973[1]
South Africa, coniferous forest	4.5E9 m3 solid vol.	Industrial wood = 4.4E6 m3, Fuel wood = 0.1E6 m3, 1973[1]
Spain, both coniferous and broadleaved forest	16.8E9 m3 solid vol.	Value is year 1973, in 1972 = 16.0E6 m3[1]
Spain, broadleaved forest	9.1E9 m3 solid vol.	Industrial wood = 3.2E6 m3, Fuel wood = 5.9E6 m3, 1973[1]
Spain, coniferous forest	7.7E9 m3 solid vol.	Industrial wood = 5.6E6 m3, Fuel wood = 2.1E6 m3, 1973[1]
Sweden, both coniferous and broadleaved forest	58.0E9 m3 solid vol.	Value is year 1973, in 1972 = 58.0E6 m3[1]
Sweden, broadleaved forest	6.3E9 m3 solid vol.	Industrial wood = 4.4E6 m3, Fuel wood = 1.9E6 m3, 1973[1]
Sweden, coniferous forest	51.7E9 m3 solid vol.	Industrial wood = 50.5E6 m3, Fuel wood = 1.2E6 m3, 1973[1]
Tanzania, both coniferous and broadleaved forest	32.7E9 m3 solid vol.	Value is year 1973, in 1972 = 32.7E6 m3[1]
Tanzania, broadleaved forest	32.6E9 m3 solid vol.	Industrial wood = 1.1E6 m3, Fuel wood = 31.5E6 m3, 1973[1]
Tanzania, coniferous forest	0.1E9 m3 solid vol.	Industrial wood = 0.1E6 m3, Fuel wood = 0.0E6 m3, 1973[1]
Total world, both coniferous and broadleaved forest	2497.5E9 m3 solid vol.	Value is year 1973, in 1972 = 2453.6E6 m3[1]
Total world, broadleaved forest	1345.4E9 m3 solid vol.	Industrial wood = 378.0E6 m3, Fuel wood = 967.4E6 m3, 1972[1]
Total world, broadleaved forest	1373.7E9 m3 solid vol.	Industrial wood = 401.5E6 m3, Fuel wood = 972.2E6 m3, 1973[1]
Total world, coniferous forest	1108.2E9 m3 solid vol.	Industrial wood = 935.3E6 m3, Fuel wood = 172.9E6 m3, 1972[1]
Total world, coniferous forest	1123.8E9 m3 solid vol.	Industrial wood = 949.7E6 m3, Fuel wood = 174.1E6 m3, 1973[1]
USA, both coniferous and broadleaved forest	355.6E9 m3 solid vol.	Value is year 1973, in 1972 = 355.7E6 m3[1]
USA, broadleaved forest	83.4E9 m3 solid vol.	Industrial wood = 72.5E6 m3, Fuel wood = 10.9E6 m3, 1973[1]
USA, coniferous forest	272.2E9 m3 solid vol.	Industrial wood = 269.7E6 m3, Fuel wood = 2.5E6 m3, 1973[1]
USSR, both coniferous and broadleaved forest	383.0E9 m3 solid vol.	Value is year 1973, in 1972 = 383.0E6 m3[1]
USSR, broadleaved forest	64.0E9 m3 solid vol.	Industrial wood = 34.1E6 m3, Fuel wood = 29.9E6 m3, 1973[1]
USSR, coniferous forest	319.0E9 m3 solid vol.	Industrial wood = 263.5E6 m3, Fuel wood = 55.5E6 m3, 1973[1]
United Kingdom, both coniferous and broadleaved forest	2.7E9 m3 solid vol.	Value is year 1973, in 1972 = 2.5E6 m3[1]
United Kingdom, broadleaved forest	0.9E9 m3 solid vol.	Industrial wood = 0.8E6 m3, Fuel wood = 0.1E6 m3, 1973[1]
United Kingdom, coniferous forest	1.8E9 m3 solid vol.	Industrial wood = 1.6E6 m3, Fuel wood = 0.2E6 m3, 1973[1]
Venezuela, both coniferous and broadleaved forest	7.4E9 m3 solid vol.	Value is year 1973, in 1972 = 7.4,7.4E6 m3[1]
Venezuela, broadleaved forest	7.4E9 m3 solid vol.	Industrial wood = 0.5E6 m3, Fuel wood = 6.9E6 m3, 1973[1]

West Germany, both coniferous and broadleaved forest	30.6E9 m3 solid vol.	Value is year 1973, in 1972 = 23.8E6 m3[1]
West Germany, broadleaved forest	7.2E9 m3 solid vol.	Industrial wood = 6.1E6 m3, Fuel wood = 1.1E6 m3, 1973[1]
West Germany, coniferous forest	23.4E9 m3 solid vol.	Industrial wood = 22.6E6 m3, Fuel wood = 0.8E6 m3, 1973[1]
Yugoslavia, both coniferous and broadleaved forest	13.2E9 m3 solid vol.	Value is year 1973, in 1972 = 13.2E6 m3[1]
Yugoslavia, broadleaved forest	9.1E9 m3 solid vol.	Industrial wood = 5.3E6 m3, Fuel wood = 3.8E6 m3, 1973[1]
Yugoslavia, coniferous forest	4.1E9 m3 solid vol.	Industrial wood = 4.1E6 m3, Fuel wood = 0.0E6 m3, 1973[1]
Zaire, both coniferous and broadleaved forest	14.7E9 m3 solid vol.	Value is year 1973, in 1972 = 14.7E6 m3[1]
Zaire, broadleaved forest	14.7E9 m3 solid vol.	Industrial wood = 1.9E6 m3, Fuel wood = 12.8E6 m3, 1973[1]

Source: 1.Skogsstyrelsen 1976. Skogstatistisj AArsbok 1974. Sveriges officiella statistik. Skogsstyrelsen. Joekoeking.

2-109 Ecosphere
Gross primary production

Species	Value	Condition
Alfalfa field	24000 kcal/(year*m**2)	Annual value, USA[1]
Algae culture	43 g dry organic matter/((m**2)*day)	Extra CO2 added, short favorable periods, mass culture[2]
Bog lake	0.3 g dry organic matter/((m**2)*day)	Phytoplankton only, Minn, average for 6 months to 1 y[2]
Boreal coniferous forest	3000 kcal/(m**2*year)	Area = 10.0E6 km2, total g. prod = 3.0E16 kcal/y[3]
Broadleaved evergreen forest	20000 kcal/(m**2*year)	Area = 14.7E6 km2, total g. prod = 29.0E16 kcal/y, tropic + sub-[3]
Coastal zones	2000 kcal/(m**2*year)	Area = 34.0E6 km2, total g. prod = 6.8E16 kcal/y, marine[4]
Coral reef	18.2 g dry organic matter/((m**2)*day)	Average for 3 pacific reefs, average for 6 months to 1 y[2]
Coral reef, Curacao	5650 g C/(m2*year)	Whole reef production, 1975[5]
Coral reef, Eniwetok Atoll	1261 g C/(m2*year)	Respirometer measurement, 1969, only algae[5]
Coral reef, Eniwetok Atoll	3241 g C/(m2*year)	Whole reef production, 1972[5]
Coral reef, Eniwetok Atoll	3504 g C/(m2*year)	Whole reef production, 1954[5]
Coral reef, Eniwetok Atoll	4380 g C/(m2*year)	Whole reef production, 1971[5]
Coral reef, Hawaii	2495 g C/(m2*year)	Respirometer measurement, 1972, only algae[5]
Coral reef, Hawaii	7358 g C/(m2*year)	Whole reef production, 1972[5]
Coral reef, Hawaii	8497 g C/(m2*year)	Whole reef production, 1971[5]
Coral reef, Kaneohe Bay	7300 g C/(m2*year)	Whole reef production, 1961[5]
Coral reef, Mokuoloe Island	2900 g C/(m2*year)	Whole reef production, 1961[5]
Coral reef, N Kapaa Reef	3000 g C/(m2*year)	Whole reef production, 1956[5]
Coral reef, Oahu Hawaii	0 g C/(m2*year)	Whole reef production, Not an autotroph reef, 1961[5]
Coral reef, Plantation Key, Florida	2303 g C/(m2*year)	Respirometer measurement, USA, 1966[5]
Coral reef, Rongelapp Atoll	1460 g C/(m2*year)	Whole reef production, N Marshall Atoll, 1948[5]
Coral reef, Rongelapp Atoll	1500 g C/(m2*year)	Whole reef production, N Marshall Isl., 1954[5]
Coral reef, reef in Arab Sea	4494 g C/(m2*year)	Whole reef production, 10 N and 72 E, 1969[5]

Cultivated land	12000 kcal/(m**2*year)	Area = 4.0E6 km2, total g. prod = 4.8E16 kcal/y, fuel subsidized[3]
Cultivated land	8000 kcal/(m**2*year)	Area = 4.9E6 km2, total g. prod = 3.9E16 kcal/y, no energy subsidy[3]
Desert and tundras	200 kcal/(m**2*year)	Area = 40.0E6 km2, total g. prod = 0.8E16 kcal/y[3]
Estuaries	20000 kcal/(m**2*year)	Area = 2.0E6 km2, total g. prod = 4.0E16 kcal/y, incl reefs, marine[6]
Estuaries	23 g dry organic matter/((m**2)*day)	Texas, short favorable periods[2]
Eutrophic lake	2.1 g dry organic matter/((m**2)*day)	Shallow lake, Japan, average for 6 months to 1 y[2]
Forest	2500 kcal/(m**2*year)	Area = 9.4E6 km2, total g. prod = 2.4E16 kcal/y, dry forest[3]
Freshwater spring	20800 kcal/(year*m**2)	Annual value, Silver Spring, FL[7]
Grasslands and pastures	2500 kcal/(m**2*year)	Area = 42.0E6 km2, total g. prod = 10.5E16 kcal/y[3]
Laguna madre	4.4 g dry organic matter/((m**2)*day)	Texas estuaries, average for 6 months to 1 y[2]
Lake Erie	1.0 g dry organic matter/((m**2)*day)	Winter, average for 6 months to 1 y[2]
Lake Erie	9.0 g dry organic matter/((m**2)*day)	Summer, average for 6 months to 1 y[2]
Land	0.172E14 kg C/yr	None[8]
Land	0.20E14 kg C/yr	None[9]
Land	0.24E14 kg C/yr	None[10]
Long Island sound	3.2 g dry organic matter/((m**2)*day)	Shallow inshore waters, average for 6 months to 1 y[2]
Marine sound	5700 kcal/(year*m**2)	Annual value, coastal, Long Island, New York[11]
Marine turtle-grass flats	34 g dry organic matter/((m**2)*day)	Florida, short favorable periods[2]
Oak - pine forest	11500 kcal/(year*m**2)	Annual value, medium aged, New York, no timber harvest[12]
Oligotrophic lake	0.7 g dry organic matter/((m**2)*day)	Clear deep lake, Lake Wisconsin, average for 6 months to 1 y[2]
Open ocean	1000 kcal/(m**2*year)	Area = 326.0E6 km2, total g. prod = 32.6E16 kcal/y, marine[4]
Pacific Ocean	0.2 g dry organic matter/((m**2)*day)	Infertile open ocean, average for 6 months to 1 y[2]
Pine forest	12200 kcal/(year*m**2)	Annual value, young plantation, mean for 0-50 y, Eng[13]
Pond	27 g dry organic matter/((m**2)*day)	Pond with untreated wastes, South Dakota, short favorable periods[2]
Pond	5.0 g dry organic matter/((m**2)*day)	Fertilized pond, NC, short favorable periods[2]
Pond	9.0 g dry organic matter/((m**2)*day)	Pond with treated sewage wastes, Denm, short favorable periods[2]
Rain forest	45000 kcal/(year*m**2)	Annual value, mature forest, Puerto Rico[14]
River	1.7 g dry organic matter/((m**2)*day)	Turbid river, NC, short favorable periods[2]
Sargasso sea	0.5 g dry organic matter/((m**2)*day)	Infertile open ocean, average for 6 months to 1 y[2]
Sea	0.235E14 kg C/yr	None[15]
Sea	0.25E14 kg C/yr	None[8]

Sea	1.24E14 kg C/yr	Range = 0.44 - 2.03[9]
Sea	1.6E14 kg C/yr	Less than value[10]
Silver springs	17.5 g dry organic matter/((m**2)*day)	Florida, average for 6 months to 1 y[2]
Silver springs	35 g dry organic matter/((m**2)*day)	Florida, short favorable periods[2]
Stream	57 g dry organic matter/((m**2)*day)	Polluted stream, Indiana, short favorable periods[2]
Temperate forest	8000 kcal/(m**2*year)	Area = 4.9E6 km2, total g. prod = 3.9E16 kcal/y, moist[3]
Total for biosphere	2000 kcal/(m**2*year)	Area = 500.0E6 km2, total g. prod = 100.0E16 kcal/y, exc. ice caps[6]
Upwelling zones	6000 kcal/(m**2*year)	Area = 0.4E6 km2, total g. prod = 0.2E16 kcal/y, marine[4]

Source: 1.Thomas, M.D. and G.R. Hill. 1949. in: Franck, J and W E. Loomis(eds). Photosynthesis in Plants, Iowa State College Press, pp 19-52.
2.Brock, T.D. 1966. Principles of Micro bial Ecology. Prentice-Hall. New Jersey.
3.Lieth, H. 1963. J. Geophys. Res. 68(13), pp 3887-3898.
4.Ryther, J.H. 1969. Science 166, pp 72-76.
5.Bolin, B., E.T. Degens, S. Kempe and P. Ketner. 1977. The Global Carbon Cycle. SCOPE and ICSU. John Wiley and Sons. NY.
6.Odum, E.P. 1971. Fundamentals of Ecol ogy. Third edition. W.B. Saunders , Philadelphia.
7.Odum, H.T. 1957. Ecol. Monogr., 27, pp 55-112.
8.Muller, D. 1960. in: Rukland, W. (ed.): Encyclopedia of Plant Physiology, 12(2), pp 934.
9.Riley, G.A. 1965. Limnol. Oceanogr., 10(suppl.), pp 202-215.
10.Fogg, G.E. 1958. Advmt. Sci., 14, pp 395.
11.Riley, G.A. 1956. Bull. Bingham Oceanogr. Coll. 15, pp 324-344.
12.Woodwell, G.M. and R. Whittaker. 1968. Amer. Zool. 8, pp 19-30.
13.Ovington, J.D. 1961 Ann. Bot. 25, pp 12-20.
14.Odum, H.T. and R.fF. Pigeon. 1970. A tropical rain forest. Nat Tech. Info. Srevice. Springfield.
15.Nielsen, E.S. 1958. Rapp. p.-v. Reun. Cons. Perm. Int. Explor. Mer., 144, pp 92.

2-110 Ecosphere
Litter

Species	Value	Condition
Bog, swamp and marsh	2500 g dry matter/m2	Total litter = 8.8E15 g DM in this ecosystem type[1]
Boreal forest	3500 g dry matter/m2	Total litter = 31.5E15 g DM in this ecosystem type[1]
Chaparral, maquis and brush land	500 g dry matter/m2	Total litter = 1.3E15 g DM in this ecosystem type[1]
Cultivated land	50 g dry matter/m2	Total litter = 0.8E15 g DM in this ecosystem type[1]
Desert and semi desert	100 g dry matter/m2	Total litter = 2.1E15 g DM in this ecosystem type[1]
Extreme desert	15 g dry matter/m2	Total litter = 0.14E15 g DM in this ecosystem type[1]
Forest	2031 g dry matter/m2	Total litter = 63.6E15 g DM in this ecosystem type[1]
Forest plantation	500 g dry matter/m2	Total litter = 0.8E15 g DM in this ecosystem type[1]
Herb lichen tundra	500 g dry matter/m2	Total litter = 1.8E15 g DM in this ecosystem type[1]
Human land	300 g dry matter/m2	Total litter = 0.4E15 g DM in this ecosystem type[1]
Mangrove forest	10000 g dry matter/m2	Total litter = 3.0E15 g DM in this ecosystem type[1]
Perpetual ice	0 g dry matter/m2	Total litter = 0E15 g DM in this ecosystem type[1]
Polar desert	30 g dry matter/m2	Total litter = 0.05E15 g DM in this ecosystem type[1]
Savanna	350 g dry matter/m2	Total litter = 7.9E15 g DM in this ecosystem type[1]
Scrub tundra	5000 g dry matter/m2	Total litter = 22E15 g DM in this ecosystem type[1]
Temperate dry grassland	325 g dry matter/m2	Total litter = 2.4E15 g DM in this ecosystem type[1]
Temperate forest	3000 g dry matter/m2	Total litter = 18E15 g DM in this ecosystem type[1]
Temperate wet grassland	500 g dry matter/m2	Total litter = 2.5E15 g DM in this ecosystem type[1]

Temperate woodland	2500 g dry matter/m2	Total litter = 5.0E15 g DM in this ecosystem type[1]
Temperated grassland	392 g dry matter/m2	Total litter = 4.9E15 g DM in this ecosystem type[1]
Total, terrestrial	795.7 g dry matter/m2	Total litter = 118.8E15 g DM in this ecosystem type[1]
Tropical rain forest	650 g dry matter/m2	Total litter = 6.5E15 g DM in this ecosystem type[1]
Tropical seasonal forest	850 g dry matter/m2	Total litter = 3.8E15 g DM in this ecosystem type[1]
Tundra, artic and alpine	250 g dry matter/m2	Total litter = 23.85E15 g DM in this ecosystem type[1]

Source: 1.Ajtay, G.L., P. Ketner and P. Duvigneaud. 1979. in: Bolin, B., E.T. Degens, S. Kempe and P. Ketner.(ds.) The Global Carbon Cycle. 129-181. SCOPE. John Wiley and Sons. NY.

2-111 Ecosphere
Litter mass

Species	Value	Condition
Boreal forest	48.0E15 g dry weight	Global value[1]
Cultivated land	1.4E15 g dry weight	Global value[1]
Desert and semi desert shrubs	0.36E15 g dry weight	Global value[1]
Extreme desert, rock and sand	0.03E15 g dry weight	Global value[1]
Lake and stream	0.0E15 g dry weight	Global value[1]
Savanna	3.0E15 g dry weight	Global value[1]
Swamp and marsh	5.0E15 g dry weight	Global value[1]
Temperate deciduous forest	14.0E15 g dry weight	Global value[1]
Temperate evergreen forest	15.0E15 g dry weight	Global value[1]
Temperate grassland	3.6E15 g dry weight	Global value[1]
Total, terrestrial	111E15 g dry weight	Global value[1]
Tropical rain forest	3.4E15 g dry weight	Global value[1]
Tropical seasonal forest	3.8E15 g dry weight	Global value[1]
Tundra and alpine	8.0E15 g dry weight	Global value[1]
Woodland and shrubland	5.1E15 g dry weight	Global value[1]

Source: 1.Whittaker, R.H. and G.E. Likens. 1973. in: WoodwellG.M. and E.V. Pecan (eds.): Carbon and Biosphere.AEC Symposium Series 30, 281-302. NTIS. Springfield. Virginia.

2-112 Ecosphere
Litterfall

Species	Value	Condition
Bog, swamp and marsh	600 g dry matter/(m2*year)	Total litterfall = 2.1E15 g DM/y in this ecosystem type[1]
Boreal closed forest	600 g dry matter/(m2*year)	Total litterfall = 3.9E15 g DM/y in this ecosystem type[1]
Boreal open forest	550 g dry matter/(m2*year)	Total litterfall = 1.4E15 g DM/y in this ecosystem type[1]
Chaparral, maquis and brush land	1000 g dry matter/(m2*year)	Total litterfall = 2.5E15 g DM/y in this ecosystem type[1]
Cultivated annual land	450 g dry matter/(m2*year)	Total litterfall = 6.8E15 g DM/y in this ecosystem type[1]
Cultivated land	438 g dry matter/(m2*year)	Total litterfall = 7.0E15 g DM/y in this ecosystem type[1]
Cultivated perennial land	150 g dry matter/(m2*year)	Total litterfall = 0.2E15 g DM/y in this ecosystem type[1]
Desert and semi desert	125 g dry matter/(m2*year)	Total litterfall = 2.6E15 g DM/y in this ecosystem type[1]
Extreme desert	15 g dry matter/(m2*year)	Total litterfall = 0.14E15 g DM/y in this ecosystem type[1]
Forest	1160 g dry matter/(m2*year)	Total litterfall = 36.3E15 g DM/y in this ecosystem type[1]
Forest plantation	875 g dry matter/(m2*year)	Total litterfall = 1.3E15 g DM/y in this ecosystem type[1]
Herb lichen tundra	145 g dry matter/(m2*year)	Total litterfall = 0.52E15 g DM/y in this ecosystem type[1]
Human land	300 g dry matter/(m2*year)	Total litterfall = 0.36E15 g DM/y in this ecosystem type[1]
Mangrove forest	600 g dry matter/(m2*year)	Total litterfall = 0.2E15 g DM/y in this ecosystem type[1]
Perpetual ice	0 g dry matter/(m2*year)	Total litterfall = 0E15 g DM/y in this ecosystem type[1]
Polar desert	20 g dry matter/(m2*year)	Total litterfall = 0.03E15 g DM/y in this ecosystem type[1]
Savanna	1390 g dry matter/(m2*year)	Total litterfall = 31.3E15 g DM/y in this ecosystem type[1]
Savanna forest	800 g dry matter/(m2*year)	Total litterfall = 2.8E15 g DM/y in this ecosystem type[1]
Savanna, grass dominated	1500 g dry matter/(m2*year)	Total litterfall = 28.5E15 g DM/y in this ecosystem type[1]
Scrub tundra	200 g dry matter/(m2*year)	Total litterfall = 0.88E15 g DM/y in this ecosystem type[1]
Temperate dry grassland	550 g dry matter/(m2*year)	Total litterfall = 4.1E15 g DM/y in this ecosystem type[1]
Temperate forest	850 g dry matter/(m2*year)	Total litterfall = 5.1E15 g DM/y in this ecosystem type[1]
Temperate wet grassland	900 g dry matter/(m2*year)	Total litterfall = 4.5E15 g DM/y in this ecosystem type[1]
Temperate woodland	1220 g dry matter/(m2*year)	Total litterfall = 2.4E15 g DM/y in this ecosystem type[1]
Temperated grassland	688 g dry matter/(m2*year)	Total litterfall = 8.6E15 g DM/y in this ecosystem type[1]
Total, terrestrial	634.3 g dry matter/(m2*year)	Total litterfall = 94.7E15 g DM/y in this ecosystem type[1]
Tropical rain forest	1850 g dry matter/(m2*year)	Total litterfall = 18.5E15 g DM/y in this ecosystem type[1]
Tropical seasonal forest	1300 g dry matter/(m2*year)	Total litterfall = 5.9E15 g DM/y in this ecosystem type[1]
Tundra, artic and alpine	150 g dry matter/(m2*year)	Total litterfall = 1.43E15 g DM/y in this ecosystem type[1]

Source: 1.Ajtay, G.L., P. Ketner and P. Duvigneaud. 1979. in: Bolin, B., E.T. Degens, S. Kempe and P. Ketner.(ds.) The Global Carbon Cycle. 129-181. SCOPE. John Wiley and Sons. NY.

2-113 Ecosphere
Masses of the earth

Species	Value	Condition
Core	1936E24 g	None[1]
Crust	24E24 g	None[1]
Limestone	0.09E24 g	7 % of total precambrian sediment mass, precambrian Limestone[1]
Limestone	0.26E24 g	14 % of phanerozoic sediment mass, phanerozoic Limestone[1]

Limestone	0.35E24 g	11 % of total sediment mass, total Limestone[1]
Mantle	4016E24 g	Crust[1]
Sandstone	0.06E24 g	4 % of total precambrian sediment mass, precambrian Sandstone[1]
Sandstone	0.20E24 g	11 % of phanerozoic sediment mass, phanerozoic Sandstone[1]
Sandstone	0.26E24 g	8 % of total sediment mass, total Sandstone[1]
Sediments and volcanics	3.2E24 g	None[1]
Shale	0.86E24 g	48 % of phanerozoic sediment mass, phanerozoic Shale[1]
Shale	0.87E24 g	62 % of total precambrian sediment mass, precambrian Shale[1]
Shale	1.73E24 g	54 % of total sediment mass, total Shale[1]
Total earth	5975E24 g	None[1]
Total phanerozoic mass	1.80E24 g	None[1]
Total precambrian mass	1.40E24 g	None[1]
Volcanics	0.38E24 g	27 % of total precambrian sediment mass, precambrian Volcanics[1]
Volcanics	0.48E24 g	27 % of phanerozoic sediment mass, phanerozoic Volcanics[1]
Volcanics	0.86E24 g	27 % of total sediment mass, total Volcanics[1]

Source: 1.Garrels, R.M. and F.T. Mackenzie. 1971. Evolution of Sedimentary Rocks. Norton and Comp. New York

2-114 Ecosphere
Mean daily productivity

Species	Value	Condition
Phytoplankton, Lake Borax	249 mg C/(day*m**2)	Ann prod = 91 g C/(m2*y), large shallow saline, 25 % of total[1]
Phytoplankton, Lake Castle	98 mg C/(day*m**2)	Ann prod = 36 g C/(m2*y), deep, alpine, Calif, rg = 6 - 317[2]
Phytoplankton, Lake Char	1.1 mg C/(day*m**2)	Ann prod = 4.1 g C/(m2*y), 80 % by benthic flora, Can, lat 74, NWT[3]
Phytoplankton, Lake Clear	438 mg C/(day*m**2)	Ann prod = 260 g C/(m2*y), very large, shallow, rg = 2 - 2440[4]
Phytoplankton, Lake Crooked	359 mg C/(day*m**2)	Ann prod = 131 g C/(m2*y), large, deep, hardwater, 1964, Ind[5]
Phytoplankton, Lake Crooked	469 mg C/(day*m**2)	Ann prod = 171 g C/(m2*y), large, deep, hardwater, 1963[5]
Phytoplankton, Lake Erie	2395 mg C/(day*m**2)	Ann prod = 310 g C/(m2*y), western stations, rg = 30 - 4760[6]
Phytoplankton, Lake Erie	790 mg C/(day*m**2)	Ann prod = 160 g C/(m2*y), eastern stations, rg = 140 - 1440[6]
Phytoplankton, Lake Erie	905 mg C/(day*m**2)	Ann prod = 210 g C/(m2*y), central stations, rg = 120 - 1690[6]
Phytoplankton, Lake Erken	285 mg C/(day*m**2)	Ann prod = 104 g C/(m2*y), large deep, nat. productive[7]
Phytoplankton, Lake Esrom	370 mg C/(day*m**2)	Ann prod = 260 g C/(m2*y), large, moderately deep, rg = 23 - 422[8]
Phytoplankton, Lake Frederiksborg slotssoe	1030 mg C/(day*m**2)	Ann prod = 376 g C/(m2*y), shallow, enriched, Denm, rg = 12 - 4160[9]
Phytoplankton, Lake Furesoe	462 mg C/(day*m**2)	Ann prod = 168 g C/(m2*y), large, deep, rg = 0 - 1380[8]
Phytoplankton, Lake Goose	729 mg C/(day*m**2)	Ann prod = 266 g C/(m2*y), small kettle, Ind, rg = 166 - 1753[10]
Phytoplankton, Lake Grane Langsoe	248 mg C/(day*m**2)	Ann prod = 91 g C/(m2*y), deep, acidic, humic stained, Denm[9]
Phytoplankton, Lake Huron	425 mg C/(day*m**2)	Ann prod = 100 g C/(m2*y), offshore stations, rg = 150 - 700[6]
Phytoplankton, Lake Lanao	1700 mg C/(day*m**2)	Ann prod = 620 g C/(m2*y), large deep tropical, Phillippines[11]
Phytoplankton, Lake Lawrence	112.6 mg C/(day*m**2)	Ann prod = 41.1 g C/(m2*y), small hardwater, 7 y mean, rg = 5 - 497[2]

Phytoplankton, Lake Little Crooked	598 mg C/(day*m**2)	Ann prod = 218 g C/(m2*y), small, deep, kettle, Ind, 1964[5]
Phytoplankton, Lake Little Crooked	618 mg C/(day*m**2)	Ann prod = 226 g C/(m2*y), small, deep, kettle, 1963, rg = 263 - 1903[5]
Phytoplankton, Lake Lunzer Untresee	123 mg C/(m**2*day)	Ann prod = 45 g C/(m2*y), small alpine[12]
Phytoplankton, Lake Martin	561 mg C/(day*m**2)	Ann prod = 205 g C/(m2*y), deep, stained marl, Ind, rg = 27 - 1708[13]
Phytoplankton, Lake Meretta	3.1 mg C/(day*m**2)	Ann prod = 11 g C/(m2*y), polluted by sewage, NWT Can, rg = 0 - 170[3]
Phytoplankton, Lake Michigan	550 mg C/(day*m**2)	Ann prod = 130 g C/(m2*y), offshore stations, rg = 70 - 1030[6]
Phytoplankton, Lake Minnetonka	820 mg C/(day*m**2)	Ann prod = 300 g C/(m2*y), large deep, complex basin, Minn[14]
Phytoplankton, Lake Olin	374 mg C/(day*m**2)	Ann prod = 137 g C/(m2*y), large deep, marl, Ind, rg = 89 - 996[13]
Phytoplankton, Lake Oliver	336 mg C/(day*m**2)	Ann prod = 123 g C/(m2*y), large deep, marl, Ind, rg = 32 - 775[13]
Phytoplankton, Lake Ontario	730 mg C/(day*m**2)	Ann prod = 180 g C/(m2*y), offshore stations, rg = 60 - 1400[6]
Phytoplankton, Lake Pretty	305 mg C/(day*m**2)	Ann prod = 111 g C/(m2*y), moderate-size, deep marl, Ind, 1964[5]
Phytoplankton, Lake Pretty	440 mg C/(day*m**2)	Ann prod = 161 g C/(m2*y), moderate-size, deep marl, Ind[5]
Phytoplankton, Lake Ranseren	44.5 mg C/(day*m**2)	Small lapplandic, Sweden, rg = 23 - 66[7]
Phytoplankton, Lake Smith Hole	194 mg C/(day*m**2)	Ann prod = 71 g C/(m2*y), shallow, humic stained, Ind, rg = 24 - 5960[13]
Phytoplankton, Lake Store Gribsoe	230 mg C/(day*m**2)	Ann prod = 84 g C/(m2*y), deep, acidic, humic stained, Denm[9]
Phytoplankton, Lake Superior	155 mg C/(day*m**2)	Most unproductive of laurentian great lakes, rg = 50 - 260[15]
Phytoplankton, Lake Sylvan	1564 mg C/(day*m**2)	Ann prod = 570 g C/(m2*y), complex basin, large, shallow, Ind, rg = 9[10]
Phytoplankton, Lake Victoria	1750 mg C/(day*m**2)	Ann prod = 640 g C/(m2*y), large, tropical, rg = 1700 - 3800[16]
Phytoplankton, Lake Walter	210 mg C/(day*m**2)	Ann prod = 77 g C/(m2*y), basin b, marl, rg = 12 - 535[13]
Phytoplankton, Lake Walter	276 mg C/(day*m**2)	Ann prod = 101 g C/(m2*y), basin c, marl, rg = 30 - 1048[13]
Phytoplankton, Lake Walter	418 mg C/(day*m**2)	Ann prod = 153 g C/(m2*y), basin a, marl, Ind, rg = 102- 1395[13]
Phytoplankton, Lake Walter	437 mg C/(day*m**2)	Ann prod = 160 g C/(m2*y), basin d, marl, rg = 98 - 1458[13]
Phytoplankton, Lake Wintergreen	1012 mg C/(day*m**2)	Ann prod = 369 g C/(m2*y), shallow, hig loading, Mich.[2]

Source: 1.Wetzel. R.G. 1964. Int. Rev. ges. Hydrobiol. 49, pp 1-64.
2.Wetzel, R.G. 1975. Limnology. Saunders. Philadelphia.
3.Kalff, J. and H.E. Welch. 1974. J. Fish. Res. bd. Can. 31, pp 621-636.
4.Goldman, C.R. and R.G. Wetzel. 1963. Ecology, 44, pp 283-294.
5.Wetzel. R.G. 1966. Verh. Int. Ver. Limnol., 16, pp 321-332.
6.Vollenweider, R.A., M. Munawar and P. Stadelmann. 1974. J. Fish. Res. bd. Canada 31, pp 739-762.
7.Rodhe, W. 1958. Verh. Int. Ver. Limnol. 13, pp 121-141.
8.Jonasson, P.M. and H. Mathiesen. 1959. Oikos 10, pp 137-167.
9.Nygaard, G. 1955. Verh. Int. Ver. Limnol. 12, pp 123-133.
10.Wetzel. R.G. 1966. Invest. Indiana Lakes Streams 7, pp 1-32.
11.Lewis, W.M. Jr. 1974. Ecol. Monogr. 44, pp 337-409.
12.Jonasson, P.M. 1972. Oikos Suppl. 14, pp 1-148.
13.Wetzel. R.G. 1973. Hydrobiol. Stud. 3, pp 91-143.
14.Megard, R.O. 1972. Limnol. Oceanogr. 17, 68-87.
15.Putnam, H.D. and T.A. Olson. 1961. Studies on the Productivity and Plankton of Lake Superior. Rep.choll Pub. Health. Univ. Minnesota.
16.Talling, J.F. 1965. Int. Rev. ges. Hydrobiol. 50, pp 1-32.

2-115 Ecosphere
Mean temperature

Species	Value	Condition
Air	286 K	Jauja, Peru, altitude = 3.4 km[1]
Air	288 K	Quito, Ecuador, altitude = 2.9 km[1]
Air	294 K	Takengoen, Indonesia altitude = 1.2 km[1]
Air	300 K	Belem, Brazil, altitude = 0 km[1]
Continental arctic air mass	227 K	Very cold, very dry, specific humidity = 0.1 g/kg[1]

Source: 1.Straheer, A.N. and A.H. Strahler. 1974. Introduction to Environmental Science, John Wiley and Sons, California.

2-116 Ecosphere
Mean temperature and humidity

Species	Value	Condition
Continental polar air mass	262 K	Cold, dry, specific humidity = 1.4 g/kg[1]
Continental tropical air mass	297 K	Warm, dry, specific humidity = 11 g/kg[1]
Maritime equatorial air mass	300 K	Warm, very moist, specific humidity = 19 g/kg[1]
Maritime polar air mass	277 K	Cool, moist, specific humidity = 4.4 g/kg[1]
Maritime tropical air mass	297 K	Warm, moist, specific humidity = 17 g/kg[1]

Source: 1.Straheer, A.N. and A.H. Strahler. 1974. Introduction to Environmental Science, John Wiley and Sons,alifornia.

2-117 Ecosphere
Net community production

Species	Value	Condition
Alfalfa field	14400 kcal/(year*m**2)	Annual value, USA[1]
Freshwater spring	2000 kcal/(year*m**2)	Annual value, Silver Spring, FL[2]
Marine sound	0.1 kcal/(year*m**2)	Annual value, coastal, Long Island, New York, very little[3]
Oak - pine forest	2000 kcal/(year*m**2)	Annual value, medium aged, New York, no timber harvest[4]
Pine forest	2900 kcal/(year*m**2)	Annual value, young plantation, mean for 0-50 y, Eng[5]
Rain forest	0 kcal/(m**2*year)	Annual value, mature forest, Puerto Rico, value equal to very little[6]

Source: 1.Thomas, M.D. and G.R. Hill. 1949. in: Franck, J and W E. Loomis(eds). Photosynthesis in Plants, Iowa State College Press, pp 19-52.
2.Odum, H.T. 1957. Ecol. Monogr., 27, pp 55-112.
3.Riley, G.A. 1956. Bull. Bingham Oceanogr. Coll. 15, pp 324-344.
4.Woodwell, G.M. and R. Whittaker. 1968. Amer. Zool. 8, pp 19-30.
5.Ovington, J.D. 1961 Ann. Bot. 25, pp 12-20.
6.Odum, H.T. and R.F. Pigeon. 1970. A tropical rain forest. Nat Tech. Info. Service. Springfield.

2-118 Ecosphere
Net primary production

Species	Value	Condition
Agricultural land	650 g dry weight/(m**2*year)	Tot npp=9.1E9 d t/y, biomass=1 dry kg/m2, tot biom.=14E9 d t[1]
Alfalfa field	15200 kcal/(year*m**2)	Annual value, USA[2]
Algal beds and reef	2500 g dry matter/(m2*year)	Rg = 500 - 4000 g dm/(m2*y)[3]
Angiosperms	0.49E15 g C/year in total	Equal to 1.07 % of total aquatic production[4]
Attached algae and estuaries	2000 g dry weight/(m**2*year)	Tot npp=4.0E9 d t/y, biomass=1 dry kg/m2, tot biom.=2.0E9 d t[1]

Benthic diatom, Danish Wadden Sea	147 g C/(m2*year)	Rg = 115-178 g C/m2[5]
Benthic diatom, Danish fjords	116 g C/(m2*year)	None[5]
Benthic diatom, False Bay	184 g C/(m2*year)	Rg = 143-226 g C/m2[5]
Benthic diatom, Isefjord	345 g C/(m2*year)	Rg = 260-430 g C/m2, Denmark[6]
Benthic diatom, Salt marsh, Georgia	200 g C/(m2*year)	None[5]
Benthic diatom, Southern New England	100 g C/(m2*year)	Rg = 60 - 140 g C/m2[5]
Benthic diatom, Ythan estuary	31 g C/(m2*year)	None[5]
Boreal forest	800 g dry weight/(m**2*year)	Tot npp=9.6E9 d t/y, biomass=20 dry kg/m2, tot biom.=240E9 d t[1]
Cold desert, grey or sandy soils	0.10 kg dry organic matter/(m2*year)	Total area = 1.0E6 km2, Total NPP = 0.05E15 g C/y, on land[7]
Cold desert, mountainous	0.15 kg dry organic matter/(m2*year)	Total area = 1.5E6 km2, Total NPP = 0.1E15 g C/y, on land[7]
Continental shelf	350 g dry weight/(m**2*year)	Tot npp=9.5E9 d t/y, biomass=0.01 dry kg/m2, tot biom.=0.3E9 d t[1]
Continental shelf	360 g dry matter/(m2*year)	Rg = 200 - 600 g dm/(m2*y)[3]
Coral reef, Curacao	2330 g C/(m2*year)	Whole reef production, 1975[4]
Coral reef, Eniwetok Atoll	2044 g C/(m2*year)	Whole reef production, 1954[4]
Coral reef, Eniwetok Atoll	2190 g C/(m2*year)	Whole reef production, 1971[4]
Coral reef, Eniwetok Atoll	2190 g C/(m2*year)	Whole reef production, 1972[4]
Coral reef, Eniwetok Atoll	946 g C/(m2*year)	Respirometer measurement, 1969, only algae[4]
Coral reef, Hawaii	2080 g C/(m2*year)	Respirometer measurement, 1972, only algae[4]
Coral reef, Hawaii	6307 g C/(m2*year)	Whole reef production, 1971[4]
Coral reef, Hawaii	6307 g C/(m2*year)	Whole reef production, 1972[4]
Coral reef, Mokuoloe Island	511 g C/(m2*year)	Whole reef production, 1961[4]
Coral reef, N Kapaa Reef	1400 g C/(m2*year)	Whole reef production, 1956[4]
Coral reef, Oahu Hawaii	0 g C/(m2*year)	Whole reef production, Not an autotroph reef, 1961[4]
Coral reef, Plantation Key, Florida	528 g C/(m2*year)	Respirometer measurement, USA, 1966[4]
Coral reef, Rongelapp Atoll	190 g C/(m2*year)	Whole reef production, N Marshall Isl., 1954[4]
Coral reef, reef in Arab Sea	3618 g C/(m2*year)	Whole reef production, 10 N and 72 E, 1969[4]
Coral reefs	0.30E15 g C/year in total	Equal to 0.65 % of total aquatic production[4]
Corn	1100 kcal/(m**2*year)	India, edible portion = 350 kcal/(m2*y)[8]
Corn	2400 kcal/(m**2*year)	World averag, edible portion = 810 kcal/(m2*y)[8]
Corn	4500 kcal/(m**2*year)	USA, edible portion = 1510 kcal/(m2*y)[8]
Costal phytoplankton, Barararia Bay, Louisiana	170 g C/(m2*year)	Coastal water, USA[6]
Costal phytoplankton, British Columbia	450 g C/(m2*year)	None[6]
Costal phytoplankton, Cochin backwater, India	124 g C/(m2*year)	Coastal water[6]

Costal phytoplankton, Columbia River	152 g C/(m2*year)	North of R, Wash.[6]
Costal phytoplankton, Columbia River	60 g C/(m2*year)	River plume, Wash[6]
Costal phytoplankton, Columbia River	88 g C/(m2*year)	River mouth, Wash.[6]
Costal phytoplankton, Columbia river	61 g C/(m2*year)	Ocean beyond river, Wash.[6]
Costal phytoplankton, Georgia	547 g C/(m2*year)	Coastal water, USA[6]
Costal phytoplankton, Hempstead Bay	395 g C/(m2*year)	State of New York, USA[6]
Costal phytoplankton, Long Island Sound	205 g C/(m2*year)	New York[6]
Costal phytoplankton, North Carolina	52 g C/(m2*year)	None[6]
Costal phytoplankton, North Sea costral water	170 g C/(m2*year)	Rg = 160-180 g C/m2, Netherlands[6]
Costal phytoplankton, Nova Scotia	190 g C/(m2*year)	Canada[6]
Costal phytoplankton, Wadden Sea	110 g C/(m2*year)	Rg = 100-120 g C/m2, Netherlands[6]
Costal phytoplankton, continental shelf	120 g C/(m2*year)	State of New York, USA[6]
Costal phytoplankton, continental slope	100 g C/(m2*year)	State of New York, USA[6]
Costal phytoplankton, shallow water off New York	160 g C/(m2*year)	State of New York, USA[6]
Cultivated land, subtropical	11.0 kg dry organic matter/(m2*year)	Total area = 0.3E6 km2, Total NPP = 1.5E15 g C/y, on land[7]
Cultivated land, temperate	1.2 kg dry organic matter/(m2*year)	Total area = 6.7E6 km2, Total NPP = 3.6E15 g C/y, on land[7]
Cultivated land, tropical	1.4 kg dry organic matter/(m2*year)	Total area = 8.5E6 km2, Total NPP = 0.3E15 g C/y, on land[7]
Estuaries	0.92E15 g C/year in total	Equal to 2.00 % of total aquatic production[4]
Estuaries	1500 g dry matter/(m2*year)	Excluding marsh, rg = 200 - 4000 g dm/(m2*y)[3]
Extreme desert, rock and ice	3 g dry weight/(m**2*year)	Tot npp=0.07E9 d t/y, biomass=0.02 dry kg/m2, tot biom.=0.5E9 d t[1]
Forest, equatorial rain	2.2 kg dry organic matter/(m2*year)	Total area = 11.0E6 km2, Total NPP = 11.0E15 g C/y, on land[7]
Forest, tropical seasonal	1.4 kg dry organic matter/(m2*year)	Total area = 5.0E6 km2, Total NPP = 3.2E15 g C/y, on land[7]
Freshwater phytoplankton	0.58E15 g C/year in total	Equal to 1.27 % of total aquatic production[4]
Freshwater spring	8800 kcal/(year*m**2)	Annual value, Silver Spring, FL[9]
Galcier	0 kg dry organic matter/(m2*year)	Total area = 13.9E6 km2, Total NPP = 0.0E15 g C/y, on land[7]
Halophytes	0.05 kg dry organic matter/(m2*year)	Total area = 0.5E6 km2, Total NPP = 0.01E15 g C/y, on land[7]
Intertidal marine angiosperm, Spartina alterniflora	1070 g C/(m2*year)	Rg = 338 - 1800 g C/m2[10]

Intertidal marine angiosperm, Spartina alterniflora	382 g C/(m2*year)	Rg = 22.5 - 855 g C/m2[11]
Intertidal marine angiosperm, Spartina alterniflora	515 g C/(m2*year)	Rg = 135 - 900 g C/m2[12]
Intertidal marine angiosperm, Spartina alterniflora	704 g C/(m2*year)	Rg = 337 - 1170 g C/m2[13]
Intertidal marine angiosperm, Spartina townsendii	337 g C/(m2*year)	None[11]
Intertidal marine angiosperm, Thalassia testudinum	599 g C/(m2*year)	Rg = 377 - 821 c G/m2[5]
Intertidal marine angiosperm, Zostera marina	270 g C/(m2*year)	Rg = 4.5 - 540 g C/m2[14]
Intertidal marine angiosperm, Zostera marina	85 g C/(m2*year)	Rg = 50 - 121 g C/m2[5]
Kelps	0.02E15 g C/year in total	Equal to 0.04 % of total aquatic production[4]
Lake and stream	0.02 kg dry organic matter/(m2*year)	Total area = 2.0E6 km2, Total NPP = 0.02E15 g C/y, on land[7]
Lake and stream	500 g dry weight/(m**2*year)	Tot npp=1.0E9 d t/y, biomass=0.02 dry kg/m2, tot biom.=0.04E9 d t[1]
Land	730 g dry weight/(m**2*year)	Tot npp=109E9 d t/y, biomass=12.5 dry kg/m2, tot biom.=1852E9 d t[1]
Mangroves	1.5 kg dry organic matter/(m2*year)	Total area = 0.3E6 km2, Total NPP = 0.2E15 g C/y, on land[7]
Marine sound	2500 kcal/(year*m**2)	Annual value, coastal, Long Island, New York[15]
Mixed forest	1.1 kg dry organic matter/(m2*year)	Total area = 1.5E6 km2, Total NPP = 0.7E15 g C/y, on land[7]
Oak - pine forest	5000 kcal/(year*m**2)	Annual value, medium aged, New York, no timber harvest[16]
Ocean	155 g dry weight/(m**2*year)	Tot npp=55E9 d t/y, biomass=0.009 dry kg/m2, tot biom.=3.3E9 d t[1]
Open ocean	125 g dry matter/(m2*year)	Rg = 2 - 400 g dm/(m2*y)[3]
Open ocean	125 g dry weight/(m**2*year)	Tot npp=41.5E9 d t/y, biomass=0.003 dry kg/m2, tot biom.=1.0E9 d[1]
Peatland	0.4 kg dry organic matter/(m2*year)	Total area = 1.5E6 km2, Total NPP = 0.3E15 g C/y, on land[7]
Phytoplankton, Lake Aranquadi	5913 g C/(m2*year)	Ethiopia[4]
Phytoplankton, Lake Baikal	113.1 g C/(m2*year)	USSR[4]
Phytoplankton, Lake Constance	164.2 g C/(m2*year)	Lower Basin, Switzerland[4]
Phytoplankton, Lake Great Bear	18.2 g C/(m2*year)	Canada[4]
Phytoplankton, Lake Great Slave	36.5 g C/(m2*year)	Canada[4]
Phytoplankton, Lake Kilotes	7811 g C/(m2*year)	Ethiopia[4]
Phytoplankton, Lake Nakuru	467.2 g C/(m2*year)	Keyna, Africa[4]
Phytoplankton, Lake Pedersborg	945.0 g C/(m2*year)	Denmark[4]
Phytoplankton, Lake Red Rock Tarn	10659 g C/(m2*year)	Australia, saline[4]
Phytoplankton, Lake Soelleroed	520.0 g C/(m2*year)	Denmark[4]
Phytoplankton, Lake Tahoe	36.1 g C/(m2*year)	Sierra Nevada, USA[4]
Phytoplankton, Lake Vanda	5.1 g C/(m2*year)	Antarctica, under 4 m ice[4]

Phytoplankton, Lake Werowrap	435 g C/(m2*year)	Australia, saline[4]
Phytoplankton, Lake Yssel	400.0 g C/(m2*year)	Netherlands[4]
Pine forest	7500 kcal/(year*m**2)	Annual value, young plantation, mean for 0-50 y, Eng[17]
Polar desert	0.05 kg dry organic matter/(m2*year)	Total area = 1.5E6 km2, Total NPP = 0.03E15 g C/y, on land[7]
Rain forest	13000 kcal/(year*m**2)	Annual value, mature forest, Puerto Rico[18]
Rice	1700 kcal/(m**2*year)	Brazil, edible portion = 580 kcal/(m2*y)[8]
Rice	2300 kcal/(m**2*year)	World average, edible portion = 760 kcal/(m2*y)[8]
Rice	5500 kcal/(m**2*year)	Japan, edible portion = 1840 kcal/(m2*y)[8]
Savanna	700 g dry weight/(m**2*year)	Tot npp=10.5E9 d t/y, biomass=4 dry kg/m2, tot biom.=60E9 d t[1]
Savanna, derived grass	2.0 kg dry organic matter/(m2*year)	Total area = 4.5E6 km2, Total NPP = 4.1E15 g C/y, on land[7]
Savanna, derived low trees	2.2 kg dry organic matter/(m2*year)	Total area = 4.5E6 km2, Total NPP = 4.5E15 g C/y, on land[7]
Savanna, spiny forest	1.2 kg dry organic matter/(m2*year)	Total area = 2.0E6 km2, Total NPP = 1.1E15 g C/y, on land[7]
Savanna, spiny shrub and tree savanna	1.3 kg dry organic matter/(m2*year)	Total area = 8.0E6 km2, Total NPP = 4.7E15 g C/y, on land[7]
Sclerophyll forest	1.0 kg dry organic matter/(m2*year)	Total area = 1.3E6 km2, Total NPP = 0.6E15 g C/y, on land[7]
Sclerophyll forest, chaparrals and macchia	0.8 kg dry organic matter/(m2*year)	Total area = 2.7E6 km2, Total NPP = 1.0E15 g C/y, on land[7]
Seas and oceans	43.50E15 g C/year in total	Equal to 94.89 % of total aquatic production[4]
Semi desert, scrub and grass, subtropical and mediterranean	0.3 kg dry organic matter/(m2*year)	Total area = 6.9E6 km2, Total NPP = 0.9E15 g C/y, on land[7]
Semi desert, scrub and grass, temperate	0.2 kg dry organic matter/(m2*year)	Total area = 2.3E6 km2, Total NPP = 0.2E15 g C/y, on land[7]
Soybeans	1400 kcal/(m**2*year)	World average, edible portion = 480 kcal/(m2*y)[8]
Soybeans	2400 kcal/(m**2*year)	Canada, edible portion = 800 kcal/(m2*y)[8]
Soybeans	780 kcal/(m**2*year)	Indonesia edible portion = 260 kcal/(m2*y)[8]
Sugar	12200 kcal/(m**2*year)	Hawaii, edible portion = 4070 kcal/(m2*y), from cane[8]
Sugar	3700 kcal/(m**2*year)	Cuba, edible portion = 1220 kcal/(m2*y), from cane[8]
Sugar	7300 kcal/(m**2*year)	Netherlands, edible portion = 2440 kcal/(m2*y), from beets[8]
Swamp and marsh	2000 g dry weight/(m**2*year)	Tot npp=4.0E9 d t/y, biomass=12 dry kg/m2, tot biom.=24E9 d t[1]
Swamp and marsh, temperate	2.5 kg dry organic matter/(m2*year)	Total area = 0.6E6 km2, Total NPP = 0.7E15 g C/y, on land[7]
Swamp and marsh, tropical	4.0 kg dry organic matter/(m2*year)	Total area = 1.4E6 km2, Total NPP = 2.6E15 g C/y, on land[7]
Sweet potatoes and yams	1100 kcal/(m**2*year)	Indonesia edible portion = 570 kcal/(m2*y)[8]
Sweet potatoes and yams	1500 kcal/(m**2*year)	World average, edible portion = 750 kcal/(m2*y)[8]
Sweet potatoes and yams	3600 kcal/(m**2*year)	Japan, edible portion = 1800 kcal/(m2*y)[8]
Taiga, middle	0.7 kg dry organic matter/(m2*year)	Total area = 6.0E6 km2, Total NPP = 1.9E15 g C/y, on land[7]
Taiga, mountainous	0.6 kg dry organic matter/(m2*year)	Total area = 5.7E6 km2, Total NPP = 1.5E15 g C/y, on land[7]

Taiga, nothern	0.5 kg dry organic matter/(m2*year)	Total area = 2.1E6 km2, Total NPP = 0.5E15 g C/y, on land[7]
Taiga, southern	0.9 kg dry organic matter/(m2*year)	Total area = 1.6E6 km2, Total NPP = 0.6E15 g C/y, on land[7]
Temperate forest	1300 g dry weight/(m**2*year)	Tot npp=23.4E9 d t/y, biomass=30 dry kg/m2, tot biom.=540E9 d t[1]
Temperate forest, coniferous plantations	1.7 kg dry organic matter/(m2*year)	Total area = 2.0E6 km2, Total NPP = 1.5E15 g C/y, on land[7]
Temperate forest, deciduous	1.3 kg dry organic matter/(m2*year)	Total area = 3.0E6 km2, Total NPP = 1.8E15 g C/y, on land[7]
Temperate forest, evergreen	1.7 kg dry organic matter/(m2*year)	Total area = 4.0E6 km2, Total NPP = 3.1E15 g C/y, on land[7]
Temperate grassland	500 g dry weight/(m**2*year)	Tot npp=4.5E9 d t/y, biomass=4 dry kg/m2, tot biom.=60E9 d t[1]
Temperate grassland, chernozems steppe	1.2 kg dry organic matter/(m2*year)	Total area = 3.5E6 km2, Total NPP = 1.9E15 g C/y, on land[7]
Temperate grassland, evergreen	1.3 kg dry organic matter/(m2*year)	Total area = 2.4E6 km2, Total NPP = 1.4E15 g C/y, on land[7]
Temperate grassland, kastanozems steppe	0.6 kg dry organic matter/(m2*year)	Total area = 2.8E6 km2, Total NPP = 0.8E15 g C/y, on land[7]
Temperate grassland, solonetz steppe	0.5 kg dry organic matter/(m2*year)	Total area = 0.3E6 km2, Total NPP = 0.07E15 g C/y, on land[7]
Total aquatic production	45.82E15 g C/year in total	Equal to 100 % of total aquatic production[4]
Total biosphere	336 g dry matter/(m2*year)	Global value[3]
Total for earth	320 g dry weight/(m**2*year)	Tot npp=164E9 d t/y, biomass=3.6 dry kg/m2, tot biom.=1855E9 d t[1]
Total marine	155 g dry matter/(m2*year)	Total marine production = 55.0E15 g dm/year[3]
Towns	0.5 kg dry organic matter/(m2*year)	Total area = 1.0E6 km2, Total NPP = 0.2E15 g C/y, on land[7]
Tropical forest	2000 g dry weight/(m**2*year)	Tot npp=40.0E9 d t/y, biomass=45 dry kg/m2, tot biom.=900E9 d t[1]
Tundra and alpine	140 g dry weight/(m**2*year)	Tot npp=1.1E9 d t/y, biomass=0.6 dry kg/m2, tot biom.=5E9 d t[1]
Tundra, grey soil	0.25 kg dry organic matter/(m2*year)	Total area = 3.8E6 km2, Total NPP = 0.4E15 g C/y, on land[7]
Tundra, mountainous	0.08 kg dry organic matter/(m2*year)	Total area = 2.9E6 km2, Total NPP = 0.1E15 g C/y, on land[7]
Upwelling zones	500 g dry matter/(m2*year)	Rg = 400 - 1000 g dm/(m2*y)[3]
Warm desert, extreme	0.001 kg dry organic matter/(m2*year)	Total area = 8.0E6 km2, Total NPP = 0.004E15 g C/y, on land[7]
Warm desert, subtropical and mediterranean	0.05 kg dry organic matter/(m2*year)	Total area = 4.2E6 km2, Total NPP = 0.1E15 g C/y, on land[7]
Warm desert, tropical	0.01 kg dry organic matter/(m2*year)	Total area = 5.6E6 km2, Total NPP = 0.03E15 g C/y, on land[7]
Weeds, excl. kelps	0.01E15 g C/year in total	Equal to 0.02 % of total aquatic production[4]
Wheat	1300 kcal/(m**2*year)	World average, edible portion = 430 kcal/(m2*y)[8]
Wheat	4400 kcal/(m**2*year)	Netherlands, edible portion = 1450 kcal/(m2*y)[8]

Wheat	900 kcal/(m**2*year)	India, edible portion = 300 kcal/(m2*y)[8]
White potatoes	1400 kcal/(m**2*year)	India, edible portion = 700 kcal/(m2*y)[8]
White potatoes	2200 kcal/(m**2*year)	World average, edible portion = 1090 kcal/(m2*y)[8]
White potatoes	4100 kcal/(m**2*year)	USA, edible portion = 2040 kcal/(m2*y)[8]
Woodland and shrubland	600 g dry weight/(m**2*year)	Tot npp=4.2E9 d t/y, biomass=6 dry kg/m2, tot biom.=42E9 d t[1]
World total C production	105E15 g C/year	Calculated by Bazilevich in 1970[3]
World total C production	115.9E15 g C/year	Calculated by Liebig in 1862[3]
World total C production	135E15 g C/year	Calculated by Bowen in 1966[3]
World total C production	24.6E15 g C/year	Calculated by Ebermayer in 1882, based on CO2 consumption[3]
World total C production	28E15 g C/year	Calculated by Fogg in 1958, conversion from gross prod by 0.5[3]
World total C production	33E15 g C/year	Calculated by Lehninger in 1965[3]
World total C production	35E15 g C/year	Calculated by Muller in 1960[3]
World total C production	36.3E15 g C/year	Calculated by Schroeder in 1919[3]
World total C production	43.7E15 g C/year	Calculated by Noddack in 1937[3]
World total C production	53E15 g C/year	Calculated by Vallentyne in 1965, rg = 44-60E15 g C/y[3]
World total C production	65.5E15 g C/year	Calculated by Golley in 1972[3]
World total C production	70.5E15 g C/year	Calculated by Lieth in 1972[3]
World total C production	73.6E15 g C/year	Calculated by Whittaker and Likens in 1973[3]
World total C production	74.5E15 g C/year	Calculated by Whittaker and Likens in 1969[3]
World total C production	74E15 g C/year	Calculated by Ryther in 1970[3]
World total C production	78.4E15 g C/year	Calculated by Whittaker and Likens in 1975[3]
World total C production	78E15 g C/year	Calculated by SCEP in 1970[3]
World total C production	83E15 g C/year	Calculated by Riley in 1944, conversion from gross prod by 0.5[3]
World total C production	90E15 g C/year	Calculated by Deevey in 1960[3]
World total C production, in sea	10.3E15 g C/year	Calculated by Muller in 1960[3]
World total C production, in sea	106E15 g C/year	Calculated by Bowen in 1966[3]
World total C production, in sea	10E15 g C/year	Calculated by Riley in 1944, concersion from gross prod. by 0.5[3]
World total C production, in sea	12E15 g C/year	Calculated by Fogg in 1958, conversion from gross by 0.5[3]
World total C production, in sea	15.1E15 g C/year	Calculated by Noddack in 1937[3]
World total C production, in sea	15.6E15 g C/year	Calculated by Duvigneaud in 1962[3]
World total C production, in sea	16.3E15 g C/year	Calculated by Schroeder in 1919[3]
World total C production, in sea	16.6E15 g C/year	Calculated by Lehninger in 1965[3]
World total C production, in sea	27E15 g C/year	Calculated by Vallentyne in 1965, rg = 22-32E15 g C/y[3]
World total C production, in sea	40.5E15 g C/year	Calculated by Golley in 1972[3]

World total C production, in sea	45.5E15 g C/year	Calculated by Lieth in 1972[3]
World total C production, in sea	45.5E15 g C/year	Calculated by Lieth in 1973[3]
World total C production, in sea	48.2E15 g C/year	Calculated by Whittaker and Likens in 1973[3]
World total C production, in sea	49.5E15 g C/year	Calculated by Whittaker and Likens in 1969[3]
World total C production, in sea	53.4E15 g C/year	Calculated by Whittaker and Likens in 1975[3]
World total C production, in sea	54E15 g C/year	Calculated by Olsson in 1970[3]
World total C production, in sea	56.4E15 g C/year	Calculated by Deevey in 1962[3]
World total C prod, in sea	56E15 g C/year	Calculated by SCEP in 1970[3]
World total C prod, in sea	78E15 g C/year	Calculated by Bazilevich et al. in 1970[3]
World total C production, on land	15E15 g C/year	Calculated by Steemann-Nielsen in 1958[3]
World total C production, on land	16.6E15 g C/year	Calculated by Lehninger in 1965[3]
World total C production, on land	16E15 g C/year	Calculated by Fogg in 1958, conversion from gross prod. by 0.5[3]
World total C production, on land	20E15 g C/year	Calculated by Ryther in 1969[3]
World total C production, on land	20E15 g C/year	Calculated by Schroeder in 1919[3]
World total C production, on land	22E15 g C/year	Calculated by SCEP in 1970[3]
World total C production, on land	23E15 g C/year	Calculated by Koblents-Mishke et al. in 1970[3]
World total C production, on land	25E15 g C/year	Calculated by Golley in 1972[3]
World total C production, on land	25E15 g C/year	Calculated by Lieth in 1972[3]
World total C production, on land	25E15 g C/year	Calculated by Muller in 1960[3]
World total C production, on land	25E15 g C/year	Calculated by Vallentyne in 1965[3]
World total C production, on land	25E15 g C/year	Calculated by Whittaker and Likens in 1969[3]
World total C production, on land	25E15 g C/year	Calculated by Whittaker and Likens in 1973[3]
World total C production, on land	25E15 g C/year	Calculated by Whittaker and Likens in 1975[3]
World total C production, on land	27E15 g C/year	Calculated by Bazilevich in 1970[3]
World total C production, on land	28.6E15 g C/year	Calculated by Noddack in 1937[3]
World total C production, on land	29E15 g C/year	Calculated by Bowen in 1966[3]
World total C production, on land	33.4E15 g C/year	Calculated by Deevey in 1960[3]

World total C production, on land	53E15 g C/year	Calculated by Ryther in 1959[3]
World total C production, on land	62E15 g C/year	Calculated by Riley in 1944, conversion from gross prod by 0.5, rg = 22-102[3]
Desert shrub	70 g dry weight/(m**2*year)	Tot npp=1.3E9 d t/y, biomass=0.7 dry kg/m2, tot biom.=13E9 d t[1]

Source: 1.Whittaker, R.H. 1970. Communities and Ecosystems. Macmillan. New York.

2.Thomas, M.D. and Hill. 1949. in: Franck, J and Loomis. Photosynthesis in Plants, Iowa State College, pp 19-52.
3.Whittaker, R.H. and G.E. Likens. 1973. in: WoodwellG.M. and E.V. Pecan (eds.): Carbon and Biosphere.AEC Symposium Series 30, 281-302. NTIS. Springfield. Virginia.
4.Bolin, B., E.T. Degens, Kempe and Ketner. 1977. The Global Carbon Cycle. SCOPE and ICSU. Johniley and Sons. NY.
5.Cadee, G.C. and J. Hegeman. 1974. Neth. J. Sea. Res. 8, 240-259.
6.Woodwell, G.M., P.H. Rich and C.A. Hall. 1973. in: Woodwell G.M. and E.V. Pecan (eds.): Carbon and te Biosphere. AEC Symp series 30, 221-239, NTIS.
7.Duvigneaud, P. 1972. La pensee et les hommes 7, 286-303.
8.FAO. United Nations. 1966. Protection Yearbook, 21.
9.Odum, H.T. 1957. Ecol. Monogr., 27, pp 55-112.
10.Odum, E.P. 1974. in: Reimond, R.J. and Queen (eds.): Ecology and Halophytes, 599-602. Academicress, London.
11.Ranwell, S.D. 1966. J. Appl. Ecol. 4, 239-256.
12.Turner, R.E., J.A. McGowan and A.W. Mantyla 1973. US Fish. Wildl. Serv. Fish. Bull. 71, 41-52.
13.Kirby, C.J. and J.G. Gosselink. 1976. Ecology 57, 1052-1059.
14.Nienhuis, P.H. and B.H.H. de Bree. 1977. Hydrobiologia 52, 55-66.
15.Riley, G.A. 1956. Bull. Bingham Oceanogr. Coll. 15, pp 324-344.
16.Woodwell, G.M. and R. Whittaker. 1968. Amer. Zool. 8, pp 19-30.
17.Ovington, J.D. 1961 Ann. Bot. 25, pp 12-20.
18.Odum, H.T. and r.f Pigeon. 1970. A tropical rain forest. Nat Tech. Info. Srevice. Springfield.

2-119 Ecosphere
Net production

Species	Value	Condition
Arable land	0.21E14 kg C/yr	In dry matter = 0.46E14 kg/yr[1]
Arable land	0.91 kg C/m**2*yr	In dry matter = 2.0 kg/m**2*yr[1]
Brown algae zone	0.009E14 kg C/yr	In dry matter = 0.026E14 kg/yr[1]
Brown algae zone	1.0 kg C/m**2*yr	In dry matter = 2.9 kg/m**2*yr[1]
Coniferous forest	0.14E14 kg C/yr	In dry matter = 0.28E14 kg/yr, val.+-50%[1]
Coniferous forest	1.4 kg C/m**2*yr	In dry matter = 2.8 kg/m**2*yr, val.+-50%[1]
Continental shelf	0.043E14 kg C/yr	In dry matter = 0.195E14 kg/yr[1]
Continental shelf	0.16 kg C/m**2*yr	In dry matter = 0.73 kg/m**2*yr[1]
Deciduous forest	0.025E14 kg C/yr	In dry matter = 0.06E14 kg/yr, val.+-50%[1]
Deciduous forest	0.54 kg C/m**2*yr	In dry matter = 1.2 kg/m**2*yr, val.+-50%[1]
Desert	0.015E14 kg C/yr	In dry matter = 0.033E14 kg/yr[1]
Desert	0.045 kg C/m**2*yr	In dry matter = 0.1 kg/m**2*yr[1]
Forest	0.078E14 kg C/yr	In dry matter=0.17E14 kg/yr, val.+-50%, other forest, excl conif,[1]
Grassland	0.245E14 kg C/yr	In dry matter = 0.54E14 kg/yr, val.+-50%[1]
Grassland	0.91 kg C/m**2*yr	In dry matter = 2.0 kg/m**2*yr, val.+-50%[1]
Land	0.071 kg C/m**2*yr	In dry matter = 1.53 kg/m**2*yr, total dry land[1]
Land	0.103E14 kg C/yr	None[2]
Land	0.151E14 kg C/yr	None[3]
Land	0.163E14 kg C/yr	None[4]
Land	0.21E14 kg C/yr	None[5]

Land	1.058E14 kg C/yr	In dry matter = 2.284E14 kg/yr, total on land[1]
Land	1.06E14 kg C/yr	None[1]
Ocean	0.072 kg C/m**2*yr	In dry matter = 0.32 kg/m**2*yr[1]
Ocean	0.24E14 kg C/yr	In dry matter = 1.070E14 kg/yr[1]
Ocean	0.292E14 kg C/yr	In dry matter = 1.291E14 kg/yr, total sea[1]
Other forest	0.54 kg C/m**2*yr	In dry matter = 1.2 kg/m**2*yr, val.+-50%[1]
Sea	0.0E14 kg C/yr	None[4]
Sea	0.15E14 kg C/yr	None[6]
Sea	0.25E14 kg C/yr	None[2]
Sea	0.286E14 kg C/yr	None[3]
Sea	0.29E14 kg C/yr	None[1]
Sea	0.32E14 kg C/yr	None[5]
Sea	0.53E14 kg C/yr	None[7]
Tropical forest	0.34E14 kg C/yr	In dry matter = 0.73E14 kg/yr, val.+-50%[1]
Tropical forest	2.3 kg C/m**2*yr	In dry matter = 5.0 kg/m**2*yr, val.+-50%[1]
Tundra	0.005E14 kg C/yr	In dry matter = 0.011E14 kg/yr[1]
Tundra	0.045 kg C/m**2*yr	In dry matter = 0.1 kg/m**2*yr[1]
Whole earth	0.265 kg C/m**2*yr	In dry matter = 0.70 kg/m**2*yr[1]
Whole earth	1.35E14 kg C/yr	In dry matter = 3.575E14 kg/yr[1]
Whole sea	0.081 kg C/m**2*yr	In dry matter = 0.36 kg/m**2*yr[1]

Source: 1.Bowen, H.J.M. 1966. Trace Elements in Biochemistry, Ac. Press.
2.Muller, D. 1960. in: Rukland, W. (ed.): Encyclopedia of Plant Physiology, 12(2), pp 934.
3.Noddack, W. 1937. Angew. Chem., 50, pp 505.
4.Schroeder, H. 1919. Naturwissenshaften, 7, pp 8.
5.Fogg, G.E. 1958. Advmt. Sci., 14, pp 395.
6.Nielsen, E.S. 1958. Rapp. p.-v. Reun. Cons. Perm. Int. Explor. Mer., 144, pp 92.
7.Ryther, J.H. 1959. Science, N.Y., 130, pp 602.

2-120 Ecosphere
Pools and fluxes within the C cycle

Species	Value	Condition
Basaltic oceanic crust carbon pool	0.32E21 g C	As non-carbonate C, Total in crust = 17.46 g C, C-content = 0.07 %[1]
Basaltic oceanic crust carbon pool	0.34E21 g C	As carbonate C, Total in crust = 48.05E21 g C, C-content = 0.20 %[1]
Basaltic oceanic crust carbon pool	0.66E21 g C	As total C, Total in crust = 65.51E21 g C, C-content = 0.27 %[1]
Biological oxidation of particulate organic carbon, POC, sediment	1.8 g C/(m2*year)	Rg = 0.5 - 3.0, oxidation rate in sediment, marine[1]
Biological oxidation of particulate organic carbon	3.0 g C/(m2*year)	Rg = 2.0 - 4.0, oxidation in water phase, marine[1]
Continental granitic crust carbon pool	1.14E21 g C	As non-carbonate C, Total in crust = 17.46 g C, C-content = 0.07 %[1]
Continental granitic crust carbon pool	7.91E21 g C	As carbonate C, Total in crust = 48.05E21 g C, C-content = 0.20 %[1]
Continental granitic crust carbon pool	9.05E21 g C	As total C, Total in crust = 65.51E21 g C, C-content = 0.27 %[1]
Continental sediment carbon pool	10.00E21 g C	As non-carbonate C, Total in crust = 17.46 g C, C-content = 0.07 %[1]

Continental sediment carbon pool	25.64E21 g C	As carbonate C, Total in crust = 48.05E21 g C, C-content = 0.20 %[1]
Continental sediment carbon pool	35.64E21 g C	As total C, Total in crust = 65.51E21 g C, C-content = 0.27 %[1]
Deep water oxidation of labile dissolved organic matter, DOC, at sea	1.5 g C/(m2*year)	Rg = 1.0 - 2.0, converted to inorganic carbon[1]
Flux of organic matter leaving the euphotic zone, sedimentation	7.1 g C/(m2*year)	Rg = 4.1 - 10.0 g C/(m2*y), marine[1]
Flux to atm. from all fluxes	2.5E15 g C/year	Man-made transfer[1]
Flux to atmosphere from firewood	0.3E15 g C/year	Man-made transfer[1]
Flux to atmosphere from forest clearing in third world	3.6E15 g C/year	Man-made transfer[1]
Flux to atmosphere from growth stimulated by CO2	-0.3E15 g C/year	Man-made transfer[1]
Flux to atmosphere from growth stimulated by NOx	-0.2E15 g C/year	Man-made transfer[1]
Flux to atmosphere from industrial use of wood	0.3E15 g C/year	Man-made transfer[1]
Flux to atmosphere from reforestation	-0.3E15 g C/year	Man-made transfer[1]
Flux to atmosphere from regrowth in temperate region	-0.5E15 g C/year	Man-made transfer[1]
Flux to atmosphere from regrowth in the tropics	-1.0E15 g C/year	Man-made transfer[1]
Flux to atmosphere from soil organic matter decomposition	0.6E15 g C/year	Man-made transfer[1]
Labile dissolved organic matter, DOC, to refractory DOC, at sea	0.5 g C/(m2*year)	Rg - 0.4 - 0.6, in water phase[1]
Net release of C to atomsphere from boreal forest	0.8E15 g C/y	10 % of total release from terrestrial plant communities[2]
Net release of C to atomsphere from detritus and humus	2.0E15 g C/y	25 % of total release from terrestrial plant communities[2]
Net release of C to atomsphere from other vegetation	0.2E15 g C/y	3 % of total release from terrestrial plant communities[2]
Net release of C to atomsphere from temperate zone forest	1.4E15 g C/y	18 % of total release from terrestrial plant communities[2]
Net release of C to atomsphere from total land	7.9E15 g C/y	100 % of total release from terrestrial plant communities[2]
Net release of C to atomsphere from total vegetation	5.9E15 g C/y	75 % of total release from terrestrial plant communities[2]
Net release of C to atomsphere from tropical forest	3.5E15 g C/y	44 % of total release from terrestrial plant communities[2]
Oceanic sediment carbon pool	14.16E21 g C	As carbonate C, Total in crust = 48.05E21 g C, C-content = 0.20 %[1]
Oceanic sediment carbon pool	20.16E21 g C	As total C, Total in crust = 65.51E21 g C, C-content = 0.27 %[1]
Oceanic sediment carbon pool	6.0E21 g C	As non-carbonate C, Total in crust = 17.46 g C, C-content = 0.07 %[1]
Particulate organic carbon, POC, trapped in sediment	0.3 g C/(m2*year)	Rg = 0.2 - 0.4, permanently trapped at sea[1]

Source: 1.Bolin, B., E.T. Degens, S. Kempe and P. Ketner. 1977. The Global Carbon Cycle. SCOPE and ICSU. Johniley and Sons. NY.
2.Woodwell, G.M. 1978. Sci. American 238, 34-43.

2-121 Ecosphere
Pools and fluxes within the Cu cycle

Species	Value	Condition
Animal biomass in seawater, pool	3.0E9 g	Average conc = 1.0 ug/g, total mass = 3E15 g[1]
Animal biomass, pool	2.4E11 g	Average conc = 60 ug/g, total mass = 2.2E18 g[1]
Assimilation of marine biomass	21E10 g/year	Global value[1]
Biomass, dead	1E17 g	Global value[1]
Biota, living	2.9E13 g	Global value[1]
Cu dissolved in oceanic pore water, pool	4.8E15 g	15 ug Cu/kg, total mass = 0.32E24 g[1]
Cu dissolved in seawater, pool	2.1E14 g	0.15 ug Cu/kg, total mass = 1.4E24[1]
Cu in atmosphere, pool	2.6E9 g	0.5 ng Cu/m3, total mass = 5.1E18 m3[1]
Cu in biomass in marshes and swamps, pool	7.2E10 g	Average conc = 12 ug/g, total mass = 6E15 g[1]
Cu in glaciers, pool	3.3E12 g	0.2 ug Cu/kg, total mass = 1.65E22[1]
Cu in lake and stream, pool	6.8E10 g	2.0 ug Cu/kg, total mass = 0.34E20 g[1]
Cu in particulate organic matter in seawater, pool	4.9E11 g	Average conc = 7.0 ug/g, total mass = 7E16 g[1]
Cu ore, reserves	4.6E14 g	Global value[1]
Cu ore, resources	1.5E16 g	Global value[1]
Deposition on earth from atmosphere	74.5E9 g /year	Stady state, global value[1]
Deposition on land from atmosphere	61.8E9 g /year	Global value[1]
Deposition on oceans from atmosphere	12.7E9 g /year	Global value[1]
Emission to atmosphere from anthropogenic sources	56E9 g /year	Global value[1]
Emission to atmosphere from fossil fuel compustion	5.4E9 g /year	Anthropogenic source[1]
Emission to atmosphere from iron and steel production	5.9E9 g /year	Anthropogenic source[1]
Emission to atmosphere from natural sources	18.5E9 g /year	Global value[1]
Emission to atmosphere from nonferrous production	21.9E9 g /year	Anthropogenic source[1]
Emission to atmosphere from other natural sources	3.9E9 g /year	Other than seasalt, exudates and dust[1]
Emission to atmosphere from other sources	22.8E9 g /year	Other than fuel, and metal production[1]
Emission to atmosphere from seasalt spray	0.08E9 g /year	Natural source[1]
Emission to atmosphere from windblown dust	12E9 g /year	Natural source[1]
Emission to atmosphere in total	74.5E9 g /year	Global value[1]
Emission to soil from dead biomass	6.2E11 g/year	Global value[1]

Emission to soil from fertilizer application	9.4E9 g/year	Global value[1]
Emission to soil from waste disposal	77E9 g/year	Global value[1]
Fossil fuel	3.4E15 g	Global value[1]
Fossil fuel deposits, coal, pool	1.7E14 g	Average conc = 17 ug/g, total mass = 10E18 g[1]
Fossil fuel deposits, crude oil, pool	3.2E10 g	Average conc = 0.14 ug/g, total mass = 0.23E18 g[1]
Fossil fuel deposits, shale oil, pool	3.2E15 g	Average conc = 70 ug/g, total mass = 46E18 g[1]
Hydrosphere pool	5E15 g	Global value[1]
Limestone in sedimentary rocks, pool	0.02E20 g	Average conc = 6 ug/g, total mass = 0.35E24 g[1]
Lithosphere, pool	14E20 g	Average conc = 24 ug/g, total mass = 57E24 g, down to 45 km[1]
Marine sedimentation	6.3E12 g/year	Global value[1]
Organic matter in soil, pool	2.4E15 g	Average conc = 350 ug/g, total mass = 6.8E18 g, down to 100 cm[1]
Outflux from marine sediment	47E10 g/year	Global value[1]
Plant biomass in seawater, pool	7.0E8 g	Average conc = 3.5 ug/g, total mass = 2E14[1]
Plant biomass, pool	2.9E13 g	Average conc = 12 ug/g, total mass = 2E16 g[1]
River run-off to oceans	6.3E12 g/year	Global value[1]
River run-off to oceans, particulate Cu	6.2E12 g/year	Global value[1]
River run-off to oceans, soluble Cu	0.6E12 g/year	Global value[1]
Sandstone in sedimantary rocks, pool	0.08E20 g	Average conc = 30 ug/g, total mass = 0.28E24 g[1]
Sedimentary rock, pool	0.75E20 g	Average conc = 30 ug/g, total mass = 2.5E24 g[1]
Sediments, pool	7.5E19 g	Global value[1]
Shale and clay in sedimentary rocks, pool	0.66E20 g	Average conc = 35 ug/g, total mass = 1.88E24[1]
Soil, pool	6.7E15 g	Average conc = 20 ug/g, total mass = 3.3E20 g, down to 100 cm[1]
Uptake of living terrestrical biomass	2E12 g/year	Global value[1]
Waste disposal in oceans	1.7E10 g/year	Global value[1]
vegetation exudates	2.5E9 g /year	Natural source[1]

Source: 1.Nriagu, J.O. 1979. Copper in the Environment. John Wiley and Sons. NY.

2-122 Ecosphere
Pools and fluxes within the H2S cycle

Species	Value	Condition
Biogenic production from terrestrial systems	110E6 tons S/y	Global value, flux[1]
Biogenic production from terrestrial systems	58E6 tons S/y	Global value, flux[2]
Biogenic production from terrestrial systems	68E6 tons S/y	Global value, flux[3]
Biogenic production from terrestrial systems	70E6 tons S/y	Global value, flux[4]
Biogenic production from terrestrial systems	90E6 tons S/y	Global value, flux[5]

Source: 1.Erikson, E. 1963. J. Geophys Res. 68, 4001-4008.
2.Friend, J.P. 1973. In: S.I. Rasool (ed.): Chemistry o the Lower Atmosphere. Plemun. NY. 177-201.
3.Robinson, E. and R.C. Robbins. 1968. Sources, Abundance and Fate of Gaseous Atmospheric Pollutants,ep. PR-6755. Stanford res inst. Menlo Park. Calif.
4.Junge, C.E. 1963. Air Chemistry and Radioactivity. Academic Press. NY. 59-74.
5.Kellog, W.W., R.D. Cadla, E.R. Allen, A.L. Lazus and E.A. Martell.1972. The Sulfur cycle. Science 17, 587-596.

2-123 Ecosphere
Pools and fluxes within the N cycle

Species	Value	Condition
Atmospheric NH3	0.07E12 g N	Over ocean areas, conc = 0.4 ppb, mixing height = 1 km, pool[1]
Atmospheric NH3	0.08E12 g N	Over other land areas, conc = 2 ppb, mixing height = 1.5 km, pool[1]
Atmospheric NH3	0.12E12 g N	In upper trophosphere, conc = 0.1 ppb, pool[1]
Atmospheric NH3	0.66E12 g N	Over tropical land areas, conc = 12 ppb, mixing height = 2 km, pool[1]
Atmospheric NH3	0.9E12 g N	In total atmosphere and upper trophosphere, pool[1]
Atmospheric NH4	0.12E12 g N	Over ocean areas, conc = 0.4 ppb, mixing height = 1 km, pool[1]
Atmospheric NH4	0.17E12 g N	Over tropical land areas, conc = 2[1]
Atmospheric NH4	0.33E12 g N	Over other land areas, conc = 3 ppb, mixing height = 1.5 km, pool[1]
Atmospheric NH4	1.16E12 g N	In upper trophosphere, conc = 0.7 ppb, pool[1]
Atmospheric NH4	1.8E12 g N	In total atmosphere and upper trophosphere, pool[1]
Atmospheric total N	0.19E12 g N	Over ocean areas, conc = 0.8 ppb, mixing height = 1 km, pool[1]
Atmospheric total N	0.41E12 g N	Over other land areas, conc = 5 ppb, mixing height = 1.5 km, pool[1]
Atmospheric total N	0.83E12 g N	Over tropical land areas, conc = 14 ppb, mixing height = 2 km, pool[1]
Atmospheric total N	1.28E12 g N	In upper trophosphere, conc = 0.8 ppb, pool[1]
Atmospheric total N	2.7E12 g N	In total atmosphere and upper trophosphere, pool[1]
Fixation of atmospheric N by agricultural ecosystems	44E12 g N/y	Global value[2]
Fixation of atmospheric N by all fluxes	264E12 g N/y	Global value, rg = 214-314E12 g N/y[2]
Fixation of atmospheric N by all terrestrial ecosystems	139E12 g N/y	Global value[2]
Fixation of atmospheric N by aquatic, pelagic biomass	70E12 g N/y	Global value, rg = 20-120E12 g N/y[2]
Fixation of atmospheric N by combustion	19E12 g N/y	Global value[2]

Fixation of atmospheric N by forest ecosystems	40E12 g N/y	Global value[2]
Fixation of atmospheric N by grassland ecosystems	45E12 g N/y	Global value[2]
Fixation of atmospheric N by industrial proces	36E12 g N/y	Global value[2]
Fixation of atmospheric N by other terrestrial ecosystems	10E12 g N/y	Global value[2]
Fixation of atmospheric N by terrestrial biomass	139E12 g N/y	Global value[2]

Source: 1.Svenson, B.H. and R. Soederlund. 1979. Ecological Bulletins No 22. N, P and S - Global Cycles. Scoperep &. Berlings. Lund.
2.Granat, L., H. Rohde and R.O. Hallberg. 1976. In: B.H. Svensson and R. Soederlund (eds.): N, P and S. SCOPE report No. 7, Ecological Bulletins, Stockholm 22, 89-134.

2-124 Ecosphere
Pools and fluxes within the Pb cycle

Species	Value	Condition
Atmispheric pool	1.8E10 gram	Pb conc in pool = 3.5 ng/kg, total pool = 5E21 g, global pool[1]
Deposition from atmosphere through rain on oceans	21E10 g/year	Oceans on the Nothern Hemisphere, 1962[2]
Deposition from atmosphere to rivers	9E10 g/year	Oceans on the Nothern Hemisphere, 1962[2]
Deposition from rain to oceans	21E10 g/year	Oceans on the Nothern Hemisphere, 1962[2]
Emission from indistrial chemicals to rivers	14E10 g/year	Oceans on the Nothern Hemisphere, 1962[2]
Emission from rivers to oceans	18E10 g/year	Oceans on the Nothern Hemisphere, 1962[2]
Emission from seasalt sprays to atmosphere	0.00002E6 kg Pb/year	Tot Prod=1000E9 kg/y, emission fac=0.00002 mg/kg, natural, global value[1]
Emission from windblown dust to atmosphere	16.0E6 kg Pb/year	Tot Prod=500E9 kg/y, emission fac=32 mg/kg, natural, global value[1]
Emission to atmosphere from Cu production	26.6E6 kg Pb/year	Tot prod=8.6E9 kg/y, emission fac=3.09 g/kg, anthropogenic, global value[1]
Emission to atmosphere from Fe and steel production	49.7E6 kg Pb/year	Tot prod=776E9 kg/y, emission fac=0.064 g/kg, anthropogenic, global value[1]
Emission to atmosphere from Fe foundries	3.90E6 kg Pb/year	Tot prod=78E9 kg/y, emission fac=0.05 g/kg, anthropogenic, global value[1]
Emission to atmosphere from Ni production	2.47E6 kg Pb/year	Tot prod=0.8E9 kg/y, emission fac=3.09 g/kg, anthropogenic, global value[1]
Emission to atmosphere from Pb milling	4.88E6 kg Pb/year	Tot prod=3.7E9 kg/y, emission fac=1.32 g/kg, anthropogenic, global value[1]
Emission to atmosphere from Pb mining	3.36E6 kg Pb/year	Tot prod=3.7E9 kg/y, emission fac=0.91 g/kg, anthropogenic, global value
Emission to atmosphere from Pb smelting and refining	26.7E6 kg Pb/year	Tot prod=4.2E9 kg/y, emission fac=6.36 g/kg, anthropogenic, global value[1]
Emission to atmosphere from Zn production	15.7E6 kg Pb/year	Tot prod=6.2E9 kg/y, emission fac=2.54 g/kg, anthropogenic, global value[1]
Emission to atmosphere from all anthropogenic sources	438E6 kg Pb/year	Global value[1]

Emission to atmosphere from all natural sources	18.55E6 kg Pb/year	Global value[1]
Emission to atmosphere from anti-knock additives	267E6 kg Pb/year	Global production = 0.42E9 kg/y, emission factor = 70 %[1]
Emission to atmosphere from cement manufacture	1.41E6 kg Pb/year	Tot prod=781E9 kg/y, emission fac=0.0018 g/kg, anthropogenic, global value[1]
Emission to atmosphere from clay production	0.66E6 kg Pb/year	Tot prod=364E9 kg/y, emission fac=0.0018 g/kg, anthropogenic, global value[1]
Emission to atmosphere from coal combustion	14.9E6 kg Pb/year	Tot prod=3300E9 kg/y, emission fac=0.0045 g/kg, anthropogenic, global value[1]
Emission to atmosphere from crushed stone	1.15E6 kg Pb/year	Tot prod=8200E9 kg/y, emission fac=0.00014 g/kg, anthropogenic, global value[1]
Emission to atmosphere from etal fabrication	3.82E6 kg Pb/year	Tot prod=2.1E9 kg/y, emission fac=1.82 g/kg, anthropogenic, global value[1]
Emission to atmosphere from ferroalloys	2.05E6 kg Pb/year	Tot prod=5.0E9 kg/y, emission fac=0.41 g/kg, anthropogenic, global value[1]
Emission to atmosphere from fertilizer and phosphate rock	0.05E6 kg Pb/year	106,0.00045[1]
Emission to atmosphere from forest fires	0.5E6 kg Pb/year	Tot Prod=36E9 kg/y, emission fac=14 mg/kg, natural, global value[1]
Emission to atmosphere from meteoric debris	0.000007E6 kg Pb/year	Tot Prod=0.0036E9 kg/y, emission fac=2.0 mg/kg, natural, global value[1]
Emission to atmosphere from mine production	400E10 g/year	Oceans on the Nothern Hemisphere, 1962, to atmosphere and ind. chemicals[2]
Emission to atmosphere from oil combustion	6.14E6 kg Pb/year	Tot prod=675E9 kg/y, emission fac=0.0091 g/kg, anthropogenic, global value[1]
Emission to atmosphere from pigments	1.31E6 kg Pb/year	Tot prod=0.36E9 kg/y, emission fac=3.63 g/kg, anthropogenic, global value[1]
Emission to atmosphere from refuse incineration	3.57E6 kg Pb/year	Tot prod=2550E9 kg/y, emission fac=0.0014 g/kg, anthropogenic, global value[1]
Emission to atmosphere from secondary Pb production	0.77E6 kg Pb/year	Tot prod=1.0E9 kg/y, emission fac=0.77 g/kg, anthropogenic, global value[1]
Emission to atmosphere from sewage sludge incineration	0.91E6 kg Pb/year	Tot prod=48E9 kg/y, emission fac=0.019 g/kg, anthropogenic, global value[1]
Emission to atmosphere from soldering	0.22E6 kg Pb/year	Tot prod=0.14E9 kg/y, emission fac=1.59 g/kg, anthropogenic, global value[1]
Emission to atmosphere from vegetation	1.65E6 kg Pb/year	Tot Prod=75E9 kg/y, emission fac=22 mg/kg, natural, global value[1]
Emission to atmosphere from volcanic particles	0.40E6 kg Pb/year	Tot Prod=25E9 kg/y, emission fac=16 mg/kg, natural, global value[1]
Emission to atmosphere from wood combustion	1.15E6 kg Pb/year	Tot prod=1260E9 kg/y, emission fac=0.00091 g/kg, anthropogenic, global value[1]
Emission to oceans from all sources	39E10 g/year	Oceans on the Nothern Hemisphere, 1962[2]
Emission to rivers from agricultural activity	3.3E10 g/year	Oceans on the Nothern Hemisphere, 1962, stimulation of denudation[2]
Freshwater biota	82.5E10 gram	Pb conc in pool = 2.5 ug/g, total pool = 3.3E17 g, global pool[1]
Glacier pools	6.1E10 gram	Pb conc in pool = 0.003 ug/kg, total pool = 2E22 g, global pool[1]
Groundwater pool	8.2E10 gram	Pb conc in pool = 20 ug/l, total pool = 4E18 g, global pool[1]
Lake and river	6.1E10 gram	Pb conc in pool = 2.0 ug/l, total pool = 3E19 g, global pool[1]

Land dead biota	210E10 gram	Pb conc in pool = 3.0 ug/g, total pool = 7.0E17 g, global pool[1]
Land living biota	8.3E10 gram	Pb conc in pool = 0.1 ug/g, total pool = 8.3E17 g, global pool[1]
Marine dead biota	250E10 gram	Pb conc in pool = 2.5 ug/g, total pool = 10E17 g, global pool[1]
Marine living biota	0.08E10 gram	Pb conc in pool = 0.5 ug/g, total pool = 1.5E15 g, global pool[1]
Natural emission to rivers	1.1E10 g/year	Oceans on the Nothern Hemisphere, 1962, preindustrial value[2]
Oceanic pool	27.4E12 gram	Pb conc in pool = 0.02 ug/kg, total pool = 13.7E23 g, global pool[1]
Sediment pool	48E18 gram	Pb conc in pool = 47 g/g, total pool = 10.2E23 g, global pool[1]
Sediment, pore water	12E15 gram	Pb conc in pool = 36 ug/l, total pool = 3.3E23 g, global pool[1]
Soil pool	4.8E15 gram	Pb conc in pool = 16 ug/g, total pool = 3E20 g, global pool[1]

Source: 1.Nriagu, J.O. 1979. The Biochemistry of Lead in The Environment. Part A. Elsevier/North Holland. Amstrdam. 2.Tatsumoto, M. and C.C. Patterson. 1963. In: J. Geiss and E.D. Goldberg (eds.): Earth Science and Metoritics. North-Holland Publ. Amsterdam. 74-89.

2-125 Ecosphere
Pools and fluxes within the Zn cycle

Species	Value	Condition
Emission to atmosphere from coal combustion, anthropogenic	14.9E9 g Zn/year	Emission factor = 4.8 g/1E6 part, total part = 3100E9 kg/y[1]
Emission to atmosphere from end use of Zn products, anthropogenic	26.0E9 g Zn/year	Global value[1]
Emission to atmosphere from ferroalloys and iron foundries	4.5E9 g Zn/year	Emission factor = 54 g/1E6 part., total proc = 83E9 kg/y[1]
Emission to atmosphere from forest fires, natural	2.1E9 g Zn/year	Emission factor = 58 g/1E6 part, total part = 36E9 kg/y[1]
Emission to atmosphere from grain handling, anthropogenic	0.18E9 g Zn/year	Emission factor = 0.5 g/1E6 part, total part = 356E9 kg/y[1]
Emission to atmosphere from iron and steel production, anthropogenic	35.1E9 g Zn/year	Emission factor = 27 g/1E6 part., total proc = 1300E9 kg/y[1]
Emission to atmosphere from mining, anthropogenic	1.6E9 g Zn/year	Emis factor=100 g/1E6 part., total proc=16E9 kg/y, mining of Zn+Cu+Pb[1]
Emission to atmosphere from oil combustion, anthropogenic	0.07E9 g Zn/year	Emission factor = 0.025 g/1E6 part, total part = 2800E9 kg/y[1]
Emission to atmosphere from phosphate fertilizers, anthropogenic	1.8E9 g Zn/year	Emission factor = 15 g/1E6 part, total part = 118E9 kg/y[1]
Emission to atmosphere from primary Al mining, anthropogenic	0.14E9 g Zn/year	Emission factor = 11 g/1E6 part., total proc = 12.3E9 kg/y[1]
Emission to atmosphere from primary Cu mining, anthropogenic	6.6E9 g Zn/year	Emission factor = 845 g/1E6 part., total proc = 7.9E9 kg/y[1]
Emission to atmosphere from primary Ni mining, anthropogenic	0.68E9 g Zn/year	Emission factor = 845 g/1E6 part., total proc = 0.8E9 kg/y[1]
Emission to atmosphere from primary Pb mining, anthropogenic	0.440E9 g Zn/year	Emission factor = 110 g/1E6 part., total proc = 4.0E9 kg/y[1]
Emission to atmosphere from primary Zn production, anthropogenic	98.6E9 g Zn/year	Emission factor = 17600 g/1E6 part., total proc = 5.6E9 kg/y[1]
Emission to atmosphere from rubber tire wear, anthropogenic	2.2E9 g Zn/year	Emission factor = 4.5 g/tire, total use = 5E8 tires[1]
Emission to atmosphere from seasalt spray, natural	0.21E9 g Zn/year	Emission factor = 0.21 g/1E6 part, total part = 1000E9 kg/y[1]
Emission to atmosphere from secondary Cu production, anthropogenic	1.1E9 g Zn/year	Emission factor = 500 g/1E6 part., total proc = 2.1E9 kg/y[1]

Emission to atmosphere from secondary Pb production, anthropogenic	0.3E9 g Zn/year	Emission factor = 300 g/1E6 part., total proc = 1.0E9 kg/y[1]
Emission to atmosphere from secondary Zn production, anthropogenic	8.1E9 g Zn/year	Emission factor = 9000 g/1E6 part., total proc = 0.9E9 kg/y[1]
Emission to atmosphere from total anthropogenic sources	315E9 g Zn/year	Global value[1]
Emission to atmosphere from total natural sources	43.7E9 g Zn/year	Global value[1]
Emission to atmosphere from vegetation, natural	9.4E9 g Zn/year	Emission factor = 125 g/1E6 part, total part = 75E9 kg/y[1]
Emission to atmosphere from volcanic particles, natural	7.0E9 g Zn/year	Emission factor = 700 g/1E6 part, total part = 10E9 kg/y[1]
Emission to atmosphere from waste inceneration, anthropogenic	75.0E9 g Zn/year	Emission factor = 500 g/1E6 part, total part = 150E9 kg/y[1]
Emission to atmosphere from windblown dust, natural	25E9 g Zn/year	Emission factor = 50 g/1E6 part, total part = 500E9 kg/y[1]
Emission to atmosphere from wood combustion, anthropogenic	37.2E9 g Zn/year	Emission factor = 58 g/1E6 part, total part = 640E9 kg/y[1]

Source: 1.Nriagu, J.O. 1980. Zinc in the Environment. Part 1: Ecological cycling. Wiley and Sons. NY.

2-126 Ecosphere
Pools within the C cycle

Species	Value	Condition
Anthropogenic emission of CO	0.55E15 g/year	Range = 0.45 - 0.64E15 g/year, global value[1]
Atmosphere	58E15 moles	As CO2[2]
Atmospheric flux of CO to the stratosphere	0.11E15 g/year	global value[1]
Biosphere	1.5E18 g	Per unit area = 0.29g/cm**2[3]
CH4 emission from C-14 free sources	0.11E15 g/year	Range = 0.016-0.21E15 g/y, global value, from mining, cars etc.[4]
CH4 emission from freshwater lakes	0.013E15 g/year	Range = 0.001-0.025E15 g/y, global value[4]
CH4 emission from oceans	0.009E15 g/year	Range = 0.001-0.018E15 g/y, global value[4]
CH4 emission from paddy fields	0.28E15 g/year	Global value[4]
CH4 emission from upland fields, forests and tundra	0.012E15 g/year	Range = 0.011-0.013E15 g/y, global value[4]
CH4, enteric fermentation of animals	0.16E15 g/year	Range = 0.10-0.22E15 g/y, global value[4]
CH4, total emission to atmosphere	0.83E15 g/year	Range = 0.6-1.07E15 g/y, global value[4]
CO, atmospheric	0.53E15 g	Based on 0.11 ppm mixing ratio[1]
CO2 consumed in photosynthesis	1.1E17 g/yr	None[3]
CO2 in ocean	0.8E18 gram C	Total on earth, as H2CO3[5]
CO2 in ocean	114.7E18 gram C	Total on earth, as HCO3(-)[5]
CO2 in ocean	14.2E18 gram C	Total on earth, as CO3(-2)[5]
CO2 in ocean water	1.4E19 g	Per unit area = 27g/cm**2[3]
CO2 in the atmosphere	2.35E18 gram C	Total on earth[5]
CO2 increase in atmosphere	5E15 g/yr	0.3 % per year[3]
CO2 produced geologically	2.5E23 g	Over 4.5 E9 years, per unit area = 5E4g/cm**2[3]

CO2 released by burning of fossil fuels	1.5E16 g/yr	None[3]
Carbonate in sediment	67000E18 gram C	Total on earth[5]
Carbonate sediments	1700000E15 moles	As carbonate[2]
Ch4 emission from swamps and marshes	0.245E15 g/year	Range = 0.19-0.30E15 g/y, global value[4]
Consumption of CH4 by oxidation by OH	1.86E15 g/year	Range = 0.86-2.86E15 g/y, global value[4]
Dead organic matter in ocean	10.0E18 gram C	Total on earth[5]
Dead organic matter on land	2.6E18 gram C	Total on earth[5]
Decaying organic matter	250E15 moles	In oceans[2]
Decaying organic matter	58E15 moles	On land[2]
Emission from forest fires and burning of CO	0.06E15 g/year	Global value, burning of agricultural waste[1]
Emission of CO from oxidation of hydrocarbon	0.06E15 g/year	Global value[1]
Emission of CO to atmosphere from degradation of chlorophyll	0.30E15 g/year	Range = 0.10-0.50E15 g/y, global value[1]
Emissions from oceans of CO to atmosphere	0.12E15 g/year	Range = 0.02 - 0.22E15 g/y, global value[1]
Flux from atmosphere of CH4 to stratosphere	0.15E15 g/year	Range = 0.10-0.21E15 g/y, global value[4]
Fossil fuel	830E15 moles	As organic carbon[2]
Fossil fuel reserves	3E18 g	Economically exploitable[3]
Input to dissolved organic carbon in oceans from resistant material	1.8E15 g C/year	Assuming 5 % of net production is resistant material[6]
Input to dissolved organic carbon in the oceans from ezcretion	3.6E15 g C/year	10 % of phytoplankton net production[6]
Input to organic carbon in the oceans by net primary production	3.6E16 g C/year	Assuming 100 g C fixed per m2 per year[6]
Input to organic carbon in the oceans from rain	2.2E14 g C/year	Assuming 1 mg C/l[6]
Input to organic carbon in the oceans from rivers	1.8E14 g C/year	Assuming 5 mg C/l[6]
Land organisms	38E15 moles	As organic carbon[2]
Living matter on land	0.3E18 gram C	Total on earth[5]
Living organic matter in ocean	0.03E18 gram C	Total on earth[5]
Loss of organic carbon in the oceans by nearshore sedimentation	2.7E12 g C/year	Flux of C[6]
Loss of organic carbon in the oceans by pelagic sedimentation	9.2E13 g C/year	Flux of C[6]
Marine organisms	0.830E15 moles	As organic matter[2]
Mass of atmospheric CO2	2.38E18 g	Per unit area = 0.45g/cm**2[3]
Non - CH4 - Carbon, anthrophgenic emission	0.065E15 g organic C/year	30 % conversion of emission to POC[7]
Non - CH4 - Carbon, emission from vegetation	0.16E15 g organic C/year	Rg = 0.01-0.35E15 g/y[7]
Non - CH4 - Carbon, pool over land as organic C	0.037E15 g	Global value, in atmosphere[7]

Non - CH4 - Carbon, pool over land of particulate organic C	0.001E15 g	Global value, in atmosphere[7]
Non - CH4 - Carbon, pool over sea as organic C	0.014E15 g	Global value, in atmosphere[7]
Non - CH4 - Carbon, pool over sea of particulate organic C	0.001E15 g	Global value, in atmosphere[7]
Non - CH4 - Carbon, total pool in atmosphere as organic C	0.051E15 g	Rg = 0.02 - 0.15E15 g C, global value[7]
Non - CH4 - Carbon, total pool of particulate organic C	0.002E15 g	Rg = 0.002 - 0.006E15 g, in atmosphere[7]
Ocean deep layers	2900E15 moles	In solution[2]
Ocean surface layers	43E15 moles	In solution[2]
Organic carbon in sediment	25000E18 gram C	Total on earth[5]
Organic carbon in sedimentary rocks	5E21 g	Per unit area = E3g/cm**2[3]
Organic carbon in the oceans in marine sediment	1.1E22 g C	Accumulated per 10E8 years[6]
Organic carbon in the oceans in plankton	3.0E15 g C	Pool of carbon in the oceans[6]
Oxidation of CO by OH	3.8E15 g/year	Range = 2.7-5.0E15 g/y, global sink of CO[1]
Oxidation of methane to CO	3.25E15 g/year	Range = 1.5 - 5E15 g/y, global value[1]
Pool of CH4 in atmosphere	4.0E15 g	Based on 1.41 ppm mixing ratio[4]
Pool of dissolved organic carbon in the oceans, DOC	1.0E18 g C	Assuming 700 ug C/l[6]
Pool of particulate organic carbon in the oceans, POC	3.0E16 g C	Assuming 20 ug C/l[6]
Total CO emission	4.5E15 g/year	Range = 2.19-6.48E15 g/y, global value[1]
Total active pools	3300E15 moles	Rounded total[2]
Total coal	15E18 g	None[3]
Total consumption of atmospheric CH4	2.01E15 g/year	Range = 0.96-03.07E15 g/y, global value[4]
Total loss of atmospheric CO	4.46E15 g/year	Range = 3.31-5.61E15 g/y, global value[1]
Total storage pools	1700000E15 moles	Rounded total[2]
Uptake of atmospheric CO by soil	0.50E15 g/year	Global value, uptake by plants is uncertain[1]

Source: 1.Warneck, P. 1974. Tellus 26, 39-46.
2.Bolin, B. 1970. Scientific American., 223(3), p 130.
3.Singer, S.F. 1975. In: S.F. Singer (ed.), The Changing Global Environment, pp 25-43. D. Reidel Publ.Comp.
4.Bolin, B., E.T. Degens, S. Kempe and P. Ketner. 1977. The Global Carbon Cycle. SCOPE and ICSU. John Wiley and Sons. NY.
5.Revelle, R. and H.E. Suess. 1957. Tellus 9, pp 18.
6.Williams, P.J. L. 1975. In: J.P. Riley and G. Skirrow. (eds.): Chemical Oceanography. 2nd ed. Vol II. Academic Press. London. 301-306.
7.Duce, R.A. 1978. Pure Appl. Geophys. 116, 244-273.

2-127 Ecosphere
Pools within the Cd cycle

Species	Value	Condition
Atmosphere	1.5E8 g	Conc in pool = 0.03 ng/m3, residence time 7 days[1]
Biosphere, freshwater biota	7E9 g	Conc in pool = 3.5 ug/g, residence time 3.5 y[1]
Biosphere, homo sapiens biomass	2E8 g	Conc in pool = 50 mg/person, residence time 1-40 years[1]
Biosphere, land animals	6E9 g	Conc in pool = 0.3 ug/g[1]
Biosphere, land plants	7.2E11 g	Conc in pool = 0.3 ug/g, residence time 20 days[1]
Biosphere, marine animals	1.2E10 g	Conc in pool = 4.0 ug/g[1]
Biosphere, marine plants	4E8 g	Conc in pool = 2.0 ug/g, residence time 18 days[1]
Fresh water, sediments	1.0E11 g	Conc in pool = 0.16 ug/g, residence time 3.6 y[1]
Freshwater, Dissolved	1.6E9 g	Conc in pool = 0.05 ug/kg[1]
Glaciers	8.2E10 g	Conc in pool = 0.005 ug/g[1]
Groundwater	4.0E8 g	Conc in pool = 0.1 ug/kg[1]
Limestone	2.8E16 g	Conc in pool = 0.08 ug/g[1]
Lithosphere	2.8E19 g	Conc in pool = 0.5 ug/g, residence time 1E9, down to 45 km[1]
Oceans, dissolved	8.4E13 g	Conc in pool = 0.06 ug/kg, residence time 2.1E4 y[1]
Oceans, particulate organic matter	3.2E11 g	Conc in pool = 4.5 ug/g, residence time 1.3 y[1]
Oceans, suspended particulates	1.4E12 g	Conc in pool = 1.0 ug/g[1]
Sandstone	2.1E16 g	Conc in pool = 0.07 ug/g[1]
Sediment pore waters	6.4E13 g	Conc in pool = 0.2 ug/kg[1]
Sedimentary rocks	2.5E18 g	Conc in pool = 1.0 ug/g[1]
Shale and clay	2.47E18 g	Conc in pool = 1.3 ug/g[1]
Soils	6.6E13 g	Conc in pool = 0.02 ug/g, up to 100 cm[1]
Soils, organic fraction	6.1E12 g	Conc in pool = 0.9 ug/g, residence time > 200 y[1]
Swamps and marshes, in biomass	3.6E9 g	Conc in pool = 0.6 ug/g[1]
Terrestrial litter	1.3E12 g	Conc in pool = 0.6 ug/g, residence time 42 y[1]

Source: 1.Mackenzie, F.T., R.J. Lantzy and V. Paterson. 1979. Global Trace Metal Cycles and Predictions. J. In. Assoc. Math. Geol.

2-128 Ecosphere
Pools within the H2O cycle

Species	Value	Condition
Atmosphere	13000 km**3	Percent of total = 0.001[1]
Average in stream channels	1250 km**3	Percent of total = 0.0001[1]
Freshwater lakes	125000 km**3	Percent of total = 0.009[1]
Ground water	4170000 km**3	Percent of total = 0.31, deep-laying[1]
Ground water	4170000 km**3	Percent of total = 0.31, within 0.8 km depth[1]
Icecaps and glaciers	29200000 km**3	Percent of total = 2.15[1]
Ocean	1322000000 km**3	Percent of total = 97.2, world oceans[1]
Saline lakes and inland seas	104000 km**3	Percent of total = 0.008[1]

Soil moisture and intermediate zone water	67000 km**3	Percent of total = 0.005[1]
Total liquid water in land areas	8637000 km**3	Percent of total = 0.635[1]

Source: 1.Raymand, L.N. 1964. U.S. Geological survey.

2-129 Ecosphere
Pools within the N cycle

Species	Value	Condition
Atmosphere	270000000E12 moles	As N2[1]
Crustal rock	1000000000E12 moles	As rock-forming minerals[1]
Decaying organic matter	120000E12 moles	As organic N[1]
Land organisms	870E12 moles	As organic N[1]
Marine organisms	69E12 moles	As organic N[1]
Ocean sediments	7100E12 moles	As inorganic N[1]
Oceans	1400000E12 moles	As N2[1]
Sedimentary rocks	29000000E12 moles	As rock-forming minerals[1]
Soils	10000E12 moles	As inorganic N[1]
Total active pools	140000E12 moles	Rounded total[1]
Total storage pools	16E20 moles	Rounded total[1]

Source: 1.Delwiche, C.C. 1970. Scientific American, 223(3), pp 137-140.

2-130 Ecosphere
Pools within the O cycle

Species	Value	Condition
Atmosphere	120E15 moles	As CO2[1]
Atmosphere	150000E15 moles	As O2[1]
Atmosphere	700E15 moles	As H2O[1]
Crustal rock	7700000E15 moles	As rock oxides[1]
Fresh surface water	6900E15 moles	As H2O[1]
Ground water	460000E15 moles	As H2O[1]
Land organisms	38E15 moles	As organic O[1]
Marine organisms	1.83E15 moles	As organic O[1]
Oceans	210E15 moles	As O2[1]
Oceans	5900E15 moles	As CO2[1]
Oceans	70000000E15 moles	As H2O[1]
Organic matter	250E15 moles	In marine sediments, as organic O[1]
Organic matter	58E15 moles	In soils, as organic O[1]
Sedimentary rocks	5100E15 moles	As carbonates[1]
Soil moisture	3700E15 moles	As H2O[1]
Total active pools	70000000E15 moles	Rounded total[1]
Total storage pools	8200000E15 moles	Rounded total[1]

Source: 1.Rubey, W.W. 1951. Geol. Soc. Amer. Bull. 62, pp 1120-1121.

2-131 Ecosphere
Primary production

Species	Value	Condition
Algae culture	72 kcal/(m**2*day)	Efficiency of primary production = 3.0 %[1]
Arctic tundra	1.8 kcal/(m**2*day)	Efficiency of primary production = 0.08 %[1]
Atlantic Ocean	10 mg C/((m**2)*day)	Danish coastal waters, Dec, min value[2]
Atlantic Ocean	100 mg C/((m**2)*day)	East.N Atlantic, 15 miles off Oporto, Sep[2]
Atlantic Ocean	110 mg C/((m**2)*day)	Equatorial Atlantic, W basin, near 20 N, Feb, rg = 60 - 160[2]
Atlantic Ocean	110 mg C/((m**2)*day)	North Sea, NE coast of Eng, Oct[2]
Atlantic Ocean	12 mg C/((m**2)*day)	Ice island t3, midsummer, rg = 0 - 24[2]
Atlantic Ocean	1300 mg C/((m**2)*day)	W.barents sea, near bear island, arctic water, May[2]
Atlantic Ocean	150 mg C/((m**2)*day)	Caribbean Sea, 10 - 20 N, rg = 100 - 200[2]
Atlantic Ocean	150 mg C/((m**2)*day)	East.N Atlantic, 200 miles off Oporto, Sep[2]
Atlantic Ocean	150 mg C/((m**2)*day)	Sargasso Sea, summer, rg = 100 - 200[2]
Atlantic Ocean	1675 mg C/((m**2)*day)	N Atlantic, Faroe-Iceland ridge, summer, rg = 650 - 2700[2]
Atlantic Ocean	175 mg C/((m**2)*day)	Equatorial Atlantic, W basin, 8 15'S, March, rg = 70 - 280[2]
Atlantic Ocean	1930 mg C/((m**2)*day)	WN Atlantic, continental shelf and inshore, spring flowering[2]
Atlantic Ocean	200 mg C/((m**2)*day)	N Atlantic, Irminger Sea, summer, rg = 150 - 250[2]
Atlantic Ocean	220 mg C/((m**2)*day)	North Sea, NE coast of Eng, May[2]
Atlantic Ocean	2250 mg C/((m**2)*day)	SW Atlantic, Walvis Bay and Benguela current, Dec, rg = 500- 4000[2]
Atlantic Ocean	235 mg C/((m**2)*day)	Equatorial Atlantic, E basin, 8 15'S, March, rg = 100 - 370[2]
Atlantic Ocean	2400 mg C/((m**2)*day)	Norwegian Sea, near Spitzbergen, N Atlantic water, June[2]
Atlantic Ocean	265 mg C/((m**2)*day)	Equatorial Atlantic, E basin, 15 45'S, Apr, rg = 90 - 420[2]
Atlantic Ocean	265 mg C/((m**2)*day)	Equatorial Atlantic, W basin, 8 15'N, Apr-May, rg = 230 - 300[2]
Atlantic Ocean	275 mg C/((m**2)*day)	W Barents sea, near Bear Island, atlantic water, May[2]
Atlantic Ocean	3 mg C/((m**2)*day)	Station Alpha, late summer, rg = 0 - 6[2]
Atlantic Ocean	300 mg C/((m**2)*day)	Danish coastal waters, March bloom[2]
Atlantic Ocean	330 mg C/((m**2)*day)	WN Atlantic, continental shelf and inshore, weighted annual mean[2]
Atlantic Ocean	35 mg C/((m**2)*day)	Mediterranean Sea, 2 miles off french coast, midsummer, rg = 30 - 40[2]
Atlantic Ocean	485 mg C/((m**2)*day)	Equatorial Atlantic, E basin, near 20 N, Feb, rg = 190 - 780[2]
Atlantic Ocean	5 mg C/((m**2)*day)	North Sea, NE coast of Eng, Feb[2]
Atlantic Ocean	500 mg C/((m**2)*day)	Norwegian Sea, near Spitzbergen, arctic water, June, rg = 400 - 600[2]
Atlantic Ocean	550 mg C/((m**2)*day)	N Atlantic, south of Greenland, summer[2]

Atlantic Ocean	560 mg C/((m**2)*day)	WN Atlantic, continental shelf and inshore, mean rate, spring[2]
Atlantic Ocean	700 mg C/((m**2)*day)	Danish coastal waters, Aug, max value[2]
Atlantic Ocean	75 mg C/((m**2)*day)	Equatorial Atlantic, W basin, 15 45'S, Apr, rg = 20 - 130[2]
Atlantic Ocean	800 mg C/((m**2)*day)	North Sea, annual range, rg = 100 - 1500[2]
Atlantic Ocean	840 mg C/((m**2)*day)	Equatorial Atlantic, E basin, 8 15'N, Apr-May, rg = 180 - 1480[2]
Atlantic Ocean	890 mg C/((m**2)*day)	Sargasso Sea, Apr bloom[2]
Atlantic Ocean	915 mg C/((m**2)*day)	N Atlantic, near Iceland, summer, rg = 530 - 1300[2]
Coral reef	95 kcal/(m**2*day)	Efficiency of primary production = 2.4 %, rg = 39 - 151[1]
Desert	0.4 kcal/(m**2*day)	Efficiency of primary production = 0.05 %[1]
Desert	1 t organic dry weight/(ha*year)	Arid, rg = 0 - 2[3]
Freshwater emergent macrophyte	38 t organic dry weight/(ha*year)	Temperate, rg = 30 - 45[3]
Freshwater emergent macrophyte	75 t organic dry weight/(ha*year)	Tropical, rg = 30 - 45[3]
Freshwater phytoplankton	2 t organic dry weight/(ha*year)	Range = 1 - 9[3]
Freshwater submersed macrophyte	17 t organic dry weight/(ha*year)	Tropical, rg = 12 - 20[3]
Freshwater submersed macrophyte	6 t organic dry weight/(ha*year)	Temperate, rg = 1 - 7[3]
Indian ocean	225 mg C/((m**2)*day)	Equatorial Indian Ocean, rg = 200 - 250[2]
Marine emergent macrophytes	30 t organic dry weight/(ha*year)	Salt marsh, rg = 25 -85[3]
Marine phytoplankton	0.2 mg C/(m**2*day)	Gulf of Mexico, 1968, n = 11, rg = 0.09 - 0.35[4]
Marine phytoplankton	1.4 mg C/(m**2*day)	Southwest Africa, n = 10, rg = 0.61 - 3.3[4]
Marine phytoplankton	2 t organic dry weight/(ha*year)	Range = 1 - 4.5[3]
Marine phytoplankton	311 mg C/(m**2*day)	Region Provencale, Mediter., chlorophyll a = 36 mg/m**2[5]
Marine phytoplankton	364 mg C/(m**2*day)	Region Cote d'azur-Corse, Mediter., chlorophyll a = 49 mg/m**2[5]
Marine phytoplankton	4.1 mg C/(m**2*day)	Peru, n = 8, rg = 1.10 - 9.2[4]
Marine submersed macrophyte	29 t organic dry weight/(ha*year)	Temperate, rg = 23 - 29[3]
Marine submersed macrophyte	29 t organic dry weight/(ha*year)	Tropical, 30 - 40[3]
Pacific Ocean	10 mg C/((m**2)*day)	Equatorial Pacific, 11 N, 115 W, autumn[2]
Pacific Ocean	175 mg C/((m**2)*day)	Equatorial Pacific, 30 N-30 S, March, rg = 100 - 250[2]
Pacific Ocean	2000 mg C/((m**2)*day)	NW Pacific, sea of Okhotsk, north of Japan, May, rg = 6 - 5100[2]
Pacific Ocean	375 mg C/((m**2)*day)	NW Pacific, Sea of Japan, Oyashio system, summer, rg = 250 - 500[2]
Pacific Ocean	605 mg C/((m**2)*day)	Equatorial Pacific, 9 N, 90 W, Costa Rica Dome, Nov, rg = 410 - 800[2]

Pacific Ocean	75 mg C/((m**2)*day)	NW Pacific, Sea of Japan, Kuroshio current, summer, rg = 50 - 100[2]
Pacific Ocean	750 mg C/((m**2)*day)	Equatorial Pacific, near coast of Ecuador, autumn, rg = 500 - 1000[2]
Rain forest	50 t organic dry weight/(ha*year)	Range = 40 - 60[3]
Subtropical deep ocean	2.9 kcal/(m**2*day)	Efficiency of primary production = 0.09 %[1]
Sugar cane field	74 kcal/(m**2*day)	Efficiency of primary production = 1.8 %[1]
Temperate annuals	22 t organic dry weight/(ha*year)	Range = 19 - 25[3]
Temperate forest	12 t organic dry weight/(ha*year)	Deciduous, rg = 9 - 15[3]
Temperate forest	28 t organic dry weight/(ha*year)	Coniferous, rg = 21 - 35[3]
Temperate herbs	20 t organic dry weight/(ha*year)	Range = 15 - 25[3]
Tropical annuals	30 t organic dry weight/(ha*year)	Range = 24 - 36[3]
Tropical forest plantation	28 kcal/(m**2*day)	Efficiency of primary production = 0.7 %[1]
Tropical marine meadow	80.5 kcal/(m**2*day)	Effi. of primary production = 2.0 %, rg = 20 - 141[1]
Tropical rainforest	131 kcal/(m**2*day)	Efficiency of primary production = 3.5 %[1]

Source: 1.Odum, H.T. 1971. Environment, Power and Society. Wiley-Interscience. New York.
2.Sisler, F.D. 1975. In: S.F. Singer (ed.), The Changing Global Environment, pp 57-69. D. Reidel Publ.Comp.
3.Westlake, D.F. 1963. Biol. Rev. 38, pp 385-425.
4.Hobson, L.A., D.W. Menzel and R.T. Barber. 1973. Mar. Biol. 19, pp 298-306.
5.Jacques, G., H.J. Minas, M. Minas and P. Nival. 1973. Mar. Biol. 23, pp 251-265.

2-132 Ecosphere Production

Species	Value	Condition
Fontinalis squamosa	92 gram/m**2	Per year assuming annual branching pattern[1]
Periphyton on Typha	170 mg dry weight/(d*m**2)	Max biomass = 1118 g[2]
Potamophylax cingulatus	4500 cal/(year*m**2)	South swedish stream[3]
Salvelinus fontinalis	167 kg wet weight/(ha*year)	Recovery year, Minn, 1969, max value[4]
Salvelinus fontinalis	50 kg wet weight/(ha*year)	Flood year, flood-damaged stream, Minn[4]
Seaweed	1750 g c/(m**2*year)	In the Sea weed zone, Nova Scotia[5]
Seaweed	648000 g c/(m shore line)	Nova Scotia[5]
Simulium sp.	1.05 kcal/(m**2*year)	Net production, lotic areas[6]

Source: 1.Dawson, F.H. 1973. Ann. Limnol. Stn. Biol. Lac Oredon, 9(3), pp 231-240.
2.Mason, C.F. and R.J. Bryant. 1975. Freshwater biol. 5(3), pp 271-277.
3.Otto, C. 1975. Oikos 26(2), pp 159-169.
4.Hanson, D.L. and T.F. Waters. 1974. Trans. Am. Fish. Soc. 103(3), pp 431-439.
5.Mann, K.H. 1972. Mar. Biol. 14, pp 199-209.
6.Neveu, A. 1973. Ann. Hydrobiol. Inst. Natl. Rech. Agron. Paris, 4(2), pp 183-199.

2-133 Ecosphere
Production of litter

Species	Value	Condition
Forest production of litter vs latitude	0.75 t dry weight/(ha*year)	Latitude = 6--65 deg, rg = 0.5-1 t/(ha*y), total litter[1]
Forest production of litter vs latitude	10 t dry weight/(ha*year)	Latitude = 0-10 deg, rg = 9-11 t/(ha*y), max = 13.5, total litter[1]
Forest production of litter vs latitude	2.5 t dry weight/(ha*year)	Latitude = 50-60 deg, rg = 1.5-3.5 t/(ha*y), max = 4.5, total litter[1]
Forest production of litter vs latitude	3 t dry weight/(ha*year)	Latitude = 40-50 deg, rg = 2-4 t/(ha*y), max = 6, total litter[1]
Forest production of litter vs latitude	5.5 t dry weight/(ha*year)	Latitude = 30-40 deg, rg = 3.5-8 t/(ha*y), total litter[1]

Source: 1.Kuhnelt, W. and N. Walker. 1976. Soil Biology. Michigan State Univ. Press. East Lansing. Michigan. 43 pp.

2-134 Ecosphere
Respiration, autotrophic

Species	Value	Condition
Alfalfa field	9200 kcal/(year*m**2)	Annual value, USA[1]
Freshwater spring	12000 kcal/(year*m**2)	Annual value, Silver Spring, FL[2]
Marine sound	3200 kcal/(year*m**2)	Annual value, coastal, Long Island, New York[3]
Oak - pine forest	6400 kcal/(year*m**2)	Annual value, medium aged, New York, no timber harvest[4]
Pine forest	4700 kcal/(year*m**2)	Annual value, young plantation, mean for 0-50 y, Eng[5]
Rain forest	32000 kcal/(year*m**2)	Annual value, mature forest, Puerto Rico[6]

Source: 1.Thomas, M.D. and G.R. Hill. 1949. in: Franck, J and W E. Loomis(eds). Photosynthesis in Plants, Iowa State College Press, pp 19-52.
2.Odum, H.T. 1957. Ecol. Monogr., 27, pp 55-112.
3.Riley, G.A. 1956. Bull. Bingham Oceanogr. Coll. 15, pp 324-344.
4.Woodwell, G.M. and R. Whittaker. 1968. Amer. Zool. 8, pp 19-30.
5.Ovington, J.D. 1961 Ann. Bot. 25, pp 12-20.
6.Odum, H.T. and R.F Pigeon. 1970. A tropical rain forest. Nat Tech. Info. Srevice. Springfield.

2-135 Ecosphere
Respiration, heterotrophic

Species	Value	Condition
Alfalfa field	800 kcal/(year*m**2)	Annual value, USA[1]
Freshwater spring	6800 kcal/(year*m**2)	Annual value, Silver Spring, FL[2]
Marine sound	2500 kcal/(year*m**2)	Annual value, coastal, Long Island, New York[3]
Oak - pine forest	3000 kcal/(year*m**2)	Annual value, medium aged, New York, no timber harvest[4]
Pine forest	4600 kcal/(year*m**2)	Annual value, young plantation, mean for 0-50 y, Eng[5]
Rain forest	13000 kcal/(year*m**2)	Annual value, mature forest, Puerto Rico[6]

Source: 1.Thomas, M.D. and G.R. Hill. 1949. in: Franck, J and W E. Loomis(eds). Photosynthesis in Plants, Iowa State College Press, pp 19-52.
2.Odum, H.T. 1957. Ecol. Monogr., 27, pp 55-112.
3.Riley, G.A. 1956. Bull. Bingham Oceanogr. Coll. 15, pp 324-344.
4.Woodwell, G.M. and R. Whittaker. 1968. Amer. Zool. 8, pp 19-30.
5.Ovington, J.D. 1961 Ann. Bot. 25, pp 12-20.
6.Odum, H.T. and R.F. Pigeon. 1970. A tropical rain forest. Nat Tech. Info. Service. Springfield.

2-136 Ecosphere
Secondary production

Species	Value	Condition
African lakes	25.2 kcal/(m**2*year)	Mans harvest, unfertilized water, mixed carnivores, rg = 0.4 - 50[1]
Fish ponds	22.5 kcal/(m**2*year)	Mans harvest, unfertilized water, stoced carnivores, rg = 9 - 34[1]
Fish ponds	268 kcal/(m**2*year)	Mans harvest, art fert. water, stoced herbivores, rg = 202 - 336,[1]
Fish ponds	51 kcal/(m**2*year)	Mans harvest, unfertilized water, stoced herbivores, rg = 22 - 80,[1]
Fish ponds	78.5 kcal/(m**2*year)	Mans harvest, artificially fert. water, stoced carnivores, rg = 45 -[1]
Great lakes	0.9 kcal/(m**2*year)	Mans harvest, unfertilized water, mixed carnivores, rg = 0.2 - 1.6[1]
Hong Kong	675 kcal/(m**2*year)	Mans harvest, outside food added, herbivores, rg = 450 - 900[1]
Lakes	18.2 kcal/(m**2*year)	Mans harvest, unfertilized water, mix carnivores, rg = 0.4 - 36, USA[1]
Malaya	785 kcal/(m**2*year)	Mans harvest, outside food added, herbivores[1]
Marine ponds	156 kcal/(m**2*year)	Mans harvest, art fert. water, stoced herbivores, rg = 112 - 202,[1]
North Sea	5.0 kcal/(m**2*year)	Mans harvest, unfertilized water, mixed carnivores[1]
One acre pond	450 kcal/(m**2*year)	Mans harvest, outside food added, carnivores, USA[1]
Peru upwelling area	335 kcal/(m**2*year)	Mans harvest, heavy natural fertilized water[1]
South China	1612 kcal/(m**2*year)	Mans harvest, outside food added, herbivores, rg = 202 - 3024[1]
World marine fishery	0.3 kcal/(m**2*year)	Mans harvest, unfertilized water, mixed carnivores[1]

Source: 1.Odum, E.P. 1971. Fundamentals of Ecology. Third edition. W.B. Saunders , Philadelphia.

2-137 Ecosphere
Stem wood

Species	Value	Condition
Africa	39.0E9 m3	Area of forest = 760E6 ha, 1973[1]
Asia, except U.S.S.R.	40.0E9 m3	Area of forest = 460E6 ha, 1973[1]
Europe, except U.S.S.R.	12.8E9 m3	Area of forest = 169E6 ha, 1973[1]
Latin America	101.1E9 m3	Area of forest = 742E6 ha, 1973[1]
North America	58.5E9 m3	Area of forest = 630E6 ha, 1973[1]
Oceania	8.5E9 m3	Area of forest = 185E6 ha, 1973[1]
Total world	338.8E9 m3	Area of forest = 3826E6 ha, 1973[1]
USSR	78.9E9 m3	Area of forest = 880E6 ha, 1973[1]

Source: 1.Skogsstyrelsen 1976. Skogstatistisj AArsbok 1974. Sveriges officiella statistik. Skogsstyrelsen. Joekoeking.

2-138 Ecosphere
Surface area of terrestrial ecosystem

Species	Value	Condition
Boreal forest	10.0E6 km2	None[1]
Boreal forest	10.0E6 km2	None[2]
Boreal forest	12.0E6 km2	None[3]
Boreal forest	12.0E6 km2	None[4]
Boreal forest	18E6 km2	None[5]
Boreal forest	19.5E6 km2	None[6]

Boreal forest	3.9E6 km2	None[7]
Boreal forest	4.0E6 km2	None[8]
Boreal forest	9.0E6 km2	None[9]
Cultivated land	10.0E6 km2	None[8]
Cultivated land	13.31E6 km2	None[7]
Cultivated land	13.31E6 km2	None[10]
Cultivated land	14.0E6 km2	None[3]
Cultivated land	14.0E6 km2	None[1]
Cultivated land	15.0E6 km2	None[4]
Cultivated land	16.0E6 km2	None[9]
Cultivated land	23.0E6 km2	None[2]
Desert and semi desert	16E6 km2	None[5]
Desert and semi desert	18.0E6 km2	None[3]
Desert and semi desert	21.0E6 km2	None[9]
Desert and semi desert	22.4E6 km2	None[7]
Desert and semi desert	25.0E6 km2	None[4]
Desert and semi desert	25.2E6 km2	None[10]
Desert and semi desert	27.1E6 km2	None[6]
Desert and semi desert	33.0E6 km2	None[2]
Desert and semi desert	52.0E6 km2	None[1]
Desert and semi desert	9.0E6 km2	None[8]
Desert, extreme	11.0E6 km2	None[2]
Desert, extreme	12.4E6 km2	None[10]
Desert, extreme	15.0E6 km2	None[4]
Desert, extreme	19.7E6 km2	None[7]
Desert, extreme	23E6 km2	None[5]
Desert, extreme	24.0E6 km2	None[3]
Desert, extreme	24.5E6 km2	None[9]
Desert, extreme	41.0E6 km2	None[8]
Desert, extreme	6.2E6 km2	None[6]
Forest	31.0E6 km2	None[9]
Forest	36.0E6 km2	None[4]
Forest	43.6E6 km2	None[1]
Forest	44.0E6 km2	None[2]
Forest	44.41E6 km2	None[7]
Forest	45.0E6 km2	None[8]
Forest	48.5E6 km2	None[4]
Forest	53.3E6 km2	None[6]
Forest	54.4E6 km2	None[10]
Forest	64E6 km2	None[5]

Grassland	12.5E6 km2	None[9]
Grassland	22.5E6 km2	None[10]
Grassland	25.0E6 km2	None[4]
Grassland	25.7E6 km2	None[1]
Grassland	27.0E6 km2	None[2]
Grassland	36.9E6 km2	None[7]
Grassland	9.0E6 km2	None[3]
Grassland	9.0E6 km2	None[6]
Grassland	9E6 km2	None[5]
Human areas	2.0E6 km2	None[9]
Lake and stream	2.0E6 km2	None[3]
Lake and stream	2.0E6 km2	None[5]
Lake and stream	2.0E6 km2	None[9]
Other forest	0.9E6 km2	None[6]
Other forest	1.5E6 km2	None[9]
Other forest	14.0E6 km2	None[1]
Other forest	14.4E6 km2	None[2]
Other forest	15.0E6 km2	None[8]
Savanna	12.9E6 km2	None[10]
Savanna	15.0E6 km2	None[3]
Savanna	15.0E6 km2	None[4]
Savanna	15E6 km2	None[5]
Savanna	22.5E6 km2	None[9]
Savanna	23.0E6 km2	None[6]
Swamp and marsh	2.0E6 km2	None[3]
Swamp and marsh	2.0E6 km2	None[9]
Swamp and marsh	3.3E6 km2	None[7]
Swamp and marsh	3.3E6 km2	None[10]
Swamp and marsh	4.0E6 km2	None[5]
Temperate deciduous forest	12E6 km2	None[5]
Temperate deciduous forest	16.6E6 km2	None[6]
Temperate deciduous forest	26.8E6 km2	None[10]
Temperate deciduous forest	3.0E6 km2	None[9]
Temperate deciduous forest	4.9E6 km2	None[1]
Temperate deciduous forest	4.9E6 km2	None[2]
Temperate deciduous forest	5.66E6 km2	None[7]
Temperate deciduous forest	7.0E6 km2	None[3]
Temperate evergreen and deciduous forest	18.0E6 km2	None[4]
Temperate evergreen forest	1.4E6 km2	None[6]

Temperate evergreen forest	14.6E6 km2	None[7]
Temperate evergreen forest	3.0E6 km2	None[9]
Temperate evergreen forest	5.0E6 km2	None[3]
Temperate evergreen forest	8.6E6 km2	None[10]
Total, terrestrial	133.6E6 km2	None[6]
Total, terrestrial	142.0E6 km2	None[8]
Total, terrestrial	148.0E6 km2	None[1]
Total, terrestrial	148.0E6 km2	None[4]
Total, terrestrial	148.5E6 km2	None[7]
Total, terrestrial	148.5E6 km2	None[10]
Total, terrestrial	149.0E6 km2	None[3]
Total, terrestrial	149.0E6 km2	None[2]
Total, terrestrial	149.3E6 km2	None[5]
Total, terrestrial	149.3E6 km2	None[9]
Tropical rain and seasonal forest	14.7E6 km2	None[1]
Tropical rain and seasonal forest	14.7E6 km2	None[2]
Tropical rain and seasonal forest	19.0E6 km2	None[10]
Tropical rain and seasonal forest	20.0E6 km2	None[8]
Tropical rain and seasonal forest	20.0E6 km2	None[4]
Tropical rain and seasonal forest	20.25E6 km2	None[7]
Tropical rain forest	10.0E6 km2	None[9]
Tropical rain forest	17.0E6 km2	None[3]
Tropical rain forest	25E6 km2	None[5]
Tropical rain forest	9.7E6 km2	None[6]
Tropical seasonal forest	4.5E6 km2	None[9]
Tropical seasonal forest	5.2E6 km2	None[6]
Tropical seasonal forest	7.5E6 km2	None[3]
Tropical seasonal forest	9E6 km2	None[5]
Tundra	10.0E6 km2	None[4]
Tundra	11.0E6 km2	None[2]
Tundra	11.2E6 km2	None[6]
Tundra	37.0E6 km2	None[8]
Tundra	8.0E6 km2	None[3]
Tundra	8.5E6 km2	None[7]
Tundra	8E6 km2	None[5]
Tundra	9.5E6 km2	None[9]
Tundra	9.9E6 km2	None[10]
Woodland and shrubland	3.8E6 km2	None[6]
Woodland and shrubland	4.5E6 km2	None[9]
Woodland and shrubland	4E6 km2	None[5]

Woodland and shrubland	7.0E6 km2	None[4]
Woodland and shrubland	8.5E6 km2	None[3]
spc6.0	6.0E6 km2	None[8]

Source: 1.Lieth, H. 1964. Versuch einer kartographischen Darstellung der Produktiviat der Pflanzendecke auf deErde. 72-80. Franz Steiner Verlag. Wiesbadeen.
2.Bowen, H.J.M. 1966. Trace Elements in Biochemistry. Academic Press. London.
3.Whittaker, R.H. and G.E. Likens. 1973. in: WoodwellG.M. and E.V. Pecan (eds.): Carbon and Biosphere.AEC Symposium Series 30, 281-302. NTIS. Springfield. Virginia.
4.Golley, F.B. 1972. Ann. Biol. Colloq. 31, 69-90.
5.Bazilevich, , N.I., L.Y. Rodin and N.N/ Rozov. 1971. Sov. Geogr. Rev. Transl. 12, 293-317.
6.Shantz, D.M. 1954. Ecology 35, 143-145.
7.Deevey, Jr. E.S. 1964. African Game Ranching. 1-75. Perganon Press. Oxford.
8.Schmitt, W.R. 1965. Ann. New York Acad. Sci. 118, 645-653.
9.Ajtay, G.L., P. Ketner and P. Duvigneaud. 1979. in: Bolin, B., E.T. Degens, S. Kempe and P. Ketner.(ds.) The Global Carbon Cycle. 129-181. SCOPE. John Wiley and Sons. NY.
10.Stamp, L.D. 1960. Our Developing World, 1-195. Faber and Faber. London.

2-139 Ecosphere
Synthesis of 1 g of biomass

Species	Value	Condition
C6H12O6	71.98E4 moles	According eq: C6H12O6 + NH3 + H2SO4 + H3PO4 <-> biomass + CO2 + H2O[1]
CO2	25.53E4 moles	According eq: C6H12O6 + NH3 + H2SO4 + H3PO4 <-> biomass + CO2 + H2O[1]
CO2	410.35E4 moles	According eq: CO2 + HNO3 + H2O + H2SO4 + H3PO4 <-> biomass + O2[1]
CO2	410.35E4 moles	According eq: CO2 + N2 + H2O + H2SO4 + H3PO4 <-> biomass + O2[1]
CO2	410.35E4 moles	According eq: CO2 + NH3 + H2O + H2SO4 + H3PO4 <-> biomass + O2[1]
H2O	175.11E4 moles	According eq: C6H12O6 + NH3 + H2SO4 + H3PO4 <-> biomass + CO2 + H2O[1]
H2O	263.77E4 moles	According eq: CO2 + NH3 + H2O + H2SO4 + H3PO4 <-> biomass + O2[1]
H2O	356.55E4 moles	According eq: CO2 + HNO3 + H2O + H2SO4 + H3PO4 <-> biomass + O2[1]
H2O	398.20E4 moles	According eq: CO2 + N2 + H2O + H2SO4 + H3PO4 <-> biomass + O2[1]
H2SO4	3.0E4 moles	According eq: C6H12O6 + NH3 + H2SO4 + H3PO4 <-> biomass + CO2 + H2O[1]
H2SO4	3.0E4 moles	According eq: CO2 + HNO3 + H2O + H2SO4 + H3PO4 <-> biomass + O2[1]
H2SO4	3.0E4 moles	According eq: CO2 + N2 + H2O + H2SO4 + H3PO4 <-> biomass + O2
H2SO4	3.0E4 moles	According eq: CO2 + NH3 + H2O + H2SO4 + H3PO4 <-> biomass + O2[1]
H3PO4	6.8E4 moles	According eq: C6H12O6 + NH3 + H2SO4 + H3PO4 <-> biomass + CO2 + H2O[1]
H3PO4	6.8E4 moles	According eq: CO2 + HNO3 + H2O + H2SO4 + H3PO4 <-> biomass + O2[1]
H3PO4	6.8E4 moles	According eq: CO2 + N2 + H2O + H2SO4 + H3PO4 <-> biomass + O2[1]
H3PO4	6.8E4 moles	According eq: CO2 + NH3 + H2O + H2SO4 + H3PO4 <-> biomass + O2[1]
HNO3	93.28E4 moles	According eq: CO2 + HNO3 + H2O + H2SO4 + H3PO4 <-> biomass + O2[1]
N2	46.64E4 moles	According eq: CO2 + N2 + H2O + H2SO4 + H3PO4 <-> biomass + O2[1]
NH3	93.28E4 moles	According eq: C6H12O6 + NH3 + H2SO4 + H3PO4 <-> biomass + CO2 + H2O[1]
NH3	93.28E4 moles	According eq: CO2 + NH3 + H2O + H2SO4 + H3PO4 <-> biomass + O2[1]
O2	434.38E4 moles	According eq: CO2 + NH3 + H2O + H2SO4 + H3PO4 <-> biomass + O2[1]
O2	506.60E4 moles	According eq: CO2 + N2 + H2O + H2SO4 + H3PO4 <-> biomass + O2[1]
O2	620.69E4 moles	According eq: CO2 + HNO3 + H2O + H2SO4 + H3PO4 <-> biomass + O2[1]

Source: 1.Morowitz, H.J., M.E. Tourtellotte, W.R. Guild, E. Castro and C. Woese. 1962. J. Mol. Boil., 4, p 93.

2-140 Ecosphere
Water balance for earth

Species	Value	Condition
Evaporation from Africa	43 cm/year	None[1]
Evaporation from Asia	31 cm/year	None[1]
Evaporation from Atlantic Ocean	124 cm/year	None[1]
Evaporation from Australia	42 cm/year	None[1]
Evaporation from Europe	39 cm/year	None[1]
Evaporation from Indian Ocean	132 cm/year	None[1]
Evaporation from North America	32 cm/year	None[1]
Evaporation from Pacific Ocean	132 cm/year	None[1]
Evaporation from South America	70 cm/year	None[1]
Evaporation from total land areas	42 cm/year	None[1]
Evaporation from world oceans	126 cm/year	None[1]
Inflow to Atlantic Ocean	12 cm/year	None[1]
Inflow to Indian Ocean	7 cm/year	None[1]
Outflow from Pacific Ocean	8 cm/year	None[1]
Outflow from world oceans	0 cm/year	None[1]
Precipitation on Africa	69 cm/year	None[1]
Precipitation on Asia	60 cm/year	None[1]
Precipitation on Atlantic Ocean	89 cm/year	None[1]
Precipitation on Australia	47 cm/year	None[1]
Precipitation on Europe	64 cm/year	None[1]
Precipitation on Indian Ocean	117 cm/year	None[1]
Precipitation on North America	66 cm/year	None[1]
Precipitation on Pacific Ocean	133 cm/year	None[1]
Precipitation on South America	163 cm/year	None[1]
Precipitation on total land areas	73 cm/year	None[1]
Precipitation on world oceans	114 cm/year	None[1]
Runoff from Africa	26 cm/year	None[1]
Runoff from Asia	29 cm/year	None[1]
Runoff from Atlantic Ocean	-23 cm/year	None[1]
Runoff from Australia	5 cm/year	None[1]
Runoff from Europe	25 cm/year	None[1]
Runoff from Indian Ocean	-8 cm/year	None[1]
Runoff from North America	34 cm/year	None[1]
Runoff from Pacific Ocean	-7 cm/year	None[1]
Runoff from South America	93 cm/year	None[1]
Runoff from total land areas	31 cm/year	None[1]
Runoff from world oceans	-12 cm/year	None[1]

Source: 1.Budyko, M.I. 1971. Climate and Life. Hydrological Publ. Co. Leningrad.

2-141 Light
Absorption

Species	Value	Condition
Monochromatic light	0.52 %	Attenuation in 1 m water, wavelength = 460 nm, extenuation coef =0.0054[1]
Monochromatic light	0.52 %	Attenuation in 1 m water, wavelength = 480 nm, extenuation coef =^{0}.00501
Monochromatic light	0.70 %	Attenuation in 1 m water, wavelength = 440 nm, extenuation coef =^{0}.00781
Monochromatic light	0.77 %	Attenuation in 1 m water, wavelength = 500 nm, extenuation coef =^{0}.00751
Monochromatic light	0.92 %	Attenuation in 1 m water, wavelength = 420 nm, extenuation coef =^{0}.00881
Monochromatic light	1.6 %	Attenuation in 1 m water, wavelength = 520 nm, extenuation coef =^{0}.0161
Monochromatic light	1.63 %	Attenuation in 1 m water, wavelength = 400 nm, extenuation coef =^{0}.01541
Monochromatic light	19.0 %	Attenuation in 1 m water, wavelength = 600 nm, extenuation coef =^{0}.2101
Monochromatic light	2.10 %	Attenuation in 1 m water, wavelength = 380 nm, extenuation coef =^{0}.02551
Monochromatic light	23.5 %	Attenuation in 1 m water, wavelength = 620 nm, extenuation coef =^{0}.2731
Monochromatic light	26.6 %	Attenuation in 1 m water, wavelength = 640 nm, extenuation coef =0.310[1]
Monochromatic light	3.0 %	Attenuation in 1 m water, wavelength = 540 nm, extenuation coef =0.030[1]
Monochromatic light	3.9 %	Attenuation in 1 m water, wavelength = 560 nm, extenuation coef =^{0}.0401
Monochromatic light	31.0 %	Attenuation in 1 m water, wavelength = 660 nm, extenuation coef =^{0}.3701
Monochromatic light	36.6 %	Attenuation in 1 m water, wavelength = 680 nm, extenuation coef =0.455[1]
Monochromatic light	45.0 %	Attenuation in 1 m water, wavelength = 700 nm, extenuation coef =0.598[1]
Monochromatic light	64.5 %	Attenuation in 1 m water, wavelength = 720 nm, extenuation coef =^{1}.041
Monochromatic light	7.0 %	Attenuation in 1 m water, wavelength = 580 nm, extenuation coef =0.078[1]
Monochromatic light	88.5 %	Attenuation in 1 m water, wavelength = 740 nm, extenuation coef =^{2}.161
Monochromatic light	89.4 %	Attenuation in 1 m water, wavelength = 800 nm, extenuation coef =^{2}.241
Monochromatic light	90.1 %	Attenuation in 1 m water, wavelength = 780 nm, extenuation coef =^{2}.311
Monochromatic light	91.1 %	Attenuation in 1 m water, wavelength = 820 nm, extenuation coef =^{2}.421
Monochromatic light	91.4 %	Attenuation in 1 m water, wavelength = 760 nm, extenuation coef =^{2}.451

Source: 1.Hutchinson, G.E. 1957. A Treatise on Limnology. I. Geography, Physics and Chemistry. John Wiley and sons. New York.

2-142 Light
Albedo

Species	Value	Condition
Bare ground	15 %	% of total incoming radiation, with diffuse reflection, rg = 10 - 20[1]
Dark cultivated soil	8.5 %	% of total incoming radiation, with diffuse reflection, rg = 7 - 10[1]
Forest, green	6.5 %	% of total incoming radiation, with diffuse reflection, rg = 3 - 10[1]
Forest, snow-covered ground	17.5 %	% of total incoming radiation, with diffuse reflection, rg = 10 - 25[1]
Fresh snow	87 %	% of total incoming radiation, with diffuse reflection[1]
Fresh snow cover	85 %	% of total incoming radiation, with diffuse reflection, rg = 75 - 95[1]
Grass, dry	20 %	% of total incoming radiation, with diffuse reflection, rg = 15 - 25[1]
Grass, high dry	32 %	% of total incoming radiation, with diffuse reflection, rg = 31 - 33[1]
Grass, high fresh	26 %	% of total incoming radiation, with diffuse reflection[1]
Light sand dunes, surf	45 %	% of total incoming radiation, with diffuse reflection, rg = 30 - 60[1]

Meadows and fields	21 %	% of total incoming radiation, with diffuse reflection, rg = 12 - 30[1]
Sandy soil	32.5 %	% of total incoming radiation, with diffuse reflection, rg = 15 - 40[1]
Snow	55 %	% of total incoming radiation, with diffuse reflection, rg = 40 - 70,[1]
Snow, old	46 %	% of total incoming radiation, with diffuse reflection[1]
Snow, several days old	70 %	% of total incoming radiation, with diffuse reflection[1]
Wood	12.5 %	% of total incoming radiation, with diffuse reflection, rg = 5 - 20[1]
Sea surface	6.5 %	% of total incoming radiation, with diffuse reflection, rg = 3 - 10[1]

Source: 1.List, R.J. 1963. Smithsonian Meteoroligical Tables. Smithsonian Inst. Wash. D.C.

2-143 Light
Penetration of light

Species	Value	Condition
Clear ice	72 %	Thickness = 43 cm, % transmission of surface insolation[1]
Clear ice with vestige snow	53 %	Thickness = 39 cm, % transmission of surface insolation[1]
Compacted old snow	3 %	Thickness = 17-20 cm, % transmission of surface insolation[1]
Ice with irregular surface	58 %	Thickness = 29 cm, % transmission of surface insolation[1]
Milky ice with bubbles	54 %	Thickness = 29 cm, % transmission of surface insolation[1]
New snow	20 %	Thickness = 5.0 cm, % transmission of surface insolation[1]
New snow	34 %	Thickness = 0.5 cm, % transmission of surface insolation[1]
New snow	7.75 %	Thickness = 17-20 cm, % transmission of surface insolation[1]
New snow	9 %	Thickness = 10 cm, % transmission of surface insolation[1]
Translucent ice	14.5 %	Thickness = 25 cm, % transmission of surface insolation, snow ice[1]
Wet ice with bubbles	41 %	Thickness = 39 cm, % transmission of surface insolation[1]

Source: 1.Albrecht, M.-L. 1964. Deutsche Fischerei-Zeitung 11, pp 371-376.

2-144 Light
Planetary radiation

Species	Value	Condition
Absorption of energy by clouds	3 % of incoming radiation	Incoming, total input equal to 263 kilolangleys/y[1]
Absorption of energy by earth, land and water surface	50 % of incoming radiation	Incoming, total input equal to 263 kilolangleys/y[1]
Absorption of energy by molecules	15 % of incoming radiation	Incoming, total input equal to 263 kilolangleys/y[1]
Diffuse reflection due to rayleight scatter and dust	5 % of incoming radiation	Incoming, total input equal to 263 kilolangleys/y[1]
Direct reflection from earths surface to space	6 % of incoming radiation	Incoming, total input equal to 263 kilolangleys/y[1]
Net outgoing radiation from atmosphere	47 % of incoming radiation	Outgoing, total input equal to 263 kilolangleys/y[1]
Net outgoing radiation from the earth- atmosphere system	68 % of incoming radiation	Outgoing, total input equal to 263 kilolangleys/y[1]
Net outgoing radiation from the earthsurface	21 % of incoming radiation	Outgoing, total input equal to 263 kilolangleys/y[1]
Radiation emitted by atmosphere	137 % of incoming radiation	Outgoing, total input equal to 263 kilolangleys/y[1]

Radiation from atmosphere lost to space	60 % of incoming radiation	Outgoing, total input equal to 263 kilolangleys/y[1]
Radiation from atmosphere returned to earth surface	77 % of incoming radiation	Outgoing, total input equal to 263 kilolangleys/y[1]
Radiation from the earth land and water surface	98 % of incoming radiation	Outgoing, total input equal to 263 kilolangleys/y[1]
Radiation from the earthabsorbed by atmosphere	90 % of incoming radiation	Outgoing, total input equal to 263 kilolangleys/y[1]
Radiation from the earthsurface lost to space directly	8 % of incoming radiation	Outgoing, total input equal to 263 kilolangleys/y[1]
Reflection from clouds to space	21 % of incoming radiation	Incoming, total input equal to 263 kilolangleys/y[1]
Total absorbed by earth - atmosphere system	68 % of incoming radiation	Incoming, total input equal to 263 kilolangleys/y[1]
Total at top of atmosphere	100 % of incoming radiation	Incoming, total input equal to 263 kilolangleys/y[1]
Total reflection from the earthto space	32 % of incoming radiation	Incoming, total input equal to 263 kilolangleys/y[1]

Source: 1.Sellers, D.D. 1965. Physical Climatology. Univ. of Chicago Press. Chicago.

2-145 Light Reflection

Species	Value	Condition
Cat, fur	40 %	Wave band = 0.3-0.7 microns, rg = 20 - 60[1]
Cat, fur	5 %	Wave band = 8-12 microns[1]
Cat, fur	65 %	Wave bond = 0.7-3.0 microns, rg = 60 - 70[1]
Fungi	10 %	Wave band = 8-12 microns[1]
Fungi	22.5 %	Wave band = 0.3-0.7 microns, rg = 10 - 35[1]
Fungi	60 %	Wave bond = 0.7-3.0 microns, rg = 35 - 80 %[1]
Homo sapiens, skin	25 %	Wave band = 0.3-0.7 microns, rg = 5 - 45[1]
Homo sapiens, skin	30 %	Wave bond = 0.7-3.0 microns, rg = 3 - 50[1]
Liverwort	4 %	Wave band = 8-12 microns[1]
Liverwort	45 %	Wave bond = 0.7-3.0 microns[1]
Liverwort	7 %	Wave band = 0.3-0.7 microns[1]
Lizard	12 %	Wave band = 0.4-0.7 microns, dorsal[1]
Lizard	30 %	Wave band = 0.7-1.1 microns, ventral[1]
Lizard	42 %	Wave band = 0.4-0.7 microns, ventral[1]
Lizard	9 %	Wave band = 0.7-1.1 microns, dorsal[1]
Pika fur, back	10 %	Wave band = 0.3-0.7 microns, rg = 5 - 15[1]
Pika fur, back	40 %	Wave bond = 0.7-3.0 microns, rg = 20 - 60[1]
Pika fur, back	5 %	Wave band = 8-12 microns[1]
Populus deltoides, leaf	35 %	Wave bond = 0.7-3.0 microns[1]
Populus deltoides, leaf	7 %	Wave band = 0.3-0.7 microns[1]
Stellar jay, back feathers	3 %	Wave band = 0.3-0.7 microns[1]
Stellar jay, back feathers	40 %	Wave bond = 0.7-3.0 microns, rg = 5 - 50 %[1]

Source: 1.Gates, D.M., H.J. Keegan and V.R. Weidner. 1966. Scientific Experiments for Manned Orbital Flight. Sience and Technology Series 4, pp 71-86.

2-146 Light
Reflection from water surface

Species	Value	Condition
Diffuse sky light	11 %	Solar elevation = 30 degrees[1]
Diffuse sky light	13 %	Solar elevation = 20 degrees[1]
Diffuse sky light	14 %	Solar elevation = 15 degrees[1]
Diffuse sky light	15 %	Solar elevation = 10 degrees[1]
Diffuse sky light	17 %	Solar elevation = 5 degrees[1]
Diffuse sky light	7 %	Solar elevation = 50 degrees[1]
Diffuse sky light	9 %	Solar elevation = 40 degrees[1]
Direct sunlight	13 %	Solar elevation = 20 degrees[1]
Direct sunlight	2 %	Solar elevation = 50 degrees[1]
Direct sunlight	21 %	Solar elevation = 15 degrees[1]
Direct sunlight	3.5 %	Solar elevation = 3.5 degrees[1]
Direct sunlight	35 %	Solar elevation = 10 degrees[1]
Direct sunlight	58 %	Solar elevation = 5 degrees[1]
Direct sunlight	6 %	Solar elevation = 30 degrees[1]

Source: 1.Golterman, H.L. 1975. Physiological Limnol., Elsevier Amsterdam.

2-147 Light
Solar energy distribution

Species	Value	Condition
Far ultraviolet light	0.02 % of solar spectrum	Wavelength = 0.1-0.2 u, energy range = 152.2-304.4 kcal/einstein[1]
Middle infrared light	2.10 % of solar spectrum	Wavelength = 3-30 u, energy range = 0.95-9.5 kcal/einstein[1]
Near infrared light	38.90 % of solar spectrum	Wavelength = 0.78-3 u, energy range = 9.5-36.7 kcal/einstein[1]
Ultraviolet	7.27 % of solar spectrum	Wavelength = 0.2-0.38 u, energy range = 75.3-152.2 kcal/einstein[1]
Visible	51.73 % of solar spectrum	Wavelength = 0.38-0.78 u, energy range = 36.7-75.3 kcal/einstein[1]

Source: 1.Johnson, F.S. 1954. J. Meteorol., 11, p 431.

2-148 Pesticides
Aerial fallout

Species	Value	Condition
DDT	359E-9 g DDT/(m**2*day)	Southern Calif, rg = 28 - 690[1]

Source: 1.Young, D.R., D.J McDermott and T. Heesen. 1976. Bull. Environ. Contam. Toxicol. 16(5), pp 604-611.

2-149 Pesticides
Degradation rate

Species	Value	Condition
2,3,5-T	21 weeks	In soil, time used to reach 70-100 % of control[1]
2,3,6-TBA	47 weeks	In soil, time used to reach 70-100 % of control[1]
2,4-D	5 weeks	In soil, time used to reach 70-100 % of control[1]
Aldrin	38 % of initial in culv	After 4 months, uncultivated:2.7-5.3 % remaining after 4months, rg =[2]

Aldrin	40 % of initial in culv soil	After 20 w, uncultivated:5 % remaining after 6w[3]
Aldrin	53 % of initial in culv soil	After 1 yr, uncultivated:6.5-13 % remaining after 1yr, rg = 44 - 62 %[4]
Aldrin	90 weeks	In soil, time used to reach 70-100 % of control[1]
Amibem	10 weeks	In soil, time used to reach 70-100 % of control[1]
Atrazine	35 weeks	In soil, time used to reach 70-100 % of control[1]
Azinphosmethyl	95 % loss	Loam soil, 303 K, 6 d[5]
BHC	140 weeks	In soil, time used to reach 70-100 % of control[1]
Barban	3 weeks	In soil, time used to reach 70-100 % of control[1]
Bensulfide	35 weeks	In soil, time used to reach 70-100 % of control[1]
CDAA	8 weeks	In soil, time used to reach 70-100 % of control[1]
Chlordane	300 weeks	In soil, time used to reach 70-100 % of control[1]
Cipc	8 weeks	In soil, time used to reach 70-100 % of control[1]
DDT	200 weeks	In soil, time used to reach 70-100 % of control[1]
Dalapon	8 weeks	In soil, time used to reach 70-100 % of control[1]
Dicamba	8 weeks	In soil, time used to reach 70-100 % of control[1]
Dichlobenil	18 weeks	In soil, time used to reach 70-100 % of control[1]
Dieldrin	140 weeks	In soil, time used to reach 70-100 % of control[1]
Diphenamid	30 weeks	In soil, time used to reach 70-100 % of control[1]
Diuron	30 weeks	In soil, time used to reach 70-100 % of control[1]
EPTC	5 weeks	In soil, time used to reach 70-100 % of control[1]
Ethion	12 % loss	Soil dust, pH = 6.9, organic matter = 0.8 %, 200 days[5]
Ethion	23 % loss	Soil dust, pH = 7.3, organic matter = 2.3 %, 200 days[5]
Ethion	36 % loss	Soil dust, pH = 7.3, organic matter = 2.1 %, 200 days[5]
Ethion	36 % loss	Soil dust, pH = 7.6, organic matter = 1.8 %, 200 days[5]
Fenuron	30 weeks	In soil, time used to reach 70-100 % of control[1]
Heptachlor	15.0 % of initialt in culv soil	After 3 - 4 months, uncultivated:3.0 % remaining after 3 - 4 months[6]
Heptachlor	38 % of initial in culv soil	After 4months, uncultivated:2.7-5.3 % remaining after 4months, rg =[2]
Heptachlor	90 weeks	In soil, time used to reach 70-100 % of control[1]
IPA	5 weeks	In soil, time used to reach 70-100 % of control[1]
Linuron	18 weeks	In soil, time used to reach 70-100 % of control[1]
MCPA	12 weeks	In soil, time used to reach 70-100 % of control[1]
Monuron	35 weeks	In soil, time used to reach 70-100 % of control[1]
Parathion	58 % loss	Soil dust, pH = 6.9, organic matter = 0.8 %, 30 days[5]
Parathion	85 % loss	Soil dust, pH = 7.3-7.6, organic matter = 1.8-2.3 %, 10 days[5]
Parathion	99 % loss	Soil dust, pH = 6.9, organic matter = 0.8 %, 50 days[5]
Phenthoate	38 % loss	Soil dust, pH = 6.9, organic matter = 0.8 %, 75 days[5]
Phenthoate	50 % loss	Soil dust, pH = 7.3-7.6, organic matter = 1.8-2.3 %, 10 days[5]
Phenthoate	95 % loss	Soil dust, pH = 7.3-7.6, organic matter = 1.8-2.3 %, 30 days[5]

Phenthoate	99 % loss	Soil dust, pH = 6.9, organic matter = 0.8 %, 150 days, more than[5]
Picloram	80 weeks	In soil, time used to reach 70-100 % of control[1]
Prometryn	10 weeks	In soil, time used to reach 70-100 % of control[1]
Propazine	80 weeks	In soil, time used to reach 70-100 % of control[1]
Simazine	40 weeks	In soil, time used to reach 70-100 % of control[1]
TCA	12 weeks	In soil, time used to reach 70-100 % of control[1]
Trifluralin	23 weeks	In soil, time used to reach 70-100 % of control[1]

Source: 1.Kearney, P.C., J.E. Oliver, C.S. Helling, A.R. Isensee and A. Kontson. 1977. J. Agric. Food. Chem. 2(5), pp 1177-1181.
2.Licktenstein, E.P., C.H. Muller, G.R. Myrdal and K.R. Schulz. 1962. J. Econ. Entomol., 55, pp 215.
3.Wheatley, G.A., J.A. Hardman and A.H. Strickland. 1962. Plant Pathol., 11, pp 81.
4.Licktenstein, E.P., G.R. Myrdal and K.R. Schulz. 1964. J. Econ. Entomol., 57(1), pp 133.
5.Iwata, I., M.E. Dusch, W.E. Westlake and F.A. Gunther. 1975. Bull. Environ. Contam. Toxicol. 14(1),p 49-56.
6.Saha, J.G. and W.W.A. Stewart. 1967. Can. J. Plant. Sci., 47, pp 79.

2-150 Pesticides
Half-life time

Species	Value	Condition
Chlorobromuron	2.5 days	In ensilage, pH = 4.5, 300 K, initial conc = 0.12 ppm[1]
Methoxychlor	24 days	Elimination from oak leaves in water, pH = 7.9[2]
Methoxychlor	26 days	Elimination from maple leaves in water, pH = 7.9[2]
Methoxychlor	7 days	Elimination from beech leaves in water, pH = 7.9[2]
Phosphamidon	2 days	Sprayed by aircraft, New Brunswich, Can, 3/6, 275-293 K, sunny, app[3]
Sevin	1.7 days	Degradation in plant tissue[4]

Source: 1.Katz, S.E., R.F. Struze and C.A. Fassbender. 1970. Bull. Environ. Contam. Toxicol., 5, pp 50-55.
2.Deer, S.K. 1974. Bull. Environ. Contam. Toxicol., 11(6), pp 500-502.
3.Varty, I.W. and W.N. Yule. 1976. Bull. Environ. Contam. Toxicol. 15(3), pp 257-264.
4.Barrett, G.W. 1968. Ecology, 49, pp 1019-1035.

2-151 Pesticides
Half-life time for photolysis

Species	Value	Condition
2,4-D BEE	1E2	Light = 3E13 photons/cm2, BEE = butoxyethyl ester of 2,4-D, lab exp.[1]
2,4-D butoxyethyl	4.0 days	In hydrocarbon, september[1]
2,4-D butoxyethyl ester	100 hours	Light int. = 3E13photons/(cm2*sec), 280-300 nm, Hg-lamp[1]
2,4-D butoxyethyl	14 days	In water, september, ester[1]
Atrazine	3000 hours	Light int. = 3E13photons/(cm2*sec), 280-300 nm, Hg-lamp[1]
Atrazine	3E3 hours	Light = 3E13 photons/cm2, 280-300 nm, lat = 30 N, lab experiment[1]
Captan	2000 hours	Light int. = 3E13photons/(cm2*sec), 280-300 nm, Hg-lamp, more than value[1]
Captan	2E3 hours	Light = 3E13 photons/cm2, 280-300 nm, lat = 30 N, lab experiment[1]
Carbaryl	102 hours	Latitude = 40 N, fall, calc for mid-season, depth < 10 cm in pure water[1]
Carbaryl	103 hours	Latitude = 30 N, winter, calc for mid-season, depth < 10 cm in pure water1
Carbaryl	121 hours	Latitude = 70 N, summer, calc for mid-season, depth < 10 cm in pure water1
Carbaryl	1760 hours	Latitude = 70 N, fall, calc for mid-season, depth < 10 cm in pure water1
Carbaryl	186 hours	Latitude = 50 N, fall, calc for mid-season, depth < 10 cm in pure water1
Carbaryl	200 hours	Latitude = 40 N, winter, calc for mid-season, depth < 10 cm in pure water1
Carbaryl	200 hours	Light int. = 3E13photons/(cm2*sec), 280-300 nm, Hg-lamp[1]

Carbaryl	292 hours	Latitude = 70 N, spring, calc for mid-season, depth < 10 cm in pure water1
Carbaryl	2E2 hours	Light = 3E13 photons/(cm2*sec), 280-300 nm, lat = 30 N, lab[1]
Carbaryl	41 hours	Latitude = 0 N, fall, calc for mid-season, depth < 10 cm in pure water1
Carbaryl	43 hours	Latitude = 0 N, spring, calc for mid-season, depth < 10 cm in pure water1
Carbaryl	43 hours	Latitude = 0 N, winter, calc for mid-season, depth < 10 cm in pure water1
Carbaryl	43800 hours	Latitude = 70 N, winter, calc for mid-season, depth < 10 cm in pure water1
Carbaryl	45 hours	Latitude = 0 N, summer, calc for mid-season, depth < 10 cm in pure water1
Carbaryl	46 hours	Latitude = 30 N, summer, calc for mid-season, depth < 10 cm in pure water1
Carbaryl	51 hours	Latitude = 30 N, spring, calc for mid-season, depth < 10 cm in pure water1
Carbaryl	52 hours	Latitude = 40 N, summer, calc for mid-season, depth < 10 cm in pure water1
Carbaryl	571 hours	Latitude = 50 N, winter, calc for mid-season, depth < 10 cm in pure water1
Carbaryl	61 hours	Latitude = 50 N, summer, calc for mid-season, depth < 10 cm in pure water1
Carbaryl	64 hours	Latitude = 40 N, spring, calc for mid-season, depth < 10 cm in pure water1
Carbaryl	68 hours	Latitude = 30 N, fall, calc for mid-season, depth < 10 cm in pure water1
Carbaryl	90 hours	Latitude = 50 N, spring, calc for mid-season, depth < 10 cm in pure water1
DDT	227000 hours	In water, direct photolysis, midday sun, summer, lat = 40 N, more than value1
DDT	227000 hours	Medium = water, latitude 40 N, larger than value[2]
DDT	280000 hours	In hydrocarbon, direct photolysis, midday sun, summer, lat = 40 N[1]
Diazinon	1000 hours	Light int. = 3E13photons/(cm2*sec), 280-300 nm, Hg-lamp[1]
Diazinon	1E3 hours	Light = 3E13 photons/cm2, 280-300 nm, lat = 30 N, lab experiment[1]
Malathion	2E4 hours	Light = 3E13 photons/cm2, 280-300 nm, lat = 30 N, lab experiment[1]
Malathion	2E4 hours	Light int. = 3E13photons/(cm2*sec), 280-300 nm, Hg-lamp[1]
Methoprene	6 hours	On glass, exposed to sunlight through glass, feb, calif.[3]
Methoxychlor	2.2 hours	Suwannee River, pH = 4.7[2]
Methoxychlor	2.2 hours	Suwannee River, pH = 4.7, midday sun, May, lat = 34 N, 40 ppb meth[1]
Methoxychlor	2.9 hours	Alabama River, pH = 7.7[2]
Methoxychlor	2.9 hours	Alabama River, pH = 7.7, midday sun, May, lat = 34 N, 40 ppb methox.[1]
Methoxychlor	30 hours	Light int. = 3E13photons/(cm2*sec), 280-300 nm, Hg-lamp[1]
Methoxychlor	300 hours	Distilled water, pH = 6.3, midday sun, May, lat = 34 N, 40 ppb meth,[1]
Methoxychlor	300 hours	Distilled water, pH = 6.3, more than value[2]
Methoxychlor	3E1 hours	Light = 3E13 photons/cm2, 280-300 nm, lat = 30 N, lab experiment[2]
Methoxychlor	4100 hours	In hydrocarbon, direct photolysis, midday sun, summer, lat = 40 N[1]
Methoxychlor	5.4 hours	Tombigbee River, pH = 7.6[2]
Methoxychlor	5.4 hours	Tombigbee River, pH = 7.6, midday sun, May, lat = 34 N, 40 ppb[1]
Methoxychlor	690 hours	In water, direct photolysis, midday sun, summer, lat = 40 N[1]
Methoxychlor	690 hours	Medium = water, latitude 40 N[2]
Methoxychlor	7.3 hours	20 ppm humic acid in dest wat, pH = 5.2, midday sun, May, 34 N, 40[1]
Methoxychlor	7.3 hours	Distlled water with 20 ppm humic acid, pH = 5.2[2]
Parathion	2000 hours	Light int. = 3E13photons/(cm2*sec), 280-300 nm, Hg-lamp, lat 30 N[1]
Toxaphene	2E4 hours	Light int. = 3E13photons/(cm2*sec), 280-300 nm, Hg-lamp, lat = 30 N,[1]

Source: 1.Wolfe, N.L., R.G. Zepp, G.L. Banghman, R.C. Fincher and J.A. Gordon. 1976. Chemical and Photochemical Transformations of selected Pesticides in Aquatic Systems, EPA-600/3-76-067.
2.Zepp, R.G., N.L. Wolfe, J.A. Gordon and R.C. Fincher. 1976. J. Agric. Food Chem., 24(4), pp 727-732.
3.Quistad, G.B., L.E. Staiger and D.A. Schooley. 1975. J. Agric. Food. Chem. 23(2), pp 299-303.

2-152 Pesticides
Half-life time for volatilization

Species	Value	Condition
2,4-D 1-butylester	1.1 days	From a 1 m deep water layer, 298 K, half-life time calculated[1]
2,4-D 1-octylester	11.5 days	From a 1 m deep water layer, 298 K, half-life time calculated[1]
2,4-D 2-buyoxyethylester	895.0 days	From a 1 m deep water layer, 298 K, half-life time calculated[1]
2,4-D methylester	22.0 days	From a 1 m deep water layer, 298 K, half-life time calculated[1]
DDT	14.3 days	o,p'-DDT, appl = 1-1.3 kg/ha, cotton[2]
DDT	15.1 days	p,p'-dde, appl = 1-1.3 kg/ha, cotton[2]
DDT	18.8 days	p,p'-DDT, appl = 1-1.3 kg/ha, cotton[2]
Toxaphene	15.1 days	Appl = 2 - 2.7 kg/ha, cotton[2]

Source: 1.Zepp, R.G., N.L. Wolfe, G.L. Baughman and J.A. Gordon. 1974. Environ. Qual. Saf. Suppl III, pp 313-39.
2.Nash, R.G., M.L. Ball, Jr. and W.L. Har ris. 1977. J. Agric. Food. Chem. 25(2), pp 336-341.

2-153 Pesticides
Half-life time in soil

Species	Value	Condition
1,3-dichloropropenes	24 days	Sandy soil, 288-293 K, isomers same rate[1]
1,3-dichloropropenes	5.5 days	Clay cont. soil, 288-293 K, rg = 3 - 8 d[1]
2,4,5-T	2 weeks	On native grasses, TX[2]
2,4,5-T	3 weeks	On litter, TX[2]
2,4-D	12 days	In soil treated with 2,4-D in several years, rg = 10 - 14[3]
2,4-D	12 days	In soil treated with 2,4-D in severel years, rg = 10 - 14[3]
2,4-D	2 weeks	On native grasses, TX[2]
2,4-D	3 days	303 K[3]
2,4-D	3 weeks	On litter, TX[2]
Akton	1 year	2 lbs/acre, granules, corn, sultan silt loam[4]
Akton	1.2 years	2 lbs/acre, spray, corn, sultan silt loam[4]
Akton	1.5 years	10 lbs/acre, spray, corn, sultan silt loam[4]
Akton	32 weeks	Lab.[4]
Aldicarb	12 days	Oxidation, value depending of soil, rg = 7 - 17[5]
Aldrin	10 days	In irrigated soil, cotten, Gezira, Sudan[6]
Aldrin	28 % remaining after 15 years	Max value[7]
Aldrin	40 % remaining after 14 years	Max value[7]
Azinphosmethyl	10 days	Clay and silt loam[8]
Azinphosmethyl	12 days	Applied as an emulsion on soil surface[8]
Azinphosmethyl	13 days	Natural soil[8]

Azinphosmethyl	135 days	Sterile soil, dry, lag period included, 298 K[9]
Azinphosmethyl	15 days	Silty clay loam[8]
Azinphosmethyl	25 days	Silt loam[8]
Azinphosmethyl	28 days	Granular form rototilling into soil[8]
Azinphosmethyl	29 days	Sterile soil, wet, lag period included, 298 K[9]
Azinphosmethyl	29 days	Wet (50 % of sat.) sterilized soil at 298k[8]
Azinphosmethyl	30 days	Sandy loam[8]
Azinphosmethyl	32 days	Natural soil, dry, lag period included, 313 K[9]
Azinphosmethyl	36 days	Sterile soil, dry, lag period included, 313 K[9]
Azinphosmethyl	484 days	Natural soil, dry, lag period included, 279 K[9]
Azinphosmethyl	484 days	Sterile soil, dry, lag period included, 279 K[9]
Azinphosmethyl	5 days	313 K, potatoes[10]
Azinphosmethyl	6 days	Sterile soil, wet, lag period included, 313 K[9]
Azinphosmethyl	65 days	Sandy loam[8]
Azinphosmethyl	70 days	Fine sandy loam[8]
Azinphosmethyl	80 days	Sandy loam and loam[8]
Azinphosmethyl	88 days	Natural soil, dry, lag period included, 298 K[9]
Azinphosmethyl	88 days	Sterile soil, wet, lag period included, 279 K[9]
BHC	10 % remaining after 14 years	Max value[7]
Benomyl	4.5 months	On turf, Delaware, rg = 3 - 6[11]
Benomyl	9 months	Bare soil, Delaware, rg = 3 - 6[11]
Bux	1.5 weeks	Hydrolysis, lab., rg = 1 - 2[5]
Carbaryl	64 days	Sandy loam, appl = 25.4 kg/ha[12]
Carbaryl	8 days	Agricultural soil[5]
Carbofuran	0.56 years	In soil, well drained, application = 2.5 ppm, clay-muck, 1973[13]
Carbofuran	0.60 years	In soil, better drained, application = 2.5 ppm, clay-muck, 1973[13]
Carbofuran	0.73 years	In soil, poorly drained, application = 2.5 ppm, 1974, clay[13]
Carbofuran	0.92 years	In soil, poorly drained, application = 2.5 ppm, clay-muck, 1973[13]
Carbofuran	30 days	Hydrolysis, formation of phenol[5]
Carbofuran	7.52 years	In soil, well drained, application = 2.5 ppm, clay-muck, 1974[13]
Chlordane	40 % remaining after 14 years	Max value[7]
Chlorobenzilate	21 days	In leon and lakeland sands[3]
Cis-chloroallylal-cohol	2 days	Clay cont. soil, 288 K, 1,3-dichloropropenes[1]
DDT	14 days	In irrigated soil, cotten, Gezira, Sudan[6]
DDT	39 % remaining after 17 years	Max value[7]
DDT	8 months	p,p'-DDT, subtropical soil, fall and winter, appl = 5 kg/ha[14]
Dicamba	2 weeks	On native grasses, TX[2]
Dicamba	3 weeks	On litter, TX[2]

Dieldrin	11 days	In irrigated soil, cotten, Gezira, Sudan[6]
Dieldrin	2.7 days	Loss by volatilization, grass, first 5 d[15]
Dieldrin	31 % remaining	Max value, after 15 years[7]
Dieldrin	7.5 months	p,p'-DDT, subtropical soil, fall and winter, appl = 5 kg/ha[14]
Dioxathion	30 days	Soil dust, pH = 7.6, organic matter = 1.8 %[8]
Dioxathion	45 days	Soil dust, pH = 7.3, organic matter = 2.3 %[8]
Dioxathion	55 days	Soil dust, pH = 7.3, organic matter = 2.1 %[8]
Endosulfan	7 days	In irrigated soil, cotten, Gezira, Sudan[6]
Endrin	41 % remaining	Max value, after 14 years[7]
Ethion	420 days	None[8]
Heptachlor	1.7 days	Loss by volatilization, grass, first 5 d[15]
Heptachlor	16 % remaining	Max value, after 14 years[7]
Malathion	3 days	Basic silty loam, Illinois, appl = 10 ppm, pH = 6.2[16]
Malathion	4.3 days	Basic silty loam, Illinois, appl = 10 ppm, pH = 7.2[16]
Malathion	7 days	Basic silty loam, Illinois, appl = 10 ppm, pH = 8.2[16]
Meobal	7 days	Field condition[5]
Methomyl	36 days	Range = 30 - 42[5]
Naproamide	102 days	Soil moisture content = 10.0 %, 287 K, initial = 4.5 kg/ha[17]
Naproamide	112 days	Soil moisture content = 7.5 %, 287 K, initial = 4.5 kg/ha[17]
Naproamide	54 days	Soil moisture content = 10.0 %, 301 K, initial = 4.5 kg/ha[17]
Naproamide	56 days	Soil moisture content = 10.0 %, 301 K, initial = 2.25 kg/ha[17]
Naproamide	63 days	Soil moisture content = 7.5 %, 301 K, initial = 4.5 kg/ha[17]
Naproamide	90 days	Soil moisture content = 3.5 %, 301 K, initial = 4.5 kg/ha[17]
Terbacil	150 days	303 K[3]
Terbacil	225 days	288 K[3]
Toxaphene	45 % remaining after 14 years	Max value[7]
Trans-dichloro-allylalcohol	1 day	Clay cont. soil, 288 K, less than value, 1,3-dichloropropenes[1]
Tsuamacide	66 % in 40 days	Half the amouth bound thigtly in soil[5]
Tsuamacide	7 days	Field condition[5]

Source: 1.Van Dijk, H. 1974. Agro-Ecosyst., 1(3), pp 193-204.
2.Morton, H.L., E.D. Robison and R.E. Meyer. 1967. Weeds 15, pp 268-271.
3.Mansell, R.S. et al. 1977. Fertilizer and Pesticide Movements from Citrus groves in Florida flatwoodSoils. EPA-600/2-77-177.
4.Getzin, L. W. 1977. Bull. Environ. Contam. Toxicol. 18(4), pp 499-502.
5.Miyamoto, J. 1975. Env. Qual. Saf. 4, pp 134-135.
6.El Zorgani, G.A. 1976. Bull. Environ. Contam. Toxicol., 15(3), pp 378-382.
7.Nash, R.G. and C.A. Woolson. 1967. Science 157, pp 924-927.
8.Iwata, I., M.E. Dusch, W.E. Westlake and F.A. Gunther. 1975. Bull. Environ. Contam. Toxicol. 14(1),p 49-56.
9.Jones, A.S. and C.S. Hodges. 1974. J. Agric. Food. Chem. 22(3), pp 435-441.
10.Yaron, B., H. Bielorai and L. Klinger. 1974. J. Environ. Qual. 3(4), pp 413-417.
11.Baude, F.J., H.L. Pease and R.F. Holt. 1974. J. agric. Fd. Chem. 22(3), pp 413-418.
12.Lafeur, K.S. 1976. J. Environ. Qual. 5(1), pp 91-92.
13.Williams, I.H., M.J. Brown and P. Whitehead. 1976. Bull. Environ. Contam. Toxicol., 15(2), pp 242-43.
14.Talekar, N.S., L.-t. Sun, E.-M. Lee and J.-S. Chen. 1977. J. Agric. Food. Chem. 25(2), pp 348-352.
15.Taylor, A.W., D.E. Glotfelty, B.C. Turner, R.E. Silver, H.P. Freeman and A. Weiss. 1977. J. Agric. Food. Chem. 25(3), pp 542-548.
16.Paschal, D.C. and M.E. Neville. 1976. J. Environ. Qual. 5(4), pp 441-443.
17.Walker, A. 1974. J. Environ. Qual. 3(4), pp 396-401.

2-154 Pesticides
Half-life time in water

Species	Value	Condition
2,4-D 1-butylester	220 days	Hydrolysis, 301 K, pH = 6[1]
2,4-D 1-butylester	5.2 hours	Hydrolysis, 301 K, pH = 9[1]
2,4-D 1-octylester	220 days	Hydrolysis, 301 K, pH = 6[1]
2,4-D 1-octylester	5.2 hours	Hydrolysis, 301 K, pH = 9[1]
2,4-D 2-butoxyethylester	0.6 hours	Hydrolysis, 301 K, pH = 9[1]
2,4-D 2-butoxyethylester	26 days	Hydrolysis, 301 K, pH = 6[1]
2,4-D 2-butoxymethyl-ethylester	180 days	Hydrolysis, 301 K, pH = 6[1]
2,4-D 2-butoxymethyl-ethylester	4.4 hours	Hydrolysis, 301 K, pH = 9[1]
2,4-D 2-octylester	1500 days	Hydrolysis, 301 K, pH = 6[1]
2,4-D 2-octylester	37.0 hours	Hydrolysis, 301 K, pH = 9[1]
2,4-D 2-propy ester	17.0 hours	Hydrolysis, 301 K, pH = 9[1]
2,4-D 2-propylester	710 days	Hydrolysis, 301 K, pH = 6[1]
2,4-D methylester	1.1 hours	Hydrolysis, 301 K, pH = 9[1]
2,4-D methylester	44 days	Hydrolysis, 301 K, pH = 6[1]
Atrazine	12.9 days	Hydrolysis, pH = 12.9[2]
Atrazine	15.2 days	Hydrolysis, pH = 11.9[2]
Atrazine	18.4 days	Hydrolysis, pH = 2.2[2]
Atrazine	5.1 days	Hydrolysis, pH = 1.3[2]
Atrazine	66.4 days	Hydrolysis, pH = 3.1[2]
Atrazine	81.1 days	Hydrolysis, pH = 11.1[2]
Captan	155 min	Tombigbee River, pH = 7.0, 301k[3]
Carbaryl	1 month	pH = 8.0, 276.5 K, seawater[4]
Carbaryl	1.0 days	pH = 8.0, 301 K, seawater[4]
Carbaryl	1.3 days	Hydrolysis, pH = 8, 300 K[3]
Carbaryl	116 min	pH = 9.2, 298 K, initial conc = 3E-3 M, dark[5]
Carbaryl	116 min	pH = 9.2, 298 K, initial conc = 3E-3 m, dark[5]
Carbaryl	13 days	Hydrolysis, pH = 7, 300 K[3]
Carbaryl	173 min	pH = 9.0, 298 K, initial conc = 3E-3 M, dark[5]
Carbaryl	173 min	pH = 9.0, 298 K, initial conc = 3E-3 m, dark[5]
Carbaryl	20 min	pH = 10.0, 298 K, initial conc = 3E-3 M, dark[5]
Carbaryl	20 min	pH = 10.0, 298 K, initial conc = 3E-3 m, dark[5]
Carbaryl	27 min	pH = 9.8, 298 K, initial conc = 3E-3 M, dark[5]
Carbaryl	27 min	pH = 9.8, 298 K, initial conc = 3E-3 m, dark[5]

Carbaryl	3.2 hours	Hydrolysis, pH = 9, 300 K[3]
Carbaryl	3.5 days	pH = 8.0, 293 K, seawater[4]
Carbaryl	3.6 years	Hydrolysis, pH = 5, 300 K[3]
Carbaryl	4.4 months	Hydrolysis, pH = 6, 300 K[3]
Carbaryl	4.8 days	pH = 8.0, 290 K, seawater[4]
Carbaryl	58 min	pH = 9.5, 298 K, initial conc = 3E-3 M, dark[5]
Carbaryl	58 min	pH = 9.5, 298 K, initial conc = 3E-3 m, dark[5]
Carbaryl	8 min	pH = 10.0, 308 K, initial conc = 3E-3 M, dark[5]
Carbaryl	8 min	pH = 10.0, 308 K, initial conc = 3E-3 m, dark[5]
Carbaryl	99 min	pH = 10.0, 285 K, initial conc = 3E-3 M, dark[5]
Carbaryl	99 min	pH = 10.0, 285 K, initial conc = 3E-3 m, dark[5]
Diazinon	0.49 days	Hydrolysis, pH = 3.1[6]
Diazinon	136 days	Hydrolysis, pH = 9.0[6]
Diazinon	185 days	Hydrolysis, pH = 7.5[6]
Diazinon	31 days	Hydrolysis, pH = 5.0[6]
Diazinon	6 days	Hydrolysis, pH = 10.4[6]
Diazoxon	0.017 day	Hydrolysis, pH = 3.1[6]
Diazoxon	0.42 days	Hydrolysis, pH = 10.4[6]
Diazoxon	1.27 days	Hydrolysis, pH = 5.0[6]
Diazoxon	18 days	Hydrolysis, pH = 9.0[6]
Diazoxon	29 days	Hydrolysis, pH = 7.5[6]
Dimilin	22.9 days	pH = 7.7, 283 K, initial = 0.1 ppm[7]
Dimilin	28.7 days	pH = 7.7, 297 K, initial = 0.1 ppm[7]
Dimilin	3.45 days	pH = 10.0, 311 K, initial = 0.1 ppm[7]
Dimilin	8.00 days	pH = 7.7, 311 K, initial = 0.09 ppm[7]
Dimilin	8.31 days	pH = 10.0, 297 K, initial = 0.1 ppm[7]
Dimilin	80.5 days	pH = 10.0, 283 K, initial = 0.1 ppm[7]
Heptachlor	23.1 hour	299.8 K, for hydrolysis, in distilled water[8]
Methoprene	30 hours	Freshwater pond, initial = 0.001 ppm, Feb, in sunlight, Calif[9]
Methoprene	40 hours	Freshwater pond, initial = 0.01 ppm, Feb, in sunlight, Calif[9]
Methoxychlor	1 hour	Hydls, conc = E-6 M, 5 % acetonitrile, H2O2 conc = 0.1 M, 338 K, less than value[3]
Methoxychlor	1.7 hour	338 K, hydrolysis, conc = E-6 M, 5 % acetonitrile, H2O2 conc = 8E-2[3]
Methoxychlor	17 hours	338 K, hydrolysis, conc = E-6 M, 5 % acetonitrile, H2O2 conc = 8E-4 M[3]
Methoxychlor	2.0 hours	338 K, hydrolysis, conc = E-6 M, 5 % acetonitrile, H2O2 conc = 8E-3 M[3]
Methoxychlor	25 hours	338 K, hydrolysis, conc = E-6 M, 5 % acetonitrile, H2O2 conc = 8E-5 M[3]
Methoxychlor	58 hours	Conc in water = E-7m, no H2O2 added, 338 K[3]
Methoxychlor	58 hours	Hydls, conc = E-6 M, 5 % acetonitrile, no H2O2 added, 338 K[3]
Parathion	0.5 hours	343 K, distilled water + dil NaOH, hydrolysis, pH = 12, less than value[10]
Parathion	15.7 hours	343 K, buffer, salt 30 g/l, hydrolysis, pH = 7.9, rg = +- 0.2[10]
Parathion	17.2 hours	343 K, seawater, salinity = 3.25 %, hydrolysis, pH = 8.0, rg = +- 0.3[10]

Parathion	180 hours	323 K, estuarine water, salinity = 2.57 %, hydrolysis, pH = 7.8[10]
Parathion	19.3 hours	343 K, seawater, salinity = 3.25 %, hydrolysis, pH = 1.0, rg = +- 0.5[10]
Parathion	21.0 hours	343 K, distilled water, 30 g NaCl/l, hydrolysis, pH = 5.5[10]
Parathion	22.7 hours	343 K, distilled water + dil NaOH, hydrolysis, pH = 8.6, rg = +- 0.6[10]
Parathion	30.0 hours	343 K, estuarine water, salinity = 2.57 %, hydrolysis, pH = 7.8[10]
Parathion	31.8 hours	343 K, buffer, salt 0.55 g/l, hydrolysis, pH = 7.5, rg = +- 0.7[10]
Parathion	39.2 hours	343 K, distilled water, hydrolysis, pH = 5.9, rg = +- 1.3[10]
Parathion	39.2 hours	343 K, distilled water, hydrolysis, pH = 5.9, rg = +-1.3[10]
Parathion	4.47 hours	363 K, estuarine water, salinity = 2.57 %, hydrolysis, pH = 7.8[10]
Parathion	5240 hours	296 K, estuarine water, salinity = 2.57 %, hydrolysis, pH = 7.8[10]
Zectran	1.5 days	pH = 9.5, 285 K, rg = 1 - 2[11]
Zectran	27 days	pH = 7.4, 285 K[11]
Zr515	100 days	Saltwater, 277.5 K, initial = 830 ppm[12]
Zr515	35 days	Freshwater, 277.5 K, initial = 763 ppm[12]

Source: 1.Zepp, R.G., N.L. Wolfe, G.L. Baughman and J.A. Gordon. 1974. Environ. Qual. Saf. Suppl III, pp 313-39.
2.Armstrong, D.E., G. Chester and R.F. Harris. 1967. Soil Sci. Soc. Amer. Proc., 31, pp 61.
3.Wolfe, N.L., R.G. Zepp, G.L. Banghman, R.C. Fincher and J.A. Gordon. 1976. Chemical and Photochemical Transformations of selected Pesticides in Aquatic Systems, EPA-600/3-76-067.
4.Karien, J.F., J.G. Lamberton, N.E. Stewart and L.C. Terriere. 1967. J. Agric. Food. Chem. 15, pp 148
5.Joergensen, S.E., J.H. Henriksen, M. Friis, L.A. Joergensen and H.F. Mejer. Handbook of Environmental Data and Ecological Parameters. 1979. ISEM. Copenhagen.
6.Faust, S.D. and H.M. Gomaa. 1972. Environ. Letters, 3, pp 171.
7.Schuth, C.K., A.R. Isensee, E.A. Woolson and P.C. Kearney. 1974. J. Agric. Food. Chem. 22(6), pp 999-1003.
8.Demayo, A. 1972. Bull. Environ. Contam. Toxicol., 8(4), pp 234-237.
9.Schooley, D.A., Bergot, L.L. Dunham and J.B. Siddall. 1975. J. Agric. Food. Chem. 23(2), pp 293-298.
10.Webber, K. 1976. Water Res. 10(3), pp 239-243.
11.Holeman, J.N. 1968. Water Res., 4, pp 737-747.
12.Pree, D.J. and D.K.R. Stewart. 1975. Bull. Environ. Contam. Toxicol. 14(1), pp 117-121.

2-155 Pesticides
Loss by runoff

Species	Value	Condition
2,4,5-T	0.02 % of application	Slope = 5 %, grass, application = foliar, soil = c, 24 months, tx[1]
2,4,5-T	0.05 % of application	Slope = -, grass, app = simulated rainfall, soil = sil, 14 months[2]
2,4,5-T	2 % of application	Slope = 3-8 %, fallow sod, app = simulated rainfall, 4 months[3]
2,4-D	0 g/ha	Deep tillage, in open drain, appl = 3 lbs/ha, 14 d, citrus[4]
2,4-D	0 g/ha	In open drain, deep tillage, appl = 3 lbs/ha, 14 d, 74[4]
2,4-D	0.04 %	1972, sandy coastal plain soil, appl = 0.56 kg/ha[5]
2,4-D	0.27 %	1971, sandy coastal plain soil, appl = 0.56 kg/ha[5]
2,4-D	0.33 %	1970, sandy coastal plain soil, appl = 0.56 kg/ha[5]
2,4-D	122.9 g/ha	Deep til. & lime, in open drain, appl = 3 lbs/ha, 14 d, citrus[4]
2,4-D	15.26 g/ha	In open drain, shallow tillage, appl = 3 lbs/ha, 14 d, 74[4]
2,4-D	19.1 g/ha	Deep tillage, in submerged drain, appl = 3 lbs/ha, 14 d, citrus[4]
2,4-D	20.55 g/ha	In open drain, deep til. & lime, appl = 3 lbs/ha, 14 d, 74[4]
2,4-D	3.19 g/ha	In submerged drain, deep tillage, appl = 3 lbs/ha, 14 d, 74[4]
2,4-D	45.8 g/ha	Shallow tillage, in submerged drain, appl = 3 lbs/ha, 14 d, citrus[4]
2,4-D	52.7 g/ha	Deep til. & lime, in submerged drain, appl = 3 lbs/ha, 14 d, citrus[4]
2,4-D	7.66 g/ha	In submerged drain, shallow tillage, appl = 3 lbs/ha, 14 d, 74[4]

2,4-D	8.81 g/ha	In submerged drain, deep til. & lime, appl = 3 lbs/ha, 14 d, 74[4]
2,4-D	91.2 g/ha	Shallow tillage, in open drain, appl = 3 lbs/ha, 14 d, citrus[4]
2,4-D amine	4 % of application	Slope = 7 %, cultivated, app = simulated rainfall, soil = sl, 48 h[6]
2,4-D amine	5 % of application	Slope = 7 %, cultivated, app = simulated rainfall, 1 h, soil = sl, GA[6]
2,4-D amine	5 % of application	Slope = 7 %, cultivated, app = simulated rainfall, soil = sl, 96 h[6]
2,4-D butylether	35 % of application	Slope = 7 %, cultivated, app = simulated rainfall, 1 h, soil = sl[6]
2,4-D isoctyl	40 % of application	Slope = 7 %, cultivated, app = simulated rainfall, 1 h, soil = sl[6]
Aldrin	5.2 % of application	Slope = 1.5 %, cultivated, application = simulated rainfall[7]
Atrazine	1.9 % of application (to sediment)	Slope = 3 %, fallow, app = incorporated sim. rainfall, soil = sl[8]
Atrazine	1.95 % of application (to sediment)	Slope = 2 %, fallow, app = incorporated sim. rainfall, soil = sl[8]
Atrazine	10.3 % of application (to water)	Slope = 4 %, fallow, app = incorporated sim. rainfall, soil = scl[8]
Atrazine	11.4 % of application (to water)	Slope = 3 %, fallow, app = incorporated sim. rainfall, soil = sl, 1[8]
Atrazine	13.3 % of application	Slope = 3 %, fallow, app = incorporated sim. rainfall, soil = sl[8]
Atrazine	16 % of application	Slope = 10-15 5, corn, app = surface, soil = sil, 58 d, IA[9]
Atrazine	2.17 % of application (to sediment)	Slope = 4 %, fallow, app = incorporated sim. rainfall, soil = scl[8]
Atrazine	2.5 % of application	Slope = 14 %, corn, application = surface, soil = sicl, 5 months, PA[10]
Atrazine	3.04 % of application (to sediment)	Slope = 6 %, fallow, app = incorporated sim. rainfall, soil = sl[8]
Atrazine	4.49 % of application (to water)	Slope = 2 %, fallow, app = incorporated sim. rainfall, soil = sl1[8]
Atrazine	4.8 % of application	Slope = 14 %, corn, app = surface, soil = sicl, 5 months, PA[10]
Atrazine	4.83 %	Appl = 4.5 kg/ha, zea mays[11]
Atrazine	5.0 % of application	Slope = 14 %, app = surface, corn, soil = sicl, 5 months, PA[10]
Atrazine	5.05 %	Appl = 2.2 kg/ha, zea mays[11]
Atrazine	6.5 % of application	Slope = 6.5 %, fallow, app = simulated rainfall, soil = sl, 1 h, ga[12]
Atrazine	7.14 % of application (to water)	Slope = 6 %, fallow, app = incorporated sim. rainfall, soil = sl1[8]
Benomyl	1.1 %	Appl = 20 lb/acre, rain = 3.5 in, 7 d, Ken[13]
Cs13529	4.97 %	Appl = 2.2 kg/ha, zea mays[11]
Cs13529	5.70 %	Appl = 2.2 kg/ha, zea mays[11]
DDT	0.81 % of application	Slope = 8 %, oats, application = surface, soil = gl, 12 months, ME[14]
DDT	1.0 % of application	Slope = 8 %, rot. potatoes, app = surface, soil = gl, 24 months, ME[14]
DDT	1.6 % of application	Slope = 8 %, cont. potatoes, app = surface, soil = gl, 24 mounths, ME[14]
DDT	2.83 % of application	Slope = 3 %, cotton, app = foliar, soil = ls & sl, 6 months, NC[15]
DDT	6.3 % of application	Slope = 1.5 %, cultivated, application = simulated rainfall[7]
Dicamba	2 % of application	Slope = 3-8 %, fallow sod, app = simulated rainfall, 4 months[8]
Dichlobenil	0.36 % of application (to sediment)	Slope = 2 %, fallow, app = incorporated sim. rainfall, soil = sl[8]
Dichlobenil	2.07 % of application (to water)	Slope = 6 %, fallow, app = incorporated sim. rainfall, soil = sl[8]

Dichlobenil	2.33 % of application (to sediment)	Slope = 3 %, fallow, app = incorporated sim. rainfall, soil = sl[8]
Dichlobenil	2.36 % of application (to water)	Slope = 2 %, fallow, app = incorporated sim. rainfall, soil = sl[8]
Dichlobenil	2.55 % of application (to sediment)	Slope = 4 %, fallow, app = incorporated sim. rainfall, soil = sl[8]
Dichlobenil	2.55 % of application (to sediment)	Slope = 6 %, fallow, app = incorparated sim. rainfall, soil = sl[8]
Dichlobenil	5.99 % of application (to water)	Slope = 4 %, fallow, app = incorporated sim. rainfall, soil = sl[8]
Dichlobenil	7.61 % of application (to water)	Slope = 3 %, fallow, app = incorporated sim. rainfall, soil = sl[8]
Dieldrin	0.007 % of application	Slope = 14 %, most corn, app = incorporated, soil = sil, 8 months[16]
Dieldrin	0.7 % of application	Slope = 14 %, most corn, app = incorporated, soil = sil, 26 months[16]
Dieldrin	4.7 % of application	Slope = 1.5 %, cultivated, app = simulated rainfall, soil = sil, KY[7]
Diuron	0.025 % of application	Appl = 0.84 kg/ha, cotton, surface runoff, rain = 9.2 cm, Miss, 1972[17]
Diuron	0.042 % of application	Appl = 0.84 kg/ha, cotton, surface runoff, rain = 9.7 cm, Miss, 1971,[17]
Diuron	0.05 % of application	Slope = 0.2 %, cotton, app = surface ponded, soil = sicl, 60 d, LA,[18]
Diuron	0.199 % of application	Appl = 0.84 kg/ha, cotton, surface runoff, rain = 15.5 cm, Miss, 1973[17]
Endosulfan	0.01 % of application	Slope = 8 %, oats, application = surface , soil = gl, 12 months, ME[1]
Endosulfan	0.25 % of application	Slope = 8 %, rot. potatoes, app = surface, soil = gl, 12 months, ME[14]
Endosulfan	0.35 % of application	Slope = 8 %, cont. potatoes, app = surface, soil = gl, 12 months, ME[14]
Endrin	0.1 % of application	Sugarcane, app = surface, soil = sicl, 9 months, LA[19]
Endrin	0.4 % of application	Sugarcane, app = surface, soil = sicl, 2 months, LA[19]
Endrin	0.6 % of application	Slope = 8 %, rot. potatoes, app = surface, soil = gl, 24 months, ME[14]
Endrin	0.9 % of application	Slope = 8 %, cont. potatoes, app = surface, soil = gl, 24 months, ME[14]
Fenac	0.004 % of application	Appl = 3.36 kg/ha, sugarcane, surface runoff, rain = 9.2 cm, Miss[17]
Fenac	2.8 % of application	Appl = 3.36 kg/ha, sugarcane, surface runoff, rain = 15.5 cm, Miss[17]
Fenac	2.87 % of application	Slope = 0.2 %, sugar cane, application = surface, soil = sicl, 60 d[18]
Gs 14254	0.02 % of application	Slope = 14 %, alfafa, app = surface, soil = sicl, 5 months, PA[20]
Gs 14254	0.03 % of application	Slope = 14 %, alfafa, application = surface, soil = sicl, 5 months[20]
Linuron	0.02 % of application	Appl = 2.24 kg/ha, soybean, surface runoff, rain = 9.2 cm, Miss, 1972[17]
Linuron	0.029 % of application	Slope = 0.2 %, cotton, app = surface ponded, soil = sicl, 60 d, LA[18]
Linuron	0.04 % of application	Appl = 2.24 kg/ha, soybean, surface runoff, rain = 9.7 cm, Miss, 1971[17]
Linuron	0.29 % of application	Appl = 2.24 kg/ha, soybean, surface runoff, rain = 15.5 cm, Miss[17]
MBC	0.63 %	Appl = 20 lb/acre, rain = 3.5 in, 7 d, Ken[1]
Methoxychlor	0.4 % of application	Grass, application = simulated rainfall, soil = sil, 14 months, Ohio[2]
Picloram	0.02 % of application	Slope = 5 %, grass, application = foliar, soil = c, 24 months, TX[1]
Picloram	3.9 % of application	Slope = 3-8 %, fallow sod, app = simulated rainfall, soil = cl, 4[3]
Prometryn	0.5 % of application	Slope = 1 %, cotton, app = surface, soil = sl, 5 months, PA[20]
Propachlor	2.3 % of application	Slope = 10-15 %, application = surface, corn, soil = sil, 8 d, IA[9]
Terbacil	1.49 g/ha	In submerged drain, deep til. & lime, appl = 4 lbs/ha, 14 d, 75[4]
Terbacil	124.4 g/ha	Deep tillage, in submerged drain, appl = 4 lbs/ha, 14 d, citrus[4]

Terbacil	144.6 g/ha	Deep tillage, in submerged drain, appl = 4 lbs/ha, 14 d, citrus[4]
Terbacil	159.5 g/ha	Deep tillage, in open drain, appl = 4 lbs/ha, 14 d, citrus[4]
Terbacil	20.81 g/ha	In submerged drain, deep til. & lime, appl = 4 lbs/ha, 14 d, 74[4]
Terbacil	221.7 g/ha	Deep til. & lime, in open drain, appl = 4 lbs/ha, 14 d, citrus[4]
Terbacil	24.19 g/ha	In submerged drain, deep tillage, appl = 4 lbs/ha, 14 d, 74[4]
Terbacil	26.88 g/ha	In open drain, deep tillage, appl = 4 lbs/ha, 14 d, 74[4]
Terbacil	35.3 g/ha	Shallow tillage, in submerged drain, appl = 4 lbs/ha, 14 d, citrus[4]
Terbacil	37.08 g/ha	In open drain, deep til. & lime, appl = 4 lbs/ha, 14 d, 74[4]
Terbacil	47.3 g/ha	Deep tillage, in submerged drain, appl = 4 lbs/ha, 14 d, citrus[4]
Terbacil	483.9 g/ha	Shallow tillage, in submerged drain, appl = 4 lbs/ha, 14 d, citrus[4]
Terbacil	494.5 g/ha	Shallow tillage, in open drain, appl = 4 lbs/ha, 14 d, citrus[4]
Terbacil	5.91 g/ha	In submerged drain, shallow tillage, appl = 4 lbs/ha, 14 d, 75[4]
Terbacil	7.94 g/ha	In submerged drain, deep til. & lime, appl = 4 lbs/ha, 14 d, 75[4]
Terbacil	8.9 g/ha	Deep til. & lime, in submerged drain, appl = 4 lbs/ha, 14 d, citrus[4]
Terbacil	80.94 g/ha	In submerged drain, shallow tillage, appl = 4 lbs/ha, 14 d, 74[4]
Terbacil	82.72 g/ha	In open drain, shallow tillage, appl = 4 lbs/ha, 14 d, 74[4]
Toxaphene	0.36 % of application	Slope = 3 %, cotton, app = foliar, soil = ls & sl, 6 months, NC[15]
Trifluralin	0 % of application	Appl = 1.40 kg/ha, soybean, surface runoff, rain = 26.5 cm, Miss[17]
Trifluralin	0.004 % of application	Appl = 1.40 kg/ha, soybean, surface runoff, rain = 13.8 cm, Miss[17]
Trifluralin	0.04 % of application	Slope = 0.2 %, cotton and Soybean, app = incorporated, soil = sicl[18]
Trifluralin	0.4 % of application	Appl = 1.40 kg/ha, soybean, surface runoff, rain = 27.5 cm, Miss[17]

Source: 1.Bovey,R.W., E. Burnett, C. Richardson, M.G. Merkle, J.R. Baur and W.G. Knisel. 1974. J. Environ. Quaity, 3, pp 61-64.
2.Edwards, W.M. and B.L. Glass. 1971. Bull. Environ. Contam. Toxicol., 6, pp 81-84.
3.Trichell, D.W., H.L. Morton and M.G. Merkle. 1968. Weed Sci., 16, pp 447-449.
4.Mansell, R.S. et al. 1977. Fertilizer and Pesticide Movements from Citrus groves in Florida flatwoodSoils. EPA-600/2-77-177.
5.White, A.W., L.E. Asmussen, E.W. Hauser and J.W. Turnbull. 1976. J. Environ. Qual. 5(4), pp 487-490.
6.Barnett, A.P., E.W. Hauser, A.W. White and J.H. Holliday. 1967. Weeds, 15, pp 133-137.
7.Haan, C.T.. Am. Soc. Agr. Eng. Trans., 14, pp 445-449
8.Bailey, G.W., A.D. Barnett, W.R. Payne, Jr and C.N. Smith. 1974. EPA-6602-74-017, U.S. Environ. Protection Agency. Washington D.C..
9.Ritter, W.F., H.P. Johnson, W.G. Lovely and M. Molnad 1974. Environ. Sci. Tech nol., 8 pp 38-42.
10.Hall, J.K. 1972. J. Environ. Quality, 1, pp 172-176.
11.Hall, J.K. 1974. J. Environ. Qual., 3(2), pp 174-180.
12.White, A.W., A.P. Barnett, B.C. Wright and J.H. Holliday. 1967. Environ. Sci Technol., 1, pp 740-744
13.Rhodes, R.C. and J.D. Long. 1974. Bull. Environ. Contam. Toxicol. 12(4), pp 385-393.
14.Epstein, E. and W.J. Grant. 1968. Soil Sci Soc. Am. Proc., 32 pp 423-426.
15.Bradley, J.R.Jr., T.J. Sheets and M.D. Jackson. 1972. J. Environ. Quality, 1, pp 102-105.
16.Caro, J.H. and A.W. Taylor. 1971. J. agric. Fd. Chem., 19, pp 379-384.
17.Willis, G.H., R.L. Rogers and L.M. Southwick. 1975. J. Environ. Qual. 4(3), pp 399-402.
18.Willis, G.H., R.L. Rogers and L.M. Southwick. 1976. J. Environ. Quality, pp 145-152.
19.Willis, G.H. and R.A. Hamilton. 1973. J. Environ.Quality 2, pp 463-466.
20.Hall, J.K. 1974. J. Environ. Quality, 3, pp 174-180.

2-156 Pesticides
Loss by volatilization

Species	Value	Condition
DDT	50 % in 5 months	Appl = 3.0 lb/acre in June, Ariz, irrigated soil[1]
DDT	67 % in 5 months	Appl = 32.0 lb/acre in June, Ariz, fallow soil[1]
DDT	80 % in 1 year	Total DDT, desert soil, appl = 1-2 lb/acre[2]
DDT	88 % in 5 months	Appl = 2.0 lb/acre in June, Ariz, fallow soil[1]
DDT	92 % in 1 year	p,p'-DDT, desert soil, appl. = 1-2 lb/acre[2]

Dieldrin	199 g/(ha*growing season)	Mean of growing season, appl = 5.6 kg/ha, 7.5 cm deep layer[3]
Dieldrin	282 g/(ha*day)	3 day after appl. with 5.6 kg/ha, farm rig, grass, observed value[4]
Dieldrin	325 g/(ha*day)	2 day after appl. with 5.6 kg/ha, farm rig, grass, observed value[4]
Dieldrin	4.0 g/(ha*day)	Incorporated, max. value, application = 5.6 kg/ha in a 7.5 cm deep[3]
Dieldrin	40.0 g/(ha*day)	9 day after appl. with 5.6 kg/ha, farm rig, grass, observed value[4]
Dieldrin	53.7 g/(ha*day)	6 day after appl. with 5.6 kg/ha, farm rig, grass, observed value[4]
Dieldrin	6.2 g/(ha*day)	23 day after appl. with 5.6 kg/ha, farm rig, grass, observed value[4]
Dieldrin	654 g/(ha*day)	1 day after appl. with 5.6 kg/ha, farm rig, grass, observed value[4]
Dieldrin	9.2 g/(ha*day)	14 day after appl. with 5.6 kg/ha, farm rig, grass, observed value[4]
Heptachlor	132 g/(ha*day)	3 day after appl. with 5.6 kg/ha, farm rig, grass, observed value[4]
Heptachlor	24.1 g/(ha*day)	9 day after appl. with 5.6 kg/ha, farm rig, grass, observed value[4]
Heptachlor	2554 g/(ha*day)	1 day after appl. with 5.6 kg/ha, farm rig, grass, observed value[4]
Heptachlor	33.5 g/(ha*day)	6 day after appl. with 5.6 kg/ha, farm rig, grass, observed value[4]
Heptachlor	335 g/(ha*day)	2 day after appl. with 5.6 kg/ha, farm rig, grass, observed value[4]
Heptachlor	383 g/(ha*growing season)	Incorporated, mean of growing season, appl = 5.6 kg/ha, 7.5 cm deep[3]
Heptachlor	5.0 g/(ha*day)	Incorporated, max. value, application = 5.6 kg/ha in a 7.5 cm deep[3]
Heptachlor	6.9 g/(ha*day)	14 day after appl. with 5.6 kg/ha, farm rig, grass, observed value[4]
Heptachlor	7.4 g/(ha*day)	23 day after appl. with 5.6 kg/ha, farm rig, grass, observed value[4]
Trifluralin	0.0517 kg/(ha*day)	Incorporated, 14 kg/ha[5]
Trifluralin	4.0 kg/(ha*day)	Surface application, 10 kg/ha[5]

Source: 1.Ware, G.W., W.P. Cahill and B.J. Esresen. 1975. Bull. Environ. Contam. Toxicol., 14(1), pp 88-97.
2.Ware, G.W., B. Estesen, W.C. Kronland and W.P. Cahill. 1977. Bull. Environ. Contam. Toxicol., 17(3),pp 317-322.
3.Taylor, A.W., D.E. Glottelty, B.L. Glass, H.P. Freeman and W.M. Edwards. 1974. J. Agric. Food. Chem.24(3), pp 625-631.
4.Tillander, M., J.K. Miettinen and I. Koivisto. 1972. in: Ruivo, M. (ed.); Marine Pollution and Sea Life. pp 303-305. Fishing Trading news (books) Ltd. London.
5.Spencer, W.F. and M.M. Cliath. 1974. J. Agric. Food. Chem. 22(6), pp 987-991.

2-157 Pesticides
Loss during application

Species	Value	Condition
Dyfonate	12.6 % of application	Appl = 3.54 lb/acre, spray from tractor, calm, 298 K, by pan analys,[1]
Dyfonate	16.7 % of application	Appl = 3.54 lb/acre, spray from tractor, calm, 298 K, by soil analys,[1]
Dyfonate	17.6 % of application	Appl = 3.54 lb/acre, spray from tractor, calm, 298 K, by pan analys,[1]
Dyfonate	24.1 % of application	Appl = 3.54 lb/acre, spray from tractor, calm, 298 K, by soil analys,[1]
Zinophos	37.3 % of application	Appl = 3.54 lb/acre, spray from tractor, calm, 298 K, by soil analys[1]
Zinophos	39.5 % of application	Appl = 3.54 lb/acre, spray from tractor, calm, 298 K, by pan analys[1]

Source: 1.Kiigemagi, U. and L.C. Terrierre. 1972. Bull. Environ. Contam. Toxicol. 7(6), pp 348-352.

2-158 Pesticides
Loss from soil

Species	Value	Condition
DDT	17 % in 5 years	Medford soil, clay adobe, cultivated, 284 K, precip = 30 inches[1]
DDT	18 % in 5 years	Hood river soil, sandy loam, no cultivation, 284 K, precip = 30 inch[1]

Source: 1.Kiigemagi, U. and L.C. Terrierre. 1972. Bull. Environ. Contam. Toxicol. 7(6), pp 348-352.

2-159 Pesticides
Quantum yield by photolysis

Species	Value	Condition
2,4-D butoxyethyl ester	0.0080	Photolysis in N-hexane, 313 nm, 301 K, acetone-sensitized[1]
2,4-D butoxyethyl ester	0.052	Photolysis in water, 313 nm, 301 K, pH = 6.6[1]
2,4-D butoxyethyl ester	0.056	Photolysis in water, 313 nm, 301 K, pH = 5.3[1]
2,4-D butoxyethyl ester	0.056	Photolysis in water, 313 nm, 301 K, pH = 7.8[1]
2,4-D butoxyethyl ester	0.17	Photolysis in N-hexadecane, 313 nm, 301 K[1]
2,4-D butoxyethyl ester	0.17	Photolysis in N-hexane, 313 nm, 301 K[1]
2,4-D methyl ester	0.031	Photolysis in water, 313 nm, 301 K, pH = 6.6[1]
2,4-D methyl ester	0.13	Photolysis in N-hexadecane, 313 nm, 301 K[1]
Carbaryl	0.0052	In water, 313 nm, 298 K, air-saturated, pH = 5.2[1]
Carbaryl	0.0055	In water, 313 nm, 298 K, air-saturated, pH = 6.9[1]
Carbaryl	0.0056	In water, 313 nm, 298 K, air-saturated, pH = 5.5[1]
Carbaryl	0.0058	In water, 313 nm, 298 K, air-saturated, pH = 6.1[1]
Carbaryl	0.016	In water, 313 nm, 298 K, degassed, pH = 5.5[1]
Methoxychlor	0.12	In N-hexane, methoxychlor conc. = 0.003m[1]
Methoxychlor	0.3	In water, methoxychlor conc. = 30ppb[1]
Methoxychlor	0.32	In 1.1 m water in acetonitrile, methoxychlor conc. = 0.003m[1]

Source: 1.Wolfe, N.L., R.G. Zepp, G.L. Banghman, R.C. Fincher and J.A. Gordon. 1976. Chemical and Photochemical Transformations of selected Pesticides in Aquatic Systems, EPA-600/3-76-067.

2-160 Pesticides
Specific sunlight absorption rate

Species	Value	Condition
2,4-D ester	194E6 1/sec	Midday sun, midsummer, latitude = 40 N[1]
Atrazine	6.6E6 1/sec	Midday sun, midsummer, latitude = 40 N[1]
Carbaryl	623E6 1/sec	Midday sun, midsummer, latitude = 40 N[1]
Diazinon	6.2E6 1/sec	Midday sun, midsummer, latitude = 40 N[1]
Malathion	1.2E6 1/sec	Midday sun, midsummer, latitude = 40 N[1]
Methoxychlor	0.87E6 1/sec	Midday sun, midsummer, latitude = 40 N[1]
Parathion	6930E6 1/sec	Midday sun, midsummer, latitude = 40 N[1]

Source: 1.Wolfe, N.L., R.G. Zepp, G.L. Banghman, R.C. Fincher and J.A. Gordon. 1976. Chemical and Photochemical Transformations of selected Pesticides in Aquatic Systems, EPA-600/3-76-067.

2-161 Pesticides
Volatility

Species	Value	Condition
Aldrin	2.3E-5 mm Hg	293-298 K, vapor pressure[1]
Camphechlor	1.0E-6 mm Hg	293-298 K, vapor pressure[1]
Chlordane	1.0E-5 mm Hg	293-298 K, vapor pressure[1]
DDT	1.9E-7 mm Hg	293-298 K, vapor pressure[1]
Dieldrin	1.8E-7 mm Hg	293-298 K, vapor pressure[1]
Endrin	2.0E-7 mm Hg	293-298 K, vapor pressure[1]
Heptachlor	3.0E-4 mm Hg	293-298 K, vapor pressure[1]
Lindane	9.4E-6 mm Hg	293-298 K, vapor pressure[1]

Source: 1.Edwards, C.A. 1975. Persistent Pesticides in the Environment, 2nd edition, CRC-Press.

2-162 Sediment
Accumulation rate

Species	Value	Condition
Freshwater sediment	0.05 cm/year	Lake Ontario, western part, from Cs-137 profile[1]
Freshwater sediment	0.05 cm/year	Lake Ontario, western part, from Sr-90 profile[1]
Freshwater sediment	0.3 cm/year	Lake Erie, western basin, from Cs-137 profile[1]
Freshwater sediment	0.3 cm/year	Lake Erie, western basin, from Sr-90 profile[1]
Freshwater sediment	0.6 cm/year	Lake Erie, central basin, from Cs-137 profile[1]
Freshwater sediment	0.6 cm/year	Lake Erie, central basin, from Sr-90 profile[1]

Source: 1.Lerman, A. and T.A. Lietzke. 1975. Limnol. Oceanogr. 20(4), pp 497-510.

2-163 Sediment
Concentration gradient

Species	Value	Condition
Sulfate	0.0 umol cm-3 cm1	difference between Porewater and overlying water, Lille Hovvatn[1]
Sulfate	1.62 10-2 umol cm-3 cm1	difference between Porewater and overlying water, Hovvatn[1]
Sulfate	2.37 10-2 umol cm-3 cm1	difference between Porewater and overlying water, Twitchell[1]
Sulfate	2.65 10-2 umol cm-3 cm1	difference between Porewater and overlying water, Lille Hovvatn[1]
Sulfate	2.87 10-2 umol cm-3 cm1	difference between Porewater and overlying water, Woods[1]
Sulfate	4.39 10-2 umol cm-3 cm1	difference between Porewater and overlying water, Big Moose[1]
Sulfate	8.25 10-3 umol cm-3 cm1	difference between Porewater and overlying water, Woods[1]
Sulfate	9.43 10-3 umol cm-3 cm1	difference between Porewater and overlying water, Hovvatn[1]

Source: 1.Rudd, J.W.M., Kelly, C.A., St.Louis, V., Hesslein, R.H. Limnol. Oceanogr., 31(6), 1267, 1986.

2-164 Sediment
Content

Species	Value	Condition
Organic carbon	1.2-0.4 % (wt./wt.)	depth 0.5-4.5 cm, Sagamore[1]
Organic carbon	14.4-11.5 % (wt./wt.)	depth 0.5-5.5 cm, Twitchell[1]
Organic carbon	15.7-7.6 % (wt./wt.)	depth 0.5-4.5 cm, Woods[1]
Organic carbon	18.1-11.4 % (wt./wt.)	depth 0.75-6.0 cm, Lille Hovvatn, water depth 2m[1]

Organic carbon	18.1-5.2 % (wt./wt.)	depth 0.75-6.0 cm, Lille Hovvatn, water depth 5m[1]
Organic carbon	19.4-13.6 % (wt./wt.)	depth 0.75-6.0 cm, Hovvatn, water depth 2m[1]
Organic carbon	2.2-0.5 % (wt./wt.)	depth 0.5-4.5 cm, Darts[1]
Organic carbon	26.8-19.4 % (wt./wt.)	depth 0.75-6.0 cm, Hovvatn, water depth 5m[1]
Organic carbon	7.3-0.7 % (wt./wt.)	depth 0.5-4.5 cm, Big Moose[1]

Source: 1.Rudd, J.W.M., Kelly, C.A. and Furutani, A. Limnol. Oceanogr., 31(6), 1281, 1986.

2-165 Sediment
Diffusion coefficient

Species	Value	Condition
Sulfate	4.3 10-6 cm2 s-1	functional diff.coeff., Darts[1]
Sulfate	4.61 10-6 cm2 s-1	functional diff.coeff., Sagamore[1]
Sulfate	4.8 10-6 cm2 s-1	functional diff.coeff., Hovvatn[1]
Sulfate	5.11 10-6 cm2 s-1	functional diff.coeff., Hovvatn[1]
Sulfate	5.26 10-6 cm2 s-1	functional diff.coeff., Big Moose[1]
Sulfate	5.42 10-6 cm2 s-1	functional diff.coeff., Lille Hovvatn[1]
Sulfate	5.42 10-6 cm2 s-1	functional diff.coeff., Twitchell[1]
Sulfate	5.84 10-6 cm2 s-1	functional diff.coeff., Lille Hovvatn[1]
Sulfate	7.30 10-6 cm2 s-1	functional diff.coeff., Woods[1]

Source: 1.Rudd, J.W.M., Kelly, C.A., St.Louis, V., Hesslein, R.H. Limnol. Oceanogr., 31(6), 1267, 1986.

2-166 Sediment
Flux

Species	Value	Condition
Sulfate	-126 ueq m-2 d-1	Darts[1]
Sulfate	-222 ueq m-2 d-1	Twitchell[1]
Sulfate	-362 ueq m-2 d-1	Woods[1]
Sulfate	-400 ueq m-2 d-1	Big Moose[1]
Sulfate	-65.6 ueq m-2 d-1	Sagamore[1]
Sulfate	143 ueq m-2 d-1	Hovvatn[1]
Sulfate	248 ueq m-2 d-1	Lille Hovvatn[1]
Sulfate	78.2 ueq m-2 d-1	Hovvatn[1]

Source: 1.Rudd, J.W.M., Kelly, C.A., St.Louis, V., Hesslein, R.H. Limnol. Oceanogr., 31(6), 1267, 1986.

2-167 Sediment
Oxygen uptake

Species	Value	Condition
Estuarine mud	1.5 gram O2/(m**2*day)	298 K, rg = 1 - 2[1]
Fiber sludge	7 gram O2/(m**2*day)	298 K, cellulosic fiber, rg = 4 - 10[1]
Freshwater sediment	0.10 mg O2/(gram sediment*hour)	Chemical uptake = 10 %, org matter = 21.3 %, 277 K, Grane Lang Lake[2]
Freshwater sediment	0.74 mg O2/(gram sediment*hour)	Chemical uptake = 65 %, org matter = 72.0 %, 281 K, Store Grib Lake[2]
Freshwater sediment	0.75 mg O2/(gram sediment*hour)	Chemical uptake = 59 %, org matter = 39.6 %, 281 K, Lake Esrom[2]
Freshwater sediment	1.14 mg O2/(gram sediment*hour)	Chemical uptake = 31 %, org matter = 39.9 %, 275 K, Lake Esrom[2]

Freshwater sediment	1.34 mg O2/(gram sediment*hour)	Chemical uptake = 54 %, org matter = 45.0 %, 283 K, , Fred Castle[2]
Freshwater sediment	1.48 mg O2/(gram sediment*hour)	Chemical uptake = 59 %, org matter = 40.0 %, 277 K, Lake Esrom[2]
Marine sediment	0 ml O2/(m**2*hour)	O2 = 0 ml/l, depth = 74 m, 282 K, total uptake, Puget Sound[3]
Marine sediment	0.5 ml O2/(m**2*hour)	Depth = 1850 m, chemical uptake = 0.0, 278 K, New England slope[4]
Marine sediment	0.7 ml O2/(m**2*hour)	Depth = 1325 m, chemical uptake = 0.4, 288 K, E Trop Pacific[4]
Marine sediment	0.8 ml O2/(m**2*hour)	Depth = 7005 m, chemical uptake = 0.2, 275 K, NE Pacific[4]
Marine sediment	0.9 ml O2/(m**2*hour)	Depth = 4145 m, chemical uptake = 0.9, 275 K, NE Pacific[4]
Marine sediment	1.2 ml O2/(m**2*hour)	Depth = 3205 m, chemical uptake = 0.8, 275 K, NE Pacific[4]
Marine sediment	1.2 ml O2/(m**2*hour)	Depth = 5720 m, chemical uptake = 1.1, 275 K, NE Pacific[4]
Marine sediment	1.5 ml O2/(m**2*hour)	Depth = 2735 m, chemical uptake = 1.4, 275 K, NE Pacific[4]
Marine sediment	1.5 ml O2/(m**2*hour)	Depth = 5830 m, chemical uptake = 1.5, 275 K, NE Pacific[4]
Marine sediment	1.7 ml O2/(m**2*hour)	Depth = 2750 m, chemical uptake not determinated, 280 K, NE Pacific[4]
Marine sediment	1.7 ml O2/(m**2*hour)	Depth = 3670 m, chemical uptake = 1.7, 275 K, NE Pacific[4]
Marine sediment	1.7 ml O2/(m**2*hour)	Depth = 4720 m, chemical uptake = 1.1, 275 K, NE Pacific[4]
Marine sediment	1.8 ml O2/(m**2*hour)	Depth = 2900 m, chemical uptake = 1.7, 288 K, E Trop Pacific[4]
Marine sediment	1.9 ml O2/(m**2*hour)	Depth = 4580 m, chemical uptake = 1.9, 275 K, NE Pacific[4]
Marine sediment	11.1 ml O2/(m**2*hour)	O2 = 2.81 ml/l, depth = 74 m, 282 K, chemical uptake, Puget Sound[3]
Marine sediment	12.3 ml O2/(m**2*hour)	O2 = 0.56 ml/l, depth = 74 m, 282 K, respiration, Puget Sound[3]
Marine sediment	15.0 ml O2/(m**2*hour)	O2 = 0.56 ml/l, depth = 74 m, 282 K, total uptake, Puget Sound[3]
Marine sediment	15.1 ml O2/(m**2*hour)	O2 = 1.05 ml/l, depth = 74 m, 282 K, respiration, Puget Sound[3]
Marine sediment	16.6 ml O2/(m**2*hour)	O2 = 2 ml/l, depth = 74 m, 282 K, respiration, Puget Sound[3]
Marine sediment	17.1 ml O2/(m**2*hour)	O2 = 2.53 ml/l, depth = 74 m, 282 K, respiration, Puget Sound[3]
Marine sediment	17.4 ml O2/(m**2*hour)	O2 = 3.44 ml/l, depth = 74 m, 282 K, chemical uptake, Puget Sound[3]
Marine sediment	18.9 ml O2/(m**2*hour)	O2 = 1.05 ml/l, depth = 74 m, 282 K, total uptake, Puget Sound[3]
Marine sediment	2.0 ml O2/(m**2*hour)	Depth = 3145 m, chemical uptake = 2.0, 275 K, NE Pacific[4]
Marine sediment	2.3 ml O2/(m**2*hour)	Depth = 2750 m, chemical uptake = 1.8, 281 K, NE Pacific[4]
Marine sediment	2.3 ml O2/(m**2*hour)	Depth = 2790 m, chemical uptake = 2.2, 281 K, NE Pacific[4]
Marine sediment	2.3 ml O2/(m**2*hour)	O2 = 0.07 ml/l, depth = 74 m, 282 K, total uptake, Puget Sound[3]
Marine sediment	2.3 ml O2/(m**2*hour)	O2 = 0.39 ml/l, depth = 74 m, 282 K, chemical uptake, Puget Sound[3]
Marine sediment	2.4 ml O2/(m**2*hour)	Depth = 1230 m, chemical uptake = 1.04, 277 K, San Diego trough[4]
Marine sediment	2.4 ml O2/(m**2*hour)	Depth = 2860 m, chemical uptake = 2.4, 281 K, NE Pacific[4]
Marine sediment	2.8 ml O2/(m**2*hour)	Depth = 5160 m, chemical uptake = 2.3, 275 K, NE Pacific[4]
Marine sediment	23.7 ml O2/(m**2*hour)	O2 = 2 ml/l, depth = 74 m, 282 K, total uptake, Puget Sound[3]
Marine sediment	26.6 ml O2/(m**2*hour)	O2 = 2.53 ml/l, depth = 74 m, 282 K, total uptake, Puget Sound[3]
Marine sediment	3.0 ml O2/(m**2*hour)	Depth = 2755 m, chemical uptake = 3.0, 281 K, NE Pacific[4]
Marine sediment	3.0 ml O2/(m**2*hour)	Depth = 5085 m, chemical uptake = 3.0, 275 K, NE Pacific[4]
Marine sediment	3.3 ml O2/(m**2*hour)	Depth = 6975 m, chemical uptake = 2.3, 275 K, NE Pacific[4]
Marine sediment	4.2 ml O2/(m**2*hour)	Depth = 2820 m, chemical uptake = 3.1, 288 K, E Tropical Pacific[4]
Marine sediment	4.3 ml O2/(m**2*hour)	O2 = 1.23 ml/l, depth = 74 m, 282 K, chemical uptake, Puget Sound[3]

Marine sediment	4.7 ml O2/(m**2*hour)	O2 = 0.14 ml/l, depth = 74 m, 282 K, total uptake, Puget Sound[3]
Marine sediment	5.2 ml O2/(m**2*hour)	Depth = 2510 m, chemical uptake = 3.7, 275 K, NE Pacific[4]
Marine sediment	5.6 ml O2/(m**2*hour)	O2 = 1.65 ml/l, depth = 74 m, 282 K, chemical uptake, Puget Sound[3]
Marine sediment	6.7 ml O2/(m**2*hour)	Depth = 180 m, O2 = 3.5 ml/l, chemical uptake = 50-59%, 282k, Puget[3]
Marine sediment	6.9 ml O2/(m**2*hour)	Depth = 23 m, O2 = 4.4-4.7 ml/l, 280 K, Puget Sound[3]
Marine sediment	7.1 ml O2/(m**2*hour)	O2 = 2.04 ml/l, depth = 74 m, 282 K, chemical uptake, Puget Sound[3]
Marine sediment	8.6 ml O2/(m**2*hour)	Depth = 23 m, O2 = 5.2-5.8 ml/l, 280 K, Puget Sound[3]
Marine sediment	9.4 ml O2/(m**2*hour)	O2 = 0.42 ml/l, depth = 74 m, 282 K, respiration, Puget Sound[3]
Mineral soils	0.07 gram O2/(m**2*day)	298 K, rg = 0.05 - 0.1[1]
Municipal sewage sludge	1.5 gram O2/(m**2*day)	298 K, aged downstream outfall, rg = 1 - 2[1]
Municipal sewage sludge	4 gram O2/(m**2*day)	298 K, outfall vicinity, rg = 2 - 10[1]
Pacific sediment	1.7 ml O2/(m**2*hour)	Depth = 2750 m, O2 = 2.4 ml/l, 280 K, NE Pacific[3]
Pacific sediment	3.0 ml O2/(m**2*hour)	Depth = 105 m, O2 = 2.2 ml/l, chemical uptake = 87-112%, 280 K, NE[3]
Pacific sediment	3.4 ml O2/(m**2*hour)	Depth = 146 m, O2 = 2.3-2.4 ml/l, chemical uptake = 66-73%, 280 K, NE[3]
Pacific sediment	6.7 ml O2/(m**2*hour)	Depth = 146 m, O2 = 1.7-2.2 ml/l, chemical uptake = 31-49%, 280 K, NE[3]
Sandy bottom	0.5 gram O2/(m**2*day)	298 K, rg = 0.2 - 1.0[1]
Sphaerotilus	7 gram O2/(m**2*day)	298 K, 10 g dry weight/m2[1]
Tropical Pacific sediment	0.7 ml O2/(m**2*hour)	Depth = 1325 m, O2 = 1.1 ml/l, chemical uptake = 42-62%, 288 K, E[3]
Tropical Pacific sediment	1.8 ml O2/(m**2*hour)	Depth = 2900 m, O2 = 1.5-1.9 ml/l, chemical uptake = 85-106%, 288 K,[3]
Tropical Pacific sediment	3.8 ml O2/(m**2*hour)	Depth = 275 m, O2 = 0.8 ml/l, chemical uptake = 71-77%, 288 K, E[3]
Tropical Pacific sediment	4.2 ml O2/(m**2*hour)	Depth = 2820 m, O2 = 1.7 ml/l, chemical uptake = 66-81%, 288 K, E[3]
Tropical Pacific	9.3 ml O2/(m**2*hour)	Depth = 85 m, O2 = 1.0 ml/l, chemical uptake = 48-101%, 288 K, sed,E[3]

Source: 1.Thomann, R.V. 1972. System Analysis and Water Quality Measurement. Environ. Research and Application. Inc. New York.
2.Hargrave, B.T. 1972. Limnol. Oceanogr., 17(4), pp 583-596.
3.Pamatmat, M.M. 1971. Limnol. Oceanogr. 16(4), pp 536-550.
4.Smith, K.L. Jr. 1974. Limnol. Oceanogr. 19(6), pp 939-944.

2-168 Sediment
Porewater concentration

Species	Value	Condition
Ammonia	0.23-50.7 umol l-1	Chubb, Sep. 82, depth 9.3 cm[1]
Ammonia	0.26-14.4 umol l-1	Plastic, Sep. 82, depth 14.8 cm[1]
Ammonia	1.07-25.9 umol l-1	Red chalk, Sep. 82, depth 5.7 cm[1]
Ammonia	10-28.6 umol l-1	Sagamore, June 84, depth 11.49 cm[1]
Ammonia	11.4-40.7 umol l-1	Twitchell, June 84, depth 9.8 cm[1]
Ammonia	2.1-60.7 umol l-1	Darts, July 83, depth 2.1 cm[1]
Ammonia	2.1-9.3 umol l-1	Little Hovvatn, Sep. 84, depth 11.9 cm[1]
Ammonia	2.86-28.6 umol l-1	Hovvatn, Sep. 84, depth 11.3 cm[1]
Ammonia	3.6-40.7 umol l-1	Darts, June 84, depth 10.3 cm[1]
Ammonia	40-130 umol l-1	Big Moose, July 83, depth 8.8 cm[1]
Ammonia	5.7-27.1 umol l-1	Hovvatn, Sep. 84, depth 10.8 cm[1]
Ammonia	5.7-8.57 umol l-1	Little Hovvatn, Sep. 84, depth 11.3 cm[1]
Nitrate	0-0.18 umol l-1	Big Moose, July 83, depth 8.8 cm[1]
Nitrate	0-1.0 umol l-1	Woods, June 84, depth 5.5 cm[1]
Nitrate	0.05-6.5 umol l-1	Darts, July 83, depth 13.0 cm[1]
Nitrate	0.14-0.36 umol l-1	Sagamore, June 84, depth 11.4 cm[1]
Nitrate	0.21-0.5 umol l-1	Darts, June 84, depth 10.3 cm[1]
Nitrate	0.3-0.5 umol l-1	Twitchell, June 84, depth 9.8 cm[1]
Nitrate	2.14-9.57 umol l-1	Lille Hovvatn, Sep. 84, depth 11.3 cm[1]
Nitrate	2.29-10.4 umol l-1	Lille Hovvatn, Sep. 84, depth 11.85 cm[1]
Nitrate	2.71-11.7 umol l-1	Hovvatn, Sep. 84, depth 10.8 cm[1]
Nitrate	2.79-11.9 umol l-1	Hovvatn, Sep. 84, depth 11.3 cm[1]
Sulfate	0.9-33.7 umol l-1	Twitchell, June 84, depth 10.8 cm[1]
Sulfate	10.1-67.0 umol l-1	Hovvatn, Sep. 84, depth 10.8 cm[1]
Sulfate	10.8-88.0 umol l-1	Red chalk, Sep. 82, depth 13.0 cm[1]
Sulfate	14.9-42.7 umol l-1	Hovvatn, Sep. 84, depth 10.8 cm[1]
Sulfate	17.7-83.3 umol l-1	Plastic, Sep. 82, depth 9.8 cm[1]
Sulfate	19.2-43.7 umol l-1	Big Moose, July 83, depth 10.1 cm[1]
Sulfate	2.21-62.9 umol l-1	Woods, June 84, depth 5.5 cm[1]
Sulfate	2.92-39.1 umol l-1	Darts, July 83, depth 12.9 cm[1]
Sulfate	25.4-38.0 umol l-1	Lille Hovvatn, Sep. 84, depth 13.9 cm[1]
Sulfate	27.9-64.0 umol l-1	Sagamore, June 84, depth 10.95 cm[1]
Sulfate	38.0-119.6 umol l-1	Lille Hovvatn, Sep. 84, depth 11.1 cm[1]
Sulfate	4.4-31.4 umol l-1	Darts, June 84, depth 13.6 cm[1]
Sulfate	6.25-86.4 umol l-1	Chubb, Sep. 82, depth 6.7 cm[1]

Source: 1.Rudd, J.W.M., Kelly, C.A., St.Louis, V., Hesslein, R.H. Limnol. Oceanogr., 31(6), 1267, 1986.

2-169 Sediment
Porewater concentrations

Species	Value	Condition
Fe (ferrous)	0.72-116 umol l-1	Hovvatn, Sep. 84, depth 10.8 cm[1]
Fe (ferrous)	0.72-180 umol l-1	Hovvatn, Sep. 84, depth 10.8 cm[1]
Fe (ferrous)	1.08-16.7 umol l-1	Big Moose, July 83, depth 10.1 cm[1]
Fe (ferrous)	10.6-225 umol l-1	Darts, June 84, depth 13.6 cm[1]
Fe (ferrous)	12.4-353 umol l-1	Lille Hovvatn, Sep. 84, depth 13.9 cm[1]
Fe (ferrous)	2.0-160 umol l-1	Darts, July 83, depth 13.0 cm[1]
Fe (ferrous)	3.04-469 umol l-1	Lille Hovvatn, Sep. 84, depth 11.1 cm[1]
Fe (ferrous)	3.6-22.7 umol l-1	Sagamore, June 84, depth 10.95 cm[1]
Fe (ferrous)	59-166 umol l-1	Twitchell, June 84, depth 10.8 cm[1]

Source: 1.Rudd, J.W.M., Kelly, C.A., St.Louis, V., Hesslein, R.H. Limnol. Oceanogr., 31(6), 1267, 1986.

2-170 Sediment
Porosity

Species	Value	Condition
Freshwater Sediment	0.645-0.55	at zero-depth calculated from regression, Sagamore[1]
Freshwater Sediment	0.65-0.53	at zero-depth calculated from regression, Darts[1]
Freshwater Sediment	0.83-0.42	at zero-depth calculated from regression, Big Moose[1]
Freshwater Sediment	0.95-0.77	at zero-depth calculated from regression, Woods[1]
Freshwater Sediment	0.97-0.95	at zero-depth calculated from regression, Twitchell[1]
Freshwater Sediment	0.98-0.95	at zero-depth calculated from regression, Lille Hovvatn[1]
Freshwater Sediment	0.98-0.96	at zero-depth calculated from regression, Hovvatn[1]

Source: 1.Rudd, J.W.M., Kelly, C.A., St.Louis, V., Hesslein, R.H. Limnol. Oceanogr., 31(6), 1267, 1986.

2-171 Sediment
Release rate

Species	Value	Condition
Sediment	0.061 g BOD/(m**2*hour)	Aerobic rate[1]
Sediment	0.1083 g NH3-N/(m**2*hour)	Aerobic rate[1]
Sediment	0.125 g PO4-P/(m**2*hour)	Aerobic rate[1]
Sediment	0.154 g COD/(m**2*hour)	Aerobic rate[1]
Sediment	0.203 g BOD/(m**2*hour)	Anaerobic rate[1]
Sediment	0.250 g PO4-P/(m**2*hour)	Anaerobic rate[1]
Sediment	0.3125 g NH3-N/(m**2*hour)	Anaerobic rate[1]
Sediment	0.495 g COD/(m**2*hour)	Anaerobic rate[1]

Source: 1.Lombardo, P.S. 1972. Hydrocomp, Inc. Mathematical Model of Water Quality in Rivers and Impoundments

2-172 Sediment
Sedimentation rate

Species	Value	Condition
Calcium	177.1 g dry wt. m-2 y-1	Lake Constance, 1981[1]
Calcium	203.5 g dry wt. m-2 y-1	Lake Constance, 1981[1]
Calcium	78.2 g dry wt. m-2 y-1	Lake Constance, 1980[1]
Particulate organic carbon	113 g m-2 y-1	in Sediment traps[2]
Particulate organic matter	128.5 g dry wt. m-2 y-1	Lake Constance, 1981[1]
Particulate organic matter	143.4 g dry wt. m-2 y-1	Lake Constance, 1982[1]
Particulate organic matter	65.1 g dry wt. m-2 y-1	Lake Constance, 1980[1]
Particulate phosphorus	2.6 g m-2 y-1	in Sediment traps[2]
Total matter	1.137 g dry wt. m-2 y-1	Lake Constance, 1981[1]
Total matter	887 g dry wt. m-2 y-1	in Sediment traps[2]
Total matter	937 g dry wt. m-2 y-1	Lake Constance, 1982[1]

Source: 1.Stabel, H.-H. Limnol. Oceanogr., 31, 1081, 1986.
2.Bloesch, J. and Uehlinger, U. Limnol. Oceanogr., 31, 1094, 1986.

2-173 Sediment
Sulfate reduction

Species	Value	Condition
Marine sediment	0.165E-9 m S/((cm**3)*day)	Bay of Kiel, Germany, 10-30 cm, 278-283 K, rg = 0.08 - 0.29[1]
Marine sediment	0E-9 m S/((cm**3)*day)	Black Sea, below 10 cm at 2000 m, 282k[2]
Marine sediment	1.4E-9 m S/((cm**3)*day)	St. Barbara Basin, Calif, upper 3 m[3]
Marine sediment	14.3E-9 M S/((cm**3)*day)	Black Sea, surface sediment at 2000 m, 282 K, rg = 0.8 - 38[2]
Marine sediment	4.9E-9 m S/((cm**3)*day)	St. Barbara Basin, Calif, upper 10 cm[4]
Marine sediment	49E-9 M S/((cm**3)*day)	Black Sea, surface sediment at 200 m, 282 K, rg = 29 - 68[2]
Marine sediment	736E-9 m S/((cm**3)*day)	Aarhus Bay, Denm, surface sediment, rg = 439 - 1013[5]
Marine sediment	8.2E-9 m S/((cm**3)*day)	Bay of Kiel, Germany, surface sediment, 278-283 K[1]
Marine sediment	95E-9 m S/((cm**3)*day)	Model system, littoral surface sediment, Limfjord, Denm, 295K[5]
Marine sediment	988E-9 m S/((cm**3)*day)	Barents Sea, littoral surface sediment, rg = 532 - 1444[6]

Source: 1.Hartmann, M. and H. Nielsen. 1969. Geol. Rdsch. 58, pp 621-655.
2.Sorokin, Y.I. 1962. Mikrobiologiya 31, pp 402-410.
3.Berner, R.A. 1964. Geochim. cosmochim. Acta, 28, pp 1497-1503.
4.Kaplan, I.R., K.O. Emery and S.C. Rittenberg. 1963. Geochim. cosmochim. Acta 27, pp 297-331.
5.Joergensen, B.B. and T. Fenchel. 1974. Mar biol. 24, pp 189-201.
6.Ivanov, M.V. 1968. Microbiological processes in the Formation of Sulfur Deposits. 298 pp. Israel Program for Scientific Translations. Jerusalem.

Chapter 3

EQUATIONS FOR BIOLOGICAL PROCESSES

Algae
Body Weight/Size
Rates and Concentrations
Zooplankton

3-001 Equation
Adsorption isotherm

Species	Value	Condition
Atrazine on ethanol - insoluble humic acid	Y = 18.50*X + 0.1411	Y = umol atra. adsor/(200 mg), X=equlibrium conc (umol/5 ml ethanol), 298 K[1]
Atrazine on ethanol - insoluble humic acid	Y = 14.90*X + 0.034	Y = umol Atra. sorb/(200 mg), X=equil. conc (umol/5 ml ethanol), 276 K[1]
2,4 - D on humic acid	Y = 1.034*X**1.034	278 K, Y = sorbed (umol/g), X = equil conc = ug/l[2]
2,4 - D on humic acid	Y = 0.869*X**0.789	298 K, Y = sorbed (umol/g), X = equil conc = ug/l[2]
2,4 - D on fulvic acid	Y = 0.815*X**0.76	278 K, Y = sorbed (umol/g), X = equil conc = ug/l[2]
2,4 - D on fulvic acid	Y = 0.716*X**0.83	298 K, Y = sorbed (umol/g), X = equil conc = ug/l[2]
Picloram on humic acid	Y = 1.132*X**0.877	278 K, Y = sorbed (umol/g), X = equil conc = ug/l[2]
Picloram on humic acid	Y = 1.046*X**0.912	298 K, Y = sorbed (umol/g), X = equil conc = ug/l[2]
2,4' - Cl2 - PCB on humic acid	Y = 3.600*X**0.860	297 K, Y = sorbed (umol/g), X = equil conc = ug/l[2]
2,4' - Cl2 - PCB on woodburn soil	0.845*X**1.180	297 K, Y = sorbed (umol/g), X = equil conc = ug/l[2]
2,4' - Cl2 - PCB on illite clay	Y = 1.301*X**0.84	297 K, Y = sorbed (umol/g), X = equil conc = ug/l[2]
2,5,2',5' - Cl4 - PCB on humic acid	Y = 4.639*X**3.820	297 K, Y = sorbed (umol/g), X = equil conc = ug/l[2]
2,5,2',5' - Cl4 - PCB on woodburn soil	Y = 1.689*X**1.0	297 K, Y = sorbed (umol/g), X = equil conc = ug/l[2]
2,5,2',5' - Cl4 - PCB on illite clay	Y = 1.326*X**0.92	297 K, Y = sorbed (umol/g), X = equil conc = ug/l[2]
2,4,5,2',4',5' - Cl6 - PCB on humic acid	Y = 9.53*X**3.78	297 K, Y = sorbed (umol/g), X = equil conc = ug/l[2]
2,4,5,2',4',5' - Cl6 - PCB on woodburn soil	Y = 2.505*X**1.25	297 K, Y = sorbed (umol/g), X = equil conc = ug/l[2]
2,4,5,2',4',5' - Cl6 - PCB on illite clay	Y = 2.301*X**1.260	297 K, Y = sorbed (umol/g), X = equil conc = ug/l[2]
Aniline on Ca - humates	Y = 1.732*X**0.820	298 K, Y = sorbed (ug/g), X = equil conc = ug/l[2]
Aniline on soignies soil - Ca	Y = 1.350*X**0.610	298 K, Y = sorbed (ug/g), X = equil conc = ug/l[2]
Dyfonate on Al(3+) - humic acid	Y = 3.330*X**0.890	273 K, Y = sorbed (ug/g), X = equil conc = ug/l[2]
Dyfonate on Al(3+) - humic acid	Y = 3.010*X**0.950	296 K, Y = sorbed (ug/g), X = equil conc = ug/l[2]
Dyfonate on Fe(3+) - humic acid	Y = 3.210*X**0.86	273 K, Y = sorbed (ug/g), X = equil conc = ug/l[2]
Dyfonate on Fe(3+) - humic acid	Y = 2.99*X**0.90	296 K, Y = sorbed (ug/g), X = equil conc = ug/l[2]
Dyfonate on Cu(2+) - humic acid	Y = 3.20*X**0.960	273 K, Y = sorbed (ug/g), X = equil conc = ug/l[2]
Dyfonate on Cu(2+) - humic acid	Y = 3.18*X**1.0	296 K, Y = sorbed (ug/g), X = equil conc = ug/l[2]
Dyfonate on Zn(2+) - humic acid	Y = 3.34*X**0.83	273 K, Y = sorbed (ug/g), X = equil conc = ug/l[2]
Dyfonate on Zn(2+) - humic acid	Y = 3.17*X**0.93	296 K, Y = sorbed (ug/g), X = equil conc = ug/l[2]

Dyfonate on Co(2+) - humic acid	Y = 3.38*X**0.89	273 K, Y = sorbed (ug/g), X = equil conc = ug/l[2]
Dyfonate on Co(2+) - humic acid	Y = 3.220*X**0.970	296 K, Y = sorbed (ug/g), X = equil conc = ug/l[2]
Dyfonate on Mn(2+) - humic acid	Y = 3.36*X**0.800	273 K, Y = sorbed (ug/g), X = equil conc = ug/l[2]
Dyfonate on Mn(2+) - humic acid	Y = 3.270*X**0.850	296 K, Y = sorbed (ug/g), X = equil conc = ug/l[2]
Dyfonate on Ni(2+) - humic acid	Y = 3.400*X**0.790	273 K, Y = sorbed (ug/g), X = equil conc = ug/l[2]
Dyfonate on Ni(2+) - humic acid	Y = 3.280*x**0.860	296 K, Y = sorbed (ug/g), X = equil conc = ug/l[2]
Dyfonate on Ca(2+) - humic acid	Y = 3.430*X**0.820	273 K, Y = sorbed (ug/g), X = equil conc = ug/l[2]
Dyfonate on Ca(2+) - humic acid	Y = 3.35*X**0.830	296 K, Y = sorbed (ug/g), X = equil conc = ug/l[2]
Dyfonate on Mg(2+) - humic acid	Y = 3.400*X**0.800	273 K, Y = sorbed (ug/g), X = equil conc = ug/l[2]
Dyfonate on Mg(2+) - humic acid	Y = 3.280*X**0.840	296 K, Y = sorbed (ug/g), X = equil conc = ug/l[2]
Dyfonate on H(+) - humic acid	Y = 3.45*X**0.85	273 K, Y = sorbed (ug/g), X = equil conc = ug/l[2]
Dyfonate on H(+) - humic acid	Y = 3.160*X**0.92	296 K, Y = sorbed (ug/g), X = equil conc = ug/l[2]

Source: 1.Li, G-C. and G.T Felbeck. 1972. Soil Sci. 113, 140.
2.Choudhry, G.G. 1982. In: O. Hutzinger: Environmental chemistry. Reactions and processes. Vol 2. PartB. Springer-verlag. Berlin. 103-128.

3-002 Equation
Algae

Species	Value	Condition
algal standing crop vs dissolved calcium	log(y) = 2.31 + 3.99*log(x)	y = algal st crop (ug/l), x=dis Ca (u eq/l), r=0.78, UK, freshwater lakes[1]
biomass vs volume	log(c) = 0.610 + 0.892*log(v + 0.1*b)	c = carbon (pico g), v=plasma vol (um**3), b=vacuole vol (um**3)[2]
cell carbon as a function of chlorophyll a	pc = 37+29*x	pc = cell carbon,x=chl.a,r=0.73,95% conf.lim.,st.1, marine phyt,n=13[3]
cell carbon as a function of chlorophyll a	pc = 2+55*x	pc = cell carbon,x=chl.a,r=0.88,95% conf.lim.,st.2, marine phyt,n=15[3]
cell carbon as a function of chlorophyll a	pc = - 1+80*x	pc = cell carbon,x=chl.a,r=0.89,95% conf.lim.,st.3, marine phyt,n=15[3]
cell carbon as a function of chlorophyll a	pc = 4+40*x	pc = cell carbon,x=chl.a,r=0.79,95% conf.lim., marine phyt,n=16[4]
cell carbon as a function of chlorophyll a	pc = 11+19*x	pc = cell carbon,x=chl.a,r=0.54,95% conf.lim., marine phyt,n=16[5]
chlorophyll versus total phosphorus	log chl = - 1.09 + 1.46*log(p - total)	143 lakes, summer level,phytoplankton, r = 0.95[6]
correlation between chlorophyll a and total iron	y = 1.782*x - 20.3	y = log(chlorophyll a),x=log(fe),r=0.842,at end of batch, as - f[7]
correlation between chlorophyll a and total iron	y = 1.171*x - 6.5	y = log(chlorophyll a),x=log(fe), haematococcus pluvialis, r=0.824[7]
correlation between chlorophyll a and total iron	y = 1.163*x - 4.8	y = log(chlorophyll a),x=log(fe),r=0.875,at end of batch, sc - q[7]
correlation between chlorophyll a and total iron	y = 1.303*x - 8.5	y = log(chlorophyll a),x=log(fe),r=0.870,at end of batch, synedra acus[7]

correlation between chlorophyll a and total phosphate	y = 1.916*x - 1.93	y = log(chlorophyll a),x=log(p),r=0.826,at end of batch, as - f[7]
correlation between chlorophyll a and total phosphate	y = 1.317*x - 6.90	y = log(chlorophyll a),x=log(p), ha - p, r=0.932,at end of batch[7]
correlation between chlorophyll a and total phosphate	y = 1.51*x - 2.90	y = log(chlorophyll a),x=log(p),r=0.872,at end of batch, sc - q[7]
correlation between chlorophyll a and total phosphate	y = 1.407*x - 9.23	y = log(chlorophyll a),x=log(p),r=0.788,at end of batch, Synedra acus[7]
dry weight as a function of optical density	y = 971 - 7206*d + 57200*d**2	y = dry weight (mg/l), d=optical density, 600 nm, spectronic 20, ch - p[8]
increase of photosynthetic efficiency with decreased light	ga = exp(0.685*iz + 0.4)	ga = photosynthetic efficiency,iz = light at depth z, phyt[9]
light inhibition as a function of uv and ir at surface(0) and depth	p = ln(uv+ir)(z) + 5)/((ln(uv+ir)(0) + 5)*(1.54))	p = ligth inh.,uv=ultra violet,ir=infrared, depth as index, phyt[9]
maximum daily production vs light intensity and Algae concentration	g24 = ca*lz/(lz + 500)*0.048	g24 = max prod (mg c/(m3*d)),ca=Algae (g(n+p)/m3),lz=avia ligth, in water[10]
maximum growth rate as a function of temperature	log(v) = 0.0275*t - 0.070	v = growth rate(doublings/day),t=temp (degress celcius),273 - 313 k, phyt[11]
Mn content in alga vs medium	y = 73.34*x+8.68	y = mn content in alga,x=mn conc.in medium,294k,light=2500 lux, pe - t[12]
mortality of phytoplankton	j = 0.005*q/t	q = biomass,t = time (h), j = flux (ug/(l*h))[9]
nutrient content per cell	q = uptake rate/cell division rate	exp.growth phase,nutrient content per cell(= q) remains const., phyt[13]
organic carbon as a function of chlorophyll a	poc = 151+81*x	poc = org c,x=chl.a,r=0.65,st.1, marine phyt,n=13[3]
organic carbon as a function of chlorophyll a	poc = 55+140*x	poc = org C,x=chl.a,r=0.70,st.2, marine phyt,n=15[3]
organic carbon as a function of chlorophyll a	poc = 23+285*x	poc = org C,x=chl.a,r=0.92,st.3, marine phyt,n=15[3]
organic carbon as a function of chlorophyll a	poc = 91+206*x	poc = org C,x=chl.a,r=0.59, marine phyt,n=16[4]
organic carbon as a function of chlorophyll a	poc = 73+32*x	poc = org C,x=chl.a,r=0.63, marine phyt,n=16[5]
particulate nitrogen as a function of ash - free dry weight	y = 0.142 + 27*x	y = part N (mg/l), x=dw (mg/l), r=0.94[14]
photosynthesis as a function of secchi depth	a/a(max) = 1.75*s**0.7	a = photosyn.,am=max of a, s=secchi d.(m), estimate, in lakes[15]
relative growth as a function of light intensity	rg = i/iopt*exp(1 - i/iopt)	i = light intensity,iopt=optimal i,rg=relative growth[16]
respiration	resp = respk*f(t)*phyt	F047[17]
respiration	resp = respk(ca/cmax)*f(t)*-phyt	F047, as c[18]
specific growth rate as a function of temperature	gr = 0.006*t - 0.035	gr = growth rate (1/h),t=water temp (cel), 279 - 301 k[19]
total organic content vs cell size	log(c) = 0.758*log(v) - 0.422	c = carbon per cell (pg c), v=volume (um**3), diatoms[20]

total organic content vs cell size	log(c) = 0.866*log(v) - 0.460	c = carbon per cell (pg c), v=volume (um**3), phyt (not diatoms)[20]

Source: 1.Gorham, E.J., J.W.G. Lund, J.E. Sanger and W.E. Dean, Jr. 1974. Limnol. Ocean ogr., 19(4), pp 601-61.
2.Booth, B.O. 1975. Limnol. Oceanogr., 20(5), pp 865-868.
3.Eppley, R.W., F.M.H. Reid and J.D.H. Strickland. 1970. Bull. Scripps Inst. Oceanogr. 17, pp 33-42.
4.Zeitschel, B. 1970. Mar. Biol. 7, pp 305-318.
5.Hobson, L.A. 1971. Invest. pesq. 35, pp 195-208.
6.Jones, J.R. and R.W. Bachmann. 1976. Water Pollut. Control, 48(9), pp 2176-2182.
7.Clasen, J. and H. Bernhardt. 1974. Water Res. 8(1), pp 31-44.
8.Nyholm, N. 1977. Biotech. Bioeng. 19, pp 467-492.
9.Richey, J.E. 1977. in: Hall, C.A.S. and J.W. Day, Jr. (eds.); Ecosystem Modeling in theory and practce. John Wiley and Sons. New York.
10.Beyer, J.E. 1976. Ecosystems, an Opera tional Research Approach. IMSOR. Tech. Univ. Denmark.
11.Eppley, R.W. 1972. Fish Bull., 70, pp 1063-1085.
12.Wettern, M., D.W. Lorch and A. Weber. 1976. Arch. Hydrobiol. 77(3), pp 267-276.
13.Parsons, T. and M. Takahashi. 1973. Biological Oceanographic Processesp. p 186 Pergamon Press.
14.Brewer, P.G. and J.C. Goldman. 1976. Limnol. Oceanogr. 21(1), pp 108-117.
15.Vollenweider, R.A. 1960. Mem. Ist. Ital. Idrobiol., 12 pp 201-244.
17.Gargas, E. 1976. A Three box eutrophica tion Model of a Mesotrophic Danish Lake. Water Quality Instiute.
18.Mann, K.H. 1972. Mar. Biol. 14, pp 199-209.
19.Lombardo, P.S. 1972. Hydrocomp, Inc. Mathematical Model of Water Quality in Rivers and Impoundsments
20.Strathmann, R.R. 1967. Limnol. Oceanogr., 12, pp 411- 418.

3-003 Equation
Body length

Species	Value	Condition
Aplodinotus grunniens	l = 57.639 + 56.348*s	l = total length (mm), s=spine radius (mm), tuttle creek, kansas[1]
Eleginops maclovinus	l = 46.57 + 65.66*r	l = tot length (mm), r=scale tot length (mm), r=0.9559, deseado r estuary[2]
Ictalurus punctatus	l = 7.50 + 87.96*s + 8.02*s**2	l = total length (mm), s=spine radius (mm), tuttle creek, kansas[3]
Loxechinus albus	y = 3.75*x**0.79	y = diameter (mm), x=height (mm),chilean coast, both sexes[4]
Pollachius virens	ln(r) = - 1.6987 + 0.6781*ln(l)	r = otolith radius (mm), l=length (cm), north and norwegian sea[5]
Anchoa mitchilli	l = 15.05*(1 - exp(- 0.18*ln(p))	l = length (mm), p=food (chlorella/l),n=29, r=0.82, 16 d after hatch[6]

Source: 1.Klaassen, H.E. and F.W. Cook, Jr. 1973. Trans. Kans. Acad. Sci. 76(3), pp 244-247.
2.Gosztonyi, A.E. 1974. Physis 33(86), pp 1-8.
3.Klaassen, H.E. and A.H. Townsend. 1973. Trans. Kans. Acad. Sci. 76(3), pp 248-253.
4.Gutierrez, M.J. and V.I. Otsu. 1975. Rev. Biol. Mar. Dep. Oceanogr. Univ. Chile 15(2), pp 179-199.
5.Cieglewicz, W. and B. Draganik. 1969. Pr. Morsk. Inst. Ryb. Gdynia, A, 15, pp 133-152.
6.Houde, E.D. 1977. Mar. Biol., 43(4), pp 333-341.

3-004 Equation
Body length vs age group

Species	Value	Condition
Chrysophrys major	lt = 484.50*(1 - exp(- 0.18*(t + 0.47)))	lt = fork l (mm) at age t (y), central waters of hiroshima prefecture[1]
Cod, North Sea	(lt - 36.5) = + - 19.25*(!t - 2.04!)**0.7225	lt = length (cm) at age t (year - class)[2]
Eleginops maclovinus	l = 108.29*t + 13.8	l = length (mm), t=time (y), deseado river estuary[3]
Gadus morhua	l = 110.8*(1 - exp(- 0.333*(t - 0.87)))	l = length (cm), t=age group (y), southern North Sea[4]
Gadus morhua	l = 118.7*(1 - exp(- 0.269*(t - 0.87)))	l = length (cm), t=age group (y), nothern North Sea[4]
Genypterus capensis	lt = 120.13*(1 - exp(- 0.1314*(t+0.1858)))	lt = length (mm) at age t (y), west coast stocks, Walvis Bay[5]
Genypterus capensis	lt = 148.51*(1 - exp(- 0.1120*(t+0.7404)))	lt = length (mm) at age t (y), southern stocks, Walvis Bay[5]
Haddock	(lt - 23.25) = + - 5.447*(!t - 1.8!)**0.79	lt = length (cm) at age t (year - class)[2]

Katsuwonus pelamis	l = 103.6*(1 - exp(- 0.302*(t - 0.016)))	l = length (cm), t=age group (y),by vertebral ring reading, taiwan[6]
Katsuwonus pelamis	l = 103.8*(1 - exp(- 0.432*(t - 0.016)))	l = length (cm), t=age group (y),by increment method, taiwan[6]
Labeo rohita	l = 1015*(1 - exp(- 0.276*(t+0.333)))	l = length (mm), t=age group (y)[7]
Macruronus magellanicus	lt = 66.8*(1 - exp(- 0.35984*(t - 0.070851)))	lt = length (cm) at age t(year - cllass)[8]
Marluccius angustimanus	lt = 32.7*(1 - exp(- 0.35*(t - 0.16)))	lt = length (cm) at age t(year - cllass)[9]
Merluccius angustimanus	lt = 32.7*(1 - exp(- 0.35(t - 0.16))	lt = length (cm) at time t (year), south golf of Calif.[10]
Panulirus laevicauda	l = 21.3*(1 - exp(- 0.708*t))	l = total length (cm), t=age group (y), ceara state, Brazil[11]
Panulirus longipes cygnus	lt = 113.47*(1 - exp(- 0.459*(t - 1.045)))	lt = length (mm) at age t (y), coastal reefs[12]
Pimelodus maculatus	l = 56.5*(1 - exp(- 0.1938*t))	l = length (cm), t=age group (y), male, Brazil[13]
Pimelodus maculatus	l = 45.5*(1 - exp(- 0.2104*t))	l = length (cm), t=age group (y), female, Brazil[13]
Placie	(lt - 39) = + - 3.802*(!t - 8.2!)**0.6776	lt = length (cm) at age t (year - class)[2]
Pollachius virens	l = 109.4*(1 - exp(- 0.160*t))	l = length (cm), t=age group (y), north vs Norwegian Sea[14]
Psenopsis anomala	l = 243.i*(1 - exp(- 0.34*(t+1.674)))	l = length (mm), t=age group (y), kii channel, Japan[15]
Pseudotolithus senegalensis	lt = 81*(1 - exp(- 0.33*(t - 0)))	lt = length (cm) at age t(year - cllass)[16]
Pseudotolithus typus	lt = 103*(1 - exp(- 0.29*(t - 0.15)))	lt = length (cm) at age t(year - cllass)[16]
Salmo salar	lt = 120.25*(1 - exp(- 0.439*(t + 0.335)))	lt = length (cm) at age t=time (year - class), baltic[17]
Salmo salar	lt = 156.03*(1 - exp(- 0.2876*(t + 0.456)))	lt = length (cm) at age t=time (year - class), baltic[18]
Scomberomorus cavalla	l = 840m(1 - exp(- 0.35m(t + 2.5)))	l = length (mm), t=age group (y), florida, male[19]
Scomberomorus maculatus	l = 515*(1 - exp(- 0.48*(t+1.12)))	l = length (mm),w\=age group (y), florida[20]
Serranus alexandrinus	(lt - 24.33) = + - 7.143*(!t - 1.93!)**0.82897	lt = length (cm) at age t (year - class)[2]
Sole, North Sea	(lt - 9.5) = + - 12.1*(!t - 1!)**0.4298	lt = length (cm) at age t (year - class)[2]
Tautogolabrus adspersus	lt = 284.77*(1 - exp(- 0.1979*(t - 0.1044)))	lt = length (mm) at age t (y), weweantic r. estuary, mass[21]
Thymallus thymallus	lt = 57.5*(1 - exp(- 0.223*(t - 0.115)))	lt = length (cm) at age t=time (year - class), llyn tegid, 1946 - 1952[22]
Thymallus thymallus	lt = 45.6*(1 - exp(- 0.386*(t - 0.222)))	lt = length (cm) at age t=time (year - class), llyn tegid, 1962 - 1968[22]
Thymallus thymallus	lt = 38.9*(1 - exp(- 0.424*(t - 0.149)))	lt = length (cm) at age t=time (year - class), uppper dee[22]
Thymallus thymallus	lt = 41.7*(1 - exp(- 0.484*(t - 0.281)))	lt = length (cm) at age t=time (year - class), corwen[22]
Thymallus thymallus	lt = 57.5*(1 - exp(- 0.233*t+0.233*0.115))	lt = length (cm) at time t (year), llyn tegid vs river dee[22]
Trachurus japonicus	lt = 424.9*(1 - exp(- 0.248*(t + 809)))	lt = fork l (cm) at age t (y), waters off cheju island[23]
Trachurus trachurus	lt = 56.41*(1 - exp(- 0.11*(t + 0.5461)))	lt = length (cm) at age t (y), cuba[24]
Trachurus trachurus	lt = 55.5*(1 - exp(- 0.11*(t - 1.96)))	lt = length (cm) at age t (y), south - west african waters[25]

Source: 1.Kuniyuki, K., M. Yano and M. Kaanishi. 1975. Bull. Nansei Reg. Fish. Res. Lab., pp 81-100.
2.Rafail, S.Z. 1971. Mar. Biol., 10, pp 13-21.
3.Gosztonyi, A.E. 1974. Physis 33(86), pp 1-8.

4.Daan, N. 1974. Neth. J. Sea Res. 8(1), pp 27-48.
5.Payne, A.I.I. 1976. Collect. Sci. Pap. ICESEAF/Recl. Doc. Sci. CIPASE/Coleco. Doc. Cient. Cipaso, v, pp 159-163.
6.Chi, K.-S. and R.-T. Yang. Acta Ocean ogr. Taiwan. 3, pp 199-222.
7.Khan, R.A. and A.Q. Siddiqui. 1973. Proc. Indian Natl. Sci. Acad., Part B. 39(5), pp 582-597.
8.Aguayo, H.M. 1974. Invest. Pesq. Inst. Fom. Pesq. Chile, (No 19), 17 p.
9.Mathews, C.P. 1975. J. Fish. Biol. 7(1), pp 83-94.
10.Mathews, C.P. 1975. J. Fish. Biol. 7(1), pp 83-94.
11.Ivo Cerrea, C.T. 1973. Arq. Cienc. Mar., 13(1), pp 19-23.
12.Chittleborough, R.G. 1976. Aust. J. Mar. Freshwater Res. 27(2), pp 279-295.
13.Fenerich, N.A., M.Y. Narahara and H.M. Godinho. 1975. Bol. Inst. Pesca, Sao Paulo, 4(1), pp 15-28.
14.Cieglewicz, W. and B. Draganik. 1969. Pr. Morsk. Inst. Ryb. Gdynia, A, 15, pp 133-152.
15.Sakamoto, T. and T. Suzuki. 1974. Bull. Jap. Soc. Sci. Fish. 40(6), pp 551-560.
16.Bayogbona, E.O. 1969. in: Proc. Symp. Oceanogr. Fish. resources in trop Atlantic. pp 349-359. Paris.Unesco.
17.Thurow, F. 1973. Ber. dt. wiss. Kommn. Meeresforsch., 22, pp 445-451.
18.Hohendorf, K. 1966. Kieler Meeres forsch., 22, pp 70-90.
19.Beaumariage, D.S. 1973. Fla. Mar. Res. Publ., 1, pp 45.
20.Powell, D. 1975. Florida Mar. Res. Publ. (no5), 21 p.
21.Serchuk, F.M. and C.F. Cole. 1974. Chea sepeake Sci. 15(4), pp 205-213.
22.Woolland, J.V. and J.W. Jones. 1975. J. Fish. Biol., 7(6), pp 749-773.
23.Ann, H.B. 1973. Bull. Fish. Res. Dev. Agency, Pusan 10, pp 73-85.
24.Terre, J.J. 1976. Collect. Sci. Pap. ICSEAF/Recl. Doc. Sci. CIPASE/Coleco. Doc. Cient. CIPASO, v. 3,pp 39-44.
25.Wengrzyn, J. 1976. Collect. Sci. Pap. ICSEAF/Recl. Doc. Sci. CIPASE/Coleco. Doc. Cient. CIPASO, v 3.pp 51-57.

3-005 Equation
Body weight

Species	Value	Condition
Armadillidium vulgare	p = p0/exp((k/0.4329)*t)	p = weight ,p0=p at birth, t=time (d), k=coefficient of growth[1]
Loxechinus albus	y = 0.0004*x**2.966	y = diameter (mm), x=weight (g), chilean coast, female[2]
Salmo trutta	y = 0.1522*x + 0.78	y = ww of gonads (g), x= ww of fish (g), female, 1967 - 1969[3]
Salmo trutta	y = 0.2378*x - 42.54	y = ww of gonads (g), x= ww of fish (g), female, 1970 - 1971[3]
Salmo trutta	y = 0.03316*x - 0.65	y = ww of gonads (g), x= ww of fish (g), male[3]
Solea solea	c = 1.49*w - 0.00393*l**3	c = total caloric cont (k cal), w=wet w (g), l=length (cm)[4]

Source: 1.Hubbel, S.P. 1971. in: Patten, B.C. (ed.); Systems Analysis and Simulation in Ecology. 1, pp 269-324Academic Press. New York.
2.Gutierrez, M.J. and V.I. Otsu. 1975. Rev. Biol. Mar. Dep. Oceanogr. Univ. Chile 15(2), pp 179-199.
3.Thorpe, J.E. 1974. Proc. R. Soc. Edinb. B. 1974, pp 299-313.
4.Bromley, P.J. 1971. J. Cons. Int. Ex plor. Mer. 34(1), pp 131-134.

3-006 Equation
Body weight vs age group

Species	Value	Condition
Chrysophrys major	w = 2368.22*(1 - exp(- 0.18*(t + 0.47)))**2.84	w = weight (g) at age t (y), central waters of hiroshima prefecture[1]
Cod, North Sea	(wt - 51) = + - 377*(!t - 1!)**1.7482	wt = weight (g) at age t (year - class[2]
Gadus morhua	w = 14150*(1 - exp(- 0.333*(t - 0.87)))	w = weight (g), t=age group (y), southern North Sea[3]
Gadus morhua	w = 17390*(1 - exp(- 0.269*(t - 0.87)))	w = weight (g), t=age group (y), nothern North Sea[3]
Haddock	(wt - 124.2) = + - 85.07*(!t - 2.03!)**1.1152	wt = weight (g) at age t (year - class[2]
Macruronus magellanicus	wt = 804.1*(1 - exp(- 0.35984*(t - 0.070851)))**3	wt = weight (g) at[4]
Placie	(wt - 920) = + - 146.3*(!t - 11.2!)**0.82617	wt = weight (g) at age t (year - class)[2]
Psenopsis anomala	w = 486.7*(1 - exp((- 0.34*(t+1.674)))*2.8912	w = weight (g), t=age group (y), kii channal, Japan[5]

Salmo salar	wt = 16504*(1 - exp(- 0.437*(t + 0.334)))	wt = weight (g) at age t=time (year - class), baltic[6]
Salmo salar	wt = 25230*(1 - exp(- 0.5397*(t - 0.082)))	wt = weight (g) at age t=time (year - class), baltic[7]
Serranus alexandrinus	(wt - 71.9) = + - 129.8*(!t - 1!)**1.4084	wt = weight (g) at age t (year - class)[2]
Sole, North Sea	(wt - 144.9) = + - 788*(!t - 3!)**0.8483	wt = weight (g) at age t (year - class)[2]
Srachurus japonicus	wt = 1.000*(1 - exp(- 0.248*(t + 809)))**3	wt = wet weight (g) at age t (y), waters off cheju island[8]
Trachurus trachurus	w = 1307.2*(1 - exp(- 0.11*(t - 1.96)))**3	w = weight (g) at age t (y), south - west african waters[9]

Source: 1.Li, G-C. and G.T Felbeck. 1972. Soil Sci. 113, 140.
2.Rafail, S.Z. 1971. Mar. Biol., 10, pp 13-21.
3.Daan, N. 1974. Neth. J. Sea Res. 8(1), pp 27-48.
4.Aguayo, H.M. 1974. Invest. Pesq. Inst. Fom. Pesq. Chile, (No 19), 17 p.
5.Sakamoto, T. and T. Suzuki. 1974. Bull. Jap. Soc. Sci. Fish. 40(6), pp 551-560.
6.Thurow, F. 1973. Ber. dt. wiss. Kommn. Meeresforsch., 22, pp 445-451.
7.Hohendorf, K. 1966. Kieler Meeres forsch., 22, pp 70-90.
8.Ann, H.B. 1973. Bull. Fish. Res. Dev. Agency, Pusan 10, pp 73-85.
9.Wengrzyn, J. 1976. Collect. Sci. Pap. ICSEAF/Recl. Doc. Sci. CIPASE/Coleco. Doc. Cient. CIPASO, v 3.pp 51-57.

3-007 Equation
Body weight vs body length

Species	Value	Condition
Acroneuria sp.	w = 0.027*l**3	Plecoptera,w = weight,l=length[1]
Agarum cribrosum	y = 0.0005*x**2.7740	y = ww (g), x=length (cm), r=0.88, depth=12 m, fox point, 1968[2]
Agarum cribrosum	y = 0.0099*x**2.2273	y = ww (g), x=length (cm), r=0.856, depth=12 m, fox point, 1970[2]
Ageneiosus valenciennesi	log(w) = 2.211 + 3.02*log(l)	w = wet w. (g), l=length (mm), Brazil, male, r=0.99[3]
Ageneiosus valenciennesi	log(w) = 3.042 + 3.042*log(l)	w = wet w. (g), l=length (mm), Brazil, female, r=0.99[3]
Ageneiosus valenciennesi	log(w) = 2.248 + 3.04*log(l)	w = wet w. (g), l=length (mm), Brazil, both sexes, r=0.99[3]
Alosa pseudoharengus	log(w) = - 5.28911 + 3.06370*log(l)	w = wet w (g), l=length (mm), claytor l., virginia[4]
Antocha sp.	w = 0.002*l**3	Diptera,w = weight,l=length[1]
Aphelochirus vittatus	w = 0.071*l**3	hemiptera,w = weight,l=length[1]
Aplodinotus grunniens	log(w) = - 5.248 + 3.123*log(l)	l = total length (mm), w=wet w. (g), tuttle creek, kansas[5]
Atherix sp.	w = 0.025*l**3	diptera,w = weight,l=length[1]
Baetis sp.	w = 0.012*l**3	ephemeroptera,w = weight,l=length[1]
Balaenoptera acutorostrata	w = 0.049574*w**2.31	w = wet weight (tonnes), l - length (m), adjusted for fluid loss[6]
Balaenoptera acutorostrata	w = 0.0076*l**3.23	w = wet weight (tonnes),l=length (m), antarctic, n=22[6]
Balaenoptera borealis	w = 0.025763*l**2.43	w = wet weight (tonnes), l - length (m), adjusted for fluid loss[6]
Balaenoptera borealis	w = 0.0242*l**2.43	w = wet weight (tonnes),l=length (m), Japan, n=16[6]
Balaenoptera brydei	w = 0.012965*w**2.74	w = wet weight (tonnes), l - length (m), adjusted for fluid loss[6]
Balaenoptera brydei	w = 0.0122*l**2.74	w = wet weight (tonnes),l=length (m), Japan, n=27[6]
Balaenoptera musculus	w = 0.002899*l**3.25	w = wet weight (tonnes), l - length (m), adjusted for fluid loss[6]
Balaenoptera musculus	w = 0.0046*l**3.09	w = wet weight (tonnes),l=length (m), antarctic, n=43[6]

Balaenoptera musculus brevicauda	w = 0.0004*l**3.97	w = wet weight (tonnes),l=length (m), antarctic, n=5[6]
Balaenoptera physalus	w = 0.007996*l**2.90	w = wet weight (tonnes), l - length (m), adjusted for fluid loss[6]
Balaenoptera physalus	w = 0.0238*l**2.53	w = wet weight (tonnes),l=length (m), antarctic, n=34[6]
Balaenoptera physalus	w = 0.0015*l**3.46	w = wet weight (tonnes),l=length (m), n. hemisphere, n=8[6]
Bosmina longirostris	w**(- 3) = 0.56*l - 0.01	w = wet wt.(mg),l=length(mm)[7]
Bosmina longirostris	w**(- 3) = 0.06*l+0.035	w = wet wt.(mg),l=length(mm)[8]
Bosmina sp.	w = 0.124*l**2.181	w = wet wt.(mg),l=length(mm)[9]
Brachycentrus sp.	w = 0.051*l**3	trichoptera,w = weight,l=length[1]
Branchinecta gigas	w = 0.108*l**3.27	w = dry weight (ug),l=length (mm), male, n=70, 14 - 60 mm[10]
Branchinecta gigas	w = 0.118*l**3.20	w = dry weight (ug),l=length (mm), female, n=65, 11.4 - 72 mm[10]
Branchinecta gigas	w = 7.56*l**1.45	w = dry weight (ug),l=length (mm), larvae, n=14, 1.1 - 4.9 mm[10]
Branchinecta mackini	w = 0.011*l**2.36	w = dry weight (ug),l=length (mm), male, n=35, 6.2 - 24.7 mm[10]
Branchinecta mackini	w = 0.003*l**2.57	w = dry weight (ug),l=length (mm), female, n=32, 6.1 - 25 mm[10]
Branchinecta mackini	w = 0.192*l**1.51	w = dry weight (ug),l=length (mm), larvae, n=26, 0.5 - 10.7 mm[10]
Carcinus mediterraneus	w = 0.1819*l**3.16	w = wet w. (g), l=length (cm), may - june, france, male[11]
Carcinus mediterraneus	w = 0.1814*l**3.08	w = wet w. (g), l=length (cm), may - june, france, female[11]
Carpiodes cyprinus	log(w) = - 4.06042 + 2.78559*log(l)	w = wet w (g), l=length (mm), four mile creek, ohio[12]
Carpiodes velifer	log(w) = - 4.05206 + 2.80837*log(l)	w = wet w (g), l=length (mm), four mile creek, ohio[12]
Cataclysta midas	w = 0.04*l**3	lepidoptera,w = weight,l=length[1]
Ceriodaphnia pulchella	w**(- 3) = 0.57*l - 0.02	w = wet wt.(mg),l=length(mm)[8]
Chironomus sp.	w = 0.0026*l**3	diptera,w = weight,l=length[1]
Chondrostoma nasus prespense	log(w) = 2.98472*log(l) - 1.75947	w = wet w. (g), l=length (mm), prespa l.[13]
Cololabis saira	w = 0.00008309*l**3.19519	w = wet w. (g), l=length (mm), usa[14]
Copepods	w**(- 3) = 0.34*l - 0.03	w = wet wt.(mg),l=length(mm)[7]
Copepods	w = 0.055*l**2.73	w = wet wt.(mg),l=length(mm)[15]
Cyclops sp.	y = 0.01610*x - 0.00080	y = dry w. (ug), x=length (mm), stonehouse pond[16]
Daphnia hyalina	w**(- 3) = 0.32*l - 0.08	w = wet wt.(mg),l=length(mm)[7]
Daphnia hyalina	w**(- 3) = 0.46*l - 0.12	w = wet wt.(mg),l=length(mm)[8]
Daphnia longispina	w**(- 3) = 0.46*l - 0.12	w = wet wt.(mg),l=length(mm)[8]
Daphnia pulicaria	log(w) = 3.14*log(l) + 0.59	w = dry weight (ug), l=length (mm)[17]
Daphnia rosea	w = 0.0053*l**2.17	w = dry wt.(mg),l=length fr.tail - base to anterior edge of eye,nongravid[18]
Daphnia rosea	w = 0.0053*l**2.17	correction fact.:add 2 ug per egg or embryo in the brood pouch[18]
Daphnia sp.	y = 0.02794*x - 0.01365	y = dry w. (ug), x=length (mm), stonehouse pond[16]

Daphnia sp.	w = 0.052*l**3.012	w = wet wt.(mg),l=length(mm)[9]
Diaphanosoma sp.	w = 0.092*l**2.449	w = wet wt.(mg),l=length(mm)[9]
Diaptomus nevadensis	w = 0.001*l**3.61	w = dry weight (ug),l=length (mm), female, n=51, 1.9 - 3.8 mm[10]
Diaptomus nevadensis	w = 0.107*l**1.79	w = dry weight (ug),l=length (mm), larvae, n=20, 0.4 - 1.9 mm[10]
Diaptomus spp.	y = 0.01629*x + 0.00061	y = dry w. (ug), x=length (mm), stonehouse pond[16]
Dinarthrodes sp.	w = 0.027*l**3	trichoptera,w = weight,l=length[1]
Ecdyonurus yoshidae	w = 0.042*l**3	ephemeroptera,w = weight,l=length[1]
Eleginops maclovinus	w = 1.25*10**(- 5)*l**3.061	w = wet w. (g), l=length (mm), r=0.9996, deseado river estuary[19]
Elmis sp.	w = 0.0062*l**3	coleoptera,larva,w = weight,l=length[1]
Elmis sp.	w = 0.07*l**3	coleoptera,imago,w = weight,l=length[1]
Elops saurus	log(w) = 5.3295 + 3.1123*log(l)	w = wet w. (g), l= fork length (mm), juvenile, louisiana coastal stream[20]
Epeorus hiemalis	w = 0.027*l**3	ephemeroptera,w = weight,l=length[1]
Ephemera lineata	w = 0.01*l**3	ephemeroptera,w = weight,l=length[1]
Ephemerella trispina	w = 0.03*l**3	ephemeroptera,w = weight,l=length[1]
Epiophlebia sp.	w = 0.029*l**3	odonata,w = weight,l=length[1]
Eriocera sp.	w = 0.026*l**3	diptera,w = weight,l=length[1]
Eschrichtius robustus	w = 0.0051*l**3.28	w = wet weight (tonnes),l=length (m), n. pacific, n=8[6]
Eubalaena glacialis sieboldii	w = 0.013200*l**3.06	w = wet weight (tonnes), l - length (m), adjusted for fluid loss[6]
Eubalaena glacialis sieboldii	w = 0.0124*l**3.06	w = wet weight (tonnes),l=length (m), n. pacific, n=21[6]
Eubranchipus bundyi	w = 0.550*l**2.50	w = dry weight (ug),l=length (mm), male, n=44, 3.1 - 10.7 mm[10]
Eubranchipus bundyi	w = 0.513*l**2.42	w = dry weight (ug),l=length (mm), female, n=37, 3.2 - 11.2 mm[10]
Eubranchipus bundyi	w = 0.001*l**3.39	w = dry weight (ug),l=length (mm), male, n=42, 1.9 - 3.3 mm[10]
Eubrianax sp.	w = 0.052*l**3	coleoptera,w = weight,l=length[1]
Fundulus heteroclitus	log(w) = 3.066*log(l) - 4.943	r = 0.968, w=wet w. (g), l=length (cm), patuxent river[21]
Fundulus majalis	log(w) = 3.076*log(l) - 5.022	r = 0.938, w=wet w. (g), l=length (cm), patuxent river[21]
Goera japonica	w = 0.028*l**3	trichoptera,w = weight,l=length[1]
Gomphus sp.	w = 0.043*l**3	odonata,w = weight,l=length[1]
Heteropneustes fossilis	log(w) = - 5.6440 + 3.1646*log(l)	w = wet w. (g), l=length (mm), darbhanga, india, juvenile, r=0.97[22]
Heteropneustes fossilis	log(w) = - 5.0817 + 2.9418*log(l)	w = wet w. (g), l=length (mm), darbhanga, india, female, r=0.975[22]
Heteropneustes fossilis	log(w) = - 4.2147 + 2.5271*log(l)	w = wet w. (g), l=length (mm), darbhanga, india, male, r=0.91[22]
Holopedium sp.	y = 0.00686*x + 0.01206	y = dry w. (ug), x=length (mm), stonehouse pond[16]
Hydropsyche sp.	w = 0.019*l**3	trichoptera,w = weight,l=length[1]
Ictalurus punctatus	log(w) = - 5.419 + 3.115*log(l)	l = total length (mm), w=wet w. (g), tuttle creek, kansas[23]
Isonychia japonica	w = 0.023*l**3	ephemeroptera,w = weight,l=length[1]

Laminaria digitata	y = 0.0041*x**2.5927	y = ww (g), x=length (cm), r=0.863, depth=5 m, Fox Point, 1968[2]
Laminaria digitata	y = 0.0128*x**2.2651	y = ww (g), x=length (cm), r=0.892, depth=12 m, Fox Point, 1968[2]
Laminaria digitata	y = 0.0187*x**2.2587	y = ww (g), x=length (cm), r=0.869, depth=5 m, fox point, 1970[2]
Laminaria digitata	y = 0.0013*x**2.8148	y = ww (g), x=length (cm), r=0.969, depth=12 m, Fox Point, 1970[2]
Laminaria longicruris	y = 0.0980*x**1.7234	y = ww (g), x=length (cm), r=0.964, depth=5 m,srrawberry isl, 1968[2]
Laminaria longicruris	y = 0.0121*x**1.9925	y = ww (g), x=length (cm), r=0.948, depth=5 m, Fox Point, 1968[2]
Laminaria longicruris	y = 0.0255*x**1.9180	y = ww (g), x=length (cm), r=0.952, depth=12 m, Fox Point, 1968[2]
Laminaria longicruris	y = 0.0038*x**2.4256	y = ww (g), x=length (cm), r=0.985, depth=5 m, Luke Isl., 1970[2]
Laminaria longicruris	y = 0.0010*x**2.7358	y = ww (g), x=length (cm), r=0.961, depth=10 m, Luke Isl., 1970[2]
Laminaria longicruris	y = 0.0018*x**2.5338	y = ww (g), x=length (cm), r=0.96, depth=5 m, Fox Point, 1970[2]
Laminaria longicruris	y = 0.0025*x**2.4150	y = ww (g), x=length (cm), r=0.923, depth=12 m, Fox Point, 1970[2]
Loxechinus albus	y = 0.0006*x**2.835	y = diameter (mm), x=weight (g), chilean coast, male[24]
Luciola cruciata	w = 0.03*l**3	coleoptera,w = weight,l=length[1]
Macronema sp.	w = 0.02*l**3	trichoptera,w = weight,l=length[1]
Megaptera novaeagliae	w = 0.016473*l**2.95	w = wet weight (tonnes), l - length (m), adjusted for fluid loss[6]
Megaptera novaeangliae	w = 0.0158*l**2.95	w = wet weight (tonnes),l=length (m), antarctic, n=17[6]
Menidia menidia	log(w) = 2.855*log(l) - 4.915	r = 0.981, w=wet w. (g), l=length (cm), patuxent river[21]
Michropterus coosae	log(w) = - 4.387 + 2.798*log(l)	w = wet w (g), l=length (mm), spring creek, tennessee[25]
Michropterus coosae	log(w) = - 4.576 + 2.870*log(l)	w = wet w (g), l=length (mm), roaring r., tennessee[25]
Micrasema sp.	w = 0.013*l**3	Trichoptera,w = weight,l=length[1]
Micropterus Salmoides	log(w) = - 4.666 + 2.9*log(l)	w = wet w. (g), l=length (mm), Moncove lake, W. Virginia, r=0.99[26]
Moina macrura	w**(- 3) = 0.66*l - 0.01	w = wet wt.(mg),l=length(mm)[8]
molanna sp.	w = 0.02*l**3	Trichoptera,w = weight,l=length[1]
Morone americanas	log(w) = 3.073*log(l) - 4.713	r = 0.967, w=wet w. (g), l=length (cm), Patuxent river[21]
Mugil capito	log(w) = - 4.5647 + 2.7612*log(l)	male, w = wet w (g), l=length (mm), Borollus lake, Egypt[27]
Mugil capito	log(w) = - 4.9029 + 2.9392*log(l)	female, w = wet w (g), l=length (mm), Borollus lake, Egypt[27]
Mugil capito	log(w) = - 4.4244 + 2.8071*log(l)	both sexes, w = wet w (g), l=length (mm), Borollus lake, Egypt[27]
Mullus barbatus	log(w) = 4.6943 + 2.88*log(l)	both sexes, w = wet w (g), l=length (mm), Abukir - rosetta, Egypt[27]
Mullus surmuletus	log(w) = - 5.3255 + 3.1635*log(l)	male, w = wet w (g), l=length (mm), Egypt mediterranian[27]
Mullus surmuletus	log(w) = - 5.3786 + 3.1941*log(l)	female, w = wet w (g), l=length (mm), Egypt mediterranian[27]

Mullus surmuletus	log(w) = - 5.3805 + 3.1924*log(l)	both sexes, w = wet w (g), l=length (mm), Egypt mediterranian[27]
Neoperla sp.	w = 0.011*l**3	plecoptera,w = weight,l=length[1]
Nephrops norvegicus	w = 0.0002188*l**3.321	w = wet w (g), l=carapace length (mm), male, moray firth[28]
Nephrops norvegicus	w = 0.0003221*l**3.204	w = wet w (g), l=carapace length (mm), male, minch[28]
Nephrops norvegicus	w = 0.0006840*l**2.963	w = wet w (g), l=carapace length (mm), female, moray firth vs minch[28]
Nephrops norvegicus	w = 0.0004523*l**3.117	w = wet w (g), l=carapace length (mm), male, ne england[29]
Nephrops norvegicus	w = 0.001076*l**2.849	w = wet w (g), l=carapace length (mm), female, minch[29]
Nephrops norvegicus	w = 0.0002245*l**3.3345	w = wet w (g), l=carapace length (mm), both sexes, irish sea[30]
Peprilus tricanthus	log(w) = - 5.1852 + 3.2646*log(l)	w = wet w (g), l=length (mm), lower york r., virginia[31]
Physeter catodon	w = 0.006648*l**3.18	w = wet weight (tonnes), l - length (m), adjusted for fluid loss[6]
Physeter catodon	w = 0.0196*l**2.74	w = wet weight (tonnes),l=length (m), antarctic + n. pacific,n=44[6]
Physeter catodon	w = 0.0029*l**3.55	w = wet weight (tonnes),l=length (m), natal, n=3[6]
Pimephales promelas	log(w) = 5.7204 + 3.5178*log(l)	w = wet w. (g), l=length (mm), north dakota saline lakes[32]
Platyphylax yokouchii	w = 0.026*l**3	trichoptera,w = weight,l=length[1]
Pollachius virens	w = 0.0699m*mm2.54	w = wet weigth (g), l = length (cm), north vs norwegian sea[33]
Polyphemus pediculus	w**(- 3) = 0.44*l - 0.02	w = wet wt.(mg),l=length(mm)[7]
Potamanthus kamonis	w = 0.014*l**3	ephemeroptera,w = weight,l=length[1]
Procopium williamsoni	log(w) = 2.1967 + 3.2108*(log(x) - 2.3710)	w = wet weigth (g), l=fork length (mm), sheep river, alberta[34]
Protothermes grandis	w = 0.034*l**3	Neuoptera,w = weight,l=length[1]
Rhyacophila brevicephala	w = 0.02*l**3	Trichoptera,w = weight,l=length[1]
Rhyacophila nigrocephala	w = 0.0072*l**3	Trichoptera,w = weight,l=length[1]
Sardinella aurita	w = 4.315*10**(- 6)*l**3.12	w = wet w. (g), l=length (cm), Egypt mediterranian[35]
Sardinella maderensis	w = 1.278*10**(- 5)*l**2.9176	w = wet w. (g), l=length (cm), Egypt mediterranian[35]
Serranus alexandrinus	log(w) = - 1.79257 + 2.92925*log(l)	w = wet weight (g), l=length (cm)[36]
Sieboldius japonica	w = 0.031*l**3	Odonata,w = weight,l=length[1]
Simocephalus vetulus	w**(- 3) = 0.65*l - 0.04	w = wet wt.(mg),l=length(mm)[8]
Simulium sp.	w = 0.01*l**3	diptera,w = weight,l=length[1]
Stenopsyche griseipennis	w = 0.013*l**3	Trichoptera,w = weight,l=length[1]
Stenopsyche sp.	w = 0.046*l**3	trichoptera,w = weight,l=length[1]
Thymallus thymallus	log(w) = - 2.288 + 3.219*log(l)	w = wet w. (g), l=length (cm), llyn tegid and river dee[37]
Tilapia zillii	log(w) = - 1.941 + 3.1849*log(l)	w = wet weight (g), l=length (cm), female, s=2.9 %[38]
Tilapia zillii	log(w) = - 1.5828 + 2.8309*log(l)	w = wet weight (g), l=length (cm), male, s=2.9 %[38]
Tilapia zillii	log(w) = - 1.825 + 3.0594*log(l)	w = wet weight (g), l=length (cm), both sex, s=2.9 %[38]
Trachurus japonicus	w = 1.318*(l**3)*10**(- 5)	w = wet w (g), l=fork length (cm), waters off Cheju Island[39]

Trachurus trachurus	w = 0.0096*l**2.95	w = wet w (g), l=length (cm), south - west African waters[40]
Trinectes maculatus	log(w) = 0.0132*log(l) + 0.034	r = 0.972, w=wet w. (g), l=length (cm), Patuxent river[21]
Tubifex tubifex	log(w) = (0.8980 + - 0.25) + (1.76 + - 0.30)*log(l)	w = wet w (mg),l=length (mm), 95 % cl., n=70[41]
Upeneus moluccensis	w = 0.0126*l**3.1526	w = wet w (g), l=fork length (cm), China Sea, female[42]
Upeneus moluccensis	w = 0.0146*l**3.0873	w = wet w (g), l=fork length (cm), China Sea, male[42]

Source: 1.Edmonson, W.T. 1971. In: W.T. Edmondson and G.G. Winberg (eds.); A Manual on Methods for the Assessmnt of Secondary Productivity in Fresh Waters. IBP Handbook No 17, p 145.
2.Mann, K.H. 1972. Mar. Biol. 14, pp 199-209.
3.Silva Bezerra, J.W. 1975. Notes Doc. Peche Piscic. (Nouv. Ser.), 11, pp 65-74.
4.Boaze, J.L. and R.T. Lackey. 1974. Prog. Fish Cult. 36(6), pp 163-164.
5.Klaassen, H.E. and F.W. Cook, Jr. 1973. Trans. Kans. Acad. Sci. 76(3), pp 244-247.
6.Lockyer, C. 1975. J. Cons. Int. Explor. Mer., 36(3), pp 259-273.
7.Shcherbakov, A.P. 1952. Dokl. Akad. Nauk SSSR, 84, pp 153-157.
8.Osmera, S. 1966. Zool. Listy 15, pp 79-83.
9.Pechen, G.A. 1965. Gidrobiol. Zh. 1, pp 19-26.
10.Daborn, G.R. 1974. Can. J. Zool., 52(III), pp 1303-1310
11.Do Chi, T. and G. Lasserre. 1974. Bull. Ecol. 5(2), pp 91-104.
12.Woodward, R.L. and T.E. Wissing. 1976. Trans. Am. Fish. Soc. 105(3), pp 411-415.
13.Sidorovski, M.N. 1974. Izd. Zavod. Ribar. SR Marked. 5(2), pp 15.
14.Sunada, J.S. 1974. Calif. Fish. Game, 60(2), pp 64-73.
15.Ambler, J.W. and B.W. Frost. 1974. Limnol. Oceanogr. 19(3), pp 446-451.
16.Ferrante, J.G. 1976. Hydrobiologia, 49(3), pp 203-204.
17.Snow, N.B. 1972. Limnol. Oceanogr. 17(6), pp 909-913.
18.McRoy, C.P. and R.J. Barsdate. 1970. Limnol. Oceanogr. 15(1) pp 6-13.
19.Gosztonyi, A.E. 1974. Physis 33(86), pp 1-8.
20.Sekavee, G.B. 1974. Trans. Am. Fish. Soc., 103(3), pp 472-476.
21.O'Connor, S.G. and A.J. McErlean. 1975. in: Cronin, L.E. (ed.); Eustarine Research. pp 494-518. Acadmic Press. New York.
22.Thakur, N.K. and N.K. das. 1975. J. In land Fish. Soc. India. 6(1974), pp 95-96.
23.Klaassen, H.E. and A.H. Townsend. 1973. Trans. Kans. Acad. Sci. 76(3), pp 248-253.
24.Gutierrez, M.J. and V.I. Otsu. 1975. Rev. Biol. Mar. Dep. Oceanogr. Univ. Chile 15(2), pp 179-199.
25.Gwinner, H.R., H.J. Cathey and F.J. Bulow. 1975. J. Tenn. Acad. Sci. 50(3), pp 102-105.
26.Montgomery, W.V. 1975. Proc. W. Va. Acad. Sci. 46(3-4), pp 265-273.
27.El-Maghraby, A.M., M.T. Hashem and H.M. El-sedfy. 1973. Bull. Inst. Oceanogr. Fish. Cairo. 3, pp 55-2.
28.Pope, J.A. and H.J. Thomas. 1967. J. Cons. perm. Int. Explor. Mer., 31, pp 265-291.
29.Symonds, D.J. 1972. Fishery Invest. Ser. II, Lond., 27(3), pp 1-35.
30.Farmer, A.S. 1974. J. Nat. Hist., 8(6), pp 605-620.
31.Du Paul, W.D. and J.D. McEachran. 1973. Cheasepeake Sci. 14(3), pp 205-207.
32.Held, J.W. and J.J. Paterka. 1974. Trans. Am. Fish. Soc., 103(4), pp 743-756.
33.Cieglewicz, W. and B. Draganik. 1969. Pr. Morsk. Inst. Ryb. Gdynia, A, 15, pp 133-152.
34.Thompson, G.E. and R.W. Davies. 1976. Trans. Am. Fish. Soc. 105(2), pp 208-219.
35.Soliman, I.A.M., G.A. Botros and A.M.El-Maghraby. 1970. Bull. Inst. Oceanogr. Fish., Cairo 1, pp 27-5.
36.Rafail, S.Z. 1971. Mar. Biol., 10, pp 13-21.
37.Woolland, J.V. and J.W. Jones. 1975. J. Fish. Biol., 7(6), pp 749-773.
38.Saleh, H.H. 1972. Mar. biol. 12, pp 255-260.
39.Ann, H.B. 1973. Bull. Fish. Res. Dev. Agency, Pusan 10, pp 73-85.
40.Wengrzyn, J. 1976. Collect. Sci. Pap. ICSEAF/Recl. Doc. Sci. CIPASE/Coleco. Doc. Cient. CIPASO, v 3.pp 51-57.
41.Kosiorek, D. 1974. Pol. Arch. Hydrobiol. 21(3-4), pp 411-422.
42.Lee, C.K.C. 1974. in: The Kuroshio 3. pp 389. National Research Council of Thailand. Bangkok.

3-008 Equation
Caloric content as a function of % organic matter

Species	Value	Condition
marine zooplankton	y = - 3370 + 136*x - 0.514*x**2	y = cal cont (cal/g dw),x=% org matter in dry plankton,r=0.94[1]
marine zooplankton	y = 1351 + 106*z - 21.6*a	z = % carbon of dw, a=ash cont (% of dw),y=cal cont (cal/g dw),r=0.98[1]

Source: 1.Platt, T., V.M. Brawn and B. Irwin. 1969. J. Fish. Res. bd. Can., 26, pp 2345-2349.

3-009 Equation
Carbon

Species	Value	Condition
Ctenophores	log(y) = 0.92*log(x) + 0.087	y = carbon (ug c/anim.), x=equatorial diameter (mm), size< 4 mm[1]
Ctenophores	log(y) = 2.35*log(x)	y = carbon (ug c/anim.), x=equatorial diameter (mm), size> 4 mm[1]

Source: 1.Mullin, M.M. and E.R. Brooks. 1970 Limnol. Oceanogr. 15(5), pp 748-755.

3-010 Equation
Cd

Species	Value	Condition
bloodplasma of dogs	cn(t) = 3100*e**(- .3308t)+830*e**(- .12166t)+20*e**(- 1025t)	0.15mg cd/kg,intrav.inject.,cn(t) = cd in blood at time t,t=minutes[1]
bloodplasma of dogs	cn(t) = 3350*e**(- .29267t)+510*e**(- .10613t)+15*e**(- .00383t)	0.15mg cd/kg,intrav.inject.,cn(t) = cd in blood at time t,t=minutes[1]
bloodplasma of dogs	cn(t) = 4150*e**(- .39346t)+900*e**(- .12552t)+30*e**(- .03401t)	0.15mg cd/kg,intrav.inject.,cn(t) = cd in blood at time t,t=minutes[1]
bloodplasma of dogs	cn(t) = 810*e**(- .46247t)+320*e**(- .12343t)+18*e**(- .00476t)	0.15mg Cd/kg,intrav.inject.,cn(t) = cd in blood at time t,t=minutes[1]
Cd content of testplant	y = 105+4.0x	r = .998,y=Cd in plant,x=cd extract by ammoniumacet.at ph=7,soil - ph=6[2]
Cd content of testplant	y = 102+2.4x	r = .992,y=Cd in plant,x=cd extract by ammoniumacet.at ph=7,soil - ph=7.2[2]
Homo sapiens, body burden	b(t) = a*(1 - k**t)/(1 - k)	b(t) = body burden at time t,a=abs.amount/d[1]
Homo sapiens, body burden	b(t) = a*(1 - k**t)/(1 - k)	k = fraction of body burden remaining after 1day[1]
Homo sapiens, body burden	b(n) = a1*(1 - k**365)/(1 - k)*(k**365)**(n - 1)+..+an*(1 - k**365)/(1 - k)	b(n) = body burden after n yrs,a1=abs.amount/d(const.dur.1yr)[1]
Homo sapiens, body burden	b(n) = a1*(1 - k**365)/(1 - k)*(k**365)**(n - 1)+..+an*(1 - k**365)/(1 - k)	k = fraction of body burden remain.after 1day[1]
Homo sapienss,smokers	accumulation; y = 0.188x+6.8	y = Cd in kidney,liver and lungs(mg),x=cigarett.packs/d*time(y) of smoking[1]
mice	y = 13.45+0.434x	r = .42,y=cd in feces(ng)/24hr,x=cd in body(ug), excr with feces[1]
mice	y = 119.0+0.274x	r = 0.36,y=cd in feces(ng)/24hr,x=cd in body(ug), excr with feces[1]
mice	y = 13.45+0.434x	inject.of 0.025mg cd/kg day(5 days/w), excr with feces[1]
mice	y = 119.0+0.274x	inject.of 0.25 mg cd/kg day(5 days/w), excr with feces[1]
mice	y = 0.112x+1.785	y = cd in urine(ng/24hr),x=body burden of cd(ug)[1]
soil - Cd extracted in ammoniumacetate	y = 17+1.9x	y = extracted cd,x=added sewage sludge betw. 0 - 58t/ha,solvent - ph=4.8[2]
soil - cd extracted in 2 m nitric acid	y = 99+2.7x	r = .995,y=extracted cd,x=amount of sewage sludge,solvent - ph=4.8[2]
urinary excretion from mice	y = 0.015+0.0002x	r = .98,y=cd in urine(ug)/24hr,x=cd in body(ug)[1]
urinary excretion from mice	y = 0.011+0.000096x	r = .93,y=cd in urine(ug)/24hr,x=cd in body(ug)[1]

urinary excretion from mice	y = 0.015+0.0002x	inject.of 0.5mg cd/kg day(5days/w)[1]
urinary excretion from mice	y = 0.011+0.000096x	inject.of 0.25mg cd/kg day(5 days/w)[1]
urinary excretion from mice	y = 0.112x+1.785	approx.,0.25mg cd/kg day(5days/w),subcut.inject.[1]

Source: 1.Friberg, L., M. Piscater, G.F. Nordberg and T. Kjellstroem. Cadmium in the Environment 2nd edition,RC-Press.
2.Walne, P.R. 1972. J. Mar. Biol. Ass. U.K., 52, pp 345-374.

3-011 Equation
Cd concentration versus PO4 - P concentration

Species	Value	Condition
Seawater	Y = - 3.6 + 34.9*X	Y = ng Cd/l, X = umol PO4, r = 0.998, off central Calif Coast, 1977[1]
Seawater	Y = - 0.032 + 0.31*X	Y = nmol Cd, X = umol PO4, r = 0.998, off central Calif Coast, 1977[1]

Source: 1.Bruland, K.W., G.A. Knauer and J.H. Martin. 1978. Limnol. Oceanogr. 23, 618-625.

3-012 Equation
Cd concentration versus fish weight

Species	Value	Condition
Merlangus merlangus	Log(Cd) = 0.3 + 0.1*log(W)	Cd in ug/g wet weight, Weight (W) in g, Severn Estuary, UK[1]
Merlangus merlangus	Log(Cd) = 0.3 + 0.8*log(L)	Cd in ug/g wet weight, Length (L) in cm, Severn Estuary, UK[1]
Boreogadus saida	Log(Cd) = 0.55 + 0.1*log(W)	Cd in ug/g dry weight, Weight (W) in g, Baffin land, r = 0.128, artic cod[2]

Source: 1.Badsha, K.S. and . Sainsbury. 1977. Mar. Pollut. Bull. 8, 164-166.
2.Bohn, A. and R.O. McElroy. 1976. J. Fish. Res. bd Canada 33, 2836-2840.

3-013 Equation
Cd concentration in liver versus concentration in kidney

Species	Value	Condition
Chicken	Y = 1.2846 + 5.4394*X	Y = kidney (ug/g ww), X = liver (ug/g ww), r=0.94, n=47[1]
Pig	Y = 1.3457 + 4.0104*X	Y = kidney (ug/g ww), X = liver (ug/g ww), r=0.92, n=23[1]
Cattle	Y = 0.3286 + 4.5427*X	Y = kidney (ug/g ww), X = liver (ug/g ww), r=0.85, n=12[1]

Source: 1.Lee, K.C., Cunningham, Paulsen, Liang, and R.B. Moore. 1976. Physiol. Plant. 36, 4-6.

3-014 Equation
Coefficient of exponential growth as a function of temperature

Species	Value	Condition
Armadillidium vulgare	log(k) = - 2.4761 + 0.03657*t	k = coef. of growth (1/d), t= cel deg., 278 - 304.5 k[1]

Source: 1.Hubbel, S.P. 1971. in: Patten, B.C. (ed.); Systems Analysis and Simulation in Ecology. 1, pp 269-324Academic Press. New York.

3-015 Equation
Cu

Species	Value	Condition
Daphina magna	y = 8*x - 0.2	y = reciprocal survival time (10**3/t),x=cuco3 activity (micro m),ph=8[1]
Daphina magna	y = 156*x - 0.2	y = reciprocal survival time (10**3/t),x=cu activity (micro m),ph=8[1]

Source: 1.Andrew, R.W., K.E. Biesinger and G.E. Glass. 1977. Water Res., 11(3), pp 309-315.

3-016 Equation
Daily metabolism as a function of body weight

Species	Value	Condition
Apodemus flavicollis	y = 2.364*w**(- 0.50)	y = metabolism (kcal/(g*d)),w=wet weight (g)[1]
Clethrionomys glareolus	y = 2.510*w**(- 0.50)	y = metabolism (kcal/(g*d)),w=wet weight (g)[1]
Crustaceans	q = 19.24*w**0.81	q = daily met (cal/(anim*d)), w=wet w (g), 293 k[2]
mammals	q = 422*w**0.75	q = daily met (cal/(anim*d)), w=wet w (g), basal metabolism[3]
poikilothermic metazoan animals	q = 16.54*w**0.75	q = daily met (cal/(anim*d)), w=wet w (g), standard metabolism[4]
Protozoa	q = 2.02w**0.755	q = daily met (cal/(anim*d)), w=wet w (g), 293 k[4]
temperate fish	q = 35.8*w**0.81	q = daily met (cal/(anim*d)), w=wet w (g), resting, 293 k[5]

Source: 1.Grodzinsky, W., B. Bobek, A. Drozdz and A. Gorecki. 1970. in: Petrusewitz, K. and L. Ryszkowsk (eds)Energy Flow through Small Mammal Populations, pp 291-298. PWN. Warszawa.
2.Winberg, G.G. 1950. Z. Obshch. Biol. 11, pp 367-380.
3.Kleiber, M. 1961. The Fire Of Life. An Introduction to Animal Energetiets. J. Wiley. New York.
4.Hemmingsen, A.M. 1960. Rep. Steno meml. Hosp. 9, part 2, pp 1-110.
5.Winberg, G.G. 1956. Fish. Res. Bd. Can. Transl. Ser., No. 194.

3-017 Equation
Degradation as a function of redox potiental and time

Species	Value	Condition
trifluralin	y = 96.4 - 17.6*x + 0.7*x**2	y = % pest. remaining, x=time (d), redox = - 50 - +50 mv, r=0.95[1]
trifluralin	y = 99.0 - 1.95*x	y = % pest. remaining, x=time (d), redox = +150 - 450 mv, r=0.86[1]

Source: 1.Willis, G.H., R.C. Wander and L.M. Southwick. 1974. J. Environ. Qual. 3(3), pp 262-265.

3-018 Equation
Dissolved oxygen saturation vs barometric and altitude

Species	Value	Condition
water	cs' = cs*((pb - pv)/(760 - pb))	cs' = O2 sat (mg/l),cs=cs' at sealevel, F049[1]

Source: 1.American Public Health Association. Standard Methods for the Examination of Water and Wastewater. 193. 13th Edition. Wash, D.C.

3-019 Equation
Dissolved oxygen saturation vs salinity

Species	Value	Condition
water	cs = cs(0)*(1 - s*10**(- 5))	cs = dissolved O2 sat (mg/l),cs(0)=cs for s=0, s=salinity (mg chlor/l)[1]

Source: 1.Fair, G.M., J.C. Geyer and D.A. Okun. 1968. Water and Wastewater Engineering, Vol 2. Water Purificaton and Wastewater Treatment and Disposal. John Wiley and Sons. New York

3-020 Equation
Dissolved oxygen saturation vs temperature

Species	Value	Condition
water	cs = 14.652+0.41022*t+0.0079910*t**2 - 0.000077774*t**3	cs = dissolved O2 sat (mg/l),t=water temp (cel),standard pressure,s=o[1]

Source: 1.Elmore, H.L. and T.W. Hayes. 1960. J. San. Eng. Div. ASCE. 86, SA4, pp 41.

3-021 Equation
Dissolved oxygen saturation vs temperature and salinity

Species	Value	Condition
seawater	cs = 14.6244 - 0.367134*t+0.0044972*t**2 - 0.09665+0.00205*s*t+0.00027*s**2	cs = O2 sat (mg/l), t=temp (cel), s=salinity (1/1000)[1]

Source: 1.Carritt, D.E. and E.J. Green. 1967. Mar. Res. 25(2)

3-022 Equation
Drift as a function of benthos density

Species	Value	Condition
Ephemerella needhami	d = 0.29x	d = drift(numb./night),x=density(no./0.03m**2),r=0.975,n=8,high food[1]
Ephemerella serrata gp.	d = 0.54x	d = drift(numb./night),x=density(no./0.03m**2),r=0.896,n=12,low food[1]
Ephemerella serrata gp.	d = 0.29x	d = drift(numb./night),x=density(no./0.03m**2),r=0.937,n=12,high food[1]
Ephemerrella needhami	d = 0.51x	d = drift(numb./night),x=density(no./0.03m**2),r=0.992,n=8,low food[1]
Hydropsychidae	d = 0.28x	d = drift(numb./night),x=density(no./0.03m**2),r=0.991,n=6[1]
Hydropsychidae	d = 0.36x	d = drift(numb./night),x=density(no./0.03m**2),r=0.998,n=8,low food[1]
Hydropsychidae	d = 0.22x	d = drift(numb./night),x=density(no./0.03m**2),r=0.980,n=7,high food[1]
Simuliidae	d = 13.45 + 0.056x	d = drift(numb./night),x=density(no./0.03m**2),r=0.950,n=8[1]
Tricorythodes sp.	d = 0.13x	d = drift(numb./night),x=density(no./0.03m**2),r=0.896,n=12,low food[1]
Tricorythodes sp.	d = 0.13x	d = drift(numb./night),x=density(no./0.03m**2),r=0.896,n=12,high food[1]

Source: 1.Hildebrand, S.G. 1974. Limnol. Oceanogr., 19(6), pp 951-957.

3-023 Equation
Eddy diffusion

Species	Value	Condition
streams	log(k*u/v*el/(v*h) = 6.45 - 0.762*log(r*v*h/p)	el = long turbulent diff coefficient, F050[1]

Source: 1.Battelle. 1974. Development of a Mathematical Water Quality Model for Gray's Harbor and the ChehalisRiver, Wash. Documentation Report. Pacific NW Lab. Richland, Wash.

3-024 Equation
Energy budget for Salmo trutta

Species	Value	Condition
change in energy content	delta(b) = c*pp - r	b = energi in body (cal), pp=energy for growth and metabol,r =resp, F054[1]
daily loss of energy in excretory products vs temp and food ratio	pu = 0.0259*t**(0.580(+ - 0.0110))*exp(- 0.299(+ - 0.0162)*c/cm))	276.8 - 293.4 k, 95 % cl.[1]
daily loss of energy in feces vs temperature and food ratio	pf = 0.212*t**(- 0.222(+ - 0.0224))*exp(0.631(+ - 0.0350)*c/cm)	276.8 - 293.4, 95 % cl.[1]
energy avaible for metabolism vs temperature and body weight	rm = 3.890*w**0.770(+ - 0.0254)*exp(0.204(+ - 0.00394)*t)	rm = energy for met (cal/d), 276.8 - 290.8 k, 95 % cl.[1]
energy avaible for metabolism vs temperature and body weight	rm = 2.215*10**7*ww**0.757(+ - 0.0208)*exp(- 0.663(+ - 0.0175)*t)	rm = energy for met (cal/d), 290.8 - 292.5 k, 95 % cl.[1]
energy avaible for metabolism vs temperature and body weight	rm = 28.833*w**0.756(+ - 0.0440)*exp(0.0325(+ - 0.0288)*t)	rm = energy for met (cal/d), 292.5 - 294.7 k, 95 % cl.[1]
energy avaible for metabolism vs temperature and body weight	rm = 3.890*w**0.770(+ - 0.0254)*exp(0.204(+ - 0.00394)*t)	rm = energy for met (cal/d), 276.8 - 290.8 k, 95 % cl.[1]
energy for growth vs energy loss in feces and excretory products	pp = 1 - pf - pu	see other part of table[1]
energy for standard metabolism vs temperature and body weight	rs = 4.126*w**0.734(+ - 0.0257)*exp(0.192(+ - 0.0125)*t)	rs = energy for standard met. (cal/d), 276.8 - 280.1 k[1]
energy for standard metabolism vs temperature and body weight	rs = 8.277*w**0.731(+ - 0.0220)*exp(0.0938(+ - 0.00407)*t)	rs = energy for standard met. (cal/d), 280.1 - 292.5 k[1]
energy in excretory products vs daily loss and total energy consumed	u = c*pu	u = energy in excretory products (cal/d),pu=fraction lost of consumated[1]
energy in feces as a function of daily loss and tatal energy consumed	f = c*pf	f = energy in feces (cal/d),pf=fraction lost of consumation=c (cal/d)[1]
energy loss in waste products vs temperature and food ration	pt = 0.00512*t**0.812*exp(0.523*(c/cmax))	pt = energy in waste (cal),cmax=max of c=food ratio,t=cel deg,277 - 280 k[1]
energy loss in waste products vs temperature and food ration	pt = 0.0477*t**0.801*exp(0.223*(c/cmax))	pt = energy in waste (cal),cmax=max of c=food ratio,t=cel deg,280 - 292 k[2]
energy reqiured for maintenance metabolism vs temp and body weight	rmain = 3.802*w**0.723(+ - 0.0998)*exp(0.245(+ - 0.0811)*t)	rmain = energy req. for main. met (cal/d), 276.8 - 280.1 k, 95 % cl.[1]
energy reqiured for maintenance metabolism vs temp and body weight	rmain = 11.866*w**0.721(+ - 0.0417)*exp(0.0915(+ - 0.0112)*t)	rmain = energy req. for main. met (cal/d), 280.1 - 288.0 k, 95 % cl.[1]
maintenance energy intake vs temperature and body weight	cmain = 6.169*w**0.716(+ - 0.107)*exp(0.244(+ - 0.102)*t)	cmain = energy intake for main. (cal/d), 276.8 - 279.6 k, 95 % cl[1]
maintenance energy intake vs temperature and body weight	cmain = 12.031*w**0.737(+ - 0.058)*exp(0.105(+ - 0.016)*t)	cmain = energy intake for main. (cal/d), 279.6 - 292.5 k, 95 % cl[1]
max daily energy content of food as a function of body weight	cm = 2.902*w**(0.762(+ - 0.027))*exp(0.418(+ - 0.035)*t)	cm = max cont. (cal/d), t=cel deg, w=wet w (g), 276.8 - 279.6 k[1]
max daily energy content of food as a function of body weight	cm = 15.018*w**(0.759(+ - 0.023))*exp(0.171(+ - 0.012)*t)	cm = max cont. (cal/d), t=cel deg, w=wet w (g), 279.6 - 286.3 k[1]
max daily energy content of food as a function of body weight	cm = 26.433*w**(0.767(+ - 0.041))*exp(0.126(+ - 0.031)*t)	cm = max cont. (cal/d), t=cel deg, w=wet w (g), 286.3 - 290.8 k[1]
max daily energy content of food as a function of body weight	cm = 3.241*10**7*(0.753(+ - 0.086))*exp(- 0.662(+ - 0.054)*t)	cm = max cont. (cal/d), t=cel deg, w=wet w (g), 290.8 - 294.7 k[1]
optimum energy intake vs temperature and body weight	copt = 2.902*w**0.762(+ - 0.027)*exp(0.418(+ - 0.035)*t)	copt = optimun energy intake (cal/d), 276.8 - 279.8 k, 95 % cl.[1]

optimum energy intake vs temperature and body weight	copt = 15.116*w**0.767(+ - 0.046)*exp(0.138(+ - 0.011)*t)	copt = optimun energy intake (cal/d), 279.8 - 288.0 k, 95 % cl.[1]
optimum energy intake vs temperature and body weight	copt = 753.9*w**0.767(+ - 0.017)*exp(- 0.118(+ - 0.006)*t)	copt = optimun energy intake (cal/d), 288.0 - 292.5 k, 95 % cl.[1]
proportion of dialy energy intake lost by feces vs temp & food ration	pf = 0.212*t**(- 0.222)*exp(0.631*(c/cmax))	pf = energy in feces (cal),cmax=max of c=food ratio,t=cel deg,n=76[1]
proportion of dialy energy intake lost in excrotory products	pu = 0.0259*t**0.580*exp(- 0.299*(c/cmax))	pu = energy in exc prod. (cal),cmax=max of c=food ratio,t=cel deg,n=80[1]

Source: 1.Elliott, J.M. 1976. J. Anim. Ecol. 45, pp 923-948.
2.Elliott, J.M. 1975. J. Anim. Ecol. 44, pp 805-821.

3-025 Equation
Energy used for maintenance as a function of assimilated energy

Species	Value	Condition
heterotherms	log(r) = 1.0733*log(p) + 0.3757	r = maintenance, p=ass. energy, units=kcal/(m2*y)[1]
homeotherms	log(r) = 1.1740*log(p) + 0.1352	r = maintenance, p=ass. energy, units=kcal/(m2*y)[1]

Source: 1.McNeill, S. and J.H. Lawton. 1970. Na ture 225, pp 472-474

3-026 Equation
Energy used for maintenence as a function of assimilated energy

Species	Value	Condition
aquatic heterotherms	y = 0.811*x	y = maintenance, x=assimilation, units=kcal/(m2*y)[1]
terrestrial heterotherms	y = 0.548*x	y = maintenance, x=assimilation, units=kcal/(m2*y)[1]

Source: 1.McNeill, S. and J.H. Lawton. 1970. Na ture 225, pp 472-474

3-027 Equation
Excretion of ammonium

Species	Value	Condition
Calanus	y = 0.659*x	food = biddulphia, y=excr,x=tot n ingested, units=g n/(anim*d)[1]
Crassostrea virginica	log(y) = 1.22*log(x) + 1.33	y = exc. (u moles nh4 - n/d), x=dry w., (g), r=0.84, ph=8, 293 k, s=3 %[2]
marine zooplankton	log(y) = 1.237 + 0.928*log(x)	y = nh4 exc (ng - atom/h*animal), x= dw. (mg/ani),n pacific,298,k,mixed[3]
Mercenaria mercenaria	log(y) = 0.94*log(x) + 1.33	y = exc. (u moles NH4 - n/d), x=dry w., (g), r=0.67, ph=8, 293 k, s=3 %[2]
Phronima sedentaria	e = 0.86*w**1.08	e = excretion (ug NH4 - n/(mg dw*d)), w=dry weight (mg), 286 k[4]
Phronima sedentaria	e = 0.57*w**1.26	e = excretion (ug tot - n/(mg dw*d)), w=dry weight (mg), 286 k[4]
zooplankton, Aequorea & Orchistoma & Cunina sp.	log(y) = - 1.64(se=+ - 0.10) + 1.49(se=+ - 0.18)*log(x)	y = excr (ug - atom n/h),x=body prot (mg),299 k, trop marine,r=0.93,n=[5]
zooplankton, Agalma okeni	log(y) = - 1.21(se=+ - 0.07) + 0.82(se=+ - 0.10)*log(x)	y = excr (ug - atom n/h),x=body prot (mg),299 k, trop marine,r=0.89,n=[5]
zooplankton, Athorybia rosacea & a. sp.	log(y) = - 0.88(se=+ - 0.08) + 0.48(se=+ - 0.09)*log(x)	y = excr (ug - atom n/h),x=body prot (mg),299 k, trop marine,r=0.88,n=[5]
zooplankton, Bathyphysa sibogae & Rhizophysa filiformis	log(y) = - 1.55(se=+ - 0.18) + 0.80(se=+ - 0.17)*log(x)	y = excr (ug - atom n/h),x=body prot (mg),299 k, trop marine,r=0.81,n=[5]

zooplankton, Corella spectabilis & Gleba cordata	log(y) = - 1.68(se=+ - 0.28) + 0.86(se=+ - 0.24)*log(x)	y = excr (ug - atom n/h),x=body prot (mg),299 k, trop marine,r=0.84,n=[5]
zooplankton, Diphyes dispar	log(y) = - 1.39(se=+ - 0.08) + 0.68(se=+ - 0.09)*log(x)	y = excr (ug - atom n/h),x=body prot (mg),299 k, trop marine,r=0.94,n=[5]
zooplankton, Eurhamphea vexilligera & Cestum sp.	log(y) = - 1.22(se=+ - 0.03) + 1.03(se=+ - 0.06)*log(x)	y = excr (ug - atom n/h),x=body prot (mg),299 k, trop marine,r=0.99,n=[5]
zooplankton, Forskalia edwardsi & F. thaloides	log(y) = - 1.20(se=+ - 0.04) + 1.10(se=+ - 0.08)*log(x)	y = excr (ug - atom n/h),x=body prot (mg),299 k, trop marine,r=0.96,n=[5]
zooplankton, Nanomia bijuga	log(y) = - 1.13(se=+ - 0.07) + 0.71(se=+ - 0.14)*log(x)	y = excr (ug - atom n/h),x=body prot (mg),299 k, trop marine,r=0.91,n=[5]
zooplankton, Pegea confederata (aggregate)	log(y) = - 1.38(se=+ - 0.08) + 1.22(se=+ - 0.17)*log(x)	y = excr (ug - atom n/h),x=body prot (mg),299 k, trop marine,r=0.89,n=[5]
zooplankton, Pterotrachea hippocampus	log(y) = - 0.76(se=+ - 0.11) + 0.77(se=+ - 0.30)*log(x)	y = excr (ug - atom n/h),x=body prot (mg),299 k, trop marine,r=0.72,n=[5]
zooplankton, Rosacea cymbiformis	log(y) = - 1.43(se=+ - 0.08) + 0.95(se=+ - 0.19)*log(x)	y = excr (ug - atom n/h),x=body prot (mg),299 k, trop marine,r=0.84,n=[5]
zooplankton, Salpa cylindrica (solitary)	log(y) = - 0.85(se=+ - 0.13) + 0.64(se=+ - 0.24)*log(x)	y = excr (ug - atom n/h),x=body prot (mg),299 k, trop marine,r=0.71,n=[5]
zooplankton, Sulculeolaria quadrivalvis	log(y) = - 0.91(se=+ - 0.13) + 0.68(se=+ - 0.09)*log(x)	y = excr (ug - atom n/h),x=body prot (mg),299 k, trop marine,r=0.77,n=[5]

Source: 1.Bohn, A. and R.O. McElroy. 1976. J. Fish. Res. bd Canada 33, 2836-2840.
2.Mayzaud, P. and S. Pallot. 1973. Mar biol. 19, pp 307-314.
3.Mullin, M.M., H.J. Perry, E.H. Renger and P.M. Evans. 1975. Mar. Sci. Commun., 1(1), pp 1-13.
4.Mayzaud, P. and S. Pallot. 1973. Mar biol. 19, pp 307-314.
5.Biggs, D.C. 1977. Limnol. Oceanogr., 22(1), pp 108-117.

3-028 Equation
Excretion of phosphorus

Species	Value	Condition
Daphnia rosea	e = 0.0286*e**(0.0387*t+0.00001*d - 3.34*p)*w**(- 0.383)	e = rel.rate(ug p/mg*hr),t=temp.(c),d=food conc.(cell/ml),[1]
Daphnia rosea	e = 0.0286*e**(0.0387*t+0.00001*d - 3.34*p)*w**(- 0.383)	p = food - p(ug p/ml),w=dry wt.(mg),food=rhodotorula[1]
epilemnetic zooplankton	r = 4.5*w**(- 0.69)	r = release(ug at.p/g*hr),w=dry wt.(ug/indiv.),293k,epilimnion[2]
epilemnetic zooplankton	r = 2.3*w**(- 0.45)	r = release(ug at.p/g*hr),w=dry wt.(ug/indiv.),278k,hypolimnion[2]
hypolemnetic zooplankton	r = 22.7*w**(- 0.92)	r = release(ug at.p/g*hr),w=dry wt.(ug/indiv.),293k,epilimnion[2]
hypolemnetic zooplankton	r = 7.2*w**(- 0.99)	r = release(ug at.p/g*hr),w=dry wt.(ug/indiv.),278k,hypolimnion[2]
marine zooplankton	log(y) = 0.405 + 0.915*log(x)	y = po4 exc (ng - atom/h*animal), x= dw. (mg/anim),n pacific,298 k,mixed[3]
zooplankton	y = - 3.1 + 0.01481*x	y = exc (ug - atom p/(g*h)), x=resp (ug - atom O2/(g*h)), marine[4]
zooplankton	e = 0.252*w**(- 0.33)	e = release rate(ug p/mg*hr),w=dry wt.(mg)[5]
zooplankton	e = 0.157*w**(- 0.383)	e = rel.rate(ug/mg*hr),w=dry wt.(mg),293k,food conc.=10**5cells/ml,[1]
zooplankton	e = 0.157*w**(- 0.383)	food - p = 0.021ug p/ml,food=rhodotorula(yeast)[1]
zooplankton	e = 0.042*w**(- 0.383)	283k,food conc. = 10**5cells/ml,food - p=0.021ug p/ml[1]

Source: 1.McRoy, C.P. and R.J. Barsdate. 1970. Limnol. Oceanogr. 15(1) pp 6-13.
2.Bishop, J.W. and J.P. Barlow. 1975. Limnol. Oceanogr., 20(1), pp 148-149.
3.Mullin, M.M., H.J. Perry, E.H. Renger and P.M. Evans. 1975. Mar. Sci. Commun., 1(1), pp 1-13.
4.Satomi, M. and L.R. Pomeroy. 1965. Ecology 46, pp 877-881.
5.Joergensen, S.E. 1976. Ecol. Model. 2, pp 147-165.

3-029 Equation
Extracellular release as a function of chlorophyll a concentration

Species	Value	Condition
Skeletonema costatum	log(y) = 2.83 - 0.016*x	y = ex. release (% of photosynthesis), x=chl a (mg/m3), r=0.95[1]

Source: 1.Ignatiades, L. and G.E. Fogg. 1973. J. mar. biol. Ass. U.K., 53, pp 937-956.

3-030 Equation
Fecundity

Species	Value	Condition
Carpiodes cyprinus	f = 7172.5 + 5089.9*t	f = no. of ova/animal, t = age (y), four mile creek, ohio[1]
Carpiodes cyprinus	f = - 15690.7 + 169.1*w - 0.1*w**2	f = no. of ova/animal, l=wet w. (g), four mile creek, ohio[1]
Carpiodes cyprinus	f = - 58755 + 2984.5*l	f = no. of ova/animal, l=length (mm), four mile creek, ohio[1]
Procopium williamsoni	y = 4347.6 + 50.7*(x - 301.5)	y = eggs/fish, x=fork length (mm), sheep river, alberta[2]
Upeneus moluccensis	f = 20040*(l - 11.43)	f = no. of eggs/anim, l=fork l. (cm), china sea[3]

Source: 1.Woodward, R.L. and T.E. Wissing. 1976. Trans. Am. Fish. Soc. 105(3), pp 411-415.
2.Thompson, G.E. and R.W. Davies. 1976. Trans. Am. Fish. Soc. 105(2), pp 208-219.
3.Lee, C.K.C. 1974. in: The Kuroshio 3. pp 389. National Research Council of Thailand. Bangkok.

3-031 Equation
Filtration rate

Species	Value	Condition
Acartia sp.	y = 0.00201*w**1.011	y = mean filt. rate (ml/(anim*h)), w=dry w. (ug), cheasepeake bay,r=.79[1]
Acartia sp.	y = 0.049*w**1.213	y = max filt. rate (ml/(anim*h)), w=dry w. (ug), cheasepeake bay,r=.91[1]
Arctica islandica	f = 1.602*w**0.66	f = ml/h, w=dry w. (g), 285 k, Algae, 0.2 - 10 g[2]
CerioDaphia reticulata	log(y) = 2.44 - 0.734*log(x)	y = filt (ml/(h*anim),x=number of part/ml (3 - 24my),298 k,anim=0.8 mm[3]
Crassostrea gigas	f = 6.714*w**0.25	f = ml/h, w=dry w. (g), 293 - 295 k,seston, 0.5 - 2 g[4]
Crassostrea gigas	f = 10.392*w**0.27	f = ml/h, w=dry w. (g), 288 k, seston, 0.5 - 10 g g[4]
Eurytemora affinis	y = - 0.0043 + 0.0171*w	y = mean filt. rate (ml/(anim*h)), w=dry w. (ug), cheasepeake bay,r=.97[1]
Eurytemora affinis	y = - 0.0077 + 0.0373*w	y = max filt. rate (ml/(anim*h)), w=dry w. (ug), cheasepeake bay,r=.90[1]
Lanice conchilega	f = 0.88*w**0.32	f = filt (ml/h), w=dry w. (mg), 285 k, r=0.96, food=40*10**6 cell/l[5]
Lanice conchilega	f = 1.07*w**0.52	f = filt (ml/h), w=dry w. (mg), 285 k, r=0.96, food=40*10**6 cell/l,max[5]
Modiolus modiolus	f = 0.929*w**0.74	f = ml/h, w=dry w. (g), 285 k, Algae, 0.43 - 4 g[2]

Mytilus californianus	f = 2.270*w**0.29	f = ml/h, w=dry w. (g), 293 k, graphite, 3 - 12 g[6]
Mytilus californianus	f = 1.162*w**0.37	f = ml/h, w=dry w. (g), 289 k, graphite, 2 - 12 g[6]
Mytilus californianus	f = 0.599*w**0.21	f = ml/h, w=dry w. (g), 283 k, graphite, 0.3 - 12 g[6]
Mytilus californianus	f = 0.604*w**0.56	f = ml/h, w=dry w. (g), 283 k, graphite, 0.3 - 2 g[6]
Mytilus californianus	f = 1.835*w**(- 0.39)	f = ml/h, w=dry w. (g), 283 k, graphite, 3 - 12 g[6]
Mytilus californianus	f = 2.069*w**0.27	f = ml/h, w=dry w. (g), 289 k, graphite, 2 - 12 g[6]
Mytilus californianus	f = 2.138*w**0.48	f = ml/h, w=dry w. (g), 294 k, graphite, 0.5 - 12 g[7]
Mytilus californianus	f = 2.156*w**0.80	f = ml/h, w=dry w. (g), 294 k, graphite, 0.5 - 12 g[7]
Mytilus californianus	f = 3.962*w**0.13	f = ml/h, w=dry w. (g), 294 k, graphite, 3 - 12 g[7]
Mytilus californianus	f = 0.527*w**0.27	f = ml/h, w=dry w. (g), 279 k, graphite, 0.5 - 12 g[7]
Mytilus californianus	f = 0.511*w**0.53	f = ml/h, w=dry w. (g), 279 k, graphite, 0.5 - 2 g[7]
Mytilus californianus	f = 0.940*w**(- 0.06)	f = ml/h, w=dry w. (g), 279 k, graphite, 3 - 12 g[7]
Mytilus edulis	f = 3.846*w**0.25	f = ml/h, w=dry w. (g), 283 k, seston, 0.5 - 4 g[4]
Mytilus edulis	f = 4.788*w**0.26	f = ml/h, w=dry w. (g), 283 k, seston, 0.5 - 2 g[4]
Mytilus edulis	f = 4.212*w**0.11	f = ml/h, w=dry w. (g), 293 - 295 k,seston, 0.5 - 2 g[4]
Mytilus edulis	f = 3.046*w**0.34	f = ml/h, w=dry w. (g), 293 - 295 k,seston, 0.2 - 3.2 g[4]
Mytilus edulis	f = 2.214*w**0.27	f = ml/h, w=dry w. (g), 0.28 - 2.9 g[2]
Mytilus edulis	f = 1.656*w**0.49	f = ml/h, w=dry w. (g), 288 k, graphite, 0.058 - 1.7 g[8]
Mytilus edulis	f = 2.410*w**0.74	f = ml/h, w=dry w. (g), 285 k, 20*10**6 cell/l, dunaliella, 0.003 - 1 g[9]
Mytilus edulis	f = 1.313*w**0.73	f = ml/h, w=dry w. (g), 285 k, 40*10**6 ceel/l, dunaliella, 0.003 - 1 g[9]
Mytilus edulis	f = 1.273*w**0.38	f = ml/h, w=dry w. (g), 285 - 288 k,silt, 0.7 - 3.1 g[10]
Ostrea edulis	f = 8.796*w**0.23	f = ml/h, w=dry w. (g), 293 - 295 k, seston, 0.5 - 2 g[4]
Pecten irradians	f = 4.742*w**0.82	f = ml/h, w=dry w. (g), 297 k,Algae, 0.66 - 4.12 g[11]
Venerupis decussata	f = 4.560*w**0.28	f = ml/h, w=dry w. (g), 293 - 295 k,seston, 0.5 - 1 g[4]
Venus mercenaria	f = 4.530*w**0.18	f = ml/h, w=dry w. (g), 293 - 295 k,seston, 0.5 - 2 g[4]

Source: 1.Allan, J.D., S. Richman, D.R. Heinle and R. Huff. 1977. Mar. Biol., 43(4), pp 317-331.
2.White, W.R. 1968. Lab. Note. cent. Elect. Res. Lab. RD/L/N 116/68, pp 1-7.
3.O'Brien, W.J. 1974. Am. Midl. Nat. 91(2), pp 509-512.
4.Walne, P.R. 1972. J. Mar. Biol. Ass. U.K., 52, pp 345-374.
5.Buhr, K.-J. 1976. Mar. Biol., 38(4), pp 373-383.
6.Rao, K.P. 1953. Biol. Bull. Mar. Biol. Lab., Woods Hole, 104, pp 171-181.
7.Segal, E., K.P. Rao and T.W. James. 1953. Nature, 172, pp 1108-1109.
8.Theede, H. 1963. Kieler Meeresforsch., 19, pp 20-41.
9.Winter, J.E. 1973. Mar. Biol., 22, pp 317-328.
10.Willemsen, J. 1952. Archs. neerls. Zool., 10, pp 152-160.
11.Chipman, W.A. and J.G. Hopkins. 1954. Biol. Bull. mar. biol.Lab., Woods Hole, 107, pp 80-91.

3-032 Equation
Filtration rate as a function of dissolved oxygen

Species	Value	Condition
Daphnia galeata mendotae	f = - 0.019(+ - 0.077) + 0.169(+ - 0.016)*x	f = filt. (ml/anim*h)),x=O2 (mg/l), r=0.98[1]
Daphnia magna	f = - 0.229 (+ - 0.091) + 0.554(+ - 0.069)*x	f = filt. (ml/anim*h)),x=O2 (mg/l), O2<3.0, small anim=1.54 mm, r=0.96[1]
Daphnia magna	f = 1.015(+ - 0.122) + 0.032(+ - 0.022)*x	f = filt. (ml/anim*h)),x=O2 (mg/l), O2>3.0, small anim=1.54 mm, r=0.63[1]
Daphnia magna	f = - 1.036(+ - 0.386) + 2.259(+ - 0.303)*x	f = filt. (ml/anim*h)),x=O2 (mg/l), O2<3.0, large anim=2.34 mm, r=0.93[1]
Daphnia magna	f = 3.482(+ - 0.772) + 0.101(+ - 0.136)*x	f = filt. (ml/anim*h)),x=O2 (mg/l), O2>3.0, large anim=2.34 mm, r=0.39[1]

Source: 1.Heisey, D. and K.G. Porter. 1977. Lim nol. Oceanogr., 22(5), pp 839-845.

3-033 Equation
Food intake

Species	Value	Condition
animals	d(r)/dt = h*f*w**2/3	r = food con., w=body weight, f=feeding level(0 - 1), h=const.[1]
Calanus and Thalassiosira	y = 12.330*(1 - exp(- 0.000843*(x - 670)))	y = ingestion (10**3 cell/(anim*h)),x=food (10**3 cell/ml)[2]
Calanus and Thalassiosira	y = 13.800*(x - 767)/(785+(x - 767)	y = ingestion (10**3 cell/(anim*h)),x=food (10**3 cell/ml)[2]
Calanus and Thalassiosira	y = 4.07*x - 622	y = ingestion (10**3 cell/(anim*h)),x=food (10**3 cell/ml), x<3300[2]
Calanus and Thalassiosira	y = - 0.154*x + 13100	y = ingestion (10**3 cell/(anim*h)),x=food (10**3 cell/ml), x>3300[2]
Calanus helgolandicus, adult	i = 22.45*(1 - exp(- 0.173*p))	i = daily ingestion, p=food conc., food= elinius modestus stage ii n[3]
Calanus on Centric sp.	y = 347*(1 - exp(- 0.0583*(x - 3.08)))	y = ingestion (10**3 cell/(anim*h)),x=food (10**3 cell/ml)[2]
Calanus on Centric sp.	y = 383*(x - 3.40)/((12.2 + (x - 3.40)))	y = ingestion (10**3 cell/(anim*h)),x=food (10**3 cell/ml)[2]
Calanus on Centric sp.	y = 9.92*x - 5.32	y = ingestion (10**3 cell/(anim*h)),x=food (10**3 cell/ml), x<40[2]
Calanus on Centric sp.	y = 0.0058*x + 331	y = ingestion (10**3 cell/(anim*h)),x=food (10**3 cell/ml), x>40[2]
Calliopius lacviusculus	i = 0.847*wc**0.748	i = ingestion (ug carbon/(anim*d)), wc=body w. (ug carbon/anim),281 k[4]
Calliopius lacviusculus	i = 0.525*wn**0.792	i = ingestion (ug nitrogen/(anim*d)), wn=body w. (ug n/anim), 281 k[4]
Calliopius lacviusculus	i = 0.877*wc**0.784	i = ingestion (ug carbon/(anim*d)), wc=body w. (ug carbon/anim),281 k[4]
Calliopius lacviusculus	i = 0.394*wn**0.888	i = ingestion (ug nitrogen/(anim*d)), wn=body w. (ug n/anim), 285 k[4]
Calliopius lacviusculus	i = 2.050*wc**0.695	i = ingestion (ug carbon/(anim*d)), wc=body w. (ug carbon/anim),281 k[4]
Calliopius lacviusculus	i = 1.031*wn**0.744	i = ingestion (ug nitrogen/(anim*d)), wn=body w. (ug n/anim), 288 k[4]

Clupea harengus	r = 1.465*w**0.744	r = food intake (g/wet weight*week),w=body wet w.(g), 277.5 k[5]
Daphnia rosea	i = f*p/w	i = ing.rate(ug p/mg*hr),f=filtrat.rate-(ml/animal*hr),[6]
Daphnia rosea	i = f*p/w	p = p - conc.in food(ug p/ml),w=dry wt.(mg),food=rhodotorula[6]
herbivores, omnivores and carnivores	r = (p+t)/u	r = food req.,1/u=effic. of assim.p=max rate of prod.,t=rate of exchah.[7]
Tortanus discaudatus	i = 38.5*(1 - e**(- 0.025*p))	food = calanus pacificus(nau.3),i=ingest.rate(prey attacked),[8]
Tortanus discaudatus	i = 23.0*(1 - e**(- 0.064*p))	food = calanus pacificus(nau.5),p=number of prey/l[8]
Tortanus discaudatus	i = 32.2*(1 - e**(- 0.046*d))	food = calanus pacificus(nau.5),d=ug prey (dry wt.)/l[8]
Tortanus discaudatus	i = 23.1*(1 - e**(- 0.064*d))	food = calanus pacificus(nau.3),d=ug prey (dry wt.)/l[8]
zooplankton	i = im*(1 - e**(- k*p))	i = ingest.rate,p=prey density,im=max.ingest.rate,-k=constant[9]

Source: 1.Beyer, J.E. and H. Lassen. 1976. Multis pecies Models of an exploited Sea. IMSOR. Tech. Univ. Denmar.
2.Frost, B.W. 1975. Limnol. Oceanogr., 29(2), pp 263-266.
3.Corner, E.D.S., R.N. Head, C.C. Kilving ton and L. Pennycuick. J. Mar. Biol. Assoc. U.K., 56(2), pp45-358.
4.Dagg, M.J. 1976. Int. Rev. Gesamt. Hydrobiol. 61(3), pp 297-357.
5.De Silva, S.S. and F. Balbontin. 1974. J. Fish. Biol., 6, pp 645-658.
6.McRoy, C.P. and R.J. Barsdate. 1970. Limnol. Oceanogr. 15(1) pp 6-13.
7.Vinogradov, M.E., V.V. Menshutkin and E.A. Shushkina. 1972. Mar. biol. 16, pp 261-268.
8.Ambler, J.W. and B.W. Frost. 1974. Limnol. Oceanogr. 19(3), pp 446-451.
9.Ivlev, V.S. 1961.

3-034 Equation
Fork length

Species	Value	Condition
Micropterus salmoides	fl = 16.615 + 0.699*s	fl = fork length, s=ant. scale radius*90, applagate, calif[1]
Micropterus salmoides	fl = 23.292 + 0.648*s	fl = fork length, s=ant. scale radius*90, potter, calif[1]
Micropterus salmoides	fl = 19.023 + 0.678*s	fl = fork length, s=ant. scale radius*90, reuter l., calif[1]
Micropterus salmoides	sl = 1.7 + 0.861*fl	sl = standard length, fl= fork length[1]
Micropterus salmoides	tl = 1.6 + 1.034*fl	tl = total length, fl=fork length, calif[1]
Trachurus japonicus	l = 149*r + 0.04	l = fork length (cm), r=ant. scale radius (mm), waters[2]

Source: 1.Schultze, R.F. and C.D. Vanicek. 1974. Calif. Fish. Game, 60(2), pp 94-96.
2.Ann, H.B. 1973. Bull. Fish. Res. Dev. Agency, Pusan 10, pp 73-85.

3-035 Equation
Freshwater and sediments

Species	Value	Condition
chlorophyll derivates vs algal standing crop	log(y) = 0.316*log(x) + 0.84	y = chl deriv (units/g org carbon),x = Algae st crop (ug dw/l), r=0.83[1]
sediments clhorophyll deriv vs chlorophyll a	log(y) = 0.59*log(x) + 0.36	y = sed chl deriv (units/g org c),x = epilimnic chl a (ug/l),r=0.81,uk[1]
total carotenoids vs algal standing crop	y = 5.01 + 4.33*x	y = carot (units/g org carbon),x = Algae st crop (ug dw/l), r=0.9, UK[1]
total carotenoids vs dissolved calcium	y = 94.9*x - 10.0	y = carot (units/g org carbon),x = calcium (u eq/l), r=0.93, UK[1]
total sedimentary sulfur vs Algae standing crop	y = 0.088 + 0.086*x	y = sed sul (% of dw),x=Algae biomass (ug dw/l),r=0.76, UK[1]
total sedimentary sulfur vs dissolved sulfate	y = 3.61*x - 0.258	y = sed sulfur (% of dry weight),x=dis sulfate(u eq/l),r=0.79, UK[1]

Source: 1.Gorham, E.J., J.W.G. Lund, J.E. Sanger and W.E. Dean, Jr. 1974. Limnol. Ocean ogr., 19(4), pp 601-61.

3-036 Equation
Freshwater sediments

Species	Value	Condition
elemental hydrogen vs organic carbon	log(y) = 0.528*log(x) - 0.333	y = hydrogen,x=org carbon,units= % of dw, UK[1]
epiphasic vs hypophasic carotenoids	log(y) = 1.04*log(x) - 0.24	y = epiphasic,x=hypophasic,units = units/g org c, r=0.98, UK[1]
loss on ignition vs organic carbon	y = 3.47 + 1.78*x	y = loss on ignition,x=org carbon,units= % of dry weight,r=0.99, UK[1]
nitrogen vs organic carbon	y = 0.044 + 0.0756*x	y = nitrogen,x=org carbon,units=% of dw,r=0.97, UK[1]
total carotenoids vs chlorophyll derivatives	log(y) = 1.34*log(x) - 0.10	y = carot.,x=chla deriv.,units = units/g org carbon, r=0.93, UK[1]
total sulfur vs carotenoids	y = 56*x - 0.8	y = carot (units/g org carbon),x = sulfur (% dw), r=0.85, UK[1]

Source: 1.Gorham, E.J., J.W.G. Lund, J.E. Sanger and W.E. Dean, Jr. 1974. Limnol. Ocean ogr., 19(4), pp 601-61.

3-037 Equation
Growth rate

Species	Value	Condition
animals	d(w)/dt = h*w**(2/3) - k*w	w = body weight, h and k = constants[1]
Calliopius lacviusculus	gr = 0.112*wc*(1 - (wc/3927.3)**0.599)	gr = ug carbon/d, wc=body weight (ug carbon/anim), 281 k[2]
Calliopius lacviusculus	gr = 0.320*wn*(1 - (wn/962.1)**0.200)	gr = ug nitrogen/d, wn=body weight (ug nitrogen/anim), 281 k[2]
Calliopius lacviusculus	gr = 0.190*wc*(1 - (wc/3266.6)**0.400)	gr = ug carbon/d, wc=body weight (ug carbon/anim), 285 k[2]
Calliopius lacviusculus	gr = 0.270*wn*(1 - (wn/736.5)**0.200)	gr = ug nitrogen/d, wn=body weight (ug nitrogen/anim), 285 k[2]
Calliopius lacviusculus	gr = 0.237*wc*(1 - (wc/2562.8)**0.599)	gr = ug carbon/d, wc=body weight (ug carbon/anim), 288 k[2]
Calliopius lacviusculus	gr = 0.182*wn*(1 - (wn/586.5)**0.578)	gr = ug nitrogen/d, wn=body weight (ug nitrogen/anim), 288 k[2]

Eichhornia crassipes	y = 1 - 0.0037*(t - 29)**2	y = growth rate (1/d), t=cel deg., in marsh[3]
Hypera postica	rt = a/(b+c*exp(- d*t(time))	rt = rate of development,t(time)=temp at time=time,a,b,c,d=constants[4]

Source: 1.Beyer, J.E. and H. Lassen. 1976. Multis pecies Models of an exploited Sea. IMSOR. Tech. Univ. Denmar.
2.Dagg, M.J. 1976. Int. Rev. Gesamt. Hydrobiol. 61(3), pp 297-357.
3.Mitsch, W.J. 1976. Ecol. Modell. 2(1), pp 69-89.
4.Shoemaker, C.A. 1977. in: Hall, C.A.S. and J.W. Day, Jr. (eds.); Ecosystem Modeling in theory and Prctice. pp 546-574. John Wiley and Sons. New York.

3-038 Equation
Half life time of parathion as a function of temperature

Species	Value	Condition
parathion	t(1/2) = 5((10**3)/k - 2.8)+1	k = kelvin degrees,sterile estuarine water, pH = 7.8, salinity = 2.57 %[1]

Source: 1.Webber, K. 1976. Water Res. 10(3), pp 239-243.

3-039 Equation
Half - rate coefficient as a function of temperature

Species	Value	Condition
sulfate reducing bacteria	ks = (1.65)*(10**(- 39))*(10**(12082/t))	t = abs temp,ks=half - vel coef.(mg/l),arrhenius Equation[1]

Source: 1.Middleton, A.C. and A.W. Lawrence. 1977. J. Water Pollut. Control, 49(7), pp 1659-1670.

3-040 Equation
Hatching time of eggs as a function of temperature

Species	Value	Condition
Calanus finmarchicus	d = 2232/((t+8)**1.536)	d = time (hours), t= temp (c), 273 - 293 k[1]

Source: 1.McLaren, I.A. 1963. J. Fish. Res. bd. Can., 20, pp 685 -727.

3-041 Equation
Hg

Species	Value	Condition
fershwater fish	c = k*w**0.81	for fish = 42cm - >1.25 mg/kg;49 - >1.5.mg/kg,c=hg,conc,w=length of fish[1]
Homo sapienss	y = 0.26x	y = hg in blood(ug/l),x=hg in urine(ug/l)[2]
Scorpaena porcus	y = 1.39*10**(- 2)*x + 1.21	y = hg conc (mg/kg w.w.), x= wet weight, r=0.88, marine, meditar.[3]

Source: 1.Vinogradov, M.E., V.V. Menshutkin and E.A. Shushkina. 1972. Mar. biol. 16, pp 261-268.
2.Friberg, L. and J. Vostal. 1974. Mercury in the Environment. CRC-Press.
3.Renzoni, A., E. Bacci and L. Falciat. 1973. Rev. Int. Oceanogr. Med. CERBOM., 31-32, pp 17-45.

3-042 Equation
Hg, methylation

Species	Value	Condition
specific methylation rate	rate = a*b**n*(Hg(total))**n (microgram/g vss day)	anaerobic,n = 0.15,a=0.61,b=ratio free Hg to total inorganic Hg,303k[1]
specific methylation rate	rate = a*b**n*(Hg(total))**n (microgram/g vss day)	aerobic,n = 0.28,a=2.,b=ratio free Hg to total inorganic Hg,303k[1]

Source: 1.Bisogni, J.J. and A.W. Lawrence. 1975. Water Pollut. Control, 47, pp 135-152.

3-043 Equation
Intracellular nitrogen conc. as a function of specific growth rate

Species	Value	Condition
Chaetoceros gracilis	in = 0.068(+ - 0.007) + 0.239(+ - 0.013)*my	in = intra cell conc (mg N/mg c),my=growth (1/hour), NH4, 298 k[1]
Chlorella pyrenoidosa	in = 0.043(+ - 0.002) + 0.239(+ - 0.016)*my	in = intra cell conc (mg N/mg dw),my=growth (1/hour), no3, 308 k[1]
Coccochloris stagina	in = 0.065(+ - 0.003) + 1.78(+ - 0.10)*my	in = intra cell conc (mg N/mg c),my=growth (1/hour), no3, 298 k[1]
Cyclotella nana	in = 0.054(+ - 0.012) + 1.21(+ - 0.23)*my	in = intra cell conc (mg N/mg c),my=growth (1/hour), no3, 298 k[1]
Dunaliella tertiolecta	in = 0.023(+ - 0.009) + 1.76(+ - 0.25)*my	in = intra cell conc (mg N/mg c),my=growth (1/hour), NH4, 298 k[1]
Monochrysis lutheri	in = 0.042(+ - 0.003) + 0.78(+ - 0.09)*my	in = intra cell conc (mg N/mg c),my=growth (1/hour), NH4, 298 k[1]
Oscillatoria agardhii	in = 0.040(+ - 0.006) + 1.45(+ - 0.24)*my	in = intra cell conc (mg N/mg dw),my=growth (1/hour), no3, 293 k[1]

Source: 1.Nyholm,N. 1977. Prog. Wat. Tech. 8(4/5), pp 347-358.

3-044 Equation
Light expression

Species	Value	Condition
phytoplankton	f(i,z) = i(0)*exp(- ee*z)/ik*exp(1 - i(0)*exp(- ee*z)/ik)	F047[1]
phytoplankton	ref = 0.08+0.02*sin(0.12083*week - 1.57079)	F047[2]
phytoplankton	i(0) = ida*cfgi/fd, ih=0.7*i(0)+3*10**5	F047, unit cal/(9m**2)*day)[3]
phytoplankton	f(i) = a/((aa+bb*phyt)*v)*ln((i(0)+ik)/(i(0)*exp((- aa - bb*phyt)*v/a)+ik)	F047, growth rate of phytoplankton as f(i)[4]
phytoplankton	f(i,z) = 2.72*fd/(ee*z)*exp(ida*exp(ee*z)/ik) - exp(ida/ik)	F047[5]
phytoplankton	fd = 0.5+0.26*sin(t - 82), ida=i(0)/fd*24	F047, lattitude 60 degrees[5]
phytoplankton	i(0) = 350+300*sin(t - 82) - 75*(1+sin(2*(t - 37))) m	F047 unit = m*wh/(cm(882)*day)[5]
phytoplankton	ee = 0.5+14*pa(1),	F047[5]
phytoplankton	i(0)(t) = ida*k*(sin(2*phi/24(t - 6)))**(1/2)	F047,[6]
phytoplankton	ida = intgrl(i(0)(t)dt	F047, k = 2.6225[6]
phytoplankton	f(i) = i/ik	F047, when i<ik[7]
phytoplankton	f(i) = 1	F047, when ik<i<ih[7]
phytoplankton	f(i) = max(0,1 - klys(i - ih))	F047, when ih<i[7]
phytoplankton	i = i(0)*exp((- aa - bb*phyt - cc*zoo - dd*detritus)*z)	F047[7]

Source: 1.Steele, J.H. 1974. The Structure of Ma rine Ecosystems. Blackwell Scientific Publications Ltd. Oxfor. pp. 74-91.
2.Patten, B.C., P.A. Egloff and T.H. Richardson. 1975. Total Ecosystem Model for a Cove in Lake Texomain: B.C. Patten (ed), Systems analysis in Ecology. Vol III. AP. N.Y. pp. 206-423.
3.Nyholm, N. 1978. Ecol. Model. 4 ,pp 172-179.
4.Joergensen, S.E. 1976. Ecol. Model. 2, pp 147-165.
5.Ahlgren, I. 1973. Scripta Limn. Upsaliensa, No 334.
6.Schofields, W.K. and R.G. Krutchoff. 1974. Bulletin 60 Stochastic Model for a Dynamic Ecosystem. VPIWRRC-Bull, 60 Virginia Water Resources Research Center.
7.Gargas, E. 1976. A Three box eutrophica tion Model of a Mesotrophic Danish Lake. Water Quality Instiute.

3-045 Equation
Longitudinal dispersion coefficient

Species	Value	Condition
estuary	el = cl*e**(1/3)*le**(4/3)	el = long. disp. coefficient (length**2/time) F053[1]
estuary	el = 0.042*abs(u)*r	simplification, F053[1]
streams	dl = 3.26*u**0.607	dl = long disp coef. (ft2/min), u=average vel (ft/min), 0.33<u<3.30[2]
streams	el = 22.6*n*u*d**0.833	el = long disp coeff., d=depth of flow, n=manning coeff., u=mean veloc.[1]

Source: 1.Feigner, K.D. and H.S. Harris. 1970. Documentation Report FWQA Dynamic Estuary Model. Federal Wateruality Administration. Wash. D.C. 248 p.
2.Gloyna, E.A. 1971. Transport of Organic and Inorganic Materials in Small-scale Ecosystems. Advancesn Chem. Series. Vol. 106.

3-046 Equation
Mean length of larvae vs food concentration

Species	Value	Condition
Achirus lineatus	l = 5.23*(1 - exp(- 0.23*ln(p)))	l = length (mm), p=food conc (no/l), 16 d after hatch, 301 k,r=0.76[1]
Anchoa mitchilli	l = 15.05*(1 - exp(- 0.18*ln(p)))	l = length (mm), p=food conc (no/l), 16 d after hatch, 299 k,r=0.82[1]

Source: 1.Houde, E.D. 1977. Mar. Biol., 43(4), pp 333-341.

3-047 Equation
Metabolism as a function of body weight

Species	Value	Condition
birds	m = 0.1856*w**0.668	m = basal metab. (kcal/d), w=wet w. (g)[1]
Lizards	m = 0.0944*w**0.62	m = basal metab. (kcal/d), w=wet w. (g)[1]
mammals	m = 0.1244*w**0.75	m = basal metab. (kcal/d), w=wet w. (g)[1]

Source: 1.Turner, F.B., R.I. Jennrich and J.D. Weintraub. 1969. Ecology 50, pp 1076-1081.

3-048 Equation
Mortality of Cu vs water hardness, effect Cu

Species	Value	Condition
Fish, Salmonaidae	Log10(Y) = 0.6942 + 0.6955*log10(X)	Y = LC50 of Cu in ug/l, 96 h, X = hardness in mg/l, r = 0.829[1]
Fish, Centachidae	Log10(Y) = 1.591 + 0.9518*log10(X)	Y = LC50 of Cu in ug/l, 96 h, X = hardness in mg/l, r = 0.977[1]
Pimephales promelas	Log10(Y) = 0.1266 + 1.143*log10(X)	Y = LC50 of Cu in ug/l, 96 h, X = hardness in mg/l, r = 0.923[1]
Fish, Cyprinids	Log10(Y) = 0.2112 + 1.041*log10(X)	Y = LC50 of Cu in ug/l, 96 h, X = hardness in mg/l, r = 0.900[1]
Zebrafish, carp and killifish	Log10(Y) = 0.2399 + 1.0567*log10(X)	Y = LC50 of Cu in ug/l, 96 h, X = hardness in mg/l, r = 0.908[1]

Source: 1.Nriagu, J.O. 1979. Copper in The Environment. Health Effects. Part II. John Wiley and sons. NY.

3-049 Equation
Oxygen uptake as a function of body protein

Species	Value	Condition
zooplankton, Aequorea & Orchistoma & Cunina sp.	log(y) = 0.81(se=+ - 0.08) + 1.27(se=+ - 0.15)*log(x)	y = O2 uptake (ml O2/h), x=prot.(mg), 299 k, trop. marine,r=0.93,n=14[1]
zooplankton, Agalma okeni	log(y) = 1.07(se=+ - 0.04) + 0.87(se=+ - 0.06)*log(x)	y = O2 uptake (ml O2/h), x=prot.(mg), 299 k, trop. marine,r=0.88,n=58[1]
zooplankton, Athorybia rosacea & a. sp.	log(y) = 1.51(se=+ - 0.05) + 0.34(se=+ - 0.06)*log(x)	y = O2 uptake (ml O2/h), x=prot.(mg), 299 k, trop. marine,r=0.9,n=13[1]
zooplankton, Bathyphysa sibogae & Rhizophysa filiformis	log(y) = 0.80(se=+ - 0.16) + 0.87(se=+ - 0.15)*log(x)	y = O2 uptake (ml O2/h), x=prot.(mg), 299 k, trop. marine,r=0.85,n=13[1]
zooplankton, Corella spectabilis & Gleba cordata	log(y) = 1.22(se=+ - 0.03) + 0.86(se=+ - 0.04)*log(x)	y = O2 uptake (ml O2/h), x=prot.(mg), 299 k, trop. marine,r=0.98,n=17[1]
zooplankton, Diphyes dispar	log(y) = 1.08(se=+ - 0.05) + 0.65(se=+ - 0.06)*log(x)	y = O2 uptake (ml O2/h), x=prot.(mg), 299 k, trop. marine,r=0.93,n=17[1]
zooplankton, Eurhamphea vexilligera & Cestum sp.	log(y) = 1.28(se=+ - 0.07) + 0.73(se=+ - 0.13)*log(x)	y = O2 uptake (ml O2/h), x=prot.(mg), 299 k, trop. marine,r=0.89,n=10[1]
zooplankton, Forskalia edwardsi & F. thaloides	log(y) = 1.26(se=+ - 0.03) + 0.89(se=+ - 0.06)*log(x)	y = O2 uptake (ml O2/h), x=prot.(mg), 299 k, trop. marine,r=0.97,n=20[1]
zooplankton, Nanomia bijuga	log(y) = 1.36(se=+ - 0.02) + 0.70(se=+ - 0.04)*log(x)	y = O2 uptake (ml O2/h), x=prot.(mg), 299 k, trop. marine,r=0.96,n=19[1]
zooplankton, Pegea confederata (aggregate)	log(y) = 1.30(se=+ - 0.04) + 1.06(se=+ - 0.09)*log(x)	y = O2 uptake (ml O2/h), x=prot.(mg), 299 k, trop. marine,r=0.94,n=21[1]
zooplankton, Pterotrachea hippocampus	log(y) = 1.32(se=+ - 0.03) + 0.86(se=+ - 0.08)*log(x)	y = O2 uptake (ml O2/h), x=prot.(mg), 299 k, trop. marine,r=0.97,n=8[1]
zooplankton, Rosacea cymbiformis	log(y) = 0.75(se=+ - 0.08) + 1.24(se=+ - 0.19)*log(x)	y = O2 uptake (ml O2/h), x=prot.(mg), 299 k, trop. marine,r=0.89,n=13[1]
zooplankton, Salpa cylindrica (solitary)	log(y) = 1.61(se=+ - 0.04) + 0.71(se=+ - 0.08)*log(x)	y = O2 uptake (ml O2/h), x=prot.(mg), 299 k, trop. marine,r=0.94,n=13[1]
zooplankton, Sulculeolaria quadrivalvis	log(y) = 1.61(se=+ - 0.08) + 0.79(se=+ - 0.15)*log(x)	y = O2 uptake (ml O2/h), x=prot.(mg), 299 k, trop. marine,r=0.88,n=10[1]

Source: 1.Biggs, D.C. 1977. Limnol. Oceanogr., 22(1), pp 108-117.

3-050 Equation
Oxygen uptake as a function of body weight

Species	Value	Condition
Apis mellifica	ox = 1.78*w**0.91	ox = O2 upt. (ul/(amin*h)), w=wet w(mg)[1]
arthropod species	log(y) = 450 - 0.31*log(x)	y = O2 uptake (ul O2/(g*h)), x=wet weight (mg)[2]
Asellus aquaticus	ox = 0.350*w**0.867	ox = O2 upt. (ul/(mg*h)), w=wet w(mg), 1.580 - 22.53 mg,first 100 d,293 k[3]
Chela bacaila	log(m) = log(0.4655) + 0.76*log(w)	m = O2 uptake (mg O2/h), w = wet w.(g), freshwater fish.[4]
Daphnia pulex	log(o) = log(0.0014) + 0.881*l0g(w)	o = O2 uptake (ul O2/(anim*h), w= dry w (mg)[5]
insectivores	y = 29.64*w**(- 0.57)	y = O2 uptake (cc O2/(g*h)), w= wet weight (g)[6]
insectivores	y = 26.80*w**(- 0.50)	y = O2 uptake (cc O2/(g*h)), x= wet weight (g), exponent forsed to .50[6]
insectivores	y = 3.087*w**(- 0.50)	y = O2 uptake (kcal/(g*d)), x= wet weight (g), exponent forsed to .50[6]
Macrocyclops albidus	ox = 1.22*w**0.84	ox = O2 uptake (ul*10**(- 3)/(anim*h)), w= wet weight (ug),all stages[7]

Macrocyclops albidus	ox = 1.25*w**0.34	ox = O2 uptake (ul*10**(- 3)/(anim*h)), w= wet weight (ug), nauplii[7]
Macrocyclops albidus	ox = 0.76*w**1.02	ox = O2 uptake (ul*10**(- 3)/(anim*h)),w=wet w (ug), adult & copepodit[7]
Macrocyclops albidus	ox = 0.372*w**0.840	ox = O2 upt. (ul/(mg*h)), w=wet w(mg), 0.001 - 0.040 mg, whole, 293 k[3]
Perlodes intricata	ox = 1.600*w**0.659	ox = O2 upt. (ul/(mg*h)), w=wet w(mg),90 - 200 mg, 2nd life cycle, 293 k[3]
Phronima sedentaria	r = 2.84*w**0.57	r = O2 uptake (ul O2/(mg dw*day), w=dry weight (mg), 286 k[8]
Pogonomyrmex badius	ox = 8.06*w**0.91	ox = O2 upt. (ul/(amin*h)), w=wet w(mg), worker[9]
Pogonomyrmex badius	ox = 1.63*w**0.91	ox = O2 upt. (ul/(amin*h)), w=wet w(mg), soldier[9]
Polydesmus complanatus	ox = 0.841*w**0.680	ox = O2 upt. (ul/(mg*h)), w=wet w(mg), 5 - 90 mg, 2nd life cycle, 293 k[3]
Rhisoglyphus echinopus	ox = 0.510*w**0.830	ox = O2 upt. (ul/(mg*h)), w=wet w(mg), 0.001 - 0.050 mg, whole, 293 k[3]
rodents	y = 19.94*w**(- 0.50)	y = O2 uptake (cc O2/(g*h)), w= wet weight (g), exponent forced to .50[6]
rodents	y = 2.297*w**(- 0.50)	y = O2 uptake (kcal/(g*d)), w= wet weight (g), exponent forced to .50[6]
rodentsvores	y = 17.76*w**(- 0.46)	y = O2 uptake (cc O2/(g*h)), w= wet weight (g)[6]
Salmo gairdneri	r = 0.192*w**0.654	r = oxygen consump (cc/h),w=wet weight (g), starvation[10]
Salmo gairdneri	r = 0.366*w**0.813	r = oxygen consump (cc/h),w=wet weight (g), maximum food level[10]
Salmo gairdneri	r = 0.229*w**0.789	r = oxygen consump (cc/h),w=wet weight (g), normal metabolism[10]
Simocephalus vetulus	ox = 0.512*w**0.874	ox = O2 upt. (ul/(mg*h)), w=wet w(mg), 0.017 - 0.350 mg, whole, 293 k[3]
Tribolium castaneum	ox = 6.653*w**0.844	ox = O2 upt. (ul/(mg*h)), w=wet w(mg), 0.020 - 3.500 mg, feeding, 293 k[3]
Tribolium castaneum	ox = 6.65*w**0.84	ox = O2 upt. (ul/(amin*h)), w=wet w(mg), larvae 2 - 7, 0.022 - 2.26 mg[11]
Tribolium castaneum	ox = 0.24*w**0.91	ox = O2 upt. (ul/(amin*h)), w=wet w(mg), ants (pupae)[12]
Tribolium castaneum	r = 13.84*w**0.93	r = O2 resp rate (ul O2/h), w=wet w (mg), mean for different stages[13]

Source: 1.Duncan, A. and R.Z. Klekowski. 1975. in: Grodzinsky, W., R.Z. Klekowski and A. Duncan (eds.); Methodfor Ecological Bioenergeticts. IBP Handbook No. 24, pp 97-147. Blackwell. Oxford.
2.Eppley, R.W., F.M.H. Reid and J.D.H. Strickland. 1970. Bull. Scripps Inst. Oceanogr. 17, pp 33-42.
3.Kamler, E. 1970. Pol. Arch. Hydrobiol. 17, pp 201-216.
4.Lohgaonker, A.L. 1974. Curr. Sci., 43(18), pp 584-585.
5.Richman, S. 1958. Ecol. Monogr. 28, pp 273-291.
6.Grodzwsky, W. 1975. in: Grodzinsky, W., R.Z. Klekowsky and A. Duncan (eds). Methods for Ecological Boenergetics IBP Handbook No. 24, pp 65-94. Blackwell, Oxford.
7.Klekowski, R.Z. and E.A. Shushkina. 1966. Verh. int. Ver. Limnol. 16, pp 399-418.
8.Mayzaud, P. and S. Pallot. 1973. Mar biol. 19, pp 307-314.
9.Golley, F.B. and J.B. Gentry. 1964. Ecology 45, pp 217-225.
10.Staples, D.J. and M. Nomura. 1976. J. Fish. Biol. 9, pp 29-43.
11.Klekowsky, R.Z., T. Prus and H. Zyromska-Rudzka. 1967. in: Petrusewitz, K. (ed). Secondary Productivty of Terrestrial Ecosystems, pp 859-879, PWN, Warszawa.
12.Winberg, G.G. 1956. Uc. Zap. Beloruss. gos. Univ., Ser. Biol. 26, pp 243-254.
13.Klekowsky, R.Z., T. Prus and H. Zyromska-Rudzka. 1967. in: Petrusewitz, K. (ed). Secondary Productivty of Terrestrial Ecosystems, pp 859-879, PWN, Warszawa.

3-051 Equation
Oxygen uptake as a function of oxygen conc. in the water

Species	Value	Condition
benthal uptake	db = k*c**0.3/h*ro**(t - 20)	db = O2 uptake,h=water depth, c=O2 conc (mg/l),ro**(t - 20)=temp factor[1]
freshwater sediment (depth = 10m)	y = 0.13*x+1.9	y = sedim.O2 uptake(ml/(m**2)*h,x=O2 sat.(%),283k,l.vidostern,r=0.99[2]
freshwater sediment (depth = 20m)	y = 0.17*x+3.1	y = sedim.O2 uptake(ml/(m**2)*h,x=O2 sat.(%),283k,l.vidostern,r=0.96[2]
freshwater sediment (depth = 30m)	y = 0.15*x+6.0	y = sedim.O2 uptake(ml/(m**2)*h,x=O2 sat.(%),283k,l.vidostern,r=0.95[2]
freshwater sediment (depth = 38m)	y = 0.14*x+6.2	y = sedim.O2 uptake(ml/(m**2)*h,x=O2 sat.(%),283k,l.vidostern,r=0.99[2]
sediments	y = 26.7*x**0.51/(z**0.36)	y = O2 uptake, x=ann. primary prod., z=mixed layer depth, sea & lake[3]

Source: 1.Lombardo, P.S. 1972. Hydrocomp, Inc. Mathematical Model of Water Quality in Rivers and Impoundsments
2.Granek, W. 1977. Vatten 3, pp 251-265.
3.Smith, K.L. Jr. 1974. Limnol. Oceanogr. 19(6), pp 939-944.

3-052 Equation
Oxygen uptake as a function of temperature

Species	Value	Condition
Bacteria	y = 1.2*x**0.1	y = O2 uptake (ml O2/m2*h),x= cel deg, freshwater sediment,shallow wat.[1]
Diaptomus	u = 0.0364*t - 0.3418*log(w) + 0.6182	u = O2 uptake (ul O2/(mg dw*h),t=temp (c), w=dry weight (mg)[2]
Hyalella azteca	ln(ox(micro l. o(2)/(amphipod*hour)) = - 12.8+ 11.13*ln(x) + 1.98*ln(c)	ox = oxygen uptake,x = dry weight(micro g), c = temp in c[3]
Mytilus edulis	y = 0.018 + 0.006084*t	y = O2 uptake (mg O2/(g ww*h)),t=temp (cel)[4]
Panterpe insignis	y = 5 - 0.417*x	y = O2 uptake (ml O2/(g*h)),x=cel deg, birds in torpor[5]
Panterpe insignis	y = 14 - 0.321*x	y = O2 uptake (ml O2/(g*h)),x=cel deg, resting birds[5]

Source: 1.Hargrave, B.T. 1969. J. Fish. Res. bd. Can. 26, pp 2003-2026.
2.Comita, G.W. 1968. Limnol. Oceanogr., 13, pp 51-57.
3.Hargrave, B.T. 1970. Limnol. Oceanogr., 15(1), pp 21-30.
4.De Vooys, C.G.N. 1976. Mar. Biol., 36, pp 25-30.
5.Wolf, L.L. and F.R. Hainsworth. 1972. Comp. Biochem. Physiol., 41, pp 167-173.

3-053 Equation
Oxygen uptake vs % organic matter

Species	Value	Condition
sediments	log(y) = - 2.15 + 0.90*log(x)	y = O2 uptake (mg O2/(g dw*h)),x= % org matter, r=0.91, UK[1]

Source: 1.Hargrave, B.T. 1972. Limnol. Oceanogr., 17(4), pp 583-596.

3-054 Equation
Oxygen uptake vs particle diameter

Species	Value	Condition
sediment detritus	log(y) = 1.45 - 0.52*log(x)	y = O2 uptake (mg O2/(gram detritus*h)),x= particle dia (um), r=0.65[1]
sediment sand	log(y) = - 0.20 - 0.85*log(x)	y = O2 uptake (mg O2/(g sand*h)),x= particle dia (um), r=0.95[1]
sediments	log(y) = 2.07 - 0.69*log(x)	y = O2 uptake (mg O2/(g org matter*h),x= particle dia (um), r=0.79[1]

Source: 1.Hargrave, B.T. 1972. Limnol. Oceanogr., 17(4), pp 583-596.

3-055 Equation
Photosynthesis

Species	Value	Condition
algal suspensions	a = a(max)*b*i/sqrt(1+(b*i)**2)	a = photosyn, i=light energy, b=constant, without inhibition[1]
algal suspensions	a = a(max)*b*i/sqrt(1+(b*i)**2)*q(ii*c)	a = photosyn, i=light energy, b,c=const.,q(i*c)=function for inhibition[1]
phytoplankton	photo = my(t)*phyt*min(i/I0,ps/ps0,ns/ns0)	F047[2]
phytoplankton	photo = my(t)*(i/(i+ik)*ns/(kn+ns)*ps/(kp+ps)),my(t)=mymax*f(t)	F047[3]
phytoplankton	photo = my(t)*i*ps*ns*phyt	F047[4]
phytoplankton	photo = mymax*phyt*(ps+ns)	F047[5]
phytoplankton	photo = my(i)*f(i)*ps/(kp+ps)*ns/(kn+ns)*cs/(kc+cs),- my(t)=mymax*f(i)	F047[6]
phytoplankton	photo = my*phyt*ps*i*t	F047[7]
phytoplankton	photo = my*phyt*f(ps)*fd*f(i)	F047[8]
phytoplankton	photo = my*phyt*f(i)*(f(ns)+f(ps)+f(cs)/3)	F047[9]
phytoplankton	photo = my*phyt*min(f(i),f(ps),f(ns)),	F047[10]
phytoplankton	up = upmax*(1/(1+pk1*pa) - 1/(1+pki*ps))	F047[11]
phytoplankton	my = mymax*f(i)*f(ps)*f(ns)*f(t)*fd*fac	F047[12]
phytoplankton	photosynthesisto = my(t)*i/(iop(t))*exp(1 - i/iop(t))*f(pa)*f(na)- F047[13]	
phytoplankton	my(t) = 0.02*exp(0.17*t), iop(t)=0.06*exp(0.22*t), f(pa)=pa/(pa - pmin)	F047[13]
phytoplankton	f(na) = na/(na - nmin), up=upmax*ps/(kp+ps), un=unmax*ns/(kn+ns)	F047[13]
phytoplankton	photosynthesisto = my*f(i)*f(na,pa), f(na,pa)=2/(1/f(pa)+1/f(na))	F047[14]
phytoplankton	f(pa) = (kpa+pamax - x)/(pamax - x)*(pa/phyt - x)/kpa+pa/phyt - x)pa+pa/phyt - p	F047,x = pamin[14]
phytoplankton	f(na) = (na/phyt - namin)/(namax - namin),F047[14]	
phytoplankton	dps = pamax*dphyt, dns=namax*dphyt, up=my*pamax, un=my*namax	F047,excess supply of nutrients[14]
phytoplankton	up and un equal to supply	F047,limiting conditions[14]
phytoplankton	photo = mymax*f(t)*f(pa)*f(na)*f(ca), f(pa)=(pa - pamin*phyt)/pa	F047[15]
phytoplankton	f(na) = (na - namin*phyt)/na, f(ca)=(ca - camin*phyt/ca	F047[15]
phytoplankton	uc = f(i)*ucmax*((camax*phyt - ca)/(camax*phyt - camin*phyt))*phyt*cs/x	F047,x = kc+cs[15]

phytoplankton	up = upmax*((pamax*phyt - pa)/(pamax*phyt - pamin*phyt))*phyt*ps/(kp+ps)	F047, un analog[15]
plants	pn = a*(1 - exp(- b*lf)) + c	pn = net photosynthesis, lf=light flux density, a,b,c=const.[16]
trees	r = a*(1 - exp(- b*(al - c))	r = relative rate of photosynthesis,al=aviable light,a,b,c=constants[17]

Source: 1.Talling, J.F. 1957. New Phytol 56, pp 29-50.
2.Broquist, S. 1971. Matematisk modell foe#r Ekosystemet i en Sjoe. Matematiske Institution, Tekniskaoegskolan, Stockholm.
3.Chen, C.W. and G.T. Orlob. 1975. in: Patten, B.C. (ed.); Systems Analysis and Simulation in Ecology,vol 3. pp 476-537. Acad. Press. N.Y.
4.Parker, R.A. 1976. Int. J. System Sci. 7. 957-962.
5.Anderson, J.M. 1973. In; D. Meadows (ed); Towards Global Equlibrium. M.I.T.-Press.
6.Dahl-Madsen, K.I. and S. Nielsen. 1974. Vand, 5, pp 24-31. Cph.
7.Jansson, B.O. 1971. Ecosystem Approach to Baltic Problem. Bulletins from the Ecological Research Comittes, no 16, Swedish Natural Science Research Council.
8.Lassen, H. and P.B. Nielsen. 1972. Simple Mathematical Model for the pp as a Function of the phosphae Conc and the incoming solar Energy applied to the North Sea. Danmarks Hav- og Fiskeri. ICES.
9.Patten, B.C., P.A. Egloff and T.H. Richardson. 1975. Total Ecosystem Model for a Cove in Lake Texomain: B.C. Patten (ed), Systems analysis in Ecology. Vol III. AP. N.Y. pp. 206-423.
10.Larsen, D.P., H.T. Mercier and K.W. Malveg. 1974. in: E.J. Middlebrooks, D.H. Falkenborg and T.E. Mooney (eds). Modeling the eutrophication Process. Ann Arbor, Mich. pp. 15-33.
11.Bierman, V.J, F.H. Verhoff, T.C. Poulson and M.W. Tenney. 1974. In; Middle brooks, E.J., D.H. Falkenorg and T.E. Moloney (eds). Modelling the eutrophication Process. Ann Arbor, Mich. pp 15-33.
12.Gargas, E. 1976. A Three box eutrophica tion Model of a Mesotrophic Danish Lake. Water Quality Instiute.
13.Cloern, J.E. 1978. Ecol. Model. 4. pp. 165-177.
14.Nyholm, N. 1978. Ecol. Model. 4 ,pp 172-179.
15.Joergensen, S.E. 1976. Ecol. Model. 2, pp 147-165.
16.Fukai, S., S. Koh and A. Kumura. 1976. J. appl. Ecol. 13(3), pp 877-887.
17.Botkin, D.B. 1977. in: Hall, C.A.S. and J.W. Day, Jr. (eds.); Ecosystem Modeling in theory and practce. pp 214-233. John Wiley and Sons. New York.

3-056 Equation
Production

Species	Value	Condition
aboveground	p = 0.502 - 0.240*log(t)	p = prod. (kcal/(m2*y), t=age of abandoned agricultural field (y)[1]
Algae	y = 0.05*x + 0.02 (mg oxygen/mg chl. *h)	x in centigrades[2]
Betula maximowicziana	p = 0.145*f	p = production (kg/(m2 ground area*year),f=leaf area (m2/m2 ground)[3]
Betula maximowicziana	y = 0.148*x	y = production (kg/(tree*year), x=leaf area (m2/tree)[3]
branch wood in trees vs stand density	y = 0.18 + 0.317*x	y = annual production (kg ww),x=area per tree(1/m2),Pinus sylvestris[4]
Cladoceran consumers and net primary productivity	y = - 250*x - 900	y = prod (g c/(m2*y), x=shannon diversity, in lakes[5]
freshwater lakes	log(p) = 2.92 - 1.81*log(rw)	p = production (mg c/(100 m3*d)), rw=renewal time (y), canada, prim. prod.[6]
plants communities	log(p) = - 1.66 + 1.66*log(ae)	p = produc. (g dw/(m2*year), ae=evapotr. (mm/year)[7]
production as a function of respiration	log(p) = - 0.4723 + 0.6081*log(r)	p = production, r=respiration, units = kcal/(ha*y)[8]
rodent	log(p) = - 2.2627 + 1.1365*log(r)	p = production, r=respiration, units = kcal/(ha*y)[8]

Source: 1.McNaughton, S.J. and L.L. Wolf. 1975. General Ecology. Holt, Rinehart and Winston, Int., New York.
2.Gauf, G.G. 1975. Ecologia (Berl.), 15, pp 17-32.
3.Satoo, T. 1970. in: Reichle, D.E. (ed.); Analysis of Temperate Forest Ecosystem. pp 55-72. Springer-erlag. Berlin.
4.Madwick, H.A.I. 1970. in: Reichle, D.E. (ed.); Analysis of Temperate Forest Ecosystem. pp 47-54. Sprnger-Verlag. Berlin.
5.Whiteside, M.C. 1970. Ecol. Monogr. 40, pp 79-118.
6.Brunskill, G.J. and D.W. Schindler. 1971. J. Fish. Res. bd. Can., 28, pp 134-155.
7.Rosenzweig, M.L. 1968. Am. Nat. 102, pp 67-74.

8.Grodzwsky, W. 1975. in: Grodzinsky, W., R.Z. Klekowsky and A. Duncan (eds). Methods for Ecological Boenergetics IBP Handbook No. 24, pp 65-94. Blackwell, Oxford.

3-057 Equation
Rate of development as a function of temperature

Species	Value	Condition
Hexarthra fennica	1/d = - 0.278 + 0.0845*t	d = duration of egg develop. (d),t=cel deg., 287 - 297 k[1]
Kellicottia bostoniensis	1/d = - 0.083 + 0.0430*t	d = duration of egg develop. (d),t=cel deg., 283 - 293 k[1]
Keratella cochlearis and Kellicottia longispina	1/d = +0.028 + 0.0361*t	d = duration of egg develop. (d),t=cel deg., 283 - 293 k[1]
Keratella quadrata	1/d = - 0.083 + 0.0528*t	d = duration of egg develop. (d),t=cel deg., 283 - 293 k[1]
Keratellploeosoma truncatum	1/d = - 0.306 + 0.0472*t	d = duration of egg develop. (d),t=cel deg., 283 - 285 k[1]
Ploesoma hudsoni	1/d = - 0.417 + 0.0722*t	d = duration of egg develop. (d),t=cel deg., 286 - 293 k[2]
Polyarthra vulgaris	1/d = - 0.083 + 0.0444*t	d = duration of egg develop. (d),t=cel deg., 278 - 293 k[2]

Source: 1.Edmonson, W.T. 1965. Ecol. Monogr., 35, pp 61-111.
2.King, C.E. 1967. Ecology 48, pp 111-128.

3-058 Equation
Respiration as a function of dissolved oxygen

Species	Value	Condition
Daphnia galeata mendotae	r = 3.998(+ - 1.207) + 1.411(+ - 0.261)*x	r = resp rate (ul O2/(mg dw*h), x=O2 conc (mg/l), r=0.83[1]
Daphnia magna	r = 2.743(+ - 0.263) + 1.177(+ - 0.239)*x	r = resp.(ul O2/(mg dw*h), x=O2 (mg/l), r=0.83[1]

Source: 1.Heisey, D. and K.G. Porter. 1977. Lim nol. Oceanogr., 22(5), pp 839-845.

3-059 Equation
Respiration as a function of temperature

Species	Value	Condition
Coregonus	y = 39.8*exp(0.085*t)	y = resp (ul cO2/h), t = ambient temp (cel)[1]
Eichhornia crassipes	y = 2**((t - 30.7)/10)	y = resp rate (1/d), t=cel deg., in marsh[2]
Eichhornia crassipes	y = 2**((t - 28.5)/10)	y = resp rate (1/d), t=cel deg., in lake[2]
Leuciscus idus	y = 16.8*exp(0.122*t)	y = resp (ul cO2/h), t = ambient temp (cel)[1]
Thymallus arcticus baicalensis	y = 51.6*exp(0.085*t)	y = resp (ul cO2/h), t = ambient temp (cel)[1]
trees	r = 0.498*t - 128	r = resp (g cO2/(m2*d)), t=kelvin deg., oak - pine,brookhaven. 1/5 - 30/9[3]
trees	r = 0.127*t - 31	r = resp (g cO2/(m2*d)), t=kelvin deg., oak - pine,brookhaven. 1/10 - 30/4[3]

Source: 1.Alikin, Y.S. 1976. Zh. Obshch. Biol. 37(1), pp 127-134.
2.Mitsch, W.J. 1976. Ecol. Modell. 2(1), pp 69-89.
3.Woodwell, G.M. and P.F. Bourdeau. 1966. Science 154, pp 1031-1034.

3-060 Equation
Salmo trutta

Species	Value	Condition
caloric content as a function of condition factor and body weight	y = 4879*exp(+0.0953*l)*-w**0.0180	y = (cal/g dw), l=condition factor, w=wet w (g), r=0.801[1]
caloric content as a function of condition factor and body weight	y = 20.41*exp(+0.0953*l)*-w**0.0180	y = (k j/g dw), l=condition factor, w=wet w (g), r=0.801[1]
caloric content as a function of condition factor and body weight	y = 751.9*exp(0.332*l)*w**0.0720	y = (cal/g ww), l=condition factor, w=wet w (g), r=0.854[1]
caloric content as a function of condition factor and body weight	y = 3.145*exp-(+0.332*l)*w**0.0720	y = (k j/g ww), l=condition factor, w=wet w (g), r=0.854[1]
caloric content as a function of condition factor and body weight	y = 5374*l**0.0907*w**0.0178	y = (cal/g dw), l=condition factor, w=wet w (g), r=0.796[1]
caloric content as a function of condition factor and body weight	y = 22.48*l**0.0907*w**0.0178	y = (k j/g dw), l=condition factor, w=wet w (g), r=0.796[1]
caloric content as a function of condition factor and body weight	y = 1053*l**0.317*w**0.0711	y = (cal/g ww), l=condition factor, w=wet w (g), r=0.850[1]
caloric content as a function of condition factor and body weight	y = 4.406*l**0.317*w**0.0711	y = (k j/g ww), l=condition factor, w=wet w (g), r=0.850[1]
caloric content as a function of fork length and body weight	y = 5471*exp(+0.00323*l)*w**(- 0.00195)	y = (cal/g dw), l=fork length (cm), w=wet w (g), r=0.753[1]
caloric content as a function of fork length and body weight	y = 22.89*exp(+0.00323*l)*w**(- 0.00195)	y = (k j/g dw), l=fork length (cm), w=wet w (g), r=0.753[1]
caloric content as a function of fork length and body weight	y = 1121*exp(+0.0113*l)*-w**(0.00193)	y = (cal/g ww), l=fork length (cm), w=wet w (g), r=0.815[1]
caloric content as a function of fork length and body weight	y = 4.690*exp(+0.0113*l)*-w**(0.00193)	y = (k j/g ww), l=fork length (cm), w=wet w (g), r=0.815[1]
caloric content as a function of fork length and body weight	y = 8148*l**(- 0.271)*w**0.108	y = (cal/g dw), l=fork length (cm), w=wet w (g), r=0.796[1]
caloric content as a function of fork length and body weight	y = 34.09*l**(- 0.271)*w**0.108 - 0.00195)	y = (k j/g dw), l=fork length (cm), w=wet w (g), r=0.796[1]
caloric content as a function of fork length and body weight	y = 4608*l**(- 0.962)*w**0.391	y = (cal/g ww), l=fork length (cm), w=wet w (g), r=0.852[1]
caloric content as a function of fork length and body weight	y = 19.28*l**(- 0.962)*w**0.391	y = cal (k j/g ww), l=fork length (cm), w=wet w (g), r=0.852[1]
consumation of food per day as a function of body weight and temp.	d = 0.742*w**0.786*exp(0.366*t)	d = food con. (mg/d), w=ww (g),t=cel ,growth=90 % of max,276.8 - 279.8 k[2]
consumation of food per day as a function of body weight and temp.	d = 3.820*w**0.772*exp(0.134*t)	d = food con. (mg/d), w=ww (g),t=cel ,growth=90 % of max,279.8 - 288 k[2]
consumation of food per day as a function of body weight and temp.	d = 5.191*w**0.794*exp(0.107*t)	d = food con. (mg/d), w=ww (g),t=cel ,growth=90 % of max,288 - 290.5 k[2]
consumation of food per day as a function of body weight and temp.	d = 0.820*w**0.763*exp(0.356*t)	d = food con. (mg/d), w=ww (g),t=cel ,growth=80 % of max,276.8 - 279.8 k[2]
consumation of food per day as a function of body weight and temp.	d = 3.446*w**0.780*exp(0.130*t)	d = food con. (mg/d), w=ww (g),t=cel ,growth=80 % of max,279.8 - 288 k[2]
consumation of food per day as a function of body weight and temp	d = 112.701*w**0.765*exp(- 0.096*t)	d = food con. (mg/d), w=ww (g),t=cel ,growth=80 % of max,288 - 290.5 k[2]
consumation of food per day as a function of body weight and temp.	d = 0.859*w**0.769*exp(0.335*t)	d = food con. (mg/d), w=ww (g),t=cel ,growth=70 % of max,276.8 - 279.8 k[2]
consumation of food per day as a function of body weight and temp.	d = 3.184*w**0.781*exp(0.129*t)	d = food con. (mg/d), w=ww (g),t=cel ,growth=70 % of max,279.8 - 288 k[2]
consumation of food per day as a function of body weight and temp.	d = 75.866*w**0.764*exp(- 0.076*t)	d = food con. (mg/d), w=ww (g),t=cel ,growth=70 % of max,288 - 290.5 k[2]

consumation of food per day as a function of body weight and temp.	d = 0.938*w**0.761*exp(0.317*t)	d = food con. (mg/d), w=ww (g),t=cel ,growth=60 % of max,276.8 - 279.8 k[2]
consumation of food per day as a function of body weight and temp.	d = 3.004*w**0.775*exp(0.129*t)	d = food con. (mg/d), w=ww (g),t=cel ,growth=60 % of max,279.8 - 288 k[2]
consumation of food per day as a function of body weight and temp.	d = 55.336*w**0.756*exp(- 0.058*t)	d = food con. (mg/d), w=ww (g),t=cel ,growth=60 % of max,288 - 290.5 k[2]
consumation of food per day as a function of body weight and temp.	d = 0.975*w**0.763*exp(0.301*t)	d = food con. (mg/d), w=ww (g),t=cel ,growth=50 % of max,276.8 - 279.8 k[2]
consumation of food per day as a function of body weight and temp.	d = 2.948*w**0.760*exp(0.130*t)	d = food con. (mg/d), w=ww (g),t=cel ,growth=50 % of max,279.8 - 288 k[2]
consumation of food per day as a function of body weight and temp.	d = 40.834*w**0.756*exp(- 0.043*t)	d = food con. (mg/d), w=ww (g),t=cel ,growth=50 % of max,288 - 290.5 k[2]
consumation of food per day as a function of body weight and temp.	d = 1.073*w**0.748*exp(0.287*t)	d = food con. (mg/d), w=ww (g),t=cel ,growth=40 % of max,276.8 - 279.8 k[2]
consumation of food per day as a function of body weight and temp.	d = 2.604*w**0.771*exp(0.130*t)	d = food con. (mg/d), w=ww (g),t=cel ,growth=40 % of max,279.8 - 288 k[2]
consumation of food per day as a function of body weight and temp.	d = 27.702*w**0.759*exp(- 0.024*t)	d = food con. (mg/d), w=ww (g),t=cel ,growth=40 % of max,288 - 290.5 k[2]
consumation of food per day as a function of body weight and temp.	d = 1.144*w**0.746*exp(0.268*t)	d = food con. (mg/d), w=ww (g),t=cel ,growth=30 % of max,276.8 - 279.8 k[2]
consumation of food per day as a function of body weight and temp.	d = 2.367*w**0.764*exp(0.133*t)	d = food con. (mg/d), w=ww (g),t=cel ,growth=30 % of max,279.8 - 288 k[2]
consumation of food per day as a function of body weight and temp.	d = 20.876*w**0.749*exp(- 0.007*t)	d = food con. (mg/d), w=ww (g),t=cel ,growth=30 % of max,288 - 290.5 k[2]
consumation of food per day as a function of body weight and temp.	d = 1.247*w**0.737*exp(0.249*t)	d = food con. (mg/d), w=ww (g),t=cel ,growth=20 % of max,276.8 - 279.8 k[2]
consumation of food per day as a function of body weight and temp.	d = 2.294*w**0.754*exp(0.132*t)	d = food con. (mg/d), w=ww (g),t=cel ,growth=20 % of max,279.8 - 288 k[2]
consumation of food per day as a function of body weight and temp.	d = 13.469*w**0.748*-exp(0.015*t)	d = food con. (mg/d), w=ww (g),t=cel ,growth=20 % of max,288 - 290.5 k[2]
consumation of food per day as a function of body weight and temp.	d = 1.235*w**0.740*exp(0.238*t)	d = food con. (mg/d), w=ww (g),t=cel ,growth=10 % of max,276.8 - 279.8 k[2]
consumation of food per day as a function of body weight and temp.	d = 1.993*w**0.755*exp(0.134*t)	d = food con. (mg/d), w=ww (g),t=cel ,growth=10 % of max,279.8 - 288 k[2]
consumation of food per day as a function of body weight and temp.	d = 8.987*w**0.749*exp(0.036*t)	d = food con. (mg/d), w=ww (g),t=cel ,growth=10 % of max,288 - 290.5 k[2]
fat content as a function of condition factor and body weight	y = 1.236*exp(+0.781*l)*w**0.159	y = % fat, l=condition factor, w=wet w (g), r=0.862[1]
fat content as a function of condition factor and body weight	y = 2.732*l**0.771*w**0.157	y = % fat, l=condition factor, w=wet w (g), r=0.862[1]
fat content as a function of fork length and body weight	y = 2.807*exp(+0.00529*l)*-w**(+0.124)	y = % fat, l=fork length (cm), w=wet w (g), r=0.802[1]
fat content as a function of fork length and body weight	y = 95.63*l**(- 2.316)*w**0.929	y = % fat, l=fork length (cm), w=wet w (g), r=0.862[1]
fat content as a function of water content	y = 51.63 - 0.613(+ - 0.0225)*x	y = % fat, x = % water, r = 0.98[1]
protein content as a function of condition factor and body weight	y = 12.53*exp(+0.148*l)*-w**0.0263	y = % protein, l=condition factor, w=wet w (g), r=0.732[1]
protein content as a function of condition factor and body weight	y = 14.53*l**0.148*w**0.0259	y = % protein, l=condition factor, w=wet w (g), r=0.725[1]
protein content as a function of fork length and body weight	y = 15.39*exp(+0.00987*l)*w**(- 0.0344)	y = % protein, l=fork length (cm), w=wet w (g), r=0.729[1]
protein content as a function of fork length and body weight	y = 27.51*l**(- 0.414)*w**0.164	y = % protein, l=fork length (cm), w=wet w (g), r=0.725[1]

protein content as a function of water content	y = 42.92 - 0.353(+ - 0.0226)*x	y = % protein, x = % water, r = 0.924[1]
specific growth rate as a function of body weight	gr = 0.622*w**(- 0.325(+ - 0.050))	gr = growth rate (% ww/d), w=wet w. (g), 95 % cl, 278.6+ - 0.5 k[3]
specific growth rate as a function of body weight	gr = 1.884*w**(- 0.332(+ - 0.032))	gr = growth rate (% ww/d), w=wet w. (g), 95 % cl, 282.5+ - 0.7 k[3]
specific growth rate as a function of body weight	gr = 2.788*w**(- 0.318(+ - 0.025))	gr = growth rate (% ww/d), w=wet w. (g), 95 % cl, 285.8+ - 0.7 k[3]
specific growth rate as a function of body weight	gr = 2.151*w**(- 0.307(+ - 0.049))	gr = growth rate (% ww/d), w=wet w. (g), 95 % cl, 288.0+ - 0.6 k[3]
specific growth rate as a function of body weight	gr = 0.029*w**(- 0.029(+ - 0.102))	gr = growth rate (% ww/d), w=wet w. (g), 95 % cl, 292.5+ - 0.8 k[3]
specific growth rate as a function of body weight	gr = 0.622*w**(- 0.325(+ - 0.050))	gr = growth rate (% ww/d), w=wet w. (g), 95 % cl, 278.6+ - 0.5 k[3]
specific growth rate as a function of body weight	gr = 1.884*w**(- 0.332(+ - 0.032))	gr = growth rate (% ww/d), w=wet w. (g), 95 % cl, 282.5+ - 0.7 k[3]
specific growth rate as a function of body weight	gr = 2.788*w**(- 0.318(+ - 0.025))	gr = growth rate (% ww/d), w=wet w. (g), 95 % cl, 285.8+ - 0.7 k[3]
specific growth rate as a function of body weight	gr = 2.151*w**(- 0.307(+ - 0.049))	gr = growth rate (% ww/d), w=wet w. (g), 95 % cl, 288.0+ - 0.6 k[3]
specific growth rate as a function of body weight	gr = 0.029*w**(- 0.029(+ - 0.102))	gr = growth rate (% ww/d), w=wet w. (g), 95 % cl, 292.5+ - 0.8 k[3]
specific growth rate as a function of body weight and temperature	gr = 100*(- 0.0100 + 0.0029(+ - 0.001)*t)*w**(- 0.3250(+ - 0.0360))	gr = growth rate (% ww/d), t=cel deg, w=wet w. (g),95% cl,276.8 - 285.8 k[3]
specific growth rate as a function of body weight and temperature	gr = 100*(+0.0820 - 0.0042(+ - 0.0002)*t)*w**(- 0.2917(+ - 0.0760))	gr = growth rate (% ww/d), t=cel deg, w=wet w. (g),95% cl,286.6 - 292.5 k[3]
specific growth rate as a function of body weight and temperature	gr = 100*(- 0.0100 + 0.0029(+ - 0.001)*t)*w**(- 0.3250(+ - 0.0360))	gr = growth rate (% ww/d), t=cel deg, w=wet w. (g),95% cl,276.8 - 285.8 k[3]
specific growth rate as a function of body weight and temperature	gr = 100*(+0.0820 - 0.0042(+ - 0.0002)*t)*w**(- 0.2917(+ - 0.0760))	gr = growth rate (% ww/d), t=cel deg, w=wet w. (g),95% cl,286.6 - 292.5 k[3]
specific growth rate as a function of temperature	gr = - 0.269 + 0.081(+ - 0.004)*t	gr = growth rate (% ww/d), t=cel deg, 95 % cl, 276.8 - 285.8 k[3]
specific growth rate as a function of temperature	gr = +2.568 - 0.131(+ - 0.006)*t	gr = growth rate (% ww/d), t=cel deg, 95 % cl, 286.5 - 292.5 k[3]
specific growth rate as a function of temperature	gr = +4.216 - 0.216(+ - 0.058)*t	gr = growth rate (% ww/d), t=cel deg, 95 % cl, 292.5 - 294.7 k[3]
specific growth rate as a function of temperature	gr = - 0.269 + 0.081(+ - 0.004)*t	gr = growth rate (% ww/d), t=cel deg, 95 % cl, 276.8 - 285.8 k[3]
specific growth rate as a function of temperature	gr = +2.568 - 0.131(+ - 0.006)*t	gr = growth rate (% ww/d), t=cel deg, 95 % cl, 286.5 - 292.5 k[3]
specific growth rate as a function of temperature	gr = +4.216 - 0.216(+ - 0.058)*t	gr = growth rate (% ww/d), t=cel deg, 95 % cl, 292.5 - 294.7 k[3]
water content as a function of condition factor and body weight	y = 81.11*l**(- 0.0705)*w**(- 0.0174)	y = % water, l=condition factor, w=wet w (g), r=0.866[1]
water content as a function of fork length and body weight	y = 79.83*exp(- 0.00283*l)*w**(- 0.0000426)	y = % water, l=fork length (cm), w=wet w (g), r=0.838[1]
water content as a function of fork length and body weight	y = 57.80*l**0.223*w**(- 0.0916)	y = % water, l=fork length (cm), w=wet w (g), r=0.869[1]

Source: 1.Elliott, J.M. 1976. J. Anim. Ecol. 45, pp 273-289.
2.Elliott, J.M. 1975, J. Anim. Ecol. 44, pp 823-842.
3.Elliott, J.M. 1975. J. Anim. Ecol. 44, pp 805-821.

3-061 Equation
Summer average chlorophyll conc vs total phosphorus concentration

Species	Value	Condition
freshwater lakes	log(c) = 1.583*log(p) - 1.134	c = chla a (mg/m3),p=spring overturn p conc (mg/m3),r=0.975, jap lakes[1]
freshwater lakes	log(c) = 1.449*log(p) - 1.136	c = chla a (mg/m3),p=spring overturn p conc (mg/m3),r=0.95, n=46[1]

Source: 1.Dillon, P.J. and F.H. Riegler. 1974. Limnol. Oceanogr., 19(5), pp 767-773.

3-062 Equation
Suspension transport vs runoff volume

Species	Value	Condition
River Rhine, Ems, Weser, Elbe, Danube and Inn	Y = - 7.19 + 7.5E - 3*X	Y = susp. transp. in tons/(km2*y), X = runoff in mm/year, 1966 - 73[1]

Source: 1.Matsuo, S. and Y. Miyake. 1966. J. Geophys. Res. 71, 5235.

3-063 Equation
Suspension concentration vs runoff volume

Species	Value	Condition
River Rhine, Ems, Weser, Elbe, Danube and Inn	Y = 32.4 + 5.8*E - 3*X	Y = conc of susp in g/m3, X = runoff (mm/year), 1966 - 73[1]

Source: 1.Matsuo, S. and Y. Miyake. 1966. J. Geophys. Res. 71, 5235.

3-064 Equation
Suspension transport vs runoff volume

Species	Value	Condition
River Rhine, Danube and Inn	Y = 148 + 0.179*X	Y = susp. transp. in tons/(km2*y), X = runoff in mm/year, 1966 - 73[1]

Source: 1.Matsuo, S. and Y. Miyake. 1966. J. Geophys. Res. 71, 5235.

3-065 Equation
Suspension concentration vs runoff volume

Species	Value	Condition
River Rhine, Danube and Inn	Y = 846 - 0.697*X	Y = conc of susp in g/m3, X = runoff (mm/year), 1966 - 73[1]

Source: 1.Matsuo, S. and Y. Miyake. 1966. J. Geophys. Res. 71, 5235.

3-066 Equation
Temperature effects upon biological rates

Species	Value	Condition
used in models	k(t) = k(20)*kot**(t - 20)	F047[1]
used in models	k(t) = k(opt)*exp(a*(t - t(opt))*(tmax - t/u)**(a*(u))	F047, u = tmax - t(opt)[2]
used in models	k(t) = k(opt)*exp(- 2.3*(t - t(opt)/(tmax - t(opt))**2)	F047, for t>t(opt)[3]
used in models	k(t) = k(opt)*exp(- 2.3*(t - t(opt)/(tmin - t(opt))**2)	F047, t<t(opt)[3]
used in models	k(t) = k(opt)*exp(- 2.3*abs((t - t(opt)/15))	F047[4]
used in models	k(t) = k(opt)*exp(k*t)	F047[4]

used in models	k(t) = k(20)*kot**(t - 20)	F047[1]
used in models	k(t) = k(opt)*exp(a*(t - t(opt))*(tmax - t/u)**(a*(u))	F047, u = tmax - t(opt)[2]
used in models	k(t) = k(opt)*exp(- 2.3*(t - t(opt)/(tmax - t(opt))**2)	F047, for t>t(opt)[3]
used in models	k(t) = k(opt)*exp(- 2.3*(t - t(opt)/(tmin - t(opt))**2)	F047, t<t(opt)[3]
used in models	k(t) = k(opt)*exp(- 2.3*abs((t - t(opt)/15))	F047[4]
used in models	k(t) = k(opt)*exp(k*t)	F047[4]

Source: 1.Edmonson, W.T. 1965. Ecol. Monogr., 35, pp 61-111.
2.Lassiter, R.R. 1975. Modeling Dynamics of Biological and Chemical Components of Aquatic Ecosystems.PA-660/3-75-012.
3.Nixon, S.W. and J.N. Kremer. 1977. in: Hall, C.A.S. and J.W. Day, Jr. (eds.); Ecosystem Modeling in theory and practice. pp 622-673. John Wiley and Sons. New York.
4.Edmonson, W.T. 1965. Ecol. Monogr., 35, pp 61-111.

3-067 Equation
Transparency as a function of maximum production rate

Species	Value	Condition
lakes	tp = 35/sqrt(g24)	tp = transparency (m), g24=max production rate (mg c/(m3*d))[1]

Source: 1.Mathiesen, H. 1970. Vatten 26, pp 146-173.

3-068 Equation
Uptake of Hg as a function of concentration in water and elimination

Species	Value	Condition
Lepomis macrochirus (1)	q(t) = (u*c/k)*(1 - exp(- k*t))	q(t) = conc in fish at time t=t, u=fraction absorbed from water at[1]
Lepomis macrochirus (2)	q(t) = ((u*c/k)+b*t)*(1 - exp(- k*t))	passage over gills,k = elim. rate,c=wat conc,b and k=turnover rate in a[1]
Lepomis macrochirus (3)	d(t) = u*c/k*(1 - exp(- k*t)) + u*c/(k - g)*(exp(- k*t) - exp(- g*t))	compartment d and h,d(t) = conc in d at time t[1]
Lepomis macrochirus (4)	q(t) = 3.84 (+ - 0.27 + 0.177 (+ - 0.002)*t	water = 0.2 ppb ch3hgcl, 281 k, fasted fish[1]
Lepomis macrochirus (5)	q(t) = 1.35 (+ - 0.32) + t/(3.96 (+ - 0.12) + 0.00367 (+ - 0.000027)*t)	water = 0.2 ppb ch3hgcl, 281 k, fasted fish, fish conc=ppb (wet)[1]
Lepomis macrochirus (6)	q(t) = c*1.37 (+ - 0.02)/0.00198 (+ - 0.00011)*(1 - exp(0.00198*t)	water = 0.2 ppb ch3hgcl, 281 k, fasted fish, fish conc=ppb (wet)in a[1]
Lepomis macrochirus (7)	q(t) = (c*1.48/0.0178 + 0.142*t)*(1 - exp(- 0.0178*t)	water = 0.2 ppb ch3hgcl, 281 k, fasted fish, fish conc=ppb (wet)[1]

Source: 1.Curtis, E.H., J.J. Beauchamp and B.C. Blaylock. 1977. Ecol. Modell., 3(4), pp 273-284.

3-069 Equation
Uptake of nitrate as a function of concentration

Species	Value	Condition
phytoplankton from North Pacific	y = - 0.08 + 0.96*x	y = specific rate of assimilation (1/d),x= conc of no3 (ug - atom/l)[1]

Source: 1.Eppley, R.W., E.H. Renger, E.L. Venrick and M.M. Mullin. 1973. Limnol. Oceanogr., 18(4), pp 534-551.

3-070 Equation
Uptake of toxaphene as a function of time

Species	Value	Condition
Gambusia affinis	y = 0.00068 + 0.00058*t	y = conc in fish (mg/kg ww), t=time (h), conc in medium = 2 ppm[1]

Source: *1.Schaper, R.A. and L.A. Crowder. 1976. Bull. Environ. Contam. Toxicol. 15(5), pp 581-587.*

3-071 Equation
Uptake of urea as a function of concentration

Species	Value	Condition
phytoplankton from North Pacific	y = - 0.003 + 0.247*x	y = specific rate of assimilation (1/d),x= conc of urea (ug - atom/l)[1]

Source: *1.Eppley, R.W., E.H. Renger, E.L. Venrick and M.M. Mullin. 1973. Limnol. Oceanogr., 18(4), pp 534-551.*

3-072 Equation
Uptake of Zn

Species	Value	Condition
Gambusia affinis	1 - up = 1.39*exp(- 0.1614*t)+16.75*exp(- 0.0316*t)+81.66*exp(- 0.003556*t)	up = uptake (%), t=time (d), 3 compartsments[1]

Source: *1.Nriagu, J.O. 1979. Copper in The Environment. Health Effects. Part II. John Wiley and sons. NY.*

3-073 Equation
Zooplankton

Species	Value	Condition
body size vs filtration rate	r = 0.164*l**2.52	r = filtration ratE(ml/animal*hr),l=length(mm),288 - 298 k, Daphnia magna[1]
body size vs filtration rate	r = 0.31*l**3.02	r = filtration ratE(ml/animal*hr),l=length(mm),293 k, Daphnia rosea[1]
carbon content vs body length	log(c) = 0.507*l + 0.190	c = carbon content (ug), l=length (mm), Daphnia pulicaria[2]
development of zooplankton copepodit as a function of temperature	log(y) = 1.57 - 0.528*log(x)	y = devel of copepodit (d), x=cel, 4 culture temp, Diaptomus pallious[3]
development of zooplankton eggs as a function of temperature	log(y) = 1.16 - 0.77*log(x)	y = devel of eggs (d), x=cel, 4 culture temp, Diaptomus pallious[3]
development of zooplankton nauplius as a function of temperature	log(y) = 1.32 - 0.455*log(x)	y = devel of nauplius (d), x=cel, 4 culture temp, Diaptomus pallious[3]
excretion of ammonium - N vs phosphate - P	y = 9.76*x+0.09	y = excr.(NH4 - n),x=excr., PO4 - p),r=0.907,n=44,22hrs,288 k, marine[4]
excretion of ammonium - N vs total - N	y = 0.75*x - 0.77	y = excr.(NH4 - n),x=excr.,(total - n),r=0.879,n=39,22hrs,288 k, marine[4]
excretion of phosphate - P vs total - P	y = 0.48*x	y = excr.(PO4 - p),x=excr.(?068)(total - p),r=0.928,n=39,22hrs,288 k, marine[4]
excretion of total - N vs total - P	y = 7.28*x+0.83	y = excr.(total - n),x=excr.(?068)(total - p),r=0.922,n=41,22hrs,288 k, marine[4]
grazing rate of zooplankton feeding on phytoplankton	f = v * (ln(c0) - ln(ct)/t	f = ml swept of zoopl.,v=vol of exp water(ml per dry weight of zooplan.[5]
grazing rate of zooplankton feeding on phytoplankton	f = v * (ln(c0) - ln(ct)/t	c0 = phyt. conc in control vol,ct=phyt in experiment bottle[5]

grazing rate, used in models	d(phyt)/dt = - k*phyt*zoo	F047[6]
grazing rate, used in models	grz = myz*zoo*((phyt - ktr)/(kz+phyt)), myz=myzmax*f(t)	F047[7]
grazing rate, used in models	grz = zoo*(phyt - ktr)	F047, e.g. g(zoo) = myz*zoo[8]
grazing rate, used in models	grz = zoo*myz*(1 - exp(dp*(phyt - ktr)), myz=myzmax*f(t)	F047[9]
grazing rate, used in models	grz = myz=(1 - zoo/ck)*zoo	F047[10]
grazing rate, used in models	grz = myz*av*(1 - zoo/ck)*zoo, av=f(phyt)*f(ox)*f(t)*f(tox)	F047, av is related to avaibability of food[11]
grazing rate, used in models	grz = myz*((phyt - ktr)/(kz+phyt))*zoo	F047, phyt>ktr[12]
grazing rate, used in models	grz = myz'*phyt/zoo,	F047, phyt<ktr[12]
mortality of freshwater zooplankton	j = 0.005*q/t	j = zoopl. death (ug/(l*h), q=biomass, t=time (h)[13]
mortality, used in models	mort = mortk*zoo	F047[14]
mortality, used in models	mort = mortk*zoo	F047[14]
respiration rate vs body weight	log(r) = - 0.169*log(w)+0.023	r = resp.rate (mic.l O2/(mg body wt.)*hr,marine, w=dry body wt.,boreal[15]
respiration rate vs body weight	log(r) = - 0.309*log(w)+0.357	r = resp.ratE(mic.l O2/(mg body wt.)*hr,marine, w=dry body wt.,temperate[15]
respiration rate vs body weight	log(r) = - 0.464*log(w)+0.874	r = resp.ratE(mic.l O2/(mg body wt.)*hr,marine, w=dry body wt.,tropical[15]
respiration vs excretion of ammonium - N	y = 0.12*x+0.05	y = resp.,x=excr.(?068)(NH4 - n),r=0.911,n=44,22hrs,288 k, marine[4]
respiration vs excretion of phosphate - P	y = 1.18*x+0.06	y = resp.,x=excr.(?068)(PO4 - p),r=0.890,n=41,22hrs,288 k, marine[4]
respiration vs excretion of total nitrogen	y = 0.08*x - 0.01	y = resp.,x=excr.(?068)(total - n),r=0.867,n=36,22hrs,288 k, marine[4]
respiration vs excretion of total phosphorus	y = 0.57*x+0.06	y = resp.,x=excr.,(total - p),r=0.897,n=40,22hrs,288 k, marine[4]
respiration, used in models	resp = respk*f(t)*zoo	F047[14]
respiration, used in models	log(resp)=0.0364*t-0.3418*log(w)+0.6182	F047[16]
respiration, used in models	resp = respk*f(t)*zoo	F047[14]
respiration, used in models	log(resp) = 0.0364*t - 0.3418*log(w) + 0.6182	F047[16]
zooplankton volume as a function of day - time	log10(y) = 1.32+0.49*cos(t)	y = zooplankton volume (ml/1000m**3),t=time of day(hr)[17]

Source: 1.Egloff. D.A. and D.S. Palmer. 1971. Limnol. Oceanogr., 16(6), pp 900-905.
2.Snow, N.B. 1972. Limnol. Oceanogr. 17(6), pp 909-913.
3.Geiling, W.T. and R.S. Campbell. 1972. Limnol. Oceanogr., 17(2), pp 304-306.
4.Le Borgne, R.P. 1973. Mar. Biol. 19, pp 249-257.
5.Martin, J.H 1970. Limnol. Oceanogr. 15(5), pp 756-761.
6.Dodson, S.I. 1975. Limnol. Oceanogr., 20(3), pp 426-433.
7.Steele, J.H. 1974. The Structure of Ma rine Ecosystems. Blackwell Scientific Publications Ltd. Oxfor. pp. 74-91.
8.Walsh,J.J. and R.C. Dugdale. 1971. In vest. Pesq.35, pp 309-330
9.O'Brien, J.J. et al. 1972. An Ecological Model of the Lower Marine Trophic Levels on the ContinentalShelf of West FL. Geophysical Fluid Dynamics. Florida Univ. Tallahassee. Fla. 170 p.
10.Lotka, A.J. 1956. Elements of Mathematical Biology. Reprinted. Dover. New York.
11.Odum, H.T. 1972. in: Patten, B.C. (ed.); Systems Analysis and Simulation in Ecology. 1, pp 139-211.cademic Press. New York.
12.Gargas, E. 1976. A Three box eutrophication Model of a Mesotrophic Danish Lake. Water Quality Instiute.
13.Richey, J.E. 1977. in: Hall, C.A.S. and J.W. Day, Jr. (eds.); Ecosystem Modeling in theory and practce. John Wiley and Sons. New York.
14.Di Toro, D.M, D.J. O'Connor and J.L. Mancini. 1975. In: Patten, B.C. (ed.); Systems Analysis and Simlation in Ecology. Ac. Press. N.Y.
15.Ikeda, T. 1970. Bull. Fac. Fish. Hokkaido Univ., 21, pp 91-112.
16.Patten, B.C., P.A. Egloff and T.H. Richardson. 1975. Total Ecosystem Model for a Cove in Lake Texomain: B.C. Patten (ed), Systems analysis in Ecology. Vol III. AP. N.Y. pp. 206-423.
17.King, J.E. and T.S. Hilda. 1954. U.S. Fish. and Wildl. Serv. Spec. Sci. Rep. Fish. 18, pp 66.

Chapter 4

ECOTOXICOLOGICAL EFFECTS OF CHEMICAL COMPOUNDS

Compounds in Alphabetic Order

4-001 Effect of Acclimation time

Parameter	Effect Value	Concentration and Condition	Species
Development time of eggs	27.22 hours at 293 K	Winter acclimated animal, 281 - 283 K	Acartia clausii[1]
Development time of eggs	28.7 hours at 293 K	Winter acclimated animal, 9 d at 293 K	Acartia clausii[1]
Development time of eggs	30.3 hours at 293 K	1. generation from winter acclimated animals, whole life at 293 K	Acartia clausii[1]
Development time of eggs	32.67 hours at 293 K	Summer acclimated animals, 291-293 K, field temp.	Acartia clausii[1]

Source: 1.Landry, M.R. 1975. Limnol. Oceanog. 20(5), pp 854-857.

4-002 Effect of Acenaphthalene

Parameter	Effect Value	Concentration and Condition	Species
Growth rate	-3 % more than control	Water = 0.3 ppm, 96 h, PAH, sporelings	Antithamnion plumula[1]

Source: 1.Boney, A.D. 1974. Mar. Pollut. Bull. 5, 185-186.

4-003 Effect of Acetaldehyde

Parameter	Concentration	Condition	Species
LD50	1.93 g pr. kg body weight	single oral dose	Rat[1]
LD50	0.64 g pr. kg body weight	subcutane dose	Rat[1]
Sublethal effect	6200 mg/l	Inhibition of cell multiplication	Microcystis aeruginosa[2]
Sublethal effect	10000 mg/l	Inhibition of cell multiplication	Pseudomonas putida[2]
TLm	70 mg/l	24 hours	Pinperch[3]
TLm	53 mg/l	96 hours	SunFish[4]

Source: 1.Henschler, D., Toxikologisch-arbeitsmedizinische Begrundung von MAH-werte, Verlag Chemie.
2.Bringmann, G. and Kuhn, R., nwf-wasser/abwasser, H.9, 1976.
3.Garret, J. T. and Dougherty, F. M. Jr., Texas J. Sci., 3, 391, 1951.
4.McKee, J. E., and Wolf, H. W., Water Quality, The resources agency of California, 1963.

4-004 Effect of Acetamide

Parameter	Concentration	Condition	Species
TLm	15.5-20 mg/l	72 hours	Gambusia affinis[1]
TLm	13.3-26.3 mg/l	24-96 hours	Mosquito Fish[2]

Source: 1.Meinck, F., Stooff, H. and Kohlschutter, H., Les eaux residuaires industrielles, 1970.2.Jones, H. R., Environmental Control in the organic and petrochemical Industries, Noyes Data Corp., 171.

4-005 Effect of Acetic acid

Parameter	Concentration	Condition	Species
LC0	100 mg/l	24 hours	Creek chub[1]
LC0	10 mg/l	long exposure, hard water	Goldfish[2]
LC100	200 mg/l	24 hours	Creek chub[1]
LD50	3.3 g pr. kg body weight	single oral dose	not specified[3]

Sublethal effect	90 mg/l	Inhibition of cell multiplication	Microcystis aeruginosa[4]
Sublethal effect	2850 mg/l	Inhibition of cell multiplication	Pseudomonas putida[4]
TLm	75 mg/l	96 hours	Bluegill[5]
TLm	100-1000 mg/l	24 hours	Lepomis macrochirus[6]
TLm	251 mg/l	24-96 hours	Mosquito Fish[7]

Source: 1.Gilette, L. A., Miller, D. L. and Redman, H. E., Sewage Ind. Wastes, 24, 1397, 1952.
2.Wallen, I. E., Greer, W. C. and Lasater, R., Sewage Ind. Wastes, 29, 695, 1957.
3.Patty, F. A., Industrial Hygiene and toxicology, 2, Interscience Publishers, New York, 1976.
4.Bringmann, G. and Kuhn, R., nwf-wasser/abwasser, H.9, 1976.
5.Price, K. S. et al., JWPCF, 46, 1974.
6.Elkins, H. F. et al., Sewage Ind. Wastes, 28, 1475, 1956.
7.Jones, H. R., Environmental Control in the organic and petrochemical Industries, Noyes Data Corp., 171.

4-006 Effect of Acetone

Parameter	Concentration	Condition	Species
Sublethal effect	300 mg/l	light irritation of mucous membrane	Man[1]
Sublethal effect	530 mg/l	Inhibition of cell multiplication	Microcystis aeruginosa[2]
Sublethal effect	1700 mg/l	Inhibition of cell multiplication	Pseudomonas putida[2]
TLm	2100 mg/l	24-48 hours	Brine shrimps[3]
TLm	10 mg/l	24-48 hours	Daphnia magna[4]
TLm	5500 mg/l	no ind.	Gammarus pulex[5]
TLm	13000 mg/l	24-96 hours	Mosquito fish[6]
TLm	1000 mg/l	protozoa	Vorticella campanula[5]

Source: 1.Henschler, D., Toxikologisch-arbeitsmedizinische Begrundung von MAH-werte, Verlag Chemie.
2.Bringmann, G. and Kuhn, R., nwf-wasser/abwasser, H.9, 1976.
3.Price, K. S. et al., JWPCF, 46, 1974.
4.Elkins, H. F. et al., Sewage Ind. Wastes, 28, 1475, 1956.
5.Meinck, F., Stooff, H. and Kohlschutter, H., Les eaux residuaires industrielles, 1970.
6.Wallen, I. E., Greer, W. C. and Lasater, R., Sewage Ind. Wastes, 29, 695, 1957.

4-007 Effect of Acetonitrile

Parameter	Concentration	Condition	Species
LD50	0.18 g / kg ww	acute oral dose	guinea pig[1]
LD50	1.7-8.5 g / kg ww	acute oral dose	not specified[1]
Sublethal effect	520 mg/l	Inhibition of cell multiplication	Microcystis aeruginosa[2]
Sublethal effect	680 mg/l	Inhibition of cell multiplication	Pseudomonas putida[2]
TLm	1000-1850 mg/l	96 hours, various species	Fish[3]

Source: 1.Patty, F. A., Industrial Hygiene and toxicology, 2, Interscience Publishers, New York, 1976.
2.Bringmann, G. and Kuhn, R., nwf-wasser/abwasser, H.9, 1976.
3.Jones, H. R., Environmental Control in the organic and petrochemical Industries, Noyes Data Corp., 171.

4-008 Effect of Acetoxime

Parameter	Concentration	Condition	Species
LC100	0.3 mg/l	not indicated	Pseudomonas[1]

Source: 1.Meinck, F., Stooff, H. and Kohlschutter, H., Les eaux residuaires industrielles, 1970.

4-009 Effect of Acetylene

Parameter	Concentration	Condition	Species
Sublethal effect	250 ppm	declination in seedling, 3 days	Sweet pea[1]
TLm	200 mg/l	33 hours	Trout[2]

Source: 1.Crocker, W., Growth and Plants, Reinhold publishing Comp., New York, 1948.
2.Meinck, F., Stooff, H. and Kohlschutter, H., Les eaux residuaires industrielles, 1970.

4-010 Effect of Acetylene tetrabromide

Parameter	Concentration	Condition	Species
LD50	0.4 g pr. kg body weight	oral dose	rabbit, guinea pig[1]

Source: 1.Patty, F. A., Industrial Hygiene and toxicology, 2, Interscience Publishers, New York, 1976.

4-011 Effect of Acid precipitation

Parameter	Effect Value	Concentration and Condition	Species
Metal concentration in lake water	400 mg/m3	Unaffected area = 13 (Blue Chalk), Clear Water, Ont.	Al[1]
Metal concentration in lake water	0.6 mg/m3	Unaffected area = 0.5 (Central Norway), SE Norway	Cd[2]
Metal concentration in lake water	0.63 mg/m3	Unaffected area = 0.23 (N. Sweden), W. Coast, Sweden	Cd[3]
Metal concentration in lake water	5 mg/m3	Unaffected area = 1 (Lake A.), Sudbury, Ont.	Cd[4]
Metal concentration in lake water	10 mg/m3	Unaffected area = 10 (Central Norway), SE Norway	Cu[2]
Metal concentration in lake water	1120 mg/m3	Unaffected area = 2 (Lake A.), Sudbury, Ont.	Cu[4]
Metal concentration in lake water	92 mg/m3	Unaffected area = 8 (Blue Chalk), Clear Water, Ont.	Cu[1]
Metal concentration in lake water	100 mg/m3	Unaffected area = 41 (Blue Chalk), Clear Water, Ont.	Fe[1]
Metal concentration in lake water	300 mg/m3	Unaffected area = 40 (Blue Chalk), Clear Water, Ont.	Mn[1]
Metal concentration in lake water	1850 mg/m3	Unaffected area = 3 (lake A.), Sudbury, Ont.	Ni[4]
Metal concentration in lake water	270 mg/m3	Unaffected area = 3 (Blue Chalk), Clear Water, Ont.	Ni[1]
Metal concentration in lake water	10 mg/m3	Unaffected area = 0.5 (Central Norway), SE Norway	Pb[2]
Metal concentration in lake water	22 mg/m3	Unaffected area = 1 (lake A.), Sudbury, Ont.	Pb[4]
Metal concentration in lake water	122 mg/m3	Unaffected area = 1 (Lake A), Sudbury, Ont.	Zn[4]
Metal concentration in lake water	122 mg/m3	Unaffected area = 13 (N. Sweden), W. Coast, Sweden,	Zn[3]
Metal concentration in lake water	35 mg/m3	Unaffected area = 17 (Central Norway), SE Norway	Zn[2]
Metal concentration in lake water	50 mg/m3	Unaffected area = 9 (Blue Chalk), Clear Water, Ont.	Zn[1]

Source: 1.Dillon, P.J., N.D. Yan, W.A. Schneider and N. Conroy. 1977. Paper presented at Symp. on Lake Metabolsm and Lake Management. Uppsala Univ. p 25.
2.Henriksen, A. and R.F. Wright. 1976. Paper presented at Nordic Hydrological Conference. Reykjavik.
3.Wright, R.F. and E.T. Gjessing. 1976. Ambio 4, pp 220-223.
4.Beamish, R.J. 1976. Water Air Soil Pollut. 6, pp 501-514.

4-012 Effect of Acrolein

Parameter	Concentration	Condition	Species
LD50	0.046 g pr. kg ww	acute oral dose	Rat[1]
Sublethal effect	0.04 mg/l	Inhibition of cell multiplication	Microcystis sp[2]
Sublethal effect	0.21 mg/l	Inhibition of cell multiplication	Pseudomonas putida[2]

Source: 1.Shell Chemie, Shell Industrie Chemicalien gids, Shell Niederland Chemie, 1975.
2.Bringmann, G. and Kuhn, R., nwf-wasser/abwasser, H.9, 1976.

4-013 Effect of Acrylamide

Parameter	Concentration	Condition	Species
LD50	2.5 g pr. kg ww	single oral dose	Rat[1]

Source: 1.Patty, F. A., Industrial Hygiene and toxicology, 2, Interscience Publishers, New York, 1976.

4-014 Effect of Acrylic acid

Parameter	Concentration	Condition	Species
Sublethal effect	0.15 mg/l	Inhibition of cell multiplication	Microcystis aeruginosa[1]
Sublethal effect	41 mg/l	Inhibition of cell multiplication	Pseudomonas putida[1]

Source: 1.Bringmann, G. and Kuhn, R., nwf-wasser/abwasser, H.9, 1976.

4-015 Effect of Acrylonitrile

Parameter	Concentration	Condition	Species
LC50	3600 mg/l	1 hour, seawater, 15 C, respiratory route	Crangon crangon[1]
LC50	25 mg/l	24 hour, seawater, 15 C, respiratory route	Crangon crangon[1]
LC50	6 mg/l	96 hour, seawater, 15 C, respiratory route	Crangon crangon[1]
LC50	1800 mg/l	1 hour, seawater, 15 C, respiratory route	Gobius minutus[1]
LC50	20 mg/l	24 hour, seawater, 15 C, respiratory route	Gobius minutus[1]
LC50	14 mg/l	96 hour, seawater, 15 C, respiratory route	Gobius minutus[1]
LD50	0.035 g pr. kg ww	acute oral dose	Mouse[2]
LD50	0.078 g pr. kg ww	acute oral dose	Rat[2]
Sublethal effect	97 ppm	Slight transitory effect	Rabbit[2]
TLm	12-33 mg/l	96 hours, various species	Fish[3]

Source: 1.Adema, D. M. M., Acute toxiciteitstoetsen met 1,2-dichlorethann, fenol, acrylonitrile, an alkylbenzeesulfonant, Delft, 1976.
2.Patty, F. A., Industrial Hygiene and toxicology, 2, Interscience Publishers, New York, 1976.
3.Jones, H. R., Environmental Control in the organic and petrochemical Industries, Noyes Data Corp., 171.

4-016 Effect of Actinomycin D

Parameter	Effect Value	Concentration and Condition	Species
Mortality	50 %	Oral dose = 0.7 mg/kg, 5.58E-7mol/kg, F007 = 6.25	Mouse[1]

Source: 1.Luckey, T.D. and B. Venugopal. 1977. Metal Toxicity in Mammals, 1. Plenum Press.

4-017 Effect of Adipic acid

Parameter	Concentration	Condition	Species
TLm	<330 mg/l	24 hours	Lepomis macrochirus[1]

Source: 1.Dowden, B. F. and Bennet, H. J., JWPCF, 37, 1310, 1965.

4-018 Effect of Adiponitrile

Parameter	Concentration	Condition	Species
TLm	720 mg/l	soft water	Bluegill[1]
TLm	1250 mg/l	hard water	Bluegill[1]

Source: 1.Pickering Q. H. and Henderson, C., JWPCF, 38, 1419, 1966.

4-019 Effect of Ag

Parameter	Effect Value	Concentration and Condition	Species
Mortality	50 %	AgNO3, water = 6 mg AgNO3/l, toxic level given, microorganisms	Azotobacter sp.[1]
Mortality	0 %	Toxicological limit = 0.01 - 0.03 mg/l, Ag-compounds	Crustacean[2]
Mortality	50 %	Ag(NO3)2, exposure time = 4days, conc in water = 30 ug/l	Daphnia magna[3]
Mortality	50 %	AgNO3, water = 6 mg AgNO3/l, toxic level given, microorganisms	Escherichia coli[1]
Mortality	0 %	Toxicological limit = 0.003 - 0.1 mg/l, Ag-compounds	Fish[2]
Mortality	50 %	AgNO3, water = 6 mg AgNO3/l, toxic level given, microorganisms	Hansenula sp.[1]
Mortality	0 %	Toxicological limit = 0.05 mg/l cronical symptoms	Homo sapiens[2]
Mortality	50 %	Colloid, oral dose = 100 mg/kg body	Mouse[4]
Mortality	50 %	Nitrate, intraperitonal, dose = 129 mg/kg body	Mouse[4]
Mortality	50 %	AgNO3, water = 6 mg AgNO3/l, toxic level given, microorganisms	Pseudomonas sp.[1]
Mortality	50 %	Oxide, oral dose = 2820 mg/kg body	Rat[4]
Mortality	50 %	AgNO3, water = 6 mg AgNO3/l, toxic level given, microorganisms	Rhodotorula sp.[1]
Mortality	50 %	AgNO3, water = 6 mg AgNO3/l, toxic level given, microorganisms	Serratia sp.[1]
Mortality	50 %	AgNO3, water = 6 mg AgNO3/l, toxic level given, microorganisms	Torula sp.[1]
Nitrification	93 % inhib.	Soil = 5 umoles/g of soil, 10 d incubation, 3 soils, lab	Soil organisms[5]

Source: 1.Hulin, N. and M. Horowitz. 1973. Phytoparasitica 1, 64.
2.Liebman, H. 1958. Handbuch der Frischwasser - und Abwasserbiologie. Band II.Munchen. Oldenbourg.
3.McKee, J.R. and H.W. Wolf. 1963. Water Quality Criteria, Publication no. 3A, Calif. State Water Res.Control Board 2nd Edition.
4.Toxic Substances List 1973, U.S. Depart ment of Health, Education and Welfare, Public Health Service National Institute for Occupational Safety and Health, Rockville Maryland, USA, June 1973.
5.Liang, C.N. and M.A. Tabatabai. 1978. Environ. Pollut. 12, 141-147.

4-020 Effect of Al

Parameter	Effect Value	Concentration and Condition	Species
Leaf symptoms	77.5 mg/kg dw in leaf	Rg = 65 - 90, Symptoms outside this range, steam sterilized soil	Tomato leaves[1]
Mortality	0 %	Toxicological limit = 136 mg/l	Crustacean[2]
Mortality	50 %	AlCl3, exposure time = 2 d, conc in water = 3900 ug/l	Daphnia magna[3]
Mortality	0 %	Toxicological limit = 1.6 mg/l, trout, Al(SO4)3	Fish[2]
Mortality	0 %	Toxicological limit = 12 g/kg, Al(SO4)3	Mammalia[2]
Mortality	50 %	Sulfate, oral dose = 770 mg/kg body	Mouse[4]
Mortality	50 %	Chloride, oral dose = 3700 mg/kg body	Rat[4]
Mortality	50 %	Nitrate, oral dose = 4300 mg/kg body	Rat[4]
Nitrification	60 % inhibation	Soil = 5 umoles/g of soil, 10 d incubation, 3 soils, lab	Soil organisms[5]

Source: 1.Harrod, M.F. 1971. In: Tech. Bull. 21, pp 176-189. Trace Elements in Soils and Crops.
2.Liebman, H. 1958. Hnadbuch der Friscwasser - und Abwasserbiologie. Band II.Munchen. Oldenbourg.
3.Biesinger, K.E. and G.M. Christensen. 1972. J. Fish. Res. bd. Can., 29(12), 1691-1700.
4.Toxic Substances List 1973, U.S. Depart ment of Health, Education and Welfare, Public Health ServiceNational Institute for Occupational Safety and Health, Rockville Maryland, USA, June 1973.
5.Liang, C.N. and M.A. Tabatabai. 1978. Environ. Pollut. 12, 141-147.

4-021 Effect of Alanap

Parameter	Effect Value	Concentration and Condition	Species
Mortality	10 %	10 ppm, 48 h, 14 g, 293 K, pH = 8.2, alk = 80 ppm, less than value	Ictalurus punctatus[1]

Source: 1.McCorle, F.M., J.E. Chambers and J.D. Yarbrough. 1977. Bull. Environ. Contam. Toxicol. 18(3), pp 267270.

4-022 Effect of Aldrin

Parameter	Effect Value	Concentration and Condition	Species
Heart beat	0 beats/min	Aldrin = 300 micro mol, after 24 h, 299 K, 5 snails	Biomphalaria glabrata[1]
Heart beat	136 beats/min	Aldrin = 1 micro mol, 24 h, 299 K	Biomphalaria glabrata[1]
Heart beat	150 beats/min	Aldrin = 1 micro mol, after 24 h, 299 K, 5 snails	Biomphalaria glabrata[1]
Heart beat	18 beats/min	Aldrin = 14 micro mol, 24 h, 299 K	Biomphalaria glabrata[1]
Heart beat	20 beats/min	Aldrin = 168 micro mol, 24 h, 299 K	Biomphalaria glabrata[1]
Heart beat	24 beats/min	Aldrin = 112 micro mol, 24 h, 299 K	Biomphalaria glabrata[1]
Heart beat	24 beats/min	Aldrin = 225 micro mol, 24 h, 299 K	Biomphalaria glabrata[1]
Heart beat	24 beats/min	Aldrin = 281 micro mol, after 24 h, 299 K, 5 snails	Biomphalaria glabrata[1]
Heart beat	24 beats/min	Aldrin = 56 micro mol, 24 h, 299 K	Biomphalaria glabrata[1]
Heart beat	25 beats/min	Aldrin = 28 micro mol, 24 h, 299 K	Biomphalaria glabrata[1]
Heart beat	29 beats/min	Aldrin = 7 micro mol, after 24 h, 299 K, 5 snails	Biomphalaria glabrata[1]
Mortality	50 %	Conc in food = 37 mg/kg(33-41), 95 % CL	Bobwhite[2]
Mortality	95 %	Conc in soil > 125 mg/kg	Earthworms[2]
Mortality	50 %	Conc in food = 34 mg/kg(28-41), 95 % CL	Japanese quail[2]
Mortality	50 %	Conc in food = 155 mg/kg(129-186), 95 % CL	Mallard[2]
Mortality	50 %	Conc in food = 14.0 mg/kg	Pheasant[3]
Mortality	50 %	Conc in food = 57 mg/kg(50-64), 95 % CL	Pheasant[2]
Mortality	50 %	Conc in food = 4.0 mg/kg	Quail[3]
Mortality	50 %	Acute dermal toxicity = 99 mg/kg	Rat[2]
Mortality	50 %	Acute oral toxicity = 32-67 mg/kg	Rat[2]
Mortality	95 %	Conc in soil > 250 mg/kg	Slugs[2]

Source: 1.Cocks, J.A. 1974. Environ. Pollut., 5, pp 149-151.
2.Edwards, C.A. 1975. Persistent Pesticides in the Environment, 2nd edition, CRC-Press.
3.Dewitt, J.B., C.M. Menzie, V.A. Adomaitis and W.L. Reickel. 1960. Trans. 25th N. Am. Wildl. Conf., p277.

4-023 Effect of Aldrin, Dieldrin or Endrin

Parameter	Effect Value	Concentration and Condition	Species
ATP level	ATP reduced	Water conc = 100000 aldrin and dieldrin ppb	Green algae[1]
Growth rate	growth completely inhibited	Water conc = < 5000 ppb, completely inhibition ppb	Blue-green algae[2]
Growth rate	growth inhibited	Water conc = 50000 endrin ppb	Blue-green algae[3]
Growth rate	growth inhibited	Water conc = 950 ppb	Blue-green algae[4]
Growth rate	growth inhibited	Water conc = 100 endrin and dieldrin ppb	Diatom[5]
Growth rate	photosynthesis prevented	Water conc = 1 dieldrin and endrin ppb	Diatom[5]

Growth rate	growth not affected	Water conc = 1000 dieldrin and endrin ppb, effect not observed	Green algae[5]
Mortality	cultures not killed	Water conc = > 15000 ppb	Blue-green algae[2]
Mortality	cultured not killed	Water conc = > 20000 ppb	Green algae[2]
Respiration	respiration not affected	Water conc = 1000 ppb, no effect observed	Green algae and Blue-green algae[2]

Source: 1.Spiridinov, Y.Y., V. S. Skhiladze and G.S. Spiridinova. 1970. Subtrop. Kultury 1, 150-155.
2.Goguadze, V.D. 1969. Subtrop. Kultury No 4, 161-167.
3.Malone, C.R. 1971. Soil Biol. Biochem. 3, 127-131.4.Micev, N. and M. Bubalov. 1972. Symp. Biol. Hung. 11, 379-384.
5.Thomas, R.E. and S.D. Rice. 1978. In: Vernberg Vernberg and Calabrese (eds.): Marine Pollution: Funcional Processes, NY. Acedemic Press.

4-024 Effect of Alkyl benzen sulfonate

Parameter	Effect Value	Concentration and Condition	Species
Growth potential	1	14 and 21 days, control = 4, initial = 20 mg surfactant/l, F008	Anabaena cylindrica[1]
Growth potential	3	7 days, control = 4, initial = 20 mg surfactant/l, F008	Anabaena cylindrica[1]
Growth potential	2	7 days, control = 4, initial = 20 mg surfactant/l, F008	Anabaena variabilis[1]
Growth potential	4	14 and 21 days, control = 4, initial = 20 mg surfactant/l, F008	Anabaena variabilis[1]
Growth potential	1	7 days, control = 4, initial = 20 mg surfactant/l, F008	Anacystis nidulans[1]
Growth potential	3	14 and 21 days, control = 4, initial = 20 mg surfactant/l, F008	Anacystis nidulans[1]
Growth potential	4	21 days, control = 4, initial = 20 mg surfactant/l, F008	Ankistrodesmus braunii[1]
Growth potential	4	7, 14 and 21 days, control = 4, initial = 20 mg surfactant/l,	Ankistrodesmus braunii[1]
Growth potential	4	7, 14 and 21 days, control = 4, initial = 20 mg surfactant/l,	Calothrix parietina[1]
Growth potential	2	7 days, control = 4, initial = 20 mg surfactant/l, F008	Chlorella pyrenoidosa[1]
Growth potential	4	14 and 21 days, control = 4, initial = 20 mg surfactant/l, F008	Chlorella pyrenoidosa[1]
Growth potential	2	7 days, control = 4, initial = 20 mg surfactant/l, F008	Chlorella vulgaris[1]
Growth potential	4	14 and 21 days, control = 4, initial = 20 mg surfactant/l, F008	Chlorella vulgaris[1]
Growth potential	1	14 days, control = 4, initial = 20 mg surfactant/l, F008	Cylindrospermum sp.[1]
Growth potential	2	21 days, control = 4, initial = 20 mg surfactant/l, F008	Cylindrospermum sp.[1]
Growth potential	3	7 days, control = 4, initial = 20 mg surfactant/l, F008	Cylindrospermum sp.[1]
Growth potential	3	7 days, control = 4, initial = 20 mg surfactant/l, F008	Gloeocapsa alpicola[1]
Growth potential	4	14 and 21 days, control = 4, initial = 20 mg surfactant/l, F008	Gloeocapsa alpicola[1]
Growth potential	3	14 days, control = 4, initial = 20 mg surfactant/l, F008	Nostoc muscorum[1]
Growth potential	4	7 and 21 days, control = 4, initial = 20 mg surfactant/l, F008	Nostoc muscorum[1]
Growth potential	3	7 and 14 days, control = 4, initial = 20 mg surfactant/l, F008	Oscillatoria borneti[1]
Growth potential	4	21 days, control = 4, initial = 20 mg surfactant/l, F008	Oscillatoria borneti[1]
Growth potential	2	7 and 14 days, control = 4, initial = 20 mg surfactant/l, F008	Oscillatoria chalybia[1]
Growth potential	3	21 days, control = 4, initial = 20 mg surfactant/l, F008	Oscillatoria chalybia[1]
Growth potential	2	7 and 14 days, control = 4, initial = 20 mg surfactant/l, F008	Oscillatoria formosa[1]
Growth potential	3	21 days, control = 4, initial conc = 20 mg surfactant/l, F008	Oscillatoria formosa[1]
Growth potential	1	14 days, control = 4, initial = 20 mg surfactant/l, F008	Oscillatoria tenuis[1]
Growth potential	2	7 days, control = 4, initial = 20 mg surfactant/l, F008	Oscillatoria tenuis[1]
Growth potential	3	21 days, control = 4, initial = 20 mg surfactant/l, F008	Oscillatoria tenuis[1]
Growth potential	2	7 days, control = 4, initial = 20 mg surfactant/l, F008	Phormidium faveolarum[1]

Growth potential	3	14 days, control = 4, initial = 20 mg surfactant/l, F008	Phormidium faveolarum[1]
Growth potential	4	21 days, control = 4, initial = 20 mg surfactant/l, F008	Phormidium faveolarum[1]
Growth potential	3	7 days, control = 4, initial = 20 mg surfactant/l, F008	Scenedesmus obliquus[1]
Growth potential	4	14 and 21 days, control = 4, initial = 20 mg surfactant/l, F008	Scenedesmus obliquus[1]
Growth potential	3	7 days, control = 4, initial = 20 mg surfactant/l, F008	Scenedesmus quadricauda[1]
Growth potential	4	14 and 21 days, control = 4, initial = 20 mg surfactant/l, F008	Scenedesmus quadricauda[1]
Growth potential	1	7 days, control = 4, initial = 20 mg surfactant/l, F008	Ulothrix fimbriata[1]
Growth potential	4	14 and 21 days, control = 4, initial = 20 mg surfactant/l, F008	Ulothrix fimbriata[1]
Mortality	50 %	ABS-comm.+enz. = 12.5 mg/l, ABS-tech. = 18 mg/l, 96 h, 294 K	Cyprinus carpio[2]
Mortality	50 %	ABS-comm.+enz. = 12.5 mg/l, ABS-tech. = 20 mg/l, 48 h, 294 K	Cyprinus carpio[2]
Mortality	50 %	ABS-comm.+enz. = 12.5 mg/l, ABS-tech. = 21 mg/l, 24 h, 294 K	Cyprinus carpio[2]
Mortality	50 %	ABS-comm.+enz. = 28 mg/l, ABS-tech. = 37 mg/l, ABS-comm. = 23 mg/l	Tilapia melanopleura[2]
Mortality	50 %	ABS-comm.+enz. = 30 mg/l, ABS-tech. = 37 mg/l, ABS-comm. = 25 mg/l	Tilapia melanopleura[2]
Mortality	50 %	ABS-comm.+enz. = 32 mg/l, ABS-tech. = 37 mg/l, ABS-comm = 25 mg/l, 24	Tilapia melanopleura[2]

Source: 1.Davis, E.M. and E.F. Gloyna. 1969. Water Pollut. Control. 41 (8), pp 1494-1504.
2.Lopez-Zavala, A. A.S. de Aluja,B. Eliasl L.B. Manjarrez, L.Buchmann, I.A. Mercado and S. Caltenco. 175. Progress Water Technol. 7(2), pp 73-82.
The F008 footnote indicates: 4 = full potiental similar to control, 3 = slight inhibitation, turbidity and chlorophyll decrease. 2 = moderate inhibitation, turbidity and chlorophyll decrease with yellowing of cells, 1 = marked inhibition, turbdity decrease, white to yellow colouration, with few living cells.

4-025 Effect of Alkyl polyethoxylate

Parameter	Effect Value	Concentration and Condition	Species
Growth potential	3	14 days, control = 4, initial = 20 mg AE7.4 /l, F008	Anabaena cylindrica[1]
Growth potential	3	7 and 14 days, control = 4, initial = 20 mg AE7.0 /l, F008	Anabaena cylindrica[1]
Growth potential	4	21 days, control = 4, initial = 20 mg AE7.0 /l, F008	Anabaena cylindrica[1]
Growth potential	4	7 and 21 days, control = 4, initial = 20 mg AE7.4 /l, F008	Anabaena cylindrica[1]
Growth potential	1	7 days, control = 4, initial = 20 mg AE7.0 /l, F008	Anabaena variabilis[1]
Growth potential	1	7 days, control = 4, initial = 20 mg AE7.4 /l, F008	Anabaena variabilis[1]
Growth potential	2	14 days, control = 4, initial = 20 mg AE7.0 /l, F008	Anabaena variabilis[1]
Growth potential	2	14 days, control = 4, initial = 20 mg AE7.4 /l, F008	Anabaena variabilis[1]
Growth potential	3	21 days, control = 4, initial = 20 mg AE7.0 /l, F008	Anabaena variabilis[1]
Growth potential	4	21 days, control = 4, initial = 20 mg AE7.4 /l, F008	Anabaena variabilis[1]
Growth potential	1	7 and 14 days, control = 4, initial = 20 mg AE7.0 /l, F008	Anacystis nidulans[1]
Growth potential	2	7 and 14 days, control = 4, initial = 20 mg AE7.4 /l, F008	Anacystis nidulans[1]
Growth potential	3	21 days, control = 4, initial = 20 mg AE7.0 /l, F008	Anacystis nidulans[1]
Growth potential	3	21 days, control = 4, initial = 20 mg AE7.4 /l, F008	Anacystis nidulans[1]
Growth potential	2	7 days, control = 4, initial = 20 mg AE7.0 /l, F008	Ankistrodesmus braunii[1]
Growth potential	2	7 days, control = 4, initial = 20 mg AE7.4 /l, F008	Ankistrodesmus braunii[1]
Growth potential	3	14 days, control = 4, initial = 20 mg AE7.0 /l, F008	Ankistrodesmus braunii[1]
Growth potential	4	14 and 21 days, control = 4, initial = 20 mg AE7.4 /l, F008	Ankistrodesmus braunii[1]
Growth potential	4	21 days, control = 4, initial = 20 mg AE7.0 /l, F008	Ankistrodesmus braunii[1]

Growth potential	4	7, 14 and 21 days, control = 4, initial = 20 mg AE7.0 /l, F008	Calothrix parietina[1]
Growth potential	4	7, 14 and 21 days, control = 4, initial = 20 mg AE7.4 /l, F008	Calothrix parietina[1]
Growth potential	1	7 days, control = 4, initial = 20 mg AE7.0 /l, F008	Chlorella pyrenoidosa[1]
Growth potential	1	7 days, control = 4, initial = 20 mg AE7.4 /l, F008	Chlorella pyrenoidosa[1]
Growth potential	3	14 days, control = 4, initial = 20 mg AE7.0 /l, F008	Chlorella pyrenoidosa[1]
Growth potential	3	14 days, control = 4, initial = 20 mg AE7.4 /l, F008	Chlorella pyrenoidosa[1]
Growth potential	4	21 days, control = 4, initial = 20 mg AE7.0 /l, F008	Chlorella pyrenoidosa[1]
Growth potential	4	21 days, control = 4, initial = 20 mg AE7.4 /l, F008	Chlorella pyrenoidosa[1]
Growth potential	1	7 and 14 days, control = 4, initial = 20 mg AE7.0 /l, F008	Chlorella vulgaris[1]
Growth potential	1	7 and 14 days, control = 4, initial = 20 mg AE7.4 /l, F008	Chlorella vulgaris[1]
Growth potential	2	21 days, control = 4, initial = 20 mg AE7.0 /l, F008	Chlorella vulgaris[1]
Growth potential	2	21 days, control = 4, initial = 20 mg AE7.4 /l, F008	Chlorella vulgaris[1]
Growth potential	2	14 days, control = 4, initial = 20 mg AE7.0 /l, F008	Cylindrospermum sp.[1]
Growth potential	2	14 days, control = 4, initial = 20 mg AE7.4 /l, F008	Cylindrospermum sp.[1]
Growth potential	3	7 and 21 days, control = 4, initial = 20 mg AE7.4 /l, F008	Cylindrospermum sp.[1]
Growth potential	3	7 days, control = 4, initial = 20 mg AE7.0 /l, F008	Cylindrospermum sp.[1]
Growth potential	4	21 days, control = 4, initial = 20 mg AE7.0 /l, F008	Cylindrospermum sp.[1]
Growth potential	2	7 and 14 days, control = 4, initial = 20 mg AE7.4 /l, F008	Gloeocapsa alpicola[1]
Growth potential	3	21 days, control = 4, initial = 20 mg AE7.4 /l, F008	Gloeocapsa alpicola[1]
Growth potential	3	7 and 14 days, control = 4, initial = 20 mg AE7.0 /l, F008	Gloeocapsa alpicola[1]
Growth potential	4	21 days, control = 4, initial = 20 mg AE7.0 /l, F008	Gloeocapsa alpicola[1]
Growth potential	4	21 days, control = 4, initial = 20 mg AE7.4 /l, F008	Nostoc muscorum[1]
Growth potential	4	7, 14 and 21 days, control = 4, initial = 20 mg AE7.0 /l, F008	Nostoc muscorum[1]
Growth potential	4	7, 14 and 21 days, control = 4, initial = 20 mg AE7.4 /l, F008	Nostoc muscorum[1]
Growth potential	2	21 days, control = 4, initial = 20 mg AE7.0 /l, F008	Oscillatoria borneti[1]
Growth potential	2	7, 14 and 21 days, control = 4, initial = 20 mg AE7.0 /l, F008	Oscillatoria borneti[1]
Growth potential	2	7, 14 and 21 days, control = 4, initial = 20 mg AE7.4 /l, F008	Oscillatoria borneti[1]
Growth potential	3	7 and 21 days, control = 4, initial = 20 mg AE7.0 /l, F008	Oscillatoria chalybia[1]
Growth potential	3	7, 14 and 21 days, control = 4, initial = 20 mg AE7.4 /l, F008	Oscillatoria chalybia[1]
Growth potential	4	14 days, control = 4, initial = 20 mg AE7.0 /l, F008	Oscillatoria chalybia[1]
Growth potential	3	7 days, control = 4, initial = 20 mg AE7.0 /l, F008	Oscillatoria formosa[1]
Growth potential	3	7 days, control = 4, initial = 20 mg AE7.4 /l, F008	Oscillatoria formosa[1]
Growth potential	4	14 and 21 days, control = 4, initial = 20 mg AE7.0 /l, F008	Oscillatoria formosa[1]
Growth potential	4	14 and 21 days, control = 4, initial = 20 mg AE7.4 /l, F008	Oscillatoria formosa[1]
Growth potential	2	7 days, control = 4, initial = 20 mg AE7.0 /l, F008	Oscillatoria tenuis[1]
Growth potential	2	7 days, control = 4, initial = 20 mg AE7.4 /l, F008	Oscillatoria tenuis[1]
Growth potential	3	14 days, control = 4, initial = 20 mg AE7.0 /l, F008	Oscillatoria tenuis[1]
Growth potential	3	14 days, control = 4, initial = 20 mg AE7.4 /l, F008	Oscillatoria tenuis[1]
Growth potential	4	21 days, control = 4, initial = 20 mg AE7.0 /l, F008	Oscillatoria tenuis[1]
Growth potential	4	21 days, control = 4, initial = 20 mg AE7.4 /l, F008	Oscillatoria tenuis[1]

Growth potential	2	7 days, control = 4, initial = 20 mg AE7.4 /l, F008	Phormidium faveolarum[1]
Growth potential	3	14 days, control = 4, initial = 20 mg AE7.4 /l, F008	Phormidium faveolarum[1]
Growth potential	4	21 days, control = 4, initial = 20 mg AE7.4 /l, F008	Phormidium faveolarum[1]
Growth potential	4	7, 14 and 21 days, control = 4, initial = 20 mg AE7.0 /l, F008	Phormidium faveolarum[1]
Growth potential	3	7 days, control = 4, initial = 20 mg AE7.0 /l, F008	Scenedesmus obliquus[1]
Growth potential	3	7 days, control = 4, initial = 20 mg AE7.4 /l, F008	Scenedesmus obliquus[1]
Growth potential	4	14 and 21 days, control = 4, initial = 20 mg AE7.0 /l, F008	Scenedesmus obliquus[1]
Growth potential	4	14 and 21 days, control = 4, initial = 20 mg AE7.4 /l, F008	Scenedesmus obliquus[1]
Growth potential	4	7, 14 and 21 days, control = 4, initial = 20 mg AE7.0 /l, F008	Scenedesmus quadricauda[1]
Growth potential	4	7, 14 and 21 days, control = 4, initial = 20 mg AE7.4 /l, F008	Scenedesmus quadricauda[1]
Growth potential	1	14 days, control = 4, initial = 20 mg AE7.0 /l, F008	Ulothrix fimbriata[1]
Growth potential	1	7 days, control = 4, initial = 20 mg AE7.4 /l, F008	Ulothrix fimbriata[1]
Growth potential	2	14 days, control = 4, initial = 20 mg AE7.4 /l, F008	Ulothrix fimbriata[1]
Growth potential	2	7 days, control = 4, initial = 20 mg AE7.0 /l, F008	Ulothrix fimbriata[1]
Growth potential	3	21 days, control = 4, initial = 20 mg AE7.0 /l, F008	Ulothrix fimbriata[1]
Growth potential	3	21 days, control = 4, initial = 20 mg AE7.4 /l, F008	Ulothrix fimbriata[1]

Source: 1.Davis, E.M. and E.F. Gloyna. 1969. Water Pollut. Control. 41 (8), pp 1494-1504.
For footnote F008 see preceding table.

4-026 Effect of Allylacetate

Parameter	Concentration	Condition	Species
LD50	0.13 g pr. kg body weight	acute oral dose	Rat[1]

Source: 1.Patty, F. A., Industrial Hygiene and toxicology, 2, Interscience Publishers, New York, 1976.

4-027 Effect of Allylalcohol

Parameter	Concentration	Condition	Species
LD50	0.096 g pr. kg body weight	single oral dose	Mouse[1]
LD50	0.071 g pr. kg body weight	single oral dose	Rabbit[1]
LD50	0.064-0.105 g pr. kg ww	single oral dose	Rat[1]
Sublethal effect	6.3 ppm	Eye irritation	Man[1]

Source: 1.Patty, F. A., Industrial Hygiene and toxicology, 2, Interscience Publishers, New York, 1976.

4-028 Effect of Allylamine

Parameter	Concentration	Condition	Species
LD50	0.106 g pr. kg body weight	acute oral dose	Rat[1]
Sublethal effect	0.35 mg/l	Inhibition of cell multiplication	Microcystis aeruginosa[2]
Sublethal effect	71 mg/l	Inhibition of cell multiplication	Pseudomonas putida[2]

Source: 1.Patty, F. A., Industrial Hygiene and toxicology, 2, Interscience Publishers, New York, 1976.
2.Bringmann, G. and Kuhn, R., nwf-wasser/abwasser, H.9, 1976.

4-029 Effect of Allylchloride

Parameter	Concentration	Condition	Species
LD50	0.7 g pr. kg body weight	oral dose	Rat[1]
Sublethal effect	50 ppm	Eye irritation	Man[1]
TLm	42-59 mg/l	24-96 hours, soft water	Bluegill[2]
TLm	21-27 mg/l	24-96 hours, soft water	Goldfish[2]
TLm	20-24 mg/l	24-96 hours, soft water	Latuhead[2]
TLm	24-26 mg/l	24-96 hours, hard water	Latuhead[2]

Source: 1.Patty, F. A., Industrial Hygiene and toxicology, 2, Interscience Publishers, New York, 1976.
2.Pickering Q. H. and Henderson, C., JWPCF, 38, 1419, 1966.

4-030 Effect of Allylglycidylether

Parameter	Concentration	Condition	Species
LD50	0.39 g pr. kg body weight	single oral dose	Mouse[1]
LD50	1.60 g pr. kg body weight	single oral dose	Rat[1]

Source: 1.Patty, F. A., Industrial Hygiene and toxicology, 2, Interscience Publishers, New York, 1976.

4-031 Effect of Ammonia

Parameter	Effect Value	Concentration and Condition	Species
Mortality	0 %	Ammonia (NH3) = 0.018 mg/l, 6 w	Chinook salmon[1]
Mortality	18 %	Ammonia (NH3) = 1.1 mg/l, 240 h, 289 K	Cyprinus carpio[1]
Mortality	15 %	Ammonia (NH3) = 0.22 mg/l, 24 h	Salmo gairdneri[1]
Mortality	5 %	Ammonia (NH3) = 0.085 mg/l, 24 h	Salmo gairdneri[1]
Mortality	50 %	Ammonia (NH3) = 0.47 mg/l, 24 h	Salmo gairdneri[1]
Mortality	50 %	Ammonia (NH3) = 0.28 mg/l, 24 h	Salmo salar[1]
Mortality	0 %	Ammonia (NH3)(unionized) = 0.025 mg/l	Salmonoids[1]
TLm	2.0-2.5 mg/l	24-96 hours	Goldfish[2]
TLm	8.2 mg/l	96 hours, hardwater	Latuhead minnows[3]

Source: 2.Gauhey, P. H. Mc, Engineering Management of Water Quality, McGraw-Hill, New York, 1968.
3.Jones, H. R., Environmental Control in the organic and petrochemical Industries, Noyes Data Corp., 171.
1.Eifac. 1970. Water Quality Criteria for European Freshwater Fish. Report on Ammonia and Inland fisheies Prepared by Eifac, Water Quality Criteria for European Freshwater Fish, FAO, Roma, no 11.

4-032 Effect of Ammoniumacetate

Parameter	ConcentratioCondition		Species
TLm	238 mg/l	24-96 hours, hardwater	Mosquito fish[1]

Source: 1.Jones, H. R., Environmental Control in the organic and petrochemical Industries, Noyes Data Corp., 171.

4-033 Effect of Aniline-hydrochloride

Parameter	ConcentratioCondition		Species
LC100	5.5 mg/l	48 hour	Goldfish[1]

Source: 1.McKee, J. E., and Wolf, H. W., Water Quality, The resources agency of California, 1963.

4-034 Effect of Antracene

Parameter	Concentration	Condition	Species
LC0	5 mg/l	24 hour	Trout[1]

Source: *1.Meinck, F., Stooff, H. and Kohlschutter, H., Les eaux residuaires industrielles, 1970.*

4-035 Effect of Anthracene

Parameter	Effect Value	Concentration and Condition	Species
Growth rate	-20 % more than control	Water = 0.3 ppm, 96 h, PAH, sporelings	Antithamnion plumula[1]
Mortality	0 % mortality	Zero level at 70 ppm, naplii sedimentated (thought dead), 1 h, S=3 %	Balanus amphitrites[2]

Source: *1.Boney, A.D. 1974. Mar. Pollut. Bull. 5, 185-186.2.Donahue, W.H., R.T. Wang, M. Welch and J.A.C. Nicol. 1977. Environ. Pollut. 13, 187-202.*

4-036 Effect of As

Parameter	Effect Value	Concentration and Condition	Species
Mortality	50 % mortality	Na2HasO4, water = 10E-4 M As, toxic level given, microorganisms	Azotobacter sp.[1]
Mortality	0 %	Toxicological limit = > 0.7 mg/l, As-compounds	Biological purification[2]
Mortality	0 %	Toxicological limit = 4 - 9.1 mg/l, As-compounds	Crustacean[3]
Mortality	50 %	Na2AsSO4, exposure time = 2 d, conc in water = 7400 ug/l	Daphnia magna[4]
Mortality	50 % mortality	48 h, water = 1800 ppm, sodium arsenite	Daphnia pulex[5]
Mortality	0 %	Toxicological limit = 1 - 23 mg/l, As-compounds	Fish[3]
Mortality	0 %	Toxi. limit = 2 mg/kg, 0.05 mg/l chronical symptoms, As-compoumd	Homo sapiens[3]
Mortality	0 %	Toxicological limit = 2 - 15 mg/kg, As-compounds	Mammalia[3]
Mortality	50 %	Oral = 9 mg/kg, 5.49E-5mol/kg, F007 = 4.26, NaH2AsSO4	Mouse[6]
Mortality	50 %	3-oxide, oral dose = 45 mg/kg body	Rat[7]
Mortality	50 %	5-oxide, oral dose = 8 mg/kg body	Rat[7]
Mortality	50 %	Acute oral toxicity = 10-40 mg/kg, calcium arsenate	Rat[8]
Mortality	50 %	Acute oral toxicity = 10-50 mg sodium arsenite/kg, pesticide	Rat[8]
Mortality	50 %	Acute oral toxicity = 138 mg/kg , As2O3, pesticide	Rat[8]
Mortality	50 %	K-arsenite, oral dose = 1140 mg/kg body	Rat[7]
Mortality	50 %	Na-arsenite, oral dose = 41 mg/kg body	Rat[7]
Mortality	50 % mortality	Arsenate, water = 2E-2 M As, toxic level given, microorganisms	Yeast sp.[9]
Nitrification	37 % inhibation	Soil = 5 umoles/g of soil, 10 d incubation, 3 soils, lab	Soil organisms[10]
Nitrification	83 % inhibation	Soil = 5 umoles/g of soil, 10 d incubation, 3 soils, lab	Soil organisms[10]

Source: *1.Zhakharyan, S.V. 1970. Symp. Biol. Hung. 11, 391-395.*
2.Jung. K.D. 1973. Wirkungskonzentration schaedigender Stoffe im Wasser. Gelenkirchen. Hygiene-Institudes Ruhrgebiets.
3.Liebman, H. 1958. Hnadbuch der Friscwasser - und Abwasserbiologie. Band II.Munchen. Oldenbourg.
4.Biesinger, K.E. and G.M. Christensen. 1972. J. Fish. Res. bd. Can., 29(12), 1691-1700.
5.Rankov, V. 1973. Pochv. Agrokhim. 8, 125-134.
6.Luckey, T.D. and B. Venugopal. 1977. Metal Toxicity in Mammals, 1. Plenum Press.
7.Toxic Substances List 1973, U.S. Depart ment of Health, Education and Welfare, Public Health ServiceNational Institute for Occupational Safety and Health, Rockville Maryland, USA, June 1973.
8.Edwards, C.A. 1975. Persistent Pesticides in the Environment, 2nd edition, CRC-Press.
9.Laskowski, D.A. and F.E. Broadbent. 1970. Proc. Soil Sci. Soc. Am. 34, 72-77.
10.Liang, C.N. and M.A. Tabatabai. 1978. Environ. Pollut. 12, 141-147.

4-037 Effect of Atrazine

Parameter	Effect Value	Concentration and Condition	Species
Sublethal effect	0.003 mg/l	Inhibition of cell multiplication	Microcystis aeruginosa[1]
Sublethal effect	>10 mg/l	Inhibition of cell multiplication	Pseudomonas putida[1]
Mortality	50 %	Acute oral toxicity = 3080 mg/kg, herbicide	Rat[2]

Source: 1.Bringmann, G. and Kuhn, R., nwf-wasser/abwasser, H.9, 1976.
2.Edwards, C.A. 1975. Persistent Pesticides in the Environment, 2nd edition, CRC-Press.

4-038 Effect of Azinphosmethyl

Parameter	Effect Value	Concentration and Condition	Species
Mortality	50 %	48 h, water = 3 mg/l	Daphnia pulex[1]

Source: 1.Rankov, V. 1973. Pochv. Agrokhim. 8, 125-134.

4-039 Effect of B

Parameter	Effect Value	Concentration and Condition	Species
Deficiency symptoms	0.5 mg/kg soil	No symptoms over this val., sandy soil, soil-pH < 6.5, air dried soil	Plant[1]
Deficiency symptoms	0.7 mg/kg soil	No symptoms over this val., heavy mineral soil, soil-pH < 6.5, air	Plant[1]
Deficiency symptoms	0.7 mg/kg soil	No symptoms over this val., sandy soil, soil-pH > 6.5, air dried soil	Plant[1]
Deficiency symptoms	1.0 mg/kg soil	No symptoms over this val., heavy mineral soil, soil-pH > 6.5, air	Plant[1]
Deficiency symptoms	1.0 mg/kg soil	No symptoms over this val., peat, soil-pH < 6.5, air dried soil	Plant[1]
Deficiency symptoms	1.5 mg/kg soil	No symptoms over this val., peat, soil-pH > 6.5, air dried soil	Plant[1]
Mortality	0 %	Toxicological limit = 1 - 10 g/l, H3BO3	Fish[2]
Mortality	50 %	Na2B4O7, exposure = 2 d, conc in water = 8200000 ug/l	Gambusia affinis[3]
Mortality	0 %	Toxicological limit = 5 - 20 g, H3BO3	Homo sapiens[2]
Mortality	0 %	Toxicological limit = 1 - 5.1 g/kg, H3BO3	Mammalia[2]
Mortality	50 %	Borax, oral dose = 41 mg/kg body	Mouse[4]
Mortality	50 %	Boron in fresh water = 113 ug/ml, 283 K, underyearling	Oncorhynchus kisutch[5]
Mortality	50 %	Boron in sea water = 12.2g/ml, 283 K, underyearling	Oncorhynchus kisutch[5]
Mortality	50 %	Borax, oral dose = 4500 mg/kg body	Rat[4]
Mortality	50 %	Boric acid, oral dose = 2660 mg/kg body	Rat[4]
Nitrification	80 % inhib.	Soil = 5 umoles/g of soil, 10 d incubation, 3 soils, lab	Soil organisms[6]
Survival time	33.4 hour	Boron = 656 microgram/ml, 20 fish, median time	Oncorhynchus kisutch[5]
Survival time	84.0 hour	Boron = 328 microgram/ml, 20 fish, median time	Oncorhynchus kisutch[5]

Source: 1.Batey, T. 1971. In: Tech. Bull. 21, pp 137-148. Trace Elements in Soils and Crops.
2.Liebman, H. 1958. Hnadbuch der Friscwasser - und Abwasserbiologie. Band II.Munchen. Oldenbourg.
3.McKee, J.R. and H.W. Wolf. 1963. Water Quality Criteria, Publication no. 3A, Calif. State Water Res.Control Board 2nd Edition.
4.Toxic Substances List 1973, U.S. Depart ment of Health, Education and Welfare, Public Health ServiceNational Institute for Occupational Safety and Health, Rockville Maryland, USA, June 1973.
5.Thompson, J.A.J. and J.C. Davis. 1976. Water Res., 10(10), pp 869-875.
6.Liang, C.N. and M.A. Tabatabai. 1978. Environ. Pollut. 12, 141-147.

4-040 Effect of BCM

Parameter	Effect Value	Concentration and Condition	Species
Growth	50 % inhibition	48 h, freshwater = 0.34 mg/l, 297 K, log phase, E7 cells/l	Chlorella pyrenoidosa[1]
Mortality	50 %	48 h, freshwater = 0.46 mg/l, 293 K, age =	Daphnia magna[1]
Mortality	50 %	48 h, freshwater = 1.8 mg/l, 288 K, age = 3 m, n = 10	Salmo gairdneri[1]

Source: 1.Canton, J.H. 1976. Bull. Environ. Contam. Toxicol. 16(2), pp 214-218.

4-041 Effect of Ba

Parameter	Effect Value	Concentration and Condition	Species
Mortality	0 %	Toxicological limit = 1000 mg/l, BaCl2	Biological purification[1]
Mortality	50 %	16000 mg/l in diet, 7 d, Ba acetate and hydroxide	Chicken[2]
Mortality	50 %	33000 mg/l in diet, Ba acetate and hydroxide, 5 d	Chicken[2]
Mortality	50 %	8000 mg/l in diet, 4 w, Ba acetate and hydroxide	Chicken[2]
Mortality	0 %	Toxicological limit = 83 - 1500 mg/l, BaCl2	Crustacean[3]
Mortality	50 %	BaCl2, exposure time = 2 d, conc in water = 14500 ug/l	Daphnia magna[4]
Mortality	0 %	Toxicological limit = 0.15 - 8.3 g/l, BaCl2	Fish[3]
Mortality	0 %	Toxicological limit = 3.3 - 8.3 mg/kg, BaCl2	Homo sapiens[3]
Mortality	0 %	Toxicological limit = 30 - 500 mg/kg, BaCl2	Mammalia[3]
Mortality	50 %	Acetate, intravenous, dose = 11 mg/kg body	Mouse[5]
Mortality	50 %	Chloride, intravenous, dose = 8 mg/kg body	Mouse[5]
Mortality	50 %	Nitrate, intravenous, dose = 8.5 mg/kg body	Mouse[5]
Mortality	50 %	Oral dose = 500 mg/kg, 2.40E-3mol/kg, F007 = 2.62, as BaCl2	Mouse[6]
Mortality	50 %	Carbonate, oral dose = 800 mg/kg body	Rat[7]
Mortality	50 %	Chloride, oral dose = 150 mg/kg body	Rat[5]
Mortality	50 %	Sulfonate, oral dose = 3000 mg/kg body	Rat[5]

Source: 1.Jung. K.D. 1973. Wirkungskonzentration schaedigender Stoffe im Wasser. Gelenkirchen. Hygiene-Institudes Ruhrgebiets.
2.Johnson, D., A.L. Mehring and H.W. TiTus. 1960. Proc. Soc. Exp. Biol. Med. 104, pp 436-438.
3.Liebman, H. 1958. Hnadbuch der Friscwasser - und Abwasserbiologie. Band II.Munchen. Oldenbourg.
4.Biesinger, K.E. and G.M. Christensen. 1972. J. Fish. Res. bd. Can., 29(12), 1691-1700.
5.Toxic Substances List 1973, U.S. Depart ment of Health, Education and Welfare, Public Health ServiceNational Institute for Occupational Safety and Health, Rockville Maryland, USA, June 1973.
6.Luckey, T.D. and B. Venugopal. 1977. Metal Toxicity in Mammals, 1. Plenum Press.
7.EPA. Water Quality Crit. Data Book. 2rd. Inorganic Pollution of Freshwater. 18010. DPU 07/71.

4-042 Effect of Barban

Parameter	Effect Value	Concentration and Condition	Species
Growth rate	7.5 % stimul.	Water conc = 1.0 mg/l, phenylcarbamate herbicide	Anabaena cylindraca[1]
Growth rate	50 % inhibition	Water conc = 2.0 mg/l, phenylcarbamate herbicide, N2-fixing conditions	Anabaena cylindrica[1]
Growth rate	50 % inhibition	Water conc = 2.8 mg/l, phenylcarbamate herbicide	Anabaena cylindrica[1]
Growth rate	50 % inhibition	Water conc = 0.3 mg/l, phenylcarbamate herbicide	Chlorella pyrenoidosa[1]
Growth rate	50 % inhibition	Water conc = 1.8 mg/l, phenylcarbamate herbicide	Tolypothrix tenuis[1]

Source: 1.Hulin, N. and M. Horowitz. 1973. Phytoparasitica 1, 64.

4-043 Effect of Be

Parameter	Effect Value	Concentration and Condition	Species
Mortality	0 %	Toxicological limit = 0.2 g/l, other than sunfish, $BeSO_4$	Fish[1]
Mortality	0 %	Toxicological limit = 1.3 g/l, sunfish, $BeSO_4$	Fish[1]
Mortality	50 %	Beryllium = 0.19 mg/l, 96 h, soft water, 296 K	Lebistes reticulatus[2]
Mortality	50 %	Beryllium = 0.43 mg/l, 48 h, soft water, 296 K	Lebistes reticulatus[2]
Mortality	50 %	Beryllium = 2.58 mg/l, 24 h, soft water, 296 K	Lebistes reticulatus[2]
Mortality	50 %	Beryllium = 20.3 mg/l, 96 h, hard water, 296 K	Lebistes reticulatus[2]
Mortality	50 %	Beryllium = 22.1 mg/l, 48 h, hard water, 296 K	Lebistes reticulatus[2]
Mortality	50 %	Beryllium = 22.3 mg/l, 24 h, hard water, 296 K	Lebistes reticulatus[2]
Mortality	50 %	Oral dose = 12 mg/kg, 1.50E-4mol/kg, F007 = 3.82, as $BeCl_2$	Mouse[3]
Mortality	50 %	$BeCl_2$, exposure time = 4 d, conc in water = 150 microgram/l, limnol	Pimephales promelas[4]
Mortality	50 %	Chloride, oral dose = 86 mg/kg body	Rat[5]
Mortality	50 %	Fluoride, oral dose = 98 mg/kg body	Rat[5]
Mortality	50 %	Hydroxide, intravenous, dose = 2.3 mg/kg body	Rat[5]
Mortality	50 %	Sulfate, oral dose = 82 mg/kg body	Rat[5]

Source: 1.Liebman, H. 1958. Handbuch der Friscwasser - und Abwasserbiologie. Band II.Munchen. Oldenbourg.
2.Canton, J.H. 1976. Bull. Environ. Contam. Toxicol. 16(2), pp 214-218.
3.Luckey, T.D. and B. Venugopal. 1977. Metal Toxicity in Mammals, 1. Plenum Press.
4.McKee, J.R. and H.W. Wolf. 1963. Water Quality Criteria, Publication no. 3A, Calif. State Water Res.Control Board 2nd Edition.
5.Toxic Substances List 1973, U.S. Depart ment of Health, Education and Welfare, Public Health ServiceNational Institute for Occupational Safety and Health, Rockville Maryland, USA, June 1973.

4-044 Effect of Benomyl

Parameter	Effect Value	Concentration and Condition	Species
Growth	50 % inhibition	48 h, freshwater = 1.4 mg/l, 297 K, log phase, E7 cells/l	Chlorella pyrenoidosa[1]
Mortality	50 %	48 h, freshwater = 0.64 mg/l, 293 K, age =	Daphnia magna[1]
Mortality	50 %	48 h, freshwater = 3.4 mg/l, 288 K, age = 3 m, n = 10	Lebistes reticulatus[1]
Mortality	50 %	Acute oral toxicity = 9590 mg/kg	Rat[2]
Mortality	50 %	48 h, freshwater = 0.48 mg/l, 288 K, age = 3 m, n = 10	Salmo gairdneri[1]

Source: 1.Canton, J.H. 1976. Bull. Environ. Contam. Toxicol. 16(2), pp 214-218.
2.Edwards, C.A. 1975. Persistent Pesticides in the Environment, 2nd edition, CRC-Press.

4-045 Effect of Bensulide

Parameter	Effect Value	Concentration and Condition	Species
Mortality	50 %	379 (327-402) ppb, 96 h, 14 g, 293 K, pH = 8.2, alk = 80 ppm, 95 % CL	Ictalurus punctatus[1]

Source: 1.McCorle, F.M., J.E. Chambers and J.D. Yarbrough. 1977. Bull. Environ. Contam. Toxicol. 18(3), pp 267270.

4-046 Effect of Benz(a)anthracene

Parameter	Effect Value	Concentration and Condition	Species
Mortality	87 % mortality	Water = 1.0 ppm, bluegill, 6 months	Lepomis macrochirus[1]

Source: 1.Brown, E.R., Sinclair, Keith, Beamer, Hazdra, Nair, and Callaghan1977. In: H.F. Kraybill, C.J. Dawe,J.C. Harshbarger and R.G. Terdiff (eds.): Aquatic Pollutants and Biological Effects with emphasis on

4-047 Effect of Benzaldehyde

Parameter	Concentration	Condition	Species
Sublethal effect	20 mg/l	Inhibition of cell multiplication	Microcystis aeruginosa[1]
Sublethal effect	132 mg/l	Inhibition of cell multiplication	Pseudomonas putida[1]
LD50	*0.05 g pr. kg ww*	*subcutane dose*	*Rabbit*[1]
LD50	5.6-5.7 g pr. kg ww	method not indicated	Rat[3]

Source: 1.Patty, F. A., Industrial Hygiene and toxicology, 2, Interscience Publishers, New York, 1976.
2.McKee, J. E., and Wolf, H. W., Water Quality, The resources agency of California, 1963.
3.Bringmann, G. and Kuhn, R., nwf-wasser/abwasser, H.9, 1976.

4-048 Effect of Benzene

Parameter	Effect Value	Concentration and Condition	Species
Mortality	100 %	Inhalation, air = 1 - 2 mg/m3	Homo sapiens and mammals[1]
Mortality	100 %	Inhalation, air = at least 0.01 - 1 g/m3, also anaemia	Homo sapiens and mammals[1]
Mortality	50 %	Water = 108 ppm, stage 1 zoeae, 96 h, Dungeness crab	Cancer magister[2]
Mortality	50 %	Water = 20 ppm, 96 h, bay shrimp	Crago franciscorum[3]
Mortality	50 %	Water = 386 ppm, 96 h	Gambusia affinis[4]
Mortality	100 %	Inhalation,	Homo sapiens and mammals[1]
Mortality	50 %	Water = 5.8 ppm, 96 h	Marone saxatilis[3]
Mortality	50 %	Water = 27 ppm, 96 h	Palaemonetes pugio[5]
Mortality	100 %	Inhalation, air = 2.8 - 5 mg/m3	Homo sapiens and mammals[6]
Mortality	100 %	Inhalation, air = 0.6-3 m/3, also mucositis and dyssomnia	Homo sapiens and mammals[6]
Mortality	100 %	Inhalation, acute lethal concentration, air = 15-65 g/m3	Homo sapiens and mammals[6]
LC0	10 mg/l	96 hours, seawater, 8 C	Coho salmon[6]
LC100	60 mg/l	24 hours	Bluegill[7]
LC100	100 mg/l	2 hours, seawater, 8 C	Coho salmon[6]
LC100	34 mg/l	24 hours	SunFish[7]
LC50	20 mg/l	24-48 hours	Bluegill[7]
LC50	20 mg/l	24-48 hours	SunFish[7]
Sublethal	>1400 mg/l	Inhibition of cell multiplication	Microcystis aeruginosa[8]
Sublethal	92 mg/l	Inhibition of cell multiplication	Pseudomonas putida[8]
TLm	22.5 mg/l	24-96 hours, softwater	Bluegill[9]
TLm	34-37 mg/l	24-96 hours, softwater	Guppie[9]
TLm	24-35 mg/l	24-96 hours, soft/hardwater	Latuhead[9]

Source: 9.Drger, Gasspurgerat. Drgerwerk AG, 24 Lubeck, PO 1339.
6.Morrow, J. E., EPA/660&3/73/018, 1974.
7.McKee, J. E., and Wolf, H. W., Water Quality, The resources agency of California, 1963.
8.Bringmann, G. and Kuhn, R., nwf-wasser/abwasser, H.9, 1976.
1.Kaiser, G. and G. Toelg. 1980. In: O. Hutzinger: The Handbook of Environmental Chemistry. Antropogenc compounds. Vol 3. Part A. pp 1-58. Springer-Verlag.
2.Caldwell, R.s., E. M. Caldarone and M.H. Mallon. 1977. D.A. Wolfe (ed.): Fate and Effects of Petrolem Hydrocarbons in Marine Ecosystems and Organisms. NY. Pergamon Press. 210-220.
3.Benville, P.E. Jr. and S. Korn. 1977. Calif. Fish and Game. 63, 204-209.
4.Wallen, I.E., W.C. Greer and R. Lasater. 1957. Sew. Indust. Wastes 29, 695-711.
5.Neff, J.M., Anderson, Cox, Laughlin, Rossi and Tatem. 1976. In: Sources, Effects and Sinks of Hydrocrbon in the Aquatic Environment. Wash DC. Amer. Inst. of Biological Science.

4-049 Effect of Benzene sulfonic acid

Parameter	Concentration	Condition	Species
LD50	0.4-3.2 g / kg ww	acute oral dose	Mouse[1]

Source: 1.Patty, F. A., Industrial Hygiene and toxicology, 2, Interscience Publishers, New York, 1976.

4-050 Effect of Benzo(a)pyrene

Parameter	Effect Value	Concentration and Condition	Species
Growth rate	49 % more than control	Water = 0.01 ppm (in methyl cellosolve), 96 h, SD = 1 %, PAH, sporelings	Antithamnion plumula[1]
Growth rate	55 % more than control	Water = 0.1 ppm (in methyl cellosolve), 96 h, SD = 5 %, PAH, sporelings	Antithamnion plumula[1]
Mortality	0 % mortality	Water = 1.0 ppm, zero level mortality in 96 h, marine	Neathes arenaceodentata[2]

Source: 1.Boney, A.D. and E.D.S. Corner. 1962. J. Mar. Biol. Assoc. UK. 43, 579-585.
2.Rossi, S.S. and J.M. Neff. 1978. Mar. Pollut. Bull. 9, 220-223.

4-051 Effect of benzonitrile

Parameter	Concentration	Condition	Species
TLm	78 mg/l	96 hours, soft water	Bluegill[1]
TLm	78-135 mg/l	96 hours, soft/hard water	Flathead minnow[2]
TLm	400 mg/l	96 hours, soft water	Guppie[2]

Source: 1.Henderson, C., Pickering, Q. H. and Lemke, A. E., Purdue Univ. Eng. Bull. Ext. Ser., 106, 120, 1960.
2.Jones, H. R., Environmental Control in the organic and petrochemical Industries, Noyes Data Corp., 171.

4-052 Effect of Benzoic acid

Parameter	Concentration	Condition	Species
Sublethal effect	146 mg/l	Immobilization, prolonged exposure	Dapnia[1]
Sublethal effect	55 mg/l	Inhibition of cell multiplication	Microcystis aeruginosa[2]
Sublethal effect	480 mg/l	Inhibition of cell multiplication	Pseudomonas putida[2]

Parameter	Concentration	Condition	Species
LD50	*1.7 g pr. kg body weight*	*acute oral dose*	*Rat*[3]
TLm	180-255 mg/l	24-96 hours	Mosquito fish[4]

Source: 3.Patty, F. A., Industrial Hygiene and toxicology, 2, Interscience Publishers, New York, 1976.
4.Jones, H. R., Environmental Control in the organic and petrochemical Industries, Noyes Data Corp., 171.
1.McKee, J. E., and Wolf, H. W., Water Quality, The resources agency of California, 1963.
2.Bringmann, G. and Kuhn, R., nwf-wasser/abwasser, H.9, 1976.

4-053 Effect of Benzonitrile

Parameter	Concentration	Condition	Species
Sublethal effect	3.4 mg/l	Inhibition of cell multiplication	Microcystis aeruginosa[1]
Sublethal effect	11 mg/l	Inhibition of cell multiplication	Pseudomonas putida[1]

Source: 1.Bringmann, G. and Kuhn, R., nwf-wasser/abwasser, H.9, 1976.

4-054 Effect of Benzylalcohol

Parameter	Concentration	Condition	Species
LC0	1000 mg/l	not indicated	Escherichia coli[2]
LD50	3.1 g pr. kg body weight	single oral dose	Rat[1]

Source: 1.Patty, F. A., Industrial Hygiene and toxicology, 2, Interscience Publishers, New York, 1976. 2.Meinck, F., Stooff, H. and Kohlschutter, H., Les eaux residuaires industrielles, 1970.

4-055 Effect of Benzylamine

Parameter	Concentration	Condition	Species
LC0	1000 mg/l	not indicated	Escherichia coli[1]

Source: 1.Meinck, F., Stooff, H. and Kohlschutter, H., Les eaux residuaires industrielles, 1970.

4-056 Effect of Benzylchloride

Parameter	Concentration	Condition	Species
Sublethal effect	17 mg/l	Paralysis	Cyprinus carpio[1]
Sublethal effect	30 mg/l	Inhibition of cell multiplication	Microcystis aeruginosa[2]
Sublethal effect	4.8 mg/l	Inhibition of cell multiplication	Pseudomonas putida[2]
Sublethal effect	10 mg/l	Paralysis	Trout[1]

Source: 1.Meinck, F., Stooff, H. and Kohlschutter, H., Les eaux residuaires industrielles, 1970. 2.Bringmann, G. and Kuhn, R., nwf-wasser/abwasser, H.9, 1976.

4-057 Effect of Bi

Parameter	Effect Value	Concentration and Condition	Species
Mortality	50 %	Dihydroxypropoxy, intravenous, dose = 13 mg/kg body	Rat[1]

Source: 1.Toxic Substances List 1973, U.S. Department of Health, Education and Welfare, Public Health ServiceNational Institute for Occupational Safety and Health, Rockville Maryland, USA, June 1973.

4-058 Effect of Biphenyl

Parameter	Effect Value	Concentration and Condition	Species
Mortality	50 % mortality	Water = 27 ppm, naplii sedimentated (thought dead), 1 h, S=3 %	Balanus amphitrites[1]

Source: 1.Donahue, W.H., R.T. Wang, M. Welch and J.A.C. Nicol. 1977. Environ. Pollut. 13, 187-202.

4-059 Effect of Bo

Parameter	Effect Value	Concentration and Condition	Species
Growth	0.001 mg/l gives deficiency symptoms	Tree, 100 d in greenhouse, conc in culture solution	Agathis australis[1]
Growth	0.500 mg/l gives normal growth	Tree, 100 d in greenhouse, conc in culture solution	Agathis australis[1]
Growth	5.000 mg/l gives toxic symptoms	Tree, 100 d in greenhouse, conc in culture solution	Agathis australis[1]

Source: 1.Peterson, P.J. 1962. N. Z. J. Sci. Technol. 5, pp 141-161.

4-060 Effect of Bordeaux mixture

Parameter	Effect Value	Concentration and Condition	Species
Mortality	50 %	Acute oral toxicity = 300 mg/kg, fungicide	Rat[1]

Source: 1.Edwards, C.A. 1975. Persistent Pesticides in the Environment, 2nd edition, CRC-Press.

4-061 Effect of Botulinal

Parameter	Effect Value	Concentration and Condition	Species
Mortality	50 %	Oral dose = 1.14E-6 mg/kg, botulinal a, F007 = 14.9	Mouse[1]
Mortality	50 %	Oral dose = 2.5E-3 mg/kg, botulinal e, F007 = 11.15	Mouse[1]
Mortality	50 %	Oral dose = 3.2E-7 mg/kg, botulinal d, F007 = 15.49	Mouse[1]

Source: 1.Luckey, T.D. and B. Venugopal. 1977. Metal Toxicity in Mammals, 1. Plenum Press.

4-062 Effect of Butylglycidylether

Parameter	Concentration	Condition	Species
LD50	1.5-2.3 g pr. kg body weight	acute intragastric dose	mouse, Rat[1]
LD50	1.5-2.3 g pr. kg body weight	acute intragastric dose	mouse, Rat[1]

Source: 1.Patty, F. A., Industrial Hygiene and toxicology, 2, Interscience Publishers, New York, 1976.

4-063 Effect of Butylglycolether

Parameter	Concentration	Condition	Species
LD50	1.2 g pr. kg ww	single oral dose	mouse, guinea pig[2]
LD50	0.32 g pr. kg ww	single oral dose	Rabbit[2]
LC50	600-1000 mg/l	48 hours	Brown shrimps[1]
LC50	550-950 mg/l	96 hours	Brown shrimps[1]

Source: 1.Blackmann, R. A. A., Marine Pollution Bull. 5, 116, 1974.
2.Patty, F. A., Industrial Hygiene and toxicology, 2, Interscience Publishers, New York, 1976.

4-064 Effect of Butylphosphate

Parameter	Concentration	Condition	Species
Sublethal effect	4.1 mg/l	Inhibition of cell multiplication	Microcystis aeruginosa[1]
Sublethal effect	>100 mg/l	Inhibition of cell multiplication	Pseudomonas putida[1]

Source: 1.Bringmann, G. and Kuhn, R., nwf-wasser/abwasser, H.9, 1976.

4-065 Effect of Butyraldehyde

Parameter	Concentration	Condition	Species
Sublethal effect	19 mg/l	Inhibition of cell multiplication	Microcystis aeruginosa[2]
Sublethal effect	100 mg/l	Inhibition of cell multiplication	Pseudomonas putida[2]
LD50	5.9 g pr. kg ww	acute oral dose	Rat[1]

Source: 1.Patty, F. A., Industrial Hygiene and toxicology, 2, Interscience Publishers, New York, 1976.
2.Bringmann, G. and Kuhn, R., nwf-wasser/abwasser, H.9, 1976.

4-066 Effect of Butyramide

Parameter	Concentration	Condition	Species
LD50	2.9-3.8 g pr. kg ww	single oral dose	Rat[1]

Source: 1.Patty, F. A., Industrial Hygiene and toxicology, 2, Interscience Publishers, New York, 1976.

4-067 Effect of Butyronitrile

Parameter	Concentration	Condition	Species
LD50	0.05-0.1 g pr. kg ww	acute oral dose	Rat[1]

Source: 1.Patty, F. A., Industrial Hygiene and toxicology, 2, Interscience Publishers, New York, 1976.

4-068 Effect of Caproic acid

Parameter	Concentration	Condition	Species
LD50	6.44 g pr. kg ww	oral dose	Rat[1]

Source: 1.Patty, F. A., Industrial Hygiene and toxicology, 2, Interscience Publishers, New York, 1976.

4-069 Effect of CCl_4

Parameter	Effect Value	Concentration and Condition	Species
Mortality	50 %	Oral dose = 4620 mg/kg, 3.0E-2mol/kg, F007 = 1.52	Mouse[1]
Sublethaleffect	5 ppm	Increased liver weight, chronic exposure	Guinea pig[2]
LD50	2.92 g pr. kg ww	oral dose	Rat[2]

Source: 2.Patty, F. A., Industrial Hygiene and toxicology, 2, Interscience Publishers, New York, 1976.
1.Luckey, T.D. and B. Venugopal. 1977. Metal Toxicity in Mammals, 1. Plenum Press.

4-070 Effect of Ca

Parameter	Effect Value	Concentration and Condition	Species
Mortality	50 %	Oral dose = 1350 mg/kg, 7.36E-3mol/kg, F007 = 2.13, as CaCl2	Mouse[1]
Mortality	50 %	Acetate, oral dose = 4280 mg/kg body	Rat[2]
Mortality	50 %	Chloride, oral dose = 4000 mg/kg body	Rat[2]

Source: 1.Luckey, T.D. and B. Venugopal. 1977. Metal Toxicity in Mammals, 1. Plenum Press.
2.Toxic Substances List 1973, U.S. Department of Health, Education and Welfare, Public Health ServiceNational Institute for Occupational Safety and Health, Rockville Maryland, USA, June 1973.

4-071 Effect of Camphechlor

Parameter	Effect Value	Concentration and Condition	Species
Mortality	50 %	Conc in food = 828 mg/kg(619-1102), 95 % CL	Bobwhite[1]
Mortality	95 %	Conc in soil > 15 mg/kg	Earthworms[1]
Mortality	50 %	Conc in food = 686 mg/kg(523-1002), 95 % CL	Japanese quail[1]
Mortality	50 %	Conc in food = 538 mg/kg(474-614), 95 % CL	Mallard[1]
Mortality	50 %	Conc in food = 450.0 mg/kg	Pheasant[2]
Mortality	50 %	Conc in food = 542 mg/kg(462-638), 95 % CL	Pheasant[1]
Mortality	50 %	Conc in food = 500.0 mg/kg	Quail[2]
Mortality	50 %	Acute dermal toxicity = 780-1075 mg/kg, insecticide	Rat[1]
Mortality	50 %	Acute oral toxicity = 40-150 mg/kg, insecticide	Rat[1]

Source: 1.Edwards, C.A. 1975. Persistent Pesticides in the Environment, 2nd edition, CRC-Press.2.Dewitt, J.B., C.M. Menzie, V.A. Adomaitis and W.L. Reickel. 1960. Trans. 25th N. Am. Wildl. Conf., p277.

4-072 Effect of Caproic acid

Parameter	Concentration	Condition	Species
TLm	22 mg/l	24 hours	Daphnia magna[1]
TLm	150-200 mg/l	24 hours	Lepomis macrochirus[2]

Source: 1.Verschuren. Handbook of Toxicology,
2.Dowden, B. F. and Bennet, H. J., JWPCF, 37, 1310, 1965.

4-073 Effect of Carbaryl

Parameter	Effect Value	Concentration and Condition	Species
Heart beat	0 beats/min	Carbaryl = 400 micro mol, after 24 h, 299 K, 5 snails	Biomphalaria glabrata[1]
Heart beat	31 beats/min	Carbaryl = 320 micro mol, after 24 h, 299 K, 5 snails	Biomphalaria glabrata[1]
Heart beat	96 beats/min	Carbaryl = 80 micro mol, after 24 h, 299 K, 5 snails	Biomphalaria glabrata[1]
Mortality	50 % mortality	48 h, water = 6 mg/l	Daphnia pulex[2]

Source: 1.Cocks, J.A. 1974. Environ. Pollut., 5, pp 149-151.2.Rankov, V. 1973. Pochv. Agrokhim. 8, 125-134.

4-074 Effect of Carbon disulfide

Parameter	Concentration	Condition	Species
TLm	135-162 mg/l	24-96 hours	Mosquito fish[1]

Source: 1.Jones, H. R., Environmental Control in the organic and petrochemical Industries, Noyes Data Corp., 171.

4-075 Effect of Cathecol

Parameter	Concentration	Condition	Species
LC0	90 mg/l	not indicated	Escherichia coli[1]
LC0	6 mg/l	not indicated	Scenedesmus[1]
LC0	1.6 mg/l	not indicated	Vorticella campanula[1]
LC100	14 mg/l	48 hours	Goldfish[2]

Source: 1.Meinck, F., Stooff, H. and Kohlschutter, H., Les eaux residuaires industrielles, 1970.
2.McKee, J. E., and Wolf, H. W., Water Quality, The resources agency of California, 1963.

4-076 Effect of Cation exchange capacity

Parameter	Effect Value	Concentration and Condition	Species
Uptake of Cd	11.0 mg Cd/kg dw	CEC=231 meq/kg soil, exch soil Cd=2.55 ppm, dw=1.34 g	Avena sativa[1]
Uptake of Cd	12.9 mg Cd/kg dw	CEC=189 meq/kg soil, exch soil Cd=2.90 ppm, dw=1.34 g	Avena sativa[1]
Uptake of Cd	13.7 mg Cd/kg dw	CEC=171 meq/kg soil, exch soil Cd=2.94 ppm, dw=0.99 g	Avena sativa[1]
Uptake of Cd	7.9 mg Cd/kg dw	CEC=305 meq/kg soil, exch soil Cd=1.95 ppm, dw=1.76 g	Avena sativa[1]
Uptake of Cd	9.5 mg Cd/kg dw	CEC =269 meq/kg,exch soil Cd=2.34 ppm, dw = 1.66 g	Avena sativa[1]

Source: 1.Haghiri, F. 1974. J. Environ. Qual., 3(2), pp 180-183.

4-077 Effect of Cd

Parameter	Effect Value	Concentration and Condition	Species
ATP level	-26 % change relative to control	Cd = 1 mM, CdSO4, 2 h, oats, excised root treatment	Avena sativa[1]
ATPase	-75 % change relative to control	Cd = 1 mM, CdSO4, oats, Mg and K stimulated	Phaseolus vulgaris[1]

Acid phosphatase	+115 % change relative to control	Cd = 1.35 uM, CdSO4, solution, soybean cv. Columbus	Glycine max[2]
Blood glucose	20.0 mg/100 ml	Cd = 0.0 mg/l, 15 d,	Pleuronectes flesus[3]
Blood glucose	21.0 mg/100 ml	Cd = 0.1 mg/l, 15 d	Pleuronectes flesus[3]
Blood glucose	24.7 mg/100 ml	Cd = 0.0 ppb, 9 w, 9-10 fish, 283 K, rg = +- 0.8	Pleuronectes flesus[3]
Blood glucose	25.9 mg/100 ml	Cd = 0.0 ppb, 4 w, 9-10 fish, 283 K, rg = +- 0.8	Pleuronectes flesus[3]
Blood glucose	27.8 mg/100 ml	Cd = 5 ppb, 4 w, 9-10 fish, 283 K, rg = +- 1.2	Pleuronectes flesus[3]
Blood glucose	28.4 mg/100 ml	Cd = 500 ppb, 9 w, 9-10 fish, 283 K, rg = +- 3.6	Pleuronectes flesus[3]
Blood glucose	29.0 mg/100 ml	Cd = 1.0 mg/l, 15 d	Pleuronectes flesus[3]
Blood glucose	29.2 mg/100 ml	Cd = 5 ppb, 9 w, 9-10 fish, 283 K, rg = +- 1.2	Pleuronectes flesus[3]
Blood glucose	31.7 mg/100 ml	Cd = 50 ppb, 4 w, 9-10 fish, 283 K, rg = +- 1.2	Pleuronectes flesus[3]
Blood glucose	34.1 mg/100 ml	Cd = 500 ppb, 4 w, 9-10 fish, 283 K, rg = +- 1.4	Pleuronectes flesus[3]
Blood glucose	35.0 mg/100 ml	Cd = 10.0 mg/l, 15 d	Pleuronectes flesus[3]
Blood glucose	41.0 mg/100 ml	Cd = 50 ppb, 9 w, 9-10 fish, 283 K, rg = +- 2.7	Pleuronectes flesus[3]
Blood lactate	1.6 mg/100 ml	Cd = 10.0 mg/l, 15 d	Pleuronectes flesus[3]
Blood lactate	3.9 mg/100 ml	Cd = 0.1 mg/l, 15 d	Pleuronectes flesus[3]
Blood lactate	4.8 mg/100 ml	Cd = 1.0 mg/l, 15 d	Pleuronectes flesus[3]
Blood lactate	5.0 mg/100 ml	Cd = 0.0 mg/100 ml, 15 d	Pleuronectes flesus[3]
Carbonic anhydrase	-18 % change relative to control	Cd = 1.35 uM, CdSO4, solution, soybean cv. Columbus	Glycine max[2]
Chlorophyll synthesis	-80 % change relative to control	Cd=10 ug/ml, CdCl2, bean cv. Red Kidney, excised embryos	Phaseolus vulgaris[4]
Corpuscular hemoglobin in blood	22.7 g/100 ml packed red cells	Cd = 0.0 ppb, 60 d, 31 fish, rg = +- 0.7	Pseudopleuronectes americanus[5]
Corpuscular hemoglobin in blood	22.7 g/100 ml packed red cells	Cd = 10 ppb, 60 d, 31 fish, rg = +- 0.7	Pseudopleuronectes americanus[5]
Corpuscular hemoglobin in blood	23.8 g/100 ml packed red cells	Cd = 5 ppb, 60 d, 31 fish, rg = +- 1.0	Pseudopleuronectes americanus[5]
Corpuscular hemoglobin in blood	26.2 picograms/cell	Cd = 0.0 ppb, 60 d, 14 fish, rg = +- 1.6	Pseudopleuronectes americanus[5]
Corpuscular hemoglobin in blood	26.2 picograms/cell	Cd = 10 ppb, 60 d, 14 fish, rg = +- 1.6	Pseudopleuronectes americanus[5]
Corpuscular hemoglobin in blood	27.9 picograms/cell	Cd = 5 ppb, 60 d, 9 fish, rg = +- 3.7	Pseudopleuronectes americanus[5]
Corpuscular volume in blood	114.0 micro**3/cell	Cd = 10 ppb, 60 d, 14 fish, rg = +- 5.0	Pseudopleuronectes americanus[5]
Corpuscular volume in blood	114.0 micro**3/cell	Cd = 5 ppb, 60 d, 9 fish, rg = +- 6.1	Pseudopleuronectes americanus[5]
Corpuscular volume in blood	121.0 micro**3/cell	Cd = 0.0 ppb, 60 d, 17 fish, rg = +- 7.0	Pseudopleuronectes americanus[5]
DNase	+300 % change relative to control	Cd = 1.35 uM, CdSO4, solution, soybean cv. Columbus	Glycine max[2]

Dry weight	-94 % change relative to control	Cd = 4 ug/g medium, CdCl2, Sand and Hoaglands sol., soybean	Glycine max[6]
Embryo curvature	-65 % change relative to control	Cd = 10 ug Cd/g medium, CdCl2, bean cv. Red Kidney	Phaseolus vulgaris[4]
Embryo curvature	0 % change relative to control	Cd = 2 ug Cd/g medium, CdCl2, bean cv. Red Kidney	Phaseolus vulgaris[4]
Embryo greening	-80 % change relative to control	Cd = 10 ug Cd/g medium, CdCl2, Bean cv. Red Kidney	Phaseolus vulgaris[4]
Embryo greening	0 % change relative to control	Cd = 2 ug Cd/g medium, CdCl2, Bean cv. Red Kidney	Phaseolus vulgaris[4]
Embryo growth	+4 % change relative to control	Cd = 2 ug/g medium, CdCl2, bean cv. Red Kidney	Phaseolus vulgaris[4]
Embryo growth	-80 % change relative to control	Cd = 10 ug/g medium, CdCl2, bean cv. Red Kidney	Phaseolus vulgaris[4]
Enzyme reaction	-100 % of control	Cd = 0.5 mM, Cd(NO3)2, spinach, isolated chloroplasts fixed in gluthrald.	DCPIP photoreduction of Spinacea oleracea[7]
Enzyme reaction	-43 % of control	Cd = 100 uM, CdCl2, spinach, H2O to methylviologen	Electron transport of Spinacea oleracea[8]
Enzyme reaction	-50 % of control	Cd = 10 mM, Cd(NO3)2, inhibits on donor side of complex	Fluorescence in Spinacea oleracea[9]
Enzyme reaction	-44 % of control	Cd = 0.01-0.5 mM, effect rg=+2- -89 %, CdCl2, isolated chloroplasts	Glycolipid biosynthesis of Spinacea oleracea[10]
Enzyme reaction	-20 % of control	Cd = 500 uM, CdCl2, isolated chloroplasts, spinach	Hill reaction of Spinacea oleracea[11]
Enzyme reaction	-55 % of control	Cd = 0.5 mM, Cd(NO3)2, spinach, isolated chloroplasts	Hill reaction of Spinacea oleracea[7]
Enzyme reaction	+120 % of control	Cd = 0.03 mM, CdCl2, isolated mitochonria, corn	Mitochondrial NADH oxidation in Zea mays[12]
Enzyme reaction	-80 % of control	Cd = 1 mM, CdCl2, isolated mitichondria, corn	Mitochondrial NADH oxidation in Zea mays[12]
Enzyme reaction	-22 % of control	Cd = 0.7 mM, Cd(NO3)2, isolated enzyme	PEP carboxylase in Spinacea oleracea[13]
Enzyme reaction	-19 % of control	Cd = 0.5 mM, isolated enzyme, tobacco	PEP carboxylase system of Nicotiana tabacu[13]
Enzyme reaction	-5 % of control	Cd = 0.5 mM, isolated enzyme, pea, var. hortense	PEP carboxylase system of Pisum sativum[13]
Enzyme reaction	-100 % of control	Cd = 0.25 mM, CdCl2, isolated mitochondria, corn	Phosphyralation , ADP / O ratio in Zea mays[12]
Enzyme reaction	-60 % of control	Cd = 100 uM, CdCl2, isolated chloroplasts, spinach	Photophosphorylation, ATP synthesis of Spinacea oleracea[8]
Enzyme reaction	-77 % of control	Cd = 10 uM, CdCl2, isolated chloroplasts, spinach	Photosynthesis c02 - 14 fixation of Spinacea oleracea[11]
Enzyme reaction	-75 % of control	Cd = 9 mM, Cd(NO3)2, isolated chloroplasts	Photosynthesis, PS II electron transport in Spinacea oleracea[9]
Enzyme reaction	-90 % of control	Cd = 0.7 mM, Cd(NO3)2, 25 min, isolated chloroplasts	Photosynthesis, PS II photoreaction in Spinacea oleracea[14]
Enzyme reaction	+15 % of control	Cd = 0.0001 mM, CdCl2, isolated mitochondria, corn	Succinate oxidation in Zea mays[12]
Enzyme reaction	-80 % of control	Cd = 0.1 mM, CdCl2, isolated mitochondria, corn	Succinate oxidation in Zea mays[12]

Glycogen in muscles	44.0 micrograms/100 mg wet weight	Cd = 0.1 mg/l, 15 d	Pleuronectes flesus[3]
Glycogen in muscles	54.0 micrograms/100 mg wet weight	Cd = 1.0 mg/l, 15 d	Pleuronectes flesus[3]
Glycogen in muscles	55.0 micrograms/100 mg wet weight	Cd = 10.0 mg/l, 15 d	Pleuronectes flesus[3]
Glycogen in muscles	74.0 micrograms/100 mg wet weight	Cd = 0.0 mg/l, 15 d	Pleuronectes flesus[3]
Glycolytic rate	0.25 umol lactate/(mg protein*hour)	Cd=537 ug Cd/g dw animal, rg=0.04, Ladye Bay, polluted, Bristol Ch.	Patella vulgata[15]
Glycolytic rate	0.27 umol lactate/(mg protein*hour)	Cd=419 ug Cd/g dw animal, rg=0.04, Weston-Mare, polluted, Bristol Ch.	Patella vulgata[15]
Glycolytic rate	0.37 umol lactate/(mg protein*hour)	Cd=257 ug Cd/g dw animal, rg=0.055, Blue Anchor, polluted, Bristol Ch.	Patella vulgata[15]
Glycolytic rate	0.45 umol lactate/(mg protein*hour)	Cd=27 ug Cd/g dw animal, rg=0.095, Sidmouth, non-polluted, Engl. Channel	Patella vulgata[15]
Glycolytic rate	0.48 umol lactate/(mg protein*hour)	Cd=116 ug Cd/g dw animal, rg=0.04, Combe Martin, polluted, Bristol Ch.	Patella vulgata[15]
Glycolytic rate	4.3 mg %	Cd=116 ug Cd/g dw animal, rg=0.6, Combe Martin, polluted, Bristol Ch.	Patella vulgata[15]
Glycolytic rate	5.1 mg %	Cd=27 ug Cd/g dw animal, rg=1.3, Sidmouth, non-polluted, Engl. Channel	Patella vulgata[15]
Glycolytic rate	6.2 mg %	Cd=257 ug Cd/g dw animal, rg=1.0, Blue Anchor, polluted, Bristol Ch.	Patella vulgata[15]
Glycolytic rate	8.0 mg %	Cd=419 ug Cd/g dw animal, rg=1.2, Weston-Mare, polluted, Bristol Ch.	Patella vulgata[15]
Glycolytic rate	8.6 mg %	Cd=537 ug Cd/g dw animal, rg=1.4, Ladye Bay, polluted, Bristol Ch.	Patella vulgata[15]
Grain yield	-25 % change relative to control	Cd=320 ug/g med., CdSO4+sludge, tis=1.8 ug Cd/g ww, Domino silt flooded	Oryza sativa[16]
Growth rate	40 % growth inhibition	Lab medium, conc = 10000 ug Cd/l	Anacystis nidulans[17]
Growth rate	50 % growth inhibition	Lab medium, growth cessation in contnuos culture, conc = 10 ug Cd/l	Asterionella formosa[18]
Growth rate	50 % growth inhibition	Lab, test tube bioassay, conc = 1000 ug Cd/l	Chlorela sorokiniana[19]
Growth rate	60 % growth inhibition	Lab, conc = 10000 ug Cd/l g Cd/l	Gleocapsa apicola[17]
Growth rate	10 % growth inhibition	Lab, conc = 10000 ug Cd/l	Merismopedia glauca f. insignis[17]
Growth rate	50 % root weight inhibition	Lab, conc = 7400 ug Cd/l	Myriophyllum spicatum[20]
Growth rate	50 % shoot weight inhibition	Lab, conc = 14600 ug Cd/l	Myriophyllum spicatum[20]
Growth rate	30 % growth inhibition	Lab, conc = 10000 ug Cd/l	Nostoc muscorum[17]
Growth rate	40 % inhibition of primary production	Lab, brachish water, conc = 100 ug/l	Phytoplankton[21]
Height	-40 % change relative to control	CdCl2 = 5 ug Cd/g medium, Sand and Hoaglands solution, silver maple	Acer saccharum[22]
Height	-88 % change relative to control	CdCl2 = 20 ug Cd/g medium, Sand and Hoaglands solution, silver maple	Acer saccharum[22]
Height	-64 % change relative to control	Cd = 4 ug/g medium, CdCl2, Sand and Hoaglands sol., soybean	Glycine max[6]

Hematocrit	15.9 %	Cd = 500 ppb, 4 w, 9-10 fish, 283 K, rg = +- 1.7	Pleuronectes flesus[3]
Hematocrit	16.0 %	Cd = 5 ppb, 9 w, 9-10 fish, 283 K, rg = +- 0.8	Pleuronectes flesus[3]
Hematocrit	16.3 %	Cd = 500 ppb, 9 w, 9-10 fish, 283 K, rg = +- 1.6	Pleuronectes flesus[3]
Hematocrit	16.5 %	Cd = 50 ppb, 9 w, 9-10 fish, 283 K, rg = +- 0.8	Pleuronectes flesus[3]
Hematocrit	17.0 %	Cd = 10.0 mg/l, 15 d, 280 K	Pleuronectes flesus[3]
Hematocrit	18 %	Cd = 1.0 mg/l, 15 d, 280 K	Pleuronectes flesus[3]
Hematocrit	18.8 %	Cd = 50 ppb, 4 w, 9-10 fish, 283 K, rg = +- 0.9	Pleuronectes flesus[3]
Hematocrit	19.0 %	Cd = 0.1 mg/l, 15 d, 280 K	Pleuronectes flesus[3]
Hematocrit	20.7 %	Cd = 5 ppb, 4 w, 9-10 fish, 283 K, rg = +- 0.9	Pleuronectes flesus[3]
Hematocrit	21.0 %	Cd = 0.0 mg/l, 15 d, 280 K	Pleuronectes flesus[3]
Hematocrit	21.8 %	Cd = 0.0 ppb, 4 w, 9-10 fish, 283 K, rg = +- 0.9	Pleuronectes flesus[3]
Hematocrit	22.1 %	Cd = 0.0 ppb, 9 w, 9-10 fish, 283 K, rg = +- 0.8	Pleuronectes flesus[3]
Hematocrit	33.0 % packed red cells	Cd = 10 ppb, 60 d, 32 fish, rg = +- 1.0	Pseudopleuronectes americanus[5]
Hematocrit	34.0 % packed red cells	Cd = 5 ppb, 60 d, 32 fish, rg = +- 1.0	Pseudopleuronectes americanus[5]
Hematocrit	35.0 % packed red cells	Cd = 0.0 ppb, 60 d, 35 fish, rg = +- 1.0	Pseudopleuronectes americanus[5]
Hemoglobin	4.4 g/100 ml	Cd = 5 ppb, 9 w, 9-10 fish, 283 K, rg = +- 0.3	Pleuronectes flesus[3]
Hemoglobin	4.5 g/100 ml	Cd = 50 ppb, 9 w, 9-10 fish, 283 K, rg = +- 0.3	Pleuronectes flesus[3]
Hemoglobin	4.6 g/100 ml	Cd = 500 ppb, 9 w, 9-10 fish, 283 K, rg = +- 0.4	Pleuronectes flesus[3]
Hemoglobin	4.8 %	Cd = 10.0 mg/l, 15 d, 280 K	Pleuronectes flesus[3]
Hemoglobin	4.9 g/100 ml	Cd = 500 ppb, 4 w, 9-10 fish, 283 K, rg = +- 0.4	Pleuronectes flesus[3]
Hemoglobin	5.2 %	Cd = 1.0 mg/l, 15 d, 280 K	Pleuronectes flesus[3]
Hemoglobin	5.5 %	Cd = 0.1 mg/l, 15 d, 280 K	Pleuronectes flesus[3]
Hemoglobin	5.5 g/100 ml	Cd = 50 ppb, 4 w, 9-10 fish, 283 K, rg = +- 0.2	Pleuronectes flesus[3]
Hemoglobin	5.8 %	Cd = 0.0 mg/l, 15 d, 280 K	Pleuronectes flesus[3]
Hemoglobin	6.1 g/100 ml	Cd = 0.0 ppb, 9 w, 9-10 fish, 283 K, rg = +- 0.2	Pleuronectes flesus[3]
Hemoglobin	6.1 g/100 ml	Cd = 5 ppb, 4 w, 9-10 fish, 283 K, rg = +- 0.3	Pleuronectes flesus[3]
Hemoglobin	6.3 g/100 ml	Cd = 0.0 ppb, 4 w, 9-10 fish, 283 K, rg = +- 0.2	Pleuronectes flesus[3]
Hemoglobin	7.4 g/100 ml whole blood	Cd = 0.0 ppb, 60 d, 32 fish, rg = +- 0.3	Pseudopleuronectes americanus[5]
Hemoglobin	7.4 g/100 ml whole blood	Cd = 10 ppb, 60 d, 32 fish, rg = +- 0.3	Pseudopleuronectes americanus[5]
Hemoglobin	8.1 g/100 ml whole blood	Cd = 5 ppb, 60 d, 31 fish, rg = +- 0.3	Pseudopleuronectes americanus[5]
K (+) influx	-50 % change relative to control	Cd = 1 mM, CdSO4, oats, 2 h	Phaseolus vulgaris[1]
Leaf lenght	-14 % relative to control	CdCl2 = 5 ug Cd/g medium, Sand and Hoaglands solution, beetroot	Beta vulgaris[22]
Leaf lenght	-48 % change relative to control	CdCl2 = 20 ug Cd/g medium, Sand and Hoaglands solution, beetroot	Beta vulgaris[22]
Leaf weight	-48 % change relative to control	CdCl2 = 5 ug Cd/g medium, Sand and Hoaglands solution, silver maple	Acer saccharum[22]

Leaf weight	-95 % change relative to control	CdCl2 = 20 ug Cd/g medium, Sand and Hoaglands solution, silver maple	Acer saccharum[22]
Leaf weight	-44 % change relative to control	Cd = 18 uM in medium, CdCl2, sand + vermiculite, soybean	Glycine max[23]
Leaf weight	-99 % change relative to control	Cd = 900 uM in medium, CdCl2, sand + vermiculite, soybean	Glycine max[23]
Liver glycogen	2.2 mg/100 mg wet weight	Cd = 0.0 mg/l, 15 d	Pleuronectes flesus[3]
Liver glycogen	3.0 mg/100 mg wet weight	Cd = 0.1 mg/l, 15 d	Pleuronectes flesus[3]
Liver glycogen	3.6 mg/100 mg wet weight	Cd = 1.0 mg/l, 15 d	Pleuronectes flesus[3]
Liver glycogen	3.8 mg/100 mg wet weight	Cd = 10.0 mg/l, 15 d	Pleuronectes flesus[3]
Maltase dehydrogenase	+69 % change relative to control	Cd = 1.35 uM, CdSO4, solution, soybean cv. Columbus	Glycine max[2]
Mchc	27.5 %	Cd = 10.0 mg/l, 15 d, 280 K	Pleuronectes flesus[3]
Mchc	28 %	Cd = 0.0 mg/l, 15 d, 280 K	Pleuronectes flesus[3]
Mchc	29 %	Cd = 0.1 mg/l, 15 d, 280 K	Pleuronectes flesus[3]
Mchc	29.5 %	Cd = 1.0 mg/l, 15 d, 280 K	Pleuronectes flesus[3]
Mortality	50 %	CdCl2 = 33 - 100 mg Cd/l, 48 h, 288 K	Agonus cataphractus[24]
Mortality	50 %	CdCl2 = 33 mg Cd/l, 96 H, 288 K	Agonus cataphractus[24]
Mortality	50 %	CdCl2 = 14 mg Cd/l, 120 h, S = 3.4 %, 292 K	Aldrichetta forsteri[25]
Mortality	50 %	CdCl2 = 16 mg Cd/l, 168 h, S = 3.4 %, 292 K	Aldrichetta forsteri[25]
Mortality	50 %	Cd = 1.0 mg/l, 240 h, CdSO4, 293 K	Artemia salina[26]
Mortality	50 %	CdCl2 = 0.8 mg Cd/l, 96 h, S = 2 %, 293 K	Asterias forbesi[27]
Mortality	50 %	CdCl2 = 12.0 mg Cd/l, 24 h, S = 2 %, 293 K	Asterias forbesi[27]
Mortality	50 %	CdCl2 = 15 - 21 mg Cd/l, 168 h, S = 3.4 %, 292 K	Atherinasoma microstoma[25]
Mortality	50 % mortality	CdCl2, water = 74-83 mg Cd/l, toxic level given, microorganisms	Azotobacter sp.[28]
Mortality	50 % mortality	CdCl2, water = E-4 M Cd, toxic level given, microorganisms	Bacterium communis[29]
Mortality	0 %	Toxicological limit = 1 - 5 mg/l, CdCl2	Biological purification[30]
Mortality	50 %	Cd = 100 mg/l, 24 h, 293 K, S = 2 %, 293 K	Carcimus maenas[27]
Mortality	50 %	Cd = 16.6 mg/l, 48 h, CdCl22, S = 2 %, 293 K	Carcinus maenas[27]
Mortality	50 %	Cd = 4.1 mg/l, 96 h, CdCl2, 293 K, S= 2 %	Carcinus maenas[27]
Mortality	50 %	CdCl2 = 10-33 mg Cd/l, 48 h, 293 K	Cardium edule[31]
Mortality	50 %	Cd = 0.32 mg/l, 96 h, 293 K, CdCl2, S = 2 %	Cragnon septemspinosa[27]
Mortality	50 %	Cd = 0.5 mg/l, 48 h, CdCl2, 293 K, S = 2 %	Cragnon septemspinosa[27]
Mortality	50 %	Cd = 2.4 mg/l, 24 h, CdCl2, 293 K, S = 2 %	Cragnon septemspinosa[27]
Mortality	50 %	Cd = 0.6 - 1.0 mg/l, CdCl2, 288 K, 96 h	Crangon crangon[24]
Mortality	50 %	Cd = 3.3 - 10 mg/l, CdCl2, 48 h, 288 K	Crangon crangon[24]
Mortality	50 %	CdCl2, exposure time = 2 d, conc in water = 1200 microgram/l,	Crangon crangon[32]

Mortality	50 %	CdCl2, exposure time = 2 d, conc in water = 3300-10000 ug/l, marine	Crangon crangon[33]
Mortality	0 %	Toxicological limit = 0.03 - 0.4 mg/l, CdCl2	Crustacean[34]
Mortality	50 %	CdCl2 = 2.5 - 5.0 mg Cd/l, 96 h	Ctenodrilus serratus[35]
Mortality	50 %	CdCl2 = 100 mg Cd/l, 24 h, S = 2 %, 293 K	Cyprinodon variegatus[27]
Mortality	50 %	CdCl2 = 50 mg Cd/l, 48 h, S = 2 %, 293 K	Cyprinodon variegatus[27]
Mortality	50 %	CdCl2 = 50 mg Cd/l, 96 h, S = 2 %, 293 K	Cyprinodon variegatus[27]
Mortality	50 %	CdCl2, exposure time = 2 d, conc in water = 65 ug/l, limnol	Daphnia magna[36]
Mortality	50 %	4000 min* mg/m**3(approx.), inhalation of Cd oxide fume for 10-30 min	Dog[37]
Mortality	0 %	Toxicological limit = 3.0 mg/l, trout, CdCl2	Fish[34]
Mortality	50 %	CdCl2 = 125 mg Cd/l, 24 h, S = 2 %, 293 K	Fundulus majalis[27]
Mortality	50 %	CdCl2 = 21 mg Cd/l, 96 h, S = 2 %, 293 K	Fundulus majalis[27]
Mortality	50 %	CdCl2 = 59 mg Cd/l, 48 h, S = 2 %, 293 K	Fundulus majalis[27]
Mortality	50 %	3500 min* mg/m**3(approx.), inhalation of Cd oxide fume for 10-30 min	Guinea pigs[37]
Mortality	0 %	Toxicological limit = 50 - 500 mg/kg, CdCl2	Homo sapiens[34]
Mortality	100 %	Lab, conc = 100 ug Cd/l	Lemna minor[38]
Mortality	0 %	Toxicological limit = 0.07 - 0.15 mg/kg, CdCl2	Mammalia[34]
Mortality	50 %	CdCl2 = 12 mg Cd/l, 48 h, S = 2 %, 293 K	Menidia menidia[39]
Mortality	50 %	CdCl2 = 13 mg Cd/l, 48 h, S = 3 %, 293 K	Menidia menidia[39]
Mortality	50 %	15000 min* mg/m**3(approx.), inhalation of Cd oxide fume for 10-30	Monkey[37]
Mortality	50 %	< 700 min* mg/m**3 (500?), inhalation of Cd oxide fume for 10-30 min	Mouse[37]
Mortality	50 %	CdCl = 25 mg Cd/l, 96 h, S = 2 %, 298 K	Mya arenaria[27]
Mortality	50 %	CdCl2 = 10.5 mg Cd/l, 96 h, S = 2 %, 298 K	Mya arenaria[27]
Mortality	50 %	CdCl2 = 2.2 mg Cd/l, 96 h, S = 2 %, 298 K	Mya arenaria[27]
Mortality	50 %	CdCl2 = 50 mg Cd/l, 48 h, S = 2 %, 298 K	Mya arenaria[27]
Mortality	50 %	CdCl2 = 6.6 mg Cd/l, 96 h, S = 2 %, 298 K	Mya arenaria[27]
Mortality	50 %	CdCl2 = > 200 mg Cd/l, 298 K, 24 h, S = 2 %	Mya arenaria[27]
Mortality	50 %	CdCl2 = 165 mg Cd/l, 48 H, S = 2 %, 298 K	Mytilus edulis[27]
Mortality	50 %	CdCl2 = > 200 mg Cd/l, 24 h, S = 2 %, 298 K	Mytilus edulis[27]
Mortality	50 %	CdCl2 = 1.6 mg Cd/l, 96 h, S = 3.3 %, 296.5 K	Mytilus edulis planulatus[40]
Mortality	50 %	CdCl2 = 2.0 - 3.3 mg Cd/l, 96 h, 293 K	Mytilus edulis planulatus[31]
Mortality	50 %	CdCl2 = 125 mg Cd/l, 48 h, S = 2 %, 298 K	Nassarius obsoletus[27]
Mortality	50 %	CdCl2 = > 200 mg Cd/l, 24 h, S = 2 %, 298 K	Nassarius obsoletus[27]
Mortality	50 %	CdCl2 = 6.4 mg Cd/l, 168 h, S = 3.3 %, 291 K	Neanthes vaali[40]
Mortality	50 %	CdSO4 = 10.0 mg Cd/l, 816 h, S = 1.75 %	Nereis diversicolor[41]
Mortality	50 %	CdSO4 = 100 mg Cd/l, 192 h, S = 1.75 %	Nereis diversicolor[41]
Mortality	50 %	CdCl2 = 11.0 mg Cd/l, 96 h, S = 2 %, 293 K	Nereis virens[27]

Mortality	50 %	CdCl2 = 25.0 mg Cd/l, 24 h, S = 2 %, 293 K	Nereis virens[27]
Mortality	50 %	Cadmium = 2.0 mg/l, 54d, hard water	Noemacheilus barbatulus[42]
Mortality	50 %	CdSO4 = 1.0 mg Cd/l, 410 h, 293 K	Ophryotrocha labronica[43]
Mortality	50 %	CdCl2 = 2.5 - 5.0 mg Cd/l, 96 h	Ophyryotrocha diadema[35]
Mortality	50 %	Cd = 6.4 - 6.8 mg/l, 290 K, CdCl2, 96 h	Palaemon sp.[40]
Mortality	50 %	Cd = 0.42 mg/l, 96 h, CdCl2, 293 K, S = 2 %	Palaemonetes vulgaris[27]
Mortality	50 %	Cd = 0.76 mg/l, 96 h, CdCl2, S = 2 %, 298 K	Palaemonetes vulgaris[44]
Mortality	50 %	Cd = 43 mg/l, 24 h, CdCl2, S = 2 %, 293 K	Palaemonetes vulgaris[27]
Mortality	50 %	Cd = 5.8 mg/l, 48 h, CdCl2, 293 K, S = 2 %	Palaemonetes vulgaris[27]
Mortality	50 %	Cd = 24.1 mg/l, 292 K, CdCl2, S = 1.75 %, 96 h	Paragrapsus gaimardii[45]
Mortality	50 %	Cd = 61.5 mg/l, 278 K, CdCl2, S = 1.75 %, 96 h	Paragrapsus gaimardii[45]
Mortality	50 %	Cd = 0.32 mg/l, 96 h, CdCl2, 293 K, S = 2 %	Pararus longicarpus[27]
Mortality	50 %	Cd = 3.7 mg/l, 48 h, 293 K, CdCl2, 293 K	Pararus longicarpus[27]
Mortality	50 %	Cd = > 200 mg/l, CdCl2, 24 h, S = 2 %, 293 K	Pararus longicarpus[27]
Mortality	50 %	CdCl2 = > 10 mg Cd/l, 96 h, S = 3.2 %, 293 K	Patiriella exigua[40]
Mortality	50 %	Cd = 3.5 mg/l, 96 h, CdCl2, 298 K, S = 2 %	Penaeus duorarum[44]
Mortality	55 %	CdSO4 = 1.6 mg/l, S = 2 %, 7 d, 293 K, flow-through, n=20, cont = 0 %	Penaeus duorarum[44]
Mortality	65 %	CdCl2 = 1.8 mg/l, S = 2 %, 7 d, 293 K, flow-through, n=20, cont = 0 %	Penaeus duorarum[44]
Mortality	75 %	Cd(NO3)2 = 1.7 mg/l, S = 2 %, 7 d, 293 K, flow-through, n=20, cont = 0 %	Penaeus duorarum[44]
Mortality	75 %	Cd-acetate = 1.6 mg/l, S = 2 %, 7 d, 293 K, flow-through, n=20, cont = 0 %	Penaeus duorarum[44]
Mortality	50 %	CdSO4 = 2.0 +- 0.1 ppm, 96 h, 25 ppm CaCO3, 293 K, pH = 7.4-7.9	Philodina sp.0[46]
Mortality	50 %	CdSO4 = 6.4 +- 0.5 ppm, 48 h, 25 ppm CaCO3, 293 K, pH = 7.4-7.9	Philodina sp.0[46]
Mortality	50 %	CdSO4 = 6.4 +- 3.1 ppm, 24 h, 25 ppm CaCO3, 293 K, pH = 7.4-7.9	Philodina sp.0[46]
Mortality	50 % mortality	CdCl2, water = 74-83 mg Cd/l, toxic level given, microorganisms	Pseudomonas sp.[28]
Mortality	50 %	2500 min* mg/m**3(approx.), inhalation of Cd oxide fume for 10-30 min	Rabbit[37]
Mortality	50 %	500 min* mg/m**3, n = 160, inhalation of Cd oxide fume for 10-30 min	Rat[37]
Mortality	50 %	Chloride, oral dose = 88 mg/kg body	Rat[47]
Mortality	50 %	Oxide, oral dose = 72 mg/kg body	Rat[47]
Mortality	50 %	Stearate, oral dose = 1225 mg/kg body	Rat[47]
Mortality	50 %	Succinate, oral dose = 660 mg/kg body	Rat[47]
Mortality	50 %	Cadmium = 0.017 mg/l, 5.5d, hard water	Salmo gairdneri[42]
Mortality	100 %	Lab, conc = 50 ug Cd/l	Salvinia natans[38]
Mortality	0 %	Toxicological limit = 0.1 mg/l, CdCl2	Self Purification[30]
Mortality	50 % mortality	CdCl2, water = 4E-6 M Cd, toxic level given, microorganisms	Tetrahymena sp.[28]

Mortality	50 %	Cd = 10.4 mg/, 96 h, CdCl2, 303 K, S = 2 %	Uca pugilator[24]
Mortality	50 %	Cd = 28 mg/l, 48 h, CdCl2, S = 2 %, 303 K	Uca pugilator[24]
Mortality	50 %	Cd = 3.5 mg/l, 240 h, CdCl2, 303 K, S = 2 %	Uca pugilator[24]
Mortality	50 %	Cd = 42 mg/l, 240 h, CdCl2, 283 K, S = 2 %	Uca pugilator[24]
Mortality	50 %	Cd = 46.6 mg/l, 96 h, 293 K, CdCl2, S = 2 %	Uca pugilator[24]
Mortality	50 %	Cd = 9.5 mg/l, 240 h, CdCl2, 293 K, S = 2 %	Uca pugilator[24]
Mortality	50 %	CdCl2 = 158 mg Cd/l, 24 h, S = 2 %, 298 K	Urosalpinx cinerea[27]
Mortality	50 %	CdCl2 = 28 mg Cd/l, 48 h, S = 2 %, 298 K	Urosalpinx cinerea[27]
New stem diameter	-12 % change relative to control	Cd=10 ug/g med., CdCl2, Drummer silty clay loam, Am. sycamore	Platanus occidentalis[48]
New stem diameter	-45 % change relative to control	Cd=100 ug/g med., CdCl2, Drummer silty clay loam, Am. sycamore	Platanus occidentalis[48]
Nitrification	77 % inhibation	Soil = 5 umoles/g of soil, 10 d incubation, 3 soils, lab	Soil organisms[49]
Nitrogen fixation	12 no of nodules/plant	Cd = 1.0 ppm in soil, control = 46	Soybean with nodule[50]
Nitrogen fixation	12 no of nodules/plant	Cd = 2.5 ppm in soil, control = 46	Soybean with nodule[50]
Nitrogen fixation	13.47 weight of one nodule in mg	Cd = 5.0	Soybean with nodule[50]
Nitrogen fixation	144 ethylene produced in nmoles/(plant*30 min)	Cd = 1.0 ppm in soil, control = 1640	Soybean with nodule[50]
Nitrogen fixation	24.58 weight of one nodule in mg	Cd = 2.5 ppm in soil, control = 62.30 mg	Soybean with nodule[50]
Nitrogen fixation	25.31 weight of one nodule in mg	Cd = 1.0 ppm in soil, control = 62.30 mg	Soybean with nodule[50]
Nitrogen fixation	3 no of nodules/plant	Cd = 5.0	Soybean with nodule[50]
Nitrogen fixation	38 ethylene produced in nmoles/(plant*30 min)	Cd = 5.0	Soybean with nodule[50]
Nitrogen fixation	87 ethylene produced in nmoles/(plant*30 min)	Cd = 2.5 ppm in soil, control = 1640	Soybean with nodule[50]
Nodule NH4 (+)	-2 % change relative to control	Cd = 18 uM, CdCl2, sand and vermicilite, soybean	Glycine max[23]
Nodule NH4 (+)	-99 % change relative to control	Cd = 900 uM, CdCl2, sand and vermicilite, soybean	Glycine max[23]
Nodule acetyle reduction	-71 % change relative to control	Cd = 18 uM, CdCl2, sand and vermiculite, soybean	Glycine max[23]
Nodule acetylene reduction	-99 % change relative to control	Cd = 900 uM, CdCl2, sand and vermiculite, soybean	Glycine max[23]
Nodule weight	-27 % change relative to control	Cd = 18 uM in medium, CdCl2, sand + vermiculite, soybean	Glycine max[23]
Nodule weight	-98 % change relative to control	Cd = 900 uM in medium, CdCl2, sand + vermiculite, soybean	Glycine max[23]
Peroxidase	+405 % change relative to control	Cd = 1.35 uM, CdSO4, solution, soybean cv. Columbus	Glycine max[2]
Photosynthesis	-100 % change relative to control	Cd = 900 uM, CdCl2, sandand vermicilite, soybean	Glycine max[23]

Photosynthesis	-17 % change relative to control	Cd = 18 uM, CdCl2, sand and vermicilite, soybean	Glycine max[23]
Photosynthesis	-10 % change relative to control	Cd = 1 ug/ml, CdCl2, Hoaglands, 4-7 d, sunflower	Hellianthus annuus[51]
Photosynthesis	-10 % change relative to control	Cd = 2 ug/ml, Excised tops, 4-5 d, Vermiculite and Hoaglands, sunflower	Hellianthus annuus[52]
Photosynthesis	-50 % change relative to control	Tissue conc = 340 ug/g ww, Hoaglands, 4-7 d, sunflower	Hellianthus annuus[51]
Photosynthesis	-50 % change relative to control	Tissue conc = 96 ug Cd/g ww, 4-5 d, vermiculite and hoaglands, sunflower	Hellianthus annuus[52]
Photosynthesis	-70 % change relative to control	Cd = 10 ug/ml, CdCl2, Hoaglands, 4-7 d, sunflower	Hellianthus annuus[51]
Photosynthesis	-80 % change relative to control	Cd = 200 ug/l, excised tops, 4-5 d, vermiculite and hoaglands, sunflower	Hellianthus annuus[52]
Photosynthesis	-25 % change relative to control	Cd = 20 ug/g soil, CdCl2, Drummer silty clay loam soil, Am. sycamore	Platanus accidentalis[48]
Photosynthesis	-5 % change relative to control	Cd = 10 ug/g soil, CdCl2, Drummer silty clay loam soil, Am. sycamore	Platanus accidentalis[48]
Photosynthesis	-50 % change relative to control	Tissue conc = 160 ug/g leaf (ww), CdCl2, nutrient solution	Zea mays[51]
Photosynthesis	-70 % change relative to control	Cd = 10 ug/ml, CdCl2, corn, nutrient solution	Zea mays[51]
Photosynthesis	0 % change relative to control	Cd = 1 ug/ml, CdCl2, corn, nutrient solution	Zea mays[51]
Plasma Ca	2.2 milli mol	Cd = 10.0 ppm, 15 d	Pleuronectes flesus[3]
Plasma Ca	2.4 milli mol	Cd = 1.0 ppm, 15 d	Pleuronectes flesus[3]
Plasma Ca	2.5 milli mol	Cd = 0.1 ppm, 15 d	Pleuronectes flesus[3]
Plasma Ca	2.9 milli-mol	Cd = 0.0 ppm, 15 d	Pleuronectes flesus[3]
Plasma K	3.1 milli mol	Cd = 1.0 ppm, 15 d	Pleuronectes flesus[3]
Plasma K	3.1 milli mol	Cd = 10.0 ppm, 15 d	Pleuronectes flesus[3]
Plasma K	3.3 milli mol	Cd = 0.1 ppm, 15 d	Pleuronectes flesus[3]
Plasma K	3.8 milli mol	Cd = 0.0 ppm, 15 d	Pleuronectes flesus[3]
Plasma Mg	0.5 milli mol	Cd = 0.0 ppm, 15 d	Pleuronectes flesus[3]
Plasma Mg	1.1 milli mol	Cd = 10.0 ppm, 15 d	Pleuronectes flesus[3]
Plasma Mg	1.2 milli mol	Cd = 0.1 ppm, 15 d	Pleuronectes flesus[3]
Plasma Mg	1.4 milli mol	Cd = 1.0 ppm, 15 d	Pleuronectes flesus[3]
Plasma PO4 - P	5.7 milli mol	Cd = 0.0 ppm, 15 d	Pleuronectes flesus[3]
Plasma PO4 - P	5.9 milli mol	Cd = 0.1 ppm, 15 d	Pleuronectes flesus[3]
Plasma PO4 - P	6.6 milli mol	Cd = 1.0 ppm, 15 d	Pleuronectes flesus[3]
Plasma PO4 - P	6.6 milli mol	Cd = 10.0 ppm, 15 d	Pleuronectes flesus[3]
Plasma osmolality	380.0 mosm/kg H2O	Cd = 5 ppb, 60 d, 22 fish, rg = +- 7.0	Pseudopleuronectes americanus[5]
Plasma osmolality	401.0 mosm/kg H2O	Cd = 0.0 ppm, 60 d, 22 fish, rg = +- 0.7	Pseudopleuronectes americanus[5]
Plasma osmolality	401.0 mosm/kg H2O	Cd = 10 ppb, 60 d, 19 fish, rg = +- 5.0	Pseudopleuronectes americanus[5]

Plasma protein	6.7 g %	Cd = 0.0 ppb, 60 d, 34 fish, rg = +- 0.2	Pseudopleuronectes americanus[5]
Plasma protein	6.7 g %	Cd = 5 ppb, 60 d, 30 fish, rg = +- 0.2	Pseudopleuronectes americanus[5]
Plasma protein	6.8 g %	Cd = 10 ppb, 60 d, 31 fish, rg = +- 0.2	Pseudopleuronectes americanus[5]
Pod weight	-35 % change relative to control	Cd = 18 uM in medium, CdCl2, sand + vermiculite, soybean	Glycine max[23]
Pod weight	-98 % change relative to control	Cd = 900 uM in medium, CdCl2, sand + vermiculite, soybean	Glycine max[23]
RNase	+226 % change relative to control	Cd = 1.35 uM, CdSO4, solution, soybean cv. Columbus	Glycine max[2]
Red blood cells	2.69E6 cells/mm**3	Cd = 5 ppb, 60 d, 10 fish, rg = +- 0.12	Pseudopleuronectes americanus[5]
Red blood cells	2.96E6 cells/mm**3	Cd = 10 ppb, 60 d, 14 fish, rg = +- 0.18	Pseudopleuronectes americanus[5]
Red blood cells	2.9E6 cells/mm**3	Cd = 0.0 ppb, 60 d, 17 fish, rg = +- 1.7	Pseudopleuronectes americanus[5]
Respiration	-35 % change relative to control	Cd = 1 mM, CdSO4, 2 h, oats, excised root treatment	Avena sativa[1]
Respiration	+53 % change relative to control	Cd = 1.35 uM, CdSO4, solution, soybean cv. Columbus	Glycine max[2]
Respiration rate	40.4 ul O2/(hour*g wet weight)	Cd = 1.00 mg/l, n=11, rg = +- 3.1 ul O2, S=2.5 %, 293 K, cont=30.5	Nassarius obsoletus[53]
Root weight	-47 % change relative to control	CdCl2 = 5 ug Cd/g medium, Sand and Hoaglands solution, silver maple	Acer saccharum[22]
Root weight	-93 % change relative to control	CdCl2 = 20 ug Cd/g medium, Sand and Hoaglands solution, silver maple	Acer saccharum[22]
Root weight	-6 % change relative to control	Cd = 18 uM in medium, CdCl2, sand + vermiculite, soybean	Glycine max[23]
Root weight	-80 % change relative to control	Cd = 900 uM in medium, CdCl2, sand + vermiculite, soybean	Glycine max[23]
Root weight	+17 % change relative to control	Cd = 0.1 ug/g medium, CdCl2, nutrient sol., lettuce	Lactuca sativa[54]
Root weight	-71 % change relative to control	Cd = 50 ug/g medium, CdCl2, nutrient sol., lettuce	Lactuca sativa[54]
Root weight	-10 % change relative to control	Cd=10 ug/g med., CdCl2, Drummer silty clay loam, Am. sycamore	Platanus occidentalis[48]
Root weight	-30 % change relative to control	Cd=100 ug/g med., CdCl2, Drummer silty clay loam, Am. sycamore	Platanus occidentalis[48]
Root weight	+9 % change relative to control	Cd = 0.5 ug/g medium, Cd(2+), Hjorth silty clay loam, radish	Raphanus sativus[55]
Root weight	-61 % change relative to control	Cd = 100 ug/g medium, Cd(2+), Hjorth silty clay loam, radish	Raphanus sativus[55]
Root weight	-46 % change relative to control	Cd = 73 ug/g med, smelting Cd complex, palmerton soil, rg = 0 - -60 %	Typha latifolia[56]
Shoot weight	+1 % change relative to control	Cd = 18 uM in medium, CdCl2, sand + vermiculite, soybean	Glycine max[23]
Shoot weight	-78 % change relative to control	Cd = 900 uM in medium, CdCl2, sand + vermiculite, soybean	Glycine max[23]
Steam dry weight	-56 % change relative to control	CdCl2 = 5 ug Cd/g medium, Sand and Hoaglands solution, silver maple	Acer saccharum[22]

Steam dry weight	-96 % change relative to control	CdCl2 = 20 ug Cd/g medium, Sand and Hoaglands solution, silver maple	Acer saccharum[22]
Stem diameter	-10 % change relative to control	Cd=10 ug/g med., CdCl2, Drummer silty clay loam, Am. sycamore	Platanus occidentalis[48]
Stem diameter	-62 % change relative to control	Cd=100 ug/g med., CdCl2, Drummer silty clay loam, Am. sycamore	Platanus occidentalis[48]
Stomatal opening	-10 % change relative to control	Cd = 1 uM, epidermal cells, 8 h, solution, sunflower	Hellianthus annuus[52]
Stomatal opening	-65 % change relative to control	Cd = 1000 uM, epidermal cells, 8 h, solution, sunflower	Hellianthus annuus[52]
Survival and reproduction	0 worms	Control = 186, Cd = 10.0 mg/l, CdCl2, numbers after 21 d, initial 40	Ctenodrilus serratus[57]
Survival and reproduction	0 worms	Control = 40, Cd = 10.0 mg/l, CdCl2, numbers after 96 h, initial 40	Ctenodrilus serratus[57]
Survival and reproduction	103 worms	Control = 186, Cd = 2.5 mg/l, CdCl2, numbers after 21 d, initial 40	Ctenodrilus serratus[57]
Survival and reproduction	12 worms	Control = 186, Cd = 5.0 mg/l, CdCl2, numbers after 21 d, initial 40	Ctenodrilus serratus[57]
Survival and reproduction	13 worms	Control = 40, Cd = 5.0 mg/l, CdCl2, numbers after 96 h, initial 40	Ctenodrilus serratus[57]
Survival and reproduction	183 worms	Control = 186, Cd = 1.0 mg/l, CdCl2, numbers after 21 d, initial 40	Ctenodrilus serratus[57]
Survival and reproduction	197 worms	Control = 186, Cd = 0.5 mg/l, CdCl2, numbers after 21 d, initial 40	Ctenodrilus serratus[57]
Survival and reproduction	239 worms	Control = 186, Cd = 0.1 mg/l, CdCl2, numbers after 21 d, initial 40	Ctenodrilus serratus[57]
Survival and reproduction	39 worms	Control = 40, Cd = 2.5 mg/l, CdCl2, numbers after 96 h, initial 40	Ctenodrilus serratus[57]
Survival and reproduction	40 worms	Control = 40, Cd = 0.1 mg/l, CdCl2, numbers after 96 h, initial 40	Ctenodrilus serratus[57]
Survival and reproduction	40 worms	Control = 40, Cd = 0.5 mg/l, CdCl2, numbers after 96 h, initial 40	Ctenodrilus serratus[57]
Survival and reproduction	40 worms	Control = 40, Cd = 1.0 mg/l, CdCl2, numbers after 96 h, initial 40	Ctenodrilus serratus[57]
Survival and reproduction	0 worms	Control = 212, Cd = 10.0 mg/l, CdCl2, numbers after 21 d, initial 40	Ophryotrocha diadema[57]
Survival and reproduction	0 worms	Control = 30, Cd = 10.0 mg/l, CdCl2, numbers after 96 h, initial 40	Ophryotrocha diadema[57]
Survival and reproduction	11 worms	Control = 30, Cd = 5.0 mg/l, CdCl2, numbers after 96 h, initial 40	Ophryotrocha diadema[57]
Survival and reproduction	218 worms	Control = 212, Cd = 0.5 mg/l, CdCl2, numbers after 21 d, initial 40	Ophryotrocha diadema[57]
Survival and reproduction	267 worms	Control = 212, Cd = 0.1 mg/l, CdCl2, numbers after 21 d, initial 40	Ophryotrocha diadema[57]
Survival and reproduction	34 worms	Control = 212, Cd = 2.5 mg/l, CdCl2, numbers after 21 d, initial 40	Ophryotrocha diadema[57]
Survival and reproduction	37 worms	Control = 30, Cd = 2.5 mg/l, CdCl2, numbers after 96 h, initial 40	Ophryotrocha diadema[57]
Survival and reproduction	39 worms	Control = 30, Cd = 0.5 mg/l, CdCl2, numbers after 96 h, initial 40	Ophryotrocha diadema[57]
Survival and reproduction	40 worms	Control = 30, Cd = 0.1 mg/l, CdCl2, numbers after 96 h, initial 40	Ophryotrocha diadema[57]

Survival and reproduction	40 worms	Control = 30, Cd = 1.0 mg/l, CdCl2, numbers after 96 h, initial 40	Ophryotrocha diadema[57]
Survival and reproduction	93 worms	Control = 212, Cd = 5.0 mg/l, CdCl2, numbers after 21 d, initial 40	Ophryotrocha diadema[57]
Survival and reproduction	95 worms	Control = 212, Cd = 1.0 mg/l, CdCl2, numbers after 21 d, initial 40	Ophryotrocha diadema[57]
Top dry weight	+4 % change relative to control	Cd = 0.1 ug/g medium, CdCl2, nutrient sol., lettuce	Lactuca sativa[54]
Top dry weight	-88 % change relative to control	Cd = 50 ug/g medium, CdCl2, nutrient sol., lettuce	Lactuca sativa[54]
Top growth	-25 % change relative to control	Cd = 1.0 ug/g medium, CdSO4, tissue = 15 ug Cd/g dw, ryegrass	Lolium sp.[58]
Top growth	-90 % change relative to control	Cd = 82 ug/g medium, CdSO4, tissue = 500 ug Cd/g dw, ryegrass	Lolium sp.[58]
Top yield	+32 % change relative to control	Cd = 0.5 ug/g medium, Cd(2+), Hjorth silty clay loam, radish	Raphanus sativus[55]
Top yield	-37 % change relative to control	Cd = 100 ug/g med, Cd(2+), Hjorth silty clay loam, radish	Raphanus sativus[55]
Top yield	-93 % change relative to control	Cd = 20 ug/g medium, CdCl2, Red podzolic soil, subterraneam clover	Trifolium sp.[59]
Top yield	-3 % change relative to control	Cd = 0.2 ug/g medium, CdCl2, Red podzolic soil	Trifolium sp., subterraneam clover[59]
Transpiration	-24 % change relative to control	Cd = 20 ug/g soil, CdCl2, Drummer silty clay loam soil, Am. sycamore	Platanus accidentalis[48]
Transpiration	-5 % change relative to control	Cd = 10 ug/g soil, CdCl2, Drummer silty clay loam soil, Am. sycamore	Platanus accidentalis[48]
Vegetative yield	+4 % change relative to control	CdCl2 = 0.01 ug Cd/g medium, Nutrient solution, tissue = 0.9 ug Cd/g dw	Beta vulgaris[60]
Vegetative yield	-38 % change relative to control	CdCl2=0.01 ug Cd/g medium, nut. sol., tissue=0.6 ug Cd/g dw, swiss chard	Beta vulgaris[60]
Vegetative yield	-53 % change relative to control	CdCl2 = 1.0 ug Cd/g medium, Nutrient solution, tissue = 2.5 ug Cd/g dw	Beta vulgaris[60]
Vegetative yield	-53 % change relative to control	CdCl2=1.0 ug Cd/g medium, nut. sol., tissue=150 ug Cd/g dw, swiss chard	Beta vulgaris[60]
Vegetative yield	-25 % change relative to control	Cd=145 ug/g med., CdSO4+sludge, tissue=43 ug Cd/g dw, domino silt loam	Cynodon dactylon[16]
Vegetative yield	-24 % change relative to control	Cd=0.01 ug/g med., CdCl2, Nut. sol., tissue=1.5 ug Cd/g dw, carrot	Daucus carota[60]
Vegetative yield	-72 % change relative to control	Cd=1.0 ug/g med., CdCl2, Nut. sol., tissue=2.2 ug Cd/g dw, carrot	Daucus carota[60]
Vegetative yield	-25 % change relative to control	Cd = 95 ug/g med., CdSO4+sludge, tissue=37 ug Cd/g dw, Domino silt loam	Festuca elatior[16]
Vegetative yield	-26 % change relative to control	Cd = 18 uM in medium, CdCl2, sand + vermiculite, soybean	Glycine max[23]
Vegetative yield	-91 % change relative to control	Cd = 900 uM in medium, CdCl2, sand + vermiculite, soybean	Glycine max[23]
Vegetative yield	+18 % change relative to control	Cd = 0.1 ug/g medium, tissue=24.3 ug Cd/g dw, nut. sol., lettuce	Lactuca sp.[60]
Vegetative yield	+26 % change relative to control	Cd = 0.01 ug/g medium, tissue=1.5 ug Cd/g dw, nut. sol., lettuce	Lactuca sp.[60]
Vegetative yield	+2 % change relative to control	Cd = 0.01 ug Cd/g med., CdCl2, tissue=1.4 ug Cd/g dw, nut. sol., tomato	Lycopersicum esculentum[60]
Vegetative yield	-53 % change relative to control	Cd=1.0 ug/g med, CdCl2, tissue=158 ug Cd/g dw, nut. sol., tomato	Lycopersicum esculentum[60]

Vegetative yield	-25 % change relative to control	Cd=30 ug/g medium, CdSO4+sludge, tissue=24 ug Cd/g dw, Domino silt loam	Medicago sativa[16]
Vegetative yield	-20 % change relative to control	Cd=10 ug/g med., CdCl2, Drummer silty clay loam, Am. sycamore	Platanus occidentalis[48]
Vegetative yield	-50 % change relative to control	Cd=100 ug/g med., CdCl2, Drummer silty clay loam, Am. sycamore	Platanus occidentalis[48]
Vegetative yield	+207 % change relative to control	Cd = 0.01 ug/g med, tissue=0.7 ug Cd/g dw, CdCl2, nut. sol., radish	Raphanus sp.[60]
Vegetative yield	+85 % change relative to control	Cd = 1.0 ug/g medium, tissue=144 ug Cd/g dw, CdCl2, nut. sol., radish	Raphanus sp.[60]
Vegetative yield	-25 % change relative to control	Cd=15 ug/g med., CdSO4+sludge, tis=9 ug Cd/g dw, Domino silt l.,sudangr.	Sorghum halepense[16]
Vegetative yield	-25 % change relative to control	Cd=40 ug/g med., CdSO4+sludge, tis=17 ug Cd/g dw, Domino silt loam	Trifolium repens[16]
Yield	18.37 dry wt	Control = 42.57 dry wt, Cd in soil = 200 ppm, 100 d growth, grains	Avenasativa[61]
Yield	18.72 dry wt	Control = 20.84 dry wt, Cd in soil = 200 ppm, 100 d growth, leaf	Avenasativa[61]
Yield	2.88 dry wt	Control = 3.75 dry wt, Cd in soil = 200 ppm, 100 d growth, root	Avenasativa[61]
Yield	21.5 dry wt	Control = 20.84 dry wt, Cd in soil = 40 ppm, 100 d growth, leaf	Avenasativa[61]
Yield	23.28 dry wt	Control = 29.90 dry wt, Cd in soil = 200 ppm, 100 d growth, stalk	Avenasativa[61]
Yield	27.28 dry wt	Control = 42.57 dry wt, Cd in soil = 40 ppm, 100 d growth, grains	Avenasativa[61]
Yield	3.18 dry wt	Control = 3.75 dry wt, Cd in soil = 40 ppm, 100 d growth, root	Avenasativa[61]
Yield	39.17 dry wt	Control = 29.90 dry wt, Cd in soil = 40 ppm, 100 d growth, stalk	Avenasativa[61]
Yield	6.85 dry wt	Control = 5.56 dry wt, Cd in soil = 200 ppm, 100 d growth, husks	Avenasativa[61]
Yield	8.31 dry wt	Control = 5.56 dry wt, Cd in soil = 40 ppm, 100 d growth, husks	Avenasativa[61]
Yield	0.17 dry wt	Control = 1.78 dry wt, Cd in soil = 200 ppm, 60 d growth, root	Brassica oleracea var. botrytis[61]
Yield	1.47 dry wt	Control = 1.78 dry wt, Cd in soil = 40 ppm, 60 d growth, root	Brassica oleracea var. botrytis[61]
Yield	3.17 dry wt	Control = 103.67 dry wt, Cd in soil = 200 ppm, 60 d growth, leaf	Brassica oleracea var. botrytis[61]
Yield	122.5 dry wt	Control = 95.93 dry wt, Cd in soil = 40 ppm, 60 d growth, leaf	Brassica oleracea var. italica[61]
Yield	2.09 dry wt	Control = 2.46 dry wt, Cd in soil = 200 ppm, 60 d growth, root	Brassica oleracea var. italica[61]
Yield	35.20 dry wt	Control = 95.93 dry wt, Cd in soil = 200 ppm, 60 d growth,leaf	Brassica oleracea var. italica[61]
Yield	4.08 dry wt	Control = 2.46 dry wt, Cd in soil = 40 ppm, 60 d growth, root	Brassica oleracea var. italica[61]
Yield	0.81 dry wt	Control = 0.96 dry wt, 130 d growth, root	Daucus carota[61]
Yield	1.5 dry wt	Control = 18.95 dry wt, Cd in soil = 200 ppm, 130 d growth, top	Daucus carota[61]
Yield	16.92 dry wt	Control = 18.95 dry wt, Cd in soil = 40 ppm, 130 d growth, top	Daucus carota[61]

Yield	38.27 dry wt	Control = 41.72 dry wt, Cd in soil = 40 ppm, 130 d growth, tubers	Daucus carota[61]
Yield	0.36 dry wt	Control = 0.91 dry wt, Cd in soil = 200 ppm, 35 d growth, root	Lactuca sativa[61]
Yield	0.99 dry wt	Control = 0.91 dry wt, Cd in soil = 40 ppm, 35 d growth, root	Lactuca sativa[61]
Yield	1.55 dry wt	Control = 17.5 dry wt, Cd in soil = 200 ppm, 35 d growth, leaf	Lactuca sativa[61]
Yield	18.82 dry wt	Control = 17.5 dry wt, Cd in soil = 40 ppm, 35 d growth, leaf	Lactuca sativa[61]
Yield	0.21 dry wt	Control = 27.45 dry wt, Cd in soil = 200 ppm, 95 d growth, seed	Pisum sativum[61]
Yield	0.56 dry wt	Control = 1.98 dry wt, Cd in soil = 200 ppm, 95 d growth, root	Pisum sativum[61]
Yield	0.84 dry wt	Control = 10.59 dry wt, Cd in soil = 200 ppm, 95 d growth, pods	Pisum sativum[61]
Yield	1.8 dry wt	Control = 1.98 dry wt, Cd in soil = 40 ppm, 95 d growth, root	Pisum sativum[61]
Yield	14.43 dry wt	Control = 19.64 dry wt, Cd in soil = 40 ppm, 95 d growth, vines	Pisum sativum[61]
Yield	16.98 dry wt	Control = 22.45 dry wt, Cd in soil = 40 ppm, 95 d growth, seed	Pisum sativum[61]
Yield	2.47 dry wt	Control = 19.64 dry wt, Cd in soil = 200 ppm, 95 d growth, vines	Pisum sativum[61]
Yield	7.45 dry wt	Control = 10.59 dry wt, Cd in soil = 40 ppm, 95 d growth, pods	Pisum sativum[61]
Yield	0.13 dry wt	Control = 0.96 dry wt, Cd in soil = 200 ppm, 130 d growth, root	Raphanus sativus[61]
Yield	0.48 dry wt	Control = 7.1 dry wt, Cd in soil = 200 ppm, 45 d growth, tubers	Raphanus sativus[61]
Yield	1.47 dry wt	Control = 8.24 dry wt, Cd in soil = 200 ppm, 45 d growth, top	Raphanus sativus[61]
Yield	1.52 dry wt	Control = 41.72 dry wt, Cd in soil = 200 ppm, 130 d growth, tubers	Raphanus sativus[61]
Yield	5.12 dry wt	Control = 7.1 dry wt, Cd in soil = 40 ppm, 45 d growth, tubers	Raphanus sativus[61]
Yield	6.25 dry wt	Control = 8.24 dry wt, Cd in soil = 40 ppm, 45 d growth, top	Raphanus sativus[61]
Yield	0 % decrease to control	Cd added to soil = 0.001 % of Cd in control	Rice[37]
Yield	31 % decrease to control	Cd added to soil = 0.1 % of Cd in control	Rice[37]
Yield	68 % decrease to control	Cd added to soil = 0.3 % of Cd in control	Rice[37]
Yield	7 % decrease to control	Cd added to soil = 0.03 % of Cd in control	Rice[37]
Yield	8 % decrease to control	Cd added to soil = 0.003 % of Cd in control	Rice[37]
Yield	8 % decrease to control	Cd added to soil = 0.01 % of Cd in control	Rice[37]
Yield	81 % decrease to control	Cd added to soil = 0.6 % of Cd in control	Rice[37]

Yield	99 % decrease to control	Cd added to soil = 1.0 % of Cd in control	Rice[37]
Yield	0.02 dry wt	Control = 0.51 dry wt, Cd in soil = 200 ppm, 55 d growth, root	Spinacia oleracea[61]
Yield	0.02 dry wt	Control = 0.51 dry wt, Cd in soil = 40 ppm, 55 d growth, root	Spinacia oleracea[61]
Yield	0.16 dry wt	Control = 10.96 dry wt, Cd in soil = 200 ppm, 55 d growth, leaf	Spinacia oleracea[61]
Yield	0.39 dry wt	Control = 10.96 dry wt, Cd in soil = 40 ppm, 55 d growth, leaf	Spinacia oleracea[61]
Yield	28 % decrease to control	Cd added to soil = 0.003 % of Cd in control	Wheat[37]
Yield	6 % increase to control	Cd added to soil = 0.001 % of Cd in control	Wheat[37]
Yield	84 % decrease to control	Cd added to soil = 0.01 % of Cd in control	Wheat[37]
Yield	87 % decrease to control	Cd added to soil = 0.03 % of Cd in control	Wheat[37]
Yield	97 % decrease to control	Cd added to soil = 0.1 % of Cd in control	Wheat[37]
Yield	97 % decrease to control	Cd added to soil = 0.3 % of Cd in control	Wheat[37]
Yield	98 % decrease to control	Cd added to soil = 0.6 % of Cd in control	Wheat[37]
Yield	99 % decrease to control	Cd added to soil = 1.0 % of Cd in control	Wheat[37]

Source: 1.Keck, R.W. 1978. Plant Physiol. 62, 94-96.
2.Lee, K.C., Cunningham, Paulsen, Liang, and R.B. Moore. 1976. Physiol. Plant. 36, 4-6.
3.Larson, Aa. The National Swedish Environment Protection Board, Brackish Wa ter. In J.H. Koeman et al(ed), Sublethal Effects of Toxic Chemicals on Aquatic Animals, Elsevier. pp 15-21.
4.Imai, I. and S.M. Siegel. 1973. Physiol. Plant. 29, 118-120.
5.Calabrise, A., M.A. Dawson, F.P. Thurberg and D.R. Wenzloff. 1975. In; Koeman, J.H. et al. (eds); Sublethal Effects of Toxic Chemicals on Aquatic Animals. Elsevier N.Y. pp 15-21.
6.Chaney, W.R., R.C. Strickland and R.J. Lamoreaux. 1977. Plant Soil 47, 275-278.
7.Zilinskas, B.A. And Govindjee. 1976. Z. Plantzenphysiol. 77, 302-314.
8.Lucero, H.A., C.S. Andreo and R.H. Vallejos. 1976. Plant Sci. Lett. 6, 309-313.
9.Van DUijvendijk-Matteoli, M.A. and G.M. Desmet. 1975. Biochim. Biophys. Acta 408, 164-169.
10.Mudd, J.B., T.T. McManus, A. Ongun and T.E. Cullogh. 1972. J. Environ. Quality 1, 288-291.
11.Hampp, R., K. Beulich and H. Ziegler. 1976. Z. Planzenphysiol. 77, 336-344.
12.Miller, R.J., J.E. Bittell and D.E. Koeppe. 1973. Physiol. Plant 28, 166-171.
13.Leblova, S. and J. Mares. 1975. Photosunthetica 9, 177-184.
14.Li. E.H. and C.D. Miles. 1975. Plant Sci. Lett. 5, 33-40.
15.Shore, R., G. Carney and T. Stygall. 1975. Mar. Pollut. Bull. 6, 187-189.
16.Bingham, F.T., A.L. Page, R.J. Mahler and T.J. Ganje. 1976. J. Environ. Quality 5, 57-60.
17.Sparling, A.B. 1968. Interaction between Blue-Green Algae and Heavy Metals. Ph D. thesis. Wash Univ.St Louis.Mo. 107 pp.
18.Conway, H.L. 1978. J. Fish. Res. bd Canada, 35, 286-294.
19.Moshe, M., N. Betzer and Y. Kott. 1972. Water Res. 6, 1165-1171.
20.Stanley, R.A. 1974. Arch. Environ. Contam. Toxicol. 2, 331-341.
21.Pietilainen, K. 1975. Intern. Conf. on Heavy Metals in the Environment. Symposium Proc., 2 (pt 2), 81-873.
22.Lamoreaux, R.J. and W.R. Chaney. 1977. J. Environ. Quality 6, 201-205.
23.Huang, C.Y., F.A. Bazzaz and L.N. Vanderhoef. 1974. Plant Physiol. 54, 122-124.
24.O'Hara, J. 1973. Fish. Bull. 71, 149-153.
25.Negilski, D.S. 1976. Aust. J. Mar. Freshwater Res. 27, 137-149.
26.Wittmann, G.T.W. and U. Forster. 1977. S. Afr. J. Sci. 72, 365-370.
27.Eisler, R. 1971. J. Fish. Res. Bd Canada 28, 1225-1234.
28.Nepomiluev, V.F. and T.I. Kuzyakina. 1967. Izv. timiryazev. selKhoz. Akad. No. 4, 84-91.
29.Ishizawa, S. and T. Matsuguchi. 1966. Bull. natn Inst. agric. Sci., Tokyo (B) 16, 1-90.
30.Jung. K.D. 1973. Wirkungskonzentration schaedigender Stoffe im Wasser. Gelenkirchen. Hygiene-Institudes Ruhrgebiets.
31.Postman, J.E. and K.W. Wilson. 1971. Shellfish Inf. Leafl. 22. Ministry of Agriculture, Fish. and Food. London.
32.Naturvaardsverkets Brackvattan Toxikologiska Lab.. Raport Jan. 1974.
33.Portmann, J.E. and K.W. Wilson. 1971. Shellfish Information Leaflet no 22.
34.Liebman, H. 1958. Handbuch der Friscwasser - und Abwasserbiologie. Band II.Munchen. Oldenbourg.
35.Reisch, D.J. and R.S. Carr. 1978. Mar. Pollut. Bull. 9, 24-27.
36.Biesinger, K.E. and G.M. Christensen. 1972. J. Fish. Res. bd. Can., 29(12), 1691-1700.
37.Friberg, L., M. Piscater, G.F. Nordberg and T. Kjellstroem. Cadmium in the Environment 2nd edition,CRC-Press.

38.Huchinson, T.C. and H. Czyrska. 1975. Verh. Int. Ver. Limnol. 19, 2102-2111.
39.Middaugh, D.P. and J.M. Dean. 1977. Bull. Environ. Cont. Toxicol. 17, 645-652.
40.James, R.O., P.J. Stiglich and T.W. Healy. 1975. Disc. Faraday Soc. 59, 142-156.
41.Bryan, G.W. 1976. in: Johnson. R. (ed.): Marine Pollution. Academic Press. NY. pp 185-302.
42.Solbe, J.F., G. Del and V.A. Cooper. 1976. Water Res. 10(6), pp 523-527.
43.Brown, B. and Ahsanullah. 1971. Mar. Pollut. Bull. 2, 182-187.
44.Nimmo, D.W.R.,D.V. Lightner and L.H. Bahner. 1977. in: Vernberg, F.J. et al. (eds.): Physiological Responses of Marine Biota to Pollutants. Academic Press. NY. pp 131-184.
45.Sullivan, J.K. 1977. Aust, J, Mar. Freshwater Res. 28, 739-743.
46.Sallivan, G.W., A.L. Buikema and J. Cairns, Jr. 1973. Va. J. Sci. 24(3), pp 120.
47.Toxic Substances List 1973, U.S. Department of Health, Education and Welfare, Public Health Service National Institute for Occupational Safety and Health, Rockville Maryland, USA, June 1973.
48.Carlson, R.W. and F.A. Bazzaz. 1977. Environ. Pollut. 12, 243-253.
49.Liang, C.N. and M.A. Tabatabai. 1978. Environ. Pollut. 12, 141-147.
50.Letchworth, M.B. and V. Blum. 1977. Environ. Pollut. 14, 303-312.
51.Carlson, R.W., F.A. Bazzaz and G.L. Rofle. 1975. Environ. Res. 10, 113-120.
52.Bazzaz, F.A., R.W. Carlson and G.L. Rofle. 1974. Environ. Pollut. 7, 241-246.
53.MacInnes, J.R., and F.P. Thurberg. 1973. Mar. Pollut. Bull. 4, 185-186.
54.John, M.K. and C.J. Van Laerhoven. 1976. Environ. Pollution 10, 163-173.
55.John, M.K. 1972. Can. J. Plant. Sci. 52, 715-719.
56.McNaughton, S.J. T.C. Folsom,Lee,Park, Price, Schmitz , Roeder. 1974. Ecology 55, 1163-1165.
58.Dijkshoorn, W., J.E.M. Lampe and A.R. Kowsolea. 1974. Neth. J. Agric. Sci. 22, 66-71.
59.Williams, C.H. and D.J. Davis. 1977. Aust. J. Soil. Res. 15, 59-68.
60.Turner, M.A. 1973. J. Environ. Quality 2, 118-119.
61.John, M.K. 1973. Environ. Pollut., 4, pp 7-15.

4-078 Effect of Cd and Cu

Parameter	Effect Value	Concentration and Condition	Species
Respiration rate	6.4 ul O2/(hour*g wet weight)	Cd=1.00 + Cu=0.25 mg/l, n=11, rg=3.0 ul O2, S=2.5 %, 293 K, cont=30.5	Nassarius obsoletus[1]

Source: 1.MacInnes, J.R., and F.P. Thurberg. 1973. Mar. Pollut. Bull. 4, 185-186.

4-079 Effect of Cd, age and salinity

Parameter	Effect Value	Concentration and Condition	Species
Mortality	50 %	CdCl2 = 12.0 mg Cd/l, 95 % CL = 9.2-15.6, S = 3 %, age = 7 d, 293 K	Fundulus heteroclitus, larvae[1]
Mortality	50 %	CdCl2 = 16.2 mg Cd/l, 95 % CL = 12.7-21.1, S = 2 %, age = 1 d, 293 K	Fundulus heteroclitus, larvae[1]
Mortality	50 %	CdCl2 = 23.0 mg Cd/l, 95 % CL = 19.2-27.6, S = 3 %, age = 1 d, 293 K	Fundulus heteroclitus, larvae[1]
Mortality	50 %	CdCl2 = 32.0 mg Cd/l, 95 % CL = 24.6-41.6, S = 2 %, age = 14 d, 293 K	Fundulus heteroclitus, larvae[1]
Mortality	50 %	CdCl2 = 7.8 mg Cd/l, 95 % CL = 5.6-10.3, S = 3 %, age = 14 d, 293 K	Fundulus heteroclitus, larvae[1]
Mortality	50 %	CdCl2 = 9.0 mg Cd/l, 95 % CL = 6.4-12.5, S = 2 %, age = 7 d, 293 K	Fundulus heteroclitus, larvae[1]
Mortality	50 %	CdCl2 = 1.6 mg Cd/l, 95 % CL = 1.1-2.4, S = 3 %, age = 14 d, 293 K	Menidia menidia, larvae[1]
Mortality	50 %	CdCl2 = 2.2 mg Cd/l, 95 % CL = 1.6-2.9, S = 2 %, age = 14 d, 293 K	Menidia menidia, larvae[1]
Mortality	50 %	CdCl2 = 3.2 mg Cd/l, 95 % CL = 1.9-4.7, S = 2 %, age = 7 d, 293 K	Menidia menidia, larvae[1]
Mortality	50 %	CdCl2 = 3.4 mg Cd/l, 95 % CL = 2.1-4.9, S = 3 %, age = 7 d, 293 K	Menidia menidia, larvae[1]
Mortality	50 %	CdCl2 = 3.8 mg Cd/l, 95 % CL = 2.7-5.3, S = 2 %, age = 1 d, 293 K	Menidia menidia, larvae[1]
Mortality	50 %	CdCl2 = 5.6 mg Cd/l, 95 % CL = 4.0-7.8, S = 3 %, age = 1 d, 293 K	Menidia menidia, larvae[1]

Source: 1.Middaugh, D.P. and J.M. Dean. 1977. Bull. Environ. Cont. Toxicol. 17, 645-652.

4-080 Effect of Cd, temperature and salinity

Parameter	Effect Value	Concentration and Condition	Species
Mortality	50 %	Cd = 100 mg/l, 48 h, S = 2.4 %, 294 K, adult	Fundulus heteroclitus[1]
Mortality	50 %	Cd = app. 27 mg/l, 96 h, 298-298 K, adult	Fundulus heteroclitus[2]
Mortality	50 %	CdCl2 = 100 mg Cd/l, 48 h, S = 1.0 %, 291 K, adult	Fundulus heteroclitus[3]
Mortality	50 %	CdCl2 = 140 mg Cd/l, 24 h, S = 2 %, 293 K, adult	Fundulus heteroclitus[1]
Mortality	50 %	CdCl2 = 15 mg Cd/l, 192 h, S = 5 %, 293 K, adult	Fundulus heteroclitus[1]
Mortality	50 %	CdCl2 = 175 - 200 mg Cd/l, 48 h, S = 2.0 %, 291 K, adult	Fundulus heteroclitus[3]
Mortality	50 %	CdCl2 = 175 - 200 mg Cd/l, S = 1.0 %, 291 K, adult, 24 h	Fundulus heteroclitus[3]
Mortality	50 %	CdCl2 = 27 - 28 mg Cd/l, S = 3.5 %, 293 K, adult, 192 h	Fundulus heteroclitus[1]
Mortality	50 %	CdCl2 = 30 mg Cd/l, 96 h, S = 3.0 %, 293 K, adult	Fundulus heteroclitus[3]
Mortality	50 %	CdCl2 = 40 mg Cd/l, 48 h, S = 3.0 %, 293 K, adult	Fundulus heteroclitus[3]
Mortality	50 %	CdCl2 = 43 mg Cd/l, S = 3.0 %, 48 h, 293 K, adult	Fundulus heteroclitus[4]
Mortality	50 %	CdCl2 = 49 - 55 mg Cd/l, S = 2 %, 293 K, adult, 96 h	Fundulus heteroclitus[1]
Mortality	50 %	CdCl2 = 50 mg Cd/l, S = 3.5 %, 278 K, adult, 192 h	Fundulus heteroclitus[1]
Mortality	50 %	CdCl2 = 60 mg Cd/l, 48 h, S = 2.0 %, 293 K, adult	Fundulus heteroclitus[4]
Mortality	50 %	CdCl2 = 73 mg Cd/l, S = 1.0 %, 96 h, 291 K, adult	Fundulus heteroclitus[3]
Mortality	50 %	CdCl2 = 75 mg Cd/l, S = 1.5 %, 278 K, adult, 192 h	Fundulus heteroclitus[1]
Mortality	50 %	CdCl2 = 78 mg Cd/l, 96 h, S = 2.0 %, 291 K, adult	Fundulus heteroclitus[3]
Mortality	50 %	CdCl2 = > 200 mg Cd/l, 24 h, S = 2.0 %, 291 K, adult	Fundulus heteroclitus[3]
Mortality	50 %	CdCl2 = > 50 mg Cd/l, 24 h, S = 3.0 %, 293 K, adult	Fundulus heteroclitus[3]
Mortality	50 %	Adult, CdCl2 = 10.4 mg Cd/l, 303 K, 96 h, S = 2 %	Uca pugilator[5]
Mortality	50 %	Adult, CdCl2 = 11.0 mg Cd/l, 303 K, 48 h, S = 1 %	Uca pugilator[5]
Mortality	50 %	Adult, CdCl2 = 11.8 mg Cd/l, 293 K, 240 h, S = 1 %	Uca pugilator[5]
Mortality	50 %	Adult, CdCl2 = 15.7 mg Cd/l, 283 K, 240 h, S = 1 %	Uca pugilator[5]
Mortality	50 %	Adult, CdCl2 = 16.5 mg Cd/l, 293 K, 192 h, S = 2 %	Uca pugilator[5]
Mortality	50 %	Adult, CdCl2 = 18.0 mg Cd/l, 293 K, 192 h, S = 1 %	Uca pugilator[5]
Mortality	50 %	Adult, CdCl2 = 2.9	Uca pugilator[5]
Mortality	50 %	Adult, CdCl2 = 2.9 mg Cd/l, 303 K, 240 h, S = 1 %	Uca pugilator[5]
Mortality	50 %	Adult, CdCl2 = 21.0 mg Cd/l, 293 K, 192 h, S = 3 %	Uca pugilator[5]
Mortality	50 %	Adult, CdCl2 = 21.3 mg Cd/l, 293 K, 144 h, S = 1 %	Uca pugilator[5]
Mortality	50 %	Adult, CdCl2 = 23.0 mg Cd/l, 293 K, 144 h, S = 2 %	Uca pugilator[5]
Mortality	50 %	Adult, CdCl2 = 23.3 mg Cd/l, 303 K, 96 h, S = 3 %	Uca pugilator[5]
Mortality	50 %	Adult, CdCl2 = 28.0 mg Cd/l, 303 K, 48 h, S = 2 %	Uca pugilator[5]
Mortality	50 %	Adult, CdCl2 = 28.5 mg Cd/l, 283 K, 192 h, S = 1 %	Uca pugilator[5]
Mortality	50 %	Adult, CdCl2 = 29.6 mg Cd/l, 293 K, 144 h, S = 3 %	Uca pugilator[5]
Mortality	50 %	Adult, CdCl2 = 3.0 mg Cd/l, 303 K, 192 h, S = 1 %	Uca pugilator[5]
Mortality	50 %	Adult, CdCl2 = 3.5	Uca pugilator[5]

Mortality	50 %	Adult, CdCl2 = 3.5 mg Cd/l, 303 K, 240 h, S = 2 %	Uca pugilator[5]
Mortality	50 %	Adult, CdCl2 = 3.7	Uca pugilator[5]
Mortality	50 %	Adult, CdCl2 = 3.7 mg Cd/l, 303 K, 192 h, S = 2 %	Uca pugilator[5]
Mortality	50 %	Adult, CdCl2 = 32.2 mg Cd/l, 293 K, 96 h, S = 1 %	Uca pugilator[5]
Mortality	50 %	Adult, CdCl2 = 33.3 mg Cd/l, 303 K, 48 h, S = 3 %	Uca pugilator[5]
Mortality	50 %	Adult, CdCl2 = 37.0 mg Cd/l, 293 K, 96 h, S = 3 %	Uca pugilator[5]
Mortality	50 %	Adult, CdCl2 = 4.0 mg Cd/l, 303 K, 144 h, S = 1 %	Uca pugilator[5]
Mortality	50 %	Adult, CdCl2 = 42.0 mg Cd/l, 283 K, 240 h, S = 2 %	Uca pugilator[5]
Mortality	50 %	Adult, CdCl2 = 46.6 mg Cd/l, 293 K, 96 h, S = 2 %	Uca pugilator[5]
Mortality	50 %	Adult, CdCl2 = 47.0 mg Cd/l, 283 K, 240 h, S = 3 %	Uca pugilator[5]
Mortality	50 %	Adult, CdCl2 = 5.2	Uca pugilator[5]
Mortality	50 %	Adult, CdCl2 = 5.2 mg Cd/l, 303 K, 144 h, S = 2 %	Uca pugilator[5]
Mortality	50 %	Adult, CdCl2 = 5.7	Uca pugilator[5]
Mortality	50 %	Adult, CdCl2 = 5.7 mg Cd/l, 303 K, 240 h, S = 3 %	Uca pugilator[5]
Mortality	50 %	Adult, CdCl2 = 51.0 mg Cd/l, 283 K, 114 h, s = 1 %	Uca pugilator[5]
Mortality	50 %	Adult, CdCl2 = 52.0 mg Cd/l, 283 K, 192 h, S = 2 %	Uca pugilator[5]
Mortality	50 %	Adult, CdCl2 = 6.50 mg Cd/l, 303 K, 192 h, S = 3 %	Uca pugilator[5]
Mortality	50 %	Adult, CdCl2 = 6.8 mg Cd/l, 303 K, 96 h, S = 1 %	Uca pugilator[5]
Mortality	50 %	Adult, CdCl2 = 7.60 mg Cd/l, 303 K, 144 h, S = 3 %	Uca pugilator[5]
Mortality	50 %	Adult, CdCl2 = 7.9	Uca pugilator[5]
Mortality	50 %	Adult, CdCl2 = 7.9 mg Cd/l, 293 K, 240 h, S = 3 %	Uca pugilator[5]
Mortality	50 %	Adult, CdCl2 = 9.5	Uca pugilator[5]
Mortality	50 %	Adult, CdCl2 = 9.5 mg Cd/l, 293 K, 240 h, S = 2 %	Uca pugilator[5]

Source: 1.Eisler, R. 1971. J. Fish. Res. Bd Canada 28, 1225-1234.
2.Jackim, E., J.M. Hamlin and S. Sonis. 1970. J. Fish Res. Bd Canada, 27, pp 383-390.
3.Stanley, R.A. 1975. Trans. Am. Fish. Soc. 104, 129-134.
4.Middaugh, D.P. and J.M. Dean. 1977. Bull. Environ. Cont. Toxicol. 17, 645-652.
5.O'Hara, J. 1973. Fish. Bull. 71, 149-153.

4-081 Effect of Ce

Parameter	Effect Value	Concentration and Condition	Species
Mortality	50 %	Nitrate, oral dose = 4200 mg/kg body	Rat[1]

Source: 1.Toxic Substances List 1973, U.S. Department of Health, Education and Welfare, Public Health ServiceNational Institute for Occupational Safety and Health, Rockville Maryland, USA, June 1973.

4-082 Effect of Chloral

Parameter	Concentration	Condition	Species
Sublethal effect	78 mg/l	Inhibition of cell multiplication	Microcystis aeruginosa[1]
LD50	0.05-0.4 g pr. kg ww	oral dose	Rat[2]
Sublethal effect	*1.6 mg/l*	*Inhibition of cell multiplication*	*Pseudomonas putida*[1]

Source: 1.Bringmann, G. and Kuhn, R., nwf-wasser/abwasser, H.9, 1976.
2.Patty, F. A., Industrial Hygiene and toxicology, 2, Interscience Publishers, New York, 1976.

4-083 Effect of Chloramben

Parameter	Effect Value	Concentration and Condition	Species
Mortality	10 %	10 ppm, 48 h, 14 g, 293 K, pH = 8.2, alk = 80 ppm, less than value	Ictalurus punctatus[1]

Source: 1.McCorle, F.M., J.E. Chambers and J.D. Yarbrough. 1977. Bull. Environ. Contam. Toxicol. 18(3), pp 267270.

4-084 Effect of Chloramine

Parameter	Effect Value	Concentration and Condition	Species
Nitrogen uptake	0.004 micro m nitrogen-N (l/h)	Control = 0.123, nitrate, NH2Cl = 0.1 mg/l, pH = 8.2, 24 h, r = 0.50	Phytoplankton[1]

Source: 1.Toetz, D., L. Varga and M. Pierce. 1977. Water Res. 11(3), pp 253-258.

4-085 Effect of Chlordane

Parameter	Effect Value	Concentration and Condition	Species
Mortality	50 %	Conc in food = 331 mg/kg(197-479), 95 % CL	Bobwhite[1]
Mortality	95 %	Conc in soil = 5 mg/kg	Earthworms[1]
Mortality	50 %	Conc in food = 350 mg/kg(305-403), 95 % CL	Japanese quail[1]
Mortality	50 %	Conc in food = 858 mg/kg(629-1241), 95 % CL	Mallard[1]
Mortality	50 %	Conc in food = 430 mg/kg(366-505), 95 % CL	Pheasant[1]
Mortality	50 %	Conc in food = 500.0 mg/kg	Pheasant[2]
Mortality	50 %	Conc in food = 250.0 mg/kg	Quail[2]
Mortality	50 %	Acute oral toxicity = 457-590 mg/kg, insecticide	Rat[1]
Mortality	95 %	Conc in soil > 10 mg/kg	Slugs[1]

Source: 1.Edwards, C.A. 1975. Persistent Pesticides in the Environment, 2nd edition, CRC-Press.
2.Dewitt, J.B., C.M. Menzie, V.A. Adomaitis and W.L. Reickel. 1960. Trans. 25th N. Am. Wildl. Conf., p277.

4-086 Effect of Chlorine

Parameter	Effect Value	Concentration and Condition	Species
Nitrogen uptake	0.007 micro m nitrogen-N (l/h)	Control = 0.326, ammonia (NH3), Cl2 = 0.1 mg/l, pH = 8.2, 24 h	Phytoplankton[1]
Nitrogen uptake	0.010 micro m nitrogen-N (l/h)	Control = 0.038, nitrate, Cl2 = 0.1 mg/l, pH = 8.2, 24 h, r = 0.99	Phytoplankton[1]
Nitrogen uptake	0.013 micro m nitrogen-N (l/h)	Control = 0.123, nitrate, Cl2 = 0.1 mg/l, r = 0.94	Phytoplankton[1]
Nitrogen uptake	0.265 micro m nitrogen-N (l/h)	Control = 0.249, nitrate, Cl2 = 0.1 mg/l, pH = 8.2, 24 h, r = 0.21	Phytoplankton[1]
Nitrogen uptake	0.822 micro m nitrogen-N (l/h)	Control = 0.581, ammonia (NH3), Cl2 = 0.1 mg/ly, r = 0.52	Phytoplankton[1]

Source: 1.Toetz, D., L. Varga and M. Pierce. 1977. Water Res. 11(3), pp 253-258.

4-087 Effect of Chloroacetaldehyde

Parameter	Concentration	Condition	Species
LD50	0.05-0.4 g pr. kg ww	oral dose	Rat[1]

Source: 1.Patty, F. A., Industrial Hygiene and toxicology, 2, Interscience Publishers, New York, 1976.

4-088 Effect of Chlorpropham

Parameter	Effect Value	Concentration and Condition	Species
Growth rate	50 % inhibition	Water conc = 12.5 mg/l, phenylcarbamate herbicide	Anabaena cylindrica[1]
Growth rate	50 % inhibition	Water conc = 8.2 mg/l, phenylcarbamate herbicide, N2-fixing conditions	Anabaena cylindrica[1]
Growth rate	50 % inhibition	Water conc = 5.5 mg/l, phenylcarbamate herbicide	Anacystis nidulans[1]
Growth rate	7.5 % stimulation	Water conc = 1.0 mg/l, phenylcarbamate herbicide	Anacystis nidulans[1]
Growth rate	50 % inhibition	Water conc = 2.7 mg/l, phenylcarbamate herbicide	Chlorella pyrenoidosa[1]
Growth rate	50 % inhibition	Water conc = 9.0 mg/l, phenylcarbamate herbicide	Gloeocapsa alpicola[1]
Growth rate	50 % inhibition	Water conc = 13.2 mg/l, phenylcarbamate herbicide	Tolypothrix tenuis[1]
Growth rate	7.5 % stimulation of growth	Water conc = 1.5 mg/l, phenylcarbamate herbicide	Tolypothrix tenuis[1]

Source: 1.Hulin, N. and M. Horowitz. 1973. Phytoparasitica 1, 64.

4-089 Effect of Chlorobenzene

Parameter	Conc.	Condition	Species
Sublethal effect	120 mg/l	Inhibition of cell multiplication	Microcystis aeruginosa[1]
Sublethal effect	17 mg/l	Inhibition of cell multiplication	Pseudomonas putida[1]
TLm	24 mg/l	24-96 hours	Bluegill[2]
TLm	29-39 mg/l	24-96 hours	Fathead minnow[2]
TLm	51-73 mg/l	24-96 hours	Goldfish[2]
TLm	45 mg/l	24-96 hours	Guppie[2]
LD50	2.83 g pr. kg ww	single oral dose	Rabbit[3]
LD50	2.2 g pr. kg ww	single oral dose	Rat[3]

Source: 1.Bringmann, G. and Kuhn, R., nwf-wasser/abwasser, H.9, 1976.
2.Jones, H. R., Environmental Control in the organic and petrochemical Industries, Noyes Data Corp., 171.
3.Henschler, D., Toxikologisch-arbeitsmedizinische Begrundung von MAH-werte, Verlag Chemie.

4-090 Effect of Chlorodimethylether

Parameter	Conc.	Condition	Species
LD50	0.5 g pr. kg ww	single oral dose	Rat[1]

Source: 1.Patty, F. A., Industrial Hygiene and toxicology, 2, Interscience Publishers, New York, 1976.

4-091 Effect of Chloroform

Parameter	Conc.	Condition	Species
Sublethal effect	185 mg/l	Inhibition of cell multiplication	Microcystis aeruginosa[1]
Sublethal effect	125 mg/l	Inhibition of cell multiplication	Pseudomonas putida[1]

Source: 1.Bringmann, G. and Kuhn, R., nwf-wasser/abwasser, H.9, 1976.

4-092 Effect of Chloroprene

Parameter	Conc.	Condition	Species
LD50	0.67 g pr. kg ww	oral dose	Rat[1]

Source: 1.Patty, F. A., Industrial Hygiene and toxicology, 2, Interscience Publishers, New York, 1976.

4-093 Effect of Chrysene

Parameter	Effect Value	Concentration and Condition	Species
Growth rate	58 % more than control	Water = 0.3 ppm, 96 h, PAH, sporelings	Antithamnion plumula[1]
Mortality	0 % mortality	Water = 1.0 ppm, zero level mortality in 96 h, marine	Neathes arenaceodentata[2]

Source: 1.Boney, A.D. 1974. Mar. Pollut. Bull. 5, 185-186.2.Rossi, S.S. and J.M. Neff. 1978. Mar. Pollut. Bull. 9, 220-223.

4-094 Effect of Citric acid

Parameter	Conc.	Condition	Species
LC0	80 mg/l	long time exposure	Daphnia magna[1]
LC0	625 mg/l	long time exposure	Goldfish[1]
LC100	120 mg/l	long time exposure	Daphnia magna[1]
LC100	894 mg/l	long time exposure	Goldfish[1]
Sublethal effect	80 mg/l	Inhibition of cell multiplication	Microcystis aeruginosa[2]
Sublethal effect	>10000 mg/l	Inhibition of cell multiplication	Pseudomonas putida[2]

Source: 1.Ellis, M. M., Detection and measurement of stream pollution, U.S. Bur. Fisheries Bull. No. 22, 365,937. 2.Bringmann, G. and Kuhn, R., nwf-wasser/abwasser, H.9, 1976.

4-095 Effect of Climate

Parameter	Effect Value	Concentration and Condition	Species
Ecosphere	5 %	Climate in city compared with surrounding contryside	Decrease in snowfall[1]
Ecosphere	12.5 %	Climate in city compared with surrounding contryside, rg = 5 - 20 %	Increase in calms[1]
Ecosphere	15 times more	Climate in city compared with surrounding contryside, rg = 5 - 25	Increase in gaseous pollutants[1]

Source: 1.Landsberg, H.E. 1970. Meteorological Monographs 11, pp 91.

4-096 Effect of Co

Parameter	Effect Value	Concentration and Condition	Species
Mortality	50 % mortality	Water = 0.7 mg CoCl2/l, toxic level given, microorganisms	Aspergillus niger[1]
Mortality	50 % mortality	Water = 1.2mg CoCl2*6H2O /l, toxic level given, microorganisms	Aspergillus niger[2]
Mortality	50 % mortality	CoCl2*6H2O, water = 10E-4 M Co, toxic level given, microorganisms	Azotobacter sp.[3]
Mortality	50 % mortality	Water = 23-26 mg Co/l, toxic level given, microorganisms, as CoCl2	Azotobacter sp.[4]
Mortality	0 %	Toxicological limit = 0.5 mg/l, protozoa, Co-compounds	Crustacean[5]
Mortality	50 %	CoCl2, exposure time = 2 d, conc in water = 1110 microgram/l, limnol.	Daphnia magna[6]
Mortality	50 % mortality	Water = 23-26 mg Co/l, toxic level given, microorganisms, as CoCl2	Escherichia coli[4]
Mortality	0 %	Toxicological limit = 0.01 - 0.1 mg/l, Co-compounds	Fish[5]
Mortality	0 %	Toxicological limit = 50 - 500 mg/kg, Co-compounds	Homo sapiens[5]
Mortality	0 %	Toxicological limit = 0.7 - 1.5 g/kg, Co-compounds	Mammalia[5]
Mortality	50 %	Chloride, oral dose = 80 mg/kg body	Mouse[7]
Mortality	50 % mortality	Water = 23-26 mg Co/l, toxic level given, microorganisms, as CoCl2	Pseudomonas sp.[4]
Mortality	50 %	Chloride, oral dose = 180 mg/kg body	Rat[7]
Mortality	50 %	Oxide, oral dose = 1700 mg/kg body	Rat[7]
Mortality	0 %	Toxicological limit = 5 mg/l, Co-compounds	Self Purification[8]
Mortality	50 % mortality	Water = 23-26 mg Co/l, toxic level given, microorganisms, as CoCl2	Serratia sp.[4]
Mortality	50 % mortality	Water = 23-26 mg Co/l, toxic level given, microorganisms, as CoCl2	Torula sp.[4]
Nitrification	45 % inhibation	Soil = 5 umoles/g of soil, 10 d incubation, 3 soils, lab	Soil organisms[9]
Oxygen consumption	0.133 ml/h	Cobalt = 0.35 mg/l, cultivation hoaglands solution, see Purvis, M.J.	Dialysis culture[10]
Oxygen consumption	0.179 ml/h	Cobalt = 0.2 mg/l, cultivation hoaglands solution, see Purvis, M.J.	Dialysis culture[10]
Oxygen consumption	0.226 ml/h	Cobalt = 0.0 mg/l, cultivation hoaglands solution, see Purvis, M.J.	Dialysis culture[10]

Source: 1.Singh, K. 1971. Pestic. India 6, 14-17.
2.Resz. A. 1968. Sonderh. Z. Plfkrankh. PflPath. PflSchutz No 4, 147-150.
3.Zhakharyan, S.V. 1970. Symp. Biol. Hung. 11, 391-395.
4.Hoffman-Kakol, I. and H. Kwarta. 1972. Zesz. nauk. Akad. Rolnicz. Szczecin. No 38, 91-103.
5.Liebman, H. 1958. Handbuch der Friscwasser - und Abwasserbiologie. Band II.Munchen. Oldenbourg.
6.Biesinger, K.E. and G.M. Christensen. 1972. J. Fish. Res. bd. Can., 29(12), 1691-1700.
7.Toxic Substances List 1973, U.S. Department of Health, Education and Welfare, Public Health ServiceNational Institute for Occupational Safety and Health, Rockville Maryland, USA, June 1973.
8.Jung. K.D. 1973. Wirkungskonzentration schaedigender Stoffe im Wasser. Gelenkirchen. Hygiene-Institudes Ruhrgebiets.
9.Liang, C.N. and M.A. Tabatabai. 1978. Environ. Pollut. 12, 141-147.
10.Sandstroem, O. and S. Soedergren. 1974. Vatten, 30, pp 229-34

4-097 Effect of Coffeine

Parameter	Effect Value	Concentration and Condition	Species
Mortality	50 %	Oral dose = 250 mg/kg, 1.29E-3mol/kg, F007 = 2.89	Mouse[1]

Source: 1.Luckey, T.D. and B. Venugopal. 1977. Metal Toxicity in Mammals, 1. Plenum Press.

4-098 Effect of Coumarin

Parameter	Concentration	Condition	Species
LD50	0.8-1.35 g pr. kg ww	single oral dose, 100 %	rat, Rabbit[1]

Source: 1.Patty, F. A., Industrial Hygiene and toxicology, 2, Interscience Publishers, New York, 1976.

4-099 Effect of Cr

Parameter	Effect Value	Concentration and Condition	Species
Acid phosphatase activity	26.94 milli-u/100 ml	Cr(+6) = 10 mg/l, 32 d, 283 K, 4*13 fish	Rutilus rutilus[1]
Acid phosphatase activity	28.46 milli-u/100 ml	Cr(+6) = 0 mg/l, 32 d, 283 K, 4*13 fish	Rutilus rutilus[1]
Acid phosphatase activity	41.18 milli-u/100 ml	Cr(+6) = 0.1 mg/l, 32 d, 283 K, 4*13 fish	Rutilus rutilus[1]
Acid phosphatase activity	42.21 milli-u/100 ml	Cr(+6) = 1.0 mg/l, 32 d, 283 K, 4*13 fish	Rutilus rutilus[1]
Acid phosphatase activity	26.94 milli-u/100 ml	Cr(+6) = 10.0 ppm, 32 d, 20 fish	Salmo gairdneri[1]
Acid phosphatase activity	28.46 milli-u/100 ml	Cr(+6) = 0.0 ppm, 32 d, 20 fish	Salmo gairdneri[1]
Acid phosphatase activity	41.18 milli-u/100 ml	Cr(+6) = 0.1 ppm, 32 d, 20 fish	Salmo gairdneri[1]
Acid phosphatase activity	42.21 milli-u/100 ml	Cr(+6) = 1.0 ppm, 32 d, 20 fish	Salmo gairdneri[1]
Destruction of intestinal epithelium	100 % effect	Cr(+6) = 100.0 mg/l, effect observed	Salmo gairdneri[2]
G-6-PDH	16.46 milli-u/mg tissue	Cr(+6) = 0.0 ppm, 32 d, 20 fish	Salmo gairdneri[1]
G-6-PDH	16.68 milli-u/mg tissue	Cr(+6) = 0.1 ppm, 32 d, 20 fish	Salmo gairdneri[1]
G-6-PDH	16.70 milli-u/mg tissue	Cr(+6) = 1.0 ppm, 32 d, 20 fish	Salmo gairdneri[1]
G-6-PDH	17.42 milli-u/mg tissue	Cr(+6) = 10.0 ppm, 32 d, 20 fish	Salmo gairdneri[1]
GOT	30.9 milli-u/mg tissue	Cr(+6) = 10.0 ppm, 32 d, 20 fish	Salmo gairdneri[1]
GOT	57.0 milli-u/mg tissue	Cr(+6) = 0.1 ppm, 32 d, 20 fish	Salmo gairdneri[1]
GOT	76.6 milli-u/mg tissue	Cr(+6) = 1.0 ppm, 32 d, 20 fish	Salmo gairdneri[1]
GOT	78.0 milli-u/mg tissue	Cr(+6) = 0.0 ppm, 32 d, 20 fish	Salmo gairdneri[1]
GPT	26.3 milli-u/mg tissue	Cr(+6) = 0.1 ppm, 32 d, 20 fish	Salmo gairdneri[1]
GPT	28.8 milli-u/mg tissue	Cr(+6) = 10.0 ppm, 32 d, 20 fish	Salmo gairdneri[1]
GPT	36.9 milli-u/mg tissue	Cr(+6) = 0.0 ppm, 32 d, 20 fish	Salmo gairdneri[1]
GPT	37.5 milli-u/mg tissue	Cr(+6) = 1.0 ppm, 32 d, 20 fish	Salmo gairdneri[1]
Glucose	30.38 mg/100 ml	Cr(+6) = 1.0 mg/l, 32 d, 283 K, 4*13 fish	Rutilus rutilus[1]
Glucose	41.69 mg/100 ml	Cr(+6) = 0.1 mg/l, 32 d, 283 K, 4*13 fish	Rutilus rutilus[1]

Glucose	42.98 mg/100 ml	Cr(+6) = 0 mg/l, 32 d, 283 K, 4*13 fish	Rutilus rutilus[1]
Glucose	57.22 mg/100 ml	Cr(+6) = 10 mg/l, 32 d, 283 K, 4*13 fish	Rutilus rutilus[1]
Glucose	3.6 millimol/l	Cr(+6) = 0 mg/l, 15-22 d, 281 K, 20 fish	Salmo gairdneri[1]
Glucose	30.38 mg/100 ml	Cr(+6) = 1.0 ppm, 32 d, 20 fish	Salmo gairdneri[1]
Glucose	41.69 mg/100 ml	Cr(+6) = 0.1 ppm, 32 d, 20 fish	Salmo gairdneri[1]
Glucose	42.98 mg/100 ml	Cr(+6) = 0.0 ppm, 32 d, 20 fish	Salmo gairdneri[1]
Glucose	57.22 mg/100 ml	Cr(+6) = 10.0 ppm, 32 d, 20 fish	Salmo gairdneri[1]
Glucose	8.4 millimol/l	Cr(+6) = 10 mg/l, 15-22 d, 281 K, 20 fish	Salmo gairdneri[1]
Gonad weight	0.84 g/100 g fish	Cr(+6) = 0.0 ppm, 32 d, 20 fish, male	Salmo gairdneri[1]
Gonad weight	0.96 g/100 g fish	Cr(+6) = 0.1 ppm, 32 d, 20 fish, male	Salmo gairdneri[1]
Gonad weight	1.23 g/100 g fish	Cr(+6) = 1.0 ppm, 32 d, 20 fish, male	Salmo gairdneri[1]
Gonad weight	1.51 g/100 g fish	Cr(+6) = 10.0 ppm, 32 d, 20 fish, male	Salmo gairdneri[1]
Gonad weight	1.82 g/100 g fish	Cr(+6) = 10.0 ppm, 32 d, 20 fish, female	Salmo gairdneri[1]
Gonad weight	1.94 g/100 g fish	Cr(+6) = 0.1 ppm, 32 d, 20 fish, female	Salmo gairdneri[1]
Gonad weight	2.41 g/100 g fish	Cr(+6) = 0.0 ppm, 32 d, 20 fish, female	Salmo gairdneri[1]
Gonad weight	2.79 g/100 g fish	Cr(+6) = 1.0 ppm, 32 d, 20 fish, female	Salmo gairdneri[1]
Growth	50 % inhibition	Cr(+6) = 0.2 mg/l, young salmon, effect observed	Salmo gairdneri[3]
Hematocrit	13.5 %	Cr(+6) = 10 mg/l, 32 d, 283 K, 4*13 fish	Rutilus rutilus[1]
Hematocrit	14.7 %	Cr(+6) = 0 mg/l, 32 d, 283 K, 4*13 fish	Rutilus rutilus[1]
Hematocrit	15.1 %	Cr(+6) = 1.0 mg/l, 32 d, 283 K, 4*13 fish	Rutilus rutilus[1]
Hematocrit	17.1 %	Cr(+6) = 0.1 mg/l, 32 d, 283 K, 4*13 fish	Rutilus rutilus[1]
Hematocrit	13.5 %	Cr(+6) = 10.0 ppm, 32 d, 20 fish	Salmo gairdneri[1]
Hematocrit	14.7 %	Cr(+6) = 0.0 ppm, 32 d, 20 fish	Salmo gairdneri[1]
Hematocrit	15.1 %	Cr(+6) = 1.0 ppm, 32 d, 20 fish	Salmo gairdneri[1]
Hematocrit	17.1 %	Cr(+6) = 0.1 ppm, 32 d, 20 fish	Salmo gairdneri[1]
Hematocrit	22.9 %	Cr(+6) = 0 mg/l, 15-22 d, 281 K, 20 fish	Salmo gairdneri[1]
Hematocrit	22.9 %	Cr(+6) = 0.0 ppm, 15-22 d, 10 fish	Salmo gairdneri[1]
Hematocrit	29.6 %	Cr(+6) = 10 mg/l, 15-22 d, 281 K, 20 fish	Salmo gairdneri[1]
Hematocrit	29.6 %	Cr(+6) = 10.0 ppm, 15-22 d, 10 fish	Salmo gairdneri[1]
Hemoglobin	2.6 g/100 ml	Cr(+6) = 1.0 mg/l, 32 d, 283 K, 4*13 fish	Rutilus rutilus[1]
Hemoglobin	3.2 g/100 ml	Cr(+6) = 0.1 mg/l, 32 d, 283 K, 4*13 fish	Rutilus rutilus[1]
Hemoglobin	3.4 g/100 ml	Cr(+6) = 10 mg/l, 32 d, 283 K, 4*13 fish	Rutilus rutilus[1]
Hemoglobin	3.5 g/100 ml	Cr(+6) = 0 mg/l, 32 d, 283 K, 4*13 fish	Rutilus rutilus[1]
Hemoglobin	12.9 micromol/l	Cr(+6) = 0 mg/l, 15-22 d, 281 K, 20 fish	Salmo gairdneri[1]
Hemoglobin	12.9 micromol/l	Cr(+6) = 0.0 ppm, 15-22 d, 10 fish	Salmo gairdneri[1]
Hemoglobin	17.6 micromol/l	Cr(+6) = 10 mg/l, 15-22 d, 281 K, 20 fish	Salmo gairdneri[1]
Hemoglobin	17.6 micromol/l	Cr(+6) = 10 ppm, 15-22 d, 10 fish	Salmo gairdneri[1]
Hemoglobin	2.8 g/100 ml	Cr(+6) = 1.0 ppm, 32 d, 20 fish	Salmo gairdneri[1]
Hemoglobin	3.2 g/100 ml	Cr(+6) = 0.1 ppm, 32 d, 20 fish	Salmo gairdneri[1]

Hemoglobin	3.4 g/100 ml	Cr(+6) = 10.0 ppm, 32 d, 20 fish	Salmo gairdneri[1]
Hemoglobin	3.5 g/100 ml	Cr(+6) = 0.0 ppm, 32 d, 20 fish	Salmo gairdneri[1]
K in blood	3.7 millieq/l	Cr(+6) = 0 mg/l, 15-22 d, 281 K, 20 fish	Salmo gairdneri[1]
K in blood	3.7 millieq/l	Cr(+6) = 0.0 ppm, 15-22 d, 10 fish	Salmo gairdneri[1]
K in blood	4.2 millieq/l	Cr(+6) = 10 mg/l, 15-22 d, 281 K, 20 fish	Salmo gairdneri[1]
K in blood	4.2 millieq/l	Cr(+6) = 10.0 ppm, 15-22 d, 10 fish	Salmo gairdneri[1]
Kidney weight	0.27 g/100 g fish	Cr(+6) = 0.1 ppm, 32 d, 20 fish	Salmo gairdneri[1]
Kidney weight	0.27 g/100 g fish	Cr(+6) = 1.0 ppm, 32 d, 20 fish	Salmo gairdneri[1]
Kidney weight	0.3 g/100 g fish	Cr(+6) = 0.0 ppm, 32 d, 20 fish	Salmo gairdneri[1]
Kidney weight	0.33 g/100 g fish	Cr(+6) = 10.0 ppm, 32 d, 20 fish	Salmo gairdneri[1]
Liver weight	0.92 g/100 g fish	Cr(+6) = 0.0 ppm, 281 K, 20 fish, 2 y	Salmo gairdneri[1]
Liver weight	0.94 g/100 g fish	Cr(+6) = 10.0 ppm, 32 d, 20 fish	Salmo gairdneri[1]
Liver weight	10.0 g/100 g fish	Cr(+6) = 1.0 ppm, 281 K, 20 fish, 2 y	Salmo gairdneri[1]
Liver weight	10.9 g/100 g fish	Cr(+6) = 0.1 ppm, 281 K, 20 fish, 2 y	Salmo gairdneri[1]
Mortality	100.0 %	Cr(+6) = 0.032-6.4 mg/l	Algae[4]
Mortality	0 %	Toxicological limit = 2 - 5 mg/l, Cr2(SO4)3	Biological purification[5]
Mortality	50 %	CrO3, exposure time = 2 d, conc in water = 100 mg/l, marine	Crangon crangon[6]
Mortality	0 %	Toxicological limit = 0.03 - 0.1 mg/l, Cr2(SO4)3	Crustacean[7]
Mortality	0 %	Toxicological limit = 0.3 - 0.7 mg/l, Cr(VI)	Crustacean[7]
Mortality	100.0 %	Cr(+6) = 0.05 mg/l	Daphnia sp.[8]
Mortality	0 %	Toxicological limit = 0.015 - 0.195 mg/l, Cr(VI)	Fish[7]
Mortality	0 %	Toxicological limit = 1.2 - 200 mg/l, Cr2(SO4)2	Fish[7]
Mortality	100.0 %	Cr(+6) = 17.-118.0 mg/l, diff. species	Fish[9]
Mortality	0 %	Toxicological limit = 0.5 - 5 g/kg, Cr(VI)	Homo sapiens[7]
Mortality	0 %	Cr(+6) = 45 mg/l, 20 d, hard water	Lepomis macrochirus[10]
Mortality	0 %	Toxicological limit = 0.45 - 11 mg/l chronical symptoms	Mammalia[7]
Mortality	100 %	Cr(+6) = 31.8 mg/l	Oncorhynchus kisutch[11]
Mortality	50 %	K2Cr2O7 = 3.1 +- 2.1 ppm, 96 h, 25 ppm CaCO3, 293 K, pH = 7.4-7.9	Philodina sp.[12]
Mortality	50 %	K2Cr2O7 = 31.2 +- 3.0 ppm, 48 h, 25 ppm CaCO3, 293 K, pH = 7.4-7.9	Philodina sp.[12]
Mortality	50 %	K2Cr2O7 = 31.6 +- 2.5 ppm, 24 h, 25 ppm CaCO3, 293 K, pH = 7.4-7.9	Philodina sp.[12]
Mortality	0.0 %	Cr(+6) = 1.0 mg/l, 96 h, hard water	Pimephales promelas[8]
Mortality	50.0 %	Cr(+6) = 33 mg/l, 96 h, hard water	Pimephales promelas[8]
Mortality	50 %	Chloride, oral dose = 1870 mg/kg body	Rat[13]
Mortality	50 %	Nitrate, oral dose = 3250 mg/kg body	Rat[13]
Mortality	0 %	Cr(+6) = 0.35 mg/l, 8m, CaCO3 = 45 mg/l, pH = 7-8	Salmo gairdneri[14]
Mortality	0.0 %	Cr(+6) = 0.3 mg/l, 96 h, soft water	Salmo gairdneri[8]
Mortality	50 %	Cr(+6) = 69 mg/l, 96 h	Salmo gairdneri[14]

Mortality	50.0 %	Cr(+6) = 69.0 mg/l, 96 h, soft water	Salmo gairdneri[8]
Mortality	0 %	Cr(+6) = 0.2 mg/l, 8 m, CaCO3 = 45 mg/l, pH = 7-8	Salvelinus fontinalis[14]
Mortality	0.0 %	Cr(+6) = 0.3 mg/l, 96 h, soft water	Salvelinus fontinalis[8]
Mortality	50 %	Cr(+6) = 59 mg/l, 96 h	Salvelinus fontinalis[14]
Mortality	50.0 %	Cr(+6) = 50 mg/l, 96 h, soft water	Salvelinus fontinalis[8]
Mortality	0 %	Toxicological limit = 0.3 mg/l, Cr2(SO4)3	Self Purification[5]
Na in blood	115.1 millieq/l	Cr(+6) = 10 mg/l, 15-22 d, 281 K, 20 fish	Salmo gairdneri[1]
Na in blood	151.0 millieq/l	Cr(+6) = 0 mg/l, 15-22 d, 281 K, 20 fish	Salmo gairdneri[1]
Na in blood	3.6 millieq/l	Cr(+6) = 0.0 ppm, 15-22 d, 10 fish	Salmo gairdneri[1]
Na in blood	8.4 millieq/l	Cr(+6) = 10.0 ppm, 15-22 d, 10 fish	Salmo gairdneri[1]
Nitrification	81 % inhibation	Soil = 5 umoles/g of soil, 10 d incubation, 3 soils, lab	Soil organisms[15]
Red blood cells	1.11E6 /microlitre	Cr(+6) = 10 mg/l, 32 d, 283 K, 4*13 fish	Rutilus rutilus[1]
Red blood cells	1.13E6 /microlitre	Cr(+6) = 1.0 mg/l, 32 d, 283 K, 4*13 fish	Rutilus rutilus[1]
Red blood cells	1.19E6 /micro-l	Cr(+6) = 0 mg/l, 32 d, 283 K, 4*13 fish	Rutilus rutilus[1]
Red blood cells	1.24E6 /microlitre	Cr(+6) = 0.1 mg/l, 32 d, 283 K, 4*13 fish	Rutilus rutilus[1]
Red blood cells	1.11E6 /microlitre	Cr(+6) = 10.0 ppm, 32 d, 20 fish	Salmo gairdneri[1]
Red blood cells	1.13E6 /microlitre	Cr(+6) = 1.0 ppm, 32 d, 20 fish	Salmo gairdneri[1]
Red blood cells	1.19E6 /microlitre	Cr(+6) = 0.0 ppm, 32 d, 20 fish	Salmo gairdneri[1]
Red blood cells	1.24E6 /microlitre	Cr(+6) = 0.1 ppm, 32 d, 20 fish	Salmo gairdneri[1]
White blood cells	82.8E3 /microlitre	Cr(+6) = 1.0 mg/l, 32 d, 283 K, 4*13 fish	Rutilus rutilus[1]
White blood cells	89.6E3 /microlitre	Cr(+6) = 10 mg/l, 32 d, 283 K, 4*13 fish	Rutilus rutilus[1]
White blood cells	95.3E3 /microlitre	Cr(+6) = 0.1 mg/l, 32 d, 283 K, 4*13 fish	Rutilus rutilus[1]
White blood cells	98.1E3 /microlitre	Cr(+6) = 0 mg/l, 32 d, 283 K, 4*13 fish	Rutilus rutilus[1]
White blood cells	82.8 mg/100 ml	Cr(+6) = 1.0 ppm, 32 d, 20 fish	Salmo gairdneri[1]
White blood cells	89.6 mg/100 ml	Cr(+6) = 10.0 ppm, 32 d, 20 fish	Salmo gairdneri[1]
White blood cells	95.3 mg/100 ml	Cr(+6) = 0.1 ppm, 32 d, 20 fish	Salmo gairdneri[1]
White blood cells	98.1 mg/100 ml	Cr(+6) = 0.0 ppm, 32 d, 20 fish	Salmo gairdneri[1]

Source: 1.Strik, J.J.T.W.A., H.H. Iongh, J.W.A. van Rijn et al. 1975. in: Koeman, J.H. et al. (eds.): SublethaEffects of Toxic Chemicals on Aquatic Animals, Elsevier. NY. pp 31-37.
2.Fromm, P.O. and R.H. Schiffman. 1959. Stream Pollut., 31 pp 205.
3.McKee, J.R. and H.W. Wolf. 1963. Water Quality Criteria, Publication no. 3A, Calif. State Water Res.Control Board 2nd Edition.
4.Hervey, R.J. 1949. Bot. Gaz., 111, pp 1.
5.Jung. K.D. 1973. Wirkungskonzentration schaedigender Stoffe im Wasser. Gelenkirchen. Hygiene-Institudes Ruhrgebiets.
6.Portmann, J.E. and K.W. Wilson. 1971. Shellfish Information Leaflet no 22.
7.Liebman, H. 1958. Handbuch der Friscwasser - und Abwasserbiologie. Band II.Munchen. Oldenbourg.
8.Committee on Biologic Effects of Atmos pheric Pollutants, Chromium, Medical and Biologic Effects ofnvironmental Pollu tants, National Academy of Sci. Wash. D.C. 1974.
9.Henderson, C. and Q.H. Pickering. 1966. Air Water Pollut. 10, pp 453.
10.Doudoroff, P. and M. Katz. 1953. Swedish. ind. Waters, 25, p 802.
11.Holland, G.A. Lasater, J.E. Neumann and W.E. Eldridge. 1960. State of Wash. Dep. of Fish. Res. Bull.5.
12.Sallivan, G.W., A.L. Buikema and J. Cairns, Jr. 1973. Va. J. Sci. 24(3), pp 120.

13.Toxic Substances List 1973, U.S. Department of Health, Education and Welfare, Public Health ServiceNational Institute for Occupational Safety and Health, Rockville Maryland, USA, June 1973.
14.Benoit, D.A. 1976. Water Res. 10(6), pp 497-500.
15.Liang, C.N. and M.A. Tabatabai. 1978. Environ. Pollut. 12, 141-147.

4-100 Effect of Croton-aldehyde

Parameter	Concentration	Condition	Species
LD50	0.16-0.3 g pr. kg ww	acute dose, 100 %	Rat[1]

Source: 1.Patty, F. A., Industrial Hygiene and toxicology, 2, Interscience Publishers, New York, 1976.

4-101 Effect of Cs

Parameter	Effect Value	Concentration and Condition	Species
Mortality	50 %	Carbonate, intravenous, dose = 940 mg/kg body	Mouse[1]
Mortality	50 %	Hydroxide, oral dose = 1026 mg/kg body	Rat[1]

Source: 1.Toxic Substances List 1973, U.S. Department of Health, Education and Welfare, Public Health ServiceNational Institute for Occupational Safety and Health, Rockville Maryland, USA, June 1973.

4-102 Effect of Cu

Parameter	Effect Value	Concentration and Condition	Species
Chromagens	10.0 % chromagens	Control = 18.8 %, Cu = 2 ppm, after 14 d, 299 K, fresh water, Lake	Bacterial colonies[1]
Chromagens	32.5 % chromagens	Control = 19.4 %, Cu = 2 ppm, after 14 d, 299 K, brackish, Clear	Bacterial colonies[1]
Diversity	0.78 shannon diversity	Soil Cu = 218.3 ppm, n = 4, control = 1.51, soil Cu = 15.0 ppm, n =	Weed[2]
Diversity	1.38 shannon diversity	Soil Cu = 63.0 ppm, n = 4, control = 1.39, soil Cu = 14.0 ppm, n = 2,	Weed[2]
Diversity	1.48 shannon diversity	Soil Cu = 140.6+-96 ppm, n = 8, control = 1.96+-0.27, soil Cu = 17.8	Weed[2]
Diversity	1.70 shannon diversity	Soil Cu = 92.6+-40 ppm, n = 8, control = 1.80+-0.09, soil Cu = 10.8	Weed[2]
Dry weight	0 % effect observed	Cu = 10 ug/l, weight of siphon, max conc without effect, marine	Tellina tenuis[3]
Dry weight	0 % effect observed	Cu = 30 ug/l, weight of siphon, max conc without effect, marine	Tellina tenuis[3]
Egg production	1 % effect observed	Cu = 20 ug/l, decrease, min conc with effect, marine	Acartia sp.[4]
Egg production	1 % decrease	Cu = > 77 ug/l, hardness = 45 mg/l, minimum conc with effect	Lepomis macrochirus[5]
Egg production	1 % decrease	Cu = 18 ug/l, hardness = 31 mg/l, minimum conc with effect	Pimephales promelas[6]
Egg production	1 % decrease	Cu = 33 ug/l, hardness = 198 mg/l, minimum conc with effect	Pimephales promelas[7]
Egg production	1 % decrease	Cu = 37 ug/l, hardness = 200 mg/l, minimum conc with effect	Pimephales promelas[8]
Egg production	1 % decrease	Cu = 66 ug/l, high organic conc, fluctaing hardness, minimum conc with effect	Pimephales promelas[9]
Egg production	1 % effect observed	Cu = 5 ug/l, decrease, min conc with effect, marine	Pseudocalanus sp., Calanus sp. and Euphausia sp.[4]

Egg production	1 % decrease	Cu = 33 ug/l, hardness = 45 mg/l, minimum conc with effect	Salvelinus fontinalis[10]
Egg production	0 % effect observed	Cu = 6.4 ug/l, max conc without effect, marine	Tigriopus japonicus[11]
Egg production	1 % effect observed	Cu = 64 ug/l, min conc with effect, marine	Tigriopus japonicus[11]
Eggs per female	0	Cu = 0.18 mg/l, control = 1185, no eggs at Cu = 0.32 and 0.57 mg/l	Pimephales promelas[12]
Eggs per female	1272	Cu = 0.033 mg/l, control = 1185	Pimephales promelas[12]
Eggs per female	735	Cu = 0.065 mg/l, control = 1185	Pimephales promelas[12]
Eggs per female	80	Cu = 0.12 mg/l, control = 1185	Pimephales promelas[12]
Eggs per spawning	0	Cu = 0.18 mg/l, control = 174, no eggs at Cu = 0.32 and 0.57 mg/l	Pimephales promelas[12]
Eggs per spawning	113	Cu = 0.065 mg/l, control = 174	Pimephales promelas[12]
Eggs per spawning	138	Cu = 0.033 mg/l, control = 174	Pimephales promelas[12]
Eggs per spawning	65	Cu = 0.12 mg/l, control = 174	Pimephales promelas[12]
Feeding rate	1 % effect observed	Cu = 5 ug/l, min conc with effect, marine	Calanus sp. and Metridia sp.[4]
Growth	0.001 mg/l gives deficiency symptoms	Tree, 100 d in greenhouse, conc in culture solution	Agathis australis[13]
Growth	0.01 mg/l gives normal growth	Tree, 100 d in greenhouse, conc in culture solution	Agathis australis[13]
Growth	0.2 gives toxic symptoms	Tree, 100 d in greenhouse, conc in culture solution	Agathis australis[13]
Growth	0.002 mm/day	Control = 0.014 mm/day, Cu = 0.1 mg/l, 10 juv. , Bredwell Brook, UK	Asellus meridianus[14]
Growth	0.005 mm/day	Control = 0.028 mm/day, Cu = 0.1 mg/l, 10 juv. , River Gannal, UK	Asellus meridianus[14]
Growth	0.006 mm/day	Control = 0.017 mm/day, Cu = 0.1 mg/l, 10 juv. , River Hayle, UK	Asellus meridianus[14]
Growth	0.006 mm/day	Control = 0.028 mm/day, Cu = 0.1 mg/l, 10 juv. , Bredwell Brook, UK	Asellus meridianus[14]
Growth	0.011 mm/day	Control = 0.033 mm/day, Cu = 0.1 mg/l, 10 juv. , River Gannal, UK	Asellus meridianus[14]
Growth	0.017 mm/day	Control = 0.033 mm/day, Cu = 0.1 mg/l, 10 juv. , River Hayle, UK	Asellus meridianus[14]
Growth	0.017 mm/day	Control = 0.044 mm/day, Cu = 0.1 mg/l, 10 juv. , River Gannal, UK	Asellus meridianus[14]
Growth	0.022 mm/day	Control = 0.022 mm/day, Cu = 0.1 mg/l, 10 juv. , River Hayle, UK	Asellus meridianus[14]
Growth	0.028 mm/day	Control = 0.033 mm/day, Cu = 0.1 mg/l, 10 juv. , River Hayle, UK	Asellus meridianus[14]
Growth	100 % effect	30 ug/l, inhibition observed	Paracentrotus lividus[15]
Growth rate	0 % decrease observed	Cu = 25 ug/l, max conc without effect, marine	Artemia salina[16]
Growth rate	1 % decrease observed	Cu = 50 ug/l, min conc with effect, marine	Artemia salina[16]
Growth rate	1 % decrease observed	Cu = 100 ug/l, no pollution adapted anim., min conc with effect, freshwater	Ascellus meridianus[17]
Growth rate	0 % decrease observed	Cu = 10 ug/l, colony hydroid, max conc without effect, marine	Campanularia flexuosa[18]

Growth rate	1 % decrease observed	Cu = 13 ug/l, colony hydroid, min conc with effect, marine	Campanularia flexuosa[18]
Growth rate	0 % decrease observed	Cu = 40 ug/l, max conc without effect, freshwater	Daphnia ambigua[19]
Growth rate	1 % decrease observed	Cu = 60 ug/l, min conc with effect, freshwater	Daphnia ambigua[19]
Growth rate	0 % decrease observed	Cu = 60 ug/l, max conc without effect, freshwater	Daphnia magna[19]
Growth rate	1 % decrease observed	Cu = 80 ug/l, min conc with effect, freshwater	Daphnia magna[19]
Growth rate	0 % decrease observed	Cu = 40 ug/l, max conc without effect, freshwater	Daphnia parvula[19]
Growth rate	1 % decrease observed	Cu = 60 ug/l, min conc with effect, freshwater	Daphnia parvula[19]
Growth rate	0 % decrease observed	Cu = 40 ug/l, max conc without effect, freshwater	Daphnia pulex[19]
Growth rate	1 % decrease observed	Cu = 60 ug/l, min conc with effect, freshwater	Daphnia pulex[19]
Growth rate	0 % decrease observed	Cu = 30 ug/l, colony hydroid, max conc without effect, marine	Eirene viridula[20]
Growth rate	1 % decrease observed	Cu = 60 ug/l, colony hydroid, min conc with effect, marine	Eirene viridula[20]
Growth rate	0 % decrease observed	Cu = 4.6 ug/l, 2. generation effected, max conc without effect, freshwater	Gammarus pseudolimnaeus[21]
Growth rate	0 % decrease observed	Cu = 8 ug/l, also survival affected, max conc without effect, freshwater	Gammarus pseudolimnaeus[21]
Growth rate	1 % decrease observed	Cu = 14.8 ug/l, also survival affected, min conc with effect, freshwater	Gammarus pseudo-limnaeus[21]
Growth rate	1 % decrease observed	Cu = 8 ug/l, 2. generation affected, min conc with effect, freshwater	Gammarus pseudolimnaeus[21]
Growth rate	1 % decrease	Cu = 19 ug/l, hardness = 187 mg/l, minimum conc with effect	Ictalurus nebulosus[22]
Growth rate	1 % decrease	Cu = 19 ug/l, hardness = 38 mg/l, minimum conc with effect	Ictalurus nebulosus[22]
Growth rate	1 % decrease	Cu = 162 ug/l, hardness = 45 mg/l, minimum conc with effect	Lepomis macrochirus[23]
Growth rate	0 % decrease observed	Cu = 4.9 ug/l, larvae, max conc without effect, marine	Mercenaria mercenaria[24]
Growth rate	1 % decrease	Cu = 20 ug/l, minimum conc with effect	Oncorhynchus tschawytscha[25]
Growth rate	1 % decrease observed	Cu = 25 ug/l, min conc with effect, marine	Ophryotrocha labronica[26]
Growth rate	1 % decrease observed	Cu = 15 ug/l, growth retarded, min conc with effect, freshwater	Orconectes rusticus[27]
Growth rate	0 % decrease observed	Cu = 8 ug/l, also survival, max conc without effect, freshwater	Physa integra[21]
Growth rate	1 % decrease observed	Cu = 14.8 ug/l, also survival, min conc with effect, freshwater	Physa integra[21]
Growth rate	1 % decrease	Cu = 95 ug/l, hardness = 198 mg/l, minimum conc with effect	Pimephales promelas[7]
Growth rate	1 % decrease	Cu = > 100 ug/l, hardness = 200 mg/l, minimum conc with effect	Pimephales promelas[8]
Growth rate	1 % decrease	Cu = 17 ug/l, hardness = 45 mg/l, minimum conc with effect	Salvelinus fontinalis[10]
Growth rate	1 % decrease	Cu = 5 ug/l, hardness = 38 mg/l, minimum conc with effect	Salvelinus fontinalis[22]
Growth rate	1 % decrease	Cu = 8 ug/l, hardness = 187 mg/l, minimum conc with effect	Salvelinus fontinalis[22]

Hatching	1 % decrease	Cu = 500 ug/l, S = 2.0 %, minimum conc with effect	Fundulus heteroclitus[28]
Hatching	1 % decrease	Cu = > 66 ug/l, hardness = 187 mg/l, minimum conc with effect	Ictalurus nebulosus[22]
Hatching	1 % decrease	Cu = > 66 ug/l, hardness = 38 mg/l, minimum conc with effect	Ictalurus nebulosus[22]
Hatching	1 % decrease	Cu = 162 ug/l, hardness = 45 mg/l, minimum conc with effect	Lepomis macrochirus[5]
Hatching	1 % decrease	Cu = 500 ug/l, S = 2.0 %, minimum conc with effect	Menidia menidia[28]
Hatching	1 % decrease	Cu = >80 ug/l, minimum conc with effect	Oncorhynchus tschawytscha[25]
Hatching	1 % decrease	Cu = 120 ug/l, high organic conc, fluctaing hardness, minimum conc with effect	Pimephales promelas[8]
Hatching	1 % decrease	Cu = > 100 ug/l, hardness = 200 mg/l, minimum conc with effect	Pimephales promelas[8]
Hatching	1 % decrease	Cu = > 11 ug/l, hardness = 31 mg/l, minimum conc with effect	Pimephales promelas[6]
Hatching	1 % decrease	Cu = > 33 ug/l, hardness = 198 mg/l, minimum conc with effect	Pimephales promelas[7]
Hatching	1 % decrease	Cu = 19 ug/l, hardness = 100 ug/l, minimum conc with effect	Salmo gairdneri[29]
Hatching	1 % decrease	Cu = 20 ug/l, hardness = 7.8 mg/l, minimum conc with effect	Salmo gairdneri[30]
Hatching	1 % decrease	Cu = 10 ug/l, hardness = 7.8 mg/l, minimum conc with effect	Salmo salar[30]
Hatching	1 % decrease	Cu = 13 ug/l, hardness = 38 mg/l, minimum conc with effect	Salvelinus fontinalis[22]
Hatching	1 % decrease	Cu = 33 ug/l, hardness = 45 mg/l, minimum conc with effect	Salvelinus fontinalis[10]
Hatching	1 % decrease	Cu = 74 ug/l, hardness = 187 mg/l, minimum conc witheffect	Salvelinus fontinalis[22]
Hatching	1 % decrease	Cu = 47 ug/l, hardness = 13 mg/l, minimum conc with effect	Stizostedion vitreum[22]
Larvae development	1 % effect observed	Cu = 10 ug/l, abnormal larvae, min conc with effect, marine	Capitella capitella[26]
Mortality	50 % mortality	Cu conc = 8300 ug/l, 96 h, stone fly, freshwater	Acroneuria lycorias[31]
Mortality	50 % mortality	Cu conc = 900 ug/l, adults, 96 h, freshwater	Amnicola sp.[32]
Mortality	50 % mortality	Cu conc = 9300 ug/l, eggs, 96 h, freshwater	Amnicola sp.[32]
Mortality	50 % mortality	Cu conc = 1000 ug/l, 110 h, larvae, marine	Artemia salina[33]
Mortality	50 % mortality	Cu conc = 1000 ug/l, 168 h, adult, marine	Artemia salina[33]
Mortality	50 % mortality	Cu conc = 2050-2554 ug/l, 24 h, marine	Artemia salina[4]
Mortality	50 % mortality	Cu conc = 1200-2500 ug/l, 48 h, freshwater	Ascellus meridianus[34]
Mortality	50 %	Cu = 1.2 mg/l, 10 juveniles, 293 K, animals from Bredwell Brook, UK	Asellus meridianus[14]
Mortality	50 %	Cu = 1.7 mg/l, 10 juveniles, 293 K, test animals from River Hayle,	Asellus meridianus[14]
Mortality	50 %	Cu = 1.9 mg/l, 10 juveniles, 293 K, test animals from River Gannel,	Asellus meridianus[14]
Mortality	50 % mortality	CuSO4*5H2O, water = E-4 M Cu, toxic level given, microorganisms	Azotobacter sp.[35]
Mortality	0 %	Toxicological limit = 1 mg/l, CuSO4	Biological purification[36]

Mortality	50 % mortality	Cu conc = 100 ug/l, 24 h, marine	Brachionus plicatilis[4]
Mortality	50 % mortality	Cu conc = 3800 ug/l, larvae, 2 h, marine	Bugula neritina[37]
Mortality	1 % mortality	Cu conc = 200 - 500 ug/l, toxicity threshold, marine	Busycon canalicalatum[38]
Mortality	50 % mortality	Cu conc = 6200 ug/l, 96 h, freshwater	Caddis fly[32]
Mortality	50 % mortality	Cu conc = 105-2778 ug/l, several spp., 24 h, marine	Calanoid copepods[4]
Mortality	50 % mortality	Cu conc = 1700 ug/l, 96 h, freshwater	Campeloma decisum[39]
Mortality	50 % mortality	Cu conc = 180 ug/l, 96 h, trochophore larvae, marine	Capitella capitella[26]
Mortality	50 % mortality	Cu conc = 200 ug/l, adult, 96 h, marine	Capitella capitella[26]
Mortality	0 % mortality	Cu = 0.025 mg/l, mortality within 4 d posthatching, 100 mg CaCO3/l	Carassius auratus, larvae[40]
Mortality	0 % mortality	Cu = 0.050 mg/l, mortality within 4 d posthatching, 100 mg CaCO3/l	Carassius auratus, larvae[40]
Mortality	0 % mortality	Cu = 0.100 mg/l, mortality within 4 d posthatching, 100 mg CaCO3/l	Carassius auratus, larvae[40]
Mortality	1 % mortality	Cu = 261 ug/l, hatchning time=3 d, 3 d exposure, 100 mg CaCO3/l	Carassius auratus, larvae[40]
Mortality	1 % mortality	Cu = 299 ug/l, hatchning time=3 d, 7 d exposure, 100 mg CaCO3/l	Carassius auratus, larvae[40]
Mortality	100 % mortality	Cu = 25.00 mg/l, mortality within 4 d posthatching, 100 mg CaCO3/l	Carassius auratus, larvae[40]
Mortality	100 % mortality	Cu = 50.00 mg/l, mortality within 4 d posthatching, 100 mg CaCO3/l	Carassius auratus, larvae[40]
Mortality	16 % mortality	Cu = 1.000 mg/l, mortality within 4 d posthatching, 100 mg CaCO3/l	Carassius auratus, larvae[40]
Mortality	50 % mortality	Cu = 5.000 mg/l, mortality within 4 d posthatching, 100 mg CaCO3/l	Carassius auratus, larvae[40]
Mortality	50 % mortality	Cu = 5.20 ug/l, hatchning time=3 d, 3 d exposure, 100 mg CaCO3/l	Carassius auratus, larvae[40]
Mortality	50 % mortality	Cu = 5.20 ug/l, hatchning time=3 d, 7 d exposure, 100 mg CaCO3/l	Carassius auratus, larvae[40]
Mortality	7 % mortality	Cu = 0.500 mg/l, mortality within 4 d posthatching, 100 mg CaCO3/l	Carassius auratus, larvae[40]
Mortality	79 % mortality	Cu = 10.00 mg/l, mortality within 4 d posthatching, 100 mg CaCO3/l	Carassius auratus, larvae[40]
Mortality	50 % mortality	Cu conc = 109000 ug/l, adult, 48 h, marine	Carcinus maenus[41]
Mortality	50 % mortality	Cu conc = 600 ug/l, larvae, 48 h, marine	Carcinus maenus[41]
Mortality	50 % mortality	Cu conc = 30 ug/l, 96 h, freshwater	Chironomous sp.[32]
Mortality	50 %	CuSO4, exposure time = 2 d, conc in water = 29.5 mg/l, marine	Crangon crangon[42]
Mortality	50 % mortality	Cu conc = 29500 ug/l, adult, 48 h, marine	Crangon crangon[41]
Mortality	50 % mortality	Cu conc = 330 ug/l, 48 h, larvae, marine	Crangon crangon[41]
Mortality	50 % mortality	Cu conc = 560 ug/l, 96 h, marine	Crassostrea gigas[43]
Mortality	50 % mortality	Cu conc = > 100 ug/l, 14 d, marine	Crassostrea gigas[43]
Mortality	1 % effect observed	Cu = 19-43 ug/l, also inhibition of spawning, min conc with effect, marine	Crassostrea virginica[44]
Mortality	3 % mortality	Cu = 0.025 mg/l, 20 w, tissue conc = 639 mg/kg ww	Crassostrea virginica[45]
Mortality	6 % mortality	Cu = 0.050 mg/l, 20 w, tissue conc = 1060 mg/kg ww	Crassostrea virginica[45]

Mortality	0 %	Toxicological limit = 0.08 - 0.8 mg/l, CuSO4	Crustacean[46]
Mortality	50 % mortality	Cu conc = 250-500 ug/l, 96 h, marine	Ctenodrilus serratus[47]
Mortality	50 % mortality	Cu conc = 2500 ug/l, 48 h, freshwater	Cyclops abyssorum[48]
Mortality	50 % mortality	Cu conc = 4600 ug/l, 96 h, freshwater	Damsel fly[32]
Mortality	50 % mortality	Cu conc = 68 ug/l, with food, 72 h, freshwater	Daphnia ambigua[49]
Mortality	50 % mortality	Cu conc = 5 ug/l, no food, 48 h, freshwater	Daphnia hyalella[48]
Mortality	50 %	CuCl2, exposure time = 2 d, conc in water = 9.8 microgram/l, limnol.	Daphnia magna[50]
Mortality	50 % mortality	Cu = 44 ug/l, 3 w, freshwater	Daphnia magna[51]
Mortality	50 % mortality	Cu conc = 60 ug/l, with food, 48 h, freshwater	Daphnia magna[52]
Mortality	50 % mortality	Cu conc = 86 ug/l, with food, 72 h, freshwater	Daphnia magna[49]
Mortality	50 % mortality	Cu conc = 9.8 ug/l, no food, 48 h, freshwater	Daphnia magna[52]
Mortality	50 % mortality	Cu conc = 72 ug/l, with food, 72 h, freshwater	Daphnia parvula[49]
Mortality	50 % mortality	Cu conc = 86 ug/l, with food, 72 h, freshwater	Daphnia pulex[49]
Mortality	50 % mortality	Cu conc = 180-200 ug/l, 14 d, freshwater	Ephmerella grandis[53]
Mortality	50 % mortality	Cu conc = 320 ug/l, 48 h, freshwater	Ephmerrella subvaria[31]
Mortality	1 % effect observed	Cu = 6 ug/l, prefeeding stages, min conc with effect, marine	Euchaeta japonica[54]
Mortality	50 % mortality	Cu conc = 500 ug/l, 48 h, freshwater	Eudiaptomus padanus[48]
Mortality	50 % mortality	Cu conc = 14-30 ug/l, 24 h, marine	Euphausia pacifica[4]
Mortality	0 %	Toxicological limit = 0.03 - 0.8 mg/l, trout, CuSO4	Fish[46]
Mortality	1 % mortality	Cu = 0.100 mg/l, mortality within 4 d posthatching, 100 mg CaCO3/l	Fowlers Toad, larvae[40]
Mortality	1 % mortality	Cu = 0.500 mg/l, mortality within 4 d posthatching, 100 mg CaCO3/l	Fowlers Toad, larvae[40]
Mortality	1 % mortality	Cu = 1436 ug/l, hatchning time=3 d, 7 d exposure, 100 mg CaCO3/l	Fowlers Toad, larvae[40]
Mortality	1 % mortality	Cu = 2415 ug/l, hatchning time=3 d, 3 d exposure, 100 mg CaCO3/l	Fowlers Toad, larvae[40]
Mortality	100 % mortality	Cu = 100.0 mg/l, mortality within 4 d posthatching, 100 mg CaCO3/l	Fowlers Toad, larvae[40]
Mortality	26 % mortality	Cu = 10.00 mg/l, mortality within 4 d posthatching, 100 mg CaCO3/l	Fowlers Toad, larvae[40]
Mortality	3 % mortality	Cu = 1.00 mg/l, mortality within 4 d posthatching, 100 mg CaCO3/l	Fowlers Toad, larvae[40]
Mortality	47 % mortality	Cu = 25.00 mg/l, mortality within 4 d posthatching, 100 mg CaCO3/l	Fowlers Toad, larvae[40]
Mortality	50 % mortality	Cu = 26.96 ug/l, hatchning time=3 d, 7 d exposure, 100 mg CaCO3/l	Fowlers Toad, larvae[40]
Mortality	50 % mortality	Cu = 35.99 ug/l, hatchning time=3 d, 3 d exposure, 100 mg CaCO3/l	Fowlers Toad, larvae[40]
Mortality	58 % mortality	Cu = 50.0 mg/l, mortality within 4 d posthatching, 100 mg CaCO3/l	Fowlers Toad, larvae[40]
Mortality	8 % mortality	Cu = 5.000 mg/l, mortality within 4 d posthatching, 100 mg CaCO3/l	Fowlers Toad, larvae[40]
Mortality	1 % of effect observed	Cu = 250 ug/l, S = 2.0 %, minimum conc with effect	Fundulus heteroclitus[28]

Mortality	50 % mortality	Cu conc = 2900 ug/l, 2 h, larvae, marine	Galeolaria caespitosa[37]
Mortality	50 % mortality	Cu conc = 20 ug/l, no food, 96 h, freshwater	Gammarus pseudolimnaeus[39]
Mortality	50 % mortality	Cu conc = 910 ug/l, 96 h, freshwater	Gammarus sp.[32]
Mortality	1 % mortality	Cu conc = 56 ug/l, 48 h, 278 and 286 K, lethal threshold, marine	Homarus americanus[55]
Mortality	50 % mortality	Cu conc = 1000 ug/l, adults, 278 K, 48 h, marine	Homarus americanus[55]
Mortality	50 % mortality	Cu conc = 560 ug/l, adults, 286 K, 48 h, marine	Homarus americanus[55]
Mortality	50 % mortality	Cu conc = 100-330 ug/l, larvae, 48 h, marine	Homarus gammarus[41]
Mortality	0 %	Toxicological limit = 8 g, CuSO4	Homo sapiens[46]
Mortality	50 % mortality	Cu conc = 32000 ug/l, 14 d, caddis fly, freshwater	Hydropsyche betteni[31]
Mortality	50 % mortality	Cu conc = > 64000 ug/l, 96 h, caddis fly, freshwater	Hydropsyche betteni[31]
Mortality	0 % mortality	Cu = 0.010 mg/l, mortality within 4 d posthatching, 100 mg CaCO3/l	Ictalurus lacustris, larvae[40]
Mortality	1 % mortality	Cu = 0.001 mg/l, mortality within 4 d posthatching, 100 mg CaCO3/l	Ictalurus lacustris, larvae[40]
Mortality	1 % mortality	Cu = 0.025 mg/l, mortality within 4 d posthatching, 100 mg CaCO3/l	Ictalurus lacustris, larvae[40]
Mortality	1 % mortality	Cu = 0.050 mg/l, mortality within 4 d posthatching, 100 mg CaCO3/l	Ictalurus lacustris, larvae[40]
Mortality	1 % mortality	Cu = 167 ug/l, hatchning time=6 d, 10 d exposure, 100 mg CaCO3/l	Ictalurus lacustris, larvae[40]
Mortality	1 % mortality	Cu = 191 ug/l, hatchning time=6 d, 6 d exposure, 100 mg CaCO3/l	Ictalurus lacustris, larvae[40]
Mortality	100 % mortality	Cu = 50.00 mg/l, mortality within 4 d posthatching, 100 mg CaCO3/l	Ictalurus lacustris, larvae[40]
Mortality	11 % mortality	Cu = 0.500 mg/l, mortality within 4 d posthatching, 100 mg CaCO3/l	Ictalurus lacustris, larvae[40]
Mortality	17 % mortality	Cu = 1.000 mg/l, mortality within 4 d posthatching, 100 mg CaCO3/l	Ictalurus lacustris, larvae[40]
Mortality	2 % mortality	Cu = 0.005 mg/l, mortality within 4 d posthatching, 100 mg CaCO3/l	Ictalurus lacustris, larvae[40]
Mortality	29 % mortality	Cu = 5.000 mg/l, mortality within 4 d posthatching, 100 mg CaCO3/l	Ictalurus lacustris, larvae[40]
Mortality	5 % mortality	Cu = 0.100 mg/l, mortality within 4 d posthatching, 100 mg CaCO3/l	Ictalurus lacustris, larvae[40]
Mortality	50 % mortality	Cu = 6.62 ug/l, hatchning time=6 d, 10 d exposure, 100 mg CaCO3/l	Ictalurus lacustris, larvae[40]
Mortality	50 % mortality	Cu = 7.56 ug/l, hatchning time=6 d, 6 d exposure, 100 mg CaCO3/l	Ictalurus lacustris, larvae[40]
Mortality	54 % mortality	Cu = 10.00 mg/l, mortality within 4 d posthatching, 100 mg CaCO3/l	Ictalurus lacustris, larvae[40]
Mortality	8 % mortality	Cu = 0.25 mg/l, mortality within 4 d posthatching, 100 mg CaCO3/l	Ictalurus lacustris, larvae[40]
Mortality	85 % mortality	Cu = 25.00 mg/l, mortality within 4 d posthatching, 100 mg CaCO3/l	Ictalurus lacustris, larvae[40]
Mortality	1 % of effect observed	Cu = 19 ug/l, hardness = 187 mg/l, minimum conc with effect	Ictalurus nebulosus[22]
Mortality	1 % of effect observed	Cu = 19 ug/l, hardness = 38 mg/l, minimum conc with effect	Ictalurus nebulosus[22]
Mortality	50 % mortality	Cu conc = 1900 ug/l, 96 h, marine	Japanese oyster[56]

Mortality	1 % of effect observed	Cu = 40 ug/l, hardness = 45 mg/l, minimum conc with effect	Lepomis macrochirus[5]
Mortality	0 %	Toxicological limit = 8 g/kg	Mammalia[46]
Mortality	0 % mortality	Cu = 0.010 mg/l, mortality within 4 d posthatching, 100 mg CaCO3/l	Marbled Salamander, larvae[40]
Mortality	1 % mortality	Cu = 21.4 ug/l, hatchning time=4 d, 8 d exposure, 100 mg CaCO3/l	Marbled Salamander, larvae[40]
Mortality	1 % mortality	Cu = 37.2 ug/l, hatchning time=4 d, 4 d exposure, 100 mg CaCO3/l	Marbled Salamander, larvae[40]
Mortality	100 % mortality	Cu = 10.00 mg/l, mortality within 4 d posthatching, 100 mg CaCO3/l	Marbled Salamander, larvae[40]
Mortality	15 % mortality	Cu = 5.000 mg/l, mortality within 4 d posthatching, 100 mg CaCO3/l	Marbled Salamander, larvae[40]
Mortality	18 % mortality	Cu = 0.100 mg/l, mortality within 4 d posthatching, 100 mg CaCO3/l	Marbled Salamander, larvae[40]
Mortality	3 % mortality	Cu = 0.050 mg/l, mortality within 4 d posthatching, 100 mg CaCO3/l	Marbled Salamander, larvae[40]
Mortality	30 % mortality	Cu = 0.500 mg/l, mortality within 4 d posthatching, 100 mg CaCO3/l	Marbled Salamander, larvae[40]
Mortality	50 % mortality	Cu = 0.77 ug/l, hatchning time=4 d, 8 d exposure, 100 mg CaCO3/l	Marbled Salamander, larvae[40]
Mortality	50 % mortality	Cu = 3.59 ug/l, hatchning time=4 d, 4 d exposure, 100 mg CaCO3/l	Marbled Salamander, larvae[40]
Mortality	55 % mortality	Cu = 1.00 mg/l, mortality within 4 d posthatching, 100 mg CaCO3/l	Marbled Salamander, larvae[40]
Mortality	1 % of effect observed	Cu = 250 ug/l, S = 2.0 %, minimum conc with effect	Menidia menidia[28]
Mortality	91 % mortality	Cu conc = 0.025 mg/l, 20 w, tissue conc = 2.2 mg/kg ww	Mercenaria mercenaria[45]
Mortality	96 % mortality	Cu conc = 0.050 mg/l, 20 w, tissue conc = 3.6 mg/kg ww	Mercenaria mercenaria[45]
Mortality	1 % mortality	Cu = 1511 ug/l, hatchning time=4 d, 4 d exposure, 100 mg CaCO3/l	Micropterus coosae, larvae[40]
Mortality	1 % mortality	Cu = 1592 ug/l, hatchning time=4 d, 8 d exposure, 100 mg CaCO3/l	Micropterus coosae, larvae[40]
Mortality	50 % mortality	Cu = 6.56 ug/l, hatchning time=4 d, 8 d exposure, 100 mg CaCO3/l	Micropterus coosae, larvae[40]
Mortality	50 % mortality	Cu = 6.97 ug/l, hatchning time=4 d, 4 d exposure, 100 mg CaCO3/l	Micropterus coosae, larvae[40]
Mortality	0 % mortality	Cu = 0.100 mg/l, mortality within 4 d posthatching, 100 mg CaCO3/l	Micropterus coosea, larvae[40]
Mortality	1 % mortality	Cu = 0.500 mg/l, mortality within 4 d posthatching, 100 mg CaCO3/l	Micropterus coosea, larvae[40]
Mortality	100 % mortality	Cu = 25.00 mg/l, mortality within 4 d posthatching, 100 mg CaCO3/l	Micropterus coosea, larvae[40]
Mortality	100 % mortality	Cu = 50.00 mg/l, mortality within 4 d posthatching, 100 mg CaCO3/l	Micropterus coosea, larvae[40]
Mortality	26 % mortality	Cu = 5.000 mg/l, mortality within 4 d posthatching, 100 mg CaCO3/l	Micropterus coosea, larvae[40]
Mortality	3 % mortality	Cu = 1.000 mg/l, mortality within 4 d posthatching, 100 mg CaCO3/l	Micropterus coosea, larvae[40]
Mortality	75 % mortality	Cu = 10.00 mg/l, mortality within 4 d posthatching, 100 mg CaCO3/l	Micropterus coosea, larvae[40]

Mortality	50 % mortality	Cu conc = 17-29 ug/l, 24 h, marine	Mnemiopsis mecradyi[4]
Mortality	1 % mortality	Cu conc = 100 - 200 ug/l, lethal threshold, marine	Mytilus edulis[57]
Mortality	1 % mortality	Cu conc = 200 - 300 ug/l, lethal threshold, marine	Mytilus edulis[58]
Mortality	50 % mortality	Cu conc = 22000 ug/l, 2 h, larvae, marine	Mytilus edulis[37]
Mortality	50 % mortality	Cu conc = 90 ug/l, 96 h, freshwater	Nais sp.[32]
Mortality	0 % mortality	Cu = 0.001 mg/l, mortality within 4 d posthatching, 100 mg CaCO3/l	Narrow-Mouthed Toad, larvae[40]
Mortality	1 % mortality	Cu = 0.6 ug/l, hatchning time=3 d, 3 d exposure, 100 mg CaCO3/l	Narrow-Mouthed Toad, larvae[40]
Mortality	1 % mortality	Cu = 0.7 ug/l, hatchning time=3 d, 7 d exposure, 100 mg CaCO3/l	Narrow-Mouthed Toad, larvae[40]
Mortality	100 % mortality	Cu = 1.00 mg/l, mortality within 4 d posthatching, 100 mg CaCO3/l	Narrow-Mouthed Toad, larvae[40]
Mortality	100 % mortality	Cu = 5.000 mg/l, mortality within 4 d posthatching, 100 mg CaCO3/l	Narrow-Mouthed Toad, larvae[40]
Mortality	34 % mortality	Cu = 0.010 mg/l, mortality within 4 d posthatching, 100 mg CaCO3/l	Narrow-Mouthed Toad, larvae[40]
Mortality	50 % mortality	Cu = 0.04 ug/l, hatchning time=3 d, 7 d exposure, 100 mg CaCO3/l	Narrow-Mouthed Toad, larvae[40]
Mortality	50 % mortality	Cu = 0.05 ug/l, hatchning time=3 d, 3 d exposure, 100 mg CaCO3/l	Narrow-Mouthed Toad, larvae[40]
Mortality	52 % mortality	Cu = 0.050 mg/l, mortality within 4 d posthatching, 100 mg CaCO3/l	Narrow-Mouthed Toad, larvae[40]
Mortality	6 % mortality	Cu = 0.005 mg/l, mortality within 4 d posthatching, 100 mg CaCO3/l	Narrow-Mouthed Toad, larvae[40]
Mortality	70 % mortality	Cu = 0.100 mg/l, mortality within 4 d posthatching, 100 mg CaCO3/l	Narrow-Mouthed Toad, larvae[40]
Mortality	87 % mortality	Cu = 0.500 mg/l, mortality within 4 d posthatching, 100 mg CaCO3/l	Narrow-Mouthed Toad, larvae[40]
Mortality	50 % mortality	Cu conc = 140 ug/l, juvenile, 28 d, marine	Neathes arenaceodentata[26]
Mortality	50 % mortality	Cu conc = 250 ug/l, adult, 28 d, marine	Neathes arenaceodentata[26]
Mortality	50 % mortality	Cu conc = 300 ug/l, 96 h, juvenile and adult, marine	Neathes arenaceodentata[26]
Mortality	50 % mortality	Cu conc = 370-480 ug/l, 1.0-3.4 % salinity, 96 h, marine	Nereis diversicolor[59]
Mortality	50 % mortality	Cu conc = 200 ug/l, 0.5 % salinity, 96 h, marine	Nereis virens[59]
Mortality	50 %	Cu = 0.26 mg/l63d, hard water	Noemacheilus barbatulus[60]
Mortality	1 % of effect observed	Cu = 20 ug/l, minimum conc with effect	Oncorhynchus tschawytscha[25]
Mortality	50 % mortality	Cu conc = 100 - 250 ug/l, 96 h, marine	Ophryotrocha diadoma[47]
Mortality	50 % mortality	Cu conc = 1000 ug/l, 4.5 h, marine	Ophryotrocha labronica[33]
Mortality	1 % mortality	Cu = 60-125 ug/l, toxicity threshold, newly hatched, 96 h, freshwater	Orconectes rusticus[61]
Mortality	1 % mortality	Cu conc = 125 ug/l, egg mortality, toxicity threshold, freshwater	Orconectes rusticus[61]
Mortality	1 % mortality	Cu conc = 500-1000 ug/l, Adults, toxicity threshold, freshwater	Orconectes rusticus[61]

Mortality	50 % mortality	Cu conc = 36 ug/l, 24 h, marine	Phialidium sp.[4]
Mortality	50 % mortality	Cu conc = 1100 ug/l, hard water, 96 h, freshwater	Philodina acuticornis[62]
Mortality	50 % mortality	Cu conc = 700 ug/l, soft water, 96 h, freshwater	Philodina acuticornis[62]
Mortality	50 %	CuSO4 = 0.4 +- 2.2 ppm, 96 h, 25 ppm CaCO3, 293 K, pH = 7.4-7.9	Philodina sp.[63]
Mortality	50 %	CuSO4 = 0.7 +- 3.3 ppm, 48 h, 25 ppm CaCO3, 293 K, pH = 7.4-7.9	Philodina sp.[63]
Mortality	50 %	CuSO4 = 1.6	Philodina sp.[63]
Mortality	1 % mortality	Cu conc = 70-80 ug/l, threshold toxicity, marine	Phyllodoce maculata[64]
Mortality	50 % mortality	Cu conc = 39 ug/l, 96 h, freshwater	Physa integra[39]
Mortality	1 % of effect observed	Cu = 11 ug/l, hardness = 31 mg/l, minimum conc with effect	Pimephales promelas[6]
Mortality	1 % of effect observed	Cu = 18 ug/l, hardness = 31 mg/l, minimum conc with effect	Pimephales promelas[6]
Mortality	1 % of effect observed	Cu = 180 ug/l, high organic conc, fluctaing hardness, minimum conc with effect	Pimephales promelas[8]
Mortality	1 % of effect observed	Cu = 95 ug/l, hardness = 198 mg/l, minimum conc with effect	Pimephales promelas[7]
Mortality	1 % of effect observed	Cu = > 100 ug/l, hardness = 200 mg/l, minimum conc with effect	Pimephales promelas[8]
Mortality	50 % mortality	Cu conc = 33 ug/l, 24 h, marine	Pleurobranchia pilens[4]
Mortality	50 % mortality	Cu conc = 500-1000 ug/l, 96 h, freshwater	Plumatella casmiana[65]
Mortality	1 % effect observed	Cu = 5 ug/l, min conc with effect, marine	Pseudocalanus sp., Calanus sp. and Euphausia sp.[4]
Mortality	50 % mortality	Water = E-3 - E-4 M Cu, P. C-1, toxic level given, microorganisms	Pseudomonas sp.[66]
Mortality	50 % mortality	Cu conc = 10000-13900 ug/l, 14 d, stone fly, freshwater	Pteronarcys californica[53]
Mortality	1 % mortality	Cu = 0.001 mg/l, mortality within 4 d posthatching, 100 mg CaCO3/l	Rana pipiens, larvae[40]
Mortality	1 % mortality	Cu = 0.4 ug/l, hatchning time=4 d, 4 d exposure, 100 mg CaCO3/l	Rana pipiens, larvae[40]
Mortality	1 % mortality	Cu = 0.5 ug/l, hatchning time=4 d, 8 d exposure, 100 mg CaCO3/l	Rana pipiens, larvae[40]
Mortality	100 % mortality	Cu = 1.00 mg/l, mortality within 4 d posthatching, 100 mg CaCO3/l	Rana pipiens, larvae[40]
Mortality	100 % mortality	Cu = 5.000 mg/l, mortality within 4 d posthatching, 100 mg CaCO3/l	Rana pipiens, larvae[40]
Mortality	13 % mortality	Cu = 0.005 mg/l, mortality within 4 d posthatching, 100 mg CaCO3/l	Rana pipiens, larvae[40]
Mortality	34 % mortality	Cu = 0.010 mg/l, mortality within 4 d posthatching, 100 mg CaCO3/l	Rana pipiens, larvae[40]
Mortality	46 % mortality	Cu = 0.050 mg/l, mortality within 4 d posthatching, 100 mg CaCO3/l	Rana pipiens, larvae[40]
Mortality	50 % mortality	Cu = 0.05 ug/l, hatchning time=4 d, 8 d exposure, 100 mg CaCO3/l	Rana pipiens, larvae[40]
Mortality	50 % mortality	Cu = 0.06 ug/l, hatchning time=4 d, 4 d exposure, 100 mg CaCO3/l	Rana pipiens, larvae[40]

Mortality	59 % mortality	Cu = 0.100 mg/l, mortality within 4 d posthatching, 100 mg CaCO3/l	Rana pipiens, larvae[40]
Mortality	82 % mortality	Cu = 0.500 mg/l, mortality within 4 d posthatching, 100 mg CaCO3/l	Rana pipiens, larvae[40]
Mortality	50 % mortality	Cu conc = 210 ug/l, 96 h, less than 0.1 % salinity, marine	Rangia cupeata[67]
Mortality	50 % mortality	Cu conc = 7000 ug/l, 96 h, 2.2 % salinity, marine	Rangia cupeata[67]
Mortality	50 % mortality	Cu conc = 8000 ug/l, 96 h, 0.55 % salinity, marine	Rangia cupeata[67]
Mortality	50 %	1-oxide, oral dose = 470 mg/kg body	Rat[68]
Mortality	50 %	Acetate, oral dose = 710 mg/kg body	Rat[68]
Mortality	50 %	Carbonate, oral dose = 159 mg/kg body	Rat[68]
Mortality	50 %	Chloride, oral dose = 140 mg/kg body	Rat[68]
Mortality	50 %	Nitrate, oral dose = 940 mg/kg body	Rat[68]
Mortality	50 %	Sulfate, oral dose = 300 mg/kg body	Rat[68]
Mortality	50 % mortality	Cu conc = 43-460 ug/l, 24 h, marine	Sagitta hispida[4]
Mortality	1 % of effect observed	Cu = 19 ug/l, hardness = 7.8 ug/l, minimum conc with effect	Salmo gairdneri[29]
Mortality	1 % of effect observed	Cu = 40 ug/l, hardness = 7.8 mg/l, minimum conc with effect	Salmo gairdneri[30]
Mortality	50 %	Cu = 0.13 mg/l, 24h, pH = 7.5	Salmo gairdneri[69]
Mortality	50 %	Cu = 0.14 mg/l, 24h, pH = 6.5	Salmo gairdneri[69]
Mortality	50 %	Cu = 0.28 mg/l, 119d, hard water	Salmo gairdneri[60]
Mortality	1 % of effect observed	Cu = 40 ug/l, hardness = 7.8 mg/l, minimum conc with effect	Salmo salar[30]
Mortality	1 % of effect observed	Cu = 17 ug/l, hardness = 45 mg/l, minimum conc with effect	Salvelinus fontinalis[10]
Mortality	1 % of effect observed	Cu = 27 ug/l, hardness = 38 mg/l, minimum conc with effect	Salvelinus fontinalis[22]
Mortality	1 % of effect observed	Cu = 49 ug/l, hardness = 187 mg/l, minimum conc with effect	Salvelinus fontinalis[22]
Mortality	1 % mortality	Cu = 3.4 ug/l, hatchning time=24 d, 28 d exposure, 100 mg CaCO3/l	Salvelinus fontinalis, larvae[40]
Mortality	1 % mortality	Cu = 4.1 ug/l, hatchning time=24 d, 24 d exposure, 100 mg CaCO3/l	Salvelinus fontinalis, larvae[40]
Mortality	100 % mortality	Cu = 1.000 mg/l, mortality within 4 d posthatching, 100 mg CaCO3/l	Salvelinus fontinalis, larvae[40]
Mortality	15 % mortality	Cu = 0.010 mg/l, mortality within 4 d posthatching, 100 mg CaCO3/l	Salvelinus fontinalis, larvae[40]
Mortality	19 % mortality	Cu = 0.005 mg/l, mortality within 4 d posthatching, 100 mg CaCO3/l	Salvelinus fontinalis, larvae[40]
Mortality	25 % mortality	Cu = 0.025 mg/l, mortality within 4 d posthatching, 100 mg CaCO3/l	Salvelinus fontinalis, larvae[40]
Mortality	39 % mortality	Cu = 0.100 mg/l, mortality within 4 d posthatching, 100 mg CaCO3/l	Salvelinus fontinalis, larvae[40]
Mortality	42 % mortality	Cu = 0.050 mg/l, mortality within 4 d posthatching, 100 mg CaCO3/l	Salvelinus fontinalis, larvae[40]
Mortality	50 % mortality	Cu = 0.11 ug/l, hatchning time=24 d, 28 d exposure, 100 mg CaCO3/l	Salvelinus fontinalis, larvae[40]
Mortality	50 % mortality	Cu = 0.12 ug/l, hatchning time=24 d, 24 d exposure, 100 mg CaCO3/l	Salvelinus fontinalis, larvae[40]

Mortality	6 % mortality	Cu = 0.001 mg/l, mortality within 4 d posthatching, 100 mg CaCO3/l	Salvelinus fontinalis, larvae[40]
Mortality	61 % mortality	Cu = 0.25 mg/l, mortality within 4 d posthatching, 100 mg CaCO3/l	Salvelinus fontinalis, larvae[40]
Mortality	85 % mortality	Cu = 0.500 mg/l, mortality within 4 d posthatching, 100 mg CaCO3/l	Salvelinus fontinalis, larvae[40]
Mortality	0 %	Toxicological limit = 0.01 mg/l, CuSO4	Self Purification[36]
Mortality	0 % mortality	Cu = 0.001 mg/l, mortality within 4 d posthatching, 100 mg CaCO3/l	Southern Gray Treefrog, larvae[40]
Mortality	1 % mortality	Cu = 0.2 ug/l, hatchning time=3 d, 3 d exposure, 100 mg CaCO3/l	Southern Gray Treefrog, larvae[40]
Mortality	1 % mortality	Cu = 0.3 ug/l, hatchning time=3 d, 7 d exposure, 100 mg CaCO3/l	Southern Gray Treefrog, larvae[40]
Mortality	10 % mortality	Cu = 0.005 mg/l, mortality within 4 d posthatching, 100 mg CaCO3/l	Southern Gray Treefrog, larvae[40]
Mortality	100 % mortality	Cu = 1.00 mg/l, mortality within 4 d posthatching, 100 mg CaCO3/l	Southern Gray Treefrog, larvae[40]
Mortality	100 % mortality	Cu = 5.000 mg/l, mortality within 4 d posthatching, 100 mg CaCO3/l	Southern Gray Treefrog, larvae[40]
Mortality	39 % mortality	Cu = 0.010 mg/l, mortality within 4 d posthatching, 100 mg CaCO3/l	Southern Gray Treefrog, larvae[40]
Mortality	49 % mortality	Cu = 0.050 mg/l, mortality within 4 d posthatching, 100 mg CaCO3/l	Southern Gray Treefrog, larvae[40]
Mortality	50 % mortality	Cu = 0.04 ug/l, hatchning time=3 d, 7 d exposure, 100 mg CaCO3/l	Southern Gray Treefrog, larvae[40]
Mortality	50 % mortality	Cu = 0.06 ug/l, hatchning time=3 d, 3 d exposure, 100 mg CaCO3/l	Southern Gray Treefrog, larvae[40]
Mortality	60 % mortality	Cu = 0.100 mg/l, mortality within 4 d posthatching, 100 mg CaCO3/l	Southern Gray Treefrog, larvae[40]
Mortality	76 % mortality	Cu = 0.500 mg/l, mortality within 4 d posthatching, 100 mg CaCO3/l	Southern Gray Treefrog, larvae[40]
Mortality	50 % mortality	Cu conc = 510 ug/l, 2 h, larvae, marine	Spirorbis lamellosa[37]
Mortality	1 % of effect observed	Cu = 13 ug/l, hardness = 13 mg/l, minimum conc with effect	Stizostedion vitreum[22]
Mortality	50 % mortality	Cu conc = 1000 ug/l, 96 h, marine	Tellina tenuis[18]
Mortality	50 % mortality	Cu conc = 210 ug/l, 34 mg/l hardness, 48 h, freshwater	Tubifex tubifex[70]
Mortality	50 % mortality	Cu conc = 6.4 ug/l, distilled water, 48 h, freshwater	Tubifex tubifex[70]
Mortality	50 % mortality	Cu conc = 890 ug/l, 261 mg/l hardness, 48 h, freshwater	Tubifex tubifex[70]
Mortality	50 % mortality	Cu conc = 570 ug/l, larvae, 2 h, marine	Watersipora cucullata[37]
Nitrification	45 % inhibation	Soil = 5 umoles/g of soil, 10 d incubation, 3 soils, lab	Soil organisms[71]
Nitrogen fixation	1509 ethylene produced in nmoles/(plant*30 min)	Cu = 1.0 ppm in soil, control = 1640	Soybean with nodule[72]
Nitrogen fixation	24 no of nodules/plant	Cu = 10.0	Soybean with nodule[72]
Nitrogen fixation	27 no of nodules/plant	Cu = 5.0 ppm in soil, control = 46	Soybean with nodule[72]
Nitrogen fixation	28 no of nodules/plant	Cu = 1.0 ppm in soil, control = 46	Soybean with nodule[72]
Nitrogen fixation	47.85 weight of one nodule in mg	Cu = 10.0	Soybean with nodule[72]

Nitrogen fixation	54.06 weight of one nodule in mg	Cu = 5.0 ppm in soil, control = 62.30 mg	Soybean with nodule[72]
Nitrogen fixation	54.09 weight of one nodule in mg	Cu = 1.0 ppm in soil, control = 62.30 mg	Soybean with nodule[72]
Nitrogen fixation	879 ethylene produced in nmoles/(plant*30 min)	Cu = 10.0	Soybean with nodule[72]
Nitrogen fixation	991 ethylene produced in nmoles/(plant*30 min)	Cu = 5.0 ppm in soil, control = 1640	Soybean with nodule[72]
Numbers of colony forming units	5.25 (log-unit)	Control = 4.62, Cu = 2 ppm, after 14 d, 299 K, brackish water, Clear	Bacterial colonies[1]
Numbers of colony forming units	6.50 (log-unit)	Control = 5.25, Cu = 2 ppm, after 14 d, 299 K, fresh water, Lake	Bacterial colonies[1]
Nutrient uptake	0.01 ppm Hg	Cu = 2.0 ppm, control = 0.02, water = 0.01 ppm Hg, 298 K, fresh wat	Bacterial cells[73]
Nutrient uptake	0.01 ppm I	Cu = 2.0 ppm, control = 0.02, water = 0.1 ppm I, 298 K, Ash River,	Bacterial cells[73]
Nutrient uptake	0.02 ppm Hg	Cu = 2.0 ppm, control = 0.02, water = 0.03 ppm Hg, 298 K, Ash River,	Bacterial cells[73]
Nutrient uptake	0.02 ppm Hg	Cu = 2.0 ppm, control = 0.08, water = 0.008 ppm Hg, 298 K, brackish,	Bacterial cells[73]
Nutrient uptake	0.05 ppm Mn	Cu = 2.0 ppm, control = 0.07, water = 0.07 ppm Mn, 298 K, Ash River,	Bacterial cells[73]
Nutrient uptake	0.06 ppm Co	Cu = 2.0 ppm, control = 0.03, water = 0.06 ppm Co, 298 K, fresh wat	Bacterial cells[73]
Nutrient uptake	0.06 ppm Co	Cu = 2.0 ppm, control = 0.08, water = 0.1 ppm Co, 298 K, Ash River,	Bacterial cells[73]
Nutrient uptake	0.08 ppm I	Cu = 2.0 ppm, control = 0.12, water = 0.05 ppm I, 298 K, fresh wat	Bacterial cells[73]
Nutrient uptake	0.1 ppm Br	Cu = 2.0 ppm, control = 0.1, water = 0.1 ppm Br, 298 K, Ash River,	Bacterial cells[73]
Nutrient uptake	0.2 ppm Co	Cu = 2.0 ppm, control = 0.04, water = 0.04 ppm Co, 298 K, brackish,	Bacterial cells[73]
Nutrient uptake	0.23 ppm Mn	Cu = 2.0 ppm, control = 0.07, water = 0.03 ppm Mn, 298 K, fresh wat	Bacterial cells[73]
Nutrient uptake	0.3 ppm Cr	Cu = 2.0 ppm, control = 0.4, water = 0.2 ppm Cr, 298 K, Ash River,	Bacterial cells[73]
Nutrient uptake	0.3 ppm Cu	Cu = 2.0 ppm, control = 0.7, water = 0.4 ppm Cu, 298 K, Ash River,	Bacterial cells[73]
Nutrient uptake	0.3 ppm Zn	Cu = 2.0 ppm, control = 0.6, water = 0.4 ppm Zn, 298 K, Ash River,	Bacterial cells[73]
Nutrient uptake	0.4 ppm Br	Cu = 2.0 ppm, control = 0.1, water = 0.2 ppm Br, 298 K, brackish,	Bacterial cells[73]
Nutrient uptake	0.5 ppm Br	Cu = 2.0 ppm, control = 1.2, water = 9.0 ppm Br, 298 K, fresh wat	Bacterial cells[73]
Nutrient uptake	0.6 ppm I	Cu = 2.0 ppm, control = 0.1, water = 0.1 ppm I, 298 K, brackish, F009	Bacterial cells[73]
Nutrient uptake	0.7 ppm Cr	Cu = 2.0 ppm, control = 0.1, water = 0.4 ppm Cr, 298 K, brackish,	Bacterial cells[73]
Nutrient uptake	0.75 ppm Cr	Cu = 2.0 ppm, control = 0.65, water = 0.25 ppm Cr, 298 K, fresh wat	Bacterial cells[73]

Nutrient uptake	1.9 ppm Ti	Cu = 2.0 ppm, control = 1.3, water = 0.9 ppm Ti, 298 K, Ash River,	Bacterial cells[73]
Nutrient uptake	12.3 ppm K	Cu = 2.0 ppm, control = 41.0, water = 80.0 ppm K, 298 K, fresh wat	Bacterial cells[73]
Nutrient uptake	13.3 ppm Mg	Cu = 2.0 ppm, control = 6.3, water = 6.2 ppm Mg, 298 K, brackish,	Bacterial cells[73]
Nutrient uptake	13.4 ppm Cl	Cu = 2.0 ppm, control = 2.3, water = 14.1 ppm Cl, 298 K, brackish,	Bacterial cells[73]
Nutrient uptake	162.0 ppm Fe	Cu = 2.0 ppm, control = 48.0, water = 16.0 ppm Fe, 298 K, brackish,	Bacterial cells[73]
Nutrient uptake	18.8 ppm Ca	Cu = 2.0 ppm, control = 16.1, water = 21.8 ppm Ca, 298 K, brackish,	Bacterial cells[73]
Nutrient uptake	2.2 ppm Mn	Cu = 2.0 ppm, control = 0.3, water = 0.09 ppm Mn, 298 K, brackish,	Bacterial cells[73]
Nutrient uptake	2.3 ppm Cl	Cu = 2.0 ppm, control = 10.7, water = 3.8 ppm Cl, 298 K, Ash River,	Bacterial cells[73]
Nutrient uptake	2.4 ppm Na	Cu = 2.0 ppm, control = 19.9, water = 7.7 ppm Na, 298 K, Ash River,	Bacterial cells[73]
Nutrient uptake	2.7 ppm Ti	Cu = 2.0 ppm, control = 2.8, water = 13.8 ppm Ti, 298 K, fresh wat	Bacterial cells[73]
Nutrient uptake	21.0 ppm K	Cu = 2.0 ppm, control = 13.8, water = 9.7 ppm K, 298 K, brackish,	Bacterial cells[73]
Nutrient uptake	22.4 ppm Mg	Cu = 2.0 ppm, control = 26.9, water = 236.0 ppm Mg, 298 K, fresh wat	Bacterial cells[73]
Nutrient uptake	23.3 ppm Fe	Cu = 2.0 ppm, control = 21.9, water = 16.9 ppm Fe, 298 K, Ash River,	Bacterial cells[73]
Nutrient uptake	29.0 ppm Ti	Cu = 2.0 ppm, control = 8.3, water = 1.1 ppm Ti, 298 K, brackish,	Bacterial cells[73]
Nutrient uptake	3.2 ppm Cu	Cu = 2.0 ppm, control = 1.2, water = 5.4 ppm Cu, 298 K, fresh wat	Bacterial cells[73]
Nutrient uptake	3.8 ppm Mg	Cu = 2.0 ppm, control = 4.9, water = 4.1 ppm Mg, 298 K, Ash River,	Bacterial cells[73]
Nutrient uptake	32.5 ppm Al-Si	Cu = 2.0 ppm, control = 16.8, water = 13.0 ppm Al, 298 K, Ash River,	Bacterial cells[73]
Nutrient uptake	33.0 ppm Fe	Cu = 2.0 ppm, control = 29.0, water = 23.0 ppm Fe, 298 K, fresh wat	Bacterial cells[73]
Nutrient uptake	4.0 ppm Cu	Cu = 2.0 ppm, control = 1.4, water = 1.2 ppm Cu, 298 K, brackish,	Bacterial cells[73]
Nutrient uptake	4.0 ppm Zn	Cu = 2.0 ppm, control = 1.4, water = 1.17 ppm Zn, 298 K, fresh wat	Bacterial cells[73]
Nutrient uptake	40.0 ppm Na	Cu = 2.0 ppm, control = 13.6, water = 19.0 ppm Na, 298 K, brackish,	Bacterial cells[73]
Nutrient uptake	43.0 ppm Al-Si	Cu = 2.0 ppm, control = 18.6, water = 13.0 ppm Al, 298 K, brackish,	Bacterial cells[73]
Nutrient uptake	44.1 ppm Ca	Cu = 2.0 ppm, control = 43.2, water = 41.0 ppm Ca, 298 K, fresh wat	Bacterial cells[73]
Nutrient uptake	5.0 ppm Ca	Cu = 2.0 ppm, control = 4.2, water = 9.2 ppm Ca, 298 K, Ash River,	Bacterial cells[73]
Nutrient uptake	6.9 ppm K	Cu = 2.0 ppm, control = 4.6, water = 6.1 ppm K, 298 K, Ash River,	Bacterial cells[73]
Nutrient uptake	8.1 ppm Al-Si	Cu = 2.0 ppm, control = 6.8, water = 3.0 ppm Al, 298 K, fresh wat	Bacterial cells[73]
Nutrient uptake	84.8 ppm Na	Cu = 2.0 ppm, control = 110.0, water = 1364.0 ppm Na, 298 K, fresh	Bacterial cells[73]

Nutrient uptake	9.5 ppm Zn	Cu = 2.0 ppm, control = 0.8, water = 0.5 ppm Zn, 298 K, brackish,	Bacterial cells[73]
Nutrient uptake	98.0 ppm Cl	Cu = 2.0 ppm, control = 123.o, water = 1434.0 ppm Cl, 298 K, fresh	Bacterial cells[73]
Photosynthesis	105 % of control	Cu = 0.001 mg/l, 3 h, Golden Horn	Aufwuchs communities[74]
Photosynthesis	38 % of control	Cu = 1.0 mg/l, 3 h, Golden Horn	Aufwuchs communities[74]
Photosynthesis	8 % of control	Cu = 10.0 mg/l, 3 h, Golden Horn	Aufwuchs communities[74]
Photosynthesis	82 % of control	Cu = 0.1 mg/l, 3 h, Golden Horn	Aufwuchs communities[74]
Photosynthesis	85 % of control	Cu = 0.01 mg/l, 3 h, Golden Horn	Aufwuchs communities[74]
Reproduction	100 % reduction	Cu = 50 ug/l, marine	Artemia salina[16]
Reproduction	0 % effect observed	Cu = 10 ug/l, larvae, max conc without effect, marine	Crassostrea virginica[24]
Reproduction	1 % effect observed	Cu = 32.8 ug/l, larvae, min conc with effect, marine	Crassostrea virginica[24]
Reproduction	0 % effect observed	Cu = 50 ug/l, max conc without effect, marine	Ctenodrilus serratus[47]
Reproduction	1 % effect observed	Cu = 100 ug/l, min conc with effect, marine	Ctenodrilus serratus[47]
Reproduction	16 % reduction	Cu = 22 ug/l, freshwater	Daphnia magna[51]
Reproduction	0 % effect observed	Cu = 50 ug/l, max conc without effect, marine	Ophryotrocha diadema[47]
Reproduction	1 % effect observed	Cu = 250 ug/l, min conc with effect, marine	Ophryotrocha diadema[47]
Reproduction	100 % reduction	Cu = 100 ug/l, marine	Ophryotrocha labronica[26]
Respiration	100 % of control	Cu = 0.001 mg/l, 3 h, Golden Horn	Aufwuchs communities[74]
Respiration	50 % of control	Cu = 10.0 mg/l, 3 h, Golden Horn	Aufwuchs communities[74]
Respiration	80 % of control	Cu = 1.0 mg/l, 3 h, Golden Horn	Aufwuchs communities[74]
Respiration	90 % of control	Cu = 0.1 mg/l, 3 h, Golden Horn	Aufwuchs communities[74]
Respiration	98 % of control	Cu = 0.01 mg/l, 3 h, Golden Horn	Aufwuchs communities[74]
Respiration rate	12.5 ul O2/(hour*g wet weight)	Cu = 0.25 mg/l, n=11, rg = +- 2.6 ul O2, S=2.5 %, 293 K, cont=30.5	Nassarius obsoletus[75]
Spawning	1 % decrease	Cu = 162 ug/l, hardness = 45 mg/l, minimum conc with effect	Lepomis macrochirus[5]
Spawning	1 % decrease	Cu = 120 ug/l, high organic conc, fluctaing hardness, minimum conc with effect	Pimephales promelas[9]
Spawning	1 % decrease	Cu = 18 ug/l, hardness = 31 mg/l, minimum conc with effect	Pimephales promelas[6]
Spawning	1 % decrease	Cu = 33 ug/l, hardness = 198 mg/l, minimum conc with effect	Pimephales promelas[7]
Spawning	1 % decrease	Cu = 61 ug/l, hardness = 200 mg/l, minimum conc with effect	Pimephales promelas[8]
Spawning	1 % decrease	Cu = > 33 ug/l, hardness = 45 mg/l, minimum conc with effect	Salvelinus fontinalis[10]
Spawnings per female	0	Cu = 0.18 mg/l, control = 174, no spawnings at Cu = 0.32 and 0.57 mg/	Pimephales promelas[12]
Spawnings per female	0.8	Cu = 0.12 mg/l, control = 6.6	Pimephales promelas[12]
Spawnings per female	6.5	Cu = 0.065 mg/l, control = 6.6	Pimephales promelas[12]
Spawnings per female	8.9	Cu = 0.033 mg/l, control = 6.6	Pimephales promelas[12]

Survival time	1.092 min	Tot Cu = 1.27, dis Cu = 1.24 + 110 micro m Na4P2O7, 291 K, Lake	Daphnia magna[76]
Survival time	120 min	Tot Cu = 1.27, dis Cu = 0.98 + 48 micro m NaHCO3, 291 K, Lake	Daphnia magna[76]
Survival time	168 min	Tot Cu = 0.8, dis Cu = 0.61 micro m, Lake Superoir, 291 K	Daphnia magna[76]
Survival time	1740 min	Tot Cu = 0.2, dis Cu = 0.2 micro m, Lake Superoir, 291 K	Daphnia magna[76]
Survival time	180 min	Tot Cu = 1.27, dis Cu = 1.02 micro m, Lake Superoir, 291 K	Daphnia magna[76]
Survival time	246 min	Tot Cu = 0.8, dis Cu = 0.6 + 48 micro m NaHCO3, 291 K, Lake Superoir	Daphnia magna[76]
Survival time	2880 min	Control, dis Cu = 0.02 micro m + 110 micro m Na4P2O7, Lake Superoir,	Daphnia magna[76]
Survival time	2880 min	Control, dis Cu = 0.02 micro m + 210 micro m Na2HPO4, Lake Superoir,	Daphnia magna[76]
Survival time	2880 min	Control, dissolved Cu = 0.02 micro m + 48 micro m NaHCO3, more than	Daphnia magna[76]
Survival time	2880 min	Control, dissolved Cu = 0.02 micro m, pH = 7.85-8.05, 291 K, Lake	Daphnia magna[76]
Survival time	2880 min	Tot Cu = 0.8, dis Cu = 0.77 + 110 micro m Na4P2O7, 291 K, Lake	Daphnia magna[76]
Survival time	2880 min	Tot Cu = 2.01, dis Cu = 1.95 + 110 micro m Na4P2O7, 291 K, Lake	Daphnia magna[76]
Survival time	316 min	Tot Cu = 2.01, dis Cu = 1.07 + 210 micro m NaHPO4, 291 K, Lake	Daphnia magna[76]
Survival time	372 min	Tot Cu = 0.51, dis Cu = 0.41 micro m, Lake Superoir, 291 K	Daphnia magna[76]
Survival time	480 min	Tot Cu = 0.32, dis Cu = 0.33 micro m, Lake Superoir, 291 K	Daphnia magna[76]
Survival time	486 min	Tot Cu = 1.27, dis Cu = 0.76 + 210 micro m NaHPO4, 291 K, Lake	Daphnia magna[76]
Survival time	618 min	Tot Cu = 0.8, dis Cu = 0.54 + 210 micro m NaHPO4, 291 K, Lake	Daphnia magna[76]
Survival time	84 min	Tot Cu = 2.01, dis Cu = 1.45 + 48 micro m NaHCO3, 291 K, Lake	Daphnia magna[76]
Survival time	84 min	Tot Cu = 2.01, dis Cu = 1.54 micro m, Lake Superoir, 291 K	Daphnia magna[76]
Survival time	18.4 days	Cu = 0.49 mg/l, 285 K, 10 fish	Noemacheilus barbatulus[60]
Survival time	3.7 days	Cu = 0.76 mg/l, 285 K, 10 fish	Noemacheilus barbatulus[60]
Survival time	35.1 days	Cu = 0.29 mg/l, 285 K, 10 fish	Noemacheilus barbatulus[60]
Survival time	4.5 days	Cu = 0.71 mg/l, 285 K, 10 fish	Noemacheilus barbatulus[60]
Survival time	64 days	0.12 mg/l, 285 K, 10 fish, more than value	Noemacheilus barbatulus[60]
Survival time	64 days	Cu = 0.002 mg/l, 285 K, 10 fish, more than value	Noemacheilus barbatulus[60]
Survival time	8.3 days	Cu = 0.39 mg/l, 285 K, 10 fish	Noemacheilus barbatulus[60]
Survival time	1000 min	Cu = 0.2 mg/l, pH = 6.5	Salmo gairdneri[69]
Survival time	260 min	Cu = 2.0 mg/l, pH = 6.5	Salmo gairdneri[69]
Survival time	330 min	Cu = 2.0 mg/l, pH = 7.5	Salmo gairdneri[69]
Survival time	340 min	Cu = 1.2 mg/l, pH = 6.5	Salmo gairdneri[69]

Survival time	400 min	Cu = 1.2 mg/l, pH = 7.5	Salmo gairdneri[69]
Survival time	500 min	Cu = 0.8 mg/l, pH = 6.5	Salmo gairdneri[69]
Survival time	560 min	Cu = 0.8 mg/l, pH = 7.5	Salmo gairdneri[69]
Survival time	620 min	Cu = 0.5 mg/l, pH = 6.5	Salmo gairdneri[69]
Survival time	640 min	Cu = 0.35 mg/l, pH = 6.5	Salmo gairdneri[69]
Survival time	680 min	Cu = 0.5 mg/l, pH = 7.5	Salmo gairdneri[69]
Survival time	740 min	Cu = 0.35 mg/l, pH = 7.5	Salmo gairdneri[69]
Survival time	900 min	Cu = 0.2 mg/l, pH = 7.5	Salmo gairdneri[69]
Yield	1.2 g wet weight/pot	Soil pH = 5.7, Cu in soil = 500 ppm	Mustard[77]
Yield	23.6 g wet weight/pot	Soil pH = 6.4, Cu in soil = 200 ppm	Mustard[77]
Yield	24.4 g wet weight/pot	Soil pH = 6.4, Cu in soil = 100 ppm	Mustard[77]
Yield	26.0 g wet weight/pot	Soil pH = 5.7, Cu in soil = 100 ppm	Mustard[77]
Yield	5.6 g wet weight/pot	Soil pH = 5.7, Cu in soil = 200 ppm	Mustard[77]
Yield	6.4 g wet weight/pot	Soil pH = 6.4, Cu in soil = 500 ppm	Mustard[77]
Yield	16.4 g wet weight/pot	Soil pH = 6.4, Cu in soil = 500 ppm	Oat[77]
Yield	17.1 g wet weight/pot	Soil pH = 5.7, Cu in soil = 200 ppm	Oat[77]
Yield	19.1 g wet weight/pot	Soil pH = 6.4, Cu in soil = 100 ppm	Oat[77]
Yield	19.4 g wet weight/pot	Soil pH = 5.7, Cu in soil = 100 ppm	Oat[77]
Yield	19.7 g wet weight/pot	Soil pH = 6.4, Cu in soil = 200 ppm	Oat[77]
Yield	7.3 g wet weight/pot	Soil pH = 5.7, Cu in soil = 500 ppm	Oat[77]
Yield	1.18 t/ha at 85 % dry matter	Added CuSO4 = 11 kg/ha, loamy sand, control = 0.70 t/ha	Spring barley[78]
Yield	1.49 t/ha at 85 % dry matter	Added CuSO4 = 45 kg/ha, loamy sand, control = 0.70 t/ha	Spring barley[78]
Yield	1.55 t/ha at 85 % dry matter	Added CuSO4 = 31-34 kg/ha, loamy sand, control = 0.70 t/ha	Spring barley[78]
Yield	1.58 t/ha at 85 % dry matter	Added CuSO4 = 22 kg/ha, loamy sand, control = 0.70 t/ha	Spring barley[78]
Yield	1.76 t/ha at 85 % dry matter	Add CuSO4 = 125-134 kg/ha, loamy sand, control = 0.70 t/ha	Spring barley[78]
Yield	1.85 t/ha at 85 % dry matter	Added CuSO4 = 63-67 kg/ha, loamy sand, control = 0.70 t/ha	Spring barley[78]
Yield	3.48 t/ha at 85 % dry matter	Added CuSO4 = 63-67 kg/ha, light peat, control = 2.55 t/ha	Spring barley[78]
Yield	2.68 t/ha at 85 % dry matter	Added CuSO4 = 31-34 kg/ha, peaty loam, control = 1.96 t/ha	Spring wheat[78]
Yield	2.75 t/ha at 85 % dry matter	Add CuSO4 = 125-134 kg/ha, peaty loam, control = 1.96 t/ha	Spring wheat[78]
Yield	2.86 t/ha at 85 % dry matter	Added CuSO4 = 63-67 kg/ha, peaty loam, control = 1.96 t/ha,	Spring wheat[78]

Source: 1.Singleton, F.L. and R.K. Guthrie. 1977. Water Res., 11(8), pp 639-642.
2.Morishima, H. and H.I Oka. 1977. Agro-Ecosyst., 3(2), pp 131-145.
3.Saward, D., A. Sterling, and G. Topping. 1975. Mar. Biol. 29, 351-361.
4.Reeve, M.R., G.D. Grice, V.R. Gibson, Walter, Darcey and Ikeda. 1976. In: A.P.M. Lockwood (ed.): Effects of Pollutants on Aquatic Organisms. Cambridge Univ. Press. Cambridge. England.
5.Benoit, D.A. 1975. Trans. Am. Fish. Soc. 104, 353.
6.Mount, D.I. and C.E. stephan. 1969. J. Fish. Res. bd Canada 26, 2449-2457.
7.Mount, D.I. 1968. Water Res. 2, 215-224.
8.Pickering, Q., W. Brungs and M. Gast. 1977. Water Res. 11, 1079-1084.

9.Brungs, W.A., J.R. Geckler and M. Gast. 1976. Water Res. 10, 37-43.
10.McKim, J.M. and D.A. Benoit. 1971. J. Fish. Res. bd Canada 28, 655-662.
11.DAgosino, A. and C. Finney. 1974. In: F.J. Vernberg and W.B. Vernberg (eds.): Pollution and Physioloy of Marine. Organisms. Symposium XIII. Academic Press. NY, 445-463.
12.Brungs, W.A., J.R. Geckler and M. Gast. 1976. Water Res. 10(1), pp 37-44.
13.Peterson, P.J. 1962. N. Z. J. Sci. Technol. 5, pp 141-161.
14.Brown, B.E. 1976. Water Res., 10(6), pp 555-559.
15.Bougis, P. 1965. C.r. hebd. Seanc. Acad. Sci. Paris 260, pp 2929-2931.
16.Saliba, L.J. and M. Ahsanullah. 1973. J. Exp. Biol. 26, 113-124.
17.Mudd, J.B., T.T. McManus, A. Ongun and T.E. Cullogh. 1972. J. Environ. Quality 1, 288-291.
18.Stirling, E.A. 1975. Mar. Pollut. Bull. 6, 122-124.
19.Kempe, S., B. Mycke and Seeger. 1977. Flussfrachten und Erosionsraten in Mitteuropa. Springer. Berli.
20.Karbe, L. 1972. Commentat. Biol. Soc. Sci. Fenn. 55, 1-19.
21.Leblova, S. and J. Mares. 1975. Photosunthetica 9, 177-184.
22.Sauter, S., K.S. Buxton and S.R. Petrocelli. 1976. Effects of Exposure to Heavy Metals in Selected Freshwater Fish. US EPA. EPA 600/3-76-088, Duluth. Minn.
23.Lett, P.F., G.J. Farmer and F.W.H. Beamish. 1973. J. Fish. Res. bd Canada 33, 1335-1342.
24.Calabrese, A., J.R. MacInnes, D.A. Nelson and J.E. Miller. 1977. Mar. Biol. 41, 179-184.
25.Hazel, G.R. and S.J. Meith. 1970. Calif Fish Game 56, 121-124.
26.Reich, D.J., J.M. Martin, F.M. Piltz and J.Q. Word. 1976. Water Res. 10, 299-302.
27.Lucero, H.A., C.S. Andreo and R.H. Vallejos. 1976. Plant Sci. Lett. 6, 309-313.
28.Gardner, G.R. and LaRoche. 1973. J. Fish. Res. bd Canada 30, 363-368.
29.Goettl, J.R. Jr., P.H. Davies and J.R. Sinley. 1976. Colorado Fisheries Research Review 1972-1975. Sate Publication Code DOW-R-R-F72-75, 68-74.
30.Grande, M. 1967. In: Advances in Water Pollution Research, Proc 3rd International Conf. Munich. Germany. Sep 1966. Vol 1. Water Poll. Cont Fed. Wash DC, 97-111.
31.Warnick, S.L. and H.L. Bell. 1969. J. Water Pollut. Control Fed. 41, 280-284.
32.Rehwoldt, R., G. Bida and B. Nerrie. 1971. Bull. Environ. Contam. Toxicol. 6, 445.
33.Brown, B. and M. Ahsanullah. 1971. Mar. Pollut. Bull. 2, 182-187.
34.Brown, B.E. 1976. Water Res. 10, 555-559.
35.Zhakharyan, S.V. 1970. Symp. Biol. Hung. 11, 391-395.
36.Jung. K.D. 1973. Wirkungskonzentration schaedigender Stoffe im Wasser. Gelenkirchen. Hygiene-Institudes Ruhrgebiets.
37.Wisely, B. and R.A.P. Blick. 1967. Aust. J. Mar. Freshwater Res. 18, 63-72.
38.Betzer, S.B. and P.P. Yevich. 1975. Biol. Bull. 148, 257-264.
39.Arthur, J.W. and E.N. Leonard. 1970. J. Fish. Res. bd Canada 27, 1277-1283.
40.Sauter, S., K.S. Buxton, K.J. Macek and S.R. Petrocelli. 1976. Effects of Exposure to Heavy Metals of Selected Freshwater Fish. Toxicity of Cu, Cd, Cr and Pb to eggs and Fry of Seven Fish species. US EPA.
41.Conner, P.M. 1972. Mar. Pollut. Bull. 3, 190-192.
42.Portmann, J.E. and K.W. Wilson. 1971. Shellfish Information Leaflet no 22.
43.Okazaki, R.K. 1976. Bull. Environ. Contam. Toxicol. 16, 658-664.
44.Mandelli, E.F. 1975. Water Res. 9, 287-295.
45.Shuster, C.N. Jr., and B.H. Pringle. 1969. In: 1968 Proc of the National Shelfish Assocation. 91-103
46.Liebman, H. 1958. Handbuch der Friscwasser - und Abwasserbiologie. Band II.Munchen. Oldenbourg.
47.Reich, D.J. and R.S. Carr. 1978. Mar. Pollut. Bull. 9, 24-27.
48.Baudouin, M.F. and P. Scoppa. 1974. Bull. Environ. Contam. Toxicol. 12, 745-751.
49.Winner, R.W. and M.P. Farrell. 1976. J. Fish. Res. bd Canada 33, 1685-1691.
50.Biesinger, K.E. and G.M. Christensen. 1972. J. Fish. Res. bd. Can., 29(12), 1691-1700.
51.Garrels, R.M. and F.T. Mackenzie. 1971. Evolution of Sedimentary Rocks. Norton and Comp. New York
52.Biesinger, K.E. and G.M. Christensen. 1975. J. Fish. Res. bd Canada 29, 1671-1700.
53.Nehring, R.B. 1976. Bull. Environ. Contam. Toxicol. 15, 147-154.
54.Lewis, A.G., P. Whitfeld and A. Ramnarine. 1973. Mar. Biol. 17, 215-221.
55.McLeese, D.W. 1974. J. Fish. Res. bd Canada 31, 1949-1952.
56.Fujiya, M. 1960. Bull. Jap. Soc. Sci. Fish. 26, 462-468.
57.Scott, D.M. and C.W. Major. 1972. Biol. Bull. 143, 679-688.
58.Delhaye, W. and D. Cornet. 1975. Comp. Biochem. Physiol. 50A, 511-518.
59.Jones, L.H., N.V. Jones and A.J. Radlett. 1976. Estuarine Coastal Mar. Sci. 4, 107-111.
60.Solbe, J.F., G. Del and V.A. Cooper. 1976. Water Res. 10(6), pp 523-527.
61.Hubschman, J.H. 1967. Crustaceana 12, 33-42.
62.Buikema, A.L. Jr., J. Jr. Cairns, and G.W. Sullivan. 1974. Water Res. Bull. 10, 648-661.
63.Sallivan, G.W., A.L. Buikema and J. Cairns, Jr. 1973. Va. J. Sci. 24(3), pp 120.
64.McLusky, D.S. and C.N.K. Phillips. 1975. Eustarine Coastal Mar. Sci. 3, 103-108.
65.Bushnell, J.H. 1974. In: C.W. Hart and S.L.H. Fuller (eds.): Pollution Ecology of Freshwater Invertebrates. Academic Press. NY. Chapter 6.
66.Hoffman-Kakol, I. and H. Kwarta. 1972. Zesz. nauk. Akad. Rolnicz. Szczecin. No 38, 105-116.
67.Olson, K.R. and R.C. Harrel. 1973. Contrib. Mar. Sci. 17, 9.
68.Toxic Substances List 1973, U.S. Department of Health, Education and Welfare, Public Health ServiceNational Institute for Occupational Safety and Health, Rockville Maryland, USA, June 1973.
69.Shaw, T.L. and V.M. Brown. 1974. Water Res., 8(6), pp 377-382.
70.Brkovic-Popovic, I. and M. Popovic. 1970. Environ. Pollut. 13, 93-98.
71.Liang, C.N. and M.A. Tabatabai. 1978. Environ. Pollut. 12, 141-147.
72.Letchworth, M.B. and V. Blum. 1977. Environ. Pollut. 14, 303-312.
73.Guthrie, R.K., F.L. Singleton and D.S. Ckerry. 1977. Water Res., 11(8), pp 643-646.
74.Orhon, D. 1975. Water Res. 9(5/6), pp 467-472.
75.MacInnes, J.R., and F.P. Thurberg. 1973. Mar. Pollut. Bull. 4, 185-186.
76.Andrew, R.W., K.E. Biesinger and G.E. Glass. 1977. Water Res., 11(3), pp 309-315.
77.Patterson, J.B.E. 1971. Tech. Bull., 21, pp 193-207.
78.Caldwell, T.H. 1971. In: Tech. Bull. 21, pp 62-87. Trace Elements in Soils and Crops.

4-103 Effect of Cu and Hg

Parameter	Effect Value	Concentration and Condition	Species
Mortality	0 % mortality	Water = 0.01 mg Cu and Hg/l, control (only Cu at 0.01 mg/l)= 1 % mort.	Carassius auratus, larvae[1]
Mortality	0 % mortality	Water = 0.025 mg Cu and Hg/l, control (only Cu at 0.025 mg/l)= 2 % mort.	Carassius auratus, larvae[1]
Mortality	100 % mortality	Water = 1.00 mg Cu and Hg/l, control (only Cu at 1.00 mg/l)= 57 % mort.	Carassius auratus, larvae[1]
Mortality	15 % mortality	Water = 0.050 mg Cu and Hg/l, control (only Cu at 0.050 mg/l)= 5 % mort.	Carassius auratus, larvae[1]
Mortality	26 % mortality	Water = 0.075 mg Cu and Hg/l, control (only Cu at 0.075 mg/l)= 17 % mort.	Carassius auratus, larvae[1]
Mortality	48 % mortality	Water = 0.10 mg Cu and Hg/l, control (only Cu at 0.100 mg/l)= 24 % mort.	Carassius auratus, larvae[1]
Mortality	50 % mortality	Water = 0.18 mg Cu and Hg/l, control (only Cu at 0.59 mg/l)= 50 % mort.	Carassius auratus, larvae[1]
Mortality	58 % mortality	Water = 0.25 mg Cu and Hg/l, control (only Cu at 0.25 mg/l)= 38 % mort.	Carassius auratus, larvae[1]
Mortality	70 % mortality	Water = 0.500 mg Cu and Hg/l, control (only Cu at 0.500 mg/l)= 47 % mort.	Carassius auratus, larvae[1]
Mortality	74 % mortality	Water = 0.750 mg Cu and Hg/l, control (only Cu at 0.750 mg/l)= 53 % mort.	Carassius auratus, larvae[1]
Mortality	0 % mortality	Water = 0.005 mg Cu and Hg/l, control (only Cu at 0.005 mg/l)= 4 % mort.	Ictalurus lacustris, larvae[1]
Mortality	1 % mortality	Water = 0.001 mg Cu and Hg/l, control (only Cu at 0.001 mg/l)= 0 % mort.	Ictalurus lacustris, larvae[1]
Mortality	100 % mortality	Water = 0.075 mg Cu and Hg/l, control (only Cu at 0.075 mg/l)= 35 % mort.	Ictalurus lacustris, larvae[1]
Mortality	100 % mortality	Water = 0.10 mg Cu and Hg/l, control (only Cu at 0.100 mg/l)= 36 % mort.	Ictalurus lacustris, larvae[1]
Mortality	100 % mortality	Water = 0.25 mg Cu and Hg/l, control (only Cu at 0.25 mg/l)= 39 % mort.	Ictalurus lacustris, larvae[1]
Mortality	100 % mortality	Water = 0.500 mg Cu and Hg/l, control (only Cu at 0.500 mg/l)= 53 % mort.	Ictalurus lacustris, larvae[1]
Mortality	100 % mortality	Water = 0.750 mg Cu and Hg/l, control (only Cu at 0.750 mg/l)= 55 % mort.	Ictalurus lacustris, larvae[1]
Mortality	100 % mortality	Water = 1.00 mg Cu and Hg/l, control (only Cu at 1.00 mg/l)= 57 % mort.	Ictalurus lacustris, larvae[1]
Mortality	16 % mortality	Water = 0.01 mg Cu and Hg/l, control (only Cu at 0.01 mg/l)= 5 % mort.	Ictalurus lacustris, larvae[1]
Mortality	50 % mortality	Water = 0.023 mg Cu and Hg/l, control (only Cu at 0.42 mg/l)= 50 % mort.	Ictalurus lacustris, larvae[1]
Mortality	53 % mortality	Water = 0.025 mg Cu and Hg/l, control (only Cu at 0.025 mg/l)= 17 % mort.	Ictalurus lacustris, larvae[1]
Mortality	80 % mortality	Water = 0.050 mg Cu and Hg/l, control (only Cu at 0.050 mg/l)= 32 % mort.	Ictalurus lacustris, larvae[1]
Mortality	0 % mortality	Water = 0.025 mg Cu and Hg/l, control (only Cu at 0.025 mg/l)= 9 % mort.	Micropterus coosae, larvae[1]
Mortality	1 % mortality	Water = 0.01 mg Cu and Hg/l, control (only Cu at 0.01 mg/l)= 7 % mort.	Micropterus coosae, larvae[1]
Mortality	100 % mortality	Water = 2.50 mg Cu and Hg/l, control (only Cu at 2.50 mg/l)= 100 % mort.	Micropterus coosae, larvae[1]

Mortality	18 % mortality	Water = 0.10 mg Cu and Hg/l, control (only Cu at 0.100 mg/l)= 16 % mort.	Micropterus coosae, larvae[1]
Mortality	27 % mortality	Water = 0.25 mg Cu and Hg/l, control (only Cu at 0.25 mg/l)= 30 % mort.	Micropterus coosae, larvae[1]
Mortality	49 % mortality	Water = 0.500 mg Cu and Hg/l, control (only Cu at 0.500 mg/l)= 43 % mort.	Micropterus coosae, larvae[1]
Mortality	50 % mortality	Water = 0.42 mg Cu and Hg/l, control (only Cu at 0.87 mg/l)= 50 % mort.	Micropterus coosae, larvae[1]
Mortality	59 % mortality	Water = 0.750 mg Cu and Hg/l, control (only Cu at 0.750 mg/l)= 51 % mort.	Micropterus coosae, larvae[1]
Mortality	7 % mortality	Water = 0.050 mg Cu and Hg/l, control (only Cu at 0.050 mg/l)= 10 % mort.	Micropterus coosae, larvae[1]
Mortality	7 % mortality	Water = 0.075 mg Cu and Hg/l, control (only Cu at 0.075 mg/l)= 15 % mort.	Micropterus coosae, larvae[1]
Mortality	84 % mortality	Water = 1.00 mg Cu and Hg/l, control (only Cu at 1.00 mg/l)= 52 % mort.	Micropterus coosae, larvae[1]
Mortality	1 % mortality	Water = 0.001 mg Cu and Hg/l, control (only Cu at 0.001 mg/l)= 9 % mort.	Salvelinus fontinetis, larvae[1]
Mortality	100 % mortality	Water = 0.10 mg Cu and Hg/l, control (only Cu at 0.100 mg/l)= 71 % mort.	Salvelinus fontinetis, larvae[1]
Mortality	18 % mortality	Water = 0.025 mg Cu and Hg/l, control (only Cu at 0.025 mg/l)= 39 % mort.	Salvelinus fontinetis, larvae[1]
Mortality	22 % mortality	Water = 0.050 mg Cu and Hg/l, control (only Cu at 0.050 mg/l)= 44 % mort.	Salvelinus fontinetis, larvae[1]
Mortality	4 % mortality	Water = 0.005 mg Cu and Hg/l, control (only Cu at 0.005 mg/l)=22 % mort.	Salvelinus fontinetis, larvae[1]
Mortality	49 % mortality	Water = 0.075 mg Cu and Hg/l, control (only Cu at 0.075 mg/l)= 54 % mort.	Salvelinus fontinetis, larvae[1]
Mortality	50 % mortality	Water = 0.018 mg Cu and Hg/l, control (only Cu at 0.015 mg/l)= 50 % mort.	Salvelinus fontinetis, larvae[1]
Mortality	6 % mortality	Water = 0.01 mg Cu and Hg/l, control (only Cu at 0.01 mg/l)=33 % mort.	Salvelinus fontinetis, larvae[1]

Source: 1.Mount, D.I. and C.E. stephan. 1969. J. Fish. Res. bd Canada 26, 2449-2457.

4-104 Effect of Cu and Zn

Parameter	Effect Value	Concentration and Condition	Species
Mortality	0 % mortality	Water = 0.050 mg Cu and Zn/l, control (only Cu at 0.050 mg/l)= 0 % mort.	Carassius auratus, larvae[1]
Mortality	0 % mortality	Water = 0.100 mg Cu and Zn/l, control (only Cu at 0.100 mg/l)= 0 % mort.	Carassius auratus, larvae[1]
Mortality	25 % mortality	Water = 1.000 mg Cu and Zn/l, control (only Cu at 1.000 mg/l)= 22 % mort.	Carassius auratus, larvae[1]
Mortality	3 % mortality	Water = 0.500 mg Cu and Zn/l, control (only Cu at 0.500 mg/l)= 3 % mort.	Carassius auratus, larvae[1]
Mortality	50 % mortality	Water = 2.63 mg Cu and Zn/l, control (only Cu at 3.99 mg/l)= 50 % mort.	Carassius auratus, larvae[1]
Mortality	68 % mortality	Water = 5.000 mg Cu and Zn/l, control (only Cu at 5.000 mg/l)= 49 % mort.	Carassius auratus, larvae[1]
Mortality	93 % mortality	Water = 10.00 mg Cu and Zn/l, control (only Cu at 10.00 mg/l)= 79 % mort.	Carassius auratus, larvae[1]
Mortality	0 % mortality	Water = 0.01 mg Cu and Zn/l, control (only Cu at 0.01 mg/l)= 0 % mort.	Ictalurus lacustris, larvae[1]

Mortality	100 % mortality	Water = 10.00 mg Cu and Zn/l, control (only Cu at 10.00 mg/l)= 75 % mort.	Ictalurus lacustris, larvae[1]
Mortality	16 % mortality	Water = 0.100 mg Cu and Zn/l, control (only Cu at 0.100 mg/l)= 17 % mort.	Ictalurus lacustris, larvae[1]
Mortality	30 % mortality	Water = 0.500 mg Cu and Zn/l, control (only Cu at 0.500 mg/l)= 33 % mort.	Ictalurus lacustris, larvae[1]
Mortality	50 % mortality	Water = 0.70 mg Cu and Zn/l, control (only Cu at 1.55 mg/l)= 50 % mort.	Ictalurus lacustris, larvae[1]
Mortality	56 % mortality	Water = 1.000 mg Cu and Zn/l, control (only Cu at 1.000 mg/l)= 54 % mort.	Ictalurus lacustris, larvae[1]
Mortality	6 % mortality	Water = 0.050 mg Cu and Zn/l, control (only Cu at 0.050 mg/l)= 0 % mort.	Ictalurus lacustris, larvae[1]
Mortality	91 % mortality	Water = 5.000 mg Cu and Zn/l, control (only Cu at 5.000 mg/l)= 67 % mort.	Ictalurus lacustris, larvae[1]
Mortality	2 % mortality	Water = 0.100 mg Cu and Zn/l, control (only Cu at 0.100 mg/l)= 1 % mort.	Micropterus coosae, larvae[1]
Mortality	23 % mortality	Water = 1.000 mg Cu and Zn/l, control (only Cu at 1.000 mg/l)= 3 % mort.	Micropterus coosae, larvae[1]
Mortality	3 % mortality	Water = 0.500 mg Cu and Zn/l, control (only Cu at 0.500 mg/l)= 1 % mort.	Micropterus coosae, larvae[1]
Mortality	44 % mortality	Water = 5.000 mg Cu and Zn/l, control (only Cu at 5.000 mg/l)= 35 % mort.	Micropterus coosae, larvae[1]
Mortality	50 % mortality	Water = 4.23 mg Cu and Zn/l, control (only Cu at 6.73 mg/l)= 50 % mort.	Micropterus coosae, larvae[1]
Mortality	79 % mortality	Water = 10.00 mg Cu and Zn/l, control (only Cu at 10.00 mg/l)= 69 % mort.	Micropterus coosae, larvae[1]

Source: 1.Mount, D.I. and C.E. Stephan. 1969. J. Fish. Res. bd Canada 26, 2449-2457.

4-105 Effect of Cu, Mo and S

Parameter	Effect Value	Concentration and Condition	Species
Daily grains	103 g/day	Food = 0.10 % SO4-S, 2 mg Mo/kg, 10 mg Cu/kg, interaction effect	Sheep[1]
Daily grains	40 g/day	Food = 0.40 % SO4-S, 2 mg Mo/kg, 40 mg Cu/kg, interaction effect	Sheep[1]
Daily grains	50 g/day	Food = 0.40 % SO4-S, 2 mg Mo/kg, 10 mg Cu/kg, interaction effect	Sheep[1]
Daily grains	61 g/day	Food = 0.40 % SO4-S, 8 mg Mo/kg, 40 mg Cu/kg, interaction effect	Sheep[1]
Daily grains	78 g/day	Food = 0.10 % SO4-S, 8 mg Mo/kg, 40 mg Cu/kg, interaction effect	Sheep[1]
Daily grains	88 g/day	Food = 0.10 % SO4-S, 8 mg Mo/kg, 10 mg Cu/kg, interaction effect	Sheep[1]
Daily grains	95 g/day	Food = 0.40 % SO4-S, 8 mg Mo/kg, 10 mg Cu/kg, interaction effect	Sheep[1]
Daily grains	99 g/day	Food = 0.10 % SO4-S, 2 mg Mo/kg, 40 mg Cu/kg, interaction effect	Sheep[1]
Hemoglobin	12.2 g/100 ml blood	Food = 0.40 % SO4-S, 2 mg Mo/kg, 10 mg Cu/kg, interaction effect	Sheep[1]
Hemoglobin	12.4 g/100 ml blood	Food = 0.10 % SO4-S, 8 mg Mo/kg, 10 mg Cu/kg, interaction effect	Sheep[1]
Hemoglobin	12.6 g/100 ml blood	Food = 0.40 % SO4-S, 8 mg Mo/kg, 40 mg Cu/kg, interaction effect	Sheep[1]
Hemoglobin	12.9 g/100 ml blood	Food = 0.40 % SO4-S, 2 mg Mo/kg, 40 mg Cu/kg, interaction effect	Sheep[1]

Hemoglobin	13.0 g/100 ml blood	Food = 0.10 % SO4-S, 8 mg Mo/kg, 40 mg Cu/kg, interaction effect	Sheep[1]
Hemoglobin	13.1 g/100 ml blood	Food = 0.10 % SO4-S, 2 mg Mo/kg, 10 mg Cu/kg, interaction effect	Sheep[1]
Hemoglobin	13.2 g/100 ml blood	Food = 0.40 % SO4-S, 8 mg Mo/kg, 10 mg Cu/kg, interaction effect	Sheep[1]
Hemoglobin	13.5 g/100 ml blood	Food = 0.10 % SO4-S, 2 mg Mo/kg, 40 mg Cu/kg, interaction effect	Sheep[1]
Liver Cu	114 ug/g dry matter	Food = 0.40 % SO4-S, 8 mg Mo/kg, 40 mg Cu/kg, interaction effect	Sheep[1]
Liver Cu	117 ug/g dry matter	Food = 0.40 % SO4-S, 2 mg Mo/kg, 10 mg Cu/kg, interaction effect	Sheep[1]
Liver Cu	118 ug/g dry matter	Food = 0.10 % SO4-S, 8 mg Mo/kg, 10 mg Cu/kg, interaction effect	Sheep[1]
Liver Cu	120 ug/g dry matter	Food = 0.40 % SO4-S, 2 mg Mo/kg, 40 mg Cu/kg, interaction effect	Sheep[1]
Liver Cu	154 ug/g dry matter	Food = 0.10 % SO4-S, 2 mg Mo/kg, 10 mg Cu/kg, interaction effect	Sheep[1]
Liver Cu	179 ug/g dry matter	Food = 0.10 % SO4-S, 8 mg Mo/kg, 40 mg Cu/kg, interaction effect	Sheep[1]
Liver Cu	268 ug/g dry matter	Food = 0.10 % SO4-S, 2 mg Mo/kg, 40 mg Cu/kg, interaction effect	Sheep[1]
Liver Cu	67 ug/g dry matter	Food = 0.40 % SO4-S, 8 mg Mo/kg, 10 mg Cu/kg, interaction effect	Sheep[1]
Plasma Cu	100 ug/100 ml	Food = 0.40 % SO4-S, 2 mg Mo/kg, 10 mg Cu/kg, interaction effect	Sheep[1]
Plasma Cu	104 ug/100 ml	Food = 0.10 % SO4-S, 8 mg Mo/kg, 10 mg Cu/kg, interaction effect	Sheep[1]
Plasma Cu	105 ug/100 ml	Food = 0.40 % SO4-S, 2 mg Mo/kg, 40 mg Cu/kg, interaction effect	Sheep[1]
Plasma Cu	107 ug/100 ml	Food = 0.10 % SO4-S, 8 mg Mo/kg, 40 mg Cu/kg, interaction effect	Sheep[1]
Plasma Cu	111 ug/100 ml	Food = 0.10 % SO4-S, 2 mg Mo/kg, 10 mg Cu/kg, interaction effect	Sheep[1]
Plasma Cu	111 ug/100 ml	Food = 0.40 % SO4-S, 8 mg Mo/kg, 10 mg Cu/kg, interaction effect	Sheep[1]
Plasma Cu	112 ug/100 ml	Food = 0.10 % SO4-S, 2 mg Mo/kg, 40 mg Cu/kg, interaction effect	Sheep[1]
Plasma Cu	116 ug/100 ml	Food = 0.40 % SO4-S, 8 mg Mo/kg, 40 mg Cu/kg, interaction effect	Sheep[1]

Source: 1.Goodrich, R.D. and A.D. Tillman. 1966. J. Nutr. 92, 237-244.

4-106 Effect of Cyanate

Parameter	Effect Value	Concentration and Condition	Species
Nitrogen fixation	100 % inhibition	0.14 mg N/l	Azotobacter sp.[1]

Source: 1.Azim, M.A. and E.R. Roberts. 1956. Biochim. Biophys. Acta, 21, pp 308-312.

4-107 Effect of Cyanide

Parameter	Effect Value	Concentration and Condition	Species
Mortality	50 %	Oral dose = 3 mg/kg, 1.11E-4mol/kg, F007 = 3.95	Mouse[1]
Numbers of eggs per female	0 eggs	88.9 mg HCN/m3, control = 3464 eggs, pH = 8.07, 256 d, 298 k	Pimephales promelas[2]
Numbers of eggs per female	0 eggs	98.1 mg HCN/m3, control = 3464 eggs, pH = 8.07, 256 d, 298 k	Pimephales promelas[2]
Numbers of eggs per female	1010 eggs	32.8 mg HCN/m3, control = 3464 eggs, pH = 8.07, 256 d, 298 k	Pimephales promelas[2]
Numbers of eggs per female	1124 eggs	40.5 mg HCN/m3, control = 3464 eggs, pH = 8.07, 256 d, 298 k	Pimephales promelas[2]
Numbers of eggs per female	1367 eggs	24.7 mg HCN/m3, control = 3464 eggs, pH = 8.07, 256 d, 298 k	Pimephales promelas[2]
Numbers of eggs per female	1468 eggs	17.9 mg HCN/m3, control = 3464 eggs, pH = 8.07, 256 d, 298 k	Pimephales promelas[2]
Numbers of eggs per female	1845 eggs	11.14 mg HCN/m3, control = 3464 eggs, pH = 8.07, 256 d, 298 k	Pimephales promelas[2]
Numbers of eggs per female	243 eggs	75.3 mg HCN/m3, control = 3464 eggs, pH = 8.07, 256 d, 298 k	Pimephales promelas[2]
Numbers of eggs per female	2512 eggs	5.9 mg HCN/m3, control = 3464 eggs, pH = 8.07, 256 d, 298 k	Pimephales promelas[2]
Numbers of eggs per female	319 eggs	66.8 mg HCN/m3, control = 3464 eggs, pH = 8.07, 256 d, 298 k	Pimephales promelas[2]
Numbers of eggs per female	72 eggs	57.5 mg HCN/m3, control = 3464 eggs, pH = 8.07, 256 d, 298 k	Pimephales promelas[2]

Source: 1.Luckey, T.D. and B. Venugopal. 1977. Metal Toxicity in Mammals, 1. Plenum Press.
2.Lind, D.T., L.L. Smith(JR) and S.J. Broderius. 1977. Water Pollut. Control, 49(2), pp 262-268.

4-108 Effect of Cyanur acid

Parameter	Concentration	Condition	Species
LD50	10 g pr. kg ww	single oral dose, 0 %	rat, Rabbit[1]

Source: 1.Patty, F. A., Industrial Hygiene and toxicology, 2, Interscience Publishers, New York, 1976.

4-109 Effect of Cyclohexane

Parameter	Concentration	Condition	Species
Sublethal effect	300 ppm	Eye irritation	Man[1]
TLm	32-43 mg/l	24-96 hours	Bluegill[2]
TLm	32-43 mg/l	24-96 hours	Fathead minnow[2]
TLm	32-43 mg/l	24-96 hours	Goldfish[2]
TLm	58 mg/l	24-96 hours	Guppie[2]
TLm	15500 mg/l	24 hours, turbid Oklahoma water	Mosquito fish[3]

Source: 1.Patty, F. A., Industrial Hygiene and toxicology, 2, Interscience Publishers, New York, 1976.
2.Pickering Q. H. and Henderson, C., JWPCF, 38, 1419, 1966.
3.Wallen, I. E., Greer, W. C. and Lasater, R., Sewage Ind. Wastes, 29, 695, 1957.

4-110 Effect of Cyclohexanol

Parameter	Concentration	Condition	Species
Sublethal effect	100 ppm	Eye irritation, 3-5 min.	Man[1]
LD50	*2.2-2.6 g pr. kg ww*	*single oral dose*	*Rabbit*[1]

Source: 1.Patty, F. A., Industrial Hygiene and toxicology, 2, Interscience Publishers, New York, 1976.

4-111 Effect of Cyclohexanone

Parameter	Concentration	Condition	Species
Sublethal effect	52 mg/l	Inhibition of cell multiplication	Microcystis aeruginosa[4]
Sublethal effect	180 mg/l	Inhibition of cell multiplication	Pseudomonas putida[4]
LC0	*1000 mg/l*	*not indicated*	*Scenedesmus*[1]
LD50	1.3-1.5 g pr. kg ww	single oral dose	Mouse[2]
LD50	1.6-1.9 g pr. kg ww	single oral dose	Rabbit[2]
LD50	1.62 g pr. kg ww	single oral dose	Rat[3]

Source: 1.Meinck, F., Stooff, H. and Kohlschutter, H., Les eaux residuaires industrielles, 1970.
2.Patty, F. A., Industrial Hygiene and toxicology, 2, Interscience Publishers, New York, 1976.
3.Henschler, D., Toxikologisch-arbeitsmedizinische Begrundung von MAH-werte, Verlag Chemie.
4.Bringmann, G. and Kuhn, R., nwf-wasser/abwasser, H.9, 1976.

4-112 Effect of Cyclohexene

Parameter	Concentration	Condition	Species
LC0	100 mg/l	96 hours, seawater, 8 C	Coho salmon[1]

Source: 1.Morrow, J. E., EPA/660&3/73/018, 1974.

4-113 Effect of Cyclohexylamine

Parameter	Concentration	Condition	Species
Sublethal effect	0.02 mg/l	Inhibition of cell multiplication	Microcystis aeruginosa[1]
Sublethal effect	10 mg/l	Inhibition of cell multiplication	Pseudomonas putida[1]
LD50	*0.4-0.8 g pr. kg ww*	*acute oral dose*	*Rat*[2]

Source: 1.Bringmann, G. and Kuhn, R., nwf-wasser/abwasser, H.9, 1976.
2.Patty, F. A., Industrial Hygiene and toxicology, 2, Interscience Publishers, New York, 1976.

4-114 Effect of Cyclopentanole

Parameter	Concentration	Condition	Species
Sublethal effect	28 mg/l	Inhibition of cell multiplication	Microcystis aeruginosa[1]
Sublethal effect	250 mg/l	Inhibition of cell multiplication	Pseudomonas putida[1]

Source: 1.Bringmann, G. and Kuhn, R., nwf-wasser/abwasser, H.9, 1976.

4-115 Effect of DDD

Parameter	Effect Value	Concentration and Condition	Species
Mortality	50 %	Conc in food = 2178 mg/kg ww (1835-2584), 95 % CL	Bobwhite[1]
Mortality	50 %	Conc in food = 3165 mg/kg ww (2534-3978), 95 % CL	Japanese quail[1]
Mortality	50 %	Conc in food = 4814 mg/kg ww (3451-7054), 95 % CL	Mallard[1]
Mortality	50 %	Conc in food = 579 mg/kg ww (499-668), 95 % CL	Pheasant[1]

Source: 1.Edwards, C.A. 1975. Persistent Pesticides in the Environment, 2nd edition, CRC-Press.

4-116 Effect of DDE

Parameter	Effect Value	Concentration and Condition	Species
Mortality	50 %	Conc in food = 825 mg/kg ww (697-976), 95 % CL	Bobwhite[1]
Mortality	50 %	Conc in food = 1355 mg/kg ww (1111-1648), 95 % CL	Japanese quail[1]
Mortality	50 %	Conc in food = 3572 mg/kg ww (2811-4669), 95 % CL	Mallard[1]
Mortality	50 %	Conc in food = 841 mg/kg ww (731-967), 95 % CL	Pheasant[1]
Thickness of egg shell	38 % decrease	pp-DDE = 30 ppm/meal, 6 m, mean egg content = 135 ppm	Anas rubripes[2]
Thickness of egg shell	10 % decrease	pp-DDE = 10 ppm/meal, more than 1 y, mean egg content = 32 ppm	Falco sparverius[2]
Thickness of egg shell	13 % decrease	pp-DDE = 40 ppm/meal, 1-2 y, mean egg content = not given	Mallard[2]
Thickness of egg shell	23 % decrease	pp-DDE = 80*150 mg/kg injections, mean egg content = 80 ppm/meal	Streptopelia risoria[2]

Source: 1.Edwards, C.A. 1975. Persistent Pesticides in the Environment, 2nd edition, CRC-Press.
2.Cooke, A.S. 1973. Environ. Pollut., 4, pp 85-148.

4-117 Effect of DDT

Parameter	Effect Value	Concentration and Condition	Species
ATP level	ATP reduced	Water conc = 100000 ppb	Green algae[1]
Body weight	10.86 g	Injection of 50 mg DDT/ kg ww, 21 d, cont = 8.96+-0.85 g, rg	Mouse[2]
Body weight	22.62 g	Injection of 50 mg DDT/ kg ww, 33 d, cont = 21.30+-2.34 g,	Mouse[2]
Body weight	7.57 g	Injection of 50 mg DDT/ kg ww, 16 d, cont = 7.25+-0.76 g, rg	Mouse[2]
Cell clumping	effect observed	Water conc = 0.3 ppb, also cytological effects	Green algae[3]
Chloroplast electron transport	transport inhibited	Water conc = 20 u M	Green algae[4]
Growth rate	growth not affected	Water conc = 99 ppb, effect not observed	Blue-green algae[5]
Growth rate	growth not affected	Water conc = 100 ppb, effect not observed	Coccolithophorid[6]
Growth rate	growth inhibited	Water conc = 100 ppb	Diatom[7]

Growth rate	growth inhibited	Water conc = 100 ppb	Diatom[8]
Growth rate	growth inhibited	Water conc = 80 ppb	Diatom[4]
Growth rate	growth not affected	Water conc = 25 ppb, effect not observed	Diatom[9]
Growth rate	growth inhibited	Water conc = 0.3 ppb	Green algae[3]
Growth rate	growth inhibited	Water conc = 100 ppb	Green algae[10]
Growth rate	growth not affected	Water conc = 1000 ppb, effect not observed	Green algae[9]
Growth rate	growth not affected	Water conc = 1000 ppb, effect not observed	Green algae[11]
Growth rate	growth not affected	Water conc = 100000 ppb, effect not observed	Green algae[12]
Growth rate	growth not affected	Water conc = 100000 ppb, effect not observed	Green algae[1]
Growth rate	growth not affected	Water conc = 20000 ppb, effect not observed	Green algae[13]
Growth rate	growth not affected	Water conc = 20000 ppb, effect not observed	Green algae[14]
Growth rate	growth not affected	Water conc = 80 ppb, effect not observed	Green algae[4]
Heart beat	0 beats/min	DDT = 137 micro mol, after 24 h, 299 K, 5 snails	Biomphalaria glabrata[15]
Heart beat	30 beats/min	DDT = 116 micro mol, after 24 h, 299 K, 5 snails	Biomphalaria glabrata[15]
Heart beat	92 beats/min	DDT = 3 micro mol, after 24 h, 299 K, 5 snails	Biomphalaria glabrata[15]
Immobilisation	100 %	DDT = 0.002 mg/l, for 50 h	Daphnia sp.[16]
Morphological changes	no morphological effects observed	Water conc = 99 ppb, effect not observed	Blue-green algae[5]
Morphological changes	no morphological effects observed	Water conc = 80 ppb, effect not observed	Diatom[4]
Mortality	50 %	Conc in food = 611 mg/kg ww (514-724), 95 % CL	Bobwhite[17]
Mortality	0.0 %	DDT = 0.005 ppm, anim = 24.8 mg, 24 h frantic behaviour, anim = 6.9 ppm	Bufo bufo[18]
Mortality	0.0 %	DDT = 0.005 ppm, anim = 25.2 mg, 48 h frantic behaviour, anim = 13.5 ppm	Bufo bufo[18]
Mortality	0.0 %	DDT = 0.02 ppm, anim = 106 mg, 24 h frantic behaviour, abnormal	Bufo bufo[18]
Mortality	0.0 %	DDT = 0.02 ppm, anim = 71.2 mg, 24 h frantic behaviour, anim = 4.8 ppm	Bufo bufo[18]
Mortality	0.0 %	DDT = 0.05 ppm, anim = 12.7 mg, 48 h frantic behaviour, 2 d obs	Bufo bufo[18]
Mortality	0.0 %	DDT = 0.05 ppm, anim = 23.0 mg, 48 h resigned behaviour, 7 d	Bufo bufo[18]
Mortality	0.0 %	DDT = 0.05 ppm, anim = 24.6 mg, 24 h frantic behaviour, anim = 65.2 ppm	Bufo bufo[18]
Mortality	0.0 %	DDT = 0.5 ppm, 21.5 mg, 24 h resigned-shuddering behaviour	Bufo bufo[18]
Mortality	0.0 %	DDT = 0.5 ppm, 64.6 mg, 24 h resigned-shuddering behaviour, 5 d	Bufo bufo[18]
Mortality	0.0 %	DDT = 0.5 ppm, anim = 101. mg, 24 h resigned-shuddering behaviour	Bufo bufo[18]
Mortality	0.0 %	DDT = 0.5 ppm, anim = 20.1 mg, 48 h shuddering behaviour, anim = 478 ppm	Bufo bufo[18]

Mortality	0.0 %	DDT=0.02 ppm, 92 mg, 48 h frantic-resigned behv, abnormal snouts	Bufo bufo[18]
Mortality	100.0 %	DDT = 0.5 ppm, 64.6 mg, 48 h treatment	Bufo bufo[18]
Mortality	100.0 %	DDT = 0.5 ppm, anim = 79 or 101 mg, 5 d after treatment	Bufo bufo[18]
Mortality	15.0 %	DDT = 0.5 ppm, anim = 9.8 or 13.1 mg, 2 d after treatment	Bufo bufo[18]
Mortality	20.0 %	DDT = 0.5 ppm, anim = 79 mg, 48 h shuddering behaviour, anim = 134	Bufo bufo[18]
Mortality	3.0 %	DDT = 0.0 ppm, anim = 9.3 mg, 24 h treatment, 2 d obs, anim = < 0.01 ppm	Bufo bufo[18]
Mortality	3.0 %	DDT = 0.02 ppm, anim = 63.7 mg, 48 h frantic behaviour, anim = 4.3 ppm	Bufo bufo[18]
Mortality	5.0 %	DDT = 0.005 ppm, anim = 24.8 or 25.2 mg, 5 d after treat, frantic	Bufo bufo[18]
Mortality	5.0 %	DDT = 0.02 ppm, anim = 63.7 or 71.6 mg, 5 d after treat, frantic	Bufo bufo[18]
Mortality	6.0 %	DDT = 0.0 ppm, anim = 13.1 mg, 48 h treatment, 2 d obs, anim < 0.01	Bufo bufo[18]
Mortality	90.0 %	DDT = 0.5 ppm, anim = 20.1 or 21.5 mg, 5 d after treat, shuddering	Bufo bufo[18]
Mortality	50 % mortality	48 h, water = 0.4 mg/l	Daphnia pulex[19]
Mortality	95 %	Conc in soil > 200 mg/kg	Earthworms[17]
Mortality	50 %	Conc in food = 568 mg/kg ww (470-687), 95 % CL	Japanese quail[17]
Mortality	50 %	Conc in food = 1869 mg/kg ww (1500-2372), 95 % CL	Mallard[17]
Mortality	50 %	Conc in food = 300.0 mg/kg ww	Pheasant[20]
Mortality	50 %	Conc in food = 311 mg/kg ww (256-374), 95 % CL	Pheasant[17]
Mortality	50 %	Conc in food = 500.0 mg/kg ww	Quail[20]
Mortality	0.0 %	DDT = 0.005 ppm, anim = 42.4 mg, 24 h frantic behaviour, 1 d obs,	Rana temporaria[18]
Mortality	0.0 %	DDT = 0.02 ppm, anim = 134 mg, 48 h resigned behv, abnormal snouts,	Rana temporaria[18]
Mortality	0.0 %	DDT = 0.02 ppm, anim = 15.9 mg, 24 h frantic behv, deformed, 2 d obs	Rana temporaria[18]
Mortality	0.0 %	DDT = 0.02 ppm, anim = 16.8 mg, 48 h frantic behaviour, 1 d obs	Rana temporaria[18]
Mortality	0.0 %	DDT = 0.02 ppm, anim = 22. mg, 24 h frantic behaviour, 1 d obs	Rana temporaria[18]
Mortality	0.0 %	DDT = 0.5 ppm, 14.6 mg, 24 h resigned behv, deformed, 2 d obs	Rana temporaria[18]
Mortality	0.0 %	DDT = 0.5 ppm, anim = 19.5 mg, 24 h resigned-moribund behaviour	Rana temporaria[18]
Mortality	0.0 %	DDT=0.02 ppm, 127 mg, 24 h frantic-resigned behv, abnormal snouts	Rana temporaria[18]
Mortality	100.0 %	DDT = 0.5 ppm, anim = 150 mg, 48 h treatment, anim = 36.6 ppm DDT	Rana temporaria[18]

Mortality	100.0 %	DDT = 0.5 ppm, anim = 347 mg, 5-8d after treat, abnormal snouts	Rana temporaria[18]
Mortality	100.0 %	DDT = 0.5 ppm, anim = 37.6 mg, 48 h treatment, anim = 53.2 ppm DDT	Rana temporaria[18]
Mortality	13.0 %	DDT = 0.5 ppm, anim = 150 mg, 24 h resigned-moribund behaviour, anim	Rana temporaria[18]
Mortality	15.0 %	DDT = 0.05 ppm, anim = 38.1 mg, 24 h treatment, 1 d obs, anim = 3.9 ppm	Rana temporaria[18]
Mortality	25.0 %	DDT = 0.5 ppm, anim = 37.6 mg, 24 h moribund behaviour, 1 d obs, anim	Rana temporaria[18]
Mortality	29.0 %	DDT = 0.02 ppm, anim = 127 or 134 mg, 5 d after treatment	Rana temporaria[18]
Mortality	5.0 %	DDT = 0.5 ppm, anim = 347 mg, 24 h frantic-resigned behv, deformed	Rana temporaria[18]
Mortality	6.0 %	DDT = 0.005 ppm, anim = 31.2 mg, 48 h treatment, 1 d obs, anim = 3.9 ppm	Rana temporaria[18]
Mortality	80.0 %	DDT = 0.05 ppm, anim = 29.4 mg, 48 h moribund behaviour, 1 d obs	Rana temporaria[18]
Mortality	80.0 %	DDT = 0.5 ppm, anim = 17.8 mg, 48 h moribund behaviour	Rana temporaria[18]
Mortality	90.0 %	DDT = 0.5 ppm, anim = 17.8 or 19.5 mg, 1 d after treat, moribund	Rana temporaria[18]
Mortality	94.0 %	DDT = 0.5 ppm, anim = 347 mg, 48 h frantic, abnormal snouts	Rana temporaria[18]
Mortality	50 %	Acute dermal toxicity = 2500-3000 mg/kg, insecticide	Rat[17]
Mortality	50 %	Acute oral toxicity = 113-250 mg/kg, insecticide	Rat[17]
Mortality	95 %	Conc in soil > 50 mg/kg	Slugs[17]
No effect	0.0 %	DDT = 0.0 ppm, anim = 105 mg, 24 h treatment, 5 d obs, anim < 0.01	Bufo bufo[18]
No effect	0.0 %	DDT = 0.0 ppm, anim = 24.1 mg, 48 h treatment, 7 d obs, anim < 0.01	Bufo bufo[18]
No effect	0.0 %	DDT = 0.0 ppm, anim = 26.3 mg, 24 h treatment, 7 d obs, anim < 0.01	Bufo bufo[18]
No effect	0.0 %	DDT = 0.0 ppm, anim = 347 mg, 48 h treatment, anim < 0.01 ppm DDT	Bufo bufo[18]
No effect	0.0 %	DDT = 0.0 ppm, anim = 72. mg, 24 h treatment, 5 d obs, anim < 0.01 ppm	Bufo bufo[18]
No effect	0.0 %	DDT = 0.0 ppm, anim = 72.2 mg, 48 h treatment, 5 d obs, anim < 0.01 ppm	Bufo bufo[18]
No effect	0.0 %	DDT = 0.0 ppm, anim = 88. mg, 48 h treatment, 5 d obs, anim < 0.01 ppm	Bufo bufo[18]
No effect	0.0 %	DDT = 0.0008 ppm, anim = 111. mg, 24 h treatment, 5 d obs,	Bufo bufo[18]
No effect	0.0 %	DDT = 0.0008 ppm, anim = 68. mg, 24 h treatment, 5 d obs, anim = 0.3 ppm	Bufo bufo[18]
No effect	0.0 %	DDT = 0.0008 ppm, anim = 68.4 mg, 48 h treatment, 5 d obs	Bufo bufo[18]
No effect	0.0 %	DDT = 0.0008 ppm, anim = 93. mg, 48 h treatment, 5 d obs, anim = 0.54 ppm	Bufo bufo[18]
No effect	0.0 %	DDT = 0.005 ppm, anim = 12.5 mg, 48 h treatment, 2 d obs, anim < 0.01 ppm	Bufo bufo[18]

No effect	0.0 %	DDT = 0.005 ppm, anim = 9.8 mg, 24 h treatment, 2 d obs, anim < 0.01 ppm	Bufo bufo[18]
No effect	0.0 %	DDT = 0.05 ppm, anim = 10. mg, 24 h treatment, 2 d obs, anim = 30 ppm	Bufo bufo[18]
No effect	0.0 %	DDT = 0.5 ppm, anim = 13.1 mg, 48 h treatment, anim = 306 ppm DDT	Bufo bufo[18]
No effect	0.0 %	DDT = 0.5 ppm, anim = 9.8 mg, 24 h treatment, anim = 82 ppm DDT	Bufo bufo[18]
No effect	0 %	DDT = 0.0 ppm, anim = 21.4 mg, 24 h treatment, 2 d obs	Rana temporaria[18]
No effect	0 %	DDT = 0.0008 ppm, anim = 14.8 mg, 24 h treatment, 2 d obs	Rana temporaria[18]
No effect	0.0 %	DDT = 0.0 ppm, anim = 15.9 mg, 24 h treatment, 2 d obs	Rana temporaria[18]
No effect	0.0 %	DDT = 0.0 ppm, anim = 152 mg, 48 h treatment, 5 d obs, anim < 0.01 ppm	Rana temporaria[18]
No effect	0.0 %	DDT = 0.0 ppm, anim = 160 mg, 24 h treatment, 5 d obs, anim = 0.23 ppm	Rana temporaria[18]
No effect	0.0 %	DDT = 0.0 ppm, anim = 27.9 mg, 48 h treatment, 2 d obs	Rana temporaria[18]
No effect	0.0 %	DDT = 0.0 ppm, anim = 347 mg, 24 h treatment, 8d obs, anim < 0.01 ppm	Rana temporaria[18]
No effect	0.0 %	DDT = 0.0 ppm, anim = 44.7 mg, 48 h treatment, 1 d obs, anim < 0.01 ppm	Rana temporaria[18]
No effect	0.0 %	DDT = 0.0 ppm, anim = 47.4 mg, 24 h treatment, 1 d obs, anim < 0.01 ppm	Rana temporaria[18]
No effect	0.0 %	DDT = 0.0008 ppm, anim = 154 mg, 48 h treatment, 5 d obs, anim = 0.29 ppm	Rana temporaria[18]
No effect	0.0 %	DDT = 0.0008 ppm, anim = 177 mg, 24 h treatment, 5 d obs, anim = 0.17 ppm	Rana temporaria[18]
No effect	0.0 %	DDT = 0.0008 ppm, anim = 23.1 mg, 24 h treatment, 2 d obs	Rana temporaria[18]
No effect	0.0 %	DDT = 0.0008 ppm, anim = 26.9 mg, 48 h treatment, 1 d obs	Rana temporaria[18]
Photosynthesis	photosynthesis not inhibited	Water conc = > 10 ppb	Coccolithophorid[6]
Photosynthesis	photosynthesis inhibited	Water conc = < 10 ppb	Diatom[21]
Photosynthesis	photosynthesis inhibited	Water conc = > 10 ppb	Diatom[6]
Photosynthesis	photosynthesis not inhibited	Water conc = > 1 ppb	Diatom[6]
Photosynthesis	effect observed	Water conc = 20000 ppb	Green algae[14]
Photosynthesis	photosynthesis inhibited	Water conc = 100 ppb	Green algae[10]
Photosynthesis	photosynthesis inhibited	Water conc = 3600-36000 ppb	Green algae[22]
Photosynthesis	photosynthesis inhibited	Water conc = < 10 ppb	Green algae[21]
Photosynthesis	photosynthesis not inhibited	Water conc = 1000 ppb	Green algae[23]

Respiration	respiration not affected	Water conc = 1000 ppb	Green algae[6]
Respiration	respiration not affected	Water conc = 1000 ppb	Green algae[13]
Thickness of egg shell	9 % decrease	pp-DDT = 10 ppm/meal, 10 w, mean egg content = not given	Chicken[24]
Thickness of egg shell	4 % decrease	pp-DDT = 30 ppm/meal, 96 d, mean egg content = not given	Colinus virginianus[24]
Thickness of egg shell	4 % decrease	op-DDT = 27 ppm, 45 d, mean egg content = 27 ppm/meal	Coturnix coturnix japonica[24]
Thickness of egg shell	6 % decrease	pp-DDT = 100 ppm, 45 d, mean egg content = 520 ppm/meal	Coturnix coturnix japonica[24]
Thickness of egg shell	7 % decrease	pp-DDT = 25 ppm/meal, 26 w, mean egg content = not given	Coturnix coturnix japonica[24]
Thickness of egg shell	17 % decrease	pp-DDT+dieldrin = 15+3 ppm/meal, 1-2 y, mean egg content = not given	Falco sparverius[24]
Thickness of egg shell	11 % decrease	pp-DDT = 100 ppm/meal, 2 m, pg = 6 ppm in yolk	Gallus domesticus[24]
Thickness of egg shell	7 % decrease	pp-DDT = 300 micrograms/d meal, 15 w, mean egg content = not given	Lonchura striata[24]
Thickness of egg shell	13 % decrease	pp-DDT = 40 ppm/meal, 1-2 y, mean egg content = not given	Mallard[24]
Thickness of egg shell	5 % decrease	pp-DDT = 30 ppm/meal, 96 d, mean egg content = not given	Mallard[24]
Thickness of egg shell	9 % decrease	pp-DDT = 1000 mg/kg d, 16 d, mean egg content = not given	Mallard[24]

Source: 1.Spiridinov, Y.Y., V. S. Skhiladze and G.S. Spiridinova. 1970. Subtrop. Kultury 1, 150-155.
2.Kihlstroem, K.I., J. Oerberg, C. Lundberg and P.B. Danielsson. 1974. Ambio. 3(6), pp 231-233.
3.Samokhvalov, A.N. 1973. Vest. Mosk. Univ. 6, 87-92.
4.Kuzyakina, T.I. 1974. Weed Abstr. 24, 820.
5. Cooke, A.S. 1973. Environ Poll. 4, 85-148.
6.Thomas, R.E. and S.D. Rice. 1978. In: Vernberg Vernberg and Calabrese (eds.): Marine Pollution: Functional Processes, NY. Acedemic Press.
7.Mikhailova, M.F. 1968. Khimiya selKhoz 6, 204-206.
8.Micev, N. and M. Bubalov. 1972. Symp. Biol. Hung. 11, 379-384.
9.Spiridinov, Y.Y., V. S. Skhiladze and G.S. Spiridinova. 1970.In: The Behavior, Conversion and Analysis of Pesticides and their Metabolites in Soil. Proc. 1st all-Union Conf. Pushino-na-OKe, Akad. Nauk.
10.Vasilev, D.S. and O.V. Enkina. 1967. Dokl. Vses. Akad. selKhoz. Nauk. 12, 10-12.
11.Teuteberg, A. 1968. Z. PflKrankh. PflSchutz 75, 72-86.
12.Spiridinov, Y.Y., V. S. Skhiladze and G.S. Spiridinova. 1970. Subtrop. Kultury 3, 152-160.
13.Goguadze, V.D. 1969. Subtrop. Kultury No 4, 161-167.
14. Cooke, A.S. 1973. Environ Poll. 4, 85-148.
15.Cocks, J.A. 1974. Environ. Pollut., 5, pp 149-151.
16.Bays, L.R. 1971. Environ. Pollut., 1(3), pp 205-234.
17.Edwards, C.A. 1975. Persistent Pesticides in the Environment, 2nd edition, CRC-Press.
18.Cooke, A.S. 1972. Environ. Pollut., 3, pp 51-68.
19.Rankov, V. 1973. Pochv. Agrokhim. 8, 125-134.
20.Dewitt, J.B., C.M. Menzie, V.A. Adomaitis and W.L. Reickel. 1960. Trans. 25th N. Am. Wildl. Conf., p277.
21.Micev, N. and M. Bubalov. 1969. Godisen Zb. zemjod.-sum. Fak. Univ. Skopje 22, 19-24.
22.Bertrand, D. and A. de Wolf. 1972. C.r. hebd. Seanc. Acad. Agric. Fr. 58, 1469-1473.
23.Micev, N. 1970. Savr. Poljopr. 18, 245-249.
24.Cooke, A.S. 1973. Environ. Pollut., 4, pp 85-148.

4-118 Effect of DNBP

Parameter	Effect Value	Concentration and Condition	Species
Mortality	50 %	118 (103-135) ppb, 96 h, 14 g, 293 K, pH = 8.2, alk = 80 ppm, 95 % CL	Ictalurus punctatus[1]

Source: 1.McCorle, F.M., J.E. Chambers and J.D. Yarbrough. 1977. Bull. Environ. Contam. Toxicol. 18(3), pp 267270.

4-119 Effect of DSMA

Parameter	Effect Value	Concentration and Condition	Species
Mortality	10 %	10 ppm, 48 h, 14 g, 293 K, pH = 8.2, alk = 80 ppm, less than value	Ictalurus punctatus[1]

Source: 1.McCorle, F.M., J.E. Chambers and J.D. Yarbrough. 1977. Bull. Environ. Contam. Toxicol. 18(3), pp 267270.

4-120 Effect of Dalapon

Parameter	Effect Value	Concentration and Condition	Species
Mortality	10 %	10 ppm, 48 h, 14 g, 293 K, pH = 8.2, alk = 80 ppm, less than value	Ictalurus punctatus[1]

Source: 1.McCorle, F.M., J.E. Chambers and J.D. Yarbrough. 1977. Bull. Environ. Contam. Toxicol. 18(3), pp 267270.

4-121 Effect of Day length

Parameter	Effect Value	Concentration and Condition	Species
Dieldrin uptake	0.11 mg dieldrin/kg wet weight	Long day, grown in sand with 10 ppm dieldrin	Secale cereale[1]
Dieldrin uptake	0.13 mg dieldrin/kg wet weight	Short day, grown in sand with 10 ppm dieldrin	Secale cereale[1]

Source: 1.Wheelen, W.B. 1970. Bull. Environ. Contam. Tixicol. 5, pp 463-467.

4-122 Effect of Degradation

Parameter	Effect Value	Concentration and Condition	Species
Litter composition	0.43 % Ca of dry weight	Change in composition initial = 0.35 % Ca, 70 % degradated	Ca[1]
Litter composition	0.25 % K of dry weight	Change in composition initial = 0.04 % K, 70 % degradated	K[1]
Litter composition	0.06 % Mg of dry weight	Change in composition initial = 0.04 % Mg, 70 % degradated	Mg[1]
Litter composition	0.79 % N of dry weight	Change in composition initial = 0.96 % N, 70 % degradated	N[1]
Litter composition	0.14 % P of dry weight	Change in composition initial = 0.04 % P, 70 % degradated	P[1]

Source: 1.Old, S.M. Ecol Monogr. 39, pp 355-383.

4-123 Effect of Detergents

Parameter	Effect Value	Concentration and Condition	Species
Mortality	100.0 %	APH-11EO = 1000.0 ppm, 24 h, 40 larvae	Aedes aegypti[1]
Mortality	100.0 %	PA-9EO = 200.0 ppm, 24 h, 40 larvae	Aedes aegypti[1]
Mortality	100.0 %	PAES = 12.0 ppm, 24 h, 40 larvae	Aedes aegypti[1]
Mortality	100.0 %	PAS 311 = 5.0 ppm, 24 h, 40 larvae	Aedes aegypti[1]
Mortality	100.0 %	Teepol ch53 = 170.0 ppm, 24 h, 40 larvae	Aedes aegypti[1]
Mortality	50.0 %	APH-11EO = 500.0 ppm, 24 h, 40 larvae	Aedes aegypti[1]
Mortality	50.0 %	PAES = 11.0 ppm, 24 h, 40 larvae	Aedes aegypti[1]
Mortality	50.0 %	PAS 311 = 4.0 ppm, 24 h, 40 larvae	Aedes aegypti[1]
Mortality	100.0 %	APH-11EO = 23.0 ppm, 24 h, 2*3 snails	Biomphalaria glabrata[1]
Mortality	100.0 %	PA-9EO = 11.0 ppm, 24 h, 2*3 snails	Biomphalaria glabrata[1]

Mortality	100.0 %	PAES = 12.0 ppm, 24 h, 2*3 snails	Biomphalaria glabrata[1]
Mortality	100.0 %	PAS 311 = 4.0 ppm, 24 h, 2*3 snails	Biomphalaria glabrata[1]
Mortality	100.0 %	Teepol ch53 = 7.0 ppm, 24 h, 2*3 snails	Biomphalaria glabrata[1]
Mortality	100.0 %	APH-11EO = 52.0 ppm, 24 h, 5 young fish	Lebistes reticulatus[1]
Mortality	100.0 %	APH-11EO = 57.0 ppm, 24 h, 3 fish, female	Lebistes reticulatus[1]
Mortality	100.0 %	APH-11EO = 64 ppm, 24 h, 3 fish, male	Lebistes reticulatus[1]
Mortality	100.0 %	PA-9EO = 11.0 ppm, 24 h, 3 fish, male	Lebistes reticulatus[1]
Mortality	100.0 %	PA-9EO = 6.0 ppm, 24 h, 5 young fish	Lebistes reticulatus[1]
Mortality	100.0 %	PA-9EO = 9.0 ppm, 24 h, 3 fish, female	Lebistes reticulatus[1]
Mortality	100.0 %	PAES = 4.0 ppm, 24 h, 5 young fish	Lebistes reticulatus[1]
Mortality	100.0 %	PAES = 5.0 ppm, 24 h, 3 fish, female	Lebistes reticulatus[1]
Mortality	100.0 %	PAES = 8.0 ppm, 24 h, 3 fish, male	Lebistes reticulatus[1]
Mortality	100.0 %	PAS 311 = 0.5 ppm, 24 h, 5 young fish	Lebistes reticulatus[1]
Mortality	100.0 %	PAS 311 = 2.0 ppm, 24 h, 3 fish, female	Lebistes reticulatus[1]
Mortality	100.0 %	PAS 311 = 2.0 ppm, 24 h, 3 fish, male	Lebistes reticulatus[1]
Mortality	100.0 %	Teepol ch53 = 4.0 ppm, 24 h, 3 fish, female	Lebistes reticulatus[1]
Mortality	100.0 %	Teepol ch53 = 4.0 ppm, 24 h, 3 fish, male	Lebistes reticulatus[1]

Source: 1.Van Emden, H.M., C.C.M. Kroon, E.N. Schoeman and H.A. van Seventer. 1974. Environ. Pollut. 6, pp 297308.

4-124 Effect of Di-2-ethylhexyl phthalate

Parameter	Effect Value	Concentration and Condition	Species
Heart beat	87 beats/min	Conc in water = 200 ppm, rg = +- 5	Goldfish[1]
Reproduction	60 % decrease	Conc in water = 3 ppb	Daphnia magna[1]
Reproduction	70 % decrease	Conc in water = 10 ppb	Daphnia magna[1]
Reproduction	83 % decrease	Conc in water = 30 ppb	Daphnia magna[1]
Reproduction	0 % decrease	Oral dose = 0.06 g/kg*day for 2 yrs	Dog[2]
Reproduction	0 % decrease	Oral dose = 0.06 g/kg*day for 2 yrs	Pig[2]

Source: 1.Jensen, A.A. 1977. Phthalater i Mil joeet. Publ. from the danish environ. agency, Miljoestyrelsen. Ch.
2.Autian, J. 1973. Environ. Helth Perspect., 4, pp 3-26.

4-125 Effect of Di-isobutylcarbinol

Parameter	Concentration	Condition	Species
LD50	3.6 g pr. kg ww	single oral dose	Rat[1]
LD50	*5.8 g pr. kg ww*	*single oral dose, Di-iso-butylketone*	*Rat*[1]

Source: 1.Patty, F. A., Industrial Hygiene and toxicology, 2, Interscience Publishers, New York, 1976.

4-126 Effect of Di-isopropylamine

Parameter	Concentration	Condition	Species
LD50	0.6-0.9 g pr. kg ww	oral dose	Rat[2]
LC0	*40 mg/l*	*24 hours*	*Creek chub*[1]
LC100	60 mg/l	24 hours	Creek chub[1]

Source: 1.Gilette, L. A., Miller, D. L. and Redman, H. E., Sewage Ind. Wastes, 24, 1397, 1952.
2.McKee, J. E., and Wolf, H. W., Water Quality, The resources agency of California, 1963.

4-127 Effect of Di-isopropylether

Parameter	Concentration	Condition	Species
LC0	125 mg/l	not indicated	Colpoda[1]
LC0	380 mg/l	24 hours	Goldfish[2]
LC0	125 mg/l	not indicated	Pseudomonas[1]
LC0	30 mg/l	not indicated	Scenedesmus[1]
LD50	*5.0-6.5 g pr. kg ww*	*single oral dose*	*Rat*[3]

Source: 1.Meinck, F., Stooff, H. and Kohlschutter, H., Les eaux residuaires industrielles, 1970.
2.Shell Chemie, Shell Industrie Chemicalien gids, Shell Niederland Chemie, 1975.
3.Patty, F. A., Industrial Hygiene and toxicology, 2, Interscience Publishers, New York, 1976.

4-128 Effect of Di-n-butylamine

Parameter	Concentration	Condition	Species
LD50	0.5 g pr. kg ww	acute oral dose	Rat[1]

Source: 1.Patty, F. A., Industrial Hygiene and toxicology, 2, Interscience Publishers, New York, 1976.

4-129 Effect of Di-n-allylamine

Parameter	Concentration	Condition	Species
Sublethal effect	20 ppm	Increased liver weight, 50 times 7 hours	Rat[1]
LD50	*0.58 g pr. kg ww*	*acute oral dose*	*Rat*[1]

Source: 1.Patty, F. A., Industrial Hygiene and toxicology, 2, Interscience Publishers, New York, 1976.

4-130 Effect of Di-n-propylamine

Parameter	Concentration	Condition	Species
LD50	0.2-0.4 g pr. kg ww	acute oral dose	Rat[1]

Source: 1.Patty, F. A., Industrial Hygiene and toxicology, 2, Interscience Publishers, New York, 1976.

4-131 Effect of Diacetonealcohol

Parameter	Concentration	Condition	Species
LC50	>5000 mg/l	24 hours	Goldfish[2]
Sublethal effect	530 mg/l	Inhibition of cell multiplication	Microcystis aeruginosa[3]
Sublethal effect	825 mg/l	Inhibition of cell multiplication	Pseudomonas putida[3]
LD50	*4.0 g pr. kg ww*	*single oral dose*	*Rat*[1]

Source: 1.Patty, F. A., Industrial Hygiene and toxicology, 2, Interscience Publishers, New York, 1976.
2.Shell Chemie, Shell Industrie Chemicalien gids, Shell Niederland Chemie, 1975.
3.Bringmann, G. and Kuhn, R., nwf-wasser/abwasser, H.9, 1976.

4-132 Effect of Diallylphtalate

Parameter	Concentration	Condition	Species
Sublethal effect	0.65 mg/l	Inhibition of cell multiplication	Microcystis aeruginosa[1]
Sublethal effect	>100 mg/l	Inhibition of cell multiplication	Pseudomonas putida[1]

Source: 1.Bringmann, G. and Kuhn, R., nwf-wasser/abwasser, H.9, 1976.

4-133 Effect of Diazinon

Parameter	Effect Value	Concentration and Condition	Species
Mortality	50 % mortality	48 h, water = 0.9 mg/l	Daphnia pulex[1]

Source: 1.Rankov, V. 1973. Pochv. Agrokhim. 8, 125-134.

4-134 Effect of Diazomethane

Parameter	Concentration	Condition	Species
Sublethal effect	-	carcinogenic	Man[1]
Sublethal effect	-	carcinogenic	Man[2]

Source: 1.Henschler, D., Toxikologisch-arbeitsmedizinische Begrundung von MAH-werte, Verlag Chemie.
2.Maximale Arbeitsplatzkonzentration 1973. BRD.

4-135 Effect of Dibenz(a)anthracene

Parameter	Effect Value	Concentration and Condition	Species
Mortality	0 % mortality	Water = 1.0 ppm, zero level mortality in 96 h, marine	Neathes arenaceodentata[1]

Source: 1.Rossi, S.S. and J.M. Neff. 1978. Mar. Pollut. Bull. 9, 220-223.

4-136 Effect of Dibenzothiophene

Parameter	Effect Value	Concentration and Condition	Species
Mortality	50 % mortality	Water = 3.18 ppm, 96 h, sheep head minnow	Cyprinodon variegatus[1]
Mortality	50 % mortality	Water = 0.28 ppm, 96 h	Palaemonetes pugio[1]

Source: 1.Neff, J.M. 1979. Polycyclic Aromatic Hydrocarbons in the Aquatic Environment. Applied Sci. Publ. London.

4-137 Effect of Dibutyl - phthalate

Parameter	Effect Value	Concentration and Condition	Species
Heart beat	105 beats/min	Conc in water = 1 ppm, rg = +- 10	Goldfish[1]
Heart beat	120 beats/min	Conc in water = 12 ppm, 200 ppm atropin added, rg = +- 5	Goldfish[1]
Heart beat	130 beats/min	Conc in water = 0.5 ppm, rg = +- 15	Goldfish[1]
Heart beat	48 beats/min	Conc in water = 12 ppm, rg = +- 10	Goldfish[1]
Heart beat	60 beats/min	Conc in water = 5 ppm, rg = +- 10	Goldfish[1]
Mortality	50 %	Oral dose = 11.9 ml/kg, 7 d	Rat[2]
Mortality	50 %	Oral dose = 13.8 ml/kg, 24 hrs	Rat[2]
Mortality	50 %	Peritoneal dose = 6.8 ml/kg, 7 d	Rat[2]
Mortality	50 %	Peritoneal dose = 8.3 ml/kg, 24 hrs	Rat[2]

Source: 1.Jensen, A.A. 1977. Phthalater i Miljoeet. Publ. from the danish environ. agency, Miljoestyrelsen. Ch.
2.Reakall, D.B. 1974. J. Environ. Qual. 52, pp 1-39.

4-138 Effect of Dichlobenil

Parameter	Effect Value	Concentration and Condition	Species
Mortality	50 % mortality	48 h, water = 3700 mg/l	Daphnia pulex[1]
Mortality	50 %	Acute oral toxicity = 3160 mg/kg, herbicide	Rat[2]

Source: 1.Rankov, V. 1973. Pochv. Agrokhim. 8, 125-134.
2.Edwards, C.A. 1975. Persistent Pesticides in the Environment, 2nd edition, CRC-Press.

4-139 Effect of Dichloroacetic acid

Parameter	Concentration	Condition	Species
LD50	5.5 g pr. kg ww	single oral dose	Mouse[1]
LD50	4.5 g pr. kg ww	single oral dose	Rat[1]

Source: 1.Patty, F. A., Industrial Hygiene and toxicology, 2, Interscience Publishers, New York, 1976.

4-140 Effect of Dichlorodiflouromethane

Parameter	Concentration	Condition	Species
LD50	>1 g pr. kg ww	single oral dose	Rat[1]

Source: 1.Henschler, D., Toxikologisch-arbeitsmedizinische Begrundung von MAH-werte, Verlag Chemie.

4-141 Effect of Dichloroisopropylether

Parameter	Concentration	Condition	Species
LD50	0.24 g pr. kg ww	single oral dose	Rat[1]

Source: 1.Patty, F. A., Industrial Hygiene and toxicology, 2, Interscience Publishers, New York, 1976.

4-142 Effect of Dieldrin

Parameter	Effect Value	Concentration and Condition	Species
Immobilisation	100 %	Dieldrin = 0.25 mg/l, immobilisation in 50 h	Daphnia sp.[1]
Mortality	50 %	Conc in food = 39 mg/kg(37-41), 95 % CL	Bobwhite[2]
Mortality	50 %	Flowing seawater = 12.5 ug/l, 96 h	Crassostrea virginica[3]
Mortality	50 %	Flowing seawater = 23.6 ug/l, 96 h	Cyprinodon variegatus[3]
Mortality	50 % mortality	48 h, water = 250 mg/l	Daphnia pulex[4]
Mortality	95 %	Conc in soil > 100 mg/kg	Earthworms[2]
Mortality	50 %	Conc in food = 57 mg/kg(53-61), 95 % CL	Japanese quail[2]
Mortality	50 %	Conc in food = 185 mg/kg(152-217), 95 % CL	Mallard[2]
Mortality	50 %	Flowing seawater = 11.4 ug/l, 96 h	Palaemonetes pugio[3]
Mortality	50 %	Flowing seawater = 0.9 ug/l, 96 h	Penaeus duorarum[3]
Mortality	50 %	Conc in food = 50.0 mg/kg	Pheasant[5]
Mortality	50 %	Conc in food = 56 mg/kg(51-60), 95 % CL	Pheasant[2]
Mortality	50 %	Conc in food = 35.0 mg/kg	Quail[5]
Mortality	50 %	Acute dermal toxicity = 10-102 mg/kg, insecticide	Rat[2]
Mortality	50 %	Acute oral toxicity = 38-87 mg/kg, insecticide	Rat[2]

Thickness of egg shell	4 % decrease	Dieldrin = 10 ppm/meal, 1-2 y, mean egg content = 44 ppm	Anas platyrhynchos[6]

Source: 1.Bays, L.R. 1971. Environ. Pollut., 1(3), pp 205-234.
2.Edwards, C.A. 1975. Persistent Pesticides in the Environment, 2nd edition, CRC-Press.
3.Mitchell, A.L. (ed.). 1974. pp 427-434. EPA, Gulf Breeze environ. Res. Lab.
4.Rankov, V. 1973. Pochv. Agrokhim. 8, 125-134.
5.Dewitt, J.B., C.M. Menzie, V.A. Adomaitis and W.L. Reickel. 1960. Trans. 25th N. Am. Wildl. Conf., p277.
6.Cooke, A.S. 1973. Environ. Pollut., 4, pp 85-148.

4-143 Effect of Diethanolamine

Parameter	Concentration	Condition	Species
LD50	1.82 g pr. kg ww	acute oral dose	Rat[1]
LC50	*800 mg/l*	*pH=9.6*	*Goldfish*[1]
LC50	>5000 mg/l	pH=7.0	Goldfish[1]
Sublethal effect	16 mg/l	Inhibition of cell multiplication	Microcystis aeruginosa[2]
Sublethal effect	17 mg/l	Inhibition of cell multiplication, neutralized	Microcystis aeruginosa[2]
Sublethal effect	16 mg/l	Inhibition of cell multiplication	Pseudomonas putida[2]
Sublethal effect	>10000 mg/l	Inhibition of cell multiplication, neutralized	Pseudomonas putida[2]
TLm	2100 mg/l	24 hours, tap water	Bluegil[3]
TLm	1800 mg/l	24 hours, turbid Oklahoma water	Mosquito fish[3]
TLm	2100 mg/l	24 hours, tap water	SunFish[4]

Source: 1.Shell Chemie, Shell Industrie Chemicalien gids, Shell Niederland Chemie, 1975.
2.Bringmann, G. and Kuhn, R., nwf-wasser/abwasser, H.9, 1976.
3.Wallen, I. E., Greer, W. C. and Lasater, R., Sewage Ind. Wastes, 29, 695, 1957.
4.Turnbull, H., DeMann, J. H. and Weston, R. F., Ind. Eng. Chem., 46, 324, 1954.

4-144 Effect of Diethylamine

Parameter	Concentration	Condition	Species
LC0	70 mg/l	24 hours	Creek chub[2]
LC0	1000 mg/l	not indicated	Escherichia coli[2]
LC100	100 mg/l	24 hours	Creek chub[2]
LD50	*0.54 g pr. kg ww*	*acute oral dose*	*Rat*[1]

Source: 1.Patty, F. A., Industrial Hygiene and toxicology, 2, Interscience Publishers, New York, 1976.
2.Gilette, L. A., Miller, D. L. and Redman, H. E., Sewage Ind. Wastes, 24, 1397, 1952.

4-145 Effect of Diethyleneglycol

Parameter	Concentration	Condition	Species
LC50	>5000 mg/l	not indicated	Goldfish[1]
Sublethal effect	1700 mg/l	Inhibition of cell multiplication	Microcystis aeruginosa[2]
Sublethal effect	8000 mg/l	Inhibition of cell multiplication	Pseudomonas putida[2]
TLm	>32000 mg/l	24 hours, turbid Oklahoma water	Mosquito fish[3]
LD50	*20.7 g pr. kg ww*	*single oral dose*	*Rat*[4]

Source: 1.Shell Chemie, Shell Industrie Chemicalien gids, Shell Niederland Chemie, 1975.
2.Bringmann, G. and Kuhn, R., nwf-wasser/abwasser, H.9, 1976.
3.Wallen, I. E., Greer, W. C. and Lasater, R., Sewage Ind. Wastes, 29, 695, 1957.
4.Patty, F. A., Industrial Hygiene and toxicology, 2, Interscience Publishers, New York, 1976.

4-146 Effect of Diethyleneglycolmonobutylether

Parameter	Concentration	Condition	Species
LD50	2.0 g pr. kg ww	single oral dose	guinea pig[2]
LD50	5.5-6.6 g pr. kg ww	single oral dose	Rat[2]
LC50	*2700 mg/l*	*24 hours*	*Goldfish*[1]

Source: 1.Shell Chemie, Shell Industrie Chemicalien gids, Shell Niederland Chemie, 1975.
2.Patty, F. A., Industrial Hygiene and toxicology, 2, Interscience Publishers, New York, 1976.

4-147 Effect of Diethyleneglycolmonoethylether

Parameter	Concentration	Condition	Species
LD50	1.0 g pr. kg ww	single oral dose, 100 %	cat[2]
LD50	3.8-5.0 g pr. kg ww	single oral dose	guinea pig[2]
LD50	8.7-9.7 g pr. kg ww	single oral dose	Rat[2]
LC50	*>5000 mg/l*	*24 hours*	*Goldfish*[1]

Source: 1.Shell Chemie, Shell Industrie Chemicalien gids, Shell Niederland Chemie, 1975.
2.Patty, F. A., Industrial Hygiene and toxicology, 2, Interscience Publishers, New York, 1976.

4-148 Effect of Diethyloxalate

Parameter	Concentration	Condition	Species
Sublethal effect	9.0 mg/l	Inhibition of cell multiplication	Microcystis aeruginosa[1]
Sublethal effect	>10000 mg/l	Inhibition of cell multiplication	Pseudomonas putida[1]

Source: 1.Bringmann, G. and Kuhn, R., nwf-wasser/abwasser, H.9, 1976.

4-149 Effect of Diethylphtalate

Parameter	Concentration	Condition	Species
Sublethal effect	15 mg/l	Inhibition of cell multiplication	Microcystis aeruginosa[1]
Sublethal effect	>400 mg/l	Inhibition of cell multiplication	Pseudomonas putida[1]

Source: 1.Bringmann, G. and Kuhn, R., nwf-wasser/abwasser, H.9, 1976.

4-150 Effect of Diglycidylether

Parameter	Concentration	Condition	Species
LD50	0.17 g pr. kg ww	acute intragrastric	Mouse[1]
LD50	0.45 g pr. kg ww	acute intragrastric	Rat[1]

Source: 1.Patty, F. A., Industrial Hygiene and toxicology, 2, Interscience Publishers, New York, 1976.

4-151 Effect of Diisopropylether

Parameter	Concentration	Condition	Species
Sublethal effect	800 ppm	Eye irritation	Man[1]

Source: 1.Patty, F. A., Industrial Hygiene and toxicology, 2, Interscience Publishers, New York, 1976.

4-152 Effect of Dimethoxymethane

Parameter	Concentration	Condition	Species
LD50	5 g pr. kg ww	subctane dose	guinea pig[1]

Source: 1.Patty, F. A., Industrial Hygiene and toxicology, 2, Interscience Publishers, New York, 1976.

4-153 Effect of Dimethylarsinic acid

Parameter	Effect Value	Concentration and Condition	Species
Mortality	50 %	Acute oral toxicity = 1350 mg/kg, pesticide	Rat[1]

Source: 1.Edwards, C.A. 1975. Persistent Pesticides in the Environment, 2nd edition, CRC-Press.

4-154 Effect of Dimethylnaphthalene

Parameter	Effect Value	Concentration and Condition	Species
Mortality	50 % mortality	Water = 6 ppm, naplii sedimentated (thought dead), 1 h, S=3 %	Balanus amphitrites[1]
Mortality	50 % mortality	Water = 0.60 ppm, 96 h, stage 1 zoeae	Cancer magister[2]
Mortality	50 % mortality	Water = 5.1 ppm, 24 h	Cyprinodon variegatus[3]
Mortality	50 % mortality	Water = 0.08 ppm, 24 h	Penaeus aztecus[3]

Source: 1.Donahue, W.H., R.T. Wang, M. Welch and J.A.C. Nicol. 1977. Environ. Pollut. 13, 187-202.
2.Caldwell, R.s., E. M. Caldarone and M.H. Mallon. 1977. D.A. Wolfe (ed.): Fate and Effects of Petroleum Hydrocarbons in Marine Ecosystems and Organisms. NY. Pergamon Press. 210-220.
3.Anderson, J.W., J.M. Neff, B.A. Cox, Tatem and Hightower. 1974. In: Vernberg and Vernberg (eds.): Pollution and Physiology of Marine Animals. NY. Academic Press. 285-310.

4-155 Effect of Dimethylsulfate

Parameter	Concentration	Condition	Species
Sublethal effect	-	Carcinogenic	Man[1]
Sublethal effect	-	Carcinogenic	Man[2]
LD50	*0.05 g pr. kg ww*	*acute oral dose, 100 %*	*Rabbit*[1]
LD50	0.44 g pr. kg ww	acute oral dose, 100 %	Rat[3]

Source: 3.Patty, F. A., Industrial Hygiene and toxicology, 2, Interscience Publishers, New York, 1976.
1.Maximale Arbeitsplatzkonzentration 1973. BRD.
2.Federal register, 39, 1974.

4-156 Effect of Dimethylteraphtalate

Parameter	Concentration	Condition	Species
LD50	>3.2 g pr. kg ww	acute oral dose	Rat[1]

Source: 1.Patty, F. A., Industrial Hygiene and toxicology, 2, Interscience Publishers, New York, 1976.

4-157 Effect of Dioctyl-phthalate

Parameter	Effect Value	Concentration and Condition	Species
Mortality	50 %	Oral dose = 33.7 ml/kg, 7 d	Rat[2]
Mortality	50 %	Oral dose = 45.1 ml/kg, 24 hrs	Rat[2]
Mortality	50 %	Peritoneal dose = 37.3 ml/kg, 7 d	Rat[2]
Mortality	50 %	Peritoneal dose = 50.5 ml/kg, 24 hrs	Rat[2]
Sublethal effect	-	Teratogenic	Rat[1]

Source: 1.Singh, A. R., Lawrence, W. H. and Autian, J., J. Pharm. Sci., 61, 51, 1972.
2.Reakall, D.B. 1974. J. Environ. Qual. 52, pp 1-39.

4-158 Effect of Diphenylether

Parameter	Concentration	Condition	Species
LD50	4.0 g pr. kg wwsingle oral dose		Rat[1]

Source: 1.Patty, F. A., Industrial Hygiene and toxicology, 2, Interscience Publishers, New York, 1976.

4-159 Effect of Dipropylenglycol

Parameter	Concentration	Condition	Species
LD50	14.8 g pr. kg ww	single oral dose	Rat[1]

Source: 1.Patty, F. A., Industrial Hygiene and toxicology, 2, Interscience Publishers, New York, 1976.

4-160 Effect of Dipropylenglycolmethylether

Parameter	Concentration	Condition	Species
LD50	5.5-7.5 g pr. kg ww	single oral dose	rat, dog[1]

Source: 1.Patty, F. A., Industrial Hygiene and toxicology, 2, Interscience Publishers, New York, 1976.

4-161 Effect of Disodium methylarsonate

Parameter	Effect Value	Concentration and Condition	Species
Mortality	50 %	Acute oral toxicity = 750 mg/kg, pesticide	Rat[1]

Source: 1.Edwards, C.A. 1975. Persistent Pesticides in the Environment, 2nd edition, CRC-Press.

4-162 Effect of Diuron

Parameter	Effect Value	Concentration and Condition	Species
Mortality	50 % mortality	48 h, water = 1400 mg/l	Daphnia pulex[1]
Mortality	10 %	10 ppm, 48 h, 14 g, 293 K, pH = 8.2, alk = 80 ppm, less than value	Ictalurus punctatus[2]
Mortality	50 %	Acute oral toxicity = 3400 mg/kg, herbicide	Rat[3]

Source: 1.Rankov, V. 1973. Pochv. Agrokhim. 8, 125-134.
2.McCorle, F.M., J.E. Chambers and J.D. Yarbrough. 1977. Bull. Environ. Contam. Toxicol. 18(3), pp 267270.
3.Edwards, C.A. 1975. Persistent Pesticides in the Environment, 2nd edition, CRC-Press.

4-163 Effect of dl-lactic acid

Parameter	Concentration	Condition	Species
LC0	170 mg/l	26-72 hours	Daphnia[1]
LD50	1.81 g pr. kg ww	single oral dose	guinea pig[2]
LD50	3.73 g pr. kg ww	single oral dose	Rat[2]
TLm	400 mg/l	not indicated	Salmo iredius[3]
TLm	100 mg/l	18 hours	Trout[3]

Source: 1.Meinck, F., Stooff, H. and Kohlschutter, H., Les eaux residuaires industrielles, 1970.
2.Patty, F. A., Industrial Hygiene and toxicology, 2, Interscience Publishers, New York, 1976.
3.Ellis, M. M., Detection and Measurement of Stream Pollution, U.S. Department of Commerce, Bureau ofisheries, 1937.

4-164 Effect of dl-tartaric acid

Parameter	Concentration	Condition	Species
LC0	200 mg/l	long time exposure, hard water	Goldfish[1]
LC0	10 mg/l	long time exposure, soft water	Goldfish[1]
LC0	250-320 mg/l	not indicated	Paramecium caudatum[2]

Source: 1.Ellis, M. M., Detection and measurement of stream pollution, U.S. Bur. Fisheries Bull. No. 22, 365,937. 2.Meinck, F., Stooff, H. and Kohlschutter, H., Les eaux residuaires industrielles, 1970.

4-165 Effect of Dy

Parameter	Effect Value	Concentration and Condition	Species
Mortality	50 %	Chloride, intraperitonal, dose = 585 mg/kg body	Mouse[1]
Mortality	50 %	Nitrate, oral dose = 3100 mg/kg body	Rat[1]

Source: 1.Toxic Substances List 1973, U.S. Department of Health, Education and Welfare, Public Health Service National Institute for Occupational Safety and Health, Rockville Maryland, USA, June 1973.

4-166 Effect of EDC-tar

Parameter	Effect Value	Concentration and Condition	Species
Survival	0 %	EDC-tar = 100.0 ppm, 20 d, 46 banacles	Balanus balanoides[1]
Survival	0 %	EDC-tar = 50.0 ppm, 20 d, 43 banacles	Balanus balanoides[1]
Survival	100 %	EDC-tar = 0.0 ppm, 20 d, 56 banacles	Balanus balanoides[1]
Survival	100 %	EDC-tar = 2.0 ppm, 20 d, 48 banacles	Balanus balanoides[1]
Survival	100 %	EDC-tar = 5.0 ppm, 20 d, 70 banacles	Balanus balanoides[1]
Survival	45 %	EDC-tar = 10.0 ppm, 20 d, 72 banacles	Balanus balanoides[1]
Survival	6 %	EDC-tar = 25.0 ppm, 20 d, 54 banacles	Balanus balanoides[1]
Survival	95 %	EDC-tar = 1.0 ppm, 20 d, 76 banacles	Balanus balanoides[1]

Source: 1.Rosenberg, R. 1972. Environ. Pollut., 3, pp 313-318.

4-167 Effect of EPTC

Parameter	Effect Value	Concentration and Condition	Species
Mortality	10 %	10 ppm, 48 h, 14 g, 293 K, pH = 8.2, alk = 80 ppm, less than value	Ictalurus punctatus[1]

Source: 1.McCorle, F.M., J.E. Chambers and J.D. Yarbrough. 1977. Bull. Environ. Contam. Toxicol. 18(3), pp 267270.

4-168 Effect of e-caprolactam

Parameter	Concentration	Condition	Species
LC0	1000 mg/l	30 days	CatFish[1]

Source: 1.Meinck, F., Stooff, H. and Kohlschutter, H., Les eaux residuaires industrielles, 1970.

4-169 Effect of Enanthic acid

Parameter	Concentration	Condition	Species
LD50	0.3 g pr. kg ww	oral dose, 2-4 days	Mouse[1]

Source: 1.Patty, F. A., Industrial Hygiene and toxicology, 2, Interscience Publishers, New York, 1976.

4-170 Effect of Endosulfan

Parameter	Effect Value	Concentration and Condition	Species
Mortality	50 %	Acute oral toxicity = 30-220 mg/kg, insecticide	Rat[1]

Source: 1.Edwards, C.A. 1975. Persistent Pesticides in the Environment, 2nd edition, CRC-Press.

4-171 Effect of Endrin

Parameter	Effect Value	Concentration and Condition	Species
Mortality	50 % mortality	48 h, water = 20 mg/l	Daphnia pulex[1]
Mortality	95 %	Conc in soil = 10 mg/kg	Earthworms[2]
Mortality	50 %	0.8 (0.5-1.2) ppb, 96 h, 14 g, 293 K, pH = 8.2, alk = 80 ppm, 95 % CL	Ictalurus punctatus[3]
Mortality	50 %	Conc in food = 14.0 mg/kg	Pheasant[4]
Mortality	50 %	Conc in food = 5.0 mg/kg	Quail[4]
Mortality	50 %	Acute dermal toxicity = 15 mg/kg, insecticide	Rat[2]
Mortality	50 %	Acute oral toxicity = 7.5-7.8 mg/kg, insecticide	Rat[2]
Mortality	95 %	Conc in soil > 20 mg/kg	Slugs[2]

Source: 1.Rankov, V. 1973. Pochv. Agrokhim. 8, 125-134.
2.Edwards, C.A. 1975. Persistent Pesticides in the Environment, 2nd edition, CRC-Press.
3.McCorle, F.M., J.E. Chambers and J.D. Yarbrough. 1977. Bull. Environ. Contam. Toxicol. 18(3), pp 267270.
4.Dewitt, J.B., C.M. Menzie, V.A. Adomaitis and W.L. Reickel. 1960. Trans. 25th N. Am. Wildl. Conf., p277.

4-172 Effect of Epichlorohydrine

Parameter	Concentration	Condition	Species
LD50	0.18 g pr. kg ww	single oral dose	guinea pig[1]
LD50	0.24 g pr. kg ww	single oral dose	Mouse[1]
LD50	0.04 g pr. kg ww	single oral dose	Rat[1]
LC50	*23 mg/l*	*24 hours*	*Goldfish*[2]
Sublethal effect	*100 ppm*	*Eye irritation*	*Man*[1]
Sublethal effect	6.0 mg/l	Inhibition of cell multiplication	Microcystis aeruginosa[3]
Sublethal effect	55 mg/l	Inhibition of cell multiplication	Pseudomonas putida[3]

Source: 2.Shell Chemie, Shell Industrie Chemicalien gids, Shell Niederland Chemie, 1975.
3.Bringmann, G. and Kuhn, R., nwf-wasser/abwasser, H.9, 1976.
1.Patty, F. A., Industrial Hygiene and toxicology, 2, Interscience Publishers, New York, 1976.

4-173 Effect of Epinephrine

Parameter	Effect Value	Concentration and Condition	Species
Mortality	50 %	Oral dose = 4 mg/kg(= 2.18E-5mol/kg), F007 = 4.66	Mouse[1]

Source: 1.Luckey, T.D. and B. Venugopal. 1977. Metal Toxicity in Mammals, 1. Plenum Press.

4-174 Effect of Er

Parameter	Effect Value	Concentration and Condition	Species
Mortality	50 %	Chloride, citratecomplex, intraperitonal, dose = 122 mg/kg body	Mouse[1]
Mortality	50 %	Nitrate, intraperitonal, dose = 230 mg/kg body	Rat[1]

Source: 1.Toxic Substances List 1973, U.S. Department of Health, Education and Welfare, Public Health ServiceNational Institute for Occupational Safety and Health, Rockville Maryland, USA, June 1973.

4-175 Effect of Ethanol

Parameter	Concentration	Condition	Species
LC0	7000 mg/l	24 hours	Creek chub[1]
LC100	9000 mg/l	24 hours	Creek chub[1]
Sublethal effect	1450 mg/l	Inhibition of cell multiplication	Microcystis aeruginosa[2]
Sublethal effect	6500 mg/l	Inhibition of cell multiplication	Pseudomonas putida[2]
LD50	*8 g pr. kg ww*	*single oral dose*	*Rabbit*[3]
LD50	9 g pr. kg ww	single oral dose	Rat[3]

Source: 3.Patty, F. A., Industrial Hygiene and toxicology, 2, Interscience Publishers, New York, 1976.
1.Gilette, L. A., Miller, D. L. and Redman, H. E., Sewage Ind. Wastes, 24, 1397, 1952.
2.Bringmann, G. and Kuhn, R., nwf-wasser/abwasser, H.9, 1976.

4-176 Effect of Ethanolamine

Parameter	Concentration	Condition	Species
LC50	190 mg/l	24 hours	Goldfish[1]
LC50	170 mg/l	96 hours	Goldfish[1]
Sublethal effect	1.6 mg/l	Inhibition of cell multiplication, neutralized	Microcystis aeruginosa[2]
Sublethal effect	2.1 mg/l	Inhibition of cell multiplication, unneutralized	Microcystis aeruginosa[2]
Sublethal effect	8 mg/l	Inhibition of cell multiplication, neutralized	Pseudomonas putida[2]
Sublethal effect	6300 mg/l	Inhibition of cell multiplication, unneutralized	Pseudomonas putida[2]
LD50	*2.7 g pr. kg ww*	*acute oral dose*	*Rat*[3]

Source: 3.Patty, F. A., Industrial Hygiene and toxicology, 2, Interscience Publishers, New York, 1976.

1.Shell Chemie, Shell Industrie Chemicalien gids, Shell Niederland Chemie, 1975.
2.Bringmann, G. and Kuhn, R., nwf-wasser/abwasser, H.9, 1976.

4-177 Effect of Ethyl-sec-amylketone

Parameter	Concentration	Condition	Species
Sublethal effect	25-100 ppm	Eye irritation, headache	Man[2]
LD50	*2.5 g pr. kg ww*	*single oral dose*	*guinea pig*[1]
LD50	3.8 g pr. kg ww	single oral dose	Mouse[1]
LD50	3.5 g pr. kg ww	single oral dose	Rat[1]

Source: 1.Patty, F. A., Industrial Hygiene and toxicology, 2, Interscience Publishers, New York, 1976.
2.Bringmann, G. and Kuhn, R., nwf-wasser/abwasser, H.9, 1976.

4-178 Effect of Ethylacetate

Parameter	Concentration	Condition	Species
Sublethal effect	90-550 mg/l	Inhibition of cell multiplication	Microcystis aeruginosa[1]
Sublethal effect	33-650 mg/l	Inhibition of cell multiplication	Pseudomonas putida[1]

Source: 1.Bringmann, G. and Kuhn, R., nwf-wasser/abwasser, H.9, 1976.

4-179 Effect of Ethylacrylate

Parameter	Concentration	Condition	Species
LD50	0.4 g pr. kg ww	acute oral dose	Rabbit[1]
LD50	1.0 g pr. kg ww	acute oral dose	Rat[1]

Source: 1.Patty, F. A., Industrial Hygiene and toxicology, 2, Interscience Publishers, New York, 1976.

4-180 Effect of Ethylamine

Parameter	Concentration	Condition	Species
LC0	30 mg/l	24 hours	Creek chub[1]
LC0	40 mg/l	not indicated	Daphnias[2]
LC0	10 mg/l	not indicated	Scenedesmus[3]
LC100	50 mg/l	24 hours	Creek chub[1]
Sublethal effect	1.1 mg/l	Inhibition of cell multiplication, neutralized	Microcystis aeruginosa[2]
Sublethal effect	1.3 mg/l	Inhibition of cell multiplication, unneutralized	Microcystis aeruginosa[3]
Sublethal effect	5.3 mg/l	Inhibition of cell multiplication, neutralized	Pseudomonas putida[3]
Sublethal effect	29 mg/l	Inhibition of cell multiplication, unneutralized	Pseudomonas putida[3]
LD50	0.4 g pr. kg ww	oral dose	Rat[4]

Source: 4.McKee, J. E., and Wolf, H. W., Water Quality, The resources agency of California, 1963.
1.Gilette, L. A., Miller, D. L. and Redman, H. E., Sewage Ind. Wastes, 24, 1397, 1952.
2.Meinck, F., Stooff, H. and Kohlschutter, H., Les eaux residuaires industrielles, 1970.
3.Bringmann, G. and Kuhn, R., nwf-wasser/abwasser, H.9, 1976.

4-181 Effect of Ethylbenzene

Parameter	Effect Value	Concentration and Condition	Species
Mortality	50 % mortality	Water = 13 ppm, 96 h, stage 1 zoeae	Cancer magister[3]
Mortality	50 % mortality	Water = 0.49 ppm, 96 h	Crago franciscorum[4]
Mortality	50 % mortality	Water = 4.3 ppm, 96 h	Marone saxatilis[4]
LC100	50 mg/l	24 hours, seawater	Coho salmon[1]
Sublethal effect	33 mg/l	Inhibition of cell multiplication	Microcystis aeruginosa[2]
Sublethal effect	12 mg/l	Inhibition of cell multiplication	Pseudomonas putida[2]

Source: 1.Morrow, J. E., EPA/660&3/73/018, 1974.
2.Bringmann, G. and Kuhn, R., nwf-wasser/abwasser, H.9, 1976.
3.Caldwell, R.s., E. M. Caldarone and M.H. Mallon. 1977. D.A. Wolfe (ed.): Fate and Effects of Petroleum Hydrocarbons in Marine Ecosystems and Organisms. NY. Pergamon Press. 210-220.
4.Benville, P.E. Jr. and S. Korn. 1977. Calif. Fish and Game. 63, 204-209.

4-182 Effect of Ethylchloride

Parameter	Concentration	Condition	Species
Sublethal effect	2000 ppm	Histopatological changes in liver and kidney, 9 hours	Guinea pig[1]

Source: *1.Patty, F. A., Industrial Hygiene and toxicology, 2, Interscience Publishers, New York, 1976.*

4-183 Effect of Ethylchloroformate

Parameter	Concentration	Condition	Species
LD50	<0.05 g pr. kg ww	acute oral dose	Rat[1]

Source: *1.Patty, F. A., Industrial Hygiene and toxicology, 2, Interscience Publishers, New York, 1976.*

4-184 Effect of Ethylcyclohexane

Parameter	Concentration	Condition	Species
LC0	100 mg/l	96 hours, seawater, 8 C	Coho salmon[1]

Source: *1.Morrow, J. E., EPA/660&3/73/018, 1974.*

4-185 Effect of Ethylcyclopentane

Parameter	Concentration	Condition	Species
LC0	100 mg/l	96 hours, seawater, 8 C	Coho salmon[1]

Source: *1.Morrow, J. E., EPA/660&3/73/018, 1974.*

4-186 Effect of Ethylene

Parameter	Concentration	Condition	Species
Sublethal effect	0.6 ppm	Growth retardation, 720 hours	Gossypium hirsutum[1]
Sublethal effect	0.2 ppm	Growth retardation, 72 hours	Lathyrus[2]
Sublethal effect	4 ppm	Growth retardation	Lillium[3]
Sublethal effect	2 ppm	Growth retardation, 72 hours	Narscissus[3]
Sublethal effect	1 ppm	Fading, 24 hours	Vanda orchid[4]

Source: *1.Heck, W. W., Pires, E. G. and Hall, W. C., JAPCA, 11, 549, 1961.*
2.Crocker, W., Zimmerman, P. W. and Hitchcock, A. E., Contrib. Boyce Thompson Inst., 4, 177, 1932.
3.Hitchcock, A. E., Crocker, W. and Zimmerman, P. W., Contrib. Boyce Thompson Inst., 4, 155, 1932.
4.Akamine, E. K. and Sakamoto, H. I., Amer. Orchid. Soc. Bull., 20, 149, 1959.

4-187 Effect of Ethylenebenzene

Parameter	Concentration	Condition	Species
TLm	32-35 mg/l	24-96 hours, soft water	Bluegill[1]
TLm	42-49 mg/l	24-96 hours, soft/hard water	Fathead minnow[1]
TLm	94-97 mg/l	24-96 hours, soft water	Goldfish[1]
TLm	94-97 mg/l	24-96 hours, soft water	Guppie[1]

Source: *1.Pickering Q. H. and Henderson, C., JWPCF, 38, 1419, 1966.*

4-188 Effect of Ethylenebromide

Parameter	Concentration	Condition	Species
LD50	0.079 g pr. kg ww	single oral dose	chicken[1]
LD50	0.079 g pr. kg ww	single oral dose	chicken[1]
LD50	0.42 g pr. kg ww	single oral dose	Female Mouse[1]
LD50	0.079 g pr. kg ww	single oral dose	Female Rabbit[1]
LD50	0.12 g pr. kg ww	single oral dose	Female Rat[1]
LD50	0.11 g pr. kg ww	single oral dose, Female	guinea pig[1]
LD50	0.11 g pr. kg ww	single oral dose	guinea pig[1]
LD50	0.15 g pr. kg ww	single oral dose	male Rat[1]
LD50	0.42 g pr. kg ww	single oral dose, Female	Mouse[1]
LD50	0.055 g pr. kg ww	single oral dose, Female	Rabbit[1]
LD50	0.15 g pr. kg ww	single oral dose, male	Rat[1]
LD50	0.12 g pr. kg ww	single oral dose, Female	Rat[1]

Source: 1.Patty, F. A., Industrial Hygiene and toxicology, 2, Interscience Publishers, New York, 1976.

4-189 Effect of Ethylenediamine

Parameter	Concentration	Condition	Species
LC0	8 mg/l	not indicated	Daphnia[1]
LC0	200 mg/l	not indicated	Escherichia coli[2]
Sublethal effect	0.08 mg/l	Inhibition of cell multiplication, neutralized	Microcystis aeruginosa[2]
Sublethal effect	0.4 mg/l	Inhibition of cell multiplication, unneutralized	Microcystis aeruginosa[2]
Sublethal effect	0.85 mg/l	Inhibition of cell multiplication, neutralized	Pseudomonas putida[2]
Sublethal effect	0.5 mg/l	Inhibition of cell multiplication, unneutralized	Pseudomonas putida[2]
LD50	1.16 g pr. kg ww	acute oral dose	Rat[3]
LD50	1.16 g pr. kg ww	acute oral dose	Rat[3]

Source: 3.Patty, F. A., Industrial Hygiene and toxicology, 2, Interscience Publishers, New York, 1976.

1.Meinck, F., Stooff, H. and Kohlschutter, H., Les eaux residuaires industrielles, 1970.
2.Bringmann, G. and Kuhn, R., nwf-wasser/abwasser, H.9, 1976.

4-190 Effect of Ethylenediaminetetraacetic acid

Parameter	Concentration	Condition	Species
Sublethal effect	76 mg/l	Inhibition of cell multiplication	Microcystis aeruginosa[1]
Sublethal effect	105 mg/l	Inhibition of cell multiplication	Pseudomonas putida[1]

Source: 1.Bringmann, G. and Kuhn, R., nwf-wasser/abwasser, H.9, 1976.

4-191 Effect of Ethylenedichloride

Parameter	Concentration	Condition	Species
LC50	2000 mg/l	3 min, seawater, 15 C	Crangon crangon[1]
LC50	345 mg/l	1 hour, seawater, 15 C	Crangon crangon[1]

LC50	125 mg/l	8 hours, seawater, 15 C	Crangon crangon[1]
LC50	65 mg/l	24-96 hours, seawater, 15 C	Crangon crangon[1]
Sublethal effect	105 mg/l	Inhibition of cell multiplication	Microcystis aeruginosa[2]
Sublethal effect	135 mg/l	Inhibition of cell multiplication	Pseudomonas putida[2]
LD50	*0.68 g pr. kg ww*	*single oral dose*	*Rat*[3]
LD50	0.68 g pr. kg ww	single oral dose	Rat[3]

Source: 3.Patty, F. A., Industrial Hygiene and toxicology, 2, Interscience Publishers, New York, 1976.
1.Adema, D. M. M., Acute toxiciteitstoetsen met 1,2-dichlorethann, fenol, acrylonitrile, an alkylbenzeesulfonant, Delft, 1976.
2.Bringmann, G. and Kuhn, R., nwf-wasser/abwasser, H.9, 1976.

4-192 Effect of Ehyleneglycol

Parameter	Concentration	Condition	Species
LD50	8.0 g pr. kg ww	single oral dose	guinea pig[1]
LD50	8.0 g pr. kg ww	single oral dose	guinea pig[1]
LD50	14.5 g pr. kg ww	single oral dose	Mouse[1]
LD50	14.5 g pr. kg ww	single oral dose	Mouse[1]
LD50	6.0 g pr. kg ww	single oral dose	Rat[1]
LD50	6.0 g pr. kg ww	single oral dose	Rat[1]

Source: 1.Patty, F. A., Industrial Hygiene and toxicology, 2, Interscience Publishers, New York, 1976.

4-193 Effect of Ethyleneglycoldiethylether

Parameter	Concentration	Condition	Species
LD50	2.44 g pr. kg ww	single oral dose	guinea pig[1]
LD50	2.44 g pr. kg ww	single oral dose	guinea pig[1]
LD50	4.39 g pr. kg ww	single oral dose	Rat[1]
LD50	4.39 g pr. kg ww	single oral dose	Rat[1]

Source: 1.Patty, F. A., Industrial Hygiene and toxicology, 2, Interscience Publishers, New York, 1976.

4-194 Effect of Ethyleneglycolmonoethylether

Parameter	Concentration	Condition	Species
LD50	1.4-2.8 g pr. kg ww	single oral dose, 100 %	guinea pig[1]
LD50	1.4-2.8 g pr. kg ww	single oral dose, 100 %	guinea pig[1]
LD50	4.3 g pr. kg ww	single oral dose, 100 %	Mouse[1]
LD50	4.3 g pr. kg ww	single oral dose, 100 %	Mouse[1]
LD50	3.1 g pr. kg ww	single oral dose, 100 %	Rabbit[1]
LD50	3.1 g pr. kg ww	single oral dose, 100 %	Rabbit[1]
LD50	3-5.5 g pr. kg ww	single oral dose, 100 %	Rat[1]
LD50	3-5.5 g pr. kg ww	single oral dose, 100 %	Rat[1]
LC50	*>5000 mg/l*	*24 hours*	*Goldfish*[2]
Sublethal effect	100 mg/l	Inhibition of cell multiplication	Microcystis aeruginosa[3]
Sublethal effect	>10000 mg/l	Inhibition of cell multiplication	Pseudomonas putida[3]

Source: 1.Shell Chemie, Shell Industrie Chemicalien gids, Shell Niederland Chemie, 1975.
2.Bringmann, G. and Kuhn, R., nwf-wasser/abwasser, H.9, 1976.
3.Patty, F. A., Industrial Hygiene and toxicology, 2, Interscience Publishers, New York, 1976.

4-195 Effect of Ethyleneglycolmonoethyletheracetate

Parameter	Concentration	Condition	Species
LD50	1.9 g pr. kg ww	single oral dose	guinea pig[1]
LD50	1.9 g pr. kg ww	single oral dose	guinea pig[1]
LD50	1.95 g pr. kg ww	single oral dose	Rabbit[1]
LD50	1.95 g pr. kg ww	single oral dose	Rabbit[1]
LD50	5.1 g pr. kg ww	single oral dose	Rat[1]
LD50	5.1 g pr. kg ww	single oral dose	Rat[1]

Source: 1.Patty, F. A., Industrial Hygiene and toxicology, 2, Interscience Publishers, New York, 1976.

4-196 Effect of Ethyleneglycolmonomethylether

Parameter	Concentration	Condition	Species
LD50	0.95 g pr. kg ww	single oral dose, 100 %	guinea pig[1]
LD50	0.95 g pr. kg ww	single oral dose, 100 %	guinea pig[1]
LD50	0.89 g pr. kg ww	single oral dose, 100 %	Rabbit[1]
LD50	0.89 g pr. kg ww	single oral dose, 100 %	Rabbit[1]
LD50	3.4 g pr. kg ww	single oral dose, 100 %	Rat[1]
LD50	3.4 g pr. kg ww	single oral dose, 100 %	Rat[1]

Source: 1.Patty, F. A., Industrial Hygiene and toxicology, 2, Interscience Publishers, New York, 1976.

4-197 Effect of Ethyleneglycolnitrate

Parameter	Concentration	Condition	Species
Sublethal effect	2 ppm	Blood changes, 1000 days	Cat[1]

Source: 1.Patty, F. A., Industrial Hygiene and toxicology, 2, Interscience Publishers, New York, 1976.

4-198 Effect of Ethyleneglycolride

Parameter	Concentration	Condition	Species
LC0	1000 mg/l	not indicated	Scenedesmus[1]

Source: 1.Meinck, F., Stooff, H. and Kohlschutter, H., Les eaux residuaires industrielles, 1970.

4-199 Effect of Ethyleneimine

Parameter	Concentration	Condition	Species
Sublethal effect	900 ppm	Eye irritation	Man[1]
Sublethal effect	-	Carcinogenic	Man[1]
Sublethal effect	-	Carcinogenic	Man[2]
Sublethal effect	0.11 mg/l	Inhibition of cell multiplication, neutralized	Microcystis aeruginosa[3]
Sublethal effect	0.12 mg/l	Inhibition of cell multiplication, unneutralized	Microcystis aeruginosa[3]
Sublethal effect	5.5 mg/l	Inhibition of cell multiplication, neutralized	Pseudomonas putida[3]
Sublethal effect	9.0 mg/l	Inhibition of cell multiplication, unneutralized	Pseudomonas putida[3]
LD50	0.008 g pr kg ww	single oral dose	cat[4]

LD50	0.008 g pr kg ww	single oral dose	cat[4]
LD50	0.004 g pr kg ww	single oral dose	Rabbit[4]
LD50	0.004 g pr kg ww	single oral dose	Rabbit[4]
LD50	0.004-0.012 g pr kg ww	single oral dose	Rat[4]
LD50	0.004-0.012 g pr kg ww	single oral dose	Rat[4]

Source: 4.Henschler, D., Toxikologisch-arbeitsmedizinische Begrundung von MAH-werte, Verlag Chemie.
1.Federal register, 39, 1974.
2.Maximale Arbeitsplatzkonzentration 1973. BRD.
3.Bringmann, G. and Kuhn, R., nwf-wasser/abwasser, H.9, 1976.

4-200 Effect of Ethyleneoxide

Parameter	Concentration	Condition	Species
LC50	90 mg/l	24 hours	Goldfish[1]
Sublethal effect	100 ppm	Anemia, 130 times 6 hours	Dog[2]

Source: 1.Shell Chemie, Shell Industrie Chemicalien gids, Shell Niederland Chemie, 1975.2.Patty, F. A., Industrial Hygiene and toxicology, 2, Interscience Publishers, New York, 1976.

4-201 Effect of Ethylformate

Parameter	Concentration	Condition	Species
Sublethal effect	330 ppm	Eye irritation	Man[1]
LD50	*<0.05 g pr. kg ww*	*acute oral dose*	*Rat*[1]
LD50	*4.3 g pr. kg ww*	*single oral dose*	*Rat*[1]

Source: 1.Patty, F. A., Industrial Hygiene and toxicology, 2, Interscience Publishers, New York, 1976.

4-202 Effect of Ethylsilicate

Parameter	Concentration	Condition	Species
Sublethal effect	250 ppm	Eye irritation	Man[1]

Source: 1.Patty, F. A., Industrial Hygiene and toxicology, 2, Interscience Publishers, New York, 1976.

4-203 Effect of Eu

Parameter	Effect Value	Concentration and Condition	Species
Mortality	50 %	Chloride, intraperitonal, dose = 500 mg/kg body	Mouse[1]
Mortality	50 %	Nitrate, intraperitonal, dose = 210 mg/kg body	Rat[1]

Source: 1.Toxic Substances List 1973, U.S. Department of Health, Education and Welfare, Public Health ServiceNational Institute for Occupational Safety and Health, Rockville Maryland, USA, June 1973.

4-204 Effect of Fe

Parameter	Effect Value	Concentration and Condition	Species
Growth	0.001 mg/l gives deficiency symptoms	Tree, 100 d in greenhouse, conc in culture solution	Agathis australis[1]
Growth	1.200 mg/l gives normal growth	Tree, 100 d in greenhouse, conc in culture solution	Agathis australis[1]
Growth	12.0 mg/l gives toxic symptoms	Tree, 100 d in greenhouse, conc in culture solution	Agathis australis[1]
Mortality	50 % mortality	FeCl3, water = E-3 M Fe, toxic level given, microorganisms	Bacterium communis[2]

Mortality	0 %	Toxicological limit = > 35 mg/l, FeSO4	Biological purification[3]
Mortality	0 %	Toxicological limit = 1.62 - 152 mg/l, FeSO4	Crustacean[4]
Mortality	0 %	Toxicological limit = 0.9 - 152 mg/l, FeSO4	Fish[4]
Mortality	0 %	Toxicological limit = 0.5 - 5 g/kg, FeSO4	Homo sapiens[4]
Mortality	0 %	Toxicological limit = 0.5 - 5 g/kg, FeSO4	Mammalia[4]
Mortality	50 %	2-sulfate, oral dose = 1170 mg/kg body	Mouse[5]
Mortality	50 %	2-chloride, oral dose = 984 mg/kg body	Rat[5]
Mortality	50 %	2-gluconate, oral dose = 4600 mg/kg body	Rat[5]
Mortality	50 %	2-sulfate, oral dose = 1480 mg/kg body	Rat[5]
Mortality	50 %	3-chloride, oral dose = 900 mg/kg body	Rat[5]
Mortality	50 %	3-nitrate, oral dose = 3250 mg/kg body	Rat[5]
Nitrification	40 % inhibation	Soil = 5 umoles/g of soil, 10 d incubation, 3 soils, lab	Soil organisms[6]
Nitrification	49 % inhibation	Soil = 5 umoles/g of soil, 10 d incubation, 3 soils, lab	Soil organisms[6]

Source: 1.Peterson, P.J. 1962. N. Z. J. Sci. Technol. 5, pp 141-161.
2.Ishizawa, S. and T. Matsuguchi. 1966. Bull. natn Inst. agric. Sci., Tokyo (B) 16, 1-90.
3.Jung. K.D. 1973. Wirkungskonzentration schaedigender Stoffe im Wasser. Gelenkirchen. Hygiene-Institudes Ruhrgebiets.
4.Liebman, H. 1958. Handbuch der Friswasser - und Abwasserbiologie. Band II.Munchen. Oldenbourg.
5.Toxic Substances List 1973, U.S. Department of Health, Education and Welfare, Public Health ServiceNational Institute for Occupational Safety and Health, Rockville Maryland, USA, June 1973.
6.Liang, C.N. and M.A. Tabatabai. 1978. Environ. Pollut. 12, 141-147.

4-205 Effect of Fenac

Parameter	Effect Value	Concentration and Condition	Species
Mortality	50 % mortality	48 h, water = 4500 mg/l	Daphnia pulex[1]
Mortality	50 %	Acute oral toxicity = 1780 mg/kg, herbicide	Rat[2]

Source: 1.Rankov, V. 1973. Pochv. Agrokhim. 8, 125-134.2.Edwards, C.A. 1975. Persistent Pesticides in the Environment, 2nd edition, CRC-Press.

4-206 Effect of Fenoprop

Parameter	Effect Value	Concentration and Condition	Species
Mortality	50 % mortality	48 h, water = 2400 mg/l	Daphnia pulex[1]

Source: 1.Rankov, V. 1973. Pochv. Agrokhim. 8, 125-134.

4-207 Effect of Fluometuron

Parameter	Effect Value	Concentration and Condition	Species
Mortality	10 %	10 ppm, 48 h, 14 g, 293 K, pH = 8.2, alk = 80 ppm, less than value	Ictalurus punctatus[1]

Source: 1.McCorle, F.M., J.E. Chambers and J.D. Yarbrough. 1977. Bull. Environ. Contam. Toxicol. 18(3), pp 267270.

4-208 Effect of Fluoranthene

Parameter	Effect Value	Concentration and Condition	Species
Mortality	50 % mortality	Water = 0.5 ppm, 96 h, marine	Neathes arenaceodentata[1]

Source: 1.Rossi, S.S. and J.M. Neff. 1978. Mar. Pollut. Bull. 9, 220-223.

4-209 Effect of Fluorene

Parameter	Effect Value	Concentration and Condition	Species
Mortality	50 % mortality	Water = 44 ppm, naplii sedimentated (thought dead), 1 h, S=3 %	Balanus amphitrites[1]
Mortality	50 % mortality	Water = 1.68 ppm, 96 h, sheep head minnow	Cyprinodon variegatus[2]
Mortality	50 % mortality	Water = 1.0 ppm, 96 h, marine	Neathes arenaceodentata[3]
Mortality	50 % mortality	Water = 0.32 ppm, 96 h	Palaemonetes pugio[2]

Source: 1.Donahue, W.H., R.T. Wang, M. Welch and J.A.C. Nicol. 1977. Environ. Pollut. 13, 187-202.
2.Neff, J.M. 1979. Polycyclic Aromatic Hydrocarbons in the Aquatic Environment. Applied Sci. Publ. London.
3.Rossi, S.S. and J.M. Neff. 1978. Mar. Pollut. Bull. 9, 220-223.

4-210 Effect of Fluoride

Parameter	Effect Value	Concentration and Condition	Species
Mortality	50 %	Na-flouride, oral dose = 125 mg/kg(= 2.98E-3mol/kg), F007 = 2.53	Mouse[1]
Mortality	0.01 %	1 mg/l, freshwater	Rainbow trout[2]

Source: 1.Luckey, T.D. and B. Venugopal. 1977. Metal Toxicity in Mammals, 1. Plenum Press.
2.Herbert, D.W.M. and D.S. Shurben. 1964. Water Waste Treat. J. 10, pp 141-142.

4-211 Effect of Fluoroacetate

Parameter	Effect Value	Concentration and Condition	Species
Mortality	50 %	Dose = 0.2 mg/kg, intravenous	Cat[1]
Mortality	50 %	Dose: > 500 mg/kg, intraperitonal, South Africa	Clawed toad[1]
Mortality	50 %	Dose = 0.06 mg/kg, intravenous	Dog[1]
Mortality	50 %	Dose = 150 mg/kg, subcutant	Frog[1]
Mortality	50 %	Dose = 0.35 mg/kg, intraperitonal	Guinea pig[1]
Mortality	50 %	Dose = 2-10 mg/kg, oral	Homo sapiens[1]
Mortality	50 %	Dose = 1.0 mg/kg, oral	Horse[1]
Mortality	50 %	Dose = 7 mg/kg, intraperitonal	Mouse[1]
Mortality	50 %	Dose = 0.5 mg/kg, intravenous	Rabbit[1]
Mortality	50 %	Dose = 5 mg/kg, intraperitonal or subcutant	Rat[1]
Mortality	50 %	Dose = 0.3 mg/kg, oral	Sheep[1]

Source: 1.Pattison, F.L.M. 1959. Toxic Aliphatic Fluorine Compounds

4-212 Effect of Formaldehyde

Parameter	Concentration	Condition	Species
Sublethal effect	0.39 mg/l	Inhibition of cell multiplication, 35 % solution	Microcystis aeruginosa[1]
Sublethal effect	14 mg/l	Inhibition of cell multiplication, 35 % solution	Pseudomonas putida[1]
TLm	50-200 mg/l	24-96 hours, soft water	Guppie[2]
LD50	*0.26 g pr. kg ww*	*single oral dose*	*guinea pig*[3]
LD50	0.1-0.8 g pr. kg ww	single oral dose	Rat[3]

Source: 3.Patty, F. A., Industrial Hygiene and toxicology, 2, Interscience Publishers, New York, 1976.

1.Bringmann, G. and Kuhn, R., nwf-wasser/abwasser, H.9, 1976.
2.Meinck, F., Stooff, H. and Kohlschutter, H., Les eaux residuaires industrielles, 1970.

4-213 Effect of Formic acid

Parameter	Concentration	Condition	Species
LC0	1000 mg/l	not indicated	Escherichia coli[1]

Source: 1.Meinck, F., Stooff, H. and Kohlschutter, H., Les eaux residuaires industrielles, 1970.

4-214 Effect of Furfural

Parameter	Concentration	Condition	Species
LC0	10 mg/l	not indicated	Gambusia affinis[1]
Sublethal effect	2.7 mg/l	Inhibition of cell multiplication	Microcystis aeruginosa[2]
Sublethal effect	16 mg/l	Inhibition of cell multiplication	Pseudomonas putida[2]
TLm	1.2-32 mg/l	24-96 hours, soft water	Bluegill[3]
TLm	1.2-32 mg/l	24-96 hours, soft water	Bluegill[4]
TLm	24 mg/l	96 hours, soft water	Gambusia affinis[1]
TLm	24-44 mg/l	24-96 hours, soft water	Mosquito fish[5]
LD50	*0.65 g pr. kg ww*	*acute oral dose*	*dog*[6]
LD50	0.5 g pr. kg ww	acute oral dose	Mouse[6]
LD50	0.05-0.1 g pr. kg ww	acute oral dose	Rat[6]

*Source: 6.Patty, F. A., Industrial Hygiene and toxicology, 2, Interscience Publishers, New York, 1976.
1.Meinck, F., Stooff, H. and Kohlschutter, H., Les eaux residuaires industrielles, 1970.
2.Bringmann, G. and Kuhn, R., nwf-wasser/abwasser, H.9, 1976.
3.Turnbull, H., DeMann, J. H. and Weston, R. F., Ind. Eng. Chem., 46, 324, 1954.
4.Jones, H. R., Environmental Control in the organic and petrochemical Industries, Noyes Data Corp., 171.5.Wallen, I. E., Greer, W. C. and Lasater, R., Sewage Ind. Wastes, 29, 695, 1957.*

4-215 Effect of Furfuryl

Parameter	Concentration	Condition	Species
LD50	0.275 g pr. kg ww	single oral dose	Rat[1]

Source: 1.Patty, F. A., Industrial Hygiene and toxicology, 2, Interscience Publishers, New York, 1976.

4-216 Effect of Furfurylalcohol

Parameter	Concentration	Condition	Species
Sublethal effect	5.2 mg/l	Inhibition of cell multiplication	Microcystis aeruginosa[1]
Sublethal effect	50 mg/l	Inhibition of cell multiplication	Pseudomonas putida[1]

Source: 1.Bringmann, G. and Kuhn, R., nwf-wasser/abwasser, H.9, 1976.

4-217 Effect of Ga

Parameter	Effect Value	Concentration and Condition	Species
Mortality	50 %	Nitrate, oral dose = 3280 mg/kg body	Guinea pig[1]
Mortality	50 %	Chloride, intraperitonal, dose = 5 mg/kg body	Mouse[1]

Source: 1.Toxic Substances List 1973, U.S. Department of Health, Education and Welfare, Public Health ServiceNational Institute for Occupational Safety and Health, Rockville Maryland, USA, June 1973.

4-218 Effect of Gd

Parameter	Effect Value	Concentration and Condition	Species
Mortality	50 %	Chloride, citratecomplex, intraperitonal, dose = 153 mg/kg body	Mouse[1]
Mortality	50 %	Nitrate, intraperitonal, dose = 225 mg/kg body	Rat[1]

Source: 1.Toxic Substances List 1973, U.S. Department of Health, Education and Welfare, Public Health ServiceNational Institute for Occupational Safety and Health, Rockville Maryland, USA, June 1973.

4-219 Effect of Ge

Parameter	Effect Value	Concentration and Condition	Species
Mortality	50 %	Oxide, intraperitonal, dose = 750 mg/kg body	Rat[1]

Source: 1.Toxic Substances List 1973, U.S. Department of Health, Education and Welfare, Public Health ServiceNational Institute for Occupational Safety and Health, Rockville Maryland, USA, June 1973.

4-220 Effect of Glutaric acid

Parameter	Concentration	Condition	Species
LC0	1000 mg/l	not indicated	Scenedesmus[1]

Source: 1.Meinck, F., Stooff, H. and Kohlschutter, H., Les eaux residuaires industrielles, 1970.

4-221 Effect of Glycerol

Parameter	Concentration	Condition	Species
LC50	>5000 mg/l	24 hours	Goldfish[1]
Sublethal effect	2900 mg/l	Inhibition of cell multiplication	Microcystis aeruginosa[2]
Sublethal effect	>10000 mg/l	Inhibition of cell multiplication	Pseudomonas putida[2]

Source: 1.Shell Chemie, Shell Industrie Chemicalien gids, Shell Niederland Chemie, 1975.
2.Bringmann, G. and Kuhn, R., nwf-wasser/abwasser, H.9, 1976.

4-222 Effect of Glycidol

Parameter	Concentration	Condition	Species
LD50	0.85 g pr. kg ww	single oral dose	Mouse[1]
LD50	0.45 g pr. kg ww	single oral dose	Rat[1]

Source: 1.Patty, F. A., Industrial Hygiene and toxicology, 2, Interscience Publishers, New York, 1976.

4-223 Effect of Glycolic acid

Parameter	Concentration	Condition	Species
LD50	1.6-3.2 g pr. kg ww	single oral dose	Rat[1]

Source: 1.Patty, F. A., Industrial Hygiene and toxicology, 2, Interscience Publishers, New York, 1976.

4-224 Effect of HCDD

Parameter	Effect Value	Concentration and Condition	Species
Mortality	50 %	Oral dose = 100 mg/kg body wt.	Rat[1]
Teratogenicity	0 % effect	Dose = 0.1 microgram/kg*day, from day 6-15 in pregnancy-periode	Rat[1]
Teratogenicity	5 % effect	Dose = 10-100 microgram/kg*day, from day 6-15 in pregnancy-periode	Rat[1]

Source: 1.Schewets, B.A. et al. 1973. Environ. health Persp., 9. pp 233-240.

4-225 Effect of Haptanal

Parameter	Concentration	Condition	Species
LD50	25 g pr. kg ww	acute oral dose	Mouse[1]

Source: 1.Duvel, W. A. and Helfgott, T., JWPCF, 47, 1975.

4-226 Effect of Heptachlor

Parameter	Effect Value	Concentration and Condition	Species
Mortality	50 %	Conc in food = 92 mg/kg ww(76-113), 95 % CL	Bobwhite[1]
Mortality	50 % mortality	48 h, water = 42 mg/l	Daphnia pulex[2]
Mortality	95 %	Conc in soil = 20 mg/kg	Earthworms[1]
Mortality	50 %	Conc in food = 93 mg/kg ww(74-116), 95 % CL	Japanese quail[1]
Mortality	50 %	Conc in food = 480 mg/kg ww(389-570), 95 % CL	Mallard[1]
Mortality	50 %	Conc in food = 150.0-400.0 mg/kg ww	Pheasant[3]
Mortality	50 %	Conc in food = 224 mg/kg ww(191-265), 95 % CL	Pheasant[1]
Mortality	50 %	Conc in food = 125.0 mg/kg ww	Quail[3]
Mortality	50 %	Acute dermal toxicity = 195 mg/kg	Rat[1]
Mortality	50 %	Acute oral toxicity = 60-100 mg/kg	Rat[1]

Source: 1.Edwards, C.A. 1975. Persistent Pesticides in the Environment, 2nd edition, CRC-Press.
2.Rankov, V. 1973. Pochv. Agrokhim. 8, 125-134.
3.Dewitt, J.B., C.M. Menzie, V.A. Adomaitis and W.L. Reickel. 1960. Trans. 25th N. Am. Wildl. Conf., p277.

4-227 Effect of Hexachloro-1,3-butadien

Parameter	Effect Value	Concentration and Condition	Species
Body weight	-7 %	HCBD = 30.0 ppm, control = 8 %, 49 d, 6 fish	Carassius auratus[1]
Body weight	1 %	HCBD = 3.0 ppm, control = 8 %, 49 d, 6 fish	Carassius auratus[1]
Body weight	11 %	HCBD = 0.3 ppm, control = 8 %, 49 d, 6 fish	Carassius auratus[1]
Body weight	27 %	HCBD = 3.0 ppm, control = 35 %, 67 d, 6 fish	Carassius auratus[1]
Body weight	28 %	HCBD = 9.6 ppm, control = 35 %, 67 d, 6 fish	Carassius auratus[1]
Body weight	35 %	HCBD = 0.3 ppm, control = 35 %, 67 d, 6 fish	Carassius auratus[1]

Body weight	4 %	HCBD = 9.6 ppm, control = 8 %, 49 d, 6 fish	Carassius auratus[1]
Body weight	6 %	HCBD = 30.0 ppm, control = 35 %, 67 d, 6 fish	Carassius auratus[1]
G-6-P	115 micro-m/(g liver*10 min)	HCBD = 3.0 ppm, control = 137 mym, 49 d, 6 fish	Carassius auratus[1]
G-6-P	130 micro-m/(g liver*10 min)	HCBD = 0.3 ppm, control = 137 mym, 49 d, 6 fish	Carassius auratus[1]
G-6-P	84 micro-m/(g liver*10 min)	HCBD = 9.6 ppm, control = 137 mym, 49 d, 6 fish	Carassius auratus[1]
G-6-PDH	1.7 micro-m/(g liver*hour)	HCBD = 0.3 ppm, control = 1.6 myh, 49 d, 6 fish	Carassius auratus[1]
G-6-PDH	2.1 micro-m/(g liver*hour)	HCBD = 3.0 ppm, control = 1.6 myh, 49 d, 6 fish	Carassius auratus[1]
G-6-PDH	2.3 micro-m/(g liver*hour)	HCBD = 9.6 ppm, control = 1.6 myh, 49 d, 6 fish	Carassius auratus[1]
Hematocrit	40.8 %	HCBD = 0.3 ppm, control = 46.3 %, 49 d, 6 fish	Carassius auratus[1]
Hematocrit	41.9 %	HCBD = 3.0 ppm, control = 46.3 %, 49 d, 6 fish	Carassius auratus[1]
Hematocrit	42.8 %	HCBD = 30.0 ppm, control = 46.3 %, 49 d, 6 fish	Carassius auratus[1]
Hematocrit	44.6 %	HCBD = 9.6 ppm, control = 46.3 %, 49 d, 6 fish	Carassius auratus[1]
Hemoglobin	8.9 g/100 ml	HCBD = 0.3 ppm, control = 10.2 g/100 ml, 49 d, 6 fish	Carassius auratus[1]
Hemoglobin	8.9 g/100 ml	HCBD = 30.0 ppm, control = 10.2 g/100 ml, 49 d, 6 fish	Carassius auratus[1]
Hemoglobin	9.1 g/100 ml	HCBD = 3.0 ppm, control = 10.2 g/100 ml, 49 d, 6 fish	Carassius auratus[1]
Hemoglobin	9.4 g/100 ml	HCBD = 9.6 ppm, control = 10.2 g/100 ml, 49 d, 6 fish	Carassius auratus[1]
Liver weight	4.1 % of ww	HCBD = 3.0 ppm, control = 4.2 %, 49 d, 6 fish	Carassius auratus[1]
Liver weight	4.2 % of ww	HCBD = 0.3 ppm, control = 4.2 %, 49 d, 6 fish	Carassius auratus[1]
Liver weight	5.4 % of ww	HCBD = 9.6 ppm, control = 4.2 %, 49 d, 6 fish	Carassius auratus[1]
Mortality	50 %	HCBD = 0.13 ppm, 96 h, 290.5 K, 6*5 aminals	Asellus aquaticus[1]
Mortality	50 %	HCBD = 0.16 ppm, 72 h, 290.5 K, 6*5 aminals	Asellus aquaticus[1]
Mortality	50 %	HCBD = 0.09 ppm, 96 h, 290.5 K, 5*5 fish	Carassius auratus[1]
Mortality	50 %	HCBD = 0.21 ppm, 24 h, 290.5 K, 5*10 specimens	Lymnaea stagnalis[1]
Mortality	50 %	HCBD = 0.21 ppm, 48 h, 290.5 K, 5*10 specimens	Lymnaea stagnalis[1]
Mortality	50 %	HCBD = 0.21 ppm, 72 h, 290.5 K, 5*10 specimens	Lymnaea stagnalis[1]
Mortality	50 %	HCBD = 0.21 ppm, 96 h, 290.5 K, 5*10 specimens	Lymnaea stagnalis[1]
Phenylalanine hydroxylase	5.5 micro-u/ml	HCBD = 0.3 ppm, control = 6.7 myu, 67 d, 6 fish	Carassius auratus[1]
Phenylalanine hydroxylase	5.8 micro-u/ml	HCBD = 30.0 ppm, control = 6.7 myu, 67 d, 6 fish	Carassius auratus[1]

Phenylalanine hydroxylase	7.9 micro-u/ml	HCBD = 3.0 ppm, control = 6.7 myu, 67 d, 6 fish	Carassius auratus[1]
Phenylalanine hydroxylase	9.1 micro-u/ml	HCBD = 9.6 ppm, control = 6.7 myu, 67 d, 6 fish	Carassius auratus[1]
Phenylalanine hydroxylase, liver	14.6 micro-m tyrosine/g of liver /h	HCBD = 0.3 ppm, control = 13.9 myt, 67 d, 6 fish	Carassius auratus[1]
Phenylalanine hydroxylase, liver	15.7 micro-m tyrosine/g of liver /h	HCBD = 3.0 ppm, control = 13.9 myt, 67 d, 6 fish	Carassius auratus[1]
Phenylalanine hydroxylase, liver	18.1 micro-m tyrosine/g of liver /h	HCBD = 9.6 ppm, control = 13.9 myt, 67 d, 6 fish	Carassius auratus[1]
Phenylalanine hydroxylase, liver	21.1 micro-m tyrosine/g of liver /h	HCBD = 30.0 ppm, control = 13.9 myt, 67 d, 6 fish	Carassius auratus[1]
Serum glutamic pyruvic transaminase	4.2 micro-u/ml	HCBD = 0.3 ppm, control = 5.5 myu, 67 d, 6 fish	Carassius auratus[1]
Serum glutamic pyruvic transaminase	4.2 micro-u/ml	HCBD = 3.0 ppm, control = 5.5 myu, 67 d, 6 fish	Carassius auratus[1]
Serum glutamic pyruvic transaminase	4.9 micro-u/ml	HCBD = 9.6 ppm, control = 5.5 myu, 67 d, 6 fish	Carassius auratus[1]
Serum glutamic pyruvic transaminase	5.0 micro-u/ml	HCBD = 30.0 ppm, control = 5.5 myu, 67 d, 6 fish	Carassius auratus[1]

Source: 1.Leeuwangh, P., H. Bult and L. Schneiders. 1975. in: Koeman, J.H. et al.(ed): Sublethal Effects of Toxic Chemicals on Aquatic Animals, Elsevier.N.Y., pp 646-658.

4-228 Effect of Hexachlorobenzene

Parameter	Effect Value	Concentration and Condition	Species
Mortality	50 %	Acute oral toxicity = 10000 mg/kg, fungicide	Rat[2]
LD50	*6-7.5 g pr. kg ww*	*single oral dose, 100 %*	*Rat, Mouse*[1]

Source: 1.Patty, F. A., Industrial Hygiene and toxicology, 2, Interscience Publishers, New York, 1976. 2.Edwards, C.A. 1975. Persistent Pesticides in the Environment, 2nd edition, CRC-Press.

4-229 Effect of Hexachlorocyclohexane

Parameter	Concentration	Condition	Species
Sublethal effect	0.3 mg/l	Inhibition of cell multiplication	Microcystis aeruginosa[1]
Sublethal effect	>5 mg/l	Inhibition of cell multiplication	Pseudomonas putida[1]

Source: 1.Bringmann, G. and Kuhn, R., nwf-wasser/abwasser, H.9, 1976.

4-230 Effect of Hexachloroethane

Parameter	Concentration	Condition	Species
LD50	0.325 g pr. kg ww	intravenous dose, 100 %	dog[1]

Source: 1.Patty, F. A., Industrial Hygiene and toxicology, 2, Interscience Publishers, New York, 1976.

4-231 Effect of Hexamethylene 1,6-diisocyanate

Parameter	Concentration	Condition	Species
LD50	1 g pr. kg ww	single oral dose	cat[1]
LD50	0.6-0.9 g pr. kg ww	single oral dose	rabbit, rat, Mouse[1]

Source: 1.Patty, F. A., Industrial Hygiene and toxicology, 2, Interscience Publishers, New York, 1976.

4-232 Effect of Hexyleneglycol

Parameter	Concentration	Condition	Species
LC50	>5000 mg/l	24 hours	Goldfish[1]
LD50	*3.5 g pr. kg ww*	*single oral dose*	*Mouse*[2]
LD50	4.4 g pr. kg ww	single oral dose	Rat[2]

Source: 2.Patty, F. A., Industrial Hygiene and toxicology, 2, Interscience Publishers, New York, 1976.
1.Shell Chemie, Shell Industrie Chemicalien gids, Shell Niederland Chemie, 1975.

4-233 Effect of Hf

Parameter	Effect Value	Concentration and Condition	Species
Mortality	50 %	Chloride, intraperi., dose=112 mg/kg ww	Mouse[1]
Mortality	50 %	Oxychloride, intraperi., dose=113 mg/kg ww	Mouse[1]
Mortality	50 %	Gluconate, intraperitonal, dose =2000 mg/kg ww	Rat[1]

Source: 1.Toxic Substances List 1973, U.S. Department of Health, Education and Welfare, Public Health ServiceNational Institute for Occupational Safety and Health, Rockville Maryland, USA, June 1973.

4-234 Effect of Hg

Parameter	Effect Value	Concentration and Condition	Species
Chromagens	12.5 % chromagens	Control = 18.8 % Hg = 2 ppm, after 14d, 299k, fresh, Lake Houston,	Bacterial colonies[1]
Chromagens	3.8 % chromagens	Control = 19.4 %, Cu = 2 ppm, after 14d, 299k, brackish, Clear Lake,	Bacterial colonies[1]
Corpuscular hemoglobin in blood	17.8 g/100 ml packed red cells	Hg = 10 ppb, 60 d, 26 fish, rg = +- 0.6	Pseudopleuronectes americanus[2]
Corpuscular hemoglobin in blood	18.9 g/100 ml packed red cell	Hg = 5 ppb, 60 d, 22 fish, rg = +- 0.4	Pseudopleuronectes americanus[2]
Corpuscular hemoglobin in blood	19.4 g/100 ml packed red cells	Hg = 0.0 ppb, 60 d, 29 fish, rg = +- 0.8	Pseudopleuronectes americanus[2]
Corpuscular hemoglobin in blood	23.5 picograms/cell	Hg = 10 ppb, 60 d, 11 fish, rg = +- 1.9	Pseudopleuronectes americanus[2]
Corpuscular hemoglobin in blood	26.5 picograms/cell	Hg = 5 ppb, 60 d, 11 fish, rg = +- 1.5	Pseudopleuronectes americanus[2]
Corpuscular hemoglobin in blood	30.7 picograms/cell	Hg = 0.0 ppb, 60 d, 12 fish, rg = +- 2.0	Pseudopleuronectes americanus[2]
Corpuscular volume in blood	127.0 U**3/cell	Hg = 10 ppb, 60 d, 10 fish, rg = +- 6.0	Pseudopleuronectes americanus[2]
Corpuscular volume in blood	145.0 U**3/cell	Hg = 5 ppb, 60 d, 12 fish, rg = +- 12.0	Pseudopleuronectes americanus[2]
Corpuscular volume in blood	151.0 U**3/cell	Hg = 0.0 ppb, 60 d, 11 fish, rg = +- 10.5	Pseudopleuronectes americanus[2]
Hematocrit	23.0 % packed red cells	Hg = 10 ppb, 60 d, 27 fish, rg = +- 1.4	Pseudopleuronectes americanus[2]
Hematocrit	28.0 % packed red cells	Hg = 0.0 ppb, 60 d, 25 fish, rg = +- 1.6	Pseudopleuronectes americanus[2]
Hematocrit	29.0 % packed red cells	Hg = 5 ppb, 60 d, 25 fish, rg = +- 1.5	Pseudopleuronectes americanus[2]
Hemoglobin	4.2 g/100 ml wholeeblood	Hg = 10 ppb, 60 d, 27 fish, rg = +- 0.2	Pseudopleuronectes americanus[2]

Hemoglobin	5.3 g/100 ml whole blood	Hg = 0.0 ppb, 60 d, 27 fish, rg = +- 0.3	Pseudopleuronectes americanus[2]
Hemoglobin	5.4 g/100 ml whole blood	Hg = 5 ppb, 60 d, 22 fish, rg = +- 0.2	Pseudopleuronectes americanus[2]
Mortality	50 % mortality	Hg metal, 0.1 mg/ml, toxic level given, microorganisms	Aspergillus niger[3]
Mortality	50 % mortality	Water = 7.4 mg HgCl2/l, toxic level given, microorganisms	Azotobacter sp.[4]
Mortality	50 % mortality	HgCl2, water = E-5 M Hg, toxic level given, microorganisms	Bacterium communis[5]
Mortality	50 %	HgCl2, expos.time = 2 d, conc in water = 3300-10000 microgram/l,	Crangon crangon[6]
Mortality	0 %	Toxicological limit = 0.03 - 0.1 mg/l, HgCl2	Crustacean[7]
Mortality	50 %	HgCl2, expos.time = 2 d, conc in water = 5 microgram/l, limnol.	Daphnia magna[8]
Mortality	50 % mortality	Water = 7.4 mg HgCl2/l, toxic level given, microorganisms	Escherichia coli[4]
Mortality	0 %	Toxicological limit = 0.15 - 0.25 mg/l, HgCl2	Fish[7]
Mortality	50 % mortality	Water = 7.4 mg HgCl2/l, toxic level given, microorganisms	Hansenula sp.[4]
Mortality	0 %	Toxicological limit = 0.1 - 1 g, HgCl2	Mammalia[7]
Mortality	50 %	HgCl2, oral dose = 5 mg/kg(= 1.84E-5mol/kg), F007 = 4.74	Mouse[9]
Mortality	50 %	Hydroxymethyl, intraperitonal, dose = 17 mg/kg body	Mouse[10]
Mortality	50 %	HgCl2 = 0.7 +- 1.9 ppm, 96 h, 25 ppm CaCO3, 293 K, pH = 7.4-7.9	Philodina sp.[11]
Mortality	50 %	HgCl2 = 1.0 +- 2.2 ppm, 24 h, 25 ppm CaCO3, 293 K, pH = 7.4-7.9	Philodina sp.[11]
Mortality	50 %	HgCl2 = 1.3 +- 2.1 ppm, 48 h, 25 ppm CaCO3, 293 K, pH = 7.4-7.9	Philodina sp.[11]
Mortality	50 % mortality	Water = 7.4 mg HgCl2/l, toxic level given, microorganisms	Pseudomonas sp.[4]
Mortality	50 %	1-chloride, oral dose = 210 mg/kg body	Rat[12]
Mortality	50 %	2-chloride, oral dose = 37 mg/kg body	Rat[12]
Mortality	50 %	Acute oral toxicity = 30 mg/kg, pesticide, ethyl mercuric chloride	Rat[13]
Mortality	50 %	Acute oral toxicity = 37 mg HgCl/kg, pesticide	Rat[13]
Mortality	50 %	Acute oral toxicity = 570 mg/kg, methoxyethylmercury	Rat[13]
Mortality	50 %	Chloroethyl, oral dose = 30 mg/kg body	Rat[12]
Mortality	50 %	Methoxyethylacetate, oral dose = 10 mg/kg body	Rat[10]
Mortality	50 %	Phenylacetate, oral dose = 22 mg/kg body	Rat[10]
Mortality	50 % mortality	Water = 7.4 mg HgCl2/l, toxic level given, microorganisms	Rhodotorula sp.[4]
Mortality	50 %	278 K, 0.40 mg Hg/l, HgCl	Salmo gairdneri[14]
Mortality	50 %	283 K, 0.28 mg Hg/l, HgCl	Salmo gairdneri[14]
Mortality	50 %	293 K, 0.22 mg Hg/l, HgCl	Salmo gairdneri[14]

Mortality	0 %	Toxicological limit = 0.018 mg/l, HgCl2	Self Purification[15]
Mortality	50 % mortality	Water = 7.4 mg HgCl2/l, toxic level given, microorganisms	Serratia sp.[4]
Mortality	50 % mortality	HgCl2, water = 7.6E-6 M Hg, toxic level given, microorganisms	Tetrahymena sp.[16]
Mortality	50 % mortality	Water = 7.4 mg HgCl2/l, toxic level given, microorganisms	Torula sp.[4]
Nitrification	96 % inhibation	Soil = 5 umoles/g of soil, 10 d incubation, 3 soils, lab	Soil organisms[17]
Numbers of colony forming units	6.25 (log-unit)	Control = 5.25, Hg = 2 ppm, after 14 d, 299 K, fresh, Lake Houston	Bacterial colonies[1]
Numbers of colony forming units	6.94 (log-unit)	Control = 4.62, Hg = 2 ppm, after 14 d, 299 K, brackish, Clear Lake	Bacterial colonies[1]
Nutrient uptake	0.03 ppm Hg	Hg = 0.04 ppm, control = 0.02, water = 0.01 ppm Hg, 298 K, fresh water	Bacterial cells[18]
Nutrient uptake	0.03 ppm I	Hg = 0.04 ppm, control = 0.02, water = 0.1 ppm I, 298 K, Ash River	Bacterial cells[18]
Nutrient uptake	0.04 ppm Hg	Hg = 0.04 ppm, control = 0.02, water = 0.03 ppm Hg, 298 K, Ash River	Bacterial cells[18]
Nutrient uptake	0.05 ppm Co	Hg = 0.04 ppm, control = 0.08, water = 0.1 ppm Co, 298 K, Ash River	Bacterial cells[18]
Nutrient uptake	0.08 ppm Co	Hg = 0.04 ppm, control = 0.04, water = 0.04 ppm Co, 298 K, brackish	Bacterial cells[18]
Nutrient uptake	0.08 ppm I	Hg = 0.04 ppm, cont = 0.12 ppm I with wat = 0.05 ppm I, 298 K, fresh	Bacterial cells[18]
Nutrient uptake	0.08 ppm Mn	Hg = 0.04 ppm, control = 0.07, water = 0.07 ppm Mn, 298 K, Ash River	Bacterial cells[18]
Nutrient uptake	0.09 ppm Br	Hg = 0.04 ppm, control = 0., water = 0.1 ppm Br, 298 K, Ash River	Bacterial cells[18]
Nutrient uptake	0.10 ppm Co	Hg = 0.04 ppm, control = 0.03, water = 0.06 ppm Co, 298 K, fresh wat,	Bacterial cells[18]
Nutrient uptake	0.2 ppm Hg	Hg = 0.04 ppm, control = 0.08, water = 0.008 ppm Hg, 298 K, brackish	Bacterial cells[18]
Nutrient uptake	0.2 ppm I	Hg = 0.04 ppm, control = 0.1, water = 0.1 ppm I, 298 K, brackish	Bacterial cells[18]
Nutrient uptake	0.3 ppm Cr	Hg = 0.04 ppm, control = 0.1, water = 0.4 ppm Cr, 298 K, brackish	Bacterial cells[18]
Nutrient uptake	0.4 ppm Br	Hg = 0.04 ppm, control = 0.1, water = 0.2 ppm Br, 298 K, brackish	Bacterial cells[18]
Nutrient uptake	0.48 ppm Mn	Hg = 0.04 ppm, control = 0.07, water = 0.03 ppm Mn, 298 K, fresh water	Bacterial cells[18]
Nutrient uptake	0.5 ppm Cr	Hg = 0.04 ppm, control = 0.4, water = 0.2 ppm Cr, 298 K, Ash River	Bacterial cells[18]
Nutrient uptake	0.5 ppm Zn	Hg = 0.04 ppm, control = 0.6, water = 0.4 ppm Zn, 298 K, Ash River	Bacterial cells[18]
Nutrient uptake	0.8 ppm Br	Hg = 0.04 ppm, cont = 1.2 ppm Br with wat = 9.0 ppm Br, 298 K, fresh water	Bacterial cells[18]
Nutrient uptake	0.9 ppm Cu	Hg = 0.04 ppm, control = 0.7, water = 0.4 ppm Cu, 298 K, Ash River	Bacterial cells[18]
Nutrient uptake	1.0 ppm Cr	Hg = 0.04 ppm, control = 0.65, water = 0.25 ppm Cr, 298 K, fresh water	Bacterial cells[18]

Nutrient uptake	1.0 ppm Mn	Hg = 0.04 ppm, control = 0.3, water = 0.09 ppm Mn, 298 K, brackish	Bacterial cells[18]
Nutrient uptake	1.3 ppm Ti	Hg = 0.04 ppm, control = 1.3, water = 0.9 ppm Ti, 298 K, Ash River	Bacterial cells[18]
Nutrient uptake	10.0 ppm Cu	Hg = 0.04 ppm, control = 1.2, water = 5.4 ppm Cu, 298 K, fresh water	Bacterial cells[18]
Nutrient uptake	10.0 ppm K	Hg = 0.04 ppm, control = 41.0, water = 80.0 ppm K, 298 K, fresh water	Bacterial cells[18]
Nutrient uptake	10.4 ppm Ca	Hg = 0.04 ppm, control = 4.2, water = 9.2 ppm Ca, 298 K, Ash River	Bacterial cells[18]
Nutrient uptake	100.0 ppm Fe	Hg = 0.04 ppm, control = 48.0, water = 16.0 ppm Fe, 298 K, brackish	Bacterial cells[18]
Nutrient uptake	11.0 ppm Mg	Hg = 0.04 ppm, control = 26.9, water = 236.0 ppm Mg, 298 K, fresh water	Bacterial cells[18]
Nutrient uptake	11.0 ppm Ti	Hg = 0.04 ppm, control = 8.3, water = 1.1 ppm Ti, 298 K, brackish	Bacterial cells[18]
Nutrient uptake	11.7 ppm Cl	Hg = 0.04 ppm, control = 10.7, water = 3.8 ppm Cl, 298 K, Ash River	Bacterial cells[18]
Nutrient uptake	13.2 ppm Na	Hg = 0.04 ppm, control = 19.9, water = 7.7 ppm Na, 298 K, Ash River	Bacterial cells[18]
Nutrient uptake	13.9 ppm Al-Si	Hg = 0.04 ppm, control = 16.8, water = 13.0 ppm Al, 298 K, Ash River	Bacterial cells[18]
Nutrient uptake	18.0 ppm Zn	Hg = 0.04 ppm, control = 0.8, water = 0.5 ppm Zn, 298 K, brackish	Bacterial cells[18]
Nutrient uptake	18.5 ppm Ca	Hg = 0.04 ppm, control = 16.1, water = 21.8 ppm Ca, 298 K, brackish	Bacterial cells[18]
Nutrient uptake	2.9 ppm Mg	Hg = 0.04 ppm, control = 4.9, water = 4.1 ppm Mg, 298 K, Ash River	Bacterial cells[18]
Nutrient uptake	22.0 ppm Na	Hg = 0.04 ppm, control = 13.6, water = 19.0 ppm Na, 298 K, brackish	Bacterial cells[18]
Nutrient uptake	23.0 ppm K	Hg = 0.04 ppm, control = 13.8, water = 9.7 ppm K, 298 K, brackish	Bacterial cells[18]
Nutrient uptake	26.0 ppm Ca	Hg = 0.04 ppm, control = 43.2, water = 41.0 ppm Ca, 298 K, fresh water	Bacterial cells[18]
Nutrient uptake	26.9 ppm Fe	Hg = 0.04 ppm, control = 21.9, water = 16.9 ppm Fe, 298 K, Ash River	Bacterial cells[18]
Nutrient uptake	27.0 ppm Cl	Hg = 0.04 ppm, cont = 123 ppm Cl with wat = 1434 ppm Cl, 298 K, fresh water	Bacterial cells[18]
Nutrient uptake	27.7 ppm Al-Si	Hg = 0.04 ppm, control = 18.6, water = 13.0 ppm Al, 298 K, brackish	Bacterial cells[18]
Nutrient uptake	28.0 ppm Al-Si	Hg = 0.04 ppm, control = 6.8, water = 3.0 ppm Al, 298 K, fresh water	Bacterial cells[18]
Nutrient uptake	3.4 ppm Cu	Hg = 0.04 ppm, control = 1.4, water = 1.2 ppm Cu, 298 K, brackish	Bacterial cells[18]
Nutrient uptake	36.0 ppm Na	Hg = 0.04 ppm, control = 110, water = 1364 ppm Na, 298 K, fresh water	Bacterial cells[18]
Nutrient uptake	4.0 ppm Ti	Hg = 0.04 ppm, control = 2.8, water = 13.8 ppm Ti, 298 K, fresh water	Bacterial cells[18]
Nutrient uptake	4.0 ppm Zn	Hg = 0.04 ppm, control = 1.4, water = 1.17 ppm Zn, 298 K, fresh water	Bacterial cells[18]
Nutrient uptake	40.8 ppm Mg	Hg = 0.04 ppm, control = 6.3, water = 6.2 ppm Mg, 298 K, brackish	Bacterial cells[18]

Nutrient uptake	5.1 ppm K	Hg = 0.04 ppm, control = 4.6, water = 6.1 ppm K, 298 K, Ash River	Bacterial cells[18]
Nutrient uptake	52.0 ppm Fe	Hg = 0.04 ppm, control = 29.0, water = 23.0 ppm Fe, 298 K, fresh water	Bacterial cells[18]
Nutrient uptake	6.6 ppm Cl	Hg = 0.04 ppm, control = 2.3, water = 14.1 ppm Cl, 298 K, brackish	Bacterial cells[18]
Photosynthesis	0 % of control	Hg = 10.0 mg/l, 3 h, Golden Horn	Aufwuchs communities[19]
Photosynthesis	16 % of control	Hg = 1.0 mg/l, 3 h, Golden Horn	Aufwuchs communities[19]
Photosynthesis	72 % of control	Hg = 0.1 mg/l, 3 h, Golden Horn	Aufwuchs communities[19]
Photosynthesis	93 % of control	Hg = 0.01 mg/l, 3 h, Golden Horn	Aufwuchs communities[19]
Photosynthesis	97 % of control	Hg = 0.001 mg/l, 3 h, Golden Horn	Aufwuchs communities[19]
Photosynthesis	100 % effect	0.1 ug/l as organomercurial fungicides, reduction observed	Phytoplankton[20]
Plasma osmolality	348.0 mosm/kg H2O	Hg = 5 ppb, 60 d, 11 fish, rg = +- 4.0	Pseudopleuronectes americanus[2]
Plasma osmolality	360.0 mosm/kg H2O	Hg = 10 ppb, 60 d, 15 fish, rg = +- 4.0	Pseudopleuronectes americanus[2]
Plasma osmolality	362.0 mosm/kg H2O	Hg = 0.0 ppb, 60 d, 12 fish, rg = +- 2.6	Pseudopleuronectes americanus[2]
Plasma protein	5.4 g/100 ml	Hg = 0.0 ppb, 60 d, 26 fish, rg = +- 0.2	Pseudopleuronectes americanus[2]
Plasma protein	6.0 g/100 ml	Hg = 10 ppb, 60 d, 27 fish, rg = +- 0.2	Pseudopleuronectes americanus[2]
Plasma protein	6.3 g/100 ml	Hg = 5 ppb, 60 d, 25 fish, rg = +- 0.2	Pseudopleuronectes americanus[2]
Red blood cells	1.78E6 cells/mm**3	Hg = 10 ppb, 60 d, 11 fish, rg = +- 0.16	Pseudopleuronectes americanus[2]
Red blood cells	1.87E6 cells/mm**3	Hg = 0.0 ppb, 60 d, 12 fish, rg = +- 0.16	Pseudopleuronectes americanus[2]
Red blood cells	2.1E6 cells/mm**3	Hg = 5 ppb, 60 d, 12 fish, rg = +- 0.18	Pseudopleuronectes americanus[2]
Respiration	10 % of control	Hg = 10.0 mg/l, 3 h, Golden Horn	Aufwuchs communities[19]
Respiration	60 % of control	Hg = 1.0 mg/l, 3 h, Golden Horn	Aufwuchs communities[19]
Respiration	70 % of control	Hg = 0.1 mg/l, 3 h, Golden Horn	Aufwuchs communities[19]
Respiration	87 % of control	Hg = 0.01 mg/l, 3 h, Golden Horn	Aufwuchs communities[19]
Respiration	94 % of control	Hg = 0.001 mg/l, 3 h, Golden Horn	Aufwuchs communities[19]

Source: 1.Singleton, F.L. and R.K. Guthrie. 1977. Water Res., 11(8), pp 639-642.
2.Calabrise, A., M.A. Dawson, F.P. Thurberg and D.R. Wenzloff. 1975. In; Koeman, J.H. et al. (eds); Sublethal Effects of Toxic Chemicals on Aquatic Animals. Elsevier N.Y. pp 15-21.
3.Singh, K. 1971. Pestic. India 6, 14-17.
4.Hoffman-Kakol, I. and H. Kwarta. 1972. Zesz. nauk. Akad. Rolnicz. Szczecin. No 38, 91-103.
5.Ishizawa, S. and T. Matsuguchi. 1966. Bull. natn Inst. agric. Sci., Tokyo (B) 16, 1-90.
6.Portmann, J.E. and K.W. Wilson. 1971. Shellfish Information Leaflet no 22.
7.Liebman, H. 1958. Handbuch der Friscwasser - und Abwasserbiologie. Band II.Munchen. Oldenbourg.
8.Biesinger, K.E. and G.M. Christensen. 1972. J. Fish. Res. bd. Can., 29(12), 1691-1700.
9.Luckey, T.D. and B. Venugopal. 1977. Metal Toxicity in Mammals, 1. Plenum Press.
10.Nordisk Hygienisk Tidsskrift, 1970. Suppl. 3, Stockholm.
11.Sallivan, G.W., A.L. Buikema and J. Cairns, Jr. 1973. Va. J. Sci. 24(3), pp 120.
12.Toxic Substances List 1973, U.S. Department of Health, Education and Welfare, Public Health ServiceNational Institute for Occupational Safety and Health, Rockville Maryland, USA, June 1973.
13.Edwards, C.A. 1975. Persistent Pesticides in the Environment, 2nd edition, CRC-Press.
14.Macleod, J.C. and E. Pessah. 1973. J. Fish. Res. bd. Can. 30(4), pp 485-492.
15.Jung. K.D. 1973. Wirkungskonzentration schaedigender Stoffe im Wasser. Gelenkirchen. Hygiene-Institudes Ruhrgebiets.
16.Nepomiluev, V.F. and T.I. Kuzyakina. 1967. Izv. timiryazev. selKhoz. Akad. No. 4, 84-91.
17.Liang, C.N. and M.A. Tabatabai. 1978. Environ. Pollut. 12, 141-147.
18.Guthrie, R.K., F.L. Singleton and D.S. Ckerry. 1977. Water Res., 11(8), pp 643-646.

19.Orhon, D. 1975. Water Res. 9(5/6), pp 467-472.
20.Harrison, F.L. 1969. Proc. 2nd Nat. Symp. on Radioecology. USAEC Conf. 670503, pp 198-220.

4-235 Effect of Hg and Cu

Parameter	Effect Value	Concentration and Condition	Species
Change in chromagens	12.5 % chromagens	Control = 19.4 %, Cu + Hg = 2 ppm, after 14 d, 299 K, brackish, Clear Lake	Bacterial colonies[1]
Change in chromagens	15.6 % chromagens	Control = 18.8 %, Cu + Hg = 2 ppm, after 14 d, 299 K, fresh, Clear Lake	Bacterial colonies[1]
Numbers of colony forming units	6.38 (log-unit)	Control = 4.62, Cu + Hg = 2 ppm, after 14 d, 299 K, brackish, Clear Lake	Bacterial colonies[1]
Numbers of colony forming units	6.70 (log-unit)	Control = 5.25, Cu + Hg = 2 ppm, after 14 d, 299 K, fresh, Clear Lake	Bacterial colonies[1]

Source: 1.Singleton, F.L. and R.K. Guthrie. 1977. Water Res., 11(8), pp 639-642.

4-236 Effect of Ho

Parameter	Effect Value	Concentration and Condition	Species
Mortality	50 %	Chloride, citratecomplex, intraperitonal, dose = 117 mg/kg body	Rat[1]
Mortality	50 %	Chloride, intraperitonal, dose = 312 mg/kg body	Rat[1]
Mortality	50 %	Nitrate, oral dose = 3000 mg/kg body	Rat[1]

Source: 1.Toxic Substances List 1973, U.S. Department of Health, Education and Welfare, Public Health ServiceNational Institute for Occupational Safety and Health, Rockville Maryland, USA, June 1973.

4-237 Effect of Humidity

Parameter	Effect Value	Concentration and Condition	Species
Dieldrin uptake	0.07 mg dieldrin/kg wet weight	High humidity, grown in sand with 10 ppm dieldrin	Secale cereale[1]
Dieldrin uptake	0.16 mg dieldrin/kg wet weight	Low humidity, grown in sand with 10 ppm dieldrin	Secale cereale[1]

Source: 1.Wheelen, W.B. 1970. Bull. Environ. Contam. Tixicol. 5, pp 463-467.

4-238 Effect of Hydrazine

Parameter	Concentration	Condition	Species
LC0	0.7 mg/l	7 months, air, respiratory route	Rabbit[1]
LC0	146 mg/l	0.5 hour	Rainbow trout[2]
LC0	0.4 mg/l	7 months, air, respiratory route	Rat[1]
LC20	1 mg/l	90 days continously, air, respiratory route	Ape[1]
LC50	570 mg/l	4 hours, air, respiratory route	Mouse[1]
LC50	252 mg/l	4 hours, air, respiratory route	Rat[1]
LC70	54 mg/l	5*6 hours, 2 weeks, air, respiratory route	Mouse[1]
LC80	225 mg/l	5*6 hours, air, respiratory route	Mouse[1]
LC96	1 mg/l	90 days continously, air, respiratory route	Rat[1]
LD50	*0.059 g pr. kg ww*	*single oral dose*	*Mouse*[3]
LD50	0.060 g pr. kg ww	single oral dose	Rat[3]

Source: 3.Patty, F. A., Industrial Hygiene and toxicology, 2, Interscience Publishers, New York, 1976.

1.Henschler, D., *Toxikologisch-arbeitsmedizinische Begrundung von MAH-werte*, Verlag Chemie.
2.McKee, J. E., and Wolf, H. W., *Water Quality*, The resources agency of California, 1963.

4-239 Effect of Hydrazoic acid

Parameter	Concentration	Condition	Species
LC100	10 mg/l	1 hour, air, respiratory route	Man[1]
LC38	1024 mg/l	1 hour, air, respiratory route	Rat[2]
LC94	1138 mg/l	1 hour, air, respiratory route	Rat[2]

Source: 1.Strafford, N., Strouts, C. R. N., and Stubbings, W. V., The Determination of Toxic Substances in airHeffer, Cambridge.
2.Patty, F. A., Industrial Hygiene and toxicology, 2, Interscience Publishers, New York, 1976.

4-240 Effect of Hydrogen cyanide

Parameter	Concentration	Condition	Species
LC0	200 mg/l	1.5 hour, air, respiratory route	Guinea pig[1]
LC100	180 mg/l	not indicated	Cat[1]
LC100	315 mg/l	1.5 hour, air, respiratory route	Guinea pig[1]
LC100	180 mg/l	10 min., air, respiratory route	Man[1]
LC50	142 mg/l	1.5 hour, air, respiratory route	Rat[1]
TLm	0.069 mg/l	24 hours	Pin perch[2]
TLm	0.05 mg/l	24 hours, seawater	Pin perch[3]
TLm	0.18 mg/l	24 hours	SunFish[4]
LD50	*0.0007-0.0035 g pr. kg ww*	*by ingestion, 100 %*	*man*[5]

Source: 5.Strafford, N., Strouts, C. R. N., and Stubbings, W. V., The Determination of Toxic Substances in airHeffer, Cambridge.
1.Patty, F. A., Industrial Hygiene and toxicology, 2, Interscience Publishers, New York, 1976.
2.Jones, H. R., Environmental Control in the organic and petrochemical Industries, Noyes Data Corp., 171.
3.Garret, J. T. and Dougherty, F. M. Jr., Texas J. Sci., 3, 391, 1951.
4.Eckenfelder, W. W., Industrial Water Pollution Control, McGraw-Hill, New York, 1966.

4-241 Effect of Hydrogen sulfide

Parameter	Concentration	Condition	Species
TLm	0.111 mg/l	96 hours	Asellus[1]
TLm	0.316 mg/l	96 hours	Ephemera[1]
TLm	0.84 mg/l	96 hours	Gammarus[1]
TLm	1.07 mg/l	96 hours	Grangonyx[1]

Source: 1.Oseid, D. M. and Smith, L. L. Jr., Water Res., 9, 739, 1974.

4-242 Effect of Hydrogencyanide

Parameter	Concentration	Condition	Species
Sublethal effect	1.8 mg/l	Immobilization, 50 % after 48 hours	Daphnia[1]

Source: 1.Eckenfelder, W. W., Industrial Water Pollution Control, McGraw-Hill, New York, 1966.

4-243 Effect of Hydrogenquinone

Parameter	Concentration	Condition	Species
LC0	0.6 mg/l	not indicated	Daphnia[1]
LC0	50 mg/l	not indicated	Escherichia coli[1]

LC0	4 mg/l	not indicated	Scenedesmus[1]
LC100	0.3 mg/l	48 hours	Goldfish[2]
LD50	*0.08 g pr. kg ww*	*single oral dose*	*cat*[3]
LD50	0.2 g pr. kg ww	single oral dose	Rabbit[3]

Source: 3.Patty, F. A., Industrial Hygiene and toxicology, 2, Interscience Publishers, New York, 1976.
1.Meinck, F., Stooff, H. and Kohlschutter, H., Les eaux residuaires industrielles, 1970.
2.McKee, J. E., and Wolf, H. W., Water Quality, The resources agency of California, 1963.

4-244 Effect of Hydrogenquinonemonobenzylether

Parameter	Concentration	Condition	Species
LC100	2.5 mg/l	48 hours	Goldfish[1]

Source: 1.McKee, J. E., and Wolf, H. W., Water Quality, The resources agency of California, 1963.

4-245 Effect of In

Parameter	Effect Value	Concentration and Condition	Species
Mortality	50 %	Nitrate, oral dose = 3350 mg/kg body	Mouse[1]
Mortality	50 %	Chloride, intraperitonal, dose = 1.8 mg/kg body	Rat[2]
Mortality	50 %	Metallic, oral dose = 4200 mg/kg body	Rat[1]

Source: 1.Toxic Substances List 1973, U.S. Department of Health, Education and Welfare, Public Health ServiceNational Institute for Occupational Safety and Health, Rockville Maryland, USA, June 1973.
2.EPA. Water Quality Crit. Data Book. 2rd. Inorganic Pollution of Freshwater. 18010. DPU 07/71.

4-246 Effect of Ionizing radiation

Parameter	Effect Value	Concentration and Condition	Species
Mortality	63 % mortality	Ionizing radiation = 120 kR, microorganism	Amoeba sp.[1]
Mortality	63 % mortality	Ionizing radiation = 2 kR, E. coli K12, microorganism	Escherichia coli[2]
Mortality	63 % mortality	Ionizing radiation = 150 kR, microorganism	Micrococcus radiodurans[3]
Mortality	63 % mortality	Ionizing radiation = 350 kR, microorganism	Paramaecium sp.[1]
Mortality	63 % mortality	Ionizing radiation = 3 kR, microorganism	Saccharomyces cerevisiae[4]
Mortality	63 % mortality	Ionizing radiation = 80 kR, microorganism	Schizosaccharomyces pombe[5]

Source: 1.Fink, R.J., O.H. Fletchall and O. Calvert. 1968. Weed Sci. 16, 104-105.
2.Pantera, H. 1972. Pam. Pulawski 51, 69-76.
3.Bakalivanov, D. 1971. Soils and Fert. 36, 607.
4.Tsevetkova, S.D. 1966. Les. Khoz. 19, 25-26.
5.Spirama Raju, K. and G. Rangaswami. 1971. Indian J. Microbiol. 11, 25-29.

4-247 Effect of Ir

Parameter	Effect Value	Concentration and Condition	Species
Mortality	50 %	As chloride, intraperitonal, dose = 56 mg/kg body	Mouse[1]

Source: 1.Toxic Substances List 1973, U.S. Department of Health, Education and Welfare, Public Health ServiceNational Institute for Occupational Safety and Health, Rockville Maryland, USA, June 1973.

4-248 Effect of Isoamylacetate

Parameter	Concentration	Condition	Species
Sublethal effect	1900 ppm	Loss of weight, 6 times 8 hours	Cat[1]

Source: 1.Patty, F. A., Industrial Hygiene and toxicology, 2, Interscience Publishers, New York, 1976.

4-249 Effect of Isobranzan

Parameter	Effect Value	Concentration and Condition	Species
Mortality	50 %	Acute oral toxicity = 4.8-5.5 mg/kg wet weight	Rat[1]

Source: 1.Edwards, C.A. 1975. Persistent Pesticides in the Environment, 2nd edition, CRC-Press.

4-250 Effect of Isobutanol

Parameter	Concentration	Condition	Species
Sublethal effect	290 mg/l	Inhibition of cell multiplication	Microcystis aeruginosa[2]
Sublethal effect	280 mg/l	Inhibition of cell multiplication	Pseudomonas putida[2]
LD50	*2.46 g pr. kg ww*	*single oral dose*	*Rat*[1]

Source: 1.Patty, F. A., Industrial Hygiene and toxicology, 2, Interscience Publishers, New York, 1976.
2.Bringmann, G. and Kuhn, R., nwf-wasser/abwasser, H.9, 1976.

4-251 Effect of Isobutylacetate

Parameter	Concentration	Condition	Species
Sublethal effect	205 mg/l	Inhibition of cell multiplication	Microcystis aeruginosa[1]
Sublethal effect	200 mg/l	Inhibition of cell multiplication	Pseudomonas putida[1]

Source: 1.Bringmann, G. and Kuhn, R., nwf-wasser/abwasser, H.9, 1976.

4-252 Effect of Isobutylamine

Parameter	Concentration	Condition	Species
LC0	20 mg/l	24 hours	Creek chub[1]
LC100	60 mg/l	24 hours	Creek chub[1]

Source: 1.Gilette, L. A., Miller, D. L. and Redman, H. E., Sewage Ind. Wastes, 24, 1397, 1952.

4-253 Effect of Isobutyric acid

Parameter	Concentration	Condition	Species
LD50	0.4-0.8 g pr. kg ww	single oral dose	?[1]

Source: 1.Patty, F. A., Industrial Hygiene and toxicology, 2, Interscience Publishers, New York, 1976.

4-254 Effect of Isooctanol

Parameter	Concentration	Condition	Species
Sublethal effect	7.3 mg/l	Inhibition of cell multiplication	Microcystis aeruginosa[1]
Sublethal effect	63 mg/l	Inhibition of cell multiplication	Pseudomonas putida[1]

Source: 1.Bringmann, G. and Kuhn, R., nwf-wasser/abwasser, H.9, 1976.

4-255 Effect of Isophorone

Parameter	Concentration	Condition	Species
Sublethal effect	25 mg/l	Eye irritation	Man[1]

Source: 1.Patty, F. A., Industrial Hygiene and toxicology, 2, Interscience Publishers, New York, 1976.

4-256 Effect of Isoprene

Parameter	Concentration	Condition	Species
TLm	42.5 mg/l	24-96 hours	Bluegill[1]
TLm	74-87 mg/l	24-96 hours	Fathead minnow[1]
TLm	180 mg/l	24-96 hours	Goldfish[1]
TLm	240 mg/l	24-96 hours	Guppie[1]

Source: 1.Cage, J.C. British J. Ind. Medicine 1970, vol 27.

4-257 Effect of Isopropanol

Parameter	Concentration	Condition	Species
LC0	900 mg/l	24 hours	Creek chub[1]
LC100	1100 mg/l	24 hours	Creek chub[1]
LC50	1400 mg/l	48 hours	Brown shrimps[2]
LC50	1150 mg/l	96 hours	Brown shrimps[2]
LC50	>5000 mg/l	24 hours	Goldfish[1]
LD50	8 g pr. kg ww	single oral dose	Rabbit[3]
LD50	5.8 g pr. kg ww	single oral dose	Rat[3]
Sublethal effect	1000 mg/l	Inhibition of cell multiplication	Microcystis aeruginosa[4]
Sublethal effect	1050 mg/l	Inhibition of cell multiplication	Pseudomonas putida[4]

Source: 1.Gilette, L. A., Miller, D. L. and Redman, H. E., Sewage Ind. Wastes, 24, 1397, 1952.2.Blackmann, R. A. A., Marine Pollution Bull. 5, 116, 1974.
3.Patty, F. A., Industrial Hygiene and toxicology, 2, Interscience Publishers, New York, 1976.
4.Bringmann, G. and Kuhn, R., nwf-wasser/abwasser, H.9, 1976.

4-258 Effect of Isopropylacetate

Parameter	Concentration	Condition	Species
Sublethal effect	200 ppm	Eye irritation	Man[1]
Sublethal effect	1400 mg/l	Inhibition of cell multiplication	Microcystis aeruginosa[2]
Sublethal effect	190 mg/l	Inhibition of cell multiplication	Pseudomonas putida[2]

Source: 1.Patty, F. A., Industrial Hygiene and toxicology, 2, Interscience Publishers, New York, 1976.2.Bringmann, G. and Kuhn, R., nwf-wasser/abwasser, H.9, 1976.

4-259 Effect of Isopropylamine

Parameter	Concentration	Condition	Species
LD50	0.82 g pr. kg ww	acute oral dose	Rat[1]

Source: 1.Patty, F. A., Industrial Hygiene and toxicology, 2, Interscience Publishers, New York, 1976.

4-260 Effect of Isopropylglycidylether

Parameter	Concentration	Condition	Species
Sublethal effect	400 ppm	Retardation of weight gain, 50 times 7 hours	Rat[1]
LD50	1.3 g pr. kg ww	acute intragastric dose	Mouse[1]
LD50	4.2 g pr. kg ww	acute intragastric dose	Rat[1]

Source: 1.Patty, F. A., Industrial Hygiene and toxicology, 2, Interscience Publishers, New York, 1976.

4-261 Effect of Isopthalic acid

Parameter	Concentration	Condition	Species
LD50	>0.5 g pr. kg ww	intraperitoneal dose	Mouse[1]

Source: 1.Patty, F. A., Industrial Hygiene and toxicology, 2, Interscience Publishers, New York, 1976.

4-262 Effect of Isovaleric acid

Parameter	Concentration	Condition	Species
LD50	<3.2 g pr. kg ww	single oral dose	Rat[1]

Source: 1.Patty, F. A., Industrial Hygiene and toxicology, 2, Interscience Publishers, New York, 1976.

4-263 Effect of K

Parameter	Effect Value	Concentration and Condition	Species
Mortality	50 %	Acetate, oral dose = 3250 mg/kg body	Rat[1]
Mortality	50 %	Carbonate, oral dose = 1870 mg/kg body	Rat[1]
Mortality	50 %	Chloride, intraperitonal, dose = 825 mg/kg body	Rat[1]
Mortality	50 %	Flouride, oral dose = 245 mg/kg body	Rat[1]
Mortality	50 %	Hydroxide, oral dose = 365 mg/kg body	Rat[1]
Nitrogen fixation	100 % inhibited	0.01 m, reversible by addition of NH4-acetate, K-orthophosphite	Azotobacter sp.[2]

Source: 1.Toxic Substances List 1973, U.S. Department of Health, Education and Welfare, Public Health ServiceNational Institute for Occupational Safety and Health, Rockville Maryland, USA, June 1973.
2.Bullen, W.A. and D.S. Frear. 1957. Arch. Biochem. Biophys., 66, pp 502-503.

4-264 Effect of l-malic acid

Parameter	Concentration	Condition	Species
LD50	1.6-3.2 g pr. kg ww	single oral dose	mouse, Rat[1]

Source: 1.Patty, F. A., Industrial Hygiene and toxicology, 2, Interscience Publishers, New York, 1976.

4-265 Effect of La

Parameter	Effect Value	Concentration and Condition	Species
Mortality	50 %	Acetate, oral dose = 4400 mg/kg body	Rat[1]
Mortality	50 %	Chloride, oral dose = 4200 mg/kg body	Rat[1]
Mortality	50 %	Metallic, intraven., dose = 3.5 mg/kg body	Rat[1]
Mortality	50 %	Nitrate, oral dose = 4500 mg/kg body	Rat[1]
Mortality	50 %	Oxide, oral dose = 10000 mg/kg body	Rat[2]
Mortality	50 %	Sulfate, intraperitonal, dose=275 mg/kg ww	Rat[1]

Source: 1.Toxic Substances List 1973, U.S. Department of Health, Education and Welfare, Public Health ServiceNational Institute for Occupational Safety and Health, Rockville Maryland, USA, June 1973.
2.EPA. Water Quality Crit. Data Book. 2rd. Inorganic Pollution of Freshwater. 18010. DPU 07/71.

4-266 Effect of Lauric acid

Parameter	Concentration	Condition	Species
LD50	35 % of diet	oral ingestion for 2 years	Rat[1]

Source: 1.Patty, F. A., Industrial Hygiene and toxicology, 2, Interscience Publishers, New York, 1976.

4-267 Effect of Lactonitrile

Parameter	Concentration	Condition	Species
TLm	0.9-4.0 mg/l	24-96 hours	Bluegill[1]
TLm	0.9 mg/l	96 hours	Fathead minnow[1]
TLm	1.4 mg/l	96 hours	Guppie[1]
TLm	0.215 mg/l	24 hours, seawater	Pin perch[2]
LD50	*0.021 g pr. kg ww*	*acute oral dose*	*Rat*[3]

Source: 3.Patty, F. A., Industrial Hygiene and toxicology, 2, Interscience Publishers, New York, 1976.
1.Jones, H. R., Environmental Control in the organic and petrochemical Industries, Noyes Data Corp., 171.
2.Garret, J. T. and Dougherty, F. M. Jr., Texas J. Sci., 3, 391, 1951.

4-268 Effect of Leadacetate

Parameter	Concentration	Condition	Species
Sublethal effect	0.45 mg/l	Inhibition of cell multiplication	Microcystis aeruginosa[1]
Sublethal effect	1.8 mg/l	Inhibition of cell multiplication	Pseudomonas putida[1]

Source: 1.Bringmann, G. and Kuhn, R., nwf-wasser/abwasser, H.9, 1976.

4-269 Effect of Li

Parameter	Effect Value	Concentration and Condition	Species
Mortality	50 %	LiCl, expos.time = 2 d, conc in water = 7200 microgram/l, marine	Daphnia magna[1]
Mortality	50 %	Flouride, oral dose = 200 mg/kg body	Guinea pig[2]
Mortality	50 %	Carbonate, oral dose = 710 mg/kg body	Rat[2]

Source: 1.McKee, J.R. and H.W. Wolf. 1963. Water Quality Criteria, Publication no. 3A, Calif. State Water Res.Control Board 2nd Edition.
2.Toxic Substances List 1973, U.S. Department of Health, Education and Welfare, Public Health ServiceNational Institute for Occupational Safety and Health, Rockville Maryland, USA, June 1973.

4-270 Effect of Light and temperature

Parameter	Effect Value	Concentration and Condition	Species
Mn content	1740 mg/kg dry weight	Cool day, upper leaves, shaded, leaf	Tomato[1]
Mn content	2000 mg/kg dry weight	Hot day, upper leaves, shaded, leaf	Tomato[1]
Mn content	2400 mg/kg dry weight	Cool day, upper leaves, unshaded, leaf	Tomato[1]
Mn content	2900 mg/kg dry weight	Cool day, lower leaves, shaded, leaf	Tomato[1]
Mn content	3090 mg/kg dry weight	Hot day, upper leaves, unshaded, leaf	Tomato[1]
Mn content	3200 mg/kg dry weight	Hot day, lower leaves, shaded, leaf	Tomato[1]
Mn content	3400 mg/kg dry weight	Cool day, lower leaves, unshaded, leaf	Tomato[1]
Mn content	3940 mg/kg dry weight	Hot day, lower leaves, unshaded, leaf	Tomato[1]

Source: 1.Harrod, M.F. 1971. In: Tech. Bull. 21, pp 176-189. Trace Elements in Soils and Crops.

4-271 Effect of Light intensity

Parameter	Effect Value	Concentration and Condition	Species
Mn content	3350 mg/kg dry weight	Shaded, Mn soil:0-15cm = 78 ppm;15-30cm = 186 ppm, toxicity symp,	Tomato[1]
Mn content	5650 mg/kg dry weight	Unshaded, Mn soil:0-15cm = 96 ppm;15-30cm = 189 ppm, severe symp,	Tomato[1]

Source: 1.Harrod, M.F. 1971. In: Tech. Bull. 21, pp 176-189. Trace Elements in Soils and Crops.

4-272 Effect of Lindane

Parameter	Effect Value	Concentration and Condition	Species
Growth rate	growth inhibited	Water conc = 100000 ppb	Blue-green algae[1]
Growth rate	growth inhibited	Water conc = 7500 ppb, not reduced in all sp. ppb	Green algae[2]
Growth rate	growth stimulated	Water conc = 1000 ppb	Green algae[2]
Immobilisation	100 %	Lindane = 0.3 mg/l, for 50 h	Daphnia sp.[3]
Mortality	50 %	Conc in food = 882 mg/kg ww (755-1041), 95 % CL	Bobwhite[4]
Mortality	50 % mortality	48 h, water = 460 mg/l	Daphnia pulex[5]
Mortality	95 %	Conc in soil > 65 mg/kg	Earthworms[4]
Mortality	50 %	Conc in food = 425 mg/kg ww (347-520), 95 % CL	Japanese quail[4]
Mortality	40 %	Conc in food = 5000 mg/kg wet weight	Mallard[4]
Mortality	50 %	Conc in food > 5000 mg/kg ww	Mallard[4]
Mortality	50 %	Conc in food = 561 mg/kg ww (445-690), 95 % CL	Pheasant[4]
Mortality	50 %	Acute dermal toxicity = 900-1000 mg/kg, insecticide	Rat[4]
Mortality	50 %	Acute oral toxicity = 88-125 mg/kg, insecticide	Rat[4]

Source: 1.Malone, C.R. 1971. Soil Biol. Biochem. 3, 127-131.
2.Teuteberg, A. 1968. Z. PflKrankh. PflSchutz 75, 72-86.
3.Bays, L.R. 1971. Environ. Pollut., 1(3), pp 205-234.
4.Edwards, C.A. 1975. Persistent Pesticides in the Environment, 2nd edition, CRC-Press.
5.Rankov, V. 1973. Pochv. Agrokhim. 8, 125-134.

4-273 Effect of Linear alkyl benzen sulfonate

Parameter	Effect Value	Concentration and Condition	Species
Growth potential	1	14 d, control = 4, initial conc = 20 mg surfactant/l, F008	Anabaena cylindrica[1]
Growth potential	3	7 and 21 d, control = 4, initial conc = 20 mg surfactant/l, F008	Anabaena cylindrica[1]
Growth potential	3	7 d, control = 4, initial conc = 20 mg surfactant/l, F008	Anabaena variabilis[1]
Growth potential	4	14 and 21 d, control = 4, initial conc = 20 mg surfactant/l, F008	Anabaena variabilis[1]
Growth potential	1	7 d, control = 4, initial conc = 20 mg surfactant/l, F008	Anacystis nidulans[1]
Growth potential	4	14 and 21 d, control = 4, initial conc = 20 mg surfactant/l, F008	Anacystis nidulans[1]
Growth potential	4	7, 14 and 21 d, control = 4, initial conc = 20 mg surfactant/l, F008	Ankistrodesmus braunii[1]
Growth potential	4	7, 14 and 21 d, control = 4, initial conc = 20 mg surfactant/l, F008	Calothrix parietina[1]
Growth potential	2	7 d, control = 4, initial conc = 20 mg surfactant/l, F008	Chlorella pyrenoidosa[1]
Growth potential	4	14 and 21 d, control = 4, initial conc = 20 mg surfactant/l, F008	Chlorella pyrenoidosa[1]
Growth potential	3	7 d, control = 4, initial conc = 20 mg surfactant/l, F008	Chlorella vulgaris[1]

Growth potential	4	14 and 21 d, control = 4, initial conc = 20 mg surfactant/l, F008	Chlorella vulgaris[1]
Growth potential	2	14 d, control = 4, initial conc = 20 mg surfactant/l, F008	Cylindrospermum sp.[1]
Growth potential	4	7 and 21 d, control = 4, initial conc = 20 mg surfactant/l, F008	Cylindrospermum sp.[1]
Growth potential	2	7 d, control = 4, initial conc = 20 mg surfactant/l, F008	Gloeocapsa alpicola[1]
Growth potential	4	14 and 21 d, control = 4, initial conc = 20 mg surfactant/l, F008	Gloeocapsa alpicola[1]
Growth potential	3	14 d, control = 4, initial conc = 20 mg surfactant/l, F008	Nostoc muscorum[1]
Growth potential	4	7 and 21 d, control = 4, initial conc = 20 mg surfactant/l, F008	Nostoc muscorum[1]
Growth potential	3	7 d, control = 4, initial conc = 20 mg surfactant/l, F008	Oscillatoria borneti[1]
Growth potential	4	14 and 21 d, control = 4, initial conc = 20 mg surfactant/l, F008	Oscillatoria borneti[1]
Growth potential	3	7 and 14 d, control = 4, initial conc = 20 mg surfactant/l, F008	Oscillatoria chalybia[1]
Growth potential	4	21 d, control = 4, initial conc = 20 mg surfactant/l, F008	Oscillatoria chalybia[1]
Growth potential	3	14 d, control = 4, initial conc = 20 mg surfactant/l, F008	Oscillatoria formosa[1]
Growth potential	4	7 and 21 d, control = 4, initial conc = 20 mg surfactant/l, F008	Oscillatoria formosa[1]
Growth potential	2	14 d, control = 4, initial conc = 20 mg surfactant/l, F008	Oscillatoria tenuis[1]
Growth potential	3	7 and 21 d, control = 4, initial conc = 20 mg surfactant/l, F008	Oscillatoria tenuis[1]
Growth potential	4	7, 14 and 21 d, control = 4, initial conc = 20 mg surfactant/l, F008	Phormidium faveolarum[1]
Growth potential	3	14 d, control = 4, initial conc = 20 mg surfactant/l, F008	Scenedesmus obliquus[1]
Growth potential	4	7 and 21 d, control = 4, initial conc = 20 mg surfactant/l, F008	Scenedesmus obliquus[1]
Growth potential	3	7 d, control = 4, initial conc = 20 mg surfactant/l, F008	Scenedesmus quadricauda[1]
Growth potential	4	14 and 21 d, control = 4, initial conc = 20 mg surfactant/l, F008	Scenedesmus quadricauda[1]
Growth potential	1	7 d, control = 4, initial conc = 20 mg surfactant/l, F008	Ulothrix fimbriata[1]
Growth potential	4	14 and 21 d, control = 4, initial conc = 20 mg surfactant/l, F008	Ulothrix fimbriata[1]
Mortality	100.0 %	LAS 345 = 8.0 ppm, 24 h, 40 larvae	Aedes aegypti[2]
Mortality	100.0 %	LAS 358 = 3.0 ppm, 24 h, 40 larvae	Aedes aegypti[2]
Mortality	50.0 %	LAS 345 = 6.0 ppm, 24 h, 40 larvae	Aedes aegypti[2]
Mortality	50.0 %	LAS 358 = 2.0 ppm, 24 h, 40 larvae	Aedes aegypti[2]
Mortality	100.0 %	LAS 345 = 3.0 ppm, 24 h, 2*3 snails	Biomphalaria glabrata[2]
Mortality	100.0 %	LAS 358 = 1.0 ppm, 24 h, 2*3 snails	Biomphalaria glabrata[2]
Mortality	50 %	LAS-comm.+enz. = 5.4 mg/l, LAS-tech = 5.0 mg/l, 96 h, 294 K	Cyprinus carpio[3]

Mortality	50 %	LAS-comm.+enz. = 5.4 mg/l, LAS-tech = 6.8 mg/l, 48 h, 294 K	Cyprinus carpio[3]
Mortality	50 %	LAS-comm.+enz. = 5.4 mg/l, LAS-tech = 6.8, 24 h,, 294 K	Cyprinus carpio[3]
Mortality	100.0 %	LAS 345 = 1.0 ppm, 24 h, 5 young fish	Lebistes reticulatus[2]
Mortality	100.0 %	LAS 345 = 2 ppm, 24 h, 3 fish, female	Lebistes reticulatus[2]
Mortality	100.0 %	LAS 345 = 2.0 ppm, 24 h, 3 fish, male	Lebistes reticulatus[2]
Mortality	100.0 %	LAS 358 = 0.5 ppm, 24 h, 5 young fish	Lebistes reticulatus[2]
Mortality	100.0 %	LAS 358 = 1.0 ppm, 24 h, 3 fish, female	Lebistes reticulatus[2]
Mortality	100.0 %	LAS 358 = 1.0 ppm, 24 h, 3 fish, male	Lebistes reticulatus[2]
Mortality	50 %	LAS-comm.+enz. = 23 mg/l, LAS-tech = 12 mg/l, LAS-comm. = 4.6 mg/l,	Tilapia melanopleura[3]
Mortality	50 %	LAS-comm.+enz. = 26 mg/l, LAS-tech = 12 mg/l, LAS-comm. = 4.8 mg/l,	Tilapia melanopleura[3]

Source: 1.Davis, E.M. and E.F. Gloyna. 1969. Water Pollut. Control. 41 (8), pp 1494-1504.
2.Van Emden, H.M., C.C.M. Kroon, E.N. Schoeman and H.A. van Seventer. 1974. Environ. Pollut. 6, pp 297308.
3.Lopez-Zavala, A. A.S. de Aluja,B. Eliasl L.B. Manjarrez, L.Buchmann, I.A. Mercado and S. Caltenco. 175. Progress Water Technol. 7(2), pp 73-82.
The F008 footnote indicates: 4 = full potiental similar to control, 3 = slight inhibitation, turbidity and chlorophyll decrease. 2 = moderate inhibitation, turbidity and chlorophyll decrease with yellowing of cells, 1 = marked inhibition, turbdity decrease, white to yellow colouration, with few living cells.

4-274 Effect of Liporic acid

Parameter	Effect Value	Concentration and Condition	Species
Nitrogen fixation	100 % inhibition	0.1 g/l, inhibition null by addition of 10 ug biotin/l	Clostridium pasteurianum[1]

Source: 1.Carnahan, J.E., L.E. Mortensen and J.E. Castle. 1960. J. Bact., 80, pp 311-315.

4-275 Effect of Lu

Parameter	Effect Value	Concentration and Condition	Species
Mortality	50 %	Chloride, citratecomplex, intraperitonal, dose = 135 mg/kg body	Mouse[1]
Mortality	50 %	Nitrate, intraperitonal, dose = 335 mg/kg body	Rat[1]

Source: 1.Toxic Substances List 1973, U.S. Department of Health, Education and Welfare, Public Health ServiceNational Institute for Occupational Safety and Health, Rockville Maryland, USA, June 1973.

4-276 Effect of MSMA

Parameter	Effect Value	Concentration and Condition	Species
Mortality	10 %	10 ppm, 48 h, 14 g, 293 K, pH = 8.2, alk = 80 ppm, less than value	Ictalurus punctatus[1]

Source: 1.McCorle, F.M., J.E. Chambers and J.D. Yarbrough. 1977. Bull. Environ. Contam. Toxicol. 18(3), pp 267-270.

4-277 Effect of Malathion

Parameter	Effect Value	Concentration and Condition	Species
Heart beat	0 beats/min	Malathion = 520 micro mol, after 24 h, 299 K, 5 snails	Biomphalaria glabrata[1]
Heart beat	100 beats/min	Malathion = 375 micro mol, after 24 h, 299 K, 5 snails	Biomphalaria glabrata[1]
Heart beat	136 beats/min	Malathion = 243 micro mol, after 24 h, 299 K, 5 snails	Biomphalaria glabrata[1]
Heart beat	32 beats/min	Malathion = 486 micro mol, after 24 h, 299 K, 5 snails	Biomphalaria glabrata[1]
Mortality	50 % mortality	48 h, water = 2 mg/l	Daphnia pulex[2]

Source: 1.Cocks, J.A. 1974. Environ. Pollut., 5, pp 149-151.
2.Rankov, V. 1973. Pochv. Agrokhim. 8, 125-134.

4-278 Effect of Maleic anhydride

Parameter	Concentration	Condition	Species
TLm	150 mg/l	24 hours, tapwater	Bluegill[1]
TLm	230-240 mg/l	24-96 hours	Mosquito fish[2]
TLm	150 mg/l	24 hours, tapwater	SunFish[1]
LD50	0.4-0.8 g pr. kg ww	single oral dose	Rat[3]

Source: 3.Patty, F. A., Industrial Hygiene and toxicology, 2, Interscience Publishers, New York, 1976.
1.Turnbull, H., DeMann, J. H. and Weston, R. F., Ind. Eng. Chem., 46, 324, 1954.
2.Jones, H. R., Environmental Control in the organic and petrochemical Industries, Noyes Data Corp., 171.

4-279 Effect of Mercaptoacetic acid

Parameter	Concentration	Condition	Species
LD50	<0.05 g pr. kg ww	single oral dose	Rat[1]

Source: 1.Patty, F. A., Industrial Hygiene and toxicology, 2, Interscience Publishers, New York, 1976.

4-280 Effect of Mesityloxide

Parameter	Concentration	Condition	Species
LD50	0.655 g pr. kg ww	acute oral dose	Rat[1]
Sublethal effect	25 ppm	Eye irritation	Man[1]
LC50	540 mg/l	24 hours	Goldfish[2]

Source: 2.Shell Chemie, Shell Industrie Chemicalien gids, Shell Niederland Chemie, 1975.
1.Patty, F. A., Industrial Hygiene and toxicology, 2, Interscience Publishers, New York, 1976.

4-281 Effect of Methanol

Parameter	Concentration	Condition	Species
LC0	1250 mg/l	not indicated	Colpoda[1]
LC0	8000 mg/l	24 hours	Creek chub[2]
LC0	10000 mg/l	48 hours	Daphnia[1]
LC0	600 mg/l	not indicated	Psudomonas[3]
LC0	10000 mg/l	not indicated	Scenedesmus[1]
LC100	17000 mg/l	24 hours	Creek chub[2]
TLm	10000 mg/l	24 hours	Brine shrimp[1]
TLm	8000 mg/l	48 hours	Trout[4]

Source: 1.Meinck, F., Stooff, H. and Kohlschutter, H., Les eaux residuaires industrielles, 1970.
2.Gilette, L. A., Miller, D. L. and Redman, H. E., Sewage Ind. Wastes, 24, 1397, 1952.
3.Bringmann, G. and Kuhn, R., nwf-wasser/abwasser, H.9, 1976.
4.Price, K. S. et al., JWPCF, 46, 1974.

4-282 Effect of Methoxychlor

Parameter	Effect Value	Concentration and Condition	Species
Growth rate	1 slight inhibition of growth	Water conc = 100 ppb	Green algae[1]
Mortality	50 %	Conc in food > 5000 mg/kg ww	Bobwhite[2]
Mortality	50 %	Methoxychlor = 0.0345 ppm, 288 K, 96 h	Catostomus commersoni[3]
Mortality	50 %	Conc in food > 5000 mg/kg ww	Japanese quail[2]
Mortality	50 %	Conc in food > 5000 mg/kg ww	Mallard[2]
Mortality	50 %	Conc in food = 25000 mg/kg ww	Pheasant[4]
Mortality	50 %	Conc in food > 5000 mg/kg ww	Pheasant[2]
Mortality	50 %	Conc in food = 22000 mg/kg ww	Quail[4]
Mortality	50 %	Acute oral toxicity = 6000 mg/kg	Rat[2]

Source: 1.Hulin, N. and M. Horowitz. 1973. Phytoparasitica 1, 64.
2.Edwards, C.A. 1975. Persistent Pesticides in the Environment, 2nd edition, CRC-Press.
3.Waiwood, K.G. and P.H. Johansen. 1974. Water Res., 8(7), pp 401-406.
4.Dewitt, J.B., C.M. Menzie, V.A. Adomaitis and W.L. Reickel. 1960. Trans. 25th N. Am. Wildl. Conf., p277.

4-283 Effect of Methyl thiophanate

Parameter	Effect Value	Concentration and Condition	Species
Mortality	50 %	Acute oral toxicity = 7500 mg/kg	Rat[1]

Source: 1.Edwards, C.A. 1975. Persistent Pesticides in the Environment, 2nd edition, CRC-Press.

4-284 Effect of Methyl-beta-ethylacrolein

Parameter	Concentration	Condition	Species
LD50	4.3 g pr. kg ww	acute oral dose	Rat[1]

Source: 1.Patty, F. A., Industrial Hygiene and toxicology, 2, Interscience Publishers, New York, 1976.

4-285 Effect of Methylacetate

Parameter	Concentration	Condition	Species
Sublethal effect	5000 ppm	Eye irritation, 20 min.	Cat[1]
Sublethal effect	6600 ppm	Weight loss, 8 times 6 hours	Cat[1]

Source: 1.Strafford, N., Strouts, C. R. N., and Stubbings, W. V., The Determination of Toxic Substances in airHeffer.

4-286 Effect of Methylacrylate

Parameter	Concentration	Condition	Species
LD50	0.2 g pr. kg ww	acute oral dose	Rabbit[1]

Source: 1.Patty, F. A., Industrial Hygiene and toxicology, 2, Interscience Publishers, New York, 1976.

4-287 Effect of Methylamine

Parameter	Concentration	Condition	Species
LC0	480 mg/l	not indicated	Daphnia[1]
LC0	1000 mg/l	not indicated	Escherichia coli[1]
LC0	4 mg/l	not indicated	Scenedesmus[1]
LD50	0.1-0.2 g pr. kg ww	acute oral dose	Rat[2]

Source: 2.Patty, F. A., Industrial Hygiene and toxicology, 2, Interscience Publishers, New York, 1976.
1.Meinck, F., Stooff, H. and Kohlschutter, H., Les eaux residuaires industrielles, 1970.

4-288 Effect ofMethylamylalcohol

Parameter	Concentration	Condition	Species
LC50	360 mg/l	24 hours	Goldfish[2]
LD50	2.6 g pr. kg ww	single oral dose	Rat[1]

Source: 1.Patty, F. A., Industrial Hygiene and toxicology, 2, Interscience Publishers, New York, 1976.
2.Shell Chemie, Shell Industrie Chemicalien gids, Shell Niederland Chemie, 1975.

4-289 Effect of Methylbenzoate

Parameter	Concentration	Condition	Species
LD50	3.4 g pr. kg ww	acute oral dose	Rat[1]

Source: 1.Patty, F. A., Industrial Hygiene and toxicology, 2, Interscience Publishers, New York, 1976.

4-290 Effect of Methylchloride

Parameter	Concentration	Condition	Species
Sublethal effect	550 mg/l	Inhibition of cell multiplication	Microcystis aeruginosa[1]
Sublethal effect	500 mg/l	Inhibition of cell multiplication	Pseudomonas putida[1]

Source: 1.Bringmann, G. and Kuhn, R., nwf-wasser/abwasser, H.9, 1976.

4-291 Effect of Methylchloroformate

Parameter	Concentration	Condition	Species
LD50	<0.05 g pr. kg body weight	acute oral dose	Rat[1]

Source: 1.Patty, F. A., Industrial Hygiene and toxicology, 2, Interscience Publishers, New York, 1976.

4-292 Effect of Methylenechloride

Parameter	Concentration	Condition	Species
LC0	1000 mg/l	not indicated	Colpoda[1]
LC0	1250 mg/l	not indicated	Daphnia[1]
LC0	1000 mg/l	not indicated	Pseudomonas[1]
LC0	125 mg/l	not indicated	Scenedesmus[1]

Source: 1.Meinck, F., Stooff, H. and Kohlschutter, H., Les eaux residuaires industrielles, 1970.

4-293 Effect of Methylethylketone

Parameter	Concentration	Condition	Species
LC0	5000 mg/l	not indicated	Colpoda[1]
LC0	2500 mg/l	not indicated	Pseudomonas[2]
LC0	12500 mg/l	not indicated	Scenedesmus[2]
LC50	5000 mg/l	24 hours	Goldfish[3]
Sublethal effect	110 mg/l	Inhibition of cell multiplication	Microcystis aeruginosa[2]
Sublethal effect	1150 mg/l	Inhibition of cell multiplication	Pseudomonas putida[2]
TLm	1690-5640 mg/l	24-96 hours, tapwater	Bluegill[4]
TLm	1690-5640 mg/l	24-96 hours, tapwater	Bluegill[5]
TLm	5600 mg/l	24-96 hours	Mosquito fish[5]
TLm	5600 mg/l	24-96 hours	Mosquito fish[4]

Source: 1.Meinck, F., Stooff, H. and Kohlschutter, H., Les eaux residuaires industrielles, 1970.
2.Bringmann, G. and Kuhn, R., nwf-wasser/abwasser, H.9, 1976.
3.Shell Chemie, Shell Industrie Chemicalien gids, Shell Niederland Chemie, 1975.
4.Turnbull, H., DeMann, J. H. and Weston, R. F., Ind. Eng. Chem., 46, 324, 1954.
5.Jones, H. R., Environmental Control in the organic and petrochemical Industries, Noyes Data Corp., 171.

4-294 Effect of Methylhydrazine

Parameter	Concentration	Condition	Species
LD50	0.033 g pr. kg body weight	acute oral dose	Rat[1]

Source: 1.Patty, F. A., Industrial Hygiene and toxicology, 2, Interscience Publishers, New York, 1976.

4-295 Effect of Methyliodide

Parameter	Concentration	Condition	Species
LD50	0.15-0.22 g pr. kg body weight	subcutane dose	Rat[1]
LD50	0.15-0.22 g pr. kg body weight	oral dose	Rat[1]

Source: 1.Patty, F. A., Industrial Hygiene and toxicology, 2, Interscience Publishers, New York, 1976.

4-296 Effect of Methylisoamylketone

Parameter	Concentration	Condition	Species
Sublethal effect	90 mg/l	Inhibition of cell multiplication	Microcystis aeruginosa[1]
Sublethal effect	115 mg/l	Inhibition of cell multiplication	Pseudomonas putida[1]

Source: 1.Bringmann, G. and Kuhn, R., nwf-wasser/abwasser, H.9, 1976.

4-297 Effect of Methylisobutylketone

Parameter	Concentration	Condition	Species
LC50	460 mg/l	24 hours	Goldfish[1]
Sublethal effect	200 ppm	Eye irritation	Man[2]
Sublethal effect	136 mg/l	Inhibition of cell multiplication	Microcystis aeruginosa[3]
Sublethal effect	275 mg/l	Inhibition of cell multiplication	Pseudomonas putida[3]
LD50	2.08 g pr. kg body weight	single oral dose	Rat[2]

Source: 1.Shell Chemie, Shell Industrie Chemicalien gids, Shell Niederland Chemie, 1975.
2.Patty, F. A., Industrial Hygiene and toxicology, 2, Interscience Publishers, New York, 1976.
3.Bringmann, G. and Kuhn, R., nwf-wasser/abwasser, H.9, 1976.

4-298 Effect of Methylisopropenylketone

Parameter	Concentration	Condition	Species
Sublethal effect	1.45 mg/l	Eye irritation	Man[1]
LD50	0.18 g pr. kg body weight	single oral dose	Rat[1]

Source: 1.Bringmann, G. and Kuhn, R., nwf-wasser/abwasser, H.9, 1976.
2.Patty, F. A., Industrial Hygiene and toxicology, 2, Interscience Publishers, New York, 1976.

4-299 Effect of Methylmercaptan

Parameter	Concentration	Condition	Species
TLm	0.55-0.9 mg/l	24-96 hours	Salmonides[1]

Source: 1.Meinck, F., Stooff, H. and Kohlschutter, H., Les eaux residuaires industrielles, 1970.

4-300 Effect of Methylmetacrylate

Parameter	Concentration	Condition	Species
Sublethal effect	120 mg/l	Inhibition of cell multiplication	Microcystis aeruginosa[1]
Sublethal effect	100 mg/l	Inhibition of cell multiplication	Pseudomonas putida[1]
TLm	232-368 mg/l	24-96 hours	Bluegill[2]
TLm	159-499 mg/l	24-96 hours	Fathead minnow[2]
TLm	277-423 mg/l	24-96 hours	Goldfish[2]
TLm	368 mg/l	24-96 hours	Guppie[2]

LD50	6-7 g pr. kg body weight	acute oral dose	Rabbit[3]
LD50	8.4 g pr. kg body weight	acute oral dose	Rat[3]

Source: 3.Patty, F. A., Industrial Hygiene and toxicology, 2, Interscience Publishers, New York, 1976.
1.Bringmann, G. and Kuhn, R., nwf-wasser/abwasser, H.9, 1976.
2.Pickering Q. H. and Henderson, C., JWPCF, 38, 1419, 1966.

4-301 Effect of Methylsalicylate

Parameter	Concentration	Condition	Species
LD50	2.1 g pr. kg body weight	acute oral dose	dog[1]
LD50	0.7 g pr. kg body weight	acute oral dose	guinea pig[1]
LD50	2.8 g pr. kg body weight	acute oral dose	Rabbit[1]

Source: 1.Patty, F. A., Industrial Hygiene and toxicology, 2, Interscience Publishers, New York, 1976.

4-302 Effect of Metribuzin

Parameter	Effect Value	Concentration and Condition	Species
Mortality	10 %	10 ppm, 48 h, 14 g, 293 K, pH = 8.2, alk = 80 ppm, less than value	Ictalurus punctatus[1]

Source: 1.McCorle, F.M., J.E. Chambers and J.D. Yarbrough. 1977. Bull. Environ. Contam. Toxicol. 18(3), pp 267270.

4-303 Effect of Mg

Parameter	Effect Value	Concentration and Condition	Species
Mortality	50 %	Acetate, oral dose = 18 mg/kg body	Rat[1]
Mortality	50 %	Chloride, intravenous, dose = 14 mg/kg body	Rat[1]
Mortality	50 %	Chloride, oral dose = 2800 mg/kg body	Rat[1]

Source: 1.Toxic Substances List 1973, U.S. Department of Health, Education and Welfare, Public Health ServiceNational Institute for Occupational Safety and Health, Rockville Maryland, USA, June 1973.

4-304 Effect of Mirex

Parameter	Effect Value	Concentration and Condition	Species
Growth rate	0 no inhibition of growth	Water conc = 0.2 ppb	Green algae[1]
Growth rate	1 slight inhibitipn of growth	Water conc = 100 ppb	Green algae[2]

Source: 1.Rankov, V. 1968. Pochv. Agrokhim. 3, 81-90.
2.Hulin, N. and M. Horowitz. 1973. Phytoparasitica 1, 64.

4-305 Effect of Mn

Parameter	Effect Value	Concentration and Condition	Species
Growth	0.001 mg/l gives deficiency symptoms	Tree, 100 d in greenhouse, conc in culture solution	Agathis australis[1]
Growth	0.50 mg/l gives normal growth	Tree, 100 d in greenhouse, conc in culture solution	Agathis australis[1]
Growth	2.50 mg/l gives toxic symptoms	Tree, 100 d in greenhouse, conc in culture solution	Agathis australis[1]
Leaf symptoms	3750 mg/kg dry weight	No symptoms at 1600 mg/kg dry wt., steam sterilized soil	Tomato leaves[2]

Mn content	1350 mg/kg dry weight	Brown spots on stem and petioles of lower leaves, added Mn = 400 ppm,	Cucumber[2]
Mn content	1920 mg/kg dry weight	Dark brown spots on stem and petioles of most leaves, leaf	Cucumber[2]
Mn content	1920 mg/kg dry weight	Veins brown and laminae chlorotic, added Mn = 600 ppm, leaf	Cucumber[2]
Mn content	600 mg/kg dry weight	No symptoms, added Mn to soil = 200 ppm, leaf	Cucumber[2]
Mn content	77 mg/kg dry weight	No symptoms, exchangeble Mn in soil = 45 ppm, leaf	Cucumber[2]
Mn content	110 mg/kg dry weight	No symptoms, exchangeble Mn = 90 ppm, pH = 7.1, leaf	Tomato[2]
Mn content	2240 mg/kg dry weight	Chlorosis lower leaf, veins purple brown, added Mn= 400 ppm, pH=6.8,	Tomato[2]
Mn content	3040 mg/kg dry weight	Added Mn = 600 ppm, pH = 6.9, leaf	Tomato[2]
Mn content	3040 mg/kg dry weight	Dark brown lesions on stem and on main veins, low.leav.chlorotic,	Tomato[2]
Mn content	740 mg/kg dry weight	No symptoms, added Mn = 200 ppm, pH = 7.0, leaf	Tomato[2]
Mortality	50 % mortality	MnCl2, water = 5E-2 M Mn, toxic level given, microorganisms	Bacterium communis[3]
Mortality	50 % mortality	MnSO4*4H2O, water = 0.1 %, mixed culture, toxic level given, microorganisms	Corynebacterium sp. and Chromobacterium sp.[4]
Mortality	0 %	Toxicological limit = 0.5 - 1 g/l, Mn-compounds	Crustacean[5]
Mortality	50 %	MnCl2, exposure time = 2 d, conc in water = 9800 microgram/l, fresh	Daphnia magna[6]
Mortality	0 %	Toxicological limit = 0.05 - 1.2 g/l, Mn-compounds	Fish[5]
Mortality	0 %	Toxicological limit = 0.5 - 5 g/kg, Mn-compounds	Homo sapiens[5]
Mortality	50 %	Sulfate, intraperitonal, dose = 44 mg/kg body	Mouse[7]
Mortality	50 %	Acetate, oral dose = 3730 mg/kg body	Rat[7]
Mortality	50 %	K-permanganate, oral dose = 1090 mg/kg body	Rat[7]
Mortality	50 % mortality	MnSO4*H20, water = > 0.05 %, toxic level given, microorganisms	Sphaerotilus discophorus[8]
Nitrification	24 % inhibation	Soil = 5 umoles/g of soil, 10 d incubation, 3 soils, lab	Soil organisms[9]
Yield	4.71 t grain/ha at 15 % moisture	Added MnSO4 = 5.6 kg/ha, org.soil, control = 3.25 t/ha, 1962, N Wales	Barley[10]
Yield	1.71 t grain/ha at 15 % moisture	Added MnSO4 = 5.6 kg/ha, org.soil, control = 1.01 t/ha, 1961, N Wales	Oat[10]
Yield	2.59 t grain/ha at 15 % moisture	Added MnSO4 = 5.6 kg/ha, org.soil, control = 1.13 t/ha, 1959, N Wales	Oat[10]
Yield	2.88 t grain/ha at 15 % moisture	Added MnSO4 = 5.6 kg/ha, org.soil, control = 0.98 t/ha, 1960, N Wales	Oat[10]
Yield	4.85 t grain/ha at 15 % moisture	Added MnSO4 = 5.6 kg/ha, control = 3.48 t/ha, 1960, N Wales	Oat[10]

Source: 1.Peterson, P.J. 1962. N. Z. J. Sci. Technol. 5, pp 141-161.
2.Harrod, M.F. 1971. In: Tech. Bull. 21, pp 176-189. Trace Elements in Soils and Crops.
3.Ishizawa, S. and T. Matsuguchi. 1966. Bull. natn Inst. agric. Sci., Tokyo (B) 16, 1-90.
4.Bone, H.T. and J.E. Kuntz. 1968. Proc. N. Cent. Weed Control Conf. 23, 34.
5.Liebman, H. 1958. Handbuch der Friscwasser - und Abwasserbiologie. Band II.Munchen. Oldenbourg.
6.Biesinger, K.E. and G.M. Christensen. 1972. J. Fish. Res. bd. Can., 29(12), 1691-1700.
7.Toxic Substances List 1973, U.S. Department of Health, Education and Welfare, Public Health ServiceNational Institute for Occupational Safety and Health, Rockville Maryland, USA, June 1973.
8.Fields, M.L., R. Der and D.D. Hemphill. 1967. Weeds 15, 195-197.
9.Liang, C.N. and M.A. Tabatabai. 1978. Environ. Pollut. 12, 141-147.
10.Batey, T. 1971. In: Tech. Bull. 21, pp 137-148. Trace Elements in Soils and Crops.

4-306 Effect of Mo

Parameter	Effect Value	Concentration and Condition	Species
Growth	0.0001 mg/l gives deficiency symptoms	Tree, 100 d in greenhouse, conc in culture solution	Agathis australis[1]
Growth	0.500 mg/l gives normal growth	Tree, 100 d in greenhouse, conc in culture solution	Agathis australis[1]
Growth	100 % solution	Tree, 100 d in greenhouse, conc in culture solution, no toxic	Agathis australis[1]
Mortality	50 %	MoO3, exposure time = 4 d, conc in water = 70000 microgram/l, fresh	Pimephales promelas[2]
Mortality	50 %	Ammoniummolybdate, oral dose = 333 mg/kg body	Rat[3]
Mortality	50 %	Trioxide, oral dose = 125 mg/kg body	Rat[4]
Nitrification	54 % inhibation	Soil = 5 umoles/g of soil, 10 d incubation, 3 soils, lab	Soil organisms[5]

Source: 1.Peterson, P.J. 1962. N. Z. J. Sci. Technol. 5, pp 141-161.
2.McKee, J.R. and H.W. Wolf. 1963. Water Quality Criteria, Publication no. 3A, Calif. State Water Res.Control Board 2nd Edition.
3.EPA. Water Quality Crit. Data Book. 2rd. Inorganic Pollution of Freshwater. 18010. DPU 07/71.
4.Toxic Substances List 1973, U.S. Department of Health, Education and Welfare, Public Health ServiceNational Institute for Occupational Safety and Health, Rockville Maryland, USA, June 1973.
5.Liang, C.N. and M.A. Tabatabai. 1978. Environ. Pollut. 12, 141-147.

4-307 Effect of Monochloroacetic acid

Parameter	Concentration	Condition	Species
LD50	0.08 g pr. kg body weight	oral dose	guinea pig[1]
LD50	0.255 g pr. kg body weight	oral dose	Mouse[1]
LD50	0.076 g pr. kg body weight	oral dose	Rat[1]

Source: 1.Patty, F. A., Industrial Hygiene and toxicology, 2, Interscience Publishers, New York, 1976.

4-308 Effect of Monuron

Parameter	Effect Value	Concentration and Condition	Species
Mortality	10 %	10 ppm, 48 h, 14 g, 293 K, pH = 8.2, alk = 80 ppm, less than value	Ictalurus punctatus[1]
Mortality	50 %	Acute oral toxicity = 3600 mg/kg, herbicide	Rat[2]

Source: 1.McCorle, F.M., J.E. Chambers and J.D. Yarbrough. 1977. Bull. Environ. Contam. Toxicol. 18(3), pp 267270.
2.Edwards, C.A. 1975. Persistent Pesticides in the Environment, 2nd edition, CRC-Press.

4-309 Effect of Morphine

Parameter	Effect Value	Concentration and Condition	Species
Mortality	50 %	Oral dose = 285 mg/kg, 9.99E-4mol/kg, F007 = 3.00	Mouse[1]

Source: 1.Luckey, T.D. and B. Venugopal. 1977. Metal Toxicity in Mammals, 1. Plenum Press.
Footnote; F007 = pT50 is shown = -log(potiental toxicity, 50 %.

4-310 Effect of Morpholine

Parameter	Concentration	Condition	Species
Sublethal effect	1.7 mg/l	Inhibition of cell multiplication	Microcystis aeruginosa[2]
Sublethal effect	310 mg/l	Inhibition of cell multiplication	Pseudomonas putida[2]
LD50	0.9 g pr. kg body weight	acute oral dose	guinea pig[1]
LD50	1.05-1.6 g pr. kg body weight	acute oral dose	Rat[1]

Source: 1.Patty, F. A., Industrial Hygiene and toxicology, 2, Interscience Publishers, New York, 1976. 2.Bringmann, G. and Kuhn, R., nwf-wasser/abwasser, H.9, 1976.

4-311 Effect of Mortality

Parameter	Effect Value	Concentration and Condition	Species
Se, DL-selenocy-stine	100 % mortality	Intraperitoneal inj, 4 mg Se/kg body weight	Rat[1]
Se, DL-seleno-methionine	50 % mortality	Intraperitoneal inj, 4.25 mg Se/kg body weight	Rat[1]
Se, dimethyl selenide	50 % mortality	Intraperitoneal inj, 1600 mg Se/kg body weight	Rat[1]
Se, diseleno-methionine	50 % mortality	Intraperitoneal inj., 25-30 mg Se/kg body weight	Rat[1]
Se, hydrogen selenide	100 % mortality	In air, All animaexposed to 0.02 mg Se/l air for 60 min died within 25 d	Rat[1]
Se, selenium oxychloride	50 % mortality	Application to skin, 4 mg compound caused death in 24 h	Rabbit[1]
Se, selenium oxychloride	50 % mortality	Application to skin, 83 mg compound caused death in 5 h	Rabbit[1]
Se, sodium selenate	50 % mortality	Intravenous injection, 2-2.5 mg Se/kg body weight	Rabbit[1]
Se, sodium selenate	50 % mortality	Intraperitoneal injection, 5.5-5.75 mg Se/kg body weight	Rat[1]
Se, sodium selenate	50 % mortality	Intravenous injection, 3 mg Se/kg body weight	Rat[1]
Se, sodium selenite	50 % mortality	Injection, 2 mg Se/kg body weight	Dogt[1]
Se, sodium selenite	50 % mortality	Injection, 0.9-1.5 mg Se/kg body weight	Rabbit[1]
Se, sodium selenite	50 % mortality	Intravenous inj., 1.5 mg Se/kg body weight	Rabbit[1]
Se, sodium selenite	50 % mortality	Injection, 3-5.7 mg Se/kg body weight	Rat[1]
Se, sodium selenite	50 % mortality	Intraperitoneal inj., 3.25-3.5 mg Se/kg body weight	Rat[1]
Se, sodium selenite	50 % mortality	Intravenous inj., 3 mg Se/kg body weight	Rat[1]
Se, trime-thylseleno-nium chloride	50 % mortality	Intraperitoneal inj., 49.4 mg Se.kg body weight	Rat[1]

Source: 1.Kiss, A. 1967. Agrokem. Talajt. 16, 11.

4-312 Effect of N-amylamine

Parameter	Concentration	Condition	Species
LC0	30 mg/l	24 hour	Creek chub[1]
LC100	50 mg/l	24 hour	Creek chub[1]

Source: 1.Gilette, L. A., Miller, D. L. and Redman, H. E., Sewage Ind. Wastes, 24, 1397, 1952.

4-313 Effect of N-dimethylnitrosamine

Parameter	Concentration	Condition	Species
Sublethal effect	-	Carcinogenic	Man[1]

Source: 1.Maximale Arbeitsplatzkonzentration 1973. BRD.

4-314 Effect of N-ethylmorpholine

Parameter	Concentration	Condition	Species
LD50	1.8 g pr. kg body weight	single oral dose	Rat[1]

Source: 1.Patty, F. A., Industrial Hygiene and toxicology, 2, Interscience Publishers, New York, 1976.

4-315 Effect of N-hexylamine

Parameter	Concentration	Condition	Species
LD50	0.39 g pr. kg body weight	acute oral dose	Rat[1]
LD50	0.67 g pr. kg body weight	oral dose	Rat[2]

Source: 1.Patty, F. A., Industrial Hygiene and toxicology, 2, Interscience Publishers, New York, 1976. 2.McKee, J. E., and Wolf, H. W., Water Quality, The resources agency of California, 1963.

4-316 Effect of Na

Parameter	Effect Value	Concentration and Condition	Species
Mortality	50 %	Oral dose = 2600 mg NaCl/kg, 4.45E-2mol/kg, F007 = 1.35	Mouse[1]
Mortality	50 %	Sulfate, intraperitonal, dose = 193 mg/kg body	Mouse[2]
Mortality	50 %	Carbonate, oral dose = 4200 mg/kg body	Rat[2]
Mortality	50 %	Chloride, oral dose = 3000 mg/kg body	Rat[2]

Source: 1.Luckey, T.D. and B. Venugopal. 1977. Metal Toxicity in Mammals, 1. Plenum Press.2.Toxic Substances List 1973, U.S. Department of Health, Education and Welfare, Public Health ServiceNational Institute for Occupational Safety and Health, Rockville Maryland, USA, June 1973.
Footnote; F007 = pT50 is shown = -log(potiental toxicity, 50 %.

4-317 Effect of Na - flouroacetate

Parameter	Effect Value	Concentration and Condition	Species
Mortality	50 %	Oral dose = 18 mg/kg, 1.80E-4mol/kg, F007 = 3.74	Mouse[1]

Source: 1.Luckey, T.D. and B. Venugopal. 1977. Metal Toxicity in Mammals, 1. Plenum Press.
Footnote; F007 = pT50 is shown = -log(potiental toxicity, 50 %.

4-318 Effect of NaCl

Parameter	Effect Value	Concentration and Condition	Species
Growth rate	100 % growth inhibition	Water = 5000 - 10000 mg NaCl/l, depended on growth medium	Anabaena variabilis[1]
Growth rate	90 % growth inhibition	Water = 20000 mg NaCl/l	Anacystis nidulans[1]
Growth rate	50 % inhibition	Water = 4800 mg NaCl/l, lake water	Ankistrodesmus falcatus[2]
Growth rate	100 % growth inhibition	Water = 29000 mg NaCl/l, if 10 mM ca(2+) present	Chlorella sorokiniana[3]
Growth rate	100 % inhibition og growth	Water = 17400 mg NaCl/l, if 10 mM Ca(2+) absent	Chlorella sorokiniana[3]
Growth rate	100 % growth inhibition	Water = 40000 mg NaCl/l, depended on growth medium	Chlorogloea fritchii[1]
Growth rate	100 % growth inhibition	Water = 10000 mg NaCl/l, depended on growth medium	Eucapsis sp.[1]
Growth rate	100 % growth inhibition	Water = 20000 - 40000 mg NaCl/l, depended on growth medium	Fisherella ambigua[1]
Growth rate	100 % growth inhibition	Water = 40000 mg NaCl/l, depended on growth medium	Tolypothrix tenuis[1]

Source: 1.Batterton, J.C. Jr and C. van Baalen. 1971. Arch. Mikrobiol. 76, 151-165.
2.Lam, D.C.L., C.K. Minns, P.V. Hodson, T.J. Simons and P. Wong. 1974. In: J.O. Nriagu (ed.): Environmental Biochemistry. Vol 2, Ann Arbor Science. Ann Arbor. Mich. 743-761.
3.Chimiklis, P.E. and E.P. Karlander. 1973. Plant. Physiol. 51, 48-56.

4-319 Effect of Naphthalene

Parameter	Effect Value	Concentration and Condition	Species
Mortality	50 %	Water = 13 ppm, naplii sedimentated (thought dead), 1 h, S=3 %	Balanus amphitrites[1]
Mortality	100 % mortality	Water = 0.008-0.012 ppm, 18-24 h, stage 1 zoeae	Cancer magister[2]
Mortality	50 % mortality	Water = > 2.0 ppm, 96 h, stage 1 zoeae	Cancer magister[3]
Mortality	50 % mortality	Water = 2.4 ppm, 24 h, sheeps head minnow	Cyprinodon variegatus[4]
Mortality	50 % mortality	Water = 2.68 ppm, 96 h, marine amphipod	Elasmopus pectenicrus[5]
Mortality	30 % mortality	Water = 0.05 ppm, 10 d	Eurytemora affinis[6]
Mortality	50 % mortality	Water = 2.5 ppm, 24 h	Gambusia affinis[4]
Mortality	50 % mortality	Water = 3.8 ppm, 96 h, marine polychaete	Neathes arenaceodentata[7]
Mortality	50 % mortality	Water = 0.92 ppm, 24 h. pink salmon fry	Oncorhynchus gorbuscha[8]
Mortality	50 % mortality	Water = 2.4 ppm, 96 h	Palaemonetes pugio[9]
Mortality	100 % mortality	Water = 0.008-0.012 ppm, 18-24 h, stage I and V spot shrimp larvae	Pandalus platyceros[2]
Mortality	50 % mortality	Water = 2.5 ppm, 24 h	Penaeus aztecus[4]
TLm	150-220 mg/l	24-96 hours, turbid Oklahoma water	Mosquito fish[10]
LC50	33 mg/l	24 hours, 20 C	Chlorella vulgaris[11]

Source: 1.Donahue, W.H., R.T. Wang, M. Welch and J.A.C. Nicol. 1977. Environ. Pollut. 13, 187-202.
2.Sanborn, H.R., and D.C. Malins. 1977. Proc. Soc. Exper. Biol. Med. 154, 151-155.
3.Caldwell, R.s., E. M. Caldarone and M.H. Mallon. 1977. D.A. Wolfe (ed.): Fate and Effects of Petroleum Hydrocarbons in Marine Ecosystems and Organisms. NY. Pergamon Press. 210-220.
4.Anderson, J.W., J.M. Neff, B.A. Cox, Tatem and Hightower. 1974. In: Vernberg and Vernberg (eds.): Pollution and Physiology of Marine Animals. NY. Academic Press. 285-310.
5.Lee, W.Y. and J.A.C. Nicol. 1978. Mar. Biol. 48, 215-222.
6.Berdugo, V., R.P. Harris and S.C. OHara. 1977. Mar. Pollut. Bull. 8, 138-143.
7.Rossi, S.S. and J.M. Neff. 1978. Mar. Pollut. Bull. 9, 220-223.
8.Thomas, R.E. and S.D. Rice. 1978. In: Vernberg Vernberg and Calabrese (eds.): Marine Pollution: Functional Processes, NY. Acedemic Press.
9.Neff, J.M., Anderson, Cox, Laughlin, Rossi and Tatem. 1976. In: Sources, Effects and Sinks of Hydrocarbon in the Aquatic Environment. Wash DC. Amer. Inst. of Biological Science.
11.Kauss, P. B. and Hutchinson, T. C., Environ Poll., 9, 1975.
10.Wallen, I. E., Greer, W. C. and Lasater, R., Sewage Ind. Wastes, 29, 695, 1957.

4-320 Effect of Nb

Parameter	Effect Value	Concentration and Condition	Species
Mortality	50 %	Chloride, oral dose = 830 mg/kg body	Mouse[1]
Mortality	50 %	Chlorid, oral dose = 41 mg/kg body	Rat[2]
Mortality	50 %	K-niobate, oral dose = 725 mg/kg body	Rat[2]

Source: 1.Toxic Substances List 1973, U.S. Department of Health, Education and Welfare, Public Health ServiceNational Institute for Occupational Safety and Health, Rockville Maryland, USA, June 1973.
2.EPA. Water Quality Crit. Data Book. 2rd. Inorganic Pollution of Freshwater. 18010. DPU 07/71.

4-321 Effect of Nd

Parameter	Effect Value	Concentration and Condition	Species
Mortality	50 %	Chloride, citratecomplex, intraperitonal, dose = 140 mg/kg body	Mouse[1]
Mortality	50 %	Chloride, intraperitonal, dose = 347 mg/kg body	Mouse[1]
Mortality	50 %	Nitrate, oral dose = 2750 mg/kg body	Rat[1]

Source: 1.Toxic Substances List 1973, U.S. Department of Health, Education and Welfare, Public Health ServiceNational Institute for Occupational Safety and Health, Rockville Maryland, USA, June 1973.

4-322 Effect of Neburon

Parameter	Effect Value	Concentration and Condition	Species
Mortality	50 %	Acute oral toxicity = 11000 mg/kg, herbicide	Rat[1]

Source: 1.Edwards, C.A. 1975. Persistent Pesticides in the Environment, 2nd edition, CRC-Press.

4-323 Effect of Ni

Parameter	Effect Value	Concentration and Condition	Species
Egg production	218 eggs	Ni = 1.6 mg/l, mal. = 6, fem. = 6, spawn. = 5, cont=(12824 egg)	Pimephales promelas[1]
Egg production	3489 eggs	Ni = 0.73 mg/l, mal. = 5, fem. = 7, spawn. = 53, cont=(12824 eggs)	Pimephales promelas[1]
Egg production	5520 eggs	Ni = 0.082 mg/l, mal. = 6, fem. = 5, spawn. = 39, cont=(12824 egg)	Pimephales promelas[1]
Egg production	9240 eggs	Ni = 0.18 mg/l, mal. = 6, fem. = 7, spawn. = 56, cont=(12824 egg)	Pimephales promelas[1]
Egg production	9787 eggs	Ni = 0.38 mg/l, mal. = 5, fem. = 7, spawn. = 60, cont = (12824 egg)	Pimephales promelas[1]
Hatchability	42.4 %	Ni = 0.73 mg/l, n = 23, control = 93.8 % (86-97), n = 10, rg = 0 - 97	Pimephales promelas, eggs[1]

Hatchability	94.2 %	Ni = 0.38 mg/l, n = 13, control = 93.8 % (86-97), n = 10	Pimephales promelas, eggs[1]
Hatchability	95.8 %	Ni = 0.18 mg/l, n = 12, control = 93.8 % (86-97), n = 10	Pimephales promelas, eggs[1]
Hatchability	95.9 %	Ni = 0.082 mg/l, n = 11, control = 93.8 % (86-97), n = 10	Pimephales promelas, eggs[1]
Mortality	50 % mortality	NiCl2, water = 0.9 mg/l, toxic level given, microorganisms	Aspergillus niger[2]
Mortality	50 % mortality	NiSO4*6H2O, water = 0.4 mg/10 ml, toxic level given, microorganisms	Aspergillus niger[3]
Mortality	50 % mortality	Ni(NO3)2*6H2O, water = E-4 M Ni, toxic level given, microorganismsAzotobacter sp.[4]	
Mortality	0 %	Toxicological limit = 6 mg/l, Ni-compounds	Biological purification[5]
Mortality	50 %	NiSO4, exposure time = 2 d, conc in water = 125000 ug/l, marine	Crangon crangon[6]
Mortality	0 %	Toxicological limit = 0.0055 - 1 g/l, Ni-compounds	Crustacean[7]
Mortality	50 %	NiCl2, exposure time = 2 d, conc in water = 510 ug/l, limnol.	Daphnia magna[8]
Mortality	50 % mortality	NiCl2*6H20, water = 23-27 mg Ni/l, toxic level given, microorganisms	Escherichia coli[9]
Mortality	0 %	Toxicological limit = 0.8 - 55 mg/l, Ni-compounds	Fish[7]
Mortality	50 % mortality	NiCl2*6H20, water = 23-27 mg Ni/l, toxic level given, microorganisms	Hansenula sp.[9]
Mortality	0 %	Toxicological limit = 50 - 500 mg/kg, Ni-compounds	Homo sapiens[7]
Mortality	50 %	Flouride, intravenous, dose = 130 mg/kg body	Mouse[10]
Mortality	50 % mortality	NiCl2*6H20, water = 23-27 mg Ni/l, toxic level given, microorganisms	Pseudomonas sp.[9]
Mortality	50 %	Nitrate, oral dose = 1620 mg/kg body	Rat[10]
Mortality	50 %	Tetracarbonyl, intravenous, dose = 22 mg/kg body	Rat[10]
Mortality	50 % mortality	NiCl2*6H20, water = 23-27 mg Ni/l, toxic level given, microorganisms	Rhodotorula sp.[9]
Mortality	0 %	Toxicological limit = 0.1 mg/l, Ni-compounds	Self Purification[5]
Nitrification	64 % inhibation	Soil = 5 umoles/g of soil, 10 d incubation, 3 soils, lab	Soil organisms[11]
Nitrogen fixation	28 no of nodules/plant	Ni = 5.0	Soybean with nodule[12]
Nitrogen fixation	38 no of nodules/plant	Ni = 2.5 ppm in soil, control = 46	Soybean with nodule[12]
Nitrogen fixation	44 no of nodules/plant	Ni = 1.0 ppm in soil, control = 46	Soybean with nodule[12]
Nitrogen fixation	47.41 weight of one nodule in mg	Ni = 5.0	Soybean with nodule[12]
Nitrogen fixation	48.33 weight of one nodule in mg	Ni = 2.5 ppm in soil, control = 62.30 mg	Soybean with nodule[12]
Nitrogen fixation	527 ethylene prod in nmoles/(plant*30 min)	Ni = 5.0	Soybean with nodule[12]
Nitrogen fixation	58.86 weight of one nodule in mg	Ni = 1.0 ppm in soil, control = 62.30 mg	Soybean with nodule[12]

Nitrogen fixation	611 ethylene prod in nmoles/(plant*30 min)	Ni = 2.5 ppm in soil, control = 1640	Soybean with nodule[12]
Nitrogen fixation	824 ethylene prod in nmoles/(plant*30 min)	Ni = 1.0 ppm in soil, control = 1640	Soybean with nodule[12]
Yield	0.4 g wet weight	pH = 5.7, Ni added to soil = 250 ppm, control = 24.8	Mustard[13]
Yield	0.8 g wet weight	pH = 5.7, Ni added to soil = 100 ppm, control = 24.8	Mustard[13]
Yield	17.2 g wet weight	pH = 5.7, Ni added to soil = 50 ppm, control = 24.8	Mustard[13]
Yield	23.2 g wet weight	pH = 6.4, Ni added to soil = 50 ppm, control = 25.6	Mustard[13]
Yield	25.8 g wet weight	pH = 6.4, Ni added to soil = 100 ppm, control = 25.6	Mustard[13]
Yield	8.0 g wet weight	pH = 6.4, Ni added to soil = 250 ppm, control = 25.6	Mustard[13]
Yield	1.0 g dry weight	pH = 5.3, added sludge = 150 t/ha, added Ni = 389 ppm, control = 1.4	Oat[13]
Yield	1.3 g dry weight	pH = 5.3, added sludge = 75 t/ha, added Ni = 194 ppm, control = 1.4Oat[13]	
Yield	1.4 g dry weight	pH = 5.3, added sludge = 37.5 t/ha, added Ni = 97 ppm, control = 1.4	Oat[13]
Yield	1.6 g dry weight	pH = 6.8, added sludge = 150 t/ha, added Ni = 389 ppm, control = 1.4	Oat[13]
Yield	1.6 g dry weight	pH = 6.8, added sludge = 37.5 t/ha, added Ni = 97 ppm, control = 1.4	Oat[13]
Yield	1.6 g dry weight	pH = 6.8, added sludge = 75 t/ha, added Ni = 194 ppm, control = 1.4	Oat[13]
Yield	13.8 g wet weight	pH = 6.4, Ni added to soil = 250 ppm, control = 19.8	Oat[13]
Yield	14.6 g wet weight	pH = 6.4, Ni added to soil = 100 ppm, control = 19.8	Oat[13]
Yield	16.8 g wet weight	pH = 6.4, Ni added to soil = 50 ppm, control = 19.8	Oat[13]
Yield	17.1 g wet weight	pH = 5.7, Ni added to soil = 50 ppm, control = 20.3	Oat[13]
Yield	2.5 g wet weight	pH = 5.7, Ni added to soil = 250 ppm, control = 20.3	Oat[13]
Yield	5.9 g wet weight	pH = 5.7, Ni added to soil = 100 ppm, control = 20.3	Oat[13]

Source: 1.Pickering, Q.H. 1974. Water Pollut. Control, 46(4), pp 760-765.
2.Singh, K. 1971. Pestic. India 6, 14-17.
3.Resz. A. 1968. Sonderh. Z. Plfkrankh. PflPath. PflSchutz No 4, 147-150.
4.Zhakharyan, S.V. 1970. Symp. Biol. Hung. 11, 391-395.
5.Jung. K.D. 1973. Wirkungskonzentration schaedigender Stoffe im Wasser. Gelenkirchen. Hygiene-Institudes Ruhrgebiets.
6.Portmann, J.E. and K.W. Wilson. 1971. Shellfish Information Leaflet no 22.
7.Liebman, H. 1958. Handbuch der Friscwasser - und Abwasserbiologie. Band II.Munchen. Oldenbourg.
8.Biesinger, K.E. and G.M. Christensen. 1972. J. Fish. Res. bd. Can., 29(12), 1691-1700.
9.Hoffman-Kakol, I. and H. Kwarta. 1972. Zesz. nauk. Akad. Rolnicz. Szczecin. No 38, 91-103.
10.Toxic Substances List 1973, U.S. Department of Health, Education and Welfare, Public Health ServiceNational Institute for Occupational Safety and Health, Rockville Maryland, USA, June 1973.
11.Liang, C.N. and M.A. Tabatabai. 1978. Environ. Pollut. 12, 141-147.
12.Letchworth, M.B. and V. Blum. 1977. Environ. Pollut. 14, 303-312.
13.Patterson, J.B.E. 1971. Tech. Bull., 21, pp 193-207.

4-324 Effect of Nicotinic acid

Parameter	Effect Value	Concentration and Condition	Species
Mortality	50 %	Oral dose = 1860 mg/kg, 1.51E-2mol/kg, F007 = 1.82	Mouse[1]

Source: 1.Luckey, T.D. and B. Venugopal. 1977. Metal Toxicity in Mammals, 1. Plenum Press.
Footnote; F007 = pT50 is shown = -log(potiental toxicity, 50 %.

4-325 Effect of Nitralin

Parameter	Effect Value	Concentration and Condition	Species
Mortality	10 %	10 ppm, 48 h, 14 g, 293 K, pH = 8.2, alk = 80 ppm, less than value	Ictalurus punctatus[1]

Source: 1.McCorle, F.M., J.E. Chambers and J.D. Yarbrough. 1977. Bull. Environ. Contam. Toxicol. 18(3), pp 267270.

4-326 Effect of Nitrilotriacetic acid

Parameter	Concentration	Condition	Species
Sublethal effect	510 mg/l	Inhibition of cell multiplication	Microcystis aeruginosa[1]
Sublethal effect	>1000 mg/l	Inhibition of cell multiplication	Pseudomonas putida[1]

Source: 1.Bringmann, G. and Kuhn, R., nwf-wasser/abwasser, H.9, 1976.

4-327 Effect of Nitrobenzene

Parameter	Concentration	Condition	Species
Sublethal effect	1.9 mg/l	Inhibition of cell multiplication	Microcystis aeruginosa[2]
Sublethal effect	7 mg/l	Inhibition of cell multiplication	Pseudomonas putida[2]
TLm	*20-24 mg/l*	*6 hours, distilled water*	*Vairon fish*[1]
TLm	90-100 mg/l	6 hours, hard water	Vairon fish[1]

Source: 1.Meinck, F., Stooff, H. and Kohlschutter, H., Les eaux residuaires industrielles, 1970.
2.Bringmann, G. and Kuhn, R., nwf-wasser/abwasser, H.9, 1976.

4-328 Effect of Nitrogen

Parameter	Effect Value	Concentration and Condition	Species
Basel area	4.48 m**2/ha	Appl = 84 kg N/ha, control = 2.8 m2/ha, Scotland, 36 y, NH4-sulphate	Pinus nigra[1]
Basel area	5.11 m**2/ha	Appl = 168 kg N/ha, control = 2.8 m2/ha, Scotland, 36 y, NH4-sulphate	Pinus nigra[1]
Basel area	5.44 m**2/ha	Appl = 336 kg N/ha, control = 2.8 m2/ha, Scotland, 36 y, NH4-sulphate	Pinus nigra[1]
Basel area	5.79 m**2/ha	Appl = 504 kg N/ha, control = 2.8 m2/ha, Scotland, 36 y, NH4-sulphate	Pinus nigra[1]
Nitrogen fixation	100 % inhibition	As nitrite, more than 1.4 mg N/l, stimulatory at lower conc	Azotobacter sp.[2]

Source: 1.Miller, H.G. and J.D. Miller. 1976. J. appl. Ecol. 13(1), pp 249-256.
2.Azim, M.A. and E.R. Roberts. 1956. Biochim. Biophys. Acta, 21, pp 308-312.

4-329 Effect of Nitroglycerine

Parameter	Concentration	Condition	Species
LC0	26 mg/l	not indicated	Daphnia[1]
LC0	2 mg/l	not indicated	Pseudomonas[1]
LC0	6.5 mg/l	not indicated	Scenedesmus[1]

Source: 1.Meinck, F., Stooff, H. and Kohlschutter, H., Les eaux residuaires industrielles, 1970.

4-330 Effect of Nitroglycol

Parameter	Concentration	Condition	Species
LC0	190 mg/l	not indicated	Daphnia[1]
LC0	190 mg/l	not indicated	Pseudomonas[1]
LC0	100 mg/l	not indicated	Scenedesmus[1]

Source: 1.Meinck, F., Stooff, H. and Kohlschutter, H., Les eaux residuaires industrielles, 1970.

4-331 Effect of Nitromethane

Parameter	Concentration	Condition	Species
LD50	0.75-1.0 g pr. kg body weight	single oral dose, 100 %	Rabbit[1]

Source: 1.Patty, F. A., Industrial Hygiene and toxicology, 2, Interscience Publishers, New York, 1976.

4-332 Effect of Nitrous Acide / oxide

Parameter	Effect Value	Concentration and Condition	Species
Nitrogen fixation	100 % inhibition	0.14 mg N/l, reversible, stimulatory at lower conc	Azotobacter sp.[1]

Source: 1.Azim, M.A. and E.R. Roberts. 1956. Biochim. Biophys. Acta, 21, pp 308-312.

4-333 Effect of Nonyl phenol polyethoxylate

Parameter	Effect Value	Concentration and Condition	Species
Growth potential	4	21 days, control = 4, initial conc = 20 mg /l, F008	Anabaena cylindrica[1]
Growth potential	4	7, 14 and 21 days, control = 4, initial conc = 20 mg /l, F008	Anabaena cylindrica[1]
Growth potential	1	7, 14 and 21 days, control = 4, initial conc = 20 mg /l, F008	Anabaena variabilis[1]
Growth potential	3	7, 14 and 21 days, control = 4, initial conc = 20 mg /l, F008	Anacystis nidulans[1]
Growth potential	4	21 days, control = 4, initial conc = 20 mg /l, F008	Anacystis nidulans[1]
Growth potential	1	7 days, control = 4, initial conc = 20 mg /l, F008	Ankistrodesmus braunii[1]
Growth potential	4	14 and 21 days, control = 4, initial conc = 20 mg /l, F008	Ankistrodesmus braunii[1]
Growth potential	4	7, 14 and 21 days, control = 4, initial conc = 20 mg /l, F008	Calothrix parietina[1]
Growth potential	1	7 days, control = 4, initial conc = 20 mg /l, F008	Chlorella pyrenoidosa[1]

Growth potential	3	14 days, control = 4, initial conc = 20 mg /l, F008	Chlorella pyrenoidosa[1]
Growth potential	4	21 days, control = 4, initial conc = 20 mg /l, F008	Chlorella pyrenoidosa[1]
Growth potential	1	7 days, control = 4, initial conc = 20 mg /l, F008	Chlorella vulgaris[1]
Growth potential	2	14 and 21 days, control = 4, initial conc = 20 mg /l, F008	Chlorella vulgaris[1]
Growth potential	3	7, 14 and 21 days, control = 4, initial conc = 20 mg /l, F008	Cylindrospermum sp.[1]
Growth potential	2	7 and 14 days, control = 4, initial conc = 20 mg /l, F008	Gloeocapsa alpicola[1]
Growth potential	3	21 days, control = 4, initial conc = 20 mg /l, F008	Gloeocapsa alpicola[1]
Growth potential	3	14 days, control = 4, initial conc = 20 mg /l, F008Nostoc muscorum[1]	
Growth potential	4	21 days, control = 4, initial conc = 20 mg /l, F008	Nostoc muscorum[1]
Growth potential	4	7 and 21 days, control = 4, initial conc = 20 mg /l, F008	Nostoc muscorum[1]
Growth potential	2	7, 14 and 21 days, control = 4, initial conc = 20 mg /l, F008	Oscillatoria borneti[1]
Growth potential	2	7, 14 and 21 days, control = 4, initial conc = 20 mg /l, F008Oscillatoria chalybia[1]	
Growth potential	2	7 days, control = 4, initial conc = 20 mg /l, F008	Oscillatoria formosa[1]
Growth potential	3	14 days, control = 4, initial conc = 20 mg /l, F008	Oscillatoria formosa[1]
Growth potential	4	21 days, control = 4, initial conc = 20 mg /l, F008	Oscillatoria formosa[1]
Growth potential	2	14 days, control = 4, initial conc = 20 mg /l, F008	Oscillatoria tenuis[1]
Growth potential	2	7 and 21 days, control = 4, initial conc = 20 mg /l, F008	Oscillatoria tenuis[1]
Growth potential	2	7 days, control = 4, initial conc = 20 mg /l, F008	Scenedesmus obliquus[1]
Growth potential	4	14 and 21 days, control = 4, initial conc = 20 mg /l, F008	Scenedesmus obliquus[1]
Growth potential	3	7 days, control = 4, initial conc = 20 mg /l, F008	Scenedesmus quadricauda[1]
Growth potential	4	14 and 21 days, control = 4, initial conc = 20 mg /l, F008Scenedesmus quadricauda[1]	
Growth potential	2	7 and 14 days, control = 4, initial conc = 20 mg /l, F008	Ulothrix fimbriata[1]
Growth potential	4	21 days, control = 4, initial conc = 20 mg /l, F008	Ulothrix fimbriata[1]

Source: 1.Davis, E.M. and E.F. Gloyna. 1969. Water Pollut. Control. 41 (8), pp 1494-1504.
The F008 footnote indicates: 4 = full potiental similar to control, 3 = slight inhibitation, turbidity and chlorophyll decrease. 2 = moderate inhibitation, turbidity and chlorophyll decrease with yellowing of cells, 1 = marked inhibition, turbdity decrease, white to yellow colouration, with few living cells.

4-334 Effect of O_3

Parameter	Effect Value	Concentration and Condition	Species
Growth rate	42 % photosynthesis of control	Plant of age = 36 d, O3 = 62 pphm, exposure time = 30 min	Barley[1]
Growth rate	62 % photosynthesis of control	Plant of age = 47 d, O3 = 60 pphm, exposure time = 60 min	Barley[1]
Growth rate	52 % net photosynthesis of control	1 week, 50 pphm O3, net rate, small chamber technique, SD=5	Black oat[2]
Growth rate	57 % net photosynthesis of control	3 weeks, 50 pphm O3, net rate, small chamber technique, SD=5	Black oat[2]
Growth rate	70 % net photosynthesis of control	1-2 days, 50 pphm O3, net rate, small chamber technique, SD=10	Black oat[2]
Growth rate	65 % photosynthesis of control	Plant of age = 51 d, O3 = 50 pphm, exposure time = 80 min	Bush bean
Growth rate	38 % photosynthesis of control	Plant of age = 77 d, O3 = 75 pphm, exposure time = 50 min	Cauliflower[1]
Growth rate	55 % photosynthesis of control	Plant of age = 78 d, O3 = 60 pphm, exposure time = 70 min	Chard[1]
Growth rate	68 % photosynthesis of control	Plant of age = 49 d, O3 = 50 pphm, exposure time = 90 min	Corn[1]
Growth rate	36 % photosynthesis of control	Plant of age = 39 d, O3 = 70 pphm, exposure time = 45 min	Lima bean[1]
Growth rate	52 % photosynthesis of control	Plant of age = 26 d, O3 = 45 pphm, exposure time = 120 min	Lima bean[1]
Growth rate	55 % photosynthesis of control	Plant of age = 47 d, O3 = 50 pphm, exposure time = 60 min	Oat[1]
Growth rate	60 % photosynthesis of control	Plant of age = 48 d, O3 = 45 pphm, exposure time = 30 min	Oat[1]
Growth rate	67 % photosynthesis of control	Plant of age = 50 d, O3 = 40 pphm, exposure time = 30 min	Oat[1]
Growth rate	69 % photosynthesis of control	Plant of age = 50 d, O3 = 60 pphm, exposure time = 60 min	Oat[1]
Growth rate	50 % photosynthesis of control	Plant of age = 56 d, O3 = 45 pphm, exposure time = 90 min	Pinto bean[1]
Growth rate	50 % photosynthesis of control	Plant of age = 39 d, O3 = 60 pphm, exposure time = 60 min	Potato[1]
Growth rate	49 % photosynthesis of control	Plant of age = 80 d, O3 = 90 pphm, exposure time = 90 min	Sugar beat[1]
Growth rate	80 % photosynthesis of control	Plant of age = 74 d, O3 = 65 pphm, exposure time = 90 min	Sugar beat[1]
Growth rate	45 % net photosynthesis of control	3 weeks, 50 pphm O3, net rate, small chamber technique, SD=11	Sugar maple[2]
Growth rate	73 % net photosynthesis of control	1 week, 50 pphm O3, net rate, small chamber technique, SD=16	Sugar maple[2]
Growth rate	79 % net photosynthesis of control	1-2 days, 50 pphm O3, net rate, small chamber technique, SD=10	Sugar maple[2]

Growth rate	22 % photosynthesis of control	Plant of age = 99 d, O3 = 40 pphm, exposure time = 90 min	Tobacco[1]
Growth rate	38 % photosynthesis of control	Plant of age = 110 d, O3 = 45 pphm, exposure time = 75 min	Tobacco[1]
Growth rate	50 % photosynthesis of control	Plant of age = 118 d, O3 = 48 pphm, exposure time = 80 min	Tobacco[1]
Growth rate	62 % photosynthesis of control	Plant of age = 70 d, O3 = 48 pphm, exposure time = 35 min	Tobacco[1]
Growth rate	67 % photosynthesis of control	Plant of age = 69 d, O3 = 50 pphm, exposure time = 45 min	Tobacco[1]
Growth rate	57 % photosynthesis of control	Plant of age = 72 d, O3 = 60 pphm, exposure time = 60 min	Tomato[1]
Growth rate	50 % photosynthesis of control	Plant of age = 56 d, O3 = 70 pphm, exposure time = 60 min	Wheat[1]
Growth rate	100 % net photo. of control	1 week, 50 pphm O3, net rate, small chamber technique, SD=15	White ash[2]
Growth rate	100 % net photo. of control	1-2 days, 50 pphm O3, net rate, small chamber technique, SD=18	White ash[2]
Growth rate	94 % net photo. of control	3 weeks, 50 pphm O3, net rate, small chamber technique, SD=12	White ash[2]
Leaf damage	0 % leaf injury	Plant of age = 36 d, O3 = 62 pphm, exposure time = 30 min	Barley[1]
Leaf damage	0 % leaf injury	Plant of age = 47 d, O3 = 60 pphm, exposure time = 60 min	Barley[1]
Leaf damage	3 % leaf injury	Plant of age = 51 d, O3 = 50 pphm, exposure time = 80 min	Bush bean[1]
Leaf damage	0 % leaf injury	Plant of age = 77 d, O3 = 75 pphm, exposure time = 50 min	Cauliflower[1]
Leaf damage	3 % leaf injury	Plant of age = 78 d, O3 = 60 pphm, exposure time = 70 min	Chard[1]
Leaf damage	3 % leaf injury	Plant of age = 49 d, O3 = 50 pphm, exposure time = 90 min	Corn[1]
Leaf damage	1 % leaf injury	Plant of age = 26 d, O3 = 45 pphm, exposure time = 120 min	Lima bean[1]
Leaf damage	5 % leaf injury	Plant of age = 39 d, O3 = 70 pphm, exposure time = 45 min	Lima bean[1]
Leaf damage	1 % leaf injury	Plant of age = 47 d, O3 = 50 pphm, exposure time = 60 min	Oat[1]
Leaf damage	1 % leaf injury	Plant of age = 48 d, O3 = 45 pphm, exposure time = 30 min	Oat[1]
Leaf damage	1 % leaf injury	Plant of age = 50 d, O3 = 40 pphm, exposure time = 30 min	Oat[1]
Leaf damage	2 % leaf injury	Plant of age = 50 d, O3 = 60 pphm, exposure time = 60 min	Oat[1]
Leaf damage	10 % leaf injury	Plant of age = 56 d, O3 = 45 pphm, exposure time = 90 min	Pinto bean[1]
Leaf damage	3 % leaf injury	Plant of age = 39 d, O3 = 60 pphm, exposure time = 60 min	Potato[1]
Leaf damage	4 % leaf injury	Plant of age = 74 d, O3 = 65 pphm, exposure time = 90 min	Sugar beat[1]

Leaf damage	7 % leaf injury	Plant of age = 80 d, O3 = 90 pphm, exposure time = 90 min	Sugar beat[1]
Leaf damage	0 % leaf injury	Plant of age = 99 d, O3 = 40 pphm, exposure time = 90 min	Tobacco[1]
Leaf damage	1 % leaf injury	Plant of age = 110 d, O3 = 45 pphm, exposure time = 75 min	Tobacco[1]
Leaf damage	1 % leaf injury	Plant of age = 118 d, O3 = 48 pphm, exposure time = 80 min	Tobacco[1]
Leaf damage	2 % leaf injury	Plant of age = 70 d, O3 = 48 pphm, exposure time = 35 min	Tobacco[1]
Leaf damage	4 % leaf injury	Plant of age = 69 d, O3 = 50 pphm, exposure time = 45 min	Tobacco[1]
Leaf damage	1 % leaf injury	Plant of age = 72 d, O3 = 60 pphm, exposure time = 60 min	Tomato[1]
Leaf damage	4 % leaf injury	Plant of age = 56 d, O3 = 70 pphm, exposure time = 60 min	Wheat[1]
Nitrogen concentration	2.79 % in root	Control = 2.66 %, 588 ug O3/m3, 2 h exposure, 1 w growth after exp.	Ladino clover[3]
Nitrogen concentration	2.87 % in root	Control = 2.66 %, 1176 ug O3/m3, 2 h exposure, 1 w growth after exp.	Ladino clover[3]
Nitrogen concentration	214.8 mg N in top/organismn	Control = 265.5, 1176 ug O3/m3, 2 h exposure, 1 w growth after exp.	Ladino clover[3]
Nitrogen concentration	247.5 mg N in top/organismn	Control = 265.5, 588 ug O3/m3, 2 h exposure, 1 w growth after exp.	Ladino clover[3]
Nitrogen concentration	3.02 % in top	Control = 2.85 %, 588 ug O3/m3, 2 h exposure, 1 w growth after exp.	Ladino clover[3]
Nitrogen concentration	3.08 % in top	Control = 2.85 %, 1176 ug O3/m3, 2 h exposure, 1 w growth after exp.	Ladino clover[3]
Nitrogen concentration	30.8 mg N in root/organismn	Control = 44.5, 1176 ug O3/m3, 2 h exposure, 1 w growth after exp.	Ladino clover[3]
Nitrogen concentration	38.0 mg N in root/organismn	Control = 44.5, 588 ug O3/m3, 2 h exposure, 1 w growth after exp.	Ladino clover[3]
Nitrogen fixation	309 no of nodules/-organismn	Control = 483 no, 1176 ug O3/m3, 2 h exposure, 1 w growth after exp.	Ladino clover[3]
Nitrogen fixation	317 no of nodules/-organismn	Control = 483 no, 588 ug O3/m3, 2 h exposure, 1 w growth after exp.	Ladino clover[3]
Root weight	1.16 g dry weight/-orgamismn	Control = 1.75 g, 1176 ug O3/m3, 2 h exposure, 1 w growth after exp.	Ladino clover[3]
Root weight	1.48 g dry weight/-orgamismn	Control = 1.75 g, 588 ug O3/m3, 2 h exposure, 1 w growth after exp.	Ladino clover[3]
Top weight	8.15 g dry weight/-orgamismn	Control = 10.37, 1176 ug O3/m3, 2 h exposure, 1 w growth after exp.	Ladino clover[3]
Top weight	9.65 g dry weight/-orgamismn	Control = 10.37, 588 ug O3/m3, 2 h exposure, 1 w growth after exp.	Ladino clover[3]
Toxic effects	Injury threshold value for field O3 fumi.	O3 = 25 ppm, Aspen+conifer comm, tree + shrub, 2 h, intermountain, USA	Abies concolor[4]
Toxic effects	Injury threshold value for field O3 fumi.	O3 = > 40 ppm, grass-oak comm, tree and shrub, 2 h, intermountain, USA	Acer grandidentatum[4]

Toxic effects	Injury threshold value for field O3 fumi.	O3 = >25 ppm, grass-oak comm, tree and shrub, 2 h, intermountain, USA	Acer negundo[4]
Toxic effects	Injury threshold value for field O3 fumi.	O3 = 25 ppm, Aspen+conifer comm, perennial forb, 2 h, intermountain, USA	Acetea arguta[4]
Toxic effects	Injury threshold value for field O3 fumi.	O3 = >30 ppm, grass-oak comm, perennial forbs, 2 h, intermountain, USA	Achillea millefolium[4]
Toxic effects	Injury threshold value for field O3 fumi.	O3 = 20 ppm, Aspen+conifer comm, perennial forb, 2 h, intermountain, USA	Agastache urticifolia[4]
Toxic effects	Injury threshold value for field O3 fumi.	O3 = >25 ppm, Aspen+conifer comm, grass, 2 h, intermountain, USA	Agropyron caninum[4]
Toxic effects	Injury threshold value for field O3 fumi.	O3 = 25 ppm, Aspen+conifer comm, perennial forb, 2 h, intermountain, USA	Allium acuminatum
Toxic effects	Injury threshold value for field O3 fumi.	O3 = >40 ppm, grass-oak comm, perennial forbs, 2 h, intermountain, USA	Ambrosia psilostachya[4]
Toxic effects	Injury threshold value for field O3 fumi.	O3 = 25 ppm, Aspen+conifer comm, tree + shrub, 2 h, intermountain, USA	Amelanchier myrsinites[4]
Toxic effects	Injury threshold value for field O3 fumi.	O3 = <25 ppm, Aspen+conifer comm, perennial forb, 2 h, intermountain, USA	Angelica pinnata[4]
Toxic effects	Injury threshold value for field O3 fumi.	O3 = 40 ppm, grass-oak comm, tree and shrub, 2 h, intermountain, USA	Artemesia tridentata[4]
Toxic effects	Injury threshold value for field O3 fumi.	O3 = 15 ppm, Aspen+conifer comm, perennial forb, 2 h, intermountain, USA	Aster engelmanni[4]
Toxic effects	Injury threshold value for field O3 fumi.	O3 = 30 ppm, grass-oak comm, grass, 2 h, intermountain, USA	Bromus brizaeformis[4]
Toxic effects	Injury threshold value for field O3 fumi.	O3 = <25 ppm, Aspen+conifer comm, grass, 2 h, intermountain, USA	Bromus carinatus[4]
Toxic effects	Injury threshold value for field O3 fumi.	O3 = 15 ppm, grass-oak comm, grass, 2 h, intermountain, USA	Bromus tectorum[4]
Toxic effects	Injury threshold value for field O3 fumi.	O3 = >40 ppm, grass-oak comm, perennial forbs, 2 h, intermountain, USA	Calochortus nuttallii[4]
Toxic effects	Injury threshold value for field O3 fumi.	O3 = 30 ppm, Aspen+conifer comm, perennial forb, 2 h, intermountain, USA	Carex siccata[4]
Toxic effects	Injury threshold value for field O3 fumi.	O3 = <25 ppm, Aspen+conifer comm, annual forb, 2 h, intermountain, USA	Chenopodium fremontii[4]
Toxic effects	Injury threshold value for field O3 fumi.	O3 = 25 ppm, Aspen+conifer comm, perennial forb, 2 h, intermountain, USA	Cichorium intybus[4]
Toxic effects	Injury threshold value for field O3 fumi.	O3 = <40 ppm, Aspen+conifer comm, perennial forb, 2 h, intermountain, USA	Cirsium arvense[4]
Toxic effects	Injury threshold value for field O3 fumi.	O3 = >40 ppm, grass-oak comm, perennial forbs, 2 h, intermountain, USA	Cirsium avense[4]
Toxic effects	Injury threshold value for field O3 fumi.	O3 = <25 ppm, Aspen+conifer comm, annual forb, 2 h, intermountain, USA	Collomia linearis[4]
Toxic effects	Injury threshold value for field O3 fumi.	O3 = >25 ppm, grass-oak comm, perennial forbs, 2 h, intermountain, USA	Conium maculatum[4]
Toxic effects	Injury threshold value for field O3 fumi.	O3 = 25 ppm, Aspen+conifer comm, annual forb, 2 h, intermountain, USA	Descurania californica[4]
Toxic effects	Injury threshold value for field O3 fumi.	O3 = 30 ppm, Aspen+conifer comm, perennial forb, 2 h, intermountain, USA	Epilobium angustifolium[4]
Toxic effects	Injury threshold value for field O3 fumi.	O3 = 30 ppm, Aspen+conifer comm, perennial forb, 2 h, intermountain, USA	Epilobium watsoni[4]
Toxic effects	Injury threshold value for field O3 fumi.	O3 = 30 ppm, Aspen+conifer comm, perennial forb, 2 h, intermountain, USA	Eriogonum heracliodes[4]
Toxic effects	Injury threshold value for field O3 fumi.	O3 = 30 ppm, Aspen+conifer comm, perennial forb, 2 h, intermountain, USA	Fragaria ovalis[4]

Toxic effects	Injury threshold value for field O3 fumi.	O3 = >30 ppm, Aspen+conifer comm, annual forb, 2 h, intermountain, USA	Galium bifolium[4]
Toxic effects	Injury threshold value for field O3 fumi.	O3 = 30 ppm, Aspen+conifer comm, annual forb, 2 h, intermountain, USA	Gayophytum racemosum[4]
Toxic effects	Injury threshold value for field O3 fumi.	O3 = >15 ppm, Aspen+conifer comm, perennial forb, 2 h, intermountain, USA	Gentiana amarella[4]
Toxic effects	Injury threshold value for field O3 fumi.	O3 = <25 ppm, Aspen+conifer comm, perennial forb, 2 h, intermountain, USA	Geranium fremontii[4]
Toxic effects	Injury threshold value for field O3 fumi.	O3 = 15 ppm, Aspen+conifer comm, perennial forb, 2 h, intermountain, USA	Geranium richardsonii[4]
Toxic effects	Injury threshold value for field O3 fumi.	O3 = 15 ppm, grass-oak comm, perennial forbs, 2 h, intermountain, USA	Hedysarum boreale[4]
Toxic effects	Injury threshold value for field O3 fumi.	O3 = >30 ppm, grass-oak comm, perennial forbs, 2 h, intermountain, USA	Helianthus anuus[4]
Toxic effects	Injury threshold value for field O3 fumi.	O3 = >25 ppm, Aspen+conifer comm, perennial forb, 2 h, intermountain, USA	Juncus sp.[4]
Toxic effects	Injury threshold value for field O3 fumi.	O3 = >25 ppm, Aspen+conifer comm, perennial forb, 2 h, intermountain, USA	Lathyrus lanzwertii[4]
Toxic effects	Injury threshold value for field O3 fumi.	O3 = 25 ppm, Aspen+conifer comm, perennial forb, 2 h, intermountain, USA	Lathyrus pauciflorus[4]
Toxic effects	Injury threshold value for field O3 fumi.	O3 = >40 ppm, grass-oak comm, tree and shrub, 2 h, intermountain, USA	Mahonia repens[4]
Toxic effects	Injury threshold value for field O3 fumi.	O3 = 25 ppm, grass-oak comm, perennial forbs, 2 h, intermountain, USA	Medicago sativa[4]
Toxic effects	Injury threshold value for field O3 fumi.	O3 = 30 ppm, Aspen+conifer comm, perennial forb, 2 h, intermountain, USA	Mertensia arizonica[4]
Toxic effects	Injury threshold value for field O3 fumi.	O3 = >25 ppm, Aspen+conifer comm, perennial forb, 2 h, intermountain, USA	Mimulus guttatus[4]
Toxic effects	Injury threshold value for field O3 fumi.	O3 = <40 ppm, Aspen+conifer comm, perennial forb, 2 h, intermountain, USA	Mimulus moschatus[4]
Toxic effects	Injury threshold value for field O3 fumi.	O3 = >30 ppm, Aspen+conifer comm, perennial forb, 2 h, intermountain, USA	Mitella stenopetala[4]
Toxic effects	Injury threshold value for field O3 fumi.	O3 = 25 ppm, Aspen+conifer comm, perennial forb, 2 h, intermountain, USA	Osmorhiza occidentalis[4]
Toxic effects	Injury threshold value for field O3 fumi.	O3 = >30 ppm, Aspen+conifer comm, tree + shrub, 2 h, intermountain, USA	Pachystima myrsinites[4]
Toxic effects	Injury threshold value for field O3 fumi.	O3 = <25 ppm, Aspen+conifer comm, perennial forb, 2 h, intermountain, USA	Phacelia heterophylla[4]
Toxic effects	Injury threshold value for field O3 fumi.	O3 = 25 ppm, grass-oak comm, grass, 2 h, intermountain, USA	Poa pratensis[4]
Toxic effects	Injury threshold value for field O3 fumi.	O3 = 30 ppm, Aspen+conifer comm, perennial forb, 2 h, intermountain, USA	Polemonium foliosissimum[4]
Toxic effects	Injury threshold value for field O3 fumi.	O3 = >25 ppm, Aspen+conifer comm, annual forb, 2 h, intermountain, USA	Polygonum douglasii[4]
Toxic effects	Injury threshold value for field O3 fumi.	O3 = 15 ppm, Aspen+conifer comm, tree + shrub, 2 h, intermountain, USA	Populus tremuloides[4]
Toxic effects	Injury threshold value for field O3 fumi.	O3 = 30 ppm, grass-oak comm, tree and shrub, 2 h, intermountain, USA	Potentilla fruticosa[4]
Toxic effects	Injury threshold value for field O3 fumi.	O3 = 25 ppm, grass-oak comm, tree and shrub, 2 h, intermountain, USA	Quercus gambelii[4]
Toxic effects	Injury threshold value for field O3 fumi.	O3 = 30 ppm, Aspen+conifer comm, tree + shrub, 2 h, intermountain, USA	Ribes hudsonianum[4]

Toxic effects	Injury threshold value for field O3 fumi.	O3 = >30 ppm, Aspen+conifer comm, tree + shrub, 2 h, intermountain, USA	Rosa woodsii[4]
Toxic effects	Injury threshold value for field O3 fumi.	O3 = <30 ppm, Aspen+conifer comm, perennial forb, 2 h, intermountain, USA	Rudbeckia occidentalis[4]
Toxic effects	Injury threshold value for field O3 fumi.	O3 = 25 ppm, grass-oak comm, perennial forbs, 2 h, intermountain, USA	Rumex crispus[4]
Toxic effects	Injury threshold value for field O3 fumi.	O3 = >25 ppm, Aspen+conifer comm, tree + shrub, 2 h, intermountain, USA	Sambucus melanocarpa[4]
Toxic effects	Injury threshold value for field O3 fumi.	O3 = <30 ppm, Aspen+conifer comm, perennial forb, 2 h, intermountain, USA	Saxifraga arguta[4]
Toxic effects	Injury threshold value for field O3 fumi.	O3 = 15 ppm, Aspen+conifer comm, perennial forb, 2 h, intermountain, USA	Senecio serra[4]
Toxic effects	Injury threshold value for field O3 fumi.	O3 = 30 ppm, Aspen+conifer comm, tree + shrub, 2 h, intermountain, USA	Symphoricarpos vaccinioides[4]
Toxic effects	Injury threshold value for field O3 fumi.	O3 = >25 ppm, Aspen+conifer comm, perennial forb, 2 h, intermountain, USA	Taraxacum officinale[4]
Toxic effects	Injury threshold value for field O3 fumi.	O3 = >25 ppm, Aspen+conifer comm, perennial forb, 2 h, intermountain, USA	Thalictrum fendleri[4]
Toxic effects	Injury threshold value for field O3 fumi.	O3 = >30 ppm, grass-oak comm, tree and shrub, 2 h, intermountain, USA	Toxicodendron radicans[4]
Toxic effects	Injury threshold value for field O3 fumi.	O3 = 30 ppm, grass-oak comm, perennial forbs, 2 h, intermountain, USA	Urtica gracilis[4]
Toxic effects	Injury threshold value for field O3 fumi.	O3 = 25 ppm, Aspen+conifer comm, perennial forb, 2 h, intermountain, USA	Veronica anagallis-aquatica[4]
Toxic effects	Injury threshold value for field O3 fumi.	O3 = >25 ppm, Aspen+conifer comm, perennial forb, 2 h, intermountain, USA	Vicia americana[4]
Toxic effects	Injury threshold value for field O3 fumi.	O3 = >40 ppm, grass-oak comm, perennial forbs, 2 h, intermountain, USA	Vicia americana[4]
Toxic effects	Injury threshold value for field O3 fumi.	O3 = >30 ppm, Aspen+conifer comm, perennial forb, 2 h, intermountain, USA	Viola adunca[4]
Photo-synthesis	0.9 mg/(g dry weight*hour)	Ozone = 0.0 pphm, 77 d, 302 K, 8 month-old seedlings	Pinus elliottii[5]
Photo-synthesis	0.9 mg/(g dry weight*hour)	Ozone = 15.0 pphm, 77 d, 302 K, 8 month-old seedlings	Pinus elliottii[5]
Photo-synthesis	1.4 mg/(g dry weight*hour)	Ozone = 15.0 pphm, 88 d, 302 K, 8 month-old seedlings	Pinus elliottii[5]
Photo-synthesis	1.6 mg/(g dry weight*hour)	Ozone = 5.0 pphm, 77 d, 302 K, 8 month-old seedlings	Pinus elliottii[5]
Photo-synthesis	1.7 mg/(g dry weight*hour)	Ozone = 5.0 pphm, 88 d, 302 K, 8 month-old seedlings	Pinus elliottii[5]
Photo-synthesis	1.9 mg/(g dry weight*hour)	Ozone = 0.0 pphm, 88 d, 302 K, 8 month-old seedlings	Pinus elliottii[5]
Photo-synthesis	2.0 mg/(g dry weight*hour)	Ozone = 15.0 pphm, 36 d, 302 K, 8 month-old seedlings	Pinus elliottii[5]
Photo-synthesis	2.6 mg/(g dry weight*hour)	Ozone = 0.0 pphm, 36 d, 302 K, 8 month-old seedlings	Pinus elliottii[5]
Photo-synthesis	3.1 mg/(g dry weight*hour)	Ozone = 5.0 pphm, 36 d, 302 K, 8 month-old seedlings	Pinus elliottii[5]
Photo-synthesis	0.9 mg/(g dry weight*hour)	Ozone = 0.0 pphm, 77 d, 302 K, 8 month-old seedlings	Pinus serotina[5]
Photo-synthesis	1.6 mg/(g dry weight*hour)	Ozone = 15.0 pphm, 77 d, 302 K, 8 month-old seedlings	Pinus serotina[5]

Photo-synthesis	1.8 mg/(g dry weight*hour)	Ozone = 5.0 pphm, 77 d, 302 K, 8 month-old seedlings	Pinus serotina[5]
Photo-synthesis	3.3 mg/(g dry weight*hour)	Ozone = 15.0 pphm, 36 d, 302 K, 8 month-old seedlings	Pinus serotina[5]
Photo-synthesis	3.3 mg/(g dry weight*hour)	Ozone = 15.0 pphm, 88 d, 302 K, 8 month-old seedlings	Pinus serotina[5]
Photo-synthesis	3.4 mg/(g dry weight*hour)	Ozone = 0.0 pphm, 88 d, 302 K, 8 month-old seedlings	Pinus serotina[5]
Photo-synthesis	3.5 mg/(g dry weight*hour)	Ozone = 0.0 pphm, 36 d, 302 K, 8 month-old seedlings	Pinus serotina[5]
Photo-synthesis	3.8 mg/(g dry weight*hour)	Ozone = 5.0 pphm, 88 d, 302 K, 8 month-old seedlings	Pinus serotina[5]
Photo-synthesis	4.2 mg/(g dry weight*hour)	Ozone = 5.0 pphm, 36 d, 302 K, 8 month-old seedlings	Pinus serotina[5]
Photo-synthesis	1.9 mg/(g dry weight*hour)	Ozone = 5.0 pphm, 77 d, 302 K, 4 month-old seedlings	Pinus strobus[5]
Photo-synthesis	2.0 mg/(g dry weight*hour)	Ozone = 15.0 pphm, 77 d, 302 K, 4 month-old seedlings	Pinus strobus[5]
Photo-synthesis	2.1 mg/(g dry weight*hour)	Ozone = 0.0 pphm, 77 d, 302 K, 4 month-old seedlings	Pinus strobus[5]
Photo-synthesis	2.5 mg/(g dry weight*hour)	Ozone = 15.0 pphm, 36 d, 302 K, 4 month-old seedlings	Pinus strobus[5]
Photo-synthesis	3.0 mg/(g dry weight*hour)	Ozone = 5.0 pphm, 36 d, 302 K, 4 month-old seedlings	Pinus strobus[5]
Photo-synthesis	3.2 mg/(g dry weight*hour)	Ozone = 0.0 pphm, 36 d, 302 K, 4 month-old seedlings	Pinus strobus[5]
Photo-synthesis	4.4 mg/(g dry weight*hour)	Ozone = 15.0 pphm, 19 d, 302 K, 4 month-old seedlings	Pinus strobus[5]
Photo-synthesis	4.8 mg/(g dry weight*hour)	Ozone = 5.0 pphm, 19 d, 302 K, 4 month-old seedlings	Pinus strobus[5]
Photo-synthesis	4.9 mg/(g dry weight*hour)	Ozone = 0.0 pphm, 19 d, 302 K, 4 month-old seedlings	Pinus strobus[5]
Photo-synthesis	1.7 mg/(g dry weight*hour)	Ozone = 15.0 pphm, 77 d, 302 K, 8 month-old seedlings	Pinus taeda[5]
Photo-synthesis	1.8 mg/(g dry weight*hour)	Ozone = 0.0 pphm, 77 d, 302 K, 8 month-old seedlings	Pinus taeda[5]
Photo-synthesis	1.8 mg/(g dry weight*hour)	Ozone = 5.0 pphm, 77 d, 302 K, 8 month-old seedlings	Pinus taeda[5]
Photo-synthesis	3.2 mg/(g dry weight*hour)	Ozone = 15.0 pphm, 36 d, 302 K, 8 month-old seedlings	Pinus taeda[5]
Photo-synthesis	3.7 mg/(g dry weight*hour)	Ozone = 0.0 pphm, 36 d, 302 K, 8 month-old seedlings	Pinus taeda[5]
Photo-synthesis	4.0 mg/(g dry weight*hour)	Ozone = 5.0 pphm, 36 d, 302 K, 8 month-old seedlings	Pinus taeda[5]
Photo-synthesis	4.0 mg/(g dry weight*hour)	Ozone = 5.0 pphm, 88 d, 302 K, 8 month-old seedlings	Pinus taeda[5]
Photo-synthesis	4.1 mg/(g dry weight*hour)	Ozone = 15.0 pphm, 88 d, 302 K, 8 month-old seedlings	Pinus taeda[5]
Photo-synthesis	4.3 mg/(g dry weight*hour)	Ozone = 0.0 pphm, 88 d, 302 K, 8 month-old seedlings	Pinus taeda[5]
Respiration	0.4 mg/(g dry weight*hour)	Ozone = 0.0 pphm, 77 d, 302 K, 8 month-old seedlings	Pinus elliottii[5]

Respiration	0.4 mg/(g dry weight*hour)	Ozone = 5.0 pphm, 77 d, 302 K, 8 month-old seedlings	Pinus elliottii[5]
Respiration	0.5 mg/(g dry weight*hour)	Ozone = 0.0 pphm, 88 d, 302 K, 8 month-old seedlings	Pinus elliottii[5]
Respiration	0.5 mg/(g dry weight*hour)	Ozone = 5.0 pphm, 88 d, 302 K, 8 month-old seedlings	Pinus elliottii[5]
Respiration	0.6 mg/(g dry weight*hour)	Ozone = 15.0 pphm, 77 d, 302 K, 8 month-old seedlings	Pinus elliottii[5]
Respiration	0.6 mg/(g dry weight*hour)	Ozone = 5.0 pphm, 36 d, 302 K, 8 month-old seedlings	Pinus elliottii[5]
Respiration	0.7 mg/(g dry weight*hour)	Ozone = 0.0 pphm, 36 d, 302 K, 8 month-old seedlings	Pinus elliottii[5]
Respiration	0.8 mg/(g dry weight*hour)	Ozone = 15.0 pphm, 88 d, 302 K, 8 month-old seedlings	Pinus elliottii[5]
Respiration	1.3 mg/(g dry weight*hour)	Ozone = 15.0 pphm, 36 d, 302 K, 8 month-old seedlings	Pinus elliottii[5]
Respiration	0.5 mg/(g dry weight*hour)	Ozone = 15.0 pphm, 77 d, 302 K, 8 month-old seedlings	Pinus serotina[5]
Respiration	0.5 mg/(g dry weight*hour)	Ozone = 5.0 pphm, 77 d, 302 K, 8 month-old seedlings	Pinus serotina[5]
Respiration	0.6 mg/(g dry weight*hour)	Ozone = 0.0 pphm, 77 d, 302 K, 8 month-old seedlings	Pinus serotina[5]
Respiration	0.9 mg/(g dry weight*hour)	Ozone = 5.0 pphm, 36 d, 302 K, 8 month-old seedlings	Pinus serotina[5]
Respiration	1.0 mg/(g dry weight*hour)	Ozone = 0.0 pphm, 36 d, 302 K, 8 month-old seedlings	Pinus serotina[5]
Respiration	1.0 mg/(g dry weight*hour)	Ozone = 0.0 pphm, 88 d, 302 K, 8 month-old seedlings	Pinus serotina[5]
Respiration	1.2 mg/(g dry weight*hour)	Ozone = 15.0 pphm, 36 d, 302 K, 8 month-old seddlings	Pinus serotina[5]
Respiration	1.2 mg/(g dry weight*hour)	Ozone = 15.0 pphm, 88 d, 302 K, 8 month-old seedlings	Pinus serotina[5]
Respiration	1.2 mg/(g dry weight*hour)	Ozone = 5.0 pphm, 88 d, 302 K, 8 month-old seedlings	Pinus serotina[5]
Respiration	0.6 mg/(g dry weight*hour)	Ozone = 0.0 pphm, 77 d, 302 K, 4 month-old seedlings	Pinus strobus[5]
Respiration	0.6 mg/(g dry weight*hour)	Ozone = 5.0 pphm, 36 d, 302 K, 4 month-old seedlings	Pinus strobus[5]
Respiration	0.6 mg/(g dry weight*hour)	Ozone = 5.0 pphm, 77 d, 302 K, 4 month-old seedlings	Pinus strobus[5]
Respiration	0.8 mg/(g dry weight*hour)	Ozone = 0.0 pphm, 36 d, 302 K, 4 month-old seedlings	Pinus strobus[5]
Respiration	0.9 mg/(g dry weight*hour)	Ozone = 15.0 pphm, 77 d, 302 K, 4 month-old seedlings	Pinus strobus[5]
Respiration	1.4 mg/(g dry weight*hour)	Ozone = 15. ppm, 36 d, 302 K, 4 month-old seedlings	Pinus strobus[5]
Respiration	0.5 mg/(g dry weight*hour)	Ozone = 15.0 pphm, 77 d, 302 K, 8 month-old seedlings	Pinus taeda[5]
Respiration	0.6 mg/(g dry weight*hour)	Ozone = 0.0 pphm, 77 d, 302 K, 8 month-old seedlings	Pinus taeda[5]
Respiration	0.6 mg/(g dry weight*hour)	Ozone = 5.0 pphm, 77 d, 302 K, 8 month-old seedlings	Pinus taeda[5]

Respiration	0.7 mg/(g dry weight*hour)	Ozone = 0.0 pphm, 36 d, 302 K, 8 month-old seedlings	Pinus taeda[5]
Respiration	0.7 mg/(g dry weight*hour)	Ozone = 5.0 pphm, 36 d, 302 K, 8 month-old seedlings	Pinus taeda[5]
Respiration	1.0 mg/(g dry weight*hour)	Ozone = 0.0 pphm, 88 d, 302 K, 8 month-old seedlings	Pinus taeda[5]
Respiration	1.0 mg/(g dry weight*hour)	Ozone = 5.0 pphm, 88 d, 302 K, 8 month-old seedlings	Pinus taeda[5]
Respiration	1.2 mg/(g dry weight*hour)	Ozone = 15.0 pphm, 36 d, 302 K, 8 month-old seedlings	Pinus taeda[5]
Respiration	1.2 mg/(g dry weight*hour)	Ozone = 15.0 pphm, 88 d, 302 K, 8 month-old seedlings	Pinus taeda[5]

Source: 5.Barnes, R. 1972. Environ. Pollut., 3, pp 133-138.
1.Hill, A.C. and N. Littlefield. 1969. Environ. Sci. Technol. 3, 52-56.
2.Keller, T. 1977. 4th International Clean Air Congress, 81-83.
3.Letchworth, M.B. and V. Blum. 1977. Environ. Pollut. 14, 303-312.
4.Threshow, M. and D. Stewart. 1973. Biol. Conserva. 5, 209-214.

4-335 Effect of OCDD

Parameter	Effect Value	Concentration and Condition	Species
Mortality	50 %	Oral dose: > 1 g/kg body wt.	Rat[1]
Teratogenicity	0 % effect	Dose = 100-500 mg/kg*day, from day 6-15 in pregnancy-periode	Rat[2]

Source: 1.Schewets, B.A. et al. 1973. Environ. health Persp., 9. pp 233-240.
2.Mercier, M.J. 1977. 2,3,7,8-Tetrachlorodibenzo-p-dioxin, An Overview. Liege. Belgium.

4-336 Effect of Oil antipollution products

Parameter	Effect Value	Concentration and Condition	Species
Mortality	50 %	BP MOD 9 = 1/160 (vol/vol), 72 h, 10 fish	Carassius auratus[1]
Mortality	50 %	Bp 1100 = 1/160 (vol/vol), 72 h, 10 fish	Carassius auratus[1]
Mortality	50 %	Dispersoc sd = 1/150 (vol/vol), 72 h, 10 fish	Carassius auratus[1]
Mortality	50 %	Drew. amer. oil spill emuls. no 1 = 1/450 (vol/vol), 72 h, 10 fish	Carassius auratus[1]
Mortality	50 %	Esso corexit 7664 = 1/10 (vol/vol), 72 h, 10 fish	Carassius auratus[1]
Mortality	50 %	Esso corexit 8666 = 1/5-1/4 (vol/vol), 72 h, 10 fish	Carassius auratus[1]
Mortality	50 %	Fina sol osr/2 = 1/40 (vol/vol), 72 h, 10 fish	Carassius auratus[1]
Mortality	50 %	Fina sol sc = 1/80 (vol/vol), 72 h, 10 fish	Carassius auratus[1]
Mortality	50 %	Gamlen oil spill remover = 1/200 (vol/vol), 72 h, 10 fish	Carassius auratus[1]
Mortality	50 %	Shell oil disperser = 1/210 (vol/vol), 72 h, 10 fish	Carassius auratus[1]
Mortality	50 %	Slip cleam m = 1/200 (vol/vol), 72 h, 10 fish	Carassius auratus[1]
Mortality	50 %	T c 66 = 1/160 (vol/vol), 72 h, 10 fish	Carassius auratus[1]
Mortality	50 %	Tar solvent es = 1/160 (vol/vol), 72 h, 10 fish	Carassius auratus[1]
Mortality	50 %	Zerol sea el = 1/300 (vol/vol), 72 h, 10 fish	Carassius auratus[1]

Source: 1.Baldini, L.E. and F. Cugurra. 1974. Water Res., 8(5), pp 323-324.

4-337 Effect of Oil pollution

Parameter	Effect Value	Concentration and Condition	Species
Numbers after pollution	12.0	Numbers 21(1956-7) 13(1966-7), Hobson Beach, Calif, USA, 1969	Plant species[1]
Numbers after pollution	12.0	Numbers before 13(1956-7) 14(1966-7), Campus Point, UCSB, Calif, USA,	Plant species[1]
Numbers after pollution	14.0	Numbers before 22(1956-7) 20(1966-7), Eucalyptus Lane, Calif, USA,	Plant species[1]
Numbers after pollution	17.0	Numbers before 27(1956-7) 26(1966-7), Loon Point, Montecito, Calif,	Plant species[1]
Numbers after pollution	2.0	Numbers before 18(1956-7) 14(1966-7), Leo Carrillo, St Beach, Calif,	Plant species[1]
Numbers after pollution	20.0	Numbers before 18(1956-7) 20(1966-7), El Capitan State Pk., Calif.	Plant species[1]
Numbers after pollution	21.0	Numbers before 29(1956-7) 40(1966-7), Santa Barbara Point, Calif,	Plant species[1]
Numbers after pollution	31.0	Numbers before 18(1956-7) 18(1966-7), Coil Oil Point, Calif, USA,	Plant species[1]
Numbers after pollution	34.0	Numbers before 14(1956-7) 29(1966-7), Carpinteria Reef, Calif, USA,	Plant species[1]
Numbers after pollution	9.0	Numbers before 19(1956-7) 17(1966-7), Palm Street, Ventura, CalifUSA,	Plant species[1]
Oil coverage	1 % damage along a transect	Coal oil point, Calif, USA, light coverrage	Chthamalus fissus[1]
Oil coverage	20 % damage along a transect	Campus point, UCBS, Calif, USA, light coverrage	Chthamalus fissus[1]
Oil coverage	15 % damage along a transect	Thalli white, Campus Point, USCB, Calif, USA, moderate coverrage	Enteromorpha intestinalis[1]
Oil coverage	20 % damage along a transect	Thalli white, Carpinteria Reef, Calif, USA, moderate coverrage	Enteromorpha intestinalis[1]
Oil coverage	3 % damage along a transect	Thalli white, Coil Oil Point, Calif, USA, light coverrage, rg = 1 - 5	Enteromorpha intestinalis[1]
Oil coverage	1.5 % damage along a transect	Loon point, Montecito, Calif, USA, light coverrage, rg = 1 - 2	Gigartina leptorynchos[1]
Oil coverage	1 % damage along a transect	Campus point, UCBS, Calif, USA, heavy coverrage	Mytilus spp.[1]
Oil coverage	40 % damage along a transect	% exposed blades damaged, Carpinteria Reef, Calif, heavy coverrage	Phyllospadix torreyi[1]
Oil coverage	55 % damage along a transect	% exposed damaged, Santa Barbara Point, Calif, USA, heavy coverrage	Phyllospadix torreyi[1]
Oil coverage	3 % damage along a transect	Campus Point, UCBS, Calif, USA, moderate coverrage, rg = 1 - 5	Pollicipes polymerus[1]
Oil coverage	3 % damage along a transect	Carpinteria Reef, Calif, USA, moderate coverrage, rg = 1 - 5	Pollicipes polymerus[1]
Oil coverage	3 % damage along a transect	Coil Oil Point, Calif, USA, moderate coverrage, rg = 1 - 5	Porphyra sp.[1]
Oil coverage	3 % damage along a transect	Thalli white, El Capitan State Pk, Calif, USA, light coverrage	Porphyra sp.[1]
Oil coverage	3 % damage along a transect	Thalli white, Hobson Beach, Calif, USA, light coverrage, rg = 1 -5	Ulva angusta[1]

Source: 1.Foster, M., M. Neushul and R. Zingmark. 1971. Environ. Pollut., 2, pp 11

4-338 Effect of Oxalic acid

Parameter	Concentration	Condition	Species
LC0	20 mg/l	long time exposure	Fish[1]
LC0	1000 mg/l	0.5 hour, pH=2.6	Goldfish[2]
LC0	200 mg/l	4 days, pH=5.3	Goldfish[2]
Sublethal effect	42 mg/l	Inhibition of cell multiplication, unneutralized	Microcystis aeruginosa[3]
Sublethal effect	80 mg/l	Inhibition of cell multiplication, neutralized	Microcystis aeruginosa[3]
Sublethal effect	41 mg/l	Inhibition of cell multiplication, unneutralized	Pseudomonas putida[3]
Sublethal effect	1550 mg/l	Inhibition of cell multiplication, neutralized	Pseudomonas putida[3]

Source: 1.Meinck, F., Stooff, H. and Kohlschutter, H., Les eaux residuaires industrielles, 1970.
2.Auer. Auergeselschaft-1000, Berlin 65 (west).
3.Bringmann, G. and Kuhn, R., nwf-wasser/abwasser, H.9, 1976.

4-339 Effect of Oxygen

Parameter	Effect Value	Concentration and Condition	Species
Growth	0 % of max growth	Oxygen conc = 1.00 mg O2/l, 278 K, weight = 400 g	Salvelinus fontinalis[1]
Growth	0 % of max growth	Oxygen conc = 1.7 mg O2/l, 278 K, weight = 5 g, 8 cm	Salvelinus fontinalis[1]
Growth	0 % of max growth	Oxygen conc = 1.74 mg O2/l, 283 K, weight = 400 g	Salvelinus fontinalis[1]
Growth	0 % of max growth	Oxygen conc = 1.8 mg O2/l, 283 K, weight = 5 g, 8 cm	Salvelinus fontinalis[1]
Growth	0 % of max growth	Oxygen conc = 2.90 mg O2/l, 288 K, weight = 400 g	Salvelinus fontinalis[1]
Growth	0 % of max growth	Oxygen conc = 3.57 mg O2/l, 293 K, weight = 400 g	Salvelinus fontinalis[1]
Growth	0 % of max growth	Oxygen conc = 4.24 mg O2/l, 288 K, weight = 5 g, 8 cm	Salvelinus fontinalis[1]
Growth	0 % of max growth	Oxygen conc = 4.33 mg O2/l, 293 K, weight = 5 g, 8 cm	Salvelinus fontinalis[1]
Growth	100 % of max growth	Oxygen conc = 11.4 mg O2/l, 283 K, weight = 5 g, 8 cm	Salvelinus fontinalis[1]
Growth	100 % of max growth	Oxygen conc = 3.13 mg O2/l, 278 K, weight = 400 g	Salvelinus fontinalis[1]
Growth	100 % of max growth	Oxygen conc = 6.05 mg O2/l, 283 K, weight = 400 g	Salvelinus fontinalis[1]
Growth	100 % of max growth	Oxygen conc = 6.9 mg O2/l, 278 K, weight = 5 g, 8 cm	Salvelinus fontinalis[1]
Growth	30 % of max growth	Oxygen conc = 1.12 mg O2/l, 278 K, weight = 400 g	Salvelinus fontinalis[1]
Growth	30 % of max growth	Oxygen conc = 1.83 mg O2/l, 283 K, weight = 400 g	Salvelinus fontinalis[1]
Growth	30 % of max growth	Oxygen conc = 2.05 mg O2/l, 278 K, weight = 5 g, 8 cm	Salvelinus fontinalis[1]
Growth	30 % of max growth	Oxygen conc = 2.23 mg O2/l, 283 K, weight = 5 g, 8 cm	Salvelinus fontinalis[1]
Growth	30 % of max growth	Oxygen conc = 3.17 mg O2/l, 288 K, weight = 400 g	Salvelinus fontinalis[1]
Growth	30 % of max growth	Oxygen conc = 4.38 mg O2/l, 293 K, weight = 400 g	Salvelinus fontinalis[1]
Growth	30 % of max growth	Oxygen conc = 4.64 mg O2/l, 288 K, weight = 5 g, 8 cm	Salvelinus fontinalis[1]

Growth	30 % of max growth	Oxygen conc = 5.36 mg O2/l, 293 K, weight = 5 g, 8 cm	Salvelinus fontinalis[1]
Growth	80 % of max growth	Oxygen conc = 2.41 mg O2/l, 278 K, weight = 400 g	Salvelinus fontinalis[1]
Growth	80 % of max growth	Oxygen conc = 3.64 mg O2/l, 283 K, weight = 400 g	Salvelinus fontinalis[1]
Growth	80 % of max growth	Oxygen conc = 4.80 mg O2/l, 278 K, weight = 5 g, 8 cm	Salvelinus fontinalis[1]
Growth	80 % of max growth	Oxygen conc = 5.09 mg O2/l, 283 K, weight = 5 g, 8 cm	Salvelinus fontinalis[1]
Growth	80 % of max growth	Oxygen conc = 6.47 mg O2/l, 288 K, weight = 400 g	Salvelinus fontinalis[1]
Growth	80 % of max growth	Oxygen conc = 9.71 mg O2/l, 288 K, weight = 5 g, 8 cm	Salvelinus fontinalis[1]

Source: 1.Madsen, B.L., P.N. Markmann and G.Rasmussen. 1977. Vand, 8(1), pp 4-8.

4-340 Effect of PCB

Parameter	Effect Value	Concentration and Condition	Species
Blood glucose	61.9 mg/100 ml	Control = 65.7, PCB = 6*1000 mg/kg ww (every sec.day), 2 w	Anguilla anguilla[1]
Blood glucose	62.5 mg/100 ml	Control = 62.6, PCB = 1*200 mg/kg ww (one time), 1 m, rg =+- 2.2	Anguilla anguilla[1]
Blood glucose	79.9 mg/100 ml	Control = 70., PCB = 1*200 mg/kg ww (one time), 4 m, rg =+- 5.5	Anguilla anguilla[1]
Blood glucose	82 mg/100 ml	Control = 82.7, PCB = 3*1000 mg/kg ww (every second day), 1 w	Anguilla anguilla[1]
Blood lactate	19.7 mg/100 ml	Control = 14.3, PCB = 3*1000 mg/kg ww (every sec.day), 1 w	Anguilla anguilla[1]
Blood lactate	6.58 mg/100 ml	Control = 5.86, PCB = 1*200 mg/kg ww (one time), 1 m, rg =+- 0.88	Anguilla anguilla[1]
Blood lactate	7.78 mg/100 ml	Control = 9.51, PCB = 1*200 mg/kg ww (one time), 4 m, rg =+- 1.35	Anguilla anguilla[1]
Blood lactate	9.0 mg/100 ml	Control = 4.9, PCB = 6*1000 mg/kg ww (every sec.day), 2 w, rg =+- 3.6	Anguilla anguilla[1]
Body weight	11.54 g	Injection of 50 mg PCB/ kg body weight, 21 d, cont = 8.96+-0.85 g	Mouse[2]
Body weight	22.76 g	Injection of 50 mg PCB/ kg body weight, 33 d, cont = 21.30+-2.34 g	Mouse[2]
Body weight	8.47 g	Injection of 50 mg PCB/ kg ww at birth, cont = 7.25+-0.76 g, 16 d	Mouse[2]
Corpuscular hemoglobin in blood	67.2 picogram/cell	Control = 70.6, PCB = 1*200 mg/kg ww (one time), 16 w, rg = +- 4.1	Anguilla anguilla[3]
Corpuscular hemoglobin in blood	79.4 picogram/cell	Control = 91.7, PCB = 1*200 mg/kg ww (one time), 4 w, rg = +- 6.7	Anguilla anguilla[3]
Corpuscular hemoglobin in blood	96.5 picogram/cell	Control = 82 , PCB = 3*1000 mg/kg ww (every sec.day), 1 w	Anguilla anguilla[3]
Corpuscular hemoglobin in blood	97.7 picogram/cell	Control = 85.4 , PCB = 6*1000 mg/kg ww (every sec.day), rg = +- 12.6	Anguilla anguilla[3]
Corpuscular hemoglobin in blood	70.9 picogram/cell	Control = 74.1, PCB = 1*200 mg/kg ww (one time), 17 d, rg = +- 3.6	Salmo gairdneri[3]

Corpuscular hemoglobin in blood	73.6 picogram/cell	Control = 83.2, PCB = 1*200 mg/kg ww (one time), 5 d, rg = +- 4.5	Salmo gairdneri[3]
Corpuscular hemoglobin in blood	81.7 picogram/cell	Control = 84.8, PCB = 1*200 mg/kg ww (one time), 10 d, rg = +- 3.8	Salmo gairdneri[3]
Corpuscular volume in blood	222.6 micro**3/cell	Control = 249.8, PCB = 1*200 mg/kg ww (one time), 16 w, rg = +- 15.2	Anguilla anguilla[3]
Corpuscular volume in blood	285.2 micro**3/cell	Control = 310.1, PCB = 1*200 mg/kg ww (one time), 4 w, rg = +- 30.6	Anguilla anguilla[3]
Corpuscular volume in blood	327.0 micro**3/cell	Control = 317.9, PCB = 6*1000 mg/kg ww (every sec.day), rg = +- 35.4	Anguilla anguilla[3]
Corpuscular volume in blood	328.4 micro**3/cell	Control = 296.9, PCB = 3*1000 mg/kg ww (every sec.day), rg = +- 58.2	Anguilla anguilla[3]
Corpuscular volume in blood	344.3 micro**3/cell	Control = 359.2, PCB = 1*200 mg/kg ww (one time), 17 d, rg = +- 21.3	Salmo gairdneri[3]
Corpuscular volume in blood	356.1 micro**3/cell	Control = 417.3, PCB = 1*200 mg/kg ww (one time), 5 d, rg = +- 21.0	Salmo gairdneri[3]
Corpuscular volume in blood	424.3 micro**3/cell	Control = 375.4, PCB = 1*200 mg/kg ww (one time), 10 d, rg = +- 21.4	Salmo gairdneri[3]
Enzyme activity in blood	182.4 ug pbg/(ml*hour)	Control = 188.9, PCB = 0.2 g/fish (one time), 21 d, rg =+- 13.2	Salmo gairdneri[4]
Enzyme activity in blood	202.0 ug pbg/(ml*hour)	Control = 180., PCB = 0.2 g/fish (one time), 7 d, rg =+- 11.1	Salmo gairdneri[4]
Enzyme activity in blood	203.7 ug pbg/(ml*hour)	Control = 238.9, PCB = 0.2 g/fish (one time), 3 d, rg =+- 11.8	Salmo gairdneri[4]
Enzyme activity in blood	215.0 ug pbg/(ml*hour)	Control = 245., PCB = 0.2 g/fish (one time), 14 d, rg =+- 16.0	Salmo gairdneri[4]
Enzyme activity in blood	5.2 ug pbg/(ml*hour)	Control = 5.6, PCB = 0.2 g/fish (one time), 3 d, rg =+- 0.2	Salmo gairdneri[4]
Enzyme activity in blood	5.3 ug pbg/(ml*hour)	Control = 4.7, PCB = 0.2 g/fish (one time), 7 d, rg =+- 0.2	Salmo gairdneri[4]
Enzyme activity in blood	5.4 ug pbg/(ml*hour)	Control = 5.1, PCB = 0.2 g/fish (one time), 21 d, rg =+- 0.2	Salmo gairdneri[4]
Enzyme activity in blood	6.4 ug pbg/(ml*hour)	Control = 5.9, PCB = 0.2 g/fish (one time), 14 d, rg =+- 0.3	Salmo gairdneri[4]
Enzyme activity in kidney	82.7 ug pbg/(gram wet weight*hour)	Control = 82.6, PCB = 0.2 g/fish (one time), 21 d, rg = +- 3.5	Salmo gairdneri[4]
Enzyme activity in kidney	88.5 ug pbg/(gram wet weight*hour)	Control = 90.1, PCB = 0.2 g/fish (one time), 7 d, rg = +- 2.0	Salmo gairdneri[4]
Enzyme activity in kidney	92.7 ug pbg/(gram wet weight*hour)	Control = 102.9, PCB = 0.2 g/fish (one time), 14 d, rg = +- 4.1	Salmo gairdneri[4]

Enzyme activity in kidney	94.7 ug pbg/(gram wet weight*hour)	Control = 94.7, PCB = 0.2 g/fish (one time), 3 d, rg =+- 6.9	Salmo gairdneri[4]
Enzyme activity in spleen	256.6 ug pbg/(gram wet weight*hour)	Control = 197.4, PCB = 0.2 g/fish (one time), 3 d, rg =+- 17.8	Salmo gairdneri[4]
Enzyme activity in spleen	339.0 ug pbg/(gram wet weight*hour)	Control = 313.1, PCB = 0.2 g/fish (one time), 7 d, rg =+- 39.5	Salmo gairdneri[4]
Enzyme activity in spleen	341.0 ug pbg/(gram wet weight*hour)	Control = 244.8, PCB = 0.2 g/fish (one time), 21 d, rg =+-25.4	Salmo gairdneri[4]
Enzyme activity in spleen	399.6 ug pbg/(gram wet weight*hour)	Control = 297.5, PCB = 0.2 g/fish (one time), 14 d, rg =+- 28.6	Salmo gairdneri[4]
Glycogen in muscles	0.125 mg/100 mg	Control = 0.147, PCB = 1*200 mg/kg ww (one time), 1 m, rg =+- 0.018	Anguilla anguilla[1]
Glycogen in muscles	0.192 mg/100 mg	Control = 0.189, PCB = 1*200 mg/kg ww (one time), 4 m, rg =+- 0.026	Anguilla anguilla[1]
Glycogen in muscles	0.2 mg/100 mg	Control = 0.26, PCB = 6*1000 mg/kg ww (every sec.day), 2 w	Anguilla anguilla[1]
Glycogen in muscles	0.22 mg/100 mg	Control = 0.2, PCB = 3*1000 mg/kg ww (every sec.day), 1 w	Anguilla anguilla[1]
Gonad somatic index	0.32	Control = 0.43, PCB = 1*200 mg/kg ww (one time), 4 m, rg =+- 0.09	Anguilla anguilla[1]
Gonad somatic index	0.35	Control = 0.38, PCB = 3*1000 mg/kg ww (every sec.day), 1 w	Anguilla anguilla[1]
Gonad somatic index	0.5	Control = 0.32, PCB = 6*1000 mg/kg ww (every sec.day), 2 w	Anguilla anguilla[1]
Gonad somatic index	0.62	Control = 0.64, PCB = 1*200 mg/kg ww (one time), 1 m, rg =+- 0.11	Anguilla anguilla[1]
Growth	15.85 %	Control = 18.04 %, PCB = 1*200 mg/kg ww (one time), 4 m, rg =+- 1.71	Anguilla anguilla[1]
Growth	5.27 %	Control = 6.73 %, PCB = 6*1000 mg/kg ww (every sec.day), 2 w	Anguilla anguilla[1]
Growth	5.92 %	Control = 5.30 %, PCB = 3*1000 mg/kg ww (every sec.day), 1 w	Anguilla anguilla[1]
Growth	8.63 %	Control = 7.36 %, PCB = 1*200 mg/kg ww (one time), 1 m, rg =+- 1.27	Anguilla anguilla[1]
Growth rate	0.74 1/day	PCB = 0 ppm, 285 K, 120 h, control = 100 %	Dunaliella tertiolecta[5]
Growth rate	0.75 1/day	PCB = 0.1 ppm, 285 K, 120 h, control = 101 %	Dunaliella tertiolecta[5]
Growth rate	0.79 1/day	PCB = 10pmm, 285 K, 120 h, control = 107 %	Dunaliella tertiolecta[5]
Growth rate	1.18 1/day	PCB = 10pmm, 290 K, 72 h, control = 97 %	Dunaliella tertiolecta[5]
Growth rate	1.22 1/day	PCB = 0 ppm, 290 K, 72 h, control = 100 %	Dunaliella tertiolecta[5]
Growth rate	1.23 1/day	PCB = 0.1 ppm, 290 K, 72 h, control = 101 %	Dunaliella tertiolecta[5]
Growth rate	1.66 1/day	PCB = 0 ppm, 298 K, 72 h, control = 100 %	Dunaliella tertiolecta[5]
Growth rate	1.66 1/day	PCB = 0.1 ppm, 298 K, 72 h, control = 100 %	Dunaliella tertiolecta[5]
Growth rate	1.69 1/day	PCB = 10pmm, 298 K, 72 h, control = 102 %	Dunaliella tertiolecta[5]
Growth rate	0.46 1/day	PCB = 1 ppm, 283 K, 192 h, control = 77 %	Rhizosolenia setigera[5]
Growth rate	0.49 1/day	PCB = 0.1 ppm, 283 K, 192 h, control = 82 %	Rhizosolenia setigera[5]
Growth rate	0.49 1/day	PCB = 10 ppm, 283 K, 192 h, control = 82 %	Rhizosolenia setigera[5]

Growth rate	0.6 1/day	PCB = 0 ppm, 283 K, 192 h, control = 100 %	Rhizosolenia setigera[5]
Growth rate	0.73 1/day	PCB = 10 ppm, 288 K, 144 h, control = 85 %	Rhizosolenia setigera[5]
Growth rate	0.75 1/day	PCB = 1 ppm, 288 K, 144 h, control = 87 %	Rhizosolenia setigera[5]
Growth rate	0.81 1/day	PCB = 0.1 ppm, 288 K, 144 h, control = 95 %	Rhizosolenia setigera[5]
Growth rate	0.86 1/day	PCB = 0 ppm, 288 K, 144 h, control = 100 %	Rhizosolenia setigera[5]
Growth rate	0.6 1/day	PCB = 10pmm, 285 K, 120 h, control = 42 %	Thalassiosira pseudonana[5]
Growth rate	1.31 1/day	PCB = 10pmm, 290 K, 72 h, control = 56 %	Thalassiosira pseudonana[5]
Growth rate	1.43 1/day	PCB = 0 ppm, 285 K, 120 h, control = 100 %	Thalassiosira pseudonana[5]
Growth rate	1.57 1/day	PCB = 0.1 ppm, 285 K, 120 h, control = 109 %	Thalassiosira pseudonana[5]
Growth rate	2.13 1/day	PCB = 10pmm, 298 K, 72 h, control = 83 %	Thalassiosira pseudonana[5]
Growth rate	2.15 1/day	PCB = 0.1 ppm, 290 K, 72 h, control = 84 %	Thalassiosira pseudonana[5]
Growth rate	2.34 1/day	PCB = 0 ppm, 290 K, 72 h, control = 100 %	Thalassiosira pseudonana[5]
Growth rate	2.57 1/day	PCB = 0 ppm, 298 K, 72 h, control = 100 %	Thalassiosira pseudonana[5]
Growth rate	2.59 1/day	PCB = 0.1 ppm, 298 K, 72 h, control = 101 %	Thalassiosira pseudonana[5]
Hematocrit	31.1 %	Control = 38. %, PCB = 6*1000 mg/kg ww (every sec.day), 2 w	Anguilla anguilla[3]
Hematocrit	31.7 %	Control = 30.4 %, PCB = 1*200 mg/kg ww (one time), 16 w, rg =+- 1.2	Anguilla anguilla[3]
Hematocrit	33.2 %	Control = 34. %, PCB = 3*1000 mg/kg ww (every sec.day), 1 w	Anguilla anguilla[3]
Hematocrit	34.1 %	Control = 36.1 %, PCB = 1*200 mg/kg ww (one time), 4 w, rg =+- 1.4	Anguilla anguilla[3]
Hematocrit	32.9 %	Control = 37.5 %, PCB = 0.2 g (one time), 21 d, rg =+- 37.5	Salmo gairdneri[4]
Hematocrit	34.4 %	Control = 41.9 %, PCB = 0.2 g (one time), 14 d, rg =+- 4.2	Salmo gairdneri[4]
Hematocrit	38.4 %	Control = 37.9 %, PCB = 0.2 g (one time), 7 d, rg =+- 2.2	Salmo gairdneri[4]
Hematocrit	39.4 %	Control = 43.9 %, PCB = 0.2 g (one time), 3 d, rg =+- 1.9	Salmo gairdneri[4]
Hematocrit	43.4 %	Control = 49.7 %, PCB = 1*200 mg/kg ww (one time), 5 d, rg =+- 4.8	Salmo gairdneri[3]
Hematocrit	48.5 %	Control = 55.2 %, PCB = 1*200 mg/kg ww (one time), 17 d, rg =+- 2.9	Salmo gairdneri[3]
Hematocrit	48.7 %	Control = 46.3 %, PCB = 1*200 mg/kg ww (one time), 10 d, rg =+- 3.2	Salmo gairdneri[3]
Hemoglobin	9.3 g/100 ml	Control = 8.6, PCB = 1*200 mg/kg ww (one time), 16 w, rg =+- 0.5	Anguilla anguilla[3]
Hemoglobin	9.6 g/100 ml	Control = 10.3, PCB = 6*1000 mg/kg ww (every sec.day), 2 w, rg =+- 0.	Anguilla anguilla[3]
Hemoglobin	9.7 g/100 ml	Control = 10.5, PCB = 1*200 mg/kg ww (one time), 4 w, rg =+- 0.4	Anguilla anguilla[3]
Hemoglobin	9.9 g/100 ml	Control = 9.3, PCB = 3*1000 mg/kg ww (every sec.day), 1 w, rg =+- 0.8	Anguilla anguilla[3]
Hemoglobin	10.0 g/100 ml	Control = 11.4, PCB = 1*200 mg/kg ww (one time), 17 d, rg =+- 0.7	Salmo gairdneri[3]
Hemoglobin	7.6 g/100 ml	Control = 8.8, PCB = 0.2 g (one time), 14 d, rg =+- 0.9	Salmo gairdneri[4]

Hemoglobin	8.0 g/100 ml	Control = 8., PCB = 0.2 g (one time), 21 d, rg =+- 0.6	Salmo gairdneri[4]
Hemoglobin	8.2 g/100 ml	Control = 8.3, PCB = 0.2 g (one time), 3 d, rg =+- 0.6	Salmo gairdneri[4]
Hemoglobin	8.5 g/100 ml	Control = 10., PCB = 1*200 mg/kg ww (one time), 5 d, rg =+- 0.4	Salmo gairdneri[3]
Hemoglobin	9.0 g/100 ml	Control = 7.8, PCB = 0.2 g (one time), 7 d, rg =+- 0.6	Salmo gairdneri[4]
Hemoglobin	9.4 g/100 ml	Control = 10.4, PCB = 1*200 mg/kg ww (one time), 10 d, rg =+- 0.6	Salmo gairdneri[3]
Liver glycogen	4.47 mg/100 mg	Control = 4.08, PCB = 1*200 mg/kg ww (one time), 4 m, rg = +- 0.33	Anguilla anguilla[1]
Liver glycogen	5.11 mg/100 mg	Control = 6.03, PCB = 6*1000 mg/kg ww (every sec.day), 2 w	Anguilla anguilla[1]
Liver glycogen	5.29 mg/100 mg	Control = 6.47, PCB = 1*200 mg/kg ww (one time), 1 m, rg = +- 0.83	Anguilla anguilla[1]
Liver glycogen	8.0 mg/100 mg	Control = 6.94, PCB = 3*1000 mg/kg ww (every sec.day), 1 w	Anguilla anguilla[1]
Liver phospholipids	2.27	Control = 2.81, PCB = 3*1000 mg/kg ww (every sec.day), 1 w	Anguilla anguilla[1]
Liver phospholipids	2.49	Control = 2.41, PCB = 6*1000 mg/kg ww (every sec.day), 2 w	Anguilla anguilla[1]
Liver protein	13.9 mg/100 mg	Control = 14.1, PCB = 3*1000 mg/kg ww (every sec.day), 1 w	Anguilla anguilla[1]
Liver protein	15.2 mg/100 mg	Control = 14.1, PCB = 6*1000 mg/kg ww (every sec.day), 2 w	Anguilla anguilla[1]
Liver protein	19.0 mg/100 mg	Control = 18.8, PCB = 1*200 mg/kg ww (one time), 1 m, rg = +- 0.5	Anguilla anguilla[1]
Liver protein	20.0 mg/100 mg	Control = 20.6, PCB = 1*200 mg/kg ww (one time), 4 m, rg = +- 1.0	Anguilla anguilla[1]
Liver somatic index	0.715	Control = 0.758, PCB = 1*200 mg/kg ww (one time), 4 m, rg = +- 0.032	Anguilla anguilla[1]
Liver somatic index	0.853	Control = 0.814, PCB = 1*200 mg/kg ww (one time), 1 m, rg = +- 0.038	Anguilla anguilla[1]
Liver somatic index	1.54	Control = 1.43, PCB = 3*1000 mg/kg ww (every sec.day), 1 w	Anguilla anguilla[1]
Liver somatic index	1.59	Control = 1.42, PCB = 6*1000 mg/kg ww (every sec.day), 2 w	Anguilla anguilla[1]
Liver triglycerides	13.0 mg/100 mg	Control = 12.6, PCB = 1*200 mg/kg ww (one time), 4 m, rg = +- 2.7	Anguilla anguilla[1]
Liver triglycerides	13.4 mg/100 mg	Control = 12.6, PCB = 1*200 mg/kg ww (one time), 1 m, rg = +- 1.0	Anguilla anguilla[1]
Liver triglycerides	6.06 mg/100 mg	Control = 5.73, PCB = 3*1000 mg/kg ww (every sec.day), 1 w	Anguilla anguilla[1]
Liver triglycerides	8.26 mg/100 mg	Control = 5.12, PCB = 6*1000 mg/kg ww (every sec.day), 2 w	Anguilla anguilla[1]
Liver water	64.7 mg/100 mg	Control = 63.2, PCB = 1*200 mg/kg ww (one time), 1 m, rg = +- 1.0	Anguilla anguilla[1]
Liver water	65.4 mg/100 mg	Control = 65.6, PCB = 1*200 mg/kg ww (one time), 4 m, rg = +- 2.0	Anguilla anguilla[1]
Liver water	66.5 mg/100 mg	Control = 68.3, PCB = 6*1000 mg/kg ww (every sec.day), 2 w	Anguilla anguilla[1]

Liver water	68.5 mg/100 mg	Control = 68.6, PCB = 3*1000 mg/kg ww (every sec.day), 1 w	Anguilla anguilla[1]
Mean weights	153.2 mg	PCB = 0.01 mg/l, control = 160.8 mg, 24 h, F022	Escherichia coli[6]
Mean weights	186.4 mg	PCB = 0.1 mg/l, control = 160.8 mg, 24 h, F022	Escherichia coli[6]
Mean weights	86.0 mg	PCB = 0.1 mg/l, control = 44.5 mg, 24 h, F022	Escherichia coli[6]
Mean weights	90.4 mg	PCB = 0.01 mg/l, control = 44.5 mg, 24 h, F022	Escherichia coli[6]
Mortality	50 %	PCB as trichlorated form (KC300) = 1.45 ppm, 96 h, fry	Cyprinus carpio[7]
Mortality	50 %	PCB as trichlorated form (KC300) = 0.90 ppm, 96 h, fry	Lebistes reticulatus[7]
Mortality	50 %	PCB as trichlorated form (KC300) = 3.2 ppm, 96 h, adult	Lebistes reticulatus[7]
Muscle phospholipids	0.51 mg/100 mg	Control = 0.54, PCB = 6*1000 mg/kg ww (every sec.day), 2 w	Anguilla anguilla[1]
Muscle phospholipids	0.66 mg/100 mg	Control = 0.63, PCB = 3*1000 mg/kg ww (every sec.day), 1 w	Anguilla anguilla[1]
Muscle protein	20.9 mg/100 mg	Control = 20.4, PCB = 6*1000 mg/kg ww (every sec.day), 2 w	Anguilla anguilla[1]
Muscle protein	21.6 mg/100 mg	Control = 21.4, PCB = 3*1000 mg/kg ww (every sec.day), 1 w	Anguilla anguilla[1]
Muscle protein	22.7 mg/100 mg	Control = 22.3, PCB = 1*200 mg/kg ww (one time), 1 m, rg =+- 0.4	Anguilla anguilla[1]
Muscle protein	23.5 mg/100 mg	Control = 23.9, PCB = 1*200 mg/kg ww (one time), 4 m, rg =+- 0.5	Anguilla anguilla[1]
Muscle triglycerides	10.35 mg/100 mg	Control = 7.23, PCB = 6*1000 mg/kg ww (every sec.day), 2 w	Anguilla anguilla[1]
Muscle triglycerides	18.9 mg/100 mg	Control = 21.6, PCB = 1*200 mg/kg ww (one time), 1 m, rg =+- 3.3	Anguilla anguilla[1]
Muscle triglycerides	21.5 mg/100 mg	Control = 18.4, PCB = 1*200 mg/kg ww (one time), 4 m, rg =+- 2.4	Anguilla anguilla[1]
Muscle triglycerides	7.02 mg/100 mg	Control = 8.66, PCB = 3*1000 mg/kg ww (every sec.day), 1 w	Anguilla anguilla[1]
Muscle water	62.9 mg/100 mg	Control = 63., PCB = 1*200 mg/kg ww (one time), 1 m, rg =+- 2.2	Anguilla anguilla[1]
Muscle water	63.7 mg/100 mg	Control = 66.1, PCB = 1*200 mg/kg ww (one time), 4 m, rg =+- 1.5	Anguilla anguilla[1]
Muscle water	69.7 mg/100 mg	Control = 73.1, PCB = 6*1000 mg/kg ww (every sec.day), 2 w	Anguilla anguilla[1]
Muscle water	70.9 mg/100 mg	Control = 69.9, PCB = 3*1000 mg/kg ww (every sec.day), 1 w	Anguilla anguilla[1]
Plasma Ca	2.67 milli-mol	Control = 2.75, PCB = 1*200 mg/kg ww (one time), 4 m, rg =+- 0.02	Anguilla anguilla[1]
Plasma Ca	2.67 milli-mol	Control = 2.85, PCB = 1*200 mg/kg ww (one time), 1 m, rg =+- 0.05	Anguilla anguilla[1]
Plasma Cl	106. milli-mol	Control = 106., PCB = 1*200 mg/kg ww (one time), 4 m, rg =+- 3.0	Anguilla anguilla[1]
Plasma Cl	123. milli-mol	Control = 114., PCB = 1*200 mg/kg ww (one time), 1 m, rg =+- 2.0	Anguilla anguilla[1]
Plasma Cl	130. milli-mol	Control = 132., PCB = 3*1000 mg/kg ww (every sec.day), 1 w	Anguilla anguilla[1]
Plasma Cl	131. milli-mol	Control = 135., PCB = 6*1000 mg/kg ww (every sec.day), 2 w	Anguilla anguilla[1]

Plasma K	2.9 milli-mol	Control = 3.1 , PCB = 1*200 mg/kg ww (one time), 4 m, rg =+- 0.1	Anguilla anguilla[1]
Plasma K	6.2 milli-mol	Control = 5.3, PCB = 1*200 mg/kg ww (one time), 1 m, rg =+- 1.0	Anguilla anguilla[1]
Plasma Mg	0.94 milli-mol	Control = 1.01 , PCB = 1*200 mg/kg ww (one time), 1 m, rg =+- 0.03	Anguilla anguilla[1]
Plasma Mg	1.07 milli-mol	Control = 1.06, PCB = 1*200 mg/kg ww (one time), 4 m, rg =+- 0.02	Anguilla anguilla[1]
Plasma Na	157. milli-mol	Control = 155., PCB = 1*200 mg/kg ww (one time), 4 m, rg =+- 2.0	Anguilla anguilla[1]
Plasma Na	161. milli-mol	Control = 159., PCB = 1*200 mg/kg ww (one time), 1 m, rg =+- 1.0	Anguilla anguilla[1]
Plasma PO4 - P	1.29 milli-mol	Control = 1.14 , PCB = 3*1000 mg/kg ww (every sec.day), 1 w	Anguilla anguilla[1]
Plasma PO4 - P	1.43 milli-mol	Control = 1.3, PCB = 1*200 mg/kg ww (one time), 4 m, rg =+- 0.04	Anguilla anguilla[1]
Plasma PO4 - P	1.43 milli-mol	Control = 1.4 , PCB = 1*200 mg/kg ww (one time), 1 m, rg =+- 0.06	Anguilla anguilla[1]
Plasma PO4 - P	1.55 milli-mol	Control = 1.81 , PCB = 6*1000 mg/kg ww (every sec.day), 2 w	Anguilla anguilla[1]
Plasma cortisol	3.8 microgram/100 ml	Control = 4., PCB = 1*200 mg/kg ww (one time), 1 m, rg =+- 1.5	Anguilla anguilla[1]
Plasma cortisol	8. microgram/100 ml	Control = 8.8, PCB = 1*200 mg/kg ww (one time), 1 m, rg =+- 1.8	Anguilla anguilla[1]
Plasma esterified cholesterol	3.13 milli-mol	Control = 3.54 , PCB = 3*1000 mg/kg ww (every sec.day), 1 w	Anguilla anguilla[1]
Plasma esterified cholesterol	3.87 milli-mol	Control = 4.06, PCB = 6*1000 mg/kg ww (every sec.day), 2 w	Anguilla anguilla[1]
Plasma free cholesterol	0.47 milli-mol	Control = 0.42, PCB = 6*1000 mg/kg ww (every sec.day), 2 w	Anguilla anguilla[1]
Plasma free cholesterol	0.79 milli-mol	Control = 0.32, PCB = 3*1000 mg/kg ww (every sec.day), 1 w	Anguilla anguilla[1]
Plasma phospholipids	7.06 milli-mol	Control = 8.1 , PCB = 3*1000 mg/kg ww (every sec.day), 1 w	Anguilla anguilla[1]
Plasma phospholipids	9.2 milli-mol	Control = 6.78, PCB = 6*1000 mg/kg ww (every sec.day), 2 w	Anguilla anguilla[1]
Plasma protein	5.75 g/100 ml	Control = 5.95, PCB = 6*1000 mg/kg ww (every sec.day), 2 w	Anguilla anguilla[1]
Plasma protein	6.34 mg/100 ml	Control = 5.98 , PCB = 1*200 mg/kg ww (one time), 4 m, rg =+- 0.95	Anguilla anguilla[1]
Plasma protein	6.42 mg/100 ml	Control = 6.48 , PCB = 1*200 mg/kg ww (one time), 1 m, rg =+- 0.18	Anguilla anguilla[1]
Plasma trigliver protein-cerides	4.66 milli-mol	Control = 4.33, PCB = 3*1000 mg/kg ww (every sec.day), 1 w	Anguilla anguilla[1]
Plasma trigliver protein-cerides	5.14 milli-mol	Control = 5.71 , PCB = 1*200 mg/kg ww (one time), 1 m, rg =+- 0.54	Anguilla anguilla[1]

Plasma trigliver proteincerides	6.28 milli-mol	Control = 5.1 , PCB = 6*1000 mg/kg ww (every sec.day), 2 w	Anguilla anguilla[1]
Plasma trigliver protein-cerides	6.95 milli-mol	Control = 7.4 , PCB = 1*200 mg/kg ww (one time), 4 m, rg =+- 1.04	Anguilla anguilla[1]
Red blood cells	1.01E6 /ul	Control = 1.26, PCB = 6*1000 mg/kg ww (every sec.day), 2 w	Anguilla anguilla[3]
Red blood cells	1.07E6 /ul	Control = 1.21, PCB = 3*1000 mg/kg ww (every sec.day), 1 w	Anguilla anguilla[3]
Red blood cells	1.29E6 /ul	Control = 1.22, PCB = 1*200 mg/kg ww (one time), 4 w, rg =+- 0.11	Anguilla anguilla[3]
Red blood cells	1.42E6 /ul	Control = 1.28, PCB = 1*200 mg/kg ww (one time), 16w, rg =+- 0.1	Anguilla anguilla[3]
Red blood cells	1.15E6 /ul	Control = 1.24 , PCB = 1*200 mg/kg ww (one time), 10d, rg =+- 0.08	Salmo gairdneri[3]
Red blood cells	1.22E6 /ul	Control = 1.23, PCB = 1*200 mg/kg ww (one time), 5 d, rg =+- 0.09	Salmo gairdneri[3]
Red blood cells	1.46E6 /ul	Control = 1.55, PCB = 1*200 mg/kg ww (one time), 17 d, rg =+- 0.11	Salmo gairdneri[3]
Spleen somatic index	0.08 (spleen weight*100)/ww	Control = 0.09, PCB = 1*200 mg/kg ww (one time), 16w, rg =+- 0.02	Anguilla anguilla[3]
Spleen somatic index	0.09 (spleen weight*100)/ww	Control = 0.07, PCB = 6*1000 mg/kg ww (every sec.day), 2 w	Anguilla anguilla[3]
Spleen somatic index	0.11 (spleen weight*100)/ww	Control = 0.07, PCB = 1*200 mg/kg ww (one time), 4 w, rg =+- 0.02	Anguilla anguilla[3]
Spleen somatic index	0.12 (spleen weight*100)/ww	Control = 0.11, PCB = 3*1000 mg/kg ww (every sec.day), 1 w	Anguilla anguilla[3]
Spleen somatic index	0.17 (spleen weight*100)/ww	Control = 0.11, PCB = 0.2g(one time), 3 d, rg =+- 0.03	Salmo gairdneri[4]
Spleen somatic index	0.2 (spleen weight*100)/ww	Control = 0.21, PCB = 0.2g(one time), 7 d, rg =+- 0.05	Salmo gairdneri[4]
Spleen somatic index	0.24 (spleen weight*100)/ww	Control = 0.13, PCB = 0.2g(one time), 21 d, rg =+- 0.05	Salmo gairdneri[4]
Spleen somatic index	0.26 (spleen weight*100)/ww	Control = 0.23, PCB = 1*200 mg/kg ww (one time), 17 d, rg =+- 0.04	Salmo gairdneri[3]
Spleen somatic index	0.29 (spleen weight*100)/ww	Control = 0.25,PCB = 1*200 mg/kg ww (one time), 1, rg =+- 0.0	Salmo gairdneri[3]
Spleen somatic index	0.3 (spleen weight*100)/ww	Control = 0.23, PCB = 0.2g(one time), 14 d, rg =+- 0.09	Salmo gairdneri[4]
Spleen somatic index	0.38 (spleen weight*100)/ww	Control = 0.23, PCB = 1*200 mg/kg ww (one time), 5 d, rg =+- 0.1	Salmo gairdneri[3]

Source: 1.Dave, G. A. Buman, M-L. Johansson Sjoebeck and U. Lidman. 1975. In; Koeman, J.H. et al. (eds); Sublehal Effects of Toxic Chemicals on Aquatic Animals, pp 197-205. Elsevier. Amsterdam.
2.Kihlstroem, K.I., J. Oerberg, C. Lund berg and P.B. Danielsson. 1974. Ambio. 3(6), pp 231-233.
3.Johansson-Sjoebeck, M-L., G. Dave and U. Lidman. 1975. In: J.H. Koeman et al.(eds.), Sublethal Effets of Toxic Chemicals on Aquatic Animals, Elsevier Scientific Publishing
4.Westman, I., M-L. Johansson-Sjoebeck and R. Faenge. 1975. in: Koeman, J.H. and J.J.W.A. Strik. (eds.: Sublethal Effects of Toxic Chemicals on Aquatic Animals, Elsevier, N.Y., pp 111-119.
5.Fisher, N.S. and C.F. Wurster. 1974. En viron. Pollut., 5, pp 205-212.
6.Keil, J.E., S.H. Sandifer, C.D. Graber and L.E. Priester. 1972 Water Res. 6(7), pp 837-842.
7.Kimura, S., H. Kumada and Y. Matida. 1973. Bull. Freshwater Fish. Res. Lab. Tokyo 23(2), pp 115-123.
Footnote: F022 = Effect of PCB on E. Coli. Effect based on laboratory experiment. Incubation time 24 hours. A volume of 500 ml is used.

4-341 Effect of Palytoxin

Parameter	Effect Value	Concentration and Condition	Species
Mortality	50 %	Oral dose = 1.5E-4 mg/kg, 4.55E-11 mol/kg, F007 = 10.34	Mouse[1]

Source: 1.Luckey, T.D. and B. Venugopal. 1977. Metal Toxicity in Mammals, 1. Plenum Press.
Footnote; F007 = pT50 is shown = -log(potiental toxicity), 50 %.

4-342 Effect of Parathion

Parameter	Effect Value	Concentration and Condition	Species
Mortality	50 %	Oral dose = 5.5 mg/kg, 1.89E-5mol/kg, F007 = 4.72	Mouse[1]
Mortality	50 %	Single oral dose = 10000 microgram/kg ww	Rat[2]

Source: 1.Luckey, T.D. and B. Venugopal. 1977. Metal Toxicity in Mammals, 1. Plenum Press.
2.Gribble, G.W. 1974. TCDD. Chemistry, 47, pp 15-18.
Footnote; F007 = pT50 is shown = -log(potiental toxicity), 50 %.

4-343 Effect of Pb

Parameter	Effect Value	Concentration and Condition	Species
Growth	0.002 mm/day	Control = 0.014 mm/day, Pb = 0.1 mg/l, 10 juv, Bredwell Brook, UK	Asellus meridianus[1]
Growth	0.006 mm/day	Control = 0.017 mm/day, Pb = 0.1 mg/l, 10 juv, River Hayle, UK	Asellus meridianus[1]
Growth	0.006 mm/day	Control = 0.028 mm/day, Pb = 0.1 mg/l, 10 juv, Bredwell Brook, UK	Asellus meridianus[1]
Growth	0.006 mm/day	Control = 0.033 mm/day, Pb = 0.1 mg/l, 10 juv, River Hayle, UK	Asellus meridianus[1]
Growth	0.022 mm/day	Control = 0.022 mm/day, Pb = 0.1 mg/l, 10 juv, River Hayle, UK	Asellus meridianus[1]
Growth	0.033 mm/day	Control = 0.028 mm/day, Pb = 0.1 mg/l, 10 juv, River Gannal, UK	Asellus meridianus[1]
Growth	0.033 mm/day	Control = 0.033 mm/day, Pb = 0.1 mg/l, 10 juv, River Hayle, UK	Asellus meridianus[1]
Growth	0.05 mm/day	Control = 0.044 mm/day, Pb = 0.1 mg/l, 10 juv, River Gannal, UK	Asellus meridianus[1]
Mortality	50 % mortality	Pb(NO3)2, water = 15-18 mg/l, toxic level given, microorganisms	Anabaena sp.[2]
Mortality	50 %	Lead = 0.28 mg/l, 10 juv, 293 K, test animals from Bredwell Brook, UK	Asellus meridianus[1]
Mortality	50 %	Lead = 1.0 mg/l, 10 juv, 293 K, test animals from River Hayle, UK	Asellus meridianus[1]
Mortality	50 %	Lead = 1.75 - 3.5 mg/l, 10 juv, 293 K, test animals from River Hayle,	Asellus meridianus[1]
Mortality	50 %	Lead = 3.5 mg/l, 10 juv, 293 K, test animals from River Gannel, UK	Asellus meridianus[1]
Mortality	50 %	Lead = 3.5 mg/l, 10 juv, 293 K, test animals from River Hayle, UK	Asellus meridianus[1]
Mortality	50 % mortality	Pb(NO3)2, water = 0.007 M, toxic level given, microorganisms	Aspergillus niger[3]
Mortality	50 % mortality	PbCl2, water = 67 mg/ml, toxic level given, microorganisms	Aspergillus niger[4]
Mortality	0 %	Toxicological limit = 5 mg/l, Pb(NO3)2	Biological purification[5]

Mortality	50 % mortality	Pb(NO3)2, water = 15-18 mg/l, toxic level given, microorganisms	Chlamydomonas sp.[2]
Mortality	50 % mortality	Pb(NO3)2, water = 5 mg/l, toxic level given, microorganisms	Cosmarium sp.[2]
Mortality	0 %	Toxicological limit = 3 - 170 mg/l, Pb(NO3)2	Crustacean[6]
Mortality	50 %	PbCl2, exposure time = 2 d, conc in water = 450 ug/l, limnol.	Daphnia magna[7]
Mortality	0 %	Toxicological limit = 0.33 - 200 mg/l, trout, Pb(NO3)2	Fish[6]
Mortality	0 %	Toxicological limit = 2 g/kg, Pb(NO3)2	Mammalia[6]
Mortality	50 % mortality	Pb(NO3)2, water = 15-18 mg/l, toxic level given, microorganisms	Navicula sp.[2]
Mortality	50 %	PbCl2 = 39.9 +- 1.5 ppm, 48 h, 25 ppm CaCO3, 293 K, pH = 7.4-7.9	Philodina sp.[8]
Mortality	50 %	PbCl2 = 40.8 +- 1.6 ppm, 96 h, 25 ppm CaCO3, 293 K, pH = 7.4-7.9	Philodina sp.[8]
Mortality	50 %	PbCl2 = 47.5 +- 1.1 ppm, 24 h, 25 ppm CaCO3, 293 K, pH = 7.4-7.9Philodina sp.[8]	
Mortality	50 % mortality	Pb(NO3)2, water = 0.1 g Pb(NO3)2/l, toxic level given, microorganisms	Pseudomonas pyocyanea sp.[9]
Mortality	50 %	2-oxide, intraperitonal, dose = 400 mg/kg body	Rat[10]
Mortality	50 %	Acetate, intravenous, dose = 120 mg/kg body	Rat[10]
Mortality	50 %	Acute oral toxicity = 800 mg/kg, pesticide, lead arsenate	Rat[11]
Mortality	50 %	Arsenate, oral dose = 100 mg/kg body	Rat[10]
Mortality	50 %	Sulfide, intraperitonal, dose = 1600 mg/kg body	Rat[10]
Mortality	50 %	Triethyl, intraperitonal, dose = 11.2 mg/kg body	Rat[10]
Nitrification	14 % inhibation	Soil = 5 umoles/g of soil, 10 d incubation, 3 soils, lab	Soil organisms[12]
Photosynthesis	20 % of control	10.0 mg Pb/l	Aufwuchs communities[13]
Photosynthesis	35 % of control	1.0 mg Pb/l, 3 h	Aufwuchs communities[13]
Photosynthesis	84 % of control	0.1 mg Pb/l, 3 h	Aufwuchs communities[13]
Photosynthesis	90 % of control	0.02 mg Pb/l, 3 h	Aufwuchs communities[13]
Photosynthesis	95 % of control	0.001 mg Pb/l, 3 h	Aufwuchs communities[13]
Reproduction	16 % decrease	30 ppb in water	Daphnia magna[14]
Reproduction	50 % decrease	100 ppb in water	Daphnia magna[14]
Respiration	104 % of control	0.001 mg Pb/l, 3 h, Golden Horn	Aufwuchs communities[13]
Respiration	17 % of control	10.0 mg Pb/l, 3 h, Golden Horn	Aufwuchs communities[13]
Respiration	40 % of control	1.0 mg Pb/l, 3 h, Golden Horn	Aufwuchs communities[13]
Respiration	70 % of control	0.1 mg Pb/l, 3 h, Golden Horn	Aufwuchs communities[13]
Respiration	84 % of control	0.01 mg Pb/l, 3 h, Golden Horn	Aufwuchs communities[13]

Source: 1.Brown, B.E. 1976. Water Res., 10(6), pp 555-559.
2.Tu, C-M. and W.B. Bollen. 1968. Weed Res. 8, 28-37.
3.Anderson, J.R. and E.A. Drew. 1976. Zbl. Bakt. abt II, 131, 247-258.
4.Singh, K. 1971. Pestic. India 6, 14-17.

5.Jung. K.D. 1973. Wirkungskonzentration schaedigender Stoffe im Wasser. Gelenkirchen. Hygiene-Institudes Ruhrgebiets.
6.Liebman, H. 1958. Handbuch der Friscwasser - und Abwasserbiologie. Band II.Munchen. Oldenbourg.
7.Biesinger, K.E. and G.M. Christensen. 1972. J. Fish. Res. bd. Can., 29(12), 1691-1700.
8.Sallivan, G.W., A.L. Buikema and J. Cairns, Jr. 1973. Va. J. Sci. 24(3), pp 120.
9.Hoffman-Kakol, I. and H. Kwarta. 1972. Zesz. nauk. Akad. Rolnicz. Szczecin. No 38, 91-103.
10.Toxic Substances List 1973, U.S. Department of Health, Education and Welfare, Public Health ServiceNational Institute for Occupational Safety and Health, Rockville Maryland, USA, June 1973.
11.Edwards, C.A. 1975. Persistent Pesticides in the Environment, 2nd edition, CRC-Press.
12.Liang, C.N. and M.A. Tabatabai. 1978. Environ. Pollut. 12, 141-147.
13.Orhon, D. 1975. Water Res. 9(5/6), pp 467-472.
14.Biesinger, K.E. and G.M. Christensen. 1972. J. Fish. Res. bd. Can., 29(12), 1691-1700.

4-344 Effect of Pentachlorophenol

Parameter	Concentration	Condition	Species
LC50	0.21 mg/l	24 hours	Fathead minnow[2]
LC50	0.21 mg/l	96 hours	Fathead minnow[2]
LD50	0.07-0.1 g pr. kg body weight	oral dose	Rabbit[1]
LD50	0.027-0.078 g pr. kg body weight	oral dose	Rat[1]

Source: 1.Patty, F. A., Industrial Hygiene and toxicology, 2, Interscience Publishers, New York, 1976.
2.Adelman, I. R., Smith, L. L. and Sieenop, G. D., Water Res., 100, 685, 1976.

4-345 Effect of Perylene

Parameter	Effect Value	Concentration and Condition	Species
Growth rate	0 % more than control	Water = 0.3 ppm, 96 h, PAH, sporelings	Antithamnion plumula[1]

Source: 1.Boney, A.D. 1974. Mar. Pollut. Bull. 5, 185-186.

4-346 Effect of Phenanthrene

Parameter	Effect Value	Concentration and Condition	Species
Mortality	50 % mortality	Water = 70 ppm, naplii sedimentated (thought dead), 1 h, S=3 %	Balanus amphitrites[1]
Mortality	50 % mortality	Water = 150 ppm, 96 h	Gambusia affinis[2]
Mortality	50 % mortality	Water = 0.6 ppm, 96 h, marine	Neathes arenaceodentata[3]
Mortality	50 % mortality	Water = 0.37 ppm, 24 h	Palaemonetes pugio[4]

Source: 1.Donahue, W.H., R.T. Wang, M. Welch and J.A.C. Nicol. 1977. Environ. Pollut. 13, 187-202.
2.EPA 1975. Water Quality Data Book. Organic Chemical Pollution of Fresh Water. Wash DC. EPA 18010 DPV
3.Rossi, S.S. and J.M. Neff. 1978. Mar. Pollut. Bull. 9, 220-223.
4.Young, G.P. 1977. Effects of Naphthalene on Palaemonetes pugio. Ms Thesis. and M Univ. Collage Statin. Texas.

4-347 Effect of Phenol

Parameter	Concentration	Condition	Species
LC0	16 mg/l	not indicated	Daphnia[6]
LC0	>1000 mg/l	not indicated	Escherichia coli[7]
LC0	40 mg/l	not indicated	Scenedesmus[7]
LC50	5600 mg/l	3 min., 15 C	Crangon crangon[8]
LC50	400 mg/l	27 min., 15 C	Crangon crangon[8]
LC50	80 mg/l	3 hours, 15 C	Crangon crangon[8]
LC50	30 mg/l	72 hours, 15 C	Crangon crangon[8]
LC50	25 mg/l	96 hours, 15 C	Crangon crangon[8]

LC50	18000 mg/l	3 min., 15 C	Gobius minutus[8]
LC50	320 mg/l	27 min., 15 C	Gobius minutus[8]
LC50	20 mg/l	3 hours, 15 C	Gobius minutus[8]
LC50	10 mg/l	48 hours, 15 C	Gobius minutus[8]
LC50	9 mg/l	96 hours, 15 C	Gobius minutus[8]
Sublethal effect	4.6 mg/l	Inhibition of cell multiplication	Microcystis aeruginosa[2]
Sublethal effect	64 mg/l	Inhibition of cell multiplication	Pseudomonas putida[2]
LD50	0.1 g pr. kg body weight	single oral dose	cat[1]
LD50	0.5 g pr. kg body weight	single oral dose	dog[1]
LD50	0.4-0.6 g pr. kg body weight	single oral dose	Rabbit[1]
LD50	0.53 g pr. kg body weight	single oral dose	Rat[1]
TLm	6-19 mg/l	24-96 hours	Bluegill[2]
TLm	25 mg/l	24 hours	Crucian carp[3]
TLm	32-39 mg/l	24-96 hours, hard water	Fathead minnow[4]
TLm	34-41 mg/l	24-96 hours, soft water	Fathead minnow[4]
TLm	45-50 mg/l	24-96 hours, soft water	Goldfish[4]
TLm	39-50 mg/l	24-96 hours, soft water	Guppie[4]
TLm	22-56 mg/l	24-96 hours	Mosquito fish[5]
TLm	15-17 mg/l	24 hours	Roach[3]
TLm	15-17 mg/l	24 hours	Tench[3]
TLm	4.0 mg/l	18 weeks	Trout[3]

Source: 1.Patty, F. A., Industrial Hygiene and toxicology, 2, Interscience Publishers, New York, 1976.
2.Turnbull, H., DeMann, J. H. and Weston, R. F., Ind. Eng. Chem., 46, 324, 1954.
3.Europian inland fishery advisory commission, Water quality criteria for european freshwater fish, Waer Res., 7, 929, 1973.
4.Pickering Q. H. and Henderson, C., JWPCF, 38, 1419, 1966.
5.Jones, H. R., Environmental Control in the organic and petrochemical Industries, Noyes Data Corp., 171.
6.Meinck, F., Stooff, H. and Kohlschutter, H., Les eaux residuaires industrielles, 1970.
7.Bringmann, G. and Kuhn, R., nwf-wasser/abwasser, H.9, 1976.
8.Adema, D. M. M., Acute toxiciteitstoetsen met 1,2-dichlorethann, fenol, acrylonitrile, an alkylbenzeesulfonant, Delft, 1976.

4-348 Effect of Phenolglycidylether

Parameter	Concentration	Condition	Species
LD50	1.4 g pr. kg body weight	acute intragastrine dose	Mouse[1]
LD50	3.85 g pr. kg body weight	acute intragastrine dose	Rat[1]

Source: 1.Patty, F. A., Industrial Hygiene and toxicology, 2, Interscience Publishers, New York, 1976.

4-349 Effect of Phenylacetate

Parameter	Concentration	Condition	Species
Sublethal effect	7.5 mg/l	Inhibition of cell multiplication	Microcystis aeruginosa[1]
Sublethal effect	115 mg/l	Inhibition of cell multiplication	Pseudomonas putida[1]

Source: 1.Bringmann, G. and Kuhn, R., nwf-wasser/abwasser, H.9, 1976.

4-350 Effect of Phenylglycidylether

Parameter	Concentration	Condition	Species
LC50	43-69 mg/l	24-96 hours	Goldfish[1]

Source: 1.Shell Chemie, Shell Industrie Chemicalien gids, Shell Niederland Chemie, 1975.

4-351 Effect of Phenylhydrazine

Parameter	Concentration	Condition	Species
LD50	0.18 g pr. kg body weight	single oral dose, 100 %	not indicated[1]

Source: 1.Patty, F. A., Industrial Hygiene and toxicology, 2, Interscience Publishers, New York, 1976.

4-352 Effect of Phenyl mercuric acetate

Parameter	Effect Value	Concentration and Condition	Species
Mortality	50 %	Acute oral toxicity = 2080 mg/kg	Rat[1]

Source: 1.Edwards, C.A. 1975. Persistent Pesticides in the Environment, 2nd edition, CRC-Press.

4-353 Effect of Phloroglucinol

Parameter	Concentration	Condition	Species
LC0	0.6 mg/l	not indicated	Daphnia[1]
LC0	>1000 mg/l	not indicated	Escherichia coli[1]
LC0	200 mg/l	not indicated	Scenedesmus[1]
LC100	630 mg/l	48 hours	Goldfish[2]
LC50	60 mg/l	24 hours	Goldfish[3]

Source: 1.Meinck, F., Stooff, H. and Kohlschutter, H., Les eaux residuaires industrielles, 1970.
2.McKee, J. E., and Wolf, H. W., Water Quality, The resources agency of California, 1963.
3.Shell Chemie, Shell Industrie Chemicalien gids, Shell Niederland Chemie, 1975.

4-354 Effect of Phosphate

Parameter	Effect Value	Concentration and Condition	Species
Nitrogen fixation	19.1E-9 m C2H2/((ug chl a)*hour)	Mainly Rhizosolenia spp., added PO4 = 50 micro mol/l	Phytoplankton[1]
Nitrogen fixation	21.8E-9 m C2H2/((ug chl a)*hour)	Mainly Rhizosolenia spp., added PO4 = 0 micro mol/l	Phytoplankton[1]
Nitrogen fixation	29.7E-9 m C2H2/((ug chl a)*hour)	Mainly Rhizosolenia spp., added PO4 = 20 micro mol/l	Phytoplankton[1]
Nitrogen fixation	37.0E-9 m C2H2/((ug chl a)*hour)	Mainly Rhizosolenia spp., added PO4 = 10 micro mol/l	Phytoplankton[1]
Nitrogen fixation	39.4E-9 m C2H2/((ug chl a)*hour)	Mainly Rhizosolenia spp., added PO4 = 2.0 micro mol/l	Phytoplankton[1]
Nitrogen fixation	41.5E-9 m C2H2/((ug chl a)*hour)	Mainly Rhizosolenia spp., added PO4 = 1.0 micro mol/l	Phytoplankton[1]
Nitrogen fixation	49.6E-9 m C2H2/((ug chl a)*hour)	Mainly Rhizosolenia spp., added PO4 = 0.5 micro mol/l	Phytoplankton[1]
Nitrogen fixation	52.8E-9 m C2H2/((ug chl a)*hour)	Mainly Rhizosolenia spp., added PO4 = 5.0 micro mol/l	Phytoplankton[1]

Source: 1.Mague, T.H., F.C. Mague and O. Holm-Hansen. 1977. Mar. Biol. 41(3), pp 213-227.

4-355 Effect of Phosphorus starvation

Parameter	Effect Value	Concentration and Condition	Species
Chlorophyll a content	0.049 mg chlorophyll a/100 mg algae	14 days in P-free media, PO4-P in algae = 0.0 mg/100 mg algae	Anabaena sp.[1]
Chlorophyll a content	0.350 mg chlorophyll a/100 mg algae	7 days in P-free media, PO4-P in algae = 0.0 mg/100 mg algae	Anabaena sp.[1]
Chlorophyll a content	2.125 mg chlorophyll a/100 mg algae	No starvation, PO4-P in algae = 0.070 mg/100 mg algae	Anabaena sp.[1]
Chlorophyll a content	0.019 mg chlorophyll a/100 mg algae	7 days in P-free media, PO4-P in algae = 0.027 mg/100 mg algae	Chlorella sp.[1]
Chlorophyll a content	2.000 mg chlorophyll a/100 mg algae	No starvation, PO4-P in algae = 0.139 mg/100 mg algae	Chlorella sp.[1]
Chlorophyll a content	0.190 mg chlorophyll a/100 mg algae	14 days in P-free media	Microcystis sp.[1]
Chlorophyll a content	0.625 mg chlorophyll a/100 mg algae	7 days in P-free media, PO4-P in algae = 0.0 mg/100 mg algae	Microcystis sp.[1]
Chlorophyll a content	1.215 mg chlorophyll a/100 mg algae	No starvation, PO4-P in algae = 0.064 mg/100 mg algae	Microcystis sp.[1]
Chlorophyll a content	0.762 mg chlorophyll a/100 mg algae	14 days in P-free media, PO4-P in algae = 0.121 mg/100 mg algae	Selenastrum sp.[1]
Chlorophyll a content	2.475 mg chlorophyll a/100 mg algae	No starvation, PO4-P in algae = 0.192 mg/100 mg algae	Selenastrum sp.[1]
Extractable soluble phosphorus	0.0 mg PO4-P/100 mg algae	7 days in P-free media, control = 0.070 mg PO4-P/100 mg algae	Anabaena sp.[1]
Extractable soluble phosphorus	0.027 mg PO4-P/100 mg algae	7 days in P-free media, control = 0.139 mg PO4-P/100 mg algae	Chlorella sp.[1]
Extractable soluble phosphorus	0.0 mg PO4-P/100 mg algae	7 days in P-free media, control = 0.064 mg PO4-P/100 mg algae	Microcystis sp.[1]
Extractable soluble phosphorus	0.121 mg PO4-P/100 mg algae	14 days in P-free media, control = 0.192 mg PO4-P/100 mg algae	Selenastrum sp.[1]

Source: 1.Gramms, L.C. and W.C. Boyle. 1972. Water Res., 6(12), pp 1433-1444.

4-356 Effect of Phthalacid ester

Parameter	Effect Value	Concentration and Condition	Species
Mortality	100 %	Dose = 1.9 mg/l, inhalation	Mouse[1]
Mortality	4.5 %	Dose = 0.43-0.61 mg/l, inhalation, n = 22	Mouse[1]
Mortality	50 %	Dose = 0.7-0.93 mg/l, inhalation, n = 22	Mouse[1]

Source: 1.Reakall, D.B. 1974. J. Environ. Qual. 52, pp 1-39.

4-357 Effect of Picloram

Parameter	Effect Value	Concentration and Condition	Species
Mortality	50 %	Acute oral toxicity = 8200 mg/kg, herbicide	Rat[1]

Source: 1.Edwards, C.A. 1975. Persistent Pesticides in the Environment, 2nd edition, CRC-Press.

4-358 Effect of Picric acid

Parameter	Concentration	Condition	Species
LC0	88 mg/l	not indicated	Daphnia[1]
LC0	1000 mg/l	not indicated	Escherichia coli[2]
LC0	240 mg/l	not indicated	Scenedesmus[2]
LC0	30 mg/l	not indicated	Fish[1]

Source: 1.Meinck, F., Stooff, H. and Kohlschutter, H., Les eaux residuaires industrielles, 1970. 2.Bringmann, G. and Kuhn, R., nwf-wasser/abwasser, H.9, 1976.

4-359 Effect of Pivalic acid

Parameter	Concentration	Condition	Species
LC50	375-400 mg/l	24-96 hours, pH=5.0	Goldfish[1]
LC50	4500 mg/l	24-96 hours, pH=7.0	Goldfish[1]

Source: 1.Shell Chemie, Shell Industrie Chemicalien gids, Shell Niederland Chemie, 1975.

4-360 Effect of Polyelectrolyte

Parameter	Effect Value	Concentration and Condition	Species
Mortality	50 %	Calcon m-500 = 0.65 mg/l, 48 h, 290-292k, 5 Daphnia sp.	Daphnia magna[1]
Mortality	50 %	Superfloc 330 = 0.34 milligram/l, 96 h, 290-292k, 5 Daphnia sp.	Daphnia magna[1]
Mortality	50 %	Calcon m-500 = 2.0 mg/l, 48 h, 279-281 K, 5 copepodes	Limnocalanus macrurus[1]
Mortality	50 %	Calcon m-500 = 2.0 mg/l, 96 h, 279-281 K, 5 copepodes	Limnocalanus macrurus[1]

Source: 1.Biesinger, K.E., A.E. Lemka , W.E. Smith and R.M. Tyo. 1976. Water Poll. Con trol, 48(1), pp 183-187

4-361 Effect of Prim-amylacetate

Parameter	Concentration	Condition	Species
Sublethal effect	63 mg/l	Inhibition of cell multiplication	Microcystis aeruginosa[1]
Sublethal effect	145 mg/l	Inhibition of cell multiplication	Pseudomonas putida[1]

Source: 1.Bringmann, G. and Kuhn, R., nwf-wasser/abwasser, H.9, 1976.

4-362 Effect of Propanil

Parameter	Effect Value	Concentration and Condition	Species
Chlorophyll synthesis	synthesis inhibited	Water conc = 0.2 ppm	Chlorococcum aplanosporum[1]
Chlorophyll synthesis	synthesis inhibited	Water conc = 8 ppm	Tolypothrix tenuis[2]
Growth rate	growth prevented	Water conc = 5 ppm	Anabaena cylindrica[3]
Growth rate	growth stimulated	Water conc = < 0.03 ppm	Anabaena cylindrica[3]
Growth rate	50 %	Water conc = 0.09 ppm	Anabaena cylindrica[3]
Growth rate	growth prevented	Water conc = 5 ppm	Anabaena variabilis[3]
Growth rate	growth inhibited	Water conc = 0.1 ppm	Calothrix brevissima[4]
Growth rate	growth prevented	Water conc = 10 ppm	Calothrix brevissima[4]
Growth rate	growth stimulated	Water conc = 0.01 ppm	Calothrix brevissima[4]

Growth rate	growth prevented	Water conc = > 2 ppm	Chlorella pyrenoidosa[3]
Growth rate	inhibition of grwoth	Water conc = 0.5-2.0 ppm, followed by full recovery	Chlorella pyrenoidosa[5]
Growth rate	50 %	Water conc = 0.18 ppm	Chlorella pyrenoidosa[3]
Growth rate	growth inhibited	Water conc = 0.2-1.0 ppm in culture	Chlorococcum aplanosporum[1]
Growth rate	growth inhibited	Water conc = 2.5 ppm in soil	Chlorococcum aplanosporum[1]
Growth rate	growth prevented	Water conc = 5 ppm	Chlorococcum aplanosporum[1]
Growth rate	inhibition of growth	Water conc = 0.5-2.0 ppm, followed by full recovery	Chlorococcum aplanosporum[1]
Growth rate	growth prevented	Water conc = 5 ppm	Gloeocapsa alpicola[3]
Growth rate	50 %	Water conc = 0.005 ppm	Gloeocapsa alpicola[3]
Growth rate	growth prevented	Water conc = 5 ppm	Nostoc entophytum[3]
Growth rate	growth stimulated	Water conc = < 0.03 ppm	Nostoc entophytum[3]
Growth rate	50 %	Water conc = 0.17 ppm	Nostoc entophytum[3]
Growth rate	growth prevented	Water conc = 5 ppm	Nostoc muscorum[3]
Growth rate	growth inhibited	Water conc = 0.1 ppm	Tolypothrix tenuis[4]
Growth rate	growth inhibited	Water conc = 0.18 ppm	Tolypothrix tenuis[2]
Growth rate	growth prevented	Water conc = 10 ppm	Tolypothrix tenuis[4]
Growth rate	growth prevented	Water conc = 18 ppm	Tolypothrix tenuis[2]
Growth rate	growth prevented	Water conc = 5 ppm	Tolypothrix tenuis[3]
Growth rate	growth stimulated	Water conc = 0.01 ppm	Tolypothrix tenuis[4]
Growth rate	50 %	Water conc = 0.2 ppm	Tolypothrix tenuis[3]
Lysis	effect observed	Water conc = up to 0.2 ppm, followed by recovery in growth	Anabaena cylindrica[3]
Nitrogen fixation	fixation inhibited	Water conc = 0.1 ppm	Calothrix brevissima[4]
Nitrogen fixation	fixation stimulated	Water conc = 0.01 ppm	Calothrix brevissima[4]
Nitrogen fixation	fixation inhibited	Water conc = 0.1 ppm	Tolypothrix tenuis[4]
Nitrogen fixation	fixation stimulated	Water conc = 0.01 ppm	Tolypothrix tenuis[4]
Photosynthesis	photosynthesis prevented	Water conc = 8 ppm	Anabaena cylindrica[3]
Photosynthesis	photosynthesis prevented	Water conc = 8 ppm	Nostoc entophytum[3]
Photosynthesis	photosynthesis prevented	Water conc = 8 ppm	Tolypothrix tenuis[3]

Source: 1.De, B.K. and S. Mukhopadhyay. 1971. Int. Rice Commn. Newsl. 20, 35-39.
2. Mc. Corle, F.M., J.E. Chambers and J.D. Yarbrough. 1977. Bull. Env. Contam. Toxicol. 18(3), 263-270.
3.Thorneburg, R.P. and J.A. Tweedy. 1973. Weed Sci. 21, 397-399.
4.Isaeva, L.I. 1967. Agrokhimiya 4, 122-125.
5.Hamdi, Y.A. and M.S. Tewfik. 1969. Acta microbiol. pol. Ser B 1(18), 53-57.

4-363 Effect of Propenoxide

Parameter	Concentration	Condition	Species
LD50	0.69 g pr. kg body weight	acute intragastric dose	guinea pig[1]
LD50	1.14 g pr. kg body weight	acute intragastric dose	Rat[1]

Source: 1.Patty, F. A., Industrial Hygiene and toxicology, 2, Interscience Publishers, New York, 1976.

4-364 Effect of Propham

Parameter	Effect Value	Concentration and Condition	Species
Growth rate	50 % inhibition	Water conc = 70 mg/l, phenylcarbamate herbicide	Anabaena cylindrica[1]
Growth rate	7.5 % stimulation of growth	Water conc = 5 mg/l, phenylcarbamate herbicide	Anabaena cylindrica[1]
Growth rate	50 % inhibition	Water conc = 52 mg/l, phenylcarbamate herbicide	Anacystis nidulans[1]
Growth rate	50 % inhibition	Water conc = 13.7 mg/l, phenylcarbamate herbicide	Chlorella pyrenoidosa[1]
Growth rate	50 % inhibition	Water conc = 41 mg/l	Gloeocapsa alpicola[1]
Growth rate	50 % inhibition	Water conc = 64 mg/l	Tolypothrix tenuis[1]
Growth rate	7.5 % stimulation of growth	Water conc = 5 mg/l, phenylcarbamate herbicide	Tolypothrix tenuis[1]

Source: 1.Hulin, N. and M. Horowitz. 1973. Phytoparasitica 1, 64.

4-365 Effect of Propionaldehyde

Parameter	Concentration	Condition	Species
LD50	0.8-1.6 g pr. kg body weight	acute oral dose	Rat[1]

Source: 1.Patty, F. A., Industrial Hygiene and toxicology, 2, Interscience Publishers, New York, 1976.

4-366 Effect of Propionic acid

Parameter	Concentration	Condition	Species
LD50	>0.4 g pr. kg body weight	single oral dose	Rat[1]
TLm	>1000 mg/l	24-48 hours	Culex[2]
TLm	130 mg/l	24 hours	Daphnia magna[3]
TLm	50 mg/l	48 hours	Daphnia magna[4]
TLm	188 mg/l	24 hours	Lepouris macrochirus[4]

Source: 1.Patty, F. A., Industrial Hygiene and toxicology, 2, Interscience Publishers, New York, 1976.
2.Dowden, B. F. and Bennet, H. J., JWPCF, 37, 1310, 1965.
3.Verschuren, K. 1977. Handbook of Environmental data on organic Chemicals. Van Nostrand Reinhold Comp.
4.Worne, H. E., Tijdschrift vand het BECEWA, Liege, Belgium.

4-367 Effect of Propylene

Parameter	Concentration	Condition	Species
Sublethal effect	1000 ppm	Declination in seedling, 3 days	Sweet pea[1]

Source: 1.Crocker, W., Growth and Plants, Reinhold publishing Comp., New York, 1948.

4-368 Effect of Propyleneglycol

Parameter	Concentration	Condition	Species
LD50	9.6 g pr. kg body weight	single oral dose	dog[1]
LD50	18.5 g pr. kg body weight	single oral dose	Rabbit[1]
LD50	32.5 g pr. kg body weight	single oral dose	Rat[1]

Source: 1.Patty, F. A., Industrial Hygiene and toxicology, 2, Interscience Publishers, New York, 1976.

4-369 Effect of Propyleneimine

Parameter	Concentration	Condition	Species
LD50	0.019 g pr. kg body weight	acute oral dose	Rat[1]

Source: 1.Patty, F. A., Industrial Hygiene and toxicology, 2, Interscience Publishers, New York, 1976.

4-370 Effect of Pyrene

Parameter	Effect Value	Concentration and Condition	Species
Growth rate	0 % more than control	Water = 0.3 ppm, 96 h, PAH, sporelings	Antithamnion plumula[1]

Source: 1.Boney, A.D. 1974. Mar. Pollut. Bull. 5, 185-186.

4-371 Effect of Pyrethrins

Parameter	Effect Value	Concentration and Condition	Species
Mortality	50 % mortality	48 h, water = 25 mg/l	Daphnia pulex[1]

Source: 1.Rankov, V. 1973. Pochv. Agrokhim. 8, 125-134.

4-372 Effect of Pyridine

Parameter	Concentration	Condition	Species
LC0	100 mg/l	not indicated	Alburnus[1]
Sublethal effect	28 mg/l	Inhibition of cell multiplication	Microcystis aeruginosa[2]
Sublethal effect	340 mg/l	Inhibition of cell multiplication	Pseudomonas putida[2]

Source: 1.Meinck, F., Stooff, H. and Kohlschutter, H., Les eaux residuaires industrielles, 1970.
2.Bringmann, G. and Kuhn, R., nwf-wasser/abwasser, H.9, 1976.

4-373 Effect of Pyrogallol

Parameter	Concentration	Condition	Species
LC0	18 mg/l	not indicated	Daphnia[1]
LC0	8 mg/l	not indicated	Scenedesmus[1]
LC100	18 mg/l	48 hours	Goldfish[2]
LD50	*1.1 g pr. kg body weight*	*oral dose, 100 %*	*Rabbit*[3]

Source: 3.Patty, F. A., Industrial Hygiene and toxicology, 2, Interscience Publishers, New York, 1976.
1.Meinck, F., Stooff, H. and Kohlschutter, H., Les eaux residuaires industrielles, 1970.
2.McKee, J. E., and Wolf, H. W., Water Quality, The resources agency of California, 1963.

4-374 Effect of Quinhydrone

Parameter	Concentration	Condition	Species
LC0	0.4 mg/l	not indicated	Daphnia[1]
LC0	5 mg/l	not indicated	Escherichia coli[1]
LC0	4 mg/l	not indicated	Scenedesmus[1]

Source: 1.Meinck, F., Stooff, H. and Kohlschutter, H., Les eaux residuaires industrielles, 1970.

4-375 Effect of Quinoline

Parameter	Concentration	Condition	Species
LC0	52 mg/l	not indicated	Daphnia[2]
LC0	1000 mg/l	not indicated	Escherichia coli[2]
LC0	140 mg/l	not indicated	Scendesmus[2]
LD50	0.46 g pr. kg body weight	single oral dose	Rat[1]

Source: 1.Patty, F. A., Industrial Hygiene and toxicology, 2, Interscience Publishers, New York, 1976.
2.Meinck, F., Stooff, H. and Kohlschutter, H., Les eaux residuaires industrielles, 1970.

4-376 Effect of Resorcinol

Parameter	Concentration	Condition	Species
LC0	0.8 mg/l	not indicated	Daphnia[1]
LC0	<1000 mg/l	not indicated	Escherichia coli[1]
LC0	60 mg/l	not indicated	Scenedesmus[1]
LC100	57.4 mg/l	48 hours	Goldfish[2]
LD50	0.37 g pr. kg body weight	oral dose, 100 %	guinea pig[3]
LD50	0.75 g pr. kg body weight	oral dose, 100 %	Rabbit[3]
LD50	0.37 g pr. kg body weight	oral dose, 100 %	Rat[3]

Source: 3.Patty, F. A., Industrial Hygiene and toxicology, 2, Interscience Publishers, New York, 1976.
1.Meinck, F., Stooff, H. and Kohlschutter, H., Les eaux residuaires industrielles, 1970.
2.McKee, J. E., and Wolf, H. W., Water Quality, The resources agency of California, 1963.

4-377 Effect of SO_2

Parameter	Effect Value	Concentration and Condition	Species
Growth rate	26 % net photosynthesis of control	3 weeks, 50 pphm SO2, net rate, small chamber technique, SD=7	Black oat[1]
Growth rate	52 % net photosynthesis of control	1 week, 50 pphm SO2, net rate, small chamber technique, SD=11	Black oat[1]
Growth rate	60 % net photosynthesis of control	1-2 days, 50 pphm SO2, net rate, small chamber technique, SD=15	Black oat[1]
Growth rate	100 % external control (CO2 upt at bud break)	Cont=100 %, 0.05 ppm SO2 in air, 3 y plant, fall, exp= 31-60 d	Norway spruce[1]
Growth rate	101 % external control (CO2 upt at bud break)	Cont=127 %, 0.1 ppm SO2 in air, 3 y plant, summer, exp= 31-60 d	Norway spruce[1]
Growth rate	107 % external control (CO2 upt at bud break)	Cont=104 %, 0.1 ppm SO2 in air, 3 y plant, spring, exp= 16-30 d	Norway spruce[1]
Growth rate	108 % external control (CO2 upt at bud break)	Cont=105 %, 0.1 ppm SO2 in air, 3 y plant, summer, exp= 1-15 d	Norway spruce[1]
Growth rate	110 % external control (CO2 upt at bud break)	Cont=116 %, 0.1 ppm SO2 in air, 3 y plant, fall, exp= 16-30 d	Norway spruce[1]
Growth rate	113 % external control (CO2 upt at bud break)	Cont=105 %, 0.05 ppm SO2 in air, 3 y plant, summer, exp= 1-15 d	Norway spruce[1]
Growth rate	113 % external control (CO2 upt at bud break)	Cont=131 %, 0.1 ppm SO2 in air, 3 y plant, summer, exp= 16-30 d	Norway spruce[1]
Growth rate	114 % external control (CO2 upt at bud break)	Cont=113 %, 0.1 ppm SO2 in air, 3 y plant, fall, exp= 1-15 d	Norway spruce[1]
Growth rate	120 % external control (CO2 upt at bud break)	Cont=127 %, 0.05 ppm SO2 in air, 3 y plant, summer, exp= 31-60 d	Norway spruce[1]

Growth rate	125 % external control (CO2 upt at bud break)	Cont=131 %, 0.05 ppm SO2 in air, 3 y plant, summer, exp= 16-30 d	Norway spruce[1]
Growth rate	133 % external control (CO2 upt at bud break)	Cont=116 %, 0.05 ppm SO2 in air, 3 y plant, fall, exp= 16-30 d	Norway spruce[1]
Growth rate	134 % external control (CO2 upt at bud break)	Cont=113 %, 0.05 ppm SO2 in air, 3 y plant, fall, exp= 1-15 d	Norway spruce[1]
Growth rate	52 % external control (CO2 upt at bud break)	Cont=56 %, 0.05 ppm SO2 in air, 3 y plant, spring, exp= 1-15 d	Norway spruce[1]
Growth rate	53 % external control (CO2 upt at bud break)	Cont=56 %, 0.2 ppm SO2 in air, 3 y plant, spring, exp= 1-15 d	Norway spruce[1]
Growth rate	58 % external control (CO2 upt at bud break)	Cont=56 %, 0.1 ppm SO2 in air, 3 y plant, spring, exp= 1-15 d	Norway spruce[1]
Growth rate	68 % external control (CO2 upt at bud break)	Cont=100 %, 0.2 ppm SO2 in air, 3 y plant, fall, exp= 31-60 d	Norway spruce[1]
Growth rate	72 % external control (CO2 upt at bud break)	Cont=110 %, 0.2 ppm SO2 in air, 3 y plant, spring, exp= 31-60 d	Norway spruce[1]
Growth rate	80 % external control (CO2 upt at bud break)	Cont=127 %, 0.2 ppm SO2 in air, 3 y plant, summer, exp= 31-60 d	Norway spruce[1]
Growth rate	84 % external control (CO2 upt at bud break)	Cont=100 %, 0.1 ppm SO2 in air, 3 y plant, fall, exp= 31-60 d	Norway spruce[1]
Growth rate	88 % external control (CO2 upt at bud break)	Cont=104 %, 0.05 ppm SO2 in air, 3 y plant, spring, exp= 16-30 d	Norway spruce[1]
Growth rate	88 % external control (CO2 upt at bud break)	Cont=110 %, 0.1 ppm SO2 in air, 3 y plant, spring, exp= 31-60 d	Norway spruce[1]
Growth rate	88 % external control (CO2 upt at bud break)	Cont=116 %, 0.2 ppm SO2 in air, 3 y plant, fall, exp= 16-30 d	Norway spruce[1]
Growth rate	89 % external control (CO2 upt at bud break)	Cont=110 %, 0.05 ppm SO2 in air, 3 y plant, spring, exp= 31-60 d	Norway spruce[1]
Growth rate	94 % external control (CO2 upt at bud break)	Cont=104 %, 0.2 ppm SO2 in air, 3 y plant, spring, exp= 16-30 d	Norway spruce[1]
Growth rate	95 % external control (CO2 upt at bud break)	Cont=105 %, 0.2 ppm SO2 in air, 3 y plant, summer, exp= 1-15 d	Norway spruce[1]
Growth rate	96 % external control (CO2 upt at bud break)	Cont=113 %, 0.2 ppm SO2 in air, 3 y plant, fall, exp= 1-15 d	Norway spruce[1]
Growth rate	96 % external control (CO2 upt at bud break)	Cont=131 %, 0.2 ppm SO2 in air, 3 y plant, summer, exp= 16-30 d	Norway spruce[1]
Growth rate	101 % external control (CO2 upt at bud break)	Cont=109 %, 0.05 ppm SO2 in air, 3 y plant, fall, exp= 16-30 d	Scotch pine[1]
Growth rate	102 % external control (CO2 upt at bud break)	Cont=149 %, 0.2 ppm SO2 in air, 3 y plant, summer, exp=16-30 d	Scotch pine[1]
Growth rate	109 % external control (CO2 upt at bud break)	Cont=109 %, 0.1 ppm SO2 in air, 3 y plant, fall, exp= 16-30 d	Scotch pine[1]
Growth rate	111 % external control (CO2 upt at bud break)	Cont=130 %, 0.2 ppm SO2 in air, 3 y plant, summer, exp= 1-15 d	Scotch pine[1]
Growth rate	113 % external control (CO2 upt at bud break)	Cont=110 %, 0.05 ppm SO2 in air, 3 y plant, fall, exp= 1-15 d	Scotch pine[1]
Growth rate	114 % external control (CO2 upt at bud break)	Cont=149 %, 0.1 ppm SO2 in air, 3 y plant, summer, exp= 16-30 d	Scotch pine[1]
Growth rate	116 % external control (CO2 upt at bud break)	Cont=110 %, 0.2 ppm SO2 in air, 3 y plant, fall, exp= 1-15 d	Scotch pine[1]
Growth rate	116 % external control (CO2 upt at bud break)	Cont=136 %, 0.05 ppm SO2 in air, 3 y plant, summer, exp= 31-60 d	Scotch pine[1]
Growth rate	118 % external control (CO2 upt at bud break)	Cont=110 %, 0.1 ppm SO2 in air, 3 y plant, fall, exp= 1-15 d	Scotch pine[1]

Growth rate	124 % external control (CO_2 upt at bud break)	Cont=130 %, 0.1 ppm SO_2 in air, 3 y plant, summer, exp= 1-15 d	Scotch pine[1]
Growth rate	132 % external control (CO_2 upt at bud break)	Cont=130 %, 0.05 ppm SO_2 in air, 3 y plant, summer, exp= 1-15 d	Scotch pine[1]
Growth rate	140 % external control (CO_2 upt at bud break)	Cont=149 %, 0.05 ppm SO_2 in air, 3 y plant, summer, exp= 16-30 d	Scotch pine[1]
Growth rate	19 mm as staminate cone length	Rg = 17-21 mm, high SO_2 incidence, near Russian Cu smelter	Scotch pine[2]
Growth rate	2.72 g as ripe cone weight	Rg = 2.39-3.03 g, high SO_2 incidence, near Russian Cu smelter	Scotch pine[2]
Growth rate	27 mm as staminate cone length	Rg = 22-31 mm, low SO_2 incidence, near Russian Cu smelter	Scotch pine[2]
Growth rate	34.8 mm as ripe cone length	Rg = 32.1 - 35.6 mm, High SO_2 incidence, near Russian Cu smelter	Scotch pine[2]
Growth rate	38 % external control (CO_2 upt at bud break)	Cont=75 %, 0.2 ppm SO_2 in air, 3 y plant, spring, exp= 1-15 d	Scotch pine[1]
Growth rate	4.86 g as ripe cone weight	Rg = 3.81-5.91 g, low SO_2 incidence, near Russian Cu smelter	Scotch pine[2]
Growth rate	41.0 mm as ripe cone length	Rg = 39.5 - 42.4 mm, low SO_2 incidence, near Russian Cu smelter	Scotch pine[2]
Growth rate	44 % external control (CO_2 upt at bud break)	Cont=136 %, 0.2 ppm SO_2 in air, 3 y plant, summer, exp= 31-60 d	Scotch pine[1]
Growth rate	49 % external control (CO_2 upt at bud break)	Cont=71 %, 0.05 ppm SO_2 in air, 3 y plant, spring, exp= 16-30 d	Scotch pine[1]
Growth rate	49 % external control (CO_2 upt at bud break)	Cont=86 %, 0.2 ppm SO_2 in air, 3 y plant, spring, exp= 31-60 d	Scotch pine[1]
Growth rate	5.43 g as weight of 1000 seeds	Rg = 4.85-5.72 g, high SO_2 incidence, near Russian Cu smelter	Scotch pine[2]
Growth rate	57 % external control (CO_2 upt at bud break)	Cont=71 %, 0.2 ppm SO_2 in air, 3 y plant, spring, exp= 16-30 d	Scotch pine[1]
Growth rate	6.07 g as weight of 1000 seeds	Rg = 5.47-6.67 g, low SO_2 incidence, near Russian Cu smelter	Scotch pine[2]
Growth rate	64 % external control (CO_2 upt at bud break)	Cont=86 %, 0.05 ppm SO_2 in air, 3 y plant, spring, exp= 31-60 d	Scotch pine[1]
Growth rate	70 % external control (CO_2 upt at bud break)	Cont=75 %, 0.1 ppm SO_2 in air, 3 y plant, spring, exp= 1-15 d	Scotch pine[1]
Growth rate	72 % external control (CO_2 upt at bud break)	Cont=100 %, 0.2 ppm SO_2 in air, 3 y plant, fall, exp= 31-60 d	Scotch pine[1]
Growth rate	77 % external control (CO_2 upt at bud break)	Cont=75 %, 0.05 ppm SO_2 in air, 3 y plant, spring, exp= 1-15 d	Scotch pine[1]
Growth rate	80 % external control (CO_2 upt at bud break)	Cont=100 %, 0.05 ppm SO_2 in air, 3 y plant, fall, exp= 31-60 d	Scotch pine[1]
Growth rate	80 % external control (CO_2 upt at bud break)	Cont=71 %, 0.1 ppm SO_2 in air, 3 y plant, spring, exp= 16-30 d	Scotch pine[1]
Growth rate	82 % external control (CO_2 upt at bud break)	Cont=100 %, 0.1 ppm SO_2 in air, 3 y plant, fall, exp= 31-60 d	Scotch pine[1]
Growth rate	83 % external control (CO_2 upt at bud break)	Cont=136 %, 0.1 ppm SO_2 in air, 3 y plant, summer, exp= 31-60 d	Scotch pine[1]
Growth rate	83 % external control (CO_2 upt at bud break)	Cont=86 %, 0.1 ppm SO_2 in air, 3 y plant, spring, exp= 31-60 d	Scotch pine[1]
Growth rate	97 % external control (CO_2 upt at bud break)	Cont=109 %, 0.2 ppm SO_2 in air, 3 y plant, fall, exp= 16-30 d	Scotch pine[1]
Growth rate	10 % external control (CO_2 upt at bud break)	Cont=40 %, 0.2 ppm SO_2 in air, 3 y plant, spring, exp= 1-15 d	Silver fir[1]

Growth rate	100 % external control (CO_2 upt at bud break)	Cont=109 %, 0.2 ppm SO_2 in air, 3 y plant, summer, exp= 1-15 d	Silver fir[1]
Growth rate	104 % external control (CO_2 upt at bud break)	Cont=109 %, 0.05 ppm SO_2 in air, 3 y plant, summer, exp= 16-30 d	Silver fir[1]
Growth rate	104 % external control (CO_2 upt at bud break)	Cont=109 %, 0.1 ppm SO_2 in air, 3 y plant, summer, exp= 16-30 d	Silver fir[1]
Growth rate	108 % external control (CO_2 upt at bud break)	Cont=109 %, 0.05 ppm SO_2 in air, 3 y plant, summer, exp= 1-15 d	Silver fir[1]
Growth rate	114 % external control (CO_2 upt at bud break)	Cont=109 %, 0.1 ppm SO_2 in air, 3 y plant, summer, exp= 1-15 d	Silver fir[1]
Growth rate	24 % external control (CO_2 upt at bud break)	Cont=93 %, 0.2 ppm SO_2 in air, 3 y plant, spring, exp= 31-60 d	Silver fir[1]
Growth rate	27 % external control (CO_2 upt at bud break)	Cont=40 %, 0.1 ppm SO_2 in air, 3 y plant, spring, exp= 1-15 d	Silver fir[1]
Growth rate	28 % external control (CO_2 upt at bud break)	Cont=40 %, 0.05 ppm SO_2 in air, 3 y plant, spring, exp= 1-15 d	Silver fir[1]
Growth rate	34 % external control (CO_2 upt at bud break)	Cont=64 %, 0.2 ppm SO_2 in air, 3 y plant, spring, exp= 16-30 d	Silver fir[1]
Growth rate	49 % external control (CO_2 upt at bud break)	Cont=93 %, 0.1 ppm SO_2 in air, 3 y plant, spring, exp= 31-60 d	Silver fir[1]
Growth rate	51 % external control (CO_2 upt at bud break)	Cont=64 %, 0.1 ppm SO_2 in air, 3 y plant, spring, exp= 16-30 d	Silver fir[1]
Growth rate	58 % external control (CO_2 upt at bud break)	Cont=64 %, 0.05 ppm SO_2 in air, 3 y plant, spring, exp= 16-30 d	Silver fir[1]
Growth rate	58 % external control (CO_2 upt at bud break)	Cont=96 %, 0.2 ppm SO_2 in air, 3 y plant, summer, exp= 31-60 d	Silver fir[1]
Growth rate	62 % external control (CO_2 upt at bud break)	Cont=100 %, 0.2 ppm SO_2 in air, 3 y plant, fall, exp= 31-60 d	Silver fir[1]
Growth rate	70 % external control (CO_2 upt at bud break)	Cont=100 %, 0.1 ppm SO_2 in air, 3 y plant, fall, exp= 31-60 d	Silver fir[1]
Growth rate	72 % external control (CO_2 upt at bud break)	Cont=100 %, 0.05 ppm SO_2 in air, 3 y plant, fall, exp= 31-60 d	Silver fir[1]
Growth rate	74 % external control (CO_2 upt at bud break)	Cont=93 %, 0.05 ppm SO_2 in air, 3 y plant, spring, exp= 31-60 d	Silver fir[1]
Growth rate	77 % external control (CO_2 upt at bud break)	Cont=96 %, 0.1 ppm SO_2 in air, 3 y plant, summer, exp= 31-60 d	Silver fir[1]
Growth rate	80 % external control (CO_2 upt at bud break)	Cont=106 %, 0.2 ppm SO_2 in air, 3 y plant, fall, exp= 16-30 d	Silver fir[1]
Growth rate	83 % external control (CO_2 upt at bud break)	Cont=106 %, 0.05 ppm SO_2 in air, 3 y plant, fall, exp= 16-30 d	Silver fir[1]
Growth rate	83 % external control (CO_2 upt at bud break)	Cont=106 %, 0.1 ppm SO_2 in air, 3 y plant, fall, exp= 16-30 d	Silver fir[1]
Growth rate	87 % external control (CO_2 upt at bud break)	Cont=122 %, 0.2 ppm SO_2 in air, 3 y plant, fall, exp= 1-15 d	Silver fir[1]
Growth rate	87 % external control (CO_2 upt at bud break)	Cont=96 %, 0.05 ppm SO_2 in air, 3 y plant, summer, exp= 31-60 d	Silver fir[1]
Growth rate	90 % external control (CO_2 upt at bud break)	Cont=109 %, 0.2 ppm SO_2 in air, 3 y plant, summer, exp= 16-30 d	Silver fir[1]
Growth rate	91 % external control (CO_2 upt at bud break)	Cont=122 %, 0.1 ppm SO_2 in air, 3 y plant, fall, exp= 1-15 d	Silver fir[1]
Growth rate	94 % external control (CO_2 upt at bud break)	Cont=122 %, 0.05 ppm SO_2 in air, 3 y plant, fall, exp= 1-15 d	Silver fir[1]

Growth rate	46 % net photosynthesis of control	1 week, 50 pphm SO2, net rate, small chamber technique, SD=9	Sugar maple[1]
Growth rate	57 % net photosynthesis of control	3 weeks, 50 pphm SO2, net rate, small chamber technique, SD=10	Sugar maple[1]
Growth rate	78 % net photosynthesis of control	1-2 days, 50 pphm SO2, net rate, small chamber technique, SD=5	Sugar maple[1]
Growth rate	73 % net photosynthesis of control	1-2 days, 50 pphm SO2, net rate, small chamber technique, SD=13	White ash[1]
Growth rate	80 % net photosynthesis of control	1 week, 50 pphm SO2, net rate, small chamber technique, SD=12	White ash[1]
Growth rate	93 % net photosynthesis of control	3 weeks, 50 pphm SO2, net rate, small chamber technique, SD=5	White ash[1]

Source: 1.Keller, T. 1977. 4th International Clean Air Congress, 81-83.
2.Mamajev, S.A. and O.D. Shkarlet. 1972. Mitteil. Forst. Bundes-Versuch. Wien 97, 443-450.

4-378 Effect of $SO_2 + O_3$

Parameter	Effect Value	Concentration and Condition	Species
Growth rate	55 % net photosynthesis of control	1-2 days, 50 SO2 + 50 pphm O3, net rate, small chamber technique, SD=8	Black oat[1]
Growth rate	56 % net photosynthesis of control	1 week, 50 SO2 + 50 pphm O3, net rate, small chamber technique, SD=9	Black oat[1]
Growth rate	26 % net photosynthesis of control	1-2 days, 50 SO2 + 50 pphm O3, net rate, small chamber technique, SD=10	Sugar maple[1]
Growth rate	59 % net photosynthesis of control	1 week, 50 SO2 + 50 pphm O3, net rate, small chamber technique, SD=10	Sugar maple[1]
Growth rate	54 % net photosynthesis of control	1-2 days, 50 SO2 + 50 pphm O3, net rate, small chamber technique, SD=8	White ash[1]
Growth rate	62 % net photosynthesis of control	1 week, 50 SO2 + 50 pphm O3, net rate, small chamber technique, SD=13	White ash[1]

Source: 1.Keller, T. 1977. 4th International Clean Air Congress, 81-83.

4-379 Effect of Sb

Parameter	Effect Value	Concentration and Condition	Species
Mortality	50 % mortality	Sb2O3SbOCl, water = E-4 M Sb, toxic level given, microorganisms	Azotobacter sp.[1]

Source: 1.Zhakharyan, S.V. 1970. Symp. Biol. Hung. 11, 391-395.

4-380 Effect of Se

Parameter	Effect Value	Concentration and Condition	Species
Mortality	0 %	Toxicological limit = 183 mg/l, protozoa, Se-compounds	Crustacean[1]
Mortality	0 %	Toxicological limit = 2 - 10.5 mg/l, Se-compounds	Fish[1]
Mortality	0 %	Toxicological limit = 0.01 mg/l chronical symptoms, Se-compounds	Homo sapiens[1]
Mortality	0 %	Toxicological limit = 5 - 10 mg/l chronical symptoms, Se-compounds	Mammalia[1]
Nitrification	91 % inhibation	Soil = 5 umoles/g of soil, 10 d incubation, 3 soils, lab	Soil organisms[2]

Source: 1.Liebman, H. 1958. Handbuch der Friscwasser - und Abwasserbiologie. Band II.Munchen. Oldenbourg.
2.Liang, C.N. and M.A. Tabatabai. 1978. Environ. Pollut. 12, 141-147.

4-381 Effect of Sn

Parameter	Effect Value	Concentration and Condition	Species
Mortality	0 %	Toxicological limit = 2 mg/l, Sn-compounds	Fish[1]
Nitrification	41 % inhibation	Soil = 5 umoles/g of soil, 10 d incubation, 3 soils, lab	Soil organisms[2]

Source: 1.Liebman, H. 1958. Handbuch der Friscwasser - und Abwasserbiologie. Band II.Munchen. Oldenbourg.2.Liang, C.N. and M.A. Tabatabai. 1978. Environ. Pollut. 12, 141-147.

4-382 Effect of Styrene

Parameter	Concentration	Condition	Species
Sublethal effect	67 mg/l	Inhibition of cell multiplication	Microcystis aeruginosa[1]
Sublethal effect	72 mg/l	Inhibition of cell multiplication	Pseudomonas putida[1]
TLm	25.1 mg/l	24-96 hours, soft water	Bluegill[2]
TLm	46-57 mg/l	24-96 hours, soft water	Fathead minnow[2]
TLm	59-63 mg/l	24-96 hours, hard water	Fathead minnow[2]
TLm	64.7 mg/l	24-96 hours, soft water	Goldfish[2]
TLm	74.8 mg/l	24-96 hours, soft water	Guppie[2]

Source: 1.Bringmann, G. and Kuhn, R., nwf-wasser/abwasser, H.9, 1976.
2.Pickering Q. H. and Henderson, C., JWPCF, 38, 1419, 1966.

4-383 Effect of Styrene oxide

Parameter	Concentration	Condition	Species
LD50	2.0 g pr. kg body weight	single oral dose, 100 %	guinea pig[1]
LD50	2.0 g pr. kg body weight	single oral dose, 100 %	Rat[1]

Source: 1.Patty, F. A., Industrial Hygiene and toxicology, 2, Interscience Publishers, New York, 1976.

4-384 Effect of Succinic acid

Parameter	Concentration	Condition	Species
LC0	125 mg/l	not indicated	Colpoda[1]
LC0	125 mg/l	not indicated	Pseudomonas[1]
LC0	1 mg/l	not indicated	Scenedesmus[1]

Source: 1.Meinck, F., Stooff, H. and Kohlschutter, H., Les eaux residuaires industrielles, 1970.

4-385 Effect of Succinonitrile

Parameter	Concentration	Condition	Species
LD50	0.45 g pr. kg body weight	acute oral dose, 100 %	Rat[1]

Source: 1.Patty, F. A., Industrial Hygiene and toxicology, 2, Interscience Publishers, New York, 1976.

4-386 Effect of Sulfanilic acid

Parameter	Concentration	Condition	Species
LD50	>3.2 g pr. kg body weight	acute oral dose	Mouse[1]

Source: 1.Patty, F. A., Industrial Hygiene and toxicology, 2, Interscience Publishers, New York, 1976.

4-387 Effect of Sulfenyl

Parameter	Effect Value	Concentration and Condition	Species
Nitrogen fixation	50 % inhibition	5 mg/l, reversible by add of Na-molybdate, trichloromethylsulphenyl	Clostridium pasteurianum[1]

Source: 1.Carnahan, J.E., L.E. Mortensen and J.E. Castle. 1960. J. Bact., 80, pp 311-315.

4-388 Effect of Sulfolane

Parameter	Concentration	Condition	Species
LC50	4800 mg/l	24 hours	Goldfish[1]

Source: 1.Shell Chemie, Shell Industrie Chemicalien gids, Shell Niederland Chemie, 1975.

4-389 Effect of TNT

Parameter	Concentration	Condition	Species
LC50	1.6 mg/l	96 hours	Fish[1]

Source: 1.Smock, L. A., Stoneburger, D. L. and Clarck, J. R., Water Research, 10, 537, 1976.

4-390 Effect of Temperature

Parameter	Effect Value	Concentration and Condition	Species
Development of zooplankton	0.8 days	From copepodit stage 4 to 5, male, 293 K, no. of groups = 6	Cyclops vicinus[1]
Development of zooplankton	1.0 days	From copepodit stage 2 to 3, 293 K, no. of groups = 6, SD = +- 0.5	Cyclops vicinus[1]
Development of zooplankton	1.3 days	From copepodit stage 3 to 4, 288 K, no. of groups = 18, SD = +- 0.5	Cyclops vicinus[1]
Development of zooplankton	1.5 days	From copepodit stage 2 to 3, 288 K, no. of groups = 18, SD = +- 0.9	Cyclops vicinus[1]
Development of zooplankton	1.6 days	From copepodit stage 1 to 2, 293 K, no. of groups = 6, SD = +- 0.6	Cyclops vicinus[1]
Development of zooplankton	1.7 days	From copepodit stage 1 to 2, 288 K, no. of groups = 18, SD = +- 1.1	Cyclops vicinus[1]
Development of zooplankton	1.7 days	From copepodit stage 3 to 4, 293 K, no. of groups = 6, SD = +- 1.3	Cyclops vicinus[1]
Development of zooplankton	1.9 days	From copepodit stage 4 to 5, male, 283 K, no. of groups = 19	Cyclops vicinus[1]
Development of zooplankton	10.4 days	From copepodit stage 4 to 5, female, 278 K, no. of groups = 15	Cyclops vicinus[1]
Development of zooplankton	10.9 days	Total copepodit development, female, 288 K, no. of groups = 14	Cyclops vicinus[1]
Development of zooplankton	12.2 days	Total copepodit development, female, 293 K, no. of groups = 6	Cyclops vicinus[1]
Development of zooplankton	13.0 days	Total copepodit development, male, 283 K, no. of groups = 19	Cyclops vicinus[1]
Development of zooplankton	13.1 days	From copepodit stage 5 to adult, female, 278 K, no. of groups = 15	Cyclops vicinus[1]
Development of zooplankton	14.5 days	From nauplius to copepodit stage 1, 283 K, no. of groups = 20	Cyclops vicinus[1]
Development of zooplankton	16.6 days	From nauplius to copepodit stage 1, 293 K, no. of groups = 6, SD = +-	Cyclops vicinus[1]

Development of zooplankton	18.8 days	From nauplius to copepodit stage 1, 288 K, no. of groups = 18, SD = +	Cyclops vicinus[1]
Development of zooplankton	18.9 days	Total copepodit development, female, 283 K, no. of groups = 19	Cyclops vicinus[1]
Development of zooplankton	2.1 days	From copepodit stage 5 to adult, male, 293 K, no. of groups = 6	Cyclops vicinus[1]
Development of zooplankton	2.2 days	From copepodit stage 4 to 5, male, 288 K, no. of groups = 12	Cyclops vicinus[1]
Development of zooplankton	2.4 days	From copepodit stage 2 to 3, 283 K, no. of groups = 20, SD = +- 0.5	Cyclops vicinus[1]
Development of zooplankton	2.4 days	From copepodit stage 4 to 5, female, 293 K, no. of groups = 6	Cyclops vicinus[1]
Development of zooplankton	2.5 days	From copepodit stage 5 to adult, male, 288 K, no. of groups = 12	Cyclops vicinus[1]
Development of zooplankton	2.6 days	From copepodit stage 3 to 4, 283 K, no. of groups = 20, SD = +- 0.7	Cyclops vicinus[1]
Development of zooplankton	2.7 days	From copepodit stage 1 to 2, 283 K, no. of groups = 20, SD = +- 0.7	Cyclops vicinus[1]
Development of zooplankton	2.8 days	From copepodit stage 4 to 5, female, 288 K, no. of groups = 14	Cyclops vicinus[1]
Development of zooplankton	25.3 days	Total copepodit development, male, 278 K, no. of groups = 18	Cyclops vicinus[1]
Development of zooplankton	3.4 days	From copepodit stage 5 to adult, female, 288 K, no. of groups = 14	Cyclops vicinus[1]
Development of zooplankton	3.5 days	From copepodit stage 5 to adult, male, 283 K, no. of groups = 19	Cyclops vicinus[1]
Development of zooplankton	3.6 days	From copepodit stage 4 to 5, male, 278 K, no. of groups = 18	Cyclops vicinus[1]
Development of zooplankton	37.3 days	From nauplius to copepodit stage 1, 278 K, no. of groups = 20	Cyclops vicinus[1]
Development of zooplankton	37.8 days	Total copepodit development, female, 278 K, no. of groups = 15	Cyclops vicinus[1]
Development of zooplankton	4.0 days	From copepodit stage 2 to 3, 278 K, no. of groups = 20, SD = +- 1.2	Cyclops vicinus[1]
Development of zooplankton	4.6 days	From copepodit stage 1 to 2, 278 K, no. of groups = 20, SD = +- 1.0	Cyclops vicinus[1]
Development of zooplankton	4.7 days	From copepodit stage 5 to adult, female, 283 K, no. of groups = 19	Cyclops vicinus[1]
Development of zooplankton	5.1 days	From copepodit stage 3 to 4, 278 K, no. of groups = 20, SD = +- 0.8	Cyclops vicinus[1]
Development of zooplankton	5.5 days	From copepodit stage 5 to adult, female, 293 K, no. of groups = 6	Cyclops vicinus[1]
Development of zooplankton	6.3 days	From copepodit stage 4 to 5, female, 283 K, no. of groups = 19	Cyclops vicinus[1]
Development of zooplankton	7.1 days	Total copepodit development, male, 293 K, no. of groups = 6	Cyclops vicinus[1]
Development of zooplankton	8.3 days	From copepodit stage 5 to adult, male, 278 K, no. of groups = 18	Cyclops vicinus[1]
Development of zooplankton	8.9 days	Total copepodit development, male, 288 K, no. of groups = 12	Cyclops vicinus[1]
Development of zooplankton	0.8 days	From copepodit stage 4 to 5, female, 293 K, no. of groups = 3	Eudiaptomus gracilis[1]

Development of zooplankton	1.8 days	From copepodit stage 4 to 5, male, 293 K, no. of groups = 3	Eudiaptomus gracilis[1]
Development of zooplankton	10.5 days	From copepodit stage 5 to adult, male, 283 K, no. of groups = 6	Eudiaptomus gracilis[1]
Development of zooplankton	12.0 days	Total copepodit development, female, 293 K, no. of groups = 2	Eudiaptomus gracilis[1]
Development of zooplankton	13.6 days	From copepodit stage 4 to 5, male, 278 K, no. of groups = 8	Eudiaptomus gracilis[1]
Development of zooplankton	15.1 days	From copepodit stage 3 to 4, 278 K, no. of groups = 13, SD = +- 5.4	Eudiaptomus gracilis[1]
Development of zooplankton	15.2 days	From nauplius to copepodit stage 1, 288 K, no. of groups = 9	Eudiaptomus gracilis[1]
Development of zooplankton	15.3 days	From copepodit stage 5 to adult, female, 283 K, no. of groups = 8	Eudiaptomus gracilis[1]
Development of zooplankton	15.5 days	Total copepodit development, male, 293 K, no. of groups = 1	Eudiaptomus gracilis[1]
Development of zooplankton	16.7 days	From copepodit stage 5 to adult, male, 278 K, no. of groups = 8	Eudiaptomus gracilis[1]
Development of zooplankton	17.3 days	From copepodit stage 5 to adult, female, 278 K, no. of groups = 7	Eudiaptomus gracilis[1]
Development of zooplankton	18.6 days	From copepodit stage 4 to 5, female, 278 K, no. of groups = 7	Eudiaptomus gracilis[1]
Development of zooplankton	18.7 days	Total copepodit development, female, 288 K, no. of groups = 6	Eudiaptomus gracilis[1]
Development of zooplankton	19.4 days	From nauplius to copepodit stage 1, 283 K, no. of groups = 19	Eudiaptomus gracilis[1]
Development of zooplankton	2.1 days	From copepodit stage 5 to adult, female, 293 K, no. of groups = 2	Eudiaptomus gracilis[1]
Development of zooplankton	2.3 days	From copepodit stage 2 to 3, 293 K, no. of groups = 3	Eudiaptomus gracilis[1]
Development of zooplankton	2.4 days	From copepodit stage 2 to 3, 288 K, no. of groups = 9, SD = +- 1.5	Eudiaptomus gracilis[1]
Development of zooplankton	2.5 days	From copepodit stage 3 to 4, 293 K, no. of groups = 3, SD = +- 1.3	Eudiaptomus gracilis[1]
Development of zooplankton	20.7 days	Total copepodit development, male, 288 K, no. of groups = 9	Eudiaptomus gracilis[1]
Development of zooplankton	3.0 days	From copepodit stage 1 to 2, 288 K, no. of groups = 9, SD = +- 1.8	Eudiaptomus gracilis[1]
Development of zooplankton	3.1 days	From copepodit stage 1 to 2, 293 K, no. of groups = 4, SD = +- 1.0	Eudiaptomus gracilis[1]
Development of zooplankton	3.3 days	From copepodit stage 3 to 4, 288 K, no. of groups = 9, SD = +- 2.1	Eudiaptomus gracilis[1]
Development of zooplankton	31.0 days	Total copepodit development, male, 283 K, no. of groups = 6	Eudiaptomus gracilis[1]
Development of zooplankton	36.2 days	From nauplius to copepodit stage 1, 278 K, no. of groups = 16	Eudiaptomus gracilis[1]
Development of zooplankton	37.9 days	Total copepodit development, female, 283 K, no. of groups = 8	Eudiaptomus gracilis[1]
Development of zooplankton	4.0 days	From copepodit stage 4 to 5, male, 288 K, no. of groups = 9	Eudiaptomus gracilis[1]
Development of zooplankton	4.0 days	From copepodit stage 5 to adult, male, 293 K, no. of groups = 1	Eudiaptomus gracilis[1]

Development of zooplankton	4.5 days	From copepodit stage 4 to 5, female, 288 K, no. of groups = 6	Eudiaptomus gracilis[1]
Development of zooplankton	4.8 days	From copepodit stage 3 to 4, 283 K, no. of groups = 11, SD = +- 2.5	Eudiaptomus gracilis[1]
Development of zooplankton	5.2 days	From copepodit stage 5 to adult, female, 288 K, no. of groups = 6	Eudiaptomus gracilis[1]
Development of zooplankton	5.3 days	From copepodit stage 1 to 2, 283 K, no. of groups = 18, SD = +- 1.8	Eudiaptomus gracilis[1]
Development of zooplankton	6.1 days	From copepodit stage 2 to 3, 283 K, no. of groups = 14, SD = +- 4.4	Eudiaptomus gracilis[1]
Development of zooplankton	60.6 days	Total copepodit development, male, 278 K, no. of groups = 8	Eudiaptomus gracilis[1]
Development of zooplankton	67.6 days	Total copepodit development, female, 278 K, no. of groups = 7	Eudiaptomus gracilis[1]
Development of zooplankton	7.8 days	From copepodit stage 4 to 5, male, 283 K, no. of groups = 6	Eudiaptomus gracilis[1]
Development of zooplankton	7.9 days	From copepodit stage 5 to adult, male, 288 K, no. of groups = 9	Eudiaptomus gracilis[1]
Development of zooplankton	8.9 days	From nauplius to copepodit stage 1, 293 K, no. of groups = 5	Eudiaptomus gracilis[1]
Development of zooplankton	9.2 days	From copepodit stage 2 to 3, 278 K, no. of groups = 14, SD = +- 1.7	Eudiaptomus gracilis[1]
Development of zooplankton	9.2 days	From copepodit stage 4 to 5, female, 283 K, no. of groups = 8	Eudiaptomus gracilis[1]
Development of zooplankton	9.8 days	From copepodit stage 1 to 2, 278 K, no. of groups = 16, SD = +- 1.4	Eudiaptomus gracilis[1]
Filtration rate	1753 ml/hr	Size = 7.6 cm, food = Chlamydomonas sp. (2E7cells/l, 10*7.5 U), 285 K	Modiolus modiolus[2]
Filtration rate	2527 ml/hr	Size = 7.6 cm, food = Chlamydomonas sp. (2E7cells/l, 10*7.5 U), 293 K	Modiolus modiolus[2]
Filtration rate	834 ml/hr	Size = 7.6 cm, food = Chlamydomonas sp. (2E7cells/l, 10*7.5 U), 277 K	Modiolus modiolus[2]
Filtration rate	106 ml/hr	Size = 4 cm, food = Platymonas suecica (5E6cells/l, 7.2*9.4 U), 303 K	Mytilus edulis[3]
Filtration rate	1100 ml/hr	Size = 4.5 cm, food = aquadag graphite(ca.2 U), 283 K	Mytilus edulis[4]
Filtration rate	1250 ml/hr	Size = 4.5 cm, food = aquadag graphite(ca.2 U), 293 K	Mytilus edulis[4]
Filtration rate	1320 ml/hr	Size = 4.5 cm, food = aquadag graphite(ca.2 U), 288 K	Mytilus edulis[4]
Filtration rate	1370 ml/hr	Size = 4.5 cm, food = aquadag graphite(ca.2 U), 290 K	Mytilus edulis[4]
Filtration rate	1468 ml/hr	Size = 4 cm, food = Platymonas suecica (5E6cells/l, 7.2*9.4 U), 283 K	Mytilus edulis[3]
Filtration rate	1751 ml/hr	Size = 4 cm, food = Platymonas suecica (5E6cells/l, 7.2*9.4 U), 288 K	Mytilus edulis[3]
Filtration rate	1796 ml/hr	Size = 4 cm, food = Platymonas suecica (5E6cells/l, 7.2*9.4 U), 293 K	Mytilus edulis[3]
Filtration rate	1798 ml/hr	Size = 4 cm, food = Platymonas suecica (5E6cells/l, 7.2*9.4 U), 298 K	Mytilus edulis[3]
Filtration rate	250 ml/hr	Size = 4.5 cm, food = aquadag graphite(ca.2 U), 298 K	Mytilus edulis[4]

Filtration rate	353 ml/hr	Size = 4 cm, food = Platymonas suecica (5E6cells/l, 7.2*9.4 U), 278 K	Mytilus edulis[3]
Filtration rate	650 ml/hr	Size = 4.5 cm, food = aquadag graphite(ca.2 U), 278 K	Mytilus edulis[4]
First mortality	15 sec	302.5 K, no mortality under 297 K	Oncorhynchus keta[5]
First mortality	2640 sec	299.8 K, no mortality under 297 K	Oncorhynchus keta[5]
First mortality	4 sec	310.9 K, no mortality under 297 K	Oncorhynchus keta[5]
First mortality	6 sec	305.3 K, no mortality under 297 K	Oncorhynchus keta[5]
First mortality	100 sec	299.8 K, no mortality under 294.2 K	Oncorhynchus tschawytscha[5]
First mortality	4 sec	305.3 K, no mortality under 294.2 K	Oncorhynchus tschawytscha[5]
First mortality	0.3 min	305 K, no mortality under 288 K	Thaleichthys pacificus[5]
First mortality	1.4 min	300 K, no mortality under 288 K	Thaleichthys pacificus[5]
First mortality	15 min	294 K, no mortality under 288 K	Thaleichthys pacificus[5]
First mortality	36 min	291 K, no mortality under 288 K	Thaleichthys pacificus[5]
First mortality	5.0 min	297 K, no mortality under 288 K	Thaleichthys pacificus[5]
Growth	0.36 mm/day	96 days old plant, 302.1 K	Pepper plants[6]
Growth	12.7 mm/day	24 days old plant, 285.7 K	Pepper plants[6]
Growth	2.3 mm/day	96 days old plant, 281.2 K	Pepper plants[6]
Growth	2.4 mm/day	96 days old plant, 298.5 K	Pepper plants[6]
Growth	2.8 mm/day	96 days old plant, 292.1 K	Pepper plants[6]
Growth	3 mm/day	24 days old plant, 281.2 K	Pepper plants[6]
Growth	4.1 mm/day	96 days old plant, 288.9 K	Pepper plants[6]
Growth	4.45 mm/day	96 days old plant, 285.7 K	Pepper plants[6]
Growth	7.24 mm/day	24 days old plant, 302.1 K	Pepper plants[6]
Growth	8 mm/day	24 days old plant, 292.1 K	Pepper plants[6]
Growth	9.0 mm/day	24 days old plant, 298.5 K	Pepper plants[6]
Mortality	0 mortality in condition	Min. temp. = 272 K, maximum temp. = 306.4 K	Ameiurus nebulosus[7]
Mortality	0 mortality in condition	Min. temp. = 276.7 K, maximum temp. = 309.5 K	Ameiurus nebulosus[7]
Mortality	274.6 K	Lower lethal temperature	Black bass[8]
Mortality	309.5 K	Upper lethal temperature	Black bass[8]
Mortality	273 K	Lower lethal temperature	Catfish[8]
Mortality	310.1 K	Upper lethal temperature	Catfish[8]
Mortality	305.5 K	Upper lethal temperature	Crappie[8]
Mortality	273 K	Average value, lower lethal temperature	Fish[8]
Mortality	307.8 K	Average value, upper lethal temperature	Fish[8]
Mortality	0 mortality in condition	Min. temp. = 273 K, maximum temp. = 303.3 K	Ictalurus lacustris[9]
Mortality	0 mortality in condition	Min. temp. = 279 K, maximum temp. = 306.5 K	Ictalurus lacustris[9]
Mortality	0 mortality in condition	Min. temp. = 275.5 K, maximum temp. = 303.7 K	Lepomis macrochirus[9]
Mortality	0 mortality in condition	Min. temp. = 278 K, maximum temp. = 304.5 K	Lepomis macrochirus[9]

Mortality	0 mortality in condition	Min. temp. = 278.5 K, maximum temp. = 305.5 K	Micropterus salmoides[9]
Mortality	0 mortality in condition	Min. temp. = 283.8 K, maximum temp. = 309.4 K	Micropterus salmoides[9]
Mortality	273 K	Lower lethal temperature	Minnow[8]
Mortality	306.4 K	Upper lethal temperature	Minnow[8]
Mortality	0 mortality in condition	Minimum temp. = none, maximum temp. = 294.3 K	Oncorhynchus gorbuscha[7]
Mortality	0 mortality in condition	Minimum temp. = none, maximum temp. = 296.9 K	Oncorhynchus gorbuscha[7]
Mortality	0 mortality in condition	Minimum temp. = 273.5 K, maximum temp. = 295.6 K	Oncorhynchus keta[7]
Mortality	0 mortality in condition	Minimum temp. = 299.5 K, maximum temp. = 296.7 K	Oncorhynchus keta[7]
Mortality	0 mortality in condition	Minimum temp. = 273.2 K, maximum temp. = 295.9 K	Oncorhynchus kisutch[7]
Mortality	0 mortality in condition	Minimum temp. = 277.5 K, maximum temp. = 298 K	Oncorhynchus kisutch[7]
Mortality	0 mortality in condition	Minimum temp. = 273 K, maximum temp. = 295.2 K	Oncorhynchus nerka[7]
Mortality	0 mortality in condition	Minimum temp. = 277.7 K, maximum temp. = 297.8 K	Oncorhynchus nerka[7]
Mortality	0 mortality in condition	Minimum temp. = 273, 8 K, maximum temp. = 297.3 K, acclimation temp.	Oncorhynchus tschawytscha[7]
Mortality	0 mortality in condition	Minimum temp. = 277.5 K, maximum temp. = 298.1 K	Oncorhynchus tschawytscha[7]
Mortality	0 mortality in condition	Minimum temp. = 274.1 K, maximum temp. = 298 K	Perca flavescens[10]
Mortality	0 mortality in condition	Minimum temp. = 276.7 K, maximum temp. = 302.7 K	Perca flavescens[10]
Mortality	273 K	Lower lethal temperature	Pickerels[8]
Mortality	307.4 K	Upper lethal temperature	Pickerels[8]
Mortality	275.5 K	Lower lethal temperature	Sunfish[8]
Mortality	308.7 K	Upper lethal temperature	Sunfish[8]
Mortality	100 %	310.5 K, 24 h, 50 animals	Thermocyclops neglectus[11]
Mortality	100 %	313 K, 2 h, 50 animals	Thermocyclops neglectus[11]
Mortality	12 %	308 K, 31 h, 50 animals	Thermocyclops neglectus[11]
Mortality	20 %	308 K, 33 h, 50 animals	Thermocyclops neglectus[11]
Mortality	22 %	308 K, 36 h, 50 animals	Thermocyclops neglectus[11]
Mortality	22 %	310.5 K, 2 h, 50 animals	Thermocyclops neglectus[11]
Mortality	4 %	308 K, 20 h, 50 animals	Thermocyclops neglectus[11]
Mortality	8 %	308 K, 24 h, 50 animals	Thermocyclops neglectus[11]
Mortality	80 %	310.5 K, 4 h, 50 animals	Thermocyclops neglectus[11]
Mortality	98 %	310.5 K, 6 h, 50 animals	Thermocyclops neglectus[11]
Mortality	273 K	Lower lethal temperature	Yellow perch[8]
Mortality	303.9 K	Upper lethal temperature	Yellow perch[8]
Optimum for growth	300 K	None	Black bass[8]
Optimum for growth	303 K	None	Catfish[8]

Optimum for growth	296 K	None	Crappie[8]
Optimum for growth	297.2 K	Average value	Fish[8]
Optimum for growth	300 K	None	Minnow[8]
Optimum for growth	298.4 K	None	Pickerels[8]
Optimum for growth	300.5 K	None	Sunfish[8]
Optimum for growth	287.2 K	None	Yellow perch[8]
Rate of reproduction	1 eggs/(female*week)	288.2 K, females/(female*w) = 0.005, time from egg to adult = 32.9	Calandra oryzae[12]
Rate of reproduction	197 eggs/(female*week)	305.3 eggs/(female*week)	Calandra oryzae[12]
Rate of reproduction	266 eggs/(female*week)	296 K, females/(female*w) = 0.43, time from egg to adult = 6.1 w	Calandra oryzae[12]
Rate of reproduction	27 eggs/(female*week)	306.5 K, females/(female*w) = 0.12, time from egg to adult = 9.4 w	Calandra oryzae[12]
Rate of reproduction	344 eggs/(female*week)	302.1 K, females/(female*w) = 0.77, time from egg to adult = 4.0 w	Calandra oryzae[12]
Rate of reproduction	384 eggs/(female*week)	298.5 K, females/(female*w) = 0.61, time from egg to adult = 4.9 w	Calandra oryzae[12]
Rate of reproduction	4 eggs/(female*week)	291.2 K, females/(female*w) = 0.14, time from egg to adult = 15.5 w	Calandra oryzae[12]
Rate of reproduction	5 eggs/(female*week)	308.0 K, females/(female*w) = 0, time from egg to adult = infinite	Calandra oryzae[12]
Rate of thermal denaturation	1.05E-6 1/sec	315 K, 9.08 % replaced in 24 h, 1 order reac., protein as biomass	Protein[13]
Rate of thermal denaturation	1.25E-7 1/sec	310 K, 1.08 % replaced in 24 h, 1 order reac., protein as biomass	Protein[13]
Rate of thermal denaturation	1.50E-6 1/sec	316 K, 13.81 % replaced in 24 h, 1 order reac., protein as biomass	Protein[13]
Rate of thermal denaturation	1.89E-7 1/sec	311 K, 1.63 % replaced in 24 h, 1 order reac., protein as biomass	Protein[13]
Rate of thermal denaturation	13.5E-6 1/sec	321 K, 116.64 % replaced in 24 h, 1 order reac., protein as biomass	Protein[13]
Rate of thermal denaturation	18.1E-6 1/sec	322 K, 161.51 % replaced in 24 h, 1 order reac., protein as biomass	Protein[13]
Rate of thermal denaturation	2.37E-6 1/sec	317 K, 20.50 % replaced in 24 h, 1 order reac., protein as biomass	Protein[13]
Rate of thermal denaturation	27.1E-6 1/sec	323 K, 234.14 % replaced in 24 h, 1 order reac., protein as biomass	Protein[13]
Rate of thermal denaturation	3.52E-6 1/sec	318 K, 28.65 % replaced in 24 h, 1 order reac., protein as biomass	Protein[13]
Rate of thermal denaturation	3.70E-7 1/sec	312 K, 3.20 % replaced in 24 h, 1 order reac., protein as biomass	Protein[13]
Rate of thermal denaturation	5.12E-7 1/sec	313 K, 4.42 % replaced in 24 h, 1 order reac., protein as biomass	Protein[13]

Rate of thermal denaturation	5.35E-6 1/sec	319 K, 46.20 % replaced in 24 h, 1 order reac., protein as biomass	Protein[13]
Rate of thermal denaturation	6.62E-7 1/sec	314 K, 5.71 % replaced in 24 h, 1 order reac., protein as biomass	Protein[13]
Rate of thermal denaturation	7.92E-6 1/sec	320 K, 68.40 % replaced in 24 h, 1 order reac., protein as biomass	Protein[13]
Respiration of zooplankton	14.9 microlitre O2/(mg dry weight*day)	283.5 K, 95 % CL = +- 0.5	Phronima sedentaria[14]
Respiration of zooplankton	16.5 microlitre O2/(mg dry weight*day)	288.5 K, 95 % CL = +- 2.7	Phronima sedentaria[14]
Respiration of zooplankton	19.4 microlitre O2/(mg dry weight*day)	292.5 K, 95 % CL = +- 1.3	Phronima sedentaria[14]
Respiration of zooplankton	9.1 microlitre O2/(mg dry weight*day)	278 K, 95 % CL = +- 1.1	Phronima sedentaria[14]
Uptake of Cd	5.55 mg Cd/kg dry weight	288.5 K, soil = 10 ppm Cd	Glycine max[15]
Uptake of Cd	6.91 mg Cd/kg dry weight	294.1 K, soil = 10 ppm Cd	Glycine max[15]
Uptake of Cd	7.90 mg Cd/kg dry weight	299.6 K, soil = 10 ppm Cd	Glycine max[15]
Uptake of Cd	9.79 mg Cd/kg dry weight	305.2 K, soil = 10 ppm Cd	Glycine max[15]
Uptake of Pb	197 mg/g dry weight	Plant roots, 40 min, 277 K	Bean[16]
Uptake of Pb	32 mg/g dry weight	Plant roots, 120 min, 297 K	Bean[16]
Uptake of Pb	38 mg/g dry weight	Plant roots, 40 min, 297 K	Bean[16]
Uptake of Pb	450 mg/g dry weight	Plant roots, 120 min, 277 K	Bean[16]
Uptake of Pb	130 mg/g dry weight	Plant roots, 120 min, 277 K	Corn[16]
Uptake of Pb	170 mg/g dry weight	Plant roots, 120 min, 297 K	Corn[16]
Uptake of Pb	59 mg/g dry weight	Plant roots, 40 min, 277 K	Corn[16]
Uptake of Pb	94 mg/g dry weight	Plant roots, 40 min, 297 K	Corn[16]
Uptake of Pb	205 mg/g dry weight	Plant roots, 40 min, 277 K	Soybean[16]
Uptake of Pb	37 mg/g dry weight	Plant roots, 120 min, 297 K	Soybean[16]
Uptake of Pb	38 mg/g dry weight	Plant roots, 40 min, 297 K	Soybean[16]
Uptake of Pb	560 mg/g dry weight	Plant roots, 120 min, 277 K	Soybean[16]
Water soluble and exchangeable Mn	10 ppm	Heated to 333 K for 30 min, unheated = 8 ppm	Soil[17]
Water soluble and exchangeable Mn	27 ppm	Heated to 355 K for 30 min, unheated = 8 ppm	Soil[17]
Water soluble and exchangeable Mn	60 ppm	Heated to 366 K for 30 min, unheated = 8 ppm	Soil[17]
Water soluble and exchangeable Mn	82 ppm	Heated to 373 K for 30 min, unheated = 8 ppm	Soil[17]

Source: 1.Munro, I.G. 1974. Oecologia 16, pp 355-367.
2.Winter, J.E. 1969. Mar. Biol., 4, pp 87-135.
3.Schultze, E.H. 1975. Mar. biol., 30(4), pp 331-341.
4.Theede, H. 1963. Kieler Meeresforsch., 19, pp 20-41.
5.Garrigan, G.A. 1977. Water Pollut. Control 49(12), pp 2380-2389.
6.Dorland, R.E. and F.W. Went. 1947. Am. J. Bot., 34, pp 393-401.
7.Brett, J.R. 1952. Bull. Fish Res. Board Can., 9, pp 265-323.
8.Leidy, G.R. and R.M. Jenkins. 1977. The development of Fishery Compartments and Population Rate Coef for Use in Resevoir Ecosystem Model. USDI Fish and Wildlife Service. US Army. COE.
9.Hart, J.S. 1952. Geographic variations of some Physiological and Morphological Characters in certainFreshwater Fish. Publ. Int. Fish Res. Lab., 72, pp 79.

10.Hart, J.S. 1947. Lethal Temparature relations of certain Fish of the Toronto. Trans. R. Soc.Can., 41, pp 57-64.
11.Martin, J.H 1970. Limnol. Oceanogr. 15(5), pp 756-761.
12.Birch, L.C. 1953. Ecology 34, pp 698-711.
13.Stearn, A.E. 1949. Advan. Enzymol. 9, pp 25.
14.Mayzaud, P. and S. Pallot. 1973. Mar biol. 19, pp 307-314.
15.Haghiri, F. 1974. J. Environ. Qual., 3(2), pp 180-183.
16.Arvik, J.H. and R.L. Zimdahl. 1974. J. Environ. Qual., 3(4), pp 374-376.
17.Harrod, M.F. 1971. In: Tech. Bull. 21, pp 176-189. Trace Elements in Soils and Crops.

4-391 Effect of Terephthalic acid

Parameter	Concentration	Condition	Species
LD50	>6.4 g pr. kg body weight	acute oral dose	Rat[1]

Source: 1.Patty, F. A., Industrial Hygiene and toxicology, 2, Interscience Publishers, New York, 1976.

4-392 Effect of Tetanus

Parameter	Effect Value	Concentration and Condition	Species
Mortality	50 %	Oral dose = 1.67E-6 mg/kg, 2.53E-14mol/kg, F007 = 13.6	Mouse[1]

Source: 1.Luckey, T.D. and B. Venugopal. 1977. Metal Toxicity in Mammals, 1. Plenum Press.

4-393 Effect of Tetracycline HCl

Parameter	Effect Value	Concentration and Condition	Species
Mortality	50 %	Oral dose = 650 mg/kg, 1.35E-3mol/kg, F007 = 2.87	Mouse[1]

Source: 1.Luckey, T.D. and B. Venugopal. 1977. Metal Toxicity in Mammals, 1. Plenum Press.
Footnote; F007 = pT50 is shown = -log(potiental toxicity, 50 %.

4-394 Effect of Tetraethyllead

Parameter	Concentration	Condition	Species
TLm	0.2-2.0 mg/l	24-96 hours	Bluegill[2]
LD50	*0.035 g pr. kg body weight*	*single oral dose*	*Rat*[1]
LD50	0.015 g pr. kg body weight	parenteral dose	Rat[1]

Source: 1.Henschler, D., Toxikologisch-arbeitsmedizinische Begrundung von MAH-werte, Verlag Chemie.
2.Jones, H. R., Environmental Control in the organic and petrochemical Industries, Noyes Data Corp., 171.

4-395 Effect of Tetrahydrofuran

Parameter	Concentration	Condition	Species
Sublethal effect	225 mg/l	Inhibition of cell multiplication	Microcystis aeruginosa[1]
Sublethal effect	580 mg/l	Inhibition of cell multiplication	Pseudomonas putida[1]

Source: 1.Bringmann, G. and Kuhn, R., nwf-wasser/abwasser, H.9, 1976.

4-396 Effect of Tetrahydrofurfurylalcohol

Parameter	Concentration	Condition	Species
LD50	0.8-1.6 g pr. kg body weight	single oral dose	guinea pig[1]
LD50	1.6-3.2 g pr. kg body weight	single oral dose	Rat[1]

Source: 1.Patty, F. A., Industrial Hygiene and toxicology, 2, Interscience Publishers, New York, 1976.

4-397 Effect of Tetramethyllead

Parameter	Concentration	Condition	Species
LD50	0.08-0.106 g pr. kg body weight	single oral dose	Rat[1]
LD50	0.088 g pr. kg body weight	intravenous dose	Rat[1]
LD50	0.08 g pr. kg body weight	intraperitoneal dose	Rat[1]

Source: 1.Henschler, D., Toxikologisch-arbeitsmedizinische Begrundung von MAH-werte, Verlag Chemie.

4-398 Effect of Tetrapropylenebenzensulphonate

Parameter	Concentration	Condition	Species
LC50	56000 mg/l	9 min., 15 C	Crangon crangon[1]
LC50	42000 mg/l	48 hours, 15 C	Crangon crangon[1]
LC50	18000 mg/l	96 hours, 15 C	Crangon crangon[1]
LC50	7000 mg/l	9 min., 15 C	Gobius minutus[1]
LC50	25 mg/l	3-96 hours, 15 C	Gobius minutus[1]

Source: 1.Adema, D. M. M., Acute toxiciteitstoetsen met 1,2-dichlorethann, fenol, acrylonitrile, an alkylbenzeesulfonant, Delft, 1976.

4-399 Effect of Tetrodotoxin

Parameter	Effect Value	Concentration and Condition	Species
Mortality	50 %	Oral dose = 1E-2 mg/kg, 3.13E-8mol/kg, F007 = 7.50	Mouse[1]

Source: 1.Luckey, T.D. and B. Venugopal. 1977. Metal Toxicity in Mammals, 1. Plenum Press.
Footnote; F007 = pT50 is shown = -log(potiental toxicity, 50 %.

4-400 Effect of Th

Parameter	Effect Value	Concentration and Condition	Species
Mortality	50 %	4-chloride, intraperitonal, dose = 114 mg/kg body	Mouse[1]
Mortality	50 %	Nitrate, intraperitonal, dose = 68 mg/kg body	Rat[1]

Source: 1.Toxic Substances List 1973, U.S. Department of Health, Education and Welfare, Public Health ServiceNational Institute for Occupational Safety and Health, Rockville Maryland, USA, June 1973.

4-401 Effect of Thiophanate-methyl

Parameter	Effect Value	Concentration and Condition	Species
Growth	50 % inhibition	48 h, freshwater = 8.5 mg/l, 297 K, log phase, E7 cells/l	Chlorella pyrenoidosa[1]
Mortality	50 %	48 h, freshwater = 16 mg/l, 293 K, age = < 1 d, n = 20	Daphnia magna[1]
Mortality	50 %	48 h, freshwater = 130 mg/l, 288 K, age = 3 m, n = 10	Lebistes reticulatus[1]
Mortality	50 %	48 h, freshwater = 7.8 mg/l, 288 K, age = 3 m, n = 10	Salmo gairdneri[1]

Source: 1.Canton, J.H. 1976. Bull. Environ. Contam. Toxicol. 16(2), pp 214-218.

4-402 Effect of Thiram

Parameter	Effect Value	Concentration and Condition	Species
Mortality	50 %	Acute oral toxicity = 275-865 mg/kg, fungicide	Rat[1]

Source: 1.Edwards, C.A. 1975. Persistent Pesticides in the Environment, 2nd edition, CRC-Press.

4-403 Effect of Tl

Parameter	Effect Value	Concentration and Condition	Species
Mortality	50 %	3-acetate, oral dose = 32 mg/kg body	Mouse[1]
Mortality	50 %	Chloride, oral dose = 24 mg/kg body	Mouse[1]
Mortality	50 %	Chloride, oral dose = 24 mg/kg, 1.0E-4mol/kg, F007 = 4.00	Mouse[2]
Mortality	50 %	Nitrate, oral dose = 32.5 mg/kg body	Mouse[1]
Mortality	50 %	2-oxide, oral dose = 43.6 mg/kg body	Rat[1]
Mortality	50 %	Sulfate, oral dose = 15.8 mg/kg body	Rat[1]

Source: 1.Toxic Substances List 1973, U.S. Department of Health, Education and Welfare, Public Health ServiceNational Institute for Occupational Safety and Health, Rockville Maryland, USA, June 1973.
2.Luckey, T.D. and B. Venugopal. 1977. Metal Toxicity in Mammals, 1. Plenum Press.

4-404 Effect of Tm

Parameter	Effect Value	Concentration and Condition	Species
Mortality	50 %	Chloride, intraperitonal, dose = 485 mg/kg body	Mouse[1]
Mortality	50 %	Nitrate, intraperitonal, dose = 255 mg/kg body	Mouse[1]

Source: 1.Toxic Substances List 1973, U.S. Department of Health, Education and Welfare, Public Health ServiceNational Institute for Occupational Safety and Health, Rockville Maryland, USA, June 1973.

4-405 Effect of Toluene

Parameter	Effect Value	Concentration and Condition	Species
Mortality	50 % mortality	Water = 28 ppm, 96 h, stage 1 zoeae	Cancer magister[1]
Mortality	50 % mortality	Water = 22.80 ppm, 96 h	Carassius auratus[2]
Mortality	50 % mortality	Water = 4.3 ppm, 96 h	Crago franciscorum[3]
Mortality	50 % mortality	Water = 3.8 ppm, 24 h, estuarine	Eurytemora affinis[4]
Mortality	50 % mortality	Water = 1180 ppm, 96 h	Gambusia affinis[5]
Mortality	50 % mortality	Water = 7.3 ppm, 96 h	Marone saxatilis[3]
Mortality	50 % mortality	Water = 5.38 ppm, 24 h	Oncorhynchus gorbuscha[6]
Mortality	50 % mortality	Water = 9.5 ppm, 96 h	Palaemonetes pugio[7]

Source: 1.Caldwell, R.s., E. M. Caldarone and M.H. Mallon. 1977. D.A. Wolfe (ed.): Fate and Effects of Petroleum Hydrocarbons in Marine Ecosystems and Organisms. NY. Pergamon Press. 210-220.
2.Brenniman, G., R. Hartung and W.J. Weber. 1976. Water Res. 10, 165-169.
3.Benville, P.E. Jr. and S. Korn. 1977. Calif. Fish and Game. 63, 204-209.
4.Ott, F.S., R.P. Harris and S.C.M. OHara. 1978. Mar. Environ. Res. 1, 49-58.
5.Wallen, I.E., W.C. Greer and R. Lasater. 1957. Sew. Indust. Wastes 29, 695-711.
6.Thomas, R.E. and S.D. Rice. 1978. In: Vernberg Vernberg and Calabrese (eds.): Marine Pollution: Funcional Processes, NY. Acedemic Press.
7.Neff, J.M., Anderson, Cox, Laughlin, Rossi and Tatem. 1976. In: Sources, Effects and Sinks of Hydrocarbon in the Aquatic Environment. Wash DC. Amer. Inst. of Biological Science.

4-406 Effect of Toluene

Parameter	Concentration	Condition	Species
LC0	10 mg/l	24-96 hours, seawater, 8 C	Coho salmon[1]
LC0	60 mg/l	not indicated	Daphnia[2]
LC0	200 mg/l	not indicated	Escherichia coli[3]
LC0	120 mg/l	not indicated	Scenedesmus[3]
LC100	50 mg/l	96 hours, seawater, 8 C	Coho salmon[1]

LC50	58 mg/l	24 hours	Fish[4]
LC90	50 mg/l	24 hours, seawater, 8 C	Coho salmon[1]
Sublethal effect	105 mg/l	Inhibition of cell multiplication	Microcystis aeruginosa[3]
Sublethal effect	29 mg/l	Inhibition of cell multiplication	Pseudomonas putida[3]
TLm	24 mg/l	24-96 hours	Bluegill[5]
TLm	34-56 mg/l	24-96 hours	Fathead minnow[5]
TLm	58 mg/l	24-96 hours	Goldfish[5]
TLm	59-63 mg/l	24-96 hours	Guppie[5]
TLm	1280-1340 mg/l	24-96 hours, turbid Oklahoma water	Mosquito fish[6]

Source: 1.Morrow, J. E., EPA/660&3/73/018, 1974.
2.Meinck, F., Stooff, H. and Kohlschutter, H., Les eaux residuaires industrielles, 1970.
3.Bringmann, G. and Kuhn, R., nwf-wasser/abwasser, H.9, 1976.
4.Shell Chemie, Shell Industrie Chemicalien gids, Shell Niederland Chemie, 1975.
5.Pickering Q. H. and Henderson, C., JWPCF, 38, 1419, 1966.
6.Wallen, I. E., Greer, W. C. and Lasater, R., Sewage Ind. Wastes, 29, 695, 1957.

4-407 Effect of Toluene-2,4-isocyanate

Parameter	Concentration	Condition	Species
LD50	5.8 g pr. kg body weight	single oral dose	Rat[1]

Source: 1.Henschler, D., Toxikologisch-arbeitsmedizinische Begrundung von MAH-werte, Verlag Chemie.

4-408 Effect of Toxaphene

Parameter	Effect Value	Concentration and Condition	Species
Mortality	50 % mortality	48 h, water = 15 mg/l	Daphnia pulex[1]

Source: 1.Rankov, V. 1973. Pochv. Agrokhim. 8, 125-134.

4-409 Effect of Traffic intensity

Parameter	Effect Value	Concentration and Condition	Species
Pb content	39.87 mg Pb/kg dry weight	Traffic = 8120 vehicles/d, VA, distance to road = 12 m	Apple[1]
Pb content	72.36 mg Pb/kg dry weight	Traffic = 21040 vehicles/d, VA, distance to road = 12 m	Apple[1]
Pb content	154.61 mg Pb/kg dry weight	Traffic = 23500 vehicles/d, VA, distance to road = 6 m	Black locust[1]
Pb content	34.82 mg Pb/kg dry weight	Traffic = 1085 vehicles/d, VA, distance to road = 6 m	Black locust[1]
Pb content	74.96 mg Pb/kg dry weight	Traffic = 21040 vehicles/d, VA, distance to road = 6 m	Black locust[1]
Pb content	24.46 mg Pb/kg dry weight	Traffic = 21040 vehicles/d, VA, distance to road = 18 m	Chicory[1]
Pb content	57.42 mg Pb/kg dry weight	Traffic = 8120 vehicles/d, VA, distance to road = 6 m	Coralberry[1]
Pb content	71.98 mg Pb/kg dry weight	Traffic = 21040 vehicles/d, VA, distance to road = 12 m	Coralberry[1]
Pb content	27.35 mg Pb/kg dry weight	Traffic = 8120 vehicles/d, VA, distance to road = 6 m	Coralberry fruit[1]
Pb content	27.43 mg Pb/kg dry weight	Traffic = 21040 vehicles/d, VA, distance to road = 18 m	Crown vetch[1]
Pb content	42.32 mg Pb/kg dry weight	Traffic = 21040 vehicles/d, VA, distance to road = 12 m	Crown vetch[1]

Pb content	74.87 mg Pb/kg dry weight	Traffic = 21040 vehicles/d, VA, distance to road = 6 m	Crown vetch[1]
Pb content	39.93 mg Pb/kg dry weight	Traffic = 21040 vehicles/d, VA, distance to road = 18 m	Dandelion[1]
Pb content	17.48 mg Pb/kg dry weight	Traffic = 8120 vehicles/d, VA, distance to road = 6 m	Japanese honeysuckle[1]
Pb content	19.97 mg Pb/kg dry weight	Traffic = 1085 vehicles/d, VA, distance to road = 6 m	Japanese honeysuckle[1]
Pb content	25.00 mg Pb/kg dry weight	Traffic = 8120 vehicles/d, VA, distance to road = 12 m	Japanese honeysuckle[1]
Pb content	29.86 mg Pb/kg dry weight	Traffic = 21040 vehicles/d, VA, distance to road = 6 m	Japanese honeysuckle[1]
Pb content	32.31 mg Pb/kg dry weight	Traffic = 1085 vehicles/d, VA, distance to road = 3 m	Japanese honeysuckle[1]
Pb content	32.39 mg Pb/kg dry weight	Traffic = 8120 vehicles/d, VA, distance to road = 3 m	Japanese honeysuckle[1]
Pb content	19.97 mg Pb/kg dry weight	Traffic = 1085 vehicles/d, VA, distance to road = 12 m	Lespedeza sericea[1]
Pb content	115.10 mg Pb/kg dry weight	Traffic = 21040 vehicles/d, VA, distance to road = 6 m	Multiflora rose[1]
Pb content	32.40 mg Pb/kg dry weight	Traffic = 1085 vehicles/d, VA, distance to road = 6 m	Multiflora rose[1]
Pb content	17.49 mg Pb/kg dry weight	Traffic = 1085 vehicles/d, VA, distance to road = 6 m	Multiflora rose fruit[1]
Pb content	27.50 mg Pb/kg dry weight	Traffic = 21040 vehicles/d, VA, distance to road = 18 m	Narrow-leaf plantain[1]
Pb content	33.50 mg Pb/kg dry weight	Traffic = 8120 vehicles/d, VA, distance to road = 12 m	Panic grass[1]
Pb content	24.95 mg Pb/kg dry weight	Traffic = 8120 vehicles/d, VA, distance to road = 6 m	Periwinkle[1]
Pb content	35.73 mg Pb/kg dry weight	Traffic = 8120 vehicles/d, VA, distance to road = 3 m	Periwinkle[1]
Pb content	12.49 mg Pb/kg dry weight	Traffic = 1085 vehicles/d, VA, distance to road = 3 m	Plant[1]
Pb content	14.97 mg Pb/kg dry weight	Traffic = 1085 vehicles/d, VA, distance to road = 12 m	Plant[1]
Pb content	17.50 mg Pb/kg dry weight	Traffic = 8120 vehicles/d, VA, distance to road = 18 m	Plant[1]
Pb content	19.95 mg Pb/kg dry weight	Traffic = 23500 vehicles/d, VA, distance to road = 18 m	Plant[1]
Pb content	19.99 mg Pb/kg dry weight	Traffic = 8120 vehicles/d, VA, distance to road = 6 m	Plant[1]
Pb content	20.97 mg Pb/kg dry weight	Traffic = 8120 vehicles/d, VA, distance to road = 3 m	Plant[1]
Pb content	24.85 mg Pb/kg dry weight	Traffic = 23500 vehicles/d, VA, distance to road = 50 m	Plant[1]
Pb content	29.88 mg Pb/kg dry weight	Traffic = 23500 vehicles/d, VA, distance to road = 12 m	Plant[1]
Pb content	29.98 mg Pb/kg dry weight	Traffic = 21040 vehicles/d, VA, distance to road = 18 m	Plant[1]
Pb content	32.44 mg Pb/kg dry weight	Traffic = 21040 vehicles/d, VA, distance to road = 6 m	Plant[1]

Pb content	69.76 mg Pb/kg dry weight	Traffic = 23500 vehicles/d, VA, distance to road = 6 m	Plant[1]
Pb content	40.84 mg Pb/kg dry weight	Traffic = 8120 vehicles/d, VA, distance to road = 12 m	Rubus[1]
Pb content	109.74 mg Pb/kg dry weight	Traffic = 8120 vehicles/d, distance from road = 3 m	Soil[1]
Pb content	126.74 mg Pb/kg dry weight	Traffic = 23500 vehicles/d, distance from road = 6 m	Soil[1]
Pb content	19.91 mg Pb/kg dry weight	Traffic = 8120 vehicles/d, distance from road = 18 m	Soil[1]
Pb content	19.92 mg Pb/kg dry weight	Traffic = 1085 vehicles/d, distance from road = 12 m	Soil[1]
Pb content	22.35 mg Pb/kg dry weight	Control area in forest, 500 m from highway	Soil[1]
Pb content	22.35 mg Pb/kg dry weight	Traffic = 23500 vehicles/d, distance from road = 50 m	Soil[1]
Pb content	22.43 mg Pb/kg dry weight	Traffic = 1085 vehicles/d, distance from road = 3 m	Soil[1]
Pb content	26.44 mg Pb/kg dry weight	Traffic = 1085 vehicles/d, distance from road = 6 m	Soil[1]
Pb content	27.37 mg Pb/kg dry weight	Traffic = 8120 vehicles/d, distance from road = 12 m	Soil[1]
Pb content	27.47 mg Pb/kg dry weight	Traffic = 1085 vehicles/d, dist. from road = 18 m	Soil[1]
Pb content	32.31 mg Pb/kg dry weight	Traffic = 23500 vehicles/d, dist. from road = 18 m	Soil[1]
Pb content	42.34 mg Pb/kg dry weight	Traffic = 23500 vehicles/d, dist. from road = 12 m	Soil[1]
Pb content	47.25 mg Pb/kg dry weight	Traffic = 8120 vehicles/d, dist. from road = 6 m	Soil[1]
Pb content	47.44 mg Pb/kg dry weight	Traffic = 21040 vehicles/d, dist. to road = 18 m	Soil[1]
Pb content	79.60 mg Pb/kg dry weight	Traffic = 21040 vehicles/d, dist. from road = 12 m	Soil[1]
Pb content	87.30 mg Pb/kg dry weight	Traffic = 21040 vehicles/d, dist. to road = 6 m	Soil[1]
Pb content	24.88 mg Pb/kg dry weight	Traffic = 8120 vehicles/d, VA, dist. to road = 6 m	Sumac[1]
Pb content	29.90 mg Pb/kg dry weight	Traffic = 21040 vehicles/d, VA, dist. to road = 6 m	Viburnum berries[1]
Pb content	72.22 mg Pb/kg dry weight	Traffic = 21040 vehicles/d, VA, dist. to road = 6 m	Viburnum sp.[1]
Pb content	49.85 mg Pb/kg dry weight	Traffic = 21040 vehicles/d, VA, dist. to road = 6 m	Virginia pine[1]

Source: 1.Goldsmith, C.D. Jr., P.F. Scancon and W.R. Pieie. 1976. Bull. Environ. Contam. Toxicol. 16(1), pp 6670.

4-410 Effect of Tri-N-allylamine

Parameter	Concentration	Condition	Species
LD50	1.3 g pr. kg body weight	acute oral dose	Rat[1]
Sublethal effect	100 ppm	Changes in liver and kidney, 50 times 7 hours	Rat[1]

Source: 1.Patty, F. A., Industrial Hygiene and toxicology, 2, Interscience Publishers, New York, 1976.

4-411 Effect of Tri-n-butylamine

Parameter	Concentration	Condition	Species
TLm	20-40 mg/l	not indicated	Gudgeon[1]

Source: 1.Meinck, F., Stooff, H. and Kohlschutter, H., Les eaux residuaires industrielles, 1970.

4-412 Effect of Tri-n-propylamine

Parameter	Concentration	Condition	Species
LC0	30 mg/l	24 hours	Creek chub[1]
LC100	70 mg/l	24 hours	Creek chub[1]

Source: 1.Gilette, L. A., Miller, D. L. and Redman, H. E., Sewage Ind. Wastes, 24, 1397, 1952.

4-413 Effect of Trichlorethylene

Parameter	Concentration	Condition	Species
LC0	100 mg/l	not indicated	Daphnia[2]
LC100	600 mg/l	40 hours	Daphnia[2]
LD50	4.8-8.0 g pr. kg body weight	oral dose, 100 %	Rat[1]

Source: 1.Patty, F. A., Industrial Hygiene and toxicology, 2, Interscience Publishers, New York, 1976.
2.Meinck, F., Stooff, H. and Kohlschutter, H., Les eaux residuaires industrielles, 1970.

4-414 Effect of Trichloroacetic acid

Parameter	Concentration	Condition	Species
Sublethal effect	250 mg/l	Inhibition of cell multiplication	Microcystis aeruginosa[1]
Sublethal effect	>1000 mg/l	Inhibition of cell multiplication	Pseudomonas putida[1]

Source: 1.Bringmann, G. and Kuhn, R., nwf-wasser/abwasser, H.9, 1976.

4-415 Effect of Trichloroethylene

Parameter	Concentration	Condition	Species
Sublethal effect	160 ppm	Eye irritation	Man[1]
Sublethal effect	63 mg/l	Inhibition of cell multiplication	Microcystis aeruginosa[2]
Sublethal effect	65 mg/l	Inhibition of cell multiplication	Pseudomonas putida[2]

Source: 1.Henschler, D., Toxikologisch-arbeitsmedizinische Begrundung von MAH-werte, Verlag Chemie.
2.Bringmann, G. and Kuhn, R., nwf-wasser/abwasser, H.9, 1976.

4-416 Effect of Triethanolamine

Parameter	Concentration	Condition	Species
LC0	160 mg/l	not indicated	Colpoda[1]
LC0	2500 mg/l	not indicated	Daphnia[1]
LC0	10000 mg/l	not indicated	Psudomonas[1]
LC0	100 mg/l	not indicated	Scenedesmus[1]
Sublethal effect	47 mg/l	Inhibition of cell multiplication, neutralized	Microcystis aeruginosa[1]
Sublethal effect	>1000 mg/l	Inhibition of cell multiplication, neutralized	Pseudomonas putida[1]
LD50	9.11 g pr. kg body weight	acute oral dose	Rat[2]

Source: 2.Patty, F. A., Industrial Hygiene and toxicology, 2, Interscience Publishers, New York, 1976.
1.Bringmann, G. and Kuhn, R., nwf-wasser/abwasser, H.9, 1976.

4-417 Effect of Triethylamine

Parameter	Concentration	Condition	Species
LC0	50 mg/l	24 hours	Creek chub[1]
LC0	200 mg/l	not indicated	Daphnia[2]
LC0	1000 mg/l	not indicated	Escherichia coli[2]
LC0	1 mg/l	not indicated	Scenedesmus[2]
LC100	80 mg/l	24 hours	Creek chub[1]
LD50	0.46 g pr. kg body weight	acute oral dose	Rat[3]

Source: 3.Patty, F. A., Industrial Hygiene and toxicology, 2, Interscience Publishers, New York, 1976.

1.Gilette, L. A., Miller, D. L. and Redman, H. E., Sewage Ind. Wastes, 24, 1397, 1952.
2.Meinck, F., Stooff, H. and Kohlschutter, H., Les eaux residuaires industrielles, 1970.

4-418 Effect of Triethyleneglycol

Parameter	Concentration	Condition	Species
Sublethal effect	3600 mg/l	Inhibition of cell multiplication, neutralized	Microcystis aeruginosa[2]
Sublethal effect	320 mg/l	Inhibition of cell multiplication, neutralized	Pseudomonas putida[2]
LD50	14.6 g pr. kg body weight	single oral dose	guinea pig[1]
LD50	22.1 g pr. kg body weight	single oral dose	Mouse[1]
LD50	10 g pr. kg body weight	single oral dose	Rabbit[1]
LD50	22.1 g pr. kg body weight	single oral dose	Rat[1]

Source: 1.Patty, F. A., Industrial Hygiene and toxicology, 2, Interscience Publishers, New York, 1976.
2.Bringmann, G. and Kuhn, R., nwf-wasser/abwasser, H.9, 1976.

4-419 Effect of Triethyleneglycolmonoethylether

Parameter	Concentration	Condition	Species
LC50	>5000 mg/l	24 hours	Goldfish[1]

Source: 1.Shell Chemie, Shell Industrie Chemicalien gids, Shell Niederland Chemie, 1975.

4-420 Effect of Trifluralin

Parameter	Effect Value	Concentration and Condition	Species
Mortality	50 % mortality	48 h, water = 240 mg/l	Daphnia pulex[1]
Mortality	50 %	417 (380-447) ppb, 96 h, 14 g, 293 K, pH = 8.2, alk = 80 ppm, 95 % CL	Ictalurus punctatus[2]

Source: 1.Rankov, V. 1973. Pochv. Agrokhim. 8, 125-134.
2.McCorle, F.M., J.E. Chambers and J.D. Yarbrough. 1977. Bull. Environ. Contam. Toxicol. 18(3), pp 267270.

4-421 Effect of Trinitrotoluene

Parameter	Effect Value	Concentration and Condition	Species
Growth	0 % inhibition	Alpha TNT = 15 mg/l, inhibition over this conc	Microcystis aeruginosa[1]
Growth	0 % inhibition	Alpha TNT = 5 mg/l, inhibition over this conc	Selenastrum capricornutum[1]
Growth	0 % inhibition	TNT = 9 mg/l, inhibition over this concentration	Selenastrum capricornutum[1]
Mortality	100 %	TNT colored complex = 50 mg/l	Microcystis aeruginosa[1]
Mortality	50 %	Alpha TNT = 2.58 mg/l	Pimephales promelas[1]
Mortality	50 %	TNT = 1.6 mg/l, pink water, deg product of TNT	Pimephales promelas[1]

Source: 1.Smock, L.A. 1976. Water Res., 10(6), pp 537-543.

4-422 Effect of Triphenylene

Parameter	Effect Value	Concentration and Condition	Species
Growth rate	-0.2 % more than control	Water = 0.3 ppm, 96 h, PAH, sporelings	Antithamnion plumula[1]

Source: 1.Boney, A.D. 1974. Mar. Pollut. Bull. 5, 185-186.

4-423 Effect of Tween 20

Parameter	Effect Value	Concentration and Condition	Species
Oxygen uptake	18.1 microlitre O2/(hr*mg protein)	Conc = 10000 ppm, pH = 1.9, 308 K, control = 71.7, sterile cont = 5.6	Thiobacillus ferrooxidans[1]
Oxygen uptake	18.2 microlitre O2/(hr*mg protein)	Conc = 1000 ppm, pH = 1.9, 308 K, control = 71.7, sterile cont = 5.6	Thiobacillus ferrooxidans[1]
Oxygen uptake	30.1 microlitre O2/(hr*mg protein)	Conc = 100 ppm, pH = 1.9, 308 K, control = 71.7, sterile cont = 5.6	Thiobacillus ferrooxidans[1]
Oxygen uptake	35.6 microlitre O2/(hr*mg protein)	Conc = 10 ppm, pH = 1.9, 308 K, control = 71.7, sterile control = 5.6	Thiobacillus ferrooxidans[1]

Source: 1.Torma, A.E. 1976. App. Sci. Publ., pp 937-945.

4-424 Effect of Tween 40

Parameter	Effect Value	Concentration and Condition	Species
Oxygen uptake	13.5 microlitre O2/(hr*mg protein)	Conc = 1000 ppm, pH = 1.9, 308 K, control = 71.7, sterile control = 5.6	Thiobacillus ferrooxidans[1]
Oxygen uptake	14.4 microlitre O2/(hr*mg protein)	Conc = 10000 ppm, pH = 1.9, 308 K, control = 71.7, sterile control = 5.6	Thiobacillus ferrooxidans[1]
Oxygen uptake	35.0 microlitre O2/(hr*mg protein)	Conc = 100 ppm, pH = 1.9, 308 K, control = 71.7, sterile control = 5.6	Thiobacillus ferrooxidans[1]
Oxygen uptake	47.0 microlitre O2/(hr*mg protein)	Conc = 10 ppm, pH = 1.9, 308 K, control = 71.7, sterile control = 5.6	Thiobacillus ferrooxidans[1]

Source: 1.Torma, A.E. 1976. App. Sci. Publ., pp 937-945.

4-425 Effect of Tween 60

Parameter	Effect Value	Concentration and Condition	Species
Oxygen uptake	14.9 microlitre O2/(hr*mg protein)	Conc = 10000 ppm, pH = 1.9, 308 K, control = 71.7, sterile control = 5.6	Thiobacillus ferrooxidans[1]
Oxygen uptake	15.3 microlitre O2/(hr*mg protein)	Conc = 1000 ppm, pH = 1.9, 308 K, control = 71.7, sterile control = 5.6	Thiobacillus ferrooxidans[1]
Oxygen uptake	36.9 microlitre O2/(hr*mg protein)	Conc = 100 ppm, pH = 1.9, 308 K, control = 71.7, sterile control = 5.6	Thiobacillus ferrooxidans[1]
Oxygen uptake	39.5 microlitre O2/(hr*mg protein)	Conc = 10 ppm, pH = 1.9, 308 K, control = 71.7, sterile control = 5.6	Thiobacillus ferrooxidans[1]

Source: 1.Torma, A.E. 1976. App. Sci. Publ., pp 937-945.

4-426 Effect of Tween 80

Parameter	Effect Value	Concentration and Condition	Species
Oxygen uptake	17.4 microlitre O2/(hr*mg protein)	Conc = 10000 ppm, pH = 1.9, 308 K, control = 71.7, sterile control = 5.6	Thiobacillus ferrooxidans[1]
Oxygen uptake	18.4 microlitre O2/(hr*mg protein)	Conc = 1000 ppm, pH = 1.9, 308 K, control = 71.7, sterile control = 5.6	Thiobacillus ferrooxidans[1]
Oxygen uptake	36.9 microlitre O2/(hr*mg protein)	Conc = 100 ppm, pH = 1.9, 308 K, control = 71.7, sterile control = 5.6	Thiobacillus ferrooxidans[1]

Oxygen uptake	45.1 microlitre O2/(hr*mg protein)	Conc = 10 ppm, pH = 1.9, 308 K, control = 71.7, sterile control = 5.6	Thiobacillus ferrooxidans[1]

Source: 1.Torma, A.E. 1976. App. Sci. Publ., pp 937-945.

4-427 Effect of U

Parameter	Effect Value	Concentration and Condition	Species
Mortality	50 % mortality	UO2(NO3)2*6H2O, water = E-4 M U, toxic level given, microorganisms	Azotobacter sp.[1]
Mortality	50 %	4-oxide, intraperitonal, dose = 6 mg/kg body	Mouse[2]
Mortality	50 %	Acetate, intraperitonal, dose = 400 mg/kg body	Mouse[2]
Mortality	50 %	Nitrate, intraperitonal, dose = 135 mg/kg body	Rat[2]

Source: 1.Zhakharyan, S.V. 1970. Symp. Biol. Hung. 11, 391-395.
2.Toxic Substances List 1973, U.S. Department of Health, Education and Welfare, Public Health ServiceNational Institute for Occupational Safety and Health, Rockville Maryland, USA, June 1973.

4-428 Effect of Ultraviolet light

Parameter	Effect Value	Concentration and Condition	Species
Mortality	63 % mortality	Ultraviolet light = 50000 ergs/mm**2, microorganism	Bodo marina[1]
Mortality	63 % mortality	Ultraviolet light = 500 ergs/mm**2, E. Coli K12, microorganism	Escherichia coli[2]
Mortality	63 % mortality	Ultraviolet light = 6000 ergs/mm**2, microorganism	Micrococcus radiodurans[3]
Mortality	63 % mortality	Ultraviolet light = 800 ergs/mm**2, microorganism	Saccharomyces cerevisiae[4]
Mortality	63 % mortality	Ultraviolet light = 1350 ergs/mm**2, microorganism	Schizosaccharomyces pombe[5]
Mortality	63 % mortality	Ultraviolet light = 1300 ergs/mm**2, microorganism	Ustiago maydis[6]

Source: 1.Prozorova, M.I. 1967. Izv. sib. Otdel. Akad. Nauk SSSR 5, 32-35.
2.Pantera, H. 1972. Pam. Pulawski 51, 69-76.
3.Bakalivanov, D. 1971. Soils and Fert. 36, 607.
4.Tsevetkova, S.D. 1966. Les. Khoz. 19, 25-26.
5.Spirama Raju, K. and G. Rangaswami. 1971. Indian J. Microbiol. 11, 25-29.
6.Bobyshev, V.G. and G.Y. Lapchenkov. 1966. Khimiya selKhoz 4, 30-31.

4-429 Effect of Undecylemic acid

Parameter	Concentration	Condition	Species
LD50	2.5 g pr. kg body weight	oral dose	Rat[1]

Source: 1.Patty, F. A., Industrial Hygiene and toxicology, 2, Interscience Publishers, New York, 1976.

4-430 Effect of V

Parameter	Effect Value	Concentration and Condition	Species
Mortality	50 % mortality	VO(SO4), water = E-4 M V, toxic level given, microorganisms	Azotobacter sp.[1]
Mortality	100 % mortality	Injection of Vanadyl sulfate equal to 35-45 mg V/kg ww	Guinea pig[2]
Mortality	100 % mortality	Injection of ammonium metavanadate equal to 1 - 2.0 mg V/kg ww	Guinea pig[2]
Mortality	100 % mortality	Injection of colloidal vanadium pentoxide equal to 20-28 mg V/kg ww	Guinea pig[2]
Mortality	100 % mortality	Injection of sodium hexavanadate equal to 40-50 mg V/kg ww	Guinea pig[2]
Mortality	100 % mortality	Injection of sodium orthovanadate equal to 1-2 mg V/kg ww	Guinea pig[2]

Mortality	100 % mortality	Injection of sodium pyrovanadate equal to 1-2 mg V/kg ww	Guinea pig[2]
Mortality	100 % mortality	Injection of sodium tetravanadate equal to 18-20 mg V/kg ww	Guinea pig[2]
Mortality	100 % mortality	Injection of sodium vanadate equal to 30-40 mg V/kg ww	Guinea pig[2]
Mortality	100 % mortality	Injection of Vanadyl sulfate equal to 125-150 mg V/kg ww	Mouse[2]
Mortality	100 % mortality	Injection of ammonium metavanadate equal to 25-30 mg V/kg ww	Mouse[2]
Mortality	100 % mortality	Injection of colloidal vanadium pentoxide equal to 87.5-117.5 mg V/kg ww	Mouse[2]
Mortality	100 % mortality	Injection of sodium hexavanadate equal to 100-150 mg V/kg ww	Mouse[2]
Mortality	100 % mortality	Injection of sodium orthovanadate equal to 50-100 mg V/kg ww	Mouse[2]
Mortality	100 % mortality	Injection of sodium pyrovanadate equal to 50-100 mg V/kg ww	Mouse[2]
Mortality	100 % mortality	Injection of sodium tetravanadate equal to 25-50 mg V/kg ww	Mouse[2]
Mortality	100 % mortality	Injection of sodium vanadate equal to 100-150 mg V/kg ww	Mouse[2]
Mortality	50 %	VSO4, exposure time = 4 d, conc in water = 4800 mg/l, limnol.	Pimephales promelas[3]
Mortality	100 % mortality	Injection of Vanadyl sulfate equal to 18-20 mg V/kg ww	Rabbit[2]
Mortality	100 % mortality	Injection of ammonium metavanadate equal to 1.5-2.0 mg V/kg wwRabbit[2]	
Mortality	100 % mortality	Injection of colloidal vanadium pentoxide equal to 1-2 mg V/kg ww	Rabbit[2]
Mortality	100 % mortality	Injection of sodium hexavanadate equal to 30-40 mg V/kg ww	Rabbit[2]
Mortality	100 % mortality	Injection of sodium orthovanadate equal to 2-3 mg V/kg ww	Rabbit[2]
Mortality	100 % mortality	Injection of sodium pyrovanadate equal to 3-4 mg V/kg ww	Rabbit[2]
Mortality	100 % mortality	Injection of sodium tetravanadate equal to 6-8 mg/kg ww	Rabbit[2]
Mortality	100 % mortality	Injection of Vanadyl sulfate equal to 158-190 mg V/kg ww	Rat[2]
Mortality	100 % mortality	Injection of ammonium metavanadate equal to 20-30 mg V/kg wwww	Rat[2]
Mortality	100 % mortality	Injection of sodium hexavanadate equal to 40-50 mg V/kg ww	Rat[2]
Mortality	100 % mortality	Injection of sodium orthovanadate equal to 50-60 mg V/kg ww	Rat[2]
Mortality	100 % mortality	Injection of sodium pyrovanadate equal to 40-50 mg V/kg ww	Rat[2]
Mortality	100 % mortality	Injection of sodium tetravanadate equal to 30-40 mg V/kg ww	Rat[2]

Mortality	100 % mortality	Injection of sodium vanadate equal to 10-20 mg V/kg ww	Rat[2]
Mortality	50 %	2-chloride, oral dose = 540 mg/kg body	Rat[4]
Mortality	50 %	3-chloride, oral dose = 350 mg/kg body	Rat[4]
Mortality	50 %	4-chloride, oral dose = 160 mg/kg body	Rat[4]
Nitrification	37 % inhibation	Soil = 5 umoles/g of soil, 10 d incubation, 3 soils, lab	Soil organisms[5]

Source: 1.Zhakharyan, S.V. 1970. Symp. Biol. Hung. 11, 391-395.
2.Faulkner-Hudson, T.G. 1964. Vanadium Toxicology and Biological Significance. NY. Elsevier.
3.McKee, J.R. and H.W. Wolf. 1963. Water Quality Criteria, Publication no. 3A, Calif. State Water Res.Control Board 2nd Edition.
4.Toxic Substances List 1973, U.S. Department of Health, Education and Welfare, Public Health ServiceNational Institute for Occupational Safety and Health, Rockville Maryland, USA, June 1973.
5.Liang, C.N. and M.A. Tabatabai. 1978. Environ. Pollut. 12, 141-147.

4-431 Effect of Vinyl-2-pyridine

Parameter	Concentration	Condition	Species
LD50	0.4-0.8 g/kg body weight	acute oral dose	Mouse[1]
LD50	0.1-0.2 g/kg body weight	acute oral dose	Rat[1]

Source: 1.Patty, F. A., Industrial Hygiene and toxicology, 2, Interscience Publishers, New York, 1976.

4-432 Effect of Vinylacetate

Parameter	Concentration	Condition	Species
TLm	18 mg/l	24-96 hours	Bluegill[1]
TLm	19-39 mg/l	24-96 hours	Fathead minnow[1]
TLm	42 mg/l	24-96 hours	Goldfish[1]
TLm	31 mg/l	24-96 hours	Guppie[1]
LD50	2.9 g pr. kg body weight	acute oral dose	Rat[2]

Source: 2.Valland, A. and Damel, R., L'hygiene et la securite dans la grande industrie chimique, Inst. nat. dereserche et de securite, 1971.
1.Pickering Q. H. and Henderson, C., JWPCF, 38, 1419, 1966.

4-433 Effect of Vinylchloride

Parameter	Concentration	Condition	Species
Sublethal effect	200 mg/l	Liver degeneration, 7 hours pr. day for 6 months	Rabbit[1]

Source: 1.Henschler, D., Toxikologisch-arbeitsmedizinische Begrundung von MAH-werte, Verlag Chemie.

4-434 Effect of W

Parameter	Effect Value	Concentration and Condition	Species
Mortality	50 %	Metallic, intraperitonal, dose = 5000 mg/kg body	Rat[1]
Mortality	50 %	Na-wolframate, oral dose = 1190 mg/kg body	Rat[1]
Nitrification	33 % inhibation	Soil = 5 umoles/g of soil, 10 d incubation, 3 soils, lab	Soil organisms[2]

Source: 1.Toxic Substances List 1973, U.S. Department of Health, Education and Welfare, Public Health ServiceNational Institute for Occupational Safety and Health, Rockville Maryland, USA, June 1973.
2.Liang, C.N. and M.A. Tabatabai. 1978. Environ. Pollut. 12, 141-147.

4-435 Effect of Xylenes

Parameter	Effect Value	Concentration and Condition	Species
Mortality	50 % mortality	Water = 16.94 ppm, 96 h	Carassius auratus[1]
Mortality	50 % mortality	Water = 7.4 ppm, 96 h	Palaemonetes pugio[2]

Source: 1.Brenniman, G., R. Hartung and W.J. Weber. 1976. Water Res. 10, 165-169.
2.Neff, J.M., Anderson, Cox, Laughlin, Rossi and Tatem. 1976. In: Sources, Effects and Sinks of Hydrocarbon in the Aquatic Environment. Wash DC. Amer. Inst. of Biological Science.

4-436 Effect of Y

Parameter	Effect Value	Concentration and Condition	Species
Mortality	50 %	3-chloride, intraperitonal, dose = 450 mg/kg body	Rat[1]
Mortality	50 %	3-nitrate, intraperitonal, dose = 350 mg/kg body	Rat[1]
Mortality	50 %	3-oxide, intraperitonal, dose = 500 mg/kg body	Rat[1]

Source: 1.Toxic Substances List 1973, U.S. Department of Health, Education and Welfare, Public Health ServiceNational Institute for Occupational Safety and Health, Rockville Maryland, USA, June 1973.

4-437 Effect of Yb

Parameter	Effect Value	Concentration and Condition	Species
Mortality	50 %	2-nitrate, oral dose = 3100 mg/kg body	Rat[1]

Source: 1.Toxic Substances List 1973, U.S. Department of Health, Education and Welfare, Public Health ServiceNational Institute for Occupational Safety and Health, Rockville Maryland, USA, June 1973.

4-438 Effect of Zn

Parameter	Effect Value	Concentration and Condition	Species
Concentration of Cd	1.35 mg/kg dry weight	Cd = 5 ug/l, S = 2.8 %, 7 d, Zn = 1000 ug/l, n=10, 283 K	Mulinia lateralis, soft part[1]
Concentration of Cd	2.30 mg/kg dry weight	Cd = 5 ug/l, S = 2.8 %, 7 d, Zn = 500 ug/l, n=10, 283 K	Mulinia lateralis, soft part[1]
Concentration of Cd	3.38 mg/kg dry weight	Cd = 5 ug/l, S = 2.8 %, 7 d, Zn = 5.0 ug/l, n=10, 283 K	Mulinia lateralis, soft part[1]
Concentration of Cd	4.46 mg/kg dry weight	Cd = 5 ug/l, S = 2.8 %, 7 d, Zn = 0.0 ug/l, n=10, 283 K	Mulinia lateralis, soft part[1]
Concentration of Cd	0.40 mg/kg dry weight	Cd = 5 ug/l, S = 2.8 %, 7 d, Zn = 1000 ug/l, n=10, 283 K	Mytilus edulis, soft part[1]
Concentration of Cd	2.16 mg/kg dry weight	Cd = 5 ug/l, S = 2.8 %, 7 d, Zn = 500 ug/l, n=10, 283 K	Mytilus edulis, soft part[1]
Concentration of Cd	8.51 mg/kg dry weight	Cd = 5 ug/l, S = 2.8 %, 7 d, Zn = 5.0 ug/l, n=10, 283 K	Mytilus edulis, soft part[1]
Concentration of Cd	9.60 mg/kg dry weight	Cd = 5 ug/l, S = 2.8 %, 7 d, Zn = 0.0 ug/l, n=10, 283 K	Mytilus edulis, soft part[1]
Growth	0.001 mg/l gives deficiency symptoms	Tree, 100 d in greenhouse, conc in culture solution	Agathis australis[2]
Growth	0.050 mg/l gives normal growth	Tree, 100 d in greenhouse, conc in culture solution	Agathis australis[2]
Growth	0.500 mg/l gives toxic symptoms	Tree, 100 d in greenhouse, conc in culture solution	Agathis australis[2]
Growth	100 % of control	Zn = 0 ppm, pH = 6.23, 288.5 K, after 96 hr	Chlorella vulgaris[3]
Growth	39.86 % of control	Zn = 1.0 ppm, pH = 6.17, 288.5 K, after 96 hr, control-pH = 6.23	Chlorella vulgaris[3]

Growth	43.87 % of control	Zn = 1.8 ppm, pH = 6.14, 288.5 K, after 96 hr, control-pH = 6.23	Chlorella vulgaris[3]
Growth	51.28 % of control	Zn = 2.4 ppm, pH = 6.12, 288.5 K, after 96 hr, control-pH = 6.23	Chlorella vulgaris[3]
Growth	58.61 % of control	Zn = 3.2 ppm, pH = 6.07, 288.5 K, after 96 hr, control-pH = 6.23	Chlorella vulgaris[3]
Growth	67.96 % of control	Zn = 10.0 ppm, pH = 5.83, 288.5 K, after 96 hr, control-pH = 6.23	Chlorella vulgaris[3]
Hemorrhaging	100 % effect	Zn = 0.2 - 6.8 mg/l	Phoxinus phoxinus[4]
Larval development	125 ug/l gives decrease in growth	Zn = 100 - 150 ug/l	Crassostrea gigas[5]
Larval development	200 ug/l gives no growth	Zn = 200 ug/l	Crassostrea gigas[5]
Larval development	50 ug/l gives normal growth	Zn = 50 ug/l	Crassostrea gigas[5]
Mortality	50 % mortality	Zn = 14.3 mg/l, 120 h, ZnSO4, hardness = 290 mg/l, pH = 7.2-7.9, 285 K	Abramis brana[6]
Mortality	50 % mortality	Zn = 14.0 mg/l, hardness = 50 ppm, pH = 7.6, 291 K, 96 h	Amnicola sp., adult[7]
Mortality	50 % mortality	Zn = 16.8 mg/l, hardness = 50 ppm, pH = 7.6, 291 K, 24 h	Amnicola sp., adult[7]
Mortality	50 % mortality	Zn = 20.2 mg/l, hardness = 50 ppm, pH = 7.6, 291 K, 96 h	Amnicola sp., eggs[7]
Mortality	50 % mortality	Zn = 28.1 mg/l, hardness = 50 ppm, pH = 7.6, 24 h, 291 K	Amnicola sp., eggs[7]
Mortality	50 % mortality	Zn = 180 mg/l, soft water, hardness = 20 ppm, ZnSO4, 293 K	Argia sp.[8]
Mortality	50 % mortality	Zn = 38.5 mg/l, 298 K, soft water, ZnSO4, hardness = 20 ppm	Asellus communis[8]
Mortality	50 % mortality	Zn = 56 mg/l, hard water, hardness = 100 ppm, 291 K, ZnSO4	Asellus communis[8]
Mortality	50 % mortality	ZnSO4*&H2O, water = 1 mg/l, toxic level given, microorganisms	Aspergillus niger[9]
Mortality	50 % mortality	Zn = 52 mg/l, 96 h, ZnCl2	Austropotamobius pallipes[10]
Mortality	0 %	Toxicological limit = 1 - 3 mg/l, ZnSO4	Biological purification[11]
Mortality	50 % mortality	Zn = 62.6 mg/l, 24 h, hardness = 50 ppm, pH = 7.6, 291 K	Caddis fly, larvae[7]
Mortality	50 % mortality	Zn = 21.5 mg/l, hardness = 50 ppm, pH = 7.6, 96 h, 291 K	Chironomus sp.[7]
Mortality	0 %	Toxicological limit = 19.4 mg/l, ZnSO4	Crustacean[12]
Mortality	50 % mortality	Zn = 5.5 mg/l, subalpine lake water, ZnSO4, 283+-0.5 K	Cyclops abyssorum[13]
Mortality	50 % mortality	Zn = 0.040 mg/l, subalpine lake water, ZnSO4, 283+-0.5 K	Daphnia hyalina[13]
Mortality	50 %	ZnCl2, exposure time = 2 d, conc in water = 100 mg/l, limnol.	Daphnia magna[14]
Mortality	50 % mortality	Zn = 10.8 mg/l, 48 h, hardness = 5.8 mg/l, soft water, ZnSO4, 293-295 K	Daphnia magna[15]
Mortality	50 % mortality	Zn = 11.6 mg/l, 24 h, soft water, hardness = 5.8 mg/l, ZnCl2, 293-295 K	Daphnia magna[15]

Mortality	50 % mortality	Zn = 14.9 mg/l, 24 h, hardness = 5.8 mg/l, soft water, ZnSO4, 293-295 K	Daphnia magna[15]
Mortality	50 % mortality	Zn = 7.4 mg/l, 96 h, hardness = 5.8 mg/l, soft water, ZnSO4, 293-295 K	Daphnia magna[15]
Mortality	50 % mortality	Zn = 7.7 mg/l, 96 h, soft water, hardness = 5.8 mg/l, ZnCl2, 293-295 K	Daphnia magna[15]
Mortality	50 % mortality	Zn = 8.1 mg/l, 72 h, hardness = 5.8 mg/l, soft water, ZnSO4, 293-295 K	Daphnia magna[15]
Mortality	50 % mortality	Zn = 8.7 mg/l, 72 h, soft water, hardness = 5.8 mg/l, ZnCl2, 293-295 K	Daphnia magna[15]
Mortality	50 % mortality	Zn = 9.8 mg/l, 48 h, soft water, hardness = 5.8 mg/l, ZnCl2, 293-295 K	Daphnia magna[15]
Mortality	50 % mortality	Zn=0.100 mg/l, with food, L Superior water, 291+-1 K, 48 h, hard=45 ppm	Daphnia magna[16]
Mortality	50 % mortality	Zn = 7.4 mg/l, 96 h, hardness - 50 pmm, 298 K, tubellaria	Dugesia tigrina[17]
Mortality	50 % mortality	Zn = 0.5 mg/l, subalpine lake water, ZnSO4, 283+-0.5 K	Eudiaptomus padanus[13]
Mortality	0 %	Toxicological limit = 1 - 5 mg/l, ZnSO4, trout	Fish[12]
Mortality	50 % mortality	Zn = 10.2 mg/l, 24 h, hardness - 50 ppm, pH=7.6, 290 K	Gammarus sp.[7]
Mortality	50 % mortality	Zn = 8.1 mg/l, 96 h, hardness = 50 ppm, pH = 7.6, 290 K	Gammarus sp.[7]
Mortality	50 % mortality	Zn = 13.5 mg/l, aerated lake water, ZnSO4, 297+-2 K, 48 h	Goniobasis livescens[18]
Mortality	50 % mortality	Zn = 18.0 mg/l, aerated lake water, ZnSO4, 297+-2 K, 24 h	Goniobasis livescens[18]
Mortality	50 % mortality	Zn = 0.87 mg/l, 96 h, soft water, hardness = 20 mg/l, ZnSO4, 285.8 K	Helisoma companulata, adult[19]
Mortality	50 % mortality	Zn = 0.96 mg/l, 72 h, soft water, hardness = 20 mg/l, ZnSO4, 285.8 K	Helisoma companulata, adult[19]
Mortality	50 % mortality	Zn = 1.27 mg/l, 72 h, Hard water, hardness = 100 mg/l, ZnSO4, 296 K	Helisoma companulata, adult[19]
Mortality	50 % mortality	Zn = 1.27 mg/l, 96 h, Hard water, hardness = 100 mg/l, ZnSO4, 296 K	Helisoma companulata, adult[19]
Mortality	50 % mortality	Zn = 1.27 mg/l, 96 h, soft water, hardness = 20 mg/l, ZnSO4, 296 K	Helisoma companulata, adult[19]
Mortality	50 % mortality	Zn = 1.48 mg/l, 72 h, soft water, hardness = 20 mg/l, ZnSO4, 296 K	Helisoma companulata, adult[19]
Mortality	50 % mortality	Zn = 1.88 mg/l, 48 h, soft water, hardness = 20 mg/l, ZnSO4, 296 K	Helisoma companulata, adult[19]
Mortality	50 % mortality	Zn = 11.07 mg/l, 24 h, hard water, hardness = 100 mg/l, ZnSO4, 285.8 K	Helisoma companulata, adult[19]
Mortality	50 % mortality	Zn = 11.07 mg/l, 24 h, soft water, hardness = 20 mg/l, ZnSO4, 285.8 K	Helisoma companulata, adult[19]
Mortality	50 % mortality	Zn = 11.07 mg/l, 48 h, hard water, hardness = 100 mg/l, ZnSO4, 285.8 K	Helisoma companulata, adult[19]
Mortality	50 % mortality	Zn = 12.66 mg/l, 24 h, soft water, hardness = 20 mg/l, ZnSO4, 296 K	Helisoma companulata, adult[19]
Mortality	50 % mortality	Zn = 3.03 mg/l, 72 h, hard water, hardness = 100 mg/l, ZnSO4, 285.8 K	Helisoma companulata, adult[19]
Mortality	50 % mortality	Zn = 3.03 mg/l, 96 h, hard water, hardness = 100 mg/l, ZnSO4, 285.8 K	Helisoma companulata, adult[19]

Mortality	50 % mortality	Zn = 5.29 mg/l, 24 h, Hard water, hardness = 100 mg/l, ZnSO4, 296 K	Helisoma companulata, adult[19]
Mortality	50 % mortality	Zn = 5.29 mg/l, 48 h, Hard water, hardness = 100 mg/l, ZnSO4, 296 K	Helisoma companulata, adult[19]
Mortality	50 % mortality	Zn = 8.70 mg/l, 48 h, soft water, hardness = 20 mg/l, ZnSO4, 285.8 K	Helisoma companulata, adult[19]
Mortality	50 % mortality	Zn = 1.9-3.6 mg/l, 96 h, synthetic water, Ca=11 and Mg=4 ppm, 303 K	Lepomis macrochirus[20]
Mortality	50 % mortality	Zn = 10.1-12.5 mg/l, 96 h, synthetic water, Ca=8700+Mg=3000 ppm, 291 K	Lepomis macrochirus[20]
Mortality	50 % mortality	Zn = 10.2-12.3 mg/l, 96 h, synthetic water, Ca=8700+Mg=3000 ppm, 303 K	Lepomis macrochirus[20]
Mortality	50 % mortality	Zn = 2.9-3.8 mg/l, 96 h, synthetic water, Ca = 11 and Mg = 4 ppm, 291 K	Lepomis macrochirus[20]
Mortality	50 % mortality	Zn = 4.15 mg/l, aerated lake water, ZnSO4, 297+-2 K, 48 h	Lymnaea emarginata[18]
Mortality	50 % mortality	Zn = 4.4 mg/l, aerated lake water, ZnSO4, 297+-2 K, 24 h	Lymnaea emarginata[18]
Mortality	0 %	Toxicological limit = 1.9 - 2.2 mg/kg, ZnSO4	Mammalia[12]
Mortality	50 % mortality	Zn = 18.4 mg/l, hardness = 50 mg/l, pH=7.6, 96 h, 290 K	Nais sp.[7]
Mortality	50 % mortality	Zn = 21.1 mg/l, hardness = 50 mg/l, pH=7.6, 24 h, 290 K	Nais sp.[7]
Mortality	50 %	Zn = 2.5 mg/l, 5d, hard water	Noemacheilus barbatulus[21]
Mortality	50 % mortality	Zn = 2.5 mg/l, 120 h, ZnSO4, hardness=240-290 mg/l, pH=7.2-7.75	Noemacheilus barbatulus[22]
Mortality	50 % mortality	Zn = 58 mg/l, 289+-1 K, 96 h, ZnCl2	Orconectes limnosus[10]
Mortality	50 % mortality	Zn = 1.1 mg/l, 96 H, ZnSO4, total hardness = 10 ppm, 288 K	Paratya tasmaniensis[23]
Mortality	50 % mortality	Zn = 14.5 mg/l, 24 h, ZnSO4, total hardness = 10 ppm, 288 K	Paratya tasmaniensis[23]
Mortality	50 % mortality	Zn = 3.3 mg/l, 72 h, ZnSO4, total hardness = 10 ppm, 288 K	Paratya tasmaniensis[23]
Mortality	50 % mortality	Zn = 8.3 mg/l, 48 h, ZnSO4, total hardness = 10 ppm, 288 K	Paratya tasmaniensis[23]
Mortality	50 % mortality	Zn = 16.0 mg/l, 120 h, ZnSO4, hardness = 290 mg/l, pH = 7.2-7.9, 285 K	Perca fluviatilis[6]
Mortality	50 % mortality	Zn = 0.9 mg/l, 48 h, soft water, hardness = 25 mg/l, 293+-2 K, ZnSO4	Philodina acuticornis[24]
Mortality	50 % mortality	Zn = 1.2 mg/l, 24 h, ZnSO4, soft water, hardness = 25 mg/l, 293+-2 K	Philodina acuticornis[24]
Mortality	50 % mortality	Zn = 1.2 mg/l, 96 h, soft water, hardness = 25 mg/l, 293+-2 K, ZnSO4	Philodina acuticornis[24]
Mortality	50 % mortality	Zn = 1.3 mg/l, 96 h, soft water, hardness = 25 mg/l, 293+-2 K, ZnCl2	Philodina acuticornis[24]
Mortality	50 % mortality	Zn = 2.0 mg/l, 48 h, hardness = 81 mg/l, 293+-1 K, ZnSO4	Philodina acuticornis[24]
Mortality	50 % mortality	Zn = 2.4 mg/l, 48 h, soft water, hardness = 25 mg/l, 293+-2 K, ZnCl2	Philodina acuticornis[24]
Mortality	50 % mortality	Zn = 2.6 mg/l, 24 h, hardness = 81 mg/l, ZnSO4, 293+-1 K, ZnSO4	Philodina acuticornis[24]
Mortality	50 % mortality	Zn = 3.7 mg/l, 24 h, soft water, hardness = 25 mg/l, 293+-2 K	Philodina acuticornis[24]

Mortality	50 %	ZnSO4 = 0.9 +- 2.3 ppm, 48 h, 25 ppm CaCO3, 293 K, pH = 7.4-7.9	Philodina sp.[25]
Mortality	50 %	ZnSO4 = 1.1 +- 2.0 ppm, 24 h, 25 ppm CaCO3, 293 K, pH = 7.4-7.9	Philodina sp.[25]
Mortality	50 %	ZnSO4 = 1.3 +- 1.8 ppm, 96 h, 25 ppm CaCO3, 293 K, pH = 7.4-7.9	Philodina sp.[25]
Mortality	50 %	Zn(NO3)2, exposure time = 4 d, conc in water = 3200 mg/l, limnol.	Phoxinus phoxinus[4]
Mortality	50 % mortality	Zn = 0.62-0.78 mg/l, 303 K, hardness = 40 mg/l, ZnCl2, soft water, 96 h	Physa heterostropha, adult[7]
Mortality	50 % mortality	Zn = 0.79-1.27 mg/l, 293 K, hardness = 40 mg/l, ZnCl2, soft water, 96 h	Physa heterostropha, adult[7]
Mortality	50 % mortality	Zn = 2.36-6.36 mg/l, 303 K, hardness = 160 mg/l, ZnCl2, 96 h, hard water	Physa heterostropha, adult[7]
Mortality	50 % mortality	Zn = 2.66-5.57 mg/l, 293 K, hardness = 160 mg/l, ZnCl2, 96 h, hard water	Physa heterostropha, adult[7]
Mortality	50 % mortality	Zn = 2.71 mg/l, 24 h, soft water, hardness = 20 mg/l, 294 K, ZnSO4	Physa heterostropha, adult[19]
Mortality	50 % mortality	Zn = 3.16 mg/l, 72 h, hard water, hardness = 100 mg/l, 294 K, ZnSO4	Physa heterostropha, adult[19]
Mortality	50 % mortality	Zn = 3.16 mg/l, 96 h, hard water, hardness = 100 mg/l, 294 K, ZnSO4	Physa heterostropha, adult[19]
Mortality	50 % mortality	Zn = 3.62 mg/l, 48 h, hard water, hardness = 100 mg/l, 294 K, ZnSO4	Physa heterostropha, adult[19]
Mortality	50 % mortality	Zn = 4.07 mg/l, 24 h, hard water, hardness = 100 mg/l, 294 K, ZnSO4	Physa heterostropha, adult[19]
Mortality	50 % mortality	Zn = 4.9 mg/l, 72 h, soft water, hardness = 20 mg/l, 294 K, ZnSO4	Physa heterostropha, adult[19]
Mortality	50 % mortality	Zn = 4.9 mg/l, 96 h, soft water, hardness = 20 mg/l, 294 K, ZnSO4	Physa heterostropha, adult[19]
Mortality	50 % mortality	Zn = 6.1 mg/l, 48 h, soft water, hardness = 20 mg/l, 294 K, ZnSO4	Physa heterostropha, adult[19]
Mortality	50 % mortality	Zn = 0.303 mg/l, 72 h, soft water, hardness = 20 mg/l, 284 K, ZnSO4	Physa heterostropha, young adult[19]
Mortality	50 % mortality	Zn = 0.303 mg/l, 96 h, soft water, hardness = 20 mg/l, 284 K, ZnSO4	Physa heterostropha, young adult[19]
Mortality	50 % mortality	Zn = 0.35 mg/l, 72 h, soft water, hardness = 20 mg/l, 305 K, ZnSO4	Physa heterostropha, young adult[19]
Mortality	50 % mortality	Zn = 0.35 mg/l, 96 h, soft water, hardness = 20 mg/l, 305 K, ZnSO4	Physa heterostropha, young adult[19]
Mortality	50 % mortality	Zn = 0.434 mg/l, 24 h, soft water, hardness = 20 mg/l, 284 K, ZnSO4	Physa heterostropha, young adult[19]
Mortality	50 % mortality	Zn = 0.434 mg/l, 48 h, hard water, hardness = 100 mg/l, 284 K, ZnSO4	Physa heterostropha, young adult[19]
Mortality	50 % mortality	Zn = 0.434 mg/l, 48 h, soft water, hardness = 20 mg/l, 284 K, ZnSO4	Physa heterostropha, young adult[19]
Mortality	50 % mortality	Zn = 0.434 mg/l, 72 h, hard water, hardness = 100 mg/l, 284 K, ZnSO4	Physa heterostropha, young adult[19]
Mortality	50 % mortality	Zn = 0.434 mg/l, 72 h, soft water, hardness = 20 mg/l, 293 K, ZnSO4	Physa heterostropha, young adult[19]
Mortality	50 % mortality	Zn = 0.434 mg/l, 96 h, hard water, hardness = 100 mg/l, 284 K, ZnSO4	Physa heterostropha, young adult[19]

Mortality	50 % mortality	Zn = 0.434 mg/l, 96 h, soft water, hardness = 20 mg/l, 293 K, ZnSO4	Physa heterostropha, young adult[19]
Mortality	50 % mortality	Zn = 0.536 mg/l, 48 h, soft water, hardness = 20 mg/l, 293 K, ZnSO4	Physa heterostropha, young adult[19]
Mortality	50 % mortality	Zn = 0.598 mg/l, 24 h, soft water, hardness = 20 mg/l, 305 K, ZnSO4	Physa heterostropha, young adult[19]
Mortality	50 % mortality	Zn = 0.598 mg/l, 48 h, soft water, hardness = 20 mg/l, 305 K, ZnSO4	Physa heterostropha, young adult[19]
Mortality	50 % mortality	Zn = 0.667 mg/l, 24 h, soft water, hardness = 20 mg/l, 293 K, ZnSO4	Physa heterostropha, young adult[19]
Mortality	50 % mortality	Zn = 0.949 mg/l, 24 h, hard water, hardness = 100 mg/l, 284 K, ZnSO4	Physa heterostropha, young adult[19]
Mortality	50 % mortality	Zn = 1.11 mg/l, 48 h, hard water, hardness = 100 mg/l, ZnSO4, 305 K	Physa heterostropha, young adult[19]
Mortality	50 % mortality	Zn = 1.11 mg/l, 72 h, hard water, hardness = 100 mg/l, ZnSO4, 305 K	Physa heterostropha, young adult[19]
Mortality	50 % mortality	Zn = 1.11 mg/l, 96 h, hard water, hardness = 100 mg/l, ZnSO4, 305 K	Physa heterostropha, young adult[19]
Mortality	50 % mortality	Zn = 1.28 mg/l, 24 h, hard water, hardness = 100 mg/l, ZnSO4, 305 K	Physa heterostropha, young adult[19]
Mortality	50 % mortality	Zn = 1.70 mg/l, 96 h, hard water, hardness = 100 mg/l, 293 K, ZnSO4	Physa heterostropha, young adult[19]
Mortality	50 % mortality	Zn = 1.96 mg/l, 72 h, hard water, hardness = 100 mg/l, 293 K, ZnSO4	Physa heterostropha, young adult[19]
Mortality	50 % mortality	Zn = 2.76 mg/l, 48 h, hard water, hardness = 100 mg/l, 293 K, ZnSO4	Physa heterostropha, young adult[19]
Mortality	50 % mortality	Zn = 3.50 mg/l, 24 h, hard water, hardness = 100 mg/l, 293 K, ZnSO4	Physa heterostropha, young adult[19]
Mortality	50 % mortality	Zn = 13.5 mg/l, aerated lake water, ZnSO4, 297+-2 K, 48 h	Physa integra[18]
Mortality	50 % mortality	Zn = 18.0 mg/l, aerated lake water, ZnSO4, 297+-2 K, 24 h	Physa integra[18]
Mortality	50 % mortality	Zn = 0.78 mg/l, 96 h, soft water	Pimephales promalas[26]
Mortality	50 %	Acetate, oral dose = 2460 mg/kg body	Rat[27]
Mortality	50 %	Chloride, oral dose = 350 mg/kg body	Rat[28]
Mortality	50 % mortality	Zn = 17.3 mg/l, 120 h, hardness = 290 mg/l, ph=7.2-7.9, 284 K, ZnSO4	Rutilus rutilus[6]
Mortality	50 %	Zn = 4.6 mg/l, 5 d, hard water	Salmo gairdneri[21]
Mortality	50 %	Zn = 4.76 mg/l, 101 h, 288.5 K, 10 fish	Salmo gairdneri[29]
Mortality	50 %	Zn = 5.3 mg/l, 24 h, 288.5 K, 10 fish	Salmo gairdneri[29]
Mortality	50 %	Zn = 6.0 mg/l, 15 h, 288.5 K, 10 fish	Salmo gairdneri[29]
Mortality	50 %	Zn = 6.6 mg/l, 11.7 h, 288.5 K, 10 fish	Salmo gairdneri[29]
Mortality	50 %	Zn = 9. mg/l, 8.3 h, 288.5 K, 10 fish	Salmo gairdneri[29]
Mortality	50 % mortality	Zn = 4.0 mg/l, 283-285 K, hardness = 240-290 mg/l, 48 h, pH=7.5-7.9	Salmo gairdneri[30]
Mortality	50 % mortality	Zn = 4.0 ppm, 48 h, well wat., Ca=120+Mg=5 ppm, fingerlings, 289 K	Salmo gairdneri[31]
Mortality	50 % mortality	Zn = 4.6 mg/l, 120 h, ZnSO4, hardness = 290 mg/l, pH = 7.2-7.9	Salmo gairdneri[6]
Mortality	50 % mortality	Zn=0.6 mg/l, 48 h, dil. well wat, Ca=4.5+Mg=0.19 ppm, fingerlings, 290 K	Salmo gairdneri[32]

Mortality	50 % mortality	Zn=2.0 Kmg/l, 48 h, dil. well wat, Ca=4.5+Mg=0.19 ppm, fingerlings, 290	Salmo gairdneri[32]
Mortality	50 % mortality	Zn=4.0 mg/l, 48 h, dil. well wat, Ca=19+Mg=0.79 ppm, fingerlings, 290 K	Salmo gairdneri[32]
Mortality	0 %	Toxicological limit = 0.1 mg/l, ZnSO4	Self Purification[11]
Mortality	50 % mortality	Zn = 2.62 mg/l, 296-298 K, ZnSO4, distilled water, 96 h, ciliata	Tetrahymena pyriformis[33]
Mortality	50 % mortality	Zn = 0.11 mg/l, 48 h, distilled water, 293 K, ZnSO4	Tubifex tubifex[34]
Mortality	50 % mortality	Zn = 0.12 mg/l, 24 h, distilled water, 293 K, ZnSO4	Tubifex tubifex[34]
Mortality	50 % mortality	Zn = 2.57 mg/l, Hard.=34 ppm+29.6 mg phosphate/l, 293 K, 48 h, ZnSO4	Tubifex tubifex[34]
Mortality	50 % mortality	Zn = 2.98 mg/l, hardness = 34.2 ppm, ZNSO4, 293 K, 48 h	Tubifex tubifex[34]
Mortality	50 % mortality	Zn = 3.64 mg/l, hard. = 34.2 ppm+29.6 mg phosphate/l, 293 K, 24 h, ZnSO4	Tubifex tubifex[34]
Mortality	50 % mortality	Zn = 4.62 mg/l, hardness = 34.2 ppm, ZnSO4, 293 K, 24 h	Tubifex tubifex[34]
Mortality	50 % mortality	Zn = 60.2 mg/l, 48 h, hardness = 261 mg/l, 293 K	Tubifex tubifex[34]
Mortality	50 % mortality	Zn = 75.8 mg/l, 24 h, hardness = 261 mg/l, 293 K	Tubifex tubifex[34]
Mortality	50 % mortality	Zn = 0.29 mg/l,	Vorticella convallaria[35]
Mortality	50 % mortality	Zn = 0.29 mg/l, Zn(NO3)2, distilled water, 12 h, ciliata	Vorticella convallaria[35]
Mortality of Zn vs temperature	1250 minutes survival	Zn = 4.4 mg/l, 286.5 K	Salmo gairdneri[36]
Mortality of Zn vs temperature	1300 minutes survival	Zn = 3.9 mg/l, 288.5 K	Salmo gairdneri[36]
Mortality of Zn vs temperature	1425 minutes survival	Zn = 3.5 mg/l, 291.5 K	Salmo gairdneri[36]
Mortality of Zn vs temperature	150 minutes survival	Zn = 14.1 mg/l, 294.5 K	Salmo gairdneri[36]
Mortality of Zn vs temperature	150 minutes survival	Zn = 19.8 mg/l, 291.5 K	Salmo gairdneri[36]
Mortality of Zn vs temperature	195 minutes survival	Zn = 23.4 mg/l, 288.5 K	Salmo gairdneri[36]
Mortality of Zn vs temperature	220 minutes survival	Zn = 27.8 mg/l, 286.5 K	Salmo gairdneri[36]
Mortality of Zn vs temperature	222 minutes survival	Zn = 11.9 mg/l, 291.5 K	Salmo gairdneri[36]
Mortality of Zn vs temperature	232 minutes survival	Zn = 14.1 mg/l, 288.5 K	Salmo gairdneri[36]
Mortality of Zn vs temperature	250 minutes survival	Zn = 8.4 mg/l, 294.5 K	Salmo gairdneri[36]
Mortality of Zn vs temperature	265 minutes survival	Zn = 15.3 mg/l, 286.5 K	Salmo gairdneri[36]
Mortality of Zn vs temperature	375 minutes survival	Zn = 6.5 mg/l, 291.5 K	Salmo gairdneri[36]
Mortality of Zn vs temperature	412 minutes survival	Zn = 7.1 mg/l, 288.5 K	Salmo gairdneri[36]

Mortality of Zn vs temperature	4200 minutes survival	Zn = 4 mg/l, 294.5 K	Salmo gairdneri[36]
Mortality of Zn vs temperature	515 minutes survival	Zn = 4.6 mg/l, 294.5 K	Salmo gairdneri[36]
Mortality of Zn vs temperature	538 minutes survival	Zn = 8.4 mg/l, 286.5 K	Salmo gairdneri[36]
Nitrification	40 % inhibation	Soil = 5 umoles/g of soil, 10 d incubation, 3 soils, lab	Soil organisms[37]
Nitrogen fixation	1252 ethylene produced in nmoles/(plant*30 min)	Zn = 10.0	Soybean with nodule[38]
Nitrogen fixation	1585 ethylene produced in nmoles/(plant*30 min)	Zn = 5.0 ppm in soil, control = 1640	Soybean with nodule[38]
Nitrogen fixation	1659 ethylene produced in nmoles/(plant*30 min)	Zn = 1.0 ppm in soil, control = 1640	Soybean with nodule[38]
Nitrogen fixation	36 no of nodules/plant	Zn = 10.0	Soybean with nodule[38]
Nitrogen fixation	38 no of nodules/plant	Zn = 5.0 ppm in soil, control = 46	Soybean with nodule[38]
Nitrogen fixation	40 no of nodules/plant	Zn = 1.0 ppm in soil, control = 46	Soybean with nodule[38]
Nitrogen fixation	52.90 weight of one nodule in mg	Zn = 10.0	Soybean with nodule[38]
Nitrogen fixation	53.69 weight of one nodule in mg	Zn = 5.0 ppm in soil, control = 62.30 mg	Soybean with nodule[38]
Nitrogen fixation	58.40 weight of one nodule in mg	Zn = 1.0 ppm in soil, control = 62.30 mg	Soybean with nodule[38]
Photosynthesis	100 % of control	Zn = 0.001 mg/l, 3 h, Golden Horn	Aufwuchs communities[39]
Photosynthesis	106 % of control	Zn = 0.1 mg/l, 3 h, Golden Horn	Aufwuchs communities[39]
Photosynthesis	60 % of control	Zn = 10.0 mg/l, 3 h, Golden Horn	Aufwuchs communities[39]
Photosynthesis	82 % of control	Zn = 1.0 mg/l, 3 h, Golden Horn	Aufwuchs communities[39]
Photosynthesis	95 % of control	Zn = 0.01 mg/l, 3 h, Golden Horn	Aufwuchs communities[39]
Respiration	104 % of control	Zn = 0.001 mg/l, 3 h, Golden Horn	Aufwuchs communities[39]
Respiration	80 % of control	Zn = 10.0 mg/l, 3 h, Golden Horn	Aufwuchs communities[39]
Respiration	88 % of control	Zn = 1.0 mg/l, 3 h, Golden Horn	Aufwuchs communities[39]
Respiration	95 % of control	Zn = 0.1 mg/l, 3 h, Golden Horn	Aufwuchs communities[39]
Respiration	98 % of control	Zn = 0.01 mg/l, 3 h, Golden Horn	Aufwuchs communities[39]
Survival time	220 min	Zn = 7 mg/l, CaCO3 = 12 mg/l, pH = 6.6-6.7	Salmo gairdneri[29]
Survival time	300 min	Zn = 7 mg/l, CaCO3 = 50 mg/l, pH = 7.-7.2	Salmo gairdneri[29]
Survival time	470 min	Zn = 7 mg/l, CaCO3 = 320 mg/l, pH = 7.6-7.8	Salmo gairdneri[29]
Survival time	660 min	Zn = 7 mg/l, CaCO3 = 504 mg/l, pH = 7.7-8.	Salmo gairdneri[29]
Uptake of Cd	4.68 mg Cd/kg dry weight	Soil = 225 ppm applied, soil = 10 ppm Cd, 288.5-305.2 K	Glycine max[40]
Uptake of Cd	5.38 mg Cd/kg dry weight	Soil = 400 ppm applied, soil = 10 ppm Cd, 288.5-305.2 K	Glycine max[40]
Uptake of Cd	6.91 mg Cd/kg dry weight	Soil = 100 ppm applied, soil = 10 ppm Cd, 288.5-305.2 K	Glycine max[40]
Uptake of Cd	7.48 mg Cd/kg dry weight	Soil = 0 ppm Zn applied, soil = 10 ppm Cd, 288.5-305.2 K	Glycine max[40]

Uptake of Cd	8.26 mg Cd/kg dry weight	Soil = 50 ppm applied, soil = 10 ppm Cd, 288.5-305.2 K	Glycine max[40]
Uptake of Cd	8.60 mg Cd/kg dry weight	Soil = 25 ppm applied, soil = 10 ppm Cd, 288.5-305.2 K	Glycine max[40]
Uptake of Cd	9.35 mg Cd/kg dry weight	Soil = 10 ppm applied, soil = 10 ppm Cd, 288.5-305.2 K	Glycine max[40]
Uptake of Cd	9.41 mg Cd/kg dry weight	Soil = 5 ppm applied, soil = 10 ppm Cd, 288.5-305.2 K	Glycine max[40]
Uptake of Cd	0.86 micro ci/(mg dry weight)	Zn = 100 ppm, Cd = 3 ppb as Cd-109, 5 d, 283-293 K, root	Phaseolus vulgaris[41]
Uptake of Cd	13.2 micro ci/(mg dry weight)	Zn = 5 ppm, Cd = 3 ppb as Cd-109, 5 d, 283-293 K, root	Phaseolus vulgaris[41]
Uptake of Cd	13.6 micro ci/(mg dry weight)	Zn = 1 ppm, Cd = 3 ppb as Cd-109, 5 d, 283-293 K, root	Phaseolus vulgaris[41]
Uptake of Cd	156.3 micro ci/(mg dry weight)	Zn = 5 ppm, Cd = 3 ppb as Cd-109, 5 d, 283-293 K, leaf	Phaseolus vulgaris[41]
Uptake of Cd	16.3 micro ci/(mg dry weight)	Zn = 0 ppm, Cd = 3 ppb as Cd-109, 5 d, 283-293 K, root	Phaseolus vulgaris[41]
Uptake of Cd	168.5 micro ci/(mg dry weight)	Zn = 100 ppm, Cd = 3 ppb as Cd-109, 5 d, 283-293 K, stem	Phaseolus vulgaris[41]
Uptake of Cd	180.6 micro ci/(mg dry weight)	Zn = 10 ppm, Cd = 3 ppb as Cd-109, 5 d, 283-293 K, leaf	Phaseolus vulgaris[41]
Uptake of Cd	195.9 micro ci/(mg dry weight)	Zn = 5 ppm, Cd = 3 ppb as Cd-109, 5 d, 283-293 K, stem	Phaseolus vulgaris[41]
Uptake of Cd	215.7 micro ci/(mg dry weight)	Zn = 10 ppm, Cd = 3 ppb as Cd-109, 5 d, 283-293 K, stem	Phaseolus vulgaris[41]
Uptake of Cd	43.7 micro ci/(mg dry weight)	Zn = 0 ppm, Cd = 3 ppb as Cd-109, 5 d, 283-293 K, leaf	Phaseolus vulgaris[41]
Uptake of Cd	50.0 micro ci/(mg dry weight)	Zn = 100 ppm, Cd = 3 ppb as Cd-109, 5 d, 283-293 K, leaf	Phaseolus vulgaris[41]
Uptake of Cd	57.2 micro ci/(mg dry weight)	Zn = 1 ppm, Cd = 3 ppb as Cd-109, 5 d, 283-293 K, stem	Phaseolus vulgaris[41]
Uptake of Cd	64.4 micro ci/(mg dry weight)	Zn = 1 ppm, Cd = 3 ppb as Cd-109, 5 d, 283-293 K, leaf	Phaseolus vulgaris[41]
Uptake of Cd	68.0 micro ci/(mg dry weight)	Zn = 0 ppm, Cd = 3 ppb as Cd-109, 5 d, 283-293 K, stem	Phaseolus vulgaris[41]
Uptake of Cd	8.3 micro ci/(mg dry weight)	Zn = 10 ppm, Cd = 3 ppb as Cd-109, 5 d, 283-293 K, root	Phaseolus vulgaris[41]
Vertebral damage	8 species	Zn = 0.2-0.3 mg/l, after 270 d, total 13 species in experiment	Fish[42]

Source: 1.Jockim, E., G. Morrison and R. Steele. 1977. Mar. Biol. 40, 303-308.
2.Peterson, P.J. 1962. N. Z. J. Sci. Technol. 5, pp 141-161.
3.Rachlin, J.W. and M. Farran. 1974. Water Res., 8(8), pp 575-577.
4.Bengtsson, B.E. 1974. Bull. Environ. Contam. Toxicol., 12, p 654.
5.Brereton, A., H. Lord, I. Thornton and J.S. Webb. 1973. Mar. biol. 19(2), pp 96-101.
6.Ball, I.R. 1967. Water Res. 1, 777-783.
7.Rehwoldt, R., L. Lasko, C. Shaw and F. Wirbowski. 1973. Bull. Environ. Contam. Toxicol. 10, 291-294.
8.Wurtz, C. B. and C.H. Bridges. 1961. Proc. Penn. Acad. Sci. 35, 51-56.
9.Resz. A. 1968. Sonderh. Z. Plfkrankh. PflPath. PflSchutz No 4, 147-150.
10.Boutet, C. and C. Chaisemartin. 1973. C.R. Seances Soc. Biol. 167, 1933-1938.
11.Jung. K.D. 1973. Wirkungskonzentration schaedigender Stoffe im Wasser. Gelenkirchen. Hygiene-Institudes Ruhrgebiets.
12.Liebman, H. 1958. Handbuch der Friscwasser - und Abwasserbiologie. Band II.Munchen. Oldenbourg.
13.Baudouin, M.F. and P. Scoppa. 1974. Bull. Environ. Contam. Toxicol. 12, 745-751.
14.Biesinger, K.E. and G.M. Christensen. 1972. J. Fish. Res. bd. Can., 29(12), 1691-1700.
15.Cabejszek, I. and M. Stasiak. 1960. Rocz. Panstw. Zakl. Hig. 533-540.
16.Biesinger, K.E. and G.M. Christensen. 1972. J. Fish. Res. bd Canada 29, 1691-1700.
17.See, C.L., A.L. Buikema and J. Cairns. 1974. ASB Bull. 21, 82.
18.Cairns, J. Jr., D.I. Messenger and W.F. Calhoun. 1976. Arch. Hydrobiol. 77, 164-175.
19.Wurtz, C.B. 1962. Nautilus 76, 53-61.
20.Cairns, J. and A. Scheier. 1958. Not. Nat. Acad. Nat. Sci. Phila. 308, 1-11.

21.Solbe, J.F., G. Del and V.A. Cooper. 1976. Water Res. 10(6), pp 523-527.
22.Solbe, J.F. de L.G. and V.A. Flook. 1975. J. Fish. Biol. 7, 631-637.
23.Thorp, V.J. and P.S. Lake. 1974. Aust. J. Mar. freshwater Res. 25, 97-104.
24.Buikema, A.L., J. Cairns and G.W. Sullivan. 1974. Water Resour. Bull. 10, 648-661.
25.Sallivan, G.W., A.L. Buikema and J. Cairns, Jr. 1973. Va. J. Sci. 24(3), pp 120.
26.Pickering, Q.H. and C. Henderson. 1966. Air Water Pollut. Int. J. 10, 453-463.
27.Toxic Substances List 1973, U.S. Department of Health, Education and Welfare, Public Health ServiceNational. Institute for Occupational Safety and Health, Rockville Maryland, USA, June 1973.
28.EPA. Water Quality Crit. Data Book. 2rd. Inorganic Pollution of Freshwater. 18010. DPU 07/71.
29.Solbe, J.F. de L.G. 1974. Water Res., 8(6), pp 389-391.
30.Brown, V.M. and D.H.M. Jordan and B.A. Tiller. 1969. J. Fish. Biol. 1, 1-9.
31.Brown, V.M. and R.A. Dalton. 1970. J. Fish. Biol. 2, 211-216.
32.Lloyd, R. 1960. Ann. Appl. Biol. 48, 84-94.
33.Carter, J.W. and I.L. Cameron. 1973. Water Res. 7, 951-961.
34.Brkovic-Popovic, I. and M. Popovic. 1977. Environ. Pollut. 13, 93-98.
35.Sartory, D.P. and B.J. LLoyd. 1976. Water Res. 10, 1123-1127.
36.Lloyd, , R.M. 1965. Factors that Affect the tolerence of fish to heavy metal Poisoning. Biol. Probles in Water Poll. 3rd Seminar 1962. US Dept. Health, Educ and Welfare. 181 pp.
37.Liang, C.N. and M.A. Tabatabai. 1978. Environ. Pollut. 12, 141-147.
38.Letchworth, M.B. and V. Blum. 1977. Environ. Pollut. 14, 303-312.
39.Orhon, D. 1975. Water Res. 9(5/6), pp 467-472.
40.Haghiri, F. 1974. J. Environ. Qual., 3(2), pp 180-183.
41.Hook, R.I. van Jr., B.G. Blaylock, E.A. Bondietti, C.W. Francis, J.W. Huckabee, D.E. Reichle, F.H. Seeton and J.P. Wintherspoon. 1976. Environ. Qual. Saf. 5, pp 167-182.
42.Bengtson, B.E. and C.H. Carlin, A.A. Larson and O. Svanberg.

4-439 Effect of Zr

Parameter	Effect Value	Concentration and Condition	Species
Mortality	50 %	4-chloride, oral dose = 1688 mg/kg body	Rat[1]
Mortality	50 %	Nitrate, oral dose = 853 mg/kg body	Rat[1]
Mortality	50 %	ZrO-acetate, oral dose = 4100 mg/kg body	Rat[1]
Mortality	50 %	ZrOCl2, oral dose = 3500 mg/kg body	Rat[1]

Source: 1.Toxic Substances List 1973, U.S. Department of Health, Education and Welfare, Public Health Service National Institute for Occupational Safety and Health, Rockville Maryland, USA, June 1973.

4-440 Effect of alpha - amanitin

Parameter	Effect Value	Concentration and Condition	Species
Mortality	50 %	Oral dose = 0.3 mg/kg, 3.28E-7mol/kg, F007 = 6.48	Mouse[1]

Source: 1.Luckey, T.D. and B. Venugopal. 1977. Metal Toxicity in Mammals, 1. Plenum Press.

4-441 Effect of alpha- butylene

Parameter	Concentration	Condition	Species
Sublethal effect	50000 ppm	Epinasty in petiole, 2 days	Tomato[1]

Source: 1.Crocker, W., Growth and Plants, Reinhold publishing Comp., New York, 1948.

4-442 Effect of alpha-chloropropionic acid

Parameter	Concentration	Condition	Species
LD50	0.98 g pr. kg body weight	oral dose, 100 %	Rat[1]

Source: 1.Patty, F. A., Industrial Hygiene and toxicology, 2, Interscience Publishers, New York, 1976.

4-443 Effect of alpha-dicyclopentadiene

Parameter	Concentration	Condition	Species
LD50	6.72 g pr. kg body weight	percutane	Rat[1]
LD50	0.41 g pr. kg body weight	single oral dose	Rat[1]
LD50	4.46 g pr. kg body weight	dermal dose	Rat[1]

Source: 1.Patty, F. A., Industrial Hygiene and toxicology, 2, Interscience Publishers, New York, 1976.

4-444 Effect of alpha-naphtoquinone

Parameter	Concentration	Condition	Species
TLm	0.3-0.6 mg/l	24-48 hours	Fish[1]

Source: 1.Meinck, F., Stooff, H. and Kohlschutter, H., Les eaux residuaires industrielles, 1970.

4-445 Effect of alpha-naphtylamine

Parameter	Concentration	Condition	Species
Sublethal effect	-	Carcinogenic	Man[1]
TLm	6 mg/l	not indicated	Gardon fish[2]

Source: 1.Federal register, 39, 1974.
2.Meinck, F., Stooff, H. and Kohlschutter, H., Les eaux residuaires industrielles, 1970.

4-446 Effect of alpha olefin sulfonate

Parameter	Effect Value	Concentration and Condition	Species
Mortality	50 %	AOS-tech. = 3.0 mg/l, 48 h, 294 K	Cyprinus carpio[1]
Mortality	50 %	AOS-tech. = 3.2 mg/l, 24 h, 294 K	Cyprinus carpio[1]
Mortality	50 %	Conc = 3.0 mg/l, 96 h, 294 K	Cyprinus carpio[1]
Mortality	50 %	Conc = 2 mg/l, 24 h, 294 K	Tilapia melanopleura[1]
Mortality	50 %	Conc = 2 mg/l, 48 h, 294 K	Tilapia melanopleura[1]
Mortality	50 %	Conc = 2 mg/l, 96 h, 294 K	Tilapia melanopleura[1]

Source: 1.Lopez-Zavala, A. A.S. de Aluja,B. Eliasl L.B. Manjarrez, L.Buchmann, I.A. Mercado and S. Caltenco. 175. Progress Water Technol. 7(2), pp 73-82.

4-447 Effect of alpha-pyridylamine

Parameter	Concentration	Condition	Species
LD50	0.8-1.6 g pr. kg body weight acute oral dose	mouse, Rat[1]	

Source: 1.Patty, F. A., Industrial Hygiene and toxicology, 2, Interscience Publishers, New York, 1976.

4-448 Effect of beta,beta'-dichloromethylether

Parameter	Concentration	Condition	Species
LD50	0.136 g pr. kg body weight	single oral dose	Mouse[1]
LD50	0.126 g pr. kg body weight	single oral dose	Rabbit[1]
LD50	0.075-0.105 g pr. kg body weight	single oral dose	Rat[1]

Source: 1.Patty, F. A., Industrial Hygiene and toxicology, 2, Interscience Publishers, New York, 1976.

4-449 Effect of beta-naphtylamine

Parameter	Concentration	Condition	Species
Sublethal effect	-	Carcinogenic	Man[1]
Sublethal effect	-	Carcinogenic	Man[2]

Source: 1.Federal register, 39, 1974.
2.Maximale Arbeitsplatzkonzentration 1973. BRD.

4-450 Effect of beta-propiolactone

Parameter	Concentration	Condition	Species
LD50	0.05-0.1 g pr. kg body weight	single oral dose	Rat[1]
Sublethal effect	-	Carcinogenic	Man[2]
Sublethal effect	-	Carcinogenic	Man[3]

Source: 1.Patty, F. A., Industrial Hygiene and toxicology, 2, Interscience Publishers, New York, 1976.
2.Maximale Arbeitsplatzkonzentration 1973. BRD.
3.Federal register, 39, 1974.

4-451 Effect of m-Xylene

Parameter	Effect Value	Concentration and Condition	Species
Mortality	50 % mortality	Water = 12.0 ppm, 96 h, dungeness crab, stage 1 zoeae	Cancer magister[1]
Mortality	50 % mortality	Water = 3.7 ppm, 96 h	Crago franciscorum[2]
Mortality	50 % mortality	Water = 9.2 ppm, 96 h	Marone saxatilis[2]
LC50	16 mg/l	24 hours	Goldfish[3]

Source: 3.Shell Chemie, Shell Industrie Chemicalien gids, Shell Niederland Chemie, 1975.
1.Caldwell, R.s., E. M. Caldarone and M.H. Mallon. 1977. D.A. Wolfe (ed.): Fate and Effects of Petroleum Hydrocarbons in Marine Ecosystems and Organisms. NY. Pergamon Press. 210-220.
2.Benville, P.E. Jr. and S. Korn. 1977. Calif. Fish and Game. 63, 204-209.

4-452 Effect of o-Xylene

Parameter	Effect Value	Concentration and Condition	Species
Mortality	50 % mortality	Water = 6.0 ppm, 96 h, stage 1 zoeae	Cancer magister[5]
Mortality	50 % mortality	Water = 1.3 ppm, 96 h	Crago franciscorum[6]
Mortality	50 % mortality	Water = 11.0 ppm, 96 h	Marone saxatilis[6]
LC0	10 mg/l	96 hours, seawater, 8 C	Coho salmon[1]
LC100	100 mg/l	96 hours, seawater, 8 C	Coho salmon[1]
LC50	55 mg/l	1 day, 20 C	Chlorella vulgaris[2]
LC50	13 mg/l	24 hours	Goldfish[3]
TLm	100-1000 mg/l	24 hours	Daphnia magna[4]

Source: 1.Morrow, J. E., EPA/660&3/73/018, 1974.
2.Kauss, P. B. and Hutchinson, T. C., Environ Poll., 9, 1975.
3.Shell Chemie, Shell Industrie Chemicalien gids, Shell Niederland Chemie, 1975.
4.Elkins, H. F. et al., Sewage Ind. Wastes, 28, 1475, 1956.
5.Caldwell, R.s., E. M. Caldarone and M.H. Mallon. 1977. D.A. Wolfe (ed.): Fate and Effects of Petroleum Hydrocarbons in Marine Ecosystems and Organisms. NY. Pergamon Press. 210-220.
6.Benville, P.E. Jr. and S. Korn. 1977. Calif. Fish and Game. 63, 204-209.

4-453 Effect of p-Xylene

Parameter	Effect Value	Concentration and Condition	Species
Mortality	50 % mortality	Water = 2.0 ppm, 96 h	Crago franciscorum[3]
Mortality	50 % mortality	Water = 2.0 ppm, 96 h	Marone saxatilis[3]
LC50	18 mg/l	24 hours	Goldfish[1]
TLm	21-24 mg/l	24-96 hours, soft water	Bluegill[2]
TLm	27-29 mg/l	24-96 hours, soft water	Fathead minnow[2]
TLm	29 mg/l	24-96 hours, hard water	Fathead minnow[2]
TLm	37 mg/l	24-96 hours, soft water	Goldfish[2]
TLm	35 mg/l	24-96 hours, soft water	Guppie[2]

Source: 1.Shell Chemie, Shell Industrie Chemicalien gids, Shell Niederland Chemie, 1975.
2.Pickering Q. H. and Henderson, C., JWPCF, 38, 1419, 1966.
3.Benville, P.E. Jr. and S. Korn. 1977. Calif. Fish and Game. 63, 204-209.

4-454 Effect of m-aminobenzoic acid

Parameter	Concentration	Condition	Species
LD50	0.25-0.50 g pr. kg body weight	intraperitoneal dose	Mouse[1]

Source: 1.Patty, F. A., Industrial Hygiene and toxicology, 2, Interscience Publishers, New York, 1976.

4-455 Effect of m-chlorobenzoic acid

Parameter	Concentration	Condition	Species
LD50	>0.5 g pr. kg body weight	acute oral dose	Rabbit[1]

Source: 1.Patty, F. A., Industrial Hygiene and toxicology, 2, Interscience Publishers, New York, 1976.

4-456 Effect of m-cresol

Parameter	Concentration	Condition	Species
LC0	28 mg/l	not indicated	Daphnia[1]
LC0	600 mg/l	not indicated	Escherichia coli[2]
LD50	1.1-2.0 g pr. kg body weight	single oral dose, 100 %	rat, Rabbit[3]
Sublethal effect	13 mg/l	Inhibition of cell multiplication	Microcystis aeruginosa[4]
TLm	10-14 mg/l	96 hours	BlueFish[5]
TLm	25 mg/l	24 hours	Crucian carp[5]
TLm	24 mg/l	24-96 hours	Mosquito fish[5]
TLm	21-23 mg/l	24 hours	Roach[5]
TLm	21-23 mg/l	24 hours	Tench[5]

Source: 1.Meinck, F., Stooff, H. and Kohlschutter, H., Les eaux residuaires industrielles, 1970.
2.Jones, H. R., Environmental Control in the organic and petrochemical Industries, Noyes Data Corp., 171.
3.Patty, F. A., Industrial Hygiene and toxicology, 2, Interscience Publishers, New York, 1976.
4.Bringmann, G. and Kuhn, R., nwf-wasser/abwasser, H.9, 1976.
5.Europian inland fishery advisory commission, Water quality criteria for european freshwater fish, Waer Res., 7, 929, 1973.

4-457 Effect of m-dinitrobenzene

Parameter	Concentration	Condition	Species
Sublethal effect	0.1-0.17 mg/l	Inhibition of cell multiplication	Microcystis aeruginosa[1]
Sublethal effect	5-14 mg/l	Inhibition of cell multiplication	Pseudomonas putida[1]

Source: 1.Bringmann, G. and Kuhn, R., nwf-wasser/abwasser, H.9, 1976.

4-458 Effect of m-hydroxybenzoic acid

Parameter	Concentration	Condition	Species
LD50	2.8 g pr. kg body weight	intraperitoneal dose	guinea pig[1]

Source: 1.Patty, F. A., Industrial Hygiene and toxicology, 2, Interscience Publishers, New York, 1976.

4-459 Effect of m-nitrobenzoic acid

Parameter	Concentration	Condition	Species
LD50	>0.2 g pr. kg body weight	acute oral dose	Rabbit[1]

Source: 1.Patty, F. A., Industrial Hygiene and toxicology, 2, Interscience Publishers, New York, 1976.

4-460 Effect of m-nitrophenol

Parameter	Concentration	Condition	Species
LC0	24 mg/l	not indicated	Daphnia[1]
LC0	300 mg/l	not indicated	Pseudomonas[1]
LC0	28 mg/l	not indicated	Scenedesmus[1]
Sublethal effect	17 mg/l	Inhibition of cell multiplication	Microcystis aeruginosa[2]
Sublethal effect	7 mg/l	Inhibition of cell multiplication	Pseudomonas putida[2]
TLm	20-22 mg/l	6 hours, hard water	Vairon fish[1]
TLm	9-10 mg/l	6 hours, distilled water	Vairon fish[1]
TLm	4-6 mg/l	6 hours, distilled water	Vairon fish[1]
TLm	30-33 mg/l	6 hours, hard water	Vairon fish[1]

Source: 1.Meinck, F., Stooff, H. and Kohlschutter, H., Les eaux residuaires industrielles, 1970.
2.Bringmann, G. and Kuhn, R., nwf-wasser/abwasser, H.9, 1976.

4-461 Effect of m-nitrotoluene

Parameter	Concentration	Condition	Species
Sublethal effect	1 mg/l	Inhibition of cell multiplication	Microcystis aeruginosa[1]
Sublethal effect	10 mg/l	Inhibition of cell multiplication	Pseudomonas putida[1]
TLm	14-18 mg/l	6 hours, distilled water	Vairon fish[2]
TLm	25-30 mg/l	6 hours, hard water	Vairon fish[2]

Source: 1.Bringmann, G. and Kuhn, R., nwf-wasser/abwasser, H.9, 1976.
2.Meinck, F., Stooff, H. and Kohlschutter, H., Les eaux residuaires industrielles, 1970.

4-462 Effect of m-toluidine

Parameter	Concentration	Condition	Species
LC0	0.6 mg/l	not indicated	Daphnia[1]
LC0	1000 mg/l	not indicated	Escherichia coli[1]
LC0	8 mg/l	not indicated	Scenedesmus[1]

Source: 1.Meinck, F., Stooff, H. and Kohlschutter, H., Les eaux residuaires industrielles, 1970.

4-463 Effect of n-butanol

Parameter	Concentration	Condition	Species
LC0	1000 mg/l	24 hours	Creek chub[1]
LC100	1400 mg/l	24 hours	Creek chub[1]
LD50	4.2-4.4 g pr. kg body weight	oral dose	rat,Rabbit[2]
Sublethal effect	25 ppm	Eye irritation	Man[2]
Sublethal effect	100 mg/l	Inhibition of cell multiplication	Microcystis aeruginosa[3]
Sublethal effect	650 mg/l	Inhibition of cell multiplication	Pseudomonas putida[3]

Source: 1.Gilette, L. A., Miller, D. L. and Redman, H. E., Sewage Ind. Wastes, 24, 1397, 1952.
2.Patty, F. A., Industrial Hygiene and toxicology, 2, Interscience Publishers, New York, 1976.
3.Bringmann, G. and Kuhn, R., nwf-wasser/abwasser, H.9, 1976.

4-464 Effect of n-butylacetate

Parameter	Concentration	Condition	Species
LC0	1000 mg/l	not indicated	Escherichia coli[1]
Sublethal effect	200-300 ppm	Eye irritation	Man[2]

Source: 1.Meinck, F., Stooff, H. and Kohlschutter, H., Les eaux residuaires industrielles, 1970.
2.Strafford, N., Strouts, C. R. N., and Stubbings, W. V., The Determination of Toxic Substances in airHeffer, Cambridge.

4-465 Effect of n-butylamine

Parameter	Concentration	Condition	Species
LD50	6.5 g pr. kg body weight	acute oral dose	Rat[1]
Sublethal effect	0.19 mg/l	Inhibition of cell multiplication, neutralized	Microcystis aeruginosa[2]
Sublethal effect	0.14 mg/l	Inhibition of cell multiplication, unneutralized	Microcystis aeruginosa[2]
Sublethal effect	800 mg/l	Inhibition of cell multiplication, neutralized	Pseudomonas putida[2]
Sublethal effect	65 mg/l	Inhibition of cell multiplication, unneutralized	Pseudomonas putida[2]

Source: 1.Patty, F. A., Industrial Hygiene and toxicology, 2, Interscience Publishers, New York, 1976.
2.Bringmann, G. and Kuhn, R., nwf-wasser/abwasser, H.9, 1976.

4-466 Effect of n-butylether

Parameter	Concentration	Condition	Species
LD50	7.4 g pr. kg body weight	acute oral dose	Rat[1]
Sublethal effect	200 ppm	Eye irritation	Man[1]

Source: 1.Patty, F. A., Industrial Hygiene and toxicology, 2, Interscience Publishers, New York, 1976.

4-467 Effect of n-butylmercaptan

Parameter	Concentration	Condition	Species
TLm	5.5-7.4 mg/l	24-48 hours	Lepomis macrochirus[1]

Source: 1.Meinck, F., Stooff, H. and Kohlschutter, H., Les eaux residuaires industrielles, 1970.

4-468 Effect of n-butyric acid

Parameter	Concentration	Condition	Species
Sublethal effect	368 mg/l	Inhibition of cell multiplication	Microcystis aeruginosa[1]
Sublethal effect	875 mg/l	Inhibition of cell multiplication	Pseudomonas putida[1]
TLm	61 mg/l	48 hours	Daphnia magna[2]
TLm	200 mg/l	24 hours	Lepomis macrochirus[3]
TLm	400 mg/l	24 hours	Salmoi iredens[4]
TLm	20-40 mg/l	24 hours	Trout[4]

Source: 1.Bringmann, G. and Kuhn, R., nwf-wasser/abwasser, H.9, 1976.
2.Dowden, B. F. and Bennet, H. J., JWPCF, 37, 1310, 1965.
3.Worne, H. E., Tijdschrift vand het BECEWA, Liege, Belgium.
4.Meinck, F., Stooff, H. and Kohlschutter, H., Les eaux residuaires industrielles, 1970.

4-469 Effect of n-heptane

Parameter	Concentration	Condition	Species
LC0	100 mg/l	96 hours, seawater, 8 C	Coho salmon[1]
LC0	5600 mg/l	not indicated	Gambusia affinis[2]
LC100	10000 mg/l	not indicated	Gambusia affinis[2]
LC100	30 mg/l	1-4 hours	Roach[3]
LC50	4 mg/l	24 hours	Goldfish[4]
TLm	5 mg/l	24-96 hours, soft water	Mosquito fish[5]

Source: 1.Morrow, J. E., EPA/660&3/73/018, 1974.
2.Meinck, F., Stooff, H. and Kohlschutter, H., Les eaux residuaires industrielles, 1970.
3.McKee, J. E., and Wolf, H. W., Water Quality, The resources agency of California, 1963.
4.Shell Chemie, Shell Industrie Chemicalien gids, Shell Niederland Chemie, 1975.
5.Wallen, I. E., Greer, W. C. and Lasater, R., Sewage Ind. Wastes, 29, 695, 1957.

4-470 Effect of n-hexane

Parameter	Concentration	Condition	Species
LC0	100 mg/l	96 hours, seawater, 8 C	Coho salmon[1]
LC0	10 mg/l	96 hours, seawater, 8 C	Macrocystis porifera[2]
LC50	4 mg/l	24 hours	Goldfish[3]

Source: 1.Morrow, J. E., EPA/660&3/73/018, 1974.
2.McKee, J. E., and Wolf, H. W., Water Quality, The resources agency of California, 1963.
3.Shell Chemie, Shell Industrie Chemicalien gids, Shell Niederland Chemie, 1975.

4-471 Effect of n-hexanol

Parameter	Concentration	Condition	Species
Sublethal effect	12 mg/l	Inhibition of cell multiplication	Microcystis aeruginosa[1]
Sublethal effect	62 mg/l	Inhibition of cell multiplication	Pseudomonas putida[1]

Source: 1.Bringmann, G. and Kuhn, R., nwf-wasser/abwasser, H.9, 1976.

4-472 Effect of n-methyl-2-pyrrolidon

Parameter	Concentration	Condition	Species
LC0	5000 mg/l	not indicated	Colpoda[1]
LC0	5000 mg/l	not indicated	Pseudomonas[1]
LC0	5000 mg/l	not indicated	Scenedesmus[1]

Source: 1.Meinck, F., Stooff, H. and Kohlschutter, H., Les eaux residuaires industrielles, 1970.

4-473 Effect of n-octane

Parameter	Concentration	Condition	Species
LC0	100 mg/l	96 hours, seawater, 8 C	Coho salmon[1]

Source: 1.Morrow, J. E., EPA/660&3/73/018, 1974.

4-474 Effect of n-pentane

Parameter	Concentration	Condition	Species
LC0	100 mg/l	96 hours, seawater, 8 C	Coho salmon[1]
LC0	60 mg/l	not indicated	Gardon[1]

Source: 1.Meinck, F., Stooff, H. and Kohlschutter, H., Les eaux residuaires industrielles, 1970.

4-475 Effect of n-propanol

Parameter	Concentration	Condition	Species
LC0	200 mg/l	24 hours	Creek chub[1]
LC100	500 mg/l	24 hours	Creek chub[1]
LD50	4.5 g pr. kg body weight	single oral dose	Mouse[2]
LD50	2.8 g pr. kg body weight	single oral dose	Rabbit[2]
LD50	1.9 g pr. kg body weight	single oral dose	Rat[2]
Sublethal effect	255 mg/l	Inhibition of cell multiplication	Microcystis aeruginosa[3]
Sublethal effect	11200 mg/l	Inhibition of cell multiplication	Pseudomonas putida[3]
TLm	200-400 mg/l	24-96 hours	Gudgeon[4]

Source: 1.Gilette, L. A., Miller, D. L. and Redman, H. E., Sewage Ind. Wastes, 24, 1397, 1952.
2.Patty, F. A., Industrial Hygiene and toxicology, 2, Interscience Publishers, New York, 1976.
3.Bringmann, G. and Kuhn, R., nwf-wasser/abwasser, H.9, 1976.
4.Meinck, F., Stooff, H. and Kohlschutter, H., Les eaux residuaires industrielles, 1970.

4-476 Effect of n-propylacetate

Parameter	Concentration	Condition	Species
Sublethal effect	530 mg/l	Inhibition of cell multiplication	Microcystis aeruginosa[1]
Sublethal effect	170 mg/l	Inhibition of cell multiplication	Pseudomonas putida[1]

Source: 1.Bringmann, G. and Kuhn, R., nwf-wasser/abwasser, H.9, 1976.

4-477 Effect of n-propylamine

Parameter	Concentration	Condition	Species
LD50	0.2-0.4 g pr. kg body weight	acute oral dose	Rat[1]

Source: 1.Patty, F. A., Industrial Hygiene and toxicology, 2, Interscience Publishers, New York, 1976.

4-478 Effect of n-propylnitrate

Parameter	Concentration	Condition	Species
LD50	5.0-7.5 g pr. kg body weight	single oral dose, 100 %	Rat[1]

Source: 1.Patty, F. A., Industrial Hygiene and toxicology, 2, Interscience Publishers, New York, 1976.

4-479 Effect of n-valeralaldehyde

Parameter	Concentration	Condition	Species
LD50	6.4-12.8 g pr. kg body weight	acute oral dose	Mouse[1]
LD50	3.2-6.4 g pr. kg body weight	acute oral dose	Rat[1]

Source: 1.Patty, F. A., Industrial Hygiene and toxicology, 2, Interscience Publishers, New York, 1976.

4-480 Effect of n-valeric acid

Parameter	Concentration	Condition	Species
LD50	>0.4 g pr. kg body weight	single oral dose	Rat[1]

Source: 1.Patty, F. A., Industrial Hygiene and toxicology, 2, Interscience Publishers, New York, 1976.

4-481 Effect of o-chlorobenzoic acid

Parameter	Concentration	Condition	Species
LD50	>0.5 g pr. kg body weight	acute oral dose	Rabbit[1]

Source: 1.Patty, F. A., Industrial Hygiene and toxicology, 2, Interscience Publishers, New York, 1976.

4-482 Effect of o-chlorophenol

Parameter	Concentration	Condition	Species
TLm	8.2-8.4 mg/l	24 hours	Bluegill[1]
TLm	11-22 mg/l	24-96 hours	Fathead minnow[2]
TLm	12-15 mg/l	24-96 hours	Goldfish[2]
TLm	20-23 mg/l	24-96 hours	Guppie[2]
TLm	8.2-8.4 mg/l	24 hours	SunFish[3]

Source: 1.Lammering, M. W. and Burbank, N. J. Jr., Purdue Univ. Eng. Bull. Ext. Ser., 106, 541, 1960. 2.Pickering Q. H. and Henderson, C., JWPCF, 38, 1419, 1966.3.McKee, J. E., and Wolf, H. W., Water Quality, The resources agency of California, 1963.

4-483 Effect of o-cresol

Parameter	Concentration	Condition	Species
LC0	16 mg/l	not indicated	Daphnia[1]
LC0	60 mg/l	not indicated	Escherichia coli[2]
LC0	40 mg/l	not indicated	Scenedesmus[2]
Sublethal effect	6.8 mg/l	Inhibition of cell multiplication	Microcystis aeruginosa[2]
Sublethal effect	33 mg/l	Inhibition of cell multiplication	Pseudomonas putida[2]
TLm	21-22 mg/l	24-96 hours	BlueFish[3]
TLm	30 mg/l	24 hours	Crucian carp[4]
TLm	13-18 mg/l	24-96 hours	Fathead minnow[3]

TLm	19-49 mg/l	24-96 hours	Goldfish[3]
TLm	18-50 mg/l	24-96 hours	Guppie[3]
TLm	15-16 mg/l	24 hours	Roach[4]
TLm	15-16 mg/l	24 hours	Tench[4]

Source: 1.Meinck, F., Stooff, H. and Kohlschutter, H., Les eaux residuaires industrielles, 1970.
2.Bringmann, G. and Kuhn, R., nwf-wasser/abwasser, H.9, 1976.
3.Pickering Q. H. and Henderson, C., JWPCF, 38, 1419, 1966.
4.Europian inland fishery advisory commission, Water quality criteria for european freshwater fish, Waer Res., 7, 929, 1973.

4-484 Effect of o-dichlorobenzene

Parameter	Concentration	Condition	Species
Sublethal effect	53 mg/l	Inhibition of cell multiplication	Microcystis aeruginosa[1]
Sublethal effect	15 mg/l	Inhibition of cell multiplication	Pseudomonas putida[1]

Source: 1.Bringmann, G. and Kuhn, R., nwf-wasser/abwasser, H.9, 1976.

4-485 Effect of o-dimethylphtalate

Parameter	Concentration	Condition	Species
LD50	2.4 g pr. kg body weight	acute oral dose	guinea pig[1]
LD50	7.2 g pr. kg body weight	acute oral dose	Mouse[1]
LD50	6.9 g pr. kg body weight	acute oral dose	Rat[1]

Source: 1.Patty, F. A., Industrial Hygiene and toxicology, 2, Interscience Publishers, New York, 1976.

4-486 Effect of o-dinitrobenzene

Parameter	Concentration	Condition	Species
Sublethal effect	0.15 mg/l	Inhibition of cell multiplication	Microcystis aeruginosa[1]
Sublethal effect	16 mg/l	Inhibition of cell multiplication	Pseudomonas putida[1]

Source: 1.Bringmann, G. and Kuhn, R., nwf-wasser/abwasser, H.9, 1976.

4-487 Effect of o-methylcyclohexanone

Parameter	Concentration	Condition	Species
LD50	1.0-1.2 g pr. kg body weight	single oral dose, 100 %	Rabbit[1]

Source: 1.Patty, F. A., Industrial Hygiene and toxicology, 2, Interscience Publishers, New York, 1976.

4-488 Effect of o-nitrobenzoic acid

Parameter	Concentration	Condition	Species
LD50	>0.2 g pr. kg body weight	acute oral dose	Rabbit[1]

Source: 1.Patty, F. A., Industrial Hygiene and toxicology, 2, Interscience Publishers, New York, 1976.

4-489 Effect of o-nitrophenol

Parameter	Concentration	Condition	Species
LC0	1000 mg/l	not indicated	Pseudomonas[1]
LC0	36 mg/l	not indicated	Scenedesmus[1]
LC0	60 mg/l	not indicated	Scenedesmus[1]
Sublethal effect	27 mg/l	Inhibition of cell multiplication	Microcystis aeruginosa[2]
Sublethal effect	0.9 mg/l	Inhibition of cell multiplication	Pseudomonas putida[2]
TLm	46-67 mg/l	24-48 hours, hard water	Bluegill[3]
TLm	46-67 mg/l	24-48 hours, hard water	Bluegill[4]
TLm	14-18 mg/l	6 hours, distilled water	Vairon fish[1]
TLm	125-130 mg/l	6 hours, hard water	Vairon fish[1]

Source: 1.Meinck, F., Stooff, H. and Kohlschutter, H., Les eaux residuaires industrielles, 1970.
2.Bringmann, G. and Kuhn, R., nwf-wasser/abwasser, H.9, 1976.
3.Lammering, M. W. and Burbank, N. J. Jr., Purdue Univ. Eng. Bull. Ext. Ser., 106, 541, 1960.
4.Jones, H. R., Environmental Control in the organic and petrochemical Industries, Noyes Data Corp., 171.

4-490 Effect of o-nitrotoluene

Parameter	Concentration	Condition	Species
Sublethal effect	3.1 mg/l	Inhibition of cell multiplication	Microcystis aeruginosa[1]
Sublethal effect	18 mg/l	Inhibition of cell multiplication	Pseudomonas putida[1]
TLm	35-40 mg/l	6 hours, hard water	Vairon fish[2]
TLm	18-20 mg/l	6 hours, distilled water	Vairon fish[2]
TLm	2-10 mg/l	24-96 hours	Fish[2]

Source: 1.Bringmann, G. and Kuhn, R., nwf-wasser/abwasser, H.9, 1976.
2.Meinck, F., Stooff, H. and Kohlschutter, H., Les eaux residuaires industrielles, 1970.

4-491 Effect of o-phenylphenol

Parameter	Concentration	Condition	Species
LD50	0.5 g pr. kg body weight	repeated oral dose, 0 %	dog[1]
LD50	2.7 g pr. kg body weight	oral dose	Rat[1]

Source: 1.Patty, F. A., Industrial Hygiene and toxicology, 2, Interscience Publishers, New York, 1976.

4-492 Effect of o-phtalic acid

Parameter	Concentration	Condition	Species
LD50	3.2 g pr. kg body weight	acute oral dose, as K-salt	Rat[1]

Source: 1.Patty, F. A., Industrial Hygiene and toxicology, 2, Interscience Publishers, New York, 1976.

4-493 Effect of o-phtalic anhydride

Parameter	Concentration	Condition	Species
LD50	0.8-1.6 g pr. kg body weight	single oral dose	Rat[1]

Source: 1.Patty, F. A., Industrial Hygiene and toxicology, 2, Interscience Publishers, New York, 1976.

4-494 Effect of o-toluic acid

Parameter	Concentration	Condition	Species
LD50	>3.2 g pr. kg body weight	acute oral dose	Rat[1]

Source: 1.Patty, F. A., Industrial Hygiene and toxicology, 2, Interscience Publishers, New York, 1976.

4-495 Effect of o-toluidine

Parameter	Concentration	Condition	Species
LC0	8 mg/l	not indicated	Daphnia[1]
LC0	1000 mg/l	not indicated	Escherichia coli[2]
LC0	10 mg/l	not indicated	Scenedesmus[2]
Sublethal effect	0.31 mg/l	Inhibition of cell multiplication	Microcystis aeruginosa[2]
Sublethal effect	16 mg/l	Inhibition of cell multiplication	Pseudomonas putida[2]
TLm	100 mg/l	not indicated	Fish[1]

Source: 1.Meinck, F., Stooff, H. and Kohlschutter, H., Les eaux residuaires industrielles, 1970.
2.Bringmann, G. and Kuhn, R., nwf-wasser/abwasser, H.9, 1976.

4-496 Effect of p-aminobenzoic acid

Parameter	Concentration	Condition	Species
LD50	1.0 g pr. kg body weight	acute oral dose	dog[1]
LD50	2.85 g pr. kg body weight	acute oral dose	Mouse[1]

Source: 1.Patty, F. A., Industrial Hygiene and toxicology, 2, Interscience Publishers, New York, 1976.

4-497 Effect of p-benzoquinone

Parameter	Concentration	Condition	Species
LD50	0.094 g pr. kg body weight	parenteral dose	Mouse[1]
LD50	0.13-0.3 g pr. kg body weight	oral dose	Rat[1]
LD50	0.025 g pr. kg body weight	parenteral dose	Rat[1]
Sublethal effect	0.1 ppm	Eye irritation	Man[1]

Source: 1.Henschler, D., Toxikologisch-arbeitsmedizinische Begrundung von MAH-werte, Verlag Chemie.

4-498 Effect of p-chlorophenol

Parameter	Concentration	Condition	Species
LD50	0.67 g pr. kg body weight	oral dose	Rat[1]

Source: 1.Patty, F. A., Industrial Hygiene and toxicology, 2, Interscience Publishers, New York, 1976.

4-499 Effect of p-cresol

Parameter	Concentration	Condition	Species
LC0	12 mg/l	not indicated	Daphnia[1]
LC0	>1000 mg/l	not indicated	Escherichia coli[1]
LC0	6 mg/l	not indicated	Scenedesmus[1]
LC50	21 mg/l	24 hours, respiratory route	Crucian carp[2]
LC50	5 mg/l	24 hours, respiratory route	Rainbow trout[3]

LC50	17 mg/l	24 hours, respiratory route	Roach[2]
LC50	16 mg/l	24 hours, respiratory route	Tench[2]
LC50	4 mg/l	24 hours, respiratory route	Trout[2]
LD50	1.1-1.8 g pr. kg body weight	single oral dose, 100 %	rat, Rabbit[4]

Source: 1.Meinck, F., Stooff, H. and Kohlschutter, H., Les eaux residuaires industrielles, 1970.
2.Europian inland fishery advisory commission, Water quality criteria for european freshwater fish, Waer Res., 7, 929, 1973.
3.Longwell and Pentelow, J. Exp. Biol., 12, 1, 1935.
4.Patty, F. A., Industrial Hygiene and toxicology, 2, Interscience Publishers, New York, 1976.

4-500 Effect of p-dichlorobenzene

Parameter	Concentration	Condition	Species
Sublethal effect	50 ppm	Eye irritation	Man[1]

Source: 1.Patty, F. A., Industrial Hygiene and toxicology, 2, Interscience Publishers, New York, 1976.

4-501 Effect of p-hydroxybenzoic acid

Parameter	Concentration	Condition	Species
LD50	3 g pr. kg body weight	intraperitoneal dose	guinea pig[1]
LD50	0.4-3.2 g pr. kg body weight	acute oral dose	Rat[1]

Source: 1.Patty, F. A., Industrial Hygiene and toxicology, 2, Interscience Publishers, New York, 1976.

4-502 Effect of p-hydroxyphenylglycine

Parameter	Concentration	Condition	Species
LC100	20 mg/l	48 hours	Goldfish[1]

Source: 1.McKee, J. E., and Wolf, H. W., Water Quality, The resources agency of California, 1963.

4-503 Effect of p-nitroaniline

Parameter	Concentration	Condition	Species
Sublethal effect	0.35 mg/l	Inhibition of cell multiplication	Microcystis aeruginosa[1]
Sublethal effect	3.8 mg/l	Inhibition of cell multiplication	Pseudomonas putida[1]

Source: 1.Bringmann, G. and Kuhn, R., nwf-wasser/abwasser, H.9, 1976.

4-504 Effect of p-nitrobenzoic acid

Parameter	Concentration	Condition	Species
LD50	>0.2 g pr. kg body weight	acute oral dose	Rabbit[1]

Source: 1.Patty, F. A., Industrial Hygiene and toxicology, 2, Interscience Publishers, New York, 1976.

4-505 Effect of p-nitrophenol

Parameter	Concentration	Condition	Species
LC0	14 mg/l	not indicated	Daphnia[1]
LC0	100 mg/l	not indicated	Pseudomonas[1]
LC0	42 mg/l	not indicated	Scenedesmus[1]
Sublethal effect	56 mg/l	Inhibition of cell multiplication	Microcystis aeruginosa[2]
Sublethal effect	4 mg/l	Inhibition of cell multiplication	Pseudomonas putida[2]

Source: 1.Meinck, F., Stooff, H. and Kohlschutter, H., Les eaux residuaires industrielles, 1970.
2.Bringmann, G. and Kuhn, R., nwf-wasser/abwasser, H.9, 1976.

4-506 Effect of p-nitrotoluene

Parameter	Concentration	Condition	Species
Sublethal effect	3.3 mg/l	Inhibition of cell multiplication	Microcystis aeruginosa[1]
Sublethal effect	26 mg/l	Inhibition of cell multiplication	Pseudomonas putida[1]
TLm	45-50 mg/l	6 hours, hard water	Vairon fish[2]
TLm	20-22 mg/l	6 hours, distilled water	Vairon fish[2]

Source: 1.Bringmann, G. and Kuhn, R., nwf-wasser/abwasser, H.9, 1976.
2.Meinck, F., Stooff, H. and Kohlschutter, H., Les eaux residuaires industrielles, 1970.

4-507 Effect of p-phenyldiamine

Parameter	Concentration	Condition	Species
LC100	5.74 mg/l	48 hours	Goldfish[1]

Source: 1.McKee, J. E., and Wolf, H. W., Water Quality, The resources agency of California, 1963.

4-508 Effect of p-tert-butylbenzoicacid

Parameter	Concentration	Condition	Species
LD50	0.57 g pr. kg body weight	acute oral dose	Mouse[1]
LD50	0.74 g pr. kg body weight	acute oral dose	Rat[1]

Source: 1.Shell Chemie, Shell Industrie Chemicalien gids, Shell Niederland Chemie, 1975.

4-509 Effect of p-tert-butylphenol

Parameter	Concentration	Condition	Species
LC0	8 mg/l	not indicated	Daphnia[1]
LC0	>100 mg/l	not indicated	Escherichia coli[1]
LC0	10 mg/l	not indicated	Scenedesmus[1]

Source: 1.Meinck, F., Stooff, H. and Kohlschutter, H., Les eaux residuaires industrielles, 1970.

4-510 Effect of p-tert-butyltoluene

Parameter	Concentration	Condition	Species
LC50	3 mg/l	24 hours	Goldfish[1]
LD50	0.014-0.028 g pr. kg body weight	percutane dose	Rabbit[2]
Sublethal effect	80 ppm	Eye irritation, 5 min.	Man[2]

Source: 1.Shell Chemie, Shell Industrie Chemicalien gids, Shell Niederland Chemie, 1975.
2.Patty, F. A., Industrial Hygiene and toxicology, 2, Interscience Publishers, New York, 1976.

4-511 Effect of p-toluene-sulfonic acid

Parameter	Concentration	Condition	Species
LC50	245 mg/l	1 day, 20 C	Chlorella vulgaris[1]

Source: 1.Kauss, P. B. and Hutchinson, T. C., Environ Poll., 9, 1975.

4-512 Effect of p-toluidine

Parameter	Concentration	Condition	Species
LC0	0.6 mg/l	not indicated	Daphnia[1]
LC0	1000 mg/l	not indicated	Escherichia coli[1]
LC0	10 mg/l	not indicated	Scenedesmus[1]

Source: 1.Meinck, F., Stooff, H. and Kohlschutter, H., Les eaux residuaires industrielles, 1970.

4-513 Effect of pH

Parameter	Effect Value	Concentration and Condition	Species
Biomass	122 microgram chl a/100 cm**2	Algae biomass, pH = 4.3 - 5.5, spring, natural stream	Biomass in streams[1]
Biomass	140 microgram chl a/100 cm**2	Algae biomass, pH = 4, spring	Biomass in streams[1]
Biomass	140 microgram chl a/100 cm**2	Algae biomass, pH = 4.3 - 5.5, summer, natural stream	Biomass in streams[1]
Biomass	186 microgram chl a/100 cm**2	Algae biomass, pH = 4, spring	Biomass in streams[1]
Biomass	36 microgram chl a/100 cm**2	Algae biomass, pH = 6, spring	Biomass in streams[1]
Biomass	43 microgram chl a/100 cm**2	Algae biomass, pH = 4.3 - 5.5, fall, natural stream	Biomass in streams[1]
Biomass	62 microgram chl a/100 cm**2	Algae biomass, pH = 4, spring	Biomass in streams[1]
Biomass	63 microgram chl a/100 cm**2	Algae biomass, pH = 6, spring	Biomass in streams[1]
Biomass	82 microgram chl a/100 cm**2	Algae biomass, pH = 6, spring	Biomass in streams[1]
Diversity	0.92 Shannon diversity	pH = 3.20, number of individual = 21,10 ft2,number of taxa = 2, strip	River community[2]
Diversity	1.43 Shannon diversity	pH = 4.90, no of ind = 250/10 ft**2, no of taxa = 15, 10 ft2 = 0.93	River community[2]
Diversity	1.94 Shannon diversity	pH = 5.00, no of ind = 359/10 ft2, no of taxa = 13	River community[2]
Diversity	1.95 Shannon diversity	pH = 4.68, no of indeveduals = 143, no of taxa = 17	River community[2]
Diversity	3.10 Shannon diversity	pH = 6, number of ind = 467,10 ft2, number of taxa = 30, control	River community[2]
Hacthing of eggs	2 %	pH = 4.0, 289 K, 253-274 eggs, 72 h after fert	Brachydanio rerio, eggs[3]
Hacthing of eggs	20 %	pH = 4.0, 289 K, 253-274 eggs, 72 h after fert	Brachydanio rerio, eggs[3]
Hacthing of eggs	39 %	pH = 7.0 - 10.0, 289 K, 253-274 eggs, 72 h after f	Brachydanio rerio, eggs[3]
Hacthing of eggs	5 %	pH = 5.5, 289 K, 253-274 eggs, 72 h after fert	Brachydanio rerio, eggs[3]
Hacthing of eggs	83 %	pH = 5.5, 289 K, 253-274 eggs, 72 h after fert	Brachydanio rerio, eggs[3]
Hacthing of eggs	89 %	pH = 10.0, 289 K, 253-274 eggs, 72 h after fert	Brachydanio rerio, eggs[3]

Hacthing of eggs	92 %	pH = 7.0, 289 K, 253-274 eggs, 72 h after fert	Brachydanio rerio, eggs[3]
Hacthing of eggs	95 %	pH = 8.5, 289 K, 253-274 eggs, 72 h after fert	Brachydanio rerio, eggs[3]
Hatching of eggs	13 % hatched of fertile eggs	pH = 4.0, 289 K, 132-133 eggs, fish 96 h after fert	Brachydanio rerio, eggs[3]
Hatching of eggs	14 % dying later than 72 hours after fert	pH = 7.0 -10.0, 289 K, 121-205 eggs, fish later 72 h after fert	Brachydanio rerio, eggs[3]
Hatching of eggs	20 % hatched of fertile eggs	pH = 5.5, 289 K, 253-274 eggs, Brachydanio rerio	Brachydanio rerio, eggs[3]
Hatching of eggs	4 % hatched of fertile eggs	pH = 4.0, 289 K, 253-274 eggs, Brachydanio rerio	Brachydanio rerio, eggs[3]
Hatching of eggs	40 % dying later than 72 hours after fert	pH = 5.5, 289 K, 121-205 eggs, fish later 72 h after fert	Brachydanio rerio, eggs[3]
Hatching of eggs	48 % hatched of fertile eggs	pH = 7.0, 289 K, 253-274 eggs, Brachydanio rerio	Brachydanio rerio, eggs[3]
Hatching of eggs	49 % hatched of fertile eggs	pH = 8.5, 289 K, 253-274 eggs, Brachydanio rerio	Brachydanio rerio, eggs[3]
Hatching of eggs	50 % hatched of fertile eggs	pH = 5.5, 289 K, 132-133 eggs, fish 96 h after fert	Brachydanio rerio, eggs[3]
Hatching of eggs	52 % hatched of fertile eggs	pH = 10.0, 289 K, 253-274 eggs, Brachydanio rerio	Brachydanio rerio, eggs[3]
Hatching of eggs	75 % hatched of fertile eggs	pH = 7.0, 289 K, 132-133 eggs, fish 96 h after fert	Brachydanio rerio, eggs[3]
Hatching of eggs	84 % hatched of fertile eggs	pH = 8.5, 289 K, 132-133 eggs, fish 96 h after fert	Brachydanio rerio, eggs[3]
Hatching of eggs	98 % hatched of fertile eggs	pH = 10.0, 289 K, 132-133 eggs, fish 96 h after fert	Brachydanio rerio, eggs[3]
Hatching of eggs	17.0 % of hatched of fertile eggs	Control = 44.4 % (pH = 6.8), pH = 5.0, after 20 d, 299 K	Jordanella floridae[4]
Hatching of eggs	51.1 % of hatched of fertile eggs	Control = 44.4 % (pH = 6.8), pH = 5.5, after 20 d, 299 K	Jordanella floridae[4]
Hatching of eggs	55.2 % of fertile eggs	Control = 90.1 % (pH = 6.8), pH = 5.0, after 20 d, 299 K	Jordanella floridae[4]
Hatching of eggs	56.5 % of hatched of fertile eggs	Control = 44.4 % (pH = 6.8), pH = 6.0, after 20 d, 299 K	Jordanella floridae[4]
Hatching of eggs	68.0 % of fertile eggs	Control = 90.1 % (pH = 6.8), pH = 5.5, after 20 d, 299 K	Jordanella floridae[4]
Hatching of eggs	78.6 % of fertile eggs	Control = 90.1 % (pH = 6.8), pH = 6.0, after 20 d, 299 K	Jordanella floridae[4]
Hatching of eggs per spawning	1 egg	Control = 2467eggs (pH = 6.8), pH = 4.5, after 20 d, 299 K	Jordanella floridae[4]
Hatching of eggs per spawning	1277 eggs per spawning	Control = 2467eggs (pH = 6.8), pH = 6.0, after 20 d, 299 K	Jordanella floridae[4]
Hatching of eggs per spawning	319 eggs per spawning	Control = 2467eggs (pH = 6.8), pH = 5.0, after 20 d, 299 K	Jordanella floridae[4]
Hatching of eggs per spawning	557 eggs per spawning	Control = 2467eggs (pH = 6.8), pH = 5.5, after 20 d, 299 K	Jordanella floridae[4]
Mortality	50 %	pH = 3.85, 30 days exposure	Acroneuria lycorias[5]
Mortality	50 %	pH = 4.42, 30 days exposure	Boyeria vinosa[5]

Mortality	50 %	pH = 2.45, 30 days exposure	Brachycentrus americanus[5]
Mortality	50 %	pH = 5.38, 30 days exposure	Ephemerella subvaria[5]
Mortality	50 %	pH = 3.38, 30 days exposure	Hydropsyche betteni[5]
Mortality	50 %	pH = 4.50, 30 days exposure	Isogenus frontalis[5]
Mortality	50 %	pH = 4.30, 30 days exposure	Ophiogomphus rupinsulensis[5]
Mortality	50 %	pH = 5.00, 30 days exposure	Pteronarcys dorsata[5]
Mortality	50 %	pH = 3.71, 30 days exposure	Taeniopteryx maura[5]
Number of spawnings	1	Control = 32 (pH = 6.8), pH = 4.5, after 20 d, 299 K	Jordanella floridae[4]
Number of spawnings	15	Control = 32 (pH = 6.8), pH = 5.5, after 20 d, 299 K	Jordanella floridae[4]
Number of spawnings	24	Control = 32 (pH = 6.8), pH = 6.0, after 20 d, 299 K	Jordanella floridae[4]
Number of spawnings	7	Control = 32 (pH = 6.8), pH = 5.0, after 20 d, 299 K	Jordanella floridae[4]
Succesful emergence	50 %	pH = 5.0, 30 days exposure	Acroneuria lycorias[5]
Succesful emergence	50 %	pH = 5.2, 30 days exposure	Boyeria vinosa[5]
Succesful emergence	50 %	pH = 4.0, 30 days exposure	Brachycentrus americanus[5]
Succesful emergence	50 %	pH = 5.9, 30 days exposure	Ephemerella subvaria[5]
Succesful emergence	50 %	pH = 4.7, 30 days exposure	Hydropsyche betteni[5]
Succesful emergence	50 %	pH = 6.6, 30 days exposure	Isogenus frontalis[5]
Succesful emergence	50 %	pH = 5.2, 30 days exposure	Ophiogomphus rupinsulensis[5]
Succesful emergence	50 %	pH = 5.8, 30 days exposure	Pteronarcys dorsata[5]
Succesful emergence	50 %	pH = 5.0, 30 days exposure	Taeniopteryx maura[5]
Uptake of Pb	1010 mg/g dry weight	Roots, 12 h, pH = 4.5, lead nitrate	Barley[6]
Uptake of Pb	1200 mg/g dry weight	Roots, 12 h, pH = 7.0, lead nitrate	Barley[6]
Uptake of Pb	1240 mg/g dry weight	Roots, 12 h, pH = 5.5, lead nitrate	Barley[6]
Uptake of Pb	1450 mg/g dry weight	Roots, pH = 4.5, 4 h, lead nitrate	Bean[6]
Uptake of Pb	1650 mg/g dry weight	Roots, pH = 5.5, 4 h, lead nitrate	Bean[6]
Uptake of Pb	1875 mg/g dry weight	Roots, 12 h, pH = 7.0, lead nitrate	Bean[6]
Uptake of Pb	2925 mg/g dry weight	Roots, 12 h, pH = 5.5, lead nitrate	Bean[6]
Uptake of Pb	3350 mg/g dry weight	Roots, pH = 7.0, 24 h, lead nitrate	Bean[6]
Uptake of Pb	4600 mg/g dry weight	Roots, pH = 4.5, 24 h, lead nitrate	Bean[6]
Uptake of Pb	5175 mg/g dry weight	Roots, pH = 5.5, 24 h, lead nitrate	Bean[6]
Uptake of Pb	837 mg/g dry weight	Roots, pH = 7.0, 4 h, lead nitrate	Bean[6]
Uptake of Pb	1200 mg/g dry weight	Roots, 12 h, pH = 7.0, lead nitrate	Corn[6]
Uptake of Pb	680 mg/g dry weight	Roots, 12 h, pH = 4.5, lead nitrate	Corn[6]

Uptake of Pb	920 mg/g dry weight	Roots, 12 h, pH = 5.5, lead nitrate	Corn[6]
Uptake of Pb	1000 mg/g dry weight	Roots, 12 h, pH = 5.5, lead nitrate	Radish[6]
Uptake of Pb	900 mg/g dry weight	Roots, 12 h, pH = 7.0, lead nitrate	Radish[6]
Uptake of Pb	910 mg/g dry weight	Roots, 12 h, pH = 4.5, lead nitrate	Radish[6]
Uptake of Pb	2950 mg/g dry weight	Roots, 12 h, pH = 4.5, lead nitrate	Soybean[6]
Uptake of Pb	3700 mg/g dry weight	Roots,12 h, pH = 4.5, lead nitrate	Soybean[6]
Uptake of Pb	4900 mg/g dry weight	Roots, 12 h, pH = 7.0, lead nitrate	Soybean[6]
Uptake of Pb	5150 mg/g dry weight	Roots, 12 h, pH = 5.5, lead nitrate	Soybean[6]
Weight	0.075 gram	Control = 0.55 gram (pH = 6.8), pH = 5.0, after 45 d, 299 k, fry	Jordanella floridae[4]
Weight	0.26 gram	Control = 0.55 gram (pH = 6.8), pH = 5.5, after 45 d, 299 k, fry	Jordanella floridae[4]
Weight	0.42 gram	Control = 0.55 gram (pH = 6.8), pH = 6.0, after 45 d, 299 k, fry	Jordanella floridae[4]

Source: 1.Hendrey, G.R., K. Baalsrud, T.S. Traaen, M. Laake and G. Raddum. 1976. Ambio. 5(5-6), pp 224-227.
2.Tomkiewicz, S.M. JR. and W.A. Dunson. 1977. Water Res., 11(4), pp 397-402.
3.Johanson, N., J.E. Kihlstroem and A. Wahlberg. 1973. Ambio. II(1-2), pp 42-43.
4.Craig, R.G. and W.F. Baksi. 1977. Water Res., 11(8), pp 621-626.
5.Bell, H.L. 1971. Water Res., 5(6), pp 313-319.
6.Arvik, J.H. and R.L. Zimdahl. 1974. J. Environ. Qual., 3(4), pp 374-376.

4-514 Effect of pH and HCO3(-)

Parameter	Effect Value	Concentration and Condition	Species
Cupric ion / total soluble Cu ratio	0.009	pH = 8.0, 5E-3m HCO3(-)	Cu[1]
Cupric ion / total soluble Cu ratio	0.02	pH = 8.0, 2E-3m HCO3(-)	Cu[1]
Cupric ion / total soluble Cu ratio	0.03	pH = 7.5, 5E-3m HCO3(-)	Cu[1]
Cupric ion / total soluble Cu ratio	0.06	pH = 7.5, 2E-3m HCO3(-)	Cu[1]
Cupric ion / total soluble Cu ratio	0.09	pH = 7.0, 5E-3m HCO3(-)	Cu[1]
Cupric ion / total soluble Cu ratio	0.11	pH = 7.5, 1E-3m HCO3(-)	Cu[1]
Cupric ion / total soluble Cu ratio	0.14	pH = 7.0, 2E-3m HCO3(-)	Cu[1]
Cupric ion / total soluble Cu ratio	0.25	pH = 7.0, 1E-3m HCO3(-)	Cu[1]

Source: 1.Tomkiewicz, S.M. JR. and W.A. Dunson. 1977. Water Res., 11(4), pp 397-402.

4-515 Effect of paraldehyde

Parameter	Concentration	Condition	Species
LD50	3.0-4.0 g pr. kg body weight	acute oral dose	dog[1]
LD50	5.0 g pr. kg body weight	acute oral dose	Rabbit[1]
LD50	1.65 g pr. kg body weight	acute oral dose	Rat[1]

Source: 1.Patty, F. A., Industrial Hygiene and toxicology, 2, Interscience Publishers, New York, 1976.

4-516 Effect of sec-butanol

Parameter	Concentration	Condition	Species
LC50	4300 mg/l	24 hours	Goldfish[1]
LD50	6.5 g pr. kg body weight	single oral dose	Rat[2]
Sublethal effect	312 mg/l	Inhibition of cell multiplication	Microcystis aeruginosa[3]
Sublethal effect	500 mg/l	Inhibition of cell multiplication	Pseudomonas putida[3]

Source: 1.Turnbull, H., DeMann, J. H. and Weston, R. F., Ind. Eng. Chem., 46, 324, 1954.
2.Patty, F. A., Industrial Hygiene and toxicology, 2, Interscience Publishers, New York, 1976.
3.Bringmann, G. and Kuhn, R., nwf-wasser/abwasser, H.9, 1976.

4-517 Effect of sec-butylbenzene

Parameter	Concentration	Condition	Species
LC50	4 mg/l	24-96 hours, pH=5	Goldfish[1]
LC50	33 mg/l	24-96 hours, pH=7	Goldfish[1]

Source: 1.Shell Chemie, Shell Industrie Chemicalien gids, Shell Niederland Chemie, 1975.

4-518 Effect of t-butanol

Parameter	Concentration	Condition	Species
LC0	3000 mg/l	24 hours	Creek chub[1]
LC100	6000 mg/l	24 hours	Creek chub[1]

Source: 1.Gilette, L. A., Miller, D. L. and Redman, H. E., Sewage Ind. Wastes, 24, 1397, 1952.

4-519 Effect of tert-butylacetate

Parameter	Concentration	Condition	Species
Sublethal effect	420 mg/l	Inhibition of cell multiplication	Microcystis aeruginosa[1]
Sublethal effect	78 mg/l	Inhibition of cell multiplication	Pseudomonas putida[1]

Source: 1.Bringmann, G. and Kuhn, R., nwf-wasser/abwasser, H.9, 1976.

4-520 Effect of 1,1,1-trichlorethane

Parameter	Concentration	Condition	Species
TLm	75-100 mg/l	not indicated	Fish[1]

Source: 1.Meinck, F., Stooff, H. and Kohlschutter, H., Les eaux residuaires industrielles, 1970.

4-521 Effect of 1,1,1-trichloroethane

Parameter	Concentration	Condition	Species
LD50	11.2 g pr. kg body weight	single oral dose	Female mouse[1]
LD50	5.66 g pr. kg body weight	single oral dose	Female rabbit[1]
LD50	9.5 g pr. kg body weight	single oral dose	male guinea pig[1]
LD50	10.3-12.3 g pr. kg body weight	single oral dose	Rat[1]

Source: 1.Patty, F. A., Industrial Hygiene and toxicology, 2, Interscience Publishers, New York, 1976.

4-522 Effect of 1,1,2-Trichloroethane

Parameter	Concentration	Condition	Species
LD50	0.1-0.2 g pr. kg body weight	single oral dose	Rat[1]

Source: 1.Patty, F. A., Industrial Hygiene and toxicology, 2, Interscience Publishers, New York, 1976.

4-523 Effect of 1,1-Dimethylhydrazine

Parameter	Concentration	Condition	Species
LD50	0.265 g pr. kg body weight	single oral dose	Mouse[1]
LD50	0.122 g pr. kg body weight	single oral dose	Rat[1]

Source: 1.Patty, F. A., Industrial Hygiene and toxicology, 2, Interscience Publishers, New York, 1976.

4-524 Effect of 1,2,3,4-Tetrachloro-1,2-difluoroethane

Parameter	Concentration	Condition	Species
LD50	2 g pr. kg body weight	oral dose	Rat[1]

Source: 1.Patty, F. A., Industrial Hygiene and toxicology, 2, Interscience Publishers, New York, 1976.

4-525 Effect of 1,2,3,4-Tetrachlorobenzene

Parameter	Concentration	Condition	Species
LD50	1.5 g pr. kg body weight	oral dose	Rabbit[1]
LD50	1.5 g pr. kg body weight	oral dose	Rat[1]

Source: 1.Stofen, D., Toxicology, 1, 187, 1973.

4-526 Effect of 1,2,3,4-Tetrachloroethane

Parameter	Concentration	Condition	Species
LD50	0.3 g pr. kg body weight	oral dose, 100 %	Rat[1]

Source: 1.Patty, F. A., Industrial Hygiene and toxicology, 2, Interscience Publishers, New York, 1976.

4-527 Effect of 1,2,4-Trimethylbenzene

Parameter	Concentration	Condition	Species
Sublethal effect	1000 mg/l	Eye irritation, 15 times 6 hours	Rat[1]

Source: 1.Cage, J.C. British J. Ind. Medicine 1970, vol 27.

4-528 Effect of 1,2,4,5-Tetramethylbenzene

Parameter	Effect Value	Concentration and Cond.	Species
Mortality	50 % mortality	Water = 2.1 ppm, 96 h, stage 1 zoeae	Cancer magister[1]

Source: 1.Caldwell, R.s., E. M. Caldarone and M.H. Mallon. 1977. D.A. Wolfe (ed.): Fate and Effects of Petrolem Hydrocarbons in Marine Ecosystems and Organisms. NY. Pergamon Press. 210-220.

4-529 Effect of 1,2,4-Trimethylbenzene

Parameter	Effect Value	Concentration and Cond.	Species
Mortality	50 % mortality	Water = 5.1 ppm, 96 h, stage 1 zoeae	Cancer magister[1]
Mortality	50 % mortality	Water = 5.1 ppm, marine amphipod, 96 h	Elasmopus pectenicrus[2]
Mortality	50 % mortality	Water = 5.4 ppm, 96 h	Palaemonetes pugio[3]

Source: 1.Caldwell, R.s., E. M. Caldarone and M.H. Mallon. 1977. D.A. Wolfe (ed.): Fate and Effects of Petrolem Hydrocarbons in Marine Ecosystems and Organisms. NY. Pergamon Press. 210-220.
2.Lee, W.Y. and J.A.C. Nicol. 1978. Mar. Biol. 48, 215-222.
3.Neff, J.M., Anderson, Cox, Laughlin, Rossi and Tatem. 1976. In: Sources, Effects and Sinks of Hydrocrbon in the Aquatic Environment. Wash DC. Amer. Inst. of Biological Science.

4-530 Effect of 1,2-Butyleneoxide

Parameter	Concentration	Condition	Species
LD50	0.5 g pr. kg body weight	single oral dose	Rat[1]

Source: 1.Duvel, W. A. and Helfgott, T., JWPCF, 47, 1975.

4-531 Effect of 1,2-Dichloropropane

Parameter	Concentration	Condition	Species
LD50	2-4 g pr. kg body weight	oral dose,100 %	guinea pig[1]

Source: 1.Patty, F. A., Industrial Hygiene and toxicology, 2, Interscience Publishers, New York, 1976.

4-532 Effect of 1,3,5-Cyclohepatriene

Parameter	Concentration	Condition	Species
LD50	0.057 g pr. kg body weight	acute oral dose	Rat[1]
LD50	0.44-0.88 g pr. kg body weight	acute dermal dose	Rat[1]
LC50	15 mg/l	24 hours	GoldFish[1]

Source: 1.Shell Chemie, Shell Industrie Chemicalien gids, Shell Niederland Chemie, 1975.

4-533 Effect of 1,3,5-Trimethylbenzene

Parameter	Effect Value	Concentration and Cond.	Species
Mortality	50 % mortality	Water = 12.52 ppm, 96 h	Carassius auratus[1]

Source: 1.Brenniman, G., R. Hartung and W.J. Weber. 1976. Water Res. 10, 165-169.

4-534 Effect of 1,3,5-Trioxane

Parameter	Concentration	Condition	Species
LC0	1 mg/l	not indicated	Colpoda[1]
LC0	5 mg/l	not indicated	Daphnia[1]
LC0	1 mg/l	not indicated	Pseudomonas[1]
LC0	1 mg/l	not indicated	Scenedesmus[1]

Source: 1.Meinck, F., Stooff, H. and Kohlschutter, H., Les eaux residuaires industrielles, 1970.

4-535 Effect of 1,3-Butadiene

Parameter	Concentration	Condition	Species
Sublethal effect	8000 ppm	Eye irritation, 8 hours	Man[1]
TLm	71.5 mg/l	24 hours	Pinperch[2]

Source: 1.Patty, F. A., Industrial Hygiene and toxicology, 2, Interscience Publishers, New York, 1976.
2.Jones, H. R., Environmental Control in the organic and petrochemical Industries, Noyes Data Corp., 171.

4-536 Effect of 1,4-Dimethoxybenzene

Parameter	Concentration	Condition	Species
LD50	8.5 g pr. kg body weight	single oral dose	Rat[1]

Source: 1.Patty, F. A., Industrial Hygiene and toxicology, 2, Interscience Publishers, New York, 1976.

4-537 Effect of 1,4-Dioxane

Parameter	Concentration	Condition	Species
LD50	3.9 g pr. kg body weight	single oral dose	guinea pig[1]
LD50	5.7 g pr. kg body weight	single oral dose	Mouse[1]
LD50	5.2 g pr. kg body weight	single oral dose	Rat[1]
Sublethal effect	300 ppm	Eye irritation	Man[1]
Sublethal effect	575 mg/l	Inhibition of cell multiplication	Microcystis aeruginosa[2]
Sublethal effect	2700 mg/l	Inhibition of cell multiplication	Pseudomonas putida[2]

Source: 1.Patty, F. A., Industrial Hygiene and toxicology, 2, Interscience Publishers, New York, 1976.
2.Bringmann, G. and Kuhn, R., nwf-wasser/abwasser, H.9, 1976.

4-538 Effect of 1,5,9-Cyclodecatriene

Parameter	Concentration	Condition	Species
LC50	4 mg/l	24 hours	Goldfish[1]

Source: 1.Shell Chemie, Shell Industrie Chemicalien gids, Shell Niederland Chemie, 1975.

4-539 Effect of 1,5-Cyclo-octadiene

Parameter	Concentration	Condition	Species
LC50	14 mg/l	24 hours	Goldfish[1]
LD50	2.7 g pr. kg body weight	acute oral dose	Rat[2]

Source: 1.Shell Chemie, Shell Industrie Chemicalien gids, Shell Niederland Chemie, 1975.
2.Patty, F. A., Industrial Hygiene and toxicology, 2, Interscience Publishers, New York, 1976.

4-540 Effect of 1-Decanol

Parameter	Concentration	Condition	Species
LD50	6.4-12.8 g pr. kg body weight	single oral dose	Mouse[1]
LD50	12.8-25.6 g pr. kg body weight	single oral dose	Rat[1]

Source: 1.Patty, F. A., Industrial Hygiene and toxicology, 2, Interscience Publishers, New York, 1976.

4-541 Effect of 1-Dodecanol

Parameter	Concentration	Condition	Species
LD50	>12.8 g pr. kg body weight	single oral dose	Rat[1]

Source: 1.Patty, F. A., Industrial Hygiene and toxicology, 2, Interscience Publishers, New York, 1976.

4-542 Effect of 1-Heptanol

Parameter	Concentration	Condition	Species
Sublethal effect	3.50 mg/l	Inhibition of cell multiplication	Microcystis aeruginosa[1]
Sublethal effect	67 mg/l	Inhibition of cell multiplication	Pseudomonas putida[1]

Source: 1.Bringmann, G. and Kuhn, R., nwf-wasser/abwasser, H.9, 1976.

4-543 Effect of 1-Hexadecanol

Parameter	Concentration	Condition	Species
LD50	3.2-12.8 g pr. kg body weight	single oral dose	mouse,Rat[1]

Source: 1.Patty, F. A., Industrial Hygiene and toxicology, 2, Interscience Publishers, New York, 1976.

4-544 Effect of 1-Methylnaphthalene

Parameter	Effect Value	Concentration and Cond.	Species
Mortality	50 % mortality	Water = 6.4 ppm, naplii sedimentated (thought dead), 1 h, S=3 %	Balanus amphitrites[1]
Mortality	50 % mortality	Water = 1.9 ppm, 96 h, stage 1 zoeae	Cancer magister[2]
Mortality	50 % mortality	Water = 3.4 ppm, 24 h	Cyprinodon variegatus[3]

Source: 1.Donahue, W.H., R.T. Wang, M. Welch and J.A.C. Nicol. 1977. Environ. Pollut. 13, 187-202. 2.Caldwell, R.s., E. M. Caldarone and M.H. Mallon. 1977. D.A. Wolfe (ed.): Fate and Effects of Petrolem Hydrocarbons in Marine Ecosystems and Organisms. NY. Pergamon Press. 210-220.3.Anderson, J.W., J.M. Neff, B.A. Cox, Tatem and Hightower. 1974. In: Vernberg and Vernberg (eds.): Polution and Physiology of Marine Animals. NY. Academic Press. 285-310.

4-545 Effect of 1-Methylphenanthrene

Parameter	Effect Value	Concentration and Cond.	Species
Mortality	50 % mortality	Water = 0.3 ppm, 96 h, marine	Neathes arenaceodentata[1]

Source: 1.Rossi, S.S. and J.M. Neff. 1978. Mar. Pollut. Bull. 9, 220-223.

4-546 Effect of 1-Nitrobutane

Parameter	Concentration	Condition	Species
LD50	0.5-0.75 g pr. kg body weight	oral dose, 100 %	Rabbit[1]

Source: 1.Patty, F. A., Industrial Hygiene and toxicology, 2, Interscience Publishers, New York, 1976.

4-547 Effect of 1-Nonanol

Parameter	Concentration	Condition	Species
LD50	6.4-12.8 g pr. kg body weight	single oral dose	Mouse[1]
LD50	3.2-6.4 g pr. kg body weight	single oral dose	Rat[1]

Source: 1.Patty, F. A., Industrial Hygiene and toxicology, 2, Interscience Publishers, New York, 1976.

4-548 Effect of 1-Octanol

Parameter	Concentration	Condition	Species
LC0	5 mg/l	24 hours	Sea lamprey[1]
Sublethal effect	1.9 mg/l	Inhibition of cell multiplication	Microcystis aeruginosa[2]
Sublethal effect	>50 mg/l	Inhibition of cell multiplication	Pseudomonas putida[2]

Source: 1.Meinck, F., Stooff, H. and Kohlschutter, H., Les eaux residuaires industrielles, 1970. 2.Bringmann, G. and Kuhn, R., nwf-wasser/abwasser, H.9, 1976.

4-549 Effect of 1-Pentanol

Parameter	Concentration	Condition	Species
LC0	350 mg/l	24 hours	Creek chub[1]
LC100	500 mg/l	24 hours	Creek chub[1]
Sublethal effect	17 mg/l	Inhibition of cell multiplication	Microcystis aeruginosa[2]
Sublethal effect	220 mg/l	Inhibition of cell multiplication	Pseudomonas putida[2]

Source: 1.Gilette, L. A., Miller, D. L. and Redman, H. E., Sewage Ind. Wastes, 24, 1397, 1952. 2.Bringmann, G. and Kuhn, R., nwf-wasser/abwasser, H.9, 1976.

4-550 Effect of 2,3,4,6-Tetrachlorophenol

Parameter	Concentration	Condition	Species
LD50	0.14 g pr. kg body weight	acute oral dose	Rat[1]

Source: 1.Patty, F. A., Industrial Hygiene and toxicology, 2, Interscience Publishers, New York, 1976.

4-551 Effect of 2,3,5-Trimethylnaphthalene

Parameter	Effect Value	Concentration and Cond.	Species
Mortality	50 % mortality	Water = 0.32 ppm, 24 h, estuarine	Eurytemora affinis[1]

Source: 1.Ott, F.S., R.P. Harris and S.C.M. OHara. 1978. Mar. Environ. Res. 1, 49-58.

4-552 Effect of 2,3,6-Trimethylnaphthalene

Parameter	Effect Value	Concentration and Cond.	Species
Mortality	50 % mortality	Water = 2.0 ppm, 96 h, marine	Neathes arenaceodentata[1]

Source: 1.Ott, F.S., R.P. Harris and S.C.M. OHara. 1978. Mar. Environ. Res. 1, 49-58.

4-553 Effect of 2,4-D

Parameter	Effect Value	Concentration and Cond.	Species
Mortality	50 % mortality	48 h, water = 3200 mg/l	Daphnia pulex[1]

Source: 1.Rankov, V. 1973. Pochv. Agrokhim. 8, 125-134.

4-554 Effect of 2,4,5-Trichlorophenol

Parameter	Concentration	Condition	Species
LC50	1.5 mg/l	not indicated	Chlorella pyrenoidosa[1]
LD50	0.82 g pr. kg body weight	oral dose	Rat[2]
LD50	0.5 g pr. kg body weight	oral dose, 100 %	Rat[3]

Source: 1.Jones, H. R., Environmental Control in the organic and petrochemical Industries, Noyes Data Corp., 171.
2.Patty, F. A., Industrial Hygiene and toxicology, 2, Interscience Publishers, New York, 1976.
3.McKee, J. E., and Wolf, H. W., Water Quality, The resources agency of California, 1963.

4-555 Effect of 2,4,5-Trichlorophenoxyacetic acid

Parameter	Concentration	Condition	Species
LC0	60 mg/l	not indicated	Gardon[1]
LC0	55 mg/l	not indicated	Perch[1]

Source: 1.Meinck, F., Stooff, H. and Kohlschutter, H., Les eaux residuaires industrielles, 1970.

4-556 Effect of 2,4,6-Trinitroresorcinol

Parameter	Concentration	Condition	Species
LC50	2.58 mg/l	96 hours	Fish[1]
Sublethal effect	0.32 mg/l	Inhibition of cell multiplication	Microcystis aeruginosa[2]
Sublethal effect	>100 mg/l	Inhibition of cell multiplication	Pseudomonas putida[2]

Source: 1.Smock, L. A., Stoneburger, D. L. and Clarck, J. R., Water Research, 10, 537, 1976.
2.Bringmann, G. and Kuhn, R., nwf-wasser/abwasser, H.9, 1976.

4-557 Effect of 2,4,6-Trinitrotoluene

Parameter	Concentration	Condition	Species
LC50	4-5 mg/l	6 hours	Fish[1]

Source: 1.Meinck, F., Stooff, H. and Kohlschutter, H., Les eaux residuaires industrielles, 1970.

4-558 Effect of 2,6-dimethylnaphthalene

Parameter	Effect Value	Concentration and Cond.	Species
Mortality	50 % mortality	Water = 0.85 ppm, 24 h, estuarine	Eurytemora affinis[1]
Mortality	50 % mortality	Water = 2.60 ppm, 96 h	Neathes arenaceodentata[2]
Mortality	50 % mortality	Water = 0.7 ppm, 96 h	Palaemonetes pugio[3]

Source: 1.Ott, F.S., R.P. Harris and S.C.M. OHara. 1978. Mar. Environ. Res. 1, 49-58.
2.Rossi, S.S. and J.M. Neff. 1978. Mar. Pollut. Bull. 9, 220-223.
3.Neff, J.M., Anderson, Cox, Laughlin, Rossi and Tatem. 1976. In: Sources, Effects and Sinks of Hydrocrbon in the Aquatic Environment. Wash DC. Amer. Inst. of Biological Science.

4-559 Effect of 2,4-Dichlorophenol

Parameter	Concentration	Condition	Species
LD50	0.58 g pr. kg body weight	single oral dose	Rat[1]
Sublethal effect	2 mg/l	Inhibition of cell multiplication	Microcystis aeruginosa[2]
Sublethal effect	6 mg/l	Inhibition of cell multiplication	Pseudomonas putida[2]

Source: 1.Patty, F. A., Industrial Hygiene and toxicology, 2, Interscience Publishers, New York, 1976.
2.Bringmann, G. and Kuhn, R., nwf-wasser/abwasser, H.9, 1976.

4-560 Effect of 2,4-Dinitrophenol

Parameter	Concentration	Condition	Species
Sublethal effect	33 mg/l	Inhibition of cell multiplication	Microcystis aeruginosa[1]
Sublethal effect	115 mg/l	Inhibition of cell multiplication	Pseudomonas putida[1]

Source: 1.Bringmann, G. and Kuhn, R., nwf-wasser/abwasser, H.9, 1976.

4-561 Effect of 2,4-Hexadienal

Parameter	Concentration	Condition	Species
LD50	0.7 g pr. kg body weight	acute oral dose	Rat[1]

Source: 1.Patty, F. A., Industrial Hygiene and toxicology, 2, Interscience Publishers, New York, 1976.

4-562 Effect of 2,4-Xylenol

Parameter	Concentration	Condition	Species
LC0	24 mg/l	not indicated	Daphnia[1]
LC0	500 mg/l	not indicated	Escherichia coli[1]
LC0	40 mg/l	not indicated	Scenedesmus[1]
TLm	30 mg/l	24 hours	Crucian carp[2]
TLm	13 mg/l	24 hours	Tench[2]

Source: 1.Meinck, F., Stooff, H. and Kohlschutter, H., Les eaux residuaires industrielles, 1970.
2.Dore, M. B. N. and Legube, B., La tribune de Cebedeau, 28, 1975.

4-563 Effect of 2,6-Di-tert-butyl-4-methylphenol

Parameter	Concentration	Condition	Species
LC0	0.4 mg/l	not indicated	Goldfish[1]

Source: 1.Shell Chemie, Shell Industrie Chemicalien gids, Shell Niederland Chemie, 1975.

4-564 Effect of 2-Aminopyridine

Parameter	Concentration	Condition	Species
Sublethal effect	5.2 ppm	Transient effect, 8 hours, absorbed through skin	Man[1]

Source: 1.Patty, F. A., Industrial Hygiene and toxicology, 2, Interscience Publishers, New York, 1976.

4-565 Effect of 2-Aminopyridine acid

Parameter	Concentration	Condition	Species
LD50	0.05 g pr. kg body weight	acute oral dose	Mouse[1]
LD50	0.2 g pr. kg body weight	acute oral dose	Rat[1]

Source: 1.Patty, F. A., Industrial Hygiene and toxicology, 2, Interscience Publishers, New York, 1976.

4-566 Effect of 2-Bromoethoxyphenol

Parameter	Concentration	Condition	Species
Sublethal effect	80 ppm	Weight loss, 13 times 5.5 hours	Rat[1]

Source: 1.Cage, J.C. British J. Ind. Medicine 1970, vol 27.

4-567 Effect of 2-Chloroethanol

Parameter	Concentration	Condition	Species
LD50	0.11 g pr. kg body weight	single oral dose	guinea pig[1]
LD50	0.072-0.095 g/kg ww	single oral dose	Rabbit[1]

Source: 1.Patty, F. A., Industrial Hygiene and toxicology, 2, Interscience Publishers, New York, 1976.

4-568 Effect of 2-Chloropropane

Parameter	Concentration	Condition	Species
LD50	3 g pr. kg body weight	single oral dose, 0 %	Rat[1]
LD50	10 g pr. kg body weight	single oral dose, 100 %	Rat[1]

Source: 1.Patty, F. A., Industrial Hygiene and toxicology, 2, Interscience Publishers, New York, 1976.

4-569 Effect of 2-Ethyl-1-butanol

Parameter	Concentration	Condition	Species
LD50	1.85 g pr. kg body weight	single oral dose	Rat[1]

Source: 1.Patty, F. A., Industrial Hygiene and toxicology, 2, Interscience Publishers, New York, 1976.

4-570 Effect of 2-Ethyl-1-hexanol

Parameter	Concentration	Condition	Species
LD50	3.2-6.4 g pr. kg body weight	single oral dose	Rat and Mouse[1]

Source: 1.Patty, F. A., Industrial Hygiene and toxicology, 2, Interscience Publishers, New York, 1976.

4-571 Effect of 2-Hexanol

Parameter	Concentration	Condition	Species
Sublethal effect	32 mg/l	Inhibition of cell multiplication	Microcystis aeruginosa[1]
Sublethal effect	63 mg/l	Inhibition of cell multiplication	Pseudomonas putida[1]

Source: 1.Bringmann, G. and Kuhn, R., nwf-wasser/abwasser, H.9, 1976.

4-572 Effect of 2-Isopropoxyethanol

Parameter	Concentration	Condition	Species
LC50	>5000 mg/l	24 hours	Goldfish[1]

Source: 1.Shell Chemie, Shell Industrie Chemicalien gids, Shell Niederland Chemie, 1975.

4-573 Effect of 2-Methoxyphenol

Parameter	Concentration	Condition	Species
LD50	1-2 g pr. kg body weight	oral dose by intubation	Rat[1]

Source: 1.Patty, F. A., Industrial Hygiene and toxicology, 2, Interscience Publishers, New York, 1976.

4-574 Effect of 2-Methyl-2-butanol

Parameter	Concentration	Condition	Species
LC0	1300 mg/l	24 hours	Creek chub[1]
LC100	2600 mg/l	24 hours	Creek chub[1]
Sublethal effect	105 mg/l	Inhibition of cell multiplication	Microcystis aeruginosa[2]
Sublethal effect	410 mg/l	Inhibition of cell multiplication	Pseudomonas putida[2]

Source: 1.Gilette, L. A., Miller, D. L. and Redman, H. E., Sewage Ind. Wastes, 24, 1397, 1952.
2.Bringmann, G. and Kuhn, R., nwf-wasser/abwasser, H.9, 1976.

4-575 Effect of 2-Methylanthracene

Parameter	Effect Value	Concentration and Cond.	Species
Growth rate	-12 % more than control	Water = 0.3 ppm, 96 h, PAH, sporelings	Antithamnion plumula[1]

Source: 1.Boney, A.D. 1974. Mar. Pollut. Bull. 5, 185-186.

4-576 Effect of 2-Methylfuran

Parameter	Concentration	Condition	Species
Sublethal effect	14 mg/l	Inhibition of cell multiplication	Microcystis aeruginosa[1]
Sublethal effect	90 mg/l	Inhibition of cell multiplication	Pseudomonas putida[1]

Source: 1.Bringmann, G. and Kuhn, R., nwf-wasser/abwasser, H.9, 1976.

4-577 Effect of 2-Methylnaphthalene

Parameter	Effect Value	Concentration and Cond.	Species
Mortality	50 % mortality	Water = 1.3 ppm, 96 h, stage 1 zoeae	Cancer magister[1]
Mortality	50 % mortality	Water = 2.0 ppm, 24 h, sheeps head minnow	Cyprinodon variegatus[2]
Mortality	50 % mortality	Water = 1.5 ppm, 24 h, estuarine	Eurytemora affinis[3]
Mortality	50 % mortality	Water = 1.1 ppm, 96 h	Palaemonetes pugio[4]
Mortality	50 % mortality	Water = 0.7 ppm, 24 h	Penaeus aztecus[2]

Source: 1.Caldwell, R.s., E. M. Caldarone and M.H. Mallon. 1977. D.A. Wolfe (ed.): Fate and Effects of Petrolem Hydrocarbons in Marine Ecosystems and Organisms. NY. Pergamon Press. 210-220.
2.Anderson, J.W., J.M. Neff, B.A. Cox, Tatem and Hightower. 1974. In: Vernberg and Vernberg (eds.): Polution and Physiology of Marine Animals. NY. Academic Press. 285-310.
3.Ott, F.S., R.P. Harris and S.C.M. OHara. 1978. Mar. Environ. Res. 1, 49-58.
4.Neff, J.M., Anderson, Cox, Laughlin, Rossi and Tatem. 1976. In: Sources, Effects and Sinks of Hydrocrbon in the Aquatic Environment. Wash DC. Amer. Inst. of Biological Science.

4-578 Effect of 2-Nitro-p-cresol

Parameter	Concentration	Condition	Species
Sublethal effect	32 mg/l	Inhibition of cell multiplication	Microcystis aeruginosa[1]
Sublethal effect	4 mg/l	Inhibition of cell multiplication	Pseudomonas putida[1]

Source: 1.Bringmann, G. and Kuhn, R., nwf-wasser/abwasser, H.9, 1976.

4-579 Effect of 2-Octanol

Parameter	Concentration	Condition	Species
LD50	>3.2 g pr. kg body weight	single oral dose	Rat[1]

Source: 1.Patty, F. A., Industrial Hygiene and toxicology, 2, Interscience Publishers, New York, 1976.

4-580 Effect of 3,4-Xylenol

Parameter	Concentration	Condition	Species
LC0	16 mg/l	not indicated	Daphnia[1]
LC0	>100 mg/l	not indicated	Escherichia coli[1]
LC0	40 mg/l	not indicated	Scenedesmus sp.[1]
TLm	21 mg/l	24 hours	Crucian carp[2]
TLm	16-18 mg/l	24 hours	Roach[2]
TLm	16-18 mg/l	24 hours	Tench[2]

Source: 1.Meinck, F., Stooff, H. and Kohlschutter, H., Les eaux residuaires industrielles, 1970.
2.Dore, M. B. N. and Legube, B., La tribune de Cebedeau, 28, 1975.

4-581 Effect of 3,5-Xylenol

Parameter	Concentration	Condition	Species
LC0	16 mg/l	not indicated	Daphnia[1]
LC0	>100 mg/l	not indicated	Escherichia coli[1]
LC0	40 mg/l	not indicated	Scenedesmus[1]
TLm	53 mg/l	24 hours	Crucian carp[2]
TLm	51 mg/l	24 hours	Tench[2]

Source: 1.Meinck, F., Stooff, H. and Kohlschutter, H., Les eaux residuaires industrielles, 1970.
2.Dore, M. B. N. and Legube, B., La tribune de Cebedeau, 28, 1975.

4-582 Effect of 3-Heptanone

Parameter	Concentration	Condition	Species
LD50	2.79 g pr. kg body weight	single oral dose	Rat[1]

Source: 1.Patty, F. A., Industrial Hygiene and toxicology, 2, Interscience Publishers, New York, 1976.

4-583 Effect of 3-Hexanol

Parameter	Concentration	Condition	Species
Sublethal effect	32 mg/l	Inhibition of cell multiplication	Microcystis aeruginosa[1]
Sublethal effect	105 mg/l	Inhibition of cell multiplication	Pseudomonas putida[1]

Source: 1.Bringmann, G. and Kuhn, R., nwf-wasser/abwasser, H.9, 1976.

4-584 Effect of 3-Hydroxybutanol

Parameter	Concentration	Condition	Species
LD50	2.2 g pr. kg body weight	acute oral dose	Rat[1]

Source: 1.Patty, F. A., Industrial Hygiene and toxicology, 2, Interscience Publishers, New York, 1976.

4-585 Effect of 3-Methylcholanthrene

Parameter	Effect Value	Concentration and Condition	Species
Growth rate	10 % more than control	Water = 2.0 ppm (in methyl cellosolve), 1 h, SD = 2 %, PAH, sporelings	Antithamnion plumula[1]
Growth rate	21 % more than control	Water = 0.003 pp (in methyl cellosolve), 96 h, SD = 3 %, sporelings	Antithamnion plumula[1]
Growth rate	48 % more than control	Water = 2.0 ppm (in methyl cellosolve), .5 h, SD = 1 %, PAH, sporelings	Antithamnion plumula[1]
Growth rate	52 % more than control	Water = 0.1 ppm (in methyl cellosolve), 96 h, SD = 3 %, PAH, sporelings	Antithamnion plumula[1]
Growth rate	53 % more than control	Water = 1.0 ppm (in methyl cellosolve), 1 h, SD = 5 %, PAH, sporelings	Antithamnion plumula[1]
Growth rate	8 % more than control	Water = 1.0 ppm (in methyl cellosolve), 0.5 h, SD = 1 %, PAH, sporelings	Antithamnion plumula[1]
Growth rate	88 % more than control	Water = 0.01 ppm (in methyl cellosolve), 96 h, SD = 5 %, PAH, sporelings	Antithamnion plumula[1]

Source: 1.Boney, A.D. and E.D.S. Corner. 1962. J. Mar. Biol. Assoc. UK. 43, 579-585.

4-586 Effect of 3-Trifluoromethyl-4-nitrophenol

Parameter	Concentration	Condition	Species
LC10	7 mg/l	not indicated	Brown trout[1]
LC10	9 mg/l	not indicated	Rainbow trout[1]
LC100	2 mg/l	larvae	Sea lamprey[1]

Source: 1.McKee, J. E., and Wolf, H. W., Water Quality, The resources agency of California, 1963.

4-587 Effect of 4-Amino-m-toluenesulfonic acid

Parameter	Concentration	Condition	Species
TLm	425 mg/l	24 hours	Mosquito Fish[1]
TLm	375 mg/l	96 hours	Mosquito Fish[1]

Source: 1.Jones, H. R., Environmental Control in the organic and petrochemical Industries, Noyes Data Corp., 171.

4-588 Effect of 4-Aminodiphenyl

Parameter	Concentration	Condition	Species
Sublethal effect	50 mg/l	Carcinogenic	Man[1]

Source: 1.Maximale Arbeitsplatzkonzentration 1973. BRD.

4-589 Effect of 4-Hydroxymethyl-2,6-di-tert-butylphenol

Parameter	Concentration	Condition	Species
LC50	9 mg/l	24-96 hours	Goldfish[1]

Source: 1.Shell Chemie, Shell Industrie Chemicalien gids, Shell Niederland Chemie, 1975.

4-590 Effect of 4-Methoxy-4-methyl-2-pentanone

Parameter	Concentration	Condition	Species
LC50	3800 mg/l	24 hours	Goldfish[1]
LD50	3.72 g pr. kg body weight	acute oral dose	Rat[1]

Source: 1.Shell Chemie, Shell Industrie Chemicalien gids, Shell Niederland Chemie, 1975.

4-591 Effect of 4-Methoxyphenol

Parameter	Concentration	Condition	Species
LC100	200 mg/l	48 hours	Goldfish[1]
LD50	1.6 g pr. kg body weight	single oral dose	Rat[2]

Source: 1.McKee, J. E., and Wolf, H. W., Water Quality, The resources agency of California, 1963.
2.Patty, F. A., Industrial Hygiene and toxicology, 2, Interscience Publishers, New York, 1976.

4-592 Effect of 4-Methylcyclohexanol

Parameter	Concentration	Condition	Species
LD50	1.75-2.0 g pr. kg ww	single oral dose	Rabbit[1]

Source: 1.Patty, F. A., Industrial Hygiene and toxicology, 2, Interscience Publishers, New York, 1976.

4-593 Effect of 4-Nitro-m-cresol

Parameter	Concentration	Condition	Species
Sublethal effect	13 mg/l	Inhibition of cell multiplication	Microcystis aeruginosa[1]
Sublethal effect	6 mg/l	Inhibition of cell multiplication	Pseudomonas putida[1]

Source: 1.Bringmann, G. and Kuhn, R., nwf-wasser/abwasser, H.9, 1976.

4-594 Effect of 5-Chloro-2-nitrophenol

Parameter	Concentration	Condition	Species
LC100	3 mg/l	larvae	Sea lamprey[1]

Source: 1.Meinck, F., Stooff, H. and Kohlschutter, H., Les eaux residuaires industrielles, 1970.

4-595 Effect of 6-Methylbenz(a)anthracene

Parameter	Effect Value	Concentration and Condition	Species
Growth rate	125 % more than control	Water = 0.1 ppm (in methyl cellosolve), 96 h, SD = 5 %, PAH, sporelings	Antithamnion plumula[1]
Growth rate	142 % more than control	Water = 0.01 ppm (in methyl cellosolve), 96 h, SD = 3 %, PAH, sporelings	Antithamnion plumula[1]

Source: 1.Boney, A.D. and E.D.S. Corner. 1962. J. Mar. Biol. Assoc. UK. 43, 579-585.

4-596 Effect of 6-Nitro-m-cresol

Parameter	Concentration	Condition	Species
Sublethal effect	34 mg/l	Inhibition of cell multiplication	Microcystis aeruginosa[1]
Sublethal effect	7 mg/l	Inhibition of cell multiplication	Pseudomonas putida[1]

Source: 1.Bringmann, G. and Kuhn, R., nwf-wasser/abwasser, H.9, 1976.

4-597 Effect of 6-Methylbenz(a)anthracene

Parameter	Effect Value	Concentration and Condition	Species
Growth rate	125 % more than control	Water = 0.1 ppm (in methyl cellosolve), 96 h, SD = 5 %, PAH, sporelings	Antithamnion plumula[1]
Growth rate	142 % more than control	Water = 0.01 ppm (in methyl cellosolve), 96 h, SD = 3 %, PAH, sporelings	Antithamnion plumula[1]

Source: 1.Boney, A.D. and E.D.S. Corner. 1962. J. Mar. Biol. Assoc. UK. 43, 579-585.

4-598 Effect of 9,10-Dimethylbenz(a)anthcene

Parameter	Effect Value	Concentration and Condition	Species
Growth rate	133 % more than control	Water = 0.01 ppm (in methyl cellosolve), 96 h, SD = 5 %, PAH, sporelings	Antithamnion plumula[1]
Growth rate	56 % more than control	Water = 0.1 ppm (in methyl cellosolve), 96 h, SD = 3 %, PAH, sporelings	Antithamnion plumula[1]

Source: 1.Boney, A.D. and E.D.S. Corner. 1962. J. Mar. Biol. Assoc. UK. 43, 579-585.

4-599 Effect of 9-Methylanthracene

Parameter	Effect Value	Concentration and Condition	Species
Growth rate	14 % more than control	Water = 0.3 ppm, 96 h, PAH, sporelings	Antithamnion plumula[1]

Source: 1.Boney, A.D. 1974. Mar. Pollut. Bull. 5, 185-186.

Chapter 5

CHEMICALS AND THE LIVING ORGANISM

Background Concentration
Concentration Factor
Concentrations
Intake, Excretion and Uptake

5-001 Background concentration
1,3-Butadiene

Item	Value	Condition
Atmospheric conc of aliphatic hydrocarbon and alkylbenzene	2 ppb	Los Angeles, 1967, STP[1]

Source: 1.Altshuller, A.P., W.A. Lonneman and S.L. Kopczynski. 1973. J. Air Pollut. Contr. Assoc. 23, 597-599.

5-002 Background concentration
2,4-Dimethylpentane

Item	Value	Condition
Atmosphere, organic gas	2.7 ppb	Rg = 0.3-5.1 ppb, mountain air in USA, 1972, STP[1]

Source: 1.Cavanagh, L.A., C.F. Sahadt and E. Robinson. 1969. Environ. Sci. Technol. 3, 251-257.

5-003 Background concentration
2-Methylbutane

Item	Value	Condition
Atmosphere, organic gas	0.5 ppb	Rg = 0 - 1.0 ppb, mountain air in USA, 1972, STP[1]

Source: 1.Cavanagh, L.A., C.F. Sahadt and E. Robinson. 1969. Environ. Sci. Technol. 3, 251-257.

5-004 Background concentration
2-Methylbutene

Item	Value	Condition
Atmospheric conc of aliphatic hydrocarbon and alkylbenzene	3 ppb	Los Angeles, 1967, STP[1]

Source: 1.Altshuller, A.P., W.A. Lonneman and S.L. Kopczynski. 1973. J. Air Pollut. Contr. Assoc. 23, 597-599.

5-005 Background concentration
ATP

Item	Value	Condition
Estuarine sediment	0.04 microgram/cm**3	Depth = 15 cm, winter, Newport Estuary[1]
Estuarine sediment	0.16 microgram/cm**3	Depth = 5 cm, winter, Newport Estuary[1]
Estuarine sediment	0.32 microgram/cm**3	Depth = 5 cm, summer, Newport Estuary[1]
Estuarine sediment	0.64 microgram/cm**3	Depth = 0 cm, winter, Newport Estuary[1]
Estuarine sediment	1.18 microgram/cm**3	Depth = 0 cm, summer, Newport Estuary[1]

Source: 1.Ferguson, R.L. and M.B. Murdoch. 1975. in: Cronin, L.E. (ed.); Estuarine Re search. Vol 1 pp 229-250Academic Press. New York.

5-006 Background concentration
Acetaldehyde

Item	Value	Condition
Atmosphere, organic gas	0.15 ppb	Rg = 0-0.3 ppb, Point Barrow, Alaska, 1968[1]

Source: 1.Cavanagh, L.A., C.F. Sahadt and E. Robinson. 1969. Environ. Sci. Technol. 3, 251-257.

5-007 Background concentration
Acetone

Item	Value	Condition
Atmosphere, organic gas	0.6 ppb	Rg = 0.3-0.9 ppb, mountain air in USA, 1972, STP[1]
Atmosphere, organic gas	1.5 ppb	Rg = 0-2.9 ppb, Point Barrow, Alaska, 1968[1]
Atmosphere, organic gas	1.6 ppb	Rg = 0.1 - 3 ppb, trop. and polar maritime air, 1972, STP[2]
Atmosphere, organic gas	240 ppb	Central Amazona jungle, 1972, STP[2]

Source: 1.Cavanagh, L.A., C.F. Sahadt and E. Robinson. 1969. Environ. Sci. Technol. 3, 251-257.
2.Robinson, E., R.A. Rasmussen, H.H. Westberg and M.W. Holdren. 1973. J. Geophys. Res. 78, 5345-5351.

5-008 Background concentration
Acetylene + Propane

Item	Value	Condition
Atmospheric conc of aliphatic hydrocarbon and alkylbenzene	76 ppb	Los Angeles, 1967, STP[1]

Source: 1.Altshuller, A.P., W.A. Lonneman and S.L. Kopczynski. 1973. J. Air Pollut. Contr. Assoc. 23, 597-599.

5-009 Background concentration
Ag

Item	Value	Condition
Anodonta sp.	0.28 mg/kg dry weight	Gastropod, L Aberffraw, 1970, freshwater[1]
Atmosphere	0.125 ng/m**3	San Francisco Bay, 9 stations, 1970, rg = 0.05 - 0.2[2]
Atmosphere	2 ng/m**3	S. Calif., 1972[3]
Carbonate sediments	0.19 mg/kg dry weight	Average value[4]
Cardium edule	0.04 mg/kg dry weight	Gastropod, Irish Sea, GB, 1970[1]
Crassostrea gigas	1.9 mg/kg dry weight	Gastropod, Knysna Estuary, R.S.A., 1975[5]
Crassostrea margaritacea	1.9 mg/kg dry weight	Gastropod, Knysna Estuary, R.S.A., 1975[6]
Crustral rock	0.000007 percent by weight	Average value[7]
Deep-sea clay sediments	0.11 mg/kg dry weight	Average value[4]
Fucus spp.	0.30 mg/kg dry weight	Range = 0.07 - 0.79 ppm, Irish Sea, North Sea, 1971[8]
Granitic rock	0.15 mg/kg dry weight	Average value[4]
Igenous rock	0.07 mg/kg dry weight	None[9]
Mytilus edulis	0.03 mg/kg dry weight	Gastropod, Irish Sea, GB, 1970[1]
Mytilus edulis	0.1 mg/kg dry weight	Gastropod, rg = 0.1 - 0.3 ppm, New Zealand, 1964[10]
Mytilus edulis	3 mg/kg dry weight	Gastropod, rg = 1 - 6 ppm, Trondheimsfjord, Norway, 1976[11]
Ostrea edulis	6.4 mg/kg dry weight	Gastropod, Knysna Estuary, R.S.A., 1975[5]
Ostrea sinuata	5.6 mg/kg dry weight	Gastropod, rg = 4.5-7.3 ppm, Tasman Bay, NZ, 1964[10]
Patella vulgata	2.7 mg/kg dry weight	Gastropod, rg = 1.3-6.6 ppm, Irish Sea, 1971[12]
Pecten novaezelandia	0.7 mg/kg dry weight	Gastropod, rg = 0.6-2.3 ppm, Tasman Bay, New Zealand, 1964[10]
River water	0.13E-3 ppm	Range = 0.1E-4-0.35E-2 ppm[13]
Sandstone sediment	0.12 mg/kg dry weight	Average value[4]
Seawater	0.0003 ppm	AgCl2(-), residence time = 2100E3 y[14]

Seawater	0.77 ug/l	Range = 0.055 - 1.5 ug/l[15]
Sedimentary rock	0.9 mg/kg dry weight	None[16]
Shale sediment	0.27 mg/kg dry weight	Average value[4]
Soil	0.1 mg/kg dry soil	Oven-dried, rg = 0.01 - 0.5[13]
Water	2.692 ug/l	Rivers, USA, rg = 0.085 - 5.3[2]

Source: 1.Segar, D.A. and R.E. Pellenbarg. 1971. J. Mar. Biol. Assoc. UK. 51, 131-136.
2.Laveskog, A., A. Lindskog and V. Sten berg. 1976. Om Metaller, Statens Naturvaardsverk, Stockholm.
3.Bruland, K.W., K. Bertine, M. Koide and E.D. Goldberg. 1974. Environ. Sci. Technol. 8, pp 425-432.
4.Turekian, K.K. and K.H. Wedepohl. 1961. Amer. Geol. Soc. Bull. 72, pp 175-192.
5.Watling, H.R. and R.J. Watling. 1976. Mar. Pollut. Bull. 7, 45-48.
6.Fourie, H.O. 1976. S. Afr. J. Sci. 72, 110-113.
7.Mason, B. 1966. Principles of Geochemistry. 3rd ed. John Wiley and Sons. New York.
8.Martin, J.H., W.W. Broenkow. 1975. Science 190, 884-885.
9.Taylor, S.R. 1964. Geochim. Cosmochim. Acta 28, pp 1273-1285.
10.Brooks, R.R., and D. Rumsey. 1974. NZJ Mar. Freshwater Res. 8, 155-166.
11.Lande, E. 1977. Environ. Pollut. 12, 187-198.
12.Preston, A., Jefferies, Dutton, Harvey and A.K. Steele. 1972. Environ. Pollut. 3, 69-82.
13.Bowen, H.J.M. 1966. Trace Elements in Biochemistry, Ac. Press.
14.Goldberg, E.D., 1963. The Sea, Interscience London, 2, pp 3.
15.Lie, R., R.G. Thomas and J.K. Scott. 1960. Health Phys. 2, pp 334-340.
16.Vinogradov, A.P. 1959. The Geochemistry of Rare and Dispersed Chemical Elements in the Soil. Transl.from Russian. Consultant Bureau Inc. NY. 2nd ed.

5-010 Background concentration Al

Item	Value	Condition
Atmosphere	0.57 ng/m**3	Southpole, Oct 1970, rg = +- 0.17[1]
Atmosphere	1125 ng/m**3	San Francisco Bay, USA, rg = 250 - 2000[1]
Atmosphere	277 ng/m**3	England, 7 stations, rg = 74 - 483[1]
Atmospheric particulates	6.7 ng/m**3	Hawaii, n > = 56, rg = +- 7.2[2]
Biosphere	4.12 moles/hectare	Average composition of living matter[3]
Crustral rock	8.1 percent by weight	Average value[4]
Freshwater, river	15 ug/l	Range 0.5 - 3 ppm, 6 rivers in California[5]
Rain	160 ug/l	England, yearly mean, 1 station[1]
River water	0.24 ppm	Range = 0.01-2.5 ppm[6]
Seawater	0.01 ppm	Residence time = 0.1E3 y[7]
Soil	71000 mg/kg dry soil	Oven-dried, rg = 10000 - 300000[6]
Water	112 ug/l	Riverin USA, rg = 11 - 213[1]

Source: 1.Laveskog, A., A. Lindskog and V. Sten berg. 1976. Om Metaller, Statens Naturvaardsverk, Stockholm.
2.Golterman, H.L. 1975. Physiological Limnol., Elsevier Amsterdam.
3.Deevey, E.S. 1970. Scientific American. 223(3), p 150.
4.Mason, B. 1966. Principles of Geochemistry. 3rd ed. John Wiley and Sons. New York.
5.Jones, B.F., V.C. Kennedy and G.W. Zellweger. 1974. Water Resour Res. 10, 791-793.
6.Bowen, H.J.M. 1966. Trace Elements in Biochemistry, Ac. Press.
7.Goldberg, E.D., 1963. The Sea, Interscience London, 2, pp 3.

5-011 Background concentration Anthanthrene

Item	Value	Condition
Atmosphere, seasonal variation	0.33 ng/m3	1. Quarter, 2. Q=0.08, 3. Q=0.14, 4. Q=0.79, LA, n=13, USA, 1975[1]

Source: 1.Gordon, R.J. 1976. Environ. Sci. Technol. 10, 370-373.

5-012 Background concentration Ar

Item	Value	Condition
Atmosphere	0.93 % by atoms	Sea-level dry atmospheric composition[1]
Atmosphere	9300 ug/m**3	293 K, 1 atm[2]
Gas in ice, glacier ice	1.08 %	Gas pres.=2.1 bar, dens=0.90 g/cm3, gas=26 cm3/kg, rg=73.2-78.3 %[3]
Gas in ice, iceberg ice	0.93 %	Gas pres.=1.3 bar, dens=0.88 g/cm3, gas=40 cm3/kg, rg=77.7-78.3 %[3]
Gas in ice, pond ice	1.0 %	Gas pres.=1.0 bar, dens=0.916 g/cm3, gas=1.0 cm3/kg, n = 1[3]
Gas in ice, sea ice	1.0 %	Gas pres.=0.9 bar, dens=91 g/cm3, gas=13 cm3/kg, rg=54.2-76.8 %[3]
Lower atmosphere	0.934 % by volume	None[4]
River water	0.6 ppm	None[5]
Seawater	0.3 ml/l	Saturated seawater, salinity = 3.43 %, 283 K[1]
Seawater	0.6 ppm	None[6]

Source: 1.Sisler, F.D. 1975. In: S.F. Singer (ed.), The Changing Global Environment, pp 57-69. D. Reidel Publ. Comp.
2.Junge, C.E. 1963. Air Chemistry and Radioactivity, Ac. Press.
3.Matsuo, S. and Y. Miyake. 1966. J. Geophys. Res. 71, 5235.
4.Gleuckauf, E. 1951. Compendium of Meteorology. Am. Meteorological Soc. Boston.
5.Bowen, H.J.M. 1966. Trace Elements in Biochemistry, Ac. Press.
6.Goldberg, E.D., 1963. The Sea, Interscience London, 2, pp 3.

5-013 Background concentration As

Item	Value	Condition
Alosa pseudoharengus	0.023 mg/kg wet weight	Freshwater fish, whole body, Lake Michigan, 1969[1]
Anarrhichas minor	78.3 mg/kg dry weight	Fish, marine, rg = 17.1 - 195 ppm, W Greenland, mining area, 1974[2]
Atmosphere	0.02 ug/m3	133 stations, rg = 0.01 - 0.75 ug/m3, USA, 1964[3]
Atmosphere	0.07 ug/m3	Prague, summer mean[4]
Atmosphere	0.56 ug/m3	Prague, winter mean[4]
Atmosphere	1.4 ug/m3	El Paso, Texas, near copper smelter, 1964[5]
Atmosphere	1.6 ug/m3	Romania, near copper smelter[5]
Atmosphere	10 ng/m**3	Less than value[6]
Blue crab	12.16 mg/kg wet weight	8 references used, SD = +-1.79, rg = 3.81 - 35.62[7]
Carbonate sediments	1 mg/kg dry weight	Average value[8]
Cardium edule	5.4 mg/kg dry weight	Gastropod, rg = 4.5-6.3 ppm, Southampton waters, GB, 1971[9]
Ceratophyllum demersum	325 mg/kg dry weight	Higher plant, Rg = 1.4-650 ppm, Waikito R, Hastings, Nz, 1976[10]
Clams	3.13 mg/kg wet weight	11 references used, SD = +- 3.85, rg = 0.90 - 12.75[7]
Cod, liver	3.4 mg/kg dry weight	Human food[6]
Cod, muscle	0.5 mg/kg dry weight	Human food[6]
Coregonus clupeaformis	0.09 mg/kg wet weight	Fish, less, Moose L, USA, 1970, animal decapiated and eviscerated[11]
Coregonus clupeaformis	0.70 mg/kg wet weight	Fish, less, L Erie, USA, 1970, animal decapiated and eviscerated[11]
Crangon crangon	0.6 mg/kg dry weight	Marine, Crustacean, S Antonio Bay, Texay, 1975, whole animal[12]

Crassostrea commercialis	1.2 mg/kg wet weight	Gastropod, rg = 0.3 - 3.4 ppm, NSW estuaries, Aust., 1974[13]
Crassostrea virginica	1.3 mg/kg dry weight	Gastropod, San Antonio B, Texas, USA, 1972[12]
Deep-sea clay sediments	13 mg/kg dry weight	Average value[8]
Elodea canadensis	156 mg/kg dry weight	Higher plant, Rg = 3.0-307 ppm, Waikito R, Hastings, Nz, 1976[10]
Esox lucius	0.05 mg/kg wet weight	Fish, less, L Erie, USA, 1970, animal decapiated and eviscerated[11]
Esox lucius	0.05 mg/kg wet weight	Fish, less, Moose L, USA, 1970, animal decapiated and eviscerated[11]
Freshwater zooplankton, effect of fertilizer	0.06 mg/kg dry weight	Area without fertilizer, Scenedesmus obliquus, Thailand, 1975[14]
Freshwater zooplankton, effect of fertilizer	2.36 mg/kg dry weight	Area with fertilizer, Scenedesmus obliquus, Thailand, 1974[14]
Freshwater, river	3.2 ug/l	Danube, Ulm, FRG, range 1 - 8 ppm[15]
Freshwater, river and lake	20 ug/l	Range 10 - 1100 ppm, USA[16]
Fucus vesiculosus	35.8 mg/kg dry weight	West Greenland, 1974[2]
Gadus morhua	2.14 mg/kg wet weight	3 references used, SD = +- 1.26, rg = 0.86 - 3.37[7]
Granitic rock	1.9 mg/kg dry weight	Average value[8]
Hippoglossoides platessiodes	188 mg/kg dry weight	Fish, marine, rg = 17 - 290 ppm, W Greenland, mining area, 1974[2]
Homarus americanus	5.7 mg/kg dry weight	Marine, Crustacean, rg =3.8-7.6 ppm, muscle, Moreton H, Newfound., 1974[17]
Ictaluridae	1.15 mg/kg wet weight	2 references used, SD = +-1.46, rg = 0.12 - 2.18[7]
Igenous rock	2 mg/kg dry weight	None[18]
Lagarosiphon major	70 mg/kg dry weight	Moss, Rg = 20-120 ppm, L Rotura, NZ, 1972[19]
Lutjanidae	0.06 mg/kg wet weight	1 references used[7]
Makaira indica	0.6 mg/kg wet weight	Fish, rg = 0.1-1.65 ppm, NE Australia, unpoll., muscle[13]
Marine zooplankton, copepods	6.0 mg/kg dry weight	West Greenland, 1974[2]
Melanogrammus aeglefinus	2.56 mg/kg wet weight	2 references used, SD = +- 2.77, rg = 0.60 - 2.18[7]
Mugil cephalus	1.0 mg/kg dry weight	Fish, less, N Atlantic, E of Gulf Stream, muscle, 1972[20]
Mytilus edulis	12.1 mg/kg dry weight	Gastropod, rg = 9.45-14.8 ppm, Southampton Waters, GB, 1971[21]
Mytilus edulis	15.4 mg/kg dry weight	Gastropod, rg = 14.1 - 16.7 ppm, W Greenland, 1974[2]
Mytilus edulis	5.3 mg/kg dry weight	Gastropod, rg = +- 5.4 ppm, Mortons Harbor, Newfoundland, 1974[17]
Notropis hudsonius	0.003 mg/kg w weight	Fish, L Michigan, 1969, whole animal[1]
Notropis hudsonius	0.0035 mg/kg wet weight	Fish, Lake Erie, USA, whole animal, 1969[1]
Oyster	9.12 mg/kg wet weight	10 references used, SD = +- 15.59, rg = 0.45 - 42.74[7]
Pandalus borealis	61.6 mg/kg dry weight	Marine, Crustacean, W Greenland, mining, muscle, rg =52-71 ppm, 1974[2]
Patella vulgata	2.7 mg/kg wet weight	Gastropod, rg = 1.0-3.9 ppm, Bristol Channel, Somerset, 1971[22]
Percidae	0.31 mg/kg wet weight	2 references used, SD = +- 0.26, rg = 0.05 - 0.57[7]
Percopsis omiscomaycus	0.007 mg/kg wet weight	Fish, rg = +- 0.002 ppm, L Superior, USA, whole animal, 1969[1]
Percopsis omiscomaycus	0.43 mg/kg wet weight	Fish, rg = +-0.012 ppm, L Michigan, 1969, whole animal[1]
Pleuronectidae	2.45 mg/kg wet weight	5 references used, SD = +- 1.92, rg = 0.11 - 4.51[7]
Pollock	2.35 mg/kg wet weight	1 reference used[7]

Potamogeton sp.	91 mg/kg dry weight	Higher plant, Rg = <6-178 ppm, Waikito R, Hastings, Nz, 1976[10]
Rain	0.65 ug/l	Range = 0.3 - 1.0[6]
River water	0.4E-3 ppm	Range <[23]
Sandstone sediment	1 mg/kg dry weight	Average value[8]
Scallops, sea	8.76 mg/kg wet weight	7 references used, SD = +- 12.80, rg = 0.39 - 33.56[7]
Seawater	0.003 ppm	AsSo4H(2-)[24]
Sedimentary rock	7 mg/kg dry weight	None[25]
Shale sediment	13 mg/kg dry weight	Average value[8]
Shellfish	105 mg/kg dry weight	Human food, rg = 40 - 170[6]
Shrimp, brown	8.71 mg/kg wet weight	8 references used, SD = +- 7.63, rg = 1.77 - 23.75[7]
Sillaginidae	0.73 mg/kg wet weight	1 reference used[7]
Soil	6 mg/kg dry soil	Oven-dried, rg = 0.1 - 40[23]
Vegetables	0.1 mg/kg dry weight	Human food[6]
Water	0.3 ug/l	Rivers, Sweden, rg = 0.2 - 0.4[6]

Source: 1.Lucas, H.F. Jr., D.N. Edgington and P.J. Colby. 1970. J. Fish. Res. Bd Canada 27, 677-684.
2.Bohn, A. 1975. Mar. Pollut. Bull. 6, 87-89
3.Sullivan, R.J. 1969. Preliminary Air Pollution Survey of As and its Compounds. NAPCA. Raleigh, N.C.
4.Vondracek, V. 1963. Cesk. Hyg. 8, pp 341-344.
5.Gabor, S. and V. Coldea. 1977. Environ. Health Perspect. 19, pp 107-108.
6.Laveskog, A., A. Lindskog and V. Stenberg. 1976. Om Metaller, Statens Naturvaardsverk, Stockholm.
7.Zook, E.G., J.J. Powell, B.M. Hackley, J.A. Emerson, J.R. Brooker and G.M. Knobl, Jr. 1976. J. Agric. Food. Chem. 24(1), pp 47-53.
8.Turekian, K.K. and K.H. Wedepohl. 1961. Amer. Geol. Soc. Bull. 72, pp 175-192.
9.Segar, D.A. and R.E. Pellenbarg. 1971. J. Mar. Biol. Assoc. UK. 51, 131-136.
10.Reay. P.F. 1972. J. Appl. Ecol. 9, 557-565.
11.Uthe, J.H. and E.G. Bligh. 1971. J. Fish. Res. bd Canada 28, 786-788.
12.Sims, R.R. and B.J. Presley. 1976. Bull. Environ. Contam. Toxicol. 16, 520-526.
13.MacKay, N.J., M.N. Kazacos, R.J. Williams and M.I. Leedow. 1975. Mar. Pollut. Bull. 6, 57-61.
14.Payer, H.D., Runkel, Schramel, Stengel, Bhumiratana and Soeder. 1976. Chemosphere 6, 413-418.
15.Quentin, K.E., and H.A. Winkler. 1974. Zentralbl. Bakt. Abt. Orig. B 158, 514-523.
16.Ferguson, J.F. and J. Gavis. 1972. Water Res. 6, 2159-2174.
17.Penrose, W.R., R. Black and M.J. Hayward. 1975. J. Fish. Res. Bd Canada 32, 1275-1281.
18.Taylor, S.R. 1964. Geochim. Cosmochim. Acta 28, pp 1273-1285.
19.Fish, G.R. 1963. N.Z.J. Bot. 1, 410-418.
20.Windom, H.L., R. Sticjney, Smith, White and F. Taylor. 1973. J. Fish. Res. Bd Canada 30, 275-279.
21.Raymont, J.E.G. 1972. Proc. R. Soc. London b 180, 451-468.
22.Peden, J.D., J.H. Crothers, C.E. Waterfall and J. Beasley. 1973. Mar. Pollut. Bull. 4, 7-9.
23.Bowen, H.J.M. 1966. Trace Elements in Biochemistry, Ac. Press.
24.Goldberg, E.D., 1963. The Sea, Interscience London, 2, pp 3.
25.Vinogradov, A.P. 1959. The Geochemistry of Rare and Dispersed Chemical Elements in the Soil. Transl. from Russian. Consultant Bureau Inc. NY. 2nd ed.

5-014 Background concentration
Au

Item	Value	Condition
Crustral rock	0.0000004 percent by weight	Average value[1]
River water	0.6E-4 ppm	Less than value[2]
Seawater	0.000011 ppm	AuCl4(-), residence time = 560E3 y[3]

Source: 1.Mason, B. 1966. Principles of Geochemistry. 3rd ed. John Wiley and Sons. New York.
2.Bowen, H.J.M. 1966. Trace Elements in Biochemistry, Ac. Press.
3.Goldberg, E.D., 1963. The Sea, Interscience London, 2, pp 3.

5-015 Background concentration
B

Item	Value	Condition
Carbonate sediments	20 mg/kg dry weight	Average value[1]
Deep-sea clay sediments	230 mg/kg dry weight	Average value[1]
Drinking water	50 ug/l	Stockholm, 1969, rg = 40 - 60[2]
Freshwater, river	201 ug/l	Range 10 - 400, Ruhr R., FGR[3]
Granitic rock	9 mg/kg dry weight	Average value[1]
Igenous rock	10 mg/kg dry weight	None[4]
Rain	10 ug/l	Florida, 1954[2]
Rain	2 ug/l	Sweden, 1953[2]
Rain	3 ug/l	France, area with no industry, rg = 2 - 4[2]
Rain	80 ug/l	England, 1968[2]
River water	0.013 ppm	Range = 0.01-1 ppm[5]
Sandstone sediment	35 mg/kg dry weight	Average value[1]
Seawater	4.6 ppm	B(OH)3[6]
Sedimentary rock	12 mg/kg dry weight	None[7]
Shale sediment	100 mg/kg dry weight	Average value[1]
Soil	10 mg/kg dry soil	Oven-dried, rg = 2 - 100[5]
Water	36 ug/l	Rivers, Sweden, rg = 5 - 67[2]
Water	635 ug/l	Trent River, UK, rg = 510 - 760[2]

Source: 1.Turekian, K.K. and K.H. Wedepohl. 1961. Amer. Geol. Soc. Bull. 72, pp 175-192.
2.Laveskog, A., A. Lindskog and V. Stenberg. 1976. Om Metaller, Statens Naturvaardsverk, Stockholm.
3.Dietz, F. 1974. Z. Wasser Abwasserforsch. 7, 74-80.
4.Taylor, S.R. 1964. Geochim. Cosmochim. Acta 28, pp 1273-1285.
5.Bowen, H.J.M. 1966. Trace Elements in Biochemistry, Ac. Press.
6.Goldberg, E.D., 1963. The Sea, Interscience London, 2, pp 3.
7.Vinogradov, A.P. 1959. The Geochemistry of Rare and Dispersed Chemical Elements in the Soil. Transl. from Russian. Consultant Bureau Inc. NY. 2nd ed.

5-016 Background concentration
Ba

Item	Value	Condition
Atmosphere	751 ng/m**3	Urban, USA, rg = 2 - 1500[1]
Atmosphere, urban	5 ng/m**3	18 USA cities, range = 0 - 1500 ng/m3, 1970[2]
Carbonate sediments	10 mg/kg dry weight	Average value[3]
Crustal rock	0.04 %	Mostly in certain rock types[2]
Deep-sea clay sediments	2300 mg/kg dry weight	Average value[3]
Freshwater	0.05 mg/l	Range = 0.007 - 15.0[2]
Granitic rock	420 mg/kg dry weight	Average value[3]
River water	0.054 ppm	Range = 0.009-0.15 ppm[4]
Sandstone sediment	190 mg/kg dry weight	Average value[3]
Seawater	0.006 mg/l	None[2]
Seawater	0.03 ppm	Ba(++), residence time = 84E3 y[5]

Shale sediment	580 mg/kg dry weight	Average value[3]
Soil	500 mg/kg dry soil	Oven-dried, rg = 100 - 3000[4]
Water	54 ug/l	Rivers, world mean[1]

Source: 1.Laveskog, A., A. Lindskog and V. Sten berg. 1976. Om Metaller, Statens Naturvaardsverk, Stockholm.
2.Schroeder, G.A., I.H. Tipton and A.P. Nason. 1972. J. Chron. Dis. 25, pp 491-517.
3.Turekian, K.K. and K.H. Wedepohl. 1961. Amer. Geol. Soc. Bull. 72, pp 175-192.
4.Bowen, H.J.M. 1966. Trace Elements in Biochemistry, Ac. Press.
5.Goldberg, E.D., 1963. The Sea, Interscience London, 2, pp 3.

5-017 Background concentration Be

Item	Value	Condition
Atmosphere	0.13 ng/m**3	Not urban, USA[1]
Atmosphere	1.5 ng/m**3	Urban, USA, rg = 1 - 2[1]
Atmosphere, urban	0.3 ng/m**3	Boston, NW = 0.5, Cleveland = 1.3, Pittsburgh = 3.0[2]
Carbonate sediments	0.2 mg/kg dry weight	Average value[3]
Deep-sea clay sediments	2.6 mg/kg dry weight	Average value[3]
Drinking water	0.355 ug/l	USA, rg = 0.01 - 0.7[1]
Granitic rock	2.0 mg/kg dry weight	Average value[3]
Igenous rock	3 mg/kg dry weight	None[4]
Lithosphere	5 mg/kg dry weight	Overall concentration[5]
River water	0.001 ppm	Range = 0.1E-3-0.001 ppm[6]
Sandstone sediment	0.3 mg/kg dry weight	Average value[3]
Seawater	0.6E-6 ppm	Residence time = 0.15E3 y[7]
Sedimentary rock	3 mg/kg dry weight	None[8]
Shale sediment	3.0 mg/kg dry weight	Average value[3]
Soil	6 mg/kg dry soil	Oven-dried, mg/kg dry soil[6]
Surface water	1 ug/l	Water, USA[1]

Source: 1.Laveskog, A., A. Lindskog and V. Stenberg. 1976. Om Metaller, Statens Naturvaardsverk, Stockholm.
2.Sterner, J.H., and M. Eisenbud. 1951. arch. Ind. Hyg. 4, pp 123-151.
3.Turekian, K.K. and K.H. Wedepohl. 1961. Amer. Geol. Soc. Bull. 72, pp 175-192.
4.Taylor, S.R. 1964. Geochim. Cosmochim. Acta 28, pp 1273-1285.
5.Nikonova, N.N. 1967. Izv. Sib. Akad. Nauk. SSSR Ser. Biol. Med. 3, pp 25-29.
6.Bowen, H.J.M. 1966. Trace Elements in Biochemistry, Ac. Press.
7.Goldberg, E.D., 1963. The Sea, Interscience London, 2, pp 3.
8.Vinogradov, A.P. 1959. The Geochemistry of Rare and Dispersed Chemical Elements in the Soil. Transl. from Russian. Consultant Bureau Inc. NY. 2nd ed.

5-018 Background concentration Benz(a)anthracene

Item	Value	Condition
Atmosphere, seasonal variation	0.30 ng/m3	1. Quarter, 2. Q=0.06, 3. Q=0.10, 4. Q=0.59, LA, n=13, USA, 1975[1]
Crassostrea virginica	10 ug/kg dry weight	Harbor of Norfolk, Vir., USA, less than value[2]
Domestic sewage	0.025 ug/liter	Dry weather condition, Hegne, GFR[3]
Domestic sewage	10.360 ug/liter	During heavy rain, Hegne, GFR[3]
Mixed sewage	1.360 ug/liter	GFR, high % industri, Stockach, Rodolfsell = 0.343 ppm[3]
Wastewater, domestic	0.191 ug/liter	Effluent conc to Rotach R, GFR[3]
Wastewater, industrial	0.167 ug/liter	Effluent conc to Rotach R, GFR[3]

Source: 1.Gordon, R.J. 1976. Environ. Sci. Technol. 10, 370-373.

2.Fazio, T. 1971. Proc. 7th Nat. Shellfish Sanit. Workshop. Wash DC. FDA, Division of Shellfish Sanitaion
3.Borneff, J., and H. Kunte. 1965. Arch. Hyg. (Berlin) 149, 226-243.

5-019 Background concentration Benzene

Item	Value	Condition
Atmosphere, organic gas	0.1 ppb	Rg = 0 - 0.2 ppb, trop. and polar maritime air, 1972, STP[1]
Atmosphere, organic gas	0.2 ppb	Rg = 0-0.4 ppb, Point Barrow, Alaska, 1968[2]
Atmosphere, organic gas	0.7 ppb	Rg = 0.3-1.1 ppb, mountain air in USA, 1972, STP[2]
Atmosphere, organic gas	10.5 ppb	Rg = 7 - 14 ppb, Central Amazona jungle, 1972, STP[1]

Source: 1.Robinson, E., R.A. Rasmussen, H.H. Westberg and M.W. Holdren. 1973. J. Geophys. Res. 78, 5345-5351.
2.Cavanagh, L.A., C.F. Sahadt and E. Robinson. 1969. Environ. Sci. Technol. 3, 251-257.

5-020 Background concentration Benzo(a)anthracene

Item	Value	Condition
Algae	2 microgram/kg wet weight	Italy[1]
Algae	60 microgram/kg wet weight	Greenland, sampled at 40 m[1]
Algae	60 microgram/kg wet weight	Greenland, sampled at seabed[1]
Cod	15 microgram/kg wet weight	Greenland[1]
Molluscs	2 microgram/kg wet weight	Italy[1]
Molluscs	60 microgram/kg wet weight	Greenland[1]
Mussel	11 microgram/kg wet weight	Italy, shell[1]
Mussel	18 microgram/kg wet weight	Greenland, shell[1]
Mussel	335 microgram/kg wet weight	Italy, body, rg = 130 - 450[1]
Mussel	55 microgram/kg wet weight	Greenland, body[1]
Plankton	13.5 microgram/kg wet weight	Italy, rg = 6 - 21[1]
Plankton	400 microgram/kg wet weight	French Channel Coast[1]
Plankton	5 microgram/kg wet weight	Greenland[1]
Sardine	65 microgram/kg wet weight	Italy[1]

Source: 1.Harrison, R.M., R. Perry and R.A. Wellings. 1975. Water Res., 9(4), pp 331-346.

5-021 Background concentration Benzo(a)pyrene

Item	Value	Condition
Atmosphere, seasonal variation	0.77 ng/m3	1. Quarter, 2. Q=0.17, 3. Q=0.26, 4. Q=1.27, LA, n=13, USA, 1975[1]
Cecarcinus planatus	119 ug/kg dry weight	Muscle, terrestrial crab, Clipperton Lagoon, Pacific, 1969[2]
Crassostrea gigas	1.3 ug/kg dry weight	Rg = 0.56 - 2.04 ppb, Coos Bay, Oregon, USA, 1976[2]
Crassostrea virginica	4 ug/kg dry weight	Harbor of Norfolk, Vir., USA, rg=2-6 ppb[3]
Domestic sewage	0.001 ug/liter	Dry weather condition, Hegne, GFR[4]
Domestic sewage	3.840 ug/liter	During heavy rain, Hegne, GFR[4]
Drinking water source	0.2 ng/liter	Ground water, USA, 1977[2]
Drinking water source	0.4 ng/liter	Well water, Engl., 1974, rg = 0.2 - 0.6 ng/l[2]

Drinking water source	0.9 ng/liter	Tap water, 9 USA cities, 1977 rg = 0.2 - 1.6 ng/l[2]
Drinking water source	3 ng/liter	Ground water, GFR, 1973, rg = 2 - 4 ng/l[2]
Drinking water source	3.6 ng/liter	Ground water, GFR, 1968, rg = 0.4 - 7.0 ng/l[2]
Drinking water source	4.8 ng/liter	Tap water, GFR, 1968, rg = 0.5 - 9.0 ng/l[2]
Drinking water source	8.5 ng/liter	Well water, GFR, 1973, rg = 2 - 15 ng/l[2]
Freshwater sediment	1.5 ug/kg dry weight	Pskov region, USSR, non-polluted, rg = 1 - 2 ppb[2]
Freshwater sediment	1.5 ug/kg dry weight	R. Sunzha, USSR, 25 km below oil-refinery discharge, rg = 0-3.1, 1968[2]
Freshwater sediment	10 ug/kg dry weight	S Beaufort Sea, 1975, rg = 0.4-20 ppb[2]
Freshwater sediment	11000 ug/kg dry weight	River Usk, UK, 1974[2]
Freshwater sediment	12500 ug/kg dry weight	R. sediment below a coke byproduct plant, USSR, 1963, rg = 8200-17000[2]
Freshwater sediment	14.1 ug/kg dry weight	R. Sunzha, USSR, 3-4 km below oil-refinery discharge, rg = 9.2-19, 1968[2]
Freshwater sediment	150 ug/kg dry weight	Benyuls, France, depth in sed = 8 m, 1963[2]
Freshwater sediment	15000 ug/kg dry weight	River Seine mud, 1959[2]
Freshwater sediment	1610 ug/kg dry weight	Grosser Ploner See, GFR, 100 m from N shore, 1974[2]
Freshwater sediment	17 ug/kg dry weight	Mouth of the Var R., France, 1963, rg = 0 - 34 ppb[2]
Freshwater sediment	1760 ug/kg dry weight	Canal dllle-Jeanty, 1962[2]
Freshwater sediment	200 ug/kg dry weight	Cerbere, France, 1963, depth in sed = 75 m[2]
Freshwater sediment	23 ug/kg dry weight	Port-Vendres, france, 1963, depth in sed = 16 m[2]
Freshwater sediment	260 ug/kg dry weight	Grosser Ploner See, GFR, 100 m from S shore, 1974[2]
Freshwater sediment	29 ug/kg dry weight	R. Seine, 40 km downstream Paris, depth in sed = 7.1 m[2]
Freshwater sediment	3.5 ug/kg dry weight	Polynesian atolls, 1974, rg = 0.3 - 6.5 ppb[2]
Freshwater sediment	3000 ug/kg dry weight	R. Seine, 40 km downstream Paris, depth in sed = 5.9 m[2]
Freshwater sediment	40 ug/kg dry weight	R. Seine, 40 km downstream Paris, depth in sed = 10 m[2]
Freshwater sediment	40 ug/kg dry weight	R. Seine, 40 km downstream Paris, depth in sed = 9 m[2]
Freshwater sediment	400 ug/kg dry weight	Port of Dunkerque, 1962[2]
Freshwater sediment	44 ug/kg dry weight	Rublevskoye Resevoir, USSR, 1971[2]
Freshwater sediment	443 ug/kg dry weight	Lake Constance, GFR, 1976[2]
Freshwater sediment	445 ug/kg dry weight	Khiminskoe Resevoir, USSR, 1971[2]
Freshwater sediment	453 ug/kg dry weight	Rhone R delta, rg = 0 - 906 ppb, 1963[2]
Freshwater sediment	470 ug/kg dry weight	Severn estuary at Aust., UK, 1977[2]
Freshwater sediment	55 ug/kg dry weight	S region of North Sea, rg = 0-122.5 ppb, 1962[2]
Freshwater sediment	750 ug/kg dry weight	Villefranche B, France, depth in sed = 4-82 m, rg = 0-1500 ppb[2]
Freshwater sediment	80 ug/kg dry weight	Near oil-refinery tanker jetty, USA, 1976, depth in sed = 10 m[2]
Freshwater, lake	0.3 ng/liter	Lake Erie at Baffalo, USA, 1977, used as drinking water source[2]
Freshwater, lake	2.3 ng/liter	Resevoirs for drinking water, Engl., 1974, rg = 0.7 - 3.8 ng/l[2]
Freshwater, lake	5.8 ng/liter	Lake Constance, drinking water source, GFR, 1963, rg = 0.2 - 11.5 ng/l[2]

Freshwater, lake	8.5 ng/liter	Used as drinking water source, Moscow, 1969, rg = 4 - 13 ng/l[2]
Freshwater, river	0.05 ng/liter	Rg = 0.01-0.1 ng/l, Pskov region, USSR, remote from human activity[2]
Freshwater, river	114 ng/liter	Rg = 78-150 ng/l, Oyster R., Conn., USA, 1970[2]
Freshwater, river	1800 ng/liter	Rg = 50-3500 ng/l, Sunzha R., USSR, below oil refinery, 1967[2]
Freshwater, river	225 ng/liter	Rg = 170-280 ng/l, Thames R., Engl., 1975[2]
Freshwater, river	23.5 ng/liter	Rg = 4 - 43 ng/l, R Aach at Stockach, GFR, 1964[2]
Freshwater, river	24 ng/liter	Rg = 1.5-48 ng/l, Severn R., Engl., 1974[2]
Freshwater, river	252 ng/liter	Rg = 5.3-504 ng/l, Trent R., Engl., 1974[2]
Freshwater, river	35 ng/liter	Rg = 10 - 60 ng/l, R Rhine at Koblenz, 1973[2]
Freshwater, river	41.1 ng/liter	Delaware R., Philadelphia, Penn., USA, 1977[2]
Freshwater, river	5.3 ng/liter	Rg = 0.6 - 10 ng/l, other Rhine and Aach in GFR, 1964[2]
Freshwater, river	5.6 ng/liter	Ohio R., Huntington, W Virginia, USA, 1977[2]
Freshwater, river	60 ng/liter	Rg = 42 - 77 ng/l, Monongahela R. at Pittsburgh, Penn., USA, 1977[2]
Freshwater, river	80 ng/liter	Rg = 50 - 100 ng/l, R Rhine at Mainz, GFR, 1963[2]
Marine sediment	0 microgram/kg dry weight	French Mediterranean Coast, depth = 58 m[5]
Marine sediment	0 microgram/kg dry weight	French Mediterranean Coast, estuary, depth = 102 m[5]
Marine sediment	1.4 microgram/kg dry weight	Italy, depth = 120 m, Naples, near volcanic pollution[5]
Marine sediment	15 microgram/kg dry weight	French Mediterranean Coast, estuary, depth = 4 m[5]
Marine sediment	1500 microgram/kg dry weight	French Mediterranean Coast, depth = 1.6 m[5]
Marine sediment	2 mg/kg dry weight	Italy, depth = 15 - 45 m, Naples, higly industrialized area, rg = 1-5[5]
Marine sediment	20 microgram/kg dry weight	French Mediterranean Coast, estuary, depth = 1 m[5]
Marine sediment	25 microgram/kg dry weight	French Mediterranean Coast, estuary, depth = 5 m[5]
Marine sediment	270 microgram/kg dry weight	Italy, depth = 2 - 65 m, Naples, highly industrialized area, rg = 10-500[5]
Marine sediment	300 microgram/kg dry weight	Italy, Naples, island affected by pollution, depth = 0 m, rg = 100-500[5]
Marine sediment	34 microgram/kg dry weight	French Mediterranean Coast, estuary, depth = 0 m[5]
Marine sediment	400 microgram/kg dry weight	French Mediterranean Coast, depth = 14 m[5]
Marine sediment	400 microgram/kg dry weight	French Mediterranean Coast, depth = 82 m[5]
Marine sediment	5 microgram/kg dry weight	Greenland, depth = 0.2 m[5]
Marine sediment	610 microgram/kg dry weight	Italy, depth = 55 m, Naples, near volcanic pollution, rg = 260 - 960[5]
Marine sediment	7.5 microgram/kg dry weight	Italy, depth = 13 m, Naples, highly industrialized area[5]
Marine sediment	75 microgram/kg dry weight	French Mediterranean Coast, depth = 48 m[5]
Mixed sewage	0.100 ug/liter	GFR, high % industri, Stockach, Rodolfsell = 0.368 ppm[4]
Mya arenaria	14 ug/kg dry weight	Rg = 1.32-26.64 ppb, Coos Bay, Oregon, USA, 1976[2]
Mytilus californeanus	0.4 ug/kg dry weight	W Coast of Vancouver Isl., BC, Canada, rg = +- 0.4 ppb, 1974[2]
Mytilus edulis	168 ug/kg dry weight	Vancouver, BC, Canada, 1974, False Creek, rg = +- 24 ppb[2]

Mytilus edulis	34 ug/kg dry weight	Rg = <0.4 - 67.4 ppb, Tillamook Bay, Oregon, USA, 1976[2]
Mytilus edulis	60.6 ug/kg dry weight	Rg = 0.48-120.8 ppb, Yaquina Bay, Oregon, USA, 1976[2]
Mytilus edulis	72 ug/kg dry weight	Vancouver, BC, Canada, 1974, wharf, marine and dock area, rg =+- 19.6[2]
Mytilus edulis	8 ug/kg dry weight	Vancover, BC, Canada, 1974, outer harbor, rg = +-1.2 ppb[2]
Rain	42 ng/liter	Used as drinking water source, Koblenz, 1973, GFR, rg = 4 - 80 ng/l[2]
Saxidomas giganteus	2.7 ug/kg dry weight	Rg = 1.16-4.20 ppb, Coos Bay, Oregon, USA, 1976[2]
Wastewater, domestic	0.038 ug/liter	Effluent conc to Rotach R, GFR[4]
Wastewater, industrial	0.100 ug/liter	Effluent conc to Rotach R, GFR[4]
Zostera marina	25 ug/kg dry weight	Bay of St. Malo, France, 1966[2]

Source: 1.Gordon, R.J. 1976. Environ. Sci. Technol. 10, 370-373.
2.Neff, J.M. 1979. Polycyclic Aromatic Hydrocarbons in the Aquatic Environment. Applied Sci. Publ. London.
3.Fazio, T. 1971. Proc. 7th Nat. Shellfish Sanit. Workshop. Wash DC. FDA, Division of Shellfish Sanitation
4.Borneff, J., and H. Kunte. 1965. Arch. Hyg. (Berlin) 149, 226-243.
5.Harrison, R.M., R. Perry and R.A. Wellings. 1975. Water Res., 9(4), pp 331-346.

5-022 Background concentration Benzo(b)fluoranthene

Item	Value	Condition
Atmosphere, seasonal variation	0.24 ng/m3	1. Quarter, 2. Q=0.24, 3. Q=0.33, 4. Q=1.30, LA, n=13, USA, 1975[1]
Crassostrea virginica	0.3 ug/kg dry weight	Aransas Bay, uncontam. area, TEX, USA, 1970, rg=0.0-0.5 ppb[2]
Crassostrea virginica	1.2 ug/kg dry weight	Galveston B., commercial oyster reef, TEX, USA, 1970, rg=0.5-1.8 ppb[2]
Crassostrea virginica	2.2 ug/kg dry weight	Galveston B., oyster reef contam. dom. sewage, TEX, rg=0.0-4.1 ppb[2]
Domestic sewage	0.039 ug/liter	Dry weather condition, Hegne, GFR[3]
Domestic sewage	10.790 ug/liter	During heavy rain, Hegne, GFR[3]
Mixed sewage	0.870 ug/liter	GFR, high % industri, Stockach, Rodolfsell = 0.525 ppm[3]
Wastewater, domestic	0.036 ug/liter	Effluent conc to Rotach R, GFR[3]
Wastewater, industrial	0.114 ug/liter	Effluent conc to Rotach R, GFR[3]

Source: 1.Gordon, R.J. 1976. Environ. Sci. Technol. 10, 370-373.
2.Fazio, T. 1971. Proc. 7th Nat. Shellfish Sanit. Workshop. Wash DC. FDA, Division of Shellfish Sanitaion
3.Borneff, J., and H. Kunte. 1965. Arch. Hyg. (Berlin) 149, 226-243.

5-023 Background concentration Benzo(e)pyrene

Item	Value	Condition
Atmosphere, seasonal variation	1.30 ng/m3	1. Quarter, 2. Q=0.42, 3. Q=0.62, 4. Q=1.96, LA, n=13, USA, 1975[1]
Crassostrea virginica	0.2 ug/kg dry weight	Aransas Bay, uncontam. area, TEX, USA, 1970, rg=0.0-0.4 ppb[2]
Crassostrea virginica	1.2 ug/kg dry weight	Galveston B., commercial oyster reef, TEX, USA, 1970, rg=0.0-4.0 ppb[2]
Crassostrea virginica	2.1 ug/kg dry weight	Galveston B., oyster reef contam. dom. sewage, TEX, rg=0.8-4.2 ppb[2]
Crassostrea virginica	20 ug/kg dry weight	Harbor of Norfolk, Vir., USA, less than value[2]

Source: 1.Gordon, R.J. 1976. Environ. Sci. Technol. 10, 370-373.
2.Fazio, T. 1971. Proc. 7th Nat. Shellfish Sanit. Workshop. Wash DC. FDA, Division of Shellfish Sanitation

5-024 Background concentration Benzo(ghi)perylene

Item	Value	Condition
Atmosphere, seasonal variation	3.80 ng/m3	1. Quarter, 2. Q=1.35, 3. Q=2.71, 4. Q=8.25, LA, n=13, USA, 1975[1]
Crassostrea virginica	3 ug/kg dry weight	Harbor of Norfolk, Vir., USA, rg=1-5 ppb[2]
Domestic sewage	0.004 ug/liter	Dry weather condition, Hegne, GFR[3]
Domestic sewage	4.180 ug/liter	During heavy rain, Hegne, GFR[3]
Mixed sewage	0.480 ug/liter	GFR, high % industri, Stockach, Rodolfsell = 0.120 ppm[3]
Wastewater, domestic	0.040 ug/liter	Effluent conc to Rotach R, GFR[3]
Wastewater, industrial	0.073 ug/liter	Effluent conc to Rotach R, GFR[3]

Source: 1.Gordon, R.J. 1976. Environ. Sci. Technol. 10, 370-373.
2.Fazio, T. 1971. Proc. 7th Nat. Shellfish Sanit. Workshop. Wash DC. FDA, Division of Shellfish Sanitation
3.Borneff, J., and H. Kunte. 1965. Arch. Hyg. (Berlin) 149, 226-243.

5-025 Background concentration Benzo(j)fluoranthene

Item	Value	Condition
Atmosphere, seasonal variation	0.26 ng/m3	1. Quarter, 2. Q=0.06, 3. Q=0.12, 4. Q=0.43, LA, n=13, USA, 1975[1]
Domestic sewage	0.057 ug/liter	Dry weather condition, Hegne, GFR[2]
Domestic sewage	9.910 ug/liter	During heavy rain, Hegne, GFR[2]
Mixed sewage	1.740 ug/liter	GFR, high % industri, Stockach, Rodolfsell = 1.100 ppm[2]
Wastewater, domestic	0.037 ug/liter	Effluent conc to Rotach R, GFR[2]
Wastewater, industrial	0.045 ug/liter	Effluent conc to Rotach R, GFR[2]

Source: 1.Gordon, R.J. 1976. Environ. Sci. Technol. 10, 370-373.
2.Borneff, J., and H. Kunte. 1965. Arch. Hyg. (Berlin) 149, 226-243.

5-026 Background concentration Benzo(k)fluoranthene

Item	Value	Condition
Atmosphere, seasonal variation	0.27 ng/m3	1. Quarter, 2. Q=0.07, 3. Q=0.15, 4. Q=0.52, LA, n=13, USA, 1975[1]
Crassostrea virginica	0.1 ug/kg dry weight	Galveston B., commercial oyster reef, TEX, USA, 1970, rg=0.0-0.6 ppb[2]
Crassostrea virginica	0.4 ug/kg dry weight	Galveston B., oyster reef contam. dom. sewage, TEX, rg=0.0-1.1 ppb[2]
Crassostrea virginica	10 ug/kg dry weight	Harbor of Norfolk, Vir., USA, rg =8-12 ppb[2]
Domestic sewage	0.022 ug/liter	Dry weather condition, Hegne, GFR[3]
Domestic sewage	1.840 ug/liter	During heavy rain, Hegne, GFR[3]
Mixed sewage	0.460 ug/liter	GFR, high % industri, Stockach, Rodolfsell = 0.336 ppm[3]
Wastewater, domestic	0.031 ug/liter	Effluent conc to Rotach R, GFR[3]
Wastewater, industrial	0.032 ug/liter	Effluent conc to Rotach R, GFR[3]

Source: 1.Gordon, R.J. 1976. Environ. Sci. Technol. 10, 370-373.
2.Fazio, T. 1971. Proc. 7th Nat. Shellfish Sanit. Workshop. Wash DC. FDA, Division of Shellfish Sanitation
3.Borneff, J., and H. Kunte. 1965. Arch. Hyg. (Berlin) 149, 226-243.

5-027 Background concentration
Bi

Item	Value	Condition
Atmosphere	0.25 ng/m**3	Not urban, USA, rg = 0.1 - 0.4[1]
Atmosphere	19.5 ng/m**3	Urban, USA, rg = 1 - 38[1]
Atmosphere, rural	0.35 ng/m**3	Range = 0.1 - 0.6 ng/m3, 1971, USA[2]
Atmosphere, urban	33 ng/m**3	Range = 1 - 66 ng/m3, 1971[3]
Cardium edule	0.025 mg/kg dry weight	Gastropod, rg = 0.01-0.04 ppm, Southampton waters, GB, 1971[4]
Crassostrea gigas	0.49 mg/kg wet weight	Gastropod, rg = +- 0.20 ppm, Saldanha B, LangeBaan B., RSA, 1975[5]
Crassostrea margaritacea	0.68 mg/kg wet weight	Gastropod, rg = +- 0.13 ppm, Saldanha B, LangeBaan B., RSA, 1975[5]
Jasus lalandii	0.81 mg/kg wet weight	Marine, Crustaceans, rg=0.53-1.1, Saldanha Bay, SA, muscle, 1975[5]
Seawater	0.17E-4 ppm	Residence time = 45E3 y[6]
Seawater	0.2 ug/l	Approximate value, 1971[7]

Source: 1.Laveskog, A., A. Lindskog and V. Stenberg. 1976. Om Metaller, Statens Naturvaardsverk, Stockholm.
2.Division of Atmospheric Surveillance. 1972. Air Quality data. Office of Air Programme, US. EPA. Res.Triangle Park. N.C.
3.Division of Atmospheric Surveillance 1972. Air Quality Data from National Surveilance Network. pp 10103. EPA. Research Triangle Park. NC.
4.Martin, J.H., W.W. Broenkow. 1975. Science 190, 884-885.
5.Fourie, H.O. 1976. S. Afr. J. Sci. 72, 110-113.
6.Goldberg, E.D., 1963. The Sea, Interscience London, 2, pp 3.
7.Environmental Studies Board. 1972. Water Quality Data. National Academy of Engineering. Wash D.C.

5-028 Background concentration
Biotin

Item	Value	Condition
Freshwater	0.002 mg/m**3	Small swiss ponds, rg = < 0.001 - 0.004[1]
Freshwater	0.00205 mg/m**3	Lindsey Pond, Conn, rg = 0.0001 - 0.004[2]
Freshwater	0.00245 mg/m**3	Particulate, Tsukui, Japan, surface, rg = 0.0005 - 0.00402[3]
Freshwater	0.00355 mg/m**3	Kasumigaura, Japan, rg = 0.0021 - 0.050[4]
Freshwater	0.0355 mg/m**3	Dissolved, Tsukui, Japan, surface, rg = 0.013 - 0.058[3]
Freshwater	0.0405 mg/m**3	Sagami, Japan, rg = 0.013 - 0.068[5]
Rain	0.0008 mg/m**3	Mean of precipatation, Mo, Nov - March[6]
Rain	0.004 mg/m**3	Mean of precipitation, Mo, April - Nov[6]

Source: 1.Clemenson, H. 1963. Schweiz. Z. Hydrol., 25, pp 157-165.
2.Benoit, R.J. 1957. Limnol. Oceanogr., 2, pp 233-240.
3.Ohwada, K., M. Otsuhata and N. Taka. 1972. Bull. Japan Soc. Sci. Fish. 38, pp 817-823.
4.Kashiwada, K., A. Kanazawa and S. Tachibanazono. 1963. Mem. Fac. Fish., Kagoshima Univ. 12, pp 153-17.
5.Ohwada, K. and N. Taga. 1972. Limnol. Oceanogr. 17, pp 315-320.
6.Parker, B.C. and M.A. Wachtel. 1971. in: Cairns, J. (ed.); The Structure and Function of Freshwatericrobial Communities. Res. Div. Monogr. Virginia Poly tech. Inst. 3, pp 195-207.

5-029 Background concentration Br

Item	Value	Condition
River water	0.021 ppm	Range = 0.005-140 ppm[1]
Seawater	65 ppm	Br(-)[2]
Soil	5 mg/kg dry soil	Oven-dried, rg = 1 - 10[1]

Source: 1.Bowen, H.J.M. 1966. Trace Elements in Biochemistry, Ac. Press.
2.Goldberg, E.D., 1963. The Sea, Interscience London, 2, pp 3.

5-030 Background concentration Bromoform

Item	Value	Condition
Drinking water	0.1 ug/m3	Typical non-polluted value, halocarbon[1]
Drinking water	40 ug/m3	High value, halocarbon[1]
Freshwater	0.05 ug/m3	Typical non-polluted value, halocarbon[1]
Freshwater	0.1 ug/m3	High value in urban area, halocarbon[1]
Groundwater	0.05 ug/m3	Typical non-polluted value, halocarbon[1]
Groundwater	2 ug/m3	High value, halocarbon[1]

Source: 1.Pearson, C.R. 1982. In: O. Hutzinger (ed.): Environmental Chemistry. Springer-Verlag. Volume 3, part B. Berlin.

5-031 Background concentration Butane

Item	Value	Condition
Atmosphere, organic gas	0.1 ppb	Rg = 0 - 0.1 ppb, Point Barrow, Alaska, 1968[1]
Atmosphere, organic gas	0.8 ppb	Rg = 0 - 1.5 ppb, trop. and polar maritime air, 1972, STP[2]
Atmosphere, organic gas	1.7 ppb	Rg = 0.1-3.6 ppb, mountain air in USA, 1972, STP[1]
Atmosphere, organic gas	73 ppb	Rg = 54 - 91 ppb, Central Amazona jungle, 1972, STP[2]
Atmospheric conc of aliphatic hydrocarbon and alkylbenzene	46 ppb	Los Angeles, 1967, STP[3]

Source: 1.Cavanagh, L.A., C.F. Sahadt and E. Robinson. 1969. Environ. Sci. Technol. 3, 251-257.
2.Robinson, E., R.A. Rasmussen, H.H. Westberg and M.W. Holdren. 1973. J. Geophys. Res. 78, 5345-5351.
3.Altshuller, A.P., W.A. Lonneman and S.L. Kopczynski. 1973. J. Air Pollut. Contr. Assoc. 23, 597-599.

5-032 Background concentration Butane (1) + Isobutane

Item	Value	Condition
Atmospheric conc of aliphatic hydrocarbon and alkylbenzene	5.5 ppb	Los Angeles, 1967, STP[1]

Source: 1.Altshuller, A.P., W.A. Lonneman and S.L. Kopczynski. 1973. J. Air Pollut. Contr. Assoc. 23, 597-599.

5-033 Background concentration Butene (2)

Item	Value	Condition
Atmospheric conc of aliphatic hydrocarbon and alkylbenzene	2 ppb	Los Angeles, 1967, STP[1]

Source: 1.Altshuller, A.P., W.A. Lonneman and S.L. Kopczynski. 1973. J. Air Pollut. Contr. Assoc. 23, 597-599.

5-034 Background concentration C

Item	Value	Condition
Atmosphere	164000 ug/m**3	293 K, 1 atm[1]
Biosphere	6560 moles/hectare	Average composition of living matter[2]
River water	11 ppm	HCO3, range = 6-19 ppm[3]
Seawater	28 ppm	CO3H(-), organic carbon[4]
Soil	20000 mg/kg dry soil	Oven-dried[3]

Source: 1.Junge, C.E. 1963. Air Chemistry and Radioactivity, Ac. Press.
2.Deevey, E.S. 1970. Scientific American. 223(3), p 150.
3.Bowen, H.J.M. 1966. Trace Elements in Biochemistry, Ac. Press.
4.Goldberg, E.D., 1963. The Sea, Interscience London, 2, pp 3.

5-035 Background concentration C-9 and C-10 alkylbenzene

Item	Value	Condition
Atmospheric conc of aliphatic hydrocarbon and alkylbenzene	12 ppb	Los Angeles, 1967, STP[1]

Source: 1.Altshuller, A.P., W.A. Lonneman and S.L. Kopczynski. 1973. J. Air Pollut. Contr. Assoc. 23, 597-599.

5-036 Background concentration CH4

Item	Value	Condition
Atmosphere	975 ug/m**3	293 K, 1 atm, residence time = 100 y, rg = 850 - 1100[1]
Lower atmosphere	0.00002 % by volume	None[2]
Seawater	15 ml/l	Saturated seawater, s = 3.43 %, 283 K, high val.only over anoxic[3]

Source: 1.Junge, C.E. 1963. Air Chemistry and Radioactivity, Ac. Press.
2.Gleuckauf, E. 1951. Compendium of Meteorology. Am. Meteorological Soc. Boston.
3.Sisler, F.D. 1975. In: S.F. Singer (ed.), The Changing Global Environment, pp 57-69. D. Reidel Publ. Comp.

5-037 Background concentration CO_2

Item	Value	Condition
Atmosphere	0.03 % by atoms	Sea-level dry atmospheric composition[1]
Atmosphere	0.45 g/cm**2	Total in atmos.(1975) = 2.25E18 g[2]
Atmosphere	4E4 g/cm**2	Total supply = 1.9E23 g**2[3]
Atmosphere	6E5 ug/m**3	293 K, 1 atm, residence time = 4 y[4]
Gas in ice, glacier ice	0.16 %	Gas pres.=2.1 bar, dens=0.90 g/cm3, gas=26 cm3/kg, rg=73.2-78.3 %[5]
Gas in ice, iceberg ice	0.08 %	Gas pres.=1.3 bar, dens=0.88 g/cm3, gas=40 cm3/kg, rg=77.7-78.3 %[5]
Gas in ice, pond ice	24.8 %	Gas pres.=1.0 bar, dens=0.916 g/cm3, gas=1.0 cm3/kg, n = 1[5]
Gas in ice, sea ice	0.9 %	Gas pres.=0.9 bar, dens=91 g/cm3, gas=13 cm3/kg, rg=54.2-76.8 %[5]
Lower atmosphere	0.033 % by volume	None[6]
Meltwater from snow	0.4 mg CO2/l	Picos, Spain, fresh meltwater[7]

Meltwater from snow	1.80 mg CO2/l	Rg = 0.85 - 1.76, fresh meltwater, Switzerland[7]
Meltwater from snow	3.14 mg CO2/l	Rg = 2.64 - 3.63 mg CO2/l, from vegetated soil, Switzerland[7]
Rain	2.2 mg CO2/l	Picos, Spain[7]
Rain	2.4 mg CO2/l	Rg = 2.2 - 2.6 mg CO2/l, Switzerland[7]
Rain	3.0 mg CO2/l	Rg = 2.5 - 3.5 mg CO2/l, Cuba[7]
Seawater	45 ml/l	Saturated seawater, salinity = 3.43 %, 283 K, rg = 34 - 56[1]

Source: 1.Sisler, F.D. 1975. In: S.F. Singer (ed.), The Changing Global Environment, pp 57-69. D. Reidel Publ.Comp.
2. Bolin, B. 1981. Carbon Cycle Modelling. SCOPE 16. John Wiley and Sons. NY.
3.Johnson, F.S. 1975. In: S.F. Singer (ed.), The Changing Global Environment, pp 49-56. D. Reichel Pub. Copm.
4.Junge, C.E. 1963. Air Chemistry and Radioactivity, Ac. Press.
5.Matsuo, S. and Y. Miyake. 1966. J. Geophys. Res. 71, 5235.
6.Gleuckauf, E. 1951. Compendium of Meteorology. Am. Meteorological Soc. Boston.
7.Bolin, B., E.T. Degens, S. Kempe and P. Ketner. 1977. The Global Carbon Cycle. SCOPE and ICSU. John Wiley and Sons. NY.

5-038 Background concentration CO_3(2-) + HCO_3

Item	Value	Condition
River, Africa	43.0 mg/l	None[1]
River, Asia	79.0 mg/l	None[1]
River, Australia	31.6 mg/l	None[1]
River, Europe	95.0 mg/l	None[1]
River, North America	68.0 mg/l	None[1]
River, South America	31.0 mg/l	None[1]
River, World	58.4 mg/l	None[1]
Seawater	17 % of total	As Na2CO3, chlorinity = 1.9 %, 298 K, pH = 8.15[2]
Seawater	67 % of total	As MgCO3, chlorinity = 1.9 %, 298 K, pH = 8.15[2]
Seawater	7 % of total	As CaCO3, chlorinity = 1.9 %, 298 K, pH = 8.15[2]
Seawater	9 % of total	As free ion, chlorinity = 1.9 %, 298 K, pH = 8.15[2]

Source: 1.Wetzel, R.G. 1975. Limnology. Saunders. Philadelphia.
2.Garrels, R.M. and M.E. Thompson. 1962. Am. J. Sci. 260., p 57.

5-039 Background concentration Ca

Item	Value	Condition
Atmospheric particulates	160 mg/m**3	Hawaii, n > = 56, rg = +- 90[1]
Biosphere	18.9 moles/hectare	Average composition of living matter[2]
River water	15 ppm	Range = 4-120 ppm[3]
River, Africa	12.5 mg/l	None[4]
River, Asia	18.4 mg/l	None[4]
River, Australia	3.9 mg/l	None[4]
River, Europe	31.1 mg/l	None[4]
River, North America	21.0 mg/l	None[4]
River, South America	7.2 mg/l	None[4]
River, World	15. mg/l	None[4]

Seawater	0.2 % of total	As $CaCO_3$(aq), chlorinity = 1.9 %, 298 K, pH = 8.15[5]
Seawater	1 % of total	As $Ca(HCO_3)_2$, chlorinity = 1.9 %, 298 K, pH = 8.15[5]
Seawater	400 ppm	Ca^{2+}, residence time = 8000E3 y[6]
Seawater	8 % of total	As $CaSO_4$, chlorinity = 1.9 %, 298 K, pH = 8.15[5]
Seawater	91 % of total	As free ion, chlorinity = 1.9 %, 298 K, pH = 8.15[5]
Soil	13700 mg/kg dry soil	Oven-dried, rg = 7000 - 500000[3]

Source: 1.Golterman, H.L. 1975. Physiological Limnol., Elsevier Amsterdam.
2.Deevey, E.S. 1970. Scientific American. 223(3), p 150.
3.Bowen, H.J.M. 1966. Trace Elements in Biochemistry, Ac. Press.
4.Wetzel, R.G. 1975. Limnology. Saunders. Philadelphia.
5.Garrels, R.M. and M.E. Thompson. 1962. Am. J. Sci. 260., p 57.
6.Goldberg, E.D., 1963. The Sea, Interscience London, 2, pp 3.

5-040 Background concentration Carbon monoxide

Item	Value	Condition
Atmosphere	10.5 ug/m**3	293 K, 1 atm, residence time = 0.3 y, rg = 1 - 20[1]

Source: 1.Junge, C.E. 1963. Air Chemistry and Radioactivity, Ac. Press.

5-041 Background concentration Carbon tetrachloride

Item	Value	Condition
Atmosphere	0.15 ug/m3	Typical value[1]
Bird, egg	16 mg/kg wet weight	Rg = 1 - 30 mg/kg wet weight, food from water organisms[2]
Bird, liver	2 mg/kg wet weight	Rg = 1 - 3 mg/kg wet weight, food from water organisms[2]
Bread	5 mg/kg wet weight	None[2]
Crustacea	1.6 mg/kg wet weight	Rg = 0.2 - 3.0 mg/kg wet weight[2]
Dairy products	5.1 mg/kg wet weight	Rg = 0.1 - 10 mg/kg wet weight[2]
Drinking water	0.5 ug/m3	High value, halocarbon[2]
Drinking water	10 ug/m3	Typical non-polluted value, halocarbon[2]
Estuarine water	0.15 ug/m3	High value in inshore water[3]
Fish, flesh	6.0 mg/kg wet weight	Rg = 2 - 10 mg/kg wet weight[2]
Fish, liver	18 mg/kg wet weight	Rg = 0.3 - 36 mg/kg wet weight[2]
Freshwater	0.1 ug/m3	Typical non-polluted value, halocarbon[2]
Freshwater	3.3 ug/m3	High value in urban area, halocarbon[2]
Fruit and vegetables	9.0 mg/kg wet weight	Rg = 3 - 15 mg/kg wet weight[2]
Groundwater	0.1 ug/m3	Typical non-polluted value, halocarbon[2]
Groundwater	0.5 ug/m3	High value, halocarbon[2]
Homo sapiens, fat	9.5 mg/kg wet weight	Rg = 1 - 20 mg/kg wet weight[2]
Homo sapiens, liver	2.5 mg/kg wet weight	Rg = 1 - 4 mg/kg wet weight[2]
Marine algae	12.5 mg/kg wet weight	Rg = 10 - 15 mg/kg wet weight[2]
Meat	8 mg/kg wet weight	Rg = 7 - 9 mg/kg wet weight[2]
Mollusca	1.1 mg/kg wet weight	Rg = 0.1 - 2.0 mg/kg wet weight[2]

Plankton	0.27 mg/kg wet weight	Rg = 0.04 - 0.5 mg/kg wet weight[2]
Rain	0.15 ug/m3	Rg = 0.001 - 0.3 ppb, halocarbon[2]
Seal, blubber	9.5 mg/kg wet weight	Rg = 4 - 15 mg/kg wet weight[2]
Seal, liver	2.1 mg/kg wet weight	Rg = 0.2 - 4 mg/kg wet weight[2]
Seawater	0.005 ug/m3	Typical open ocean[3]
Sediment	0.5 ug/m3	Typical conc in inshore sediment[3]
Snow	0.0017 ug/m3	Rg = 0.0003 - 0.003 ppb, halocarbon[2]
Vegetable oil	9.0 mg/kg wet weight	Rg = 0.05 - 18 mg/kg wet weight[2]

Source: 1.Lovelock, J.E. and P. Simmons. 1980. In: Sugden (ed.): The Aerosol Controversy. Sci-Horwood.
2.Pearson, C.R. 1982. In: O. Hutzinger (ed.): Environmental Chemistry. Springer-Verlag. Volume 3, partB. Berlin.
3.National Academy of Science. 1978. Chloroform, Carbon tetrachloride, and other halomethanes. An Envionmental Assessment. US NAS.

5-042 Background concentration Cd

Item	Value	Condition
Anodonta sp.	1.2 mg/kg dry weight	Gastropod, L Aberffraw, 1970, freshwater[4]
Arripis trutta, gill	0.32 mg/kg wet weight	Commercial sea fish, New Zealand, 1973, marine[5]
Arripis trutta, gonad	0.08 mg/kg wet weight	Commercial sea fish, New Zealand, 1973, marine[5]
Arripis trutta, kidney	2.27 mg/kg wet weight	Commercial sea fish, New Zealand, 1973, marine[5]
Arripis trutta, liver	0.40 mg/kg wet weight	Commercial sea fish, New Zealand, 1973, marine[5]
Arripis trutta, muscle	0.002 mg/kg wet weight	Commercial sea fish, New Zealand, 1973, marine[5]
Arripis trutta, spleen	0.28 mg/kg wet weight	Commercial sea fish, New Zealand, 1973, marine[5]
Asellus aquaticus	1.21 mg/kg dry weight	Freshwater, Crustaceans, rg = 0.47-2.97 ppm, Elsenz R, 1976[6]
Atmosphere	0.003 ng/m3	Rg = 0.0003 - 0.62, N. Atlantic - 30 deg N[8]
Atmosphere	120 ng/m3	El Paso, USA[11]
Bacon, human food	0.040 mg/kg dry weight	Rg = 0 - 1.555 ppm, SD = +- 0.111 ppm, FDA, USA, 1974[30]
Bread, white, human food	0.036 mg/kg dry weight	Rg = 0 - 0.439 ppm, SD = +- 0.063 ppm, FDA, USA, 1974[30]
Butter, human food	0.032 mg/kg dry weight	Rg = 0 - 0.68 ppm, SD = +- 0.071 ppm, FDA, USA, 1974[30]
Cancer pagurus	0.36 mg/kg wet weight	Marine, Crustacean, rg=+-0.42, Northcumberland C, GB, 1976, whole anim[37]
Carcinus maenas	1 mg/kg dry weight	Marine, Crustacean, less, muscle, Hardanger, Norway, 1976[3]
Cardium edule	0.45 mg/kg dry weight	Gastropod, rg = 0.3-0.6 ppm, SW Spain and Portugal, Medit., 1974[39]
Cardium edule	1.5 mg/kg dry weight	Gastropod, Irish Sea, GB, 1970[4]
Cladophora glomerata	3.9 mg/kg dry weight	Algae, Deadman Bay, Kingston, Ontario, 1975[43]
Cladophora rivularis	0.45 mg/kg dry weight	Algae, Rg = 0.08-1.37 ppm, Waikito R., NZ, 1971[44]
Clams, all species	0.22 mg/kg wet weight	12 references used, SD = +- 0.17, rg = 0.03 - 0.58[33]
Coal fly ash	25 mg/kg dry weight	Range = 0.1 - 50 ug/g dw[26]
Crangon crangon	0.4 mg/kg dry weight	Marine, Crustacean, S Antonio Bay, Texas, whole animal, 1975, less than[46]

Crangon vulgaris	3.5 mg/kg wet weight	Marine, Crustacean, whole animal, Northumberland C, GB, rg =+- 2.1, 1975[37]
Crassostrea commercialis	0.2 mg/kg wet weight	Gastropod, rg = 0.1-1.0 ppm, NSW estuaries, Aust., 1974[47]
Crassostrea gigas	33.18 mg/kg dry weight	Gastropod, rg = 7.6 ppm, Tamar R Est., Tasmania, 1973[51]
Cyprinus carpio	0.035 mg/kg wet weight	Fish, rg=0.011-0.069 ppm, Illinois R, muscle, 1973[40]
Dairy products, human food	0.27 mg/kg wet weight	Rg = 0.03 - 0.56[24]
Drinking water	0.55 ug/l	Sweden, rg = 0.1 - 1[17]
Freshwater zooplankton, effect of fertilizer	0.35 mg/kg dry weight	Area without fertilizer, Scenedesmus obliquus, Thailand, 1974[56]
Freshwater zooplankton, effect of fertilizer	2.46 mg/kg dry weight	Area with fertilizer, Scenedesmus obliquus, FRG, 1974[56]
Freshwater, river	0.06 ug/l	Range 0.02 - 0.12 ppm, 3 rivers in Australia[57]
Freshwater, river	220 ug/l	Neckar River, FRG, max value[58]
Fucus sp.	2.5 mg/kg dry weight	Range = 1.7 - 2.3 ppm, Portugal and Spain, 1974, Atlantic[39]
Fucus vesiculosus	11.2 mg/kg dry weight	Range = 3.82 - 19.5 ppm, Bristol Channel, GB, 1973[68]
Gadus morrhua	0.015 mg/kg dry weight	Fish, rg = +-0.005, Trondsheimfj, Norway, muscle, 1976[3]
Littorina littorea	27 mg/kg dry weight	Gastropod, rg = 8-75 ppm, Bristol Channel, 1971, marine[71]
Littorina littorea	68 mg/kg dry weight	Gastropod, Severn Estuary, Bristol Ch., 1971, rg = 15-210 ppm, marine[72]
Marine phytoplankton	10.7 mg/kg dry weight	Range = 4.8 - 16.5 ppm, Monterey Bay, Baja Calif., 1975, pure diatoms[73]
Marine sediment, polluted	1.1 mg/kg dry weight	Baltic Sea, Sweden, rg = <0.01-8.1 ppm, HCl, 1976[24]
Marine zooplankton, copepods	10.6 mg/kg wet weight	Range = 6.0 - 15.2 ppm, Monterey Bay, Calif, 1974, pure copepods[74]
Meats, human food	0.042 mg/kg dry weight	Rg = 0.02 - 0.052 ppm, Japan[41]
Mercenaria mercenaria	0.19 mg/kg wet weight	Gastropod, Atlantic coast, USA, 1967[52]
Mya arenaria	0.27 mg/kg wet weight	Gastropod, Atlantic coast, USA, 1967[52]
Myriophyllum spicatum	1.06 mg/kg dry weight	Higher plant, Rg = 0.48 - 3.04 ppm, Lago Maggiore, Italy, 1972[70]
Mytilus edulis	0.63 mg/kg wet weight	Gastropod, rg = 0.26-1.6 ppm, New Zealand, 1964[36]
Mytilus edulis	24.6 mg/kg dry weight	Gastropod, rg = +-21.8 ppm, Port Phillip, 1975[85]
Oyster	1.88 mg/kg wet weight	11 references used, SD = +- 1.29, rg = 0.11 - 3.60[33]
Patella vulgata	10 mg/kg dry weight	Gastropod, rg = 2-22 ppm, Mortons Harbor, Newfoundl., 1974[87]
Patella vulgata	220 mg/kg dry weight	Gastropod, rg = 67-440 ppm, Bristol Channel, N. Somerset, 1973[84]
Pecten maximus	13 mg/kg dry weight	Gastropod, Irish Sea, GB, 1970[4]
Platichthys flesus	1.4 mg/kg dry weight	Rg=1.1-1.7 ppm, North Devon, Engl., age effects, marine, fish[90]
Platichthys flesus	4.6 mg/kg dry weight	Rg=3.4-7.3 ppm, Severn Estuary, age and seasonal effects, marine, fish[90]
Pleuronectes platessa	1.44 mg/kg wet weight	Fish, marine, Northumberl, power station, muscle, 1975[37]

Potamogeton pectinatus	0.67 mg/kg dry weight	Higher plant, Rg = 0.49 - 0.85 ppm, Leine R, FRG, 1976[92]
Rain	0.6 ug/l	14 July-7 Nov 1971, New Hampshire, USA, bulk precipitation[94]
River water	0.08 ppm	None[96]
Rutilus ruthilus	0.015 mg/kg wet weight	Fish, rg = 0.002-0.055 ppm, Neckar R, muscle, 1977, FRG[97]
Seawater, dissolved Cd	3.9 ng/l	Filt., 0.5 m depth, Pacific (8 N - 8 S to 155 W), DPASV on MFE[79]
Seawater, dissolved Cd	390 ng/kg	Filt., C Verde, NE Atlantic, Depth=10-2500 m, rg=70-710, Chelating resin+AAS[79]
Sediment, non-polluted	0.14 mg/kg dry weight	Jubilee Creek, Ill., USA, rg=0.08-0.23, tot sed, HNO3/HClO4, SD=0.07[100]
Sediment, non-polluted	0.7 mg/kg dry weight	Undeleloped tributaries around Great Lakes, n=25, tot sed., HNO3/HClO4[107]
Sediment, polluted, freshwater	0.68 mg/kg dry weight	Bodensee, FRG, ref background = 0.21, max = 1.57, HNO3/HCl[108]
Sediment, polluted, freshwater	7.8 mg/kg dry weight	Lake Illawarra, Aust., max = 45 ppm, HNO3/HClO4, ref background = 0.02[116]
Sediment, polluted, river	12.2 mg/kg dry weight	Blies, Saar, Frg, rg = 0.5 - 24, < 63 um, HNO3/HCl[118]
Shrimp, brown	0.09 mg/kg wet weight	8 references used, SD = +- 0.06, rg = 0.04 - 0.24[33]
Soil	0.06 mg/kg dry soil	Oven-dried, rg = 0.01 - 0.7[96]
Solea solea	2.1 mg/kg dry weight	Fish, marine, Sagres, S Portugal, muscle, 1974[31]
Ulva lactuca	1.25 mg/kg dry weight	Range = 0.5 - 2.0 ppm, Portugal and Spain, 1974, Atlantic[84]
Water	1 ug/l	729 Riverand lakes, USA, less than[17]

Source: 1.Lucas, H.F. Jr., D.N. Edgington and P.J. Colby. 1970. J. Fish. Res. Bd Canada 27, 677-684.
3.Lande, E. 1977. Environ. Pollut. 12, 187-198.
4.Segar, D.A. and R.E. Pellenbarg. 1971. J. Mar. Biol. Assoc. UK. 51, 131-136.
5.Brooks, R.R. and M.G. Rumsby. 1974. N.Z.J. Mar. Freshwat. Res. 8, 155-166.
6.Prosi, F. 1977. Schwermetallbelastung in den sedimenten der Elsenz und ihre Auswerkung auf Limnische Organismen. Diss. Univ. Heidelberg.
8.Duce, R.A. 1976. in: Windom, H.L. and R.A. Duce. (eds): Marine Pollutant Transfer. D.C. Health, Lexigton, Mass. pp 77-119.
11.Commission Europ Communities. 1978. Criteria (Dose/effect relationships) for Cd. Pergamon. Oxford.
17.Laveskog, A., A. Lindskog and V. Stenberg. 1976. Om Metaller, Statens Naturvaardsverk, Stockholm.
24.Nriagu, J.O. 1980. Cadmium in the Environment. Part I: Ecological Cycling. John Wiley and Sons. New York.
26.Yost, K.J. 1978. in: Cadmium 77: Proceedings of the 1st International Cd Conference. Metal Bulletin Ltd. London. pp 147-166.
30.Food and Drug Administration. 1975. Compliance Program Evaluation. FY. 1974. Heavy Metals in Food Survey. US Dep. Health.
31.Martin, J.B.M. 1974. Mar. Biol. 28, 245-251.
33.Zook, E.G., J.J. Powell, B.M. Hackley, J.A. Emerson, J.R. Brooker and G.M. Knobl, Jr. 1976. J. Agric. Food. Chem. 24(1), pp 47-53.
36.Heydt, G. 1977. Schwermetallgehalte von Wasser, Wasserpflanzen, Chironomidae und Mollusca der Elsenz. Dipl. Arbeit Univ. Heidelberg. 143 pp.
37.Wright, D.A. 1976. Mar. Pollut. Bull. 7, 36-38.
39.Foster, P. 1976. Environ. Pollut. 10, 45-53.
40.Mathis, B.J. and N.R. Kevern. 1973. Distribution of Hg, Cd, Pb and Th in a eutrophic lake. Int. Water Res. Michigan State Univ., Technol. Dep. 34, 23 pp.
41.Yamagata, N. and I. Shigematsu. 1970. Bull. Jap. Inst. Pub. Health 19, 1-27.
43.Keeney, W.L., W.G. Breck, G.W. Van Loon and J.A. Page 1976. Water Res. 10, 981-984.
44.Reay. P.F. 1972. J. Appl. Ecol. 9, 557-565.
46.Sims, R.R. and B.J. Presley. 1976. Bull. Environ. Contam. Toxicol. 16, 520-526.
47.MacKay, N.J., M.N. Kazacos, R.J. Williams and M.I. Leedow. 1975. Mar. Pollut. Bull. 6, 57-61.
51.Ayling, G.M. 1974. Water Res. 8, 729-738.
52.Pringle, B.H., D.E. Hissong, E.L. Katz and S.T. Mulawka 1968. J. Sanit. Eng. Div. Am. Soc. Civ. Eng. 94, 455-475.
56.Payer, H.D., Runkel, Schramel, Stengel, Bhumiratana and Soeder. 1976. Chemosphere 6, 413-418.
57.Doolan, K.J. and L.E. Smythe. 1973. Search 4, 162-163.
58.Foerster, U. 1980. In: O. Hutzinger: The Handbook of Environmental Chemistry. Antropogenic compounds, Vol 3. Part A. pp 59-107. Springer-Verlag.
68.Fuge, R. and K.H. James. 1974. Mar. Pollut. Bull. 5, 9-12.
70.Ravera, O., R. Gommes and H. Muntau. 1973. In: Problems of the Contamination of Man and His Environmnt by Hg and Cd. Luxembourg. Eur. Comm. Coll. pp 317-331.

71.Nickless, G., R. Stenner and N. Terrile. 1972. Mar. Pollut. Bull. 3, 188-190.
72.Butterworth, J., P. Lester and G. Nickless. 1972. Mar. Pollut. Bull. 3, 72-74.
73.Martin, J.H., W.W. Broenkow. 1975. Science 190, 884-885.
74.Van As, D. H.O. Fourie and C.M. Vleggaar. 1975. S. Afr. J. Sci. 71, 151-154.
79.Martin, J.H., G.A. Knauer and A.R. Flegal 1980. in: Nriagu, J.O. (ed); Cadmium in the Environment. Prt 1. Ecological Cycling. John Wiley and sons. pp 141-236.
84.Stenner, R.D. and G. Nickless. 1975. Mar. Pollut. Bull. 6, 89-92.
85.Talbot, V.W., R.J. Magee and M. Hassain. 1976. Mar. Pollut. Bull. 84-86.
87.Penrose, W.R., R. Black and M.J. Hayward. 1975. J. Fish. Res. Bd Canada 32, 1275-1281.
90.Hardisty, M.W., R.J. Huggins, S. Kartar and M. Sainsbury. 1974. Mar. Pollut. Bull. 5, 61-63.
92.Abo-Rady, M.D.K. 1977. Die Belastung der oberen Leine mit Schwermetallen durch kommunale und industrelle Abwasser. Diss. Univ. Gottingen. 120 p.
94.Schlesinger, W.H., W.A. Reiners and D.S. Knopman. 1973. Environ. Pollut., 6, pp 39-47.
96.Bowen, H.J.M. 1966. Trace Elements in Biochemistry, Ac. Press.
97.Muller, G. and F. Prosi. 1977. Naturwissenschaften 64, 530.
100.Enk, M.D. and B.J. Mathis. 1977. Hydrobiologia 52, 153-158.
107.Fitchko, J. and T.C. Hutchinson. 1975. J. Great Lakes res. 1, 46-78.
108.Forster, U. and G. Muller. 1974. Tschermacks Min. Petr. Mitt. 21, 145-163.
116.Depers, A.M. 1978. 10th International Congress on Sedimentology, Jerusalem. July 9-14, 171-172.
117.Wittmann, G.T.W. and U. Forster. 1977. S. Afr. J. Sci. 72, 365-370.
119.Farrah, H. and W.F. Pickering. 1977. Water, Air, Soil Pollut. 8, 189-197.

5-043 Background concentration
Ce

Species	Value	Condition
Seawater	0.007 microgram/l	Soluble concentration, Japan, Cl = 1.873 %[1]
Seawater	0.058 microgram/l	Total concentration, Japan, Cl = 1.873 %[1]
Seawater	0.4E-3 ppm	Residence time = 6.1E3 y[2]
Soil	50 mg/kg dry soil	Oven-dried[3]

Source: 1.Suzuki, H., T. Koganage and M. Saki. 1975. Impacts of Nuclear Releases into the Aquatic Environment. IAEA SM-198/19, pp 77-89. Vienna.
2.Goldberg, E.D., 1963. The Sea, Interscience London, 2, pp 3.
3.Bowen, H.J.M. 1966. Trace Elements in Biochemistry, Ac. Press.

5-044 Background concentration
Chlorodibromomethane

Species	Value	Condition
Drinking water	1 ug/m3	Typical non-polluted value, halocarbon[1]
Drinking water	10 ug/m3	High value, halocarbon[1]
Freshwater	0.05 ug/m3	Typical non-polluted value, halocarbon[1]
Freshwater	0.5 ug/m3	High value in urban area, halocarbon[1]
Groundwater	0.05 ug/m3	Typical non-polluted value, halocarbon[1]
Groundwater	0.5 ug/m3	High value, halocarbon[1]

Source: 1.Pearson, C.R. 1982. In: O. Hutzinger (ed.): Environmental Chemistry. Springer-Verlag. Volume 3, partB. Berlin.

5-045 Background concentration
Chloroform

Species	Value	Condition
Atmosphere	0.03 ug/m3	Typical value[1]
Bird, egg	15 mg/kg wet weight	Rg = 0.7 - 29 mg/kg wet weight, food from water organisms[2]
Bird, liver	9.3 mg/kg wet weight	Rg = 1.3 - 17.3 mg/kg wet weight, food from water organisms[2]
Bread	2 mg/kg wet weight	None[2]
Crustacea	90 mg/kg wet weight	Rg = 0.05 - 180 mg/kg wet weight[2]
Dairy products	17 mg/kg wet weight	Rg = 1 - 33 mg/kg wet weight[2]

Drinking water	0.1 ug/m3	Typical non-polluted value, halocarbon[2]
Drinking water	100 ug/m3	High value, halocarbon[2]
Estuarine water	0.1 ug/m3	High value in inshore water[3]
Fish, flesh	57 mg/kg wet weight	Rg = 5 - 110 mg/kg wet weight[2]
Fish, liver	12 mg/kg wet weight	Rg = 6 - 18 mg/kg wet weight[2]
Freshwater	0.3 ug/m3	Typical non-polluted value, halocarbon[2]
Freshwater	6.0 ug/m3	High value in urban area, rg = 2.0 - 10 ppb, halocarbon[2]
Fruit and vegetables	9.0 mg/kg wet weight	Rg = 0.05 - 18 mg/kg wet weight[2]
Groundwater	0.3 ug/m3	Typical non-polluted value, halocarbon[2]
Groundwater	2 ug/m3	High value, halocarbon[2]
Homo sapiens, fat	37 mg/kg wet weight	Rg = 5 - 68 mg/kg wet weight[2]
Homo sapiens, liver	5.5 mg/kg wet weight	Rg = 1 - 10 mg/kg wet weight[2]
Meat	2.5 mg/kg wet weight	Rg = 1 - 4 mg/kg wet weight[2]
Mollusca	75 mg/kg wet weight	Rg = 0.05 - 150 mg/kg wet weight[2]
Plankton	2.5 mg/kg wet weight	Rg = 0.02 - 5 mg/kg wet weight[2]
Rain	0.1 ug/m3	Rg = 0.005 - 0.2 ppb, halocarbon[2]
Seal, blubber	14.8 mg/kg wet weight	Rg = 7.6 - 22 mg/kg wet weight[2]
Seal, liver	6 mg/kg wet weight	Rg = 0.01 - 12 mg/kg wet weight[2]
Seawater	0.015 ug/m3	Typical open ocean[3]
Sediment	0.5 ug/m3	Typical conc in inshore sediment[3]
Snow	0.06 ug/m3	Rg = 0.002 - 0.12 ppb, halocarbon[2]
Vegetable oil	5.0 mg/kg wet weight	Rg = 0.05 - 10 mg/kg wet weight[2]

Source: 1.Lovelock, J.E. and P. Simmons. 1980. In: Sugden (ed.): The Aerosol Controversy. Sci-Horwood. 2.Pearson, C.R. 1982. In: O. Hutzinger (ed.): Environmental Chemistry. Springer-Verlag. Volume 3, partB. Berlin. 3.National Academy of Science. 1978. Chloroform, Carbon tetrachloride, and other halomethanes. An Envionmental Assessment. US NAS.

5-046 Background concentration Chlorophyll a

Species	Value	Condition
Estuarine sediment	0.60 microgram/cm**3	Depth = 15 cm, winter, Newport Estuary[1]
Estuarine sediment	2.00 microgram/cm**3	Depth = 5 cm, winter, Newport Estuary[1]
Estuarine sediment	2.01 microgram/cm**3	Depth = 5 cm, summer, Newport Estuary[1]
Estuarine sediment	3.77 microgram/cm**3	Depth = 0 cm, summer, Newport Estuary[1]
Estuarine sediment	7.45 microgram/cm**3	Depth = 0 cm, winter, Newport Estuary[1]

Source: 1.Ferguson, R.L. and M.B. Murdoch. 1975. in: Cronin, L.E. (ed.); Estuarine Research. Vol 1 pp 229-250. Academic Press. New York.

5-047 Background concentration Chrysene

Species	Value	Condition
Atmosphere, seasonal variation	0.70 ng/m3	1. Quarter, 2. Q=0.26, 3. Q=0.44, 4. Q=1.57, LA, n=13, USA, 1975[1]
Crassostrea virginica	0.2 ug/kg dry weight	Galveston B., commercial oyster reef, TEX, USA, 1970, rg=0.0-1.7 ppb[2]
Crassostrea virginica	0.3 ug/kg dry weight	Aransas Bay, uncontam. area, TEX, USA, 1970, rg=0.3-0.6 ppb[2]

Crassostrea virginica	0.6 ug/kg dry weight	Galveston B., oyster reef contam. dom. sewage, TEX, rg=0.0-2.0 ppb[2]

Source: 1.Gordon, R.J. 1976. Environ. Sci. Technol. 10, 370-373.
2.Fazio, T. 1971. Proc. 7th Nat. Shellfish Sanit. Workshop. Wash DC. FDA, Division of Shellfish Sanitation

5-048 Background concentration
Cl

Species	Value	Condition
Biosphere	2.80 moles/hectare	Average composition of living matter[1]
River water	7.8 ppm	Range = 5-35 ppm[2]
River, Africa	12.1 mg/l	None[3]
River, Asia	8.7 mg/l	None[3]
River, Australia	10. mg/l	None[3]
River, Europe	6.9 mg/l	None[3]
River, North America	8. mg/l	None[3]
River, South America	4.9 mg/l	None[3]
River, World	7.8 mg/l	None[3]
Seawater	100 % of total	As free ion, chlorinity = 1.9 %, 298 K, pH = 8.15[4]
Seawater	19000 ppm	Cl(-)[5]
Soil	100 mg/kg dry soil	Oven-dried[2]

Source: 1.Deevey, E.S. 1970. Scientific American. 223(3), p 150.
2.Bowen, H.J.M. 1966. Trace Elements in Biochemistry, Ac. Press.
3.Wetzel, R.G. 1975. Limnology. Saunders. Philadelphia.
4.Garrels, R.M. and M.E. Thompson. 1962. Am. J. Sci. 260., p 57.
5.Goldberg, E.D., 1963. The Sea, Interscience London, 2, pp 3.

5-049 Background concentration
Co

Species	Value	Condition
Alosa pseudoharengus	0.023 mg/kg wet weight	Freshwater fish, whole body, Lake Michigan, 1969, rg = 0.025 ppm[1]
Amblema plicata	0.7 mg/kg wet weight	Gastropod, rg = 0.4-1.2 ppm, Illinois R., 1972, freshwater[2]
Anodonta sp.	0.42 mg/kg dry weight	Gastropod, L Aberffraw, 1970, freshwater[3]
Atmosphere	0.00084 ng/m**3	Southpole, 1970, rg = +- 0.00027[4]
Atmosphere	2.04 ng/m**3	England, 7 stations, rg = 0.08 - 4.0[4]
Atmosphere	6 ng/m**3	Urban, USA[4]
Atmosphere, urban	11.5 ng/m**3	Range = 0.3 - 23 ng/m3, Chicago 1969[5]
Carbonate sediments	0.1 mg/kg dry weight	Average value[6]
Cardium edule	7.1 mg/kg dry weight	Gastropod, Irish Sea, GB, 1970[3]
Choromytilus meridionalis	0.038 mg/kg wet weight	Gastropod, rg = +- 0.006 ppm, Cape, W Coast, S Africa, 1974[7]
Crassostrea gigas	2.8 mg/kg dry weight	Gastropod, rg = 2.2-3.5 ppm, Hinkley Pt. Lower Station, Severn Est., 73[8]
Crassostrea virginica	0.1 mg/kg wet weight	Gastropod, Atlantic Coast, USA, 1967[9]
Crustral rock	0.0025 percent by weight	Average value[10]
Deep-sea clay sediments	74 mg/kg dry weight	Average value[6]
Dreissena ploymorpha	2.0 mg/kg dry weight	Rg = 1.2-2.8 ppm, Elbe Est., transplanted from Schaal L, 1973, freshwater[11]

Dreissena polymorpha	0.7 mg/kg dry weight	Gastropod, Schaal L, to be transplanted to Elbe Est., 1973, freshwater[11]
Esox lucius	0.07 mg/kg wet weight	Fish, rg = 0.03-0.11 ppm, Illinois R, USA, muscle, 1972[12]
Food	0.18 mg/kg dry weight	Vegetables[4]
Food	0.2 mg/kg dry weight	Animal, rg = 0.1 - 0.3[4]
Forest foliage	6.2 mg/kg ash weight	Uncontaminated value for USA, rg = <1-10000 ppm[13]
Freshwater sediment	17 mg/kg dry weight	Range of 90 % = 3 - 40 ppm, 74 lakes, pelitic[14]
Freshwater zooplankton, effect of fertilizer	0.43 mg/kg dry weight	Area without fertilizer, Scenedesmus obliquus, Thailand, 1974[15]
Freshwater zooplankton, effect of fertilizer	0.83 mg/kg dry weight	Area with fertilizer, Scenedesmus obliquus, FRG, 1974[15]
Freshwater, river and lake	5.0 ug/l	Range = 1 - 4500 ppm, 720 lakes and rivers, USA[16]
Fusconaria flava	1.2 mg/kg wet weight	Gastropod, rg= 0.6-1.6 ppm, Illinois R., 1972, freshwater[2]
Granitic rock	7 mg/kg dry weight	Average value[6]
Igenous rock	25 mg/kg dry weight	None[17]
Jasus lalandii	0.004 mg/kg wet weight	Marine, Crusta., rg =+-0.002, Melkbosch C, W Coast, SA, muscle, 1975[7]
Marine phytoplankton	0.094 mg/kg wet weight	Range = 35 - 44 ppm, Cape W Coast, S Africa, 1973[7]
Mercenaria mercenaria	0.2 mg/kg wet weight	Gastropod, Atlantic coast, USA, 1967[9]
Merluccius capensis	0.004 mg/kg wet weight	Fish, marine, rg =+-0.001 ppm, Cape W Coast, S Africa, muscle, 1974[7]
Micropterus dolomieu	0.15 mg/kg wet weight	Fish, rg = 0.14-0.16 ppm, Illinois R, USA, muscle, 1972[12]
Micropterus salmoides	0.09 mg/kg wet weight	Fish, rg = 0.06-0.18 ppm, Illinois R, USA, muscle, 1972[12]
Mya arenaria	0.1 mg/kg wet weight	Gastropod, Atlantic coast, USA, 1967[9]
Mytilus edulis	1.6 mg/kg dry weight	Gastropod, Irish Sea, GB, 1970[3]
Mytilus edulis	2.8 mg/kg dry weight	Gastropod, Medit., NW Coast, France + Italy, 1975[18]
Notropis hudsonius	0.025 mg/kg wet weight	Fish, rg = +- 0.004 ppm, Lake Erie, whole animal, 1969[1]
Notropis hudsonius	0.042 mg/kg wet weight	Fish, rg = +- 0.15 ppm, L Michigan, 1969, whole animal[1]
Nucella lapillus	0.3 mg/kg dry weight	Gastropod, Irish Sea, GB, 1970, marine[3]
Patella vulgata	0.4 mg/kg dry weight	Gastropod, marine, Irish Sea, 1971[3]
Patella vulgata	2 mg/kg dry weight	Gastropod, rg = 1 - 4 ppm, Mortons Harbor, Newfoundl., 1974[19]
Pecten maximus	8.5 mg/kg dry weight	Gastropod, Irish Sea, GB, 1970[3]
Percopsis omiscomaycus	0.022 mg/kg wet weight	Fish, rg = +- 0.005 ppm, L Superior, USA, whole animal, 1969[1]
Percopsis omiscomaycus	0.024 mg/kg wet weight	Fish, rg = +-0.002 ppm, L Michigan, 1969, whole animal[1]
Phosphate rocks	3 mg/kg dry weight	Rock from Kola, USSR[20]
Phosphate rocks	3 mg/kg dry weight	Rock from Morocco[20]
Phosphate rocks	5 mg/kg dry weight	Rock from Florida, USA[20]
Polyphosphate	0.1 mg/kg dry weight	Rock from Florida, USA[20]
Quadrula quadrula	0.8 mg/kg wet weight	Gastropod, rg = 0.5-1.3 ppm, Illinois R., 1972, freshwater[2]
Rain	0.01 ug/l	Sweden, 3 stations[4]
Rain	0.25 ug/l	England, yearly mean[4]
River water	0.9E-3 ppm	Range < 0.006 ppm[21]

Sandstone sediment	0.3 mg/kg dry weight	Average value[6]
Scomber japonicus	0.02 mg/kg wet weight	Fish, marine, rg =+-0.017, Cape W Coast, S Africa, muscle, 1974[7]
Seawater	0.27E-3 ppm	Co(2+), residence time = 18E3 y[22]
Sedimentary rock	23 mg/kg dry weight	None[23]
Seriola pappei	0.008 mg/kg wet weight	Fish, marine, rg =+- 0.005 ppm, Cape W Coast, S Africa, muscle, 1974[7]
Shale	19 mg/kg dry weight	Average value[6]
Shale sediment	19 mg/kg dry weight	Average value[6]
Soil	8 mg/kg dry soil	Oven-dried, rg = 1 - 40[21]
Superphosphate fertilizer	0.8 mg/kg dry weight	Rock from Florida, USA[20]
Ulva sp.	0.038 mg/kg wet weight	Range = +- 0.013 ppm, Cape W Coast, S Africa, marine, 1974[7]
Water	0.50 ug/l	Baltic Sea, Gotlandsdybet:250 m, 1970-71[4]
Water	18.5 ug/l	Rivers, USA, rg = 1 - 36[4]
Water	3.75 ug/l	Lakes, Sweden[4]

Source: 1.Lucas, H.F. Jr., D.N. Edgington and P.J. Colby. 1970. J. Fish. Res. Bd Canada 27, 677-684.
2.Mathis, B.J. and T.F. Cummings. 1973. J. Water Pollut Cont. Fed. 45, 1573-1583.
3.Segar, D.A. and R.E. Pellenbarg. 1971. J. Mar. Biol. Assoc. UK. 51, 131-136.
4.Laveskog, A., A. Lindskog and V. Stenberg. 1976. Om Metaller, Statens Naturvaardsverk, Stockholm.
5.Brar, S.S., D.M. Nelson, J.R. Kline, P.F. Gustafson, E.L. Kanabrocki, E.E. Moore and D.M. Hattori. 170. J. Geochem. Res.75, pp 2939-2945.
6.Turekian, K.K. and K.H. Wedepohl. 1961. Amer. Geol. Soc. Bull. 72, pp 175-192.
7.Van As, D. H.O. Fourie and C.M. Vleggaar. 1975. S. Afr. J. Sci. 71, 151-154.
8.Romeril, M.G. 1971. Mar. Biol. 9, 347-354.
9.Pringle, B.H., D.E. Hissong, E.L. Katz and S.T. Mulawka 1968. J. Sanit. Eng. Div. Am. Soc. Civ. Eng., 94, 455-475.
10.Mason, B. 1966. Principles of Geochemis try. 3rd ed. John Wiley and Sons. New York.
11.Karbe, L., N. Antonacopoulos and C. Schnier. 1975. Verh. Int. Ver. Limnol. 19, 2094-2101.
12.Mathis, B.J. and N.R. Kevern. 1973. Distribution of Hg, Cd, Pb and Th in a eutrophic lake. Int. Water Res. Michigan State Univ., Technol. Dep. 34, 23 pp.
13.Shacklette, H.T., J.A. Erdman, T.F. Harms and C.S.E. Pupp. 1978. In: F.W. Oeheme (ed.): Toxicity of Heavy Metals in the Environment. Dekkker. NY. 25-68.
14.Foerstner, V. 1977. Arch. Hydrobiol., 80(2), pp 172-191.
15.Payer, H.D., Runkel, Schramel, Stengel, Bhumiratana and Soeder. 1976. Chemosphere 6, 413-418.
16.Durum, W.H., J.D. Hem and S.G. Heidel. 1971. US Geol. Surv. Circ. 643, 49.
17.Taylor, S.R. 1964. Geochim. Cosmochim. Acta 28, pp 1273-1285.
18.Fowler, S.W. and B. Oregioni. 1976. Mar. Pollut. Bull. 7, 26-29.
19.Penrose, W.R., R. Black and M.J. Hayward. 1975. J. Fish. Res. Bd Canada 32, 1275-1281.
20.Langmyhr, F.J. R. Solberg and Y. Thomassen. 1977. Anal. Chim. Acta 92, pp 105-109.
21.Bowen, H.J.M. 1966. Trace Elements in Biochemistry, Ac. Press.
22.Goldberg, E.D., 1963. The Sea, Interscience London, 2, pp 3.
23.Vinogradov, A.P. 1959. The Geochemistry of Rare and Dispersed Chemical Elements in the Soil. Transl. from Russian. Consultant Bureau Inc. NY. 2nd ed.

5-050 Background concentration Coronene

Species	Value	Condition
Atmosphere, seasonal variation	2.49 ng/m3	1. Quarter, 2. Q=1.13, 3. Q=1.66, 4. Q=4.44, LA, n=13, USA, 1975[1]

Source: 1.Gordon, R.J. 1976. Environ. Sci. Technol. 10, 370-373.

5-051 Background concentration Cr

Species	Value	Condition
Alosa pseudoharengus	1.1 mg/kg wet weight	Freshwater fish, whole body, Lake Michigan, 1969[1]
Amblema plicata	4.4 mg/kg wet weight	Gastropod, rg = 0.6-9.9 ppm, Illinois R., 1972, freshwater[2]
Anarrhichas lupus	1 mg/kg dry weight	Fish, marine, Trondheimfj. Norway, 1976, muscle[3]
Anodonta sp.	0.84 mg/kg dry weight	Gastropod, L Aberffraw, 1970, freshwater[4]

Atmosphere	0.0053 ng/m**3	Southpole, 1970, rg = +- 0.003[5]
Atmosphere	27 ng/m**3	Urban, USA, rg = 1 - 55[5]
Atmosphere	305 ng/m**3	Sweden, mean from 8 stations, rg = 130 - 480[5]
Atmosphere	6.15 ng/m**3	England, 7 stations, rg = 0.8 - 11.5[5]
Atmosphere	6.5 ng/m**3	Not urban, USA, rg = 1 - 12[5]
Atmosphere, rural	10 ng/m**3	Annual mean of 10 ng/m3 seldom reached[6]
Atmosphere, urban	25 ng/m**3	Range = 10 - 50 ng/m**3[6]
Blue crab	0.14 mg/kg wet weight	2 references used, SD = +- 0.06, rg = 0.10 - 0.17[7]
Carbonate sediments	11 mg/kg dry weight	Average value[8]
Carcinus maenas	8 mg/kg dry weight	Marine, Crustacean, muscle, Hardanger, Norway, 1976[3]
Choromytilus meridionalis	1.05 mg/kg wet weight	Gastropod, rg = +- 0.29 ppm, Saldanha B, LangeBaan B., RSA, 1975[9]
Clams, all species	0.33 mg/kg wet weight	6 references used, SD = +- 0.12, rg = 0.17 - 0.44[7]
Coregonus clupeaformis	0.017 mg/kg wet weight	Fish, less, L Erie, USA, 1970, animal decapiated and eviscerated[10]
Coregonus clupeaformis	0.033 mg/kg wet weight	Fish, less, Moose L, USA, 1970, animal decapiated and eviscerated[10]
Crassostrea gigas	0.54 mg/kg wet weight	Gastropod, Saldanha B, LangeBaan B., RSA, 1975[9]
Crassostrea gigas	12.45 mg/kg dry weight	Gastropod, rg = +- 7.5 ppm, Tamar R Est., Tasmania, 1973[11]
Crassostrea virginica	0.40 mg/kg wet weight	Gastropod, Atlantic Coast, USA, 1967[12]
Crustral rock	0.010 percent by weight	Average value[13]
Cyprinus carpio	0.16 mg/kg wet weight	Fish, rg=0.02-0.46 ppm, Illinois R, muscle, 1973[14]
Deep-sea clay sediments	90 mg/kg dry weight	Average value[8]
Dreissena ploymorpha	34.7 mg/kg dry weight	Rg = 9.4-60 ppm, Elbe Est., transplanted from Schaal L, 1973, freshwater[15]
Dreissena ploymorpha	6.6 mg/kg dry weight	Gastropod, Schaal L, to be transplanted to Elbe Est., 1973, freshwater[15]
Esox lucius	0.031 mg/kg wet weight	Fish, less, L Erie, USA, 1970, animal decapiated and eviscerated[10]
Esox lucius	0.035 mg/kg wet weight	Fish, less, Moose L, USA, 1970, animal decapiated and eviscerated[10]
Esox lucius	0.13 mg/kg wet weight	Fish, rg = 0.02-0.22 ppm, Illinois R, USA, muscle, 1972[14]
Food	0.6 mg/kg dry weight	Vegetables, rg = 0.2 - 1.0[5]
Forest foliage	8.0 mg/kg ash weight	Uncontaminated value for USA, rg = <2-150 ppm[16]
Freshwater	2.5 ug/l	Range < 0.1 - 5 ug/l[6]
Freshwater	5.5 ug/l	Range = 1 - 10 ug/l, river and lakes[6]
Freshwater sediment	82 mg/kg dry weight	Range of 90 % = 25 - 180ppm, 74 lakes, pelitic[17]
Freshwater, lake	5.7 ug/l	Range = 2 - 10 ppm, Lake Constance, Lindau[18]
Freshwater, river	3.0 ug/l	Range 0.3 - 69, streams to Lake Michigan[19]
Freshwater, river	7.2 ug/l	Range = 3 - 18 ppm, Danube, Ulm[18]
Freshwater, river and lake	1 ug/l	Less than value, rg = 0.7 - 69, 720 river and lakes, USA[20]
Fucus vesiculosus	4.5 mg/kg dry weight	Menai Strait, Anglesey, 1975[21]
Fusconaria flava	7.7 mg/kg wet weight	Gastropod, rg = 1.1-11.6 ppm, Illinois R., 1972, freshwater[2]
Gadus morrhua	1.5 mg/kg dry weight	Fish, rg = +- 1, Trondsheimfj, Norway, muscle, 1976[3]

Granitic rock	22 mg/kg dry weight	Average value[8]
Ictaluridae	0.23 mg/kg wet weight	1 reference used[7]
Ictiobus cyprinellus	0.13 mg/kg wet weight	Fish, rg = 0.02-0.53 ppm, Illinois R, USA, muscle, 1972[14]
Igenous rock	100 mg/kg dry weight	None[22]
Jasus lalandii	0.03 mg/kg wet weight	Marine, Crustaceans, Saldanha Bay, SA, muscle, 1975[9]
Jasus lalandii	0.08 mg/kg wet weight	Marine, Crustaceans, rg =+-0.08, Melkbosch C, W Coast, SA, muscle, 1975[23]
Lutjanidae	0.12 mg/kg wet weight	1 reference used[7]
Mactra clabrata	0.54 mg/kg wet weight	Gastropod, rg = +- 0.17 ppm, Saldanha B, Langebaan B, RSA, 1975[9]
Marine phytoplankton	0.54 mg/kg wet weight	Cape W Coast, S Africa, 1973[23]
Marine zooplankton	1300 mg/kg wet weight	Range = 2.7 - 2900 ppm, Cape W Coast, S Africa, 1974[23]
Marine zooplankton	9.0 mg/kg wet weight	Range = 0.03 - 18 ppm, Cape W Coast, S Africa, 1974[23]
Melanogrammus aeglefinus	0.11 mg/kg wet weight	4 references used, SD = +- 0.12, rg = 0.02 - 0.30[7]
Mercenaria mercenaria	0.31 mg/kg wet weight	Gastropod, Atlantic coast, USA, 1967[12]
Merluccius capensis	0.26 mg/kg wet weight	Fish, marine, rg =+-0.44 ppm, Cape W Coast, S Africa, muscle, 1974[23]
Merluccius sp.	0.09 mg/kg wet weight	2 references used, SD = +- 0.00, rg = 0.09 - 0.09[7]
Micropterus dolomieu	0.16 mg/kg wet weight	Fish, rg = 0.04-0.27 ppm, Illinois R, USA, muscle, 1972[14]
Micropterus salmoides	0.11 mg/kg wet weight	Fish, rg = 0.04-0.24 ppm, Illinois R, USA, muscle, 1972[14]
Mya arenaria	0.52 mg/kg wet weight	Gastropod, Atlantic coast, USA, 1967[12]
Mytilus edulis	1.5 mg/kg dry weight	Gastropod, Irish Sea, GB, 1970[4]
Mytilus edulis	16 mg/kg dry weight	Gastropod, rg = 9 - 24 ppm, New Zealand, 1964[24]
Mytilus edulis	20 mg/kg dry weight	Gastropod, rg = 4 - 49 ppm, Trondheimsfjord, Norway, 1976[3]
Mytilus edulis	7.5 mg/kg dry weight	Gastropod, rg = 0.5-288 ppm, Medit., NW Coast, France + Italy, 1975[25]
Notropis hudsonius	0.9 mg/kg wet weight	Fish, rg = +- 0.5 ppm, L Michigan, 1969, whole animal[1]
Notropis hudsonius	10 mg/kg wet weight	Fish, Lake Erie, USA, whole animal, 1969[1]
Ostrea sinuata	3 mg/kg dry weight	Gastropod, rg = 2 - 6 ppm, Tasman Bay, NZ, 1964[24]
Oyster	0.12 mg/kg wet weight	2 references used, SD = +- 0.04, rg = 0.09 - 0.14[7]
Patella vulgata	10 mg/kg dry weight	Gastropod, rg = 7-17 ppm, Mortons Harbor, Newfoundl., 1974[26]
Pecten novaezelandia	10 mg/kg dry weight	Gastropod, rg = 3-23 ppm, Tasman Bay, New Zealand, 1964[24]
Percidae	0.07 mg/kg wet weight	1 references used[7]
Percopsis omiscomaycus	1.6 mg/kg wet weight	Fish, rg = +-0.2 ppm, L Michigan, 1969, whole animal[1]
Percopsis omiscomaycus	3 mg/kg wet weight	Fish, less, L Superior, USA, whole animal, 1969[1]
Phosphate rocks	2 mg/kg dry weight	Rock from Kola, USSR[27]
Phosphate rocks	306 mg/kg dry weight	Rock from Morocco[27]
Phosphate rocks	69 mg/kg dry weight	Rock from Florida, USA[27]
Pleuronectidae	0.09 mg/kg wet weight	5 references used, SD = +- 0.08, rg = 0.01 - 0.18[7]
Pollock	0.08 mg/kg wet weight	1 reference used[7]
Polyphosphate	5 mg/kg dry weight	Rock from Florida, USA[27]
Quadrula quadrula	4.7 mg/kg wet weight	Gastropod, rg = 1.8-8.3 ppm, Illinois R., 1972, freshwater[2]
Rain	0.25 ug/l	Sweden, 3 stations, rg = 0.1 - 0.4[5]

Rain	2.9 ug/l	England, yearly mean, 1 station[5]
River	42.3 ppb	15 Rivers, North America, Cr(+6), rg = 0.7 - 84.0[28]
River water	0.18E-3 ppm	Range = 0.1E-3-0.06 ppm[29]
Sandstone sediment	35 mg/kg dry weight	Average value[8]
Scallops, sea	0.09 mg/kg wet weight	5 references used, SD = +- 0.02, rg = 0.05 - 0.11[7]
Scomber japonicus	0.65 mg/kg wet weight	Fish, marine, rg =+-0.85 ppm, Cape W Coast, S Africa, muscle, 1974[23]
Seawater	0.5E-4 ppm	Residence time = 0.35E3 y[30]
Sedimentary rock	160 mg/kg dry weight	None[31]
Seriola grandis	0.02 mg/kg wet weight	Fish, marine, rg =+-0.01 ppm, North NZ, muscle, 1973[24]
Seriola pappei	1.0 mg/kg wet weight	Fish, marine, rg =+- 1.8 ppm, Cape W Coast, S Africa, muscle, 1974[23]
Shale	90 mg/kg dry weight	Average value[8]
Shale sediment	90 mg/kg dry weight	Average value[8]
Shrimp, brown	0.07 mg/kg wet weight	7 references used, SD = +- 0.05, rg = 0.01 - 0.12[7]
Soil	100 mg/kg dry soil	Oven-dried, rg =5 - 3000[29]
Stormwater	27 g/m3	Lodi, New Jersey, USA, rain = 1.4 ug/l, Saddle R = 3.0 ug/l[32]
Superphosphate fertilizer	119 mg/kg dry weight	Rock from Florida, USA[27]
Surface water	48.5 ppb	Holland, Cr(+6), rg = 9.0 - 88.0[33]
Trigla kumu	0.01 mg/kg wet weight	Fish, rg = 0.01 -0.02 ppm, N Island, NZ, muscle, 1973[24]
Ulva sp.	0.45 mg/kg wet weight	Range = +- 0.38 ppm, Cape W Coast, S Africa, marine, 1974[34]
Wastewater, Bakery	330 ug/l	Average value, New York industries[35]
Wastewater, Fat rendering	210 ug/l	Average value, New York industries[35]
Wastewater, Fish processing	230 ug/l	Average value, New York industries[35]
Wastewater, Meat processing	150 ug/l	Average value, New York industries[35]
Wastewater, Textile dyeing	820 ug/l	Average value, New York industries[35]
Wastewater, brewery	60 ug/l	Average value, New York industries[35]
Wastewater, car wash	140 ug/l	Average value, New York industries[35]
Wastewater, fur dressing and dyeing	20140 ug/l	Average value, New York industries[35]
Wastewater, ice cream	50 ug/l	Average value, New York industries[35]
Wastewater, laundry	1220 ug/l	Average value, New York industries[35]
Wastewater, miscellaneous chemical indistries	280 ug/l	Average value, New York industries[35]
Wastewater, miscellaneous foods	150 ug/l	Average value, New York industries[35]
Wastewater, soft drinks and flavorings	1800 ug/l	Average value, New York industries[35]

Source: 1.Lucas, H.F. Jr., D.N. Edgington and P. J. Colby. 1970. J. Fish. Res. Bd Canada 27, 677-684.
2.Mathis, B.J. and T.F. Cummings. 1973. J. Water Pollut Cont. Fed. 45, 1573-1583.
3.Lande, E. 1977. Environ. Pollut. 12, 187-198.
4.Segar, D.A. and R.E. Pellenbarg. 1971. J. Mar. Biol. Assoc. UK. 51, 131-136.
5.Laveskog, A., A. Lindskog and V. Stenberg. 1976. Om Metaller, Statens Naturvaardsverk, Stockholm.
6.NAS 1973. Chromium. Committee on Biological Effects of Atmospheric Pollutants. National Acedemy of Sience. Wash. D.C.
7.Zook, E.G., J.J. Powell, B.M. Hackley, J.A. Emerson, J.R. Brooker and G.M. Knobl, Jr. 1976. J. Agric. Food. Chem. 24(1), pp 47-53.
8.Turekian, K.K. and K.H. Wedepohl. 1961. Amer. Geol. Soc. Bull. 72, pp 175-192.

9.Fourie, H.O. 1976. S. Afr. J. Sci. 72, 110-113.
10.Uthe, J.H. and E.G. Bligh. 1971. J. Fish. Res. bd. Canada 28, 786-788.
11.Ayling, G.M. 1974. Water Res. 8, 729-738.
12.Pringle, B.H., D.E. Hissong, E.L. Katz and S.T. Mulawka 1968. J. Sanit. Eng. Div. Am. Soc. Civ. Eng.94, 455-475.
13.Mason, B. 1966. Principles of Geochemis try. 3rd ed. John Wiley and Sons. New York.
14.Mathis, B.J. and N.R. Kevern. 1973. Distribution of Hg, Cd, Pb and Th in a eutrophic lake. Int. Water Res. Michigan State Univ., Technol. Dep. 34, 23 pp.
15.Karbe, L., N. Antonacopoulos and C. Schnier. 1975. Verh. Int. Ver. Limnol. 19, 2094-2101.
16.Shacklette, H.T., J.A. Erdman, T.F. Harms and C.S.E. Pupp. 1978. In: F.W. Oeheme (ed.): Toxicity of Heavy Metals in the Environment. Dekkker. NY. 25-68.
17.Foerstner, V. 1977. Arch. Hydrobiol., 80(2), pp 172-191.
18.Quentin, K.E., and H.A. Winkler. 1974. Zentralbl. Bakt. Abt. Orig. B 158, 514-523.
19.Robbins, J.A., E. Landstroem and M. Wahlgren. 1972. Proc 15th Conf. Great Lakes Res. Int. Assoc. Gret Lakes, 270-290.
20.Durum, W.H., J.D. Hem and S.G. Heidel. 1971. US Geol. Surv. Circ. 643, 49.
21.Foster, P. 1976. Environ. Pollut. 10, 45-53.
22.Taylor, S.R. 1964. Geochim. Cosmochim. Acta 28, pp 1273-1285.
23.Van As, D. H.O. Fourie and C.M. Vleggaar. 1975. S. Afr. J. Sci. 71, 151-154.
24.Brooks, R.R., and D. Rumsey. 1974. NZJ Mar. Freshwater Res. 8, 155-166.
25.Fowler, S.W. and B. Oregioni. 1976. Mar. Pollut. Bull. 7, 26-29.
26.Penrose, W.R., R. Black and M.J. Hayward. 1975. J. Fish. Res. Bd Canada 32, 1275-1281.
27.Langmyhr, F.J. R. Solberg and Y. Thomassen. 1977. Anal. Chim. Acta 92, pp 105-109.
28.Haffty, J. 1960. Residu Method for common minor Elements, U.S. Geological Survey Water Supply Paper 50-A, U.S. Goverment Printing Office, Wash. D.C., pp 9.
29.Bowen, H.J.M. 1966. Trace Elements in Biochemistry, Ac. Press.
30.Goldberg, E.D., 1963. The Sea, Interscience London, 2, pp 3.
31.Vinogradov, A.P. 1959. The Geochemistry of Rare and Dispersed Chemical Elements in the Soil. Transl. from Russian. Consultant Bureau Inc. NY. 2nd ed.
32.Nriagu, J.O. 1980. Zinc in the Environment. Part 1: Ecological cycling. Wiley and Sons. NY.
33.Strik, J.J.T.W.A., H.H. Iongh, J.W.A. van Rijn et al. 1975. in: Koeman, J.H. et al. (eds.): Sublethal Effects of Toxic Chemicals on Aquatic Animals, Elsevier. NY. pp 31-37.
34.Martin, J.H., W.W. Broenkow. 1975. Science 190, 884-885.
35.Klein, L.A., M. Land, N. Nash and S.L. Kirschner. 1974. J. Wat. Poll. Cont. Fed. 46, pp 2653-2662.

5-052 Background concentration
Cs

Species	Value	Condition
River water	0.2E-3 ppm	Range = 0.5E-4-0.2E-3 ppm[1]
Seawater	0.5E-3 ppm	Cs(+), residence time = 40E3 y[2]
Soil	6 mg/kg dry soil	Oven-dried, rg = 0.3 - 25[1]

Source: 1.Bowen, H.J.M. 1966. Trace Elements in Biochemistry, Ac. Press.
2.Goldberg, E.D., 1963. The Sea, Interscience London, 2, pp 3.

5-053 Background concentration
Cu

Species	Value	Condition
Alosa pseudoharengus	0.800 mg/kg wet weight	Freshwater fish, whole body, Lake Michigan, 1969, rg = 0.15 ppm[2]
Asellus aquaticus	12.99 mg/kg dry weight	Freshwater, Crustaceans, rg = 8.6-19.9 ppm, Elsenz R, 1976[6]
Atmosphere, non-polluted	0.01 ng/m3	Chacaltaty, Bolivia, 1975, enrichment factor (Al) = 8[9]
Atmosphere, non-polluted	12 ng/m3	N Indian Ocean, 1968-69, enrichment factor (Al) = 150[9]
Aufwuchs communities	34.2 mg/kg dry weight	Rg = 22.7-45.5 ppm, Elsenz R, FRG, 1976[11]
Belone belone	4.1 mg/kg dry weight	Fish, marine, rg = +- 1 ppm, NE Atlantic, Cornwall, muscle, 1974[13]
Cancer pagurus	6.80 mg/kg wet weight	Marine, Crustacean, Northcumberland C, GB, 1976, whole anim.[15]
Carcinus maenas	175 mg/kg dry weight	Marine, Crustacean, muscle, Hardanger, Norway, 1976[4]
Carcinus maenas	6.00 mg/kg wet weight	Marine, Crustacean, muscle, Northumberland C, GB, rg =+-3.0, 1975[15]
Cardium edule	11 mg/kg dry weight	Gastropod, Irish Sea, GB, 1970[5]
Cladophora glomerata	16.1 mg/kg dry weight	Algae, Rg = 9.1 - 23.0 ppm, Leine R., FRG, 1976[20]

Cladophora glomerata	6.4 mg/kg dry weight	Algae, Deadman Bay, Kingston, Ontario, 1975[21]
Clupea sprattus	4.61 mg/kg wet weight	Fish, marine, rg =+-4.1, Northumberl, power station, muscle, 1975[15]
Crangon crangon	34 mg/kg dry weight	Marine, Crustacean, S Antonio Bay, Texas, whole animal, 1975[24]
Crangon vulgaris	25.9 mg/kg wet weight	Marine, Crustacean, Northumberland Coast, GB, 1975, muscle[15]
Crassostrea gigas	3.7 mg/kg wet weight	Gastropod, rg = +- 1.0 ppm, Saldanha B, LangeBaan B., RSA, 1975[19]
Crassostrea gigas	4100 mg/kg dry weight	Gastropod, rg = 1760-6480 ppm, Hinkley Pt. Lower Station,Severn Est., 73[27]
Cyprinus carpio	0.24 mg/kg wet weight	Fish, rg = 0.12-0.41 ppm, Illinois R, muscle, 1973[18]
Food	1.17 mg/kg dry weight	Vegetables, USA[7]
Food	1.49 mg/kg dry weight	Fish and shellfish excl oyster, USA[7]
Food	1.76 mg/kg dry weight	Milk products, USA[7]
Food	14.82 mg/kg dry weight	Nuts, USA[7]
Food	3.92 mg/kg dry weight	Meat incl liver, USA, liver = 11 mg/kg dry weight[7]
Food	4.63 mg/kg dry weight	Fat excl lecithin, USA[7]
Forest foliage	128 mg/kg ash weight	Uncontaminated value for USA, rg = <10-3000 ppm[33]
Freshwater	5.3 mg/m3	Median value, rg = 0.83 - 105 mg/m3[34]
Freshwater sediment	46 mg/kg dry weight	Range of 90 % = 20 - 80 ppm, 74 lakes, pelitic[35]
Freshwater, river	10 ug/l	Range = 5 - 26 ppm, Danube, Austria[39]
Freshwater, river	40 ug/l	Range = <5 - 120 ppm, Kansas, 12 rivers[42]
Fucus vesiculosus	71 mg/kg dry weight	Range = 49 - 97 ppm, Afon Goch, Gb, 1975, Anglesey[17]
Fucus vesiculosus	9 mg/kg dry weight	Range = 7.4 - 10 ppm, Menai Strait, Anglesey, 1975[17]
Gadus morrhua	3.5 mg/kg dry weight	Fish, rg = +-1 ppm, Trondsheimfj, Norway, muscle, 1976[4]
Mya arenaria	5.8 mg/kg wet weight	Gastropod, Atlantic coast, USA, 1967[30]
Mytilus edulis	0.73 mg/kg wet weight	Gastropod, rg = 0.47-1.45 ppm, Port Phillips B, W Australia, 1975[48]
Mytilus edulis	24 mg/kg dry weight	Gastropod, rg = 5 - 88 ppm, Trondheimsfjord, Norway, 1976[4]
Nephrops norvegicus	4.8 mg/kg dry weight	Marine, Crustaceans, Faro, Portugal, muscle, 1974[12]
Nephrops norvegicus	41 mg/kg dry weight	Marine, Crustaceans, muscle, Guadiana Estu., SW Spain, 1974[12]
Ostrea edulis	38 mg/kg dry weight	Gastropod, Knysna Estuary, R.S.A., 1975[26]
Pecten maximus	3.3 mg/kg dry weight	Gastropod, Irish Sea, GB, 1970[5]
Potamogeton pectinatus	19.7 mg/kg dry weight	Higher plant, Rg = 5.1 -34.3 ppm, Elsenz R, FRG, 1976[14]
Rain	23 ug/l	England, 1971, rel uncontaminated, Cu in atm. = 22 ng/m**3[7]
Rain	6.75 ug/l	Sweden, rg = 3 - 8.5[7]
Seawater	0.3E-2 ppm	Cu(2+), residence time = 50E3 y[62]
Soil	15 mg/kg dry weight	Cretaceous, Chalk, Wiltshire, rg=7-28 ppm, n=128, Engl[9]
Soil	620 mg/kg dry weight	Polluted, Tamar V, SW Engl., rg=35-2000 ppm, n=12, alluvial soil, mining[9]
Solea solea	4.4 mg/kg dry weight	Fish, marine, Sagres, S Portugal, muscle, 1974[12]
Superphosphate fertilizer	16 mg/kg dry weight	Rock from Florida, USA[57]
Tellina tenuis	50 mg/kg dry weight	Gastropod, less, Loch Ewe, Scotland[65]
Ulva lactuca	16 mg/kg dry weight	Range = 5.5 - 26 ppm, Portugal and Spain, 1974, Atlantic[49]

Wastewater, fur dressing and dyeing	7040 ug/l	Average value, New York industries[66]
Water	5.3 ug/l	Rivers, USA, rg = 0.83 - 105[7]

Source:2.Lucas, H.F. Jr., D.N. Edgington and P.J. Colby. 1970. J. Fish. Res. Bd Canada 27, 677-684.
4.Lande, E. 1977. Environ. Pollut. 12, 187-198.
5.Segar, D.A. and R.E. Pellenbarg. 1971. J. Mar. Biol. Assoc. UK. 51, 131-136.
6.Prosi, F. 1977. Schwermetallbelastung in den sedimenten der Elsenz und ihre Auswerkung auf Limnische Organismen. Diss. Univ. Heidelberg.
7.Laveskog, A., A. Lindskog and V. Stenberg. 1976. Om Metaller, Statens Naturvaardsverk, Stockholm.
9.Nriagu, J.O. 1979. Copper in the Environment. John Wiley and Sons. NY.
11.Bibo, 1977. Schwermetalluntersuchungen an Wasser, Schwebstoffen, Aufwuchs und Cladophora rivularis des Elsenz. Diplom-Arbeit. Univ. Heidelberg. 111 p.
12.Martin, J.B.M. 1974. Mar. Biol. 28, 245-251.
13.Stevens, J.D. and B.E. Brown. 1974. Mar. Biol. 26, 287-293.
14.Heydt, G. 1977. Schwermetallgehalte von Wasser, Wasserpflanzen, Chironomidae und Mollusca der Elsenz. Dipl. Arbeit Univ. Heidelberg. 143 pp.
15.Wright, D.A. 1976. Mar. Pollut. Bull. 7, 36-38.
17.Foster, P. 1976. Environ. Pollut. 10, 45-53.
18.Mathis, B.J. and N.R. Kevern. 1973. Distribution of Hg, Cd, Pb and Th in a eutrophic lake. Int. Water Res. Michigan State Univ., Technol. Dep. 34, 23 pp.
19.Fourie, H.O. 1976. S. Afr. J. Sci. 72, 110-113.
20.Abo-Rady, M.D.K. 1977. Die Belastung der oberen Leine mit Schwermetallen durch kommunale und industrelle Abwasser. Diss. Univ. Gottingen. 120 p.
21.Keeney, W.L., W.G. Breck, G.W. Van Loon and J.A. Page 1976. Water Res. 10, 981-984.
22.Reay. P.F. 1972. J. Appl. Ecol. 9, 557-565.
24.Sims, R.R. and B.J. Presley. 1976. Bull. Environ. Contam. Toxicol. 16, 520-526.
26.Watling, H.R. and R.J. Watling. 1976. Mar. Pollut. Bull. 7, 45-48.
27.Romeril, M.G. 1971. Mar. Biol. 9, 347-354.
30.Pringle, B.H., D.E. Hissong, E.L. Katz and S.T. Mulawka 1968. J. Sanit. Eng. Div. Am. Soc. Civ. Eng.94, 455-475.
34.Duran, W.H., J.H. Hem and S.G. Heidel. 1971. Geological Survey Circular 642. U.S. Dept. of the Intero. Wash. DC.
35.Foerstner, V. 1977. Arch. Hydrobiol., 80(2), pp 172-191.
39.Kempf, T. 1973. Wasser Boden Lufthyg. 40, 149-153.
42.Angino, E.E., L.M. Magnuson, T.C. Waugh, Galle and Bredfeldt 1974. Water Resour. Res. 10, 1187-1191.
45.Preston, A., Jefferies, Dutton, Harvey and A.K. Steele. 1972. Environ. Pollut. 3, 69-82.
48.Philips, D.J.H. 1976. Mar. Biol. 38, 71-80.
49.Stenner, R.D. and G. Nickless. 1975. Mar. Pollut. Bull. 6, 89-92.
57.Langmyhr, F.J. R. Solberg and Y. Thomassen. 1977. Anal. Chim. Acta 92, pp 105-109.
62.Goldberg, E.D., 1963. The Sea, Interscience London, 2, pp 3.
65.Saward, D., A. Stirling and G. Topping. 1975. Mar. Biol. 29, 351-361.
66.Klein, L.A., M. Land, N. Nash and S.L. Kirschner. 1974. J. Wat. Poll. Cont. Fed. 46, pp 2653-2662.

5-054 Background concentration DDT

Species	Value	Condition
Algae in open ocean	0.1 mg/kg dry weight	Total DDT = 1.0E8 tons, dry biomass = 1.0E9 tons[1]
Algae on continental shelf	1.0 mg/kg dry weight	Total DDT = 3.0E8 tons, dry biomass = 0.3E9 tons[1]
Animals in oceans other than fish and mammalia	0.1 mg/kg dry weight	Total DDT = 3.02E8 tons, dry biomass = 3.02E9 tons[1]
Attached algae in ocean	1.0 mg/kg dry weight	Total DDT = 20.0E8 tons, dry biomass = 2.0E9 tons[1]
Bird	1.0 mg/kg dry weight	Total DDT = 0.002E8 tons, dry biomass = 0.00024E9 tons[1]
Domestic mammalia on land	1.0 mg/kg dry weight	Total DDT = 1.7E8 tons, dry biomass = 0.17E9 tons[1]
Feral mammalia on land	1.0 mg/kg dry weight	Total DDT = 0.09E8 tons, dry biomass = 0.009E9 tons[1]
Fish in oceans	1.0 mg/kg dry weight	Total DDT = 6.5E8 tons, dry biomass = 0.65E9 tons[1]
Homo sapiens	1.0 mg/kg dry weight	Total DDT = 3.0E8 tons, dry biomass = 0.30E9 tons[1]
Mammalia in oceans	1.0 mg/kg dry weight	Total DDT = 0.55E8 tons, dry biomass = 0.055E9 tons[1]
Plant in lakes and streams	0.010 mg/kg dry weight	Total DDT = 0.004E8 tons, dry biomass = 0.04E9 tons[1]
Plant in swamps and marshes	0.001 mg/kg dry weight	Total DDT = 0.240E8 tons, dry biomass = 24E9 tons[1]
Plant in terrestrial vegetation	0.0001 mg/kg dry weight	Total DDT = 1.814E8 tons, dry biomass = 1814E9 tons[1]
Plant on agricultural land	0.1 mg/kg dry weight	Total DDT = 14.000E8 tons, dry biomass = 14E9 tons[1]

Source: 1.Woodwell. G.M., P.P. Graig and H.A. Johnson. 1975. in: Singer, S.F. (ed.); The Changing Global Environment. pp 295-309. D. Reidel Publ. Comp. New York.

5-055 Background concentration Dichlorobromomethane

Species	Value	Condition
Drinking water	40 ug/m3	High value, halocarbon[1]
Drinking water	5 ug/m3	Typical non-polluted value, halocarbon[1]
Freshwater	0.1 ug/m3	Typical non-polluted value, halocarbon[1]
Freshwater	0.5 ug/m3	High value in urban area, halocarbon[1]
Groundwater	0.1 ug/m3	Typical non-polluted value, halocarbon[1]
Groundwater	0.5 ug/m3	High value, halocarbon[1]

Source: 1.Pearson, C.R. 1982. In: O. Hutzinger (ed.): Environmental Chemistry. Springer-Verlag. Volume 3, part B. Berlin.

5-056 Background concentration Dissolved organic carbon

Species	Value	Condition
East China Sea	1.25 mg C/l	May 1968, n = 20, rg = 0.96 - 1.42[1]
North Pacific	0.72 mg C/l	Calif. - Samoa, April-May 1967, n = 6, rg = 0.50 - 1.04[2]
North Pacific	1.09 mg C/l	Hawaii - Codiac, Jan 1966, n = 10, rg = 0.72 - 2.00[2]
North Pacific	2.06 mg C/l	June 1969 - June 1970, n = 8, rg = 1.10 - 4.10[3]
North-equatorial Pacific	1.12 mg C/l	Apia - Tokyo, Aug-Nov 1968, n = 59, rg = 0.88 - 1.49[4]
North-equatorial Pacific	1.14 mg C/l	155 deg W, 50 deg N - 15 deg S, Aug-Nov 1969, n = 15, rg = 0.81 -[4]
Sagami bay	1.24 mg C/l	Jan - Dec 1972, n = 34, rg = 0.73 - 2.31[5]
Sagami bay	1.31 mg C/l	Oct 1971, n = 20, rg = 1.04 - 1.69[3]
Sagami bay	1.37 mg C/l	Jan - Dec 1971, n = 30, rg = 0.85 - 1.97[5]
Sagami bay	1.55 mg C/l	Jan - Dec 1973, n = 12, rg = 0.93 - 2.48[3]
Sagami bay	1.56 mg C/l	Oct 1970, n = 30, rg = 1.22 - 1.96[3]
South East Pacific	0.91 mg C/l	Peru current, April 1966, n = 12, rg = 0.75 - 1.20[6]
South East Pacific	2.09 mg C/l	Off Peru, March-April 1966, n = 12, rg = 1.05 - 4.30[6]
Tokyo Bay	1.59 mg C/l	Nov - Dec 1971, n = 13, rg = 0.94 - 2.32[3]
Tokyo Bay	1.73 mg C/l	Dec 1973, n = 32, rg = 0.88 - 5.27[3]
Tokyo Bay	1.94 mg C/l	Sept 1971, n = 10, rg = 1.29 - 3.01[3]
Tokyo Bay	2.53 mg C/l	June 1972, n = 9, rg = 1.24 - 3.90[3]
Tokyo Bay	3.07 mg C/l	Aug 1971, n = 7, rg = 2.40 - 3.70[3]
Tokyo Bay	3.61 mg C/l	July - Aug 1971, n = 11, rg = 2.26 - 5.32[3]
Tokyo Bay	3.63 mg C/l	June 1973, n = 9, rg = 1.63 - 4.36[3]
Tokyo Bay	3.88 mg C/l	Aug 1973, n = 30, rg = 2.79 - 5.66[3]
Tokyo Bay	4.07 mg C/l	Aug 1973, n = 27, rg = 2.32 - 7.00[3]
Western North Pacific	0.88 mg C/l	Jume 1967, n = 17, rg = 0.73 - 1.13[7]
Western North Pacific	1.10 mg C/l	June 1966 - May 1968, n = 22, rg = 0.83 - 1.68[8]

Source: 1.Ogura, N. 1970. Limnol. Oceanogr., 15, pp 476-479.
2.Williams, P.M. 1971. In S.J. Faust and Hunter. (eds.): Organic Compounds in Aquatic Env., N.Y. Marcel Dekker In.
3.Ogura, N. 1975. Mar. Biol., 31(2), pp 101-111.
4.Ogura, N. 1970. Nature. 227, pp 1335-1336.
5.Ogura, N., N. Nakamoto, M. Funakoshi, A. Kamatani and S. Iwata. 1975. J. Oceanogr. Soc. Japan, 31, p43-47.
6.Barber, R.T. 1968. Nature, 220, pp 274-275.
7.Ogura, N. 1970. Deep-Sea Res., 17, pp 221-231.
8.Ogura, N. 1972. In K. Sugawara(ed), the Kuroshio II, Tokyo Saikon., pp 201-205.

5-057 Background concentration EDTA

Species	Value	Condition
Sewage, variation during the course of a day	110 ug/l	Time (hr) = 14.15, Secondary domestic sewage, Stevenage[1]
Sewage, variation during the course of a day	150 ug/l	Time (hr) = 12.35, Secondary domestic sewage, Stevenage[1]
Sewage, variation during the course of a day	180 ug/l	Time (hr) = 15.40, Secondary domestic sewage, Stevenage[1]
Sewage, variation during the course of a day	180 ug/l	Time (hr) = 17.00, Secondary domestic sewage, Stevenage[1]
Sewage, variation during the course of a day	190 ug/l	Time (hr) = 11.15, Secondary domestic sewage, Stevenage[1]
Sewage, variation during the course of a day	210 ug/l	Time (hr) = 08.45, Secondary domestic sewage, Stevenage[1]
Sewage, variation during the course of a day	210 ug/l	Time (hr) = 09.55, Secondary domestic sewage, Stevenage[1]

Source: 1.Gardiner, J. 1976. Water Res. 10, 507-514.

5-058 Background concentration Ethane + Ethylene

Species	Value	Condition
Atmosphere, organic gas	0.05 ppb	Rg = 0 - 0.1 ppb, Point Barrow, Alaska, 1968[1]
Atmosphere, organic gas	32 ppb	Rg = 19-44 ppb, Central Amazona jungle, 1972, STP[2]
Atmospheric conc of aliphatic hydrocarbon and alkylbenzene	102 ppb	Los Angeles, 1967, STP[3]

Source: 1.Cavanagh, L.A., C.F. Sahadt and E. Robinson. 1969. Environ. Sci. Technol. 3, 251-257.
2.Robinson, E., R.A. Rasmussen, H.H. Westberg and M.W. Holdren. 1973. J. Geophys. Res. 78, 5345-5351.
3.Altshuller, A.P., W.A. Lonneman and S.L. Kopczynski. 1973. J. Air Pollut. Contr. Assoc. 23, 597-599.

5-059 Background concentration Ethylbenzene

Species	Value	Condition
Atmospheric conc of aliphatic hydrocarbon and alkylbenzene	5 ppb	Los Angeles, 1967, STP[1]

Source: 1.Altshuller, A.P., W.A. Lonneman and S.L. Kopczynski. 1973. J. Air Pollut. Contr. Assoc. 23, 597-599.

5-060 Background concentration Ethyltoluene

Species	Value	Condition
Atmospheric conc of aliphatic hydrocarbon and alkylbenzene	7.5 ppb	Los Angeles, 1967, STP[1]

Source: 1.Altshuller, A.P., W.A. Lonneman and S.L. Kopczynski. 1973. J. Air Pollut. Contr. Assoc. 23, 597-599.

5-061 Background concentration Eu

Species	Value	Condition
Seawater	0.0003 microgram/l	Soluble concentration, Japan, Cl = 1.873 %[1]
Seawater	0.0016 microgram/l	Total concentration, Japan, Cl = 1.873 %[1]

Source: 1.Suzuki, H., T. Koganage and M. Saki. 1975. Impacts of Nuclear Releases into the Aquatic Environment.IAEA SM-198/19, pp 77-89. Vienna.

5-062 Background concentration F

Species	Value	Condition
Igneous rock	625 mg/kg dry weight	None[1]
River water	0.09 ppm	None[2]
Seawater	1.3 ppm	None[3]
Sedimentary rock	500 mg/kg dry weight	None[4]
Soil	200 mg/kg dry soil	Oven-dried, rg = 30 - 300[2]
Soil	200 mg/kg dry weight	None[4]

Source: 1.Taylor, S.R. 1964. Geochim. Cosmochim. Acta 28, pp 1273-1285.
2.Bowen, H.J.M. 1966. Trace Elements in Biochemistry, Ac. Press.
3.Goldberg, E.D., 1963. The Sea, Interscience London, 2, pp 3.
4.Vinogradov, A.P. 1959. The Geochemistry of Rare and Dispersed Chemical Elements in the Soil. Transl. from Russian. Consultant Bureau Inc. NY. 2nd ed.

5-063 Background concentration Fe

Species	Value	Condition
Anarrhichas lupus	29 mg/kg dry weight	Fish, marine, Trondheinfj. Norway, 1976, muscle[1]
Anodonta sp.	5500 mg/kg dry weight	Gastropod, L Aberffraw, 1970, freshwater[2]
Atmosphere	0.84 ng/m**3	Southpole, 1970, rg = +- 0.21[3]
Atmosphere	2250 ng/m**3	Urban, USA, rg = 300 - 4200[3]
Atmosphere	370 ng/m**3	Engl, 7 stations, rg = 90 - 650[3]
Atmosphere	390 ng/m**3	Not urban, USA, rg = 80 - 700[3]
Atmosphere	5600 ng/m**3	Sweden, mean from 12 stations, rg = 200 - 11000[3]
Atmosphere, rural	0.39 ug/m3	Rg = 0.08 - 0.7 ug/m3, 29 stations, USA[4]
Atmospheric particulates	12 ng/m**3	Hawaii, n > = 56, rg = +- 9[5]
Biosphere	1.38 moles/hectare	Average composition of living matter[6]
Cancer irroratus	60.0 mg/kg wet weight	Marine, Crustacean, rg =+-28, Terrence B, Nova Scotia, whole, 1973[7]
Carcinus maenas	666 mg/kg dry weight	Marine, Crustacean, muscle, Hardanger, Norway, 1976[1]
Cardium edule	590 mg/kg dry weight	Gastropod, Irish Sea, GB, 1970[2]
Choromytilus meridionalis	14.5 mg/kg wet weight	Gastropod, rg = +- 8.16 ppm, Saldanha B, LangeBaan B., RSA, 1975[8]
Crassostrea gigas	12 mg/kg wet weight	Gastropod, rg = +- 5 ppm, Saldanha B, LangeBaan B., RSA, 1975[8]
Crassostrea gigas	128 mg/kg dry weight	Gastropod, Knysna Estuary, R.S.A., 1975[9]
Crassostrea gigas	315 mg/kg dry weight	Gastropod, rg = 214-422 ppm, Hinkley Pt. Lower Station, Severn Est., 73[10]

Crassostrea margaritacea	4.7 mg/kg wet weight	Gastropod, rg = +- 2.5 ppm, Saldanha B, LangeBaan B., RSA, 1975[8]
Crassostrea margaritacea	57 mg/kg dry weight	Gastropod, Knysna Estuary, R.S.A., 1975[9]
Crassostrea virginica	67 mg/kg wet weight	Gastropod, Atlantic Coast, USA, 1967[11]
Crustral rock	5.0 percent by weight	Average value[12]
Dreissena ploymorpha	250 mg/kg dry weight	Gastropod, Schaal L, to be transplanted to Elbe Est., 1973, freshwater[13]
Dreissena ploymorpha	724 mg/kg dry weight	Rg = 531-916 ppm, Elbe Est., transplanted from Schaal L, 1973, freshwater[13]
Freshwater sediment	40100 mg/kg dry weight	Range of 90 % = 12000 - 69000 ppm, 74 lakes, pelitic[14]
Freshwater, lake	32 ug/l	Range = 5 - 53 ppm, Lake Constance, Lindau[15]
Freshwater, river	10 ug/l	Less than value, rg = 0.7 - 8.8 ppm, Calif Coast Stream[16]
Freshwater, river	74 ug/l	Range = 12 - 130 ppm, Danube, Ulm[15]
Fucus spp.	750 mg/kg dry weight	Range = 56 - 1517 ppm, Irish Sea, North Sea, 1971[17]
Fucus vesiculosus	218 mg/kg dry weight	Range = 146 - 360 ppm, Menai Straits, Anglesey, 1975[18]
Fucus vesiculosus	75 mg/kg dry weight	Range = 41 - 168 ppm, Afon Goch, Anglesey, GB, 1975[18]
Gadus morrhua	29 mg/kg dry weight	Fish, rg = +-2 ppm, Trondsheimfj, Norway, muscle, 1976[1]
Jasus lalandii	1.11 mg/kg wet weight	Marine, Crustaceans, rg = 0.62-1.6, Saldanha Bay, SA, muscle, 1975[8]
Jasus lalandii	2.7 mg/kg wet weight	Marine, Crusta., rg=+-1.3, Melkbosch C, W Coast, SA, muscle, 19, 1975[19]
Mactra clabrata	41 mg/kg wet weight	Gastropod, rg = +- 14 ppm, Saldanha B, Langebaan B, RSA, 1975[8]
Marine phytoplankton	120 mg/kg wet weight	Range = 73 - 170 ppm, Cape W Coast, S Africa, 1973[19]
Mercenaria mercenaria	30 mg/kg wet weight	Gastropod, Atlantic coast, USA, 1967[11]
Merluccius capensis	4.3 mg/kg wet weight	Fish, marine, rg =+- 3.3 ppm, Cape W Coast, S Africa, muscle, 1974[19]
Mya arenaria	405 mg/kg wet weight	Gastropod, Atlantic coast, USA, 1967[11]
Mytilus edulis	1700 mg/kg dry weight	Gastropod, Irish Sea, GB, 1970[2]
Mytilus edulis	1960 mg/kg dry weight	Gastropod, rg = 960-2640 ppm, New Zealand, 1964[20]
Mytilus edulis	363 mg/kg dry weight	Gastropod, rg = 112-1623 ppm, Trondheimsfjord, Norway, 1976[1]
Mytilus edulis	443 mg/kg dry weight	Gastropod, rg = 149-2220 ppm, Medit., NW Coast, France + Italy, 1975[21]
Nucella lapillus	65 mg/kg dry weight	Gastropod, Irish Sea, GB, 1970, marine[2]
Ostrea edulis	167 mg/kg dry weight	Gastropod, Knysna Estuary, R.S.A., 1975[9]
Ostrea sinuata	682 mg/kg dry weight	Gastropod, rg = 630-750 ppm, Tasman Bay, NZ, 1964[20]
Patella vulgata	150 mg/kg dry weight	Gastropod, marine, Irish Sea, 1971[2]
Patella vulgata	1817 mg/kg dry weight	Gastropod, rg = 1285-2505 ppm, Mortons Harbor, Newfoundl., 1974[22]
Patella vulgata	2060 mg/kg dry weight	Gastropod,rg =1380-3270 ppm, Irish Sea, 1971[17]
Pecten maximus	170 mg/kg dry weight	Gastropod, Irish Sea, GB, 1970[2]
Pecten novaezelandia	2915 mg/kg dry weight	Gastropod, rg = 1140-6900 ppm, Tasman Bay, New Zealand, 1964[20]
Pomatomus saltatrix	7.0 mg/kg wet weight	Fish, marine, Cape Hatteras, NC, USA, 1972, muscle[23]
Rain	200 ug/l	Engl, yearly mean[3]
River water	0.67 ppm	Range = 0.01-1.4 ppm[24]

River, Africa	1.3 mg/l	None[25]
River, Asia	0.01 mg/l	None[25]
River, Australia	0.3 mg/l	None[25]
River, Europe	0.8 mg/l	None[25]
River, North America	0.16 mg/l	None[25]
River, South America	1.4 mg/l	None[25]
River, World	0.67 mg/l	None[25]
Scomber japonicus	17 mg/kg wet weight	Fish, marine, rg =+- 2 ppm, Cape W Coast, S Africa, muscle, 1974[19]
Seawater	0 nmol kg-1	Depth 10 m, ferrous Fe, offshore waters of Peru[26]
Seawater	0.01 ppm	Fe(OH)3, residence time = 0.14E3 y[27]
Seawater	122 nmol kg-1	Depth 38 m, total Fe, offshore waters of Peru[26]
Seawater	13 nmol kg-1	Depth 0 m, ferrous and ferric Fe, offshore waters of Peru[26]
Seawater	166 nmol kg-1	Depth 74 m, total Fe, offshore waters of Peru[26]
Seawater	171 nmol kg-1	Depth 79 m, total Fe, offshore waters of Peru[26]
Seawater	2 nmol kg-1	Depth 10 m, ferrous and ferric Fe, offshore waters of Peru[26]
Seawater	2 nmol kg-1	Depth 19 m, ferrous and ferric Fe, offshore waters of Peru[26]
Seawater	2 nmol kg-1	Depth 5 m, ferrous Fe, offshore waters of Peru[26]
Seawater	214 nmol kg-1	Depth 68 m, total Fe, offshore waters of Peru[26]
Seawater	224 nmol kg-1	Depth 63 m, total Fe, offshore waters of Peru[26]
Seawater	230 nmol kg-1	Depth 49 m, total Fe, offshore waters of Peru[26]
Seawater	34 nmol kg-1	Depth 74 m, ferrous Fe, offshore waters of Peru[26]
Seawater	35 nmol kg-1	Depth 5 m, total Fe,offshore waters of Peru[26]
Seawater	37 nmol kg-1	Depth 74 m, ferrous and ferric Fe, offshore waters of Peru[26]
Seawater	38 nmol kg-1	Depth 0 m, total Fe[26]
Seawater	38 nmol kg-1	Depth 0 m, total Fe, offshore waters of Peru[26]
Seawater	4 nmol kg-1	Depth 28 m, ferrous Fe, offshore waters of Peru[26]
Seawater	43 nmol kg-1	Depth 79 m, ferrous Fe, offshore waters of Peru[26]
Seawater	47 nmol kg-1	Depth 63 m, ferrous Fe, offshore waters of Peru[26]
Seawater	6 nmol kg-1	Depth 0 m, ferrous Fe, offshore waters of Peru[26]
Seawater	6 nmol kg-1	Depth 13 m, total Fe, offshore waters of Peru[26]
Seawater	6 nmol kg-1	Depth 19 m, total Fe, offshore waters of Peru[26]
Seawater	6 nmol kg-1	Depth 28 m, ferrous and ferric Fe, offshore waters of Peru[26]
Seawater	6 nmol kg-1	Depth 49 m, ferrous Fe, offshore waters of Peru[26]
Seawater	61 nmol kg-1	Depth 79 m, ferrous and ferric Fe, offshore waters of Peru[26]
Seawater	79 nmol kg-1	Depth 63 m, ferrous and ferric Fe, offshore waters of Peru[26]
Seawater	8 nmol kg-1	Depth 5 m, ferrous and ferric Fe, offshore waters of Peru[26]
Seawater	9 nmol kg-1	Depth 10 m, total Fe, offshore waters of Peru[26]
Seawater	9 nmol kg-1	Depth 28 m, total Fe, offshore waters of Peru[26]
Seawater	9 nmol kg-1	Depth 49 m, ferrous and ferric Fe, offshore waters of Peru[26]
Seriola grandis	6.8 mg/kg wet weight	Fish, marine, rg =2-42 ppm, North NZ, muscle, 1973[20]

Seriola pappei	14 mg/kg wet weight	Fish, marine, rg =+-3 ppm, Cape W Coast, S Africa, muscle, 1974[19]
Shale	46700 mg/kg dry weight	Average value[28]
Soil	38000 mg/kg dry soil	Oven-dried, rg = 7000 - 550000[24]
Trigla kumu	4.2 mg/kg wet weight	Fish, rg = 2.1 - 13 ppm, N Island, NZ, muscle, 1973[20]
Ulva sp.	39 mg/kg wet weight	Range = +- 13 ppm, Cape W Coast, S Africa, marine, 1974[19]
Water	31.9 ug/l	Baltic Sea, 8 stations, 1970-71, rg = 0.8 - 63[3]
Water	852 ug/l	Rivers, USA, rg = 31 - 1670[3]

Source: 1.Lande, E. 1977. Environ. Pollut. 12, 187-198.
2.Segar, D.A. and R.E. Pellenbarg. 1971. J. Mar. Biol. Assoc. UK. 51, 131-136.
3.Laveskog, A., A. Lindskog and V. Stenberg. 1976. Om Metaller, Statens Naturvaardsverk, Stockholm.
4.Zoller, W.H., E.S. Gladney and R.A. Duce. 1974. Science 183, pp 198-200.
5.Golterman, H.L. 1975. Physiological Limnol., Elsevier Amsterdam.
6.Deevey, E.S. 1970. Scientific American. 223(3), p 150.
7.Martin, J.B.M. 1974. Mar. Biol. 28, 245-251.
8.Fourie, H.O. 1976. S. Afr. J. Sci. 72, 110-113.
9.Watling, H.R. and R.J. Watling. 1976. Mar. Pollut. Bull. 7, 45-48.
10.Romeril, M.G. 1971. Mar. Biol. 9, 347-354.
11.Pringle, B.H., D.E. Hissong, E.L. Katz and S.T. Mulawka 1968. J. Sanit. Eng. Div. Am. Soc. Civ. Eng. 94, 455-475.
12.Mason, B. 1966. Principles of Geochemistry. 3rd ed. John Wiley and Sons. New York.
13.Karbe, L., N. Antonacopoulos and C. Schnier. 1975. Verh. Int. Ver. Limnol. 19, 2094-2101.
14.Foerstner, V. 1977. Arch. Hydrobiol., 80(2), pp 172-191.
15.Quentin, K.E., and H.A. Winkler. 1974. Zentralbl. Bakt. Abt. Orig. B 158, 514-523.
16.Jones, B.F., V.C. Kennedy and G.W. Zellweger. 1974. Water Resour Res. 10, 791-793.
17.Preston, A., Jefferies, Dutton, Harvey and A.K. Steele. 1972. Environ. Pollut. 3, 69-82.
18.Foster, P. 1976. Environ. Pollut. 10, 45-53.
19.Van As, D. H.O. Fourie and C.M. Vleggaar. 1975. S. Afr. J. Sci. 71, 151-154.
20.Brooks, R.R., and D. Rumsey. 1974. NZJ Mar. Freshwater Res. 8, 155-166.
21.Fowler, S.W. and B. Oregioni. 1976. Mar. Pollut. Bull. 7, 26-29.
22.Penrose, W.R., R. Black and M.J. Hayward. 1975. J. Fish. Res. Bd Canada 32, 1275-1281.
23.Cross, F.A., L.H. Hardy, N.Y. Jones and R.T. Barber. 1973. J. Fish. Res. bd Canada 30, 1287-1291.
24.Bowen, H.J.M. 1966. Trace Elements in Biochemistry, Ac. Press.
25.Wetzel, R.G. 1975. Limnology. Saunders. Philadelphia.
26.Hong, H. and Kester, D.R. Limnol. Oceanogr., 31, 512, 1986.
27.Goldberg, E.D., 1963. The Sea, Interscience London, 2, pp 3.
28.Turekian, K.K. and K.H. Wedepohl. 1961. Amer. Geol. Soc. Bull. 72, pp 175-192.

5-064 Background concentration
Fluoranthene

Species	Value	Condition
Atmosphere, seasonal variation	0.38 ng/m3	1. Quarter, 2. Q=0.15, 3. Q=0.24, 4. Q=0.68, LA, n=13, USA, 1975[1]
Crassostrea virginica	1.7 ug/kg dry weight	Aransas Bay, uncontam. area, TEX, USA, 1970[2]
Crassostrea virginica	3.0 ug/kg dry weight	Galveston B., commercial oyster reef, TEX, USA, 1970, rg=1.6-5.6 ppb[2]
Crassostrea virginica	7.8 ug/kg dry weight	Galveston B., oyster reef contam. dom. sewage, TEX, rg=2.8-14 ppb[2]
Crassostrea virginica	800 ug/kg dry weight	Harbor of Norfolk, Vir., USA, rg=600-1000 ppb[2]
Domestic sewage	0.352 ug/liter	Dry weather condition, Hegne, GFR[3]
Domestic sewage	16.350 ug/liter	During heavy rain, Hegne, GFR[3]
Mixed sewage	3.420 ug/liter	GFR, high % industri, Stockach, Rodolfsell = 2.660 ppm[3]
Wastewater, domestic	0.273 ug/liter	Effluent conc to Rotach R, GFR[3]
Wastewater, industrial	2.198 ug/liter	Effluent conc to Rotach R, GFR[3]

Source: 1.Gordon, R.J. 1976. Environ. Sci. Technol. 10, 370-373.
2.Fazio, T. 1971. Proc. 7th Nat. Shellfish Sanit. Workshop. Wash DC. FDA, Division of Shellfish Sanitation
3.Borneff, J., and H. Kunte. 1965. Arch. Hyg. (Berlin) 149, 226-243.

5-065 Background concentration
Folic acid

Species	Value	Condition
Freshwater	0.142 mg/m**3	Kasumigaura, Japan, rg = 0.040 - 0.244[1]
Freshwater	0.24 mg/m**3	Small swiss ponds, rg = < 0.01 - 0.48[2]

Source: 1.Kashiwada, K., A. Kanazawa and S. Tachibanazono. 1963. Mem. Fac. Fish., Kagoshima Univ. 12, pp 153-17.
2.Clemenson, H. 1963. Schweiz. Z. Hydrol., 25, pp 157-165.

5-066 Background concentration
Ga

Species	Value	Condition
River water	0.001 ppm	Less than value[1]
Seawater	0.3E-4 ppm	Residence time = 1.4E3 y[2]
Soil	30 mg/kg dry soil	Oven-dried, rg = 0.4 - 300[1]

Source: 1.Bowen, H.J.M. 1966. Trace Elements in Biochemistry, Ac. Press.
2.Goldberg, E.D., 1963. The Sea, Interscience London, 2, pp 3.

5-067 Background concentration
Ge

Species	Value	Condition
Earth crust	1.5 mg/kg dry weight	None[1]
Seawater	0.7E-4 ppm	Ge(OH)4, residence time = 7E3 y[2]
Soil	0.95 mg/kg dry weight	Rg= 0.6 - 13 ppm, USA[3]
Soil	1 mg/kg dry soil	Oven-dried, rg = 1 - 50[4]

Source: 1.Fyfe, W.S. 1974.Geochemistry, p 8, Claredon Press, Oxford.
2.Goldberg, E.D., 1963. The Sea, Interscience London, 2, pp 3.
3.Schroeder, H.A.and J.J. Balassa. 1967. J. Chron. Dis. 20, pp 211-224.
4.Bowen, H.J.M. 1966. Trace Elements in Biochemistry, Ac. Press.

5-068 Background concentration
H

Species	Value	Condition
Atmosphere	308 ug/m**3	293 K, 1 atm, exclusive H2O, rg = 250 - 365[1]
Atmosphere	5E-5 % by atoms	Sea-level dry atmospheric composition[2]
Atmosphere	63 ug/m**3	293 K, 1 atm, rg = 36 - 90[1]
Biosphere	13100 moles/hectare	Average composition of living matter[3]
Lower atmosphere	0.00005 % by volume	None[4]
Seawater	108000 ppm	H2O[5]
Seawater	3E-3 ml/l	Sat.seawater, salinity = 3.43 %, temp. = 283k[2]

Source: 1.Junge, C.E. 1963. Air Chemistry and Radioactivity, Ac. Press.
2.Sisler, F.D. 1975. In: S.F. Singer (ed.), The Changing Global Environment, pp 57-69. D. Reidel Publ.Comp.
3.Deevey, E.S. 1970. Scientific American. 223(3), p 150.
4.Gleuckauf, E. 1951. Compendium of Meteorology. Am. Meteorological Soc. Boston.
5.Goldberg, E.D., 1963. The Sea, Interscience London, 2, pp 3.

5-069 Background concentration
H_2O

Species	Value	Condition
Atmosphere	151E5 ug/m**3	293 K, 1 atm, residence time = 0.027 y, rg = 3E4 - 3E7[1]
Atmosphere	2 g/cm**2	Total in atmos.(1975) = 10**19 g[2]
Atmosphere	3E5 g/cm**2	Total supply = 1.5E24 g[2]

Source: 1.Junge, C.E. 1963. Air Chemistry and Radioactivity, Ac. Press.
2.Johnson, F.S. 1975. In: S.F. Singer (ed.), The Changing Global Environment, pp 49-56. D. Reichel Pub. Copm.

5-070 Background concentration
HCO3

Species	Value	Condition
Seawater	19 % of total	As Mg(HCO3)2, chlorinity = 1.9 %, 298 K, pH = 8.15[1]
Seawater	4 % of total	As Ca(HCO3)2, chlorinity = 1.9 %, 298 K, pH = 8.15[1]
Seawater	69 % of total	As free ion, chlorinity = 1.9 %, 298 K, pH = 8.15[1]
Seawater	8 % of total	As NaHCO3, chlorinity = 1.9 %, 298 K, pH = 8.15[1]

Source: 1.Garrels, R.M. and M.E. Thompson. 1962. Am. J. Sci. 260., p 57.

5-071 Background concentration
He

Species	Value	Condition
Atmosphere	5.2 ug/m**3	293 K, 1 atm, residence time = 2E6 y[1]
Atmosphere	5E-4 % by atoms	Sea-level dry atmospheric composition[2]
Lower atmosphere	0.00053 % by volume	None[3]
Seawater	0.69E-5 ppm	He, residence time = 20000E3 y[4]

Source: 1.Junge, C.E. 1963. Air Chemistry and Radioactivity, Ac. Press.
2.Sisler, F.D. 1975. In: S.F. Singer (ed.), The Changing Global Environment, pp 57-69. D. Reidel Publ. Comp.
3.Gleuckauf, E. 1951. Compendium of Meteorology. Am. Meteorological Soc. Boston.
4.Goldberg, E.D., 1963. The Sea, Interscience London, 2, pp 3.

5-072 Background concentration
Hf

Species	Value	Condition
Seawater	0.8E-5 ppm	Less than value[1]
Soil	6 mg/kg dry soil	Oven-dried[2]

Source: 1.Goldberg, E.D., 1963. The Sea, Interscience London, 2, pp 3.
2.Bowen, H.J.M. 1966. Trace Elements in Biochemistry, Ac. Press.

5-073 Background concentration
Hg

Species	Value	Condition
Acerina cernua	1.50 mg/kg wet weight	Fish, Elbe estuary, muscle, 1974[1]
Alosa fallax	0.76 mg/kg wet weight	Fish, Elbe estuary, muscle, 1974[1]
Anarrhichas lupus	0.2 mg/kg wet weight	Fish, marine, Trondheinfj. Norway, 1976, muscle[1]
Anguilla anguilla	0.93 mg/kg wet weight	Fish, max = 2.02 ppm, Elbe estuary, muscle, 1974[1]
Anguilla anguilla	1.62 mg/kg wet weight	Fish, rg = 0.81 - 2.03 ppm, Swedish Sound, 1969, muscle[2]

Anodonta anatina	0.022 mg/kg wet weight	Gastropod, Schaal L, to be transplanted to Elbe Est., 1973, freshwater[3]
Anodonta anatina	0.055 mg/kg wet weight	Elbe Est., Transplanted from Schaal L, rg = 0.034-0.075 ppm, 1973, freshwater[3]
Asellus aquaticus	0.05 mg/kg wet weight	Freshwater, Crustaceans, Lake Aasjoen, Sweden, 1973[4]
Asellus aquaticus	0.11 mg/kg wet weight	Elbe estuary, limnic and brackish, crustaceans[5]
Asellus aquaticus	0.15 mg/kg wet weight	Freshwater, Crustaceans, Lake Kyrksjoen, rg = 0.1 - 0.2 ppm[5]
Asellus aquaticus	0.16 mg/kg wet weight	Freshwater, Crustaceans, Sweden, above paper mill, 1965[6]
Asellus aquaticus	0.35 mg/kg wet weight	Freshwater, Crustaceans, Elbe Estuary, 1976, rg = 0.22 - 0.56 ppm[5]
Asellus aquaticus	1.9 mg/kg wet weight	Freshwater, Crustaceans, Sweden, below paper mill, 1965[6]
Atmosphere	25 ng/m**3	Range = 0.001 - 50[7]
Atmosphere	5 ng/m3	Rg - 1 - 10 ng/m3, global average[8]
Aufwuchs communities	12.6 mg/kg dry weight	Rg = 0.5 - 25 ppm, Rhine R, 1976[9]
Blue crab	0.16 mg/kg wet weight	6 references used, SD = +- 0.15, rg = 0.07 - 0.48[10]
Brevoortia tyrannus	0.095 mg/kg wet weight	Fish, marine, viscera, Core Sound, W Atlantic, 1972[11]
Brevoortia tyrannus	0.166 mg/kg wet weight	Fish, marine, viscera, Chesapeake Bay, Md, USA, 1972[11]
Carbonate sediments	0.04 mg/kg dry weight	Average value[12]
Cardium edule	0.035 mg/kg wet weight	Gastropod, rg = +-0.005 ppm, Elbe R, 1976[13]
Cardium edule	0.48 mg/kg dry weight	Gastropod, rg = 0.16-0.80 ppm, Southampton waters, GB, 1971[14]
Cellana exerata	0.02 mg/kg wet weight	Gastropod, rg = +- 0.03 ppm, coastal waters of Hawaii, 1975, marine[15]
Cetorrhinus maximus	2.1 mg/kg wet weight	Fish, marine, rg = 0.9-4.0 ppm, N Atlantic, 1974, muscle[16]
Cladophora glomerata	0.61 mg/kg dry weight	Algae, Rg = 0.53 - 0.68 ppm, Leine R., FRG, 1976[17]
Clams, all species	0.08 mg/kg wet weight	22 references used, SD = +- 0.05, rg = 0.01 - 0.12[10]
Clupea harengus	0.09 mg/kg wet weight	Fish, marine, mean of N Atlantic and N Sea, 1974, muscle[16]
Coregonus clupeaformis	0.07 mg/kg wet weight	Fish, Moose L, USA, 1970, animal decapiated and eviscerated[18]
Coregonus clupeaformis	0.09 mg/kg wet weight	Fish, rg = 0.06-0.14 ppm, La Grande R, Canada, 1974, muscle[19]
Coregonus clupeaformis	0.17 mg/kg wet weight	Fish, Moose L, USA, 1970, animal decapiated and eviscerated[18]
Crangon crangon	0.02 mg/kg dry weight	Marine, Crustacean, S Antonio Bay, Texas, 1975, whole animal, less than[20]
Crangon crangon	7.5 mg/kg wet weight	Marine, Crustacean, Elbe estuary, whole animal, rg = 0.03 - 0.12 ppm[21]
Crassostrea virginica	0.05 mg/kg dry weight	Gastropod, San Antonio B, Texas, USA, 1972[20]
Crassostrea virginica	0.47 mg/kg wet weight	Gastropod, Mispilion R, Delaware, USA, 1972[22]
Crustral rock	0.000008 percent by weight	Average value[23]
Cyprinus carpio	0.25 mg/kg wet weight	Fish, rg = 0.14-0.26 ppm, Lake Oahe, S Dakota, 1973[24]
Cyprinus carpio	0.25 mg/kg wet weight	Fish, rg = 0.18-0.34 ppm, Lake Powell, N M, USA, 1974[25]
Deep-sea clay sediments	0.2 mg/kg dry weight	Average value, rg = 0.001 - 0.4[12]
Dreissena polymorpha	0.08 mg/kg dry weight	Gastropod, Schaal L, to be transplanted to Elbe Est., 1973, freshwater[3]
Dreissena polymorpha	0.020 mg/kg wet weight	Gastropod, Schaal L, to be transplanted to Elbe Est., 1973, freshwater[3]
Dreissena polymorpha	0.088 mg/kg wet weight	Elbe Est., transplanted from Schaal L., rg = 0.04-0.135 ppm, 1973, freshwater[3]

Dreissena polymorpha	0.44 mg/kg dry weight	Rg = 0.16-0.67 ppm, Elbe Est., transplanted from Schaal L, 1973, freshwater[3]
Engraulis mordax	0.04 mg/kg wet weight	Fish, marine, rg=0.01-0.08 ppm, Monterey B, Calif, 1971, muscle[26]
Esox lucius	0.11 mg/kg wet weight	Fish, Moose L, USA, 1970, animal decapiated and eviscerated[18]
Esox lucius	0.3 mg/kg wet weight	Fish, Tidoe-Lindoe, Maelaren, muscle, 1966[27]
Esox lucius	0.37 mg/kg wet weight	Fish, rg = 0.20 - 0.52 ppm, Lake Oahe, S Dakota, USA, muscle[24]
Esox lucius	0.49 mg/kg wet weight	Fish, Moose L, USA, 1970, animal decapiated and eviscerated[18]
Esox lucius	0.57 mg/kg wet weight	Fish, rg = 0.38-0.75 ppm, Skotfoss, S Norway, 1972, whole animal[4]
Esox lucius	0.76 mg/kg wet weight	Fish, rg = 0.45 - 1.21 ppm, La Grande R, Canada, 1973, muscle[19]
Esox lucius	1.14 mg/kg wet weight	Fish, Lake AAsjoen, Sweden, 1973, muscle[28]
Esox lucius	2.3 mg/kg wet weight	Fish, Braaviken, Oestergoetland, 1966, muscle, 1966[27]
Esox lucius	2.3 mg/kg wet weight	Fish, Stockholm archipelogo, 1966, muscle[27]
Esox lucius	5.9 mg/kg wet weight	Fish, rg = 1.8-9.9 ppm, Lake St. Clair, Canada, 1973, muscle[29]
Fish	0.1 ppm	Marine, excl.large predators[30]
Fish	0.5 ppm	Freshwater, predators, great variation[30]
Fontinalis antipyr.	0.25 mg/kg dry weight	Moss, Rg = 0.12-0.36 ppm, Ruhr R, FRG, 1971[31]
Food	0.055 mg/kg wet weight	Sweden, fish from sea, uncontaminated, rg = 0.01 - 0.1[7]
Food	0.125 mg/kg wet weight	Sweden, fish from lakes, uncontaminated, rg = 0.05 - 0.2[7]
Forest foliage	25 mg/kg ash weight	Uncontaminated value for USA, rg = <25-50 ppm[32]
Freshwater sediment	0.49 mg/kg dry weight	Range of 90 % = 0.15 - 1.20 ppm, 74 lakes, pelitic[33]
Freshwater zooplankton, effect of fertilizer	0.09 mg/kg dry weight	Area with fertilizer, Scenedesmus obliquus, FRG, 1974[34]
Freshwater, lake	0.07 ug/l	Range = 0.02 - 0.13 ppm, Clay Lake, Ont, 1972[35]
Freshwater, lake	0.18 ug/l	Range = 0.03 - 0.47 ppm, Lake Constance. Lindau, 1973[36]
Freshwater, river	0.08 ug/l	Range = 0.02 - 0.38 ppm, Danube, Ulm/Regensburg, 1973[36]
Freshwater, river	0.18 ug/l	Range = 0.07 - 0.75 ppm, Danube, Austria, n = 11, 1972[37]
Freshwater, river and lake	0.055 ug/l	Range = 0.02 - 2.8 ppm, 67 rivers and lakes, NE USA , 1971[38]
Gadus morrhua	0.15 mg/kg wet weight	Fish, rg = 0.04 - 0.45 ppm, Iceland, Faroer, Bornholm, Elbe, muscle, 74[39]
Gadus morrhua	0.2 mg/kg wet weight	Fish, Trondsheimfj, Norway, muscle, 1976[1]
Gadus morrhua	1.06 mg/kg wet weight	Fish, rg = 0.12-3.18 ppm, Kilsfjord, S Norway, muscle[40]
Gadus virens	0.11 mg/kg wet weight	Fish, rg = 0.06 - 0.15 ppm, Iceland, Faroer, Norway, 1974, muscle[16]
Granitic rock	0.08 mg/kg dry weight	Average value[12]
Greenland ice	0.052 ug/l	Before 1946, rg = 0.030-0.075 ppm[41]
Greenland ice	0.158 ug/l	After 1946, rg = 0.087 - 0.230 ppm[41]
Grey seal, liver	30 ppm	Range = 10 - 50[42]
Hygroamblystegium sp.	0.25 mg/kg dry weight	Moss, Rg = 0.03-0.64 ppm, Ruhr R, FRG, 1971[31]
Ictaluridae	0.17 mg/kg wet weight	11 references used, SD = +- 0.13, rg = 0.04 - 0.50[10]
Igneous rock	0.08 mg/kg dry weight	None[43]
Lemna nasus	2.2 mg/kg wet weight	Fish, marine, rg = 1.5-2.0 ppm, N Atlantic, 1974, muscle[16]
Lepomis macrochirus	0.09 mg/kg wet weight	Fish, rg = 0.06-0.13 ppm, Lake Powell, NM, USA, 1974, muscle[25]

Littorina littorea	0.105 mg/kg wet weight	Gastropod, rg = 0.06-0.156 ppm, Elbe Estuary, FRG, 1976, marine[21]
Lucioperca sandra	0.065 mg/kg wet weight	Fish, rg = 0.051-0.080 ppm, Swedish Sound, 1969, muscle[2]
Lutjanidae	0.44 mg/kg wet weight	2 references used, SD = +- 0.23, rg = 0.28 - 0.60[10]
Macoma baltica	0.05 mg/kg wet weight	Gastropod, rg = +- 0.02 ppm, Elbe Estuary, FRG, 1976[21]
Maia squinada	0.03 mg/kg dry weight	Marine, Crustacean, Quarteira, S Portugal, 1974, muscle[44]
Makaira indica	7.3 mg/kg wet weight	Fish, rg = 0.5-16.5, NE Australia, unpoll., muscle[45]
Makaira nigricans	2.06 mg/kg wet weight	Fish, rg = 0.3-8.35 ppm, Hawaiian waters, 1975, muscle[46]
Marine phytoplankton	0.207 mg/kg dry weight	Range = 0.104 - 0.509, Monterey submarine canyon, < 76 um[26]
Marine phytoplankton	0.41 mg/kg wet weight	Range = 0.115 - 0.713 ppm, Hawaii-Montery transect, 1971, < 76 um[26]
Marine zooplankton	0.130 mg/kg dry weight	Range = 0.039 - 0.448 ppm, Hawaii - Monterey transect, 1971[26]
Marine zooplankton	0.15 mg/kg dry weight	Range = 0.11 - 0.19 ppm, 3 sites at USA coast, NC + NY + Md[47]
Marine zooplankton	0.5 mg/kg dry weight	Range = 0.06 - 5.3 ppm, Cape Hatteras - Cape Cod, NC, USA, 1972[47]
Marine zooplankton, euphausids, ctenophores	0.119 mg/kg dry weight	Range = 0.051 - 0.290 ppm, Monterey submarine canyon, 1971[26]
Marine zooplankton, euphausids, ctenophores	0.188 mg/kg dry weight	Range = 0.122 - 0.166 ppm, Oregon Coast, 1971[26]
Melanogrammus aeglefinus	0.07 mg/kg wet weight	9 references used, SD = +- 0.06, rg = 0.02 - 0.17[10]
Melanogrammus aeglefinus	0.08 mg/kg wet weight	Fish, rg = 0.05-0.12 ppm, Iceland, Faroer, Bornholm, Elbe, 1974, muscle[16]
Merluccius sp.	0.25 mg/kg wet weight	10 references used, SD = +- 0.13, rg = 0.09 - 0.45[10]
Micropterus dolomieu	0.31 mg/kg wet weight	Fish, rg = 0.19-0.69 ppm, Lake Powell, NM, USA[25]
Micropterus salmoides	3.2 mg/kg wet weight	Fish, rg = 1.87-4.49 ppm, Lake Jocasse, USA, 1976, muscle[48]
Molva byrkelange	0.34 mg/kg wet weight	Fish, rg = 0.27-0.52 ppm, Iceland, faroer, Norway, 1974, muscle[16]
Moss	0.4 mg/kg dry weight	No pollution, Great Smoky Mountains national park[49]
Mugil cephalus	0.1 mg/kg dry weight	Fish, N Atlantic, E of Gulf Stream, muscle, 1972[47]
Myriophyllum spicatum	0.20 mg/kg dry weight	Higher plant, Rg = 0.09 - 0.33 ppm, Ruhr R, FRG, 1971[31]
Mytilus edulis	0.09 mg/kg wet weight	Gastropod, rg = 0.05-0.16 ppm, English Channel, France, 1974[50]
Mytilus edulis	0.11 mg/kg wet weight	Gastropod, rg = 0.06-0.2 ppm, Jutland, 1974[50]
Mytilus edulis	0.16 mg/kg wet weight	Gastropod, rg = 0.05-0.29 ppm, Atlantic, France, 1974[50]
Mytilus edulis	0.18 mg/kg wet weight	Gastropod, rg = 0.06-0.83 ppm, North Sea, Germany, France, 1974[50]
Mytilus edulis	0.29 mg/kg wet weight	Gastropod, rg = 0.08-0.7 ppm, English Channel, GB, 1974[50]
Mytilus edulis	0.29 mg/kg wet weight	Gastropod, rg = 0.11-1.65 ppm, North Sea, GB, 1974[50]
Mytilus edulis	0.39 mg/kg dry weight	Gastropod, rg = 0.16-0.629 ppm, St. Lawrence estuary, Canada, 1975[51]
Mytilus edulis	1.15 mg/kg dry weight	Gastropod, rg = 0.43-1.86 ppm, Southampton Waters, GB, 1971[14]
Mytilus edulis	4.2 mg/kg dry weight	Gastropod, rg = 0.4-13 ppm, Derwent Estuary, Tasmania, 1976[52]
Nerita picea	0.03 mg/kg wet weight	Gastropod, rg = +-0.01 ppm, coastal waters of Hawaii, 1975, marine[15]
Nuphar luteum	0.11 mg/kg dry weight	Higher plant, Rg = 0.03-0.43 ppm, Ruhr R, FRG, 1971[31]
Osmerus eperlanus	0.54 mg/kg wet weight	Fish, Elbe estuary, muscle, 1974[1]
Oyster	0.10 mg/kg wet weight	9 references used, SD = +- 0.06, rg = 0.03 - 0.23[10]

Patella vulgata	0.07 mg/kg wet weight	Gastropod, rg = 0.02-0.31 ppm, Bristol Channel, Somerset, 1971[53]
Pelagic organisms	0.22 mg/kg dry weight	Azores, rg = 0.06 - 0.38[54]
Perca flavescens	0.11 mg/kg wet weight	Fish, rg = 0.06-0.24 ppm, muscle, 1973, Millstaetter Lake, Austria[25]
Perca flavescens	0.19 mg/kg wet weight	Fish, rg = 0.04-0.32 ppm, Lake Oahe, S Dakota, muscle, 1973[24]
Perca flavescens	0.30 mg/kg wet weight	Fish, rg = 0.13-0.59 ppm, above paper mill, Kammerfoss, Sweden, muscle[55]
Perca flavescens	2.8 mg/kg wet weight	Fish, rg = 1.39- 4.16 ppm, Swedish Sound, 1969, muscle[2]
Perca flavescens	4.93 mg/kg wet weight	Fish, rg = 3.86-6.08 ppm, below paper mill, Kammerfoss, Sweden, muscle[55]
Percidae	1.18 mg/kg wet weight	56 references used, SD = +- 2.32, rg = 0.08 - 15.70[10]
Periphyton	0.032 mg/kg wet weight	Algae, Lake Powell, Colorado R, USA, 1974[56]
Plant	0.2 mg/kg air dried weight	No pollution, Great Smoky Mountains national park[49]
Platichthyes flesus	0.46 mg/kg wet weight	Fish, marine, Swedish Sound, 1969, rg = 0.05 - 0.86 ppm[2]
Pleuronectes platessa	1.53 mg/kg wet weight	Fish, marine, Swedish Sound, 1969, rg = 0.07 - 3.1 ppm[2]
Pleuronectidae	0.22 mg/kg wet weight	24 references used, SD = +- 0.20, rg = 0.02 - 0.80[10]
Pollock	0.12 mg/kg wet weight	4 references used, SD = +- 0.08, rg = 0.05 - 0.23[10]
Pomatomus saltatrix	0.37 mg/kg wet weight	Fish, marine, Cape Hatteras, NC, USA, 1972, muscle[57]
Potamogeton pectinatus	0.42 mg/kg dry weight	Higher plant, Rg = 0.37-0.46 ppm, Leine R, FRG, 1976[17]
Radix peregra	0.09 mg/kg wet weight	Gastropod, rg = 0.08-0.09 ppm, Elbe, Upstream Hamburg, 1975, freshwater[21]
Radix peregra	0.33 mg/kg wet weight	Gastropod, rg = 0.25-0.40 ppm, Elbe, downstream Hamburg, 1975, freshwater[21]
Rain	0.06 microgram/l	30 July - 7 Nov 1971, New Hamshire, USA, bulk precipation[58]
Rain	0.1 ug/l	Sweden, 60 ug Hg/m**2*yr[7]
Raja clavata	0.53 mg/kg dry weight	Fish, marine, Sagress, S Portugal, 1975[44]
River water	0.8E-4 ppm	None[59]
Rutilus ruthilus	0.43 mg/kg wet weight	Fish, rg = 0.22-0.65 ppm, muscle, Lake Aasjon, Sweden, 1973[60]
Sagittaria sagittifolia	0.15 mg/kg dry weight	Higher plant, Rg = 0.08 - 0.33 ppm, Ruhr R, FRG, 1971[31]
Salmo gairdneri	0.08 mg/kg wet weight	Fish, Lake Powell, NM, USA, muscle[25]
Salmo trutta	0.21 mg/kg wet weight	Fish, rg = 0.09-0.47 ppm, above paper mill, Kilsfjord, Norway, muscle[55]
Salmo trutta	4.35 mg/kg wet weight	Fish, rg = 2.65-7.38 ppm, below paper mill, Kilsfjord, Norway, muscle[55]
Salvelinus namaycush	0.43 mg/kg dry weight	Fish, rg = 0.19-0.66 ppm, Lake Cayuga, NY, USA, whole animal[61]
Sandstone sediment	0.03 mg/kg dry weight	Average value[12]
Scallops, sea	0.04 mg/kg wet weight	5 references used, SD = +- 0.008, rg = 0.03 - 0.05[10]
Scardinius erythrophthalmus	0.12 mg/kg wet weight	Fish, rg = 0.03-0.21 ppm, muscle, Worther L, Austria, 1973[25]
Scomber scombrus	0.15 mg/kg wet weight	Fish, marine, mean of N Atlantic and N Sea, 1974, muscle[16]
Seawater	0.3E-4 ppm	HgCl4(2-), residence time = 42E3 y[62]
Seawater	79.5 ng/l	Azores, highest conc in deep water, rg = 17 - 142 ng/l[54]
Sebastes marinus	0.14 mg/kg wet weight	Fish, rg = 0.07 - 0.22 ppm, Iceland, Faroer, Shetland, 1974, muscle[16]

Sebastes mentella	0.24 mg/kg wet weight	Fish, rg = 0.07-0.54 ppm, Iceland, Faroer, Shetland, 1974, muscle[16]
Sedimentary rock	0.4 mg/kg dry weight	None[63]
Shale	0.4 mg/kg dry weight	Average value[12]
Shale sediment	0.4 mg/kg dry weight	Average value[12]
Shrimp, brown	0.70 mg/kg wet weight	13 references used, SD = +- 2.02, rg = 0.02 - 7.36[10]
Soil	0.01 mg/kg dry weight	None[63]
Solea solea	0.17 mg/kg dry weight	Fish, marine, Sagres, S Portugal, muscle, 1974[44]
Squalus acanthias	0.24 mg/kg wet weight	Fish, marine, N Atlantic, 1974[16]
Stizostedion v. vitreum	0.43 mg/kg wet weight	Fish, Lake Powell, NM, USA, muscle, 1974[25]
Stizostedion vitreum	0.30 mg/kg wet weight	Fish, rg = 0.12-0.39 ppm, L Oahe, S Dakota, muscle, 1973[24]
Thalamita crenata	0.061 mg/kg wet weight	Marine, Crustacean, rg =+- 0.02, muscle, Hawaii, Ala Wai Canal, 1975[64]
Thalamita crenata	0.09 mg/kg wet weight	Marine, Crustacean, rg =+-0.1, Hawaii, Kanuai Oahu, 1975, whole ani.[15]
Thunnus sp.	0.51 mg/kg wet weight	35 references used, SD = +- 0.30, rg = 0.12 - 1.39[10]
Unio tumidus	0.035 mg/kg wet weight	Gastropod, Schaal L, to be transplanted to Elbe Est., 1973, freshwater[3]
Unio tumidus	0.087 mg/kg wet weight	Rg = 0.040-0.135 ppm, Elbe Est., transplanted from Schaal L, 1973, freshwater[3]
Volcanic exhalation	20000 ng/m3	Rg = 730 - 40000 ng/m3, Hawaii[8]
Volcanic exhalation	5000 ng/m3	Rg = 100 - 9600 ng/m3, Russia[8]
Water	0.045 ug/l	Groundwater, rg = 0.02 - 0.07[7]
Water	0.055 ug/l	Normal value for rivers and lakes, rg = 0.01 - 0.1[7]
Water	0.12 ug/l	Baltic Sea, surface water[7]
Atmosphere	10.2 ng/m3	Rg = 0.4 - 20 ng/m3, atlantic, 1977[8]
Atmosphere	1000 ng/m3	Rg = 30 - 1E6 ng/m3, air over deposits mines and geysers, Japan[8]
Atmosphere	13 ng/m3	Rg = 1 - 25 ng/m3, San francisco Bay, winter[8]
Atmosphere	14 ng/m3	Less than value, Japan, non industry[8]
Atmosphere	19 ng/m3	Rg = 2 - 37 ng/m3, FRG, 1977[8]
Atmosphere	1E4 ng/m3	Japan, air over agricultural area spreyed with fungicides[8]
Atmosphere	26 ng/m3	Rg = 2 - 50 ng/m3, San Francisco Bay, summer[8]

Source: 1.Lande, E. 1977. Environ. Pollut. 12, 187-198.
2.Ackefors, H. 1971. Proc. R. Soc. London B 177, 365-387.
3.Karbe, L., N. Antonacopoulos and C. Schnier. 1975. Verh. Int. Ver. Limnol. 19, 2094-2101.
4.Havre, G.N., B. Underdal and C. Christiansen. 1973. Oikos 24, 155-157.
5.Fourie, H.O. 1976. S. Afr. J. Sci. 72, 110-113.
6.Walker, G., P.S. Rainbow, P. Foster and D.J. Crisp. 1975. Mar. Biol. 30, 57-65.
7.Laveskog, A., A. Lindskog and V. Stenberg. 1976. Om Metaller, Statens Naturvaardsverk, Stockholm.
8.Kaiser, G. and G. Toelg. 1980. In: O. Hutzinger: The Handbook of Environmental Chemistry. Antropogenic compounds. Vol 3. Part A. pp 1-58. Springer-Verlag.
9.Vogt, G. and F. Kittelberger. 1977. in: Reichenbach-Klinke, H. (ed.): Fish und Umwelt. Vol 3. Fisher Stuttgart.
10.Zook, E.G., J.J. Powell, B.M. Hackley, J.A. Emerson, J.R. Brooker and G.M. Knobl, Jr. 1976. J. Agric. Food. Chem. 24(1), pp 47-53.
11.Cocoros, G., P.H. Cahn and W. Siler. 1973. J. Fish. Biol. 5, 641-647.
12.Turekian, K.K. and K.H. Wedepohl. 1961. Amer. Geol. Soc. Bull. 72, pp 175-192.
13.Segar, D.A. and R.E. Pellenbarg. 1971. J. Mar. Biol. Assoc. UK. 51, 131-136.
14.Raymont, J.E.G. 1972. Proc. R. Soc. London b 180, 451-468.
15.Klemmer, H.W., C.S. Unniayer and W.I. Okubo. 1976. Bull. Environ. Contam. Toxicol. 15, 454-457.
16.Kruger, K.E., L. Nieper and H.-J. Auslitz. 1975. Mitt. Arch. Lebensmittelhyg. 26, 201-207.
17.Abo-Rady, M.D.K. 1977. Die Belastung der oberen Leine mit Schwermetallen durch kommunale und industrelle Abwasser. Diss. Univ. Gottingen. 120 p.
18.Uthe, J.H. and E.G. Bligh. 1971. J. Fish. Res. bd Canada 28, 786-788.

19.Smith, E.G., F. Berkes and J.A. Spence. 1975. Bull. Environ. Contam. Toxicol. 13, 673-677.
20.Sims, R.R. and B.J. Presley. 1976. Bull. Environ. Contam. Toxicol. 16, 520-526.
21.Zauke, G.-P. 1977. Helgol. Wiss. Meeresunters 29, 358-374.
22.Cunninghan, P.A. and M.R. Tripp. 1973. Mar. Biol. 20, 14-19.
23.Mason, B. 1966. Principles of Geochemistry. 3rd ed. John Wiley and Sons. New York.24.Walter, C.M., H.G. Brown and C.P. Hensley. 1974. Water Res. 8, 413-418.
25.Krocza, W., P. Glantschig and W. Stoeckl. 1974. Wien. Tieraerzl. Monatsschr. 61, 169-178.
26.Knauer, G.A. and J.H. Martin. 1972. Limnol. Oceanogr. 17, 868-876.
27.Johnels, A.G. T. Westermark, W. Berg, P. Persson and Sjoestrand.1967. Oikos 18, 323-333.
28.Hasselrot, T.B. and A. Goethberg. 1974. Proc. Int. Conf. Persist. Chemicals in Aquatic Eco. Ottawa III, 37-43.
29.Bishop, J.N. and B.P. Neary. 1974. Proc. Inf. Conf. Transp. Persist. Chemicals in Aquatic Ecosystems Ottawa III, 25-29.
30.Therkelsen, O. 1977. Vatten 3, pp 299-317.
31.Dietz, F. 1974. Z. Wasser Abwasserforsch. 7, 74-80.
32.Shacklette, H.T., J.A. Erdman, T.F. Harms and C.S.E. Pupp. 1978. In: F.W. Oeheme (ed.): Toxicity of Heavy Metals in the Environment. Dekkker. NY. 25-68.
33.Foerstner, V. 1977. Arch. Hydrobiol., 80(2), pp 172-191.
34.Payer, H.D., Runkel, Schramel, Stengel, Bhumiratana and Soeder. 1976. Chemosphere 6, 413-418.
35.Armstrong, F.A.J. and A.L. Hamilton. 1973. in P.C. Singer. Trace Metals and Metals and Metal-Organic Interactions in Natural Waters. pp 131-156.
36.Quentin, K.E., and H.A. Winkler. 1974. Zentralbl. Bakt. Abt. Orig. B 158, 514-523.
37.Dworsky, R., F. Ebner, H. Gams and L.J. Ottendorfer. 1973. Oesterr. Abwasserrundsch. 18, 22-27.
38.Klein, D.H. 1972. In: R. Hartung and B.D. Dinman (eds.): Environmental Hg Contamination. An Arbor Sc. Publ. 25-29.
39.Brooks, R.R., and D. Rumsey. 1974. NZJ Mar. Freshwater Res. 8, 155-166.
40.Underwood, E.J. 1971. Trace elements in human and animal nutrition. 3rd ed. New York. Academic Press
41.Weiss, H.V., M. Koide and E.D. Goldberg. 1971. Science 174, 692-693.
42.Holden, A.V. 1973. J. Food. Tech. 8, pp 1-25.
43.Taylor, S.R. 1964. Geochim. Cosmochim. Acta 28, pp 1273-1285.
44.Martin, J.B.M. 1974. Mar. Biol. 28, 245-251.
45.MacKay, N.J., M.N. Kazacos, R.J. Williams and M.I. Leedow. 1975. Mar. Pollut. Bull. 6, 57-61.
46.Schultz, C.D., D. Crear, J.E. Pearson, J.B. Rivers and Hylim. 1976. Bull. Environ. Contam. Toxicol.5, 230-234.
47.Windom, H.L., R. Sticjney, Smith, White and F. Taylor. 1973. J. Fish. Res. Bd Canada 30, 275-279.
48.Abernathy, A.R. and P.M. Cumbie. 1977. Bull. Environ. Contam. Toxicol. 17, 595-602.
49.Huckabee, J.W. and B.G. Blaylock. 1972. in: Dhar, S.K. (ed.); Metal Ions in Biological Systems. pp 15-160. Plenum Publ. Corp. New York.
50.Wolf, P. de. 1975. Mar. Pollut. Bull. 6, 61-63.
51.Bourget, E. and D. Cossa. 1976. Mar. Pollut. Bull. 7, 237-240.
52.Bloom, H. and G.M. Ayling. 1977. Environ. Geol. 2, 3-22.
53.Peden, J.D., J.H. Crothers, C.E. Waterfall and J. Beasley. 1973. Mar. Pollut. Bull. 4, 7-9.
54.Leatherland, T.M., J.D. Burton, F. Cul kin, M.J. McCartney and R.J. Morris. 1973. Deep-Sea Res. 20,p 679-685.
55.Ui, J. and S. Kitamura. 1971. Mar. Pollut. Bull. 2, 56-58.
56.Potter, L., D. Kidd and D. Standiford. 1975. Environ. Sci. Technol. 9, 41-46.
57.Cross, F.A., L.H. Hardy, N.Y. Jones and R.T. Barber. 1973. J. Fish. Res. bd Canada 30, 1287-1291.
58.Schlesinger, W.H., W.A. Reiners and D.S. Knopman. 1973. Environ. Pollut., 6, pp 39-47.
59.Bowen, H.J.M. 1966. Trace Elements in Biochemistry, Ac. Press.
60.Mathis, B.J. and N.R. Kevern. 1973. Distribution of Hg, Cd, Pb and Th in a eutrophic lake. Int. Water Res. Michigan State Univ., Technol. Dep. 34, 23 pp.
61.Bache, C.A., W.H. Gutemann and D.J. Lisk. 1971. Science 172, 951-952.
62.Goldberg, E.D., 1963. The Sea, Interscience London, 2, pp 3.
63.Vinogradov, A.P. 1959. The Geochemistry of Rare and Dispersed Chemical Elements in the Soil. Transl. from Russian. Consultant Bureau Inc. NY. 2nd ed.
64.Luoma, S.N. 1976. Bull. Environ. Contam. Toxicol. 16, 719-723.

5-074 Background concentration
Hydrogen sulfide

Species	Value	Condition
Atmosphere	16.5 ug/m**3	293 K, 1 atm, residence time = 0.11 y, rg = 3 - 30[1]
Seawater	11 ml/l	Sat.seawater, S = 3.43 %, 283 K, high val.only over anoxic basin, rg=[2]

Source: 1.Junge, C.E. 1963. Air Chemistry and Radioactivity, Ac. Press.
2.Sisler, F.D. 1975. In: S.F. Singer (ed.), The Changing Global Environment, pp 57-69. D. Reidel Publ. Comp.

5-075 Background concentration
I

Species	Value	Condition
River water	0.002 ppm	None[1]
Seawater	0.06 ppm	I(-), IO3(-)?[2]
Soil	5 mg/kg dry soil	Oven-dried[1]

Source: 1.Bowen, H.J.M. 1966. Trace Elements in Biochemistry, Ac. Press.
2.Goldberg, E.D., 1963. The Sea, Interscience London, 2, pp 3.

5-076 Background concentration In

Species	Value	Condition
Atmosphere	0.0525 ng/m**3	England, less than, rg = 0.025 - 0.08[1]
Earth crust	1E-5 %	None[2]
Rain	0.59 ug/l	England[3]
Rain	0.59 ug/l	England, yearly mean, less than[1]
Seawater	0.02 ppm	Less than value[4]
Seawater	0.1 mg/l	Atlantic Ocean, Neutron activation, 1969[5]

Source: 1.Laveskog, A., A. Lindskog and V. Stenberg. 1976. Om Metaller, Statens Naturvaardsverk, Stockholm.
2.Sunderman, D.N. and T. Hasegawa. 1960. The Radiochemistry of Indium. USAEC Division of Tech Inform. Extentions. Oak Ridge. Temm. p 46.
3.Peirson, D.H., P.A. Cawse, L. Salmon and R.S. Cambray. 1973. Nature 241, pp 252-256.
4.Goldberg, E.D., 1963. The Sea, Interscience London, 2, pp 3.
5.Mattews, A.D. and J.P. Riley. 1970. Anal. Chim. Acta 51, pp 287-294.

5-077 Background concentration Indenol(1,2,3-cd)pyrene

Species	Value	Condition
Atmosphere, seasonal variation	1.79 ng/m3	1. Quarter, 2. Q=0.68, 3. Q=1.00, 4. Q=2.64, LA, n=13, USA, 1975[1]

Source: 1.Gordon, R.J. 1976. Environ. Sci. Technol. 10, 370-373.

5-078 Background concentration Indenol(1,2,3-cd_pyrene)

Species	Value	Condition
Domestic sewage	0.017 ug/liter	Dry weather condition, Hegne, GFR[1]
Domestic sewage	4.980 ug/liter	During heavy rain, Hegne, GFR[1]
Mixed sewage	0.930 ug/liter	GFR, high % industri, Stockach, Rodolfsell = 0.476 ppm[1]
Wastewater, domestic	0.022 ug/liter	Effluent conc to Rotach R, GFR[1]
Wastewater, industrial	0.057 ug/liter	Effluent conc to Rotach R, GFR[1]

Source: 1.Borneff, J., and H. Kunte. 1965. Arch. Hyg. (Berlin) 149, 226-243.

5-079 Background concentration Isobutane

Species	Value	Condition
Atmosphere, organic gas	0.8 ppb	Rg = 0 - 1.7 ppb, mountain air in USA, 1972, STP[1]
Atmospheric conc of aliphatic hydrocarbon and alkylbenzene	12 ppb	Los Angeles, 1967, STP[2]

Source: 1.Cavanagh, L.A., C.F. Sahadt and E. Robinson. 1969. Environ. Sci. Technol. 3, 251-257.
2.Altshuller, A.P., W.A. Lonneman and S.L. Kopczynski. 1973. J. Air Pollut. Contr. Assoc. 23, 597-599.

5-080 Background concentration Isopentane

Species	Value	Condition
Atmospheric conc of aliphatic hydrocarbon and alkylbenzene	35 ppb	Los Angeles, 1967, STP[1]

Source: 1.Altshuller, A.P., W.A. Lonneman and S.L. Kopczynski. 1973. J. Air Pollut. Contr. Assoc. 23, 597-599.

5-081 Background concentration K

Species	Value	Condition
Atmospheric particulates	120 ng/m**3	Hawaii, n > = 56, rg = +- 60[1]
Biosphere	11.7 moles/hectare	Average composition of living matter[2]
River water	2.3 ppm	Range = 1.4-10 ppm[3]
River, Asia	3.8 mg/l	None[4]
River, Australia	1.4 mg/l	None[4]
River, Europe	1.7 mg/l	None[4]
River, North America	1.4 mg/l	None[4]
River, South America	2.0 mg/l	None[4]
River, World	2.3 mg/l	None[4]
Seawater	1 % of total	As K2SO4, chlorinity = 1.9 %, 298 K, pH = 8.15[5]
Seawater	380 ppm	K(+), residence time = 11000E3 y[6]
Seawater	99 % of total	As free ion, chlorinity = 1.9 %, 298 K, pH = 8.15[5]
Soil	14000 mg/kg dry soil	Oven-dried, rg = 400 - 30000[3]

Source: 1.Golterman, H.L. 1975. Physiological Limnol., Elsevier Amsterdam.
2.Deevey, E.S. 1970. Scientific American. 223(3), p 150.
3.Bowen, H.J.M. 1966. Trace Elements in Biochemistry, Ac. Press.
4.Wetzel, R.G. 1975. Limnology. Saunders. Philadelphia.
5.Garrels, R.M. and M.E. Thompson. 1962. Am. J. Sci. 260., p 57.
6.Goldberg, E.D., 1963. The Sea, Interscience London, 2, pp 3.

5-082 Background concentration Kr

Species	Value	Condition
Atmosphere	1.0E-4 % by atoms	Sea-level dry atmospheric composition[1]
Atmosphere	1.1 ug/m**3	293 K, 1 atm[2]
Lower atmosphere	0.00012 % by volume	None[3]
Seawater	0.25E-2 ppm	Kr[4]

Source: 1.Sisler, F.D. 1975. In: S.F. Singer (ed.), The Changing Global Environment, pp 57-69. D. Reidel Publ. Comp.
2.Junge, C.E. 1963. Air Chemistry and Radioactivity, Ac. Press.
3.Gleuckauf, E. 1951. Compendium of Meteorology. Am. Meteorological Soc. Boston.
4.Goldberg, E.D., 1963. The Sea, Interscience London, 2, pp 3.

5-083 Background concentration La

Species	Value	Condition
Seawater	0.12E-4 ppm	Residence time = 0.44E3 y[1]
Soil	30 mg/kg dry soil	Oven-dried, rg = 1 - 5000[2]

Source: 1.Goldberg, E.D., 1963. The Sea, Interscience London, 2, pp 3.
2.Bowen, H.J.M. 1966. Trace Elements in Biochemistry, Ac. Press.

5-084 Background concentration Li

Species	Value	Condition
Carbonate sediments	5 mg/kg dry weight	Average value[1]
Deep-sea clay sediments	57 mg/kg dry weight	Average value[1]
Freshwater sediment	57 mg/kg dry weight	Range of 90 % = 15 - 200 ppm, 74 lakes, pelitic[2]
Freshwater, lake	220000 ug/l	Tuz Goelue, Turkey[3]
Freshwater, river	10 ug/l	Range = 1 - 25 ppm, Illinois River[4]
Freshwater, river	10 ug/l	Range = 4.1 - 11 ppm, Rhine River[5]
Granitic rock	24 mg/kg dry weight	Average value[1]
River water	0.11E-2 ppm	Range = 0.7E-4-0.04 ppm[6]
Sandstone sediment	15 mg/kg dry weight	Average value[1]
Seawater	0.18 ppm	Li(+), residence time = 20000E3 y[7]
Shale	66 mg/kg dry weight	Average value[1]
Shale sediment	66 mg/kg dry weight	Average value[1]
Soil	30 mg/kg dry soil	Oven-dried, rg = 7 - 200[6]
Water	1 ug/l	Rivers, rg = 0.1 - 400[8]

Source: 1.Turekian, K.K. and K.H. Wedepohl. 1961. Amer. Geol. Soc. Bull. 72, pp 175-192.
2.Foerstner, V. 1977. Arch. Hydrobiol., 80(2), pp 172-191.
3.Irion, G. 1972. Naturwissenschaften 59, 467.
4.Biggs, R.B., J.C. Miller and M.J. Otley. 1972. Trace metals in several Delaware watersheds - a progress report. Newark. Water Resources Center. Univ. Delaware.
5.Reichert, J.K. 1973. Vom Wasser 39, 137-146.
6.Bowen, H.J.M. 1966. Trace Elements in Biochemistry, Ac. Press.
7.Goldberg, E.D., 1963. The Sea, Interscience London, 2, pp 3.
8.Laveskog, A., A. Lindskog and V. Stenberg. 1976. Om Metaller, Statens Naturvaardsverk, Stockholm.

5-085 Background concentration Lu

Species	Value	Condition
Seawater	0.0003 microgram/l	Soluble concentration, Japan, Cl = 1.873 %[1]
Seawater	0.0007 microgram/l	Total concentration, Japan, Cl = 1.873 %[1]

Source: 1.Suzuki, H., T. Koganage and M. Saki. 1975. Impacts of Nuclear Releases into the Aquatic Environment. IAEA SM-198/19, pp 77-89. Vienna.

5-086 Background concentration Methane

Species	Value	Condition
Atmospheric conc of aliphatic hydrocarbon and alkylbenzene	2400 ppb	Los Angeles, 1967, STP[1]

Source: 1.Altshuller, A.P., W.A. Lonneman and S.L. Kopczynski. 1973. J. Air Pollut. Contr. Assoc. 23, 597-599.

5-087 Background concentration Methanol + Ethanol

Species	Value	Condition
Atmosphere, organic gas	0.6 ppb	Rg = 0 - 1.2 ppb, Point Barrow, Alaska, 1968[1]

Source: 1.Cavanagh, L.A., C.F. Sahadt and E. Robinson. 1969. Environ. Sci. Technol. 3, 251-257.

5-088 Background concentration Methyl bromide

Species	Value	Condition
Atmosphere	0.05 ug/m3	Typical value[1]
Estuarine water	0.005 ug/m3	High value in inshore water[2]
Seawater	0.001 ug/m3	Typical open ocean[2]
Sediment	0.5 ug/m3	Typical conc in inshore sediment[2]

Source: 1.Lovelock, J.E. and P. Simmons. 1980. In: Sugden (ed.): The Aerosol Controversy. Sci-Horwood. 2.National Academy of Science. 1978. Chloroform, Carbon tetrachloride, and other halomethanes. An Envionmental Assessment. US NAS.

5-089 Background concentration Methyl chloride

Species	Value	Condition
Atmosphere	0.6 ug/m3	Typical value[1]
Seawater	0.015 ug/m3	Typical open ocean[2]

Source: 1.Lovelock, J.E. and P. Simmons. 1980. In: Sugden (ed.): The Aerosol Controversy. Sci-Horwood. 2.National Academy of Science. 1978. Chloroform, Carbon tetrachloride, and other halomethanes. An Envionmental Assessment. US NAS.

5-090 Background concentration Methyl chloroform

Species	Value	Condition
Atmosphere	0.10 ug/m3	Typical value[1]
Estuarine water	0.15 ug/m3	High value in inshore water[2]
Seawater	0.010 ug/m3	Typical open ocean[2]
Sediment	0.25 ug/m3	Typical conc in inshore sediment[2]

Source: 1.Lovelock, J.E. and P. Simmons. 1980. In: Sugden (ed.): The Aerosol Controversy. Sci-Horwood. 2.National Academy of Science. 1978. Chloroform, Carbon tetrachloride, and other halomethanes. An Envionmental Assessment. US NAS.

5-091 Background concentration Methyl iodide

Species	Value	Condition
Atmosphere	0.01 ug/m3	Typical value[1]
Estuarine water	0.2 ug/m3	High value in inshore water[2]
Seawater	0.080 ug/m3	Typical open ocean[2]

Source: 1.Lovelock, J.E. and P. Simmons. 1980. In: Sugden (ed.): The Aerosol Controversy. Sci-Horwood. 2.National Academy of Science. 1978. Chloroform, Carbon tetrachloride, and other halomethanes. An Envionmental Assessment. US NAS.

5-092 Background concentration Methylchloroform

Species	Value	Condition
Drinking water	0.016 ug/m3	Typical non-polluted value, halocarbon[1]
Drinking water	0.1 ug/m3	High value, halocarbon[1]
Freshwater	0.1 ug/m3	Typical non-polluted value, halocarbon[1]
Freshwater	5.5 ug/m3	High value in urban area, rg = 1.0 - 10 ppb, halocarbon[1]
Groundwater	0.016 ug/m3	High value, halocarbon[1]
Groundwater	0.2 ug/m3	Typical non-polluted value, halocarbon[1]
Rain	0.047 ug/m3	Rg = 0.005 - 0.09 ppb, halocarbon[1]
Snow	0.016 ug/m3	Rg = 0.001 - 0.03 ppb, halocarbon[1]

Source: 1.Pearson, C.R. 1982. In: O. Hutzinger (ed.): Environmental Chemistry. Springer-Verlag. Volume 3, part B. Berlin.

5-093 Background concentration Methylene chloride

Species	Value	Condition
Drinking water	20 ug/m3	Typical non-polluted value, halocarbon[1]
Freshwater	0.05 ug/m3	Typical non-polluted value, halocarbon[1]
Freshwater	0.5 ug/m3	High value in urban area, halocarbon[1]
Groundwater	5 ug/m3	High value, halocarbon[1]

Source: 1.Pearson, C.R. 1982. In: O. Hutzinger (ed.): Environmental Chemistry. Springer-Verlag. Volume 3, part B. Berlin.

5-094 Background concentration Mg

Species	Value	Condition
Atmospheric particulates	410 ng/m**3	Hawaii, n > = 56, rg = +- 240[1]
Biosphere	8.18 moles/hectare	Average composition of living matter[2]
Crustral rock	2.1 percent by weight	Average value[3]
River water	4.1 ppm	Range = 1.5-6 ppm[4]
River, Africa	3.8 mg/l	None[5]
River, Asia	5.6 mg/l	None[5]
River, Australia	2.7 mg/l	None[5]
River, Europe	5.6 mg/l	None[5]
River, North America	5.0 mg/l	None[5]
River, South America	1.5 mg/l	None[5]
River, World	4.1 mg/l	None[5]
Seawater	0.3 % of total	As MgCO3(aq), chlorinity = 1.9 %, 298 K, pH = 8.15[6]
Seawater	1 % of total	As Mg(HCO3)2, chlorinity = 1.9 %, 298 K, pH = 8.15[6]
Seawater	11 % of total	As MgSO4, chlorinity = 1.9 %, 298 K, pH = 8.15[6]
Seawater	1350 ppm	Mg(2+), residence time = 45000E3 y[7]
Seawater	87 % of total	As free ion, chlorinity = 1.9 %, 298 K, pH = 8.15[6]
Soil	5000 mg/kg dry soil	Oven-dried, rg = 600 - 6000[4]

Source: 1.Golterman, H.L. 1975. Physiological Limnol., Elsevier Amsterdam.
2.Deevey, E.S. 1970. Scientific American. 223(3), p 150.
3.Mason, B. 1966. Principles of Geochemistry. 3rd ed. John Wiley and Sons. New York.
4.Bowen, H.J.M. 1966. Trace Elements in Biochemistry, Ac. Press.
5.Wetzel, R.G. 1975. Limnology. Saunders. Philadelphia.
6.Garrels, R.M. and M.E. Thompson. 1962. Am. J. Sci. 260., p 57.
7.Goldberg, E.D., 1963. The Sea, Interscience London, 2, pp 3.

5-095 Background concentration Mn

Species	Value	Condition
Anarrhichas lupus	4 mg/kg dry weight	Fish, marine, Trondheinfj. Norway, 1976, muscle[1]
Anodonta sp.	2100 mg/kg dry weight	Gastropod, L Aberffraw, 1970, freshwater[2]
Atmosphere	0.01 ng/m**3	Southpole, 1970, rg = +- 0.005[3]
Atmosphere	100 ng/m**3	USA, mean from 103 stations, 1960-65, max = 9980 ng/m**3[3]
Atmosphere	1720 ng/m**3	Sweden, mean from 6 stations, rg = 440 - 3000[3]
Atmosphere	22 ng/m**3	England, 7 stations, rg = 3 - 41[3]
Atmosphere	26 ng/m**3	Not urban, USA, rg = 5 - 47[3]
Atmosphere	2660 ng/m**3	USA, 1974, rg = 220 - 300, area with foundaries[4]
Atmosphere	50 ng/m**3	USA, 1974, rg = 30 - 70, area without foundries[4]
Atmosphere	725 ng/m**3	Urban, USA, rg = 10 - 1440[3]
Atmosphere, urban	250 ng/m**3	24 h measurements over 3 y, n=15, rg = 50 - 450 ng/m**3[5]
Atmospheric particulates	0.19 ng/m**3	Hawaii, n > = 56, rg = +- 0.14[6]
Biosphere	0.765 moles/hectare	Average composition of living matter[7]
Cancer irroratus	10.6 mg/kg wet weight	Marine, Crustacean, rg =+- 4.2, Terrence B, Nova Scotia, whole, 1973[8]
Carbonate sediments	1100 mg/kg dry weight	Average value[9]
Cardium edule	6.3 mg/kg dry weight	Gastropod, Irish Sea, GB, 1970[2]
Choromytilus meridionalis	1.83 mg/kg wet weight	Gastropod, rg = +- 0.46 ppm, Saldanha B, LangeBaan B., RSA, 1975[10]
Coregonus clupeaformis	0.66 mg/kg wet weight	Fish, L Erie, USA, 1970, animal decapiated and eviscerated[11]
Coregonus clupeaformis	0.69 mg/kg wet weight	Fish, Moose L, USA, 1970, animal decapiated and eviscerated[11]
Crassostrea gigas	1.3 mg/kg wet weight	Gastropod, rg = +- 0.5 ppm, Saldanha B, LangeBaan B., RSA, 1975[10]
Crassostrea gigas	16 mg/kg dry weight	Gastropod, Knysna Estuary, R.S.A., 1975[12]
Crassostrea gigas	31 mg/kg dry weight	Gastropod, rg = 16-46 ppm, Hinkley Pt. Lower Station, Severn Est., 1973[13]
Crassostrea margaritacea	0.55 mg/kg wet weight	Gastropod, rg = +- 3.6 ppm, Saldanha B, LangeBaan B., RSA, 1975[10]
Crassostrea margaritacea	2 mg/kg dry weight	Gastropod, Knysna Estuary, R.S.A., 1975[12]
Crassostrea virginica	4.3 mg/kg wet weight	Gastropod, Atlantic Coast, USA, 1967[14]
Crustral rock	0.10 percent by weight	Average value[15]
Deep-sea clay sediments	6700 mg/kg dry weight	Average value[9]
Esox lucius	0.93 mg/kg wet weight	Fish, L Erie, USA, 1970, animal decapiated and eviscerated[11]
Esox lucius	2.98 mg/kg wet weight	Fish, Moose L, USA, 1970, animal decapiated and eviscerated[11]
Food	1.0 mg/kg dry weight	Vegetables, rg = 0.5 - 1.5[3]

Food	170 mg/kg dry weight	Tea, rg = 150 - 170[3]
Food	96 ug/l	Milk, USA, rg = 63 - 129[3]
Freshwater sediment	882 mg/kg dry weight	Range of 90 % = 100 - 1800 ppm, 74 lakes, pelitic[16]
Freshwater, lake	5.4 ug/l	Range = 3 - 10 ppm, Lake Constance, Lindau, 1973[17]
Freshwater, river	12 ug/l	Range = 0.3 - 130 ppm, streams to Lake Michigan[18]
Freshwater, river	29 ug/l	Range = 4 - 64 ppm, Danube, Ulm/Regensburg, 1973[17]
Freshwater, river	34 ug/l	Range = <4 - 170 ppm, Kansas, 12 rivers, 1973[19]
Freshwater, river and lake	10 ug/l	Range = 0.8 - 28 ppm, Wales[20]
Fucus spp.	64 mg/kg dry weight	Range = 33 - 190 ppm, Irish Sea, North Sea, 1971[21]
Fucus vesiculosus	103 mg/kg dry weight	Range = 7.1 - 8.9 ppm, Menai Straits, Anglesey, 1975[22]
Fucus vesiculosus	63.7 mg/kg dry weight	Range = 38.7 - 89.2 ppm, Bristol Channel, GB, 1973[23]
Fucus vesiculosus	71 mg/kg dry weight	Range = 52 - 97 ppm, Afon Goch, Anglesey, GB, 1975[22]
Gadus morrhua	4.5 mg/kg dry weight	Fish, rg = +- 1 ppm, Trondsheimfj, Norway, muscle, 1976[1]
Granitic rock	540 mg/kg dry weight	Average value[9]
Igenous rock	950 mg/kg dry weight	None[24]
Jasus lalandii	0.27 mg/kg wet weight	Marine, Crusta., rg=+-0.06, Melkbosch C, W Coast, SA, muscle, 1975[25]
Jasus lalandii	0.33 mg/kg wet weight	Marine, Crustaceans, rg =+-0.9, Saldanha Bay, SA, muscle, 1975[10]
Mactra clabrata	0.34 mg/kg wet weight	Gastropod, rg = +- 0.13 ppm, Saldanha B, Langebaan B, RSA, 1975[10]
Marine phytoplankton	1.03 mg/kg wet weight	Range = 0.95 - 1.1 ppm, Cape W Coast, S Africa, 1973[25]
Marine zooplankton	3.97 mg/kg wet weight	Range = 0.14 - 7.8 ppm, Cape W Coast, S Africa, 1974[25]
Mercenaria mercenaria	5.8 mg/kg wet weight	Gastropod, Atlantic coast, USA, 1967[14]
Merluccius capensis	0.22 mg/kg wet weight	Fish, marine, rg =+- 0.11 ppm, Cape W Coast, S Africa, muscle, 1974[25]
Mya arenaria	6.7 mg/kg wet weight	Gastropod, Atlantic coast, USA, 1967[14]
Mytilus edulis	21.1 mg/kg dry weight	Gastropod, rg = 3.3-69.8 ppm, Medit., NW Coast, France + Italy, 1975[26]
Mytilus edulis	27 mg/kg dry weight	Gastropod, rg = 12-38 ppm, New Zealand, 1964[27]
Mytilus edulis	3.5 mg/kg dry weight	Gastropod, Irish Sea, GB, 1970[2]
Nucella lapillus	12 mg/kg dry weight	Gastropod, Irish Sea, GB, 1970, marine[2]
Ostrea edulis	6 mg/kg dry weight	Gastropod, Knysna Estuary, R.S.A., 1975[12]
Ostrea sinuata	8 mg/kg dry weight	Gastropod, rg = 1 - 11 ppm, Tasman Bay, NZ, 1964[27]
Patella vulgata	13 mg/kg dry weight	Gastropod, marine, Irish Sea, 1971[2]
Patella vulgata	42 mg/kg dry weight	Gastropod, rg = 23 - 88 ppm, Irish Sea, 1971[21]
Pecten maximus	140 mg/kg dry weight	Gastropod, Irish Sea, GB, 1970[2]
Pecten novaezelandia	111 mg/kg dry weight	Gastropod, rg = 12-306 ppm, Tasman Bay, New Zealand, 1964[27]
Phosphate rocks	10 mg/kg dry weight	Rock from Morocco[28]
Phosphate rocks	266 mg/kg dry weight	Rock from Kola, USSR[28]
Phosphate rocks	771 mg/kg dry weight	Rock from Florida, USA[28]
Polyphosphate	10 mg/kg dry weight	Rock from Florida, USA[28]
Pomatomus saltatrix	0.22 mg/kg wet weight	Fish, marine, Cape Hatteras, NC, USA, 1972, muscle[29]

Rain	11.5 ug/l	Sweden, 3 stations, rg = 6 - 17[3]
Rain	8.1 ug/l	England, yearly mean, 1 station[3]
River water	0.012 ppm	Range = 0.2E-4-0.13 ppm[30]
Sandstone sediment	390 mg/kg dry weight	Average value[9]
Scomber japonicus	0.28 mg/kg wet weight	Fish, marine, rg =+-0.14 ppm, Cape W Coast, S Africa, muscle, 1974[25]
Seawater	0.002 ppm	Mn(2+), residence time = 1.4E3 y[31]
Seawater	2 ug/l	None[32]
Sedimentary rock	670 mg/kg dry weight	None[33]
Seriola grandis	0.34 mg/kg wet weight	Fish, marine, rg =0.12-1.1 ppm, North NZ, muscle, 1973[27]
Seriola pappei	0.29 mg/kg wet weight	Fish, marine, rg =+-0.08 ppm, Cape W Coast, S Africa, muscle, 1974[25]
Shale	850 mg/kg dry weight	Average value[9]
Shale sediment	850 mg/kg dry weight	Average value[9]
Soil	850 mg/kg dry soil	Oven-dried, rg = 100 - 4000[30]
Superphosphate fertilizer	225 mg/kg dry weight	Rock from Florida, USA[28]
Trigla kumu	1.15 mg/kg wet weight	Fish, rg = 2.1 - 3.7 ppm, N Island, NZ, muscle, 1973[27]
Ulva sp.	4.1 mg/kg wet weight	Range = +- 2.8 ppm, Cape W Coast, S Africa, marine, 1974[25]
Water	80 ug/l	Ground- and surface water, USA, 1961, rg = 0 - 80[3]

Source: 1.Lande, E. 1977. Environ. Pollut. 12, 187-198.
2.Segar, D.A. and R.E. Pellenbarg. 1971. J. Mar. Biol. Assoc. UK. 51, 131-136.
3.Laveskog, A., A. Lindskog and V. Stenberg. 1976. Om Metaller, Statens Naturvaardsverk, Stockholm.
4.EPA 1975. Scientific and Technical Report on Manganese. EPA-600/6-7-002. U.S. EPA, Wash DC.
5.Greger, J.L., P. Baligar, R.P. Abernathy, Bennet and Peterson. 1978. Am. J. Clin. Nutr. 31, pp 117-11.
6.Golterman, H.L. 1975. Physiological Limnol., Elsevier Amsterdam.
7.Deevey, E.S. 1970. Scientific American. 223(3), p 150.
8.Martin, J.B.M. 1974. Mar. Biol. 28, 245-251.
9.Turekian, K.K. and K.H. Wedepohl. 1961. Amer. Geol. Soc. Bull. 72, pp 175-192.
10.Fourie, H.O. 1976. S. Afr. J. Sci. 72, 110-113.
11.Uthe, J.H. and E.G. Bligh. 1971. J. Fish. Res. bd Canada 28, 786-788.
12.Watling, H.R. and R.J. Watling. 1976. Mar. Pollut. Bull. 7, 45-48.
13.Romeril, M.G. 1971. Mar. Biol. 9, 347-354.
14.Pringle, B.H., D.E. Hissong, E.L. Katz and S.T. Mulawka 1968. J. Sanit. Eng. Div. Am. Soc. Civ. Eng. 94, 455-475.
15.Mason, B. 1966. Principles of Geochemis try. 3rd ed. John Wiley and Sons. New York.
16.Foerstner, V. 1977. Arch. Hydrobiol., 80(2), pp 172-191.
17.Quentin, K.E., and H.A. Winkler. 1974. Zentralbl. Bakt. Abt. Orig. B 158, 514-523.
18.Robbins, J.A., E. Landstroem and M. Wahlgren. 1972. Proc 15th Conf. Great Lakes Res. Int. Assoc. Gret Lakes, 270-290.
19.Angino, E.E., L.M. Magnuson, T.C. Waugh, Galle and Bredfeldt 1974. Water Resour. Res. 10, 1187-1191.
20.Abdullah, M.I. and L.G. Royle. 1974. Cd in some British coastal and fresh water Environments. Proc.nt. Symp. Problems of the Contamination of Man and his Environment by Hg and Cd. Luxembourg.
21.Preston, A., Jefferies, Dutton, Harvey and A.K. Steele. 1972. Environ. Pollut. 3, 69-82.
22.Foster, P. 1976. Environ. Pollut. 10, 45-53.
23.Fuge, R. and K.H. James. 1974. Mar. Pollut. Bull. 5, 9-12.
24.Taylor, S.R. 1964. Geochim. Cosmochim. Acta 28, pp 1273-1285.
25.Van As, D. H.O. Fourie and C.M. Vleggaar. 1975. S. Afr. J. Sci. 71, 151-154.
26.Fowler, S.W. and B. Oregioni. 1976. Mar. Pollut. Bull. 7, 26-29.
27.Brooks, R.R., and D. Rumsey. 1974. NZJ Mar. Freshwater Res. 8, 155-166.
28.Langmyhr, F.J. R. Solberg and Y. Thomassen. 1977. Anal. Chim. Acta 92, pp 105-109.
29.Cross, F.A., L.H. Hardy, N.Y. Jones and R.T. Barber. 1973. J. Fish. Res. bd Canada 30, 1287-1291.
30.Bowen, H.J.M. 1966. Trace Elements in Biochemistry, Ac. Press.
31.Goldberg, E.D., 1963. The Sea, Interscience London, 2, pp 3.
32.Shigematsu, T., M. Matsui, O. Fujino and K. Kinoshita. 1975. Anal. Chim. Acta 76, pp 329-336.
33.Vinogradov, A.P. 1959. The Geochemistry of Rare and Dispersed Chemical Elements in the Soil. Transl. from Russian. Consultant Bureau Inc. NY. 2nd ed.

5-096 Background concentration Mo

Species	Value	Condition
Atmosphere	1.65 ng/m**3	Not urban, USA, rg = 0.1 - 3.2[1]
Atmosphere	20 ng/m**3	Urban, USA, rg = 10 - 30[1]
Bean sprouts	0.61 mg/kg dry weight	Normal leaves[2]
Cabbage, leaf	0.09 mg/kg dry weight	Deficient leaves[2]
Cabbage, leaf	0.42 mg/kg dry weight	Normal leaves[2]
Carbonate sediments	0.4 mg/kg dry weight	Average value[3]
Cauliflower, leaves	0.05 mg/kg dry weight	Deficient leaves, high Mn in leaves, rg = 0.02 - 0.08[2]
Cauliflower, leaves	0.45 mg/kg dry weight	Deficient leaves, Mn toxicity complicating factor, rg = 0.4 - 0.5[2]
Cauliflower, leaves	1.0 mg/kg dry weight	Normal leaves, Mn toxicity complicating factor[2]
Cauliflower, leaves	1.55 mg/kg dry weight	Normal leaves, high Mn in leaves, rg = 0.10 - 3.0[2]
Citrus, leaves	0.025 mg/kg dry weight	Deficient leaves, severely affected leaves, in water culture[2]
Citrus, leaves	0.055 mg/kg dry weight	Deficient leaves, symptoms pronounced below soil pH = 5.6, rg = 0.03[2]
Crustral rock	0.00015 percent by weight	Average value[4]
Deep-sea clay sediments	27 mg/kg dry weight	Average value[3]
Drinking water	15 ug/l	Schweiz, rg = 1 - 29[1]
Food	0.14 mg/kg dry weight	Chicken[1]
Food	0.21 mg/kg dry weight	Potato[1]
Food	0.44 mg/kg dry weight	Rice[1]
Food	3 mg/kg dry weight	Sheep, rg = 2 - 4[1]
Freshwater, river	0.1 ug/l	Less than value, Rhine, FRG, 1971[5]
Freshwater, river	7.0 ug/l	Streams to Lake Michigan, 1971[6]
Granitic rock	1.0 mg/kg dry weight	Average value[3]
Hops, leaves	0.02 mg/kg dry weight	Deficient leaves, white interveinal breakdown of leaves[2]
Hops, leaves	0.04 mg/kg dry weight	Normal leaves[2]
Igenous rock	1.5 mg/kg dry weight	None[7]
Lettuce, leaves	0.06 mg/kg dry weight	Deficient leaves, very large yield increase[2]
Lettuce, leaves	0.10 mg/kg dry weight	Normal leaves, very large yield increase[2]
River water	0.35E-4 ppm	Range < 0.007 ppm[8]
Sandstone sediment	0.2 mg/kg dry weight	Average value[3]
Seawater	0.01 ppm	MoO4(2-), residence time = 500E3 y[9]
Shale sediment	2.6 mg/kg dry weight	Average value[3]
Soil	2 mg/kg dry soil	Oven-dried, rg = 0.2 - 5[8]
Water	0.68 ug/l	Surface water, USA, mean[1]

Source: 1.Laveskog, A., A. Lindskog and V. Stenberg. 1976. Om Metaller, Statens Naturvaardsverk, Stockholm.
2.Williams, J.H. 1971. In: Tech. Bull. 21, pp 119-135. Trace Elements in Soils and Crops.
3.Turekian, K.K. and K.H. Wedepohl. 1961. Amer. Geol. Soc. Bull. 72, pp 175-192.
4.Mason, B. 1966. Principles of Geochemistry. 3rd ed. John Wiley and Sons. New York.
5.Reichert, J.K. 1973. Vom Wasser 39, 137-146.

6.Robbins, J.A., E. Landstroem and M. Wahlgren. 1972. Proc 15th Conf. Great Lakes Res. Int. Assoc. Great Lakes, 270-290.
7.Taylor, S.R. 1964. Geochim. Cosmochim. Acta 28, pp 1273-1285.
8.Bowen, H.J.M. 1966. Trace Elements in Biochemistry, Ac. Press.
9.Goldberg, E.D., 1963. The Sea, Interscience London, 2, pp 3.

5-097 Background concentration N

Species	Value	Condition
Ambient concentration	0.2 ug/m**3	Compound = NO3, atmospheric mass as N = 0.2E6 tons[1]
Ambient concentration	0.25 ppm	Compound = N2O, atmospheric mass as N = 1500E6 tons[1]
Ambient concentration	1 ppb	Compound = NO/NO2, ocean, atmospheric mass as N = 3E6 tons[1]
Ambient concentration	1.0 ug/m**3	Compound = NH4, atmospheric mass as N = 4.1E6 tons[1]
Ambient concentration	4 ppb as NO2	Compound = NO/NO2, land, atmospheric mass as N = 12E6 tons[1]
Ambient concentration	6 ppb	Compound = NH3, atmospheric mass as N = 30E6 tons[1]
Atmosphere	3 ug/m**3	293 K, 1 atm, residence time = short, NO2[2]
Atmosphere	7.5 ug/m**3	293 K, 1 atm, residence time = short, NH3, rg = 0 - 15[2]
Atmosphere	78.09 % by atoms	Sea-level dry atmospheric composition[3]
Atmosphere	850 ug/m**3	293 K, 1 atm, residence time = 4 y, N2O, rg = 500 - 1200[2]
Atmosphere	8E2 g/cm**2	Total in atmosphere (1975) = 4E21 g,N2[4]
Atmosphere	9.73E10 ug/m**3	293 K, 1 atm, N2[2]
Biosphere	71.6 moles/hectare	Average composition of living matter[5]
Lower atmosphere	0.00005 % by volume	N2o[6]
Lower atmosphere	78.084 % by volume	N2[6]
River water	0.23 ppm	NO3(-), range = 0.01-0.8 ppm[7]
River, Africa	0.8 mg/l	NO3(-)[8]
River, Asia	0.7 mg/l	NO3(-)[8]
River, Australia	0.05 mg/l	NO3(-)[8]
River, Europe	3.7 mg/l	NO3(-)[8]
River, North America	1.0 mg/l	NO3(-)[8]
River, South America	0.7 mg/l	NO3(-)[8]
River, World	1.0 mg/l	NO3(-)[8]
Seawater	0.5 ppm	Organic N, NO3(-), NH4(+), residence time = 2.5E3 y[9]
Seawater	0.6 umol kg-1	Depth 10 m, nitrite, offshore waters of Peru[10]
Seawater	0.8 umol kg-1	Depth 13 m, nitrite, offshore waters of Peru[10]
Seawater	1.2 umol kg-1	Depth 0 m, nitrite, offshore waters of Peru[10]
Seawater	1.3 umol kg-1	Depth 5 m, nitrite, offshore waters of Peru[10]
Seawater	11.56 ml/l	Saturated seawater, salinity = 3.43 %, 283 K, N2[3]
Seawater	2.5 umol kg-1	Depth 19 m, nitrite, offshore waters of Peru[10]
Seawater	2.9 umol kg-1	Depth 38 m, nitrite, offshore waters of Peru[10]
Seawater	3.5 umol kg-1	Depth 49 m, nitrite, offshore waters of Peru[10]
Seawater	3.7 umol kg-1	Depth 74 m, nitrite, offshore waters of Peru[10]
Seawater	4.2 umol kg-1	Depth 79 m, nitrite, offshore waters of Peru[10]

Seawater	4.3 umol kg-1	Depth 28 m, nitrite, offshore waters of Peru[10]
Seawater	4.6 umol kg-1	Depth 68 m, nitrite, offshore waters of Peru[10]
Seawater	4.9 umol kg-1	Depth 63 m, nitrite, offshore waters of Peru[10]
Soil	1000 mg/kg dry soil	Oven-dried, rg = 200 - 2500[7]

Source: 1.Robinson, E. and R.C. Robbins, 1975. In: S.F. Singer (ed.), The Changing Global Environment, pp 111-23. D. Reidel Publ. Comp.
2.Junge, C.E. 1963. Air Chemistry and Radioactivity, Ac. Press.
3.Sisler, F.D. 1975. In: S.F. Singer (ed.), The Changing Global Environment, pp 57-69. D. Reidel Publ.Comp.
4.Johnson, F.S. 1975. In: S.F. Singer (ed.), The Changing Global Environment, pp 49-56. D. Reidel Pub. Comp.
5.Deevey, E.S. 1970. Scientific American. 223(3), p 150.
6.Gleuckauf, E. 1951. Compendium of Meteorology. Am. Meteorological Soc. Boston.
7.Bowen, H.J.M. 1966. Trace Elements in Biochemistry, Ac. Press.
8.Wetzel, R.G. 1975. Limnology. Saunders. Philadelphia.
9.Goldberg, E.D., 1963. The Sea, Interscience London, 2, pp 3.
10.Hong, H. and Kester, D.R. Limnol. Oceanogr., 31, 512, 1986.

5-098 Background concentration
N2

Species	Value	Condition
Gas in ice, glacier ice	75.75 %	Gas pres.=2.1 bar, dens=0.90 g/cm3, gas=26 cm3/kg, rg=73.2-78.3 %[1]
Gas in ice, iceberg ice	77.9 %	Gas pres.=1.3 bar, dens=0.88 g/cm3, gas=40 cm3/kg, rg=77.7-78.3 %[1]
Gas in ice, pond ice	52.2 %	Gas pres.=1.0 bar, dens=0.916 g/cm3, gas=1.0 cm3/kg, n = 1[1]
Gas in ice, sea ice	69 %	Gas pres.=0.9 bar, dens=91 g/cm3, gas=13 cm3/kg, rg=54.2-76.8 %[1]

Source: 1.Matsuo, S. and Y. Miyake. 1966. J. Geophys. Res. 71, 5235.

5-099 Background concentration
Na

Species	Value	Condition
Atmospheric particulates	3500 ng/m**3	Hawaii, n > = 56, rg = +- 2000[1]
Biosphere	1.65 moles/hectare	Average composition of living matter[2]
River water	6.3 ppm	Range = 3-25 ppm[3]
River, Africa	11.0 mg/l	None[4]
River, Asia	5.5 mg/l	None[4]
River, Australia	2.9 mg/l	None[4]
River, Europe	5.4 mg/l	None[4]
River, North America	9.0 mg/l	None[4]
River, South America	4.0 mg/l	None[4]
River, World	6.3 mg/l	None[4]
Seawater	1 % of total	As Na2SO4, chlorinity = 1.9 %, 298 K, pH = 8.15[5]
Seawater	10500 ppm	Na(+), residence time = 260000E3 y[6]
Seawater	99 % of total	As free ion, chlorinity = 1.9 %, 298 K, pH = 8.15[5]
Soil	6300 mg/kg dry soil	Oven-dried, rg = 750 - 7500[3]

Source: 1.Golterman, H.L. 1975. Physiological Limnol., Elsevier Amsterdam.
2.Deevey, E.S. 1970. Scientific American. 223(3), p 150.
3.Bowen, H.J.M. 1966. Trace Elements in Biochemistry, Ac. Press.
4.Wetzel, R.G. 1975. Limnology. Saunders. Philadelphia.
5.Garrels, R.M. and M.E. Thompson. 1962. Am. J. Sci. 260., p 57.
6.Goldberg, E.D., 1963. The Sea, Interscience London, 2, pp 3.

5-100 Background concentration Nb

Species	Value	Condition
Seawater	0.1E-4 ppm	Residence time = 0.3E3 y[1]

Source: 1.Goldberg, E.D., 1963. The Sea, Interscience London, 2, pp 3.

5-101 Background concentration Ne

Species	Value	Condition
Atmosphere	18 ug/m**3	293 K, 1 atm[1]
Atmosphere	2E-3 % by atoms	Sea-level dry atmospheric composition[2]
Lower atmosphere	0.00182 % by volume	None[3]
Seawater	0.14E-3 ppm	Ne[4]
Seawater	1.7E-4 ml/l	Saturated seawater, salinity = 3.43 %, 283 K[2]

Source: 1.Junge, C.E. 1963. Air Chemistry and Radioactivity, Ac. Press.
2.Sisler, F.D. 1975. In: S.F. Singer (ed.), The Changing Global Environment, pp 57-69. D. Reidel Publ. Comp.
3.Gleuckauf, E. 1951. Compendium of Meteorology. Am. Meteorological Soc. Boston.
4.Goldberg, E.D., 1963. The Sea, Interscience London, 2, pp 3.

5-102 Background concentration Ni

Species	Value	Condition
Amblema plicata	1.1 mg/kg wet weight	Gastropod, rg = 0.4-2.3 ppm, Illinois R., 1972, freshwater[1]
Anodonta sp.	0.35 mg/kg dry weight	Gastropod, L Aberffraw, 1970, freshwater[2]
Atmosphere	190 ng/m**3	Sweden, mean from 2 stations with rg = 80 - 300, max = 300-800 ng/[3]
Atmosphere	23 ng/m**3	England, 7 stations, rg = 1 - 45[3]
Atmosphere	59.5 ng/m**3	Urban, USA, rg = 1 - 118[3]
Atmosphere	6.3 ng/m**3	Not urban, USA, rg = 0.6 - 12[3]
Atmosphere, rural	6 ng/m**3	Non-urban area, USA[4]
Atmosphere, urban	25 ng/m**3	During cold seson, warm seson = 17, USA[4]
Atmospheric particulates	150 ng/m**3	Yearly average, New York[4]
Cancer pagurus	9.80 mg/kg wet weight	Marine, Crustacean, Northcumberland C, GB, 1976, whole anim.[5]
Carbonate sediments	20 mg/kg dry weight	Average value[6]
Carcinus maenas	11 mg/kg dry weight	Marine, Crustacean, muscle, Hardanger, Norway, 1976[7]
Carcinus maenas	6.20 mg/kg wet weight	Marine, Crustacean, muscle, Northumberland C, GB, 1975[5]
Carcinus maenas	6.50 mg/kg wet weight	Marine, Crustacean, whole animal, Northcumberland C, GB, 1975[5]
Cardium edule	7.9 mg/kg dry weight	Gastropod, Irish Sea, GB, 1970[2]
Carpiodes cyprinus	0.18 mg/kg wet weight	Fish, rg = 0.15-0.45 ppm, Illinois R, muscle, 1973[8]
Choromytilus meridionalis	0.26 mg/kg wet weight	Gastropod, rg = +- 0.13 ppm, Saldanha B, LangeBaan B., RSA, 1975[9]
Cladophora glomerata	17.8 mg/kg dry weight	Algae, Rg = 11.9 - 23.8 ppm, Leine R., FRG, 1976[10]
Clupea sprattus	7.2 mg/kg wet weight	Fish, marine, rg =+-4.8, Northumberl, power station, muscle, 1975[5]
Coregonus clupeaformis	0.2 mg/kg wet weight	Fish, less, L Erie, USA, 1970, animal decapiated and eviscerated[11]

Coregonus clupeaformis	0.2 mg/kg wet weight	Fish, less, Moose L, USA, 1970, animal decapiated and eviscerated[11]
Crangon vulgaris	8.8 mg/kg wet weight	Marine, Crustacean, Northumberland Coast, GB, 1975, muscle[5]
Crassostrea gigas	0.25 mg/kg wet weight	Gastropod, rg = +- 0.65 ppm, Saldanha B, LangeBaan B., RSA, 1975[9]
Crassostrea gigas	1.6 mg/kg dry weight	Gastropod, Knysna Estuary, R.S.A., 1975[12]
Crassostrea gigas	5.15 mg/kg dry weight	Gastropod, rg = 3.8-6.5 ppm, Hinkley Pt. Lower Station, Severn Est., 73[13]
Crassostrea margaritacea	0.20 mg/kg wet weight	Gastropod, rg = +- 0.05 ppm, Saldanha B, LangeBaan B., RSA, 1975[9]
Crassostrea margaritacea	1.6 mg/kg dry weight	Gastropod, Knysna Estuary, R.S.A., 1975[12]
Crassostrea virginica	0.19 mg/kg wet weight	Gastropod, Atlantic Coast, USA, 1967[14]
Crustral rock	0.0075 percent by weight	Average value[15]
Cyprinus carpio	0.19 mg/kg wet weight	Fish, rg=0.04-0.28 ppm, Illinois R, muscle, 1973[8]
Deep-sea clay sediments	225 mg/kg dry weight	Average value[6]
Dorosoma cepedianum	0.28 mg/kg wet weight	Fish, rg = 0.06-0.52 ppm, Illinois R, USA, muscle, 1972[8]
Drinking water	1.1 ug/l	USA, rg = 0.8 - 1.5[3]
Drinking water	5.3 ug/l	Germany, rg = 0.8 - 34[3]
Esox lucius	0.15 mg/kg wet weight	Fish, rg = 0.08-0.19 ppm, Illinois R, USA, muscle, 1972[8]
Esox lucius	0.2 mg/kg wet weight	Fish, less, L Erie, USA, 1970, animal decapiated and eviscerated[11]
Esox lucius	0.2 mg/kg wet weight	Fish, less, Moose L, USA, 1970, animal decapiated and eviscerated[11]
Food	1.8 mg/kg dry weight	Vegetables, rg = 0.3 - 3.3[3]
Forest foliage	37 mg/kg ash weight	Uncontaminated value for USA, rg = <2-1300 ppm[16]
Freshwater sediment	103 mg/kg dry weight	Range of 90 % = 30 - 250 ppm, 74 lakes, pelitic[17]
Freshwater, lake	5.0 ug/l	Range = 3 - 7 ppm, Lake Constance, Lindau, 1973[18]
Freshwater, river	10 ug/l	Range = 2 - 17 ppm, Danube, Ulm/Regensburg, 1973[18]
Freshwater, river	27 ug/l	Range = <10 - 89 ppm, Kansas, 92 rivers, 1972[19]
Freshwater, river	4.6 ug/l	Range = 3.2 - 6.9 ppm, St. Lawrence River, 1975[20]
Freshwater, river and lake	1.5 ug/l	Range = 0.5 - 3.5 ppm, Wales[18]
Fucus spp.	9.9 mg/kg dry weight	Range = 1.8 - 18.0 ppm, Irish Sea, North Sea, 1971[21]
Fucus vesiculosus	7.3 mg/kg dry weight	Range = 7.0 - 7.6 ppm, Afon Goch, Anglesey, GB, 1975[22]
Fusconaria flava	2.1 mg/kg wet weight	Gastropod, rg = 0.7-3 ppm, Illinois R., 1972, freshwater[1]
Granitic rock	15 mg/kg dry weight	Average value[6]
Ictiobus cyprinellus	0.10 mg/kg wet weight	Fish, rg = 0.02-0.18 ppm, Illinois R, USA, muscle, 1972[8]
Jasus lalandii	0.21 mg/kg wet weight	Marine, Crustaceans, rg =+-0.5, Saldanha Bay, SA, muscle, 1975[9]
Mercenaria mercenaria	0.24 mg/kg wet weight	Gastropod, Atlantic coast, USA, 1967[14]
Micropterus dolomieu	0.13 mg/kg wet weight	Fish, rg = 0.08-0.19 ppm, Illinois R, USA, muscle, 1972[8]
Micropterus salmoides	0.11 mg/kg wet weight	Fish, rg = 0.05-0.23 ppm, Illinois R, USA, muscle, 1972[8]
Mya arenaria	0.27 mg/kg wet weight	Gastropod, Atlantic coast, USA, 1967[14]
Mytilus edulis	12 mg/kg dry weight	Gastropod, rg = 5 - 19 ppm, Berkeley Marine Calif, 1975[23]
Mytilus edulis	15 mg/kg dry weight	Gastropod, rg = 6 - 43 ppm, Trondheimsfjord, Norway, 1976[7]

Mytilus edulis	3.7 mg/kg dry weight	Gastropod, Irish Sea, GB, 1970[2]
Mytilus edulis	4.3 mg/kg dry weight	Gastropod, rg = 0.9-14.1 ppm, Medit., NW Coast, France + Italy, 1975[24]
Mytilus edulis	7 mg/kg dry weight	Gastropod, rg = 1 - 17 ppm, New Zealand, 1964[25]
Nucella lapillus	2.4 mg/kg dry weight	Gastropod, Irish Sea, GB, 1970, marine[2]
Ostrea edulis	1.7 mg/kg dry weight	Gastropod, Knysna Estuary, R.S.A., 1975[12]
Ostrea sinuata	2 mg/kg dry weight	Gastropod, rg = 1 - 3 ppm, Tasman Bay, NZ, 1964[25]
Patella vulgata	2.5 mg/kg dry weight	Gastropod, marine, Irish Sea, 1971[2]
Patella vulgata	7 mg/kg dry weight	Gastropod, rg = 4-11 ppm, Mortons Harbor, Newfoundl., 1974[26]
Patella vulgata	7.0 mg/kg dry weight	Gastropod, rg = 4.5-9.9 ppm, Irish Sea, 1971[21]
Pecten maximus	49 mg/kg dry weight	Gastropod, Irish Sea, GB, 1970[2]
Pecten novaezelandia	6 mg/kg dry weight	Gastropod, rg = 2 - 17 ppm, Tasman Bay, New Zealand, 1964[25]
Periphyton	155 mg/kg dry weight	Rg = 46 - 265 ppm, Vermillion R, Ill, 1973[27]
Phosphate rocks	2 mg/kg dry weight	Rock from Kola, USSR[28]
Phosphate rocks	39 mg/kg dry weight	Rock from Florida, USA[28]
Phosphate rocks	69 mg/kg dry weight	Rock from Morocco[28]
Pleuronectes platessa	2.8 mg/kg wet weight	Fish, marine, Northumberl, power station, muscle, 1975[5]
Polyphosphate	18 mg/kg dry weight	Rock from Florida, USA[28]
Potamogeton pectinatus	12.5 mg/kg dry weight	Higher plant, Rg = 9.2 - 15.8 ppm, Leine R, FRG, 1976[10]
Pseudogobio brevirostris	0.3 mg/kg wet weight	Fish, Ta-Tu R, Taiwan, 1973, whole animal[29]
Quadrula quadrula	0.9 mg/kg wet weight	Gastropod, rg = 0.4-1.6 ppm, Illinois R., 1972, freshwater[1]
Rain	3 ug/l	Sweden, 3 stations[3]
Rain	6 ug/l	England, 1 station, yearly mean[3]
Rhinogobio similis	0.3 mg/kg wet weight	Fish, Ta-Tu R, Taiwan, 1973, whole animal[29]
River water	0.01 ppm	Range = 0.2E-3-0.02 ppm[30]
Sandstone sediment	2 mg/kg dry weight	Average value[6]
Seawater	0.3 ug/l	Range = 0.1 - 0.5 ug/l[4]
Seawater	0.54E-2 ppm	Ni(2+), residence time = 18E3 y[31]
Seriola grandis	0.03 mg/kg wet weight	Fish, marine, rg =0.02-0.08 ppm, North NZ, muscle, 1973[25]
Shale	68 mg/kg dry weight	Average value[6]
Shale sediment	68 mg/kg dry weight	Average value[6]
Soil	40 mg/kg dry soil	Oven-dried, rg = 10 - 1000[30]
Soil	500 mg/kg dry weight	Rg = 3 - 1000 ppm, wide range dependend of mineral content, world value[4]
Stormwater	57 g/m3	Lodi, New Jersey, USA, rain = 3 ug/l, Saddle R = 16 ug/l[32]
Superphosphate fertilizer	34 mg/kg dry weight	Rock from Florida, USA[28]
Trigla kumu	0.02 mg/kg wet weight	Fish, rg = 0.01 - 0.03 ppm, N Island, NZ, muscle, 1973[25]
Wastewater, Bakery	430 ug/l	Average value, New York industries[33]
Wastewater, Fat rendering	280 ug/l	Average value, New York industries[33]
Wastewater, Fish processing	140 ug/l	Average value, New York industries[33]
Wastewater, Meat processing	70 ug/l	Average value, New York industries[33]

Wastewater, Textile dyeing	250 ug/l	Average value, New York industries[33]
Wastewater, brewery	40 ug/l	Average value, New York industries[33]
Wastewater, car wash	190 ug/l	Average value, New York industries[33]
Wastewater, fur dressing and dyeing	740 ug/l	Average value, New York industries[33]
Wastewater, ice cream	1100 ug/l	Average value, New York industries[33]
Wastewater, laundry	100 ug/l	Average value, New York industries[33]
Wastewater, miscellaneous chemical indistries	100 ug/l	Average value, New York industries[33]
Wastewater, miscellaneous foods	110 ug/l	Average value, New York industries[33]
Wastewater, soft drinks and flavorings	2200 ug/l	Average value, New York industries[33]
Water	0.2 ug/l	Baltic Sea, Gotlandsdybet : 230 m, 1970-71[3]
Water	30.5 ug/l	Rivers, USA, rg = 5 - 56[3]
Water	7.75 ug/l	Baltic Sea, Fladen : 50 m, 1970-71[3]

Source: 1.Mathis, B.J. and T.F. Cummings. 1973. J. Water Pollut Cont. Fed. 45, 1573-1583.
2.Segar, D.A. and R.E. Pellenbarg. 1971. J. Mar. Biol. Assoc. UK. 51, 131-136.
3.Laveskog, A., A. Lindskog and V. Stenberg. 1976. Om Metaller, Statens Naturvaardsverk, Stockholm.
4.NAS 1975. Nickel. National Research Council, National Academy of Science, Wash. DC.
5.Wright, D.A. 1976. Mar. Pollut. Bull. 7, 36-38.
6.Turekian, K.K. and K.H. Wedepohl. 1961. Amer. Geol. Soc. Bull. 72, pp 175-192.
7.Lande, E. 1977. Environ. Pollut. 12, 187-198.
8.Mathis, B.J. and N.R. Kevern. 1973. Distribution of Hg, Cd, Pb and Th in a eutrophic lake. Int. Water Res. Michigan State Univ., Technol. Dep. 34, 23 pp.
9.Fourie, H.O. 1976. S. Afr. J. Sci. 72, 110-113.
10.Abo-Rady, M.D.K. 1977. Die Belastung der oberen Leine mit Schwermetallen durch kommunale und industrelle Abwasser. Diss. Univ. Gottingen. 120 p.
11.Uthe, J.H. and E.G. Bligh. 1971. J. Fish. Res. bd Canada 28, 786-788.
12.Watling, H.R. and R.J. Watling. 1976. Mar. Pollut. Bull. 7, 45-48.
13.Romeril, M.G. 1971. Mar. Biol. 9, 347-354.
14.Pringle, B.H., D.E. Hissong, E.L. Katz and S.T. Mulawka 1968. J. Sanit. Eng. Div. Am. Soc. Civ. Eng. 94, 455-475.
15.Mason, B. 1966. Principles of Geochemis try. 3rd ed. John Wiley and Sons. New York.
16.Shacklette, H.T., J.A. Erdman, T.F. Harms and C.S.E. Pupp. 1978. In: F.W. Oeheme (ed.): Toxicity of Heavy Metals in the Environment. Dekkker. NY. 25-68.
17.Foerstner, V. 1977. Arch. Hydrobiol., 80(2), pp 172-191.
18.Quentin, K.E., and H.A. Winkler. 1974. Zentralbl. Bakt. Abt. Orig. B 158, 514-523.
19.Angino, E.E., L.M. Magnuson, T.C. Waugh, Galle and Bredfeldt 1974. Water Resour. Res. 10, 1187-1191.
20.Subramanian, V. and R.J. Gibbs. 1972. Abstr. Annu. Meet. Geol. Soc. Amer. Trace metal adsorption clay minerals.
21.Preston, A., Jefferies, Dutton, Harvey and A.K. Steele. 1972. Environ. Pollut. 3, 69-82.
22.Foster, P. 1976. Environ. Pollut. 10, 45-53.
23.Bohn, A. 1975. Mar. Pollut. Bull. 6, 87-89
24.Fowler, S.W. and B. Oregioni. 1976. Mar. Pollut. Bull. 7, 26-29.
25.Brooks, R.R., and D. Rumsey. 1974. NZJ Mar. Freshwater Res. 8, 155-166.
26.Penrose, W.R., R. Black and M.J. Hayward. 1975. J. Fish. Res. Bd Canada 32, 1275-1281.
27.Leland, H.V. and J.M. McNerney. 1974. Lead Transport in a river ecosystem. Proc. Int. Conf. Transp. Persist. Chem. Aquatic Ecosystem. Ottawa III 17-23.
28.Langmyhr, F.J. R. Solberg and Y. Thomassen. 1977. Anal. Chim. Acta 92, pp 105-109.
29.Chung, I.H. and S.S. Jeng. 1974. Bull. Inst. Acad. Sin. 13, 69-73.
30.Bowen, H.J.M. 1966. Trace Elements in Biochemistry, Ac. Press.
31.Goldberg, E.D., 1963. The Sea, Interscience London, 2, pp 3.
32.Nriagu, J.O. 1980. Zinc in the Environment. Part 1: Ecological cycling. Wiley and Sons. NY.
33.Klein, L.A., M. Land, N. Nash and S.L. Kirschner. 1974. J. Wat. Poll. Cont. Fed. 46, pp 2653-2662.

5-103 Background concentration
Niacin

Species	Value	Condition
Freshwater	0.52 mg/m**3	Lindsey Pond, Conn, nicotinic acidamide, rg = 0.15 - 0.89[1]
Freshwater	1.5 mg/m**3	Small swiss ponds, nicotinic acidamide, rg = < 0.001 - 3.0[2]
Freshwater	1.8 mg/m**3	Kasumigaura, Japan, nicotinic acidamide, rg = 0.30 - 3.3[3]
Rain	0.42 mg/m**3	Mean of precipatation, MO, Nov-March, nicotinic acidamide[4]

Rain	2.0 mg/m**3	Mean of precipitation, MO, April-Nov., nicotinic acidamide[4]

Source: 1.Benoit, R.J. 1957. Limnol. Oceanogr., 2, pp 233-240.
2.Clemenson, H. 1963. Schweiz. Z. Hydrol., 25, pp 157-165.
3.Kashiwada, K., A. Kanazawa and S. Tachibanazono. 1963. Mem. Fac. Fish., Kagoshima Univ. 12, pp 153-17.
4.Parker, B.C. and M.A. Wachtel. 1971. in: Cairns, J. (ed.); The Structure and Function of Freshwater Microbial Communities. Res. Div. Monogr. Virginia Polytech. Inst. 3, pp 195-207.

5-104 Background concentration
Nonane + Decane

Species	Value	Condition
Atmospheric conc of aliphatic hydrocarbon and alkylbenzene	5 ppb	Los Angeles, 1967, STP[1]

Source: 1.Altshuller, A.P., W.A. Lonneman and S.L. Kopczynski. 1973. J. Air Pollut. Contr. Assoc. 23, 597-599.

5-105 Background concentration
O

Species	Value	Condition
Biosphere	6540 moles/hectare	Average composition of living matter[1]

Source: 1.Deevey, E.S. 1970. Scientific American. 223(3), p 150.

5-106 Background concentration
O_2

Species	Value	Condition
Atmosphere	2.99E10 ug/m**3	293 K, 1 atm[1]
Atmosphere	20.95 % by atoms	Sea-level dry atmospheric composition[2]
Atmosphere	2E2 g/cm**2	Total in atmosphere (1975) = E21 g[3]
Atmosphere	5E3 g/cm**2	Total supply = 2.7E22 g[4]
Gas in ice, glacier ice	23.0 %	Gas pres.=2.1 bar, dens=0.90 g/cm3, gas=26 cm3/kg, rg=73.2-78.3 %[5]
Gas in ice, iceberg ice	21.0 %	Gas pres.=1.3 bar, dens=0.88 g/cm3, gas=40 cm3/kg, rg=77.7-78.3 %[5]
Gas in ice, pond ice	22.0 %	Gas pres.=1.0 bar, dens=0.916 g/cm3, gas=1.0 cm3/kg, n = 1[5]
Gas in ice, sea ice	26.0	Gas pres.=0.9 bar, dens=91 g/cm3, gas=13 cm3/kg, rg=54.2-76.8 %[5]
Lower atmosphere	20.946 % by volume	None[6]
Seawater	6.44 ml/l	Saturated seawater, salinity = 3.43 %, 283 K[2]
Seawater	857000 ppm	H2O, O2, SO4(2-)[7]
Soil	490000 mg/kg dry soil	Oven-dried[8]

Source: 1.Junge, C.E. 1963. Air Chemistry and Radioactivity, Ac. Press.
2.Sisler, F.D. 1975. In: S.F. Singer (ed.), The Changing Global Environment, pp 57-69. D. Reidel Publ. Comp.
3.Shannon, L.R. 1977. Bull. Environ. Contam. Toxicol. 17(6), pp 637-644.
4.Johnson, F.S. 1975. In: S.F. Singer (ed.), The Changing Global Environment, pp 49-56. D. Reidel Pub. Comp.
5.Matsuo, S. and Y. Miyake. 1966. J. Geophys. Res. 71, 5235.
6.Gleuckauf, E. 1951. Compendium of Meteorology. Am. Meteorological Soc. Boston.
7.Goldberg, E.D., 1963. The Sea, Interscience London, 2, pp 3.
8.Bowen, H.J.M. 1966. Trace Elements in Biochemistry, Ac. Press.

5-107 Background concentration
O3

Species	Value	Condition
Atmosphere	1.0E-6 % by atoms	Sea-level dry atmospheric composition[1]
Atmosphere	50 ug/m**3	293 K, 1 atm, residence time = 2 y, rg = 0 - 100[2]

Source: 1.Sisler, F.D. 1975. In: S.F. Singer (ed.), The Changing Global Environment, pp 57-69. D. Reidel Publ. Comp.
2.Junge, C.E. 1963. Air Chemistry and Radioactivity, Ac. Press.

5-108 Background concentration
P

Species	Value	Condition
Biosphere	3.35 moles/hectare	Average composition of living matter[1]
River water	0.005 ppm	Range = 0.001-0.3E-12 ppm[2]
Seawater	0.07 ppm	HPO4(2-)[3]
Seawater	2E-9 ppm	None[3]
Sediment	0.30 10E-4 mg P mg-1 sed	extractable P, stream draining wooded area, org.cont. 2.15 %[4]
Sediment	0.35 ug l-1	equlibrium conc., stream draining agricultural area, org.cont. 3.54 %[4]
Sediment	0.39 10E-4 mg P mg-1 sed	extractable P, stream draining wooded area, org.cont. 1.80 %[4]
Sediment	0.55 ug l-1	equlibrium conc., stream draining wooded area, org.cont. 1.80 %[4]
Sediment	0.99 10E-4 mg P mg-1 sed	extractable P, stream draining agricultural area, org.cont. 2.97 %[4]
Sediment	1.26 ug l-1	equlibrium conc., stream draining agricultural area, org.cont. 2.97 %[4]
Sediment	1.46 10E-4 mg P mg-1 sed	extractable P, stream draining agricultural area, org.cont. 3.54 %[4]
Sediment	2.25 ug l-1	equlibrium conc., stream draining wooded area, org.cont. 2.15 %[4]
Soil	650 mg/kg dry soil	Oven-dried[2]

Source: 1.Deevey, E.S. 1970. Scientific American. 223(3), p 150.
2.Bowen, H.J.M. 1966. Trace Elements in Biochemistry, Ac. Press.
3.Goldberg, E.D., 1963. The Sea, Interscience London, 2, pp 3.
4.Klotz, R.L., Limnol.Oceanogr., 30(3), 543, 1985.

5-109 Background concentration
Pantothenic acid

Species	Value	Condition
Freshwater	0.017 mg/m**3	Small swiss ponds, rg = < 0.01 - 0.034[1]
Freshwater	0.135 mg/m**3	Kasumigaura, Japan[2]

Source: 1.Clemenson, H. 1963. Schweiz. Z. Hydrol., 25, pp 157-165.
2.Kashiwada, K., A. Kanazawa and S. Tachibanazono. 1963. Mem. Fac. Fish., Kagoshima Univ. 12, pp 153-17.

5-110 Background concentration
Pb

Species	Value	Condition
Amblema plicata	2.7 mg/kg wet weight	Gastropod, rg = 1.1-7.6 ppm, Illinois R., 1972, freshwater[1]
Amphiboles, rock forming mineral	15 mg/kg dry weight	Rg = 1 - 70 ppm, n = 85[2]
Andesite, igneous rock	8.3 mg/kg dry weight	Rg = 0.6 - 36 ppm, n =[2]
Andesite, igneous rock	8.3 mg/kg dry weight	Rg = 0.6 - 36 ppm, n = 166[2]
Anodonta sp.	1.2 mg/kg dry weight	Gastropod, L Aberffraw, 1970, freshwater[3]
Apatite, rock forming mineral	102 mg/kg dry weight	Rg = 5 - 2000 ppm, n = 58[2]
Arsenopyrite, rock mineral	118 mg/kg dry weight	Rg = 78 - 160 ppm, n = 3[2]
Asellus aquaticus	6.80 mg/kg dry weight	Freshwater, Crustaceans, rg = 2.69-16.5 ppm, Elsenz R, 1976[4]
Atlantic deep-sea carbonate ooze sediment	12.6 mg/kg dry weight	N = 52[5]
Atlantic deep-sea clay sediment	48 mg/kg dry weight	N = 487[5]

Callitriche palustris	17.4 mg/kg dry weight	Higher plant, Rg = 13.5-21.3 ppm, Elsenz R, FRG, 1976[15]
Cardium edule	0.76 mg/kg dry weight	Gastropod, Irish Sea, GB, 1970[3]
Carpiodes cyprinus	0.64 mg/kg wet weight	Fish, rg = 0.09-1.3 ppm, Illinois R, muscle, 1973[17]
Choromytilus meridionalis	0.15 mg/kg wet weight	Gastropod, rg = +- 0.14 ppm, Saldanha B, LangeBaan B., RSA, 1975[18]
Cladophora glomerata	27.2 mg/kg dry weight	Algae, rg = 5.2 - 49.2 ppm, Leine R., FRG, 1976[20]
Cladophora sp.	181 mg/kg dry weight	Algae, Rg = 14.9 - 347 ppm, Vermillion R., Ill., 1973[22]
Coal	11.6 mg/kg dry weight	Belgium, n = 44, coal ash = 317 ppm Pb[5]
Coal	99 mg/kg dry weight	East Germany, n = 378. coal ash = 1030 ppm[5]
Coregonus clupeaformis	0.5 mg/kg wet weight	Fish, less, Moose L, USA, 1970, animal decapiated and eviscerated[24]
Crangon crangon	0.2 mg/kg dry weight	Marine, Crustacean, less than, S Antonio Bay, Texas, whole anim., 1975[25]
Crassostrea commercialis	0.8 mg/kg wet weight	Gastropod, rg = 0.1-1.3 ppm, NSW estuaries, Aust., 1974[26]
Crassostrea gigas	0.82 mg/kg dry weight	Gastropod, rg = +- 1.4 ppm, Tamar R Est., Tasmania, 1973[27]
Crassostrea gigas	16 mg/kg dry weight	Gastropod, rg = 15-17 ppm, Hinkley Pt. Lower Station, Severn Est., 1973[28]
Crude oil	0.04 mg/kg dry weight	USA, n = 96, crude oil ash = 735 ppm Pb[5]
Cyprinus carpio	0.56 mg/kg wet weight	Fish, rg=0.15-2.13 ppm, Illinois R, muscle, 1973[17]
Deep-sea clay sediments	80 mg/kg dry weight	Average value[16]
Drinking water	50 ug/l	Max value, rg = 0 - 50 ug/l[32]
Esox lucius	0.5 mg/kg wet weight	Fish, less, L Erie, USA, 1970, animal decapiated and eviscerated[24]
Fontinalis antipyr.	1320 mg/kg dry weight	Moss, Rg = 450 - 2180 ppm, Ruhr R, FRG, 1971[33]
Fontinalis antipyr.	137 mg/kg dry weight	Moss, Rg = 56.3-217.5 ppm, Elsenz R, FRG, 1976[15]
Food	1.5 mg/kg dry weight	Liver, rg = 0.2 - 2.8[7]
Food	2.95 mg/kg dry weight	Kidney, rg = 0.2 - 5.7[7]
Forest foliage	135 mg/kg ash weight	Uncontaminated value for USA, rg = <10-3000 ppm[34]
Freshwater	8.5 ug/l	Rg = 7 - 10 ug/l, Midi-Pyrenees, dissolved Pb[35]
Freshwater river sediment	13 mg/kg dry weight	24 tributaries to Lake Superior[5]
Freshwater river sediment	850 mg/kg dry weight	R Rhine, estuarine, 1972[5]
Freshwater zooplankton, effect of fertilizer	34.8 mg/kg dry weight	Area with fertilizer, Scenedesmus obliquus, FRG, 1974[37]
Freshwater zooplankton, effect of fertilizer	6.03 mg/kg dry weight	Area without fertilizer, Scenedesmus obliquus, Thailand, 1974[37]
Freshwater, river	0.9 ug/l	Streams to Cayuga Lake[38]
Freshwater, river	9.0 ug/l	Range = 5 - 50 ppm, Danube, Austria[43]
Fucus sp.	9.0 mg/kg dry weight	Range = 5 - 13 ppm, Portugal and Spain, Atlantic, 1974[30]
Gadus morhua	0.28 mg/kg wet weight	3 references used, SD = +- 0.07, rg = 0.20 - 0.34[14]
Littorina littorea	0.6 mg/kg dry weight	Gastropod, Severn Estuary, Bristol Ch., 1971, rg = 0.1-3 ppm, marine[48]
Littorina littorea	7 mg/kg dry weight	Gastropod, rg = 1-19 ppm, Bristol Channel, 1971, marine[49]
Marine	36 mg Pb/m**3	Dissolved Pb, 300 km from Calif[23]
Marine sediment	152 mg/kg dry weight	North Sea, German Bay, 1974[5]
Marine sediment	9.2 mg/kg dry weight	Mediterranean Sea, rg = 8.4 - 10 ppm, 1976[5]

Mercenaria mercenaria	0.52 mg/kg wet weight	Gastropod, Atlantic coast, USA, 1967[29]
Micropterus salmoides	0.59 mg/kg wet weight	Fish, rg = 0.36 - 1.13 ppm, Illinois R, USA, muscle, 1972[17]
Mya arenaria	0.7 mg/kg wet weight	Gastropod, Atlantic coast, USA, 1967[29]
Myriophyllum spicatum	30 mg/kg dry weight	Higher plant, Ruhr R, FRG, 1971[33]
Mytilus edulis	199 mg/kg dry weight	Gastropod, rg = 3 - 352 ppm, Derwent Estuary, Tasmania, 1976[54]
Mytilus edulis	3.3 mg/kg dry weight	Gastropod, rg = 3.1 - 3.5 ppm, North Sea, Weser Estuary, 1973[56]
Nephrops norvegicus	3.5 mg/kg dry weight	Marine, Crustaceans, Faro, Portugal, muscle, 1974[50]
Nuphar luteum	20 mg/kg dry weight	Higher plant, Ruhr R, FRG, 1971[33]
Oyster	24.92 mg/kg wet weight	12 references used, SD = +- 38.97, rg = 0.20 - 100[14]
Peats	38 mg/kg dry weight	N Sweden, n = 1253[5]
Percidae	0.55 mg/kg wet weight	2 references used, SD = +- 0.07, rg = 0.50 - 0.60[14]
Physa sp.	42 mg/kg dry weight	Gastropod, rg = +- 25 ppm, Illinois R., 1972, freshwater[1]
Platichthyes flesus	21.5 mg/kg dry weight	Fish, marine, Oldbury, Severn Estu., power station, rg = +- 7.5 ppm[60]
Potamogeton crispus	11.1 mg/kg dry weight	Higher plant, Rg = 7.4 -14.8 ppm, Elsenz R, FRG, 1976[15]
Rain	13.4 ug/l	14 July - 7 Nov 1971, New Hampshire, USA[62]
Rain	30 ug/l	Japan, Business area, in residential area = 8 ug/l[63]
Rain	90 ug/l	Sweden, 1972[7]
Sardina pilchardus	0.2 mg/kg dry weight	Fish, marine, less, NE Atlantic, Cornwall, muscle, 1974[13]
Scallops, sea	0.32 mg/kg wet weight	6 references used, SD = +- 0.12, rg = 0.15 - 0.47[14]
Seawater	0.035 ug/l	North Atlantic, subsurface, rg = 0.01 - 0.06 ppm, 1971[68]
Seawater	0.14 ug/l	Rg = 0.09 - 0.19 ppm, New York Bight, Subsurface, 1971[68]
Seawater	2.5 ug/l	Rg = 2 - 3 ppm, Black Sea, to 2000 m, 1971[75]
Shrimp, brown	0.46 mg/kg wet weight	4 references used, SD = +- 0.06, rg = 0.39 - 0.52[14]
Soil, non-polluted	0.2 mg/kg dry weight	Solonetz[5]
Soil, non-polluted	44 mg/kg dry weight	Organic soil[5]
Soil, polluted	100 mg/kg dry weight	Rg = 5 - 1400 ppm, Anvil, Yukon[5]
Soil, urban	834 mg/kg dry weight	USA, New York, Central Park[5]
Solea solea	2 mg/kg dry weight	Fish, marine, Sagres, S Portugal, muscle, 1974[50]
Street dust	2000 mg/kg dry weight	Rg = 1600 - 2400, 77 stations in Mid-westerrn cities, USA[79]
Street dust	7930 mg/kg dry weight	Birmingham, around Pb factory, rg = 136-470000 ppm, internal surfaces[81]
Superphosphate fertilizer	0.7 mg/kg dry weight	Rock from Florida, USA[59]
Ulva lactuca	14 mg/kg dry weight	Range = 10 - 14 ppm, Portugal and Spain, 1974, Atlantic, marine[57]
Water	20 ug/l	Waste water from Stockholm, rg = 19 - 21[7]

Source: 1.Mathis, B.J. and T.F. Cummings. 1973. J. Water Pollut Cont. Fed. 45, 1573-1583.
3.Segar, D.A. and R.E. Pellenbarg. 1971. J. Mar. Biol. Assoc. UK. 51, 131-136.
5.Nriagu, J.O. 1979. The Biochemistry of Lead in The Environment. Part A. Elsevier/North Holland. Amsterdam.
6.Egorov, V.V. T.N. Zhigalowskaja and S.G. Malakhov. 1970. J. Geophys. Res. 75, pp 3650-3656.
7.Laveskog, A., A. Lindskog and V. Stenberg. 1976. Om Metaller, Statens Naturvaardsverk, Stockholm.
11.Vogt, G. and F. Kittelberger. 1977. in: Reichenbach-Klinke, H. (ed.): Fish und Umwelt. Vol 3. Fisher Stuttgart.
12.Bibo, 1977. Schwermetalluntersuchungen an Wasser, Schwebstoffen, Aufwuchs und Cladophora rivularis ds Elsenz. Diplom-Arbeit. Univ. Heidelberg. 111 p.
13.Stevens, J.D. and B.E. Brown. 1974. Mar. Biol. 26, 287-293.
14.Zook, E.G., J.J. Powell, B.M. Hackley, J.A. Emerson, J.R. Brooker and G.M. Knobl, Jr. 1976. J. Agric. Food. Chem. 24(1), pp 47-53.

15.Heydt, G. 1977. Schwermetallgehalte von Wasser, Wasserpflanzen, Chironomidae und Mollusca der Elsenz. Dipl. Arbeit Univ. Heidelberg. 143 pp.
17.Mathis, B.J. and N.R. Kevern. 1973. Distribution of Hg, Cd, Pb and Th in a eutrophic lake. Int. WateRes. Michigan State Univ., Technol. Dep. 34, 23 pp.
18.Fourie, H.O. 1976. S. Afr. J. Sci. 72, 110-113.
20.Abo-Rady, M.D.K. 1977. Die Belastung der oberen Leine mit Schwermetallen durch kommunale und industrelle Abwasser. Diss. Univ. Gottingen. 120 p.
22.Leland, H.V. and J.M. McNerney. 1974. Lead Transport in a river ecosystem. Proc. Int. Conf. Transp. Persist. Chem. Aquatic Ecosystem. Ottawa III 17-23.
23.Harriss, R.C. et al. 1972. Mar. Poll. Bull. 3(12), pp 182-183.
24.Uthe, J.H. and E.G. Bligh. 1971. J. Fish. Res. bd Canada 28, 786-788.
26.MacKay, N.J., M.N. Kazacos, R.J. Williams and M.I. Leedow. 1975. Mar. Pollut. Bull. 6, 57-61.
28.Romeril, M.G. 1971. Mar. Biol. 9, 347-354.
29.Pringle, B.H., D.E. Hissong, E.L. Katz and S.T. Mulawka 1968. J. Sanit. Eng. Div. Am. Soc. Civ. Eng. 94, 455-475.
30.Foster, P. 1976. Environ. Pollut. 10, 45-53.
32.NAS 1972. Airborne Lead in Perspective. National Research Council, National Acedemy of Science, WashC.
33.Dietz, F. 1974. Z. Wasser Abwasserforsch. 7, 74-80.
34.Shacklette, H.T., J.A. Erdman, T.F. Harms and C.S.E. Pupp. 1978. In: F.W. Oeheme (ed.): Toxicity of Heavy Metals in the Environment. Dekkker. NY. 25-68.
35.Servant, J. 1973. In: Environmental Health Aspects of Lead, pp 155-165. Center for information and Documentation CID, Luxembourg.
37.Payer, H.D., Runkel, Schramel, Stengel, Bhumiratana and Soeder. 1976. Chemosphere 6, 413-418.
38.Potter, L., D. Kidd and D. Standiford. 1975. Environ. Sci. Technol. 9, 41-46.
43.Ebner, F. and H. Gams. 1973. Oesterr. Abwasser Rundsch. 18, 47-48.
48.Butterworth, J., P. Lester and G. Nickless. 1972. Mar. Pollut. Bull. 3, 72-74.
49.Nickless, G., R. Stenner and N. Terrile. 1972. Mar. Pollut. Bull. 3, 188-190.
50.Martin, J.B.M. 1974. Mar. Biol. 28, 245-251.
54.Bloom, H. and G.M. Ayling. 1977. Environ. Geol. 2, 3-22.
56.Schulz-Baldes, M. 1974. Mar. Biol. 25, 177-193.
57.Stenner, R.D. and G. Nickless. 1975. Mar. Pollut. Bull. 6, 89-92.
59.Langmyhr, F.J. R. Solberg and Y. Thomassen. 1977. Anal. Chim. Acta 92, pp 105-109.
60.Hardisty, M.W., R.J. Huggins, S. Kartar and M. Sainsbury. 1974. Mar. Pollut. Bull. 5, 61-63.
62.Schlesinger, W.H., W.A. Reiners and D.S. Knopman. 1973. Environ. Pollut., 6, pp 39-47.
63.Kobayashi, Y. 1972. Nippon Eiseigaku Zasshi 28, pp 253-260.
68.Piotrowicz, S.R., B.J. Ray, G.L. Hoffman and R.A. Duce. 1972. J. Geophys Res. 77, 5243-5254.
75.Vinogradova, Z.A., N.Z. Eremina and G.M. Kogan. 1972. Dokl. Akad. Nauk SSSR 204, 704-706.
79.Kennedy, V.C. 1960. U.S. Geol. Survey Bull. p 1098 A.
81.Archer, A. and R.S. Barret. 1976. Sci. Tot. Environ. 6, 275-286.

5-111 Background concentration
Pentane

Species	Value	Condition
Atmosphere, organic gas	0.15 ppb	Rg = 0 - 0.3 ppb, Point Barrow, Alaska, 1968[1]
Atmosphere, organic gas	0.15 ppb	Rg = 0 - 0.3 ppb, trop. and polar maritime air, 1972, STP[2]
Atmosphere, organic gas	0.25 ppb	Rg = 0 - 0.5 ppb, mountain air in USA, 1972, STP[1]
Atmosphere, organic gas	202 ppb	Rg = 135 - 269 ppb, Central Amazona jungle, 1972, STP[2]
Atmospheric conc of aliphatic hydrocarbon and alkylbenzene	21 ppb	Los Angeles, 1967, STP[3]

Source: 1.Cavanagh, L.A., C.F. Sahadt and E. Robinson. 1969. Environ. Sci. Technol. 3, 251-257.
2.Robinson, E., R.A. Rasmussen, H.H. Westberg and M.W. Holdren. 1973. J. Geophys. Res. 78, 5345-5351.
3.Altshuller, A.P., W.A. Lonneman and S.L. Kopczynski. 1973. J. Air Pollut. Contr. Assoc. 23, 597-599.

5-112 Background concentration
Perchloroethylene

Species	Value	Condition
Atmosphere	0.005 ug/m3	Typical value[1]
Bird, egg	20.2 mg/kg wet weight	Rg = 1.4 - 39 mg/kg wet weight, food from water organisms[2]
Bird, liver	2.3 mg/kg wet weight	Rg = 1.5 - 3.1 mg/kg wet weight, food from water organisms[2]
Bread	1 mg/kg wet weight	None[2]
Crustacea	8.5 mg/kg wet weight	Rg = 2 - 15 mg/kg wet weight[2]
Dairy products	6.5 mg/kg wet weight	Rg = 0.05 - 13 mg/kg wet weight[2]

Drinking water	0.1 ug/m3	High value, halocarbon[2]
Drinking water	0.1 ug/m3	Typical non-polluted value, halocarbon[2]
Estuarine water	0.15 ug/m3	High value in inshore water[3]
Fish, flesh	5.7 mg/kg wet weight	Rg = 0.3 - 11 mg/kg wet weight[2]
Fish, liver	21 mg/kg wet weight	Rg = 1 - 41 mg/kg wet weight[2]
Freshwater	0.5 ug/m3	Typical non-polluted value, halocarbon[2]
Freshwater	11 ug/m3	High value in urban area, rg = 1.5 - 20 ppb, halocarbon[2]
Fruit and vegetables	0.64 mg/kg wet weight	Rg = 0.05 - 1.2 mg/kg wet weight[2]
Groundwater	0.2 ug/m3	Typical non-polluted value, halocarbon[2]
Groundwater	5 ug/m3	High value, halocarbon[2]
Homo sapiens, fat	2.4 mg/kg wet weight	Rg = 0.5 - 4.3 mg/kg wet weight[2]
Homo sapiens, liver	14.7 mg/kg wet weight	Rg = 0.4 - 29 mg/kg wet weight[2]
Marine algae	16.5 mg/kg wet weight	Rg = 13 - 20 mg/kg wet weight[2]
Meat	2.9 mg/kg wet weight	Rg = 0.9 - 5 mg/kg wet weight[2]
Mollusca	3.2 mg/kg wet weight	Rg = 0.05 - 6.4 mg/kg wet weight[2]
Plankton	1.7 mg/kg wet weight	Rg = 0.05 - 2.3 mg/kg wet weight[2]
Rain	0.08 ug/m3	Rg = 0.001 - 0.15 ppb, halocarbon[2]
Seal, blubber	9.8 mg/kg wet weight	Rg = 0.6 - 19 mg/kg wet weight[2]
Seal, liver	1.6 mg/kg wet weight	Rg = 0.05 - 3.2 mg/kg wet weight[2]
Seawater	0.010 ug/m3	Typical open ocean[3]
Sediment	1.000 ug/m3	Typical conc in inshore sediment[3]
Snow	0.011 ug/m3	Rg = 0.001 - 0.02 ppb, halocarbon[2]
Vegetable oil	3.5 mg/kg wet weight	Rg = 0.01 - 7 mg/kg wet weight[2]

Source: 1.Lovelock, J.E. and P. Simmons. 1980. In: Sugden (ed.): The Aerosol Controversy. Sci-Horwood. 2.Pearson, C.R. 1982. In: O. Hutzinger (ed.): Environmental Chemistry. Springer-Verlag. Volume 3, partB. Berlin. 3.National Academy of Science. 1978. Chloroform, Carbon tetrachloride, and other halomethanes. An Environmental Assessment. US NAS.

5-113 Background concentration
Perylene

Species	Value	Condition
Atmosphere, seasonal variation	0.22 ng/m3	1. Quarter, 2. Q=0.22, 3. Q=0.06, 4. Q=0.22, LA, n=13, USA, 1975[1]
Crassostrea virginica	0.1 ug/kg dry weight	Galveston B., oyster reef contam. dom. sewage, TEX, rg=0.0-1.5 ppb[2]
Crassostrea virginica	0.7 ug/kg dry weight	Galveston B., commercial oyster reef, TEX, USA, 1970, rg=0.0-3.5 ppb[2]
Marine sediment	160.7 ug/kg dry weight	Depth in sed=478[3]
Marine sediment	160.7 ug/kg dry weight	Depth in sed=478 cm, water depth=5168 m, aerobic, 10 54 N, 44 03 W[3]
Marine sediment	1675 ug/kg dry weight	Depth in sed=17[3]
Marine sediment	1675 ug/kg dry weight	Depth in sed=17 cm, water depth=200 m, anaerobic, 48 30 N, 123 30 W[3]
Marine sediment	184.6 ug/kg dry weight	Depth in sed=100[3]
Marine sediment	184.6 ug/kg dry weight	Depth in sed=100 cm, water depth=5168 m, anaerobic, 10 54 N, 44 03 W[3]

Marine sediment	2370.0 ug/kg dry weight	Depth in sed=34[3]
Marine sediment	2370.0 ug/kg dry weight	Depth in sed=34 cm, water depth=200 m, anaerobic, 48 30 N, 123 30 W[3]
Marine sediment	244 ug/kg dry weight	Depth in sed=1.2[3]
Marine sediment	244 ug/kg dry weight	Depth in sed=1.2 cm, water depth=580 m, anaerobic, 34 11 N, 120 03 W[3]
Marine sediment	260 ug/kg dry weight	Depth in sed=1.3[3]
Marine sediment	260 ug/kg dry weight	Depth in sed=1.3 cm, water depth=600 m, anaerobic, 34 13 N, 120 02 W[3]
Marine sediment	3010 ug/kg dry weight	Glaston peat, Avalon, Engl. 51 09 N, 2 45 W, anaerobic[3]
Marine sediment	331.1 ug/kg dry weight	Depth in sed=230 cm, water depth=5168 m, anaerobic, 10 54 N, 44 03 W[3]
Marine sediment	343.2 ug/kg dry weight	Depth in sed=0.05 cm, water depth=200 m, anaerobic, 48 30 n, 123 30 W[3]
Marine sediment	68 ug/kg dry weight	Depth in sed=0.65 cm, water depth=600 m, anaerobic, 34 14N, 120 02 W[3]
Marine sediment	741 ug/kg dry weight	Depth in sed=2.0[3]
Marine sediment	741 ug/kg dry weight	Depth in sed=2.0 cm, water depth=200 m, anaerobic, 48 30 N, 123 30 W[3]
Marine sediment	78 ug/kg dry weight	Depth in sed=2.5[3]
Marine sediment	78 ug/kg dry weight	Depth in sed=2.5 cm, water depth=600 m, anaerobic, 34 14N, 120 01 W[3]

Source: 1.Gordon, R.J. 1976. Environ. Sci. Technol. 10, 370-373.
2.Fazio, T. 1971. Proc. 7th Nat. Shellfish Sanit. Workshop. Wash DC. FDA, Division of Shellfish Sanitation
3.Aizenshat, Z. 1973. Geochim. Cosmoschim. Acta 37, 559-567.

5-114 Background concentration
Phenanthrene

Species	Value	Condition
Crassostrea virginica	2.2 ug/kg dry weight	Galveston B., oyster reef contam. dom. sewage, TEX, rg=0.0-6.9 ppb[1]

Source: 1.Fazio, T. 1971. Proc. 7th Nat. Shellfish Sanit. Workshop. Wash DC. FDA, Division of Shellfish Sanitation

5-115 Background concentration
Polynuclear aromatic hydrocarbons

Species	Value	Condition
Lake water	10 ng/l	Lake Constance, centrifuged water[1]
River water	800 ng/l	River Main, centrifuged water[1]
Tap water	7 ng/l	Mainz[1]
Well water	25 ng/l	South of Frankfurt[1]

Source: 1.White, J.D. and J.A. Dracup. 1977. Water Pollut. Control 49(10), pp 2179-2189.

5-116 Background concentration
Propane

Species	Value	Condition
Atmosphere, organic gas	215 ppb	Rg = 183-247 ppb, Central Amazona jungle, 1972, STP[1]

Source: 1.Robinson, E., R.A. Rasmussen, H.H. Westberg and M.W. Holdren. 1973. J. Geophys. Res. 78, 5345-5351.

5-117 Background concentration Propylbenzene

Species	Value	Condition
Atmospheric conc of aliphatic hydrocarbon and alkylbenzene	4.5 ppb	Los Angeles, 1967, STP[1]

Source: 1.Altshuller, A.P., W.A. Lonneman and S.L. Kopczynski. 1973. J. Air Pollut. Contr. Assoc. 23, 597-599.

5-118 Background concentration Propylene

Species	Value	Condition
Atmospheric conc of aliphatic hydrocarbon and alkylbenzene	10.5 ppb	Los Angeles, 1967, STP[1]

Source: 1.Altshuller, A.P., W.A. Lonneman and S.L. Kopczynski. 1973. J. Air Pollut. Contr. Assoc. 23, 597-599.

5-119 Background concentration Pt

Species	Value	Condition
Crustral rock	0.000001 percent by weight	Average value[1]

Source: 1.Mason, B. 1966. Principles of Geochemistry. 3rd ed. John Wiley and Sons. New York.

5-120 Background concentration Pyrene

Species	Value	Condition
Atmosphere, seasonal variation	0.58 ng/m3	1. Quarter, 2. Q=0.23, 3. Q=0.25, 4. Q=1.24, LA, n=13, USA, 1975[1]
Crassostrea virginica	0.9 ug/kg dry weight	Aransas Bay, uncontam. area, TEX, USA, 1970, rg=0.8-1.0 ppb[2]
Crassostrea virginica	1.8 ug/kg dry weight	Galveston B., commercial oyster reef, TEX, USA, 1970, rg=1.3-5.9 ppb[2]
Crassostrea virginica	30 ug/kg dry weight	Harbor of Norfolk, Vir., USA, rg=20-40 ppb[2]
Crassostrea virginica	6.5 ug/kg dry weight	Galveston B., oyster reef contam. dom. sewage, TEX, rg=3.8-13 ppb[2]
Domestic sewage	0.254 ug/liter	Dry weather condition, Hegne, GFR[3]
Domestic sewage	16.050 ug/liter	During heavy rain, Hegne, GFR[3]
Mixed sewage	3.120 ug/liter	GFR, high % industri, Stockach, Rodolfsell = 2.560 ppm[3]
Wastewater, industrial	1.957 ug/liter	Effluent conc to Rotach R, GFR[3]

Source: 1.Gordon, R.J. 1976. Environ. Sci. Technol. 10, 370-373.
2.Fazio, T. 1971. Proc. 7th Nat. Shellfish Sanit. Workshop. Wash DC. FDA, Division of Shellfish Sanitation
3.Borneff, J., and H. Kunte. 1965. Arch. Hyg. (Berlin) 149, 226-243.

5-121 Background concentration Ra

Species	Value	Condition
River water	3.9E-10 ppm	None[1]
Seawater	6E-11 ppm	None[2]
Soil	8E-7 mg/kg dry soil	Oven-dried, rg = 3 - 20 E-7[1]

Source: 1.Bowen, H.J.M. 1966. Trace Elements in Biochemistry, Ac. Press.
2.Goldberg, E.D., 1963. The Sea, Interscience London, 2, pp 3.

5-122 Background concentration Rb

Species	Value	Condition
River water	0.15E-2 ppm	Range < 0.008[1]
Seawater	0.12 ppm	Rb(+), residence time = 270E3 y[2]
Soil	100 mg/kg dry soil	Oven-dried, rg = 20 - 600[1]

Source: 1.Bowen, H.J.M. 1966. Trace Elements in Biochemistry, Ac. Press.
2.Goldberg, E.D., 1963. The Sea, Interscience London, 2, pp 3.

5-123 Background concentration Rn

Species	Value	Condition
Atmosphere	6E-8 % by atoms	Sea-level dry atmospheric composition[1]
River water	1.7E-15 ppm	None[2]
Seawater	6E-16 ppm	Rn[3]

Source: 1.Sisler, F.D. 1975. In: S.F. Singer (ed.), The Changing Global Environment, pp 57-69. D. Reidel Publ.Comp.
2.Bowen, H.J.M. 1966. Trace Elements in Biochemistry, Ac. Press.
3.Goldberg, E.D., 1963. The Sea, Interscience London, 2, pp 3.

5-124 Background concentration S

Species	Value	Condition
Atmosphere	0.006 ug/m3	Indian Ocean, 40 deg S, 1973, SO4-S[1]
Atmosphere	0.03 ug/m3	Before Agung eruption 1963, Stratosphere, rg = 0.01 - 0.05 ug/m3, SO4-S[1]
Atmosphere	0.03 ug/m3	S Pacific, 0-20 deg S, 1973, SO4-S[1]
Atmosphere	0.05 ug/m3	S Pacific Ocean, average value, 1973, reduced SO2-S[1]
Atmosphere	0.09 ug/m3	N Atlantic, Faroe Islands, 1974, reduced SO2-S[1]
Atmosphere	0.14 ug/m3	N Atlantic, Faroe Islands, n = 4, 1974, SO4-S[1]
Atmosphere	0.17 ug/m3	Typical value in non-urban area, 1962, SO4-S[1]
Atmosphere	0.35 ug/m3	S Pacific Ocean, Rg = 0.2 - 0.5 ug/m3, 1973, total S, SO4-S[1]
Atmosphere	0.4 ug/m3	21 samples of 36 less than value, Antarctic, 1967, reduced SO2-S[1]
Atmosphere	0.4 ug/m3	Brazil, 1973, reduced SO2-S[1]
Atmosphere	0.5 ug/m3	N Atlantic, Total S, n=4, 1973, SO4-S[1]
Atmosphere	0.6 ug/m3	S hemisphere, vary variable, average value, 1973, SO4-S[1]
Atmosphere	0.9 ug/m3	Rg = 0.4 - 1.4 ug/m3, Panama, 1973, reduced SO2-S[1]
Atmosphere	1.1 ug/m3	Rg = 0.5 - 1.7 ug/m3, Caribbean, 1973, reduced SO2-S[1]
Atmosphere	1.5 ug/m3	N hemisphere, vary variable, average, 1973, SO4-S[1]
Atmosphere	26.3 ug/m**3	293 K, 1 atm, residence time = 0.014 y, SO2, rg = 0 - 50[2]
Biosphere	4.44 moles/hectare	Average composition of living matter[3]
Peat	1.8 umol g-1 dry wt.	inorganic S0-S in Big Run Bog, oven-dried samples[4]
Peat	1.8 umol g-1 dry wt.	inorganic SO4-S in Big Run Bog, oven-dried samples[4]
Peat	1.9 umol g-1 dry wt.	inorganic S0-S in Big Run Bog, oven-dried samples[4]
Peat	10.5 umol g-1 dry wt.	inorganic, FeS2- S in Big Run Bog, oven-dried samples[4]

Peat	116.2 umol g-1 dry wt.	organic S in Big Run Bog, oven-dried samples[4]
Peat	116.3 umol g-1 dry wt.	organic S in Big Run Bog, moist samples[4]
Peat	135.7 umol g-1 dry wt.	total S in Big Run Bog, oven-dried samples[4]
Peat	136.6 umol g-1 dry wt.	total S in Big Run Bog, moist samples[4]
Peat	14.2 umol g-1 dry wt.	inorganic FeS2-S in Big Run Bog, oven-dried samples[4]
Peat	19.5 umol g-1 dry wt.	inorganic S in Big Run Bog, oven-dried samples[4]
Peat	2.4 umol g-1 dry wt.	inorganic FeS-S in Big Run Bog, oven-dried samples[4]
Peat	20.3 umol g-1 dry wt.	inorganic S in Big Run Bog, oven-dried samples[4]
Peat	3.4 umol g-1 dry wt.	inorganic FeS-S in Big Run Bog, oven-dried samples[4]
Peat	30.8 umol g-1 dry wt.	organic, ester-sulfate S in Big Run Bog, oven-dried samples[4]
Peat	4.7 umol g-1 dry wt.	inorganic SO4-S in Big Run Bog, oven-dried samples[4]
Peat	56.0 umol g-1 dry wt.	organic, carbon-bonded S in Big Run Bog, oven-dried samples[4]
Peat	60.2 umol g-1 dry wt.	organic, ester-sulfate S in Big Run Bog, oven-dried samples[4]
Peat	85.5 umol g-1 dry wt.	organic, carbon-bonded S in Big Run Bog, oven-dried samples[4]
River water	3.7 ppm	Range = 0.9-30 ppm[5]
River, Africa	13.5 mg/l	As SO4(2-)[6]
River, Asia	8.4 mg/l	As SO4(2-)[6]
River, Australia	2.6 mg/l	As SO4(2-)[6]
River, Europe	24.0 mg/l	As SO4(2-)[6]
River, North America	20.0 mg/l	As SO4(2-)[6]
River, South America	4.8 mg/l	As SO4(2-)[6]
River, World	11.2 mg/l	As SO4(2-)[6]
Seawater	0.5 % of total	As K2S, chlorinity = 1.9 %, 298 K, pH = 8.15, of SO4 pool[7]
Seawater	21 % of total	As Na2S, chlorinity = 1.9 %, 298 K, pH = 8.15, of SO4 pool[7]
Seawater	22 % of total	As MgS, chlorinity = 1.9 %, 298 K, pH = 8.15, of SO4 pool[7]
Seawater	3 % of total	As CaS, chlorinity = 1.9 %, 298 K, pH = 8.15, of SO4 pool[7]
Seawater	54 % of total	As free ion, chlorinity = 1.9 %, 298 K, pH = 8.15, of SO4 pool[7]
Seawater	885 ppm	SO4(-2)[8]
Troposphere	0.7 ug SO4-S/m**3	As SO4: 2 ug/m**3[9]

Source: 1.Keckes, S. and J.K. Miettinen. 1970. Proc. FAO Conf. Rome. Dec 9-18.
2.Junge, C.E. 1963. Air Chemistry and Radioactivity, Ac. Press.
3.Deevey, E.S. 1970. Scientific American. 223(3), p 150.
4.Wieder, R.K., Lang, G.E. and Granus, V.A., Limnol.Oceanogr., 30(5), 1109, 1985.
5.Bowen, H.J.M. 1966. Trace Elements in Biochemistry, Ac. Press.
6.Wetzel, R.G. 1975. Limnology. Saunders. Philadelphia.
7.Garrels, R.M. and M.E. Thompson. 1962. Am. J. Sci. 260., p 57.
8.Goldberg, E.D., 1963. The Sea, Interscience London, 2, pp 3.
9.Robinson, E. and R.C. Robbins, 1975. In: S.F. Singer (ed.), The Changing Global Environment, pp 111-23. D. Reidel Publ. Comp.

5-125 Background concentration Sb

Species	Value	Condition
Atmosphere	1 ng/m**3	N Engl, 1971[1]
Atmosphere	1.5 ng/m**3	Not urban, USA, rg = 1 - 2[1]
Atmosphere	1.7 pg/m**3	Southpole, 1970, rg = 0.6[1]

Atmosphere	63.5 ng/m**3	Urban, USA, rg = 42 - 85[1]
Atmosphere, urban	32 ng/m**3	Range = 1.4 - 55 ng/m3, Chicago[2]
Carbonate sediments	0.2 mg/kg dry weight	Average value[3]
Choromytilus meridionalis	0.20 mg/kg wet weight	Gastropod, Cape, W Coast, S Africa, 1974[4]
Choromytilus meridionalis	2.3 mg/kg wet weight	Gastropod, rg = +- 0.69 ppm, Saldanha B, LangeBaan B., RSA, 1975[5]
Crassostrea gigas	1.4 mg/kg wet weight	Gastropod, rg = +- 0.4 ppm, Saldanha B, LangeBaan B., RSA, 1975[5]
Crassostrea margaritacea	1.8 mg/kg wet weight	Gastropod, rg = +- 0.13 ppm, Saldanha B, LangeBaan B., RSA, 1975[5]
Crustral rock	0.00002 percent by weight	Average value[6]
Deep-sea clay sediments	1.0 mg/kg dry weight	Average value[3]
Freshwater	0.1 ug/l	River Rhine, 1969[7]
Granitic rock	0.2 mg/kg dry weight	Average value[3]
Jasus lalandii	0.12 mg/kg wet weight	Marine, Crusta., rg =+-0.07, Melkbosch C, W Coast, SA, muscle, 1975[4]
Jasus lalandii	2.75 mg/kg wet weight	Marine, Crustaceans, rg=0.5, Saldanha Bay, SA, muscle, 1975[5]
Marine zooplankton	0.26 mg/kg wet weight	Range = 0.004 - 0.52 ppm, Cape W Coast, S Africa, 1974[4]
Mytilus edulis	0.04 mg/kg dry weight	Gastropod, Southampton Waters, GB, 1971[8]
Pelagic organisms	0.1 mg/kg dry weight	Azores[9]
Rain	0.15 ug/l	Sweden, rg = 0.1 - 0.2[1]
Rain	1.8 ug/l	N Engl[1]
Sandstone sediment	0.04 mg/kg dry weight	Average value[3]
Seawater	0.2 ug/l	Northeastern Pacific, 1968[7]
Seawater	0.33E-3 ppm	Residence time = 350E3 y[10]
Shale sediment	1.5 mg/kg dry weight	Average value[3]
Soil	6 ppm of dry soil	Oven-dried, rg = 2 - 10[11]
Ulva sp.	0.16 mg/kg dry weight	Range = +- 0.09 ppm, Cape W Coast, S Africa, marine, 1974[4]
Water	0.065 ug/l	Rivers, Sweden, rg = 0.05 - 0.08[1]

Source: 1.Laveskog, A., A. Lindskog and V. Stenberg. 1976. Om Metaller, Statens Naturvaardsverk, Stockholm.
2.Dams, R. J.A. Robbins, K.A. Rahn and J.W. Winchester. 1970. Anal. Chem 42, pp 861-867.
3.Turekian, K.K. and K.H. Wedepohl. 1961. Amer. Geol. Soc. Bull. 72, pp 175-192.
4.Van As, D. H.O. Fourie and C.M. Vleggaar. 1975. S. Afr. J. Sci. 71, 151-154.
5.Fourie, H.O. 1976. S. Afr. J. Sci. 72, 110-113.
6.Mason, B. 1966. Principles of Geochemistry. 3rd ed. John Wiley and Sons. New York.
7.Spencer, D.W., D.E. Robertson, K.K. Turekian and T.R. Folsom 1970. J. Geophys. Res. 75, pp 7688-7696
8.Raymont, J.E.G. 1972. Proc. R. Soc. London b 180, 451-468.
9.Leatherland, T.M., J.D. Burton, F. Culkin, M.J. McCartney and R.J. Morris. 1973. Deep-Sea Res. 20,p 679-685.
10.Goldberg, E.D., 1963. The Sea, Interscience London, 2, pp 3.
11.Bowen, H.J.M. 1966. Trace Elements in Biochemistry, Ac. Press.

5-126 Background concentration
Sc

Species	**Value**	**Condition**
Seawater	0.4E-5 ppm	Residence time = 5.6E3 y, less than value[1]
Soil	7 mg/kg dry soil	Oven-dried, rg = 10 - 25[2]

Source: 1.Goldberg, E.D., 1963. The Sea, Interscience, London, 2, pp 3.
2.Bowen, H.J.M. 1966. Trace Elements in Biochemistry, Ac. Press.

5-127 Background concentration Se

Species	Value	Condition
Atmosphere	0.0056 ng/m**3	Southpole, 1972, rg = +- 0.0012[1]
Atmosphere	1 ng/m**3	USA, 1964-65[1]
Atmosphere	1.74 ng/m**3	England, 7 cities, 1972, rg = 0.48 - 3.0[1]
Atmosphere	2 ng/m**3	app value[2]
Carbonate sediments	0.08 mg/kg dry weight	Average value[3]
Deep-sea clay sediments	0.17 mg/kg dry weight	Average value[3]
Drinking water	0.061 ug/l	Stockholm, Sweden[1]
Earth crust	0.07 mg/kg dry weight	Rg = 0.05 - 0.07, average value[4]
Freshwater	200 ug/l	Rg = 0.1 - 200, also ground water[5]
Freshwater, river	1.3 ug/l	Range = < 1 - 4 ppm, Lake Constance, Lindau, 1973[6]
Freshwater, river	2.4 ug/l	Range = < 1 - 8 ppm, Danube, Ulm/Regensburg, 1973[6]
Granitic rock	0.05 mg/kg dry weight	Average value[3]
Ground water	6000 ug/l	Max value, dependent on geology[7]
Igenous rock	0.05 mg/kg dry weight	None[8]
Rain	0.34 ug/l	England, 1971, rel.uncontaminated, Se in atm. = 1 ng/m**3[1]
River water	0.02 ppm	Less than value[9]
Sandstone sediment	0.05 mg/kg dry weight	Average value[3]
Seawater	0.01-1.06 nmol kg-1	as org. Se, varying with depth[10]
Seawater	0.02-1.03 nmol kg-1	as selenite, varying with depth[10]
Seawater	0.13-1.48 nmol kg-1	as selenate, varying with depth[10]
Seawater	0.95-2.44 nmol kg-1	total Se, varying with depth[10]
Seawater	0.9E-4 ppm	None[11]
Sedimentary rock	0.6 mg/kg dry weight	None[12]
Shale sediment	0.6 mg/kg dry weight	Average value[3]
Soil	0.01 mg/kg dry weight	None[12]
Soil	0.1 mg/kg dry weight	NZ, Se-deficient areas[4]
Soil	0.2 mg/kg dry soil	Oven-dried, rg = 0.01 - .0[9]
Soil	1200 mg/kg dry weight	Ireland, seleniferous area[13]
Water	0.125 ug/l	Sweden, water-spring, rg = 0.010 - 0.015[1]

Source: 1.Laveskog, A., A. Lindskog and V. Stenberg. 1976. Om Metaller, Statens Naturvaardsverk, Stockholm.
2.John, W., R. Kaiper, K. Rahm and J.J. Westowshki. 1973. Atmos. Environ. 7, pp 107-118.
3.Turekian, K.K. and K.H. Wedepohl. 1961. Amer. Geol. Soc. Bull. 72, pp 175-192.
4.Wells, N. 1967. N.Z.J. Geol. Geophys. 10, pp 198-208.
5.Lindberg, P. 1968. Acta Vet. Scand. Suppl. 23.
6.Quentin, K.E., and H.A. Winkler. 1974. Zentralbl. Bakt. Abt. Orig. B 158, 514-523.
7.Cannon, H.G. 1964. M.S. Geol. Survey Bull. No. 1176.
8.Taylor, S.R. 1964. Geochim. Cosmochim. Acta 28, pp 1273-1285.
9.Bowen, H.J.M. 1966. Trace Elements in Biochemistry, Ac. Press.
10.Cutter, G.A. and Bruland, K.W., Limnol.Oceanogr., 29(6), 1179, 1984.
11.Goldberg, E.D., 1963. The Sea, Interscience, London, 2, pp 3.
12.Vinogradov, A.P. 1959. The Geochemistry of Rare and Dispe ed Chemical Elements in the Soil. Transl. from Russian. Consultant Bureau Inc. NY. 2nd ed.
13.Fleming, G.A. 1962. Soil Sci. 94, pp 28-35.

5-128 Background concentration Si

Species	Value	Condition
Biosphere	8.62 moles/hectare	Average composition of living matter[1]
River water	6.5 ppm	Range = 2-12 ppm , as Si[2]
River, Africa	23.2 mg/l	As SiO2[3]
River, Asia	11.7 mg/l	As SiO2[3]
River, Australia	3.9 mg/l	As SiO2[3]
River, Europe	7.5 mg/l	As SiO2[3]
River, North America	9.0 mg/l	As SiO2[3]
River, South America	11.9 mg/l	As SiO2[3]
River, World	13.1 mg/l	As SiO2[3]
Seawater	3 ppm	Si(OH)4, residence time = 8E3 y , as Si[4]
Soil	330000 mg/kg dry soil	Oven-dried, as Si, rg = 250000 - 350000[2]

Source: 1.Deevey, E.S. 1970. Scientific American. 223(3), p 150.
2.Bowen, H.J.M. 1966. Trace Elements in Biochemistry, Ac. Press.
3.Wetzel, R.G. 1975. Limnology. Saunders. Philadelphia.
4.Goldberg, E.D., 1963. The Sea, Interscience London, 2, pp 3.

5-129 Background concentration Sn

Species	Value	Condition
Atmosphere	1.0 ng/m**3	Not urban, USA, rg = 0.2 - 1.8[1]
Atmosphere	100 ng/m**3	1972, Heidelberg, W. Germany, Neutron activation[2]
Atmosphere	20 ng/m**3	Urban, USA, rg = 10 - 30[1]
Atmosphere	4.1 ug/m3	1973, Japan, 700 m from industrial plant[3]
Crustral rock	0.00020 percent by weight	Average value[4]
Igneous rock	2 mg/kg dry weight	None[5]
Rain	10 ug/l	Sweden, 3 stations[1]
River water	0.4E-4 ppm	None[6]
Seawater	0.003 ppm	Residence time = 100E3 y[7]
Sedimentary rock	30 mg/kg dry weight	None[8]
Sewage sludge	160 mg/kg dry weight	Rg = 40 - 700 ppm[9]
Soil	10 mg/kg dry soil	Oven-dried, rg = 2 - 200[6]
Water	5.5 ug/l	Rivers, USA, rg = 0.3 - 11[1]

Source: 1.Laveskog, A., A. Lindskog and V. Stenberg. 1976. Om Metaller, Statens Naturvaardsverk, Stockholm.
2.Friberg, L. G.N. Nordberg and V.B. Vouk. 1979. Handbook on the Toxicology of Metals. Elseveir.
3.Environmental Agency. Japan. 1974. Japanese bacground paper No. 2. Effects on Health and Specific Air Pollutants from Industrial Emission. Geneva 4-9 Nov 1974. WHO meeting.
4.Mason, B. 1966. Principles of Geochemistry. 3rd ed. John Wiley and Sons. New York.
5.Taylor, S.R. 1964. Geochim. Cosmochim. Acta 28, pp 1273-1285.
6.Bowen, H.J.M. 1966. Trace Elements in Biochemistry, Ac. Press.
7.Goldberg, E.D., 1963. The Sea, Interscience London, 2, pp 3.
8.Vinogradov, A.P. 1959. The Geochemistry of Rare and Dispersed Chemical Elements in the Soil. Transl. from Russian. Consultant Bureau Inc. NY. 2nd ed.
9.Berrow, M.L. and J. Webber. 1972. J. Sci. Food Agric. 23, pp 93-100.

5-130 Background concentration Sr

Species	Value	Condition
Atmospheric particulates	2.4 ng/m**3	Hawaii, n > = 56, rg = +- 1.5[1]
Carbonate sediments	610 mg/kg dry weight	Average value[2]
Deep-sea clay sediments	180 mg/kg dry weight	Average value[2]
Food	0.50 ug/g	USA, milk[3]
Food	1.20 ug/g	USA, meat[3]
Food	2.46 ug/g	USA, vegetables[3]
Food	9.56 ug/g	USA, fish and shellfish[3]
Freshwater sediment	216 mg/kg dry weight	Range of 90 % = 55 - 600 ppm, 74 lakes, pelitic[4]
Granitic rock	440 mg/kg dry weight	Average value[2]
River water	0.08 ppm	Range = 0.003-0.8 ppm[5]
Sandstone sediment	20 mg/kg dry weight	Average value[2]
Seawater	8.1 ppm	Sr(9+), residence time = 19000E3 y[6]
Shale	300 mg/kg dry weight	Average value[2]
Shale sediment	300 mg/kg dry weight	Average value[2]
Soil	300 ppm dry soil	Oven-dried, rg = 50 - 1000[5]
Water	405 ug/l	Rivers and lakes, USA, 1958-67, rg = 11 - 802[3]

Source: 1.Golterman, H.L. 1975. Physiological Limnol., Elsevier Amsterdam.
2.Turekian, K.K. and K.H. Wedepohl. 1961. Amer. Geol. Soc. Bull. 72, pp 175-192.
3.Laveskog, A., A. Lindskog and V. Stenberg. 1976. Om Metaller, Statens Naturvaardsverk, Stockholm.
4.Foerstner, V. 1977. Arch. Hydrobiol., 80(2), pp 172-191.
5.Bowen, H.J.M. 1966. Trace Elements in Biochemistry, Ac. Press.
6.Goldberg, E.D., 1963. The Sea, Interscience London, 2, pp 3.

5-131 Background concentration Ta

Species	Value	Condition
Seawater	0.25E-5 ppm	Less than value[1]

Source: 1.Goldberg, E.D., 1963. The Sea, Interscience London, 2, pp 3.

5-132 Background concentration Tb

Species	Value	Condition
Seawater	0.0008 microgram/l	Soluble concentration, Japan, Cl = 1.873 %[1]
Seawater	0.0042 microgram/l	Total concentration, Japan, Cl = 1.873 %[1]

Source: 1.Suzuki, H., T. Koganage and M. Saki. 1975. Impacts of Nuclear Releases into the Aquatic Environment. IAEA SM-198/19, pp 77-89. Vienna.

5-133 Background concentration Te

Species	Value	Condition
Atmosphere	2 ug/m3	2 km from electrolytic Cu refining plant[1]

Source: 1.Seljankina, K.P. and R.R. Brooks. 1971. Gig. Sanit. 35, pp 95-96.

5-134 Background concentration Th

Species	Value	Condition
Forest foliage	4 mg/kg ash weight	Uncontaminated value for USA, rg = 2-100 ppm[1]
Marine algae	1E-7 g Th/g ash weight	Adriatic Sea, filtred water, rg = 1E-6 - 1E-8[2]
River water	0.0E-4 ppm	None[3]
Seawater	0.5E-4 ppm	Residence time = 0.35E3 y[4]
Seawater	8E-10 g Th/l	Adriatic Sea, unfiltred water[2]
Seawater	8E-13 g Th/l	Adriatic Sea, filtred water[2]
Soil	5 ppm dry soil	Oven-dried, rg = 0.1 - 12[3]

Source: 1.Shacklette, H.T., J.A. Erdman, T.F. Harms and C.S.E. Pupp. 1978. In: F.W. Oeheme (ed.): Toxicity of Heavy Metals in the Environment. Dekkker. NY. 25-68.
2.Strohal, P. and T. Pinter. 1973. Limnol. Oceanogr. 18(2), pp 250-253.
3.Bowen, H.J.M. 1966. Trace Elements in Biochemistry, Ac. Press.
4.Goldberg, E.D., 1963. The Sea, Interscience London, 2, pp 3.

5-135 Background concentration Thiamine

Species	Value	Condition
Freshwater	0.0425 mg/m**3	Lindsey Pond, Conn, rg = 0.008 - 0.077[1]
Freshwater	0.095 mg/m**3	Particulate, Tsukui, Japan, surface, rg = 0.031 - 0.159[2]
Freshwater	0.17 mg/m**3	Sagami, Japan, rg = 0.001 - 0.38[3]
Freshwater	0.25 mg/m**3	Dissolved, Tsukui, Japan, surface, rg = 0.075 - 0.436[2]
Freshwater	0.5 mg/m**3	Small swiss ponds, rg = < 0.001 - 1.0[4]
Freshwater	6 mg/m**3	Northern german lakes, rg = 0.05 - 12[5]

Source: 1.Benoit, R.J. 1957. Limnol. Oceanogr., 2, pp 233-240.
2.Ohwada, K., M. Otsuhata and N. Taka. 1972. Bull. Japan Soc. Sci. Fish. 38, pp 817-823.
3.Ohwada, K. and N. Taga. 1972. Limnol. Oceanogr. 17, pp 315-320.
4.Clemenson, H. 1963. Schweiz. Z. Hydrol., 25, pp 157-165.
5.Hagedorn, H. 1971. Arch. Hydrobiol., 68, pp 382-399.

5-136 Background concentration Ti

Species	Value	Condition
Atmosphere	0.1 ug/m3	Annual average, urban air, lower in rural air[1]
Atmosphere	0.5 ug/m3	Industrialized area, Japan[1]
Atmosphere	10.5 ng/m**3	Not urban, USA, rg = 3 - 18[2]
Atmosphere	30 ng/m**3	Sweden, TiO2[2]
Atmosphere	40 ng/m**3	Urban, USA, rg = 10 - 70[2]
Crustral rock	0.44 percent by weight	Average value[3]
Drinking water	2 ug/l	Rg = 0.5 - 15 ug/l, USA[4]
Food	1.2 mg/kg dry weight	Cabbage[2]
Food	1.8 mg/kg dry weight	Potato[2]
Food	12 mg/kg dry weight	Spinach[2]
Food	3 mg/kg dry weight	Sheep, horse and hare, rg = 2 - 4[2]
Freshwater	54.5 ug/l	Rg = 2 - 107, Canada and USA[5]

River water	0.86E-2 ppm	Range < 0.11 ppm[6]
Seawater	0.001 ppm	Residence time = 0.16E3 y[7]
Seawater	0.8 ug/l	Rg = 0.6 - 1 ug/l, values up to 9 ug/l have been found[8]
Soil	3 %	Rg = 0.3 - 6 %, up to 10 %, enrichnent near coal combustion plants[9]
Soil	5000 ppm dry soil	Oven-dried, rg = 1000 - 10000[6]

Source: 1.EPA, USA. 1973. Air Quality Data for Metals, 1968-69. National Air Surveilance Network. No. APTD-146, pp 13/1-13/13.
2.Laveskog, A., A. Lindskog and V. Stenberg. 1976. Om Metaller, Statens Naturvaardsverk, Stockholm.
3.Mason, B. 1966. Principles of Geochemistry. 3rd ed. John Wiley and Sons. New York.
4.Friberg, L. G.N. Nordberg and V.B. Vouk. 1979. Handbook on the Toxicology of Metals. Elsevier.
5.Durum, W.H. and J. Haffty. 1961. U.S. Geol. Surv. Circ. 445, pp 1-11.
6.Bowen, H.J.M. 1966. Trace Elements in Biochemistry, Ac. Press.
7.Goldberg, E.D., 1963. The Sea, Interscience, London, 2, pp 3.
8.Silvay, W.D. 1967. U.S. Geol. Surv. Water Supply Pap. 1535-L, L1-L25.
9.Vinogradov, A.P. 1959. In: The Geochemistry of Rare and Dispersed Chemical Elements in Soils, 2ed, Consultants bureau. NY, pp 95-209.

5-137 Background concentration Tl

Species	Value	Condition
Atmosphere	0.24 ng/m**3	Rg = 0.04 - 0.48 ng/m3, Nebraska[1]
Freshwater	7 ug/l	Range = 0.01 - 14 ug/l[1]
Seawater	0.01 ug/l	App value[1]
Seawater	0.1E-4 ppm	Tl(+), less than value[2]
Soil	0.1 ppm dry soil	Oven-dried[3]

Source: 1.Carson, B.L. and I.C. Smith. 1977. Thallium: An Appraisal of Environmental Exposure. Technical Report no. 5. NIEHS, Research Triangle Park, N.C.
2.Goldberg, E.D., 1963. The Sea, Interscience, London, 2, pp 3.
3.Bowen, H.J.M. 1966. Trace Elements in Biochemistry, Ac. Press.

5-138 Background concentration Toluene

Species	Value	Condition
Atmosphere, organic gas	0.1 ppb	Rg = 0 - 0.2 ppb, trop. and polar maritime air, 1972, STP[1]
Atmosphere, organic gas	0.6 ppb	Rg = 0.1-1.0 ppb, mountain air in USA, 1972, STP[2]
Atmosphere, organic gas	8 ppb	Rg = 1 - 15 ppb, Central Amazona jungle, 1972, STP[1]
Atmospheric conc of aliphatic hydrocarbon and alkylbenzene	30 ppb	Los Angeles, 1967, STP[3]

Source: 1.Robinson, E., R.A. Rasmussen, H.H. Westberg and M.W. Holdren. 1973. J. Geophys. Res. 78, 5345-5351.
2.Cavanagh, L.A., C.F. Sahadt and E. Robinson. 1969. Environ. Sci. Technol. 3, 251-257.
3.Altshuller, A.P., W.A. Lonneman and S.L. Kopczynski. 1973. J. Air Pollut. Contr. Assoc. 23, 597-599.

5-139 Background concentration Total hydrocarbons

Species	Value	Condition
Seawater	0.00 ug/liter	Sargasso Sea, Bermudo, 1974-76, depth = 1200 m[1]
Seawater	0.00 ug/liter	Sargasso Sea, Bermudo, 1974-76, depth = 2000 m[1]
Seawater	0.26 ug/liter	Sargasso Sea, Bermudo, 1974-76, depth = 300 m, rg = +- 0.16 ug/l[1]
Seawater	0.32 ug/liter	Sargasso Sea, Bermudo, 1974-76, depth = 1 m, rg = +- 0.12 ug/l[1]
Seawater	0.47 ug/liter	Sargasso Sea, Bermudo, 1974-76, depth = 30 m, rg = +- 0.25 ug/l[1]
Seawater	13.9 ug/liter	Baltic Sea, 1973, depth = 1 m, rg = +- 3.8 ug/l[1]

Seawater	16.0 ug/liter	Mediterranean Sea, 1975, depth = 1 m[1]
Seawater	3.5 ug/liter	Nova Scotia to Gulf Stream, 1973, depth = 10 m, rg = +- 1.8 ug/l[1]
Seawater	3.8 ug/liter	Baltic Sea, 1973, depth = 1 m above sed., rg = +- 1.4 ug/l[1]
Seawater	4.7 ug/liter	Nova Scotia to Gulf Stream, 1973, depth = 1 m, rg = +- 1.3 ug/l[1]
Seawater	5.4 ug/liter	Baltic Sea, 1973, depth = 10 - 50 m, rg = +- 1.8 ug/l[1]
Seawater	6.1 ug/liter	Nova Scotia to Gulf Stream, 1973, depth = 25 m, rg = +- 1.8 ug/l[1]

Source: 1.Zsolnay, A. 1977. Mar. Chem. 5, 465-475.

5-140 Background concentration Total non - CH4 carbon

Species	Value	Condition
Atmosphere, organic gas	10 ug organic carbon/m3	More than value ppb, Point Barrow, Alaska, 1968, excluding n-Butanol[1]
Atmosphere, organic gas	1400 ug organic carbon/m3	Rg = 1200 - 1600 ppb, Central Amazona jungle, 1972, STP[2]
Atmosphere, organic gas	17 ug organic carbon/m3	Rg = 6 - 28 ppb, trop. and polar maritime air, 1972, STP[2]
Atmosphere, organic gas	18 ug organic carbon/m3	Rg = > 5 - 30 ppb, mountain air in USA, 1972, STP[1]
Atmospheric conc of aliphatic hydrocarbon and alkylbenzene	950 ug organic carbon/m3	Los Angeles, 1967, STP[3]

Source: 1.Cavanagh, L.A., C.F. Sahadt and E. Robinson. 1969. Environ. Sci. Technol. 3, 251-257.
2.Robinson, E., R.A. Rasmussen, H.H. Westberg and M.W. Holdren. 1973. J. Geophys. Res. 78, 5345-5351.
3.Altshuller, A.P., W.A. Lonneman and S.L. Kopczynski. 1973. J. Air Pollut. Contr. Assoc. 23, 597-599.

5-141 Background concentration Total polyaromatic hydrocarbons, PAH

Species	Value	Condition
Domestic sewage	0.075 ug/liter	Dry weather condition, Hegne, GFR, unidentified PAH[1]
Domestic sewage	0.771 ug/liter	Dry weather condition, Hegne, GFR, identified PAH[1]
Domestic sewage	0.846 ug/liter	Dry weather condition, Hegne, GFR[1]
Domestic sewage	78.300 ug/liter	During heavy rain, Hegne, GFR, identified PAH[1]
Domestic sewage	87.500 ug/liter	During heavy rain, Hegne, GFR[1]
Domestic sewage	9.200 ug/liter	During heavy rain, Hegne, GFR, unidentified PAH[1]
Mixed sewage	12.48 ug/liter	GFR, high % industri, Stockach, identified PAH, Rodolfsell = 8.488 ppm[1]
Mixed sewage	14.98 ug/liter	GFR, high % industri, Stockach, Rodolfsell = 8.558 ppm[1]
Mixed sewage	2.500 ug/liter	GFR, high % industri, Stockach, unidentified PAH, Rodolfsell = 0.070 ppm[1]
Wastewater, domestic	0.130 ug/liter	Effluent conc to Rotach R, GFR, unidentified PAH[1]
Wastewater, domestic	0.668 ug/liter	Effluent conc to Rotach R, GFR, identified PAH[1]
Wastewater, domestic	0.798 ug/liter	Effluent conc to Rotach R, GFR[1]
Wastewater, industrial	0.390 ug/liter	Effluent conc to Rotach R, GFR, unidentified PAH[1]
Wastewater, industrial	4.743 ug/liter	Effluent conc to Rotach R, GFR, identified PAH[1]
Wastewater, industrial	5.133 ug/liter	Effluent conc to Rotach R, GFR[1]

Source: 1.Borneff, J., and H. Kunte. 1965. Arch. Hyg. (Berlin) 149, 226-243.

5-142 Background concentration Total polycyclic aromatic hydrocarbon, PAH

Species	Value	Condition
Atmosphere, seasonal var.	13.19 ng/m3	1. Quarter, 2. Q=4.90, 3. Q=8.08, 4. Q=25.9, LA, n=13, USA, 1975[1]
Drinking water source	150 ng/liter	Ground water, GFR, 1973, rg = 100 - 200 ng/l[2]
Drinking water source	4.7 ng/liter	Well water, Engl., 1974, rg = 3.6-5.8 ng/l[2]
Drinking water source	425 ng/liter	Well water, GFR, 1973, rg = 100 - 700 ng/l[2]
Drinking water source	67.3 ng/liter	Ground water, GFR, 1968, rg = 10.9-123.5 ng/l[2]
Drinking water source	7.8 ng/liter	Tap water, 9 USA cities, 1977 rg = 0.9 - 14.9 ng/l[2]
Drinking water source	77.2 ng/liter	Tap water, GFR, 1968, rg = 29.2 - 125.5 ng/l[2]
Drinking water source	8.3 ng/liter	Ground water, USA, 1977[2]
Freshwater, lake	130 ng/liter	Lake Constance, drinking water source, GFR, 1963, rg = 25 - 234 ng/l[2]
Freshwater, lake	26.2 ng/liter	Resevoirs for drinking water, Engl., 1974, rg = 9.1 - 43.2 ng/l[2]
Freshwater, lake	4.7 ng/liter	Lake Erie at Baffalo, USA, 1977, used as drinking water source[2]
Freshwater, river	1140 ng/liter	Rg = 730 - 1500 ng/l, R Rhine at Mainz, GFR, 1963[2]
Freshwater, river	143 ng/liter	Rg = 20 - 266 ng/l, Severn R., Engl., 1974[2]
Freshwater, river	1575 ng/liter	Rg = 800 - 2350 ng/l, Thames R., Engl., 1975[2]
Freshwater, river	1750 ng/liter	Rg = 500 -3000 ng/l, R Rhine at Koblenz, 1973[2]
Freshwater, river	1900 ng/liter	Rg = 25 - 3790 ng/l, Trent R., Engl., 1974[2]
Freshwater, river	2270 ng/liter	Rg = 1440-3100 ng/l, R Aach at Stockach, GFR, 1964[2]
Freshwater, river	351.8 ng/liter	Delaware R., Philadelphia, Penn., USA, 1977[2]
Freshwater, river	57.9 ng/liter	Ohio R., Huntington, W Virginia, USA, 1977[2]
Freshwater, river	600 ng/liter	Rg = 200 - 1000 ng/l, other Rhine and Aach in GFR, 1964[2]
Freshwater, river	632 ng/liter	Rg = 600 - 663 ng/l, Monongahela R. at Pittsburgh, Penn., USA, 1977[2]
Rain	2100 ng/liter	Used as drinking water source, Koblenz, 1973, GFR, rg = 200 - 4000 ng/l[2]
Seawater	0.00 ng/liter	Sargasso Sea, Bermudo, 1974-76, depth = 2000 m[3]
Seawater	1 ng/liter	Sargasso Sea, Bermudo, 1974-76, depth = 30 m, rg = +- 0 ng/l[3]
Seawater	1 ng/liter	Sargasso Sea, Bermudo, 1974-76, depth = 300 m, rg = +- 1 ng/l[3]
Seawater	148 ng/liter	Mediterranean Sea, 1975, depth = 1 m, rg = +- 36 ng/l[3]
Seawater	16.8 ng/liter	Nova Scotia to Gulf Stream, 1973, depth = 10 m, rg = +- 4 ng/l[3]
Seawater	277 ng/liter	Baltic Sea, 1973, depth = 1 m, rg = +- 121 ng/l[3]
Seawater	30.8 ng/liter	Nova Scotia to Gulf Stream, 1973, depth = 1 m, rg = +- 10 ng/l[3]
Seawater	31 ng/liter	Sargasso Sea, Bermudo, 1974-76, depth = 1 m, rg = +- 10 ng/l[3]
Seawater	4 ng/liter	Sargasso Sea, Bermudo, 1974-76, depth = 1200 m[3]
Seawater	47 ng/liter	Baltic Sea, 1973, depth = 1 m above sed., rg = +- 13 ng/l[3]
Seawater	52 ng/liter	Baltic Sea, 1973, depth = 10 - 50 m, rg = +- 9 ng/l[3]
Seawater	8.6 ng/liter	Nova Scotia to Gulf Stream, 1973, depth = 25 m, rg = +- 9 ng/l[3]

Source: 1.Gordon, R.J. 1976. Environ. Sci. Technol. 10, 370-373.
2.Neff, J.M. 1979. Polycyclic Aromatic Hydrocarbons in the Aquatic Environment. Applied Sci. Publ. London.
3.Zsolnay, A. 1977. Mar. Chem. 5, 465-475.

5-143 Background concentration Trichloroethane

Species	Value	Condition
Bird, egg	16.5 mg/kg wet weight	Rg = 3 - 30 mg/kg wet weight, food from water organisms[1]
Bird, liver	2.5 mg/kg wet weight	Rg = 1 - 4 mg/kg wet weight, food from water organisms[1]
Bread	2 mg/kg wet weight	None[1]
Crustacea	17 mg/kg wet weight	Rg = 0.7 - 34 mg/kg wet weight[1]
Dairy products	5.1 mg/kg wet weight	Rg = 0.1 - 10 mg/kg wet weight[1]
Fish, flesh	2.8 mg/kg wet weight	Rg = 0.7 - 5 mg/kg wet weight[1]
Fish, liver	8.0 mg/kg wet weight	Rg = 1 - 15 mg/kg wet weight[1]
Fruit and vegetables	2.5 mg/kg wet weight	Rg = 1 - 4 mg/kg wet weight[1]
Homo sapiens, fat	13 mg/kg wet weight	Rg = 1 - 25 mg/kg wet weight[1]
Homo sapiens, liver	3.0 mg/kg wet weight	Rg = 1 - 5 mg/kg wet weight[1]
Marine algae	17.5 mg/kg wet weight	Rg = 10 - 25 mg/kg wet weight[1]
Meat	4.5 mg/kg wet weight	Rg = 3 - 6 mg/kg wet weight[1]
Mollusca	5.5 mg/kg wet weight	Rg = 0.05 - 10 mg/kg wet weight[1]
Plankton	5.4 mg/kg wet weight	Rg = 0.03 - 10.7 mg/kg wet weight[1]
Seal, blubber	16 mg/kg wet weight	Rg = 8 - 24 mg/kg wet weight[1]
Seal, liver	2.1 mg/kg wet weight	Rg = 0.2 - 4 mg/kg wet weight[1]
Vegetable oil	5.7 mg/kg wet weight	Rg = 0.5 - 10 mg/kg wet weight[1]

Source: 1.Pearson, C.R. 1982. In: O. Hutzinger (ed.): Environmental Chemistry. Springer-Verlag. Volume 3, part B. Berlin.

5-144 Background concentration Trichloroethylene

Species	Value	Condition
Atmosphere	0.005 ug/m3	Typical value[1]
Bird, egg	17.7 mg/kg wet weight	Rg = 2.4 - 33 mg/kg wet weight, food from water organisms[2]
Bird, liver	4.1 mg/kg wet weight	Rg = 2.1 - 6 mg/kg wet weight, food from water organisms[2]
Bread	7 mg/kg wet weight	None[2]
Crustacea	9.3 mg/kg wet weight	Rg = 2.6 - 16 mg/kg wet weight[2]
Dairy products	5.2 mg/kg wet weight	Rg = 0.3 - 10 mg/kg wet weight[2]
Drinking water	0.1 ug/m3	High value, halocarbon[2]
Drinking water	0.1 ug/m3	Typical non-polluted value, halocarbon[2]
Estuarine water	0.3 ug/m3	High value in inshore water[3]
Fish, flesh	5.9 mg/kg wet weight	Rg = 0.8 - 11 mg/kg wet weight[2]
Fish, liver	29 mg/kg wet weight	Rg = 2 - 56 mg/kg wet weight[2]
Freshwater	0.5 ug/m3	Typical non-polluted value, halocarbon[2]
Freshwater	10 ug/m3	High value in urban area, rg = 1.0 - 20 ppb, halocarbon[2]
Fruit and vegetables	0.27 mg/kg wet weight	Rg = 0.05 - 0.5 mg/kg wet weight[2]
Groundwater	0.2 ug/m3	Typical non-polluted value, halocarbon[2]
Groundwater	5 ug/m3	High value, halocarbon[2]

Homo sapiens, fat	17.7 mg/kg wet weight	Rg = 1.4 - 32 mg/kg wet weight[2]
Homo sapiens, liver	3.9 mg/kg wet weight	Rg = 2 - 5.8 mg/kg wet weight[2]
Marine algae	19 mg/kg wet weight	Rg = 16 - 22 mg/kg wet weight[2]
Meat	17 mg/kg wet weight	Rg = 12 - 22 mg/kg wet weight[2]
Mollusca	6.0 mg/kg wet weight	Rg = 0.05 - 12 mg/kg wet weight[2]
Plankton	0.47 mg/kg wet weight	Rg = 0.05 - 0.9 mg/kg wet weight[2]
Rain	0.08 ug/m3	Rg = 0.005 - 0.15 ppb, halocarbon[2]
Seal, blubber	4.9 mg/kg wet weight	Rg = 2.5 - 7.2 mg/kg wet weight[2]
Seal, liver	4.6 mg/kg wet weight	Rg = 3 - 6.2 mg/kg wet weight[2]
Seawater	0.010 ug/m3	Typical open ocean[3]
Sediment	0.500 ug/m3	Typical conc in inshore sediment[3]
Snow	0.021 ug/m3	Rg = 0.001 - 0.004 ppb, halocarbon[2]
Vegetable oil	4.5 mg/kg wet weight	Rg = 0.05 - 9 mg/kg wet weight[2]

Source: 1.Lovelock, J.E. and P. Simmons. 1980. In: Sugden (ed.): The Aerosol Controversy. Sci-Horwood. 2.Pearson, C.R. 1982. In: O. Hutzinger (ed.): Environmental Chemistry. Springer-Verlag. Volume 3, partB. Berlin. 3.National Academy of Science. 1978. Chloroform, Carbon tetrachloride, and other halomethanes. An Environmental Assessment. US NAS.

5-145 Background concentration U

Species	Value	Condition
Atmosphere	0.02 ng/m**3	UK, 1965[1]
Atmosphere	3.5E-9 ng/m**3	Air over Atlantic Ocean[2]
Carbonate sediments	2.2 mg/kg dry weight	Average value[3]
Crustral rock	0.00018 percent by weight	Average value[4]
Deep-sea clay sediments	1.3 mg/kg dry weight	Average value[3]
Drinking water	32 ng/l	New York, Japan = 4.8 - 11.4 ng/l[5]
Freshwater sediment	0.6 mg U3O8/kg dry weight	Gyttja, Germany, n=8, Halterner Stausee, rg=0.5-0.7 ppm U3O8[6]
Freshwater sediment	1.0 mg U3O8/kg dry weight	Gyttja, Germany, n=6, Baldeneysee, n=6, rg=0.5-1.5 ppm U3O8[6]
Freshwater sediment	1.6 mg U3O8/kg dry weight	Gyttja, Germany, Hengsteysee, n=6, rg=0.7-2.5 ppm U3O8[6]
Freshwater sediment	2.8 mg U3O8/kg dry weight	Gyttja, Germany, Selenter See, n=16, rg = 0.5-5.0 ppm U3O8[6]
Freshwater sediment	2.8 mg U3O8/kg dry weight	Gyttja, Schlkenmehrer Maar, n=11, rg=0.6-5.0, germany[6]
Freshwater sediment	3.0 mg U3O8/kg dw	Gyttja, Laacher See, n=15, rg = 1-5.0 ppm U3O8, Germany[6]
Freshwater sediment	3.5 mg U3O8/kg dry weight	Gyttja, Germany, Pulvermaar, n=13, rg=1.0-6.0 ppm U3O8[6]
Freshwater sediment	6.8 mg U3O8/kg dry weight	Gyttja, Germany, n=17, Meerfelder Maar, rg=1.5-13.0 ppm U3O8[6]
Granitic rock	3.0 mg/kg dry weight	Average value[3]
Groundwater	14 mg U3O8/l	Lauenburg and Avendorf, germany, rg=7-21 ppm U3O8, n=23[6]
Groundwater	2.0 mg U3O8/l	Elbe and weser alluvium, rg =0-4.0 ppm U3O8, germany, n=23[6]
River water	0.001 ppm	Range = 0.2E-4-0.05 ppm[7]
Sandstone sediment	0.45 mg/kg dry weight	Average value[3]
Seawater	0.003 ppm	UO2(CO3)3(4-), residence time = 500E3 y[8]
Seawater	3.3 ug/l	None[9]
Shale sediment	3.7 mg/kg dry weight	Average value[3]

Soil	0.5 mg U3O8/kg dry weight	Spain, vertisols, terrestrical soil, n=9, rg=0.0-1.0 ppm U3O8[6]
Soil	0.7 mg U3O8/kg dry weight	Portugal, vertisols, terrestrical soil, rg = 0-1.5 ppm U3O8, n=19[6]
Soil	1 ppm dry soil	Oven-dried, rg = 0.9 - 9[7]
Soil	1.0 mg U3O8/kg dry weight	Ditmarschen, Elbe and Weser Marsh, n=33, saline, rg=0-2.0 ppm U3O8[6]
Soil	1.3 mg U3O8/kg dry weight	Rhein Ruhr, n-28, Grey brown podsol, A-horizon, rg=0.5-2.0 ppm U3O8[6]
Soil	2.0 mg U3O8/kg dry weight	W-coast, Bornholm, Denmark, lignite[6]
Soil	5 mg U3O8/kg dry weight	Volcanic and Maar Eifel, n=27, rg = 1-10 ppm U3O8[6]

Source: 1.Friberg, L. G.N. Nordberg and V.B. Vouk. 1979. Handbook on the Toxicology of Metals. Elsevier.
2.Hamilton, E.I. 1970. Health Phys. 19, pp 511-520.
3.Turekian, K.K. and K.H. Wedepohl. 1961. Amer. Geol. Soc. Bull. 72, pp 175-192.
4.Mason, B. 1966. Principles of Geochemistry. 3rd ed. John Wiley and Sons. New York.
5.Welford, G.A. and R. Baird. 1967. Heath Phys. 13, pp 1321-1324.
6.Colwell, R.R., G.S. Sayler, J.D. Nelson and A. Justice. 1976. In: J.O. Nriagu (ed.): Environmental Biochemistry, Vol 2. Ann Arbor Science. Ann Arbor. Mich. 473-487.
7.Bowen, H.J.M. 1966. Trace Elements in Biochemistry, Ac. Press.
8.Goldberg, E.D., 1963. The Sea, Interscience London, 2, pp 3.
9.Keen, N.J. 1968. J. Brit. Nucl. Energy Soc. 7, pp 178-183.

5-146 Background concentration
V

Species	Value	Condition
Angiosperms	1.6 mg/kg dry weight	USA[1]
Annelida	1.2 mg/kg dry weight	USA[1]
Atmosphere	0.0015 ng/m**3	Southpole, 1970, rg = 0.0006[2]
Atmosphere	12.25 ng/m**3	Not urban, USA, rg = 0.5 - 24[2]
Atmosphere	18.2 ng/m**3	Engl, 7 cities, 1972, rg = 4.2 - 32.2[2]
Atmosphere	23.5 ng/m**3	Urban, USA, rg = 1 - 46[2]
Atmosphere	60 ng/m**3	Chicago, annual mean[3]
Atmospheric particulates	0.16 ng/m**3	Hawaii, n > = 56, rg = +- 0.11[4]
Brown algae	2 mg/kg dry weight	USA[1]
Bryophytes	2.3 mg/kg dry weight	USA[1]
Carbonate sediments	20 mg/kg dry weight	Average value[5]
Coelenterates	2.3 mg/kg dry weight	USA[1]
Crustacenas	0.4 mg/kg dry weight	USA[1]
Crustral rock	0.014 percent by weight	Average value[6]
Deep-sea clay sediments	120 mg/kg dry weight	Average value[5]
Drinking water	0.5 ug/l	Sweden[2]
Echinodermata	1.9 mg/kg dry weight	USA[1]
Ferns	0.13 mg/kg dry weight	USA[1]
Fish	0.14 mg/kg dry weight	USA[1]
Food	0.0001 mg/kg wet weight	Milk, Sweden[2]
Food	0.00082 mg/kg wet weight	Potato, Sweden[2]
Food	0.01 mg/kg wet weight	Calfliver, Sweden[2]
Food	0.23 mg/kg dry weight	USA, polished rice[2]

Food	0.82 mg/kg dry weight	Japan, polished rice[2]
Forest foliage	7.7 mg/kg ash weight	Uncontaminated value for USA, rg = <5-70 ppm[7]
Fungi	0.67 mg/kg dry weight	USA[1]
Granitic rock	88 mg/kg dry weight	Average value[5]
Gymnosperms	0.69 mg/kg dry weight	USA[1]
Insecta	0.15 mg/kg dry weight	USA[1]
Mammalia	0.4 mg/kg dry weight	USA, less than value[1]
Mollusca	0.7 mg/kg dry weight	USA[1]
Plankton	5 mg/kg dry weight	USA[1]
Rain	4.1 ug/l	Engl, 1971, relative uncontaminated, V in atmosphere = 9.6 ng/m**3[2]
River water	0.001 ppm	Range < 0.007 ppm[8]
Sandstone sediment	20 mg/kg dry weight	Average value[5]
Seawater	0.002 ppm	VO5H3(2-), residence time = 10E3 y[9]
Shale sediment	130 mg/kg dry weight	Average value[5]
Soil	100 ppm dry soil	Oven-dried, rg = 20 - 500[8]
Water	1 ug/l	Surface water, mean[2]

Source: 1.National Academy of Science. 1977. Effects of Atmospheric Pollutants, Vanadium. Wash DC. Nat. Aca. Sc., p 117, 104.
2.Laveskog, A., A. Lindskog and V. Stenberg. 1976. Om Metaller, Statens Naturvaardsverk, Stockholm.
3.Lee, R.E., S.S. Goranson, R.E. Enrione and G.B. Morgan. 1972. Environ. Sci. Technol. 6, pp 1025-1030
4.Golterman, H.L. 1975. Physiological Limnol., Elsevier Amsterdam.
5.Turekian, K.K. and K.H. Wedepohl. 1961. Amer. Geol. Soc. Bull. 72, pp 175-192.
6.Mason, B. 1966. Principles of Geochemistry. 3rd ed. John Wiley and Sons. New York.
7.Shacklette, H.T., J.A. Erdman, T.F. Harms and C.S.E. Pupp. 1978. In: F.W. Oeheme (ed.): Toxicity ofeavy Metals in the Environment. Dekkker. NY. 25-68.
8.Bowen, H.J.M. 1966. Trace Elements in Biochemistry, Ac. Press.
9.Goldberg, E.D., 1963. The Sea, Interscience London, 2, pp 3.

5-147 Background concentration Vitamin b12

Species	Value	Condition
Freshwater	0.0024 mg/m**3	Dissolved, Tsukui, Japan, surface, rg = 0.0005 - 0.0042[1]
Freshwater	0.0111 mg/m**3	Particulate, Tsukui, Japan, surface, rg = 0.0019 - 0.0203[1]
Freshwater	0.0165 mg/m**3	Kasumigaura, Japan, rg = 0.005 - 0.028[2]
Freshwater	0.067 mg/m**3	Lindsey Pond, Conn, rg = 0.06 - 0.075[3]
Freshwater	0.426 mg/m**3	Sagami, Japan, rg = 0.005 - 0.85[4]
Rain	0.0004 mg/m**3	Mean of precipatation, MO, Nov-March[5]
Rain	0.001 mg/m**3	Mean of precipitation, MO, April-Nov.[5]

Source: 1.Ohwada, K., M. Otsuhata and N. Taka. 1972. Bull. Japan Soc. Sci. Fish. 38, pp 817-823.
2.Kashiwada, K., A. Kanazawa and S. Tachibanazono. 1963. Mem. Fac. Fish., Kagoshima Univ. 12, pp 153-17.
3.Benoit, R.J. 1957. Limnol. Oceanogr., 2, pp 233-240.
4.Ohwada, K. and N. Taga. 1972. Limnol. Oceanogr. 17, pp 315-320.
5.Parker, B.C. and M.A. Wachtel. 1971. in: Cairns, J. (ed.); The Structure and Function of Freshwater Microbial Communities. Res. Div. Monogr. Virginia Polytech. Inst. 3, pp 195-207.

5-148 Background concentration
W

Species	Value	Condition
Atmosphere	0.0015 ng/m**3	Southpole, 1970, rg = +- 0.0006[1]
Atmosphere	0.4 ng/m**3	Not urban, USA, 1969[1]
Atmosphere	1 ng/m**3	Not urban, Engl, 1971[1]
Atmosphere	1 ug/l	Less than value, UK, 1972-1973[2]
Atmosphere	5.8 ng/m**3	Chicago, 1969[1]
Carbonate sediments	0.6 mg/kg dry weight	Average value[3]
Crustral rock	0.00015 percent by weight	Average value[4]
Deep-sea clay sediments	4.5 mg/kg dry weight	Average value[3]
Drinking water	0.05 ug/l	Rg = 0.03 - 0.1 ug/l, Sweden, neutron activation[5]
Granitic rock	1.3 mg/kg dry weight	Average value[3]
Rain	0.35 ug/l	Not urban, Engl, 1971, W in atm. = 1 ng/m**3[1]
Sandstone sediment	1.6 mg/kg dry weight	Average value[3]
Seawater	0.1E-3 ppm	WO4(2-), residence time = 1E3 y[6]
Shale sediment	1.8 mg/kg dry weight	Average value[3]
Water	0.03 ug/l	River, Sweden, 1965[1]

Source: 1.Laveskog, A., A. Lindskog and V. Stenberg. 1976. Om Metaller, Statens Naturvaardsverk, Stockholm.
2.Cawse, P.A. 1974. A Survey of Atmospheric Trace Elements in the UK. 1972-73. AERER R 7669. Her Majestys office. London.
3.Turekian, K.K. and K.H. Wedepohl. 1961. Amer. Geol. Soc. Bull. 72, pp 175-192.
4.Mason, B. 1966. Principles of Geochemis try. 3rd ed. John Wiley and Sons. New York.
5.Bostroem, H. and P.O. Wester. 1967. Acta Med. Scand. 181, pp 465-473.
6.Goldberg, E.D., 1963. The Sea, Interscience London, 2, pp 3.

5-149 Background concentration
Xe

Species	Value	Condition
Atmosphere	0.086 ug/m**3	293 K, 1 atm[1]
Atmosphere	8E-6 % by atoms	Sea-level dry atmospheric composition[2]
Lower atmosphere	0.00009 % by volume	None[3]
Seawater	0.52E-4 ppm	Xe[4]

Source: 1.Junge, C.E. 1963. Air Chemistry and Radioactivity, Ac. Press.
2.Sisler, F.D. 1975. In: S.F. Singer (ed.), The Changing Global Environment, pp 57-69. D. Reidel Publ. Comp.
3.Gleuckauf, E. 1951. Compendium of Meteorology. Am. Meteorological Soc. Boston.
4.Goldberg, E.D., 1963. The Sea, Interscience London, 2, pp 3.

5-150 Background concentration
Xylenes

Species	Value	Condition
Atmospheric conc of aliphatic hydrocarbon and alkylbenzene	23.5 ppb	Los Angeles, 1967, STP[1]

Source: 1.Altshuller, A.P., W.A. Lonneman and S.L. Kopczynski. 1973. J. Air Pollut. Contr. Assoc. 23, 597-599.

5-151 Background concentration Y

Species	Value	Condition
Seawater	0.3E-3 ppm	Residence time = 7.5E3 y[1]
Soil	50 ppm dry soil	Oven-dried, rg = 25 - 250[2]

Source: 1.Goldberg, E.D., 1963. The Sea, Interscience London, 2, pp 3. 2.Bowen, H.J.M. 1966. Trace Elements in Biochemistry, Ac. Press.

5-152 Background concentration Yb

Species	Value	Condition
Seawater	0.0021 microgram/l	Soluble concentration, Japan, Cl = 1.873 %[1]
Seawater	0.0056 microgram/l	Total concentration, Japan, Cl = 1.873 %[1]

Source: 1.Suzuki, H., T. Koganage and M. Saki. 1975. Impacts of Nuclear Releases into the Aquatic Environment. IAEA SM-198/19, pp 77-89. Vienna.

5-153 Background concentration Zn

Species	Value	Condition
Anodonta sp.	120 mg/kg dry weight	Gastropod, L Aberffraw, 1970, freshwater[4]
Asellus aquaticus	181.9 mg/kg dry weight	Freshwater, Crustaceans, rg = 118-203 ppm, Elsenz R, 1976[5]
Atmosphere	0.03 ng/m**3	Southpole, 1970a, rg = +- 0.01[6]
Atmosphere	130.9 ng/m3	Pasadena, Calif. mass median dia = 3.4 um, May 1975[1]
Atmosphere	1700 ng/m3	Cincinnati, Ohio, mass median dia = 1.1 um, 1970[1]
Atmosphere	2820 ng/m3	Liege, Belgium, mass median dia = 1.5 um, 1972[1]
Belone belone	48 mg/kg dry weight	Fish, marine, rg - +- 6, NE Atlantic, Cornwall, muscle, 1974[11]
Bulk precipitation	108 ug/l	Lake Ontario Basin, winter, 1969-1975, fallout = 5.8 ug/(cm2*y)[12]
Bulk precipitation	92 ug/l	Total Zn, 899 stations, N Ont., rg = 62-243 ug/l[12]
Cancer irroratus	30.5 mg/kg wet weight	Marine, Crustacean, rg =+- 10.2, Terrence B, Nova Scotia, whole, 1973[8]
Cardium edule	150 mg/kg dry weight	Gastropod, rg = 100-200 ppm, SW Spain and Portugal, Medit., 1974[16]
Carpiodes cyprinus	3.4 mg/kg wet weight	Fish, rg = 2.1-5.5 ppm, Illinois R, muscle, 1973[17]
Chthamalus stellatus	129 mg/kg dry weight	Marine, Crustaceans, rg = 26, soft parts, SW Spain, Guadiana E., 1975[8]
Cladophora glomerata	126 mg/kg dry weight	Algae, Rg = 62 - 190 ppm, Leine R., FRG, 1976[20]
Cladophora glomerata	23.7 mg/kg dry weight	Algae, Deadman Bay, Kingston, Ontario, 1975[21]
Cladophora glomerata	8.2 mg/kg dry weight	Algae, Main Duck, L Ontario, 1975[21]
Crangon crangon	14 mg/kg dry weight	Marine, Crustacean, S Antonio Bay, Texas, 1975, whole animal[24]
Crassostrea gigas	1510 mg/kg dry weight	Gastropod, rg = 1350-2670 ppm, Helford, Colne, Poole Est., GB, 1974[26]
Crassostrea gigas	21 mg/kg wet weight	Gastropod, rg = +- 5 ppm, Saldanha B, LangeBaan B., RSA, 1975[18]
Crassostrea gigas	22500 mg/kg dry weight	Gastropod, rg = 9860-35120 ppm, Hinkley Pt. Lower Sta, Severn Est., 73[27]
Crassostrea gigas	396 mg/kg dry weight	Gastropod, Knysna Estuary, R.S.A., 1975[28]

Crassostrea gigas	7227 mg/kg dry weight	Gastropod, rg = +- 4010 ppm, Tamar R Est., Tasmania, 1973[29]
Crustral rock	0.0070 percent by weight	Average value[31]
Cyprinus carpio	10.2 mg/kg wet weight	Fish, rg = 4.1-16.1 ppm, Illinois R, muscle, 1973[17]
Deep-sea clay sediments	165 mg/kg dry weight	Average value[15]
Dreissena polymorpha	174 mg/kg dry weight	Rg = 140-208 ppm, Elbe Est., transplanted from Schaal L, 1973, freshwater[32]
Esox lucius	11 mg/kg wet weight	Fish, L Erie, USA, 1970, animal decapiated and eviscerated[23]
Esox lucius	19 mg/kg wet weight	Fish, Moose L, USA, 1970, animal decapiated and eviscerated[23]
Esox lucius	2.6 mg/kg wet weight	Fish, rg = 2.3 - 2.8 ppm, Illinois R, USA, muscle, 1972[17]
Food	125 mg/kg dry weight	Animals, rg = 50 - 200[6]
Food	55 mg/kg dry weight	Vegetables, rg = 10 - 100[6]
Freshwater, lake	20 ug/l	Range = 19 - 95 ppm, Maryrnong, Yarra, Australia, 1974[36]
Freshwater, river	70 ug/l	Range = 11 - 120 ppm, Danube, Ulm/Regensburg, 1973[42]
Freshwater, river and lake	20 ug/l	Range = < 12 - 42000 ppm, 720 river and lakes, 1970[44]
Fucus sp.	227 mg/kg dry weight	Range = 110 - 345 ppm, Portugal and Spain, Atlantic, 1974[16]
Gadus morrhua	43 mg/kg dry weight	Fish, rg = 38-48 ppm, Trondsheimfj, Norway, muscle, 1976[3]
Glacier ice	0.38 ug/l	Greenland, before 1900[12]
Glacier ice	1.05 ug/l	Greenland ice, 1966-1971[12]
Granitic rock	60 mg/kg dry weight	Average value[15]
Hypnum cupressiforme, non-polluted area	74 mg/kg dry weight	Wales, 1970, Bryophyte[1]
Hypnum cupressiforme, non-polluted area	95 mg/kg dry weight	Rg = 90 - 100 ppm, Denmark, Bryophyte[1]
Hypnum cupressiforme, polluted area	187 mg/kg dry weight	Rg = 47-328 ppm, Consett, England, Fe and steelworks, 1975, Bryophyte[1]
Hypnum cupressiforme, polluted area	345 mg/kg dry weight	Max. value, Swansea, Wales, 1970, smelter district, Bryophyte[1]
Littorina littorea	117 mg/kg dry weight	Gastropod, rg = 60-210 ppm, Bristol Channel, 1971, marine[49]
Littorina littorea	219 mg/kg dry weight	Gastropod, Severn Estuary, Bristol Ch., 1971, rg = 100-520 ppm, marine[50]
Marine phytoplankton	39.5 mg/kg wet weight	Range = 35 - 44 ppm, Cape W Coast, S Africa, 1973[19]
Marine zooplankton, copepods, sagitta, euphausids	4.45 mg/kg wet weight	Range = 1.9 - 7.0 ppm, Cape W Coast, S Africa, 1974[19]
Mytilus edulis	14 mg/kg wet weight	Gastropod, rg = 3.8-26 ppm, New Zealand, 1964[13]
Mytilus edulis	147 mg/kg dry weight	Gastropod, rg = 62-250 ppm, Bristol Channel, GB, 1971[49]
Phosphate rocks	222 mg/kg dry weight	Rock from Morocco[60]
Phosphate rocks	24 mg/kg dry weight	Rock from Kola, USSR[60]
Phosphate rocks	94 mg/kg dry weight	Rock from Florida, USA[60]
Platichthyes flesus	126 mg/kg dry weight	Fish, marine, Oldbury, Severn Estu., power station, rg=+-49 ppm[61]
Pleuronectes platessa	30.3 mg/kg wet weight	Fish, marine, Northumberl, power station, muscle, 1975[14]
Potamogeton pectinatus	175 mg/kg dry weight	Higher plant, Rg = 137-213 ppm, Leine R, FRG, 1976[20]
Rain	49.5 ug/l	Sweden, 3 stations, rg = 15.5 - 84[6]
Rutilus ruthilus	8.4 mg/kg wet weight	Fish, rg = 6.1-11.9 ppm, Elsenz R, muscle, 1977, FRG[65]

Seawater	0.01 ppm	Zn(2+), residence time = 180E3 y[66]
Soil	50 ppm dry soil	Oven-dried, rg = 10 - 300[64]
Stormwater	470 g/m3	Lodi, New Jersey, USA, rain = 44 ug/l, Saddle R = 47 ug/l[1]
Superphosphate fertilizer	363 mg/kg dry weight	Rock from Florida, USA[60]
Ulva lactuca	110 mg/kg dry weight	Range = 59 - 160 ppm, Portugal and Spain, 1974, Atlantic[52]

Source: 1.Nriagu, J.O. 1980. Zinc in the Environment. Part 1: Ecological cycling. Wiley and Sons. NY.
3.Lande, E. 1977. Environ. Pollut. 12, 187-198.
4.Segar, D.A. and R.E. Pellenbarg. 1971. J. Mar. Biol. Assoc. UK. 51, 131-136.
5.Prosi, F. 1977. Schwermetallbelastung in den sedimenten der Elsenz und ihre Auswerkung auf Limnische Organismen. Diss. Univ. Heidelberg.
6.Laveskog, A., A. Lindskog and V. Sten berg. 1976. Om Metaller, Statens Naturvaardsverk, Stockholm.
8.Martin, J.B.M. 1974. Mar. Biol. 28, 245-251.
11.Stevens, J.D. and B.E. Brown. 1974. Mar. Biol. 26, 287-293.
12.Bohn, H.L. 1978. On organic soil carbon and CO2. Tellus 30.
13.Heydt, G. 1977. Schwermetallgehalte von Wasser, Wasserpflanzen, Chironomidae und Mollusca der Elsenz. Dipl. Arbeit Univ. Heidelberg. 143 pp.
14.Wright, D.A. 1976. Mar. Pollut. Bull. 7, 36-38.
15.Turekian, K.K. and K.H. Wedepohl. 1961. Amer. Geol. Soc. Bull. 72, pp 175-192.
16.Foster, P. 1976. Environ. Pollut. 10, 45-53.
17.Mathis, B.J. and N.R. Kevern. 1973. Distribution of Hg, Cd, Pb and Th in a eutrophic lake. Int. Water Res. Michigan State Univ., Technol. Dep. 34, 23 pp.
18.Fourie, H.O. 1976. S. Afr. J. Sci. 72, 110-113.
19.Van As, D. H.O. Fourie and C.M. Vleggaar. 1975. S. Afr. J. Sci. 71, 151-154.
20.Abo-Rady, M.D.K. 1977. Die Belastung der oberen Leine mit Schwermetallen durch kommunale und industrelle Abwasser. Diss. Univ. Gottingen. 120 p.
21.Keeney, W.L., W.G. Breck, G.W. Van Loon and J.A. Page 1976. Water Res. 10, 981-984.
23.Uthe, J.H. and E.G. Bligh. 1971. J. Fish. Res. bd Canada 28, 786-788.
24.Sims, R.R. and B.J. Presley. 1976. Bull. Environ. Contam. Toxicol. 16, 520-526.
26.Thornton, I., H. Watling and A. Darracott. 1975. Sci. Tot. Environ. 4, 325-345.
27.Romeril, M.G. 1971. Mar. Biol. 9, 347-354.
28.Watling, H.R. and R.J. Watling. 1976. Mar. Pollut. Bull. 7, 45-48.
29.Ayling, G.M. 1974. Water Res. 8, 729-738.
31.Mason, B. 1966. Principles of Geochemistry. 3rd ed. John Wiley and Sons. New York.
32.Karbe, L., N. Antonacopoulos and C. Schnier. 1975. Verh. Int. Ver. Limnol. 19, 2094-2101.
36.Phillip, A.T., R.W. Pettis, J.E. Harris, Fabris, Boar and Bone. 1975. Proc. R. Austr. Chem. Inst. 42,209-215.
42.Quentin, K.E., and H.A. Winkler. 1974. Zentralbl. Bakt. Abt. Orig. B 158, 514-523.
44.Durum, W.H., J.D. Hem and S.G. Heidel. 1971. US Geol. Surv. Circ. 643, 49.
49.Nickless, G., R. Stenner and N. Terrile. 1972. Mar. Pollut. Bull. 3, 188-190.
50.Butterworth, J., P. Lester and G. Nickless. 1972. Mar. Pollut. Bull. 3, 72-74.
52.Stenner, R.D. and G. Nickless. 1975. Mar. Pollut. Bull. 6, 89-92.
60.Langmyhr, F.J. R. Solberg and Y. Thomassen. 1977. Anal. Chim. Acta 92, pp 105-109.
61.Hardisty, M.W., R.J. Huggins, S. Kartar and M. Sainsbury. 1974. Mar. Pollut. Bull. 5, 61-63.
64.Bowen, H.J.M. 1966. Trace Elements in Biochemistry, Ac. Press.
65.Muller, G. and F. Prosi. 1977. Naturwissenschften 64, 530.
66.Goldberg, E.D., 1963. The Sea, Interscience London, 2, pp 3.

5-154 Background concentration Zr

Species	Value	Condition
River water	0.26E-2 ppm	Range = 0.5E-4-0.022 ppm[1]
Seawater	0.22E-4 ppm	None[2]
Soil	300 ppm dry soil	Oven-dried, rg = 60 - 2000[1]

Source: 1.Bowen, H.J.M. 1966. Trace Elements in Biochemistry, Ac. Press.
2.Goldberg, E.D., 1963. The Sea, Interscience, London, 2, pp 3.

5-155 Background concentration n-Butanol

Species	Value	Condition
Atmosphere, organic gas	88 ppb	Rg = 51-126 ppb, Point Barrow, Alaska, 1968[1]

Source: 1.Cavanagh, L.A., C.F. Sahadt and E. Robinson. 1969. Environ. Sci. Technol. 3, 251-257.

5-156 Concentration factor
1-Methylnaphthalene

Species	Value	Condition
Oncorhynchus kisutch, muscle	100 based on wet weight	Water sol fraction = 4 ppb, 288 K, 5 w, flow through, prudhoe B crude oil[1]

Source: 1.Roubal, W.T., D.H. Bovee, T.K. Collier and S.I. Stranahan. 1977. Proc of 1977 Oil Spill Conf. (Prevention, Behavior, Control, Cleanup). Wash DC. American Petroleum Institute. 551-556.

5-157 Concentration factor
2,4,2(a)chlorobiphenyl

Species	Value	Condition
Salmo salar	0.414 based on wet weight	Uptake from food, steady state cf., 288 K, 48 h, 6.24 g fish[1]
Salmo salar	389 based on wet weight	Uptake from water, steady state cf., 288 K, 48 h, 6.24 g fish[1]

Source: 1.Zitko, V. and O. Hutzinger. 1976. Bull. Environ. Contam. Toxicol. 16(6), pp 665-673.

5-158 Concentration factor
2,4,5-T

Species	Value	Condition
Scenedesmus acutus	3.3 based on dry weight	Water = 0.5E-7 mol, 303 K, pH = 7-8, 8 h[1]

Source: 1.Boehm, H.H. and H. Muller. 1976. Naturwisenschaftens 63(3), pp 296.

5-159 Concentration factor
2,4-D

Species	Value	Condition
Scenedesmus acutus	2.2 based on dry weight	Water = 0.5E-7 mol, 303 K, pH = 7-8, 8 h[1]

Source: 1.Boehm, H.H. and H. Muller. 1976. Naturwissenschaftens 63(3), pp 296.

5-160 Concentration factor
2,4-Di-bromobiphenyl

Species	Value	Condition
Salmo salar	0.318 based on wet weight	Uptake from food, steady state cf., 288 K, 48 h, 6.24 g fish[1]
Salmo salar	1343 based on wet weight	Uptake from water, steady state cf., 288 K, 48 h, 6.24 g fish[1]

Source: 1.Zitko, V. and O. Hutzinger. 1976. Bull. Environ. Contam. Toxicol. 16(6), pp 665-673.

5-161 Concentration factor
2,4-Di-chlorobiphenyl

Species	Value	Condition
Salmo salar	0.054 based on wet weight	Uptake from food, steady state cf., 288 K, 48 h, 6.24 g fish[1]
Salmo salar	894 based on wet weight	Uptake from water, steady state cf., 288 K, 48 h, 6.24 g fish[1]

Source: 1.Zitko, V. and O. Hutzinger. 1976. Bull. Environ. Contam. Toxicol. 16(6), pp 665-673.

5-162 Concentration factor
2,5,3(b)chlorobiphenyl

Species	Value	Condition
Salmo salar	0.586 based on wet weight	Uptake from food, steady state cf., 288 K, 48 h, 6.24 g fish[1]
Salmo salar	389 based on wet weight	Uptake from water, steady state cf., 288 K, 48 h, 6.24 g fish[1]

Source: 1.Zitko, V. and O. Hutzinger. 1976. Bull. Environ. Contam. Toxicol. 16(6), pp 665-673.

5-163 Concentration factor
2,5,4(c)chlorobiphenyl

Species	Value	Condition
Salmo salar	0.494 based on wet weight	Uptake from food, steady state cf., 288 K, 48 h, 6.24 g fish[1]
Salmo salar	385 based on wet weight	Uptake from water, steady state cf., 288 K, 48 h, 6.24 g fish[1]

Source: 1.Zitko, V. and O. Hutzinger. 1976. Bull. Environ. Contam. Toxicol. 16(6), pp 665-673.

5-164 Concentration factor
2,6-di-bromobiphenyl

Species	Value	Condition
Salmo salar	0.179 based on wet weight	Uptake from food, steady state cf., 288 K, 48 h, 6.24 g fish[1]
Salmo salar	1267 based on wet weight	Uptake from water, steady state cf., 288 K, 48 h, 6.24 g fish[1]

Source: 1.Zitko, V. and O. Hutzinger. 1976. Bull. Environ. Contam. Toxicol. 16(6), pp 665-673.

5-165 Concentration factor
2,6-di-chlorobiphenyl

Species	Value	Condition
Salmo salar	0.137 based on wet weight	Uptake from food, steady state cf., 288 K, 48 h, 6.24 g fish[1]
Salmo salar	919 based on wet weight	Uptake from water, steady state cf., 288 K, 48 h, 6.24 g fish[1]

Source: 1.Zitko, V. and O. Hutzinger. 1976. Bull. Environ. Contam. Toxicol. 16(6), pp 665-673.

5-166 Concentration factor
2,6-di-chlorobiphenyl1

Species	Value	Condition
Salmo salar	0.449 based on wet weight	Uptake from food, steady state cf., 288 K, 48 h, 6.24 g fish[1]
Salmo salar	425 based on wet weight	Uptake from water, steady state cf., 288 K, 48 h, 6.24 g fish[1]

Source: 1.Zitko, V. and O. Hutzinger. 1976. Bull. Environ. Contam. Toxicol. 16(6), pp 665-673.

5-167 Concentration factor
2,6-di-chlorobiphenyl2

Species	Value	Condition
Salmo salar	0.587 based on wet weight	Uptake from food, steady state cf., 288 K, 48 h, 6.24 g fish[1]
Salmo salar	314 based on wet weight	Uptake from water, steady state cf., 288 K, 48 h, 6.24 g fish[1]

Source: 1.Zitko, V. and O. Hutzinger. 1976. Bull. Environ. Contam. Toxicol. 16(6), pp 665-673.

5-168 Concentration factor 2,6-di-chlorobiphenyl3

Species	Value	Condition
Salmo salar	0.571 based on wet weight	Uptake from food, steady state cf., 288 K, 48 h, 6.24 g fish[1]
Salmo salar	111 based on wet weight	Uptake from water, steady state cf., 288 K, 48 h, 6.24 g fish[1]

Source: 1.Zitko, V. and O. Hutzinger. 1976. Bull. Environ. Contam. Toxicol. 16(6), pp 665-673.

5-169 Concentration factor 2-Methylnaphthalene

Species	Value	Condition
Oncorhynchus kisutch, muscle	140 based on wet weight	Water sol fraction = 4 ppb, 288 K, 5 w, flow through, prudhoe B crude oil[1]

Source: 1.Roubal, W.T., D.H. Bovee, T.K. Collier and S.I. Stranahan. 1977. Proc of 1977 Oil Spill Conf. (Prevetion, Behavior, Control, Cleanup). Wash DC. American Petroleum Institute. 551-556.

5-170 Concentration factor 3,4-di-bromobiphenyl

Species	Value	Condition
Salmo salar	0 based on wet weight	Uptake from food, steady state cf., 288 K, 48 h, 6.24 g fish[1]
Salmo salar	1343 based on wet weight	Uptake from water, steady state cf., 288 K, 48 h, 6.24 g fish[1]

Source: 1.Zitko, V. and O. Hutzinger. 1976. Bull. Environ. Contam. Toxicol. 16(6), pp 665-673.

5-171 Concentration factor 3,4-di-chlorobiphenyl

Species	Value	Condition
Salmo salar	0 based on wet weight	Uptake from food, steady state cf., 288 K, 48 h, 6.24 g fish[1]
Salmo salar	358 based on wet weight	Uptake from water, steady state cf., 288 K, 48 h, 6.24 g fish[1]

Source: 1.Zitko, V. and O. Hutzinger. 1976. Bull. Environ. Contam. Toxicol. 16(6), pp 665-673.

5-172 Concentration factor Ag

Species	Value	Condition
Atmospheric particulates	830	US urban air, relative to earths crust, Fe standard, F200[1]
Brown algae	240 based on wet weight	None[2]
Daphnia magna	11 based on wet weight	Ag = 5.0 mg/l, 7 w[3]
Daphnia magna	20 based on wet weight	Ag = 5.0 mg/l, 3 w[3]
Daphnia magna	24 based on wet weight	Ag = 0.5 mg/l, 7 w[3]
Daphnia magna	26 based on wet weight	Ag = 0.5 mg/l, 10 w[3]
Daphnia magna	28 based on wet weight	Ag = 5.0 mg/l, 5 w[3]
Daphnia magna	32 based on wet weight	Ag = 0.5 mg/l, 3 w[3]
Daphnia magna	36 based on wet weight	Ag = 0.5 mg/l, 5 w[3]
Daphnia magna	5 based on wet weight	Ag = 5.0 mg/l, 1 w[3]
Daphnia magna	9 based on wet weight	Ag = 5.0 mg/l, 10 w[3]
Daphnia magna	9.8 based on wet weight	Ag = 0.5 mg/l, 1 w[3]

Fucus vesiculosus	5E3 based on dry weight	18 obs., all areas around the British Isles[4]
Ligumia sp.	0.5 based on wet weight	Ag = 5.0 mg/l, 1 w, body weight = 6.70 g, also Margaritifera sp.[3]
Ligumia sp.	1.9 based on wet weight	Ag = 5.0 mg/l, 3 w, body weight = 2.67 g, also Margaritifera sp.[3]
Ligumia sp.	13.2 based on wet weight	Ag = 5.0 mg/l, 5 w, body weight = 2.06 g, also Margaritifera sp.[3]
Ligumia sp.	22.0 based on wet weight	Ag = 5.0 mg/l, 7 w, body weight = 5.16 g, also Margaritifera sp.[3]
Ligumia sp.	3 based on wet weight	Ag = 0.5 mg/l, 1 w, body weight = 5.67 g, also Margaritifera sp.[3]
Ligumia sp.	3.4 based on wet weight	Ag = 0.5 mg/l, 3 w, body weight = 3.05 g, also Margaritifera sp.[3]
Ligumia sp.	6.6 based on wet weight	Ag = 0.5 mg/l, 10 w, body weight = 3.12 g, also Margaritifera sp.[3]
Ligumia sp.	8.0 based on wet weight	Ag = 0.5 mg/l, 7 w, body weight = 4.60 g, also Margaritifera sp.[3]
Ligumia sp.	9.8 based on wet weight	Ag = 5.0 mg/l, 10 w, body weight = 4.24 g, also Margaritifera sp.[3]
Ligumia sp. and Margaritefera sp.	3.6 based on wet weight	Ag = 0.5 mg/l, 5 w, body weight = 4.11 g, also Margaritifera sp.[3]
Patella vulgata	3.1E4 based on dry weight	17 obs., all areas around the British Isles[4]
Phytoplankton	7810 based on wet weight	Range = 620 - 15000[5]
Pimephales promelas	0.7 based on wet weight	Ag = 5.0 mg/l, 1 w, body weight = 0.89 g[3]
Pimephales promelas	101 based on wet weight	0.5 mg Ag/l, 7 w, body weight = 0.84 g[3]
Pimephales promelas	2.4 based on wet weight	Ag = 0.5 mg/l, 5 w, body weight = 1.16 g[3]
Pimephales promelas	2.4 based on wet weight	Ag = 0.5 mg/l, 7 w, body weight = 0.85 g[3]
Pimephales promelas	2.7 based on wet weight	Ag = 5.0 mg/l, 3 w, body weight = 0.94 g[3]
Pimephales promelas	3.2 based on wet weight	Ag = 0.5 mg/l, 1 w, body weight = 1.14 g[3]
Pimephales promelas	3.2 based on wet weight	Ag = 0.5 mg/l, 3 w, body weight = 1.03 g[3]
Pimephales promelas	4.0 based on wet weight	Ag = 0.5 mg/l, 10 w, body weight = 1.20 g[3]
Pimephales promelas	4.2 based on wet weight	Ag = 5.0 mg/l, 7 w, body weight = 0.87 g[3]
Pimephales promelas	4.2 based on wet weight	Ag = 5.0 mg/l, Hg = 5.0 mg Hg/m3, body weight = 0.80 g[3]
Pimephales promelas	6.2 based on wet weight	Ag = 5.0 mg/l, 10 w, body weight = 1.19 g[3]
Plankton	210 based on wet weight	None[2]
Porphyra umbilicalis	1.9E3 based on dry weight	20 obs., all areas around the British Isles[4]
Scenedesmus sp.	10 based on wet weight	Ag = 5.0 mg/l, 1 w[3]
Scenedesmus sp.	150 based on wet weight	Ag = 5.0 mg/l, 10 w[3]
Scenedesmus sp.	182 based on dry weight	Ag = 0.5 mg/l, 3 w[3]
Scenedesmus sp.	200 based on wet weight	Ag = 0.5 mg/l, 7 w[3]
Scenedesmus sp.	240 based on w weight	Ag = 0.5 mg/l, 5 w[3]
Scenedesmus sp.	30 based on wet weight	Ag = 5.0 mg/l, 3 w[3]
Scenedesmus sp.	50 based on wet weight	Ag = 5.0 mg/l, 5 w[3]
Scenedesmus sp.	52 based on dry weight	Ag = 0.5 mg/l, 1 w[3]
Scenedesmus sp.	72 based on wet weight	Ag = 5.0 mg/l, 7 w[3]
Scenedesmus sp.	96 based on wet weight	Ag = 0.5 mg/l, 10 w[3]

Source: 1.Foestner, U. and G.T.W. Wittmann. 1979. Metal Pollution in the Aquatic Environment. Springer-Verlag.Berlin.
2.Bowen, H.J.M. 1966. Trace Elements in Biochemistry, Ac. Press.
3.Terhaar, C.J., W.S. Ewell, S.P. Dziuba, W.W. White and P.J. Murphy. 1977. Water Res. 11(1), pp 101-11.
4.Preston, A., D.R. Jefferies , J.W.R. Harvey and A.K. Steele. 1972. Environ. Pollut., 3, pp 183-214.
5.Cushing, C.E. and L.A. Rancitelli. 1972. Northwest Science, 46(2), pp 115-121.

5-173 Concentration factor
Al

Species	Value	Condition
Atmospheric particulates	0.5	US urban air, relative to earths crust, Fe standard, F200[1]
Atmospheric particulates	1.0	N Atlantic Westerlies, relative to earths crust, Al standard, F200[2]
Atmospheric particulates	1.0	South Pole, relative to earths crust, Al standard, F200[3]
Brown algae	1550 based on wet weight	None[4]
Plankton	25000 based on wet weight	None[4]

Source: 1.Foestner, U. and G.T.W. Wittmann. 1979. Metal Pollution in the Aquatic Environment. Springer-Verlag.Berlin.
2.Chester, R. and J.H. Stoner. 1973. Nature (london) 245, pp 27-28.
3.Duce, R.A., J. Quinn, C.M. Olney, S. Piotrowits, B. Ray and Wade.1972. Science 176, pp 161-163.
4.Bowen, H.J.M. 1966. Trace Elements in Biochemistry, Ac. Press.

5-174 Concentration factor
Aldrin

Species	Value	Condition
Alfalfa	0.011 based on wet weight	Conc in soil = 0.84 mg/kg dry wt.[1]
Beetle	3.83 based on wet weight	Conc in soil = 0.06 mg/kg dry wt., USA, sites = 2[1]
Buffalofish	30000 based on wet weight	Conc in water = 0.007 ug/l[1]
Buffalofish	3913 based on wet weight	Conc in water = 0.023 ug/l[1]
Carrot, root	0.001 based on wet weight	Conc in muck soil = 8.36 mg/kg dry wt.[1]
Carrot, root	0.021 based on wet weight	Conc in clay soil = 0.48 mg/kg dry wt.[1]
Carrot, root	0.340 based on wet weight	Conc in soil = 0.94 mg/kg dry wt.[1]
Carrot, root	0.396 based on wet weight	Conc in silt loam soil = 1.34 mg/kg dry wt.[1]
Catfish	1428 based on wet weight	Conc in water = 0.007 ug/l[1]
Catfish	1590 based on wet weight	Conc in water = 0.044 ug/l[1]
Catfish	4444 based on wet weight	Conc in water = 0.009 ug/l[1]
Catfish	952 based on wet weight	Conc in water = 0.021 ug/l[1]
Crayfish	84286 based on wet weight	Conc in water = 0.014 ug/l, USA[1]
Cucumber, fruit	0.031 based on wet weight	Conc in soil = 3.7 mg/kg dry wt.[1]
Earthworms	0.38 based on wet weight	Conc in soil = 0.24 mg/kg dry wt., GB, sites = 2[1]
Earthworms	0.57 based on wet weight	Conc in arable soil = 0.37 mg/kg dry wt., GB, sites = 10[1]
Earthworms	0.6 based on wet weight	Conc in arable soil = 0.10 mg/kg dry wt., GB, sites = 7[1]
Earthworms	0.91 based on wet weight	Conc in soil = 0.72 mg/kg dry wt., GB, sites = 1, excl Lumbricus[1]
Earthworms	1.17 based on wet weight	Conc in soil = 0.06 mg/kg dry wt., USA, sites = 1[1]
Earthworms	3.33 based on dry weight	Conc in soil = 0.03 mg/kg dry wt., USA, sites = 67[1]
Earthworms	4.0 based on dry weight	Conc in soil = 0.008 mg/kg dry wt., USA, sites = 1[1]
Gryllus assimilis	0.17 based on wet weight	Conc in soil = 0.06 mg/kg dry wt., USA, sites = 2[1]
Lettuce, foliage	0.013 based on wet weight	Conc in silt loam soil = 2.3 mg/kg dry wt.[1]
Lumbricus terrestris	0.07 based on wet weight	Conc in soil = 0.72 mg/kg dry wt., GB, sites = 1[1]
Oyster	10 based on wet weight	Conc in water: < 1.0 ug/l, conc in sediment < 0.001 mg/kg, USA[1]

Oyster	10 based on wet weight	Conc in water: < 1.0 ug/l, conc in sediment < 0.01 mg/kg, USA[1]
Pea, foliage	0.083 based on wet weight	Conc in sand = 2.3 mg/kg dry wt.[1]
Pea, root	22.39 based on wet weight	Conc in sand = 2.3 mg/kg dry wt.[1]
Potato, tubers	0.074 based on wet weight	Conc in soil = 0.94 mg/kg dry wt.[1]
Slugs	16.67 based on wet weight	Conc in soil = 0.006 mg/kg dry wt., USA, sites = 4[1]
Soybean	6.0 based on wet weight	Conc in soil < 0.0001 mg/kg dry wt., less than value[1]
Sugar beet, root	0.057 based on wet weight	Conc in silt loam soil = 2.3 mg/kg dry wt.[1]
Sugar beet, root	0.284 based on wet weight	Conc in loam soil = 0.35 mg/kg dry wt.[1]
Turnips	0.022 based on wet weight	Conc in silt loam soil = 2.3 mg/kg dry wt.[1]
Wheat, foliage	0.008 based on wet weight	Conc in silt loam soil = 1.8 mg/kg dry wt.[1]

Source: 1.Edwards, C.A. 1975. Persistent Pesticides in the Environment, 2nd edition, CRC-Press.

5-175 Concentration factor
Am

Species	Value	Condition
Barley, grain	1.4E-6 based on dry plant and dry soil	Nevada, lab. site 11241 am, 293-301 K[1]
Barley, grain	3E-6 based on dry plant and dry soil	Nevada, lab. site 13241 am, 293-301 K[1]
Barley, lower vegetation	17E-6 based on dry plant and dry soil	Nevada, lab. site 11, 239, 249 Pu, 293-301 K[1]
Barley, lower vegetation	58E-6 based on dry plant and dry soil	Nevada, lab. site 13241 am, 293-301 K[1]
Barley, upper vegetation	156E-6 based on dry plant and dry soil	Nevada, lab. site 11241 am, 293-301 K[1]
Barley, upper vegetation	93E-6 based on dry plant and dry soil	Nevada, lab. site 13241 am, 293-301 K[1]
Nereis diversicolor	0.0003 conc in anim/(conc in sediment)	Wet weight both sed. and anim., Bravo Crater, SD = +- 0.0004[2]
Nereis diversicolor	0.0004 conc in anim/(conc in sediment)	Dry weight both sed. and anim., Bikini Atoll, n = 9, 5 d[2]
Nereis diversicolor	0.0006 conc in anim/(conc in sediment)	Wet weight both sed. and anim., Windscale, SD = +- 0.0006[2]
Nereis diversicolor	0.0007 conc in anim/(conc in sediment)	Dry weight both sed. and anim., Bikini Atoll, n = 10, 15 d[2]
Nereis diversicolor	0.0010 conc in anim/(conc in sediment)	Dry weight both sed. and anim., Bikini Atoll, n = 31, 40 d[2]
Nereis diversicolor	0.0012 conc in anim/(conc in sediment)	Dry weight both sed. and anim., Windscale, n = 22, 5 d[2]
Nereis diversicolor	0.0013 conc in anim/(conc in sediment)	Dry weight both sed. and anim., Bikini Atoll, n = 6, 40 d[2]
Nereis diversicolor	0.0020 conc in anim/(conc in sediment)	Dry weight both sed. and anim., Windscale, n = 25, 15 d[2]
Nereis diversicolor	0.0024 conc in anim/(conc in sediment)	Dry weight both sed. and anim., Windscale, n = 23, 40 d[2]
Nereis diversicolor	0.0043 conc in anim/(conc in sediment)	Dry weight both sed. and anim., Windscale, n = 1, 225 d[2]

Source: 1.Schultz, R.R., G.A. Tompkin, L. Leven thal and K.L. Babcock. 1976. J. Environ. Qual. 5(4), pp 406-41.
2.Beasley, T.M. and S.W. Fowler. 1976. Mar. Biol., 38(2), pp 95-100.

5-176 Concentration factor
As

Species	Value	Condition
Atmospheric particulates	310	US urban air, relative to earths crust, Fe standard, F200[1]
Atmospheric particulates	40	N Atlantic Westerlies, relative to earths crust, Al standard, F200[2]
Atmospheric particulates	69	South Pole, relative to earths crust, Al standard, F200[3]
Brown algae	2500 based on wet weight	None[4]
Brown algae	3100 based on wet weight	Range = 200 - 6000[5]
Ceratophyllum demersum	16500 based on dry weight	Waikato River, Japan, 524 - 971 mg As/kg dw, rg = 13000 - 20000[6]
Compsopogon hookeri	7000 based on dry weight	Waikato River, Japan, 550 mg As/kg dw[6]
Enteromorpha compressa	4000 based on dry weight	England, 12 mg As/kg dw[6]
Enteromorpha nana	1700 based on dry weight	Waikato River, Japan, 20 - 40 mg As/kg dw, rg = 800 - 2800[6]
Fucus serratus	600 based on dry weight	England, 1.7 mg As/kg dw[6]
Ictalurus punctatus	0.4 based on dry weight	296 K, lab. system, 8 d, willacy sandy loam[7]
Lagarosiphon major	4950 based on dry weight	Waikato River, Japan, 32-405 mg As/kg dw, rg = 4600 - 5300[6]
Lemna minor	3 based on dry weight	296 K, lab. system, 20 d, willacy sandy loam[7]
Macrocystis pyrifera	19000 based on dry weight	New Zealand, 58 mg As/kg dw[6]
Oedogonium cardiacum	17 based on dry weight	296 K, lab. system, 20 d, hidalgo clay loam[7]
Oedogonium cardiacum	3 based on dry weight	296 K, lab. system, 20 d, willacy sandy loam[7]
Oedogonium cardiacum	5 based on dry weight	296 K, lab. system, 20 d, laredo silt loam[7]
Physa sp.	2 based on dry weight	296 K, lab. system, 20 d, willacy sandy loam[7]
Physa sp.	21 based on dry weight	296 K, lab. system, 20 d, laredo silt loam[7]
Physa sp.	23.0 based on dry weight	296 K, lab. system, 8 d, hidalgo clay loam[7]
Physa sp.	7 based on dry weight	296 K, lab. system, 8 d, willacy sandy loam[7]
Procambarus clarki, chitin	0.1 based on dry weight	296 K, lab. system, 20 d, hidalgo clay loam[7]
Procambarus clarki, chitin	0.7 based on dry weight	296 K, lab. system, 8 d, laredo silt loam[7]
Procambarus clarki, chitin	0.9 based on dry weight	296 K, lab. system, 20 d, willacy sandy loam[7]
Procambarus clarki, chitin	2.5 based on dry weight	296 K, lab. system, 20 d, laredo silt loam[7]
Procambarus clarki, chitin	5.3 based on dry weight	296 K, lab. system, 8 d, willacy sandy loam[7]
Procambarus clarki, soft tissue	16.0 based on dry weight	296 K, lab. system, 20 d, laredo silt loam[7]
Procambarus clarki, soft tissue	4.2 based on dry weight	296 K, lab. system, 8 d, hidalgo clay loam[7]
Procambarus clarki, soft tissue	4.7 based on dry weight	296 K, lab. system, 8 d, willacy sandy loam[7]
Procambarus clarki, soft tissue	5.2 based on dry weight	296 K, lab. system, 8 d, laredo silt loam[7]
Salmo gairdneri, egg	0.007 based on wet weight	10.0 mg As(5+)/l, 160-200 eggs, 279.5 K, 15 d[8]
Salmo gairdneri, egg	0.01 based on wet weight	5.0 mg As(5+)/l, 160-200 eggs, 279.5 K, 15 d[8]

Salmo gairdneri, egg	0.02 based on wet weight	50.0 mg As(5+)/l, 160-200 eggs, 279.5 K, 15 d[8]
Salmo gairdneri, egg	0.03 based on wet weight	50.0 mg As(3+)/l, 160-200 eggs. 279.5 K, 15 d[8]
Salmo gairdneri, egg	0.05 based on wet weight	10.0 mg As(3+)/l, 160-200 eggs, 279.5 K, 15 d[8]
Salmo gairdneri, egg	0.08 based on wet weight	1.0 mg As(5+)/l, 160-200 eggs, 279.5 K, 33 d[8]
Salmo gairdneri, egg	0.1 based on wet weight	5.0 mg As(3+)/l, 160-200 eggs, 279.5 K, 15 d[8]
Salmo gairdneri, egg	0.2 based on wet weight	1.0 mg As(3+)/l, 160-200 eggs, 279.5 K, 33 d[8]
Salmo gairdneri, egg	0.8 based on wet weight	0.5 mg As(5+)/l, 160-200 eggs, 279.5 K, 33 d[8]
Salmo gairdneri, egg	0.9 based on wet weight	0.5 mg As(3+)/l, 160-200 eggs, 279.5 K, 33 d[8]
Salmo gairdneri, egg	18.5 based on wet weight	0.05 mg As(5+)/l, 160-200 eggs, 279.5 K, 33 d[8]
Salmo gairdneri, egg	2.7 based on wet weight	0.05 mg As(3+)/l, 160-200 eggs, 279.5 K, 33 d[8]
Salmo gairdneri, egg	4.0 based on wet weight	0.1 mg As(5+)/l, 160-200 eggs, 279.5 K, 33 d[8]
Salmo gairdneri, egg	5.5 based on wet weight	0.1 mg As(3+)/l, 160-200 eggs, 279.5 K, 33 d[8]
Typha orientalis	100 based on dry weight	Waikato River, Japan, 8 mg As/kg dw[6]

Source: 1.Foestner, U. and G.T.W. Wittmann. 1979. Metal Pollution in the Aquatic Environment. Springer-Verlag.Berlin.
2.Chester, R. and J.H. Stoner. 1973. Nature (london) 245, pp 27-28.
3.Duce, R.A., J. Quinn, C.M. Olney, S. Piotrowits, B. Ray and Wade.1972. Science 176, pp 161-163.
4.Bowen, H.J.M. 1966. Trace Elements in Biochemistry, Ac. Press.
5.Mauchline, J. and W.L Templeton. 1964. Oceanogr. Mar. Biol. Ass. Rev., 2, pp 229.
6.Reay, P.F. 1972. J. appl. Ecol. 9, pp 557-565.
7.Schuth, C.K., A.R. Isensee, E.A. Woolson and P.C. Kearney. 1974. J. Agric. Food. Chem. 22(6), pp 9991003.
8.Dabrowski, K.R.. 1976. Water Res., 10(9), pp 793-796.

5-177 Concentration factor
Atrazine

Species	Value	Condition
Gambusia affinis	17 based on wet weight	Model ecosystem, organism = 197 ppb, water = 5.0-3.8 ppb, 18 d[1]
Helisoma sp.	8 based on wet weight	Model ecosystem, organism = 98 ppb, water = 5.0-3.8 ppb, 18 d[1]
Oedogonium cardiacum	9 based on wet weight	Model ecosystem, organism = 104 ppb, water = 5.0-3.8 ppb, 18 d[1]
Scenedesmus acutus	10 based on dry weight	Water = 4.65E-6 mol/l, 303 K, pH = 7-8, 8 h, living cells[2]
Scenedesmus acutus	12 based on dry weight	Water = 4.65E-9 mol/l, 303 K, pH = 7-8, 8 h, dead cells[2]
Scenedesmus acutus	31.8 based on dry weight	Water = 0.5E-7 mol, 303 K, pH = 7-8, 8 h[2]
Scenedesmus acutus	35 based on dry weight	Water = 4.65E-7 mol/l, 303 K, pH = 7-8, 8 h, living cells[2]
Scenedesmus acutus	60 based on dry weight	Water = 4.65E-8 mol/l, 303 K, pH = 7-8, 8 h, living cells[2]
Scenedesmus acutus	9 based on dry weight	Water = 4.65E-6 mol/l, 303 K, pH = 7-8, 8 h, dead cells[2]
Scenedesmus acutus	9 based on dry weight	Water = 4.65E-7 mol/l, 303 K, pH = 7-8, 8 h, dead cells[2]
Scenedesmus acutus	9 based on dry weight	Water = 4.65E-8 mol/l, 303 K, pH = 7-8, 8 h, dead cells[2]
Scenedesmus acutus	90 based on dry weight	Water = 4.65E-9 mol/l, 303 K, pH = 7-8, 8 h, living cells[2]

Source: 1.Kearney, P.C., J.E. Oliver, C.S. Helling, A.R. Isensee and A. Kontson. 1977. J. Agric. Food. Chem. 2(5), pp 1177-1181.
2.Boehm, H.H. and H. Muller. 1976. Naturwissenschaftens 63(3), pp 296.

5-178 Concentration factor
Au

Species	Value	Condition
Brown algae	270 based on wet weight	None[1]

Source: 1.Bowen, H.J.M. 1966. Trace Elements in Biochemistry, Ac. Press.

5-179 Concentration factor
Au - 198

Species	Value	Condition
Anodonta nuttalliana, soft parts	7 based on wet weight	No food, 292 K, pH = 7.5, freshwat = 2.3 part & 16 solub p ci/ml, 7 d[1]
Anodonta nuttalliana, soft parts	7 based on wet weight	No food, 292 K, pH = 7.5, freshwat = 8.9 part & 25 solub p ci/ml, 21d[1]
Astacus sp., muscle	1 based on wet weight	No food, 292 K, pH = 7.5, freshwat = 2.3 part & 16 solub p ci/ml, 7d[1]
Astacus sp., muscle	1 based on wet weight	No food, 292 K, pH = 7.5, freshwat = 8.9 part & 25 solub p ci/ml, 21d[1]
Astacus sp., soft parts	11 based on wet weight	No food, 292 K, pH = 7.5, freshwat = 2.3 part & 16 solub p ci/ml, 7 d[1]
Astacus sp., soft parts	12 based on wet weight	No food, 292 K, pH = 7.5, freshwat = 8.9 part & 25 solub p ci/ml, 21d[1]
Astacus sp., viscera	14 based on wet weight	No food, 292 K, pH = 7.5, freshwat = 2.3 part & 16 solub p ci/ml, 7 d[1]
Astacus sp., viscera	14 based on wet weight	No food, 292 K, pH = 7.5, freshwat = 8.9 part & 25 solub p ci/ml, 21d[1]
Cancer productus, muscle	4 based on wet weight	No food, 285 K, pH = 8.2, seawater = 3.3 part & 12 solub p ci/ml, 20d[1]
Cancer productus, sort parts	12 based on wet weight	No food, 285 K, pH = 8.2, seawater = 3.3 part & 12 solub p ci/ml, 20d[1]
Cancer productus, viscera	16 based on wet weight	No food, 285 K, pH = 8.2, seawater = 3.3 part & 12 solub p ci/ml, 20d[1]
Cancer productus, viscera	20 based on wet weight	No food, 285 K, pH = 8.2, seawater = 3.5 part & 6.9 solub p ci/ml, 6d[1]
Carassius auratus, muscle	1 based on wet weight	No food, 292 K, pH = 7.5, freshwat = 2.3 part & 16 solub p ci/ml, 7d[1]
Carassius auratus, nonmuscle	9 based on wet weight	No food, 292 K, pH = 7.5, freshwat = 2.3 part & 16 solub p ci/ml, 7 d[1]
Carassius auratus, whole body	9 based on wet weight	No food, 292 K, pH = 7.5, freshwat = 8.9 part & 25 solub p ci/ml, 21d[1]
Gillichthys mirabilis, muscle	1 based on wet weight	No food, 285 K, pH = 8.2, seawater = 3.3 part & 12 solub p ci/ml, 20d[1]
Gillichthys mirabilis, nonmuscle	1 based on wet weight	No food, 285 K, pH = 8.2, seawater = 3.3 part & 12 solub p ci/ml, 20d[1]
Gillichthys mirabilis, nonmuscle	2 based on wet weight	No food, 285 K, pH = 8.2, seawater = 3.3 part & 12 solub p ci/ml, 20d[1]
Gillichthys mirabilis, whole body	1 based on wet weight	No food, 285 K, pH = 8.2, seawater = 3.5 part . and 6.9 solub p ci/ml[1]
Mytilus edulis, soft parts	11 based on wet weight	No food, 285 K, pH = 8.2, seawater = 3.5 part & 6.9 solub p ci/ml, 6d[1]
Saxidonus giganteus, soft parts	3 based on wet weight	No food, 285 K, pH = 8.2, seawater = 3.3 part & 12 solub p ci/ml, 20d[1]

Source: 1.Harrison, F.L. 1973. Health Physics, 24, pp 331-343.

5-180 Concentration factor
B

Species	Value	Condition
Brown algae	6.6 based on wet weight	None[1]
Coral	1.50E-2 based on Ca content	Shallow coastal ocean, mean of 13 determinations, F023[2]
Coral	1.72E-2 based on Ca content	Deep ocean coral, mean of 13 determinations, F023[2]
Coral	2.03E-2 based on Ca content	Shallow open ocean coral, mean of 8 determinations, F023[2]

Source: 1.Bowen, H.J.M. 1966. Trace Elements in Biochemistry, Ac. Press.
2.Livingston, H.D. and G. Thompson. 1971. Limnol. Oceanogr. 16(5), pp 786-796.

5-181 Concentration factor
BHC

Species	Value	Condition
Cabbage, leaf	0.218 based on wet weight	Conc in sandy loam soil = 1.70 mg/kg dry wt.[1]
Carrot	0.021 based on wet weight	Conc in muck soil = 0.34 mg/kg dry wt.[1]
Carrot	0.263 based on wet weight	Conc in sandy soil = 0.09 mg/kg dry wt.[1]
Carrot, root	3.53 based on wet weight	Conc in sandy loam soil = 1.70 mg/kg dry wt.[1]
Crassostrea virginica	18 based on wet weight	Conc in water = 0.33 ug/l, USA[1]
Crayfish	33 based on wet weight	Conc in water = 0.03 ug/l, USA[1]
Earthworms	3.75 based on wet weight	Conc in soil(orchard) = 0.08 mg/kg dry wt., GB, sites = 2[1]
Earthworms	4.25 based on wet weight	Conc in soil = 0.004 mg/kg dry wt., GB, sites = 1, excl Lumbricus[1]
Earthworms	5.0 based on wet weight	Conc in arable soil = 0.01 mg/kg dry wt., GB, sites = 10[1]
Feral pigeon, tissue	2.0 based on wet weight	Conc in food = 72 mg/kg[1]
Gyrinidae	33 based on wet weight	Conc in water = 0.03 ug/l, USA[1]
Lumbricus terrestris	1.60 based on wet weight	Conc in soil = 0.004 mg/kg dry wt., GB, sites = 1[1]
Oyster	10 based on wet weight	Conc in water: < 1.0 ug/l, conc in sediment < 0.001 mg/kg, USA[1]
Oyster	10 based on wet weight	Conc in water: < 1.0 ug/l, conc in sediment < 0.01, USA[1]
Pea, foliage	36.94 based on wet weight	Conc in sand = 0.62 mg/kg dry wt.[1]
Pea, root	133.23 based on wet weight	Conc in sand = 0.62 mg/kg dry wt.[1]
Pea, vines	0.894 based on wet weight	Conc in sandy loam soil = 1.70 mg/kg dry wt.[1]
Plankton	2500 based on wet weight	Conc in water = 0.040 ug/l, USA[1]
Plankton	7 based on wet weight	Conc in water = 0.03 ug/l, USA[1]
Potato, tubers	0.207 based on wet weight	Conc in sandy loam soil = 3.0 mg/kg dry wt.[1]
Soybean	1.0 based on wet weight	Conc in soil = 0.006 mg/kg dry wt.[1]
Tendipendidae, larvae	33 based on wet weight	Conc in water = 0.03 ug/l, USA[1]

Source: 1.Edwards, C.A. 1975. Persistent Pesticides in the Environment, 2nd edition, CRC-Press.

5-182 Concentration factor
Ba

Species	Value	Condition
Brown algae	260 based on wet weight	None[1]
Coral	1.11 based on Ca content	Shallow open ocean coral, mean of 8 determinations, F023[2]
Coral	1.34 based on Ca content	Deep ocean coral, mean of 13 determinations, F023[2]
Coral	2.38 based on Ca content	Shallow coastal ocean, mean of 13 determinations, F023[2]
Plankton	120 based on wet weight	None[1]

Source: 1.Bowen, H.J.M. 1966. Trace Elements in Biochemistry, Ac. Press.
2.Livingston, H.D. and G. Thompson. 1971. Limnol. Oceanogr. 16(5), pp 786-796.

5-183 Concentration factor
Ba - 140 / La - 140

Species	Value	Condition
Anodonta nuttalliana, soft parts	1.6 based on wet weight	No food, 292 K, pH = 7.5, freshwat = 0.69 part & 110 solub p ci/ml, 7d[1]
Anodonta nuttalliana, soft parts	3.2 based on wet weight	No food, 292 K, pH = 7.5, freshwat = 3.5 part & 280 solub p ci/ml, 21d[1]
Astacus sp., muscle	0.4 based on wet weight	No food, 292 K, pH = 7.5, freshwat = 0.69 part & 110 solub p ci/ml, 7d[1]
Astacus sp., muscle	1.7 based on wet weight	No food, 292 K, pH = 7.5, freshwat = 3.5 part & 280 solub p ci/ml, 21d[1]
Astacus sp., soft parts	25 based on wet weight	No food, 292 K, pH = 7.5, freshwat = 90 part & 0 solub p ci/l, 7 d[1]
Astacus sp., soft parts	9 based on wet weight	No food, 292 K, pH = 7.5, freshwat = 0.69 part & 110 solub p ci/ml, 7d[1]
Astacus sp., viscera	10 based on wet weight	No food, 292 K, pH = 7.5, freshwat = 0.69 part & 110 solub p ci/ml, 7d[1]
Astacus sp., viscera	30 based on wet weight	No food, 292 K, pH = 7.5, freshwat = 3.5 part & 280 solub p ci/ml, 21d[1]
Cancer productus, muscle	1.3 based on w weight	No food, 285 K, pH = 8.2, seawater = 0.67 part & 230 solub p ci/ml[1]
Cancer productus, sort parts	7 based on wet weight	No food, 285 K, pH = 8.2, seawater = 0.67 part & 230 solub p ci/ml[1]
Cancer productus, viscera	5 based on wet weight	No food, 285 K, pH = 8.2, seawater = 1.10 part & 220 solub p ci/ml, 6d[1]
Cancer productus, viscera	9 based on wet weight	No food, 285 K, pH = 8.2, seawater = 0.67 part & 230 solub p ci/ml[1]
Carassius auratus, muscle	1 based on wet weight	No food, 292 K, pH = 7.5, freshwat = 0.69 part & 110 solub p ci/ml, 7d[1]
Carassius auratus, nonmuscle	3.3 based on wet weight	No food, 292 K, pH = 7.5, freshwat = 0.69 part & 110 solub p ci/ml, 7d[1]
Carassius auratus, whole body	3 based on wet weight	No food, 292 K, pH = 7.5, freshwat = 3.5 part & 280 solub p ci/ml, 21d[1]
Gillichthys mirabilis, muscle	0.9 based on wet weight	No food, 285 K, pH = 8.2, seawater = 0.67 part & 230 solub p ci/ml[1]
Gillichthys mirabilis, muscle	1 based on wet weight	No food, 285 K, pH = 8.2, seawater = 1.10 part & 220 solub p ci/ml, 6d[1]
Gillichthys mirabilis, nonmuscle	0.8 based on wet weight	No food, 285 K, pH = 8.2, seawater = 0.67 part & 230 solub p ci/ml[1]

Gillichthys mirabilis, nonmuscle	1.3 based on wet weight	No food, 285 K, pH = 8.2, seawater = 1.10 part & 220 solub p ci/ml, 6d[1]
Gillichthys mirabilis, whole body	0.8 based on wet weight	No food, 285 K, pH = 8.2, seawater = 120 part & 0 solub p ci/l, 6 d[1]
Gillichthys mirabilis, whole body	1.2 based on wet weight	No food, 285 K, pH = 8.2, seawater = 0.67 part & 230 solub p ci/ml[1]
Mytilus edulis, soft parts	0.9 based on wet weight	No food, 285 K, pH = 8.2, seawater = 1.10 part & 220 solub p ci/ml, 6d[1]
Saxidonus giganteus, soft parts	0.7 based on wet weight	No food, 285 K, pH = 8.2, seawater = 0.67 part & 230 solub p ci/ml[1]
Suspended solids	530 based on wet weight	Danube, 1973, nuclear power plant since 1967, rg = 320 - 780[2]

Source: 1.Harrison, F.L. 1973. Health Physics, 24, pp 331-343.
2.Herrmann, H., M. Ruf, K. Hubel and W. Lunsmann. 1975. Impacts of Nuclear Releases into the Aquatic Evironment. IAEA-SM-198/14. pp 461-470. Vienna.

5-184 Concentration factor
Be

Species	Value	Condition
Brown algae	1500 based on wet weight	None[1]

Source: 1.Mauchline, J. and W.L Templeton. 1964. Oceanogr. Mar. Biol. Ass. Rev., 2, pp 229.

5-185 Concentration factor
Benzo(a)pyrene

Species	Value	Condition
Macoma inquinata	0.09 based on sediment concentration	Sediment=2000 ug prudhoe B crude oil/g, 7 d, sed=0.64 ppm, water=0.043 ppb[1]
Macoma inquinata	861 based on wet weight	Sediment=2000 ug prudhoe B crude oil/g, 7 d, sed=0.64 ppm, water=0.043 ppb[1]
Rangia cuneata	8.7 based on wet weight	Estuarine, water = 0.052 ppm, tissue = 0.45+-0.1 ppm[2]

Source: 1.Roesijadi, G., J.W. Anderson and J.W. Blaylock. 1978. J. Fish. Res. bd Canada 35, 608-614.
2.Neff, J.M., Anderson, Cox, Laughlin, Rossi and Tatem. 1976. In: Sources, Effects and Sinks of Hydrocarbon in the Aquatic Environment. Wash DC. Amer. Inst. of Biological Science.

5-186 Concentration factor
Br

Species	Value	Condition
Brown algae	2.8 based on wet weight	None[1]

Source: 1.Bowen, H.J.M. 1966. Trace Elements in Biochemistry, Ac. Press.

5-187 Concentration factor
C2 - Benzenes

Species	Value	Condition
Oncorhynchus kisutch, muscle	1.39 based on wet weight	Water sol fraction = 352 ppb, 288 K, 5 w, flow through, prudhoe B crude oil[1]

Source: 1.Roubal, W.T., D.H. Bovee, T.K. Collier and S.I. Stranahan. 1977. Proc of 1977 Oil Spill Conf. (Prevention, Behavior, Control, Cleanup). Wash DC. American Petroleum Institute. 551-556.

5-188 Concentration factor
C2 - Naphthalenes

Species	Value	Condition
Oncorhynchus kisutch, muscle	85.0 based on wet weight	Water sol fraction = 10 ppb, 288 K, 5 w, flow through, prudhoe B crude oil[1]

Source: 1.Roubal, W.T., D.H. Bovee, T.K. Collier and S.I. Stranahan. 1977. Proc of 1977 Oil Spill Conf. (Prevention, Behavior, Control, Cleanup). Wash DC. American Petroleum Institute. 551-556.

5-189 Concentration factor
C3 - Benzene

Species	Value	Condition
Oncorhynchus kisutch, muscle	37.5 based on wet weight	Water sol fraction = 40 ppb, 288 K, 5 w, flow through, prudhoe B crude oil[1]

Source: 1.Roubal, W.T., D.H. Bovee, T.K. Collier and S.I. Stranahan. 1977. Proc of 1977 Oil Spill Conf. (Prevention, Behavior, Control, Cleanup). Wash DC. American Petroleum Institute. 551-556.

5-190 Concentration factor
C3 - Naphthalene

Species	Value	Condition
Oncorhynchus kisutch, muscle	113.3 based on wet weight	Water sol fraction = 6 ppb, 288 K, 5 w, flow through, prudhoe B crude oil[1]

Source: 1.Roubal, W.T., D.H. Bovee, T.K. Collier and S.I. Stranahan. 1977. Proc of 1977 Oil Spill Conf. (Prevention, Behavior, Control, Cleanup). Wash DC. American Petroleum Institute. 551-556.

5-191 Concentration factor
C4 - Benzenes and C5 - Benzenes

Species	Value	Condition
Oncorhynchus kisutch, muscle	458.3 based on wet weight	Water sol fraction = 12 ppb, 288 K, 5 w, flow through, prudhoe B crude oil[1]

Source: 1.Roubal, W.T., D.H. Bovee, T.K. Collier and S.I. Stranahan. 1977. Proc of 1977 Oil Spill Conf. (Prevention, Behavior, Control, Cleanup). Wash DC. American Petroleum Institute. 551-556.

5-192 Concentration factor
Ca

Species	Value	Condition
Brassica oleracea	13 based on wet weight	Conc in soil solution = 750ppm[1]
Brown algae	5.25 based on wet weight	Range = 0.5 - 10[2]
Brown algae	7.2 based on wet weight	None[1]
Cancer irroratus	128.0 based on wet weight	Seawater conc = 400 ppm, Nova Scotia, AAS, SD = +- 4.1[3]
Crustacea	40 based on dry weight	Marine, rg = 0.5 - 250, 12 values used[4]
Daphnia magna	6.9 based on dry weight	On dry wt basis relative to the algae consumed[5]
Fish	1.9 based on dry weight	Marine, rg = 0.5 - 7.6, 20 values used[4]
Fish	70 based on dry weight	Freshwater, rg = 0.5 - 470, 18 values used[4]
Molluscs	16.5 based on dry weight	Marine, rg = 0.2 - 112, 20 values used[4]
Phytoplankton	48.5 based on wet weight	Range = 12 - 85[6]
Plankton	10 based on wet weight	None[2]
Plankton	5 based on wet weight	None[1]
Plants	10 based on dry weight	Marine, rg = 1.8 - 31, 20 values used[4]

Plants	265 based on dry weight	Algae, bryophytes and vascular hydrophytes, freshwater, 32 sp[7]
Plants	350 based on dry weight	Freshwater, rg = 64 - 720, 5 values used[4]
Zooplankton	10.5 based on wet weight	Seawater concentration 400.0 mg/m3[8]

Source: 1.Bowen, H.J.M. 1966. Trace Elements in Biochemistry, Ac. Press.
2.Mauchline, J. and W.L Templeton. 1964. Oceanogr. Mar. Biol. Ass. Rev., 2, pp 229.
3.Martin, J.-L.M. 1974. Mar. biol., 28(4), pp 245-251.
4.Jinks, S.M. and M. Eisenbud. 1972. Radiat. Data Rep. 13, pp 243-247.
5.Verduin, J. 1975. Limnol. Oceanogr. 20(6), pp 1052.
6.Cushing, C.E. and L.A. Rancitelli. 1972. Northwest Science, 46(2), pp 115-121.
7.Timofeeva-Resovskaya, E.A., T-R. N.V. and E.A. Gileva. 1961. Dokl. Akad. Nauk. SSSR, 140, pp 1437.
8.Martin, J.H 1970. Limnol. Oceanogr. 15(5), pp 756-761.

5-193 Concentration factor
Cacodylic acid

Species	Value	Condition
Daphnia magna	147 based on dry weight	296 K, lab. system, 8 d, laredo silt loam[1]
Daphnia magna	51 based on dry weight	296 K, lab. system, 8 d, hidalgo clay loam[1]
Ictalurus punctatus	0.7 based on dry weight	296 K, lab. system, 8 d, willacy sandy loam[1]
Ictalurus punctatus	0.8 based on dry weight	296 K, lab. system, 8 d, hidalgo clay loam[1]
Ictalurus punctatus	2.1 based on dry weight	296 K, lab. system, 20 d, willacy sandy loam[1]
Ictalurus punctatus	2.8 based on dry weight	296 K, lab. system, 8 d, laredo silt loam[1]
Ictalurus punctatus	275 based on dry weight	296 K, lab. system, 20 d, laredo silt loam[1]
Ictalurus punctatus	8.5 based on dry weight	296 K, lab. system, 20 d, hidalgo clay loam[1]
Lemna minor	1244 based on dry weight	296 K, lab. system, 20 d, hidalgo clay loam[1]
Lemna minor	3466 based on dry weight	296 K, lab. system, 20 d, laredo silt loam[1]
Lemna minor	52 based on dry weight	296 K, lab. system, 8 d, hidalgo clay loam[1]
Lemna minor	58 based on dry weight	296 K, lab. system, 8 d, willacy sandy loam[1]
Lemna minor	88 based on dry weight	296 K, lab. system, 8 d, laredo silt loam[1]
Lemna minor	89 based on dry weight	296 K, lab. system, 20 d, willacy sandy loam[1]
Oedogonium cardiacum	101 based on dry weight	296 K, lab. system, 8 d, willacy sandy loam[1]
Oedogonium cardiacum	163 based on dry weight	296 K, lab. system, 20 d, willacy sandy loam[1]
Oedogonium cardiacum	2483 based on dry weight	296 K, lab. system, 20 d, hidalgo clay loam[1]
Oedogonium cardiacum	26959 based on dry weight	296 K, lab. system, 20 d, laredo silt loam[1]
Oedogonium cardiacum	298 based on dry weight	296 K, lab. system, 8 d, laredo silt loam[1]
Oedogonium cardiacum	80 based on dry weight	296 K, lab. system, 8 d, hidalgo clay loam[1]
Physa sp.	1050 based on dry weight	296 K, lab. system, 20 d, laredo silt loam[1]
Physa sp.	22 based on dry weight	296 K, lab. system, 8 d, willacy sandy loam[1]
Physa sp.	25 based on dry weight	296 K, lab. system, 8 d, laredo silt loam[1]
Physa sp.	26 based on dry weight	296 K, lab. system, 8 d, hidalgo clay loam[1]
Physa sp.	4.3 based on dry weight	296 K, lab. system, 20 d, willacy sandy loam[1]
Procambarus clarki, soft tissue	0.5 based on dry weight	296 K, lab. system, 8 d, hidalgo clay loam[1]
Procambarus clarki, soft tissue	13.6 based on dry weight	296 K, lab. system, 20 d, laredo silt loam[1]

Procambarus clarki, soft tissue	2.0 based on dry weight	296 K, lab. system, 8 d, laredo silt loam[1]
Procambarus clarki, soft tissue	2.0 based on dry weight	296 K, lab. system, 8 d, willacy sandy loam[1]
Procambarus clarki, soft tissue	3.0 based on dry weight	296 K, lab. system, 20 d, willacy sandy loam[1]
Procambarus clarki, soft tissue	5.9 based on dry weight	296 K, lab. system, 20 d, hidalgo clay loam[1]

Source: 1.Schuth, C.K., A.R. Isensee, E.A. Woolson and P.C. Kearney. 1974. J. Agric. Food. Chem. 22(6), pp 999-1003.

5-194 Concentration factor Camphechlor

Species	Value	Condition
Aquatic plant	18 based on wet weight	Conc in water = 4.8 ng/l[1]
Aquatic plant	512 based on wet weight	Conc in water = 0.41 ng/l[1]
Crayfish	20 based on wet weight	Conc in water: < 1.0 ug/l, USA[1]
Fish	2250 based on wet weight	Conc in water = 1.0-4.0 ug/l, rg = 2000 - 2500[1]
Freshwater invertebrates	2270 based on wet weight	Conc in water = 0.63 ug/l, USA[1]
Gyrinidae	20 based on wet weight	Conc in water: < 1.0 ug/l, USA[1]
Plankton	5 based on wet weight	Conc in water: < 1.0 ug/l, USA[1]
Salmo gairdneri	18829 based on wet weight	Conc in water = 0.41 ug/l[1]
Tendipendidae, larvae	20 based on wet weight	Conc in water: < 1.0 ug/l, USA[1]

Source: 1.Edwards, C.A. 1975. Persistent Pesticides in the Environment, 2nd edition, CRC-Press.

5-195 Concentration factor Carbaryl

Species	Value	Condition
Ictalurus punctatus	140 based on dry weight	Model ecosystem, 22 d, 296 K, water = 1.66-9.45 ppb, soil = 2.43 ppm[1]
Lemna minor	3600 based on dry weight	Model ecosystem, 22 d, 296 K, water = 1.66-9.45 ppb, soil = 2.43 ppm[1]
Oedogonium cardiacum	4000 based on dry weight	Model ecosystem, 22 d, 296 K, water = 1.66-9.45 ppb, soil = 2.43 ppm[1]
Physa sp.	300 based on dry weight	Model ecosystem, 22 d, 296 K, water = 1.66-9.45 ppb, soil = 2.43 ppm[1]
Procambarus sp.	260 based on dry weight	Model ecosystem, 22 d, 296 K, water = 1.66-9.45 ppb, soil = 2.43 ppm[1]

Source: 1.Kanazawa, J., A.R. Isensee and P.C. Kearney. 1975. J. Agric. Food. Chem. 23(4), pp 760-763.

5-196 Concentration factor Cd

Species	Value	Condition
Adsorption on fish fecals	200 based on dry weight	Added Cd = 2.1 mg/m3, 2 h, rg = 100 - 300[1]
Adsorption on fish fecals	600 based on dry weight	Added Cd = 2.1 mg/m3, 24 h, rg = 200 - 1000[1]
Adsorption on humic acid	14000 based on dry weight	Added Cd = 2.1 - 4.3 mg/m3, 2 h, rg = +- 2000[1]

Adsorption on humic acid	18000 based on dry weight	Added Cd = 2.1 - 4.3 mg/m3, 24 h, rg = +- 3000[1]
Adsorption on kaolin	250 based on dry weight	Added Cd = 7.3 mg/m3, 2 h, rg = +- 30[1]
Adsorption on kaolin	380 based on dry weight	Added Cd = 7.3 mg/m3, 24 h, rg = +- 50[1]
Adsorption on plant material	1000 based on dry weight	Added Cd = 2.1 mg/m3, 24 h[1]
Adsorption on plant material	800 based on dry weight	Added Cd = 2.1 mg/m3, 2 h[1]
Adsorption on silica	1000 based on dry weight	Added Cd = 7.3 mg/m3, 24 h, rg = +- 100[1]
Adsorption on silica	430 based on dry weight	Added Cd = 7.3 mg/m3, 2 h, rg = +- 40[1]
Adsorption on suspended solids	1200 based on dry weight	Added Cd = 177 mg/m3, suspended solids = 34 mg/l, 1 h[1]
Adsorption on suspended solids	13000 based on dry weight	Added Cd = 5.5 mg/m3, suspended solids = 34 mg/l, 1 h[1]
Adsorption on suspended solids	14000 based on dry weight	Added Cd = 2.1 mg/m3, 2 h, Pix Brook[1]
Adsorption on suspended solids	15500 based on dry weight	Added Cd = 2.1 mg/m3, 2 h, Stevenage Brook, rg = 6000 - 25000[1]
Adsorption on suspended solids	17000 based on dry weight	Added Cd = 2.1 mg/m3, 2 h, River Ivel[1]
Adsorption on suspended solids	20000 based on dry weight	EDTA = 2.8 mg/m3, added Cd = 2.1 mg/m3, suspended solids = 88 mg/l[1]
Adsorption on suspended solids	24000 based on dry weight	EDTA = 27 mg/m3, added Cd = 2.1 mg/m3, suspended solids = 88 mg/l[1]
Adsorption on suspended solids	25000 based on dry weight	Added Cd = 2.1 mg/m3, suspended solids = 34 mg/l, 1 h[1]
Adsorption on suspended solids	2600 based on dry weight	Added Cd = 56 mg/m3, suspended solids = 34 mg/l, 1 h[1]
Adsorption on suspended solids	26000 based on dry weight	Added Cd = 2.1 mg/m3, 24 h, Pix Brook[1]
Adsorption on suspended solids	3000 based on dry weight	Added Cd = 2.1 mg/m3, 2 h, River Hiz[1]
Adsorption on suspended solids	40000 based on dry weight	Added Cd = 2.1 mg/m3, 24 h, River Ivel[1]
Adsorption on suspended solids	42000 based on dry weight	EDTA = 0 mg/m3, added Cd = 2.1 mg/m3, suspended solids = 88 mg/l[1]
Adsorption on suspended solids	6000 based on dry weight	Added Cd = 2.1 mg/m3, 24 h, River Hiz[1]
Adsorption on suspended solids	6900 based on dry weight	EDTA = 300 mg/m3, added Cd = 2.1 mg/m3, suspended solids = 88 mg/l[1]
Adsorption on suspended solids	8000 based on dry weight	Added Cd = 13.0 mg/m3, suspended solids = 34 mg/l, 1 h[1]
Atmospheric particulates	1900	US urban air, relative to earths crust, Fe standard, F200[2]
Atmospheric particulates	300	N Atlantic Westerlies, relative to earths crust, Al standard, F200[3]
Brown algae	890 based on wet weight	None[4]
Carassius auratus	136 based on wet weight	Water = 3 ug/l[5]
Carassius auratus	400 based on wet weight	Water = 1 ug/l[5]
Cladophora glomerata	18E3 based on dry weight	Water from main Duck Isl, Lake Ontario, water = 0.1 ng/ml[6]
Crustacean	250 based on wet weight	Marine, Cd-115m, greater than value[7]

Echinoderm	110 based on wet weight	Marine, Cd-115m[7]
Fish, viscera	8 based on wet weight	Marine, Cd-115m, rg = 6 - 10[7]
Fucus vesiculosus	2.7E3 based on dry weight	18 obs., all areas around the British Isles[8]
Humic acid	14000 based on dry weight	After 2 h, added Cd = 2.1 - 4.3 ug/l, rg = +- 2000[9]
Humic acid	18000 based on dry weight	After 24 h, added Cd = 2.1 - 4.3 ug/l, rg = +- 3000[9]
Kaolin	250 based on dry weight	After 2 h, added Cd = 7.3 ug/l, rg = +- 20[9]
Kaolin	380 based on dry weight	After 24 h, added Cd = 7.3 ug/l, rg = +- 50[9]
Lysmata seticaudata	120 based on dry weight	Steady state, 281 K, n=6, Cd-109, whole shrimp[10]
Lysmata seticaudata	230 based on dry weight	Steady state, 286 K, n=6, Cd-109, whole shrimp[10]
Lysmata seticaudata	500 based on dry weight	Steady state, 295 K, n=6, Cd-109, whole shrimp[10]
Lysmata seticaudata	600 based on dry weight	Uptake from water, 2 m, Cd-109[11]
Mytilus edulis	220 based on wet weight	Steady state cf, water = 0.050 mg Cd/l[12]
Mytilus galloprovincialis	130 based on dry weight	Uptake from water, 2 m, Cd-109[11]
Mytilus galloprovincialis	400 based on wet weight	Water = 4.34 ppm[13]
Patella vulgata	2.5E4 based on dry weight	17 obs., all areas around the British Isles[8]
Plankton	910 based on wet weight	None[4]
Plant material	1000 based on dry weight	After 24 h, added Cd = 2.1 ug/l, watercress[9]
Plant material	800 based on dry weight	After 2 h, added Cd = 2.1 ug/l, watercress[9]
Plants	1620 based on dry weight	Algae, bryophytes and vascular hydrophytes, freshwater, 32 sp[14]
Porphyra umbilicalis	6.6E2 based on dry weight	20 obs., all areas around the British Isles[8]
River sediment	10000 based on dry weight	After 2 h, added Cd = 2.1 ug/l, rg = 3000 - 17000[9]
River sediment	23000 based on dry weight	After 24 h, added Cd = 2.1 ug/l, rg = 6000 - 40000[9]
Silica	1000 based on dry weight	After 24 h, added Cd = 7.3 ug/l, rg = +- 100[9]
Silica	430 based on dry weight	After 2 h, added Cd = 7.3 ug/l, rg = +- 40[9]
Zooplankton	6000 based on wet weight	Seawater concentration 0.0001 mg/m3[15]

Source: 1.Gardiner, J. 1974. Water Res., 8(3), pp 157-164.
2.Foestner, U. and G.T.W. Wittmann. 1979. Metal Pollution in the Aquatic Environment. Springer-Verlag.Berlin.
3.Chester, R. and J.H. Stoner. 1973. Nature (london) 245, pp 27-28.
4.Bowen, H.J.M. 1966. Trace Elements in Biochemistry, Ac. Press.
5.Myttenaere, C., M. Merlini, R.Bittel, P. Dabin, J.M. Mousny and G. Pozzi. 1975. Impacts of Nuclear releases into the Aquatic Environment. IAEA-SM-198/16, pp 29-47. Vienna.
6.Keeney, W.L., W.G. Breck , G.W. Vanloon and J.A. Page. 1976. Water Res., 10(11)., pp 981-984.
7.Hiyama, Y. and M. Shimizu. 1964. Rec. Oceanogr. Works. Jap. 7(2), pp 43-77.
8.Preston, A., D.R. Jefferies , J.W.R. Harvey and A.K. Steele. 1972. Environ. Pollut., 3, pp 183-214.
9.Sparling, A.B. 1968. Interaction between Blue-Green Algae and Heavy Metals. Ph D. thesis. Wash Univ.St Louis. Mo. 107 pp.
10.Fowler, S.W. and G. Benayoun. 1974. In: Comparative Studies of Food and Environmental Contamination.IAEA. Vienna. 159-178.
11.Fowler, S.W. and G. Benayoun. 1974. in: Comparative Studies of Food and Environmental Contamination.IAEA, pp 159-178. Vienna.
12.Calapaj, G.G. 1974. Ig. Mod. 67(2), pp 136-145.
13.Majori, L. and F. Petronio. 1973. Rev. Int. Oceanogr. Med. CERBOM., 31-32, pp 55-90.
14.Timofeeva-Resovskaya, E.A., T-R. N.V. and E.A. Gileva. 1961. Dokl. Akad. Nauk. SSSR, 140, pp 1437.
15.Martin, J.H 1970. Limnol. Oceanogr. 15(5), pp 756-761.

5-197 Concentration factor
Ce

Species	Value	Condition
Crustacea	600 based on dry weight	Freshwater, rg = 300 - 1000, 2 values used[1]
Crustacea	88 based on dry weight	Marine, rg = 5 - 220, 4 values used[1]
Fish	81 based on dry weight	Freshwater, rg = 2 - 160, 2 values used[1]
Fish	99 based on dry weight	Marine, rg = 0.3 - 538, 6 values used[1]
Hizikia fusiforme, whole body	2740 based on wet weight	Rel to soluble conc, tot conc = 58 ng/l, solub = 7 ng/l, seawat,[2]
Meretrix lusoria, soft part	520 based on wet weight	Rel to total conc, tot conc = 58 ng/l, solub = 7 ng/l, seawat, Japan[2]
Molluscs	1100 based on dry weight	Freshwater, rg = 400 - 1500, 4 values used[1]
Molluscs	240 based on dry weight	Marine, rg = 100 - 350, 3 values used[1]
Paralichthys olivaceus, muscle	200 based on wet weight	Rel to soluble conc, tot conc = 58 ng/l, solub = 7 ng/l, seawat[2]
Plants	1610 based on dry weight	Marine, rg = 120 - 4500, 13 values used[1]
Plants	3180 based on dry weight	Freshwater, rg = 200 - 35000, 12 values used[1]
Plants	7100 based on dry weight	Algae, bryophytes and vascular hydrophytes, freshwater, 32 sp[3]
Sargassum fulvellum, whole body	290 based on wet weight	Rel to total conc, tot conc = 58 ng/l, solub = 7 ng/l, seawat, Japan[2]
Seriola quinqueradiata, muscle	340 based on wet weight	Rel to soluble conc, tot conc = 58 ng/l, solub = 7 ng/l, seawat,[2]
Ulva pertusa, whole body	520 based on wet weight	Rel to total conc, tot conc = 58 ng/l, solub = 7 ng/l, seawat, Japan[2]
Undaria pinnatifida, whole body	530 based on wet weight	Rel to soluble conc, tot conc = 58 ng/l, solub = 7 ng/l, seawat,[2]

Source: 1.Jinks, S.M. and M. Eisenbud. 1972. Radiat. Data Rep. 13, pp 243-247.
2.Suzuki, H., T. Koganage and M. Saki. 1975. Impacts of Nuclear Releases into the Aquatic Environment. IAEA SM-198/19, pp 77-89. Vienna.
3.Timofeeva-Resovskaya, E.A., T-R. N.V. and E.A. Gileva. 1961. Dokl. Akad. Nauk. SSSR, 140, pp 1437.

5-198 Concentration factor
Ce - 141

Species	Value	Condition
Anodonta nuttalliana, soft parts	3 based on wet weight	No food, 292 K, pH = 7.5, freshwat = 370 part & 0 solub p ci/l, 21 d[1]
Anodonta nuttalliana, soft parts	6 based on wet weight	No food, 292 K, pH = 7.5, freshwat = 90 part & 0 solub p ci/l, 7 d[1]
Astacus sp., muscle	1 based on wet weight	No food, 292 K, pH = 7.5, freshwat = 370 part & 0 solub p ci/l, 21 d,[1]
Astacus sp., muscle	1 based on wet weight	No food, 292 K, pH = 7.5, freshwat = 90 part & 0 solub p ci/l, 7 d,[1]
Astacus sp., soft parts	110 based on wet weight	No food, 292 K, pH = 7.5, freshwat = 90 part & 0 solub p ci/l, 7 d[1]
Astacus sp., soft parts	38 based on wet weight	No food, 292 K, pH = 7.5, freshwat = 2.3 part & 1200 solub p ci/ml, 7d[1]
Astacus sp., viscera	130 based on wet weight	No food, 292 K, pH = 7.5, freshwat = 90 part & 0 solub p ci/l, 7 d[1]

Astacus sp., viscera	46 based on wet weight	No food, 292 K, pH = 7.5, freshwat = 370 part & 0 solub p ci/l, 21 d[1]
Cancer productus, muscle	1 based on wet weight	No food, 285 K, pH = 8.2, seawater = 30 part & 0 solub p ci/l, 20 d,[1]
Cancer productus, sort parts	140 based on wet weight	No food, 285 K, pH = 8.2, seawater = 30 part & 0 solub p ci/l, 20 d[1]
Cancer productus, viscera	200 based on wet weight	No food, 285 K, pH = 8.2, seawater = 30 part & 0 solub p ci/l, 20 d[1]
Cancer productus, viscera	61 based on wet weight	No food, 285 K, pH = 8.2, seawater = 120 part & 0 solub p ci/l, 6 d[1]
Carassius auratus, muscle	1 based on wet weight	No food, 292 K, pH = 7.5, freshwat = 90 part and 0 solub p ci/l, 7 d[1]
Carassius auratus, nonmuscle	7 based on wet weight	No food, 292 K, pH = 7.5, freshwat = 90 part and 0 solub p ci/l, 7 d[1]
Carassius auratus, whole body	7 based on wet weight	No food, 292 K, pH = 7.5, freshwat = 370 part and 0 solub p ci/l, 21[1]
Fish, muscle	150 based on wet weight	Danube, 1973, nuclear power plant since 1967, rg = 30 - 560[2]
Freshwater sediment	750 based on wet weight	Danube, 1973, nuclear power plant since 1967, rg = 290 - 1080[2]
Gillichthys mirabilis, muscle	1 based on wet weight	No food, 285 K, pH = 8.2, seawater = 120 part and 0 solub p ci/l, 6d[1]
Gillichthys mirabilis, muscle	1 based on wet weight	No food, 285 K, pH = 8.2, seawater = 30 part and 0 solub p ci/l, 20d[1]
Gillichthys mirabilis, nonmuscle	0.6 based on wet weight	No food, 285 K, pH = 8.2, seawater = 120 part and 0 solub p ci/l, 6 d[1]
Gillichthys mirabilis, nonmuscle	57 based on wet weight	No food, 285 K, pH = 8.2, seawater = 30 part and 0 solub p ci/l, 20 d[1]
Gillichthys mirabilis, whole body	5 based on wet weight	No food, 285 K, pH = 8.2, seawater = 30 part and 0 solub p ci/l, 20 d[1]
Gillichthys mirabilis, whole body	52 based on wet weight	No food, 285 K, pH = 8.2, seawater = 34 part and 1300 solub p ci/ml[1]
Mytilus edulis, soft parts	5 based on wet weight	No food, 285 K, pH = 8.2, seawater = 120 part and 0 solub p ci/l, 6 d[1]
Saxidonus giganteus, soft parts	60 based on wet weight	No food, 285 K, pH = 8.2, seawater = 30 part and 0 solub p ci/l, 20 d[1]
Suspended solids	1020 based on wet weight	Danube, 1973, nuclear power plant since 1967, rg = 340 - 1860[2]
Water plant	270 based on wet weight	Danube, 1973, nuclear power plant since 1967, rg = 50 - 810[2]

Source: 1.Harrison, F.L. 1973. Health Physics, 24, pp 331-343.
2.Herrmann, H., M. Ruf, K. Hubel and W. Lunsmann. 1975. Impacts of Nuclear Releases into the Aquatic Environment. IAEA-SM-198/14. pp 461-470. Vienna.

5-199 Concentration factor
Ce - 144

Species	Value	Condition
Algae	100 based on wet weight	Marine, greater than[1]
Algae	6000 based on wet weight	Lab value, rg = 2000 - 10000[2]
Corallina officinalis	15000 based on dry weight	Seawater = 1 - 2 p ci/l, rg = 10000 - 20000[3]
Crustacean	200 based on wet weight	Edible part, rg = 100 - 300, marine[4]
Crustacean, exoskeleton	2 based on wet weight	Marine[1]
Crustacean, muscle	25 based on wet weight	Marine[1]
Crustacean, viscera	200 based on wet weight	Marine[1]

Echinoderm, digestive tract	250 based on wet weight	Marine, greater than[1]
Echinoderm, gonad	2.5 based on wet weight	Marine[1]
Fish	200 based on wet weight	Lab value, less[2]
Fish, bone	390 based on wet weight	Danube, 1973, nuclear power plant since 1967, rg = 140 - 760[5]
Fish, gill	70 based on wet weight	Marine, greater than[1]
Fish, muscle	0.3 based on wet weight	Marine[1]
Fish, muscle	60 based on wet weight	Danube, 1973, nuclear power plant since 1967, rg = 20 - 200[5]
Freshwater sediment	1010 based on wet weight	Danube, 1973, nuclear power plant since 1967, rg = 260 - 2180[5]
Invertebrate	1250 based on wet weight	Lab value, rg = 500 - 2000[2]
Lampsilis radiata	900 based on wet weight	91 days, freshwater[6]
Molluscs	200 based on wet weight	Edible part, rg = 100 - 300, marine[4]
Molluscs, adductor muscle	5 based on wet weight	Marine[1]
Molluscs, viscera	40 based on wet weight	Marine, greater than[1]
Nitella opaca	26000 based on wet weight	Equilibrium after 100 days[7]
Phytoplankton	1650 based on wet weight	30 w, 6 species, rg = 300 - 3000, marine[8]
Suspended silt	86700 based on dry weight	Seawater, Bombay, rg = 16900 - 132000[9]
Suspended solids	1330 based on wet weight	Danube, 1973, nuclear power plant since 1967, rg = 60 - 8160[5]
Water plant	310 based on wet weight	Danube, 1973, nuclear power plant since 1967, rg = 50 - 630[5]

Source: 1.Hiyama, Y. and M. Shimizu. 1964. Rec. Oceanogr. Works. Jap. 7(2), pp 43-77.
2.Ancellin, J. and A. Vilquin. 1968. Radioprot. 3(3), pp 185-213.
3.Ancellin, J. and A. Vilquin. 1966. in: Disposal of Radioactive Wastes into Seas, Ocean and Surface Water. pp 583-604. IAEA. Vienna.
4.LeGall, P. and J. Ancellin. 1971. Available from Nat. Tech. Inf. Ser., Springfield, Va. as CEA-N-148, pp 1-17.
5.Herrmann, H., M. Ruf, K. Hubel and W. Lunsmann. 1975. Impacts of Nuclear Releases into the Aquatic Environment. IAEA-SM-198/14. pp 461-470. Vienna.
6.Harvey, R.S. 1969. Health Physics, 17, pp 149-154.
7.Hampson, M.A. 1967. J. Exp. Bot. 18, pp 17-53.
8.Rice, T.R. and V.M. Willis. 1959. Limnol. Oceanogr. 4, pp 227-290.
9.Pillai, K.C., N.N. Dey, E. Mathew and B.U. Kothari. 1975. Impacts of Nuclear releases into the Aquatic Environment. IAEA-SM-198/16, pp 277-299. Vienna.

5-200 Concentration factor Chlordane

Species	Value	Condition
Alfalfa, foliage	302.4 based on wet weight	Conc in soil = 0.25 mg/kg dry wt[1]
Algae	2 based on wet weight	Conc in water = 6.6 ng/l[1]
Chub	1.2 based on wet weight	Conc in water = 6.6 ug/l[1]
Earthworms	5.0 based on wet weight	Conc in soil = 0.02 mg/kg dry wt, USA, sites = 67[1]
Gonidea sp.	1000 based on wet weight	Conc in water = 0.007 ug/l, USA[1]
Kormophytes	0.46 based on wet weight	Conc in water = 6.6 ng/l[1]
Oyster	10 based on wet weight	Conc in water: < 1.0 ug/l, cf: < 0.001 mg/kg, USA[1]
Oyster	10 based on wet weight	Conc in water: < 1.0 ug/l, conc in sediment < 0.01 mg/kg, USA[1]
Soybean	0.005 based on wet weight	Conc in soil = 0.020 mg/kg dry wt[1]
Sugar beet, root	0.182 based on wet weight	Conc in loam soil = 1.233 mg/kg dry wt[1]
Sweet potato, tubers	0.004 based on wet weight	Conc in soil = 0.28 mg/kg dry wt[1]

Source: 1.Edwards, C.A. 1975. Persistent Pesticides in the Environment, 2nd edition, CRC-Press.

5-201 Concentration factor Chlorinated naphthalene

Species	Value	Condition
Chlorococcum sp.	110 based on wet weight	Conc in water = 10ppb, 24 h, halowax 1014, F026[1]
Chlorococcum sp.	110 based on wet weight	Conc in water = 50 ppb, 24 h, halowax 1014, F026[1]
Chlorococcum sp.	120 based on wet weight	Conc in water = 100ppb, 24 h, halowax 1013, F025[1]
Chlorococcum sp.	120 based on wet weight	Conc in water = 100ppb, 24 h, halowax 1014, F026[1]
Chlorococcum sp.	124 based on wet weight	Conc in water = 25 ppb, 24 h, halowax 1014, F026[1]
Chlorococcum sp.	140 based on wet weight	Conc in water = 5 ppb, 24 h, halowax 1014, F026[1]
Chlorococcum sp.	25 based on wet weight	Conc in water = 50 ppb, 24 h, halowax 1000, F024[1]
Chlorococcum sp.	32 based on wet weight	Conc in water = 100ppb, 24 h, halowax 1000, F024[1]
Chlorococcum sp.	60 based on wet weight	Conc in water = 10ppb, 24 h, halowax 1013, F025[1]
Chlorococcum sp.	60 based on wet weight	Conc in water = 5 ppb, 24 h, halowax 1013, F025[1]
Chlorococcum sp.	84 based on wet weight	Conc in water = 25 ppb, 24 h, halowax 1013, F025[1]
Chlorococcum sp.	86 based on wet weight	Conc in water = 50 ppb, 24 h, halowax 1013, F025[1]

Source: 1.Knaver, G.A. and J.H. Martin. 1972. Limnol. Oceanogr. 17(6), pp 868-876.

5-202 Concentration factor Chrysene

Species	Value	Condition
Macoma inquinata	0.04 based on sediment concentration	Sediment=2000 ug prudhoe B crude oil/g, 7 d, sed=8.37 ppm, water=0.43 ppb[1]
Macoma inquinata	694 based on wet weight	Sediment=2000 ug prudhoe B crude oil/g, 7 d, sed=8.37 ppm, water=0.43 ppb[1]
Rangia cuneata	8.2 based on wet weight	Estuarine, water = 0.066 ppm, tissue = 0.54+-0.3 ppm[2]

Source: 1.Roesijadi, G., J.W. Anderson and J.W. Blaylock. 1978. J. Fish. Res. bd Canada 35, 608-614.
2.Neff, J.M., Anderson, Cox, Laughlin, Rossi and Tatem. 1976. In: Sources, Effects and Sinks of Hydrocarbon in the Aquatic Environment. Wash DC. Amer. Inst. of Biological Science.

5-203 Concentration factor Cl

Species	Value	Condition
Brassica oleracea	38 based on wet weight	Conc in soil solution = 29ppm[1]
Brown algae	0.062 based on wet weight	None[1]
Daphnia magna	17.5 based on dry weight	On dry wt basis relative to the algae consumed[2]
Plankton	1 based on wet weight	None[1]

Source: 1.Bowen, H.J.M. 1966. Trace Elements in Biochemistry, Ac. Press.
2.Verduin, J. 1975. Limnol. Oceanogr. 20(6), pp 1052.

5-204 Concentration factor Co

Species	Value	Condition
Agarum cribrosum	200 based on dry weight	Seawater = 5E-7 g/l, Vostok Bay, Sea of Japan[1]
Ahnfeltia paradoxa	290 based on wet weight	Seawater = 0.14 ug/l, Japan, Co-60[2]
Ahnfeltia paradoxa	910 based on wet weight	Seawater = 0.14 ug/l, Japan[2]

Atmospheric particulates	0.45	Low level marine part, relative to earths crust, Fe standard, F200[3]
Atmospheric particulates	2	US urban air, relative to earths crust, Fe standard, F200[4]
Atmospheric particulates	3.5	N Atlantic Westerlies, relative to earths crust, Al standard, F200[5]
Atmospheric particulates	4.7	South Pole, relative to earths crust, Al standard, F200[6]
Bivalves	880 based on wet weight	Bombay, rg = 30 - 1730[7]
Brown algae	450 based on wet weight	None[8]
Brown algae	650 based on wet weight	None[9]
Chondrus ocellatus	833 based on wet weight	Seawater = 0.14 ug/l, Japan, Co-60[2]
Chondrus ocellatus	860 based on wet weight	Seawater = 0.14 ug/l, Japan[2]
Chondrus yendoi	150 based on dry weight	Seawater = 5E-7 g/l, Vostok Bay, Sea of Japan[1]
Chorda filum	540 based on dry weight	Seawater = 5E-7 g/l, Vostok Bay, Sea of Japan[1]
Clibanarius virescens, whole body	520 based on dry weight	15 - 30 d, marine, Co-60[10]
Codium yessoensis	60 based on dry weight	Seawater = 5E-7 g/l, Vostok Bay, Sea of Japan[1]
Costaria costata	100 based on dry weight	Seawater = 5E-7 g/l, Vostok Bay, Sea of Japan[1]
Crassostrea gigas, body fluid	20 based on wet weight	Steady state cf, initial cf = 20, Co-60[11]
Crassostrea gigas, digestive gland + stomach	140 based on wet weight	Steady state cf, initial cf = 60, Co-60[11]
Crassostrea gigas, gills + labial palps	96 based on wet weight	Steady state cf, initial cf = 42, Co-60[11]
Crassostrea gigas, intact animal	300 based on wet weight	Steady state cf, initial cf = 90, Co-60[11]
Crassostrea gigas, mantle	82 based on wet weight	Steady state cf, initial cf = 38, Co-60[11]
Crassostrea gigas, muscle	41 based on wet weight	Steady state cf, initial cf = 33, Co-60[11]
Crassostrea gigas, shell	610 based on wet weight	Steady state cf, initial cf = 200, Co-60[11]
Crassostrea gigas, soft tissues	51 based on wet weight	Steady state cf, initial cf = 30, Co-60[11]
Crassostrea gigas, visceral remains	38 based on wet weight	Steady state cf, initial cf = 19, Co-60[11]
Crustacea	1700 based on dry weight	Marine, rg = 300 - 4000, 3 values used[12]
Eisenia bicyclis	200 based on wet weight	Seawater = 0.14 ug/l, Japan[2]
Eisenia bicyclis	34 based on wet weight	Seawater = 0.14 ug/l, Japan, Co-60[2]
Enteromorpha prolifera	1400 based on dry weight	Seawater = 5E-7 g/l, Vostok Bay, Sea of Japan[1]
Fish	1615 based on dry weight	Freshwater, rg = 60 - 3450, 5 values used[12]
Fish	50 based on wet weight	Marine organisms, typical value[13]
Fish	650 based on dry weight	Marine, rg = 20 - 5000, 9 values used[12]
Fish	90 based on wet weight	Freshwater organisms, typical value[14]
Fish, bone	220 based on wet weight	Danube, 1973, nuclear power plant since 1967, rg = 100 - 310, Co-58[15]

Fish, bone	310 based on wet weight	Danube, 1973, nuclear power plant since 1967, rg = 140 - 420, Co-60[15]
Fish, muscle	130 based on wet weight	Danube, 1973, nuclear power plant since 1967, rg = 10 - 380, Co-60[15]
Fish, muscle	130 based on wet weight	Danube, 1973, nuclear power plant since 1967, rg = 30 - 310, Co-58[15]
Freshwater plankton	37000 based on wet weight	Conc in water = 0.03 ug/l, Schoehsee[16]
Freshwater sediment	2830 based on wet weight	Danube, 1973, nuclear power plant since 1967, rg = 540 - 6070, Co-60[15]
Freshwater sediment	890 based on wet weight	Danube, 1973, nuclear power plant since 1967, rg = 310 - 2420, Co-58[15]
Hizikia fusiforme	120 based on wet weight	Seawater = 0.14 ug/l, Japan[2]
Hizikia fusiforme	127 based on wet weight	Seawater = 0.14 ug/l, Japan, Co-60[2]
Laminaria chichorioides	500 based on dry weight	Seawater = 5E-7 g/l, Vostok Bay, Sea of Japan[1]
Laminaria japonica	138 based on wet weight	Seawater = 0.14 ug/l, Japan, Co-60[2]
Laminaria japonica	70 based on wet weight	Seawater = 0.14 ug/l, Japan[2]
Lampsilis radiata	790 based on wet weight	91 days, freshwater, Co-60[17]
Leander pacificus, carapace	18 based on dry weight	15 - 30 d, marine, Co-60[10]
Leander pacificus, flesh	7 based on dry weight	15 - 30 d, marine, Co-60[10]
Macrophytes	1367 based on wet weight	Freshwater organisms, typical value[14]
Molluscs	166 based on dry weight	Marine, rg = 1 - 210, 6 values used[12]
Molluscs	300 based on wet weight	Freshwater organisms, typical value[18]
Molluscs	32408 based on dry weight	Freshwater, rg = 300 - 85000, 5 values used[12]
Mytilus edulis, byssus	1500 based on dry weight	Lab., marine, Co-58[19]
Mytilus edulis, shell	190 based on dry weight	15 - 30 d, marine, Co-60[10]
Mytilus edulis, shell	190 based on dry weight	Lab., marine, Co-58[19]
Mytilus edulis, soft parts	140 based on dry weight	15 - 30 d, marine, Co-60[10]
Pelvetia wrightii	300 based on dry weight	Seawater = 5E-7 g/l, Vostok Bay, Sea of Japan[1]
Phyllospadix iwatensis	460 based on dry weight	Seawater = 5E-7 g/l, Vostok Bay, Sea of Japan[1]
Phytoplankton	100000 based on dry weight	Marine[20]
Phytoplankton	83950 based on dry weight	Range = 15900 - 152000[21]
Plankton	4600 based on wet weight	None[9]
Plants	4425 based on dry weight	Algae, bryophytes and vascular hydrophytes, freshwater, 32 sp[22]
Plants	553 based on dry weight	Marine, rg = 60 - 1400, 5 values used[12]
Plants	6760 based on dry weight	Freshwater, rg = 300 - 30000, 11 values used[12]
Polysiphonia japonica	740 based on dry weight	Seawater = 5E-7 g/l, Vostok Bay, Sea of Japan[1]
Ptilota filicina	900 based on dry weight	Seawater = 5E-7 g/l, Vostok Bay, Sea of Japan[1]
Rhodomela larix	750 based on dry weight	Seawater = 5E-7 g/l, Vostok Bay, Sea of Japan[1]
Sargassum pallidum	300 based on dry weight	Seawater = 5E-7 g/l, Vostok Bay, Sea of Japan[1]
Sargassum ringgoldianum	260 based on wet weight	Seawater = 0.14 ug/l, Japan[2]
Sargassum thunbergii	1890 based on wet weight	Seawater = 0.14 ug/l, Japan[2]

Sargassum thunbergii	2574 based on wet weight	Seawater = 0.14 ug/l, Japan, Co-60[2]
Sediment	16 based on dry weight	Sand sediment, marine, Urazoko Bay, Japan, Co-60[2]
Sediment	169 based on dry weight	Silty sand sediment, marine, Urazoko Bay, Japan, Co-60[2]
Sediment	20 based on dry weight	Sand sediment, marine, Urazoko Bay, Japan, Co-60[2]
Sediment	231 based on dry weight	Sandy silt sediment, marine, Urazoko Bay, Japan, Co-60[2]
Sediment	26 based on dry weight	Sand sediment, marine, Urazoko Bay, Japan, Co-60[2]
Sediment	36 based on dry weight	Sand sediment, marine, Urazoko Bay, Japan, Co-60[2]
Seston	3.8 based on wet weight	Depth = 17 m, Schoehsee[16]
Seston	4.2 based on wet weight	Depth = 8 m, Schoehsee[16]
Styela plicata, test	400 based on dry weight	15 - 30 d, marine, Co-60[10]
Styela plicata, viscera	850 based on dry weight	15 - 30 d, marine, Co-60[10]
Suspended solids	2860 based on wet weight	Danube, 1973, nuclear power plant since 1967, rg = 290 - 15400, Co-58[15]
Suspended solids	5380 based on wet weight	Danube, 1973, nuclear power plant since 1967, rg = 470 - 12040, Co-60[15]
Ulva fenestrata	200 based on dry weight	Seawater = 5E-7 g/l, Vostok Bay, Sea of Japan[1]
Ulva pertusa	340 based on wet weight	Seawater = 0.14 ug/l, Japan[2]
Ulva pertusa	439 based on wet weight	Seawater = 0.14 ug/l, Japan, Co-60[2]
Undaria pinnatifida	190 based on wet weight	Seawater = 0.14 ug/l, Japan[2]
Undaria pinnatifida	210 based on wet weight	Seawater = 0.14 ug/l, Japan, Co-60[2]
Water plant	220 based on wet weight	Danube, 1973, nuclear power plant since 1967, rg = 25 - 520, Co-60[15]
Water plant	530 based on wet weight	Danube, 1973, nuclear power plant since 1967, rg = 15 - 1800, Co-58[15]
Zooplankton	3200 based on wet weight	Seawater concentration 0.0005 mg/m3[23]
Zostera asiatica	770 based on dry weight	Seawater = 5E-7 g/l, Vostok Bay, Sea of Japan[1]

Source: 1.Saenko, G.N., M.D. Koryakov, V.F. Makienko and I.G. Dobrosmy. 1976. Mar. biol., 34(2), pp 169-176.
2.Gilat, Y., Laichter and N.H. Shafrir. 1975. Impacts of Nuclear releases into the Aquatic Environment. IAEA-SM-198/18, pp 63-75. Vienna.
3.Zoller, W.H., E.S. Gladey and R.A. Duce. 1974. Science 183, pp 200.
4.Foestner, U. and G.T.W. Wittmann. 1979. Metal Pollution in the Aquatic Environment. Springer-Verlag.Berlin.
5.Chester, R. and J.H. Stoner. 1973. Nature (london) 245, pp 27-28.
6.Duce, R.A., J. Quinn, C.M. Olney, S. Piotrowits, B. Ray and Wade.1972. Science 176, pp 161-163.
7.Bhatt, Y.M., V.N. Sastry, S.M. Shah and T.M. Krishnamoorthy. 1968. Proc. Nat. Inst. Sci., India, pt, Biol. Sec. 34(B,6), pp 283-287.
8.Mauchline, J. and W.L Templeton. 1964. Oceanogr. Mar. Biol. Ass. Rev., 2, pp 229.
9.Bowen, H.J.M. 1966. Trace Elements in Biochemistry, Ac. Press.
10.Shimizu, M., T. Kajihara and Y. Hiyama. 1970. Rec. Oceanogr. Works. Jap. 10(2), pp 137-145.
11.Cranmore, G. and F.L. Harrison. 1975. Health Phys., 28, pp 319-333.
12.Jinks, S.M. and M. Eisenbud. 1972. Radiat. Data Rep. 13, pp 243-247.
13.Goldberg, E.D. 1962. Limnol Oceanogr., 7, pp 72.
14.Merlini, M., C. Rigliocca, A. Berg and G. Pozzi. 1970. IAEA/ SM, 142/27.
15.Herrmann, H., M. Ruf, K. Hubel and W. Lunsmann. 1975. Impacts of Nuclear Releases into the Aquatic Evironment. IAEA-SM-198/14. pp 461-470. Vienna.
16.Groth, P. 1971. Arch. Hydrobiol. 68(3), pp 305-375.
17.Harvey, R.S. 1969. Health Physics, 17, pp 149-154.
18.Mathis, B.J. and T.F. Cummins. 1973. Water Pollut. Control 45 pp 1573.
19.Shimizu, M., T. Kajihara, I. Suyama and Y. Hiyama. 1971. J. Rad. Res. 12(1), pp 17-28.
20.Lowman, F.G. 1963. in: Schultz, V. and A.W. Klement, Jr. (eds.); Radioecology. pp 561-567. Reinhold publ. Corp. New York.
21.Cushing, C.E. and L.A. Rancitelli. 1972. Northwest Science, 46(2), pp 115-121.
22.Timofeeva-Resovskaya, E.A., T-R. N.V. and E.A. Gileva. 1961. Dokl. Akad. Nauk. SSSR, 140, pp 1437.
23.Martin, J.H 1970. Limnol. Oceanogr. 15(5), pp 756-761.

5-205 Concentration factor Cr

Species	Value	Condition
Agarum cribrosum	590000 based on dry weight	Seawater = 2.0E-9 g/l, Vostok Bay, Sea of Japan[1]
Atmospheric particulates	1.2	Low level marine part, relative to earths crust, Fe standard, F200[2]
Atmospheric particulates	11	US urban air, relative to earths crust, Fe standard, F200[3]
Brown algae	5150 based on wet weight	Range = 300 - 10000[4]
Brown algae	6500 based on wet weight	None[5]
Chondrus yendoi	450000 based on dry weight	Seawater = 2.0E-9 g/l, Vostok Bay, Sea of Japan[1]
Chorda filum	310000 based on dry weight	Seawater = 2.0E-9 g/l, Vostok Bay, Sea of Japan[1]
Codium yessoensis	1095000 based on dry weight	Seawater = 2.0E-9 g/l, Vostok Bay, Sea of Japan[1]
Costaria costata	175000 based on dry weight	Seawater = 2.0E-9 g/l, Vostok Bay, Sea of Japan[1]
Enteromorpha prolifera	1170000 based on dry weight	Seawater = 2.0E-9 g/l, Vostok Bay, Sea of Japan[1]
Fish	10 based on wet weight	Freshwater organisms, typical value[6]
Hermione hystrix	15 based on wet weight	19 d, Cr-51[7]
Heterochordaria abietina	435000 based on dry weight	Seawater = 2.0E-9 g/l, Vostok Bay, Sea of Japan[1]
Laminaria chichorioides	335000 based on dry weight	Seawater = 2.0E-9 g/l, Vostok Bay, Sea of Japan[1]
Laminaria japonica	145000 based on dry weight	Seawater = 2.0E-9 g/l, Vostok Bay, Sea of Japan[1]
Lampsilis radiata	440 based on wet weight	91 days, freshwater, Cr-51[8]
Macrophytes	2880 based on wet weight	Marine organisms, typical value[9]
Molluscs	21800 based on wet weight	Marine organisms, typical value[10]
Molluscs	267 based on wet weight	Freshwater organisms, typical value[11]
Pelvetia wrightii	325000 based on dry weight	Seawater = 2.0E-9 g/l, Vostok Bay, Sea of Japan[1]
Phyllospadix iwatensis	440000 based on dry weight	Seawater = 2.0E-9 g/l, Vostok Bay, Sea of Japan[1]
Phytoplankton	49450 based on wet weight	Range = 17900 - 81000[12]
Phytoplankton	7800 based on wet weight	Marine organisms, typical value[13]
Plankton	17000 based on wet weight	None[5]
Plants	695 based on dry weight	Algae, bryophytes and vascular hydrophytes, freshwater, 32 sp[14]
Polysiphonia japonica	1120000 based on dry weight	Seawater = 2.0E-9 g/l, Vostok Bay, Sea of Japan[1]
Ptilota filicina	300000 based on dry weight	Seawater = 2.0E-9 g/l, Vostok Bay, Sea of Japan[1]
Rhodomela larix	785000 based on dry weight	Seawater = 2.0E-9 g/l, Vostok Bay, Sea of Japan[1]
Sargassum pallidum	195000 based on dry weight	Seawater = 2.0E-9 g/l, Vostok Bay, Sea of Japan[1]
Ulva fenestrata	172000 based on dry weight	Seawater = 2.0E-9 g/l, Vostok Bay, Sea of Japan[1]
Ulvaria spendens	55000 based on dry weight	Seawater = 2.0E-9 g/l, Vostok Bay, Sea of Japan[1]
Zostera asiatica	215000 based on dry weight	Seawater = 2.0E-9 g/l, Vostok Bay, Sea of Japan[1]

Source: 1.Saenko, G.N., M.D. Koryakov, V.F. Makienko and I.G. Dobrosmy. 1976. Mar. biol., 34(2), pp 169-176.
2.Zoller, W.H., E.S. Gladey and R.A. Duce. 1974. Science 183, pp 200.
3.Foestner, U. and G.T.W. Wittmann. 1979. Metal Pollution in the Aquatic Environment. Springer-Verlag.Berlin.
4.Mauchline, J. and W.L Templeton. 1964. Oceanogr. Mar. Biol. Ass. Rev., 2, pp 229.
5.Bowen, H.J.M. 1966. Trace Elements in Biochemistry, Ac. Press.
6.Merlini, M., C. Rigliocca, A. Berg and G. Pozzi. 1970. IAEA/ SM, 142/27.
7.Chipman, W.A. 1967. Proc. Int. Symp. Radioecol. Concent. Progress, 1966. Stockholm, pp 931-941.

8.Harvey, R.S. 1969. Health Physics, 17, pp 149-154.
9.Haegerhaell, B. 1973. Bot. Mar., 16, p 53.
10.Pringle, B.H., D.E. Hissong, E.L. Katz and S.T. Mulawka. 1968. J. San. Eng. Div. Am. Soc. Civ. Eng., 94(5970), pp 455-475.
11.Mathis, B.J. and T.F. Cummins. 1973. Water Pollut. Control 45 pp 1573.
12.Cushing, C.E. and L.A. Rancitelli. 1972. Northwest Science, 46(2), pp 115-121.
13.Martin, J.H. and G.A. Knauer. 1973. Geochim. Cosmochim. Acta., 37, p 1639.
14.Timofeeva-Resovskaya, E.A., T-R. N.V. and E.A. Gileva. 1961. Dokl. Akad. Nauk. SSSR, 140, pp 1437.

5-206 Concentration factor
Cs

Species	Value	Condition
32 freshwater plant	480 based on dry weight	Algae, bryophytes and vascular hydrophytes, freshwater, 32 sp[1]
Ameiurus nebulosus	212 based on wet weight	Wintergreen Lake, stable Cs, May 1971[2]
Bowfin	268 based on wet weight	Wintergreen Lake, stable Cs, May 1971[2]
Brown algae	33 based on wet weight	None[3]
Brown algae	50.5 based on wet weight	Range = 1 - 100[4]
Ceratophyllum sp.	420 based on wet weight	Wintergreen Lake, stable Cs, May 1971[2]
Crustacea	18 based on dry weight	Marine, rg = 0.5 - 26, 5 values used[5]
Fish	3680 based on dry weight	Freshwater, rg = 120 - 22000, 15 values used[5]
Fish	48 based on dry weight	Marine, rg = 5 - 244, 10 values used[5]
Hybrid sunfish	152 based on wet weight	Wintergreen Lake, stable Cs, May 1971[2]
Hybrid sunfish	333 based on wet weight	Wintergreen Lake, stable Cs, Sep 1970[2]
Lake chubsucker	262 based on wet weight	Wintergreen Lake, stable Cs, Sep 1970[2]
Lake chubsucker	292 based on wet weight	Wintergreen Lake, stable Cs, may 1971[2]
Largemouth bass	3188 based on wet weight	Wintergreen Lake, stable Cs, may 1971[2]
Largemouth bass	348 based on wet weight	Wintergreen Lake, stable Cs, may 1971[2]
Largemouth bass	723 based on wet weight	Wintergreen Lake, stable Cs, Sep 1970[2]
Molluscs	15 based on dry weight	Marine, rg = 3 - 28, 7 values used[5]
Nuphar sp.	416 based on wet weight	Wintergreen Lake, stable Cs, may 1971[2]
Phytoplankton	112 based on wet weight	Wintergreen Lake, stable Cs, may 1971[2]
Phytoplankton	1595 based on wet weight	Wintergreen Lake, stable Cs, Sep 1970[2]
Phytoplankton	22000 based on wet weight	Range = 2000 - 42000[6]
Plankton	3 based on wet weight	Range = 1 - 5[4]
Plants	51 based on dry weight	Marine, rg = 17 - 240, 10 values used[5]
Plants	907 based on dry weight	Freshwater, rg = 80 - 4000, 9 values used[5]
Salmo trutta	1.02E3 based on wet weight	Whole fish, stable Cs, fish = 6.6 g, soft water area[7]
Salmo trutta	6.80E3 based on wet weight	Whole fish, stable Cs, fish = 354 g, soft water area[7]
Salmo trutta	7.80E3 based on wet weight	Whole fish, stable Cs, fish = 850 g, soft water area[7]
Salmo trutta	7.8E3 based on wet weight	Whole fish, stable Cs, fish = 850 g, soft water area[7]
Yellow perch	1016 based on wet weight	Wintergreen Lake, stable Cs, may 1971[2]
Yellow perch	2009 based on wet weight	Wintergreen Lake, stable Cs, Sep 1970[2]
Zooplankton	248 based on wet weight	Wintergreen Lake, stable Cs, Sep 1970[2]

Source: 1.Timofeeva-Resovskaya, E.A., T-R. N.V. and E.A. Gileva. 1961. Dokl. Akad. Nauk. SSSR, 140, pp 1437.
2.Eyman, L.D. and T.R. Trabalka. 1977. Health Phys., 32(6), pp 475-478.
3.Bowen, H.J.M. 1966. Trace Elements in Biochemistry, Ac. Press.
4.Mauchline, J. and W.L Templeton. 1964. Oceanogr. Mar. Biol. Ass. Rev., 2, pp 229.
5.Jinks, S.M. and M. Eisenbud. 1972. Radiat. Data Rep. 13, pp 243-247.
6.Cushing, C.E. and L.A. Rancitelli. 1972. Northwest Science, 46(2), pp 115-121.
7.Hewett, C.J. and D.F. Jefferies. 1976. J. Fish. Biol. 9, pp 479-489.

5-207 Concentration factor
Cs - 133

Species	Value	Condition
Gammarus lacustris	1030 based on wet weight	393 d after start, East Twin Lake, rel. to filtered water[1]
Gammarus lacustris	184 based on wet weight	393 d after start, East Twin Lake, rel. to total conc.[1]
Gammarus lacustris	24 based on wet weight	0.06 d after start, East Twin Lake, rel. to filtered water[1]
Gammarus lacustris	243 based on wet weight	3 d after start, East Twin Lake, rel. to filtered water[1]
Gammarus lacustris	994 based on wet weight	39 d after start, East Twin Lake, rel. to filtered water[1]
Lemna trisulca	1264 based on wet weight	393 d after start, East Twin Lake, rel. to filtered water[1]
Lemna trisulca	140 based on wet weight	3 d after start, East Twin Lake, rel. to filtered water[1]
Lemna trisulca	228 based on wet weight	393 d after start, East Twin Lake, rel. to total conc.[1]
Lemna trisulca	324 based on wet weight	39 d after start, East Twin Lake, rel. to filtered water[1]
Nuphar polysepalum	101 based on wet weight	39 d after start, East Twin Lake, rel. to filtered water[1]
Nuphar polysepalum	18 based on wet weight	3 d after start, East Twin Lake, rel. to filtered water[1]
Nuphar polysepalum	280 based on wet weight	393 d after start, East Twin Lake, rel. to filtered water[1]
Nuphar polysepalum	50 based on wet weight	393 d after start, East Twin Lake, rel. to total conc.[1]
Salmo gairdneri	2447 based on wet weight	393 d after start, East Twin Lake, rel. to filtered water[1]
Salmo gairdneri	430 based on wet weight	393 d after start, East Twin Lake, rel. to total conc.[1]
Salmo gairdneri	85 based on wet weight	3 d after start, East Twin Lake, rel. to filtered water[1]
Salmo gairdneri	873 based on wet weight	39 d after start, East Twin Lake, rel. to filtered water[1]
Sediment	1357 ml/cm**2 based on surface of sediment	393 d after start, East Twin Lake, rel. to total conc.[1]
Sediment	1945 ml/cm**2 based on surface of sediment	39 d after start, East Twin Lake, rel. to filtered water[1]
Sediment	639 ml/cm**2 based on surface of sediment	3 d after start, East Twin Lake, rel. to filtered water[1]
Sediment	7570 ml/cm**2 based on surface of sediment	393 d after start, East Twin Lake, rel. to filtered water[1]
Sediment	86 ml/cm**2 based on surface of sediment	0.06 d after start, East Twin Lake, rel. to filtered water[1]
Zooplankton	112 based on wet weight	393 d after start, East Twin Lake, rel. to filtered water[1]
Zooplankton	20 based on wet weight	393 d after start, East Twin Lake, rel. to total conc.[1]
Zooplankton	228 based on wet weight	39 d after start, East Twin Lake, rel. to filtered water[1]
Zooplankton	63 based on wet weight	3 d after start, East Twin Lake, rel. to filtered water[1]
Zooplankton	9.3 based on wet weight	0.06 d after start, East Twin Lake, rel. to filtered water[1]

Source: 1.Hakonson, T.E. and F.W. Whicker. 1975. Health Physics., 28, pp 699-706.

5-208 Concentration factor Cs - 134

Species	Value	Condition
Crustacean, soft parts	15 based on wet weight	4 d, marine[1]
Fish, bone	240 based on wet weight	Danube, 1973, nuclear power plant since 1967, rg = 60 - 500[2]
/Fish, muscle	200 based on wet weight	Danube, 1973, nuclear power plant since 1967, rg = 30 - 380[2]
Freshwater phytoplankton	8500 based on dry weight	Equilibrium after 36 h, rg = 4000 - 13000[3]
Freshwater sediment	3040 based on wet weight	Danube, 1973, nuclear power plant since 1967, rg = 840 - 5360[2]
Marine algae	1.6 based on wet weight	1 d, rg = 1.2 - 2.0[1]
Rooted aquatic plant	400 based on dry weight	Equilibrium after 36 h[3]
Suspended solids	13950 based on wet weight	Danube, 1973, nuclear power plant since 1967, rg = 100 - 58210[2]
Water plant	320 based on wet weight	Danube, 1973, nuclear power plant since 1967, rg = 10 - 1350[2]

Source: 1.Hiyama, Y. and M. Shimizu. 1964. Rec. Oceanogr. Works. Jap. 7(2), pp 43-77.
2.Herrmann, H., M. Ruf, K. Hubel and W. Lunsmann. 1975. Impacts of Nuclear Releases into the Aquatic Environment. IAEA-SM-198/14. pp 461-470. Vienna.
3.Gloyna, E.F., Y.A. Yosef and T.J. Padden. 1971. U.S. Atom. Ener. Comm. Contract AT(11-1)-490, Tech.ept. 21, CRWR-75, EHE-71-1, ORO-21, pp 1-66.

5-209 Concentration factor Cs - 137

Species	Value	Condition
Algae	40 based on wet weight	Lab value, less[1]
Ameiurus nebulosus	842 based on wet weight	Wintergreen Lake, may 1971[2]
Astropecten aranciacus	8 based on dry weight	Eastern Mediterranean, dry weight = 43.4 g[3]
Boops boops	208 based on dry weight	Eastern Mediterranean, dry weight = 91.7 g[3]
Bowfin	440 based on wet weight	Wintergreen Lake, may 1971[2]
Brissopsis lyrifera	261 based on dry weight	Eastern Mediterranean, dry weight = 333.0 g[3]
Brissopsis lyrifera	339 based on dry weight	Eastern Mediterranean, dry weight = 153.0 g[3]
Ceratophyllum sp.	302 based on wet weight	Wintergreen Lake, may 1971[2]
Charydis longicollis	57 based on dry weight	Eastern Mediterranean, dry weight = 90.5 g[3]
Crassostrea gigas, body fluid	3.1 based on wet weight	Steady state cf, initial cf = 3.1[4]
Crassostrea gigas, digestive gland + stomach	11.3 based on wet weight	Steady state cf, initial cf = 10.7[4]
Crassostrea gigas, gills + labial palps	10.1 based on wet weight	Steady state cf, initial cf = 9.5[4]
Crassostrea gigas, intact animal	7.7 based on wet weight	Steady state cf, initial cf = 5.3[4]
Crassostrea gigas, mantle	9.4 based on wet weight	Steady state cf, initial cf = 8.6[4]
Crassostrea gigas, muscle	6.6 based on wet weight	Steady state cf, initial cf = 5.4[4]
Crassostrea gigas, shell	9.2 based on wet weight	Steady state cf, initial cf = 7.2[4]
Crassostrea gigas, soft tissues	9.4 based on wet weight	Steady state cf, initial cf = 8.8[4]

Crassostrea gigas, visceral remains	6.9 based on wet weight	Steady state cf, initial cf = 6.4[4]
Crustacean	7 based on wet weight	Edible part, marine, rg = 4 - 10[5]
Enteromorpha sp.	19 based on wet weight	Conc. in body = 4900 mg/kg ww[6]
Eskmeals silt	449.0 based on wet weight	Yearly mean, Cumberland coastral area, 1969[7]
Fish	40 based on wet weight	Lab value, less[1]
Fish, bone	210 based on wet weight	Danube, 1973, nuclear power plant since 1967, rg = 20 - 640[8]
Fish, muscle	170 based on wet weight	Danube, 1973, nuclear power plant since 1967, rg = 35 - 350[8]
Freshwater sediment	10000 based on dry weight	Lake Erie, central basin[9]
Freshwater sediment	10000 based on dry weight	Lake Ontario, western part[9]
Freshwater sediment	3160 based on wet weight	Danube, 1973, nuclear power plant since 1967, rg = 1340 - 5950[8]
Freshwater sediment	3500 based on dry weight	Lake Erie, western basin[9]
Fucus serratus	32 based on wet weight	None[10]
Fucus serratus	46 based on wet weight	Conc. in body = 10550 mg/kg ww[6]
Fucus serratus	53 based on wet weight	None[11]
Fucus vesiculosus	111 based on wet weight	None[12]
Fucus vesiculosus	27 based on wet weight	None[13]
Fucus vesiculosus	46 based on wet weight	Conc. in body = 10140 mg/kg ww[6]
Fucus vesiculosus	52 based on wet weight	Range = 28 - 78[14]
Fucus vesiculosus	62.5 based on wet weight	Range = 51 - 74[15]
Fucus vesiculosus	74 based on wet weight	None[10]
Hybrid sunfish	429 based on wet weight	Wintergreen Lake, may 1971[2]
Hybrid sunfish	525 based on wet weight	Wintergreen Lake, Sep 1970[2]
Inshore sand	36.3 based on wet weight	Yearly mean, Cumberland coastral area, 1969[7]
Invertebrate	40 based on wet weight	Lab value, less[1]
Lake chubsucker	642 based on wet weight	Wintergreen Lake, may 1971[2]
Lake chubsucker	875 based on wet weight	Wintergreen Lake, Sep 1970[2]
Lampsilis radiata	220 based on wet weight	91 days, freshwater[16]
Largemouth bass	1426 based on wet weight	Wintergreen Lake, may 1971[2]
Largemouth bass	1671 based on wet weight	Wintergreen Lake, Sep 1970[2]
Largemouth bass	701 based on wet weight	Wintergreen Lake, may 1971[2]
Littorina littorea, shell	13 based on wet weight	None[6]
Littorina littorea, soft parts	43 based on wet weight	Conc. in body = 3320 mg/kg ww[6]
Littorina littorea, soft parts	45 based on wet weight	Range = 18 - 72[10]
Marine algae	10 based on wet weight	Green algae[17]
Marine algae	138 based on dry weight	Eastern Mediterranean, dry weight = 185.4 g[3]
Marine algae	30 based on wet weight	Brown algae[17]
Marine algae	53 based on dry weight	Eastern Mediterranean, dry weight = 200.0 g[3]
Metapenaeus monoceros	57 based on dry weight	Eastern Mediterranean, dry weight = 127 g[3]
Molluscs	7 based on wet weight	Edible part, marine, rg = 4 - 10[5]

Mugil capito	106 based on dry weight	Eastern Mediterranean, dry weight = 238.0 g[3]
Mugil cephalus	106 based on dry weight	Eastern Mediterranean, dry weight = 238.0 g[3]
Mytilus edulis, shell	9 based on wet weight	None[18]
Mytilus edulis, soft parts	30 based on wet weight	Range = 12 - 44[19]
Mytilus edulis, soft parts	47 based on wet weight	Conc. in body = 3040 mg/kg ww[18]
Nephtys sp.	3.5 based on wet weight	Yearly mean, Cumberland coastal area, 1969[7]
Nucella lapillus	57 based on wet weight	None[19]
Nucella lapillus, shell	31 based on wet weight	None[18]
Nucella lapillus, soft parts	99 based on wet weight	Conc. in body = 2890 mg/kg ww[18]
Nucula sp.	14.3 based on wet weight	Yearly mean, Cumberland coastal area, 1969[7]
Nuphar sp.	333 based on wet weight	Wintergreen Lake, may 1971[2]
Parapenaeus longirostris	102 based on dry weight	Eastern Mediterranean, dry weight = 50.0 g[3]
Penaeus japonicus	35 based on dry weight	Eastern Mediterranean, dry weight = 133.0 g[3]
Penaeus japonicus	57 based on dry weight	Eastern Mediterranean, dry weight = 208.0 g[3]
Pennatula rubra	71 based on dry weight	Eastern Mediterranean, dry weight = 97.2 g[3]
Phytoplankton	151 based on wet weight	Wintergreen Lake, may 1971[2]
Phytoplankton	833 based on wet weight	Wintergreen Lake, Sep 1970[2]
Pleuronectes platessa, flesh	26.7 based on wet weight	Yearly mean, Cumberland coastal area, 1969[7]
Pleuronectes platessa, gut content	39.0 based on wet weight	Yearly mean, Cumberland coastal area, 1969[7]
Salmo trutta	1.8 based on wet weight	Whole fish, fish = 600 -820.0 g, uptake from water = 1 u ci/l[20]
Salmo trutta	3.6 based on wet weight	Whole fish, fish = 75.0-280.0 g, uptake from water = 1 u ci/l[20]
Salmo trutta	3.90E3 based on wet weight	Whole fish, soft water area[20]
Salmo trutta	4.4 based on wet weight	Whole fish, fish = 8.0 - 40.0 g, uptake from water = 1 u ci/l[20]
Salmo trutta	5.5 based on wet weight	Whole fish, fish = 1.5 - 16.5 g, uptake from water = 1 u ci/l[20]
Salmo trutta, blood cells	1.99 based on wet weight	Steady state cf, uptake from water = 1 u ci/l, 420 d, alk = 224 ppm[20]
Salmo trutta, blood plasma	0.37 based on wet weight	Steady state cf, uptake from water = 1 u ci/l, 420 d, alk = 224 ppm[20]
Salmo trutta, bone	3.4 based on wet weight	Steady state cf, uptake from water = 1 u ci/l, 420 d, alk = 224 ppm[20]
Salmo trutta, gill filament	4.9 based on wet weight	Steady state cf, uptake from water = 1 u ci/l, 420 d, alk = 224 ppm[20]
Salmo trutta, gonad	8.0 based on wet weight	Steady state cf, uptake from water = 1 u ci/l, 420 d, alk = 224 ppm[20]
Salmo trutta, gut	8.2 based on wet weight	Steady state cf, uptake from water = 1 u ci/l, 420 d, alk = 224 ppm[20]
Salmo trutta, heart	10.9 based on wet weight	Steady state cf, uptake from water = 1 u ci/l, 420 d, alk = 224 ppm[20]
Salmo trutta, kidney	5.5 based on wet weight	Steady state cf, uptake from water = 1 u ci/l, 420 d, alk = 224 ppm[20]
Salmo trutta, liver	4.2 based on wet weight	Steady state cf, uptake from water = 1 u ci/l, 420 d, alk = 224 ppm[20]

Salmo trutta, muscle	8.2 based on wet weight	Steady state cf, uptake from water = 1 u ci/l, 420 d, alk = 224 ppm[20]
Salmo trutta, nerve cord	7.0 based on wet weight	Steady state cf, uptake from water = 1 u ci/l, 420 d, alk = 224 ppm[20]
Salmo trutta, skin	2.9 based on wet weight	Steady state cf, uptake from water = 1 u ci/l, 420 d, alk = 224 ppm[20]
Salmo trutta, spleen	5.8 based on wet weight	Steady state cf, uptake from water = 1 u ci/l, 420 d, alk = 224 ppm[20]
Salmo trutta, whole blood	1.08 based on wet weight	Steady state cf, uptake from water = 1 u ci/l, 420 d, alk = 224, ppm[20]
Salmo trutta, whole fish	5.1 based on wet weight	Steady state cf, uptake from water = 1 u ci/l, 420 d, alk = 224 ppm[20]
Sardina pilchardus	110 based on dry weight	Eastern Mediterranean, dry weight = 127.0 g[3]
Sardinella aurita	133 based on dry weight	Eastern Mediterranean, dry weight = 113.9 g[3]
Sardinella aurita	133 based on dry weight	Eastern Mediterranean, dry weight = 204.6 g[3]
Sardinella aurita	51 based on dry weight	Eastern Mediterranean, dry weight = 198.2 g[3]
Sardinella aurita	61 based on dry weight	Eastern Mediterranean, dry weight = 198.3 g[3]
Scomber japonicus	102 based on dry weight	Eastern Mediterranean, dry weight = 238.6 g[3]
Sediment	15 based on dry weight	Sand sediment, marine, Urazoko Bay, Japan[3]
Sediment	16 based on dry weight	Sand sediment, marine, Urazoko Bay, Japan[3]
Sediment	32 based on dry weight	Sand sediment, marine, Urazoko Bay, Japan[3]
Sediment	48 based on dry weight	Silty sand sediment, marine, Urazoko Bay, Japan[3]
Sediment	5 based on dry weight	Sand sediment, marine, Urazoko Bay, Japan[3]
Sediment	66 based on dry weight	Sandy silt sediment, marine, Urazoko Bay, Japan[3]
Sepia offcinalis	47 based on dry weight	Eastern Mediterranean, dry weight = 245.5 g, excluding shell[3]
Sepia offcinalis	65 based on dry weight	Eastern Mediterranean, dry weight = 245.5 g, shell[3]
Siganus luridus	41 based on dry weight	Eastern Mideterranean, dry weight = 88.4 g[3]
Sphyraena sp.	184 based on dry weight	Eastern Mideterranean, dry weight = 100.2 g[3]
Suspended silt	870 based on dry weight	Seawater, Bombay, rg = 422 - 1570[21]
Suspended solids	18780 based on wet weight	Danube, 1973, nuclear power plant since 1967, rg = 1210 - 93140[8]
Trifolium perenne, aerial part	0.44 based on dry weight	Flowering, culture sol. = 4E-1 u ci/ml, soil only, rg = 0.12 - 0.76[22]
Trifolium perenne, aerial part	0.53 based on dry weight	Sown, culture sol. = 4E-1 u ci/ml, rg = 0.02 - 1.04[22]
Trifolium perenne, aerial part	0.76 based on dry weight	Flowering, culture sol. = 4E-2 u ci/ml, overhead irrig[22]
Trifolium perenne, aerial part	0.89 based on dry weight	Flowering, culture sol. = 4E-1 u ci/ml, overhead irrig[22]
Trifolium perenne, aerial part	1.08 based on dry weight	Flowering, culture sol. = 4E-2 u ci/ml, soil only, rg = 0.07 - 2.09[22]
Trifolium perenne, aerial part	1.88 based on dry weight	Shoeting, culture sol. = 4E-1 u ci/ml, soil only, rg = 0.04 - 3.74[22]
Trifolium perenne, aerial part	1.89 based on dry weight	Shoeting, culture sol. = 4E-2 u ci/ml, soil only, rg = 0.06 - 3.73[22]
Trifolium perenne, aerial part	13.4 based on dry weight	Shoeting, culture sol. = 4E-4 u ci/ml, soil only, rg = 0.51 - 26.36[22]

Trifolium perenne, aerial part	14.7 based on dry weight	Flowering, culture sol. = 4E-4 u ci/ml, overhead irrig[22]
Trifolium perenne, aerial part	2.14 based on dry weight	Shoeting, culture sol. = 4E-1 u ci/ml, irrigated, rg = 0.01 - 4.27[22]
Trifolium perenne, aerial part	2.53 based on dry weight	Sown, culture sol. = 4E-2 u ci/ml, rg = 0.07 - 5.10[22]
Trifolium perenne, aerial part	3.13 based on dry weight	Shoeting, culture sol. = 4E-4 u ci/ml, irrigated, rg = 0.15 - 7.50[22]
Trifolium perenne, aerial part	8.42 based on dry weight	Shoeting, culture sol. = 4E-2 u ci/ml, irrigated, rg = 0.02 - 16.82[22]
Trifolium perenne, aerial part	83.4 based on dry weight	Sown, culture sol. = 4E-4 u ci/ml, rg = 0.54 - 166.22[22]
Trifolium perenne, aerial part	9.62 based on dry weight	Flowering, culture sol. = 4E-4 u ci/ml, soil only, rg = 5.36 - 13.88[22]
Trifolium perenne, root	0.34 based on dry weight	Shoeting, culture sol. = 4E-2 u ci/ml, irrigated, rg = 0.04 - 0.63[22]
Trifolium perenne, root	0.61 based on dry weight	Shoeting, culture sol. = 4E-1 u ci/ml, irrigated, rg = 0.07 - 1.14[22]
Trifolium perenne, root	0.67 based on dry weight	Flowering, culture sol. = 4E-2 u ci/ml, irrigated, rg = 0.27 - 1.06[22]
Trifolium perenne, root	0.89 based on dry weight	Flowering, culture sol. = 4E-1 u ci/ml, irrigated, rg = 0.04 - 1.74[22]
Trifolium perenne, root	1.06 based on dry weight	Sown, culture sol. = 4E-2 u ci/ml, rg = 0.14 - 1.98[22]
Trifolium perenne, root	1.68 based on dry weight	Flowering, culture sol. = 4E-2 u ci/ml, soil only, rg = 0.08 - 3.28[22]
Trifolium perenne, root	11.2 based on dry weight	Shoeting, culture sol. = 4E-4 u ci/ml, soil only, rg = 1.83 - 20.51[2]
Trifolium perenne, root	2.12 based on dry weight	Shoeting, culture sol. = 4E-4 u ci/ml, irrigated, rg = 0.56 - 3.67[22]
Trifolium perenne, root	2.14 based on dry weight	Flowering, culture sol. = 4E-1 u ci/ml, soil only, rg = 0.36 - 3.92[22]
Trifolium perenne, root	2.19 based on dry weight	Sown, culture sol. = 4E-1 u ci/ml, rg = 0.02 - 5.10[22]
Trifolium perenne, root	2.21 based on dry weight	Shoeting, culture sol. = 4E-1 u ci/ml, soil only, rg = 0.28 - 4.11[22]
Trifolium perenne, root	2.83 based on dry weight	Shoeting, culture sol. = 4E-2 u ci/ml, soil only, rg = 0.15 - 5.10[22]
Trifolium perenne, root	29.9 based on dry weight	Sown, culture sol. = 4E-4 u ci/ml, rg = 1.47 - 58.34[22]
Trifolium perenne, root	4.71 based on dry weight	Flowering, culture sol. = 4E-4 u ci/ml, irrigated, rg = 1.18 - 8.23[22]
Trifolium perenne, root	70.8 based on dry weight	Flowering, culture sol. = 4E-4 u ci/ml, soil only, rg = 2.11 - 139.55[22]
Upeneus moluccensis	37 based on dry weight	Red Sea, dry weight = 121.0 g[3]
Water plant	180 based on wet weight	Danube, 1973, nuclear power plant since 1967, rg = 10 - 1000[8]
Yellow perch	1043 based on wet weight	Wintergreen Lake, Sep 1970[2]
Yellow perch	1182 based on wet weight	Wintergreen Lake, May 1971[2]
Zooplankton	845 based on wet weight	Wintergreen Lake, Sep 1970[2]

Source: 1.Ancellin, J. and A. Vilquin. 1968. Radioprot. 3(3), pp 185-213.
2.Eyman, L.D. and T.R. Trabalka. 1977. Health Phys., 32(6), pp 475-478.
3.Gilat, Y., Laichter and N.H. Shafrir. 1975. Impacts of Nuclear releases into the Aquatic Environment.IAEA-SM-198/18, pp 63-75. Vienna.
4.Cranmore, G. and F.L. Harrison. 1975. Health Phys., 28, pp 319-333.
5.LeGall, P. and J. Ancellin. 1971. Available from Nat. Tech. Inf. Ser., Springfield, Va. as CEA-N-148, pp 1-17.
6.Vinogradov, M.E., V.V. Menshutkin and E.A. Shushkina. 1972. Mar. biol. 16, pp 261-268.

7.Pentreath, R.J. and D.F. Jefferies. 1971. J. mar. biol. Ass. U.K. 51, pp 963-976.
8.Herrmann, H., M. Ruf, K. Hubel and W. Lunsmann. 1975. Impacts of Nuclear Releases into the Aquatic Environment. IAEA-SM-198/14. pp 461-470. Vienna.
9.Lerman, A. and T.A. Lietzke. 1975. Limnol. Oceanogr. 20(4), pp 497-510.
10.Bryan, G.W., A. Preston and W.L. Templeton. 1966. In: Disposal of Radioactive Wastes into Seas, Ocens and Surface Waters, pp 623-667.
11.Templeton, W.L. 1962. In Proc. Seminar on Aspects of Radioactive Contamination in normal and emergeny situations, Rome, FAO, pp 42-64.
12.Mitchell, N.T. 1969. Radioactivity in Surface and Coastal Waters of the British Isles. 168. M.A.F.F.FRL 5, Lowestoft, Suffolk.
13.Mitchell, N.T. 1971. Radioactivity in Surface and Coastal Water of the British Isles, 1970 M.A.F.F.RL 8, Lowertoft, Suffolk.
14.Mitchell, N.T. 1968. Radioactivity in Surface and Coastal Waters of the British Isles. 1968. M.A.F.FFRL 2, Lowestoft, Suffolk.
15.Mitchell, N.T. 1971. Radioactivity in Surface and Coastal Waters of the British Isles . 1969. M.A.F.FFRL 7, Lowestoft, Suffolk.
16.Harvey, R.S. 1969. Health Physics, 17, pp 149-154.
17.Polikarpov, G.G. 1966. Radioecology of Aquatic Organisms. N.Y. Reinhold.
18.Bryan, G.W. 1963. J. Mar. Biol. Ass. U.K., 43, pp 519.
19.Mauchline, J. 1963. The Biological and Geographical distribution in the Irish Sea of Radioactive Effluent from Windscale Works, 1959-1960. UKAEA AHSB(RP) R.27,
20.Hewett, C.J. and D.F. Jefferies. 1976. J. Fish. Biol. 9, pp 479-489.
21.Pillai, K.C., N.N. Dey, E. Mathew and B.U. Kothari. 1975. Impacts of Nuclear releases into the Aquatic Environment. IAEA-SM-198/16, pp 277-299. Vienna.
22.Bergamini, P.G., G. Palmas, F. Piantelli and M. Rigato. 1970. Health Phys., 19(4), pp 521-528.

5-210 Concentration factor
Cu

Species	Value	Condition
Agarum cribrosum	900 based on dry weight	Seawater = 2.5E-7 g/l, Vostok Bay, Sea of Japan[1]
Ascophyllum nodusum	8600 based on wet weight	Marine, field collection[2]
Atmospheric particulates	3.4	Low level marine part, relative to earths crust, Fe standard, F200[3]
Atmospheric particulates	83	US urban air, relative to earths crust, Fe standard, F200[4]
Atmospheric particulates	84	N Atlantic Westerlies, relative to earths crust, Al standard, F200[5]
Blidingia minima	1.62E4 based on dry weight	Rel to dissolved metal concentration in seawater, rg = 0.71 - 1.83E4[6]
Brassica oleracea	275 based on wet weight	Conc in soil solution = 0.04 ppm[7]
Brown algae	100 based on wet weight	None[8]
Brown algae	920 based on wet weight	None[7]
Busycon canaliculatum	3800 based on wet weight	Marine, field collection[9]
Cancer irroratus	42275 based on wet weight	Seawater conc = 0.47E-3 ppm, Nova Scotia, AAS, Sd = +- 7232[10]
Chlamys opercularis, digestive gland	22000 based on wet weight	Marine, field collection[11]
Chlamys opercularis, kidney	540000 based on wet weight	Marine, field collection[11]
Chlamys opercularis, soft parts	4600 based on wet weight	Marine, field collection[11]
Chlorella regularis	1.5E3 based on dry weight	Lab, pH = 4.6, Cu conc = 0.5-2.0 mg/l[12]
Chlorella vulgaris	0.85E3 based on dry weight	Rg = 0.7-1.0E3, lab, Cu tolerent algae[2]
Chlorella vulgaris	2.5E3 based on dry weight	Rg = 0.4-4.6E3, non-Cu Tolerent algae[2]
Chondrus yendoi	2300 based on dry weight	Seawater = 2.5E-7 g/l, Vostok Bay, Sea of Japan[1]
Chorda filum	560 based on dry weight	Seawater = 2.5E-7 g/l, Vostok Bay, Sea of Japan[1]
Cladophora glomerata	1.9E3 based on dry weight	Water from main Duck Isl., Lake Ontario = 2.9+-0.6, (conc in ng/ml+-SD[13]

Cladophora glomerata	1E3 based on dry weight	Lake Erie, eutrophic, field collection[14]
Cladophora glomerata	2.2E3 based on dry weight	Water from Deadman Bay , Lake Ontario = 5.7+-1.1, (conc in ng/ml+-SD)[13]
Cladophora glomerata	2.5E3 based on dry weight	Spokane River, industrial pollution, field collection[15]
Cladophora sp.	1.8E3 based on dry weight	Alive, Lower Swansea V., distant from Zn smelter, field algae bloom[16]
Cladophora sp.	3.5E3 based on dry weight	Dead, Lower Swansea V., distant from Zn smelter, field algae bloom[16]
Codium yessoensis	500 based on dry weight	Seawater = 2.5E-7 g/l, Vostok Bay, Sea of Japan[1]
Coral	1.79 based on Ca content	Shallow open ocean coral, mean of 8 determinations, F023[17]
Coral	2.22 based on Ca content	Shallow coastal ocean, mean of 13 determinations, F023[17]
Coral	2.76 based on Ca content	Deep ocean coral, mean of 13 determinations, F023[17]
Crassostrea virginica	14800 based on wet weight	Marine, field collection[9]
Enteromorpha linza	0.675E4 based on dry weight	Rel to dissolved metal concentration in seawater, rg = 0.56 - 0.77E4[6]
Enteromorpha prolifera	2800 based on dry weight	Seawater = 2.5E-7 g/l, Vostok Bay, Sea of Japan[1]
Fish	127 based on wet weight	Marine organisms, typical value[18]
Fish	60 based on wet weight	Freshwater organisms, typical value[19]
Freshwater plankton	60000 based on dry weight	Conc in water = 1.0 ug/l, Schoehsee[20]
Fucus vesiculosus	0.25E4 based on dry weight	Rel to dissolved metal concentration in seawater[6]
Fucus vesiculosus	0.45E4 based on dry weight	Rel to dissolved metal concentration in seawater[6]
Fucus vesiculosus	0.55E4 based on dry weight	Rel to dissolved metal concentration in seawater, rg = 0.36 - 0.74E4[6]
Fucus vesiculosus	1.56E4 based on dry weight	Rel to dissolved metal concentration in seawater, rg = 0.45 - 2.7E4[6]
Fucus vesiculosus	15200 based on wet weight	Rg = 3700 - 27000, marine, field collection[2]
Fucus vesiculosus	4.5E3 based on dry weight	18 obs., all areas around the British Isles[21]
Fucus vesiculosus	55000 based on dry weight	Rg = 36000-74000, marine, field collection[22]
Helix aspersa	2 based on dry weight	Relative to conc in dry soil, terrestrial gastropod, rg=20-100 ppm Cu[23]
Heterochordaria abietina	800 based on dry weight	Seawater = 2.5E-7 g/l, Vostok Bay, Sea of Japan[1]
Isopoda	5 based on dry weight	Relative to conc in dry soil, terrestrial isopod, rg=100-500 ppm Cu[23]
Laminaria chichorioides	1600 based on dry weight	Seawater = 2.5E-7 g/l, Vostok Bay, Sea of Japan[1]
Laminaria digitata, lamina	170 based on dry weight	Cu = 0.05 mg/l, exposure 4 d, postexp 6 d, control = 1.7 ppm[24]
Laminaria digitata, lamina	226 based on dry weight	Cu = 0.20 mg/l, exposure 4 d, postexp 6 d, control = 1.7 ppm[24]
Laminaria digitata, lamina	522 based on dry weight	Cu = 0.5 mg/l, exposure 4 d, postexp 6 d, control = 1.7 ppm[24]
Laminaria japonica	105000 based on wet weight	Marine, field collection, rg = 10000-200000[25]
Laminaria japonica	360 based on dry weight	Seawater = 2.5E-7 g/l, Vostok Bay, Sea of Japan[1]
Lumbricus rubellus	0.05 based on dry weight	Relative to conc in dry soil, terrestrial earthworm, rg=10-20 ppm Cu[23]
Macrophytes	158 based on wet weight	Freshwater organisms, typical value[19]

Macrophytes	2890 based on wet weight	Marine organisms, typical value[26]
Mercenaria mercenaria	900 based on wet weight	Marine, field collection[9]
Microspora sp.	1.6E4 based on dry weight	Lower Swansea V., near zinc smelter waste, field algal bloom[16]
Molluscs	1500 based on wet weight	Freshwater organisms, typical value[27]
Molluscs	3800 based on wet weight	Marine organisms, typical value[28]
Mya arenaria	1 based on wet weight	Water=0.01 ug Cu/l, 1 d, spring, 289-295 K, control (0 mg Cu/l)=0.7 ppm[29]
Mya arenaria	132 based on wet weight	Water=0.01 ug Cu/l, 1 d, spring, 289-295 K, control (0 mg Cu/l)=0.7 ppm[29]
Mya arenaria	1600 based on wet weight	273 - 283 K, winter, Cu = 0.000 mg/l, control value[29]
Mya arenaria	200 based on wet weight	Marine, field collection[9]
Mya arenaria	250 based on wet weight	Water=0.05 ug Cu/l, 7 d, spring, 289-295 K, control (0 mg Cu/l)=0.7 ppm[29]
Mya arenaria	34 based on wet weight	Water=0.05 ug Cu/l, 1 d, spring, 289-295 K, control (0 mg Cu/l)=0.7 ppm[29]
Mya arenaria	373 based on wet weight	Water=0.01 ug Cu/l, 14 d, spring, 289-295 K, control (0 mg Cu/l)=0.7 ppm[29]
Mya arenaria	398 based on wet weight	273 - 283 K, winter, Cu = 0.010 mg/l, 6 w[29]
Mya arenaria	464 based on wet weight	273 - 283 K, winter, Cu = 0.050 mg/l, 6 w[29]
Mya arenaria	560 based on wet weight	273 - 283 K, winter, Cu = 0.010 mg/l, 16 w[29]
Mytilus edulis	1150 based on wet weight	Marine, field collection[9]
Mytilus galloprovincialis	800 based on wet weight	Water = 9.25 ppm[30]
Nereis diversicolor	0.45 ratio organism/sediment	Both in wet weight concentration, marine, field collection[31]
Patella vulgata	7.6E3 based on dry weight	17 obs., all areas around the British Isles[21]
Pecten maximus, digestive gland	36000 based on wet weight	Marine, field collection[11]
Pecten maximus, kidney	7800 based on wet weight	Marine, field collection[11]
Pecten maximus, soft parts	2400 based on wet weight	Marine, field collection[11]
Pelvetia wrightii	440 based on dry weight	Seawater = 2.5E-7 g/l, Vostok Bay, Sea of Japan[1]
Phyllospadix iwatensis	1000 based on dry weight	Seawater = 2.5E-7 g/l, Vostok Bay, Sea of Japan[1]
Phytoplankton	2800 based on wet weight	Marine organisms, typical value[32]
Phytoplankton	4830 based on wet weight	Range = 360 - 9300[33]
Plankton	17000 based on wet weight	None[7]
Polysiphonia japonica	2800 based on dry weight	Seawater = 2.5E-7 g/l, Vostok Bay, Sea of Japan[1]
Pomatomus saltatrix, white muscle	200 based on wet weight	Marine, field collection[34]
Porphyra umbilicalis	6.3E3 based on dry weight	20 obs., all areas around the British Isles[21]
Ptilota filicina	5200 based on dry weight	Seawater = 2.5E-7 g/l, Vostok Bay, Sea of Japan[1]
Rhodomela larix	1800 based on dry weight	Seawater = 2.5E-7 g/l, Vostok Bay, Sea of Japan[1]
Sargassum fulvellum	22800 based on wet weight	Marine, field collection[35]
Sargassum pallidum	1700 based on dry weight	Seawater = 2.5E-7 g/l, Vostok Bay, Sea of Japan[1]
Scenedesmus acuminatus	4.0E3 based on dry weight	Lab, 20 min uptake, chemostat[36]

Scenedesmus acutiformis	1.2E4 based on dry weight	Lab, 8 d growth, batch, Cu tolerent algae, rg = 0.87-1.5E3[23]
Seston	0.8 based on wet weight	Depth = 17 m, Schoehsee[20]
Seston	1.0 based on wet weight	Depth = 8 m, Schoehsee[20]
Spisula solidissima	450 based on wet weight	Marine, field collection[11]
Tribonema sp.	1.3E4 based on dry weight	Alive, adjacent to Zn smelter, Lower Swansea V., field algal bloom[16]
Tribonema sp.	8.3E4 based on dry weight	Dead, adjacent to Zn smelter, Lower Swansea V., field algal bloom[16]
Ulva fenestrata	8400 based on dry weight	Seawater = 2.5E-7 g/l, Vostok Bay, Sea of Japan[1]
Ulva sp.	0.12E4 based on dry weight	Rel to dissolved metal concentration in seawater[6]
Ulva sp.	0.625E4 based on dry weight	Rel to dissolved metal concentration in seawater, rg = 0.47 - 0.86E4[6]
Ulvaria spendens	5700 based on dry weight	Seawater = 2.5E-7 g/l, Vostok Bay, Sea of Japan[1]
Undaria pinnatifida, blade	15000 based on wet weight	Marine, field collection[35]
Undaria pinnatifida, sporophyll	12800 based on wet weight	Marine, field collection[35]
Zooplankton	1630 based on wet weight	Marine, field collection[37]
Zooplankton	1630 based on wet weight	Seawater concentration 1630 mg/m3[38]
Zooplankton	1800 based on wet weight	Marine organisms, typical value[32]
Zostera asiatica	600 based on dry weight	Seawater = 2.5E-7 g/l, Vostok Bay, Sea of Japan[1]

Source: 1.Saenko, G.N., M.D. Koryakov, V.F. Makienko and I.G. Dobrosmy. 1976. Mar. biol., 34(2), pp 169-176.
3.Zoller, W.H., E.S. Gladey and R.A. Duce. 1974. Science 183, pp 200.
4.Foestner, U. and G.T.W. Wittmann. 1979. Metal Pollution in the Aquatic Environment. Springer-Verlag.Berlin.
5.Chester, R. and J.H. Stoner. 1973. Nature (london) 245, pp 27-28.
6.Seelinger, U. and P. Edwards. 1977. Mar. Pollut. Bull. 8(1), pp 16-19.
7.Bowen, H.J.M. 1966. Trace Elements in Biochemistry, Ac. Press.
8.Mauchline, J. and W.L Templeton. 1964. Oceanogr. Mar. Biol. Ass. Rev., 2, pp 229.
9.Pringle, B.H., D.E. Hissong, E.L. Katz and S.T. Mulawka. 1972. Environ. Pollut. 3, 69-82.
10.Martin, J.-L.M. 1974. Mar. biol., 28(4), pp 245-251.
11.Bryan, G.W. 1973. J. Mar. Biol. Assoc. UK. 53, 145-166.
12.Sakaguchi, T., T. Horikashi and A. Nakajima. 1977. J. Agric. Chem. Jap. 51, 497-505.
13.Keeney, W.L., W.G. Breck , G.W. Vanloon and J.A. Page. 1976. Water Res., 10(11)., pp 981-984.
14.Taft, C.E. and W.J. Kisher. 1973. Water Resour. Cent. Ohio Sate Univ. Proj. Completion Rep. 332X, 33X.
15.Funk, W.H., F.W. Rabe, R. Fibly, G. Bailey, Bennett and Shah. 1975. An Integrated Study on the Impact of Metallic Trace Element Pollution on the Coeur DAlene-Spokane Rivers-Lake Drainage System. Wash. S
16.Trollope, D.R. and B. Evans. 1976. Environ. Pollut. 11, 109-116.
17.Livingston, H.D. and G. Thompson. 1971. Limnol. Oceanogr. 16(5), pp 786-796.
18.Goldberg, E.D. 1962. Limnol Oceanogr., 7, pp 72.
19.Merlini, M., C. Rigliocca, A. Berg and G. Pozzi. 1970. IAEA/ SM, 142/27.
20.Groth, P. 1971. Arch. Hydrobiol. 68(3), pp 305-375.
21.Preston, A., D.R. Jefferies , J.W.R. Harvey and A.K. Steele. 1972. Environ. Pollut., 3, pp 183-214.
22.Seelinger, U. and P. Edwards. 1977. Mar. Pollut. Bull. 8, 16-19.
23.Nriagu, J.O. 1979. Copper in the Environment. John Wiley and Sons. NY.
24.Bryan, G.W. 1969. J. Mar. Biol. Assoc. UK, 49, 225-243.
25.Yoshimura, A., H. Tada, M. Sakai, T. Harada and K. Oishi. 1976. Bull. Jap. Soc. Sci. Fish. 42, 661-64.
26.Haegerhaell, B. 1973. Bot. Mar., 16, p 53.
27.Mathis, B.J. and T.F. Cummins. 1973. Water Pollut. Control 45 pp 1573.
28.Pringle, B.H., D.E. Hissong, E.L. Katz and S.T. Mulawka. 1968. J. San. Eng. Div. Am. Soc. Civ. Eng., 94(5970), pp 455-475.
29.Eisler, R. 1977. Mar. Biol. 43, 265-276.
30.Majori, L. and F. Petronio. 1973. Rev. Int. Oceanogr. Med. CERBOM., 31-32, pp 55-90.
31.Bryan, G.W. and L.G. Hummerstone. 1971. J. Mar. Biol. Assoc. UK. 51, 845-863.
32.Martin, J.H. and G.A. Knauer. 1973. Geochim. Cosmochim. Acta., 37, p 1639.
33.Cushing, C.E. and L.A. Rancitelli. 1972. Northwest Science, 46(2), pp 115-121.
34.Cross, F.A., L.H. Hardy, N.Y. Jones and R.T. Barber. 1973. J. Fish. Res. bd Canada 30, 1287-1291.
35.Kim, C.Y. and J.H. Won. 1974. Bull. Korean Fish. Soc. 7, 169-178.
36.Mierle, G.M. and P.M. Stokes. 1976. In: D.D. Hemphill (ed.): Trace Substances in Environmentent. Elsevier.
37.Martin, J.H. 1970. Limnol. Oceanogr. 15, 756-761.
38.Martin, J.H 1970. Limnol. Oceanogr. 15(5), pp 756-761.

5-211 Concentration factor DDD

Species	Value	Condition
Ameiurus nebulosus	1439 based on wet weight	Conc in water = 14 ug/l, rg = 1107 - 1771[1]
Black crappie	4300 based on wet weight	Conc in water = 14 ug/l, rg = 386 - 8214[1]
Bluegill	593 based on wet weight	Conc in water = 14 ug/l, rg = 471 - 714[1]
Largemouth bass	1596 based on wet weight	Conc in water = 14 ug/l, rg = 1407 - 1785[1]
Sacramento blackfish	1018 based on wet weight	Conc in water = 14 ug/l, rg = 778 - 1257[1]
White catfish	5693 based on wet weight	Conc in water = 14 ug/l, rg = 2172 - 9214[1]

Source: 1.Edwards, C.A. 1975. Persistent Pesticides in the Environment, 2nd edition, CRC-Press.

5-212 Concentration factor DDE

Species	Value	Condition
Mytilus spp.	310000 based on wet weight	p,p-DDE, seawater = 5.1 ng/l, Calif, Royal Palms, 19/9 72[1]
Mytilus spp.	45000 based on wet weight	p,p-DDE, seawater = 0.11 ng/m3, Calif, Frisco Bay, 1975[1]

Source: 1.Risebrough, R.W., B.W. De Lappe and T.T. Schmidt. 1976. Mar. Pollut. Bull. 7(12), pp 225-228.

5-213 Concentration factor DDT

Species	Value	Condition
Agriolimax reticulatus	2.33 based on wet weight	Soil = 17.20 ppm, terrestrial gastropod[1]
Agriolimax reticulatus	8.26 based on wet weight	Soil = 2.30 ppm, terrestrial gastropod[1]
Alfalfa, foliage	0.081 based on wet weight	Conc in soil = 1.39 mg/kg dry wt.[2]
Algae	0.06 based on wet weight	Conc in water = 1000.0 ng/l[2]
Algae	0.35 based on wet weight	Conc in water = 5.80 ng/l[2]
Algae	126400 based on wet weight	Conc in water = 0.20 ng/l[2]
Algae	21250 based on wet weight	Conc in water = 0.004 ng/l[2]
Algae	3.3E1 based on wet weight	Max value[3]
Algae	30 based on wet weight	Conc in water = 0.33 ng/l[2]
Algae	500 based on wet weight	Conc in water = 0.016 ng/l[2]
Amphidinium carteri	1.2E5 based on wet weight	Marine algae, cell C = 961 ug, equilibrium = 0.8 ppt[4]
Amphidinium carteri	1.4E5 based on wet weight	Marine algae, cell C = 961 ug, equilibrium = 0.5 ppt[4]
Amphidinium carteri	2.2E5 based on wet weight	Marine algae, cell C = 1226 ug, equilibrium = 0.5 ppt[4]
Amphidinium carteri	9600 based on wet weight	Medium = 0.7 ug DDT/m3, metabolites found = DDE[5]
Amphipoda	65000 based on wet weight	Conc in water = 0.20 ug/l, USA[2]
Anabaena cylindrica	268 based on wet weight	Medium = 1000 ug DDT/m3, no metabolits found[6]
Anacystis nidulans	0.07 based on wet weight	Conc in water = 1000.0 ng/l[2]
Anacystis nidulans	849 based on wet weight	Medium = 1000 ug DDT/m3, no metabolits found[7]
Ankistrodesmus amalloides	21300 based on wet weight	30 min, 100 cells/ml, water conc = 0.72 ppb[8]

Ankistrodesmus amalloides	43600 based on wet weight	60 min, 100 cells/ml, water conc = 0.72 ppb[8]
Ankistrodesmus amalloides	53100 based on wet weight	120 min, 100 cells/ml, water conc = 0.72 ppb[8]
Ankistrodesmus amalloides	616 based on wet weight	Max value, medium = 0.72 ug DDT/m3, metabolites found = DDE and TDE[9]
Ankistrodesmus amalloides	61600 based on wet weight	180 min, 100 cells/ml, water conc = 0.72 ppb[8]
Aplysia sp.	178000 based on wet weight	Conc in water = 0.01 ug/l, USA[2]
Aquatic plant	1E5 based on wet weight	Max value[3]
Aquatic vegetation	326 based on wet weight	Conc in water = 20.0 ng/l[2]
Aquatic vegetation	375 based on wet weight	Conc in water = 200.0 ng/l[2]
Aquipecten irradians	27000 based on wet weight	Conc in water = 0.001(estim.), USA[2]
Arion sp.	3.17 based on wet weight	Soil = 2.30 ppm, terrestrial gastropod[10]
Bald eagle, muscle	0.36 based on wet weight	Conc in food = 800 mg/kg[2]
Beetle	0.43 based on wet weight	Conc in soil(orchard) = 9.75 mg/kg dry wt., GB, sites = 2[2]
Beetle	0.52 based on wet weight	Conc in soil(orchard) = 9.75 mg/kg dry wt., GB, sites = 2[2]
Beetle	1.0 based on wet weight	Conc in arable soil = 0.31 mg/kg dry wt., GB, sites = 10[2]
Beetle	2.14 based on wet weight	Conc in arable soil = 0.58 mg/kg dry wt., GB, sites = 10[2]
Beetle	4.33 based on wet weight	Conc in soil = 0.06 mg/kg dry wt., USA, sites = 2[2]
Beetle larvae	35.33 based on wet weight	Conc in soil = 0.015 mg/kg dry wt., USA, sites = 2[2]
Black bullhead	155 based on wet weight	Conc in water = 20 ug/l[2]
Bobwhite quail, tissue	0.03 based on wet weight	Conc in food = 50 mg/kg[2]
Bobwhite quail, tissue	0.04 based on wet weight	Conc in food = 100 mg/kg[2]
Bobwhite quail, tissue	0.08 based on wet weight	Conc in food = 250 mg/kg[2]
Brachidontes demissus plicatulus	157000 based on wet weight	Conc in water = 0.001 ug/l(estim.), USA[2]
Brachidontes recurvus	24000 based on wet weight	Conc in water = 1.0 ug/l, USA[2]
Bullhead trout	150 based on wet weight	Conc in water = 20 ug/l, rg = 100 - 200[2]
Callinectes sapidus	79000 based on wet weight	Conc in water = 0.001 ug/l(estim.), USA[2]
Campetoma sp.	43 based on wet weight	Conc in water = 3.28 ug/l, Canada[2]
Carrot	0.0003 based on wet weight	Conc in muck soil = 12.61 mg/kg dry wt.[2]
Carrot	0.007 based on wet weight	Conc in sandy soil = 4.99 mg/kg dry wt.[2]
Carrot, root	0.128 based on wet weight	Conc in sandy loam soil = 24.8 mg/kg dry wt.[2]
Ceratophyllum demersum	13550 based on wet weight	Conc in water = 0.20 ng/l[2]
Ceratophyllum demersum	2013 based on wet weight	Conc in water = 0.77 ng/l[2]
Chara sp.	13050 based on wet weight	Conc in water = 0.20 ng/l[2]
Chara sp.	1636 based on wet weight	Conc in water = 0.77 ng/l[2]
Chicken, muscle	0.16 based on wet weight	Conc in food = 250.00 mg/kg[2]
Chicken, muscle	0.21 based on wet weight	Conc in food = 10.00 mg/kg[2]
Chicken, muscle	0.21 based on wet weight	Conc in food = 50.00 mg/kg[2]
Chironomidae	26800 based on wet weight	Conc in water = 0.20 ug/l, USA[2]
Chub	5 based on wet weight	Conc in water = 5.8 ug/l[2]

Cladium jamaicensis	471 based on wet weight	Conc in water = 0.017 ng/l[2]
Cladophora gracilis	1660 based on wet weight	Conc in water = 0.05 ng/l[2]
Cladophora sp.	12701 based on wet weight	Conc in water = 0.77 ng/l[2]
Cowbird, tissue	0.25 based on wet weight	Conc in food = 500 mg/kg[2]
Crab	1.44E2 based on wet weight	Max value[3]
Crab	144 based on wet weight	Conc in water = 50.0 ug/l, USA[2]
Crassostrea gigas	20000 based on wet weight	Conc in water = 1.0 ug/l, USA[2]
Crassostrea gigas	70000 based on wet weight	Conc in water = 0.1 ug/l, USA[2]
Crassostrea virginica	15100 based on wet weight	Conc in water = 10.0 ug/l, USA[2]
Crassostrea virginica	26000 based on wet weight	Conc in water = 1.0 ug/l, USA[2]
Crassostrea virginica	30000 based on wet weight	Conc in water = 1.0 ug/l, USA[2]
Crassostrea virginica	70000 based on wet weight	Conc in water = 0.001 ug/l(estim.), USA[2]
Crassostrea virginica	70000 based on wet weight	Conc in water = 0.10 ug/l, USA[2]
Crayfish	25 based on wet weight	Conc in water = 0.08 ug/l, USA[2]
Crayfish	9.7E1 based on wet weight	Max value[3]
Croakers	20000 based on wet weight	Conc in water = 0.1 ug/l[2]
Cyclotella nana	58100 based on wet weight	Max value, medium = 0.7 ug DDT/m3, metabolites found = DDE[5]
Cylindrotheca closterium	190 based on wet weight	Medium = 100 ug DDT/m3, metabolites found = DDE[11]
Deroceras sp. and Limax sp.	17.93 based on dry weight	Soil = 2.94 ppm, terrestrial gastropod[12]
Deroceras sp. and Limax sp.	2.25 based on dry weight	Soil = 15.30 ppm, terrestrial gastropod[12]
Deroceras sp. and Limax sp.	2.95 based on dry weight	Soil = 15.30 ppm, terrestrial gastropod[12]
Deroceras sp. and Limax sp.	5.53 based on dry weight	Soil = 3.39 ppm, terrestrial gastropod[12]
Earthworms	0.02 based on wet weight	Conc in soil = 12.50 mg/kg dry wt., USA, sites = 1[2]
Earthworms	0.15 based on wet weight	Conc in soil(forest) = 1.23 mg/kg dry wt., Canada[2]
Earthworms	0.60 based on wet weight	Conc in soil = 117.33 mg/kg dry wt., USA, sites = 1, excl Lumbricus[2]
Earthworms	0.64 based on wet weight	Conc in soil = 17.50 mg/kg dry wt., New Zealand, sites = 1[2]
Earthworms	0.71 based on wet weight	Conc in soil = 11.43 mg/kg dry wt., GB, sites = 1[2]
Earthworms	0.78 based on wet weight	Conc in soil = 83.2 mg/kg dry wt., USA, sites = 1, excl Lumbricus[2]
Earthworms	0.79 based on wet weight	Conc in soil = 0.94 mg/kg dry wt., GB, sites = 1, excl Lumbricus[2]
Earthworms	1.58 based on wet weight	Conc in soil = 7.77 mg/kg dry wt., GB, sites = 2[2]
Earthworms	18.90 based on dry weight	USA, sites = 3[2]
Earthworms	2.01 based on wet weight	Conc in soil(orchard) = 9.75 mg/kg dry wt., GB, sitea = 2[2]
Earthworms	2.16 based on dry weight	Conc in soil = 0.06 mg/kg dry wt., USA, sites = 1[2]
Earthworms	2.40 based on wet weight	Conc in soil = 10.65 mg/kg dry wt., USA, sites = 2, excl Lumbricus[2]
Earthworms	2.73 based on wet weight	Conc in arable soil = 0.32 mg/kg dry wt., GB, sites = 7[2]
Earthworms	2.73 based on wet weight	Conc in soil = 1.96 mg/kg dry wt., GB, sites = 1[2]

Earthworms	2.73 based on wet weight	Conc in soil(orchard) = 9.75 mg/kg dry wt., GB, sites = 2[2]
Earthworms	3.16 based on wet weight	Conc in soil = 3.74 mg/kg dry wt., GB, sites = 1[2]
Earthworms	5.86 based on wet weight	Conc in arable soil = 0.58 mg/kg dry wt., GB, sites = 10[2]
Earthworms	6.0 based on dry weight	Conc in soil = 0.01 mg/kg dry wt., USA, sites = 20[2]
Earthworms	9.04 based on dry weight	Conc in soil = 1.36 mg/kg dry wt., USA, sites = 67[2]
Earthworms	9.44 based on wet weight	Conc in soil = 1.80 mg/kg dry wt., USA, sites = 3[2]
Earthworms	9.8 based on dry weight	Conc in soil = 0.98 mg/kg dry wt., USA, sites = 67[2]
Eleocharis cellulosa	533 based on wet weight	Conc in water = 0.015 ng/l[2]
Ensis directus	29000 based on wet weight	Conc in water = 0.001 ug/l(estim.), USA[2]
Euglena gracilis	0.14 based on wet weight	Conc in water = 1000.0 ng/l[2]
Euglena gracilis	99 based on wet weight	Medium = 1000 ug DDT/m3, no metabolits found[7]
Feral pigeon, muscle	0.45 based on wet weight	Conc in food = 72 mg/kg[2]
Fish	12333 based on wet weight	Conc in water = 0.30 ug/l, rg = 3333 - 21333[2]
Fish	35000 based on wet weight	Conc in water = 0.11 ug/l[2]
Fish	55 based on wet weight	Conc in water = 10 ug/l, rg = 10 - 100[2]
Fish	790 based on wet weight	5 spp., conc in water = 30-40 ug/l, rg = 130 - 1450[2]
Fish	8.29E5 based on wet weight	Max value[3]
Fish	829300 based on wet weight	Conc in water = 0.015 ug/l[2]
Gonidea sp.	1640000 based on wet weight	Conc in water = 0.005 ug/l, USA[2]
Gryllus assimilis	1.0 based on wet weight	Conc in soil = 0.06 mg/kg dry weight, USA, sites = 2[2]
Gyrinidae	11488 based on wet weight	Conc in water = 0.08 ug/l, USA[2]
Hibiscus palustris	1403 based on wet weight	Conc in water = 0.77 ng/l[2]
Hirudinea	31400 based on wet weight	Conc in water = 0.20 ug/l, USA[2]
Homarus americanus	216000 based on wet weight	Conc in water = 0.001 ug/l(estim.), USA[2]
Kestrel, tissue	18.38 based on wet weight	Conc in food = 2.8 mg/kg, died[2]
Kestrel, tissue	25.0 based on wet weight	Conc in food = 2.8 mg/kg, sacrificed[2]
Lemna minor	13550 based on wet weight	Conc in water = 0.20 ng/l[2]
Lemna sp.	2974 based on wet weight	Conc in water = 0.77 ng/l[2]
Lumbricus terrestris	0.67 based on wet weight	Conc in soil = 11.43 mg/kg dry weight, GB., sites = 1[2]
Lumbricus terrestris	0.69 based on wet weight	Conc in soil = 83.2 mg/kg dry weight, USA, sites = 1[2]
Lumbricus terrestris	1.17 based on wet weight	Conc in soil = 0.94 mg/kg dry weight, GB., sites = 1[2]
Lumbricus terrestris	1.24 based on wet weight	Conc in soil = 10.65 mg/kg dry weight, USA, sites = 2[2]
Lumbricus terrestris	2.03 based on wet weight	Conc in soil = 31.06 mg/kg dry weight, USA, sites = 1[2]
Lumbricus terrestris	2.53 based on wet weight	Conc in soil = 6.88 mg/kg dry weight, GB., sites = 1[2]
Marsh snail	19000 based on wet weight	Conc in water = 0.001 ug/l(estim.), USA[2]
Mercenaria mercenaria	26000 based on wet weight	Conc in water = 0.001 ug/l(estim.), USA[2]
Mercenaria mercenaria	3000 based on wet weight	Conc in water = 1.0 ug/l, USA[2]
Mercenaria mercenaria	6000 based on wet weight	Conc in water = 1.0 ug/l, USA[2]
Mercenaria mercenaria	70000 based on wet weight	Conc in water = 0.10 ug/l, USA[2]

Mercenaria mercenaria	8400 based on wet weight	Conc in water = 0.05 ug/l, conc in sediment = 0.14 mg/kg(estim.), USA[2]
Microcystis aeruginosa	230 based on wet weight	Medium = 1000 ug DDT/m3, no metabolits found[6]
Moss	30 based on wet weight	Conc in water = 0.33 ng/l[2]
Mya arenaria	62000 based on wet weight	Conc in water = 0.001 ug/l(estim.), USA[2]
Myriophyllum sp.	8350 based on wet weight	Conc in water = 0.20 ng/l[2]
Myriophyllum sp.	948 based on wet weight	Conc in water = 0.77 ng/l[2]
Mytilus edulis	98000 based on wet weight	Conc in water = 0.001 ug/l(estim.), USA[2]
Mytilus spp.	210000 based on wet weight	p,p-DDT, seawater = 0.25 ng/m3, France, Carro, 11/7 74[13]
Mytilus spp.	40000 based on wet weight	p,p-DDT, seawater = 0.35 ng/m3, France, Stes-Maries, 10/7 74[13]
Mytilus spp.	690000 based on wet weight	p,p-DDT, seawater = 1.3 ng/m3, France, Marseille, 12/7 74[13]
Mytilus spp.	90000 based on wet weight	p,p-DDT, seawater = 0.54 ng/l, France, Valras-Plage, 9/7 74[13]
Nassarius obsoletus	5200 based on wet weight	Conc in water = 0.05 ug/l, USA[2]
Nassarius obsoletus	76000 based on wet weight	Conc in water = 0.001 ug/l(estim.), USA[2]
Notonectidae	16000 based on wet weight	Conc in water = 0.20 ug/l, USA[2]
Oedogonium sp.	270 based on wet weight	Medium = 1000 ug DDT/m3, no metabolits found[9]
Ostrea edulis	15000 based on wet weight	Conc in water = 1.0 ug/l, USA[2]
Ostrea equestris	23000 based on wet weight	Conc in water = 1.0 ug/l, USA[2]
Oyster	14670 based on wet weight	Conc in water = 1.0 ug/l, USA[2]
Oyster	40 based on wet weight	Conc in water: < 1.0 ug/l, conc in sediment = 0.01 mg/kg, USA[2]
Oyster	450 based on wet weight	Conc in water: < 1.0 ug/l, conc in sediment = 0.082 mg/kg, USA[2]
Oyster	70000 based on wet weight	Conc in water = 0.1 ug/l, USA[2]
Oyster	75 based on wet weight	Conc in water = 0.28 ug/l, USA[2]
Oyster	7E4 based on wet weight	Max value, eastern oyster and eastern clam[3]
Palaemonetes vulgaris	61000 based on wet weight	Conc in water = 0.001 ug/l(estim.), USA[2]
Pheasant, muscle	0.07 based on wet weight	Conc in food = 50 mg/kg, sacrificed[2]
Pheasant, muscle	0.48 based on wet weight	Conc in food = 50 mg/kg, died[2]
Pheasant, tissue	2.91 based on wet weight	Conc in food = 10 mg/kg[2]
Pinfish	12000 based on wet weight	Conc in water = 1.0 ug/l[2]
Pinfish	40000 based on wet weight	Conc in water = 0.1 ug/l[2]
Plankton	1.67E4 based on wet weight	Max value[3]
Plankton	130 based on wet weight	Conc in water = 0.115 ug/l, USA[2]
Plankton	16666 based on wet weight	Conc in water = 0.30 ug/l, USA[2]
Plankton	250 based on wet weight	Conc in water = 20.0 ug/l, USA[2]
Plankton	63 based on wet weight	Conc in water = 0.08 ug/l, USA[2]
Plankton	70 based on wet weight	Conc in water = 1.0 ug/l(estim.), USA[2]
Plankton	800 based on wet weight	Conc in water = 0.05 ug/l, USA[2]

Planorbidae	5150 based on wet weight	Conc in water = 0.20 ug/l, USA[2]
Pontederia cordata	636 based on wet weight	Conc in water = 0.77 ng/l[2]
Pontederia lanceolata	240 based on wet weight	Conc in water = 0.025 ng/l[2]
Potamogeton crispus	961 based on wet weight	Conc in water = 0.77 ng/l[2]
Potamogeton foliosus	1727 based on wet weight	Conc in water = 0.77 ng/l[2]
Potamogeton pectinatus	1338 based on wet weight	Conc in water = 0.77 ng/l[2]
Potamogeton pectinatus	326 based on wet weight	Conc in water = 20.0 ng/l[2]
Potamogeton pectinatus	9300 based on wet weight	Conc in water = 0.20 ng/l[2]
Potamogeton sp.	160 based on wet weight	Conc in water = 10.0 ng/l[2]
Potato, tubers	0.066 based on wet weight	Conc in sandy loam soil = 24.8 mg/kg dry weight[2]
Rutabaga	0.036 based on wet weight	Conc in sandy loam soil = 24.8 mg/kg dry weight[2]
Sagittaria latifolia	6273 based on wet weight	Conc in water = 0.77 ng/l[2]
Salmo gairdneri	207 based on wet weight	Conc in water = 20 ug/l[2]
Salvelinus fontinalis	710 based on wet weight	Conc in water = 24 ug/l[2]
Scenedesmus obliquus	0.05 based on wet weight	Conc in water = 1000.0 ng/l[2]
Scenedesmus obliquus	134 based on wet weight	Medium = 1000 ug DDT/m3, no metabolits found[6]
Scenedesmus obliquus	626 based on wet weight	Medium = 1000 ug DDT/m3, no metabolites detected[7]
Scirpus sp.	3650 based on wet weight	Conc in water = 0.20 ng/l[2]
Scirpus validus	675 based on wet weight	Conc in water = 0.77 ng/l[2]
Sea hare	1.78E5 based on wet weight	Max value[3]
Sea squid	1E6 based on wet weight	Max value[3]
Sea squirt	1600000 based on wet weight	Conc in water = 0.01 ug/l, USA[2]
Sea squirt	200 based on wet weight	Conc in water = 100.0 ug/l, USA[2]
Sea squirt	200000 based on wet weight	Conc in water = 0.10 ug/l, USA[2]
Shrimp	2.8E3 based on wet weight	Max value[3]
Shrimp	280 based on wet weight	Conc in water = 0.5 ug/l, USA[2]
Shrimp	3200 based on wet weight	Conc in water = 0.05 ug/l, USA[2]
Shrimp	478 based on wet weight	Conc in water = 0.115 ug/l, USA[2]
Skeletonema costatum	38400 based on wet weight	Max value, medium = 0.7 ug DDT/m3, metabolites found = DDE[5]
Slugs	0.30 based on wet weight	Conc in grass root = 20.90 mg/kg dry weight, GB., sites = 1[2]
Slugs	11.69 based on wet weight	Conc in soil = 7.10 mg/kg dry weight, GB., sites = 1[2]
Slugs	2.25 based on wet weight	Conc in soil(orchard) = 9.75 mg/kg dry weight, GB., sites = 2[2]
Slugs	2.41 based on wet weight	Conc in soil(orchard) = 9.75 mg/kg dry weight, GB., sites = 2[2]
Slugs	3.31 based on wet weight	Conc in soil = 1.96 mg/kg dry weight, GB., sites = 1[2]
Slugs	3.86 based on wet weight	Conc in soil = 3.74 mg/kg dry weight, GB., sites = 1[2]
Slugs	6.33 based on dry weight	Conc in soil = 5.53 mg/kg dry weight, USA, sites = 4[2]
Slugs	6.54 based on wet weight	Conc in soil = 6.88 mg/kg dry weight, GB., sites = 1[2]
Snail	0.14 based on dry weight	Soil = 15.30 ppm, terrestrial gastropod[12]

Snail	0.83 based on dry weight	Soil = 2.94 ppm, terrestrial gastropod[12]
Snail	1.05 based on dry weight	Soil = 2.94 ppm, terrestrial gastropod[12]
Snail	1.48E3 based on wet weight	Max value[3]
Snail	1480 based on wet weight	Conc in water = 50.0 ug/l, USA[2]
Soybean	0.074 based on wet weight	Conc in soil = 0.57 mg/kg dry weight[2]
Sparganium eurycarpum	6450 based on wet weight	Conc in water = 0.20 ng/l[2]
Sparganium eurycarpum	740 based on wet weight	Conc in water = 0.77 ng/l[2]
Spartina sp., mat	4549 based on wet weight	Conc in water = 1.44 ng/l(approx.)[2]
Spirogyra sp.	325 based on wet weight	Conc in water = 0.77 ng/l[2]
Sugar beet, root	0.067 based on wet weight	Conc in loam soil = 1.537 mg/kg dry weight[2]
Sweet potato, tubers	0.008 based on wet weight	Conc in soil = 1.19 mg/kg dry weight[2]
Syracosphaera carteri	2.3E5 based on wet weight	Marine algae, cell C = 301 ug, equilibrium = 2.0 ppt[4]
Syracosphaera carteri	2.9E5 based on wet weight	Marine algae, cell C = 151 ug, equilibrium = 2.6 ppt[4]
Tendipendidae, larvae	11488 based on wet weight	Conc in water = 0.08 ug/l, USA[2]
Tetraselmis chuii	6300 based on wet weight	Medium = 0.7 ug DDT/m3, metabolites found = DDE[5]
Thalassiosira fluviatilis	1.4E5 based on wet weight	Marine algae, cell C = 730 ug, equilibrium = 2.4 ppt[4]
Thalassiosira fluviatilis	16E5 based on wet weight	Marine algae, cell C = 365 ug, equilibrium = 3.3 ppt[4]
Trout	200 based on wet weight	Conc in water = 20 ug/l[2]
Typha latifolia	1300 based on wet weight	Conc in water = 0.20 ng/l[2]
Uca sp.	68000 based on wet weight	Conc in water = 0.001 ug/l(estim.), USA[2]
Utricularia vulgaris	29200 based on wet weight	Conc in water = 0.20 ng/l[2]
Utricularia vulgaris	3299 based on wet weight	Conc in water = 0.77 ng/l[2]
Vascular plant	0.52 based on wet weight	Conc in water = 5.8 ng/l[2]

Source: 1.Davis, B.N.K. and M.E. French. 1969. Soil Biol. Biochem. 1, p 45.
2.Edwards, C.A. 1975. Persistent Pesticides in the Environment, 2nd edition, CRC-Press.
3.Leonard, R.A., G.W. Baily and R.R. Swank(JR). 1976. Land Application of Water Materials, pp 48-78.
4.Cox, J.L. 1970. Bull Environ. Contam. Toxicol., 5, pp 218-221.
5.Akopyan, E.A. and A.G. Agaronyan. 1968. Soils and Fert. 33, 3133.
6.Goguadze, V.D. 1969. Subtrop. Kultury No 4, 161-167.
7.Milkowska, A. and A. Gorzelak. 1966. Sylwan 110, 13-22.
8.Neudorf, S. and M.A.Q. Khan. 1975. Bull. Environ. Contam. Toxicol. 31, pp 443-450.
9.Kulinska. 1967. Roczn. Nauk. roln. (A) 93, 229-262.
10.Davis, B.N.K. 1968. Am. Appl. Biol. 61, pp 45.
11.Balasubbramaniam, A. and R. Sidaramappa. 1974. Mysore J. agric. Sci. 8, 214-219.
12.Gish, C.D. 1970. Pestic. Monit. J. 3, pp 241.
13.Risebrough, R.W., B.W. De Lappe and T.T. Schmidt. 1976. Mar. Pollut. Bull. 7(12), pp 225-228.

5-214 Concentration factor DDT analog

SpeciesValue		Condition
Fish	140 based on wet weight	DDT analog with CH3- and CH3 as radicals in aromatic rings[1]
Fish	1400 based on wet weight	DDT analog with Cl- and CH3- as radicals in aromatic rings[1]
Fish	1536 based on wet weight	DDT analog with C2H5O and C2H5O- as radicals in aromatic rings[1]
Fish	1545 based on wet weight	DDT analog with CH3O- and CH3O- as radicals in aromatic rings[1]
Fish	310 based on wet weight	DDT analog with CH3O- and CH3S- as radicals in aromatic rings[1]
Fish	400 based on wet weight	DDT analog with CH3- and C2H5_- as radicals in aromatic rings[1]
Fish	5.5 based on wet weight	DDT analog with CH3S- and CH3- as radicals in aromatic rings[1]

Fish	84500 based on wet weight	DDT analog with Cl- and Cl- as radicals in aromatic rings[1]
Snail	120000 based on wet weight	DDT analog with CH3O- and CH3O- as radicals in aromatic rings[1]
Snail	120270 based on wet weight	DDT analog with CH3- and CH3 as radicals in aromatic rings[1]
Snail	21000 based on wet weight	DDT analog with Cl- and CH3- as radicals in aromatic rings[1]
Snail	300 based on wet weight	DDT analog with CH3S- and CH3- as radicals in aromatic rings[1]
Snail	3400 based on wet weight	DDT analog with CH3O- and CH3S- as radicals in aromatic rings[1]
Snail	34500 based on wet weight	DDT analog with Cl- and Cl- as radicals in aromatic rings[1]
Snail	42000 based on wet weight	DDT analog with CH3- and C2H5_- as radicals in aromatic rings[1]
Snail	97645 based on wet weight	DDT analog with C2H5O and C2H5O- as radicals in aromatic rings[1]

Source: 1.Kapoor, I.P. et al. 1973. J. Agric. fd Chem. 21, 310.

5-215 Concentration factor Dieldrin

Species	Value	Condition
Alfalfa	0.481 based on wet weight	Conc in loam soil = 0.054 mg/kg dry weight[1]
Algae	4091 based on wet weight	Conc in water = 0.011 ng/l[1]
Algae	7480 based on wet weight	300 K, freshwater, Metcalf model ecosystem[2]
Amphidinium carteri	982 based on dry weight	1.7 ppb dieldrin, 2 h, 2.3 % removed from sol., 1.67E6 cell/ml[3]
Ankistrodesmus amalloides	11700 based on wet weight	30 min, 100 cells/ml, water conc = 0.72 ppb[4]
Ankistrodesmus amalloides	19100 based on wet weight	60 min, 100 cells/ml, water conc = 0.72 ppb[4]
Ankistrodesmus amalloides	27000 based on wet weight	120 min, 100 cells/ml, water conc = 0.72 ppb[4]
Ankistrodesmus amalloides	32000 based on wet weight	180 min, 100 cells/ml, water conc = 0.72 ppb[4]
Beetle	0.30 based on wet weight	Conc in arable soil = 0.37 mg/kg dry weight, GB., sites = 10[1]
Beetle	0.50 based on wet weight	Conc in arable soil = 0.15 mg/kg dry weight, GB., sites = 10[1]
Beetle	20.64 based on wet weight	Conc in soil = 0.25 mg/kg dry weight, USA, sites = 2[1]
Buffalofish	30000 based on wet weight	Conc in water = 0.007 ug/l[1]
Buffalofish	3913 based on wet weight	Conc in water = 0.023 ug/l[1]
Callinectes sapidus	13000 based on wet weight	Conc in water = 0.001 ug/l(estim.), USA[1]
Carrot	0.003 based on wet weight	Conc in muck soil = 1.53 mg/kg dry weight[1]
Carrot	0.039 based on wet weight	Conc in sandy soil = 1.165 mg/kg dry weight[1]
Carrot, root	0.005 based on wet weight	Conc in muck soil = 3.90 mg/kg dry weight[1]
Carrot, root	0.229 based on wet weight	Conc in clay soil = 0.48 mg/kg dry weight[1]
Catfish	1428 based on wet weight	Conc in water = 0.007 ug/l[1]
Catfish	1590 based on wet weight	Conc in water = 0.044 ug/l[1]
Catfish	4444 based on wet weight	Conc in water = 0.009 ug/l[1]
Catfish	952 based on wet weight	Conc in water = 0.021 ug/l[1]
Cheumatopsyche sp., larvae	5133 based on wet weight	Conc in water = 11.5 ug/l, USA[1]
Chicken, tissue	0.05 based on wet weight	Conc in food = 20.00 mg/kg[1]
Chicken, tissue	0.06 based on wet weight	Conc in food = 10.00 mg/kg[1]
Chicken, tissue	40.80 based on wet weight	Conc in food = 0.25 mg/kg[1]
Chicken, tissue	47.60 based on wet weight	Conc in food = 0.75 mg/kg[1]

Corbicula manilensis	1015 based on wet weight	300 K, freshwater, Metcalf model ecosystem[2]
Corn, root	11.13 based on wet weight	Conc in soil = 0.84 mg/kg dry weight[1]
Corn, root	13.45 based on wet weight	Conc in soil = 4.12 mg/kg dry weight[1]
Corn, shoot	0.201 based on wet weight	Conc in soil = 4.12 mg/kg dry weight[1]
Corn, shoot	0.286 based on wet weight	Conc in soil = 0.84 mg/kg dry weight[1]
Corydalis cornutus, larvae	68 based on wet weight	Conc in water = 17.0 ug/l, USA[1]
Crassostrea virginica	3500 based on wet weight	Conc in water = 1.0 ug/l, USA[1]
Crayfish	18 based on wet weight	Conc in water = 29.0 ug/l, USA[1]
Cucumber, fruit	0.031 based on wet weight	Conc in soil = 1.4 mg/kg dry weight[1]
Cucumber, fruit	0.031 based on wet weight	Conc in soil = 3.7 mg/kg dry weight[1]
Cutworms	2.83 based on wet weight	GB, sites = 1[1]
Cyclotella nana	4810 based on dry weight	1.7 ppb dieldrin, 2 h, 13.0 % removed from sol., 0.83E6 cell/ml[3]
Daphnia sp.	2145 based on wet weight	300 K, freshwater, Metcalf model ecosystem[2]
Earthworms	1.16 based on wet weight	Conc in arable soil = 0.37 mg/kg dry weight, GB., sites = 10[1]
Earthworms	1.97 based on wet weight	Conc in soil = 0.63 mg/kg dry weight, USA, sites = 4[1]
Earthworms	126.60 based on dry weight	Conc in soil = 0.006 mg/kg dry weight, USA, sites = 67[1]
Earthworms	4.64 based on wet weight	Conc in soil = 0.28 mg/kg dry weight, GB., sites = 2[1]
Earthworms	4.82 based on wet weight	Conc in soil = 0.64 mg/kg dry weight, GB., sites = 1, excl Lumbricus[1]
Earthworms	4.86 based on wet weight	Conc in arable soil = 0.07 mg/kg dry weight, GB., sites = 7[1]
Earthworms	5.68 based on dry weight	Conc in soil = 0.25 mg/kg dry weight, USA, sites = 1[1]
Earthworms	5.85 based on dry weight	Conc in soil = 0.103 mg/kg dry weight, USA, sites = 1[1]
Earthworms	7.50 based on dry weight	Conc in soil = 0.40 mg/kg dry weight, USA, sites = 2[1]
Earthworms	9.9 based on dry weight	Conc in soil = 0.10 mg/kg dry weight, USA, sites = 67[1]
Elodea sp.	1280 based on wet weight	300 K, freshwater, Metcalf model ecosystem[2]
Ensis directus	4000 based on wet weight	Conc in water = 0.001 ug/l(estim.), USA[1]
Fish	55 based on wet weight	Conc in water = 10 ug/l, rg = 10 - 100[1]
Fish	6145 based on wet weight	300 K, freshwater, Metcalf model ecosystem[2]
Gryllus assimilis	0.88 based on wet weight	Conc in soil = 0.25 mg/kg dry weight, USA, sites = 2[1]
Homarus americanus	12000 based on wet weight	Conc in water = 0.001 ug/l(estim.), USA[1]
Hydropsyche, larvae	5133 based on wet weight	Conc in water = 11.5 ug/l, USA[1]
Isochrysis galbana	8238 based on dry weight	1.7 ppb dieldrin, 2 h, 15.5 % removed from sol., 1.40E6 cell/ml[3]
Lamb, tissue	0.70 based on wet weight	Conc in food = 2.25 mg/kg[1]
Lamb, tissue	0.80 based on wet weight	Conc in food = 0.75 mg/kg[1]
Lamb, tissue	1.60 based on wet weight	Conc in food = 0.25 mg/kg[1]
Legume foliage	0.05 based on wet weight	Conc in soil = 0.05 mg/kg dry weight[1]
Leiostomus xanthurus	6000 based on wet weight	Uptake from water[5]
Lettuce, foliage	0.013 based on wet weight	Conc in silt loam soil = 2.3 mg/kg dry weight[1]
Lumbricus terrestris	2.50 based on wet weight	Conc in soil = 0.64 mg/kg dry weight, GB., sites = 1[1]
Nassarius obsoletus	11000 based on wet weight	Conc in water = 0.001 ug/l(estim.), USA[1]
Olisthodiscus luteus	4900 based on dry weight	1.7 ppb dieldrin, 2 h, 13.0 % removed from sol., 0.12E6 cell/ml[3]

Palaemonetes vulgaris	236000 based on wet weight	Conc in water = 0.001 ug/l(estim.), USA[1]
Pheasant, muscle	0.05 based on wet weight	Conc in food = 50 mg/kg[1]
Physa sp.	114935 based on wet weight	300 K, freshwater, Metcalf model ecosystem[2]
Physa sp.	7978 based on wet weight	Conc in water = 6.0 ug/l, USA[1]
Pig, tissue	1.60 based on wet weight	Conc in food = 0.25 mg/kg[1]
Pig, tissue	1.90 based on wet weight	Conc in food = 2.25 mg/kg[1]
Pig, tissue	3.70 based on wet weight	Conc in food = 0.75 mg/kg[1]
Plankton	278 based on wet weight	Conc in water = 0.018 ug/l, USA[1]
Rangia cuneata	54 based on dry weight	Uptake from food = 11.16 ug/g ww in algae, 48 h, n = 6[6]
Rat, tissue	15.10 based on wet weight	Conc in food = 1.0 mg/kg[1]
Rat, tissue	6.70 based on wet weight	Conc in food = 10 mg/kg[1]
Shrimp	278 based on wet weight	Conc in water = 0.018 ug/l, USA[1]
Simulium vittatum, larvae	1786 based on wet weight	Water = 11.5 ug/l, USA[1]
Skeletonema costatum	15882 based on dry weight	1.7 ppb dieldrin, 2 h, 42.0 % removed from sol., 1.19E6 cell/ml[3]
Slugs	0.81 based on wet weight	Conc in arable soil = 0.37 mg/kg dry weight, GB, sites = 10[1]
Slugs	571.11 based on dry weight	Conc in soil = 0.009 mg/kg dry weight, USA, sites = 4[1]
Slugs	6.75 based on wet weight	Conc in soil = 0.20 mg/kg dry weight, GB, sites = 1[1]
Slugs	7.50 based on wet weight	Conc in arable soil = 0.04 mg/kg dry weight, GB, sites = 1[1]
Steer, tissue	3.20 based on wet weight	Conc in food = 0.25 mg/kg[1]
Steer, tissue	3.90 based on wet weight	Conc in food = 2.25 mg/kg[1]
Steer, tissue	4.70 based on wet weight	Conc in food = 0.75 mg/kg[1]
Sugar beet, root	0.057 based on wet weight	Conc in silt loam soil = 2.3 mg/kg dry weight[1]
Sugar beet, root	0.284 based on wet weight	Conc in loam soil = 0.35 mg/kg dry weight[1]
Sweet clover	0.481 based on wet weight	Conc in loam soil = 0.054 mg/kg dry weight[1]
Sweet potato, tubers	0.024 based on wet weight	Conc in soil = 0.17 mg/kg dry weight[1]
Tetraselmis chuii	8588 based on dry weight	1.7 ppb dieldrin, 2 h, 16.0 % removed from sol., 0.13E6 cell/ml[3]
Trout	3300 based on wet weight	Water = 2.3 ug/l[1]
Tubificid worms	4563 based on wet weight	Water = 80.0 ug/l, USA[1]
Turnips	0.022 based on wet weight	Conc in silt loam soil = 2.3 mg/kg dry weight[1]
Uca ininax	247 based on wet weight	300 K, freshwater, Metcalf model ecosystem[2]
Uca sp.	15000 based on wet weight	Water = 0.001 ug/l(estim.), USA[1]
Wheat	0.038 based on wet weight	Conc in soil = 28.39 mg/kg dry weight[1]
Wheat	0.15 based on wet weight	Conc in soil = 1.13 mg/kg dry weight[1]
Wheat, foliage	0.008 based on wet weight	Conc in silt loam soil = 1.8 mg/kg dry weight[1]
White - tailed deer, muscle	0.02 based on wet weight	Conc in food = 5 mg/kg[1]
White - tailed deer, muscle	0.05 based on wet weight	Conc in food = 25 mg/kg[1]

Source: 1.Edwards, C.A. 1975. Persistent Pesticides in the Environment, 2nd edition, CRC-Press.
2.Sanborn, J.R. and C.-C. Yu. 1973. Bull. Environ. Contam. Toxicol. 10(6), pp 340-346.
3.Rice, C.P. and H.C. Sikka. 1973. Bull. Environ. Contam. Toxicol. 9, pp 116-123.
4.Neudorf, S. and M.A.Q. Khan. 1975. Bull. Environ. Contam. Toxicol. 31, pp 443-450.
5.Mitchell, A.L. (ed.). 1974. pp 427-434. EPA, Gulf Breeze environ. Res. Lab.
6.Petrocelli, S.K., J.W. Anderson and A.R. Hanks. 1975. Mar. Biol., 31, pp 215-218.

5-216 Concentration factor Dimethylbenz(a)anthrecene

Species	Value	Condition
Macoma inquinata	0.06 based on sediment concentration	Sediment=2000 ug prudhoe B crude oil/g, 7 d, sed=4.63 ppm, water=0.63 ppb[1]
Macoma inquinata	1349 based on wet weight	Sediment=2000 ug prudhoe B crude oil/g, 7 d, sed=4.63 ppm, water=0.63 ppb[1]

Source: 1.Roesijadi, G., J.W. Anderson and J.W. Blaylock. 1978. J. Fish. Res. bd. Canada 35, 608-614.

5-217 Concentration factor EDC-tar

Species	Value	Condition
Leander adspersus	159 based on wet weight	Seawater, s = 3.3-3.4 %, 283-287 K, 16 d, n = 14, water = 1 ppm[1]
Leander adspersus	220 based on wet weight	Seawater, s = 3.3-3.4 %, 283-287 K, 16 d, n = 14, water = 0.1 ppm[1]
Leander adspersus	2900 based on wet weight	Seawater, s = 3.3-3.4 %, 283-287 K, 16 d, n = 14, water = 0.01 ppm[1]
Mytilus edulis	1030 based on wet weight	Seawater, s = 3.3-3.4 %, 283-287 K, 16 d, n = 14, water = 0.1 ppm[1]
Mytilus edulis	37 based on wet weight	Seawater, s = 3.3-3.4 %, 283-287 K, 16 d, n = 14, water = 10 ppm[1]
Mytilus edulis	59 based on wet weight	Seawater, s = 3.3-3.4 %, 283-287 K, 16 d, n = 14, water = 1 ppm[1]
Pleuronectes platessa	15 based on wet weight	Seawater, s = 3.3-3.4 %, 283-287 K, 16 d, n = 14, water = 1 ppm[1]

Source: 1.Jerneloev, A., R. Rosenberg and S. Jensen. 1972. Water Res. 6(10), pp 1181-1191.

5-218 Concentration factor Endrin

Species	Value	Condition
Algae	0.67 based on wet weight	Water = 10.5 ng/l[1]
Aquatic vegetation	27 based on wet weight	Water = 7.4 ng/l[1]
Chub	4.7 based on wet weight	Water = 10.5 ug/l[1]
Conidea sp.	1182 based on wet weight	Water = 0.011 ug/l, USA[1]
Earthworms	3.57 based on wet weight	Conc in soil = 0.49 mg/kg dry weight, USA, sites = 67[1]
Earthworms	5.8 based on wet weight	Conc in soil = 0.12 mg/kg dry weight, USA, sites = 2[1]
Fish	70000 based on wet weight	Water = 0.10 ug/l[1]
Oyster	10 based on wet weight	Water < 1.0 ug/l, conc in sediment < 0.001 mg/kg, USA[1]
Oyster	10 based on wet weight	Water < 1.0 ug/l, conc in sediment < 0.01 mg/kg, USA[1]
Pimephales promelas	10000 based on wet weight	Water = 0.015 ug/l[1]
Slugs	40.62 based on dry weight	Conc in soil = 1.20 mg/kg dry weight, USA, sites = 4[1]
Vascular plant	0.57 based on wet weight	Water = 10.5 ng/l[1]
Wheat, foliage	0.020 based on wet weight	Conc in silt loam soil = 1.918 mg/kg dry weight[1]

Source: 1.Edwards, C.A. 1975. Persistent Pesticides in the Environment, 2nd edition, CRC-Press.

5-219 Concentration factor Eu

Species	Value	Condition
Engraulis japonicus, whole body	730 based on wet weight	Rel to soluble conc, total = 1.6 ng/l, solub = 0.3 ng/l, marine, Japan[1]
Engraulis japonicus, whole body	92 based on wet weight	Rel to total conc, total = 1.6 ng/l, solub = 0.3 ng/l, marine, Japan[1]
Hizikia fusiforme, whole body	100 based on wet weight	Rel to total conc, total = 1.6 ng/l, solub = 0.3 ng/l, marine, Japan[1]
Hizikia fusiforme, whole body	800 based on wet weight	Rel to soluble conc, total = 1.6 ng/l, solub = 0.3 ng/l, marine,[1]
Meretrix lusoria, soft part	6870 based on wet weight	Rel to soluble conc, total = 1.6 ng/l, solub = 0.3 ng/l, marine,[1]
Meretrix lusoria, soft part	860 based on wet weight	Rel to total conc, total = 1.6 ng/l, solub = 0.3 ng/l, marine, Japan[1]
Paralichthys olivaceus, muscle	130 based on wet weight	Rel to soluble conc, total = 1.6 ng/l, solub = 0.3 ng/l, marine,[1]
Paralichthys olivaceus, muscle	17 based on wet weight	Rel to total conc, total = 1.6 ng/l, solub = 0.3 ng/l, marine, Japan[1]
Sargassum fulvellum, whole body	1800 based on wet weight	Rel to soluble conc, total = 1.6 ng/l, solub = 0.3 ng/l, marine,[1]
Sargassum fulvellum, whole body	230 based on wet weight	Rel to total conc, total = 1.6 ng/l, solub = 0.3 ng/l, marine, Japan[1]
Seriola quinqueradiata, muscle	470 based on wet weight	Rel to soluble conc, total = 1.6 ng/l, solub = 0.3 ng/l, marine,[1]
Seriola quinqueradiata, muscle	58 based on wet weight	Rel to total conc, total = 1.6 ng/l, solub = 0.3 ng/l, marine, Japan[1]
Ulva pertusa, whole body	2570 based on wet weight	Rel to soluble conc, total = 1.6 ng/l, solub = 0.3 ng/l, marine,[1]
Ulva pertusa, whole body	320 based on wet weight	Rel to total conc, total = 1.6 ng/l, solub = 0.3 ng/l, marine, Japan[1]
Undaria pinnatifida, whole body	300 based on wet weight	Rel to soluble conc, total = 1.6 ng/l, solub = 0.3 ng/l, marine,[1]
Undaria pinnatifida, whole body	38 based on wet weight	Rel to total conc, total = 1.6 ng/l, solub = 0.3 ng/l, marine, Japan[1]

Source: 1.Suzuki, H., T. Koganage and M. Saki. 1975. Impacts of Nuclear Releases into the Aquatic Environment. IAEA SM-198/19, pp 77-89. Vienna.

5-220 Concentration factor Fe

Species	Value	Condition
Agarum cribrosum	2900 based on dry weight	Seawater = 37.5E-6 g/l, Vostok Bay, Sea of Japan[1]
Algae	150 based on dry weight	EDTA added, Fe-59, marine benthic, rg = 100 - 300[2]
Atmospheric particulates	1.0	Low level marine part, relative to earths crust, Fe standard, F200[3]
Atmospheric particulates	1.0	US urban air, relative to earths crust, Fe standard, F200[4]
Atmospheric particulates	1.1	N Atlantic Westerlies, relative to earths crust, Al standard, F200[5]
Atmospheric particulates	2.1	South Pole, relative to earths crust, Al standard, F200[6]
Brassica oleracea	120 based on wet weight	Conc in soil solution = 0.25ppm[7]

Brown algae	0.86 based on wet weight	None[7]
Brown algae	17000 based on wet weight	None[7]
Brown algae	27500 based on wet weight	Range = 20000 - 35000[8]
Cancer irroratus	5165 based on wet weight	Seawater conc = 11.62E-3 ppm, Nova Scotia, AAS, SD = +- 726[9]
Chondrus yendoi	1600 based on dry weight	Seawater = 37.5E-6 g/l, Vostok Bay, Sea of Japan[1]
Chorda filum	100 based on dry weight	Seawater = 37.5E-6 g/l, Vostok Bay, Sea of Japan[1]
Codium yessoensis	1000 based on dry weight	Seawater = 37.5E-6 g/l, Vostok Bay, Sea of Japan[1]
Costaria costata	250 based on dry weight	Seawater = 37.5E-6 g/l, Vostok Bay, Sea of Japan[1]
Crustacea	2000 based on dry weight	Marine, rg = 1000 - 4000, 2 values used[10]
Crustacea	930 based on dry weight	Freshwater, rg = 60 - 1800, 3 values used[10]
Enteromorpha prolifera	1300 based on dry weight	Seawater = 37.5E-6 g/l, Vostok Bay, Sea of Japan[1]
Fish	1800 based on dry weight	Marine, rg = 600 - 3000, 6 values used[10]
Fish	190 based on wet weight	Freshwater organisms, typical value[11]
Fish	191 based on dry weight	Freshwater, rg = 0.1 - 1225, 8 values used[10]
Fish, muscle	540 based on wet weight	Danube, 1973, nuclear power plant since 1967, Fe-59[12]
Fucus vesiculosus	2.4E4 based on dry weight	18 obs., all areas around the British Isles[13]
Heterochordaria abietina	2000 based on dry weight	Seawater = 37.5E-6 g/l, Vostok Bay, Sea of Japan[1]
Laminaria chichorioides	460 based on dry weight	Seawater = 37.5E-6 g/l, Vostok Bay, Sea of Japan[1]
Laminaria japonica	70 based on dry weight	Seawater = 37.5E-6 g/l, Vostok Bay, Sea of Japan[1]
Macrophytes	3642 based on wet weight	Freshwater organisms, typical value[11]
Molluscs	14400 based on wet weight	Marine organisms, typical value[14]
Molluscs	25170 based on dry weight	Freshwater, rg = 20 - 80000, 4 values used[10]
Molluscs	7600 based on dry weight	Marine, rg = 1000 - 13000, 6 values used[10]
Patella vulgata	2.8E5 based on dry weight	17 obs., all areas around the British Isles[13]
Pelvetia wrightii	940 based on dry weight	Seawater = 37.5E-6 g/l, Vostok Bay, Sea of Japan[1]
Phyllospadix iwatensis	3500 based on dry weight	Seawater = 37.5E-6 g/l, Vostok Bay, Sea of Japan[1]
Phytoplankton	28300 based on wet weight	Marine organisms, typical value[15]
Phytoplankton	284350 based on wet weight	Range = 76700 - 492000[16]
Phytoplankton	90000 based on dry weight	Marine[17]
Plankton	63000 based on wet weight	Water = 15 ug/l, Schoehsee, freshwater[18]
Plankton	721000 based on wet weight	Range = 2000 - 140000[8]
Plankton	87000 based on wet weight	None[7]
Plants	2260 based on dry weight	Marine, rg = 300 - 6000, 7 values used[10]
Plants	4935 based on dry weight	Algae, bryophytes and vascular hydrophytes, freshwater, 32 sp[19]
Plants	6675 based on dry weight	Freshwater, rg = 40 - 45000, 10 values used[10]
Polysiphonia japonica	24000 based on dry weight	Seawater = 37.5E-6 g/l, Vostok Bay, Sea of Japan[1]
Porphyra umbilicalis	3.6E4 based on dry weight	20 obs., all areas around the British Isles[13]

Ptilota filicina	6000 based on dry weight	Seawater = 37.5E-6 g/l, Vostok Bay, Sea of Japan[1]
Rhodomela larix	7300 based on dry weight	Seawater = 37.5E-6 g/l, Vostok Bay, Sea of Japan[1]
Sargassum pallidum	300 based on dry weight	Seawater = 37.5E-6 g/l, Vostok Bay, Sea of Japan[1]
Sediment	660 based on wet weight	Danube, 1973, nuclear power plant since 1967, Fe-59[12]
Seston	6.1 based on wet weight	Depth = 8 m, Schoehsee[18]
Seston	6.5 based on wet weight	Depth = 17 m, Schoehsee[18]
Ulva fenestrata	2200 based on dry weight	Seawater = 37.5E-6 g/l, Vostok Bay, Sea of Japan[1]
Ulvaria spendens	4500 based on dry weight	Seawater = 37.5E-6 g/l, Vostok Bay, Sea of Japan[1]
Zooplankton	114600 based on wet weight	Marine organisms, typical value[15]
Zooplankton	144000 based on wet weight	Seawater concentration 0.01 mg/m3[20]
Zostera asiatica	230 based on dry weight	Seawater = 37.5E-6 g/l, Vostok Bay, Sea of Japan[1]

Source: 1.Saenko, G.N., M.D. Koryakov, V.F. Makienko and I.G. Dobrosmy. 1976. Mar. biol., 34(2), pp 169-176.
2.Taylor, W.R. and E.P. Odum. 1960. Biol. Bull. 119, p 343.
3.Zoller, W.H., E.S. Gladey and R.A. Duce. 1974. Science 183, pp 200.
4.Foestner, U. and G.T.W. Wittmann. 1979. Metal Pollution in the Aquatic Environment. Springer-Verlag.Berlin.
5.Chester, R. and J.H. Stoner. 1973. Nature (london) 245, pp 27-28.
6.Duce, R.A., J. Quinn, C.M. Olney, S. Piotrowits, B. Ray and Wade.1972. Science 176, pp 161-163.
7.Bowen, H.J.M. 1966. Trace Elements in Biochemistry, Ac. Press.
8.Mauchline, J. and W.L Templeton. 1964. Oceanogr. Mar. Biol. Ass. Rev., 2, pp 229.
9.Martin, J.-L.M. 1974. Mar. biol., 28(4), pp 245-251.
10.Jinks, S.M. and M. Eisenbud. 1972. Radiat. Data Rep. 13, pp 243-247.
11.Merlini, M., C. Rigliocca, A. Berg and G. Pozzi. 1970. IAEA/ SM, 142/27.
12.Herrmann, H., M. Ruf, K. Hubel and W. Lunsmann. 1975. Impacts of Nuclear Releases into the Aquatic Environment. IAEA-SM-198/14. pp 461-470. Vienna.
13.Preston, A., D.R. Jefferies , J.W.R. Harvey and A.K. Steele. 1972. Environ. Pollut., 3, pp 183-214.
14.Pringle, B.H., D.E. Hissong, E.L. Katz and S.T. Mulawka. 1968. J. San. Eng. Div. Am. Soc. Civ. Eng., 94(5970), pp 455-475.
15.Martin, J.H. and G.A. Knauer. 1973. Geochim. Cosmochim. Acta., 37, p 1639.
16.Cushing, C.E. and L.A. Rancitelli. 1972. Northwest Science, 46(2), pp 115-121.
17.Lowman, F.G. 1963. in: Schultz, V. and A.W. Klement, Jr. (eds.); Radioecology. pp 561-567. Reinhold publ. Corp. New York.
18.Groth, P. 1971. Arch. Hydrobiol. 68(3), pp 305-375.
19.Timofeeva-Resovskaya, E.A., T-R. N.V. and E.A. Gileva. 1961. Dokl. Akad. Nauk. SSSR, 140, pp 1437.
20.Martin, J.H 1970. Limnol. Oceanogr. 15(5), pp 756-761.

5-221 Concentration factor
Ga

Species	Value	Condition
Brown algae	4200 based on wet weight	None[1]
Plankton	12000 based on wet weight	None[1]

Source: 1.Bowen, H.J.M. 1966. Trace Elements in Biochemistry, Ac. Press.

5-222 Concentration factor
Ge

Species	Value	Condition
Brown algae	108 based on wet weight	Range = 15 - 200[1]
Plants	305 based on dry weight	Algae, bryophytes and vascular hydrophytes, freshwater, 32 sp[2]

Source: 1.Mauchline, J. and W.L Templeton. 1964. Oceanogr. Mar. Biol. Ass. Rev., 2, pp 229.
2.Timofeeva-Resovskaya, E.A., T-R. N.V. and E.A. Gileva. 1961. Dokl. Akad. Nauk. SSSR, 140, pp 1437.

5-223 Concentration factor HCH

Species	Value	Condition
Salmo gairdneri	0.1 based on wet weight	Uptake from food, conc in fish fat, food: HCH = 10 mg/kg, 2 w[1]
Salmo gairdneri	0.26 based on wet weight	Uptake from food, conc in fish fat, food: HCH = 50 mg/kg, 2 w[1]
Salmo gairdneri	0.34 based on wet weight	Uptake from food, conc in fish fat, food: HCH = 1250 mg/kg, 4 w[1]
Salmo gairdneri	0.35 based on wet weight	Uptake from food, conc in fish fat, food: HCH = 250 mg/kg, 4 w[1]
Salmo gairdneri	0.37 based on wet weight	Uptake from food, conc in fish fat, food: HCH = 50 mg/kg, 12 w[1]
Salmo gairdneri	0.38 based on wet weight	Uptake from food, conc in fish fat, food: HCH = 250 mg/kg, 2 w[1]
Salmo gairdneri	0.39 based on wet weight	Uptake from food, conc in fish fat, food: HCH = 50 mg/kg, 8 w[1]
Salmo gairdneri	0.40 based on wet weight	Uptake from food, conc in fish fat, food: HCH = 250 mg/kg, 8 w[1]
Salmo gairdneri	0.41 based on wet weight	Uptake from food, conc in fish fat, food: HCH = 1250 mg/kg, 12 w[1]
Salmo gairdneri	0.42 based on wet weight	Uptake from food, conc in fish fat, food: HCH = 50 mg/kg, 4 w[1]
Salmo gairdneri	0.43 based on wet weight	Uptake from food, conc in fish fat, food: HCH = 1250 mg/kg, 8 w[1]
Salmo gairdneri	0.44 based on wet weight	Uptake from food, conc in fish fat, food: HCH = 10 mg/kg, 8 w[1]
Salmo gairdneri	0.47 based on wet weight	Uptake from food, conc in fish fat, food: HCH = 10 mg/kg, 4 w[1]
Salmo gairdneri	0.54 based on wet weight	Uptake from food, conc in fish fat, food: HCH = 10 mg/kg, 12 w[1]
Salmo gairdneri	0.70 based on wet weight	Uptake from food, conc in fish fat, food: HCH = 250 mg/kg, 12 w[1]

Source: 1.Canton, J.H., G.J. van Fisch and P.A. Greve, 1977. In: S.H. Jenkins, Vol. 2(1), pp 111-115.

5-224 Concentration factor Heptachlor

Species	Value	Condition
Alfalfa	0.036 based on wet weight	Conc in soil = 0.78 mg/kg dry weight[1]
Alfalfa, foliage	2.863 based on wet weight	Conc in soil = 0.234 mg/kg dry weight[1]
Alfalfa, root	1.752 based on wet weight	Conc in soil = 0.234 mg/kg dry weight[1]
Bluegill	1130 based on wet weight	Water = 50 ug/l[1]
Bluegill	314 based on wet weight	Water = 50 ug/l[1]
Carrot	0.003 based on wet weight	Conc in muck soil = 1.40 mg/kg dry weight[1]
Carrot	0.089 based on wet weight	Conc in sandy soil = 0.44 mg/kg dry weight[1]
Carrot, root	0.735 based on wet weight	Conc in soil = 0.49 mg/kg dry weight[1]
Carrot, root	0.737 based on wet weight	Conc in silt loam soil = 1.33 mg/kg dry weight[1]
Chicken, liver	0.13 based on wet weight	Conc in food = 0.05 mg/kg[1]
Cucumber, fruit	0.721 based on wet weight	Conc in soil = 3.8 mg/kg dry weight[1]
Earthworms	3.50 based on dry weight	Conc in soil = 0.02 mg/kg dry weight, USA, sites = 67[1]
Lettuce	0.016 based on wet weight	Conc in silt loam soil = 2.5 mg/kg dry weight[1]
Oat, seed	0.02 based on wet weight	Conc in soil = 1.0 mg/kg dry weight[1]
Potato, tubers	0.102 based on wet weight	Conc in soil = 0.49 mg/kg dry weight[1]
Rutabaga, root	0.075 based on wet weight	Conc in soil = 0.320 mg/kg dry weight[1]
Soybean	0.11 based on wet weight	Conc in soil = 1.0 mg/kg dry weight[1]
Sugar beet, root	0.056 based on wet weight	Conc in silt loam soil = 2.5 mg/kg dry weight[1]

Sweet potato, tubers	0.02 based on wet weight	Conc in soil = 0.02 mg/kg dry weight[1]
Turnips	0.036 based on wet weight	Conc in silt loam soil = 2.5 mg/kg dry weight[1]
Wheat	0.045 based on wet weight	Conc in soil = 9.74 mg/kg dry weight[1]
Wheat	0.057 based on wet weight	Conc in soil = 1.94 mg/kg dry weight[1]
Woodcock, tissue	2.62 based on wet weight	Conc in food = 0.65 mg/kg[1]
Woodcock, tissue	4.55 based on wet weight	Conc in food = 2.86 mg/kg[1]

Source: 1.Edwards, C.A. 1975. Persistent Pesticides in the Environment, 2nd edition, CRC-Press.

5-225 Concentration factor
Heptachlor and heptaepoxide

Species	Value	Condition
Crassostrea virginica	7 based on wet weight	Water = 0.28 ug/l, USA[1]
Crayfish	17 based on wet weight	Water = 0.06 ug/l, USA[1]
Gyrinidae	5200 based on wet weight	Water = 0.06 ug/l, USA[1]
Oyster	10 based on wet weight	Water < 1.0 ug/l, conc in sediment < 0.002 mg/kg, USA[1]
Oyster	20 based on wet weight	Water < 1.0 ug/l, conc in sediment < 0.02 mg/kg, USA[1]
Plankton	18 based on wet weight	Water = 0.06 ug/l, USA[1]
Tendipendidae, larvae	5200 based on wet weight	Water = 0.06 ug/l, USA[1]

Source: 1.Edwards, C.A. 1975. Persistent Pesticides in the Environment, 2nd edition, CRC-Press.

5-226 Concentration factor
Hexabromobenzene

Species	Value	Condition
Salmo salar	1 based on wet weight	Uptake from food, steady state cf, 288 K, 48 h, 6.24 g fish[1]
Salmo salar	1 based on wet weight	Uptake from water, steady state cf, 288 K, 48 h, 6.24 g fish[1]

Source: 1.Zitko, V. and O. Hutzinger. 1976. Bull. Environ. Contam. Toxicol. 16(6), pp 665-673.

5-227 Concentration factor
Hexabromobiphenyl

Species	Value	Condition
Salmo salar	0.114 based on wet weight	Juvenile, uptake from food, 48 h, 5.29 g fish[1]
Salmo salar	1.73 based on wet weight	Juvenile, uptake from water, 48 h, 5.29 g fish[1]

Source: 1.Zitko, V. 1977. Bull. Environ. Contam. Toxicol. 17(3), pp 285-292.

5-228 Concentration factor
Hexachlorobenzene

Species	Value	Condition
Salmo salar	0.174 based on wet weight	Uptake from food, steady state cf, 288 K, 48 h, 6.24 g fish[1]
Salmo salar	690 based on wet weight	Uptake from water, steady state cf, 288 K, 48 h, 6.24 g fish[1]

Source: 1.Zitko, V. and O. Hutzinger. 1976. Bull. Environ. Contam. Toxicol. 16(6), pp 665-673.

5-229 Concentratiorfactor Hg

Species	Value	Condition
Aedes aegypti	102.1 based on wet weight	300 K, 203-Hg-nitrate, 72 h exp, larval age = 7 d, F027, SD = +- 3.1[1]
Aedes aegypti	114.3 based on wet weight	300 K, 203-Hg-nitrate, 96 h exp, larval age = 5 d, F027, SD = +- 4.8[1]
Aedes aegypti	130.2 based on wet weight	300 K, 203-Hg-nitrate, 96 h exp, larval age = 7 d, F027, SD = +- 3.3[1]
Aedes aegypti	165.0 based on wet weight	300 K, phenyl-203-Hg-acetate, 96 h exp, larval age = 3 d, F027[1]
Aedes aegypti	167.3 based on wet weight	300 K, phenyl-203-Hg-acetate, 24 h exp, larval age = 3 d, F027[1]
Aedes aegypti	20.5 based on wet weight	300 K, 203-Hg-nitrate, 24 h exp, larval age = 3 d, F027, SD = +- 1.2[1]
Aedes aegypti	238.1 based on wet weight	300 K, phenyl-203-Hg-acetate, 48 h exp, larval age = 3 d, F027[1]
Aedes aegypti	24.1 based on wet weight	300 K, 203-Hg-nitrate, 48 h exp, larval age = 3 d, F027, SD = +- 2.1[1]
Aedes aegypti	291.9 based on wet weight	300 K, phenyl-203-Hg-acetate, 24 h exp, larval age = 5 d, F027[1]
Aedes aegypti	325.0 based on wet weight	300 K, phenyl-203-Hg-acetate, 96 h exp, larval age = 5 d, F027[1]
Aedes aegypti	334.5 based on wet weight	300 K, phenyl-203-Hg-acetate, 48 h exp, larval age = 5 d, F027[1]
Aedes aegypti	35.1 based on wet weight	300 K, 203-Hg-nitrate, 24 h exp, larval age = 5 d, F027, SD = +- 1.8[1]
Aedes aegypti	372.1 based on wet weight	300 K, phenyl-203-Hg-acetate, 96 h exp, larval age = 7 d, F027[1]
Aedes aegypti	381.2 based on wet weight	300 K, phenyl-203-Hg-acetate, 72 h exp, larval age = 3 d, F027[1]
Aedes aegypti	385.3 based on wet weight	300 K, phenyl-203-Hg-acetate, 24 h exp, larval age = 7 d, F027[1]
Aedes aegypti	42.7 based on wet weight	300 K, 203-Hg-nitrate, 72 h exp, larval age = 3 d, F027, SD = +- 1.7[1]
Aedes aegypti	447.2 based on wet weight	300 K, phenyl-203-Hg-acetate, 48 h exp, larval age = 7 d, F027[1]
Aedes aegypti	46.7 based on wet weight	300 K, 203-Hg-nitrate, 48 h exp, larval age = 5 d, F027, SD = +- 2.8[1]
Aedes aegypti	491.8 based on wet weight	300 K, phenyl-203-Hg-acetate, 72 h exp, larval age = 5 d, F027[1]
Aedes aegypti	50.1 based on wet weight	300 K, 203-Hg-nitrate, 24 h exp, larval age = 7 d, F027, SD = +- 3.2[1]
Aedes aegypti	579.1 based on wet weight	300 K, phenyl-203-Hg-acetate, 48 h exp, larval age = 7 d, F027[1]
Aedes aegypti	63.1 based on wet weight	300 K, 203-Hg-nitrate, 72 h exp, larval age = 5 d, F027, SD = +- 4.3[1]
Aedes aegypti	79.8 based on wet weight	300 K, 203-Hg-nitrate, 48 h exp, larval age = 7 d, F027, SD = +- 6.1[1]
Aedes aegypti	95.1 based on wet weight	300 K, 203-Hg-nitrate, 96 h exp, larval age = 3 d, F027, SD = +- 4.7[1]
Algae	700 based on dry weight	Range = 200 - 1200[2]
Atmospheric particulates	1100	US urban air, relative to earths crust, Fe standard, F200[3]
Bacillus licheniformis	13.6 based on wet weight	310 K, 203-Hg-nitrate, 5 h exp, F028, SD = +- 2.8[1]
Bacillus licheniformis	16.8 based on wet weight	310 K, 203-Hg-nitrate, 96 h exp, F028, SD = +- 7.2[1]
Bacillus licheniformis	17.7 based on wet weight	296 K, 203-Hg-nitrate, 24 h exp, F028, SD = +- 3.1[1]
Bacillus licheniformis	19.0 based on wet weight	310 K, phenyl-203-Hg-acetate, 96 h exp, F028, SD = +- 3.2[1]
Bacillus licheniformis	22.4 based on wet weight	318 K, phenyl-203-Hg-acetate, 96 h exp, F028, SD = +- 3.9[1]
Bacillus licheniformis	24.0 based on wet weight	318 K, 203-Hg-nitrate, 96 h exp, F028, SD = +- 3.5[1]
Bacillus licheniformis	25.7 based on wet weight	296 K, 203-Hg-nitrate, 48 h exp, F028, SD = +- 5.2[1]
Bacillus licheniformis	28.6 based on wet weight	296 K, 203-Hg-nitrate, 96 h exp, F028, SD = +- 5.7[1]
Bacillus licheniformis	31.3 based on wet weight	310 K, phenyl-203-Hg-acetate, 24 h exp, F028, SD = +- 3.5[1]
Bacillus licheniformis	33.6 based on wet weight	296 K, 203-Hg-nitrate, 72 h exp, F028, SD = +- 5.3[1]

Bacillus licheniformis	34.7 based on wet weight	318 K, 203-Hg-nitrate, 5 h exp, F028, SD = +- 6.5[1]
Bacillus licheniformis	39.4 based on wet weight	310 K, phenyl-203-Hg-acetate, 72 h exp, F028, SD = +- 10.6[1]
Bacillus licheniformis	4.3 based on wet weight	310 K, phenyl-203-Hg-acetate, 5 h exp, F028, SD = +- 0.7[1]
Bacillus licheniformis	42.2 based on wet weight	318 K, 203-Hg-nitrate, 72 h exp, F028, SD = +- 5.8[1]
Bacillus licheniformis	42.8 based on wet weight	318 K, phenyl-203-Hg-acetate, 24 h exp, F028, SD = +- 6.3[1]
Bacillus licheniformis	51.4 based on wet weight	318 K, phenyl-203-Hg-acetate, 48 h exp, F028, SD = +- 9.0[1]
Bacillus licheniformis	51.9 based on wet weight	318 K, 203-Hg-nitrate, 48 h exp, F028, SD = +- 3.0[1]
Bacillus licheniformis	53.9 based on wet weight	310 K, phenyl-203-Hg-acetate, 48 h exp, F028, SD = +- 10.7[1]
Bacillus licheniformis	55.5 based on wet weight	318 K, phenyl-203-Hg-acetate, 72 h exp, F028, SD = +- 5.6[1]
Bacillus licheniformis	6.7 based on wet weight	296 K, phenyl-203-Hg-acetate, 5 h exp, F028, SD = +- 0.5[1]
Bacillus licheniformis	61.6 based on wet weight	310 K, 203-Hg-nitrate, 24 h exp, F028, SD = +- 3.7[1]
Bacillus licheniformis	69.9 based on wet weight	318 K, 203-Hg-nitrate, 24 h exp, F028, SD = +- 10.8[1]
Bacillus licheniformis	70.9 based on wet weight	310 K, 203-Hg-nitrate, 48 h exp, F028, SD = +- 6.7[1]
Bacillus licheniformis	78.0 based on wet weight	310 K, 203-Hg-nitrate, 72 h exp, F028, SD = +- 2.0[1]
Bacillus licheniformis	8.1 based on wet weight	296 K, phenyl-203-Hg-acetate, 24 h exp, F028, SD = +- 1.3[1]
Bacillus licheniformis	8.2 based on wet weight	296 K, phenyl-203-Hg-acetate, 48 h exp, F028, SD = +- 1.8[1]
Bacillus licheniformis	8.8 based on wet weight	296 K, phenyl-203-Hg-acetate, 96 h exp, F028, SD = +- 1.9[1]
Bacillus licheniformis	8.8 based on wet weight	318 K, phenyl-203-Hg-acetate, 5 h exp, F028, SD = +- 0.5[1]
Bacillus licheniformis	9.1 based on wet weight	296 K, 203-Hg-nitrate, 5 h exp, F028, SD = +- 2.0[1]
Bacillus licheniformis	9.4 based on wet weight	296 K, phenyl-203-Hg-acetate, 72 h exp, F028, SD = +- 0.4[1]
Brown algae	250 based on wet weight	None[4]
Crassostrea virginica	2100 based on wet weight	Water = 10 - 100 ug/l, 45 d, rg = 1400 - 2800[5]
Daphnia magna	20 based on wet weight	5.0 mg Hg/m3, 10 w[6]
Daphnia magna	25 based on wet weight	5.0 mg Hg/m3, 7 w[6]
Daphnia magna	33 based on wet weight	2.0 mg Hg/m3, 3 w[6]
Daphnia magna	38 based on wet weight	5.0 mg Hg/m3, 5 w[6]
Daphnia magna	45 based on wet weight	5.0 mg Hg/m3, 1 w[6]
Daphnia magna	46 based on wet weight	5.0 mg Hg/m3, 3 w[6]
Daphnia magna	50 based on wet weight	2.0 mg Hg/m3, 10 w[6]
Daphnia magna	55 based on wet weight	2.0 mg Hg/m3, 5 w[6]
Daphnia magna	55 based on wet weight	2.0 mg Hg/m3, 7 w[6]
Daphnia magna	65 based on wet weight	2.0 mg Hg/m3, 1 w[6]
Esox lucius, axial muscle	3000 based on wet weight	Min value, Sweden[7]
Fish	3000 based on wet weight	Freshwater, methyl-Hg(+)[8]
Lebistes reticulatus	108.4 based on wet weight	203-Hg-nitrate, 24 h exp[1]
Lebistes reticulatus	169.1 based on wet weight	203-Hg-nitrate, 96 h exp, based on wet weight[1]
Lebistes reticulatus	169.1 based on wet weight	Phenyl-203-Hg-acetate, 72 h exp, based on wet weight[1]
Lebistes reticulatus	18.3 based on wet weight	Phenyl-203-Hg-acetate, 96 h exp, based on wet weight[1]
Lebistes reticulatus	184.1 based on wet weight	203-Hg-nitrate, 72 h exp, based on wet weight[1]

Lebistes reticulatus	282.2 based on wet weight	203-Hg-nitrate, 48 h exp, based on wet weight[1]
Lebistes reticulatus	33.3 based on wet weight	Phenyl-203-Hg-acetate, 24 h exp, based on wet weight[1]
Lebistes reticulatus	91.6 based on wet weight	Phenyl-203-Hg-acetate, 48 h exp, based on wet weight[1]
Ligumia sp.	10.0 based on wet weight	5.0 mg Hg/m3, 5 w, body weight = 4.27 g, also Margaritifera sp.[6]
Ligumia sp.	12.5 based on wet weight	2.0 mg Hg/m3, 1 w, body weight = 6.17 g, also Margaritifera sp.[6]
Ligumia sp.	160 based on wet weight	5.0 mg Hg/m3, 1 w, body weight = 0.75 g, also Margaritifera sp.[6]
Ligumia sp.	20.0 based on wet weight	5.0 mg Hg/m3, 1 w, body weight = 9.61 g, also Margaritifera sp.[6]
Ligumia sp.	30.0 based on wet weight	2.0 mg Hg/m3, 5 w, body weight = 4.31 g, also Margaritifera sp.[6]
Ligumia sp.	30.0 based on wet weight	5.0 mg Hg/m3, 7 w, body weight = 3.42 g, also Margaritifera sp.[6]
Ligumia sp.	35 based on wet weight	5.0 mg Hg/m3, 10 w, body weight = 5.09 g, also Margaritifera sp.[6]
Ligumia sp.	37.5 based on wet weight	2.0 mg Hg/m3, 3 w, body weight = 4.13 g, also Margaritifera sp.[6]
Ligumia sp.	55.0 based on wet weight	2.0 mg Hg/m3, 10 w, body weight = 4.61 g, also Margaritifera sp.[6]
Ligumia sp.	67.0 based on wet weight	5.0 mg Hg/m3, 3 w, body weight = 2.86 g, also Margaritifera sp.[6]
Ligumia sp.	67.5 based on wet weight	2.0 mg Hg/m3, 7 w, body weight = 4.43 g, also Margaritifera sp.[6]
Longfin smelt	3400 based on wet weight	Size = 9.3 cm, Lake Washington, water conc = 0.020 mg/m3, muscle[9]
Mytilus galloprovincialis	15625 based on wet weight	Water = 41.2 ppm[10]
Oryzias latipes	1600 based on wet weight	Water = 10 ppb, measured on eggs, pH = 6.9, 299 K, HgCl[11]
Oryzias latipes	1900 based on wet weight	Water = 15 ppb, measured on eggs, pH = 6.9, 299 K, HgCl[11]
Oryzias latipes	1900 based on wet weight	Water = 30 ppb, measured on eggs, pH = 6.9, 299 K, HgCl[11]
Oryzias latipes	2700 based on wet weight	Water = 20 ppb, measured on eggs, pH = 6.9, 299 K, HgCl[11]
Palaemonetes vulgaris, exoskeleton	11.3 based on wet weight	Methylmercury cloride, water = 1.5 ppb, 72 h, 293 K[12]
Palaemonetes vulgaris, exoskeleton	7.54 based on wet weight	Methylmercury cloride, water = 1.5 ppb, 24 h, 293 K[12]
Palaemonetes vulgaris, exoskeleton	8.51 based on wet weight	Mercuric cloride, water = 1.5 ppb, 24 h, 293 K[12]
Palaemonetes vulgaris, foregut	4.30 based on wet weight	Mercuric cloride, water = 1.5 ppb, 24 h, 293 K[12]
Palaemonetes vulgaris, foregut	5.58 based on wet weight	Methylmercury cloride, water = 1.5 ppb, 24 h, 293 K[12]
Palaemonetes vulgaris, foregut	8.7 based on wet weight	Methylmercury cloride, water = 1.5 ppb, 72 h, 293 K[12]
Palaemonetes vulgaris, gills	147.54 based on wet weight	Mercuric cloride, water = 1.5 ppb, 24 h, 293 K[12]
Palaemonetes vulgaris, gills	16.3 based on wet weight	Methylmercury cloride, water = 1.5 ppb, 72 h, 293 K[12]
Palaemonetes vulgaris, gills	24.44 based on wet weight	Methylmercury cloride, water = 1.5 ppb, 24 h, 293 K[12]
Palaemonetes vulgaris, remainder	2.19 based on wet weight	Methylmercury cloride, water = 1.5 ppb, 24 h, 293 K[12]
Palaemonetes vulgaris, remainder	2.41 based on wet weight	Mercuric cloride, water = 1.5 ppb, 24 h, 293 K[12]
Palaemonetes vulgaris, remainder	5.3 based on wet weight	Methylmercury cloride, water = 1.5 ppb, 72 h, 293 K[12]
Palaemonetes vulgaris, ventral nerve cord	113.22 based on wet weight	Methylmercury cloride, water = 1.5 ppb, 24 h, 293 K[12]

Palaemonetes vulgaris, ventral nerve cord	29.23 based on wet weight	Mercuric cloride, water = 1.5 ppb, 24 h, 293 K[12]
Palaemonetes vulgaris, ventral nerve cord	7.6 based on wet weight	Methylmercury cloride, water = 1.5 ppb, 72 h, 293 K[12]
Palaemonetes vulgaris, whole bodies	5.60 based on wet weight	Methylmercury cloride, water = 1.5 ppb, 24 h, 293 K[12]
Palaemonetes vulgaris, whole bodies	5.70 based on wet weight	Mercuric cloride, water = 1.5 ppb, 24 h, 293 K[12]
Palaemonetes vulgaris, whole bodies	7.90 based on wet weight	Methylmercury cloride, water = 1.5 ppb, 72 h, 293 K[12]
Phytoplankton	2435 based on wet weight	Range = 70 - 4800[13]
Pimephales promelas	100 based on wet weight	2.0 mg Hg/m3, 3 w, body weight = 0.72 g[6]
Pimephales promelas	126 based on wet weight	5.0 mg Hg/m3, 5 w, body weight = 1.01 g[6]
Pimephales promelas	138 based on wet weight	2.0 mg Hg/m3, 7 w, body weight = 0.85 g[6]
Pimephales promelas	145 based on wet weight	5.0 mg Hg/m3, 10 w, body weight = 0.88 g[6]
Pimephales promelas	146 based on wet weight	5.0 mg Hg/m3, 3 w, body weight = 0.86 g[6]
Pimephales promelas	158 based on wet weight	2.0 mg Hg/m3, 10 w, body weight = 1.03 g[6]
Pimephales promelas	65 based on wet weight	2.0 mg Hg/m3, 1 w, body weight = 0.84 g[6]
Plants	5915 based on dry weight	Algae, bryophytes and vascular hydrophytes, freshwater, 32 sp[14]
Ptychocheilus oregonensis	40000 based on wet weight	Size = 50 cm, Lake Washington, water conc = 0.020 mg/m3, muscle[9]
Ptychocheilus oregonensis	8500 based on wet weight	Size = 30 cm, Lake Washington, water conc = 0.020 mg/m3, muscle[9]
Salmo gairdneri	10 based on wet weight	Water = 0.1 mg Hg/l as HgCl, 283 k[15]
Salmo gairdneri	22 based on wet weight	Water = 0.1 mg Hg/l as HgCl, 293 K[15]
Salmo gairdneri	4 based on wet weight	Water = 0.1 mg Hg/l as HgCl, 278 k[15]
Salmo gairdneri	90 based on wet weight	293 K, 0.1 mg Hg/l as phenylmercuric acetate, rg = 80 - 100[15]
Scenedesmus sp.	100 based on wet weight	5.0 mg/m3, 3 w[6]
Scenedesmus sp.	113 based on wet weight	2.0 mg/m3, 5 w[6]
Scenedesmus sp.	120 based on wet weight	5.0 mg/m3, 5 w[6]
Scenedesmus sp.	144 based on wet weight	5.0 mg/m3, 7 w[6]
Scenedesmus sp.	146 based on wet weight	5.0 mg/m3, 10 w[6]
Scenedesmus sp.	180 based on wet weight	2.0 mg/m3, 10 w[6]
Scenedesmus sp.	180 based on wet weight	2.0 mg/m3, 7 w[6]
Scenedesmus sp.	43 based on wet weight	2.0mg/m3, 1 week[6]
Scenedesmus sp.	66 based on wet weight	5.0 mg/m3, 1 w[6]
Scenedesmus sp.	73 based on wet weight	2.0 mg/m3, 3 w[6]
Sockeye salmon	1750 based on wet weight	Size = 9.0 cm, Lake Washington, water conc = 0.020 mg/m3, muscle[9]
Zooplankton	650 based on wet weight	Lake Washington, water conc = 0.020 mg/m3[9]

Source: 1.Zitko, V. and O. Hutzinger. 1976. Bull. Environ. Contam. Toxicol. 16(6), pp 665-673.
2.Hannerz, L. 1968. Rep. Inst. Freshwat. Res. Drottningholm 48, pp 120-176.
3.Foestner, U. and G.T.W. Wittmann. 1979. Metal Pollution in the Aquatic Environment. Springer-Verlag.Berlin.
4.Bowen, H.J.M. 1966. Trace Elements in Biochemistry, Ac. Press.
5.Cunningham, P.A. and M.R.Tripp. 1973. Mar. biol. 20, pp 14-19.
6.Terhaar, C.J., W.S. Ewell, S.P. Dziuba, W.W. White and P.J. Murphy. 1977. Water Res. 11(1), pp 101-11.
7.Johnels, A.G., T. Westermark, W. Berg., P.I. Persson and B. Sjoestrand. 1967. Oikos 18, pp 323-333.
8.Nordisk Hygienisk Tidsskrift, 1970. Suppl. 3, Stockholm.
9.Schell, W.R. and R.S. Barnes. 1974. in: Rubin, A.J. (ed.): Aqueous-Environmental Chemistry of Metal. pp 155-158.
10.Majori, L. and F. Petronio. 1973. Rev. Int. Oceanogr. Med. CERBOM., 31-32, pp 55-90.

11.Heisinger, J.F. and W. Green. 1975. Bull. Environ. Contam. Toxicol. 14(6), pp 665-673.
12.Ray, G.L. and M.R. Tripp. 1976. J. Environ. Qual. 5(2), pp 193-197.
13.Cushing, C.E. and L.A. Rancitelli. 1972. Northwest Science, 46(2), pp 115-121.
14.Timofeeva-Resovskaya, E.A., T-R. N.V. and E.A. Gileva. 1961. Dokl. Akad. Nauk. SSSR, 140, pp 1437.
15.Macleod, J.C. and E. Pessah. 1973. J. Fish. Res. bd. Can. 30(4), pp 485-492.

5-230 Concentration factor
I

Species	Value	Condition
Anodonta nuttalliana, soft parts	0.5 based on wet weight	No food, 292 K, pH = 7.5, freshwat = 1.7 part & 250 solub p ci/ml, 7d[1]
Anodonta nuttalliana, soft parts	0.7 based on wet weight	No food, 292 K, pH = 7.5, freshwat = 5.7 part & 280 p ci/ml, 21 d[1]
Astacus sp., muscle	0.1 based on wet weight	No food, 292 K, pH = 7.5, freshwat = 1.7 part & 250 solub p ci/ml, 7d[1]
Astacus sp., muscle	1 based on wet weight	No food, 292 K, pH = 7.5, freshwat = 5.7 part & 280 p ci/ml, 21 d[1]
Astacus sp., soft parts	15 based on wet weight	No food, 292 K, pH = 7.5, freshwat = 0.69 part & 110 solub p ci/ml, 7d[1]
Astacus sp., viscera	17 based on wet weight	No food, 292 K, pH = 7.5, freshwat = 5.7 part & 280 p ci/ml, 21 d[1]
Brown algae	6200 based on wet weight	None[2]
Brown algae	6500 based on wet weight	Range = 3000 - 10000[3]
Cancer productus, muscle	0.5 based on wet weight	No food, 285 K, pH = 8.2, seawater = 1.3 part & 450 p ci/ml, 20 d[1]
Cancer productus, sort parts	19 based on wet weight	No food, 285 K, pH = 8.2, seawater = 1.3 part & 450 p ci/ml, 20 d[1]
Cancer productus, viscera	2 based on wet weight	No food, 285 K, pH = 8.2, seawater = 1.1 part & 440 solub p ci/ml, 6d[1]
Cancer productus, viscera	21 based on wet weight	No food, 285 K, pH = 8.2, seawater = 1.3 part & 450 p ci/ml, 20 d[1]
Carassius auratus, muscle	0.2 based on wet weight	No food, 292 K, pH = 7.5, freshwat = 1.7 part & 250 solub p ci/ml, 7d[1]
Carassius auratus, nonmuscle	1 based on wet weight	No food, 292 K, pH = 7.5, freshwat = 1.7 part & 250 solub p ci/ml, 7d[1]
Carassius auratus, whole body	0.9 based on wet weight	No food, 292 K, pH = 7.5, freshwat = 5.7 part & 280 p ci/ml, 21 d[1]
Crustacea	31 based on dry weight	Marine, rg = 20 - 48, 4 values used[4]
Fish	10 based on dry weight	Marine, rg = 3 - 15, 5 values used[4]
Fish	9 based on dry weight	Freshwater, rg = 0.5 - 25, 9 values used[4]
Fish, muscle	520 based on wet weight	Danube, 1973, nuclear power plant since 1967, I-131, rg = 250 - 780[5]
Gillichthys mirabilis, muscle	1 based on wet weight	No food, 285 K, pH = 8.2, seawater = 1.3 part & 450 p ci/ml, 20 d[1]
Gillichthys mirabilis, nonmuscle	0.7 based on wet weight	No food, 285 K, pH = 8.2, seawater = 1.1 part & 440 solub p ci/ml, 6d[1]
Gillichthys mirabilis, nonmuscle	0.9 based on wet weight	No food, 285 K, pH = 8.2, seawater = 1.3 part & 450 p ci/ml, 20 d[1]
Gillichthys mirabilis, whole body	0.7 based on wet weight	No food, 285 K, pH = 8.2, seawater = 1.3 part & 450 p ci/ml, 20 d[1]
Gillichthys mirabilis, whole body	0.8 based on wet weight	No food, 285 K, pH = 8.2, seawater = 1.10 part & 220 solub p ci/ml, 6d[1]
Molluscs	320 based on dry weight	Freshwater, rg = 60 - 1000, 4 values used[4]
Molluscs	5010 based on dry weight	Marine, rg = 20 - 20000, 4 values used[4]

Mytilus edulis, soft parts	0.8 based on wet weight	No food, 285 K, pH = 8.2, seawater = 1.1 part & 440 solub p ci/ml, 6d[1]
Plankton	1200 based on wet weight	None[2]
Plants	1065 based on dry weight	Marine, rg = 30 - 6800, 8 values used[4]
Plants	370 based on dry weight	Algae, bryophytes and vascular hydrophytes, freshwater, 32 sp[6]
Plants	69 based on dry weight	Freshwater, rg = 10 - 200, 8 values used[4]
Sediment	6800 based on wet weight	Danube, 1973, nuclear power plant since 1967, I-131, rg = 510 - 13800[5]

Source: 1.Harrison, F.L. 1973. Health Physics, 24, pp 331-343.
2.Bowen, H.J.M. 1966. Trace Elements in Biochemistry, Ac. Press.
3.Mauchline, J. and W.L Templeton. 1964. Oceanogr. Mar. Biol. Ass. Rev., 2, pp 229.
4.Jinks, S.M. and M. Eisenbud. 1972. Radiat. Data Rep. 13, pp 243-247.
5.Herrmann, H., M. Ruf, K. Hubel and W. Lunsmann. 1975. Impacts of Nuclear Releases into the Aquatic Environment. IAEA-SM-198/14. pp 461-470. Vienna.
6.Timofeeva-Resovskaya, E.A., T-R. N.V. and E.A. Gileva. 1961. Dokl. Akad. Nauk. SSSR, 140, pp 1437.

5-231 Concentration factor Insecticides

Species	Value	Condition
Slugs	50.5 based on wet weight	Cf = conc in anim (wet wt)/conc in soil (dry wt), rg = 1 - 100[1]

Source: 1.Edwards, C.A. 1975. Persistent Pesticides in the Environment, 2nd edition, CRC-Press.

5-232 Concentration factor K

Species	Value	Condition
Brassica oleracea	115 based on wet weight	Conc in soil solution = 5.4 ppm[1]
Brown algae	26.5 based on wet weight	Range = 3 - 50[2]
Brown algae	34 based on wet weight	None[1]
Cancer irroratus	7.76 based on wet weight	Seawater conc = 8.0 ppm, Nova Scotia, AAS, SD = +- 0.81[3]
Crustacea	12 based on dry weight	Marine, rg = 8 - 19, 10 values used[4]
Fish	16 based on dry weight	Marine, rg = 6.7 - 34, 10 values used[4]
Fish	4400 based on dry weight	Freshwater, rg = 340 - 18000, 6 values used[4]
Fish, bone	1560 based on wet weight	Danube, 1973, nuclear power plant since 1967, K-40, rg = 970 - 2450[5]
Fish, muscle	700 based on wet weight	Danube, 1973, nuclear power plant since 1967, K-40, rg = 570 - 1580[5]
Molluscs	8 based on dry weight	Marine, rg = 3.5 - 10, 10 values used[4]
Phytoplankton	4300 based on wet weight	Range = 1400 - 7200[6]
Plants	13 based on dry weight	Marine, rg = 4 - 31, 10 values used[4]
Sediment	1300 based on wet weight	Danube, 1973, nuclear power plant since 1967, K-40, rg = 650 - 2080[5]
Suspended solids	1070 based on wet weight	Danube, 1973, nuclear power plant since 1967, K-40, rg = 520 - 2010[5]
Water plant	320 based on wet weight	Danube, 1973, nuclear power plant since 1967, K-40, rg = 220 - 490[5]

Source: 1.Bowen, H.J.M. 1966. Trace Elements in Biochemistry, Ac. Press.
2.Mauchline, J. and W.L Templeton. 1964. Oceanogr. Mar. Biol. Ass. Rev., 2, pp 229.
3.Martin, J.-L.M. 1974. Mar. biol., 28(4), pp 245-251.
4.Jinks, S.M. and M. Eisenbud. 1972. Radiat. Data Rep. 13, pp 243-247.
5.Herrmann, H., M. Ruf, K. Hubel and W. Lunsmann. 1975. Impacts of Nuclear Releases into the Aquatic Environment. IAEA-SM-198/14. pp 461-470. Vienna.
6.Cushing, C.E. and L.A. Rancitelli. 1972. Northwest Science, 46(2), pp 115-121.

5-233 Concentration factor La

Species	Value	Condition
Brown algae	8300 based on wet weight	None[1]

Source: 1.Bowen, H.J.M. 1966. Trace Elements in Biochemistry, Ac. Press.

5-234 Concentration factor Li

Species	Value	Condition
Brown algae	8 based on wet weight	None[1]
Coral	0.63 based on Ca content	Shallow open ocean coral, mean of 8 determinations, F023[2]
Coral	0.64 based on Ca content	Shallow coastal ocean, mean of 13 determinations, F023[2]
Coral	0.83 based on Ca content	Deep ocean coral, mean of 13 determinations, F023[2]

Source: 1.Bowen, H.J.M. 1966. Trace Elements in Biochemistry, Ac. Press.
2.Livingston, H.D. and G. Thompson. 1971. Limnol. Oceanogr. 16(5), pp 786-796.

5-235 Concentration factor Lu

Species	Value	Condition
Engraulis japonicus, whole body	200 based on wet weight	Rel to soluble conc, total = 0.7 ng/l, solub = 0.3 ng/l, marine, Japan[1]
Engraulis japonicus, whole body	75 based on wet weight	Rel to total conc, total = 0.7 ng/l, solub = 0.3 ng/l, marine, Japan[1]
Hizikia fusiforme, whole body	280 based on wet weight	Rel to total conc, total = 0.7 ng/l, solub = 0.3 ng/l, marine, Japan[1]
Hizikia fusiforme, whole body	730 based on wet weight	Rel to soluble conc, total = 0.7 ng/l, solub = 0.3 ng/l, marine, Japan[1]
Meretrix lusoria, soft part	300 based on wet weight	Rel to total conc, total = 0.7 ng/l, solub = 0.3 ng/l, marine, Japan[1]
Meretrix lusoria, soft part	770 based on wet weight	Rel to soluble conc, total = 0.7 ng/l, solub = 0.3 ng/l, marine, Japan[1]
Paralichthys olivaceus, muscle	13 based on wet weight	Rel to total conc, total = 0.7 ng/l, solub = 0.3 ng/l, marine, Japan[1]
Paralichthys olivaceus, muscle	33 based on wet weight	Rel to soluble conc, total = 5.6 ng/l, solub = 2.1 ng/l, marine, Japan[1]
Sargassum fulvellum, whole body	350 based on wet weight	Rel to total conc, total = 0.7 ng/l, solub = 0.3 ng/l, marine, Japan[1]
Sargassum fulvellum, whole body	930 based on wet weight	Rel to soluble conc, total = 0.7 ng/l, solub = 0.3 ng/l, marine, Japan1
Seriola quinqueradiata, muscle	13 based on wet weight	Rel to total conc, total = 0.7 ng/l, solub = 0.3 ng/l, marine, Japan[1]
Seriola quinqueradiata, muscle	33 based on wet weight	Rel to soluble conc, total = 5.6 ng/l, solub = 2.1 ng/l, marine, Japan1
Ulva pertusa, whole body	300 based on wet weight	Rel to total conc, total = 0.7 ng/l, solub = 0.3 ng/l, marine, Japan[1]
Ulva pertusa, whole body	800 based on wet weight	Rel to soluble conc, total = 0.7 ng/l, solub = 0.3 ng/l, marine, Japan1
Undaria pinnatifida, whole body	110 based on wet weight	Rel to total conc, total = 0.7 ng/l, solub = 0.3 ng/l, marine, Japan[1]

Undaria pinnatifida, whole body	300 based on wet weight	Rel to soluble conc, total = 0.7 ng/l, solub = 0.3 ng/l, marine, Japan[1]

Source: 1.Suzuki, H., T. Koganage and M. Saki. 1975. Impacts of Nuclear Releases into the Aquatic Environment. IAEA SM-198/19, pp 77-89. Vienna.

5-236 Concentration factor Methoxychlor

Species	Value	Condition
Chicken, muscle and tissue	0.01 based on wet weight	Conc in food = 1000 mg/kg, mean value[1]
Chicken, tissue	0.006 based on wet weight	Conc in food = 100.00 mg/kg[1]
Earthworms	4.17 based on wet weight	Conc in soil = 2.66 mg/kg dry wt., USA, sites = 2, excl Lumbricus[1]
Lumbricus terrestris	2.11 based on wet weight	Conc in soil = 2.66 mg/kg dry wt., USA, sites = 2[1]
Oyster	10 based on wet weight	Conc in water: < 1.0 ug/l, conc in sediment < 0.01 mg/kg, USA[1]

Source: 1.Edwards, C.A. 1975. Persistent Pesticides in the Environment, 2nd edition, CRC-Press.

5-237 Concentration factor Mg

Species	Value	Condition
Brassica oleracea	17 based on wet weight	Conc in soil solution = 22 ppm[1]
Brown algae	0.96 based on wet weight	None[1]
Cancer irroratus	2.24 based on wet weight	Seawater conc = 1350 ppm, Nova Scotia, AAS, SD = +- 0.103[2]
Phytoplankton	1104 based on wet weight	Range = 230 - 660[3]
Plankton	0.59 based on wet weight	None[1]
Zooplankton	1.3 based on wet weight	Seawater concentration 1350.0 mg/m3[4]

Source: 1.Bowen, H.J.M. 1966. Trace Elements in Biochemistry, Ac. Press.
2.Martin, J.-L.M. 1974. Mar. biol., 28(4), pp 245-251.
3.Cushing, C.E. and L.A. Rancitelli. 1972. Northwest Science, 46(2), pp 115-121.
4.Martin, J.H 1970. Limnol. Oceanogr. 15(5), pp 756-761.

5-238 Concentration factor Mirex

Species	Value	Condition
Chlorococcum sp.	10000 based on wet weight	Lab., water = 10 pptr[1]

Source: 1.Hollister, T.A., G.E. Walsh and J. Forester. 1975. Bull. Environ. Contam. Toxicol. 14(6), pp 753-759

5-239 Concentration factor Mn

Species	Value	Condition
Agarum cribrosum	86600 based on dry weight	Seawater = 6.4E-7 g/l, Vostok Bay, Sea of Japan[1]
Brassica oleracea	460 based on wet weight	Conc in soil solution = 0.072 ppm[2]
Brown algae	6500 based on wet weight	None[2]
Brown algae	6500 based on wet weight	None[3]
Cancer irroratus	12699 based on wet weight	Seawater conc = 0.84E-3 ppm, Nova Scotia, AAS, SD = +- 1502[4]
Chondrus yendoi	10000 based on dry weight	Seawater = 6.4E-7 g/l, Vostok Bay, Sea of Japan[1]
Chorda filum	2200 based on dry weight	Seawater = 6.4E-7 g/l, Vostok Bay, Sea of Japan[1]
Codium yessoensis	24000 based on dry weight	Seawater = 6.4E-7 g/l, Vostok Bay, Sea of Japan[1]

Costaria costata	7000 based on dry weight	Seawater = 6.4E-7 g/l, Vostok Bay, Sea of Japan[1]
Crustacea	125000 based on dry weight	Freshwater, rg = 1700 - 250000, 6 values used[5]
Crustacea	2270 based on dry weight	Marine, rg = 600 - 7500, 8 values used[5]
Enteromorpha prolifera	71000 based on dry weight	Seawater = 6.4E-7 g/l, Vostok Bay, Sea of Japan[1]
Fish	23 based on wet weight	Freshwater organisms, typical value[6]
Fish	363 based on dry weight	Marine, rg = 35 - 1800, 9 values used[5]
Fish	373 based on wet weight	Marine organisms, typical value[7]
Fish	81 based on dry weight	Freshwater, rg = 0.1 - 400, 19 values used[5]
Fish, muscle	230 based on wet weight	Danube, 1973, nuclear power plant since 1967, Mn-54, rg = 60 - 440[8]
Fucus vesiculosus	2.3E4 based on dry weight	18 obs., all areas around the British Isles[9]
Heterochordaria abietina	4600 based on dry weight	Seawater = 6.4E-7 g/l, Vostok Bay, Sea of Japan[1]
Laminaria chichorioides	8300 based on dry weight	Seawater = 6.4E-7 g/l, Vostok Bay, Sea of Japan[1]
Laminaria japonica	1700 based on dry weight	Seawater = 6.4E-7 g/l, Vostok Bay, Sea of Japan[1]
Lampsilis radiata	2400 based on wet weight	91 days, freshwater, Mn-54[10]
Macrophytes	1450 based on wet weight	Freshwater organisms, typical value[6]
Molluscs	22080 based on dry weight	Marine, rg = 170 - 150000, 11 values used[5]
Molluscs	2300 based on wet weight	Marine organisms, typical value[11]
Molluscs	300000 based on dry weight	Freshwater, rg = 1100 - 1600000, 9 values used[5]
Patella vulgata	1.1E4 based on dry weight	17 obs., all areas around the British Isles[9]
Pelvetia wrightii	4600 based on dry weight	Seawater = 6.4E-7 g/l, Vostok Bay, Sea of Japan[1]
Phyllospadix iwatensis	8000 based on dry weight	Seawater = 6.4E-7 g/l, Vostok Bay, Sea of Japan[1]
Phytoplankton	3800 based on wet weight	Marine organisms, typical value[12]
Plankton	29000 based on wet weight	Conc in water = 4.5 ug/l, Schoehsee, freshwater[13]
Plankton	750 based on wet weight	None[3]
Plankton	9400 based on wet weight	None[2]
Plants	150000 based on dry weight	Freshwater, rg = 1300 - 600000, 9 values used[5]
Plants	5230 based on dry weight	Marine, rg = 2000 - 20000, 10 values used[5]
Polysiphonia japonica	9000 based on dry weight	Seawater = 6.4E-7 g/l, Vostok Bay, Sea of Japan[1]
Porphyra umbilicalis	7.E3 based on dry weight	20 obs., all areas around the britsh isles[9]
Ptilota filicina	178000 based on dry weight	Seawater = 6.4E-7 g/l, Vostok Bay, Sea of Japan[1]
Rhodomela larix	37000 based on dry weight	Seawater = 6.4E-7 g/l, Vostok Bay, Sea of Japan[1]
Sargassum pallidum	7000 based on dry weight	Seawater = 6.4E-7 g/l, Vostok Bay, Sea of Japan[1]
Sediment	1520 based on wet weight	Danube, 1973, nuclear power plant since 1967, Mn-54, rg = 570 - 2450[8]
Seston	1.6 based on wet weight	Depth = 17 m, Schoehsee[13]
Seston	5.4 based on wet weight	Depth = 8 m, Schoehsee[13]
Suspended solids	1030 based on wet weight	Danube, 1973, nuclear power plant since 1967, Mn-54, rg = +- 500 -[8]

Ulva fenestrata	44000 based on dry weight	Seawater = 6.4E-7 g/l, Vostok Bay, Sea of Japan[1]
Ulvaria spendens	47000 based on dry weight	Seawater = 6.4E-7 g/l, Vostok Bay, Sea of Japan[1]
Unio mancus	100000 based on dry weight	Lake Maggiore[14]
Unio mancus	300000 based on dry weight	Lake Maggiore, Mn-54[14]
Unio mancus	40000 based on dry weight	Lake Varsee, Mn-54[14]
Unio mancus	80000 based on dry weight	Lake Varsee[14]
Water plant	240 based on wet weight	Danube, 1973, nuclear power plant since 1967, Mn-54, rg = 65 - 430[8]
Zooplankton	1650 based on wet weight	Seawater concentration 0.002 mg/m3[15]
Zooplankton	3900 based on wet weight	Marine organisms, typical value[12]
Zostera asiatica	18600 based on dry weight	Seawater = 6.4E-7 g/l, Vostok Bay, Sea of Japan[1]

Source: 1.Saenko, G.N., M.D. Koryakov, V.F. Makienko and I.G. Dobrosmy. 1976. Mar. biol., 34(2), pp 169-176.
2.Bowen, H.J.M. 1966. Trace Elements in Biochemistry, Ac. Press.
3.Mauchline, J. and W.L Templeton. 1964. Oceanogr. Mar. Biol. Ass. Rev., 2, pp 229.
4.Martin, J.-L.M. 1974. Mar. biol., 28(4), pp 245-251.
5.Jinks, S.M. and M. Eisenbud. 1972. Radiat. Data Rep. 13, pp 243-247.
6.Merlini, M., C. Rigliocca, A. Berg and G. Pozzi. 1970. IAEA/ SM, 142/27.
7.Goldberg, E.D. 1962. Limnol Oceanogr., 7, pp 72.
8.Herrmann, H., M. Ruf, K. Hubel and W. Lunsmann. 1975. Impacts of Nuclear Releases into the Aquatic Environment. IAEA-SM-198/14. pp 461-470. Vienna.
9.Preston, A., D.R. Jefferies , J.W.R. Harvey and A.K. Steele. 1972. Environ. Pollut., 3, pp 183-214.
10.Harvey, R.S. 1969. Health Physics, 17, pp 149-154.
11.Pringle, B.H., D.E. Hissong, E.L. Katz and S.T. Mulawka. 1968. J. San. Eng. Div. Am. Soc. Civ. Eng., 94(5970), pp 455-475.
12.Martin, J.H. and G.A. Knauer. 1973. Geochim. Cosmochim. Acta., 37, p 1639.
13.Groth, P. 1971. Arch. Hydrobiol. 68(3), pp 305-375.
14.Merlini, M. 1967. Proc. Int. Symp. Radioecol. Concent. Process., 1966, Stockholm, pp 977-982.
15.Martin, J.H 1970. Limnol. Oceanogr. 15(5), pp 756-761.

5-240 Concentration factor
Mo

Species	Value	Condition
Agarum cribrosum	60 based on dry weight	Seawater = 11.0E-7 g/l, Vostok Bay, Sea of Japan[1]
Brown algae	11 based on wet weight	None[2]
Chondrus yendoi	100 based on dry weight	Seawater = 11.0E-7 g/l, Vostok Bay, Sea of Japan[1]
Chorda filum	200 based on dry weight	Seawater = 11.0E-7 g/l, Vostok Bay, Sea of Japan[1]
Costaria costata	20 based on dry weight	Seawater = 11.0E-7 g/l, Vostok Bay, Sea of Japan[1]
Enteromorpha prolifera	250 based on dry weight	Seawater = 11.0E-7 g/l, Vostok Bay, Sea of Japan[1]
Heterochordaria abietina	10 based on dry weight	Seawater = 11.0E-7 g/l, Vostok Bay, Sea of Japan[1]
Laminaria chichorioides	10 based on dry weight	Seawater = 11.0E-7 g/l, Vostok Bay, Sea of Japan[1]
Laminaria japonica	70 based on dry weight	Seawater = 11.0E-7 g/l, Vostok Bay, Sea of Japan[1]
Pelvetia wrightii	100 based on dry weight	Seawater = 11.0E-7 g/l, Vostok Bay, Sea of Japan[1]
Phyllospadix iwatensis	20 based on dry weight	Seawater = 11.0E-7 g/l, Vostok Bay, Sea of Japan[1]
Plankton	20000 based on wet weight	Conc in water = 0.21 ug/l, Schoehsee[3]
Plankton	25 based on wet weight	None[2]
Polysiphonia japonica	270 based on dry weight	Seawater = 11.0E-7 g/l, Vostok Bay, Sea of Japan[1]
Ptilota filicina	350 based on dry weight	Seawater = 11.0E-7 g/l, Vostok Bay, Sea of Japan[1]
Rhodomela larix	80 based on dry weight	Seawater = 11.0E-7 g/l, Vostok Bay, Sea of Japan[1]

Sargassum pallidum	100 based on dry weight	Seawater = 11.0E-7 g/l, Vostok Bay, Sea of Japan[1]
Seston	0.7 based on wet weight	Depth = 17 m, Schoehsee[3]
Seston	0.8 based on wet weight	Depth = 8 m, Schoehsee[3]
Ulva fenestrata	50 based on dry weight	Seawater = 11.0E-7 g/l, Vostok Bay, Sea of Japan[1]
Ulvaria spendens	50 based on dry weight	Seawater = 11.0E-7 g/l, Vostok Bay, Sea of Japan[1]
Zostera asiatica	180 based on dry weight	Seawater = 11.0E-7 g/l, Vostok Bay, Sea of Japan[1]

Source: 1.Saenko, G.N., M.D. Koryakov, V.F. Makienko and I.G. Dobrosmy. 1976. Mar. biol., 34(2), pp 169-176.
2.Bowen, H.J.M. 1966. Trace Elements in Biochemistry, Ac. Press.
3.Groth, P. 1971. Arch. Hydrobiol. 68(3), pp 305-375.

5-241 Concentration factor
N

Species	Value	Condition
Brassica oleracea	49 based on wet weight	Conc in soil solution = 215 ppm[1]
Brown algae	7500 based on wet weight	None[1]
Plankton	19000 based on wet weight	None[1]

Source: 1.Bowen, H.J.M. 1966. Trace Elements in Biochemistry, Ac. Press.

5-242 Concentration factor
N-nitrosoatrazine

Species	Value	Condition
Gambusia affinis	16 based on wet weight	Model ecosystem, organism = 186 ppb, water = 5.0-3.8 ppb, 18 d[1]
Gambusia affinis	22 based on wet weight	Model ecosystem, organism = 188 ppb, water = 5.0-3.8 ppb, 18 d[1]
Gambusia affinis	31 based on wet weight	Model ecosystem, organism = 301 ppb, water = 5.0-3.8 ppb, 18 d[1]

Source: 1.Kearney, P.C., J.E. Oliver, C.S. Helling, A.R. Isensee and A. Kontson. 1977. J. Agric. Food. Chem. 2(5), pp 1177-1181.

5-243 Concentration factor
Na

Species	Value	Condition
Brassica oleracea	19 based on wet weight	Conc in soil solution = 39 ppm[1]
Brown algae	0.78 based on wet weight	None[1]
Brown algae	1 based on wet weight	None[2]
Cancer irroratus	0.84 based on wet weight	Seawater conc = 10500 ppm, Nova Scotia, AAS, SD = +- 0.052[3]
Daphnia magna	4.5 based on dry weight	On dry wt basis relative to the algae consumed[4]
Phytoplankton	1302 based on wet weight	Range = 105 - 2600[5]
Plankton	0.14 based on wet weight	None[1]
Pseudemys sp.	300 based on dry weight	In-vitro, freshwater = 575 mg/l[6]

Source: 1.Bowen, H.J.M. 1966. Trace Elements in Biochemistry, Ac. Press.
2.Mauchline, J. and W.L Templeton. 1964. Oceanogr. Mar. Biol. Ass. Rev., 2, pp 229.
3.Martin, J.-L.M. 1974. Mar. biol., 28(4), pp 245-251.
4.Verduin, J. 1975. Limnol. Oceanogr. 20(6), pp 1052.
5.Cushing, C.E. and L.A. Rancitelli. 1972. Northwest Science, 46(2), pp 115-121.
6.Dunson, W.A. 1969. Proc. 2nd Nat. Symp. Radioecology, USAEC. Conf. 670503, pp 191-197.

5-244 Concentration factor Naphthalene

Species	Value	Condition
Oncorhynchus kisutch, muscle	60 based on wet weight	Water sol fraction = 4 ppb, 288 K, 5 w, flow through, prudhoe B crude oil[1]
Rangia cuneata	17.1 based on wet weight	Conc in water = 0.24 ppm, 24 h, dimethylnaphthalene[2]
Rangia cuneata	2.3 based on wet weight	Conc in water = 0.84 ppm, 24 h[2]
Rangia cuneata	26.7 based on wet weight	Conc in water = 0.03 ppm, 24 h, trimethylnaphthalene[2]
Rangia cuneata	6.1 based on wet weight	Estuarine, water = 0.071 ppm, tissue = 0.43+-0.1 ppm[3]
Rangia cuneata	8.1 based on wet weight	Conc in water = 0.48 ppm, 24 h, 2-methylnaphthalene[2]
Rangia cuneata	8.5 based on wet weight	Conc in water = 0.34 ppm, 24 h, 1-methylnaphthalene[2]

Source: 1.Roubal, W.T., D.H. Bovee, T.K. Collier and S.I. Stranahan. 1977. Proc of 1977 Oil Spill Conf. (Prevention, Behavior, Control, Cleanup). Wash DC. American Petroleum Institute. 551-556.
2.Neff, J.M., B.A. Lox, D. Dixit and J.W. Anderson. 1976. Mar. Biol. 38(3), pp 279-289.
3.Neff, J.M., Anderson, Cox, Laughlin, Rossi and Tatem. 1976. In: Sources, Effects and Sinks of Hydrocarbon in the Aquatic Environment. Wash DC. Amer. Inst. of Biological Science.

5-245 Concentration factor Nb

Species	Value	Condition
Brown algae	725 based on wet weight	Range = 450 - 1000[1]
Fish, muscle	180 based on wet weight	Danube, 1973, nuclear power plant since 1967, Nb-95, rg = +- 40 - 460[2]
Plants	7640 based on dry weight	Algae, bryophytes and vascular hydrophytes, freshwater, 32 sp[3]
Sediment	1850 based on wet weight	Danube, 1973, nuclear power plant since 1967, Nb-95, rg = +- 710 -[2]
Suspended solids	870 based on wet weight	Danube, 1973, nuclear power plant since 1967, Nb-95, rg = +- 80 -[2]
Water plant	100 based on wet weight	Danube, 1973, nuclear power plant since 1967, Nb-95, rg = +- 20 - 250[2]

Source: 1.Mauchline, J. and W.L Templeton. 1964. Oceanogr. Mar. Biol. Ass. Rev., 2, pp 229.
2.Herrmann, H., M. Ruf, K. Hubel and W. Lunsmann. 1975. Impacts of Nuclear Releases into the Aquatic Environment. IAEA-SM-198/14. pp 461-470. Vienna.
3.Timofeeva-Resovskaya, E.A., T-R. N.V. and E.A. Gileva. 1961. Dokl. Akad. Nauk. SSSR, 140, pp 1437.

5-246 Concentration factor Ni

Species	Value	Condition
Agarum cribrosum	320 based on dry weight	Seawater = 7.6E-7 g/l, Vostok Bay, Sea of Japan[1]
Atmospheric particulates	1.5	Low level marine part, relative to earths crust, Fe standard, F200[2]
Atmospheric particulates	12	US urban air, relative to earths crust, Fe standard, F200[3]
Brown algae	140 based on wet weight	None[4]
Brown algae	500 based on wet weight	None[5]
Chondrus yendoi	20 based on dry weight	Seawater = 7.6E-7 g/l, Vostok Bay, Sea of Japan[1]
Chorda filum	50 based on dry weight	Seawater = 7.6E-7 g/l, Vostok Bay, Sea of Japan[1]
Codium yessoensis	30 based on dry weight	Seawater = 7.6E-7 g/l, Vostok Bay, Sea of Japan[1]
Costaria costata	110 based on dry weight	Seawater = 7.6E-7 g/l, Vostok Bay, Sea of Japan[1]

Enteromorpha prolifera	30 based on dry weight	Seawater = 7.6E-7 g/l, Vostok Bay, Sea of Japan[1]
Fish	235 based on wet weight	Marine organisms, typical value[6]
Fish	85 based on wet weight	Freshwater organisms, typical value[7]
Fucus vesiculosus	2.8E3 based on dry weight	18 obs., all areas around the British Isles[8]
Heterochordaria abietina	40 based on dry weight	Seawater = 7.6E-7 g/l, Vostok Bay, Sea of Japan[1]
Laminaria chichorioides	60 based on dry weight	Seawater = 7.6E-7 g/l, Vostok Bay, Sea of Japan[1]
Laminaria japonica	100 based on dry weight	Seawater = 7.6E-7 g/l, Vostok Bay, Sea of Japan[1]
Macrophytes	1050 based on wet weight	Marine organisms, typical value[9]
Molluscs	4000 based on wet weight	Marine organisms, typical value[10]
Molluscs	650 based on wet weight	Freshwater organisms, typical value[11]
Patella vulgata	3.6E3 based on dry weight	17 obs., all areas around the British Isles[8]
Pelvetia wrightii	30 based on dry weight	Seawater = 7.6E-7 g/l, Vostok Bay, Sea of Japan[1]
Phyllospadix iwatensis	40 based on dry weight	Seawater = 7.6E-7 g/l, Vostok Bay, Sea of Japan[1]
Phytoplankton	570 based on wet weight	Marine organisms, typical value[12]
Plankton	1700 based on wet weight	None[4]
Porphyra umbilicalis	1.1E3 based on dry weight	20 obs., all areas around the British Isles[8]
Ptilota filicina	200 based on dry weight	Seawater = 7.6E-7 g/l, Vostok Bay, Sea of Japan[1]
Rhodomela larix	350 based on dry weight	Seawater = 7.6E-7 g/l, Vostok Bay, Sea of Japan[1]
Ulva fenestrata	200 based on dry weight	Seawater = 7.6E-7 g/l, Vostok Bay, Sea of Japan[1]
Ulvaria spendens	780 based on dry weight	Seawater = 7.6E-7 g/l, Vostok Bay, Sea of Japan[1]
Zooplankton	2500 based on wet weight	Seawater concentration 0.002 mg/m3[13]
Zooplankton	560 based on wet weight	Marine organisms, typical value[12]
Zostera asiatica	230 based on dry weight	Seawater = 7.6E-7 g/l, Vostok Bay, Sea of Japan[1]

Source: 1.Saenko, G.N., M.D. Koryakov, V.F. Makienko and I.G. Dobrosmy. 1976. Mar. biol., 34(2), pp 169-176.
2.Zoller, W.H., E.S. Gladey and R.A. Duce. 1974. Science 183, pp 200.
3.Foestner, U. and G.T.W. Wittmann. 1979. Metal Pollution in the Aquatic Environment. Springer-Verlag.Berlin.
4.Bowen, H.J.M. 1966. Trace Elements in Biochemistry, Ac. Press.
5.Mauchline, J. and W.L Templeton. 1964. Oceanogr. Mar. Biol. Ass. Rev., 2, pp 229.
6.Goldberg, E.D. 1962. Limnol Oceanogr., 7, pp 72.
7.Merlini, M., C. Rigliocca, A. Berg and G. Pozzi. 1970. IAEA/ SM, 142/27.
8.Preston, A., D.R. Jefferies , J.W.R. Harvey and A.K. Steele. 1972. Environ. Pollut., 3, pp 183-214.
9.Haegerhaell, B. 1973. Bot. Mar., 16, p 53.
10.Pringle, B.H., D.E. Hissong, E.L. Katz and S.T. Mulawka. 1968. J. San. Eng. Div. Am. Soc. Civ. Eng., 94(5970), pp 455-475.
11.Mathis, B.J. and T.F. Cummins. 1973. Water Pollut. Control 45 pp 1573.
12.Martin, J.H. and G.A. Knauer. 1973. Geochim. Cosmochim. Acta., 37, p 1639.
13.Martin, J.H 1970. Limnol. Oceanogr. 15(5), pp 756-761.

5-247 Concentration factor Organochlorines

Species	Value	Condition
Aquatic plant	101000 based on wet weight	Conc in water = 0.30 ng/l[1]
Aquatic plant	2222 based on wet weight	Conc in water = 0.45 ng/l[1]
Aquatic plant	3143 based on wet weight	Conc in water = 0.35 ng/l[1]
Aquatic plant	3478 based on wet weight	Conc in water = 0.23 ng/l[1]

Source: 1.Edwards, C.A. 1975. Persistent Pesticides in the Environment, 2nd edition, CRC-Press.

5-248 Concentration factor P

Species	Value	Condition
Brassica oleracea	1600 based on wet weight	Conc in soil solution = 0.70 ppm[1]
Brown algae	10000 based on wet weight	None[1]
Brown algae	10000 based on wet weight	None[2]
Plankton	15000 based on wet weight	None[1]
Plants	5480 based on dry weight	Algae, bryophytes and vascular hydrophytes, freshwater, 32 sp[3]
Zostera marina	1011 based on dry weight	Leaf base, 3000 min, added to lower compartment[4]
Zostera marina	1806 based on dry weight	Leaf tip, light, 3000 min, added to upper compartment[4]
Zostera marina	213 based on dry weight	Leaf tip, dark, 3000 min, added to lower compartment[4]
Zostera marina	215 based on dry weight	Leaf base, dark, 3000 min, added to lower compartmen[4]
Zostera marina	382 based on dry weight	Leaf middle, light, 3000 min, added to upper compartment[4]
Zostera marina	455 based on dry weight	Leaf base, dark, 3000 min, added to upper compartment[4]
Zostera marina	47 based on dry weight	Stem, dark, 3000 min, added to upper compartment[4]
Zostera marina	4823 based on dry weight	Root, light, 3000 min, added to upper compartment[4]
Zostera marina	743 based on dry weight	Root, 3000 min, light, added to lower compartment[4]
Zostera marina	87 based on dry weight	3000 min, rhizome, light, added to lower compartment[4]
Zostera marina	96 based on dry weight	Stem, 3000 min, dark, added to lower compartment[4]

Source: 1.Bowen, H.J.M. 1966. Trace Elements in Biochemistry, Ac. Press.
2.Mauchline, J. and W.L Templeton. 1964. Oceanogr. Mar. Biol. Ass. Rev., 2, pp 229.
3.Timofeeva-Resovskaya, E.A., T-R. N.V. and E.A. Gileva. 1961. Dokl. Akad. Nauk. SSSR, 140, pp 1437.
4.McRoy, C.P. and R.J. Barsdate. 1970. Limnol. Oceanogr. 15(1) pp 6-13.

5-249 Concentration factor PCB

Species	Value	Condition
Ambloplites rupestris	117E3 based on wet weight	Water = 3.0 ppb, ratio tissue/sed = 0.12 on ww, Hudson River[1]
Chaoborus punctipennis	22000 based on dry weight	Water = 1.3 ppb, 1 d, 294 K, pH = 7.4, uptake from water, no food[2]
Chaoborus punctipennis	23000 based on dry weight	Water = 1.3 ppb, 4 d, 294 K, pH = 7.4, uptake from water, no food[2]
Chaoborus punctipennis	23800 based on dry weight	Water = 1.3 ppb. 7 d, 294 K, pH = 7.4, uptake from water, no food[2]
Chaoborus punctipennis	24800 based on dry weight	Water = 1.3 ppb, 14 d, 294 K, pH = 7.4, uptake from water, no food[2]
Corydalus cornutus	1400 based on dry weight	Water = 1.1 ppb, 1 d, 294 K, pH = 7.4, uptake from water, no food[2]
Corydalus cornutus	4600 based on dry weight	Water = 1.1 ppb, 4 d, 294 K, pH = 7.4, uptake from water, no food[2]
Corydalus cornutus	5700 based on dry weight	Water = 1.1 ppb, 7 d, 294 K, pH = 7.4, uptake from water, no food[2]
Corydalus cornutus	6600 based on dry weight	Water = 1.1 ppb, 14 d, 294 K, pH = 7.4, uptake from water, no food[2]
Corydalus cornutus	6800 based on dry weight	Water = 1.1 ppb, 21 d, 294 K, pH = 7.4, uptake from water, no food[2]
Culex tarsalis	12600 based on dry weight	Water = 1.5 ppb, 1 d, 294 K, pH = 7.4, uptake from water, no food[2]
Culex tarsalis	18000 based on dry weight	Water = 1.5 ppb, 4 d, 294 K, pH = 7.4, uptake from water, no food[2]

Culex tarsalis	20000 based on dry weight	Water = 1.5 ppb, 7 d, 294 K, pH = 7.4, uptake from water, no food[2]
Daphnia magna	24700 based on dry weight	Water = 1.1 ppb, 1 d, 294 K, pH = 7.4, uptake from water, no food[2]
Daphnia magna	47000 based on dry weight	Water = 1.1 ppb, 4 d, 294 K, pH = 7.4, uptake from water, no food[2]
Gammarus pseudolimnaeus	17000 based on dry weight	Water = 1.6 ppb, 1 d, 294 K, pH = 7.4, uptake from water, no food[2]
Gammarus pseudolimnaeus	24000 based on dry weight	Water = 1.6 ppb, 4 d, 294 K, pH = 7.4, uptake from water, no food[2]
Gammarus pseudolimnaeus	26000 based on dry weight	Water = 1.6 ppb, 7 d, 294 K, pH = 7.4, uptake from water, no food[2]
Gammarus pseudolimnaeus	27000 based on dry weight	Water = 1.6 ppb, 21 d, 294 K, pH = 7.4, uptake from water, no food[2]
Gammarus pseudolimnaeus	27500 based on dry weight	Water = 1.6 ppb, 14 d, 294 K, pH = 7.4, uptake from water, no food[2]
Mytilus spp.	190000 based on wet weight	Seawater = 1.2 ng/l, Calif, Royal Palms, 19/9 72[3]
Mytilus spp.	270000 based on wet weight	Seawater = 0.94 ng/m3, France, Valras-Plage, 9/7 74[3]
Mytilus spp.	300000 based on wet weight	Seawater = 0.68 ng/m3, France, Carro, 11/7 74[3]
Mytilus spp.	470000 based on wet weight	Seawater = 0.95 ng/m3, France, Stes-Maries, 10/7 74[3]
Mytilus spp.	69000 based on wet weight	Seawater = 0.98 ng/m3, Calif, Frisco Bay, 1975[3]
Mytilus spp.	690000 based on wet weight	Seawater = 1.6 ng/m3, France, Marseille, 12/7 74[3]
Notropis cornutus	7.0E3 based on wet weight	Water = 1.0 ppb, ratio tissue/sed = 0.98 on ww, Hudson River[1]
Notropis cornutus	78E3 based on wet weight	Water = 1.0 ppb, ratio tissue/sed = 4.09 on ww, Hudson River[1]
Orconectes nais	1700 based on dry weight	Water = 1.2 ppb, 4 d, 294 K, pH = 7.4, uptake from water, no food[2]
Orconectes nais	3400 based on dry weight	Water = 1.2 ppb, 7 d, 294 K, pH = 7.4, uptake from water, no food[2]
Orconectes nais	4500 based on dry weight	Water = 1.2 ppb, 14 d, 294 K, pH = 7.4, uptake from water, no food[2]
Orconectes nais	5100 based on dry weight	Water = 1.2 ppb, 21 d, 294 K, pH = 7.4, uptake from water, no food[2]
Orconectes nais	570 based on dry weight	Water = 1.2 ppb, 1 d, 294 K, pH = 7.4, uptake from water, no food[2]
Palaemonetes kadiakensis	10300 based on dry weight	Water = 1.3 ppb, 1 d, 294 K, pH = 7.4, uptake from water, no food[2]
Palaemonetes kadiakensis	12300 based on dry weight	Water = 1.3 ppb, 4 d, 294 K, pH = 7.4, uptake from water, no food[2]
Palaemonetes kadiakensis	13700 based on dry weight	Water = 1.3 ppb, 7 d, 294 K, pH = 7.4, uptake from water, no food[2]
Palaemonetes kadiakensis	14200 based on dry weight	Water = 1.3 ppb, 14 d, 294 K, pH = 7.4, uptake from water, no food[2]
Palaemonetes kadiakensis	16600 based on dry weight	Water = 1.3 ppb, 21 d, 294 K, pH = 7.4, uptake from water, no food[2]
Palaemonetes pugio	11000 based on wet weight	Aroclor 1254, flowing seawat = 2.3 ug/l, s = 2.6 %, 7 d, 290-301 K[4]
Palaemonetes pugio	14000 based on wet weight	Seawater = 1.3 ug/l, 290-301 K, 16 d, mortality = 40 %[4]
Palaemonetes pugio	3000 based on wet weight	Aroclor 1254, flowing seawat = 5.3 ug/l, s = 2.6 %, 7 d, 290-301 K[4]
Palaemonetes pugio	3200 based on wet weight	Aroclor 1254, flowing seawat = 1.0 ug/l, s = 2.6 %, 7 d, 290-301 K[4]
Palaemonetes pugio	3700 based on wet weight	Seawater = 12.5 ug/l, 290-301 K, 16 d, mortality = 55 %[4]
Palaemonetes pugio	4800 based on wet weight	Aroclor 1254, flowing seawat = 3.2 ug/l, s = 2.6 %, 7 d, 290-301 K[4]
Palaemonetes pugio	5600 based on wet weight	Aroclor 1254, flowing seawat = 5.2 ug/l, s = 2.6 %, 7 d, 290-301 K[4]

Palaemonetes pugio	5700 based on wet weight	Aroclor 1254, flowing seawat = 5.3 ug/l, s = 2.6 %, 7 d, 290-301 K[4]
Palaemonetes pugio	6700 based on wet weight	Seawater = 4.0 ug/l, 290-301 K, 16 d, mortality = 45 %[4]
Palaemonetes pugio	7000 based on wet weight	Aroclor 1254, flowing seawat = 2.7 ug/l, s = 2.6 %, 7 d, 290-301 K[4]
Palaemonetes pugio	7100 based on wet weight	Aroclor 1254, flowing seawat = 9.1 ug/l, s = 2.6 %, 7 d, 290-301 K[4]
Palaemonetes pugio	7100 based on wet weight	Seawater = 9.1 ug/l, 290-301 K, 7 d, mortality = 60 %[4]
Palaemonetes pugio	7600 based on wet weight	Aroclor 1254, flowing seawat = 0.17 ug/l, s = 2.6 %, 7 d, 290-301 K[4]
Palaemonetes pugio	7600 based on wet weight	Seawater = 0.17 ug/l, 290-301 K, 7 d, mortality = 8 %[4]
Palaemonetes pugio	8100 based on wet weight	Aroclor 1254, flowing seawat = 3.2 ug/l, s = 2.6 %, 7 d, 290-301 K[4]
Palaemonetes pugio	8700 based on wet weight	Aroclor 1254, flowing seawat = 0.62 ug/l, s = 2.6 %, 7 d, 290-301 K[4]
Palaemonetes pugio	8700 based on wet weight	Seawater = 0.62 ug/l, 290-301 K, 7 d, mortality = 4 %[4]
Physa sp.	1.9E3 based on wet weight	Water = 1.0 ppb, ratio tissue/sed = 3.63 on ww, Hudson River[1]
Physa sp.	27E3 based on wet weight	Water = 1.0 ppb, ratio tissue/sed = 11.82 on ww, Hudson River[1]
Physa sp.	45E3 based on wet weight	Water = 2.2 ppb, ratio tissue/sed = 6.52 on ww, Hudson River[1]
Pteronarcys dorsata	2100 based on dry weight	Water = 2.8 ppb, 1 d, 294 K, pH = 7.4, uptake from water, no food[2]
Pteronarcys dorsata	2500 based on dry weight	Water = 2.8 ppb, 4 d, 294 K, pH = 7.4, uptake from water, no food[2]
Pteronarcys dorsata	2800 based on dry weight	Water = 2.8 ppb, 21 d, 294 K, pH = 7.4, uptake from water, no food[2]
Pteronarcys dorsata	2800 based on dry weight	Water = 2.8 ppb, 7 d, 294 K, pH = 7.4, uptake from water, no food[2]
Pteronarcys dorsata	2900 based on dry weight	Water = 2.8 ppb, 14 d, 294 K, pH = 7.4, uptake from water, no food[2]
Salmo salar	0.358 based on wet weight	Juvenile, uptake from food, 48 h, 5.29 g fish, aroclor 1254[5]
Salmo salar	282 based on wet weight	Juvenile, uptake from water, 48 h, 5.29 g fish, aroclor 1254[5]
Yellow perch	17.0E3 based on wet weight	Water = 1.0 ppb, ratio tissue/sed = 0.41 on ww, Hudson River[1]

Source: 1.Nadeau, P.J. and R.A. Davis. 1976. Bull. Environ. Contam. Toxicol. 16(4), pp 436-444.
2.Sanders, H.O. and J.H. Chandler. 1972. Bull. Environ. Contam. Toxicol. 7(5), pp 257-263.
3.Risebrough, R.W., B.W. De Lappe and T.T. Schmidt. 1976. Mar. Pollut. Bull. 7(12), pp 225-228.
4.Nimmo, D.R., J. Forester, P.T. Heitmuller and G.H. Cook. 1974. Bull. Environ. Contam., 11, pp 303-30.
5.Zitko, V. 1977. Bull. Environ. Contam. Toxicol. 17(3), pp 285-292.

5-250 Concentration factor
Pb

Species	Value	Condition
Atmospheric particulates	2300	N Atlantic Westerlies, relative to earths crust, Al standard, F200[1]
Atmospheric particulates	2300	US urban air, relative to earths crust, Fe standard, F200[2]
Atmospheric particulates	2500	South Pole, relative to earths crust, Al standard, F200[3]
Atmospheric particulates	49	Low level marine part, relative to earths crust, Fe standard, F200[4]
Blidingia minima	5.45E4 based on dry weight	Rel to dissolved metal concentration in seawater[5]
Brown algae	70000 based on wet weight	None[6]
Cladophora glomerata	16E3 based on dry weight	Water from Deadman Bay, Lake Ontario = 4.1+-1.9 ng/ml (SD)[7]
Cladophora glomerata	20E3 based on dry weight	Water from Main Duck Isl, Lake Ontario = 0.5+-0.3 ng/ml (SD)[7]
Crassostrea gigas	75 based on wet weight	Big Reef, water conc = 1.3 mg/m3[8]
Enteromorpha linza	3.25E4 based on dry weight	Rel to dissolved metal concentration in seawater, rg = 2.0 - 4.5E4[5]

Fucus vesiculosus	0.087E4 based on dry weight	Rel to dissolved metal concentration in seawater[5]
Fucus vesiculosus	0.23E4 based on dry weight	Rel to dissolved metal concentration in seawater[5]
Fucus vesiculosus	0.24E4 based on dry weight	Rel to dissolved metal concentration in seawater[5]
Fucus vesiculosus	1.85E4 based on dry weight	Rel to dissolved metal concentration in seawater, rg = 1.3 - 2.4E4[5]
Fucus vesiculosus	2.4E3 based on dry weight	13 obs., all areas around the British Isles[9]
Longfin smelt	75 based on wet weight	Size = 4 - 12 cm, Lake Washington, water conc = 2 mg/m3, muscle[8]
Mytilus edulis	150 based on wet weight	Big Reef, water conc = 1.3 mg/m3[8]
Mytilus galloprovincialis	323 based on wet weight	Water = 6.40 ppm[10]
Patella vulgata	5.5E3 based on dry weight	12 obs, all areas around the British Isles[9]
Plankton	41000 based on wet weight	None[6]
Porphyra umbilicalis	2.E3 based on dry weight	15 obs, all areas around the British Isles[9]
Ptychocheilus oregonensis	1700 based on wet weight	Size = 30 cm, Lake Washington, water conc = 2 mg/m3, muscle[8]
Ptychocheilus oregonensis	650 based on wet weight	Size = 50 cm, Lake Washington, water conc = 2 mg/m3, muscle[8]
Sockeye salmon	65 based on wet weight	Size = 5 - 17 cm, Lake Washington, water conc = 2 mg/m3, muscle[8]
Ulva sp.	3.3E4 based on dry weight	Rel to dissolved metal concentration in seawater, rg = 1.7 - 4.9E4[5]
Zooplankton	1500 based on wet weight	Lake Washington, water conc = 2 mg/m3[8]
Zooplankton	197000 based on wet weight	Seawater concentration 0.000003 mg/m3[11]

Source: 1.Chester, R. and J.H. Stoner. 1973. Nature (london) 245, pp 27-28.
2.Foestner, U. and G.T.W. Wittmann. 1979. Metal Pollution in the Aquatic Environment. Springer-Verlag.Berlin.
3.Duce, R.A., J. Quinn, C.M. Olney, S. Piotrowits, B. Ray and Wade.1972. Science 176, pp 161-163.
4.Zoller, W.H., E.S. Gladey and R.A. Duce. 1974. Science 183, pp 200.
5.Seelinger, U. and P. Edwards. 1977. Mar. Pollut. Bull. 8(1), pp 16-19.
6.Bowen, H.J.M. 1966. Trace Elements in Biochemistry, Ac. Press.
7.Keeney, W.L., W.G. Breck , G.W. Vanloon and J.A. Page. 1976. Water Res., 10(11)., pp 981-984.
8.Schell, W.R. and R.S. Barnes. 1974. in: Rubin, A.J. (ed.): Aqueous-Environmental Chemistry of Metal. pp 155-158.
9.Preston, A., D.R. Jefferies , J.W.R. Harvey and A.K. Steele. 1972. Environ. Pollut., 3, pp 183-214.
10.Majori, L. and F. Petronio. 1973. Rev. Int. Oceanogr. Med. CERBOM., 31-32, pp 55-90.
11.Martin, J.H 1970. Limnol. Oceanogr. 15(5), pp 756-761.

5-251 Concentration factor Phenanthrene

Species	Value	Condition
Macoma inquinata	0.20 based on sediment concentration	Sediment=2000 ug prudhoe B crude oil/g, 7 d, sed=0.49 ppm, water=3.7 ppb[1]
Macoma inquinata	10.3 based on wet weight	Sediment=2000 ug prudhoe B crude oil/g, 7 d, sed=0.49 ppm, water=3.7 ppb[1]
Rangia cuneata	32.0 based on wet weight	Estuarine, water = 0.089 ppm, tissue = 2.85+-1.1 ppm[2]

Source: 1.Roesijadi, G., J.W. Anderson and J.W. Blaylock. 1978. J. Fish. Res. bd Canada 35, 608-614.
2.Neff, J.M., Anderson, Cox, Laughlin, Rossi and Tatem. 1976. In: Sources, Effects and Sinks of Hydrocarbon in the Aquatic Environment. Wash DC. Amer. Inst. of Biological Science.

5-252 Concentration factor Photo dieldrin

Species	Value	Condition
Ankistrodesmus amalloides	3800 based on wet weight	60 min, 100 cells/ml, water conc = 0.72 ppb[1]
Ankistrodesmus amalloides	4700 based on wet weight	180 min, 100 cells/ml, water = 0.72 ppb[1]
Ankistrodesmus amalloides	4700 based on wet weight	30 min, 100 cells/ml, water conc = 0.72 ppb[1]
Ankistrodesmus amalloides	4900 based on wet weight	120 min, 100 cells/ml, water = 0.72 ppb[1]

Source: 1.Neudorf, S. and M.A.Q. Khan. 1975. Bull. Environ. Contam. Toxicol. 31, pp 443-450.

5-253 Concentration factor Po

Species	Value	Condition
Algae	322 based on wet weight	Ecalgrain Bay, marine, primary producers[1]
Ascidians	785 based on wet weight	Ecalgrain Bay, marine, primary consumers[1]
Asterina gibbosa	452 based on wet weight	Ecalgrain Bay, marine, secondary consumers[1]
Balanus balanoides	503 based on wet weight	Ecalgrain Bay, marine, primary consumers[1]
Benthic motile predators	47 based on wet weight	Ecalgrain Bay, marine, tertiary consumers[1]
Burrowing predaceous species	315 based on wet weight	Ecalgrain Bay, marine, secondary consumers[1]
Celery	2.9E-5 based on dry weight	Based on dry plant and dry soil conc at end of exp., mature crop,[2]
Corn	0.35E-5 based on dry weight	Based on dry plant and dry soil conc at end of exp., mature crop,[2]
Fucus serratus	523 based on wet weight	Ecalgrain Bay, marine, primary producers[1]
Lichens	3756 based on wet weight	Ecalgrain Bay, marine, primary producers[1]
Littorina littoralis	205 based on wet weight	Ecalgrain Bay, marine, primary consumers[1]
Motile predaceous species	63 based on wet weight	Ecalgrain Bay, marine, secondary consumers[1]
Nucella lapillus	70 based on wet weight	Ecalgrain Bay, marine, secondary consumers[1]
Onions	5.1E-5 based on dry weight	Based on dry plant and dry soil conc at end of exp., mature crop,[2]
Pea	3.5E-5 based on dry weight	Based on dry plant and dry soil conc at end of exp., mature crop,[2]
Pelagic predators	8 based on wet weight	Ecalgrain Bay, marine, tertiary consumers[1]
Potato	62E-5 based on dry weight	Based on dry plant and dry soil conc at end of exp., mature crop,[2]
Primary consumers	772 based on wet weight	Ecalgrain Bay, marine, rg = 48 - 1495[1]
Radish	58E-5 based on dry weight	Based on dry plant and dry soil conc at end of exp., mature crop,[2]
Secondary consumers	245 based on wet weight	Ecalgrain Bay, marine, rg = 38 - 452[1]
Sedentary herbivorous species	119 based on wet weight	Ecalgrain Bay, marine, primary consumers[1]
Sedentary limnivorous species	119 based on wet weight	Ecalgrain Bay, marine, primary consumers[1]
Sedentary predaceous species	261 based on wet weight	Ecalgrain Bay, marine, secondary consumers[1]
Sesile macrophagous species	165 based on wet weight	Ecalgrain Bay, marine, primary consumers[1]
Sesile microphagous species	1082 based on wet weight	Ecalgrain Bay, marine, primary consumers[1]
Spinach	80E-5 based on dry weight	Based on dry plant and dry soil conc at end of exp., mature crop,[2]

Sponges	1370 based on wet weight	Ecalgrain Bay, marine, primary consumers[1]
Tertiary consumers	37 based on wet weight	Ecalgrain Bay, marine, rg = 1 - 73[1]
Tomato	0.12E-5 based on dry weight	Based on dry plant and dry soil conc at end of exp., mature crop,[2]
Wheat	0.9E-5 based on dry weight	Based on dry plant and dry soil conc at end of exp., mature crop,[2]

Source: 1.Guary, J.C. and A. Fraizier. 1977. Health Phys., 32(1), pp 21-28.
2.Watters, R.L. and W.R. Hansen. 1970. Health Physics. 18(4), pp. 409-413.

5-254 Concentration factor
Polybrominated biphenyl

Species	Value	Condition
Salmo salar	1.0 based on wet weight	Juvenile, uptake from food, 48 h, 5.29 g fish, bp-6[1]
Salmo salar	48 based on wet weight	Juvenile, uptake from water, 48 h, 5.29 g fish, bp-6[1]

Source: 1.Zitko, V. 1977. Bull. Environ. Contam. Toxicol. 17(3), pp 285-292.

5-255 Concentration factor
Pu

Species	Value	Condition
Algae	10000 based on wet weight	At Rocky Flats plant, USA, Pu-239[1]
Barley, grain	0.2E-6 based on dry plant and dry soil	Nevada, lab. site 11, 239 and 249-Pu, 293-301 K[2]
Barley, grain	0.4E-6 based on dry plant and dry soil	Nevada, lab. site 13, 239 and 249-Pu, 293-301 K[2]
Barley, lower vegetation	63E-6 based on dry plant and dry soil	Nevada, lab. site 11241 am, 293-301 K[2]
Barley, lower vegetation	9.2E-6 based on dry plant and dry soil	Nevada, lab. site 13, 239 and 249-Pu, 293-301 K[2]
Barley, root	110000 based on dry plant and dry soil	Soil = 0.05 u ci/g soil, ritzville silt loam[3]
Barley, root	36000 based on dry plant and dry soil	Soil = 10 u ci/g soil, ritzville silt loam[3]
Barley, root	60000 based on dry plant and dry soil	Soil = 0.5 u ci/g soil, ritzville silt loam[3]
Barley, shoots	20000 based on dry plant and dry soil	Soil = 0.5 u ci/g soil, ritzville silt loam, rg = 19000 - 21000[3]
Barley, shoots	35000 based on dry plant and dry soil	Soil = 0.05, rg = 14000 - 56000[3]
Barley, shoots	5500 based on dry plant and dry soil	Soil = 10 u ci/g soil, ritzville silt loam[3]
Barley, upper vegetation	23E-6 based on dry plant and dry soil	Nevada, lab. site 11, 239 and 249-Pu, 293-301 K[2]
Barley, upper vegetation	8.8E-6 based on dry plant and dry soil	Nevada, lab. site 13, 239 and 249-Pu, 293-301 K[2]
Bass, backbone	160 based on wet weight	Marine vertebrate, Pu-239[4]
Bass, gut	36 based on wet weight	Marine vertebrate, Pu-239[4]
Bass, liver	14 based on wet weight	Marine vertebrate, Pu-239[4]
Bass, muscle	4 based on wet weight	Marine vertebrate, Pu-239[4]
Blue fin tuna, backbone	90 based on wet weight	Marine vertebrate, Pu-239[4]

Blue fin tuna, muscle	2 based on wet weight	Marine vertebrate, Pu-239[4]
Blue fish, backbone	570 based on wet weight	Marine vertebrate, Pu-239[4]
Blue fish, liver	26 based on wet weight	Marine vertebrate, Pu-239[4]
Blue fish, muscle	5 based on wet weight	Marine vertebrate, Pu-239[4]
Blue shark, backbone	21 based on wet weight	Marine vertebrate, Pu-239[4]
Blue shark, liver	175 based on wet weight	Marine vertebrate, Pu-239[4]
Blue shark, muscle	4 based on wet weight	Marine vertebrate, Pu-239[4]
Bonito, whole animal	3 based on wet weight	Marine vertebrate, Pu-239[4]
Bullhead minnow	4.19 based on wet weight	At Rocky Flats plant, USA, rg = 0.89 - 7.5, Pu-239[1]
Crassostrea gigas	170 based on wet weight	Water = 0.0119 p ci/l, entire animal, St Vaast-la-Hougue, France,[5]
Crassostrea gigas	210 based on wet weight	Water = 0.0119 p ci/l, shell, St Vaast-la-Hougue, France, F029[5]
Crassostrea gigas	30 based on wet weight	Water = 0.0119 p ci/l, soft parts, St Vaast-la-Hougue, France, F029[5]
Crayfish	1000 based on wet weight	At Rocky Flats plant, USA, Pu-239[1]
Crepidula fornicata	1580 based on wet weight	Water = 0.0119 p ci/l, entire animal, St Vaast-la-Hougue, France,[5]
Crepidula fornicata	2225 based on wet weight	Water = 0.0119 p ci/l, shell, St Vaast-la-Hougue, France, F029[5]
Crepidula fornicata	30 based on wet weight	Water = 0.0119 p ci/l, soft parts, St Vaast-la-Hougue, France, F029[5]
Dusky shark, backbone	49 based on wet weight	Marine vertebrate, Pu-239[4]
Ensis ensis	110 based on wet weight	Water = 0.0119 p ci/l, shell, St Vaast-la-Hougue, France, F029[5]
Ensis ensis	20 based on wet weight	Water = 0.0119 p ci/l, soft parts, St Vaast-la-Hougue, France, F029[5]
Ensis ensis	60 based on wet weight	Water = 0.0119 p ci/l, entire animal, St Vaast-la-Hougue, France,[5]
Fish	2.6 based on dry weight	Marine, rg = 0.1 - 5, 2 values used[6]
Flounder, backbone	50 based on wet weight	Marine vertebrate, Pu-239[4]
Flounder, gut	1060 based on wet weight	Marine vertebrate, Pu-239[4]
Flounder, liver	57 based on wet weight	Marine vertebrate, Pu-239[4]
Flounder, muscle	1 based on wet weight	Marine vertebrate, Pu-239[4]
Limanda limanda	30 based on wet weight	Organism from Irish Sea, Pu-239[7]
Littorina littoralis	70 based on wet weight	Water = 0.0119 p ci/l, soft parts, St Vaast-la-Hougue, France, F029[5]
Littorina littoralis	80 based on wet weight	Water = 0.0119 p ci/l, entire animal, St Vaast-la-Hougue, France,[5]
Littorina littoralis	80 based on wet weight	Water = 0.0119 p ci/l, shell, St Vaast-la-Hougue, France, F029[5]
Littorina littorea	2E3 based on wet weight	Organism from Irish Sea, Pu-239[7]
Lysmata seticaudata	127 based on wet weight	7 - 17 days, rg = 94 - 160, Pu-237[8]
Mako shark, backbone	29 based on wet weight	Marine vertebrate, Pu-239[4]
Mako shark, muscle	13 based on wet weight	Marine vertebrate, Pu-239[4]
Molluscs	286 based on dry weight	Marine[6]
Mytilus edulis	190 based on wet weight	Water = 0.0119 p ci/l, entire animal, St Vaast-la-Hougue, France,[5]

Mytilus edulis	2E3 based on wet weight	Organism from Irish Sea, Pu-239[7]
Mytilus edulis	330 based on wet weight	Water = 0.0119 p ci/l, shell, St Vaast-la-Hougue, France, F029[5]
Mytilus edulis	40 based on wet weight	Water = 0.0119 p ci/l, soft parts, St Vaast-la-Hougue, France, F029[5]
Nereis diversicolor	0.0008 conc in anim/(conc in sediment)	Dry weight in sed and anim., Bikini Atoll, n = 25, 15 d[9]
Nereis diversicolor	0.0012 conc in anim/(conc in sediment)	Dry weight in sed and anim., Bikini Atoll, n = 9, 5 d, SD = +- 0.0001[9]
Nereis diversicolor	0.0012 conc in anim/(conc in sediment)	Wet weight in sed and anim., Windscale, SD = +- 0.0001[9]
Nereis diversicolor	0.0015 conc in anim/(conc in sediment)	Wet weight in sed and anim., Bravo Crater, SD = +- 0.0001[9]
Nereis diversicolor	0.0018 conc in anim/(conc in sediment)	Dry weight in sed and anim., Windscale, n = 22, 5 d, SD = +- 0.0002[9]
Nereis diversicolor	0.0021 conc in anim/(conc in sediment)	Dry weight in sed and anim., Bikini Atoll, n = 13, 40 d[9]
Nereis diversicolor	0.0028 conc in anim/(conc in sediment)	Dry weight in sed and anim., Bikini Atoll, n = 6, 40 d[9]
Nereis diversicolor	0.0035 conc in anim/(conc in sediment)	Dry weight in sed and anim., Windscale, n = 23, 40 d, SD = +- 0.0005[9]
Nereis diversicolor	0.0035 conc in anim/(conc in sediment)	Dry weight in sed and anim., Windscale, n = 25, 15 d, SD = +- 0.0003[9]
Nereis diversicolor	0.0047 conc in anim/(conc in sediment)	Dry weight in sed and anim., Bikini Atoll, n = 10, 15 d[9]
Nereis diversicolor	0.0049 conc in anim/(conc in sediment)	Dry weight in sed and anim., Bikini Atoll, n = 31, 40 d[9]
Nereis diversicolor	0.0060 conc in anim/(conc in sediment)	Dry weight in sed and anim., Windscale, n = 1, 225 d, SD = +- 0.0010[9]
Nucella lapillus	400 based on wet weight	Water = 0.0119 p ci/l, entire animal, St Vaast-la-Hougue, France[5]
Nucella lapillus	440 based on wet weight	Water = 0.0119 p ci/l, shell, St Vaast-la-Hougue, France, F029[5]
Nucella lapillus	50 based on wet weight	Water = 0.0119 p ci/l, soft parts, St Vaast-la-Hougue, France, F029[5]
Ostrea edulis	250 based on wet weight	Water = 0.0119 p ci/l, entire animal, St Vaast-la-Hougue, France,[5]
Ostrea edulis	290 based on wet weight	Water = 0.0119 p ci/l, shell, St Vaast-la-Hougue, France, F029[5]
Ostrea edulis	40 based on wet weight	Water = 0.0119 p ci/l, soft parts, St Vaast-la-Hougue, France, F029[5]
Patella vulgata	120 based on wet weight	Water = 0.0119 p ci/l, entire animal, St Vaast-la-Hougue, France,[5]
Patella vulgata	180 based on wet weight	Water = 0.0119 p ci/l, shell, St Vaast-la-Hougue, France, F029[5]
Patella vulgata	50 based on wet weight	Water = 0.0119 p ci/l, soft parts, St Vaast-la-Hougue, France, F029[5]
Plants	1000 based on dry weight	Marine[6]
Pleuronectes platessa	30 based on wet weight	Organism from Irish Sea, Pu-239[7]
Porphyra sp.	3E3 based on wet weight	Organism from Irish Sea, Pu-239[7]
Salmo salar	1 based on wet weight	Organism from Irish Sea, Pu-239[7]
Sativum leguminosae	0.458 based on dry weight	N = 4, colloid Pu size = 0.6-1.0 u, Pu in 8 cm depth[10]
Sativum leguminosae	0.517 based on dry weight	N = 4, colloid Pu size = 0.2-0.6 u, Pu in 8 cm depth[10]
Sativum leguminosae	0.532 based on dry weight	N = 4, colloid Pu size = > 1.00 u, Pu in 8 cm depth[10]

Sativum leguminosae	0.546 based on dry weight	N = 4, colloid Pu size = 0.2-0.6 u, Pu in 2 cm depth[10]
Sativum leguminosae	0.797 based on dry weight	N = 4, colloid Pu size = 0.6-1.0 u, Pu in 2 cm depth[10]
Sativum leguminosae	2.480 based on dry weight	N = 4, colloid Pu size = > 1.00 u, Pu in 2 cm depth[10]
Sativum leguminosae	6.20E-4 based on dry weight	N = 4, colloid Pu size = > 1.00 u, Pu in 2 cm depth, unchealated[10]
Sativum leguminosae	6.98E-4 based on dry weight	N = 4, colloid Pu size = 0.6-1.0 u, Pu in 8 cm depth, unchealated,[10]
Sativum leguminosae	7.43E-4 based on dry weight	N = 4, colloid Pu size = 0.2-0.6 u, Pu in 8 cm depth, unchealated,[10]
Sediment	0.1E4 based on dry weight	Water = 0.040 p ci/l, 110 km from outfall, Irish Sea, Pu-239[7]
Sediment	0.6E4 based on dry weight	Water = 0.035 p ci/l, 75 km from outfall, Irish Sea, Pu-239[7]
Sediment	0.7E4 based on dry weight	Water = 0.05 p ci/l, 90 km from outfall, Irish Sea, Pu-239[7]
Sediment	2.0E4 based on dry weight	Water = 1.02 p ci/l, 9 km from outfall, Irish Sea, Pu-239[7]
Sediment	22E4 based on dry weight	Water = 0.37 p ci/l, 4.5 km from outfall, Irish Sea, Pu-239[7]
Sediment	2E4 based on dry weight	Water = 0.03 p ci/l, 110 km from outfall, Irish Sea, Pu-239[7]
Sediment	3.5E4 based on dry weight	Water = 0.65 p ci/l, 4.5 km from outfall, Irish Sea, Pu-239[7]
Sediment	3E4 based on dry weight	Water = 0.31 p ci/l, 9 km from outfall, Irish Sea, Pu-239[7]
Sediment	3E4 based on dry weight	Water = 0.46 p ci/l, 9 km from outfall, Irish Sea, Pu-239[7]
Sediment	7E4 based on dry weight	Water = 0.48 p ci/l, 1 km from outfall, Irish Sea, Pu-239[7]
Sediment	9E4 based on dry weight	Water = 0.31 p ci/l, 9 km from outfall, Irish Sea, Pu-239[7]
Seston	55000 based on wet weight	At Rocky Flats plant, USA, rg = 10000 - 100000, Pu-239[1]
Suspended silt	89600 based on dry weight	Seawater, Bombay, rg = 48400 - 131000, Pu-239[11]
Tautog, backbone	510 based on wet weight	Marine vertebrate, Pu-239[4]
Tautog, gut	730 based on wet weight	Marine vertebrate, Pu-239[4]
Tautog, liver	64 based on wet weight	Marine vertebrate, Pu-239[4]
Vascular plant	0.055 based on wet weight	At Rocky Flats plant, USA, rg = 0.1 - 0.01, Pu-239[1]
Venerupis decussata	120 based on wet weight	Water = 0.0119 p ci/l, entire animal, St Vaast-la-Hougue, France,[5]
Venerupis decussata	160 based on wet weight	Water = 0.0119 p ci/l, shell, St Vaast-la-Hougue, France, F029[5]
Venerupis decussata	50 based on wet weight	Water = 0.0119 p ci/l, soft parts, St Vaast-la-Hougue, France, F029[5]
Worm	0.17 conc in anim/(conc in sediment)	Wet weight in sed and anim., Mass coast[9]
Zooplankton	1000 based on wet weight	At Rocky Flats plant, USA, Pu-239[1]

Source: 1.Thompson, M.A. 1975. Impacts of Nuclear Releases into the Aquatic Environment. IAEA-SM-198/38, pp 21-225. Vienna.
2.Schultz, R.R., G.A. Tompkin, L. Leventhal and K.L. Babcock. 1976. J. Environ. Qual. 5(4), pp 406-41.
3.Wildung, R.E. and T.R. Garland. 1974. J. Agric. Food. Chem. 22(5), pp 836-838.
4.Noshkin, V.E. 1972. Health Physics. 21(6), pp 537-549.
5.Guary, J.C. and A. Fraizier. 1977. Mar. Biol. 41(3), pp 263-267.
6.Jinks, S.M. and M. Eisenbud. 1972. Radiat. Data Rep. 13, pp 243-247.
7.Hethering, J.A., D.F. Jefferies and M.B. Lovett. 1975. Impacts of Nuclear Releases into the Aquatic Environment. IAEA-SM-198/29, pp 193-210. vienna.
8.Fowler, S.W. and G. Benayoun. 1974. in: Comparative Studies of Food and Environmental Contamination.IAEA, pp 159-178. Vienna.
9.Beasley, T.M. and S.W. Fowler. 1976. Mar. Biol., 38(2), pp 95-100.
10.Lipton, W.V. and A.S. Goldin. 1976. Health Physics. 31(5), pp 425-430.
11.Pillai, K.C., N.N. Dey, E. Mathew and B.U. Kothari. 1975. Impacts of Nuclear releases into the Aquatic Environment. IAEA-SM-198/16, pp 277-299. Vienna.

5-256 Concentration factor
Ra

Species	Value	Condition
Brown algae	100 based on wet weight	None[1]
Brown algae	370 based on wet weight	None[2]
Fish, bone	510 based on wet weight	Danube, 1973, nuclear power plant since 1967, rg = 170 - 810, Ra-226[3]
Fish, muscle	70 based on wet weight	Danube, 1973, nuclear power plant since 1967, rg = 25 - 110, Ra-226[3]
Plankton	2750 based on wet weight	None[1]
Plankton	4500 based on wet weight	None[2]
Sediment	2070 based on wet weight	Danube, 1973, nuclear power plant since 1967, rg = 1390 - 2570,[3]
Suspended solids	2410 based on wet weight	Danube, 1973, nuclear power plant since 1967, rg = 480 - 10160,[3]
Water plant	190 based on wet weight	Danube, 1973, nuclear power plant since 1967, rg = 25 - 690, Ra-226[3]

Source: 1.Mauchline, J. and W.L Templeton. 1964. Oceanogr. Mar. Biol. Ass. Rev., 2, pp 229.
2.Bowen, H.J.M. 1966. Trace Elements in Biochemistry, Ac. Press.
3.Herrmann, H., M. Ruf, K. Hubel and W. Lunsmann. 1975. Impacts of Nuclear Releases into the Aquatic Environment. IAEA-SM-198/14. pp 461-470. Vienna.

5-257 Concentration factor
Rb

Species	Value	Condition
Brown algae	15 based on wet weight	None[1]
Brown algae	27.5 based on wet weight	Range = 5 - 50[2]
Crustacea	13 based on dry weight	Marine, rg = 8.9 - 17.3, 3 values used[3]
Fish	17 based on dry weight	Marine, rg = .7.6 - 23.8, 6 values used[3]
Molluscs	17 based on dry weight	Marine, rg = 11 - 27, 3 values used[3]
Phytoplankton	21000 based on wet weight	Range = 6600 - 35400[4]
Plant	1230 based on dry weight	Algae, bryophytes and vascular hydrophytes, freshwater, 32 sp[5]
Plants	23 based on dry weight	Marine, rg = 12 - 42, 3 values used[3]

Source: 1.Bowen, H.J.M. 1966. Trace Elements in Biochemistry, Ac. Press.
2.Mauchline, J. and W.L Templeton. 1964. Oceanogr. Mar. Biol. Ass. Rev., 2, pp 229.
3.Jinks, S.M. and M. Eisenbud. 1972. Radiat. Data Rep. 13, pp 243-247.
4.Cushing, C.E. and L.A. Rancitelli. 1972. Northwest Science, 46(2), pp 115-121.
5.Timofeeva-Resovskaya, E.A., T-R. N.V. and E.A. Gileva. 1961. Dokl. Akad. Nauk. SSSR, 140, pp 1437.

5-258 Concentration factor
Ru

Species	Value	Condition
Algae	150 based on wet weight	Lab value, rg = 100 - 200, Ru-106[1]
Anodonta nuttalliana, soft parts	0.4 based on wet weight	No food, 292 K, pH = 7.5, freshwat = 40 part & 3100 solub p ci/l, 7d[2]
Anodonta nuttalliana, soft parts	1.1 based on wet weight	No food, 292 K, pH = 7.5, freshwat = 330 part & 5500 solub p ci/l, 21 d[2]
Astacus sp., muscle	1 based on wet weight	No food, 292K, pH=7.5, freshwat = 40 part & 3100 solub p ci/l, 7 d[2]
Astacus sp., muscle	1 based on wet weight	No food, 292K, pH=7.5, freshwat=330 part & 5500 solub p ci/l, 21 d[2]
Astacus sp., soft parts	6 based on wet weight	No food, 292 K, pH = 7.5, freshwat = 0 part & 3600 solub p ci/l, 7 d[2]

Astacus sp., viscera	7 based on wet weight	No food, 292 K, pH = 7.5, freshwat = 330 part & 5500 solub p ci/l, 21 d[2]
Brown algae	1008 based on wet weight	Range = 15 - 2000[3]
Cancer productus, muscle	0.3 based on wet weight	No food, 285 K, pH = 8.2, seawater = 40 part & 4000 solub p ci/l, 20 d[2]
Cancer productus, sort parts	3 based on wet weight	No food, 285 K, pH = 8.2, seawater = 40 part & 4000 solub p ci/l, 20 d[2]
Cancer productus, viscera	3 based on wet weight	No food, 285 K, pH = 8.2, seawater = 40 part & 4000 solub p ci/l, 20 d[2]
Cancer productus, viscera	3 based on wet weight	No food, 285 K, pH = 8.2, seawater = 90 part & 2800 solub p ci/l, 6 d[2]
Carassius auratus, muscle	1 based on wet weight	No food, 292 K, pH = 7.5, fresh = 40 part & 3100 solub p ci/l, 7 d[2]
Carassius auratus, nonmuscle	0.8 based on wet weight	No food, 292 K, pH = 7.5, freshwat = 40 part & 3100 solub p ci/l, 7 d[2]
Carassius auratus, whole body	0.7 based on wet weight	No food, 292 K, pH = 7.5, freshwat = 330 part & 5500 solub p ci/l, 21 d[2]
Corallina officinalis	150 based on dry weight	Seawater = 1 - 2 p ci/l, rg = 100 - 200, Ru-106[4]
Crustacea	38 based on dry weight	Marine, rg = 1 - 100, 4 values used[5]
Crustacean	15.5 based on wet weight	Shell, rg = 3 - 28, Ru-106, marine[6]
Crustacean	6 based on wet weight	Hepatopancreas, rg = 2 - 10, Ru-106, marine[6]
Crustacean	7.5 based on wet weight	Gills, rg = 4 - 11, Ru-106, marine[6]
Eskmeals silt	7529.9 based on wet weight	Yearly mean, Cumberland coastal area, 1969, Ru-106[7]
Fish	1 based on wet weight	Lab value, Ru-106[1]
Fish, bone	580 based on wet weight	Danube, 1973, nuclear power plant since 1967, rg = 300 - 1430, Ru-106[8]
Fish, muscle	170 based on wet weight	Danube, 1973, nuclear power plant since 1967, rg = 40 - 270, Ru-103[8]
Fish, muscle	70 based on wet weight	Danube, 1973, nuclear power plant since 1967, rg = 15 - 170, Ru-106[8]
Gillichthys mirabilis, muscle	1 based on wet weight	No food, 285 K, pH = 8.2, sea = 40 part & 4000 solub p ci/l, 20 d[2]
Gillichthys mirabilis, muscle	1 based on wet weight	No food, 285 K, pH = 8.2, sea = 90 part & 2800 solub p ci/l, 6 d[2]
Gillichthys mirabilis, nonmuscle	0.5 based on wet weight	No food, 285 K, pH = 8.2, seawater = 40 part & 4000 solub p ci/l, 20 d[2]
Gillichthys mirabilis, nonmuscle	0.8 based on wet weight	No food, 285 K, pH = 8.2, seawater = 90 part & 2800 solub p ci/l, 6 d2
Gillichthys mirabilis, whole body	0.4 based on wet weight	No food, 285 K, pH = 8.2, seawater = 0 part & 7500 solub p ci/l, 6 d[2]
Gillichthys mirabilis, whole body	0.7 based on wet weight	No food, 285 K, pH = 8.2, seawater = 40 part & 4000 solub p ci/l, 20 d[2]
Invertebrate	35 based on wet weight	Lab value, rg = 20 - 50, Ru-106[1]
Mollusca	15.5 based on wet weight	Shell, rg = 3 - 28, Ru-106, marine[6]
Mollusca	6 based on wet weight	Hepatopancreas, rg = 2 - 10, Ru-106, marine[6]
Mollusca	7.5 based on wet weight	Gills, rg = 4 - 11, Ru-106, marine[6]
Molluscs	2.2 based on dry weight	Marine, rg = 1 - 3.6, 3 values used[5]
Mytilus edulis, soft parts	1.8 based on wet weight	No food, 285 K, pH = 8.2, seawater = 90 part & 2800 solub p ci/l, 6 d2
Nephtys sp.	507.2 based on wet weight	Yearly mean, Cumberland coastral area, 1969, Ru-106[7]

Nucula sp.	3247.8 based on wet weight	Yearly mean, Cumberland coastral area, 1969, Ru-106[7]
Plankton	1800 based on wet weight	Range = 600 - 3000[3]
Plant	1700 based on dry weight	Algae, bryophytes and vascular hydrophytes, freshwater, 32 sp[9]
Plants	448 based on dry weight	Marine, rg = 15 - 2000, 8 values used[5]
Pleuronectes platessa, gut content	663.7 based on wet weight	Yearly mean, Cumberland coastral area, 1969, Ru-106[7]
Sand	1217.0 based on wet weight	Yearly mean, Cumberland coastal area, 1969, Ru-106, inshore sand[7]
Saxidonus giganteus, soft parts	0.4 based on wet weight	No food, 285 K, pH = 8.2, seawater = 40 part & 4000 solub p ci/l, 20[2]
Sediment	1190 based on wet weight	Danube, 1973, nuclear power plant since 1967, rg = 260 - 2790, Ru-106[8]
Sediment	450 based on wet weight	Danube, 1973, nuclear power plant since 1967, rg = 40 - 1190, Ru-103[8]
Suspended silt	4860 based on dry weight	Seawater, Bombay, rg = 1750 - 8720, Ru-106[10]
Suspended solids	4730 based on wet weight	Danube, 1973, nuclear power plant since 1967, rg = 140 - 10760,[8]
Suspended solids	870 based on wet weight	Danube, 1973, nuclear power plant since 1967, rg = 360 - 2420, Ru-103[8]
Water plant	100 based on wet weight	Danube, 1973, nuclear power plant since 1967, rg = 25 - 250, Ru-103[8]
Water plant	30 based on wet weight	Danube, 1973, nuclear power plant since 1967, rg = 10 - 90, Ru-106[8]

Source: 1.Ancellin, J. and A. Vilquin. 1968. Radioprot. 3(3), pp 185-213.
2.Harrison, F.L. 1973. Health Physics, 24, pp 331-343.
3.Mauchline, J. and W.L Templeton. 1964. Oceanogr. Mar. Biol. Ass. Rev., 2, pp 229.
4.Ancellin, J. and A. Vilquin. 1966. in: Disposal of Radioactive Wastes into Seas, Ocean and Surface Water. pp 583-604. IAEA. Vienna.
5.Jinks, S.M. and M. Eisenbud. 1972. Radiat. Data Rep. 13, pp 243-247.
6.LeGall, P. and J. Ancellin. 1971. Available from Nat. Tech. Inf. Ser., Springfield, Va. as CEA-N-148, pp 1-17.
7.Pentreath, R.J. and D.F. Jefferies. 1971. J. mar. biol. Ass. U.K. 51, pp 963-976.
8.Herrmann, H., M. Ruf, K. Hubel and W. Lunsmann. 1975. Impacts of Nuclear Releases into the Aquatic Environment. IAEA-SM-198/14. pp 461-470. Vienna.
9.Timofeeva-Resovskaya, E.A., T-R. N.V. and E.A. Gileva. 1961. Dokl. Akad. Nauk. SSSR, 140, pp 1437.
10.Pillai, K.C., N.N. Dey, E. Mathew and B.U. Kothari. 1975. Impacts of Nuclear releases into the Aquatic Environment. IAEA-SM-198/16, pp 277-299. Vienna.

5-259 Concentration factor
S

Species	Value	Condition
Brassica oleracea	21 based on wet weight	Conc in soil solution = 205 ppm[1]
Brown algae	10 based on wet weight	None[2]
Brown algae	3.4 based on wet weight	None[1]
Lampsilis radiata	240 based on wet weight	91 days, freshwater, S-35[3]
Plankton	1.7 based on wet weight	None[1]
Plant	165 based on dry weight	Algae, bryophytes and vascular hydrophytes, freshwater, 32 sp[4]

Source: 1.Bowen, H.J.M. 1966. Trace Elements in Biochemistry, Ac. Press.
2.Mauchline, J. and W.L Templeton. 1964. Oceanogr. Mar. Biol. Ass. Rev., 2, pp 229.
3.Harvey, R.S. 1969. Health Physics, 17, pp 149-154.
4.Timofeeva-Resovskaya, E.A., T-R. N.V. and E.A. Gileva. 1961. Dokl. Akad. Nauk. SSSR, 140, pp 1437.

5-260 Concentration factor
Sb

Species	Value	Condition
Anodonta nuttalliana, soft parts	1.7 based on wet weight	No food, 292 K, pH = 7.5, freshwat = 0 part & 3600 solub p ci/l, 7 d[1]
Anodonta nuttalliana, soft parts	1.7 based on wet weight	No food, 292 K, pH = 7.5, freshwat = 0 part & 8800 solub p ci/l, 21d[1]
Astacus sp., muscle	1 based on wet weight	No food, 292 K, pH = 7.5, fresh = 0 part & 3600 solub p ci/l, 7 d[1]
Astacus sp., muscle	1 based on wet weight	No food, 292 K, pH = 7.5, fresh = 0 part & 8800 solub p ci/l, 21 d[1]
Astacus sp., soft parts	1 based on wet weight	No food, 292 K, pH=7.5, freshwat = 0 part & 3600 solub p ci/l, 7 d[1]
Astacus sp., soft parts	1.2 based on wet weight	No food, 292 K, pH = 7.5, freshwat = 1.7 part & 250 solub p ci/ml, 7d[1]
Astacus sp., viscera	1 based on wet weight	No food, 292 K, pH=7.5, freshwat=0 part & 3600 solub p ci/l, 7 d,[1]
Astacus sp., viscera	1.4 based on wet weight	No food, 292 K, pH = 7.5, freshwat = 0 part & 8800 solub p ci/l, 21 d[1]
Atmospheric particulates	1300	South Pole, relative to earths crust, Al standard, F200[2]
Atmospheric particulates	2800	US urban air, relative to earths crust, Fe standard, F200[3]
Atmospheric particulates	3000	N Atlantic Westerlies, relative to earths crust, Al standard, F200[4]
Cancer productus, muscle	1 based on wet weight	No food, 285 K, pH=8.2, seawater=70 part & 13000 solub p ci/l, 20 d[1]
Cancer productus, sort parts	0.7 based on wet weight	No food, 285 K, pH = 8.2, seawater = 70 part & 13000 solub p ci/l, 20 d[1]
Cancer productus, viscera	0.7 based on wet weight	No food, 285 K, pH = 8.2, seawater = 0 part & 7500 solub p ci/l, 6 d[1]
Cancer productus, viscera	0.8 based on wet weight	No food, 285 K, pH = 8.2, seawater = 70 part & 13000 solub p ci/l, 20 d[1]
Carassius auratus, muscle	1 based on wet weight	No food, 292 K, pH=7.5, freshwat = 0 part & 3600 solub p ci/l, 7 d[1]
Carassius auratus, nonmuscle	1.7 based on wet weight	No food, 292 K, pH = 7.5, freshwat = 0 part & 3600 solub p ci/l, 7 d[1]
Carassius auratus, whole body	1.5 based on wet weight	No food, 292 K, pH = 7.5, freshwat = 0 part & 8800 solub p ci/l, 21 d[1]
Fish, bone	470 based on wet weight	Danube, 1973, nuclear power plant since 1967, Sb-125[5]
Fish, bone	810 based on wet weight	Danube, 1973, nuclear power plant since 1967, rg = 350 - 1100, Sb-124[5]
Fish, muscle	100 based on wet weight	Danube, 1973, nuclear power plant since 1967, rg = 30 - 170, Sb-125[5]
Fish, muscle	90 based on wet weight	Danube, 1973, nuclear power plant since 1967, rg = 70 - 120, Sb-124[5]
Gillichthys mirabilis, muscle	1 based on wet weight	No food, 285 K, pH=8.2, seawater = 0 part & 7500 solub p ci/l, 6 d[1]
Gillichthys mirabilis, muscle	1 based on wet weight	No food, 285K, pH=8.2, seawater=70 part & 13000 solub p ci/l, 20 d[1]
Gillichthys mirabilis, nonmuscle	0.4 based on wet weight	No food, 285 K, pH = 8.2, seawater = 70 part & 13000 solub p ci/l, 20 d[1]

Gillichthys mirabilis, nonmuscle	0.9 based on wet weight	No food, 285 K, pH = 8.2, seawater = 0 part & 7500 solub p ci/l, 6 d[1]
Gillichthys mirabilis, whole body	0.4 based on wet weight	No food, 285 K, pH = 8.2, seawater = 1.1 part & 440 solub p ci/ml, 6 d[1]
Gillichthys mirabilis, whole body	0.9 based on wet weight	No food, 285 K, pH = 8.2, seawater = 70 part & 13000 solub p ci/l, 20 d[1]
Mytilus edulis, soft parts	0.7 based on wet weight	No food, 285 K, pH = 8.2, seawater = 0 part & 7500 solub p ci/l, 6 d[1]
Phytoplankton	1740 based on wet weight	Range = 280 - 3200[6]
Plankton	50 based on wet weight	None[7]
Saxidonus giganteus, soft parts	1 based on wet weight	No food, 285K, pH=8.2, seawater=70 part & 13000 solub p ci/l, 20 d[1]
Sediment	220 based on wet weight	Danube, 1973, nuclear power plant since 1967, freshwater[5]
Sediment	560 based on wet weight	Danube, 1973, nuclear power plant since 67, freshwater[5]
Suspended solids	4300 based on wet weight	Danube, 1973, nuclear power plant since 1967, rg = 480 - 28000,[5]
Suspended solids	780 based on wet weight	Danube, 1973, nuclear power plant since 1967, rg = 760 - 800, Sb-124[5]
Water plant	140 based on wet weight	Danube, 1973, nuclear power plant since 1967, rg = 60 - 330, Sb-125[5]
Water plant	30 based on wet weight	Danube, 1973, nuclear power plant since 1967, Sb-124[5]

Source: 1.Harrison, F.L. 1973. Health Physics, 24, pp 331-343.
2.Duce, R.A., J. Quinn, C.M. Olney, S. Piotrowits, B. Ray and Wade.1972. Science 176, pp 161-163.
3.Foestner, U. and G.T.W. Wittmann. 1979. Metal Pollution in the Aquatic Environment. Springer-Verlag.Berlin.
4.Chester, R. and J.H. Stoner. 1973. Nature (London) 245, pp 27-28.
5.Herrmann, H., M. Ruf, K. Hubel and W. Lunsmann. 1975. Impacts of Nuclear Releases into the Aquatic Environment. IAEA-SM-198/14. pp 461-470. Vienna.
6.Cushing, C.E. and L.A. Rancitelli. 1972. Northwest Science, 46(2), pp 115-121.
7.Mauchline, J. and W.L Templeton. 1964. Oceanogr. Mar. Biol. Ass. Rev., 2, pp 229.

5-261 Concentration factor
Sc

Species	Value	Condition
Brown algae	2050 based on wet weight	Range = 1500 - 2600[1]
Phytoplankton	527500 based on wet weight	Range = 160000 - 995000[2]

Source: 1.Mauchline, J. and W.L Templeton. 1964. Oceanogr. Mar. Biol. Ass. Rev., 2, pp 229.
2.Cushing, C.E. and L.A. Rancitelli. 1972. Northwest Science, 46(2), pp 115-121.

5-262 Concentration factor
Se

Species	Value	Condition
Atmospheric particulates	16000	N Atlantic Westerlies, relative to earths crust, Al standard, F200[1]
Atmospheric particulates	18000	South Pole, relative to earths crust, Al standard, F200[2]
Atmospheric particulates	2500	US urban air, relative to earths crust, Fe standard, F200[3]
Lysmata seticaudata, exoskeleton	30.0 based on wet weight	51 d, seawater, uptake from water, Se-75[4]
Lysmata seticaudata, eyes	15.2 based on wet weight	51 d, seawater, uptake from water, Se-75[4]
Lysmata seticaudata, muscle	17.8 based on wet weight	51 d, seawater, uptake from water, Se-75[4]
Lysmata seticaudata, viscera	25.7 based on wet weight	51 d, seawater, uptake from water, Se-75[4]

Lysmata seticaudata, whole body	21.9 based on wet weight	51 d, seawater, uptake from water, Se-75[4]
Mytilus galloprovincialis	13 based on wet weight	Wet weight = 21.8 g, uptake of se(+4), 13 d, n = 4, SD = +- 2.5, Se-75[5]
Mytilus galloprovincialis	29 based on wet weight	Wet weight = 13.2 g, uptake of se(+4), 13 d, n = 5, SD = +- 8, Se-75[5]
Mytilus galloprovincialis	46 based on wet weight	Wet weight = 2.1 g, uptake of se(+4), 13 d, n = 3, SD = +- 10, Se-75[5]
Mytilus galloprovincialis, gills	284 based on wet weight	65 d, seawater, uptake from water, Se-75[4]
Mytilus galloprovincialis, mantle	80.7 based on wet weight	65 d, seawater, uptake from water, Se-75[4]
Mytilus galloprovincialis, muscle	186 based on wet weight	65 d, seawater, uptake from water, Se-75[4]
Mytilus galloprovincialis, shell	127 based on wet weight	65 d, seawater, uptake from water, Se-75[4]
Mytilus galloprovincialis, viscera	494 based on wet weight	65 d, seawater, uptake from water, Se-75[4]
Mytilus galloprovincialis, whole body	91.2 based on wet weight	65 d, seawater, uptake from water, Se-75[4]
Phytoplankton	3200 based on wet weight	Range = 900 - 5500[6]

Source: 1.Chester, R. and J.H. Stoner. 1973. Nature (london) 245, pp 27-28.
2.Duce, R.A., J. Quinn, C.M. Olney, S. Piotrowits, B. Ray and Wade.1972. Science 176, pp 161-163.
3.Foestner, U. and G.T.W. Wittmann. 1979. Metal Pollution in the Aquatic Environment. Springer-Verlag.Berlin.
4.Fowler, S.W. and G. Benayoun. 1976. Bull. Environ. Contam. Toxicol., 16(3), pp 339-346.
5.Fowler, S.W. and G. Benayoun. 1976. Mar. Biol., 37, pp 59-68.
6.Cushing, C.E. and L.A. Rancitelli. 1972. Northwest Science, 46(2), pp 115-121.

5-263 Concentration factor
Si

Species	Value	Condition
Brown algae	120 based on wet weight	None[1]
Plankton	17000 based on wet weight	None[1]

Source: 1.Bowen, H.J.M. 1966. Trace Elements in Biochemistry, Ac. Press.

5-264 Concentration factor
Sn

Species	Value	Condition
Brown algae	92 based on wet weight	None[1]
Plankton	2900 based on wet weight	None[1]

Source: 1.Bowen, H.J.M. 1966. Trace Elements in Biochemistry, Ac. Press.

5-265 Concentration factor
Sr

Species	Value	Condition
Algae	6.0 based on wet weight	1 d, rg = 5.3 - 6.7, marine, Sr-89[1]
Brown algae	20.5 based on wet weight	Range = 1 - 40[2]
Brown algae	44 based on wet weight	None[3]
Cancer irroratus	84.2 based on wet weight	Seawater = 380 ppm, Nova Scotia, AAS, SD = +- 2.65[4]
Carp	33 based on wet weight	Semimigratory fish, Caspian and Azov Seas, Sr-90[5]
Coral	1.10 based on Ca content	Shallow coastal ocean, mean of 13 determinations, F023[6]
Coral	1.22 based on Ca content	Shallow open ocean coral, mean of 8 determinations, F023[6]

Coral	1.25 based on Ca content	Deep ocean coral, mean of 13 determinations, F023[6]
Crustacea	0.6 based on dry weight	Marine, rg = 0.1 - 1.1, 8 values used[7]
Crustacean, exoskeleton	15 based on wet weight	4 d, marine, Sr-89[1]
Detritus	1000 based on dry weight	Marine detritus, Indian Ocean, less, Sr-90[8]
Echinoderm	0.9 based on wet weight	1 d, marine, Sr-89[1]
Fish	0.43 based on dry weight	Marine, rg = 0.1 - 1.5, 9 values used[7]
Fish	14 based on dry weight	Freshwater, rg = 0.85 - 90, 13 values used[7]
Fish, bone	360 based on wet weight	Danube, 1973, nuclear power plant since 1967, rg = 19 - 1090, Sr-90[9]
Fish, bone	40 based on wet weight	Danube, 1973, nuclear power plant since 1967, rg = 10 - 60, Sr-89[9]
Fish, muscle	20 based on wet weight	Danube, 1973, nuclear power plant since 1967, rg = 10 - 60, Sr-89[9]
Fish, muscle	20 based on wet weight	Danube, 1973, nuclear power plant since 1967, rg = 5 - 40, Sr-90[9]
Kutum	2.5 based on wet weight	Migratory fish, Caspian and Azov Seas, Sr-90[5]
Molluscs	1.7 based on dry weight	Marine, rg = 0.1 - 10, 10 values used[7]
Phytoplankton	450 based on dry weight	Equilibrium after 36 h, rg = 200 - 700, freshwater, Sr-85[10]
Plankton	8 based on wet weight	None[3]
Plankton	9 based on wet weight	None[2]
Plant	475 based on dry weight	Algae, bryophytes and vascular hydrophytes, freshwater, 32 sp[11]
Plants	200 based on dry weight	Freshwater, rg = 80 - 410, 5 values used[7]
Plants	21 based on dry weight	Marine, rg = 0.2 - 82, 21 values used[7]
Roach	40 based on wet weight	Semimigratory fish, Caspian and Azov Seas, Sr-90[5]
Rooted aquatic plant	600 based on dry weight	Equilibrium after 36 h, Sr-85[10]
Sediment	120 based on dry weight	Lake Ontario, western part, freshwater, Sr-90[12]
Sediment	140 based on wet weight	Danube, 1973, nuclear power plant since 1967, freshwater[9]
Sediment	45 based on dry weight	Lake Erie, western basin, freshwater, Sr-90[12]
Sediment	80 based on dry weight	Lake Erie, central basin, freshwater, Sr-90[12]
Sediment	810 based on wet weight	Danube, 1973, nuclear power plant since 1967, freshwater[9]
Suspended silt	15.8 based on dry weight	Seawater, Bombay, rg = 4.3 - 31.0, Sr-90[13]
Suspended solids	1250 based on wet weight	Danube, 1973, nuclear power plant since 1967, rg = 1 - 8340, Sr-89[9]
Suspended solids	2460 based on wet weight	Danube, 1973, nuclear power plant since 1967, rg = 10 - 10560, Sr-90[9]
Ulva lactuca	0.32 based on wet weight	Equilibrium after 3 h, pH = 8, carrier free, Sr-90[14]
Water plant	130 based on wet weight	Danube, 1973, nuclear power plant since 1967, rg = 2 - 510, Sr-89[9]
Water plant	20 based on wet weight	Danube, 1973, nuclear power plant since 1967, rg = 3 - 50, Sr-90[9]
Zooplankton	5.1 based on wet weight	Seawater concentration 8.0 mg/m3[15]

Source: 1.Hiyama, Y. and M. Shimizu. 1964. Rec. Oceanogr. Works. Jap. 7(2), pp 43-77.
2.Mauchline, J. and W.L Templeton. 1964. Oceanogr. Mar. Biol. Ass. Rev., 2, pp 229.
3.Bowen, H.J.M. 1966. Trace Elements in Biochemistry, Ac. Press.
4.Martin, J.-L.M. 1974. Mar. biol., 28(4), pp 245-251.
5.Timoshuck, V.I. 1970. Available from Nat. Tech. Inf. Ser. Springfield, Va. as Part of Aec-tr-7299, p185-196.
6.Livingston, H.D. and G. Thompson. 1971. Limnol. Oceanogr. 16(5), pp 786-796.
7.Jinks, S.M. and M. Eisenbud. 1972. Radiat. Data Rep. 13, pp 243-247.
8.Patin, S.A. and N.I. Popov. 1966. Proc. Symp. Int. Atom. Energ. Agency, pp 443-449.
9.Herrmann, H., M. Ruf, K. Hubel and W. Lunsmann. 1975. Impacts of Nuclear Releases into the Aquatic Environment. IAEA-SM-198/14. pp 461-470. Vienna.
10.Gloyna, E.F., Y.A. Yosef and T.J. Padden. 1971. U.S. Atom. Ener. Comm. Contract AT(11-1)-490, Tech.ept. 21, CRWR-75, EHE-71-1, ORO-21, pp 1-66.

11.Timofeeva-Resovskaya, E.A., T-R. N.V. and E.A. Gileva. 1961. Dokl. Akad. Nauk. SSSR, 140, pp 1437.
12.Lerman, A. and T.A. Lietzke. 1975. Limnol. Oceanogr. 20(4), pp 497-510.
13.Pillai, K.C., N.N. Dey, E. Mathew and B.U. Kothari. 1975. Impacts of Nuclear releases into the Aquatic Environment. IAEA-SM-198/16, pp 277-299. Vienna.
14.Hampson, M.A. 1967. J. Exp. Bot. 18, pp 17-53.
15.Martin, J.H 1970. Limnol. Oceanogr. 15(5), pp 756-761.

5-266 Concentration factor TCDD

Species	Value	Condition
Algae	15000 based on wet weight	Greather than or equal to value[1]
Daphnia magna	18100 based on dry weight	Water = 0.26 ppt, soil = 0.001 ppm, 4 d, Metcalf model, cf to wat.[2]
Daphnia magna	22900 based on dry weight	Water = 7.13 ppt, soil = 0.10 ppm, 4 d, Metcalf model, cf to wat.[2]
Daphnia magna	26800 based on dry weight	Water = 0.66 ppt, soil = 0.01 ppm, 4 d, Metcalf model, cf to wat.[2]
Daphnia magna	31200 based on dry weight	Water = 239 ppt, soil = 3.17 ppm, 4 d, Metcalf model, cf to wat. conc[2]
Daphnia magna	48000 based on dry weight	Water = 0.05 ppt, soil = 0.0001 ppm, 4 d, Metcalf model, cf to wat.[2]
Daphnia magna	7800 based on dry weight	Water = 1330 ppt, soil = 7.45 ppm, 4 d, Metcalf model, cf to wat.[2]
Fish	15000 based on wet weight	Greather than or equal to value[1]
Gambusia affinis	1000 based on dry weight	Water = 1330 ppt, soil = 7.45 ppm, 4 d, Metcalf model, cf to wat.[2]
Gambusia affinis	11300 based on dry weight	Water = 48 ppt, soil = 0.53 ppm, 4 d, Metcalf model, cf to wat. conc[2]
Gambusia affinis	12900 based on dry weight	Water = 7 ppt, soil = 0.15 ppm, 4 d, Metcalf model, cf to wat. conc[2]
Gambusia affinis	22700 based on dry weight	Water = 0.26 ppt, soil = 0.001 ppm, 4 d, Metcalf model, cf to wat.[2]
Gambusia affinis	23300 based on dry weight	Water = 18 ppt, soil = 0.29 ppm, 4 d, Metcalf model, cf to wat. conc[2]
Gambusia affinis	24000 based on dry weight	Water = 0.05 ppt, soil = 0.0001 ppm, 4 d, Metcalf model, cf to wat.[2]
Gambusia affinis	61600 based on dry weight	Water = 7.13 ppt, soil = 0.10 ppm, 4 d, Metcalf model, cf to wat.[2]
Gambusia affinis	63300 based on dry weight	Water = 0.66 ppt, soil = 0.01 ppm, 4 d, Metcalf model, cf to wat.[2]
Gambusia affinis	9200 based on dry weight	Water = 239 ppt, soil = 3.17 ppm, 4 d, Metcalf model, cf to wat. conc[2]
Ictalurus punctatus	11400 based on dry weight	Water = 7 ppt, soil = 0.15 ppm, 4 d, Metcalf model, cf to wat. conc[2]
Ictalurus punctatus	14400 based on dry weight	Water = 7.13 ppt, soil = 0.10 ppm, 4 d, Metcalf model, cf to wat.[2]
Ictalurus punctatus	2000 based on dry weight	Water = 0.05 ppt, soil = 0.0001 ppm, 4 d, Metcalf model, cf to wat.[2]
Ictalurus punctatus	2300 based on dry weight	Water = 48 ppt, soil = 0.53 ppm, 4 d, Metcalf model, cf to wat. conc[2]
Ictalurus punctatus	27900 based on dry weight	Water = 0.66 ppt, soil = 0.01 ppm, 4 d, Metcalf model, cf to wat.[2]
Ictalurus punctatus	3000 based on dry weight	Water = 239 ppt, soil = 3.17 ppm, 4 d, Metcalf model, cf to wat. conc[2]
Ictalurus punctatus	4600 based on dry weight	Water = 0.26 ppt, soil = 0.001 ppm, 4 d, Metcalf model, cf to wat.[2]
Ictalurus punctatus	6700 based on dry weight	Water = 18 ppt, soil = 0.29 ppm, 4 d, Metcalf model, cf to wat. conc[2]
Insecta	15000 based on wet weight	Greather than or equal to value[1]
Lemna minor	1200 based on dry weight	Water = 0.26 ppt, soil = 0.001 ppm, 4 d, Metcalf model, cf to wat.[2]
Lemna minor	4000 based on dry weight	Water = 0.05 ppt, soil = 0.0001 ppm, 4 d, Metcalf model, cf to wat.[2]
Lemna minor	4300 based on dry weight	Water = 7.13 ppt, soil = 0.10 ppm, 4 d, Metcalf model, cf to wat.[2]
Lemna minor	5000 based on dry weight	Water = 0.66 ppt, soil = 0.01 ppm, 4 d, Metcalf model, cf to wat.[2]

Oedogonium cardiacum	10500 based on dry weight	Water = 239 ppt, soil = 3.17 ppm, 4 d, Metcalf model, cf to wat. conc[2]
Oedogonium cardiacum	11100 based on dry weight	Water = 7.13 ppt, soil = 0.10 ppm, 4 d, Metcalf model, cf to wat.[2]
Oedogonium cardiacum	12800 based on dry weight	Water = 48 ppt, soil = 0.53 ppm, 4 d, Metcalf model, cf to wat. conc[2]
Oedogonium cardiacum	18600 based on dry weight	Water = 7 ppt, soil = 0.15 ppm, 4 d, Metcalf model, cf to wat. conc[2]
Oedogonium cardiacum	2000 based on dry weight	Water = 0.05 ppt, soil = 0.0001 ppm, 4 d, Metcalf model, cf to wat.[2]
Oedogonium cardiacum	5000 based on dry weight	Water = 1330 ppt, soil = 7.45 ppm, 4 d, Metcalf model, cf to wat.[2]
Oedogonium cardiacum	5400 based on dry weight	Water = 0.26 ppt, soil = 0.001 ppm, 4 d, Metcalf model, cf to wat.[2]
Oedogonium cardiacum	7600 based on dry weight	Water = 0.66 ppt, soil = 0.01 ppm, 4 d, Metcalf model, cf to wat.[2]
Oedogonium cardiacum	8100 based on dry weight	Water = 48 ppt, soil = 0.53 ppm, 4 d, Metcalf model, cf to wat. conc[2]
Physa sp.	11600 based on dry weight	Water = 239 ppt, soil = 3.17 ppm, 4 d, Metcalf model, cf to wat. conc[2]
Physa sp.	1400 based on dry weight	Water = 1330 ppt, soil = 7.45 ppm, 4 d, Metcalf model, cf to wat.[2]
Physa sp.	14700 based on dry weight	Water = 0.66 ppt, soil = 0.01 ppm, 4 d, Metcalf model, cf to wat.[2]
Physa sp.	16100 based on dry weight	Water = 18 ppt, soil = 0.29 ppm, 4 d, Metcalf model, cf to wat. conc[2]
Physa sp.	17500 based on dry weight	Water = 7.13 ppt, soil = 0.10 ppm, 4 d, Metcalf model, cf to wat.[2]
Physa sp.	24000 based on dry weight	Water = 0.05 ppt, soil = 0.0001 ppm, 4 d, Metcalf model, cf to wat.[2]
Physa sp.	41000 based on dry weight	Water = 48 ppt, soil = 0.53 ppm, 4 d, Metcalf model, cf to wat. conc[2]
Physa sp.	47100 based on dry weight	Water = 7 ppt, soil = 0.15 ppm, 4 d, Metcalf model, cf to wat. conc[2]
Physa sp.	5400 based on dry weight	Water = 0.26 ppt, soil = 0.001 ppm, 4 d, Metcalf model, cf to wat.[2]
Snail	15000 based on wet weight	Greather than or equal to value[1]

Source: 1.Isensee, A.R. 1977. Bioaccumulation of TCDD. Draft Subject to Revision, Conf. paper. Stockholm. 2.Isensee, A.R. and G.E. Jones. 1975. Environ. Sci. Technol. 9(7), pp 668-672.

5-267 Concentration factor Tb

Species	Value	Condition
Engraulis japonicus, whole body	1040 based on wet weight	Rel to soluble conc., total = 4.2 ng/l, solub = 0.8 ng/l, marine, Japan[1]
Engraulis japonicus, whole body	150 based on wet weight	Rel to total conc., total = 4.2 ng/l, solub = 0.8 ng/l, marine, Japan[1]
Hizikia fusiforme, whole body	110 based on wet weight	Rel to total conc., total = 4.2 ng/l, solub = 0.8 ng/l, marine, Japan[1]
Hizikia fusiforme, whole body	780 based on wet weight	Rel to soluble conc., total = 4.2 ng/l, solub = 0.8 ng/l, marine, Japan[1]
Meretrix lusoria, soft part	3180 based on wet weight	Rel to soluble conc., total = 4.2 ng/l, solub = 0.8 ng/l, marine, Japan[1]
Meretrix lusoria, soft part	460 based on wet weight	Rel to total conc., total = 4.2 ng/l, solub = 0.8 ng/l, marine, Japan[1]
Paralichthys olivaceus, muscle	38 based on wet weight	Rel to soluble conc., total = 4.2 ng/l, solub = 0.8 ng/l, marine, Japan1
Paralichthys olivaceus, muscle	5 based on wet weight	Rel to total conc., total = 4.2 ng/l, solub = 0.8 ng/l, marine, Japan[1]
Sargassum fulvellum, whole body	2230 based on wet weight	Rel to soluble conc., total = 4.2 ng/l, solub = 0.8 ng/l, marine, Japan[1]

Sargassum fulvellum, whole body	320 based on wet weight	Rel to total conc., total = 4.2 ng/l, solub = 0.8 ng/l, marine, Japan[1]
Seriola quinqueradiata, muscle	13 based on wet weight	Rel to total conc., total = 4.2 ng/l, solub = 0.8 ng/l, marine, Japan[1]
Seriola quinqueradiata, muscle	883 based on wet weight	Rel to soluble conc., total = 4.2 ng/l, solub = 0.8 ng/l, marine, Japan[1]
Ulva pertusa, whole body	1210 based on wet weight	Rel to soluble conc., total = 4.2 ng/l, solub = 0.8 ng/l, marine, Japan[1]
Ulva pertusa, whole body	180 based on wet weight	Rel to total conc., total = 4.2 ng/l, solub = 0.8 ng/l, marine, Japan1
Undaria pinnatifida, whole body	380 based on wet weight	Rel to soluble conc., total = 4.2 ng/l, solub = 0.8 ng/l, marine, Japan[1]
Undaria pinnatifida, whole body	54 based on wet weight	Rel to total conc., total = 4.2 ng/l, solub = 0.8 ng/l, marine, Japan[1]

Source: 1.Suzuki, H., T. Koganage and M. Saki. 1975. Impacts of Nuclear Releases into the Aquatic Environment.IAEA SM-198/19, pp 77-89. Vienna.

5-268 Concentration factor
Tc

Species	Value	Condition
Haliotis rufescens, columnar muscle	0.5 based on wet weight	Seawater with 50 n ci/l, n = 2-4, 284 K, 3 d, Tc-95[1]
Haliotis rufescens, columnar muscle	0.8 based on wet weight	Seawater with 50 n ci/l, n = 2-4, 284 K, 10 d, Tc-95[1]
Haliotis rufescens, columnar muscle	1.3 based on wet weight	Seawater with 50 n ci/l, n = 2-4, 284 K, 25 d, Tc-95[1]
Haliotis rufescens, digestive gland	271 based on wet weight	Seawater with 50 n ci/l, n = 2-4, 284 K, 25 d, Tc-95[1]
Haliotis rufescens, digestive gland	28 based on wet weight	Seawater with 50 n ci/l, n = 2-4, 284 K, 3 d, Tc-95[1]
Haliotis rufescens, digestive gland	80 based on wet weight	Seawater with 50 n ci/l, n = 2-4, 284 K, 10 d, Tc-95[1]
Haliotis rufescens, gill	117 based on wet weight	Seawater with 50 n ci/l, n = 2-4, 284 K, 3 d, Tc-95[1]
Haliotis rufescens, gill	253 based on wet weight	Seawater with 50 n ci/l, n = 2-4, 284 K, 25 d, Tc-95[1]
Haliotis rufescens, gill	309 based on wet weight	Seawater with 50 n ci/l, n = 2-4, 284 K, 10 d, Tc-95[1]
Haliotis rufescens, gonad	1.5 based on wet weight	Seawater with 50 n ci/l, n = 2-4, 284 K, 3 d, Tc-95[1]
Haliotis rufescens, gonad	3 based on wet weight	Seawater with 50 n ci/l, n = 2-4, 284 K, 10 d, Tc-95[1]
Haliotis rufescens, gonad	5 based on wet weight	Seawater with 50 n ci/l, n = 2-4, 284 K, 25 d, Tc-95[1]
Haliotis rufescens, heart	2.2 based on wet weight	Seawater with 50 n ci/l, n = 2-4, 284 K, 10 d, Tc-95[1]
Haliotis rufescens, heart	4 based on wet weight	Seawater with 50 n ci/l, n = 2-4, 284 K, 25 d, Tc-95[1]
Haliotis rufescens, heart	6 based on wet weight	Seawater with 50 n ci/l, n = 2-4, 284 K, ww d, Tc-95[1]
Haliotis rufescens, kidney	147 based on wet weight	Seawater with 50 n ci/l, n = 2-4, 284 K, 25 d, Tc-95[1]
Haliotis rufescens, kidney	28 based on wet weight	Seawater with 50 n ci/l, n = 2-4, 284 K, 3 d, Tc-95[1]
Haliotis rufescens, kidney	67 based on wet weight	Seawater with 50 n ci/l, n = 2-4, 284 K, 10 d, Tc-95[1]

Source: 1.Spies, R.B. 1975. Health Physics, 29, pp 695-699.

5-269 Concentration factor
Th

Species	Value	Condition
Brown algae	10 based on wet weight	None[1]
Fish, bone	190 based on wet weight	Danube, 1973, nuclear power plant since 1967, Tc-232[2]
Fish, muscle	80 based on wet weight	Danube, 1973, nuclear power plant since 1967, rg = 40 - 110, Tc-232[2]

Mya arenaria	17.6 based on wet weight	Water = 100 mg/m3, 80 d, thallium(I) sulfate[3]
Mya arenaria	18.6 based on wet weight	Water = 50 mg/m3, 80 d, thallium(I) sulfate[3]
Mytilus edulis	10.9 based on wet weight	Water = 100 mg/m3, 40 d, thallium(I) sulfate[3]
Mytilus edulis	12.4 based on wet weight	Water = 50 mg/m3, 40 d, thallium(I) sulfate[3]
Sediment	3160 based on wet weight	Danube, 1973, nuclear power plant since 67, freshwater, rg = 810 -[2]
Suspended solids	1900 based on wet weight	Danube, 1973, nuclear power plant since 1967, rg = 280 - 5000, Tc-232[2]
Water plant	125 based on wet weight	Danube, 1973, nuclear power plant since 1967, rg = 90 - 180, Tc-232[2]

Source: 1.Mauchline, J. and W.L Templeton. 1964. Oceanogr. Mar. Biol. Ass. Rev., 2, pp 229.
2.Herrmann, H., M. Ruf, K. Hubel and W. Lunsmann. 1975. Impacts of Nuclear Releases into the Aquatic Environment. IAEA-SM-198/14. pp 461-470. Vienna.
3.Zitko, V. and W.V. Carson. 1975. Bull. Environ. Contam. Toxicol. 14(5), pp 530-533.

5-270 Concentration factor
Ti

Species	Value	Condition
Agarum cribrosum	23000 based on dry weight	Seawater = 1.0E-7 g/l, Vostok Bay, Sea of Japan[1]
Brown algae	3000 based on wet weight	None[2]
Chondrus yendoi	500 based on dry weight	Seawater = 1.0E-7 g/l, Vostok Bay, Sea of Japan[1]
Chorda filum	1200 based on dry weight	Seawater = 1.0E-7 g/l, Vostok Bay, Sea of Japan[1]
Codium yessoensis	3200 based on dry weight	Seawater = 1.0E-7 g/l, Vostok Bay, Sea of Japan[1]
Costaria costata	900 based on dry weight	Seawater = 1.0E-7 g/l, Vostok Bay, Sea of Japan[1]
Enteromorpha prolifera	1900 based on dry weight	Seawater = 1.0E-7 g/l, Vostok Bay, Sea of Japan[1]
Heterochordaria abietina	6000 based on dry weight	Seawater = 1.0E-7 g/l, Vostok Bay, Sea of Japan[1]
Laminaria chichorioides	900 based on dry weight	Seawater = 1.0E-7 g/l, Vostok Bay, Sea of Japan[1]
Laminaria japonica	200 based on dry weight	Seawater = 1.0E-7 g/l, Vostok Bay, Sea of Japan[1]
Pelvetia wrightii	800 based on dry weight	Seawater = 1.0E-7 g/l, Vostok Bay, Sea of Japan[1]
Phyllospadix iwatensis	15700 based on dry weight	Seawater = 1.0E-7 g/l, Vostok Bay, Sea of Japan[1]
Phytoplankton	2700 based on wet weight	Marine organisms, typical value[3]
Plankton	20000 based on wet weight	None[2]
Polysiphonia japonica	37000 based on dry weight	Seawater = 1.0E-7 g/l, Vostok Bay, Sea of Japan[1]
Ptilota filicina	44000 based on dry weight	Seawater = 1.0E-7 g/l, Vostok Bay, Sea of Japan[1]
Rhodomela larix	35000 based on dry weight	Seawater = 1.0E-7 g/l, Vostok Bay, Sea of Japan[1]
Sargassum pallidum	5300 based on dry weight	Seawater = 1.0E-7 g/l, Vostok Bay, Sea of Japan[1]
Ulva fenestrata	9400 based on dry weight	Seawater = 1.0E-7 g/l, Vostok Bay, Sea of Japan[1]
Ulvaria spendens	18000 based on dry weight	Seawater = 1.0E-7 g/l, Vostok Bay, Sea of Japan[1]
Zostera asiatica	11000 based on dry weight	Seawater = 1.0E-7 g/l, Vostok Bay, Sea of Japan[1]

Source: 1.Saenko, G.N., M.D. Koryakov, V.F. Makienko and I.G. Dobrosmy. 1976. Mar. biol., 34(2), pp 169-176.
2.Bowen, H.J.M. 1966. Trace Elements in Biochemistry, Ac. Press.
3.Martin, J.H. and G.A. Knauer. 1973. Geochim. Cosmochim. Acta., 37, p 1639.

5-271 Concentration factor
Toxaphene

Species	Value	Condition
Aspergillus sp.	1.7E4 based on wet weight	Equilibrium cf., 301 K, 24 h[1]
Bacillus subtilis	3.4E3 based on wet weight	Equilibrium cf., 301 K, 24 h[1]
Chlorella pyrenoidosa	1.7E4 based on wet weight	Equilibrium cf., 288 K, 200 h, 170 ftc[1]
Flavobacterium harrisonii	5.2E3 based on wet weight	Equilibrium cf., 301 K, 24 h[1]
Seston	6.6E3 based on wet weight	Equilibrium cf., 301 K, 24 h[1]

Source: 1.Paris, D.F., D.L. Lewis and J.T. Barnett. 1977. Bull. Environ. Contam. Toxicol. 17(5), pp 564-572.

5-272 Concentration factor
U

Species	Value	Condition
Brown algae	10 based on wet weight	None[1]
Coral	1.08 based on Ca content	Shallow coastal ocean, mean of 10 determinations, F023[2]
Coral	1.35 based on Ca content	Deep ocean coral, mean of 8 determinations, F023[2]
Coral	1.36 based on Ca content	Shallow open ocean coral, mean of 8 determinations, F023[2]

Source: 1.Mauchline, J. and W.L Templeton. 1964. Oceanogr. Mar. Biol. Ass. Rev., 2, pp 229.
2.Livingston, H.D. and G. Thompson. 1971. Limnol. Oceanogr. 16(5), pp 786-796.

5-273 Concentration factor
V

Species	Value	Condition
Agarum cribrosum	1100 based on dry weight	Seawater = 4.4E-7 g/l, Vostok Bay, Sea of Japan[1]
Atmospheric particulates	1.4	Low level marine part, relative to earths crust, Fe standard, F200[2]
Atmospheric particulates	1.4	South Pole, relative to earths crust, Al standard, F200[3]
Atmospheric particulates	13	N Atlantic Westerlies, relative to earths crust, Al standard, F200[4]
Atmospheric particulates	42	US urban air, relative to earths crust, Fe standard, F200[5]
Brown algae	250 based on wet weight	None[6]
Chondrus yendoi	800 based on dry weight	Seawater = 4.4E-7 g/l, Vostok Bay, Sea of Japan[1]
Chorda filum	370 based on dry weight	Seawater = 4.4E-7 g/l, Vostok Bay, Sea of Japan[1]
Codium yessoensis	330 based on dry weight	Seawater = 4.4E-7 g/l, Vostok Bay, Sea of Japan[1]
Heterochordaria abietina	300 based on dry weight	Seawater = 4.4E-7 g/l, Vostok Bay, Sea of Japan[1]
Laminaria chichorioides	80 based on dry weight	Seawater = 4.4E-7 g/l, Vostok Bay, Sea of Japan[1]
Laminaria japonica	120 based on dry weight	Seawater = 4.4E-7 g/l, Vostok Bay, Sea of Japan[1]
Pelvetia wrightii	300 based on dry weight	Seawater = 4.4E-7 g/l, Vostok Bay, Sea of Japan[1]
Phyllospadix iwatensis	1100 based on dry weight	Seawater = 4.4E-7 g/l, Vostok Bay, Sea of Japan[1]
Plankton	620 based on wet weight	None[6]
Polysiphonia japonica	3500 based on dry weight	Seawater = 4.4E-7 g/l, Vostok Bay, Sea of Japan[1]
Ptilota filicina	1300 based on dry weight	Seawater = 4.4E-7 g/l, Vostok Bay, Sea of Japan[1]

Rhodomela larix	600 based on dry weight	Seawater = 4.4E-7 g/l, Vostok Bay, Sea of Japan[1]
Sargassum pallidum	130 based on dry weight	Seawater = 4.4E-7 g/l, Vostok Bay, Sea of Japan[1]
Ulva fenestrata	120 based on dry weight	Seawater = 4.4E-7 g/l, Vostok Bay, Sea of Japan[1]
Ulvaria spendens	400 based on dry weight	Seawater = 4.4E-7 g/l, Vostok Bay, Sea of Japan[1]

Source: 1.Saenko, G.N., M.D. Koryakov, V.F. Makienko and I.G. Dobrosmy. 1976. Mar. biol., 34(2), pp 169-176.
2.Zoller, W.H., E.S. Gladey and R.A. Duce. 1974. Science 183, pp 200.
3.Duce, R.A., J. Quinn, C.M. Olney, S. Piotrowits, B. Ray and Wade.1972. Science 176, pp 161-163.
4.Chester, R. and J.H. Stoner. 1973. Nature (London) 245, pp 27-28.
5.Foestner, U. and G.T.W. Wittmann. 1979. Metal Pollution in the Aquatic Environment. Springer-Verlag.Berlin.
6.Bowen, H.J.M. 1966. Trace Elements in Biochemistry, Ac. Press.

5-274 Concentration factor
W

Species	Value	Condition
Anodonta nuttalliana, soft parts	0.5 based on wet weight	No food, 292 K, pH = 7.5, freshwat = 2.3 part & 1200 solub p ci/ml, 7 d[1]
Anodonta nuttalliana, soft parts	0.8 based on wet weight	No food, 292 K, pH = 7.5, freshwat = 34 part & 1900 solub p ci/ml, 21 d[1]
Astacus sp., muscle	1 based on wet weight	No food, 292 K, ph=7.5, freshwat=2.3 part & 1200 solub p ci/ml, 7 d[1]
Astacus sp., muscle	1 based on wet weight	No food, 292 K, ph=7.5, freshwat=34 part & 1900 solub p ci/ml, 21 d[1]
Astacus sp., soft parts	0.7 based on wet weight	No food, 292 K, pH = 7.5, freshwat = 2.3 part & 1200 solub p ci/ml, 7 d[1]
Astacus sp., soft parts	4 based on wet weight	No food, 292 K, pH = 7.5, freshwat = 2.3 part. and 16 solub p ci/ml[1]
Astacus sp., viscera	0.9 based on wet weight	No food, 292 K, pH = 7.5, freshwat = 2.3 part & 1200 solub p ci/ml, 7 d[1]
Astacus sp., viscera	5 based on wet weight	No food, 292 K, pH = 7.5, freshwat = 34 part & 1900 solub p ci/ml, 21 d[1]
Brown algae	87 based on wet weight	None[2]
Cancer productus, muscle	0.8 based on wet weight	No food, 285 K, pH = 8.2, seawater = 20 part & 1500 solub p ci/ml, 20 d[1]
Cancer productus, sort parts	10 based on wet weight	No food, 285 K, pH = 8.2, seawater = 20 part & 1500 solub p ci/ml, 20 d[1]
Cancer productus, viscera	0.5 based on wet weight	No food, 285 K, pH = 8.2, seawater = 34 part & 1300 solub p ci/ml, 6 d[1]
Cancer productus, viscera	15 based on wet weight	No food, 285 K, pH = 8.2, seawater = 20 part & 1500 solub p ci/ml, 20 d[1]
Carassius auratus, muscle	1 based on wet weight	No food, 292 K, ph=7.5, freshwat=2.3 part & 1200 solub p ci/ml, 7 d[1]
Carassius auratus, nonmuscle	0.7 based on wet weight	No food, 292 K, pH = 7.5, freshwat = 2.3 part & 1200 solub p ci/ml, 7 d[1]
Carassius auratus, whole body	0.6 based on wet weight	No food, 292 K, pH = 7.5, freshwat = 34 part & 1900 solub p ci/ml, 21 d[1]
Gillichthys mirabilis, muscle	0.2 based on wet weight	No food, 285 K, pH = 8.2, seawater = 20 part & 1500 solub p ci/ml, 20 d[1]
Gillichthys mirabilis, muscle	1 based on wet weight	No food, 285 K, ph=8.2, seawater=34 part & 1300 solub p ci/ml, 6 d[1]
Gillichthys mirabilis, nonmuscle	0.5 based on wet weight	No food, 285 K, pH = 8.2, seawater = 20 part & 1500 solub p ci/ml, 20 d[1]
Gillichthys mirabilis, nonmuscle	0.8 based on wet weight	No food, 285 K, pH = 8.2, seawater = 34 part & 1300 solub p ci/ml, 6 d[1]

Gillichthys mirabilis, whole body	0.5 based on wet weight	No food, 285 K, pH = 8.2, seawater = 3.5 part & 6.9 solub p ci/ml, 6 d[1]
Gillichthys mirabilis, whole body	0.6 based on wet weight	No food, 285 K, pH = 8.2, seawater = 20 part & 1500 solub p ci/ml, 20 d[1]
Mytilus edulis, soft parts	1.3 based on wet weight	No food, 285 K, pH = 8.2, seawater = 34 part & 1300 solub p ci/ml, 6 d[1]
Plant	10 based on wet weight	Conc in soil = 2.7 ug/g, rg = 2 - 18[3]
Saxidonus giganteus, soft parts	0.4 based on wet weight	No food, 285 K, pH = 8.2, seawater = 20 part & 1500 solub p ci/ml, 20 d[4]

Source: 1.Harrison, F.L. 1973. Health Physics, 24, pp 331-343.
2.Bowen, H.J.M. 1966. Trace Elements in Biochemistry, Ac. Press.
3.Wedepoki, K.H.(ed). 1969. Handbook of Geochemistry, Springer-Verlag Berlin.
4.Kearney, P.C., J.E. Oliver, C.S. Helling, A.R. Isensee and A. Kontson. 1977. J. Agric. Food. Chem. 2(5), pp 1177-1181.

5-275 Concentration factor XMC

Species	Value	Condition
Ictalurus punctatus	550 based on dry weight	Model ecosystem, 22 d, 296 K, water = 2.0-35.8 ppb, soil = 2.78 ppm[1]
Lemna minor	2300 based on dry weight	Model ecosystem, 22 d, 296 K, water = 2.0-35.8 ppb, soil = 2.78 ppm[1]
Oedogonium cardiacum	3050 based on dry weight	Model ecosystem, 22 d, 296 K, water = 2.0-35.8 ppb, soil = 2.78 ppm[1]
Physa sp.	260 based on dry weight	Model ecosystem, 22 d, 296 K, water = 2.0-35.8 ppb, soil = 2.78 ppm[1]
Procambarus sp.	104 based on dry weight	Model ecosystem, 22 d, 296 K, water = 2.0-35.8 ppb, soil = 2.78 ppm[1]

Source: 1.Kanazawa, J., A.R. Isensee and P.C. Kearney. 1975. J. Agric. Food. Chem. 23(4), pp 760-763.

5-276 Concentration factor Y

Species	Value	Condition
Brown algae	550 based on wet weight	Range = 100 - 1000[1]
Plant	6880 based on dry weight	Algae, bryophytes and vascular hydrophytes, freshwater, 32 sp[2]

Source: 1.Mauchline, J. and W.L Templeton. 1964. Oceanogr. Mar. Biol. Ass. Rev., 2, pp 229.
2.Timofeeva-Resovskaya, E.A., T-R. N.V. and E.A. Gileva. 1961. Dokl. Akad. Nauk. SSSR, 140, pp 1437.

5-277 Concentration factor Yb

Species	Value	Condition
Engraulis japonicus, whole body	210 based on wet weight	Rel to total conc., total = 5.6 ng/l, solub = 2.1 ng/l, marine, Japan[1]
Engraulis japonicus, whole body	710 based on wet weight	Rel to soluble conc., total = 5.6 ng/l, solub = 2.1 ng/l, marine, Japan[1]
Hizikia fusiforme, whole body	320 based on wet weight	Rel to soluble conc., total = 5.6 ng/l, solub = 2.1 ng/l, marine, Japan[1]
Hizikia fusiforme, whole body	94 based on wet weight	Rel to total conc., total = 5.6 ng/l, solub = 2.1 ng/l, marine, Japan[1]
Meretrix lusoria, soft part	2590 based on wet weight	Rel to soluble conc., total = 5.6 ng/l, solub = 2.1 ng/l, marine, Japan[1]
Meretrix lusoria, soft part	750 based on wet weight	Rel to total conc., total = 5.6 ng/l, solub = 2.1 ng/l, marine, Japan[1]
Paralichthys olivaceus, muscle	130 based on wet weight	Rel to soluble conc., total = 5.6 ng/l, solub = 2.1 ng/l, marine, Japan[1]
Paralichthys olivaceus, muscle	39 based on wet weight	Rel to total conc., total = 5.6 ng/l, solub = 2.1 ng/l, marine, Japan[1]

Sargassum fulvellum, whole body	140 based on wet weight	Rel to total conc., total = 5.6 ng/l, solub = 2.1 ng/l, marine, Japan[1]
Sargassum fulvellum, whole body	460 based on wet weight	Rel to soluble conc., total = 5.6 ng/l, solub = 2.1 ng/l, marine, Japan[1]
Seriola quinqueradiata, muscle	190 based on wet weight	Rel to soluble conc., total = 5.6 ng/l, solub = 2.1 ng/l, marine, Japan[1]
Seriola quinqueradiata, muscle	54 based on wet weight	Rel to total conc., total = 5.6 ng/l, solub = 2.1 ng/l, marine, Japan[1]
Ulva pertusa, whole body	230 based on wet weight	Rel to total conc., total = 5.6 ng/l, solub = 2.1 ng/l, marine, Japan[1]
Ulva pertusa, whole body	800 based on wet weight	Rel to soluble conc., total = 5.6 ng/l, solub = 2.1 ng/l, marine, Japan1
Undaria pinnatifida, whole body	290 based on wet weight	Rel to soluble conc., total = 5.6 ng/l, solub = 2.1 ng/l, marine, Japan[1]
Undaria pinnatifida, whole body	85 based on wet weight	Rel to total conc., total = 5.6 ng/l, solub = 2.1 ng/l, marine, Japan[1]

Source: 1.Suzuki, H., T. Koganage and M. Saki. 1975. Impacts of Nuclear Releases into the Aquatic Environment.IAEA SM-198/19, pp 77-89. Vienna.

5-278 Concentration factor
Yt

Species	Value	Condition
Ulva lactuca	550 based on wet weight	Equilibrium after 3 h, pH = 8, carrier free, Yt-90[1]

Source: 1.Hampson, M.A. 1967. J. Exp. Bot. 18, pp 17-53.

5-279 Concentration factor
Zn

Species	Value	Condition
Agarum cribrosum	200 based on dry weight	Seawater = 66.6E-7 g/l, Vostok Bay, Sea of Japan[1]
Algae	290 based on wet weight	Marine, Zn-65[2]
Arthrobacter marinus	200 based on dry weight	Total concentration = 350 ppb, added Zn = 0 ppb, 293 K[3]
Arthrobacter marinus	202 based on dry weight	Total concentration = 18350 ppb, added Zn = 19614 ppb, 293 K[3]
Arthrobacter marinus	48 based on dry weight	Total concentration = 6960 ppb, added Zn = 6538 ppb, 293 K[3]
Arthrobacter marinus	55 based on dry weight	Total concentration = 3650 ppb, added Zn = 3269 ppb, 293 K[3]
Arthrobacter marinus	72 based on dry weight	Total concentration = 13390 ppb, added Zn = 13076 ppb, 293 K[3]
Arthrobacter marinus	97 based on dry weight	Total concentration = 1050 ppb, added Zn = 654 ppb, 293 K[3]
Atmospheric particulates	13	Low level marine part, relative to earths crust, Fe standard, F200[4]
Atmospheric particulates	270	US urban air, relative to earths crust, Fe standard, F200[5]
Atmospheric particulates	93	South Pole, relative to earths crust, Al standard, F200[6]
Brassica oleracea	41 based on wet weight	Conc in soil solution = 0.2 ppm[7]
Brown algae	3400 based on wet weight	None[7]
Brown algae	6550 based on wet weight	Range = 100 - 13000[8]
Cancer irroratus	16599 based on wet weight	Seawater = 1.77E-3 ppm, Nova Scotia, AAS, SD = +- 1368[9]
Carassius auratus	75 based on wet weight	Water = 0.67 - 0.22 ppm[10]
Carcinus sp.	130 based on wet weight	n = 4, uptake from water = 23.3 - 117.1 ug/l. 96 d, Zn-65[11]
Carcinus sp.	210 based on wet weight	n = 5, uptake from water and food, water = 23.3-117.1 ug/l, 96 d,[11]

Carcinus sp.	450 based on wet weight	n = 6, uptake from water and food, water = 23.3-117.1 ug/l, 96 d[11]
Carcinus sp.	520 based on wet weight	n = 6, uptake from water = 23.3 - 117.1 ug/l. 96 d[11]
Chironomid larvae	1700 based on wet weight	Lake Malar, Sweden, water = 3E-6 u ci/ml, Zn-65[12]
Chondrus yendoi	260 based on dry weight	Seawater = 66.6E-7 g/l, Vostok Bay, Sea of Japan[1]
Chorda filum	60 based on dry weight	Seawater = 66.6E-7 g/l, Vostok Bay, Sea of Japan[1]
Cladophora glomerata	1E3 based on dry weight	Main Duck Isl., Lake Ontario = 1.7+-1.2 ug/l +- SD[13]
Cladophora glomerata	2.9E3 based on dry weight	Deadman Bay , Lake Ontario = 7.7+-3.1 ug/l +- SD[13]
Cladophora glomerata	49E3 based on dry weight	Deadman Bay , Lake Ontario = 0.6+-0.2 ug/l +- SD[13]
Codium yessoensis	540 based on dry weight	Seawater = 66.6E-7 g/l, Vostok Bay, Sea of Japan[1]
Coral	0.40 based on Ca content	Shallow open ocean coral, mean of 13 determinations, F023[14]
Coral	0.71 based on Ca content	Shallow open ocean coral, mean of 13 determinations, F023[14]
Coral	5.04 based on Ca content	Shallow open ocean coral, mean of 8 determinations, F023[14]
Costaria costata	60 based on dry weight	Seawater = 66.6E-7 g/l, Vostok Bay, Sea of Japan[1]
Crassostrea gigas	0.88E4 based on dry weight	13 mounths, water = 1.04-19.63E-2 p ci/ml, 1962, Zn-65[15]
Crassostrea gigas	0.98E4 based on dry weight	13 mounths, water = 1.04-19.63E-2 p ci/ml, 1964, Zn-65[15]
Crassostrea gigas	14000 based on wet weight	Willapa Bay, rg = +- 9700[16]
Crassostrea gigas	9900 based on wet weight	Willapa Bay, Zn-65, rg = +- 2700[16]
Crassostrea gryphoides	9250 based on wet weight	Bombay, rg = 5500 - 13000[17]
Crustacea	1800 based on dry weight	Freshwater, rg = 300 - 4000, 3 values used[18]
Crustacea	5300 based on dry weight	Marine, rg = 1700 - 15000, 6 values used[18]
Crustacean, viscera	500 based on wet weight	Marine, Zn-65, greater than[2]
Echinoderm, digestive tract	200 based on wet weight	Marine, Zn-65, greater than[2]
Enteromorpha prolifera	500 based on dry weight	Seawater = 66.6E-7 g/l, Vostok Bay, Sea of Japan[1]
Fish	11 based on wet weight	42 d, Zn-65, rg = 2 - 20[19]
Fish	1744 based on dry weight	Freshwater, rg = 10 - 7600, 10 values used[18]
Fish	228 based on wet weight	Freshwater organisms, typical value[20]
Fish	3400 based on dry weight	Marine, rg = 280 - 15500, 12 values used[18]
Fish	533 based on wet weight	Marine organisms, typical value[21]
Fish, muscle	510 based on wet weight	Danube, 1973, nuclear power plant since 1967, Zn-65, rg = 190 - 780[22]
Fish, viscera	22 based on wet weight	Marine, Zn-65[2]
Fucus serratus	1.1E4 based on wet weight	Steady state cf., 283 K, Zn = 5-25 mg/m3, March[23]
Fucus serratus	7.2E3 based on wet weight	Steady state cf., 283 K, Zn = 5-25 mg/m3, Nov[23]
Fucus serratus	7.4E3 based on wet weight	Steady state cf., 283 K, Zn = 5-25 mg/m3, July[23]
Fucus vesiculosus	2E4 based on dry weight	18 obs., all areas around the British Isles[24]
Gobius sp.	10 based on wet weight	n = 4, uptake from water = 23.3 - 117.1 ug/l. 96 d, Zn-65[11]
Gobius sp.	1400 based on wet weight	n = 5, uptake from water = 23.3 - 117.1 ug/l. 96 d[11]
Gobius sp.	25 based on wet weight	n = 4, uptake from water and food, water = 23.3-117.1 ug/l, 96 d,[11]

Gobius sp.	900 based on wet weight	n = 4, uptake from water and food, water = 23.3-117.1 ug/l, 96 d[11]
Heterochordaria abietina	150 based on dry weight	Seawater = 66.6E-7 g/l, Vostok Bay, Sea of Japan[1]
Laminaria chichorioides	150 based on dry weight	Seawater = 66.6E-7 g/l, Vostok Bay, Sea of Japan[1]
Laminaria japonica	70 based on dry weight	Seawater = 66.6E-7 g/l, Vostok Bay, Sea of Japan[1]
Lampsilis radiata	4100 based on wet weight	91 days, freshwater, Zn-65[25]
Leeches	400 based on wet weight	Lake Malar, Sweden, water = 3E-6 u ci/ml, Zn-65[12]
Lysmata seticaudata	2020 based on wet weight	n = 6, uptake from water = 23.3 - 117.1 ug/l. 96 d[11]
Lysmata seticaudata	3200 based on wet weight	n = 6, uptake from water and food, water = 23.3-117.1 ug/l, 96 d[11]
Lysmata seticaudata	380 based on wet weight	n = 2, uptake from water = 23.3 - 117.1 ug/l. 96 d, Zn-65[11]
Lysmata seticaudata	380 based on wet weight	n = 4, uptake from water and food, water = 23.3-117.1 ug/l, 96 d[11]
Macrophytes	318 based on wet weight	Freshwater organisms, typical value[20]
Macrophytes	7000 based on wet weight	Marine organisms, typical value[26]
Marine algae	7100 based on dry weight	Max cf. at 500 flc, 291-298 K, not related to water conc., Zn-65[27]
Mollusca	2258 based on wet weight	Freshwater organisms, typical value[28]
Mollusca	27300 based on wet weight	Marine organisms, typical value[29]
Mollusca, vicera	68 based on wet weight	Marine, Zn-65[2]
Molluscs	33544 based on dry weight	Freshwater, rg = 30 - 140000, 5 values used[18]
Molluscs	47000 based on dry weight	Marine, rg = 2100 - 330000, 11 values used[18]
Mytilus edulis	75 based on wet weight	Steady state cf., water = 0.750 mg zn/l[30]
Mytilus galloprovincialis	170 based on wet weight	n = 6, uptake from water = 23.3 - 117.1 ug/l. 96 d, Zn-65[11]
Mytilus galloprovincialis	580 based on wet weight	n = 6, uptake from water = 23.3 - 117.1 ug/l. 96 d[11]
Patella vulgata	1.9E4 based on dry weight	17 obs., all areas around the British Isles[24]
Pelvetia wrightii	20 based on dry weight	Seawater = 66.6E-7 g/l, Vostok Bay, Sea of Japan[1]
Phyllospadix iwatensis	200 based on dry weight	Seawater = 66.6E-7 g/l, Vostok Bay, Sea of Japan[1]
Phytoplankton	42250 based on wet weight	Range = 8900 - 75600[31]
Phytoplankton	5500 based on wet weight	Marine organisms, typical value[32]
Pike	1250 based on wet weight	Lake Malar, Sweden, water = 3E-6 u ci/ml, Zn-65[12]
Plankton	1000 based on wet weight	None[8]
Plankton	61000 based on wet weight	Conc in water = 1.8 ug/l, Schoehsee, freshwater[33]
Plankton	65000 based on wet weight	None[7]
Plant	4600 based on dry weight	Algae, bryophytes and vascular hydrophytes, freshwater, 32 sp[34]
Plants	3155 based on dry weight	Freshwater, rg = 140 - 15000, 13 values used[18]
Plants	900 based on dry weight	Marine, rg = 80 - 2500, 4 values used[18]
Polysiphonia japonica	300 based on dry weight	Seawater = 66.6E-7 g/l, Vostok Bay, Sea of Japan[1]
Porphyra umbilicalis	1E4 based on dry weight	20 obs., all areas around the British Isles[24]
Ptilota filicina	480 based on dry weight	Seawater = 66.6E-7 g/l, Vostok Bay, Sea of Japan[1]
Rhodomela larix	230 based on dry weight	Seawater = 66.6E-7 g/l, Vostok Bay, Sea of Japan[1]
Sargassum pallidum	40 based on dry weight	Seawater = 66.6E-7 g/l, Vostok Bay, Sea of Japan[1]

Sediment	20000 based on wet weight	Lake Malar, Sweden, water = 3E-6 u ci/ml, Zn-65[12]
Sediment	480 based on wet weight	Danube, 1973, nuclear power plant since 1967, Zn-65, rg = 230 - 740[22]
Seston	1.3 based on wet weight	Depth = 17 m, Schoehsee[33]
Seston	1.5 based on wet weight	Depth = 8 m, Schoehsee[33]
Snail	590 based on wet weight	Lake Malar, Sweden, water = 3E-6 u ci/ml, Zn-65[12]
Suspended solids	3030 based on wet weight	Danube, 1973, nuclear power plant since 1967, Zn-65, rg = 90 - 8820[22]
Ulva fenestrata	30 based on dry weight	Seawater = 66.6E-7 g/l, Vostok Bay, Sea of Japan[1]
Ulvaria spendens	100 based on dry weight	Seawater = 66.6E-7 g/l, Vostok Bay, Sea of Japan[1]
Water plant	150 based on wet weight	Danube, 1973, nuclear power plant since 1967, Zn-65[22]
Zooplankton	5100 based on wet weight	Seawater concentration 0.01 mg/m3[35]
Zooplankton	8800 based on wet weight	Marine organisms, typical value[32]

Source: 1.Saenko, G.N., M.D. Koryakov, V.F. Makienko and I.G. Dobrosmy. 1976. Mar. biol., 34(2), pp 169-176.
2.Hiyama, Y. and M. Shimizu. 1964. Rec. Oceanogr. Works. Jap. 7(2), pp 43-77.
3.Jones, G.E., L.G. Royle and L. Mur ray.1976. Applied Science Publ., 5(12), pp 889-895.
4.Zoller, W.H., E.S. Gladey and R.A. Duce. 1974. Science 183, pp 200.
5.Foestner, U. and G.T.W. Wittmann. 1979. Metal Pollution in the Aquatic Environment. Springer-Verlag.Berlin.
6.Duce, R.A., J. Quinn, C.M. Olney, S. Piotrowits, B. Ray and Wade.1972. Science 176, pp 161-163.
7.Bowen, H.J.M. 1966. Trace Elements in Biochemistry, Ac. Press.
8.Mauchline, J. and W.L Templeton. 1964. Oceanogr. Mar. Biol. Ass. Rev., 2, pp 229.
9.Martin, J.-L.M. 1974. Mar. biol., 28(4), pp 245-251.
10.Myttenaere, C., M. Merlini, R.Bittel, P. Dabin, J.M. Mousny and G. Pozzi. 1975. Impacts of Nuclear releases into the Aquatic Environment. IAEA-SM-198/16, pp 29-47. Vienna.
11.Renfro, W.C., S.W. Fowler, M. Heyraud and J. La Rosa. 1975. J. Fish. Res. bd. Canada. 32(8), pp 13391345.
12.Hannerz, L. 1966. Proc. 1st Int. Congr. Radiat. Prot., pp 417.
13.Keeney, W.L., W.G. Breck , G.W. Vanloon and J.A. Page. 1976. Water Res., 10(11)., pp 981-984.
14.Livingston, H.D. and G. Thompson. 1971. Limnol. Oceanogr. 16(5), pp 786-796.
15.Salo, E.O. and W.L. Leet. 1969. Proc. 2nd Nat. Symp. on Radioecology. USAEC. Conf. 670503, pp 363-37.
16.Seymour, A. 1966. Disposal of Radioactive Wastes into Seas, Oceans and Surface Waters, IAEA, Vienna. pp 605-619.
17.Bhatt, Y.M., V.N. Sastry, S.M. Shah and T.M. Krishnamoorthy. 1968. Proc. Nat. Inst. Sci., India, pt, Biol. Sec. 34(B,6), pp 283-287.
18.Jinks, S.M. and M. Eisenbud. 1972. Radiat. Data Rep. 13, pp 243-247.
19.Kameda, K., M. Shimizu and Y. Hiyama. 1968. J. Rad. Res. 9, pp 50-62.
20.Merlini, M., C. Rigliocca, A. Berg and G. Pozzi. 1970. IAEA/ SM, 142/27.
21.Goldberg, E.D. 1962. Limnol Oceanogr., 7, pp 72.
22.Herrmann, H., M. Ruf, K. Hubel and W. Lunsmann. 1975. Impacts of Nuclear Releases into the Aquatic Environment. IAEA-SM-198/14. pp 461-470. Vienna.
23.Young, M.L. 1975. J. Mar. Biol. Ass. U.K., 55, pp 583-610.
24.Preston, A., D.R. Jefferies , J.W.R. Harvey and A.K. Steele. 1972. Environ. Pollut., 3, pp 183-214.
25.Harvey, R.S. 1969. Health Physics, 17, pp 149-154.
26.Haegerhaell, B. 1973. Bot. Mar., 16, p 53.
27.Bedrosian, P.H. 1962. Dissertation Abstracts 22, pp 3146.
28.Mathis, B.J. and T.F. Cummins. 1973. Water Pollut. Control 45 pp 1573.
29.Pringle, B.H., D.E. Hissong, E.L. Katz and S.T. Mulawka. 1968. J. San. Eng. Div. Am. Soc. Civ. Eng., 94(5970), pp 455-475.
30.Calapaj, G.G. 1974. Ig. Mod. 67(2), pp 136-145.
31.Cushing, C.E. and L.A. Rancitelli. 1972. Northwest Science, 46(2), pp 115-121.
32.Martin, J.H. and G.A. Knauer. 1973. Geochim. Cosmochim. Acta., 37, p 1639.
33.Groth, P. 1971. Arch. Hydrobiol. 68(3), pp 305-375.
34.Timofeeva-Resovskaya, E.A., T-R. N.V. and E.A. Gileva. 1961. Dokl. Akad. Nauk. SSSR, 140, pp 1437.
35.Martin, J.H 1970. Limnol. Oceanogr. 15(5), pp 756-761.

5-280 Concentration factor
Zr

Species	Value	Condition
Brown algae	675 based on wet weight	Range = 350 - 1000[1]
Fish, bone	500 based on wet weight	Danube, 1973, nuclear power plant since 1967, rg = 180 - 1220, Zr-95[2]
Fish, muscle	160 based on wet weight	Danube, 1973, nuclear power plant since 1967, rg = 40 - 440, Zr-95[2]
Phytoplankton	2250 based on wet weight	Range = 1500 - 3000[3]
Plankton	2250 based on wet weight	Range = 1500 - 3000[1]
Plant	6230 based on dry weight	Algae, bryophytes and vascular hydrophytes, freshwater, 32 sp[4]
Sediment	1430 based on wet weight	Danube, 1973, nuclear power plant since 1967, rg = 340 - 3570, Zr-95[2]
Suspended solids	1250 based on wet weight	Danube, 1973, nuclear power plant since 1967, rg = 180 - 3570, Zr-95[2]
Water plant	150 based on wet weight	Danube, 1973, nuclear power plant since 1967, rg = 65 - 460, Zr-95[2]

Source: 1.Mauchline, J. and W.L Templeton. 1964. Oceanogr. Mar. Biol. Ass. Rev., 2, pp 229.
2.Herrmann, H., M. Ruf, K. Hubel and W. Lunsmann. 1975. Impacts of Nuclear Releases into the Aquatic Environment. IAEA-SM-198/14. pp 461-470. Vienna.
3.Cushing, C.E. and L.A. Rancitelli. 1972. Northwest Science, 46(2), pp 115-121.
4.Timofeeva-Resovskaya, E.A., T-R. N.V. and E.A. Gileva. 1961. Dokl. Akad. Nauk. SSSR, 140, pp 1437.

5-281 Concentration factor
Zr - 95 / Nb - 95

Species	Value	Condition
Eskmeals silt	34637.7 based on wet weight	Yearly mean, Cumberland coastral area, 1969, Zr-95[1]
Inshore sand	5268.7 based on wet weight	Yearly mean, Cumberland coastral area, 1969, Zr-95[1]
Nephtys sp.	733.8 based on wet weight	Yearly mean, Cumberland coastral area, 1969, Zr-95[1]
Nucula sp.	15886.0 based on wet weight	Yearly mean, Cumberland coastral area, 1969, Zr-95[1]
Pleuronectes platessa, gut content	2670.6 based on wet weight	Yearly mean, Cumberland coastral area, 1969, Zr-95[1]

Source: 1.Pentreath, R.J. and D.F. Jefferies. 1971. J. mar. biol. Ass. U.K. 51, pp 963-976.

5-282 Concentration factor
Zr / Nb

Species	Value	Condition
Crustacea	51 based on dry weight	Marine, rg = 1 - 100, 4 values used[1]
Fish	86 based on dry weight	Marine, rg = 0.05 - 247, 5 values used[1]
Molluscs	81 based on dry weight	Marine, rg = 8 - 165, 4 values used[1]
Plants	1119 based on dry weight	Marine, rg = 170 - 2900, 12 values used[1]

Source: 1.Jinks, S.M. and M. Eisenbud. 1972. Radiat. Data Rep. 13, pp 243-247.

5-283 Concentrations
1,3-Dimethylnaphthalene

Species	Value	Condition
Fundulus heteroclitus	0.36 mg/kg wet weight	570 l S. Louisiana crude oil in 810 m2 shoreline area, 6 h after spill[1]
Fundulus heteroclitus	0.40 mg/kg wet weight	570 l S. Louisiana crude oil in 810 m2 shoreline area, 216 h after spill[1]
Fundulus heteroclitus	1.18 mg/kg wet weight	570 l S. Louisiana crude oil in 810 m2 shoreline area, 31 h after spill[1]
Fundulus heteroclitus	1.54 mg/kg wet weight	570 l S. Louisiana crude oil in 810 m2 shoreline area, 76 h after spill[1]

Source: 1.Bieri, R.H., V.C. Stamoudis and M.K. Cueman. 1977. Proc. of 1977 Oil spill Conference (Prevention, Behavior, Control, Cleanup). Wash. DC. American Petrolium Institute. 511-516.

5-284 Concentrations
1,5-Dimethylnaphthalene

Species	Value	Condition
Fundulus heteroclitus	0.13 mg/kg wet weight	570 l S. Louisiana crude oil in 810 m2 shoreline area, 6 h after spill[1]
Fundulus heteroclitus	0.16 mg/kg wet weight	570 l S. Louisiana crude oil in 810 m2 shoreline area, 216 h after spill[1]
Fundulus heteroclitus	0.37 mg/kg wet weight	570 l S. Louisiana crude oil in 810 m2 shoreline area, 31 h after spill[1]
Fundulus heteroclitus	0.47 mg/kg wet weight	570 l S. Louisiana crude oil in 810 m2 shoreline area, 76 h after spill[1]

Source: 1.Bieri, R.H., V.C. Stamoudis and M.K. Cueman. 1977. Proc. of 1977 Oil spill Conference (Prevention, Behavior, Control, Cleanup). Wash. DC. American Petrolium Institute. 511-516.

5-285 Concentrations
1-Methylnaphthalene

Species	Value	Condition
Crassostrea virginica	4.7 mg/kg wet weight	Upt in 8 h from 302 ppm No. 2 fuel, 0.82 % 1-MN, 120 h excr, init=8.7 ppm[1]
Crassostrea virginica	5.1 mg/kg wet weight	Upt in 8 h from 302 ppm No. 2 fuel, 0.82 % 1-MN, 6 h excr, init=8.7 ppm[1]
Crassostrea virginica	5.7 mg/kg wet weight	Upt 8 h from 302 ppm No. 2 fuel, 0.82 % 1-MN, 24 h excr, init=8.7 ppm[1]
Crassostrea virginica	8.4 mg/kg wet weight	Upt in 8 h from 302 ppm No. 2 fuel, 0.82 % 1-MN, 3 h excr, init=8.7 ppm[1]
Crassostrea virginica	8.7 mg/kg wet weight	302 ppm dispersed No. 2 fuel oil with 0.82 % 1-MN, flow through, 8 h[1]
Fundulus heteroclitus	0.13 mg/kg wet weight	570 l S. Louisiana crude oil in 810 m2 shoreline area, 216 h after spill[2]
Fundulus heteroclitus	0.34 mg/kg wet weight	570 l S. Louisiana crude oil in 810 m2 shoreline area, 6 h after spill[2]
Fundulus heteroclitus	0.98 mg/kg wet weight	570 l S. Louisiana crude oil in 810 m2 shoreline area, 76 h after spill[2]
Fundulus heteroclitus	1.26 mg/kg wet weight	570 l S. Louisiana crude oil in 810 m2 shoreline area, 31 h after spill[2]
Marine sediment near oil separator platform	200 ug/kg wet weight	Trinity B., TEX, USA, sample 15 m from brine effluent = 370 ug/l[3]
Seawater near oil separator platform	0.20 ug/liter	Trinity B., TEX, USA, sample 15 m from brine effluent = 370 ug/l[3]

Source: 1.Neff, J.M., B.A. Cox, D. Dixit and J.W. Anderson. 1976. Mar. Biol. 38, 279-289.
2.Bieri, R.H., V.C. Stamoudis and M.K. Cueman. 1977. Proc. of 1977 Oil spill Conference (Prevention, Behavior, Control, Cleanup). Wash. DC. American Petrolium Institute. 511-516.
3.Armstrong, H.W., K. Fucik, J.W. Anderson and J.M. Neff. 1977. Effects of oilfield brine effluent on benthic organisms in Trinity Bay. American Petrolium Institute Publ. No. 4291. Wash. DC.

5-286 Concentrations
1-Naphthol

Species	Value	Condition
Oncorhynchus kisutch, brain	32 % of total	Distribution 24 h after intraperi. inj. of 2.5 uCi naphthalene-1-14C[1]
Oncorhynchus kisutch, flesh	47 % of total	Distribution 24 h after intraperi. inj. of 2.5 uCi naphthalene-1-14C[1]
Oncorhynchus kisutch, gall bladder	7 % of total	Distribution 24 h after intraperi. inj. of 2.5 uCi naphthalene-1-14C[1]
Oncorhynchus kisutch, heart	61 % of total	Distribution 24 h after intraperi. inj. of 2.5 uCi naphthalene-1-14C[1]
Oncorhynchus kisutch, liver	40 % of total	Distribution 24 h after intraperi. inj. of 2.5 uCi naphthalene-1-14C[1]

Source: 1.Roubal, W.T., D.H. Bovee, T.K. Collier and S.I. Stranahan. 1977. Proc of 1977 Oil Spill Conf. (Prevention, Behavior, Control, Cleanup). Wash DC. American Petroleum Institute. 551-556.

5-287 Concentrations
1-Naphthyl glycuronic acid

Species	Value	Condition
Oncorhynchus kisutch, brain	35 % of total	Distribution 24 h after intraperi. inj. of 2.5 uCi naphthalene-1-14C[1]
Oncorhynchus kisutch, flesh	14 % of total	Distribution 24 h after intraperi. inj. of 2.5 uCi naphthalene-1-14C[1]
Oncorhynchus kisutch, gall bladder	75 % of total	Distribution 24 h after intraperi. inj. of 2.5 uCi naphthalene-1-14C[1]
Oncorhynchus kisutch, heart	13 % of total	Distribution 24 h after intraperi. inj. of 2.5 uCi naphthalene-1-14C[1]
Oncorhynchus kisutch, liver	25 % of total	Distribution 24 h after intraperi. inj. of 2.5 uCi naphthalene-1-14C[1]

Source: 1.Roubal, W.T., D.H. Bovee, T.K. Collier and S.I. Stranahan. 1977. Proc of 1977 Oil Spill Conf. (Prevention, Behavior, Control, Cleanup). Wash DC. American Petroleum Institute. 551-556.

5-288 Concentrations
2,3,5-Trimethylnaphthalene

Species	Value	Condition
Fundulus heteroclitus	0.07 mg/kg wet weight	570 l S. Louisiana crude oil in 810 m2 shoreline area, 6 h after spill[1]
Fundulus heteroclitus	0.08 mg/kg wet weight	570 l S. Louisiana crude oil in 810 m2 shoreline area, 216 h after spill[1]
Fundulus heteroclitus	0.13 mg/kg wet weight	570 l S. Louisiana crude oil in 810 m2 shoreline area, 31 h after spill[1]
Fundulus heteroclitus	0.25 mg/kg wet weight	570 l S. Louisiana crude oil in 810 m2 shoreline area, 76 h after spill[1]

Source: 1.Bieri, R.H., V.C. Stamoudis and M.K. Cueman. 1977. Proc. of 1977 Oil spill Conference (Prevention, Behavior, Control, Cleanup). Wash. DC. American Petrolium Institute. 511-516.

5-289 Concentrations
2,3-Dimethylnaphthalene

Species	Value	Condition
Fundulus heteroclitus	0.03 mg/kg wet weight	570 l S. Louisiana crude oil in 810 m2 shoreline area, 76 h after spill[1]
Fundulus heteroclitus	0.04 mg/kg wet weight	570 l S. Louisiana crude oil in 810 m2 shoreline area, 6 h after spill[1]
Fundulus heteroclitus	0.05 mg/kg wet weight	570 l S. Louisiana crude oil in 810 m2 shoreline area, 216 h after spill[1]
Fundulus heteroclitus	0.05 mg/kg wet weight	570 l S. Louisiana crude oil in 810 m2 shoreline area, 31 h after spill[1]

Source: 1.Bieri, R.H., V.C. Stamoudis and M.K. Cueman. 1977. Proc. of 1977 Oil spill Conference (Prevention, Behavior, Control, Cleanup). Wash. DC. American Petrolium Institute. 511-516.

5-290 Concentrations
2-Methylnaphthalene

Species	Value	Condition
Crassostrea virginica	0.1 mg/kg wet weight	Upt in 8 h from 302 ppm No. 2 fuel, 1.89 % 2-MN, 672 h excr, init=15.0 ppm[1]
Crassostrea virginica	12.0 mg/kg wet weight	Upt in 8 h from 302 ppm No. 2 fuel, 1.89 % 2-MN, 3 h excr, init=15.0 ppm[1]
Crassostrea virginica	15.0 mg/kg wet weight	302 ppm dispersed No. 2 fuel oil with 1.89 % 2-MN,flow through, 8 h[1]
Crassostrea virginica	6.8 mg/kg wet weight	Upt in 8 h from 302 ppm No. 2 fuel, 1.89 % 2-MN, 120 h excr, init=15.0 ppm[1]
Crassostrea virginica	7.3 mg/kg wet weight	Upt in 8 h from 302 ppm No. 2 fuel, 1.89 % 2-MN, 6 h excr, init=15.0 ppm[1]
Crassostrea virginica	7.6 mg/kg wet weight	Upt in 8 h from 302 ppm No. 2 fuel, 1.89 % 2-MN, 24 h excr, init=15.0 ppm[1]
Fundulus heteroclitus	0.34 mg/kg wet weight	570 l S. Louisiana crude oil in 810 m2 shoreline area, 6 h after spill[2]
Fundulus heteroclitus	1.12 mg/kg wet weight	570 l S. Louisiana crude oil in 810 m2 shoreline area, 76 h after spill[2]
Fundulus heteroclitus	1.56 mg/kg wet weight	570 l S. Louisiana crude oil in 810 m2 shoreline area, 31 h after spill[2]
Marine sediment near oil separator platform	300 ug/kg wet weight	Trinity B., TEX, USA, sample 15 m from brine effluent = 500 ug/l[3]
Seawater near oil separator platform	0.60 ug/liter	Trinity B., TEX, USA, sample 15 m from brine effluent = 500 ug/l[3]

Source: 1.Neff, J.M., B.A. Cox, D. Dixit and J.W. Anderson. 1976. Mar. Biol. 38, 279-289.
2.Bieri, R.H., V.C. Stamoudis and M.K. Cueman. 1977. Proc. of 1977 Oil spill Conference (Prevention, Behavior, Control, Cleanup). Wash. DC. American Petrolium Institute. 511-516.
3.Armstrong, H.W., K. Fucik, J.W. Anderson and J.M. Neff. 1977. Effects of oilfield brine effluent on benthic organisms in Trinity Bay. American Petrolium Institute Publ. No. 4291. Wash. DC.

5-291 Concentrations
3-Methylnaphthalene, 4-Methylnaphthalene, Naphthal.

Species	Value	Condition
Fundulus heteroclitus	0.10 mg/kg wet weight	570 l S. Louisiana crude oil in 810 m2 shoreline area, 6 h after spill[1]
Fundulus heteroclitus	0.14 mg/kg wet weight	570 l S. Louisiana crude oil in 810 m2 shoreline area, 216 h after spill[1]
Fundulus heteroclitus	0.27 mg/kg wet weight	570 l S. Louisiana crude oil in 810 m2 shoreline area, 31 h after spill[1]
Fundulus heteroclitus	0.56 mg/kg wet weight	570 l S. Louisiana crude oil in 810 m2 shoreline area, 76 h after spill[1]

Source: 1.Bieri, R.H., V.C. Stamoudis and M.K. Cueman. 1977. Proc. of 1977 Oil spill Conference (Prevention, Behavior, Control, Cleanup). Wash. DC. American Petrolium Institute. 511-516.

5-292 Concentrations
Ag

Species	Value	Condition
Zooplankton	0.96 mg/kg dry weight	n = 19, NY, conc in algae: < 0.61 mg/kg dw, mean, Long Isl, less than value[1]

Source: 1.Greig, R.A., A. Adams and D.R. Wenzloff. 1977. Bull. Environ. Contam. and Toxic., 18(1), pp 3-8.

5-293 Concentrations
Aldrin

Species	Value	Condition
Sediment	0.001 mg/kg wet weight	Conc in water: < 1.0 ug/l, USA, less than value[1]
Sediment	0.01 mg/kg wet weight	Conc in water: < 1.0 ug/l, USA, less than value[1]
Sediment	0.17 mg/kg wet weight	Conc in water = 0.014 ug/l, USA[1]

Source: 1.Edwards, C.A. 1975. Persistent Pesticides in the Environment, 2nd edition, CRC-Press.

5-294 Concentrations
Anthracene

Species	Value	Condition
Crassostrea virginica	2.5 mg/kg dry weight	Exposure = 8 d, water = 1 ug/l, crude oil, lab[1]
Crassostrea virginica	5.6 mg/kg dry weight	Exposure = 2 d, water = 13 ug/l, crude oil, lab[1]

Source: 1.Lee, R.F., W.S. Gardner, J.W. Anderson, Balylock and B-Clarke. 1978. Environ. Sci. Technol. 12, 832-38.

5-295 Concentrations
As

Species	Value	Condition
Sediment	17.5 mg/kg dry weight	Water conc = < 10 mg/m3, geologigal age = permian, rg = < 5 - 30[1]

Source: 1.Aston, S.R., I. Thornton and J.S. Webb. 1975. Sci. Total Environ., 4(4), pp 345-358.

5-296 Concentrations
Atrazine

Species	Value	Condition
Oat, root	0.195 mg/kg dry weight	Soil, added atrazine = 0.5 ppm, lab, 293-303 K, 2 w, hydroxy-atrazine[1]
Oat, root	0.312 mg/kg dry weight	Soil, added atrazine = 4.5 kg/ha, 293-303 K, 2 w, hydroxy - atrazine[1]
Oat, root	0.464 mg/kg dry weight	Soil, added atrazine = 4.5 kg/ha, 293-303 K, 2 w, hydroxy-atrazine[1]
Oat, root	0.564 mg/kg wet weight	Soil, added atrazine = 4.5 kg/ha, 293-303 K, 2 w, hydroxy-atrazine[1]
Oat, root	0.634 mg/kg dry weight	Soil, added atrazine = 4.5 kg/ha, 293-303 K, 2 w, hydroxy-atrazine[1]
Oat, root	0.653 mg/kg dry weight	Soil, added atrazine = 4.5 kg/ha, 293-303 K, 2 w, hydroxy - atrazine[1]
Oat, root	0.660 mg/kg dry weight	Soil, added atrazine = 4.5 kg/ha, 293-303 K, 2 w, hydroxy - atrazine[1]
Oat, root	0.670 mg/kg dry weight	Soil, added atrazine = 1.0 ppm, lab, 293-303 K, 2 w, hydroxy-atrazine[1]
Oat, root	0.950 mg/kg dry weight	Soil, added atrazine = 2.0 ppm, lab, 293-303 K, 2 w, hydroxy-atrazine[1]
Oat, root	1.030 mg/kg dry weight	Soil, added atrazine = 4.5 ppm, lab, 293-303 K, 2 w, hydroxy-atrazine[1]
Oat, shoots	0.193 mg/kg wet weight	Soil, added atrazine = 4.5 kg/ha, 293-303 K, 2 w, hydroxy-atrazine[1]
Oat, shoots	0.236 mg/kg wet weight	Soil, added atrazine = 0.5 ppm, lab, 293-303 K, 2 w, hydroxy-atrazine[1]
Oat, shoots	0.330 mg/kg wet weight	Soil, added atrazine = 1.0 ppm, lab, 293-303 K, 2 w, hydroxy-atrazine[1]
Oat, shoots	0.392 mg/kg wet weight	Soil, add atrazine = 4.5 ppm, lab, 293-303 K, 2 w, hydroxy-atrazine[1]
Oat, shoots	0.409 mg/kg wet weight	Soil, added atrazine = 4.5 kg/ha, 293-303 K, 2 w, hydroxy-atrazine[1]
Oat, shoots	0.514 mg/kg wet weight	Soil, added atrazine = 2.0 ppm, lab, 293-303 K, 2 w, hydroxy-atrazine[1]

Source: 1.Khan, S.U. and P.B. Marriage. 1977. J. Agric. Food. Chem. 25(6), pp 1408-1413.

5-297 Concentrations
B

Species	Value	Condition
Cabbage, plant	21. mg/kg dry weight	15 plants, rural area[1]
Cabbage, plant	28.8 mg/kg dry weight	15 plants, urban area[1]
Oncorrhynchus kisutch	224 mg/kg dry weight	Boron in water = 328 mg/l, 84 h, 20 fish[2]
Oncorrhynchus kisutch	354 mg/kg dry weight	Boron in water = 656 mg/l, 33.4 h, 20 fish[2]

Oncorrhynchus kisutch	391 mg/kg dry weight	Boron in water = 656 mg/l, 33.4 h, 20 fish[2]

Source: *1.Purves, D. 1972. Environ. Pollut., 3, pp 17-24.*
2.Thompson, J.A.J. and J.C. Davis. 1976. Water Res., 10(10), pp 869-875.

5-298 Concentrations BHC

Species	Value	Condition
Chicken, brain	0.50 mg/kg wet weight	Conc in food = 10.00 mg/kg wet weight[1]
Chicken, fat	1.18 mg/kg wet weight	Conc in food = 1.00 mg/kg wet weight[1]
Chicken, fat	1.20 mg/kg wet weight	Conc in food = 1.00 mg/kg wet weight[1]
Chicken, fat	20.38 mg/kg wet weight	Conc in food = 10.00 mg/kg wet weight[1]
Chicken, fat	21.0 mg/kg wet weight	Conc in food = 10.00 mg/kg wet weight[1]
Chicken, liver	0.026 mg/kg wet weight	Conc in food = 0.05 mg/kg wet weight[1]
Chicken, liver	0.039 mg/kg wet weight	Conc in food = 0.15 mg/kg wet weight[1]
Chicken, liver	0.082 mg/kg wet weight	Conc in food = 0.45 mg/kg wet weight[1]
Feral pigeon, tissue	144.0 mg/kg wet weight	Conc in food = 72 mg/kg wet weight[1]
Sediment	0.0002 mg/kg wet weight	Conc in water = 0.03 ug/l, USA, less than value[1]
Sediment	0.001 mg/kg wet weight	Conc in water: < 1.0 ug/l, USA, less than value[1]
Sediment	0.01 mg/kg wet weight	Conc in water: < 1.0 ug/l, USA, less than value[1]
Sediment	0.027 mg/kg wet weight	Conc in water = 0.040 ug/l, USA[1]
Soil	0.1 mg/kg wet weight	Joydabpur(b), Dacca, gamma BHC, 10 g[2]
Soil	0.1 mg/kg wet weight	Mirpur, Dacca, gamma BHC, 10 g[2]
Soil	0.43 mg/kg wet weight	Demra(c), Dacca, gamma BHC 10 g[2]
Soil	0.45 mg/kg wet weight	Demra(a), Dacca, gamma BHC, 10 g[2]
Soil	0.5 mg/kg wet weight	Demra(b), Dacca, gamma BHC, 10 g[2]
Soil	0.5 mg/kg wet weight	Joydabpur(a), Dacca, gamma BHC, 10 g[2]

Source: 1.Edwards, C.A. 1975. Persistent Pesticides in the Environment, 2nd edition, CRC-Press.
2.Nasim, A.L., M.M.H. Paig and K.A. Lord. 1971. Environ. Pollut., 2, pp 7-12.

5-299 Concentrations BOD

Species	Value	Condition
Agricultural cropland	7 mg/l	General characteristics, runoff[1]
Animal feedlot, runoff	6000 mg/l	General characteristics, rg = 1000 - 11000[1]
Municipal sewage	135 mg/l	Typical treated, primary, rg = 70 - 200[1]
Municipal sewage	200 mg/l	Typical untreated, rg = 100 - 300[1]
Municipal sewage	25 mg/l	Typical treated, secondary, rg = 15 - 45[1]
Precipitation	12.5 mg/l	General characteristics, rg = 12 - 13[1]
Urban land, drainage	86 mg/l	General characteristics, rg = 12 - 160[1]

Source: 1.Donigian, A.S. and N.H. Crawford. 1976. Modeling nonpoint Pollution from the Land Surface, EPA-600/376-083. July.

5-300 Concentrations
Be

Species	Value	Condition
Rat, lung	2 mg/kg dry weight	In equilibrium with air of 1.75 ug/m3, 36 w[1]

Source: 1.Reeves, A.L. 1967. Cancer Res. 27, pp 1895-1899.

5-301 Concentrations
Benz(a)anthracene

Species	Value	Condition
Crassostrea virginica	1.8 mg/kg dry weight	Exposure = 8 d, water = 0.1 ug/l, crude oil, lab[1]
Crassostrea virginica	2.8 mg/kg dry weight	Exposure = 2 d, water = 5.3 ug/l, crude oil, lab[1]

Source: 1.Lee, R.F., W.S. Gardner, J.W. Anderson, Balylock and B-Clarke. 1978. Environ. Sci. Technol. 12, 832-38.

5-302 Concentrations
Benzene

Species	Value	Condition
Seawater near oil separator platform	1.5 ug/liter	Trinity B., TEX, USA, sample 15 m from brine effluent = 3300 ug/l[1]

Source: 1.Armstrong, H.W., K. Fucik, J.W. Anderson and J.M. Neff. 1977. Effects of oilfield brine effluent on benthic organisms in Trinity Bay. American Petrolium Institute Publ. No. 4291. Wash. DC.

5-303 Concentrations
Benzo(a)pyrene

Species	Value	Condition
Aquatic vascular plants	376 ug/kg dry weight	Clipperton lagoon, eq Pacific, seawater = 1.6 ppb, inside lagoon, 1969[1]
Benthic animals, muscle	110 ug/kg dry weight	Dunkerque, France, marine, sediment = 220 ppb[2]
Benthic animals, muscle	20 ug/kg dry weight	Dunkerque, France, marine, sediment = 20 ppb[2]
Benthic animals, muscle	20 ug/kg dry weight	Seine estuary, France, marine, sediment = 80 ppb[2]
Benthic animals, muscle	30 ug/kg dry weight	Seine estuary, France, marine, sediment = 100 ppb[2]
Benthic animals, muscle	40 ug/kg dry weight	Seine estuary, France, marine, sediment = 130 ppb[2]
Benthic animals, muscle	45 ug/kg dry weight	Dunkerque, France, marine, sediment = 180 ppb, rg = 40-50 ppb[2]
Callinectes sapidus, blood	12E-4 ug total in tissue	2 d upt. from wat=2.5 ug BaP/l, init=30, juv, marine, n=3, 2 d excr[3]
Callinectes sapidus, blood	30E-4 ug total in tissue	Water = 2.5 ug BaP/l, juvenile, marine, n=3, 2 d exp+- 85E-4[3]
Callinectes sapidus, blood	50E-4 ug total in tissue	Water = 2.5 ug BaP/l, juvenile, marine, n=3, 1 d exp+- 70E-4[3]
Callinectes sapidus, gill	110E-4 ug total in tissue	Water = 2.5 ug BaP/l, juvenile, marine, n=3, 1 d exp+- 70E-4[3]
Callinectes sapidus, gill	290E-4 ug total in tissue	Water = 2.5 ug BaP/l, juvenile, marine, n=3, 2 d exp+- 85E-4[3]
Callinectes sapidus, gill	70E-4 ug total in tissue	2 d upt. from wat=2.5 ug BaP/l, init=290, juv, marine, n=3, 2 d excr[3]
Callinectes sapidus, hepatopancreas	160E-4 ug total in tissue	Water = 2.5 ug BaP/l, juvenile, marine, n=3, 1 d exp+- 70E-4[3]
Callinectes sapidus, hepatopancreas	280E-4 ug total in tissue	Water = 2.5 ug BaP/l, juvenile, marine, n=3, 2 d exp+- 85E-4[3]
Callinectes sapidus, hepatopancreas	40E-4 ug total in tissue	2 d upt. from wat=2.5 ug BaP/l, init=280, juv, marine, n=3, 2 d excr[3]
Callinectes sapidus, muscle	2E-4 ug total in tissue	2 d upt. from wat=2.5 ug BaP/l, init=2, juv, marine, n=3, 2 d excr[3]

Callinectes sapidus, muscle	2E-4 ug total in tissue	Water = 2.5 ug BaP/l, juvenile, marine, n=3, 2 d exp+- 85E-4[3]
Callinectes sapidus, muscle	4E-4 ug total in tissue	Water = 2.5 ug BaP/l, juvenile, marine, n=3, 1 d exp+- 70E-4[3]
Callinectes sapidus, stomach	1E-4 ug total in tissue	2 d upt. from wat=2.5 ug BaP/l, init=2, juv, marine, n=3, 2 d excr[3]
Callinectes sapidus, stomach	2E-4 ug total in tissue	Water = 2.5 ug BaP/l, juvenile, marine, n=3, 2 d exp+- 85E-4[3]
Callinectes sapidus, stomach	3E-4 ug total in tissue	Water = 2.5 ug BaP/l, juvenile, marine, n=3, 1 d exp+- 70E-4[3]
Caranx lugibris, integument	13.6 ug/kg dry weight	Clipperton lagoon, eq Pacific, muscle = 0.5 ppb, seawater=1.6 ppb, 1969[1]
Caranx lugibris, intestine	5.5 ug/kg dry weight	Clipperton lagoon, eq Pacific, muscle = 0.5 ppb, seawater=1.6 ppb, 1969[1]
Caranx lugibris, viscera	2.4 ug/kg dry weight	Clipperton lagoon, eq Pacific, muscle = 0.5 ppb, seawater=1.6 ppb, 1969[1]
Cenchus sp. and Eragrostis sp.	71.5 ug/kg dry weight	Clipperton lagoon, eq Pacific, seawater = 1.6 ppb, shore of lag., 1969[1]
Crassostrea virginica	0.30 mg/kg dry weight	Exposure = 8 d, water = 0.1 ug/l, crude oil, lab[4]
Crassostrea virginica	0.36 mg/kg dry weight	Exposure = 2 d, water = 1.9 ug/l, crude oil, lab[4]
Kulhia taeneura, muscle	4.0 ug/kg dry weight	Clipperton Lagoon, Eq. Pacific, water=1.6 ppb, 1969, integument = 26 ppb[1]
Lygia sp.	536 ug/kg dry weight	Clipperton lagoon, eq Pacific, seawater = 1.6 ppb, inside lagoon, 1969[1]
Mugil chelo, muscle	17.3 ug/kg dry weight	Scales conc = 22.0 ppb, Bay of St. Malo, France, 1966[5]
Mugil chelo, viscera	155.0 ug/kg dry weight	Scales conc = 22.0 ppb, Bay of St. Malo, France, 1965[5]
Phytoplankton and zooplankton	7.3 ug/kg dry weight	Clipperton lagoon, eq Pacific, seawater = 1.6 ppb, inside lagoon, 1969[1]

Source: 1.Niaussat, P. and C. Auger. 1970. C. R. Acad. Sci. (Paris) Ser. D. 270, 2702-2705.
2.Perdriau, J. 1964. Cah. Oceanogr. 16, 205-229.
3.Lee, R.F., C. Ryan and M.L. Neuhauser. 1976. Mar. Biol. 37, 363-370.
4.Lee, R.F., W.S. Gardner, J.W. Anderson, Balylock and B-Clarke. 1978. Environ. Sci. Technol. 12, 832-38.
5.Neff, J.M. 1979. Polycyclic Aromatic Hydrocarbons in the Aquatic Environment. Applied Sci. Publ. London.

5-304 Concentrations
Biphenyl

Species	Value	Condition
Crassostrea virginica	0.1 mg/kg wet weight	Upt in 8 h from 302 ppm No. 2 fuel, 0.16 % B, 120 h excr, init=9.1 ppm[1]
Crassostrea virginica	0.1 mg/kg wet weight	Upt in 8 h from 302 ppm No. 2 fuel, 0.16 % B, 6 h excr, init=9.1 ppm[1]
Crassostrea virginica	0.2 mg/kg wet weight	Upt in 8 h from 302 ppm No. 2 fuel, 0.16 % B, 24 h excr, init=9.1 ppm[1]
Crassostrea virginica	0.3 mg/kg wet weight	302 ppm dispersed No. 2 fuel oil with 0.16 % B, flow through, 8 h[1]
Crassostrea virginica	0.3 mg/kg wet weight	Upt in 8 h from 302 ppm No. 2 fuel, 0.16 % B, 3 h excr, init=9.1 ppm[1]

Source: 1.Neff, J.M., B.A. Cox, D. Dixit and J.W. Anderson. 1976. Mar. Biol. 38, 279-289.

5-305 Concentrations
Biphenyl and 2,6-Dimethylnaphthalene

Species	Value	Condition
Fundulus heteroclitus	0.25 mg/kg wet weight	570 l S. Louisiana crude oil in 810 m2 shoreline area, 216 h after spill[1]
Fundulus heteroclitus	0.25 mg/kg wet weight	570 l S. Louisiana crude oil in 810 m2 shoreline area, 6 h after spill[1]
Fundulus heteroclitus	0.81 mg/kg wet weight	570 l S. Louisiana crude oil in 810 m2 shoreline area, 31 h after spill[1]
Fundulus heteroclitus	0.99 mg/kg wet weight	570 l S. Louisiana crude oil in 810 m2 shoreline area, 76 h after spill[1]

Source: 1.Bieri, R.H., V.C. Stamoudis and M.K. Cueman. 1977. Proc. of 1977 Oil spill Conference (Prevention, Behavior, Control, Cleanup). Wash. DC. American Petrolium Institute. 511-516.

5-306 Concentrations
Br

Species	Value	Condition
Stigeoclonium pachyderum	72.8 % of the amount added to medium	Medium = 1 mg/l, 45 min, 298 K, 40 w[1]

Source: 1.Vance, B.D. and A.W. Maki. 1976. Bull. Environ. Contam. Toxicol. 15(5), pp 601-607.

5-307 Concentrations
C12 - C24 n-paraffins

Species	Value	Condition
Crassostrea virginica	10 mg/kg wet weight	Upt in 8 h from 302 ppm No. 2 fuel, 7.38 % n-P, 120 h excr, init=235 ppm[1]
Crassostrea virginica	156 mg/kg wet weight	Upt in 8 h from 302 ppm No. 2 fuel, 7.38 % n-P, 3 h excr, init=235 ppm[1]
Crassostrea virginica	18 mg/kg wet weight	Upt in 8 h from 302 ppm No. 2, 7.38 % n-P, 24 h excr, init=235 ppm[1]
Crassostrea virginica	235 mg/kg wet weight	302 ppm dispersed No. 2 fuel oil with 7.38 % n-P, flow through, 8 h[1]
Crassostrea virginica	68 mg/kg wet weight	Upt in 8 h from 302 ppm No. 2 fuel, 7.38 % n-P, 6 h excr, init=235 ppm[1]

Source: 1.Neff, J.M., B.A. Cox, D. Dixit and J.W. Anderson. 1976. Mar. Biol. 38, 279-289.

5-308 Concentrations
C12 - C28 n-paraffins

Species	Value	Condition
Crassostrea virginica	0.10 mg/kg wet weight	Exposed 0[1]
Crassostrea virginica	0.10 mg/kg wet weight	Exposed 0 d to sediment with 887.4 ug total oil/g sed, prudhoe B crude oil[1]
Crassostrea virginica	0.69 mg/kg wet weight	Exposed 40 d to sediment with 887.4 ug total oil/g sed, prudhoe B crude oil[1]
Crassostrea virginica	2.08 mg/kg wet weight	Exposed 60 d to sediment with 887.4 ug total oil/g sed, prudhoe B crude oil[1]
Mytilus edulis	0.10 mg/kg wet weight	Exposed 0[1]
Mytilus edulis	0.10 mg/kg wet weight	Exposed 0 d to sediment with 887.4 ug total oil/g sed, prudhoe B crude oil[1]
Mytilus edulis	1.32 mg/kg wet weight	Exposed 40 d to sediment with 887.4 ug total oil/g sed, prudhoe B crude oil[1]
Mytilus edulis	1.48 mg/kg wet weight	Exposed 60 d to sediment with 887.4 ug total oil/g sed, prudhoe B crude oil[1]
Nephthys incisa	0.10 mg/kg wet weight	Exposed 0[1]
Nephthys incisa	0.10 mg/kg wet weight	Exposed 0 d to sediment with 887.4 ug total oil/g sed, prudhoe B crude oil[1]
Nephthys incisa	0.10 mg/kg wet weight	Exposed 40 d to sediment with 887.4 ug total oil/g sed, prudhoe B crude oil[1]
Nephthys incisa	0.10 mg/kg wet weight	Exposed 60 d to sediment with 887.4 ug total oil/g sed, prudhoe B crude oil[1]

Source: 1.Roesijadi, G., J.W. Anderson and J.W. Blaylock. 1978. J. Fish. Res. bd Canada 35, 608-614.

5-309 Concentrations
C3 - Benzene

Species	Value	Condition
Marine sediment near oil separator platform	600 ug/kg wet weight	Trinity B., TEX, USA, sample 15 m from brine effluent = 650 ug/l[1]
Seawater near oil separator platform	0.80 ug/liter	Trinity B., TEX, USA, sample 15 m from brine effluent = 650 ug/l[1]

Source: 1.Armstrong, H.W., K. Fucik, J.W. Anderson and J.M. Neff. 1977. Effects of oilfield brine effluent on benthic organisms in Trinity Bay. American Petrolium Institute Publ. No. 4291. Wash. DC.

5-310 Concentrations
C3 - Naphthalene

Species	Value	Condition
Fundulus heteroclitus	0.09 mg/kg wet weight	570 l S. Louisiana crude oil in 810 m2 shoreline area, 6 h after spill[1]
Fundulus heteroclitus	0.12 mg/kg wet weight	570 l S. Louisiana crude oil in 810 m2 shoreline area, 216 h after spill[1]
Fundulus heteroclitus	0.20 mg/kg wet weight	570 l S. Louisiana crude oil in 810 m2 shoreline area, 31 h after spill[1]
Fundulus heteroclitus	0.42 mg/kg wet weight	570 l S. Louisiana crude oil in 810 m2 shoreline area, 76 h after spill[1]

Source: 1.Bieri, R.H., V.C. Stamoudis and M.K. Cueman. 1977. Proc. of 1977 Oil spill Conference (Prevention, Behavior, Control, Cleanup). Wash. DC. American Petrolium Institute. 511-516.

5-311 Concentrations
C3 - Naphthalene and C2 - Biphenyl

Species	Value	Condition
Fundulus heteroclitus	0.03 mg/kg wet weight	570 l S. Louisiana crude oil in 810 m2 shoreline area, 216 h after spill[1]
Fundulus heteroclitus	0.03 mg/kg wet weight	570 l S. Louisiana crude oil in 810 m2 shoreline area, 6 h after spill[1]
Fundulus heteroclitus	0.09 mg/kg wet weight	570 l S. Louisiana crude oil in 810 m2 shoreline area, 31 h after spill[1]
Fundulus heteroclitus	0.22 mg/kg wet weight	570 l S. Louisiana crude oil in 810 m2 shoreline area, 76 h after spill[1]

Source: 1.Bieri, R.H., V.C. Stamoudis and M.K. Cueman. 1977. Proc. of 1977 Oil spill Conference (Prevention, Behavior, Control, Cleanup). Wash. DC. American Petrolium Institute. 511-516.

5-312 Concentrations
C4 - Naphthalenes, C2 - Biphenyl and C3 - Biphenyl

Species	Value	Condition
Fundulus heteroclitus	0.03 mg/kg wet weight	570 l S. Louisiana crude oil in 810 m2 shoreline area, 216 h after spill[1]
Fundulus heteroclitus	0.03 mg/kg wet weight	570 l S. Louisiana crude oil in 810 m2 shoreline area, 6 h after spill[1]
Fundulus heteroclitus	0.04 mg/kg wet weight	570 l S. Louisiana crude oil in 810 m2 shoreline area, 31 h after spill[1]
Fundulus heteroclitus	0.11 mg/kg wet weight	570 l S. Louisiana crude oil in 810 m2 shoreline area, 76 h after spill[1]

Source: 1.Bieri, R.H., V.C. Stamoudis and M.K. Cueman. 1977. Proc. of 1977 Oil spill Conference (Prevention, Behavior, Control, Cleanup). Wash. DC. American Petrolium Institute. 511-516.

5-313 Concentrations
CCl3F

Species	Value	Condition
Troposphere	50 ppt	1950-71, global production = 1.48E6 tons, load = 1.01E6 tons[1]
Troposphere	60 ppt	1972, global production = 0.27E6 tons, load = 1.21E6 tons[1]
Troposphere	80 ppt	1974, global production = 0.27E6 tons, load = 1.62E6 tons[1]

Source: 1.Wilkniss, P.E., J.W. Swinnerton, R.A. Lamontague and D.J. Bressan. 1975. Science 187, pp 832.

5-314 Concentrations COD

Species	Value	Condition
Agricultural cropland	80 mg/l	General characteristics, runoff[1]
Animal feedlot	22050 mg/l	General characteristics, runoff, rg = 3100 - 41000[1]
Municipal sewage	330 mg/l	Typical treated, primary, rg = 165 - 500[1]
Municipal sewage	500 mg/l	Typical untreated, rg = 250 - 750[1]
Municipal sewage	55 mg/l	Typical treated, secondary, rg = 25 - 80[1]
Precipitation	12.5 mg/l	General characteristics, rg = 9 - 16[1]
Urban land, drainage	97.5 mg/l	General characteristics, rg = 85 - 110[1]

Source: 1.Donigian, A.S. and N.H. Crawford. 1976. Modeling nonpoint Pollution from the Land Surface, EPA-600/376-083. July.

5-315 Concentrations Ca

Species	Value	Condition
Different species in freshwater, Ca bound with Humic	0.7 mol %	298 K, analytic data, 1 atm, freshwater, L. Celyn[1]
Different species in freshwater, Ca(2+)	97.8 mol %	298 K, analytic data, 1 atm, freshwater, L. Celyn[1]
Different species in freshwater, CaHCO3(-)	0.9 mol %	298 K, analytic data, 1 atm, freshwater, L. Celyn[1]
Different species in freshwater, CaSO4	0.6 mol %	298 K, analytic data, 1 atm, freshwater, L. Celyn[1]
Different species in seawater, Ca(2+)	90.6 mol %	298 K, pH = 8.0, 1 atm, seawater, model distribution[1]
Different species in seawater, CaCO3	0.2 mol %	298 K, pH = 8.0, 1 atm, seawater, model distribution[1]
Different species in seawater, CaSO4	9.2 mol %	298 K, pH = 8.0, 1 atm, seawater, model distribution[1]

Source: 1.Mantoura, R.F.C., A. Dickson and J.P. Riley. 1978. Estuarine Coastal Mar. Sci. 6, 387-408.

5-316 Concentrations Camphechlor

Species	Value	Condition
Sediment	0.01 mg/kg wet weight	Conc in water: < 1.0 ug/l, USA, less than value[1]
Sediment	0.65 mg/kg wet weight	Conc in water = 0.63 ug/l, USA[1]

Source: 1.Edwards, C.A. 1975. Persistent Pesticides in the Environment, 2nd edition, CRC-Press.

5-317 Concentrations Cd

Species	Value	Condition
Alfalfa, root	10.51 mg/kg dry weight	Cd added = 5.0 ug/g soil, Grenville loam, control = 0.52 ppm[1]
Alfalfa, root	7.99 mg/kg dry weight	Cd added = 2.5 ug/g soil, Grenville loam, control = 0.52 ppm[1]
Alfalfa, top	1.34 mg/kg dry weight	Cd added = 2.5 ug/g soil, Grenville loam, control = 0.28 ppm[1]
Alfalfa, top	1.72 mg/kg dry weight	Cd added = 5.0 ug/g soil, Grenville loam, control = 0.28 ppm[1]
Aquipecten irradians	19 % more than control	Seawater, 10 ug Cd/l added as CdCl2, 21 d[2]
Avena sativa, grains	20.8 mg/kg dry weight	Cd in control plants = 3.9 ppm, Cd in soil = 40 ppm[3]
Avena sativa, grains	33.6 mg/kg dry weight	Cd in control plants = 3.9 ppm, Cd in soil = 200 ppm[3]
Avena sativa, husks	37.8 mg/kg dry weight	Cd in control plants = 4.5 ppm, Cd in soil = 40 ppm[3]

Avena sativa, husks	96 mg/kg dry weight	Cd in control plants = 4.5 ppm, Cd in soil = 200 ppm[3]
Avena sativa, leaf	177 mg/kg dry weight	Cd in control plants = 3.9 ppm, Cd in soil = 200 ppm[3]
Avena sativa, leaf	45.4 mg/kg dry weight	Cd in control plants = 3.9 ppm, Cd in soil = 40 ppm[3]
Avena sativa, root	144.8 mg/kg dry weight	Cd in control plants = 3.6 ppm, Cd in soil = 40 ppm[3]
Avena sativa, root	663.2 mg/kg dry weight	Cd in control plants = 3.6 ppm, Cd in soil = 200 ppm[3]
Avena sativa, stalk	116.5 mg/kg dry weight	Cd in control plants = 3.9 ppm, Cd in soil = 200 ppm[3]
Avena sativa, stalk	47.4 mg/kg dry weight	Cd in control plants = 3.9 ppm, Cd in soil = 40 ppm[3]
Beech and spruce forest soil	0.056 mg/kg dry weight soil	Central German, soil depth = 0-10 cm[4]
Beech and spruce forest soil	0.060 mg/kg dry weight soil	Central German, soil depth = 10-20 cm, 71 g/ha[4]
Beech and spruce forest soil	0.074 mg/kg dry weight soil	Central German, soil depth = 30-40 cm, 96 g/ha[4]
Beech and spruce forest soil	0.096 mg/kg dry weight soil	Central German, soil depth = 40-50 cm, 142 g/ha[4]
Beverages	0.040 mg/kg wet weight	USA, 1968-1970, n = 30, AAS and/or polarography, max value[5]
Bound in stream	95 % of added	Cd released in a stream as CdCl2, bound in 100 first m[6]
Brachydanio rerio	12.7 mg/kg wet weight	Food = 10 ppm Cd as acetate, female, 300 K, alk = 25 mg Ca/l[7]
Brachydanio rerio	5.10 mg/kg wet weight	Food = 10 ppm Cd as acetate, male, 300 K, alk = 25 mg Ca/l[7]
Brassica oleracea var. botrytis, root	1356.7 mg/kg dry weight	Cd in control plants = 1.8 ppm, Cd in soil = 200 ppm[3]
Brassica oleracea var. botrytis, root	202.5 mg/kg dry weight	Cd in control plants = 1.8 ppm, Cd in soil = 40 ppm[3]
Brassica oleracea var. italica, leaves	268.5 mg/kg dry weight	Cd in control plants = 2.7 ppm, Cd in soil = 200 ppm[3]
Brassica oleracea var. italica, leaves	36 mg/kg dry weight	Cd in control plants = 2.7 ppm, Cd in soil = 40 ppm[3]
Brassica oleracea var. italica, root	153 mg/kg dry weight	Cd in control plants = 6.5 ppm, Cd in soil = 40 ppm[3]
Brassica oleracea var. italica, root	1647.4 mg/kg dry weight	Cd in control plants = 6.5 ppm, Cd in soil = 200 ppm[3]
Brassica oleracea var. oleracea, leaf	18.5 mg/kg dry weight	Cd in control plants = 4.8 ppm, Cd in soil = 40 ppm[3]
Brassica oleracea var. oleracea, leaf	198.6 mg/kg dry weight	Cd in control plants = 4.8 ppm, Cd in soil = 200 ppm[3]
Chesnut oak	0.8 mg/kg dry weight soil	E Tennessee, soil litter layer O1, g/ha = 15[8]
Callinectes sapidus, carapace	14.4 mg/kg wet weight	CdCl2 = 11140 ug/l, 8 d, S = 3 %, cont = 2.4 ppm, 293 K[9]
Callinectes sapidus, carapace	2.8 mg/kg wet weight	CdCl2 = 0.1 ug/l, 8 d, S = 3 %, cont = 2.4 ppm, 293 K[9]
Callinectes sapidus, carapace	2.8 mg/kg wet weight	CdCl2 = 110 ug/l, 8 d, S = 3 %, cont = 2.4 ppm, 293 K[9]
Callinectes sapidus, claw	0.3 mg/kg wet weight	CdCl2 = 0.1 ug/l, 8 d, S = 3 %, cont = 0.3 ppm, 293 K[9]
Callinectes sapidus, claw	0.4 mg/kg wet weight	CdCl2 = 110 ug/l, 8 d, S = 3 %, cont = 0.3 ppm, 293 K[9]
Callinectes sapidus, claw	3.5 mg/kg wet weight	CdCl2 = 11140 ug/l, 8 d, S = 3 %, cont = 0.3 ppm, 293 K[9]
Callinectes sapidus, eyestalk	0.3 mg/kg wet weight	CdCl2 = 0.1 ug/l, 8 d, S = 3 %, cont = 0.7 ppm, 293 K[9]
Callinectes sapidus, eyestalk	1.0 mg/kg wet weight	CdCl2 = 110 ug/l, 8 d, S = 3 %, cont = 0.7 ppm, 293 K[9]
Callinectes sapidus, eyestalk	6.4 mg/kg wet weight	CdCl2 = 11140 ug/l, 8 d, S = 3 %, cont = 0.7 ppm, 293 K[9]

Callinectes sapidus, gastric mill	0.3 mg/kg wet weight	CdCl2 = 0.1 ug/l, 8 d, S = 3 %, cont = 3.1 ppm, 293 K[9]
Callinectes sapidus, gastric mill	0.3 mg/kg wet weight	CdCl2 = 110 ug/l, 8 d, S = 3 %, cont = 3.1 ppm, 293 K[9]
Callinectes sapidus, gastric mill	11.7 mg/kg wet weight	CdCl2 = 11140 ug/l, 8 d, S = 3 %, cont = 0.6 ppm, 293 K[9]
Callinectes sapidus, gill	0.3 mg/kg wet weight	CdCl2 = 0.1 ug/l, 8 d, S = 3 %, cont = 0.2 ppm, 293 K[9]
Callinectes sapidus, gills	1.0 mg/kg wet weight	CdCl2 = 110 ug/l, 8 d, S = 3 %, cont = 0.2 ppm, 293 K[9]
Callinectes sapidus, gills	56.6 mg/kg wet weight	CdCl2 = 11140 ug/l, 8 d, S = 3 %, cont = 0.4 ppm, 293 K[9]
Callinectes sapidus, green gland	3.9 mg/kg wet weight	CdCl2 = 11140 ug/l, 8 d, S = 3 %, cont = 0 ppm, 293 K[9]
Callinectes sapidus, heart	0 mg/kg wet weight	CdCl2 = 0.1 ug/l, 8 d, S = 3 %, cont = 0.6 ppm, 293 K[9]
Callinectes sapidus, heart	0 mg/kg wet weight	CdCl2 = 110 ug/l, 8 d, S = 3 %, cont = 0.6 ppm, 293 K[9]
Callinectes sapidus, heart	3.2 mg/kg wet weight	CdCl2 = 11140 ug/l, 8 d, S = 3 %, cont = 0.2 ppm, 293 K[9]
Callinectes sapidus, hepatopancreas	1.4 mg/kg wet weight	CdCl2 = 110 ug/l, 8 d, S = 3 %, cont = 0.7 ppm, 293 K[9]
Callinectes sapidus, hepatopancreas	1.5 mg/kg wet weight	CdCl2 = 0.1 ug/l, 8 d, S = 3 %, cont = 0.7 ppm, 293 K[9]
Callinectes sapidus, hepatopancreas	41.1 mg/kg wet weight	CdCl2 = 11140 ug/l, 8 d, S = 3 %, cont = 3.1 ppm, 293 K[9]
Carrot, root	2.53 mg/kg dry weight	Cd added = 2.5 ug/g soil, Grenville loam, control = 0.24 ppm[1]
Carrot, root	2.65 mg/kg dry weight	Cd added = 5.0 ug/g soil, Grenville loam, control = 0.24 ppm[1]
Carrot, root	57.5 mg/kg dry weight	Flowing solution, Cd = 10 ug/l, CdCl2, 7 d uptake, fibrous root[10]
Carrot, shoot	4.1 mg/kg dry weight	Flowing solution, Cd = 10 ug/l, CdCl2, 7 d uptake[10]
Carrot, top	5.66 mg/kg dry weight	Cd added = 2.5 ug/g soil, Grenville loam, control = 0.46 ppm[1]
Carrot, top	7.70 mg/kg dry weight	Cd added = 5.0 ug/g soil, Grenville loam, control = 0.46 ppm[1]
Cd(2+)	89 %	pH = 7.2, Observed, E = 1.5 mV, WPRL, England[11]
Cd(2+)	92 %	pH = 7.2, Calculated, E = 1.5 mV, WPRL, England[11]
CdCO3	3.9 %	pH = 7.2, Calculated, E = 1.5 mV, WPRL, England[11]
CdCl(+)	1.8 %	pH = 7.2, Calculated, E = 1.5 mV, WPRL, England[11]
CdSO4	0.6 %	pH = 7.2, Calculated, E = 1.5 mV, WPRL, England[11]
Cocksfoot, root	145.3 mg/kg dry weight	Flowing solution, Cd = 10 ug/l, CdCl2, 7 d uptake, fibrous root[10]
Cocksfoot, shoot	4.9 mg/kg dry weight	Flowing solution, Cd = 10 ug/l, CdCl2, 7 d uptake[10]
Corn, top	1.84 mg/kg dry weight	Cd added = 2.5 ug/g soil, Grenville loam, control = Corn, top ppm[1]
Corn, top	10.47 mg/kg dry weight	Cd added = 2.5 ug/g soil, Grenville loam, control = 0.73 ppm[1]
Corn, top	17.02 mg/kg dry weight	Cd added = 5.0 ug/g soil, Grenville loam, control = 0.73 ppm[1]
Corn, top	2.68 mg/kg dry weight	Cd added = 5.0 ug/g soil, Grenville loam, control = Corn, top ppm[1]
Crassostrea virginica	104.91 mg Cd/kg dry weight	40 w, 5 ppb in unfilt. seawater, control = 12.55 ppm, 283 K[12]
Crassostrea virginica	11.83 mg Cd/kg dry weight	0 w, 5 ppb in unfilt. seawater, control = 11.83 ppm, 283 K[12]
Crassostrea virginica	21.75 mg Cd/kg dry weight	4 w, 5 ppb in unfilt. seawater, control = 13.10 ppm, 283 K[12]

Crassostrea virginica	22.00 mg Cd/kg dry weight	8 w, 5 ppb in unfilt. seawater, control = 15.57 ppm, 283 K[12]
Crassostrea virginica	24.90 mg Cd/kg dry weight	12 w, 5 ppb in unfilt. seawater, control = 19.35 ppm, 283 K[12]
Crassostrea virginica	27.59 mg Cd/kg dry weight	20 w, 5 ppb in unfilt. seawater, control = 13.60 ppm, 283 K[12]
Crassostrea virginica	28.55 mg Cd/kg dry weight	16 w, 5 ppb in unfilt. seawater, control = 13.42 ppm, 283 K[12]
Crassostrea virginica	33.77 mg Cd/kg dry weight	24 w, 5 ppb in unfilt. seawater, control = 16.75 ppm, 283 K[12]
Crassostrea virginica	352 % more than control	Seawater, 10 ug Cd/l added as CdCl2, 21 d[2]
Crassostrea virginica	39.22 mg Cd/kg dry weight	28 w, 5 ppb in unfilt. seawater, control = 10.85 ppm, 283 K[12]
Crassostrea virginica	63.27 mg Cd/kg dry weight	32 w, 5 ppb in unfilt. seawater, control = 12.19 ppm, 283 K[12]
Crassostrea virginica	85.83 mg Cd/kg dry weight	36 w, 5 ppb in unfilt. seawater, control = 17.47 ppm, 283 K[12]
Crataegus monogyna, leaf	6.04 mg/kg dry weight	Avonmouth, Bristol, GB, unwashed leaves[13]
Dairy products	0.050 mg/kg wet weight	USA, 1968-1970, n = 30, AAS and/or polarography, max value[5]
Daucus carota, root	1737.5 mg/kg dry weight	Cd in control plants = 8.5 ppm, Cd in soil = 40 ppm[3]
Daucus carota, tops	79.3 mg/kg dry weight	Cd in control plants = 6.6 ppm, Cd in soil = 40 ppm[3]
Daucus carota, tubers	26.8 mg/kg dry weight	Cd in control plants = 2.4 ppm, Cd in soil = 40 ppm[3]
Different Cd species in filt., settled sewage water, Humic Cd complex	39 %	pH = 7.2, Calculated, E = 12.3 mV, WPRL, England[11]
Different Cd species in filtered river water, Cd(2+)	42 %	pH = 7.7, Observed, E = 7.9 mV, R. Trent, Swarkeston, Engl.[11]
Different Cd species in filtered river water, Cd(2+)	55 %	pH = 7.7, Calculated, E = 6.9 mV, Stevenage Br., WPRL, Eng.[11]
Different Cd species in filtered river water, Cd(2+)	58 %	pH = 7.7, Calculated, E = 7.9 mV, R. Trent, Swarkeston, Engl.[11]
Different Cd species in filtered river water, Cd(2+)	63 %	pH = 7.7, Observed, E = 6.9 mV, Stevenage Br., WPRL, Eng.[11]
Different Cd species in filtered river water, CdCO3	3.9 %	pH = 7.7, Calculated, E = 7.9 mV, R. Trent, Swarkeston, Engl.[11]
Different Cd species in filtered river water, CdCO3	6.1 %	pH = 7.7, Calculated, E = 6.9 mV, Stevenage Br., WPRL, Eng.[11]
Different Cd species in filtered river water, CdCl(+)	3.5 %	pH = 7.7, Calculated, E = 7.9 mV, R. Trent, Swarkeston, Engl.[11]
Different Cd species in filtered river water, CdCl(+)	4.6 %	pH = 7.7, Calculated, E = 6.9 mV, Stevenage Br., WPRL, Eng.[11]
Different Cd species in filtered river water, CdOH(+)	2.6 %	pH = 7.7, Calculated, E = 7.9 mV, R. Trent, Swarkeston, Engl.[11]
Different Cd species in filtered river water, CdOH(+)	2.8 %	pH = 7.7, Calculated, E = 6.9 mV, Stevenage Br., WPRL, Eng.[11]
Different Cd species in filtered river water, CdSO4	7.2 %	pH = 7.7, Calculated, E = 6.9 mV, Stevenage Br., WPRL, Eng.[11]
Different Cd species in filtered river water, CdSO4	7.2 %	pH = 7.7, Calculated, E = 7.9 mV, R. Trent, Swarkeston, Engl.[11]
Different Cd species in filtered river water, Humic Cd complex	24 %	pH = 7.7, Calculated, E = 6.9 mV, Stevenage Br., WPRL, Eng.[11]

Different Cd species in filtered river water, Humic Cd complex	24 %	pH = 7.7, Calculated, E = 7.9 mV, R. Trent, Swarkeston, Engl.[11]
Different Cd species in filtered, settled sewage water, Cd(2+)	38 %	pH = 7.7, Observed, E = 12.3 mV, WPRL, England[11]
Different Cd species in filtered, settled sewage water, Cd(2+)	41 %	pH = 7.2, Calculated, E = 12.3 mV, WPRL, England[11]
Different Cd species in filtered, settled sewage water, CdCO3	9.0 %	pH = 7.2, Calculated, E = 12.3 mV, WPRL, England[11]
Different Cd species in filtered, settled sewage water, CdCl(+)	5.3 %	pH = 7.2, Calculated, E = 12.3 mV, WPRL, England[11]
Different Cd species in filtered, settled sewage water, CdSO4	3.1 %	pH = 7.2, Calculated, E = 12.3 mV, WPRL, England[11]
Different Cd species in filtered, settled sewage, CdOH(+)	1.8 %	pH = 7.2, Calculated, E = 12.3 mV, WPRL, England[11]
Different Cd species in fresh ground water, CdOH(+)	1.4 %	pH = 7.2, Calculated, E = 1.5 mV, WPRL, England[11]
Different Cd species in percolating filter effluent, Cd(2+)	29 %	pH = 8.0, Observed, E = 15.9 mV, WPRL, England, good quality[11]
Different Cd species in percolating filter effluent, Cd(2+)	32 %	pH = 8.0, Observed, E = 14.4 mV, WPRL, England, poor quality[11]
Different Cd species in percolating filter effluent, Cd(2+)	35 %	pH = 8.0, Calculated, E = 14.4 mV, WPRL, England, poor quality[11]
Different Cd species in percolating filter effluent, Cd(2+)	38 %	pH = 8.0, Calculated, E = 15.9 mV, WPRL, England, good quality[11]
Different Cd species in percolating filter effluent, CdCO3	12 %	pH = 8.0, Calculated, E = 15.9 mV, WPRL, England, good quality[11]
Different Cd species in percolating filter effluent, CdCO3	15 %	pH = 8.0, Calculated, E = 14.4 mV, WPRL, England, poor quality[11]
Different Cd species in percolating filter effluent, CdCl(+)	5.2 %	pH = 8.0, Calculated, E = 14.4 mV, WPRL, England, poor quality[11]
Different Cd species in percolating filter effluent, CdCl(+)	6.2 %	pH = 8.0, Calculated, E = 15.9 mV, WPRL, England, good quality[11]
Different Cd species in percolating filter effluent, CdOH(+)	3.2 %	pH = 8.0, Calculated, E = 14.4 mV, WPRL, England, poor quality[11]
Different Cd species in percolating filter effluent, CdOH(+)	3.6 %	pH = 8.0, Calculated, E = 15.9 mV, WPRL, England, good quality[11]
Different Cd species in percolating filter effluent, CdSO4	2.5 %	pH = 8.0, Calculated, E = 14.4 mV, WPRL, England, poor quality[11]
Different Cd species in percolating filter effluent, CdSO4	3.0 %	pH = 8.0, Calculated, E = 15.9 mV, WPRL, England, good quality[11]

Different Cd species in percolating filter effluent, Humic Cd complex	37 %	pH = 8.0, Calculated, E = 15.9 mV, WPRL, England, good quality[11]
Different Cd species in percolating filter effluent, Humic Cd complex	38 %	pH = 8.0, Calculated, E = 14.4 mV, WPRL, England, poor quality[11]
Different Cd species in polluted sediments, CaCO3	0 % of total Cd	Ashtabula, Lake Erie, Total Cd = 4.1 ppm[14]
Different Cd species in polluted sediments, CaCO3	21 % of total Cd	German Bight, Total Cd = 1.2 ppm[15]
Different Cd species in polluted sediments, CaCO3	4 % of total Cd	Bridgeport, Conn, USA, Total Cd = 15.2 ppm[14]
Different Cd species in polluted sediments, CaCO3	43 % of total Cd	Lower Rhine, Total Cd = 28.4 ppm[16]
Different Cd species in polluted sediments, CaCO3	5 % of total Cd	Los Angeles Harbor, Total Cd = 2.2 ppm[17]
Different Cd species in polluted sediments, cation exchangable	1 % of total Cd	Ashtabula, Lake Erie, Total Cd = 4.1 ppm[14]
Different Cd species in polluted sediments, cation exchangable	1 % of total Cd	Bridgeport, Conn, USA, Total Cd = 15.2 ppm[14]
Different Cd species in polluted sediments, cation exchangable	1 % of total Cd	Los Angeles Harbor, Total Cd = 2.2 ppm[17]
Different Cd species in polluted sediments, cation exchangable	30 % of total Cd	German Bight, Total Cd = 1.2 ppm[15]
Different Cd species in polluted sediments, cation exchangable	4 % of total Cd	Lower Rhine, Total Cd = 28.4 ppm[16]
Different Cd species in polluted sediments, easily reducible	0 % of total Cd	German Bight, Total Cd = 1.2 ppm[15]
Different Cd species in polluted sediments, easily reducible	0 % of total Cd	Lower Rhine, Total Cd = 28.4 ppm[16]
Different Cd species in polluted sediments, easily reducible	1 % of total Cd	Bridgeport, Conn, USA, Total Cd = 15.2 ppm[14]
Different Cd species in polluted sediments, easily reducible	2 % of total Cd	Ashtabula, Lake Erie, Total Cd = 4.1 ppm[14]
Different Cd species in polluted sediments, easily reducible	9 % of total Cd	Los Angeles Harbor, Total Cd = 2.2 ppm[17]
Different Cd species in polluted sediments, moderately reducible	1 % of total Cd	Ashtabula, Lake Erie, Total Cd = 4.1 ppm[14]
Different Cd species in polluted sediments, moderately reducible	1 % of total Cd	Bridgeport, Conn, USA, Total Cd = 15.2 ppm[14]
Different Cd species in polluted sediments, moderately reducible	11 % of total Cd	German Bight, Total Cd = 1.2 ppm[15]
Different Cd species in polluted sediments, moderately reducible	34 % of total Cd	Los Angeles Harbor, Total Cd = 2.2 ppm[17]

Different Cd species in polluted sediments, moderately reducible	6 % of total Cd	Lower Rhine, Total Cd = 28.4 ppm[16]
Different Cd species in polluted sediments, organic sulfides	10 % of total Cd	German Bight, Total Cd = 1.2 ppm[15]
Different Cd species in polluted sediments, organic sulfides	38 % of total Cd	Los Angeles Harbor, Total Cd = 2.2 ppm[17]
Different Cd species in polluted sediments, organic sulfides	47 % of total Cd	Lower Rhine, Total Cd = 28.4 ppm[16]
Different Cd species in polluted sediments, organic sulfides	76 % of total Cd	Ashtabula, Lake Erie, Total Cd = 4.1 ppm[14]
Different Cd species in polluted sediments, organic sulfides	96 % of total Cd	Bridgeport, Conn, USA, Total Cd = 15.2 ppm[14]
Different Cd species in polluted sediments, residual	1 % of total Cd	Bridgeport, Conn, USA, Total Cd = 15.2 ppm[14]
Different Cd species in polluted sediments, residual	1 % of total Cd	Lower Rhine, Total Cd = 28.4 ppm[16]
Different Cd species in polluted sediments, residual	14 % of total Cd	Los Angeles Harbor, Total Cd = 2.2 ppm[17]
Different Cd species in polluted sediments, residual	22 % of total Cd	Ashtabula, Lake Erie, Total Cd = 4.1 ppm[14]
Different Cd species in polluted sediments, residual	28 % of total Cd	German Bight, Total Cd = 1.2 ppm[15]
Different Cd species in seawater, Cd(2+)	1.8 %	Model distribution[18]
Different Cd species in seawater, Cd(2+)	2.5 %	Model distribution[19]
Different Cd species in seawater, Cd(2+)	3 %	Model distribution[20]
Different Cd species in seawater, CdCl(+)	29 %	Model distribution[18]
Different Cd species in seawater, CdCl(+)	34 %	Model distribution[20]
Different Cd species in seawater, CdCl(+)	40 %	Model distribution[19]
Different Cd species in seawater, CdCl2	38 %	Model distribution[18]
Different Cd species in seawater, CdCl2	50 %	Model distribution[19]
Different Cd species in seawater, CdCl2	51 %	Model distribution[20]
Different Cd species in seawater, CdCl3(-)	12 %	Model distribution[20]
Different Cd species in seawater, CdCl3(-)	28 %	Model distribution[18]
Different Cd species in seawater, CdCl3(-)	6 %	Model distribution[19]
Different Cd species in seawater, Wattamolla Beach	11 %	Dissociable by Chelex-100 and determinable by ASV[21]
Different Cd species in seawater, Wattamolla Beach	11 %	Nondissociable by Chelex-100 and nondeterminable by ASV, organic[21]

Different Cd species in seawater, Wattamolla Beach	7 %	Nondissociable by Chelex-100 and nondeterminable by ASV, inorganic[21]
Different Cd species in seawater, Wattamolla Beach	75 %	Nondissociable by Chelex-100 and determinable by ASV[21]
Different species in freshwater, Cd bound with Humic	2.7 mol %	298 K, analytic data, 1 atm, freshwater, L. Celyn[22]
Different species in freshwater, Cd(2+)	92.8 mol %	298 K, analytic data, 1 atm, freshwater, L. Celyn[22]
Different species in freshwater, CdCO3	0.2 mol %	298 K, analytic data, 1 atm, freshwater, L. Celyn[22]
Different species in freshwater, CdCl2	0.7 mol %	298 K, analytic data, 1 atm, freshwater, L. Celyn[22]
Different species in freshwater, CdHCO3(-)	3.0 mol %	298 K, analytic data, 1 atm, freshwater, L. Celyn[22]
Different species in freshwater, CdOH(+)	0.1 mol %	298 K, analytic data, 1 atm, freshwater, L. Celyn[22]
Different species in freshwater, CdSO4	0.5 mol %	298 K, analytic data, 1 atm, freshwater, L. Celyn[22]
Different species in seawater, Cd(2+)	0.9 mol %	298 K, pH = 8.0, 1 atm, seawater, model distribution[22]
Different species in seawater, CdCO3	6.4 mol %	298 K, pH = 8.0, 1 atm, seawater, model distribution[22]
Different species in seawater, CdCl(+)	16.2 mol %	298 K, pH = 8.0, 1 atm, seawater, model distribution[22]
Different species in seawater, CdCl2	45.9 mol %	298 K, pH = 8.0, 1 atm, seawater, model distribution[22]
Different species in seawater, CdCl3(-)	30.2 mol %	298 K, pH = 8.0, 1 atm, seawater, model distribution[22]
Different species in seawater, CdSO4	0.3 mol %	298 K, pH = 8.0, 1 atm, seawater, model distribution[22]
Fish and poultry, meat	0.045 mg/kg wet weight	USA, 1968-1970, n = 30, AAS and/or polarography, max value[5]
Fodder beet, root	151.1 mg/kg dry weight	Flowing solution, Cd = 10 ug/l, CdCl2, 7 d uptake, fibrous root[10]
Fodder beet, shoot	21.1 mg/kg dry weight	Flowing solution, Cd = 10 ug/l, CdCl2, 7 d uptake[10]
Fruits	0.225 mg/kg wet weight	USA, 1968-1970, n = 30, AAS and/or polarography, max value[5]
Garden fruits	0.070 mg/kg wet weight	USA, 1968-1970, n = 30, AAS and/or polarography, max value[5]
Glycine max, leaf	0.4 mg/kg dry weight	Diagnostic part, new leaf, soil = 0.0 ug Cd/g, pH = 7.7, Zn = 47 ppm[23]
Glycine max, leaf	16.0 mg/kg dry weight	Diagnostic part, new leaf, soil = 10.0 ug Cd/g, pH = 7.7, Zn = 47 ppm[23]
Glycine max, leaf	22.0 mg/kg dry weight	Diagnostic part, new leaf, soil = 20.0 ug Cd/g, pH = 7.7, Zn = 47 ppm[23]
Glycine max, leaf	29.0 mg/kg dry weight	Diagnostic part, new leaf, soil = 40.0 ug Cd/g, pH = 7.7, Zn = 47 ppm[23]
Glycine max, leaf	3.7 mg/kg dry weight	Diagnostic part, new leaf, soil = 2.5 ug Cd/g, pH = 7.7, Zn = 47 ppm[23]
Glycine max, leaf	35.0 mg/kg dry weight	Diagnostic part, new leaf, soil = 160.0 ug Cd/g, pH = 7.7, Zn = 47[23]
Glycine max, leaf	36.0 mg/kg dry weight	Diagnostic part, new leaf, soil = 80.0 ug Cd/g, pH = 7.7, Zn = 47 ppm[23]
Glycine max, leaf	5.5 mg/kg dry weight	Diagnostic part, new leaf, soil = 5.0 ug Cd/g, pH = 7.7, Zn = 47 ppm[23]

Glycine max, seed	11.0 mg/kg dry weight	Ediable part, soil = 10.0 ug Cd/g, pH = 7.7, Zn = 47 ppm[23]
Glycine max, seed	17.0 mg/kg dry weight	Ediable part, soil = 20.0 ug Cd/g, pH = 7.7, Zn = 47 ppm[23]
Glycine max, seed	21.0 mg/kg dry weight	Ediable part, soil = 160.0 ug Cd/g, pH = 7.7, Zn = 47 ppm[23]
Glycine max, seed	29.0 mg/kg dry weight	Ediable part, soil = 40.0 ug Cd/g, pH = 7.7, Zn = 47 ppm[23]
Glycine max, seed	29.0 mg/kg dry weight	Ediable part, soil = 80.0 ug Cd/g, pH = 7.7, Zn = 47 ppm[23]
Glycine max, seed	3.9 mg/kg dry weight	Ediable part, soil = 2.5 ug Cd/g, pH = 7.7, Zn = 47 ppm[23]
Glycine max, seed	6.8 mg/kg dry weight	Ediable part, soil = 5.0 ug Cd/g, pH = 7.7, Zn = 47 ppm[23]
Grain and cereal products	0.070 mg/kg wet weight	USA, 1968-1970, n = 30, AAS and/or polarography, max value[5]
Guinea pig, renal cortex	10.8 mg/kg dry weight	1 month of cigarette smoke, n = 22,control = 8.4, n = 15, SD = 2.8[5]
Homarus americanus	25 % more than control	Seawater, 10 ug Cd/l added as CdCl2, 21 d[2]
Homo sapiens, blood of exposed workers	27 ng/g wet weight	Sweden, presently exposed, n = 22, spectrographic, 1971[5]
Homo sapiens, blood of exposed workers	41 ng/g wet weight	W Germany, presently exposed, n = 22, AAS after extraction[5]
Homo sapiens, blood of exposed workers	8.6 ng/g wet weight	Sweden, not exposed for 10-20 years, n = 35, spectrographic, 1971[5]
Homo sapiens, blood of not exposed persons	2 ng/g wet weight	Sweden, n = 150, spectrographic, 1971, less than[5]
Homo sapiens, blood of not exposed persons	3.5 ng/g wet weight	W Germany, n = 18, AAS after extraction, 1972[5]
Homo sapiens, blood of not exposed persons	4.9 ng/g wet weight	UK, children, n = 89, AAS after extraction, 1972[5]
Homo sapiens, blood of not exposed persons	5 ng/g wet weight	Sweden, n = 6, neutron activation, 1961, rg = 2 - 8[5]
Homo sapiens, blood of not exposed persons	5 ng/g wet weight	USA, n = 243, AAS after extraction, 1968, less than[5]
Homo sapiens, blood of not exposed persons	9.5 ng/g wet weight	Belgium, n = 24, AAS after extraction, 1972, SE = +- 2.1[5]
Homo sapiens, serum of exposed workers	1.9 ng/g wet weight	W Germany, presently exposed, n = 17, AAS after extraction, 1969[5]
Homo sapiens, serum of exposed workers	13.7 ng/g wet weight	W Germany, presently exposed, n = 22, AAS after extraction[5]
Homo sapiens, serum of not exposed persons	12 ng/g wet weight	USA, n = 27, AAS , 1970[5]
Homo sapiens, serum of not exposed persons	3.3 ng/g wet weight	W Germany, n = 15, AAS after extraction, 1968[5]
Humic Cd complex	0 %	pH = 7.2, Calculated, E = 1.5 mV, WPRL, England[11]
Kale, root	18.9 mg/kg dry weight	Flowing solution, Cd = 10 ug/l, CdCl2, 7 d uptake, fibrous root[10]
Kale, shoot	3.3 mg/kg dry weight	Flowing solution, Cd = 10 ug/l, CdCl2, 7 d uptake[10]
Lactuca sativa, leaf	51.1 mg/kg dry weight	Cd in control plants = 4.8 ppm, Cd in soil = 40 ppm[3]
Lactuca sativa, leaf	667.7 mg/kg dry weight	Cd in control plants = 4.8, Cd in soil = 200 ppm[3]
Lactuca sativa, root	1628.2 mg/kg dry weight	Cd in control plants = 2.3 ppm, Cd in soil = 200 ppm[3]
Lactuca sativa, root	294.7 mg/kg dry weight	Cd in control plants = 2.3 ppm, Cd in soil = 40 ppm[3]
Leafy vegetables	0.110 mg/kg wet weight	USA, 1968-1970, n = 30, AAS and/or polarography, max value[5]
Legume vegetables	0.035 mg/kg wet weight	USA, 1968-1970, n = 30, AAS and/or polarography, max value[5]
Lettuce, root	2.96 mg/kg dry weight	Cd added = 2.5 ug/g soil, Grenville loam, control = 0.40 ppm[1]

Lettuce, root	48.6 mg/kg dry weight	Flowing solution, Cd = 10 ug/l, CdCl2, 7 d uptake, fibrous root[10]
Lettuce, root	5.60 mg/kg dry weight	Cd added = 5.0 ug/g soil, Grenville loam, control = 0.40 ppm[1]
Lettuce, shoot	15.0 mg/kg dry weight	Flowing solution, Cd = 10 ug/l, CdCl2, 7 d uptake[10]
Lettuce, top	10.36 mg/kg dry weight	Cd added = 5.0 ug/g soil, Grenville loam, control = 0.66 ppm[1]
Lettuce, top	7.72 mg/kg dry weight	Cd added = 2.5 ug/g soil, Grenville loam, control = 0.66 ppm[1]
Lucerne, root	33.4 mg/kg dry weight	Flowing solution, Cd = 10 ug/l, CdCl2, 7 d uptake, fibrous root[10]
Lucerne, shoot	3.1 mg/kg dry weight	Flowing solution, Cd = 10 ug/l, CdCl2, 7 d uptake[10]
Lysmata seticaudata, exoskeleton	23.2 % of total body burden in compartme	63 d uptake, compartment = 21.8 % of wet shrimp, Cd-109[24]
Lysmata seticaudata, exoskeleton	23.2 % of total body burden in compartment	63 d uptake, compartment = 21.8 %[24]
Lysmata seticaudata, exoskeleton	26.2 % of total body burden in compartme	After 245 d cleansing, compartment = 21.8 % of wet shrimp, Cd-109[24]
Lysmata seticaudata, exoskeleton	26.2 % of total body burden in compartment	After 245 d cleansing, compartment = 21.8 %[24]
Lysmata seticaudata, eyes	0.3 % of total body burden in compartmen	63 d uptake, compartment = 1.4 % of wet shrimp, Cd-109[24]
Lysmata seticaudata, eyes	0.3 % of total body burden in compartment	63 d uptake, compartment = 1.4 %[24]
Lysmata seticaudata, eyes	0.6 % of total body burden in compartmen	After 245 d cleansing, compartment = 1.4 % of wet shrimp, Cd-109[24]
Lysmata seticaudata, eyes	0.6 % of total body burden in compartment	After 245 d cleansing, compartment = 1.4 %[24]
Lysmata seticaudata, hemolymph	0.6 % of total body burden in compartmen	63 d uptake, compartment = 19.9 % of wet shrimp, Cd-109[24]
Lysmata seticaudata, hemolymph	0.6 % of total body burden in compartment	63 d uptake, compartment = 19.9 %[24]
Lysmata seticaudata, hemolymph	1.5 % of total body burden in compartmen	After 245 d cleansing, compartment = 19.9 % of wet shrimp, Cd-109[24]
Lysmata seticaudata, hemolymph	1.5 % of total body burden in compartment	After 245 d cleansing, compartment = 19.9 %[24]
Lysmata seticaudata, muscle	49.7 % of total body burden in compartme	After 245 d cleansing, compartment = 50.3 % of wet shrimp, Cd-109[24]
Lysmata seticaudata, muscle	49.7 % of total body burden in compartment	After 245 d cleansing, compartment = 50.3 %[24]
Lysmata seticaudata, muscle	54.2 % of total body burden in compartme	63 d uptake, compartment = 50.3 % of wet shrimp, Cd-109[24]
Lysmata seticaudata, muscle	54.2 % of total body burden in compartment	63 d uptake, compartment = 50.3 %[24]
Lysmata seticaudata, viscera	21.7 % of total body burden in compartme	63 d uptake, compartment = 6.6 % of wet shrimp, Cd-109[24]
Lysmata seticaudata, viscera	21.7 % of total body burden in compartment	63 d uptake, compartment = 6.6 %[24]
Lysmata seticaudata, viscera	21.9 % of total body burden in compartme	After 245 d cleansing, compartment = 6.6 % of wet shrimp, Cd-109[24]
Lysmata seticaudata, viscera	21.9 % of total body burden in compartment	After 245 d cleansing, compartment = 6.6 %[24]
Maize, root	57.5 mg/kg dry weight	Flowing solution, Cd = 10 ug/l, CdCl2, 7 d uptake, fibrous root[10]
Maize, shoot	11.5 mg/kg dry weight	Flowing solution, Cd = 10 ug/l, CdCl2, 7 d uptake[10]

Meadow fescue, root	108.1 mg/kg dry weight	Flowing solution, Cd = 10 ug/l, CdCl2, 7 d uptake, fibrous root[10]
Meadow fescue, shoot	5.8 mg/kg dry weight	Flowing solution, Cd = 10 ug/l, CdCl2, 7 d uptake[10]
Milk	0.0028 mg/kg wet weight	USA, rg = 0.0015 - 0.004[5]
Milk	0.003 mg/kg wet weight	Japan, nonpolluted area[5]
Milk	0.009 mg/kg wet weight	W Germany[5]
Milk	0.01 mg/kg wet weight	Czechoslovakia[5]
Mouse, blood	14.9 ng/g wet weight	6 w, injection of 0.025 mg Cd/kg*d (5 d/w) during w 1-3, n = 4[5]
Mouse, blood	23.8 ng/g wet weight	3 w, injection of 0.025 mg Cd/kg*d (5 d/w) during w 1-3, n = 4[5]
Mouse, blood	25.9 ng/g wet weight	24 w, injection of 0.025 mg Cd/kg*d (5 d/w) during w 7-21, n = 8[5]
Mouse, blood	48.4 ng/g wet weight	21 w, injection of 0.025 mg Cd/kg*d (5 d/w) during w 7-21, n = 4[5]
Mouse, kidney	162 mg/kg wet weight	Injection of 0.25 mg Cd/kg*d (5 d/w) for 6 m, n = 4, rg = 141 - 172[5]
Mouse, kidney	172 mg/kg wet weight	Injection of 0.5 mg Cd/kg*d (5 d/w) for 6 m, n = 4, rg = 158 - 194[5]
Mouse, liver	258 mg/kg wet weight	Injection of 0.5 mg Cd/kg*d (5 d/w) for 6 m, n = 4, rg = 190 - 288[5]
Mouse, liver	281 mg/kg wet weight	Injection of 0.25 mg Cd/kg*d (5 d/w) for 6 m, n = 4, rg = 271 - 288[5]
Mouse, pancreas	114 mg/kg wet weight	Injection of 0.5 mg Cd/kg*d (5 d/w) for 6 m, n = 4, rg = 111 - 121[5]
Mouse, pancreas	71 mg/kg wet weight	Injection of 0.25 mg Cd/kg*d (5 d/w) for 6 m, n = 4, rg = 67 - 74[5]
Mouse, spleen	10.1 mg/kg wet weight	Injection of 0.25 mg Cd/kg*d (5 d/w) for 6 m, n = 4, rg = 9.5 - 10.6[5]
Mouse, spleen	19.5 mg/kg wet weight	Injection of 0.5 mg Cd/kg*d (5 d/w) for 6 m, n = 4, rg = 18.4 - 21.4[5]
Mouse, whole body	11.0 ug/animal	3 w, injection of 0.025 mg Cd/kg*d (5 d/w) during w 1-3, n = 4[5]
Mouse, whole body	11.9 ug/animal	6 w, injection of 0.025 mg Cd/kg*d (5 d/w) during w 1-3, n = 4[5]
Mouse, whole body	61.1 ug/animal	21 w, injection of 0.025 mg Cd/kg*d (5 d/w) during w 7-21, n = 4[5]
Mouse, whole body	65.2 ug/animal	24 w, injection of 0.025 mg Cd/kg*d (5 d/w) during w 7-21, n = 8[5]
Mulinia lateralis	10.33 ug Cd/g dry weight	Uptake from water = 20 ug/l, 283 K, 14 d, s = 3 %, n = 10[25]
Mulinia lateralis	20.46 ug Cd/g dry weight	Uptake from water = 20 ug/l, 293 K, 14 d, s = 3 %, n = 10[25]
Mulinia lateralis	3.6 ug Cd/g dry weight	Uptake from water = 5 ug/l, 283 K, 14 d, s = 3 %, n = 10[25]
Mulinia lateralis	8.9 ug Cd/g dry weight	Uptake from water = 5 ug/l, 293 K, 14 d, s = 3 %, n = 10, SD = +-[25]
Mulinia lateralis, soft part	10.3 mg/kg dry weight	Cd = 20 ug/l, 283 K, SD = 10.3 ppm, n = 10, Cd-109, 14 d, S=3.0 %[26]
Mulinia lateralis, soft part	20.46 mg/kg dry weight	Cd = 20 ug/l, 293 K, n = 10, Cd-109, 14 d, S=3.0 %[26]
Mulinia lateralis, soft part	24.35 mg/kg dry weight	Cd = 20 ug/l, 283 K, n = 10, CdCl2, 21 d, S=3.0 %[26]
Mulinia lateralis, soft part	3.6 mg/kg dry weight	Cd = 5 ug/l, 283 K, SD = 0.21 ppm, n = 10, Cd-109, 14 d, S=3.0 %[26]

Mulinia lateralis, soft part	36.91 mg/kg dry weight	Cd = 20 ug/l, 293 K, n = 10, CdCl2, 21 d, S=2.0 %[26]
Mulinia lateralis, soft part	52.27 mg/kg dry weight	Cd = 20 ug/l, 283 K, n = 10, CdCl2, 21 d, S=2.0 %[26]
Mulinia lateralis, soft part	8.785 mg/kg dry weight	Cd = 20 ug/l, 293 K, n = 10, CdCl2, 21 d, S=3.0 %[26]
Mulinia lateralis, soft part	8.9 mg/kg dry weight	Cd = 5 ug/l, 293 K, SD = 0.21 ppm, n = 10, Cd-109, 14 d, S=3.0 %[26]
Mussel	19 mg/kg dry weight	Conc in sed = 2.5 mg/kg dw, filt seawater = 0.8 ppb, Derwent Est, Tasmania[27]
Mussel	20 mg/kg dry weight	Conc in sed = 8.0 mg/kg dw, filt seawater = 0.6 ppb, Derwent Est, Tasmania[27]
Mussel	22 mg/kg dry weight	Conc in sed = 13 mg/kg dw, filt seawater = <0.5 ppb, Derwent Est, Tasmania[27]
Mussel	23 mg/kg dry weight	Conc in sed = 40 mg/kg dw, filt seawater = 1.6 ppb, Derwent Est, Tasmania[27]
Mussel	25 mg/kg dry weight	Conc in sed = 50 mg/kg dw, filt seawater = 1.0 ppb, Derwent Est, Tasmania[27]
Mussel	29 mg/kg dry weight	Conc in sed = 15 mg/kg dw, filt seawater = 2.1 ppb, Derwent Est, Tasmania[27]
Mussel	38 mg/kg dry weight	Conc in sed = 50 mg/kg dw, filt seawater = <0.5 ppb, Derwent Est, Tasmania[27]
Mya arenaria	16.76 ug Cd/g dry weight	Uptake from water = 20 ug/l, 283 K, 14 d, s = 3 %, n = 10[25]
Mya arenaria	2.2 ug Cd/g dry weight	Uptake from water = 5 ug/l, 283 K, 14 d, s = 3 %, n = 10[25]
Mya arenaria	29.04 ug Cd/g dry weight	Uptake from water = 20 ug/l, 293 K, 14 d, s = 3 %, n = 10[25]
Mya arenaria	4.2 ug Cd/g dry weight	Uptake from water = 5 ug/l, 293 K, 14 d, s = 3 %, n = 10[25]
Mya arenaria, soft part	16.8 mg/kg dry weight	Cd = 20 ug/l, 283 K, SD = 1.23 ppm, n = 10, Cd-109, 14 d, S=3.0 %[26]
Mya arenaria, soft part	2.2 mg/kg dry weight	Cd = 5 ug/l, 283 K, SD = 0.31 ppm, n = 10, Cd-109, 14 d, S=3.0 %[26]
Mya arenaria, soft part	29.0 mg/kg dry weight	Cd = 20 ug/l, 293 K, SD = 29.0 ppm, n = 10, Cd-109, 14 d, S=3.0 %[26]
Mya arenaria, soft part	4.2 mg/kg dry weight	Cd = 5 ug/l, 293 K, SD = 0.31 ppm, n = 10, Cd-109, 14 d, S=3.0 %[26]
Mytilus edulis	50.55 ug Cd/g dry weight	Uptake from water = 20 ug/l, 283 K, 14 d, s = 3 %, n = 10[25]
Mytilus edulis	60.35 ug Cd/g dry weight	Uptake from water = 20 ug/l, 293 K, 14 d, s = 3 %, n = 10[25]
Mytilus edulis	9.3 ug Cd/g dry weight	Uptake from water = 5 ug/l, 283 K, 14 d, s = 3 %, n = 10[25]
Mytilus edulis	9.4 ug Cd/g dry weight	Uptake from water = 5 ug/l, 293 K, 14 d, s = 3 %, n = 10[25]
Mytilus edulis, soft part	0.99 mg/kg dry weight	Cd=40 ug/l, n=10, cont.=0.51 ppm, 6 d exposure, 283 K, SD=0.27, S=3.5 %[28]
Mytilus edulis, soft part	1.01 mg/kg dry weight	Cd=40 ug/l, n=10, cont.=0.47 ppm, 6 d exposure, 291 K, SD=0.41, S=3.5 %[28]
Mytilus edulis, soft part	1.22 mg/kg dry weight	Cd=40 ug/l, n=10, cont.=0.47 ppm, 14 d exposure, 291 K, SD=0.57, S=3.5 %[28]
Mytilus edulis, soft part	1.42 mg/kg dry weight	Cd=40 ug/l, n=10, cont.=0.51 ppm, 14 d exposure, 283 K, SD=0.41, S=3.5 %[28]
Mytilus edulis, soft part	1.55 mg/kg dry weight	Cd=40 ug/l, n=10, cont.=0.52 ppm, 6 d exposure, 283 K, SD=0.34, S=1.5 %[28]
Mytilus edulis, soft part	108.12 mg/kg dry weight	Cd = 20 ug/l, 293 K, n = 10, CdCl2, 21 d, S=2.0 %[26]
Mytilus edulis, soft part	2.23 mg/kg dry weight	Cd=40 ug/l, n=10, cont.=0.49 ppm, 6 d exposure, 291 K, SD=0.4, S=1.5 %[28]
Mytilus edulis, soft part	2.25 mg/kg dry weight	Cd=40 ug/l, n=10, cont.=0.52 ppm, 14 d exposure, 283 K, SD=0.66, S=1.5 %[28]

Mytilus edulis, soft part	32.08 mg/kg dry weight	Cd = 20 ug/l, 283 K, n = 10, CdCl2, 21 d, S=3.0 %[26]
Mytilus edulis, soft part	50.5 mg/kg dry weight	Cd = 20 ug/l, 283 K, SD = 8.57 ppm, n = 10, Cd-109, 14 d, S=3.0 %[26]
Mytilus edulis, soft part	6.52 mg/kg dry weight	Cd=40 ug/l, n=10, cont.=0.49 ppm, 14 d exposure, 291 K, SD=2.04, S=1.5 %[28]
Mytilus edulis, soft part	60.3 mg/kg dry weight	Cd = 20 ug/l, 293 K, SD = 11.8 ppm, n = 10, Cd-109, 14 d, S=3.0 %[26]
Mytilus edulis, soft part	83.16 mg/kg dry weight	Cd = 20 ug/l, 283 K, n = 10, CdCl2, 21 d, S=2.0 %[26]
Mytilus edulis, soft part	86.62 mg/kg dry weight	Cd = 20 ug/l, 293 K, n = 10, CdCl2, 21 d, S=3.0 %[26]
Mytilus edulis, soft part	9.3 mg/kg dry weight	Cd = 5 ug/l, 283 K, SD = 2.48 ppm, n = 10, Cd-109, 14 d, S=3.0 %[26]
Mytilus edulis, soft part	9.4 mg/kg dry weight	Cd = 5 ug/l, 293 K, SD = 2.48 ppm, n = 10, Cd-109, 14 d, S=3.0 %[26]
Nucula proxima	2.0 ug Cd/g dry weight	Uptake from water = 20 ug/l, 283 K, 14 d, s = 3 %, n = 10[25]
Nucula proxima	4.56 ug Cd/g dry weight	Uptake from water = 20 ug/l, 293 K, 14 d, s = 3 %, n = 10[25]
Nucula proxima, soft part	0.61 mg/kg dry weight	Cd = 20 ug/l, 283 K, n = 10, CdCl2, 21 d, S=3.0 %[26]
Nucula proxima, soft part	2.08 mg/kg dry weight	Cd = 20 ug/l, 283 K, n = 10, CdCl2, 21 d, S=2.0 %[26]
Nucula proxima, soft part	2.61 mg/kg dry weight	Cd = 20 ug/l, 293 K, n = 10, CdCl2, 21 d, S=3.0 %[26]
Nucula proxima, soft part	5.35 mg/kg dry weight	Cd = 20 ug/l, 293 K, n = 10, CdCl2, 21 d, S=2.0 %[26]
Oak - Hickory	0.2 mg/kg dry weight soil	E Tennessee, soil litter layer O1, g/ha = 4[8]
Oak forest soil	0.3 mg/kg dry weight soil	E Chicago, USA, urban, horizon = B, 14-25 cm[29]
Oak forest soil	10.4 mg/kg dry weight soil	E Chicago, USA, urban, horizon = A1, 0-2.5 cm[29]
Oak forest soil	3.8 mg/kg dry weight soil	E Chicago, USA, urban, horizon = A1, 2.5-14 cm[29]
Oak forest soil	4.7 mg/kg dry weight soil	E Chicago, USA, urban, horizon = O , surface litter[29]
Oats, grain	1.50 mg/kg dry weight	Cd added = 2.5 ug/g soil, Grenville loam, control = 0.21 ppm[1]
Oats, grain	2.07 mg/kg dry weight	Cd added = 2.5 ug/g soil, Grenville loam, control = 0.21 ppm[1]
Oats, root	5.06 mg/kg dry weight	Cd added = 2.5 ug/g soil, Grenville loam, control = 0.81 ppm[1]
Oats, root	8.72 mg/kg dry weight	Cd added = 5.0 ug/g soil, Grenville loam, control = 0.81 ppm[1]
Oats, straw	2.30 mg/kg dry weight	Cd added = 2.5 ug/g soil, Grenville loam, control = 0.29 ppm[1]
Oats, straw	3.70 mg/kg dry weight	Cd added = 5.0 ug/g soil, Grenville loam, control = 0.29 ppm[1]
Oil, fats and shortening	0.085 mg/kg wet weight	USA, 1968-1970, n = 30, AAS and/or polarography, max value[5]
Oryza sativa, grain	0.1 mg/kg dry weight	Edible part, soil = 0.0 ug Cd/g, pH = 7.7, Zn = 47 ppm[23]
Oryza sativa, grain	0.1 mg/kg dry weight	Edible part, soil = 2.5 ug Cd/g, pH = 7.7, Zn = 47 ppm[23]
Oryza sativa, grain	0.2 mg/kg dry weight	Edible part, soil = 10.0 ug Cd/g, pH = 7.7, Zn = 47 ppm[23]
Oryza sativa, grain	0.2 mg/kg dry weight	Edible part, soil = 5.0 ug Cd/g, pH = 7.7, Zn = 47 ppm[23]
Oryza sativa, grain	0.3 mg/kg dry weight	Edible part, soil = 20.0 ug Cd/g, pH = 7.7, Zn = 47 ppm[23]
Oryza sativa, grain	0.6 mg/kg dry weight	Edible part, soil = 40.0 ug Cd/g, pH = 7.7, Zn = 47 ppm[23]
Oryza sativa, grain	0.7 mg/kg dry weight	Edible part, soil = 80.0 ug Cd/g, pH = 7.7, Zn = 47 ppm[23]
Oryza sativa, grain	1.1 mg/kg dry weight	Edible part, soil = 160.0 ug Cd/g, pH = 7.7, Zn = 47 ppm[23]
Oryza sativa, leaf	0.1 mg/kg dry weight	Diagnostic part, 2 top leaves, soil = 0.0 ug Cd/g, pH = 7.7, Zn = 47 ppm[23]
Oryza sativa, leaf	0.1 mg/kg dry weight	Diagnostic part, 2 top leaves, soil = 10.0 ug Cd/g, pH = 7.7, Zn = 47 ppm[23]

Oryza sativa, leaf	0.1 mg/kg dry weight	Diagnostic part, 2 top leaves, soil = 2.5 ug Cd/g, pH = 7.7, Zn = 47 ppm[23]
Oryza sativa, leaf	0.1 mg/kg dry weight	Diagnostic part, 2 top leaves, soil = 20.0 ug Cd/g, pH = 7.7, Zn = 47 ppm[23]
Oryza sativa, leaf	0.1 mg/kg dry weight	Diagnostic part, 2 top leaves, soil = 40.0 ug Cd/g, pH = 7.7, Zn = 47 ppm[23]
Oryza sativa, leaf	0.1 mg/kg dry weight	Diagnostic part, 2 top leaves, soil = 5.0 ug Cd/g, pH = 7.7, Zn = 47 ppm[23]
Oryza sativa, leaf	0.3 mg/kg dry weight	Diagnostic part, 2 top leaves, soil = 80.0 ug Cd/g, pH = 7.7, Zn = 47 ppm[23]
Oryza sativa, leaf	0.7 mg/kg dry weight	Diagnostic part, 2 top leaves, soil = 160.0 ug Cd/g, pH = 7.7, Zn = 47 ppm[23]
PINE	0.6 mg/kg dry weight soil	E Tennessee, soil litter layer O1, g/ha = 9[8]
Parsnip, root	163.8 mg/kg dry weight	Flowing solution, Cd = 10 ug/l, CdCl2, 7 d uptake, fibrous root[10]
Parsnip, shoot	2.5 mg/kg dry weight	Flowing solution, Cd = 10 ug/l, CdCl2, 7 d uptake[10]
Penaeus duorarum, exoskeleton	10.2 mg/kg wet weight	Cd = 1 mg/l, 96 h, S = 2 %, 298 K, CdCl2, control = 1.0 ppm[30]
Penaeus duorarum, exoskeleton	32 mg/kg wet weight	Cd = 3 mg/l, 96 h, S = 2 %, 298 K, CdCl2, control = 1.0 ppm[30]
Penaeus duorarum, gills	320 mg/kg wet weight	Cd = 3 mg/l, 96 h, S = 2 %, 298 K, CdCl2, control = 3.4 ppm[30]
Penaeus duorarum, gills	82 mg/kg wet weight	Cd = 1 mg/l, 96 h, S = 2 %, 298 K, CdCl2, control = 3.4 ppm[30]
Penaeus duorarum, hepatopancreas	210 mg/kg wet weight	Cd = 1 mg/l, 96 h, S = 2 %, 298 K, CdCl2, control = 3.9 ppm[30]
Penaeus duorarum, hepatopancreas	270 mg/kg wet weight	Cd = 3 mg/l, 96 h, S = 2 %, 298 K, CdCl2, control = 3.9 ppm[30]
Penaeus duorarum, muscle	1.8 mg/kg wet weight	Cd = 1 mg/l, 96 h, S = 2 %, 298 K, CdCl2, control = 2.0 ppm[30]
Penaeus duorarum, muscle	8.3 mg/kg wet weight	Cd = 3 mg/l, 96 h, S = 2 %, 298 K, CdCl2, control = 2.0 ppm[30]
Perennial ryegrass, root	102.8 mg/kg dry weight	Flowing solution, Cd = 10 ug/l, CdCl2, 7 d uptake, fibrous root[10]
Perennial ryegrass, shoot	6.1 mg/kg dry weight	Flowing solution, Cd = 10 ug/l, CdCl2, 7 d uptake[10]
Phaseolus vulgaris, bean	0.05 mg/kg dry weight	Edible part, soil = 0.0 ug Cd/g, pH = 7.7, Zn = 47 ppm[23]
Phaseolus vulgaris, bean	0.2 mg/kg dry weight	Edible part, soil = 2.5 ug Cd/g, pH = 7.7, Zn = 47 ppm[23]
Phaseolus vulgaris, bean	0.4 mg/kg dry weight	Edible part, soil = 5.0 ug Cd/g, pH = 7.7, Zn = 47 ppm[23]
Phaseolus vulgaris, bean	0.7 mg/kg dry weight	Edible part, soil = 10.0 ug Cd/g, pH = 7.7, Zn = 47 ppm[23]
Phaseolus vulgaris, bean	1.0 mg/kg dry weight	Edible part, soil = 20.0 ug Cd/g, pH = 7.7, Zn = 47 ppm[23]
Phaseolus vulgaris, bean	1.7 mg/kg dry weight	Edible part, soil = 40.0 ug Cd/g, pH = 7.7, Zn = 47 ppm[23]
Phaseolus vulgaris, bean	2.3 mg/kg dry weight	Edible part, soil = 80.0 ug Cd/g, pH = 7.7, Zn = 47 ppm[23]
Phaseolus vulgaris, bean	3.0 mg/kg dry weight	Edible part, soil = 160.0 ug Cd/g, pH = 7.7, Zn = 47 ppm[23]
Phaseolus vulgaris, leaf	0.6 mg/kg dry weight	Diagnostic part, new leaf, soil = 0.0 ug Cd/g, pH = 7.7, Zn = 47 ppm[23]
Phaseolus vulgaris, leaf	1.8 mg/kg dry weight	Diagnostic part, new leaf, soil = 2.5 ug Cd/g, pH = 7.7, Zn = 47 ppm[23]
Phaseolus vulgaris, leaf	10.0 mg/kg dry weight	Diagnostic part, new leaf, soil = 10.0 ug Cd/g, pH = 7.7, Zn = 47 ppm[23]

Phaseolus vulgaris, leaf	12.0 mg/kg dry weight	Diagnostic part, new leaf, soil = 20.0 ug Cd/g, pH = 7.7, Zn = 47 ppm[23]
Phaseolus vulgaris, leaf	14.0 mg/kg dry weight	Diagnostic part, new leaf, soil = 40.0 ug Cd/g, pH = 7.7, Zn = 47 ppm[23]
Phaseolus vulgaris, leaf	19.0 mg/kg dry weight	Diagnostic part, new leaf, soil = 160.0 ug Cd/g, pH = 7.7, Zn = 47 ppm[23]
Phaseolus vulgaris, leaf	19.0 mg/kg dry weight	Diagnostic part, new leaf, soil = 80.0 ug Cd/g, pH = 7.7, Zn = 47 ppm[23]
Phaseolus vulgaris, leaf	4.3 mg/kg dry weight	Diagnostic part, new leaf, soil = 5.0 ug Cd/g, pH = 7.7, Zn = 47 ppm[23]
Phaseolus vulgaris, leaves and stems	29.8 p ci/mg dry weight	225 ppm N, Cd-109 = 3 ppb, 7 d, pH = 6.5, 293 K, tot prod = 468 mg dw[31]
Phaseolus vulgaris, leaves and stems	49.0 p ci/mg dry weight	225 ppm N, Cd-109 = 3 ppb, 7 d, pH = 6.5, 293 K, tot prod = 444 mg dw[31]
Phaseolus vulgaris, leaves and stems	72.9 p ci/mg dry weight	225 ppm N, Cd-109 = 3 ppb, 7 d, pH = 6.5, 293 K, tot prod = 577 mg dw[31]
Phaseolus vulgaris, lower leaves	10.0 p ci/mg dry weight	1463 ppm N, Cd-109 = 3 ppb, 7 d, pH = 6.5, 293 K, prod = 110 mg dw[31]
Phaseolus vulgaris, lower leaves	29.1 p ci/mg dry weight	338 ppm N, Cd-109 = 3 ppb, 7 d, pH = 6.5, 293 K, prod = 98 mg dw[31]
Phaseolus vulgaris, lower leaves	48.4 p ci/mg dry weight	Wat = 225 ppm N, Cd-109 = 3 ppb, 7 d, pH = 6.5, 293 K, prod = 106 mg[31]
Phaseolus vulgaris, root	3707 p ci/mg dry weight	Wat = 225 ppm N, Cd-109 = 3 ppb, 7 d, pH = 6.5, 293 K, prod = 5.1 mg[31]
Phaseolus vulgaris, root	4325 p ci/mg dry weight	1463 ppm N, Cd-109 = 3 ppb, 7 d, pH = 6.5, 293 K, prod = 62.0 mg dw[31]
Phaseolus vulgaris, root	4553 p ci/mg dry weight	338 ppm N, Cd-109 = 3 ppb, 7 d, pH = 6.5, 293 K, prod = 65.1 mg dw[31]
Phaseolus vulgaris, stems	116.2 p ci/mg dry weight	Wat = 225 ppm N, Cd-109 = 3 ppb, 7 d, pH = 6.5, 293 K, prod = 135 mg[31]
Phaseolus vulgaris, stems	72.6 p ci/mg dry weight	1463 ppm N, Cd-109 = 3 ppb, 7 d, pH = 6.5, 293 K, prod = 113 mg dw[31]
Phaseolus vulgaris, stems	93.1 p ci/mg dry weight	338 ppm N, Cd-109 = 3 ppb, 7 d, pH = 6.5, 293 K, prod = 104 mg dw[31]
Phaseolus vulgaris, top leaves	25.5 p ci/mg dry weight	1463 ppm N, Cd-109 = 3 ppb, 7 d, pH = 6.5, 293 K, prod = 82 mg dw[31]
Phaseolus vulgaris, top leaves	40.5 p ci/mg dry weight	338 ppm N, Cd-109 = 3 ppb, 7 d, pH = 6.5, 293 K, prod = 80 mg dw[31]
Phaseolus vulgaris, top leaves	68.3 p ci/mg dry weight	Wat = 225 ppm N, Cd-109 = 3 ppb, 7 d, pH = 6.5, 293 K, prod = 145 mg[31]
Pisum sativum, pods	28.2 mg/kg dry weight	Cd in control plants = 5.7 ppm, Cd in soil = 200 ppm[3]
Pisum sativum, pods	9.5 mg/kg dry weight	Cd in control plants = 5.7 ppm, Cd in soil = 40 ppm[3]
Pisum sativum, root	1571.2 mg/kg dry weight	Cd in control plants = 8.6 ppm, Cd in soil = 200 ppm[3]
Pisum sativum, root	409.7 mg/kg dry weight	Cd in control plants = 8.6 ppm, Cd in soil = 40 ppm[3]
Pisum sativum, seed	10.1 mg/kg dry weight	Cd in control plants = 5.4 ppm, Cd in soil = 40 ppm[3]
Pisum sativum, seed	19.7 mg/kg dry weight	Cd in control plants = 5.4 ppm, Cd in soil = 200 ppm[3]
Pisum sativum, vines	116.9 mg/kg dry weight	Cd in control plants = 7.4 ppm, Cd in soil = 200 ppm[3]
Pisum sativum, vines	37.2 mg/kg dry weight	Cd in control plants = 7.4 ppm, Cd in soil = 40 ppm[3]
Potato	0.001 mg/kg wet weight	USA[5]
Potato	0.017 mg/kg wet weight	Roumania[5]

Potato	0.038 mg/kg wet weight	Japan, nonpolluted area[5]
Potato	0.039 mg/kg wet weight	W Germany[5]
Potato	0.08 mg/kg wet weight	UK[5]
Potato	0.080 mg/kg wet weight	USA, 1969-70, n = 30, AAS and/or polarography, max value[5]
Potato	0.09 mg/kg wet weight	Czechoslovakia[5]
Potato, top	3.46 mg/kg dry weight	Cd added = 2.5 ug/g soil, Grenville loam, control = 0.58 ppm[1]
Potato, top	7.35 mg/kg dry weight	Cd added = 5.0 ug/g soil, Grenville loam, control = 0.58 ppm[1]
Potato, tuber	0.89 mg/kg dry weight	Cd added = 2.5 ug/g soil, Grenville loam, control = 0.18 ppm[1]
Potato, tuber	1.09 mg/kg dry weight	Cd added = 5.0 ug/g soil, Grenville loam, control = 0.18 ppm[1]
Pseudopleuronectes americanus, blood	0.2 mg/kg wet weight	Cd = 10 ppb, 60 d, 3.6 fish, less than[32]
Pseudopleuronectes americanus, blood	0.2 mg/kg wet weight	Cd = 5 ppb, 60 d, 3-6 fish, less than[32]
Pseudopleuronectes americanus, blood	0.3 mg/kg wet weight	Cd = 0 ppb, 60 d, 3-6 fish, less than[32]
Pseudopleuronectes americanus, gill	0.3 mg/kg wet weight	Cd = 0 ppb, 60 d, 3-6 fish, less than[32]
Pseudopleuronectes americanus, gill	0.3 mg/kg wet weight	Cd = 10 ppb, 60 d, 3-6 fish, less than[32]
Pseudopleuronectes americanus, gill	0.3 mg/kg wet weight	Cd = 5 ppb, 60 d, 3-6 fish, less than[32]
Quercus robur, leaf	6.82 mg/kg dry weight	Avonmouth, Bristol, GB, unwashed leaves[13]
Radish, root	33.6 mg/kg dry weight	Flowing solution, Cd = 10 ug/l, CdCl2, 7 d uptake, fibrous root[10]
Radish, shoot	1.8 mg/kg dry weight	Flowing solution, Cd = 10 ug/l, CdCl2, 7 d uptake[10]
Raphanus sativus, tops	264.7 mg/kg dry weight	Cd in control plants = 9.8 ppm, Cd in soil = 40 ppm[3]
Raphanus sativus, tops	398 mg/kg dry weight	Cd in control plants = 9.8 ppm, Cd in soil = 40 ppm[3]
Raphanus sativus, tubers	54.6 mg/kg dry weight	Cd in control plants = 3.6 ppm, Cd in soil = 40 ppm[3]
Rat, kidney	1.6 % of total intracellular amount	Microsomes, 10 anim, 24 h after injec of 277 ug Cd(+2)/275 g[33]
Rat, kidney	2.1 % of total intracellular amount	Lysosomes, 10 anim, 24 h after injec of 277 ug Cd(+2)/275 g[33]
Rat, kidney	2.2 % of total intracellular amount	Mitochondria, 10 anim, 24 h after injec of 277 ug Cd(+2)/275 g[33]
Rat, kidney	4.4 % of total intracellular amount	Nuclei, 10 anim, 24 h after injec of 277 ug Cd(+2)/275 g[33]
Rat, kidney	89.7 % of total intracellular amount	Soluble fraction, 10 anim, 24 h after injec of 277 ug Cd(+2)/275 g[33]
Rat, liver	1 % of total intracellular amount	Lysosomes, 10 anim, 24 h after injec of 277 ug Cd(+2)/275 g[33]
Rat, liver	2.3 % of total intracellular amount	Microsomes, 10 anim, 24 h after injec of 277 ug Cd(+2)/275 g[33]
Rat, liver	2.7 % of total intracellular amount	Mitochondria, 10 anim, 24 h after injec of 277 ug Cd(+2)/275 g[33]
Rat, liver	3.5 % of total intracellular amount	Nuclei, 10 anim , 24 h after injection of 277 micro g Cd(+2)[33]
Rat, liver	90.5 % of total intracellular amount	Soluble fraction, 10 anim, 24 h after injec of 277 ug Cd(+2)/275 g[33]

Red clover, root	61.8 mg/kg dry weight	Flowing solution, Cd = 10 ug/l, CdCl2, 7 d uptake, fibrous root[10]
Red clover, shoot	8.3 mg/kg dry weight	Flowing solution, Cd = 10 ug/l, CdCl2, 7 d uptake[10]
Root vegetables	0.080 mg/kg wet weight	USA, 1968-1970, n = 30, AAS and/or polarography, max value[5]
Sainfoin, root	4.8 mg/kg dry weight	Flowing solution, Cd = 10 ug/l, CdCl2, 7 d uptake, fibrous root[10]
Sainfoin, shoot	2.3 mg/kg dry weight	Flowing solution, Cd = 10 ug/l, CdCl2, 7 d uptake[10]
Salix alba, leaf	7.85 mg/kg dry weight	Avonmouth, Bristol, GB, unwashed leaves[13]
Seawater	0.11 ug/l unfiltered	Filtered conc = 0.09 ug/l, labile content, depth=10 m, Baltic[34]
Seawater	0.13 ug/l unfiltered	Filtered conc = 0.05 ug/l, labile content, depth=58 m, Baltic[34]
Seawater	0.13 ug/l unfiltered	Filtered conc = 0.08 ug/l, labile content, depth=30 m, Baltic[34]
Seawater	0.23 ug/l unfiltered	Filtered conc = 0.14 ug/l, UV-labile content, depth=58 m, Baltic[34]
Seawater	0.25 ug/l unfiltered	Filtered conc = 0.18 ug/l, labile content, depth=0.5 m, Baltic[34]
Seawater	0.56 ug/l unfiltered	Filtered conc = 0.22 ug/l, UV-labile content, depth=0.5 m, Baltic[34]
Seawater	0.68 ug/l unfiltered	Filtered conc = 0.42 ug/l, UV-labile content, depth=30 m, Baltic[34]
Seawater	0.80 ug/l unfiltered	Filtered conc = 0.15 ug/l, UV-labile content, depth=10 m, Baltic[34]
Sediment	0.37 mg/kg dry weight	Unpolluted area, estuary, Urr water, rg = 0.2 - 0.7[35]
Sediment	0.4 mg/kg dry weight	Unpolluted area, River, Illinois River, rg = 0.3 - 0.5[36]
Sediment	0.9 mg/kg dry weight	Unpolluted area, estuary, Urr water, rg = 0.8 - 1.3[37]
Sediment	1.0 mg/kg dry weight	Unpolluted area, estuary, Urr water, rg = 0.7 - 1.2[37]
Sediment	1.4 mg/kg dry weight	Range 0.4-2.4 ppm, River Rheidol, Cardiganshire, GB[38]
Sediment	12 mg/kg dry weight	Max water = 6 microgram/l, max conc, Gannel River, GB[39]
Sediment	12 mg/kg dry weight	Water conc = 3 microgram/l, yearly mean, Gannel River, GB[39]
Sediment	2 mg/kg dry weight	Max water = 1 microgram/l, 1 years measurement, Carnon River, GB[39]
Sediment	2 mg/kg dry weight	Max water = 1 microgram/l, 1 years measurement, Fowey River, GB[39]
Sediment	2 mg/kg dry weight	Max water = 1 microgram/l, 1 years measurement, Red River, GB[39]
Sediment	2 mg/kg dry weight	Max water = 2 microgram/l, 1 years measurement, Fowey River, GB[39]
Sediment	2 mg/kg dry weight	Max water = 2 microgram/l, max conc, Gannel River, GB[39]
Sediment	2 mg/kg dry weight	Water conc = 1 microgram/l, yearly mean, Gannel River, GB[39]
Sediment	2 mg/kg dry weight	Water conc = 1 microgram/l, yearly mean, Red River, GB[39]
Sediment	2 mg/kg dry weight	Water conc = < 1 microgram/l, yearly mean, Carnon River, GB[39]
Sediment	2 mg/kg dry weight	Water conc = < 1 microgram/l, yearly mean, Fowey River, GB[39]
Sediment	5 mg/kg dry weight	Max water = 4 microgram/l, 1 years measurement, Red River, GB[39]
Sediment	5 mg/kg dry weight	Water conc = 2 microgram/l, yearly mean, Red River, GB[39]
Sediment	6.7 mg/kg dry weight	Max water = 195 microgram/l, 1 years measurement, Carnon River, GB[39]

Sorghum sp., root	59.6 mg/kg dry weight	Flowing solution, Cd = 10 ug/l, CdCl2, 7 d uptake, fibrous root[10]
Sorghum sp., shoot	17.5 mg/kg dry weight	Flowing solution, Cd = 10 ug/l, CdCl2, 7 d uptake[10]
Soybean, grain	1.88 mg/kg dry weight	Cd added = 2.5 ug/g soil, Grenville loam, control = 0.29 ppm[1]
Soybean, grain	2.51 mg/kg dry weight	Cd added = 5.0 ug/g soil, Grenville loam, control = 0.29 ppm[1]
Soybean, root	11.77 mg/kg dry weight	Cd added = 5.0 ug/g soil, Grenville loam, control = 0.99 ppm[1]
Soybean, root	6.09 mg/kg dry weight	Cd added = 2.5 ug/g soil, Grenville loam, control = 0.99 ppm[1]
Soybean, vegetative part	3.95 mg/kg dry weight	Cd added = 2.5 ug/g soil, Grenville loam, control = 0.71 ppm[1]
Soybean, vegetative part	4.88 mg/kg dry weight	Cd added = 5.0 ug/g soil, Grenville loam, control = 0.71 ppm[1]
Spinacia oleracea, leaves	207.5 mg/kg dry weight	Cd in control plants = 12.2 ppm, Cd in soil = 40 ppm[3]
Spinacia oleracea, leaves	239.3 mg/kg dry weight	Cd in control plants = 12.2 ppm, Cd in soil = 200 ppm[3]
Spinacia oleracea, root	213.9 mg/kg dry weight	Cd in control plants = 8.5 ppm, Cd in soil = 40 ppm[3]
Spinacia oleracea, root	490.3 mg/kg dry weight	Cd in control plants = 8.5 ppm, Cd in soil = 200 ppm[3]
Spruce forest soil	24 mg/kg dry weight soil	Central Sweden, needle litter, O1[40]
Spruce forest soil	44 mg/kg dry weight soil	Central Sweden, Humus, O2[40]
Sugar and adjuncts	0.055 mg/kg wet weight	USA, 1968-1970, n = 30, AAS and/or polarography, max value[5]
Sunflower, root	83.7 mg/kg dry weight	Flowing solution, Cd = 10 ug/l, CdCl2, 7 d uptake, fibrous root[10]
Sunflower, shoot	7.3 mg/kg dry weight	Flowing solution, Cd = 10 ug/l, CdCl2, 7 d uptake[10]
Tick bean, root	36.9 mg/kg dry weight	Flowing solution, Cd = 10 ug/l, CdCl2, 7 d uptake, fibrous root[10]
Tick bean, shoot	4.2 mg/kg dry weight	Flowing solution, Cd = 10 ug/l, CdCl2, 7 d uptake[10]
Timothy, root	138.7 mg/kg dry weight	Flowing solution, Cd = 10 ug/l, CdCl2, 7 d uptake, fibrous root[10]
Timothy, root	15.28 mg/kg dry weight	Cd added = 5.0 ug/g soil, Grenville loam, control = 0.61 ppm[1]
Timothy, root	8.97 mg/kg dry weight	Cd added = 2.5 ug/g soil, Grenville loam, control = 0.61 ppm[1]
Timothy, shoot	5.2 mg/kg dry weight	Flowing solution, Cd = 10 ug/l, CdCl2, 7 d uptake[10]
Timothy, top	1.04 mg/kg dry weight	Cd added = 2.5 ug/g soil, Grenville loam, control = 0.21 ppm[1]
Timothy, top	1.41 mg/kg dry weight	Cd added = 5.0 ug/g soil, Grenville loam, control = 0.21 ppm[1]
Tobacco, leaf	11.57 mg/kg dry weight	Cd added = 5.0 ug/g soil, Grenville loam, control = 0.49 ppm[1]
Tobacco, leaf	5.41 mg/kg dry weight	Cd added = 2.5 ug/g soil, Grenville loam, control = 0.49 ppm[1]
Tobacco, root	2.78 mg/kg dry weight	Cd added = 2.5 ug/g soil, Grenville loam, control = 0.42 ppm[1]
Tobacco, root	5.54 mg/kg dry weight	Cd added = 5.0 ug/g soil, Grenville loam, control = 0.42 ppm[1]
Tobacco, stem	2.42 mg/kg dry weight	Cd added = 2.5 ug/g soil, Grenville loam, control = 0.28 ppm[1]
Tobacco, stem	4.80 mg/kg dry weight	Cd added = 5.0 ug/g soil, Grenville loam, control = 0.28 ppm[1]
Tomato	0 mg/kg wet weight	USA[5]
Tomato	0.013 mg/kg wet weight	Rumania[5]
Tomato	0.015 mg/kg wet weight	W Germany[5]
Tomato	0.032 mg/kg wet weight	Japan, nonpolluted area[5]
Tomato	0.06 mg/kg wet weight	UK[5]
Tomato, root	10.35 mg/kg dry weight	Cd added = 5.0 ug/g soil, Grenville loam, control = 0.59 ppm[1]
Tomato, root	121.6 mg/kg dry weight	Flowing solution, Cd = 10 ug/l, CdCl2, 7 d uptake, fibrous root[10]

Tomato, root	5.08 mg/kg dry weight	Cd added = 2.5 ug/g soil, Grenville loam, control = 0.59 ppm[1]
Tomato, shoot	5.6 mg/kg dry weight	Flowing solution, Cd = 10 ug/l, CdCl2, 7 d uptake[10]
Tomato, vegetative part	0.99 mg/kg dry weight	Cd added = 2.5 ug/g soil, Grenville loam, control = 0.23 ppm[1]
Tomato, vegetative part	1.03 mg/kg dry weight	Cd added = 5.0 ug/g soil, Grenville loam, control = 0.23 ppm[1]
Tomato, vegetative part	5.26 mg/kg dry weight	Cd added = 2.5 ug/g soil, Grenville loam, control = 0.51 ppm[1]
Tomato, vegetative part	6.46 mg/kg dry weight	Cd added = 5.0 ug/g soil, Grenville loam, control = 0.51 ppm[1]
Triticum aestivum, grain	0.1 mg/kg dry weight	Edible part, soil = 0.0 ug Cd/g, pH = 7.7, Zn = 47 ppm[23]
Triticum aestivum, grain	1.5 mg/kg dry weight	Edible part, soil = 2.5 ug Cd/g, pH = 7.7, Zn = 47 ppm[23]
Triticum aestivum, grain	10.0 mg/kg dry weight	Edible part, soil = 40.0 ug Cd/g, pH = 7.7, Zn = 47 ppm[23]
Triticum aestivum, grain	13.0 mg/kg dry weight	Edible part, soil = 80.0 ug Cd/g, pH = 7.7, Zn = 47 ppm[23]
Triticum aestivum, grain	14.0 mg/kg dry weight	Edible part, soil = 160.0 ug Cd/g, pH = 7.7, Zn = 47 ppm[23]
Triticum aestivum, grain	2.8 mg/kg dry weight	Edible part, soil = 5.0 ug Cd/g, pH = 7.7, Zn = 47 ppm[23]
Triticum aestivum, grain	5.8 mg/kg dry weight	Edible part, soil = 10.0 ug Cd/g, pH = 7.7, Zn = 47 ppm[23]
Triticum aestivum, grain	8.7 mg/kg dry weight	Edible part, soil = 20.0 ug Cd/g, pH = 7.7, Zn = 47 ppm[23]
Triticum aestivum, leaf	0.1 mg/kg dry weight	Diagnostic part, 2 top leaves, soil = 0.0 ug Cd/g, pH = 7.7, Zn = 47 ppm[23]
Triticum aestivum, leaf	12.0 mg/kg dry weight	Diagnostic part, 2 top leaves, soil = 10.0 ug Cd/g, pH = 7.7, Zn = 47 ppm[23]
Triticum aestivum, leaf	2.7 mg/kg dry weight	Diagnostic part, 2 top leaves, soil = 2.5 ug Cd/g, pH = 7.7, Zn = 47 ppm[23]
Triticum aestivum, leaf	24.0 mg/kg dry weight	Diagnostic part, 2 top leaves, soil = 20.0 ug Cd/g, pH = 7.7, Zn = 47 ppm[23]
Triticum aestivum, leaf	31.0 mg/kg dry weight	Diagnostic part, 2 top leaves, soil = 40.0 ug Cd/g, pH = 7.7, Zn = 47 ppm[23]
Triticum aestivum, leaf	41.0 mg/kg dry weight	Diagnostic part, 2 top leaves, soil = 80.0 ug Cd/g, pH = 7.7, Zn = 47 ppm[23]
Triticum aestivum, leaf	48.0 mg/kg dry weight	Diagnostic part, 2 top leaves, soil = 160.0 ug Cd/g, pH = 7.7, Zn = 47 ppm[23]
Triticum aestivum, leaf	6.4 mg/kg dry weight	Diagnostic part, 2 top leaves, soil = 5.0 ug Cd/g, pH = 7.7, Zn = 47 ppm[23]
Uca pugilator, gill and hepatopancreas	0.09 mg/kg dry weight	Cd = 1 mg/l, 278 K, 72 h, S = 3.0 %[41]
Uca pugilator, gill and hepatopancreas	0.24 mg/kg dry weight	Cd = 1 mg/l, 278 K, 72 h, S = 0.5 %[41]
Uca pugilator, gill and hepatopancreas	0.37 mg/kg dry weight	Cd = 1 mg/l, 298 K, 72 h, S = 3.0 %[41]
Uca pugilator, gill and hepatopancreas	0.53 mg/kg dry weight	Cd = 1 mg/l, 298 K, 72 h, S = 0.5 %[41]
Uca pugilator, gill and hepatopancreas	0.74 mg/kg dry weight	Cd = 1 mg/l, 308 K, 72 h, S = 3.0 %[41]
Uca pugilator, gill and hepatopancreas	1.00 mg/kg dry weight	Cd = 1 mg/l, 308 K, 72 h, S = 0.5 %[41]
Watercress, root	77.5 mg/kg dry weight	Flowing solution, Cd = 10 ug/l, CdCl2, 7 d uptake, fibrous root[10]
Watercress, shoot	21.0 mg/kg dry weight	Flowing solution, Cd = 10 ug/l, CdCl2, 7 d uptake[10]
Wheat, flour	0.02 mg/kg wet weight	Czechoslovakia[5]

Wheat, flour	0.025 mg/kg wet weight	Japan, nonpolluted area[5]
Wheat, flour	0.047 mg/kg wet weight	W Germany[5]
Wheat, flour	0.07 mg/kg wet weight	USA[5]
Wheat, flour	0.075 mg/kg wet weight	Canada, rg = 0.05 - 0.10[5]
White clover, root	82.0 mg/kg dry weight	Flowing solution, Cd = 10 ug/l, CdCl2, 7 d uptake, fibrous root[10]
White clover, shoot	8.5 mg/kg dry weight	Flowing solution, Cd = 10 ug/l, CdCl2, 7 d uptake[10]
Winter barley, root	48.8 mg/kg dry weight	Flowing solution, Cd = 10 ug/l, CdCl2, 7 d uptake, fibrous root[10]
Winter barley, shoot	2.6 mg/kg dry weight	Flowing solution, Cd = 10 ug/l, CdCl2, 7 d uptake[10]
Winter oat, root	44.2 mg/kg dry weight	Flowing solution, Cd = 10 ug/l, CdCl2, 7 d uptake, fibrous root[10]
Winter oat, shoot	6.6 mg/kg dry weight	Flowing solution, Cd = 10 ug/l, CdCl2, 7 d uptake[10]
Winter wheat, root	58.0 mg/kg dry weight	Flowing solution, Cd = 10 ug/l, CdCl2, 7 d uptake, fibrous root[10]
Winter wheat, shoot	8.1 mg/kg dry weight	Flowing solution, Cd = 10 ug/l, CdCl2, 7 d uptake[10]
Yellow poplar	1.0 mg/kg dry weight soil	E Tennessee, soil litter layer O1, g/ha = 8[8]
Zea mays, grain	0.05 mg/kg dry weight	Edible part, soil = 0.0 ug Cd/g, pH = 7.7, Zn = 47 ppm[23]
Zea mays, grain	0.05 mg/kg dry weight	Edible part, soil = 2.5 ug Cd/g, pH = 7.7, Zn = 47 ppm[23]
Zea mays, grain	0.5 mg/kg dry weight	Edible part, soil = 5.0 ug Cd/g, pH = 7.7, Zn = 47 ppm[23]
Zea mays, grain	1.4 mg/kg dry weight	Edible part, soil = 10.0 ug Cd/g, pH = 7.7, Zn = 47 ppm[23]
Zea mays, grain	2.4 mg/kg dry weight	Edible part, soil = 20.0 ug Cd/g, pH = 7.7, Zn = 47 ppm[23]
Zea mays, grain	3.2 mg/kg dry weight	Edible part, soil = 80.0 ug Cd/g, pH = 7.7, Zn = 47 ppm[23]
Zea mays, grain	3.4 mg/kg dry weight	Edible part, soil = 40.0 ug Cd/g, pH = 7.7, Zn = 47 ppm[23]
Zea mays, grain	6.6 mg/kg dry weight	Edible part, soil = 160.0 ug Cd/g, pH = 7.7, Zn = 47 ppm[23]
Zea mays, leaf	11.0 mg/kg dry weight	Diagnostic part, 3+4 leaf from top, soil = 2.5 ug Cd/g, pH = 7.7, Zn[23]
Zea mays, leaf	23.0 mg/kg dry weight	Diagnostic part, 3+4 leaf from top, soil = 5.0 ug Cd/g, pH = 7.7, Zn = 47 ppm[23]
Zea mays, leaf	27.0 mg/kg dry weight	Diagnostic part, 3+4 leaf from top, soil = 10.0 ug Cd/g, pH = 7.7, Zn = 47 ppm[23]
Zea mays, leaf	3.9 mg/kg dry weight	Diagnostic part, 3+4 leaf from top, soil = 0.0 ug Cd/g, pH = 7.7, Zn = 47 ppm[23]
Zea mays, leaf	40.0 mg/kg dry weight	Diagnostic part, 3+4 leaf from top, soil = 20.0 ug Cd/g, pH = 7.7, Zn = 47 ppm[23]
Zea mays, leaf	58.0 mg/kg dry weight	Diagnostic part, 3+4 leaf from top, soil = 40.0 ug Cd/g, pH = 7.7, Zn = 47 ppm[23]
Zea mays, leaf	71.0 mg/kg dry weight	Diagnostic part, 3+4 leaf from top, soil = 80.0 ug Cd/g, pH = 7.7, Zn = 47 ppm[23]
Zea mays, leaf	94.0 mg/kg dry weight	Diagnostic part, 3+4 leaf from top, soil = 160.0 ug Cd/g, pH = 7.7, Zn = 47 ppm[23]
Zooplankton	2.08 mg/kg dry weight	19 st, NY, conc in algae = 1.6 mg/kg dw, mean value, Long Isl[42]

Source: 1.MacLean, A.J. 1976. Can. J. Soil Sci. 56, 129-138.
2.Eisler, R. G.E. Zaaroogian and R.J. Hennekey. 1972. J. Fish. Res. bd. Canada 29, pp 1367-1369.
3.John, M.K. 1973. Environ. Pollut., 4, pp 7-15.
4.Heinrichs, H. and R. Meyer. 1977. J. Environ. Qual. 6, 402-407.
5.Friberg, L., M. Piscater, G.F. Nordberg and T. Kjellstroem. Cadmium in the Environment 2nd edition,CRC-Press.
6.Huckabee, J.W. and B.G. Blaylock. 1972. in: Dhar, S.K. (ed.); Metal Ions in Biological Systems. pp 15-160. Plenum Publ. Corp. New York.

7.Rehwoldt, R. and D. Karimian-Teherani. 1976. Bull. Environ. Contam. Toxicol. 15(4), pp 442-446.
8.Van Hook, R.I., W.F. Harris and G.S. Henderson. 1977. Ambio 6, 281-286.
9.Hutcheson, M.S. 1974. Chesapeake Sci. 15, 237-241.
10.Jarvis, S.C., L.H.P. Jones and M.J. Hopper. 1976. Plant Soil 44, 179-191.
11.Gardiner, J. 1974. Water Res. 8, 23-30.
12.Zaroogian, G.E. and S. Cheer. 1976. Nature 261, pp 408-410.
13.Little, P. 1974. Environ. Pollut., 5, pp 159-172.
14.Brannon, J.M., R.M. Engler, J.R. Rose, P.G. Hunt and I. Smith. 1976. in: Proceedings of a special Conference on Dredging and Its Environmental Effects, Mobile, Ala, Jan. 26-28 1976. ASCE, pp 455-495.
15.Patchineelam, S.R. and U. Forstner. 1977. Senckenberg. Marit. 9, 75-104.
16.Patchineelam, S.R. 1975. Untersuchugen uber die Hauptbindindungsarten und die Mobiisierbarkeit von Shwermetallen in Fluviatilen Sedimenten. Diss. Univ Heidelberg. Germany. 136 pp.
17.Gupta, S.K. and K.Y. Chen. 1975. Environ. Lett. 10, 129-158.
18.Dyrssen, D. and M. Wedborg. 1973. in: Goldberg, E.D. (ed.): The Sea. Vol 5. Wiley-Intersciece. New Yrk. pp 181-195.
19.Zirino, A. and S. Yamamoto. 1972. Limnol. Oceanogr. 17, 661-671.
20.Ahrland, S. 1975. in: Goldberg, E.D. (ed.): The Nature of Sea Water. Dahlem Konferenzen. Berlin. pp19-243.
21.Batley, G.E. and T.M. Florence. 1976. Mar. Chem. 4, 347-363.
22.Mantoura, R.F.C., A. Dickson and J.P. Riley. 1978. Eustarine Coastral Mar. Sci. 6, 387-408.
23.Bingham, F.T., A.L. Page, R.J. Mahler and T.J. Ganje. 1976. J. Environ. Qual., 5(4), pp 501-513.
24.Fowler, S.W. and B. Oregioni. 1976. Mar. Pollut. Bull. 7, 26-29.
25.Amiard-Triquet, C. 1974. Mar. Biol. 26, pp 161-165.
26.Jockim, E., G. Morrison and R. Steele. 1977. Mar. Biol. 40, 303-308.
27.Bloom, H. and G.M. Ayling. 1977. Environ. Geol. 2, pp 3-22.
28.Phillips, D.J.H. 1976. Mar. Biol. 38, 59-69.
29.Tyler, G. 1972. Ambio 1, 53-59.
30.Nimmo, D.W.R., D.V. Lightner and L.H. Bahner. 1977. In: F.J. Vernberg, A. Calabrese, F.P. Thurberg ad W.B. Vernberg (eds.): Physiological Responses of Marine Biota to Pollutants. Academic Press. NY. 1
31.Hook, R.I. van Jr., B.G. Blaylock, E.A. Bondietti, C.W. Francis, J.W. Huckabee, D.E. Reichle, F.H. Seeton and J.P. Wintherspoon. 1976. Environ. Qual. Saf. 5, pp 167-182.
32.Calabrise, A., M.A. Dawson, F.P. Thurberg and D.R. Wenzloff. 1975. In; Koeman, J.H. et al. (eds); Sulethal Effects of Toxic Chemicals on Aquatic Animals. Elsevier N.Y. pp 15-21.
33.Sabbioni, E. and F. Girardi. 1977. Sci Total Environ., 7(2), pp 145-179.
34.Bruegmann, L. 1974. Acta Hydrochim. Hydrobiol. 2, 123-138.
35.Taylor, D. 1976. Estuarine Coastal Mar. Sci. 2, pp 417.
36.Mathis, B.J. and T.F. Cummins. 1973. Water Pollut. Control 45 pp 1573.
37.Taylor, D. 1976. Sci. Toatal Environ. 6(3), pp 259-264.
38.Davies, B.E. and J. Lewin. 1974. Envi ron. Pollut., 6, pp 49-57.
39.Altshuller, A.P. and J.J. Bufalini. 1971. Environ. Sci. Technol., 5(1), pp 39-64.
40.Inman, R.E., R.B. Ingersoll and E.A. Levy. 1971. Science 172, 1229-1231.
41.Vernberg, W.B., P.J. DeCoursey and J. OHara. 1974. In: F.J. Vernberg and W.B. Vernberg (eds.): Polltion Physiology of Marine Organisms. Academic Press. NY, pp 381-425.
42.Greig, R.A., A. Adams and D.R. Wenzloff. 1977. Bull. Environ. Contam. and Toxic., 18(1), pp 3-8.

5-318 Concentrations
Cd and salinity

Species	Value	Condition
Clupea harengus, egg	100 mg/kg dry weight	Cd = 1.0 mg/l, salinity = 2.5 %, control (Cd=0 mg/l) = 32 ppm[1]
Clupea harengus, egg	124 mg/kg dry weight	Cd = 0.5 mg/l, salinity = 0.5 %, control (Cd=0 mg/l) = 32 ppm[1]
Clupea harengus, egg	152 mg/kg dry weight	Cd = 1.0 mg/l, salinity = 1.6 %, control (Cd=0 mg/l) = 32 ppm[1]
Clupea harengus, egg	191 mg/kg dry weight	Cd = 1.0 mg/l, salinity = 0.5 %, control (Cd=0 mg/l) = 32 ppm[1]
Clupea harengus, egg	196 mg/kg dry weight	Cd = 5.0 mg/l, salinity = 3.2 %, control (Cd=0 mg/l) = 32 ppm[1]
Clupea harengus, egg	213 mg/kg dry weight	Cd = 5.0 mg/l, salinity = 2.5 %, control (Cd=0 mg/l) = 32 ppm[1]
Clupea harengus, egg	263 mg/kg dry weight	Cd = 5.0 mg/l, salinity = 1.6 %, control (Cd=0 mg/l) = 32 ppm[1]
Clupea harengus, egg	396 mg/kg dry weight	Cd = 5.0 mg/l, salinity = 0.5 %, control (Cd=0 mg/l) = 32 ppm[1]
Clupea harengus, egg	46 mg/kg dry weight	Cd = 0.5 mg/l, salinity = 3.2 %, control (Cd=0 mg/l) = 32 ppm[1]
Clupea harengus, egg	52 mg/kg dry weight	Cd = 0.5 mg/l, salinity = 2.5 %, control (Cd=0 mg/l) = 32 ppm[1]

Clupea harengus, egg	83 mg/kg dry weight	Cd = 0.5 mg/l, salinity = 1.6 %, control (Cd=0 mg/l) = 32 ppm[1]
Clupea harengus, egg	93 mg/kg dry weight	Cd = 1.0 mg/l, salinity = 3.2 %, control (Cd=0 mg/l) = 32 ppm[1]
Pleuronectes flesus, egg	13 mg/kg dry weight	Cd = 1.0 mg/l, salinity = 4.2 %, control (Cd=0 mg/l) = 9 ppm[1]
Pleuronectes flesus, egg	17 mg/kg dry weight	Cd = 1.0 mg/l, salinity = 3.2 %, control (Cd=0 mg/l) = 9 ppm[1]
Pleuronectes flesus, egg	23 mg/kg dry weight	Cd = 2.0 mg/l, salinity = 4.2 %, control (Cd=0 mg/l) = 9 ppm[1]
Pleuronectes flesus, egg	30 mg/kg dry weight	Cd = 2.0 mg/l, salinity = 2.5 %, control (Cd=0 mg/l) = 9 ppm[1]
Pleuronectes flesus, egg	35 mg/kg dry weight	Cd = 1.0 mg/l, salinity = 1.6 %, control (Cd=0 mg/l) = 9 ppm[1]
Pleuronectes flesus, egg	39 mg/kg dry weight	Cd = 3.0 mg/l, salinity = 3.2 %, control (Cd=0 mg/l) = 9 ppm[1]
Pleuronectes flesus, egg	48 mg/kg dry weight	Cd = 5.0 mg/l, salinity = 4.2 %, control (Cd=0 mg/l) = 9 ppm[1]
Pleuronectes flesus, egg	52 mg/kg dry weight	Cd = 3.0 mg/l, salinity = 2.5 %, control (Cd=0 mg/l) = 9 ppm[1]
Pleuronectes flesus, egg	54 mg/kg dry weight	Cd = 5.0 mg/l, salinity = 3.2 %, control (Cd=0 mg/l) = 9 ppm[1]
Pleuronectes flesus, egg	61 mg/kg dry weight	Cd = 2.0 mg/l, salinity = 1.6 %, control (Cd=0 mg/l) = 9 ppm[1]
Pleuronectes flesus, egg	65 mg/kg dry weight	Cd = 3.0 mg/l, salinity = 2.5 %, control (Cd=0 mg/l) = 9 ppm[1]
Pleuronectes flesus, egg	70 mg/kg dry weight	Cd = 5.0 mg/l, salinity = 2.5 %, control (Cd=0 mg/l) = 9 ppm[1]
Pleuronectes flesus, egg	74 mg/kg dry weight	Cd = 3.0 mg/l, salinity = 1.6 %, control (Cd=0 mg/l) = 9 ppm[1]
Pleuronectes flesus, egg	85 mg/kg dry weight	Cd = 5.0 mg/l, salinity = 2.5 %, control (Cd=0 mg/l) = 9 ppm[1]
Pleuronectes flesus, egg	93 mg/kg dry weight	Cd = 5.0 mg/l, salinity = 1.6 %, control (Cd=0 mg/l) = 9 ppm[1]

Source: 1.Phillips, D.J.H. 1976. Mar. Biol. 38, 59-69.

5-319 Concentrations
Cd in liver / Cd in kidney

Species	Value	Condition
Dog	0.17	0.5 ug Cd/g water, after 4 y[1]
Dog	0.19	2.5 ug Cd/g water, after 4 y[1]
Dog	0.24	5.0 ug Cd/g water, after 4 y[1]
Dog	0.30	10.0 ug Cd/g water, after 4 y[1]
Rat	0.14	0.1 ug Cd/g water, after 12 m[1]
Rat	0.19	0.5 ug Cd/g water, after 12 m[1]
Rat	0.20	0.5 ug Cd/g water, after 6 m[1]
Rat	0.24	2.5 ug Cd/g water, after 12 m[1]
Rat	0.31	5.0 ug Cd/g water, after 12 m[1]
Rat	0.38	2.5 ug Cd/g water, after 6 m[1]
Rat	0.43	5.0 ug Cd/g water, after 6 m[1]

Rat	0.47	10.0 ug Cd/g water, after 12 m[1]
Rat	0.67	10.0 ug Cd/g water, after 6 m[1]

Source: 1.Friberg, L., M. Piscater, G.F. Nordberg and T. Kjellstroem. Cadmium in the Environment 2nd edition,CRC-Press.

5-320 Concentrations
Chlordane

Species	Value	Condition
Sediment	0.001 mg/kg wet weight	Conc in water: < 1.0 ug/l, USA, less than[1]
Sediment	0.01 mg/kg wet weight	Conc in water: < 1.0 ug/l, USA, less than[1]

Source: 1.Edwards, C.A. 1975. Persistent Pesticides in the Environment, 2nd edition, CRC-Press.

5-321 Concentrations
Cl

Species	Value	Condition
Forest soil	150 mg/kg dry weight soil	E Tennessee, various forests, depth = 120 - 160 cm[1]
Forest soil	150 mg/kg dry weight soil	E Tennessee, various forests, depth = 80 - 120 cm[1]
Forest soil	50 mg/kg dry weight soil	E Tennessee, various forests, depth = 40 - 80 cm[1]
Forest soil	50 mg/kg dry weight soil	E Tennessee, various forests, depth = > 160 cm[1]
Forest soil	50 mg/kg dry weight soil	E Tennessee, various forests, depth = litter[1]
Forest soil	55 mg/kg dry weight soil	E Tennessee, various forests, depth = 0 - 40 cm[1]

Source: 1.Reiners, W.A., R.H. Marks and P.M. Vitousek. 1975. Oikos 26, 264-275.

5-322 Concentrations
Co

Species	Value	Condition
Beech and spruce forest soil	11 mg/kg dry weight	Central German, depth = 10-20 cm, 13.1 kg/ha[1]
Beech and spruce forest soil	12 mg/kg dry weight	Central German, depth = 30-40 cm, 15.6 kg/ha[1]
Beech and spruce forest soil	13 mg/kg dry weight	Central German, depth = 40-50 cm, 19.2 kg/ha[1]
Beech and spruce forest soil	7 mg/kg dry weight	Central German, depth = 0-10 cm, 8.1 kg/ha[1]
Cocksfoot	0.10 mg/kg dry weight	Added CoSO4 = 2.2 kg/ha, pH = 7.0, control = 0.02 mg/kg dry weight[2]
Cocksfoot	0.15 mg/kg dry weight	Added CoSO4 = 4.4 kg/ha, pH = 7.0, control = 0.02 mg/kg dry weight[2]
Cocksfoot	0.16 mg/kg dry weight	Added CoSO4 = 2.2 kg/ha, pH = 5.6, control = 0.03 mg/kg dry weight[2]
Cocksfoot	0.25 mg/kg dry weight	Added CoSO4 = 4.4 kg/ha, pH = 5.6, control = 0.03 mg/kg dry weight[2]
Hypolimnion	0.07 ug/l	Conc in epilimnion = 0.03 ug/l, concentration factor = 2.3, Schoehsee[3]
Particle fractions	1.6 % of total	Size = 2.0-0.5 mm, weight = 3.2 %, area = 0.2 %, marine, Japan, Co-60[4]
Particle fractions	10.5 % of total	Size = 0.125-0.075 mm, weight = 16.9 %, area = 10.4 %, marine, Japan, Co-60[4]
Particle fractions	10.8 % of total	Size = 0.25-0.125 mm, weight = 23 %, area = 10.1 %, marine, Japan, Co-60[4]
Particle fractions	3.2 % of total	Size = 0.5-0.25 mm, weight = 8.5 %, area = 1.4 %, marine, Japan, Co-60[4]

Particle fractions	73.9 % of total	Size below 0.075 mm, weight = 48.4 %, area = 77.9 %, marine, Japan, Co-60[4]
Red clover	0.17 mg/kg dry weight	Added $CoSO4$ = 2.2 kg/ha, pH = 7.0, control = 0.04 mg/kg dry weight[2]
Red clover	0.31 mg/kg dry weight	Added $CoSO4$ = 2.2 kg/ha, pH = 5.6, control = 0.06 mg/kg dry weight[2]
Red clover	0.38 mg/kg dry weight	Added $CoSO4$ = 4.4 kg/ha, pH = 7.0, control = 0.04 mg/kg dry weight[2]
Red clover	0.53 mg/kg dry weight	Added $CoSO4$ = 4.4 kg/ha, pH = 5.6, control = 0.06 mg/kg dry weight[2]
Sediment	5.7 mg/kg dry weight	Unpolluted area, estuary, Urr water, rg = 3.8 - 7.5[5]
Sediment	6 mg/kg dry weight	Unpolluted area, River, Illinois River, rg = 4 - 8[6]
Sediment	6.6 mg/kg dry weight	Unpolluted area, estuary, Urr water[7]
Sediment	7.9 mg/kg dry weight	Unpolluted area, estuary, Urr water, rg = 5.3 - 11.9[5]
Sediment	8.3 mg/kg dry weight	Conc in epilimnion = 0.03 ug/l, concentration factor = 280000,[3]
Timothy	0.08 mg/kg dry weight	Added limestone = 15.0 t/ha, soil-pH = 6.0, N Wales[8]
Timothy	0.10 mg/kg dry weight	Added limestone = 5.0 t/ha, soil-pH = 5.2, N Wales[8]
Timothy	0.11 mg/kg dry weight	Added limestone = 2.5 t/ha, soil-pH = 5.0, N Wales[8]
Timothy	0.19 mg/kg dry weight	Added limestone = 0.63 t/ha, soil-pH = 4.7, N Wales[8]
Vegetation	0.16 mg/kg dry weight	Conc in podzol:f = 1, h = 1, a1a2 = 2, b1 = 2, b2 = 8, b3 = 5, F030[2]
White clover	0.29 mg/kg dry weight	Added limestone = 15.0 t/ha, soil-pH = 6.0, N Wales[8]
White clover	0.30 mg/kg dry weight	Added limestone = 5.0 t/ha, soil-pH = 5.2, N Wales[8]
White clover	0.36 mg/kg dry weight	Added limestone = 2.5 t/ha, soil-pH = 5.0, N Wales[8]
White clover	0.40 mg/kg dry weight	Added limestone = 0.63 t/ha, soil-pH = 4.7, N Wales[8]

Source: 1.Heinrichs, H. and R. Meyer. 1977. J. Environ. Qual. 6, 402-407.
2.Mitchell, R. and C. Chamberlin. 1978. in: Berg. G. (ed.); Indicators of Enteric Contamination in Natural Waters.
3.Groth, P. 1971. Arch. Hydrobiol. 68(3), pp 305-375.
4.Nagaya, Y. and K. Nakamura. 1975. Im pacts of Nuclear releases into the Aquatic Environment. IAEA-SM198/21, pp 313-328. Vienna.
5.Taylor, D. 1976. Sci. Toatal Environ. 6(3), pp 259-264.
6.Mathis, B.J. and T.F. Cummins. 1973. Water Pollut. Control 45 pp 1573.
7.Taylor, D. 1976. Estuarine Coastal Mar. Sci. 2, pp 417.
8.Archer, F.C. 1971. In: Tech. Bull. 21, pp 150-158. Trace Elements in Soils and Crops.

5-323 Concentrations
Compounds in dry biomass

Species	Value	Condition
Adenine	1.40E-4 moles/g dry biomass	None[1]
Alanine	4.54E-4 moles/g dry biomass	None[1]
Arginine	2.52E-4 moles/g dry biomass	None[1]
Asparagine	1.01E-4 moles/g dry biomass	None[1]
Aspartic acid	2.01E-4 moles/g dry biomass	None[1]
Compounds in dry biomass, other than mentioned	2.08E-4 moles/g dry biomass	None[1]

Cysteine	1.01E-4 moles/g dry biomass	None[1]
Cytosine	1.40E-4 moles/g dry biomass	None[1]
Deoxyribose	0.96E-4 moles/g dry biomass	None[1]
Fatty acid	2.80E-4 moles/g dry biomass	None[1]
Glutamic acid	3.53E-4 moles/g dry biomass	None[1]
Glutamine	2.01E-4 moles/g dry biomass	None[1]
Glycerin	1.40E-4 moles/g dry biomass	None[1]
Glycine	4.03E-4 moles/g dry biomass	None[1]
Guanine	1.40E-4 moles/g dry biomass	None[1]
Hexose	10.26E-4 moles/g dry biomass	None[1]
Histidine	0.50E-4 moles/g dry biomass	None[1]
Isoleucine	2.52E-4 moles/g dry biomass	None[1]
Leucine	4.03E-4 moles/g dry biomass	None[1]
Lysine	4.03E-4 moles/g dry biomass	None[1]
Methionine	2.01E-4 moles/g dry biomass	None[1]
PO4 - P	6.80E-4 moles/g dry biomass	None[1]
Phenylalanine	1.51E-4 moles/g dry biomass	None[1]
Proline	2.52E-4 moles/g dry biomass	None[1]
Ribose	4.47E-4 moles/g dry biomass	None[1]
Serine	3.02E-4 moles/g dry biomass	None[1]
Threonine	2.52E-4 moles/g dry biomass	None[1]
Thymine	0.24E-4 moles/g dry biomass	None[1]
Tryptophan	0.50E-4 moles/g dry biomass	None[1]
Tyrosine	1.01E-4 moles/g dry biomass	None[1]
Uracil	1.15E-4 moles/g dry biomass	None[1]
Valine	3.02E-4 moles/g dry biomass	None[1]

Source: 1.Morowitz, H.J., M.E. Tourtellotte, W.R. Guild, E. Castro and C. Woese. 1962. J. Mol. Boil., 4, p 93.

5-324 Concentrations
Cr

Species	Value	Condition
Alisma plantago aquatica, old leaves	1.48 mg/kg dry weight	150 m below pollution entry I, water = 0.030 mg/l, processed leaves[1]
Alisma plantago aquatica, old leaves	2.07 mg/kg dry weight	800 and 50 m below pollution I and II, water = 0.050 mg/l, processed leaves[1]
Alisma plantago aquatica, old leaves	2.45 mg/kg dry weight	800 and 50 m below pollution I and II, water = 0.050 mg/l, whole leaves[1]
Alisma plantago aquatica, old leaves	2.71 mg/kg dry weight	800 m above pollution entry I, water = not detect., processed leaves[1]
Alisma plantago aquatica, old leaves	3.36 mg/kg dry weight	150 m below pollution entry I, water = 0.030 mg/l, whole leaves[1]
Alisma plantago aquatica, old leaves	4.92 mg/kg dry weight	800 m above pollution entry I, water = not detect., whole leaves[1]
Alisma plantago aquatica, young leaves	1.90 mg/kg dry weight	800 m above pollution entry I, water = not detect., processed leaves[1]
Alisma plantago aquatica, young leaves	14.10 mg/kg dry weight	150 m below pollution entry I, water = 0.030 mg/l, whole leaves[1]
Alisma plantago aquatica, young leaves	2.65 mg/kg dry weight	800 m above pollution entry I, water = not detect., whole leaves[1]
Alisma plantago aquatica, young leaves	5.03 mg/kg dry weight	800 and 50 m below pollution I and II, water = 0.050 mg/l, processed leaves[1]
Alisma plantago aquatica, young leaves	5.90 mg/kg dry weight	800 and 50 m below pollution I and II, water = 0.050 mg/l, whole leaves[1]
Alisma plantago aquatica, young leaves	7.93 mg/kg dry weight	150 m below pollution entry I, water = 0.030 mg/l, processed leaves[1]
Barley, leaves	2.1 mg/kg dry weight	Soil-pH = 5.6, added Cr-alum. = 220 ppm[2]
Barley, leaves	2.2 mg/kg dry weight	Soil-pH = 5.6, added K2Cr2O7 = 200 ppm[2]
Barley, leaves	2.3 mg/kg dry weight	Soil-pH = 6.4, added K2Cr2O7 = 200 ppm, control = 0.8 mg/kg dry[2]
Barley, leaves	2.6 mg/kg dry weight	Soil-pH = 7.8, added Cr-alum. = 200 ppm, control = 0.5 mg/kg dry[2]
Barley, leaves	2.7 mg/kg dry weight	Soil-pH = 6.4, added Cr-alum. = 200 ppm, control = 0.8 mg/kg dry[2]
Barley, leaves	3.2 mg/kg dry weight	Soil-pH = 7.8, added K2Cr2O7 = 200 ppm, control = 0.5 mg/kg dry[2]
Barley, root	115 mg/kg dry weight	Soil-pH = 6.4, added K2Cr2O7 = 200 ppm, control = 2.6 mg/kg dry[2]
Barley, root	137 mg/kg dry weight	Soil-pH = 5.6, added K2Cr2O7 = 200 ppm[2]
Barley, root	146 mg/kg dry weight	Soil-pH = 7.8, added Cr-alum. = 200 ppm, control = 1.9 mg/kg dry[2]
Barley, root	168 mg/kg dry weight	Soil-pH = 7.8, added K2Cr2O7 = 200 ppm, control = 1.9 mg/kg dry[2]
Barley, root	46 mg/kg dry weight	Soil-pH = 5.6, added Cr-alum. = 200 ppm[2]
Barley, root	78 mg/kg dry weight	Soil-pH = 6.4, added Cr-alum. = 200 ppm, control = 2.6 mg/kg dry[2]
Beech and spruce forest soil	51 mg/kg dry weight soil	Central German, soil depth = 40-50 cm, 76E3 g/ha[3]
Beech and spruce forest soil	52 mg/kg dry weight soil	Central German, soil depth =[3]
Beech and spruce forest soil	56 mg/kg dry weight soil	Central German, soil depth = 30-40 cm, 73E3 g/ha[3]

Beech and spruce forest soil	59 mg/kg dry weight soil	Central German, soil depth = 10-20 cm, 70E3 g/ha[3]
Dunaliella sp.	2.6 % of total	Lab, 24 h, insoluble fraction, without algae = 6.4 %, Cr-51[4]
Dunaliella sp.	3.4 % of total	Lab, 1 h, insoluble fraction, without algae = 5.8 %, Cr-51[4]
Dunaliella sp.	94.2 % of total	Lab, 1 h, soluble fraction, without algae = 94.2 %, Cr-51[4]
Dunaliella sp.	97.4 % of total	Lab, 24 h, soluble fraction, without algae = 93.6 %, Cr-51[4]
Forest soil	17 mg/kg dry weight soil	E Tennessee, various forests, depth = 40 - 80 cm[5]
Forest soil	2 mg/kg dry weight soil	E Tennessee, various forests, depth = 80 - 120 cm[5]
Forest soil	2 mg/kg dry weight soil	E Tennessee, various forests, depth = litter[5]
Forest soil	5 mg/kg dry weight soil	E Tennessee, various forests, depth = 0 - 40 cm[5]
Forest soil	5 mg/kg dry weight soil	E Tennessee, various forests, depth = 120 - 160 cm[5]
Forest soil	50 mg/kg dry weight soil	E Tennessee, various forests, depth = > 160 cm[5]
Living spruce needles	0.85 mg/kg dry weight	Litter = 2.3-6.7, underlaying soil = 23-81 mg/kg dw, S Sweden[6]
Sediment	10.43 mg/kg dry weight	800 m above pollution entry I, water = not detect., rg = 5.2 - 14.0[1]
Sediment	3.7 mg/kg dry weight	Unpolluted area, estuary, Urr water, rg = 1.6 - 5.7[7]
Sediment	5.7 mg/kg dry weight	Unpolluted area, estuary, Urr water, rg = 3.9 - 8.4[7]
Sediment	53.00 mg/kg dry weight	150 m below pollution entry I, water = 0.030 mg/l, rg = 43.58 - 59.0[1]
Sediment	6 mg/kg dry weight	Unpolluted area, River, Illinois River, rg = 3 - 7[8]
Sediment	64.74 mg/kg dry weight	800 and 50 m below pollution I and II, water = 0.050 mg/l, rg = 62.67[1]
Sediment	9.9 mg/kg dry weight	Unpolluted area, estuary, Urr water[9]
Vegetation	1.3 mg/kg dry weight	Conc in podzol:f = 5, h = 8, a1a2 = 80, b1 = 80, b2 = 100, b3 = 100[10]
Zooplankton	7.92 mg/kg dry weight	19 st, NY, conc in algae: < 2.88 mg/kg dw, mean, less, Long Isl[11]

Source: 1.Patrick, F.M. and M.W. Loutit. 1977. Water Res., 11(8), pp 699-703.
2.Patterson, J.B.E. 1971. Tech. Bull., 21, pp 193-207.
3.Heinrichs, H. and R. Meyer. 1977. J. Environ. Qual. 6, 402-407.
4.Fraizer, A. and J.C. Ancellin. 1975. Impacts of Nuclear releases into the Aquatic Environment. IAEA-m-198/8, pp 49-61. Vienna.
5.Reiners, W.A., R.H. Marks and P.M. Vitousek. 1975. Oikos 26, 264-275.
6.Nilsson, I. 1972. Oikos, 22, pp 366.
7.Taylor, D. 1976. Sci. Total Environ. 6(3), pp 259-264.
8.Mathis, B.J. and T.F. Cummins. 1973. Water Pollut. Control 45 pp 1573.
9.Taylor, D. 1976. Estuarine Coastal Mar. Sci. 2, pp 417.
10.Mitchell, R. and C. Chamberlin. 1978. in: Berg. G. (ed.); Indicators of Enteric Contamination in Natural Waters.
11.Greig, R.A., A. Adams and D.R. Wenzloff. 1977. Bull. Environ. Contam. and Toxic., 18(1), pp 3-8.

5-325 Concentrations
Cs - 137

Species	Value	Condition
Particle fractions	0.3 % of total	Size = 2.0-0.5 mm, weight = 3.2 %, area = 0.2 %, marine, Japan[1]
Particle fractions	15.5 % of total	Size = 0.125-0.075 mm, weight = 16.9 %, area = 10.4 %, marine, Japan[1]
Particle fractions	16.2 % of total	Size = 0.25-0.125 mm, weight = 23 %, area = 10.1 %, marine, Japan[1]
Particle fractions	3.9 % of total	Size = 0.5-0.25 mm, weight = 8.5 %, area = 1.4 %, marine, Japan[1]

Particle fractions	64.1 % of total	Size below 0.075 mm, weight = 48.4 %, area = 77.9 %, marine, Japan[1]

Source: 1.Nagaya, Y. and K. Nakamura. 1975. Im pacts of Nuclear releases into the Aquatic Environment. IAEA-SM198/21, pp 313-328. Vienna.

5-326 Concentrations Cu

Species	Value	Condition
Alisma plantago aquatica, old leaves	113.90 mg/kg dry weight	800 and 50 m below pollution I and II, water = 0.085 mg/l, whole leaves[1]
Alisma plantago aquatica, old leaves	129.58 mg/kg dry weight	800 m above pollution entry I, water = 0.032 mg/l, whole leaves[1]
Alisma plantago aquatica, old leaves	51.54 mg/kg dry weight	150 m below pollution entry I, water = 0.113 mg/l, processed leaves[1]
Alisma plantago aquatica, old leaves	68.66 mg/kg dry weight	800 m above pollution entry I, water = 0.032 mg/l, processed leaves[1]
Alisma plantago aquatica, old leaves	77.10 mg/kg dry weight	150 m below pollution entry I, water = 0.113 mg/l, whole leaves[1]
Alisma plantago aquatica, old leaves	81.58 mg/kg dry weight	800 and 50 m below pollution I and II, water = 0.085 mg/l, processed leaves[1]
Alisma plantago aquatica, young leaves	128.51 mg/kg dry weight	150 m below pollution entry I, water = 0.113 mg/l, whole leaves[1]
Alisma plantago aquatica, young leaves	136.31 mg/kg dry weight	800 and 50 m below pollution I and II, water = 0.085 mg/l, whole leaves[1]
Alisma plantago aquatica, young leaves	155.67 mg/kg dry weight	800 m above pollution entry I, water = 0.032 mg/l, whole leaves[1]
Alisma plantago aquatica, young leaves	70.05 mg/kg dry weight	800 m above pollution entry I, water = 0.032 mg/l, processed leaves[1]
Alisma plantago aquatica, young leaves	75.67 mg/kg dry weight	150 m below pollution entry I, water = 0.113 mg/l, processed leaves[1]
Alisma plantago aquatica, young leaves	77.60 mg/kg dry weight	800 and 50 m below pollution I and II, water = 0.085 mg/l, processed leaves[1]
Ameiurus nebdosus, gill	1.8	Wanapitei R, Can, 1974, wat=9+-2 ppb, polluted, n=9[2]
Ameiurus nebdosus, kidney	2.5	Wanapitei R, Can, 1974, wat=9+-2 ppb, polluted, n=9[2]
Ameiurus nebdosus, liver	30.3	Wanapitei R, Can, 1974, wat=9+-2 ppb, polluted, n=9[2]
Ameiurus nebdosus, muscle	1.3	Wanapitei R, Can, 1974, wat=9+-2 ppb, polluted, n=9[2]
Anabaena variabilis	3.4 ug/cell	Added = 0.0 mg Cu(+2) and Ag(+)/l, inoculum = 5x, Cu in media = 11.0[3]
Anabaena variabilis	4.7 ug/cell	Added = 0.0 mg Cu(+2) and Ag(+)/l, inoculum = 1x, Cu in media = 8.4[3]
Anabaena variabilis	7.2 ug/cell	Added = 0.1 mg Cu(+2) and Ag(+)/l, inoculum = 5x, Cu in media = 29.0[3]
Anabaena variabilis	9.5 ug/cell	Added = 0.1 mg Cu(+2) and Ag(+)/l, inoculum = 1x, Cu in media = 29.0[3]
Anarchis canadensis	175 mg/kg wet weight	Water = 9+-2 ppb, pH=7.3, Wanapitei R, Can, silt-ladden and poll.[2]
Anarchis canadensis	18 mg/kg wet weight	Water = 4+-1 ppb, pH=7.2, Main outlet (French R) to L. Huron, Can[2]
Anarchis canadensis	4 mg/kg wet weight	Water = 2+-2 ppb, pH=7.2, East outlet (French R) to L. Huron, Can[2]
Anarchis canadensis	5 mg/kg wet weight	Water = 2+-2 ppb, pH=7.2, Pickerel R, Can, non-poll.[2]
Anarchis canadensis	55 mg/kg wet weight	Water = 6+-2 ppb, pH=7.3, W Wanapitei B, Can, 1974[2]

Anarchis canadensis	9 mg/kg wet weight	Water = 3+-2 ppb, pH=7.1, French R, Can, non-poll.[2]
Anguilla anguilla, liver	37.2 mg/kg dry weight	River water, power station cooling water, fish[4]
Anguilla anguilla, muscle	8.97 mg/kg dry weight	River water, power station cooling water, fish[4]
Beech and spruce forest soil	17 mg/kg dry weight soil	Central German, soil depth = 40-50 cm, 25E3 g/ha[5]
Beech and spruce forest soil	18 mg/kg dry weight soil	Central German, soil depth = 10-20 cm, 21E3[5]
Beech and spruce forest soil	23 mg/kg dry weight soil	Central German, soil depth = 30-40 cm, 30E3[5]
Beech and spruce forest soil	24 mg/kg dry weight soil	Central German, soil depth =[5]
Busycon canaliculatum, digestive gland	650 mg/kg wet weight	Summer, Narragansett Bay, rg = 400 - 900[6]
Busycon canaliculatum, digestive gland	92.5 mg/kg wet weight	Winter, Narragansett Bay, rg = 65 - 120[6]
Cabbage, plant	3.6 mg/kg dry weight	15 plants, urban area[7]
Cabbage, plant	4.1 mg/kg dry weight	15 plants, rural area[7]
Catostomus commersoni, gill	1.3	Wanapitei R, Can, 1974, wat=9+-2 ppb, polluted, n=7[2]
Catostomus commersoni, kidney	1.7	Wanapitei R, Can, 1974, wat=9+-2 ppb, polluted, n=7[2]
Catostomus commersoni, liver	15.0	Wanapitei R, Can, 1974, wat=9+-2 ppb, polluted, n=7[2]
Catostomus commersoni, muscle	1.4	Wanapitei R, Can, 1974, wat=9+-2 ppb, polluted, n=7[2]
Clam	10 mg/kg wet weight	Water = 2+-2 ppb, pH=7.2, Pickerel R, Can, non-poll.[2]
Clam	20 mg/kg wet weight	Water = 2+-2 ppb, pH=7.2, East outlet (French R) to L. Huron, Can[2]
Clam	20 mg/kg wet weight	Water = 9+-2 ppb, pH=7.3, Wanapitei R, Can, silt-ladden and poll.[2]
Clam	30 mg/kg wet weight	Water = 4+-1 ppb, pH=7.2, Main outlet (French R) to L. Huron, Can[2]
Clam	5 mg/kg wet weight	Water = 3+-2 ppb, pH=7.1, French R, Can, non-poll.[2]
Clam	68 mg/kg wet weight	Water = 6+-2 ppb, pH=7.3, W Wanapitei B, Can, 1974[2]
Clam	74 mg/kg wet weight	Water = 5+-2 ppb, pH=7.2, Western outlet (French R) to L. Huron, Can[2]
Clam	91 mg/kg wet weight	Water = 7+-2 ppb, pH=7.2, Hartley Bay, Canada, 1974[2]
Coniferous trees, needles	130 mg/kg ash weight	Calcareous rock = 34+-7, soil = 42+-10 ppm, Finland, n = 9[8]
Coniferous trees, needles	240 mg/kg ash weight	Silicic rock = 50+-4, soil = 110+-10 ppm, Finland, n = 58, SD = +-26[8]
Coniferous trees, needles	59 mg/kg ash weight	Ultrabasic rock = 27+-4, soil = 22+-5 ppm, Finland, n = 10[8]
Coniferous trees, twigs	290 mg/kg ash weight	Calcareous rock = 34+-7, soil = 42+-10 ppm, Finland, n = 9[8]
Coniferous trees, twigs	510 mg/kg ash weight	Silicic rock = 50+-4, soil = 110+-10 ppm, Finland, n = 58, SD = +-71[8]
Coniferous trees, twigs	82 mg/kg ash weight	Ultrabasic rock = 27+-4, soil = 22+-5 ppm, Finland, n = 10[8]
Crassostrea virginica	1020 mg/kg wet weight	Cu conc = 0.050 mg/l, 16 w, rg = 931 - 1109 ppm[9]
Crassostrea virginica	1035 mg/kg wet weight	Cu conc = 0.050 mg/l, 20 w, rg = 944 - 1125 ppm[9]
Crassostrea virginica	194 mg/kg wet weight	Cu conc = 0.025 mg/l, 4 w, rg = 144 - 250 ppm[9]
Crassostrea virginica	263 mg/kg wet weight	Cu conc = 0.025 mg/l, 8 w, rg = 220 - 306 ppm[9]
Crassostrea virginica	267 mg/kg wet weight	Cu conc = 0.050 mg/l, 4 w, rg = 184 - 350 ppm[9]

Crassostrea virginica	421 mg/kg wet weight	Cu conc = 0.025 mg/l, 12 w, rg = 280 - 563 ppm[9]
Crassostrea virginica	535 mg/kg wet weight	Cu conc = 0.050 mg/l, 8 w, rg = 397 - 670 ppm[9]
Crassostrea virginica	608 mg/kg wet weight	Cu conc = 0.025 mg/l, 16 w, rg = 478 - 737 ppm[9]
Crassostrea virginica	683 mg/kg wet weight	Cu conc = 0.025 mg/l, 20 w, rg = 650 - 716 ppm[9]
Crassostrea virginica	715 mg/kg wet weight	Cu conc = 0.050 mg/l, 12 w, rg = 584 - 833 ppm[9]
Crassostrea virginica	86 mg/kg wet weight	Cu conc = 0.025 mg/l, 0 w, rg = 71 - 121 ppm[9]
Crassostrea virginica	86 mg/kg wet weight	Cu conc = 0.050 mg/l, 0 w, rg = 71 - 121 ppm[9]
Crayfish	153 mg/kg wet weight	Water = 2+-2 ppb, pH=7.2, East outlet (French R) to L. Huron, Can[2]
Crayfish	212 mg/kg wet weight	Water = 4+-1 ppb, pH=7.2, Main outlet (French R) to L. Huron, Can[2]
Crayfish	23 mg/kg wet weight	Water = 5+-2 ppb, pH=7.2, Western outlet (French R) to L. Huron, Can[2]
Crayfish	265 mg/kg wet weight	Water = 3+-2 ppb, pH=7.1, French R, Can, non-poll.[2]
Crayfish	352 mg/kg wet weight	Water = 9+-2 ppb, pH=7.3, Wanapitei R, Can, silt-ladden and poll.[2]
Crayfish	87 mg/kg wet weight	Water = 6+-2 ppb, pH=7.3, W Wanapitei B, Can, 1974[2]
Deciduous trees and shrubs, leaves	150 mg/kg ash weight	Silicic rock = 50+-4, soil = 110+-10 ppm, Finland, n = 27, SD = +-21[8]
Deciduous trees and shrubs, leaves	94 mg/kg ash weight	Ultrabasic rock = 27+-4, soil = 22+-5 ppm, Finland, n = 14[8]
Deciduous trees and shrubs, leaves	99 mg/kg ash weight	Calcareous rock = 34+-7, soil = 42+-10 ppm, Finland, n = 16[8]
Deciduous trees and shrubs, twigs	130 mg/kg ash weight	Ultrabasic rock = 27+-4, soil = 22+-5 ppm, Finland, n = 14[8]
Deciduous trees and shrubs, twigs	270 mg/kg ash weight	Silicic rock = 50+-4, soil = 110+-10 ppm, Finland, n = 27, SD = +-47[8]
Deciduous trees and shrubs, twigs	96 mg/kg ash weight	Calcareous rock = 34+-7, soil = 42+-10 ppm, Finland, n = 16[8]
Equibibrium distribution in seawater, Cu(2+)	0.1 % of total inorganic Cu	298 K, seawater = 3.5 % S, pH = 8.0-8.2[10]
Equibibrium distribution in seawater, Cu(2+)	0.6 % of total inorganic Cu	298 K, seawater = 3.5 % S, pH = 8.0-8.2[11]
Equibibrium distribution in seawater, Cu(2+)	0.7 % of total inorganic Cu	298 K, seawater = 3.5 % S, pH = 8.0-8.2[12]
Equibibrium distribution in seawater, Cu(2+)	1.4 % of total inorganic Cu	298 K, seawater = 3.5 % S, pH = 8.0-8.2[13]
Equibibrium distribution in seawater, Cu(2+)	17 % of total inorganic Cu	298 K, seawater = 3.5 % S, pH = 8.0-8.2[14]
Equibibrium distribution in seawater, Cu(2+)	3 % of total inorganic Cu	298 K, seawater = 3.5 % S, pH = 8.0-8.2[15]
Equibibrium distribution in seawater, Cu(OH)2	0.3 % of total inorganic Cu	298 K, seawater = 3.5 % S, pH = 8.0-8.2[11]
Equibibrium distribution in seawater, Cu(OH)2	90.0 % of total inorganic Cu	298 K, seawater = 3.5 % S, pH = 8.0-8.2[13]
Equibibrium distribution in seawater, Cu(OH)2	90.8 % of total inorganic Cu	298 K, seawater = 3.5 % S, pH = 8.0-8.2[10]
Equibibrium distribution in seawater, CuCO3	21.6 % of total inorganic Cu	298 K, seawater = 3.5 % S, pH = 8.0-8.2, sum of CuCO3 and Cu(CO3)2(2-)[12]
Equibibrium distribution in seawater, CuCO3	3.2 % of total inorganic Cu	298 K, seawater = 3.5 % S, pH = 8.0-8.2, sum of CuCO3 and Cu(CO3)2(2-)[11]

Equibibrium distribution in seawater, CuCO3	49 % of total inorganic Cu	298 K, seawater = 3.5 % S, pH = 8.0-8.2, sum of CuCO3 and Cu(CO3)2(2-)[14]
Equibibrium distribution in seawater, CuCO3	6.2 % of total inorganic Cu	298 K, seawater = 3.5 % S, pH = 8.0-8.2, sum of CuCO3 and Cu(CO3)2(2-)[10]
Equibibrium distribution in seawater, CuCO3	6.3 % of total inorganic Cu	298 K, seawater = 3.5 % S, pH = 8.0-8.2, sum of CuCO3 and Cu(CO3)2(2-)[13]
Equibibrium distribution in seawater, CuCO3	84 % of total inorganic Cu	298 K, seawater = 3.5 % S, pH = 8.0-8.2, sum of CuCO3 and Cu(CO3)2(2-)[15]
Equibibrium distribution in seawater, CuCl(+)	0.1 % of total inorganic Cu	298 K, seawater = 3.5 % S, pH = 8.0-8.2[13]
Equibibrium distribution in seawater, CuCl(+)	0.5 % of total inorganic Cu	298 K, seawater = 3.5 % S, pH = 8.0-8.2[10]
Equibibrium distribution in seawater, CuCl(+)	1.6 % of total inorganic Cu	298 K, seawater = 3.5 % S, pH = 8.0-8.2[15]
Equibibrium distribution in seawater, CuCl(+)	10 % of total inorganic Cu	298 K, seawater = 3.5 % S, pH = 8.0-8.2[14]
Equibibrium distribution in seawater, CuCl(+)	5.8 % of total inorganic Cu	298 K, seawater = 3.5 % S, pH = 8.0-8.2[12]
Equibibrium distribution in seawater, CuCl(+)	95.5 % of total inorganic Cu	298 K, seawater = 3.5 % S, pH = 8.0-8.2[11]
Equibibrium distribution in seawater, CuCl2	0.1 % of total inorganic Cu	298 K, seawater = 3.5 % S, pH = 8.0-8.2[10]
Equibibrium distribution in seawater, CuCl2	0.1 % of total inorganic Cu	298 K, seawater = 3.5 % S, pH = 8.0-8.2[11]
Equibibrium distribution in seawater, CuCl2	0.1 % of total inorganic Cu	298 K, seawater = 3.5 % S, pH = 8.0-8.2[14]
Equibibrium distribution in seawater, CuCl2	0.1 % of total inorganic Cu	298 K, seawater = 3.5 % S, pH = 8.0-8.2[13]
Equibibrium distribution in seawater, CuCl2	0.3 % of total inorganic Cu	298 K, seawater = 3.5 % S, pH = 8.0-8.2[15]
Equibibrium distribution in seawater, CuCl2	1.6 % of total inorganic Cu	298 K, seawater = 3.5 % S, pH = 8.0-8.2[12]
Equibibrium distribution in seawater, CuOH(+)	0.1 % of total inorganic Cu	298 K, seawater = 3.5 % S, pH = 8.0-8.2[11]
Equibibrium distribution in seawater, CuOH(+)	0.5 % of total inorganic Cu	298 K, seawater = 3.5 % S, pH = 8.0-8.2[15]
Equibibrium distribution in seawater, CuOH(+)	1.2 % of total inorganic Cu	298 K, seawater = 3.5 % S, pH = 8.0-8.2[13]
Equibibrium distribution in seawater, CuOH(+)	2.0 % of total inorganic Cu	298 K, seawater = 3.5 % S, pH = 8.0-8.2[10]
Equibibrium distribution in seawater, CuOH(+)	22 % of total inorganic Cu	298 K, seawater = 3.5 % S, pH = 8.0-8.2[14]
Equibibrium distribution in seawater, CuOH(+)	3.7 % of total inorganic Cu	298 K, seawater = 3.5 % S, pH = 8.0-8.2[12]
Equibibrium distribution in seawater, CuOHCl	65.2 % of total inorganic Cu	298 K, seawater = 3.5 % S, pH = 8.0-8.2[12]
Equibibrium distribution in seawater, CuSO4	0.1 % of total inorganic Cu	298 K, seawater = 3.5 % S, pH = 8.0-8.2[11]
Equibibrium distribution in seawater, CuSO4	0.1 % of total inorganic Cu	298 K, seawater = 3.5 % S, pH = 8.0-8.2[12]
Equibibrium distribution in seawater, CuSO4	0.1 % of total inorganic Cu	298 K, seawater = 3.5 % S, pH = 8.0-8.2[13]
Equibibrium distribution in seawater, CuSO4	0.3 % of total inorganic Cu	298 K, seawater = 3.5 % S, pH = 8.0-8.2[10]

Equibibrium distribution in seawater, CuSO4	2 % of total inorganic Cu	298 K, seawater = 3.5 % S, pH = 8.0-8.2[14]
Equisetum palustre, root	18 mg/kg wet weight	Water = 3+-2 ppb, pH=7.1, French R, Can, non-poll.[2]
Equisetum palustre, root	218 mg/kg wet weight	Water = 6+-2 ppb, pH=7.3, W Wanapitei B, Can, 1974[2]
Equisetum palustre, root	26 mg/kg wet weight	Water = 2+-2 ppb, pH=7.2, Pickerel R, Can, non-poll.[2]
Equisetum palustre, root	34 mg/kg wet weight	Water = 2+-2 ppb, pH=7.2, East outlet (French R) to L. Huron, Can[2]
Equisetum palustre, root	45 mg/kg wet weight	Water = 5+-2 ppb, pH=7.2, Western outlet (French R) to L. Huron, Can[2]
Equisetum palustre, root	54 mg/kg wet weight	Water = 4+-1 ppb, pH=7.2, Main outlet (French R) to L. Huron, Can[2]
Equisetum palustre, root	61 mg/kg wet weight	Water = 7+-2 ppb, pH=7.2, Hartley Bay, Canada, 1974[2]
Equisetum palustre, root	88 mg/kg wet weight	Water = 9+-2 ppb, pH=7.3, Wanapitei R, Can, silt-ladden and poll.[2]
Equisetum palustre, stem	15 mg/kg wet weight	Water = 2+-2 ppb, pH=7.2, East outlet (French R) to L. Huron, Can[2]
Equisetum palustre, stem	16 mg/kg wet weight	Water = 5+-2 ppb, pH=7.2, Western outlet (French R) to L. Huron, Can[2]
Equisetum palustre, stem	21 mg/kg wet weight	Water = 6+-2 ppb, pH=7.3, W Wanapitei B, Can, 1974[2]
Equisetum palustre, stem	22 mg/kg wet weight	Water = 4+-1 ppb, pH=7.2, Main outlet (French R) to L. Huron, Can[2]
Equisetum palustre, stem	28 mg/kg wet weight	Water = 7+-2 ppb, pH=7.2, Hartley Bay, Canada, 1974[2]
Equisetum palustre, stem	28 mg/kg wet weight	Water = 9+-2 ppb, pH=7.3, Wanapitei R, Can, silt-ladden and poll.[2]
Equisetum palustre, stem	6 mg/kg wet weight	Water = 3+-2 ppb, pH=7.1, French R, Can, non-poll.[2]
Equisetum palustre, stem	7 mg/kg wet weight	Water = 2+-2 ppb, pH=7.2, Pickerel R, Can, non-poll.[2]
Esox sp., gill	1.9	Wanapitei R, Can, 1974, wat=9+-2 ppb, polluted, n=4, Northern pike[2]
Esox sp., kidney	2.6	Wanapitei R, Can, 1974, wat=9+-2 ppb, polluted, n=4, Northern pike[2]
Esox sp., liver	10.9	Wanapitei R, Can, 1974, wat=9+-2 ppb, polluted, n=4, Northern pike[2]
Esox sp., muscle	1.4	Wanapitei R, Can, 1974, wat=9+-2 ppb, polluted, n=4, Northern pike[2]
Fundulus heteroclitus, whole body	121 mg/kg ash weight	Ini Cu conc = 6.0 mg/l, control = 93 ppm, exposure = 2 d + 2 d postexp.[16]
Fundulus heteroclitus, whole body	125 mg/kg ash weight	Ini Cu conc = 6.0 mg/l, control = 93 ppm, exposure = 1 d + 1 d postexp.[16]
Fundulus heteroclitus, whole body	255 mg/kg ash weight	Initial Cu conc = 6.0 mg/l, control = 93 ppm, exposure = 2 d[16]
Fundulus heteroclitus, whole body	294 mg/kg ash weight	Initial Cu conc = 6.0 mg/l, control = 93 ppm, exposure = 1 d[16]
Fundulus heteroclitus, whole body	31 mg/kg ash weight	Initial Cu conc = 8.0 mg/l, control = 93 ppm, exposure = 4 d[16]
Fundulus heteroclitus, whole body	66 mg/kg ash weight	Initial Cu conc = 1.0 mg/l, control = 93 ppm, exposure = 4 d[16]
Fundulus heteroclitus, whole body	87 mg/kg ash weight	Initial Cu conc = 2.0 mg/l, control = 93 ppm, exposure = 4 d[16]
Fundulus heteroclitus, whole body	92 mg/kg ash weight	Initial Cu conc = 3.0 mg/l, control = 93 ppm, exposure = 4 d[16]
Haliotis spp., digestive gland	18.1 mg/kg dry weight	Cu conc = 0.032 mg/l, control = 12 ppm, 96 h, rg = 13.2 - 22.9 ppm[17]

Haliotis spp., digestive gland	9.7 mg/kg dry weight	Cu conc = 0.01 mg/l, control = 12 ppm, 96 h, rg = 9.0 - 10.3 ppm[17]
Haliotis spp., gill	17.3 mg/kg dry weight	Cu conc = 0.01 mg/l, control = 12 ppm, 96 h[17]
Haliotis spp., gill	35.3 mg/kg dry weight	Cu conc = 0.32 mg/l, control = 12 ppm, 96 h, rg = 30.9 - 39.7 ppm[17]
Helix aspersa collar muscle	70 mg/kg dry weight	Avon, Engl., rg =25-114 ppm, whole body = 68+-10 ppm, terrestrial[18]
Helix aspersa, albumen gland	38 mg/kg dry weight	Avon, Engl., rg = 34-42 ppm, whole body = 68+-10 ppm, terrestrial[18]
Helix aspersa, crop	32 mg/kg dry weight	Avon, Engl., rg = 21-42 ppm, whole body = 68+-10 ppm, terrestrial[18]
Helix aspersa, foot	68 mg/kg dry weight	Avon, Engl., rg = 41-96 ppm, whole body = 68+-10 ppm, terrestrial[18]
Helix aspersa, hepatopancreas	45 mg/kg dry weight	Avon, Engl., rg = 31-60 ppm, whole body = 68+-10 ppm, terrestrial[18]
Helix pomatia, albumen gland	202 mg/kg dry weight	Germany, Ausust, conc in hepatoancreas=109-505 ppm, rg = 57-147 ppm, terrestrial[18]
Helix pomatia, albumen gland	202 mg/kg dry weight	Germany, Ausust, conc in hepatoancreas=109-505 ppm, rg = 57-147 ppm, terrestrial[18]
Helix pomatia, albumen gland	82 mg/kg dry weight	Tirol, rg=+-6[18]
Helix pomatia, albumen gland	82 mg/kg dry weight	Tirol, rg=+-6 ppm, conc in hepatopancreas=48+-24 ppm, winter, terrestrial[18]
Helix pomatia, albumen gland	82 mg/kg dry weight	Tirol, rg=+-8[18]
Helix pomatia, albumen gland	82 mg/kg dry weight	Tirol, rg=+-8 ppm, conc in hepatopancreas=30+-12 ppm, summer, terrestrial[18]
Helix pomatia, foot	120 mg/kg dry weight	Tirol, rg=+-45 ppm, conc in hepatopancreas=30+-12 ppm, summer, terrestrial[18]
Helix pomatia, foot	85 mg/kg dry weight	Tirol, rg=+-34 ppm, conc in hepatopancreas=48+-24 ppm, winter, terrestrial[18]
Helix pomatia, intestine	114 mg/kg dry weight	Tirol, rg=+-5 ppm, conc in hepatopancreas=30+-12 ppm, summer, terrestrial[18]
Helix pomatia, intestine	40 mg/kg dry weight	Tirol, rg=+-21 ppm, conc in hepatopancreas=48+-24 ppm, winter, terrestrial[18]
Helix pomatia, spermoviduct	77 mg/kg dry weight	Germany, Ausust, conc in hepatoancreas=109-505 ppm, rg = 52-102 ppm, terrestrial[18]
Helix pomatia, spermoviduct	77 mg/kg dry weight	Germany, Ausust, conc in hepatoancreas=109-505 ppm, rg = 52-102 ppm, terrestrial[18]
Herbage	10.1 mg/kg dry weight	No Cu added, S Wales[19]
Herbage	11.0 mg/kg dry weight	Added CuSO4 = 11.2kg/ha, S Wales[19]
Herbage	12.0 mg/kg dry weight	Added Cu-oxide = 22.4kg/ha, S Wales[19]
Herbage	12.0 mg/kg dry weight	Added CuSO4 = 22.4kg/ha, S Wales[19]
Herbage	12.4 mg/kg dry weight	Added Cu-oxide = 11.2kg/ha, S Wales[19]
Homo sapiens, blood	0.65 uM	Cu in tap water, 1st flush = 0.10, running wat = 0.04 Um, n = 30[20]
Hypolimnion	0.9 ug/l	Conc in epilimnion = 1.0 ug/l, concentration factor = 0.9, Schoehsee[21]
Ictalurus nebulosus, gill	6.9 mg/kg dry weight	Lab, Cu conc = 0.027 mg/l, 20 month, control = 3.8 ppm, fish[22]
Ictalurus nebulosus, gill	9.4 mg/kg dry weight	Lab, Cu conc = 0.027 mg/l, pH = 7.2-8.2, 30 d, control = 5.7 ppm, fish[22]
Ictalurus nebulosus, kidney	10 mg/kg dry weight	Lab, Cu conc = 0.027 mg/l, 20 month, control = 7.8 ppm, fish[22]

Ictalurus nebulosus, kidney	10 mg/kg dry weight	Lab, Cu conc = 0.027 mg/l, pH = 7.2-8.2, 30 d, control = 8 ppm, fish[22]
Ictalurus nebulosus, liver	11 mg/kg dry weight	Lab, Cu conc = 0.027 mg/l, 20 month, control = 29 ppm, fish[22]
Ictalurus nebulosus, liver	33 mg/kg dry weight	Lab, Cu conc = 0.027 mg/l, pH = 7.2-8.2, 30 d, control = 24 ppm, fish[22]
Ictalurus sp., gill	1.8 mg/kg dry weight	French River, Cu = 2-9 mg/m3, fish[2]
Ictalurus sp., kidney	2.5 mg/kg dry weight	French River, Cu = 2-9 mg/m3, fish[2]
Ictalurus sp., liver	30.3 mg/kg dry weight	French River, Cu = 2-9 mg/m3, fish[2]
Ictalurus sp., muscle	1.3 mg/kg dry weight	French River, Cu = 2-9 mg/m3, fish[2]
Living spruce needles	2.45 mg/kg dry weight	Litter = 5.1-13.5, underlaying soil = 37-48 mg/kg dw, S Sweden[23]
Micropterus sp., gill	2.1	Wanapitei R, Can, 1974, wat=9+-2 ppb, polluted, n=8, N rock bass[2]
Micropterus sp., kidney	3.0	Wanapitei R, Can, 1974, wat=9+-2 ppb, polluted, n=8, N rock bass[2]
Micropterus sp., liver	4.9	Wanapitei R, Can, 1974, wat=9+-2 ppb, polluted, n=8, N rock bass[2]
Micropterus sp., muscle	1.7	Wanapitei R, Can, 1974, wat=9+-2 ppb, polluted, n=8, N rock bass[2]
Moxostoma sp., gill	1.9	Wanapitei R, Can, 1974, wat=9+-2 ppb, polluted, n=9, redhorse sucker[2]
Moxostoma sp., kidney	2.0	Wanapitei R, Can, 1974, wat=9+-2 ppb, polluted, n=9, redhorse sucker[2]
Moxostoma sp., liver	12.8	Wanapitei R, Can, 1974, wat=9+-2 ppb, polluted, n=9, redhorse sucker[2]
Moxostoma sp., muscle	1.5	Wanapitei R, Can, 1974, wat=9+-2 ppb, polluted, n=9, redhorse sucker[2]
Mya arenaria	200 mg/kg wet weight	Cu conc = 0.05 mg/l, 8 d, 303.5 K, initial tissue = 35 ppm[24]
Mya arenaria	79 mg/kg wet weight	Cu conc = 0.1 mg/l, 10 d, 298 K, initial tissue = 23 ppm[24]
Mya arenaria	85 mg/kg wet weight	Cu conc = 0.2 mg/l, 25 d, 298 K, initial tissue = 15 ppm[24]
Mytilus edulis	20 mg/kg dry weight	Exposure 30 d, Cu=0.010 mg/l, control=20 ppm for 3 d[25]
Mytilus edulis	35 mg/kg dry weight	Exposure 30 d, Cu=0.012 mg/l, control=20 ppm for 3 d[25]
Mytilus edulis	60 mg/kg dry weight	Exposure 10 d, Cu=0.025 mg/l, control=20 ppm for 3 d[25]
Mytilus edulis	70 mg/kg dry weight	Exposure 3 d, Cu=0.045 mg/l, control=20 ppm for 3 d, rg=50-90 ppm[25]
Mytilus edulis	80 mg/kg dry weight	Exposure 30 d, Cu=0.017 mg/l, control=20 ppm for 3 d[25]
Neomacheilus barbatulus, gill	164 mg/kg dry weight	Lab, Cu conc = 0.76 mg/l, pH = 8.6, 64 d, control = 26.4 ppm, fish[26]
Neomacheilus barbatulus, liver	115 mg/kg dry weight	Lab, Cu conc = 0.76 mg/l, pH = 8.6, 64 d, control = 53 ppm, fish[26]
Neomacheilus barbatulus, muscle	8.5 mg/kg dry weight	Lab, Cu conc = 0.76 mg/l, pH = 8.6, 64 d, control = 5.1 ppm, fish[26]
Nereis virens	16 mg/kg dry weight	Initial Cu = 0.02 mg/l, 96 h[27]
Nereis virens	250 mg/kg dry weight	Some deaths, initial Cu = 0.16 mg/l, 96 h[27]
Nereis virens	38 mg/kg dry weight	Initial Cu = 0.04 mg/l, 96 h[27]
Nereis virens	63 mg/kg dry weight	Initial Cu = 0.08 mg/l, 96 h[27]
Nereis virens, body wall	113 mg/kg dry weight	Cu = 0.2 mg/l, 22 h, control = 32 ppm[27]
Nereis virens, body wall	140 mg/kg dry weight	Cu = 0.2 mg/l, 48 h, control = 27 ppm[27]
Nereis virens, body wall	179 mg/kg dry weight	Cu = 0.2 mg/l, 72 h, control = 28 ppm[27]
Nereis virens, body wall	266 mg/kg dry weight	Cu = 0.2 mg/l, 96 h, control = 25 ppm[27]

Nereis virens, gut	117 mg/kg dry weight	Cu = 0.2 mg/l, 72 h, control = 17 ppm[27]
Nereis virens, gut	128 mg/kg dry weight	Cu = 0.2 mg/l, 96 h, control = 16 ppm[27]
Nereis virens, gut	62 mg/kg dry weight	Cu = 0.2 mg/l, 22 h, control = 15 ppm[27]
Nereis virens, gut	87 mg/kg dry weight	Cu = 0.2 mg/l, 48 h, control = 19 ppm[27]
Noemacheilus barbatulus	10.3 mg/kg dry weight	Cu = 0.38 mg/l, control = 3.mg/l, 285 K, 10 fish[28]
Noemacheilus barbatulus	3.7 mg/kg dry weight	Cu = 0.12 mg/l, control = 3.mg/l, 285 K, 10 fish[28]
Noemacheilus barbatulus	5.2 mg/kg dry weight	Cu = 0.71 mg/l, control = 3.mg/l, 285 K, 10 fish[28]
Noemacheilus barbatulus	6.2 mg/kg dry weight	Cu = 0.49 mg/l, control = 3.mg/l, 285 K, 10 fish[28]
Noemacheilus barbatulus	7.8 mg/kg dry weight	Cu = 0.29 mg/l, control = 3.mg/l, 285 K, 10 fish[28]
Noemacheilus barbatulus	7.9 mg/kg dry weight	Cu = 0.76 mg/l, control = 3.mg/l, 285 K, 10 fish[28]
Noemacheilus barbatulus, brain	12.1 mg/kg dry weight	Cu = 0.71 mg/l, no control, 285 K, 10 fish[28]
Noemacheilus barbatulus, brain	25.9 mg/kg dry weight	Cu = 0.29 mg/l, no control, 285 K, 10 fish[28]
Noemacheilus barbatulus, brain	4.2 mg/kg dry weight	Cu = 0.49 mg/l, no control, 285 K, 10 fish[28]
Noemacheilus barbatulus, brain	43.8 mg/kg dry weight	Cu = 0.38 mg/l, no control, 285 K, 10 fish[28]
Noemacheilus barbatulus, brain	53 mg/kg dry weight	Cu = 0.76 mg/l, no control, 285 K, 10 fish[28]
Noemacheilus barbatulus, eye	18.6 mg/kg dry weight	Cu = 0.76 mg/l, control = 19.6 mg/l, 285 K, 10 fish[28]
Noemacheilus barbatulus, eye	20.8 mg/kg dry weight	Cu = 0.29 mg/l, control = 19.6 mg/l, 285 K, 10 fish[28]
Noemacheilus barbatulus, eye	21.8 mg/kg dry weight	Cu = 0.71 mg/l, control = 19.6 mg/l, 285 K, 10 fish[28]
Noemacheilus barbatulus, eye	3.4 mg/kg dry weight	Cu = 0.12 mg/l, control = 19.6 mg/l, 285 K, 10 fish[28]
Noemacheilus barbatulus, eye	36.7 mg/kg dry weight	Cu = 0.38 mg/l, control = 19.6 mg/l, 285 K, 10 fish[28]
Noemacheilus barbatulus, eye	44.8 mg/kg dry weight	Cu = 0.49 mg/l, control = 19.6 mg/l, 285 K, 10 fish[28]
Noemacheilus barbatulus, gill	157 mg/kg dry weight	Cu = 0.38 mg/l, control = 26.4 mg/l, 285 K, 10 fish[28]
Noemacheilus barbatulus, gill	164 mg/kg dry weight	Cu = 0.76 mg/l, control = 26.4 mg/l, 285 K, 10 fish[28]
Noemacheilus barbatulus, gill	208 mg/kg dry weight	Cu = 0.49 mg/l, control = 26.4 mg/l, 285 K, 10 fish[28]
Noemacheilus barbatulus, gill	21.4 mg/kg dry weight	Cu = 0.12 mg/l, control = 26.4 mg/l, 285 K, 10 fish[28]
Noemacheilus barbatulus, gill	489 mg/kg dry weight	Cu = 0.29 mg/l, control = 26.4 mg/l, 285 K, 10 fish[28]
Noemacheilus barbatulus, gill	92 mg/kg dry weight	Cu = 0.71 mg/l, control = 26.4 mg/l, 285 K, 10 fish[28]
Noemacheilus barbatulus, liver	115 mg/kg dry weight	Cu = 0.76 mg/l, control = 53 mg/l, 285 K, 10 fish[28]
Noemacheilus barbatulus, liver	153 mg/kg dry weight	Cu = 0.38 mg/l, control = 53 mg/l, 285 K, 10 fish[28]
Noemacheilus barbatulus, liver	163 mg/kg dry weight	Cu = 0.29 mg/l, control = 53 mg/l, 285 K, 10 fish[28]
Noemacheilus barbatulus, liver	56 mg/kg dry weight	Cu = 0.71 mg/l, control = 53 mg/l, 285 K, 10 fish[28]
Noemacheilus barbatulus, liver	77 mg/kg dry weight	Cu = 0.12 mg/l, control = 53 mg/l, 285 K, 10 fish[28]
Noemacheilus barbatulus, liver	92 mg/kg dry weight	Cu = 0.49 mg/l, control = 53 mg/l, 285 K, 10 fish[28]
Noemacheilus barbatulus, muscle	10.1 mg/kg dry weight	Cu = 0.12 mg/l, control = 5.1 mg/l, 285 K, 10 fish[28]
Noemacheilus barbatulus, muscle	33.5 mg/kg dry weight	Cu = 0.49 mg/l, control = 5.1 mg/l, 285 K, 10 fish[28]
Noemacheilus barbatulus, muscle	45.7 mg/kg dry weight	Cu = 0.71 mg/l, control = 5.1 mg/l, 285 K, 10 fish[28]

Noemacheilus barbatulus, muscle	53 mg/kg dry weight	Cu = 0.38 mg/l, control = 5.1 mg/l, 285 K, 10 fish[28]
Noemacheilus barbatulus, muscle	8.2 mg/kg dry weight	Cu = 0.29 mg/l, control = 5.1 mg/l, 285 K, 10 fish[28]
Noemacheilus barbatulus, muscle	8.5 mg/kg dry weight	Cu = 0.76 mg/l, control = 5.1 mg/l, 285 K, 10 fish[28]
Noemacheilus barbatulus, operculum	21.6 mg/kg dry weight	Cu = 0.76 mg/l, no control, 285 K, 10 fish[28]
Noemacheilus barbatulus, operculum	31.5 mg/kg dry weight	Cu = 0.29 mg/l, no control, 285 K, 10 fish[28]
Noemacheilus barbatulus, operculum	44 mg/kg dry weight	Cu = 0.38 mg/l, no control, 285 K, 10 fish[28]
Noemacheilus barbatulus, operculum	69 mg/kg dry weight	Cu = 0.71 mg/l, no control, 285 K, 10 fish[28]
Noemacheilus barbatulus, operculum	85 mg/kg dry weight	Cu = 0.49 mg/l, no control, 285 K, 10 fish[28]
Nuphar variegatum, leaf	10 mg/kg wet weight	Water = 5+-2 ppb, pH=7.2, Western outlet (French R) to L. Huron, Can[2]
Nuphar variegatum, leaf	12 mg/kg wet weight	Water = 7+-2 ppb, pH=7.2, Hartley Bay, Canada, 1974[2]
Nuphar variegatum, leaf	18 mg/kg wet weight	Water = 9+-2 ppb, pH=7.3, Wanapitei R, Can, silt-ladden and poll.[2]
Nuphar variegatum, leaf	4 mg/kg wet weight	Water = 2+-2 ppb, pH=7.2, East outlet (French R) to L. Huron, Can[2]
Nuphar variegatum, leaf	4 mg/kg wet weight	Water = 2+-2 ppb, pH=7.2, Pickerel R, Can, non-poll.[2]
Nuphar variegatum, leaf	7 mg/kg wet weight	Water = 4+-1 ppb, pH=7.2, Main outlet (French R) to L. Huron, Can[2]
Nuphar variegatum, leaf	9 mg/kg wet weight	Water = 6+-2 ppb, pH=7.3, W Wanapitei B, Can, 1974[2]
Nuphar variegatum, peduncle	11 mg/kg wet weight	Water = 9+-2 ppb, pH=7.3, Wanapitei R, Can, silt-ladden and poll.[2]
Nuphar variegatum, peduncle	12 mg/kg wet weight	Water = 6+-2 ppb, pH=7.3, W Wanapitei B, Can, 1974[2]
Nuphar variegatum, peduncle	12 mg/kg wet weight	Water = 7+-2 ppb, pH=7.2, Hartley Bay, Canada, 1974[2]
Nuphar variegatum, peduncle	2 mg/kg wet weight	Water = 2+-2 ppb, pH=7.2, Pickerel R, Can, non-poll.[2]
Nuphar variegatum, peduncle	4 mg/kg wet weight	Water = 2+-2 ppb, pH=7.2, East outlet (French R) to L. Huron, Can[2]
Nuphar variegatum, peduncle	6 mg/kg wet weight	Water = 4+-1 ppb, pH=7.2, Main outlet (French R) to L. Huron, Can[2]
Nuphar variegatum, petiole	12 mg/kg wet weight	Water = 6+-2 ppb, pH=7.3, W Wanapitei B, Can, 1974[2]
Nuphar variegatum, petiole	18 mg/kg wet weight	Water = 7+-2 ppb, pH=7.2, Hartley Bay, Canada, 1974[2]
Nuphar variegatum, petiole	2 mg/kg wet weight	Water = 2+-2 ppb, pH=7.2, Pickerel R, Can, non-poll.[2]
Nuphar variegatum, petiole	23 mg/kg wet weight	Water = 9+-2 ppb, pH=7.3, Wanapitei R, Can, silt-ladden and poll.[2]
Nuphar variegatum, petiole	4 mg/kg wet weight	Water = 5+-2 ppb, pH=7.2, Western outlet (French R) to L. Huron, Can[2]
Nuphar variegatum, petiole	5 mg/kg wet weight	Water = 4+-1 ppb, pH=7.2, Main outlet (French R) to L. Huron, Can[2]
Nuphar variegatum, petiole	6 mg/kg wet weight	Water = 2+-2 ppb, pH=7.2, East outlet (French R) to L. Huron, Can[2]
Nuphar variegatum, root	20 mg/kg wet weight	Water = 9+-2 ppb, pH=7.3, Wanapitei R, Can, silt-ladden and poll.[2]
Nuphar variegatum, root	5 mg/kg wet weight	Water = 2+-2 ppb, pH=7.2, Pickerel R, Can, non-poll.[2]
Nuphar variegatum, root	5 mg/kg wet weight	Water = 6+-2 ppb, pH=7.3, W Wanapitei B, Can, 1974[2]
Nuphar variegatum, root	5 mg/kg wet weight	Water = 7+-2 ppb, pH=7.2, Hartley Bay, Canada, 1974[2]

Nuphar variegatum, root	6 mg/kg wet weight	Water = 2+-2 ppb, pH=7.2, East outlet (French R) to L. Huron, Can[2]
Nuphar variegatum, root	8 mg/kg wet weight	Water = 4+-1 ppb, pH=7.2, Main outlet (French R) to L. Huron, Can[2]
Oak forest soil	119 mg/kg dry weight soil	E Chicago, USA, urban, horizon = A1, 0-2.5 cm[29]
Oak forest soil	3 mg/kg dry weight soil	E Chicago, USA, urban, horizon = B, 14-25 cm[29]
Oak forest soil	30 mg/kg dry weight soil	E Chicago, USA, urban, horizon = A1, 2.5-14 cm[29]
Oak forest soil	76 mg/kg dry weight soil	E Chicago, USA, urban, horizon = O , surface litter[29]
Perca flavescens, liver	23.7 mg/kg dry weight	Rg = 2.8-44.5 ppm, DAlene-Spokane River, metal-polluted, fish[30]
Perca flavescens, muscle	0.73 mg/kg dry weight	Rg = 0.4-1.06 ppm, DAlene-Spokane River, metal-polluted, fish[30]
Periphyton	123 mg/kg wet weight	Water = 6+-2 ppb, pH=7.3, W Wanapitei B, Can, 1974[2]
Periphyton	151 mg/kg wet weight	Water = 7+-2 ppb, pH=7.2, Hartley Bay, Canada, 1974[2]
Periphyton	159 mg/kg wet weight	Water = 9+-2 ppb, pH=7.3, Wanapitei R, Can, silt-ladden and poll.[2]
Periphyton	20 mg/kg wet weight	Water = 3+-2 ppb, pH=7.1, French R, Can, non-poll.[2]
Periphyton	25 mg/kg wet weight	Water = 2+-2 ppb, pH=7.2, Pickerel R, Can, non-poll.[2]
Periphyton	42 mg/kg wet weight	Water = 2+-2 ppb, pH=7.2, East outlet (French R) to L. Huron, Can[2]
Periphyton	62 mg/kg wet weight	Water = 4+-1 ppb, pH=7.2, Main outlet (French R) to L. Huron, Can[2]
Periphyton	75 mg/kg wet weight	Water = 5+-2 ppb, pH=7.2, Western outlet (French R) to L. Huron, Can[2]
Pig, feces	135.0 mg/kg dry weight	Feed conc = 11.4, weight of pig = 64 kg[31]
Pig, feces	1421.8 mg/kg dry weight	Feed conc = 328.1 mg/kg dw, weight of pig = 45 kg, Cu suppl. in feed[31]
Pig, feces	150.5 mg/kg dry weight	Feed conc = 25.9 mg/kg dw, weight of pig = 73 kg[31]
Pig, feces	625.0 mg/kg dry weight	Feed conc = 41.3 mg/kg dw, weight of pig = 61 kg[31]
Pig, feces	82.5 mg/kg dry weight	Feed conc = 13.6 mg/kg dw, weight of pig = 76 kg[31]
Pig, feces	93.8 mg/kg dry weight	Feed conc = 11.4 mg/kg dw, weight of pig = 53 kg[31]
Pig, urine	0.125 mg/kg dry weight	Feed conc = 11.4 mg/kg dw, weight of pig = 64 kg[31]
Pig, urine	0.13 mg/kg dry weight	Feed conc = 25.9 mg/kg dw, weight of pig = 73 kg[31]
Pig, urine	0.2 mg/kg dry weight	Feed conc = 11.4 mg/kg dw, weight of pig = 53 kg[31]
Pig, urine	0.2 mg/kg dry weight	Feed conc = 41.3 mg/kg dw, weight of pig = 61 kg[31]
Pig, urine	0.25 mg/kg dry weight	Feed conc = 328.1 mg/kg dw, weight of pig = 45 kg, Cu suppl. in feed[31]
Pig, urine	0.78 mg/kg dry weight	Feed conc = 13.6 mg/kg dw, weight of pig = 76 kg[31]
Pleuronectes platessa, viscera	147 mg/kg dry weight	Cu = 0.03 mg/l, control = 30 ppm, 100 d[32]
Pleuronectes platessa, viscera	567 mg/kg dry weight	Cu = 0.10 mg/l, control = 30 ppm, 100 d[32]
Pleuronectes platessa, viscera	71 mg/kg dry weight	Cu = 0.01 mg/l, control = 30 ppm, 100 d[32]
Porcellio scaber, hepatopanceras	14300 mg/kg dry weight	Cornwall, UK, conc in body wall = 50 ppm, terrestrial[18]
Porcellio scaber, intestine	760 mg/kg dry weight	Cornwall, UK, conc in body wall = 50 ppm, terrestrial[18]
Porcellio scaber, whole animal	1020 mg/kg dry weight	Cornwall, UK, conc in body wall = 50 ppm, terrestrial[18]

Potamogeton sp., leaf	18 mg/kg wet weight	Water = 3+-2 ppb, pH=7.1, French R, Can, non-poll.[2]
Potamogeton sp., leaf	21 mg/kg wet weight	Water = 2+-2 ppb, pH=7.2, East outlet (French R) to L. Huron, Can[2]
Potamogeton sp., leaf	27 mg/kg wet weight	Water = 2+-2 ppb, pH=7.2, Pickerel R, Can, non-poll.[2]
Potamogeton sp., leaf	48 mg/kg wet weight	Water = 5+-2 ppb, pH=7.2, Western outlet (French R) to L. Huron, Can[2]
Potamogeton sp., leaf	81 mg/kg wet weight	Water = 9+-2 ppb, pH=7.3, Wanapitei R, Can, silt-ladden and poll.[2]
Potamogeton sp., leaf	85 mg/kg wet weight	Water = 6+-2 ppb, pH=7.3, W Wanapitei B, Can, 1974[2]
Potamogeton sp., leaf	93 mg/kg wet weight	Water = 7+-2 ppb, pH=7.2, Hartley Bay, Canada, 1974[2]
Potamogeton sp., stem	15 mg/kg wet weight	Water = 2+-2 ppb, pH=7.2, Pickerel R, Can, non-poll.[2]
Potamogeton sp., stem	15 mg/kg wet weight	Water = 3+-2 ppb, pH=7.1, French R, Can, non-poll.[2]
Potamogeton sp., stem	25 mg/kg wet weight	Water = 5+-2 ppb, pH=7.2, Western outlet (French R) to L. Huron, Can[2]
Potamogeton sp., stem	37 mg/kg wet weight	Water = 6+-2 ppb, pH=7.3, W Wanapitei B, Can, 1974[2]
Potamogeton sp., stem	48 mg/kg wet weight	Water = 7+-2 ppb, pH=7.2, Hartley Bay, Canada, 1974[2]
Potamogeton sp., stem	48 mg/kg wet weight	Water = 9+-2 ppb, pH=7.3, Wanapitei R, Can, silt-ladden and poll.[2]
Potamogeton sp., stem	5 mg/kg wet weight	Water = 2+-2 ppb, pH=7.2, East outlet (French R) to L. Huron, Can[2]
Red clover	10.6 mg/kg dry weight	Added CuSO4 = 22kg/ha, pH = 5.4, control = 7.7 mg/kg dry weight[33]
Red clover	12.1 mg/kg dry weight	Added CuSO4 = 22kg/ha, pH = 6.2, control = 3.8 mg/kg dry weight[33]
Salvelinus fontinalis, gill	5.0 mg/kg dry weight	Lab, Cu conc = 0.003 mg/l, fish[34]
Salvelinus fontinalis, gill	6.2 mg/kg dry weight	Lab, Cu conc = 0.009 mg/l, fish[34]
Salvelinus fontinalis, kidney	16.5 mg/kg dry weight	Lab, Cu conc = 0.009 mg/l, fish[34]
Salvelinus fontinalis, kidney	19.7 mg/kg dry weight	Lab, Cu conc = 0.003 mg/l, fish[34]
Salvelinus fontinalis, liver	238 mg/kg dry weight	Lab, Cu conc = 0.009 mg/l, fish[34]
Salvelinus fontinalis, liver	239 mg/kg dry weight	Lab, Cu conc = 0.003 mg/l, fish[34]
Salvelinus fontinalis, muscle	15.3 mg/kg dry weight	Lab, Cu conc = 0.003 mg/l, fish[34]
Salvelinus fontinalis, muscle	17.0 mg/kg dry weight	Lab, Cu conc = 0.009 mg/l, fish[34]
Sediment	109 mg/kg wet weight	Water = 9+-2 ppb, pH=7.3, Wanapitei R, Can, silt-ladden and poll., SD=55[2]
Sediment	11 mg/kg wet weight	Water = 2+-2 ppb, pH=7.2, Pickerel R, Can, non-poll., SD=7[2]
Sediment	11.2 mg/kg dry weight	Unpolluted area, estuary, Urr water, rg = 9.2 - 16.8[35]
Sediment	1100 mg/kg dry weight	Max water = 39 microgram/l, 1 years measurement, Red River, GB[36]
Sediment	1100 mg/kg dry weight	Water conc = 30 microgram/l, yearly mean, Red River, GB[36]
Sediment	112 mg/kg wet weight	Water = 5+-2 ppb, pH=7.2, Western outlet (French R) to L. Huron, Can, SD=38[2]
Sediment	140.26 mg/kg dry weight	800 m above pollution entry I, water = 0.032 mg/l, rg = 126.2 - 185.7[1]
Sediment	19.0 mg/kg dry weight	Range 5-13 ppm, River Rheidol, Cardiganshire, GB[37]
Sediment	244.31 mg/kg dry weight	150 m below pollution entry I, water = 0.113 mg/l, rg = 221.5 - 290.2[1]
Sediment	26 mg/kg dry weight	Max water = 2 microgram/l, 1 years measurement, Fowey River, GB[36]

Sediment	26 mg/kg dry weight	Water conc = 2 microgram/l, yearly mean, fowey River, GB[36]
Sediment	286.34 mg/kg dry weight	800 and 50 m below pollution I and II, water = 0.085 mg/l, rg = 274.2[1]
Sediment	291 mg/kg dry weight	Max water = 4 microgram/l, 1 years measurement, Carnon River, GB[36]
Sediment	291 mg/kg dry weight	Water conc = 3 microgram/l, yearly mean, Carnon River, GB[36]
Sediment	32 mg/kg dry weight	Max water = 3 microgram/l, max conc, Gannel River, GB[36]
Sediment	32 mg/kg dry weight	Water conc = 2 microgram/l, yearly mean, Gannel River, GB[36]
Sediment	333 mg/kg dry weight	Max water = 15 microgram/l, max conc, Gannel River, GB[36]
Sediment	333 mg/kg dry weight	Water conc = 11 microgram/l, yearly mean, Gannel River, GB[36]
Sediment	374 mg/kg dry weight	Max water = 8 microgram/l, 1 years measurement, Red River, GB[36]
Sediment	374 mg/kg dry weight	Water conc = 5 microgram/l, yearly mean, Red River, GB[36]
Sediment	39 mg/kg wet weight	Water = 7+-2 ppb, pH=7.2, Hartley Bay, Canada, 1974, SD=15[2]
Sediment	4.2 mg/kg dry weight	Unpolluted area, estuary, Urr water[38]
Sediment	45 mg/kg wet weight	Water = 6+-2 ppb, pH=7.3, W Wanapitei B, Can, 1974, SD=10[2]
Sediment	57 mg/kg wet weight	Water = 4+-1 ppb, pH=7.2, Main outlet (French R) to L. Huron, Can, SD=18[2]
Sediment	6.9 mg/kg dry weight	Unpolluted area, estuary, Urr water, rg = 1.9 - 12.2[35]
Sediment	66 mg/kg dry weight	Max water = 12 microgram/l, 1 years measurement, Fowey River, GB[36]
Sediment	66 mg/kg dry weight	Water conc = 7 microgram/l, yearly mean, fowey River, GB[36]
Sediment	7 mg/kg wet weight	Water = 3+-2 ppb, pH=7.1, French R, Can, non-poll.[2]
Sediment	7.7 mg/kg dry weight	Unpolluted area, River, Illinois River, rg = 3.5 - 11.2[39]
Sediment	8 mg/kg wet weight	Water = 2+-2 ppb, pH=7.2, East outlet (French R) to L. Huron, Can[2]
Sediment	89 mg/kg dry weight	Max water = 8 microgram/l, 1 years measurement, Carnon River, GB[36]
Sediment	89 mg/kg dry weight	Water conc = 4 microgram/l, yearly mean, Carnon River, GB[36]
Sediment	95 mg/kg dry weight	Conc in epilimnion = 1.0 ug/l, concentration factor = 95000,[21]
Sediment in suspension	26 mg Cu/g wet weight	200 mg sediment/l, pH = 6.8, release : 7.0 mg Ca/l and 0.50 mg Mg/l,[40]
Sediment in suspension	28 mg Cu/g wet weight	100 mg sediment/l, pH = 6.7, release : 4.5 mg Ca/l and 0.30 mg Mg/l,[40]
Sediment in suspension	28 mg Cu/g wet weight	150 mg sediment/l, pH = 6.7, release : 5.7 mg Ca/l and 0.43 mg Mg/l,[40]
Sediment in suspension	44 mg Cu/g wet weight	50 mg sediment/l, pH = 6.7, release : 5.5 mg Ca/l and 0.28 mg Mg/l,[40]
Size distribution of airborne particulates	11 % of total particulate Cu	Trebanos, Glamorgs., Wales[41]
Size distribution of airborne particulates	11 % of total particulate Cu	Trebanos, Glamorgs., Wales, diameter = 1.1-2.3[41]
Size distribution of airborne particulates	11 % of total particulate Cu	Trebanos, Glamorgs., Wales, diameter = 2.3-3.1[41]
Size distribution of airborne particulates	11 % of total particulate Cu	Trebanos, Glamorgs., Wales, diameter = >11[41]
Size distribution of airborne particulates	13 % of total particulate Cu	Missouri, 0.8 km from smelter[41]

Size distribution of airborne particulates	13 % of total particulate Cu	Missouri, 0.8 km from smelter, diameter = 3.3-4.7, control = 13 %[41]
Size distribution of airborne particulates	15 % of total particulate Cu	Missouri, 0.8 km from smelter[41]
Size distribution of airborne particulates	15 % of total particulate Cu	Missouri, 0.8 km from smelter, diameter = 4.7-7, control = 9.5 %[41]
Size distribution of airborne particulates	15 % of total particulate Cu	Trebanos, Glamorgs., Wales[41]
Size distribution of airborne particulates	15 % of total particulate Cu	Trebanos, Glamorgs., Wales, diameter = 0.68-1.1[41]
Size distribution of airborne particulates	21 % of total particulate Cu	Missouri, 0.8 km from smelter[41]
Size distribution of airborne particulates	21 % of total particulate Cu	Missouri, 0.8 km from smelter, diameter = 0.43-0.65, control = 13 %[41]
Size distribution of airborne particulates	22 % of total particulate Cu	Missouri, 0.8 km from smelter[41]
Size distribution of airborne particulates	22 % of total particulate Cu	Missouri, 0.8 km from smelter, diameter = > 11, control = 8.4 %[41]
Size distribution of airborne particulates	22 % of total particulate Cu	Trebanos, Glamorgs., Wales[41]
Size distribution of airborne particulates	22 % of total particulate Cu	Trebanos, Glamorgs., Wales, diameter = < 0.47[41]
Size distribution of airborne particulates	4.1 % of total particulate Cu	Missouri, 0.8 km from smelter[41]
Size distribution of airborne particulates	4.1 % of total particulate Cu	Missouri, 0.8 km from smelter, diameter = 0.65-1.1[41]
Size distribution of airborne particulates	6 % of total particulate Cu	Trebanos, Glamorgs., Wales[41]
Size distribution of airborne particulates	6 % of total particulate Cu	Trebanos, Glamorgs., Wales, diameter = 3.1-5.0[41]
Size distribution of airborne particulates	6 % of total particulate Cu	Trebanos, Glamorgs., Wales, diameter = 7.5-11[41]
Size distribution of airborne particulates	7.7 % of total particulate Cu	Missouri, 0.8 km from smelter[41]
Size distribution of airborne particulates	7.7 % of total particulate Cu	Missouri, 0.8 km from smelter, diameter = 2.1-3.3, control = 6.3 %[41]
Size distribution of airborne particulates	8.2 % of total particulate Cu	Missouri, 0.8 km from smelter[41]
Size distribution of airborne particulates	8.2 % of total particulate Cu	Missouri, 0.8 km from smelter, diameter = 1.1-2.1, control = 8.2 %[41]
Size distribution of airborne particulates	9 % of total particulate Cu	Trebanos, Glamorgs., Wales[41]
Size distribution of airborne particulates	9 % of total particulate Cu	Trebanos, Glamorgs., Wales, diameter = 0.47-0.68[41]
Size distribution of airborne particulates	9 % of total particulate Cu	Trebanos, Glamorgs., Wales, diameter = 5.0-7.5[41]
Size distribution of airborne particulates	9.3 % of total particulate Cu	Missouri, 0.8 km from smelter[41]
Size distribution of airborne particulates	9.3 % of total particulate Cu	Missouri, 0.8 km from smelter, diameter = 7-11, control = 38 %[41]
Spruce forest soil	260 mg/kg dry weight soil	Central Sweden, needle litter, O1[42]
Spruce forest soil	660 mg/kg dry weight soil	Central Sweden, Humus, O2[42]

Stizostedion vitreum, gill	1.8 mg/kg dry weight	French River, Cu = 2-9 mg/m3, fish[2]
Stizostedion vitreum, kidney	6.0 mg/kg dry weight	French River, Cu = 2-9 mg/m3, fish[2]
Stizostedion vitreum, liver	2.8 mg/kg dry weight	French River, Cu = 2-9 mg/m3, fish[2]
Stizostedion vitreum, muscle	1.4 mg/kg dry weight	French River, Cu = 2-9 mg/m3, fish[2]
Strophocheilus sp., albumen gland	35 mg/kg dry weight	Brazil, conc in hepatopancereas = 234 ppm, terrestrial[18]
Strophocheilus sp., intstine	144 mg/kg dry weight	Brazil, conc in hepatopancereas = 234 ppm, terrestrial[18]
Strophocheilus sp., mucus	225 mg/kg dry weight	Brazil, conc in hepatopancereas = 234 ppm, terrestrial[18]
Strophocheilus sp., spermoviduct	18 mg/kg dry weight	Brazil, conc in hepatopancereas = 234 ppm, terrestrial[18]
Suspended solids in water	0.43 fraction of total Cu	Soluble Cu = 31 mg/m3, Pix Brook[43]
Suspended solids in water	0.47 fraction of total Cu	Soluble Cu = 85 mg/m3, River Tame[43]
Suspended solids in water	0.57 fraction of total Cu	Soluble Cu = 13 mg/m3, Stevenage Brook[43]
Suspended solids in water	0.74 fraction of total Cu	Soluble Cu = 54 mg/m3, River Tame[43]
Suspended solids in water	0.76 fraction of total Cu	Soluble Cu = 66 mg/m3, River Tame[43]
Suspended solids in water	0.77 fraction of total Cu	Soluble Cu = 48 mg/m3, River Tame[43]
Suspended solids in water	0.77 fraction of total Cu	Soluble Cu = 56 mg/m3, River Tame[43]
Suspended solids in water	0.83 fraction of total Cu	Soluble Cu = 200 mg/m3, River Tame[43]
Suspended solids in water	0.88 fraction of total Cu	Soluble Cu = 20 mg/m3, River Lee[43]
Tellina tenuis	1100 mg/kg dry weight	Cu conc = 0.100 mg/l, control = 50 ppm, 100 d[32]
Tellina tenuis	270 mg/kg dry weight	Cu conc = 0.010 mg/l, control = 50 ppm, 100 d[32]
Tellina tenuis	470 mg/kg dry weight	Cu conc = 0.030 mg/l, control = 50 ppm, 100 d[32]
Timothy	1.8 mg/kg dry weight	Added limestone = 2.5 t/ha, soil-pH = 5.0, N Wales[19]
Timothy	1.9 mg/kg dry weight	Added limestone = 0.63 t/ha, soil-pH = 4.7, N Wales[19]
Timothy	2.0 mg/kg dry weight	Added limestone = 15.0 t/ha, soil-pH = 6.0, N Wales[19]
Timothy	2.4 mg/kg dry weight	Added limestone = 5.0 t/ha, soil-pH = 5.2, N Wales[19]
Vegetation	5.0 mg/kg dry weight	Conc in podzol:f = 10, h = 6, a1a2: < 10, b1 = 20, b2 = 10, b3 = 10,[33]
White clover	2.5 mg/kg dry weight	Added limestone = 0.63 t/ha, soil-pH = 4.7, N Wales[19]
White clover	2.5 mg/kg dry weight	Added limestone = 5.0 t/ha, soil-pH = 5.2, N Wales[19]
White clover	2.6 mg/kg dry weight	Added limestone = 15.0 t/ha, soil-pH = 6.0, N Wales[19]
White clover	2.6 mg/kg dry weight	Added limestone = 2.5 t/ha, soil-pH = 5.0, N Wales[19]
Yellow pickerel, gill	1.8	Wanapitei R, Can, 1974, wat=9+-2 ppb, polluted, n=8[2]
Yellow pickerel, kidney	6.0	Wanapitei R, Can, 1974, wat=9+-2 ppb, polluted, n=8[2]
Yellow pickerel, liver	2.8	Wanapitei R, Can, 1974, wat=9+-2 ppb, polluted, n=8[2]
Yellow pickerel, muscle	1.4	Wanapitei R, Can, 1974, wat=9+-2 ppb, polluted, n=8[2]
Zooplankton	12 mg/kg wet weight	Water = 3+-2 ppb, pH=7.1, French R, Can, non-poll.[2]
Zooplankton	15 mg/kg wet weight	Water = 2+-2 ppb, pH=7.2, Pickerel R, Can, non-poll.[2]

Zooplankton	16 mg/kg wet weight	Water = 2+-2 ppb, pH=7.2, East outlet (French R) to L. Huron, Can[2]
Zooplankton	16.22 mg/kg dry weight	19 st, NY, conc in algae = 9.07 mg/kg dw, mean, Long Isl, less[44]
Zooplankton	22 mg/kg wet weight	Water = 4+-1 ppb, pH=7.2, Main outlet (French R) to L. Huron, Can[2]
Zooplankton	24 mg/kg wet weight	Water = 5+-2 ppb, pH=7.2, Western outlet (French R) to L. Huron, Can[2]
Zooplankton	31 mg/kg wet weight	Water = 9+-2 ppb, pH=7.3, Wanapitei R, Can, silt-ladden and poll.[2]
Zooplankton	34 mg/kg wet weight	Water = 6+-2 ppb, pH=7.3, W Wanapitei B, Can, 1974[2]
Zooplankton	36 mg/kg wet weight	Water = 7+-2 ppb, pH=7.2, Hartley Bay, Canada, 1974[2]

Source: 1.Patrick, F.M. and M.W. Loutit. 1977. Water Res., 11(8), pp 699-703.
2.Hutchinson, T.C., Fedorenko, Fitchko, Kuja, Loon and Lichwa. 1974. In: J.O. Nriagu (ed.): Environmenal Biochemistry. Ann Arbor Sci. Publ. Ann Arbor, Mich, 565-585.
3.Young, R.G. and D.J. Lisk. 1972. Water Pollut. Control, 44(8), pp 1643-1647.
4.Romeril, M.G. and M.H. Davis. 1976. Aquaculture 8, 139-149.
5.Heinrichs, H. and R. Meyer. 1977. J. Environ. Qual. 6, 402-407.
6.Betzer, S.B. and M.E.Q. Pilson. 1974. Biol. Bull. 146(2), pp 165-175.
7.Purves, D. 1972. Environ. Pollut., 3, pp 17-24.
8.Lounamaa, J. 1967. in: Kvauheim, A. (ed.); Geochemical Prospecting in Fennoscandia. pp 287-317. Interscience Publ. New York.
9.Shuster, C.N. Jr., and B.H. Pringle. 1969. In: 1968 Proc of the National Shelfish Assocation. 91-103
10.Long, D.T. and E.E. Angino. 1977. Geochim. Cosmochim. Acta. 41, 1183
11.Morgan, J.J. and T.H. Sibley. 1975. Proceedings, Civil Engineering in the Oceans, Vol III, American Society of Chemical Engineers, p 1332.
12.Dyressen, P. and M. Wedborg. 1974. In: E.D. Goldberg (ed.): The Sea. Vol 5. Marine Chemistry Wiley-Interscience. NY.
13.Zirino, A. and S. Yamamoto. 1972. Limnol. Oceanogr. 17, 661-665.
14.Ahrland, S. 1975. In: E.D. Goldberg (ed.): Dahlem Kanferenzen (Nature of Seawater). Berlin.
15.Stumm, W. and P.A. Brauner. 1975. In: J.P. Riley and G. Skirrow (eds.): Chemical Oceanography. Vol 1. Academic Press. New York.
16.Eisler, R. and G.R. Gardner. 1973. J. Fish. Biol. 5, 111-142.
17.Martin, J.H., M.D. Stephenson and J.H. Martin. 1977. Calif. Fish. Game 63, 95-100.
18.Nriagu, J.O. 1979. Copper in the Environment. John Wiley and Sons. NY.
19.Archer, F.C. 1971. In: Tech. Bull. 21, pp 150-158. Trace Elements in Soils and Crops.
20.Moore, M.E. 1977. Sci. Total Environ., 7(2), pp 109-115.
21.Groth, P. 1971. Arch. Hydrobiol. 68(3), pp 305-375.
22.Brungs, W.A., E.N. Leonard and J.M. McKim. 1973. J. Fish. Res. bd Canada 30, 583-586.
23.Nilsson, I. 1972. Oikos, 22, pp 366.
24.Pringle, B.H., D.E. Hissong, E.L. Katz and S.T. Mulawka. 1972. Environ. Pollut. 3, 69-82.
25.Adema, D.M.M., S.I. De Swaaf-Mooy and P. Bais. 1972. TNO-Nieuws, 27, 482-487.
26.De, J.F., J.F. Solbe and V.A. Copper. 1976. Water Res. 10, 523-527.
27.Raymont, J.E.G. and J. Shields. 1963. Int. J. Air. Water Poll. 7, 435-443.
28.Solbe, J.F., G. Del and V.A. Cooper. 1976. Water Res. 10(6), pp 523-527.
29.Tyler, G. 1972. Ambio 1, 53-59.
30.Funk, W.H., F.W. Rabe, R. Fibly, G. Bailey, Bennett and Shah. 1975. An Integrated Study on the Impact of Metallic Trace Element Pollution on the Coeur D'Alene-Spokane Rivers-Lake Drainage System. Wash. S
31.Robinson, K., S.R. Draper and A.L. Gelman. 1971. Environ. Pollut., 2, pp 49-56.
32.Saward, D., A. Sterling and G. Topping. 1975. Mar. Biol. 29, 351-356.
33.Mitchell, R. and C. Chamberlin. 1978. in: Berg. G. (ed.); Indicators of Enteric Contamination in Natural Waters.
34.Mckim, J. and D.A. Benoit. 1974. J. Fish Res. Bd Canada 31, 449-452.
35.Taylor, D. 1976. Sci. Toatal Environ. 6(3), pp 259-264.
36.Altshuller, A.P. and J.J. Bufalini. 1971. Environ. Sci. Technol., 5(1), pp 39-64.
37.Davies, B.E. and J. Lewin. 1974. Environ. Pollut., 6, pp 49-57.
38.Taylor, D. 1976. Estuarine Coastal Mar. Sci. 2, pp 417.
39.Mathis, B.J. and T.F. Cummins. 1973. Water Pollut. Control 45 pp 1573.
40.Sauchez, I. and G.F. Lee. 1973. Water Res. 7(4), pp 587-593.
41.Dorn, C.R., J.O. Pierce, G.R. Chase and P.E. Phillpis. 1976. Atmos. Environ. 10, 443-446.
42.Inman, R.E., R.B. Ingersoll and E.A. Levy. 1971. Science 172, 1229-1231.
43.Stiff, M.J. 1971. Water Res., 5(8), pp 585-600.
44.Greig, R.A., A. Adams and D.R. Wenzloff. 1977. Bull. Environ. Contam. and Toxic., 18(1), pp 3-8.

5-327 Concentrations
DBT

Species	Value	Condition
Wild fish	0 mg/kg wet weight	After 35 d in aquarium, DBT in water = 0.01 ppm, 32 d, 0. ppm, 3 d[1]
Wild fish	0.17 mg/kg wet weight	After 32 d in aquarium, DBT in water = 0.01 ppm[1]

Wild fish	0.35 mg/kg wet weight	After 8 d in aquarium, DBT in water = 0.01 ppm[1]
Wild fish	0.37 mg/kg wet weight	After 1 d in aquarium, DBT in water = 0.01 ppm[1]

Source: 1.Jensen, S. and O. Petterson. 1971. Environ. Pollut., 2, pp 145-155.

5-328 Concentrations DDD

Species	Value	Condition
Escherichia coli	1.5 mg/kg wet weight	DDT in medium = 0.01 mg/l, 24 h, lab[1]
Escherichia coli	6.0 mg/kg wet weight	DDT in medium = 0.1 mg/l, 24 h, lab[1]

Source: 1.Keil, J.E., S.H. Sandifer, C.D. Graber and L.E. Priester. 1972 Water Res. 6(7), pp 837-842.

5-329 Concentrations DDD and DDT

Species	Value	Condition
Casmerodius albus, brain	2.1 mg/kg wet weight	5 obs., % lipid = 6.8, dieldrin = 4.36, endrin < 0.18, Audubon canyon[1]
Casmerodius albus, breast muscle	3.7 mg/kg wet weight	5 obs., % lipid = 2.4, Audubon canyon ranch, Calif, USA[1]
Casmerodius albus, liver	6.1 mg/kg wet weight	5 obs., % lipid = 3.5, hg = 6.08, Audubon canyon ranch, Calif, USA[1]

Source: 1.Faber, R.B., R.W. Risenbough and H.M. Pratt. 1972. Environ. Pollut., 3, pp 111-122.

5-330 Concentrations DDE

Species	Value	Condition
Bird	94.0 mg/kg wet weight	24 April 1970, % lipid = 100, Audubon canyon ranch, Calif, USA, egg[1]
Casmerodius albus, brain	59.7 mg/kg wet weight	5 obs., % lipid = 6.8, dieldrin = 4.36, endrin < 0.18, Audubon cayon[1]
Casmerodius albus, breast muscle	93.9 mg/kg wet weight	5 obs., % lipid = 2.4, Audubon canyon ranch, Calif, USA[1]
Casmerodius albus, liver	124.3 mg/kg wet weight	5 obs., % lipid = 3.5, hg = 6.08, Audubon canyon ranch, Calif, USA[1]
Egret	0.55 mg/kg wet weight	7 May 1970, % lipid = 1.2, Audubon canyon ranch, Calif, USA, 1 w[1]
Egret	1.1 mg/kg wet weight	11 May 1970, % lipid = 1.2, Audubon canyon ranch, Calif, USA[1]
Egret	1.1 mg/kg wet weight	13 May 1970, % lipid = 1.2, Audubon canyon ranch, Calif, USA[1]
Egret	2.1 mg/kg wet weight	11 May 1970, % lipid = 2.1, Audubon canyon ranch, Calif, USA[1]
Egret	248.0 mg/kg wet weight	No date, Audubon canyon ranch, Calif, USA, egg yolk lipid[1]
Egret	3.2 mg/kg wet weight	15-1969-70, % lipid = 1.7, Audubon canyon ranch, Calif, USA, 1 - 2 w[1]
Egret	4.0 mg/kg wet weight	11 May 1970, % lipid = 2.2, Audubon canyon ranch, Calif, USA, 5 - 6 d[1]
Escherichia coli	0.5 mg/kg wet weight	DDT in medium = 0.01 mg/l, 24 h, lab[2]
Escherichia coli	2.3 mg/kg wet weight	DDT in medium = 0.1 mg/l, 24 h, lab[2]
Heron	0.16 mg/kg wet weight	24 April 1970, % lipid = 1.2, Audubon canyon ranch, Calif, USA, 4.5 w[1]
Heron	0.49 mg/kg wet weight	24 April 1970, % lipid = 1.6, Audubon canyon ranch, Calif, USA, 2 w[1]

Heron	1.18 mg/kg wet weight	24 April 1970, % lipid = 1., Audubon canyon ranch, Calif, USA, 3 w[1]
Heron	1.2 mg/kg wet weight	No date, % lipid = 3.6, Audubon canyon ranch, Calif, USA, embryo[1]
Heron	108.0 mg/kg wet weight	20 April 1970, % lipid = 100, Audubon canyon ranch, Calif, USA, egg[1]
Heron	287.0 mg/kg wet weight	30 March 1970, % lipid = 100, Audubon canyon ranch, Calif, USA, egg[1]
Pagophilus groenlandicus, blubber	1.10 mg/kg wet weight	83, 0 % fat, 15 samples, 1973, Gulf of St Lawrence, SD = +- 0.43[3]
Pagophilus groenlandicus, blubber	2.44 mg/kg wet weight	97.8 % fat, 10 samples, 1971, Gulf of St Lawrence, SD = +- 1.45[3]
Pagophilus groenlandicus, blubber	3.18 mg/kg wet weight	95.5 % fat, 4 animals, St Lawrence, Canada, SD = +- 1.12[4]
Pagophilus groenlandicus, brain	0.01 mg/kg wet weight	6.6 % fat, 14 samples, 1973, Gulf of St Lawrence, SD = +- 0.01[3]
Pagophilus groenlandicus, brain	0.03 mg/kg wet weight	11.5 % fat, 3 animals, St Lawrence, Canada, SD = +- 0.03[4]
Pagophilus groenlandicus, brain	7.5 mg/kg wet weight	7.5 % fat, 10 samples, 1971, Gulf of St Lawrence, SD = +- 1.3[3]
Pagophilus groenlandicus, kidney	14.4 mg/kg wet weight	14.4 % fat, 13 samples, 1973, Gulf of St Lawrence, SD = +- 9.2[3]
Pagophilus groenlandicus, liver	0.02 mg/kg wet weight	1.8 % fat, 4 ainmals, Gulf of St Lawrence, Canada, SD = +- 0.01[4]
Pagophilus groenlandicus, liver	0.11 mg/kg wet weight	4.4 % fat, 16 samples, 1973, Gulf of St Lawrence, SD = +- 0.14[3]
Pagophilus groenlandicus, liver	0.15 mg/kg wet weight	7.0 % fat, 11 samples, 1971, Gulf of St Lawrence, SD = +- 0.19[3]
Pagophilus groenlandicus, muscle	0.01 mg/kg wet weight	1.1 % fat, 4 animals, St Lawrence, Canada, SD = +- 0.00[4]

Source: 1.Faber, R.B., R.W. Risenbough and H.M. Pratt. 1972. Environ. Pollut., 3, pp 111-122.
2.Keil, J.E., S.H. Sandifer, C.D. Graber and L.E. Priester. 1972 Water Res. 6(7), pp 837-842.
3.Frank, R., K. Ronald and H.E. Braun. 1973. Bull. Fish Res. Board Can., 30, pp 1053.
4.Jones, D., K. Ronald, D.H. Lavinge, R. Frank, M. Holdrinet and J.F. Uthe. 1976. Sci. Total Environ.,5(2), pp 181-196.

5-331 Concentrations DDT

Species	Value	Condition
Bald eagle, brain	0.70 mg/kg wet weight	Conc in food = 5 mg/kg wet weight[1]
Bald eagle, brain	80.0 mg/kg wet weight	Conc in food = 800 mg/kg wet weight[1]
Bald eagle, fat	35.70 mg/kg wet weight	Conc in food = 5 mg/kg wet weight[1]
Bald eagle, liver	1.90 mg/kg wet weight	Conc in food = 5 mg/kg wet weight[1]
Bald eagle, muscle	291.0 mg/kg wet weight	Conc in food = 800 mg/kg wet weight[1]
Bird, egg yolk lipid	108.0 mg/kg wet weight	24 April 1970, % lipid = 100, Audubon canyon, Calif, Total DDT[2]
Bivalves	0.23 mg/kg wet weight	Initial water = 1 ug/l, 4 h, small pond, Finland, rg = +- 0.05[3]
Bivalves	1.02 mg/kg wet weight	Initial water = 1 ug/l, 8 d, small pond, Finland, rg = +- 0.09[3]
Bladderwort	0.40 mg/kg wet weight	Initial water = 1 ug/l, 4 h, small pond, Finland, rg = +- 0.02[3]
Bladderwort	1.35 mg/kg wet weight	Initial water = 1 ug/l, 8 d, small pond, Finland, rg = +- 0.04[3]

Bobwhite quail, brain	0.40 mg/kg wet weight	Conc in food = 25 mg/kg wet weight[1]
Bobwhite quail, brain	0.57 mg/kg wet weight	Conc in food = 50 mg/kg wet weight[1]
Bobwhite quail, brain	1.02 mg/kg wet weight	Conc in food = 100 mg/kg wet weight[1]
Bobwhite quail, brain	1.94 mg/kg wet weight	Conc in food = 200 mg/kg wet weight[1]
Bobwhite quail, brain	14.20 mg/kg wet weight	Conc in food = 800 mg/kg wet weight[1]
Bobwhite quail, brain	8.75 mg/kg wet weight	Conc in food = 400 mg/kg wet weight[1]
Bobwhite quail, tissue	1.60 mg/kg wet weight	Conc in food = 50 mg/kg wet weight[1]
Bobwhite quail, tissue	19.80 mg/kg wet weight	Conc in food = 250 mg/kg wet weight[1]
Bobwhite quail, tissue	4.10 mg/kg wet weight	Conc in food = 100 mg/kg wet weight[1]
Brevoortia tyrannus	0.343 mg/kg wet weight	Long Isl, marine, estimated DDT conc in water = E-6 ng/l, Total DDT[4]
Brevoortia tyrannus, body	1.530 mg/kg wet weight	Long Isl, marine, estimated DDT conc in water = E-6 ng/l, Total DDT[4]
Brevoortia tyrannus, head and body cavity	2.330 mg/kg wet weight	Long Isl, marine, estimated DDT conc in water = E-6 ng/l, Total DDT[4]
Caddisfly larvae	0.27 mg/kg wet weight	Initial water = 1 ug/l, 4 h, small pond, Finland, rg = +- 0.05[3]
Caddisfly larvae	1.33 mg/kg wet weight	Initial water = 1 ug/l, 8 d, small pond, Finland, rg = +- 0.08[3]
Callinectes sapidus	0.073 mg/kg wet weight	Long Isl, marine, estimated DDT conc in water = E-6 ng/l, Total DDT,[4]
Chicken, brain	0.60 mg/kg wet weight	Conc in food = 10.00 mg/kg wet weight[1]
Chicken, brain	18.30 mg/kg wet weight	Conc in food = 250.00 mg/kg wet weight[1]
Chicken, brain	3.70 mg/kg wet weight	Conc in food = 50.00 mg/kg wet weight[1]
Chicken, fat	2670.0 mg/kg wet weight	Conc in food = 250.00 mg/kg wet weight[1]
Chicken, fat	518.0 mg/kg wet weight	Conc in food = 50.00 mg/kg wet weight[1]
Chicken, fat	73.0 mg/kg wet weight	Conc in food = 10.00 mg/kg wet weight[1]
Chicken, fat	80.73 mg/kg wet weight	Conc in food = 10.00 mg/kg wet weight[1]
Chicken, liver	0.16 mg/kg wet weight	Conc in food = 0.05 mg/kg wet weight[1]
Chicken, liver	0.41 mg/kg wet weight	Conc in food = 0.15 mg/kg wet weight[1]
Chicken, liver	0.53 mg/kg wet weight	Conc in food = 0.45 mg/kg wet weight[1]
Chicken, liver	198.0 mg/kg wet weight	Conc in food = 150.00 mg/kg wet weight[1]
Chicken, liver	2.12 mg/kg wet weight	Conc in food = 10.00 mg/kg wet weight[1]
Chicken, liver	4.80 mg/kg wet weight	Conc in food = 10.00 mg/kg wet weight[1]
Chicken, liver	46.0 mg/kg wet weight	Conc in food = 50.00 mg/kg wet weight[1]
Chicken, muscle	10.50 mg/kg wet weight	Conc in food = 50.00 mg/kg wet weight[1]
Chicken, muscle	2.10 mg/kg wet weight	Conc in food = 10.00 mg/kg wet weight[1]
Chicken, muscle	39.80 mg/kg wet weight	Conc in food = 250.00 mg/kg wet weight[1]
Circus hudsonius, eggs	38.200 mg/kg wet weight	Long Isl, food = Rallus, Sterna and Rynchops, Total DDT[4]

Cormorant	26.4 mg/kg wet weight	Water = 0.00005 ppm, total DDT, whole organisns basis, estaury[5]
Cowbird, brain	66.0 mg/kg wet weight	Conc in food = 500 mg/kg wet weight[1]
Cowbird, tissue	123.0 mg/kg wet weight	Conc in food = 500 mg/kg wet weight[1]
Crucian carp, gills	0.16 mg/kg wet weight	Initial water = 1 ug/l, 4 h, small pond, Finland, rg = +- 0.03[3]
Crucian carp, gills	0.39 mg/kg wet weight	Initial water = 1 ug/l, 8 d, small pond, Finland, rg = +- 0.04[3]
Crucian carp, liver	0.06 mg/kg wet weight	Initial water = 1 ug/l, 4 h, small pond, Finland, rg = +- 0.02[3]
Crucian carp, liver	0.06 mg/kg wet weight	Initial water = 1 ug/l, 8 d, small pond, Finland, rg = +- 0.01[3]
Crucian carp, muscle	0.08 mg/kg wet weight	Initial water = 1 ug/l, 8 d, small pond, Finland, rg = +- 0.01[3]
Crucian carp, muscle	0.15 mg/kg wet weight	Initial water = 1 ug/l, 4 h, small pond, Finland, rg = +- 0.03[3]
Daphnia magna	1150 mg/kg wet weight	Water = 50.0 ppb, 26 h, 294.1 K, total DDT, first instar, 9 ug animal[6]
Daphnia magna	1260 mg/kg wet weight	Water = 100 ppb, 26 h, 294.1 K, total DDT, first instar, 9 ug animal[6]
Daphnia magna	128 mg/kg wet weight	Water = 8.0 ppb, 26 h, 294.1 K, total DDT, first instar, 9 ug animal[6]
Daphnia magna	2364 mg/kg wet weight	Water = 200 ppb, 26 h, 294.1 K, total DDT, first instar, 9 ug animal[6]
Daphnia magna	266 mg/kg wet weight	Water = 16.0 ppb, 26 h, 294.1 K, total DDT, first instar, 9 ug animal[6]
Daphnia magna	4280 mg/kg wet weight	Water = 1100 ppb, 26 h, 294.1 K, total DDT, first instar, 9 ug animal[6]
Egret	0.59 mg/kg wet weight	7 May 1970, % lipid = 1.2, Audubon canyon, Calif, Total DDT, 1 w[2]
Egret	1.2 mg/kg wet weight	11 May 1970, % lipid = 1.2, Audubon canyon, Calif, Total DDT, 2 w[2]
Egret	1.2 mg/kg wet weight	13 May 1970, % lipid = 1.2, Audubon canyon, Calif, Total DDT,[2]
Egret	2.2 mg/kg wet weight	11 May 1970, % lipid = 2.1, Audubon canyon, Calif, Total DDT, 2.5 w[2]
Egret	3.3 mg/kg wet weight	15 May 1970, % lipid = 1.7, Audubon canyon, Calif, Total DDT, 1.5 w[2]
Egret	4.1 mg/kg wet weight	11 May 1970, % lipid = 2.2, Audubon canyon, Calif, Total DDT, 6 d[2]
Egret, egg yolk lipid	285.0 mg/kg wet weight	No date, % lipid = 100, Audubon canyon, Calif, Total DDT[2]
Ensis directus	0.029 mg/kg wet weight	Long Isl, marine, estimated DDT conc in water = E-6 ng/l, Total DDT[4]
Escherichia coli	0 mg/kg wet weight	DDT in medium = 0.01 mg/l, 24 h, lab[7]
Escherichia coli	11.2 mg/kg wet weight	DDT in medium = 0.1 mg/l, 24 h, lab, total DDT[7]
Escherichia coli	2.1 mg/kg wet weight	DDT in medium = 0.1 mg/l, 24 h, lab[7]
Escherichia coli	2.2 mg/kg wet weight	DDT in medium = 0.01 mg/l, 24 h, lab[7]
Feral pigeon, liver	32.80 mg/kg wet weight	Conc in food = 72 mg/kg wet weight[1]
Feral pigeon, muscle	32.33 mg/kg wet weight	Conc in food = 72 mg/kg wet weight[1]
Fish hawk, egg	13.8 mg/kg wet weight	Water = 0.00005 ppm, total DDT, whole organisns basis, estaury[5]
Fundulus majalis	0.412 mg/kg wet weight	Long Isl, marine, estimated DDT conc in water = E-6 ng/l, Total DDT[4]

Gambusia affinis	20 microgram/kg wet weight	Fish = 380 mg, initial water = 41 ppt, 48 h, 293 k[8]
Gambusia affinis	28 microgram/kg wet weight	Fish = 220 mg, initial water = 41 ppt, 48 h, 293 k[8]
Gambusia affinis	34 microgram/kg wet weight	Fish = 150 mg, initial water = 41 ppt, 48 h, 293 k[8]
Gambusia affinis	36 microgram/kg wet weight	Fish = 70.0 mg, initial water = 41 ppt, 48 h, 293 k[8]
Gambusia affinis	9 microgram/kg wet weight	Fish = 900 mg, initial water = 41 ppt, 48 h, 293 k[8]
Guinea pig, fat	1.59 mg/kg wet weight	Conc in food = 10 mg/kg wet weight[1]
Heron	0.18 mg/kg wet weight	24 April 1970, % lipid = 1.2, Audubon canyon, Calif, Total DDT, 5 w[2]
Heron	0.56 mg/kg wet weight	24 April 1970, % lipid = 1.6, Audubon canyon, Calif, Total DDT, 2 w[2]
Heron	1.33 mg/kg wet weight	24 April 1970, % lipid = 1., Audubon canyon, Calif, Total DDT, 3 w[2]
Heron	3.57 mg/kg wet weight	Water = 0.00005 ppm, total DDT, whole organisns basis, estaury[5]
Heron, egg yolk lipid	122.0 mg/kg wet weight	20 April 1970, % lipid = 100, Audubon canyon, Calif, Total DDT[2]
Heron, egg yolk lipid	320 mg/kg wet weight	30 March 1970, % lipid = 100, Audubon canyon, Calif, Total DDT[2]
Heron, embryo	1.4 mg/kg wet weight	No date, % lipid = 3.6, Audubon canyon, Calif, Total DDT[2]
Herring gull	6.00 mg/kg wet weight	Water = 0.00005 ppm, total DDT, whole organisns basis, estaury[5]
Homarus americanus	0.216 mg/kg wet weight	Long Isl, marine, estimated DDT conc in water = E-6 ng/l, Total DDT[4]
Homo sapiens	100 mg/kg wet weight	Intake from food = 3.5 mg p,p-DDT/(man*d), steady state[1]
Homo sapiens	353 mg/kg wet weight	Intake from food = 35 mg p,p-DDT/(man*d), steady state[1]
Kestrel, brain	14.90 mg/kg wet weight	Conc in food = 2.8 mg/kg, sacrificed[1]
Kestrel, brain	256.80 mg/kg wet weight	Conc in food = 2.8 mg/kg, died[1]
Kestrel, liver	190.80 mg/kg wet weight	Conc in food = 2.8 mg/kg, died[1]
Kestrel, liver	24.0 mg/kg wet weight	Conc in food = 2.8 mg/kg, sacrificed[1]
Kestrel, tissue	51.45 mg/kg wet weight	Conc in food = 2.8 mg/kg, died[1]
Kestrel, tissue	70.0 mg/kg wet weight	Conc in food = 2.8 mg/kg, sacrificed[1]
Melampus bidenatus	0.019 mg/kg wet weight	Long Isl, marine, estimated DDT conc in water = E-6 ng/l, Total DDT[4]
Menidia menidia	0.149 mg/kg wet weight	Long Isl, marine, estimated DDT conc in water = E-6 ng/l, Total DDT[4]
Merganser	22.8 mg/kg wet weight	Water = 0.00005 ppm, total DDT, whole organisms basis, estaury[5]
Moss	0.36 mg/kg wet weight	Initial water = 1 ug/l, 4 h, small pond, Finland, rg = +- 0.01[3]
Moss	0.73 mg/kg wet weight	Initial water = 1 ug/l, 8 d, small pond, Finland, rg = +- 0.02[3]
Nassarius obsoletus	0.076 mg/kg wet weight	Long Isl, marine, estimated DDT conc in water = E-6 ng/l, Total DDT[4]
Needlefish	2.07 mg/kg wet weight	Water = 0.00005 ppm, total DDT, whole organisns basis, estaury[5]
Newt	0.07 mg/kg wet weight	Initial water = 1 ug/l, 4 h, small pond, Finland, rg = +- 0.02[3]
Newt	1.17 mg/kg wet weight	Initial water = 1 ug/l, 8 d, small pond, Finland, rg = +- 0.05[3]

Notonecta sp.	0.08 mg/kg wet weight	Initial water = 1 ug/l, 4 h, small pond, Finland, rg = +- 0.01[3]
Notonecta sp.	0.21 mg/kg wet weight	Initial water = 1 ug/l, 8 d, small pond, Finland, rg = +- 0.02[3]
Opsanus tau	0.123 mg/kg wet weight	Long Isl, marine, estimated DDT conc in water = E-6 ng/l, Total DDT[4]
Pagophilus groenlandicus, blubber	0.19 mg/kg wet weight	83, 0 % fat, 15 samples, 1973, Gulf of St Lawrence, SD = +- 0.14[9]
Pagophilus groenlandicus, blubber	0.50 mg/kg wet weight	97.8 % fat, 10 samples, 1971, Gulf of St Lawrence, SD = +- 0.32[9]
Pagophilus groenlandicus, blubber	0.85 mg/kg wet weight	95.5 % fat, 4 animals, St Lawrence, Canada, SD = +- 0.12[10]
Pagophilus groenlandicus, blubber	1.34 mg/kg wet weight	83, 0 % fat, 15 samples, 1973, Gulf of St Lawrence, SD = +- 0.59,[9]
Pagophilus groenlandicus, blubber	1.41 mg/kg wet weight	97.8 % fat, 10 samples, 1971, Gulf of St Lawrence, SD = +- 0.66,[9]
Pagophilus groenlandicus, blubber	4.40 mg/kg wet weight	95.5 % fat, 4 animals, St Lawrence, Canada, SD = +- 1.27, Total DDT[10]
Pagophilus groenlandicus, brain	0 mg/kg wet weight	11.5 % fat, 3 animals, St Lawrence, Canada[10]
Pagophilus groenlandicus, brain	0 mg/kg wet weight	6.6 % fat, 14 samples, 1973, Gulf of St Lawrence[9]
Pagophilus groenlandicus, brain	0.02 mg/kg wet weight	6.6 % fat, 14 samples, 1973, Gulf of St Lawrence, SD = +- 0.02, Total[9]
Pagophilus groenlandicus, brain	0.06 mg/kg wet weight	7.5 % fat, 10 samples, 1971, Gulf of St Lawrence, SD = +- 0.19[9]
Pagophilus groenlandicus, brain	0.09 mg/kg wet weight	7.5 % fat, 10 samples, 1971, Gulf of St Lawrence, SD = +- 0.17, Total[9]
Pagophilus groenlandicus, kidney	0.03 mg/kg wet weight	14.4 % fat, 13 samples, 1973, Gulf of St Lawrence, SD = +- 0.03[9]
Pagophilus groenlandicus, kidney	0.49 mg/kg wet weight	14.4 % fat, 13 samples, 1973, Gulf of St Lawrence, SD = +- 0.82,[9]
Pagophilus groenlandicus, liver	0 mg/kg wet weight	1.8 % fat, 4 ainmals, Gulf of St Lawrence, Canada[10]
Pagophilus groenlandicus, liver	0.01 mg/kg wet weight	4.4 % fat, 16 samples, 1973, Gulf of St Lawrence, SD = +- 0.01[9]
Pagophilus groenlandicus, liver	0.07 mg/kg wet weight	7.0 % fat, 11 samples, 1971, Gulf of St Lawrence, SD = +- 0.05[9]
Pagophilus groenlandicus, liver	0.16 mg/kg wet weight	4.4 % fat, 16 samples, 1973, Gulf of St Lawrence, SD = +- 0.21, Total[9]
Pagophilus groenlandicus, liver	0.28 mg/kg wet weight	7.0 % fat, 11 samples, 1971, Gulf of St Lawrence, SD = +- 0.17, Total[9]
Pagophilus groenlandicus, muscle	0.01 mg/kg wet weight	1.1 % fat, 4 animals, St Lawrence, Canada, SD = +- 0.00, Total DDT[10]
Palaemonetes vulgaris	0.061 mg/kg wet weight	Long Isl, marine, estimated DDT conc in water = E-6 ng/l, Total DDT[4]
Perch, gills	1.67 mg/kg wet weight	Initial water = 1 ug/l, 4 h, small pond, Finland, rg = +- 0.06[3]
Perch, gills	1.72 mg/kg wet weight	Initial water = 1 ug/l, 8 d, small pond, Finland, rg = +- 0.06[3]
Perch, liver	2.00 mg/kg wet weight	Initial water = 1 ug/l, 8 d, small pond, Finland, rg = +- 0.06[3]
Perch, liver	5.63 mg/kg wet weight	Initial water = 1 ug/l, 4 h, small pond, Finland, rg = +- 0.16[3]
Perch, mes. adip.	0.39 mg/kg wet weight	Initial water = 1 ug/l, 4 h, small pond, Finland, rg = +- 0.09[3]
Perch, mes. adip.	16.80 mg/kg wet weight	Initial water = 1 ug/l, 8 d, small pond, Finland, rg = +- 0.40[3]

Perch, muscle	0.19 mg/kg wet weight	Initial water = 1 ug/l, 4 h, small pond, Finland, rg = +- 0.03[3]
Perch, muscle	0.22 mg/kg wet weight	Initial water = 1 ug/l, 8 d, small pond, Finland, rg = +- 0.02[3]
Pheasant, muscle	23.90 mg/kg wet weight	Conc in food = 50 mg/kg, died[1]
Pheasant, muscle	3.60 mg/kg wet weight	Conc in food = 50 mg/kg, sacrificed[1]
Pheasant, tissue	29.10 mg/kg wet weight	Conc in food = 10 mg/kg wet weight[1]
Pickerel	1.33 mg/kg wet weight	Water = 0.00005 ppm, total DDT, whole organisns basis, estaury[5]
Plankton	0.04 mg/kg wet weight	Water = 0.00005 ppm, total DDT, whole organisns basis, estaury[5]
Pomatomus saltatrix	0.199 mg/kg wet weight	Long Isl, marine, estimated DDT conc in water = E-6 ng/l, Total DDT[4]
Pseudopleuronectes americanus	0.059 mg/kg wet weight	Long Isl, marine, estimated DDT conc in water = E-6 ng/l, Total DDT[4]
Rabbit, fat	18.78 mg/kg wet weight	Conc in food = 10 mg/kg wet weight[1]
Rabbit, liver	0.77 mg/kg wet weight	Conc in food = 10 mg/kg wet weight[1]
Rallus longirostris	1.430 mg/kg wet weight	Long Isl, food = Melampus sp. and Uca sp., Total DDT[4]
Rallus longirostris, egg	1.400 mg/kg wet weight	Long Isl, food = Melampus sp. and Uca sp., Total DDT[4]
Rat, fat	56.43 mg/kg wet weight	Conc in food = 10 mg/kg wet weight[1]
Rat, liver	1.96 mg/kg wet weight	Conc in food = 10 mg/kg wet weight[1]
Roccus americanus	1.990 mg/kg wet weight	Long Isl, marine, estimated DDT conc in water = E-6 ng/l, Total DDT[4]
Rynchops nigra nigra	13.900 mg/kg wet weight	Long Isl, food = Menidia sp. and Fundulus sp., Total DDT[4]
Rynchops nigra nigra, eggs	5.400 mg/kg wet weight	Long Isl, food = Menidia sp. and Fundulus sp., Total DDT[4]
Scomber scombrus	0.177 mg/kg wet weight	Long Isl, marine, estimated DDT conc in water = E-6 ng/l, Total DDT[4]
Sediment	0.0022 mg/kg wet weight	Conc in water = 0.08 ug/l, USA[1]
Sediment	0.01 mg/kg wet weight	Conc in water: < 1.0 ug/l, USA[1]
Sediment	0.047 mg/kg wet weight	Conc in water = 3.28 ug/l, Canada[1]
Sediment	0.082 mg/kg wet weight	Conc in water: < 1.0 ug/l, USA[1]
Sediment	0.14 mg/kg wet weight	Conc in water = 0.05 ug/l, USA[1]
Sediment	0.20 mg/kg wet weight	Conc in water = 0.115 ug/l, USA[1]
Sediment	0.74 mg/kg wet weight	Initial water = 1 ug/l, 4 h, small pond, Finland, rg = +- 0.03[3]
Sediment	0.94 mg/kg wet weight	Initial water = 1 ug/l, 8 d, small pond, Finland, rg = +- 0.03[3]
Sediment	1.10 mg/kg wet weight	Conc in water = trace, USA[1]
Sediment	1.80 mg/kg wet weight	Conc in water = 0.30 ug/l, USA[1]
Seston	2.9 mg/kg wet weight	Initial water = 1 ug/l, 8 d, small pond, Finland, rg = +- 1.3, 8 U[3]
Seston	3.7 mg/kg wet weight	Initial water = 1 ug/l, 8 d, small pond, Finland, rg = +- 17.7, 3 U[3]
Seston	36.1 mg/kg wet weight	Initial water = 1 ug/l, 4 h, small pond, Finland, rg = +- 62.7, 3 U[3]
Seston	9.0 mg/kg wet weight	Initial water = 1 ug/l, 4 h, small pond, Finland, rg = +- 4.0, 8 U[3]

Sheep, fat	8.13 mg/kg wet weight	Conc in food = 10 mg/kg wet weight[1]
Sheep, liver	0.44 mg/kg wet weight	Conc in food = 10 mg/kg wet weight[1]
Sheephead minnow	0.94 mg/kg wet weight	Water = 0.00005 ppm, total DDT, whole organisns basis, estaury[5]
Silverside minnow	0.23 mg/kg wet weight	Water = 0.00005 ppm, total DDT, whole organisns basis, estaury[5]
Soil	0.1 mg/kg wet weight	Joudabpur(a), Dacca, 10 g, o,p-DDT[11]
Soil	0.3 mg/kg wet weight	Denra(a), Dacca, 10 g, o,p-DDT[11]
Soil	0.4 mg/kg wet weight	Demra(c), Dacca, 10 g, o,p-DDT[11]
Soil	0.5 mg/kg wet weight	Demra(b), Dacca, 10 g, o,p-DDT[11]
Soil	0.5 mg/kg wet weight	Joydabpur(b), Dacca, 10 g, o,p-DDT[11]
Soil	1.0 mg/kg wet weight	Mirpur, Dacca, 10 g, o,p-DDT[11]
Soil	1.3 mg/kg wet weight	Demra(a), Dacca, 10 g, p,p-DDT[11]
Soil	1.6 mg/kg wet weight	Demra(b), Dacca, 10 g, p,p-DDT[11]
Soil	1.8 mg/kg wet weight	Joydabpur(a), Dacca, 10 g, p,p-DDT[11]
Soil	2.0 mg/kg wet weight	Demra(c), Dacca, 10 g, p,p-DDT[11]
Soil	2.0 mg/kg wet weight	Joydabpur(b), Dacca, 10 g, p,p-DDT[11]
Soil	3.0 mg/kg wet weight	Mirpur, Dacca, 10 g, p,p-DDT[11]
Sterna hirundo	3.760 mg/kg wet weight	Long Isl, food = Menidia sp. and Fundulus sp., Total DDT[4]
Sterna hirundo, egg	5.700 mg/kg wet weight	Long Isl, food = Menidia sp. and Fundulus sp., Total DDT[4]
Tern	3.91 mg/kg wet weight	Water = 0.00005 ppm, total DDT, whole organisns basis, estaury[5]
Uca sp.	0.054 mg/kg wet weight	Long Isl, marine, estimated DDT conc in water = E-6 ng/l, Total DDT[4]
Uca sp.	0.082 mg/kg wet weight	Long Isl, marine, estimated DDT conc in water = E-6 ng/l, Total DDT,[4]

Source: 1.Edwards, C.A. 1975. Persistent Pesticides in the Environment, 2nd edition, CRC-Press.
2.Faber, R.B., R.W. Risenbough and H.M. Pratt. 1972. Environ. Pollut., 3, pp 111-122.
3.Salonen, L. and H.A. Vaajakorpi. 1976. Environ. Qual. Saf. 5, pp 130-140.
4.Foehrenbach, J. 1972. Water Pollut. Control, 44(4), pp 619-624.
5.Woodwell, G.M., C.F. Wurster and P.A. Isaacson. 1967. Science 156, pp 821-824.
6.Crosby, D.G. and R.K. Tucker. 1971. Environ. Sci. Technol., 5(8), pp 714-716.
7.Keil, J.E., S.H. Sandifer, C.D. Graber and L.E. Priester. 1972 Water Res. 6(7), pp 837-842.
8.Murphy, P.G. 1971. Bull. Environ. Contam. Toxicol. 6(1), pp 20-23.
9.Frank, R., K. Ronald and H.E. Braun. 1973. Bull. Fish Res. Board Can., 30, pp 1053.
10.Jones, D., K. Ronald, D.H. Lavinge, R. Frank, M. Holdrinet and J.F. Uthe. 1976. Sci. Total Environ.,5(2), pp 181-196.
11.Nasim, A.L., M.M.H. Paig and K.A. Lord. 1971. Environ. Pollut., 2, pp 7-12.

5-332 Concentrations Di-2-ethylhexyl phthalate

Species	Value	Condition
Pimephales promelas	1.2 mg/kg wet weight	1.24 g, 56 d, uptake from water, conc = 1.9 ug/l[1]
Pimephales promelas	2.2 mg/kg wet weight	1.24 g, 56 d, uptake from water, conc = 2.5 ug/l[1]
Pimephales promelas	3.6 mg/kg wet weight	1.24 g, 56 d, uptake from water, conc = 8.1 ug/l[1]
Pimephales promelas	4.1 mg/kg wet weight	1.24 g, 56 d, uptake from water, conc = 4.6 ug/l[1]
Pimephales promelas	5.0 mg/kg wet weight	1.24 g, 56 d, uptake from water, conc = 14 ug/l[1]
Pimephales promelas	8.6 mg/kg wet weight	1.24 g, 56 d, uptake from water, conc = 30 ug/l[1]
Pimephales promelas	9.6 mg/kg wet weight	1.24 g, 56 d, uptake from water, conc = 62 ug/l[1]
Sediment	200 ng/g wet weight	Conc in water = 300 ng/g, Lake Superior, Ontario[2]

Walleye	800 ng/g wet weight	Conc in water = 300 ng/g, Lake Superior, Ontario[2]

Source: 1.Mayer, 1976. J. Fish. Res. Bd Can. 33, pp 2610-2613.
2.Reakall, D.B. 1974. J. Environ. Qual. 52, pp 1-39.

5-333 Concentrations
Dibenzothiophene

Species	Value	Condition
Crassostrea virginica	0.1 mg/kg wet weight	Upt in 8 h from 302 ppm No. 2 fuel, 0.07 % DBT, 120 h excr, init=0.3 ppm[1]
Crassostrea virginica	0.1 mg/kg wet weight	Upt in 8 h from 302 ppm No. 2 fuel, 0.07 % DBT, 6 h excr, init=0.3 ppm[1]
Crassostrea virginica	0.2 mg/kg wet weight	Upt in 8 h from 302 ppm No. 2 fuel, 0.07 % DBT, 24 h excr, init=0.3 ppm[1]
Crassostrea virginica	0.3 mg/kg wet weight	302 ppm dispersed No. 2 fuel oil with 0.07 % DBT, flow through, 8 h[1]
Crassostrea virginica	0.3 mg/kg wet weight	Upt in 8 h from 302 ppm No. 2 fuel, 0.07 % DBT, 3 h excr, init=0.3 ppm[1]

Source: 1.Neff, J.M., B.A. Cox, D. Dixit and J.W. Anderson. 1976. Mar. Biol. 38, 279-289.

5-334 Concentrations
Dibenzothiophene, C4 - Biphenyl and C5 - Naphthalene

Species	Value	Condition
Fundulus heteroclitus	0.01 mg/kg wet weight	570 l S. Louisiana crude oil in 810 m2 shoreline area, 6 h after spill[1]
Fundulus heteroclitus	0.02 mg/kg wet weight	570 l S. Louisiana crude oil in 810 m2 shoreline area, 31 h after spill[1]

Source: 1.Bieri, R.H., V.C. Stamoudis and M.K. Cueman. 1977. Proc. of 1977 Oil spill Conference (Prevention, Behavior, Control, Cleanup). Wash. DC. American Petrolium Institute. 511-516.

5-335 Concentrations
Dieldrin

Species	Value	Condition
Callinectes sapidus	25.3 microgram/kg wet weight	Food = 3 g with 103 ug/kg/d, 10 d, pH = 8.2, 294 K, cont = 7.5 ppm,[1]
Callinectes sapidus	32.8 microgram/kg wet weight	Food = 3 g with 193 ug/kg/d, 10 d, pH = 8.2, 294 K, cont = 3.4 ppm,[1]
Callinectes sapidus	47 % of intake	Uptake from food = 103 ug/kg, water = 0 ppb[1]
Callinectes sapidus	68 % of intake	Uptake from food = 181 ug/kg, water = 0 ppb[1]
Callinectes sapidus	78 % of intake	Uptake from food = 193 ug/kg, water = 0 ppb[1]
Callinectes sapidus	96.6 microgram/kg wet weight	Food = 3 g with 181 ug/kg/d, 10 d, pH = 8.2, 294 K, cont = 4.7 ppm,[1]
Chicken, liver	0.10 mg/kg wet weight	Conc in food = 0.05 mg/kg wet weight[2]
Chicken, liver	0.12 mg/kg wet weight	Conc in food = 0.05 mg/kg wet weight[2]
Chicken, liver	0.20 mg/kg wet weight	Conc in food = 0.15 mg/kg wet weight[2]
Chicken, liver	0.27 mg/kg wet weight	Conc in food = 0.15 mg/kg wet weight[2]
Chicken, liver	0.35 mg/kg wet weight	Conc in food = 0.45 mg/kg wet weight[2]
Chicken, liver	0.50 mg/kg wet weight	Conc in food = 0.45 mg/kg wet weight[2]
Chicken, tissue	0.60 mg/kg wet weight	Conc in food = 10.00 mg/kg wet weight[2]
Chicken, tissue	1.0 mg/kg wet weight	Conc in food = 20.0 mg/kg wet weight[2]

Chicken, tissue	10.20 mg/kg wet weight	Conc in food = 0.25 mg/kg wet weight[2]
Chicken, tissue	35.70 mg/kg wet weight	Conc in food = 0.75 mg/kg wet weight[2]
Coturnix quail, brain	0.72 mg/kg wet weight	Conc in food = 2 mg/kg wet weight[2]
Coturnix quail, brain	10.13 mg/kg wet weight	Conc in food = 10 mg/kg wet weight[2]
Coturnix quail, brain	12.49 mg/kg wet weight	Conc in food = 250 mg/kg wet weight[2]
Coturnix quail, brain	13.19 mg/kg wet weight	Conc in food = 50 mg/kg wet weight[2]
Coturnix quail, liver	22.02 mg/kg wet weight	Conc in food = 10 mg/kg wet weight[2]
Coturnix quail, liver	23.08 mg/kg wet weight	Conc in food = 250 mg/kg wet weight[2]
Coturnix quail, liver	59.83 mg/kg wet weight	Conc in food = 50 mg/kg wet weight[2]
Dunaliella peircei	0.77 mg/g dry weight	96 h, water = 10 ug/l, n = 10[3]
Dunaliella peircei	0.84 mg/g dry weight	72 h, water = 10 ug/l, n = 10[3]
Dunaliella peircei	0.91 mg/g dry weight	48 h, water = 10 ug/l, n = 10[3]
Dunaliella peircei	1.27 mg/g dry weight	24 h, water = 10 ug/l, n = 10[3]
Ictalurus punctatus, dorsal muscle	1199 microgram/kg wet weight	Uptake from food = 2 ppm and water = 75 ng/l, 21 d, 350-400 mm[4]
Ictalurus punctatus, dorsal muscle	1243 microgram/kg wet weight	Uptake from food = 2 ppm, 28 d, 350-400 mm[4]
Ictalurus punctatus, dorsal muscle	175 microgram/kg wet weight	Uptake from water = 75 ng/l, 28 d, 150-225 mm[4]
Ictalurus punctatus, dorsal muscle	179 microgram/kg wet weight	Uptake from food = 2 ppm, 7 d, 150-225 mm[4]
Ictalurus punctatus, dorsal muscle	179 microgram/kg wet weight	Uptake from water = 75 ng/l, 21 d, 350-400 mm[4]
Ictalurus punctatus, dorsal muscle	2418 microgram/kg wet weight	Uptake from food = 2 ppm and water = 75 ng/l, 28 d, 350-400 mm[4]
Ictalurus punctatus, dorsal muscle	246 microgram/kg wet weight	Uptake from food = 2 ppm, 7 d, 350-400 mm[4]
Ictalurus punctatus, dorsal muscle	274 microgram/kg wet weight	Uptake from water = 75 ng/l, 28 d, 350-400 mm[4]
Ictalurus punctatus, dorsal muscle	280 microgram/kg wet weight	Uptake from food = 2 ppm and water = 75 ng/l, 7 d, 350-400 mm[4]
Ictalurus punctatus, dorsal muscle	303 microgram/kg wet weight	Uptake from food = 2 ppm and water = 75 ng/l, 7 d, 150-225 mm[4]
Ictalurus punctatus, dorsal muscle	45 microgram/kg wet weight	Uptake from water = 75 ng/l, 7 d, 150-225 mm[4]
Ictalurus punctatus, dorsal muscle	506 microgram/kg wet weight	Uptake from food = 2 ppm, 21 d, 150-225 mm[4]
Ictalurus punctatus, dorsal muscle	544 microgram/kg wet weight	Uptake from food = 2 ppm, 28 d, 150-225 mm[4]
Ictalurus punctatus, dorsal muscle	606 microgram/kg wet weight	Uptake from food = 2 ppm and water = 75 ng/l, 21 d, 150-225 mm[4]
Ictalurus punctatus, dorsal muscle	74 microgram/kg wet weight	Uptake from water = 75 ng/l, 7 d, 350-400 mm[4]

Ictalurus punctatus, dorsal muscle	796 microgram/kg wet weight	Uptake from food = 2 ppm, 21 d, 350-400 mm[4]
Ictalurus punctatus, dorsal muscle	81 microgram/kg wet weight	Uptake from water = 75 ng/l, 21 d, 150-225 mm[4]
Ictalurus punctatus, dorsal muscle	898 microgram/kg wet weight	Uptake from food = 2 ppm and water = 75 ng/l, 28 d, 150-225 mm[4]
Lamb, tissue	0.40 mg/kg wet weight	Conc in food = 0.25 mg/kg wet weight[2]
Lamb, tissue	0.60 mg/kg wet weight	Conc in food = 0.75 mg/kg wet weight[2]
Lamb, tissue	1.70 mg/kg wet weight	Conc in food = 2.25 mg/kg wet weight[2]
Pagophilus groenlandicus, blubber	0 mg/kg wet weight	97.8 % fat, 10 samples, 1971, Gulf of St Lawrence[5]
Pagophilus groenlandicus, blubber	0.050 mg/kg wet weight	83.0 % fat, 15 samples, 1973, Gulf of St Lawrence, SD = +- 0.046[5]
Pagophilus groenlandicus, blubber	0.150 mg/kg wet weight	95.5 % fat, 4 animals, St Lawrence, Canada, SD = +- 0.025[6]
Pagophilus groenlandicus, brain	0 mg/kg wet weight	6.6 % fat, 14 samples, 1973, Gulf of St Lawrence[5]
Pagophilus groenlandicus, brain	0 mg/kg wet weight	7.5 % fat, 10 samples, 1971, Gulf of St Lawrence[5]
Pagophilus groenlandicus, brain	0.001 mg/kg wet weight	11.5 % fat, 3 animals, St Lawrence, Canada, SD = +- 0.001[6]
Pagophilus groenlandicus, kidney	0.012 mg/kg wet weight	14.4 % fat, 13 samples, 1973, Gulf of St Lawrence, SD = +- 0.019[5]
Pagophilus groenlandicus, liver	0 mg/kg wet weight	7.0 % fat, 11 samples, 1971, Gulf of St Lawrence[5]
Pagophilus groenlandicus, liver	0.001 mg/kg wet weight	1.8 % fat, 4 ainmals, SD = +- 0.001[6]
Pagophilus groenlandicus, liver	0.002 mg/kg wet weight	4.4 % fat, 16 samples, 1973, Gulf of St Lawrence, SD = +- 0.002[5]
Pagophilus groenlandicus, muscle	0.001 mg/kg wet weight	1.1 % fat, 4 animals, St Lawrence, Canada, SD = +- 0.001[6]
Pheasant, muscle	2.70 mg/kg wet weight	Conc in food = 50 mg/kg wet weight[2]
Pig, tissue	0.40 mg/kg wet weight	Conc in food = 0.25 mg/kg wet weight[2]
Pig, tissue	2.80 mg/kg wet weight	Conc in food = 0.75 mg/kg wet weight[2]
Pig, tissue	4.30 mg/kg wet weight	Conc in food = 2.25 mg/kg wet weight[2]
Rat, blood	0.254 mg/kg wet weight	Uptake from food with 50 ppm dieldrin, steady state, 180 d[2]
Rat, brain	0.24 mg/kg wet weight	Conc in food = 10 mg/kg wet weight[2]
Rat, fat	15.85 mg/kg wet weight	Conc in food = 10 mg/kg wet weight[2]
Rat, fat	183 mg/kg wet weight	Uptake from food with 50 ppm dieldrin, steady state, 180 d[2]
Rat, liver	0.85 mg/kg wet weight	Conc in food = 10 mg/kg wet weight[2]
Rat, liver	7.85 mg/kg wet weight	Uptake from food with 50 ppm dieldrin, steady state, 180 d[2]
Rat, tissue	15.10 mg/kg wet weight	Conc in food = 1.0 mg/kg wet weight[2]
Rat, tissue	67.50 mg/kg wet weight	Conc in food = 10 mg/kg wet weight[2]
Rhizophora mangle, hypocotyl	0.01 mg/kg wet weight	Application = 0.06 kg/ha, pH = 6.3-6.7, s = 2.5 %, 6 w[7]
Rhizophora mangle, hypocotyl	0.02 mg/kg wet weight	Application = 0.11 kg/ha, pH = 6.3-6.7, s = 2.5 %, 6 w[7]

Rhizophora mangle, hypocotyl	0.02 mg/kg wet weight	Application = 0.28 kg/ha, pH = 6.3-6.7, s = 2.5 %, 6 w[7]
Rhizophora mangle, hypocotyl	0.04 mg/kg wet weight	Application = 0.56 kg/ha, pH = 6.3-6.7, s = 2.5 %, 6 w[7]
Rhizophora mangle, hypocotyl	0.08 mg/kg wet weight	Application = 1.12 kg/ha, pH = 6.3-6.7, s = 2.5 %, 6 w[7]
Rhizophora mangle, hypocotyl	0.08 mg/kg wet weight	Application = 2.80 kg/ha, pH = 6.3-6.7, s = 2.5 %, 6 w[7]
Rhizophora mangle, hypocotyl	0.08 mg/kg wet weight	Application = 5.60 kg/ha, pH = 6.3-6.7, s = 2.5 %, 6 w[7]
Rhizophora mangle, hypocotyl	0.16 mg/kg wet weight	Application = 11.20 kg/ha, pH = 6.3-6.7, s = 2.5 %, 6 w[7]
Rhizophora mangle, leaves	0.07 mg/kg wet weight	Application = 0.06 kg/ha, pH = 6.3-6.7, s = 2.5 %, 6 w[7]
Rhizophora mangle, leaves	0.07 mg/kg wet weight	Application = 0.56 kg/ha, pH = 6.3-6.7, s = 2.5 %, 6 w[7]
Rhizophora mangle, leaves	0.07 mg/kg wet weight	Application = 2.80 kg/ha, pH = 6.3-6.7, s = 2.5 %, 6 w[7]
Rhizophora mangle, leaves	0.09 mg/kg wet weight	Application = 5.60 kg/ha, pH = 6.3-6.7, s = 2.5 %, 6 w[7]
Rhizophora mangle, leaves	0.10 mg/kg wet weight	Application = 11.20 kg/ha, pH = 6.3-6.7, s = 2.5 %, 6 w[7]
Rhizophora mangle, leaves	0.11 mg/kg wet weight	Application = 0.11 kg/ha, pH = 6.3-6.7, s = 2.5 %, 6 w[7]
Rhizophora mangle, leaves	0.11 mg/kg wet weight	Application = 0.28 kg/ha, pH = 6.3-6.7, s = 2.5 %, 6 w[7]
Rhizophora mangle, leaves	0.11 mg/kg wet weight	Application = 1.12 kg/ha, pH = 6.3-6.7, s = 2.5 %, 6 w[7]
Rhizophora mangle, root	0.03 mg/kg wet weight	Application = 1.12 kg/ha, pH = 6.3-6.7, s = 2.5 %, 6 w[7]
Rhizophora mangle, root	0.04 mg/kg wet weight	Application = 2.80 kg/ha, pH = 6.3-6.7, s = 2.5 %, 6 w[7]
Rhizophora mangle, root	0.04 mg/kg wet weight	Application = 5.60 kg/ha, pH = 6.3-6.7, s = 2.5 %, 6 w[7]
Rhizophora mangle, root	0.06 mg/kg wet weight	Application = 11.20 kg/ha, pH = 6.3-6.7, s = 2.5 %, 6 w[7]
Sediment	0.0002 mg/kg wet weight	Conc in water = trace, USA[2]
Sediment	0.089 mg/kg wet weight	Conc in water = 0.018 ug/l, USA[2]
Sediment	0.38 mg/kg wet weight	Conc in water = 29.0 ug/l, USA[2]
Sediment	0.89 mg/kg wet weight	Conc in water = 0.018 ug/l, USA[2]
Steer, tissue	0.80 mg/kg wet weight	Conc in food = 0.25 mg/kg wet weight[2]
Steer, tissue	3.50 mg/kg wet weight	Conc in food = 0.75 mg/kg wet weight[2]
Steer, tissue	8.70 mg/kg wet weight	Conc in food = 2.25 mg/kg wet weight[2]
White - tailed deer, brain	0.20 mg/kg wet weight	Conc in food = 5 mg/kg wet weight[2]
White - tailed deer, brain	0.35 mg/kg wet weight	Conc in food = 5 mg/kg wet weight[2]
White - tailed deer, brain	0.40 mg/kg wet weight	Conc in food = 15 mg/kg wet weight[2]
White - tailed deer, brain	2.50 mg/kg wet weight	Conc in food = 45 mg/kg wet weight[2]
White - tailed deer, brain	4.41 mg/kg wet weight	Conc in food = 25 mg/kg wet weight[2]
White - tailed deer, liver	0.98 mg/kg wet weight	Conc in food = 5 mg/kg wet weight[2]
White - tailed deer, liver	9.13 mg/kg wet weight	Conc in food = 25 mg/kg wet weight[2]
White - tailed deer, muscle	0.10 mg/kg wet weight	Conc in food = 5 mg/kg wet weight[2]
White - tailed deer, muscle	1.19 mg/kg wet weight	Conc in food = 25 mg/kg wet weight[2]

Source: 1.Petrocelli, S.R. and J.W. Anderson. 1975. Bull. Environ. Contam. Toxicol. 31, pp 108-116.
2.Edwards, C.A. 1975. Persistent Pesticides in the Environment, 2nd edition, CRC-Press.
3.Petrocelli, S.K., J.W. Anderson and A.R. Hanks. 1975. Mar. Biol., 31, pp 215-218.
4.Shannon, L.R. 1977. Bull. Environ. Contam. Toxicol. 17(6), pp 637-644.
5.Frank, R., K. Ronald and H.E. Braun. 1973. Bull. Fish Res. Board Can., 30, pp 1053.
6.Jones, D., K. Ronald, D.H. Lavinge, R. Frank, M. Holdrinet and J.F. Uthe. 1976. Sci. Total Environ.,5(2), pp 181-196.
7.Walch, G.E., T.A. Hollister and J. Forester. 1974. Bull. Environ. Contam. Toxicol. 12(2), pp 129-135

5-336 Concentrations
Dimethylbiphenyls

Species	Value	Condition
Marine sediment near oil separator platform	800 ug/kg wet weight	Trinity B., TEX, USA, sample 15 m from brine effluent = 30 ug/l[1]

Source: 1.Armstrong, H.W., K. Fucik, J.W. Anderson and J.M. Neff. 1977. Effects of oilfield brine effluent on benthic organisms in Trinity Bay. American Petrolium Institute Publ. No. 4291. Wash. DC.

5-337 Concentrations
Dimethylfluorene

Species	Value	Condition
Marine sediment near oil separator platform	900 ug/kg wet weight	Trinity B., TEX, USA, sample 15 m from brine effluent = 42 ug/l[1]

Source: 1.Armstrong, H.W., K. Fucik, J.W. Anderson and J.M. Neff. 1977. Effects of oilfield brine effluent on benthic organisms in Trinity Bay. American Petrolium Institute Publ. No. 4291. Wash. DC.

5-338 Concentrations
Dimethylnaphthalene

Species	Value	Condition
Crassostrea virginica	0.01 mg/kg wet weight	Exposed 0[1]
Crassostrea virginica	0.01 mg/kg wet weight	Exposed 0 d to sediment with 887.4 ug total oil/g sed, prudhoe B crude oil[1]
Crassostrea virginica	0.89 mg/kg wet weight	Exposed 40 d to sediment with 887.4 ug total oil/g sed, prudhoe B crude oil[1]
Crassostrea virginica	2.19 mg/kg wet weight	Exposed 60 d to sediment with 887.4 ug total oil/g sed, prudhoe B crude oil[1]
Crassostrea virginica	72 mg/kg dry weight	Exposure = 8 d, water = 2 ug/l, crude oil, lab[2]
Crassostrea virginica	84 mg/kg dry weight	Exposure = 2 d, water = 10 ug/l, crude oil, lab[2]
Marine sediment near oil separator platform	8000 ug/kg wet weight	Trinity B., TEX, USA, sample 15 m from brine effluent = 190 ug/l[3]
Mytilus edulis	0.01 mg/kg wet weight	Exposed 0[1]
Mytilus edulis	0.01 mg/kg wet weight	Exposed 0 d to sediment with 887.4 ug total oil/g sed, prudhoe B crude oil[1]
Mytilus edulis	0.18 mg/kg wet weight	Exposed 60 d to sediment with 887.4 ug total oil/g sed, prudhoe B crude oil[1]
Mytilus edulis	0.38 mg/kg wet weight	Exposed 40 d to sediment with 887.4 ug total oil/g sed, prudhoe B crude oil[1]
Nephthys incisa	0.01 mg/kg wet weight	Exposed 0[1]
Nephthys incisa	0.01 mg/kg wet weight	Exposed 0 d to sediment with 887.4 ug total oil/g sed, prudhoe B crude oil[1]
Nephthys incisa	0.01 mg/kg wet weight	Exposed 40 d to sediment with 887.4 ug total oil/g sed, prudhoe B crude oil[1]
Nephthys incisa	0.16 mg/kg wet weight	Exposed 60 d to sediment with 887.4 ug total oil/g sed, prudhoe B crude oil[1]
Seawater near oil separator platform	0.70 ug/liter	Trinity B., TEX, USA, sample 15 m from brine effluent = 190 ug/l[3]

Source: 1.Roesijadi, G., J.W. Anderson and J.W. Blaylock. 1978. J. Fish. Res. bd Canada 35, 608-614.
2.Lee, R.F., W.S. Gardner, J.W. Anderson, Balylock and B-Clarke. 1978. Environ. Sci. Technol. 12, 832-38.
3.Armstrong, H.W., K. Fucik, J.W. Anderson and J.M. Neff. 1977. Effects of oilfield brine effluent on benthic organisms in Trinity Bay. American Petrolium Institute Publ. No. 4291. Wash. DC.

5-339 Concentrations Dimethylnaphthalenes

Species	Value	Condition
Crassostrea virginica	0.5 mg/kg wet weight	Upt in 8 h from 302 ppm No. 2 fuel, 3.11 % DMN, 672 h excr, init=15.0 ppm[1]
Crassostrea virginica	13.2 mg/kg wet weight	Upt in 8 h from 302 ppm No. 2 fuel, 3.11 % DMN, 6 h excr, init=15.0 ppm[1]
Crassostrea virginica	13.4 mg/kg wet weight	Upt in 8 h from 302 ppm No. 2 fuel, 3.11 % DMN, 120 h excr, init=15.0 ppm[1]
Crassostrea virginica	14.8 mg/kg wet weight	Upt 8 h from 302 ppm No. 2 fuel, 3.11 % DMN, 24 h excr, init=15.0 ppm[1]
Crassostrea virginica	21.8 mg/kg wet weight	302 ppm dispersed No. 2 fuel oil with 3.11 % DMN, flow through, 8 h[1]
Crassostrea virginica	22.7 mg/kg wet weight	Upt in 8 h from 302 ppm No. 2 fuel, 3.11 % DMN, 3 h excr, init=15.0 ppm[1]
Penaeus aztecus, tissue	14700 ug/kg wet weight	Water = 15.2 ppb after 24 h, tissue after 72 h, fuel oil No. 2, exp[2]
Sesarma cinereum, tissue	24700 ug/kg wet weight	Water = 15.2 ppb after 24 h, tissue after 72 h, fuel oil No. 2, exp[2]
Uca minax, tissue	16800 ug/kg wet weight	Water = 15.2 ppb after 24 h, tissue after 72 h, fuel oil No. 2, exp[2]

Source: 1.Neff, J.M., B.A. Cox, D. Dixit and J.W. Anderson. 1976. Mar. Biol. 38, 279-289.
2.Cox, B.A., J.W. Anderson and J.C. Parker. 1975. Proc. 1975 Conference on Prevention and Control of Oil Pollution. American Petrolium Institute. Wash DC. 607-612.

5-340 Concentrations Dimethylphenanthrenes

Species	Value	Condition
Crassostrea virginica	0.1 mg/kg wet weight	Upt in 8 h from 302 ppm No. 2 fuel, 0.53 % DMP, 6 h excr, init=0.3 ppm[1]
Crassostrea virginica	0.2 mg/kg wet weight	Upt in 8 h from 302 ppm No. 2 fuel, 0.53 % DMP, 120 h excr, init=0.3 ppm[1]
Crassostrea virginica	0.2 mg/kg wet weight	Upt in 8 h from 302 ppm No. 2 fuel, 0.53 % DMP, 3 h excr, init=0.3 ppm[1]
Crassostrea virginica	0.3 mg/kg wet weight	302 ppm dispersed No. 2 fuel oil with 0.53 % DMP, flow through, 8 h[1]
Crassostrea virginica	0.3 mg/kg wet weight	Upt in 8 h from 302 ppm No. 2 fuel, 0.53 % DMP, 24 h excr, init=0.3 ppm[1]

Source: 1.Neff, J.M., B.A. Cox, D. Dixit and J.W. Anderson. 1976. Mar. Biol. 38, 279-289.

5-341 Concentrations Dimethypheanthrene

Species	Value	Condition
Marine sediment near oil separator platform	1000 ug/kg wet weight	Trinity B., TEX, USA, sample 15 m from brine effluent = 20 ug/l[1]

Source: 1.Armstrong, H.W., K. Fucik, J.W. Anderson and J.M. Neff. 1977. Effects of oilfield brine effluent on benthic organisms in Trinity Bay. American Petrolium Institute Publ. No. 4291. Wash. DC.

5-342 Concentrations
Dinitramine

Species	Value	Condition
Carp, bile	22.7 mg/kg wet weight	12 h, uptake from water = 1 mg/l, 285 K, pH = 7.8, Ca = 220 mg/l[1]
Carp, muscle	3.96 mg/kg wet weight	12 h, uptake from water = 1 mg/l, 285 K, pH = 7.8, Ca = 220 mg/l[1]
Carp, plasma	24.4 mg/kg wet weight	12 h, uptake from water = 1 mg/l, 285 K, pH = 7.8, Ca = 220 mg/l[1]
Channel catfish, bile	13.4 mg/kg wet weight	12 h, uptake from water = 1 mg/l, 285 K, pH = 7.8, Ca = 220 mg/l[1]
Channel catfish, muscle	46.1 mg/kg wet weight	12 h, uptake from water = 1 mg/l, 285 K, pH = 7.8, Ca = 220 mg/l[1]
Channel catfish, plasma	58.5 mg/kg wet weight	12 h, uptake from water = 1 mg/l, 285 K, pH = 7.8, Ca = 220 mg/l[1]

Source: 1.Olson, L.E., J.L. Allen. and W.L. Mauck. 1975. J. Agric. Food. Chem. 23(3), pp 437-439.

5-343 Concentrations
Diol

Species	Value	Condition
Oncorhynchus kisutch, brain	5 % of total	Distribution 24 h after intraperi. inj. of 2.5 uCi naphthalene-1-14C[1]
Oncorhynchus kisutch, flesh	7 % of total	Distribution 24 h after intraperi. inj. of 2.5 uCi naphthalene-1-14C[1]
Oncorhynchus kisutch, gall bladder	7 % of total	Distribution 24 h after intraperi. inj. of 2.5 uCi naphthalene-1-14C[1]
Oncorhynchus kisutch, heart	26 % of total	Distribution 24 h after intraperi. inj. of 2.5 uCi naphthalene-1-14C[1]
Oncorhynchus kisutch, liver	11 % of total	Distribution 24 h after intraperi. inj. of 2.5 uCi naphthalene-1-14C[1]

Source: 1.Roubal, W.T., D.H. Bovee, T.K. Collier and S.I. Stranahan. 1977. Proc of 1977 Oil Spill Conf. (Prevention, Behavior, Control, Cleanup). Wash DC. American Petroleum Institute. 551-556.

5-344 Concentrations
Distribution of benzo(a)pyrene and hexadecane

Species	Value	Condition
Gadus morrhua, aquarium water	0.0 % of total activity	After 48 h, C-14 labbelled BaP and hexadecane in food = squid[1]
Gadus morrhua, aquarium water	0.12 % of total activity	After 96 h, C-14 labbelled BaP and hexadecane in food = squid[1]
Gadus morrhua, aquarium water	0.2 % of total activity	After 72 h, C-14 labbelled BaP and hexadecane in food = squid[1]
Gadus morrhua, bile fluid	0.6 % of total activity	After 48 h, C-14 labbelled BaP and hexadecane in food = squid[1]
Gadus morrhua, bile fluid	1.9 % of total activity	After 96 h, C-14 labbelled BaP and hexadecane in food = squid[1]
Gadus morrhua, bile fluid	12.5 % of total activity	After 72 h, C-14 labbelled BaP and hexadecane in food = squid[1]
Gadus morrhua, blood	0.0 % of total activity	After 48 h, C-14 labbelled BaP and hexadecane in food = squid[1]
Gadus morrhua, blood	0.04 % of total activity	After 96 h, C-14 labbelled BaP and hexadecane in food = squid[1]
Gadus morrhua, blood	0.06 % of total activity	After 72 h, C-14 labbelled BaP and hexadecane in food = squid[1]
Gadus morrhua, feces	0.5 % of total activity	After 48 h, C-14 labbelled BaP and hexadecane in food = squid, in water[1]

Gadus morrhua, feces	20.7 % of total activity	After 96 h, C-14 labbelled BaP and hexadecane in food = squid, in water[1]
Gadus morrhua, feces	41.4 % of total activity	After 72 h, C-14 labbelled BaP and hexadecane in food = squid, in water[1]
Gadus morrhua, gills	0.2 % of total activity	After 48 h, C-14 labbelled BaP and hexadecane in food = squid[1]
Gadus morrhua, gills	0.4 % of total activity	After 96 h, C-14 labbelled BaP and hexadecane in food = squid[1]
Gadus morrhua, gills	0.5 % of total activity	After 72 h, C-14 labbelled BaP and hexadecane in food = squid[1]
Gadus morrhua, intestine content	12.9 % of total activity	After 48 h, C-14 labbelled BaP and hexadecane in food = squid[1]
Gadus morrhua, intestine content	36.3 % of total activity	After 96 h, C-14 labbelled BaP and hexadecane in food = squid[1]
Gadus morrhua, intestine content	8.6 % of total activity	After 72 h, C-14 labbelled BaP and hexadecane in food = squid[1]
Gadus morrhua, liver	2.2 % of total activity	After 48 h, C-14 labbelled BaP and hexadecane in food = squid[1]
Gadus morrhua, liver	3.4 % of total activity	After 96 h, C-14 labbelled BaP and hexadecane in food = squid[1]
Gadus morrhua, liver	3.6 % of total activity	After 72 h, C-14 labbelled BaP and hexadecane in food = squid[1]
Gadus morrhua, plasma	0.03 % of total activity	After 48 h, C-14 labbelled BaP and hexadecane in food = squid[1]
Gadus morrhua, plasma	0.05 % of total activity	After 96 h, C-14 labbelled BaP and hexadecane in food = squid[1]
Gadus morrhua, plasma	0.08 % of total activity	After 72 h, C-14 labbelled BaP and hexadecane in food = squid[1]
Gadus morrhua, spleen	0.06 % of total activity	After 48 h, C-14 labbelled BaP and hexadecane in food = squid[1]
Gadus morrhua, spleen	0.1 % of total activity	After 72 h, C-14 labbelled BaP and hexadecane in food = squid[1]
Gadus morrhua, spleen	0.1 % of total activity	After 96 h, C-14 labbelled BaP and hexadecane in food = squid[1]
Gadus morrhua, stomach	32.9 % of total activity	After 72 h, C-14 labbelled BaP and hexadecane in food = squid[1]
Gadus morrhua, stomach	37.1 % of total activity	After 96 h, C-14 labbelled BaP and hexadecane in food = squid[1]
Gadus morrhua, stomach	83.5 % of total activity	After 48 h, C-14 labbelled BaP and hexadecane in food = squid[1]
Gadus morrhua, urine	0.0 % of total activity	After 48 h, C-14 labbelled BaP and hexadecane in food = squid[1]
Gadus morrhua, urine	0.0 % of total activity	After 72 h, C-14 labbelled BaP and hexadecane in food = squid[1]
Gadus morrhua, urine	0.03 % of total activity	After 96 h, C-14 labbelled BaP and hexadecane in food = squid[1]

Source: 1.McNaughton, S.J. T.C. Folsom,Lee,Park, Price, Schmitz , Roeder. 1974. Ecology 55, 1163-1165.

5-345 Concentrations
EDC-tar

Species	Value	Condition
Leander adspersus	160 mg/kg wet weight	EDC-tar in water = 1 mg/l, 283-287 K, 16 d of exposure[1]
Leander adspersus	3 mg/kg wet weight	EDC-tar in water = 0.01 mg/l, 283-287 K, 16 d of exposure[1]
Leander adspersus	8 mg/kg wet weight	EDC-tar in water = 0.1 mg/l, 283-287 K, 16 d of exposure[1]
Mytilus edulis	13 mg/kg wet weight	EDC-tar in water = 1 mg/l, 283-287 K, 16 d of exposure[1]
Mytilus edulis	26 mg/kg wet weight	EDC-tar in water = 10 mg/l, 283-287 K, 16 d of exposure[1]
Mytilus edulis	270 mg/kg wet weight	EDC-tar in water = 10 mg/l, 283-287 K, 16 d of exposure[1]
Mytilus edulis	6 mg/kg wet weight	EDC-tar in water = 0.1 mg/l, 283-287 K, 16 d of exposure[1]

Source: 1.Jerneloev, A., R. Rosenberg and S. Jensen. 1972. Water Res. 6(10), pp 1181-1191.

5-346 Concentrations Endosulfan

Species	Value	Condition
Chalamys opercularis, adductor muscle	0.88 mg/kg wet weight	Water = 0.1 mg/l, 2 w[1]
Chalamys opercularis, digestive gland	6.1 mg/kg wet weight	Water = 0.1 mg/l, 2 w[1]
Chalamys opercularis, gill	0.46 mg/kg wet weight	Water = 0.1 mg/l, 2 w[1]
Chalamys opercularis, gonad	3.8 mg/kg wet weight	Water = 0.1 mg/l, 2 w[1]
Chalamys opercularis, mantle	1.34 mg/kg wet weight	Water = 0.1 mg/l, 2 w[1]
Mytilus edulis, digestive gland	6.2 mg/kg wet weight	Water = 0.1 mg/l, 2 w[1]
Mytilus edulis, foot	0.75 mg/kg wet weight	Water = 0.1 mg/l, 2 w[1]
Mytilus edulis, gill	2.1 mg/kg wet weight	Water = 0.1 mg/l, 2 w[1]
Mytilus edulis, gonad	0.66 mg/kg wet weight	Water = 0.1 mg/l, 2 w[1]
Mytilus edulis, mantle	0.66 mg/kg wet weight	Water = 0.1 mg/l, 2 w[1]
Mytilus edulis, posterior adductor muscle	0.75 mg/kg wet weight	Water = 0.1 mg/l, 2 w[1]

Source: 1.Roberts, D. 1975. Bull. Environ. Contam. Toxicol. 31, pp 170-176.

5-347 Concentrations Endrin

Species	Value	Condition
Sediment	0.0006 mg/kg wet weight	Conc in water = trace, USA[1]
Sediment	0.001 mg/kg wet weight	Conc in water: < 1.0 ug/l, USA, less than value[1]
Sediment	0.01 mg/kg wet weight	Conc in water: < 1.0 ug/l, USA, less than value[1]

Source: 1.Edwards, C.A. 1975. Persistent Pesticides in the Environment, 2nd edition, CRC-Press.

5-348 Concentrations Fe

Species	Value	Condition
Alisma plantago aquatica, old leaves	549.81 mg/kg dry weight	800 and 50 m below pollution I and II, water = 0.888 mg/l, processed leaves[1]
Alisma plantago aquatica, old leaves	596.23 mg/kg dry weight	800 m above pollution entry I, water = 0.708 mg/l, processed leaves[1]
Alisma plantago aquatica, old leaves	599.09 mg/kg dry weight	150 m below pollution entry I, water = 0.893 mg/l, processed leaves[1]
Alisma plantago aquatica, old leaves	795.83 mg/kg dry weight	800 and 50 m below pollution I and II, water = 0.888 mg/l, whole leaves[1]
Alisma plantago aquatica, old leaves	837.54 mg/kg dry weight	150 m below pollution entry I, water = 0.893 mg/l, whole leaves[1]
Alisma plantago aquatica, old leaves	869.07 mg/kg dry weight	800 m above pollution entry I, water = 0.708 mg/l, whole leaves[1]
Alisma plantago aquatica, young leaves	1115.78 mg/kg dry weight	150 m below pollution entry I, water = 0.893 mg/l, whole leaves[1]

Alisma plantago aquatica, young leaves	1145.85 mg/kg dry weight	800 m above pollution entry I, water = 0.708 mg/l, whole leaves[1]
Alisma plantago aquatica, young leaves	728.12 mg/kg dry weight	150 m below pollution entry I, water = 0.893 mg/l, processed leaves[1]
Alisma plantago aquatica, young leaves	863.99 mg/kg dry weight	800 m above pollution entry I, water = 0.708 mg/l, processed leaves[1]
Alisma plantago aquatica, young leaves	878.56 mg/kg dry weight	800 and 50 m below pollution I and II, water = 0.888 mg/l, processed leaves[1]
Alisma plantago aquatica, young leaves	966.42 mg/kg dry weight	800 and 50 m below pollution I and II, water = 0.888 mg/l, whole leaves[1]
Beech and spruce forest soil	1.6E5 mg/kg dry weight	Central German, depth = 0-10 cm, 18.4 t/ha[2]
Beech and spruce forest soil	1.9E5 mg/kg dry weight	Central German, depth = 10-20 cm, 22.6 t/ha[2]
Beech and spruce forest soil	1.9E5 mg/kg dry weight	Central German, depth = 30-40 cm, 24.7 t/ha[2]
Beech and spruce forest soil	2.0E5 mg/kg dry weight	Central German, depth = 40-50 cm, 29.6 t/ha[2]
Hypolimnion	425 ug/l	Conc in epilimnion = 15 ug/l, concentration factor = 28, Schoehsee[3]
Living spruce needles	70 mg/kg dry weight	Litter = 574-2850, underlaying soil = 19000-24000 mg/kg dw, S Sweden,[4]
Sediment	1593.91 mg/kg dry weight	800 and 50 m below pollution I and II, water = 0.888 mg/l[1]
Sediment	2.6 mg/kg dry weight	Max water = 45 microgram/l, 1 y measurement, Fowey River[5]
Sediment	2.6 mg/kg dry weight	Water conc = 33 microgram/l, yearly mean, Fowey River[5]
Sediment	3.0 mg/kg dry weight	Max water = 26 microgram/l, 1 y measurement, Carnon River[5]
Sediment	3.0 mg/kg dry weight	Water conc = 14 microgram/l, yearly mean, Carnon River[5]
Sediment	3.6 mg/kg dry weight	Max water = 28 microgram/l, 1 y measurement, Fowey River[5]
Sediment	3.6 mg/kg dry weight	Water conc = 20 microgram/l, yearly mean, Fowey River[5]
Sediment	4.2 mg/kg dry weight	Max water = 120 microgram/l, max conc, Gannel River[5]
Sediment	4.2 mg/kg dry weight	Water conc = 79 microgram/l, yearly mean, Gannel River[5]
Sediment	5.1 mg/kg dry weight	Max water = 50 microgram/l, 1 y measurement, Red River[5]
Sediment	5.1 mg/kg dry weight	Water conc = 22 microgram/l, yearly mean, Red River[5]
Sediment	58000 mg/kg dry weight	Conc in epilimnion = 15 ug/l, concentration factor = 3.9E6, Schoehsee[3]
Sediment	6.7 mg/kg dry weight	Water conc = 91 microgram/l, yearly mean, Carnon River[5]
Sediment	649.31 mg/kg dry weight	800 m above pollution entry I, water = 0.708 mg/l, rg = 597 - 683[1]
Sediment	683.96 mg/kg dry weight	150 m below pollution entry I, water = 0.893 mg/l, rg = 629 - 714[1]
Sediment	7.1 mg/kg dry weight	Max water = 485 microgram/l, max conc, Gannel River[5]
Sediment	7.1 mg/kg dry weight	Water conc = 380 microgram/l, yearly mean, Gannel River[5]
Sediment	7.7 mg/kg dry weight	Max water = 140 microgram/l, 1 y measurement, Red River[5]
Sediment	7.7 mg/kg dry weight	Water conc = 83 microgram/l, yearly mean, Red River[5]
Sediment	727 mg/kg dry weight	Max water = 280 microgram/l, 1 y measurement, Carnon River[5]
Sediment	727 mg/kg dry weight	Water conc = 167 microgram/l, yearly mean, Carnon River[5]

Source: 1.Patrick, F.M. and M.W. Loutit. 1977. Water Res., 11(8), pp 699-703.
2.Heinrichs, H. and R. Meyer. 1977. J. Environ. Qual. 6, 402-407.
3.Groth, P. 1971. Arch. Hydrobiol. 68(3), pp 305-375.
4.Nilsson, I. 1972. Oikos, 22, pp 366.
5.Altshuller, A.P. and J.J. Bufalini. 1971. Environ. Sci. Technol., 5(1), pp 39-64.

5-349 Concentrations
Fluoranthene

Species	Value	Condition
Crassostrea virginica	4.0 mg/kg dry weight	Exposure = 8 d, water = 0.4 ug/l, crude oil, lab[1]
Crassostrea virginica	5.0 mg/kg dry weight	Exposure = 2 d, water = 7.2 ug/l, crude oil, lab[1]

Source: 1.Lee, R.F., W.S. Gardner, J.W. Anderson, Balylock and B-Clarke. 1978. Environ. Sci. Technol. 12, 832-38.

5-350 Concentrations
Fluorene

Species	Value	Condition
Crassostrea virginica	0.2 mg/kg wet weight	Upt in 8 h from 302 ppm No. 2 fuel, 0.36 % F, 120 h excr, init=0.5 ppm[1]
Crassostrea virginica	0.4 mg/kg wet weight	Upt in 8 h from 302 ppm No. 2 fuel, 0.36 % F, 6 h excr, init=0.5 ppm[1]
Crassostrea virginica	0.5 mg/kg wet weight	Upt in 8 h from 302 ppm No. 2 fuel, 0.36 % F, 24 h excr, init=0.5 ppm[1]
Crassostrea virginica	0.7 mg/kg wet weight	Upt in 8 h from 302 ppm No. 2 fuel, 0.36 % F, 3 h excr, init=0.5 ppm[1]
Crassostrea virginica	1.0 mg/kg wet weight	302 ppm dispersed No. 2 fuel oil with 0.36 % F, flow through, 8 h[1]
Marine sediment near oil separator platform	100 ug/kg wet weight	Trinity B., TEX, USA, sample 15 m from brine effluent = 14 ug/l[2]

Source: 1.Neff, J.M., B.A. Cox, D. Dixit and J.W. Anderson. 1976. Mar. Biol. 38, 279-289.
2.Armstrong, H.W., K. Fucik, J.W. Anderson and J.M. Neff. 1977. Effects of oilfield brine effluent on benthic organisms in Trinity Bay. American Petrolium Institute Publ. No. 4291. Wash. DC.

5-351 Concentrations
Fluorene, C4 - C5 Naphthalene, C2 - Biphenyl

Species	Value	Condition
Fundulus heteroclitus	0.04 mg/kg wet weight	570 l S. Louisiana crude oil in 810 m2 shoreline area, 6 h after spill[1]
Fundulus heteroclitus	0.05 mg/kg wet weight	570 l S. Louisiana crude oil in 810 m2 shoreline area, 216 h after spill[1]
Fundulus heteroclitus	0.11 mg/kg wet weight	570 l S. Louisiana crude oil in 810 m2 shoreline area, 31 h after spill[1]
Fundulus heteroclitus	0.45 mg/kg wet weight	570 l S. Louisiana crude oil in 810 m2 shoreline area, 76 h after spill[1]

Source: 1.Bieri, R.H., V.C. Stamoudis and M.K. Cueman. 1977. Proc. of 1977 Oil spill Conference (Prevention, Behavior, Control, Cleanup). Wash. DC. American Petrolium Institute. 511-516.

5-352 Concentrations
Glycoside and sulfate

Species	Value	Condition
Oncorhynchus kisutch, brain	17 % of total	Distribution 24 h after intraperi. inj. of 2.5 uCi naphthalene-1-14C[1]
Oncorhynchus kisutch, flesh	26 % of total	Distribution 24 h after intraperi. inj. of 2.5 uCi naphthalene-1-14C[1]

Oncorhynchus kisutch, gall bladder	2 % of total	Distribution 24 h after intraperi. inj. of 2.5 uCi naphthalene-1-14C[1]
Oncorhynchus kisutch, heart	0 % of total	Distribution 24 h after intraperi. inj. of 2.5 uCi naphthalene-1-14C[1]
Oncorhynchus kisutch, liver	2 % of total	Distribution 24 h after intraperi. inj. of 2.5 uCi naphthalene-1-14C[1]

Source: 1.Roubal, W.T., D.H. Bovee, T.K. Collier and S.I. Stranahan. 1977. Proc of 1977 Oil Spill Conf. (Prevention, Behavior, Control, Cleanup). Wash DC. American Petroleum Institute. 551-556.

5-353 Concentrations Heptachlor

Species	Value	Condition
Chicken, liver	0.15 mg/kg wet weight	Conc in food = 0.15 mg/kg wet weight[1]
Chicken, liver	0.31 mg/kg wet weight	Conc in food = 0.45 mg/kg wet weight[1]
Woodcock, tissue	1.70 mg/kg wet weight	Conc in food = 0.65 mg/kg wet weight[1]
Woodcock, tissue	13.0 mg/kg wet weight	Conc in food = 2.86 mg/kg wet weight[1]

Source: 1.Edwards, C.A. 1975. Persistent Pesticides in the Environment, 2nd edition, CRC-Press.

5-354 Concentrations Heptachlor and heptaepoxide

Species	Value	Condition
Sediment	0.0008 mg/kg wet weight	Conc in water = 0.06 ug/l, USA[1]
Sediment	0.002 mg/kg wet weight	Conc in water: < 1.0 ug/l, USA, less than[1]
Sediment	0.02 mg/kg wet weight	Conc in water: < 1.0 ug/l, USA, less than[1]

Source: 1.Edwards, C.A. 1975. Persistent Pesticides in the Environment, 2nd edition, CRC-Press.

5-355 Concentrations Hexachlorophene

Species	Value	Condition
Cambarinae	1338 microgram/kg wet weight	Sediment = 9.3-377.0 ug/kg dw, immature, Buffalo Creek[1]
Hydrophilus triangularis	1500 microgram/kg wet weight	Sediment = 9.3-377.0 ug/kg dw, Buffalo Creek[1]
Sediment	106.8 microgram/kg dry weight	Surface water = 18.0 ppm, bottom water = 22.0 ppm, Buffalo Creek[1]
Sediment	110.9 microgram/kg dry weight	Surface water = 31.2 ppm, bottom water = 21.6 ppm, Buffalo Creek[1]
Sediment	159.2 microgram/kg dry weight	Surface water = 28.9 ppm, bottom water = 48.3 ppm, Buffalo Creek[1]
Sediment	19.5 microgram/kg dry weight	Surface water = 16.4 ppm, bottom water = 21.2 ppm, Buffalo Creek[1]
Sediment	25.4 microgram/kg dry weight	Surface water = 41.5 ppm, bottom water = 19.2 ppm, Buffalo Creek[1]
Sediment	27.1 microgram/kg dry weight	Surface water = 21.6 ppm, bottom water = 26.8 ppm, Buffalo Creek[1]
Sediment	30.5 microgram/kg dry weight	Surface water = 37.8 ppm, bottom water = 41.4 ppm, Buffalo Creek[1]
Sediment	377.0 microgram/kg dry weight	Surface water = 44.3 ppm, bottom water = 30.0 ppm, Buffalo Creek[1]

Sediment	44.9 microgram/kg dry weight	Surface water = 37.6 ppm, bottom water = 15.2 ppm, Buffalo Creek[1]
Sediment	81.4 microgram/kg dry weight	Surface water = 21.2 ppm, bottom water = 22.0 ppm, Buffalo Creek[1]
Sediment	84.8 microgram/kg dry weight	Surface water = 34.8 ppm, bottom water = 16.4 ppm, Buffalo Creek[1]
Sediment	9.3 microgram/kg dry weight	Surface water = 20.0 ppm, bottom water = 20.8 ppm, Buffalo Creek[1]
Sigara sp.	27800 microgram/kg wet weight	Sediment = 9.3-377.0 ug/kg dw, Buffalo Creek[1]
Tubifex tubifex	335 microgram/kg wet weight	Sediment = 9.3-377.0 ug/kg dw, Buffalo Creek[1]

Source: 1.Sims, J.L. and F.K. Pfaender. 1975. Bull. Environ. Contam. Toxicol. 14(2), pp 214-220.

5-356 Concentrations
Hg

Species	Value	Condition
Asellus aquaticus	1 mg/kg dry weight	Sediment = 1 ppm dw, Elbe delta, brackish[1]
Asellus aquaticus	12 mg/kg dry weight	Sediment = 14 ppm dw, Elbe delta, freshwater[1]
Asellus aquaticus	4 mg/kg dry weight	Sediment = 33 ppm dw, Elbe delta, brackish[1]
Asellus aquaticus	7 mg/kg dry weight	Sediment = 5 ppm dw, Elbe delta, freshwater[1]
Avena sativa, grains	0.013 mg/kg dry weight	Control = 0.009, soil = 4 ppm, pH = 5.1, silt loam, CEC = .383 meq/g[2]
Avena sativa, grains	0.020 mg/kg dry weight	Control = 0.009, soil = 20 ppm, pH = 5.1, silt loam, CEC = .383 meq/g[2]
Avena sativa, husks	0.229 mg/kg dry weight	Control = 0.107, soil = 4 ppm, pH = 5.1, silt loam, CEC = .383 meq/g[2]
Avena sativa, husks	0.266 mg/kg dry weight	Control = 0.107, soil = 20 ppm, pH = 5.1, silt loam, CEC = .383 meq/g[2]
Avena sativa, leaves	0.193 mg/kg dry weight	Control = 0.176, soil = 4 ppm, pH = 5.1, silt loam, CEC = .383 meq/g[2]
Avena sativa, leaves	0.199 mg/kg dry weight	Control = 0.176, soil = 20 ppm, pH = 5.1, silt loam, CEC = .383 meq/g[2]
Avena sativa, root	0.157 mg/kg dry weight	Control = 0.151, soil = 4 ppm, pH = 5.1, silt loam, CEC = .383 meq/g[2]
Avena sativa, root	0.426 mg/kg dry weight	Control = 0.151, soil = 20 ppm, pH = 5.1, silt loam, CEC = .383 meq/g[2]
Avena sativa, stalks	0.008 mg/kg dry weight	Control = 0.011, soil = 4 ppm, pH = 5.1, silt loam, CEC = .383 meq/g[2]
Avena sativa, stalks	0.026 mg/kg dry weight	Control = 0.011, soil = 20 ppm, pH = 5.1, silt loam, CEC = .383 meq/g[2]
Beech and spruce forest soil	0.02 mg/kg dry weight	Central German, depth = 40-50 cm, 27 g/ha[3]
Beech and spruce forest soil	0.05 mg/kg dry weight	Central German, depth = 30-40 cm, 65 g/ha[3]
Beech and spruce forest soil	0.08 mg/kg dry weight	Central German, depth = 10-20 cm, 100 g/ha[3]
Beech and spruce forest soil	0.12 mg/kg dry weight	Central German, depth = 0-10 cm, 138 g/ha[3]
Brassica oleracea var. botrytis, leaves	0.061 mg/kg dry weight	Control = 0.079, soil = 20 ppm, pH = 5.1, silt loam, CEC = .383 meq/g[2]
Brassica oleracea var. botrytis, leaves	0.068 mg/kg dry weight	Control = 0.079, soil = 4 ppm, pH = 5.1, silt loam, CEC = .383 meq/g[2]
Brassica oleracea var. botrytis, root	0.197 mg/kg dry weight	Control = 0.019, soil = 4 ppm, pH = 5.1, silt loam, CEC = .383 meq/g[2]

Brassica oleracea var. botrytis, root	2.447 mg/kg dry weight	Control = 0.019, soil = 20 ppm, pH = 5.1, silt loam, CEC = .383 meq/ g[2]
Brassica oleracea var. italica, leaves	0.029 mg/kg dry weight	Control = 0.063, soil = 20 ppm, pH = 5.1, silt loam, CEC = .383 meq/g[2]
Brassica oleracea var. italica, leaves	0.078 mg/kg dry weight	Control = 0.063, soil = 4 ppm, pH = 5.1, silt loam, CEC = .383 meq/g[2]
Brassica oleracea var. italica, root	0.505 mg/kg dry weight	Control = 0.171, soil = 4 ppm, pH = 5.1, silt loam, CEC = .383 meq/g[2]
Brassica oleracea var. italica, root	1.870 mg/kg dry weight	Control = 0.171, soil = 20 ppm, pH = 5.1, silt loam, CEC = .383 meq/g[2]
Calf, brain	12 mg/kg wet weight	47 mg Hg/kg day, oral, ethyl-Hg-p-toluene, 9 d, n = 1[4]
Calf, brain	29 mg/kg wet weight	4.7 mg Hg/kg day, oral, ethyl-Hg-p-toluene, 38 d, n = 1[4]
Calf, brain	3.2 mg/kg wet weight	120 mg Hg/kg, oral, ethyl-Hg-p-toluene, 3 d, n = 1[4]
Calf, kidney	29 mg/kg wet weight	120 mg Hg/kg, oral, ethyl-Hg-p-toluene, 3 d, n = 1[4]
Calf, kidney	60 mg/kg wet weight	4.7 mg Hg/kg day, oral, ethyl-Hg-p-toluene, 38 d, n = 1[4]
Calf, kidney	62 mg/kg wet weight	47 mg Hg/kg day, oral, ethyl-Hg-p-toluene, 9 d, n = 1[4]
Calf, liver	46 mg/kg wet weight	120 mg Hg/kg, oral, ethyl-Hg-p-toluene, 3 d, n = 1[4]
Calf, liver	50 mg/kg wet weight	4.7 mg Hg/kg day, oral, ethyl-Hg-p-toluene, 38 d, n = 1[4]
Calf, liver	58 mg/kg wet weight	47 mg Hg/kg day, oral, ethyl-Hg-p-toluene, 9 d, n = 1[4]
Calf, whole blood	23 mg/kg wet weight	47 mg Hg/kg day, oral, ethyl-Hg-p-toluene, 9 d, n = 1[4]
Calf, whole blood	5 mg/kg wet weight	120 mg Hg/kg, oral, ethyl-Hg-p-toluene, 3 d, n = 1[4]
Calf, whole blood	7 mg/kg wet weight	4.7 mg Hg/kg day, oral, ethyl-Hg-p-toluene, 38 d, n = 1[4]
Calidris alpina, liver	1.7 mg/kg dry weight	6 adult birds, 1 first winter, 25-27/10 1972, British Isles[5]
Calidris alpina, liver	1.9 mg/kg dry weight	3 first winter birds, 3 adult, 20/9 1971 - 10/10 1972, British Isles[5]
Calidris alpina, liver	10.5 mg/kg dry weight	1 adult bird, 19/3 1972, British Isles[5]
Calidris alpina, liver	3.3 mg/kg dry weight	2 first winter birds, 1 adult, 9/11 1972, British Isles[5]
Calidris alpina, liver	6.9 mg/kg dry weight	9 birds, 10/1 1970, British Isles[5]
Calidris canutus, liver	1 mg/kg dry weight	2 first winter birds, 29/8 1972, British Isles[5]
Calidris canutus, liver	1.3 mg/kg dry weight	3 first winter birds, 1 adult, 26/9 - 10/10 1972, British Isles[5]
Calidris canutus, liver	14.4 mg/kg dry weight	6 adult birds, 19/3 1972, British Isles[5]
Calidris canutus, liver	18.5 mg/kg dry weight	2 first winter birds, 26/3 1972, British Isles[5]
Calidris canutus, liver	7.3 mg/kg dry weight	2 adult birds, 11/1 1970, British Isles[5]
Calidris canutus, liver	9.8 mg/kg dry weight	7 adult birds, 19/2 1972, British Isles[5]
Cat, brain	28 mg/kg wet weight	1 mg Hg/kg day, oral, methyl-Hg-OH, 37-45 d, n = 2[4]
Cat, brain	9 mg/kg wet weight	0.2-0.4 mg Hg/kg day, oral, methyl-Hg-OH, 76-126 d, n = 2[4]
Cat, brain	9.2 mg/kg wet weight	1.5 mg Hg/kg day, oral, methyl-Hg-SH-methyl, 22 d, n = 4[4]
Cat, kidney	19 mg/kg wet weight	0.2-0.4 mg Hg/kg day, oral, methyl-Hg-OH, 76-126 d, n = 2[4]
Cat, kidney	20 mg/kg wet weight	1.5 mg Hg/kg day, oral, methyl-Hg-SH-methyl, 22 d, n = 4[4]
Cat, kidney	62 mg/kg wet weight	1 mg Hg/kg day, oral, methyl-Hg-OH, 37-45 d, n = 2[4]
Cat, liver	100 mg/kg wet weight	1 mg Hg/kg day, oral, methyl-Hg-OH, 37-45 d, n = 2[4]
Cat, liver	31 mg/kg wet weight	0.2-0.4 mg Hg/kg day, oral, methyl-Hg-OH, 76-126 d, n = 2[4]
Cat, liver	62 mg/kg wet weight	1.5 mg Hg/kg day, oral, methyl-Hg-SH-methyl, 22 d, n = 4[4]

Cat, whole blood	13 mg/kg wet weight	1.5 mg Hg/kg day, oral, methyl-Hg-SH-methyl, 22 d, n = 4[4]
Cat, whole blood	19 mg/kg wet weight	0.2-0.4 mg Hg/kg day, oral, methyl-Hg-OH, 76-126 d, n = 2[4]
Cat, whole biood	61 mg/kg wet weight	1 mg Hg/kg day, oral, methyl-Hg-OH, 37-45 d, n = 2[4]
Crassostrea sp., adductor muscle	2.5 % of total Hg	Water = 10 ug HgCl2/l, 4 d, Hg-203, 288 K, control exp, rg=+-2.5[6]
Crassostrea sp., adductor muscle	2.7 % of total Hg	Water = 10 ug HgCl2/l, 4 d, Hg-203, 288 K, Hg reducing bac, rg=+-2.5[6]
Crassostrea sp., adductor muscle	2.9 % of total Hg	Water = 10 ug HgCl2/l, 4 d, Hg-203, 288 K, Hg accumulat. bac, rg=+-2.5[6]
Crassostrea sp., gills	44.2 % of total Hg	Water = 10 ug HgCl2/l, 4 d, Hg-203, 288 K, Hg reducing bac, rg=+-9.0[6]
Crassostrea sp., gills	50.4 % of total Hg	Water = 10 ug HgCl2/l, 4 d, Hg-203, 288 K, control exp, rg=+-9.0[6]
Crassostrea sp., gills	65.0 % of total Hg	Water = 10 ug HgCl2/l, 4 d, Hg-203, 288 K, Hg accumulat. bac, rg=+-9.0[6]
Crassostrea sp., mantle	13.3 % of total Hg	Water = 10 ug HgCl2/l, 4 d, Hg-203, 288 K, control exp, rg=+-7.6[6]
Crassostrea sp., mantle	16.4 % of total Hg	Water = 10 ug HgCl2/l, 4 d, Hg-203, 288 K, Hg accumulat. bac, rg=+-7.6[6]
Crassostrea sp., mantle	26.6 % of total Hg	Water = 10 ug HgCl2/l, 4 d, Hg-203, 288 K, Hg reducing bac, rg=+-7.6[6]
Crassostrea sp., mantle fluid	0.9 % of total Hg	Water = 10 ug HgCl2/l, 4 d, Hg-203, 288 K, control exp, rg=+-0.8[6]
Crassostrea sp., mantle fluid	1.5 % of total Hg	Water = 10 ug HgCl2/l, 4 d, Hg-203, 288 K, Hg accumulat. bac, rg=+-0.8[6]
Crassostrea sp., mantle fluid	2.8 % of total Hg	Water = 10 ug HgCl2/l, 4 d, Hg-203, 288 K, Hg reducing bac, rg=+-0.8[6]
Crassostrea sp., viscera	15.7 % of total Hg	Water = 10 ug HgCl2/l, 4 d, Hg-203, 288 K, Hg accumulat. bac, rg=+-11.2[6]
Crassostrea sp., viscera	26.3 % of total Hg	Water = 10 ug HgCl2/l, 4 d, Hg-203, 288 K, control exp, rg=+-11.2[6]
Crassostrea sp., viscera	30.3 % of total Hg	Water = 10 ug HgCl2/l, 4 d, Hg-203, 288 K, Hg reducing bac, rg=+-11.2[6]
Crassostrea sp., whole	12.4 % of total Hg	Water = 10 ug HgCl2/l, 4 d, Hg-203, 288 K, Hg accumulat. bac, time=0 d, rg=+-1[6]
Crassostrea sp., whole	3.1 % of total Hg	Water = 10 ug HgCl2/l, 4 d, Hg-203, 288 K, control exp, time=0 d, rg=+-1[6]
Crassostrea sp., whole	6.7 % of total Hg	Water = 10 ug HgCl2/l, 4 d, Hg-203, 288 K, Hg reducing bac, time=0 d, rg=+-1[6]
Daucus carota var. sativa, root	0.428 mg/kg dry weight	Control = 0.163, soil = 4 ppm, pH = 5.1, silt loam, CEC = .383 meq/g[2]
Daucus carota var. sativa, root	1.058 mg/kg dry weight	Control = 0.163, soil = 20 ppm, pH = 5.1, silt loam, CEC = .383 meq/g[2]
Daucus carota var. sativa, tops	0.061 mg/kg dry weight	Control = 0.044, soil = 4 ppm, pH = 5.1, silt loam, CEC = .383 meq/g[2]
Daucus carota var. sativa, tops	0.072 mg/kg dry weight	Control = 0.044, soil = 20 ppm, pH = 5.1, silt loam, CEC = .383 meq/g[2]
Daucus carota var. sativa, tubers	0.039 mg/kg dry weight	Control = 0.044, soil = 20 ppm, pH = 5.1, silt loam, CEC = .383 meq/g[2]
Daucus carota var. sativa, tubers	0.053 mg/kg dry weight	Control = 0.044, soil = 4 ppm, pH = 5.1, silt loam, CEC = .383 meq/g[2]
Distribution of Hg in lab experiment, feces	0.3 % of total added	Control exp, 10 ug HgCl2/l, Hg-203, 4 d, 288 K, rg=+-0.2[6]
Distribution of Hg in lab experiment, feces	2.8 % of total added	Hg reducing bac. present, 10 ug HgCl2/l, Hg-203, 4 d, 288 K, rg=+-0.9[6]

Distribution of Hg in lab experiment, feces	3.9 % of total added	Hg accumulation bac. pres., 10 ug HgCl2/l, Hg-203, 4 d, 288 K, rg=+-2.6[6]
Distribution of Hg in lab experiment, microparticulate	3.8 % of total added	Control exp, 10 ug HgCl2/l, Hg-203, 4 d, 288 K, rg=+-3.8[6]
Distribution of Hg in lab experiment, microparticulate	6.6 % of total added	Hg accumulation bac. pres., 10 ug HgCl2/l, Hg-203, 4 d, 288 K, rg=+-3.5[6]
Distribution of Hg in lab experiment, microparticulate	6.6 % of total added	Hg reducing bac. present, 10 ug HgCl2/l, Hg-203, 4 d, 288 K, rg=+-0.9[6]
Distribution of Hg in lab experiment, oyster uptake	19.5 % of total added	Control exp, 10 ug HgCl2/l, Hg-203, 4 d, 288 K, rg=+-9.0[6]
Distribution of Hg in lab experiment, oyster uptake	28.2 % of total added	Hg reducing bac. present, 10 ug HgCl2/l, Hg-203, 4 d, 288 K, rg=+-15.8[6]
Distribution of Hg in lab experiment, oyster uptake	49.6 % of total added	Hg accumulation bac. pres., 10 ug HgCl2/l, Hg-203, 4 d, 288 K, rg=+-14[6]
Distribution of Hg in lab experiment, total recovery	36.9 % of total added	Control exp, 10 ug HgCl2/l, Hg-203, 4 d, 288 K, rg=+-2.2[6]
Distribution of Hg in lab experiment, total recovery	53.1 % of total added	Hg reducing bac. present, 10 ug HgCl2/l, Hg-203, 4 d, 288 K, rg=+-6.4[6]
Distribution of Hg in lab experiment, total recovery	78.9 % of total added	Hg accumulation bac. pres., 10 ug HgCl2/l, Hg-203, 4 d, 288 K, rg=+-28[6]
Distribution of Hg in lab experiment, volatile	14.6 % of total added	Hg reducing bac. present, 10 ug HgCl2/l, Hg-203, 4 d, 288 K, rg=+-8.3[6]
Distribution of Hg in lab experiment, volatile	18.0 % of total added	Hg accumulation bac. pres., 10 ug HgCl2/l, Hg-203, 4 d, 288 K, rg=+-14[6]
Distribution of Hg in lab experiment, volatile	7.3 % of total added	Control exp, 10 ug HgCl2/l, Hg-203, 4 d, 288 K, rg=+-7.3[6]
Distribution of Hg in lab experiment, water phase	0.8 % of total added	Hg accumulation bac. pres., 10 ug HgCl2/l, Hg-203, 4 d, 288 K, rg=+-0.4[6]
Distribution of Hg in lab experiment, water phase	0.9 % of total added	Hg reducing bac. present, 10 ug HgCl2/l, Hg-203, 4 d, 288 K, rg=+-0.1[6]
Distribution of Hg in lab experiment, water phase	3.8 % of total added	Control exp, 10 ug HgCl2/l, Hg-203, 4 d, 288 K, rg=+-2.2[6]
Fish	0.27 mg/kg dry weight	Hg sediment < 0.04-0.62 dry wt, Hg water < 0.2 microg/l, 225 fish[7]
Fish, non-stationary	0.030 mg/kg wet weight	Conc in sediment = 0.240 mg/kg wet weight[8]
Fish, non-stationary	0.046 mg/kg wet weight	Conc in sediment = 1 mg/kg wet weight[8]
Fish, non-stationary	0.068 mg/kg wet weight	Conc in sediment = 1.67 mg/kg wet weight[8]
Fish, non-stationary	0.122 mg/kg wet weight	Conc in sediment = 3.00 mg/kg wet weight[8]
Fish, stationary	0.137 mg/kg wet weight	Conc in sediment = 0.240 mg/kg wet weight[8]
Fish, stationary	0.213 mg/kg wet weight	Conc in sediment = 1 mg/kg wet weight[8]
Fish, stationary	0.289 mg/kg wet weight	Conc in sediment = 1.67 mg/kg wet weight[8]
Fish, stationary	0.425 mg/kg wet weight	Conc in sediment = 3.0 mg/kg wet weight[8]
Fish, stationary	0.547 mg/kg wet weight	Conc in sediment = 4.67 mg/kg wet weight[8]
Homo sapiens	0.02 ug methyl-Hg/g brain	ADI value intake = 0.03 mg methyl Hg/day[9]

Homo sapiens	0.02 ug methyl-Hg/g brain	Estimated intake in USA = 0.02 mg methyl Hg/day[9]
Homo sapiens	12 ug methyl-Hg/g brain	Intake of methyl-Hg = 1.2 mg/day[9]
Homo sapiens	2.0 ug methyl-Hg/g brain	Intake of methyl-Hg = 0.2 mg/day, 500 g food with 0.5 ug Hg/g[9]
Homo sapiens	5.0 ug methyl-Hg/g brain	Intake of methyl-Hg = 0.5 mg/day[9]
Homo sapiens	8.0 ug methyl-Hg/g brain	Intake of methyl-Hg = 0.75 mg/day, fish eaters Niigata district, Japan[9]
Lactuca sativa var. crispa, leaves	0.017 mg/kg dry weight	Control = 0.031, soil = 4 ppm, pH = 5.1, silt loam, CEC = .383 meq/g[2]
Lactuca sativa var. crispa, leaves	0.045 mg/kg dry weight	Control = 0.031, soil = 20 ppm, pH = 5.1, silt loam, CEC = .383 meq/g[2]
Lactuca sativa var. crispa, root	0.175 mg/kg dry weight	Control = 0.112, soil = 4 ppm, pH = 5.1, silt loam, CEC = .383 meq/g[2]
Lactuca sativa var. crispa, root	0.387 mg/kg dry weight	Control = 0.112, soil = 20, pH = 5.1, silt loam, CEC = .383 meq/g[2]
Lebistes reticulatus, body	5.2 % of total added	Water = 1E-7 m Hg-ethylCl, 3-5 d, 298 K, in solution = 77.2 %[10]
Lebistes reticulatus, fin	0.8 % of total added	Water = 1E-7 m Hg-ethylCl, 3-5 d, 298 K, in solution = 77.2 %[10]
Lebistes reticulatus, head	5.6 % of total added	Water = 1E-7 m Hg-ethylCl, 3-5 d, 298 K, in solution = 77.2 %[10]
Lebistes reticulatus, internal organs	9.8 % of total added	Water = 1E-7 m Hg-ethylCl, 3-5 d, 298 K, in solution = 77.2 %[10]
Lebistes reticulatus, tail	1.4 % of total added	Water = 1E-7 m Hg-ethylCl, 3-5 d, 298 K, in solution = 77.2 %[10]
Mouse, brain	0.02 mg/kg wet weight	0.03 mg Hg/kg, intravenous, methyl-Hg-OH, 6 d, n = 10[4]
Mouse, brain	0.12 mg/kg wet weight	3 mg Hg/kg, subcutanous, n-amyl-Hg-acetate, 7 d, n = 3[4]
Mouse, brain	0.14 mg/kg wet weight	3 mg Hg/kg, subcutanous, iso-propyl-Hg-acetate, 7 d, n = 3[4]
Mouse, brain	0.2 mg/kg wet weight	0.3 mg Hg/kg, intravenous, methyl-Hg-OH, 6 d, n = 10[4]
Mouse, brain	0.2 mg/kg wet weight	3 mg Hg/kg, subcutanous, n-propyl-Hg-acetate, 7 d, n = 3[4]
Mouse, brain	0.23 mg/kg wet weight	3 mg Hg/kg, subcutanous, n-butyl-Hg-acetate, 7 d, n = 3[4]
Mouse, brain	0.24 mg/kg wet weight	3 mg Hg/kg, subcutanous, iso-butyl-Hg-acetate, 7 d, n = 3[4]
Mouse, brain	0.37 mg/kg wet weight	1.0 mg Hg/kg, intravenous, methyl-Hg-Cl, 22 d, n = 2[4]
Mouse, brain	0.6 mg/kg wet weight	1.0 mg Hg/kg, intravenous, methyl-Hg-OH, 6 d, n = 10[4]
Mouse, brain	3.3 mg/kg wet weight	5.0 mg Hg/kg, intravenous, methyl-Hg-OH, 6 d, n = 10[4]
Mouse, brain	4 mg/kg wet weight	2.5 mg Hg/kg day, subcutanous, methyl-Hg-acetate, 10 d, n = 3[4]
Mouse, brain	6 mg/kg wet weight	5 mg Hg/kg day, subcutanous, methyl-Hg-acetate, 10 d, n = 3[4]
Mouse, kidney	0.02 mg/kg wet weight	0.03 mg Hg/kg, intravenous, methyl-Hg-OH, 6 d, n = 10[4]
Mouse, kidney	0.2 mg/kg wet weight	0.3 mg Hg/kg, intravenous, methyl-Hg-OH, 6 d, n = 10[4]
Mouse, kidney	0.6 mg/kg wet weight	1.0 mg Hg/kg, intravenous, methyl-Hg-OH, 6 d, n = 10[4]
Mouse, kidney	17 mg/kg wet weight	3 mg Hg/kg, subcutanous, n-amyl-Hg-acetate, 7 d, n = 3[4]
Mouse, kidney	19 mg/kg wet weight	3 mg Hg/kg, subcutanous, n-butyl-Hg-acetate, 7 d, n = 3[4]
Mouse, kidney	2.7 mg/kg wet weight	1.0 mg Hg/kg, intravenous, methyl-Hg-Cl, 22 d, n = 2[4]
Mouse, kidney	21 mg/kg wet weight	3 mg Hg/kg, subcutanous, iso-butyl-Hg-acetate, 7 d, n = 3[4]
Mouse, kidney	22 mg/kg wet weight	3 mg Hg/kg, subcutanous, iso-propyl-Hg-acetate, 7 d, n = 3[4]
Mouse, kidney	32 mg/kg wet weight	3 mg Hg/kg, subcutanous, n-propyl-Hg-acetate, 7 d, n = 3[4]

Mouse, kidney	4.0 mg/kg wet weight	5.0 mg Hg/kg, intravenous, methyl-Hg-OH, 6 d, n = 10[4]
Mouse, kidney	40 mg/kg wet weight	2.5 mg Hg/kg day, subcutanous, methyl-Hg-acetate, 10 d, n = 3[4]
Mouse, kidney	60 mg/kg wet weight	5 mg Hg/kg day, subcutanous, methyl-Hg-acetate, 10 d, n = 3[4]
Mouse, liver	0.08 mg/kg wet weight	0.03 mg Hg/kg, intravenous, methyl-Hg-OH, 6 d, n = 10[4]
Mouse, liver	0.6 mg/kg wet weight	0.3 mg Hg/kg, intravenous, methyl-Hg-OH, 6 d, n = 10[4]
Mouse, liver	0.7 mg/kg wet weight	1.0 mg Hg/kg, intravenous, methyl-Hg-Cl, 22 d, n = 2[4]
Mouse, liver	11 mg/kg wet weight	3 mg Hg/kg, subcutanous, n-butyl-Hg-acetate, 7 d, n = 3[4]
Mouse, liver	15 mg/kg wet weight	3 mg Hg/kg, subcutanous, n-propyl-Hg-acetate, 7 d, n = 3[4]
Mouse, liver	16 mg/kg wet weight	3 mg Hg/kg, subcutanous, iso-butyl-Hg-acetate, 7 d, n = 3[4]
Mouse, liver	16 mg/kg wet weight	3 mg Hg/kg, subcutanous, iso-propyl-Hg-acetate, 7 d, n = 3[4]
Mouse, liver	2.1 mg/kg wet weight	1.0 mg Hg/kg, intravenous, methyl-Hg-OH, 6 d, n = 10[4]
Mouse, liver	20 mg/kg wet weight	2.5 mg Hg/kg day, subcutanous, methyl-Hg-acetate, 10 d, n = 3[4]
Mouse, liver	30 mg/kg wet weight	5 mg Hg/kg day, subcutanous, methyl-Hg-acetate, 10 d, n = 3[4]
Mouse, liver	6.3 mg/kg wet weight	3 mg Hg/kg, subcutanous, n-amyl-Hg-acetate, 7 d, n = 3[4]
Mouse, liver	9.8 mg/kg wet weight	5.0 mg Hg/kg, intravenous, methyl-Hg-OH, 6 d, n = 10[4]
Mouse, whole blood	0.02 mg/kg wet weight	0.03 mg Hg/kg, intravenous, methyl-Hg-OH, 6 d, n = 10[4]
Mouse, whole blood	0.2 mg/kg wet weight	0.3 mg Hg/kg, intravenous, methyl-Hg-OH, 6 d, n = 10[4]
Mouse, whole blood	0.33 mg/kg wet weight	3 mg Hg/kg, subcutanous, iso-propyl-Hg-acetate, 7 d, n = 3[4]
Mouse, whole blood	0.46 mg/kg wet weight	3 mg Hg/kg, subcutanous, n-amyl-Hg-acetate, 7 d, n = 3[4]
Mouse, whole blood	0.5 mg/kg wet weight	1.0 mg Hg/kg, intravenous, methyl-Hg-Cl, 22 d, n = 2[4]
Mouse, whole blood	0.5 mg/kg wet weight	1.0 mg Hg/kg, intravenous, methyl-Hg-OH, 6 d, n = 10[4]
Mouse, whole blood	0.80 mg/kg wet weight	3 mg Hg/kg, subcutanous, iso-butyl-Hg-acetate, 7 d, n = 3[4]
Mouse, whole blood	0.86 mg/kg wet weight	3 mg Hg/kg, subcutanous, n-butyl-Hg-acetate, 7 d, n = 3[4]
Mouse, whole blood	1.1 mg/kg wet weight	3 mg Hg/kg, subcutanous, n-propyl-Hg-acetate, 7 d, n = 3[4]
Mouse, whole blood	3.7 mg/kg wet weight	5.0 mg Hg/kg, intravenous, methyl-Hg-OH, 6 d, n = 10[4]
Mouse, whole blood	5 mg/kg wet weight	2.5 mg Hg/kg day, subcutanous, methyl-Hg-acetate, 10 d, n = 3[4]
Mouse, whole blood	9 mg/kg wet weight	5 mg Hg/kg day, subcutanous, methyl-Hg-acetate, 10 d, n = 3[4]
Mussel	12 mg/kg dry weight	Conc in sed = 5.8 mg/kg dw, filt seawater = 16 ppb, Derwent Est, Tasmania[11]
Mussel	13 mg/kg dry weight	Conc in sed = 13 mg/kg dw, filt seawater = 0.5 ppb, Derwent Est, Tasmania[11]
Mussel	3.6 mg/kg dry weight	Conc in sed = 12 mg/kg dw, filt seawater = 0.6 ppb, Derwent Est, Tasmania[11]
Mussel	4.8 mg/kg dry weight	Conc in sed = 12 mg/kg dw, filt seawater = 0.1 ppb, Derwent Est, Tasmania[11]
Mussel	5.0 mg/kg dry weight	Conc in sed = 10.5 mg/kg dw, filt seawater = 0.2 ppb, Derwent Est, Tasmania[11]
Mussel	5.1 mg/kg dry weight	Conc in sed = 1.6 mg/kg dw, filt seawater = 0.6 ppb, Derwent Est, Tasmania[11]
Mussel	7.7 mg/kg dry weight	Conc in sed = 3.6 mg/kg dw, filt seawater = 0.1 ppb, Derwent Est, Tasmania[11]
Pagophilus groenlandicus, adrenal	10.40 mg/kg wet weight	Methyl-Hg, oral dose = 25.0 mg/kg*d, 26 d, F031, immature[12]
Pagophilus groenlandicus, adrenal	10.75 mg/kg wet weight	Methyl-Hg, oral dose = 0.25 mg/kg*d, 60 d, F031, adult[12]

Pagophilus groenlandicus, adrenal	14.20 mg/kg wet weight	Total Hg, oral dose = 0.25 mg/kg*d, 60 d, F031, adult[12]
Pagophilus groenlandicus, adrenal	21.70 mg/kg wet weight	Methyl-Hg, oral dose = 0.25 mg/kg*d, 90 d, F031, control = n.d,[12]
Pagophilus groenlandicus, adrenal	25.00 mg/kg wet weight	Total Hg, oral dose = 0.25 mg/kg*d, 90 d, F031, control = n.d,[12]
Pagophilus groenlandicus, adrenal	26.00 mg/kg wet weight	Total Hg, oral dose = 25.0 mg/kg*d, 20 d, F031, immature[12]
Pagophilus groenlandicus, adrenal	34.90 mg/kg wet weight	Total Hg, oral dose = 25.0 mg/kg*d, 26 d, F031, immature[12]
Pagophilus groenlandicus, adrenal	7.50 mg/kg wet weight	Methyl-Hg, oral dose = 25.0 mg/kg*d, 20 d, F031, immature[12]
Pagophilus groenlandicus, blood	0.07 mg/l	Total Hg, oral dose = 0.25 mg/kg*d, 90 d, F031, control = 0.09[12]
Pagophilus groenlandicus, blood	0.07 mg/l	Total Hg, oral dose = 25.0 mg/kg*d, 60 d, F031, control = 0.11[12]
Pagophilus groenlandicus, blood	0.08 mg/l	Total Hg, oral dose = 25.0 mg/kg*d, 90 d, F031, control = 0.09[12]
Pagophilus groenlandicus, blood	0.09 mg/l	Methyl-Hg, oral dose = 0.25 mg/kg*d, 90 d, F031, control = 0.10[12]
Pagophilus groenlandicus, blood	0.10 mg/l	Methyl-Hg, oral dose = 25.0 mg/kg*d, 60 d, F031, control = 0.15[12]
Pagophilus groenlandicus, blood	0.10 mg/l	Methyl-Hg, oral dose = 25.0 mg/kg*d, 90 d, F031, control = 0.10[12]
Pagophilus groenlandicus, blood	0.14 mg/l	Tot Hg, oral dose = 0.25 mg/kg*d, 60 d, F031, cont = 0.11+-0.004[12]
Pagophilus groenlandicus, blood	0.16 mg/l	CH3Hg, oral dose = 0.25 mg/kg*d, 60 d, F031, cont = 0.15+-0.06,[12]
Pagophilus groenlandicus, blood	12.50 mg/kg wet weight	Methyl-Hg, oral dose = 0.25 mg/kg*d, 90 d, F031, control = 0.10 ppm[12]
Pagophilus groenlandicus, blood	13.10 mg/kg wet weight	Total Hg, oral dose = 0.25 mg/kg*d, 90 d, F031, control = 0.10 ppm[12]
Pagophilus groenlandicus, blood	19.00 mg/kg wet weight	Methyl-Hg, oral dose = 25.0 mg/kg*d, 20 d, F031, immature[12]
Pagophilus groenlandicus, blood	23.70 mg/kg wet weight	Methyl-Hg, oral dose = 25.0 mg/kg*d, 26 d, F031, immature[12]
Pagophilus groenlandicus, blood	28.80 mg/kg wet weight	Total Hg, oral dose = 25.0 mg/kg*d, 20 d, F031, immature[12]
Pagophilus groenlandicus, blood	30.30 mg/kg wet weight	Total Hg, oral dose = 25.0 mg/kg*d, 26 d, F031, immature[12]
Pagophilus groenlandicus, blood	8.85 mg/kg wet weight	Methyl-Hg, oral dose = 0.25 mg/kg*d, 60 d, F031, control = 0.10 ppm,[12]
Pagophilus groenlandicus, blood	9.93 mg/kg wet weight	Total Hg, oral dose = 0.25 mg/kg*d, 60 d, F031, control = 0.12 ppm[12]
Pagophilus groenlandicus, blubber	0.20 mg/kg wet weight	Total Hg, oral dose = 0.25 mg/kg*d, 90 d, F031, control = n.d[12]
Pagophilus groenlandicus, blubber	0.28 mg/kg wet weight	Total Hg, oral dose = 0.25 mg/kg*d, 60 d, F031, control = n.d, adult[12]
Pagophilus groenlandicus, blubber	0.55 mg/kg wet weight	Methyl-Hg, oral dose = 0.25 mg/kg*d, 90 d, F031, control = n.d[12]
Pagophilus groenlandicus, blubber	0.58 mg/kg wet weight	Methyl-Hg, oral dose = 0.25 mg/kg*d, 60 d, F031, control = 0.05 ppm[12]
Pagophilus groenlandicus, blubber	0.59 mg/kg wet weight	Methyl-Hg, oral dose = 25.0 mg/kg*d, 20 d, F031, immature[12]

Pagophilus groenlandicus, blubber	0.94 mg/kg wet weight	Total Hg, oral dose = 25.0 mg/kg*d, 26 d, F031, immature[12]
Pagophilus groenlandicus, blubber	1.00 mg/kg wet weight	Methyl-Hg, oral dose = 25.0 mg/kg*d, 26 d, F031, immature[12]
Pagophilus groenlandicus, blubber	1.71 mg/kg wet weight	Total Hg, oral dose = 25.0 mg/kg*d, 20 d, F031, immature[12]
Pagophilus groenlandicus, brain	14.80 mg/kg wet weight	Total Hg, oral dose = 0.25 mg/kg*d, 60 d, F031, control = 0.45 ppm[12]
Pagophilus groenlandicus, brain	21.80 mg/kg wet weight	Total Hg, oral dose = 0.25 mg/kg*d, 90 d, F031, control = 0.36 ppm[12]
Pagophilus groenlandicus, brain	23.80 mg/kg wet weight	Total Hg, oral dose = 25.0 mg/kg*d, 26 d, F031, immature[12]
Pagophilus groenlandicus, brain	42.90 mg/kg wet weight	Total Hg, oral dose = 25.0 mg/kg*d, 20 d, F031, immature[12]
Pagophilus groenlandicus, claws	21.90 mg/kg wet weight	Total Hg, oral dose = 0.25 mg/kg*d, 60 d, F031, control = 5.10 ppm[12]
Pagophilus groenlandicus, claws	22.30 mg/kg wet weight	Total Hg, oral dose = 0.25 mg/kg*d, 90 d, F031, control = 2.36 ppm[12]
Pagophilus groenlandicus, claws	38.80 mg/kg wet weight	Total Hg, oral dose = 25.0 mg/kg*d, 26 d, F031, immature[12]
Pagophilus groenlandicus, claws	81.80 mg/kg wet weight	Total Hg, oral dose = 25.0 mg/kg*d, 20 d, F031, immature[12]
Pagophilus groenlandicus, gonad	10.08 mg/kg wet weight	Methyl-Hg, oral dose = 0.25 mg/kg*d, 60 d, F031, control = n.d, adult[12]
Pagophilus groenlandicus, gonad	11.00 mg/kg wet weight	Methyl-Hg, oral dose = 0.25 mg/kg*d, 90 d, F031, immature[12]
Pagophilus groenlandicus, gonad	13.00 mg/kg wet weight	Total Hg, oral dose = 0.25 mg/kg*d, 60 d, F031, control = n.d, adult[12]
Pagophilus groenlandicus, gonad	21.50 mg/kg wet weight	Methyl-Hg, oral dose = 25.0 mg/kg*d, 20 d, F031, immature[12]
Pagophilus groenlandicus, gonad	27.00 mg/kg wet weight	Total Hg, oral dose = 25.0 mg/kg*d, 20 d, F031, immature[12]
Pagophilus groenlandicus, gonad	29.10 mg/kg wet weight	Methyl-Hg, oral dose = 25.0 mg/kg*d, 26 d, F031, immature[12]
Pagophilus groenlandicus, gonad	33.00 mg/kg wet weight	Total Hg, oral dose = 25.0 mg/kg*d, 26 d, F031, immature[12]
Pagophilus groenlandicus, hair	1.66 mg/kg wet weight	Total Hg, oral dose = 0.25 mg/kg*d, 90 d, F031, control = 1.26 ppm[12]
Pagophilus groenlandicus, hair	322.00 mg/kg wet weight	Total Hg, oral dose = 25.0 mg/kg*d, 20 d, F031, immature[12]
Pagophilus groenlandicus, hair	350.00 mg/kg wet weight	Methyl-Hg, oral dose = 25.0 mg/kg*d, 26 d, F031, immature[12]
Pagophilus groenlandicus, hair	4.10 mg/kg wet weight	Methyl-Hg, oral dose = 0.25 mg/kg*d, 90 d, F031, control = 1.50 ppm[12]
Pagophilus groenlandicus, hair	428.00 mg/kg wet weight	Methyl-Hg, oral dose = 25.0 mg/kg*d, 20 d, F031, immature[12]
Pagophilus groenlandicus, hair	434.00 mg/kg wet weight	Total Hg, oral dose = 25.0 mg/kg*d, 26 d, F031, immature[12]
Pagophilus groenlandicus, hair	7.30 mg/kg wet weight	Methyl-Hg, oral dose = 0.25 mg/kg*d, 60 d, F031, control = 1.27 ppm[12]
Pagophilus groenlandicus, hair	8.21 mg/kg wet weight	Total Hg, oral dose = 0.25 mg/kg*d, 60 d, F031, control = 9.17 ppm[12]
Pagophilus groenlandicus, heart	12.90 mg/kg wet weight	Total Hg, oral dose = 0.25 mg/kg*d, 60 d, F031, control = 0.25 ppm[12]

Pagophilus groenlandicus, heart	15.90 mg/kg wet weight	Total Hg, oral dose = 0.25 mg/kg*d, 90 d, F031, control = 0.38 ppm[12]
Pagophilus groenlandicus, heart	18.04 mg/kg wet weight	Methyl-Hg, oral dose = 0.25 mg/kg*d, 60 d, F031, adult[12]
Pagophilus groenlandicus, heart	40.20 mg/kg wet weight	Methyl-Hg, oral dose = 25.0 mg/kg*d, 20 d, F031, immature[12]
Pagophilus groenlandicus, heart	41.50 mg/kg wet weight	Methyl-Hg, oral dose = 25.0 mg/kg*d, 26 d, F031, immature[12]
Pagophilus groenlandicus, heart	46.30 mg/kg wet weight	Total Hg, oral dose = 25.0 mg/kg*d, 20 d, F031, immature[12]
Pagophilus groenlandicus, heart	50.90 mg/kg wet weight	Total Hg, oral dose = 25.0 mg/kg*d, 26 d, F031, immature[12]
Pagophilus groenlandicus, kidney	10.50 mg/kg wet weight	Methyl-Hg, oral dose = 0.25 mg/kg*d, 90 d, F031, control = 0.20 ppm[12]
Pagophilus groenlandicus, kidney	11.70 mg/kg wet weight	Methyl-Hg, oral dose = 25.0 mg/kg*d, 20 d, F031, immature[12]
Pagophilus groenlandicus, kidney	20.50 mg/kg wet weight	Methyl-Hg, oral dose = 25.0 mg/kg*d, 26 d, F031, immature[12]
Pagophilus groenlandicus, kidney	48.30 mg/kg wet weight	Total Hg, oral dose = 25.0 mg/kg*d, 20 d, F031, immature[12]
Pagophilus groenlandicus, kidney	50.60 mg/kg wet weight	Total Hg, oral dose = 0.25 mg/kg*d, 90 d, F031, control = 0.92 ppm[12]
Pagophilus groenlandicus, kidney	51.60 mg/kg wet weight	Methyl-Hg, oral dose = 0.25 mg/kg*d, 60 d, F031, control = 0.27 ppm[12]
Pagophilus groenlandicus, kidney	62.50 mg/kg wet weight	Total Hg, oral dose = 25.0 mg/kg*d, 26 d, F031, immature[12]
Pagophilus groenlandicus, kidney	69.50 mg/kg wet weight	Total Hg, oral dose = 0.25 mg/kg*d, 60 d, F031, control = 5.98 ppm[12]
Pagophilus groenlandicus, liver	125.00 mg/kg wet weight	Methyl-Hg, oral dose = 25.0 mg/kg*d, 26 d, F031, immature[12]
Pagophilus groenlandicus, liver	127.00 mg/kg wet weight	Methyl-Hg, oral dose = 25.0 mg/kg*d, 20 d, F031, immature[12]
Pagophilus groenlandicus, liver	134.00 mg/kg wet weight	Total Hg, oral dose = 25.0 mg/kg*d, 20 d, F031, immature[12]
Pagophilus groenlandicus, liver	142.00 mg/kg wet weight	Total Hg, oral dose = 25.0 mg/kg*d, 26 d, F031, immature[12]
Pagophilus groenlandicus, liver	18.48 mg/kg wet weight	Methyl-Hg, oral dose = 0.25 mg/kg*d, 60 d, F031, control = 0.16 ppm[12]
Pagophilus groenlandicus, liver	64.00 mg/kg wet weight	Total Hg, oral dose = 0.25 mg/kg*d, 60 d, F031, control = 25.8 ppm[12]
Pagophilus groenlandicus, liver	75.70 mg/kg wet weight	Methyl-Hg, oral dose = 0.25 mg/kg*d, 90 d, F031, control = 0.20 ppm[12]
Pagophilus groenlandicus, liver	82.50 mg/kg wet weight	Total Hg, oral dose = 0.25 mg/kg*d, 90 d, F031, control = 1.72 ppm[12]
Pagophilus groenlandicus, lung	10.60 mg/kg wet weight	Methyl-Hg, oral dose = 0.25 mg/kg*d, 90 d, F031, control = n.d[12]
Pagophilus groenlandicus, lung	10.60 mg/kg wet weight	Total Hg, oral dose = 0.25 mg/kg*d, 60 d, F031, control = 0.12 ppm[12]
Pagophilus groenlandicus, lung	18.20 mg/kg wet weight	Total Hg, oral dose = 0.25 mg/kg*d, 90 d, F031, control = 0.26 ppm[12]
Pagophilus groenlandicus, lung	24.80 mg/kg wet weight	Total Hg, oral dose = 25.0 mg/kg*d, 20 d, F031, immature[12]
Pagophilus groenlandicus, lung	30.80 mg/kg wet weight	Total Hg, oral dose = 25.0 mg/kg*d, 26 d, F031, immature[12]

Pagophilus groenlandicus, lung	5.70 mg/kg wet weight	Methyl-Hg, oral dose = 25.0 mg/kg*d, 26 d, F031, immature[12]
Pagophilus groenlandicus, lung	7.30 mg/kg wet weight	Methyl-Hg, oral dose = 25.0 mg/kg*d, 20 d, F031, immature[12]
Pagophilus groenlandicus, muscle	115.00 mg/kg wet weight	Methyl-Hg, oral dose = 25.0 mg/kg*d, 26 d, F031, immature[12]
Pagophilus groenlandicus, muscle	125.00 mg/kg wet weight	Total Hg, oral dose = 25.0 mg/kg*d, 26 d, F031, immature[12]
Pagophilus groenlandicus, muscle	27.60 mg/kg wet weight	Total Hg, oral dose = 0.25 mg/kg*d, 60 d, F031, control = 0.78 ppm[12]
Pagophilus groenlandicus, muscle	35.20 mg/kg wet weight	Methyl-Hg, oral dose = 0.25 mg/kg*d, 60 d, F031, control = 0.56 ppm[12]
Pagophilus groenlandicus, muscle	40.20 mg/kg wet weight	Methyl-Hg, oral dose = 0.25 mg/kg*d, 90 d, F031, control = 0.50 ppm[12]
Pagophilus groenlandicus, muscle	42.70 mg/kg wet weight	Total Hg, oral dose = 0.25 mg/kg*d, 90 d, F031, control = 0.54 ppm[12]
Pagophilus groenlandicus, muscle	91.50 mg/kg wet weight	Methyl-Hg, oral dose = 25.0 mg/kg*d, 20 d, F031, immature[12]
Pagophilus groenlandicus, muscle	96.70 mg/kg wet weight	Total Hg, oral dose = 25.0 mg/kg*d, 20 d, F031, immature[12]
Pagophilus groenlandicus, small intestine	0.50 mg/kg wet weight	Methyl-Hg, oral dose = 0.25 mg/kg*d, 60 d, F031, control = 0.09 ppm[12]
Pagophilus groenlandicus, small intestine	0.50 mg/kg wet weight	Methyl-Hg, oral dose = 0.25 mg/kg*d, 90 d, F031, control = n.d[12]
Pagophilus groenlandicus, small intestine	1367.00 mg/kg wet weight	Methyl-Hg, oral dose = 25.0 mg/kg*d, 26 d, F031, immature[12]
Pagophilus groenlandicus, small intestine	1660.00 mg/kg wet weight	Total Hg, oral dose = 25.0 mg/kg*d, 20 d, F031, immature[12]
Pagophilus groenlandicus, small intestine	17.10 mg/kg wet weight	Total Hg, oral dose = 0.25 mg/kg*d, 90 d, F031, control = 0.37 ppm[12]
Pagophilus groenlandicus, small intestine	17.20 mg/kg wet weight	Total Hg, oral dose = 0.25 mg/kg*d, 60 d, F031, control = 0.44 ppm[12]
Pagophilus groenlandicus, small intestine	1730.00 mg/kg wet weight	Total Hg, oral dose = 25.0 mg/kg*d, 26 d, F031, immature[12]
Pagophilus groenlandicus, spleen	14.50 mg/kg wet weight	Methyl-Hg, oral dose = 0.25 mg/kg*d, 90 d, F031, control = n.d[12]
Pagophilus groenlandicus, spleen	20.00 mg/kg wet weight	Total Hg, oral dose = 0.25 mg/kg*d, 90 d, F031, control = 0.33 ppm[12]
Pagophilus groenlandicus, spleen	20.40 mg/kg wet weight	Total Hg, oral dose = 0.25 mg/kg*d, 60 d, F031, control = 9.92 ppm[12]
Pagophilus groenlandicus, spleen	51.50 mg/kg wet weight	Methyl-Hg, oral dose = 25.0 mg/kg*d, 26 d, F031, immature[12]
Pagophilus groenlandicus, spleen	60.30 mg/kg wet weight	Methyl-Hg, oral dose = 25.0 mg/kg*d, 20 d, F031, immature[12]
Pagophilus groenlandicus, spleen	68.60 mg/kg wet weight	Total Hg, oral dose = 25.0 mg/kg*d, 26 d, F031, immature[12]
Pagophilus groenlandicus, spleen	81.80 mg/kg wet weight	Total Hg, oral dose = 25.0 mg/kg*d, 20 d, F031, immature[12]
Pig, brain	13 mg/kg wet weight	5.0 mg Hg/kg day, intramuscular, methyl-Hg-acetate, 7 d, n = 1[4]
Pig, brain	3.7 mg/kg wet weight	1.0 mg Hg/kg day, intramuscular, methyl-Hg-acetate, 7 d, n = 1[4]
Pig, kidney	12 mg/kg wet weight	1.0 mg Hg/kg day, intramuscular, methyl-Hg-acetate, 7 d, n = 1[4]
Pig, kidney	57 mg/kg wet weight	5.0 mg Hg/kg day, intramuscular, methyl-Hg-acetate, 7 d, n = 1[4]

Pig, liver	22 mg/kg wet weight	1.0 mg Hg/kg day, intramuscular, methyl-Hg-acetate, 7 d, n = 1[4]
Pig, liver	50 mg/kg wet weight	5.0 mg Hg/kg day, intramuscular, methyl-Hg-acetate, 7 d, n = 1[4]
Pig, whole blood	2.0 mg/kg wet weight	1.0 mg Hg/kg day, intramuscular, methyl-Hg-acetate, 7 d, n = 1[4]
Pig, whole blood	7.0 mg/kg wet weight	5.0 mg Hg/kg day, intramuscular, methyl-Hg-acetate, 7 d, n = 1[4]
Pisum sativum, pods	0.011 mg/kg dry weight	Control = 0.005, soil = 4 ppm, pH = 5.1, silt loam, CEC = .383 meq/g[2]
Pisum sativum, pods	0.042 mg/kg dry weight	Control = 0.005, soil = 20 ppm, pH = 5.1, silt loam, CEC = .383 meq/g[2]
Pisum sativum, root	0.160 mg/kg dry weight	Control = 0.011, soil = 4 ppm, pH = 5.1, silt loam, CEC = .383 meq/g[2]
Pisum sativum, root	1.415 mg/kg dry weight	Control = 0.011, soil = 20 ppm, pH = 5.1, silt loam, CEC = .383 meq/g[2]
Pisum sativum, seeds	0.002 mg/kg dry weight	Control = 0.001, soil = 4 ppm, pH = 5.1, silt loam, CEC = .383 meq/g[2]
Pisum sativum, seeds	0.003 mg/kg dry weight	Control = 0.001, soil = 20 ppm, pH = 5.1, silt loam, CEC = .383 meq/g[2]
Pisum sativum, vines	0.085 mg/kg dry weight	Control = 0.110, soil = 20 ppm, pH = 5.1, silt loam, CEC = .383 meq/g[2]
Pisum sativum, vines	0.187 mg/kg dry weight	Control = 0.110, soil = 4 ppm, pH = 5.1, silt loam, CEC = .383 meq/g[2]
Pseudopleuronectes americanus, blood	0.4 mg/kg wet weight	Hg = 0 ppb, 60 d, 3-6 fish, less than[13]
Pseudopleuronectes americanus, blood	2.9 mg/kg wet weight	Hg = 5 ppb, 60 d, 3-6 fish[13]
Pseudopleuronectes americanus, blood	3.8 mg/kg wet weight	Hg = 10 ppb, 60 d, 3-6 fish[13]
Pseudopleuronectes americanus, gill	0.14 mg/kg wet weight	Hg = 0 ppb, 60 d, 3-6 fish, less than[13]
Pseudopleuronectes americanus, gill	20.6 mg/kg wet weight	Hg = 5 ppb, 60 d, 3-6 fish[13]
Pseudopleuronectes americanus, gill	42.8 mg/kg wet weight	Hg = 10 ppb, 60 d, 3-6 fish[13]
Rabbit, brain	0.01 mg/kg wet weight	2 mg Hg/kg, subcutanous, phenyl-Hg-acetate, 40 d, n = 2[4]
Rabbit, kidney	6 mg/kg wet weight	2 mg Hg/kg, subcutanous, phenyl-Hg-acetate, 40 d, n = 2[4]
Rabbit, liver	0.6 mg/kg wet weight	2 mg Hg/kg, subcutanous, phenyl-Hg-acetate, 40 d, n = 2[4]
Rabbit, whole blood	0.01 mg/kg wet weight	2 mg Hg/kg, subcutanous, phenyl-Hg-acetate, 40 d, n = 2[4]
Raphanus sativus, tops	0.218 mg/kg dry weight	Control = 0.237, soil = 4 ppm, pH = 5.1, silt loam, CEC = .383 meq/g[2]
Raphanus sativus, tops	0.585 mg/kg dry weight	Control = 0.237, soil = 20 ppm, pH = 5.1, silt loam, CEC = .383 meq/g[2]
Raphanus sativus, tubers	0.026 mg/kg dry weight	Control = 0.013, soil = 4 ppm, pH = 5.1, silt loam, CEC = .383 meq/g[2]
Raphanus sativus, tubers	0.663 mg/kg dry weight	Control = 0.013, soil = 20 ppm, pH = 5.1, silt loam, CEC = .383 meq/g[2]
Rat, brain	0.002 mg/kg wet weight	0.03 mg Hg/kg, intravenous, methoxyethyl-Hg-OH, 12 d, n = 3[4]
Rat, brain	0.004 mg/kg wet weight	0.05 mg Hg/kg, intravenous, phenyl-Hg-OH, 4 d, n = 3[4]
Rat, brain	0.007 mg/kg wet weight	0.3 mg Hg/kg, intravenous, methoxyethyl-Hg-OH, 12 d, n = 3[4]

Rat, brain	0.009 mg/kg wet weight	0.05 mg Hg/kg, intravenous, methoxyethyl-Hg-OH, 4 d, n = 4[4]
Rat, brain	0.009 mg/kg wet weight	0.5 mg Hg/kg*d, subcutanous, methoxyethyl-Hg-OH, 12 d, n = 4[4]
Rat, brain	0.01 mg/kg wet weight	0.04 mg Hg/kg, intravenous, CH3HgOH, 6 d, n = 3[4]
Rat, brain	0.017 mg/kg wet weight	0.5 mg Hg/kg, intravenous, phenyl-Hg-OH, 4 d, n = 3[4]
Rat, brain	0.018 mg/kg wet weight	0.5 mg Hg/kg, intravenous, methoxyethyl-Hg-OH, 4 d, n = 4[4]
Rat, brain	0.018 mg/kg wet weight	0.5 mg Hg/kg, phenyl-Hg-OH, 16 d, n = 3[4]
Rat, brain	0.02 mg/kg wet weight	0.1 mg Hg/kg, intravenous, methyl-Hg-Cl, 32 d, n = 10[4]
Rat, brain	0.022 mg/kg wet weight	0.5 mg Hg/kg, intravenous, methoxyethyl-Hg-OH, 4 d, n = 4[4]
Rat, brain	0.027 mg/kg wet weight	0.03 mg Hg/kg, intravenous, methoxyethyl-Hg-OH, 3 d, n = 3[4]
Rat, brain	0.03 mg/kg wet weight	0.05 mg Hg/kg*d, subcutanous, phenyl-Hg-OH, 18 d, n = 4[4]
Rat, brain	0.036 mg/kg wet weight	0.3 mg Hg/kg, intravenous, methoxyethyl-Hg-OH, 3 d, n = 3[4]
Rat, brain	0.04 mg/kg wet weight	0.1 mg Hg/kg, intravenous, CH3HgOH, 4 d, n = 5[4]
Rat, brain	0.06 mg/kg wet weight	3 mg Hg/kg, oral, phenyl-Hg-acetate, 2 d, n = 2[4]
Rat, brain	0.076 mg/kg wet weight	3 mg Hg/kg, intravenous, methoxyethyl-Hg-OH, 12 d, n = 3[4]
Rat, brain	0.13 mg/kg wet weight	0.5 mg Hg/kg, intravenous, CH3HgOH, 4 d, n = 5[4]
Rat, brain	0.14 mg/kg wet weight	0.4 mg Hg/kg, intravenous, CH3HgOH, 6 d, n = 3[4]
Rat, brain	0.15 mg/kg wet weight	3 mg Hg/kg, intravenous, methoxyethyl-Hg-OH, 3 d, n = 3[4]
Rat, brain	0.16 mg/kg wet weight	10 mg Hg/kg, subcutanous, phenyl-HgCl, 8 d, n = 3[4]
Rat, brain	0.17 mg/kg wet weight	0.5 mg Hg/kg, intravenous, CH3HgOH, 16 d, n = 3[4]
Rat, brain	0.19 mg/kg wet weight	0.05 mg Hg/kg*d, subcutanous, CH3HgOH, 13 d, n = 4[4]
Rat, brain	0.2 mg/kg wet weight	20 mg Hg/kg, subcutanous, methoxyethyl-Hg-OH, 3 d, n = 20[4]
Rat, brain	0.47 mg/kg wet weight	5 mg Hg/kg, intravenous, phenyl-Hg-OH, 3 d, n = 3[4]
Rat, brain	0.5 mg/kg wet weight	25 mg Hg/kg, subcutanous, phenyl-Hg-OH, 3 d, n = 20[4]
Rat, brain	0.51 mg/kg wet weight	0.2 mg Hg/kg, intravenous, phenyl-Hg-acetate, 1 d, n = 4[4]
Rat, brain	0.7 mg/kg wet weight	3 mg Hg/kg, intramuscular, ethyl-HgCl, 7 d, n = 6[4]
Rat, brain	1 mg/kg wet weight	0.65 mg Hg/kg*d, subcutanous, phenyl-Hg-acetate, 42 d, n = 6[4]
Rat, brain	1.4 mg/kg wet weight	10 mg Hg/kg, subcutanous, ethyl-Hg-cysteine, 8 d, n = 3[4]
Rat, brain	1.4 mg/kg wet weight	10 mg Hg/kg, subcutanous, ethyl-HgCl, 8 d, n = 3[4]
Rat, brain	1.5 mg/kg wet weight	4.0 mg Hg/kg, intravenous, CH3HgOH, 6 d, n = 3[4]
Rat, brain	21 mg/kg wet weight	40 mg Hg/kg, subcutanous, CH3HgOH, 3 d, n = 20[4]
Rat, brain	4 mg/kg wet weight	0.65 mg Hg/kg*d, subcutanous, methyl-Hg-dicyandiamide, 42 d, n = 6[4]
Rat, kidney	0.26 mg/kg wet weight	0.1 mg Hg/kg, intravenous, methyl-Hg-Cl, 32 d, n = 10[4]
Rat, kidney	0.42 mg/kg wet weight	0.03 mg Hg/kg, intravenous, methoxyethyl-Hg-OH, 12 d, n = 3[4]
Rat, kidney	0.49 mg/kg wet weight	0.04 mg Hg/kg, intravenous, CH3HgOH, 6 d, n = 3[4]
Rat, kidney	0.59 mg/kg wet weight	0.1 mg Hg/kg, intravenous, CH3HgOH, 4 d, n = 5[4]

Rat, kidney	0.73 mg/kg wet weight	3 mg Hg/kg, intravenous, methoxyethyl-Hg-OH, 12 d, n = 3[4]
Rat, kidney	1.2 mg/kg wet weight	0.03 mg Hg/kg, intravenous, methoxyethyl-Hg-OH, 3 d, n = 3[4]
Rat, kidney	1.7 mg/kg wet weight	0.05 mg Hg/kg, intravenous, phenyl-Hg-OH, 4 d, n = 3[4]
Rat, kidney	1.7 mg/kg wet weight	0.5 mg Hg/kg, intravenous, CH3HgOH, 4 d, n = 5[4]
Rat, kidney	110 mg/kg wet weight	10 mg Hg/kg, subcutanous, ethyl-HgCl, 8 d, n = 3[4]
Rat, kidney	12 mg/kg wet weight	0.5 mg Hg/kg, intravenous, methoxyethyl-Hg-OH, 4 d, n = 4[4]
Rat, kidney	15 mg/kg wet weight	0.5 mg Hg/kg, intravenous, phenyl-Hg-OH, 4 d, n = 3[4]
Rat, kidney	16 mg/kg wet weight	0.5 mg Hg/kg, phenyl-Hg-OH, 16 d, n = 3[4]
Rat, kidney	17 mg/kg wet weight	0.5 mg Hg/kg, intravenous, methoxyethyl-Hg-OH, 4 d, n = 4[4]
Rat, kidney	18 mg/kg wet weight	10 mg Hg/kg, subcutanous, phenyl-HgCl, 8 d, n = 3[4]
Rat, kidney	19 mg/kg wet weight	3 mg Hg/kg, intravenous, methoxyethyl-Hg-OH, 3 d, n = 3[4]
Rat, kidney	2.2 mg/kg wet weight	0.4 mg Hg/kg, intravenous, CH3HgOH, 6 d, n = 3[4]
Rat, kidney	2.4 mg/kg wet weight	0.05 mg Hg/kg, intravenous, methoxyethyl-Hg-OH, 4 d, n = 4[4]
Rat, kidney	2.6 mg/kg wet weight	0.5 mg Hg/kg, intravenous, CH3HgOH, 16 d, n = 3[4]
Rat, kidney	22 mg/kg wet weight	4.0 mg Hg/kg, intravenous, CH3HgOH, 6 d, n = 3[4]
Rat, kidney	27 mg/kg wet weight	0.5 mg Hg/kg*d, subcutanous, methoxyethyl-Hg-OH, 12 d, n = 4[4]
Rat, kidney	27 mg/kg wet weight	3 mg Hg/kg, oral, phenyl-Hg-acetate, 2 d, n = 2[4]
Rat, kidney	29 mg/kg wet weight	20 mg Hg/kg, subcutanous, methoxyethyl-Hg-OH, 3 d, n = 20[4]
Rat, kidney	3.4 mg/kg wet weight	0.3 mg Hg/kg, intravenous, methoxyethyl-Hg-OH, 12 d, n = 3[4]
Rat, kidney	33 mg/kg wet weight	0.05 mg Hg/kg*d, subcutanous, phenyl-Hg-OH, 18 d, n = 4[4]
Rat, kidney	4.6 mg/kg wet weight	0.05 mg Hg/kg*d, subcutanous, CH3HgOH, 13 d, n = 4[4]
Rat, kidney	42 mg/kg wet weight	5 mg Hg/kg, intravenous, phenyl-Hg-OH, 3 d, n = 3[4]
Rat, kidney	51 mg/kg wet weight	0.65 mg Hg/kg*d, subcutanous, methyl-Hg-dicyandiamide, 42 d, n = 6[4]
Rat, kidney	52 mg/kg wet weight	25 mg Hg/kg, subcutanous, phenyl-Hg-OH, 3 d, n = 20[4]
Rat, kidney	69 mg/kg wet weight	3 mg Hg/kg, intramuscular, ethyl-HgCl, 7 d, n = 6[4]
Rat, kidney	7.6 mg/kg wet weight	0.3 mg Hg/kg, intravenous, methoxyethyl-Hg-OH, 3 d, n = 3[4]
Rat, kidney	9.1 mg/kg wet weight	0.2 mg Hg/kg, intravenous, phenyl-Hg-acetate, 1 d, n = 4[4]
Rat, kidney	90 mg/kg wet weight	0.65 mg Hg/kg*d, subcutanous, phenyl-Hg-acetate, 42 d, n = 6[4]
Rat, kidney	95 mg/kg wet weight	10 mg Hg/kg, subcutanous, ethyl-Hg-cysteine, 8 d, n = 3[4]
Rat, kidney	98 mg/kg wet weight	40 mg Hg/kg, subcutanous, CH3HgOH, 3 d, n = 20[4]
Rat, liver	0.005 mg/kg wet weight	0.03 mg Hg/kg, intravenous, methoxyethyl-Hg-OH, 12 d, n = 3[4]
Rat, liver	0.021 mg/kg wet weight	0.3 mg Hg/kg, intravenous, methoxyethyl-Hg-OH, 12 d, n = 3[4]
Rat, liver	0.04 mg/kg wet weight	0.04 mg Hg/kg, intravenous, CH3HgOH, 6 d, n = 3[4]
Rat, liver	0.047 mg/kg wet weight	0.03 mg Hg/kg, intravenous, methoxyethyl-Hg-OH, 3 d, n = 3[4]
Rat, liver	0.061 mg/kg wet weight	0.05 mg Hg/kg, intravenous, methoxyethyl-Hg-OH, 4 d, n = 4[4]
Rat, liver	0.08 mg/kg wet weight	0.1 mg Hg/kg, intravenous, methyl-Hg-Cl, 32 d, n = 10[4]
Rat, liver	0.081 mg/kg wet weight	0.05 mg Hg/kg, intravenous, phenyl-Hg-OH, 4 d, n = 3[4]

Rat, liver	0.083 mg/kg wet weight	0.5 mg Hg/kg, phenyl-Hg-OH, 16 d, n = 3[4]
Rat, liver	0.17 mg/kg wet weight	0.1 mg Hg/kg, intravenous, CH3HgOH, 4 d, n = 5[4]
Rat, liver	0.17 mg/kg wet weight	0.3 mg Hg/kg, intravenous, methoxyethyl-Hg-OH, 3 d, n = 3[4]
Rat, liver	0.24 mg/kg wet weight	0.5 mg Hg/kg, intravenous, methoxyethyl-Hg-OH, 4 d, n = 4[4]
Rat, liver	0.25 mg/kg wet weight	0.5 mg Hg/kg*d, subcutanous, methoxyethyl-Hg-OH, 12 d, n = 4[4]
Rat, liver	0.30 mg/kg wet weight	0.5 mg Hg/kg, intravenous, methoxyethyl-Hg-OH, 4 d, n = 4[4]
Rat, liver	0.33 mg/kg wet weight	0.5 mg Hg/kg, intravenous, phenyl-Hg-OH, 4 d, n = 3[4]
Rat, liver	0.36 mg/kg wet weight	0.05 mg Hg/kg*d, subcutanous, phenyl-Hg-OH, 18 d, n = 4[4]
Rat, liver	0.41 mg/kg wet weight	0.4 mg Hg/kg, intravenous, CH3HgOH, 6 d, n = 3[4]
Rat, liver	0.48 mg/kg wet weight	0.5 mg Hg/kg, intravenous, CH3HgOH, 16 d, n = 3[4]
Rat, liver	0.51 mg/kg wet weight	3 mg Hg/kg, intravenous, methoxyethyl-Hg-OH, 12 d, n = 3[4]
Rat, liver	0.52 mg/kg wet weight	0.2 mg Hg/kg, intravenous, phenyl-Hg-acetate, 1 d, n = 4[4]
Rat, liver	0.52 mg/kg wet weight	0.5 mg Hg/kg, intravenous, CH3HgOH, 4 d, n = 5[4]
Rat, liver	0.58 mg/kg wet weight	3 mg Hg/kg, oral, phenyl-Hg-acetate, 2 d, n = 2[4]
Rat, liver	0.92 mg/kg wet weight	0.05 mg Hg/kg*d, subcutanous, CH3HgOH, 13 d, n = 4[4]
Rat, liver	1.3 mg/kg wet weight	0.65 mg Hg/kg*d, subcutanous, phenyl-Hg-acetate, 42 d, n = 6[4]
Rat, liver	1.3 mg/kg wet weight	10 mg Hg/kg, subcutanous, phenyl-HgCl, 8 d, n = 3[4]
Rat, liver	11 mg/kg wet weight	10 mg Hg/kg, subcutanous, ethyl-HgCl, 8 d, n = 3[4]
Rat, liver	11 mg/kg wet weight	5 mg Hg/kg, intravenous, phenyl-Hg-OH, 3 d, n = 3[4]
Rat, liver	12 mg/kg wet weight	10 mg Hg/kg, subcutanous, ethyl-Hg-cysteine, 8 d, n = 3[4]
Rat, liver	16 mg/kg wet weight	0.65 mg Hg/kg*d, subcutanous, methyl-Hg-dicyandiamide, 42 d, n = 6[4]
Rat, liver	19 mg/kg wet weight	25 mg Hg/kg, subcutanous, phenyl-Hg-OH, 3 d, n = 20[4]
Rat, liver	4.0 mg/kg wet weight	3 mg Hg/kg, intravenous, methoxyethyl-Hg-OH, 3 d, n = 3[4]
Rat, liver	4.2 mg/kg wet weight	4.0 mg Hg/kg, intravenous, CH3HgOH, 6 d, n = 3[4]
Rat, liver	44 mg/kg wet weight	20 mg Hg/kg, subcutanous, methoxyethyl-Hg-OH, 3 d, n = 20[4]
Rat, liver	5.3 mg/kg wet weight	3 mg Hg/kg, intramuscular, ethyl-HgCl, 7 d, n = 6[4]
Rat, liver	77 mg/kg wet weight	40 mg Hg/kg, subcutanous, CH3HgOH, 3 d, n = 20[4]
Rat, whole blood	0.001 mg/kg wet weight	0.03 mg Hg/kg, intravenous, methoxyethyl-Hg-OH, 12 d, n = 3[4]
Rat, whole blood	0.004 mg/kg wet weight	0.3 mg Hg/kg, intravenous, methoxyethyl-Hg-OH, 12 d, n = 3[4]
Rat, whole blood	0.006 mg/kg wet weight	0.05 mg Hg/kg, intravenous, phenyl-Hg-OH, 4 d, n = 3[4]
Rat, whole blood	0.009 mg/kg wet weight	0.05 mg Hg/kg, intravenous, methoxyethyl-Hg-OH, 4 d, n = 4[4]
Rat, whole blood	0.033 mg/kg wet weight	0.5 mg Hg/kg*d, subcutanous, methoxyethyl-Hg-OH, 12 d, n = 4[4]
Rat, whole blood	0.039 mg/kg wet weight	0.05 mg Hg/kg*d, subcutanous, phenyl-Hg-OH, 18 d, n = 4[4]
Rat, whole blood	0.04 mg/kg wet weight	0.5 mg Hg/kg, intravenous, phenyl-Hg-OH, 4 d, n = 3[4]
Rat, whole blood	0.044 mg/kg wet weight	0.03 mg Hg/kg, intravenous, methoxyethyl-Hg-OH, 3 d, n = 3[4]

Rat, whole blood	0.047 mg/kg wet weight	0.5 mg Hg/kg, phenyl-Hg-OH, 16 d, n = 3[4]
Rat, whole blood	0.05 mg/kg wet weight	0.1 mg Hg/kg, intravenous, methyl-Hg-Cl, 32 d, n = 10[4]
Rat, whole blood	0.054 mg/kg wet weight	0.5 mg Hg/kg, intravenous, methoxyethyl-Hg-OH, 4 d, n = 4[4]
Rat, whole blood	0.066 mg/kg wet weight	3 mg Hg/kg, intravenous, methoxyethyl-Hg-OH, 12 d, n = 3[4]
Rat, whole blood	0.068 mg/kg wet weight	0.5 mg Hg/kg, intravenous, methoxyethyl-Hg-OH, 4 d, n = 4[4]
Rat, whole blood	0.082 mg/kg wet weight	0.3 mg Hg/kg, intravenous, methoxyethyl-Hg-OH, 3 d, n = 3[4]
Rat, whole blood	0.14 mg/kg wet weight	0.04 mg Hg/kg, intravenous, CH3HgOH, 6 d, n = 3[4]
Rat, whole blood	0.22 mg/kg wet weight	0.2 mg Hg/kg, intravenous, phenyl-Hg-acetate, 1 d, n = 4[4]
Rat, whole blood	0.29 mg/kg wet weight	10 mg Hg/kg, subcutanous, phenyl-HgCl, 8 d, n = 3[4]
Rat, whole blood	0.43 mg/kg wet weight	0.1 mg Hg/kg, intravenous, CH3HgOH, 4 d, n = 5[4]
Rat, whole blood	0.5 mg/kg wet weight	0.65 mg Hg/kg*d, subcutanous, phenyl-Hg-acetate, 42 d, n = 6, app[4]
Rat, whole blood	0.74 mg/kg wet weight	3 mg Hg/kg, oral, phenyl-Hg-acetate, 2 d, n = 2[4]
Rat, whole blood	1.1 mg/kg wet weight	3 mg Hg/kg, intravenous, methoxyethyl-Hg-OH, 3 d, n = 3[4]
Rat, whole blood	1.7 mg/kg wet weight	0.4 mg Hg/kg, intravenous, CH3HgOH, 6 d, n = 3[4]
Rat, whole blood	1.8 mg/kg wet weight	0.5 mg Hg/kg, intravenous, CH3HgOH, 16 d, n = 3[4]
Rat, whole blood	11 mg/kg wet weight	20 mg Hg/kg, subcutanous, methoxyethyl-Hg-OH, 3 d, n = 20[4]
Rat, whole blood	14 mg/kg wet weight	3 mg Hg/kg, intramuscular, ethyl-HgCl, 7 d, n = 6[4]
Rat, whole blood	19 mg/kg wet weight	10 mg Hg/kg, subcutanous, ethyl-HgCl, 8 d, n = 3[4]
Rat, whole blood	19 mg/kg wet weight	4.0 mg Hg/kg, intravenous, CH3HgOH, 6 d, n = 3[4]
Rat, whole blood	2.2 mg/kg wet weight	0.5 mg Hg/kg, intravenous, CH3HgOH, 4 d, n = 5[4]
Rat, whole blood	23 mg/kg wet weight	10 mg Hg/kg, subcutanous, ethyl-Hg-cysteine, 8 d, n = 3[4]
Rat, whole blood	290 mg/kg wet weight	40 mg Hg/kg, subcutanous, CH3HgOH, 3 d, n = 20[4]
Rat, whole blood	3 mg/kg wet weight	0.05 mg Hg/kg*d, subcutanous, CH3HgOH, 13 d, n = 4[4]
Rat, whole blood	4.4 mg/kg wet weight	5 mg Hg/kg, intravenous, phenyl-Hg-OH, 3 d, n = 3[4]
Rat, whole blood	40 mg/kg wet weight	0.65 mg Hg/kg*d, subcutanous, methyl-Hg-dicyandiamide, 42 d, n=6[4]
Rat, whole blood	6 mg/kg wet weight	25 mg Hg/kg, subcutanous, phenyl-Hg-OH, 3 d, n = 20[4]
Saimiri sciureus, brain	14 mg/kg wet weight	0.3-0.7 mg Hg/kg*d, oral, CH3HgOH, 28 d, n = 1, killed 4 d , F082[4]
Saimiri sciureus, brain	18 mg/kg wet weight	0.3-0.7 mg Hg/kg*d, oral, CH3HgOH, 35-36 d, n = 2, killed 1 & 9 d , F082[4]
Saimiri sciureus, brain	2.5 mg/kg wet weight	0.3-0.7 mg Hg/kg*d, oral, CH3HgOH, 21 d, n = 2, killed 85 d , F082[4]
Saimiri sciureus, kidney	12 mg/kg wet weight	0.3-0.7 mg Hg/kg*d, oral, CH3HgOH, 28 d, n = 1, killed 4 d , F082[4]
Saimiri sciureus, kidney	4.9 mg/kg wet weight	0.3-0.7 mg Hg/kg*d, oral, CH3HgOH, 21 d, n = 2, killed 85 d , F082[4]
Saimiri sciureus, kidney	6.0 mg/kg wet weight	0.3-0.7 mg Hg/kg*d, oral, CH3HgOH, 35-36 d, n = 2, killed 1 & 9 d , F082[4]
Saimiri sciureus, liver	2.7 mg/kg wet weight	0.3-0.7 mg Hg/kg*d, oral, CH3HgOH, 21 d, n = 2, killed 85 d , F082[4]

Saimiri sciureus, liver	24 mg/kg wet weight	0.3-0.7 mg Hg/kg*d, oral, CH3HgOH, 28 d, n = 1, killed 4 d , F082[4]
Saimiri sciureus, liver	9.8 mg/kg wet weight	0.3-0.7 mg Hg/kg*d, oral, CH3HgOH, 35-36 d, n = 2, killed 1 & 9 d , F082[4]
Saimiri sciureus, whole blood	0.3 mg/kg wet weight	0.3-0.7 mg Hg/kg*d, oral, CH3HgOH, 21 d, n = 2, killed 85 d , F082[4]
Saimiri sciureus, whole blood	1.5 mg/kg wet weight	0.3-0.7 mg Hg/kg*d, oral, CH3HgOH, 35-36 d, n = 2, killed 1 & 9 d , F082[4]
Saimiri sciureus, whole blood	1.9 mg/kg wet weight	0.3-0.7 mg Hg/kg*d, oral, CH3HgOH, 28 d, n = 1, killed 4 d , F082[4]
Sarcosaprophagous flies	0.62 mg/kg dry weight	Food = fish = 0.47 mg/kg dry weight, full-grown fly larvae[14]
Sarcosaprophagous flies	6.90 mg/kg dry weight	Food = fish = 1.90 mg/kg dry weight, full-grown fly larvae[14]
Sediment	0.06 mg/kg dry weight	Unpolluted area, estuary, Urr water, rg = 0.02 - 0.17[15]
Sediment	0.07 mg/kg dry weight	Unpolluted area, estuary, Urr water, rg = 0.02 - 0.33[16]
Sediment	0.07 mg/kg dry weight	Unpolluted area, estuary, Urr water, rg = 0.03 - 0.17[15]
Spinacia oleracea, leaves	0.339 mg/kg dry weight	Control = 0.094, soil = 4 ppm, pH = 5.1, silt loam, CEC = .383 meq/g[2]
Spinacia oleracea, leaves	0.695 mg/kg dry weight	Control = 0.094, soil = 20 ppm, pH = 5.1, silt loam, CEC = .383 meq/g[2]
Spinacia oleracea, root	1.022 mg/kg dry weight	Control = 0.095, soil = 4 ppm, pH = 5.1, silt loam, CEC = .383 meq/g[2]
Spinacia oleracea, root	1.067 mg/kg dry weight	Control = 0.095, soil = 20 ppm, pH = 5.1, silt loam, CEC = .383 meq/g2
Tringa totanus, liver	1.6 mg/kg dry weight	3 birds(killed), 29.8.72, British Isles[5]
Tringa totanus, liver	12.3 mg/kg dry weight	5 birds(2 killed, 3 found dead), 5.2-3.3-72, British Isles[5]
Tringa totanus, liver	2.7 mg/kg dry weight	3 birds(found dead), 4-29.11.72, British Isles[5]
Uca pugilator, gill and hepatopancreas	0.40 mg/kg dry weight	Hg = 0.18 mg/l, 308 K, 72 h, S = 3.0 %[17]
Uca pugilator, gill and hepatopancreas	0.41 mg/kg dry weight	Hg = 0.18 mg/l, 278 K, 72 h, S = 3.0 %[17]
Uca pugilator, gill and hepatopancreas	0.44 mg/kg dry weight	Hg = 0.18 mg/l, 298 K, 72 h, S = 0.5 %[17]
Uca pugilator, gill and hepatopancreas	0.44 mg/kg dry weight	Hg = 0.18 mg/l, 298 K, 72 h, S = 3.0 %[17]
Uca pugilator, gill and hepatopancreas	0.47 mg/kg dry weight	Hg = 0.18 mg/l, 278 K, 72 h, S = 0.5 %[17]
Uca pugilator, gill and hepatopancreas	0.47 mg/kg dry weight	Hg = 0.18 mg/l, 308 K, 72 h, S = 0.5 %[17]

Source: 1.Foerster, U. and G. Muller. 1975. GWF-Wasser Abwasser 116, pp 74-80.
2.John, M.K. 1972. Bull. Environ. Contam. Toxicol. 8(2), pp 77-80.
3.Heinrichs, H. and R. Meyer. 1977. J. Environ. Qual. 6, 402-407.
4.Friberg, L. and J. Vostal. 1974. Mercury in the Environment. CRC-Press.
5.Parslow, J.L.E. 1974. Environ. Pollut., 5, pp 295-304.
6.Colwell, R.R., G.S. Sayler, J.D. Nelson and A. Justice. 1976. In: J.O. Nriagu (ed.): Environmental Biochemistry, Vol 2. Ann Arbor Science. Ann Arbor. Mich. 473-487.
7.Walter, C.M., H.G. Brown and C.P. Hensley. 1974. Water Res., 8(7), pp 413-418.
8.Foestner, U. and G.T.W. Wittmann. 1979. Metal Pollution in the Aquatic Environment. Springer-Verlag.Berlin.
9.Commission European Communities. 1978. Criteria (Dose/effect relationships) for Cd. Pergamob. Oxford.
10.Fang, S.C. 1974. Environ. Res., 8(1), pp 112-118.
11.Bloom, H. and G.M. Ayling. 1977. Environ. Geol. 2, pp 3-22.
12.Ronald, K., S.V. Tessaro, J.F. Uthe, H.C. Freeman and R. Frank. 1977. Sci. Total Environ. 8, pp 1-11
13.Calabrise, A., M.A. Dawson, F.P. Thurberg and D.R. Wenzloff. 1975. In; Koeman, J.H. et al. (eds); Sublethal Effects of Toxic Chemicals on Aquatic Animals. Elsevier N.Y. pp 15-21.
14.Nuorteva, P. and E. Hasanen. 1972. Ann. Zool. Fenn. 9, pp 23-27.
15.Taylor, D. 1976. Sci. Total Environ. 6(3), pp 259-264.
16.Taylor, D. 1976. Estuarine Coastal Mar. Sci. 2, pp 417.
17.Vernberg, W.B., P.J. DeCoursey and J. OHara. 1974. In: F.J. Vernberg and W.B. Vernberg (eds.): Pollution Physiology of Marine Organisms. Academic Press. NY, pp 381-425.

5-357 Concentrations Humic

Species	Value	Condition
Different species in freshwater, Free Humic	75.2 mol %	298 K, analytic data, 1 atm, freshwater, L. Celyn[1]
Different species in freshwater, Humic bound with Ca	15.0 mol %	298 K, analytic data, 1 atm, freshwater, L. Celyn[1]
Different species in freshwater, Humic bound with Cd	0.1 mol %	298 K, analytic data, 1 atm, freshwater, L. Celyn, less than value[1]
Different species in freshwater, Humic bound with Cu	0.8 mol %	298 K, analytic data, 1 atm, freshwater, L. Celyn[1]
Different species in freshwater, Humic bound with Mg	6.4 mol %	298 K, analytic data, 1 atm, freshwater, L. Celyn[1]
Different species in freshwater, Humic bound with Mn	1.5 mol %	298 K, analytic data, 1 atm, freshwater, L. Celyn[1]
Different species in freshwater, Humic bound with Ni	0.1 mol %	298 K, analytic data, 1 atm, freshwater, L. Celyn[1]
Different species in freshwater, Humic bound with Zn	1.0 mol %	298 K, analytic data, 1 atm, freshwater, L. Celyn[1]

Source: 1.Mantoura, R.F.C., A. Dickson and J.P. Riley. 1978. Eustarine Coastal Mar. Sci. 6, 387-408.

5-358 Concentrations Humus

Species	Value	Condition
C	55 % of dry weight	Roughly estimate, rg = 50 - 60[1]
H	5 % of dry weight	Roughly estimate[1]
Nitrogen	4 % of dry weight	Roughly estimate, rg = 3 - 5[1]
Oxygen	35 % of dry weight	Roughly estimate, rg = 30 - 40[1]

Source: 1.Golterman, H.L. 1975. Physiological Limnol., Elsevier Amsterdam.

5-359 Concentrations Hydrocarbons

Species	Value	Condition
Distribution in aerosols above seawater	13 % as aromatic hydrocarbons	E Trop. Atl., height = 12 m, surface sample, tot hydrocarbons = 9 ug/m3[1]
Distribution in aerosols above seawater	2 % as n-alkanes	E Trop. Atl., height = 12 m, surface sample, tot hydrocarbons = 9 ug/m3[1]
Distribution in aerosols above seawater	87 % as aliphatic hydrocarbons	E Trop. Atl., height = 12 m, surface sample, tot hydrocarbons = 9 ug/m3[1]
Distribution in seawater	33 % as aromatic hydrocarbons	E Tropical Atlantic Ocean, surface sample, tot hydrocarbons = 10 ug/l[1]
Distribution in seawater	6 % as n-alkanes	E Tropical Atlantic Ocean, surface sample, tot hydrocarbons = 10 ug/l[1]
Distribution in seawater	67 % as aliphatic hydrocarbons	E Tropical Atlantic Ocean, surface sample, tot hydrocarbons = 10 ug/l[1]

Distribution in seawater surface microlayer	3 % as n-alkanes	E Tropical Atlantic Ocean, surface sample, tot hydrocarbons = 39 ug/l[1]
Distribution in seawater surface microlayer	45 % as aliphatic hydrocarbons	E Tropical Atlantic Ocean, surface sample, tot hydrocarbons = 39 ug/l[1]
Distribution in seawater surface microlayer	55 % as aromatic hydrocarbons	E Tropical Atlantic Ocean, surface sample, tot hydrocarbons = 39 ug/l[1]
Mytilus edulis	225 mg/kg wet weight	Uptake from water = 302 ug/l, 30 days, extrac = n-hexane, Nov-Dec[2]
Mytilus edulis	353 mg/kg wet weight	Uptake from water = 302 ug/l, 80 days, extrac = n-hexane, June-Aug[2]
Mytilus edulis	407 mg/kg wet weight	Uptake from water = 302 ug/l, 55 days, extrac = n-hexane, March-May[2]

Source: 1.Marthy, J.-C., A. Saliot and M.J. Tissier. 1977. C.R. Acad. Sci. (Paris) ser. D 286, 833-836.
2.Fossato, V.U. and W.J. Canzonier. 1976. Mar Biol., 36, pp 243-250.

5-360 Concentrations
Hydroxybenzo(a)pyrene

Species	Value	Condition
Callinectes sapidus, blood	12E-4 ug total in tissue	Water = 2.5 ug BaP/l, juvenile, marine, n=3, 2 d exp,+-85E-4[1]
Callinectes sapidus, blood	21E-4 ug total in tissue	2 d upt. from wat=2.5 ug BaP/l, init=12, juv, marine, n=3, 2 d excr[1]
Callinectes sapidus, blood	27E-4 ug total in tissue	Water = 2.5 ug BaP/l, juvenile, marine, n=3, 1 d exp, +-70E-4[1]
Callinectes sapidus, gill	10E-4 ug total in tissue	Water = 2.5 ug BaP/l, juvenile, marine, n=3, 1 d exp,+-70E-4[1]
Callinectes sapidus, gill	15E-4 ug total in tissue	Water = 2.5 ug BaP/l, juvenile, marine, n=3, 2 d exp, +-85E-4[1]
Callinectes sapidus, gill	5E-4 ug total in tissue	2 d upt. from wat=2.5 ug BaP/l, init=15, juv, marine, n=3, 2 d excr[1]
Callinectes sapidus, hepatopancreas	20E-4 ug total in tissue	Water = 2.5 ug BaP/l, juvenile, marine, n=3, 1 d exp, +-70E-4[1]
Callinectes sapidus, hepatopancreas	84E-4 ug total in tissue	Water = 2.5 ug BaP/l, juvenile, marine, n=3, 2 d exp, +-85E-4[1]
Callinectes sapidus, hepatopancreas	90E-4 ug total in tissue	2 d upt. from wat=2.5 ug BaP/l, init=84, juv, marine, n=3, 2 d excr[1]
Callinectes sapidus, muscle	1E-4 ug total in tissue	2 d upt. from wat=2.5 ug BaP/l, init=2, juv, marine, n=3, 2 d excr[1]
Callinectes sapidus, muscle	2E-4 ug total in tissue	Water = 2.5 ug BaP/l, juvenile, marine, n=3, 1 d exp, +-70E-4[1]
Callinectes sapidus, muscle	2E-4 ug total in tissue	Water = 2.5 ug BaP/l, juvenile, marine, n=3, 2 d exp, +-85E-4[1]
Callinectes sapidus, stomach	2E-4 ug total in tissue	2 d upt. from wat=2.5 ug BaP/l, init=2, juv, marine, n=3, 2 d excr[1]
Callinectes sapidus, stomach	2E-4 ug total in tissue	Water = 2.5 ug BaP/l, juvenile, marine, n=3, 1 d exp, +-70E-4[1]
Callinectes sapidus, stomach	2E-4 ug total in tissue	Water = 2.5 ug BaP/l, juvenile, marine, n=3, 2 d exp, +-85E-4[1]

Source: 1.Lee, R.F., C. Ryan and M.L. Neuhauser. 1976. Mar. Biol. 37, 363-370.

5-361 Concentrations
Insecticides

Species	Value	Condition
Buffalo	0.032 mg/kg dry weight	S Dakota, freshwater, water = 0.230 mg/m3, total insect., n = 10[1]
Bullhead	0.031 mg/kg dry weight	S Dakota, freshwater, water = 0.230 mg/m3, total insecticides, n = 10[1]

Carp	0.092 mg/kg dry weight	S Dakota, freshwater, water = 0.230 mg/m3, total insecticides, n = 10[1]
Crappie	0.016 mg/kg dry weight	S Dakota, freshwater, water = 0.230 mg/m3, total insecticides, n = 10[1]
Sediment	0.0016 mg/kg wet weight	S Dakota, freshwater, water = 0.230 mg/m3, total insecticides, n = 10[1]

Source: 1.Greichus, Y.A., A. Greichus and R.J. Emerick. 1973. Bull. Environ. Contam. Toxicol. 9, pp 321-327.

5-362 Concentrations K

Species	Value	Condition
Particle fractions	2.0 % of total	Size = 2.0-0.5 mm, weight = 3.2 %, area = 0.2 %, marine, Japan, K-40[1]
Particle fractions	20.1 % of total	Size = 0.125-0.075 mm, weight = 16.9 %, area = 10.4 %, marine, Japan, K-40[1]
Particle fractions	23.1 % of total	Size = 0.25-0.125 mm, weight = 23 %, area = 10.1 %, marine, Japan, K-40[1]
Particle fractions	46.3 % of total	Size below 0.075 mm, weight = 48.4 %, area = 77.9 %, marine, Japan, K-40[1]
Particle fractions	8.5 % of total	Size = 0.5-0.25 mm, weight = 8.5 %, area = 1.4 %, marine, Japan, K-40[1]

Source: 1.Nagaya, Y. and K. Nakamura. 1975. Im pacts of Nuclear releases into the Aquatic Environment. IAEA-SM198/21, pp 313-328. Vienna.

5-363 Concentrations Mercapturic acid

Species	Value	Condition
Oncorhynchus kisutch, brain	11 % of total	Distribution 24 h after intraperi. inj. of 2.5 uCi naphthalene-1-14C[1]
Oncorhynchus kisutch, flesh	6 % of total	Distribution 24 h after intraperi. inj. of 2.5 uCi naphthalene-1-14C[1]
Oncorhynchus kisutch, gall bladder	9 % of total	Distribution 24 h after intraperi. inj. of 2.5 uCi naphthalene-1-14C[1]
Oncorhynchus kisutch, heart	0 % of total	Distribution 24 h after intraperi. inj. of 2.5 uCi naphthalene-1-14C[1]
Oncorhynchus kisutch, liver	22 % of total	Distribution 24 h after intraperi. inj. of 2.5 uCi naphthalene-1-14C[1]

Source: 1.Roubal, W.T., D.H. Bovee, T.K. Collier and S.I. Stranahan. 1977. Proc of 1977 Oil Spill Conf. (Prevention, Behavior, Control, Cleanup). Wash DC. American Petroleum Institute. 551-556.

5-364 Concentrations Methoxychlor

Species	Value	Condition
Chicken, fat	0.10 mg/kg wet weight	Conc in food = 8.00 mg/kg wet weight[1]
Chicken, fat	1.26 mg/kg wet weight	Conc in food = 100.00 mg/kg wet weight[1]
Chicken, fat	1.40 mg/kg wet weight	Conc in food = 100 mg/kg wet weight[1]
Chicken, fat	54.10 mg/kg wet weight	Conc in food = 1000 mg/kg wet weight[1]
Chicken, liver	0.68 mg/kg wet weight	Conc in food = 1000 mg/kg wet weight[1]
Chicken, muscle	0.18 mg/kg wet weight	Conc in food = 1000 mg/kg wet weight[1]
Chicken, tissue	0.56 mg/kg wet weight	Conc in food = 100.00 mg/kg wet weight[1]

Chicken, tissue	18.90 mg/kg wet weight	Conc in food = 1000 mg/kg wet weight[1]
Rhizophora mangle, hypocotyl	0.03 mg/kg wet weight	Application = 0.28 kg/ha, pH = 6.3-6.7, s = 2.5 %, 6 w[2]
Rhizophora mangle, hypocotyl	0.26 mg/kg wet weight	Application = 0.56 kg/ha, pH = 6.3-6.7, s = 2.5 %, 6 w[2]
Rhizophora mangle, hypocotyl	0.26 mg/kg wet weight	Application = 1.12 kg/ha, pH = 6.3-6.7, s = 2.5 %, 6 w[2]
Rhizophora mangle, hypocotyl	0.29 mg/kg wet weight	Application = 2.80 kg/ha, pH = 6.3-6.7, s = 2.5 %, 6 w[2]
Rhizophora mangle, hypocotyl	0.31 mg/kg wet weight	Application = 5.60 kg/ha, pH = 6.3-6.7, s = 2.5 %, 6 w[2]
Rhizophora mangle, hypocotyl	0.49 mg/kg wet weight	Application = 11.20 kg/ha, pH = 6.3-6.7, s = 2.5 %, 6 w[2]
Rhizophora mangle, root	0.02 mg/kg wet weight	Application = 2.80 kg/ha, pH = 6.3-6.7, s = 2.5 %, 6 w[2]
Rhizophora mangle, root	0.06 mg/kg wet weight	Application = 5.60 kg/ha, pH = 6.3-6.7, s = 2.5 %, 6 w[2]
Rhizophora mangle, root	0.07 mg/kg wet weight	Application = 11.20 kg/ha, pH = 6.3-6.7, s = 2.5 %, 6 w[2]
Sediment	0.01 mg/kg wet weight	Conc in water: < 1.0 ug/l, USA, less than[1]

Source: 1.Edwards, C.A. 1975. Persistent Pesticides in the Environment, 2nd edition, CRC-Press.
2.Walch, G.E., T.A. Hollister and J. Forester. 1974. Bull. Environ. Contam. Toxicol. 12(2), pp 129-135

5-365 Concentrations
Methylbiphenyl

Species	Value	Condition
Crassostrea virginica	0.1 mg/kg wet weight	Upt in 8 h from 302 ppm No. 2 fuel, 0.16 % MN, 120 h excr, init=0.3 ppm[1]
Crassostrea virginica	0.2 mg/kg wet weight	Upt in 8 h from 302 ppm No. 2 fuel, 0.16 % MN, 24 h excr, init=0.3 ppm[1]
Crassostrea virginica	0.2 mg/kg wet weight	Upt in 8 h from 302 ppm No. 2 fuel, 0.16 % MN, 6 h excr, init=0.3 ppm[1]
Crassostrea virginica	0.4 mg/kg wet weight	Upt in 8 h from 302 ppm No. 2 fuel, 0.16 % MN, 3 h excr, init=0.3 ppm[1]
Crassostrea virginica	0.5 mg/kg wet weight	302 ppm dispersed No. 2 fuel oil with 0.16 % MN, flow through, 8 h[1]
Marine sediment near oil separator platform	200 ug/kg wet weight	Trinity B., TEX, USA, sample 15 m from brine effluent = 11 ug/l[2]

Source: 1.Neff, J.M., B.A. Cox, D. Dixit and J.W. Anderson. 1976. Mar. Biol. 38, 279-289.
2.Armstrong, H.W., K. Fucik, J.W. Anderson and J.M. Neff. 1977. Effects of oilfield brine effluent on benthic organisms in Trinity Bay. American Petrolium Institute Publ. No. 4291. Wash. DC.

5-366 Concentrations
Methylbiphenyl and C3 - Naphthalene

Species	Value	Condition
Fundulus heteroclitus	0.07 mg/kg wet weight	570 l S. Louisiana crude oil in 810 m2 shoreline area, 6 h after spill[1]
Fundulus heteroclitus	0.09 mg/kg wet weight	570 l S. Louisiana crude oil in 810 m2 shoreline area, 216 h after spill[1]
Fundulus heteroclitus	0.17 mg/kg wet weight	570 l S. Louisiana crude oil in 810 m2 shoreline area, 31 h after spill[1]
Fundulus heteroclitus	0.44 mg/kg wet weight	570 l S. Louisiana crude oil in 810 m2 shoreline area, 76 h after spill[1]

Source: 1.Bieri, R.H., V.C. Stamoudis and M.K. Cueman. 1977. Proc. of 1977 Oil spill Conference (Prevention, Behavior, Control, Cleanup). Wash. DC. American Petrolium Institute. 511-516.

5-367 Concentrations
Methylfluorene

Species	Value	Condition
Marine sediment near oil separator platform	700 ug/kg wet weight	Trinity B., TEX, USA, sample 15 m from brine effluent = 63 ug/l[1]

Source: 1.Armstrong, H.W., K. Fucik, J.W. Anderson and J.M. Neff. 1977. Effects of oilfield brine effluent on benthic organisms in Trinity Bay. American Petrolium Institute Publ. No. 4291. Wash. DC.

5-368 Concentrations
Methylfluorene, C4 - Naphthalene and C3 - Biphenyl

Species	Value	Condition
Fundulus heteroclitus	0.11 mg/kg wet weight	570 l S. Louisiana crude oil in 810 m2 shoreline area, 6 h after spill[1]
Fundulus heteroclitus	0.13 mg/kg wet weight	570 l S. Louisiana crude oil in 810 m2 shoreline area, 216 h after spill[1]
Fundulus heteroclitus	0.23 mg/kg wet weight	570 l S. Louisiana crude oil in 810 m2 shoreline area, 31 h after spill[1]
Fundulus heteroclitus	0.54 mg/kg wet weight	570 l S. Louisiana crude oil in 810 m2 shoreline area, 76 h after spill[1]

Source: 1.Bieri, R.H., V.C. Stamoudis and M.K. Cueman. 1977. Proc. of 1977 Oil spill Conference (Prevention, Behavior, Control, Cleanup). Wash. DC. American Petrolium Institute. 511-516.

5-369 Concentrations
Methylfluorenes

Species	Value	Condition
Crassostrea virginica	0.1 mg/kg wet weight	Upt in 8 h from 302 ppm No. 2 fuel, 0.36 % MF, 120 h excr, init=1.0 ppm[1]
Crassostrea virginica	0.2 mg/kg wet weight	Upt in 8 h from 302 ppm No. 2 fuel, 0.36 % MF, 6 h excr, init=1.0 ppm[1]
Crassostrea virginica	0.7 mg/kg wet weight	Upt in 8 h from 302 ppm No. 2 fuel, 0.36 % MF, 24 h excr, init=1.0 ppm[1]
Crassostrea virginica	0.7 mg/kg wet weight	Upt in 8 h from 302 ppm No. 2 fuel, 0.36 % MF, 3 h excr, init=1.0 ppm[1]
Crassostrea virginica	1.2 mg/kg wet weight	302 ppm dispersed No. 2 fuel oil with 0.36 % MF, flow through, 8 h[1]

Source: 1.Neff, J.M., B.A. Cox, D. Dixit and J.W. Anderson. 1976. Mar. Biol. 38, 279-289.

5-370 Concentrations
Methylnaphthalene

Species	Value	Condition
Crassostrea virginica	0.005 mg/kg wet weight	Exposed 0[1]
Crassostrea virginica	0.005 mg/kg wet weight	Exposed 0 d to sediment with 887.4 ug total oil/g sed, prudhoe B crude oil[1]
Crassostrea virginica	0.06 mg/kg wet weight	Exposed 40 d to sediment with 887.4 ug total oil/g sed, prudhoe B crude oil[1]
Crassostrea virginica	0.27 mg/kg wet weight	Exposed 60 d to sediment with 887.4 ug total oil/g sed, prudhoe B crude oil[1]
Crassostrea virginica	36 mg/kg dry weight	Exposure = 8 d, water = 3 ug/l, crude oil, lab[2]
Crassostrea virginica	58 mg/kg dry weight	Exposure = 2 d, water = 8 ug/l, crude oil, lab[2]
Mytilus edulis	0.005 mg/kg wet weight	Exposed 0[1]

Mytilus edulis	0.005 mg/kg wet weight	Exposed 0 d to sediment with 887.4 ug total oil/g sed, prudhoe B crude oil[1]
Mytilus edulis	0.06 mg/kg wet weight	Exposed 60 d to sediment with 887.4 ug total oil/g sed, prudhoe B crude oil[1]
Mytilus edulis	0.12 mg/kg wet weight	Exposed 40 d to sediment with 887.4 ug total oil/g sed, prudhoe B crude oil[1]
Nephthys incisa	0.005 mg/kg wet weight	Exposed 0 days1
Nephthys incisa	0.005 mg/kg wet weight	Exposed 0 d to sediment with 887.4 ug total oil/g sed, prudhoe B crude oil[1]
Nephthys incisa	0.005 mg/kg wet weight	Exposed 40 d to sediment with 887.4 ug total oil/g sed, prudhoe B crude oil[1]
Nephthys incisa	0.02 mg/kg wet weight	Exposed 60 d to sediment with 887.4 ug total oil/g sed, prudhoe B crude oil[1]
Penaeus aztecus, tissue	3610 ug/kg wet weight	Water = 15.4 ppb after 24 h, tissue after 72 h, fuel oil No. 2, exp[3]
Sesarma cinereum, tissue	6050 ug/kg wet weight	Water = 15.4 ppb after 24 h, tissue after 72 h, fuel oil No. 2, exp[3]
Uca minax, tissue	4520 ug/kg wet weight	Water = 15.4 ppb after 24 h, tissue after 72 h, fuel oil No. 2, exp[3]

Source: 1.Roesijadi, G., J.W. Anderson and J.W. Blaylock. 1978. J. Fish. Res. bd Canada 35, 608-614.
2.Lee, R.F., W.S. Gardner, J.W. Anderson, Balylock and B-Clarke. 1978. Environ. Sci. Technol. 12, 832-38.
3.Cox, B.A., J.W. Anderson and J.C. Parker. 1975. Proc. 1975 Conference on Prevention and Control of Oil Pollution. American Petrolium Institute. Wash DC. 607-612.

5-371 Concentrations Methylphenanthrene

Species	Value	Condition
Crassostrea virginica	0.4 mg/kg wet weight	Upt in 8 h from 302 ppm No. 2 fuel, 0.53 % MP, 120 h excr, init=1.9 ppm[1]
Crassostrea virginica	0.6 mg/kg wet weight	Upt in 8 h from 302 ppm No. 2 fuel, 0.53 % MP, 6 h excr, init=1.9 ppm[1]
Crassostrea virginica	1.3 mg/kg wet weight	Upt in 8 h from 302 ppm No. 2 fuel, 0.53 % MP, 24 h excr, init=1.9 ppm[1]
Crassostrea virginica	1.3 mg/kg wet weight	Upt in 8 h from 302 ppm No. 2 fuel, 0.53 % MP, 3 h excr, init=1.9 ppm[1]
Crassostrea virginica	1.9 mg/kg wet weight	302 ppm dispersed No. 2 fuel oil with 0.53 % MP, flow through, 8 h[1]
Marine sediment near oil separator platform	600 ug/kg wet weight	Trinity B., TEX, USA, sample 15 m from brine effluent = 63 ug/l[2]

Source: 1.Neff, J.M., B.A. Cox, D. Dixit and J.W. Anderson. 1976. Mar. Biol. 38, 279-289.
2.Armstrong, H.W., K. Fucik, J.W. Anderson and J.M. Neff. 1977. Effects of oilfield brine effluent on benthic organisms in Trinity Bay. American Petrolium Institute Publ. No. 4291. Wash. DC.

5-372 Concentrations Mg

Species	Value	Condition
Different Mg species in seawater, Mg(2+)	87.4 mol %	298 K, pH = 8.0, 1 atm, seawater, model distribution[1]
Different species in freshwater, Mg bound with Humic	0.8 mol %	298 K, analytic data, 1 atm, freshwater, L. Celyn[1]
Different species in freshwater, Mg(2+)	98.2 mol %	298 K, analytic data, 1 atm, freshwater, L. Celyn[1]

Different species in freshwater, MgHCO3(-)	0.5 mol %	298 K, analytic data, 1 atm, freshwater, L. Celyn[1]
Different species in freshwater, MgSO4	0.5 mol %	298 K, analytic data, 1 atm, freshwater, L. Celyn[1]
Different species in seawater, MgCO3	0.3 mol %	298 K, pH = 8.0, 1 atm, seawater, model distribution[1]
Different species in seawater, MgF(+)	1.9 mol %	298 K, pH = 8.0, 1 atm, seawater, model distribution[1]
Different species in seawater, MgSO4	10.4 mol %	298 K, pH = 8.0, 1 atm, seawater, model distribution[1]

Source: 1.Mantoura, R.F.C., A. Dickson and J.P. Riley. 1978. Eustarine Coastal Mar. Sci. 6, 387-408.

5-373 Concentrations
Mn

Species	Value	Condition
Alisma plantago aquatica, old leaves	124.31 mg/kg dry weight	150 m below pollutionentry I, water = 0.075 mg/l, processed leaves[1]
Alisma plantago aquatica, old leaves	127.00 mg/kg dry weight	150 m below pollutionentry I, water = 0.075 mg/l, whole leaves[1]
Alisma plantago aquatica, old leaves	2553.65 mg/kg dry weight	800 m above pollutionentry I, water = 0.038 mg/l, processed leaves[1]
Alisma plantago aquatica, old leaves	2591.60 mg/kg dry weight	800 m above pollutionentry I, water = 0.038 mg/l, whole leaves[1]
Alisma plantago aquatica, old leaves	95.92 mg/kg dry weight	800 and 50 m below pollution I and II, water = 0.099 mg/l, processed leaves[1]
Alisma plantago aquatica, old leaves	99.42 mg/kg dry weight	800 and 50 m below pollution I and II, water = 0.099 mg/l, whole leaves[1]
Alisma plantago aquatica, young leaves	1306.32 mg/kg dry weight	800 m above pollutionentry I, water = 0.038 mg/l, whole leaves[1]
Alisma plantago aquatica, young leaves	1513.98 mg/kg dry weight	800 m above pollutionentry I, water = 0.038 mg/l, processed leaves[1]
Alisma plantago aquatica, young leaves	160.20 mg/kg dry weight	800 and 50 m below pollution I and II, water = 0.099 mg/l, processed leaves[1]
Alisma plantago aquatica, young leaves	165.00 mg/kg dry weight	150 m below pollutionentry I, water = 0.075 mg/l, whole leaves[1]
Alisma plantago aquatica, young leaves	171.99 mg/kg dry weight	150 m below pollutionentry I, water = 0.075 mg/l, processed leaves[1]
Alisma plantago aquatica, young leaves	188.12 mg/kg dry weight	800 and 50 m below pollution I and II, water = 0.099 mg/l, whole leaves[1]
Beech and spruce forest soil	330 mg/kg dry weight soil	Central German, soil depth = 0-10 cm[2]
Beech and spruce forest soil	700 mg/kg dry weight soil	Central German, soil depth = 10-20 cm, 833E3 g/ha[2]
Beech and spruce forest soil	730 mg/kg dry weight soil	Central German, soil depth = 30-40 cm, 949E3 g/ha[2]
Beech and spruce forest soil	740 mg/kg dry weight soil	Central German, soil depth = 40-50 cm, 110E3 g/ha[2]
Coniferous trees, needles	14000 mg/kg ash weight	Silicic rock = 610+-59, soil = 1500+-150 ppm, Finland, n = 64[3]
Coniferous trees, needles	16000 mg/kg ash weight	Ultrabasic rock = 1400+-110, soil = 1200+-150 ppm, Finland, n = 10[3]
Coniferous trees, needles	7800 mg/kg ash weight	Calcareous rock = 2300+-350, soil = 1800+-370 ppm, Finland, n = 9[3]

Coniferous trees, twigs	10000 mg/kg ash weight	Silicic rock = 610+-59, soil = 1500+-150 ppm, Finland, n = 64[3]
Coniferous trees, twigs	5200 mg/kg ash weight	Ultrabasic rock = 1400+-110, soil = 1200+-150 ppm, Finland, n = 10[3]
Coniferous trees, twigs	6100 mg/kg ash weight	Calcareous rock = 2300+-350, soil = 1800+-370 ppm, Finland, n = 9[3]
Deciduous trees and shrubs, leaves	12000 mg/kg ash weight	Silicic rock = 610+-59, soil = 1500+-150 ppm, Finland, n = 80[3]
Deciduous trees and shrubs, leaves	15000 mg/kg ash weight	Ultrabasic rock = 1400+-110, soil = 1200+-150 ppm, Finland, n = 14[3]
Deciduous trees and shrubs, leaves	5000 mg/kg ash weight	Calcareous rock = 2300+-350, soil = 1800+-370 ppm, Finland, n = 23[3]
Deciduous trees and shrubs, twigs	11000 mg/kg ash weight	Ultrabasic rock = 1400+-110, soil = 1200+-150 ppm, Finland, n = 14[3]
Deciduous trees and shrubs, twigs	7400 mg/kg ash weight	Calcareous rock = 2300+-350, soil = 1800+-370 ppm, Finland, n = 23[3]
Deciduous trees and shrubs, twigs	9000 mg/kg ash weight	Silicic rock = 610+-59, soil = 1500+-150 ppm, Finland, n = 80[3]
Forest soil	2000 mg/kg dry weight soil	E Tennessee, various forests, depth = > 160 cm[4]
Forest soil	2000 mg/kg dry weight soil	E Tennessee, various forests, depth = litter[4]
Forest soil	25 mg/kg dry weight soil	E Tennessee, various forests, depth = 120 - 160 cm[4]
Forest soil	700 mg/kg dry weight soil	E Tennessee, various forests, depth = 0 - 40 cm[4]
Forest soil	700 mg/kg dry weight soil	E Tennessee, various forests, depth = 40 - 80 cm[4]
Forest soil	75 mg/kg dry weight soil	E Tennessee, various forests, depth = 80 - 120 cm[4]
Hypolimnion	590 ug/l	Conc in epilimnion = 4.5 ug/l, concentration factor = 130, Schoehsee[5]
Living spruce needles	581 mg/kg dry weight	Litter = 131-318, underlaying soil = 290-323 mg/kg dw, S Sweden[6]
Particle fractions	12.5 % of total	Size = 0.125-0.075 mm, weight = 16.9 %, area = 10.4 %, marine, Japan[7]
Particle fractions	19.6 % of total	Size = 0.25-0.125 mm, weight = 23 %, area = 10.1 %, marine, Japan[7]
Particle fractions	2.2 % of total	Size = 2.0-0.5 mm, weight = 3.2 %, area = 0.2 %, marine, Japan, Mn-54[7]
Particle fractions	3.2 % of total	Size = 0.5-0.25 mm, weight = 8.5 %, area = 1.4 %, marine, Japan[7]
Particle fractions	62.9 % of total	Size below 0.075 mm, weight = 48.4 %, area = 77.9 %, marine, Japan[7]
Pediastrum tetras	27.40 mg/kg dry weight	Mn in medium = 0.2 ppm, 294 K, light = 2500 lux[8]
Pediastrum tetras	39.99 mg/kg dry weight	Mn in medium = 0.4 ppm, 294 K, light = 2500 lux[8]
Pediastrum tetras	45.05 mg/kg dry weight	Mn in medium = 0.5 ppm, 294 K, light = 2500 lux[8]
Pediastrum tetras	7.61 mg/kg dry weight	Mn in medium = 0.0 ppm, 294 K, light = 2500 lux[8]
Sediment	128.3 mg/kg dry weight	Unpolluted area, estuary, Urr water, rg = 81.2 - 265.9[9]
Sediment	130.11 mg/kg dry weight	800 and 50 m below pollution I and II, water = 0.099 mg/l, rg = 113.7[1]
Sediment	1600 mg/kg dry weight	Conc in epilimnion = 4.5 ug/l, concentration factor = 355000,[5]
Sediment	1688 mg/kg dry weight	Max water = 3000 microgram/l, 1 y measurement, Red River, GB[10]

Sediment	1688 mg/kg dry weight	Water conc = 781 microgram/l, yearly mean, Red River, GB[10]
Sediment	245 mg/kg dry weight	Max water = 15 microgram/l, 1 y measurement, Fowey River, GB[10]
Sediment	245 mg/kg dry weight	Water conc = 9 microgram/l, yearly mean, fowey River, GB[10]
Sediment	3245 mg/kg dry weight	Max water = 515 microgram/l, 1 y measurement, Carnon River, GB[10]
Sediment	3245 mg/kg dry weight	Water conc = 244 microgram/l, yearly mean, Carnon River, GB[10]
Sediment	333 mg/kg dry weight	Unpolluted area, estuary, Urr water, rg = 221 - 587[11]
Sediment	3970 mg/kg dry weight	Max water = 614 microgram/l, max conc, Gannel River, GB[10]
Sediment	3970 mg/kg dry weight	Water conc = 367 microgram/l, yearly mean, Gannel River, GB[10]
Sediment	45.29 mg/kg dry weight	800 m above pollutionentry l, water = 0.038 mg/l, rg = 36.9 - 51.3[1]
Sediment	490 mg/kg dry weight	Max water = 9 microgram/l, 1 y measurement, Carnon River, GB[10]
Sediment	490 mg/kg dry weight	Water conc = 6 microgram/l, yearly mean, Carnon River, GB[10]
Sediment	50.39 mg/kg dry weight	150 m below pollutionentry l, water = 0.075 mg/l, rg = 42.2 - 59.5[1]
Sediment	520 mg/kg dry weight	Max water = 11 microgram/l, 1 y measurement, Fowey River, GB[10]
Sediment	520 mg/kg dry weight	Water conc = 8 microgram/l, yearly mean, fowey River, GB[10]
Sediment	527 mg/kg dry weight	Max water = 22 microgram/l, max conc, Gannel River, GB[10]
Sediment	527 mg/kg dry weight	Water conc = 16 microgram/l, yearly mean, Gannel River, GB[10]
Sediment	601 mg/kg dry weight	Unpolluted area, estuary, Urr water, rg = 81.2 - 265.9[11]
Sediment	635 mg/kg dry weight	Max water = 200 microgram/l, 1 y measurement, Red River, GB[10]
Sediment	635 mg/kg dry weight	Water conc = 69 microgram/l, yearly mean, Red River, GB[10]
Tomato, leaf	185 mg Mn/kg dry weight	5 ppm Mn, 3.02 meq KH2PO4/l culture solution, effect PO4[12]
Tomato, leaf	19.0 mg Mn/kg dry weight	0.50 ppm Mn, 3.02 meq KH2PO4/l culture solution, effect PO4[12]
Tomato, leaf	21.5 mg Mn/kg dry weight	0.50 ppm Mn, 9.07 meq KH2PO4/l culture solution, effect PO4[12]
Tomato, leaf	5000 mg Mn/kg dry weight	Added superphosphate = 9.6 kg/m**3 soil, 350 ppm exchangeable Mn[12]
Tomato, leaf	5300 mg Mn/kg dry weight	Added superphosphate = 2.4 kg/m**3 soil, 350 ppm exchangeable Mn[12]
Tomato, leaf	762 mg Mn/kg dry weight	25 ppm Mn, 9.07 meq KH2PO4/l culture solution, effect PO4[12]
Tomato, leaf	876 mg Mn/kg dry weight	25 ppm Mn, 3.02 meq KH2PO4/l culture solution, effect PO4[12]
Tomato, root	2371 mg Mn/kg dry weight	25 ppm Mn, 9.07 meq KH2PO4/l culture solution, effect PO4[12]
Tomato, root	3066 mg Mn/kg dry weight	25 ppm Mn, 3.02 meq KH2PO4/l culture solution, effect PO4[12]
Tomato, root	95.9 mg Mn/kg dry weight	0.50 ppm Mn, 9.07 meq KH2PO4/l culture solution, effect PO4[12]
Tomato, root	99.7 mg Mn/kg dry weight	0.50 ppm Mn, 3.02 meq KH2PO4/l culture solution, effect PO4[12]
Tomato, root	990 mg Mn/kg dry weight	5 ppm Mn, 3.02 meq KH2PO4/l culture solution, effect PO4[12]
Tomato, shoot	177 mg Mn/kg dry weight	5 ppm Mn, 3.02 meq KH2PO4/l culture solution, effect PO4[12]

Tomato, shoot	19.9 mg Mn/kg dry weight	0.50 ppm Mn, 3.02 meq KH2PO4/l culture solution, effect PO4[12]
Tomato, shoot	23.5 mg Mn/kg dry weight	0.50 ppm Mn, 9.07 meq KH2PO4/l culture solution, effect PO4[12]
Tomato, shoot	738 mg Mn/kg dry weight	25 ppm Mn, 9.07 meq KH2PO4/l culture solution, effect PO4[12]
Tomato, shoot	832 mg Mn/kg dry weight	25 ppm Mn, 3.02 meq KH2PO4/l culture solution, effect PO4[12]
Tomato, whole plant	1153 mg Mn/kg dry weight	25 ppm Mn, 9.07 meq KH2PO4/l culture solution, effect PO4[12]
Tomato, whole plant	1320 mg Mn/kg dry weight	25 ppm Mn, 3.02 meq KH2PO4/l culture solution, effect PO4[12]
Tomato, whole plant	339 mg Mn/kg dry weight	5 ppm Mn, 3.02 meq KH2PO4/l culture solution, effect PO4[12]
Tomato, whole plant	37.1 mg Mn/kg dry weight	0.50 ppm Mn, 3.02 meq KH2PO4/l culture solution, effect PO4[12]
Tomato, whole plant	38.3 mg Mn/kg dry weight	0.50 ppm Mn, 9.07 meq KH2PO4/l culture solution, effect PO4[12]
Vegetation	380 mg/kg dry weight	Conc in soil: f = 150, h = 100, a1a2 = 100, b1 = 500, b2 = 2000[13]

Source: 1.Patrick, F.M. and M.W. Loutit. 1977. Water Res., 11(8), pp 699-703.
2.Heinrichs, H. and R. Meyer. 1977. J. Environ. Qual. 6, 402-407.
3.Lounamaa, J. 1967. in: Kvauheim, A. (ed.); Geochemical Prospecting in Fennoscandia. pp 287-317. Interscience Publ. New York.
4.Reiners, W.A., R.H. Marks and P.M. Vitousek. 1975. Oikos 26, 264-275.
5.Groth, P. 1971. Arch. Hydrobiol. 68(3), pp 305-375.
6.Nilsson, I. 1972. Oikos, 22, pp 366.
7.Nagaya, Y. and K. Nakamura. 1975. Impacts of Nuclear releases into the Aquatic Environment. IAEA-SM198/21, pp 313-328. Vienna.
8.Wettern, M., D.W. Lorch and A. Weber. 1976. Arch. Hydrobiol. 77(3), pp 267-276.
9.Taylor, D. 1976. Estuarine Coastal Mar. Sci. 2, pp 417.
10.Altshuller, A.P. and J.J. Bufalini. 1971. Environ. Sci. Technol., 5(1), pp 39-64.
11.Taylor, D. 1976. Sci. Toatal Environ. 6(3), pp 259-264.
12.Harrod, M.F. 1971. In: Tech. Bull. 21, pp 176-189. Trace Elements in Soils and Crops.
13.Mitchell, R. and C. Chamberlin. 1978. in: Berg. G. (ed.); Indicators of Enteric Contamination in Natural Waters.

5-374 Concentrations
Mo

Species	Value	Condition
Coniferous trees, needles	15 mg/kg ash weight	Ultrabasic rock = 3.0 ppm, soil = 3.0 ppm, Finland, n = 10, SD = +- 8[1]
Coniferous trees, needles	2.2 mg/kg ash weight	Calcareous rock = 3.2 ppm, soil = 3.0 ppm, Finland, n = 9, SD = +- 1[1]
Coniferous trees, needles	2.3 mg/kg ash weight	Silicic rock = 3.3, soil = 3.4+-0.2 ppm, Finland, n = 64, SD = +- 0.6[1]
Coniferous trees, needles	7.4 mg/kg ash weight	Calcareous rock = 3.2 ppm, soil = 3.0 ppm, Finland, n = 9, SD = +- 4[1]
Coniferous trees, twigs	1.6 mg/kg ash weight	Silicic rock = 3.3, soil = 3.4+-0.2 ppm, Finland, n = 64, SD = +- 0.3[1]
Coniferous trees, twigs	12 mg/kg ash weight	Ultrabasic rock = 3.0 ppm, soil = 3.0 ppm, Finland, n = 10, SD = +- 6[1]
Deciduous trees and shrubs, leaves	28 mg/kg ash weight	Ultrabasic rock = 3.0 ppm, soil = 3.0 ppm, Finland, n = 14, SD = +- 9[1]
Deciduous trees and shrubs, leaves	5.4 mg/kg ash weight	Silicic rock = 3.3, soil = 3.4+-0.2 ppm, Finland, n = 75, SD = +- 1[1]
Deciduous trees and shrubs, leaves	7.7 mg/kg ash weight	Calcareous rock = 3.2 ppm, soil = 3.0 ppm, Finland, n = 23, SD = +-[1]

Deciduous trees and shrubs, twigs	14.0 mg/kg ash weight	Calcareous rock = 3.2 ppm, soil = 3.0 ppm, Finland, n = 23, SD = +- 4[1]
Deciduous trees and shrubs, twigs	23 mg/kg ash weight	Ultrabasic rock = 3.0 ppm, soil = 3.0 ppm, Finland, n = 14, SD = +- 6[1]
Deciduous trees and shrubs, twigs	6.8 mg/kg ash weight	Silicic rock = 3.3, soil = 3.4+-0.2 ppm, Finland, n = 75, SD = +- 1.5[1]
Hypolimnion	0.30 ug/l	Conc in epilimnion = 0.21 ug/l, concentration factor = 1.4, Schoehsee[2]
Nitella flexilis	130 ug/kg wet weight	Water = 3.3 mg/l, lab., Mo-99[3]
Nitella flexilis	8.8 ug/kg wet weight	Water = 0.014 ug/l, uptake from water, Fern Lake, Mo-99[3]
Red clover	18.6 mg/kg dry weight	Added NaMoO4 = 1.1 kg/ha, pH = 6.1, control = 0.7 mg/kg dry weight[4]
Red clover	35.6 mg/kg dry weight	Added NaMoO4 = 1.1 kg/ha, pH = 6.5, control = 3.4 mg/kg dry weight[4]
Rye grass	2.0 mg/kg dry weight	Added NaMoO4 = 1.1 kg/ha, pH = 6.1, control = 0.8 mg/kg dry weight[4]
Rye grass	3.4 mg/kg dry weight	Added NaMoO4 = 1.1 kg/ha, pH = 6.5, control = 1.1 mg/kg dry weight[4]
Sediment	1.4 mg/kg dry weight	Conc in epilimnion = 0.21 ug/l, concentration factor = 7000,[2]
White clover	0.4 mg/kg dry weight	Added limestone = 0.63 t/ha, soil-pH = 4.7, N Wales[5]
White clover	0.4 mg/kg dry weight	Added limestone = 2.5 t/ha, soil-pH = 5.0, N Wales[5]
White clover	0.6 mg/kg dry weight	Added limestone = 5.0 t/ha, soil-pH = 5.2, N Wales[5]
White clover	1.0 mg/kg dry weight	Added limestone = 15.0 t/ha, soil-pH = 6.0, N Wales[5]
White clover	2.3 mg/kg dry weight	Added limestone = 0.63 t/ha and Na2MoO4 = 280 g/ha, soil-pH = 4.7, N[5]
White clover	2.3 mg/kg dry weight	Added limestone = 2.5 t/ha and Na2MoO4 = 280 g/ha, soil-pH = 5.0, N[5]
White clover	4.0 mg/kg dry weight	Added limestone = 5.0 t/ha and Na2MoO4 = 280 g/ha, soil-pH = 5.2, N[5]
White clover	6.3 mg/kg dry weight	Added limestone = 15.0 t/ha and Na2MoO4 = 280 g/ha, soil-pH = 6.0, N[5]

Source: 1.Lounamaa, J. 1967. in: Kvauheim, A. (ed.); Geochemical Prospecting in Fennoscandia. pp 287-317. Interscience Publ. New York.
2.Groth, P. 1971. Arch. Hydrobiol. 68(3), pp 305-375.
3.Short, Z.F., P.R. Olson, R.F. Palumbo, J.R. Donaldson and F.G. Lowman. 1971. Radionuclides in Ecosystems, Proc. Third Nat. Symp. Radioecology 1, pp 474-485.
4.Mitchell, R. and C. Chamberlin. 1978. in: Berg. G. (ed.); Indicators of Enteric Contamination in Natural Waters.
5.Archer, F.C. 1971. In: Tech. Bull. 21, pp 150-158. Trace Elements in Soils and Crops.

5-375 Concentrations
Naphthalene

Species	Value	Condition
Calanus helgolandicus	26 picogram naphth/animal	24 h, water = 1 ug/l, 289 K, copepod, C-14 marked, marine[1]
Calanus helgolandicus	56 picogram naphth/animal	24 h, water = 1 ug/l, 286 K, copepod, C-14 marked, marine[2]
Calanus helgolandicus	58 picogram naphth/animal	24 h, water = 1 ug/l, 282 K, copepod, C-14 marked, marine[2]
Calanus helgolandicus	79 picogram naphth/animal	24 h, water = 1 ug/l, 278 K, copepod, C-14 marked, marine[3]
Crassostrea virginica	0.005 mg/kg wet weight	Exposed 0[4]

Crassostrea virginica	0.005 mg/kg wet weight	Exposed 0 d to sediment with 887.4 ug total oil/g sed, prudhoe B crude oil[4]
Crassostrea virginica	0.005 mg/kg wet weight	Exposed 40 d to sediment with 887.4 ug total oil/g sed, prudhoe B crude oil[4]
Crassostrea virginica	0.02 mg/kg wet weight	Exposed 60 d to sediment with 887.4 ug total oil/g sed, prudhoe B crude oil[4]
Crassostrea virginica	12 mg/kg dry weight	Exposure = 8 d, water = 3.0 ug/l, crude oil, lab[5]
Crassostrea virginica	12.0 mg/kg wet weight	Upt in 8 h from 302 ppm No. 2 fuel, 0.4 % N, 3 h excr, init=14.7 ppm[6]
Crassostrea virginica	14.7 mg/kg wet weight	302 ppm dispersed No. 2 fuel oil with 0.4 % N, flow through, 8 h[6]
Crassostrea virginica	30 mg/kg dry weight	Exposure = 2 d, water = 5.0 ug/l, crude oil, lab[5]
Crassostrea virginica	4.7 mg/kg wet weight	Upt in 8 h from 302 ppm No. 2 fuel, 0.4 % N, 120 h excr, init=14.7 ppm[6]
Crassostrea virginica	6.5 mg/kg wet weight	Upt in 8 h from 302 ppm No. 2 fuel, 0.4 % N, 24 h excr, init=14.7 ppm[6]
Crassostrea virginica	7.3 mg/kg wet weight	Upt in 8 h from 302 ppm No. 2 fuel, 0.4 % N, 6 h excr, init=14.7 ppm[6]
Fundulus heteroclitus	0.14 mg/kg wet weight	570 l S. Louisiana crude oil in 810 m2 shoreline area, 76 h after spill[7]
Fundulus heteroclitus	0.25 mg/kg wet weight	570 l S. Louisiana crude oil in 810 m2 shoreline area, 6 h after spill[7]
Fundulus heteroclitus	0.68 mg/kg wet weight	570 l S. Louisiana crude oil in 810 m2 shoreline area, 31 h after spill[7]
Marine sediment near oil separator platform	200 ug/kg wet weight	Trinity B., TEX, USA, sample 15 m from brine effluent = 300 ug/l[8]
Mytilus edulis	0.005 mg/kg wet weight	Exposed 0[4]
Mytilus edulis	0.005 mg/kg wet weight	Exposed 0 d to sediment with 887.4 ug total oil/g sed, prudhoe B crude oil[4]
Mytilus edulis	0.005 mg/kg wet weight	Exposed 40 d to sediment with 887.4 ug total oil/g sed, prudhoe B crude oil[4]
Mytilus edulis	0.01 mg/kg wet weight	Exposed 60 d to sediment with 887.4 ug total oil/g sed, prudhoe B crude oil[4]
Nephthys incisa	0.005 mg/kg wet weight	Exposed 0[4]
Nephthys incisa	0.005 mg/kg wet weight	Exposed 0 d to sediment with 887.4 ug total oil/g sed, prudhoe B crude oil[4]
Nephthys incisa	0.005 mg/kg wet weight	Exposed 40 d to sediment with 887.4 ug total oil/g sed, prudhoe B crude oil[4]
Nephthys incisa	0.005 mg/kg wet weight	Exposed 60 d to sediment with 887.4 ug total oil/g sed, prudhoe B crude oil[4]
Penaeus aztecus, tissue	450 ug/kg wet weight	Water = 2.3 ppb after 24 h, tissue after 72 h, fuel oil No. 2, exp[9]
Seawater near oil separator platform	0.40 ug/liter	Trinity B., TEX, USA, sample 15 m from brine effluent = 300 ug/l[8]
Sesarma cinereum, tissue	930 ug/kg wet weight	Water = 2.3 ppb after 24 h, tissue after 72 h, fuel oil No. 2, exp[9]
Uca minax, tissue	750 ug/kg wet weight	Water = 2.3 ppb after 24 h, tissue after 72 h, fuel oil No. 2, exp[9]

Source: 1.Bieri, R.H., and C.V. Stamoudis. 1977. In: Wolfe, D.A. (ed.): Fate and Effects of Petrolium Hydrocarbons in Marine Ecosystems and Organisms. NY. Pergamon Press. 332-344.
2.Harris, R.P., V. Berdugo, E.D.S. Corner, Kilvington and Ohara. 1977. In: D.A. Wolfe (ed.): Fate and Effects of Petroleum Hydrocarbons in Marine Ecosystems and Organisms. NY. Pergamon Press.
3.Corner, E.D.S., R.P. Harris, K.J. Whittle and P.R. Mackie. 1976. In: A.P.M. Lockwood (ed.): Effects of Pollutants on Aquatic Organisms. Cambridge Univ. Press. Cambridge.
4.Roesijadi, G., J.W. Anderson and J.W. Blaylock. 1978. J. Fish. Res. bd Canada 35, 608-614.
5.Lee, R.F., W.S. Gardner, J.W. Anderson, Balylock and B-Clarke. 1978. Environ. Sci. Technol. 12, 832-38.
6.Neff, J.M., B.A. Cox, D. Dixit and J.W. Anderson. 1976. Mar. Biol. 38, 279-289.

7.Bieri, R.H., V.C. Stamoudis and M.K. Cueman. 1977. Proc. of 1977 Oil spill Conference (Prevention, Behavior, Control, Cleanup). Wash. DC. American Petrolium Institute. 511-516.
8.Armstrong, H.W., K. Fucik, J.W. Anderson and J.M. Neff. 1977. Effects of oilfield brine effluent on benthic organisms in Trinity Bay. American Petrolium Institute Publ. No. 4291. Wash. DC.
9.Cox, B.A., J.W. Anderson and J.C. Parker. 1975. Proc. 1975 Conference on Prevention and Control of Ol Pollution. American Petrolium Institute. Wash DC. 607-612.

5-376 Concentrations
Naphthalene, methylnaphthalene, dimethylnaphthalene

Species	Value	Condition
Crassostrea virginica	0.02 mg/kg wet weight	Exposed 0[1]
Crassostrea virginica	0.02 mg/kg wet weight	Exposed 0 d to sediment with 887.4 ug total oil/g sed, prudhoe B crude oil[1]
Crassostrea virginica	0.95 mg/kg wet weight	Exposed 40 d to sediment with 887.4 ug total oil/g sed, prudhoe B crude oil[1]
Crassostrea virginica	2.46 mg/kg wet weight	Exposed 60 d to sediment with 887.4 ug total oil/g sed, prudhoe B crude oil[1]
Mytilus edulis	0.02 mg/kg wet weight	Exposed 0[1]
Mytilus edulis	0.02 mg/kg wet weight	Exposed 0 d to sediment with 887.4 ug total oil/g sed, prudhoe B crude oil[1]
Mytilus edulis	0.25 mg/kg wet weight	Exposed 60 d to sediment with 887.4 ug total oil/g sed, prudhoe B crude oil[1]
Mytilus edulis	0.50 mg/kg wet weight	Exposed 40 d to sediment with 887.4 ug total oil/g sed, prudhoe B crude oil[1]
Nephthys incisa	0.02 mg/kg wet weight	Exposed 0[1]
Nephthys incisa	0.02 mg/kg wet weight	Exposed 0 d to sediment with 887.4 ug total oil/g sed, prudhoe B crude oil[1]
Nephthys incisa	0.02 mg/kg wet weight	Exposed 40 d to sediment with 887.4 ug total oil/g sed, prudhoe B crude oil[1]
Nephthys incisa	0.18 mg/kg wet weight	Exposed 60 d to sediment with 887.4 ug total oil/g sed, prudhoe B crude oil[1]

Source: 1.Roesijadi, G., J.W. Anderson and J.W. Blaylock. 1978. J. Fish. Res. bd Canada 35, 608-614.

5-377 Concentrations
Ni

Species	Value	Condition
Ameiurus nebdosus, gill	11.1	Wanapitei R, Can, 1974, wat=42+-22 ppb, polluted, n=9[1]
Ameiurus nebdosus, kidney	11.8	Wanapitei R, Can, 1974, wat=42+-22 ppb, polluted, n=9[1]
Ameiurus nebdosus, liver	10.7	Wanapitei R, Can, 1974, wat=42+-22 ppb, polluted, n=9[1]
Ameiurus nebdosus, muscle	9.5	Wanapitei R, Can, 1974, wat=42+-22 ppb, polluted, n=9[1]
Anarchis canadensis	153 mg/kg wet weight	Water = 10+-4 ppb, pH=7.2, Main outlet (French R) to L. Huron, Can[1]
Anarchis canadensis	18 mg/kg wet weight	Water = 2+-1 ppb, pH=7.2, Pickerel R, Can, non-poll.[1]
Anarchis canadensis	210 mg/kg wet weight	Water = 29+-12 ppb, pH=7.3, W Wanapitei B, Can, 1974[1]
Anarchis canadensis	23 mg/kg wet weight	Water = 3+-1 ppb, pH=7.2, East outlet (French R) to L. Huron, Can[1]
Anarchis canadensis	23 mg/kg wet weight	Water = 3+-3 ppb, pH=7.1, French R, Can, non-poll.[1]
Anarchis canadensis	510 mg/kg wet weight	Water = 42+-22 ppb, pH=7.3, Wanapitei R, Can, silt-ladden and poll.[1]
Catostomus commersoni, gill	12.6	Wanapitei R, Can, 1974, wat=42+-22 ppb, polluted, n=7[1]

Catostomus commersoni, kidney	14.0	Wanapitei R, Can, 1974, wat=42+-22 ppb, polluted, n=7[1]
Catostomus commersoni, liver	16.5	Wanapitei R, Can, 1974, wat=42+-22 ppb, polluted, n=7[1]
Catostomus commersoni, muscle	13.2	Wanapitei R, Can, 1974, wat=42+-22 ppb, polluted, n=7[1]
Clam	11 mg/kg wet weight	Water = 42+-22 ppb, pH=7.3, Wanapitei R, Can, silt-ladden and poll.[1]
Clam	12 mg/kg wet weight	Water = 10+-4 ppb, pH=7.2, Main outlet (French R) to L. Huron, Can[1]
Clam	16 mg/kg wet weight	Water = 3+-1 ppb, pH=7.2, East outlet (French R) to L. Huron, Can[1]
Clam	20 mg/kg wet weight	Water = 15+-11 ppb, pH=7.2, Western outlet (French R) to L. Huron, Can[1]
Clam	21 mg/kg wet weight	Water = 29+-12 ppb, pH=7.3, W Wanapitei B, Can, 1974[1]
Clam	29 mg/kg wet weight	Water = 32+-18 ppb, pH=7.2, Hartley Bay, Canada, 1974[1]
Clam	4 mg/kg wet weight	Water = 2+-1 ppb, pH=7.2, Pickerel R, Can, non-poll.[1]
Clam	6 mg/kg wet weight	Water = 3+-3 ppb, pH=7.1, French R, Can, non-poll.[1]
Crayfish	11 mg/kg wet weight	Water = 3+-1 ppb, pH=7.2, East outlet (French R) to L. Huron, Can[1]
Crayfish	13 mg/kg wet weight	Water = 29+-12 ppb, pH=7.3, W Wanapitei B, Can, 1974[1]
Crayfish	39 mg/kg wet weight	Water = 42+-22 ppb, pH=7.3, Wanapitei R, Can, silt-ladden and poll.[1]
Crayfish	5 mg/kg wet weight	Water = 15+-11 ppb, pH=7.2, Western outlet (French R) to L. Huron, Can[1]
Crayfish	8 mg/kg wet weight	Water = 10+-4 ppb, pH=7.2, Main outlet (French R) to L. Huron, Can[1]
Crayfish	9 mg/kg wet weight	Water = 3+-3 ppb, pH=7.1, French R, Can, non-poll.[1]
Equisetum palustre, root	125 mg/kg wet weight	Water = 15+-11 ppb, pH=7.2, Western outlet (French R) to L. Huron, Can[1]
Equisetum palustre, root	193 mg/kg wet weight	Water = 42+-22 ppb, pH=7.3, Wanapitei R, Can, silt-ladden and poll.[1]
Equisetum palustre, root	24 mg/kg wet weight	Water = 2+-1 ppb, pH=7.2, Pickerel R, Can, non-poll.[1]
Equisetum palustre, root	263 mg/kg wet weight	Water = 29+-12 ppb, pH=7.3, W Wanapitei B, Can, 1974[1]
Equisetum palustre, root	27 mg/kg wet weight	Water = 3+-1 ppb, pH=7.2, East outlet (French R) to L. Huron, Can[1]
Equisetum palustre, root	290 mg/kg wet weight	Water = 32+-18 ppb, pH=7.2, Hartley Bay, Canada, 1974[1]
Equisetum palustre, root	48 mg/kg wet weight	Water = 3+-3 ppb, pH=7.1, French R, Can, non-poll.[1]
Equisetum palustre, root	84 mg/kg wet weight	Water = 10+-4 ppb, pH=7.2, Main outlet (French R) to L. Huron, Can[1]
Equisetum palustre, stem	13 mg/kg wet weight	Water = 2+-1 ppb, pH=7.2, Pickerel R, Can, non-poll.[1]
Equisetum palustre, stem	165 mg/kg wet weight	Water = 32+-18 ppb, pH=7.2, Hartley Bay, Canada, 1974[1]
Equisetum palustre, stem	18 mg/kg wet weight	Water = 3+-3 ppb, pH=7.1, French R, Can, non-poll.[1]
Equisetum palustre, stem	30 mg/kg wet weight	Water = 10+-4 ppb, pH=7.2, Main outlet (French R) to L. Huron, Can[1]
Equisetum palustre, stem	36 mg/kg wet weight	Water = 29+-12 ppb, pH=7.3, W Wanapitei B, Can, 1974[1]
Equisetum palustre, stem	40 mg/kg wet weight	Water = 15+-11 ppb, pH=7.2, Western outlet (French R) to L. Huron, Can[1]
Equisetum palustre, stem	79 mg/kg wet weight	Water = 42+-22 ppb, pH=7.3, Wanapitei R, Can, silt-ladden and poll.[1]

Equisetum palustre, stem	8 mg/kg wet weight	Water = 3+-1 ppb, pH=7.2, East outlet (French R) to L. Huron, Can[1]
Esox sp., gill	17.3	Wanapitei R, Can, 1974, wat=42+-22 ppb, polluted, n=4, Northern pike[1]
Esox sp., kidney	27.7	Wanapitei R, Can, 1974, wat=42+-22 ppb, polluted, n=4, Northern pike[1]
Esox sp., liver	15.4	Wanapitei R, Can, 1974, wat=42+-22 ppb, polluted, n=4, Northern pike[1]
Esox sp., muscle	13.3	Wanapitei R, Can, 1974, wat=42+-22 ppb, polluted, n=4, Northern pike[1]
Living spruce needles	1.9 mg/kg dry weight	Litter = 2.7-5.6, underlaying soil = 18-28 mg/kg dw, S Sweden[2]
Micropterus sp., gill	31.7	Wanapitei R, Can, 1974, wat=42+-22 ppb, polluted, n=8, N rock bass[1]
Micropterus sp., kidney	17.3	Wanapitei R, Can, 1974, wat=42+-22 ppb, polluted, n=8, N rock bass[1]
Micropterus sp., liver	17.0	Wanapitei R, Can, 1974, wat=42+-22 ppb, polluted, n=8, N rock bass[1]
Micropterus sp., muscle	12.5	Wanapitei R, Can, 1974, wat=42+-22 ppb, polluted, n=8, N rock bass[1]
Moxostoma sp., gill	12.2	Wanapitei R, Can, 1974, wat=42+-22 ppb, polluted, n=9, redhorse sucker[1]
Moxostoma sp., kidney	14.1	Wanapitei R, Can, 1974, wat=42+-22 ppb, polluted, n=9, redhorse sucker[1]
Moxostoma sp., liver	14.5	Wanapitei R, Can, 1974, wat=42+-22 ppb, polluted, n=9, redhorse sucker[1]
Moxostoma sp., muscle	12.9	Wanapitei R, Can, 1974, wat=42+-22 ppb, polluted, n=9, redhorse sucker[1]
Nuphar variegatum, leaf	13 mg/kg wet weight	Water = 3+-1 ppb, pH=7.2, East outlet (French R) to L. Huron, Can[1]
Nuphar variegatum, leaf	17 mg/kg wet weight	Water = 15+-11 ppb, pH=7.2, Western outlet (French R) to L. Huron, Can[1]
Nuphar variegatum, leaf	20 mg/kg wet weight	Water = 10+-4 ppb, pH=7.2, Main outlet (French R) to L. Huron, Can[1]
Nuphar variegatum, leaf	41 mg/kg wet weight	Water = 32+-18 ppb, pH=7.2, Hartley Bay, Canada, 1974[1]
Nuphar variegatum, leaf	47 mg/kg wet weight	Water = 42+-22 ppb, pH=7.3, Wanapitei R, Can, silt-ladden and poll.[1]
Nuphar variegatum, leaf	62 mg/kg wet weight	Water = 29+-12 ppb, pH=7.3, W Wanapitei B, Can, 1974[1]
Nuphar variegatum, leaf	8 mg/kg wet weight	Water = 2+-1 ppb, pH=7.2, Pickerel R, Can, non-poll.[1]
Nuphar variegatum, peduncle	3 mg/kg wet weight	Water = 2+-1 ppb, pH=7.2, Pickerel R, Can, non-poll.[1]
Nuphar variegatum, peduncle	4 mg/kg wet weight	Water = 32+-18 ppb, pH=7.2, Hartley Bay, Canada, 1974[1]
Nuphar variegatum, peduncle	5 mg/kg wet weight	Water = 3+-1 ppb, pH=7.2, East outlet (French R) to L. Huron, Can[1]
Nuphar variegatum, peduncle	5 mg/kg wet weight	Water = 42+-22 ppb, pH=7.3, Wanapitei R, Can, silt-ladden and poll.[1]
Nuphar variegatum, peduncle	7 mg/kg wet weight	Water = 29+-12 ppb, pH=7.3, W Wanapitei B, Can, 1974[1]
Nuphar variegatum, peduncle	9 mg/kg wet weight	Water = 10+-4 ppb, pH=7.2, Main outlet (French R) to L. Huron, Can[1]
Nuphar variegatum, petiole	13 mg/kg wet weight	Water = 10+-4 ppb, pH=7.2, Main outlet (French R) to L. Huron, Can[1]
Nuphar variegatum, petiole	13 mg/kg wet weight	Water = 15+-11 ppb, pH=7.2, Western outlet (French R) to L. Huron, Can[1]

Nuphar variegatum, petiole	20 mg/kg wet weight	Water = 29+-12 ppb, pH=7.3, W Wanapitei B, Can, 1974[1]
Nuphar variegatum, petiole	25 mg/kg wet weight	Water = 32+-18 ppb, pH=7.2, Hartley Bay, Canada, 1974[1]
Nuphar variegatum, petiole	35 mg/kg wet weight	Water = 42+-22 ppb, pH=7.3, Wanapitei R, Can, silt-ladden and poll.[1]
Nuphar variegatum, petiole	5 mg/kg wet weight	Water = 3+-1 ppb, pH=7.2, East outlet (French R) to L. Huron, Can[1]
Nuphar variegatum, petiole	7 mg/kg wet weight	Water = 2+-1 ppb, pH=7.2, Pickerel R, Can, non-poll.[1]
Nuphar variegatum, root	14 mg/kg wet weight	Water = 42+-22 ppb, pH=7.3, Wanapitei R, Can, silt-ladden and poll.[1]
Nuphar variegatum, root	5 mg/kg wet weight	Water = 29+-12 ppb, pH=7.3, W Wanapitei B, Can, 1974[1]
Nuphar variegatum, root	5 mg/kg wet weight	Water = 3+-1 ppb, pH=7.2, East outlet (French R) to L. Huron, Can[1]
Nuphar variegatum, root	6 mg/kg wet weight	Water = 10+-4 ppb, pH=7.2, Main outlet (French R) to L. Huron, Can[1]
Nuphar variegatum, root	6 mg/kg wet weight	Water = 2+-1 ppb, pH=7.2, Pickerel R, Can, non-poll.[1]
Nuphar variegatum, root	7 mg/kg wet weight	Water = 32+-18 ppb, pH=7.2, Hartley Bay, Canada, 1974[1]
Periphyton	34 mg/kg wet weight	Water = 3+-3 ppb, pH=7.1, French R, Can, non-poll.[1]
Periphyton	342 mg/kg wet weight	Water = 15+-11 ppb, pH=7.2, Western outlet (French R) to L. Huron, Can[1]
Periphyton	385 mg/kg wet weight	Water = 10+-4 ppb, pH=7.2, Main outlet (French R) to L. Huron, Can[1]
Periphyton	43 mg/kg wet weight	Water = 2+-1 ppb, pH=7.2, Pickerel R, Can, non-poll.[1]
Periphyton	55 mg/kg wet weight	Water = 3+-1 ppb, pH=7.2, East outlet (French R) to L. Huron, Can[1]
Periphyton	693 mg/kg wet weight	Water = 29+-12 ppb, pH=7.3, W Wanapitei B, Can, 1974[1]
Periphyton	769 mg/kg wet weight	Water = 32+-18 ppb, pH=7.2, Hartley Bay, Canada, 1974[1]
Periphyton	826 mg/kg wet weight	Water = 42+-22 ppb, pH=7.3, Wanapitei R, Can, silt-ladden and poll.[1]
Potamogeton sp., leaf	258 mg/kg wet weight	Water = 15+-11 ppb, pH=7.2, Western outlet (French R) to L. Huron, Can[1]
Potamogeton sp., leaf	365 mg/kg wet weight	Water = 29+-12 ppb, pH=7.3, W Wanapitei B, Can, 1974[1]
Potamogeton sp., leaf	39 mg/kg wet weight	Water = 2+-1 ppb, pH=-Tetrachlorophenol[3]
Potamogeton sp., leaf	39 mg/kg wet weight	Water = 2+-1 ppb, pH=7.2, Pickerel R, Can, non-poll.[1]
Potamogeton sp., leaf	480 mg/kg wet weight	Water = 42+-22 ppb, pH=7.3, Wanapitei R, Can, silt-ladden and poll.[1]
Potamogeton sp., leaf	50 mg/kg wet weight	Water = 3+-3 ppb, pH=7.1, French R, Can, non-poll.[1]
Potamogeton sp., leaf	60 mg/kg wet weight	Water = 3+-1 ppb, pH=7.2, East outlet (French R) to L. Huron, Can[1]
Potamogeton sp., leaf	690 mg/kg wet weight	Water = 32+-18 ppb, pH=7.2, Hartley Bay, Canada, 1974[1]
Potamogeton sp., stem	13 mg/kg wet weight	Water = 3+-1 ppb, pH=7.2, East outlet (French R) to L. Huron, Can[1]
Potamogeton sp., stem	150 mg/kg wet weight	Water = 32+-18 ppb, pH=7.2, Hartley Bay, Canada, 1974[1]
Potamogeton sp., stem	18 mg/kg wet weight	Water = 3+-3 ppb, pH=7.1, French R, Can, non-poll.[1]
Potamogeton sp., stem	255 mg/kg wet weight	Water = 42+-22 ppb, pH=7.3, Wanapitei R, Can, silt-ladden and poll.[1]
Potamogeton sp., stem	62 mg/kg wet weight	Water = 15+-11 ppb, pH=7.2, Western outlet (French R) to L. Huron, Can[1]
Potamogeton sp., stem	7 mg/kg wet weight	Water = 2+-1 ppb, pH=7.2, Pickerel R, Can, non-poll.[1]

Potamogeton sp., stem	78 mg/kg wet weight	Water = 29+-12 ppb, pH=7.3, W Wanapitei B, Can, 1974[1]
Raphanus sativus, root	1150 mg/kg dry weight	Soil Ni = 2330 ppm, pH=2.96, 7.4 km from Caniston smelter, Sudbury, lab[4]
Raphanus sativus, shoot	430 mg/kg dry weight	Soil Ni = 2330 ppm, pH=2.96, 7.4 km from Caniston smelter, Sudbury, lab[4]
Sediment	116 mg/kg wet weight	Water = 32+-18 ppb, pH=7.2, Hartley Bay, Canada, 1974, SD=32[1]
Sediment	118 mg/kg wet weight	Water = 10+-4 ppb, pH=7.2, Main outlet (French R) to L. Huron, Can, SD=21[1]
Sediment	13 mg/kg wet weight	Water = 2+-1 ppb, pH=7.2, Pickerel R, Can, non-poll., SD=3[1]
Sediment	147 mg/kg wet weight	Water = 15+-11 ppb, pH=7.2, Western outlet (French R) to L. Huron, Can, SD=39[1]
Sediment	15 mg/kg wet weight	Water = 3+-3 ppb, pH=7.1, French R, Can, non-poll., SD=1[1]
Sediment	157 mg/kg wet weight	Water = 29+-12 ppb, pH=7.3, W Wanapitei B, Can, 1974, SD=29[1]
Sediment	16.0 mg/kg dry weight	Unpolluted area, river, Illinois River, rg = 10 - 22[5]
Sediment	18.7 mg/kg dry weight	Unpolluted area, estuary, Urr water, rg = 9.0 - 34.5[6]
Sediment	224 mg/kg wet weight	Water = 42+-22 ppb, pH=7.3, Wanapitei R, Can, silt-ladden and poll., rg=110[1]
Sediment	23 mg/kg wet weight	Water = 3+-1 ppb, pH=7.2, East outlet (French R) to L. Huron, Can[1]
Sediment	7.2 mg/kg dry weight	Unpolluted area, estuary, Urr water, rg = 4.2 - 15.0[7]
Sediment	9.2 mg/kg dry weight	Unpolluted area, estuary, Urr water, rg = 6.1 - 14.3[6]
Spring wheat	0.5 mg/kg dry weight	Soil pH = 7.5, Ni added to soil = 10 ppm[8]
Spring wheat	0.5 mg/kg dry weight	Soil pH = 7.5, Ni added to soil = 5 ppm[8]
Spring wheat	0.75 mg/kg dry weight	Soil pH = 6.5, Ni added to soil = 5 ppm[8]
Spring wheat	0.75 mg/kg dry weight	Soil pH = 7.5, Ni added to soil = 40 ppm[8]
Spring wheat	1.0 mg/kg dry weight	Soil pH = 6.5, Ni added to soil = 0 ppm[8]
Spring wheat	1.25 mg/kg dry weight	Soil pH = 7.5, Ni added to soil = 80 ppm[8]
Spring wheat	10.0 mg/kg dry weight	Soil pH = 5.1, Ni added to soil = 40 ppm[8]
Spring wheat	105 mg/kg dry weight	Soil pH = 5.5, Ni added to soil = 160 ppm[8]
Spring wheat	17.2 mg/kg dry weight	Soil pH = 5.5, Ni added to soil = 80 ppm[8]
Spring wheat	2.0 mg/kg dry weight	Soil pH = 6.5, Ni added to soil = 20 ppm[8]
Spring wheat	2.2 mg/kg dry weight	Soil pH = 5.5, Ni added to soil = 0 ppm[8]
Spring wheat	2.2 mg/kg dry weight	Soil pH = 6.5, Ni added to soil = 10 ppm[8]
Spring wheat	2.5 mg/kg dry weight	Soil pH = 5.1, Ni added to soil = 0 ppm[8]
Spring wheat	2.5 mg/kg dry weight	Soil pH = 5.5, Ni added to soil = 5 ppm[8]
Spring wheat	2.75 mg/kg dry weight	Soil pH = 6.5, Ni added to soil = 40 ppm[8]
Spring wheat	3.0 mg/kg dry weight	Soil pH = 5.1, Ni added to soil = 10 ppm[8]
Spring wheat	3.0 mg/kg dry weight	Soil pH = 6.5, Ni added to soil = 80 ppm[8]
Spring wheat	3.0 mg/kg dry weight	Soil pH = 7.5, Ni added to soil = 160 ppm[8]
Spring wheat	3.7 mg/kg dry weight	Soil pH = 5.5, Ni added to soil = 10 ppm[8]
Spring wheat	4.5 mg/kg dry weight	Soil pH = 5.1, Ni added to soil = 5 ppm[8]
Spring wheat	4.7 mg/kg dry weight	Soil pH = 5.5, Ni added to soil = 20 ppm[8]
Spring wheat	6.5 mg/kg dry weight	Soil pH = 5.5, Ni added to soil = 40 ppm[8]

Spring wheat	74.0 mg/kg dry weight	Soil pH = 5.1, Ni added to soil = 80 ppm[8]
Spring wheat	8.0 mg/kg dry weight	Soil pH = 5.1, Ni added to soil = 20 ppm[8]
Spring wheat	8.25 mg/kg dry weight	Soil pH = 6.5, Ni added to soil = 160 ppm[8]
Spruce forest soil	26 mg/kg dry weight soil	Central Sweden, needle litter, O1[9]
Spruce forest soil	36 mg/kg dry weight soil	Central Sweden, Humus, O2[9]
Vegetation	1.1 mg/kg dry weight	Conc in podzol:f = 8, h = 5, a1a2 = 20, b1 = 15, b2 = 20, b3 = 30,[10]
Yellow pickerel, gill	16.0	Wanapitei R, Can, 1974, wat=42+-22 ppb, polluted, n=8[1]
Yellow pickerel, kidney	51.6	Wanapitei R, Can, 1974, wat=42+-22 ppb, polluted, n=8[1]
Yellow pickerel, liver	14.4	Wanapitei R, Can, 1974, wat=42+-22 ppb, polluted, n=8[1]
Yellow pickerel, muscle	13.8	Wanapitei R, Can, 1974, wat=42+-22 ppb, polluted, n=8[1]
Zooplankton	11 mg/kg wet weight	Water = 10+-4 ppb, pH=7.2, Main outlet (French R) to L. Huron, Can[1]
Zooplankton	17 mg/kg wet weight	Water = 15+-11 ppb, pH=7.2, Western outlet (French R) to L. Huron, Can[1]
Zooplankton	18 mg/kg wet weight	Water = 29+-12 ppb, pH=7.3, W Wanapitei B, Can, 1974[1]
Zooplankton	2.75 mg/kg dry weight	19 st, NY, conc in algae = 2.62 mg/kg dw, mean, Long Island[11]
Zooplankton	27 mg/kg wet weight	Water = 42+-22 ppb, pH=7.3, Wanapitei R, Can, silt-ladden and poll.[1]
Zooplankton	3 mg/kg wet weight	Water = 3+-3 ppb, pH=7.1, French R, Can, non-poll.[1]
Zooplankton	51 mg/kg wet weight	Water = 32+-18 ppb, pH=7.2, Hartley Bay, Canada, 1974[1]
Zooplankton	7 mg/kg wet weight	Water = 2+-1 ppb, pH=7.2, Pickerel R, Can, non-poll.[1]
Zooplankton	8 mg/kg wet weight	Water = 3+-1 ppb, pH=7.2, East outlet (French R) to L. Huron, Can[1]

Source: 1.Hutchinson, T.C., Fedorenko, Fitchko, Kuja, Loon and Lichwa. 1974. In: J.O. Nriagu (ed.): Environmenal Biochemistry. Ann Arbor Sci. Publ. Ann Arbor, Mich, 565-585.
2.Nilsson, I. 1972. Oikos, 22, pp 366.
3.Tyunyaeva, G.N., A.K. Minenko and L.A. Penkov. 1974. Agrokhimiya No 6, 110-114.
4.Whitby, L., P.M. Stokes, T.C. Hutchinson and G. Myslik. 1976. Can. Mineral. 14, 47-57.
5.Mathis, B.J. and T.F. Cummins. 1973. Water Pollut. Control 45 pp 1573.
6.Taylor, D. 1976. Sci. Total Environ. 6(3), pp 259-264.
7.Taylor, D. 1976. Estuarine Coastal Mar. Sci. 2, pp 417.
8.Patterson, J.B.E. 1971. Tech. Bull., 21, pp 193-207.
9.Inman, R.E., R.B. Ingersoll and E.A. Levy. 1971. Science 172, 1229-1231.
10.Mitchell, R. and C. Chamberlin. 1978. in: Berg. G. (ed.); Indicators of Enteric Contamination in Natural Waters.
11.Greig, R.A., A. Adams and D.R. Wenzloff. 1977. Bull. Environ. Contam. and Toxic., 18(1), pp 3-8.

5-378 Concentrations
Nitrate - N

Species	Value	Condition
Agricultural cropland	0.4 mg/l water	General characteristics, runoff[1]
Animal feedlot	16.5 mg/l water	General characteristics, runoff, rg = 10 - 23[1]
Forested land	0.7 mg/l water	General characteristics, runoff, rg = 0.1 - 1.3[1]
Precipitation	0.62 mg/l water	General characteristics, runoff, rg = 0.14 - 1.1[1]

Source: 1.Donigian, A.S. and N.H. Crawford. 1976. Modeling nonpoint Pollution from the Land Surface, EPA-600/376-083. July.

5-379 Concentrations
Octachlorodibenzo-p-dioxin

Species	Value	Condition
Rat, adipose tissue	0.04 microgram	Ingested = 120.7 ug, 2 w, 106 g anim, SE = +- 0.04[1]
Rat, feces	13.9 microgram	Ingested = 22.7 ug, 2 w, 106 g anim, SE = +- 0.7[1]
Rat, feces	45.2 microgram	Ingested = 120.7 ug, 2 w, 106 g anim, SE = +- 1.7[1]
Rat, liver	0.46 microgram	Ingested = 22.7 ug, 2 w, 106 g anim, SE = +- 0.09[1]
Rat, liver	1.35 microgram	Ingested = 120.7 ug, 2 w, 106 g anim, SE = +- 0.19[1]

Source: 1.Williams, D.T., H.M. Cunningham and B.J. Blanchfield. 1972. Bull. Environ. Contam. Toxicol. 7(1), pp 57-62.

5-380 Concentrations
Organic Hg

Species	Value	Condition
Anchovy	100 % of total Hg	Monterey Bay, Calif, total Hg = 47 ng/g ww, 4. year class[1]
Anchovy	76 % of total Hg	Monterey Bay, Calif, total = 38 ng/g ww, first year class[1]
Anchovy	88 % of total Hg	Monterey Bay, Calif, total Hg = 75 ng/g ww, 3. year class[1]
Phytoplankton	12 % of total Hg	Monterey Bay, Calif, 28 June, total = 15.2 ng/g ww[1]
Phytoplankton	34 % of total Hg	Monterey Bay, Calif, 8 June, total = 5.9 ng/g ww[1]
Phytoplankton	42 % of total Hg	Monterey Bay, Calif, 15 Aug, total = 10.6 ng/g ww[1]
Phytoplankton	67 % of total Hg	Monterey Bay, Calif, 24 May, total = 5.7 ng/g ww[1]

Source: 1.Knaver, G.A. and J.H. Martin. 1972. Limnol. Oceanogr. 17(6), pp 868-876.

5-381 Concentrations
P

Species	Value	Condition
Tomato, leaf	0.44 % of dry weight	Added superphosphate = 2.4 kg/m**3 soil[1]
Tomato, leaf	0.96 % of dry weight	Added superphosphate = 9.6 kg/m**3 soil[1]

Source: 1.Harrod, M.F. 1971. In: Tech. Bull. 21, pp 176-189. Trace Elements in Soils and Crops.

5-382 Concentrations
PCB

Species	Value	Condition
Bird, egg yolk lipid	241. mg/kg dry weight	24 April 1970, % lipid = 100, Audubon Canyon ranch, Calif, USA[1]
Buffalo	0.06 mg/kg dry weight	S Dakota, freshwater, water = 0.5 mg/m3, n = 10[2]
Bullhead	0.11 mg/kg dry weight	S Dakota, freshwater, water = 0.5 mg/m3, n = 10[2]
Carp	0.11 mg/kg dry weight	S Dakota, freshwater, water = 0.5 mg/m3, n = 10[2]
Casmerodius albus, brain	9.7 mg/kg wet weight	5 obs, % lipid = 6.8, dieldrin = 4.36, endrin < 0.18, Audubon Canyon,[1]
Casmerodius albus, breast muscle	16. mg/kg wet weight	5 obs., % lipid = 2.4, Audubon Canyon ranch, Calif, USA[1]
Casmerodius albus, liver	37.3 mg/kg wet weight	5 obs., % lipid = 3.5, Hg = 6.08, Audubon Canyon ranch, Calif, USA[1]
Chlorella pyrenoidosa	16 mg/kg wet weight	Water initial = 10 ppb, 1 h, 295 K, 165 ftc, octachloro-PCB[3]
Chlorella pyrenoidosa	32 mg/kg wet weight	Water initial = 10 ppb, 1 h, 295 K, 165 ftc, tetrachloro-PCB[3]
Chlorella pyrenoidosa	52 mg/kg wet weight	Water initial = 10 ppb, 1 h, 295 K, 165 ftc, decachloro-PCB[3]

Chlorella pyrenoidosa	70 mg/kg wet weight	Water initial = 10 ppb, 1 h, 295 K, 165 ftc, hexachloro-PCB[3]
Chlorella pyrenoidosa, dead cells	156 mg/kg wet weight	Water initial = 10 ppb, 1 h, 295 K, 165 ftc, hexachloro-PCB[3]
Chlorella pyrenoidosa, dead cells	62 mg/kg wet weight	Water initial = 10 ppb, 1 h, 295 K, 165 ftc, tetrachloro-PCB[3]
Crappie	0.05 mg/kg dry weight	S Dakota, freshwater, water = 0.5 mg/m3, n = 10[2]
Egret	0.7 mg/kg wet weight	11 May 1970, % lipid = 2.1, Audubon Canyon ranch, Calif, USA[1]
Egret	1.3 mg/kg wet weight	11 May 1970, % lipid = 2.2, Audubon Canyon ranch, Calif, USA[1]
Egret	1.7 mg/kg wet weight	11 May 1970, % lipid = 1.2, Audubon Canyon ranch, Calif, USA[1]
Egret	1.75 mg/kg wet weight	7 May 1970, % lipid = 1.2, Audubon Canyon ranch, Calif, USA[1]
Egret	1.8 mg/kg wet weight	15 May 1970, % lipid = 1.7, Audubon Canyon ranch, Calif, USA[1]
Egret	4.6 mg/kg wet weight	13 May 1970, % lipid = 1.2, Audubon Canyon ranch, Calif, USA[1]
Egret, egg yolk lipid	40.3 mg/kg dry weight	No date, % lipid = 100, Audubon Canyon ranch, Calif, USA[1]
Escherichia coli	13.6 mg/kg wet weight	PCB in medium = 0.1 mg/l, 24 h, lab[4]
Escherichia coli	4.4 mg/kg wet weight	PCB in medium = 0.01 mg/kg, 24 h, lab[4]
Heron	0.49 mg/kg wet weight	24 April 1970, % lipid = 1.2, Audubon Canyon ranch, Calif, USA[1]
Heron	1.35 mg/kg wet weight	24 April 1970, % lipid = 1.6, Audubon Canyon ranch, Calif, USA[1]
Heron	1.82 mg/kg wet weight	24 April 1970, % lipid = 1., Audubon Canyon ranch, Calif, USA[1]
Heron, egg yolk lipid	50.6 mg/kg dry weight	20 April 1970, % lipid = 100, Audubon Canyon ranch, Calif, USA[1]
Heron, embryo	1.2 mg/kg wet weight	No date, % lipid = 3.6, Audubon Canyon ranch, Calif, USA[1]
Mytilus edulis	10 ng/g wet weight	Conc in sediment = 5.7 ng/g dw, S Pudet Sound[5]
Mytilus edulis	11 ng/g wet weight	Conc in sediment = 1.8 ng/g dw, S Pudet Sound[5]
Mytilus edulis	11 ng/g wet weight	Conc in sediment = 3.1 ng/g dw, S Pudet Sound[5]
Mytilus edulis	14 ng/g wet weight	Conc in sediment = 0.7 ng/g dw, S Pudet Sound[5]
Mytilus edulis	14 ng/g wet weight	Conc in sediment = 10 ng/g dw, S Pudet Sound[5]
Mytilus edulis	16 ng/g wet weight	Conc in sediment = 0.7 ng/g dw, S Pudet Sound[5]
Mytilus edulis	16 ng/g wet weight	Conc in sediment = 1.5 ng/g dw, S Pudet Sound[5]
Mytilus edulis	210 ng/g wet weight	Conc in sediment = 330 ng/g dw, S Pudet Sound[5]
Mytilus edulis	24 ng/g wet weight	Conc in sediment = 1.5 ng/g dw, S Pudet Sound[5]
Mytilus edulis	27 ng/g wet weight	Conc in sediment = 1.1 ng/g dw, S Pudet Sound[5]
Mytilus edulis	30 ng/g wet weight	Conc in sediment = 2.1 ng/g dw, S Pudet Sound[5]
Mytilus edulis	31 ng/g wet weight	Conc in sediment = 1.0 ng/g dw, S Pudet Sound[5]
Mytilus edulis	38 ng/g wet weight	Conc in sediment = 17 ng/g dw, S Pudet Sound[5]
Mytilus edulis	40 ng/g wet weight	Conc in sediment = 7.3 ng/g dw, S Pudet Sound[5]
Mytilus edulis	50 ng/g wet weight	Conc in sediment = 2.6 ng/g dw, S Pudet Sound[5]
Mytilus edulis	72 ng/g wet weight	Conc in sediment = 6.3 ng/g dw, S Pudet Sound[5]
Mytilus edulis	85 ng/g wet weight	Conc in sediment = 1.8 ng/g dw, S Pudet Sound[5]
Mytilus edulis	95 ng/g wet weight	Conc in sediment = 70 ng/g dw, S Pudet Sound[5]
Pagophilus groenlandicus, blubber	1.08 mg/kg wet weight	83.0 % fat, 15 samples, 1973, Gulf of St Lawrence, SD = +- 0.53[6]

Pagophilus groenlandicus, blubber	2.46 mg/kg wet weight	97.8 % fat, 10 samples, 1971, Gulf of St Lawrence, SD = +- 1.38[6]
Pagophilus groenlandicus, blubber	6.05 mg/kg wet weight	95.5 % fat, 4 animals, St Lawrence, Canada, SD = +- 2.05[7]
Pagophilus groenlandicus, brain	0.03 mg/kg wet weight	6.6 % fat, 14 samples, 1973, Gulf of St Lawrence, SD = +- 0.04[6]
Pagophilus groenlandicus, brain	0.05 mg/kg wet weight	7.5 % fat, 10 samples, 1971, Gulf of St Lawrence, SD = +- 0.13[6]
Pagophilus groenlandicus, brain	0.09 mg/kg wet weight	11.5 % fat, 3 animals, St Lawrence, Canada, SD = +- 0.0[7]
Pagophilus groenlandicus, kidney	0.43 mg/kg wet weight	14.4 % fat, 13 samples, 1973, Gulf of St Lawrence, SD = +- 0.44[6]
Pagophilus groenlandicus, liver	0.07 mg/kg wet weight	1.8 % fat, 4 ainmals, SD = +- 0.01[7]
Pagophilus groenlandicus, liver	0.14 mg/kg wet weight	7.0 % fat, 11 samples, 1971, Gulf of St Lawrence, SD = +- 0.10[6]
Pagophilus groenlandicus, liver	0.16 mg/kg wet weight	4.4 % fat, 16 samples, 1973, Gulf of St Lawrence, SD = +- 0.18[6]
Pagophilus groenlandicus, muscle	0.45 mg/kg wet weight	1.1 % fat, 4 animals, St Lawrence, Canada, SD = +- 0.01[7]
Rhizophora mangle, hypocotyl	0.35 mg/kg wet weight	Application = 0.56 kg/ha, pH = 6.3-6.7, s = 2.5 %, 6 w, aroclor 1242[8]
Rhizophora mangle, hypocotyl	0.53 mg/kg wet weight	Application = 5.60 kg/ha, pH = 6.3-6.7, s = 2.5 %, 6 w, aroclor 1242[8]
Rhizophora mangle, hypocotyl	0.79 mg/kg wet weight	Application = 1.12 kg/ha, pH = 6.3-6.7, s = 2.5 %, 6 w, aroclor 1242[8]
Rhizophora mangle, hypocotyl	0.92 mg/kg wet weight	Application = 2.80 kg/ha, pH = 6.3-6.7, s = 2.5 %, 6 w, aroclor 1242[8]
Rhizophora mangle, hypocotyl	1.50 mg/kg wet weight	Application = 11.20 kg/ha, pH = 6.3-6.7, s = 2.5 %, 6 w, aroclor 1242[8]
Rhizophora mangle, leaves	0.46 mg/kg wet weight	Application = 0.56 kg/ha, pH = 6.3-6.7, s = 2.5 %, 6 w, aroclor 1242[8]
Rhizophora mangle, leaves	0.80 mg/kg wet weight	Application = 5.60 kg/ha, pH = 6.3-6.7, s = 2.5 %, 6 w, aroclor 1242[8]
Rhizophora mangle, leaves	0.90 mg/kg wet weight	Application = 11.20 kg/ha, pH = 6.3-6.7, s = 2.5 %, 6 w, aroclor 1242[8]
Rhizophora mangle, leaves	0.92 mg/kg wet weight	Application = 2.80 kg/ha, pH = 6.3-6.7, s = 2.5 %, 6 w, aroclor 1242[8]
Rhizophora mangle, leaves	1.11 mg/kg wet weight	Application = 1.12 kg/ha, pH = 6.3-6.7, s = 2.5 %, 6 w, aroclor 1242[8]
Rhizophora mangle, root	0.10 mg/kg wet weight	Application = 11.20 kg/ha, pH = 6.3-6.7, s = 2.5 %, 6 w, aroclor 1242[8]
Rhizophora mangle, root	0.10 mg/kg wet weight	Application = 5.60 kg/ha, pH = 6.3-6.7, s = 2.5 %, 6 w, aroclor 1242[8]
Sediment	0.0064 mg/kg wet weight	S Dakota, freshwater, water = 0.5 mg/m3, n = 10[2]

Source: 1.Faber, R.B., R.W. Risenbough and H.M. Pratt. 1972. Environ. Pollut., 3, pp 111-122.
2.Greichus, Y.A., A. Greichus and R.J. Emerick. 1973. Bull. Environ. Contam. Toxicol. 9, pp 321-327.
3.Urey, J.C., J.C. Kricher and J.M. Boy lan. 1976. Bull. Environ. Contam. Toxicol. 16(1), pp 81-85.
4.Keil, J.E., S.H. Sandifer, C.D. Graber and L.E. Priester. 1972 Water Res. 6(7), pp 837-842.
5.Mowrer, J., J. Calambokidis, N. Musgrove, B. Drager, M.W. Beug and S.G. Herman. 1977. Bull. Environ.Contam. Toxicol. 18(5), pp 588-594.
6.Frank, R., K. Ronald and H.E. Braun. 1973. Bull. Fish Res. Board Can., 30, pp 1053.
7.Jones, D., K. Ronald, D.H. Lavinge, R. Frank, M. Holdrinet and J.F. Uthe. 1976. Sci. Total Environ., 5(2), pp 181-196.
8.Walch, G.E., T.A. Hollister and J. Forester. 1974. Bull. Environ. Contam. Toxicol. 12(2), pp 129-135

5-383 Concentrations Pb

Species	Value	Condition
Alisma plantago aquatica, old leaves	21.72 mg/kg dry weight	150 m below pollution entry I, water = 0.068 mg/l water, processed leaves[1]
Alisma plantago aquatica, old leaves	3.03 mg/kg dry weight	800 m above pollution entry I, water = not detect., processed leaves[1]
Alisma plantago aquatica, old leaves	38.10 mg/kg dry weight	150 m below pollution entry I, water = 0.068 mg/l water, whole leaves[1]
Alisma plantago aquatica, old leaves	4.72 mg/kg dry weight	800 and 50 m below pollution I and II, water = 0.060 mg/l water[1]
Alisma plantago aquatica, old leaves	4.92 mg/kg dry weight	800 m above pollution entry I, water = not detect., whole leaves[1]
Alisma plantago aquatica, old leaves	7.36 mg/kg dry weight	800 and 50 m below pollution I and II, water = 0.060 mg/l water[1]
Alisma plantago aquatica, young leaves	14.25 mg/kg dry weight	800 m above pollution entry I, water = not detect., processed leaves[1]
Alisma plantago aquatica, young leaves	14.37 mg/kg dry weight	800 m above pollution entry I, water = not detect., whole leaves[1]
Alisma plantago aquatica, young leaves	18.25 mg/kg dry weight	800 and 50 m below pollution I and II, water = 0.060 mg/l water[1]
Alisma plantago aquatica, young leaves	18.77 mg/kg dry weight	150 m below pollution entry I, water = 0.068 mg/l water, processed leaves[1]
Alisma plantago aquatica, young leaves	23.61 mg/kg dry weight	150 m below pollution entry I, water = 0.068 mg/l water, whole leaves[1]
Alisma plantago aquatica, young leaves	9.81 mg/kg dry weight	800 and 50 m below pollution I and II, water = 0.060 mg/l water[1]
Bryum pseudotriquetrum	1079 mg/kg dry weight	Pb in soil = 1150 ppm, Sand Haw mine, W Yorkshire, GB[2]
Bryum pseudotriquetrum	16670 mg/kg dry weight	Pb in soil = 11104 ppm, Blea Beck Junction, W Yorkshire, GB[2]
Bryum pseudotriquetrum	3806 mg/kg dry weight	Pb in soil = 1640 ppm, Slater mine, Foolow, Derbyshire, GB[2]
Buccinum undatum	41 mg/kg dry weight	Sediment = 182 ppm, Firth of Clyde[3]
Buccinum undatum	5.4 mg/kg dry weight	Sediment = 50 ppm, Irish Sea[3]
Cabbage, plant	0.73 mg/kg dry weight	15 plants, urban area[4]
Cabbage, plant	0.81 mg/kg dry weight	15 plants, rural area[4]
Camptothecium lutescens	164 mg/kg dry weight	Pb in soil = 80 ppm, Blea Beck Head, W Yorkshire, GB[2]
Corn, old leaves	10000 mg/kg dry weight	Hydroponic solution conc = 100 mg/l water, 6 d, phosphate deficient[5]
Corn, old leaves	2000 mg/kg dry weight	Hydroponic solution conc = 100 mg/l water, 6 d, adequate phosphorus[5]
Corn, root	10600 mg/kg dry weight	Hydroponic solution conc = 100 mg/l water, 3 d[5]
Corn, top	300 mg/kg dry weight	Soil conc = 200 mg/kg dw, 4 w, rg = 200 - 400[5]
Corn, top	3900 mg/kg dry weight	Hydroponic solution conc = 100 mg/l water, 3 d[5]
Corn, young leaves	6700 mg/kg dry weight	Hydroponic solution conc = 100 mg/l water, 6 d, phosphate deficit[5]
Corn, young leaves	900 mg/kg dry weight	Hydroponic solution conc = 100 mg/l water, 6 d, adequate phosphorus[5]

Cornicularia muricata	12045 mg/kg dry weight	Pb in soil = 2570 ppm, Blea Beck Head, W Yorkshire, GB[2]
Crangon allmani	17 mg/kg dry weight	Sediment = 50 ppm, Irish Sea[3]
Crangon allmani	32 mg/kg dry weight	Sediment = 182 ppm, Firth of Clyde[3]
Crataegus monogyna	509 mg/kg dry weight	Avonmouth, Bristol, GB, unwashed leaves[6]
Crataegus monogyna, leaf	1 % removed of soluble metal	Washed with deionised water, Bristol, GB[6]
Crataegus monogyna, leaf	1 % removed of total metal	Washed with deionised water, Bristol, GB[6]
Crataegus monogyna, leaf	1.07 % removed of soluble metal	Boiled, Bristol, GB[6]
Crataegus monogyna, leaf	1.15 % removed of total metal	Boiled, Bristol, GB[6]
Crataegus monogyna, leaf	1.51 total/soluble metal removed	Boiled, Bristol, GB[6]
Crataegus monogyna, leaf	12.17 total/soluble metal removed	Washed with deionised water, Bristol, GB[6]
Crataegus monogyna, leaf	3.08 % removed of total metal	Washed with a detergent, Bristol, GB[6]
Crataegus monogyna, leaf	4.53 total/soluble metal removed	Washed with a detergent, Bristol, GB[6]
Dicranella varia	10460 mg/kg dry weight	Pb in soil = 7625 ppm, Slater mine, Foolow, Derbyshire, GB[2]
Dicranella varia	11950 mg/kg dry weight	Pb in soil = 8490 ppm, Pasryremine, Foolow, Derbyshire, GB[2]
Dicranella varia	1325 mg/kg dry weight	Pb in soil = 975 ppm, Speed mine, Foolow, Derbyshire, GB[2]
Dicranella varia	42300 mg/kg dry weight	Pb in soil = 1870 ppm, Slater mine, Foolow, Derbyshire, GB[2]
Dicranella varia	540 mg/kg dry weight	Pb in soil = 1108 ppm, Blythe Wood, alport, Derbyshire, GB[2]
Dicranella varia	60450 mg/kg dry weight	Pb in soil = 2400 ppm, Pasryremine, Foolow, Derbyshire, GB[2]
Dicranella varia	7 mg/kg dry weight	Pb in soil = 4 ppm, Botanic Garden, Cottingham, W Yorkshire, GB[2]
Different Pb species in freshwater, bed sediments	110 mg/l	Copper Ridge District, Tenn, sample no 2, less than 5 um fraction[7]
Different Pb species in freshwater, bed sediments	50 mg/l	Copper Ridge District, Tenn, sample no 1, less than 5 um fraction[7]
Different Pb species in freshwater, bed sediments	69 mg/l	Copper Ridge District, Tenn, sample no 3, less than 5 um fraction[7]
Different Pb species in freshwater, coarse particulates	124 mg/l	Copper Ridge District, Tenn, sample no 1, suspended solids[7]
Different Pb species in freshwater, coarse particulates	213 mg/l	Copper Ridge District, Tenn, sample no 2, suspended solids[7]
Different Pb species in freshwater, coarse particulates	653 mg/l	Copper Ridge District, Tenn, sample no 3, suspended solids[7]
Different Pb species in freshwater, colloidal material	2820 mg/l	Copper Ridge District, Tenn, sample no 2, suspended solids[7]
Different Pb species in freshwater, colloidal material	2820 mg/l	Copper Ridge District, Tenn, sample no 3, suspended solids[7]

Different Pb species in freshwater, colloidal material	62 mg/l	Copper Ridge District, Tenn, sample no 1, suspended solids[7]
Different Pb species in freshwater, dissolved solids	75 mg/l	Copper Ridge District, Tenn, sample no 2[7]
Different Pb species in freshwater, dissolved solids	77 mg/l	Copper Ridge District, Tenn, sample no 1[7]
Different Pb species in freshwater, dissolved solids	84 mg/l	Copper Ridge District, Tenn, sample no 3[7]
Different Pb species in freshwater, total bed sediments	106 mg/l	Copper Ridge District, Tenn, sample no 1[7]
Different Pb species in freshwater, total bed sediments	130 mg/l	Copper Ridge District, Tenn, sample no 2[7]
Different Pb species in freshwater, total bed sediments	916 mg/l	Copper Ridge District, Tenn, sample no 3[7]
Different Pb species in freshwater, water phase	0.011 mg/l	Copper Ridge District, Tenn, sample no 3[7]
Different Pb species in freshwater, water phase	0.015 mg/l	Copper Ridge District, Tenn, sample no 2[7]
Different Pb species in freshwater, water phase	0.019 mg/l	Copper Ridge District, Tenn, sample no 1[7]
Different Pb species in marine sediment	0.006 mg/kg dry weight	Silty sand, LA harbor, USA, 1974, extraction with distilled water[8]
Different Pb species in marine sediment	0.007 mg/kg dry weight	Sandy silt, LA harbor, USA, 1974, extraction with distilled water[8]
Different Pb species in marine sediment	0.008 mg/kg dry weight	Silty clay, LA harbor, USA, 1974, extraction with distilled water[8]
Different Pb species in marine sediment	0.1 mg/kg dry weight	Silty sand, LA harbor, USA, 1974, extrac. with Na dithionite+Na citrate[8]
Different Pb species in marine sediment	0.25 mg/kg dry weight	Sandy silt, LA harbor, USA, 1974, extrac. with Na dithionite+Na citrate[8]
Different Pb species in marine sediment	0.4 mg/kg dry weight	Silty clay, LA harbor, USA, 1974, extrac. with Na dithionite+Na citrate[8]
Different Pb species in marine sediment	0.5 mg/kg dry weight	Silty sand, LA harbor, USA, 1974, extraction with H2O2+0.01 M HNO3[8]
Different Pb species in marine sediment	1.0 mg/kg dry weight	Silty clay, LA harbor, USA, 1974, extraction with acetic acid[8]
Different Pb species in marine sediment	1.2 mg/kg dry weight	Sandy silt, LA harbor, USA, 1974, extraction with H2O2+0.01 M HNO3[8]
Different Pb species in marine sediment	11.0 mg/kg dry weight	Silty clay, LA harbor, USA, 1974, extraction with H2O2+0.01 M HNO3[8]
Different Pb species in marine sediment	145 mg/kg dry weight	Silty clay, LA harbor, 1974, extrac. with 25 % Acetic+0.04 M NH2OH HCl[8]
Different Pb species in marine sediment	17.0 mg/kg dry weight	Sandy silt, LA harbor, USA, 1974, extraction with NH2OH HCl[8]
Different Pb species in marine sediment	2.0 mg/kg dry weight	Silty sand, LA harbor, USA, 1974, extraction with H2O2+ammonium acetate[8]
Different Pb species in marine sediment	20.0 mg/kg dry weight	Silty clay, LA harbor, USA, 1974, extraction with NH2OH HCl[8]
Different Pb species in marine sediment	21.0 mg/kg dry weight	Sandy silt, LA harbor, 1974, extrac. with 25 % Acetic+0.04 M NH2OH HCl[8]

Different Pb species in marine sediment	42 mg/kg dry weight	Silty clay, LA harbor, USA, 1974, extraction with H2O2+ammonium acetate[8]
Different Pb species in marine sediment	5.0 mg/kg dry weight	Sandy silt, LA harbor, USA, 1974, extraction with H2O2+ammonium acetate[8]
Different Pb species in marine sediment	6.8 mg/kg dry weight	Silty sand, LA harbor, USA, 1974, extraction with acetic acid[8]
Different Pb species in marine sediment	7.2 mg/kg dry weight	Sandy silt, LA harbor, USA, 1974, extraction with acetic acid[8]
Different Pb species in marine sediment	8.0 mg/kg dry weight	Silty sand, LA harbor, USA, 1974, extraction with NH2OH HCl[8]
Different Pb species in marine sediment	9.1 mg/kg dry weight	Silty sand, LA harbor, 1974, extrac. with 25 % Acetic+0.04 M NH2OH HCl[8]
Different Pb species in marine sediment, biogenous Pb	2.0 mg/kg dry weight	Silty sand, LA harbor, USA, 1974[8]
Different Pb species in marine sediment, biogenous Pb	42 mg/kg dry weight	Silty clay, LA harbor, USA, 1974[8]
Different Pb species in marine sediment, biogenous Pb	5.0 mg/kg dry weight	Sandy silt, LA harbor, USA, 1974[8]
Different Pb species in marine sediment, interstitial Pb	0.0003 mg/kg dry weight	Silty sand, LA harbor, USA, 1974, interstitial water conc[8]
Different Pb species in marine sediment, interstitial Pb	0.0004 mg/kg dry weight	Sandy silt, LA harbor, USA, 1974, interstitial water conc[8]
Different Pb species in marine sediment, interstitial Pb	0.00045 mg/kg dry weight	Silty clay, LA harbor, USA, 1974, interstitial water conc[8]
Different Pb species in marine sediment, lithogenous Pb	165 mg/kg dry weight	Silty clay, LA harbor, USA, 1974, lithogenous fraction[8]
Different Pb species in marine sediment, lithogenous Pb	17 mg/kg dry weight	Silty sand, LA harbor, USA, 1974, lithogenous fraction[8]
Different Pb species in marine sediment, lithogenous Pb	24 mg/kg dry weight	Sandy silt, LA harbor, USA, 1974, lithogenous fraction[8]
Different Pb species in marine sediment, nodular Pb	134 mg/kg dry weight	Silty clay, LA harbor, USA, 1974, nodular hydrogenous fraction[8]
Different Pb species in marine sediment, nodular Pb	15 mg/kg dry weight	Silty sand, LA harbor, USA, 1974, nodular hydrogenous fraction[8]
Different Pb species in marine sediment, nodular Pb	34.0 mg/kg dry weight	Sandy silt, LA harbor, USA, 1974, nodular hydrogenous fraction[8]
Different Pb species in marine sediment, non-nodular Pb	1.0 mg/kg dry weight	Silty clay, LA harbor, USA, 1974, non-nodular hydrogenous fraction[8]
Different Pb species in marine sediment, non-nodular Pb	6.8 mg/kg dry weight	Silty sand, LA harbor, USA, 1974, non-nodular hydrogenous fraction[8]
Different Pb species in marine sediment, non-nodular Pb	7.2 mg/kg dry weight	Sandy silt, LA harbor, USA, 1974, non-nodular hydrogenous fraction[8]
Different Pb species in marine sediment, non-residual Pb	177 mg/kg dry weight	Silty clay, LA harbor, USA, 1974[8]
Different Pb species in marine sediment, non-residual Pb	24 mg/kg dry weight	Silty sand, LA harbor, USA, 1974[8]

Different Pb species in marine sediment, non-residual Pb	46 mg/kg dry weight	Sandy silt, LA harbor, USA, 1974[8]
Different Pb species in marine sediment, total Pb	332 mg/kg dry weight	Silty clay, LA harbor, USA, 1974[8]
Different Pb species in marine sediment, total Pb	47 mg/kg dry weight	Silty sand, LA harbor, USA, 1974[8]
Different Pb species in marine sediment, total Pb	67 mg/kg dry weight	Sandy silt, LA harbor, USA, 1974[8]
Different Pb species in seawater, Pb(2+)	1 % of inorganic Pb	Model calculations, natural seawater, includes 18 organic ligands[9]
Different Pb species in seawater, Pb(2+)	2 % of inorganic Pb	Model calculations, natural seawater[10]
Different Pb species in seawater, Pb(2+)	4.5 % of inorganic Pb	Model calculations, natural seawater, includes glycine[11]
Different Pb species in seawater, Pb(OH)2	2 % of inorganic Pb	Model calculations, natural seawater[12]
Different Pb species in seawater, PbBr(+)	0.15 % of inorganic Pb	Model calculations, natural seawater, includes glycine[11]
Different Pb species in seawater, PbCO3	0.4 % of inorganic Pb	Model calculations, natural seawater, includes glycine[11]
Different Pb species in seawater, PbCO3	40 % of inorganic Pb	Model calculations, natural seawater[12]
Different Pb species in seawater, PbCO3	75 % of inorganic Pb	Model calculations, natural seawater[10]
Different Pb species in seawater, PbCl(+)	12 % of inorganic Pb	Model calculations, natural seawater[10]
Different Pb species in seawater, PbCl(+)	15 % of inorganic Pb	Model calculations, natural seawater[12]
Different Pb species in seawater, PbCl(+)	18.9 % of inorganic Pb	Model calculations, natural seawater, includes glycine[11]
Different Pb species in seawater, PbCl(+)	97 % of inorganic Pb	Model calculations, natural seawater, includes 18 organic ligands[9]
Different Pb species in seawater, PbCl2	42.3 % of inorganic Pb	Model calculations, natural seawater, includes glycine[11]
Different Pb species in seawater, PbCl2	5 % of inorganic Pb	Model calculations, natural seawater[10]
Different Pb species in seawater, PbCl2	8 % of inorganic Pb	Model calculations, natural seawater[12]
Different Pb species in seawater, PbCl3(-)	5 % of inorganic Pb	Model calculations, natural seawater[10]
Different Pb species in seawater, PbCl3(-)	9.2 % of inorganic Pb	Model calculations, natural seawater, includes glycine[11]
Different Pb species in seawater, PbCl3(2-)	1 % of inorganic Pb	Model calculations, natural seawater[10]
Different Pb species in seawater, PbCl3(_)	5 % of inorganic Pb	Model calculations, natural seawater[12]
Different Pb species in seawater, PbCl4(2-)	3.6 % of inorganic Pb	Model calculations, natural seawater, includes glycine[11]
Different Pb species in seawater, PbHCO3(+)	1 % of inorganic Pb	Model calculations, natural seawater[10]
Different Pb species in seawater, PbHCO3(+)	1.4 % of inorganic Pb	Model calculations, natural seawater, includes glycine[11]

Different Pb species in seawater, PbOH(+)	1 % of inorganic Pb	Model calculations, natural seawater[10]
Different Pb species in seawater, PbOH(+)	1 % of inorganic Pb	Model calculations, natural seawater, includes 18 organic ligands[9]
Different Pb species in seawater, PbOH(+)	10.2 % of inorganic Pb	Model calculations, natural seawater, includes glycine[11]
Different Pb species in seawater, PbOH(+)	30 % of inorganic Pb	Model calculations, natural seawater[12]
Different Pb species in seawater, PbOHCl	8.8 % of inorganic Pb	Model calculations, natural seawater, includes glycine[11]
Different Pb species in seawater, PbSO4	0.5 % of inorganic Pb	Model calculations, natural seawater, includes glycine[11]
Different Pb species in seawater, PbSO4	1 % of inorganic Pb	Model calculations, natural seawater[10]
Encalypta vulgaris	2423 mg/kg dry weight	Pi in soil = 2423 ppm, Alport Cupola Salg Heap, Derbyshire, GB[2]
Homo sapiens, blood	1.46 micro m	Pb in tap water, 1st flush = 3.16, running = 1.86 micro m n = 27, GB,[13]
Mussel	103 mg/kg dry weight	Conc in sed = 13 mg/kg dw, filt seawater = 6 ppb, Derwent Est, Tasmania[14]
Mussel	121 mg/kg dry weight	Conc in sed = 1820 mg/kg dw, filt seawater = 6 ppb, Derwent Est, Tasmania[14]
Mussel	127 mg/kg dry weight	Conc in sed = 13 mg/kg dw, filt seawater = 9 ppb, Derwent Est, Tasmania[14]
Mussel	16 mg/kg dry weight	Conc in sed = 36 mg/kg dw, filt seawater = 10 ppb, Derwent Est, Tasmania[14]
Mussel	171 mg/kg dry weight	Conc in sed = 2500 mg/kg dw, filt seawater = 36 ppb, Derwent Est, Tasmania[14]
Mussel	226 mg/kg dry weight	Conc in sed = 384 mg/kg dw, filt seawater = 10 ppb, Derwent Est, Tasmania[14]
Mussel	228 mg/kg dry weight	Conc in sed = 4090 mg/kg dw, filt seawater = 46 ppb, Derwent Est, Tasmania[14]
Mussel	239 mg/kg dry weight	Conc in sed = 59 mg/kg dw, filt seawater = 11 ppb, Derwent Est, Tasmania[14]
Mussel	255 mg/kg dry weight	Conc in sed = 678 mg/kg dw, filt seawater = 11 ppb, Derwent Est, Tasmania[14]
Mussel	291 mg/kg dry weight	Conc in sed = 266 mg/kg dw, filt seawater = 11 ppb, Derwent Est, Tasmania[14]
Mussel	376 mg/kg dry weight	Conc in sed = 5 mg/kg dw, filt seawater = 17 ppb, Derwent Est, Tasmania[14]
Mussel	378 mg/kg dry weight	Conc in sed = 1000 mg/kg dw, filt seawater = 16 ppb, Derwent Est, Tasmania[14]
Mussel	452 mg/kg dry weight	Conc in sed = 600 mg/kg dw, filt seawater = 16 ppb, Derwent Est, Tasmania[14]
Mussel	532 mg/kg dry weight	Conc in sed = 950 mg/kg dw, filt seawater = 10 ppb, Derwent Est, Tasmania[14]
Mussel	58 mg/kg dry weight	Conc in sed = 830 mg/kg dw, filt seawater = 6 ppb, Derwent Est, Tasmania[14]
Mussel	64 mg/kg dry weight	Conc in sed = 1000 mg/kg dw, filt seawater = 10 ppb, Derwent Est, Tasmania[14]
Nucula sulcata	50 mg/kg dry weight	Sediment = 182 ppm, Firth of Clyde[3]
Pandalus montagui	24 mg/kg dry weight	Sediment = 50 ppm, Irish Sea[3]
Pandalus montagui	31 mg/kg dry weight	Sediment = 182 ppm, Firth of Clyde[3]

Pediastrum tetras	2.650 mg/kg dry weight	Pb in medium = 0.2 ppm, 294 K, light = 2500 lux[15]
Pediastrum tetras	2.749 mg/kg dry weight	Pb in medium = 0.0 ppm, 294 K, light = 2500 lux[15]
Pediastrum tetras	3.336 mg/kg dry weight	Pb in medium = 1.0 ppm, 294 K, light = 2500 lux[15]
Pediastrum tetras	3.747 mg/kg dry weight	Pb in medium = 1.5 ppm, 294 K, light = 2500 lux[15]
Pediastrum tetras	4.753 mg/kg dry weight	Pb in medium = 0.5 ppm, 294 K, light = 2500 lux[15]
Peltigera canina	5385 mg/kg dry weight	Pb in soil = 2950 ppm, Alport Cupola Salg Heap, Derbyshire, GB[2]
Peltigera canina	9943 mg/kg dry weight	Pb in soil = 1654 ppm, Blea Beck Head, W Yorkshire, GB[2]
Philonotis fontana	10015 mg/kg dry weight	Pb in soil = 940 ppm, Blea Beck Junction, W Yorkshire, GB, old plant[2]
Philonotis fontana	237 mg/kg dry weight	Pb in soil = 1640 ppm, Slater mine, Foolow, Derbyshire, GB, new plant[2]
Philonotis fontana	432 mg/kg dry weight	Pb in soil = 940 ppm, Blea Beck Junction, W Yorkshire, GB, new plant[2]
Philonotis fontana	8750 mg/kg dry weight	Pb in soil = 1640 ppm, Slater mine, Foolow, Derbyshire, GB, old plant[2]
Quercus robur	753 mg/kg dry weight	Avonmouth, Bristol, GB, unwashed laeves[6]
Quercus robur, leaf	1 % removed of soluble metal	Boiled, Bristol, GB[6]
Quercus robur, leaf	1 % removed of total metal	Boiled, Bristol, GB[6]
Quercus robur, leaf	1.25 % removed of total metal	Washed with deionised water, Bristol, GB[6]
Quercus robur, leaf	1.43 total/soluble metal removed	Boiled, Bristol, GB[6]
Quercus robur, leaf	2.2 % removed of soluble metal	Washed with deionised water, Bristol, GB[6]
Quercus robur, leaf	2.91 % removed of total metal	Washed with a detergent, Bristol, GB[6]
Quercus robur, leaf	6.847 total/soluble metal removed	Washed with deionised water, Bristol, GB[6]
Radish, root	0.7 mg/kg dry weight	Air = 2.32 ug/m3, soil = 82 ppm, HNO3 extraction[16]
Radish, root	0.8 mg/kg dry weight	Air = 1.07 ug/m3, soil = 27 ppm, HNO3 extraction[16]
Radish, root	0.8 mg/kg dry weight	Air = 1.71 ug/m3, soil = 45 ppm, HNO3 extraction[16]
Radish, root	1.8 mg/kg dry weight	Air = 0.96 ug/m3, soil = 27 ppm, HNO3 extraction[16]
Radish, root	1.9 mg/kg dry weight	Air = 0.09 ug/m3, soil = 17 ppm, HNO3 extraction[16]
Radish, root	19.8 mg/kg dry weight	From air, control in greenhouse = 19.3 ppm, Pb in soil = 29.9 ppm[17]
Radish, root	2.0 mg/kg dry weight	Air = 0.03 ug/m3, soil = 27 ppm, HNO3 extraction[16]
Radish, root	2.0 mg/kg dry weight	Air = 1.45 ug/m3, soil = 17 ppm, HNO3 extraction[16]
Radish, root	28.8 mg/kg dry weight	From air, control in greenhouse = 25.5 ppm, Pb in soil = 165 ppm[17]
Radish, root	35.5 mg/kg dry weight	From air, control in greenhouse = 31.8 ppm, Pb in soil = 299 ppm[17]
Radish, top	25.5 mg/kg dry weight	From air, control in greenhouse = 12.9 ppm, Pb in soil = 29.9 ppm[17]
Radish, top	30.4 mg/kg dry weight	From air, control in greenhouse = 17.8 ppm, Pb in soil = 165 ppm[17]
Radish, top	38.8 mg/kg dry weight	From air, control in greenhouse = 24.2 ppm, Pb in soil = 299 ppm[17]

Rat, kidney	26.1 % of total intracellular amount	Mitichondria, 36 rats, 24 h after injection of 18 ug Pb(+2)/275 g[18]
Rat, kidney	32.1 % of total intracellular amount	Soluble frac., 36 rats, 24 h after injection of 18 ug Pb(+2)/275 g[18]
Rat, kidney	34.2 % of total intracellular amount	Nuclei, 36 rats, 24 h after injection of 18 ug Pb(+2)/275 g[18]
Rat, kidney	5.0 % of total intracellular amount	Microsomes, 36 rats, 24 h after injection of 18 ug Pb(+2)/275 g[18]
Rat, kidney	5.3 % of total intracellular amount	Lysosomes, 36 rats, 24 h after injection of 18 ug Pb(+2)/275 g[18]
Rat, liver	10.2 % of total intracellular amount	Lysosomes, 36 rats, 24 h after injection of 18 ug Pb(+2)/275 g[18]
Rat, liver	10.9 % of total intracellular amount	Microsomes, 36 rats, 24 h after injection of 18 ug Pb(+2)/275 g[18]
Rat, liver	28.3 % of total intracellular amount	Nuclei, 36 rats, 24 h after injection of 18 micro g Pb(+2)/275 g[18]
Rat, liver	39.2 % of total intracellular amount	Soluble frac., 36 rats, 24 h after injection of 18 ug Pb(+2)/275 g[18]
Rat, liver	9.5 % of total intracellular amount	Mitochondria, 36 rats, 24 h after injection of 18 ug Pb(+2)/275 g[18]
Salix alba	260 mg/kg dry weight	Avonmouth, Bristol, GB, unwashed leaf[6]
Salix alba, leaf	1 % removed of total metal	Washed with a detergent, Bristol, GB[6]
Salix alba, leaf	1.15 % removed of total metal	Washed with deionised water, Bristol, GB[6]
Salix alba, leaf	1.37 total/soluble metal removed	Boiled, Bristol, GB[6]
Salix alba, leaf	1.59 % removed of total metal	Boiled, Bristol, GB[6]
Salix alba, leaf	1.66 % removed of soluble metal	Boiled, Bristol, GB[6]
Salix alba, leaf	1.72 % removed of soluble metal	Washed with deionised water, Bristol, GB[6]
Salix alba, leaf	8.0 total/soluble metal removed	Washed with deionised water, Bristol, GB[6]
Sediment	17 mg/kg dry weight	Unpolluted area, river, Illinois River, rg = 13 - 27[19]
Sediment	21.5 mg/kg dry weight	Unpolluted area, estuary, Urr water, rg = 11.5 - 34.1[20]
Sediment	23.2 mg/kg dry weight	Unpolluted area, estuary, Urr water, rg = 12.9 - 32.9[20]
Sediment	2750 mg/kg dry weight	Max water conc = 2000 microgram/l, 1 y measurement, Carnon River[21]
Sediment	2750 mg/kg dry weight	Water = 1107 microgram/l, yearly mean, Carnon River[21]
Sediment	31.2 mg/kg dry weight	Unpolluted area, estuary, Urr water, rg = 21.3 - 65.7[22]
Sediment	367 mg/kg dry weight	Max water conc = 15 microgram/l, 1 y measurement, Red River[21]
Sediment	367 mg/kg dry weight	Water = 7 microgram/l, yearly mean, Red River[21]
Sediment	42. mg/kg dry weight	Range 12-72 ppm, River Rhediol, Cardiganshire, GB[23]
Sediment	4360 mg/kg dry weight	Max water conc = 200 microgram/l, max conc, Gannel River[21]
Sediment	4360 mg/kg dry weight	Water = 73 microgram/l, yearly mean, Gannel River[21]
Sediment	46 mg/kg dry weight	Max water conc = 4 microgram/l, 1 y measurement, Fowey River[21]
Sediment	46 mg/kg dry weight	Water = 2 microgram/l, yearly mean, Fowey River[21]

Sediment	53.08 mg/kg dry weight	800 and 50 m below pollution I and II, water = 0.060 mg/l water[1]
Sediment	65 mg/kg dry weight	Max water conc = 3 microgram/l, 1 y measurement, Red River[21]
Sediment	65 mg/kg dry weight	Water = 2 microgram/l, yearly mean, Red River[21]
Sediment	72.11 mg/kg dry weight	800 m above pollution entry I, water = not detect., rg = 64.4 - 78.7[1]
Sediment	79.93 mg/kg dry weight	150 m below pollution entry I, water = 0.068 mg/l water[1]
Sediment	82 mg/kg dry weight	Max water conc = 8 microgram/l, max conc, Gannel River[21]
Sediment	82 mg/kg dry weight	Water = 3 microgram/l, yearly mean, Gannel River[21]
Sediment	89 mg/kg dry weight	Max water conc = 1.5 microgram/l, 1 y measurement, Carnon River[21]
Sediment	89 mg/kg dry weight	Water = < 1 microgram/l, yearly mean, Carnon River[21]
Sediment	98 mg/kg dry weight	Max water conc = 3 microgram/l, 1 y measurement, Fowey River[21]
Sediment	98 mg/kg dry weight	Water = 2 microgram/l, yearly mean, Fowey River[21]
Size distribution of Pb in street dust	1310 mg/kg dry weight	Size frac. = .105-0.15 mm, 14000 cars/d, size of frac.= 57.4 g/kg street dust[24]
Size distribution of Pb in street dust	202 mg/kg dry weight	Size frac. = 0.85-1.20 mm, 14000 cars/d, size of frac.= 87.1 g/kg street dust[24]
Size distribution of Pb in street dust	229 mg/kg dry weight	Size frac. = 1.20-1.70 mm, 14000 cars/d, size of frac.= 32.7 g/kg street dust[24]
Size distribution of Pb in street dust	2550 mg/kg dry weight	Size frac. = 0-0.050 mm, 14000 cars/d, size of frac.= 5.8 g/kg street dust[24]
Size distribution of Pb in street dust	2710 mg/kg dry weight	Size frac. = 0.075-.105 mm, 14000 cars/d, size of frac.= 14.3 g/kg street dust[24]
Size distribution of Pb in street dust	336 mg/kg dry weight	Size frac. = > 1.70 mm, 14000 cars/d, size of frac.= 0.5 g/kg street dust[24]
Size distribution of Pb in street dust	3470 mg/kg dry weight	Size frac. = 0.05-0.075 mm, 14000 cars/d, size of frac.= 7.3 g/kg street dust[24]
Size distribution of Pb in street dust	445 mg/kg dry weight	Size frac. = 0.3-0.42 mm, 14000 cars/d, size of frac.= 172.7 g/kg street dust[24]
Size distribution of Pb in street dust	464 mg/kg dry weight	Size frac. = 0.42-0.60 mm, 14000 cars/d, size of frac.= 166.2 g/kg street dust[24]
Size distribution of Pb in street dust	485 mg/kg dry weight	Size frac. = 0.21-0.30 mm, 14000 cars/d, size of frac.= 212.3 g/kg street dust[24]
Size distribution of Pb in street dust	580 mg/kg dry weight	Size frac. = 0.6-0.850 mm, 14000 cars/d, size of frac.= 100.1 g/kg street dust[24]
Size distribution of Pb in street dust	650 mg/kg dry weight	Size frac. = 0.15-0.21 mm, 14000 cars/d, size of frac.= 134.8 g/kg street dust[24]
Street dust	1030 mg/kg dry weight	Road type = side street roads, gutter, SD = +- 1030[25]
Street dust	1040 mg/kg dry weight	Edgware, SD = +- 450[25]
Street dust	1040 mg/kg dry weight	Wimbledon, midurban, n = 16, SD = +- 300[25]
Street dust	1330 mg/kg dry weight	Islington, central, n = 15, SD = +- 660[25]
Street dust	1460 mg/kg dry weight	Road type = main sreet, roads, pavement, SD = +- 1460[25]
Street dust	1530 mg/kg dry weight	Road type = main sreet, roads, gutter, SD = +- 600[25]
Street dust	1840 mg/kg dry weight	Lambeth, central, n = 15, SD = +- 510[25]
Street dust	35 mg/kg dry weight	Walton on the Hill, rural, n = 7, SD = +- 15[25]
Street dust	900 mg/kg dry weight	Road type = side street roads, pavement, SD = +- 420[25]
Street dust	920 mg/kg dry weight	Harold Wood, midurban, n = 18, SD = +- 500[25]

Tomato, root	418 mg/kg dry weight	Substrate = sand = 1 ppm Pb[26]
Tomato, root	51 mg/kg dry weight	Soil = 164 mg/kg dry weight[26]
Vegetation	7.6 mg/kg dry weight	Conc in podzol:f = 200, h = 100, a1a2 = 20, b1 = 15, b2 = 20[27]
Zalophus californianus, aorta	2.2 mg/kg dry weight	6 sea lions from San Luis Obispo County, Calif, USA[28]
Zalophus californianus, cerabral hemisphere	3.8 mg/kg dry weight	6 sea lions from San Luis Obispo County, Calif, USA[28]
Zalophus californianus, cerebellum	2.7 mg/kg dry weight	6 sea lions from San Luis Obispo County, Calif, USA[28]
Zalophus californianus, fat	0.3 mg/kg dry weight	6 sea lions from San Luis Obispo County, Calif, USA[28]
Zalophus californianus, femur	20.6 mg/kg dry weight	6 sea lions from San Luis Obispo County, Calif, USA[28]
Zalophus californianus, gonad	3 mg/kg dry weight	6 sea lions from San Luis Obispo County, Calif, USA[28]
Zalophus californianus, hair-skin	1.6 mg/kg dry weight	6 sea lions from San Luis Obispo County, Calif, USA[28]
Zalophus californianus, humerrus	34.2 mg/kg dry weight	6 sea lions from San Luis Obispo County, Calif, USA[28]
Zalophus californianus, kidney cortex	2.4 mg/kg dry weight	6 sea lions from San Luis Obispo County, Calif, USA[28]
Zalophus californianus, kidney medulla	2.0 mg/kg dry weight	6 sea lions from San Luis Obispo County, Calif, USA[28]
Zalophus californianus, large intestine	1.5 mg/kg dry weight	6 sea lions from San Luis Obispo county, Calif, USA[28]
Zalophus californianus, liver	1.3 mg/kg dry weight	6 sea lions from San Luis Obispo county, Calif, USA[28]
Zalophus californianus, lung	2.0 mg/kg dry weight	6 sea lions from San Luis Obispo county, Calif, USA[28]
Zalophus californianus, muscle	1.1 mg/kg dry weight	6 sea lions from San Luis Obispo county, Calif, USA[28]
Zalophus californianus, rib	8.7 mg/kg dry weight	6 sea lions from San Luis Obispo county, Calif, USA[28]
Zalophus californianus, small intestine	3.4 mg/kg dry weight	6 sea lions from San Luis Obispo county, Calif, USA[28]
Zalophus californianus, spleen	1.3 mg/kg dry weight	6 sea lions from San Luis Obispo county, Calif, USA[28]
Zalophus californianus, stomac	0.5 mg/kg dry weight	6 sea lions from San Luis Obispo county, Calif, USA[28]
Zalophus californianus, tooth	12.9 mg/kg dry weight	6 sea lions from San Luis Obispo county, Calif, USA[28]
Zooplankton	25.03 mg/kg dry weight	19 st, NY, conc in algae = 16.8 mg/kg dw, mean value, Long Island[29]

Source: 1.Patrick, F.M. and M.W. Loutit. 1977. Water Res., 11(8), pp 699-703.
2.Shimwell, D.W. and A.E. Laurie. 1972. Environ. Pollut., 3, pp 291-301.
3.Halcrow, W. et al. 1972. Mar. Poll. Bull., 3(1), pp 7-10.
4.Purves, D. 1972. Environ. Pollut., 3, pp 17-24.
5.Koeppe, D.E. 1977. Sci. Total Environ. 7(3), pp 197-206.
6.Little, P. 1974. Environ. Pollut., 5, pp 159-172.
7.Perhac, R.M. and C.J. Whelan. 1972. J. Geochem. Explor. 1, 47-53.
8.Gupta, S.K. and K.Y. Chen. 1975. Environ. Lett. 10, 129-158.
9.Morel, F. and J. Morgan. 1972. Environ. Sci. Tech. 6, 58-67.
10.Zirino, A. and S. Yamamoto. 1972. Limnol. Oceanogr. 17, 661-665.
11.Dyressen, P. and M. Wedborg. 1974. In: E.D. Goldberg (ed.): The Sea. Vol 5. Marine Chemistry Wiley-Interscience. NY.
12.Stumm, W. and H. Bilinski. 1972. In: S.H. Jenkins (ed.): Advances in Water Pollution Research. Proc. 6th Intl. Conf. Jerusalem. 39-52.
13.Moore, M.E. 1977. Sci. Total Environ., 7(2), pp 109-115.
14.Bloom, H. and G.M. Ayling. 1977. Environ. Geol. 2, pp 3-22.
15.Wettern, M., D.W. Lorch and A. Weber. 1976. Arch. Hydrobiol. 77(3), pp 267-276.
16.Dedolph,R. et al. 1970. Environ. Sci. Technol., 4(3), pp 217-223.
17.Lagerwerff, J.V. 1972. in: Morvedt, Giordano and Lindsay (eds.) ; Micronutrients in Agriculture. Maison.
18.Sabbioni, E. and F. Girardi. 1977. Sci Total Environ., 7(2), pp 145-179.
19.Mathis, B.J. and T.F. Cummins. 1973. Water Pollut. Control 45 pp 1573.
20.Taylor, D. 1976. Sci. Total Environ. 6(3), pp 259-264.
21.Altshuller, A.P. and J.J. Bufalini. 1971. Environ. Sci. Technol., 5(1), pp 39-64.
22.Taylor, D. 1976. Estuarine Coastal Mar. Sci. 2, pp 417.

23.Davies, B.E. and J. Lewin. 1974. Environ. Pollut., 6, pp 49-57.
24.Rameau, J.T.L.B. 1973: In: Environmental Health Aspects of Lead. CEC. Luxembourg 189-197.
25.Duggan, M.J. and S. Willams. 1977. Sci. Total Environ., 7(11), pp 91-97.
26.Motto, H.L. et al. 1970. Environ. Sci. Technol. 4(3), pp 231-237.
27.Mitchell, R. and C. Chamberlin. 1978. in: Berg. G. (ed.); Indicators of Enteric Contamination in Natural Waters.
28.Braham, H.W. 1974. Environ. Pollut., 5, pp 253-258.
29.Greig, R.A., A. Adams and D.R. Wenzloff. 1977. Bull. Environ. Contam. and Toxic., 18(1), pp 3-8.

5-384 Concentrations
Perylene

Species	Value	Condition
Aquatic vascular plants	71.6 ug/kg dry weight	Clipperton lagoon, eq Pacific, seawater = 3.05 ppb, inside lagoon, 1969[1]
Caranx lugibris, integument	25 ug/kg dry weight	Clipperton lagoon, eq Pacific, seawater = 3.05 ppb, inside lagoon, 1969[1]
Caranx lugibris, intestine	88 ug/kg dry weight	Clipperton lagoon, eq Pacific, seawater = 3.05 ppb, inside lagoon, 1969[1]
Caranx lugibris, muscle	1.7 ug/kg dry weight	Clipperton lagoon, eq Pacific, seawater = 3.05 ppb, inside lagoon, 1969[1]
Caranx lugibris, viscera	10.5 ug/kg dry weight	Clipperton lagoon, eq Pacific, seawater = 3.05 ppb, inside lagoon, 1969[1]
Cenchus sp. and Eragrostis sp.	219 ug/kg dry weight	Clipperton lagoon, eq Pacific, seawater = 3.05 ppb, shore of lag, 1969[1]
Kulhia taeneura, integument	30 ug/kg dry weight	Clipperton lagoon, eq Pacific, seawater = 3.05 ppb, inside lagoon, 1969[1]
Kulhia taeneura, muscle	5.0 ug/kg dry weight	Clipperton lagoon, eq Pacific, seawater = 3.05 ppb, inside lagoon, 1969[1]
Lygia sp.	865 ug/kg dry weight	Clipperton lagoon, eq Pacific, seawater = 3.05 ppb, inside lagoon, 1969[1]
Phytoplankton and zoopankton	2.7 ug/kg dry weight	Clipperton lagoon, eq Pacific, seawater = 3.05 ppb, inside lagoon, 1969[1]

Source: 1.Niaussat, P. and C. Auger. 1970. C. R. Acad. Sci. (Paris) Ser. D. 270, 2702-2705.

5-385 Concentrations
Phenanthrene

Species	Value	Condition
Crassostrea virginica	0.4 mg/kg wet weight	Upt in 8 h from 302 ppm No. 2 fuel, 0.53 % P, 120 h excr, init=1.2 ppm[1]
Crassostrea virginica	0.6 mg/kg wet weight	Upt in 8 h from 302 ppm No. 2 fuel, 0.53 % P, 6 h excr, init=1.2 ppm[1]
Crassostrea virginica	1.2 mg/kg wet weight	Upt in 8 h from 302 ppm No. 2 fuel, 0.53 % P, 24 h excr, init=1.2 ppm[1]
Crassostrea virginica	1.3 mg/kg wet weight	Upt in 8 h from 302 ppm No. 2 fuel, 0.53 % P, 3 h excr, init=1.2 ppm[1]
Crassostrea virginica	1.9 mg/kg wet weight	302 ppm dispersed No. 2 fuel oil with 0.53 % P, flow through, 8 h[1]
Fundulus heteroclitus	0.08 mg/kg wet weight	570 l S. Louisiana crude oil in 810 m2 shoreline area, 216 h after spill[2]
Fundulus heteroclitus	0.08 mg/kg wet weight	570 l S. Louisiana crude oil in 810 m2 shoreline area, 6 h after spill[2]
Fundulus heteroclitus	0.11 mg/kg wet weight	570 l S. Louisiana crude oil in 810 m2 shoreline area, 31 h after spill[2]
Fundulus heteroclitus	0.26 mg/kg wet weight	570 l S. Louisiana crude oil in 810 m2 shoreline area, 76 h after spill[2]

Marine sediment near oil separator platform	100 ug/kg wet weight	Trinity B., TEX, USA, sample 15 m from brine effluent = 36 ug/l[3]

Source: 1.Neff, J.M., B.A. Cox, D. Dixit and J.W. Anderson. 1976. Mar. Biol. 38, 279-289.
2.Bieri, R.H., V.C. Stamoudis and M.K. Cueman. 1977. Proc. of 1977 Oil spill Conference (Prevention, Behavior, Control, Cleanup). Wash. DC. American Petrolium Institute. 511-516.
3.Armstrong, H.W., K. Fucik, J.W. Anderson and J.M. Neff. 1977. Effects of oilfield brine effluent on benthic organisms in Trinity Bay. American Petrolium Institute Publ. No. 4291. Wash. DC.

5-386 Concentrations
Po

Species	Value	Condition
Cystoseria osmundacea	50 p ci/kg wet weight	1.6 f ci/cm2, wat = 23 f ci/kg, P Loma, Calif, 73, Po-210[1]
Cystoseria osmundacea	70 p ci/kg wet weight	1.2 f ci/cm2, wat = 23 f ci/kg, P Loma, Calif, 73, Po-210[1]
Eisenia arborea	58 p ci/kg wet weight	2.0 f ci/cm2, wat = 23 f ci/kg, P Loma, Calif, 73, Po-210[1]
Eisenia arborea	66 p ci/kg wet weight	2.5 f ci/cm2, wat = 23 f ci/kg, P Loma, Calif, 73, Po-210[1]
Laminaria saccharina	122 p ci/kg wet weight	5.5 f ci/cm2, wat = 23 f ci/kg, P Loma, Calif, 73, Po-210[1]
Laminaria saccharina	52 p ci/kg wet weight	2.4 f ci/cm2, wat = 23 f ci/kg, P Loma, Calif, 73, Po-210[1]
Macrocystis pyrifera	105 p ci/kg wet weight	3 cm from tip, 2.5 f ci/cm2, wat = 23 f ci/kg, P Loma, Calif, Po-210[1]
Macrocystis pyrifera	105 p ci/kg wet weight	Tip, 1.9 f ci/cm2, wat = 23 f ci/kg, P Loma, Calif, Po-210[1]
Macrocystis pyrifera	126 p ci/kg wet weight	1 cm from tip, 2.4 f ci/cm2, wat = 23 f ci/kg, P Loma, Calif, Po-210[1]
Macrocystis pyrifera	142 p ci/kg wet weight	4 cm from tip, 3.2 f ci/cm2, wat = 23 f ci/kg, P Loma, Calif, Po-210[1]
Macrocystis pyrifera	173 p ci/kg wet weight	5 cm from tip, 5.1 f ci/cm2, wat = 23 f ci/kg, P Loma, Calif, Po-210[1]
Macrocystis pyrifera	183 p ci/kg wet weight	6 cm from tip, 4.8 f ci/cm2, wat = 23 f ci/kg, P Loma, Calif, Po-210[1]
Macrocystis pyrifera	260 p ci/kg wet weight	7 cm from tip, 7.9 f ci/cm2, wat = 23 f ci/kg, P Loma, Calif, Po-210[1]
Macrocystis pyrifera	283 p ci/kg wet weight	10 cm from tip, 7.0 f ci/cm2, wat = 23 f ci/kg, P Loma, Calif[1]
Macrocystis pyrifera	320 p ci/kg wet weight	8 cm from tip, 7.6 f ci/cm2, wat = 23 f ci/kg, P Loma, Calif, Po-210[1]
Macrocystis pyrifera	410 p ci/kg wet weight	9 cm from tip, 8.9 f ci/cm2, wat = 23 f ci/kg, P Loma, Calif, Po-210[1]
Macrocystis pyrifera	60 p ci/kg wet weight	1.7 f ci/cm2, wat = 23 f ci/kg, P Loma, Calif, 73, new bla, Po-210[1]
Macrocystis pyrifera	64 p ci/kg wet weight	1.8 f ci/cm2, wat = 23 f ci/kg, P Loma, Calif, 73, new bla, Po-210[1]
Macrocystis pyrifera	75 p ci/kg wet weight	2 cm from tip, 1.8 f ci/cm2, wat = 23 f ci/kg, P Loma, Calif, Po-210[1]
Pterygophora californica	138 p ci/kg wet weight	2.3 f ci/cm2, wat = 23 f ci/kg, P Loma, Calif, 73, Po-210[1]
Pterygophora californica	96 p ci/kg wet weight	2.6 f ci/cm2, wat = 23 f ci/kg, P Loma, Calif, 73, Po-210[1]
Schizymenia epiphytica	144 p ci/kg wet weight	2.2 f ci/cm2, wat = 23 f ci/kg, P Loma, Calif, 73, Po-210[1]
Schizymenia epiphytica	290 p ci/kg wet weight	7.1 f ci/cm2, wat = 23 f ci/kg, P Loma, Calif, 73, Po-210[1]

Source: 1.Hodge, V.F., F.L. Hoffman and T.R. Folson. 1974. Health Phys., 27, pp 29-35.

5-387 Concentrations
Polar metabolits of BaP

Species	Value	Condition
Callinectes sapidus, blood	10E-4 ug total in tissue	Water = 2.5 ug BaP/l, juvenile, marine, n=3, 1 d exp, +-70E-4[1]
Callinectes sapidus, blood	26E-4 ug total in tissue	Water = 2.5 ug BaP/l, juvenile, marine, n=3, 2 d exp, +-85E-4[1]
Callinectes sapidus, blood	32E-4 ug total in tissue	2 d upt. from wat=2.5 ug BaP/l, init=12, juv, marine, n=3, 2 d excr[1]
Callinectes sapidus, gill	12E-4 ug total in tissue	2 d upt. from wat=2.5 ug BaP/l, init=3 juv, marine, n=3, 2 d excr[1]
Callinectes sapidus, gill	15E-4 ug total in tissue	Water = 2.5 ug BaP/l, juvenile, marine, n=3, 1 d exp, +-70E-4[1]
Callinectes sapidus, gill	3E-4 ug total in tissue	Water = 2.5 ug BaP/l, juvenile, marine, n=3, 2 d exp, +-85E-4[1]
Callinectes sapidus, hepatopancreas	140E-4 ug total in tissue	2 d upt. from wat=2.5 ug BaP/l, init=84, juv, marine, n=3, 2 d excr[1]
Callinectes sapidus, hepatopancreas	210E-4 ug total in tissue	Water = 2.5 ug BaP/l, juvenile, marine, n=3, 2 d exp, +-85E-4[1]
Callinectes sapidus, hepatopancreas	89E-4 ug total in tissue	Water = 2.5 ug BaP/l, juvenile, marine, n=3, 1 d exp, +-70E-4[1]
Callinectes sapidus, muscle	2E-4 ug total in tissue	Water = 2.5 ug BaP/l, juvenile, marine, n=3, 1 d exp, +-70E-4[1]
Callinectes sapidus, muscle	4E-4 ug total in tissue	2 d upt. from wat=2.5 ug BaP/l, init=2, juv, marine, n=3, 2 d excr[1]
Callinectes sapidus, muscle	8E-4 ug total in tissue	Water = 2.5 ug BaP/l, juvenile, marine, n=3, 2 d exp, +-85E-4[1]
Callinectes sapidus, stomach	6E-4 ug total in tissue	Water = 2.5 ug BaP/l, juvenile, marine, n=3, 2 d exp, +-85E-4[1]
Callinectes sapidus, stomach	7E-4 ug total in tissue	2 d upt. from wat=2.5 ug BaP/l, init=2, juv, marine, n=3, 2 d excr[1]
Callinectes sapidus, stomach	8E-4 ug total in tissue	Water = 2.5 ug BaP/l, juvenile, marine, n=3, 1 d exp, +-70E-4[1]

Source: 1.Lee, R.F., C. Ryan and M.L. Neuhauser. 1976. Mar. Biol. 37, 363-370.

5-388 Concentrations
Pu

Species	Value	Condition
Macrocystis pyrifera	0.32 p ci/kg wet weight	Tip, 7 a ci/cm2, wat = 250 a ci/kg, P Loma, Calif, Pu-239[1]
Macrocystis pyrifera	0.44 p ci/kg wet weight	8 cm from tip, 26 a ci/cm2, wat = 250 a ci/kg, P Loma, Calif, Pu-239[1]
Macrocystis pyrifera	0.48 p ci/kg wet weight	1 cm from tip, 11 a ci/cm2, wat = 250 a ci/kg, P Loma, Calif, Pu-239[1]
Macrocystis pyrifera	0.50 p ci/kg wet weight	3 cm from tip, 13 a ci/cm2, wat = 250 a ci/kg, P Loma, Calif, Pu-239[1]
Macrocystis pyrifera	0.54 p ci/kg wet weight	2 cm from tip, 15 a ci/cm2, wat = 250 a ci/kg, P Loma, Calif, Pu-239[1]

Macrocystis pyrifera	0.60 p ci/kg wet weight	4 cm from tip, 15 a ci/cm2, wat = 250 a ci/kg, P Loma, Calif, Pu-239[1]
Macrocystis pyrifera	0.61 p ci/kg wet weight	6 cm from tip, 19 a ci/cm2, wat = 250 a ci/kg, P Loma, Calif, Pu-239[1]
Macrocystis pyrifera	0.68 p ci/kg wet weight	5 cm from tip, 19 a ci/cm2, wat = 250 a ci/kg, P Loma, Calif, Pu-239[1]
Macrocystis pyrifera	1.00 p ci/kg wet weight	7 cm from tip, 31 a ci/cm2, wat = 250 a ci/kg, P Loma, Calif, Pu-239[1]
Macrocystis pyrifera	1.44 p ci/kg wet weight	9 cm from tip, 36 a ci/cm2, wat = 250 a ci/kg, P Loma, Calif, Pu-239[1]
Macrocystis pyrifera	1.90 p ci/kg wet weight	10 cm from tip, 40 a ci/cm2, wat = 250 a ci/kg, P Loma, Calif Pu-239[1]
River water	34 %	Retained by a 1.2 um filter, Pu(+6), particulate Pu, Pu-237[2]
River water	48 %	Retained by a 0.45 um filter, Pu(+6), particulate Pu, Pu-237[2]
River water	56 %	Retained by a 0.10 um filter, Pu(+6), particulate Pu, Pu-237[2]
River water	57 %	Retained by a 0.22 um filter, Pu(+6), particulate Pu, Pu-237[2]
Seawater	34 %	Retained by a 1.2 um filter, Pu(+6), particulate Pu, Pu-237[2]
Seawater	40 %	Retained by a 0.10 um filter, Pu(+6), particulate Pu, Pu-237[2]
Seawater	40 %	Retained by a 0.45 um filter, Pu(+6), particulate Pu, Pu-237[2]
Seawater	41 %	Retained by a 0.22 um filter, Pu(+6), particulate Pu, Pu-237[2]

Source: 1.Hodge, V.F., F.L. Hoffman and T.R. Folson. 1974. Health Phys., 27, pp 29-35.
2.Murray, C.N. and R. Fucai. 1975. Impacts of Nuclear releases into the Aquatic Environment. IAEA-SM-18/25, pp 179-191. Vienna.

5-389 Concentrations
Relative maximum concentration in measuring period

Species	Value	Condition
Any compound	0.5 any unit	Average over 1 month[1]
Any compound	1.0 any unit	Average over 24 h[1]
Any compound	1.2 any unit	Average over 8 h[1]
Any compound	1.8 any unit	Average over 2 h[1]
Any compound	2.2 any unit	Average over 1 h[1]
Any compound	2.4 any unit	Average over 30 min[1]
Any compound	2.7 any unit	Average over 15 min[1]
Any compound	3.3 any unit	Just one measurement[1]

Source: 1.Stern, A.C. 1964. J. Air Pollut. Cont. Ass. 14, 5-15.

5-390 Concentrations
Se

Species	Value	Condition
Astragalus bisulcatus	2050 mg/kg dry weight	Soil = 2 ppm, wyoming, USA[1]
Astragalus bisulcatus	2620 mg/kg dry weight	Soil = 1.6 ppm, N Dakota, USA[1]
Astragalus missouriensis	25 mg/kg dry weight	Soil = 0.7 ppm, NE USA[1]
Astragalus missouriensis	25 mg/kg dry weight	Soil = 5 ppm, NE USA[1]
Astragalus missouriensis	5 mg/kg dry weight	Soil = 2 ppm, Wyoning, USA[1]
Astragalus missouriensis	5 mg/kg dry weight	Soil = 2-14 ppm, NE USA[1]
Astragalus racemosus	400 mg/kg dry weight	Soil = 0.7 ppm, NE USA[1]

Astragalus racemosus	450 mg/kg dry weight	Soil = 2-14 ppm, NE USA[1]
Astragalus racemosus	5560 mg/kg dry weight	Soil = 5 ppm, NE USA[1]
Corn	0.06 microgram/g dry weight	Added to field: 120 g Se/ha; control = 0.01 ug/g, rg = +- 5.8 %[2]
Corn	0.1 microgram/g dry weight	Addition of 75 g Se/ha*yr for 5 yrs; control = 0.01 ug/g dry weight[2]
Rat, kidney	10.2 % of total intracellular amount	Mitochondria, 9 rats, 24 h after injection of 6 ug Se/275 g[3]
Rat, kidney	10.6 % of total intracellular amount	Lysosomes, 9 rats, 24 h after injection of 6 ug Se/275 g[3]
Rat, kidney	16.8 % of total intracellular amount	Soluble fraction, 9 rats, 24 h after injection of 6 ug Se/275 g[3]
Rat, kidney	18.3 % of total intracellular amount	Microsomes, 9 rats, 24 h after injection of 6 ug Se/275 g[3]
Rat, kidney	44.1 % of total intracellular amount	Nuclei, 9 rats, 24 h after injection of 6 ug Se/275 g[3]
Rat, liver	13.4 % of total intracellular amount	Lysosomes, 9 rats, 24 h after injection of 6 ug Se/275 g[3]
Rat, liver	15.6 % of total intracellular amount	Soluble fraction, 9 rats, 24 h after injection of 6 ug Se/275 g[3]
Rat, liver	21 % of total intracellular amount	Microsomes, 9 rats, 24 h after injection of 6 ug Se/275 g[3]
Rat, liver	41.8 % of total intracellular amount	Nuclei, 9 rats on 250-300 g, 12 h after injection of 6 micro g SeO3[3]
Rat, liver	8.2 % of total intracellular amount	Mitochondria, 9 rats, 24 h after injection of 6 ug Se/275 g[3]

Source: 1.Baird, R.B., S. Pourian and S.M. Gabrielian. 1972. Analyt. Chem. 36, 1359.
2.Gissel-Nielsen, G. 1973. Ambio, 2, pp 114-117.
3.Sabbioni, E. and F. Girardi. 1977. Sci Total Environ., 7(2), pp 145-179.

5-391 Concentrations
Suspended solids

Species	Value	Condition
Municipal sewage	15 mg/l	Typical treated, secondary, rg = 10 - 30[1]
Municipal sewage	200 mg/l	Typical untreated, rg = 100 - 350[1]
Municipal sewage	80 mg/l	Typical treated, primary, rg = 40 - 120[1]
Precipitation	12 mg/l	General characteristics, rg = 11 - 13[1]
Urban land, drainage	3500 mg/l	General characteristics, rg = 5 - 7340[1]

Source: 1.Donigian, A.S. and N.H. Crawford. 1976. Modeling nonpoint Pollution from the Land Surface, EPA-600/376-083. July.

5-392 Concentrations
TDE

Species	Value	Condition
Pagophilus groenlandicus, blubber	0.13 mg/kg wet weight	83.0 % fat, 15 samples, 1973, Gulf of St Lawrence, SD = +- 0.12 1
Pagophilus groenlandicus, blubber	0.18 mg/kg wet weight	95.5 % fat, 4 animals, St Lawrence, Canada, SD = +- 0.12[2]
Pagophilus groenlandicus, blubber	0.50 mg/kg wet weight	97.8 % fat, 10 samples, 1971, Gulf of St Lawrence, SD = +- 0.32[1]
Pagophilus groenlandicus, brain	0 mg/kg wet weight	11.5 % fat, 3 animals, St Lawrence, Canada[2]

Pagophilus groenlandicus, brain	0.01 mg/kg wet weight	6.6 % fat, 14 samples, 1973, Gulf of St Lawrence, SD = +- 0.01[1]
Pagophilus groenlandicus, brain	0.01 mg/kg wet weight	7.5 % fat, 10 samples, 1971, Gulf of St Lawrence[1]
Pagophilus groenlandicus, kidney	0.04 mg/kg wet weight	14.4 % fat, 13 samples, 1973, Gulf of St Lawrence, SD = +- 0.06[1]
Pagophilus groenlandicus, liver	0.01 mg/kg wet weight	1.8 % fat, 4 ainmals, Gulf of St Lawrence, Canada, SD = +- 0.01[2]
Pagophilus groenlandicus, liver	0.01 mg/kg wet weight	4.4 % fat, 16 samples, 1973, Gulf of St Lawrence, SD = +- 0.01[1]
Pagophilus groenlandicus, liver	0.03 mg/kg wet weight	7.0 % fat, 11 samples, 1971, Gulf of St Lawrence, SD = +- 0.13[1]
Pagophilus groenlandicus, muscle	0 mg/kg wet weight	1.1 % fat, 4 animals, St Lawrence, Canada[2]

Source: 1.Frank, R., K. Ronald and H.E. Braun. 1973. Bull. Fish Res. Board Can., 30, pp 1053. 2.Jones, D., K. Ronald, D.H. Lavinge, R. Frank, M. Holdrinet and J.F. Uthe. 1976. Sci. Total Environ.,5(2), pp 181-196.

5-393 Concentrations Ta

Species	Value	Condition
Ascidiella aspersa	0.19 mg/kg dry weight	Assumed seawater conc = 0.02 mg/m3, Southampton water[1]
Buccinum undatum	0.019 mg/kg dry weight	Assumed seawater conc = 0.02 mg/m3, Whitstable, Thames Estuary[1]
Carcinus maenas, mantle	0.030 mg/kg dry weight	Assumed seawater conc = 0.02 mg/m3, Southampton water[1]
Carcinus maenas, muscle	0.024 mg/kg dry weight	Assumed seawater conc = 0.02 mg/m3, Southampton water[1]
Cardium edule	0.043 mg/kg dry weight	Assumed seawater conc = 0.02 mg/m3, Whitstable, Thames Estuary[1]
Ciona intestinalis, with test	0.24 mg/kg dry weight	Assumed seawater conc = 0.02 mg/m3, Southampton water[1]
Ciona intestinalis, without test	0.096 mg/kg dry weight	Assumed seawater conc = 0.02 mg/m3, Southampton water[1]
Crassostrea angulata	0.009 mg/kg dry weight	Assumed seawater conc = 0.02 mg/m3, Whitstable, Thames Estuary[1]
Holothuria forskali	0.032 mg/kg dry weight	Assumed seawater conc = 0.02 mg/m3, Engl channal, off Plymouth[1]
Mercenaria mercenaria, digestive gland	0.02 mg/kg dry weight	Assumed seawater conc = 0.02 mg/m3, Southampton water[1]
Mercenaria mercenaria, foot	0.19 mg/kg dry weight	Assumed seawater conc = 0.02 mg/m3, Southampton water[1]
Mytilus edulis, mantle	0.01 mg/kg dry weight	Assumed seawater conc = 0.02 mg/m3, Southampton water[1]
Mytilus edulis, muscle	0.01 mg/kg dry weight	Assumed seawater conc = 0.02 mg/m3, Southampton water, less[1]
Neomysis integer	2.3 mg/kg dry weight	Assumed seawater conc = 0.02 mg/m3, Totton marshes, Test River, Engl[1]
Porania pulvillus	0.010 mg/kg dry weight	Assumed seawater conc = 0.02 mg/m3, Engl channal, off Plymouth[1]

Source: 1.Burton, J.D. and K.S. Massie. 1971. J. mar. biol. Ass. U.K. 51, pp 675-683.

5-394 Concentrations Tc

Species	Value	Condition
Soybean	1.4 mg/kg dry weight	Soil = 0.01 ug/g dw, Tc-99[1]
Soybean	380 mg/kg dry weight	Soil = 1.0 ug/g dw, Tc-99[1]
Soybean	380 mg/kg dry weight	Soil = 5.0 ug/g dw, Tc-99[1]
Soybean	6.7 mg/kg dry weight	Soil = 0.1 ug/g dw, Tc-99[1]

Soybean, bud	0.00534 mg/kg dry weight	Soil = 0.001 ug/g dw, Tc-99[1]
Soybean, bud	0.0134 mg/kg dry weight	Soil = 0.01 ug/g dw, Tc-99[1]
Soybean, bud	0.255 mg/kg dry weight	Soil = 0.1 ug/g dw, Tc-99[1]
Soybean, cotyledon	0.38 mg/kg dry weight	Soil = 0.001 ug/g dw, Tc-99[1]
Soybean, cotyledon	20.7 mg/kg dry weight	Soil = 0.01 ug/g dw, Tc-99[1]
Soybean, cotyledon	342 mg/kg dry weight	Soil = 5.0 ug/g dw, Tc-99[1]
Soybean, cotyledon	484 mg/kg dry weight	Soil = 1.0 ug/g dw, Tc-99[1]
Soybean, cotyledon	50.4 mg/kg dry weight	Soil = 0.1 ug/g dw, Tc-99[1]
Soybean, leaves	0.186 mg/kg dry weight	Soil = 0.001 ug/g dw, Tc-99[1]
Soybean, leaves	0.964 mg/kg dry weight	Soil = 0.01 ug/g dw, Tc-99[1]
Soybean, leaves	8.46 mg/kg dry weight	Soil = 0.1 ug/g dw, Tc-99[1]
Soybean, stem	0.0306 mg/kg dry weight	Soil = 0.001 ug/g dw, Tc-99[1]
Soybean, stem	0.230 mg/kg dry weight	Soil = 0.01 ug/g dw, Tc-99[1]
Soybean, stem	1.25 mg/kg dry weight	Soil = 0.1 ug/g dw, Tc-99[1]
Soybean, stem	202 mg/kg dry weight	Soil = 1.0 ug/g dw, Tc-99[1]
Soybean, stem	640 mg/kg dry weight	Soil = 5.0 ug/g dw, Tc-99[1]
Wheat	0.34 mg/kg dry weight	Soil = 0.001 ug/g dw, Tc-99[1]
Wheat	1.5 mg/kg dry weight	Soil = 0.01 ug/g dw, Tc-99[1]
Wheat	17 mg/kg dry weight	Soil = 0.1 ug/g dw, Tc-99[1]
Wheat, blade	0.476 mg/kg dry weight	Soil = 0.001 ug/g dw, Tc-99[1]
Wheat, blade	3.12 mg/kg dry weight	Soil = 0.01 ug/g dw, Tc-99[1]
Wheat, blade	39.7 mg/kg dry weight	Soil = 0.1 ug/g dw, Tc-99[1]
Wheat, culm	0.00279 mg/kg dry weight	Soil = 0.001 ug/g dw, Tc-99[1]
Wheat, culm	0.0235 mg/kg dry weight	Soil = 0.01 ug/g dw, Tc-99[1]
Wheat, culm	0.130 mg/kg dry weight	Soil = 0.1 ug/g dw, Tc-99[1]
Wheat, head	0.000437 mg/kg dry weight	Soil = 0.001 ug/g dw, Tc-99[1]
Wheat, head	0.0201 mg/kg dry weight	Soil = 0.01 ug/g dw, Tc-99[1]
Wheat, head	0.0230 mg/kg dry weight	Soil = 0.1 ug/g dw, Tc-99[1]
Wheat, head	1.19 mg/kg dry weight	Soil = 0.001 ug/g dw, Tc-99[1]

Source: 1.Wildung, R.E., T.R. Garland and D.A. Cataldo. 1977. Health Physics. 32(4), pp 314-316.

5-395 Concentrations Telodrin

Species	Value	Condition
Coturnix quail, brain	1.40 mg/kg wet weight	Conc in food = 10 mg/kg wet weight[1]
Coturnix quail, brain	1.60 mg/kg wet weight	Conc in food = 2 mg/kg wet weight[1]

Source: 1.Edwards, C.A. 1975. Persistent Pesticides in the Environment, 2nd edition, CRC-Press.

5-396 Concentrations
Th

Species	Value	Condition
Beech and spruce forest soil	0.2 mg/kg dry weight	Central German, depth = 30-40 cm, 247 g/ha[1]
Beech and spruce forest soil	0.2 mg/kg dry weight	Central German, depth = 40-50 cm, 340 g/ha[1]
Beech and spruce forest soil	0.3 mg/kg dry weight	Central German, depth = 10-20 cm, 381 g/ha[1]
Beech and spruce forest soil	0.5 mg/kg dry weight	Central German, depth = 0-10 cm, 610 g/ha[1]

Source: 1.Heinrichs, H. and R. Meyer. 1977. J. Environ. Qual. 6, 402-407.

5-397 Concentrations
Ti

Species	Value	Condition
Vegetation	21 mg/kg dry weight	Conc soil :f=250, h = 500, a1a2 = 8000, b1 = 5000, b2 = 8000, b3 =[1]

Source: 1.Mitchell, R. and C. Chamberlin. 1978. in: Berg. G. (ed.); Indicators of Enteric Contamination in Natural Waters.

5-398 Concentrations
Toluene

Species	Value	Condition
Seawater near oil separator platform	3.2 ug/liter	Trinity B., TEX, USA, sample 15 m from brine effluent = 3500 ug/l[1]

Source: 1.Armstrong, H.W., K. Fucik, J.W. Anderson and J.M. Neff. 1977. Effects of oilfield brine effluent on benthic organisms in Trinity Bay. American Petrolium Institute Publ. No. 4291. Wash. DC.

5-399 Concentrations
Total N

Species	Value	Condition
Agricultural cropland	9 mg/l	General characteristics[1]
Animal feedlot	1510 mg/l	General characteristics, rg = 920 - 2100[1]
Forested land	1.05 mg/l	General characteristics, rg = 0.3 - 1.8[1]
Municipal sewage	30 mg/l	Typical treated, secondary[1]
Municipal sewage	35 mg/l	Typical treated, primary[1]
Municipal sewage	40 mg/l	Typical untreated[1]
Precipitation	1.25 mg/l	General characteristics, rg = 1.2 - 1.3[1]
Urban land, drainage	3 mg/l	General characteristics[1]

Source: 1.Donigian, A.S. and N.H. Crawford. 1976. Modeling nonpoint Pollution from the Land Surface, EPA-600/376-083. July.

5-400 Concentrations
Total P

Species	Value	Condition
Agricultural cropland	0.86 mg/l	General characteristics, rg = 0.02 - 1.7[1]
Animal feedlot	325 mg/l	General characteristics, rg = 290 - 360[1]
Forested land	0.055 mg/l	General characteristics, rg = 0.01 - 0.11[1]
Municipal sewage	10 mg/l	Typical untreated[1]
Municipal sewage	5.0 mg/l	Typical treated, secondary[1]

Municipal sewage	7.5 mg/l	Typical treated, primary[1]
Precipitation	0.03 mg/l	General characteristics, rg = 0.02 - 0.04[1]
Urban land, drainage	0.65 mg/l	General characteristics, rg = 0.2 - 1.1[1]

Source: 1.Donigian, A.S. and N.H. Crawford. 1976. Modeling nonpoint Pollution from the Land Surface, EPA-600/376-083. July.

5-401 Concentrations
Total aromatic hydrocarbons

Species	Value	Condition
Marine sediment near oil separator platform	34200 ug/kg wet weight	Trinity B., TEX, USA, sample 15 m from brine effluent = 11764 ug/l[1]
Seawater near oil separator platform	10.50 ug/liter	Trinity B., TEX, USA, sample 15 m from brine effluent = 11764 ug/l[1]

Source: 1.Armstrong, H.W., K. Fucik, J.W. Anderson and J.M. Neff. 1977. Effects of oilfield brine effluent on benthic organisms in Trinity Bay. American Petrolium Institute Publ. No. 4291. Wash. DC.

5-402 Concentrations
Total petroleum hydrocarbons

Species	Value	Condition
Crassostrea virginica	108.9 mg/kg wet weight	Upt in 8 h from 302 ppm No. 2 fuel, 16.6 % TOT, 6 h excr, init=312 ppm[1]
Crassostrea virginica	227.1 mg/kg wet weight	Upt in 8 h from 302 ppm No. 2 fuel, 16.6 % TOT, 3 h excr, init=312 ppm[1]
Crassostrea virginica	311.7 mg/kg wet weight	302 ppm dispersed No. 2 fuel oil with 16.6 % TOT, flow through, 8 h[1]
Crassostrea virginica	46.1 mg/kg wet weight	Upt in 8 h from 302 ppm No. 2 fuel, 16.6 % TOT, 120 h excr, init=312 ppm[1]
Crassostrea virginica	66.7 mg/kg wet weight	Upt in 8 h from 302 ppm No. 2 fuel, 16.6 % TOT, 24 h excr, init=312 ppm[1]

Source: 1.Neff, J.M., B.A. Cox, D. Dixit and J.W. Anderson. 1976. Mar. Biol. 38, 279-289.

5-403 Concentrations
Total polycyclic aromatic hydrocarbon, PAH

Species	Value	Condition
Callinectes sapidus, blood	80E-4 ug total in tissue	2 d upt. from wat=2.5 ug BaP/l, init=85, juv, marine, n=3, 2 d excr[1]
Callinectes sapidus, blood	85E-4 ug total in tissue	Water = 2.5 ug BaP/l, juvenile, marine, n=3, 2 d exp[1]
Callinectes sapidus, blood	90E-4 ug total in tissue	Water = 2.5 ug BaP/l, juvenile, marine, n=3, 1 d exp[1]
Callinectes sapidus, gill	100E-4 ug total in tissue	2 d upt. from wat=2.5 ug BaP/l, init=410, juv, marine, n=3, 2 d excr[1]
Callinectes sapidus, gill	150E-4 ug total in tissue	Water = 2.5 ug BaP/l, juvenile, marine, n=3, 1 d exp[1]
Callinectes sapidus, gill	410E-4 ug total in tissue	Water = 2.5 ug BaP/l, juvenile, marine, n=3, 2 d exp[1]
Callinectes sapidus, hepatopancreas	270E-4 ug total in tissue	Water = 2.5 ug BaP/l, juvenile, marine, n=3, 1 d exp[1]
Callinectes sapidus, hepatopancreas	300E-4 ug total in tissue	2 d upt. from wat=2.5 ug BaP/l, init=590, juv, marine, n=3, 2 d excr[1]
Callinectes sapidus, hepatopancreas	590E-4 ug total in tissue	Water = 2.5 ug BaP/l, juvenile, marine, n=3, 2 d exp[1]

Callinectes sapidus, muscle	10E-4 ug total in tissue	Water = 2.5 ug BaP/l, juvenile, marine, n=3, 1 d exp[1]
Callinectes sapidus, muscle	18E-4 ug total in tissue	Water = 2.5 ug BaP/l, juvenile, marine, n=3, 2 d exp[1]
Callinectes sapidus, muscle	8E-4 ug total in tissue	2 d upt. from wat=2.5 ug BaP/l, init=18, juv, marine, n=3, 2 d excr[1]
Callinectes sapidus, stomach	10E-4 ug total in tissue	2 d upt. from wat=2.5 ug BaP/l, init=11, juv, marine, n=3, 2 d excr[1]
Callinectes sapidus, stomach	11E-4 ug total in tissue	Water = 2.5 ug BaP/l, juvenile, marine, n=3, 2 d exp[1]
Callinectes sapidus, stomach	14E-4 ug total in tissue	Water = 2.5 ug BaP/l, juvenile, marine, n=3, 1 d exp[1]

Source: 1.Lee, R.F., C. Ryan and M.L. Neuhauser. 1976. Mar. Biol. 37, 363-370.

5-404 Concentrations Total solids

Species	Value	Condition
Urban land, drainage	4407 mg/l	General characteristics, rg = 194 - 8620[1]

Source: 1.Donigian, A.S. and N.H. Crawford. 1976. Modeling nonpoint Pollution from the Land Surface, EPA-600/376-083. July.

5-405 Concentrations Trimethylnaphthalene

Species	Value	Condition
Crassostrea virginica	0.9 mg/kg wet weight	Upt in 8 h from 302 ppm No. 2 fuel, 1.84 % TMN, 672 h excr, init=21.8 ppm[1]
Crassostrea virginica	10.8 mg/kg wet weight	Upt in 8 h from 302 ppm No. 2 fuel, 1.84 % TMN, 3 h excr, init=21.8 ppm[1]
Crassostrea virginica	4.9 mg/kg wet weight	Upt in 8 h from 302 ppm No. 2 fuel, 1.84 % TMN, 120 h excr, init=21.8 ppm[1]
Crassostrea virginica	5.7 mg/kg wet weight	Upt in 8 h from 302 ppm No. 2 fuel, 1.84 % TMN, 6 h excr, init=21.8 ppm[1]
Crassostrea virginica	9.1 mg/kg wet weight	302 ppm dispersed No. 2 fuel oil with 1.84 % TMN, flow through, 8 h[1]
Crassostrea virginica	9.5 mg/kg wet weight	Upt 8 h from 302 ppm No. 2 fuel, 1.84 % TMN, 24 h excr, init=21.8 ppm[1]
Marine sediment near oil separator platform	10000 ug/kg wet weight	Trinity B., TEX, USA, sample 15 m from brine effluent = 260 ug/l[2]

Source: 1.Neff, J.M., B.A. Cox, D. Dixit and J.W. Anderson. 1976. Mar. Biol. 38, 279-289.
2.Armstrong, H.W., K. Fucik, J.W. Anderson and J.M. Neff. 1977. Effects of oilfield brine effluent on benthic organisms in Trinity Bay. American Petrolium Institute Publ. No. 4291. Wash. DC.

5-406 Concentrations
U

Species	Value	Condition
Particle fractions	10.6 % of total	Size = 0.5-0.25 mm, weight = 8.5 %, area = 1.4 %, marine, Japan,[1]
Particle fractions	17.5 % of total	Size = 0.125-0.075 mm, weight = 16.9 %, area = 10.4 %, marine, Japan,[1]
Particle fractions	19.4 % of total	Size = 0.25-0.125 mm, weight = 23 %, area = 10.1 %, marine, Japan,[1]
Particle fractions	47.4 % of total	Size below 0.075 mm, weight = 48.4 %, area = 77.9 %, marine, Japan,[1]
Particle fractions	5.1 % of total	Size = 2.0-0.5 mm, weight = 3.2 %, area = 0.2 %, marine, Japan, U-238[1]

Source: 1.Nagaya, Y. and K. Nakamura. 1975. Im pacts of Nuclear releases into the Aquatic Environment. IAEA-SM198/21, pp 313-328. Vienna.

5-407 Concentrations
V

Species	Value	Condition
Beech and spruce forest soil	55 mg/kg dry weight	Central German, depth = 0-10 cm, 63 kg/ha[1]
Beech and spruce forest soil	62 mg/kg dry weight	Central German, depth = 10-20 cm, 74 kg/ha[1]
Beech and spruce forest soil	62 mg/kg dry weight	Central German, depth = 30-40 cm, 81 kg/ha[1]
Beech and spruce forest soil	63 mg/kg dry weight	Central German, depth = 40-50 cm, 93 kg/ha[1]
Rat, kidney	28.1 % of total intracellular amount	Soluble frac., 12 rats, 24 h after injection of 10 ug V(4+)/275 g[2]
Rat, kidney	29.6 % of total intracellular amount	Mitochondria, 12 rats, 24 h after injection of 10 ug V(4+)/275 g[2]
Rat, kidney	31.5 % of total intracellular amount	Nuclei, 12 rats, 24 h after injection of 10 ug V(4+)/275 g[2]
Rat, kidney	4.3 % of total intracellular amount	Microsomes, 12 rats, 24 h after injection of 10 ug V(4+)/275 g[2]
Rat, kidney	6.6 % of total intracellular amount	Lysosomes, 12 rats, 24 h after injection of 10 ug V(4+)/275 g[2]
Rat, liver	21.4 % of total intracellular amount	Mitochondria, 24 h after adm., 12 rats on 250-300 g, 70 u ci V(4+)/275 g[2]
Rat, liver	23.1 % of total intracellular amount	Soluble fraction, 12 rats, 24 h after admin.70 micro ci V(4+)/275 g[2]
Rat, liver	4 % of total intracellular amount	Microsomes, 12 rats on 250-300g, 24 h after admin. of 70 u ci V(4+)/275 g[2]
Rat, liver	44.3 % of total intracellular amount	Nuclei, 24 h after injection, 12 rats on 250-300 g, 70 u ci V(4+)/275 g[2]
Rat, liver	7.2 % of total intracellular amount	Lysosomes, 24 h after injection, 12 rats on 250-300 g, 70 u ci V(4+)/275g[2]
Vegetation	1.0 mg/kg dry weight	Conc soil :f = 20, h = 10, a1a2 = 150, b1 = 200, b2 = 100, b3 = 100[3]

Source: 1.Heinrichs, H. and R. Meyer. 1977. J. Environ. Qual. 6, 402-407.
2.Sabbioni, E. and F. Girardi. 1977. Sci Total Environ., 7(2), pp 145-179.
3.Mitchell, R. and C. Chamberlin. 1978. in: Berg. G. (ed.); Indicators of Enteric Contamination in Natural Waters.

5-408 Concentrations
Zn

Species	Value	Condition
Alisma plantago aquatica, old leaves	117.63 mg/kg dry weight	800 and 50 m below pollution I and II, water = 0.107 mg/l, whole leaf[1]
Alisma plantago aquatica, old leaves	52.64 mg/kg dry weight	150 m below pollution entry I, water = 0.190 mg/l, processed leaves[1]
Alisma plantago aquatica, old leaves	87.99 mg/kg dry weight	150 m below pollution entry I, water = 0.190 mg/l, whole leaves[1]
Alisma plantago aquatica, old leaves	97.46 mg/kg dry weight	800 m above pollution entry I, water = 0.056 mg/l, processed leaves[1]
Alisma plantago aquatica, old leaves	98.22 mg/kg dry weight	800 and 50 m below pollution I and II, water = 0.107 mg/l, processed leaves[1]
Alisma plantago aquatica, young leaves	119.36 mg/kg dry weight	800 m above pollution entry I, water = 0.056 mg/l, whole leaves[1]
Alisma plantago aquatica, young leaves	119.82 mg/kg dry weight	150 m below pollution entry I, water = 0.190 mg/l, processed leaves[1]
Alisma plantago aquatica, young leaves	145.87 mg/kg dry weight	800 and 50 m below pollution I and II, water = 0.107 mg/l, processed leaves[1]
Alisma plantago aquatica, young leaves	192.45 mg/kg dry weight	800 and 50 m below pollution I and II, water = 0.107 mg/l, whole leaf[1]
Alisma plantago aquatica, young leaves	199.36 mg/kg dry weight	150 m below pollution entry I, water = 0.190 mg/l, whole leaves[1]
Alisma plantago aquatica, young leaves	98.25 mg/kg dry weight	800 m above pollution entry I, water = 0.056 mg/l, processed leaves[1]
Alpine tundra forest soil	29 mg/kg dry weight soil	White Mountains, NH, all sites max value = 169 ppm, organic layer[2]
Anarchis canadensis	101 mg/kg wet weight	Water = 8+-7 ppb, pH=7.2, Pickerel R, Can, non-poll.[3]
Anarchis canadensis	106 mg/kg wet weight	Water = 7+-6 ppb, pH=7.3, Wanapitei R, Can, silt-ladden and poll.[3]
Anarchis canadensis	59 mg/kg wet weight	Water = 7+-5 ppb, pH=7.2, East outlet (French R) to L. Huron, Can[3]
Anarchis canadensis	77 mg/kg wet weight	Water = 6+-6 ppb, pH=7.1, French R, Can, non-poll.[3]
Anarchis canadensis	80 mg/kg wet weight	Water = 5+-5 ppb, pH=7.3, W Wanapitei B, Can, 1974[3]
Anarchis canadensis	90 mg/kg wet weight	Water = 7+-6 ppb, pH=7.2, Main outlet (French R) to L. Huron, Can[3]
Arthrobacter marinus	286.4 microgram/kg wet weight	Total concentration = 13390 ppb, added Zn = 13076 ppb, 293 K[4]
Arthrobacter marinus	31.7 microgram/kg wet weight	Total concentration = 350 ppb, added Zn = 0 ppb, 293 K[4]
Arthrobacter marinus	40.9 microgram/kg wet weight	Total concentration = 1050 ppb, added Zn = 654 ppb, 293 K[4]
Arthrobacter marinus	480.8 microgram/kg wet weight	Total concentration = 18350 ppb, added Zn = 19614 ppb, 293 K[4]
Arthrobacter marinus	62.2 microgram/kg wet weight	Total concentration = 3650 ppb, added Zn = 3269 ppb, 293 K[4]
Arthrobacter marinus	90.9 microgram/kg wet weight	Total concentration = 6960 ppb, added Zn = 6538 ppb, 293 K[4]
Beech and spruce forest soil	38 mg/kg dry weight soil	Central German, soil depth =[5]
Beech and spruce forest soil	49 mg/kg dry weight soil	Central German, soil depth = 40-50 cm, 73E3 g/ha[5]

Beech and spruce forest soil	50 mg/kg dry weight soil	Central German, soil depth = 10-20 cm, 60E3 g/ha[5]
Beech and spruce forest soil	55 mg/kg dry weight soil	Central German, soil depth = 30-40 cm, 72E3 g/ha[5]
Bryum pseudotriquetrum	14045 mg/kg dry weight	Zn in soil = 1840 ppm, Slater mine, Foolow, Derbyshire, GB[6]
Bryum pseudotriquetrum	1544 mg/kg dry weight	Zn in soil = 1910 ppm, Blea Beck Junction, W Yorkshire, GB[6]
Bryum pseudotriquetrum	435 mg/kg dry weight	Zn in soil = 815 ppm, Sand Haw mine, W Yorkshire, GB[6]
Chestnut oak	110 mg/kg dry weight soil	E Tennessee, soil litter layer O2, g/ha = 2000[7]
Chestnut oak	42 mg/kg dry weight soil	E Tennessee, soil litter layer O1, g/ha = 290[7]
Cabbage, plant	20.9 mg/kg dry weight	15 plants, rural area[8]
Cabbage, plant	30.2 mg/kg dry weight	15 plants, urban area[8]
Camptothecium lutescens	4220 mg/kg dry weight	Zn in soil = 13220 ppm, Blea Beck Head, W Yorkshire, GB[6]
Clam	135 mg/kg wet weight	Water = 5+-5 ppb, pH=7.3, W Wanapitei B, Can, 1974[3]
Clam	151 mg/kg wet weight	Water = 7+-6 ppb, pH=7.3, Wanapitei R, Can, silt-ladden and poll.[3]
Clam	157 mg/kg wet weight	Water = 8+-7 ppb, pH=7.2, Pickerel R, Can, non-poll.[3]
Clam	185 mg/kg wet weight	Water = 7+-6 ppb, pH=7.2, Main outlet (French R) to L. Huron, Can[3]
Clam	201 mg/kg wet weight	Water = 5+-3 ppb, pH=7.2, Hartley Bay, Canada, 1974[3]
Clam	226 mg/kg wet weight	Water = 5+-5 ppb, pH=7.2, Western outlet (French R) to L. Huron, Can[3]
Clam	264 mg/kg wet weight	Water = 7+-5 ppb, pH=7.2, East outlet (French R) to L. Huron, Can[3]
Clam	86 mg/kg wet weight	Water = 6+-6 ppb, pH=7.1, French R, Can, non-poll.[3]
Coniferous trees, needles	1500 mg/kg ash weight	Silicic rock = 180+-17, soil = 320+-50 ppm, Finland, n = 34[9]
Coniferous trees, needles	860 mg/kg ash weight	Calcareous rock = 230+-44, soil = 220+-64 ppm, Finland, n = 8[9]
Coniferous trees, needles	950 mg/kg ash weight	Ultrabasic rock = 190+-29, soil = 270+-49 ppm, Finland, n = 10[9]
Coniferous trees, twigs	1200 mg/kg ash weight	Calcareous rock = 230+-44, soil = 220+-64 ppm, Finland, n = 8[9]
Coniferous trees, twigs	1200 mg/kg ash weight	Ultrabasic rock = 190+-29, soil = 270+-49 ppm, Finland, n = 10[9]
Coniferous trees, twigs	1600 mg/kg ash weight	Silicic rock = 180+-17, soil = 320+-50 ppm, Finland, n = 34[9]
Cornicularia muricata	5640 mg/kg dry weight	Zn in soil = 13220 ppm, Blea Beck Head, W Yorkshire, GB[6]
Crataegus monogyna, leaf	0.99 total/soluble metal removed	Boiled, Bristol, GB[10]
Crataegus monogyna, leaf	1 % removed of soluble metal	Washed with a detergent, Bristol, GB[10]
Crataegus monogyna, leaf	1 % removed of total metal	Washed with a detergent, Bristol, GB[10]
Crataegus monogyna, leaf	1.34 % removed of soluble metal	Washed with deionised water, Bristol, GB[10]
Crataegus monogyna, leaf	1.34 % removed of total metal	Boiled, Bristol, GB[10]
Crataegus monogyna, leaf	1.75 % removed of soluble metal	Boiled, Bristol, GB[10]
Crataegus monogyna, leaf	1.96 % removed of total metal	Washed with deionised water, Bristol, GB[10]

Crataegus monogyna, leaf	2.57 total/soluble metal removed	Washed with deionised water, Bristol, GB[10]
Crataegus monogyna, leaf	3.42 total/soluble metal removed	Washed with a detergent, Bristol, GB[10]
Crataegus monogyna, unwashed leaves	1300 mg/kg dry weight	Avonmouth, bristol, GB[10]
Crayfish	101 mg/kg wet weight	Water = 5+-5 ppb, pH=7.3, W Wanapitei B, Can, 1974[3]
Crayfish	114 mg/kg wet weight	Water = 6+-6 ppb, pH=7.1, French R, Can, non-poll.[3]
Crayfish	168 mg/kg wet weight	Water = 7+-6 ppb, pH=7.3, Wanapitei R, Can, silt-ladden and poll.[3]
Crayfish	176 mg/kg wet weight	Water = 7+-5 ppb, pH=7.2, East outlet (French R) to L. Huron, Can[3]
Crayfish	232 mg/kg wet weight	Water = 7+-6 ppb, pH=7.2, Main outlet (French R) to L. Huron, Can[3]
Crayfish	89 mg/kg wet weight	Water = 5+-5 ppb, pH=7.2, Western outlet (French R) to L. Huron, Can[3]
Daphnia magna	1 % of total Zn available	Uptake from water[11]
Deciduous trees and shrubs, leaves	1000 mg/kg ash weight	Calcareous rock = 230+-44, soil = 220+-64 ppm, Finland, n = 14[9]
Deciduous trees and shrubs, leaves	1200 mg/kg ash weight	Ultrabasic rock = 190+-29, soil = 270+-49 ppm, Finland, n = 12[9]
Deciduous trees and shrubs, leaves	2100 mg/kg ash weight	Silicic rock = 180+-17, soil = 320+-50 ppm, Finland, n = 36[9]
Deciduous trees and shrubs, twigs	1700 mg/kg ash weight	Calcareous rock = 230+-44, soil = 220+-64 ppm, Finland, n = 14[9]
Deciduous trees and shrubs, twigs	2700 mg/kg ash weight	Ultrabasic rock = 190+-29, soil = 270+-49 ppm, Finland, n = 12[9]
Deciduous trees and shrubs, twigs	4100 mg/kg ash weight	Silicic rock = 180+-17, soil = 320+-50 ppm, Finland, n = 36[9]
Dicranella varia	18 mg/kg dry weight	Zn in soil = 20 ppm, Botanic Garden, Cottingham, W Yorkshire, GB[6]
Dicranella varia	2470 mg/kg dry weight	Zn in soil = 2700 ppm, Pastyremine, Foolow, Derbyshire, GB[6]
Dicranella varia	3600 mg/kg dry weight	Zn in soil = 2700 ppm, Pastyremine, Foolow, Derbyshire, GB[6]
Dicranella varia	4540 mg/kg dry weight	Zn in soil = 2645 ppm, Blythe Wood, Alport, Derbyshire, GB[6]
Dicranella varia	5250 mg/kg dry weight	Zn in soil = 2810 ppm, Slater mine, Foolow, Derbyshire, GB[6]
Dicranella varia	5985 mg/kg dry weight	Zn in soil = 2816 ppm, Speed mine, Foolow, Derbyshire, GB[6]
Dicranella varia	6150 mg/kg dry weight	Zn in soil = 2810 ppm, Slater mine, Foolow, Derbyshire, GB[6]
Different Zn species in seawater, Zn(2+)	12.5 % of total Zn concentration	Environmental model[12]
Different Zn species in seawater, Zn(2+)	16.1 % of total Zn concentration	Environmental model[13]
Different Zn species in seawater, Zn(2+)	17 % of total Zn concentration	Environmental model[14]
Different Zn species in seawater, Zn(2+)	25 % of total Zn concentration	Environmental model[15]
Different Zn species in seawater, Zn(2+)	5.7 % of total Zn concentration	Environmental model[16]
Different Zn species in seawater, Zn(2+)	55 % of total Zn concentration	Environmental model[17]
Different Zn species in seawater, Zn(OH)2	62 % of total Zn concentration	Environmental model[14]

Different Zn species in seawater, Zn(OH)2	71.7 % of total Zn concentration	Environmental model[16]
Different Zn species in seawater, ZnCO3	2.4 % of total Zn concentration	Environmental model[16]
Different Zn species in seawater, ZnCO3	3.3 % of total Zn concentration	Environmental model[13]
Different Zn species in seawater, ZnCO3	6 % of total Zn concentration	Environmental model[14]
Different Zn species in seawater, ZnCO3	6 % of total Zn concentration	Environmental model[17]
Different Zn species in seawater, ZnCl(+)	31 % of total Zn concentration	Environmental model[17]
Different Zn species in seawater, ZnCl(+)	44.3 % of total Zn concentration	Environmental model[13]
Different Zn species in seawater, ZnCl(+)	6 % of total Zn concentration	Environmental model[14]
Different Zn species in seawater, ZnCl(+)	79 % of total Zn concentration	Environmental model, including all chlorides[12]
Different Zn species in seawater, ZnCl(+)	9.9 % of total Zn concentration	Environmental model[16]
Different Zn species in seawater, ZnCl2	15.4 % of total Zn concentration	Environmental model[13]
Different Zn species in seawater, ZnCl2	16 % of total Zn concentration	Environmental model, including all chlorides[15]
Different Zn species in seawater, ZnCl2	4 % of total Zn concentration	Environmental model[14]
Different Zn species in seawater, ZnCl2	5.4 % of total Zn concentration	Environmental model[16]
Different Zn species in seawater, ZnCl3(-)	1 % of total Zn concentration	Environmental model[14]
Different Zn species in seawater, ZnCl3(-)	1.5 % of total Zn concentration	Environmental model[16]
Different Zn species in seawater, ZnCl3(-)	1.7 % of total Zn concentration	Environmental model[13]
Different Zn species in seawater, ZnCl4(2-)	0.4 % of total Zn concentration	Environmental model[14]
Different Zn species in seawater, ZnCl4(2-)	1.0 % of total Zn concentration	Environmental model[16]
Different Zn species in seawater, ZnCl4(2-)	2.3 % of total Zn concentration	Environmental model[13]
Different Zn species in seawater, ZnHCO3(+)	0.2 % of total Zn concentration	Environmental model[16]
Different Zn species in seawater, ZnHCO3(+)	0.3 % of total Zn concentration	Environmental model[13]
Different Zn species in seawater, ZnHCO3(+)	0.7 % of total Zn concentration	Environmental model[14]
Different Zn species in seawater, ZnHCO3(+)	1.6 % of total Zn concentration	Environmental model[12]
Different Zn species in seawater, ZnOH(+)	0.1 % of total Zn concentration	Environmental model[16]
Different Zn species in seawater, ZnOH(+)	0.2 % of total Zn concentration	Environmental model[14]
Different Zn species in seawater, ZnOH(+)	0.6 % of total Zn concentration	Environmental model, including all hydroxides[12]

Different Zn species in seawater, ZnOH(+)	2 % of total Zn concentration	Environmental model[17]
Different Zn species in seawater, ZnOH(+)	2.3 % of total Zn concentration	Environmental model[13]
Different Zn species in seawater, ZnOH(+)	63 % of total Zn concentration	Environmental model, including all hydroxides[15]
Different Zn species in seawater, ZnOHCl	12.5 % of total Zn concentration	Environmental model[13]
Different Zn species in seawater, ZnSO4	1.6 % of total Zn concentration	Environmental model[12]
Different Zn species in seawater, ZnSO4	1.9 % of total Zn concentration	Environmental model[13]
Different Zn species in seawater, ZnSO4	2.2 % of total Zn concentration	Environmental model[16]
Different Zn species in seawater, ZnSO4	4 % of total Zn concentration	Environmental model[14]
Different Zn species in seawater, ZnSO4	4 % of total Zn concentration	Environmental model[17]
Encalypta vulgaris	1667 mg/kg dry weight	Zn in soil = 3960 ppm, Alport Cupola Slag Heap, Derbyshire, GB[6]
Equisetum palustre, root	101 mg/kg wet weight	Water = 8+-7 ppb, pH=7.2, Pickerel R, Can, non-poll.[3]
Equisetum palustre, root	46 mg/kg wet weight	Water = 7+-6 ppb, pH=7.3, Wanapitei R, Can, silt-ladden and poll.[3]
Equisetum palustre, root	68 mg/kg wet weight	Water = 7+-6 ppb, pH=7.2, Main outlet (French R) to L. Huron, Can[3]
Equisetum palustre, root	76 mg/kg wet weight	Water = 5+-5 ppb, pH=7.2, Western outlet (French R) to L. Huron, Can[3]
Equisetum palustre, root	81 mg/kg wet weight	Water = 5+-5 ppb, pH=7.3, W Wanapitei B, Can, 1974[3]
Equisetum palustre, root	81 mg/kg wet weight	Water = 7+-5 ppb, pH=7.2, East outlet (French R) to L. Huron, Can[3]
Equisetum palustre, root	85 mg/kg wet weight	Water = 5+-3 ppb, pH=7.2, Hartley Bay, Canada, 1974[3]
Equisetum palustre, root	91 mg/kg wet weight	Water = 6+-6 ppb, pH=7.1, French R, Can, non-poll.[3]
Equisetum palustre, stem	59 mg/kg wet weight	Water = 5+-5 ppb, pH=7.3, W Wanapitei B, Can, 1974[3]
Equisetum palustre, stem	59 mg/kg wet weight	Water = 6+-6 ppb, pH=7.1, French R, Can, non-poll.[3]
Equisetum palustre, stem	59 mg/kg wet weight	Water = 7+-6 ppb, pH=7.3, Wanapitei R, Can, silt-ladden and poll.[3]
Equisetum palustre, stem	61 mg/kg wet weight	Water = 5+-3 ppb, pH=7.2, Hartley Bay, Canada, 1974[3]
Equisetum palustre, stem	61 mg/kg wet weight	Water = 5+-5 ppb, pH=7.2, Western outlet (French R) to L. Huron, Can[3]
Equisetum palustre, stem	68 mg/kg wet weight	Water = 7+-6 ppb, pH=7.2, Main outlet (French R) to L. Huron, Can[3]
Equisetum palustre, stem	70 mg/kg wet weight	Water = 7+-5 ppb, pH=7.2, East outlet (French R) to L. Huron, Can[3]
Equisetum palustre, stem	71 mg/kg wet weight	Water = 8+-7 ppb, pH=7.2, Pickerel R, Can, non-poll.[3]
Fir forest soil	90 mg/kg dry weight soil	White Mountains, NH, all sites max value = 169 ppm, organic layer[2]
Fir krummholz forest soil	73 mg/kg dry weight soil	White Mountains, NH, all sites max value = 169 ppm, organic layer[2]
Glycine max, leaf	28 mg/kg dry weight	Diagnostic part, new leaf, soil = 160.0 ug Cd/g, pH = 7.7, Zn = 47[18]
Glycine max, leaf	36 mg/kg dry weight	Diagnostic part, new leaf, soil = 80.0 ug Cd/g, pH = 7.7, Zn = 47 ppm[18]

Glycine max, leaf	38 mg/kg dry weight	Diagnostic part, new leaf, soil = 40.0 ug Cd/g, pH = 7.7, Zn = 47 ppm[18]
Glycine max, leaf	48 mg/kg dry weight	Diagnostic part, new leaf, soil = 20.0 ug Cd/g, pH = 7.7, Zn = 47 ppm[18]
Glycine max, leaf	49 mg/kg dry weight	Diagnostic part, new leaf, soil = 5.0 ug Cd/g, pH = 7.7, Zn = 47 ppm[18]
Glycine max, leaf	53 mg/kg dry weight	Diagnostic part, new leaf, soil = 10.0 ug Cd/g, pH = 7.7, Zn = 47 ppm[18]
Glycine max, leaf	57 mg/kg dry weight	Diagnostic part, new leaf, soil = 0.0 ug Cd/g, pH = 7.7, Zn = 47 ppm[18]
Glycine max, leaf	58 mg/kg dry weight	Diagnostic part, new leaf, soil = 2.5 ug Cd/g, pH = 7.7, Zn = 47 ppm[18]
Glycine max, leaf and petiole	130 mg/kg wet weight	Total Zn in soil = 339-136 mg/kg ww, sludge = 25.4 mm[19]
Glycine max, seed	37 mg/kg dry weight	Edible part, soil = 160.0 ug Cd/g, pH = 7.7, Zn = 47 ppm[18]
Glycine max, seed	38 mg/kg dry weight	Diagnostic, 3+4 leaf from top, soil = 0.0 ug Cd/g, pH = 7.7, Zn = 47[18]
Glycine max, seed	54 mg/kg dry weight	Edible part, soil = 40.0 ug Cd/g, pH = 7.7, Zn = 47 ppm[18]
Glycine max, seed	54 mg/kg dry weight	Edible part, soil = 80.0 ug Cd/g, pH = 7.7, Zn = 47 ppm[18]
Glycine max, seed	55 mg/kg dry weight	Edible part, soil = 20.0 ug Cd/g, pH = 7.7, Zn = 47 ppm[18]
Glycine max, seed	61 mg/kg dry weight	Edible part, soil = 0.0 ug Cd/g, pH = 7.7, Zn = 47 ppm[18]
Glycine max, seed	63 mg/kg dry weight	Edible part, soil = 5.0 ug Cd/g, pH = 7.7, Zn = 47 ppm[18]
Glycine max, seed	68 mg/kg dry weight	Edible part, soil = 10.0 ug Cd/g, pH = 7.7, Zn = 47 ppm[18]
Glycine max, seed	72 mg/kg dry weight	Edible part, soil = 2.5 ug Cd/g, pH = 7.7, Zn = 47 ppm[18]
Glycine max, total seed	49 mg/kg dry weight	Total Zn in soil = 69-13 mg/kg dw, slugde = 0.0 mm[19]
Glycine max, total seed	51 mg/kg dry weight	Total Zn in soil = 74-12 mg/kg dw, slugde = 0.0 mm[19]
Glycine max, total seed	54 mg/kg dry weight	Total Zn in soil = 71-12 mg/kg dw, slugde = 0.0 mm[19]
Glycine max, total seed	55 mg/kg dry weight	Total Zn in soil = 80-13 mg/kg dw, slugde = 0.0 mm[19]
Glycine max, total seed	59 mg/kg dry weight	Total Zn in soil = 178-69 mg/kg dw, slugde = 6.4 mm[19]
Glycine max, total seed	62 mg/kg dry weight	Total Zn in soil = 232-145 mg/kg dw, slugde = 12.7 mm[19]
Glycine max, total seed	63 mg/kg dry weight	Total Zn in soil = 136-63 mg/kg dw, slugde = 6.4 mm[19]
Glycine max, total seed	65 mg/kg dry weight	Total Zn in soil = 203-126 mg/kg dw, slugde = 12.4 mm[19]
Glycine max, total seed	66 mg/kg dry weight	Total Zn in soil = 339-236 mg/kg dw, slugde = 25.4 mm[19]
Glycine max, total seed	71 mg/kg dry weight	Total Zn in soil = 175-74 mg/kg dw, slugde = 6.4 mm[19]
Glycine max, total seed	72 mg/kg dry weight	Total Zn in soil = 143-70 mg/kg dw, slugde = 6.4 mm[19]
Glycine max, total seed	78 mg/kg dry weight	Total Zn in soil = 248-144 mg/kg dw, slugde = 12.7 mm[19]
Glycine max, total seed	78 mg/kg dry weight	Total Zn in soil = 370-269 mg/kg dw, slugde = 25.4 mm[19]
Glycine max, total seed	82 mg/kg dry weight	Total Zn in soil = 265-193 mg/kg dw, slugde = 12.7 mm[19]
Glycine max, total seed	85 mg/kg dry weight	Total Zn in soil = 336-298 mg/kg dw, slugde = 25.4 mm[19]
Glycine max, total seed	86 mg/kg dry weight	Total Zn in soil = 356-243 mg/kg dw, slugde = 25.4 mm[19]
Glycine max, total seed	94 mg/kg dry weight	Total Zn in soil = 390-316 mg/kg dw, slugde = 25.4 mm[19]
Guinea pig, renal cortex	151 mg/kg dry weight	1 m of cigarette smoke, n = 22, control = 133+-9 ug/g dw, n = 15, SD[20]
Hardwood forest soil	100 mg/kg dry weight soil	White Mountains, NH, all sites max value = 169 ppm, organic layer[2]

Hypolimnion	1.9 ug/l	Conc in epilimnion = 1.9 ug/l, concentration factor = 1.1, Schoehsee[21]
Living spruce needles	107.5 mg/kg dry weight	Litter = 60-126, underlaying soil = 28-41 mg/kg dw, S Sweden, rg = 27[22]
Min value in forest soil	19 mg/kg dry weight soil	White Mountains, NH, all sites max value = 169 ppm, organic layer[2]
Mussel	1350 mg/kg dry weight	Conc in sed = 3900 mg/kg dw, filt seawater = 230 ppb, Derwent Est, Tasmania[23]
Mussel	180 mg/kg dry weight	Conc in sed = 9120 mg/kg dw, filt seawater = 78 ppb, Derwent Est, Tasmania[23]
Mussel	296 mg/kg dry weight	Conc in sed = 172 mg/kg dw, filt seawater = 6 ppb, Derwent Est, Tasmania[23]
Mussel	322 mg/kg dry weight	Conc in sed = 50 mg/kg dw, filt seawater = 210 ppb, Derwent Est, Tasmania[23]
Mussel	354 mg/kg dry weight	Conc in sed = 250 mg/kg dw, filt seawater = 70 ppb, Derwent Est, Tasmania[23]
Mussel	401 mg/kg dry weight	Conc in sed = 50 mg/kg dw, filt seawater = 120 ppb, Derwent Est, Tasmania[23]
Mussel	556 mg/kg dry weight	Conc in sed = 1100 mg/kg dw, filt seawater = 64 ppb, Derwent Est, Tasmania[23]
Mussel	572 mg/kg dry weight	Conc in sed = 169 mg/kg dw, filt seawater = 232 ppb, Derwent Est, Tasmania[23]
Mussel	699 mg/kg dry weight	Conc in sed = 100 mg/kg dw, filt seawater = 144 ppb, Derwent Est, Tasmania[23]
Mussel	702 mg/kg dry weight	Conc in sed = 1810 mg/kg dw, filt seawater = 232 ppb, Derwent Est, Tasmania[23]
Mussel	725 mg/kg dry weight	Conc in sed = 2000 mg/kg dw, filt seawater = 112 ppb, Derwent Est, Tasmania[23]
Mussel	744 mg/kg dry weight	Conc in sed = 3180 mg/kg dw, filt seawater = 230 ppb, Derwent Est, Tasmania[23]
Mussel	854 mg/kg dry weight	Conc in sed = 851 mg/kg dw, filt seawater = 128 ppb, Derwent Est, Tasmania[23]
Nereis diversicolor	240 mg/kg dry weight	Range in sedimennt = 100 - 300 ug/g dw, Engl, rg = 130 - 350[24]
Nuphar variegatum, leaf	17 mg/kg wet weight	Water = 5+-3 ppb, pH=7.2, Hartley Bay, Canada, 1974[3]
Nuphar variegatum, leaf	20 mg/kg wet weight	Water = 5+-5 ppb, pH=7.3, W Wanapitei B, Can, 1974[3]
Nuphar variegatum, leaf	29 mg/kg wet weight	Water = 7+-6 ppb, pH=7.3, Wanapitei R, Can, silt-ladden and poll.[3]
Nuphar variegatum, leaf	35 mg/kg wet weight	Water = 7+-5 ppb, pH=7.2, East outlet (French R) to L. Huron, Can[3]
Nuphar variegatum, leaf	38 mg/kg wet weight	Water = 8+-7 ppb, pH=7.2, Pickerel R, Can, non-poll.[3]
Nuphar variegatum, leaf	44 mg/kg wet weight	Water = 5+-5 ppb, pH=7.2, Western outlet (French R) to L. Huron, Can[3]
Nuphar variegatum, leaf	56 mg/kg wet weight	Water = 7+-6 ppb, pH=7.2, Main outlet (French R) to L. Huron, Can[3]
Nuphar variegatum, peduncle	12 mg/kg wet weight	Water = 5+-3 ppb, pH=7.2, Hartley Bay, Canada, 1974[3]
Nuphar variegatum, peduncle	18 mg/kg wet weight	Water = 7+-6 ppb, pH=7.3, Wanapitei R, Can, silt-ladden and poll.[3]
Nuphar variegatum, peduncle	24 mg/kg wet weight	Water = 7+-6 ppb, pH=7.2, Main outlet (French R) to L. Huron, Can[3]
Nuphar variegatum, peduncle	25 mg/kg wet weight	Water = 7+-5 ppb, pH=7.2, East outlet (French R) to L. Huron, Can[3]
Nuphar variegatum, peduncle	29 mg/kg wet weight	Water = 8+-7 ppb, pH=7.2, Pickerel R, Can, non-poll.[3]
Nuphar variegatum, peduncle	6 mg/kg wet weight	Water = 5+-5 ppb, pH=7.3, W Wanapitei B, Can, 1974[3]

Nuphar variegatum, petiole	16 mg/kg wet weight	Water = 5+-3 ppb, pH=7.2, Hartley Bay, Canada, 1974[3]
Nuphar variegatum, petiole	20 mg/kg wet weight	Water = 5+-5 ppb, pH=7.2, Western outlet (French R) to L. Huron, Can[3]
Nuphar variegatum, petiole	20 mg/kg wet weight	Water = 5+-5 ppb, pH=7.3, W Wanapitei B, Can, 1974[3]
Nuphar variegatum, petiole	25 mg/kg wet weight	Water = 7+-6 ppb, pH=7.3, Wanapitei R, Can, silt-ladden and poll.[3]
Nuphar variegatum, petiole	30 mg/kg wet weight	Water = 8+-7 ppb, pH=7.2, Pickerel R, Can, non-poll.[3]
Nuphar variegatum, petiole	33 mg/kg wet weight	Water = 7+-6 ppb, pH=7.2, Main outlet (French R) to L. Huron, Can[3]
Nuphar variegatum, petiole	35 mg/kg wet weight	Water = 7+-5 ppb, pH=7.2, East outlet (French R) to L. Huron, Can[3]
Nuphar variegatum, root	15 mg/kg wet weight	Water = 5+-3 ppb, pH=7.2, Hartley Bay, Canada, 1974[3]
Nuphar variegatum, root	17 mg/kg wet weight	Water = 5+-5 ppb, pH=7.3, W Wanapitei B, Can, 1974[3]
Nuphar variegatum, root	24 mg/kg wet weight	Water = 7+-6 ppb, pH=7.2, Main outlet (French R) to L. Huron, Can[3]
Nuphar variegatum, root	33 mg/kg wet weight	Water = 7+-5 ppb, pH=7.2, East outlet (French R) to L. Huron, Can[3]
Nuphar variegatum, root	34 mg/kg wet weight	Water = 7+-6 ppb, pH=7.3, Wanapitei R, Can, silt-ladden and poll.[3]
Nuphar variegatum, root	37 mg/kg wet weight	Water = 8+-7 ppb, pH=7.2, Pickerel R, Can, non-poll.[3]
Oak - Hickory	125 mg/kg dry weight soil	E Tennessee, soil litter layer O2, g/ha = 2200[7]
Oak - Hickory	48 mg/kg dry weight soil	E Tennessee, soil litter layer O1, g/ha = 420[7]
Oak forest soil	103 mg/kg dry weight soil	E Chicago, USA, urban, horizon = B, 14-25 cm[25]
Oak forest soil	2456 mg/kg dry weight soil	E Chicago, USA, urban, horizon = A1, 0-2.5 cm[25]
Oak forest soil	711 mg/kg dry weight soil	E Chicago, USA, urban, horizon = A1, 2.5-14 cm[25]
Oak forest soil	915 mg/kg dry weight soil	E Chicago, USA, urban, horizon = O , surface litter[25]
Oryza sativa, grain	21 mg/kg dry weight	Edible part, soil = 160.0 ug Cd/g, pH = 7.7, Zn = 47 ppm[18]
Oryza sativa, grain	24 mg/kg dry weight	Edible part, soil = 80.0 ug Cd/g, pH = 7.7, Zn = 47 ppm[18]
Oryza sativa, grain	27 mg/kg dry weight	Edible part, soil = 20.0 ug Cd/g, pH = 7.7, Zn = 47 ppm[18]
Oryza sativa, grain	28 mg/kg dry weight	Edible part, soil = 40.0 ug Cd/g, pH = 7.7, Zn = 47 ppm[18]
Oryza sativa, grain	29 mg/kg dry weight	Edible part, soil = 10.0 ug Cd/g, pH = 7.7, Zn = 47 ppm[18]
Oryza sativa, grain	37 mg/kg dry weight	Edible part, soil = 5.0 ug Cd/g, pH = 7.7, Zn = 47 ppm[18]
Oryza sativa, grain	39 mg/kg dry weight	Edible part, soil = 0.0 ug Cd/g, pH = 7.7, Zn = 47 ppm[18]
Oryza sativa, grain	43 mg/kg dry weight	Edible part, soil = 2.5 ug Cd/g, pH = 7.7, Zn = 47 ppm[18]
Oryza sativa, leaf	13 mg/kg dry weight	Diagnostic part, 2 top leaves, soil = 160.0 ug Cd/g, pH = 7.7, Zn = 47 ppm[18]
Oryza sativa, leaf	14 mg/kg dry weight	Diagnostic part, 2 top leaves, soil = 20.0 ug Cd/g, pH = 7.7, Zn = 47 ppm[18]
Oryza sativa, leaf	15 mg/kg dry weight	Diagnostic part, 2 top leaves, soil = 0.0 ug Cd/g, pH = 7.7, Zn = 47 ppm[18]
Oryza sativa, leaf	16 mg/kg dry weight	Diagnostic part, 2 top leaves, soil = 10.0 ug Cd/g, pH = 7.7, Zn = 47 ppm[18]
Oryza sativa, leaf	16 mg/kg dry weight	Diagnostic part, 2 top leaves, soil = 2.5 ug Cd/g, pH = 7.7, Zn = 47 ppm[18]

Oryza sativa, leaf	16 mg/kg dry weight	Diagnostic part, 2 top leaves, soil = 40.0 ug Cd/g, pH = 7.7, Zn = 47 ppm[18]
Oryza sativa, leaf	17 mg/kg dry weight	Diagnostic part, 2 top leaves, soil = 5.0 ug Cd/g, pH = 7.7, Zn = 47 ppm[18]
Oryza sativa, leaf	20 mg/kg dry weight	Diagnostic part, 2 top leaves, soil = 80.0 ug Cd/g, pH = 7.7, Zn = 47 ppm[18]
Pine	56 mg/kg dry weight soil	E Tennessee, soil litter layer O1, g/ha = 610[7]
Pine	59 mg/kg dry weight soil	E Tennessee, soil litter layer O2, g/ha = 930[7]
Peltigera canina	452 mg/kg dry weight	Zn in soil = 1800 ppm, Blea Beck Head, W Yorkshire, GB[6]
Peltigera canina	5650 mg/kg dry weight	Zn in soil = 3960 ppm, Alport Cupola Slap Heap, Derbyshire, GB[6]
Periphyton	108 mg/kg wet weight	Water = 5+-5 ppb, pH=7.3, W Wanapitei B, Can, 1974[3]
Periphyton	108 mg/kg wet weight	Water = 7+-6 ppb, pH=7.3, Wanapitei R, Can, silt-ladden and poll.[3]
Periphyton	111 mg/kg wet weight	Water = 7+-6 ppb, pH=7.2, Main outlet (French R) to L. Huron, Can[3]
Periphyton	118 mg/kg wet weight	Water = 5+-3 ppb, pH=7.2, Hartley Bay, Canada, 1974[3]
Periphyton	77 mg/kg wet weight	Water = 6+-6 ppb, pH=7.1, French R, Can, non-poll.[3]
Periphyton	89 mg/kg wet weight	Water = 7+-5 ppb, pH=7.2, East outlet (French R) to L. Huron, Can[3]
Periphyton	90 mg/kg wet weight	Water = 8+-7 ppb, pH=7.2, Pickerel R, Can, non-poll.[3]
Periphyton	94 mg/kg wet weight	Water = 5+-5 ppb, pH=7.2, Western outlet (French R) to L. Huron, Can[3]
Phaseolus vulgaris, bean	12 mg/kg dry weight	Edible part, soil = 160.0 ug Cd/g, pH = 7.7, Zn = 47 ppm[18]
Phaseolus vulgaris, bean	13 mg/kg dry weight	Edible part, soil = 80.0 ug Cd/g, pH = 7.7, Zn = 47 ppm[18]
Phaseolus vulgaris, bean	15 mg/kg dry weight	Edible part, soil = 40.0 ug Cd/g, pH = 7.7, Zn = 47 ppm[18]
Phaseolus vulgaris, bean	20 mg/kg dry weight	Edible part, soil = 20.0 ug Cd/g, pH = 7.7, Zn = 47 ppm[18]
Phaseolus vulgaris, bean	21 mg/kg dry weight	Edible part, soil = 10.0 ug Cd/g, pH = 7.7, Zn = 47 ppm[18]
Phaseolus vulgaris, bean	21 mg/kg dry weight	Edible part, soil = 5.0 ug Cd/g, pH = 7.7, Zn = 47 ppm[18]
Phaseolus vulgaris, bean	23 mg/kg dry weight	Edible part, soil = 0.0 ug Cd/g, pH = 7.7, Zn = 47 ppm[18]
Phaseolus vulgaris, bean	25 mg/kg dry weight	Edible part, soil = 2.5 ug Cd/g, pH = 7.7, Zn = 47 ppm[18]
Phaseolus vulgaris, leaf	21 mg/kg dry weight	Diagnostic part, new leaf, soil = 160.0 ug Cd/g, pH = 7.7, Zn = 47 ppm[18]
Phaseolus vulgaris, leaf	21 mg/kg dry weight	Diagnostic part, new leaf, soil = 80.0 ug Cd/g, pH = 7.7, Zn = 47 ppm[18]
Phaseolus vulgaris, leaf	29 mg/kg dry weight	Diagnostic part, new leaf, soil = 20.0 ug Cd/g, pH = 7.7, Zn = 47 ppm[18]
Phaseolus vulgaris, leaf	29 mg/kg dry weight	Diagnostic part, new leaf, soil = 40.0 ug Cd/g, pH = 7.7, Zn = 47 ppm[18]
Phaseolus vulgaris, leaf	31 mg/kg dry weight	Diagnostic part, new leaf, soil = 10.0 ug Cd/g, pH = 7.7, Zn = 47 ppm[18]
Phaseolus vulgaris, leaf	39 mg/kg dry weight	Diagnostic part, new leaf, soil = 5.0 ug Cd/g, pH = 7.7, Zn = 47 ppm[18]
Phaseolus vulgaris, leaf	44 mg/kg dry weight	Diagnostic part, new leaf, soil = 2.5 ug Cd/g, pH = 7.7, Zn = 47 ppm[18]
Phaseolus vulgaris, leaf	47 mg/kg dry weight	Diagnostic part, new leaf, soil = 0.0 ug Cd/g, pH = 7.7, Zn = 47 ppm[18]
Philonotis fontana	204 mg/kg dry weight	Zn in soil = 1840 ppm, Slater mine, Foolow, Derbyshire, GB, new plant[6]

Philonotis fontana	2420 mg/kg dry weight	Zn in soil = 1350 ppm, Blea Beck Junction, W Yorkshire, GB, old leaf[6]
Philonotis fontana	297 mg/kg dry weight	Zn in soil = 1350 ppm, Slater mine, Foolow, Derbyshire, GB, new plant[6]
Philonotis fontana	8750 mg/kg dry weight	Zn in soil = 1840 ppm, Slater mine, Foolow, Derbyshire, GB, old leaf[6]
Plantago aquatica	125.11 mg/kg dry weight	800 m above pollution entry I, water = 0.056 mg/l, whole leaves, old[1]
Potamogeton sp., leaf	147 mg/kg wet weight	Water = 5+-5 ppb, pH=7.2, Western outlet (French R) to L. Huron, Can[3]
Potamogeton sp., leaf	161 mg/kg wet weight	Water = 6+-6 ppb, pH=7.1, French R, Can, non-poll.[3]
Potamogeton sp., leaf	202 mg/kg wet weight	Water = 5+-5 ppb, pH=7.3, W Wanapitei B, Can, 1974[3]
Potamogeton sp., leaf	209 mg/kg wet weight	Water = 7+-5 ppb, pH=7.2, East outlet (French R) to L. Huron, Can[3]
Potamogeton sp., leaf	241 mg/kg wet weight	Water = 5+-3 ppb, pH=7.2, Hartley Bay, Canada, 1974[3]
Potamogeton sp., leaf	244 mg/kg wet weight	Water = 8+-7 ppb, pH=7.2, Pickerel R, Can, non-poll.[3]
Potamogeton sp., leaf	62 mg/kg wet weight	Water = 7+-6 ppb, pH=7.3, Wanapitei R, Can, silt-ladden and poll.[3]
Potamogeton sp., stem	37 mg/kg wet weight	Water = 5+-5 ppb, pH=7.2, Western outlet (French R) to L. Huron, Can[3]
Potamogeton sp., stem	38 mg/kg wet weight	Water = 7+-5 ppb, pH=7.2, East outlet (French R) to L. Huron, Can[3]
Potamogeton sp., stem	39 mg/kg wet weight	Water = 5+-3 ppb, pH=7.2, Hartley Bay, Canada, 1974[3]
Potamogeton sp., stem	45 mg/kg wet weight	Water = 7+-6 ppb, pH=7.3, Wanapitei R, Can, silt-ladden and poll.[3]
Potamogeton sp., stem	57 mg/kg wet weight	Water = 6+-6 ppb, pH=7.1, French R, Can, non-poll.[3]
Potamogeton sp., stem	72 mg/kg wet weight	Water = 5+-5 ppb, pH=7.3, W Wanapitei B, Can, 1974[3]
Potamogeton sp., stem	93 mg/kg wet weight	Water = 8+-7 ppb, pH=7.2, Pickerel R, Can, non-poll.[3]
Quercus robur	1617 mg/kg dry weight	Avonmouth, Bristol, GB, unwashed laaves[10]
Quercus robur, leaf	1 % removed of soluble metal	Boiled, Bristol, GB[10]
Quercus robur, leaf	1 % removed of total metal	Boiled, Bristol, GB[10]
Quercus robur, leaf	1.34 total/soluble metal removed	Boiled, Bristol, GB[10]
Quercus robur, leaf	1.44 total/soluble metal removed	Washed with deionised water, Bristol, GB[10]
Quercus robur, leaf	1.95 total/soluble metal removed	Washed with a detergent, Bristol, GB[10]
Quercus robur, leaf	3.3 % removed of total metal	Washed with a detergent, Bristol, GB[10]
Quercus robur, leaf	4.12 % removed of soluble metal	Washed with a detergent, Bristol, GB[10]
Quercus robur, leaf	5.38 % removed of total metal	Washed with deionised water, Bristol, GB[10]
Quercus robur, leaf	6.49 % removed of soluble metal	Washed with deionised water, Bristol, GB[10]
Red clover	44 mg/kg dry weight	Added ZnSO4 = 22 kg/ha, pH = 6.5, control = 29 mg/kg dw, effect pH[26]
Red clover	75 mg/kg dry weight	Added ZnSO4 = 22 kg/ha, pH = 6.1, control = 33 mg/kg dw, effect pH[26]
Salix alba	1699 mg/kg dry weight	Avonmouth, Bristol, GB, unwashed leaves[10]

Salix alba, leaf	1 % removed of soluble metal	Washed with deionised water, Bristol, GB[10]
Salix alba, leaf	1 % removed of total metal	Washed with deionised water, Bristol, GB[10]
Salix alba, leaf	1.02 total/soluble metal removed	Boiled, Bristol, GB[10]
Salix alba, leaf	1.25 total/soluble metal removed	Washed with a detergent, Bristol, GB[10]
Salix alba, leaf	1.54 % removed of total metal	Boiled, Bristol, GB[10]
Salix alba, leaf	1.61 % removed of total metal	Washed with a detergent, Bristol, GB[10]
Salix alba, leaf	1.64 % removed of soluble metal	Boiled, Bristol, GB[10]
Salix alba, leaf	1.7 % removed of soluble metal	Washed with a detergent, Bristol, GB[10]
Salix alba, leaf	1.87 total/soluble metal removed	Washed with deionised water, Bristol, GB[10]
Sediment	100.29 mg/kg dry weight	800 and 50 m below pollution I and II, water = 0.107 mg/l. rg = 96 -[1]
Sediment	105.84 mg/kg dry weight	800 m above pollution entry I, water = 0.056 mg/l, rg = 96.3 - 112.6[1]
Sediment	1235 mg/kg dry weight	Max water conc = 198 microgram/l, 1 y measurement, Red River[27]
Sediment	1235 mg/kg dry weight	Water conc = 110 microgram/l, yearly mean, Red River[27]
Sediment	129.0 mg/kg dry weight	Range 66-192 ppm, River Rheidol, Cardiganshire, GB[28]
Sediment	161 mg/kg dry weight	Max water conc = 8 microgram/l, 1 y measurement, Fowey River[27]
Sediment	161 mg/kg dry weight	Water conc = 3 microgram/l, yearly mean, fowey River[27]
Sediment	24.8 mg/kg dry weight	Unpolluted area, estuary, Urr water, rg = 17.2 - 42.0[29]
Sediment	247 mg/kg dry weight	Max water conc = 35 microgram/l, max conc, Gannel River[27]
Sediment	247 mg/kg dry weight	Water conc = 10 microgram/l, yearly mean, Gannel River[27]
Sediment	2610 mg/kg dry weight	Max water conc = 1000 microgram/l, max conc, Gannel River[27]
Sediment	2610 mg/kg dry weight	Water conc = 338 microgram/l, yearly mean, Gannel River[27]
Sediment	3 mg/kg dry weight	Max water conc = 5 microgram/l, 1 y measurement, Carnon River[27]
Sediment	3 mg/kg dry weight	Water conc = 3 microgram/l, yearly mean, Carnon River[27]
Sediment	30 mg/kg dry weight	Unpolluted area, river, Illinois River, rg = 18 - 41[30]
Sediment	320 mg/kg dry weight	Max water conc = 12 microgram/l, 1 y measurement, Carnon River[27]
Sediment	320 mg/kg dry weight	Water conc = 10 microgram/l, yearly mean, Carnon River[27]
Sediment	345.77 mg/kg dry weight	150 m below pollution entry I, water = 0.190 mg/l, rg = 328 - 365[1]
Sediment	350 mg/kg dry weight	Conc in epilimnion = 1.8 ug/l, concentration factor = 195000,[21]
Sediment	41.2 mg/kg dry weight	Unpolluted area, estuary, Urr water, rg = 24.8 - 65.0[31]
Sediment	445 mg/kg dry weight	Max water conc = 105 microgram/l, 1 y measurement, Red River[27]
Sediment	445 mg/kg dry weight	Water conc = 33 microgram/l, yearly mean, Red River[27]
Sediment	48 mg/kg wet weight	Water = 7+-5 ppb, pH=7.2, East outlet (French R) to L. Huron, Can[3]
Sediment	57 mg/kg wet weight	Water = 6+-6 ppb, pH=7.1, French R, Can, non-poll.[3]

Sediment	64 mg/kg wet weight	Water = 5+-5 ppb, pH=7.2, Western outlet (French R) to L. Huron, Can[3]
Sediment	65 mg/kg dry weight	Max water conc = 12 microgram/l, 1 y measurement, Fowey River[27]
Sediment	65 mg/kg dry weight	Water conc = 6 microgram/l, yearly mean, fowey River[27]
Sediment	66 mg/kg wet weight	Water = 8+-7 ppb, pH=7.2, Pickerel R, Can, non-poll.[3]
Sediment	67.0 mg/kg dry weight	Unpolluted area, estuary, Urr water, rg = 43.0 - 97.5[31]
Sediment	68 mg/kg wet weight	Water = 7+-6 ppb, pH=7.3, Wanapitei R, Can, silt-ladden and poll., SD=17[3]
Sediment	74 mg/kg wet weight	Water = 5+-3 ppb, pH=7.2, Hartley Bay, Canada, 1974, SD=5[3]
Sediment	82 mg/kg wet weight	Water = 7+-6 ppb, pH=7.2, Main outlet (French R) to L. Huron, Can, SD=5[3]
Sediment	827 mg/kg wet weight	Water = 5+-5 ppb, pH=7.3, W Wanapitei B, Can, 1974, SD=22[3]
Spruce fir forest soil	53 mg/kg dry weight soil	White Mountains, NH, all sites max value = 169 ppm, organic layer[2]
Spruce forest soil	2300 mg/kg dry weight soil	Central Sweden, Humus, O2[32]
Spruce forest soil	530 mg/kg dry weight soil	Central Sweden, needle litter, O1[32]
Triticum aestivum, grain	29 mg/kg dry weight	Edible part, soil = 160.0 ug Cd/g, pH = 7.7, Zn = 47 ppm[18]
Triticum aestivum, grain	32 mg/kg dry weight	Edible part, soil = 80.0 ug Cd/g, pH = 7.7, Zn = 47 ppm[18]
Triticum aestivum, grain	41 mg/kg dry weight	Edible part, soil = 20.0 ug Cd/g, pH = 7.7, Zn = 47 ppm[18]
Triticum aestivum, grain	42 mg/kg dry weight	Edible part, soil = 5.0 ug Cd/g, pH = 7.7, Zn = 47 ppm[18]
Triticum aestivum, grain	44 mg/kg dry weight	Edible part, soil = 40.0 ug Cd/g, pH = 7.7, Zn = 47 ppm[18]
Triticum aestivum, grain	45 mg/kg dry weight	Edible part, soil = 0.0 ug Cd/g, pH = 7.7, Zn = 47 ppm[18]
Triticum aestivum, grain	47 mg/kg dry weight	Edible part, soil = 2.5 ug Cd/g, pH = 7.7, Zn = 47 ppm[18]
Triticum aestivum, grain	48 mg/kg dry weight	Edible part, soil = 10.0 ug Cd/g, pH = 7.7, Zn = 47 ppm[18]
Triticum aestivum, leaf	24 mg/kg dry weight	Diagnostic part, 3 top leaf, soil = 160.0 ug Cd/g, pH = 7.7, Zn = 47 ppm[18]
Triticum aestivum, leaf	31 mg/kg dry weight	Diagnostic part, 3 top leaf, soil = 80.0 ug Cd/g, pH = 7.7, Zn = 47 ppm[18]
Triticum aestivum, leaf	34 mg/kg dry weight	Diagnostic part, 3 top leaf, soil = 20.0 ug Cd/g, pH = 7.7, Zn = 47 ppm[18]
Triticum aestivum, leaf	35 mg/kg dry weight	Diagnostic part, 3 top leaf, soil = 5.0 ug Cd/g, pH = 7.7, Zn = 47 ppm[18]
Triticum aestivum, leaf	37 mg/kg dry weight	Diagnostic part, 3 top leaf, soil = 40.0 ug Cd/g, pH = 7.7, Zn = 47 ppm[18]
Triticum aestivum, leaf	38 mg/kg dry weight	Diagnostic part, 3 top leaf, soil = 2.5 ug Cd/g, pH = 7.7, Zn = 47 ppm[18]
Triticum aestivum, leaf	41 mg/kg dry weight	Diagnostic part, 3 top leaf, soil = 10.0 ug Cd/g, pH = 7.7, Zn = 47 ppm[18]
Triticum aestivum, leaf	44 mg/kg dry weight	Diagnostic part, 3 top leaf, soil = 0.0 ug Cd/g, pH = 7.7, Zn = 47 ppm[18]
Yellow poplar	130 mg/kg dry weight soil	E Tennessee, soil litter layer O2, g/ha = 980[7]
Yellow poplar	58 mg/kg dry weight soil	E Tennessee, soil litter layer O1, g/ha = 380[7]
Zea mays, grain	32 mg/kg dry weight	Edible part, soil = 0.0 ug Cd/g, pH = 7.7, Zn = 47 ppm[18]
Zea mays, grain	38 mg/kg dry weight	Edible part, soil = 2.5 ug Cd/g, pH = 7.7, Zn = 47 ppm[18]

Zea mays, grain	41 mg/kg dry weight	Edible part, soil = 80.0 ug Cd/g, pH = 7.7, Zn = 47 ppm[18]
Zea mays, grain	45 mg/kg dry weight	Edible part, soil = 20.0 ug Cd/g, pH = 7.7, Zn = 47 ppm[18]
Zea mays, grain	47 mg/kg dry weight	Edible part, soil = 10.0 ug Cd/g, pH = 7.7, Zn = 47 ppm[18]
Zea mays, grain	47 mg/kg dry weight	Edible part, soil = 5.0 ug Cd/g, pH = 7.7, Zn = 47 ppm[18]
Zea mays, grain	53 mg/kg dry weight	Edible part, soil = 160.0 ug Cd/g, pH = 7.7, Zn = 47 ppm[18]
Zea mays, grain	54 mg/kg dry weight	Edible part, soil = 40.0 ug Cd/g, pH = 7.7, Zn = 47 ppm[18]
Zea mays, leaf	22 mg/kg dry weight	Diagnostic part, 3+4 leaf from top, soil = 160.0 ug Cd/g, pH = 7.7, Zn = 47 ppm[18]
Zea mays, leaf	24 mg/kg dry weight	Diagnostic part, 3+4 leaf from top, soil = 80.0 ug Cd/g, pH = 7.7, Zn = 47 ppm[18]
Zea mays, leaf	26 mg/kg dry weight	Diagnostic part, 3+4 leaf from top, soil = 20.0 ug Cd/g, pH = 7.7, Zn = 47 ppm[18]
Zea mays, leaf	27 mg/kg dry weight	Diagnostic part, 3+4 leaf from top, soil = 40.0 ug Cd/g, pH = 7.7, Zn = 47 ppm[18]
Zea mays, leaf	33 mg/kg dry weight	Diagnostic part, 3+4 leaf from top, soil = 10.0 ug Cd/g, pH = 7.7, Zn = 47 ppm[18]
Zea mays, leaf	39 mg/kg dry weight	Diagnostic part, 3+4 leaf from top, soil = 5.0 ug Cd/g, pH = 7.7, Zn = 47 ppm[18]
Zea mays, leaf	42 mg/kg dry weight	Diagnostic part, 3+4 leaf from top, soil = 2.5 ug Cd/g, pH = 7.7, Zn = 47 ppm[18]
Zooplankton	103 mg/kg wet weight	Water = 5+-5 ppb, pH=7.3, W Wanapitei B, Can, 1974[3]
Zooplankton	110 mg/kg wet weight	Water = 7+-6 ppb, pH=7.3, Wanapitei R, Can, silt-ladden and poll.[3]
Zooplankton	112 mg/kg wet weight	Water = 7+-6 ppb, pH=7.2, Main outlet (French R) to L. Huron, Can[3]
Zooplankton	116 mg/kg wet weight	Water = 5+-5 ppb, pH=7.2, Western outlet (French R) to L. Huron, Can[3]
Zooplankton	122 mg/kg wet weight	Water = 7+-5 ppb, pH=7.2, East outlet (French R) to L. Huron, Can[3]
Zooplankton	137 mg/kg wet weight	Water = 8+-7 ppb, pH=7.2, Pickerel R, Can, non-poll.[3]
Zooplankton	139 mg/kg wet weight	Water = 6+-6 ppb, pH=7.1, French R, Can, non-poll.[3]
Zooplankton	157 mg/kg wet weight	Water = 5+-3 ppb, pH=7.2, Hartley Bay, Canada, 1974[3]
Zooplankton	162.53 mg/kg dry weight	19 st, NY, conc in algae = 63.93 mg/kg dw, mean, Long Island[33]

Source: 1.Patrick, F.M. and M.W. Loutit. 1977. Water Res., 11(8), pp 699-703.
2.Reiners, W.A., R.H. Marks and P.M. Vitousek. 1975. Oikos 26, 264-275.
3.Hutchinson, T.C., Fedorenko, Fitchko, Kuja, Loon and Lichwa. 1974. In: J.O. Nriagu (ed.): Environmenal Biochemistry. Ann Arbor Sci. Publ. Ann Arbor, Mich, 565-585.
4.Jones, G.E., L.G. Royle and L. Mur ray.1976. Applied Science Publ., 5(12), pp 889-895.
5.Heinrichs, H. and R. Meyer. 1977. J. Environ. Qual. 6, 402-407.
6.Shimwell, D.W. and A.E. Laurie. 1972. Environ. Pollut., 3, pp 291-301.
7.Van Hook, R.I., W.F. Harris and G.S. Henderson. 1977. Ambio 6, 281-286.
8.Purves, D. 1972. Environ. Pollut., 3, pp 17-24.
9.Lounamaa, J. 1967. in: Kvauheim, A. (ed.); Geochemical Prospecting in Fennoscandia. pp 287-317. Interscience Publ. New York.
10.Little, P. 1974. Environ. Pollut., 5, pp 159-172.
11.Bachmann, R.W. 1963. in: Schutz, V. and A.W. Klement, Jr. (eds.); Radioecology. pp 485-496. Reinhold Publ. Corp. New York.
12.Morgan, J.J. and T.H. Sibley. 1975. Proc. of the American Soc. Civ. Eng. Conf. on Ocean Engineering.Univ. Delaware, Newark.
13.Dyrssen, D. and M. Wedborg. 1973. in: Goldberg, E.D. (ed.): The Sea. Vol 5. Wiley-Intersciece. New York. pp 181-195.
14.Zirino, A. and S. Yamamoto. 1972. Limnol. Oceanogr. 17, 661-671.
15.Stumm, W. and P.A. Brauner. 1972. In: J.P. Riley and G. Skirrow (eds.): Chemical Oceanography. 2nd e. Academic Press. NY, 173-239.
16.Florence, T.M. and G.E. Batley. 1976. Talanta 23, 179-186.
17.Ahrland, S. 1975. in: Goldberg, E.D. (ed.): The Nature of Sea Water. Dahlem Konferenzen. Berlin. pp19-243.
18.Bingham, F.T., A.L. Page, R.J. Mahler and T.J. Ganje. 1976. J. Environ. Qual., 5(4), pp 501-513.
19.Hinesly, T.D. R.L. Jones , J.J. Tyler and E.L. Ziegler. 1976. Water Pollut. Control., 48(9), pp 21372152.
20.Friberg, L., M. Piscater, G.F. Nordberg and T. Kjellstroem. Cadmium in the Environment 2nd edition, CRC-Press.

21.Groth, P. 1971. Arch. Hydrobiol. 68(3), pp 305-375.
22.Nilsson, I. 1972. Oikos, 22, pp 366.
23.Bloom, H. and G.M. Ayling. 1977. Environ. Geol. 2, pp 3-22.
24.Bryan, G.W. and L.G. Hummerstone. 1973. J. mar. biol. Ass. U.K. 53, pp 839-857.
25.Tyler, G. 1972. Ambio 1, 53-59.
26.Mitchell, R. and C. Chamberlin. 1978. in: Berg. G. (ed.); Indicators of Enteric Contamination in Natural Waters.
27.Altshuller, A.P. and J.J. Bufalini. 1971. Environ. Sci. Technol., 5(1), pp 39-64.
28.Davies, B.E. and J. Lewin. 1974. Envi ron. Pollut., 6, pp 49-57.
29.Taylor, D. 1976. Estuarine Coastal Mar. Sci. 2, pp 417.
30.Mathis, B.J. and T.F. Cummins. 1973. Water Pollut. Control 45 pp 1573.
31.Taylor, D. 1976. Sci. Toatal Environ. 6(3), pp 259-264.
32.Inman, R.E., R.B. Ingersoll and E.A. Levy. 1971. Science 172, 1229-1231.
33.Greig, R.A., A. Adams and D.R. Wenzloff. 1977. Bull. Environ. Contam. and Toxic., 18(1), pp 3-8.

5-409 Daily intake from, Atmosphere

Species	Value	Condition
Ti	3 ug/day	Urban air, rg = 1 - 5 ug[1]

Source: 1.Schroeder, H.A., J.J. Balassa and I.H. Tipton. 1963. J. Chron. Dis. 16, pp 55-69.

5-410 Daily intake from, Food

Species	Value	Condition
Ag	8.5 mg/day	Neutron activation, rg = 1 - 16 mg, 1970[1]
Al	27 mg/day	Normal intake, rg = 18 - 36[2]
B	20.7 mg/day	Mostly vegetables, rg = 0.3 - 41[2]
Ba	0.735 mg/day	USA[2]
Be	0.012 mg/day	USA[2]
Be	20 ug/day	Homo sapiens, only a minor fraction from inhalation[3]
Bi	0.02 mg/day	USA[2]
Bi	5 ug/day	UK, Homo sapiens, less than value, 1971[4]
Cd	0.05 mg/day	None[2]
Cd	32 ug/day	Homo sapiens, USA, dithizone, 1961, rg = 4 - 60[5]
Cd	38 ug/day	Homo sapiens, USA, 1970, dithizone or AAS[5]
Cd	47 ug/day	Homo sapiens Japan, non polluted area, 1972[5]
Cd	48 ug/day	Homo sapiens, W Germany, AAS after extraction, 1969[5]
Cd	50 ug/day	Homo sapiens, USA, 1969, dithizone or AAS[5]
Cd	51 ug/day	Homo sapiens, Roumania, dithizone, 1970, rg = 38 - 64[5]
Cd	59 ug/day	Homo sapiens Japan, non polluted area, dithizone or AAS[5]
Cd	60 ug/day	Homo sapiens, Czcheoslovakia, dithizone, isotope dilution or AAS[5]
Cd	92 ug/day	Homo sapiens, USA, AAS, 1971[5]
Co	0.39 mg/day	USA, inclusive drinking water[2]
Co	25 ug/day	Homo sapiens, rg = 5 - 45 ug/day[6]
Co	41 ug/day	None[7]
Cr	0.065 mg/day	Range = 0.03 - 0.1 mg/day[8]
Cr	0.245 mg/day	USA[2]
Cu	2 mg/day	Range = 1 - 3 mg/day[9]
Fe	15 mg/day	Sweden[2]
Ge	367 ug/day	UK, rg= +-159 ug[10]
Hg	0.015 mg/day	USA, rg = 0.01 - 0.02[2]
Li	2 mg/day	None[2]
Mn	3.4 mg/day	Rg = 2.7 - 4.1, NZ, USA, Canada, W Germany[11]
Mn	4.40 mg/day	Inclusive water, USA[2]
Mn	8.1 mg/day	Rg = 5.8 - 12.4, India[12]
Mo	0.335 mg/day	Inclusive water, USA[2]
Ni	0.4 mg/day	USA, 1961, rg = 0.3 - 0.5 mg/day[13]
Ni	0.6 mg/day	USA, inclusive water[2]

Ni	289 ug/day	USSR, students, amonth in feces = 258 ug/day[14]
Pb	250 ug/day	Range = 200 - 300 ug/day[15]
Sb	0.1 mg/day	USA[2]
Sb	10 ug/day	Sweden, 1973[16]
Se	0.105 mg/day	USA, rg = 0.06 - 0.15[2]
Se	326 ug/day	Venezuela[17]
Se	56 ug/day	NZ[17]
Sn	1 mg/day	1963[18]
Sn	187 ug/day	UK, 1970[10]
Sn	5.2 mg/day	Long term study, n=4[19]
Te	100 ug	1966, USA[20]
Ti	1.4 mg/day	USA[2]
Ti	1.65 mg/day	Rg = 0.3 - 2 mg, USA[21]
Tl	2 ug/day	Sparce data[22]
U	1 ug/day	UK, 1971[23]
U	1.3 ug/day	USA, 1966[24]
U	1.5 ug/day	Japan, urban area areas, 1969[25]
U	30 ug/day	Sovjet, 1968[26]
U	4.55 ug/day	Japan, near uranium mine, 1973[27]
V	0.116 mg/day	USA[2]
V	10 ug/day	Great variation, low value[28]
V	2 mg/day	Great variation, high value[29]
W	10.5 ug/day	Rg = 8 - 13 ug/d, n=8, 1973[30]
Zn	14.5 mg/day	Inclusive water, USA[2]

Source: 1.Wester, P.O. 1971. Acta Med. Scand. 190, pp 155-161.
2.Laveskog, A., A. Lindskog and V. Stenberg. 1976. Om Metaller, Statens Naturvaardsverk, Stockholm.
3.Petzow, G. and P. Zorn. 1974. Chemiker-Ztg 98, pp 236-241.
4.Hamilton, E.J. and M.J. Minski. 1972/1973. Sci. Total Environ. 1, pp 375-394.
5.Friberg, L., M. Piscater, G.F. Nordberg and T. Kjellstroem. Cadmium in the Environment 2nd edition, CRC-Press.
6.Schleuts, R. 1977. J. Radional. Chem. 37, pp 539-548.
7.Smith,T., C.J. Edmonds and C.F. Barnaby. 1972. Health Phys. 22, pp 359-367.
8.Schettwein-Gsell, D and S. Mommsen-Straub. 1973. Int. Z. Vitam. Ernaehrungsforch. Beiheft 13.
9.Klevay, L.M. 1975. Nutr. Rep. Int. 11, pp 237-242.
10.Hamilton, E.I. and M.J. Minski. 1972/1973. Sci. Tot. Environ. 1, pp 341-374.
11.Greger, J.L., P. Baligar, R.P. Abernathy, Bennet and Peterson. 1978. Am. J. Clin. Nutr. 31, pp 117-11.
12.Soman, S.D., V.K. Panday, K.T. Joseph and S.J. Raut. 1969. Health Phys. 17, pp 35-40.
13.Schroeder, H.A., J.J. Balassa, and I.H. Tipton. 1962. J. Chron. Dis. 15, pp 51-65.
14.Nodiya, R.I. 1972. Gig. Sanit. 37, pp 108-109.
15.WHO 1977. Lead. Environmental Health Criteria 3. World Health Organzation. Geneva.
16.Wester, P.O. 1974. Atherosclerosis 20, pp 207-215.
17.Levander, O.A. 1976. In: Selenium-tellurium in the Environment, pp 26-53. Industrial Health Found. Pittsburgh.
18.Schroeder, H.A., J.J. Balassa and I.H. Tipton. 1964. J. Chron. Dis. 17, pp 483-502.
19.Tipton, I.H., P.L. Stewart and J. Dickson. 1969. Health Phys. 16, pp 455-462.
20.Nason, A.P. and H.A. Schroeder. 1967. J. Chron. Dis. 20, p 671.
21.Tipton, I.H. and P.L. Stewart. 1969. In: D.D. Hemphill (ed.): Trace Substances in Environmental Healh. Columbia.
22.Carson, B.L. and I.C. Smith. 1977. Thallium: An Appraisal of Environmental Exposure. Technicl Reportno. 5. NIEHS, Research Triangle Park, N.C.
23.Hamilton, E.I. 1970. Health Phys. 19, pp 511-520.
24.Welford, G.A. and R. Baird. 1967. Heath Phys. 13, pp 1321-1324.
25.Nozaki, T., M. Ichikawa, T. Sasuga and M. Inarida. 1970. J. Radioanal. Chem. 6, pp 33-40.
26.Prister, B.S. 1969. GKIAE Report by Atmoizdat, Moscow.
27.Yamamoto, T., E. Yunoki, M. Yamakawa, M. Shimizu and Nukada. 1974. J. Radiat. Res. 15, pp 156-162.
28.Byrne, A.R. and L. Kosta. 1978. Sci. Tot. Environ. 10, pp 17-30.
29.ICRP 1960. Report of Committiee II on Permissible Dose for Internal Radiation. ICRP 2. Pergamon Press. Oxf.
30.Wester, P.O. 1974. Atheroscherosis 20, pp 207-215.

5-411 Excretion, Ag

Species	Value	Condition
Dog	90 % of adsorbed amount	With feces, mainly by biliary excretion[1]
Dog, lung	30 %	During day 2 - 7, inh of monodisperse Ag-coated teflon, 4 um part[2]
Dog, lung	30 %	During the first day, inh of monodisperse Ag-coated teflon, 4 um part[2]
Dog, lung	59 %	During 1.7 days, inh of monodisperse Ag-coated teflon, 4 um part[2]
Monkey	90 % of adsorbed amount	With feces, mainly by biliary excretion[1]
Mouse	90 % of adsorbed amount	With feces, mainly by biliary excretion[1]
Rat	90 % of adsorbed amount	With feces, mainly by biliary excretion[1]

Source: 1.Furchner, J.E., C.R. Richmond and G.A. Drake. 1968. Health Phys. 15, pp 505-514.
2.Camner, P. M. Lundborg and P.A. Hellstroem. 1974. Arch. Environ. Health 29, pp 211-213

5-412 Excretion Al

Species	Value	Condition
Homo sapiens	84 % of dose	With urine, oral intake = 18 mg/day, 30 d[1]
Homo sapiens	94.5 % of dose	With feces, oral intake = 18 mg/day, 30 d[1]

Source: 1.Tipton, I.H., P.L. Stewart and P.G. Mar tin. 1966. Health Phys., 12, pp 1683-1689.

5-413 Excretion Anthracene

Species	Value	Condition
Crassostrea virginica	0.1 mg/kg dry weight	Exposure=8 d, 7 d release, wat=1 ug/l, crude oil, lab, init conc=2.5 ppm[1]
Crassostrea virginica	0.4 mg/kg dry weight	Exposure=8 d, 23 d release, wat=1 ug/l, crude oil, lab, init conc=2.5 ppm[1]
Crassostrea virginica	1.2 mg/kg dry weight	Exposure=2 d, 7 d release, wat=13 ug/l, crude oil, lab, init. conc=5.6 ppm[1]

Source: 1.Lee, R.F., W.S. Gardner, J.W. Anderson, Balylock and B-Clarke. 1978. Environ. Sci. Technol. 12, 832-38.

5-414 Excretion As

Species	Value	Condition
Homo sapiens	2 % of absorbed	With feces, one oral dosis, inorganic As[1]
Homo sapiens	70 % per day	With urine, continuously intake, inorganic As[2]
Homo sapiens	75 % in 2 days	With urine, organic As in seafood, rg = 70 - 80 %[3]

Source: 1.Mealy, J. Jr., G.L. Brownell and W.H. Sweet. 1959. Arch. Neurol. Psychiatr. 81, pp 310-320.
2.Mappes, R. 1977. Int. Arch. Occup. Environ. Health 40, pp 267-272.
3.Westoeoe, G. and M. Rydaelv. 1972. Vaer foeda 24, pp 21-40.

5-415 Excretion, Ba

Species	Value	Condition
Homo sapiens	0.26 ug/day	With urine, non exposed person, rg = 0.22 - 0.33[1]
Homo sapiens	3 %	With urine, in equilibrium[2]
Homo sapiens	6 %	With sweat, in equilibrium[2]
Homo sapiens	91 %	With feces, in equilibrium[2]
Rat	20 % of dose	With feces, injection of $BaSO_4$, 24 h[3]

Rat	7 % of dose	With urine, injection of BaSO4, 24 h[3]

Source: 1.Harrison, C.E., T.E.F. Carr and A. Sutton. 1967. Int. J. Radiat. Biol. 13, pp 235-247.
2.Schroeder, G.A., I.H. Tipton and A.P. Nason. 1972. J. Chron. Dis. 25, pp 491-517.
3.Patty, F.A.(ed). 1963. Industrial Hygiene and Toxicology. II, Interscience Publ., N.Y.

5-416 Excretion, Be

Species	Value	Condition
Animal	1.6 % of dose	With urine, oral intake[1]
Animal	94 % of dose	With feces, oral intake[1]

Source: 1.Browning, E. 1969. Toxicity of Industrial Metals. Butterwoths. London.

5-417 Excretion, Benz(a)anthracene

Species	Value	Condition
Crassostrea virginica	0.3 mg/kg dry weight	Exposure=8 d, 7 d release, wat=0.1 ug/l, crude oil, lab, init conc=1.8 ppm[1]
Crassostrea virginica	1.0 mg/kg dry weight	Exposure=8 d, 23 d release, wat=0.1 ug/l, crude oil, lab, init conc=1.8 ppm[1]
Crassostrea virginica	1.9 mg/kg dry weight	Exposure=2 d, 7 d release, wat=5.3 ug/l, crude oil, lab, init. conc=2.8 ppm[1]

Source: 1.Lee, R.F., W.S. Gardner, J.W. Anderson, Balylock and B-Clarke. 1978. Environ. Sci. Technol. 12, 832-38.

5-418 Excretion, Benzo(a)pyrene

Species	Value	Condition
Callinectes sapidus, blood	1E-4 ug total in tissue	2 d upt. from wat=2.5 ug BaP/l, init=30, juv, marine, n=3, 10 d excr[1]
Callinectes sapidus, blood	1E-4 ug total in tissue	2 d upt. from wat=2.5 ug BaP/l, init=30, juv, marine, n=3, 18 d excr[1]
Callinectes sapidus, blood	8E-4 ug total in tissue	2 d upt. from wat=2.5 ug BaP/l, init=30, juv, marine, n=3, 6 d excr[1]
Callinectes sapidus, gill	0.5E-4 ug total in tissue	2 d upt. from wat=2.5 ug BaP/l, init=290, juv, marine, n=3, 18 d excr[1]
Callinectes sapidus, gill	10E-4 ug total in tissue	2 d upt. from wat=2.5 ug BaP/l, init=290, juv, marine, n=3, 10 d excr[1]
Callinectes sapidus, gill	22E-4 ug total in tissue	2 d upt. from wat=2.5 ug BaP/l, init=290, juv, marine, n=3, 6 d excr[1]
Callinectes sapidus, hepatopancreas	21E-4 ug total in tissue	2 d upt. from wat=2.5 ug BaP/l, init=280, juv, marine, n=3, 6 d excr[1]
Callinectes sapidus, hepatopancreas	2E-4 ug total in tissue	2 d upt. from wat=2.5 ug BaP/l, init=280, juv, marine, n=3, 18 d excr[1]
Callinectes sapidus, hepatopancreas	32E-4 ug total in tissue	2 d upt. from wat=2.5 ug BaP/l, init=280, juv, marine, n=3, 10 d excr[1]
Callinectes sapidus, muscle	0.5E-4 ug total in tissue	2 d upt. from wat=2.5 ug BaP/l, init=2, juv, marine, n=3, 10 d excr[1]
Callinectes sapidus, muscle	0.5E-4 ug total in tissue	2 d upt. from wat=2.5 ug BaP/l, init=2, juv, marine, n=3, 18 d excr[1]
Callinectes sapidus, muscle	1E-4 ug total in tissue	2 d upt. from wat=2.5 ug BaP/l, init=2, juv, marine, n=3, 6 d excr[1]
Callinectes sapidus, stomach	0.5E-4 ug total in tissue	2 d upt. from wat=2.5 ug BaP/l, init=2, juv, marine, n=3, 10 d excr[1]
Callinectes sapidus, stomach	0.5E-4 ug total in tissue	2 d upt. from wat=2.5 ug BaP/l, init=2, juv, marine, n=3, 18 d excr[1]
Callinectes sapidus, stomach	0.5E-4 ug total in tissue	2 d upt. from wat=2.5 ug BaP/l, init=2, juv, marine, n=3, 6 d excr[1]

Crassostrea virginica	0.12 mg/kg dry weight	Exposure=8 d, 7 d release, wat=0.1 ug/l, crude oil, lab, init conc=0.3 ppm[2]
Crassostrea virginica	0.20 mg/kg dry weight	Exposure=8 d, 23 d release, wat=0.1 ug/l, crude oil, lab, init conc=0.3 ppm[2]
Crassostrea virginica	0.40 mg/kg dry weight	Exposure=2 d, 7 d release, wat=1.9 ug/l, crude oil, lab, init. conc=0.36 ppm[2]
Rangia cuneata	16 % in 24 hours	Estuarine, water = 0.052 ppm, init. conc = 0.45+-0.1 ppm[3]

Source: 1.Lee, R.F., C. Ryan and M.L. Neuhauser. 1976. Mar. Biol. 37, 363-370.
2.Lee, R.F., W.S. Gardner, J.W. Anderson, Balylock and B-Clarke. 1978. Environ. Sci. Technol. 12, 832-38.
3.Neff, J.M., Anderson, Cox, Laughlin, Rossi and Tatem. 1976. In: Sources, Effects and Sinks of Hydrocarbon in the Aquatic Environment. Wash DC. Amer. Inst. of Biological Science.

5-419 Excretion, Bi

Species	Value	Condition
Homo sapiens	1.6 ug	With urine, Ingestion 20 ug[1]
Homo sapiens	18 ug	With feces, ingestion 20 ug, only little absorption[1]

Source: 1.ICRP. 1975. Report of the Task group on Reference Man. p 365. No 23. Pergamon Press.

5-420 Excretion, Ca

Species	Value	Condition
Fucus vesiculosus	90 %	In first few hours, 295 k, Ca-45[1]

Source: 1.Swift, E. and W.R. Taylor. 1960. Biol. Bull. 119, pp 342.

5-421 Excretion Cd

Species	Value	Condition
Acheta domesticus	0.09 1/day	Grassland food chain, predator, Cd oxide, slow phase = 17 %, rg = +-[1]
Acheta domesticus	0.11 1/day	Grassland food chain, predator, Cd(NO3)2, slow phase = 17 %, rg = +-[1]
Acheta domesticus	0.8 1/day	Grassland food chain, predator, Cd(NO3)2, fast phase = 83 %, rg = +-[1]
Acheta domesticus	0.9 1/day	Grassland food chain, predator, Cd oxide, fast phase = 83 %, rg = +-[1]
Animal	94 % of dose	With feces, oral intake, rg = 90 - 98[2]
Goat, feces	5.6 % of dose	Single intravenous injection of Cd-109, after 5 d[2]
Goat, urine	0.002 % of dose	Single intravenous injection of Cd-109, 3-7 d after injection[2]
Goat, urine	0.004 % of dose	Single intravenous injection of Cd-109, 2 d after injection[2]
Goat, urine	0.025 % of dose	Single intravenous injection of Cd-109, 1 d after injection[2]
Goat, urine	0.1 % of absorbed amount	Single oral dose of Cd-109, 3-7 d after exposure[2]
Goat, urine	0.5 % of absorbed amount	Single oral dose of Cd-109, 1 d after exposure[2]
Goat, urine	1.0 % of absorbed amount	Single oral dose of Cd-109, 2 d after exposure[2]
Homo sapiens, kidney	0.005 % body burden per day	Biological half-life time = 38 y[2]
Homo sapiens, kidney	0.01 % body burden per day	Biological half-life time = 19 y[2]
Homo sapiens, kidney	0.02 % body burden per day	Biological half-life time = 9.5 y[2]
Homo sapiens, kidney	0.05 % body burden per day	Biological half-life time = 3.8 y[2]
Homo sapiens, urine of exposed workers	22.5 ug/l	Engl, n = 10, CdO dust, not directly exposed, dithizone, 1957[2]

Homo sapiens, urine of exposed workers	220.5 ug/l	Engl, n = 27, CdO fume, dithizone, 1957, rg = 16 - 425[2]
Homo sapiens, urine of exposed workers	221 ug/l	Engl, n = 26, CdO dust, dithizone, 1957, rg = 14 - 428[2]
Homo sapiens, urine of exposed workers	250 ug/l	Engl, n = 12, CdO dust, dithizone, 1955, rg = 12 - 487[2]
Homo sapiens, urine of exposed workers	3.5 ug/g creatinine	W Germany, n = 18, CdO fume cds, AAS, mibk-apdc extraction, 1969[2]
Homo sapiens, urine of exposed workers	71.5 ug/l	Japan, n = 13, CdO fume, dithizone, 1967, rg = 3 - 140[2]
Homo sapiens, urine of exposed workers	8 ug/l	Japan, n = 34, Cd stearate, dithizone, 1965, rg = 0 - 15.9[2]
Homo sapiens, urine of exposed workers	825 ug/day	France, n = 2, CdO dust, nephelometric, 1954[2]
Homo sapiens, urine of exposed workers	84 ug/l	Engl, n = 56, CdO dust, dithizone, 1969, rg = 0 - 168[2]
Homo sapiens, urine of not exposed persons	0.21 ug/g creatinine	Sweden, 15-16 = age, n = 10, AAS, after extraction, rg= 0.1 - 0.3[2]
Homo sapiens, urine of not exposed persons	0.39 ug/day	Sweden, 20-47 = age, n = 10, AAS, after mibk-apdc extraction[2]
Homo sapiens, urine of not exposed persons	0.47 ug/l	Japan, Gifu, 4-6 y, n = 46, AAS, after extraction, 1971, SD = +- 0.25[2]
Homo sapiens, urine of not exposed persons	0.62 ug/g creatinine	Sweden, 50-59 = age, n = 88, AAS, after extraction, 1973[2]
Homo sapiens, urine of not exposed persons	0.65 ug/l	Japan, Gifu, 9-10 y, n = 40, AAS, after extraction, 1971[2]
Homo sapiens, urine of not exposed persons	0.72 ug/l	Japan, Gifu, 14-15 y, n = 41, AAS, after extraction, 1971[2]
Homo sapiens, urine of not exposed persons	0.95 ug/g creatinine	Belgium, adults, n = 44, AAS, after extraction, 1972, SD = +- 0.8[2]
Homo sapiens, urine of not exposed persons	0.98 ug/day	W Germany, adults, n = 15, AAS, after mibk-apdc extraction, 1968[2]
Homo sapiens, urine of not exposed persons	0.99 ug/l	Japan, Gifu, 20-29 y, n = 56, AAS, after extraction, 1971[2]
Homo sapiens, urine of not exposed persons	1.0 ug/l	W Germany, adults, n = 14, dithizone, 1969[2]
Homo sapiens, urine of not exposed persons	1.13 ug/l	Japan, Gifu, 30-39 y, n = 37, AAS, after extraction, 1971[2]
Homo sapiens, urine of not exposed persons	1.25 ug/g creatinine	W Germany, incl.children, n = 169, AAS, after mibk-apdc extract.[2]
Homo sapiens, urine of not exposed persons	1.59 ug/l	USA, adult, n = 154, spectrogr.after dithizone extraction, 1963[2]
Homo sapiens, urine of not exposed persons	1.7 ug/g creatinine	Sweden, 34-63 y, contaminated area, n = 10, AAS, extraction, 1972[2]

Homo sapiens, urine of not exposed persons	1.75 ug/l	Japan, Gifu, 50-59 y, n = 43, AAS, after extraction, 1971[2]
Homo sapiens, urine of not exposed persons	1.76 ug/l	Japan, Gifu, 40-49 y, n = 40, AAS, after extraction, 1971[2]
Homo sapiens, urine of not exposed persons	3.1 ug/l	Japan, adult, n = 30, dithizone, 1965, rg = 0 - 15.9[2]
Lycosa sp.	0.007 1/day	Grassland food chain, predator[1]
Mouse, urine	0.80 ng/day	3 w, injection of 0.025 mg Cd/kg*d (5 d/w) during w 1-3, n = 4[2]
Mouse, urine	0.80 ng/day	6 w, injection of 0.025 mg Cd/kg*d (5 d/w) during w 1-3, n = 4[2]
Mouse, urine	4.65 ng/day	21 w, injection of 0.025 mg Cd/kg*d (5 d/w) during w 7-21, n = 4[2]
Mouse, urine	6.34 ng/day	24 w, injection of 0.025 mg Cd/kg*d (5 d/w) during w 7-21, n = 8[2]
Rabbit, feces	1.6 % of dose/day	0.25 mg Cd/kg day, 5 d/w for 29 w, subcutanous injection, after 11 w[2]
Rabbit, feces	11.2 % of dose/day	0.25 mg Cd/kg day, 5 d/w for 29 w, subcutanous injection, after 23 w[2]
Rabbit, feces	2.8 % of dose/day	0.25 mg Cd/kg day, 5 d/w for 29 w, subcutanous injection, after 17 w[2]
Rabbit, feces	6.6 % of dose/day	0.25 mg Cd/kg day, 5 d/w for 29 w, subcutanous injection, after 29 w[2]
Rabbit, urine	1 % of dose	0.65 mg Cd-115/kg day, subcutanous injection, 0-7 w, less than value[2]
Rabbit, urine	100 times the value for week 0 - 7	0.65 mg Cd-115/kg day, subcutanous injection, 8-10 ws[2]
Rat, biles	0.005 % of dose/hr	0.5 mg Cd-109/kg, subcutanous injection, during 3 d[2]
Rat, biles	5 % of fecal excretion/day	0.5 mg Cd-109/kg, subcutanous injection, during 3 d[2]
Rat, feces	18.5 % of dose	0.63 mg Cd-115/kg, single intravenous injection, after 72 h[2]
Rat, feces	5 % of dose	2.2 ug Cd-109, single subcutanous injection, after 4 d[2]
Rat, feces	7.3 % of dose	0.63 mg Cd-115/kg, single intravenous injection, after 24 h[2]
Rat, urine	0.005 % of dose/day	Single subcutanous injection, after 4 d, rg = 0.003 - 0.007[2]
Rat, urine	0.65 % of absorbed amount	Single oral dose, 6 d after exposure[2]
Rat, urine	1.25 % of absorbed amount	Single oral dose, 1 d after exposure[2]

Source: 1.Hook, R.I. van and A.J. Yates. 1975. Environ. Res. 9(2), pp 76-83.
2.Friberg, L., M. Piscater, G.F. Nordberg and T. Kjellstroem. Cadmium in the Environment 2nd edition, CRC-Press.

5-422 Excretion
Ce

Species	Value	Condition
Meganyctiphanes norvegica	99 %	Loss of total body burden by 1 molt, uptake from water, Ce-141[1]

Source: 1.Fowler, S.W., M. Heyraud, L.F. Small and G. Benayoun. 1973. Mar. Biol., 21, pp 317-325.

5-423 Excretion, Chrysene

Species	Value	Condition
Rangia cuneata	26 % in 24 hours	Estuarine, water = 0.066 ppm, init. conc = 0.54+-0.3 ppm[1]

Source: 1.Neff, J.M., Anderson, Cox, Laughlin, Rossi and Tatem. 1976. In: Sources, Effects and Sinks of Hydrocarbon in the Aquatic Environment. Wash DC. Amer. Inst. of Biological Science.

5-424 Excretion, Co

Species	Value	Condition
Dog	5 %	In 72 h, biliary excretion, CoCl2 intravenous[1]
Homo sapiens	15 % of dose	With feces, oral intake, 23 women, 3 w, rg = 3 - 27[2]
Homo sapiens	67 % of dose	With urine, oral intake, 23 women, 3 w[2]

Source: 1.Sheline, G.E., I.L. Chaikoff and M.L. Montgomery. 1946. Am. J. Physiol. 145, pp 285-290.
2.Harp, M.J. and F.I. Scoular. 1952. J. of Natr., 47, pp 67-72.

5-425 Excretion Cr

Species	Value	Condition
Homo sapiens	4.5 %	With urine, daily intake of 41 ug[1]
Homo sapiens	6 ug/l	Daili intake = 30 - 100 ug/l, with urine, rg = 2 - 10[2]
Homo sapiens	93.5 %	With feces, daily intake of 41 ug[1]
Rat	2 %	8 h after injection, Cr(III), with urine[3]
Rat	20 %	In 4 days, with urine, Cr(III) injection[4]

Source: 1.Smith,T., C.J. Edmonds and C.F. Barnaby. 1972. Health Phys. 22, pp 359-367.
2.Friberg, L. G.N. Nordberg and V.B. Vouk. 1979. Handbook on the Toxicology of Metals. Elseveir.
3.Hopkins, L.L. Jr. 1965. Am. J. Physiol. 209, pp 731-735.
4.Visek, W.J., I.B. Whitney, U.S.G. Kuhn and C.L. Comar. 1962. Can. J. Biochem. Physiol. 40, pp 1677-183.

5-426 Excretion Cu

Species	Value	Condition
Animal	1 % of dose	With urine, oral intake, less than value[1]
Animal	70 % of dose	With feces, oral intake[1]
Homo sapiens	1 %	With urine, in 72 h, injection of Cu-64 acetate[2]
Homo sapiens	9 %	With feces, in 72 h, injection of Cu-64 acetate[2]
Mercenaria mercenaria	1.9 mg Cu/(kg wet weight*day)	298 K, inital tissue = 33 ppm, final = 20 ppm, 7 d[3]
Mercenaria mercenaria	1.9 mg Cu/(kg wet weight*day)	298 K, inital tissue = 52 ppm, final = 24 ppm, 15 d[3]
Mercenaria mercenaria	12.5 mg Cu/(kg wet weight*day)	298 K, inital tissue = 124 ppm, final = 36 ppm, 7[3]
Mercenaria mercenaria	12.5 mg Cu/(kg wet weight*day)	298 K, inital tissue = 124 ppm, final = 36 ppm, 7 d[3]
Mercenaria mercenaria	2.6 mg Cu/(kg wet weight*day)	298 K, inital tissue = 58 ppm, final = 20 ppm, 15 d[3]
Mercenaria mercenaria	5.1 mg Cu/(kg wet weight*day)	298 K, inital tissue = 96 ppm, final = 34 ppm, 11 d[3]
Mercenaria mercenaria	5.9 mg Cu/(kg wet weight*day)	298 K, inital tissue = 68 ppm, final = 9 ppm, 10 d[3]
Mercenaria mercenaria	6.5 mg Cu/(kg wet weight*day)	298 K, inital tissue = 48 ppm, final = 22 ppm, 4 d[3]
Mercenaria mercenaria	7.0 mg Cu/(kg wet weight*day)	298 K, inital tissue = 62 ppm, final = 34 ppm, 4 d[3]

Source: 1.Florkin and Stutz (ed). 1972. Comprehensive Biochemistry, Elsevier Publ. Comp., 21(2).
2.Dekaban, A. R. Aamodt, W.F. Rumble, G.S. Johnston and O Reilly. 1975. Arch. Neurol. 32, pp 672-675.
3.Pringle, B.H., D.E. Hissong, E.L. Katz and S.T. Mulawka. 1972. Environ. Pollut. 3, 69-82.

5-427 Excretion
Dieldrin

Species	Value	Condition
Ictalurus punctatus, dorsal muscle	108 ug/kg wet weight	Uptake from food = 2 ppm, initial = 287 ppm, 28 d, 150-225 mm[1]
Ictalurus punctatus, dorsal muscle	109 ug/kg wet weight	Uptake from food = 2 ppm, initial = 302 ppm, 28 d, 350-400 mm[1]
Ictalurus punctatus, dorsal muscle	142 ug/kg wet weight	Uptake from food = 2 ppm, initial = 302 ppm, 21 d, 350-400 mm[1]
Ictalurus punctatus, dorsal muscle	152 ug/kg wet weight	Uptake from food = 2 ppm, initial = 287 ppm, 21 d, 150-225 mm[1]
Ictalurus punctatus, dorsal muscle	152 ug/kg wet weight	Uptake from water = 75 ng/l, initial = 176 ppm, 7 d, 150-225 mm[1]
Ictalurus punctatus, dorsal muscle	162 ug/kg wet weight	Uptake from water = 75 ng/l, initial = 186 ppm, 7 d, 350-400 mm[1]
Ictalurus punctatus, dorsal muscle	168 ug/kg wet weight	Uptake from food = 2 ppm, initial = 287 ppm, 14 d, 150-225 mm[1]
Ictalurus punctatus, dorsal muscle	170 ug/kg wet weight	From food = 2 ppm and water = 75 ng/l, initial = 634 ppm, 28 d[1]
Ictalurus punctatus, dorsal muscle	172 ug/kg wet weight	From food = 2 ppm and water = 75 ng/l, initial = 365 ppm, 28 d[1]
Ictalurus punctatus, dorsal muscle	184 ug/kg wet weight	Uptake from food = 2 ppm, initial = 302 ppm, 14 d, 350-400 mm[1]
Ictalurus punctatus, dorsal muscle	194 ug/kg wet weight	From food = 2 ppm and water = 75 ng/l, initial = 634 ppm, 21 d[1]
Ictalurus punctatus, dorsal muscle	195 ug/kg wet weight	From food = 2 ppm and water = 75 ng/l, initial = 365 ppm, 21 d[1]
Ictalurus punctatus, dorsal muscle	217 ug/kg wet weight	From food = 2 ppm and water = 75 ng/l, initial = 365 ppm, 14 d,[1]
Ictalurus punctatus, dorsal muscle	217 ug/kg wet weight	Uptake from food = 2 ppm, initial = 287 ppm, 7 d, 150-225 mm[1]
Ictalurus punctatus, dorsal muscle	221 ug/kg wet weight	From food = 2 ppm and water = 75 ng/l, initial = 634 ppm, 14 d,[1]
Ictalurus punctatus, dorsal muscle	229 ug/kg wet weight	Uptake from food = 2 ppm, initial = 302 ppm, 7 d, 350-400 mm[1]
Ictalurus punctatus, dorsal muscle	24 ug/kg wet weight	Uptake from water = 75 ng/l, initial = 176 ppm, 28 d, 150-225 mm[1]
Ictalurus punctatus, dorsal muscle	241 ug/kg wet weight	From food = 2 ppm and water = 75 ng/l, initial = 365 ppm, 7 d,[1]
Ictalurus punctatus, dorsal muscle	28 ug/kg wet weight	Uptake from water = 75 ng/l, initial = 186 ppm, 28 d, 350-400 mm[1]
Ictalurus punctatus, dorsal muscle	358 ug/kg wet weight	From food = 2 ppm and water = 75 ng/l, initial = 634 ppm, 7 d,[1]
Ictalurus punctatus, dorsal muscle	47 ug/kg wet weight	Uptake from water = 75 ng/l, initial = 176 ppm, 21 d, 150-225 mm[1]
Ictalurus punctatus, dorsal muscle	59 ug/kg wet weight	Uptake from water = 75 ng/l, initial = 186 ppm, 21 d, 350-400 mm[1]
Ictalurus punctatus, dorsal muscle	78 ug/kg wet weight	Uptake from water = 75 ng/l, initial = 176 ppm, 14 d, 150-225 mm[1]
Ictalurus p., dor. m.	84 ug/kg wet weight	From water = 75 ng/l, init = 186 ppm, 14 d, 350-400 mm[1]

Source: 1.Shannon, L.R. 1977. Bull. Environ. Contam. Toxicol. 17(6), pp 637-644.

5-428 Excretion
Dimethylnaphthalene

Species	Value	Condition
Crassostrea virginica	0 mg/kg dry weight	Exposure=8 d, 7 d release, wat=2 ug/l, crude oil, lab, init conc=72 ppm[1]
Crassostrea virginica	4 mg/kg dry weight	Exposure=8 d, 23 d release, wat=2 ug/l, crude oil, lab, init conc=72 ppm[1]
Crassostrea virginica	8 mg/kg dry weight	Exposure=2 d, 7 d release, wat=10 ug/l, crude oil, lab, init. conc=84 ppm[1]

Source: 1.Lee, R.F., W.S. Gardner, J.W. Anderson, Balylock and B-Clarke. 1978. Environ. Sci. Technol. 12, 832-38.

5-429 Excretion
Dinitramine

Species	Value	Condition
Carp, bile	13.8 mg/kg wet weight	24 h after transference to pure water, initial = 22.7 mg/kg wet[1]
Carp, muscle	4.98 mg/kg wet weight	24 h after transference to pure water, initial = 39.6 mg/kg wet[1]
Carp, plasma	5.1 mg/kg wet weight	24 h after transference to pure water, initial = 24.4 mg/kg wet[1]
Channel catfish, bile	14.5 mg/kg wet weight	24 h after transference to pure water, initial = 13.4 mg/kg wet[1]
Channel catfish, muscle	20.0 mg/kg wet weight	24 h after transference to pure water, initial = 46.1 mg/kg wet[1]
Channel catfish, plasma	19.6 mg/kg wet weight	24 h after transference to pure water, initial = 58.5 mg/kg wet[1]

Source: 1.Olson, L.E., J.L. Allen. and W.L. Mauck. 1975. J. Agric. Food. Chem. 23(3), pp 437-439.

5-430 Excretion
Ethylenthiourea

Species	Value	Condition
Guinea pig	0 % of administrated	48 - 72 h after dose, excretion with urine, n = 6[1]
Guinea pig	17.75 % of administrated	0 - 6 h after dose, excretion with urine, n = 6, SE = +- 1.72[1]
Guinea pig	2.48 % of administrated	48 - 72 h after dose, excretion with urine, n = 6, SE = +- 0.43[1]
Guinea pig	27.18 % of administrated	6 - 24 h after dose, excretion with urine, n = 6, SE = +- 0.66[1]
Rat	0 % of administrated	72 - 96 h after dose, excretion with urine, n = 6[1]
Rat	0.42 % of administrated	48 -72 h after dose, excretion with urine, n = 6, SE = +- 0.12[1]
Rat	18.60 % of administrated	0 - 6 h after dose, excretion with urine, n = 6, SE = +- 2.38[1]
Rat	3.65 % of administrated	24 - 48 h after dose, excretion with urine, n = 6, SE = +- 0.50[1]
Rat	42.68 % of administrated	6 - 24 h after dose, excretion with urine, n = 6, SE = +- 2.36[1]

Source: 1.Newsome, W.H. 1974. Bull. Environ. Contam.,11, pp 174-176.

5-431 Excretion
Fe

Species	Value	Condition
Homo sapiens	85 % of dose	With feces, oral intake, person with Fe-deficit, rg = 80 - 90[1]
Homo sapiens	92.5 % of dose	With feces, oral intake, normal person, rg = 90 - 95[1]

Source: 1.Florkin and Stutz (ed). 1972. Comprehensive Biochemistry, Elsevier Publ. Comp., 21(2).

5-432 Excretion, Fluoranthene

Species	Value	Condition
Crassostrea virginica	0.4 mg/kg dry weight	Exposure=8 d, 7 d release, wat=0.4 ug/l, crude oil, lab, init conc=4.0 ppm[1]
Crassostrea virginica	1.4 mg/kg dry weight	Exposure=8 d, 23 d release, wat=0.4 ug/l, crude oil, lab, init conc=4.0 ppm[1]
Crassostrea virginica	1.7 mg/kg dry weight	Exposure=2 d, 7 d release, wat=7.2 ug/l, crude oil, lab, init. conc=5.0 ppm[1]

Source: 1.Lee, R.F., W.S. Gardner, J.W. Anderson, Balylock and B-Clarke. 1978. Environ. Sci. Technol. 12, 832-38.

5-433 Excretion Ge

Species	Value	Condition
Dog	90 % in 72 hours	With urine, injection of 3.4 mg Ge/kg, GeO2[1]
Dog	90 % of dose	With urine, intravenous injection, 72 h[2]
Rabbit	68.5 %	With urine, rg = 60 - 77 %, injection of 13.5 mg GeO2[1]
Rabbit	75 % of dose	With urine, intravenous injection[2]
Rabbit	9 %	With feces, rg = 6 - 12 %, injection of 13.5 mg GeO2[1]
Rabbit	9 % of dose	With feces, intravenous injection[2]
Rat	64.8 %	With urine, in 1 day, intraperitonal NaGeSO4[3]
Rat	78.8 %	With urine, in 1 day, intraperitoneal, NaGeSO4[3]
Rat, lung	48 % in 24 h	Deposited elemental GE part., mean size = 1.4 um[4]
Rat, lung	79 % in 24 h	Deposited GeO2 part., mean size 0.4 um[4]

Source: 1.Dudly, H.C. and E.J. Wallace. 1952. AMA Arc. Ind. Hyg. Occup. Med. 6, pp 263-270.
2.Browning, E. 1969. Toxicity of Industrial Metals. Butterwoths. London.
3.Rosenfeld, G. 1954. Arch. Biochem Biophys. 48, pp 84-94.
4.Dudley, H.C. 1952. AMA Arch. Ind. Hyg. Occup. Med. 6, pp 263-270.

5-434 Excretion Hg

Species	Value	Condition
Homo sapiens, feces	0.09 mg/day	0.05-0.1 mg Hg-vapor/m**3, n = 15 workers, respiratory uptake[1]
Homo sapiens, feces	0.14 mg/day	0.11-0.2 mg Hg-vapor/m**3, n = 10 workers, respiratory uptake[1]
Homo sapiens, feces	1.9 % of dose/day	Oral dose of methyl-Hg, during w 1[1]
Homo sapiens, feces	34 % of dose	Oral dose of methyl-Hg, after 49 d[1]
Homo sapiens, feces	90 % of total elimination	Oral dose of methyl-Hg[1]
Homo sapiens, urine	0.01 % of dose/day	Oral dose of methyl-Hg, during w 1[1]
Homo sapiens, urine	0.12 mg/day	0.05-0.1 mg Hg-vapor/m**3, n = 15 workers, respiratory uptake[1]
Homo sapiens, urine	0.19 mg/day	0.11-0.2 mg Hg-vapor/m**3, n = 10 workers, respiratory uptake[1]
Homo sapiens, urine	3 % of dose	Oral dose of methyl-Hg, after 49 d[1]
Mouse	4 ratio feces/urine	Intravenous injection of methyl-Hg-salt, during 21 d[1]
Pig, feces	10 % of dose	Injectionof methyl-Hg[1]
Pig, urine	2 % of dose	Injectionof methyl-Hg, after 14 d[1]
Rabbit, feces	27 % of dose	Intravenous injection of 1 mg HgCl2/kg, after 14 d[1]

Rabbit, urine	39 % of dose	Intravenous injection of 1 mg HgCl2/kg, after 14 d[1]
Rat	20 % of injected amount	Intravenous injection, Hg vapor, after 30sec.[1]
Rat	3 ratio feces/urine	Respiratory intake of elemental Hg (1.4 mg/m**3)for 5 h, after 2 d,[1]
Rat	4 ratio feces/urine	Respiratory intake of elemental Hg (1.4 mg/m**3)for 5 h, during the[1]
Rat, feces	10 % of dose	Intravenous injection of Hg acetate[1]
Rat, feces	18 % of dose	Intravenous injection of 0.5 mg Hg-acetate/kg, after 48 h[1]
Rat, feces	30 % of dose	Intravenous injection of phenyl-Hg-acetate[1]
Rat, feces	30 % of dose	Subcutanous injection of 0.5 mg HgCl2/kg day, 7 d/w, after 14 d[1]
Rat, feces	80 % of dose	Oral dose of 0.5 mg Hg-acetate/kg, after 48 h[1]
Rat, urine	1 % of dose	Oral dose of 0.5 mg Hg-acetate/kg, after 48 h[1]
Rat, urine	10 % of dose	Intravenous injection of 0.5 mg Hg-acetate/kg, after 48 h[1]
Rat, urine	70 % of dose	Subcutanous injection of 0.5 mg HgCl2/kg day, 7 d/w, after 14 d[1]

Source: 1.Friberg, L. and J. Vostal. 1974. Mercury in the Environment. CRC-Press.

5-435 Excretion, Hydroxybenzo(a)pyrene

Species	Value	Condition
Callinectes sapidus, blood	2E-4 ug total in tissue	2 d upt. from wat=2.5 ug BaP/l, init=12, juv, marine, n=3, 18 d excr[1]
Callinectes sapidus, blood	6E-4 ug total in tissue	2 d upt. from wat=2.5 ug BaP/l, init=12, juv, marine, n=3, 10 d excr[1]
Callinectes sapidus, blood	9E-4 ug total in tissue	2 d upt. from wat=2.5 ug BaP/l, init=12, juv, marine, n=3, 6 d excr[1]
Callinectes sapidus, gill	0.5E-4 ug total in tissue	2 d upt. from wat=2.5 ug BaP/l, init=15, juv, marine, n=3, 18 d excr[1]
Callinectes sapidus, gill	11E-4 ug total in tissue	2 d upt. from wat=2.5 ug BaP/l, init=15, juv, marine, n=3, 6 d excr[1]
Callinectes sapidus, gill	7E-4 ug total in tissue	2 d upt. from wat=2.5 ug BaP/l, init=15, juv, marine, n=3, 10 d excr[1]
Callinectes sapidus, hepatopancreas	24E-4 ug total in tissue	2 d upt. from wat=2.5 ug BaP/l, init=84, juv, marine, n=3, 10 d excr[1]
Callinectes sapidus, hepatopancreas	40E-4 ug total in tissue	2 d upt. from wat=2.5 ug BaP/l, init=84, juv, marine, n=3, 6 d excr[1]
Callinectes sapidus, hepatopancreas	5E-4 ug total in tissue	2 d upt. from wat=2.5 ug BaP/l, init=84, juv, marine, n=3, 18 d excr[1]
Callinectes sapidus, muscle	0.5E-4 ug total in tissue	2 d upt. from wat=2.5 ug BaP/l, init=2, juv, marine, n=3, 10 d excr[1]
Callinectes sapidus, muscle	0.5E-4 ug total in tissue	2 d upt. from wat=2.5 ug BaP/l, init=2, juv, marine, n=3, 18 d excr[1]
Callinectes sapidus, muscle	0.5E-4 ug total in tissue	2 d upt. from wat=2.5 ug BaP/l, init=2, juv, marine, n=3, 6 d excr[1]
Callinectes sapidus, stomach	0.5E-4 ug total in tissue	2 d upt. from wat=2.5 ug BaP/l, init=2, juv, marine, n=3, 18 d excr[1]
Callinectes sapidus, stomach	0.5E-4 ug total in tissue	2 d upt. from wat=2.5 ug BaP/l, init=2, juv, marine, n=3, 6 d excr[1]
Callinectes sapidus, stomach	1E-4 ug total in tissue	2 d upt. from wat=2.5 ug BaP/l, init=2, juv, marine, n=3, 10 d excr[1]

Source: 1.Lee, R.F., C. Ryan and M.L. Neuhauser. 1976. Mar. Biol. 37, 363-370.

5-436 Excretion
In

Species	Value	Condition
Animal	100 % of dose	Oral intake, 4 d[1]
Animal	40 % of dose	Injection, urinary and fecal excretion, 30 d[1]
Mouse	52 %	Oral dose, ionic In, with urine[2]
Mouse	53 %	Oral dose, colloidal In, with feces[2]
Rat, lung	60 %	In 16 days, intratracheal injection[1]

Source: 1.Smith, G.A., R.G. Thomas and J.K. Scott. 1960. Health Phys., 4, pp 101-108.
2.Castronovo, F.P. and H.N. Wagner. 1973. J. Nucl. Med. 14, pp 677-682.

5-437 Excretion
Li

Species	Value	Condition
Animal	60 % of amount in body	With urine, 24 h, rg = 45 - 75[1]

Source: 1.Noyes, R. 1969. Dis. Nerv. Syst., 30, pp 318-321.

5-438 Excretion
Methylnaphthalene

Species	Value	Condition
Crassostrea virginica	0 mg/kg dry weight	Exposure=8 d, 7 d release, wat=3 ug/l, crude oil, lab, init conc=36 ppm[1]
Crassostrea virginica	1 mg/kg dry weight	Exposure=2 d, 7 d release, wat=8 ug/l, crude oil, lab, init. conc=58 ppm[1]
Crassostrea virginica	2 mg/kg dry weight	Exposure=8 d, 23 d release, wat=3 ug/l, crude oil, lab, init conc=36 ppm[1]

Source: 1.Lee, R.F., W.S. Gardner, J.W. Anderson, Balylock and B-Clarke. 1978. Environ. Sci. Technol. 12, 832-38.

5-439 Excretion
Mirex

Species	Value	Condition
Japanese quail	0.09 % of oral dosis	Cumulativ release with white, female, 3 d after dosis, in body = 76 %[1]
Japanese quail	0.14 % of oral dosis	Cumulativ release with white, female, 7 d after dosis, in body = 38 %[1]
Japanese quail	10 % of oral dosis	Cumulativ release with feces, female, 1 d after dosis, in body = 86 %[1]
Japanese quail	11 % of oral dosis	Cumulativ release with feces, female, 3 d after dosis, in body = 76 %[1]
Japanese quail	12 % of oral dosis	Cumulativ release with feces, female, 28 d after dosis, in body = 20[1]
Japanese quail	12 % of oral dosis	Cumulativ release with feces, female, 7 d after dosis, in body = 38 %[1]
Japanese quail	12 % of oral dosis	Cumulativ release with feces, female, 84 d after dosis, in body = 2 %[1]
Japanese quail	12 % of oral dosis	Cumulativ release with yolk, female, 28 d after dosis, in body = 20 %[1]
Japanese quail	13 % of oral dosis	Cumulativ release with yolk, female, 3 d after dosis, in body = 76 %[1]
Japanese quail	25 % of oral dosis	Cumulativ release with feces, male, 1 d after dosis, in body = 73 %[1]
Japanese quail	26 % of oral dosis	Cumulativ release with feces, male, 28 d after dosis, in body = 52 %[1]
Japanese quail	26 % of oral dosis	Cumulativ release with feces, male, 3 d after dosis, in body = 71 %[1]
Japanese quail	26 % of oral dosis	Cumulativ release with feces, male, 7 d after dosis, in body = 68 %[1]
Japanese quail	26 % of oral dosis	Cumulativ release with feces, male, 84 d after dosis, in body = 53 %[1]
Japanese quail	48 % of oral dosis	Cumulativ release with yolk, female, 7 d after dosis, in body = 38 %[1]

Japanese quail	85 % of oral dosis	Cumulativ release with yolk, female, 84 d after dosis, in body = 2 %[1]

Source: 1.Ivie, G.W., H.W. Dorough and H.E. Bryant. 1974. Bull. Environ. Contam., 11, pp 129-135.

5-440 Excretion
Mn

Species	Value	Condition
Rat	0.01 %	Steady state intake of 10 mg, with urine[1]
Rat	0.1 %	With urine, total in 5 d after intravenous dose[1]
Rat	90 % of dose	With feces, parenteral, no uptake through intestine[2]
Rat	99 %	With feces, total in 5 d after intravenous dose[1]

Source: 1.Miller, E., Kanabrocki, Case, Graham, Fields, Oester and Kaplan. 1967. J. Nucl. Med. 8, pp 891-895.
2.Greenberg, D.H. and W.W. Campbell. 1940. Proc. National Acad. Sciences Wash., 24.

5-441 Excretion
Mo

Species	Value	Condition
Homo sapiens	26.5 % of dose	With urine, injection, 10 d, rg = 24 - 29[1]
Homo sapiens	3.9 % of dose	With feces, injection, 10 d, rg = 1 - 6.9[1]

Source: 1.Odell, B.L., B.J. Cambell. 1972. Trace Elements: Metabolism and Metabolic Function Ch. 2. In Comprehensive Chemistry., vol.21, Elsevier.

5-442 Excretion
N

Species	Value	Condition
Calanus sp.	26.8 % of daily ration	Feeding on spring diatoms, available for growth[1]
Calanus sp.	35.7 % of daily ration	Feeding on spring diatoms, lost as soluble excretion[1]
Calanus sp.	37.5 % of daily ration	Feeding on spring diatoms, lost as fecal pellets[1]

Source: 1.Chamberlin, C. and R. Mitchell. 1978. in: Mitchell, R. (ed.); Water Pollution Microbiology, Vol 2.

5-443 Excretion
Naphthalene

Species	Value	Condition
Crassostrea virginica	0 mg/kg dry weight	Exposure=8 d, 7 d release, wat=3.0 ug/l, crude oil, lab, init conc=12 ppm[1]
Crassostrea virginica	1 mg/kg dry weight	Exposure=2 d, 7 d release, wat=5.0 ug/l, crude oil, lab, init. conc=30 ppm[1]
Crassostrea virginica	2 mg/kg dry weight	Exposure=8 d, 23 d release, wat=3.0 ug/l, crude oil, lab, init conc=12 ppm[1]
Rangia cuneata	32 % in 24 h	Tissue conc = 4.1 ppm after 24 h exposure, dimethylnaphthalene[2]
Rangia cuneata	34 % in 24 h	Tissue conc = 2.9 ppm after 24 h exposure, 1-methylnaphthalene[2]
Rangia cuneata	50 % in 24 h	Tissue conc = 0.8 ppm after 24 h exposure, trimethylnaphthalene[2]
Rangia cuneata	51 % in 24 h	Tissue conc = 3.9 ppm after 24 h exposure, 2-methylnaphthalene[2]
Rangia cuneata	66 % in 24 hours	Estuarine, water = 0.071 ppm, init. conc = 0.43+-0.1 ppm[3]
Rangia cuneata	79 % in 24 h	Tissue conc = 1.9 ppm after 24 h exposure[2]

Source: 1.Lee, R.F., W.S. Gardner, J.W. Anderson, Balylock and B-Clarke. 1978. Environ. Sci. Technol. 12, 832-38.
2.Neff, J.M., B.A. Lox, D. Dixit and J.W. Anderson. 1976. Mar. Biol. 38(3), pp 279-289.
3.Neff, J.M., Anderson, Cox, Laughlin, Rossi and Tatem. 1976. In: Sources, Effects and Sinks of Hydrocarbon in the Aquatic Environment. Wash DC. Amer. Inst. of Biological Science.

5-444 Excretion
Ni

Species	Value	Condition
Rat	98 %	With urine, injection of Ni[1]

Source: 1.Sunderman, F.W. and C.E. Selin. 1968. Toxicol. Appl. Pharmacol. 12, pp 207-218.

5-445 Excretion
P

Species	Value	Condition
Acartia sp.	1.0 mg P/((g body weight)*d)	Summer, C V-VI, soluble P[1]
Acartia sp.	1.3 mg P/((g body weight)*d)	Summer, C II-IV, soluble P[1]
Acartia sp.	1.4 mg P/((g body weight)*d)	Summer, nauplii, soluble P[1]
Acartia spp.	11 mg P/((g body weight)*d)	Spring, soluble P[1]
Acartia spp.	2.4 mg P/((g body weight)*d)	Spring, soluble P[1]
Acartia spp.	6.6 mg P/((g body weight)*d)	Autumn, soluble P[1]
Calanus finmarchicus	0.6 mg P/((g body weight)*d)	Autumn, soluble P[1]
Calanus finmarchicus	2.2 mg P/((g body weight)*d)	Spring, soluble P[1]
Calanus sp.	17.2 % of daily ration	Feeding on spring diatoms, available for growth[1]
Calanus sp.	23.0 % of daily ration	Feeding on spring diatoms, lost as fecal pellets[1]
Calanus sp.	59.8 % of daily ration	Feeding on spring diatoms, lost as soluble excretion[1]
Oithona similis	0.9 mg P/((g body weight)*d)	Summer, soluble P[1]
Pseudocalanus minutus	1.5 mg P/((g body weight)*d)	Summer, soluble P[1]
Sagitta hispida	2.4 mg P/((g body weight)*d)	Soluble p[1]
Temora longicornis	1.3 mg P/((g body weight)*d)	Summer, soluble P[1]
Zooplankton	3.7 mg P/((g body weight)*d)	Summer, soluble P, mixed zooplankton[1]

Source: 1.Chamberlin, C. and R. Mitchell. 1978. in: Mitchell, R. (ed.); Water Pollution Microbiology, Vol

5-446 Excretion
Pb

Species	Value	Condition
Homo sapiens	16 %	Through gastrointestinal secretion[1]
Homo sapiens	65 %	Oral intake, for a long period, rg = 50 - 80 %[2]
Homo sapiens	76 %	With urine[1]
Homo sapiens	8 %	With hair, nails and sweet[1]
Homo sapiens	99 % of total excreted	Inorganic pb[3]
Rabbit	92.5 % in 60 days	Oral intake, rg = 90 - 95 %[2]
Sheep	7.5 % in 10 days	Oral intake, rg = 5 - 10 %[2]

Source: 1.Rabinowitz, M.B., G.W. Wetherhill and J.D. Kopple. 1973. Science 182, pp 725-727.
2.Thompson, J.A. 1971. Brit. J. Industr. Med. 27, pp 189-194.
3.Waldron, H.A. and D. Stoeten. 1974. Subclinical Lead Poisoning.

5-447 Excretion
Phenanthrene

Species	Value	Condition
Rangia cuneata	13 % in 24 hours	Estuarine, water = 0.089 ppm, init. conc = 2.85+-1.1 ppm[1]

Source: 1.Neff, J.M., Anderson, Cox, Laughlin, Rossi and Tatem. 1976. In: Sources, Effects and Sinks of Hydrocarbon in the Aquatic Environment. Wash DC. Amer. Inst. of Biological Science.

5-448 Excretion
Polar metabolits of BaP

Species	Value	Condition
Callinectes sapidus, blood	10E-4 ug total in tissue	2 d upt. from wat=2.5 ug BaP/l, init=26, juv, marine, n=3, 10 d excr[1]
Callinectes sapidus, blood	2E-4 ug total in tissue	2 d upt. from wat=2.5 ug BaP/l, init=26, juv, marine, n=3, 18 d excr[1]
Callinectes sapidus, blood	36E-4 ug total in tissue	2 d upt. from wat=2.5 ug BaP/l, init=26, juv, marine, n=3, 6 d excr[1]
Callinectes sapidus, gill	0.5E-4 ug total in tissue	2 d upt. from wat=2.5 ug BaP/l, init=3 juv, marine, n=3, 18 d excr[1]
Callinectes sapidus, gill	11E-4 ug total in tissue	2 d upt. from wat=2.5 ug BaP/l, init=3 juv, marine, n=3, 10 d excr[1]
Callinectes sapidus, gill	17E-4 ug total in tissue	2 d upt. from wat=2.5 ug BaP/l, init=3 juv, marine, n=3, 6 d excr[1]
Callinectes sapidus, hepatopancreas	130E-4 ug total in tissue	2 d upt. from wat=2.5 ug BaP/l, init=210, juv, marine, n=3, 6 d excr[1]
Callinectes sapidus, hepatopancreas	224E-4 ug total in tissue	2 d upt. from wat=2.5 ug BaP/l, init=210, juv, marine, n=3, 10 d excr[1]
Callinectes sapidus, hepatopancreas	29E-4 ug total in tissue	2 d upt. from wat=2.5 ug BaP/l, init=210, juv, marine, n=3, 18 d excr[1]
Callinectes sapidus, muscle	1E-4 ug total in tissue	2 d upt. from wat=2.5 ug BaP/l, init=8, juv, marine, n=3, 10 d excr[1]
Callinectes sapidus, muscle	1E-4 ug total in tissue	2 d upt. from wat=2.5 ug BaP/l, init=8, juv, marine, n=3, 18 d excr[1]
Callinectes sapidus, muscle	2E-4 ug total in tissue	2 d upt. from wat=2.5 ug BaP/l, init=8, juv, marine, n=3, 6 d excr[1]
Callinectes sapidus, stomach	1E-4 ug total in tissue	2 d upt. from wat=2.5 ug BaP/l, init=6, juv, marine, n=3, 18 d excr[1]
Callinectes sapidus, stomach	2E-4 ug total in tissue	2 d upt. from wat=2.5 ug BaP/l, init=6, juv, marine, n=3, 10 d excr
Callinectes sapidus, stomach	2E-4 ug total in tissue	2 d upt. from wat=2.5 ug BaP/l, init=6, juv, marine, n=3, 6 d excr[1]

Source: 1.Lee, R.F., C. Ryan and M.L. Neuhauser. 1976. Mar. Biol. 37, 363-370.

5-449 Excretion
Sb

Species	Value	Condition
Hamster	15 % in 24 h	With feces, intraperitional dose, Sb(V)[1]
Hamster	50 % in 24 h	Intraperitoneal dose of 0.43 mg, fecal excretion, oral dose of Sb-K-tartrate[2]
Hamster	50 % in 24 h	With feces, intraperitional dose, Sb(III)[1]
Homo sapiens	20 %	With urine, intravenous dose, Sb(II)-K-tartrate[3]
Homo sapiens	80 %	With feces, intravenous dose, Sb(III)-K-tartrate[3]

Mouse	55 % in 6 hours	Intraperitional dose, with urine[4]
Mouse	70 % in 48 hours	Intraperitional dose, with urine[4]
Rat	95.5 % of dose	With feces, oral intake of 609 ug $SbCl_3$, rg = 93 - 98[5]

Source: 1.Abdallah, A. and A.Badran. 1962. In: G.E.W. Wolstenholme and M. O'Connor (eds.): Bilharziais. Ciba.p 287-309. Chuchill. London.
2.Gellhorn, A., M.T. Khayyal, E. Mc Connel and J. Norton. 1972. East Afr. Med. J. 47, pp 576-581.
3.Bartter, F.C., D.B. Cowie, H. Most, A.T. Ness and S. Forbush. 1947. J. Trop. Med. 27, pp 403-416.
4.Goodwin, L.G. and J.E. Page. 1943. Biochem. J. 37, pp 198-209.
5.Leopold, G. E. Furakawa, W. Forth and W. Rummel. 1969. Naunym Schmiedeberg Arch. Pharm., 263, pp 275-276.

5-450 Excretion Se

Species	Value	Condition
Pig	40 % of dose	With feces, oral intake, rg = 15 - 65[1]
Rat, urine	50 % of food intake	Dependent of intake, low intake -> low urine excretion[2]

Source: 1.Florkin and Stutz (ed). 1972. Comprehensive Biochemistry, Elsevier Publ. Comp., 21(2).
2.Burk, R.F., R.J. Seely and K.W. Kiker. 1973. Proc. Soc. Exp. Biol. Med. 142, pp 214-216.

5-451 Excretion Sn

Species	Value	Condition
Dog	5.6 %	With feces, inj of Sn2Cl, 3 days[1]
Dog	62.8 %	With urine, inj of Sn2Cl, 3 days[1]
Mouse	10.6 %	With feces, inj of Sn2Cl, 3 days[1]
Mouse	19.8 %	With urine, inj of Sn2Cl, 3 days[1]
Rat	1 %	With feces, 1 day, inj of Sn[2]
Rat	20 %	With feces, injection of Sn, total in 30 d[2]
Rat	33 %	Injection of Sn(II) or Sn(IV), in 48 h, with urine[3]
Rat	50 %	With urine, 1 day, inj of Sn[2]
Rat	99 % of dose	With feces, oral intake, 48 h[4]

Source: 1.Furchner J.E. and G.A. Drake. 1976. Health Phys. 31, pp 219-224.
2.Durbin, P.W., K.G. Scott and J.G. Hamilton. 1957. Univ. California Publ. Pharmacol. 3, pp 1-34.
3.Hiles, R.A. 1974. Toxicol. Appl. Pharmacol. 27, pp 366-379.
4.Durbin, P.W., K.G. Scott and J.G. Hamil ton. 1957. The Distribution of Radioiso topes of some heavy Metals in Rats. Univ. of Calif. Publ. in Pharmacology.

5-452 Excretion Sr

Species	Value	Condition
Homo sapiens	0.24 mg/day	With urine, woman, oral intake = 1.37 mg/day for 30 d[1]
Homo sapiens	0.81 mg/day	With feces, oral intake = 1.37 mg/day for 30 d, woman[1]
Tilapia mossambica	50 %	Remaining after 1 d, uptake by eating capsules, Sr-89[2]

Source: 1.Tipton, I.H., P.L. Stewart and P.G. Mar tin. 1966. Health Phys., 12, pp 1683-1689.
2.Boroughs, H., S.J. Townsley and R.W. Hiatt. 1956. Biol. Bull. 111, pp 336-351.

5-453 Excretion
Te

Species	Value	Condition
Animal	13.6 % of dose	Intravenous injection, 1 h, rg = 10.7 - 16.4[1]
Animal	75 % of dose	With feces, oral intake[1]
Cow	0.5 %	With milk, 6 days[2]
Dog	13.5 %	Billary excretion, injection of sodium tellurite[3]
Dog	6 %	With urine, female, rg = 3-9 %, 6-10 d, injec. of NaTe and TeO2[4]
Homo sapiens	0.1 %	Exhaled, expose of elemental Te or Te(IV)[5]
Pig	18.5 %	1-2 d, with urine, injection of NaTe and TeO2[4]
Pig	70 %	1 d, injec. of NaTe and TeO2, with feces, rg = 60-80 %[4]
Rat	20.5 %	With urine, in 24 h, injected sodium tellurite[5]
Rat	33 %	With urine, one week, injected sodium tellurite[5]
Rat	6 %	6 - 10 d, inj. of NaTe and TeO2, with urine, rg = 3-9 %[4]
Rat	70 %	1 d, inj. of NaTe and TeO2, with feces, rg = 60-80 %[4]
Sheep	0.4 %	With milk, rg = 0.3 - 0.5 %, Te-132, 4 days[2]

Source: 1.Browning, E. 1969. Toxicity of Industrial Metals. Butterwoths. London.
2.Casey, H.W., A.C. Case, R.V. McClellan and L.K. Bustad. 1963. Health Phys. 3, pp 189-200.
3.De Meio, R.H. and F.C. Henriques, Jr. 1947. J. Biol. Chem. 169, pp 609-623.
4.Chertok, R.J. and S. Lake. 1971. Health Phys. 20, pp 317-324.
5.Hollins, J.G. 1969. Health Phys. 17, pp 497-505.

5-454 Excretion
Ti

Species	Value	Condition
Animal	97 % of dose	With feces, oral intake[1]
Homo sapiens	0.44 mg/day	Normal adult, with urine[2]

Source: 1.Browning, E. 1969. Toxicity of Industrial Metals. Butterwoths. London.
2.Tipton, I.H., P.L. Stewart and P.G. Mar tin. 1966. Health Phys., 12, pp 1683-1689.

5-455 Excretion
Tl

Species	Value	Condition
Homo sapiens	3.2 % per day	With urine, TINO3 or TISO4, oral dose[1]
Rat	0.5 %	With urine, independent of adm route, day 10[2]
Rat	12 %	With feces, independent of adm route, first day[2]
Rat	2.5 %	With feces, independent of adm route, day 10[2]
Rat	6 %	With urine, independent of adm route, first day[2]

Source: 1.Barclay, R.K., W.C. Peacock and D.A. Karnofsky. 1953. J. Pharmacol. Exp. Ther. 107, pp 178-187.
2.Lie, R., R.G. Thomas and J.K. Scott. 1960. Health Phys. 2, pp 334-340.

5-456 Excretion
Total polycyclic aromatic hydrocarbon, PAH

Species	Value	Condition
Callinectes sapidus, blood	20E-4 ug total in tissue	2 d upt. from wat=2.5 ug BaP/l, init=85, juv, marine, n=3, 10 d excr[1]
Callinectes sapidus, blood	65E-4 ug total in tissue	2 d upt. from wat=2.5 ug BaP/l, init=85, juv, marine, n=3, 6 d excr[1]
Callinectes sapidus, blood	6E-4 ug total in tissue	2 d upt. from wat=2.5 ug BaP/l, init=85, juv, marine, n=3, 18 d excr[1]
Callinectes sapidus, gill	1E-4 ug total in tissue	2 d upt. from wat=2.5 ug BaP/l, init=410, juv, marine, n=3, 18 d excr[1]
Callinectes sapidus, gill	35E-4 ug total in tissue	2 d upt. from wat=2.5 ug BaP/l, init=410, juv, marine, n=3, 10 d excr[1]
Callinectes sapidus, gill	70E-4 ug total in tissue	2 d upt. from wat=2.5 ug BaP/l, init=410, juv, marine, n=3, 6 d excr[1]
Callinectes sapidus, hepatopancreas	210E-4 ug total in tissue	2 d upt. from wat=2.5 ug BaP/l, init=590, juv, marine, n=3, 6 d excr[1]
Callinectes sapidus, hepatopancreas	320E-4 ug total in tissue	2 d upt. from wat=2.5 ug BaP/l, init=590, juv, marine, n=3, 10 d excr[1]
Callinectes sapidus, hepatopancreas	40E-4 ug total in tissue	2 d upt. from wat=2.5 ug BaP/l, init=590, juv, marine, n=3, 18 d excr[1]
Callinectes sapidus, muscle	2E-4 ug total in tissue	2 d upt. from wat=2.5 ug BaP/l, init=18, juv, marine, n=3, 10 d excr[1]
Callinectes sapidus, muscle	2E-4 ug total in tissue	2 d upt. from wat=2.5 ug BaP/l, init=18, juv, marine, n=3, 18 d excr[1]
Callinectes sapidus, muscle	4E-4 ug total in tissue	2 d upt. from wat=2.5 ug BaP/l, init=18, juv, marine, n=3, 6 d excr[1]
Callinectes sapidus, stomach	2E-4 ug total in tissue	2 d upt. from wat=2.5 ug BaP/l, init=11, juv, marine, n=3, 18 d excr[1]
Callinectes sapidus, stomach	4E-4 ug total in tissue	2 d upt. from wat=2.5 ug BaP/l, init=11, juv, marine, n=3, 6 d excr[1]
Callinectes sapidus, stomach	6E-4 ug total in tissue	2 d upt. from wat=2.5 ug BaP/l, init=11, juv, marine, n=3, 10 d excr[1]

Source: 1.Lee, R.F., C. Ryan and M.L. Neuhauser. 1976. Mar. Biol. 37, 363-370.

5-457 Excretion
V

Species	Value	Condition
Dog	11 %	Inj of sodium metavanadate, with feces, 1 day[1]
Dog	61 %	Inj of sodium metavanadate, with urine, 1 day[1]
Homo sapiens	99.9 % of dose	With feces, oral intake[2]

Source: 1.Roscin, A.V. 1967. in: Izaraelson, Z.I. (ed.): Toxicology of Rare Metals, pp 52-60. AEC-tr-, Wash. DC. 2.Patty, F.A.(ed). 1963. Industrial Hygiene and Toxicology. II, Interscience Publ., N.Y..

5-458 Excretion, W

Species	Value	Condition
Dog	91 %	With urine, 1 day, inj of sodium W-181 tungstate[1]
Rat	40 %	With urine, 1 day, gast uptake, tungstate[2]
Rat	7 % of dose	With feces, intramuscular injection, 24 h, rg = 5 - 9[3]
Rat	88 % of dose	With urine, intramuscular injection, 24 h, rg = 86 - 90[3]

Source: 1.Aamodt, R.L. 1973. Health Phys. 24, pp 519-524.
2.Kaye, S.V. 1968. Health Phys. 15, pp 398-417.
3.Durbin, P.W., K.G. Scott and J.G. Hamil ton. 1957. The Distribution of Radioiso topes of some heavy Metals in Rats. Univ. of Calif. Publ. in Pharmacology.

5-459 Excretion
Y

Species	Value	Condition
Tilapia mossambica	2 %	Remaining after 2 d, uptake by eating capsules, Y-91[1]

Source: 1.Boroughs, H., S.J. Townsley and R.W. Hiatt. 1956. Biol. Bull. 111, pp 352-357.

5-460 Excretion
Zn

Species	Value	Condition
Homo sapiens	42 % of dose	With feces, oral intake = 13.2 mg/d[1]
Homo sapiens	8 % of dose	With urine, oral intake = 13.2 mg/d[1]
Mouse	20 %	Inj of 6.5 ug ZnCl2, with feces, 1 w, intravenous inj[2]
Mouse	4 %	Biliary excretion, 48 h[2]
Mouse	50 %	Inj of 0.3 ug ZnCl2, with feces, 1 w, intravenous inj[2]

Source: 1.Florkin and Stutz (ed). 1972. Comprehensive Biochemistry, Elsevier Publ. Comp., 21(2).
2.Barrowman, J.A., R. Bouett and P.J. Dryden. 1973. Biochem. Soc. Trans. 1, pp 988-989.

5-461 Excretion
Zr

Species	Value	Condition
Porphyra umbilicalis	50 % of initial	Remaining in item after 6 d, Zr-95[1]
Porphyra umbilicalis	96 % of initial	Remaining in item after 65 d, Zr-95[1]

Source: 1.Foreman, E.E. and W.L. Templeton. 1958. U.S. Atomic Energy Authority, R and DB(W) TN-187, pp 16.

5-462 Uptake
Ag

Species	Value	Condition
Abies grandis	0.12 kg/(acre*100 years)	Used for growth in 100 y[1]
Anguilla japonica	2.7E-7 mg/hour	Through gills, rg = 1.2 - 4.2E-7[2]
Dog	10 %	Gast, less than value, radioactive Ag[3]
Dog, lung	17 %	Inhalation of 0.5 um spherial Ag part.[4]
Monkey	10 %	Gast, less than value, radioactive Ag[3]
Mouse	10 %	Gast, less than value, radioactive Ag[3]
Rat	10 %	Gast, less than value, radioactive Ag[3]

Source: 1.Fortescue, J.A.C. and G.G. Marten. 1970. in: Richle, D.E. (ed.); Analysis of Temperate Forest Ecosystems. pp 173-198. Springer-Verlag. Berlin.
2.Hibiya, T. and M. Oguri. 1961. Bull. Jap. Soc. Sci. Fish. 27, pp 996-1000.
3.Furchner, J.E., C.R. Richmond and G.A. Drake. 1968. Health Phys. 15, pp 505-514.
4.Phalen, R.F. and P.E. Morrow. 1973. Health Phys. 24, pp 509-518.

5-463 Uptake
Al

Species	Value	Condition
Abies grandis	9.0 kg/(acre*100 years)	Used for growth in 100 y[1]
Homo sapiens	15 %	Uremic patients, gast, great variation[2]

Source: 1.Fortescue, J.A.C. and G.G. Marten. 1970. in: Richle, D.E. (ed.); Analysis of Temperate Forest Ecosystems. pp 173-198. Springer-Verlag. Berlin.
2.Clarkso, E.M., V.A. Luch, W.V. Hynson, R.R. Bailey, J.B. Eastwood, J.S. Woodhead, V.R. Clements, O'Brian, and DeWerdener.1972. Clin. Sci. 43, pp 519-531.

5-464 Uptake
As

Species	Value	Condition
Homo sapiens	80 %	Oral intake, gast, diss inorganic As(III), As(VI) or organic As[1]

Source: 1.Charbonneau, S.M., G.K.H. Tam, F. Bryce and B. Collins. 1978. Proceedings of the 12 Annual Conference of Trace Substances in Environmental Health. June 1978. Missouri, Columbia. pp 1-11.

5-465 Uptake
B

Species	Value	Condition
Abies grandis	0.180 kg/(acre*100 years)	Used for growth in 100 y[1]

Source: 1.Fortescue, J.A.C. and G.G. Marten. 1970. in: Richle, D.E. (ed.); Analysis of Temperate Forest Ecosystems. pp 173-198. Springer-Verlag. Berlin.

5-466 Uptake
Ba

Species	Value	Condition
Abies grandis	7.4 kg/(acre*100 years)	Used for growth in 100 y[1]
Hamster	20 %	Gast, BaCl2, rg = 10 - 30 %[2]
Homo sapiens	70 % in 4 hours	Nasal absorption, BaCl2, rg = 60 - 80 %[2]

Source: 1.Fortescue, J.A.C. and G.G. Marten. 1970. in: Richle, D.E. (ed.); Analysis of Temperate Forest Ecosystems. pp 173-198. Springer-Verlag. Berlin.
2.Cuddihy, and J.A. Ozog. 1973. Health Phys. 26, pp 405-416.

5-467 Uptake
Be

Species	Value	Condition
Rat	20 %	Gast, 0.6 - 6.6 ug be/day in drinking water, BeSO4[1]

Source: 1.Reeves, A.L. 1967. Cancer Res. 27, pp 1895-1899.

5-468 Uptake
Biphenyl

Species	Value	Condition
Rat	98.2 % retained	Oral dose = 100 mg/kg[1]
Rat	98.8 % retained	Oral dose = 50 mg/kg[1]
Rat	99.5 % retained	Oral dose = 5 mg/kg[1]

Source: 1.Hutzinger, O., S. Safe and V. Zitko. 1974. The Chemistry of PCBs, CRC-Press.

5-469 Uptake
CL2

Species	Value	Condition
Alfalfa sp.	12.4 liter/(min*m**2)	Equivalent deposition velocity = 2.07 cm/sec, 1 pphm[1]

Source: 1.Hill, A.C. and E.M. Chamberlain. 1974. Atmospheric-Surface Exchange of Particulate and Gaseous Pollutants Symp. Richland. Wash. Sept 4-6, 1974.

5-470 Uptake
CO

Species	Value	Condition
Alfalfa sp.	0.0 liter/(min*m**2)	Equivalent deposition velocity = 0.0 cm/sec, 1 pphm[1]

Source: 1.Hill, A.C. and E.M. Chamberlain. 1974. Atmospheric-Surface Exchange of Particulate and Gaseous Pollutants Symp. Richland. Wash. Sept 4-6, 1974.

5-471 Uptake
CO from atmosphere

Species	Value	Condition
Model forest hectare	2.2 t/(ha*y)	Pollutant removal flux[1]
Soil in agricultural land, total world	86.0E1 tons CO/(year*mi**2)	Area = 4.60E6 mi**2[2]
Soil in all areas in USA	505.12E6 t CO/y in USA	Area = 2977 mi**2[2]
Soil in all areas, total world	14250E6 t CO/year	Area = 56.96E6 mi**2[2]
Soil in almond weeds	2.65 mg CO/(hour*m**2 of soil)	Uptake from atmosphere with 80-130 ppm CO, 298 K, Calif, test[3]
Soil in appalachian forest in USA	313.7 tons CO/(year*mi**2)	Area = 265519 mi**2[2]
Soil in bluestem prairie	2.3 mg CO/(hour*m**2 of soil)	Uptake from atmosphere with 100 ppm CO, 298 K, field, KA, USA[2]
Soil in boreal forest	4.2 mg CO/(hour*m**2 of soil)	Uptake from atmosphere with 100 ppm CO, 298 K, field, Alberta, Can[2]
Soil in boreal forest	6.8 mg CO/(hour*m**2 of soil)	Uptake from atmosphere with 100 ppm CO, 298 K, field, Alberta, Can[2]
Soil in chaparral	4.31 mg CO/(hour*m**2 of soil)	Uptake from atmosphere with 80-130 ppm CO, 298 K, Calif, test[3]
Soil in chaparral	6.46 mg CO/(hour*m**2 of soil)	Uptake from atmosphere with 80-130 ppm CO, 298 K, Calif, test[3]
Soil in coast redwoods	14.39 mg CO/(hour*m**2 of soil)	Uptake from atmosphere with 80-130 ppm CO, 298 K, Calif, test[3]
Soil in coast redwoods	16.99 mg CO/(hour*m**2 of soil)	Uptake from atmosphere with 80-130 ppm CO, 298 K, Calif, test[3]
Soil in coastal forest	14.1 mg CO/(hour*m**2 of soil)	Uptake from atmosphere with 100 ppm CO, 298 K, field, OR, USA[2]
Soil in coastal forest in USA	200.7 tons CO/(year*mi**2)	Area = 329390 mi**2[2]
Soil in cotten	2.82 mg CO/(hour*m**2 of soil)	Uptake from atmosphere with 80-130 ppm CO, 298 K, Calif, test[3]
Soil in covered area in USA	0 tons CO/(year*mi**2)	Area = 26500 mi**2[2]
Soil in cropland in USA	86.0 tons CO/(year*mi**2)	Area = 468000 mi**2[2]

Soil in deciduous forest	26.3 mg CO/(hour*m**2 of soil)	Uptake from atmosphere with 100 ppm CO, 298 K, field, NY, USA[2]
Soil in deciduous forest in USA	254.8 tons CO/(year*mi**2)	Area = 339485 mi**2[2]
Soil in desert scrub in USA	52.2 tons CO/(year*mi**2)	Area = 220723 mi**2[2]
Soil in desert, total world	52.2E1 tons CO/(year*mi**2)	Area = 10.86E6 mi**2[2]
Soil in flood plain forest	26.2 mg CO/(hour*m**2 of soil)	Uptake from atmosphere with 100 ppm CO, 298 K, field, LA, USA[2]
Soil in forest	9.90 mg CO/(hour*m**2 of soil)	Uptake from atmosphere with 80-130 ppm CO, 298 K, Calif, test[3]
Soil in grapefruit plantation	11.48 mg CO/(hour*m**2 of soil)	Uptake from atmosphere with 80-130 ppm CO, 298 K, Calif, test[3]
Soil in grass - legume pasture	11.94 mg CO/(hour*m**2 of soil)	Uptake from atmosphere with 80-130 ppm CO, 298 K, Calif, test[3]
Soil in grass meadow	10.52 mg CO/(hour*m**2 of soil)	Uptake from atmosphere with 80-130 ppm CO, 298 K, Calif, test[3]
Soil in grassland	11.9 mg CO/(hour*m**2 of soil)	Uptake from atmosphere with 100 ppm CO, 298 K, field, TX, USA[2]
Soil in grassland	2.7 mg CO/(hour*m**2 of soil)	Uptake from atmosphere with 100 ppm CO, 298 K, field, Saskatchewan[2]
Soil in grassland	3.6 mg CO/(hour*m**2 of soil)	Uptake from atmosphere with 100 ppm CO, 298 K, field, KA, USA[2]
Soil in lakes, rivers etc. in USA	0 tons CO/(year*mi**2)	Area = 78267 mi**2[2]
Soil in mixed and broadleaf forest, total world	254.8E1 tons CO/(year*mi**2)	Area = 1.32E6 mi**2[2]
Soil in mixed deciduous forest	21.3 mg CO/(hour*m**2 of soil)	Uptake from atmosphere with 100 ppm CO, 298 K, field, OH, USA[2]
Soil in model forest	1.9E4 ug(m**2 soil surface*hour)	Gaseous pollutant flux rate for dry soil[1]
Soil in model forest	2.6E3 ug(m**2 vegetative surface*hour)	Gaseous pollutant flux rate for dry soil[1]
Soil in montane forest in USA	242.0 tons CO/(year*mi**2)	Area = 87150 mi**2[2]
Soil in montane forest, total world	175.7E1 tons CO/(year*mi**2)	Area = 4.66E6 mi**2[2]
Soil in oak	15.92 mg CO/(hour*m**2 of soil)	Uptake from atmosphere with 80-130 ppm CO, 298 K, Calif, test[3]
Soil in oak - hickory forest	10.2 mg CO/(hour*m**2 of soil)	Uptake from atmosphere with 100 ppm CO, 298 K, field, MO, USA[2]
Soil in oak - hickory forest	21.3 mg CO/(hour*m**2 of soil)	Uptake from atmosphere with 100 ppm CO, 298 K, field, OH, USA[2]
Soil in oak stubble	6.23 mg CO/(hour*m**2 of soil)	Uptake from atmosphere with 80-130 ppm CO, 298 K, Calif, test[3]
Soil in pasture in USA	179.3 tons CO/(year*mi**2)	Area = 616674 mi**2[2]
Soil in pasture, total world	179.3E1 tons CO/(year*mi**2)	Area = 1.84E6 mi**2[2]
Soil in paved roads in USA	0 tons CO/(year*mi**2)	Area = 28100 mi**2[2]

Soil in ponderosa pine	13.89 mg CO/(hour*m**2 of soil)	Uptake from atmosphere with 80-130 ppm CO, 298 K, Calif, test[3]
Soil in raods, ice, water etc., total world	0E1 tons CO/(year*mi**2)	Area = 9.2E6 mi**2, total CO uptake = 0E6 tons/y[2]
Soil in southern flood plain forest in USA	595.0 tons CO/(year*mi**2)	Area = 58782 mi**2[2]
Soil in southern mixed forest in USA	86.5 tons CO/(year*mi**2)	Area = 48644 mi**2[2]
Soil in southern pine forest, total world	505.8E1 tons CO/(year*mi**2)	Area = 0.29E6 mi**2[2]
Soil in steppe, total world	76.7E1 tons CO/(year*mi**2)	Area = 3.45E6 mi**2[2]
Soil in taiga forest, total world	127.5E1 tons CO/(year*mi**2)	Area = 4.71E6 mi**2[2]
Soil in temperate grassland, total world	86.5E1 tons CO/(year*mi**2)	Area = 3.10E6 mi**2[2]
Soil in tropical deciduous forest, total world	1105.9E1 tons CO/(year*mi**2)	Area = 1.78E6 mi**2[2]
Soil in tropical grassland, total world	886.9E1 tons CO/(year*mi**2)	Area = 3.45E6 mi**2[2]
Soil in tropical rain forest, total world	805.6E1 tons CO/(year*mi**2)	Area = 6.55E6 mi**2[2]
Soil in tundra, total world	123E1 tons CO/(year*mi**2)	Area = 1.15E6 mi**2, total CO uptake = 141.3 t/y[2]
Soil in white fir	3.48 mg CO/(hour*m**2 of soil)	Uptake from atmosphere with 80-130 ppm CO, 298 K, Calif, test[3]

Source: 1.EPA 1976. Open space as an Air Resource Management Measure. Vol II. EPA-450/3/76-028b, Research Tringle Park, NC.
2.Ingersoll, R.B., R.E. Inman and W.R. Fisher. 1974. Tellus 26, 151-158.
3.Inman, R.E., R.B. Ingersoll and E.A. Levy. 1971. Science 172, 1229-1231.

5-472 Uptake
CO2

Species	Value	Condition
Alfalfa sp.	2.0 liter/(min*m**2)	Equivalent deposition velocity = 0.33 cm/sec, 1 ppm[1]

Source: 1.Hill, A.C. and E.M. Chamberlain. 1974. Atmospheric-Surface Exchange of Particulate and Gaseous Polluants Symp. Richland. Wash. Sept 4-6, 1974.

5-473 Uptake
Ca

Species	Value	Condition
Abies grandis	333.0 kg/(acre*100 years)	Used for growth in 100 y[1]

Source: 1.Fortescue, J.A.C. and G.G. Marten. 1970. in: Richle, D.E. (ed.); Analysis of Temperate Forest Ecosystems. pp 173-198. Springer-Verlag. Berlin.

5-474 Uptake
Cd

Species	Value	Condition
Acer saccharum, leaf and twig	60. mg/tree	Single growing season, CD from atm.[1]
Corn, root	1200 mg/kg dw	Solution conc = 100 mg/l water, 12 d[2]
Corn, root	3 times the conc in the shoot	Grown in solution, 12 d[2]
Corn, shoot	48 mg/kg dw	Loamy sand soil, 10 mg/kg dw, 4 w[2]

Corn, shoot	500 mg/kg dw	Solution conc = 100 mg/l water, 12 d[2]
Corn, shoot	8.4 mg/kg dw	Silty clay loam soil, 10 mg/kg dw, 4 w[2]
Homo sapiens, non-smokers	0.108 ug/day	Rg = 0.0005-0.215 ug/d, rural areas[3]
Homo sapiens, non-smokers	1.75 ug/day	Rg = 0.01 - 3.5 ug/d, urban areas[3]
Homo sapiens, non-smokers	13 ug/day	Rg = 0.05-25 ug/d, industrial areas[3]
Homo sapiens, smokers	15.5 ug/day	Rg = 3.05 - 28 ug/d, 40 cigarettes/d, industrial areas[3]
Homo sapiens, smokers	3.105 ug/day	Rg = 3.0005 - 3.215 ug/d, 40 cigarettes/d, rural area[3]
Homo sapiens, smokers	4.5 ug/day	Rg = 3.01 - 6.5 ug/d, 40 cigarettes/d, urban areas[3]
Lycosa sp.	0.08 1/day	Grassland food chain, predator[4]
Pteronarcella badia	0.005 microgram/(g*day)	Water conc = 9.0 mg/l water, day 1 - 4[5]
Pteronarcella badia	19.0 microgram/(g*day)	Water conc = 9.0 mg/l water, first 24 h[5]
Pteronarcys californica	0.001 microgram/(g*day)	Water conc = 5.0 mg/l water, day 1 - 4[5]
Pteronarcys californica	0.06 microgram/(g*day)	Water conc = 3.0 mg/l water, day 1 - 4[5]
Pteronarcys californica	1.45 microgram/(g*day)	Water conc = 3.0 mg/l water, first 24 h[5]
Pteronarcys californica	1.90 microgram/(g*day)	Water conc = 5.0 mg/l water, first 24 h[5]
Pteronemobius fasciatus	0.55 1/day	Grassland food chain, primary consumer[4]
Rice, bran	0.44 mg/kg dry weight	Cd added to soil = 0 % of Cd in control, pH-dependent[6]
Rice, bran	0.79 mg/kg dry weight	Cd added to soil = 0.001 % of Cd in control, pH-dependent[6]
Rice, bran	0.84 mg/kg dry weight	Cd added to soil = 0.003 % of Cd in control, pH-dependent[6]
Rice, bran	1.60 mg/kg dry weight	Cd added to soil = 0.01 % of Cd in control, pH-dependent[6]
Rice, bran	2.68 mg/kg dry weight	Cd added to soil = 0.03 % of Cd in control, pH-dependent[6]
Rice, bran	2.94 mg/kg dry weight	Cd added to soil = 0.1 % of Cd in control, pH-dependent[6]
Rice, bran	3.19 mg/kg dry weight	Cd added to soil = 0.3 % of Cd in control, pH-dependent[6]
Rice, bran	3.94 mg/kg dry weight	Cd added to soil = 0.6 % of Cd in control, pH-dependent[6]
Rice, polished	0.16 mg/kg dry weight	Cd added to soil = 0 % of Cd in control, pH-dependent[6]
Rice, polished	0.28 mg/kg dry weight	Cd added to soil = 0.001 % of Cd in control, pH-dependent[6]
Rice, polished	0.40 mg/kg dry weight	Cd added to soil = 0.003 % of Cd in control, pH-dependent[6]
Rice, polished	0.59 mg/kg dry weight	Cd added to soil = 0 % of Cd in control, pH-dependent[6]
Rice, polished	0.78 mg/kg dry weight	Cd added to soil = 0.01 % of Cd in control, pH-dependent[6]
Rice, polished	1.37 mg/kg dry weight	Cd added to soil = 0.03 % of Cd in control, pH-dependent[6]
Rice, polished	1.62 mg/kg dry weight	Cd added to soil = 0.1 % of Cd in control, pH-dependent[6]
Rice, polished	1.73 mg/kg dry weight	Cd added to soil = 0.6 % of Cd in control, pH-dependent[6]
Rice, polished	1.94 mg/kg dry weight	Cd added to soil = 0.3 % of Cd in control, pH-dependent[6]
Rice, unpolished	4.98 mg/kg dry weight	Cd added to soil = 1.0 % of Cd in control, pH-dependent[6]
Sediment	220.0 microgram/(g*day)	Water conc = 3.0 mg/l water, first 24 h[5]
Sediment	320.0 microgram/(g*day)	Water conc = 5.0 mg/l water, first 24 h[5]

Sediment	64.0 microgram/(g*day)	Water conc = 3.0 mg/l water, day 1 - 4[5]
Sediment	84.0 microgram/(g*day)	Water conc = 5.0 mg/l water, day 1 - 4[5]
Soil	9.4 mg Cd/kg soil	Pots placed 500 meters from a CD emission source, aerosol fallout,[2]
Sugar maple	60 mg/(tree*year)	Tree = 30 cm diameter, urban air, New Haven[7]
Wheat, whole grain	139.0 mg/kg dry weight	Cd added to soil = 1.0 % of Cd in control, pH-dependent[6]
Wheat, whole grain	15.5 mg/kg dry weight	Cd added to soil = 0.003 % of Cd in control, pH-dependent[6]
Wheat, whole grain	29.9 mg/kg dry weight	Cd added to soil = 0.01 % of Cd in control, pH-dependent[6]
Wheat, whole grain	41.4 mg/kg dry weight	Cd added to soil = 0.03 % of Cd in control, pH-dependent[6]
Wheat, whole grain	48.6 mg/kg dry weight	Cd added to soil = 0.3 % of Cd in control, pH-dependent[6]
Wheat, whole grain	60.7 mg/kg dry weight	Cd added to soil = 0.1 % of Cd in control, pH-dependent[6]
Wheat, whole grain	90.8 mg/kg dry weight	Cd added to soil = 0.6 % of Cd in control, pH-dependent[6]

Source: 1.Smith, W.H. 1973. Metal contamination of Urban Woody Plants, Science
2.Koeppe, D.E. 1977. Sci. Total Environ. 7(3), pp 197-206.
3.Foerster, U. 1980. In: O. Hutzinger: The Handbook of Environmental Chemistry. Antropogenic compounds. Vol 3. Part A. pp 59-107. Springer-Verlag.
4.Hook, R.I. van and A.J. Yates. 1975. Environ. Res. 9(2), pp 76-83.
5.Clubb, R.W., A.R. Gaufin and J.L. Lords. 1975. Environ. Res., 9(3), pp 332-341.
6.Friberg, L., M. Piscater, G.F. Nordberg and T. Kjellstroem. Cadmium in the Environment 2nd edition, CRC-Press.
7.Smith, W.H. 1974. Environ. Pollut. 6, 111-129.

5-475 Uptake
Co

Species	Value	Condition
Abies grandis	0.007 kg/(acre*100 years)	Used for growth in 100 y[1]
Hamster	0.5 %	Less than value, oral dose = 5 mg of CoO[2]
Hamster	33 %	Inhalation of CoO[2]
Homo sapiens	25 %	Gast, oral intake[3]
Rat	30 %	Oral dose of CoCl[4]
Soil	7.38E4 kg/(km**2*year)	Co from atm.[5]

Source: 1.Fortescue, J.A.C. and G.G. Marten. 1970. in: Richle, D.E. (ed.); Analysis of Temperate Forest Ecosystems. pp 173-198. Springer-Verlag. Berlin.
2.Wehner, A.P. and D.K. Craig. 1972. Am. J. Ind. Hyg. Assoc. 33, pp 146-155.
3.Smith,T., C.J. Edmonds and C.F. Barnaby. 1972. Health Phys. 22, pp 359-367.
4.Tayor, D.M. 1962. Phys. Med. Biol. 6, pp 445-451.
5.Inman, R.E., R.B. Ingersoll and F.A. Levy. 1971. Science, N.Y., 172, pp 1220-1231.

5-476 Uptake
Co - 60

Species	Value	Condition
Haliotis discus	25.5 %	Co - 60, from food, food = Eisenia, 2.1 g food, 2 d, n = 4[1]
Haliotis discus	30.5 %	Co - 60, from food, food = Undularia, 2.7 g food, 2 d, n = 6[1]
Haliotis discus	46.7 %	Co - 60, from food, food = Ulva, 1.2 g food, 2 d, n = 4, SD = +- 6.0[1]
Haliotis discus	47.1 %	Co - 60, from food, food = Laminaria, 1.75 g food, 2 d, n = 6[1]
Rat	0.60 %	Uptake from food[2]

Source: 1.Gilat, Y., Laichter and N.H. Shafrir. 1975. Impacts of Nuclear releases into the Aquatic Environment. IAEA-SM-198/18, pp 63-75. Vienna.
2.Amiard, J.-C. 1976. Health Phys., 31(4), pp 371-373.

5-477 Uptake
Cr

Species	Value	Condition
Abies grandis	0.015 kg/(acre*100 years)	Used for growth in 100 y[1]
Acer saccharum, leaf and twig	140. mg/tree	Single growing season, Cr from atm.[2]
Alisma plantago aquatica, old plant	15.51 %	800 and 50 m below pollution I and II, removed from surface, F046[3]
Alisma plantago aquatica, old plant	44.92 %	800 m above pollution entry I, removed from surface, F046[3]
Alisma plantago aquatica, old plant	55.95 %	150 m below pollution entry I, removed from surface, F046[3]
Alisma plantago aquatica, young plant	14.75 %	800 and 50 m below pollution I and II, removed from surface, F046[3]
Alisma plantago aquatica, young plant	28.30 %	800 m above pollution entry I, removed from surface, F046[3]
Alisma plantago aquatica, young plant	43.76 %	150 m below pollution entry I, removed from surface, F046[3]
Anguilla japonica	01.35E-8 mg/hour	Through gills, rg = 0.8 - 1.9E-8[4]
Homo sapiens	1 %	Gast, Less than value, Cr(III)[5]
Homo sapiens	2 %	Gast, Cr(VI)[6]
Rat	2 %	Gast, Cr(VI), rg = 3 - 6 %[7]
Sugar maple	140 mg/(tree*year)	Tree = 30 cm diameter, urban air, New Haven[8]

Source: 1.Fortescue, J.A.C. and G.G. Marten. 1970. in: Richle, D.E. (ed.); Analysis of Temperate Forest Ecosysems. pp 173-198. Springer-Verlag. Berlin.
2.Smith, W.H. 1973. Metal contamination of Urban Woody Plants, Science
3.Patrick, F.M. and M.W. Loutit. 1977. Water Res., 11(8), pp 699-703.
4.Hibiya, T. and M. Oguri. 1961. Bull. Jap. Soc. Sci. Fish. 27, pp 996-1000.
5.Mertz, W. 1969. Physiol. Rev. 49, pp 163-239.
6.Donaldson, R.M. and R.F. Barreras. 1966. J. Lab. Clin. Med. 68, pp 484-493.
7.Mertz, W. 1975. Nutr. Rev. 33, 129-135.
8.Smith, W.H. 1974. Environ. Pollut. 6, 111-129.

5-478 Uptake
Cs

Species	Value	Condition
Pleuronectes platessa	50 %	% of body burden obtained from food, min value[1]
Raja clavata	80 %	% of body burden obtained from food, min value[1]

Source: 1.Jefferies, D.F. and C.J. Hewett. 1971. J. mar. biol. Ass. U.K. 51, pp 411-422.

5-479 Uptake
Cu

Species	Value	Condition
Abies grandis	0.760 kg/(acre*100 years)	Used for growth in 100 y[1]
Alisma plantago aquatica, old plant	28.42 %	800 and 50 m below pollution I and II, removed from surface, F046[2]
Alisma plantago aquatica, old plant	33.15 %	150 m below pollution entry I, removed from surface, F046[2]
Alisma plantago aquatica, old plant	47.01 %	800 m above pollution entry I, removed from surface, F046[2]

Alisma plantago aquatica, young plant	41.12 %	150 m below pollution entry I, removed from surface, F046[2]
Alisma plantago aquatica, young plant	43.07 %	800 and 50 m below pollution I and II, removed from surface, F046[2]
Alisma plantago aquatica, young plant	55.00 %	800 m above pollution entry I, removed from surface, F046[2]
Chlorella pyrenoidosa	54 % in 1 hour	Initial Cu = 0.05 mg/l, pH = 8, 1E7 cells per liter[3]
Chlorella pyrenoidosa	6 % in 1 hour	Initial Cu = 0.05 mg/l, pH = 5, 1E8 cells per liter[3]
Chlorella pyrenoidosa	7 % in 1 hour	Initial Cu = 0.05 mg/l, pH = 5, 1E7 cells per liter[3]
Chlorella pyrenoidosa	80 % in 1 hour	Initial Cu = 0.05 mg/l, pH = 8, 1E8 cells per liter[3]
Homo sapiens	50 %	Gast, oral dosis = 1 ug, rel. smaller uptake with larger dose[4]
Menhaden	9 %	Assimilation efficiency, ingested = 1.0 ug/d, juvenile[5]
Pinfish	2.9 %	Assimilation efficiency, ingested = 0.14 ug/d, juvenile[5]
Spot	1 %	Assimilation efficiency, ingested = 5.14 ug/d, juvenile[5]

Source: 1.Fortescue, J.A.C. and G.G. Marten. 1970. in: Richle, D.E. (ed.); Analysis of Temperate Forest Ecosysems. pp 173-198. Springer-Verlag. Berlin.
2.Patrick, F.M. and M.W. Loutit. 1977. Water Res., 11(8), pp 699-703.
3.Steeman-Nielsen, E., L. Kamp-Nielsen and S. Wium-Andersen. 1970. Mar. Biol. 6, 93-97.
4.Owen, C.A. Jr. 1964. Am. J. Physiol. 207, pp 1203-1206.
5.Cross, F.A. J.N. Willis, L.H. Hardy, N.Y. Jones and J.M. Lewis 1975. in: Cronin, L.E. (ed.); Estuarine Research. Vol 1 pp 45-63. Academic Press. New York.

5-480 Uptake
DDT

Species	Value	Condition
Amphidinium carteri	93.2 % gross	max value, medium = 80 ug DDT/m3, no metabolits found[1]
Cyclotella nana	96.6 % gross	Max value, medium = 80 ug DDT/m3, metabolites found = DDE[1]
Dunaliella tertiolecta	97.8 % gross	Max value, medium = 80 ug DDT/m3, metabolites found = DDE[1]
Homo sapiens	1.0 % of total intake	From sugars and adjuncts[2]
Homo sapiens	1.5 % of total intake	From root vegetables[2]
Homo sapiens	13.4 % of total intake	From fruits[2]
Homo sapiens	14.4 % of total intake	From dairy products[2]
Homo sapiens	2.0 % of total intake	From legume vegetables[2]
Homo sapiens	2.0 % of total intake	From potatoes[2]
Homo sapiens	3.1 % of total intake	From oils and fats[2]
Homo sapiens	3.3 % of total intake	From leafy vegetables[2]
Homo sapiens	40.8 % of total intake	From meats, fish and poultry[2]
Homo sapiens	7.8 % of total intake	From grains-cerals[2]
Homo sapiens	9.9 % of total intake	From garden fruits[2]
Skeletonema costatum	95.7 % gross	Medium = 80 ug DDT/m3, metabolites found = DDE[1]
Synechococcus elongatus	84 %	Medium = 99 ug DDT/m3, metabolites found = DDE[3]
Thalassiosira fluviatilis	94.4 % gross	Max value, medium = 80 ug DDT/m3, metabolites found = DDE[1]

Source: 1.Kuzyakina, T.I. 1974. Weed Abstr. 24, 820.
2.Duggan, R.E. and J.R.. Weatherway. 1967. Science, 157, pp 1006-1010.

5-481 Uptake
Dieldrin

Species	Value	Condition
Leptidius floridanus	0.010 mg/(kg dry weight*day)	From food = 213 ppm, seawater[1]
Leptidius floridanus	0.191 mg/(kg dry weight*day)	Only from water = 0.5 ppb, seawater[1]

Source: 1.Epifano, C.E. 1973. Mar. Biol., 19, pp 320-322.

5-482 Uptake
Endrin

Species	Value	Condition
Gambusia affinis, brain	198.6 mg/(kg ww*h)	Water = 1000 ppb, uptake from water, not resistant fish, fish died[1]
Gambusia affinis, brain	28.76 mg/(kg ww*h)	Water = 1000 ppb, uptake from water, resistant fish[1]
Gambusia affinis, liver	176.7 mg/(kg ww*h)	Water = 1000 ppb, uptake from water, resistant fish[1]
Gambusia affinis, liver	213.6 mg/(kg ww*h)	Water = 1000 ppb, uptake from water, not resistant fish, fish died[1]

Source: 1.Fabacher, D.L. and H. Chambers. 1976. Bull. Environ. Contam. Toxicol., 16(2), pp 203-207.

5-483 Uptake
Fe

Species	Value	Condition
Abies grandis	1.8 kg/(acre*100 years)	Used for growth in 100 y[1]
Alisma plantago aquatica, old plant	28.46 %	150 m below pollution entry I, removed from surface, F046[2]
Alisma plantago aquatica, old plant	30.91 %	800 and 50 m below pollution entry I and II, removed from surface,[2]
Alisma plantago aquatica, old plant	31.39 %	800 m above pollution entry I, removed from surface, F046[2]
Alisma plantago aquatica, young plant	24.60 %	800 m above pollution entry I, removed from surface, F046[2]
Alisma plantago aquatica, young plant	34.74 %	150 m below pollution entry I, removed from surface, F046[2]
Alisma plantago aquatica, young plant	9.09 %	800 and 50 m below pollution entry I and II, removed from surface,[2]
Menhaden	1 %	Assimilation efficiency, ingested = 920 ug/d, juvenile[3]
Pinfish	0.6 %	Assimilation efficiency, ingested = 240 ug/d, juvenile[3]
Spot	0.3 %	Assimilation efficiency, ingested = 1400 ug/d, juvenile[3]

Source: 1.Fortescue, J.A.C. and G.G. Marten. 1970. in: Richle, D.E. (ed.); Analysis of Temperate Forest Ecosystems. pp 173-198. Springer-Verlag. Berlin.
2.Patrick, F.M. and M.W. Loutit. 1977. Water Res., 11(8), pp 699-703.
3.Cross, F.A. J.N. Willis, L.H. Hardy, N.Y. Jones and J.M. Lewis 1975. in: Cronin, L.E. (ed.); Estuarine Research. Vol 1 pp 45-63. Academic Press. New York.

5-484 Uptake
Ge

Species	Value	Condition
Rat	73.4 % in 4 hours	Neutralized GeO2, stomach intubation[1]
Rat	96.4 % in 8 hours	Neutralized GeO2, stomach intubation[1]

Source: 1.Rosenfeld, G. 1954. Arch. Biochem Biophys. 48, pp 84-94.

5-485 Uptake
Hf

Species	Value	Condition
Alfalfa sp.	22.6 liter/(min*m**2)	Equivalent deposition velocity = 3.77 cm/sec, 1 pphm[1]

Source: 1.Hill, A.C. and E.M. Chamberlain. 1974. Atmospheric-Surface Exchange of Particulate and Gaseous Pollutants Symp. Richland. Wash. Sept 4-6, 1974.

5-486 Uptake
Hg

Species	Value	Condition
Anguilla japonica	1.75E-6 mg/hour	Through gills, rg = 0.3 - 3.2E-6[1]
Cat	90 % of dose	Oral, methyl-Hg, uptake through intestine[2]
Guinea pig	6 % of dose	16 mg HgCl2/ml, after 5 h, uptake through skin, max value[2]
Homo sapiens	100 %	Oral, methyl-Hg-salt and -proteinate, uptake through intestine, app[2]
Homo sapiens	15 % of dose	Peroral intake of protein-Hg-nitrate, uptake through intestine, n = 7[2]
Homo sapiens	8 % of dose	Oral, HgCl2, after 20 min - 1 h, uptake through intestine, min value[2]
Homo sapiens	82.5 % of inhaled amount	Respiratory intake of elemental Hg (50-350 ug/m**3), rg = 78 - 85[2]
Monkey	90 % of dose	Oral, methyl-Hg, uptake through intestine[2]
Mouse	2 % of daily intake	Oral, HgCl2, uptake through intestine[2]
Mouse	89.5 % of inhaled amount	Respiratory intake of elemental Hg, rg = 75 - 100[2]
Rat	0.01 % of dose	Oral, elemental Hg, uptake through intestine, less than[2]
Rat	20 % of dose	Oral, 0.5 mg Hg-acetate/kg, after 48 h, uptake through intestine[2]
Rat	40 % of dose	Oral, 0.5 mg phenyl-Hg-acetate/kg, 48 h, uptake through intestine[2]
Rat	75 % of inhaled amount	Respiratory intake of elemental Hg, rg = 50 - 100[2]
Rat	90 % of dose	Oral, methyl-Hg, uptake through intestine[2]

Source: 1.Hibiya, T. and M. Oguri. 1961. Bull. Jap. Soc. Sci. Fish. 27, pp 996-1000.
2.Friberg, L. and J. Vostal. 1974. Mercury in the Environment. CRC-Press.

5-487 Uptake
In

Species	Value	Condition
Rat	0.5 %	Oral dose, gast[1]
Rat	4.5 %	Inhalation, 1 hour[2]

Source: 1.Smith, G.A., R.G. Thomas and J.K. Scott. 1960. Health Phys., 4, pp 101-108.
2.Morrow, P.E., F.R. Gibb, R. Cloutier, L.J. Casarett and Scott. 1958. Univ. Rochester Atomic Energy Poj. No. UR508. Univ. Roch. NY.

5-488 Uptake
K

Species	Value	Condition
Abies grandis	148.0 kg/(acre*100 years)	Used for growth in 100 y[1]

Source: 1.Fortescue, J.A.C. and G.G. Marten. 1970. in: Richle, D.E. (ed.); Analysis of Temperate Forest Ecosystems. pp 173-198. Springer-Verlag. Berlin.

5-489 Uptake
Li

Species	Value	Condition
Abies grandis	0.082 kg/(acre*100 years)	Used for growth in 100 y[1]

Source: 1.Fortescue, J.A.C. and G.G. Marten. 1970. in: Richle, D.E. (ed.); Analysis of Temperate Forest Ecosystems. pp 173-198. Springer-Verlag. Berlin.

5-490 Uptake
Mg

Species	Value	Condition
Abies grandis	64.0 kg/(acre*100 years)	Used for growth in 100 y[1]

Source: 1.Fortescue, J.A.C. and G.G. Marten. 1970. in: Richle, D.E. (ed.); Analysis of Temperate Forest Ecosystems. pp 173-198. Springer-Verlag. Berlin.

5-491 Uptake
Mn

Species	Value	Condition
Abies grandis	8.0 kg/(acre*100 years)	Used for growth in 100 y[1]
Alisma plantago aquatica, old plant	1.46 %	800 m above pollution entry I, removed from surface, F046[2]
Alisma plantago aquatica, old plant	2.12 %	150 m below pollution entry I, removed from surface, F046[2]
Alisma plantago aquatica, old plant	3.52 %	800 and 50 m below pollution entry I and II, removed from surface,[2]
Alisma plantago aquatica, young plant	14.84 %	800 and 50 m below pollution entry I and II, removed from surface,[2]
Homarus vulgaris	0.4 microgram/day	500 g animal, uptake from seawater[3]
Homo sapiens	3 %	Gast, n = 1, dose = 200 ug MnCl[4]
Menhaden	3 %	Assimilation efficiency, ingested = 11 ug/d, juvenile[5]
Pinfish	7 %	Assimilation efficiency, ingested = 2.3 ug/d, juvenile[5]
Rat	2 %	Rg = 1 - 3.5 %, gast[6]
Spot	0.7 %	Assimilation efficiency, ingested = 29 ug/d, juvenile[5]

Source: 1.Fortescue, J.A.C. and G.G. Marten. 1970. in: Richle, D.E. (ed.); Analysis of Temperate Forest Ecosystems. pp 173-198. Springer-Verlag. Berlin.
2.Patrick, F.M. and M.W. Loutit. 1977. Water Res., 11(8), pp 699-703.
3.Bryan, G.W. and E. Ward. 1965. J. mar. biol. Ass. U. K. 45, pp 65-95.
4.Mena, I., K. Horichi, K. Burke and G.C. Cotzias. 1969. Neurology 19, pp 1000-1006.
5.Cross, F.A. J.N. Willis, L.H. Hardy, N.Y. Jones and J.M. Lewis 1975. in: Cronin, L.E. (ed.); Estuarie Research. Vol 1 pp 45-63. Academic Press. New York.
6.Pollack, S., George, Reba, R.C. Kaufman and W.H. Crosby. 1965. J. Clin. Invest. 44, pp 1470-1473.

5-492 Uptake
Mo

Species	Value	Condition
Abies grandis	0.003 kg/(acre*100 years)	Used for growth in 100 y[1]

Source: 1.Fortescue, J.A.C. and G.G. Marten. 1970. in: Richle, D.E. (ed.); Analysis of Temperate Forest Ecosystems. pp 173-198. Springer-Verlag. Berlin.

5-493 Uptake
NH_3

Species	Value	Condition
Soybean	2.E3 kg/(km**2*year)	NH3 from atmosphere[1]

Source: 1.Hutchinson, J.C., R.J. Millington and D.B. Peters. 1972. Science, N.Y., 175, pp 771-772.

5-494 Uptake
NO

Species	Value	Condition
Alfalfa sp.	0.6 liter/(min*m**2)	Equivalent deposition velocity = 0.10 cm/sec, 1 pphm[1]

Source: 1.Hill, A.C. and E.M. Chamberlain. 1974. Atmospheric-Surface Exchange of Particulate and Gaseous Pollutants Symp. Richland. Wash. Sept 4-6, 1974.

5-495 Uptake
NO2

Species	Value	Condition
Alfalfa	3.58E4 kg/(km**2*year)	NO2 from atmosphere[1]
Alfalfa sp.	11.4 liter/(min*m**2)	Equivalent deposition velocity = 1.90 cm/sec, 1 pphm[2]

Source: 1.Hill, A.C. 1971. J.Air Pollut. Control, 21, pp 341-346.
2.Hill, A.C. and E.M. Chamberlain. 1974. Atmospheric-Surface Exchange of Particulate and Gaseous Pollutants Symp. Richland. Wash. Sept 4-6, 1974.

5-496 Uptake
NOx from atmosphere

Species	Value	Condition
Model forest hectare	0.38 t/(ha*y)	Pollutant removal flux[1]
Soil in model forest	2.0E2 ug(m**2 soil surface*hour)	Gaseous pollutant flux rate for dry soil[1]
Soil in model forest	2.3E3 ug(m**2 vegetative surface*hour)	Gaseous pollutant flux rate for dry soil[1]

Source: 1.EPA 1976. Open space as an Air Resource Management Measure. Vol II. EPA-450/3/76-028b, Research Tringle Park, NC.

5-497 Uptake, Na

Species	Value	Condition
Abies grandis	3.0 kg/(acre*100 years)	Used for growth in 100 y[1]

Source: 1.Fortescue, J.A.C. and G.G. Marten. 1970. in: Richle, D.E. (ed.); Analysis of Temperate Forest Ecosystems. pp 173-198. Springer-Verlag. Berlin.

5-498 Uptake, Ni

Species	Value	Condition
Abies grandis	0.028 kg/(acre*100 years)	Used for growth in 100 y[1]
Acer saccharum, leaf and twig	820. mg/tree	Single growing season, Ni from atmosphere[2]
Homo sapiens	10 %	Gast, steady state, oral intake = 289 ug/day[3]
Rat	50 %	Inhalation of Ni carbonyl of 200 mg/l in 15 min[4]
Sugar maple	820 mg/(tree*year)	Tree = 30 cm diameter, urban air, New Haven[5]

Source: 1.Fortescue, J.A.C. and G.G. Marten. 1970. in: Richle, D.E. (ed.); Analysis of Temperate Forest Ecosystems. pp 173-198. Springer-Verlag. Berlin.
2.Smith, W.H. 1973. Metal contamination of Urban Woody Plants, Science
3.Nodiya, R.I. 1972. Gig. Sanit. 37, pp 108-109.

4.Sunderman, F.W. and C.E. Selin. 1968. Toxicol. Appl. Pharmacol. 12, pp 207-218.
5.Smith, W.H. 1974. Environ. Pollut. 6, 111-129.

5-499 Uptake, O3 from atmosphere

Species	Value	Condition
Model forest hectare	9.6E4 t/(ha*y)	Pollutant removal flux[1]
Soil in model forest	1.0E9 ug(m**2 soil surface*hour)	Gaseous pollutant flux rate for dry soil[1]
Soil in model forest	6.2E4 ug(m**2 vegetative surface*hour)	Gaseous pollutant flux rate for dry soil[1]

Source: 1.EPA 1976. Open space as an Air Resource Management Measure. Vol II. EPA-450/3/76-028b, Research Tringle Park, NC.

5-500 Uptake, Octachlorodibenzo-p-dioxin

Species	Value	Condition
Rat	7 % of dose	Uptake through stomach-intestine, oral dose = 100 ug/day for 21 d[1]

Source: 1.Norbac, D.H. and J.R. Allen. 1973. Environ. health Persp., 9, 233-240.

5-501 Uptake, O

Species	Value	Condition
Alfalfa sp.	10.0 liter/(min*m**2)	Equivalent deposition velocity = 1.67 cm/sec, 1 pphm[1]

Source: 1.Hill, A.C. and E.M. Chamberlain. 1974. Atmospheric-Surface Exchange of Particulate and Gaseous Pollutants Symp. Richland. Wash. Sept 4-6, 1974.

5-502 Uptake, PAN

Species	Value	Condition
Alfalfa sp.	3.8 liter/(min*m**2)	Equivalent deposition velocity = 0.63 cm/sec, 1 pphm[1]

Source: 1.Hill, A.C. and E.M. Chamberlain. 1974. Atmospheric-Surface Exchange of Particulate and Gaseous Pollutants Symp. Richland. Wash. Sept 4-6, 1974.

5-503 Uptake, Pb

Species	Value	Condition
Abies grandis	1.3 kg/(ha*y)	Used for growth in 100 y[1]
Acer saccharum, leaf and twig	5800 mg/tree	Single growing season, Pb from atmosphere[2]
Alisma plantago aquatica, old plant	35.87 %	800 and 50 m below pollution entry I and II, removed from surface[3]
Alisma plantago aquatica, old plant	38.42 %	800 m above pollution entry I, removed from surface, F046[3]
Alisma plantago aquatica, old plant	42.99 %	150 m below pollution entry I, removed from surface, F046[3]
Alisma plantago aquatica, young plant	0.86 %	800 m above pollution entry I, removed from surface, F046[3]
Al/isma plantago aquatica, young plant	20.50 %	150 m below pollution entry I, removed from surface, F046[3]
Alisma plantago aquatica, young plant	46.25 %	800 and 50 m below pollution entry I and II, removed from surface,[3]
Homo sapiens	10 %	Gast, adult[4]
Homo sapiens	11 % of intake	Oral intake, rg = 5 - 17[5]
Homo sapiens	3 %	Gast, max value for children, age = 1/4 - 8 year[6]
Homo sapiens, lung	10.5 %	Rg = 5 - 16 %, Pb(II) oxide part., dia = 0.9 - 2.9 um 1.35 l air/min[7]
Homo sapiens, lung	10.5 %	Rg = 5 - 16 %, Pb(III) oxide part., dia = 0.9 - 2.9 um, 1.35 l air/min[7]

Homo sapiens, lung	52.9 %	Rg = 42.5 - 63.2 %, Pb(II) oxide part., dia = 0.05 um, 1.35 l air/min[7]
Homo sapiens, lung	52.9 %	Rg=42.5-63.2 %, Pb(III) oxide part., dia = 0.9 - 2.9 um, 1.35 l air/min[7]
Plant, leaf	0.01 % of intake	Uptake from air fallout on leaves[8]
Rabbit	0.9 % of intake	Oral intake, rg = 0.8 - 1.0[5]
Sheep	1.3 % of intake	Oral intake[5]
Sugar maple	5800 mg/(tree*year)	Tree = 30 cm diameter, urban air, New Haven[9]

Source: 1.Fortescue, J.A.C. and G.G. Marten. 1970. in: Richle, D.E. (ed.); Analysis of Temperate Forest Ecosystems. pp 173-198. Springer-Verlag. Berlin.
2.Smith, W.H. 1973. Metal contamination of Urban Woody Plants, Science
3.Patrick, F.M. and M.W. Loutit. 1977. Water Res., 11(8), pp 699-703.
4.Rabinowits, M.B., G.W. Wetherhill and J.D. Kopple. 1974. Environ. Health Perspect. 7, pp 145-153.
5.Hichs, R.M. 1972. Chem. Biol. Interact. 5, pp 361-390.
6.WHO 1977. Lead. Environmental Health Criteria 3. World Health Organzation. Geneva.
7.Nozaki, K. 1966. Ind. Health 4, pp 118-128.
8.Koeppe, D.E. 1977. Sci. Total Environ. 7(3), pp 197-206.
9.Smith, W.H. 1974. Environ. Pollut. 6, 111-129.

5-504 Uptake, Peroxyacetylnitrate from atmosphere

Species	Value	Condition
Model forest hectare	0.17 t/(ha*y)	Pollutant removal flux[1]
Soil in model forest	1.2E3 ug(m**2 vegetative surface*hour)	Gaseous pollutant flux rate for dry soil[1]

Source: 1.EPA 1976. Open space as an Air Resource Management Measure. Vol II. EPA-450/3/76-028b, Research Tringle Park, NC.

5-505 Uptake, Rb

Species	Value	Condition
Abies grandis	0.89 kg/(acre*100 years)	Used for growth in 100 y[1]

Source: 1.Fortescue, J.A.C. and G.G. Marten. 1970. in: Richle, D.E. (ed.); Analysis of Temperate Forest Ecosystems. pp 173-198. Springer-Verlag. Berlin.

5-506 Uptake SO_2 from atmosphere

Species	Value	Condition
Alfalfa	1.27E6 kg/(km**2*year)	SO2 from atmosphere[4]
Alfalfa sp.	17.0 liter/(min*m**2)	Equivalent deposition velocity = 2.83 cm/sec, 1 pphm[5]
Azalea	0.044 mg SO2/(dm**2 leaf*hour)	1.0 ppm SO2, foliar sorption, 1 h, 300 K, 51 % RH, 1300 ftc[1]
Azalea	0.072 mg SO2/(g leaf*hour)	1.0 ppm SO2, foliar sorption, 1 h, 300 K, 51 % RH, 1300 ftc[1]
Coliseum maple	0.502 mg O3/(dm**2 leaf*hour)	0.20 ppm O3, few hours, 299 K, 45 % RH, 2100 ftc, foliar sorption[2]
Coliseum maple	0.991 mg O3/(g leaf*hour)	0.20 ppm O3, few hours, 299 K, 45 % RH, 2100 ftc, foliar sorption[2]
Firethorn	0.072 mg SO2/(dm**2 leaf*hour)	1.0 ppm SO2, foliar sorption, 1 h, 300 K, 51 % RH, 1300 ftc[1]
Firethorn	0.213 mg SO2/(g leaf*hour)	1.0 ppm SO2, foliar sorption, 1 h, 300 K, 51 % RH, 1300 ftc[1]
Model forest hectare	748 t/(ha*y)	Pollutant removal flux[3]
Ohio buckeye	0.362 mg O3/(dm**2 leaf*hour)	0.20 ppm O3, few hours, 299 K, 45 % RH, 2100 ftc, foliar sorption[2]
Ohio buckeye	0.927 mg O3/(g leaf*hour)	0.20 ppm O3, few hours, 299 K, 45 % RH, 2100 ftc, foliar sorption[2]

Privet	0.068 mg SO2/(dm**2 leaf*hour)	1.0 ppm SO2, foliar sorption, 1 h, 300 K, 51 % RH, 1300 ftc[1]
Privet	0.134 mg SO2/(g leaf*hour)	1.0 ppm SO2, foliar sorption, 1 h, 300 K, 51 % RH, 1300 ftc[1]
Red maple	0.088 mg SO2/(dm**2 leaf*hour)	1.0 ppm SO2, foliar sorption, 1 h, 300 K, 51 % RH, 1300 ftc[1]
Red maple	0.260 mg SO2/(g leaf*hour)	1.0 ppm SO2, foliar sorption, 1 h, 300 K, 51 % RH, 1300 ftc[1]
Red maple	0.272 mg O3/(dm**2 leaf*hour)	0.20 ppm O3, few hours, 299 K, 45 % RH, 2100 ftc, foliar sorption[2]
Red maple	0.555 mg O3/(g leaf*hour)	0.20 ppm O3, few hours, 299 K, 45 % RH, 2100 ftc, foliar sorption[2]
Redvein maple	0.285 mg O3/(dm**2 leaf*hour)	0.20 ppm O3, few hours, 299 K, 45 % RH, 2100 ftc, foliar sorption[2]
Redvein maple	0.911 mg O3/(g leaf*hour)	0.20 ppm O3, few hours, 299 K, 45 % RH, 2100 ftc, foliar sorption[2]
Rhododendron	0.056 mg SO2/(dm**2 leaf*hour)	1.0 ppm SO2, foliar sorption, 1 h, 300 K, 51 % RH, 1300 ftc[1]
Rhododendron	0.079 mg SO2/(g leaf*hour)	1.0 ppm SO2, foliar sorption, 1 h, 300 K, 51 % RH, 1300 ftc[1]
Soil in model forest	4.1E4 ug(m**2 vegetative surface*hour)	Gaseous pollutant flux rate for dry soil[3]
Soil in model forest	7.7E6 ug(m**2 soil surface*hour)	Gaseous pollutant flux rate for dry soil[3]
Sugar maple	0.371 mg O3/(dm**2 leaf*hour)	0.20 ppm O3, few hours, 299 K, 45 % RH, 2100 ftc, foliar sorption[2]
Sugar maple	0.863 mg O3/(g leaf*hour)	0.20 ppm O3, few hours, 299 K, 45 % RH, 2100 ftc, foliar sorption[2]
Sweetgum	0.074 mg SO2/(dm**2 leaf*hour)	1.0 ppm SO2, foliar sorption, 1 h, 300 K, 51 % RH, 1300 ftc[1]
Sweetgum	0.267 mg SO2/(g leaf*hour)	1.0 ppm SO2, foliar sorption, 1 h, 300 K, 51 % RH, 1300 ftc[1]
Sweetgum	0.278 mg O3/(dm**2 leaf*hour)	0.20 ppm O3, few hours, 299 K, 45 % RH, 2100 ftc, foliar sorption[2]
Sweetgum	0.854 mg O3/(g leaf*hour)	0.20 ppm O3, few hours, 299 K, 45 % RH, 2100 ftc, foliar sorption[2]
White ash	0.046 mg SO2/(dm**2 leaf*hour)	1.0 ppm SO2, foliar sorption, 1 h, 300 K, 51 % RH, 1300 ftc[1]
White ash	0.118 mg SO2/(g leaf*hour)	1.0 ppm SO2, foliar sorption, 1 h, 300 K, 51 % RH, 1300 ftc[1]
White ash	0.239 mg O3/(dm**2 leaf*hour)	0.20 ppm O3, few hours, 299 K, 45 % RH, 2100 ftc, foliar sorption[2]
White ash	0.562 mg O3/(g leaf*hour)	0.20 ppm O3, few hours, 299 K, 45 % RH, 2100 ftc, foliar sorption[2]
White birch	0.086 mg SO2/(dm**2 leaf*hour)	1.0 ppm SO2, foliar sorption, 1 h, 300 K, 51 % RH, 1300 ftc[1]
White birch	0.268 mg SO2/(g leaf*hour)	1.0 ppm SO2, foliar sorption, 1 h, 300 K, 51 % RH, 1300 ftc[1]
White birch	0.536 mg O3/(dm**2 leaf*hour)	0.20 ppm O3, few hours, 299 K, 45 % RH, 2100 ftc, foliar sorption[2]
White birch	2.347 mg O3/(g leaf*hour)	0.20 ppm O3, few hours, 299 K, 45 % RH, 2100 ftc, foliar sorption[2]
White oak	0.635 mg O3/(dm**2 leaf*hour)	0.20 ppm O3, few hours, 299 K, 45 % RH, 2100 ftc, foliar sorption[2]
White oak	1.318 mg O3/(g leaf*hour)	0.20 ppm O3, few hours, 299 K, 45 % RH, 2100 ftc, foliar sorption[2]

Source: 1.Roberts, B.R. and C.R. Krause. 1976. Hort. Sci. 11, 111-112.
2.Townsend, A.M. 1974. J. Am. Soc. Hort. Sci. 99, 206-208.
3.EPA 1976. Open space as an Air Resource Management Measure. Vol II. EPA-450/3/76-028b, Research Tringle Park, NC.
4.Hill, A.C. 1971. J.Air Pollut. Control, 21, pp 341-346.
5.Hill, A.C. and E.M. Chamberlain. 1974. Atmospheric-Surface Exchange of Particulate and Gaseous Pollutants Symp. Richland. Wash. Sept 4-6, 1974.

5-507 Uptake
Sb

Species	Value	Condition
Mouse	15 %	By recovering in urine and tissue[1]

Source: 1.Waitz, J.A. R.E. Ober, J.E. Meisenhelder and P.E. Thompson. 1965. WHO Bull. 27, pp 38-45.

5-508 Uptake
Se

Species	Value	Condition
Homo sapiens	80 %	Gast, of daily selenium intake[1]
Homo sapiens	93 %	Gast, selenite[2]
Rat	97.5 % of intake	Gast, independent of dose, soluble Se, dosis = 0.02 - 4.02 mg/kg[3]

Source: 1.Stewart, R.D.H., N.M. Griffiths, C.D. Thomson and M.F. Robinson. 1978. Brit. J. Nutr. 40, pp 45-54.
2.Thompson, C.D. 1977. Trace Elements in Human Health and Animal Nutrition, 4th d. Academic Press. NY.
3.Brown, D.G., R.F. Burk, R.J. Seely and K.W. Kwiker. 1972. Int. J. Vitam. Nutr. Res. 42, pp 588-591.

5-509 Uptake
Si

Species	Value	Condition
Abies grandis	24.0 kg/(acre*100 years)	Used for growth in 100 y[1]

Source: 1.Fortescue, J.A.C. and G.G. Marten. 1970. in: Richle, D.E. (ed.); Analysis of Temperate Forest Ecosystems. pp 173-198. Springer-Verlag. Berlin.

5-510 Uptake
Sn

Species	Value	Condition
Abies grandis	0.003 kg/(acre*100 years)	Used for growth in 100 y[1]
Cow	10 %	Gast, triphenyl-Sn Acetate[2]
Homo sapiens	2 %	Gast[3]
Rat	0.1 %	Gast, tetractyl-sn, less than value[2]
Rat	0.6 %	Gast, Sn(IV)[4]
Rat	10 %	Gast, triphenyl-Sn Acetate[2]
Rat	2 %	Gast, tricyclohexyl-Sn chloride, less than value[2]
Rat	2.8 %	Gast, Sn(II)[4]
Rat	4 %	Gast, dibutyl-Sn chloride[2]
Rat	8 %	Gast, monoethyl-Sn chloride[5]

Source: 1.Fortescue, J.A.C. and G.G. Marten. 1970. in: Richle, D.E. (ed.); Analysis of Temperate Forest Ecosystems. pp 173-198. Springer-Verlag. Berlin.
2.Bruggemann, J., K. Barth and K.-H. Niesar. 1964. Zentralbl. Vet. Med. Reihe A. VII, pp 4-19.
3.Calloway, D.H. and J.J. McMullen. 1966. Am. J. Clin. Nutr. 18, pp 1-6.
4.Hiles, R.A. 1974. Toxicol. Appl. Pharmacol. 27, pp 366-379.
5.Bridges, J.W., D.S. Davis and R.T. Williams. 1967. Biochem. J. 105, pp 1261-1267.

5-511 Uptake
Sr

Species	Value	Condition
Abies grandis	6.7 kg/(acre*100 years)	Used for growth in 100 y[1]

Source: 1.Fortescue, J.A.C. and G.G. Marten. 1970. in: Richle, D.E. (ed.); Analysis of Temperate Forest Ecosystems. pp 173-198. Springer-Verlag. Berlin.

5-512 Uptake
Te

Species	Value	Condition
Rat	12.5 %	Rg = 10 - 15 %, gast[1]
Rat	25 %	Rg = 10 - 15 %, gast[2]

Source: 1.Hollins, J.G. 1969. Health Phys. 17, pp 497-505.
2.Slouka, V. and J. Hradil. 1970. Sb. Ved. Pr. VLVDU Hradce Kralove 47, pp 21-43.

5-513 Uptake
Ti

Species	Value	Condition
Abies grandis	1.1 kg/(acre*100 years)	Used for growth in 100 y[1]

Source: 1.Fortescue, J.A.C. and G.G. Marten. 1970. in: Richle, D.E. (ed.); Analysis of Temperate Forest Ecosystems. pp 173-198. Springer-Verlag. Berlin.

5-514 Uptake
Tl

Species	Value	Condition
Rat	100 %	As nitrate, 204-Tl, independent of administration route[1]

Source: 1.Lie, R., R.G. Thomas and J.K. Scott. 1960. Health Phys. 2, pp 334-340.

5-515 Uptake
U

Species	Value	Condition
Homo sapiens	12 %	Steady state, gast, calculated from urine concentration, USA[1]
Homo sapiens	2.7 %	Gast, U nitrate hexahydrate, dose = 11 mg[1]
Homo sapiens	31 %	Steady state, calculated from urine concentration, UK[2]

Source: 1.Welford, G.A. and R. Baird. 1967. Heath Phys. 13, pp 1321-1324.
2.Hamilton, E.I. 1970. Health Phys. 19, pp 511-520.

5-516 Uptake
V

Species	Value	Condition
Abies grandis	0.0002 kg/(acre*100 years)	Used for growth in 100 y[1]
Carrot, peeled	0.91E-5 % on wet weight basis	Uptake in % per gram ww of plant compared with total amount in soil[2]
Carrot, peeled	6.02E-5 % on dry weight basis	Uptake in % per gram dw of plant compared with total amount in soil[2]
Carrot, peeled	6.55E-4 % on ash weight basis	Uptake in % per gram ash w of plant compared with total amount in soil[2]
Dill	1.05E-4 % on ash weight basis	Uptake in % per gram ash w of plant compared with total amount in soil[2]
Dill	2.98E-6 % on wet weight basis	Uptake in % per gram ww of plant compared with total amount in soil[2]
Dill	3.16E-5 % on dry weight basis	Uptake in % per gram dw of plant compared with total amount in soil[2]

Homo sapiens	2 %	Gast, soluble compound[3]
Homo sapiens	25 %	Inhalation[3]
Lettuce	0.211E-4 % on ash weight basis	Uptake in % per gram ash w of plant compared with total amount in soil[2]
Lettuce	0.68E-6 % on wet weight basis	Uptake in % per gram ww of plant compared with total amount in soil[2]
Lettuce	1.01E-5 % on dry weight basis	Uptake in % per gram dw of plant compared with total amount in soil[2]
Parsley	1.45E-4 % on wet weight basis	Uptake in % per gram ww of plant compared with total amount in soil[2]
Parsley	1.50E-3 % on dry weight basis	Uptake in % per gram dw of plant compared with total amount in soil[2]
Parsley	5.66E-3 % on ash weight basis	Uptake in % per gram ash w of plant compared with total amount in soil[2]
Potato, leaf	0.30E-5 % on wet weight basis	Uptake in % per gram ww of plant compared with total amount in soil[2]
Potato, leaf	2.29E-4 % on ash weight basis	Uptake in % per gram ash w of plant compared with total amount in soil[2]
Potato, leaf	4.41E-5 % on dry weight basis	Uptake in % per gram dw of plant compared with total amount in soil[2]
Potato, peeled	0.34E-4 % on wet weight basis	Uptake in % per gram ww of plant compared with total amount in soil[2]
Potato, peeled	2.27E-4 % on dry weight basis	Uptake in % per gram dw of plant compared with total amount in soil[2]
Potato, peeled	7.87E-4 % on ash weight basis	Uptake in % per gram ash w of plant compared with total amount in soil[2]
Radish, leaf	0.14E-4 % on wet weight basis	Uptake in % per gram ww of plant compared with total amount in soil[2]
Radish, leaf	1.49E-4 % on dry weight basis	Uptake in % per gram dw of plant compared with total amount in soil[2]
Radish, leaf	6.28E-4 % on ash weight basis	Uptake in % per gram ash w of plant compared with total amount in soil[2]
Radish, peeled	0.21E-4 % on wet weight basis	Uptake in % per gram ww of plant compared with total amount in soil[2]
Radish, peeled	1.58E-3 % on ash weight basis	Uptake in % per gram ash w of plant compared with total amount in soil[2]
Radish, peeled	2.53E-4 % on dry weight basis	Uptake in % per gram dw of plant compared with total amount in soil[2]
Spinach	0.595E-4 % on ash weight basis	Uptake in % per gram ash w of plant compared with total amount in soil[2]
Spinach	1.72E-6 % on wet weight basis	Uptake in % per gram ww of plant compared with total amount in soil[2]
Spinach	2.09E-5 % on dry weight basis	Uptake in % per gram dw of plant compared with total amount in soil[2]

Source: 1.Fortescue, J.A.C. and G.G. Marten. 1970. in: Richle, D.E. (ed.); Analysis of Temperate Forest Ecosystems. pp 173-198. Springer-Verlag. Berlin.
2.Soremark, R. 1967. J. Nutr. 92, 183.
3.ICRP 1960. Report of Committiee II on Permissible Dose for Internal Radiation. ICRP Publ 2. PergamonPress. Oxford.

5-517 Uptake
W

Species	Value	Condition
Dog	60 %	Inhalation, W-181 oxide aerosols[1]
Homo sapiens	47 %	From gast to blood, oral dose[2]

Source: 1.Cawse, P.A. 1974. A Survey of Atmospheric Trace Elements in the UK. 1972-73. AERER R 7669. Her Majesys office. London.
2.Ballow, J.E. 1960. Metabolism of W-185 in the Rat. USAEC Document HW-64112.

5-518 Uptake
Zn

Species	Value	Condition
Abies grandis	0.549 kg/(acre*100 years)	Used for growth in 100 y[1]
Alisma plantago aquatica, old plant	16.50 %	800 and 50 m below pollution entry I and II, removed from surface,[2]
Alisma plantago aquatica, old plant	22.10 %	800 m above pollution entry I, removed from surface, F046[2]
Alisma plantago aquatica, old plant	40.18 %	150 m below pollution entry I, removed from surface, F046[2]
Alisma plantago aquatica, young plant	17.69 %	800 m above pollution entry I, removed from surface, F046[2]
Alisma plantago aquatica, young plant	24.20 %	800 and 50 m below pollution entry I and II, removed from surface,[2]
Alisma plantago aquatica, young plant	39.90 %	150 m below pollution entry I, removed from surface, F046[2]
Anguilla japonica	2.4E-6 mg/hour	Through gills, rg = 0.7 - 4.1[3]
Menhaden	36 %	Assimilation efficiency, ingested = 4.6 ug/d, juvenile[4]
Pinfish	19 %	Assimilation efficiency, ingested = 3.5 ug/d, juvenile[4]
Spot	2 %	Assimilation efficiency, ingested = 34 ug/d, juvenile[4]

Source: 1.Fortescue, J.A.C. and G.G. Marten. 1970. in: Richle, D.E. (ed.); Analysis of Temperate Forest Ecosystems. pp 173-198. Springer-Verlag. Berlin.
2.Patrick, F.M. and M.W. Loutit. 1977. Water Res., 11(8), pp 699-703.
3.Hibiya, T. and M. Oguri. 1961. Bull. Jap. Soc. Sci. Fish. 27, pp 996-1000.
4.Cross, F.A. J.N. Willis, L.H. Hardy, N.Y. Jones and J.M. Lewis 1975. in: Cronin, L.E. (ed.); Estuarie Research. Vol 1 pp 45-63. Academic Press. New York.

5-519 Uptake
Zr

Species	Value	Condition
Abies grandis	0.12 kg/(acre*100 years)	Used for growth in 100 y[1]

Source: 1.Fortescue, J.A.C. and G.G. Marten. 1970. in: Richle, D.E. (ed.); Analysis of Temperate Forest Ecosystems. pp 173-198. Springer-Verlag. Berlin.

5-520 Uptake from soil with added sludge

	Species	Value	Condition
Al	Allium cepa	18 mg/kg dry weight	pH = 7.1, 2 y,control = 7 ppm, soil = 39 g/kg, sludge = 42 g/kg, F085[1]
Al	Brassica oleracea var. capitata	24 mg/kg dry weight	pH = 7.1, 2 y, control = 24 ppm, soil = 39 g/kg, sludge = 42 g/kg[1]
Al	Daucus carota var. sativa	35 mg/kg dry weight	pH = 7.1, 2 y, control = 14 ppm, soil = 39 g/kg, sludge = 42 g/kg[1]
Al	Echinochloa crusgalli var. frumentacea	28 mg/kg dry weight	pH = 7.1, 2 y, control missing, soil = 39 g/kg, sludge = 42 g/kg,[1]
Al	Lycopersicon esculentum	15 mg/kg dry weight	pH = 7.1, 2 y, control = 13 ppm, soil = 39 g/kg, sludge = 42 g/kg[1]
Al	Phaseolus vulgaris	19 mg/kg dry weight	pH = 7.1, 2 y, control = 17 ppm, soil = 39 g/kg, sludge = 42 g/kg[1]
Al	Solanum tuberosum	11 mg/kg dry weight	pH = 7.1, 2 y, control = 10 ppm, soil = 39 g/kg, sludge = 42 g/kg[1]
As	Fodder rape	61 (% increase in plant)/(% increase in soil)	Background level = 12.3 ppm, sludge level = 6.6 ppm[1]
B	Allium cepa	24 mg/kg dry weight	pH = 7.1, 2 y, control = 15 ppm, soil = 14 , sludge = 63 ppm, F085[1]
B	Allium cepa	9 mg/kg dry weight	pH = 7.1, 1 y, control = 6 ppm, soil = 14 ppm, sludge = 63 ppm, F085[1]
B	Brassica oleracea var. capitata	25 mg/kg dry weight	pH = 7.1, 1 y, control = 23 ppm, soil = 14 ppm, sludge = 63 ppm, F085[1]
B	Brassica oleracea var. capitata	35 mg/kg dry weight	pH = 7.1, 2 y, control = 35 ppm, soil = 14 , sludge = 63 ppm, F085[1]
B	Daucus carota var. sativa	22 mg/kg dry weight	pH = 7.1, 2 y, control = 1.1 ppm, soil = 14 , sludge = 63 ppm, F085[1]
B	Echinochloa crusgalli var. frumentacea	12 mg/kg dry weight	pH = 7.1, 2 y, control = 10 ppm, soil = 14 , sludge = 63 ppm, F085[1]
B	Echinochloa crusgalli var. frumentacea	9 mg/kg dry weight	pH = 7.1, 1 y, control = 17 ppm, soil = 14 ppm, sludge = 63 ppm, F085[1]
B	Fodder rape	0.8 (% increase in plant)/(% increase in soil)	Background level = 0.59 ppm, sludge level = 7.3 ppm[8]
B	Lycopersicon esculentum	12 mg/kg dry weight	pH = 7.1, 2 y, control = 9 ppm, soil = 14 , sludge = 63 ppm, F085[1]
B	Phaseolus vulgaris	28 mg/kg dry weight	pH = 7.1, 1 y, control = 16 ppm, soil = 14 ppm, sludge = 63 ppm, F085[1]
B	Phaseolus vulgaris	41 mg/kg dry weight	pH = 7.1, 2 y, control = 18 ppm, soil = 14 , sludge = 63 ppm, F085[1]
B	Solanum tuberosum	10 mg/kg dry weight	pH = 7.1, 2 y, control = 6 ppm, soil = 14 , sludge = 63 ppm, F085[1]
B	Solanum tuberosum	9 mg/kg dry weight	pH = 7.1, 1 y, control = 4 ppm, soil = 14 ppm, sludge = 63 ppm, F085[1]
Br	Allium cepa	11.7 mg/kg dry weight	pH = 7.1, 1 y, control = 5.0 ppm, soil = 3.5, sludge = 10 ppm, F085[1]
Br	Allium cepa	3.0 mg/kg dry weight	pH = 7.1, 2 y, control = 2.9 ppm, soil = 3.5 ppm, sludge = 10 ppm,[1]
Br	Allium cepa	5.4 mg/kg dry weight	pH = 7.1, 1 y, control = 1.8 ppm, soil = 3.5, sludge = 10 ppm, F085[1]
Br	Brassica oleracea var. capitata	11 mg/kg dry weight	pH = 7.1, 1 y, control = 6.5 ppm, soil = 3.5, sludge = 10 ppm, F085[1]

Br	Brassica oleracea var. capitata	11.1 mg/kg dry weight	pH = 7.1, 2 y, control = 8.4 ppm, soil = 3.5 ppm, sludge = 10 ppm[1]
Br	Daucus carota var. sativa	4.5 mg/kg dry weight	pH = 7.1, 1 y, control = 4.9 ppm, soil = 3.5, sludge = 10 ppm, F085[1]
Br	Daucus carota var. sativa	6.0 mg/kg dry weight	pH = 7.1, 2 y, control = 4.0 ppm, soil = 3.5 ppm, sludge = 10 ppm[1]
Br	Echinochloa crusgalli var. frumentacea	10 mg/kg dry weight	pH = 7.1, 1 y, control = 13 ppm, soil = 3.5, sludge = 10 ppm, F085[1]
Br	Echinochloa crusgalli var. frumentacea	8.2 mg/kg dry weight	pH = 7.1, 2 y, control = 8.8 ppm, soil = 3.5 ppm, sludge = 10 ppm[1]
Br	Lycopersicon esculentum	5.6 mg/kg dry weight	pH = 7.1, 2 y, control = 3.7 ppm, soil = 3.5 ppm, sludge = 10 ppm[1]
Br	Phaseolus vulgaris	3.5 mg/kg dry weight	pH = 7.1, 2 y, control = 5.1 ppm, soil = 3.5 ppm, sludge = 10 ppm[1]
Br	Phaseolus vulgaris	6.2 mg/kg dry weight	pH = 7.1, 1 y, control = 5.5 ppm, soil = 3.5, sludge = 10 ppm, F085[1]
Br	Solanum tuberosum	5.2 mg/kg dry weight	pH = 7.1, 2 y, control = 3.3 ppm, soil = 3.5 ppm, sludge = 10 ppm[1]
Br	Solanum tuberosum	6.1 mg/kg dry weight	pH = 7.1, 1 y, control = 6.2 ppm, soil = 3.5, sludge = 10 ppm, F085[1]
Ca	Allium cepa	13200 mg/kg dry weight	pH = 7.1, 1 y, control = 4210 ppm, soil = 3.4 g/kg, sl = 71 g/kg[1]
Ca	Allium cepa	6360 mg/kg dry weight	pH = 7.1, 2 y, control = 1920 ppm, soil = 3.4 g/kg, sl = 71 g/kg[1]
Ca	Brassica oleracea var. capitata	19500 mg/kg dry weight	pH = 7.1, 2 y, control = 7700 ppm, soil = 3.4 g/kg, sl = 71 g/kg[1]
Ca	Brassica oleracea var. capitata	7630 mg/kg dry weight	pH = 7.1, 1 y, control = 4500 ppm, soil = 3.4 g/kg, sl = 71 g/kg[1]
Ca	Daucus carota var. sativa	2430 mg/kg dry weight	pH = 7.1, 1 y, control = 1650 ppm, soil = 3.4 g/kg, sl = 71 g/kg[1]
Ca	Daucus carota var. sativa	2710 mg/kg dry weight	pH = 7.1, 2 y, control = 2060 ppm, soil = 3.4 g/kg, sludge = 71 g/kg[1]
Ca	Echinochloa crusgalli var. frumentacea	2650 mg/kg dry weight	pH = 7.1, 1 y, control = 3700 ppm, soil = 3.4 g/kg, sludge = 71 g/kg[1]
Ca	Echinochloa crusgalli var. frumentacea	3950 mg/kg dry weight	pH = 7.1, 2 y, control = 3780 ppm, soil = 3.4 g/kg, sludge = 71 g/kg[1]
Ca	Lycopersicon esculentum	1780 mg/kg dry weight	pH = 7.1, 2 y, control = 786 ppm, soil = 3.4 g/kg, sludge = 71 g/kg[1]
Ca	Lycopersicon esculentum	600 mg/kg dry weight	pH = 7.1, 1 y, control = 780 ppm, soil = 3.4 g/kg, sludge = 71 g/kg[1]
Ca	Phaseolus vulgaris	6000 mg/kg dry weight	pH = 7.1, 2 y, control = 4610 ppm, soil = 3.4 g/kg, sludge = 71 g/kg[1]
Ca	Phaseolus vulgaris	7050 mg/kg dry weight	pH = 7.1, 1 y, control = 2670 ppm, soil = 3.4 g/kg, sludge = 71 g/kg[1]
Ca	Solanum tuberosum	342 mg/kg dry weight	pH = 7.1, 2 y, control = 83 ppm, soil = 3.4 g/kg, sludge = 71 g/kg[1]
Ca	Solanum tuberosum	426 mg/kg dry weight	pH = 7.1, 1 y, control = 237 ppm, soil = 3.4 g/kg, sludge = 71 g/kg[1]
Cd	Allium cepa	0.9 mg/kg dry weight	pH = 7.1, 2 y, control = 0.3 ppm, soil = 0.1 ppm, sludge = 44 ppm[1]
Cd	Allium cepa	2.2 mg/kg dry weight	pH = 7.1, 1 y, control = 0.6 ppm, soil = 0.1 ppm, sludge = 44 ppm[1]
Cd	Brassica oleracea var. capitata	0.7 mg/kg dry weight	pH = 7.1, 1 y, control = 0.2 ppm, soil = 0.1 ppm, sludge = 44 ppm[1]

Cd	Brassica oleracea var. capitata	0.9 mg/kg dry weight	pH = 7.1, 2 y, control = 0.3 ppm, soil = 0.1 ppm, sludge = 44 ppm,[1]
Cd	Corn plant, grain	0.08 mg/kg dry weight	Cd added = 0.2 kg Cd/ha as sludge, control = 0.09 ppm[2]
Cd	Corn plant, grain	0.09 mg/kg dry weight	Cd added = 1.07 kg Cd/ha as sludge, control = 0.09 ppm[2]
Cd	Corn plant, grain	0.09 mg/kg dry weight	Cd added = 2.15 kg Cd/ha as sludge, control = 0.09 ppm[2]
Cd	Corn plant, grain	0.10 mg/kg dry weight	Cd added = 0.53 kg Cd/ha as sludge, control = 0.09 ppm[2]
Cd	Corn plant, grain	0.10 mg/kg dry weight	Cd added = 4.30 kg Cd/ha as sludge, control = 0.09 ppm[2]
Cd	Corn plant, grain	0.21 mg/kg dry weight	Cd added = 12.9 kg Cd/ha as sludge, control = 0.10 ppm[3]
Cd	Corn plant, grain	0.57 mg/kg dry weight	Cd added = 25.7 kg Cd/ha as sludge, control = 0.10 ppm[3]
Cd	Corn plant, grain	0.96 mg/kg dry weight	Cd added = 51.4 kg Cd/ha as sludge, control = 0.10 ppm[3]
Cd	Corn plant, leaf	0.12 mg/kg dry weight	Cd added = 1.3 kg Cd/ha as sludge, control = 0.04 ppm[4]
Cd	Corn plant, leaf	0.19 mg/kg dry weight	Cd added = 2.6 kg Cd/ha as sludge, control = 0.04 ppm[4]
Cd	Corn plant, leaf	0.25 mg/kg dry weight	Cd added = 5.1 kg Cd/ha as sludge, control = 0.04 ppm[4]
Cd	Corn plant, leaf	0.45 mg/kg dry weight	Cd added = 2.2 kg Cd/ha as sludge, control = 0.05 ppm[5]
Cd	Corn plant, leaf	1.07 mg/kg dry weight	Cd added = 4.5 kg Cd/ha as sludge, control = 0.05 ppm[5]
Cd	Corn plants, leaf	0.08 mg/kg dry weight	Cd = 0.2 kg Cd/ha, cont = 0.04 ppm[6]
Cd	Corn plants, leaf	0.09 mg/kg dry weight	Cd = 1.07kg Cd/ha, cont = 0.04 ppm[6]
Cd	Corn plants, leaf	0.09 mg/kg dry weight	Cd = 2.15kg Cd/ha, cont = 0.04 ppm[6]
Cd	Corn plants, leaf	0.10 mg/kg dry weight	Cd = 0.53kg Cd/ha, cont = 0.04 ppm[6]
Cd	Corn plants, leaf	0.10 mg/kg dry weight	Cd = 4.30kg Cd/ha, cont = 0.04 ppm[6]
Cd	Corn plants, leaf	0.12 mg/kg dry weight	Cd = 1.3 kg Cd/ha, cont = 0.04 ppm[6]
Cd	Corn plants, leaf	0.19 mg/kg dry weight	Cd = 2.6 kg Cd/ha, cont = 0.04 ppm[6]
Cd	Corn plants, leaf	0.21 mg/kg dry weight	Cd = 12.9kg Cd/ha, cont = 0.04 ppm[6]
Cd	Corn plants, leaf	0.25 mg/kg dry weight	Cd = 5.1 kg Cd/ha, cont = 0.04 ppm[6]
Cd	Corn plants, leaf	0.45 mg/kg dry weight	Cd = 2.2 kg Cd/ha, cont = 0.05 ppm[7]
Cd	Corn plants, leaf	0.57 mg/kg dry weight	Cd = 25.7kg Cd/ha, cont = 0.04 ppm[6]
Cd	Corn plants, leaf	0.96 mg/kg dry weight	Cd = 51.4kg Cd/ha, cont = 0.04 ppm[6]
Cd	Corn plants, leaf	1.07 mg/kg dry weight	Cd = 4.5 kg Cd/ha, cont = 0.05 ppm[7]
Cd	Daucus carota var. sativa	0.5 mg/kg dry weight	pH = 7.1, 1 y, control = 1.1 ppm, soil = 0.1 ppm, sludge = 44 ppm[1]
Cd	Daucus carota var. sativa	0.8 mg/kg dry weight	pH = 7.1, 2 y, control = 1.9 ppm, soil = 0.1 ppm, sludge = 44 ppm[1]
Cd	Echinochloa crusgalli var. frumentacea	0.4 mg/kg dry weight	pH = 7.1, 1 y, control = 0.2 ppm, soil = 0.1 ppm, sludge = 44 ppm[1]
Cd	Echinochloa crusgalli var. frumentacea	0.4 mg/kg dry weight	pH = 7.1, 2 y, control = 0.2 ppm, soil = 0.1 ppm, sludge = 44 ppm[1]
Cd	Fodder rape	0.0 (% increase in plant)/(% increase in soil)	Background level = 1.2 ppm, sludge level = 11.0 ppm[8]
Cd	Fodder rape	10.9 ug/pot	Added Cd in sewage sludge = 2291 ug/pot, pH = 7.2[9]
Cd	Fodder rape	1021 ng/g dry weight	Added sewage sludge = 58 t dw/ha, soil-pH = 4.8[9]
Cd	Fodder rape	105 ng/g dry weight	Added sewage sludge = 0 t dw/ha, soil-pH = 7.2[9]
Cd	Fodder rape	11.2 ug/pot	Added Cd in sewage sludge = 0 ug/pot, pH = 4.8[9]
Cd	Fodder rape	110 ng/g dry weight	Added sewage sludge = 6.5 t dw/ha, soil-pH = 7.2[9]

Cd	Fodder rape	121 ng/g dry weight	Added sewage sludge = 0 t dw/ha, soil-pH = 6.0[9]
Cd	Fodder rape	160 ng/g dry weight	Added sewage sludge = 19 t dw/ha, soil-pH = 7.2[9]
Cd	Fodder rape	17.3 ug/pot	Added Cd in sewage sludge = 6874 ug/pot, pH = 4.8[9]
Cd	Fodder rape	17.6 ug/pot	Added Cd in sewage sludge = 2291 ug/pot, pH = 6.0[9]
Cd	Fodder rape	17.7 ug/pot	Added Cd in sewage sludge = 255 ug/pot, pH = 4.8[9]
Cd	Fodder rape	171 ng/g dry weight	Added sewage sludge = 6.5 t dw/ha, soil-pH = 6.0[9]
Cd	Fodder rape	19.2 ug/pot	Added Cd in sewage sludge = 6874 ug/pot, pH = 7.2[9]
Cd	Fodder rape	20.0 ug/pot	Added Cd in sewage sludge = 764 ug/pot, pH = 4.8[9]
Cd	Fodder rape	220 weight	Added sewage sludge = 19 t dw/ha, soil-pH = 6.0[9]
Cd	Fodder rape	23.1 ug/pot	Added Cd in sewage sludge = 2291 ug/pot, pH = 4.8[9]
Cd	Fodder rape	236 ng/g dry weight	Added sewage sludge = 58 t dw/ha, soil-pH = 7.2[9]
Cd	Fodder rape	3.0 ug/pot	Added Cd in sewage sludge = 0 ug/pot, pH = 7.2[9]
Cd	Fodder rape	3.4 ug/pot	Added Cd in sewage sludge = 0 ug/pot, pH = 6.0[9]
Cd	Fodder rape	3.7 ug/pot	Added Cd in sewage sludge = 255 ug/pot, pH = 7.2[9]
Cd	Fodder rape	30.4 ug/pot	Added Cd in sewage sludge = 6874 ug/pot, pH = 6.0[9]
Cd	Fodder rape	384 ng/g dry weight	Added sewage sludge = 175 t dw/ha, soil-pH = 7.2[9]
Cd	Fodder rape	412 ng/g dry weight	Added sewage sludge = 58 t dw/ha, soil-pH = 6.0[9]
Cd	Fodder rape	5.5 ug/pot	Added Cd in sewage sludge = 255 ug/pot, pH = 6.0[9]
Cd	Fodder rape	5.8 ug/pot	Added Cd in sewage sludge = 764 ug/pot, pH = 7.2[9]
Cd	Fodder rape	604 ng/g dry weight	Added sewage sludge = 175 t dw/ha, soil-pH = 6.0[9]
Cd	Fodder rape	637 ng/g dry weight	Added sewage sludge = 0 t dw/ha, soil-pH = 4.8[9]
Cd	Fodder rape	7.9 ug/pot	Added Cd in sewage sludge = 764 ug/pot, pH = 6.0[9]
Cd	Fodder rape	719 ng/g dry weight	Added sewage sludge = 175 t dw/ha, soil-pH = 4.8[9]
Cd	Fodder rape	845 ng/g dry weight	Added sewage sludge = 6.5 t dw/ha, soil-pH = 4.8[9]
Cd	Fodder rape	906 ng/g dry weight	Added sewage sludge = 19 t dw/ha, soil-pH = 4.8[9]
Cd	Glycine max, leaf and petiole	0.06 mg/kg dry weight	Total Cd in soil = 0.66-0.1 mg/kg dw, slugde = 0.0 mm, less than[10]
Cd	Glycine max, leaf and petiole	0.19 mg/kg dry weight	Total Cd in soil = 0.51-0.16 mg/kg dw, slugde = 0.0 mm[10]
Cd	Glycine max, leaf and petiole	0.22 mg/kg dry weight	Total Cd in soil < 0.25-0.2 mg/kg dw, slugde = 0.0 mm[10]
Cd	Glycine max, leaf and petiole	0.25 mg/kg dry weight	Total Cd in soil < 0.25-0.15 mg/kg dw, slugde = 0.0 mm[10]
Cd	Glycine max, leaf and petiole	0.27 mg/kg dry weight	Total Cd in soil = 0.57-0.19 mg/kg dw, slugde = 0.0 mm[10]
Cd	Glycine max, leaf and petiole	0.63 mg/kg dry weight	Total Cd in soil = 2.67-2.94 mg/kg dw, slugde = 6.4 mm[10]
Cd	Glycine max, leaf and petiole	0.64 mg/kg dry weight	Total Cd in soil = 4.56-4.43 mg/kg dw, slugde = 6.4 mm[10]
Cd	Glycine max, leaf and petiole	0.85 mg/kg dry weight	Total Cd in soil = 4.91-3.86 mg/kg dw, slugde = 6.4 mm[10]
Cd	Glycine max, leaf and petiole	0.89 mg/kg dry weight	Total Cd in soil = 3.96-4.03 mg/kg dw, slugde = 6.4 mm[10]
Cd	Glycine max, leaf and petiole	1.22 mg/kg dry weight	Total Cd in soil = 9.3-8.74 mg/kg dw, slugde = 12.7 mm[10]

Cd	Glycine max, leaf and petiole	1.76 mg/kg dry weight	Total Cd in soil = 10.10-8.51 mg/kg dw, slugde = 12.7 mm[10]
Cd	Glycine max, leaf and petiole	1.87 mg/kg dry weight	Total Cd in soil = 4.37-3.63 mg/kg dw, slugde = 6.4 mm[10]
Cd	Glycine max, leaf and petiole	13.6 mg/kg dry weight	Total Cd in soil = 18.13-15.27 mg/kg dw, slugde = 25.4 mm[10]
Cd	Glycine max, leaf and petiole	2.46 mg/kg dry weight	Total Cd in soil = 14.56-14.13 mg/kg dw,slugde = 25.4 mm[10]
Cd	Glycine max, leaf and petiole	2.73 mg/kg dry weight	Total Cd in soil = 7.57-8.14 mg/kg dw, slugde = 12.7 mm[10]
Cd	Glycine max, leaf and petiole	2.89 mg/kg dry weight	Total Cd in soil = 6.94-8.11 mg/kg dw, slugde = 12.7 mm[10]
Cd	Glycine max, leaf and petiole	3.85 mg/kg dry weight	Total Cd in soil = 16.7-13.66 mg/kg dw, slugde = 25.4 mm[10]
Cd	Glycine max, leaf and petiole	4.19 mg/kg dry weight	Total Cd in soil = 11.11-9.63 mg/kg dw, slugde = 12.7 mm[10]
Cd	Glycine max, leaf and petiole	5.78 mg/kg dry weight	Total Cd in soil = 16.77-14.69 mg/kg dw, slugde = 25.4 mm[10]
Cd	Glycine max, leaf and petiole	6.28 mg/kg dry weight	Total Cd in soil = 11.35-13.33 mg/kg dw, slugde = 25.4 mm[10]
Cd	Glycine max, seed	0.6 mg/kg dry weight	Ediable part, soil = 0.0 ug Cd/g, pH = 7.7, Zn = 47 ppm[11]
Cd	Glycine max, seed total	0.027 mg/kg dry weight	Total Cd in soil = 0.66-0.1 mg/kg, slugde = 0.0 mm[10]
Cd	Glycine max, seed total	0.13 mg/kg dry weight	Total Cd in soil < 0.25-0.15 mg/kg, slugde = 0.0 mm[10]
Cd	Glycine max, seed total	0.18 mg/kg dry weight	Total Cd in soil = 0.51-0.16 mg/kg, slugde = 0.0 mm[10]
Cd	Glycine max, seed total	0.23 mg/kg dry weight	Total Cd in soil = 4.56-4.43 mg/kg, slugde = 6.4 mm[10]
Cd	Glycine max, seed total	0.29 mg/kg dry weight	Total Cd in soil = 0.57-0.19 mg/kg, slugde = 0.0 mm[10]
Cd	Glycine max, seed total	0.31 mg/kg dry weight	Total Cd in soil < 0.25-0.2 mg/kg, slugde = 0.0 mm[10]
Cd	Glycine max, seed total	0.31 mg/kg dry weight	Total Cd in soil = 3.96-4.03 mg/kg, slugde = 6.4 mm[10]
Cd	Glycine max, seed total	0.31 mg/kg dry weight	Total Cd in soil = 4.91-3.86 mg/kg, slugde = 6.4 mm[10]
Cd	Glycine max, seed total	0.35 mg/kg dry weight	Total Cd in soil = 14.56-14.13 mg/kg,slugde = 25.4m[10]
Cd	Glycine max, seed total	0.35 mg/kg dry weight	Total Cd in soil = 9.3-8.74 mg/kg, slugde = 12.7 mm[10]
Cd	Glycine max, seed total	0.41 mg/kg dry weight	Total Cd in soil = 2.67-2.94 mg/kg, slugde = 6.4 mm[10]
Cd	Glycine max, seed total	0.49 mg/kg dry weight	Total Cd in soil = 4.37-3.63 mg/kg, slugde = 6.4 mm[10]
Cd	Glycine max, seed total	0.55 mg/kg dry weight	Total Cd in soil = 10.10-8.51 mg/kg, slugde = 12.7m[10]
Cd	Glycine max, seed total	0.57 mg/kg dry weight	Total Cd in soil = 7.57-8.14 mg/kg, slugde = 12.7 mm[10]
Cd	Glycine max, seed total	0.75 mg/kg dry weight	Total Cd in soil = 16.7-13.66 mg/kg, slugde = 25.4m[10]
Cd	Glycine max, seed total	0.90 mg/kg dry weight	Total Cd in soil = 6.94-8.11 mg/kg, slugde = 12.7 mm[10]
Cd	Glycine max, seed total	0.92 mg/kg dry weight	Total Cd in soil = 16.77-14.69 mg/kg, slugde = 25.4[10]
Cd	Glycine max, seed total	0.94 mg/kg dry weight	Total Cd in soil = 11.11-9.63 mg/kg, slugde = 12.7m[10]
Cd	Glycine max, seed total	1.55 mg/kg dry weight	Total Cd in soil = 11.35-13.33 mg/kg, slugde = 25.4[10]
Cd	Glycine max, seed total	3.0 mg/kg dry weight	Total Cd in soil = 18.13-15.27 mg/kg, slugde = 25.4[10]
Cd	Goat, intestine	2 % of dose	Oral dose of Cd-109[12]
Cd	Lycopersicon esculentum	0.3 mg/kg dry weight	pH = 7.1, 1 y, control = 0.1 ppm, soil = 0.1 ppm, sludge = 44 ppm[1]
Cd	Lycopersicon esculentum	0.3 mg/kg dry weight	pH = 7.1, 2 y, control = 0.1 ppm, soil = 0.1 ppm, sludge = 44 pp-m[1]

Cd	Phaseolus vulgaris	0.1 mg/kg dry weight	pH = 7.1, 1 y, control = 0.1 ppm, soil = 0.1 ppm, sludge = 44 ppm[1]
Cd	Phaseolus vulgaris	0.1 mg/kg dry weight	pH = 7.1, 2 y, control = 0.1 ppm, soil = 0.1 ppm, sludge = 44 ppm[1]
Cd	Solanum tuberosum	0.3 mg/kg dry weight	pH = 7.1, 2 y, control = 0.3 ppm, soil = 0.1 ppm, sludge = 44 ppm[1]
Cd	Solanum tuberosum	0.5 mg/kg dry weight	pH = 7.1, 1 y, control = 0.3 ppm, soil = 0.1 ppm, sludge = 44 ppm[1]
Cd	Wheat	0.029 mg/kg dry weight	Sludge = 0 t/ha, added Cd in soil = 0 ug/g, pH = 7.4[12]
Cd	Wheat	0.033 mg/kg dry weight	Sludge = 6.5 t/ha, added Cd in soil = 0.031 ug/g, pH = 7.2[12]
Cd	Wheat	0.045 mg/kg dry weight	Sludge = 0 t/ha, added Cd in soil = 0 ug/g, pH = 6.1[12]
Cd	Wheat	0.050 mg/kg dry weight	Sludge = 19 t/ha, added Cd in soil = 0.094 ug/g, pH = 7.2[12]
Cd	Wheat	0.067 mg/kg dry weight	Sludge = 0 t/ha, added Cd in soil = 0 ug/g, pH = 4.8[12]
Cd	Wheat	0.086 mg/kg dry weight	Sludge = 58 t/ha, added Cd in soil = 0.28 ug/g, pH = 6.8[12]
Cd	Wheat	0.086 mg/kg dry weight	Sludge = 6.5 t/ha, added Cd in soil = 0.031 ug/g, pH = 6.1[12]
Cd	Wheat	0.119 mg/kg dry weight	Sludge = 6.5 t/ha, added Cd in soil = 0.031 ug/g, pH = 4.8[12]
Cd	Wheat	0.122 mg/kg dry weight	Sludge = 175 t/ha, added Cd in soil = 0.84 ug/g, pH = 6.5[12]
Cd	Wheat	0.123 mg/kg dry weight	Sludge = 19 t/ha, added Cd in soil = 0.094 ug/g, pH = 6.2[12]
Cd	Wheat	0.124 mg/kg dry weight	Sludge = 175 t/ha, added Cd in soil = 0.84 ug/g, pH = 5.8[12]
Cd	Wheat	0.134 mg/kg dry weight	Sludge = 58 t/ha, added Cd in soil = 0.28 ug/g, pH = 6.2[12]
Cd	Wheat	0.147 mg/kg dry weight	Sludge = 175 t/ha, added Cd in soil = 0.84 ug/g, pH = 5.9[12]
Cd	Wheat	0.170 mg/kg dry weight	Sludge = 19 t/ha, added Cd in soil = 0.094 ug/g, pH = 5.1[12]
Cd	Wheat	0.257 mg/kg dry weight	Sludge = 58 t/ha, added Cd in soil = 0.28 ug/g, pH = 5.3[12]
Cd	Wheat, whole grain	8.27 mg/kg dry weight	Cd added to soil = 0.001 % of Cd in control, pH-dependent[12]
Cl	Allium cepa	1120 mg/kg dry weight	pH = 7.1, 2 y, control = 1260 ppm, soil = 82 ppm, sludge = 5670 ppm[1]
Cl	Allium cepa	4280 mg/kg dry weight	pH = 7.1, 1 y, control = 1650 ppm, soil = 82 ppm, sludge = 5670 ppm[1]
Cl	Brassica oleracea var. capitata	3220 mg/kg dry weight	pH = 7.1, 1 y, control = 3210 ppm, soil = 82 ppm, sludge = 5670 ppm[1]
Cl	Brassica oleracea var. capitata	4400 mg/kg dry weight	pH = 7.1, 2 y, control = 4000 ppm, soil = 82 ppm, sludge = 5670 ppm[1]
Cl	Daucus carota var. sativa	2040 mg/kg dry weight	pH = 7.1, 1 y, control = 3220 ppm, soil = 82 ppm, sludge = 5670 ppm[1]
Cl	Daucus carota var. sativa	2500 mg/kg dry weight	pH = 7.1, 2 y, control = 1880 ppm, soil = 82 ppm, sludge = 5670 ppm[1]
Cl	Echinochloa crusgalli var. frumentacea	5780 mg/kg dry weight	pH = 7.1, 1 y, control = 5260 ppm, soil = 82 ppm, sludge = 5670 ppm[1]

Cl	Echinochloa crusgalli var. frumentacea	6230 mg/kg dry weight	pH = 7.1, 2 y, control = 4100 ppm, soil = 82 ppm, sludge = 5670 ppm[1]
Cl	Lycopersicon esculentum	2470 mg/kg dry weight	pH = 7.1, 2 y, control = 2390 ppm, soil = 82 ppm, sludge = 5670 ppm[1]
Cl	Lycopersicon esculentum	3160 mg/kg dry weight	pH = 7.1, 1 y, control = 3200 ppm, soil = 82 ppm, sludge = 5670 ppm[1]
Cl	Phaseolus vulgaris	1970 mg/kg dry weight	pH = 7.1, 2 y, control = 1980 ppm, soil = 82 ppm, sludge = 5670 ppm[1]
Cl	Phaseolus vulgaris	2460 mg/kg dry weight	pH = 7.1, 1 y, control = 2270 ppm, soil = 82 ppm, sludge = 5670 ppm[1]
Cl	Solanum tuberosum	2620 mg/kg dry weight	pH = 7.1, 2 y, control = 2800 ppm, soil = 82 ppm, sludge = 5670 ppm[1]
Cl	Solanum tuberosum	2750 mg/kg dry weight	pH = 7.1, 1 y, control = 870 ppm, soil = 82 ppm, sludge = 5670 ppm[1]
Co	Fodder rape	5.7 (% increase in plant)/(% increase in soil)	Background level = 14.2 ppm, sludge level = 14.6 ppm[1]
Cr	Allium cepa	0.2 mg/kg dry weight	pH = 7.1, 2 y, control = 0.1 ppm, soil = 17.3 ppm, sludge = 302 ppm[1]
Cr	Allium cepa	5.9 mg/kg dry weight	pH = 7.1, 1 y, control = 1.1 ppm, soil = 17.3 ppm, sludge = 302 ppm[1]
Cr	Brassica oleracea var. capitata	0.2 mg/kg dry weight	pH = 7.1, 2 y, control = 0.2 ppm, soil = 17.3 ppm, sludge = 302 ppm[1]
Cr	Brassica oleracea var. capitata	3.5 mg/kg dry weight	pH = 7.1, 1 y, control = 0.2 ppm, soil = 17.3 ppm, sludge = 302 ppm[1]
Cr	Daucus carota var. sativa	0.4 mg/kg dry weight	pH = 7.1, 2 y, control = 0.1 ppm, soil = 17.3 ppm, sludge = 302 ppm[1]
Cr	Daucus carota var. sativa	0.9 mg/kg dry weight	pH = 7.1, 1 y, control = 2.8 ppm, soil = 17.3 ppm, sludge = 302 ppm[1]
Cr	Echinochloa crusgalli var. frumentacea	0.2 mg/kg dry weight	pH = 7.1, 1 y, control = 0.5 ppm, soil = 17.3 ppm, sludge = 302 ppm[1]
Cr	Echinochloa crusgalli var. frumentacea	0.2 mg/kg dry weight	pH = 7.1, 2 y, control = 0.3 ppm, soil = 17.3 ppm, sludge = 302 ppm[1]
Cr	Fodder rape	0.8 (% increase in plant)/(% increase in soil)	Background level = 36.1 ppm, sludge level = 61.0 ppm[2]
Cr	Lycopersicon esculentum	0.2 mg/kg dry weight	pH = 7.1, 1 y, control = 0.2 ppm, soil = 17.3 ppm, sludge = 302 ppm[1]
Cr	Lycopersicon esculentum	0.3 mg/kg dry weight	pH = 7.1, 2 y, control = 0.1 ppm, soil = 17.3 ppm, sludge = 302 ppm[1]
Cr	Phaseolus vulgaris	0.2 mg/kg dry weight	pH = 7.1, 2 y, control = 0.2 ppm, soil = 17.3 ppm, sludge = 302 ppm[1]
Cr	Phaseolus vulgaris	0.6 mg/kg dry weight	pH = 7.1, 1 y, control = 3.5 ppm, soil = 17.3 ppm, sludge = 302 ppm[1]
Cr	Solanum tuberosum	0.11 mg/kg dry weight	pH = 7.1, 1 y, control = 0.07 ppm, soil = 17.3 ppmsl = 302 ppm, F085[1]
Cr	Solanum tuberosum	0.3 mg/kg dry weight	pH = 7.1, 2 y, control = 0.2 ppm, soil = 17.3 ppm, sludge = 302 ppm[1]
Cu	Allium cepa	1.5 mg/kg dry weight	pH = 7.1, 2 y, control = 1.0 ppm, soil = 47 ppm, sludge = 734 ppm[1]
Cu	Allium cepa	1.7 mg/kg dry weight	pH = 7.1, 1 y, control = 3.4 ppm, soil = 47 ppm, sludge = 734 ppm[1]
Cu	Brassica oleracea var. capitata	2.2 mg/kg dry weight	pH = 7.1, 2 y, control = 0.6 ppm, soil = 47 ppm, sludge = 734 ppm[1]

Cu	Brassica oleracea var. capitata	4.2 mg/kg dry weight	pH = 7.1, 1 y, control = 3.0 ppm, soil = 47 ppm, sludge = 734 ppm[1]
Cu	Daucus carota var. sativa	2.6 mg/kg dry weight	pH = 7.1, 2 y, control = 1.3 ppm, soil = 47 ppm, sludge = 734 ppm[1]
Cu	Echinochloa crusgalli var. frumentacea	2.0 mg/kg dry weight	pH = 7.1, 2 y, control = 1.2 ppm, soil = 47 ppm, sludge = 734 ppm[1]
Cu	Echinochloa crusgalli var. frumentacea	3.9 mg/kg dry weight	pH = 7.1, 1 y, control = 2.4 ppm, soil = 47 ppm, sludge = 734 ppm[1]
Cu	Fodder rape	0.4 (% increase in plant)/(% increase in soil)	Background level = 25.5 ppm, sludge level = 90.5 ppm[2]
Cu	Lycopersicon esculentum	1.7 mg/kg dry weight	pH = 7.1, 2 y, control = 1.6 ppm, soil = 47 ppm, sludge = 734 ppm[1]
Cu	Lycopersicon esculentum	2.5 mg/kg dry weight	pH = 7.1, 1 y, control = 2.2 ppm, soil = 47 ppm, sludge = 734 ppm[1]
Cu	Phaseolus vulgaris	2.2 mg/kg dry weight	pH = 7.1, 2 y, control = 1.8 ppm, soil = 47 ppm, sludge = 734 ppm[1]
Cu	Phaseolus vulgaris	3.4 mg/kg dry weight	pH = 7.1, 1 y, control = 3.2 ppm, soil = 47 ppm, sludge = 734 ppm[1]
Cu	Solanum tuberosum	4.6 mg/kg dry weight	pH = 7.1, 2 y, control = 2.5 ppm, soil = 47 ppm, sludge = 734 ppm[1]
Cu	Solanum tuberosum	5.7 mg/kg dry weight	pH = 7.1, 1 y, control = 3.1 ppm, soil = 47 ppm, sludge = 734 ppm[1]
Fe	Allium cepa	264 mg/kg dry weight	pH = 7.1, 1 y, control = 100 ppm, soil = 25 g/kg, sludge = 61 g/kg[1]
Fe	Allium cepa	96 mg/kg dry weight	pH = 7.1, 2 y, control = 40 ppm, soil = 25 g/kg, sludge = 61 g/kg[1]
Fe	Brassica oleracea var. capitata	177 mg/kg dry weight	pH = 7.1, 1 y, control = 78 ppm, soil = 25 g/kg, sludge = 61 g/kg[1]
Fe	Brassica oleracea var. capitata	90 mg/kg dry weight	pH = 7.1, 2 y, control = 136 ppm, soil = 25 g/kg, sludge = 61 g/kg[1]
Fe	Daucus carota var. sativa	146 mg/kg dry weight	pH = 7.1, 1 y, control = 143 ppm, soil = 25 g/kg, sludge = 61 g/kg[1]
Fe	Daucus carota var. sativa	34 mg/kg dry weight	pH = 7.1, 2 y, control = 71 ppm, soil = 25 g/kg, sludge = 61 g/kg[1]
Fe	Echinochloa crusgalli var. frumentacea	100 mg/kg dry weight	pH = 7.1, 1 y, control = 49 ppm, soil = 25 g/kg, sludge = 61 g/kg[1]
Fe	Echinochloa crusgalli var. frumentacea	164 mg/kg dry weight	pH = 7.1, 2 y, control = 74 ppm, soil = 25 g/kg, sludge = 61 g/kg[1]
Fe	Lycopersicon esculentum	170 mg/kg dry weight	pH = 7.1, 1 y, control = 173 ppm, soil = 25 g/kg, sludge = 61 g/kg[1]
Fe	Lycopersicon esculentum	85 mg/kg dry weight	pH = 7.1, 2 y, control = 20 ppm, soil = 25 g/kg, sludge = 61 g/kg[1]
Fe	Phaseolus vulgaris	111 mg/kg dry weight	pH = 7.1, 2 y, control = 84 ppm, soil = 25 g/kg, sludge = 61 g/kg[1]
Fe	Phaseolus vulgaris	97 mg/kg dry weight	pH = 7.1, 1 y, control = 108 ppm, soil = 25 g/kg, sludge = 61 g/kg[1]
Fe	Solanum tuberosum	56 mg/kg dry weight	pH = 7.1, 2 y, control = 35 ppm, soil = 25 g/kg, sludge = 61 g/kg[1]
Fe	Solanum tuberosum	70 mg/kg dry weight	pH = 7.1, 1 y, control = 49 ppm, soil = 25 g/kg, sludge = 61 g/kg[1]
Hg	Allium cepa	0.2 mg/kg dry weight	pH = 7.1, 2 y, control = 0.1 ppm, soil = 0.1 ppm, sludge = 11.7 ppm[1]

Hg	Allium cepa	0.5 mg/kg dry weight	pH = 7.1, 1 y, control = 0.3 ppm, soil = 0.1 ppm, sludge = 11.7 ppm[1]
Hg	Brassica oleracea var. capitata	0.2 mg/kg dry weight	pH = 7.1, 2 y, control = 0.3 ppm, soil = 0.1 ppm, sludge = 11.7 ppm[1]
Hg	Brassica oleracea var. capitata	0.4 mg/kg dry weight	pH = 7.1, 1 y, control = 0.1 ppm, soil = 0.1 ppm, sludge = 11.7 ppm[1]
Hg	Daucus carota var. sativa	0.04 mg/kg dry weight	pH = 7.1, 2 y, control = 0.1 ppm, soil = 0.1 ppm, sludge = 11.7 ppm[1]
Hg	Daucus carota var. sativa	0.1 mg/kg dry weight	pH = 7.1, 1 y, control = 0.1 ppm, soil = 0.1 ppm, sludge = 11.7 ppm[1]
Hg	Echinochloa crusgalli var. frumentacea	0.5 mg/kg dry weight	pH = 7.1, 2 y, control = 0.4 ppm, soil = 0.1 ppm, sludge = 11.7 ppm[1]
Hg	Echinochloa crusgalli var. frumentacea	1.0 mg/kg dry weight	pH = 7.1, 1 y, control = 0.7 ppm, soil = 0.1 ppm, sludge = 11.7 ppm[1]
Hg	Fodder rape	0.01 (% increase in plant)/(% increase in soil)	Background level = 0.018 ppm, sludge level = 12.0 ppm[2]
Hg	Lycopersicon esculentum	0.2 mg/kg dry weight	pH = 7.1, 2 y, control = 0.1 ppm, soil = 0.1 ppm, sludge = 11.7 ppm[1]
Hg	Lycopersicon esculentum	0.4 mg/kg dry weight	pH = 7.1, 1 y, control = 0.1 ppm, soil = 0.1 ppm, sludge = 11.7 ppm[1]
Hg	Phaseolus vulgaris	0.1 mg/kg dry weight	pH = 7.1, 1 y, control = 0.2 ppm, soil = 0.1 ppm, sludge = 11.7 ppm[1]
Hg	Phaseolus vulgaris	0.1 mg/kg dry weight	pH = 7.1, 2 y, control = 0.3 ppm, soil = 0.1 ppm, sludge = 11.7 ppm[1]
Hg	Solanum tuberosum	0.1 mg/kg dry weight	pH = 7.1, 1 y, control = 0.1 ppm, soil = 0.1 ppm, sludge = 11.7 ppm[1]
Hg	Solanum tuberosum	0.1 mg/kg dry weight	pH = 7.1, 2 y, control = 0.1 ppm, soil = 0.1 ppm, sludge = 11.7 ppm[1]
I	Allium cepa	1.3 mg/kg dry weight	pH = 7.1, 1 y, control = 0.7 ppm, soil = 1.2 ppm, F085[1]
I	Brassica oleracea var. capitata	2.9 mg/kg dry weight	pH = 7.1, 1 y, control = 0.3 ppm, soil = 1.2 ppm, F085[1]
I	Daucus carota var. sativa	1.0 mg/kg dry weight	pH = 7.1, 1 y, control = 0.9 ppm, soil = 1.2 ppm, F085[1]
I	Echinochloa crusgalli var. frumentacea	11.7 mg/kg dry weight	pH = 7.1, 1 y, control missing, soil = 1.2 ppm, F085[1]
I	Lycopersicon esculentum	1.7 mg/kg dry weight	pH = 7.1, 1 y, control = 0.4 ppm, soil = 1.2 ppm, F085[1]
I	Phaseolus vulgaris	0.7 mg/kg dry weight	pH = 7.1, 1 y, control = 0.7 ppm, soil = 1.2 ppm, F085[1]
I	Solanum tuberosum	0.5 mg/kg dry weight	pH = 7.1, 1 y, control = 0.04 ppm, soil = 1.2 ppm, F085[1]
K	Allium cepa	11000 mg/kg dry weight	pH = 7.1, 2 y, control = 14000 ppm, soil = 14 g/kg, sludge = 2.18 g/[1]
K	Brassica oleracea var. capitata	28400 mg/kg dry weight	pH = 7.1, 2 y, control = 17400 ppm, soil = 14 g/kg, sludge = 2.18 g/[1]
K	Daucus carota var. sativa	19300 mg/kg dry weight	pH = 7.1, 2 y, control = 18400 ppm, soil = 14 g/kg, sludge = 2.18 g/[1]
K	Echinochloa crusgalli var. frumentacea	15500 mg/kg dry weight	pH = 7.1, 2 y, control = 13100 ppm, soil = 14 g/kg, sludge = 2.18 g/[1]
K	Lycopersicon esculentum	21900 mg/kg dry weight	pH = 7.1, 2 y, control = 17000 ppm, soil = 14 g/kg, sludge = 2.18 g/[1]
K	Phaseolus vulgaris	20600 mg/kg dry weight	pH = 7.1, 2 y, control = 24600 ppm, soil = 14 g/kg, sludge = 2.18 g/[1]
K	Solanum tuberosum	31400 mg/kg dry weight	pH = 7.1, 2 y, control = 20000 ppm, soil = 14 g/kg, sludge = 2.18 g/[1]

Mg	Allium cepa	630 mg/kg dry weight	pH = 7.1, 2 y, control = 720 ppm, soil = 7.47 g/kg, sludge = 8.57 g/[1]
Mg	Brassica oleracea var. capitata	1050 mg/kg dry weight	pH = 7.1, 2 y, control = 2850 ppm, soil = 7.47 g/kg, sludge = 8.57 g/[1]
Mg	Daucus carota var. sativa	880 mg/kg dry weight	pH = 7.1, 2 y, control = 1260 ppm, soil = 7.47 g/kg, sludge = 8.57 g/[1]
Mg	Echinochloa crusgalli var. frumentacea	4650 mg/kg dry weight	pH = 7.1, 2 y, control = 2100 ppm, soil = 7.47 g/kg, sludge = 8.57 g/[1]
Mg	Lycopersicon esculentum	960 mg/kg dry weight	pH = 7.1, 2 y, control = 900 ppm, soil = 7.47 g/kg, sludge = 8.57 g/[1]
Mg	Phaseolus vulgaris	1670 mg/kg dry weight	pH = 7.1, 2 y, control = 2360 ppm, soil = 7.47 g/kg, sludge = 8.57 g/[1]
Mg	Solanum tuberosum	870 mg/kg dry weight	pH = 7.1, 2 y, control = 640 ppm, soil = 7.47 g/kg, sludge = 8.57 g/[1]
Mn	Fodder rape	18 (% increase in plant)/(% increase in soil)	Background level = 476 ppm, sludge level = 480 ppm[2]
Mo	Allium cepa	1.8 mg/kg dry weight	pH = 7.1, 1 y, control = 0.7 ppm, sludge = 21 ppm, F085[1]
Mo	Brassica oleracea var. capitata	0.4 mg/kg dry weight	pH = 7.1, 1 y, control = 1.0 ppm, sludge = 21 ppm, F085[1]
Mo	Brassica oleracea var. capitata	4.9 mg/kg dry weight	pH = 7.1, 2 y, control = 0.8 ppm, sludge = 21 ppm, F085[1]
Mo	Daucus carota var. sativa	0.2 mg/kg dry weight	pH = 7.1, 1 y, control = 0.2 ppm, sludge = 21 ppm, F085[1]
Mo	Daucus carota var. sativa	0.6 mg/kg dry weight	pH = 7.1, 2 y, control = 0.2 ppm, sludge = 21 ppm, F085[1]
Mo	Echinochloa crusgalli var. frumentacea	2.3 mg/kg dry weight	pH = 7.1, 1 y, control = 0.3 ppm, sludge = 21 ppm, F085[1]
Mo	Fodder rape	1.6 (% increase in plant)/(% increase in soil)	Background level = 0.53 ppm, sludge level = 7.4 ppm[2]
Mo	Lycopersicon esculentum	0.9 mg/kg dry weight	pH = 7.1, 1 y, control = 0.5 ppm, sludge = 21 ppm, F085[1]
Mo	Phaseolus vulgaris	13 mg/kg dry weight	pH = 7.1, 1 y, control = 1.0 ppm, sludge = 21 ppm, F085[1]
Mo	Phaseolus vulgaris	30 mg/kg dry weight	pH = 7.1, 2 y, control = 2.0 ppm, sludge = 21 ppm, F085[1]
Mo	Solanum tuberosum	0.8 mg/kg dry weight	pH = 7.1, 1 y, control = 0.2 ppm, sludge = 21 ppm, F085[1]
Ni	Allium cepa	1.4 mg/kg dry weight	pH = 7.1, 2 y, control = 0.6 ppm, soil = 17 ppm, sludge = 191 ppm[1]
Ni	Brassica oleracea var. capitata	1.9 mg/kg dry weight	pH = 7.1, 2 y, control = 2.1 ppm, soil = 17 ppm, sludge = 191 ppm[1]
Ni	Brassica oleracea var. capitata	4.8 mg/kg dry weight	pH = 7.1, 1 y, control = 1.9 ppm, soil = 17 ppm, sludge = 191 ppm[1]
Ni	Daucus carota var. sativa	1.1 mg/kg dry weight	pH = 7.1, 2 y, control = 1.2 ppm, soil = 17 ppm, sludge = 191 ppm[1]
Ni	Echinochloa crusgalli var. frumentacea	2.3 mg/kg dry weight	pH = 7.1, 2 y, control = 1.8 ppm, soil = 17 ppm, sludge = 191 ppm[1]
Ni	Echinochloa crusgalli var. frumentacea	3.1 mg/kg dry weight	pH = 7.1, 1 y, control = 1.4 ppm, soil = 17 ppm, sludge = 191 ppm[1]
Ni	Fodder rape	1.6 (% increase in plant)/(% increase in soil)	Background level = 28.2 ppm, sludge level = 343.3 ppm[2]
Ni	Lycopersicon esculentum	0.9 mg/kg dry weight	pH = 7.1, 2 y, control = 0.4 ppm, soil = 17 ppm, sludge = 191 ppm[1]
Ni	Lycopersicon esculentum	1.4 mg/kg dry weight	pH = 7.1, 1 y, control = 0.5 ppm, soil = 17 ppm, sludge = 191 ppm[1]

Ni	Oat	120 mg/kg dry weight	Soil-pH = 5.3, added sludge = 75 t/ha, added Ni = 194 ppm[8]
Ni	Oat	210 mg/kg dry weight	Soil-pH = 5.3, added sludge = 150 t/ha, added Ni = 389 ppm[8]
Ni	Oat	28 mg/kg dry weight	Soil-pH = 6.8, added sludge = 37.5 t/ha, added Ni = 97 ppm[8]
Ni	Oat	4 mg/kg dry weight	Soil-pH = 6.8, no Ni added[8]
Ni	Oat	50 mg/kg dry weight	Soil-pH = 6.8, added sludge = 75 t/ha, added Ni = 194 ppm[8]
Ni	Oat	70 mg/kg dry weight	Soil-pH = 6.8, added sludge = 150 t/ha, added Ni = 389 ppm[8]
Ni	Oat	8 mg/kg dry weight	Soil-pH = 5.3, no Ni added[8]
Ni	Oat	90 mg/kg dry weight	Soil-pH = 5.3, added sludge = 37.5 t/ha, added Ni = 97 ppm[8]
Ni	Phaseolus vulgaris	5.5 mg/kg dry weight	pH = 7.1, 2 y, control = 4.4 ppm, soil = 17 ppm, sludge = 191 ppm[1]
Ni	Phaseolus vulgaris	6.7 mg/kg dry weight	pH = 7.1, 1 y, control = 3.9 ppm, soil = 17 ppm, sludge = 191 ppm[1]
Ni	Solanum tuberosum	0.9 mg/kg dry weight	pH = 7.1, 2 y, control = 1.0 ppm, soil = 17 ppm, sludge = 191 ppm[1]
Ni	Solanum tuberosum	1.1 mg/kg dry weight	pH = 7.1, 1 y, control = 0.6 ppm, soil = 17 ppm, sludge = 191 ppm[1]
Pb	Fodder rape	0.7 (% increase in plant)/(% increase in soil)	Background level = 25.7 ppm, sludge level = 293 ppm[1]
Pb	Top soil concentration	1015 mg/kg dry weight	Sludge = 37.4 t/(ha*y), control = 61.3, long term appl[2]
Pb	Top soil concentration	11 mg/kg dry weight	Sludge = 44 t/(ha*y), control = 6.6, long term appl[2]
Pb	Top soil concentration	17 mg/kg dry weight	Sludge = 88 t/(ha*y), control = 6.6, long term appl[2]
Pb	Top soil concentration	30 mg/kg dry weight	Sludge = 166 t/(ha*y), control = 6.6, long term appl[2]
Pb	Top soil concentration	4.1 mg/kg dry weight	Sludge = 8 t/(ha*y), control = 3.12 t/(ha*y), acetic-acid Pb, long term appl[2]
Pb	Top soil concentration	43.9 mg/kg dry weight	Sludge = 84 t/(ha*y), control = 25.7, long term appl[2]
Pb	Top soil concentration	48 mg/kg dry weight	Sludge = 88 t/(ha*y), control = 44, long term appl[2]
Pb	Top soil concentration	5.0 mg/kg dry weight	Sludge = 66 t/(ha*y), control = 1.2 t/(ha*y), acetic-acid Pb, long term appl[2]
Pb	Top soil concentration	53 mg/kg dry weight	Sludge = 166 t/(ha*y), control = 44, long term appl[2]
Sb	Allium cepa	11.5 mg/kg dry weight	pH = 7.1, 1 y, control = 0.8 ppm, soil = 0.8 ppm, sludge = 2.9 pp-m[1]
Sb	Brassica oleracea var. capitata	0.9 mg/kg dry weight	pH = 7.1, 1 y, control = 1.4 ppm, soil = 0.8 ppm, sludge = 2.9 ppm[1]
Sb	Daucus carota var. sativa	0.9 mg/kg dry weight	pH = 7.1, 1 y, control = 0.9 ppm, soil = 0.8 ppm, sludge = 2.9 ppm[1]
Sb	Echinochloa crusgalli var. frumentacea	0.3 mg/kg dry weight	pH = 7.1, 1 y, control = 1.1 ppm, soil = 0.8 ppm, sludge = 2.9 ppm[1]
Sb	Lycopersicon esculentum	0.5 mg/kg dry weight	pH = 7.1, 1 y, control = 2.5 ppm, soil = 0.8 ppm, sludge = 2.9 ppm[1]
Sb	Phaseolus vulgaris	1.0 mg/kg dry weight	pH = 7.1, 1 y, control = 0.4 ppm, soil = 0.8 ppm, sludge = 2.9 ppm[1]
Sb	Solanum tuberosum	0.7 mg/kg dry weight	pH = 7.1, 1 y, control = 0.4 ppm, soil = 0.8 ppm, sludge = 2.9 ppm[1]
Sc	Allium cepa	0.02 mg/kg dry weight	pH = 7.1, 1 y, control = 0.01 ppm, soil = 8.7 ppm, sludge = 3.2 ppm[1]
Sc	Brassica oleracea var. capitata	0.01 mg/kg dry weight	pH = 7.1, 1 y, control = 0.03 ppm, soil = 8.7 ppm, sludge = 3.2 ppm[1]

Sc	Daucus carota var. sativa	0.00 mg/kg dry weight	pH = 7.1, 1 y, control = 0.01 ppm, soil = 8.7 ppm, sludge = 3.2 ppm[1]
Sc	Echinochloa crusgalli var. frumentacea	0.02 mg/kg dry weight	pH = 7.1, 1 y, control = 0.00 ppm, soil = 8.7 ppm, sludge = 3.2 ppm[1]
Sc	Lycopersicon esculentum	0.01 mg/kg dry weight	pH = 7.1, 1 y, control = 0.03 ppm, soil = 8.7 ppm, sludge = 3.2 ppm[1]
Sc	Phaseolus vulgaris	0.02 mg/kg dry weight	pH = 7.1, 1 y, control = 0.01 ppm, soil = 8.7 ppm, sludge = 3.2 ppm[1]
Sc	Solanum tuberosum	0.01 mg/kg dry weight	pH = 7.1, 1 y, control = 0.004 ppm, soil = 8.7 ppm, sludge = 3.2 ppm[1]
Se	Allium cepa	0.00 mg/kg dry weight	pH = 7.1, 1 y, control = 0.00 ppm, soil = 0.3 ppm, sludge = 1.4 ppm[1]
Se	Allium cepa	0.02 mg/kg dry weight	pH = 7.1, 2 y, control = 0.02 ppm, soil = 0.3 ppm, sludge = 1.4 ppm[1]
Se	Brassica oleracea var. capitata	0.03 mg/kg dry weight	pH = 7.1, 1 y, control = 0.01 ppm, soil = 0.3 ppm, sludge = 1.4 ppm[1]
Se	Brassica oleracea var. capitata	0.04 mg/kg dry weight	pH = 7.1, 2 y, control = 0.02 ppm, soil = 0.3 ppm, sludge = 1.4 ppm[1]
Se	Daucus carota var. sativa	0.01 mg/kg dry weight	pH = 7.1, 1 y, control = 0.00 ppm, soil = 0.3 ppm, sludge = 1.4 ppm[1]
Se	Daucus carota var. sativa	0.04 mg/kg dry weight	pH = 7.1, 2 y, control = 0.01 ppm, soil = 0.3 ppm, sludge = 1.4 ppm[1]
Se	Echinochloa crusgalli var. frumentacea	0.02 mg/kg dry weight	pH = 7.1, 1 y, control = 0.02 ppm, soil = 0.3 ppm, sludge = 1.4 ppm[1]
Se	Echinochloa crusgalli var. frumentacea	0.03 mg/kg dry weight	pH = 7.1, 2 y, control = 0.03 ppm, soil = 0.3 ppm, sludge = 1.4 ppm[1]
Se	Fodder rape	0 (% increase in plant)/(% increase in soil)	Background level = 0.238, sludge level7.3 ppm[2]
Se	Lycopersicon esculentum	0.01 mg/kg dry weight	pH = 7.1, 1 y, control = 0.01 ppm, soil = 0.3 ppm, sludge = 1.4 ppm[1]
Se	Lycopersicon esculentum	0.02 mg/kg dry weight	pH = 7.1, 2 y, control = 0.02 ppm, soil = 0.3 ppm, sludge = 1.4 ppm[1]
Se	Phaseolus vulgaris	0.02 mg/kg dry weight	pH = 7.1, 2 y, control = 0.01 ppm, soil = 0.3 ppm, sludge = 1.4 ppm[1]
Se	Phaseolus vulgaris	0.04 mg/kg dry weight	pH = 7.1, 1 y, control = 0.02 ppm, soil = 0.3 ppm, sludge = 1.4 ppm[1]
Se	Solanum tuberosum	0.02 mg/kg dry weight	pH = 7.1, 1 y, control = 0.01 ppm, soil = 0.3 ppm, sludge = 1.4 ppm[1]
Se	Solanum tuberosum	0.03 mg/kg dry weight	pH = 7.1, 2 y, control = 0.02 ppm, soil = 0.3 ppm, sludge = 1.4 ppm[1]
Ta	Allium cepa	0.6 mg/kg dry weight	pH = 7.1, 1 y, control = 0.3 ppm, soil = 0.6 ppm, sludge = 0.6 ppm[1]
Ta	Brassica oleracea var. capitata	0.06 mg/kg dry weight	pH = 7.1, 1 y, control = 0.03 ppm, soil = 0.6 ppm, sludge = 0.6 ppm[1]
Ta	Daucus carota var. sativa	0.06 mg/kg dry weight	pH = 7.1, 1 y, control = 0.07 ppm, soil = 0.6 ppm, sludge = 0.6 ppm[1]
Ta	Echinochloa crusgalli var. frumentacea	0.03 mg/kg dry weight	pH = 7.1, 1 y, control = 0.07 ppm, soil = 0.6 ppm, sludge = 0.6 ppm[1]
Ta	Lycopersicon esculentum	0.01 mg/kg dry weight	pH = 7.1, 1 y, control = 0.01 ppm, soil = 0.6 ppm, sludge = 0.6 ppm[1]
Ta	Solanum tuberosum	0.06 mg/kg dry weight	pH = 7.1, 1 y, control = 0.04 ppm, soil = 0.6 ppm, sludge = 0.6 ppm[1]

Ti	Allium cepa	3.9 mg/kg dry weight	pH = 7.1, 2 y, control = 0.5 ppm, soil = 3292 ppm, sludge = 2270 ppm[1]
Ti	Brassica oleracea var. capitata	6.7 mg/kg dry weight	pH = 7.1, 2 y, control = 3.2 ppm, soil = 3292 ppm, sludge = 2270 ppm[1]
Ti	Daucus carota var. sativa	6.1 mg/kg dry weight	pH = 7.1, 2 y, control missing, soil = 3292 ppm, sludge = 2270 ppm[1]
Ti	Lycopersicon esculentum	3.1 mg/kg dry weight	pH = 7.1, 2 y, control = 2.0 ppm, soil = 3292 ppm, sludge = 2270 ppm[1]
Ti	Phaseolus vulgaris	1.6 mg/kg dry weight	pH = 7.1, 2 y, control = 12 ppm, soil = 3292 ppm, sludge = 2270 ppm[1]
Ti	Solanum tuberosum	2.5 mg/kg dry weight	pH = 7.1, 2 y, control = 3.0 ppm, soil = 3292 ppm, sludge = 2270 ppm[1]
V	Allium cepa	0.4 mg/kg dry weight	pH = 7.1, 1 y, control = 0.2 ppm, soil = 67 ppm, sludge = 29 ppm[1]
V	Brassica oleracea var. capitata	0.3 mg/kg dry weight	pH = 7.1, 1 y, control = 0.4 ppm, soil = 67 ppm, sludge = 29 ppm[1]
V	Daucus carota var. sativa	0.1 mg/kg dry weight	pH = 7.1, 1 y, control = 0.4 ppm, soil = 67 ppm, sludge = 29 ppm[1]
V	Echinochloa crusgalli var. frumentacea	0.2 mg/kg dry weight	pH = 7.1, 1 y, control = 0.04 ppm, soil = 67 ppm, sludge = 29 ppm[1]
V	Lycopersicon esculentum	0.02 mg/kg dry weight	pH = 7.1, 1 y, control = 0.12 ppm, soil = 67 ppm, sludge = 29 ppm[1]
V	Phaseolus vulgaris	0.1 mg/kg dry weight	pH = 7.1, 1 y, control = 0.06 ppm, soil = 67 ppm, sludge = 29 ppm[1]
V	Solanum tuberosum	0.04 mg/kg dry weight	pH = 7.1, 1 y, control = 4.8 ppm, soil = 67 ppm, sludge = 29 ppm[1]
W	Allium cepa	0.4 mg/kg dry weight	pH = 7.1, 1 y, control = 0.2 ppm, soil = 2.1 ppm, sludge = 4 ppm[1]
W	Brassica oleracea var. capitata	1.0 mg/kg dry weight	pH = 7.1, 1 y, control = 0.4 ppm, soil = 2.1 ppm, sludge = 4 ppm[1]
W	Daucus carota var. sativa	0.4 mg/kg dry weight	pH = 7.1, 1 y, control = 0.4 ppm, soil = 2.1 ppm, sludge = 4 ppm[1]
W	Echinochloa crusgalli var. frumentacea	0.3 mg/kg dry weight	pH = 7.1, 1 y, control = 0.3 ppm, soil = 2.1 ppm, sludge = 4 ppm[1]
W	Lycopersicon esculentum	0.5 mg/kg dry weight	pH = 7.1, 1 y, control = 0.4 ppm, soil = 2.1 ppm, sludge = 4 ppm[1]
W	Phaseolus vulgaris	0.5 mg/kg dry weight	pH = 7.1, 1 y, control = 0.2 ppm, soil = 2.1 ppm, sludge = 4 ppm[1]
W	Solanum tuberosum	0.4 mg/kg dry weight	pH = 7.1, 1 y, control = 0.04 ppm, soil = 2.1 ppm, sludge = 4 ppm[1]
Yb	Allium cepa	0.2 mg/kg dry weight	pH = 7.1, 1 y, control = 0.2 ppm, soil = 2.0 ppm, sludge = 0.2 ppm[1]
Yb	Brassica oleracea var. capitata	0.1 mg/kg dry weight	pH = 7.1, 1 y, control = 0.01 ppm, soil = 2.0 ppm, sludge = 0.2 ppm[1]
Yb	Daucus carota var. sativa	0.1 mg/kg dry weight	pH = 7.1, 1 y, control = 0.01 ppm, soil = 2.0 ppm, sludge = 0.2 ppm[1]
Yb	Echinochloa crusgalli var. frumentacea	0.1 mg/kg dry weight	pH = 7.1, 1 y, control = 0.1 ppm, soil = 2.0 ppm, sludge = 0.2 ppm[1]
Yb	Lycopersicon esculentum	0.2 mg/kg dry weight	pH = 7.1, 1 y, control = 0.1 ppm, soil = 2.0 ppm, sludge = 0.2 ppm[1]
Yb	Phaseolus vulgaris	0.1 mg/kg dry weight	pH = 7.1, 1 y, control = 0.04 ppm, soil = 2.0 ppm, sludge = 0.2 ppm[1]
Yb	Solanum tuberosum	0.1 mg/kg dry weight	pH = 7.1, 1 y, control missing, soil = 2.0 ppm, sludge = 0.2 ppm[1]
Zn	Fodder rape	0.8 (% increase in plant)/(% increase in soil)	Background level = 97.9 ppm, sludge level = 368.8 ppm[1]

Zn	Glycine max, leaf and petiole	103 mg/kg wet weight	Total Zn in soil = 136-63 mg/kg ww, sludge = 6.4 mm^2
Zn	Glycine max, leaf and petiole	106 mg/kg wet weight	Total Zn in soil = 232-145 mg/kg ww, sludge = 12.7 mm^2
Zn	Glycine max, leaf and petiole	107 mg/kg wet weight	Total Zn in soil = 132-41 mg/kg ww, sludge = 6.4 mm^2
Zn	Glycine max, leaf and petiole	120 mg/kg wet weight	Total Zn in soil = 248-144 mg/kg ww, sludge = 12.7 mm^2
Zn	Glycine max, leaf and petiole	132 mg/kg wet weight	Total Zn in soil = 203-126 mg/kg ww, sludge = 12.7 mm^2
Zn	Glycine max, leaf and petiole	143 mg/kg wet weight	Total Zn in soil = 370-269 mg/kg ww, sludge = 25.4 mm^2
Zn	Glycine max, leaf and petiole	146 mg/kg wet weight	Total Zn in soil = 143-70 mg/kg ww, sludge = 6.4 mm^2
Zn	Glycine max, leaf and petiole	162 mg/kg wet weight	Total Zn in soil = 239-148 mg/kg ww, sludge = 12.7 mm^2
Zn	Glycine max, leaf and petiole	188 mg/kg wet weight	Total Zn in soil = 356-243 mg/kg ww, sludge = 25.4 mm^2
Zn	Glycine max, leaf and petiole	203 mg/kg wet weight	Total Zn in soil = 336-298 mg/kg ww, sludge = 25.4 mm^2
Zn	Glycine max, leaf and petiole	222 mg/kg wet weight	Total Zn in soil = 265-193 mg/kg ww, sludge = 12.7 mm^2
Zn	Glycine max, leaf and petiole	42 mg/kg wet weight	Total Zn in soil = 80-13 mg/kg ww, sludge = 0.0 mm^2
Zn	Glycine max, leaf and petiole	444 mg/kg wet weight	Total Zn in soil = 390-316 mg/kg ww, sludge = 25.4 mm^2
Zn	Glycine max, leaf and petiole	49 mg/kg wet weight	Total Zn in soil = 53-13 mg/kg ww, sludge = 0.0 mm^2
Zn	Glycine max, leaf and petiole	49 mg/kg wet weight	Total Zn in soil = 69-13 mg/kg ww, sludge = 0.0 mm^2
Zn	Glycine max, leaf and petiole	50 mg/kg wet weight	Total Zn in soil = 74-12 mg/kg ww, sludge = 0.0 mm^2
Zn	Glycine max, leaf and petiole	56 mg/kg wet weight	Total Zn in soil = 71-12 mg/kg ww, sludge = 0.0 mm^2
Zn	Glycine max, leaf and petiole	81 mg/kg wet weight	Total Zn in soil = 175-74 mg/kg ww, sludge = 6.4 mm^2
Zn	Glycine max, leaf and petiole	85 mg/kg wet weight	Total Zn in soil = 178-69 mg/kg ww, sludge = 6.4 mm^2
Zn	Glycine max, total seed	58 mg/kg wet weight	Total Zn in soil = 53-13 mg/kg ww, sludge = 0.0 mm^2
Zn	Glycine max, total seed	72 mg/kg wet weight	Total Zn in soil = 132-41 mg/kg ww, sludge = 6.4 mm^2
Zn	Glycine max, total seed	81 mg/kg wet weight	Total Zn in soil = 239-148 mg/kg ww, sludge = 12.7 mm^2

Source: 1.Furr, A.K., W.C. Kelly, C.A. Bache. W.H. Gutenmann and D.J. Lisk. 1976. Bull. Environ. Contam. Toxicol., 16(6), pp 756-763.
3.Kelling, K.A., D.R. Keeney, L.M. Walsh and J.A. Ryan. 1976. J. Environ. Quality 6, 352-358.
2.Hinesly, T.D., Chaney and P.W. Simson. 1976. Application of Sewage Sludge to Cropland. US EPA. Rep MD 33, p 51.
4.Clapp, C.E., R.H. Dowdy and W.E. Larson. 1976. Application of Sewage Sludge to Croplnd. US EPA. RepCD-3.
5.Baker, D.E., M.C. Amcher and W.T. Doty. 1976. In: Land as a Waste Management Alternative. Proc. 8th Annual Waste Management Conference, Rochester, NY.
6.Clapp, C.E., R.H. Dowdy and W.E. Larson. 1976 . in: Application of Sewage Sludge to Cropland: Appraial of Potential Hazards of the Heavy Metals To Plants and Animals. US EPA. Rep MCD-33.
7.Baker, D.E., M.C. Amacher and W.T. Doty. 1976. in: Land as a Waste Management Alternative. Proc. ofhe 8th Annual Waste Management Conf. Rochester NY.
8.Anderson, A. and K.O. Nilsson. 1972. Ambio, 1(5), pp 176-179.
9.Andersson, A. and K.O. Nilsson. 1974. Ambio, 3(5), pp 198-199.
10.Hinesly, T.D. R.L. Jones , J.J. Tyler and E.L. Ziegler. 1976. Water Pollut. Control., 48(9), pp 21372152.
11.Bingham, F.T., A.L. Page, R.J. Mahler and T.J. Ganje. 1976. J. Environ. Qual., 5(4), pp 501-513.
12.Friberg, L., M. Piscater, G.F. Nordberg and T. Kjellstroem. Cadmium in the Environment 2nd edition, CRC-Press.

Chapter 6

PROCESSES IN THE ENVIRONMENT

Deposition and Emission
Equilibrium
Rate Constant

6-001 Atmosphere
Deposition

Species	Value	Condition
Cd, atmospheric precipitation	0.47 mg/(m2*year)	33 stations, Ontario, solub Cd, rain conc=0.8 ug/l, rg=0.11-1.6, 1975[1]
Cd, atmospheric precipitation	0.51 mg/(m2*year)	33 stations, Ontario, total Cd, rain conc=0.8 ug/l, rg=0.15-1.9, 1975[1]
Cd, bulk precipation	0.58 ug/l	Rg = 0.1-1.0 ug/l, Gottigen, Germany[2]
Cd, bulk precipation	0.6 ug/l	Mt. Moosilauke, fallout rate = 0.88 mg/(m2*y), New Hampshire, USA, 1973[2]
Cd, bulk precipation	0.8 ug/l	Rg = 0.01-50 ug/l, fallout rate = 0.51 mg/(m2*y), Ontario, 1975[2]
Cd, bulk precipation	1.0 ug/l	Lower Great Lakes, fallout rate = 0.55 mg/(m2*y), 1977[2]
Cd, bulk precipation	37 ug/l	Rg = 5-50 ug/l, fallout = 4.94 mg/(m2*y), Heathrow , Engl, land side, 1974[2]
Cd, bulk precipation	6 ug/l	Rg = 1-11 ug/l, fallout rate = 2.47 mg/(m2*y), London SW 7, England[2]
Cd, bulk precipation	6 ug/l	Rg = 1-12 ug/l, fallout rate = 2.47 mg/(m2*y), Hyde Park, London, Engl.[2]
Cd, bulk precipation	7 ug/l	Rg = 2-14 ug/l, fallout = 2.72 mg/(m2*y), Heathrow, Engl. air side, 1974[2]
Cd, dustfall	0.18 ug/l	S coast of Calif., USA, 1977[2]
Cd, dustfall	0.29 ug/l	Richmond, Calif., USA, 1974[2]
Cd, dustfall	0.5 mg/(m2*year)	Midwest USA cities, residential areas, 1971[2]
Cd, dustfall	0.69 mg/(m2*year)	Delaware watersheds, USA, 1972[2]
Cd, dustfall	0.76 ug/l	Conc = 13 mg/kg particulates, Cincinnati, near roadways, 1972, USA[2]
Cd, dustfall	0.8 mg/(m2*year)	Midwest USA cities, Commercial areas, 1971[2]
Cd, dustfall	0.9 mg/(m2*year)	Midwest USA cities, indistrial areas, 1971[2]
Cd, dustfall	1.1 ug/l	SE Missouri, 75 km from smelter, 1972[2]
Cd, dustfall	2.9 mg/(m2*year)	Southwest Manhattan, NY, 1975[2]
Cd, dustfall	52 ug/l	SE Missouri, near smelter, 1972[2]
Cd, dustfall	7.2 mg/(m2*year)	New York Harbor, USA, 1976[2]
Cu, fallout rate, dustfall	0.2 ug Cu/(cm2*year)	Leiston, Suffolk, Engl., less than value[3]
Cu, fallout rate, dustfall	0.4 ug Cu/(cm2*year)	Less than value, Chilton, Engl.[3]
Cu, fallout rate, dustfall	0.64 ug Cu/(cm2*year)	New York, Riverhead, Cu conc = 805 ppm[3]
Cu, fallout rate, dustfall	0.8 ug Cu/(cm2*year)	Shetland Islands, less than value[3]
Cu, fallout rate, dustfall	15.9 ug Cu/(cm2*year)	New York, Bronx, Cu conc = 2943 ppm[3]
Cu, fallout rate, dustfall	20.4 ug Cu/(cm2*year)	New York, Lower Manhattan[3]
Cu, fallout rate, dustfall	3.0 ug Cu/(cm2*year)	New York, Queens, Cu conc = 1087 ppm[3]
Cu, fallout rate, dustfall	5.2 ug Cu/(cm2*year)	Trebanos, Glamorgs., Wales[3]
Cu, fallout rate, dustfall near smelter	0.46 ug Cu/(cm2*year)	85 km from smelter, Missouri, 1974[3]
Cu, fallout rate, dustfall near smelter	2.8 ug Cu/(cm2*year)	0.8 km from smelter, Missouri, 1974[3]
Cu, fallout rate, moss sample	0.36 ug Cu/(cm2*year)	S New England Coast[3]

Cu, fallout rate, precipitation only	0.18 ug Cu/(cm2*year)	Chedron Neb., USA, rain conc = 4.5, snow conc = 4.0 mg Cu/l[3]
Cu, fallout rate, precipitation only	0.48 ug Cu/(cm2*year)	Bishop, Calif., USA[3]
Cu, fallout rate, precipitation only	1.1 ug Cu/(cm2*year)	Mauna Loa, Hawaii[3]
Cu, fallout rate, precipitation only	1.6 ug Cu/(cm2*year)	Tampa, Fl., USA[3]
Cu, fallout rate, precipitation only	6.5 ug Cu/(cm2*year)	OHare Airport, Chicago, USA[3]
Cu, fallout rate, residue bulk	0.54 ug Cu/(cm2*year)	Gottingen, Germany, Cu conc = 490 ppm[3]
Cu, fallout rate, snowfall	0.007 ug Cu/(cm2*year)	Dye-3 Camp, Greenland, before 1900, Cu conc = 0.15 g/kg[3]
Cu, fallout rate, snowfall	0.01 ug Cu/(cm2*year)	Camp century, Greenland, before 1900, Cu conc = 0.290 g/kg[3]
Cu, fallout rate, snowfall	0.02 ug Cu/(cm2*year)	Camp century, Greenland, after 1945, Cu conc = 0.58 g/kg[3]
Cu, fallout rate, snowfall	0.039 ug Cu/(cm2*year)	Dye-3 Camp, Greenland, after 1966, Cu conc = 0.85 g/kg[3]
Cu, fallout rate, soluble bulk	0.52 ug Cu/(cm2*year)	Gottingen, Germany, Cu conc = 6.8 mg/l[3]
Cu, fallout rate, soluble bulk	1.8 ug Cu/(cm2*year)	Average for 32 stations in USA, Cu conc = 21 mg/l, 1967[3]
Cu, fallout rate, total bulk	1.1 ug Cu/(cm2*year)	Gottingen, Germany, Cu conc = 41 mg/l[3]
Cu, fallout rate, total bulk	1.2 ug Cu/(cm2*year)	Chilton, Engl., Cu conc = 20 mg/l[3]
Cu, fallout rate, total bulk	1.2 ug Cu/(cm2*year)	Leiston, Suffolk, Engl., Cu conc = 28 mg/l[3]
Cu, fallout rate, total bulk	2.3 ug Cu/(cm2*year)	Walker Branch Watershed, Cu conc = 14 mg/l[3]
Cu, fallout rate, total bulk	2.6 ug Cu/(cm2*year)	Shetland Islands, Cu conc = 22 mg/l[3]
Cu, fallout rate, total bulk	3.6 ug Cu/(cm2*year)	Central Swansea, Wales, Cu conc = 57 mg/l[3]
Cu, fallout rate, total bulk	4.0 ug Cu/(cm2*year)	Switzerland, n = 7[3]
Cu, fallout rate, total bulk	5.2 ug Cu/(cm2*year)	Trebanos, Glamores, Wales, Cu conc = 44 mg/l[3]
Cu, fallout rate, total bulk	9.8 ug Cu/(cm2*year)	New York City, Cu conc = 67 mg/l[3]
Pb from bulk precipitation	0.9 mg/(m2*year)	High Sierra, Calif, 1973[4]
Pb from bulk precipitation	11.0 mg/(m2*year)	L Erie Basin, summer, 1975, pb conc = 17 ug/l[5]
Pb from bulk precipitation	123 mg/(m2*year)	Sudbury, 1.7 km from smelter, 1975[6]
Pb from bulk precipitation	17.4 mg/(m2*year)	L Ontario Basin, winter, 1975, pb conc = 59 ug/l[5]
Pb from bulk precipitation	18 mg/(m2*year)	Leiston, Suffolk, Engl, 1973[7]
Pb from bulk precipitation	19.4 mg/(m2*year)	L Ontario Basin, 1975, summer, pb conc = 35 ug/l[5]
Pb from bulk precipitation	19.6 mg/(m2*year)	Mt. Moosilauke, New Hampshire, 1973, pb conc = 13.4 ug/l[8]
Pb from bulk precipitation	25 mg/(m2*year)	Cambridgeshire, 1974, pb conc = 25 ug/l[9]
Pb from bulk precipitation	25 mg/(m2*year)	Madison, Wisconsin, 1972, pb conc = 20 ug/l[10]
Pb from bulk precipitation	26 mg/(m2*year)	Wraymires, Lancs, Engl, 1973[7]
Pb from bulk precipitation	34 mg/(m2*year)	Shetland Isl, 1973[7]
Pb from bulk precipitation	35 mg/(m2*year)	Chilton, Berks, Engl, 1973,[7]
Pb from bulk precipitation	350 mg/(m2*year)	New York, 1970[11]
Pb from bulk precipitation	40 ug/(m2*year)	Greenland, 1968[12]
Pb from bulk precipitation	54 mg/(m2*year)	London, 6 sites, 1974, pb conc = 190 ug/l[9]
Pb from bulk precipitation	55 mg/(m2*year)	Styrrups, Notts, Engl, 1973[7]
Pb from bulk precipitation	8.9 mg/(m2*year)	L Erie Basin, winter, 1975, pb conc = 27 ug/l[5]

Pb from bulk precipitation	93 ug/(m2*year)	Eastern Pacific, 1973[13]
Pb from dustfall	1265 mg/(m2*year)	Missouri, 800 m from Pb smelter, 1974[14]
Pb from dustfall	165 mg/(m2*year)	Cincinnati, 100 ft from roadway, 1971, Pb conc = 2942 ppm in part.[15]
Pb from dustfall	24 mg/(m2*year)	La Jolla, Calif, 1969[16]
Pb from dustfall	27.3 mg/(m2*year)	Southern California Bight, 1974[17]
Pb from dustfall	304 mg/(m2*year)	Cincinnati, 25 ft from roadway, 1971, Pb conc = 3743 ppm in part.[15]
Pb from dustfall	350 mg/(m2*year)	San Diego, Calif, 1969[16]
Pb from dustfall	98 mg/(m2*year)	Missouri, 85 km from Pb smelter, 1974[14]
Pb from rainfall	10 mg/(m2*year)	WesTerj Michigaj, 1966, Pb cojc = 18*9 Ug/l[18]
Pb from rainfall	10 mg/(m2*year)	Western Michigan, 1966, pb conc = 18.9 ug/l[18]
Pb from rainfall	116 mg/(m2*year)	Sudbury, 0.8 km from smelter, 1975[6]
Pb from rainfall	13.4 mg/(m2*year)	Wraymires, Lancs, Engl, 1973, pb conc = 15.5 ug/l[7]
Pb from rainfall	14.6 mg/(m2*year)	France, 11 stations, 1972, Pb conc = 14.5 ug/l[19]
Pb from rainfall	196 mg/(m2*year)	New York, 1970, pb conc = 147 ug/l[11]
Pb from rainfall	22.4 mg/(m2*year)	Chilton, Berks, Engl, 1973, pb conc = 58.4 ug/l[7]
Pb from rainfall	23.8 mg/(m2*year)	Shetland Isl, 1973, pb conc = 28.5 ug/l[7]
Pb from rainfall	28.9 mg/(m2*year)	Styrrups, Notts, Engl, 1973, pb conc = 98.5 ug/l[7]
Pb from rainfall	37.8 mg/(m2*year)	Sudbury, 100 km from smelter, 1975[6]
Pb from rainfall	4.0 mg/(m2*year)	Gottingen, Germany, 1974, Pb conc = 5.4 ug/l[20]
Pb from rainfall	6.8 mg/(m2*year)	Southern California Bight, 1974[17]
Pb from rainfall	8.2 mg/(m2*year)	Leiston, Suffolk, Engl, 1973, pb conc = 40.5 ug/l[7]
Pb from rainfall and snow	1.9 mg/(m2*year)	Chedron, Nebraska, 1975, pb conc = 4.8 ug/l[21]
Pb from residue in rainwater	19.0 mg/(m2*year)	Gottingen, Germany, 1974, Pb residue = 1410 ppm[20]
W	2 ug/cm2*year	Less than value, UK, 1972-1973[22]

Source: 1.Lee, K.C., Cunningham, Paulsen, Liang, and R.B. Moore. 1976. Physiol. Plant. 36, 4-6.
2.Nriagu, J.O. 1980. Cadmium in the Environment. Part I: Ecological Cycling. John Wiley and Sons. New York.
3.Nriagu, J.O. 1979. Copper in the Environment. John Wiley and Sons. NY.
4.Hirao, Y. and C.C. Patterson. 1974. Science 184, 989-992.
5.Nriagu, J.O. 1979. The Biochemistry of Lead in The Environment. Part A. Elsevier/North Holland. Amsterdam.
6.Hutchinson, T.C. and L.M. Whitby. 1976. In: Proc. 1st Intl. Symp. on Acid Precp. and Forest Ecosyste. L.S. Dochinger and Seliga (eds.). USDA, Upper Darby, Penna. Dept. No. NE-23, 745-765.
7.Cawse, P.A. 1972. A Survey of Atmospheric trace Elements in the U.K. (1972-1973). AERE Report No. R-669, HMSO, London, 84 pp.
8.Schlesinger, W.H., W.A. Reiners and D.S. Knopman. 1974. Environ. Pollut. 6, 39-47.
9.Harrison, R.M., R. Perry and R.A. Wellings. 1975. J. Air Pollut. Control Assoc. 25, 627-630.
10.Peterson, J.O. 1973. A Study of Lead in Lake Sediments. Thesis. Univ. Wisconsion. Madison. 246 pp.
11.Volchok, H.L. and D. Bogden. 1971. H.A.S.L. Report No. 281, April 1974.
12.Murozumi, M., T.J. Chow and C.C. Patterson. 1969. Geochim. Cosmoschim. Acta 33, 1247-1294.
13.Chester, R., S.R. Aston, J.H. Stoner and D. Bruty. 1974. J. Res. Atmosph. 777-789.
14.Dorn, C.R., J.O. Pierce, G.R. Chase and P.E. Phillips. 1975. Environ. res. 9, 159-172.
15.Creason, J.P., McNulty, Heiderscheit, Swanson and Buechley. 1972. Trace Substances in Environ. Health 5, 129-142.
16.Chow, T.J. and J.L. Earl. 1970. Science 178, 401-402.
17.Huntzicker, J.J., S.K. Friedlander and C.I. Davidson. 1975. Environ. Sci. Technol. 9, 448-457.
18.Ter Haar, G.L., R.B. Holtzman and H.F. Lucas, Jr. 1967. Nature 216, 353-355.
19.Servant, J. 1975. In: Environmental Health Aspects of Lead. CEC. Amsterdam. 155-165.
20.Ruppert, H. 1975. Water, Air, Soil Pollut. 4, 447-460.
21.Struempler, A.W. 1976. Atmosphjeric Environ. 9, 990-999.
22.Cawse, P.A. 1974. A Survey of Atmospheric Trace Elements in the UK. 1972-73. AERER R 7669. Her Majestys office. London.

6-002 Atmosphere
Dry deposition rate

Species	Value	Condition
Pb particles	0.13 cm/sec	Foothill site, Calif, 1970[1]
Pb particles	0.16 cm/sec	Norway, on snow surface, 1975[2]
Pb particles	0.17 cm/sec	Leiston, Suffolk, Engl, 1973[3]
Pb particles	0.19 cm/sec	Cincinnati, industrial, 1971[4]
Pb particles	0.22 cm/sec	Styrrup, Notts, 1973[5]
Pb particles	0.24 cm/sec	Freeway location, Calif, 1970[1]
Pb particles	0.25 cm/sec	Chilton, Engl, 1973[3]
Pb particles	0.26 cm/sec	Cincinnati, suburban, 1971[4]
Pb particles	0.30 cm/sec	Trebanos, Glamorgs, 1973[3]
Pb particles	0.31 cm/sec	Cincinnati, residential, 1971[4]
Pb particles	0.36 cm/sec	Lake Michigan, 1974[6]
Pb particles	0.38 cm/sec	Cincinnata, commercial, 1971[4]
Pb particles	0.46 cm/sec	Wraymires, Westmoreland, 1973[7]
Pb particles	0.5 cm/sec	Downtown San Diego, 1970[1]
Pb particles	0.5 cm/sec	New York Bight, 1974[8]
Pb particles	0.65 cm/sec	New York City, 1974[8]
Pb particles	0.69 cm/sec	Plynlimen, Montgoms, 1973[3]
Pb particles	1.4 cm/sec	Shetland Isl, 1973[3]
Pb particles	1.4 cm/sec	Toronto, near highways, rg = 0.7-2.0 cm/s, 1973[5]
Pb particles	11 cm/sec	Near Smelters, Toronto, rg = 8 - 14 cm/sec, 1973[5]
Pb particles	2.0 cm/sec	Near smelters, Missouri, 11974[9]

Source: 1.Robinson, E. And R.C. Robbins. 1971. Emissions, concentrations and fate of particulate atmospheric pollutants. Stanford. Res. Inst. Rep. SCC-8507, 108 pp.
2.Dorland, H. and N.S. Dorfman. 1972. Economics of the Environment. NY. W.W. Norton and Co. 426 pp.
3.Cawse, P.A. 1972. A Survey of Atmospheric trace Elements in the U.K. (1972-1973). AERE Report No. R-669, HMSO, London, 84 pp.
4.Creason, J.P., O. McNullty, Heiderscheit, Swanson and Buechley. 1972. Health 5, 129-142.
5.Roberts, T.M., Paciga, Hutchinson, Jervis, Chattopadhyay, Loon. 1974. Pbul. No. EE-1, Inst. Environ.Studies, Univ. Toronto. 74 pp.
6.Gatz, D.F. 1975. Atmosph. Environ. 5, 239-251.
7.Peirson, D.H., P.A. Cawse, L. Salmon and R.S. Cambray. 1973. Nature 241, 252-256.
8.Duce, R.A., G.T. Wallace and B.J. Ray. 1975. Univ. of Rhode Isl. Final Report, NOAA-MESA Grant, No.14-022-36, 32 p.
9.Dorn, C.R., J.O. Pierce, G.R. Chase and P.E. Phillips. 1975. Environ. res. 9, 159-172.

6-003 Atmosphere
Emission

Species	Value	Condition
Anthanthrene from coal combustion	290 ug PAH/million BTU heat input	Airborne PAH, underfeed stoker[1]
Anthanthrene from coal combustion	90000 ug PAH/million BTU heat input	Airborne PAH, hand stoked coal[1]
Anthanthrene from gas combustion	200 ug PAH/million BTU heat input	Airborne PAH, premix burner[1]
Anthracene from coal combustion	370 ug PAH/million BTU heat input	Airborne PAH, pulverized coal[1]

Anthracene from coal combustion	400000 ug PAH/million BTU heat input	Airborne PAH, hand stoked coal[1]
Anthracene from coal combustion	850 ug PAH/million BTU heat input	Airborne PAH, underfeed stoker[1]
Anthracene from oil combustion	3900 ug PAH/million BTU heat input	Airborne PAH, low pressure atomized oil[1]
As to atmosphere from fossil-fuel burning	10 t/year	Global value, from oil, fossil-fuel mobilization, 1971[2]
As to atmosphere from fossil-fuel burning	3200 t/year	Global value, from cement production, 1971[2]
As to atmosphere from fossil-fuel burning	5000 t/year	Global value, from coal, fossil-fuel mobilization, 1971[2]
Benz(a)anthracene and chrysene from municipal incinerator	0.72 mg/1000 m3	Stack gas, flow = 655 m3 gas/min, 293 K, STP, 340 mg/d total[3]
Benz(a)anthracene from 2,2,4-trimethylpentane and methylbutane comb.	32.3 ug BaA/g fuel	1 minute run, 25 % trimethylp. + 25 % 2-methylnbutane + 50 % o-xylene[4]
Benz(a)anthracene from 2,2,4-trimethylpentane combustion	0.8 ug BaA/g fuel	1 minute run, hydrocarbon fuel[4]
Benz(a)anthracene from 2,4,4-trimethyl-1-pentane combustion	0.4 ug BaA/g fuel	1 minute run, hydrocarbon fuel[4]
Benz(a)anthracene from coal combustion	19 ug PAH/million BTU heat input	Airborne PAH, pulverized coal[1]
Benz(a)anthracene from coal combustion	2000 ug PAH/million BTU heat input	Airborne PAH, rg = 560-3900, underfeed stoker[1]
Benz(a)anthracene from gasoline combustion	17.3 ug BaA/g fuel	1 minute run, hydrocarbon fuel[4]
Benz(a)anthracene from o-xylene and benzene combustion	56.3 ug BaA/g fuel	1 minute run, hydrocarbon fuel, 50 % of each fuel[4]
Benz(a)anthracene from oil combustion	27 ug PAH/million BTU heat input	Airborne PAH, stream atomized oil[1]
Benzo(a)pyrene and benzo(e)pyrene from municipal incinerator	0.02 mg/1000 m3	Stack gas, flow = 655 m3 gas/min, 293 K, STP, 9 mg/d in total[3]
Benzo(a)pyrene from 2,2,4-trimethylpentane and methylbutane	9.0 ug BaP/g fuel	1 minute run, 25 % trimethylp. + 25 % 2-methylnbutane + 50 % o-xylene[4]
Benzo(a)pyrene from 2,2,4-trimethylpentane combustion	2.6 ug BaP/g fuel	1 minute run, hydrocarbon fuel[4]
Benzo(a)pyrene from 2,4,4-trimethyl-1-pentane combustion	0.7 ug BaP/g fuel	1 minute run, hydrocarbon fuel[4]
Benzo(a)pyrene from all sources, total	1283 t/year	Period 1966-1969, USA[5]
Benzo(a)pyrene from all sources, total	3761 t/year	Period 1966-1969, Worldwide (excl. USA)[5]
Benzo(a)pyrene from coal combustion	26 ug PAH/million BTU heat input	Airborne PAH, rg = 19-32, pulverized coal[1]
Benzo(a)pyrene from coal combustion	26 ug PAH/million BTU heat input	Airborne PAH, spreader stoker[1]
Benzo(a)pyrene from coal combustion	300000 ug PAH/million BTU heat input	Airborne PAH, hand stoked coal[1]
Benzo(a)pyrene from coal combustion	37 ug PAH/million BTU heat input	Airborne PAH, chain grate stoker[1]

Benzo(a)pyrene from coal combustion	3800 ug PAH/million BTU heat input	Airborne PAH, rg = 120-10000, underfeed stoker[1]
Benzo(a)pyrene from coal combustion	400000 ug PAH/million BTU heat input	Airborne PAH, hand stoked coal[1]
Benzo(a)pyrene from coal combustion	92 ug PAH/million BTU heat input	Airborne PAH, pulverized coal[1]
Benzo(a)pyrene from enclosed industrial inceneration	23 t/year	Period 1966-1969, USA[5]
Benzo(a)pyrene from enclosed industrial inceneration	46 t/year	Period 1966-1969, Worldwide (excl. USA)[5]
Benzo(a)pyrene from enclosed non-industrial inceneration	11 t/year	Period 1966-1969, USA[5]
Benzo(a)pyrene from enclosed non-industrial inceneration	22 t/year	Period 1966-1969, Worldwide (excl. USA)[5]
Benzo(a)pyrene from gas combustion	100 ug PAH/million BTU heat input	Airborne PAH, rg = <20-200, premix burner[1]
Benzo(a)pyrene from gas combustion	1800 ug PAH/million BTU heat input	Airborne PAH, premix burner[1]
Benzo(a)pyrene from gasoline combustion	9.65 ug BaP/g fuel	1 minute run, hydrocarbon fuel[4]
Benzo(a)pyrene from glucose pyrolysis	1200 ug BaP/g fuel	Pyrolysis of bioorganic material in N2 at 1113 K[6]
Benzo(a)pyrene from glucose pyrolysis	288.8 ug BaP/g fuel	Pyrolysis of bioorganic material in N2 at 1113 K[6]
Benzo(a)pyrene from glucose pyrolysis	3130 ug BaP/g fuel	Pyrolysis of bioorganic material in N2 at 1113 K[6]
Benzo(a)pyrene from glucose pyrolysis	3750 ug BaP/g fuel	Pyrolysis of bioorganic material in N2 at 1113 K[6]
Benzo(a)pyrene from glucose pyrolysis	47.5 ug BaP/g fuel	Pyrolysis of bioorganic material in N2 at 1113 K[6]
Benzo(a)pyrene from glucose pyrolysis	98.4 ug BaP/g fuel	Pyrolysis of bioorganic material in N2 at 1113 K[6]
Benzo(a)pyrene from heat and power generation, total	2129 t/year	Period 1966-1969, Worldwide (excl. USA)[5]
Benzo(a)pyrene from heat and power generation, total	475 t/year	Period 1966-1969, USA[5]
Benzo(a)pyrene from heating and power generation using coal	1945 t/year	Period 1966-1969, USA[5]
Benzo(a)pyrene from heating and power generation using coal	2376 t/year	Period 1966-1969, Worldwide[5]
Benzo(a)pyrene from heating and power generation using coal	431 t/year	Period 1966-1969, USA[5]
Benzo(a)pyrene from heating and power generation using gas	1 t/year	Period 1966-1969, Worldwide (excl. USA)[5]
Benzo(a)pyrene from heating and power generation using gas	2 t/year	Period 1966-1969, USA[5]
Benzo(a)pyrene from heating and power generation using oil	2 t/year	Period 1966-1969, USA[5]
Benzo(a)pyrene from heating and power generation using oil	3 t/year	Period 1966-1969, Worldwide (excl. USA)[5]
Benzo(a)pyrene from heating and power generation using wood	180 t/year	Period 1966-1969, Worldwide (excl. USA)[5]
Benzo(a)pyrene from heating and power generation using wood	40 t/year	Period 1966-1969, USA[5]

Benzo(a)pyrene from industrial processes, catalytic cracking	6 t/year	Period 1966-1969, USA[5]
Benzo(a)pyrene from industrial processes, catalytic cracking	6 t/year	Period 1966-1969, Worldwide (excl. USA)[5]
Benzo(a)pyrene from industrial processes, coke production	192 t/year	Period 1966-1969, USA[5]
Benzo(a)pyrene from industrial processes, coke production	841 t/year	Period 1966-1969, Worldwide (excl. USA)[5]
Benzo(a)pyrene from industrial processes, total	198 t/year	Period 1966-1969, USA[5]
Benzo(a)pyrene from industrial processes, total	847 t/year	Period 1966-1969, Worldwide (excl. USA)[5]
Benzo(a)pyrene from o-xylene and benzene combustion	25.8 ug BaP/g fuel	1 minute run, hydrocarbon fuel, 50 % of each fuel[4]
Benzo(a)pyrene from oil combustion	100 ug PAH/million BTU heat input	Airborne PAH, less than value, vaporized oil[1]
Benzo(a)pyrene from oil combustion	300 ug PAH/million BTU heat input	Airborne PAH, low pressure atomized oil[1]
Benzo(a)pyrene from oil combustion	35 ug PAH/million BTU heat input	Airborne PAH, rg = <20 - 47, stream atomized oil[1]
Benzo(a)pyrene from oil combustion	50 ug PAH/million BTU heat input	Airborne PAH, rg = <40 - <60, centrifugal atomized oil[1]
Benzo(a)pyrene from oil combustion	900 ug PAH/million BTU heat input	Airborne PAH, low pressure atomized oilg[1]
Benzo(a)pyrene from open burning, total	588 t/year	Period 1966-1969, USA[5]
Benzo(a)pyrene from open burning, total	762 t/year	Period 1966-1969, Worldwide (excl. USA)[5]
Benzo(a)pyrene from open coal burning	340 t/year	Period 1966-1969, USA[5]
Benzo(a)pyrene from open coal burning	340 t/year	Period 1966-1969, Worldwide (excl. USA)[5]
Benzo(a)pyrene from open forest and agriculture burning	140 t/year	Period 1966-1969, USA[5]
Benzo(a)pyrene from open forest and agriculture burning	280 t/year	Period 1966-1969, Worldwide (excl. USA)[5]
Benzo(a)pyrene from other open burning	74 t/year	Period 1966-1969, other than coal, forest and agriculture, USA[5]
Benzo(a)pyrene from other open burning	74 t/year	Period 1966-1969, other than coal, forest, worldwide (excl. USA)[5]
Benzo(a)pyrene from vechiles, trucks and busses	12 t/year	Period 1966-1969, USA[5]
Benzo(a)pyrene from vechiles, trucks and busses	17 t/year	Period 1966-1969, Worldwide (excl. USA)[5]
Benzo(a)pyrene from vehichles, automobiles	10 t/year	Period 1966-1969, USA[5]
Benzo(a)pyrene from vehichles, automobiles	6 t/year	Period 1966-1969, Worldwide (excl. USA)[5]
Benzo(a)pyrene from vehicles, total	22 t/year	Period 1966-1969, USA[5]
Benzo(a)pyrene from vehicles, total	23 t/year	Period 1966-1969, Worldwide (excl. USA)[5]
Benzo(b)fluoranthene from municipal incinerator	0.32 mg/1000 m3	Stack gas= 655 m3 min, 293 K, STP, also -(k) and (j), 151 mg/d in total[3]

Benzo(e)pyrene from coal combustion	100000 ug PAH/million BTU heat input	Airborne PAH, hand stoked coal[1]
Benzo(e)pyrene from coal combustion	350 ug PAH/million BTU heat input	Airborne PAH, spreader stoker[1]
Benzo(e)pyrene from coal combustion	390 ug PAH/million BTU heat input	Airborne PAH, chain grate stoker[1]
Benzo(e)pyrene from coal combustion	5400 ug PAH/million BTU heat input	Airborne PAH, rg = 230-7900, underfeed stoker[1]
Benzo(e)pyrene from gas combustion	250 ug PAH/million BTU heat input	Airborne PAH, rg = 18-490, premix burner[1]
Benzo(ghi)perylene from coal combustion	2500 ug PAH/million BTU heat input	Airborne PAH, rg = 580-4500, underfeed stoker[1]
Benzo(ghi)perylene from municipal incinerator	0.42 mg/1000 m3	Stack gas, flow = 655 m3 gas/min, 293 K, STP, 198 mg/d in total[3]
Benzopyrene from all sources, total	5044 t/year	Period 1966-1969, Worldwide[5]
Benzopyrene from enclosed industrial inceneration	69 t/year	Period 1966-1969, Worldwide[5]
Benzopyrene from enclosed non-industrial inceneration	33 t/year	Period 1966-1969, Worldwide[5]
Benzopyrene from heat and power generation, total	2604 t/year	Period 1966-1969, Worldwide[5]
Benzopyrene from heating and power generation using gas	3 t/year	Period 1966-1969, Worldwide[5]
Benzopyrene from heating and power generation using oil	5 t/year	Period 1966-1969, Worldwide[5]
Benzopyrene from heating and power generation using wood	220 t/year	Period 1966-1969, Worldwide[5]
Benzopyrene from industrial processes, catalytic cracking	12 t/year	Period 1966-1969, Worldwide[5]
Benzopyrene from industrial processes, coke production	1033 t/year	Period 1966-1969, Worldwide[5]
Benzopyrene from industrial processes, total	1045 t/year	Period 1966-1969, Worldwide[5]
Benzopyrene from open burning, total	1350 t/year	Period 1966-1969, Worldwide[5]
Benzopyrene from open coal burning	680 t/year	Period 1966-1969, Worldwide[5]
Benzopyrene from open forest and agriculture burning	420 t/year	Period 1966-1969, Worldwide[5]
Benzopyrene from other open burning	148 t/year	Period 1966-1969, other than coal, forest and agriculture, Worldwide[5]
Benzopyrene from vechiles, trucks and busses	29 t/year	Period 1966-1969, Worldwide[5]
Benzopyrene from vehichles, automobiles	16 t/year	Period 1966-1969, Worldwide[5]
Benzopyrene from vehicles, total	45 t/year	Period 1966-1969, Worldwide[5]
CO from douglas fir and western larch slash fire	129 kg CO/tons of fuel	Emission factor, Lolo national forest, Montana[7]
CO from forest wildfires and managed burning	2.3E6 tons/year	Total for USA in 1975[8]
CO from forest wildfires and managed burning	2.9E6 tons/year	Total for USA in 1973[8]

CO from forest wildfires and managed burning	3.5E6 tons/year	Total for USA in 1970[8]
CO from forest wildfires and managed burning	3.6E6 tons/year	Total for USA in 1972[8]
CO from forest wildfires and managed burning	3.9E6 tons/year	Total for USA in 1974[8]
CO from forest wildfires and managed burning	4.8E6 tons/year	Total for USA in 1976[8]
CO from forest wildfires and managed burning	5.1E6 tons/year	Total for USA in 1971[8]
CO from ponderosa pine slash fire	66 kg CO/tons of fuel	Emission factor, San Bernardino, California[9]
CO from woody brush in artificial burning	29 kg CO/tons of fuel	Emission factor, green bush, Calif[10]
CO from woody brush in artificial burning	61 kg CO/tons of fuel	Emission factor, green bush, Calif[10]
CO to atmosphere from anthropogenic sources	1000E6 tons CO/y	Min value, global value[11]
CO to atmosphere from anthropogenic sources	540E6 tons CO/y	Min value, nothern hemisphere[11]
CO to atmosphere from anthropogenic sources	600E6 tons CO/y	Min value, global value[11]
CO to atmosphere from anthropogenic sources	900E6 tons CO/y	Min value, nothern hemisphere[11]
CO to atmosphere from chlorophyll degradation	200E6 tons CO/y	Natural source, min value, nothern hemisphere[11]
CO to atmosphere from chlorophyll degradation	300E6 tons CO/y	Natural source, min value, global value[11]
CO to atmosphere from chlorophyll degradation	500E6 tons CO/y	Natural source, max value, nothern hemisphere[11]
CO to atmosphere from chlorophyll degradation	700E6 tons CO/y	Natural source, max value, global value[11]
CO to atmosphere from methane oxidation	2000E6 tons CO/y	Natural source, max value, nothern hemisphere[11]
CO to atmosphere from methane oxidation	200E6 tons CO/y	Natural source, min value, nothern hemisphere[11]
CO to atmosphere from methane oxidation	4000E6 tons CO/y	Natural source, max value, global value[11]
CO to atmosphere from methane oxidation	400E6 tons CO/y	Natural source, min value, global value[11]
CO to atmosphere from oceans	100E6 tons CO/y	Natural source, min value, global value[11]
CO to atmosphere from oceans	220E6 tons CO/y	Natural source, max value, global value[11]
CO to atmosphere from oceans	40E6 tons CO/y	Natural source, min value, nothern hemisphere[11]
CO to atmosphere from oceans	90E6 tons CO/y	Natural source, max value, nothern hemisphere[11]
CO to atmosphere from terrestrial plants	140E6 tons CO/y	Natural source, max value, nothern hemisphere[11]
CO to atmosphere from terrestrial plants	14E6 tons CO/y	Natural source, min value, nothern hemisphere[11]
CO to atmosphere from terrestrial plants	200E6 tons CO/y	Natural source, max value, global value[11]
CO to atmosphere from terrestrial plants	20E6 tons CO/y	Natural source, min value, global value[11]

CO2 from douglas fir and western larch slash fire	1130 kg CO2/tons of fuel	Emission factor, Lolo national forest, Montana[7]
CO2 from woody brush in artificial burning	1168 kg CO2/tons of fuel	Emission factor, green bush, Calif[10]
CO2 from woody brush in artificial burning	693 kg CO2/tons of fuel	Emission factor, green bush, Calif[10]
CO2 to atmosphere from all changes in terrestrical ecosystems	1.0E9 tons C/year	From human perturbation of terrestrial ecosystem, since 1900, 0.6[12]
CO2 to atmosphere from all changes in terrestrical ecosystems	70E9 tons C in period	From human perturbation of terrestrial ecosystem, since 1900, +-30[12]
CO2 to atmosphere from change of organic matter in soil	0.3E9 tons C/year	From human perturbation of terrestrial ecosystem, since 1900, +-0.2[12]
CO2 to atmosphere from change of organic matter in soil	24E9 tons C in period	From human perturbation of terrestrial ecosystem, since 1900, +-15[12]
CO2 to atmosphere from reduction of forest in developed areas	0E9 tons C/year	From human perturbation of terrestrial ecosystem, since 1900, +-0.1[12]
CO2 to atmosphere from reduction of forest in developing areas	-0.3E9 tons C/year	From perturbation of terrestrial ecosystem, since 1900, forestation[12]
CO2 to atmosphere from reduction of forest in developing areas	0.8E9 tons C/year	From perturbation of terrestrial ecosystem, since 1900, deforestation[12]
CO2 to atmosphere from reduction of forest in developing areas	45E9 tons C in period	From human perturbation of terrestrial ecosystem, since 1900, +-15[12]
CO2 to atmosphere from use of fuel wood in developing areas	0.3E9 tons C/year	From human perturbation of terrestrial ecosystem, since 1900, +-0.2[12]
Carbon from dust from soil, POC	1.3E12 g/year	Particulate organic carbon, diameter < 1 um[1]
Carbon from dust from soil, POC	10.8E12 g/year	Particulate organic carbon, in total[13]
Carbon from dust from soil, POC	9.5E12 g/year	Particulate organic carbon, diameter > 1 um[13]
Carbon from forest fires	0.5E12 g/year	Particulate organic carbon, diameter > 1 um[13]
Carbon from forest fires	1.1E12 g/year	Particulate organic carbon, diameter < 1 um[13]
Carbon from forest fires	1.6E12 g/year	Particulate organic carbon, in total[13]
Carbon from ocean derived aerosols, POC	1.46E12 g/year	Particulate organic carbon, diameter < 1 um[13]
Carbon from ocean derived aerosols, POC	12.6E12 g/year	Particulate organic carbon, diameter > 1 um[13]
Carbon from ocean derived aerosols, POC	14.0E12 g/year	Particulate organic carbon, in total[13]
Carbon from total anthropogenic sources	13.9E12 g/year	Particulate organic carbon, diameter > 1 um[13]
Carbon from total anthropogenic sources	15.9E12 g/year	Particulate organic carbon, diameter < 1 um[13]
Carbon from total anthropogenic sources	29.8E12 g/year	Particulate organic carbon, in total[13]
Carbon from total both natural and man made	19.7E12 g/year	Particulate organic carbon, diameter < 1 um[13]
Carbon from total both natural and man made	36.5E12 g/year	Particulate organic carbon, diameter > 1 um[13]
Carbon from total both natural and man made	56.2E12 g/year	Particulate organic carbon, in total[13]
Carbon from total natural sources	22.6E12 g/year	Particulate organic carbon, diameter > 1 um[13]
Carbon from total natural sources	26.4E12 g/year	Particulate organic carbon, in total[13]

Carbon from total natural sources	3.8E12 g/year	Particulate organic carbon, diameter < 1 um[13]
Cd from Cd extraction in production	111E6 g/year	Emission factor 6500 mg/kg, Annual handling = 0.0017E9 kg, 1976, world[14]
Cd from Cu metal melting production	1580E6 g/year	Emission factor 200 mg/kg, Annual handling = 7.9E9 kg, 1976, world[14]
Cd from Fe and steel production	72E6 g/year	Emission factor 0.055 mg/kg, Annual handling = 1300E9 kg, 1976, world[14]
Cd from Pb metal melting production	230E6 g/year	Emission factor 50 mg/kg, Annual handling = 4.6E9 kg, 1976, world[14]
Cd from Zn metal production	2800E6 g/year	Emission factor 500 mg/kg, Annual handling = 5.6E9 kg, 1976, world[14]
Cd from Zn mining operations	2.8E6 g/year	Emission factor 0.5 mg/kg, Annual handling = 5.6E9 kg, 1976, world[14]
Cd from alloys and batteries	6.4E6 g/year	1976, world[14]
Cd from coal combustion	62E6 g/year	Emission factor 0.02 mg/kg, Annual handling = 3100E9 kg, 1976, world[14]
Cd from electroplating shops	8.5E6 g/year	1976, world[14]
Cd from forest fires	12E6 g Cd/year	Global prod = 36E9 kg/y, emission factor = 0.32 ug/g, global, natural source[15]
Cd from oil combustion	2.8E6 g/year	Emission factor 0.001 mg/kg, Annual handling = 2800E9 kg, 1976, world[14]
Cd from phosphate fertilizers	118E6 g/year	Emission factor 1.0 mg/kg, Annual handling = 118E9 kg, 1976, world[14]
Cd from pigments and plastics	38E6 g/year	1976, world[14]
Cd from rubber tiree wear	10E6 g/year	1976, world[14]
Cd from seasalt sprays	1.0E6 g Cd/year	Global prod = 1000E9 kg/y, emission factor = 0.04 ug/kg, global, natural source[15]
Cd from seundary nonferrous metal production	595E6 g/year	1976, world[14]
Cd from total natural source sources	843E6 g Cd/year	Global value, natural source[15]
Cd from vegetation	210E6 g Cd/year	Global prod = 75E9 kg/y, emission factor = 2.8 ug/g, global, natural source[15]
Cd from volcanic particles	520E6 g Cd/year	Global prod = 25E9 kg/y, emission factor = 2.8 ug/g, global, natural source[15]
Cd from waste incineration	1350E6 g/year	Emission factor 9.0 mg/kg, Annual handling = 150E9 kg, 1976, world[14]
Cd from windblown dust	100E6 g Cd/year	Global prod = 500E9 kg/y, emission factor = 0.2 ug/g, global, natural source[15]
Cd from wood combustion	200E6 g/year	Emission factor 3.0 mg/kg, Annual handling = 640E9 kg, 1976, world[14]
Cd to atmosphere from fossil-fuel burning	140 t/year	Global value, from coal, fossil-fuel mobilization, 1971[2]
Cd to atmosphere from fossil-fuel burning	2 t/year	Global value, from oil, fossil-fuel mobilization, 1971[2]
Cd to atmosphere from fossil-fuel burning	80 t/year	Global value, from cement production, 1971[2]
Cd, Anthropogenic sources	7.3E9 g/year	None[16]
Cd, Fertilizer manufacture	0.2E9 g/year	None[16]
Cd, Mixed anthropogenic sources	0.4E9 g/year	Excluding metal prod, incineration and fertilizer manufacture[16]
Cd, Natural sources	0.8E9 g/year	None[16]

Cd, Total emission	8.1E9 g/year	None[16]
Cd, Vegetation exudates	0.2E9 g/year	None[16]
Cd, Volcanic emission	0.5E9 g/year	None[16]
Cd, Waste incineration	1.4E9 g/year	None[16]
Cd, Windblown dusts	0.1E9 g/year	None[16]
Cd, mixed natural sources	0.01E9 g/year	Excluding dusts, volcanes and exudates[16]
Cd, non-ferrous metal production	5.3E9 g/year	None[16]
Co to atmosphere from fossil-fuel burning	30 t/year	Global value, from oil, fossil-fuel mobilization, 1971[2]
Co to atmosphere from fossil-fuel burning	700 t/year	Global value, from coal, fossil-fuel mobilization, 1971[2]
Coronene from coal combustion	26 ug PAH/million BTU heat input	Airborne PAH, spreader stoker[1]
Coronene from coal combustion	30000 ug PAH/million BTU heat input	Airborne PAH, hand stoked coal[1]
Coronene from coal combustion	330 ug PAH/million BTU heat input	Airborne PAH, rg = 330-1200, underfeed stoker[1]
Coronene from gas combustion	2000 ug PAH/million BTU heat input	Airborne PAH, premix burner[1]
Coronene from municipal incinerator	0.04 mg/1000 m3	Stack gas, flow = 655 m3 gas/min, 293 K, STP, 19 mg/d in total[3]
Coronene from oil combustion	2100 ug PAH/million BTU heat input	Airborne PAH, low pressure atomized oil[1]
Cr to atmosphere from fossil-fuel burning	1400 t/year	Global value, from coal, fossil-fuel mobilization, 1971[2]
Cr to atmosphere from fossil-fuel burning	50 t/year	Global value, from oil, fossil-fuel mobilization, 1971[2]
Cu from Fe and steel foundries	0.37E6 kg/year	Emission factor = 2.3 g/t fabricated, Handling = 4.0E6 kg/y[17]
Cu from Fe and steel production	5.9E6 kg/year	Emission factor = 4.5 g/t produced, Handling = 1304E6 kg/y[17]
Cu from Pb production	0.29E6 kg/year	Emission factor = 73 g/t Pb produced, Handling = 5.6E6 kg/y[17]
Cu from Zn production	0.78E6 kg/year	Emission factor = 0.14 kg/t Zn produced, Handling = 5.6E6 kg/y[17]
Cu from all anthropogenic sources	56.3E6 kg/year	Global value[17]
Cu from all natural sources	18.5E6 kg/year	Global value[17]
Cu from coal combustion	5.6E6 kg/year	Emission factor = 1.8 g/t coal burned, Handling = 3100E6 kg/y[17]
Cu from copper metalurgical processing	19.8E6 kg/year	Emission factor = 2.5 kg/t Cu produced, Handling = 7.9E6 kg/y[17]
Cu from copper mining	0.75E6 kg/year	Emission factor = 0.1 kg/t Cu mined, Handling = 7.5E6 kg/y[17]
Cu from forest fires	0.3E6 kg/year	Emission factor = 8 ug/g, total production = 38E9 kg/y[17]
Cu from fuel oil combustion	0.74E6 kg/year	Emission factor = 0.7 g/t oil burned, handling = 1060E6 kg/y[17]
Cu from grain handling	0.03E6 kg/year	Emission factor = 0.08 g/t grain handled, Handling = 330E6 kg/y[17]
Cu from industrial applications	4.9E6 kg/year	Global value[17]
Cu from secondary copper processing	0.33E6 kg/year	Emission factor = 0.15 kg/t Cu produced, Handling = 2.2E6 kg/y[17]
Cu from vegetation	2.5E6 kg/year	Emission factor = 0.082 ug/g, total production = 1000E9 kg/y[17]

Cu from volcanic particles	3.6E6 kg/year	Emission factor = 360 ug/g, total production = 10E9 kg/y[17]
Cu from waste incineration	5.3E6 kg/year	Emission factor = 3.5 ug/g waste, Handling = 1500E6 kg/y[17]
Cu from windblown dust	12.0E6 kg/year	Emission factor = 24 ug/g dust, total production = 500E9 kg/y[17]
Cu from wood combustion	11.5E6 kg/year	Emission factor = 18 g/t wood burned, Handling = 640E6 kg/y[17]
Cu to atmosphere from fossil-fuel burning	2100 t/year	Global value, from coal, fossil-fuel mobilization, 1971[2]
Cu to atmosphere from fossil-fuel burning	23 t/year	Global value, from oil, fossil-fuel mobilization, 1971[2]
Dimethyl sulfide to atmosphere from all sources	3.8E12 g S/y	Global value, rg = 2.1-5.5E12 g S/y[18]
Dimethyl sulfide to atmosphere from fresh leaves	0.01E12 g S/y	Global value[18]
Dimethyl sulfide to atmosphere from marine algae	0.05E12 g S/y	Global value[18]
Dimethyl sulfide to atmosphere from senescent leaves	0.53E12 g S/y	Global value[18]
Dimethyl sulfide to atmosphere from soil	3.2E12 g S/y	Global value, rg = 1.5-4.9E12 g S/y[18]
Dimethylsulphide	0.01E15 g/y	Natural source, from fresh leaves[19]
Dimethylsulphide	0.05E15 g/y	Natural source, from marine algae[19]
Dimethylsulphide	0.53E15 g/y	Natural source, from senescent leaves[19]
Dimethylsulphide	3.20E15 g/y	Natural source, from soils[19]
Dimethylsulphide	3.80E15 g/y	Natural source, total value[19]
Fluoranthene from coal combustion	1000000 ug PAH/million BTU heat input	Airborne PAH, hand stoked coal[1]
Fluoranthene from coal combustion	360 ug PAH/million BTU heat input	Airborne PAH, rg = 180-550, pulverized coal[1]
Fluoranthene from coal combustion	360 ug PAH/million BTU heat input	Airborne PAH, spreader stoker[1]
Fluoranthene from coal combustion	38000 ug PAH/million BTU heat input	Airborne PAH, rg = 3200-47000, underfeed stoker[1]
Fluoranthene from coal combustion	680 ug PAH/million BTU heat input	Airborne PAH, chain grate stoker[1]
Fluoranthene from gas combustion	320 ug PAH/million BTU heat input	Airborne PAH, rg = 100-3200, premix burner[1]
Fluoranthene from municipal incinerator	0.58 mg/1000 m3	Stack gas, flow = 655 m3 gas/min, 293 K, STP, 274 mg/d total[3]
Fluoranthene from oil combustion	15000 ug PAH/million BTU heat input	Airborne PAH, vaporized oil[1]
Fluoranthene from oil combustion	165 ug PAH/million BTU heat input	Airborne PAH, rg = 56-270, stream atomized oil[1]
Fluoranthene from oil combustion	1900 ug PAH/million BTU heat input	Airborne PAH, low pressure atomized oil[1]
Fluoranthene from oil combustion	5000 ug PAH/million BTU heat input	Airborne PAH, centrifugal atomized oil[1]
Hg to atmosphere from fossil-fuel burning	100 t/year	Global value, from cement production, 1971[2]
Hg to atmosphere from fossil-fuel burning	1600 t/year	Global value, from oil, fossil-fuel mobilization, 1971[2]

Hg to atmosphere from fossil-fuel burning	400 t/year	Global value, from coal, fossil-fuel mobilization, 1971[2]
Hydrocarbons from forest wildfires and managed burning	0.4E6 tons/year	Total for USA in 1975[8]
Hydrocarbons from forest wildfires and managed burning	0.6E6 tons/year	Total for USA in 1973[8]
Hydrocarbons from forest wildfires and managed burning	0.6E6 tons/year	Total for USA in 1974[8]
Hydrocarbons from forest wildfires and managed burning	0.7E6 tons/year	Total for USA in 1970[8]
Hydrocarbons from forest wildfires and managed burning	0.7E6 tons/year	Total for USA in 1972[8]
Hydrocarbons from forest wildfires and managed burning	0.8E6 tons/year	Total for USA in 1976[8]
Hydrocarbons from forest wildfires and managed burning	1.0E6 tons/year	Total for USA in 1971[8]
Hydrocarbons from ponderosa pine slash fire	3.6 kg HC/tons of fuel	Emission factor, San Bernardino, Calif, 15-40 % methane and ethylene[9]
Hydrocarbons from woody brush in artificial burning	12 kg HC/tons of fuel	Emission factor, green bush, Calif[10]
Hydrocarbons from woody brush in artificial burning	2 kg HC/tons of fuel	Emission factor, green bush, Calif[10]
Indeno(1,2,3-cd)pyrene from municipal incinerator	0.18 mg/1000 m3	Stack gas, flow = 655 m3 gas/min, 293 K, STP, 85 mg/d in total[3]
NO2 from douglas fir and western larch slash fire	7.7 kg NO2/tons of fuel	Emission factor, Lolo national forest, Montana[7]
Net CO2 to atmosphere from fuel wood burning	0E15 g C/y	Area burned = 1.2E10 m2/y, boreal forest[20]
Net CO2 to atmosphere from fuel wood burning	0E15 g C/y	Area burned = 1.3E10 m2/y, boreal nonwoodland forest[20]
Net CO2 to atmosphere from fuel wood burning	0E15 g C/y	Area burned = 300E10 m2/y, existing tropical shifting cultivation[20]
Net CO2 to atmosphere from fuel wood burning	0E15 g C/y	Area burned = 4.6E10 m2/y, temperate forest[20]
Net CO2 to atmosphere from fuel wood burning	0E15 g C/y	Area burned = 4.6E10 m2/y, temperate nonwooded land[20]
Net CO2 to atmosphere from fuel wood burning	1.4E15 g C/y	Area burned = 7.5E10 m2/y, due to increase in rural population[20]
Net CO2 to atmosphere from fuel wood burning	1.5E15 g C/y	Area burned = 24.0E10 m2/y, tropical forest new clearing[20]
Ni to atmosphere from fossil-fuel burning	1600 t/year	Global value, from oil, fossil-fuel mobilization, 1971[2]
Ni to atmosphere from fossil-fuel burning	2100 t/year	Global value, from coal, fossil-fuel mobilization, 1971[2]
Nitrogen oxides from forest wildfires and managed burning	0.1E6 tons/year	Total for USA in 1970[8]
Nitrogen oxides from forest wildfires and managed burning	0.1E6 tons/year	Total for USA in 1972[8]
Nitrogen oxides from forest wildfires and managed burning	0.1E6 tons/year	Total for USA in 1973[8]
Nitrogen oxides from forest wildfires and managed burning	0.1E6 tons/year	Total for USA in 1974[8]

Nitrogen oxides from forest wildfires and managed burning	0.1E6 tons/year	Total for USA in 1975[8]
Nitrogen oxides from forest wildfires and managed burning	0.2E6 tons/year	Total for USA in 1971[8]
Nitrogen oxides from forest wildfires and managed burning	0.2E6 tons/year	Total for USA in 1976[8]
Particulates from douglas fir and western larch slash fire	6.3 kg Part./tons of fuel	Emission factor, Lolo national forest, Montana[7]
Particulates from douglas fir slash fire	2.8 kg part./tons of fuel	Emission factor, 24 h value, Flathead Nat. forest, MO, needles and twigs[21]
Particulates from douglas fir slash fire	3.3 kg part./tons of fuel	Emission factor, 24 h value, Flathead Nat. forest, MO, twigs 1-2.5 cm[21]
Particulates from douglas fir slash fire	3.8 kg part./tons of fuel	Emission factor, 24 h value, Flathead Nat. forest, MO, twigs 0-1 cm[21]
Particulates from douglas fir slash fire	4.8 kg part./tons of fuel	Emission factor, 24 h value, Flathead Nat. forest, MO, needles[21]
Particulates from duff fir slash fire	7.5 kg part./tons of fuel	Emission factor, 24 h value, Flathead Nat. forest, MO, duff[21]
Particulates from forest wildfires and managed burning	0.3E6 tons/year	Total for USA in 1975[8]
Particulates from forest wildfires and managed burning	0.4E6 tons/year	Total for USA in 1973[8]
Particulates from forest wildfires and managed burning	0.5E6 tons/year	Total for USA in 1970[8]
Particulates from forest wildfires and managed burning	0.5E6 tons/year	Total for USA in 1972[8]
Particulates from forest wildfires and managed burning	0.5E6 tons/year	Total for USA in 1974[8]
Particulates from forest wildfires and managed burning	0.6E6 tons/year	Total for USA in 1976[8]
Particulates from forest wildfires and managed burning	0.7E6 tons/year	Total for USA in 1971[8]
Particulates from larch slash fire	3.3 kg part. /tons of fuel	Emission factor, 24 h value, Flathead Nat. forest, MO, twigs 1-2.5 cm[21]
Particulates from larch slash fire	5.1 kg part. /tons of fuel	Emission factor, 24 h value, Flathead Nat. forest, MO, needles and twigs[21]
Particulates from larch slash fire	5.8 kg part./tons of fuel	Emission factor, 24 h value, Flathead Nat. forest, MO, needles[21]
Particulates from larch slash fire	9.3 kg part. /tons of fuel	Emission factor, 24 h value, Flathead Nat. forest, MO, twigs 0-1 cm[21]
Particulates from ponderosa pine slash fire	66 kg Part. /tons of fuel	Emission factor, San Bernardino, California[9]
Particulates from spruce slash fire	3.7 kg part./tons of fuel	Emission factor, 24 h value, Flathead Nat. forest, MO, spruce[21]
Pb to atmosphere from fossil-fuel burning	30000 t/year	Global value, from cement production, 1971[2]
Pb to atmosphere from fossil-fuel burning	3500 t/year	Global value, from coal, fossil-fuel mobilization, 1971[2]
Pb to atmosphere from fossil-fuel burning	50 t/year	Global value, from oil, fossil-fuel mobilization, 1971[2]
Pb-210 from atmospheric natural lead	7E4 kCi/year	Source of Pb-210 in atmosphere, natural source[22]

Pb-210 from dispersion of gypsum by-products	0.01 kCi/year	Source of Pb-210 in atmosphere, anthropogenic source[23]
Pb-210 from dispersion of phosphate fertilizers	0.15 kCi/year	Source of Pb-210 in atmosphere, anthropogenic source, rg = 0.1-0.2[23]
Pb-210 from fossil fuel burning	0.2 kCi/year	Source of Pb-210 in atmosphere, anthropogenic source[24]
Pb-210 from large sea salt particles	0.01 kCi/year	Source of Pb-210 in atmosphere, natural source[25]
Pb-210 from meteoric and volcanic activity	6E-6 kCi/year	Source of Pb-210 in atmosphere, natural source[26]
Pb-210 from nuclear detonations	500 kCi	Source of Pb-210 in atmosphere. 1958-1959, anthropogenic source[24]
Pb-210 from nuclear detonations	900 kCi	Source of Pb-210 in atmosphere, 1961-1962, anthropogenic source[24]
Pb-210 from plant exudates	0.9 kCi/year	Source of Pb-210 in atmosphere, natural source[27]
Pb-210 from radon emanation from land	690 kCi/year	Source of Pb-210 in atmosphere, natural source[24]
Pb-210 from radon emanation from world oceans	14 kCi/year	Source of Pb-210 in atmosphere, natural source[28]
Pb-210 from resuspension of soils	9 kCi/year	Source of Pb-210 in atmosphere, natural source[23]
Pb-210 from tetramethyl Pb combustion	0.03 kCi/year	Source of Pb-210 in atmosphere, anthropogenic source[24]
Perylene from coal combustion	1600 ug PAH/million BTU heat input	Airborne PAH, underfeed stoker[1]
Perylene from coal combustion	60000 ug PAH/million BTU heat input	Airborne PAH, hand stoked coal[1]
Perylene from municipal incinerator	0.18 mg/1000 m3	Stack gas, flow = 655 m3 gas/min, 293 K, STP, 85 mg/d in total[3]
Phenanthrene from coal combustion	10000 ug PAH/million BTU heat input	Airborne PAH, rg = 1000-29000, underfeed stoker[1]
Phenanthrene from coal combustion	1000000 ug PAH/million BTU heat input	Airborne PAH, hand stoked coal[1]
Phenanthrene from gas combustion	77 ug PAH/million BTU heat input	Airborne PAH, premix burner[1]
Phenanthrene from oil combustion	1800 ug PAH/million BTU heat input	Airborne PAH, stream atomized oil[1]
Phenanthrene from oil combustion	3500 ug PAH/million BTU heat input	Airborne PAH, low pressure atomized oil[1]
Phenanthrene from oil combustion	8900 ug PAH/million BTU heat input	Airborne PAH, centrifugal atomized oil[1]
Pyrene from coal combustion	195 ug PAH/million BTU heat input	Airborne PAH, rg = 150-240, pulverized coal[1]
Pyrene from coal combustion	390 ug PAH/million BTU heat input	Airborne PAH, chain grate stokerg[1]
Pyrene from coal combustion	590 ug PAH/million BTU heat input	Airborne PAH, spreader stokerg[1]
Pyrene from coal combustion	600000 ug PAH/million BTU heat input	Airborne PAH, hand stoked coal[1]
Pyrene from coal combustion	7700 ug PAH/million BTU heat input	Airborne PAH, rg = 230-7900, underfeed stoker[1]
Pyrene from gas combustion	160 ug PAH/million BTU heat input	Airborne PAH, rg = 120-18000, premix burner[1]

Pyrene from municipal incinerator	1.58 mg/1000 m3	Stack gas, flow = 655 m3 gas/min, 293 K, STP, 754 mg/d total[3]
Pyrene from oil combustion	1200 ug PAH/million BTU heat input	Airborne PAH, vaporized oil[1]
Pyrene from oil combustion	175 ug PAH/million BTU heat input	Airborne PAH, rg = 49-300, stream atomized oil[1]
Pyrene from oil combustion	6100 ug PAH/million BTU heat input	Airborne PAH, low pressure atomized oilg[1]
Pyrene from oil combustion	900 ug PAH/million BTU heat input	Airborne PAH, rg = 15-1800, centrifugal atomized oil[1]
Se	100 tons/year	1969, total of USA[29]
Se	20 % of total	Se from glass manufacturing, USA[29]
Se	64 % of total	Se from coal combustion, USA[29]
Se	7 % of total	Se from fuel oil, USA[29]
Se	9 % of total	Se from metallurgy, USA[29]
Se to atmosphere from fossil-fuel burning	30 t/year	Global value, from oil, fossil-fuel mobilization, 1971[2]
Se to atmosphere from fossil-fuel burning	420 t/year	Global value, from coal, fossil-fuel mobilization, 1971[2]
Se to atmosphere from fossil-fuel burning	700 t/year	Global value, from cement production, 1971[2]
Total PAH from municipal incinerator	1906 mg/day	Stack gas, flow = 655 m3 gas/min, 293 K, STP[3]
V to atmosphere from fossil-fuel burning	3500 t/year	Global value, from coal, fossil-fuel mobilization, 1971[2]
V to atmosphere from fossil-fuel burning	8200 t/year	Global value, from oil, fossil-fuel mobilization, 1971[2]
Zn to atmosphere from fossil-fuel burning	30000 t/year	Global value, from cement production, 1971[2]
Zn to atmosphere from fossil-fuel burning	40 t/year	Global value, from oil, fossil-fuel mobilization, 1971[2]
Zn to atmosphere from fossil-fuel burning	7000 t/year	Global value, from coal, fossil-fuel mobilization, 1971[2]

Source: 1.Hangebrauck, R.P., D.J. von Lehmden and J.E. Meeker. 1964. J Air Pollut. Contr. Assoc. 14, 267-278.
2.Bertine, K.K. and E.D. Goldberg. 1971. Science 178, pp 233-235.
3.Davies, I.W.R., Harrison, Perry, Ratnayaka and R.A. Wellings. 1976. Environ. Sci. Technol. 10, 451-43.
4.Brunnemann, K.D. and D. Hoffmann. 1976. In: R. Freudenthal and P.W. Jones (eds.): Carcinogenesis-Comprehensive Survey. Vol 1. Polynuclear Aromatic Hydrocarbons. NY. Raven Press. 283-298.
5.Suess, M.J. 1972. Water Res. 6, 981-985.
6.Schmelts, N. and D. Hoffman. 1976. In: R. Freudenthal and P.W. Jones (eds.): Carcinogenes - Comprehensive Survey. Vol 1. Polynuclear Aromatic Hydrocarbones. NY. Raven Press.
7.Malte, P.C. 1975. Pollutant production from forest slash burns. Bull 339. Coll of Eng. Wash. State Univ. Pullman. Wash.
8.EPA. 1977. National Air Quality and Emissions Trends Report 1976. USEPA, EPA-450/1-77-002, Research Triangle Park. NC.
9.Sandberg, D.V., S.G. Pickford and E.F. Darley. 1975. J. Air Pollut. Control Asssoc. 21, 629-635.
10.Darley, E.F., F.R. Burleson, E.H. Malte, J. Middleton and Osterli 1966. J. Air Pollut. Control Assoc11, 685-690.
11.Nozhevnikova, A.N. and L.N. Yuganov. 1978. Adv. Microbial Ecol. 2, 203-244.
12.Bolin, B. 1977. Science 196, 613-615.
13.Bolin, B., E.T. Degens, S. Kempe and P. Ketner. 1977. The Global Carbon Cycle. SCOPE and ICSU. Johniley and Sons. NY.
14.U.S. Bureau of Mines. 1976. Minerals Yearbook. U.S. Department of the Interior, Wash. DC.
15.Nriagu, J.O. 1980. Cadmium in the Environment. Part I: Ecological Cycling. John Wiley and Sons. New York.
16.Nriagu, J.O. 1979. in: Nriagu, J.O. (ed): Copper in the Environment. Part 1A. Elsevier. Amsterdam. p137-183.
17.Nriagu, J.O. 1979. Copper in the Environment. John Wiley and Sons. NY.
18.Granat, L., H. Rohde and R.O. Hallberg. 1976. In: B.H. Svensson and R. Soederlund (eds.): N, P and S. SCOPE report No. 7, Ecological Bulletins, Stockholm 22, 89-134.
19.Zepp, R.G. and D.M. Cline. 1977. Environ. Sci. Technol. 11, 359.
20.Wong, C.S. 1978. Science 200, 197-200.
21.Adams, D.F., R.K. Koppe and E. Robinson.. 1976. In: Air Quality and Smoke from Urban and Forest Fire. National Academy of Science. Wash DC, 105-147.
22.Jaworowski, Z., J. Bilkiewicz, L. Kownacka and S. Wlodek. 1972. In: Adams, Lowsder and Gesell (eds.)The Natural

Radiation Environment. Vol 2. Proc. 2nd Int. Conf. Houston. Tex. CONF-720805-P2, 809-818
23.Moore, H.E., E.A. Martell and S.E. Poet. 1976. Environ. Sci. Technol. 10, 586-591.
24.Nriagu, J.O. 1979. The Biochemistry of Lead in The Environment. Part A. Elsevier/North Holland. Amsterdam.
25.Junge, C. 1963. Air Chemistry and Radioactivity. Academic Press. New York 382 pp.
26.Burton, W.M. and N.G. Stewart. 1960. Nature 186, 584-589.
27.Gardner, L.R. 1974. Geochim. Cosmochim. Acta 38, 1297-1302.
28.Wilkening, M.H. and W.F. Clements. 1975. J. Geophys. Res. 80, 3823-3830.
29.Davies, W.E. 1972. National Inventory of Sources and Emissions: Ba, B, Cu, S, Zn, and Se. Section IV. Report for EPA. Kansas.

6-004 Deposition
Atmosphere

Species	Value	Condition
Zn found in Leucobryum glaucum	10.1 mg/(m2*year)	Rg = +- 1.6, Coastal Maine, USA, 1975[1]
Zn found in Leucobryum glaucum	11.0 mg/(m2*year)	N New Hampshire, USA, rg = +- 1.9, 1975[1]
Zn found in Leucobryum glaucum	13.5 mg/(m2*year)	Rg = +- 1.2, S New England coast, USA, 1975[1]
Zn found in Leucobryum glaucum	14.9 mg/(m2*year)	Rg = +- 1.9, Inland Mass., USA, 1975[1]
Zn found in Leucobryum glaucum	5.8 mg/(m2*year)	Cape Cod, Mass., USA, 1975[1]
Zn found in Sphagnum fuscum	13.5 mg/(m2*year)	S Finland, 1975[1]
Zn found in Sphagnum magellanicum	43 mg/(m2*year)	Rg = 24.8-61.2, S Sweden, 1970[1]
Zn found in precipitation	120 mg/(m2*year)	Albany, NY, USA, 1969[1]
Zn found in precipitation	63.6 mg/(m2*year)	Caribou, Me, USA, 1969[1]
Zn found in precipitation	75.6 mg/(m2*year)	Nantucket, Mass., USA, 1969[1]

Source: 1.Nriagu, J.O. 1980. Zinc in the Environment. Part 1: Ecological cycling. Wiley and Sons. NY.

6-005 Emission
Solid waste composition

Species	Value	Condition
Food waste	18.2 % of dry weight	Unseasonal state, USA, Fl., municipal refuse, ann. average, 1968[1]
Food waste	19.5 % of dry weight	Semiseasonal state, USA, Ala., municipal refuse, ann. average, 1968[1]
Food waste	21.1 % of dry weight	Seasonal state, Mass, USA, municipal refuse, ann. average, 1968[1]
Glass	7.6 % of dry weight	Unseasonal state, USA, Fl., municipal refuse, ann. average, 1968[1]
Glass	8.1 % of dry weight	Semiseasonal state, USA, Ala., municipal refuse, ann. average, 1968[1]
Glass	8.8 % of dry weight	Seasonal state, Mass, USA, municipal refuse, ann. average, 1968[1]
Leather and rubber	1.3 % of dry weight	Unseasonal state, USA, Fl., municipal refuse, ann. average, 1968[1]
Leather and rubber	1.4 % of dry weight	Semiseasonal state, USA, Ala., municipal refuse, ann. average, 1968[1]
Leather and rubber	1.5 % of dry weight	Seasonal state, Mass, USA, municipal refuse, ann. average, 1968[1]
Metal	7.5 % of dry weight	Unseasonal state, USA, Fl., municipal refuse, ann. average, 1968[1]
Metal	8.1 % of dry weight	Semiseasonal state, USA, Ala., municipal refuse, ann. average, 1968[1]
Metal	8.7 % of dry weight	Seasonal state, Mass, USA, municipal refuse, ann. average, 1968[1]
Miscellaneous solid waste	1.6 % of dry weight	Unseasonal state, USA, Fl., municipal refuse, ann. average, 1968[1]
Miscellaneous solid waste	1.7 % of dry weight	Semiseasonal state, USA, Ala., municipal refuse, ann. average, 1968[1]
Miscellaneous solid waste	1.8 % of dry weight	Seasonal state, Mass, USA, municipal refuse, ann. average, 1968[1]
Paper	32.6 % of dry weight	Unseasonal state, USA, Fl., municipal refuse, ann. average, 1968
Paper	35.1 % of dry weight	Semiseasonal state, USA, Ala., municipal refuse, ann. average, 1968[1]
Paper	38.2 % of dry weight	Seasonal state, Mass, USA, municipal refuse, ann. average, 1968[1]

Plastics	1.0 % of dry weight	Unseasonal state, USA, Fl., municipal refuse, ann. average, 1968[1]
Plastics	1.1 % of dry weight	Seasonal state, Mass, USA, municipal refuse, ann. average, 1968[1]
Plastics	1.1 % of dry weight	Semiseasonal state, USA, Ala., municipal refuse, ann. average, 1968[1]
Textiles	1.8 % of dry weight	Unseasonal state, USA, Fl., municipal refuse, ann. average, 1968[1]
Textiles	2.0 % of dry weight	Seasonal state, Mass, USA, municipal refuse, ann. average, 1968[1]
Textiles	91.9 % of dry weight	Semiseasonal state, USA, Ala., municipal refuse, ann. average, 1968[1]
Wood	2.3 % of dry weight	Unseasonal state, USA, Fl., municipal refuse, ann. average, 1968[1]
Wood	2.4 % of dry weight	Semiseasonal state, USA, Ala., municipal refuse, ann. average, 1968[1]
Wood	2.7 % of dry weight	Seasonal state, Mass, USA, municipal refuse, ann. average, 1968[1]
Yard waste	14.1 % of dry weight	Seasonal state, Mass, USA, municipal refuse, ann. average, 1968[1]
Yard waste	20.7 % of dry weight	Semiseasonal state, USA, Ala., municipal refuse, ann. average, 1968[1]
Yard waste	26.1 % of dry weight	Unseasonal state, USA, Fl., municipal refuse, ann. average, 1968[1]

Source: 1.Baum, B. and C.H. Parker. 1973. Solid Waste Disposal. Vol 1. Incinerarion and Landfill, Ann Arbor Science. Ann Arbor, Michigan.

6-006 Equilibrium Adsorption of Zn

Species	Value	Condition
Zn sorption on clay, bentonite	10 % fixed of added Zn	30 d wet inc., add =500 meq/kg clay, pH=6.2, desorp = 1 M NH4OAc (pH=7)[1]
Zn sorption on clay, bentonite	11 % fixed of added Zn	30 d wet inc., add=1000 meq/kg clay, pH=8.8, desorp = 1 M NH4OAc (pH=7)[1]
Zn sorption on clay, bentonite	12 % fixed of added Zn	30 d wet inc., add =125 meq/kg clay, pH=6.2, desorp = 1 M NH4OAc (pH=7)[1]
Zn sorption on clay, bentonite	13 % fixed of added Zn	30 d wet inc., add =500 meq/kg clay, pH=7.6, desorp = 1 M NH4OAc (pH=7)[1]
Zn sorption on clay, bentonite	13 % fixed of added Zn	Wet and dry cycles, add=1000 meq/kg clay, pH=4.9, desorp = 1 M NH4OAc (pH=7)[1]
Zn sorption on clay, bentonite	14 % fixed of added Zn	30 d wet inc., add =250 meq/kg clay, pH=6.2, desorp = 1 M NH4OAc (pH=7)[1]
Zn sorption on clay, bentonite	14 % fixed of added Zn	Wet and dry cycles, add =125 meq/kg clay, pH=4.9, desorp = 1 M NH4OAc (pH=7)[1]
Zn sorption on clay, bentonite	14 % fixed of added Zn	Wet and dry cycles, add =250 meq/kg clay, pH=4.9, desorp = 1 M NH4OAc (pH=7)[1]
Zn sorption on clay, bentonite	14 % fixed of added Zn	Wet and dry cycles, add =500 meq/kg clay, pH=4.9, desorp = 1 M NH4OAc (pH=7)[1]
Zn sorption on clay, bentonite	14 % fixed of added Zn	Wet and dry cycles, add=1000 meq/kg clay, pH=6.2, desorp = 1 M NH4OAc (pH=7)[1]
Zn sorption on clay, bentonite	17 % fixed of added Zn	Wet and dry cycles, add =500 meq/kg clay, pH=6.2, desorp = 1 M NH4OAc (pH=7)[1]
Zn sorption on clay, bentonite	18 % fixed of added Zn	30 d wet inc., add =250 meq/kg clay, pH=7.6, desorp = 1 M NH4OAc (pH=7)[1]
Zn sorption on clay, bentonite	20 % fixed of added Zn	Wet and dry cycles, add =500 meq/kg clay, pH=7.6, desorp = 1 M NH4OAc (pH=7)[1]
Zn sorption on clay, bentonite	20 % fixed of added Zn	Wet and dry cycles, add=1000 meq/kg clay, pH=7.6, desorp = 1 M NH4OAc (pH=7)[1]
Zn sorption on clay, bentonite	21 % fixed of added Zn	30 d wet inc., add =125 meq/kg clay, pH=7.6, desorp = 1 M NH4OAc (pH=7)[1]
Zn sorption on clay, bentonite	22 % fixed of added Zn	Wet and dry cycles, add =250 meq/kg clay, pH=6.2, desorp = 1 M NH4OAc (pH=7)[1]

Zn sorption on clay, bentonite	23 % fixed of added Zn	Wet and dry cycles, add=1000 meq/kg clay, pH=8.8, desorp = 1 M NH4OAc (pH=7)[1]
Zn sorption on clay, bentonite	26 % fixed of added Zn	30 d wet inc., add =250 meq/kg clay, pH=8.8, desorp = 1 M NH4OAc (pH=7)[1]
Zn sorption on clay, bentonite	27 % fixed of added Zn	Wet and dry cycles, add =500 meq/kg clay, pH=8.8, desorp = 1 M NH4OAc (pH=7)[1]
Zn sorption on clay, bentonite	29 % fixed of added Zn	Wet and dry cycles, add =250 meq/kg clay, pH=7.6, desorp = 1 M NH4OAc (pH=7)[1]
Zn sorption on clay, bentonite	34 % fixed of added Zn	Wet and dry cycles, add =125 meq/kg clay, pH=6.2, desorp = 1 M NH4OAc (pH=7)[1]
Zn sorption on clay, bentonite	38 % fixed of added Zn	30 d wet inc., add =125 meq/kg clay, pH=8.8, desorp = 1 M NH4OAc (pH=7)[1]
Zn sorption on clay, bentonite	39 % fixed of added Zn	Wet and dry cycles, add =125 meq/kg clay, pH=7.6, desorp = 1 M NH4OAc (pH=7)[1]
Zn sorption on clay, bentonite	39 % fixed of added Zn	Wet and dry cycles, add =250 meq/kg clay, pH=8.8, desorp = 1 M NH4OAc (pH=7)[1]
Zn sorption on clay, bentonite	4 % fixed of added Zn	30 d wet inc., add=1000 meq/kg clay, pH=4.9, desorp = 1 M NH4OAc (pH=7)[1]
Zn sorption on clay, bentonite	46 % fixed of added Zn	Wet and dry cycles, add =125 meq/kg clay, pH=8.8, desorp = 1 M NH4OAc (pH=7)[1]
Zn sorption on clay, bentonite	6 % fixed of added Zn	30 d wet inc., add =500 meq/kg clay, pH=4.9, desorp = 1 M NH4OAc (pH=7)[1]
Zn sorption on clay, bentonite	7 % fixed of added Zn	30 d wet inc., add =125 meq/kg clay, pH=4.9, desorp = 1 M NH4OAc (pH=7)[1]
Zn sorption on clay, bentonite	7 % fixed of added Zn	30 d wet inc., add =250 meq/kg clay, pH=4.9, desorp = 1 M NH4OAc (pH=7)[1]
Zn sorption on clay, bentonite	7 % fixed of added Zn	30 d wet inc., add=1000 meq/kg clay, pH=6.2, desorp = 1 M NH4OAc (pH=7)[1]
Zn sorption on clay, bentonite	9 % fixed of added Zn	30 d wet inc., add=1000 meq/kg clay, pH=7.6, desorp = 1 M NH4OAc (pH=7)[1]
Zn sorption on clay, illite	10 % fixed of added Zn	30 d wet inc., add =320 meq/kg clay, pH=6.2, desorp = 1 M NH4OAc (pH=7)[1]
Zn sorption on clay, illite	13 % fixed of added Zn	30 d wet inc., add = 40 meq/kg clay, pH=4.9, desorp = 1 M NH4OAc (pH=7)[1]
Zn sorption on clay, illite	13 % fixed of added Zn	30 d wet inc., add = 80 meq/kg clay, pH=4.9, desorp = 1 M NH4OAc (pH=7)[1]
Zn sorption on clay, illite	14 % fixed of added Zn	30 d wet inc., add =320 meq/kg clay, pH=7.6, desorp = 1 M NH4OAc (pH=7)[1]
Zn sorption on clay, illite	15 % fixed of added Zn	30 d wet inc., add = 40 meq/kg clay, pH=6.2, desorp = 1 M NH4OAc (pH=7)[1]
Zn sorption on clay, illite	15 % fixed of added Zn	30 d wet inc., add =160 meq/kg clay, pH=6.2, desorp = 1 M NH4OAc (pH=7)[1]
Zn sorption on clay, illite	16 % fixed of added Zn	30 d wet inc., add =320 meq/kg clay, pH=8.8, desorp = 1 M NH4OAc (pH=7)[1]
Zn sorption on clay, illite	19 % fixed of added Zn	30 d wet inc., add = 80 meq/kg clay, pH=6.2, desorp = 1 M NH4OAc (pH=7)[1]
Zn sorption on clay, illite	22 % fixed of added Zn	30 d wet inc., add =160 meq/kg clay, pH=7.6, desorp = 1 M NH4OAc (pH=7)[1]
Zn sorption on clay, illite	22 % fixed of added Zn	Wet and dry cycles, add =320 meq/kg clay, pH=4.9, desorp = 1 M NH4OAc (pH=7)[1]
Zn sorption on clay, illite	23 % fixed of added Zn	Wet and dry cycles, add = 40 meq/kg clay, pH=4.9, desorp = 1 M NH4OAc (pH=7)[1]

Zn sorption on clay, illite	24 % fixed of added Zn	Wet and dry cycles, add =160 meq/kg clay, pH=4.9, desorp = 1 M NH4OAc (pH=7)[1]
Zn sorption on clay, illite	25 % fixed of added Zn	30 d wet inc., add = 40 meq/kg clay, pH=7.6, desorp = 1 M NH4OAc (pH=7)[1]
Zn sorption on clay, illite	25 % fixed of added Zn	Wet and dry cycles, add = 80 meq/kg clay, pH=4.9, desorp = 1 M NH4OAc (pH=7)[1]
Zn sorption on clay, illite	26 % fixed of added Zn	Wet and dry cycles, add =320 meq/kg clay, pH=6.2, desorp = 1 M NH4OAc (pH=7)[1]
Zn sorption on clay, illite	28 % fixed of added Zn	30 d wet inc., add = 80 meq/kg clay, pH=7.6, desorp = 1 M NH4OAc (pH=7)[1]
Zn sorption on clay, illite	28 % fixed of added Zn	30 d wet inc., add =160 meq/kg clay, pH=8.8, desorp = 1 M NH4OAc (pH=7)[1]
Zn sorption on clay, illite	29 % fixed of added Zn	Wet and dry cycles, add = 80 meq/kg clay, pH=6.2, desorp = 1 M NH4OAc (pH=7)[1]
Zn sorption on clay, illite	34 % fixed of added Zn	Wet and dry cycles, add =160 meq/kg clay, pH=6.2, desorp = 1 M NH4OAc (pH=7)[1]
Zn sorption on clay, illite	35 % fixed of added Zn	Wet and dry cycles, add = 40 meq/kg clay, pH=6.2, desorp = 1 M NH4OAc (pH=7)[1]
Zn sorption on clay, illite	38 % fixed of added Zn	30 d wet inc., add = 40 meq/kg clay, pH=8.8, desorp = 1 M NH4OAc (pH=7)[1]
Zn sorption on clay, illite	38 % fixed of added Zn	30 d wet inc., add = 80 meq/kg clay, pH=8.8, desorp = 1 M NH4OAc (pH=7)[1]
Zn sorption on clay, illite	39 % fixed of added Zn	Wet and dry cycles, add =320 meq/kg clay, pH=7.6, desorp = 1 M NH4OAc (pH=7)[1]
Zn sorption on clay, illite	4 % fixed of added Zn	30 d wet inc., add =320 meq/kg clay, pH=4.9, desorp = 1 M NH4OAc (pH=7)[1]
Zn sorption on clay, illite	40 % fixed of added Zn	Wet and dry cycles, add = 40 meq/kg clay, pH=7.6, desorp = 1 M NH4OAc (pH=7)[1]
Zn sorption on clay, illite	43 % fixed of added Zn	Wet and dry cycles, add = 80 meq/kg clay, pH=7.6, desorp = 1 M NH4OAc (pH=7)[1]
Zn sorption on clay, illite	44 % fixed of added Zn	Wet and dry cycles, add =320 meq/kg clay, pH=8.8, desorp = 1 M NH4OAc (pH=7)[1]
Zn sorption on clay, illite	45 % fixed of added Zn	Wet and dry cycles, add = 40 meq/kg clay, pH=8.8, desorp = 1 M NH4OAc (pH=7)[1]
Zn sorption on clay, illite	48 % fixed of added Zn	Wet and dry cycles, add =160 meq/kg clay, pH=7.6, desorp = 1 M NH4OAc (pH=7)[1]
Zn sorption on clay, illite	55 % fixed of added Zn	Wet and dry cycles, add = 80 meq/kg clay, pH=8.8, desorp = 1 M NH4OAc (pH=7)[1]
Zn sorption on clay, illite	60 % fixed of added Zn	Wet and dry cycles, add =160 meq/kg clay, pH=8.8, desorp = 1 M NH4OAc (pH=7)[1]
Zn sorption on clay, illite	8 % fixed of added Zn	30 d wet inc., add =160 meq/kg clay, pH=4.9, desorp = 1 M NH4OAc (pH=7)[1]
Zn sorption on clay, kaolinite	0 % fixed of added Zn	30 d wet inc., add = 40 meq/kg clay, pH=4.9, desorp = 1 M NH4OAc (pH=7)[1]
Zn sorption on clay, kaolinite	0 % fixed of added Zn	30 d wet inc., add = 40 meq/kg clay, pH=6.2, desorp = 1 M NH4OAc (pH=7)[1]
Zn sorption on clay, kaolinite	0 % fixed of added Zn	Wet and dry cycles, add = 40 meq/kg clay, pH=4.9, desorp = 1 M NH4OAc (pH=7)[1]
Zn sorption on clay, kaolinite	0 % fixed of added Zn	Wet and dry cycles, add = 40 meq/kg clay, pH=6.2, desorp = 1 M NH4OAc (pH=7)[1]
Zn sorption on clay, kaolinite	0 % fixed of added Zn	Wet and dry cycles, add = 80 meq/kg clay, pH=4.9, desorp = 1 M NH4OAc (pH=7)[1]

Zn sorption on clay, kaolinite	0 % fixed of added Zn	Wet and dry cycles, add = 80 meq/kg clay, pH=6.2, desorp = 1 M NH4OAc (pH=7)[1]
Zn sorption on clay, kaolinite	0 % fixed of added Zn	Wet and dry cycles, add =160 meq/kg clay, pH=4.9, desorp = 1 M NH4OAc (pH=7)[1]
Zn sorption on clay, kaolinite	0 % fixed of added Zn	Wet and dry cycles, add =160 meq/kg clay, pH=6.2, desorp = 1 M NH4OAc (pH=7)[1]
Zn sorption on clay, kaolinite	1 % fixed of added Zn	Wet and dry cycles, add = 40 meq/kg clay, pH=7.6, desorp = 1 M NH4OAc (pH=7)[1]
Zn sorption on clay, kaolinite	10 % fixed of added Zn	30 d wet inc., add =160 meq/kg clay, pH=4.9, desorp = 1 M NH4OAc (pH=7)[1]
Zn sorption on clay, kaolinite	11 % fixed of added Zn	30 d wet inc., add =160 meq/kg clay, pH=6.2, desorp = 1 M NH4OAc (pH=7)[1]
Zn sorption on clay, kaolinite	11 % fixed of added Zn	30 d wet inc., add =160 meq/kg clay, pH=7.6, desorp = 1 M NH4OAc (pH=7)[1]
Zn sorption on clay, kaolinite	11 % fixed of added Zn	30 d wet inc., add =160 meq/kg clay, pH=8.8, desorp = 1 M NH4OAc (pH=7)[1]
Zn sorption on clay, kaolinite	11 % fixed of added Zn	Wet and dry cycles, add =160 meq/kg clay, pH=8.8, desorp = 1 M NH4OAc (pH=7)[1]
Zn sorption on clay, kaolinite	17 % fixed of added Zn	30 d wet inc., add =500 meq/kg clay, pH=8.8, desorp = 1 M NH4OAc (pH=7)[1]
Zn sorption on clay, kaolinite	4 % fixed of added Zn	Wet and dry cycles, add = 80 meq/kg clay, pH=7.6, desorp = 1 M NH4OAc (pH=7)[1]
Zn sorption on clay, kaolinite	7 % fixed of added Zn	30 d wet inc., add = 80 meq/kg clay, pH=4.9, desorp = 1 M NH4OAc (pH=7)[1]
Zn sorption on clay, kaolinite	7 % fixed of added Zn	Wet and dry cycles, add = 80 meq/kg clay, pH=8.8, desorp = 1 M NH4OAc (pH=7)[1]
Zn sorption on clay, kaolinite	8 % fixed of added Zn	30 d wet inc., add = 40 meq/kg clay, pH=7.6, desorp = 1 M NH4OAc (pH=7)[1]
Zn sorption on clay, kaolinite	8 % fixed of added Zn	30 d wet inc., add = 40 meq/kg clay, pH=8.8, desorp = 1 M NH4OAc (pH=7)[1]
Zn sorption on clay, kaolinite	8 % fixed of added Zn	30 d wet inc., add = 80 meq/kg clay, pH=6.2, desorp = 1 M NH4OAc (pH=7)[1]
Zn sorption on clay, kaolinite	8 % fixed of added Zn	Wet and dry cycles, add = 40 meq/kg clay, pH=8.8, desorp = 1 M NH4OAc (pH=7)[1]
Zn sorption on clay, kaolinite	8 % fixed of added Zn	Wet and dry cycles, add =160 meq/kg clay, pH=7.6, desorp = 1 M NH4OAc (pH=7)[1]
Zn sorption on clay, kaolinite	9 % fixed of added Zn	30 d wet inc., add = 80 meq/kg clay, pH=7.6, desorp = 1 M NH4OAc (pH=7)[1]
Zn sorption on clay, kaolinite	9 % fixed of added Zn	30 d wet inc., add = 80 meq/kg clay, pH=8.8, desorp = 1 M NH4OAc (pH=7)[1]

Source: 1.Reddy, M.R. and H.F. Perkins. 1974. Soil Sci. Soc. Am. Proc. 38, 229-231.

6-007 Equilibrium constant Ca

Species	Value	Condition
Humic compound and Ca	3.4 = log(K)	Soil as humic comp., overall stability constant, 0.1 M Ionic, pH=5.0[1]
Humic compound and Ca	3.60 = log(K)	Seawater as humic comp., overall stability const., 0.02 M Ionic, pH=8.0[2]
Humic compound and Ca	3.65 = log(K)	Lochs as humic comp., overall stability constant, 0.02 M Ionic, pH=8.0[2]
Humic compound and Ca	3.65 = log(K)	Peat as humic comp., overall stability const. 0.02 M Ionic, pH = 8.0[2]

Humic compound and Ca	3.95 = log(K)	Lakes as humic comp., overall stability constant, 0.02 M Ionic, pH=8.0[2]
Humic compound and Ca	4.65 = log(K)	Sediment as humic comp., overall stability onst. , 0.02 M Ionic, pH=8.0[2]

Source: 1.Schnitzer, M. and S.U. Khan. 1972. in: Humic Substances in the Environment. Marcel Dekker. N. pp 203-251.
2.Mantoura, R.F.C., A. Dickson and J.P. Riley. 1978. Eustarine Coastral Mar. Sci. 6, 387-408.

6-008 Equilibrium constant
Cd

Species	Value	Condition
Cd(2+) + NTA(3-) <-> Cd-NTA(-)	3.96 = log(K)	Operational K, total Cd = 1E-7 M, pH = 8.01, Adriatic seawater[1]
Cd(2+) + NTA(3-) <-> Cd-NTA(-)	3.96 = log(K)	Operational K, total Cd = 1E-7 M, pH = 8.29, Artificial seawater[1]
Cd(2+) + NTA(3-) <-> Cd-NTA(-)	5.16 = log(K)	Operational K, total Cd = 1E-7 M, pH = 7.46, 0.0536 M MgCl2[1]
Cd(2+) + NTA(3-) <-> Cd-NTA(-)	5.55 = log(K)	Operational K, total Cd = 1E-7 M, pH = 7.38, 0.01 M CaCl2[1]
Cd(2+) + NTA(3-) <-> Cd-NTA(-)	6.46 = log(K)	Operational K, total Cd = 1E-7 M, pH = 7.83, 0.59 M NaCl[1]
Cd(2+) + NTA(3-) <-> Cd-NTA(-)	7.05 = log(K)	Operational K, total Cd = 1E-7 M, pH = 7.35, 0.7 M NaClO4[1]
Cd(2+) + NTA(3-) <-> Cd-NTA(-)	9.33 = log(K)	Common conditional K, total Cd = 1E-7 M, pH = 7.46, 0.0536 M MgCl2[1]
Cd(2+) + NTA(3-) <-> Cd-NTA(-)	9.44 = log(K)	Common conditional K, total Cd = 1E-7 M, pH = 8.01, Adriatic seawater[1]
Cd(2+) + NTA(3-) <-> Cd-NTA(-)	9.44 = log(K)	Common conditional K, total Cd = 1E-7 M, pH = 8.29, Artificial seawater[1]
Cd(2+) + NTA(3-) <-> Cd-NTA(-)	9.53 = log(K)	Common conditional K, total Cd = 1E-7 M, pH = 7.38, 0.01 M CaCl2[1]
Cd(2+) + NTA(3-) <-> Cd-NTA(-)	9.60 = log(K)	Common conditional K, total Cd = 1E-7 M, pH = 7.35, 0.7 M NaClO4[1]
Cd(2+) + NTA(3-) <-> Cd-NTA(-)	9.90 = log(K)	Common conditional K, total Cd = 1E-7 M, pH = 7.83, 0.59 M NaCl[1]
Humic compound and Cd	4.57 = log(K)	Lakes as humic comp., overall stability constant, 0.02 M Ionic, pH=8.0[2]
Humic compound and Cd	4.57 = log(K)	Peat as humic comp., overall stability constant, 0.02 M Ionic, pH=8.0[2]
Humic compound and Cd	4.95 = log(K)	Lochs as humic comp., overall stability constant, 0.02 M Ionic, pH=8.0[2]
Humic compound and Cd	5.08 = log(K)	Soil as humic comp., overall stability constant, 0.1 M Ionic, pH=4-5[3]
Metal ligand system, Cd(+2) + CO3	1.21E5	Second constant, ionic medium, infinite sol., n=34, conc = 1E-6 M[4]
Metal ligand system, Cd(+2) + CO3	4.46E4	Second constant, ionic medium, H(+) -> zero, n=34, conc = 1E-6 M[4]
Metal ligand system, Cd(+2) + CO3	4.67E2	First constant, ionic medium, H(+) -> zero, n=34, conc = 1E-6 M[4]
Metal ligand system, Cd(+2) + CO3	7.98E2	First constant, ionic medium, infinite sol., n=34, conc = 1E-6 M[4]
Metal ligand system,Cd(+2) + HCO3	1.54	First constant, ionic medium, infinite sol., n=34, conc = 1E-6 M[4]
Metal ligand system,Cd(+2) + HCO3	1.84	First constant, ionic medium, H(+) -> zero, n=34, conc = 1E-6 M[4]
Metal ligand system,Cd(+2) + HCO3	22.8	Second constant, ionic medium, H(+) -> zero, n=34, conc = 1E-6 M[4]
Metal ligand system,Cd(+2) + HCO3	24.92	Second constant, ionic medium, infinite sol., n=34, conc = 1E-6 M[4]

Source: 1.Raspor,B., P. Valenta, H.W. Nurnberg and M. Branica. 1977. Sci. Total Env. 9, pp 87-109.
2.Mantoura, R.F.C., A. Dickson and J.P. Riley. 1978. Eustarine Coastal Mar. Sci. 6, 387-408.
3.Cheam, V. and D.S. Gamble. 1974. Mar. Chem. 2, 17-32.
4.Sipos, L., P. Valenta, H.W. Nurnberg and M. Branica. 1979. J. Electronal. Chem. 87, 347-352.

6-009 Equilibrium constant Co

Species	Value	Condition
Humic compound and Co	4.2 = log(K)	Soil as humic comp., overall stability constant, 0.1 M Ionic, pH=5.0[1]
Humic compound and Co	4.29 = log(K)	Lochs as humic comp., overall stability constant, 0.02 M Ionic, pH=8.0[2]
Humic compound and Co	4.51 = log(K)	Peat as humic comp., overall stability constant, 0.02 M Ionic, pH=8.0[2]
Humic compound and Co	4.83 = log(K)	Lakes as humic comp., overall stability constant, 0.02 M Ionic, pH=8.0[2]
Humic compound and Co	4.83 = log(K)	Seawater as humic comp., overall stability const., 0.02 M Ionic, pH=8.0[2]
Humic compound and Co	4.91 = log(K)	Sediment as humic comp., overall stability const., 0.02 M Ionic, pH=8.0[2]
Humic compound and Co	6.6 = log(K)	Soil as humic comp., overall stability constant, 0.1 M Ionic, pH=6.0[3]

Source: 1.Schnitzer, M. and S.U. Khan. 1972. in: Humic Substances in the Environment. Marcel Dekker. N. pp 203-251.
2.Mantoura, R.F.C., A. Dickson and J.P. Riley. 1978. Eustarine Coastal Mar. Sci. 6, 387-408.
3.Malcolm, E.L. 1972. in: Nelson, E.D. (ed): Environmental Framework of Coastal Plain Estuaries. Geol.Survey of Wash.

6-010 Equilibrium constant Cu

Species	Value	Condition
1-Methylimidazole + Cu(2+)	4.2 = log(K)	298 K, fresh water, organic complex formation[1]
1/2 Cu2(OH)2(CO3) (s) + 2 H(+) <-> Cu(2+) + 3/2 H2O + 1/2 CO2 (g)	6.5 = log(K)	298 K, fresh water[2]
1/3 Cu3(OH)2(CO3)2 (s) + 2 H(+) <-> Cu(2+) + 4/3 H2O + 2/3 CO2 (g)	6.5 = log(K)	298 K, fresh water[2]
2 Cu(2+) + 2 OH(-) <-> Cu2(OH)2(2+)	17.7 = log(K)	298 K, fresh water[3]
8-Hydrixyquinoline + Cu(2+)	12.2 = log(K)	298 K, fresh water, organic complex formation[1]
Acetic acid + Cu(2+)	1.87 = log(K)	298 K, fresh water, organic complex formation[1]
Adenine + Cu(2+)	8.9 = log(K)	298 K, fresh water, organic complex formation[1]
Alanine + Cu(2+)	8.5 = log(K)	298 K, fresh water, organic complex formation[1]
Argimine + Cu(2+)	9.5 = log(K)	298 K, fresh water, organic complex formation[1]
Aspargine + Cu(2+)	7.8 = log(K)	298 K, fresh water, organic complex formation[1]
Aspartic acid + Cu(2+)	8.6 = log(K)	298 K, fresh water, organic complex formation[1]
Catechol + Cu(2+)	14.1 = log(K)	298 K, fresh water, organic complex formation[1]
Citric acid + Cu(2+)	13.2 = log(K)	298 K, fresh water, organic complex formation[1]
Cu(+) + 2 Cl(-) <-> CuCl2(-)	5.5 = log(K)	298 K, fresh water[3]
Cu(+) + 2 NH3 <-> Cu(NH3)2(+)	10.3 = log(K)	298 K, fresh water[3]
Cu(+) + 3 Cl(-) <-> CuCl3(2-)	5.7 = log(K)	298 K, fresh water[3]
Cu(+) + NH3 <-> Cu(NH3)(+)	5.5 = log(K)	298 K, fresh water[3]
Cu(2+) + 2 CO3(2-) <-> Cu(CO3)2(2-)	9.9 = log(K)	298 K, fresh water[3]
Cu(2+) + 2 NH3 + 2 OH(-) <-> Cu(NH3)2(OH)2	22.5 = log(K)	298 K, fresh water[3]
Cu(2+) + 2 NH3 <-> Cu(NH3)2(2+)	10.7 = log(K)	298 K, fresh water[3]
Cu(2+) + 2 OH(-) <-> Cu2(OH)2	14.3 = log(K)	298 K, fresh water[3]
Cu(2+) + 3 HS(-) <-> Cu(HS)3(-)	26.5 = log(K)	298 K, fresh water[3]

Cu(2+) + 3 NH3 + OH(-) <-> Cu(NH3)3(OH)(+)	21.1 = log(K)	298 K, fresh water[3]
Cu(2+) + 3 NH3 <-> Cu(NH3)3(2+)	14.7 = log(K)	298 K, fresh water[3]
Cu(2+) + 3 OH(-) <-> Cu(OH)3(-)	14.2 = log(K)	298 K, fresh water[3]
Cu(2+) + 4 NH3 <-> Cu(NH3)4(2+)	17.6 = log(K)	298 K, fresh water[3]
Cu(2+) + CO3(2-) <-> CuCO3	6.1 = log(K)	298 K, fresh water[3]
Cu(2+) + F(-) <-> CuF(+)	1.23 = log(K)	298 K, fresh water[3]
Cu(2+) + H(+) + PO4(3-) <-> Cu(HPO4)	16.6 = log(K)	298 K, fresh water[3]
Cu(2+) + NH3 <-> Cu(NH3)(2+)	5.8 = log(K)	298 K, fresh water[3]
Cu(2+) + OH(-) <-> Cu(OH)(-)	6.0 = log(K)	298 K, fresh water[3]
Cu(2+) + SO4(2-) <-> CuSO4	2.3 = log(K)	298 K, fresh water[3]
Cu(2+) 4 HS(-) <-> CuS(HS)3(3-) + H(+)	18.12 = log(K)	298 K, fresh water[3]
Cu(OH)2 (s) <-> Cu(2+) + 2 OH(-)	19.36 = log(K)	298 K, fresh water[2]
Cu2O (s) + H2O <-> 2 Cu(2+) + 2 OH(-)	23.6 = log(K)	298 K, fresh water[2]
Cu2S (s) <-> Cu(+) + S(2-)	48.1 = log(K)	298 K, fresh water[2]
Cu3(PO4)2 (s) <-> 3 Cu(2+) + 2 PO4(3-)	37.7 = log(K)	298 K, fresh water[2]
CuCO3 (s) <-> Cu(2+) + CO3(2-)	9.6 = log(K)	298 K, fresh water[2]
CuCl (s) <-> Cu(+) + Cl(-)	6.7 = log(K)	298 K, fresh water[2]
CuO(s) + H2O <-> Cu(2+) + 2 OH(-)	20.4 = log(K)	298 K, fresh water[2]
CuOH (s) <-> Cu(+) + OH(-)	28.7 = log(K)	298 K, fresh water[2]
CuS (s) <-> Cu(2+) + S(2-)	36.2 = log(K)	298 K, fresh water[2]
Ethylendiamine + Cu(2+)	10.6 = log(K)	298 K, fresh water, organic complex formation[1]
Ethylendiamine tetraacetic acid + Cu(2+)	18.8 = log(K)	298 K, fresh water, organic complex formation[1]
Formic acid + Cu(2+)	1.98 = log(K)	298 K, fresh water, organic complex formation[1]
Glutamic acid + Cu(2+)	7.9 = log(K)	298 K, fresh water, organic complex formation[1]
Glycollic acid + Cu(2+)	2.5 = log(K)	298 K, fresh water, organic complex formation[1]
Glycylglycine + Cu(2+)	0 = log(K)	298 K, fresh water, organic complex formation[1]
Histiamine + Cu(2+)	9.5 = log(K)	298 K, fresh water, organic complex formation[1]
Histidine + Cu(2+)	10.6 = log(K)	298 K, fresh water, organic complex formation[1]
Humic compound and Cu	11.37 = log(K)	Sediment as humic comp., overall stability const., 0.02 M Ionic, pH=8.0[4]
Humic compound and Cu	4.0 = log(K)	Soil as humic comp., overall stability constant, 0.1 M Ionic, pH=5.0[5]
Humic compound and Cu	7.8 = log(K)	Soil as humic comp., overall stability constant, 0.1 M Ionic, pH=6.8[6]
Humic compound and Cu	7.85 = log(K)	Peat as humic comp., overall stability constant, 0.02 M Ionic, pH=8.0[4]
Humic compound and Cu	8.89 = log(K)	Lochs as humic comp., overall stability constant, 0.02 M Ionic, pH=8.0[4]
Humic compound and Cu	8.89 = log(K)	Seawater as humic comp., overall stability const., 0.02 M Ionic, pH=8.0[4]
Humic compound and Cu	9.48 = log(K)	Rivers as humic comp., overall stability constant, 0.02 M Ionic, pH=8.0[4]

Humic compound and Cu	9.83 = log(K)	Lakes as humic comp., overall stability constant, 0.02 M Ionic, pH=8.0[4]
Hydroxamic acid + Cu(2+)	7.0 = log(K)	298 K, humic-type material, conditional stability, I -> 0[7]
Hydroxyproline + Cu(2+)	6.9 = log(K)	298 K, fresh water, organic complex formation[1]
Imidazole + Cu(2+)	4.3 = log(K)	298 K, fresh water, organic complex formation[1]
Lake water fluvic acid + Cu(2+)	8.42 = log(K)	298 K, humic-type material, conditional stability, pH = 8.0, I = 0.02[7]
Leucine + Cu(2+)	7.6 = log(K)	298 K, fresh water, organic complex formation[1]
Loch water humic acid + Cu(2+)	9.6 = log(K)	298 K, humic-type material, conditional stability, pH = 8.0, I = 0.02[7]
Lycine + Cu(2+)	8.12 = log(K)	298 K, fresh water, organic complex formation[1]
Nitrilotriacetic acid + Cu(2+)	13.0 = log(K)	298 K, fresh water, organic complex formation[1]
Oxalic acid + Cu(2+)	6.2 = log(K)	298 K, fresh water, organic complex formation[1]
Peat fulvic acid + Cu(2+)	7.85 = log(K)	298 K, humic-type material, conditional stability, overall K[7]
Peat humic acid + Cu(2+)	8.29 = log(K)	298 K, humic-type material, conditional stability, pH = 8.0, I = 0.02[7]
Peat humic acid + Cu(2+)	9.4 = log(K)	298 K, humic-type mat., conditional stability, rg=4.0-14.8, pH = 6.8[7]
Phthalic acid + Cu(2+)	3.46 = log(K)	298 K, fresh water, organic complex formation[1]
Picolinic acid + Cu(2+)	8.0 = log(K)	298 K, fresh water, organic complex formation[1]
Protocatechuic acid + Cu(2+)	14.5 = log(K)	298 K, fresh water, organic complex formation[1]
Purine + Cu(2+)	6.9 = log(K)	298 K, fresh water, organic complex formation[1]
Pyridine + Cu(2+)	2.6 = log(K)	298 K, fresh water, organic complex formation[1]
Pyruvic acid + Cu(2+)	2.2 = log(K)	298 K, fresh water, organic complex formation[1]
River water humic acid + Cu(2+)	9.54 = log(K)	298 K, humic-type material, conditional stability, pH = 8.0, I = 0.02[7]
Salicylic acid + Cu(2+)	10.6 = log(K)	298 K, fresh water, organic complex formation[1]
Seawater humic acid + Cu(2+)	9.3 = log(K)	298 K, humic-type material, conditional stability, pH = 8.0, I = 0.02[7]
Sediment humic acid + Cu(2+)	10.5 = log(K)	298 K, humic-type material, conditional stability, pH = 8.0, I = 0.02[7]
Serine + Cu(2+)	7.9 = log(K)	298 K, fresh water, organic complex formation[1]
Soil fulvic acid + Cu(2+)	3.3 = log(K)	298 K, humic-type material, conditional stability, pH = 5[7]
Soil fulvic acid + Cu(2+)	4.0 = log(K)	298 K, humic-type material, conditional stability, pH=5, I -> 0[7]
Soil fulvic acid + Cu(2+)	4.7 = log(K)	298 K, humic-type material, conditional stability, pH = 3, I = 0.1[7]
Soil fulvic acid + Cu(2+)	5.78 = log(K)	298 K, humic-type material, conditional stability, pH = 3.5[7]
Soil fulvic acid + Cu(2+)	8.69 = log(K)	298 K, humic-type material, conditional stability, pH = 5.0[7]
Succinic acid + Cu(2+)	3.3 = log(K)	298 K, fresh water, organic complex formation[1]
Thereonine + Cu(2+)	8.4 = log(K)	298 K, fresh water, organic complex formation[1]
n - Valeric acid + Cu(2+)	2.1 = log(K)	298 K, fresh water, organic complex formation[1]

Source: 1.Sillen, L.G. and A.E. Martell. 1971. Stability Constants of Metal-Ion Complexes. Chem. Soc. PSec. Pul. 25. London.
2.Wagma, D.D. et al. 1971. Selected Values of Chemical Thermodynamics Properties. US Natl. Bur. Stand.Tech. Note 270. Parts 3-6, Wash DC.
3.Bilinski, H., R. Huston and W. Stumm. 1976. Anal. Chim. Acta 84, 157.
4.Mantoura, R.F.C., A. Dickson and J.P. Riley. 1978. Eustarine Coastal Mar. Sci. 6, 387-408.
5.Schnitzer, M. and S.U. Khan. 1972. in: Humic Substances in the Environment. Marcel Dekker. N. pp 203-251.
6.Ernst, R., H.E. Allan and K.H. Mancy. 1975. Water Res. 9, 969-979.
7.Mantoura, R.F.C., A. Dickson and J.P. Riley. 1978. Estuar. Coastal Mar. Sci. 6, 387-408.

6-011 Equilibrium constant
Hg

Species	Value	Condition
Humic compound and Hg	18.1 = log(K)	Seawater as humic comp., overall stability const., 0.02 M Ionic, pH=8.0[1]
Humic compound and Hg	18.3 = log(K)	Peat as humic comp., overall stability constant, 0.02 M Ionic, pH=8.0[1]
Humic compound and Hg	19.4 = log(K)	Lakes as humic comp., overall stability constant, 0.02 M Ionic, pH=8.0[1]
Humic compound and Hg	19.7 = log(K)	Rivers as humic comp., overall stability constant, 0.02 M Ionic, pH=8.0[1]
Humic compound and Hg	20.6 = log(K)	Lochs as humic comp., overall stability constant, 0.02 M Ionic, pH=8.0[1]
Humic compound and Hg	21.3 = log(K)	Sediment as humic comp., overall stability const., 0.02 M Ionic, pH=8.0[1]
Humic compound and Hg	5.2 = log(K)	Soil as humic comp., overall stability constant, 0.02 M Ionic, pH=8.0[2]

Source: 1.Mantoura, R.F.C., A. Dickson and J.P. Riley. 1978. Eustarine Coastal Mar. Sci. 6, 387-408.
2.Strohal, P. and D. Huljev. 1971. in: Nuclear Techniques in Environmental Science. IAEA. Vienna.

6-012 Equilibrium constant
Mg

Species	Value	Condition
Humic compound and Mg	2.2 = log(K)	Soil as humic comp., overall stability constant, 0.1 M Ionic, pH=5.0[1]
Humic compound and Mg	3.50 = log(K)	Lochs as humic comp., overall stability constant, 0.02 M Ionic, pH=8.0[2]
Humic compound and Mg	3.50 = log(K)	Seawater as humic comp., overall stability const., 0.02 M Ionic, pH=8.0[2]
Humic compound and Mg	3.81 = log(K)	Peat as humic comp., overall stability constant, 0.02 M Ionic, pH=8.0[2]
Humic compound and Mg	4.00 = log(K)	Lakes as humic comp., overall stability constant, 0.02 M Ionic, pH=8.0[2]
Humic compound and Mg	4.09 = log(K)	Sediment as humic comp., overall stability const., 0.02 M Ionic, pH=8.0[2]

Source: 1.Schnitzer, M. and S.U. Khan. 1972. in: Humic Substances in the Environment. Marcel Dekker. N. pp 203251.
2.Mantoura, R.F.C., A. Dickson and J.P. Riley. 1978. Eustarine Coastal Mar. Sci. 6, 387-408.

6-013 Equilibrium constant
Mn

Species	Value	Condition
Humic compound and Mn	3.7 = log(K)	Soil as humic comp., overall stability constant, 0.1 M Ionic, pH=5.0[1]
Humic compound and Mn	4.17 = log(K)	Peat as humic comp., overall stability constant, 0.02 M Ionic, pH=8.0[2]
Humic compound and Mn	4.45 = log(K)	Seawater as humic comp., overall stability const., 0.02 M Ionic, pH=8.0[2]
Humic compound and Mn	4.85 = log(K)	Lakes as humic comp., overall stability constant, 0.02 M Ionic, pH=8.0[2]

Source: 1.Schnitzer, M. and S.U. Khan. 1972. in: Humic Substances in the Environment. Marcel Dekker. N. pp 203-251.
2.Mantoura, R.F.C., A. Dickson and J.P. Riley. 1978. Eustarine Coastal Mar. Sci. 6, 387-408.

6-014 Equilibrium constant
Ni

Species	Value	Condition
Humic compound and Ni	4.2 = log(K)	Soil as humic comp., overall stability constant, 0.1 M Ionic, pH=5.0[1]
Humic compound and Ni	4.35 = log(K)	Soil as humic comp., overall stability constant, 0.1 M Ionic, pH=4-5[2]
Humic compound and Ni	4.98 = log(K)	Peat as humic comp., overall stability constant, 0.02 M Ionic, pH=8.0[3]
Humic compound and Ni	5.14 = log(K)	Lakes as humic comp., overall stability constant, 0.02 M Ionic, pH=8.0[3]
Humic compound and Ni	5.31 = log(K)	Lochs as humic comp., overall stability constant, 0.02 M Ionic, pH=8.0[3]
Humic compound and Ni	5.41 = log(K)	Seawater as humic comp., overall stability const., 0.02 M Ionic, pH=8.0[3]

Source: 1.Schnitzer, M. and S.U. Khan. 1972. in: Humic Substances in the Environment. Marcel Dekker. N. pp 203-251.
2.Cheam, V. and D.S. Gamble. 1974. Mar. Chem. 2, 17-32.
3.Mantoura, R.F.C., A. Dickson and J.P. Riley. 1978. Eustarine Coastal Mar. Sci. 6, 387-408.

6-015 Equilibrium constant
Pb

Species	Value	Condition
Pb2Cu(PO4)	2.99 = log(K)	Pb2Cu(PO4)(OH)3*3H20 + 4 H(+)<->2 Pb(2+) + Cu(2+) + HPO4(2-) + 6 H2O298 K, natural reaction[1]
Pb(4+)	28.6 = log(pe0)	298 K, natural system, stability constant[1]
Pb(CO3)2(2-)	9.8 = log(beta2)	298 K, natural system, stability constant, ionic strength = 1[1]
Pb(CSN)2	1.15 = log(beta2)	298 K, natural system, stability constant[1]
Pb(ClO3)2	-0.61 = log(beta2)	298 K, natural system, stability constant[1]
Pb(HPO4)2(2-)	2.37 = log(beta2)	298 K, natural sys., stability constant, internally inconstistent value[1]
Pb(NO3)(+)	1.07 = log(K1)	298 K, natural system, stability constant[1]
Pb(OH)2	10.8 = log(beta2)	298 K, natural system, stability constant[1]
Pb(OH)3(-)	13.9 = log(beta3)	298 K, natural system, stability constant[1]
Pb(P2O7)(2-)	11.2 = log(K1)	298 K, natural system, stability constant[1]
Pb3(OH)4(2+)	32.7 = log(beta4 3)	298 K, natural system, stability constant, 0.3 M NaClO4[1]
Pb4(OH)4(4+)	36.7 = log(beta4 4)	298 K, natural system, stability constant, 0.3 M NaClO4[1]
Pb6(OH)8(4+)	69.4 = log(beta8 6)	298 K, natural system, stability constant, 0.3 M NaClO4[1]
PbBr(+)	1.8 = log(K1)	298 K, natural system, stability constant[1]
PbBr2	1.44 = log(beta2)	298 K, natural system, stability constant, very uncertain value[1]
PbBr3(-)	1.19 = log(beta3)	298 K, natural system, stability constant, very uncertain value[1]
PbCO3	6.4 = log(K1)	298 K, natural system, stability constant[1]
PbCSN(+)	0.89 = log(K1)	298 K, natural system, stability constant[1]
PbCl(+)	1.61 = log(K1)	298 K, natural system, stability constant[1]
PbCl2	1.80 = log(beta2)	298 K, natural system, stability constant[1]
PbCl3(-)	1.7 = log(beta3)	298 K, natural system, stability constant[1]
PbCl4(2+)	1.4 = log(beta4)	298 K, natural system, stability constant[1]
PbClO3(+)	-0.32 = log(K1)	298 K, natural system, stability constant[1]
PbF(+)	1.25 = log(K1)	298 K, natural system, stability constant, very uncertain value[1]
PbF2	2.56 = log(beta2)	298 K, natural system, stability constant, very uncertain value[1]
PbH2PO4(+)	1.5 = log(K1)	298 K, natural system, stability constant[1]
PbHPO4	3.1 = log(K1)	298 K, natural system, stability constant[1]
PbI(+)	2.0 = log(K1)	298 K, natural system, stability constant[1]
PbI2	3.15 = log(beta2)	298 K, natural system, stability constant[1]
PbI3(-)	3.92 = log(beta3)	298 K, natural system, stability constant[1]
PbI4(2-)	4.47 = log(beta4)	298 K, natural system, stability constant[1]
PbOH(+)	7.83 = log(K1)	298 K, natural system, stability constant[1]
PbSO4	2.7 = log(K1)	293 K, natural system, stability constant[1]

Source: 1.Wagman, D.D., W.H. Evans, Parker, Hallow, Bailey and Schumm. 1969. Selected values of chemical thermodynamic properties. Nat. Bur. Stand. Tech. Note, 270-273.

6-016 Equilibrium constant
Pb as organic Pb in anoxic seawater

Species	Value	Condition
Acetate	16.6 = -log(activity)	Eff. 100 % of the complexed Pb (=80 % of tot Pb) as cysteine complex[1]
Acetate	2.6 = -log(free ligand activity)	Eff. 100 % of the complexed Pb (=80 % of tot Pb) as cysteine complex[1]
Alanine	20.0 = -log(activity)	Eff. 100 % of the complexed Pb (=80 % of tot Pb) as cysteine complex[1]
Alanine	8.1 = -log(free ligand activity)	Eff. 100 % of the complexed Pb (=80 % of tot Pb) as cysteine complex[1]
Arginine	20.1 = -log(activity)	Eff. 100 % of the complexed Pb (=80 % of tot Pb) as cysteine complex[1]
Arginine	8.1 = -log(free ligand activity)	Eff. 100 % of the complexed Pb (=80 % of tot Pb) as cysteine complex[1]
Aspartate	17.7 = -log(activity)	Eff. 100 % of the complexed Pb (=80 % of tot Pb) as cysteine complex[1]
Aspartate	7.5 = -log(free ligand activity)	Eff. 100 % of the complexed Pb (=80 % of tot Pb) as cysteine complex[1]
Citrate	17.2 = -log(activity)	Eff. 100 % of the complexed Pb (=80 % of tot Pb) as cysteine complex[1]
Citrate	5.5 = -log(free ligand activity)	Eff. 100 % of the complexed Pb (=80 % of tot Pb) as cysteine complex[1]
Cysteine	11.7 = -log(free ligand activity)	Eff. 100 % of the complexed Pb (=80 % of tot Pb) as cysteine complex[1]
Cysteine	15.8 = -log(activity)	Eff. 100 % of the complexed Pb (=80 % of tot Pb) as cysteine complex[1]
Glutamate	19.2 = -log(activity)	Eff. 100 % of the complexed Pb (=80 % of tot Pb) as cysteine complex[1]
Glutamate	7.7 = -log(free ligand activity)	Eff. 100 % of the complexed Pb (=80 % of tot Pb) as cysteine complex[1]
Glycine	17.7 = -log(activity)	Eff. 100 % of the complexed Pb (=80 % of tot Pb) as cysteine complex[1]
Glycine	6.7 = -log(free ligand activity)	Eff. 100 % of the complexed Pb (=80 % of tot Pb) as cysteine complex[1]
Glycollate	19.9 = -log(activity)	Eff. 100 % of the complexed Pb (=80 % of tot Pb) as cysteine complex[1]
Glycollate	6.0 = -log(free ligand activity)	Eff. 100 % of the complexed Pb (=80 % of tot Pb) as cysteine complex[1]
Histidine	17.1 = -log(activity)	Eff. 100 % of the complexed Pb (=80 % of tot Pb) as cysteine complex[1]
Histidine	7.9 = -log(free ligand activity)	Eff. 100 % of the complexed Pb (=80 % of tot Pb) as cysteine complex[1]
Lactate	20.1 = -log(activity)	Eff. 100 % of the complexed Pb (=80 % of tot Pb) as cysteine complex[1]
Lactate	6.0 = -log(free ligand activity)	Eff. 100 % of the complexed Pb (=80 % of tot Pb) as cysteine complex[1]
Methionine	19.8 = -log(activity)	Eff. 100 % of the complexed Pb (=80 % of tot Pb) as cysteine complex[1]
Methionine	8.1 = -log(free ligand activity)	Eff. 100 % of the complexed Pb (=80 % of tot Pb) as cysteine complex[1]
Pb(2+)	16.1 = -log(activity)	Eff. 100 % of the complexed Pb (=80 % of tot Pb) as cysteine complex[1]
Total Pb	15.6 = -log(activity)	Eff. 100 % of the complexed Pb (=80 % of tot Pb) as cysteine complex[1]
Valine	20.3 = -log(activity)	Eff. 100 % of the complexed Pb (=80 % of tot Pb) as cysteine complex[1]
Valine	8.2 = -log(free ligand activity)	Eff. 100 % of the complexed Pb (=80 % of tot Pb) as cysteine complex[1]

Source: 1.Gardner, L.R. 1974. Geochim. Cosmochim. Acta 38, 1297-1302.

6-017 Equilibrium constant
Pb organic complexes

Species	Value	Condition
Acetate and Pb	2.19 = log(K)	1.98 M acetic acid, 298 K, stability constant[1]
Acetate and Pb	2.91 = log(beta2)	1.98 M acetic acid, 298 K, stability constant[1]
Acetate and Pb	3.52 = log(beta3)	1.98 M acetic acid, 298 K, stability constant[1]
Adenosine and Pb	-0.5 = log(K)	1.0 M NaNO3, 293 K, stability constant[1]
Adrenaline and Pb	21.06 = log(beta PbHL)	0.5 M NaNO3, 293 K, stability constant[2]

Alanine and Pb	5.0 = log(K)	Ionic strength = 0, 298 K, stability constant[1]
Alanine and Pb	8.24 = log(beta2)	Ionic strength = 0, 298 K, stability constant[1]
Arginine and Pb	4.06 = log(K)	Ionic strength = 0, 298 K, stability constant[1]
Arginine and Pb	7.42 = log(beta2)	Ionic strength = 0, 298 K, stability constant[1]
Asparagine and Pb	4.91 = log(K)	3.0 M NaClO4, 298 K, stability constant[3]
Aspargine and Pb	7.82 = log(beta2)	3.0 M NaClO4, 298 K, stability constant[3]
Aspargine and Pb	8.82 = log(beta3)	3.0 M NaClO4, 298 K, stability constant[3]
Aspartate and Pb	6.02 = log(K)	1.0 M KNO3, 303 K, stability constant[1]
Aspartate and Pb	8.18 = log(beta2)	1.0 M KNO3, 303 K, stability constant[1]
Citrate and Pb	10.85 = log(beta PbH2L)	2.0 M NaClO4, 303 K, stability constant[4]
Citrate and Pb	14.95 = log(beta PbH2L2)	2.0 M NaClO4, 303 K, stability constant[4]
Citrate and Pb	4.08 = log(K)	2.0 M NaClO4, 303 K, stability constant[4]
Citrate and Pb	6.06 = log(beta2)	2.0 M NaClO4, 303 K, stability constant[4]
Citrate and Pb	8.15 = log(beta PbHL)	2.0 M NaClO4, 303 K, stability constant[4]
Cysteine and Pb	11.39 = log(K)	0.1 M KNO3, 298 K, stability constant[3]
Cysteine and Pb	13.16 = log(K)	3.0 M NaClO4, 298 K, stability constant[3]
Cysteine and Pb	19.20 = log(beta2)	3.0 M NaClO4, 298 K, stability constant[3]
Cysteine and Pb	22.47 = log(beta3)	3.0 M NaClO4, 298 K, stability constant[3]
D-penicillamine and Pb	13.0 = log(K)	0.15 M KNO3, 298 K, stability constant[1]
DOPA and Pb	5.56 = log(K)	0.37 M NaNO3, 293 K, stability constant[2]
Diglycolate and Pb	4.95 = log(K)	Ionic strength = 0, 298 K, stability constant[1]
Diglycolate and Pb	7.40 = log(K)	Ionic strength = 0, 298 K, stability constant[1]
Dithiocarbamate and Pb	15.5 = log(beta2)	0.1 M KNO3, 293 K, stability constant[1]
Dithiodiacetate and Pb	2.4 = log(K)	0.1 M KClO4, 298 K, stability constant[1]
Dopamine and Pb	22.23 = log(beta PbHL)	0.37 M NaNO3, 293 K, stability constant[2]
EDTA and Pb	10.61 = log(K PbHL)	0.1 M KNO3, 293 K, stability constant[1]
EDTA and Pb	18.04 = log(K)	0.1 M KNO3, 293 K, stability constant[1]
Formate and Pb	0.48 = log(K)	Ionic strength = 0, 298 K, stability constant[5]
Formate and Pb	0.71 = log(beta4)	Ionic strength = 0, 298 K, stability constant[5]
Formate and Pb	0.72 = log(beta2)	Ionic strength = 0, 298 K, stability constant[5]
Formate and Pb	0.77 = log(beta3)	Ionic strength = 0, 298 K, stability constant[5]
Fulvate and Pb	3.21 = log(K)	Ionic strength = 0, 298 K, stability constant[6]
Glutamate colinate + Pb	4.48 = log(K)	Ionic strength = 0, 298 K, stability constant[7]
Glutamate colinate + Pb	4.60 = log(K)	1.0 M KNO3, 303 K, stability constant[7]
Glutamate colinate and Pb	6.22 = log(beta2)	1.0 M KNO3, 303 K, stability constant[7]
Glutamate colinate and Pb	6.22 = log(beta2)	Ionic strength = 0, 298 K, stability constant[7]
Glutamate colinate and Pb	8.70 = log(beta3)	Ionic strength = 0, 298 K, stability constant[7]
Glutamine and Pb	10.12 = log(beta3)	3.0 M NaClO4, 298 K, stability constant[3]
Glutamine and Pb	4.70 = log(K)	3.0 M NaClO4, 298 K, stability constant[3]
Glutamine and Pb	8.36 = log(beta2)	3.0 M NaClO4, 298 K, stability constant[3]

Glutathione and Pb	10.6 = log(K)	0.1 M KNO3, 298 K, stability constant[1]
Glycine and Pb	3.23 = log(K)	Ionic strength = 0, 298 K, stability constant[1]
Glycine and Pb	5.47 = log(K)	Ionic strength = 0, 298 K, stability constant[8]
Glycine and Pb	8.86 = log(beta2)	Ionic strength = 0, 298 K, stability constant[8]
Glycolate and Pb	1.90 = log(K)	1.0 M KNO3, 298 K, stability constant[1]
Glycolate and Pb	3.16 = log(beta2)	1.0 M KNO3, 298 K, stability constant[1]
Glycylglycine and Pb	1.5 = log(K PbHL)	0.1 M NaClO4, 298 K, stability constant[1]
Glycylglycine and Pb	3.23 = log(K)	Ionic strength = 0, 298 K, stability constant[1]
Glycylglycine and Pb	3.32 = log(K)	0.1 M NaCLO4, 298 K, stability constant[8]
Glycylglycine and Pb	5.35 = log(beta2)	0.1 M NaClO4, 298 K, stability constant[8]
Glycylglycine and Pb	5.93 = log(beta2)	Ionic strength = 0, 298 K, stability constant[1]
Glycylglycylcine and Pb	3.02 = log(K)	3.0 M NaClO4, 298 K, stability constant[1]
Glycylglycylcine and Pb	5.75 = log(beta2)	Ionic strength = 0, 298 K, stability constant[1]
Histidine and Pb	6.84 = log(K)	0.15 M KNO3, 298 K, stability constant[3]
Histidine and Pb	6.93 = log(K)	3.0 M NaClO4, 298 K, stability constant[3]
Histidine and Pb	9.81 = log(beta2)	3.0 M NaClO4, 298 K, stability constant[3]
Humate and Pb	8.5 = log(beta PbL2)	Ionic strength = 0, 298 K, stability constant[9]
Lactate and Pb	1.98 = log(K)	3.0 M NaClO4, 298 K, stability constant[1]
Lactate and Pb	2.98 = log(beta2)	3.0 M NaClO4, 298 K, stability constant[1]
Methionine and Pb	4.38 = log(K)	0.1 M KNO3, 298 K, stability constant[1]
Methionine and Pb	8.62 = log(beta2)	0.1 M KNO3, 298 K, stability constant[1]
NTA and Pb	11.47 = log(K)	0.1 M KNO3, 293 K, stability constant[1]
Noradrenaline and Pb	20.47 = log(beta PbHL)	0.5 NaNO3, 293 K, stability constant[2]
Oxalate and Pb	3.32 = log(K)	1.5 M KNO3, 303 K, stability constant[1]
Oxalate and Pb	5.03 = log(beta2)	1.5 M KNO3, 303 K, stability constant[1]
Phenylanine and Pb	4.63 = log(K)	3.0 M NaClO4, 298 K, stability constant[3]
Phenylanine and Pb	8.35 = log(beta2)	3.0 M NaClO4, 298 K, stability constant[3]
Pyrodoxine and Pb	1.02 = log(K)	0.1 M KNO3, vitamin B6, 298 K, stability constant[10]
Pyrodoxine and Pb	1.81 = log(beta2)	0.1 M KNO3, vitamin B6, 298 K, stability constant[10]
Serine and Pb	5.05 = log(K)	3.0 M NaClO4, 298 K, stability constant[3]
Serine and Pb	8.26 = log(beta2)	3.0 M NaClO4, 298 K, stability constant[3]
Serine and Pb	9.96 = log(beta3)	3.0 M NaClO4, 298 K, stability constant[3]
Tartarate and Pb	2.92 = log(K)	0.1 M KClO4, 293 K, stability constant[1]
Tetraglycine and Pb	3.02 = log(K)	Ionic strength = 0, 298 K, stability constant[1]
Triglycine and Pb	3.02 = log(K)	Ionic strength = 0, 298 K, stability constant[1]
Tryptophan and Pb	4.89 = log(K)	3.0 M NaClO4, 298 K, stability constant[3]
Uracil ribonucleoside and Pb	3.4 = log(K)	1.0 M NaNO3, 293 K, stability constant[1]
Valine and Pb	4.02 = log(K)	1.0 M INO3, 303 K, stability constant[1]
Valine and Pb	5.89 = log(beta2)	1.0 M INO3, 303 K, stability constant[1]

Source: 1.Sillen, L.G. and A.E. Martell. 1971. Stability Constants of Metal-Ion Complexes. Chem. Soc. PSec. Pul.

25. London.
2.Grgas-Kuznar, B., V.L. Simeon and A.O. Weber. 1974. J. Inorg. Nucl. Chem. 36, 2151-2154.
3.Corrie, A.M., M.C.D. Touche and D.R. Williams. 1973. J. Chem. Soc. (London) Dalton Proc. pp 2561-256. Part 14
4.Bottari, E. and M. Vicedomini. 1973. J. Inorg. Nucl. Chem. 35, 1269-1278.
5.Milhailov, M.H., V.T. Mihailova and V.A. Khalkin. 1974. J. Inorg. Nucl. Chem. 36, 115-126.
6.Ramamoorthy, S. and P.G. Manning. 1974. J. Inorg. Nucl. Chem. 36, 695-698.
7.Ringen, J.C. and R.E. Kirby. 1974. J. Inorg. Nucl. Chem. 36, 151-161.
8.Nag, K. and P. Banerjee. 1974. J. Inorg. Nucl. Chem. 36, 2145-2149.
9.Stevenson, F.J. 1976. In: J.O. Nriagu (ed.): Environmental Biochemistry. Michigan. Ann Arbor Science 519-540.
10.Chaturvedi, D.W. and C.M. Gupta. 1974. J. Inorg. Nucl. Chem. 36, 2155-2157.

6-018 Equilibrium constant
Pb reaction

Species	Value	Condition
Pb(OH)2 + 2 H(+) <-> Pb(2+) + 2 H2O	7.34 = log(K)	278 K, natural reaction[1]
Pb(OH)2 + 2 H(+) <-> Pb(2+) + 2 H2O	7.56 = log(K)	298 K, natural reaction[1]
Pb(OH)2 + 2 H(+) <-> Pb(2+) + 2 H2O	7.72 = log(K)	313 K, natural reaction[1]
Pb(OH)2*PbCl2 + 2 H(+) <-> 2 Pb(2+) + 2 Cl(-) + 2 H2O	14.2 = log(K)	298 K, natural reaction[1]
Pb(UO2)2(PO4)2*4 H2O + 2 H(+) <-> Pb(2+) 2 UO2(2+) + 2 HPO4(2-) + ww	-23.4 = log(K)	298 K, natural reaction, ww = 4 H2O[1]
Pb2(CO3)2(OH)2 + 4 H(+) <-> 3 Pb(2+) + 2 HCO3(-) + 2 H2O	3.66 = log(K)	298 K, natural reaction[1]
Pb2(UO2)3(PO4)2(OH)4*3 H2O+6 H(+)<->2 Pb(2+)+3 UO2(2+)+2 HPO4(2-)+w	-10.7 = log(K)	298 K, natural reaction, w = 7 H2O[1]
Pb2Cu(PO4)(OH)3*3H20 + 4 H(+)<->2 Pb(2+) + Cu(2+) + HPO4(2-) + 6 H2O	2.99 = log(K)	298 K, natural reaction[1]
Pb2SiO4 + 4 H(+) <-> 2 Pb(2+) + H4SiO4	19.7 = log(K)	278 K, natural reaction[1]
Pb2SiO4 + 4 H(+) <-> 2 Pb(2+) + H4SiO4	19.8 = log(K)	298 K, natural reaction[1]
Pb2SiO4 + 4 H(+) <-> 2 Pb(2+) + H4SiO4	19.8 = log(K)	313 K, natural reaction[1]
Pb2UO2(PO4)2*2H2O + 2 H(+)<->2 Pb(2+) + UO2(2+) + 2 HPO4(2-) + 2 H2O	-21.2 = log(K)	298 K, natural reaction[1]
Pb3(PO4)2 + 2 H(+) <-> 3 Pb(2+) + 2 HPO4(2-)	-18.3 = log(K)	278 K, natural reaction[1]
Pb3(PO4)2 + 2 H(+) <-> 3 Pb(2+) + 2 HPO4(2-)	-19.7 = log(K)	298 K, natural reaction[1]
Pb3(PO4)2 + 2 H(+) <-> 3 Pb(2+) + 2 HPO4(2-)	-20.7 = log(K)	313 K, natural reaction[1]
Pb3(PO4)2 + 4 H(+) <-> 3 Pb(2+) + 2 H2PO4(-)	-4.76 = log(K)	278 K, natural reaction[1]
Pb3(PO4)2 + 4 H(+) <-> 3 Pb(2+) + 2 H2PO4(-)	-5.26 = log(K)	298 K, natural reaction[1]
Pb3(PO4)2 + 4 H(+) <-> 3 Pb(2+) + 2 H2PO4(-)	-5.63 = log(K)	313 K, natural reaction[1]
Pb4O(PO4)2 + 4 H(+) <-> 4 Pb(2+) + 2 HPO4(2-) + H2O	-10.9 = log(K)	278 K, natural reaction[1]
Pb4O(PO4)2 + 4 H(+) <-> 4 Pb(2+) + 2 HPO4(2-) + H2O	-12.2 = log(K)	298 K, natural reaction[1]
Pb4O(PO4)2 + 4 H(+) <-> 4 Pb(2+) + 2 HPO4(2-) + H2O	-13.2 = log(K)	313 K, natural reaction[1]
Pb4O(PO4)2 + 6 H(+) <-> 4 Pb(2+) + 2 H2PO4(-) + H2O	1.90 = log(K)	313 K, natural reaction[1]
Pb4O(PO4)2 + 6 H(+) <-> 4 Pb(2+) + 2 H2PO4(-) + H2O	2.24 = log(K)	298 K, natural reaction[1]
Pb4O(PO4)2 + 6 H(+) <-> 4 Pb(2+) + 2 H2PO4(-) + H2O	2.68 = log(K)	278 K, natural reaction[1]
Pb5(PO4)3Cl + 3 H(+) <-> 5 Pb(2+) + 3 HPO4(-) + Cl(-)	-47.4 = log(K)	298 K, natural reaction[1]
Pb5(PO4)3Cl + 6 H(+) <-> 5 Pb(2+) + 3 H2PO4(-) + Cl(-)	-25.8 = log(K)	298 K, natural reaction[1]
PbAl3(PO4)(OH)6*SO4 + 7 H(+) <-> Pb(2+) 3 Al(3+) + HPO4(2-) + 6 H2O	-2.71 = log(K)	298 K, natural reaction[1]
PbAl3(PO4)2(OH)5*H2O+7 H(+)<->Pb(2+)+3 Al(3+)+2 HPO4(2-)+6 H2O	-4.65 = log(K)	298 K, natural reaction[1]
PbBr2 <-> Pb(2+) + 2 Br(-)	-5.15 = log(K)	313 K, natural reaction[1]
PbBr2 <-> Pb(2+) + 2 Br(-)	-5.18 = log(K)	298 K, natural reaction[1]

PbBr2 <-> Pb(2+) + 2 Br(-)	-5.23 = log(K)	278 K, natural reaction[1]
PbCO3 + H(+) <-> Pb(2+) + HCO3(-)	-2.40 = log(K)	278 K, natural reaction[1]
PbCO3 + H(+) <-> Pb(2+) + HCO3(-)	-2.50 = log(K)	298 K, natural reaction[1]
PbCO3 + H(+) <-> Pb(2+) + HCO3(-)	-2.58 = log(K)	313 K, natural reaction[1]
PbCO3*PbCl2 + H(+) <-> 2 Pb(2+) + HCO3(-) + 2 Cl(-)	-9.35 = log(K)	278 K, natural reaction[1]
PbCO3*PbCl2 + H(+) <-> 2 Pb(2+) + HCO3(-) + 2 Cl(-)	-9.50 = log(K)	298 K, natural reaction[1]
PbCO3*PbCl2 + H(+) <-> 2 Pb(2+) + HCO3(-) + 2 Cl(-)	-9.61 = log(K)	313 K, natural reaction[1]
PbCO3*PbO + 3 H(+) <-> 2 Pb(2+) + HCO3(-) + H2O	9.77 = log(K)	313 K, natural reaction[1]
PbCO3*PbO + 3 H(+) <-> 2 Pb(2+) + HCO3(-) + H2O	9.83 = log(K)	298 K, natural reaction[1]
PbCO3*PbO + 3 H(+) <-> 2 Pb(2+) + HCO3(-) + H2O	9.91 = log(K)	278 K, natural reaction[1]
PbCl2 <-> Pb(2+) + 2 Cl(-)	-4.73 = log(K)	278 K, natural reaction[1]
PbCl2 <-> Pb(2+) + 2 Cl(-)	-4.77 = log(K)	298 K, natural reaction[1]
PbCl2 <-> Pb(2+) + 2 Cl(-)	-4.81 = log(K)	313 K, natural reaction[1]
PbClOH + H(+) <-> Pb(2+) + Cl(-) + H2O	-15.2 = log(K)	313 K, natural reaction[1]
PbClOH + H(+) <-> Pb(2+) + Cl(-) + H2O	-15.3 = log(K)	298 K, natural reaction[1]
PbClOH + H(+) <-> Pb(2+) + Cl(-) + H2O	-15.4 = log(K)	278 K, natural reaction[1]
PbF2 <-> Pb(2+) + 2 F(-)	-5.70 = log(K)	278 K, natural reaction[1]
PbF2 <-> Pb(2+) + 2 F(-)	-6.15 = log(K)	298 K, natural reaction[1]
PbF2 <-> Pb(2+) + 2 F(-)	-6.49 = log(K)	313 K, natural reaction[1]
PbFe3(3+)(PO4)(OH)6*SO4+7 H(+)<->Pb(2+)+3 Fe(3+)+HPO4(2-)+SO4(2-)+w	-16.3 = log(K)	298 K, natural reaction, w = 6 H2O[1]
PbHPO4 + H(+) <-> Pb(2+) + H2PO4(-)	-4.25 = log(K)	298 K, natural reaction[1]
PbHPO4 <-> Pb(2+) + HPO4(2-)	-11.4 = log(K)	298 K, natural reaction[1]
PbO (red) + 2 H(+) <-> Pb(2+) + H2O	12.7 = log(K)	278 K, natural reaction[1]
PbO (red) + 2 H(+) <-> Pb(2+) + H2O	12.7 = log(K)	298 K, natural reaction[1]
PbO (red) + 2 H(+) <-> Pb(2+) + H2O	12.8 = log(K)	313 K, natural reaction[1]
PbO (yellow) + 2 H(+) <-> Pb(2+) + H2O	12.9 = log(K)	278 K, natural reaction[1]
PbO (yellow) + 2 H(+) <-> Pb(2+) + H2O	12.9 = log(K)	298 K, natural reaction[1]
PbO (yellow) + 2 H(+) <-> Pb(2+) + H2O	13.0 = log(K)	313 K, natural reaction[1]
PbS + 2 H(+) <-> Pb(2+) H2S	-8.04 = log(K)	313 K, natural reaction[1]
PbS + 2 H(+) <-> Pb(2+) H2S	-8.14 = log(K)	298 K, natural reaction[1]
PbS + 2 H(+) <-> Pb(2+) H2S	-8.29 = log(K)	278 K, natural reaction[1]
PbS + H(+) <-> Pb(2+) + HS(-)	-15.1 = log(K)	278 K, natural reaction[1]
PbS + H(+) <-> Pb(2+) + HS(-)	-15.1 = log(K)	298 K, natural reaction[1]
PbS + H(+) <-> Pb(2+) + HS(-)	-15.2 = log(K)	313 K, natural reaction[1]
PbSO4 <-> Pb(2+) + SO4(2-)	-7.31 = log(K)	278 K, natural reaction[1]
PbSO4 <-> Pb(2+) + SO4(2-)	-7.72 = log(K)	298 K, natural reaction[1]
PbSO4 <-> Pb(2+) + SO4(2-)	-8.03 = log(K)	313 K, natural reaction[1]
PbSO4*2 NH3 + 2 H(+) <-> Pb(2+) SO4(2-) + 2 NH4(+)	16.3 = log(K)	313 K, natural reaction[1]
PbSO4*2 NH3 + 2 H(+) <-> Pb(2+) SO4(2-) + 2 NH4(+)	16.5 = log(K)	298 K, natural reaction[1]
PbSO4*2 NH3 + 2 H(+) <-> Pb(2+) SO4(2-) + 2 NH4(+)	16.8 = log(K)	278 K, natural reaction[1]

PbSO4*2 PbO + 4 H(+) <-> 3 Pb(2+) + SO4(2-) + H2O	10.7 = log(K)	313 K, natural reaction[1]
PbSO4*2 PbO + 4 H(+) <-> 3 Pb(2+) + SO4(2-) + H2O	10.9 = log(K)	298 K, natural reaction[1]
PbSO4*2 PbO + 4 H(+) <-> 3 Pb(2+) + SO4(2-) + H2O	11.2 = log(K)	278 K, natural reaction[1]
PbSO4*3 PbO + 6 H(+) <-> 4 Pb(2+) SO4(2-) + 3 H2O	22.0 = log(K)	313 K, natural reaction[1]
PbSO4*3 PbO + 6 H(+) <-> 4 Pb(2+) SO4(2-) + 3 H2O	22.1 = log(K)	298 K, natural reaction[1]
PbSO4*3 PbO + 6 H(+) <-> 4 Pb(2+) SO4(2-) + 3 H2O	22.4 = log(K)	278 K, natural reaction[1]
PbSO4*PbO + 2 H(+) <-> 2 Pb(2+) + SO4(2-) + H2O	-0.26 = log(K)	298 K, natural reaction[1]
PbSO4*PbO + 2 H(+) <-> 2 Pb(2+) + SO4(2-) + H2O	-0.51 = log(K)	313 K, natural reaction[1]
PbSO4*PbO + 2 H(+) <-> 2 Pb(2+) + SO4(2-) + H2O	0.08 = log(K)	278 K, natural reaction[1]
PbSiO3 + H(+) <-> Pb(2+) + H2SiO3	7.28 = log(K)	278 K, natural reaction[1]
PbSiO3 + H(+) <-> Pb(2+) + H2SiO3	7.31 = log(K)	298 K, natural reaction[1]
PbSiO3 + H(+) <-> Pb(2+) + H2SiO3	7.33 = log(K)	313 K, natural reaction[1]

Source: 1.Wagman, D.D., W.H. Evans, Parker, Hallow, Bailey and Schumm. 1969. Selected values of chemical thermodynamic properties. Nat. Bur. Stand. Tech. Note, 270-273.

6-019 Equilibrium constant Precipitation with Fe / Mn Hydroxides

Species	Value	Condition
Co	25	Metal = 10 ug/l, Precp 22 Fe+28 Mn (mg/l), lake water+30 mg EDTA/l[1]
Co	47	Metal = 10 ug/l, Precp 22 Fe+28 Mn (mg/l), distilled water[1]
Co	67	Metal = 10 ug/l, Precp 22 Fe+28 Mn (mg/l), lake water, no mixture added[1]
Co	75	Metal = 10 ug/l, Precp 22 Fe+28 Mn (mg/l), lake water+2.6 mg humic acid[1]
Cu	95	Metal = 10 ug/l, Precp 22 Fe+28 Mn (mg/l), distilled water[1]
Cu	97	Metal = 10 ug/l, Precp 22 Fe+28 Mn (mg/l), lake water+2.6 mg humic acid[1]
Cu	98	Metal = 10 ug/l, Precp 22 Fe+28 Mn (mg/l), lake water+30 mg EDTA/l[1]
Cu	98	Metal = 10 ug/l, Precp 22 Fe+28 Mn (mg/l), lake water, no mixture added[1]
Fe	99	Metal = 10 ug/l, Precp 22 Fe+28 Mn (mg/l), lake water+2.6 mg humic acid[1]
Fe	99	Metal = 10 ug/l, Precp 22 Fe+28 Mn (mg/l), lake water+30 mg EDTA/l[1]
Fe	99	Metal = 10 ug/l, Precp 22 Fe+28 Mn (mg/l), lake water, no mixture added[1]
Mn	25	Metal = 10 ug/l, Precp 22 Fe+28 Mn (mg/l), lake water+2.6 mg humic acid[1]
Mn	25	Metal = 10 ug/l, Precp 22 Fe+28 Mn (mg/l), lake water+30 mg EDTA/l[1]
Mn	25	Metal = 10 ug/l, Precp 22 Fe+28 Mn (mg/l), lake water, no mixture added[1]
Mn	80	Metal = 10 ug/l, Precp 22 Fe+28 Mn (mg/l), distilled water[1]
Zn	14	Metal = 10 ug/l, Precp 22 Fe+28 Mn (mg/l), distilled water[1]
Zn	18	Metal = 10 ug/l, Precp 22 Fe+28 Mn (mg/l), lake water+30 mg EDTA/l[1]
Zn	86	Metal = 10 ug/l, Precp 22 Fe+28 Mn (mg/l), lake water, no mixture added[1]
Zn	88	Metal = 10 ug/l, Precp 22 Fe+28 Mn (mg/l), lake water+2.6 mg humic acid[1]
Fe	98	Metal = 10 ug/l, Precp 22 Fe+28 Mn (mg/l), distilled water[1]

Source: 1.Groth, P. 1971. Arch. Hydrobiol. 68(3), pp 305-375.

6-020 Equilibrium constant
Redox reactions

Species	Value	Condition
Chlorophyll a	0.43 V	Chloroplast component, pH = 7, shift to reduced form given in value[1]
Cycochrome b6	-0.07 V	Chloroplast component, pH = 7, shift to reduced form given in value[1]
Cytochrome f	0.37 V	Chloroplast component, pH = 7, shift to reduced form given in value[1]
Ferredoxin	-0.43 V	Chloroplast component, pH = 7, shift to reduced form given in value[1]
H(+) <-> H2	-0.42 V	Chloroplast component, pH = 7, shift to reduced form given in value[1]
H2O <-> O2	0.81 V	Chloroplast component, pH = 7, shift to reduced form given in value[1]
NADPH2	-0.34 V	Chloroplast component, pH = 7, shift to reduced form given in value[1]
Plastocyanin	0.37 V	Chloroplast component, pH = 7, shift to reduced form given in value[1]
Plastoquinone	0.0 V	Chloroplast component, pH = 7, shift to reduced form given in value[1]

Source: 1.Hall, D.O. and K.K. Rao. 1978. Phytosynthesis. 2nd ed. Camelot Press. Southampton.

6-021 Equilibrium constant
Zn

Species	Value	Condition
Humic compound and Zn	3.04 = log(K)	Soil as humic comp., overall stability constant, 0.1 M Ionic, pH=4-5[1]
Humic compound and Zn	3.7 = log(K)	Soil as humic comp., overall stability constant, 0.1 M Ionic, pH=5.0[2]
Humic compound and Zn	4.83 = log(K)	Peat as humic comp., overall stability constant, 0.02 M Ionic, pH=8.0[3]
Humic compound and Zn	5.14 = log(K)	Lakes as humic comp., overall stability constant, 0.02 M Ionic, pH=8.0[3]
Humic compound and Zn	5.27 = log(K)	Seawater as humic comp., overall stability const., 0.02 M Ionic, pH=8.0[3]
Humic compound and Zn	5.36 = log(K)	Rivers as humic comp., overall stability constant, 0.02 M Ionic, pH=8.0[3]
Humic compound and Zn	5.87 = log(K)	Sediment as humic comp., overall stability const., 0.02 M Ionic, pH=8.0[3]

Source: 1.Cheam, V. and D.S. Gamble. 1974. Mar. Chem. 2, 17-32.
2.Schnitzer, M. and S.U. Khan. 1972. in: Humic Substances in the Environment. Marcel Dekker. N. pp 203-251.
3.Mantoura, R.F.C., A. Dickson and J.P. Riley. 1978. Eustarine Coastal Mar. Sci. 6, 387-408.

6-022 Equilibrium constant
pH value for 50 % precipitation

Species	Value	Condition
Cellulose and Cd	6.9 pH units	Metal ion = 10E-4 M, rg = +-0.2, metal hydroxy species[1]
Cellulose and Cu	6.4 pH units	Metal ion = 10E-4 M, rg = +-0.2, metal hydroxy species[1]
Cellulose and Pb	6.0 pH units	Metal ion = 10E-4 M, rg = +-0.2, metal hydroxy species[1]
Cellulose and Zn	6.8 pH units	Metal ion = 10E-4 M, rg = +-0.2, metal hydroxy species[1]
Geothite, FeOOH, and Cd	6.8 pH units	Metal ion = 10E-4 M, rg = +-0.2, metal hydroxy species[2]
Goethite, FeOOH, and Cd	7.7 pH units	Metal ion = 10E-4 M, rg = +-0.2, metal hydroxy species[3]
Goethite, FeOOH, and Cu	5.2 pH units	Metal ion = 10E-4 M, rg = +-0.2, metal hydroxy species[3]
Goethite, FeOOH, and Pb	5.6 pH units	Metal ion = 10E-4 M, rg = +-0.2, metal hydroxy species[3]
Hydrous Al oxide and Cd	6.6 pH units	Metal ion = 10E-4 M, rg = +-0.2, metal hydroxy species[4]
Hydrous Al oxide and Cu	4.8 pH units	Metal ion = 10E-4 M, rg = +-0.2, metal hydroxy species[4]
Hydrous Al oxide and Pb	5.2 pH units	Metal ion = 10E-4 M, rg = +-0.2, metal hydroxy species[4]
Hydrous Al oxide and Zn	5.6 pH units	Metal ion = 10E-4 M, rg = +-0.2, metal hydroxy species[4]

Hydrous Fe oxide and Cd	5.8 pH units	Metal ion = 10E-4 M, rg = +-0.2, metal hydroxy species[4]
Hydrous Fe oxide and Cu	4.4 pH units	Metal ion = 10E-4 M, rg = +-0.2, metal hydroxy species[4]
Hydrous Fe oxide and Pb	3.1 pH units	Metal ion = 10E-4 M, rg = +-0.2, metal hydroxy species[4]
Hydrous Fe oxide and Zn	5.4 pH units	Metal ion = 10E-4 M, rg = +-0.2, metal hydroxy species[4]
Illite and Cd	7.3 pH units	Metal ion = 10E-4 M, rg = +-0.2, metal hydroxy species[5]
Illite and Cu	5.8 pH units	Metal ion = 10E-4 M, rg = +-0.2, metal hydroxy species[5]
Illite and Pb	5.9 pH units	Metal ion = 10E-4 M, rg = +-0.2, metal hydroxy species[5]
Illite and Zn	6.6 pH units	Metal ion = 10E-4 M, rg = +-0.2, metal hydroxy species[5]
Kaolin and Cd	7.2 pH units	Metal ion = 10E-4 M, rg = +-0.2, metal hydroxy species[5]
Kaolin and Cu	6.1 pH units	Metal ion = 10E-4 M, rg = +-0.2, metal hydroxy species[5]
Kaolin and Pb	5.9 pH units	Metal ion = 10E-4 M, rg = +-0.2, metal hydroxy species[5]
Kaolin and Zn	6.8 pH units	Metal ion = 10E-4 M, rg = +-0.2, metal hydroxy species[5]
Montmorillonite and Cd	7.5 pH units	Metal ion = 10E-4 M, rg = +-0.2, metal hydroxy species[5]
Montmorillonite and Cu	5.9 pH units	Metal ion = 10E-4 M, rg = +-0.2, metal hydroxy species[5]
Montmorillonite and Pb	6.1 pH units	Metal ion = 10E-4 M, rg = +-0.2, metal hydroxy species[5]
Montmorillonite and Zn	6.6 pH units	Metal ion = 10E-4 M, rg = +-0.2, metal hydroxy species[5]
TiO2 and Cd	6.0 pH units	Metal ion = 10E-4 M, rg = +-0.2, metal hydroxy species[2]

Source: 1.Farrah, H. and W.F. Pickering. 1978. Aust. J. Chem. 31, 1501-1509.
2.Sparling, A.B. 1968. Interaction between Blue-Green Algae and Heavy Metals. Ph D. thesis. Wash Univ.St Louis. Mo. 107 pp.
3.Forbes, E.A., A.M. Posner and J.P. Quirk. 1976. Anal. Chem. J. Soil. Sci. 27, 154-166.
4.Kinniburgh, D.G., M.L. Jackson and J.K. Syers. 1976. Soil Sci. Soc. Am. Proc. 40, 796-799.
5.Farrah, H. and W.F. Pickering. 1977. Water, Air, Soil Pollut. 8, 189-197.

6-023 Rate constant
Cd

Species	Value	Condition
Formation of Cd(2+) - EDTA	unmeasurably fast	Elektrolyte 0.59 M NaCl, EDTA and Cd = 5E-7[1]
Formation of Cd(2+) - EDTA	2.8E3 mol/(l*sec)	Elektrolyte = 0.0536 MgCl2, EDTA and Cd = 5E-7[1]
Formation of Cd(2+) - EDTA	3.3E2 mol/(l*sec)	Elektrolyte = artificial seawater, EDTA and Cd = 5E-7[1]
Formation of Cd(2+) - EDTA	3.6E2 mol/(l*sec)	Elektrolyte = Adriatic seawater, EDTA and Cd = 5E-7[1]
Formation of Cd(2+) - EDTA	4.2E2 mol/(l*sec)	Elektrolyte = 0.01 CaCl2, EDTA and Cd = 5E-7[1]

Source: 1.Raspor, B., H.W. Nurnberg, P. Valenta and M. Branica. 1980. Sci. Tot. Environ. Kinetics and Mecanism of Trace Metal Chelation in Seawater. 11.

6-024 Rate constant
Degradation by photolysis

Species	Value	Condition
Anthracene	0.55 relative value	Sunlamp light, relative to deg. in summer sun = 1[1]
Anthracene	0.57 relative value	Medium pressure Hg lamp, relative to deg. in summer sun = 1[1]
Anthracene	0.94 relative value	Xenon lamp, relative to deg. in summer sun = 1[1]
Anthracene	1 relative value	Summer sunlight, relative to deg. in summer sun = 1[1]
Anthracene	1.02 relative value	Winter sunlight, midday, latitude 40 deg, relative to deg. in summer sun=1[1]
Anthracene	1.13 relative value	Blacklight, relative to deg. in summer sun = 1[1]
Naphthacene	0.043 relative value	Blacklight, relative to deg. in summer sun = 1[1]

Naphthacene	0.12 relative value	Sunlamp light, relative to deg. in summer sun = 1[1]
Naphthacene	0.31 relative value	Medium pressure Hg lamp, relative to deg. in summer sun = 1[1]
Naphthacene	0.65 relative value	Xenon lamp, relative to deg. in summer sun = 1[1]
Naphthacene	1 relative value	Summer sunlight, relative to deg. in summer sun = 1[1]
Naphthacene	1.25 relative value	Winter sunlight, midday, latitude 40 deg, relative to deg. in summer sun=1[1]
Naphthalene	0.17 relative value	Blacklight, relative to deg. in summer sun = 1[1]
Naphthalene	0.55 relative value	Winter sunlight, midday, latitude 40 deg, relative to deg. in summer sun=1[1]
Naphthalene	1 relative value	Summer sunlight, relative to deg. in summer sun = 1[1]
Naphthalene	11.6 relative value	Sunlamp light, relative to deg. in summer sun = 1[1]
Naphthalene	12.7 relative value	Medium pressure Hg lamp, relative to deg. in summer sun = 1[1]
Naphthalene	2.6 relative value	Xenon lamp, relative to deg. in summer sun = 1[1]

Source: 1.Zepp, R.G. and D.M. Cline. 1977. Environ. Sci. Technol. 11, 359.

6-025 Rate constant
Degradation in water

Species	Value	Condition
2,4-D butoxyethyl ester	5.42E-10 l/(orgamism (bac.)*hour)	Rg = +- 0.97E-10, n=31, second order reaction rate[1]
2,4-D butoxyethyl ester and sediment	2.3E-10 l/(orgamism (bac.) * hour)	Second order rate, fraction sorbed on sed= 0, rg=+-0.8[2]
2-Methylnaphthalene	0E3 ug/(liter*day)	Estuarine water, initial = 30 ug/l, Gulf Stream, USA[3]
2-Methylnaphthalene	0E3 ug/(liter*day)	Estuarine water, initial = 30 ug/l, coastal, W. coast, USA[3]
2-Methylnaphthalene	250E3 ug/(liter*day)	Estuarine water, initial = 30 ug/l, estuarine, W. coast, USA, rg=+-60[3]
Benzo(a)pyrene	0 ug/(liter*day)	Initial=16 ug/l, samp depth=5-10 m, 48 h incubation, 0 d, Lab, No. 2 fuel oil[4]
Benzo(a)pyrene	1 ug/(liter*day)	Initial=16 ug/l, samp depth=5-10 m, 24 h incubation, 3 d, Lab, No. 2 fuel oil[4]
Chlorpropham	1.42E-13 l/(orgamism (bac.) * hour)	Second order rate, fraction sorbed on sed= 0, rg =+-0.6[2]
Chlorpropham	2.29E-13 l/(orgamism (bac.) * hour)	Second order rate, fraction sorbed on sed= 0.90, rg=+-0.6, app. value[2]
Chlorpropham	2.63E-14 l/(orgamism (bac.)*hour)	Rg = +- 0.72E-10, n=11, second order reaction rate[1]
Chlorpropham	3.45E-13 l/(orgamism (bac.) * hour)	Second order rate, fraction sorbed on sed= 0.15, rg=+-0.2, app. value[2]
Chlorpropham	4.15E-13 l/(orgamism (bac.) * hour)	Second order rate, fraction sorbed on sed= 0.15, rg=+-0.7[2]
Chlorpropham	7.4E-15 l/(orgamism (bac.) * hour)	Second order rate, fraction sorbed on sed= 0.90, rg=+-1.2[2]
Di(n-octyl)-phthalate	2.1E-13 l/(orgamism (bac.) * hour)	Second order rate, fraction sorbed on sed= 0.15, rg=1.2[2]
Di(n-octyl)-phthalate	3.6E-13 l/(orgamism (bac.) * hour)	Second order rate, fraction sorbed on sed= 0.15, rg=0.7, app. value[2]
Di(n-octyl)-phthalate	3.7E-13 l/(orgamism (bac.) * hour)	Second order rate, fraction sorbed on sed= 0, rg=+-0.6[2]
Di(n-octyl)phthalate	3.1E-15 l/(orgamism (bac.) * hour)	Second order rate, fraction sorbed on sed= 0.9, rg=+-0.4[2]

Di(n-octyl)phthalate	6.9E-13 l/(orgamism (bac.) * hour)	Second order rate, fraction sorbed on sed= 0.9, rg=+-2.3, app. value[2]
Dibutylphthalate	2.6E-11 l/(orgamism (bac.) * hour)	Second order rate, fraction sorbed on sed= 0.9, rg=+-0.2, app. value[2]
Dibutylphthalate	2.7E-11 l/(orgamism (bac.) * hour)	Second order rate, fraction sorbed on sed= 0.15, rg=+-1.5[2]
Dibutylphthalate	3.1E-11 l/(orgamism (bac.) * hour)	Second order rate, fraction sorbed on sed= 0, rg=+-0.8[2]
Dibutylphthalate	3.5E-11 l/(orgamism (bac.) * hour)	Second order rate, fraction sorbed on sed= 0.15, rg=+-1.7, app. value[2]
Dibutylphthalate	6.1E-13 l/(orgamism (bac.) * hour)	Second order rate, fraction sorbed on sed= 0.9, rg=+-0.5[2]
Fluorene	0 ug/(liter*day)	Initial=30 ug/l, samp depth=5-10 m, 28 h incubation, 0 d, Lab, No. 2 fuel oil[4]
Fluorene	0 ug/(liter*day)	Initial=30 ug/l, samp depth=5-10 m, 48 h incubation, 3 d, Lab, No. 2 fuel oil[4]
Malathion	4.0E-11 l/(orgamism (bac.) * hour)	Second order rate, fraction sorbed on sed= 0, rg =;-0.9[2]
Malathion	4.54E-11 l/(orgamism (bac.)*hour)	Rg = +- 0.74E-10, n=14, second order reaction rate[1]
Methoxychlor	3.0E-14 l/(orgamism (bac.) * hour)	Second order rate, fraction sorbed on sed= 0.15, rg=0.5, app. value[2]
Methoxychlor	3.2E-14 l/(orgamism (bac.) * hour)	Second order rate, fraction sorbed on sed= 0.9, rg =+-1.2, app. value[2]
Methoxychlor	5.2E-14 l/(orgamism (bac.) * hour)	Second order rate, fraction sorbed on sed= 0.15, rg=2.3[2]
Methoxychlor	6.1E-16 l/(orgamism (bac.) * hour)	Second order rate, fraction sorbed on sed= 0.9, rg =+-2.8[2]
Methoxychlor	6.7E-14 l/(orgamism (bac.) * hour)	Second order rate, fraction sorbed on sed= 0, rg=1.2[2]
Methylnaphthalene	10 ug/(liter*day)	Initial=50 ug/l, samp depth=0-5 m, 24 h incubation, 0 d, Lab, No. 2 fuel oil[4]
Methylnaphthalene	26 ug/(liter*day)	Initial=50 ug/l, samp depth=0-5 m, 24 h incubation, 3 d, Lab, No. 2 fuel oil[4]
Mytilus edulis	10 ug/(liter*day)	Initial=50 ug/l, samp depth=0-5 m, 24 h incubation, 0 d, Lab, No. 2 fuel oil[4]
Mytilus edulis	100 ug/(liter*day)	Initial=50 ug/l, samp depth=0-5 m, 10 h incubation, 4 d, Lab, No. 2 fuel oil[4]
Mytilus edulis	250 ug/(liter*day)	Initial=50 ug/l, samp depth=0-5 m, 16 h incubation, 3 d, Lab, No. 2 fuel oil[4]
Mytilus edulis	500 ug/(liter*day)	Initial=50 ug/l, samp depth=5-10 m, 10 h incubation, 4 d, Lab, No. 2 fuel oil[4]
Naphthalene	12E3 ug/(liter*day)	Estuarine water, initial = 30 ug/l, Gulf Stream, rg = +- 250[3]
Naphthalene	330E3 ug/(liter*day)	Estuarine water, initial = 30 ug/l, Coastal, W coast, USA, rg=+-40[3]
Naphthalene	870E3 ug/(liter*day)	Estuarine water, initial = 30 ug/l, estuarine, W coast, USA, rg=+-20[3]
bis(2-Ethylhexyl)-phthalate	1.8E-15 l/(orgamism (bac.) * hour)	Second order rate, fraction sorbed on sed= 0.15, rg=0.3[2]
bis(2-Ethylhexyl)-phthalate	3.6E-15 l/(orgamism (bac.) * hour)	Second order rate, fraction sorbed on sed= 0.15, rg=1.3, app. value[2]
bis(2-Ethylhexyl)-phthalate	4.2E-15 l/(orgamism (bac.) * hour)	Second order rate, fraction sorbed on sed= 0, rg=+-0.7[2]

bis(2-Ethylhexyl)-phthalate	5.7E-17 l/(orgamism (bac.) * hour)	Second order rate, fraction sorbed on sed= 0.9, rg=1.7[2]
bis(2-Ethylhexyl)-phthalate	6.1E-15 l/(orgamism (bac.) * hour)	Second order rate, fraction sorbed on sed= 0.9, rg=2.2, app. value[2]

Source: 1.Zepp, R.G. and D.M. Cline. 1977. Environ. Sci. Technol. 11, 359.
2.Steen, W.C., D.F. Paris and G.L. Baugman. 1980. In: Contaminants and Sediments. Vol 1. Ann Arbor Inc. 477-482.
3.Reish, D.J., F. Piltz, J.M. Martin and J.Q. Word. 1974. Mar. Pollut. Bull. 5, 125-126.
4.Lee, R.F. and M. Takahashi. 1977. In: A.D. McIntyre and K.J. Whittle (eds.): Petroleum Hydrocarbons in the Marine Environment., Cons. Intern. Explor. Mer. Vol 171, 150-156.

6-026 Rate constant
Degradation rate

Species	Value	Condition
Birch forest	689 g C/(m2*year)	Location = Norway, 1974[1]
Desert	10 g C/(m2*year)	Location = Not given[1]
Forest in temperate zone	0.125 g C/(m2*hour)	Location = Rg = 0.05 - 2.0, 1966[1]
Lichen heath	94 g C/(m2*year)	Location = Norway, 1974[1]
Lowland rain forest	535 g C/(m2*year)	Location = N Borneo, 1969[1]
Meadow, dry	457 g C/(m2*year)	Location = Norway, 1974[1]
Meadow, wet	253 g C/(m2*year)	Location = Norway, 1974[1]
Mountain rain forest	800 g C/(m2*year)	Location = Tjibodas, 1969[1]
Secondary forest, woodland and shrubland	0.7 g C/(m2*hour)	Location = Costa Rica, 1966[1]
Semi desert	285 g C/(m2*year)	Location = Not given[1]
Snow bed	177 g C/(m2*year)	Location = Norway, 1974[1]
Subartic mire	71 g C/(m2*year)	Location = Sweden, 1974[1]
Temperate coniferous forest	0.043 g C/(m2*year)	Location = Oak Ridge, 1965, leaf litter removed[1]
Temperate coniferous forest	108 g C/(m2*year)	Location = Belgium, 1976[1]
Temperate coniferous forest	8 g C/(m2*year)	Location = Tennessee, 1965, litter only[1]
Temperate deciduous forest	0.048 g C/(m2*year)	Location = Tennesse, 1965, leaf litter removed[1]
Temperate deciduous forest	48.2 g C/(m2*year)	Location = Holland, 1965[1]
Temperate deciduous forest	84.1 g C/(m2*year)	Location = Tennessee, 1965, litter only[1]
Temperate heathland forest	86.3 g C/(m2*hour)	Location = Holland, 1968[1]
Temperate mixed forest	150 g C/(m2*year)	Location = Belgium, 1970[1]
Temperate mixed forest	233 g C/(m2*year)	Location = Belgium, 1976[1]
Temperate wet grassland	675 g C/(m2*year)	Location = S Sweden, 1969[1]
Temperate woodland forest	450 g C/(m2*year)	Location = Sweden, 1969[1]
Tropical deciduous dry forest	0.10 g C/(m2*hour)	Location = Costa Rica, 1966[1]
Tropical gallery forest	0.37 g C/(m2*hour)	Location = Costa Rica, 1966[1]
Tropical rain forest	0.43 g C/(m2*hour)	Location = Costa Rica, 1966[1]
Tropical savanna	0.09 g C/(m2*hour)	Location = Costa Rica, 1966[1]
Tundra, herb and lichen	60 g C/(m2*year)	Location = Canada, 1973[1]

Source: 1.Ajtay, G.L., P. Ketner and P. Duvigneaud. 1979. in: Bolin, B., E.T. Degens, S. Kempe and P. Ketner.(ds.) The Global Carbon Cycle. 129-181. SCOPE. John Wiley and Sons. NY.

6-027 Rate constant
Hydrolysis in water

Species	Value	Condition
Atrazine	1.04E-3	285 K, 0.04 M acetate buffer as catalyst, hydrolysis, init atra=60 umol/l[1]
Atrazine	1.42E-2	303 K, hydrogen ion as catalyst, pH=4, hydrolysis, init atra=60 umol/l[1]
Atrazine	2.02E-2	303 K, 0.04 M acetate buffer as catalyst, hydrolysis, init atra=60 umol/l[1]
Atrazine	2.29E-3	285 K, 0.08 M acetate buffer as catalyst, hydrolysis, init atra=60 umol/l[1]
Atrazine	2.84E-3	298 K, hydrogen ion as catalyst, pH=4, hydrolysis, init atra=60 umol/l[1]
Atrazine	3.76E-2	303 K, 0.08 M acetate buffer as catalyst, hydrolysis, init atra=60 umol/l[1]
Atrazine	4.01E-1	313 K, 0.2 g humic acid as catalyst, hydrolysis, init atra=60 umol/l[1]
Atrazine	5.02E-3	298 K, 0.04 M acetate buffer as catalyst, hydrolysis, init atra=60 umol/l[1]
Atrazine	5.12E-4	285 K, hydrogen ion as catalyst, pH=4, hydrolysis, init atra=60 umol/l[1]
Atrazine	6.11E-2	313 K, 0.2 g humic acid as catalyst, hydrolysis, init atra=60 umol/l[1]
Atrazine	9.70E-3	298 K, 0.08 M acetate buffer as catalyst, hydrolysis, init atra=60 umol/l[1]
Atrazine	9.90E-1	313 K, 0.2 g humic acid as catalyst, hydrolysis, init atra=60 umol/l[1]

Source: 1.Li, G-C. and G.T Felbeck. 1972. Soil Sci. 113, 140.

6-028 Rate constant
Photolysis rate in water

Species	Value	Condition
7H-dibenzo(c,g)carbazole	0.122E-3 1/sec	Irradiation = 366 nm, 56 % reaction, some humic acid, 11.5 ng/ml[1]
7H-dibenzo(c,g)carbazole	1.52E-3 1/sec	Irradiation = 366 nm, 84 % reaction, 0.0 ug humic/ml, 9.1 ng/ml[1]
9H-Carbazole	2.46E-5 1/sec	Irradiation = 313 nm, 35 % reaction, 8.0 ug humic/ml, 1.0 ug/ml[1]
9H-Carbazole	7.50E-5 1/sec	Irradiation = 313 nm, 85 % reaction, 0.0 ug humic/ml, 1.0 ug/ml[1]
Benz(a)anthracene	1.98E-5 1/sec	Irradiation = 313 nm, 38 % reaction, 0.0 ug humic/ml, 18.0 ng/ml[1]
Benz(a)anthracene	12.3E-5 1/sec	Irradiation = 366 nm, 67 % reaction, 0.0 ug humic/ml, 9.8 ng/ml[1]
Benz(a)anthracene	3.9E-5 1/sec	Irradiation = 366 nm, 57 % reaction, 8.0 ug humic/ml, 17.3 ng/m3[1]
Benzo(a)pyrene	0.690E-4 1/sec	Irradiation = 366 nm, 43 % reaction, some humic acid, 13.0 ng/ml[1]
Benzo(a)pyrene	4.37E-4 1/sec	Irradiation = 366 nm, 87 % reaction, 0.0 ug humic/ml, 13 ng/ml[1]
Benzo(b)thiopene	0.851E-6 1/sec	Irradiation = 313 nm, 22 % reaction, 9.5 ug humic/ml, 1.0 ug/ml[1]
Benzo(b)thiopene	3.50E-6 1/sec	Irradia.=borosilicate filter, 39 % reaction, 0.0 ug humic/ml, 1.0 ug/ml[1]
Benzo(f)quinoline	0.275E-5 1/sec	Irradiation = 313 nm, 14 % reaction, 8.0 ug humic/ml, 12.1 ug/ml[1]
Benzo(f)quinoline	1.73E-5 1/sec	Irradiation = 313 nm, 39 % reaction, 0.0 ug humic/ml, 7.3 ug/ml[1]
Dibenzothiophene	1.93E-6 1/sec	Irradiation = 313 nm, 55 % reaction, 8.0 ug humic/ml, 0.5 ug/ml[1]
Dibenzothiophene	2.04E-6 1/sec	Irradiation = solar, March, 35 % reaction, 0.0 ug humic/ml, 0.5 ug/ml
Dibenzothiophene	2.32E-6 1/sec	Irradiation = 313 nm, 60 % reaction, 0.0 ug humic/ml, 0.5 ug/ml[1]
Dibenzothiophene	4.60E-6 1/sec	Irradiation = solar, March, 63 % reaction, 8.0 ug humic/ml, 0.5 ug/ml[1]
Methyl parathion	32.3E-7 1/sec	Irradiation = 313 nm, 64 % reaction, 0.0 ug humic/ml, 26 ug/ml[1]
Methyl parathion	36.2E-7 1/sec	Irradiation = 313 nm, 59 % reaction, 9.5 ug humic/ml, 26 ug/ml[1]

Quinoline	0.770E-6 1/sec	Irradiation = solar, late June, 0.0 ug humic/ml, 0.82 ug/ml[1]
Quinoline	2.60E-6 1/sec	Irradiation = 313 nm, 60 % reaction, 9.5 ug humic/ml 0.67 ug/ml[1]
Quinoline	3.04E-6 1/sec	Irradiation = solar, late June, 9.5 ug humic/ml, 0.67 ug/ml[1]
Quinoline	6.07E-6 1/sec	Irradiation = 313 nm, 60 % reaction, 0.0 ug humic/ml, 0.82 ug/ml[1]
p-Cresol	0.68E-6 1/sec	Irradiation = solar, april, 32 % reaction, 0.0 ug humic/ml, 1.0 ug/ml[1]
p-Cresol	1.84E-6 1/sec	Irradiation = 313 nm, 40 % reaction, 9.5 ug humic/ml, 1.0 ug/ml[1]
p-Cresol	8.0E-6 1/sec	Irradiation = solar, April, 93 % reaction, 9.5 ug humic/ml, 1.0 ug/ml[1]
p-Cresol	0.72E-6 1/sec	Irradiation = 313 nm, 28 % reaction, 0.0 ug humic/ml, 1.0 ug/ml[1]

Source: 1.Smith, J.H. et al. 1978. Environmental Pathways of selected chemicals in freshwater systems, part 2.EPA-600/7-78-074.

6-029 Rate constant Photooxidation

Species	Value	Condition
3-Carene	1.4E4 1/sec	7 ppb NOx and 10 ppb HC[1]
Alpha-Phellandrene	12E4 1/sec	7 ppb NOx and 10 ppb HC[1]
Carvomethene	2.1E4 1/sec	7 ppb NOx and 10 ppb HC[1]
Dihydromyrcene	3.0E4 1/sec	7 ppb NOx and 10 ppb HC[1]
Isobutene	0.84E4 1/sec	7 ppb NOx and 10 ppb HC[1]
Isoprene	1.3E4 1/sec	7 ppb NOx and 10 ppb HC[1]
Limonene	2.4E4 1/sec	7 ppb NOx and 10 ppb HC[1]
Mycene	3.8E4 1/sec	7 ppb NOx and 10 ppb HC[1]
Oximene	5.3E4 1/sec	7 ppb NOx and 10 ppb HC[1]
Terpinolene	11E4 1/sec	7 ppb NOx and 10 ppb HC[1]
alpha-Pinene	1.3E4 1/sec	7 ppb NOx and 10 ppb HC[1]
alpha-Terpinene	80E4 1/sec	7 ppb NOx and 10 ppb HC, rg = 55 - 100E4/sec[1]
beta-Phellandrene	1.9E4 1/sec	7 ppb NOx and 10 ppb HC[1]
beta-Pinene	1.1E4 1/sec	7 ppb NOx and 10 ppb HC[1]
p-Cymene	0.25E4 1/sec	7 ppb NOx and 10 ppb HC[1]
p-Methane	0.11E4 1/sec	7 ppb NOx and 10 ppb HC[1]

Source: 1.Grimsrud, E.P., H.H. Westberg and R.A. Rasmussen. 1975. J. Chem. Kinet. Symp. 1, 183-195.

6-030 Rate constant Sedimentation of Pb

Species	Value	Condition
Deposition in freshwater Lake Devils	1.4 ug/(cm2*y)	Rg = 0.5 - 2.3 ug/(cm2*y), N Wisconsin[1]
Deposition in freshwater Lake Erie	11 ug/(cm2*y)	Pre-industrial flux = 0.7 ug/(cm2*u)[1]
Deposition in freshwater Lake Huron	2.2 ug/(cm2*y)	Pre-industrial flux = 0.7 ug/(cm2*u)[1]
Deposition in freshwater Lake Mendola	8 ug/(cm2*y)	Rg = 6 - 10 ug/(cm2*y), S Wisconsin[1]
Deposition in freshwater Lake Michigan	1.3 ug/(cm2*y)	Pre-industrial flux = 0.16 ug/(cm2*u)[1]
Deposition in freshwater Lake Minocqua	1.5 ug/(cm2*y)	N Wisconsin[1]
Deposition in freshwater Lake Monona	25 ug/(cm2*y)	Rg = 20 - 30 ug/(cm2*y), S Wisconsin[1]
Deposition in freshwater Lake Ontario	7.0 ug/(cm2*y)	Pre-industrial flux = 1.0 ug/(cm2*u)[1]

Deposition in freshwater Lake Trout	2 ug/(cm2*y)	N Wisconsin[1]
Deposition to marine sediment at Baltic Sea	2.2 ug/(cm2*y)	1973[1]
Deposition to marine sediment at San Clemente	0.04 ug/(cm2*y)	1976[1]
Deposition to marine sediment at San Piedro	1.7 ug/(cm2*y)	Pre-industrial flux = 0.26 ug/(cm2*u)[1]
Deposition to marine sediment at Santa Barbara	2.1 ug/(cm2*y)	Pre-industrial flux = 1.0 ug/(cm2*u)[1]
Deposition to marine sediment at Santa Monica	0.9 ug/(cm2*y)	Pre-industrial flux = 0.24 ug/(cm2*u)[1]

Source: 1.Nriagu, J.O. 1979. The Biochemistry of Lead in The Environment. Part A. Elsevier/North Holland. Amsterdam.

6-031 Rate of degradation Bromophenol in soil

Species	Value	Condition
2-Bromophenol	14 days	Days to complete disapperance, aromatic herbicide[1]
3-Bromophenol	72 days	Days to complete disapperance, aromatic herbicide, more than value[1]
4-Bromophenol	16 days	Days to complete disapperance, aromatic herbicide[1]

Source: 1.Tyunyaeva, G.N., A.K. Minenko and L.A. Penkov. 1974. Agrokhimiya No 6, 110-114.

6-032 Rate of degradation Chlorophenol in soil

Species	Value	Condition
2,3,4,6-Tetrachlorophenol	72 days	Days to complete disapperance, aromatic herbicide, more than value[1]
2,4,5-Trichlorophenol	72 days	Days to complete disapperance, aromatic herbicide, more than value[1]
2,4,6-Trichlorophenol	5 days	Days to complete disapperance, aromatic herbicide[1]
2,4-Dichlorophenol	9 days	Days to complete disapperance, aromatic herbicide[1]
2,5-Dichlorophenol	72 days	Days to complete disapperance, aromatic herbicide, more than value[1]
2-Chlorophenol	14 days	Days to complete disapperance, aromatic herbicide[1]
3-Chlorophenol	72 days	Days to complete disapperance, aromatic herbicide, more than value[1]
4-Chlorophenol	9 days	Days to complete disapperance, aromatic herbicide[1]
Pentachlorophenol	72 days	Days to complete disapperance, aromatic herbicide, more than value[1]
Phenol	2 days	Days to complete disapperance, aromatic herbicide[1]

Source: 1.Tyunyaeva, G.N., A.K. Minenko and L.A. Penkov. 1974. Agrokhimiya No 6, 110-114.

6-033 Soil
Emission

Species	Value	Condition
Benz(a)anthracene and chrysene, sludge from municipal incinerator	171 ug/kg dry weight	23 % dry matter, 30 t/d, 4010 mg/d in total[1]
Benzo(a)pyrene and benzo(e)pyrene, sludge from municipal incinerator	147 ug/kg dry weight	23 % dry matter, 30 t/d, 3450 mg/d in total[1]
Benzo(b)fluoranthene in sludge from municipal incinerator	292 ug/kg dry weight	23 % dry matter, 30 t/d, 6850 mg/d in total[1]
Benzo(ghi)perylene in sludge from municipal incinerator	47 ug/kg dry weight	23 % dry matter, 30 t/d, 1100 mg/d in total[1]
Coronene in sludge from municipal incinerator	20 ug/kg dry weight	23 % dry matter, 30 t/d, <470 mg/d in total[1]
Fluoranthene in sludge from municipal incinerator	58 ug/kg dry weight	23 % dry matter, 30 t/d, 1360 mg/d in total[1]
Indeno(1,2,3-cd)pyrene in sludge from municipal incinerator	10 ug/kg dry weight	23 % dry matter, 30 t/d, <230 mg/d in total[1]
Perylene in sludge from municipal incinerator	82 ug/kg dry weight	23 % dry matter, 30 t/d, 1920 mg/d in total[1]
Pyrene in sludge from municipal incinerator	49 ug/kg dry weight	23 % dry matter, 30 t/d, 1150 mg/d in total[1]
Total PAH in sludge from municipal incinerator	19840 mg/day	23 % dry matter, 30 t/d, no concentration[1]

Source: 1.Davies, I.W.R., Harrison, Perry, Ratnayaka and R.A. Wellings. 1976. Environ. Sci. Technol. 10, 451-43.

6-034 Waste water treatment
Removal by activated sludge plant

Species	Value	Condition
Cd	100 %	Toronto Humber, 1970, inflow =0.04 mg Cd/l, outflow = 0 mg Cd/l, Ont., Canada[1]
Cr	100 %	Chatham, 1970, inflow = 1.5 mg Cr/l, outflow = 0.0 mg Cr/l, Ont., Canada[1]
Cr	100 %	Stratford, 1970, inflow = 0.2 mg Cr/l, outflow = 0 mg Cr/l, Ont., Canada[1]
Cr	87 %	Kitchener, 1970, inflow = 3.0 mg Cr/l, outflow = 0.4 mg Cr/l, Ont., Canada[1]
Cr	95 %	Toronto Humber, 1970, inflow = 1.3 mg Cr/l, outflow = 0.06 mg Cr/l, Ont., Canad[1]
Cr	95 %	Toronto Humber, 1970, inflow = 1.3 mg Cr/l, outflow = 0.06 mg Cr/l, Ont., Canada[1]
Cu	100 %	Stratford, 1970, inflow =0.29 mg Cu/l, outflow = 0 mg Cu/l, Ont., Canada[1]
Cu	89 %	Toronto Humber, 1970, inflow =0.55 mg Cu/l, outflow = 0.06 mg Cu/l, Ont., Canad[1]
Cu	89 %	Toronto Humber, 1970, inflow =0.55 mg Cu/l, outflow = 0.06 mg Cu/l, Ont., Canada[1]
Cu	97 %	Chatham, 1970, inflow = 0.3 mg Cu/l, outflow = .01 mg Cu/l, Ont., Canada[1]
Ni	100 %	Stratford, 1970, inflow = 0.4 mg Ni/l, outflow = 0 mg Ni/l, Ont., Canada[1]
Ni	52 %	Chatham, 1970, inflow = 4.2 mg Ni/l, outflow = 2.0 mg Ni/l, Ont., Canada[1]
Ni	58 %	Kitchener, 1970, inflow =0.48 mg Ni/l, outflow = 0.2 mg Ni/l, Ont., Canada[1]
Ni	64 %	Toronto Humber, 1970, inflow = 0.9 mg Ni/l, outflow = 0.32 mg Ni/l, Ont., Canad[1]
Ni	64 %	Toronto Humber, 1970, inflow = 0.9 mg Ni/l, outflow = 0.32 mg Ni/l, Ont., Canada[1]
Pb	100 %	Kitchener, 1970, inflow =0.16 mg Pb/l, outflow = 0.0 mg Pb/l, Ont., Canada[1]
Zn	72 %	Toronto Humber, 1970, inflow =1.25 mg Zn/l, outflow =00.35 mg Zn/l, Ont., Canad[1]
Zn	72 %	Toronto Humber, 1970, inflow =1.25 mg Zn/l, outflow =00.35 mg Zn/l, Ont., Canada[1]

Zn	85 %	Kitchener, 1970, inflow =0.48 mg Zn/l, outflow = .07 mg Zn/l, Ont., Canada[1]
Zn	85 %	Stratford, 1970, inflow = 0.2 mg Zn/l, outflow =.03mg Zn/l, Ont., Canada[1]
Zn	91 %	Chatham, 1970, inflow = 0.8 mg Zn/l, outflow = .07 mg Zn/l, Ont., Canada[1]

Source: 1.Abbott, R.A. 1977. In: Waste Handling, Disposal and Recovery in the Metal Finishing Industry. Fisheres and Environment of Canada, Econ. Tech. Rev. Rep. EPS-3 WP-77-3, Ottawa, Ont, 20-104.

6-035 Waste water treatment
Removal by primary treatment plant

Species	Value	Condition
Cd	42 %	Wnidsor, 1970, inflow = 0.12 mg Cd/l, outflow = 0.07 mg Cd/l, Ont., Canada[1]
Cr	20 %	Wnidsor, 1970, inflow = 0.56 mg Cr/l, outflow = 0.45 mg Cr/l, Ont., Canada[1]
Cu	77 %	Wnidsor, 1970, inflow = 0.26 mg Cu/l, outflow = 0.06 mg Cu/l, Ont., Canada[1]
Ni	64 %	Wnidsor, 1970, inflow = 0.44 mg Ni/l, outflow = 0.17 mg Ni/l, Ont., Canada[1]
Zn	52 %	Wnidsor, 1970, inflow = 1.74 mg Zn/l, outflow = 0.84 mg Zn/l, Ont., Canada[1]

Source: 1.Abbott, R.A. 1977. In: Waste Handling, Disposal and Recovery in the Metal Finishing Industry. Fisheres and Environment of Canada, Econ. Tech. Rev. Rep. EPS-3 WP-77-3, Ottawa, Ont, 20-104.

6-036 Water
Emission

Species	Value	Condition
Benz(a)anthracene and chrysene, output from municipal incinerator	0.64 ug/liter	Discharge of 25000 l/d,[1]
Benz(a)anthracene and chrysene, output from municipal incinerator	0.64 ug/liter	Discharge of 25000 l/d, 16.0 mg/d in total[1]
Benzo(a)pyrene and benzo(e)pyrene, output from municipal incinerator	0.14 ug/liter	Discharge of 25000 l/d,[1]
Benzo(a)pyrene and benzo(e)pyrene, output from municipal incinerator	0.14 ug/liter	Discharge of 25000 l/d, 3.5 mg/d in total[1]
Benzo(a)pyrene from all sources, total	697 t/year	Global value[2]
Benzo(a)pyrene from biosynthesis	25 t/year	Global value[2]
Benzo(a)pyrene from domestic and industrial waste	29 t/year	Global value[2]
Benzo(a)pyrene from fallout and rainout	500 t/year	Global value[2]
Benzo(a)pyrene from petroleum spillage	25 t/year	Global value[2]
Benzo(a)pyrene from surface runoff from land	118 t/year	Global value[2]
Benzo(b)fluoranthene in output water from municipal incinerator	0.14 ug/liter	Discharge of 25000 l/d,[1]
Benzo(b)fluoranthene in output water from municipal incinerator	0.14 ug/liter	Discharge of 25000 l/d, 3.5 mg/d in total[1]
Benzo(ghi)perylene in output water from municipal incinerator	0.03 ug/liter	Discharge of 25000 l/d,[1]
Benzo(ghi)perylene in output water from municipal incinerator	0.03 ug/liter	Discharge of 25000 l/d, 0.8 mg/d in total[1]
Coronene in output water from municipal incinerator	0.011 ug/liter	Discharge of 25000 l/d,[1]
Coronene in output water from municipal incinerator	0.011 ug/liter	Discharge of 25000 l/d, <0.3 mg/d in total[1]
Fluoranthene in output water from municipal incinerator	0.62 ug/liter	Discharge of 25000 l/d,[1]
Fluoranthene in output water from municipal incinerator	0.62 ug/liter	Discharge of 25000 l/d, 15.5 mg/d in total[1]
Indeno(1,2,3-cd)pyrene in output water from municipal incinerator	0.01 ug/liter	Discharge of 25000 l/d,[1]

Indeno(1,2,3-cd)pyrene in output water from municipal incinerator	0.01 ug/liter	Discharge of 25000 l/d, <0.3 mg/d in total[1]
Perylene in output water from municipal incinerator	0.13 ug/liter	Discharge of 25000 l/d,[1]
Perylene in output water from municipal incinerator	0.13 ug/liter	Discharge of 25000 l/d, 3.5 mg/d in total[1]
Pyrene in output water from municipal incinerator	0.54 ug/liter	Discharge of 25000 l/d,[1]
Pyrene in output water from municipal incinerator	0.54 ug/liter	Discharge of 25000 l/d, 13.5 mg/d in total[1]
Total PAH from all sources, in total	230040 t/year	Global value[2]
Total PAH from biosynthesis	2700 t/year	Global value[2]
Total PAH from domestic and industrial waste	4400 t/year	Global value[2]
Total PAH from fallout and rainout	50000 t/year	Global value[2]
Total PAH from petroleum spillage	170000 t/year	Global value[2]
Total PAH from surface runoff from land	2940 t/year	Global value[2]
Total PAH in output water from municipal incinerator	56.1 mg/day	Discharge of 25000 l/d,[1]
Total PAH in output water from municipal incinerator	56.1 mg/day	Discharge of 25000 l/d, no concentration[1]

Source: 1.Davies, I.W.R., Harrison, Perry, Ratnayaka and R.A. Wellings. 1976. Environ. Sci. Technol. 10, 451-43. 2.Neff, J.M. 1979. Polycyclic Aromatic Hydrocarbons in the Aquatic Environment. Applied Sci. Publ. London.

6-037 Degradation in organism DDT analog

Species	Value	Condition
Fish	0.015 ratio of polar/nonpolar metabolites	DDT analog with Cl- and Cl- as radicals in aromatic rings[1]
Fish	0.94 ratio of polar/nonpolar metabolites	DDT analog with CH3O- and CH3O- as radicals in aromatic rings[1]
Fish	1.20 ratio of polar/nonpolar metabolites	DDT analog with CH3- and C2H5_- as radicals in aromatic rings[1]
Fish	2.69 ratio of polar/nonpolar metabolites	DDT analog with C2H5O and C2H5O- as radicals in aromatic rings[1]
Fish	2.75 ratio of polar/nonpolar metabolites	DDT analog with CH3O- and CH3S- as radicals in aromatic rings[1]
Fish	3.43 ratio of polar/nonpolar metabolites	DDT analog with Cl- and CH3- as radicals in aromatic rings[1]
Fish	47 ratio of polar/nonpolar metabolites	DDT analog with CH3S- and CH3- as radicals in aromatic rings[1]
Fish	7.14 ratio of polar/nonpolar metabolites	DDT analog with CH3- and CH3 as radicals in aromatic rings[1]
Snail	0.045 ratio of polar/nonpolar metabolites	DDT analog with Cl- and Cl- as radicals in aromatic rings[1]
Snail	0.08 ratio of polar/nonpolar metabolites	DDT analog with CH3- and CH3 as radicals in aromatic rings[1]
Snail	0.13 ratio of polar/nonpolar metabolites	DDT analog with CH3O- and CH3O- as radicals in aromatic rings[1]
Snail	0.25 ratio of polar/nonpolar metabolites	DDT analog with CH3- and C2H5_- as radicals in aromatic rings[1]
Snail	0.39 ratio of polar/nonpolar metabolites	DDT analog with C2H5O and C2H5O- as radicals in aromatic rings[1]
Snail	0.77 ratio of polar/nonpolar metabolites	DDT analog with CH3S- and CH3- as radicals in aromatic rings[1]
Snail	105 ratio of polar/nonpolar metabolites	DDT analog with CH3O- and CH3S- as radicals in aromatic rings[1]

Source: 1.Kapoor, I.P. et al. 1973. J. Agric. fd Chem. 21, 310.

Chapter 7

ECOTOXICOLOGICAL EFFECTS OF PESTICIDES

Pesticides in Alphabetic Order

7-001 Effect of Additives of Gramoxone

Parameter	Effect Value	Concentration and Condition	Species
Growth rate and numbers	Slight decreases immediately after dose, no serve effects	Calcereous loam, pH=8, 7.5 % OM, field and lab, 5.7 and 57 kg/ha, in situ	Soil bacteria[1]
Nitrification and nitrifiers	No significant effect	Calcareous loam, pH=8, 7.5 % OM, 5.6 and 56 kg/ha	Soil bacteria[2]

Source: 1.Anderson, J.R. and E.A. Drew. 1976. Zbl. Bakt. abt II, 131, 247-258.
2.Anderson, J.R. and E.A. Drew. 1976. Zbl. Bakt. Abt. II, 131, 136-147.

7-002 Effect of Agelon

Parameter	Effect Value	Concentration and Condition	Species
Growth rate and numbers	No effect	Alluvial, 3 kg/ha, in situ	Soil bacteria[1]
Nitrogen fixation in soil	Annual applications had little effect	Heavy clay degraded chernozem, >2 % OM, 2.5 kg/ha, 7 y	Soil bacteria, free living[2]

Source: 1.Micev, N. and M. Bubalov. 1972. Symp. Biol. Hung. 11, 379-384.
2.Husarova, M. 1972. Ved. Pr. vyzk. Ust. Kukur. Trnave, No 6, 45-69.

7-003 Effect of Alachlor

Parameter	Effect Value	Concentration and Condition	Species
Growth rate and numbers	Decrease in populations	Heavy clay, 7 kg/ha, 7 y appl., in situ	Soil bacteria[2]
Nitrification and nitrifiers	No effect	Silt loam, pH=6, 2 % OM, 0.6-1.2 kg/ha, recommended field rate	Soil bacteria[1]

Source: 1.Thorneburg, R.P. and J.A. Tweedy. 1973. Weed Sci. 21, 397-399.
2.Husarova, M. 1972. Ved. Pr. vyzk. Ust. Kukur. Trnave, No 6, 45-69.

7-004 Effect of Aldicarb

Parameter	Effect Value	Concentration and Condition	Species
Nitrification and nitrifiers	20 % - 30 % inhibition of activity	Sandy loam, pH=7.4, 0.5 kg/ha	Soil organisms[1]
Soil respiration	Little effect on CO2-evolution	Sandy loam, pH=7.4, 0.5 and 50 kg/ha	Soil organisms[1]

Source: 1.Kuseske, D.W., B.R. Funke and J.T. Schultz. 1974. Pl. Soil 41, 255-269.

7-005 Effect of Aldrin

Parameter	Effect Value	Concentration and Condition	Species
Nitrification and nitrifiers	Inhibited activity, NO2(-) accumulation	Sandy loam, 2.5 and 10 and 100 kg/ha	Soil organisms affected by insecticides[1]
Nitrification and nitrifiers	No effect	Sandy loam, 0.1 and 1.0 kg/ha	Soil organisms[1]
Soil respiration	Low rates, no effect. Higher rates decreased CO2-evolution	Alluvial, 0.1 - 100 kg/ha	Soil organisms[1]

Source: 1.Bardiya, M.C. and A.C. Gaur. 1970. Zbl. Bakt. Abt. II, 124, 552-555.

7-006 Effect of Alipur

Parameter	Effect Value	Concentration and Condition	Species
Growth rate and numbers	No effect	Pale and dark chestnut loams, 2.4 kg/ha, in situ	Soil bacteria[1]
Nitrification and nitrifiers	Initially decreased then increased activity and population	Medium podzol loam, 1.4-2.5 % OM, 0.6 kg/ha	Soil bacteria[2]
Nitrification and nitrifiers	No effect	Light loam podzol, 4 kg/ha	Soil bacteria[3]
Nitrification and nitrifiers	No effect	Light loamy medium podzol, pH=5.6-6, 2-4 kg/ha	Soil bacteria[3]
Nitrification and nitrifiers	Slightly decreased activity in wet conditions	Pale chestnut and medium loams, 2.4 kg/ha	Soil bacteria[4]
Nitrification and nitrifiers	Transiently decreased then increased activity	Sod-podzolic loam, 0.6 and 1.8 kg/ha	Soil bacteria[5]
Nitrogen fixation in soil	Decreased Azotobacter population	Light loam pale chestnut, 1.7 % OM, 1.2-2 kg/ha	Soil bacteria, free living[6]
Nitrogen fixation in soil	Increased Azotobacter population	Leached medium loam chernozem, 1.2 kg/ha	Soil bacteria, free living[7]
Nitrogen fixation in soil	Initially decreased numbers of Azotobacter chroococcum	Light loam podzol, 4 kg/ha	Soil bacteria, free living[3]
Soil microbial ammonification	No effect	Light loam pale chestnut, 1.7 % OM, 1.2-2 kg/ha	Soil organisms[6]
Soil microbial ammonification	No effect	Loess, 4 and 40 kg/ha	Soil organisms[8]

Source: 1.Chulakov, S.A. and S.U. Zharasov. 1971. Vest. Akad. Nauk kazakh. SSR 27, 35-40.
2.Isaeva, L.I. 1967. Agrokhimiya 4, 122-125.
3.Nepomiluev, V.F., I.P. Grechin and T.I. Kuzyakina. 1966. Izv. timiryazev sel-khoz Akad. No 2., 128-16.
4.Zharasov, S.U. 1971. Vest. selKhoz Nauki, Kazakhskoi 14, 41-43.
5.Isaeva, L.I. and E.D. Novogrudskaya. 1966. Vest selKhozPNauki, Mosk. 11, 57-59.
6.Lazareva, N.V. and V.G. Blokhin. 1969. Khimiya selKhoz 7, 609-611.
7.Matsneva, N.G. and O.A. Semikhatova. 1973. Khimiya selKhoz 11, 222-224.
8.Uziak, S., K. Leonaik and K. Gostkowska. 1971. Ann. Univ. Marie Curie-Sklodowska, C, 26, 57-72.

7-007 Effect of Alipur + PCP + endothal

Parameter	Effect Value	Concentration and Condition	Species
Growth rate and numbers	Decreased population	Six soils, Alipur=1.2 + PCP=4 + endothal=4 kg/ha, in situ	Soil bacteria[1]

Source: 1.Au, F.H.F. 1968. Effect of endothal and certain other selective herbicides on microbial activity in six different soils, Ph. D. Thesis, Oregon state Univ.

7-008 Effect of Alipur + dalapon

Parameter	Effect Value	Concentration and Condition	Species
Growth rate and numbers	No effect	Pale and dark chestnut loam, Alipur=2.4 and dalapon=6.0 kg/ha, in situ	Soil bacteria[1]

Source: 1.Chulakov, S.A. and S.U. Zharasov. 1971. Vest. Akad. Nauk kazakh. SSR 27, 35-40.

7-009 Effect of Allidochlor

Parameter	Effect Value	Concentration and Condition	Species
Nitrification and nitrifiers	Temporarily decreased then increased population and activity	Medium podzol loam, 1.9-2.5 % OM, 7 kg/ha	Soil bacteria[1]

Source: 1.Isaeva, L.I. 1967. Agrokhimiya 4, 122-125.

7-010 Effect of Ametryne

Parameter	Effect Value	Concentration and Condition	Species
Nitrification and nitrifiers	Inhibited activity	Acid volcanic ash, silt loam and neutral alluvium, 10 kg/ha	Soil bacteria[1]
Nitrification and nitrifiers	Inhibited activity	Marine sand, 10 kg/ha	Soil bacteria[2]
Nitrification and nitrifiers	No significant effect	Sandy loam, 2-4 kg/ha	Soil bacteria[3]
Soil respiration activity	Small to moderate increased activity	Sandy clay loam and 2 clay soils, 1.0 and 10 and 25 kg/ha	Soil organisms[4]

Source: 1.Mikhailova, M.F. 1968. Khimiya selKhoz 6, 204-206.
2.Dubey, H.D. 1969. Proc. Soil Soc. Am. 33, 893-896.
3.Leiderman, J., J.M. Hinojo and F.A. Fogliata. 1971. Revta ind. agric. Tucuman 48, 7-14.
4.Lui, L.C. and H.R. Cibes-Viade. 1972. J. Agric. Univ. P. Rico 56, 417-425.

7-011 Effect of Aminotriazole and simazine and MCPA

Parameter	Effect Value	Concentration and Condition	Species
Soil microbial ammonification	Increased activity	Sandy clay chernozem, 1-1.6 % OM, tot=$0 kg/ha, 38 % + 20 % + 16 %	Soil organisms[1]

Source: 1.Peshakov, G., E. Raikov and D. Tsvetanov. 1969. Pochv. Agrokhim 4, 89-94.

7-012 Effect of Aminotriazole, simazine and MCPA

Parameter	Effect Value	Concentration and Condition	Species
Nitrification and nitrifiers	Decreased activity	Sandy clay chernozem, 1.3 % OM, tot 10 kg/ha, Amino(38)+sim.(20)+MCPA(10 %)	Soil bacteria[1]
Nitrogen fixation in soil	No effect in medium to heavy leached chernozem	1-1.6 % OM, tot=10 kg/ha, amino. (38 %)+simazine(20 %) + MCPA(16 %)	Soil bacteria, free living[1]

Source: 1.Peshakov, G., E. Raikov and D. Tsvetanov. 1969. Pochv. Agrokhim 4, 89-94.

7-013 Effect of Amitrole-T

Parameter	Effect Value	Concentration and Condition	Species
Nitrification and nitrifiers	Inhibited activity	Sandy loam, 65 % MHC, 87.5 and 875 kg/ha	Soil bacteria[1]
Nitrification and nitrifiers	Transient decreased activity	Sandy loam, 65 % MHC, 8.75 kg/ha	Soil bacteria[1]
Soil respiration activity	Decreased CO2-evolution	Sandy loam polder, 35 and 350 and 3500 kg/ha	Soil organisms[1]

Source: 1.Schreven, D.A. van, D.J. Lindenbergh abd A. Koridon. 1970. Pl. Soil 33, 513-532.

7-014 Effect of Atraton

Parameter	Effect Value	Concentration and Condition	Species
Growth rate and numbers	Decreased population	Leached medium loam chernozem, pH=6.0, 4 % OM, 3 kg/ha, in situ	Soil bacteria[1]
Nitrification and nitrifiers	Increased activity	Leached medium loam chernozem, pH=6, 4 % OM, 3 kg/ha	Soil bacteria[1]
Nitrogen fixation in soil	Increased Azotobacter population	Leached medium loam chernozem, 3 kg/ha, pH=6, 4 % OM	Soil bacteria, free living[1]

Source: 1.Mikhailova, M.F. 1968. Khimiya selKhoz 6, 204-206.

7-015 Effect of Atrazine

Parameter	Effect Value	Concentration and Condition	Species
Denitrification in soil	Increased activity and population	Medium loam meadow soil, 1.2-2.2 % OM, 3 kg/ha	Soil organisms[1]
Denitrification in soil	No effect	Medium to heavy sandy clay leached chernozem, 1-1.6 % OM, 10 kg/ha	Soil organisms[2]
Growth rate and numbers	Decreased population	Chernozem (calcerous), 0-20 kg/ha, in situ	Soil bacteria[3]
Growth rate and numbers	Decreased population	Heavy degraded chernozem, >2 % OM, 7 y annual appl., in situ	Soil bacteria[4]
Growth rate and numbers	Decreased population	Red sandy loam, 0.5 kg/ha, in situ	Soil bacteria[5]
Growth rate and numbers	Decreased population	Calcerous heavy loam, pH=6.8-7.2, 3-4 kg/ha, selective effect, in situ	Soil bacteria[6]
Growth rate and numbers	Decreased population	Peaty-bog, pH=5.5, 47 % OM, dry years, 4.5 and 10 kg/ha, in situ	Soil bacteria[7]
Growth rate and numbers	Increased population	Weak to medium podzol, 7 % OM, 10 and 20 kg/ha, in situ	Soil bacteria[8]
Growth rate and numbers	Increased population when N added	Loam with and without N fert., 5.2 kg/ha, in situ	Soil bacteria[9]
Growth rate and numbers	Increased then decreased population	Peaty-bog, pH=5.5, 47 % OM, wet years, 4.5 and 10 kg/ha, in situ	Soil bacteria[7]
Growth rate and numbers	Initial increase then return to normal population	Forest soil, 1.0 and 3 and 5 kg/ha, in situ	Soil bacteria[10]
Growth rate and numbers	No effect	Medium clay loam, pH=5.3, 2 % humus, 2 and 6 kg/ha, in situ	Soil bacteria[11]
Growth rate and numbers	No effect	2-10 kg/ha, slightly increased population, in situ	Soil bacteria[12]
Growth rate and numbers	Sometimes increased population	Peaty-bog, pH=5.5, 47 % OM, wet years, 2nd appl., 4.5 and 10 kg/ha, in situ	Soil bacteria[7]
Growth rate and numbers	Strong inhibition	Liquid media=100 ppm, in culture	Soil bacteria[13]
Growth rate and numbers	Decrease in population	Alluvial and terra rossa soils, 8 kg/ha, in situ	Soil bacteria, non-spore formers[14]
Growth rate and numbers	Decreased population	20-100 kg/ha, in situ	Soil bacteria, non-spore formers[12]
Growth rate and numbers	Decrease in population	Alluvial and terra rossa soils, 8 kg/ha, in situ	Soil bacteria, spore formers[14]
Growth rate and numbers	Especially decreased population	20-100 kg/ha, in situ	Soil bacteria, spore formers[12]
Growth rate and numbers	Inhibition of growth rate	Solid+liquid media=5000-10000 ppm, non-spore formers less sensetive, in culture	Soil bacteria, spore formers[15]
Morphological change	Morphological change observed	Liquid media=1000 ppm, in culture	Arthrobacter sp.[16]
Morphological change	Morphological change observed	Liquid media=1000 ppm, in culture	Pseudomonas fluorescens[16]
Morphological change	Morphological change observed	Liquid media=1000 ppm, in culture	apco[16]
Mortality	Toxic effects	Liquid media=>10000 ppm, surfactants increase effect, in culture	Rhizosphere bacteria[17]

Nitrification and nitrifiers	Annual applications temporarily decreased population	Loamy sand, pH=5.3, 2.4 % humus, 4 kg/ha, 13 y	Soil bacteria[18]
Nitrification and nitrifiers	Decreased activity	Black clay loam, 1.0 and 10 kg/ha	Soil bacteria[19]
Nitrification and nitrifiers	Decreased activity	Medium loam meadow bog, 4.0 kg/ha	Soil bacteria[20]
Nitrification and nitrifiers	Decreased activity	Heavy loam chernozem, low OM, 3 kg/ha	Soil bacteria[21]
Nitrification and nitrifiers	Decreased activity and population	Black allivial and sandy soils, 30-40 kg/ha, recomm. field rate	Soil bacteria[22]
Nitrification and nitrifiers	Increased activity	Weak to medium podzol, 7 % OM, 10 and 20 kg/ha	Soil bacteria[8]
Nitrification and nitrifiers	Increased activity	Fine sand, 3 kg/ha	Soil bacteria[23]
Nitrification and nitrifiers	Increased activity	Podzol, 2 % humus, 1.0-4.0 kg/ha	Soil bacteria[11]
Nitrification and nitrifiers	Increased activity	Calcareous chernozem, 3 kg/ha	Soil bacteria[20]
Nitrification and nitrifiers	Increased activity	Medium loam meadow bog, 1.0 and 2 kg/ha	Soil bacteria[24]
Nitrification and nitrifiers	Increased activity	Alluvial sandy loam, pH=7.9, 0.5-2.0 kg/ha	Soil bacteria[1]
Nitrification and nitrifiers	Increased activity	Heavy loam, pH=6.3, 8-12 kg/ha	Soil bacteria[25]
Nitrification and nitrifiers	Increased activity	Medium loam chernozem, 0.2-3 kg/ha, with gamma-BHC and/or NPK	Soil bacteria[26]
Nitrification and nitrifiers	Increased activity and population	Leach black calcareous clay, 15 kg/ha	Soil bacteria[27]
Nitrification and nitrifiers	Little effetc	Peaty bog, pH=5.5, 47 % OM, 4.5 and 10 kg/ha	Soil bacteria[28]
Nitrification and nitrifiers	Low rate no effect, high rates inhibited activity	Leached chernozem, 1-40 kg/ha, recomm. field rate	Soil bacteria[2]
Nitrification and nitrifiers	No effect	Fine medium loam, 3 kg/ha	Soil bacteria[23]
Nitrification and nitrifiers	No effect	Loamy clay, 2 and 4.5 kg/ha	Soil bacteria[29]
Nitrification and nitrifiers	No effect	Leached sandy clay, 1.0-1.6 % OM, 10 kg/ha	Soil bacteria[2]
Nitrification and nitrifiers	No effect	Heavy loam chernozem, 4-5 % OM, 4 and 10 kg/ha	Soil bacteria[30]
Nitrification and nitrifiers	No effect	Heavy loam chernozem, low OM, 2 kg/ha	Soil bacteria[21]
Nitrification and nitrifiers	No effect	Normal chernozem, 1-4 kg/ha	Soil bacteria[21]
Nitrification and nitrifiers	No effect in field	Medium loam meadow, 1.2-2.2 % OM, 3 kg/ha, decreased activity in lab	Soil bacteria[31]
Nitrification and nitrifiers	No significant effect	Silt loam, pH=6, 2 % OM, 1.2 and 5 kg/ha	Soil bacteria[32]
Nitrification and nitrifiers	No significant effects	Sandy loam, 1-4 kg/ha, recomm. field rate	Soil bacteria[33]
Nitrogen fixation in soil	Annual application decreased soil population	Light loam, 20 kg/ha, 5 y	Soil bacteria, free living[3]
Nitrogen fixation in soil	Annual application increased soil population	Loamy sand, pH=5.3, 4 kg/ha	Soil bacteria, free living[18]

Nitrogen fixation in soil	Annual application increased soil population	Heavy clay degraded chernozem, >2 % OM, 4 kg/ha	Soil bacteria, free living[4]
Nitrogen fixation in soil	Decreased soil population	Medium loam meadow soil, 3 kg/ha	Soil bacteria, free living[1]
Nitrogen fixation in soil	Decreased soil population	Medium to heavy sandy clay, 1-1.6 % OM, 10 kg/ha	Soil bacteria, free living[2]
Nitrogen fixation in soil	Increased population	Alluvial, 8 kg/ha	Soil bacteria, free living[14]
Nitrogen fixation in soil	Increased Azotobacter	Red sandy loam, 0.5 kg/ha	Soil bacteria, free living[5]
Nitrogen fixation in soil	Increased Azotobacter population	Calcareous heavy loam, pH=6.8-7.2, 3-4 kg/ha	Soil bacteria, free living[6]
Nitrogen fixation in soil	No effect	Chernozem, 3 kg/ha	Soil bacteria, free living[34]
Nitrogen fixation in soil	No effect on numbers of Azotobacter sp. and Clostridium sp.	Leached meadow-cinnamon and alluvial soils, 1-4 kg/ha, recomm. field rate	Soil bacteria, free living[35]
Numbers and activity of soil algae	Chlorella sp. sensitive	Sandy soils, low OM, 0-0.05 kg/ha	Herbicide applied to soil in situ[36]
Numbers and activity of soil algae	Decreased activity and population	Alluvial black and sandy, 10-40 kg/ha, recomm. field rate	Herbicide applied to soil in situ[22]
Numbers and activity of soil algae	Decreased population but recovery within 3 months	Light loam, 17.4 kg/ha	Herbicide applied to soil in situ[37]
Numbers and activity of soil algae	Decreased population for up to 5 months	Sandy clay, 3-50 kg/ha	Herbicide applied to soil in situ[38]
Numbers and activity of soil algae	Decreased populations, especially blue-green algae	Gleyed heavy loam, >8 % OM; peaty-podzol, 35 % OM, 5 kg/ha	Herbicide applied to soil in situ[39]
Numbers and activity of soil algae	Some diatoms and Chlorococcum sp. tolerant, others decreased	Light sandy loam, >3 % OM, 5 kg/ha	Herbicide applied to soil in situ[39]
Soil microbial ammonification	Annual apllications decreased population	Heavy clay degraded chernozem, 4 kg/ha, 7 y	Soil organisms affected by herbicides[4]
Soil microbial ammonification	Annual applications increased population	Loamy sand, pH=5.3, 2.4 % humus, 4 kg/ha, 13 y	Soil organisms[18]
Soil microbial ammonification	Decreased activity	Sandy clay leached chernozem, 1-1.6 % OM, 10 kg/ha	Soil organisms[2]
Soil microbial ammonification	Decreased activity	Medium loam, 1.2-2.2 % OM, 3 kg/ha	Soil organisms[31]
Soil microbial ammonification	Little effect	Red-brown loamy clay, 2 and 4.5 kg/ha	Soil organisms[29]
Soil microbial ammonification	Little effect	Peaty-bog, pH=5.5, 47 % OM, 4.5 and 10 kg/ha	Soil organisms[40]
Soil respiration activity	Decreased CO2-evolution	Heavy clay degraded dernozem, 4 kg/ha	Soil organisms[4]
Soil respiration activity	Decreased CO2-evolution from sclerotium rolfsii	Autoclaved sandy loam, 1.2 kg/ha, effect dependend on C/N ratio	Soil organisms[41]
Soil respiration activity	Decreased O2-uptake	Fine sandy loam, 88 kg/ha	Soil organisms[42]

Soil respiration activity	Increased activity	Calcareous chernozem, 1-4 kg/ha	Soil organisms[43]
Soil respiration activity	Increased activity	Heavy and light loam chernozem, 2 kg/ha	Soil organisms[21]
Soil respiration activity	Inhibited CO2-evolution	Chernozem medium clay, pH=7.2, 1.5 % OM; sand, pH=5.8, 0.9 % OM, 50 kg/ha	Soil organisms[44]
Soil respiration activity	No effect	Heavy and light loam dernozem, 3 kg/ha	Soil organisms[21]
Soil respiration activity	No effect	Chernozem medium clay, pH=7.2, 1.5 % OM; sand, pH=5.8, 0.9 % OM, 2 kg/ha	Soil organisms[44]
Soil respiration activity	Small to moderate increased activity	Sandy clay loam and 2 clay soils, 1.0 and 10 and 25 kg/ha	Soil organisms[45]

Source: 1.Joshi, O.P., M.S. Sachdev, K.L. Sahrawat and B.N. Kohli. 1976. Pl. Soil 44, 367-375.
2.Peshakov, G., E. Raikov and D. Tsvetanov. 1969. Pochv. Agrokhim 4, 89-94.
3.Pantera, H. 1972. Pam. Pulawski 51, 69-76.
4.Husarova, M. 1972. Ved. Pr. vyzk. Ust. Kukur. Trnave, No 6, 45-69.
5.Spirama Raju, K. and G. Rangaswami. 1971. Indian J. Microbiol. 11, 25-29.
6.Bobyshev, V.G. and G.Y. Lapchenkov. 1966. Khimiya selKhoz 4, 30-31.
7.Nepomiluev, V.F. and T.I. Kuzyakina. 1967. Izv. timiryazev. selKhoz. Akad. No. 4, 84-91.
8.Prozorova, M.I. 1967. Izv. sib. Otdel. Akad. Nauk SSSR 5, 32-35.
9.Fink, R.J., O.H. Fletchall and O. Calvert. 1968. Weed Sci. 16, 104-105.
10.Milkowska, A. and A. Gorzelak. 1966. Sylwan 110, 13-22.
11.Tsevetkova, S.D. 1966. Les. Khoz. 19, 25-26.
12.Simon, L. 1971. Folia Microbiol. 16, 516.
13.Krezel, Z. and M. Musial. 1969. Acta microbiol. pol. Ser B 1, 93-87.
14.Goguadze, V.D. 1969. Subtrop. Kultury No 4, 161-167.
15.Kozlova, E.I., A.A. Belousova and V. Vandareva. 1964. Agrobiologiya 2, 271-277.
16.Sobieszczanski, J. 1969. Acta microbiol. pol. Ser B, 1(18), 99-104.
17.Balicka, N., J. Sobieszczanski and T. Niewiadoma. 1969. Acta microbiiol. pol.1(18), 11-14.
18.Voets, J.P., M. Meerschman and W. Verstraete. 1974. Soil Biol. Biochem. 6, 149-152.
19.Setty, R.A., V.C. Baligar and S.V. Patil. 1970. Mysore J. agric Sci. 4, 111-113.
20.Spiridinov, Y.Y., V. S. Skhiladze and G.S. Spiridinova. 1970. Subtrop. Kultury 3, 152-160.
21.Fisyunov, A.V. 1969. Agrkhimiya 2, 122-126.
22.Hauke-Pacewiczowa, T. 1971. Pam. Pulawski 46, 5-48.
23.Nederost, J. 1967. Sb. vys. Sk. zemed. Ceskych. Budejovicich 13, 1-19.
24.Tarlapan, M.I. and V.P. Zakhariya. 1966. Agrokhimiya 11, 115.
25.Zavarzin, V.I. 1966. Agrokhimiya 4, 121-124.
26.Kudzin, Y.K., A.Y. Fisyunov, N.A. Chernyavskaya and A.Y. Makarova.1973. Dokl. Vses. Akad. SelKhoz. Nuk. imeni V.I. Lenina 9, 13-15.
27.Andreeva-Fetvadzhieva, N. and B. Kolcheva. 1969. Nauchni Trud. vissh selkostop. Inst. Georgi Dimitro 20, 49-69.
28.Zavarzin, V.I. and T.V. Belyaeva. 1966. Khimiya selKhoz 4, 34-36.
29.Budoi, G. and I. Budoi. 1967. Lucr. stiint. Inst. Agron. Nicolae Balcescu (ser. A) 10, 369-379.
30.Pilipets, S.G. and I.A. Litvinov. 1972. Trudy kharkov selKhoz Inst. No 12, 142-146.
31.Tulabaev, B.D. and N. Azimbegov. 1967. Khimiya selKhoz 5, 43-44.
32.Thorneburg, R.P. and J.A. Tweedy. 1973. Weed Sci. 21, 397-399.
33.Leiderman, J., J.M. Hinojo and F.A. Fogliata. 1971. Revta ind. agric. Tucuman 48, 7-14.
34.Yurkevich, I.V. and N.Z. Tolkachev. 1972. Khimiya selKhoz 10, 696-698.
35.Rankov, V., E. Elenkov, P. Surlekov and B. Velev. 1966. Agrokhimiya 4, 115-120.
36.Atkins, C. and Y.T. Tchan. 1967. Pl. Soil 27, 432-442.
37.Kiss, A. 1967. Agrokem. Talajt. 16, 11.
38.Pantera, H. 1970. Meded. Fac. Landb. Rijksuniv. Gent. 35, 847-854.
39.Mikhailova, E.I. and Kruglov. 1973. Pochvovedenie 8, 81-85.
40.Krykhanov, L.I. and Z.V. Pashkovskaya. 1967. Trudy ural. nauchnoissled. Inst. selKhoz 7, 66-74.
41.Curl, E.A., R. Rodriguez-Kabana and H. Funderburk. 1966. Phytopathology 56, 874.
42.Withworth, J.W., W. Garner and B.C. Williams. 1965. Res. Prog. Rept. West. Weed Contr. Conf. 1965, 17-118.
43.Bakalivanov, D. 1971. Soils and Fert. 36, 607.
44.Kulimska, D. 1967. Roczn. Nauk. roln. (A) 93, 125-130.
45.Lui, L.C. and H.R. Cibes-Viade. 1972. J. Agric. Univ. P. Rico 56, 417-425.

7-016 Effect of Atrazine + prometryne

Parameter	Effect Value	Concentration and Condition	Species
Growth rate and numbers	Decreased population	Heavy clay degraded chernozem, atraz=1.66 + prom=0.8 kg/ha, 7 y, in situ	Soil bacteria[1]

Source: 1.Husarova, M. 1972. Ved. Pr. vyzk. Ust. Kukur. Trnave, No 6, 45-69.

7-017 Effect of Atrazine and mecoprop and 2,4,5-T

Parameter	Effect Value	Concentration and Condition	Species
Numbers and activity of soil algae	Decreased population, but recovery within 3 months	Light loam, tot=13.9 kg/ha, atraz(30 %)+meco.(11.5 %)+2,4,5-T(5.2 %)	Herbicide applied to soil in situ[1]

Source: 1.Kiss, A. 1967. Agrokem. Talajt. 16, 11.

7-018 Effect of Atrazine and prometryne

Parameter	Effect Value	Concentration and Condition	Species
Soil microbial ammonification	Annual apllication decreased population	Heavy clay degraded chernozem, atra.(1.6)+prom.(0.8 kg/ha)	Soil organisms[1]
Soil respiration activity	Decreased CO2-evolution	Heavy degraded chernozem, >2 % OM, atrazine(1.6)+prom.(0.8 kg/ha)	Soil organisms[1]

Source: 1.Husarova, M. 1972. Ved. Pr. vyzk. Ust. Kukur. Trnave, No 6, 45-69.

7-019 Effect of Barban

Parameter	Effect Value	Concentration and Condition	Species
Growth rate and numbers	Increased populations	Sandy loam, pH=5, 7.1 % OM, 20 kg/ha, in situ	Soil bacteria[1]
Nitrification and nitrifiers	Completely inhibited for 18 weeks	20 Barban and 140.8 Carbyne (kg/ha), sandy loam, pH=5, 7.1 % OM	Soil bacteria[1]
Soil respiration activity	Initial decrease in CO2-evolution	Sandy loam, pH=5, 7.1 % OM, 20 kg/ha	Soil organisms[1]

Source: 1.Anderson, J.R. 1978. In: I.R. Hill and S.J.L. Wright (eds.): Pesticide Microbiology. Academic Press.London. 313-533.

7-020 Effect of Benefin

Parameter	Effect Value	Concentration and Condition	Species
Growth rate	Slight inhibition of growth	Solid and liquid media = 200 ppm, herbicid, soil bacteria in culture	Bdellovibrio bacteriovorus[1]

Source: 1.Wehr, N.B. and D.A. Klein. 1971. Soil Biol. Biochem. 3, 143-149.

7-021 Effect of Bensulfide

Parameter	Effect Value	Concentration and Condition	Species
Soil respiration activity	Decreased activity	Fine sandy loam, 24 and 96 kg/ha	Soil organisms[1]
Soil respiration activity	Increased activity	Fine sandy loam, 4.8 kg/ha	Soil organisms[1]

Source: 1.Withworth, J.W., W. Garner and B.C. Williams. 1965. Res. Prog. Rept. West. Weed Contr. Conf. 1965, 17-118.

7-022 Effect of Bentazon

Parameter	Effect Value	Concentration and Condition	Species
Soil respiration activity	No effect	Loam and loamy sand, 4 kg/ha	Soil organisms[1]

Source: 1.Jung, J. 1972. Z. PflErnaehr. Dung. Bodenk. 133, 18-23.

7-023 Effect of Bromacil

Parameter	Effect Value	Concentration and Condition	Species
Growth rate and numbers	Slightly inhibition of growth	Solid media = 200 ppm, recomm. filed rate, in culture	Bdellovibrio bacteriovorus[1]
Nitrification and nitrifiers	Decreased activity	Sandy, 10 kg/ha	Soil bacteria[2]
Nitrification and nitrifiers	Increased activity	Alluvial, 2 % OM, 4 and 8 kg/ha	Soil bacteria[3]
Nitrification and nitrifiers	Inhibited activity	Clay, 10 and 100 kg/ha	Soil bacteria[4]
Nitrification and nitrifiers	No effect	Clay, 0.1 kg/ha	Soil bacteria[4]

Source: 1.Wehr, N.B. and D.A. Klein. 1971. Soil Biol. Biochem. 3, 143-149.
2.Pancholy, S.K. and J.a. Lynd. 1969. Weed Sci. 17, 460-463.
3.Horowitz, M., T. Blumenfield, G. Herzlinger and N. Hulin. 1974. Weed Res. 14, 97-109.4.Hulin, N. and M. Horowitz. 1973. Phytoparasitica 1, 64.

7-024 Effect of Cacodylic acid

Parameter	Effect Value	Concentration and Condition	Species
Soil respiration activity	Decreased CO2-evolution from forest litter	Forest soil, 10 and 100 kg/ha	Soil organisms[1]

Source: 1.Bollen, W.R., L.A. Norris and K.L. Stowers. 1974. Weed Sci. 22, 557-562.

7-025 Effect of Camphechlor

Parameter	Effect Value	Concentration and Condition	Species
Soil respiration	Slightly stimulated activity	Loam, pH=7.2, 1.49 % C, 600 and 6000 kg/ha	Soil organisms[1]

Source: 1.Naumann, K. 1971. Pedobiologia 11, 286-295.

7-026 Effect of Carbofuran

Parameter	Effect Value	Concentration and Condition	Species
Nitrification and nitrifiers	Indirect, slightly stimulatory effect	Silt loam, 0.75 and 10 and 50 kg/ha	Soil organisms[1]

Source: 1.Kuseske, D.W., B.R. Funke and J.T. Schultz. 1974. Pl. Soil 41, 255-269.

7-027 Effect of Carbyne

Parameter	Effect Value	Concentration and Condition	Species
Growth rate and numbers	Increased populations	Sandy loam, pH=5, 7.1 % OM, 140.8 kg/ha, solvents of carbyne, in situ	Soil bacteria[1]
Growth rate and numbers	Increased populations	Sandy loam, pH=5, 7.1 % OM, 140.8 kg/ha, in situ	Soil bacteria[1]
Nitrification and nitrifiers	Transiently inhibited	Solvents of Carbyne, sandy loam, pH=5, 7.1 % OM, 140.8 kg/ha	Nitrobacter spp.[1]
Nitrification and nitrifiers	Temporary decrease in activity	Solvents of Carbyne, Sandy loam, pH=5, 7.1 % OM, 140.8 kg/ha	Soil bacteria[1]
Soil respiration activity	Depressed CO2-evolution in later stages	Sandy loam, pH=5, 7.1 % OM, 140.8 kg/ha	Soil organisms[1]
Soil respiration activity	Initial decrease in CO2-evolution	Sandy loam, pH=5, 7.1 % OM, 140.8 kg/ha	Soil organisms[1]

Source: 1.Anderson, J.R. 1978. In: I.R. Hill and S.J.L. Wright (eds.): Pesticide Microbiology. Academic Press.London. 313-533.

7-028 Effect of Carbyne formulation additives

Parameter	Effect Value	Concentration and Condition	Species
Soil respiration activity	Initial decrease in CO2-evolution	Sandy loam, pH=5, 7.1 % OM, 140.8 kg/ha	Soil organisms[1]

Source: 1.Anderson, J.R. 1978. In: I.R. Hill and S.J.L. Wright (eds.): Pesticide Microbiology. Academic Press.London. 313-533.

7-029 Effect of Chloramben

Parameter	Effect Value	Concentration and Condition	Species
Nitrogen fixation in soil	Slightly decreased azotobacters	Irrigated light chestnut, 2 % OM, 5 kg/ha	Soil bacteria, free living[1]

Source: 1.Shkola, A.I. t1970. Vest. selKhoz Nauki, Alma-Ata 13, 34-37.

7-030 Effect of Chlorazine

Parameter	Effect Value	Concentration and Condition	Species
Growth rate and numbers	Decreased population	Peaty-bog, pH=5.5, 47 % OM, dry years, 4.5 and 10 kg/ha, in situ	Soil bacteria[1]
Growth rate and numbers	Increased then decreased population	Peaty-bog, pH=5.5, 47 % OM, wet years, 4.5 and 10 kg/ha, in situ	Soil bacteria[1]
Growth rate and numbers	Sometimes increased population	Peaty-bog, pH=5.5, 47 % OM, wet years, 2nd appl., 4.5 and 10 kg/ha, in situ	Soil bacteria[1]
Nitrification and nitrifiers	No effect	Normal chernozem, 2-6 kg/ha	Soil bacteria[2]
Nitrification and nitrifiers	No effect	Light loamy medium podzol, pH=5.6-6, 4-6 kg/ha	Soil bacteria[3]
Soil respiration activity	Increased activity	Heavy and light loam dernozem, 2 kg/ha	Soil organisms[2]
Soil respiration activity	No effect	Heavy and light loam dernozem, 3 kg/ha	Soil organisms[2]

Source: 1.Nepomiluev, V.F. and T.I. Kuzyakina. 1967. Izv. timiryazev. selKhoz. Akad. No. 4, 84-91.
2.Fisyunov, A.V. 1969. Agrkhimiya 2, 122-126.
3.Nepomiluev, V.F., I.P. Grechin and T.I. Kuzyakina. 1966. Izv. timiryazev sel-khoz Akad. No 2., 128-16.

7-031 Effect of Chlorbromuron

Parameter	Effect Value	Concentration and Condition	Species
Growth rate and numbers	Growth decreased initially	Liquid media=10 ppm, revover rapid and complete, in culture	Bacillus cereus[1]

Source: 1.Macek, J. and L. Milharcic. 1972. Fragm. herb. Croatica No 16, 1-6.

7-032 Effect of Chlordecone

Parameter	Effect Value	Concentration and Condition	Species
Growth rate and numbers	Numbers decreased, then increased and reverted to normal	Sandy loam, calcareous loam and clay, 22 kg/ha	Soil bacteria[1]
Soil respiration	CO2-evolution decreased, then increased and returned to normal	Sandy loam, calcareous clay and clay, 22 kg/ha	Soil organisms[1]

Source: 1.Gawaad, A.A.A., M.H. Hammad and F.H. El-Gayar. 1973. Agrokem. Talajt. 22, 161-168.

7-033 Effect of Chlorfenac

Parameter	Effect Value	Concentration and Condition	Species
Growth rate and numbers	Retarded populations in 20 d	Loam, 3.5 % org mat., 1.5 kg/ha,, in situ	Soil bacteria[1]

Source: 1.Geshtovt, Y.N., S.U. Zharasov and N. Adekenov. 1970. Vest. SelKhoz. Nauki. Kazakhskoi SSR. 13, 35-38

7-034 Effect of Chlorfenvinphos

Parameter	Effect Value	Concentration and Condition	Species
Nitrification and nitrifiers	Inhibited activity	Alluvial meadow, 100 kg/ha	Soil organisms[1]
Nitrification and nitrifiers	No effect	Alluvial meadow, 1.0 and 10 kg/ha	Soil organisms[1]

Source: 1.Helweg, A. 1972. Tidsskr. Plante Avl 76, 519-527.

7-035 Effect of Chloroxuron

Parameter	Effect Value	Concentration and Condition	Species
Growth rate and numbers	Isolates strongly inhibited	Liquid media=100 ppm, in culture	Soil bacteria[1]
Nitrification and nitrifiers	Inhibited activity	Sandy loam soils, pH=6.1-7.4, 1.3-1.6 % OM, 50 kg/ha	Soil bacteria[2]
Soil microbial ammonification	Increased activity	Alluvial meadow and leached cinnamon, 5 kg/ha	Soil organisms[3]
Soil respiration activity	Initially increased, then decreased CO2-evolution	Sandy loam, pH=7.4, 1.5 % OM; 50 kg/ha, coarse sandy loam, pH=6.1, 1.6 % OM	Soil organisms[2]
Soil respiration activity	No effect, of slight increase in CO2-evolution	Three clay soils woth pH=5.2, 6.2, 6.5, 0.1-10 kg/ha	Soil organisms[4]

Source: 1.Krezel, Z. and M. Musial. 1969. Acta microbiol. pol. Ser B 1, 93-87.
2.Grossbard, E. and J.A.P. Marsh. 1974. Pestic Sci. 5, 609-623.
3.Rankov, V. 1972. Prvi Nat. Kongr. Pochv. Bulg. 1969, 227-232.
4.Odu, C.T.I. and M.A. Horsfall. 1971. Pestic. Sci. 2, 122-125.

7-036 Effect of Chlorpropham

Parameter	Effect Value	Concentration and Condition	Species
Growth rate and numbers	Growth retarded	Liquid media=10-50 ppm, increase in glutamic acid, in culture	Arthrobacter globiformis[1]
Growth rate and numbers	Growth temporarily retarded	Liquid media=10-100 ppm, Pseudomonas sp. unaffected, in culture	Bacillus mesentericus[2]
Growth rate and numbers	Increased populations	Volcanic ash, 285 kg/ha, in situ	Soil bacteria[3]
Growth rate and numbers	Inhibation of growth rate	Liquid media=0.2-1000 ppm, effect depending of surfactants, in culture	Soil bacteria[4]
Nitrification and nitrifiers	Decreased activity	Black alluvial and sandy soils, 100 kg/ha, recomm. field rate	Soil bacteria[5]
Nitrification and nitrifiers	Increased activity	Sandy clay loam, 10 kg/ha	Soil bacteria[6]
Nitrification and nitrifiers	Inhibited activity	Volcanic clay loam, 10 and 100 kg/ha	Soil bacteria[3]
Nitrification and nitrifiers	No effect	Light alluvial meadow, pH=7, 1.2 kg/ha, low OM	Soil bacteria[7]
Nitrification and nitrifiers	No effect	Heavy leached meadow-cinnamon, pH=7, 2 % OM, 1.2 kg/ha	Soil bacteria[7]

Numbers and activity of soil algae	Decreased population and activity	Alluvial black or sandy soils, 100-120 kg/ha, recomm. field rate	Herbicide applied to soil in situ[4]
Soil respiration activity	Decreased activity	Silty medium loam derno-podzol, pH=5.9, 12 kg/ha	Soil organisms[9]

Source: 1.Kosinkiewicz, B. 1973. Acta microbiol. pol. Ser B, 5, 145-150.
2.Balicka, N. and Z. Krezel. 1969. Weed Res. 9, 37-42.
3.Matsuguchi, T. and S. Ishizawa. 1969. J. Sci. Soil Manure, Tokyo 40, 20-25.
4.Balicka, N. 1969. Acta microbiol. pol. Ser B, 1(18), 43-45.
5.Hauke-Pacewiczowa, T. 1971. Pam. Pulawski 46, 5-48.
6.Taha, S.M., S.A.Z. Mahoud, A.M. Abdel-hafez and A.S. Hamed. 1972. Egypt. J. Microbiol 7, 53-61.
7.Rankov, V. 1968. Pochv. Agrokhim. 3, 81-90.
8.Rankov, V., E. Elenkov, P. Surlekov and B. Velev. 1966. Agrokhimiya 4, 115-120.
9.Zinchenko, V.A. and T.V. Osinskaya. 1969. Agrokhimiya 9, 94-101.
????

7-037 Effect of Chlorpyrifos

Parameter	Effect Value	Concentration and Condition	Species
Growth rate and numbers	Numbers decreased initially then recovered	Sandy loam, 1.0 and 10 kg/ha	Soil bacteria[1]
Nitrification and nitrifiers	No effect or slight inhibition	Sandy loam, 1.0 and 10 kg/ha	Soil organisms[1]
Soil respiration	Increased activity	Sandy loam, 1.0 and 10 kg/ha	Soil organisms[1]

Source: 1.Tu, C.-M. 1969. Bact. Proc. , 1-3.

7-038 Effect of Chlorthal-dimethyl

Parameter	Effect Value	Concentration and Condition	Species
Denitrification in soil	Slight decrease in population but rapidly recovering	Serozem-meadow soil, 1.5-3 kg/ha	Soil organisms[1]
Nitrification and nitrifiers	Increased activity	Alluvial soil, 2 % OM, 7.5 and 15 kg/ha	Soil bacteria[2]
Nitrification and nitrifiers	Inhibited activity	Serozem-meadow, 3 kg/ha	Soil bacteria[1]
Nitrification and nitrifiers	No effect	Serozem-meadow, 1.5 kg/ha	Soil bacteria[1]
Nitrogen fixation in soil	Inhibited Azotobacter sp.	Serozem-meadow, 3.0 kg/ha	Soil bacteria, free living[1]
Nitrogen fixation in soil	Temporarily decreased Azotobacter popul.	Serozem-meadow, 1.5 kg/ha	Soil bacteria, free living[1]
Soil microbial ammonification	High rate inhibited activity	Serozem-meadow, 1.5 and 3 kg/ha	Soil organisms[1]
Soil respiration activity	No effect	Sandy clay loam, 4.5 and 8.2 and 50 kg/ha	Soil organisms[3]

Source: 3.Kosinkiewicz, B. 1970. Meded Fac. Landb Rijksuniv. Gent 35, 673-680.

Source: 1.Tulabaev, B.G. 1972. Khimiya selKhoz 10, 776-780.
2.Horowitz, M., T. Blumenfield, G. Herzlinger and N. Hulin. 1974. Weed Res. 14, 97-109.

7-039 Effect of Chlorthaldimethyl

Parameter	Effect Value	Concentration and Condition	Species
Growth rate and numbers	No effect	Silt loam and silty clay loam, 11.4 kg/ha, in situ	Soil bacteria[1]
Growth rate and numbers	No effect	4.5 and 13.5 kg/ha, in situ	Soil bacteria[2]

Source: 1.Fields, M.L., R. Der and D.D. Hemphill. 1967. Weeds 15, 195-197.
2.Bone, H.T. and J.E. Kuntz. 1968. Proc. N. cent. Weed Control Conf. 23, 34.

7-040 Effect of Chlorthiamid

Parameter	Effect Value	Concentration and Condition	Species
Nitrification and nitrifiers	Decreased activity	Soil, pH = 6.7, 1.0 and 10 and 100 kg/ha	Soil bacteria[1]
Soil microbial ammonification	No effect	Alluvial meadow, 2.8 and 28 and 280 kg/ha	Soil organisms[1]
Soil respiration activity	Decreased CO2-evolution	Calcareous and shale soils, 10-20 kg/ha, recomm. field rate	Soil organisms[2]

Source: 1.Helweg, A. 1972. Tidsskr. Plante Avl 76, 145-155.
2.Walter, B. 1970. Sonderh. Z. PflKrankh. PflPath. PflSchutz 5, 29-31.

7-041 Effect of Chlortoluron

Parameter	Effect Value	Concentration and Condition	Species
Growth rate and numbers	Growth decreased initially	Liquid media=0.005-5 ppm, revover rapid and complete, in culture	Bacillus cereus[1]

Source: 1.Macek, J. and L. Milharcic. 1972. Fragm. herb. Croatica No 16, 1-6.

7-042 Effect of Cl

Parameter	Effect Value	Concentration and Condition	Species
Foliar disease	Chlorosis, upper surface fleck similar to O3	Air, 0.5-1.5 ppm for 0.5-3 h, chlorine	Angiosperms[1]
Foliar disease	Distal necrosis	Air, 0.5-1.5 ppm for 0.5-3 h, chlorine	Gymnosperms[1]

Source: 1.Smith, W.H. 1981. Air Pollution and Forests. Springer Verlag. NY. 384 pp.

7-043 Effect of Clorthiamid

Parameter	Effect Value	Concentration and Condition	Species
Soil respiration activity	Decreased activity	Limestone, 3.2 % OM, 10-20 kg/ha, recomm. field rate	Soil organisms[1]

Source: 1.Berthold, W. 1970. 7th Int. Congr. Pl. Protection, Paris 1970, 295-297.

7-044 Effect of Cotoran

Parameter	Effect Value	Concentration and Condition	Species
Soil microbial ammonification	Increased activity and population	Leached chernozem, 0.5-4 kg/ha, recomm. field rate	Soil organisms[1]

Source: 1.Voinova, Z., S. Vlakhova and P. Petkova. 1970. Pochv. Agrokhim 4, 103-110.

7-045 Effect of Counter

Parameter	Effect Value	Concentration and Condition	Species
Nitrification and nitrifiers	No inhibition	Three soils, pH=5.7;6.3 and 7.0, 0.1-10 kg/ha	Soil organisms[1]
Soil respiration	No effect on CO2-evolution	Three soils, pH=5.7;6.3 and 7.0, 0.1-10 kg/ha	Soil organisms[1]

Source: 1.Laveglia, J. and P.A. Dahm. 1974. Environ. Ent. 33, 528-533.

7-046 Effect of DCA

Parameter	Effect Value	Concentration and Condition	Species
Growth rate and numbers	Inhibition in growth	Clay soil, anaerobic, 1.0-10 kg/ha, CaCO3 reduces effect, in situ	Sb, methane producing[1]
Soil respiration activity	CO2 and CH4-evolution inhibited, especially at high rate	Anaerobic clay, 1.0 and 10 kg/ha	Soil organisms[1]

Source: 1.Laskowski, D.A. and F.E. Broadbent. 1970. Proc. Soil Sci. Soc. Am. 34, 72-77.

7-047 Effect of DCU

Parameter	Effect Value	Concentration and Condition	Species
Denitrification in soil	No effect	Leached chernozem, 5 kg/ha	Soil organisms[1]
Denitrification in soil	Temporarily decreased population	Leached chernozem, 15 kg/ha	Soil organisms[1]
Growth rate and numbers	Increased population	Pale chestnut light loam, pH=8, 2 % OM, 18 kg/ha, in situ	Soil bacteria[2]
Nitrification and nitrifiers	Decreased then increased activity	Leached chernozem, low OM, plae chestnut light loam, pH=8, 2 % OM, 15 kg/ha	Soil bacteria[1]
Nitrification and nitrifiers	Increased activity	Pale chestnut light loam, pH=8, 2 % OM, 18 kg/ha	Soil bacteria[3]
Nitrification and nitrifiers	No effect	Leached chernozem, low OM, pale chestnut light loam, pH=8, 2 % OM, 5 kg/ha	Soil bacteria[1]
Nitrogen fixation in soil	Decreased Azotobacter population	Leached chernozem, 15 kg/ha	Soil bacteria, free living[1]
Nitrogen fixation in soil	No effect	Leached chernozem, 5 kg/ha	Soil bacteria, free living[1]
Nitrogen fixation in soil	No effect	Pale chestnut light loam, pH=8, 2 % OM, 18 kg/ha	Soil bacteria, free living[3]

Source: 1.Lobanov, V.E. and L.P. Poddubnaya. 1967. Khimiya selKhoz. 5, 32-34.
2.Chulakov, S.A. and S.U. Zharasov. 1971. Vest. Akad. Nauk kazakh. SSR 27, 35-40.
3.Chulakov, S.A. and S.U. Zharasov. 1967. Vest. selKhoz. Nauki, Alma-Ata 10, 122-124.

7-048 Effect of DDT

Parameter	Effect Value	Concentration and Condition	Species
Nitrification and nitrifiers	Decreased activity	Silt loam, 10 and 50 kg/ha	Soil organisms[1]
Nitrification and nitrifiers	No effect	Silt loam, 0.75 kg/ha	Soil organisms[1]
Nitrification and nitrifiers	No effect on rate of nitrification, but numbers of nitrifiers decreas	Alluvial sandy loam, 0.5-10 kg/ha	Soil organisms[2]

Source: 1.Ross, D.J. 1974. N.Z. Jl agric. Res. 17, 9-17.
2.Gaur, A.C. and R.P. Pareek. 1971. Indian J. Ent. 33, 368-370.

7-049 Effect of DNBP

Parameter	Effect Value	Concentration and Condition	Species
Growth rate and numbers	Inhibition of growth rate	Solid+liquid media=200 ppm, recomm. field app. rate, in culture	Bdellovibrio bacteriovorus[1]

Source: 1.Wehr, N.B. and D.A. Klein. 1971. Soil Biol. Biochem. 3, 143-149.

7-050 Effect of DNOC

Parameter	Effect Value	Concentration and Condition	Species
Growth rate and numbers	No effect on growth	Loam, 11 y annual appl., 5-10 kg/ha, in situ	Soil bacteria[1]
Nitrification and nitrifiers	Annual applications increased activity	Light loam dernopodzol, 2-10 kg/ha, recomm. field rate	Soil bacteria[2]
Nitrification and nitrifiers	Decreased population	Light alluvial meadow, pH=7, 10 kg/ha	Soil bacteria[3]
Nitrification and nitrifiers	Decreased population	Heavy leached meadow-cinnamon, pH=7, 10 kg/ha	Soil bacteria[3]

Nitrogen fixation in soil	Decreased azotobacter population	Leached meadow-cinnamon and alluvial soils, 15 kg/ha, DNOC 50 %	Soil bacteria, free living[4]

Source: 1.Resz. A. 1968. Sonderh. Z. Plfkrankh. PflPath. PflSchutz No 4, 147-150.
2.Chunderova, A.I., T.P. Zuberts and A.M. Sofinskii. 1971. Khimiya selKhoz 9, 527-530.
3.Rankov, V. 1968. Pochv. Agrokhim. 3, 81-90.
4.Rankov, V., E. Elenkov, P. Surlekov and B. Velev. 1966. Agrokhimiya 4, 115-120.

7-051 Effect of Dalapon

Parameter	Effect Value	Concentration and Condition	Species
Growth rate and numbers	Inhibition in growth	Clay soil, anaerobic, 1.0-10 kg/ha, CaCO3 reduces effect, in situ	Sb, methane producing[1]
Growth rate and numbers	Decreased population	Red clay, 0.15-0.75 kg/ha, in situ	Soil bacteria[2]
Growth rate and numbers	Increase in populations	10 kg/ha, in situ	Soil bacteria[3]
Growth rate and numbers	Initially decreased population	Irrigated brown soil, pH=7-8, 20 kg/ha, recover after 30 d, in situ	Soil bacteria[4]
Growth rate and numbers	Temporarily decreased populations	Meadow-type medium loam, 1.5 % OM, 8 kg/ha, in situ	Soil bacteria[5]
Growth rate and numbers	Decrease in population	Light brown soil, 0.5-360 kg/ha, in situ	Soil bacteria, spore formers[6]
Nitrification and nitrifiers	Decreased activity	Meadow type medium loam, 1.5 % OM, 8 kg/ha	Soil bacteria[5]
Nitrification and nitrifiers	No effect	Sandy loam, 0.8-20 kg/ha, recomm. field rate	Soil bacteria[7]
Nitrification and nitrifiers	Silghtly decreased activity	Sandy loam, 65 % MHC, 127.5 kg/ha	Soil bacteria[8]
Nitrification and nitrifiers	Slightly increased activity	Medium podzol loam, 1.9-2.5 % OM, 10 kg/ha	Soil bacteria[9]
Nitrification and nitrifiers	Strongly decreased activity	Sandy loam, 65 % MHC, 1275 kg/ha	Soil bacteria[8]
Nitrogen fixation in soil	Decrease in azotobacter population	Red clay, 0.45 and 0.6 and 0.75 kg/ha	Soil bacteria, free living[2]
Nitrogen fixation in soil	Decreased azotobacter population	Polder soil, 65 % MHC, 0.9 and 90 kg/ha, recomm. field rate	Soil bacteria, free living[8]
Nitrogen fixation in soil	Decreased azotobacter population	Meadow-type medium loam, 1.5 % OM	Soil bacteria, free living[5]
Nitrogen fixation in soil	No effect	Red clay, 0.15 and 0.3 kg/ha	Soil bacteria, free living[2]
Soil microbial ammonification	Decreased population	Meadow type medium loam, 1.5 % OM, 8 kg/ha	Soil organisms[5]
Soil microbial ammonification	Increased activity	Brown heavy loam, pH=6.3, 10-15 kg/ha	Soil organisms[10]
Soil respiration activity	CO2 and CH4-evolution inhibited, especially at high rate	Anaerobic clay, 1.0 and 10 kg/ha	Soil organisms[1]
Soil respiration activity	Decreased CO2-evolution, especially at 2 highest rates	Sandy loam, 12.8 and 128 and 1280 kg/ha	Soil organisms[8]
Soil respiration activity	Decreased activity	Silty medium loam derno-podzol, pH=5.9, 20 kg/ha	Soil organisms[11]

Source: 1.Laskowski, D.A. and F.E. Broadbent. 1970. Proc. Soil Sci. Soc. Am. 34, 72-77.
2.Sharma, L.N. and S.N. Saxena. 1972. Andhra agric. J. 18, 74-78.
3.Krykha, hanov, L.I. and Z.V. Pashkovskaya. 1967. Trudy ural. nauchnoissled. Inst. selKhoz. 7, 66-74.
4.Akropyan, E.A., B.P. Avakyan and A.A. Kalantarov. 1974. Khimiya selKhoz. 12, 538-539.
5.Tulabaev, B.D. 1970. Uzbek. biol. Zh. 14, 11-14.
6.Zhakharyan, S.V. 1970. Symp. Biol. Hung. 11, 391-395.

7.Leiderman, J., J.M. Hinojo and F.A. Fogliata. 1971. Revta ind. agric. Tucuman 48, 7-14.
8.Schreven, D.A. van, D.J. Lindenbergh abd A. Koridon. 1970. Pl. Soil 33, 513-532.
9.Isaeva, L.I. 1967. Agrokhimiya 4, 122-125.
10.Zavarzin, V.I. and T.V. Belyaeva. 1966. Khimiya selKhoz 4, 34-36.
11.Zinchenko, V.A. and T.V. Osinskaya. 1969. Agrokhimiya 9, 94-101.

7-052 Effect of Dalapon and 2,4-D

Parameter	Effect Value	Concentration and Condition	Species
Nitrification and nitrifiers	Increased activity	Derno-podzolic light loam, dalapon (20) + 2,4-D(2 kg/ha)	Soil bacteria[1]
Soil respiration activity	Decreased activity	Dalapon(20)+2,4-D(2 kg/ha), podzolic light loam, moderate loam	Soil organisms[2]
Soil respiration activity	Decreased activity	Dalapon(20)+2,4-D(2 kg/ha), podzolic compact; heavy loam	Soil organisms[2]
Soil respiration activity	Increased activity	Dalapon(60)+2,4-D(6 kg/ha), podzolic light loam; moderate loam	Soil organisms[2]
Soil respiration activity	Increased activity	Dalapon(60)+2,4-D(6 kg/ha), podzolic compact; heavy loam	Soil organisms[2]

Source: 1.Bliev, Y.K. and L.M. Kozlova. 1972. Khimiya selKhoz 10, 850-853.
2.Bliev, O.K. 1973. Pochvovedenie 7, 61-68.

7-053 Effect of Dalapon and 2,4-DA

Parameter	Effect Value	Concentration and Condition	Species
Nitrogen fixation in soil	Increased N-fixation during first 10 days	Dalapon(60)+2,4-DA(6 kg/ha), podzolic light loam; moderate loam	Soil bacteria, free living[1]
Nitrogen fixation in soil	Little to no effect	Alluvial, dalapon(20) + 2,4-DA(2 kg/ha)	Soil bacteria, free living[1]
Nitrogen fixation in soil	Increased N-fixation during first 10 days	Dalapon(60)+2,4-DA(6 kg/ha), Podzolic compact; thick heavy loam	Soil bacteria, free living[1]

Source: 1.Bliev, O.K. 1973. Pochvovedenie 7, 61-68.

7-054 Effect of Diazinon

Parameter	Effect Value	Concentration and Condition	Species
Nitrification and nitrifiers	No effect or slight inhibition	Sandy loam, 1.0 and 10 kg/ha	Soil organisms[1]
Soil respiration	Increased respiration	Sandy loam, 1.0 and 10 kg/ha	Soil organisms[1]

Source: 1.Tu, C.-M. 1969. Bact. Proc., 1-3.

7-055 Effect of Diazon

Parameter	Effect Value	Concentration and Condition	Species
Growth rate and numbers	Numbers decreased initially then recovered	Sandy loam, 1.0 and 10 kg/ha	Soil bacteria[1]

Source: 1.Tu, C.-M. 1969. Bact. Proc., 1-3.

7-056 Effect of Dicamba

Parameter	Effect Value	Concentration and Condition	Species
Nitrification and nitrifiers	Inhibited activity	Derno-podzolic medium loam, 0.12 kg/ha	Soil bacteria[1]

Source: 1.Bezuglov, V.G., A.K. Minenko and E.P. Shelestov. 1973. Khimiya selKhoz 11, 854-856.

7-057 Effect of Dicamba dimethylamine

Parameter	Effect Value	Concentration and Condition	Species
Growth rate and numbers	Increase in growth rate	Light loam dernopodzol, 0.12 kg/ha, in situ	Clostridium acetobutylicum[1]
Growth rate and numbers	Decrease in growth rate	Light loam dernopodzol, 0.12 kg/ha, in situ	Clostridium butylicum[1]
Growth rate and numbers	Decrease in growth rate	Light loam dernopodzol, 0.12 kg/ha, in situ	Clostridium pasteurianum[1]

Source: 1.Kozyrev, M.A. and A.A. Laptev. 1972. Khimiya selKhoz. 10, 614-617.

7-058 Effect of Dichlobenil

Parameter	Effect Value	Concentration and Condition	Species
Growth rate and numbers	Inhibited populations for 90 days	Chernozemsmolnitsa, 7.5 kg/ha, in situ	Soil bacteria[1]
Nitrification and nitrifiers	Decreased activity	Soil, pH = 6.7, 10 and 100 kg/ha	Soil bacteria[2]
Nitrogen fixation in soil	No effect	Chernozem-smolnitsa, 7.5 kg/ha	Soil bacteria, free living[1]
Soil microbial ammonification	No effect	Alluvial meadow, 2.8 and 28 and 280 kg/ha	Soil organisms[2]
Soil respiration activity	Decreased CO2-evolution	Calcareous and shale soils, 2.8-13 kg/ha, recomm. field rate	Soil organisms[3]
Soil respiration activity	Decreased activity	Clay shale, 4.4 % OM, 2.8-13 kg/ha, recomm. field rate	Soil organisms[4]

Source: 1.Nikolova, G. and D. Bakalivanov. 1972. Symp. Biol. Hung. 11, 385-389.
2.Helweg, A. 1972. Tidsskr. Plante Avl 76, 145-155.
3.Walter, B. 1970. Sonderh. Z. PflKrankh. PflPath. PflSchutz 5, 29-31.
4.Berthold, W. 1970. 7th Int. Congr. Pl. Protection, Paris 1970, 295-297.

7-059 Effect of Dichlorprop

Parameter	Effect Value	Concentration and Condition	Species
Nitrification and nitrifiers	Slightly decreased activity	Sandy loam, 65 % MHC, 2.6 kg/ha	Soil bacteria[1]
Nitrification and nitrifiers	Strongly decreased activity	Sandy loam, 65 % MHC, 255 kg/ha	Soil bacteria[1]
Nitrogen fixation in soil	Decreased azotobacter population	Polder soil, 65 % MHC, 2.5 and 250 kg/ha, recomm. field rate	Soil bacteria, free living[1]
Soil respiration activity	Decreased CO2-evolution	Sandy loam, 2.5 and 25 and 250 kg/ha	Soil organisms[1]

Source: 1.Schreven, D.A. van, D.J. Lindenbergh abd A. Koridon. 1970. Pl. Soil 33, 513-532.

7-060 Effect of Dieldrin

Parameter	Effect Value	Concentration and Condition	Species
Growth rate and numbers	Decreased numbers in rhizosphere of maize	Chernozem, 150 kg/ha	Soil bacteria[1]
Growth rate and numbers	Increased, then decreased numbers	Clay soil + 2 % compost, 2.5 and 50 kg/ha	Soil bacteria[2]
Nitrification and nitrifiers	Inhibited activity, NO2(-) accumulation	Sandy loam, 2.5 and 10 and 100 kg/ha	Soil organisms[3]
Nitrification and nitrifiers	No effect	Sandy loam, 0.1 and 1.0 kg/ha	Soil organisms[3]

Nitrification and nitrifiers	Numbers of nitrifying bacteria severely depressed	Clay + 2 % compost, 2.5 and 50 kg/ha	Soil organisms[2]
Soil respiration	Low rates, no effect. Higher rates decreased CO2-evolution	Alluvial, 0.1 - 100 kg/ha	Soil organisms[3]

Source: 1.Tsirkov, Y.I. 1971. Bulg: Akad. Nauk. (1971), 267-272.
2.Mahmond, S.A.Z., K.G. Selim and T. El-Mokadem. 1970. Zbl. Bakt. Abt II 125, 134-149.
3.Bardiya, M.C. and A.C. Gaur. 1970. Zbl. Bakt. Abt. II, 124, 552-555.

7-061 Effect of Dinitramine

Parameter	Effect Value	Concentration and Condition	Species
Growth rate and numbers	Increased populations	Silty loam, 0.1-0.4 kg/ha, in situ	Soil bacteria[1]
Nitrification and nitrifiers	No effect	Sandy loam, 0.1 kg/ha	Soil bacteria[1]

Source: 1.Belles, W.S. 1972. Proc. N. cent. Weed Control Conf. 27, 50-51.

7-062 Effect of Dinoseb

Parameter	Effect Value	Concentration and Condition	Species
Nitrification and nitrifiers	No effect	Light loamy medium podzol, pH=5.6-6.0, 2-4 kg/ha, dinoseb 20 %	Soil bacteria[1]
Soil respiration activity	Decreased activity	Silty medium loam derno-podzol, pH=5.9, 5 kg/ha	Soil organisms[2]

Source: 1.Nepomiluev, V.F., I.P. Grechin and T.I. Kuzyakina. 1966. Izv. timiryazev sel-khoz Akad. No 2., 128-16.
2.Zinchenko, V.A. and T.V. Osinskaya. 1969. Agrokhimiya 9, 94-101.

7-063 Effect of Dinoseb acetate

Parameter	Effect Value	Concentration and Condition	Species
Growth rate and numbers	Slightly decrease in population	Sandy loam, selective effect, 1.2-2 kg/ha, in situ	Soil bacteria[1]
Growth rate and numbers	Slightly increased population	Peaty bog, pH=5.5, 47 % OM, 1.5 kg/ha, recomm. filed rate, in situ	Soil bacteria[2]
Nitrification and nitrifiers	Decreased activity	Sandy loam, 1.2 - 2 kg/ha	Soil bacteria[1]
Nitrification and nitrifiers	Decreased activity and population	Black alluvial and soils, 15-25 kg/ha, recomm. field rate	Soil bacteria[3]
Nitrification and nitrifiers	Inhibited activity	Four soils, 1.5-2.5 kg/ha, less effect in heavier textured soils	Soil bacteria[4]
Nitrogen fixation in soil	No effect on numbers of Azotobacter chroococcum or Clostridium sp.	Leached meadow-cinnamon and alluvial soils, 1.5-2.5 kg/ha, recomm. field	Soil bacteria, free living[5]
Nitrogen fixation in soil	Temporarily decreased population	Sandy loam, 1.2 - 2 kg/ha	Soil bacteria, free living[6]
Numbers and activity of soil algae	Decreased activity and population	Alluvial black and sandy soils, 15-25 kg/ha, recomm. field rate	Herbicide applied to soil in situ[3]
Soil microbial ammonification	Inhibited activity	Alluvial meadow, 2-1000 kg/ha	Soil organisms[7]
Soil microbial ammonification	Less than normal inhibition	Heavier textured, high OM, 2-1000 kg/ha	Soil organisms[7]
Soil microbial ammonification	No effect	Loess, 5 and 50 kg/ha	Soil organisms[8]

Source: 1.Hoffman-Kakol, I. and H. Kwarta. 1972. Zesz. nauk. Akad. Rolnicz. Szczecin. No 38, 91-103.
2.Nepomiluev, V.F. and T.I. Kuzyakina. 1967. Izv. timiryazev. selKhoz. Akad. No. 4, 84-91.
3.Hauke-Pacewiczowa, T. 1971. Pam. Pulawski 46, 5-48.
4.Rankov, V. 1973. Pochv. Agrokhim. 8, 125-134.
5.Rankov, V., E. Elenkov, P. Surlekov and B. Velev. 1966. Agrokhimiya 4, 115-120.

6.Hoffman-Kakol, I. and H. Kwarta. 1972. Zesz. nauk. Akad. Rolnicz. Szczecin. No 38, 105-116.
7.Rankov, V. 1971. Blarskata Akad. Nauk (1971), 253-257.
8.Uziak, S., K. Leonaik and K. Gostkowska. 1971. Ann. Univ. Marie Curie-Sklodowska, C, 26, 57-72.

7-064 Effect of Dinoseb triethanolamine

Parameter	Effect Value	Concentration and Condition	Species
Growth rate and numbers	Slightly increased population	Peaty bog, pH=5.5, 47 % OM, 1.5 kg/ha, recomm. filed rate, in situ	Soil bacteria[1]

Source: 1.Nepomiluev, V.F. and T.I. Kuzyakina. 1967. Izv. timiryazev. selKhoz. Akad. No. 4, 84-91.

7-065 Effect of Diphenamid

Parameter	Effect Value	Concentration and Condition	Species
Growth rate and numbers	Depressed in growth	Red and alluvial soil, 6-8 kg/ha, in situ	Soil bacteria, non-spore formers[1]
Nitrification and nitrifiers	Decreased activity for 60 days	Red soil and alluvial, 6 and 8 kg/ha	Soil bacteria[1]
Nitrification and nitrifiers	Increased activity in field	Alluvial soil, 2 % OM, no effect in lab, 2.4 and 4.8 kg/ha	Soil bacteria[2]
Nitrification and nitrifiers	Initially inhibited then increased activity	Clay, 0.1-4 kg/ha, recommended field rate	Soil bacteria[3]
Nitrification and nitrifiers	No effect	Light alluvial meadow, pH=7.0, low OM, 0.6 kg/ha	Soil bacteria[4]
Nitrogen fixation in soil	No effect	Red and alluvial, 6 and 8 kg/ha	Soil bacteria, free living[1]
Nitrogen fixation in soil	Noo on population	Leached meadow-cinnamon and alluvial soils, 4-6 kg/ha, recomm. field rate	Soil bacteria, free living[5]

Source: 1.Mickovski, M.D. 1972. Symp. Biol. Hung. 11, 397-400.
2.Horowitz, M., T. Blumenfield, G. Herzlinger and N. Hulin. 1974. Weed Res. 14, 97-109.
3.Hulin, N. and M. Horowitz. 1973. Phytoparasitica 1, 64.
4.Rankov, V. 1968. Pochv. Agrokhim. 3, 81-90.
5.Rankov, V., E. Elenkov, P. Surlekov and B. Velev. 1966. Agrokhimiya 4, 115-120.

7-066 Effect of Diquat

Parameter	Effect Value	Concentration and Condition	Species
Denitrification in soil	No effect	Lake sediment, 1-5 kg/ha, recomm. field rate	Soil organisms[1]
Nitrogen fixation in soil	Fixation inhibited	Lake sediment, 1.1-2 kg/ha, recomm. field rate	Soil bacteria, free living[1]

Source: 1.Atkinson, G. 1973. Proc. Polln. Res. Conf., June 1973. Wairakei, NZ, DSIR Information Series 97, 529539.

7-067 Effect of Disulfoton

Parameter	Effect Value	Concentration and Condition	Species
Growth rate and numbers	Decreased numbers	Volcanic ash soil, 60 kg/ha	Soil bacteria[1]
Nitrification and nitrifiers	Increased activity	Volcanic ash, 60 kg/ha	Soil organisms[2]

Source: 1.Kobryashi, T. and S. Katsura. 1968. Jap. J. Appl. Ent. Zool. 12, 53-63.
2.Kobayashi, T. and S. Katsura. 1968. Jap. J. Appl. Ent. Zool. 12, 53-63.

7-068 Effect of Diuron

Parameter	Effect Value	Concentration and Condition	Species
Growth rate and numbers	Moderately inhibition	Solid media = 200 ppm, recomm. filed rate, in culture	Bdellovibrio bacteriovorus[1]

Growth rate and numbers	4 fold decrease in population	Irrigated brown soil, 8 kg/ha, in situ	Soil bacteria[2]
Growth rate and numbers	Decreased population	Meadow bog, monthly applications, 8-16 kg/ha, in situ	Soil bacteria[3]
Growth rate and numbers	Decreased population	Red krasnozem, monthly applications, 12-20 kg/ha, in situ	Soil bacteria[3]
Growth rate and numbers	Decreased population	Red sandy loam, 0.5 kg/ha, in situ	Soil bacteria[4]
Growth rate and numbers	Increased population	0.15-0.25 kg/ha, higher appl. no effect, in situ	Soil bacteria[5]
Growth rate and numbers	No effect	Light clay, 3.1-3.5 % OM, 8 kg/ha, in situ	Soil bacteria[6]
Growth rate and numbers	Some species decreased	Medium loam red soil, 12-20 kg/ha, in situ	Soil bacteria[7]
Nitrification and nitrifiers	Decreased population in field	Irrigated brown soil, 8 kg/ha, no effect in lab	Soil bacteria[2]
Nitrification and nitrifiers	Increased activity	Brown heavy loam, pH=6.3, 6-12 kg/ha	Soil bacteria[8]
Nitrification and nitrifiers	Increased activity	Alluvial, 2 % OM, 0.8 and 1.6 kg/ha	Soil bacteria[9]
Nitrification and nitrifiers	Inhibited activity	Loam soil, pH=7-7.4, 10-15 kg/ha	Soil bacteria[10]
Nitrification and nitrifiers	Inhibited activity	Sandy loam soils, pH=6.1-7.4, 1.3-1.6 % OM, 50 kg/ha	Soil bacteria[11]
Nitrification and nitrifiers	Inhibited activity	Acid volcanic ash, silt loam and neutral alluvium, 2 kg/ha	Soil bacteria[12]
Nitrification and nitrifiers	Inhibited activity	Marine sand, 2 kg/ha	Soil bacteria[12]
Nitrification and nitrifiers	Little effect	Meadow-type chernozem, 1.5 kg/ha	Soil bacteria[13]
Nitrification and nitrifiers	No significant effect	Clay loam, 0.1-0.4 kg/ha	Soil bacteria[14]
Nitrogen fixation in soil	Decreased soil population	Irrigated brown soil, 8 kg/ha	Soil bacteria, free living[2]
Nitrogen fixation in soil	Increased soil population	Red sandy loam, 0.5 kg/ha	Soil bacteria, free living[4]
Nitrogen fixation in soil	Inhibition of soil	Acid soil, pH=5, 6-8 kg/ha	Soil bacteria, free living[15]
Nitrogen fixation in soil	No effect	Meadow soil, 1.5 kg/ha	Soil bacteria, free living[13]
Nitrogen fixation in soil	No effect	Alluvial kranozem, pH=7, 6-8 kg/ha	Soil bacteria, free living[15]
Numbers and activity of soil algae	Decreased population, but recovery within 3 months	Light loam, 5.2 kg/ha	Herbicide applied to soil in situ[16]
Soil microbial ammonification	Inhibited activity	Medium loam meadow soil, 1.2-2.0 % OM, 1.5 kg/ha	Soil organisms[17]
Soil microbial ammonification	No effect	Silt, 60 kg/ha	Soil organisms[18]
Soil respiration activity	Decreased CO2-evolution	Silt loam, 10 kg/ha, recomm. field rate	Soil organisms[19]
Soil respiration activity	Decreased O2-uptake	Fine sandy loam, 88 kg/ha	Soil organisms[20]

Soil respiration activity	Decreased activity for 28 d, then recovered, high OM decreased inhibi	Nine soil types, variable OM, 5-100 kg/ha	Soil organisms[21]
Soil respiration activity	Increased activity	Sandy clay loam and 2 clay soils, 1.0 and 10 kg/ha	Soil organisms[22]
Soil respiration activity	Initially increased, then decreased CO2-evolution	Sandy loam, pH=7.4, 1.5 % OM; 50 kg/ha, coarse sandy loam, pH=6.1, 1.6 % OM	Soil organisms[11]
Soil respiration activity	No effect	Sandy clay loam and 2 clay soils, 25 kg/ha	Soil organisms[22]
Soil respiration activity	Sometimes increased activity	Silty medium loam derno-podzol, pH=5.9, 5 kg/ha	Soil organisms[23]

Source: 1.Wehr, N.B. and D.A. Klein. 1971. Soil Biol. Biochem. 3, 143-149.
2.Akopyan, E.A. and A.G. Agaronyan. 1968. Soils and Fert. 33, 3133.
3.Spiridinov, Y.Y., V. S. Skhiladze and G.S. Spiridinova. 1970.In: The Behavior, Conversion and Analyss of Pesticides and their Metabolites in Soil. Proc. 1st all-Union Conf. Pushino-na-OKe, Akad. Nauk.
4.Spirama Raju, K. and G. Rangaswami. 1971. Indian J. Microbiol. 11, 25-29.
5.Sharma, L.N. and S.N. Saxena. 1972. Andhra agric. J. 18, 74-78.
6.Spiridinov, Y.Y., V. S. Skhiladze and G.S. Spiridinova. 1970. Subtrop. Kultury 1, 150-155.
7.Spiridinov, Y.Y., V. S. Skhiladze and G.S. Spiridinova. 1970. Subtrop. Kultury 3, 152-160.
8.Kudzin, Y.K., A.Y. Fisyunov, N.A. Chernyavskaya and A.Y. Makarova.1973. Dokl. Vses. Akad. SelKhoz. Nuk. imeni V.I. Lenina 9, 13-15.
9.Horowitz, M., T. Blumenfield, G. Herzlinger and N. Hulin. 1974. Weed Res. 14, 97-109.
10.Corke, C.T. and F.R. Thompson. 1970. Can. J. Microbiol. 16, 567-571.
11.Grossbard, E. and J.A.P. Marsh. 1974. Pestic Sci. 5, 609-623.
12.Dubey, H.D. 1969. Proc. Soil Soc. Am. 33, 893-896.
13.Tulabaev, B.D. and S. Tamikaev. 196. Uzbek. biol. Zh. 2, 14-17.
14.Kuramato, M., H.P. Haag and J.R. Sarruge. 1970. An. Esc. Sup. Agric. Luiz de Queroz 27, 117-123.
15.Goguadze, V.D. 1968. Agrokhmiya 5, 99-103.
16.Kiss, A. 1967. Agrokem. Talajt. 16, 11.
17.Tulabaev, B.D. and N. Azimbegov. 1967. Khimiya selKhoz 5, 43-44.
18.Geltser, Y.G., V.A. Geptner and L.D.Stonov. 1972. Agrokhimiya 9, 119-123.
19.Doxtader, K.G. 1968. Bact. Proc. 4, A21.
20.Withworth, J.W., W. Garner and B.C. Williams. 1965. Res. Prog. Rept. West. Weed Contr. Conf. 1965, 17-118.
21.Chandra, P., W.R. Furtick and W.B. Bollen. 1960. Weeds 8, 589-598.
22.Lui, L.C. and H.R. Cibes-Viade. 1972. J. Agric. Univ. P. Rico 56, 417-425.
23.Zinchenko, V.A. and T.V. Osinskaya. 1969. Agrokhimiya 9, 94-101.

7-069 Effect of EPTC

Parameter	Effect Value	Concentration and Condition	Species
Growth rate and numbers	Moderate inhibition of growth	Solid+liquid media=200 ppm, in culture	Bdellovibrio bacteriovorus[1]
Growth rate and numbers	Decrease in populations	Loamy chernozem, 5-7 % OM, 3-4 w, 6 kg/ha, in situ	Soil bacteria[2]
Growth rate and numbers	Decrease in populations	Dark chestnut, heavy loam, 4 % OM, 4.5 kg/ha, in situ	Soil bacteria[3]
Growth rate and numbers	Increase in populations	Dark chestnut, heavy loam, 4 % OM, 3 kg/ha, in situ	Soil bacteria[3]
Growth rate and numbers	Increased populations	Leached medium loam chernozem, 5 kg/ha, in situ	Soil bacteria[4]
Growth rate and numbers	No effect on growth	Loamy chernozem, 5-7 % OM, 3-4 w, 4-5 kg/ha, in situ	Soil bacteria[5]
Growth rate and numbers	Stimulated growth	Pale chestnut light loam, pH=8, 2 % OM, 4.5 kg/ha, in situ	Soil bacteria[6]
Growth rate and numbers	Retarding populations	Pale and dark chestnut, 3-4.5 kg/ha, in situ	Soil bacteria, spore formers[3]
Nitrification and nitrifiers	Increased activity	1.5 kg/ha	Soil bacteria[7]
Nitrification and nitrifiers	Increased activity	Derno-podzol light or medium loams, pH=6, 2-3 % OM, 4 kg/ha	Soil bacteria[8]

Nitrification and nitrifiers	Increased activity	Pale chestnut light loam, pH=8, 2 % OM, 4.5 kg/ha	Soil bacteria[9]
Nitrification and nitrifiers	Increased activity	Dark chestnut heavy loam, 4 % OM, 3 and 4.5 kg/ha	Soil bacteria[3]
Nitrification and nitrifiers	No effect	Meadow type medium loam, 1.5 % OM, 8 kg/ha	Soil bacteria[10]
Nitrification and nitrifiers	No effect	Pale chestnut medium loam, 3 and 4.5 kg/ha	Soil bacteria[3]
Numbers and activity of soil algae	Adversely affected N2-fixation	Paddy soil, 3-6 kg/ha, recomm. field rate	Herbicide applied to soil in situ[11]
Numbers and activity of soil algae	Decreased population after 6 days, Later blue-green algae were stimul	Gleyed heavy loam, >8 % OM; peaty-podzol, 35 % OM, 5 kg/ha	Herbicide applied to soil in situ[12]
Numbers and activity of soil algae	Decreased population after 6 days, later blue-green algae were stimul	Light sandy loam, 5 kg/ha	Herbicide applied to soil in situ[12]
Soil microbial ammonification	Increased activity	Light or medium loam, 4 kg/ha	Soil organisms[13]

Source: 1.Wehr, N.B. and D.A. Klein. 1971. Soil Biol. Biochem. 3, 143-149.
2.Balkova, E.N. and O.A. Semikhatova. 1969. Khimiya selKhoz. 7, 846-849.
3.Zharasov, S.U. 1972. Vest. selKhoz. Nauki. Kazakhskoi 15, 25-29.
4.Matsneva, N.G. and O.A. Semikhatova. 1973. Khimiya selKhoz 11, 222-224.
5.Chulakov, S.A. and S.U. Zharasov. 1971. Vest. Akad. Nauk kazakh. SSR 27, 35-40.
6.Chukalov, S.A. and S.U. Zharasov. 1971. Vest. selKhoz. Nauki. Alma-Ata 27, 35-40.
7.Zhukova, P.A. and Z.I. Konoshevich. 1967. Puti Povysh. Urozh. ovoshchn., Minsk 1967, 138-150.
8.Chunderova, A.I., A.M. Sofinskii and T.P. Zubets. 1968. Khimiya selKhoz 6, 280-282.
9.Chulakov, S.H. and S.U. Zharasov. 1973. Izv. Akad. Nauk. kazakh SSR, ser. Biol 11, 7-13.
10.Tulabaev, B.D. 1970. Uzbek. biol. Zh. 14, 11-14.
11.Ibrahim, A.N. 1972. Symp. Biol. Hung. 111, 445-448.
12.Mikhailova, E.I. and Kruglov. 1973. Pochvovedenie 8, 81-85.
13.Chunderova, A.I., T.P. Zuberts and A.M. Sofinskii. 1971. Khimiya selKhoz 9, 527-530.

7-070 Effect of EPTC + prometryne

Parameter	Effect Value	Concentration and Condition	Species
Growth rate and numbers	Increased population especially Bacillus spp.	Leached medium loam, EPTC=5 and prometryne=2 kg/ha, in situ	Soil bacteria[1]

Source: 1.Matsneva, N.G. and O.A. Semikhatova. 1973. Khimiya selKhoz 11, 222-224.

7-071 Effect of EPTC and sulfallate

Parameter	Effect Value	Concentration and Condition	Species
Nitrification and nitrifiers	Increased activity	Medium podzolic loam, 1.9-2.5 % OM, 4.5 EPTC+ 6 sulf. kg/ha	Soil bacteria[1]

Source: 1.Isaeva, L.I. 1967. Agrokhimiya 4, 122-125.

7-072 Effect of Endothal

Parameter	Effect Value	Concentration and Condition	Species
Growth rate and numbers	No effect on growth	Six soils, 10-20 kg/ha, in situ	Soil bacteria[1]
Nitrification and nitrifiers	No effect	Six soils, 2 and 20 kg/ha	Soil bacteria[1]
Nitrification and nitrifiers	Temporarily decreased then increased population and activity	Medium podzol loam, 1.9-2.5 % OM, 8 kg/ha	Soil bacteria[2]
Soil microbial ammonification	No effect	Six soils, 2 and 200 kg/ha	Soil organisms[1]
Soil respiration activity	Bimodal inhibition of O2-uptake	Sandy, 2, 20 and 200 kg/ha	Soil organisms[1]

Source: 1.Au, F.H.F. 1968. Effect of endothal and certain other selective herbicides on microbial activity in six different soils, Ph. D. Thesis, Oregon state Univ.
2.Isaeva, L.I. 1967. Agrokhimiya 4, 122-125.

7-073 Effect of Endrin

Parameter	Effect Value	Concentration and Condition	Species
Growth rate and numbers	No appreciable effect	Sandy loam, 1.5 kg/ha, recomm. field rate	Soil bacteria[1]
Growth rate and numbers	No effect	Rice soils, 0.45-2.25 kg/ha	Soil bacteria[2]
Nitrification and nitrifiers	No effect	Sandy loam, 1.8 kg/ha, recomm. field rate	Soil organisms[1]

Source: 1.Bollen, W.B. and C.M. Tu. 1971. U.S. Dept. Agr. Forest Serv. Res. Papar PNW-114, 4p.
2.Nair, K.S., C. Ramakrishnan and S. Sothanantham. 1973. Madras agric. J. 60, 441-443.

7-074 Effect of Eptam

Parameter	Effect Value	Concentration and Condition	Species
Nitrogen fixation in soil	No effect	Pale chestnut light loam, ph=8, 2 % OM, 4.5 kg/ha	Soil bacteria, free living[1]
Nitrogen fixation in soil	No effect	Meadow-type medium loam, 1.5 % OM, 8 kg/ha	Soil bacteria, free living[2]
Nitrogen fixation in soil	No effect	Dark chestnut heavy loam, 4 % OM, 3 and 4.5 kg/ha	Soil bacteria, free living[3]

Source: 1.Chulakov, S.A. and S.U. Zharasov. 1967. Vest. selKhoz. Nauki, Alma-Ata 10, 122-124.
2.Tulabaev, B.D. 1970. Uzbek. biol. Zh. 14, 11-14.
3.Zharasov, S.U. 1972. Vest. selKhoz. Nauki. Kazakhskoi 15, 25-29.

7-075 Effect of Eptam and prometryne

Parameter	Effect Value	Concentration and Condition	Species
Nitrogen fixation in soil	Increased Azotobacter population	Leached medium loam chernozem, eptam(5) + prometryne(2 kg/ha)	Soil bacteria, free living[1]

Source: 1.Husarova, M. 1972. Ved. Pr. vyzk. Ust. Kukur. Trnave, No 6, 45-69.

7-076 Effect of Ethylene

Parameter	Effect Value	Concentration and Condition	Species
Foliar disease	Chlorosis, necrosis and abscission	Air, Variable and undetermined concentrations	Angiosperms[1]
Foliar disease	Dwarfing and premature defoliation	Air, Variable and undetermined concentrations	Gymnosperms[1]

Source: 1.Smith, W.H. 1981. Air Pollution and Forests. Springer Verlag. NY. 384 pp.

7-077 Effect of F

Parameter	Effect Value	Concentration and Condition	Species
Foliar disease	Tip and margin necrosis	Air, less than 100 ug/g dw in basis foliar	Angiosperms[1]
Foliar disease	Distal necrosis	Air, Less than 100 ug F/g dw in basis foliar	Gymnosperms[1]
Photosynthesis	Threshold dose for photosynthesis suppression	30 ug F/g dw in basis foliar tissue, various pines, tree seedling	Pine[1]

Source: 1.Smith, W.H. 1981. Air Pollution and Forests. Springer Verlag. NY. 384 pp.

7-078 Effect of Fenitrothion

Parameter	Effect Value	Concentration and Condition	Species
Nitrification and nitrifiers	Initial decrease, but subsequently inc. activity	Silt loam, 10 and 50 kg/ha	Soil organisms[1]
Nitrification and nitrifiers	No effect	Silt loam, 0.75 kg/ha	Soil organisms[1]

Source: 1.Ross, D.J. 1974. N.Z. Jl agric. Res. 17, 9-17.

7-079 Effect of Fenuron

Parameter	Effect Value	Concentration and Condition	Species
Denitrification in soil	Increased activity and population	Medium loam meadow soil, 1.2-2.2 % OM, 1.5 kg/ha	Soil organisms[1]
Nitrification and nitrifiers	Decreased population	Meadow-type chernozem, 1.5 kg/ha	Soil bacteria[2]
Nitrification and nitrifiers	Persistent decreased in activity	Medium loam meadow, 1.2-2 % OM, 1.5 kg/ha	Soil bacteria[3]
Nitrogen fixation in soil	Decreased Azotobacter populations	Meadow soil, 1.5 kg/ha	Soil bacteria, free living[2]
Soil microbial ammonification	Inhibited activity	Medium loam meadow soil, 1.2-2.0 % OM, 1.5 kg/ha	Soil organisms[3]
Soil respiration activity	Decreased CO2-evolution	Silt loam, 10 kg/ha, recomm. field rate	Soil organisms[4]

Source: 1.Joshi, O.P., M.S. Sachdev, K.L. Sahrawat and B.N. Kohli. 1976. Pl. Soil 44, 367-375.
2.Tulabaev, B.D. and S. Tamikaev. 196. Uzbek. biol. Zh. 2, 14-17.
3.Tulabaev, B.D. and N. Azimbegov. 1967. Khimiya selKhoz 5, 43-44.
4.Doxtader, K.G. 1968. Bact. Proc. 4, A21.

7-080 Effect of Fluometuron

Parameter	Effect Value	Concentration and Condition	Species
Denitrification in soil	Decreased activity and population	Serozem-meadow soil, 1.5-3 kg/ha	Soil organisms[1]
Nitrification and nitrifiers	Decreased activity	Sandy loam, 10 and 100 kg/ha	Soil bacteria[2]
Nitrification and nitrifiers	Decreased activity	Serozem-medow soil, 3.0 kg/ha	Soil bacteria[1]
Nitrification and nitrifiers	Increased activity	Alluvial, 2 % OM, 1.6 and 3.2 kg/ha	Soil bacteria[3]
Nitrification and nitrifiers	Inhibited activity	Leached chernozem, 1.0 - 2.0 kg/ha	Soil bacteria[4]
Nitrification and nitrifiers	Inhibited activity	Sandy loam soils, pH=6.1-7.4, 1.3-.16 % OM, 50 kg/ha	Soil bacteria[5]
Nitrification and nitrifiers	No effect	Sandy loam, 0.1 and 1.0 kg/ha	Soil bacteria[2]
Nitrification and nitrifiers	No effect	Serozem-meadow soil, 1.5 kg/ha	Soil bacteria[1]
Nitrogen fixation in soil	Inhibited Azotobacter	Serozem-meadow, 3 kg/ha	Soil bacteria, free living[1]
Nitrogen fixation in soil	Temporarily decreased Azotobacter population	Serozem-meadow, 1.5 kg/ha, later recovered	Soil bacteria, free living[1]
Soil microbial ammonification	Decreased population	Serozem-meadow soil, 1.5 kg/ha	Soil organisms[1]
Soil microbial ammonification	Inhibited activity	Serozem-meadow soil, 3 kg/ha	Soil organisms[1]

Soil respiration activity	Increased CO2-evolution	Leached chernozem, 15-60 kg/ha	Soil organisms[4]
Soil respiration activity	Increased activity	Sandy clay loam and 2 clay soils, 1.0 and 10 kg/ha	Soil organisms[6]
Soil respiration activity	Initially increased, then decreased CO2-evolution	Sandy loam, pH=7.4, 1.5 % OM; 50 kg/ha, coarse sandy loam, pH=6.1, 1.6 % OM	Soil organisms[5]
Soil respiration activity	No effect	Sandy clay loam and 2 clay soils, 25 kg/ha	Soil organisms[6]
Soil respiration	Variable effects	Sandy loam, 0.1 - 100 kg/ha	Soil organisms[7]

Source: 1.Tulabaev, B.G. 1972. Khimiya selKhoz 10, 776-780.
2.Bozarth, G.A., H.H. Funderburk and E.A. Curl. 1969. Abstr. Meet. Weed Sci. Soc. Am. 1969, 236.
3.Horowitz, M., T. Blumenfield, G. Herzlinger and N. Hulin. 1974. Weed Res. 14, 97-109.
4.Voinova, Z., S. Vlakhova and P. Petkova. 1970. Pochv. Agrokhim 4, 103-110.
5.Grossbard, E. and J.A.P. Marsh. 1974. Pestic Sci. 5, 609-623.
6.Lui, L.C. and H.R. Cibes-Viade. 1972. J. Agric. Univ. P. Rico 56, 417-425.
7.Bozarth, G.A., H.H. Funderburk and E.A. Curl. 1968. Proc. Sth. Weed Conf. 21, 340.

7-081 Effect of Fonofos

Parameter	Effect Value	Concentration and Condition	Species
Growth rate and numbers	Numbers decreased, then increased	Sandy loam, calcareous clay and clay, 11 kg/ha, later again normal	Soil bacteria[1]
Soil respiration	CO2-evolution decreased, then increased	Sandy loam, calcareous clay and clay, 11 kg/ha, later again normal	Soil organisms[1]

Source: 1.Gawaad, A.A.A., M.H. Hammad and F.H. El-Gayar. 1973. Agrokem. Talajt. 22, 161-168.

7-082 Effect of Gesaprim

Parameter	Effect Value	Concentration and Condition	Species
Nitrogen fixation in soil	Little to no effect	Alluvial, 3 kg/ha	Soil bacteria, free living[1]
Soil microbial ammonification	Decreased population initially, increased population later	Rhizosphere soil of maize, 3 kg/ha	Soil organisms[1]

Source: 1.Micev, N. and M. Bubalov. 1969. Godisen Zb. zemjod.-sum. Fak. Univ. Skopje 22, 19-24.

7-083 Effect of Gesaran

Parameter	Effect Value	Concentration and Condition	Species
Nitrification and nitrifiers	No effect	Alluvial or bog-alluvial, 2.25 and 4.5 kg/ha	Soil bacteria[1]
Nitrification and nitrifiers	Transient decreased activity	Alluvial or bog-alluvial, 6.75 kg/ha	Soil bacteria[1]
Nitrogen fixation in soil	No effect	Alluvial or bog-alluvial soils, 2.25 and 4.5 kg/ha	Soil bacteria, free living[1]

Source: 1.Mickovski, M.D. 1967. Ann. Fac. Agric. Univ. Skopje 20, 43-54.

7-084 Effect of Gramoxone

Parameter	Effect Value	Concentration and Condition	Species
Growth rate and numbers	Slight decreases immediately after dose, no serve effects	Calcereous loam, pH=8, 7.5 % OM, field and lab, 5.7 and 57 kg/ha, in situ	Soil bacteria[1]
Nitrification and nitrifiers	No significant effect	Calcareous loam, pH=8, 7.5 % OM, 5.6 and 56 kg/ha	Soil bacteria[2]
Nitrogen fixation in soil	No effect for first 3 weeks	Calcareous loam, pH=8, 7.5 % OM, 5.6 and 56 kg/ha	Soil bacteria, free living[2]
Numbers and activity of soil algae	Decrease in numbers during first 24 hr then recovery	Calcareous loam, pH=8, 7.5 % OM, 5.6 and 56 kg/ha	Herbicide applied to soil in situ[2]

Soil respiration activity	Minor and irregular changes in CO2-evolution	Calcareous loam, 5.6 and 56 kg/ha, pH=8, 7.5 % OM	Soil organisms[2]

Source: 1.Anderson, J.R. and E.A. Drew. 1976. Zbl. Bakt. abt II, 131, 247-258.
2.Anderson, J.R. and E.A. Drew. 1976. Zbl. Bakt. Abt. II, 131, 136-147.

7-085 Effect of Gramoxone formulation additives

Parameter	Effect Value	Concentration and Condition	Species
Soil respiration activity	Minor and irregular changes in CO2-evolution	Calcareous loam, pH=8, 7.5 % OM, 5.6 and 56 kg/ha	Soil organisms[1]

Source: 1.Anderson, J.R. and E.A. Drew. 1976. Zbl. Bakt. Abt. II, 131, 136-147.

7-086 Effect of H_2S

Parameter	Effect Value	Concentration and Condition	Species
Foliar disease	Interveinal necrotic blotches	Air, 100 ppm for 5 h	Angiosperms[1]
Foliar disease	Distal necrosis	Air, 100 ppm for 5 h	Gymnosperms[1]

Source: 1.Smith, W.H. 1981. Air Pollution and Forests. Springer Verlag. NY. 384 pp.

7-087 Effect of Heptachlor

Parameter	Effect Value	Concentration and Condition	Species
Growth rate and numbers	Decreased numbers in rhizosphere of maize	Chernozem, 150 kg/ha	Soil bacteria[1]
Growth rate and numbers	Numbers increased	Silt loam, 1.1 and 7.9 and 55.3 kg/ha	Soil bacteria[2]

Source: 1.Tsirkov, Y.I. 1971. Bulg: Akad. Nauk. (1971), 267-272.
2.Shamiyeh, N.B. and L.F. Johnson. 1973. Soil Biol. Biochem. 5, 309-314.

7-088 Effect of IPC

Parameter	Effect Value	Concentration and Condition	Species
Growth rate and numbers	Moderate inhibition of growth	Solid+liquid media=200 ppm, in culture	Bdellovibrio bacteriovorus[1]

Source: 1.Wehr, N.B. and D.A. Klein. 1971. Soil Biol. Biochem. 3, 143-149.

7-089 Effect of Ioxynil

Parameter	Effect Value	Concentration and Condition	Species
Nitrification and nitrifiers	Inhibited activity	Sandy loam, 1.32 and 13.2 kg/ha	Soil bacteria[1]
Nitrification and nitrifiers	No effect	Podzolic loam, pH=7, 1.5 kg/ha	Soil bacteria[2]
Nitrogen fixation in soil	Decreased azobacter population	Polder soil, 65 % MHC, 0.5 and 50 kg/ha, recomm. field rate	Soil bacteria, free living[1]
Nitrogen fixation in soil	No effect on anaerobic forms but azotobacters increased	Podzolic loam, pH=7, 1.5 kg/ha	Soil bacteria, free living[2]
Soil microbial ammonification	No effect	Podzolic loam, pH=7.0, 1.5 kg/ha	Soil organisms[2]
Soil respiration activity	Decreased CO2-evolution	Sandy loam, 1.32 and 13.2 kg/ha, high effect at 132 kg/ha	Soil organisms[1]

Source: 1.Schreven, D.A. van, D.J. Lindenbergh abd A. Koridon. 1970. Pl. Soil 33, 513-532.
2.Hauke-Pacewiczowa, T. and M. Krolowa. 1968. Pam. Pulawski 31, 17-28.

7-090 Effect of Isobenzan

Parameter	Effect Value	Concentration and Condition	Species
Nitrification and nitrifiers	Inhibited activity	0.4 and 0.75 kg/ha	Soil organisms[1]

Source: 1.Laveglia, J. and P.A. Dahm. 1974. Environ. Ent. 33, 528-533.

7-091 Effect of Lenacil

Parameter	Effect Value	Concentration and Condition	Species
Growth rate and numbers	Increased population, but some species decreased	Chernozem, 3.0 % OM, 1.0 kg/ha, in situ	Soil bacteria[1]
Growth rate and numbers	No effect	Dark chestnut heavy loam, 1.2-1.7 kg/ha, in situ	Soil bacteria[2]
Growth rate and numbers	No effect	Pale chestnut light loam, 1.2-1.7 kg/ha, in situ	Soil bacteria[2]
Growth rate and numbers	Increased population	Chernozem, lenacil=1 + DCU=5 kg/ha, in situ	Soil bacteria[1]
Growth rate and numbers	Increased population	Chernozem, lenacil=1 + TCA=5 kg/ha, in situ	Soil bacteria[1]
Growth rate and numbers	Increased population	Chernozem, lenacil=1 + pebulate=4 kg/ha, in situ	Soil bacteria[1]
Soil microbial ammonification	Decreased population	Chernozem, 3 % OM, lenacil(1) + TCA(5 kg/ha)	Soil organisms[1]
Denitrification in soil	Increased activity	Chernozem, 3 % OM, lenacil(1)+pebulate(4 kg/ha)	Soil organisms[1]
Soil microbial ammonification	Decreased population	Chernozem, 3 % OM, lenacil(1) + pebulate(4 kg/ha)	Soil organisms [1]
Soil microbial ammonification	Decreased population	Chernozem, 3 % OM, lenacil(1) + DCU(5 kg/ha)	Soil organisms[3]

Source: 3.Karki, A.B., L. Coupin, P. Kaiser and M. Moussin. 1973. Rev. Ecol. Biol. Sol 10, 3-11.
1.Barona, V.P. 1974. Agrokhimiya 11, 131-134.
2.Chulakov, S.H. and S.U. Zharasov. 1973. Izv. Akad. Nauk. kazakh SSR, ser. Biol 11, 7-13.

7-092 Effect of Lindane

Parameter	Effect Value	Concentration and Condition	Species
Growth rate and numbers	Decreased numbers in rhizosphere of maize	Chernozem, 30 kg/ha	Soil bacteria[1]
Growth rate and numbers	Increased numbers	Clay soil + 2 % compost, 0.22 and 4.4 kg/ha	Soil bacteria[2]
Growth rate and numbers	No effect	Rice soils, 0.45-2.25 kg/ha	Soil bacteria[3]
Nitrification and nitrifiers	Inhibited activity	Sandy loam, 2.5 and 10 and 100 kg/ha	Soil organisms[4]
Nitrification and nitrifiers	Inhibited activity, temperature affected degree of inhibition	Arable loam, pH=6.2, 1.7 % OM; meadow loam, pH=6.7, 3.1 % OM, 1.5-3 kg/ha	Soil organisms[5]
Nitrification and nitrifiers	No effect	Sandy loam, 0.1 and 1.0 kg/ha	Soil organisms[4]
Nitrification and nitrifiers	No effect	Arable loam, pH=5.8-6.6, 1.7 % OM; meadow loam, pH=6.3-7, 3 % OM, .3 kg/ha	Soil organisms[5]
Nitrification and nitrifiers	Numbers of nitrifying bacteria severely depressed	Clay + 2 % compost, 0.22 and 4.4 kg/ha	Soil organisms[2]

Nitrification and nitrifiers	Slight stimulation in moist soil, numbers depressed in drier soil	Loam, pH=7.2, 1.49 % C, 500 kg/ha	Soil organisms[6]
Soil bacteria, growth rate and numbers	Numbers decreased, then increased and reverted to normal	Sandy loam, calcareous clay and clay, 22 kg/ha	Soil bacteria[7]
Soil respiration	CO2-evolution decreased, then increased and returned to normal	Sandy loam, calcareous clay and clay, 22 kg/ha	Soil organisms[7]
Soil respiration	Increased O2-uptake	Loamy sand, pH=6.9, 0.64 % C, 1.0 - 100 kg/ha	Soil organisms[8]
Soil respiration	Increased O2-uptake	Sandy loam, pH=7.5, 3.42 % C; clay loam, pH=7.7, 2.3 % C, 1.0-100 kg/ha	Soil organisms[8]
Soil respiration	Low rates, no effect. Higher rates decreased CO2-evolution	Alluvial, 0.1 - 100 kg/ha	Soil organisms[6]
Soil respiration	Slightly stimulated activity	Loam, pH=7.2, 1.49 % C, 6000 kg/ha	Soil organisms[6]

Source: 1.Tsirkov, Y.I. 1971. Bulg: Akad. Nauk. (1971), 267-272.
2.Mahmond, S.A.Z., K.G. Selim and T. El-Mokadem. 1970. Zbl. Bakt. Abt II 125, 134-149.
3.Nair, K.S., C. Ramakrishnan and S. Sothanantham. 1973. Madras agric. J. 60, 441-443.
4.Bardiya, M.C. and A.C. Gaur. 1970. Zbl. Bakt. Abt. II, 124, 552-555.
5.Verstraeten, L.M. J. and K. Vlassak. 1973. Pl. Soil 39, 15-28.
6.Naumann, K. 1971. Pedobiologia 11, 286-295.
7.Gawaad, A.A.A., M.H. Hammad and F.H. El-Gayar. 1973. Agrokem. Talajt. 22, 161-168.
8.Tu, C.-M. 1969. Bact. Proc. , 1-3.

7-093 Effect of Linuron

Parameter	Effect Value	Concentration and Condition	Species
Antibiotic production	Antibiotic secretion inhibited	Liquid media=10 ppm, in culture	Bacillus brevis[1]
Antibiotic production	Increased antibiotic secretion	Liquid media=10 ppm, in culture	Bacillus oligonitrophilus[1]
Antibiotic production	No effect on secretion	Liquid media=10 ppm, in culture	Bacillus polymyxa[1]
Denitrification in soil	Decreased activity and population	Medium to heavy sandy clay leached chernozem, 1-1.6 % OM, 0.5-2 kg/ha	Soil organisms[2]
Denitrification in soil	No effect	Sandy clay and silty clay, 2 and 20 kg/ha	Soil organisms[3]
Growth rate and numbers	Growth ratarded	Liquid media=1.5-4.0 ppm, increase in glutamic acid, in culture	Arthrobacter globiformis[4]
Growth rate and numbers	No effect on growth rate	Liquid media=10 ppm, in culture	Bacillus brevis[1]
Growth rate and numbers	Inhibition of growth	Liquid media=0.2-100 ppm, in culture	Bacillus mycoides[5]
Growth rate and numbers	Growth decreased	Liquid media=10 ppm, in culture	Bacillus polymyxa[1]
Growth rate and numbers	Inhibition of growth	Liquid media=10-100 ppm, in culture	Bacillus sp.[6]
Growth rate and numbers	Sensitive at acid pH	Liquid media=5-50 ppm, in culture	Bacillus spp.[7]
Growth rate and numbers	Inhibition of growth	Liquid media=12-16 ppm, in culture	Bdellovibrio bacteriovorus[8]
Growth rate and numbers	Resistant at all pH values	Liquid media=5-50 ppm, in culture	Pseudomonas sp.[7]
Growth rate and numbers	No effect on growth	Liquid media=10-100 ppm, in culture	Pseudomonas spp.[6]

Growth rate and numbers	Isolates strongly inhibited	Liquid media=10 ppm, in culture	Soil bacteria[9]
Growth rate and numbers	Marked inhibition of isolates	Liquid media=10 ppm, in culture	Soil bacteria[5]
Mortality	Toxic effects	Liquid media=1000 ppm, in culture	Bacillus spp.[10]
Mortality	Toxic effects	Liquid media=1000 ppm, in culture	Corynebacterium sp.[10]
Mortality	Toxic effects	Liquid media=1000 ppm, in culture	Arthrobacter sp.[10]
Nitrification and nitrifiers	Decreased activity	Clay loam, pH=5.35, 0.5-2 kg/ha, recomm. field rate	Soil bacteria[11]
Nitrification and nitrifiers	Decreased activity	Sandy and silty clay loams, 20 kg/ha	Soil bacteria[3]
Nitrification and nitrifiers	Decreased activity and population	Black alluvial and sandy soil, 5-20 kg/ha, recomm. field rate	Soil bacteria[12]
Nitrification and nitrifiers	Decreased population	Light alluvial meadow, pH=7, 0.2 kg/ha	Soil bacteria[13]
Nitrification and nitrifiers	Decreased population	Heavy leached meadow-cinnamon, pH=7, 2 % OM, 0.2 kg/ha	Soil bacteria[13]
Nitrification and nitrifiers	Inhibited activity	Leached chernozem, 0.5-2 kg/ha, recomm. field rate	Soil bacteria[14]
Nitrification and nitrifiers	Inhibited activity	Loam soil, pH=7-7.4, 10-15 kg/ha	Soil bacteria[15]
Nitrification and nitrifiers	Inhibited activity	Sandy loam soils, pH=6.1-7.4, 1.3-1.6 % OM, 50 kg/ha	Soil bacteria[16]
Nitrification and nitrifiers	Initially decreased, then increased activity	Sandy loam, 0.5-2 kg/ha, recomm. field rate	Soil bacteria[17]
Nitrification and nitrifiers	No effect	Sandy and silty clay loams, 2 kg/ha	Soil bacteria[3]
Nitrification and nitrifiers	No effect	Light loam podzol, 1.5 kg/ha	Soil bacteria[18]
Nitrification and nitrifiers	Retarded activity	Sandy clay chernozem, 1.0-1.6 % OM, 2 kg/ha	Soil bacteria[19]
Nitrogen fixation in soil	Decreased no. of Azo. chroococcum, but later increased	Light loam podzol, 1.5 kg/ha	Soil bacteria, free living[20]
Nitrogen fixation in soil	Increased soil population	Medium to heavy sandy clay chernozem, 1-1.6 % OM, 0.5-2 kg/ha	Soil bacteria, free living[2]
Nitrogen fixation in soil	No effect on numbers of Azotobacter sp. and Clostridium sp.	Leached meadow-cinnamon and alluvial, 0.5-2.5, recom field rate	Soil bacteria, free living[21]
Numbers and activity of soil algae	Decreased activity and population	Alluvial black and sandy, 5-25 kg/ha, recomm. field rate	Herbicide applied to soil in situ[12]
Numbers and activity of soil algae	Decreased population	Clay loam, pH=5.4, 0.5-2.5 kg/ha, recomm. field rate	Herbicide applied to soil in situ[11]
Numbers and activity of soil algae	Decreased population	Light and sandy clay, 3-50 kg/ha	Herbicide applied to soil in situ[22]
Soil microbial ammonification	Increased activity	Sandy clay leached chernozem, 1-1.6 % OM, 0.5-2.0 kg/ha	Soil organisms[23]
Soil microbial ammonification	Increased activity	Alluvial meadow and leached cinnamon, 2 kg/ha	Soil organisms[2]

Soil microbial ammonification	Increased activity and population	Clay loam, pH=5.4, 0.5-2.5 kg/ha, recomm. field rate	Soil organisms[24]
Soil microbial ammonification	Increased activity and population	Leached chernozem, 0.5-2.5 kg/ha, recomm. field rate	Soil organisms[14]
Soil microbial ammonification	Slightly decreased population	Sandy, 3 and 30 kg/ha	Soil organisms[25]
Soil respiration activity	Applied annually, increased and than decreased CO2-evolution	Sandy loams, 5 kg/ha	Soil organisms[26]
Soil respiration activity	Decreased CO2-evolution	Sandy loams, 130 kg/ha	Soil organisms[27]
Soil respiration activity	Decreased CO2-evolution	Sandy loam, pH=7.4, 1.5 % OM, 5 and 50 kg/ha	Soil organisms[16]
Soil respiration activity	Decreased CO2-evolution	Coarse sandy loam, pH=6.1, 1.6 % OM, 5 and 50 kg/ha	Soil organisms[16]
Soil respiration activity	Decreased activity	Sandy, 3 kg/ha	Soil organisms[28]
Soil respiration activity	Increased CO2-evolution	Leached chernozem, 15-60 kg/ha	Soil organisms[14]
Soil respiration activity	Increased activity	Sandy clay loam and 2 clay soils, 10 and 25 kg/ha	Soil organisms[29]
Soil respiration activity	No effect	Sandy, 0.3 kg/ha	Soil organisms[28]
Soil respiration activity	No effect	Loess, 0.3 and 3 kg/ha	Soil organisms[28]
Soil respiration activity	No effect	Sandy clay loam and 2 clay soils, 1.0 kg/ha	Soil organisms[29]

Source: 1.Kosinkiewicz, B. 1970. Meded Fac. Landb Rijksuniv. Gent 35, 673-680.
2.Peshakov, G., E. Raikov and D. Tsvetanov. 1969. Pochv. Agrokhim 4, 89-94.
3.Torstenson, L. 1974. Swedish J. agric. Res. 4, 151-160.
4.Kosinkiewicz, B. 1973. Acta microbiol. pol. Ser B, 5, 145-150.
5.Balicka, N., J. Sobieszczanski and T. Niewiadoma. 1969. Acta microbiiol. pol.1(18), 11-14.
6.Balicka, N. and Z. Krezel. 1969. Weed Res. 9, 37-42.
7.Balicka, N. 1969. Acta microbiol. pol. Ser B, 1(18), 43-45.
8.Wehr, N.B. and D.A. Klein. 1971. Soil Biol. Biochem. 3, 143-149.
9.Krezel, Z. and M. Musial. 1969. Acta microbiol. pol. Ser B 1, 93-87.
10.Sobieszczanski, J. 1969. Acta microbiol. pol. Ser B, 1(18), 99-104.
11.Sivasithamparam, K. 1970. Riso 19, 339-346
12.Hauke-Pacewiczowa, T. 1971. Pam. Pulawski 46, 5-48.
13.Mezharaupe, V.A. 1967. Mikroorg. Rast. Trudy Inst. Mikrobiol. Akad. Nauk. SSR 65-83.
14.Voinova, Z., S. Vlakhova and P. Petkova. 1970. Pochv. Agrokhim 4, 103-110.
15.Corke, C.T. and F.R. Thompson. 1970. Can. J. Microbiol. 16, 567-571.
16.Grossbard, E. and J.A.P. Marsh. 1974. Pestic Sci. 5, 609-623.
17.Beckmann, E.O. and W. Pestemer. 1975. Landw. Forsch. 28, 24-33.
18.Kozaczenko, H. and W. Sobieraj. 1973. Biul. Warzywn. 14, 161-178.
19.Tulabaev, B.D. and N. Azimbegov. 1967. Khimiya selKhoz 5, 43-44.
20.Nepomiluev, V.F., I.P. Grechin and T.I. Kuzyakina. 1966. Izv. timiryazev sel-khoz Akad. No 2., 128-16.
21.Rankov, V., E. Elenkov, P. Surlekov and B. Velev. 1966. Agrokhimiya 4, 115-120.
22.Pantera, H. 1970. Meded. Fac. Landb. Rijksuniv. Gent. 35, 847-854.
23.Rankov, V. 1972. Prvi Nat. Kongr. Pochv. Bulg. 1969, 227-232.
24.Fields, M.L., R. Der and D.D. Hemphill. 1967. Weeds 15, 195-197.
25.Uziak, S., K. Leonaik and K. Gostkowska. 1971. Ann. Univ. Marie Curie-Sklodowska, C, 26, 57-72.
26.Grossbard, E. 1971. Weed Res. 11, 263-275.
27.Marsh, J.A.P., H.A. Davies and E. Grossbard. 1974. Proc. 12th Br. Weed Control. Conf. 53-58.
28.Manninger, E. and J. Szava. 1972. Symp. Biol. Hung. 11, 369-372.
29.Lui, L.C. and H.R. Cibes-Viade. 1972. J. Agric. Univ. P. Rico 56, 417-425.

7-094 Effect of Linuron, monolinuron and chloroxuron

Parameter	Effect Value	Concentration and Condition	Species
Growth rate and numbers	Inhibition of growth	All soil conc = 0.2-1000 ppm, in culture	Soil bacteria[1]

Source: 1.Balicka, N., J. Sobieszczanski and T. Niewiadoma. 1969. Acta microbiiol. pol.1(18), 11-14.

7-095 Effect of Lumeton

Parameter	Effect Value	Concentration and Condition	Species
Growth rate and numbers	Increased aerobic population	Rhizosphere soil of maize, 3 kg/ha, in situ	Soil bacteria, spore formers[1]
Nitrogen fixation in soil	No effect	Alluvial or bog-alluvial soils, 3 and 6 kg/ha	Soil bacteria, free living[2]
Nitrification and nitrifiers	No effect	Alluvial or bog-alluvial, 3 and 6 kg/ha	Soil bacteria[2]
Nitrification and nitrifiers	Transient decreased activity	Alluvial or bog-alluvial, 9 kg/ha	Soil bacteria[2]

Source: 1.Micev, N. 1970. Savr. Poljopr. 18, 245-249.
2.Mickovski, M.D. 1967. Ann. Fac. Agric. Univ. Skopje 20, 43-54.

7-096 Effect of MCA

Parameter	Effect Value	Concentration and Condition	Species
Growth rate and numbers	Inhibition in growth	Clay soil, anaerobic, 1.0-10 kg/ha, CaCO3 reduces effect, in situ	Sb, methane producing[1]
Soil respiration activity	CO2 and CH4-evolution inhibited, especially at high rate	Anaerobic clay, 1.0 and 10 kg/ha	Soil organisms[1]

Source: 1.Laskowski, D.A. and F.E. Broadbent. 1970. Proc. Soil Sci. Soc. Am. 34, 72-77.

7-097 Effect of MCPA

Parameter	Effect Value	Concentration and Condition	Species
Denitrification in soil	Large decreased in population and activity	Paddy soil, 2.25 kg/ha	Soil organisms[1]
Denitrification in soil	No effect	Sandy clay and silty clay, 4 and 40 kg/ha	Soil organisms[2]
Growth rate and numbers	Moderate inhibition	Solid+liquid media=200 ppm, recomm. filed rate, in culture	Bdellovibrio bacteriovorus[3]
Growth rate and numbers	Increase in numbers	Light loam dernopodzol, 1.0 kg/ha, in situ	Clostridium pasteurianum[4]
Growth rate and numbers	No effect on growth	Sandy loam, 1.0-1.5 kg/ha, in situ	Soil bacteria[5]
Nitrification and nitrifiers	Decreased activity	Sandy clay and silty clay, 40 kg/ha	Soil bacteria[2]
Nitrification and nitrifiers	Decreased activity	Sandy loam, 1.0-1.5 kg/ha	Soil bacteria[5]
Nitrification and nitrifiers	Decreased population	West Bengal paddy soil, 2.0 and 3.8 kg/ha	Soil bacteria[1]
Nitrification and nitrifiers	Decreased population and activity	Black alluvial and sandy soils, 20-25 kg/ha, recomm. field rate	Soil bacteria[6]

Nitrification and nitrifiers	No effect	Sandy clay and silty clay, 4 kg/ha	Soil bacteria[2]
Nitrification and nitrifiers	No effect	Podzolic loam, pH=7, 5 kg/ha	Soil bacteria[7]
Nitrification and nitrifiers	No effect from annual applications	Light loam dernopodzol, 5 y, 2.5 kg/ha	Soil bacteria[8]
Nitrogen fixation in soil	Decreased azotobacter population	Polder soil, 65 % MHC, 30-255 kg/ha, recomm. field rate	Soil bacteria, free living[9]
Nitrogen fixation in soil	Increased population of azotobacter	Podzolic loam, pH=7, 5 kg/ha	Soil bacteria, free living[7]
Nitrogen fixation in soil	No effect on Azotobacter chroococcum or Clostridium sp. numbers	Leached meadow-cinnamon and alluvial soils, 0.3-2.25 kg/ha, recomm. field	Soil bacteria, free living[10]
Numbers and activity of soil algae	Decreased activity and population	Alluvial black and sandy, 3-23 kg/ha, recomm. field rate	Herbicide applied to soil in situ[11]
Soil microbial ammonification	Increased activity	Alluvial, 1-2 kg/ha	Soil organisms[12]
Soil respiration activity	Increased activity	Alluvial, 1-2 kg/ha	Soil organisms[13]
Soil respiration activity	No effect on O2-uptake in lab	Sandy loam, 6.7 kg/ha	Soil organisms[14]

Source: 1.De, B.K. and S. Mukhopadhyay. 1971. Int. Rice Commn. Newsl. 20, 35-39.
2.Torstenson, L. 1974. Swedish J. agric. Res. 4, 151-160.
3.Wehr, N.B. and D.A. Klein. 1971. Soil Biol. Biochem. 3, 143-149.
4.Kozyrev, M.A. and A.A. Laptev. 1972. Khimiya selKhoz. 10, 614-617.
5.Hoffman-Kakol, I. and H. Kwarta. 1972. Zesz. nauk. Akad. Rolnicz. Szczecin. No 38, 105-116.
6.Hauke-Pacewiczowa, T. 1971. Pam. Pulawski 46, 5-48.
7.Hauke-Pacewiczowa, T. and M. Krolowa. 1968. Pam. Pulawski 31, 17-28.
8.Chunderova, A.I., T.P. Zuberts and A.M. Sofinskii. 1971. Khimiya selKhoz 9, 527-530.
9.Schreven, D.A. van, D.J. Lindenbergh abd A. Koridon. 1970. Pl. Soil 33, 513-532.
10.Rankov, V., E. Elenkov, P. Surlekov and B. Velev. 1966. Agrokhimiya 4, 115-120.
11.Anderson, J.R. and E.A. Drew. 1976. Zbl. Bakt. Abt. II, 131, 136-147.
12.Lazareva, N.V. and V.G. Blokhin. 1969. Khimiya selKhoz 7, 609-611.
13.Bakalivanov, D. and G. Nikolova. 1971. In: I. Pashev (ed.): Vtori Kongres po Mikrobihologiya. IV. Chst. Sofia. 1969. Bulg. Akad. Nauk 1971. 2>3-266.
14.Grossbard, E. 1971. Weed Res. 11, 263-275.

7-098 Effect of MCPA and dichlorprop

Parameter	Effect Value	Concentration and Condition	Species
Nitrification and nitrifiers	Sligthly decreased activity	Sandy loam, 65 % MHC, total 23.5 kg/ha	Soil bacteria[1]
Nitrification and nitrifiers	Strongly inhibited activity	Sandy loam, 65 % MHC, total 235 kg/ha	Soil bacteria[1]

Source: 1.Schreven, D.A. van, D.J. Lindenbergh and A. Koridon. 1970. Pl. Soil 33, 513-532.

7-099 Effect of MCPB

Parameter	Effect Value	Concentration and Condition	Species
Nitrification and nitrifiers	No effect from annual applications	Light loam dernopodzol, 5 y, 2.0 kg/ha	Soil bacteria[1]

Source: 1.Chunderova, A.I., T.P. Zuberts and A.M. Sofinskii. 1971. Khimiya selKhoz 9, 527-530.

7-100 Effect of MSMAp

Parameter	Effect Value	Concentration and Condition	Species
Soil respiration activity	Decreased CO2-evolution from litter and soil	Forest soil, 10 and 100 kg/ha	Soil organisms[1]

Source: 1.Bollen, W.R., L.A. Norris and K.L. Stowers. 1974. Weed Sci. 22, 557-562.

7-101 Effect of Malathion

Parameter	Effect Value	Concentration and Condition	Species
Growth rate and numbers	Decreased numbers	4.2 and 5 and 6 kg/ha	Soil bacteria[1]
Growth rate and numbers	Increased numbers	0.2 kg/ha	Soil bacteria[2]

Source: 1.Swaminathan, R. and S.B. Sullia. 1969. Curr. Sci. 38, 282-284.2.Yurovskaya, Y.M. and V.A. Zhulinskaya. 1974. Khimiya selKhoz 12, 38-41.

7-102 Effect of Mecoprop

Parameter	Effect Value	Concentration and Condition	Species
Nitrification and nitrifiers	Slightly decreased activity	Sandy loam, 65 % MHC, 2.6 kg/ha	Soil bacteria[1]
Nitrification and nitrifiers	Strongly decreased activity	Sandy loam, 65 % MHC, 100 kg/ha	Soil bacteria[1]
Nitrogen fixation in soil	Decreased azotobacter population	Polder soil, 65 % MHC, 2 and 200 kg/ha, recomm. field rate	Soil bacteria, free living[1]
Soil respiration activity	Decreased CO2-evolution	Sandy loam, 1.0 and 10 and 100 kg/ha	Soil organisms[1]

Source: 1.Schreven, D.A. van, D.J. Lindenbergh abd A. Koridon. 1970. Pl. Soil 33, 513-532.

7-103 Effect of Methan-Na

Parameter	Effect Value	Concentration and Condition	Species
Numbers and activity of soil algae	Chlorococcum sp. and Phormidium sp. were tolerent, others suppressed	Light sandy loam, 500 kg/ha	Herbicide applied to soil in situ[1]
Numbers and activity of soil algae	Suppressed all growth	Gleyed heavy loam, >8 % OM; peaty-podzol, 35 % OM, 500 kg/ha	Herbicide applied to soil in situ[1]

Source: 1.Mikhailova, E.I. and Kruglov. 1973. Pochvovedenie 8, 81-85.

7-104 Effect of Methoprometryne

Parameter	Effect Value	Concentration and Condition	Species
Nitrification and nitrifiers	Inhibited activity	Alluvial soil, 2.5 kg/ha, methoprometryne 25 %	Soil bacteria[1]

Source: 1.Micev, N. and M. Bubalov. 1972. Symp. Biol. Hung. 11, 379-384.

7-105 Effect of Methoprotryne

Parameter	Effect Value	Concentration and Condition	Species
Growth rate and numbers	Increased aerobic spore formers	Alluvial, 2.5 kg/ha, in situ	Soil bacteria[1]
Soil microbial ammonification	Increased activity	Alluvial, methoprotryne 25 % = 2.5 kg/ha	Soil organisms[1]

Source: 1.Micev, N. and M. Bubalov. 1972. Symp. Biol. Hung. 11, 379-384.

7-106 Effect of Methurin

Parameter	Effect Value	Concentration and Condition	Species
Denitrification in soil	Decreased activity and population	Serozem-meadow soil, 1.5-3 kg/ha	Soil organisms[1]
Nitrogen fixation in soil	Inhibited soil population	Serozem-meadow, 3 kg/ha	Soil bacteria, free living[1]
Nitrogen fixation in soil	Temporarily decreased soil population	Serozem-meadow, 1.5 kg/ha	Soil bacteria, free living[1]
Soil respiration activity	Decreased activity	Silty medium loam derno-podzol, pH=5.9, 3 kg/ha	Soil organisms[2]

Source: 1.Tulabaev, B.G. 1972. Khimiya selKhoz 10, 776-780.
2.Zinchenko, V.A. and T.V. Osinskaya. 1969. Agrokhimiya 9, 94-101.

7-107 Effect of Metobromuron

Parameter	Effect Value	Concentration and Condition	Species
Growth rate and numbers	Growth decreased initially	Liquid media=0.003-30 ppm, revover rapid and complete, in culture	Bacillus cereus[1]

Source: 1.Macek, J. and L. Milharcic. 1972. Fragm. herb. Croatica No 16, 1-6.

7-108 Effect of Metoxuron

Parameter	Effect Value	Concentration and Condition	Species
Denitrification in soil	No significant effect	Calcareous soil, 100-400 kg/ha, recomm. field rate	Soil organisms[1]
Nitrification and nitrifiers	Inhibited activity	Sandy loam soils, pH=6.1-7.4, 1.3-1.6 % OM, 50 kg/ha	Soil bacteria[2]
Nitrification and nitrifiers	Slightly decreased activity	Calcareous soil, 100-200 kg/ha	Soil bacteria[1]
Soil respiration activity	Decreased CO2-evolution	Sandy loam, pH=7.4, 1.5 % OM, 50 kg/ha, coarse sandy loam, pH=6.1, 1.6 % OM	Soil organisms[2]
Soil respiration activity	No effect	Sandy loam, pH=7.4, 1.5 % OM, 0.5 and 5 kg/ha	Soil organisms[2]
Soil respiration activity	No effect	Coarse sandy loam, pH=6.1, 1.6 % OM, 0.5 and 5 kg/ha	Soil organisms[2]

Source: 1.Simon, J.C., P. Jamet, {.M. Lemaire and B. Jouan. 1973. Sci. Agron. Rennes, Ecole Nat. Sup. Agron. 21-244.
2.Grossbard, E. and J.A.P. Marsh. 1974. Pestic Sci. 5, 609-623.

7-109 Effect of Mirex

Parameter	Effect Value	Concentration and Condition	Species
Growth rate and numbers	No effect	Sandy, 0.98 % OM; forest sandy, 4.94 % OM; sand, 0.63 % OM, 20-2000 kg/ha	Soil bacteria[1]

Source: 1.Jonez, A.S. and C.S. Hodges. 1974. J. agric. Fd Chem. 22, 435-439.

7-110 Effect of Molinate

Parameter	Effect Value	Concentration and Condition	Species
Numbers and activity of soil algae	No adverse effect	Paddy soil, 2-4 kg/ha, recomm. field rate	Herbicide applied to soil in situ[1]

Source: 1.Ibrahim, A.N. 1972. Symp. Biol. Hung. 111, 445-448.

7-111 Effect of Monolinuron

Parameter	Effect Value	Concentration and Condition	Species
Denitrification in soil	Decreased activity and population	Medium to heavy sandy clay leached chernozem, 1-1.6 % OM, 0.5-2 kg/ha	Soil organisms[1]
Growth rate and numbers	Isolates strongly inhibited	Liquid media=100 ppm, in culture	Soil bacteria[2]
Nitrification and nitrifiers	Inhibited activity	Leached chernozem, 1.0-3 kg/ha, recomm. field rate	Soil bacteria[3]
Nitrification and nitrifiers	No effect	Light loam podzol, 1.5 kg/ha	Soil bacteria[4]
Nitrification and nitrifiers	No effect	Light alluvial meadow, pH=7, 0.2 kg/ha	Soil bacteria[5]
Nitrification and nitrifiers	No effect	Heavy leached meadow-cinnamon, pH=7, 2 % OM, 0.2 kg/ha	Soil bacteria[5]
Nitrification and nitrifiers	Retarded activity	Sandy clay chernozem, 1.0-1.6 % OM, 10 kg/ha	Soil bacteria[6]
Nitrogen fixation in soil	Decreased numbers of Azotobacter chroococcum, but later increased	Light loam podzol, 1.5 kg/ha	Soil bacteria, free living[7]
Nitrogen fixation in soil	Incresed Azobacter ch. population	Medium to heavy sandy clay chernozem, 1.-1.6 % OM, 0.5-2 kg/ha	Soil bacteria, free living[1]
Soil microbial ammonification	Increased activity	Alluvial meadow and leached cinnamon, 1-2 kg/ha	Soil organisms[8]
Soil microbial ammonification	Increased activity and population	Leached	Soil organisms[3]
Soil respiration activity	Decreased activity	Sandy, 3 kg/ha	Soil organisms[9]
Soil respiration activity	Increased CO2-evolution	Leached chernozem, 15-60 kg/ha	Soil organisms[3]
Soil respiration activity	No effect	Sandy, 0.3 kg/ha	Soil organisms[9]
Soil respiration activity	No effect	Loess, 0.3 and 3 kg/ha	Soil organisms[9]

Source: 1.Peshakov, G., E. Raikov and D. Tsvetanov. 1969. Pochv. Agrokhim 4, 89-94.
2.Krezel, Z. and M. Musial. 1969. Acta microbiol. pol. Ser B 1, 93-87.
3.Voinova, Z., S. Vlakhova and P. Petkova. 1970. Pochv. Agrokhim 4, 103-110.
4.Kozaczenko, H. and W. Sobieraj. 1973. Biul. Warzywn. 14, 161-178.
5.Mezharaupe, V.A. 1967. Mikroorg. Rast. Trudy Inst. Mikrobiol. Akad. Nauk. SSR 65-83.
6.Tulabaev, B.D. and N. Azimbegov. 1967. Khimiya selKhoz 5, 43-44.
7.Nepomiluev, V.F., I.P. Grechin and T.I. Kuzyakina. 1966. Izv. timiryazev sel-khoz Akad. No 2., 128-16.
8.Rankov, V. 1972. Prvi Nat. Kongr. Pochv. Bulg. 1969, 227-232.
9.Manninger, E. and J. Szava. 1972. Symp. Biol. Hung. 11, 369-372.

7-112 Effect of Monurun

Parameter	Effect Value	Concentration and Condition	Species
Growth rate and numbers	No effect	Light clay, 3.1-3.5 % OM, 12-16 kg/ha, in situ	Soil bacteria[1]
Growth rate and numbers	Some species decreased	Medium loam red soil, 12-20 kg/ha, in situ	Soil bacteria[2]
Nitrification and nitrifiers	Inhibited activity	Sandy loam soils, pH=6.1-7.4, 1.3-1.6 % OM, 50 kg/ha	Soil bacteria[3]
Nitrification and nitrifiers	Little effect	Meadow-type chernozem, 1.5 kg/ha	Soil bacteria[4]

Nitrogen fixation in soil	Decreased soil population	Irrigated brown soil, 8 kg/ha	Soil bacteria, free living[5]
Nitrogen fixation in soil	Increased population	Alluvial, 8 kg/ha	Soil bacteria, free living[6]
Nitrogen fixation in soil	Inhibition of soil	Acid soil, 8-10 kg/ha, pH=5	Soil bacteria, free living[7]

Source: 1.Spiridinov, Y.Y., V. S. Skhiladze and G.S. Spiridinova. 1970. Subtrop. Kultury 1, 150-155.
2.Spiridinov, Y.Y., V. S. Skhiladze and G.S. Spiridinova. 1970. Subtrop. Kultury 3, 152-160.
3.Grossbard, E. and J.A.P. Marsh. 1974. Pestic Sci. 5, 609-623.
4.Tulabaev, B.D. and S. Tamikaev. 196. Uzbek. biol. Zh. 2, 14-17.
5.Akopyan, E.A. and A.G. Agaronyan. 1968. Soils and Fert. 33, 3133.
6.Goguadze, V.D. 1969. Subtrop. Kultury No 4, 161-167.
7.Goguadze, V.D. 1968. Agrokhmiya 5, 99-103.

7-113 Effect of Monuron

Parameter	Effect Value	Concentration and Condition	Species
Growth rate and numbers	50 % decrease in population	Irrigated brown soil, 8 kg/ha, in situ	Soil bacteria[1]
Growth rate and numbers	Decreased population	Meadow bog, monthly applications, 8-16 kg/ha, in situ	Soil bacteria[2]
Growth rate and numbers	Decreased population	Red krasnozem, monthly applications, 12-20 kg/ha, in situ	Soil bacteria[2]
Growth rate and numbers	Total population decreased	Alluvial, 8 kg/ha, in situ	Soil bacteria[3]
Growth rate and numbers	Increased population	Alluvial, 8 kg/ha, in situ	Soil bacteria, spore formers[4]
Nitrification and nitrifiers	Annual application increased activity	Light loam dernopodzol, 1.5-5 kg/ha, recomm. field rate, 5 y	Soil bacteria[5]
Nitrification and nitrifiers	Decreased activity	Heavy loam soils, 12 kg/ha	Soil bacteria[6]
Nitrification and nitrifiers	Decreased population in field	Irrigated brown soil, 8 kg/ha, no effect in lab	Soil bacteria[1]
Nitrification and nitrifiers	Increased activity	Heavy loam soils, 6 kg/ha	Soil bacteria[6]
Nitrification and nitrifiers	Increased activity	Brown heavy loam, pH=6.3, 4 kg/ha	Soil bacteria[7]
Nitrification and nitrifiers	Second application decreased activity	Heavy loam soils, 6 kg/ha each time	Soil bacteria[6]
Nitrogen fixation in soil	No effect	Meadow soil, 1.5 kg/ha	Soil bacteria, free living[8]
Nitrogen fixation in soil	No effect	Alluvial krasnozem, pH=7, 8-10 kg/ha	Soil bacteria, free living[9]
Numbers and activity of soil algae	Decreased population, but recovery within 3 months	Light loam, 5.2 kg/ha	Herbicide applied to soil in situ[10]
Numbers and activity of soil algae	Decreased population, effect lasted 4 years	Light and sandy clay, 10 and 20 kg/ha	Herbicide applied to soil in situ[11]
Numbers and activity of soil algae	Decreased populations, except diatom Navicula atomus, very persistent	Gleyed heavy loam, >8 % OM; peaty-podzol, 35 % OM, 5-10 kg/ha	Herbicide applied to soil in situ[12]
Numbers and activity of soil algae	Decreased populations, except diatom Navicula atomus, very persistent	Light sandy loam, > 3 % OM, 5-10 kg/ha, recomm. field rate	Herbicide applied to soil in situ[12]
Soil microbial ammonification	Decreased activity	Silt, 60 kg/ha	Soil organisms[13]

Soil microbial ammonification	Increased activity then decreased activity	Heavy loam, 6-12 kg/ha	Soil organisms[6]
Soil respiration activity	Decreased CO2-evolution	Silt loam, 10 kg/ha, recomm. field rate	Soil organisms[14]
Soil respiration activity	Decreased activity for 28 d, then recovered, high OM decreased inhibi	Nine soil types, variable OM, 5-100 kg/haSoil organisms[15]	
Soil respiration activity	Initially increased, then decreased CO2-evolution	Sandy loam, pH=7.4, 1.5 % OM; 50 kg/ha, coarse sandy loam, pH=6.1, 1.6 % OM	Soil organisms[16]

Source: 1.Akopyan, E.A. and A.G. Agaronyan. 1968. Soils and Fert. 33, 3133.
2.Spiridinov, Y.Y., V. S. Skhiladze and G.S. Spiridinova. 1970.In: The Behavior, Conversion and Analyss of Pesticides and their Metabolites in Soil. Proc. 1st all-Union Conf. Pushino-na-OKe, Akad. Nauk.
3.Goguadze, V.D. 1969. Subtrop. Kultury No 4, 161-167.
4.Micev, N. and M. Bubalov. 1972. Symp. Biol. Hung. 11, 379-384.
5.Chunderova, A.I., T.P. Zuberts and A.M. Sofinskii. 1971. Khimiya selKhoz 9, 527-530.
6.Zavarzin, V.I. and T.V. Belyaeva. 1966. Khimiya selKhoz 4, 34-36.
7.Kudzin, Y.K., A.Y. Fisyunov, N.A. Chernyavskaya and A.Y. Makarova.1973. Dokl. Vses. Akad. SelKhoz. Nuk. imeni V.I. Lenina 9, 13-15.
8.Tulabaev, B.D. and S. Tamikaev. 196. Uzbek. biol. Zh. 2, 14-17.
9.Goguadze, V.D. 1968. Agrokhmiya 5, 99-103.
10.Kiss, A. 1967. Agrokem. Talajt. 16, 11.
11.Pantera, H. 1970. Meded. Fac. Landb. Rijksuniv. Gent. 35, 847-854.
12.Mikhailova, E.I. and Kruglov. 1973. Pochvovedenie 8, 81-85.
13.Geltser, Y.G., V.A. Geptner and L.D.Stonov. 1972. Agrokhimiya 9, 119-123.
14.Doxtader, K.G. 1968. Bact. Proc. 4, A21.
15.Chandra, P., W.R. Furtick and W.B. Bollen. 1960. Weeds 8, 589-598.
16.Grossbard, E. and J.A.P. Marsh. 1974. Pestic Sci. 5, 609-623.

7-114 Effect of Murbetol

Parameter	Effect Value	Concentration and Condition	Species
Denitrification in soil	Decreased activity and population	Sod-podzol loam, 9.1 kg/ha	Soil organisms[1]
Denitrification in soil	No effect	Sod-podzol loam, 7.6 kg/ha	Soil organisms[1]
Denitrification in soil	No effect	Leached chernozem, 40 and 80 kg/ha	Soil organisms[2]
Nitrification and nitrifiers	Decreased activity	Sod-podzolic loam, 7.6 and 9.1 kg/ha	Soil bacteria[1]
Nitrification and nitrifiers	Decreased activity	Leached chernozem, low OM, 80 kg/ha	Soil bacteria[2]
Nitrification and nitrifiers	No effect	Leached chernozem, low OM, 40 kg/ha	Soil bacteria[2]
Nitrification and nitrifiers	Slightly decreased activity in wet conditions	Pale chestnut and medium loams, 22.2 kg/ha	Soil bacteria[3]
Nitrogen fixation in soil	Increased Azobacter population and numbers of Clostridium pasteurianum	Leached chernozem, 40 and 80 kg/ha	Soil bacteria, free living[2]
Soil microbial ammonification	Increased population	Derno-podzol light or medium loam, pH=6, 2-3 % OM, 10 and 17.5 kg/ha	Soil organisms[4]

Source: 1.Isaeva, L.I. and E.D. Novogrudskaya. 1966. Vest selKhozPNauki, Mosk. 11, 57-59.
2.Lobanov, V.E. and L.P. Poddubnaya. 1967. Khimiya selKhoz. 5, 32-34.
3.Zharasov, S.U. 1971. Vest. selKhoz Nauki, Kazakhskoi 14, 41-43.
4.Chunderova, A.I., T.P. Zuberts and A.M. Sofinskii. 1971. Khimiya selKhoz 9, 527-530.

7-115 Effect of NH_3

Parameter	Effect Value	Concentration and Condition	Species
Foliar disease	Interveinal necrotic blotches similar to SO2 injury	Air, 55 ppm for 1 h	Angiosperms[1]
Foliar disease	Distal necrosis	Air, 55 pppm for 1 h	Gymnosperms[1]

Source: 1.Smith, W.H. 1981. Air Pollution and Forests. Springer Verlag. NY. 384 pp.

7-116 Effect of NO_x

Parameter	Effect Value	Concentration and Condition	Species
Foliar disease	Interveinal necrotic blotches similar to SO2 injury	Air, 20 ppm for 1 h or 1.6-2.6 ppm for 48 h or 1 ppm for 100 h	Angiosperms[1]
Foliar disease	Redbrown distal necrosis	Air, 20 ppm for 1 h or 1.6-2.6 ppm for 48 h or 1 ppm for 100 h	Gymnosperms[1]

Source: 1.Smith, W.H. 1981. Air Pollution and Forests. Springer Verlag. NY. 384 pp.

7-117 Effect of Naptalam

Parameter	Effect Value	Concentration and Condition	Species
Growth rate and numbers	No effect	Medium loam, 1.5 % OM, 5 kg/ha, in situ	Soil bacteria[1]
Nitrification and nitrifiers	No effect	Meadow-type medium loam, 1.5 % OM, 5 kg/ha	Soil bacteria[1]

Source: 1.Tulabaev, B.D. 1970. Uzbek. biol. Zh. 14, 11-14.

7-118 Effect of Neburon

Parameter	Effect Value	Concentration and Condition	Species
Nitrification and nitrifiers	Annual application increased activity	Light loam dernopodzol, 2-3 kg/ha	Soil bacteria[1]
Nitrification and nitrifiers	Decreased activity	Heavy loam soils, 8 kg/ha	Soil bacteria[2]
Nitrification and nitrifiers	Increased activity	Heavy loam soils, 4 kg/ha	Soil bacteria[2]
Nitrification and nitrifiers	Increased activity	Alluvial, 2 % OM, 0.9 and 1.8 kg/ha	Soil bacteria[3]
Nitrification and nitrifiers	Second application increased activity	Heavy loam soils, 8 kg/ha each time	Soil bacteria[2]
Soil microbial ammonification	Increased activity then decreased activity	Heavy loam, 4-8 kg/ha	Soil organisms[2]

Source: 1.Chunderova, A.I., T.P. Zuberts and A.M. Sofinskii. 1971. Khimiya selKhoz 9, 527-530.
2.Zavarzin, V.I. and T.V. Belyaeva. 1966. Khimiya selKhoz 4, 34-36.
3.Horowitz, M., T. Blumenfield, G. Herzlinger and N. Hulin. 1974. Weed Res. 14, 97-109.

7-119 Effect of Nitrofen

Parameter	Effect Value	Concentration and Condition	Species
Soil respiration activity	No effect	Sandy, 1-3 kg/ha, recomm. field rate	Soil organisms[1]

Source: 1.Gowda, T.K.S. and R.B. Patil. 1972. Biochem. J. 128, 56-57.

7-120 Effect of Noruron

Parameter	Effect Value	Concentration and Condition	Species
Nitrification and nitrifiers	Increased activity	Serozem-meadow soil, 1.5 and 3 kg/ha	Soil bacteria[1]
Nitrogen fixation in soil	Inhibited soil population	Serozem-meadow, 3 kg/ha	Soil bacteria, free living[1]
Nitrogen fixation in soil	Temporarily decreased soil population, but recovered	Serozem-meadow, 1.5 kg/ha	Soil bacteria, free living[1]
Soil microbial ammonification	Decreased population	Serozem-meadow soil, 1.5 kg/ha	Soil organisms[1]

Soil microbial ammonification	Inhibited activity	Serozem-meadow soil, 3 kg/ha	Soil organisms[1]

Source: 1.Tulabaev, B.G. 1972. Khimiya selKhoz 10, 776-780.

7-121 Effect of O_3

Parameter	Effect Value	Concentration and Condition	Species
Foliar disease	Upper surface flecks	Air, 0.20-0.30 ppm for 2-4 h	Angiosperms[1]
Foliar disease	Distal necrosis and stunted needles	Air, 0.20-0.30 ppm for 2-4 h, some conifers 0.08 ppm for 12-13 h	Gymnosperms[1]
Photosynthesis	Threshold dose for photosynthesis suppression	Air = 0.5 ppm, 7-11 h, 1-2 d treatment, tree sapling	Black oak[1]
Photosynthesis	Threshold dose for photosynthesis suppression	Air = 0.15 ppm, continuous, 84 days	Loblolly pine[1]
Photosynthesis	Threshold dose for photosynthesis suppression	Air = 0.15 ppm, continuous, 84 days	Pond pine[1]
Photosynthesis	Threshold dose for photosynthesis suppression	Air = 0.30 ppm, 9 h/day, 10 days, tree seedling	Ponderosa pine[1]
Photosynthesis	Threshold dose for photosynthesis suppression	Air = 0.15 ppm, continuous, 84 days, tree seedling	Slash pine[1]
Photosynthesis	Threshold dose for photosynthesis suppression	Air = 0.5 ppm, 7-11 h, 1-2 d exposure, tree sapling	Sugar maple[1]
Photosynthesis	Threshold dose for photosynthesis suppression	Air = 0.15 ppm, continuous, 19 days, tree seedling	White pine[1]
Photosynthesis	Threshold dose for photosynthesis suppression	Air = 0.5 ppm, 4 h, single treatment, tree sapling	White pine[1]

Source: 1.Smith, W.H. 1981. Air Pollution and Forests. Springer Verlag. NY. 384 pp.

7-122 Effect of PCP

Parameter	Effect Value	Concentration and Condition	Species
Denitrification in soil	Increased population	Leached chernozem, 1.5 kg/ha	Soil organisms[1]
Denitrification in soil	Increased population	Clay loam, 1.3 % C; alluvial, 1.1 % C, 0.8 and 1.6 and 3 kg/ha	Soil organisms[2]
Denitrification in soil	Increased population	Volcanic	Soil organisms[2]
Growth rate and numbers	Increased populations and relative more than other bacteria	Alluvial clay loam, pH=5.7, 2.7 % C, 12-120 kg/ha, in situ	Pseudomonos spp.[3]
Growth rate and numbers	Decrease in population	Six soils, 4 kg/ha, in situ	Soil bacteria[4]
Growth rate and numbers	Increase in population	Leached cherzonem, low OM, 1.5 kg/ha, in situ	Soil bacteria[1]
Growth rate and numbers	Increased populations	Alluvial clay loam, pH=5.7, 2.7 % C, 12-120 kg/ha, in situ	Soil bacteria[3]
Growth rate and numbers	Increased populations	Alluvial clay loam, pH=5.7, 2.7 % C, 12-120 kg/ha, in situ	Soil bacteria, anaerobes[3]
Nitrification and nitrifiers	Decreased activity	Alluvial clay loam, 20 kg/ha	Soil bacteria[2]
Nitrification and nitrifiers	Inhibited activity	Volcanic clay loam, 100 kg/ha	Soil bacteria[5]
Nitrification and nitrifiers	No effect	Alluvial clay loam, 2 kg/ha	Soil bacteria[2]

Numbers and activity of soil algae	Decreased populations of diatoms, blue-green and filamentous forms	Alluvial clay loam, clay loam, volcanic light clay, alluvial, 0.8-2 kg/ha	Herbicide applied to soil in situ[3]
Soil microbial ammonification	Inhibited activity	Alluvial clay loam, pH=5.7, 2.7 % C, 12-120 kg/ha	Soil organisms[2]
Soil respiration activity	Decreased activity	Six soil types, 4 kg/ha	Soil organisms[4]
Soil respiration activity	Increased activity	Clay loam, 1-3 kg/ha, recomm. field rate	Soil organisms[5]

Source: 1.Tyagny-Ryadno, M.G. 1967. Trudy Kamenets-podolsk. selKhoz Inst. 9, 43-48.
2.Ishizawa, S. and T. Motsuguchi. 1966. Bull. natn. Inst. agric. Sci., Tokyo (B) 16, 1-90.
3.Ishizawa, S. and T. Matsuguchi. 1966. Bull. natn Inst. agric. Sci., Tokyo (B) 16, 1-90.
4.Au, F.H.F. 1968. Effect of endothal and certain other selective herbicides on microbial activity in six different soils, Ph. D. Thesis, Oregon state Univ.
5.Matsuguchi, T. and S. Ishizawa. 1969. J. Sci. Soil Manure, Tokyo 40, 20-25.

7-123 Effect of PCP and endothal

Parameter	Effect Value	Concentration and Condition	Species
Nitrification and nitrifiers	Decreased activity	Six soils, PCP (4) + endothal (2 kg/ha)	Soil bacteria[1]
Soil microbial ammonification	Decreased population	Six soils, PCP(4) + endothal(2 kg/ha)	Soil organisms[1]
Soil respiration activity	Decreased activity	Six soil types, PCP(4)+endothal(2 kg/ha)	Soil organisms[1]

Source: 1.Au, F.H.F. 1968. Effect of endothal and certain other selective herbicides on microbial activity in six different soils, Ph. D. Thesis, Oregon state Univ.

7-124 Effect of Paraquat

Parameter	Effect Value	Concentration and Condition	Species
Growth rate and numbers	Almost completely inhibited	Liquid media=10-12 ppm, revover if medium diluted, in culture	Escherichia coli[1]
Growth rate and numbers	Growth reversely depressed	Liquid media=1-2 ppm, recover if medium diluted, in culture	Escherichia coli[1]
Growth rate and numbers	Count increased	Solid media = 5-10 ppm, in culture	Soil bacteria[1]
Growth rate and numbers	Decreased population	Silt loam, pH=6.3, 0.11 % OM, 0.57 kg/ha, in situ	Soil bacteria[2]
Growth rate and numbers	Slight decreases immediately after dose, no serve effects	Calcereous loam, pH=8, 7.5 % OM, field and lab, 1.14 and 11.4 kg/ha, in situ	Soil bacteria[3]
Mortality	90 % mortality	Liquid media=1800-2000 ppm, after 60 min, in culture	Escherichia coli[1]
Nitrification and nitrifiers	No significant effect	Silt loam, pH=6.3, 0.11 % OM, 0.6 and 6 kg/ha	Soil bacteria[4]
Nitrification and nitrifiers	No significant effect	Calcareous loam, pH=8, 7.5 % OM, 1.14 and 11.4 kg/ha	Soil bacteria[5]
Nitrification and nitrifiers	No significant effect	Silt loam, pH=6, 2 % OM, 0.6 and 1.2 kg/ha	Soil bacteria[6]
Numbers and activity of soil algae	Decrease in numbers during first 24 hr then recovery	Calcareous loam, pH=8, 7.5 % OM, 1.14 and 11.4 kg/ha	Herbicide applied to soil in situ[5]
Soil microbial ammonification	Decreased activity	Silt loam, pH=6.4, 0.11 % OM, 0.6 kg/ha	Soil organisms[4]
Soil respiration activity	Decreased CO2-evolution by Sclerotium rolfsii	Sandy loam, 1.25-100 kg/ha	Soil organisms[7]

Soil respiration activity	Minor and irregular changes in CO2-evolution	Calcareous loam, 1.14 and 11.4 kg/ha, pH=8, 7.5 % OM	Soil organisms[5]
Nitrogen fixation in soil	No effect for first 3 weeks	Calcareous loam, pH=8, 7.5 % OM, 1.14 and 11.4 kg/ha	Soil bacteria, free living[3]

Source: 1.Davidson, C.L. and B. Papirmeister. 1971. Proc. Soc. exp. Biol. Med. 136, 359-364.
2.Tu, C-M. and W.B. Bollen. 1968. Weed Res. 8, 28-37.
3.Anderson, J.R. and E.A. Drew. 1976. Zbl. Bakt. abt II, 131, 247-258.
4.Tu, C.-M. and W.B. Bollen. 1968. Weed Res. 8, 28-37.
5.Anderson, J.R. and E.A. Drew. 1976. Zbl. Bakt. Abt. II, 131, 136-147.
6.Thorneburg, R.P. and J.A. Tweedy. 1973. Weed Sci. 21, 397-399.
7.Curl, E.A., R. Rodriguez-Kabana and H.H. Funderburk. 1967. wPhytopathology 57, 7.

7-125 Effect of Parathion-methyl

Parameter	Effect Value	Concentration and Condition	Species
Growth rate and numbers	Increased numbers	0.3 kg/ha	Soil bacteria[1]
Nitrification and nitrifiers	Decreased numbers	Loam, pH=7.2, 1.49 % C, 0.15 and 1.5 kg/ha	Soil organisms[2]
Nitrification and nitrifiers	Decreased numbers	Loam, pH=7.2, 1.49 % C, 300 kg/ha, higher temp greater effect	Soil organisms[3]
Nitrification and nitrifiers	Initially decreased then increased numbers	Loam, pH=7.2, 1.49 % C, 15 and 150 and 300 kg/ha	Soil organisms[2]

Source: 1.Yurovskaya, Y.M. and V.A. Zhulinskaya. 1974. Khimiya selKhoz 12, 38-41.
2.Naumann, K. 1970. Zbl. Bakt. Abt. II 124, 755-765.
3.Bardiya, M.C. and A.C. Gaur. 1970. Zbl. Bakt. Abt. II, 124, 552-555.

7-126 Effect of Pb

Parameter	Effect Value	Concentration and Condition	Species
Photosynthesis	Threshold dose for photosynthesis suppression	<10 ug/g dw in basis foliar tissue, tree seedling	American sycamore[1]

Source: 1.Smith, W.H. 1981. Air Pollution and Forests. Springer Verlag. NY. 384 pp.

7-127 Effect of Pebulate

Parameter	Effect Value	Concentration and Condition	Species
Growth rate and numbers	Increased soil populations	4 kg/ha, but no. of C. pasteurianum decreased, in situ	Soil bacteria[1]
Morphological change	Morphological changes observed	Liquid media=1000 ppm, in culture	Arthrobacter sp.[2]
Morphological change	Morphological changes observed	Liquid media=1000 ppm, in culture	Corynebacterium sp.[2]
Morphological change	Morphological changes observed	Liquid media=1000 ppm, in culture	Pseudomonas fluorens[2]
Nitrification and nitrifiers	Increased activity	4 - 6 kg/ha	Soil bacteria[3]

Source: 1.Barona, V.P. 1974. Agrokhimiya 11, 131-134.
2.Sobieszczanski, J. 1969. Acta microbiol. pol. Ser B, 1(18), 99-104.
3.Balica, N. and J. Sobieszczanski. 1969. Acta. microbiol. pol. 1(18), 3-6.

7-128 Effect of Peroxyacetylnitrate

Parameter	Effect Value	Concentration and Condition	Species
Foliar disease	Lower surface bronzing	Air, 0.20-0.80 ppm for 8 h	Angiosperms[1]
Foliar disease	Chlorosis and early senescence	Air, 0.20-0.80 ppm for 8 h	Gymnosperms[1]

Source: 1.Smith, W.H. 1981. Air Pollution and Forests. Springer Verlag. NY. 384 pp.

7-129 Effect of Phenazone

Parameter	Effect Value	Concentration and Condition	Species
Growth rate and numbers	Inhibited populations	Leached dernocarbonate or dernopodzol, 8-18 kg/ha, in situ	Soil bacteria[1]
Growth rate and numbers	No effect on growth	Leached dernocarbonate or dernopodzol, 3-6 kg/ha, in situ	Soil bacteria[1]
Nitrification and nitrifiers	Inhibited activity	Leached dernocarbonate and dernopodzol, 3 - 18 kg/ha	Soil bacteria[1]
Nitrogen fixation in soil	Inhibited azobacters	Leached derno-carbonate, 8-18 kg/ha	Soil bacteria, free living[2]
Nitrogen fixation in soil	No effect	Leached derno-carbonate, 3-6 kg/ha	Soil bacteria, free living[2]
Nitrogen fixation in soil	No effect	Derno-podzol, 12 kg/ha	Soil bacteria, free living[2]
Soil microbial ammonification	Inhibited activity	Leached derno-carbonate and derno-podzol, 8-18 kg/ha	Soil organisms[2]
Soil microbial ammonification	No effect	Leached derno-carbonate and derno-podzol, 3-6 kg/ha	Soil organisms[2]

Source: 1.Mezharaupe, V.A. 1967. Mikroorg. Rast. Trudy Inst. Mikrobiol. Akad. Nauk. SSR 65-83.
2.Mezharupe, V.A. 1967. Mikroorg. Rast. Trudy Inst. Mikrobiol. Akad. Nauk. SSR 65-83.

7-130 Effect of Phenobenzuron

Parameter	Effect Value	Concentration and Condition	Species
Growth rate and numbers	No effect	Sandy loam, 1.4 % OM or limestone soil with 9.0 % OM, 10 kg/ha, in situ	Soil bacteria[1]
Growth rate and numbers	No effect on growth rate	Liquid media=100 ppm, in culture	Soil bacteria[1]
Nitrification and nitrifiers	No effect	Sandy loam, 1.4 % OM, limestone soil, 10 and 100 kg/ha	Soil bacteria[1]

Source: 1.Catroux, G. and J.-C. Fournier. 1971. Conf. Commte Francais de Lutte contre les Mauvaises Herbes (Couma) 1971, 69-78.

7-131 Effect of Phytar 560

Parameter	Effect Value	Concentration and Condition	Species
Growth rate and numbers	Increased population	Sandy loam, 300 kg/ha, in situ	Soil bacteria[1]
Growth rate and numbers	No effect	Sandy loam, 100 kg/ha, in situ	Soil bacteria[1]

Source: 1.Khikmatov, A., A. Rakhimov and Z. Muslimov. 1969. Agrokhimiya 6, 111-112.

7-132 Effect of Picloram

Parameter	Effect Value	Concentration and Condition	Species
Growth rate and numbers	Slightly inhibited growth rate	Solid+liquid media=200 ppm, recomm. filed rate, in culture	Bdellovibrio bacteriovorus[1]
Nitrification and nitrifiers	Decreased activity, especially in dry conditions	Derno-podzol medium loam, 0.07 kg/ha	Soil bacteria[2]
Nitrification and nitrifiers	Increased population	Alluvial soil, 2 kg/ha	Soil bacteria[3]
Nitrification and nitrifiers	Inhibited acticity	Acid volcanic ash, silt loam and neutral alluvium, 0.2 - 10 kg/ha	Soil bacteria[4]

Nitrification and nitrifiers	Inhibited activity	Marine sand, 0.2 - 10 kg/ha	Soil bacteria[4]
Nitrification and nitrifiers	No effect	Sandy loam, 65 % MHC, 0.48 and 4.8 and 48 kg/ha	Soil bacteria[5]
Nitrification and nitrifiers	No effect	Heavy clay, 0.5 kg/ha	Soil bacteria[6]
Nitrification and nitrifiers	No significant effect	Silt loam, pH=6, 2 % OM, 0.6 and 1.2 kg/ha	Soil bacteria[7]
Nitrogen fixation in soil	Decreased soil population	Polder soil, 65 % MHC, 100-300 kg/ha, recomm. field rate	Soil bacteria, free living[5]
Nitrogen fixation in soil	Increased population	Alluvial, 2 kg/ha	Soil bacteria, free living[3]
Soil microbial ammonification	Increased activity	Silty clay loam and silt loam, 0.1-1.0 kg/ha	Soil organisms[8]
Soil microbial ammonification	Increased population	Alluvial, 2 kg/ha	Soil organisms[3]
Soil respiration activity	CO2-evolution and O2-uptake not affected	Heavy clay, 0.03-0.12 kg/ha	Soil organisms[6]
Soil respiration activity	Increased CO2-evolution	Sandy loam polder soil, 5 kg/ha	Soil organisms[5]
Soil respiration activity	Increased activity	Fine sandy loam, 24 and 48 kg/ha	Soil organisms[9]
Soil respiration activity	No effect	Sandy loam polder soil, 0.5 and 50 kg/ha	Soil organisms[5]

Source: 1.Wehr, N.B. and D.A. Klein. 1971. Soil Biol. Biochem. 3, 143-149.
2.Bezuglov, V.G., A.K. Minenko and E.P. Shelestov. 1973. Khimiya selKhoz 11, 854-856.
3.Goguadze, V.D., E.A. Gobronidze and L.G. Gogiya. 1972. Subtrop Kultury 3, 157-159.
4.Dubey, H.D. 1969. Proc. Soil Soc. Am. 33, 893-896.
5.Schreven, D.A. van, D.J. Lindenbergh abd A. Koridon. 1970. Pl. Soil 33, 513-532.
6.Grover, R. 1972. Weed Res. 12, 112-114.
7.Thorneburg, R.P. and J.A. Tweedy. 1973. Weed Sci. 21, 397-399.
8.Tu, C.-M. and W.B. Bollen. 1968. Weed Res. 8, 28-37.
9.Withworth, J.W., W. Garner and B.C. Williams. 1965. Res. Prog. Rept. West. Weed Contr. Conf. 1965, 17-118.

7-133 Effect of Picloram and 2,4-D

Parameter	Effect Value	Concentration and Condition	Species
No. and activity of soil algae	No effect	Clay loam, pH=8.2, 2.5 % OM,	Herbicide applied to soil in situ[1]

Source: 1.Arvik, J.H. D.L. Willson and L.C. Darlington. 1971. Weed Sci. 19, 276-278.

7-134 Effect of Pirimiphos-ethyl

Parameter	Effect Value	Concentration and Condition	Species
Soil respiration	CO2-evolution decreased, then increased and returned to normal	Sandy loam, calcareous c{ay and clay, 11 kg/ha	Soil organisms[1]

Source: 1.Gawaad, A.A.A., M.H. Hammad and F.H. El-Gayar. 1973. Agrokem. Talajt. 22, 161-168.

7-135 Effect of Printazol

Parameter	Effect Value	Concentration and Condition	Species
Growth rate and numbers	Decreased then increased population	Rich loam, pH=7, 3.3 % OM, 0.25-2.0 kg/ha, in situ	Soil bacteria[1]
Soil microbial ammonification	Up to 1 kg/ha, retarded then increased activity, but 2 kg/ha inhibite	Rich loam, pH=7, 3.3 % OM, 0.25-2 kg/ha	Soil organisms[1]

Source: 1.Bertrand, D. and A. de Wolf. 1972. C.r. hebd. Seanc. Acad. Agric. Fr. 58, 1469-1473.

7-136 Effect of Prometryne

Parameter	Effect Value	Concentration and Condition	Species
Growth rate and numbers	Inhibition of growth arte	Liquid media=10-100 ppm, in culture	Bacillus sp.[1]
Growth rate and numbers	Sencivite at acid pH values	Liquid media=5-100 ppm, in culture	Bacillus sp.[2]
Growth rate and numbers	Increased population	Light loam dernopodzol, 2 kg/ha, in situ	Clostridium acetobutylicum[3]
Growth rate and numbers	No effect on growth rate	Liquid media=10-100 ppm, in culture	Pseudomonas sp.[1]
Growth rate and numbers	Resistant at all pH values	Liquid media=5-100 ppm, in culture	Pseudomonas sp.[2]
Growth rate and numbers	Decreased population	Loam, 3.5 % OM, 20 d, 1.0 kg/ha, in situ	Soil bacteria[4]
Growth rate and numbers	Decreased population	Leached medium loam chernozem, pH=6.0, 4 % OM, 2-3 kg/ha, in situ	Soil bacteria[5]
Growth rate and numbers	Increased population	Armenian vineyard, 2 kg/ha, in situ	Soil bacteria[6]
Growth rate and numbers	Increased population	Leached medium loam chernozem, 2 kg/ha, in situ	Soil bacteria[7]
Growth rate and numbers	Increased population	Derno-podzoloc loam, lab exp., 0.004-10000 kg/ha, in situ	Soil bacteria[8]
Growth rate and numbers	Increased population	Leached chernozem, 60 % MHC, 25 and 25 kg/ha, in situ	Soil bacteria[9]
Growth rate and numbers	No effect	0.5-9 kg/ha, slightly increased population, in situ	Soil bacteria[10]
Growth rate and numbers	No effect	Derno-podzolic loam, field exp., 2 kg/ha, in situ	Soil bacteria[8]
Growth rate and numbers	No effect	Normal chernozem, 2.0-25 kg/ha, in situ	Soil bacteria[9]
Growth rate and numbers	Strong inhibition	Liquid media=100 ppm, in culture	Soil bacteria[11]
Growth rate and numbers	Increased then decreased population	Light loam dernopodzol, 2 kg/ha, in situ	Soil bacteria, butyric acid bacteria[3]
Growth rate and numbers	Decreased population	10-50 kg/ha, in situ	Soil bacteria, non-spore formers[10]
Growth rate and numbers	Especially decreased population	10-50 kg/ha, in situ	Soil bacteria, spore formers[10]
Mortality	Toxic effects	Liquid media=1000 ppm, surfactants increase effect, in culture	Rhizosphere bacteria[12]
Nitrification and nitrifiers	Annual applications increased activity	Light loam dernopodzol, 2.5 kg/ha, 5 y	Soil bacteria[13]
Nitrification and nitrifiers	Decreased activity	Light alluvial and heavy leached meadow cinnamom, pH=7, 2 % OM, 4 kg/ha	Soil bacteria[14]
Nitrification and nitrifiers	Increased activity	Leached medium loam chernozem, pH=6, 4 % OM, 2-3 kg/ha	Soil bacteria[15]
Nitrification and nitrifiers	Increased activity and population	Leach black calcareous clay, 15 kg/ha	Soil bacteria[16]

Nitrification and nitrifiers	Increased activity and population	Heavy cis-Caucasian chernozem, 1.0-2.5 kg/ha	Soil bacteria[17]
Nitrification and nitrifiers	Increased activity and population, high rate delayed increase	Leached chernozem and chernozem, 2.5 and 25 kg/ha	Soil bacteria[9]
Nitrification and nitrifiers	Inhibited activity	Leached chernozem, 1.0-2.5 kg/ha	Soil bacteria[18]
Nitrification and nitrifiers	Inhibited activity	Acid volcanis ash, silt loam and neutral alluvium, 10 kg/ha	Soil bacteria[15]
Nitrification and nitrifiers	Inhibited activity	Marine sand, 10 kg/ha	Soil bacteria[15]
Nitrification and nitrifiers	No effect	Clay soil, 0.1 kg/ha	Soil bacteria[19]
Nitrification and nitrifiers	Transiently decreased activity	Clay soil, 1.0 and 5 kg/ha	Soil bacteria[19]
Nitrogen fixation in soil	Annual application decreased soil population	Light loam, 20 kg/ha, 5 y	Soil bacteria, free living[20]
Nitrogen fixation in soil	Decreased soil population	Irrigated light chestnut,, 2 % OM, 2.5 kg/ha	Soil bacteria, free living[21]
Nitrogen fixation in soil	Decreased soil population	Light loam pale chestnut, 1.7 % OM, 2-4 kg/ha	Soil bacteria, free living[22]
Nitrogen fixation in soil	Decreased soil population, but no loss of N-fixation	Cinnamon, alluvial and skeletal soils, 5-7 kg/ha	Soil bacteria, free living[23]
Nitrogen fixation in soil	Increased soil population	Leached medium loam chernozem, pH=6, 4 % OM, 2-3 kg/ha	Soil bacteria, free living[5]
Nitrogen fixation in soil	Increased soil population	Leached medium loam chernozem, 2 kg/ha	Soil bacteria, free living[7]
Nitrogen fixation in soil	No effect on numbers of Azotobacter sp. and Clostridium sp.	Leached meadow-cinnamon and alluvial soils, .5-1.5 kg/ha, recom field rate	Soil bacteria, free living[24]
Nitrogen fixation in soil	No effect on numbers of Azotobacter sp. and Clostridium sp.	Leached meadow-cinnamon and alluvial, .5-1.5 kg/ha, recom field rate	Soil bacteria, free living[24]
Numbers and activity of soil algae	Decreased population for up to 5 months	Sandy clay, 3-50 kg/ha	Herbicide applied to soil in situ[25]
Numbers and activity of soil algae	No effect	Leached medium loam chernozem, 2 kg/ha	Herbicide applied to soil in situ[26]
Soil microbial ammonification	Increased activity	Chernozems, 2.5 and 25 kg/ha	Soil organisms[27]
Soil microbial ammonification	Increased activity and population	Leached chernozem, 0.5-1.5 kg/ha, recomm. field rate	Soil organisms[18]
Soil microbial ammonification	Increased population 4-5 fold	Chernozem, 1-2.5 kg/ha	Soil organisms[17]
Soil microbial ammonification	No effect	Light loam pale chestnut, 1.7 % OM, 2-4 kg/ha	Soil organisms[22]
Soil respiration activity	Increased CO2-evolution	Leached chernozem, 15-60 kg/ha	Soil organisms[18]

Source: 1.Balicka, N. and Z. Krezel. 1969. Weed Res. 9, 37-42.
2.Balicka, N. 1969. Acta microbiol. pol. Ser B, 1(18), 43-45.
3.Kozyrev, M.A. and A.A. Laptev. 1972. Khimiya selKhoz. 10, 614-617.
4.Geshtovt, Y.N., S.U. Zharasov and N. Adekenov. 1970. Vest. SelKhoz. Nauki. Kazakhskoi SSR. 13, 35-38
5.Mikhailova, M.F. 1968. Khimiya selKhoz 6, 204-206.
6.Krykha, hanov, L.I. and Z.V. Pashkovskaya. 1967. Trudy ural. nauchnoissled. Inst. selKhoz. 7, 66-74.
7.Matsneva, N.G. and O.A. Semikhatova. 1973. Khimiya selKhoz 11, 222-224.
8.Samokhvalov, A.N. 1973. Vest. Mosk. Univ. 6, 87-92.

9.Vasilev, D.S. and O.V. Enkina. 1967. Dokl. Vses. Akad. selKhoz. Nauk. 12, 10-12.
10.Simon, L. 1971. Folia Microbiol. 16, 516.
11.Krezel, Z. and M. Musial. 1969. Acta microbiol. pol. Ser B 1, 93-87.
12.Balicka, N., J. Sobieszczanski and T. Niewiadoma. 1969. Acta microbiiol. pol.1(18), 11-14.
13.Chunderova, A.I., T.P. Zuberts and A.M. Sofinskii. 1971. Khimiya selKhoz 9, 527-530.
14.Rankov, V. 1968. Pochv. Agrokhim. 3, 81-90.
15.Dubey, H.D. 1969. Proc. Soil Soc. Am. 33, 893-896.
16.Andreeva-Fetvadzhieva, N. and B. Kolcheva. 1969. Nauchni Trud. vissh selkostop. Inst. Georgi Dimitro20, 49-69.
17.Yaroslavskaya, P.M., D.S. Vasilev and N.T. Agarkova. 1969. Agrokhimiya 1, 97-101.
18.Voinova, Z., S. Vlakhova and P. Petkova. 1970. Pochv. Agrokhim 4, 103-110.
19.Hulin, N. and M. Horowitz. 1973. Phytoparasitica 1, 64.
20.Pantera, H. 1972. Pam. Pulawski 51, 69-76.
21.Shkola, A.I. t1970. Vest. selKhoz Nauki, Alma-Ata 13, 34-37.
22.Lazareva, N.V. and V.G. Blokhin. 1969. Khimiya selKhoz 7, 609-611.
23.Goguadze, V.D. and L.N. Gogiya. 1971. Subtrop Kultury 4, 161-167.
24.Rankov, V., E. Elenkov, P. Surlekov and B. Velev. 1966. Agrokhimiya 4, 115-120.
25.Pantera, H. 1970. Meded. Fac. Landb. Rijksuniv. Gent. 35, 847-854.
26.Platonova, V.P. 1967. Trudy kirov. selKhoz. Inst 20, 215-221.
27.Anderson, J.R. 1978. In: I.R. Hill and S.J.L. Wright (eds.): Pesticide Microbiology. Academic Press.London. 313-533.

7-137 Effect of Prometryne + 2,4-D

Parameter	Effect Value	Concentration and Condition	Species
Growth rate and numbers	Increased population	Loam, 3.5 % OM, prometryne=1.0 followed by 2,4-D=0.9 kg/ha, in situ	Soil bacteria[1]

Source: 1.Geshtovt, Y.N., S.U. Zharasov and N. Adekenov. 1970. Vest. SelKhoz. Nauki. Kazakhskoi SSR. 13, 35-38

7-138 Effect of Prometryne and simazine

Parameter	Effect Value	Concentration and Condition	Species
Nitrification and nitrifiers	No effect	Light alluvial meadow, pH=7, tot 2 kg/ha, prom.(40 %)+simazine(15 %)	Soil bacteria[1]
Nitrification and nitrifiers	No effect	Heavy leached meadow cinnamon, pH=7, tot 2 kg/ha, prom.(40)+simazine(15 %)	Soil bacteria[1]

Source: 1.Rankov, V. 1968. Pochv. Agrokhim. 3, 81-90.

7-139 Effect of Pronanil

Parameter	Effect Value	Concentration and Condition	Species
Numbers and activity of soil algae	No adverse effect on N2-fixation	Paddy soil, 1-4 kg/ha, recomm. field rate	Herbicide applied to soil in situ[1]

Source: 1.Ibrahim, A.N. 1972. Symp. Biol. Hung. 111, 445-448.

7-140 Effect of Propachlor

Parameter	Effect Value	Concentration and Condition	Species
Growth rate and numbers	Decrease in populations	Degraded chernozem 7 kg/ha, 7 y annual appl., in situ	Soil bacteria[1]
Soil microbial ammonification	Annual application decreased numbers, especially in dry years	Heavy clay degraded chernozem, 1.0 kg/ha, 7 y	Soil organisms[1]

Source: 1.Husarova, M. 1972. Ved. Pr. vyzk. Ust. Kukur. Trnave, No 6, 45-69.

7-141 Effect of Propanil

Parameter	Effect Value	Concentration and Condition	Species
Denitrification in soil	Large decreased in population and activity	Paddy soil, 10.5 kg/ha	Soil organisms[1]
Nitrification and nitrifiers	Decreased population	West Bengal paddy soil, 3.14 and 5.3 kg/ha	Soil bacteria[1]
Numbers and activity of soil algae	Inhibited Chlorococcum aplanosporum and decreased numbers	Sandy and sandy loam, 0.1 and 0.25 kg/ha	Herbicide applied to soil in situ[2]

Source: 1.De, B.K. and S. Mukhopadhyay. 1971. Int. Rice Commn. Newsl. 20, 35-39.
2.Sharabi, N. El-Din. 1969. Influence of Propanil and related compounds on growth of the algae Chlorocccum aplanosporium in soil and in solution culture. Ph.D. Thesis. Rutgers Univ. New Brunswick. USA.

7-142 Effect of Propazine

Parameter	Effect Value	Concentration and Condition	Species
Nitrification and nitrifiers	Annual applications increased activity	Light loam dernopodzol, 2 kg/ha, 5 y	Soil bacteria[1]
Soil respiration activity	No effect	Heavy and light loam dernozem, 2 and 3 kg/ha	Soil organisms[2]

Source: 1.Chunderova, A.I., T.P. Zuberts and A.M. Sofinskii. 1971. Khimiya selKhoz 9, 527-530.2.Fisyunov, A.V. 1969. Agrkhimiya 2, 122-126.

7-143 Effect of Propham and diuron

Parameter	Effect Value	Concentration and Condition	Species
Soil respiration activity	No effect	Chernozem medium clay, pH=7.2, 1.5 % OM, recomm. field rate	Soil organisms[1]

Source: 1.Kulimska, D. 1967. Roczn. Nauk. roln. (A) 93, 125-130.

7-144 Effect of Propoxur

Parameter	Effect Value	Concentration and Condition	Species
Nitrification and nitrifiers	20 % - 30 % inhibition of activity	Sandy loam, pH=7.4, 0.5 kg/ha	Soil organisms[1]
Nitrification and nitrifiers	Inhibited activity, NO2(-) accumulation	Sandy silt-clay, pH=8.3, 0.67 % OM, 2.5 and 12.5 and 125 kg/ha	Soil organisms[2]
Soil respiration	Decreased CO2-evolution, after initial increase	Sandy silt-clay, pH=8.3, 0.67 % OM, 125 kg/ha	Soil organisms[2]
Soil respiration	Decreased CO2-evolution, but recovery	Sandy silt-clay, pH=8.3, 0.67 % OM, 2.5 and 12.5 kg/ha	Soil organisms[2]
Soil respiration	Increased O2-uotake due to presence of organic carrier	Sandy loam, pH=7.4, 0.5 and 50 kg/ha, little insecticide effect	Soil organisms[1]

Source: 1.Kuseske, D.W., B.R. Funke and J.T. Schultz. 1974. Pl. Soil 41, 255-269.
2.Gupta, K.G., R.K. Sud, P.K. Aggarwal and J.C. Aggarwal. 1975. Pl Soil 42, 317-325.

7-145 Effect of Pyrazon

Parameter	Effect Value	Concentration and Condition	Species
Denitrification in soil	Increased activity and population	Medium loam derno-podzol, 1.5 % OM, 4 kg/ha	Soil organisms[1]
Growth rate and numbers	Toxic effects observed	Solid media = 25000 ppm, in culture	Bacillus spp.[2]
Growth rate and numbers	Slight inhibition of growth rate	Solid media = 50000-100000 ppm, in culture	Pseudomonas sp.[2]

Growth rate and numbers	Decreased populations	Dark chestnut heavy loam, 4 % OM, 4.5 kg/ha, in situ	Soil bacteria[3]
Growth rate and numbers	Increased populations	Dark chestnut heavy loam, 4 % OM, 3 kg/ha, in situ	Soil bacteria[3]
Growth rate and numbers	Increased populations	Light loam dernopodzol, 3 kg/ha, 5 y annual appl., in situ	Soil bacteria[4]
Growth rate and numbers	No effect on growth	1.5 - 3.5 kg/ha, in situ	Soil bacteria[5]
Growth rate and numbers	Decrease in populations	15 - 35 kg/ha, in situ	Soil bacteria, spore formers[5]
Nitrification and nitrifiers	Annual application had no adverse effect	Light loam dernopodzol, 3 kg/ha, 5 years appl.	Soil bacteria[6]
Nitrification and nitrifiers	Decreased activity	Dark chestnut heavy loam, 4 % OM, 12-36 kg/ha, lab	Soil bacteria[3]
Nitrification and nitrifiers	Decreased activity	Pale chestnut medium loam, 4 % OM, 12-36 kg/ha, lab	Soil bacteria[3]
Nitrification and nitrifiers	Decreased population and zctivity	Black alluvial or sandy soils, 30-40 kg/ha, recomm. field rate	Soil bacteria[7]
Nitrification and nitrifiers	Increased activity	Medium podzolic loam, 1.9-2.5 % OM, 4 kg/ha	Soil bacteria[8]
Nitrification and nitrifiers	Increased activity	Derno-podzol light or medium soils, pH=6, 2-3 % OM, 4 kg/ha	Soil bacteria[9]
Nitrification and nitrifiers	Increased activity	Pale chest medium loam, 4 % OM, 3 and 4.5 kg/ha, field	Soil bacteria[3]
Nitrification and nitrifiers	Increased activity in field	Alluvial, 2 % OM, 1.6 and 3.2 kg/ha, no effect in lab	Soil bacteria[10]
Nitrification and nitrifiers	Increased activity in field	Dark chestnut heavy loam, 4 % OM, 3 and 4.5 kg/ha	Soil bacteria[3]
Nitrogen fixation in soil	No effect	Dark chestnut heavy loam, 4 % OM, 3 and 4.5 kg/ha	Soil bacteria, free living[3]
Numbers and activity of soil algae	Decreased activity and population	Alluvial black and sandy soils, 15-30 kg/ha, recomm. field rate	Herbicide applied to soil in situ[7]
Soil microbial ammonification	Decreased population	Podzolized light loam, 4.8 kg/ha	Soil organisms[11]
Soil microbial ammonification	Increased activity	Light or medium loam, 4 kg/ha	Soil organisms[6]
Soil respiration activity	Decreased activity	Silty medium loam derno-podzol, pH=5.9, 4 kg/ha	Soil organisms[12]
Soil respiration activity	No effect	Chernozem medium clay, pH=7.2, 1.5 % OM, 1.5-3 kg/ha	Soil organisms[13]

Source: 1.Karasavich, E.K. and L.I. Isaeva. 1969. Khimiya selKhoz 7, 452-453.
2.Zharasov, S.U. 1969. Vest. sel-Khoz. Nauki, Alma-Ata 12, 94-98.
3.Zharasov, S.U. 1972. Vest. selKhoz. Nauki. Kazakhskoi 15, 25-29.
4.Chunderova, A.I., T.P. Zubets and A.M. Sofinskii. 1971. Khimiya selKhoz. 9, 527-530.
5.Simon, L. 1971. Folia Microbiol. 16, 516.
6.Chunderova, A.I., T.P. Zuberts and A.M. Sofinskii. 1971. Khimiya selKhoz 9, 527-530.
7.Hauke-Pacewiczowa, T. 1971. Pam. Pulawski 46, 5-48.
8.Zhukova, P.S. and A.M. Botineva. 1974. Khimizat. sel'Khoz. 12, 53-56.
9.Chunderova, A.I., A.M. Sofinskii and T.P. Zubets. 1968. Khimiya selKhoz 6, 280-282.
10.Horowitz, M., T. Blumenfield, G. Herzlinger and N. Hulin. 1974. Weed Res. 14, 97-109.
11.Pranaitene, E.V. 1970. Mat. Pribalt. Sov. Zashch. Rast. Elgava 8-12.
12.Zinchenko, V.A. and T.V. Osinskaya. 1969. Agrokhimiya 9, 94-101.
13.Kulimska, D. 1967. Roczn. Nauk. roln. (A) 93, 125-130.

7-146 Effect of SO_2

Parameter	Effect Value	Concentration and Condition	Species
Foliar disease	Interveinnal necrotic blotches	0.70 ppm for 1 h or 0.18 ppm for 8 h or 0.008-0.017 for growing season	Angiosperms[1]
Foliar disease	Redbrown dieback or banding	0.70 ppm for 1 h or 0.18 ppm for 8 h or 0.008-0.017 for growing season	Gymnosperms[1]
Photosynthesis	Threshold dose for photosynthesis suppression	Air = 0.5 ppm, 7-11 h, 1-2 d exposure, tree sapling	Black oat[1]
Photosynthesis	Threshold dose for photosynthesis suppression	Air = 0.2 ppm, continuous, 2 weeks, tree seedling	Norway spruce and Scotch pine[1]
Photosynthesis	Threshold dose for photosynthesis suppression	Air = 1 ppm, 2-4 h, single treatment, tree seedling	Quaking aspen[1]
Photosynthesis	Threshold dose for photosynthesis suppression	Air = 6 ppm, 4-6 h, single treatment, tree seedling	Red maple[1]
Photosynthesis	Threshold dose for photosynthesis suppression	Air = 1 ppm, 30 min, single treatment, tree sapling	Silver maple[1]
Photosynthesis	Threshold dose for photosynthesis suppression	Air = 0.5 ppm, 7-11 h, 1-2 d exposure, tree sapling	Sugar maple[1]
Photosynthesis	Threshold dose for photosynthesis suppression	Air = 0.5 ppm, 7-11 h, 1-2 d exposure, tree sapling	White ash[1]
Photosynthesis	Threshold dose for photosynthesis suppression	Air = 0.1 ppm, continuous, 2 weeks, tree seedling	White fir[1]

Source: 1.Smith, W.H. 1981. Air Pollution and Forests. Springer Verlag. NY. 384 pp.

7-147 Effect of Sevidol

Parameter	Effect Value	Concentration and Condition	Species
Growth rate and numbers	No effect	Rice soils, 0.45-2.25 kg/ha	Soil bacteria[1]

Source: 1.Nair, K.S., C. Ramakrishnan and S. Sothananthan. 1973. Madras agric. J. 60, 441-443.

7-148 Effect of Siduron

Parameter	Effect Value	Concentration and Condition	Species
Nitrogen fixation in soil	Annual applications decreased Azotobacter population	Turfed soil, 11.25 kg/ha, 3 y	Soil bacteria, free living[1]
Numbers and activity of soil algae	Annual applications had no effect	Turfed soils, 3 y, 11.3 kg/ha	Herbicide applied to soil in situ[2]

Source: 1.Voets, J.P., M. Meerschman and W. Verstraete. 1974. Soil Biol. Biochem. 6, 149-152.
2.Fields, M.L. and D.D. Hemphill. 1968. Weed Sci. 16, 417-420.

7-149 Effect of Simazine

Parameter	Effect Value	Concentration and Condition	Species
Denitrification in soil	Annual application decreased activity	Orchard soil, 10 kg/ha, 3 y	Soil organisms[1]
Denitrification in soil	Decreased activity and population	Derno-podzol, 100 kg/ha	Soil organisms[2]
Denitrification in soil	Increased activity and population	Medium loam meadow soil, 1.2-2.2 % OM, 3 kg/ha	Soil organisms[3]
Denitrification in soil	Increased population	Leached chernozem, 1.5 kg/ha	Soil organisms[4]

Denitrification in soil	Increased population	Sandy clay and silty clay, 2 and 20 kg/ha	Soil organisms[5]
Denitrification in soil	No effect	Derno-podzol, 2-80 kg/ha	Soil organisms[2]
Growth rate and numbers	Especially increased population	Light loam dernopodzol, 0.4 kg/ha, in situ	Clostridium pasteurianum[6]
Growth rate and numbers	Decrease followed by increased population	Sandy soil, pH=7.2, 0.9 % OM, 50 kg/ha, in situ	Soil bacteria[7]
Growth rate and numbers	Decreased population	Red sandy loam, 0.5 kg/ha, in situ	Soil bacteria[8]
Growth rate and numbers	Decreased population	Calcerous heavy loam, pH=6.8-7.2, 3-4 kg/ha, selective effect, in situ	Soil bacteria[9]
Growth rate and numbers	Decreased population	Peaty-bog, pH=5.5, 47 % OM, dry years, 4.5 and 10 kg/ha, in situ	Soil bacteria[10]
Growth rate and numbers	Decreased population	Chernozem medium clay, pH=7.2, 1.5 % OM, 50 kg/ha, in situ	Soil bacteria[7]
Growth rate and numbers	Decreased population	30 and 40 kg/ha, in situ	Soil bacteria[11]
Growth rate and numbers	Decreased population	Red sandy loam, 1.0 and 10 kg/ha, in situ	Soil bacteria[12]
Growth rate and numbers	Increased population when N added	Loam with and without N fert., 5.2 kg/ha, in situ	Soil bacteria[13]
Growth rate and numbers	Increased population, especially Clostridium pasteurianum	Leached chernozem, low OM, 1.5 kg/ha, in situ	Soil bacteria[4]
Growth rate and numbers	Increased then decreased population	Peaty-bog, pH=5.5, 47 % OM, wet years, 4.5 and 10 kg/ha, in situ	Soil bacteria[10]
Growth rate and numbers	Initial increase then return to normal population	Forest soil, 1.0 and 3 and 5 kg/ha, in situ	Soil bacteria[14]
Growth rate and numbers	No effect	Medium clay loam, pH=5.3, 2 % humus, 2 and 6 kg/ha, in situ	Soil bacteria[15]
Growth rate and numbers	No effect	5-40 kg/ha, slightly increased population, in situ	Soil bacteria[16]
Growth rate and numbers	No effect	Chernozem medium clay, pH=7.2, 1.5 % OM, 2 kg/ha, in situ	Soil bacteria[7]
Growth rate and numbers	No effect	Sandy soil, pH=5.8, 0.9 % OM, 2 kg/ha, in situ	Soil bacteria[7]
Growth rate and numbers	No effect	1.1 kg/ha, in situ	Soil bacteria[11]
Growth rate and numbers	Sometimes increased population	Peaty-bog, pH=5.5, 47 % OM, wet years, 2nd appl., 4.5 and 10 kg/ha, in situ	Soil bacteria[10]
Growth rate and numbers	Strong inhibition	Liquid media=100 ppm, in culture	Soil bacteria[17]
Growth rate and numbers	Decreased population	Derno-podzol, 2-6 kg/ha (and higher up to 100), in situ	Soil bacteria, aerobic[2]
Growth rate and numbers	Increased population	Light loam dernopodzol, 0.4 kg/ha, in situ	Soil bacteria, butyric acid bacteria[6]
Growth rate and numbers	Decrease in population	Alluvial and terra rossa soils, 4-10 kg/ha, in situ	Soil bacteria, non-spore formers[18]

Growth rate and numbers	Decreased population	50-250 kg/ha, in situ	Soil bacteria, non-spore formers[16]
Growth rate and numbers	Decreased population	Irrigated brown soil, 8 kg/ha, in situ	Soil bacteria, non-spore formers[19]
Growth rate and numbers	No reduction in growth rate	Sod-podzol, 1.5-2.5 kg/ha, in situ	Soil bacteria, non-spore formers[20]
Growth rate and numbers	Decrease in population	Alluvial and terra rossa soils, 8 kg/ha, in situ	Soil bacteria, spore formers[18]
Growth rate and numbers	Decreased population	Sod-podzol, 1.5-2.5 kg/ha, in situ	Soil bacteria, spore formers[20]
Growth rate and numbers	Especially decreased population	50-250 kg/ha, in situ	Soil bacteria, spore formers[16]
Growth rate and numbers	Increased population	Irrigated brown soil, 8 kg/ha, in situ	Soil bacteria, spore formers[19]
Growth rate and numbers	Inhibition of growth rate	Solid+liquid media=5000-10000 ppm, non-spore formers less sensetive, in culture	Soil bacteria, spore formers[20]
Mortality	Toxic effects	Liquid media=>10000 ppm, surfactants increase effect, in culture	Rhizosphere bacteria[21]
Nitrification and nitrifiers	Decreased activity	Medium loam meadow bog, 4.0 kg/ha	Soil bacteria[22]
Nitrification and nitrifiers	Decreased activity	Moderate loam carbonate chernozem, 6 and 8 kg/ha	Soil bacteria[23]
Nitrification and nitrifiers	Decreased activity	Derno-podzolic loam, 100 kg/ha	Soil bacteria[2]
Nitrification and nitrifiers	Decreased activity and population	Black alluvial and sandy soils, 30-40 kg/ha, recomm. field rate	Soil bacteria[24]
Nitrification and nitrifiers	Decreased activity in lab	Medium loam meadow, 1.2-2.2 % OM, 3 kg/ha, no efftct in field	Soil bacteria[25]
Nitrification and nitrifiers	Increased activity	Podzol, 2 % humus, 1.0-4.0 kg/ha	Soil bacteria[15]
Nitrification and nitrifiers	Increased activity	Calcareius chernozem, 3 kg/ha	Soil bacteria[22]
Nitrification and nitrifiers	Increased activity	Medium loam meadow bog, 1.0 and 2 kg/ha	Soil bacteria[22]
Nitrification and nitrifiers	Increased activity	Alluvial sandy loam, pH=7.9, 0.5-2.0 kg/ha	Soil bacteria[3]
Nitrification and nitrifiers	Increased activity	Heavy loam, pH=6.3, 4-12 kg/ha	Soil bacteria[26]
Nitrification and nitrifiers	Increased activity	Medium loam chernozem, 0.2-3 kg/ha, with gamma-BHC and/or NPK	Soil bacteria[27]
Nitrification and nitrifiers	Increased activity	Red sandy loam, 1.0 kg/ha	Soil bacteria[12]
Nitrification and nitrifiers	Increased activity	Heavy loam, 8-12 kg/ha	Soil bacteria[28]
Nitrification and nitrifiers	Increased activity	Medium clay chernozem, pH=7.2, 1.5 % OM, 2 kg/ha	Soil bacteria[7]
Nitrification and nitrifiers	Increased activity	Derno-podzolic light loam, 2 kg/ha	Soil bacteria[29]
Nitrification and nitrifiers	Increased activity	Red sandy loam, pH=5.3, 1.0 and 10 kg/ha	Soil bacteria[30]

Nitrification and nitrifiers	Increased activity and population	Leach black calcareous clay, 15 kg/ha	Soil bacteria[31]
Nitrification and nitrifiers	Increased activity and population	Silty medium loam, pH=5.3, 4 % OM, 2 kg/ha	Soil bacteria[32]
Nitrification and nitrifiers	Increased heterotrophic nitrification	Loam, pH=4.9, 1.0 kg/ha	Soil bacteria[33]
Nitrification and nitrifiers	Inhibited activity	Sandy loam, 1.0 kg/ha	Soil bacteria[34]
Nitrification and nitrifiers	Inhibited activity	Loam, 3 and 6.9 kg/ha	Soil bacteria[35]
Nitrification and nitrifiers	Inhibited activity	Medium clay chernozem, pH=7.2, 1.5 % OM, 50 kg/ha	Soil bacteria[7]
Nitrification and nitrifiers	Little effect	Peaty bog, pH=5.5, 47 % OM, 4.5 and 10 kg/ha	Soil bacteria[28]
Nitrification and nitrifiers	No effect	Heavy loam chernozem, low OM, 2-3 kg/ha	Soil bacteria[36]
Nitrification and nitrifiers	No effect	Normal chernozem, 1-4 kg/ha	Soil bacteria[36]
Nitrification and nitrifiers	No effect	Sandy loam, 0.25 kg/ha	Soil bacteria[34]
Nitrification and nitrifiers	No effect	Red sandy loam, 10 kg/ha	Soil bacteria[12]
Nitrification and nitrifiers	No effect	Sod-podzol, 1.5-5 kg/ha	Soil bacteria[20]
Nitrification and nitrifiers	No effect	Loam, 1.0 kg/ha	Soil bacteria[35]
Nitrification and nitrifiers	No effect	Sandy and silty clay loams, 2 and 20 kg/ha	Soil bacteria[5]
Nitrification and nitrifiers	No effect	Derno-podzolic loam, 2-80 kg/ha	Soil bacteria[2]
Nitrification and nitrifiers	No effect	Light alluvial and heavy leached meadow cinnamom, pH=7, 2 % OM, 2 kg/ha	Soil bacteria[37]
Nitrification and nitrifiers	No significant effect	Silt loam, pH=6, 2 % OM, 1.2 and 5 kg/ha	Soil bacteria[38]
Nitrification and nitrifiers	No significant effect	Irrigated brown soil, 8 kg/ha	Soil bacteria[19]
Nitrification and nitrifiers	No significant effects	Sandy loam, 1-4 kg/ha, recomm. field rate	Soil bacteria[39]
Nitrification and nitrifiers	Temporarily decreased activity	Derno-podzolic light loam, 6 kg/ha	Soil bacteria[29]
Nitrogen fixation in soil	Decreased population in soil, but increased in rhizosphere of crop	Alluvial, 3 kg/ha	Soil bacteria, free living[40]
Nitrogen fixation in soil	Decreased soil population	Medium loam meadow soil, 3 kg/ha	Soil bacteria, free living[3]
Nitrogen fixation in soil	Decreased soil population	Alluvial, 3.0 kg/ha	Soil bacteria, free living[41]
Nitrogen fixation in soil	Decreased soil population	Chernozem, 200 kg/ha	Soil bacteria, free living[42]
Nitrogen fixation in soil	Decreased soil population	Irrigated brown soil, 8 kg/ha	Soil bacteria, free living[19]
Nitrogen fixation in soil	Increased N-fixing capacity of Beijerinckia spp.	Alluvial krasnozem, pH=7, 12-14 kg/ha	Soil bacteria, free living[43]

Nitrogen fixation in soil	Increased numbers and N-fixing capacity of Azotobacter	Smonitsa, 4.0 kg/ha	Soil bacteria, free living[44]
Nitrogen fixation in soil	Increased population	Alluvial, 10 kg/ha	Soil bacteria, free living[18]
Nitrogen fixation in soil	Increased zzz population	Calcareous heavy loam, pH=6.8-7.2, 3-4 kg/ha	Soil bacteria, free living[9]
Nitrogen fixation in soil	Increased Azotobacter sp. population	Chernozem, 4 kg/ha	Soil bacteria, free living[42]
Nitrogen fixation in soil	Inhibited Azotobacter sp.	Red sandy loam, 0.5 kg/ha	Soil bacteria, free living[8]
Nitrogen fixation in soil	Inhibited Azotobacter sp.	Acid soil, pH=5, 12-14 kg/ha	Soil bacteria, free living[43]
Nitrogen fixation in soil	Inhibited Azotobacter sp.	Chernozem, 4 kg/ha	Soil bacteria, free living[42]
Nitrogen fixation in soil	No effect	Chernozem, 3 kg/ha	Soil bacteria, free living[45]
Nitrogen fixation in soil	No effect on numbers of Azotobacter sp. and Clostridium sp.	Leached meadow-cinnamon and alluvial soils, 1-4 kg/ha, recomm. field rate	Soil bacteria, free living[46]
Nitrogen fixation in soil	No effect on numbers of Azotobacter sp. and Clostridium sp.	Leached meadow-cinnamon and alluvial soil, 1-4 kg/ha, recom field rate	Soil bacteria, free living[46]
Numbers and activity of soil algae	Decreased activity and population	Alluvial black and sandy, 10-40 kg/ha, recomm. field rate	Herbicide applied to soil in situ[24]
Numbers and activity of soil algae	Decreased population for up to 5 months	Sandy clay, 3-50 kg/ha	Herbicide applied to soil in situ[47]
Numbers and activity of soil algae	Decreased population, especially blue-green algae	Gleyed heavy loam, >8 % OM; peaty-podzol, 35 % OM, 5 kg/ha	Herbicide applied to soil in situ[48]
Numbers and activity of soil algae	No effect	Leached medium loam chernozem, 4 kg/ha	Herbicide applied to soil in situ[49]
Numbers and activity of soil algae	Some diatoms and Chlorococcum sp. tolerent, others decreased	Light sandy loam, >3 % OM, 5 kg/ha	Herbicide applied to soil in situ[48]
Soil microbial ammonification	Decreased activity	Medium loam, 1.2-2.2 % OM, 3 kg/ha	Soil organisms[10]
Soil microbial ammonification	Decreased activity after 10 months	Heavy loam, 4-8 and 8-12 kg/ha	Soil organisms[50]
Soil microbial ammonification	Increased activity	Silty medium loam, pH=5.3, 4 % OM	Soil organisms[20]
Soil microbial ammonification	Increased activity	Derno-podzol light or medium loams, pH=6, 2-3 % OM, 2 kg/ha	Soil organisms[51]
Soil microbial ammonification	Inhibited activity	Brown heavy loam, pH=6.3, 12 kg/ha	Soil organisms[28]
Soil microbial ammonification	Little effect	Peaty-bog, pH=5.5, 47 % OM, 4.5 and 10 kg/ha	Soil organisms[52]
Soil microbial ammonification	No effect	Sod-podzol, 1.5-2.5 kg/ha	Soil organisms[28]
Soil microbial ammonification	No effect	Derno-podzol, 2-30 kg/ha	Soil organisms[51]
Soil microbial ammonification	Temporarily decreased activity	Derno-podzol, 30-100 kg/ha	Soil organisms[51]

Soil respiration activity	Annual application decreased CO2-evolution	Sandy loam, 7.5 kg/ha, 7 y	Soil organisms[53]
Soil respiration activity	Decreased CO2-evolution	Sandy loams, 133 kg/ha	Soil organisms[54]
Soil respiration activity	Decreased activity	Silty medium loam derno-podzol, pH=5.9, 5 kg/ha	Soil organisms[55]
Soil respiration activity	Decreased activity	Red sandy loam, 1.0 and 10 kg/ha	Soil organisms[12]
Soil respiration activity	Increased activity	Heavy and light loam dernozem, 2 kg/ha	Soil organisms[36]
Soil respiration activity	Inhibited CO2-evolution, high organic matter decreased inhibition	Nine soils, variable OM, 0.5-10 kg/ha	Soil organisms[56]
Soil respiration activity	No effect	Heavy and light loam dernozem, 3 kg/ha	Soil organisms[36]
Soil respiration activity	No effect	Chernozem medium clay, pH=7.2, 1.5 % OM; sand, pH=5.8, 0.9 % OM, 2-50 kg/ha	Soil organisms[57]
Soil respiration activity	Transient decreased activity	Sandy, 1-4 kg/ha, recomm. field rate	Soil organisms[58]

Source: 1.Kurydina, T.I. 1965/ Sb. nauch. Rab. Inst. Sadov I.V. Michurina II, 113-116.
2.Kuzyakina, T.I. 1974. Weed Abstr. 24, 820.
3.Joshi, O.P., M.S. Sachdev, K.L. Sahrawat and B.N. Kohli. 1976. Pl. Soil 44, 367-375.
4.Tyagny-Ryadno, M.G. 1967. Trudy Kamenets-podolsk. selKhoz Inst. 9, 43-48.
5.Torstenson, L. 1974. Swedish J. agric. Res. 4, 151-160.
6.Kozyrev, M.A. and A.A. Laptev. 1972. Khimiya selKhoz. 10, 614-617.
7.Kulinska. 1967. Roczn. Nauk. roln. (A) 93, 229-262.
8.Spirama Raju, K. and G. Rangaswami. 1971. Indian J. Microbiol. 11, 25-29.
9.Bobyshev, V.G. and G.Y. Lapchenkov. 1966. Khimiya selKhoz 4, 30-31.
10.Nepomiluev, V.F. and T.I. Kuzyakina. 1967. Izv. timiryazev. selKhoz. Akad. No. 4, 84-91.
11.Singh, K. 1971. Pestic. India 6, 14-17.
12.Balasubbramaniam, A. and R. Sidaramappa. 1974. Mysore J. agric. Sci. 8, 214-219.
13.Fink, R.J., O.H. Fletchall and O. Calvert. 1968. Weed Sci. 16, 104-105.
14.Milkowska, A. and A. Gorzelak. 1966. Sylwan 110, 13-22.
15.Tsevetkova, S.D. 1966. Les. Khoz. 19, 25-26.
16.Simon, L. 1971. Folia Microbiol. 16, 516.
17.Krezel, Z. and M. Musial. 1969. Acta microbiol. pol. Ser B 1, 93-87.
18.Goguadze, V.D. 1969. Subtrop. Kultury No 4, 161-167.
19.Akopyan, E.A. and A.G. Agaronyan. 1968. Soils and Fert. 33, 3133.
20.Nepomiluev, V.F., I.P. Grechin and T.I. Kuzyakina. 1966. Izv. timiryazev sel-khoz Akad. No 2., 128-16.
21.Balicka, N., J. Sobieszczanski and T. Niewiadoma. 1969. Acta microbiiol. pol.1(18), 11-14.
22.Spiridinov, Y.Y., V. S. Skhiladze and G.S. Spiridinova. 1970. Subtrop. Kultury 3, 152-160.
23.Ivanov, A.I. 1974. Agrokhimiya 11, 113-115.
24.Hauke-Pacewiczowa, T. 1971. Pam. Pulawski 46, 5-48.
25.Tulabaev, B.D. and N. Azimbegov. 1967. Khimiya selKhoz 5, 43-44.
26.Zavarzin, V.I. 1966. Agrokhimiya 4, 121-124.
27.Kudzin, Y.K., A.Y. Fisyunov, N.A. Chernyavskaya and A.Y. Makarova.1973. Dokl. Vses. Akad. SelKhoz. Nuk. imeni V.I. Lenina 9, 13-15.
28.Zavarzin, V.I. and T.V. Belyaeva. 1966. Khimiya selKhoz 4, 34-36.
29.Bliev, Y.K. and L.M. Kozlova. 1972. Khimiya selKhoz 10, 850-853.
30.Balasubramanian, A., R. Sideramappa and G. Oblisami. 1973. Pesticides 7, 14.
31.Andreeva-Fetvadzhieva, N. and B. Kolcheva. 1969. Nauchni Trud. vissh selkostop. Inst. Georgi Dimitro20, 49-69.
32.Boiko, V.S., M.G. Degtyareva and I.P. Kazakova. 1969. Khimiya selKhoz 7, 203-207.
33.Smith, M.S. and C.S. Weeraratna. 1974. Trans. 10th Int. Congr. Soil Sci III, 173-178.
34.Nayyar, V.K., N.S. Randhawa and S.L. Chopra. 1970. Indian J. agric. Sci. 40, 445-451.
35.Farmer, F.H., R.E. Benoit and W.E. Chappell. 1965. Proc. N.E. Weed Control Conf. 19, 350-354.
36.Fisyunov, A.V. 1969. Agrkhimiya 2, 122-126.
37.Rankov, V. 1968. Pochv. Agrokhim. 3, 81-90.
38.Thorneburg, R.P. and J.A. Tweedy. 1973. Weed Sci. 21, 397-399.
39.Leiderman, J., J.M. Hinojo and F.A. Fogliata. 1971. Revta ind. agric. Tucuman 48, 7-14.
40.Micev, N. and M. Bubalov. 1969. Agrokem^. Talajt 18, 71-75.
41.Micev, N. and M. Bubalov. 1972. Symp. Biol. Hung. 11, 379-384.
42.Taha, S.M., S.A.Z. Mahmoud, A.M. Abdel-Hafez and A.S. Hamed. 1972. Egypt. J. Microbiol. 7, 53-61.
43.Goguadze, V.D. 1968. Agrokhmiya 5, 99-103.
44.Bakalivanov, D. and G. Nikolova. 1971. In: I. Pashev (ed.): Vtori Kongres po Mikrobi{ologiya. IV. Chst. Sofia. 1969. Bulg. Akad. Nauk 1971. 2>3-266.
45.Yurkevich, I.V. and N.Z. Tolkachev. 1972. Khimiya selKhoz 10, 696-698.
46.Rankov, V., E. Elenkov, P. Surlekov and B. Velev. 1966. Agrokhimiya 4, 115-120.
47.Pantera, H. 1970. Meded. Fac. Landb. Rijksuniv. Gent. 35, 847-854.
48.Mikhailova, E.I. and Kruglov. 1973. Pochvovedenie 8, 81-85.
49.Platonova, V.P. 1967. Zap. Voronezh. selKhoz Inst. 34, 41-44.
50.Chunderova, A.I., T.P. Zuberts and A.M. Sofinskii. 1971. Khimiya selKhoz 9, 527-530.

51.Pancholy, S.K. and J.a. Lynd. 1969. Weed Sci. 17, 460-463.
52.Boiku, V.S., M.G. Degtyareva and I.P. Kazakova. 1969. Khimiya selKhoz 7, 203-207.
53.Grossbard, E. 1971. Weed Res. 11, 263-275.
54.Marsh, J.A.P., H.A. Davies and E. Grossbard. 1974. Proc. 12th Br. Weed Control. Conf. 53-58.
55.Zinchenko, V.A. and T.V. Osinskaya. 1969. Agrokhimiya 9, 94-101.
56.Chandra, P., W.R. Furtick and W.B. Bollen. 1960. Weeds 8, 589-598.
57.Kulimska, D. 1967. Roczn. Nauk. roln. (A) 93, 125-130.
58.Gowda, T.K.S. and R.B. Patil. 1972. Biochem. J. 128, 56-57.

7-150 Effect of Simazine and 2,3,6-TBA

Parameter	Effect Value	Concentration and Condition	Species
Nitrogen fixation in soil	Increased soil population	Pale chestnut soils, simazine(3)+TBA(1 kg/ha)	Soil bacteria, free living[1]

Source: 1.Malanina, E.I. and P.D. Nikitin. 1971. Vest. selKhoz Nauki 10, 122-124.

7-151 Effect of Simazine and MCPA and aminotriazole

The target organism and reference number are in the 4. colu

Parameter	Effect Value	Concentration and Condition	Species
Soil respiration activity	Decreased CO2-evolution	Orchard soils, total = 15 kg/ha	Soil organisms[1]

Source: 1.Teuteberg, A. 1968. Z. PflKrankh. PflSchutz 75, 72-86.

7-152 Effect of Simazine and aminotriazole

Parameter	Effect Value	Concentration and Condition	Species
Numbers and activity of soil algae	Decreased population, but recovery within 3 months	Light loam, tot=13.9 kg/ha, simazine(13.9 %)+amino.(34 %)	Herbicide applied to soil in situ[1]
Soil respiration activity	Decreased CO2-evolution	Orchard soils, total = 10 kg/ha	Soil organisms[2]
Growth rate and numbers	*Decreased population*	*Apple orchard soil, 10 kg/ha, in situ*	*Soil bacteria*[3]

Source: 3.Teuteberg, A. 1968. Z. PflKrankh. PflSchutz 75, 72-86.
Source: 1.Kiss, A. 1967. Agrokem. Talajt. 16, 11.
2.Husarova, M. 1972. Ved. Pr. vyzk. Ust. Kukur. Trnave, No 6, 45-69.

7-153 Effect of Simazine and terbutryne

Parameter	Effect Value	Concentration and Condition	Species
Nitrogen fixation in soil	Slight inhibition of population in soil, but increase in rhizosphere of crop	Alluvial, simazine(1.1)+terbutryne(1.5 kg/ha)	Soil bacteria, free living[1]

Source: 1.Micev, N. and M. Bubalov. 1969. Agrokem. Talajt 18, 71-75.

7-154 Effect of Sodium cacodylate

Parameter	Effect Value	Concentration and Condition	Species
Growth rate and numbers	Temporarily increased population	Alluvial loam, 10.2 and 30.7 kg/ha, in situ	Soil bacteria[1]

Source: 1.Malone, C.R. 1971. Soil Biol. Biochem. 3, 127-131.

7-155 Effect of Sodium chlorate

Parameter	Effect Value	Concentration and Condition	Species
Denitrification in soil	Large decreased in population	Acid sandy soil, low OM, 67.5 kg/ha	Soil organisms[1]
Denitrification in soil	No effect	Fallow soil or turfed soil, 150 kg/ha	Soil organisms[2]
Denitrification in soil	Retarded or inhibited activity	Acidic sandy soil, low OM, 6.75 + 20.3 + 67.5 kg/ha	Soil organisms[3]
Denitrification in soil	Small decreased in population	Acid sandy soil, low OM, 6.75 and 20.3 kg/ha	Soil organisms[1]
Nitrification and nitrifiers	Inhibited Nitrobacter sp.	Various soils, lab, 6.75-150 kg/ha	Soil bacteria[1]
Nitrification and nitrifiers	No effect	Sandy loam, field study, 150 kg/ha	Soil bacteria[2]
Soil microbial ammonification	No effect	Grassed or fallow acid sandy soils, 150 kg/ha	Soil organisms[4]
Soil respiration activity	Decreased CO2-evolution	Fallow or turfed acid sandy soil, 150 kg/ha	Soil organisms[2]
Soil respiration activity	Increased CO2-evolution	Acid sandy soil + lucerne powder, 67.5 kg/ha	Soil organisms[3]

Source: 1.Karki, A.B., P. Kaiser and J.J. Pochon. 1972. C.R. hebd. Seanc. Acad. Sci., Paris, Sci. Nat. 274D, 209-2814.
2.Karki, A.B., L. Coupin, P. Kaiser and M. Moussin. 1973. Rev. Ecol. Biol. Sol 10, 3-11.
3.Karki, A.B. and P. Kaiser. 1974. Rev. Ecol. Biol. Sol 11, 477-498.
4.Barona, V.P. 1974. Agrokhimiya 11, 131-134.

7-156 Effect of TCA

Parameter	Effect Value	Concentration and Condition	Species
Denitrification in soil	Inhibited activity	Leached chernozem, 15 kg/ha	Soil organisms[1]
Denitrification in soil	No effect	Leached chernozem, 5 kg/ha	Soil organisms[1]
Growth rate and numbers	No effect no growth	Solid media = 500-50000 ppm, in culture	Pseudomonas spp.[2]
Growth rate and numbers	Inhibition in growth	Clay soil, anaerobic, 1.0-10 kg/ha, CaCO3 reduces effect, in situ	Sb, methane producing[3]
Growth rate and numbers	Increase in population	Chernozem, 3 % OM, 5 kg/ha, drop in No. of C pasteurianum, in situ	Soil bacteria[4]
Growth rate and numbers	Increase in population	Pale chestnut light loam, pH=8, 2 % OM, 18 kg/ha, in situ	Soil bacteria[5]
Growth rate and numbers	No effect on growth	15-30 kg/ha, slightly increase for low dose, in situ	Soil bacteria[6]
Growth rate and numbers	Decrease in population	Light brown soil, 30-360 kg/ha, in situ	Soil bacteria, spore formers[7]
Growth rate and numbers	Decrease in population	150 - 300 kg/ha, recomm. filed rate, in situ	Soil bacteria, spore formers[6]
Nitrification and nitrifiers	Decreased then increased population	Leached chernozem, low OM, 15 kg/ha	Soil bacteria[8]
Nitrification and nitrifiers	Increased activity	Pale chestnut light loam, pH=8, 2 % OM, 18 kg/ha	Soil bacteria[5]
Nitrification and nitrifiers	Inhibited activity	Pale chestnut medium loam, 198 kg/ha	Soil bacteria[9]

Nitrification and nitrifiers	No effect	Sandy loam, 15 - 30 kg/ha, recomm. field rate	Soil bacteria[10]
Nitrification and nitrifiers	No effect	Leached chernozem, low OM, 5 kg/ha	Soil bacteria[8]
Nitrification and nitrifiers	No effect	Pale chestnut medium loam, 18 kg/ha	Soil bacteria[9]
Nitrogen fixation in soil	Decreased population of azotobacters	Leached dernozem, 15 kg/ha	Soil bacteria, free living[1]
Nitrogen fixation in soil	No effect	Chernozem, 6 kg/ha	Soil bacteria, free living[11]
Nitrogen fixation in soil	No effect	Leached dernozem, 5 kg/ha	Soil bacteria, free living[1]
Nitrogen fixation in soil	No effect	Pale chestnut light loam, pH=8, 18 kg/ha, 2 % OM	Soil bacteria, free living[12]
Numbers and activity of soil algae	Decreased population	Light clay, 50 and 75 kg/ha	Herbicide applied to soil in situ[13]
Soil microbial ammonification	Increased population	Podzolic light loam, 25 kg/ha	Soil organisms[14]
Soil respiration activity	CO2 and CH4-evolution inhibited, especially at high rate	Anaerobic clay, 1.0 and 10 kg/haSoil organisms[3]	

Source: 1.Lobanov, V.E. and L.P. Poddubnaya. 1967. Khimiya selKhoz. 5, 32-34.
2.Zharasov, S.U. 1969. Vest. sel-Khoz. Nauki, Alma-Ata 12, 94-98.
3.Laskowski, D.A. and F.E. Broadbent. 1970. Proc. Soil Sci. Soc. Am. 34, 72-77.
4.Barona, V.P. 1974. Agrokhimiya 11, 131-134.
5.Chukalov, S.A. and S.U. Zharasov. 1971. Vest. selKhoz. Nauki. Alma-Ata 27, 35-40.
6.Simon, L. 1971. Folia Microbiol. 16, 516.
7.Zhakharyan, S.V. 1970. Symp. Biol. Hung. 11, 391-395.
8.Anderson, J.R. 1978. In: I.R. Hill and S.J.L. Wright (eds.): Pesticide Microbiology. Academic Press.London. 313-533.
9.Zharasov, S.U. 1972. Vest. selKhoz. Nauki. Kazakhskoi 15, 25-29.
10.Leiderman, J., J.M. Hinojo and F.A. Fogliata. 1971. Revta ind. agric. Tucuman 48, 7-14.
11.Yurkevich, I.V. and N.Z. Tolkachev. 1972. Khimiya selKhoz 10, 696-698.
12.Chulakov, S.A. and S.U. Zharasov. 1967. Vest. selKhoz. Nauki, Alma-Ata 10, 122-124.
13.Pantera, H. 1970. Meded. Fac. Landb. Rijksuniv. Gent. 35, 847-854.
14.Pranaitene, E.V. 1970. Mat. Pribalt. Sov. Zashch. Rast. Elgava 8-12.

7-157 Effect of Tendrin

Parameter	Effect Value	Concentration and Condition	Species
Growth rate and numbers	Numbers decreased then increased and reverted to normal	Sandy loam, calcareous clay and clay, 22 kg/ha	Soil bacteria[1]

Source: 1.Gawaad, A.A.A., M.H. Hammad and F.H. El-Gayar. 1973. Agrokem. Talajt. 22, 161-168.

7-158 Effect of Terbacil

Parameter	Effect Value	Concentration and Condition	Species
Soil respiration activity	Small to moderate increased activity	Sandy loam and 2 clay soils, 1.0 and 10 and 25 kg/ha	Soil organisms[1]

Source: 1.Lui, L.C. and H.R. Cibes-Viade. 1972. J. Agric. Univ. P. Rico 56, 417-425.

7-159 Effect of Terbutryne

Parameter	Effect Value	Concentration and Condition	Species
Growth rate and numbers	Increased aerobic spore formers	Alluvial, 2.5 kg/ha, in situ	Soil bacteria[1]
Nitrification and nitrifiers	No effect	Alluvial or bog-alluvial, 2.25 - 6.75 kg/ha	Soil bacteria[2]
Nitrogen fixation in soil	No effect	Alluvial or bog-alluvial soils, 2.25 and 4.5 kg/ha	Soil bacteria, free living[2]
Nitrogen fixation in soil	Sometimes inhibited soil population	Alluvial or bog-alluvial soils, 6.75 kg/ha	Soil bacteria, free living[2]
Soil microbial ammonification	Inhibited activity	Alluvial, terbutryne 50 % = 2.5 kg/ha	Soil organisms[1]

Source: 1.Micev, N. and M. Bubalov. 1972. Symp. Biol. Hung. 11, 379-384.
2.Mickovski, M.D. 1967. Ann. Fac. Agric. Univ. Skopje 20, 43-54.

7-160 Effect of Thionazin

Parameter	Effect Value	Concentration and Condition	Species
Nitrification and nitrifiers	No effect or slight inhibition	Sandy loam, 1.0 and 10 kg/ha	Soil organisms[1]
Soil bacteria, growth rate and numbers	Numbers decreased initially then recovered	Sandy loam, 1.0 and 10 kg/ha	Soil bacteria[1]
Soil respiration	Increased activity	Sandy loam, 1.0 and 10 kg/ha	Soil organisms[1]

Source: 1.Tu, C.-M. 1969. Bact. Proc. , 1-3.

7-161 Effect of Toxaphene

Parameter	Effect Value	Concentration and Condition	Species
Nitrification and nitrifiers	Slight stimulation in moist soil, but numbers depressed in drier soil	Loam, pH=7.2, 600-6000 kg/ha	Soil organisms[1]

Source: 1.Naumann, K. 1971. Pedobiologia 11, 286-295.

7-162 Effect of Trace metals

Parameter	Effect Value	Concentration and Condition	Species
Foliar disease	Interveinal chlorosis, tip and margin necrosis	Air, Variable or undetermined air concentration	Angiosperms[1]

Source: 1.Smith, W.H. 1981. Air Pollution and Forests. Springer Verlag. NY. 384 pp.

7-163 Effect of Trichloronate

Parameter	Effect Value	Concentration and Condition	Species
Growth rate and numbers	Numbers decreased initially then recovered	Sandy loam, 1.0 and 10 kg/ha	Soil bacteria[1]
Nitrification and nitrifiers	No effect or slight inhibition	Sandy loam, 1.0 and 10 kg/ha	Soil organisms[1]

Source: 1.Tu, C.-M. 1969. Bact. Proc. , 1-3.

7-164 Effect of Trichlorphon

Parameter	Effect Value	Concentration and Condition	Species
Growth rate and numbers	Increased numbers	1.3 kg/ha	Soil bacteria[1]

Source: 1.Yurovskaya, Y.M. and V.A. Zhulinskaya. 1974. Khimiya selKhoz 12, 38-41.

7-165 Effect of Trietazien

Parameter	Effect Value	Concentration and Condition	Species
Growth rate and numbers	Decreased population	Leached medium loam chernozem, pH=6.0, 4 % om, 5 kg/ha, in situ	Soil bacteria[1]

Source: 1.Mikhailova, M.F. 1968. Khimiya selKhoz 6, 204-206.

7-166 Effect of Trietazine

Parameter	Effect Value	Concentration and Condition	Species
Nitrification and nitrifiers	Increased activity	Leached medium loam chernozem, pH=6, 4 % OM, 5 kg/ha	Soil bacteria[1]
Nitrogen fixation in soil	Increased Azotobacter sp. population	Leached medium loam chernozem, pH=6, 4 % OM, 5 kg/ha	Soil bacteria, free living[1]
Soil microbial ammonification	Inhibited activity	Brown heavy loam, pH=6.3, 8 kg/ha	Soil organisms[2]

Source: 1.Mikhailova, M.F. 1968. Khimiya selKhoz 6, 204-206.
2.Zavarzin, V.I. and T.V. Belyaeva. 1966. Khimiya selKhoz 4, 34-36.

7-167 Effect of Trifluralin

Parameter	Effect Value	Concentration and Condition	Species
Growth rate and numbers	Decrease in populations	Sandy loam, 1.1 kg/ha, 5 y annual appl., in situ	Soil bacteria[1]
Nitrification and nitrifiers	Decreased activity	Leached smonitza, 0.45-1.8 kg/ha	Soil bacteria[2]
Nitrification and nitrifiers	Decreased activity	Bog soil, 3 kg/ha	Soil bacteria[3]
Nitrification and nitrifiers	Decreased activity, but less in heavy textured soils	0.45-1.8 kg/ha, alluvial meadow, leached meadow cinnamon, humus CaCO3	Soil bacteria[2]
Nitrification and nitrifiers	Increased activity	Alluvial meadow, 0.45-1.8 kg/ha	Soil bacteria[2]
Nitrification and nitrifiers	Increased activity in field	Alluvial soil, 2 % OM, 0.67 and 1.34 kg/ha, no effect in lab	Soil bacteria[4]
Nitrification and nitrifiers	No effect	Silt loam, pH=6.0, 2 % OM, 0.6 and 1.2 kg/ha	Soil bacteria[5]
Nitrification and nitrifiers	Severely decreased activity	Calcareous loam,	Soil bacteria[6]
Nitrification and nitrifiers	Toxic to nitrifiers	Podzolic loamy chernozem, 1.0-6.0 kg/ha	Soil bacteria[7]
Nitrogen fixation in soil	No effect	Loamy soil, 1-6 kg/ha	Soil bacteria, free living[7]
Soil microbial ammonification	Increased activity	Alluvial meadow, 1-4 kg/ha	Soil organisms[8]
Soil microbial ammonification	Inhibited activity	Alluvial meadow, 2-1000 kg/ha	Soil organisms[9]
Soil microbial ammonification	Less than normal inhibition	Heavy textured, high OM, 2-1000 kg/ha	Soil organisms[9]
Soil microbial ammonification	No effect	Calcareous loam, 11300 kg/ha	Soil organisms[6]
Soil respiration activity	Decreased O2-uptake	Fine sandy loam, 15 kg/ha	Soil organisms[10]
Soil respiration activity	Decreased O2-uptake	Clay, 1.0 kg/ha	Soil organisms[11]

Soil respiration activity	Inhibition	Calcareous loam, 11300 kg/ha, deposition of herbicide	Soil organisms[6]
Soil respiration activity	Respiration of sclerotium rolfsii decreased	Sandy loam, 0.6 kg/ha	Soil organisms[12]

Source: 1.Breazeale, F.W. and N.D. Camper. 1970. Appl. Microbiol. 19, 379-380.
2.Rankov, V. 1973. Pochv. Agrokhim. 8, 125-134.
3.Tyunyaeva, G.N., A.K. Minenko and L.A. Penkov. 1974. Agrokhimiya No 6, 110-114.
4.Horowitz, M., T. Blumenfield, G. Herzlinger and N. Hulin. 1974. Weed Res. 14, 97-109.
5.Thorneburg, R.P. and J.A. Tweedy. 1973. Weed Sci. 21, 397-399.
6.Naidu, S.M. 1972. Diss. Abstr. Int. B, 33, 2894.
7.Hamdi, Y.A. and M.S. Tewfik. 1969. Acta microbiol. pol. Ser B 1(18), 53-57.
8.Rankov, V. and E. Elenkov. 1970. Pochv. Agrokhim. 5, 127-136.
9.Rankov, V. 1971. Blarskata Akad. Nauk (1971), 253-257.
10.Withworth, J.W., W. Garner and B.C. Williams. 1965. Res. Prog. Rept. West. Weed Contr. Conf. 1965, 17-118.
11.Lui, L.C. and H.R. Cibes-Viade. 1972. J. Agric. Univ. P. Rico 56, 417-425.
12.Curl, E.A., R. Rodriguez-Kabana and H.H. Funderburk. 1967. wPhytopathology 57, 7.

7-168 Effect of VCS 438 and dicamba

Parameter	Effect Value	Concentration and Condition	Species
Nitrogen fixation in soil	Little to no effect	Alluvial, VCS 438 (1)+dicamba(0.5 kg/ha)	Soil bacteria, free living[1]
Soil microbial ammonification	Decreased population initially, increased population later	Rhizosphere soil of maize, VCS438(1) + dicamba(0.5 kg/ha)	Soil organisms[1]
Growth rate and numbers	Inhibited population	Rhizosphere soil of maize, VCS-438=1 and dicamba=0.5 kg/ha, in situ	Soil bacteria, spore formers[1]

Source: 1.Micev, N. and M. Bubalov. 1969. Godisen Zb. zemjod.-sum. Fak. Univ. Skopje 22, 19-24.

7-169 Effect of Zectran

Parameter	Effect Value	Concentration and Condition	Species
Soil respiration	No effect on CO2-evolution	Forest soil with or without litter, 0.17 kg/ha	Soil organisms[1]

Source: 1.Bollen, W.B., K.C. Lu and R.F. Terrant. 1970. U.S. Dept. Agr. Forest Serv. Res. Note PNW-124, 10 p

7-170 Effect of Zn

Parameter	Effect Value	Concentration and Condition	Species
Growth rate	Minimum concentration with toxic effects	Soil = 100 ppm Zn, isolate from non-Zn-contaminated soil	Actinomycetes[1]
Growth rate	Minimum concentration with toxic effects	Soil = 600-2000 ppm Zn, isolate from Zn-contaminated soil	Actinomycetes[1]
Growth rate	Minimum concentration with toxic effects	Soil = 200 ppm Zn, isolate from non-Zn-contaminated soil	Bacillus sp.[1]
Growth rate	Minimum concentration with toxic effects	Soil = 100-200 ppm Zn, isolate from Zn-contaminated soil	Bacillus sp.[1]
Growth rate	Minimum concentration with toxic effects	Soil = 100-1000 ppm Zn, isolate from non-Zn-contaminated soil	Fungi[1]
Growth rate	Minimum concentration with toxic effects	Soil = 100-2000 ppm Zn, isolate from Zn-contaminated soil	Fungi[1]
Growth rate	Minimum concentration with toxic effects	Soil = 100-200 ppm Zn, isolate from non-Zn-contaminated soil	Non-spore-forming bacteria[1]
Growth rate	Minimum concentration with toxic effects	Soil = 600-2000 ppm Zn, isolate from Zn-contaminated soil	Non-spore-forming bacteria[1]

Source: 1.Jordan, N.G. and M.P. Lechevalier. 1975. Can. J. Microbiol. 21, 1855-1865.

7-171 Effect of p-Chloroformanilide

Parameter	Effect Value	Concentration and Condition	Species
Growth rate and numbers	Increase of populations	Grey serozem, MHC=65 %, 20-50 kg/ha, in situ	Soil bacteria[1]
Growth rate and numbers	No effect	Soil conc up to 33 ppm, in culture	Soil bacteria[2]
Growth rate and numbers	No effect on growth	Grey serozen, MHC=65 %, 50-100 kg/ha, in situ	Soil bacteria[1]

Source: 1.Khikmatov, A., A. Rakhimov and Z. Muslimov. 1969. Agrokhimiya 6, 111-112.
2.Khikmatov, A., A. Rakhimov and Z. Muslimov. 1969. Agrokhimiya 6, 111-112.

7-172 Effect of 2,4,5-T

Parameter	Effect Value	Concentration and Condition	Species
Denitrification in soil	No effect	Sandy clay and silty clay, 4 and 40 kg/ha	Soil organisms[1]
Growth rate and numbers	Moderate inhibition	Solid+liquid media=200 ppm, recomm. filed rate, in culture	Bdellovibrio bacteriovorus[2]
Nitrification and nitrifiers	Decreased activity	Calcerous loam, dispocal of herbicide, 11300 kg/ha	Soil bacteria[3]
Nitrification and nitrifiers	Decreased activity	Sandy clay and silty clay, 40 kg/ha	Soil bacteria[1]
Nitrification and nitrifiers	No effect	Sandy and clay soils, 2-20 kg/ha, recomm. field rate	Soil bacteria[4]
Nitrification and nitrifiers	No effect	Sandy clay and silty clay, 4 kg/ha	Soil bacteria[1]
Soil microbial ammonification	No effect	Calcareous loam, 11300 kg/ha	Soil organisms[5]
Soil respiration activity	Inhibited activity	Calcareous loam, 11300 kg/ha, deposition of herbicide	Soil organisms[3]

Source: 1.Torstenson, L. 1974. Swedish J. agric. Res. 4, 151-160.
2.Wehr, N.B. and D.A. Klein. 1971. Soil Biol. Biochem. 3, 143-149.
3.Naidu, S.M. 1972. Diss. Abstr. Int. B, 33, 2894.
4.Lozano-Calle, J.M. 1970. Meded. Fac. Landb. Rijsuniv. Gent 35, 599-614.
5.Malanina, E.I. and P.D. Nikitin. 1971. Vest. selKhoz Nauki 10, 122-124.

7-173 Effect of 2,4-D

Parameter	Effect Value	Concentration and Condition	Species
Denitrification in soil	Increased population	Leached chernozem, 1.5 kg/ha	Soil organisms[1]
Growth rate and numbers	Unaffected until stationary phase	Liquid media=2.2-3.3 ppm, decrease in No in stationary phase, in culture	Aerobacter sp.[2]
Growth rate and numbers	Sensitive to compound	Liquid media=0.2-1.1 ppm, aerobic and anaerobic test, in culture	Bacillus megaterium[3]
Growth rate and numbers	Sensitive to compound	Liquid media=0.2-1.1 ppm, aerobic and anaerobic test, in culture	Bacillus subtilus[3]
Growth rate and numbers	Moderate inhibition	Solid+liquid media=200 ppm, recomm. filed rate, in culture	Bdellovibrio bacteriovorus[4]
Growth rate and numbers	Unaffected until stationary phase	Liquid media=2.2-3.3 ppm, decrease in No in stationary phase, in culture	Clostridium sp.[2]
Growth rate and numbers	Sensitive to compound	Liquid media=0.2-1.1 ppm, aerobic and anaerobic test, in culture	Corynebacterium fasciens[3]
Growth rate and numbers	Unaffected until stationary phase	Liquid media=2.2-3.3 ppm, decrease in No in stationary phase, in culture	Escherichia coli[2]

Growth rate and numbers	Decrease in population	Sandy loam, 5 appl of 4.5 kg/ha in 4 y, in situ	Soil bacteria[5]
Growth rate and numbers	Decreased population	30 and 40 kg/ha, in situ	Soil bacteria[6]
Growth rate and numbers	Decreased spore formers and bacterial activity	Peaty-bog, pH=5.5, 47 % OM, 1-2 kg/ha, in situ	Soil bacteria[7]
Growth rate and numbers	No effect on growth	Leached chernozem, low OM, 1.5 kg/ha, in situ	Soil bacteria[1]
Growth rate and numbers	No effect on growth	1.1 kg/ha, in situ	Soil bacteria[6]
Growth rate and numbers	No permanent effects	Loam, 1-2 kg/ha, 11 y annual appl., in situ	Soil bacteria[8]
Growth rate and numbers	Slightly increased population	Light loam dernopodzol, 0.5-0.75 kg/ha, in situ	Soil bacteria[9]
Nitrification and nitrifiers	Increased activity	Normal chernozem, 1.5 kg/ha	Soil bacteria[10]
Nitrification and nitrifiers	Increased activity	Silty medium loam, pH=5.3, 4 % OM, 2 kg/ha	Soil bacteria[11]
Nitrification and nitrifiers	Increased activity	Leached chernozem, 1.5 kg/ha	Soil bacteria[1]
Nitrification and nitrifiers	Increased activity	Heavy loam, 1.5 kg/ha	Soil bacteria[12]
Nitrification and nitrifiers	Increased activity	Silty medium loam, pH=5.3, 4 % OM, 0.75 kg/ha	Soil bacteria[11]
Nitrification and nitrifiers	Increased population	Derno-podzol, 1.5-15 kg/ha	Soil bacteria[13]
Nitrification and nitrifiers	No effect	Sandy loam, 1.0 - 2.0 kg/ha	Soil bacteria[14]
Nitrification and nitrifiers	No effect	Medium loam chernozem, 0.2 kg/ha	Soil bacteria[15]
Nitrification and nitrifiers	No effect	Sandy and clay soils, 1.5-15 kg/ha, recomm. field rate	Soil bacteria[16]
Nitrification and nitrifiers	No effect from annual applications	Light loam dernopodzol, 5 y, 0.5-0.75 kg/ha	Soil bacteria[17]
Nitrogen fixation in soil	Decreased population of Azotobacter sp.	Chernozem, 100 kg/ha	Soil bacteria, free living[18]
Nitrogen fixation in soil	Increased population of Azotobacter sp.	Chernozem, 2 kg/ha	Soil bacteria, free living[18]
Nitrogen fixation in soil	Increased Azotobacter sp. population	Calcareous heavy loam, pH=6.8-7.2, 0.7 kg/ha	Soil bacteria, free living[19]
Nitrogen fixation in soil	No effect	Chernozem, 1.5 kg/ha	Soil bacteria, free living[20]
Numbers and activity of soil algae	Decreased population	Leached medium loam chernozem, 5 kg/ha	Herbicide applied to soil in situ[21]
Numbers and activity of soil algae	No effect	Leached medium loam chernozem, 2 kg/ha	Herbicide applied to soil in situ[22]
Soil microbial ammonification	Increased activity	Silty medium loam, pH=5.3, 4 % OM, 0.75 kg/ha	Soil organisms[23]
Soil microbial ammonification	Increased activity	Alluvial, 1-2 kg/ha	Soil organisms[24]

Soil microbial ammonification	Increased population	Derno-podzol, 1.5-15 kg/ha	Soil organisms[13]
Soil microbial ammonification	No effect	Leached chernozem, low om, 1.5 kg/ha	Soil organisms[1]
Soil respiration activity	Decreased O2-uptake	Clay, 1.0 kg/ha	Soil organisms[25]
Soil respiration activity	Decreased activity	Silty, medium loam derno-podzol, pH=5.9, 2 kg/ha	Soil organisms[26]
Soil respiration activity	Increased activity	Alluvial, 1-2 kg/ha	Soil organisms[24]

Source: 1.Tyagny-Ryadno, M.G. 1967. Trudy Kamenets-podolsk. selKhoz Inst. 9, 43-48.
2.Lamartinere, C.A., L.T. Hart and A.D. Larson. 1969. Bull. Environ. Contam. Toxicol. 4, 113-119.
3.Hart, L.T. and A.D. Larson. 1966. Bull. Environ. Contam. Toxicol. 1, 108-120.
4.Wehr, N.B. and D.A. Klein. 1971. Soil Biol. Biochem. 3, 143-149.
5.Khikmatov, A., A. Rakhimov and Z. Muslimov. 1969. Agrokhimiya 6, 111-112.
6.Singh, K. 1971. Pestic. India 6, 14-17.
7.Hoffman-Kakol, I. and H. Kwarta. 1972. Zesz. nauk. Akad. Rolnicz. Szczecin. No 38, 91-103.
8.Resz. A. 1968. Sonderh. Z. Plfkrankh. PflPath. PflSchutz No 4, 147-150.
9.Chunderova, A.I., T.P. Zubets and A.M. Sofinskii. 1971. Khimiya selKhoz. 9, 527-530.
10.Fisyunov, A.V. 1969. Agrkhimiya 2, 122-126.
11.Boiko, V.S., M.G. Degtyareva and I.P. Kazakova. 1969. Khimiya selKhoz 7, 203-207.
12.Fisyunov, A.V. 1969. Agrokhimiya 3, 112-116.
13.Abueva, A.A. 1970. Khimiya selKhoz 8, 601-604.
14.Leiderman, J., J.M. Hinojo and F.A. Fogliata. 1971. Revta ind. agric. Tucuman 48, 7-14.
15.Kudzin, Y.K., A.Y. Fisyunov, N.A. Chernyavskaya and A.Y. Makarova.1973. Dokl. Vses. Akad. SelKhoz. Nuk. imeni V.I. Lenina 9, 13-15.
16.Lozano-Calle, J.M. 1970. Meded. Fac. Landb. Rijsuniv. Gent 35, 599-614.
17.Chunderova, A.I., T.P. Zuberts and A.M. Sofinskii. 1971. Khimiya selKhoz 9, 527-530.
18.Taha, S.M., S.A.Z. Mahmoud, A.M. Abdel-Hafez and A.S. Hamed. 1972. Egypt. J. Microbiol. 7, 53-61.
19.Bobyshev, V.G. and G.Y. Lapchenkov. 1966. Khimiya selKhoz 4, 30-31.
20.Yurkevich, I.V. and N.Z. Tolkachev. 1972. Khimiya selKhoz 10, 696-698.
21.Hauke-Pacewiczowa, T. 1971. Pam. Pulawski 46, 5-48.
22.Platonova, V.P. 1967. Trudy kirov. selKhoz. Inst 20, 215-221.
23.Boiku, V.S., M.G. Degtyareva and I.P. Kazakova. 1969. Khimiya selKhoz 7, 203-207.
24.Bakalivanov, D. and G. Nikolova. 1971. In: I. Pashev (ed.): Vtori Kongres po Mikrobiologiya. IV. Chst. Sofia. 1969. Bulg. Akad. Nauk 1971. 2>3-266.
25.Lui, L.C. and H.R. Cibes-Viade. 1972. J. Agric. Univ. P. Rico 56, 417-425.
26.Zinchenko, V.A. and T.V. Osinskaya. 1969. Agrokhimiya 9, 94-101.

7-174 Effect of 2,4-D + dicamba dimethylamine

Parameter	Effect Value	Concentration and Condition	Species
Growth rate and numbers	Decreased population	Light loam dernopodzol, 2,4-D=0.8 + dicamba di=0.75 kg/ha, in situ	Soil bacteria[1]
Growth rate and numbers	No effect	Light loam dernopodzol, 2,4-D=0.8 + dicamba di=0.75 kg/ha, in situ	Soil bacteria, anaerobic[1]

Source: 1.Kozyrev, M.A. and A.A. Laptev. 1972. Khimiya selKhoz. 10, 614-617.

7-175 Effect of 2,4-D amine

Parameter	Effect Value	Concentration and Condition	Species
Growth rate and numbers	Decreased population for 20 days	Loam, 3.5 % OM, 0.9 kg/ha, in situ	Soil bacteria[1]

Source: 1.Geshtovt, Y.N., S.U. Zharasov and N. Adekenov. 1970. Vest. SelKhoz. Nauki. Kazakhskoi SSR. 13, 35-38

7-176 Effect of 2,4-D ester

Parameter	Effect Value	Concentration and Condition	Species
Nitrification and nitrifiers	No effect	1.0 - 2.5 kg/ha, recomm. field rate	Soil bacteria[1]

Source: 1.Deshmukh, V.A. and J.G. Shrikhande 1974. J. Indian Soc. Soil Sci. 22, 36-42.

7-177 Effect of 2,4-D, 2,4-DB, MCPA, MCPB

Parameter	Effect Value	Concentration and Condition	Species
Nitrification and nitrifiers	Decreased activity from annual application of this mixture	Light loam dernopodzol, 5 y, 2,4-D(0.6) 2,4-DB(0.6) MCPA(2.5) MCPB(2 kg/ha)	Soil bacteria[1]

Source: 1.Chunderova, A.I., T.P. Zuberts and A.M. Sofinskii. 1971. Khimiya selKhoz 9, 527-530.

7-178 Effect of 2,4-DB

Parameter	Effect Value	Concentration and Condition	Species
Nitrification and nitrifiers	Decreased activity	Brown soil, 1.5-3.0 kg/ha, recomm. field rate	Soil bacteria[1]
Nitrification and nitrifiers	No effect from annual applications	Light loam dernopodzol, 5 y, 0.5-0.75 kg/ha	Soil bacteria[2]

Source: 1.Chandra, P. 1964. Weed Res. 4, 54-63.
2.Chunderova, A.I., T.P. Zuberts and A.M. Sofinskii. 1971. Khimiya selKhoz 9, 527-530.

7-179 Effect of 2,4-DEP

Parameter	Effect Value	Concentration and Condition	Species
Nitrification and nitrifiers	Decreased activity	Heavy loam, 30 kg/ha	Soil bacteria[1]
Nitrification and nitrifiers	Increased activity	Heavy loam, 6 kg/ha	Soil bacteria[1]
Nitrogen fixation in soil	No effect	Light loam pale chestnut, 1.7 % OM, 0-8 kg/ha	Soil bacteria, free living[2]
Soil microbial ammonification	No effect	Heavy loam, 6-30 kg/ha	Soil organisms[3]

Source: 1.Zavarzin, V.I. 1966. Agrokhimiya 4, 121-124.
2.Lazareva, N.V. and V.G. Blokhin. 1969. Khimiya selKhoz 7, 609-611.
3.Zavarzin, V.I. and T.V. Belyaeva. 1966. Khimiya selKhoz 4, 34-36.

7-180 Effect of 2,4-DES

Parameter	Effect Value	Concentration and Condition	Species
Nitrification and nitrifiers	Decreased activity	Heavy loam, 10 kg/ha	Soil bacteria[1]
Nitrification and nitrifiers	Increased activity	Heavy loam, 5 kg/ha	Soil bacteria[1]
Soil microbial ammonification	Increased activity after 10 months	Heavy loam, 5-10 kg/ha	Soil organisms[2]

Source: 1.Zavarzin, V.I. 1966. Agrokhimiya 4, 121-124.
2.Zavarzin, V.I. and T.V. Belyaeva. 1966. Khimiya selKhoz 4, 34-36.

7-181 Effect of 4-CPA isopropyl ester and atrazine

Parameter	Effect Value	Concentration and Condition	Species
Soil respiration activity	Decreased activity	Sandy soil, 3 kg/ha, 4-CPA(40 %)+atrazine(30 %)	Soil organisms[1]
Soil respiration activity	NO effetc, applied after linuron	Sandy soil, 0.3 kg/ha, 4-CPA(40 %)+atrazine(30 %)	Soil organisms[1]
Soil respiration activity	No effect, applied after linuron	Loess soil, 0.3 and 3 kg/ha, 4-CPA(40 %)+atrazine(30 %)	Soil organisms[1]

Source: 1.Manninger, E. and J. Szava. 1972. Symp. Biol. Hung. 11, 369-372.

ABBREVIATIONS
FOOTNOTES
INDEX

ABBREVIATIONS

+-	indicating interval of determination
*	Multiplication
**	iteration
---	missing value, value not measured
AAS	Atomic absorption spectophotometer
Ab.ground	above ground
Abs	absorption
Abs	Alkyl benzen sulfonate
Abs	Kelvin degrees
Absorp	absorption
Acc	acceleration
Ae7.0	Alkyl polyethoxylate
Ae7.4	Alkyl polyethoxylate
Alk	Alkalinity
Anim	Animal
Ann	Annual
Aos	alpha olefin sulfonate
appl	application, applied
approx	approxiamtely
assim	assimilation
att	attenuation
av	average
avia	aviable
B.C.	British Colombia
Bact	bacteria
begin	beginning
bel	below
betw	between
bhc	benxen hexachloride
bi	two generations pr. year
biol	biological
biom	biomass
br.h.	breast height
cal	calories
cal/cm2*min	experimental conditions of light intensity
calc	calculated
Calif	California
Can	Canada
CAP	Phenylalanine hycroxylase
Carot	carotenoids
cc	cm3 or millilitre
Cdo	control condition for Cadmium experiment
Cec	Cation exchange capacity
Cel	Celcius
Cell	cellular
Cfl	confidence limit
Ch-p	Chlorella pyrenoidosa
Chem	chemically
Chemo	chemostat
Chl	chlorophyll a
Chla	chlorophyll a
Chlor	chloride
Ci	Curie
Cii	Copepodit stage 2
Cl	confidence limit
Clean	cleaning
Coef	coefficient
Coeff	coefficient
Col	colony
Comb	combustion
Comm	community
Con	consumation
conc	concentration
cond	conductivity
conp	consumption
const	constant
consump	consumption
cont	content
cont	continous
cont	counted
contamin	contaminated
cr	creek
cult	cultivable
culture: xx K	xx Kelvin was used as temperature for cultivation i culture
cum	cumulative
c1	copepodit stage 1
d	(in condition) no. of days)measuring period

d	days
def	deficit
deg	degrees
Denm	Denmark
deriv	derivative
develop	development
dg	degrees days
dia	diameter
diam	diameter
dif	different
diff	different
diff	diffusion
dis	dissolved
disapp	disappearence
disp	dispersion
diss	dissolved
dissol	dissolved
DK	Denmark
DM	dry matter
DOM	dissolved organic matter
dom	dominating
dur	during
E	East
eff	effective
effic	efficiency
elim	elimination
Engl	England
Enz	enzyme
Equilibr	equilibrium
Erodibil	Erodibilae
Erupt	eruptions (volcanoes)
Est	estimate
Estu	estuarie
Evap	evaporation
Exch	exchange
Excl	exclusive
Excr	excretion
Exp	experimental
Exp. xx K conditions	xx Kelvin was used as experimental
Ext	extension
Ext	extinction
F	females
Fact	Factor
Fert	fertilization
Field	field data
Filt	filtering
Filtrat	filtration
Fl	Florida
Form	formation
Fps	feet per second
FR	France
fr	from
Fract	fraction
Freund	Freundlic isotherm, se footnote F057
Ftc	foot light candles
Funct	function
GB	Great Brittain
G-6-P-D-H	Glucose-6-phophate dehydrogenase
G-6-PDH	do.
G/G	gram per gram (used about ratios)
GOT	glutamate oxaloacetate transaminase
GPT	glucose phosphate transferase
GPT	glucose pyruvate transaminase
Gr	growth rate
Gr.rate	growth rate
H	hour
Hb	hemoglobine
Hrs	hours
Humid	humidity
Ik	linear saturation point of photosynthesis curve
Imp	imperrious
In	inch
Inch	inches
Ind	Indiana
Ingest	ingestion
Inh	inhibition

Init	initial concentration in experiment
Inject	injection
Inorg	inorganic
Insol	insolation
Interv	interval
Intrav	intravenous
Jap	Japan, Japanese
Julian C.D.	Julian calendar day
Juv	Juveniles
K	Kelvin
Kcal	kilo calories
Kj	kilo joules
Klux	kilo lux
L	lake
L	litre
LA	Louisiana
Lab	experiment done in laboratorium
LC50	concentration at which 50% of experiment organisms dies
Le	less
Long	longitudinal
Ly	langleys
M	males
M	meter
M	milli
M	molar
M	month
M/s	meter per second
Main	maintenance
Mat	matter
Matc	maximum acceptable toxicant concentration
Max	maximum
Mch	mean corpuscle hemoglobin (in blood)
Mcv	mean corpuscle volume
Mean of x	mean of x determinations
Met	metabolism
Metab	metabolism
Mic.g	microgram
MICH	Michigan
Min	minute
MINN	Minnesota
Miss	missing
Mosm	milli osmotic pressure
Multi	more than two generations per year
Mv	millivolt
Mw	milliwatt
Myh	micro M/g liver/hour
Mym	micro M/g liver/10 minutes
Myt	Micro M tyrosine/g liver/hour
Myu	micr units/ml
N	nauplius
N	nitrogen
N	North
N=x	number of samples
NC	North Carolina
Ng	nanogram
No.	number
Np	net production
NV	North west
NWT	North western territories in Canada
NZ	New Zealand
Obs	Observation period
OKL	Oklahoma
Opt	optimum conditions
Org	organic
Org	organism
Ova	ovaries
O2	oxygen
P	pico
P	production
P/R	production/respiration, respiratory coeff
Part	particle, particulate
Pe-t	Pediastrum tetras
Photosynth	photosynthesis
Phyt	phytoplankton
Pollut	pollutant, polluted
Pot	potential
ppb	parts per 10**9
ppm	parts per 10**6
Prob	probable
Prod	production
Prot	protein
Pt-50	-log (potential toxicity)
Pyr	Pyridine
R	regression coefficient
R	river
Rad	radiation
RBC	red blood cells
Ref	reference
Rel	relation
Rel	relative
Rel	release
Remain	remaining
Reprod	reproduction
Req	required
Req	requirements
Resid	residence
Resp	respiration
Rest	resistance
S	salinity
S	second
S	south
SAP	phenyl alanine hydroxylase
Sat	saturation
SC	South Carolina
SD	Standard deviation
SE	standard error of mean
Se-p	Selenastrum capricornutum
Sec	second
Sed	sediment
Sedim	sediment
SGPT	serum glutamic pyruvic transaminase
South	southern
Sp	number of spawns
Sp	species
Sp	specific
Spec	Spectronic 20
Sph	sphere
SSI	spleen somatic ration (g spleen/g body weight)
St	standing
Stage C1	Copepodit stage 1
Strato	Stratosphere
SW	South West
T	tons
Tech	technice
Temp	temperature
Tenn	Tennesee
Thr	through
Tot	total
Transp	transport
Trop	tropical
Tween	see footnotes F010, F018, F019 and F020
U	unit
U	10**-6 m
Ug	microgram
UK	United Kingdom
Um	micromol
Uni	one generation per year
Upt	uptake
Util	utilization
Val	value
Vel	velocity
Veloc	velocity
Vert	vertical
vol	volume
Voltinism	number of generations per year, see Uni, Bi and Multi
vs	versus
W	weeks
W	West
Wat	water
Wbc	White blood cells
Wisc	Wisconsin
Wt	Weight
WW	wet weight
Y	years
Yr	year

FOOTNOTES

F001 No growth of algae without addition of nitrogen, and no growth with a mixture of a amino acids in dark.

F002 The uptake rate of nutrients is a modified equation, so that the uptake rate becomes zero for a positive value of nutrient concentration. the concentration of nutrient is given by SI.

F003 See table

F004 see table

F005 Fecal pellets from Meganyctipphanus norvegia, Ephausia krohnii and Nematoscelis magalops

F006 Food is juvenile Artemia sp. or Phaeodactylum tricornatum.

F007 pt50 = -log(potential toxicity)50

F008 Numbers indicate comparison with growth in control cultures:
4: full potential, similar to control
3: slight inhibition, turbidity and chlorophyll decrease
2: moderate inhibition, turbidity and chlorophyll decrease with yellowing of cells
1: marked inhibition, turbidity decrease, white to yellow coloration with few living cells

F009 Bacterial cells: Bacillus sp., Sarcina sp., Achromobacter sp., Flavobacterium sp., Brevibacterium sp., Chromobacterium sp., Pseudomonas sp., Azotobacter sp. and Streptomyces sp.

F010 TWEEN 80 = polyoxyethylene (80) monoleate

F011 Bacterial colonies: Clear lake: Pseudomonas sp., Flavobacterium sp., Brevibacterium sp., Enterobacter sp., Achromobacter sp., Escherichia sp., Sarcina sp., Caulobacter sp., Proteus sp., Micrococcus sp. and Streptomyces sp.
Lake Houston: Pseudomonas sp., Brevibacterium sp., Bacillus sp., Flavobacterium sp., Sarcina sp., Achromobacterium sp., Micrococcus sp. and Enterobacter sp.

F012 Weed, unpolluted soil. Eclipta prostrata, Gnaphalium affine, Lapsana apogonoides, Dopatrium junceum, Vandellia procumbens, Mazus japonicus, Rotala indica, Vicia angustifolia, Stellaria alsine, Chenopodium ambrosiodes, Cyperus difformis, Fimbristylis miliacea, Echinochloa crusgalli, Setaria viridis and Digitaria ascendens.

F013 The first value in the column for values gives the mortality in %, the second value gives the concentration in the animal (if only one value is indicated, it is the mortality).

F014 Pseudomonas sp., Xanthomonas sp., Achromobacter sp. and flavobacterium sp.

F015 Bidwellii sp., Gloeosporium sp., etc.

F016 Numbers of isolates per genus espressed as % of total numbers of colonies isolated.

F017 ZY CL = earth type

F018 TWEEN 20 = polyoxyethylene (20) sorbitan monolaurate

F019 TWEEN 40 = polyoxyethylene (40) monopalmitate

F020 TWEEN 60 = polyoxyethylene (60) monostearate

F021 The range given in conditions shows the temperature for zero mortality for the first 20 observations. The lower and upper limit is given. Outside this range there is a mortality from the temperature stress.

F022 This table shows the effect of PCB on E. coli. The effect is based on laboratory experiments. The incubation time is 24 hours and a volume of 500 ml is used. The mean harvest weight is determined from this volume.

F023 Concentration factor is base on:
CF(element) = ((mol % conc of element in coral)/(mol % conc of Ca in coral))/((mol % conc of element in sea water)/(mol % conc of Ca in seawater)).

F024 Halowax 1000 = 60 % mono- and 40% dichloro naphtalene, 26% chlorine (weight).

F025 Halowax 1013 = 10% tri-, 50% tetra- plus 40% pentachloronaphtalene, 56% chlorine (weight)

F026 Halowax 1014 = 20% tetra-, 40% penta- plus 40% hexachloronaphtalene, 62% chlorine (weight)

F027 Conc factor = specific activity of larvae (cpm/g)/specific activity of water (cpm/g). Initial 203-Hg(+) 6.89*10**5 cpm/g

F028 Conc factor = specific activity of bacterial cells (cpm/g)/specific activity of media (cpm/g). Initial 203-Hg(+) 1.63*10**5 cpm/g.

F029 Conc factor = conc in org (p Ci/kg wet wt)/ conc in water (p Ci/l).

F030 Podzol - Horizons: F = 2.5 - 5 cm, H = 10 - 20 cm, A1A2 = 25.5 - 33 cm, B1(pan) = 33 cm, B2 = 33 - 43 cm, B3 = 61 - 71 cm.

F031 Background sources of Hg: water = 0.028 ppb, fish = 0.102 ppb, food suppl. = 0.4 ug/day, for Pagophilus groenlandicus.

F032 Food = Asparagus, broccoli, brussels sprouts, cabbage, cucumber, eggplant, horseradish, onions, parsnips, peppers, pimentoes, radish and Raddish tops.

F033 Food = potatoes, sweet potatoes, rutabages, turnips, sugar beet and radish.

F034 Food = asparagus, broccoli, Brussels sprouts, cabbage, celery, cauliflower, mustard greens, spinach.

F035 Food = beans, peas, eggplant, tomatoes, collards (=coleworts).

F036 Food = wheat, rye, oats, rice (polished), maize, popcorn, sorghum.

F037 Food = cantalopes, cucumbers, pumpkins, squash, water melon.

F038 Food = Almonds, bananas, figs, filberts, guavas, mangos, olives, passionfruit, papayas, pecans, pomegranates, pineapples, strawberries and walnuts.

F039 Food = Citrus, pome and stone fruits.

F040 Food = crude soybean and linseseed oils, crude cottonseed oils, edible cottonseed oil, edible soybean oil.

F041 Apples, pears, peachea, apricots, small fruits (except strawberries), vegetables (except root vegetables), meat and poultry (on fat basis).

F042 Nuts (shelled) strawberries and root vegetables.

F043 Cherries, plums, citrus and tropical fruit.

F044 Edible cottonseed and maize oil.

F045 Apples, wheat, barley, sorghum, rice (husked and/or polished).

F046 % = ((conc in whole plant) - (conc in pressed plant))/(conc in whole plant)*100

F047 This general footnote indicates as follows:
a = a constant; A = area; Asse = assimilability of food; AV = availability of food; CA = concentration phytoplankton carbon, g/m**3; CFGI = conversion factor from global irradiance to light, active photosynthesis; CK = carrying capacity, g/m**3; CS = conc of inorganic carbon, g/m**3; D = diffusion constan; Dp = constant; Denit = rate of denitrification, g/(m**3*d); Denitk = rate coefficient of denitrification, 1/d; Derv = detritus mineralization rate, g/(m***d); Det = detritus concentration, g/m**3; FD = relative day length; Fmax = maximum fractional reduction in daily specific growth rate over euphotic depth; Gek = gas exchange coefficient, m/s; I = irradiance, I(0) = irradiance , surface intensity; Ida = irradiance, daily average; Iop(t) = optimum irradiance, IH light saturation parameter, Ik light saturation parameter; K = constant, unspecific; K20 = constant or coefficient at 20 celsius; K(t) = constant or coefficient at t celsius; Kopt = constant at optimum temperature; Kc = half saturation constant of Cs uptake, g/m**3; Kdet = saturation constant of detritus mineralization, g/m**3;

Klys = constant for light inhibition; Kn = half saturation constant for Ns uptake, g/m**3; Knfix = nitrogen fixation coefficient, 1/d; Knit = coefficient for nitrification, g/m**3; Kot = temperature coefficient; Kp = half saturation constant for Ps uptake, g/m**3; Kpa = saturation constant for intracellular phosphorus; Ktp = treshold concentration for grazing, g/m**3; Kz = half saturation constant for grazing, g/m**3; L0 = constant; M = concentration of denitrification microorganisms, g/m**3; M(0) = concentration of denitrification microorganisms at time = 0, g/m**3; Mixr(z) = mixing ratio at depth; Mort = mortality; Mortk = mortality coefficient, 1/d; My = growth rate of phytoplankton, 1/d; Mydet = detritus mineralization rate coefficient, 1/d; Mym = growth rate coefficient of denitrification microorganisms, 1/d; Myz = growth rate of zooplankton; Na = concentration of nitrogen in phytoplankton, g/m**3; Nh4-ns = ammonia-nitrogen concentration, g/m**3; Nit = concentration of nitrate-nitrogen, g/m**3; Nit(0) = concentration of nitrate-nitrogen at time = 0, g/m**3; Nitrik = nitrification rate, g/(m**3*d); Nitrik = nitrification rate coefficient, 1/d; Ns = concentration of soluble inorganic nitrogen, g/m**3; Ns0 = constant; Ox = oxygen concentration, g/m**3; Pa = concentration of phosphorus in phytoplankton, g/m**3; Photo = photosynthetic rate, g/(m**3*d); Phyt = phytoplankton concentration, g/m**3; Phyt(n) = concentration of phytoplankton with nitrogen fixation, g/m**3; Pki = equilibrium constant for reactions between phosphorus and carrier, liter/mol; Ps = concentration of soluble inorganic phosphorus, g/m**3; Pse= soluble inorganic phosphorus in epilimnion, g/m**3; Psh = soluble inorganic phosphorus in the hypolimnion, g/m**3; Ps0 = constant, Ref = reflected light; R(t) = total daily radiation; Relk = release coefficient; Respv = respiration coefficient, g/(m**3*d); Resp = respiration, 1/d; Sedt = settling rate of detritus, m/d; Seph = settling rate of phytoplankton, m/d; Settl = removed by settling, 1/d; Sev(0) = surface overflow rate, m/d; Sev(p) = settling rate of particles, m/d; Sn = Nitrogen in upper layer of sediment, g/l; Sse = average concentration of solute s in epilimnion, g/m**3; Ssj = average concentration of solute s in depth interval j, g/m**3; Sto = stochiometric constant mg bacterial mass per mg substrate; t = time, days; T = temperature, celsius; Ta = temperature, Kelvin; Tmax,Tmin = temperature, maximum and minimum, rate coefficient = 0 at Tmax and Tmin, celsius; Top = temperature, optimum, celsius; Tr = temperature reference; Tox = concentration of toxic material, g/m**3; U = mass flow, m/d; Uj = mass flow from the j'th deoht interval; Uc = uptake rate of inorganic carbon, g/(m**3*d); Ucmax = maximum uptake rate of inorganic carbon, g/(m**3*d); Ue = mass transport due to epilimnion expansion, kg/d; Un = nitrogen uptake rate, g/(m**3*d); Unmax = maximum nitrogen uptake rate, g/(m**3*d); Up = phosphorus uptake rate, g/(m**3*d); Upmax = maximum phosphorus uptake rate, g/(m**3*d); V = volume, m**3; Delta(v) = change of volume, m**3; Ve = volume of epilimnion, m**3; W = weight, kg; Wis = wind speed, m/sec; Delta(x) = distance between centres of two boxes, m; Z = depth, m; Zoo = concentration of zooplankton, g/m**3; Aa = attenuation of water, 1/m; Bb = specific attenuation of phytoplankton, g/m**2; Cc = specific attenuation of zooplankton, g/m**2; Dd = specific attenuation coefficient of detritus, g/m**2; EE = attenuation coefficent, 1/m.

F048 The Bowen ratio is used when calculating the heat loss from a water surface; and it is defined by Qc = R * Qe, where R is the Bowen ratio, Qc = the convective heat transfer and Qe = heat loss due to evaporation (kcal/(m**2*sec)). The symbols used are T = water surface temperature (celsius), Ta = dry bulb air temperature (Celsius), Es = saturation vapour pressure at the water surface temperature (mb), Ea = vapour pressure at the overlaying atmosphere (mb).

F049 Cs = saturation value at sea level, at the water temperature (mg/l), Cs' = corrected value at the altitude of the river (mg/l), Pb = barometric pressure at the river altitude (mm/Hg), Pv = saturation vapour pressure of water at the river temperature (mm/Hg).

F050 K = regional dispersion factor (equal to 1 for Big Blue River), U = average velocity, V = x/tp, where x = reach length and tp = time to peak arrival of constituent concentration at sampling station, h = A/B, Where A = cross-sectional area and B = top width of the flow.

F051 DG = degrees days calculated as the number of days of the year for which the mean temperature exceeds 40 F.

F052 From the table can the expression for the light saturation curve be derived. I(t) = 0 = 1/C * ln((B(t) = A(t)/B(t).

F053 El = longitudinal dispersion coefficient, (length**2/time), E = rate of energy dissipation per unit mass, Le = mean size of Eddies participating in mixing process, Cl = function of relative channel roughness.

F054 Se all or more than one of the condition lines, to get an explanation of the parameters.

F055 All expressions given to base e and for 20 celcius unless otherwise noted. Definition of symbols: Dm: molecular diffusion coefficient (8.1*10**(-5) hour (at 20 celsius). U: stream velocity H: average stram depth. Delta(h): change in stream elevation between two points. T = travel time between two points for which delta(h) is measured. S = slope (change in water surface elevation divided by distance). Q: Specific discharge (cfs per square mile of drainage area). Dl = longitudinal dispersion coefficient Ux = Bed shear velocity.

F056 I = ionic strength = 1/2*sigma(Ci*z**2), A = 2*10**6*(e*t)**(2/3), e = dielectric constant, A = 0.5 for water at 25 celsius, z = charge of ion, B = 0.33 in water at 25 celsius, c = adjustable parameter (Ångstroem units) corresponding to size of ion.

F057 Langmuir isotherms: SP = pcs*ce(/k+ce) - nap. Sp: sorbed amount of sorbant (mg/g dry wt), Pcs: sorbant sorbtion capacity (mg/g dry wt), Ce: equilibrium concentration of sorbant (mg/l). K half saturation constant (mg/l), Nap: native amount of adsorbed sorbant.
Freundlig isotherm: Sp = Kc*ce**p - nap
Sp: sorbed amount of sorbant (mg/g dry wt), Ce: equilibrium concentration of sorbant (mg/l), Kc and p are constants.

F058 Experiment 1: oxidized mud and demineralized water.

F059 Experiment 2: oxidized mud and lake water.

F060 Experiment 3: reduced mud and demineralized water.

F061 El = effective mixing coefficent (ft**2/sec), Cl = tidally induced mixing coefficient (no dimension), Y = tidally averaged depth, U = tidally averaged absolute value of velocity, Ru = standard deviation of the velocity, Ry = standard deviation of the depth, C = density-induced mixing coefficient (ft**3/sec/mg/l-salinity).

F062 SI = incident solar radiation, Ta = air temperature and Tw = water temperature (both in F degrees), E = evaporation (inches/hour) = 0.000073*u(rw-ra) Rw = water pressure for water (mb), Ra = the vapour presure for air (mb).

F063 Lw = latent heat of vaporization (kcal/(m**2*sec)), E = evaporation rate (m/sec). E is approximated by E = (a+B*v)*(ES-Ea), where A and B are empirical coefficients, V = wind speed over water surface (m/sec), Es = saturation vapour pressure at the surface water temperature (mb), Ea = vapour pressure of the overlying atmosphere. A is often zero and B about 10**(-9).

F064 Wind velocity U is given in cm/sec, the width of diameter of organisms (w) in cm, T in celsius degrees and HC is in 10**(-3) cal/(cm**2*min(cel deg)).

F065 Pt = penetration of radiation, Lai = leaf area index, Sp = shadow projection of leaf axis of orientation, Is = solar altitude.

F066 Ra = resistance (s/m), O = foliage density = ratio of leaf plan to leaf frontal area, here 4.0, D = sqrt(leaf area), U = wind speed (m/s).

F067 F = flow, N2 - N1 = concentration gradient, z = thickness of layer, B = 200 cm**2/sec in the upper layer from the upper surface to the boundary of the thermocline; in the thermocline B = 20 cm**2/sec, and in the waters below the thermocline B = 150 cm**2/sec.

F068 Ks = stomata conductance (cm/s), De = vapour pressure deficit, 1 year old needles, temperature > 288 K, light > 500 uE/(m**2*sec).

F069 Respiration measured in ml O2/(mg dry wt*day).

F070 Excretion measured in ug atom N or P/(mg dry wt*day).

F071 Ke = extinction coefficient (1/m), Ao = light extinction coefficient of water (1/m), R(I) = self shading coefficient of constituent I, S(I) = concentration of suspended matter, I, mg/l (detritus, sediment or algae). Sigma denotes the summation over the different I's.

F072 The constant M is species specific. M has the values of 207, 115 and 55 kg/(ha*day) for sorghum, wheat and alfalfa respectively.

F073 Qat = net long wave atmospheric radiation (kcal/(m**2*sec)), Cloud = fraction of cloud cover, Ta = dry bulb air temperature in celsius.

F074 The coefficient of bed activity in a river is defined according to Kd = K1 + n*(V/D), where n is this coefficient. Kd is the deoxygenation constant in the river, K1 is the BOD (biological oxygen demand) measured in the laboratory, V is the stream velocity (ft/sec) and D is the stream depth. The formula and the table enables one to transfer laboratory values to the field.

F075 The system has some, but not all the properties of seawater. These properties are total metal concentration and total ligand concentration, values given in condition lines.

F076 Used in models with densimetric flow in vertical direction. D = half the thickness of the withdrawal of the inflow flow field (m), Q = inflow or outflow rate (m**3/sec), W = width of the reservoir at inflow or withdrawal depth (m), B = density gradient at the point of inflow or withdrawal (kg/m**4).

F077 Uptake equation is a modified Michaelis Menten kinetic, where actual uptake rate equals V=Vmax*((Si-S0)/(KSi+(Si-S0)))

F078 Detritus composed of Dunalilla marina in a concentration of 40*10**6 cells/liter.

F080 See also tables 1-351, 1-361, 1-362, 1-365, 1-368, 1-369, 1-373, 1-385, 1-387, 1-388, 1-394, 1-396, 1-397, 1-398, 1-399.

F081 Sorption determined by amount remaining in filtrate after addition of 20 mg/l at beginning of runs.

F082 Symptoms of poisoning in all or some animals.

F085 Element concentrations in edible parts, crops grown in pots.

F200 Based on Aluminium concentrations in atmospheric particles and soil both relative to iron.

INDEX